Instructor's Solutions Manual
Part II

Thomas/Finney

Calculus and Analytic Geometry

9TH Edition

Maurice D. Weir
U.S. Naval Postgraduate School

Addison-Wesley Publishing Company

Reading, Massachusetts • Menlo Park, California • New York
Don Mills, Ontario • Harlow, United Kingdom• Amsterdam • Bonn
Sydney • Singapore • Tokyo • Madrid • San Juan • Milan • Paris

Reproduced by Addison-Wesley from camera-ready copy supplied by the author.

ISBN 0-201-53178-X
6 7 8 9 10- CRS-99

PREFACE TO THE INSTRUCTOR

This Instructor's Solutions Manual contains the solutions to every exercise in the 9th Edition of CALCULUS AND ANALYTIC GEOMETRY by Ross L. Finney and George B. Thomas, Jr., including the Computer Algebra System (CAS) exercises. The corresponding Student's Solutions Manual omits the solutions to the even-numbered exercises as well as the solutions to the CAS exercises (because the CAS command templates would give them all away).

In addition to including the solutions to all of the new exercises in this edition of Thomas/Finney, we have carefully revised or rewritten every solution which appeared in previous solutions manuals to ensure that each solution

- conforms exactly to the methods, procedures and steps presented in the text
- is mathematically correct
- includes all of the steps necessary so a typical calculus student can follow the logical argument and algebra
- includes a graph or figure whenever called for by the exercise
- is formatted in an appropriate style to aid in its understanding

Every CAS exercise is solved in both the MAPLE and MATHEMATICA computer algebra systems. A template showing an example of the CAS commands needed to execute the solution is provided for each exercise type. Similar exercises within the text grouping require a change only in the input function or other numerical input parameters associated with the problem (such as the interval endpoints or the number of iterations).

Acknowledgments

We are grateful to the following individuals who contributed solutions to this manual:

David Canright, Naval Postgraduate School
Thomas Cochran, Belleville Area College
Leonard Klosinski, Santa Clara University
Jeffrey Oldham, Stanford University
Michael Schneider, Belleville Area College
Kirby Smith, Texas A&M University
Steven Szydlik, University of Wisconsin at Madison

We also thank Robert Lande for his excellent job of word processing the final manuscript.

TABLE OF CONTENTS

14 Integration in Vector Fields 1247

CHAPTER 8 INFINITE SERIES

8.1 LIMITS OF SEQUENCES OF NUMBERS

1. $a_1 = \frac{1-1}{1^2} = 0$, $a_2 = \frac{1-2}{2^2} = -\frac{1}{4}$, $a_3 = \frac{1-3}{3^2} = -\frac{2}{9}$, $a_4 = \frac{1-4}{4^2} = -\frac{3}{16}$

2. $a_1 = \frac{1}{1!} = 1$, $a_2 = \frac{1}{2!} = \frac{1}{2}$, $a_3 = \frac{1}{3!} = \frac{1}{6}$, $a_4 = \frac{1}{4!} = \frac{1}{24}$

3. $a_1 = \frac{(-1)^2}{2-1} = 1$, $a_2 = \frac{(-1)^3}{4-1} = -\frac{1}{3}$, $a_3 = \frac{(-1)^4}{6-1} = \frac{1}{5}$, $a_4 = \frac{(-1)^5}{8-1} = -\frac{1}{7}$

4. $a_1 = 2 + (-1)^1 = 1$, $a_2 = 2 + (-1)^2 = 3$, $a_3 = 2 + (-1)^3 = 1$, $a_4 = 2 + (-1)^4 = 3$

5. $a_1 = \frac{2}{2^2} = \frac{1}{2}$, $a_2 = \frac{2^2}{2^3} = \frac{1}{2}$, $a_3 = \frac{2^3}{2^4} = \frac{1}{2}$, $a_4 = \frac{2^4}{2^5} = \frac{1}{2}$

6. $a_1 = \frac{2-1}{2} = \frac{1}{2}$, $a_2 = \frac{2^2-1}{2^2} = \frac{3}{4}$, $a_3 = \frac{2^3-1}{2^3} = \frac{7}{8}$, $a_4 = \frac{2^4-1}{2^4} = \frac{15}{16}$

7. $a_1 = 1$, $a_2 = 1 + \frac{1}{2} = \frac{3}{2}$, $a_3 = \frac{3}{2} + \frac{1}{2^2} = \frac{7}{4}$, $a_4 = \frac{7}{4} + \frac{1}{2^3} = \frac{15}{8}$, $a_5 = \frac{15}{8} + \frac{1}{2^4} = \frac{31}{16}$, $a_6 = \frac{63}{32}$,
   $a_7 = \frac{127}{64}$, $a_8 = \frac{255}{128}$, $a_9 = \frac{511}{256}$, $a_{10} = \frac{1023}{512}$

8. $a_1 = 1$, $a_2 = \frac{1}{2}$, $a_3 = \frac{\left(\frac{1}{2}\right)}{3} = \frac{1}{6}$, $a_4 = \frac{\left(\frac{1}{6}\right)}{4} = \frac{1}{24}$, $a_5 = \frac{\left(\frac{1}{24}\right)}{5} = \frac{1}{120}$, $a_6 = \frac{1}{720}$, $a_7 = \frac{1}{5040}$, $a_8 = \frac{1}{40{,}320}$,
   $a_9 = \frac{1}{362{,}880}$, $a_{10} = \frac{1}{3{,}628{,}800}$

9. $a_1 = 2$, $a_2 = \frac{(-1)^2(2)}{2} = 1$, $a_3 = \frac{(-1)^3(1)}{2} = -\frac{1}{2}$, $a_4 = \frac{(-1)^4\left(-\frac{1}{2}\right)}{2} = -\frac{1}{4}$, $a_5 = \frac{(-1)^5\left(-\frac{1}{4}\right)}{2} = \frac{1}{8}$,
   $a_6 = \frac{1}{16}$, $a_7 = -\frac{1}{32}$, $a_8 = -\frac{1}{64}$, $a_9 = \frac{1}{128}$, $a_{10} = \frac{1}{256}$

10. $a_1 = -2$, $a_2 = \frac{1 \cdot (-2)}{2} = -1$, $a_3 = \frac{2 \cdot (-1)}{3} = -\frac{2}{3}$, $a_4 = \frac{3 \cdot \left(-\frac{2}{3}\right)}{4} = -\frac{1}{2}$, $a_5 = \frac{4 \cdot \left(-\frac{1}{2}\right)}{5} = -\frac{2}{5}$, $a_6 = -\frac{1}{3}$,
    $a_7 = -\frac{2}{7}$, $a_8 = -\frac{1}{4}$, $a_9 = -\frac{2}{9}$, $a_{10} = -\frac{1}{5}$

11. $a_1 = 1$, $a_2 = 1$, $a_3 = 1 + 1 = 2$, $a_4 = 2 + 1 = 3$, $a_5 = 3 + 2 = 5$, $a_6 = 8$, $a_7 = 13$, $a_8 = 21$, $a_9 = 34$, $a_{10} = 55$

12. $a_1 = 2$, $a_2 = -1$, $a_3 = -\frac{1}{2}$, $a_4 = \frac{\left(-\frac{1}{2}\right)}{-1} = \frac{1}{2}$, $a_5 = \frac{\left(\frac{1}{2}\right)}{\left(-\frac{1}{2}\right)} = -1$, $a_6 = -2$, $a_7 = 2$, $a_8 = -1$, $a_9 = -\frac{1}{2}$, $a_{10} = \frac{1}{2}$

13. $a_n = (-1)^{n+1}$, $n = 1, 2, \ldots$

14. $a_n = (-1)^n$, $n = 1, 2, \ldots$

15. $a_n = (-1)^{n+1}n^2$, $n = 1, 2, \ldots$

16. $a_n = \frac{(-1)^{n+1}}{n^2}$, $n = 1, 2, \ldots$

17. $a_n = n^2 - 1$, $n = 1, 2, \ldots$

18. $a_n = n - 4$, $n = 1, 2, \ldots$

19. $a_n = 4n - 3$, $n = 1, 2, \ldots$

20. $a_n = 4n - 2$, $n = 1, 2, \ldots$

21. $a_n = \frac{1 + (-1)^{n+1}}{2}$, $n = 1, 2, \ldots$

22. $a_n = \frac{n - \frac{1}{2} + (-1)^n\left(\frac{1}{2}\right)}{2} = \lfloor \frac{n}{2} \rfloor$, $n = 1, 2, \ldots$

23. $\left|\sqrt[n]{0.5} - 1\right| < 10^{-3} \Rightarrow -\frac{1}{1000} < \left(\frac{1}{2}\right)^{1/n} - 1 < \frac{1}{1000} \Rightarrow \left(\frac{999}{1000}\right)^n < \frac{1}{2} < \left(\frac{1001}{1000}\right)^n \Rightarrow n > \frac{\ln\left(\frac{1}{2}\right)}{\ln\left(\frac{999}{1000}\right)} \Rightarrow n > 692.8$
$\Rightarrow N = 692$; $a_n = \left(\frac{1}{2}\right)^{1/n}$ and $\lim_{n\to\infty} a_n = 1$

24. $\left|\sqrt[n]{n} - 1\right| < 10^{-3} \Rightarrow -\frac{1}{1000} < n^{1/n} - 1 < \frac{1}{1000} \Rightarrow \left(\frac{999}{1000}\right)^n < n < \left(\frac{1001}{1000}\right)^n \Rightarrow n > 9123 \Rightarrow N = 9123$;
$a_n = \sqrt[n]{n} = n^{1/n}$ and $\lim_{n\to\infty} a_n = 1$

25. $(0.9)^n < 10^{-3} \Rightarrow n \ln(0.9) < -3 \ln 10 \Rightarrow n > \frac{-3 \ln 10}{\ln(0.9)} \approx 65.54 \Rightarrow N = 65$; $a_n = \left(\frac{9}{10}\right)^n$ and $\lim_{n\to\infty} a_n = 0$

26. $\frac{2^n}{n!} < 10^{-7} \Rightarrow n! > 2^n 10^7$ and by calculator experimentation, $n > 14 \Rightarrow N = 14$; $a_n = \frac{2^n}{n!}$ and $\lim_{n\to\infty} a_n = 0$

27. (a) $f(x) = x^2 - a \Rightarrow f'(x) = 2x \Rightarrow x_{n+1} = x_n - \frac{x_n^2 - a}{2x_n} \Rightarrow x_{n+1} = \frac{2x_n^2 - \left(x_n^2 - a\right)}{2x_n} = \frac{x_n^2 + a}{2x_n} = \frac{\left(x_n + \frac{a}{x_n}\right)}{2}$

(b) $x_1 = 2$, $x_2 = 1.75$, $x_3 = 1.732142857$, $x_4 = 1.73205081$, $x_5 = 1.732050808$; we are finding the positive number where $x^2 - 3 = 0$; that is, where $x^2 = 3$, $x > 0$, or where $x = \sqrt{3}$.

28. $x_1 = 1.5$,, $x_2 = 1.416666667$, $x_3 = 1.414215686$, $x_4 = 1.414213562$, $x_5 = 1.414213562$; we are finding the positive number $x^2 - 2 = 0$; that is, where $x^2 = 2$, $x > 0$, or where $x = \sqrt{2}$.

29. $x_1 = 1$, $x_2 = 1 + \cos(1) = 1.540302306$, $x_3 = 1.540302306 + \cos(1 + \cos(1)) = 1.570791601$,
$x_4 = 1.570791601 + \cos(1.570791601) = 1.570796327 = \frac{\pi}{2}$ to 9 decimal places. After a few steps, the arc$\left(x_{n-1}\right)$ and line segment $\cos\left(x_{n-1}\right)$ are nearly the same as the quarter circle.

30. (a) $S_1 = 6.815$, $S_2 = 6.4061$, $S_3 = 6.021734$, $S_4 = 5.66042996$, $S_5 = 5.320804162$, $S_6 = 5.001555913$, $S_7 = 4.701462558$, $S_8 = 4.419374804$, $S_9 = 4.154212316$, $S_{10} = 3.904959577$, $S_{11} = 3.670662003$, $S_{12} = 3.450422282$ so it will take Ford about 12 years to catch up

(b) $3.5 = 7.25(0.94)^n \Rightarrow (0.94)^n = \frac{3.5}{7.25}$

$\Rightarrow n \ln(0.94) = \ln \frac{3.5}{7.25} \Rightarrow n = \frac{\ln\left(\frac{3.5}{7.25}\right)}{\ln(0.94)}$

$\Rightarrow n \approx 11.764 \approx 12$

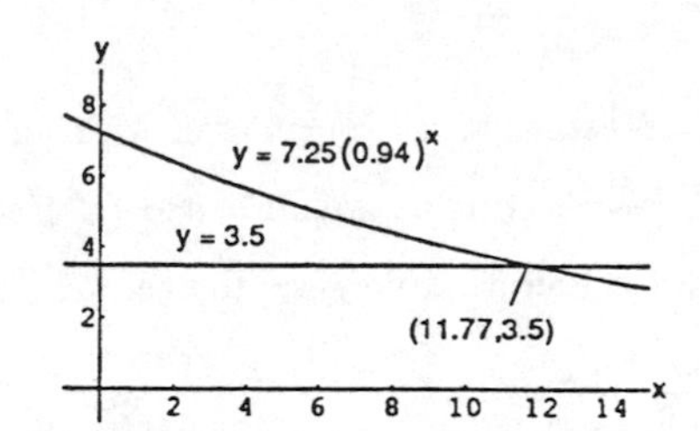

31. $a_{n+1} \ge a_n \Rightarrow \frac{3(n+1)+1}{(n+1)+1} > \frac{3n+1}{n+1} \Rightarrow \frac{3n+4}{n+2} > \frac{3n+1}{n+1} \Rightarrow 3n^2+3n+4n+4 > 3n^2+6n+n+2$ $\Rightarrow 4 > 2$; the steps are reversible so the sequence is nondecreasing; $\frac{3n+1}{n+1} < 3 \Rightarrow 3n+1 < 3n+3$ $\Rightarrow 1 < 3$; the steps are reversible so the sequence is bounded above by 3

32. $a_{n+1} \ge a_n \Rightarrow \frac{(2(n+1)+3)!}{((n+1)+1)!} > \frac{(2n+3)!}{(n+1)!} \Rightarrow \frac{(2n+5)!}{(n+2)!} > \frac{(2n+3)!}{(n+1)!} \Rightarrow \frac{(2n+5)!}{(2n+3)!} > \frac{(n+2)!}{(n+1)!}$ $\Rightarrow (2n+5)(2n+4) > n+2$; the steps are reversible so the sequence is nondecreasing; the sequence is not bounded since $\frac{(2n+3)!}{(n+1)!} = (2n+3)(2n+2)\cdots(n+2)$ can become as large as we please

33. $a_{n+1} \le a_n \Rightarrow \frac{2^{n+1}3^{n+1}}{(n+1)!} \le \frac{2^n3^n}{n!} \Rightarrow \frac{2^{n+1}3^{n+1}}{2^n3^n} \le \frac{(n+1)!}{n!} \Rightarrow 2\cdot 3 \le n+1$ which is true for $n \ge 5$; the steps are reversible so the sequence is decreasing after a_5, but it is not nondecreasing for all its terms; $a_1 = 6$, $a_2 = 18$, $a_3 = 36$, $a_4 = 54$, $a_5 = \frac{324}{5} = 64.8 \Rightarrow$ the sequence is bounded from above by 64.8

34. $a_{n+1} \ge a_n \Rightarrow 2 - \frac{2}{n+1} - \frac{1}{2^{n+1}} \ge 2 - \frac{2}{n} - \frac{1}{2^n} \Rightarrow \frac{2}{n} - \frac{2}{n+1} \ge \frac{1}{2^{n+1}} - \frac{1}{2^n} \Rightarrow \frac{2}{n(n+1)} \ge -\frac{1}{2^{n+1}}$; the steps are reversible so the sequence is nondecreasing; $2 - \frac{2}{n} - \frac{1}{2^n} \le 2 \Rightarrow$ the sequence is bounded from above

35. $a_n = 1 - \frac{1}{n}$ converges because $\frac{1}{n} \to 0$ by Example 2; also it is a nondecreasing sequence bounded above by 1

36. $a_n = n - \frac{1}{n}$ diverges because $n \to \infty$ and $\frac{1}{n} \to 0$ by Example 2, so the sequence is unbounded

37. $a_n = \frac{2^n - 1}{2^n} = 1 - \frac{1}{2^n}$ and $0 < \frac{1}{2^n} < \frac{1}{n}$; since $\frac{1}{n} \to 0$ (by Example 2) $\Rightarrow \frac{1}{2^n} \to 0$, the sequence converges; also it is a nondecreasing sequence bounded above by 1

38. $a_n = \frac{2^n - 1}{3^n} = \left(\frac{2}{3}\right)^n - \frac{1}{3^n}$; $0 < \left(\frac{2}{3}\right)^{n+1} < \left(\frac{2}{3}\right)^n$ and $0 < \frac{1}{3^n} < \frac{1}{n} \Rightarrow$ the sequence converges by definition of convergence

39. $a_n = ((-1)^n + 1)\left(\frac{n+1}{n}\right)$ diverges because $a_n = 0$ for n odd, while for n even $a_n = 2\left(1 + \frac{1}{n}\right)$ converges to 2; it diverges by definition of divergence

40. $x_n = \max\{\cos 1, \cos 2, \cos 3, \ldots, \cos n\}$ and $x_{n+1} = \max\{\cos 1, \cos 2, \cos 3, \ldots, \cos(n+1)\} \ge x_n$ with $x_n \le 1$ so the sequence is nondecreasing and bounded above by $1 \Rightarrow$ the sequence converges.

41. If $\{a_n\}$ is nonincreasing with lower bound M, then $\{-a_n\}$ is a nondecreasing sequence with upper bound $-M$. By Theorem 1, $\{-a_n\}$ converges and hence $\{a_n\}$ converges. If $\{a_n\}$ has no lower bound, then $\{-a_n\}$ has no upper bound and therefore diverges. Hence, $\{a_n\}$ also diverges.

42. $a_n \geq a_{n+1} \Leftrightarrow \frac{n+1}{n} \geq \frac{(n+1)+1}{n+1} \Leftrightarrow n^2+2n+1 \geq n^2+2n \Leftrightarrow 1 \geq 0$ and $\frac{n+1}{n} \geq 1$; thus the sequence is nonincreasing and bounded below by $1 \Rightarrow$ it converges

43. $a_n \geq a_{n+1} \Leftrightarrow \frac{1+\sqrt{2n}}{\sqrt{n}} \geq \frac{1+\sqrt{2(n+1)}}{\sqrt{n+1}} \Leftrightarrow \sqrt{n+1}+\sqrt{2n^2+2n} \geq \sqrt{n}+\sqrt{2n^2+2n} \Leftrightarrow \sqrt{n+1} \geq \sqrt{n}$ and $\frac{1+\sqrt{2n}}{\sqrt{n}} \geq \sqrt{2}$; thus the sequence is nonincreasing and bounded below by $\sqrt{2} \Rightarrow$ it converges

44. $a_n \geq a_{n+1} \Leftrightarrow \frac{1-4^n}{2^n} \geq \frac{1-4^{n+1}}{2^{n+1}} \Leftrightarrow 2^{n+1}-2^{n+1}4^n \geq 2^n - 2^n4^{n+1} \Leftrightarrow 2^{n+1}-2^n \geq 2^{n+1}4^n - 2^n4^{n+1}$ $\Leftrightarrow 2-1 \geq 2\cdot 4^n - 4^{n+1} \Leftrightarrow 1 \geq 4^n(2-4) \Leftrightarrow 1 \geq (-2)\cdot 4^n$; thus the sequence is nonincreasing. However, $a_n = \frac{1}{2^n} - \frac{4^n}{2^n} = \frac{1}{2^n} - 2^n$ which is not bounded below so the sequence diverges

45. $\frac{4^{n+1}+3^n}{4^n} = 4+\left(\frac{3}{4}\right)^n$ so $a_n \geq a_{n+1} \Leftrightarrow 4+\left(\frac{3}{4}\right)^n \geq 4+\left(\frac{3}{4}\right)^{n+1} \Leftrightarrow \left(\frac{3}{4}\right)^n \geq \left(\frac{3}{4}\right)^{n+1} \Leftrightarrow 1 \geq \frac{3}{4}$ and $4+\left(\frac{3}{4}\right)^n \geq 4$; thus the sequence is nonincreasing and bounded below by $4 \Rightarrow$ it converges

46. $a_1 = 1$, $s_2 = 2-3$, $a_3 = 2(2-3)-3 = 2^2-2\cdot 3$, $a_4 = 2(2^2-2\cdot 3)-3 = 2^3-(2^3-1)3$, $a_5 = 2[2^3-(2^3-1)3]-3 = 2^4-(2^4-1)3, \ldots, a_n = 2^{n-1}-(2^{n-1}-1)3 = 2^{n-1}-3\cdot 2^{n-1}+3$ $= 2^{n-1}(1-3)+3 = -2^n+3$; $a_n \geq a_{n+1} \Leftrightarrow -2^n+3 \geq -2^{n+1}+3 \Leftrightarrow -2^n \geq -2^{n+1} \Leftrightarrow 1 \leq 2$ so the sequence is nonincreasing but not bounded below and therefore diverges

47. Let $0 < M < 1$ and let N be an integer greater than $\frac{M}{1-M}$. Then $n > N \Rightarrow n > \frac{M}{1-M} \Rightarrow n - nM > M$ $\Rightarrow n > M + nM \Rightarrow n > M(n+1) \Rightarrow \frac{n}{n+1} > M$.

48. Since M_1 is a least upper bound and M_2 is an upper bound, $M_1 \leq M_2$. Since M_2 is a least upper bound and M_1 is an upper bound, $M_2 \leq M_1$. We conclude that $M_1 = M_2$ so the least upper bound is unique.

49. The sequence $a_n = 1+\frac{(-1)^n}{2}$ is the sequence $\frac{1}{2}, \frac{3}{2}, \frac{1}{2}, \frac{3}{2}, \ldots$. This sequence is bounded above by $\frac{3}{2}$, but it clearly does not converge, by definition of convergence.

50. Let L be the limit of the convergent sequence $\{a_n\}$. Then by definition of convergence, for $\frac{\epsilon}{2}$ there corresponds an N such that for all m and n, $m > N \Rightarrow |a_m - L| < \frac{\epsilon}{2}$ and $n > N \Rightarrow |a_n - L| < \frac{\epsilon}{2}$. Now

$|a_m - a_n| = |a_m - L + L - a_n| \le |a_m - L| + |L - a_n| < \frac{\epsilon}{2} + \frac{\epsilon}{2} = \epsilon$ whenever $m > N$ and $n > N$.

51. Given an $\epsilon > 0$, by definition of convergence there corresponds an N such that for all $n > N$, $|L_1 - a_n| < \epsilon$ and $|L_2 - a_n| < \epsilon$. Now $|L_2 - L_1| = |L_2 - a_n + a_n - L_1| \le |L_2 - a_n| + |a_n - L_1| < \epsilon + \epsilon = 2\epsilon$. $|L_2 - L_1| < 2\epsilon$ says that the difference between two fixed values is smaller than any positive number 2ϵ. The only nonnegative number smaller than every positive number is 0, so $|L_1 - L_2| = 0$ or $L_1 = L_2$.

52. Let k(n) and i(n) be two order-preserving functions whose domains are the set of positive integers and whose ranges are a subset of the positive integers. Consider the two subsequences $a_{k(n)}$ and $a_{i(n)}$, where $a_{k(n)} \to L_1$, $a_{i(n)} \to L_2$ and $L_1 \ne L_2$. Given an $\epsilon > 0$ there corresponds an N_1 such that for $k(n) > N_1$, $|a_{k(n)} - L_1| < \epsilon$, and an N_2 such that for $i(n) > N_2$, $|a_{i(n)} - L_2| < \epsilon$. Let $N = \max\{N_1, N_2\}$. Then for $n > N$, we have that $|a_n - L_1| < \epsilon$ and $|a_n - L_2| < \epsilon$. This implies $a_n \to L_1$ and $a_n \to L_2$ where $L_1 \ne L_2$. Since the limit of a sequence is unique (by Exercise 51), a_n does not converge and hence diverges.

53. $a_{2k} \to L \Leftrightarrow$ given an $\epsilon > 0$ there corresponds an N_1 such that $\left[2k > N_1 \Rightarrow |a_{2k} - L| < \epsilon\right]$. Similarly, $a_{2k+1} \to L \Leftrightarrow \left[2k+1 > N_2 \Rightarrow |a_{2k+1} - L| < \epsilon\right]$. Let $N = \max\{N_1, N_2\}$. Then $n > N \Rightarrow |a_n - L| < \epsilon$ whether n is even or odd, and hence $a_n \to L$.

54. Assume $a_n \to 0$. This implies that given an $\epsilon > 0$ there corresponds an N such that $n > N \Rightarrow |a_n - 0| < \epsilon \Rightarrow |a_n| < \epsilon \Rightarrow ||a_n|| < \epsilon \Rightarrow ||a_n| - 0| < \epsilon \Rightarrow |a_n| \to 0$. On the other hand, assume $|a_n| \to 0$. This implies that given an $\epsilon > 0$ there corresponds an N such that for $n > N$, $||a_n| - 0| < \epsilon \Rightarrow ||a_n|| < \epsilon \Rightarrow |a_n| < \epsilon \Rightarrow |a_n - 0| < \epsilon \Rightarrow a_n \to 0$.

55-66. Example CAS Commands:

Maple:

```
a:= n -> (n)^(1/n);
j:= 9400: k:= 9800: A:= plot(a(n), n=j..k, style=POINT, symbol=CIRCLE):
f:= x -> 0.999: g:= x -> 1.001:
B:= plot({f(x), g(x)}, x=j..k):
with(plots): display({A,B});
```

Mathematica:

```
Clear[a,i,n]
a[n_] = n^(1/n)
atab = Table[ a[i], {i,25} ] // N;
ListPlot[ atab ]
L = Limit[ a[n], n -> Infinity ]

Note: for this a[n], the first n for which |a[n]-L|<0.001 is n = 1!  Let's
find the next...

a[1] - L

First check several orders of magnitude, then zoom in by trial & error:
```

```
Table[ {i, N[a[10^i] - L]}, {i,10} ]
N[a[9000] - L]
N[a[9200] - L]
N[a[9123] - L]
N[a[9124] - L]
```

This is the first n for which |a[n] − L| < 0.001; for 0.0001, we get the rough estimate:

```
N[a[120000] - L]
```

67. Example CAS Commands:

Maple:

```
n:='n':
recur:= proc(f,a0,n) local i,j;
a(0):= evalf(a0);
for i from 1 to n do
a(i):= evalf(f(a(i - 1)))
od;
[[j,a(j)] $j=1..n];
end;
a:= 'a': f:= a -> (1 + r/m)*a + b;
r:= 0.02015; m:= 12; b:= 50;
recur(f,1000,100):
plot(",style=POINT,symbol=CIRCLE):
a(60);
```

Mathematica:

```
Clear[a,r,m,b]
a[n_] := (1 + r/m) a[n - 1] + b
(a)
a[0] = 1000; r = 0.02015; m = 12; b = 50;
atab = Table[ a[i], {i,0,50} ] // N;
ListPlot[ atab ]
a[60]
a[0] = 1000; r = 0.02015; m = 12; b = 50;
ak[n_] := (1 + r/m)^n (a[0] + m b/r) - m b/r
atab = Table[ {a[i],ak[i]}, {i,0,50} ] // N
ak[n + 1] == (1 + r/m) ak[n] + b // Simplify
```

68. Example CAS Commands:

Maple:

```
n:= 'n':
iterate:= proc(f,a0,n) local i,j;
   a(0):= evalf(a0);
   for i from 1 to n do
   a(i):= evalf(f(a(i - 1)))
   od;
   [[j, a(j)] $j= 1..n];
   end;
a:= 'a': f:= a -> r*a*(1 - a);
r:= 3.75;
iterate(f, 0.301, 300):
plot(", style=POINT, symbol=CIRCLE, title='LOGISTIC PLOT, r = 3.75, a = .301');
```

Mathematica:

```
Note: We could define a[n] recursively, but here we need only the first
several values so it's easier to use an iterated function:

Clear[a,r,n,i]
iter[ an_ ] = r an (1 - an)
r = 3/4;
atab = NestList[ iter, 0.3, 100 ];
ListPlot[ atab ]

To plot several lists together:

<< Graphics'MultipleListPlot'
r = 3.65;
MultipleListPlot[
 NestList[ iter, 0.3, 300 ],
 NestList[ iter, 0.301, 300] ]
r = 3.75;
MultipleListPlot[
 NestList[ iter, 0.3, 300 ],
 NestList[ iter, 0.301, 300 ] ]
```

8.2 THEOREMS FOR CALCULATING LIMITS OF SEQUENCES

1. $\lim\limits_{n\to\infty} 2+(0.1)^n = 2 \Rightarrow$ converges (Table 8.1, #4)

2. $\lim\limits_{n\to\infty} \frac{n+(-1)^n}{n} = \lim\limits_{n\to\infty} 1+\frac{(-1)^n}{n} = 1 \Rightarrow$ converges

3. $\lim\limits_{n\to\infty} \frac{1-2n}{1+2n} = \lim\limits_{n\to\infty} \frac{\left(\frac{1}{n}\right)-2}{\left(\frac{1}{n}\right)+2} = \lim\limits_{n\to\infty} \frac{-2}{2} = -1 \Rightarrow$ converges

4. $\lim\limits_{n\to\infty} \frac{2n+1}{1-3\sqrt{n}} = \lim\limits_{n\to\infty} \frac{2+\left(\frac{1}{n}\right)}{\left(\frac{1}{n}-\frac{3}{\sqrt{n}}\right)} = \infty \Rightarrow$ diverges

5. $\lim\limits_{n\to\infty} \frac{1-5n^4}{n^4+8n^3} = \lim\limits_{n\to\infty} \frac{\left(\frac{1}{n^4}\right)-5}{1+\left(\frac{8}{n}\right)} = -5 \Rightarrow$ converges

6. $\lim\limits_{n\to\infty} \frac{n+3}{n^2+5n+6} = \lim\limits_{n\to\infty} \frac{n+3}{(n+3)(n+2)} = \lim\limits_{n\to\infty} \frac{1}{n+2} = 0 \Rightarrow$ converges

7. $\lim\limits_{n\to\infty} \frac{n^2-2n+1}{n-1} = \lim\limits_{n\to\infty} \frac{(n-1)(n-1)}{n-1} = \lim\limits_{n\to\infty} (n-1) = \infty \Rightarrow$ diverges

8. $\lim\limits_{n\to\infty} \frac{1-n^3}{70-4n^2} = \lim\limits_{n\to\infty} \frac{\left(\frac{1}{n^2}\right)-n}{\left(\frac{70}{n^2}\right)-4} = \infty \Rightarrow$ diverges

9. $\lim\limits_{n\to\infty} (1+(-1)^n)$ does not exist $\Rightarrow$ diverges

10. $\lim\limits_{n\to\infty} (-1)^n\left(1-\frac{1}{n}\right)$ does not exist $\Rightarrow$ diverges

11. $\lim\limits_{n\to\infty} \left(\frac{n+1}{2n}\right)\left(1-\frac{1}{n}\right) = \lim\limits_{n\to\infty} \left(\frac{1}{2}+\frac{1}{2n}\right)\left(1-\frac{1}{n}\right) = \frac{1}{2} \Rightarrow$ converges

12. $\lim\limits_{n\to\infty} \left(2-\frac{1}{2^n}\right)\left(3+\frac{1}{2^n}\right) = 6 \Rightarrow$ converges

13. $\lim\limits_{n\to\infty} \frac{(-1)^{n+1}}{2n-1} = 0 \Rightarrow$ converges

14. $\lim\limits_{n\to\infty} \left(-\frac{1}{2}\right)^n = \lim\limits_{n\to\infty} \frac{(-1)^n}{2^n} = 0 \Rightarrow$ converges

15. $\lim\limits_{n\to\infty} \sqrt{\frac{2n}{n+1}} = \sqrt{\lim\limits_{n\to\infty} \frac{2n}{n+1}} = \sqrt{\lim\limits_{n\to\infty} \left(\frac{2}{1+\frac{1}{n}}\right)} = \sqrt{2} \Rightarrow$ converges

16. $\lim\limits_{n\to\infty} \frac{1}{(0.9)^n} = \lim\limits_{n\to\infty} \left(\frac{10}{9}\right)^n = \infty \Rightarrow$ diverges

17. $\lim\limits_{n\to\infty} \sin\left(\frac{\pi}{2}+\frac{1}{n}\right) = \sin\left(\lim\limits_{n\to\infty} \left(\frac{\pi}{2}+\frac{1}{n}\right)\right) = \sin\frac{\pi}{2} = 1 \Rightarrow$ converges

18. $\lim\limits_{n\to\infty} n\pi\cos(n\pi) = \lim\limits_{n\to\infty} (n\pi)(-1)^n$ does not exist $\Rightarrow$ diverges

19. $\lim\limits_{n\to\infty} \frac{\sin n}{n} = 0$ because $-\frac{1}{n} \le \frac{\sin n}{n} \le \frac{1}{n} \Rightarrow$ converges by the Sandwich Theorem for sequences

20. $\lim\limits_{n\to\infty} \frac{\sin^2 n}{2^n} = 0$ because $0 \le \frac{\sin^2 n}{2^n} \le \frac{1}{2^n} \Rightarrow$ converges by the Sandwich Theorem for sequences

21. $\lim\limits_{n\to\infty} \frac{n}{2^n} = \lim\limits_{n\to\infty} \frac{1}{2^n \ln 2} = 0 \Rightarrow$ converges (using l'Hôpital's rule)

22. $\lim\limits_{n\to\infty} \frac{3^n}{n^3} = \lim\limits_{n\to\infty} \frac{3^n \ln 3}{3n^2} = \lim\limits_{n\to\infty} \frac{3^n(\ln 3)^2}{6n} = \lim\limits_{n\to\infty} \frac{3^n(\ln 3)^3}{6} = \infty \Rightarrow$ diverges (using l'Hôpital's rule)

23. $\lim\limits_{n\to\infty} \frac{\ln(n+1)}{\sqrt{n}} = \lim\limits_{n\to\infty} \frac{\left(\frac{1}{n+1}\right)}{\left(\frac{1}{2\sqrt{n}}\right)} = \lim\limits_{n\to\infty} \frac{2\sqrt{n}}{n+1} = \lim\limits_{n\to\infty} \frac{\left(\frac{2}{\sqrt{n}}\right)}{1+\left(\frac{1}{n}\right)} = 0 \Rightarrow$ converges

24. $\lim\limits_{n\to\infty} \frac{\ln n}{\ln 2n} = \lim\limits_{n\to\infty} \frac{\left(\frac{1}{n}\right)}{\left(\frac{2}{2n}\right)} = 1 \Rightarrow$ converges

25. $\lim\limits_{n\to\infty} 8^{1/n} = 1 \Rightarrow$ converges (Table 8.1, #3)

26. $\lim\limits_{n\to\infty} (0.03)^{1/n} = 1 \Rightarrow$ converges (Table 8.1, #3)

27. $\lim\limits_{n\to\infty} \left(1+\frac{7}{n}\right)^n = e^7 \Rightarrow$ converges (Table 8.1, #5)

28. $\lim\limits_{n\to\infty} \left(1 - \frac{1}{n}\right)^n = \lim\limits_{n\to\infty} \left[1 + \frac{(-1)}{n}\right]^n = e^{-1} \Rightarrow$ converges (Table 8.1, #5)

29. $\lim\limits_{n\to\infty} \sqrt[n]{10n} = \lim\limits_{n\to\infty} 10^{1/n} \cdot n^{1/n} = 1 \cdot 1 = 1 \Rightarrow$ converges (Table 8.1, #3 and #2)

30. $\lim\limits_{n\to\infty} \sqrt[n]{n^2} = \lim\limits_{n\to\infty} \left(\sqrt[n]{n}\right)^2 = 1^2 = 1 \Rightarrow$ converges (Table 8.1, #2)

31. $\lim\limits_{n\to\infty} \left(\frac{3}{n}\right)^{1/n} = \frac{\lim\limits_{n\to\infty} 3^{1/n}}{\lim\limits_{n\to\infty} n^{1/n}} = \frac{1}{1} = 1 \Rightarrow$ converges (Table 8.1, #3 and #2)

32. $\lim\limits_{n\to\infty} (n+4)^{1/(n+4)} = \lim\limits_{x\to\infty} x^{1/x} = 1 \Rightarrow$ converges; (let $x = n + 4$, then use Table 8.1, #2)

33. $\lim\limits_{n\to\infty} \frac{\ln n}{n^{1/n}} = \frac{\lim\limits_{n\to\infty} \ln n}{\lim\limits_{n\to\infty} n^{1/n}} = \frac{\infty}{1} = \infty \Rightarrow$ diverges (Table 8.1, #2)

34. $\lim\limits_{n\to\infty} [\ln n - \ln(n+1)] = \lim\limits_{n\to\infty} \ln\left(\frac{n}{n+1}\right) = \ln\left(\lim\limits_{n\to\infty} \frac{n}{n+1}\right) = \ln 1 = 0 \Rightarrow$ converges

35. $\lim\limits_{n\to\infty} \sqrt[n]{4^n n} = \lim\limits_{n\to\infty} 4 \sqrt[n]{n} = 4 \cdot 1 = 4 \Rightarrow$ converges (Table 8.1, #2)

36. $\lim\limits_{n\to\infty} \sqrt[n]{3^{2n+1}} = \lim\limits_{n\to\infty} 3^{2+(1/n)} = \lim\limits_{n\to\infty} 3^2 \cdot 3^{1/n} = 9 \cdot 1 = 9 \Rightarrow$ converges (Table 8.1, #3)

37. $\lim\limits_{n\to\infty} \frac{n!}{n^n} = \lim\limits_{n\to\infty} \frac{1 \cdot 2 \cdot 3 \cdots (n-1)(n)}{n \cdot n \cdot n \cdots n \cdot n} \le \lim\limits_{n\to\infty} \left(\frac{1}{n}\right) = 0$ and $\frac{n!}{n^n} \ge 0 \Rightarrow \lim\limits_{n\to\infty} \frac{n!}{n^n} = 0 \Rightarrow$ converges

38. $\lim\limits_{n\to\infty} \frac{(-4)^n}{n!} = 0 \Rightarrow$ converges (Table 8.1, #6)

39. $\lim\limits_{n\to\infty} \frac{n!}{10^{6n}} = \lim\limits_{n\to\infty} \frac{1}{\left(\frac{(10^6)^n}{n!}\right)} = \infty \Rightarrow$ diverges (Table 8.1, #6)

40. $\lim\limits_{n\to\infty} \frac{n!}{2^n 3^n} = \lim\limits_{n\to\infty} \frac{1}{\left(\frac{6^n}{n!}\right)} = \infty \Rightarrow$ diverges (Table 8.1, #6)

41. $\lim\limits_{n\to\infty} \left(\frac{1}{n}\right)^{1/(\ln n)} = \lim\limits_{n\to\infty} \exp\left(\frac{1}{\ln n} \ln\left(\frac{1}{n}\right)\right) = \lim\limits_{n\to\infty} \exp\left(\frac{\ln 1 - \ln n}{\ln n}\right) = e^{-1} \Rightarrow$ converges

42. $\lim\limits_{n\to\infty} \ln\left(1 + \frac{1}{n}\right)^n = \ln\left(\lim\limits_{n\to\infty} \left(1 + \frac{1}{n}\right)^n\right) = \ln e = 1 \Rightarrow$ converges (Table 8.1, #5)

43. $\lim_{n\to\infty} \left(\frac{3n+1}{3n-1}\right)^n = \lim_{n\to\infty} \exp\left(n \ln\left(\frac{3n+1}{3n-1}\right)\right) = \lim_{n\to\infty} \exp\left(\frac{\ln(3n+1) - \ln(3n-1)}{\frac{1}{n}}\right)$

$= \lim_{n\to\infty} \exp\left(\frac{\frac{3}{3n+1} - \frac{3}{3n-1}}{\left(-\frac{1}{n^2}\right)}\right) = \lim_{n\to\infty} \exp\left(\frac{6n^2}{(3n+1)(3n-1)}\right) = \exp\left(\frac{6}{9}\right) = e^{2/3} \Rightarrow$ converges

44. $\lim_{n\to\infty} \left(\frac{n}{n+1}\right)^n = \lim_{n\to\infty} \exp\left(n \ln\left(\frac{n}{n+1}\right)\right) = \lim_{n\to\infty} \exp\left(\frac{\ln n - \ln(n+1)}{\left(\frac{1}{n}\right)}\right) = \lim_{n\to\infty} \exp\left(\frac{\frac{1}{n} - \frac{1}{n+1}}{\left(-\frac{1}{n^2}\right)}\right)$

$= \lim_{n\to\infty} \exp\left(-\frac{n^2}{n(n+1)}\right) = e^{-1} \Rightarrow$ converges

45. $\lim_{n\to\infty} \left(\frac{x^n}{2n+1}\right)^{1/n} = \lim_{n\to\infty} x\left(\frac{1}{2n+1}\right)^{1/n} = x \lim_{n\to\infty} \exp\left(\frac{1}{n} \ln\left(\frac{1}{2n+1}\right)\right) = x \lim_{n\to\infty} \exp\left(\frac{-\ln(2n+1)}{n}\right)$

$= x \lim_{n\to\infty} \exp\left(\frac{-2}{2n+1}\right) = xe^0 = x,\ x > 0 \Rightarrow$ converges

46. $\lim_{n\to\infty} \left(1 - \frac{1}{n^2}\right)^n = \lim_{n\to\infty} \exp\left(n \ln\left(1 - \frac{1}{n^2}\right)\right) = \lim_{n\to\infty} \exp\left(\frac{\ln\left(1 - \frac{1}{n^2}\right)}{\left(\frac{1}{n}\right)}\right) = \lim_{n\to\infty} \exp\left[\frac{\left(\frac{2}{n^3}\right)/\left(1 - \frac{1}{n^2}\right)}{\left(-\frac{1}{n^2}\right)}\right]$

$= \lim_{n\to\infty} \exp\left(\frac{-2n}{n^2-1}\right) = e^0 = 1 \Rightarrow$ converges

47. $\lim_{n\to\infty} \frac{3^n \cdot 6^n}{2^{-n} \cdot n!} = \lim_{n\to\infty} \frac{36^n}{n!} = 0 \Rightarrow$ converges (Table 8.1, #6)

48. $\lim_{n\to\infty} \frac{\left(\frac{10}{11}\right)^n}{\left(\frac{9}{10}\right)^n + \left(\frac{11}{12}\right)^n} = \lim_{n\to\infty} \frac{\left(\frac{12}{11}\right)^n\left(\frac{10}{11}\right)^n}{\left(\frac{12}{11}\right)^n\left(\frac{9}{10}\right)^n + \left(\frac{12}{11}\right)^n\left(\frac{11}{12}\right)^n} = \lim_{n\to\infty} \frac{\left(\frac{120}{121}\right)^n}{\left(\frac{108}{110}\right)^n + 1} = 0 \Rightarrow$ converges

(Table 8.1, #4)

49. $\lim_{n\to\infty} \tanh n = \lim_{n\to\infty} \frac{e^n - e^{-n}}{e^n + e^{-n}} = \lim_{n\to\infty} \frac{e^{2n} - 1}{e^{2n} + 1} = \lim_{n\to\infty} \frac{2e^{2n}}{2e^{2n}} = \lim_{n\to\infty} 1 = 1 \Rightarrow$ converges

50. $\lim_{n\to\infty} \sinh(\ln n) = \lim_{n\to\infty} \frac{e^{\ln n} - e^{-\ln n}}{2} = \lim_{n\to\infty} \frac{n - \left(\frac{1}{n}\right)}{2} = \infty \Rightarrow$ diverges

51. $\lim_{n\to\infty} \frac{n^2 \sin\left(\frac{1}{n}\right)}{2n-1} = \lim_{n\to\infty} \frac{\sin\left(\frac{1}{n}\right)}{\left(\frac{2}{n} - \frac{1}{n^2}\right)} = \lim_{n\to\infty} \frac{-\left(\cos\left(\frac{1}{n}\right)\right)\left(\frac{1}{n^2}\right)}{\left(-\frac{2}{n^2} + \frac{2}{n^3}\right)} = \lim_{n\to\infty} \frac{-\cos\left(\frac{1}{n}\right)}{-2 + \left(\frac{2}{n}\right)} = \frac{1}{2} \Rightarrow$ converges

52. $\lim\limits_{n\to\infty} n\left(1-\cos\frac{1}{n}\right) = \lim\limits_{n\to\infty} \frac{\left(1-\cos\frac{1}{n}\right)}{\left(\frac{1}{n}\right)} = \lim\limits_{n\to\infty} \frac{\left[\sin\left(\frac{1}{n}\right)\right]\left(\frac{1}{n^2}\right)}{\left(\frac{1}{n^2}\right)} = \lim\limits_{n\to\infty} \sin\left(\frac{1}{n}\right) = 0 \Rightarrow$ converges

53. $\lim\limits_{n\to\infty} \tan^{-1} n = \frac{\pi}{2} \Rightarrow$ converges

54. $\lim\limits_{n\to\infty} \frac{1}{\sqrt{n}} \tan^{-1} n = 0 \cdot \frac{\pi}{2} = 0 \Rightarrow$ converges

55. $\lim\limits_{n\to\infty} \left(\frac{1}{3}\right)^n + \frac{1}{\sqrt{2^n}} = \lim\limits_{n\to\infty} \left(\left(\frac{1}{3}\right)^n + \left(\frac{1}{\sqrt{2}}\right)^n\right) = 0 \Rightarrow$ converges (Table 8.1, #4)

56. $\lim\limits_{n\to\infty} \sqrt[n]{n^2+n} = \lim\limits_{n\to\infty} \exp\left[\frac{\ln(n^2+n)}{n}\right] = \lim\limits_{n\to\infty} \exp\left(\frac{2n+1}{n^2+n}\right) = e^0 = 1 \Rightarrow$ converges

57. $\lim\limits_{n\to\infty} \frac{(\ln n)^{200}}{n} = \lim\limits_{n\to\infty} \frac{200(\ln n)^{199}}{n} = \lim\limits_{n\to\infty} \frac{200 \cdot 199(\ln n)^{198}}{n} = \ldots = \lim\limits_{n\to\infty} \frac{200!}{n} = 0 \Rightarrow$ converges

58. $\lim\limits_{n\to\infty} \frac{(\ln n)^5}{\sqrt{n}} = \lim\limits_{n\to\infty} \left[\frac{\left(\frac{5(\ln n)^4}{n}\right)}{\left(\frac{1}{2\sqrt{n}}\right)}\right] = \lim\limits_{n\to\infty} \frac{10(\ln n)^4}{\sqrt{n}} = \lim\limits_{n\to\infty} \frac{80(\ln n)^3}{\sqrt{n}} = \ldots = \lim\limits_{n\to\infty} \frac{3840}{\sqrt{n}} = 0 \Rightarrow$ converges

59. $\lim\limits_{n\to\infty} \left(n-\sqrt{n^2-n}\right) = \lim\limits_{n\to\infty} \left(n-\sqrt{n^2-n}\right)\left(\frac{n+\sqrt{n^2-n}}{n+\sqrt{n^2-n}}\right) = \lim\limits_{n\to\infty} \frac{n}{n+\sqrt{n^2-n}} = \lim\limits_{n\to\infty} \frac{1}{1+\sqrt{1-\frac{1}{n}}}$
$= \frac{1}{2} \Rightarrow$ converges

60. $\lim\limits_{n\to\infty} \frac{1}{\sqrt{n^2-1}-\sqrt{n^2+n}} = \lim\limits_{n\to\infty} \left(\frac{1}{\sqrt{n^2-1}-\sqrt{n^2+n}}\right)\left(\frac{\sqrt{n^2-1}+\sqrt{n^2+n}}{\sqrt{n^2-1}+\sqrt{n^2+n}}\right) = \lim\limits_{n\to\infty} \frac{\sqrt{n^2-1}+\sqrt{n^2+n}}{-1-n}$
$= \lim\limits_{n\to\infty} \frac{\sqrt{1-\frac{1}{n^2}}+\sqrt{1+\frac{1}{n}}}{\left(-\frac{1}{n}-1\right)} = -2 \Rightarrow$ converges

61. $\lim\limits_{n\to\infty} \frac{1}{n} \int_1^n \frac{1}{x}\,dx = \lim\limits_{n\to\infty} \frac{\ln n}{n} = \lim\limits_{n\to\infty} \frac{1}{n} = 0 \Rightarrow$ converges (Table 8.1, #1)

62. $\lim\limits_{n\to\infty} \int_1^n \frac{1}{x^p}\,dx = \lim\limits_{n\to\infty} \left[\frac{1}{1-p}\frac{1}{x^{p-1}}\right]_1^n = \lim\limits_{n\to\infty} \frac{1}{1-p}\left(\frac{1}{n^{p-1}}-1\right) = \frac{1}{p-1} \Rightarrow$ converges when $p > 1$;
diverges for $p \le 1$

63. $1, 1, 2, 4, 8, 16, 32, \ldots = 1, 2^0, 2^1, 2^2, 2^3, 2^4, 2^5, \ldots \Rightarrow x_1 = 1$ and $x_n = 2^{n-2}$ for $n \ge 2$

64. (a) $1^2 - 2(1)^2 = -1$, $3^2 - 2(2)^2 = 1$; let $f(a,b) = (a+2b)^2 - 2(a+b)^2 = a^2 + 4ab + 4b^2 - 2a^2 - 4ab - 2b^2$ $= 2b^2 - a^2$; $a^2 - 2b^2 = -1 \Rightarrow f(a,b) = 2b^2 - a^2 = 1$; $a^2 - 2b^2 = 1 \Rightarrow f(a,b) = 2b^2 - a^2 = -1$

(b) $r_n^2 - 2 = \left(\frac{a+2b}{a+b}\right)^2 - 2 = \frac{a^2 + 4ab + 4b^2 - 2a^2 - 4ab - 2b^2}{(a+b)^2} = \frac{-(a^2 - 2b^2)}{(a+b)^2} = \frac{\pm 1}{y_n^2} \Rightarrow r_n = \sqrt{2 \pm \left(\frac{1}{y_n}\right)^2}$

In the first and second fractions, $y_n \geq n$. Let $\frac{a}{b}$ represent the $(n-1)$th fraction where $\frac{a}{b} \geq 1$ and $b \geq n-1$ for n a positive integer ≥ 3. Now the nth fraction is $\frac{a+2b}{a+b}$ and $a + b \geq 2b \geq 2n - 2 \geq n \Rightarrow y_n \geq n$. Thus, $\lim_{n\to\infty} r_n = \sqrt{2}$.

65. (a) $f(x) = x^2 - 2$; the sequence converges to $1.414213562 \approx \sqrt{2}$

(b) $f(x) = \tan(x) - 1$; the sequence converges to $0.7853981635 \approx \frac{\pi}{4}$

(c) $f(x) = e^x$; the sequence 1, 0, −1, −2, −3, −4, −5, ... diverges

66. (a) $\lim_{n\to\infty} nf\left(\frac{1}{n}\right) = \lim_{\Delta x \to 0^+} \frac{f(\Delta x)}{\Delta x} = \lim_{\Delta x \to 0^+} \frac{f(0 + \Delta x) - f(0)}{\Delta x} = f'(0)$, where $\Delta x = \frac{1}{n}$

(b) $\lim_{n\to\infty} n \tan^{-1}\left(\frac{1}{n}\right) = f'(0) = \frac{1}{1+0^2} = 1$, $f(x) = \tan^{-1} x$

(c) $\lim_{n\to\infty} n\left(e^{1/n} - 1\right) = f'(0) = e^0 = 1$, $f(x) = e^x$

(d) $\lim_{n\to\infty} n \ln\left(1 + \frac{2}{n}\right) = f'(0) = \frac{2}{1 + 2(0)} = 2$, $f(x) = \ln(1 + 2x)$

67. (a) If $a = 2n + 1$, then $b = \lfloor \frac{a^2}{2} \rfloor = \lfloor \frac{4n^2 + 4n + 1}{2} \rfloor = \lfloor 2n^2 + 2n + \frac{1}{2} \rfloor = 2n^2 + 2n$, $c = \lceil \frac{a^2}{2} \rceil = \lceil 2n^2 + 2n + \frac{1}{2} \rceil$ $= 2n^2 + 2n + 1$ and $a^2 + b^2 = (2n+1)^2 + \left(2n^2 + 2n\right)^2 = 4n^2 + 4n + 1 + 4n^4 + 8n^3 + 4n^2$ $= 4n^4 + 8n^3 + 8n^2 + 4n + 1 = \left(2n^2 + 2n + 1\right)^2 = c^2$.

(b) $\lim_{a\to\infty} \frac{\lfloor \frac{a^2}{2} \rfloor}{\lceil \frac{a^2}{2} \rceil} = \lim_{a\to\infty} \frac{2n^2 + 2n}{2n^2 + 2n + 1} = 1$ or $\lim_{a\to\infty} \frac{\lfloor \frac{a^2}{2} \rfloor}{\lceil \frac{a^2}{2} \rceil} = \lim_{a\to\infty} \sin\theta = \lim_{\theta \to \pi/2} \sin\theta = 1$

68. (a) $\lim_{n\to\infty} (2n\pi)^{1/(2n)} = \lim_{n\to\infty} \exp\left(\frac{\ln 2n\pi}{2n}\right) = \lim_{n\to\infty} \exp\left(\frac{\left(\frac{2\pi}{2n\pi}\right)}{2}\right) = \lim_{n\to\infty} \exp\left(\frac{1}{2n}\right) = e^0 = 1$;

$n! \approx \left(\frac{n}{e}\right)^n \sqrt{2n\pi}$, Stirlings approximation $\Rightarrow \sqrt[n]{n!} \approx \left(\frac{n}{e}\right)(2n\pi)^{1/(2n)} \approx \frac{n}{e}$ for large values of n

(b)

n	$\sqrt[n]{n!}$	$\frac{n}{e}$
40	15.76852702	14.71517765
50	19.48325423	18.39397206
60	23.19189561	22.07276647

69. (a) $\lim_{n\to\infty} \frac{\ln n}{n^c} = \lim_{n\to\infty} \frac{\left(\frac{1}{n}\right)}{cn^{c-1}} = \lim_{n\to\infty} \frac{1}{cn^c} = 0$

(b) For all $\epsilon > 0$, there exists an $N = e^{-(\ln \epsilon)/c}$ such that $n > e^{-(\ln \epsilon)/c} \Rightarrow \ln n > -\frac{\ln \epsilon}{c} \Rightarrow \ln n^c > \ln\left(\frac{1}{\epsilon}\right)$

$\Rightarrow n^c > \frac{1}{\epsilon} \Rightarrow \frac{1}{n^c} < \epsilon \Rightarrow \left|\frac{1}{n^c} - 0\right| < \epsilon \Rightarrow \lim_{n\to\infty} \frac{1}{n^c} = 0$

70. Let $\{a_n\}$ and $\{b_n\}$ be sequences both converging to L. Define $\{c_n\}$ by $c_{2n} = b_n$ and $c_{2n-1} = a_n$, where $n = 1, 2, 3, \ldots$. For all $\epsilon > 0$ there exists N_1 such that when $n > N_1$ then $|a_n - L| < \epsilon$ and there exists N_2 such that when $n > N_2$ then $|b_n - L| < \epsilon$. If $n > \max\{N_1, N_2\}$, then both inequalities hold and hence $|c_n - L| < \epsilon$, so $\{c_n\}$ converges to L.

71. $\lim_{n\to\infty} n^{1/n} = \lim_{n\to\infty} \exp\left(\frac{1}{n} \ln n\right) = \lim_{n\to\infty} \exp\left(\frac{1}{n}\right) = e^0 = 1$

72. $\lim_{n\to\infty} x^{1/n} = \lim_{n\to\infty} \exp\left(\frac{1}{n} \ln x\right) = e^0 = 1$, because x remains fixed while n gets large

73. Assume the hypotheses of the theorem and let ϵ be a positive number. For all ϵ there exists a N_1 such that when $n > N_1$ then $|a_n - L| < \epsilon \Rightarrow -\epsilon < a_n - L < \epsilon \Rightarrow L - \epsilon < a_n$, and there exists a N_2 such that when $n > N_2$ then $|c_n - L| < \epsilon \Rightarrow -\epsilon < c_n - L < \epsilon \Rightarrow c_n < L + \epsilon$. If $n > \max\{N_1, N_2\}$, then $L - \epsilon < a_n \le b_n \le c_n < L + \epsilon \Rightarrow |b_n - L| < \epsilon \Rightarrow \lim_{n\to\infty} b_n = L$.

74. $|a_n - L| < \delta \Rightarrow |f(a_n) - f(L)| < \epsilon \Rightarrow f(a_n) \to f(L)$

75. $g(x) = \sqrt{x}$; $2 \to 1.00000132$ in 20 iterations; $.1 \to 0.9999956$ in 20 iterations; a root is 1

76. $g(x) = x^2$; $x_0 = .5 \to 0.0000152$ in 5 iterations; $-.5 \to 0.0000152$ in 5 iterations; a root is 0

77. $g(x) = -\cos x$; $x_0 = .1 \to 0.73908456$ in 35 iterations

78. $g(x) = \cos x - 1$; $x_0 = .1 \to 0$ in 4 iterations

79. $g(x) = 0.1 + \sin x$; $x_0 = -2 \to 0.853748068$ in 43 iterations

80. $g(x) = \left(4 - \sqrt{1+x}\right)^2$; $x_0 = 3.5 \to 3.51562548$ in 85 iterations

81. $x_0 =$ initial guess $> 0 \Rightarrow x_1 = \sqrt{x_0} = (x_0)^{1/2} \Rightarrow x_2 = \sqrt{x_0^{1/2}} = x_0^{1/4}, \ldots \Rightarrow x_n = x_0^{1/(2n)} \Rightarrow x_n \to 1$ as $n \to \infty$

82. $x_0 =$ initial guess $\Rightarrow x_1 = x_0^2 \Rightarrow x_2 = \left(x_0^2\right)^2 = x_0^4, \ldots \Rightarrow x_n = x_0^{2n}$; $|x_0| < 1 \Rightarrow x_n \to 0$ as $n \to \infty$; $|x_0| > 1 \Rightarrow x_n \to \infty$ as $n \to \infty$

83. $g(x) = 2x + 3 \Rightarrow g^{-1}(x) = \frac{x-3}{2}$ and when the iterative method is applied to $g^{-1}(x)$ we have $x_0 = 2$ $\to -2.99999881$ in 23 iterations $\Rightarrow -3$ is the fixed point

84. $g(x) = 1 - 4x \Rightarrow g^{-1}(x) = \frac{1-x}{4}$ and when the iterative method is applied to $g^{-1}(x)$ we have $x_0 = 2$ $\to 0.199999571$ in 12 iterations $\Rightarrow 0.2$ is the fixed point

8.3 INFINITE SERIES

1. $s_n = \frac{a(1-r^n)}{(1-r)} = \frac{2\left(1-\left(\frac{1}{3}\right)^n\right)}{1-\left(\frac{1}{3}\right)} \Rightarrow \lim_{n\to\infty} s_n = \frac{2}{1-\left(\frac{1}{3}\right)} = 3$

2. $s_n = \frac{a(1-r^n)}{(1-r)} = \frac{\left(\frac{9}{100}\right)\left(1-\left(\frac{1}{100}\right)^n\right)}{1-\left(\frac{1}{100}\right)} \Rightarrow \lim_{n\to\infty} s_n = \frac{\left(\frac{9}{100}\right)}{1-\left(\frac{1}{100}\right)} = \frac{1}{11}$

3. $s_n = \frac{a(1-r^n)}{(1-r)} = \frac{1-\left(-\frac{1}{2}\right)^n}{1-\left(-\frac{1}{2}\right)} \Rightarrow \lim_{n\to\infty} s_n = \frac{1}{\left(\frac{3}{2}\right)} = \frac{2}{3}$

4. $s_n = \frac{1-(-2)^n}{1-(-2)}$, a geometric series where $|r| > 1 \Rightarrow$ divergence

5. $\frac{1}{(n+1)(n+2)} = \frac{1}{n+1} - \frac{1}{n+2} \Rightarrow s_n = \left(\frac{1}{2}-\frac{1}{3}\right)+\left(\frac{1}{3}-\frac{1}{4}\right)+\ldots+\left(\frac{1}{n+1}-\frac{1}{n+2}\right) = \frac{1}{2}-\frac{1}{n+2} \Rightarrow \lim_{n\to\infty} s_n = \frac{1}{2}$

6. $\frac{5}{n(n+1)} = \frac{5}{n} - \frac{5}{n+1} \Rightarrow s_n = \left(5-\frac{5}{2}\right)+\left(\frac{5}{2}-\frac{5}{3}\right)+\left(\frac{5}{3}-\frac{5}{4}\right)+\ldots+\left(\frac{5}{n-1}-\frac{5}{n}\right)+\left(\frac{5}{n}-\frac{5}{n+1}\right) = 5-\frac{5}{n+1}$

$\Rightarrow \lim_{n\to\infty} s_n = 5$

7. $1-\frac{1}{4}+\frac{1}{16}-\frac{1}{64}+\ldots$, the sum of this geometric series is $\frac{1}{1-\left(-\frac{1}{4}\right)} = \frac{1}{1+\left(\frac{1}{4}\right)} = \frac{4}{5}$

8. $\frac{1}{16}+\frac{1}{64}+\frac{1}{256}+\ldots$, the sum of this geometric series is $\frac{\left(\frac{1}{16}\right)}{1-\left(\frac{1}{4}\right)} = \frac{1}{12}$

9. $\frac{7}{4}+\frac{7}{16}+\frac{7}{64}+\ldots$, the sum of this geometric series is $\frac{\left(\frac{7}{4}\right)}{1-\left(\frac{1}{4}\right)} = \frac{7}{3}$

10. $5-\frac{5}{4}+\frac{5}{16}-\frac{5}{64}+\ldots$, the sum of this geometric series is $\frac{5}{1-\left(-\frac{1}{4}\right)} = 4$

11. $(5+1)+\left(\frac{5}{2}+\frac{1}{3}\right)+\left(\frac{5}{4}+\frac{1}{9}\right)+\left(\frac{5}{8}+\frac{1}{27}\right)+\ldots$, is the sum of two geometric series; the sum is

$\frac{5}{1-\left(\frac{1}{2}\right)} + \frac{1}{1-\left(\frac{1}{3}\right)} = 10+\frac{3}{2} = \frac{23}{2}$

12. $(5-1)+\left(\frac{5}{2}-\frac{1}{3}\right)+\left(\frac{5}{4}-\frac{1}{9}\right)+\left(\frac{5}{8}-\frac{1}{27}\right)+\ldots$, is the difference of two geometric series; the sum is

$\frac{5}{1-\left(\frac{1}{2}\right)} - \frac{1}{1-\left(\frac{1}{3}\right)} = 10-\frac{3}{2} = \frac{17}{2}$

13. $(1+1)+\left(\frac{1}{2}-\frac{1}{5}\right)+\left(\frac{1}{4}+\frac{1}{25}\right)+\left(\frac{1}{8}-\frac{1}{125}\right)+\ldots$, is the sum of two geometric series; the sum is

$$\frac{1}{1-\left(\frac{1}{2}\right)}+\frac{1}{1+\left(\frac{1}{5}\right)}=2+\frac{5}{6}=\frac{17}{6}$$

14. $2+\frac{4}{5}+\frac{8}{25}+\frac{16}{125}+\ldots=2\left(1+\frac{2}{5}+\frac{4}{25}+\frac{8}{125}+\ldots\right)$; the sum of this geometric series is $2\left(\frac{1}{1-\left(\frac{2}{5}\right)}\right)=\frac{10}{3}$

15. $\frac{4}{(4n-3)(4n+1)}=\frac{1}{4n-3}-\frac{1}{4n+1} \Rightarrow s_n=\left(1-\frac{1}{5}\right)+\left(\frac{1}{5}-\frac{1}{9}\right)+\left(\frac{1}{9}-\frac{1}{13}\right)+\ldots+\left(\frac{1}{4n-7}-\frac{1}{4n-3}\right)$

$+\left(\frac{1}{4n-3}-\frac{1}{4n+1}\right)=1-\frac{1}{4n+1} \Rightarrow \lim\limits_{n\to\infty} s_n=\lim\limits_{n\to\infty}\left(1-\frac{1}{4n+1}\right)=1$

16. $\frac{6}{(2n-1)(2n+1)}=\frac{A}{2n-1}+\frac{B}{2n+1}=\frac{A(2n+1)+B(2n-1)}{(2n-1)(2n+1)} \Rightarrow A(2n+1)+B(2n-1)=6$

$\Rightarrow (2A+2B)n+(A-B)=6 \Rightarrow \begin{cases} 2A+2B=0 \\ A-\ \ B=6 \end{cases} \Rightarrow \begin{cases} A+B=0 \\ A-B=6 \end{cases} \Rightarrow 2A=6 \Rightarrow A=3$ and $B=-3$. Hence,

$\sum\limits_{n=1}^{k} \frac{6}{(2n-1)(2n+1)}=3\sum\limits_{n=1}^{k}\left(\frac{1}{2n-1}-\frac{1}{2n+1}\right)=3\left(\frac{1}{1}-\frac{1}{3}+\frac{1}{3}-\frac{1}{5}+\frac{1}{5}-\frac{1}{7}+\ldots-\frac{1}{2(k-1)+1}+\frac{1}{2k-1}-\frac{1}{2k+1}\right)$

$=3\left(1-\frac{1}{2k+1}\right) \Rightarrow$ the sum is $\lim\limits_{k\to\infty} 3\left(1-\frac{1}{2k+1}\right)=3$

17. $\frac{40n}{(2n-1)^2(2n+1)^2}=\frac{A}{(2n-1)}+\frac{B}{(2n-1)^2}+\frac{C}{(2n+1)}+\frac{D}{(2n+1)^2}$

$=\frac{A(2n-1)(2n+1)^2+B(2n+1)^2+C(2n+1)(2n-1)^2+D(2n-1)^2}{(2n-1)^2(2n+1)^2}$

$\Rightarrow A(2n-1)(2n+1)^2+B(2n+1)^2+C(2n+1)(2n-1)^2+D(2n-1)^2=40n$

$\Rightarrow A\left(8n^3+4n^2-2n-1\right)+B\left(4n^2+4n+1\right)+C\left(8n^3-4n^2-2n+1\right)=D\left(4n^2-4n+1\right)=40n$

$\Rightarrow (8A+8C)n^3+(4A+4B-4C+4D)n^2+(-2A+4B-2C-4D)n+(-A+B+C+D)=40n$

$\Rightarrow \begin{cases} 8A+8C=\ 0 \\ 4A+4B-4C+4D=\ 0 \\ -2A+4B-2C-4D=40 \\ -A+\ B+\ C+\ D=\ 0 \end{cases} \Rightarrow \begin{cases} 8A+8C=\ 0 \\ A+\ B-C+\ D=\ 0 \\ -A+2B-C-2D=20 \\ -A+\ B+C+\ D=\ 0 \end{cases} \Rightarrow \begin{cases} B+\ D=\ 0 \\ 2B-2D=20 \end{cases} \Rightarrow 4B=20 \Rightarrow B=5$ and

$D=-5 \Rightarrow \begin{cases} A+C=0 \\ -A+5+C-5=0 \end{cases} \Rightarrow C=0$ and $A=0$. Hence, $\sum\limits_{n=1}^{k}\left[\frac{40n}{(2n-1)^2(2n+1)^2}\right]$

$=5\sum\limits_{n=1}^{k}\left[\frac{1}{(2n-1)^2}-\frac{1}{(2n+1)^2}\right]=5\left(\frac{1}{1}-\frac{1}{9}+\frac{1}{9}-\frac{1}{25}+\frac{1}{25}-\ldots-\frac{1}{(2(k-1)+1)^2}+\frac{1}{(2k-1)^2}-\frac{1}{(2k+1)^2}\right)$

$=5\left(1-\frac{1}{(2k+1)^2}\right) \Rightarrow$ the sum is $\lim\limits_{n\to\infty} 5\left(1-\frac{1}{(2k+1)^2}\right)=5$

18. $\frac{2n+1}{n^2(n+1)^2} = \frac{1}{n^2} - \frac{1}{(n+1)^2} \Rightarrow s_n = \left(1 - \frac{1}{4}\right) + \left(\frac{1}{4} - \frac{1}{9}\right) + \left(\frac{1}{9} - \frac{1}{16}\right) + \ldots + \left[\frac{1}{(n-1)^2} - \frac{1}{n^2}\right] + \left[\frac{1}{n^2} - \frac{1}{(n+1)^2}\right]$

$\Rightarrow \lim_{n\to\infty} s_n = \lim_{n\to\infty} \left[1 - \frac{1}{(n+1)^2}\right] = 1$

19. $s_n = \left(1 - \frac{1}{\sqrt{2}}\right) + \left(\frac{1}{\sqrt{2}} - \frac{1}{\sqrt{3}}\right) + \left(\frac{1}{\sqrt{3}} - \frac{1}{\sqrt{4}}\right) + \ldots + \left(\frac{1}{\sqrt{n-1}} + \frac{1}{\sqrt{n}}\right) + \left(\frac{1}{\sqrt{n}} - \frac{1}{\sqrt{n+1}}\right) = 1 - \frac{1}{\sqrt{n+1}}$

$\Rightarrow \lim_{n\to\infty} s_n = \lim_{n\to\infty} \left(1 - \frac{1}{\sqrt{n+1}}\right) = 1$

20. $s_n = \left(\frac{1}{2} - \frac{1}{2^{1/2}}\right) + \left(\frac{1}{2^{1/2}} - \frac{1}{2^{1/3}}\right) + \left(\frac{1}{2^{1/3}} - \frac{1}{2^{1/4}}\right) + \ldots + \left(\frac{1}{2^{1/(n-1)}} - \frac{1}{2^{1/n}}\right) + \left(\frac{1}{2^{1/n}} - \frac{1}{2^{1/(n+1)}}\right) = \frac{1}{2} - \frac{1}{2^{1/(n+1)}}$

$\Rightarrow \lim_{n\to\infty} s_n = \frac{1}{2} - \frac{1}{1} = -\frac{1}{2}$

21. $s_n = \left(\frac{1}{\ln 3} - \frac{1}{\ln 2}\right) + \left(\frac{1}{\ln 4} - \frac{1}{\ln 3}\right) + \left(\frac{1}{\ln 5} - \frac{1}{\ln 4}\right) + \ldots + \left(\frac{1}{\ln(n+1)} - \frac{1}{\ln n}\right) + \left(\frac{1}{\ln(n+2)} - \frac{1}{\ln(n+1)}\right)$

$= -\frac{1}{\ln 2} + \frac{1}{\ln(n+2)} \Rightarrow \lim_{n\to\infty} s_n = -\frac{1}{\ln 2}$

22. $s_n = \left[\tan^{-1}(1) - \tan^{-1}(2)\right] + \left[\tan^{-1}(2) - \tan^{-1}(3)\right] + \ldots + \left[\tan^{-1}(n-1) - \tan^{-1}(n)\right]$

$+ \left[\tan^{-1}(n) - \tan^{-1}(n+1)\right] = \tan^{-1}(1) - \tan^{-1}(n+1) \Rightarrow \lim_{n\to\infty} s_n = \tan^{-1}(1) - \frac{\pi}{2} = \frac{\pi}{4} - \frac{\pi}{2} = -\frac{\pi}{4}$

23. convergent geometric series with sum $\frac{1}{1 - \left(\frac{1}{\sqrt{2}}\right)} = \frac{\sqrt{2}}{\sqrt{2} - 1} = 2 + \sqrt{2}$

24. divergent geometric series with $|r| = \sqrt{2} > 1$

25. convergent geometric series with sum $\frac{\left(\frac{3}{2}\right)}{1 - \left(-\frac{1}{2}\right)} = 1$

26. $\lim_{n\to\infty} (-1)^{n+1} n \neq 0 \Rightarrow$ diverges

27. $\lim_{n\to\infty} \cos(n\pi) = \lim_{n\to\infty} (-1)^n \neq 0 \Rightarrow$ diverges

28. $\cos(n\pi) = (-1)^n \Rightarrow$ convergent geometric series with sum $\frac{1}{1 - \left(-\frac{1}{5}\right)} = \frac{5}{6}$

29. convergent geometric series with sum $\frac{1}{1 - \left(\frac{1}{e^2}\right)} = \frac{e^2}{e^2 - 1}$

30. $\lim_{n\to\infty} \ln \frac{1}{n} = -\infty \neq 0 \Rightarrow$ diverges

31. convergent geometric series with sum $\frac{2}{1 - \left(\frac{1}{10}\right)} - 2 = \frac{20}{9} - \frac{18}{9} = \frac{2}{9}$

32. convergent geometric series with sum $\frac{1}{1 - \left(\frac{1}{x}\right)} = \frac{x}{x - 1}$

33. difference of two geometric series with sum $\frac{1}{1-\left(\frac{2}{3}\right)} - \frac{1}{1-\left(\frac{1}{3}\right)} = 3 - \frac{3}{2} = \frac{3}{2}$

34. $\lim\limits_{n\to\infty} \left(1-\frac{1}{n}\right)^n = \lim\limits_{n\to\infty} \left(1+\frac{-1}{n}\right)^n = e^{-1} \neq 0 \Rightarrow$ diverges

35. $\lim\limits_{n\to\infty} \frac{n!}{1000^n} = \infty \neq 0 \Rightarrow$ diverges

36. $\lim\limits_{n\to\infty} \frac{n^n}{n!} = \lim\limits_{n\to\infty} \frac{n \cdot n \cdots n}{1 \cdot 2 \cdots n} > \lim\limits_{n\to\infty} n = \infty \Rightarrow$ diverges

37. $\sum\limits_{n=1}^{\infty} \ln\left(\frac{n}{n+1}\right) = \sum\limits_{n=1}^{\infty} [\ln(n) - \ln(n+1)] \Rightarrow s_n = [\ln(1) - \ln(2)] + [\ln(2) - \ln(3)] + [\ln(3) - \ln(4)] + \ldots$
$+ [\ln(n-1) - \ln(n)] + [\ln(n) - \ln(n+1)] = \ln(1) - \ln(n+1) = -\ln(n+1) \Rightarrow \lim\limits_{n\to\infty} s_n = -\infty, \Rightarrow$ diverges

38. $\lim\limits_{n\to\infty} a_n = \lim\limits_{n\to\infty} \ln\left(\frac{n}{2n+1}\right) = \ln\left(\frac{1}{2}\right) \neq 0 \Rightarrow$ diverges

39. convergent geometric series with sum $\frac{1}{1-\left(\frac{e}{\pi}\right)} = \frac{\pi}{\pi - e}$

40. divergent geometric series with $|r| = \frac{e^\pi}{\pi^e} \approx \frac{23.141}{22.459} > 1$

41. $\sum\limits_{n=0}^{\infty} (-1)^n x^n = \sum\limits_{n=0}^{\infty} (-x)^n$; $a = 1$, $r = -x$; converges to $\frac{1}{1-(-x)} = \frac{1}{1+x}$ for $|x| < 1$

42. $\sum\limits_{n=0}^{\infty} (-1)^n x^{2n} = \sum\limits_{n=0}^{\infty} \left(-x^2\right)^n$; $a = 1$, $r = -x^2$; converges to $\frac{1}{1+x^2}$ for $|x| < 1$

43. $a = 3$, $r = \frac{x-1}{2}$; converges to $\frac{3}{1-\left(\frac{x-1}{2}\right)} = \frac{6}{3-x}$ for $-1 < \frac{x-1}{2} < 1$ or $-1 < x < 3$

44. $\sum\limits_{n=0}^{\infty} \frac{(-1)^n}{2}\left(\frac{1}{3+\sin x}\right)^n = \sum\limits_{n=0}^{\infty} \frac{1}{2}\left(\frac{-1}{3+\sin x}\right)^n$; $a = \frac{1}{2}$, $r = \frac{-1}{3+\sin x}$; converges to $\frac{\left(\frac{1}{2}\right)}{1-\left(\frac{-1}{3+\sin x}\right)}$
$= \frac{3+\sin x}{2(4+\sin x)} = \frac{3+\sin x}{8+2\sin x}$ for all x $\left(\text{since } \frac{1}{4} \leq \frac{1}{3+\sin x} \leq \frac{1}{2} \text{ for all } x\right)$

45. $a = 1$, $r = 2x$; converges to $\frac{1}{1-2x}$ for $|2x| < 1$ or $|x| < \frac{1}{2}$

46. $a = 1$, $r = -\frac{1}{x^2}$; converges to $\frac{1}{1-\left(\frac{-1}{x^2}\right)} = \frac{x^2}{x^2+1}$ for $\left|x^2\right| < 1$ or $|x| < 1$

47. $a = 1$, $r = -(x+1)^n$; converges to $\frac{1}{1+(x+1)} = \frac{1}{2+x}$ for $|x+1| < 1$ or $-2 < x < 0$

48. $a = 1$, $r = \frac{3-x}{2}$; converges to $\frac{1}{1-\left(\frac{3-x}{2}\right)} = \frac{2}{x-1}$ for $\left|\frac{3-x}{2}\right| < 1$ or $1 < x < 5$

49. $a = 1$, $r = \sin x$; converges to $\frac{1}{1-\sin x}$ for $x \neq (2k+1)\frac{\pi}{2}$, k an integer

50. $a = 1$, $r = \ln x$; converges to $\frac{1}{1 - \ln x}$ for $|\ln x| < 1$ or $e^{-1} < x < e$

51. $0.\overline{23} = \sum_{n=0}^{\infty} \frac{23}{100}\left(\frac{1}{10^2}\right)^n = \frac{\left(\frac{23}{100}\right)}{1 - \left(\frac{1}{100}\right)} = \frac{23}{99}$

52. $0.\overline{234} = \sum_{n=0}^{\infty} \frac{234}{1000}\left(\frac{1}{10^3}\right)^n = \frac{\left(\frac{234}{1000}\right)}{1 - \left(\frac{1}{1000}\right)} = \frac{234}{999}$

53. $0.\overline{7} = \sum_{n=0}^{\infty} \frac{7}{10}\left(\frac{1}{10}\right)^n = \frac{\left(\frac{7}{10}\right)}{1 - \left(\frac{1}{10}\right)} = \frac{7}{9}$

54. $0.\overline{d} = \sum_{n=0}^{\infty} \frac{d}{10}\left(\frac{1}{10}\right)^n = \frac{\left(\frac{d}{10}\right)}{1 - \left(\frac{1}{10}\right)} = \frac{d}{9}$

55. $0.0\overline{6} = \sum_{n=0}^{\infty} \left(\frac{1}{10}\right)\left(\frac{6}{10}\right)\left(\frac{1}{10}\right)^n = \frac{\left(\frac{6}{100}\right)}{1 - \left(\frac{1}{10}\right)} = \frac{6}{90} = \frac{1}{15}$

56. $1.\overline{414} = 1 + \sum_{n=0}^{\infty} \frac{414}{1000}\left(\frac{1}{10^3}\right)^n = 1 + \frac{\left(\frac{414}{1000}\right)}{1 - \left(\frac{1}{1000}\right)} = 1 + \frac{414}{999} = \frac{1413}{999}$

57. $1.24\overline{123} = \frac{124}{100} + \sum_{n=0}^{\infty} \frac{123}{10^5}\left(\frac{1}{10^3}\right)^n = \frac{124}{100} + \frac{\left(\frac{123}{10^5}\right)}{1 - \left(\frac{1}{10^3}\right)} = \frac{124}{100} + \frac{123}{10^5 - 10^2} = \frac{124}{100} + \frac{123}{99{,}900} = \frac{123{,}753}{99{,}900} = \frac{41{,}251}{33{,}300}$

58. $3.\overline{142857} = 3 + \sum_{n=0}^{\infty} \frac{142{,}857}{10^6}\left(\frac{1}{10^6}\right)^n = 3 + \frac{\left(\frac{142{,}857}{10^6}\right)}{1 - \left(\frac{1}{10^6}\right)} = 3 + \frac{142{,}857}{10^6 - 1} = \frac{2{,}857{,}140}{999{,}999} = \frac{317{,}460}{111{,}111}$

59. (a) $\sum_{n=-2}^{\infty} \frac{1}{(n+4)(n+5)}$ (b) $\sum_{n=0}^{\infty} \frac{1}{(n+2)(n+3)}$ (c) $\sum_{n=5}^{\infty} \frac{1}{(n-3)(n-2)}$

60. (a) $\sum_{n=-1}^{\infty} \frac{5}{(n+2)(n+3)}$ (b) $\sum_{n=3}^{\infty} \frac{5}{(n-2)(n-1)}$ (c) $\sum_{n=20}^{\infty} \frac{5}{(n-19)(n-18)}$

61. (a) one example is $\frac{1}{2} + \frac{1}{4} + \frac{1}{8} + \frac{1}{16} + \ldots = \frac{\left(\frac{1}{2}\right)}{1 - \left(\frac{1}{2}\right)} = 1$

(b) one example is $-\frac{3}{2} - \frac{3}{4} - \frac{3}{8} - \frac{3}{16} - \ldots = \frac{\left(-\frac{3}{2}\right)}{1 - \left(\frac{1}{2}\right)} = -3$

(c) one example is $1 - \frac{1}{2} - \frac{1}{4} - \frac{1}{8} - \frac{1}{16} - \ldots$; the series $\frac{k}{2} + \frac{k}{4} + \frac{k}{8} + \ldots = \frac{\left(\frac{k}{2}\right)}{1 - \left(\frac{1}{2}\right)} = k$ where k is any positive or negative number.

62. The term-by-term sum of the divergent series $\sum_{n=1}^{\infty} (1)$ and the divergent series $\sum_{n=1}^{\infty} (-1)$ is $\sum_{n=1}^{\infty} 0 = 0$.

63. Let $a_n = b_n = \left(\frac{1}{2}\right)^n$. Then $\sum_{n=1}^{\infty} a_n = \sum_{n=1}^{\infty} b_n = \sum_{n=1}^{\infty} \left(\frac{1}{2}\right)^n = 1$, while $\sum_{n=1}^{\infty} \left(\frac{a_n}{b_n}\right) = \sum_{n=1}^{\infty} (1)$ diverges.

64. Let $a_n = b_n = \left(\frac{1}{2}\right)^n$. Then $\sum_{n=1}^{\infty} a_n = \sum_{n=1}^{\infty} b_n = \sum_{n=1}^{\infty} \left(\frac{1}{2}\right)^n = 1$, while $\sum_{n=1}^{\infty} (a_n b_n) = \sum_{n=1}^{\infty} \left(\frac{1}{4}\right)^n = \frac{1}{3}$.

65. Let $a_n = \left(\frac{1}{4}\right)^n$ and $b_n = \left(\frac{1}{2}\right)^n$. Then $A = \sum_{n=1}^{\infty} a_n = \frac{1}{3}$, $B = \sum_{n=1}^{\infty} b_n = 1$ and $\sum_{n=1}^{\infty} \left(\frac{a_n}{b_n}\right) = \sum_{n=1}^{\infty} \left(\frac{1}{2}\right)^n = 1 \neq \frac{A}{B}$.

66. Yes: $\sum \left(\frac{1}{a_n}\right)$ diverges. The reasoning: $\sum a_n$ converges $\Rightarrow a_n \to 0 \Rightarrow \frac{1}{a_n} \to \infty \Rightarrow \sum \left(\frac{1}{a_n}\right)$ diverges by the nth-Term Test.

67. Since the sum of a finite number of terms is finite, adding or subtracting a finite number of terms from a series that diverges does not change the divergence of the series.

68. Let $A_n = a_1 + a_2 + \ldots + a_n$ and $\lim_{n\to\infty} A_n = A$. Assume $\sum (a_n + b_n)$ converges to S. Let $S_n = (a_1 + b_1) + (a_2 + b_2) + \ldots + (a_n + b_n) \Rightarrow S_n = (a_1 + a_2 + \ldots + a_n) + (b_1 + b_2 + \ldots + b_n)$ $\Rightarrow b_1 + b_2 + \ldots + b_n = S_n - A_n \Rightarrow \lim_{n\to\infty} (b_1 + b_2 + \ldots + b_n) = S - A \Rightarrow \sum b_n$ converges. This contradicts the assumption that $\sum b_n$ diverges; therefore, $\sum (a_n + b_n)$ diverges.

69. (a) $\frac{2}{1-r} = 5 \Rightarrow \frac{2}{5} = 1 - r \Rightarrow r = \frac{3}{5}$; $2 + 2\left(\frac{3}{5}\right) + 2\left(\frac{3}{5}\right)^2 + \ldots$

(b) $\frac{\left(\frac{13}{2}\right)}{1-r} = 5 \Rightarrow \frac{13}{10} = 1 - r \Rightarrow r = -\frac{3}{10}$; $\frac{13}{2} - \frac{13}{2}\left(\frac{3}{10}\right) + \frac{13}{2}\left(\frac{3}{10}\right)^2 - \frac{13}{2}\left(\frac{3}{10}\right)^3 + \ldots$

70. $1 + e^b + e^{2b} + \ldots = \frac{1}{1 - e^b} = 9 \Rightarrow \frac{1}{9} = 1 - e^b \Rightarrow e^b = \frac{8}{9} \Rightarrow b = \ln\left(\frac{8}{9}\right)$

71. $s_n = 1 + 2r + r^2 + 2r^3 + r^4 + 2r^5 + \ldots + r^{2n} + 2r^{2n+1}$, $n = 0, 1, \ldots$

$\Rightarrow s_n = \left(1 + r^2 + r^4 + \ldots + r^{2n}\right) + \left(2r + 2r^3 + 2r^5 + \ldots + 2r^{2n+1}\right) \Rightarrow \lim_{n\to\infty} s_n = \frac{1}{1-r^2} + \frac{2r}{1-r^2}$
$= \frac{1+2r}{1-r^2}$, if $|r^2| < 1$ or $|r| < 1$

72. $L - s_n = \frac{a}{1-r} - \frac{a\left(1 - r^n\right)}{1-r} = \frac{ar^n}{1-r}$

73. distance $= 4 + 2\left[(4)\left(\frac{3}{4}\right) + (4)\left(\frac{3}{4}\right)^2 + \ldots\right] = 4 + 2\left(\frac{3}{1 - \left(\frac{3}{4}\right)}\right) = 28$ m

74. time $= \sqrt{\frac{4}{4.9}} + 2\sqrt{\left(\frac{4}{4.9}\right)\left(\frac{3}{4}\right)} + 2\sqrt{\left(\frac{4}{4.9}\right)\left(\frac{3}{4}\right)^2} + 2\sqrt{\left(\frac{4}{4.9}\right)\left(\frac{3}{4}\right)^3} + \ldots = \sqrt{\frac{4}{4.9}} + 2\sqrt{\frac{4}{4.9}}\left[\sqrt{\frac{3}{4}} + \sqrt{\left(\frac{3}{4}\right)^2} + \ldots\right]$

$$= \frac{2}{\sqrt{4.9}} + \left(\frac{4}{\sqrt{4.9}}\right)\left[\frac{\sqrt{\frac{3}{4}}}{1-\sqrt{\frac{3}{4}}}\right] = \frac{2}{\sqrt{4.9}} + \left(\frac{4}{\sqrt{4.9}}\right)\left(\frac{\sqrt{3}}{2-\sqrt{3}}\right) = \frac{(4-2\sqrt{3})+4\sqrt{3}}{\sqrt{4.9}(2-\sqrt{3})} = \frac{4+2\sqrt{3}}{\sqrt{4.9}(2-\sqrt{3})} \approx 12.58 \text{ sec}$$

75. area $= 2^2 + \left(\sqrt{2}\right)^2 + (1)^2 + \left(\frac{1}{\sqrt{2}}\right)^2 + \ldots = 4 + 2 + 1 + \frac{1}{2} + \ldots = \frac{4}{1-\frac{1}{2}} = 8 \text{ m}^2$

76. area $= 2\left[\frac{\pi\left(\frac{1}{2}\right)^2}{2}\right] + 4\left[\frac{\pi\left(\frac{1}{4}\right)^2}{2}\right] + 8\left[\frac{\pi\left(\frac{1}{8}\right)^2}{2}\right] + \ldots = \pi\left(\frac{1}{4} + \frac{1}{8} + \frac{1}{16} + \ldots\right) = \pi\left(\frac{\left(\frac{1}{4}\right)}{1-\left(\frac{1}{2}\right)}\right) = \frac{\pi}{2}$

77. (a) $L_1 = 3,\ L_2 = 3\left(\frac{4}{3}\right),\ L_3 = 3\left(\frac{4}{3}\right)^2,\ \ldots,\ L_n = 3\left(\frac{4}{3}\right)^{n-1} \Rightarrow \lim_{n\to\infty} L_n = \lim_{n\to\infty} 3\left(\frac{4}{3}\right)^{n-1} = \infty$

(b) $A_1 = \frac{1}{2}(1)\left(\frac{\sqrt{3}}{2}\right) = \frac{\sqrt{3}}{4},\ A_2 = A_1 + 3\left(\frac{1}{2}\right)\left(\frac{1}{3}\right)\left(\frac{\sqrt{3}}{6}\right) = \frac{\sqrt{3}}{4} + \frac{\sqrt{3}}{12},\ A_3 = A_2 + 12\left(\frac{1}{2}\right)\left(\frac{1}{9}\right)\left(\frac{\sqrt{3}}{18}\right)$

$$= \frac{\sqrt{3}}{4} + \frac{\sqrt{3}}{12} + \frac{\sqrt{3}}{27},\ A_4 = A_3 + 48\left(\frac{1}{2}\right)\left(\frac{1}{27}\right)\left(\frac{\sqrt{3}}{54}\right),\ \ldots,\ A_n = \frac{\sqrt{3}}{4} + \frac{27\sqrt{3}}{64}\left(\frac{4}{9}\right)^2 + \frac{27\sqrt{3}}{64}\left(\frac{4}{9}\right)^3 + \ldots$$

$$= \frac{\sqrt{3}}{4} + \sum_{n=2}^{\infty} \frac{27\sqrt{3}}{64}\left(\frac{4}{9}\right)^n = \frac{\sqrt{3}}{4} + \frac{\left(\frac{27\sqrt{3}}{64}\right)\left(\frac{4}{9}\right)^2}{1-\left(\frac{4}{9}\right)} = \frac{\sqrt{3}}{4} + \frac{\left(\frac{27\sqrt{3}}{64}\right)\left(\frac{16}{9}\right)}{9-4} = \frac{\sqrt{3}}{4} + \frac{3\sqrt{3}}{4\cdot 5} = \frac{5\sqrt{3}+3\sqrt{3}}{20} = \frac{2\sqrt{3}}{5}$$

78. Each term of the series $\sum_{n=1}^{\infty} \frac{1}{n^2}$ represents the area of one of the squares shown in the figure, and all of the squares lie inside the rectangle of width 1 and length $\sum_{n=0}^{\infty} \left(\frac{1}{2}\right)^n = \frac{1}{1-\frac{1}{2}} = 2$. Since the squares do not fill the rectangle completely, and the area of the rectangle is 2, we have $\sum_{n=1}^{\infty} \frac{1}{n^2} < 2$.

8.4 THE INTEGRAL TEST FOR SERIES OF NONNEGATIVE TERMS

1. converges; a geometric series with $r = \frac{1}{10} < 1$

2. converges; a geometric series with $r = \frac{1}{e} < 1$

3. diverges; by the nth-Term Test for Divergence, $\lim_{n\to\infty} \frac{n}{n+1} = 1 \neq 0$

4. diverges by the Integral Test; $\int_1^n \frac{5}{x+1}\,dx = \ln(n+1) - \ln 2 \Rightarrow \int_1^\infty \frac{5}{x+1}\,dx \to \infty$

5. diverges; $\sum_{n=1}^{\infty} \frac{3}{\sqrt{n}} = 3\sum_{n=1}^{\infty} \frac{1}{\sqrt{n}}$, which is a divergent p-series

6. converges; $\sum_{n=1}^{\infty} \frac{-2}{n\sqrt{n}} = -2 \sum_{n=1}^{\infty} \frac{1}{n^{3/2}}$, which is a convergent p-series

7. converges; a geometric series with $r = \frac{1}{8} < 1$

8. diverges by the nonzero constant multiple rule since $\sum_{n=1}^{\infty} \frac{1}{n}$ diverges

9. diverges by the Integral Test: $\int_2^n \frac{\ln x}{x}\,dx = \frac{1}{2}\left(\ln^2 n - \ln 2\right) \Rightarrow \int_2^{\infty} \frac{\ln x}{x}\,dx \to \infty$

10. diverges by the Integral Test: $\int_2^{\infty} \frac{\ln x}{\sqrt{x}}\,dx;\ \begin{bmatrix} t = \ln x \\ dt = \frac{dx}{x} \\ dx = e^t\,dt \end{bmatrix} \to \int_{\ln 2}^{\infty} te^{t/2}\,dt = \lim_{b\to\infty} \left[2te^{t/2} - 4e^{t/2}\right]_{\ln 2}^{b}$

$= \lim_{b\to\infty} \left[2e^{b/2}(b-2) - 2e^{(\ln 2)/2}(\ln 2 - 2)\right] = \infty$

11. converges; a geometric series with $r = \frac{2}{3} < 1$

12. diverges; $\lim_{n\to\infty} \frac{5^n}{4^n + 3} = \lim_{n\to\infty} \frac{5^n \ln 5}{4^n \ln 4} = \lim_{n\to\infty} \left(\frac{\ln 5}{\ln 4}\right)\left(\frac{5}{4}\right)^n \neq 0$

13. diverges; $\sum_{n=0}^{\infty} \frac{-2}{n+1} = -2 \sum_{n=0}^{\infty} \frac{1}{n+1}$, which diverges by the Integral Test

14. diverges by the Integral Test: $\int_1^n \frac{dx}{2x-1} = \frac{1}{2}\ln(2n-1) \to \infty$ as $n \to \infty$

15. diverges; $\lim_{n\to\infty} a_n = \lim_{n\to\infty} \frac{2^n}{n+1} = \lim_{n\to\infty} \frac{2^n \ln 2}{1} = \infty \neq 0$

16. diverges by the Integral Test: $\int_1^n \frac{dx}{\sqrt{x}\left(\sqrt{x}+1\right)};\ \begin{bmatrix} u = \sqrt{x}+1 \\ du = \frac{dx}{\sqrt{x}} \end{bmatrix} \to \int_2^{\sqrt{n}+1} \frac{du}{u} = \ln\left(\sqrt{n}+1\right) - \ln 2$

$\to \infty$ as $n \to \infty$

17. diverges; $\lim_{n\to\infty} \frac{\sqrt{n}}{\ln n} = \lim_{n\to\infty} \frac{\left(\frac{1}{2\sqrt{n}}\right)}{\left(\frac{1}{n}\right)} = \lim_{n\to\infty} \frac{\sqrt{n}}{2} = \infty \neq 0$

18. diverges; $\lim_{n\to\infty} a_n = \lim_{n\to\infty} \left(1 + \frac{1}{n}\right)^n = e \neq 0$

19. diverges; a geometric series with $r = \frac{1}{\ln 2} \approx 1.44 > 1$

20. converges; a geometric series with $r = \frac{1}{\ln 3} \approx 0.91 < 1$

21. converges by the Integral Test: $\int_3^\infty \frac{\left(\frac{1}{x}\right)}{(\ln x)\sqrt{(\ln x)^2 - 1}}\, dx;\ \begin{bmatrix} u = \ln x \\ du = \frac{1}{x}\, dx \end{bmatrix} \to \int_{\ln 3}^\infty \frac{1}{u\sqrt{u^2-1}}\, du$

$= \lim_{b\to\infty} \left[\sec^{-1}|u|\right]_{\ln 3}^b = \lim_{b\to\infty} \left[\sec^{-1} b - \sec^{-1}(\ln 3)\right] = \lim_{b\to\infty} \left[\cos^{-1}\left(\frac{1}{b}\right) - \sec^{-1}(\ln 3)\right]$

$= \cos^{-1}(0) - \sec^{-1}(\ln 3) = \frac{\pi}{2} - \sec^{-1}(\ln 3) \approx 1.1439$

22. converges by the Integral Test: $\int_1^\infty \frac{1}{x(1+\ln^2 x)}\, dx = \int_1^\infty \frac{\left(\frac{1}{x}\right)}{1+(\ln x)^2}\, dx;\ \begin{bmatrix} u = \ln x \\ du = \frac{1}{x}\, dx \end{bmatrix} \to \int_0^\infty \frac{1}{1+u^2}\, du$

$= \lim_{b\to\infty} \left[\tan^{-1} u\right]_0^b = \lim_{b\to\infty} \left(\tan^{-1} b - \tan^{-1} 0\right) = \frac{\pi}{2} - 0 = \frac{\pi}{2}$

23. diverges by the nth-Term Test for divergence; $\lim_{n\to\infty} n \sin\left(\frac{1}{n}\right) = \lim_{n\to\infty} \frac{\sin\left(\frac{1}{n}\right)}{\left(\frac{1}{n}\right)} = \lim_{x\to 0} \frac{\sin x}{x} = 1 \neq 0$

24. diverges by the nth-Term Test for divergence; $\lim_{n\to\infty} n \tan\left(\frac{1}{n}\right) = \lim_{n\to\infty} \frac{\tan\left(\frac{1}{n}\right)}{\left(\frac{1}{n}\right)} = \lim_{n\to\infty} \frac{\left(-\frac{1}{n^2}\right)\sec^2\left(\frac{1}{n}\right)}{\left(-\frac{1}{n^2}\right)}$

$= \lim_{n\to\infty} \sec^2\left(\frac{1}{n}\right) = \sec^2 0 = 1 \neq 0$

25. converges by the Integral Test: $\int_1^\infty \frac{e^x}{1+e^{2x}}\, dx;\ \begin{bmatrix} u = e^x \\ du = e^x\, dx \end{bmatrix} \to \int_e^\infty \frac{1}{1+u^2}\, du = \lim_{n\to\infty} \left[\tan^{-1} u\right]_e^b$

$= \lim_{b\to\infty} \left(\tan^{-1} b - \tan^{-1} e\right) = \frac{\pi}{2} - \tan^{-1} e \approx 0.35$

26. converges by the Integral Test: $\int_1^\infty \frac{2}{1+e^x}\, dx;\ \begin{bmatrix} u = e^x \\ du = e^x\, dx \\ dx = \frac{1}{u}\, du \end{bmatrix} \to \int_e^\infty \frac{2}{u(1+u)}\, du = \int_e^\infty \left(\frac{2}{u} - \frac{2}{u+1}\right) du$

$= \lim_{b\to\infty} \left[2 \ln \frac{u}{u+1}\right]_e^b = \lim_{b\to\infty} 2 \ln\left(\frac{b}{b+1}\right) - 2 \ln\left(\frac{e}{e+1}\right) = 2 \ln 1 - 2 \ln\left(\frac{e}{e+1}\right) = -2 \ln\left(\frac{e}{e+1}\right)$

27. converges by the Integral Test: $\int_1^\infty \frac{8 \tan^{-1} x}{1+x^2}\, dx;\ \begin{bmatrix} u = \tan^{-1} x \\ du = \frac{dx}{1+x^2} \end{bmatrix} \to \int_{\pi/4}^{\pi/2} 8u\, du = \left[4u^2\right]_{\pi/4}^{\pi/2} = 4\left(\frac{\pi^2}{4} - \frac{\pi^2}{16}\right) = \frac{3\pi^2}{4}$

28. diverges by the Integral Test: $\int_1^\infty \frac{x}{x^2+1}\,dx$; $\left[\begin{array}{l} u = x^2+1 \\ du = 2x\,dx \end{array}\right] \to \frac{1}{2}\int_2^\infty \frac{du}{4} = \lim\limits_{b\to\infty} \left[\frac{1}{2}\ln u\right]_2^b$

$= \lim\limits_{b\to\infty} \frac{1}{2}(\ln b - \ln 2) = \infty$

29. converges by the Integral Test: $\int_1^\infty \operatorname{sech} x\,dx = 2 \lim\limits_{b\to\infty} \int_1^b \frac{e^x}{1+(e^x)^2}\,dx = 2 \lim\limits_{b\to\infty} \left[\tan^{-1} e^x\right]_1^b$

$= 2 \lim\limits_{b\to\infty} \left(\tan^{-1} e^b - \tan^{-1} e\right) = \pi - 2\tan^{-1} e$

30. converges by the Integral Test: $\int_1^\infty \operatorname{sech}^2 x\,dx = \lim\limits_{b\to\infty} \int_1^b \operatorname{sech}^2 x\,dx = \lim\limits_{b\to\infty} [\tanh x]_1^b = \lim\limits_{b\to\infty} (\tanh b - \tanh 1)$

$= 1 - \tanh 1$

31. $\int_1^\infty \left(\frac{a}{x+2} - \frac{1}{x+4}\right) dx = \lim\limits_{b\to\infty} \left[a \ln|x+2| - \ln|x+4|\right]_1^b = \lim\limits_{b\to\infty} \ln \frac{(b+2)^a}{b+4} - \ln\left(\frac{3^a}{5}\right)$;

$\lim\limits_{b\to\infty} \frac{(b+2)^a}{b+4} = a \lim\limits_{b\to\infty} (b+2)^{a-1} = \begin{cases} \infty, & a > 1 \\ 1, & a = 1 \end{cases}$ $\Rightarrow$ the series converges to $\ln\left(\frac{5}{3}\right)$ if $a = 1$ and diverges to ∞ if $a > 1$. If $a < 1$, the terms of the series eventually become negative and the Integral Test does not apply. From that point on, however, the series behaves like a negative multiple of the harmonic series, and so it diverges.

32. $\int_3^\infty \left(\frac{1}{x-1} - \frac{2a}{x+1}\right) dx = \lim\limits_{b\to\infty} \left[\ln\left|\frac{x-1}{(x+1)^{2a}}\right|\right]_3^b = \lim\limits_{b\to\infty} \ln \frac{b-1}{(b+1)^{2a}} - \ln\left(\frac{2}{4^{2a}}\right)$; $\lim\limits_{b\to\infty} \frac{b-1}{(b+1)^{2a}}$

$= \lim\limits_{b\to\infty} \frac{1}{2a(b+1)^{2a-1}} = \begin{cases} 1, & a = \frac{1}{2} \\ \infty, & a < \frac{1}{2} \end{cases}$ $\Rightarrow$ the series converges to $\ln\left(\frac{4}{2}\right) = \ln 2$ if $a = \frac{1}{2}$ and diverges to ∞ if

if $a < \frac{1}{2}$. If $a > \frac{1}{2}$, the terms of the series eventually become negative and the Integral Test does not apply. From that point on, however, the series behaves like a negative multiple of the harmonic series, and so it diverges.

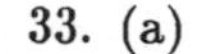

33. (a)

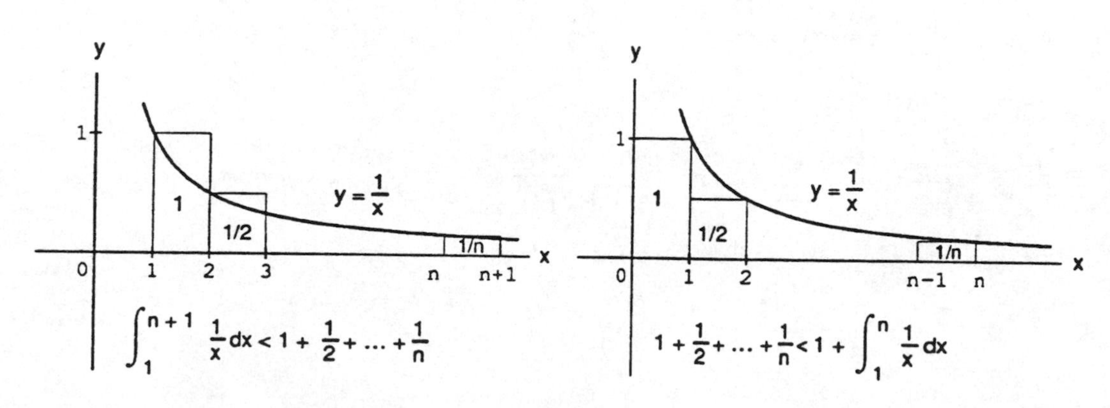

(b) There are $(13)(365)(24)(60)(60)(10^9)$ seconds in 13 billion years; by part (a) $s_n \le 1 + \ln n$ where $n = (13)(365)(24)(60)(60)(10^9) \Rightarrow s_n \le 1 + \ln\left((13)(365)(24)(60)(60)(10^9)\right)$
$= 1 + \ln(13) + \ln(365) + \ln(24) + 2\ln(60) + 9\ln(10) \approx 41.55$

34. No, because $\sum_{n=1}^{\infty} \frac{1}{nx} = \frac{1}{x}\sum_{n=1}^{\infty} \frac{1}{n}$ and $\sum_{n=1}^{\infty} \frac{1}{n}$ diverges

35. Yes. If $\sum_{n=1}^{\infty} a_n$ is a divergent series of positive numbers, then $\left(\frac{1}{2}\right)\sum_{n=1}^{\infty} a_n = \sum_{n=1}^{\infty} \left(\frac{a_n}{2}\right)$ also diverges and $\frac{a_n}{2} < a_n$.

There is no "smallest" divergent series of positive numbers: for any divergent series $\sum_{n=1}^{\infty} a_n$ of positive numbers $\sum_{n=1}^{\infty} \left(\frac{a_n}{2}\right)$ has smaller terms and still diverges.

36. No, if $\sum_{n=1}^{\infty} a_n$ is a convergent series of positive numbers, then $2\sum_{n=1}^{\infty} a_n = \sum_{n=1}^{\infty} 2a_n$ also converges, and $2a_n \ge a_n$.
There is no "largest" convergent series of positive numbers.

37. Let $A_n = \sum_{k=1}^{n} a_k$ and $B_n = \sum_{k=1}^{n} 2^k a_{(2^k)}$, where $\{a_k\}$ is a nonincreasing sequence of positive terms converging to 0. Note that $\{A_n\}$ and $\{B_n\}$ are nondecreasing sequences of positive terms. Now,

$$B_n = 2a_2 + 4a_4 + 8a_8 + \ldots + 2^n a_{(2^n)} = 2a_2 + (2a_4 + 2a_4) + (2a_8 + 2a_8 + 2a_8 + 2a_8) + \ldots$$
$$+ \underbrace{\left(2a_{(2^n)} + 2a_{(2^n)} + \ldots + 2a_{(2^n)}\right)}_{2^{n-1} \text{ terms}} \le 2a_1 + 2a_2 + (2a_3 + 2a_4) + (2a_5 + 2a_6 + 2a_7 + 2a_8) + \ldots$$
$$+ \left(2a_{(2^{n-1})} + 2a_{(2^{n-1}+1)} + \ldots + 2a_{(2^n)}\right) = 2A_{(2^n)} \le 2\sum_{k=1}^{\infty} a_k.$$

Therefore if $\sum a_k$ converges, then $\{B_n\}$ is bounded above $\Rightarrow \sum 2^k a_{(2^k)}$ converges. Conversely,

$$A_n = a_1 + (a_2 + a_3) + (a_4 + a_5 + a_6 + a_7) + \ldots + a_n < a_1 + 2a_2 + 4a_4 + \ldots + 2^n a_{(2^n)} = a_1 + B_n < a_1 + \sum_{k=1}^{\infty} 2^k a_{(2^k)}.$$

Therefore, if $\sum_{k=1}^{\infty} 2^k a_{(2^k)}$ converges, then $\{A_n\}$ is bounded above and hence converges.

38. (a) $a_{(2^n)} = \frac{1}{2^n \ln(2^n)} = \frac{1}{2^n \cdot n(\ln 2)} \Rightarrow \sum_{n=2}^{\infty} 2^n a_{(2^n)} = \sum_{n=2}^{\infty} 2^n \frac{1}{2^n \cdot n(\ln 2)} = \frac{1}{\ln 2}\sum_{n=2}^{\infty} \frac{1}{n}$, which diverges $\Rightarrow \sum_{n=2}^{\infty} \frac{1}{n \ln n}$ diverges.

(b) $a_{(2^n)} = \frac{1}{2^{np}} \Rightarrow \sum_{n=1}^{\infty} 2^n a_{(2^n)} = \sum_{n=1}^{\infty} 2^n \cdot \frac{1}{2^{np}} = \sum_{n=1}^{\infty} \frac{1}{(2^n)^{p-1}} = \sum_{n=1}^{\infty} \left(\frac{1}{2^{p-1}}\right)^n$, a geometric series that converges if $\frac{1}{2^{p-1}} < 1$ or $p > 1$, but diverges if $p \ge 1$.

39. (a) $\int_2^{\infty} \frac{dx}{x(\ln x)^p}; \left[\begin{array}{l} u = \ln x \\ du = \frac{dx}{x} \end{array}\right] \to \int_{\ln 2}^{\infty} u^{-p}\, du = \lim_{b\to\infty} \left[\frac{u^{-p+1}}{-p+1}\right]_{\ln 2}^{b} = \lim_{b\to\infty} \left(\frac{1}{1-p}\right)[b^{-p+1} - (\ln 2)^{-p+1}]$

$= \begin{cases} \frac{1}{p-1}(\ln 2)^{-p+1}, & p > 1 \\ \infty, & p < 1 \end{cases}$ $\Rightarrow$ the improper integral converges if $p > 1$ and diverges

if $p < 1$. For $p = 1$: $\int_2^{\infty} \frac{dx}{x \ln x} = \lim_{b\to\infty} \left[\ln(\ln x)\right]_2^b = \lim_{b\to\infty} \left[\ln(\ln b) - \ln(\ln 2)\right] = \infty$, so the improper integral diverges if $p = 1$.

(b) Since the series and the integral converge or diverge together, $\sum_{n=2}^{\infty} \frac{1}{n(\ln n)^p}$ converges if and only if $p > 1$.

40. (a) $p = 1 \Rightarrow$ the series diverges

(b) $p = 1.01 \Rightarrow$ the series converges

(c) $\sum_{n=2}^{\infty} \frac{1}{n(\ln n^3)} = \frac{1}{3} \sum_{n=2}^{\infty} \frac{1}{n(\ln n)}$; $p = 1 \Rightarrow$ the series diverges

(d) $p = 3 \Rightarrow$ the series converges

41. (a) From Fig. 8.13 in the text with $f(x) = \frac{1}{x}$ and $a_k = \frac{1}{k}$, we have $\int_1^{n+1} \frac{1}{x}\, dx \le 1 + \frac{1}{2} + \frac{1}{3} + \ldots + \frac{1}{n}$

$\le 1 + \int_1^n f(x)\, dx \Rightarrow \ln(n+1) \le 1 + \frac{1}{2} + \frac{1}{3} + \ldots + \frac{1}{n} \le 1 + \ln n \Rightarrow 0 \le \ln(n+1) - \ln n$

$\le \left(1 + \frac{1}{2} + \frac{1}{3} + \ldots + \frac{1}{n}\right) - \ln n \le 1$. Therefore the sequence $\left\{\left(1 + \frac{1}{2} + \frac{1}{3} + \ldots + \frac{1}{n}\right) - \ln n\right\}$ is bounded above by 1 and below by 0.

(b) From the graph in Fig. 8.13(a) with $f(x) = \frac{1}{x}$, $\frac{1}{n+1} < \int_n^{n+1} \frac{1}{x}\, dx = \ln(n+1) - \ln n$

$\Rightarrow 0 > \frac{1}{n+1} - [\ln(n+1) - \ln n] = \left(1 + \frac{1}{2} + \frac{1}{3} + \ldots + \frac{1}{n+1} - \ln(n+1)\right) - \left(1 + \frac{1}{2} + \frac{1}{3} + \ldots + \frac{1}{n} - \ln n\right)$.

If we define $a_n = 1 + \frac{1}{2} = \frac{1}{3} + \frac{1}{n} - \ln n$, then $0 > a_{n+1} - a_n \Rightarrow a_{n+1} < a_n \Rightarrow \{a_n\}$ is a decreasing sequence of nonnegative terms.

42. $e^{-x^2} \le e^{-x}$ for $x \ge 1$, and $\int_1^{\infty} e^{-x}\, dx = \lim_{b\to\infty} \left[-e^{-x}\right]_1^b = \lim_{b\to\infty} \left(-e^{-b} + e^{-1}\right) = e^{-1} \Rightarrow \int_1^{\infty} e^{-x^2}\, dx$ converges by the Comparison Test for improper integrals $\Rightarrow \sum_{n=0}^{\infty} e^{-n^2} = 1 + \sum_{n=1}^{\infty} e^{-n^2}$ converges by the Integral Test.

8.5 COMPARISON TESTS FOR SERIES OF NONNEGATIVE TERMS

1. diverges by the Limit Comparison Test (part 1) when compared with $\sum_{n=1}^{\infty} \frac{1}{\sqrt{n}}$, a divergent p-series:

$$\lim_{n\to\infty} \frac{\left(\frac{1}{2\sqrt{n}+\sqrt[3]{n}}\right)}{\left(\frac{1}{\sqrt{n}}\right)} = \lim_{n\to\infty} \frac{\sqrt{n}}{2\sqrt{n}+\sqrt[3]{n}} = \lim_{n\to\infty} \left(\frac{1}{2+n^{-1/6}}\right) = \frac{1}{2}$$

2. diverges by the Direct Comparison Test since $n+n+n > n+\sqrt{n}+0 \Rightarrow \frac{3}{n+\sqrt{n}} > \frac{1}{n}$, which is the nth term of the divergent series $\sum_{n=1}^{\infty} \frac{1}{n}$

3. converges by the Direct Comparison Test; $\frac{\sin^2 n}{2^n} \le \frac{1}{2^n}$, which is the nth term of a convergent geometric series

4. converges by the Direct Comparison Test; $\frac{1+\cos n}{n^2} \le \frac{2}{n^2}$ and the p-series $\sum \frac{1}{n^2}$ converges

5. diverges since $\lim_{n\to\infty} \frac{2n}{3n-1} = \frac{2}{3} \ne 0$

6. converges by the Limit Comparison Test (part 1) with $\frac{1}{n^{3/2}}$, the nth term of a convergent p-series:

$$\lim_{n\to\infty} \frac{\left(\frac{n+1}{n^2\sqrt{n}}\right)}{\left(\frac{1}{n^{3/2}}\right)} = \lim_{n\to\infty} \left(\frac{n+1}{n}\right) = 1$$

7. converges by the Direct Comparison Test; $\left(\frac{n}{3n+1}\right)^n < \left(\frac{n}{3n}\right)^n < \left(\frac{1}{3}\right)^n$, the nth term of a convergent geometric series

8. converges by the Limit Comparison Test (part 1) with $\frac{1}{n^{3/2}}$, the nth term of a convergent p-series:

$$\lim_{n\to\infty} \frac{\left(\frac{1}{n^{3/2}}\right)}{\left(\frac{1}{\sqrt{n^3+2}}\right)} = \lim_{n\to\infty} \sqrt{\frac{n^3+2}{n^3}} = \lim_{n\to\infty} \sqrt{1+\frac{2}{n^3}} = 1$$

9. diverges by the Direct Comparison Test; $n > \ln n \Rightarrow \ln n > \ln \ln n \Rightarrow \frac{1}{\ln n} < \frac{1}{\ln (\ln n)}$ and the series $\sum_{n=3}^{\infty} \frac{1}{n}$ diverges

10. diverges by the Limit Comparison Test (part 3) when compared with $\sum_{n=2}^{\infty} \frac{1}{n}$, a divergent p-series:

$$\lim_{n\to\infty} \frac{\left(\frac{1}{(\ln n)^2}\right)}{\left(\frac{1}{n}\right)} = \lim_{n\to\infty} \frac{n}{(\ln n)^2} = \lim_{n\to\infty} \frac{1}{2(\ln n)\left(\frac{1}{n}\right)} = \frac{1}{2}\lim_{n\to\infty} \frac{n}{\ln n} = \frac{1}{2}\lim_{n\to\infty} \frac{1}{\left(\frac{1}{n}\right)} = \frac{1}{2}\lim_{n\to\infty} n = \infty$$

11. converges by the Limit Comparison Test (part 2) when compared with $\sum_{n=1}^{\infty} \frac{1}{n^2}$, a convergent p-series:

$$\lim_{n\to\infty} \frac{\left[\frac{(\ln n)^2}{n^3}\right]}{\left(\frac{1}{n^2}\right)} = \lim_{n\to\infty} \frac{(\ln n)^2}{n} = \lim_{n\to\infty} \frac{2(\ln n)\left(\frac{1}{n}\right)}{1} = 2 \lim_{n\to\infty} \frac{\ln n}{n} = 0 \qquad \text{(Table 8.1)}$$

12. converges by the Limit Comparison Test (part 2) when compared with $\sum_{n=1}^{\infty} \frac{1}{n^2}$, a convergent p-series:

$$\lim_{n\to\infty} \frac{\left[\frac{(\ln n)^3}{n^3}\right]}{\left(\frac{1}{n^2}\right)} = \lim_{n\to\infty} \frac{(\ln n)^3}{n} = \lim_{n\to\infty} \frac{3(\ln n)^2\left(\frac{1}{n}\right)}{1} = 3 \lim_{n\to\infty} \frac{(\ln n)^2}{n} = 3 \lim_{n\to\infty} \frac{2(\ln n)\left(\frac{1}{n}\right)}{1} = 6 \lim_{n\to\infty} \frac{\ln n}{n}$$

$$= 6 \cdot 0 = 0 \qquad \text{(Table 8.1)}$$

13. diverges by the Limit Comparison Test (part 3) with $\frac{1}{n}$, the nth term of the divergent harmonic series:

$$\lim_{n\to\infty} \frac{\left[\frac{1}{\sqrt{n}\ln n}\right]}{\left(\frac{1}{n}\right)} = \lim_{n\to\infty} \frac{\sqrt{n}}{\ln n} = \lim_{n\to\infty} \frac{\left(\frac{1}{2\sqrt{n}}\right)}{\left(\frac{1}{n}\right)} = \lim_{n\to\infty} \frac{\sqrt{n}}{2} = \infty$$

14. converges by the Limit Comparison Test (part 2) with $\frac{1}{n^{5/4}}$, the nth term of a convergent p-series:

$$\lim_{n\to\infty} \frac{\left[\frac{(\ln n)^2}{n^{3/2}}\right]}{\left(\frac{1}{n^{5/4}}\right)} = \lim_{n\to\infty} \frac{(\ln n)^2}{n^{1/4}} = \lim_{n\to\infty} \frac{\left(\frac{2\ln n}{n}\right)}{\left(\frac{1}{4n^{3/4}}\right)} = 8 \lim_{n\to\infty} \frac{\ln n}{n^{1/4}} = 8 \lim_{n\to\infty} \frac{\left(\frac{1}{n}\right)}{\left(\frac{1}{4n^{3/4}}\right)} = 32 \lim_{n\to\infty} \frac{1}{n^{1/4}} = 32 \cdot 0 = 0$$

15. diverges by the Limit Comparison Test (part 3) with $\frac{1}{n}$, the nth term of the divergent harmonic series:

$$\lim_{n\to\infty} \frac{\left(\frac{1}{1+\ln n}\right)}{\left(\frac{1}{n}\right)} = \lim_{n\to\infty} \frac{n}{1+\ln n} = \lim_{n\to\infty} \frac{1}{\left(\frac{1}{n}\right)} = \lim_{n\to\infty} n = \infty$$

16. diverges by the Limit Comparison Test (part 3) with $\frac{1}{n}$, the nth term of the divergent harmonic series:

$$\lim_{n\to\infty} \frac{\left(\frac{1}{(1+\ln n)^2}\right)}{\left(\frac{1}{n}\right)} = \lim_{n\to\infty} \frac{n}{(1+\ln n)^2} = \lim_{n\to\infty} \frac{1}{\left[\frac{2(1+\ln n)}{n}\right]} = \lim_{n\to\infty} \frac{n}{2(1+\ln n)} = \lim_{n\to\infty} \frac{1}{\left(\frac{2}{n}\right)} = \lim_{n\to\infty} \frac{n}{2} = \infty$$

17. diverges by the Integral Test: $\int_2^{\infty} \frac{\ln(x+1)}{x+1}\,dx = \int_{\ln 3}^{\infty} u\,du = \lim_{b\to\infty} \left[\frac{1}{2}u^2\right]_{\ln 3}^{b} = \lim_{b\to\infty} \frac{1}{2}\left(b^2 - \ln^2 3\right) = \infty$

18. diverges by the Limit Comparison Test (part 3) with $\frac{1}{n}$, the nth term of the divergent harmonic series:

$$\lim_{n\to\infty} \frac{\left(\frac{1}{1+\ln^2 n}\right)}{\left(\frac{1}{n}\right)} = \lim_{n\to\infty} \frac{n}{1+\ln^2 n} = \lim_{n\to\infty} \frac{1}{\left(\frac{2\ln n}{n}\right)} = \lim_{n\to\infty} \frac{n}{2\ln n} = \lim_{n\to\infty} \frac{1}{\left(\frac{2}{n}\right)} = \lim_{n\to\infty} \frac{n}{2} = \infty$$

19. converges by the Direct Comparison Test with $\frac{1}{n^{3/2}}$, the nth term of a convergent p-series: $n^2 - 1 > n$ for

$n \geq 2 \Rightarrow n^2(n^2-1) > n^3 \Rightarrow n\sqrt{n^2-1} > n^{3/2} \Rightarrow \frac{1}{n^{3/2}} > \frac{1}{n\sqrt{n^2-1}}$

20. converges by the Direct Comparison Test with $\frac{1}{n^{3/2}}$, the nth term of a convergent p-series: $n^2+1 > n^2$

$\Rightarrow n^2 + 1 > \sqrt{n}n^{3/2} \Rightarrow \frac{n^2+1}{\sqrt{n}} > n^{3/2} \Rightarrow \frac{\sqrt{n}}{n^2+1} < \frac{1}{n^{3/2}}$

21. converges because $\sum\limits_{n=1}^{\infty} \frac{1-n}{n2^n} = \sum\limits_{n=1}^{\infty} \frac{1}{n2^n} + \sum\limits_{n=1}^{\infty} \frac{-1}{2^n}$ which is the sum of two convergent series:

$\sum\limits_{n=1}^{\infty} \frac{1}{n2^n}$ converges by the Direct Comparison Test since $\frac{1}{n2^n} < \frac{1}{2^n}$, and $\sum\limits_{n=1}^{\infty} \frac{-1}{2^n}$ is a convergent geometric series

22. converges by the Direct Comparison Test: $\sum\limits_{n=1}^{\infty} \frac{n+2^n}{n^2 2^n} = \sum\limits_{n=1}^{\infty} \left(\frac{1}{n2^n} + \frac{1}{n^2}\right)$ and $\frac{1}{n2^n} + \frac{1}{n^2} \leq \frac{1}{2^n} + \frac{1}{n^2}$, the sum of the nth terms of a convergent geometric series and a convergent p-series

23. converges by the Direct Comparison Test: $\frac{1}{3^{n-1}+1} < \frac{1}{3^{n-1}}$, which is the nth term of a convergent geometric series

24. diverges; $\lim\limits_{n\to\infty} \left(\frac{3^{n-1}+1}{3^n}\right) = \lim\limits_{n\to\infty} \left(\frac{1}{3} + \frac{1}{3^n}\right) = \frac{1}{3} \neq 0$

25. diverges by the Limit Comparison Test (part 1) with $\frac{1}{n}$, the nth term of the divergent harmonic series:

$$\lim_{n\to\infty} \frac{\left(\sin\frac{1}{n}\right)}{\left(\frac{1}{n}\right)} = \lim_{x\to 0} \frac{\sin x}{x} = 1$$

26. diverges by the Limit Comparison Test (part 1) with $\frac{1}{n}$, the nth term of the divergent harmonic series:

$$\lim_{n\to\infty} \frac{\left(\tan\frac{1}{n}\right)}{\left(\frac{1}{n}\right)} = \lim_{n\to\infty} \left(\frac{1}{\cos\frac{1}{n}}\right) \frac{\left(\sin\frac{1}{n}\right)}{\left(\frac{1}{n}\right)} = \lim_{x\to 0} \left(\frac{1}{\cos x}\right)\left(\frac{\sin x}{x}\right) = 1 \cdot 1 = 1$$

27. converges by the Limit Comparison Test (part 1) with $\frac{1}{n^2}$, the nth term of a convergent p-series:

$$\lim_{n\to\infty} \frac{\left(\frac{10n+1}{n(n+1)(n+2)}\right)}{\left(\frac{1}{n^2}\right)} = \lim_{n\to\infty} \frac{10n^2+n}{n^2+3n+2} = \lim_{n\to\infty} \frac{20n+1}{2n+3} = \lim_{n\to\infty} \frac{20}{2} = 10$$

28. converges by the Limit Comparison Test (part 1) with $\frac{1}{n^2}$, the nth term of a convergent p-series:

$$\lim_{n\to\infty} \frac{\left(\frac{5n^3-3n}{n^2(n-2)(n^2+5)}\right)}{\left(\frac{1}{n^2}\right)} = \lim_{n\to\infty} \frac{5n^3-3n}{n^3-2n^2+5n-10} = \lim_{n\to\infty} \frac{15n^2-3}{3n^2-4n+5} = \lim_{n\to\infty} \frac{30n}{6n-4} = 5$$

29. converges by the Direct Comparison Test: $\frac{\tan^{-1} n}{n^{1.1}} < \frac{\frac{\pi}{2}}{n^{1.1}}$ and $\sum_{n=1}^{\infty} \frac{\frac{\pi}{2}}{n^{1.1}} = \frac{\pi}{2} \sum_{n=1}^{\infty} \frac{1}{n^{1.1}}$ is the product of a convergent p-series and a nonzero constant

30. converges by the Direct Comparison Test: $\sec^{-1} n < \frac{\pi}{2} \Rightarrow \frac{\sec^{-1} n}{n^{1.3}} < \frac{\left(\frac{\pi}{2}\right)}{n^{1.3}}$ and $\sum_{n=1}^{\infty} \frac{\left(\frac{\pi}{2}\right)}{n^{1.3}} = \frac{\pi}{2} \sum_{n=1}^{\infty} \frac{1}{n^{1.3}}$ is the product of a convergent p-series and a nonzero constant

31. converges by the Limit Comparison Test (part 1) with $\frac{1}{n^2}$: $\lim_{n\to\infty} \frac{\left(\frac{\coth n}{n^2}\right)}{\left(\frac{1}{n^2}\right)} = \lim_{n\to\infty} \coth n = \lim_{n\to\infty} \frac{e^n+e^{-n}}{e^n-e^{-n}}$
$= \lim_{n\to\infty} \frac{1+e^{-2n}}{1-e^{-2n}} = 1$

32. converges by the Limit Comparison Test (part 1) with $\frac{1}{n^2}$: $\lim_{n\to\infty} \frac{\left(\frac{\tanh n}{n^2}\right)}{\left(\frac{1}{n^2}\right)} = \lim_{n\to\infty} \tanh n = \lim_{n\to\infty} \frac{e^n-e^{-n}}{e^n+e^{-n}}$
$= \lim_{n\to\infty} \frac{1-e^{-2n}}{1+e^{-2n}} = 1$

33. diverges: $f(x) = \sqrt[x]{x} \Rightarrow f'(x) > 0$ when $1 < x < e$ and $f'(x) < 0$ when $x > e \Rightarrow \sqrt[e]{e} > \sqrt[n]{n}$ for all $n \geq 3$; also $3^e > e \Rightarrow 3 > \sqrt[e]{e}$. Consequently, $3n > n\sqrt[n]{n} \Rightarrow \frac{1}{3n} < \frac{1}{n\sqrt[n]{n}} \Rightarrow \sum_{n=1}^{\infty} \frac{1}{n\sqrt[n]{n}}$ diverges by the Direct Comparison Test

34. converges by the Limit Comparison Test (part 1) with $\frac{1}{n^2}$: $\lim_{n\to\infty} \frac{\left(\frac{\sqrt[n]{n}}{n^2}\right)}{\left(\frac{1}{n^2}\right)} = \lim_{n\to\infty} \sqrt[n]{n} = 1$ (Table 8.1)

35. $\frac{1}{1+2+3+\ldots+n} = \frac{1}{\left(\frac{n(n+1)}{2}\right)} = \frac{2}{n(n+1)} \leq \frac{1}{n^2} \Rightarrow$ the series converges by the Direct Comparison Test

36. $\frac{1}{1+2^2+3^2+\ldots+n^2} = \frac{1}{\frac{n(n+1)(2n+1)}{6}} = \frac{6}{n(n+1)(2n+1)} \le \frac{6}{n^3} \Rightarrow$ the series converges by the Direct Comparison Test

37. (a) If $\lim_{n\to\infty} \frac{a_n}{b_n} = 0$, then there exists an integer N such that for all $n > N$, $\left|\frac{a_n}{b_n} - 0\right| < 1 \Rightarrow -1 < \frac{a_n}{b_n} < 1$ $\Rightarrow a_n < b_n$. Thus, if $\sum b_n$ converges, then $\sum a_n$ converges by the Direct Comparison Test.

(b) If $\lim_{n\to\infty} \frac{a_n}{b_n} = \infty$, then there exists an integer N such that for all $n > N$, $\frac{a_n}{b_n} > 1 \Rightarrow a_n > b_n$. Thus, if $\sum b_n$ diverges, then $\sum a_n$ diverges by the Direct Comparison Test.

38. Yes, $\sum_{n=1}^{\infty} \frac{a_n}{n}$ converges by the Direct Comparison Test because $\frac{a_n}{n} < a_n$

39. $\lim_{n\to\infty} \frac{a_n}{b_n} = \infty \Rightarrow$ there exists an integer N such that for all $n > N$, $\frac{a_n}{b_n} > 1 \Rightarrow a_n > b_n$. If $\sum a_n$ converges, then $\sum b_n$ converges by the Direct Comparison Test

40. $\sum a_n$ converges $\Rightarrow \lim_{n\to\infty} a_n = 0 \Rightarrow$ there exists an integer N such that for all $n > N$, $0 \le a_n < 1 \Rightarrow a_n^2 < a_n$ $\Rightarrow \sum a_n^2$ converges by the Direct Comparison Test

41. Example CAS commands:

Maple:

```
s:= k -> sum(1/(n^3*(sin^2)(n)), n=1..k);
limit(s(k), k=infinity);
plot(s(k), k=1..100, style=POINT, symbol=CIRCLE);
plot(s(k), k=1..200, style=POINT, symbol=CIRCLE);
plot(s(k), k=1..400, style=POINT, symbol=CIRCLE);
evalf(355/113);
```

Mathematica:

```
Clear[a,k,n,s]
a[n_] = 1/ (n^3 Sin[n]^2)
s[k_] = Sum[ a[n], {n,1,k} ]

Note: To make Mathematica smart about limits, load the package:

<< Calculus`Limit`
Limit[ s[k], k -> Infinity ]

But Mathematica still cannot find the limit...

Note: For plotting many partial sums, it is far more efficient to do the
calculations numerically rather than exactly. $o we redefine s[k] (where
the "s[k_] := s[k] = ..." causes Mathematica to remember previous results)

Clear[s]
s[k_] := s[k] = s[k-1] + N[a[k] ]
```

```
s[1] = N[a[1]]
ListPlot[ Table[ s[k], {k,100} ] ]
ListPlot[ Table[ s[k], {k,200} ] ]
ListPlot[ Table[ s[k], {k,400} ] ]
```

Note: Change PlotRange so Mathematica does not cut off the jump.

```
Show[ %, PlotRange -> All ]
N[ 355/113 ]
N[ Pi - 355/113 ]
Sin[ 355 ] // N
a[ 355 ] // N
```

8.6 THE RATIO AND ROOT TESTS FOR SERIES WITH NONNEGATIVE TERMS

1. converges by the Ratio Test: $\lim\limits_{n\to\infty} \frac{a_{n+1}}{a_n} = \lim\limits_{n\to\infty} \frac{\left[\frac{(n+1)^{\sqrt{2}}}{2^{n+1}}\right]}{\left[\frac{n^{\sqrt{2}}}{2^n}\right]} = \lim\limits_{n\to\infty} \frac{(n+1)^{\sqrt{2}}}{2^{n+1}} \cdot \frac{2^n}{n^{\sqrt{2}}}$
$= \lim\limits_{n\to\infty} \left(1+\frac{1}{n}\right)^{\sqrt{2}} \left(\frac{1}{2}\right) = \frac{1}{2} < 1$

2. converges by the Ratio Test: $\lim\limits_{n\to\infty} \frac{a_{n+1}}{a_n} = \lim\limits_{n\to\infty} \frac{\left(\frac{(n+1)^2}{e^{n+1}}\right)}{\left(\frac{n^2}{e^n}\right)} = \lim\limits_{n\to\infty} \frac{(n+1)^2}{e^{n+1}} \cdot \frac{e^n}{n^2} = \lim\limits_{n\to\infty} \left(1+\frac{1}{n}\right)^2 \left(\frac{1}{e}\right) = \frac{1}{e} < 1$

3. diverges by the Ratio Test: $\lim\limits_{n\to\infty} \frac{a_{n+1}}{a_n} = \lim\limits_{n\to\infty} \frac{\left(\frac{(n+1)!}{e^{n+1}}\right)}{\left(\frac{n!}{e^n}\right)} = \lim\limits_{n\to\infty} \frac{(n+1)!}{e^{n+1}} \cdot \frac{e^n}{n!} = \lim\limits_{n\to\infty} \frac{n+1}{e} = \infty$

4. diverges by the Ratio Test: $\lim\limits_{n\to\infty} \frac{a_{n+1}}{a_n} = \lim\limits_{n\to\infty} \frac{\left(\frac{(n+1)!}{10^{n+1}}\right)}{\left(\frac{n!}{10^n}\right)} = \lim\limits_{n\to\infty} \frac{(n+1)!}{10^{n+1}} \cdot \frac{10^n}{n!} = \lim\limits_{n\to\infty} \frac{n}{10} = \infty$

5. converges by the Ratio Test: $\lim\limits_{n\to\infty} \frac{a_{n+1}}{a_n} = \lim\limits_{n\to\infty} \frac{\left(\frac{(n+1)^{10}}{10^{n+1}}\right)}{\left(\frac{n^{10}}{10^n}\right)} = \lim\limits_{n\to\infty} \frac{(n+1)^{10}}{10^{n+1}} \cdot \frac{10^n}{n^{10}} = \lim\limits_{n\to\infty} \left(1+\frac{1}{n}\right)^{10} \left(\frac{1}{10}\right)$
$= \frac{1}{10} < 1$

6. diverges; $\lim\limits_{n\to\infty} a_n = \lim\limits_{n\to\infty} \left(\frac{n-2}{n}\right)^n = \lim\limits_{n\to\infty} \left(1+\frac{-2}{n}\right)^n = e^{-2} \neq 0$

7. converges by the Direct Comparison Test: $\frac{2+(-1)^n}{(1.25)^n} = \left(\frac{4}{5}\right)^n [2+(-1)^n] \le \left(\frac{4}{5}\right)^n (3)$

8. converges; a geometric series with $|r| = \left|-\frac{2}{3}\right| < 1$

9. diverges; $\lim\limits_{n\to\infty} a_n = \lim\limits_{n\to\infty} \left(1 - \frac{3}{n}\right)^n = \lim\limits_{n\to\infty} \left(1 + \frac{-3}{n}\right)^n = e^{-3} \approx 0.05 \neq 0$

10. diverges; $\lim\limits_{n\to\infty} a_n = \lim\limits_{n\to\infty} \left(1 - \frac{1}{3n}\right)^n = \lim\limits_{n\to\infty} \left(1 + \frac{\left(-\frac{1}{3}\right)}{n}\right)^n = e^{-1/3} \approx 0.72 \neq 0$

11. converges by the Direct Comparison Test: $\frac{\ln n}{n^3} < \frac{n}{n^3} = \frac{1}{n^2}$ for $n \geq 2$

12. converges by the nth-Root Test: $\lim\limits_{n\to\infty} \sqrt[n]{a_n} = \lim\limits_{n\to\infty} \sqrt[n]{\frac{(\ln n)^n}{n^n}} = \lim\limits_{n\to\infty} \frac{((\ln n)^n)^{1/n}}{(n^n)^{1/n}} = \lim\limits_{n\to\infty} \frac{\ln n}{n}$

$= \lim\limits_{n\to\infty} \frac{\left(\frac{1}{n}\right)}{1} = 0 < 1$

13. diverges by the Direct Comparison Test: $\frac{1}{n} - \frac{1}{n^2} = \frac{n-1}{n^2} > \frac{1}{2}\left(\frac{1}{n}\right)$ for $n > 2$

14. converges by the nth-Root Test: $\lim\limits_{n\to\infty} \sqrt[n]{a_n} = \lim\limits_{n\to\infty} \sqrt[n]{\left(\frac{1}{n} - \frac{1}{n^2}\right)^n} = \lim\limits_{n\to\infty} \left(\left(\frac{1}{n} - \frac{1}{n^2}\right)^n\right)^{1/n}$

$= \lim\limits_{n\to\infty} \left(\frac{1}{n} - \frac{1}{n^2}\right) = 0 < 1$

15. diverges by the Direct Comparison Test: $\frac{\ln n}{n} > \frac{1}{n}$ for $n \geq 3$

16. converges by the Ratio Test: $\lim\limits_{n\to\infty} \frac{a_{n+1}}{a_n} = \lim\limits_{n\to\infty} \frac{(n+1)\ln(n+1)}{2^{n+1}} \cdot \frac{2^n}{n \ln(n)} = \frac{1}{2} < 1$

17. converges by the Ratio Test: $\lim\limits_{n\to\infty} \frac{a_{n+1}}{a_n} = \lim\limits_{n\to\infty} \frac{(n+2)(n+3)}{(n+1)!} \cdot \frac{n!}{(n+1)(n+2)} = 0 < 1$

18. converges by the Ratio Test: $\lim\limits_{n\to\infty} \frac{a_{n+1}}{a_n} = \lim\limits_{n\to\infty} \frac{(n+1)^3}{e^{n+1}} \cdot \frac{e^n}{n^3} = \frac{1}{e} < 1$

19. converges by the Ratio Test: $\lim\limits_{n\to\infty} \frac{a_{n+1}}{a_n} = \lim\limits_{n\to\infty} \frac{(n+4)!}{3!\,(n+1)!\,3^{n+1}} \cdot \frac{3!\,n!\,3^n}{(n+3)!} = \lim\limits_{n\to\infty} \frac{n+4}{3(n+1)} = \frac{1}{3} < 1$

20. converges by the Ratio Test: $\lim\limits_{n\to\infty} \frac{a_{n+1}}{a_n} = \lim\limits_{n\to\infty} \frac{(n+1)2^{n+1}(n+2)!}{3^{n+1}(n+1)!} \cdot \frac{3^n n!}{n2^n(n+1)!}$

$= \lim\limits_{n\to\infty} \left(\frac{n+1}{n}\right)\left(\frac{2}{3}\right)\left(\frac{n+2}{n+1}\right) = \frac{2}{3} < 1$

21. converges by the Ratio Test: $\lim_{n\to\infty} \frac{a_{n+1}}{a_n} = \lim_{n\to\infty} \frac{(n+1)!}{(2n+3)!} \cdot \frac{(2n+1)!}{n!} = \lim_{n\to\infty} \frac{n+1}{(2n+3)(2n+2)} = 0 < 1$

22. converges by the Ratio Test: $\lim_{n\to\infty} \frac{a_{n+1}}{a_n} = \lim_{n\to\infty} \frac{(n+1)!}{(n+1)^{n+1}} \cdot \frac{n^n}{n!} = \lim_{n\to\infty} \left(\frac{n}{n+1}\right)^n = \lim_{n\to\infty} \frac{1}{\left(\frac{n+1}{n}\right)^n}$

$= \lim_{n\to\infty} \frac{1}{\left(1+\frac{1}{n}\right)^n} = \frac{1}{e} < 1$

23. converges by the Root Test: $\lim_{n\to\infty} \sqrt[n]{a_n} = \lim_{n\to\infty} \sqrt[n]{\frac{n}{(\ln n)^n}} = \lim_{n\to\infty} \frac{\sqrt[n]{n}}{\ln n} = \lim_{n\to\infty} \frac{1}{\ln n} = 0 < 1$

24. converges by the Root Test: $\lim_{n\to\infty} \sqrt[n]{a_n} = \lim_{n\to\infty} \sqrt[n]{\frac{n}{(\ln n)^{n/2}}} = \lim_{n\to\infty} \frac{\sqrt[n]{n}}{\sqrt{\ln n}} = \frac{\lim_{n\to\infty} \sqrt[n]{n}}{\lim_{n\to\infty} \sqrt{\ln n}} = 0 < 1$

$\left(\lim_{n\to\infty} \sqrt[n]{n} = 1\right)$

25. converges by the Direct Comparison Test: $\frac{n! \ln n}{n(n+2)!} = \frac{\ln n}{n(n+1)(n+2)} < \frac{n}{n(n+1)(n+2)} = \frac{1}{(n+1)(n+2)} < \frac{1}{n^2}$ which is the nth-term of a convergent p-series

26. converges by the Ratio Test: $\lim_{n\to\infty} \frac{a_{n+1}}{a_n} = \lim_{n\to\infty} \frac{3(n+1)}{(n+1)^3 2^{n+1}} \cdot \frac{n^3 2^n}{3n} = \lim_{n\to\infty} \frac{n^2}{(n+1)^2}\left(\frac{1}{2}\right) = \frac{1}{2} < 1$

27. converges by the Ratio Test: $\lim_{n\to\infty} \frac{a_{n+1}}{a_n} = \lim_{n\to\infty} \frac{\left(\frac{1+\sin n}{n}\right)a_n}{a_n} = 0 < 1$

28. converges by the Ratio Test: $\lim_{n\to\infty} \frac{a_{n+1}}{a_n} = \lim_{n\to\infty} \frac{\left(\frac{1+\tan^{-1} n}{n}\right)a_n}{a_n} = \lim_{n\to\infty} \frac{1+\tan^{-1} n}{n} = 0$ since the numerator approaches $1+\frac{\pi}{2}$ while the denominator tends to ∞

29. diverges by the Ratio Test: $\lim_{n\to\infty} \frac{a_{n+1}}{a_n} = \lim_{n\to\infty} \frac{\left(\frac{3n-1}{2n+1}\right)a_n}{a_n} = \lim_{n\to\infty} \frac{3n-1}{2n+1} = \frac{3}{2} > 1$

30. diverges; $a_{n+1} = \frac{n}{n+1} a_n \Rightarrow a_{n+1} = \left(\frac{n}{n+1}\right)\left(\frac{n-1}{n} a_{n-1}\right) \Rightarrow a_{n+1} = \left(\frac{n}{n+1}\right)\left(\frac{n-1}{n}\right)\left(\frac{n-2}{n-1} a_{n-2}\right)$

$\Rightarrow a_{n+1} = \left(\frac{n}{n+1}\right)\left(\frac{n-1}{n}\right)\left(\frac{n-2}{n-1}\right)\cdots\left(\frac{1}{2}\right)a_1 \Rightarrow a_{n+1} = \frac{a_1}{n+1} \Rightarrow a_{n+1} = \frac{3}{n+1}$, which is a constant times the general term of the diverging harmonic series

31. converges by the Ratio Test: $\lim_{n\to\infty} \frac{a_{n+1}}{a_n} = \lim_{n\to\infty} \frac{\left(\frac{2}{n}\right)a_n}{a_n} = \lim_{n\to\infty} \frac{2}{n} = 0 < 1$

32. converges by the Ratio Test: $\lim_{n\to\infty} \frac{a_{n+1}}{a_n} = \lim_{n\to\infty} \frac{\left(\frac{\sqrt[n]{n}}{2}\right)a_n}{a_n} = \lim_{n\to\infty} \frac{\sqrt[n]{n}}{n} = \frac{1}{2} < 1$

33. converges by the Ratio Test: $\lim\limits_{n\to\infty} \frac{a_{n+1}}{a_n} = \lim\limits_{n\to\infty} \frac{\left(\frac{1+\ln n}{n}\right)a_n}{a_n} = \lim\limits_{n\to\infty} \frac{1+\ln n}{n} = \lim\limits_{n\to\infty} \frac{1}{n} = 0 < 1$

34. $\frac{n+\ln n}{n+10} > 0$ and $a_1 = \frac{1}{2} \Rightarrow a_n > 0$; $\ln n > 10$ for $n > e^{10} \Rightarrow n + \ln n > n + 10 \Rightarrow \frac{n+\ln n}{n+10} > 1$
$\Rightarrow a_{n+1} = \frac{n+\ln n}{n+10}\, a_n > a_n$; thus $a_{n+1} > a_n \geq \frac{1}{2} \Rightarrow \lim\limits_{n\to\infty} a_n \neq 0$, so the series diverges by the nth-Term Test

35. diverges by the nth-Term Test: $a_1 = \frac{1}{3}$, $a_2 = \sqrt[2]{\frac{1}{3}}$, $a_3 = \sqrt[3]{\sqrt[2]{\frac{1}{3}}} = \sqrt[6]{\frac{1}{3}}$, $a_4 = \sqrt[4]{\sqrt[3]{\sqrt[2]{\frac{1}{3}}}} = \sqrt[4!]{\frac{1}{3}}, \ldots,$
$a_n = \sqrt[n!]{\frac{1}{3}} \Rightarrow \lim\limits_{n\to\infty} a_n = 1$ because $\left\{\sqrt[n!]{\frac{1}{3}}\right\}$ is a subsequence of $\left\{\sqrt[n]{\frac{1}{3}}\right\}$ whose limit is 1 by Table 8.1

36. converges by the Direct Comparison Test: $a_1 = \frac{1}{2}$, $a_2 = \left(\frac{1}{2}\right)^2$, $a_3 = \left(\left(\frac{1}{2}\right)^2\right)^3 = \left(\frac{1}{2}\right)^6$, $a_4 = \left(\left(\frac{1}{2}\right)^6\right)^4 = \left(\frac{1}{2}\right)^{24}, \ldots$
$\Rightarrow a_n = \left(\frac{1}{2}\right)^{n!} < \left(\frac{1}{2}\right)^n$ which is the nth-term of a convergent geometric series

37. converges by the Ratio Test: $\lim\limits_{n\to\infty} \frac{a_{n+1}}{a_n} = \lim\limits_{n\to\infty} \frac{2^{n+1}(n+1)!\,(n+1)!}{(2n+2)!} \cdot \frac{(2n)!}{2^n n!\, n!} = \lim\limits_{n\to\infty} \frac{2(n+1)(n+1)}{(2n+2)(2n+1)}$
$= \lim\limits_{n\to\infty} \frac{n+1}{2n+1} = \frac{1}{2} < 1$

38. diverges by the Ratio Test: $\lim\limits_{n\to\infty} \frac{a_{n+1}}{a_n} = \lim\limits_{n\to\infty} \frac{(3n+3)!}{(n+1)!\,(n+2)!\,(n+3)!} \cdot \frac{n!\,(n+1)!\,(n+2)!}{(3n)!}$
$= \lim\limits_{n\to\infty} \frac{(3n+3)(3+2)(3n+1)}{(n+1)(n+2)(n+3)} = \lim\limits_{n\to\infty} 3\left(\frac{3n+2}{n+2}\right)\left(\frac{3n+1}{n+3}\right) = 3 \cdot 3 \cdot 3 = 27 > 1$

39. diverges by the Root Test: $\lim\limits_{n\to\infty} \sqrt[n]{a_n} \equiv \lim\limits_{n\to\infty} \sqrt[n]{\frac{(n!)^n}{(n^n)^2}} = \lim\limits_{n\to\infty} \frac{n!}{n^2} = \infty > 1$

40. converges by the Root Test: $\lim\limits_{n\to\infty} \sqrt[n]{\frac{(n!)^n}{n^{n^2}}} = \lim\limits_{n\to\infty} \sqrt[n]{\frac{(n!)^n}{(n^n)^n}} = \lim\limits_{n\to\infty} \frac{n!}{n^n} = \lim\limits_{n\to\infty} \left(\frac{1}{n}\right)\left(\frac{2}{n}\right)\left(\frac{3}{n}\right)\cdots\left(\frac{n-1}{n}\right)\left(\frac{n}{n}\right)$
$\leq \lim\limits_{n\to\infty} \left(\frac{n-1}{n}\right)^{n-1} = \lim\limits_{n\to\infty} \frac{\left(\frac{n-1}{n}\right)^n}{\left(\frac{n-1}{n}\right)} = \lim\limits_{n\to\infty} \frac{\left(1-\frac{1}{n}\right)^n}{\left(1-\frac{1}{n}\right)} = \frac{e^{-1}}{1} = \frac{1}{e} < 1$

41. converges by the Root Test: $\lim\limits_{n\to\infty} \sqrt[n]{a_n} = \lim\limits_{n\to\infty} \sqrt[n]{\frac{n^n}{2^{n^2}}} = \lim\limits_{n\to\infty} \frac{n}{2^n} = \lim\limits_{n\to\infty} \frac{1}{2^n \ln 2} = 0 < 1$

42. diverges by the Root Test: $\lim\limits_{n\to\infty} \sqrt[n]{a_n} = \lim\limits_{n\to\infty} \sqrt[n]{\frac{n^n}{(2^n)^2}} = \lim\limits_{n\to\infty} \frac{n}{4} = \infty > 1$

43. converges by the Ratio Test: $\lim_{n\to\infty} \frac{a_{n+1}}{a_n} = \lim_{n\to\infty} \frac{1\cdot 3\cdot \cdots \cdot(2n-1)(2n+1)}{4^{n+1}2^{n+1}(n+1)!}\cdot\frac{4^n 2^n n!}{1\cdot 3\cdot\cdots\cdot(2n-1)}$

$= \lim_{n\to\infty} \frac{2n+1}{(4\cdot 2)(n+1)} = \frac{1}{4} < 1$

44. converges by the Ratio Test: $a_n = \frac{1\cdot 3\cdots(2n-1)}{(2\cdot 4\cdots 2n)(3^n+1)} = \frac{1\cdot 2\cdot 3\cdot 4\cdots(2n-1)(2n)}{(2\cdot 4\cdots 2n)^2(3^n+1)} = \frac{(2n)!}{(2^n n!)^2(3^n+1)}$

$\Rightarrow \lim_{n\to\infty} \frac{(2n+2)!}{[2^{n+1}(n+1)!]^2(3^{n+1}+1)}\cdot\frac{(2^n n!)^2(3^n+1)}{(2n)!} = \lim_{n\to\infty} \frac{(2n+1)(2n+2)(3^n+1)}{2^2(n+1)^2(3^{n+1}+1)}$

$= \lim_{n\to\infty} \left(\frac{4n^2+6n+2}{4n^2+8n+4}\right)\frac{(1+3^{-n})}{(3+3^{-n})} = 1\cdot\frac{1}{3} = \frac{1}{3} < 1$

45. Ratio: $\lim_{n\to\infty} \frac{a_{n+1}}{a_n} = \lim_{n\to\infty} \frac{1}{(n+1)^p}\cdot\frac{n^p}{1} = \lim_{n\to\infty} \left(\frac{n}{n+1}\right)^p = 1^p = 1 \Rightarrow$ no conclusion

Root: $\lim_{n\to\infty} \sqrt[n]{a_n} = \lim_{n\to\infty} \sqrt[n]{\frac{1}{n^p}} = \lim_{n\to\infty} \frac{1}{(\sqrt[n]{n})^p} = \frac{1}{(1)^p} = 1 \Rightarrow$ no conclusion

46. Ratio: $\lim_{n\to\infty} \frac{a_{n+1}}{a_n} = \lim_{n\to\infty} \frac{1}{(\ln(n+1))^p}\cdot\frac{(\ln n)^p}{1} = \left[\lim_{n\to\infty} \frac{\ln n}{\ln(n+1)}\right]^p = \left[\lim_{n\to\infty} \frac{\left(\frac{1}{n}\right)}{\left(\frac{1}{n+1}\right)}\right]^p = \left(\lim_{n\to\infty} \frac{n+1}{n}\right)^p$

$= (1)^p = 1 \Rightarrow$ no conclusion

Root: $\lim_{n\to\infty} \sqrt[n]{a_n} = \lim_{n\to\infty} \sqrt[n]{\frac{1}{(\ln n)^p}} = \frac{1}{\left(\lim_{n\to\infty} (\ln n)^{1/n}\right)^p}$; let $f(n) = (\ln n)^{1/n}$, then $\ln f(n) = \frac{\ln(\ln n)}{n}$

$\Rightarrow \lim_{n\to\infty} \ln f(n) = \lim_{n\to\infty} \frac{\ln(\ln n)}{n} = \lim_{n\to\infty} \frac{\left(\frac{1}{n \ln n}\right)}{1} = \lim_{n\to\infty} \frac{1}{n \ln n} = 0 \Rightarrow \lim_{n\to\infty} (\ln n)^{1/n}$

$= \lim_{n\to\infty} e^{\ln f(n)} = e^0 = 1$; therefore $\lim_{n\to\infty} \sqrt[n]{a_n} = \frac{1}{\left(\lim_{n\to\infty} (\ln n)^{1/n}\right)^p} = \frac{1}{(1)^p} = 1 \Rightarrow$ no conclusion

47. $a_n \le \frac{n}{2^n}$ for every n and the series $\sum_{n=1}^{\infty} \frac{n}{2^n}$ converges by the Ratio Test since $\lim_{n\to\infty} \frac{(n+1)}{2^{n+1}}\cdot\frac{2^n}{n} = \frac{1}{2} < 1$

$\Rightarrow \sum_{n=1}^{\infty} a_n$ converges by the Direct Comparison Test

8.7 ALTERNATING SERIES, ABSOLUTE AND CONDITIONAL CONVERGENCE

1. converges absolutely $\Rightarrow$ converges by the Absolute Convergence Test since $\sum_{n=1}^{\infty} |a_n| = \sum_{n=1}^{\infty} \frac{1}{n^2}$ which is a convergent p-series

2. converges absolutely $\Rightarrow$ converges by the Absolute Convergence Test since $\sum_{n=1}^{\infty} |a_n| = \sum_{n=1}^{\infty} \frac{1}{n^{3/2}}$ which is a convergent p-series

3. diverges by the nth-Term Test since for $n > 10 \Rightarrow \frac{n}{10} > 1 \Rightarrow \lim_{n\to\infty} \left(\frac{n}{10}\right)^n \neq 0 \Rightarrow \sum_{n=1}^{\infty} (-1)^{n+1}\left(\frac{n}{10}\right)^n$ diverges

4. diverges by the nth-Term Test since $\lim_{n\to\infty} \frac{10^n}{n^{10}} = \lim_{n\to\infty} \frac{10^n(\ln 10)^{10}}{10!} = \infty$ (after 10 applications of L'Hôpital's rule)

5. converges by the Alternating Series Test because $f(x) = \ln x$ is an increasing function of $x \Rightarrow \frac{1}{\ln x}$ is decreasing $\Rightarrow u_n \geq u_{n+1}$ for $n \geq 1$; also $u_n \geq 0$ for $n \geq 1$ and $\lim_{n\to\infty} \frac{1}{\ln n} = 0$

6. converges by the Alternating Series Test since $f(x) = \frac{\ln x}{x} \Rightarrow f'(x) = \frac{1 - \ln x}{x^2} < 0$ when $x > e \Rightarrow f(x)$ is decreasing $\Rightarrow u_n \geq u_{n+1}$; also $u_n \geq 0$ for $n \geq 1$ and $\lim_{n\to\infty} u_n = \lim_{n\to\infty} \frac{\ln n}{n} = \lim_{n\to\infty} \frac{\left(\frac{1}{n}\right)}{1} = 0$

7. diverges by the nth-Term Test since $\lim_{n\to\infty} \frac{\ln n}{\ln n^2} = \lim_{n\to\infty} \frac{\ln n}{2 \ln n} = \lim_{n\to\infty} \frac{1}{2} = \frac{1}{2} \neq 0$

8. converges by the Alternating Series Test since $f(x) = \ln\left(1 + x^{-1}\right) \Rightarrow f'(x) = \frac{-1}{x(x+1)} < 0$ for $x > 0 \Rightarrow f(x)$ is decreasing $\Rightarrow u_n \geq u_{n+1}$; also $u_n \geq 0$ for $n \geq 1$ and $\lim_{n\to\infty} u_n = \lim_{n\to\infty} \ln\left(1 + \frac{1}{n}\right) = \ln\left(\lim_{n\to\infty} \left(1 + \frac{1}{n}\right)\right) = \ln 1 = 0$

9. converges by the Alternating Series Test since $f(x) = \frac{\sqrt{x}+1}{x+1} \Rightarrow f'(x) = \frac{1 - x - 2\sqrt{x}}{2\sqrt{x}\,(x+1)^2} < 0 \Rightarrow f(x)$ is decreasing $\Rightarrow u_n \geq u_{n+1}$; also $u_n \geq 0$ for $n \geq 1$ and $\lim_{n\to\infty} u_n = \lim_{n\to\infty} \frac{\sqrt{n}+1}{n+1} = 0$

10. diverges by the nth-Term Test since $\lim_{n\to\infty} \frac{3\sqrt{n+1}}{\sqrt{n}+1} = \lim_{n\to\infty} \frac{3\sqrt{1+\frac{1}{n}}}{1+\left(\frac{1}{\sqrt{n}}\right)} = 3 \neq 0$

11. converges absolutely since $\sum_{n=1}^{\infty} |a_n| = \sum_{n=1}^{\infty} \left(\frac{1}{10}\right)^n$ a convergent geometric series

12. converges absolutely by the Direct Comparison Test since $\left|\frac{(-1)^{n+1}(0.1)^n}{n}\right| = \frac{1}{(10)^n n} < \left(\frac{1}{10}\right)^n$ which is the nth term of a convergent geometric series

13. converges conditionally since $\frac{1}{\sqrt{n}} > \frac{1}{\sqrt{n+1}} > 0$ and $\lim_{n\to\infty} \frac{1}{\sqrt{n}} = 0 \Rightarrow$ convergence; but $\sum_{n=1}^{\infty} |a_n| = \sum_{n=1}^{\infty} \frac{1}{n^{1/2}}$ is a divergent p-series

14. converges conditionally since $\frac{1}{1+\sqrt{n}} > \frac{1}{1+\sqrt{n+1}} > 0$ and $\lim_{n\to\infty} \frac{1}{1+\sqrt{n}} = 0 \Rightarrow$ convergence; but $\sum_{n=1}^{\infty} |a_n| = \sum_{n=1}^{\infty} \frac{1}{1+\sqrt{n}}$ is a divergent series since $\frac{1}{1+\sqrt{n}} \ge \frac{1}{2\sqrt{n}}$ and $\sum_{n=1}^{\infty} \frac{1}{n^{1/2}}$ is a divergent p-series

15. converges absolutely since $\sum_{n=1}^{\infty} |a_n| = \sum_{n=1}^{\infty} \frac{n}{n^3+1}$ and $\frac{n}{n^3+1} < \frac{1}{n^2}$ which is the nth-term of a converging p-series

16. diverges by the nth-Term Test since $\lim_{n\to\infty} \frac{n!}{2^n} = \infty$ (Table 8.1)

17. converges conditionally since $\frac{1}{n+3} > \frac{1}{(n+1)+3} > 0$ and $\lim_{n\to\infty} \frac{1}{n+3} = 0 \Rightarrow$ convergence; but $\sum_{n=1}^{\infty} |a_n|$ $= \sum_{n=1}^{\infty} \frac{1}{n+3}$ diverges because $\frac{1}{n+3} \ge \frac{1}{4n}$ and $\sum_{n=1}^{\infty} \frac{1}{n}$ is a divergent series

18. converges absolutely because the series $\sum_{n=1}^{\infty} \left|\frac{\sin n}{n^2}\right|$ converges by the Direct Comparison Test since $\left|\frac{\sin n}{n^2}\right| \le \frac{1}{n^2}$

19. diverges by the nth-Term Test since $\lim_{n\to\infty} \frac{3+n}{5+n} = 1 \ne 0$

20. converges conditionally since $f(x) = \ln x$ is an increasing function of $x \Rightarrow \frac{1}{3 \ln x} = \frac{1}{\ln(x^3)}$ is decreasing $\Rightarrow \frac{1}{3 \ln n} > \frac{1}{3 \ln (n+1)} > 0$ for $n \ge 2$ and $\lim_{n\to\infty} \frac{1}{3 \ln n} = 0 \Rightarrow$ convergence; but $\sum_{n=2}^{\infty} |a_n| = \sum_{n=2}^{\infty} \frac{1}{\ln(n^3)}$ $= \sum_{n=2}^{\infty} \frac{1}{3 \ln n}$ diverges because $\frac{1}{3 \ln n} > \frac{1}{3n}$ and $\sum_{n=2}^{\infty} \frac{1}{n}$ diverges

21. converges conditionally since $f(x) = \frac{1}{x^2} + \frac{1}{x} \Rightarrow f'(x) = -\left(\frac{2}{x^3} + \frac{1}{x^2}\right) < 0 \Rightarrow f(x)$ is decreasing and hence $u_n > u_{n+1} > 0$ for $n \ge 1$ and $\lim_{n\to\infty} \left(\frac{1}{n^2} + \frac{1}{n}\right) = 0 \Rightarrow$ convergence; but $\sum_{n=1}^{\infty} |a_n| = \sum_{n=1}^{\infty} \frac{1+n}{n^2}$ $= \sum_{n=1}^{\infty} \frac{1}{n^2} + \sum_{n=1}^{\infty} \frac{1}{n}$ is the sum of a convergent and divergent series, and hence diverges

22. converges absolutely by the Direct Comparison Test since $\left|\frac{(-2)^{n+1}}{n+5^n}\right| = \frac{2^{n+1}}{n+5^n} < 2\left(\frac{2}{5}\right)^n$ which is the nth term of a convergent geometric series

23. converges absolutely by the Ratio Test: $\lim_{n\to\infty} \left(\frac{u_{n+1}}{u_n}\right) = \lim_{n\to\infty} \left[\frac{(n+1)^2\left(\frac{2}{3}\right)^{n+1}}{n^2\left(\frac{2}{3}\right)^n}\right] = \frac{2}{3} < 1$

24. diverges by the nth-Term Test since $\lim_{n\to\infty} a_n = \lim_{n\to\infty} 10^{1/n} = 1 \neq 0$ (Table 8.1)

25. converges absolutely by the Integral Test since $\int_1^{\infty} (\tan^{-1} x)\left(\frac{1}{1+x^2}\right) dx = \lim_{b\to\infty} \left[\frac{(\tan^{-1} x)^2}{2}\right]_1^b$

$= \lim_{b\to\infty} \left[(\tan^{-1} b)^2 - (\tan^{-1} 1)^2\right] = \frac{1}{2}\left[\left(\frac{\pi}{2}\right)^2 - \left(\frac{\pi}{4}\right)^2\right] = \frac{3\pi^2}{32}$

26. converges conditionally since $f(x) = \frac{1}{x \ln x} \Rightarrow f'(x) = -\frac{[\ln(x) + 1]}{(x \ln x)^2} < 0 \Rightarrow f(x)$ is decreasing

$\Rightarrow u_n > u_{n+1} > 0$ for $n \geq 2$ and $\lim_{n\to\infty} \frac{1}{n \ln n} = 0 \Rightarrow$ convergence; but by the Integral Test,

$\int_2^{\infty} \frac{dx}{x \ln x} = \lim_{b\to\infty} \int_2^b \left(\frac{\left(\frac{1}{x}\right)}{\ln x}\right) dx = \lim_{b\to\infty} [\ln(\ln x)]_2^b = \lim_{b\to\infty} [\ln(\ln b) - \ln(\ln 2)] = \infty$

$\Rightarrow \sum_{n=1}^{\infty} |a_n| = \sum_{n=1}^{\infty} \frac{1}{n \ln n}$ diverges

27. diverges by the nth-Term Test since $\lim_{n\to\infty} \frac{n}{n+1} = 1 \neq 0$

28. converges conditionally since $f(x) = \frac{\ln x}{x - \ln x} \Rightarrow f'(x) = \frac{\left(\frac{1}{x}\right)(x - \ln x) - (\ln x)\left(1 - \frac{1}{x}\right)}{(x - \ln x)^2}$

$= \frac{1 - \left(\frac{\ln x}{x}\right) - \ln x + \left(\frac{\ln x}{x}\right)}{(x - \ln x)^2} = \frac{1 - \ln x}{(x - \ln x)^2} < 0 \Rightarrow u_n \geq u_{n+1} > 0$ when $n > e$ and $\lim_{n\to\infty} \frac{\ln n}{n - \ln n}$

$= \lim_{n\to\infty} \frac{\left(\frac{1}{n}\right)}{1 - \left(\frac{1}{n}\right)} = 0 \Rightarrow$ convergence; but $n - \ln n < n \Rightarrow \frac{1}{n - \ln n} > \frac{1}{n} \Rightarrow \frac{\ln n}{n - \ln n} > \frac{1}{n}$ so that

$\sum_{n=1}^{\infty} |a_n| = \sum_{n=1}^{\infty} \frac{\ln n}{n - \ln n}$ diverges by the Direct Comparison Test

29. converges absolutely by the Ratio Test: $\lim_{n\to\infty} \left(\frac{u_{n+1}}{u_n}\right) = \lim_{n\to\infty} \frac{(100)^{n+1}}{(n+1)!} \cdot \frac{n!}{(100)^n} = \lim_{n\to\infty} \frac{100}{n+1} = 0 < 1$

30. converges absolutely since $\sum_{n=1}^{\infty} |a_n| = \sum_{n=1}^{\infty} \left(\frac{1}{5}\right)^n$ is a convergent geometric series

31. converges absolutely by the Direct Comparison Test since $\sum_{n=1}^{\infty} |a_n| = \sum_{n=1}^{\infty} \frac{1}{n^2 + 2n + 1}$ and

$\frac{1}{n^2 + 2n + 1} < \frac{1}{n^2}$ which is the nth-term of a convergent p-series

32. converges absolutely since $\sum_{n=1}^{\infty} |a_n| = \sum_{n=1}^{\infty} \left(\frac{\ln n}{\ln n^2}\right)^n = \sum_{n=1}^{\infty} \left(\frac{\ln n}{2 \ln n}\right)^n = \sum_{n=1}^{\infty} \left(\frac{1}{2}\right)^n$ is a convergent geometric series

33. converges absolutely since $\sum_{n=1}^{\infty} |a_n| = \sum_{n=1}^{\infty} \left|\frac{(-1)^n}{n\sqrt{n}}\right| = \sum_{n=1}^{\infty} \frac{1}{n^{3/2}}$ is a convergent p-series

34. converges conditionally since $\sum_{n=1}^{\infty} \frac{\cos n\pi}{n} = \sum_{n=1}^{\infty} \frac{(-1)^n}{n}$ is the convergent alternating harmonic series, but $\sum_{n=1}^{\infty} |a_n| = \sum_{n=1}^{\infty} \frac{1}{n}$ diverges

35. converges absolutely by the Root Test: $\lim_{n\to\infty} \sqrt[n]{|a_n|} = \lim_{n\to\infty} \left(\frac{(n+1)^n}{(2n)^n}\right)^{1/n} = \lim_{n\to\infty} \frac{n+1}{2n} = \frac{1}{2}$

36. converges absolutely by the Ratio Test: $\lim_{n\to\infty} \left|\frac{a_{n+1}}{a_n}\right| = \lim_{n\to\infty} \frac{((n+1)!)^2}{((2n+2)!)} \cdot \frac{(2n)!}{(n!)^2} = \lim_{n\to\infty} \frac{(n+1)^2}{(2n+2)(2n+1)} = \frac{1}{4} < 1$

37. diverges by the nth-Term Test since $\lim_{n\to\infty} |a_n| = \lim_{n\to\infty} \frac{(2n)!}{2^n n!\, n} = \lim_{n\to\infty} \frac{(n+1)(n+2)\cdots(2n)}{2^n n}$ $= \lim_{n\to\infty} \frac{(n+1)(n+2)\cdots(n+(n-1))}{2^{n-1}} > \lim_{n\to\infty} \left(\frac{n+1}{2}\right)^{n-1} = \infty \neq 0$

38. converges absolutely by the Ratio Test: $\lim_{n\to\infty} \left|\frac{a_{n+1}}{a_n}\right| = \lim_{n\to\infty} \frac{(n+1)!\,(n+1)!\,3^{n+1}}{(2n+3)!} \cdot \frac{(2n+1)!}{n!\,n!\,3^n}$ $= \lim_{n\to\infty} \frac{(n+1)^2 3}{(2n+2)(2n+3)} = \frac{3}{4} < 1$

39. converges conditionally since $\frac{\sqrt{n+1} - \sqrt{n}}{1} \cdot \frac{\sqrt{n+1} + \sqrt{n}}{\sqrt{n+1} + \sqrt{n}} = \frac{1}{\sqrt{n+1} + \sqrt{n}}$ and $\left\{\frac{1}{\sqrt{n+1} + \sqrt{n}}\right\}$ is a decreasing sequence of positive terms which converges to 0 $\Rightarrow \sum_{n=1}^{\infty} \frac{(-1)^n}{\sqrt{n+1} + \sqrt{n}}$ converges; but $n > \frac{1}{3} \Rightarrow 3n > 1$ $\Rightarrow 4n > n+1 \Rightarrow 2\sqrt{n} > \sqrt{n+1} \Rightarrow 3\sqrt{n} > \sqrt{n+1} + \sqrt{n} \Rightarrow \frac{1}{3\sqrt{n}} < \frac{1}{\sqrt{n+1} + \sqrt{n}} \Rightarrow \sum_{n=1}^{\infty} \frac{1}{\sqrt{n+1} + \sqrt{n}}$ diverges by the Direct Comparison Test

40. diverges by the nth-Term Test since $\lim_{n\to\infty} \left(\sqrt{n^2+n} - n\right) = \lim_{n\to\infty} \left(\sqrt{n^2+n} - n\right) \cdot \left(\frac{\sqrt{n^2+n} + n}{\sqrt{n^2+n} + n}\right)$ $= \lim_{n\to\infty} \frac{n}{\sqrt{n^2+n} + n} = \lim_{n\to\infty} \frac{1}{\sqrt{1 + \frac{1}{n}} + 1} = \frac{1}{2} \neq 0$

41. diverges by the nth-Term Test since $\lim_{n\to\infty}\left(\sqrt{n+\sqrt{n}}-\sqrt{n}\right)=\lim_{n\to\infty}\left[\left(\sqrt{n+\sqrt{n}}-\sqrt{n}\right)\left(\frac{\sqrt{n+\sqrt{n}}+\sqrt{n}}{\sqrt{n+\sqrt{n}}+\sqrt{n}}\right)\right]$

$=\lim_{n\to\infty}\frac{\sqrt{n}}{\sqrt{n+\sqrt{n}}+\sqrt{n}}=\lim_{n\to\infty}\frac{1}{\sqrt{1+\frac{1}{\sqrt{n}}}+1}=\frac{1}{2}\neq 0$

42. converges conditionally since $\left\{\frac{1}{\sqrt{n}+\sqrt{n+1}}\right\}$ is a decreasing sequence of positive terms converging to 0

$\Rightarrow \sum_{n=1}^{\infty}\frac{(-1)^n}{\sqrt{n}+\sqrt{n+1}}$ converges; but $\lim_{n\to\infty}\frac{\left(\frac{1}{\sqrt{n}+\sqrt{n+1}}\right)}{\left(\frac{1}{\sqrt{n}}\right)}=\lim_{n\to\infty}\frac{\sqrt{n}}{\sqrt{n}+\sqrt{n+1}}=\lim_{n\to\infty}\frac{1}{1+\sqrt{1+\frac{1}{n}}}=1$

so that $\sum_{n=1}^{\infty}\frac{1}{\sqrt{n}+\sqrt{n+1}}$ diverges by the Limit Comparison Test with $\sum_{n=1}^{\infty}\frac{1}{\sqrt{n}}$ which is a divergent p-series

43. converges absolutely by the Direct Comparison Test since $\operatorname{sech}(n)=\frac{2}{e^n+e^{-n}}=\frac{2e^n}{e^{2n}+1}<\frac{2e^n}{e^{2n}}=\frac{2}{e^n}$ which is the nth term of a convergent geometric series

44. converges absolutely by the Integral Test since $\int_1^{\infty}\operatorname{csch} x\,dx=\int_1^{\infty}\left(\frac{2}{e^x-e^{-x}}\cdot\frac{e^x}{e^x}\right)dx=-2\int_1^{\infty}\frac{e^x}{1-(e^x)^2}\,dx$

$=-2\lim_{b\to\infty}\int_1^b\frac{e^x}{1-(e^x)^2}\,dx=-2\lim_{b\to\infty}\left[\coth^{-1}e^x\right]_1^b=-2\lim_{b\to\infty}\left[\coth^{-1}(e^b)-\coth^{-1}e\right]$

$=-2\lim_{b\to\infty}\left[\frac{1}{2}\ln\left(\frac{e^b+1}{e^n-1}\right)-\frac{1}{2}\ln\left(\frac{e+1}{e-1}\right)\right]=\ln\left(\frac{e+1}{e-1}\right)-\ln\left(\lim_{n\to\infty}\left(\frac{e^b+1}{e^b-1}\right)\right)=\ln\left(\frac{e+1}{e-1}\right)-\ln 1\approx 0.77$

$\Rightarrow \sum_{n=1}^{\infty}|a_n|=\sum_{n=1}^{\infty}\operatorname{csch} n$ converges

45. $|\text{error}|<\left|(-1)^6\left(\frac{1}{5}\right)\right|=0.2$

46. $|\text{error}|<\left|(-1)^6\left(\frac{1}{10^5}\right)\right|=0.00001$

47. $|\text{error}|<\left|(-1)^6\frac{(0.01)^5}{5}\right|=2\times10^{-11}$

48. $|\text{error}|<\left|(-1)^4t^4\right|=t^4<1$

49. $\frac{1}{(2n)!}<\frac{5}{10^6}\Rightarrow(2n)!>\frac{10^6}{5}=200{,}000\Rightarrow n\geq5\Rightarrow 1-\frac{1}{2!}+\frac{1}{4!}-\frac{1}{6!}+\frac{1}{8!}\approx0.54030$

50. $\frac{1}{n!}<\frac{5}{10^6}\Rightarrow\frac{10^6}{5}<n!\Rightarrow n\geq9\Rightarrow 1-1+\frac{1}{2!}-\frac{1}{3!}+\frac{1}{4!}-\frac{1}{5!}+\frac{1}{6!}-\frac{1}{7!}+\frac{1}{8!}\approx0.367881944$

51. (a) $a_n\geq a_{n+1}$ fails since $\frac{1}{3}<\frac{1}{2}$

(b) Since $\sum_{n=1}^{\infty} |a_n| = \sum_{n=1}^{\infty} \left[\left(\frac{1}{3}\right)^n + \left(\frac{1}{2}\right)^n\right] = \sum_{n=1}^{\infty} \left(\frac{1}{3}\right)^n + \sum_{n=1}^{\infty} \left(\frac{1}{2}\right)^n$ is the sum of two absolutely convergent series, we can rearrange the terms of the original series to find its sum:

$$\left(\frac{1}{3} + \frac{1}{9} + \frac{1}{27} + \ldots\right) - \left(\frac{1}{2} + \frac{1}{4} + \frac{1}{8} + \ldots\right) = \frac{\left(\frac{1}{3}\right)}{1 - \left(\frac{1}{3}\right)} - \frac{\left(\frac{1}{2}\right)}{1 - \left(\frac{1}{2}\right)} = \frac{1}{2} - 1 = -\frac{1}{2}$$

52. $s_{20} = 1 - \frac{1}{2} + \frac{1}{3} - \frac{1}{4} + \ldots + \frac{1}{19} - \frac{1}{20} \approx 0.6687714032 \Rightarrow s_{20} + \frac{1}{2} \cdot \frac{1}{21} \approx 0.692580927$

53. The unused terms are $\sum_{j=n+1}^{\infty} (-1)^{j+1} a_j = (-1)^{n+1}(a_{n+1} - a_{n+2}) + (-1)^{n+3}(a_{n+3} - a_{n+4}) + \cdots$ $= (-1)^{n+1}[(a_{n+1} - a_{n+2}) + (a_{n+3} - a_{n+4}) + \cdots]$. Each grouped term is positive, so the remainder has the same sign as $(-1)^{n+1}$, which is the sign of the first unused term.

54. $s_n = \frac{1}{1 \cdot 2} + \frac{1}{2 \cdot 3} + \frac{1}{3 \cdot 4} + \ldots + \frac{1}{n(n+1)} = \sum_{k=1}^{n} \frac{1}{k(k+1)} = \sum_{k=1}^{n} \left(\frac{1}{k} - \frac{1}{k+1}\right)$
$= \left(1 - \frac{1}{2}\right) + \left(\frac{1}{2} - \frac{1}{3}\right) + \left(\frac{1}{3} - \frac{1}{4}\right) + \left(\frac{1}{4} - \frac{1}{5}\right) + \ldots + \left(\frac{1}{n} - \frac{1}{n+1}\right)$ which are the first 2n terms of the first series, hence the two series are the same. Yes, for

$$s_n = \sum_{k=1}^{n} \left(\frac{1}{k} - \frac{1}{k+1}\right) = \left(1 - \frac{1}{2}\right) + \left(\frac{1}{2} - \frac{1}{3}\right) + \left(\frac{1}{3} - \frac{1}{4}\right) + \left(\frac{1}{4} - \frac{1}{5}\right) + \ldots + \left(\frac{1}{n-1} - \frac{1}{n}\right) + \left(\frac{1}{n} - \frac{1}{n+1}\right) = 1 - \frac{1}{n+1}$$

$\Rightarrow \lim_{n\to\infty} s_n = \lim_{n\to\infty} \left(1 - \frac{1}{n+1}\right) = 1 \Rightarrow$ both series converge to 1. The sum of the first $2n+1$ terms of the first series is $\left(1 - \frac{1}{n+1}\right) + \frac{1}{n+1} = 1$. Their sum is $\lim_{n\to\infty} s_n = \lim_{n\to\infty} \left(1 - \frac{1}{n+1}\right) = 1$.

55. Using the Direct Comparison Test, since $|a_n| \geq a_n$ and $\sum_{n=1}^{\infty} a_n$ diverges we must have that $\sum_{n=1}^{\infty} |a_n|$ diverges.

56. $|a_1 + a_2 + \ldots + a_n| \leq |a_1| + |a_2| + \ldots + |a_n|$ for all n; then $\sum_{n=1}^{\infty} |a_n|$ converges $\Rightarrow \sum_{n=1}^{\infty} a_n$ converges and these imply that $\left|\sum_{n=1}^{\infty} a_n\right| \leq \sum_{n=1}^{\infty} |a_n|$

57. (a) $\sum_{n=1}^{\infty} |a_n + b_n|$ converges by the Direct Comparison Test since $|a_n + b_n| \leq |a_n| + |b_n|$ and hence $\sum_{n=1}^{\infty} (a_n + b_n)$ converges absolutely

(b) $\sum_{n=1}^{\infty} |b_n|$ converges $\Rightarrow \sum_{n=1}^{\infty} -b_n$ converges absolutely; since $\sum_{n=1}^{\infty} a_n$ converges absolutely and $\sum_{n=1}^{\infty} -b_n$ converges absolutely, we have $\sum_{n=1}^{\infty} [a_n + (-b_n)] = \sum_{n=1}^{\infty} (a_n - b_n)$ converges absolutely by part (a)

(c) $\sum_{n=1}^{\infty} |a_n|$ converges $\Rightarrow |k| \sum_{n=1}^{\infty} |a_n| = \sum_{n=1}^{\infty} |ka_n|$ converges $\Rightarrow \sum_{n=1}^{\infty} ka_n$ converges absolutely

58. If $a_n = b_n = (-1)^n \frac{1}{\sqrt{n}}$, then $\sum_{n=1}^{\infty} (-1)^n \frac{1}{\sqrt{n}}$ converges, but $\sum_{n=1}^{\infty} a_n b_n = \sum_{n=1}^{\infty} \frac{1}{n}$ diverges

59. $s_1 = -\frac{1}{2}$, $s_2 = -\frac{1}{2} + 1 = \frac{1}{2}$,

$s_3 = -\frac{1}{2} + 1 - \frac{1}{4} - \frac{1}{6} - \frac{1}{8} - \frac{1}{10} - \frac{1}{12} - \frac{1}{14} - \frac{1}{16} - \frac{1}{18} - \frac{1}{20} - \frac{1}{22} \approx -0.5099$,

$s_4 = s_3 + \frac{1}{3} \approx -0.1766$,

$s_5 = s_4 - \frac{1}{24} - \frac{1}{26} - \frac{1}{28} - \frac{1}{30} - \frac{1}{32} - \frac{1}{34} - \frac{1}{36} - \frac{1}{38} - \frac{1}{40} - \frac{1}{42} - \frac{1}{44} \approx -0.512$,

$s_6 = s_5 + \frac{1}{5} \approx -0.312$,

$s_7 = s_6 - \frac{1}{46} - \frac{1}{48} - \frac{1}{50} - \frac{1}{52} - \frac{1}{54} - \frac{1}{56} - \frac{1}{58} - \frac{1}{60} - \frac{1}{62} - \frac{1}{64} - \frac{1}{66} \approx -0.51106$

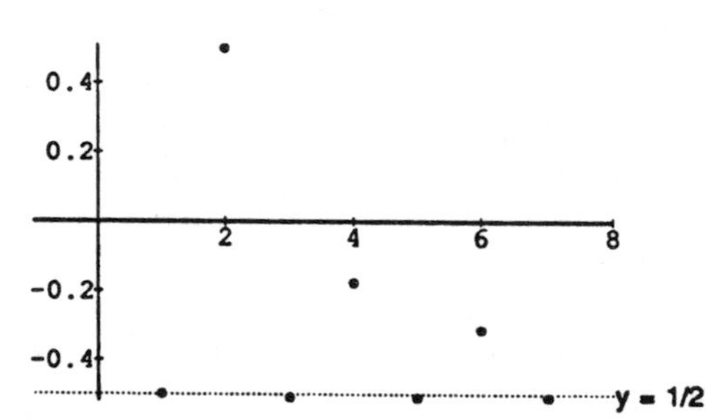

60. (a) Since $\sum |a_n|$ converges, say to M, for $\epsilon > 0$ there is an integer N_1 such that $\left| \sum_{n=1}^{N_1-1} |a_n| - M \right| < \frac{\epsilon}{2}$

$\Leftrightarrow \left| \sum_{n=1}^{N_1-1} |a_n| - \left(\sum_{n=1}^{N_1-1} |a_n| + \sum_{n=N_1}^{\infty} |a_n| \right) \right| < \frac{\epsilon}{2} \Leftrightarrow \left| -\sum_{n=N_1}^{\infty} |a_n| \right| < \frac{\epsilon}{2} \Leftrightarrow \sum_{n=N_1}^{\infty} |a_n| < \frac{\epsilon}{2}$. Also, $\sum a_n$ converges to L $\Leftrightarrow$ for $\epsilon > 0$ there is an integer N_2 (which we can choose greater than or equal to N_1) such that $\left| s_{N_2} - L \right| < \frac{\epsilon}{2}$. Therefore, $\sum_{n=N_1}^{\infty} |a_n| < \frac{\epsilon}{2}$ and $\left| s_{N_2} - L \right| < \frac{\epsilon}{2}$.

(b) The series $\sum_{n=1}^{\infty} |a_n|$ converges absolutely, say to M. Thus, there exists N_1 such that $\left| \sum_{n=1}^{k} |a_n| - M \right| < \epsilon$ whenever $k > N_1$. Now all of the terms in the sequence $\{|b_n|\}$ appear in $\{|a_n|\}$. Sum together all of the terms in $\{|b_n|\}$, in order, until you include all of the terms $\{|a_n|\}_{n=1}^{N_1}$, and let N_2 be the largest index in the sum $\sum_{n=1}^{N_2} |b_n|$ so obtained. Then $\left| \sum_{n=1}^{N_2} |b_n| - M \right| < \epsilon$ as well $\Rightarrow \sum_{n=1}^{\infty} |b_n|$ converges to M.

61. (a) If $\sum_{n=1}^{\infty} |a_n|$ converges, then $\sum_{n=1}^{\infty} a_n$ converges and $\frac{1}{2}\sum_{n=1}^{\infty} a_n + \frac{1}{2}\sum_{n=1}^{\infty} |a_n| = \sum_{n=1}^{\infty} \frac{a_n + |a_n|}{2}$

converges where $b_n = \frac{a_n + |a_n|}{2} = \begin{cases} a_n, & \text{if } a_n \ge 0 \\ 0, & \text{if } a_n < 0 \end{cases}$.

(b) If $\sum_{n=1}^{\infty} |a_n|$ converges, then $\sum_{n=1}^{\infty} a_n$ converges and $\frac{1}{2}\sum_{n=1}^{\infty} a_n - \frac{1}{2}\sum_{n=1}^{\infty} |a_n| = \sum_{n=1}^{\infty} \frac{a_n - |a_n|}{2}$

converges where $c_n = \frac{a_n - |a_n|}{2} = \begin{cases} 0, & \text{if } a_n \ge 0 \\ a_n, & \text{if } a_n < 0 \end{cases}$.

62. The terms in this conditionally convergent series were not added in the order given.

63. Here is an example figure when $N = 5$. Notice that $u_3 > u_2 > u_1$ and $u_3 > u_5 > u_4$, but $u_n \ge u_{n+1}$ for $n \ge 5$.

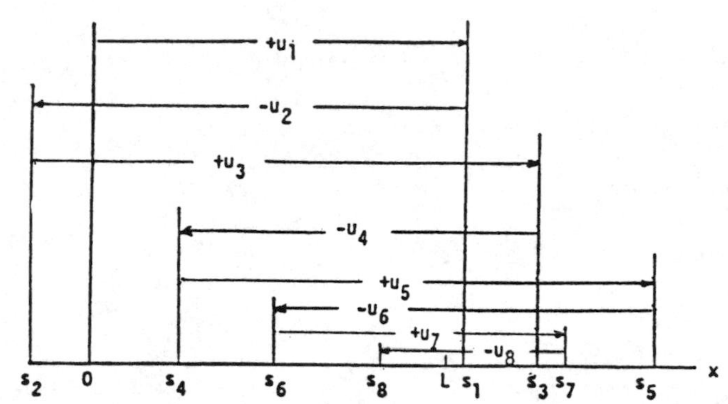

8.8 POWER SERIES

1. $\lim_{n\to\infty} \left|\frac{u_{n+1}}{u_n}\right| < 1 \Rightarrow \lim_{n\to\infty} \left|\frac{x^{n+1}}{x^n}\right| < 1 \Rightarrow |x| < 1 \Rightarrow -1 < x < 1$; when $x = -1$ we have $\sum_{n=1}^{\infty} (-1)^n$, a divergent series; when $x = 1$ we have $\sum_{n=1}^{\infty} 1$, a divergent series

(a) the radius is 1; the interval of convergence is $-1 < x < 1$

(b) the interval of absolute convergence is $-1 < x < 1$

(c) there are no values for which the series converges conditionally

2. $\lim_{n\to\infty} \left|\frac{u_{n+1}}{u_n}\right| < 1 \Rightarrow \lim_{n\to\infty} \left|\frac{(x+5)^{n+1}}{(x+5)^n}\right| < 1 \Rightarrow |x+5| < 1 \Rightarrow -6 < x < -4$; when $x = -6$ we have $\sum_{n=1}^{\infty} (-1)^n$, a divergent series; when $x = -4$ we have $\sum_{n=1}^{\infty} 1$, a divergent series

(a) the radius is 1; the interval of convergence is $-6 < x < -4$

(b) the interval of absolute convergence is $-6 < x < -4$

(c) there are no values for which the series converges conditionally

3. $\lim_{n\to\infty} \left|\frac{u_{n+1}}{u_n}\right| < 1 \Rightarrow \lim_{n\to\infty} \left|\frac{(4x+1)^{n+1}}{(4x+1)^n}\right| < 1 \Rightarrow |4x+1| < 1 \Rightarrow -1 < 4x+1 < 1 \Rightarrow -\frac{1}{2} < x < 0$; when $x = -\frac{1}{2}$ we have $\sum_{n=1}^{\infty} (-1)^n(-1)^n = \sum_{n=1}^{\infty} (-1)^{2n} = \sum_{n=1}^{\infty} 1^n$, a divergent series; when $x = 0$ we have $\sum_{n=1}^{\infty} (-1)^n(1)^n = \sum_{n=1}^{\infty} (-1)^n$, a divergent series

(a) the radius is $\frac{1}{4}$; the interval of convergence is $-\frac{1}{2} < x < 0$

(b) the interval of absolute convergence is $-\frac{1}{2} < x < 0$

(c) there are no values for which the series converges conditionally

4. $\lim_{n\to\infty} \left|\frac{u_{n+1}}{u_n}\right| < 1 \Rightarrow \lim_{n\to\infty} \left|\frac{(3x-2)^{n+1}}{n+1} \cdot \frac{n}{(3x-2)^n}\right| < 1 \Rightarrow |3x-2| \lim_{n\to\infty} \left(\frac{n}{n+1}\right) < 1 \Rightarrow |3x-2| < 1$ $\Rightarrow -1 < 3x-2 < 1 \Rightarrow \frac{1}{3} < x < 1$; when $x = \frac{1}{3}$ we have $\sum_{n=1}^{\infty} \frac{(-1)^n}{n}$ which is the alternating harmonic series and is conditionally convergent; when $x = 1$ we have $\sum_{n=1}^{\infty} \frac{1}{n}$, the divergent harmonic series

(a) the radius is $\frac{1}{3}$; the interval of convergence is $\frac{1}{3} \le x < 1$

(b) the interval of absolute convergence is $\frac{1}{3} < x < 1$

(c) the series converges conditionally at $x = \frac{1}{3}$

5. $\lim_{n\to\infty} \left|\frac{u_{n+1}}{u_n}\right| < 1 \Rightarrow \lim_{n\to\infty} \left|\frac{(x-2)^{n+1}}{10^{n+1}} \cdot \frac{10^n}{(x-2)^n}\right| < 1 \Rightarrow \frac{|x-2|}{10} < 1 \Rightarrow |x-2| < 10 \Rightarrow -10 < x-2 < 10$ $\Rightarrow -8 < x < 12$; when $x = -8$ we have $\sum_{n=1}^{\infty} (-1)^n$, a divergent series; when $x = 12$ we have $\sum_{n=1}^{\infty} 1$, a divergent series

(a) the radius is 10; the interval of convergence is $-8 < x < 12$

(b) the interval of absolute convergence is $-8 < x < 12$

(c) there are no values for which the series converges conditionally

6. $\lim_{n\to\infty} \left|\frac{u_{n+1}}{u_n}\right| < 1 \Rightarrow \lim_{n\to\infty} \left|\frac{(2x)^{n+1}}{(2x)^n}\right| < 1 \Rightarrow \lim_{n\to\infty} |2x| < 1 \Rightarrow |2x| < 1 \Rightarrow -\frac{1}{2} < x < \frac{1}{2}$; when $x = -\frac{1}{2}$ we have $\sum_{n=1}^{\infty} (-1)^n$, a divergent series; when $x = \frac{1}{2}$ we have $\sum_{n=1}^{\infty} 1$, a divergent series

(a) the radius is $\frac{1}{2}$; the interval of convergence is $-\frac{1}{2} < x < \frac{1}{2}$

(b) the interval of absolute convergence is $-\frac{1}{2} < x < \frac{1}{2}$

(c) there are no values for which the series converges conditionally

7. $\lim_{n\to\infty} \left|\frac{u_{n+1}}{u_n}\right| < 1 \Rightarrow \lim_{n\to\infty} \left|\frac{(n+1)x^{n+1}}{(n+3)} \cdot \frac{(n+2)}{nx^n}\right| < 1 \Rightarrow |x| \lim_{n\to\infty} \frac{(n+1)(n+2)}{(n+3)(n)} < 1 \Rightarrow |x| < 1$

$\Rightarrow -1 < x < 1$; when $x = -1$ we have $\sum_{n=1}^{\infty} (-1)^n \frac{n}{n+2}$, a divergent series by the nth-term Test; when $x = 1$ we have $\sum_{n=1}^{\infty} \frac{n}{n+2}$, a divergent series

(a) the radius is 1; the interval of convergence is $-1 < x < 1$

(b) the interval of absolute convergence is $-1 < x < 1$

(c) there are no values for which the series converges conditionally

8. $\lim_{n\to\infty} \left|\frac{u_{n+1}}{u_n}\right| < 1 \Rightarrow \lim_{n\to\infty} \left|\frac{(x+2)^{n+1}}{n+1} \cdot \frac{n}{(x+2)^n}\right| < 1 \Rightarrow |x+2| \lim_{n\to\infty} \left(\frac{n}{n+1}\right) < 1 \Rightarrow |x+2| < 1$

$\Rightarrow -1 < x+2 < 1 \Rightarrow -3 < x < -1$; when $x = -3$ we have $\sum_{n=1}^{\infty} \frac{1}{n}$, a divergent series; when $x = -1$ we have $\sum_{n=1}^{\infty} \frac{(-1)^n}{n}$, a convergent series

(a) the radius is 1; the interval of convergence is $-3 < x \le -1$

(b) the interval of absolute convergence is $-3 < x < -1$

(c) the series converges conditionally at $x = -1$

9. $\lim_{n\to\infty} \left|\frac{u_{n+1}}{u_n}\right| < 1 \Rightarrow \lim_{n\to\infty} \left|\frac{x^{n+1}}{(n+1)\sqrt{n+1}\,3^{n+1}} \cdot \frac{n\sqrt{n}\,3^n}{x^n}\right| < 1 \Rightarrow \frac{|x|}{3}\left(\lim_{n\to\infty} \frac{n}{n+1}\right)\left(\sqrt{\lim_{n\to\infty} \frac{n}{n+1}}\right) < 1$

$\Rightarrow \frac{|x|}{3}(1)(1) < 1 \Rightarrow |x| < 3 \Rightarrow -3 < x < 3$; when $x = -3$ we have $\sum_{n=1}^{\infty} \frac{(-1)^n}{n^{3/2}}$, an absolutely convergent series; when $x = 3$ we have $\sum_{n=1}^{\infty} \frac{1}{n^{3/2}}$, a convergent p-series

(a) the radius is 3; the interval of convergence is $-3 \le x \le 3$

(b) the interval of absolute convergence is $-3 \le x \le 3$

(c) there are no values for which the series converges conditionally

10. $\lim_{n\to\infty} \left|\frac{u_{n+1}}{u_n}\right| < 1 \Rightarrow \lim_{n\to\infty} \left|\frac{(x-1)^{n+1}}{\sqrt{n+1}} \cdot \frac{\sqrt{n}}{(x-1)^n}\right| < 1 \Rightarrow |x-1| \sqrt{\lim_{n\to\infty} \frac{n}{n+1}} < 1 \Rightarrow |x-1| < 1$

$\Rightarrow -1 < x-1 < 1 \Rightarrow 0 < x < 2$; when $x = 0$ we have $\sum_{n=1}^{\infty} \frac{(-1)^n}{n^{1/2}}$, a conditionally convergent series; when $x = 2$ we have $\sum_{n=1}^{\infty} \frac{1}{n^{1/2}}$, a divergent series

(a) the radius is 1; the interval of convergence is $0 \le x < 2$

(b) the interval of absolute convergence is $0 < x < 2$

(c) the series converges conditionally at $x = 0$

11. $\lim_{n\to\infty} \left|\frac{u_{n+1}}{u_n}\right| < 1 \Rightarrow \lim_{n\to\infty} \left|\frac{x^{n+1}}{(n+1)!} \cdot \frac{n!}{x^n}\right| < 1 \Rightarrow |x| \lim_{n\to\infty} \left(\frac{1}{n+1}\right) < 1$ for all x

(a) the radius is ∞; the series converges for all x

(b) the series converges absolutely for all x

(c) there are no values for which the series converges conditionally

12. $\lim_{n\to\infty} \left|\frac{u_{n+1}}{u_n}\right| < 1 \Rightarrow \lim_{n\to\infty} \left|\frac{3^{n+1} x^{n+1}}{(n+1)!} \cdot \frac{n!}{3^n x^n}\right| < 1 \Rightarrow 3|x| \lim_{n\to\infty} \left(\frac{1}{n+1}\right) < 1$ for all x

(a) the radius is ∞; the series converges for all x

(b) the series converges absolutely for all x

(c) there are no values for which the series converges conditionally

13. $\lim_{n\to\infty} \left|\frac{u_{n+1}}{u_n}\right| < 1 \Rightarrow \lim_{n\to\infty} \left|\frac{x^{2n+3}}{(n+1)!} \cdot \frac{n!}{x^{2n+1}}\right| < 1 \Rightarrow x^2 \lim_{n\to\infty} \left(\frac{1}{n+1}\right) < 1$ for all x

(a) the radius is ∞; the series converges for all x

(b) the series converges absolutely for all x

(c) there are no values for which the series converges conditionally

14. $\lim_{n\to\infty} \left|\frac{u_{n+1}}{u_n}\right| < 1 \Rightarrow \lim_{n\to\infty} \left|\frac{(2x+3)^{2n+3}}{(n+1)!} \cdot \frac{n!}{(2x+3)^{2n+1}}\right| < 1 \Rightarrow (2x+3)^2 \lim_{n\to\infty} \left(\frac{1}{n+1}\right) < 1$ for all x

(a) the radius is ∞; the series converges for all x

(b) the series converges absolutely for all x

(c) there are no values for which the series converges conditionally

15. $\lim_{n\to\infty} \left|\frac{u_{n+1}}{u_n}\right| < 1 \Rightarrow \lim_{n\to\infty} \left|\frac{x^{n+1}}{\sqrt{(n+1)^2+3}} \cdot \frac{\sqrt{n^2+3}}{x^n}\right| < 1 \Rightarrow |x| \sqrt{\lim_{n\to\infty} \frac{n^2+3}{n^2+2n+4}} < 1 \Rightarrow |x| < 1$

$\Rightarrow -1 < x < 1$; when $x = -1$ we have $\sum_{n=1}^{\infty} \frac{(-1)^n}{\sqrt{n^2+3}}$, a conditionally convergent series; when $x = 1$ we have $\sum_{n=1}^{\infty} \frac{1}{\sqrt{n^2+3}}$, a divergent series

(a) the radius is 1; the interval of convergence is $-1 \le x < 1$

(b) the interval of absolute convergence is $-1 < x < 1$

(c) the series converges conditionally at $x = -1$

16. $\lim_{n\to\infty} \left|\frac{u_{n+1}}{u_n}\right| < 1 \Rightarrow \lim_{n\to\infty} \left|\frac{x^{n+1}}{\sqrt{(n+1)^2+3}} \cdot \frac{\sqrt{n^2+3}}{x^n}\right| < 1 \Rightarrow |x| \sqrt{\lim_{n\to\infty} \frac{n^2+3}{n^2+2n+4}} < 1 \Rightarrow |x| < 1$

$\Rightarrow -1 < x < 1$; when $x = -1$ we have $\sum_{n=1}^{\infty} \frac{1}{\sqrt{n^2+3}}$, a divergent series; when $x = 1$ we have $\sum_{n=1}^{\infty} \frac{(-1)^n}{n^2+3}$, a conditionally convergent series

(a) the radius is 1; the interval of convergence is $-1 < x \le 1$

(b) the interval of absolute convergence is $-1 < x < 1$

(c) the series converges conditionally at $x = 1$

17. $\lim\limits_{n\to\infty} \left|\frac{u_{n+1}}{u_n}\right| < 1 \Rightarrow \lim\limits_{n\to\infty} \left|\frac{(n+1)(x+3)^{n+1}}{5^{n+1}} \cdot \frac{5^n}{n(x+3)^n}\right| < 1 \Rightarrow \frac{|x+3|}{5} \lim\limits_{n\to\infty} \left(\frac{n+1}{n}\right) < 1 \Rightarrow \frac{|x+3|}{5} < 1$

$\Rightarrow |x+3| < 5 \Rightarrow -5 < x+3 < 5 \Rightarrow -8 < x < 2$; when $x = -8$ we have $\sum\limits_{n=1}^{\infty} \frac{n(-5)^n}{5^n} = \sum\limits_{n=1}^{\infty} (-1)^n n$, a divergent series; when $x = 2$ we have $\sum\limits_{n=1}^{\infty} \frac{n5^n}{5^n} = \sum\limits_{n=1}^{\infty} n$, a divergent series

(a) the radius is 5; the interval of convergence is $-8 < x < 2$

(b) the interval of absolute convergence is $-8 < x < 2$

(c) there are no values for which the series converges conditionally

18. $\lim\limits_{n\to\infty} \left|\frac{u_{n+1}}{u_n}\right| < 1 \Rightarrow \lim\limits_{n\to\infty} \left|\frac{(n+1)x^{n+1}}{4^{n+1}(n^2+2n+2)} \cdot \frac{4^n(n^2+1)}{nx^n}\right| < 1 \Rightarrow \frac{|x|}{4} \lim\limits_{n\to\infty} \left|\frac{(n+1)(n^2+1)}{n(n^2+2n+2)}\right| < 1 \Rightarrow |x| < 4$

$\Rightarrow -4 < x < 4$; when $x = -4$ we have $\sum\limits_{n=1}^{\infty} \frac{n(-1)^n}{n^2+1}$, a conditionally convergent series; when $x = 4$ we have $\sum\limits_{n=1}^{\infty} \frac{n}{n^2+1}$, a divergent series

(a) the radius is 4; the interval of convergence is $-4 \le x < 4$

(b) the interval of absolute convergence is $-4 < x < 4$

(c) the series converges conditionally at $x = -4$

19. $\lim\limits_{n\to\infty} \left|\frac{u_{n+1}}{u_n}\right| < 1 \Rightarrow \lim\limits_{n\to\infty} \left|\frac{\sqrt{n+1}\,x^{n+1}}{3^{n+1}} \cdot \frac{3^n}{\sqrt{n}\,x^n}\right| < 1 \Rightarrow \frac{|x|}{3} \sqrt{\lim\limits_{n\to\infty} \left(\frac{n+1}{n}\right)} < 1 \Rightarrow \frac{|x|}{3} < 1 \Rightarrow |x| < 3$

$\Rightarrow -3 < x < 3$; when $x = -3$ we have $\sum\limits_{n=1}^{\infty} (-1)^n \sqrt{n}$, a divergent series; when $x = 3$ we have $\sum\limits_{n=1}^{\infty} \sqrt{n}$, a divergent series

(a) the radius is 3; the interval of convergence is $-3 < x < 3$

(b) the interval of absolute convergence is $-3 < x < 3$

(c) there are no values for which the series converges conditionally

20. $\lim\limits_{n\to\infty} \left|\frac{u_{n+1}}{u_n}\right| < 1 \Rightarrow \lim\limits_{n\to\infty} \left|\frac{\sqrt[n+1]{n+1}\,(2x+5)^{n+1}}{\sqrt[n]{n}\,(2x+5)^n}\right| < 1 \Rightarrow |2x+5| \lim\limits_{n\to\infty} \left(\frac{\sqrt[n+1]{n+1}}{\sqrt[n]{n}}\right) < 1$

$\Rightarrow |2x+5| \left(\frac{\lim\limits_{t\to\infty} \sqrt[t]{t}}{\lim\limits_{n\to\infty} \sqrt[n]{n}}\right) < 1 \Rightarrow |2x+5| < 1 \Rightarrow -1 < 2x+5 < 1 \Rightarrow -3 < x < -2$; when $x = -3$ we have

$\sum_{n=1}^{\infty} (-1)^n \sqrt[n]{n}$, a divergent series since $\lim_{n\to\infty} \sqrt[n]{n} = 1$; when $x = -2$ we have $\sum_{n=1}^{\infty} \sqrt[n]{n}$, a divergent series

(a) the radius is $\frac{1}{2}$; the interval of convergence is $-3 < x < -2$

(b) the interval of absolute convergence is $-3 < x < -2$

(c) there are no values for which the series converges conditionally

21. $\lim_{n\to\infty} \left|\frac{u_{n+1}}{u_n}\right| < 1 \Rightarrow \lim_{n\to\infty} \left|\frac{\left(1+\frac{1}{n+1}\right)^{n+1} x^{n+1}}{\left(1+\frac{1}{n}\right)^n x^n}\right| < 1 \Rightarrow |x| \left(\frac{\lim_{t\to\infty}\left(1+\frac{1}{t}\right)^t}{\lim_{n\to\infty}\left(1+\frac{1}{n}\right)^n}\right) < 1 \Rightarrow |x|\left(\frac{e}{e}\right) < 1 \Rightarrow |x| < 1$

$\Rightarrow -1 < x < 1$; when $x = -1$ we have $\sum_{n=1}^{\infty} (-1)^n \left(1+\frac{1}{n}\right)^n$, a divergent series by the nth-Term Test since

$\lim_{n\to\infty} \left(1+\frac{1}{n}\right)^n = e \neq 0$; when $x = 1$ we have $\sum_{n=1}^{\infty} \left(1+\frac{1}{n}\right)^n$, a divergent series

(a) the radius is 1; the interval of convergence is $-1 < x < 1$

(b) the interval of absolute convergence is $-1 < x < 1$

(c) there are no values for which the series converges conditionally

22. $\lim_{n\to\infty} \left|\frac{u_{n+1}}{u_n}\right| < 1 \Rightarrow \lim_{n\to\infty} \left|\frac{\ln(n+1)x^{n+1}}{x^n \ln n}\right| < 1 \Rightarrow |x| \lim_{n\to\infty} \left|\frac{\left(\frac{1}{n+1}\right)}{\left(\frac{1}{n}\right)}\right| < 1 \Rightarrow |x| \lim_{n\to\infty} \left(\frac{n}{n+1}\right) < 1 \Rightarrow |x| < 1$

$\Rightarrow -1 < x < 1$; when $x = -1$ we have $\sum_{n=1}^{\infty} (-1)^n \ln n$, a divergent series by the nth-Term Test since

$\lim_{n\to\infty} \ln n \neq 0$; when $x = 1$ we have $\sum_{n=1}^{\infty} \ln n$, a divergent series

(a) the radius is 1; the interval of convergence is $-1 < x < 1$

(b) the interval of absolute convergence is $-1 < x < 1$

(c) there are no values for which the series converges conditionally

23. $\lim_{n\to\infty} \left|\frac{u_{n+1}}{u_n}\right| < 1 \Rightarrow \lim_{n\to\infty} \left|\frac{(n+1)^{n+1}x^{n+1}}{n^n x^n}\right| < 1 \Rightarrow |x| \left(\lim_{n\to\infty} \left(1+\frac{1}{n}\right)^n\right)\left(\lim_{n\to\infty} (n+1)\right) < 1$

$\Rightarrow e|x| \lim_{n\to\infty} (n+1) < 1 \Rightarrow$ only $x = 0$ satisfies this inequality

(a) the radius is 0; the series converges only for $x = 0$

(b) the series converges absolutely only for $x = 0$

(c) there are no values for which the series converges conditionally

24. $\lim_{n\to\infty} \left|\frac{u_{n+1}}{u_n}\right| < 1 \Rightarrow \lim_{n\to\infty} \left|\frac{(n+1)!\,(x-4)^{n+1}}{n!\,(x-4)^n}\right| < 1 \Rightarrow |x-4| \lim_{n\to\infty} (n+1) < 1 \Rightarrow$ only $x = 4$ satisfies this inequality

(a) the radius is 0; the series converges only for $x = 4$

(b) the series converges absolutely only for $x = 4$

(c) there are no values for which the series converges conditionally

25. $\lim_{n\to\infty} \left|\frac{u_{n+1}}{u_n}\right| < 1 \Rightarrow \lim_{n\to\infty} \left|\frac{(x+2)^{n+1}}{(n+1)\,2^{n+1}} \cdot \frac{n2^n}{(x+2)^n}\right| < 1 \Rightarrow \frac{|x+2|}{2} \lim_{n\to\infty} \left(\frac{n}{n+1}\right) < 1 \Rightarrow \frac{|x+2|}{2} < 1 \Rightarrow |x+2| < 2$

$\Rightarrow -2 < x+2 < 2 \Rightarrow -4 < x < 0$; when $x = -4$ we have $\sum_{n=1}^{\infty} \frac{-1}{n}$, a divergent series; when $x = 0$ we have $\sum_{n=1}^{\infty} \frac{(-1)^{n+1}}{n}$, the alternating harmonic series which converges conditionally

(a) the radius is 2; the interval of convergence is $-4 < x \le 0$

(b) the interval of absolute convergence is $-4 < x < 0$

(c) the series converges conditionally at $x = 0$

26. $\lim_{n\to\infty} \left|\frac{u_{n+1}}{u_n}\right| < 1 \Rightarrow \lim_{n\to\infty} \left|\frac{(-2)^{n+1}(n+2)(x-1)^{n+1}}{(-2)^n(n+1)(x-1)^n}\right| < 1 \Rightarrow 2\,|x-1| \lim_{n\to\infty} \left(\frac{n+2}{n+1}\right) < 1 \Rightarrow 2\,|x-1| < 1$

$\Rightarrow |x-1| < \frac{1}{2} \Rightarrow -\frac{1}{2} < x-1 < \frac{1}{2} \Rightarrow \frac{1}{2} < x < \frac{3}{2}$; when $x = \frac{1}{2}$ we have $\sum_{n=1}^{\infty} (n+1)$, a divergent series; when $x = \frac{3}{2}$ we have $\sum_{n=1}^{\infty} (-1)^n(n+1)$, a divergent series

(a) the radius is $\frac{1}{2}$; the interval of convergence is $\frac{1}{2} < x < \frac{3}{2}$

(b) the interval of absolute convergence is $\frac{1}{2} < x < \frac{3}{2}$

(c) there are no values for which the series converges conditionally

27. $\lim_{n\to\infty} \left|\frac{u_{n+1}}{u_n}\right| < 1 \Rightarrow \lim_{n\to\infty} \left|\frac{x^{n+1}}{(n+1)(\ln(n+1))^2} \cdot \frac{n(\ln n)^2}{x^n}\right| < 1 \Rightarrow |x| \left(\lim_{n\to\infty} \frac{n}{n+1}\right)\left(\lim_{n\to\infty} \frac{\ln n}{\ln(n+1)}\right)^2 < 1$

$\Rightarrow |x|(1)\left(\lim_{n\to\infty} \frac{\left(\frac{1}{n}\right)}{\left(\frac{1}{n+1}\right)}\right)^2 < 1 \Rightarrow |x|\left(\lim_{n\to\infty} \frac{n+1}{n}\right)^2 < 1 \Rightarrow |x| < 1 \Rightarrow -1 < x < 1$; when $x = -1$ we have $\sum_{n=1}^{\infty} \frac{(-1)^n}{n(\ln n)^2}$ which converges absolutely; when $x = 1$ we have $\sum_{n=1}^{\infty} \frac{1}{n(\ln n)^2}$ which converges

(a) the radius is 1; the interval of convergence is $-1 \le 1 \le 1$

(b) the interval of absolute convergence is $-1 \le x \le 1$

(c) there are no values for which the series converges conditionally

28. $\lim_{n\to\infty} \left|\frac{u_{n+1}}{u_n}\right| < 1 \Rightarrow \lim_{n\to\infty} \left|\frac{x^{n+1}}{(n+1)\ln(n+1)} \cdot \frac{n\ln(n)}{x^n}\right| < 1 \Rightarrow |x| \left(\lim_{n\to\infty} \frac{n}{n+1}\right)\left(\lim_{n\to\infty} \frac{\ln(n)}{\ln(n+1)}\right) < 1$

$\Rightarrow |x|(1)(1) < 1 \Rightarrow |x| < 1 \Rightarrow -1 < x < 1$; when $x = -1$ we have $\sum_{n=2}^{\infty} \frac{(-1)^n}{n\ln n}$, a convergent alternating series; when $x = 1$ we have $\sum_{n=2}^{\infty} \frac{1}{n\ln n}$ which diverges by Exercise 39, Section 8.4

(a) the radius is 1; the interval of convergence is $-1 \le x < 1$

(b) the interval of absolute convergence is $-1 < x < 1$

(c) the series converges conditionally at $x = -1$

29. $\lim_{n\to\infty}\left|\frac{u_{n+1}}{u_n}\right|<1 \Rightarrow \lim_{n\to\infty}\left|\frac{(4x-5)^{2n+3}}{(n+1)^{3/2}}\cdot\frac{n^{3/2}}{(4x-5)^{2n+1}}\right|<1 \Rightarrow (4x-5)^2\left(\lim_{n\to\infty}\frac{n}{n+1}\right)^{3/2}<1 \Rightarrow (4x-5)^2<1$

$\Rightarrow |4x-5|<1 \Rightarrow -1<4x-5<1 \Rightarrow 1<x<\frac{3}{2}$; when $x=1$ we have $\sum_{n=1}^{\infty}\frac{(-1)^{2n+1}}{n^{3/2}}=\sum_{n=1}^{\infty}\frac{-1}{n^{3/2}}$ which is absolutely convergent; when $x=\frac{3}{2}$ we have $\sum_{n=1}^{\infty}\frac{(1)^{2n+1}}{n^{3/2}}$, a convergent p-series

(a) the radius is $\frac{1}{4}$; the interval of convergence is $1\le x\le\frac{3}{2}$

(b) the interval of absolute convergence is $1\le x\le\frac{3}{2}$

(c) there are no values for which the series converges conditionally

30. $\lim_{n\to\infty}\left|\frac{u_{n+1}}{u_n}\right|<1 \Rightarrow \lim_{n\to\infty}\left|\frac{(3x+1)^{n+2}}{2n+4}\cdot\frac{2n+2}{(3x+1)^{n+1}}\right|<1 \Rightarrow |3x+1|\lim_{n\to\infty}\left(\frac{2n+2}{2n+4}\right)<1 \Rightarrow |3x+1|<1$

$\Rightarrow -1<3x+1<1 \Rightarrow -\frac{2}{3}<x<0$; when $x=-\frac{2}{3}$ we have $\sum_{n=1}^{\infty}\frac{(-1)^{n+1}}{2n+1}$, a conditionally convergent series; when $x=0$ we have $\sum_{n=1}^{\infty}\frac{(1)^{n+1}}{2n+1}=\sum_{n=1}^{\infty}\frac{1}{2n+1}$, a divergent series

(a) the radius is $\frac{1}{3}$; the interval of convergence is $-\frac{2}{3}\le x<0$

(b) the interval of absolute convergence is $-\frac{2}{3}<x<0$

(c) the series converges conditionally at $x=-\frac{2}{3}$

31. $\lim_{n\to\infty}\left|\frac{u_{n+1}}{u_n}\right|<1 \Rightarrow \lim_{n\to\infty}\left|\frac{(x+\pi)^{n+1}}{\sqrt{n+1}}\cdot\frac{\sqrt{n}}{(x+\pi)^n}\right|<1 \Rightarrow |x+\pi|\lim_{n\to\infty}\left|\sqrt{\frac{n}{n+1}}\right|<1$

$\Rightarrow |x+\pi|\sqrt{\lim_{n\to\infty}\left(\frac{n}{n+1}\right)}<1 \Rightarrow |x+\pi|<1 \Rightarrow -1<x+\pi<1 \Rightarrow -1-\pi<x<1-\pi$;

when $x=-1-\pi$ we have $\sum_{n=1}^{\infty}\frac{(-1)^n}{\sqrt{n}}=\sum_{n=1}^{\infty}\frac{(-1)^n}{n^{1/2}}$, a conditionally convergent series; when $x=1-\pi$ we have $\sum_{n=1}^{\infty}\frac{1^n}{\sqrt{n}}=\sum_{n=1}^{\infty}\frac{1}{n^{1/2}}$, a divergent p-series

(a) the radius is 1; the interval of convergence is $(-1-\pi)\le x<(1-\pi)$

(b) the interval of absolute convergence is $-1-\pi<x<1-\pi$

(c) the series converges conditionally at $x=-1-\pi$

32. $\lim_{n\to\infty}\left|\frac{u_{n+1}}{u_n}\right|<1 \Rightarrow \lim_{n\to\infty}\left|\frac{\left(x-\sqrt{2}\right)^{2n+3}}{2^{n+1}}\cdot\frac{2^n}{\left(x-\sqrt{2}\right)^{2n+1}}\right|<1 \Rightarrow \frac{\left(x-\sqrt{2}\right)^2}{2}\lim_{n\to\infty}|1|<1$

$\Rightarrow \frac{\left(x-\sqrt{2}\right)^2}{2}<1 \Rightarrow \left(x-\sqrt{2}\right)^2<2 \Rightarrow \left|x-\sqrt{2}\right|<\sqrt{2} \Rightarrow -\sqrt{2}<x-\sqrt{2}<\sqrt{2} \Rightarrow 0<x<2\sqrt{2}$; when $x=0$ we have $\sum_{n=1}^{\infty}\frac{\left(-\sqrt{2}\right)^{2n+1}}{2^n}=-\sum_{n=1}^{\infty}\frac{2^{n+1/2}}{2^n}=-\sum_{n=1}^{\infty}\sqrt{2}$ which diverges since $\lim_{n\to\infty}a_n\ne 0$; when $x=2\sqrt{2}$ we have $\sum_{n=1}^{\infty}\frac{\left(\sqrt{2}\right)^{2n+1}}{2^n}=\sum_{n=1}^{\infty}\frac{2^{n+1/2}}{2^n}=\sum_{n=1}^{\infty}\sqrt{2}$, a divergent series

(a) the radius is $\sqrt{2}$; the interval of convergence is $0 < x < 2\sqrt{2}$

(b) the interval of absolute convergence is $0 < x < 2\sqrt{2}$

(c) there are no values for which the series converges conditionally

33. $\lim_{n\to\infty} \left|\frac{u_{n+1}}{u_n}\right| < 1 \Rightarrow \lim_{n\to\infty} \left|\frac{(x-1)^{2n+2}}{4^{n+1}} \cdot \frac{4^n}{(x-1)^{2n}}\right| < 1 \Rightarrow \frac{(x-1)^2}{4} \lim_{n\to\infty} |1| < 1 \Rightarrow (x-1)^2 < 4 \Rightarrow |x-1| < 2$
$\Rightarrow -2 < x-1 < 2 \Rightarrow -1 < x < 3$; at $x = -1$ we have $\sum_{n=0}^{\infty} \frac{(-2)^{2n}}{4^n} = \sum_{n=0}^{\infty} \frac{4^n}{4^n} = \sum_{n=0}^{\infty} 1$, which diverges; at $x = 3$ we have $\sum_{n=0}^{\infty} \frac{2^{2n}}{4^n} = \sum_{n=0}^{\infty} \frac{4^n}{4^n} = \sum_{n=0}^{\infty} 1$, a divergent series; the interval of convergence is $-1 < x < 3$; the series $\sum_{n=0}^{\infty} \frac{(x-1)^{2n}}{4^n} = \sum_{n=0}^{\infty} \left(\left(\frac{x-1}{2}\right)^2\right)^n$ is a convergent geometric series when $-1 < x < 3$ and the sum is

$$\frac{1}{1-\left(\frac{x-1}{2}\right)^2} = \frac{1}{\left[\frac{4-(x-1)^2}{4}\right]} = \frac{4}{4-x^2+2x-1} = \frac{4}{3+2x-x^2}$$

34. $\lim_{n\to\infty} \left|\frac{u_{n+1}}{u_n}\right| < 1 \Rightarrow \lim_{n\to\infty} \left|\frac{(x+1)^{2n+2}}{9^{n+1}} \cdot \frac{9^n}{(x+1)^{2n}}\right| < 1 \Rightarrow \frac{(x+1)^2}{9} \lim_{n\to\infty} |1| < 1 \Rightarrow (x+1)^2 < 9 \Rightarrow |x+1| < 3$
$\Rightarrow -3 < x+1 < 3 \Rightarrow -4 < x < 2$; when $x = -4$ we have $\sum_{n=0}^{\infty} \frac{(-3)^{2n}}{9^n} = \sum_{n=0}^{\infty} 1$ which diverges; at $x = 2$ we have $\sum_{n=0}^{\infty} \frac{3^{2n}}{9^n} = \sum_{n=0}^{\infty} 1$ which also diverges; the interval of convergence is $-4 < x < 2$; the series $\sum_{n=0}^{\infty} \frac{(x+1)^{2n}}{9^n} = \sum_{n=0}^{\infty} \left(\left(\frac{x+1}{3}\right)^2\right)^n$ is a convergent geometric series when $-4 < x < 2$ and the sum is

$$\frac{1}{1-\left(\frac{x+1}{3}\right)^2} = \frac{1}{\left[\frac{9-(x+1)^2}{9}\right]} = \frac{9}{9-x^2-2x-1} = \frac{9}{8-2x-x^2}$$

35. $\lim_{n\to\infty} \left|\frac{u_{n+1}}{u_n}\right| < 1 \Rightarrow \lim_{n\to\infty} \left|\frac{\left(\sqrt{x}-2\right)^{n+1}}{2^{n+1}} \cdot \frac{2^n}{\left(\sqrt{x}-2\right)^n}\right| < 1 \Rightarrow \left|\sqrt{x}-2\right| < 2 \Rightarrow -2 < \sqrt{x}-2 < 2 \Rightarrow 0 < \sqrt{x} < 4$
$\Rightarrow 0 < x < 16$; when $x = 0$ we have $\sum_{n=0}^{\infty} (-1)^n$, a divergent series; when $x = 16$ we have $\sum_{n=0}^{\infty} (1)^n$, a divergent series; the interval of convergence is $0 < x < 16$; the series $\sum_{n=0}^{\infty} \left(\frac{\sqrt{x}-2}{2}\right)^n$ is a convergent geometric series when $0 < x < 16$ and its sum is $\frac{1}{1-\left(\frac{\sqrt{x}-2}{2}\right)} = \frac{1}{\left(\frac{2-\sqrt{x}+2}{2}\right)} = \frac{2}{4-\sqrt{x}}$

36. $\lim_{n\to\infty} \left|\frac{u_{n+1}}{u_n}\right| < 1 \Rightarrow \lim_{n\to\infty} \left|\frac{(\ln x)^{n+1}}{(\ln x)^n}\right| < 1 \Rightarrow |\ln x| < 1 \Rightarrow -1 < \ln x < 1 \Rightarrow e^{-1} < x < e$; when $x = e^{-1}$ or e we obtain the series $\sum_{n=0}^{\infty} 1^n$ and $\sum_{n=0}^{\infty} (-1)^n$ which both diverge; the interval of convergence is $e^{-1} < x < e$;

$\sum_{n=0}^{\infty} (\ln x)^n = \frac{1}{1 - \ln x}$ when $e^{-1} < x < e$

37. $\lim_{n\to\infty} \left|\frac{u_{n+1}}{u_n}\right| < 1 \Rightarrow \lim_{n\to\infty} \left|\left(\frac{x^2+1}{3}\right)^{n+1} \cdot \left(\frac{3}{x^2+1}\right)^n\right| < 1 \Rightarrow \frac{(x^2+1)}{3} \lim_{n\to\infty} |1| < 1 \Rightarrow \frac{x^2+1}{3} < 1 \Rightarrow x^2 < 2$ $\Rightarrow |x| < \sqrt{2} \Rightarrow -\sqrt{2} < x < \sqrt{2}$; at $x = \pm\sqrt{2}$ we have $\sum_{n=0}^{\infty} (1)^n$ which diverges; the interval of convergence is $-\sqrt{2} < x < \sqrt{2}$; the series $\sum_{n=0}^{\infty} \left(\frac{x^2+1}{3}\right)^n$ is a convergent geometric series when $-\sqrt{2} < x < \sqrt{2}$ and its sum is $\frac{1}{1 - \left(\frac{x^2+1}{3}\right)} = \frac{1}{\left(\frac{3 - x^2 - 1}{3}\right)} = \frac{3}{2 - x^2}$

38. $\lim_{n\to\infty} \left|\frac{u_{n+1}}{u_n}\right| < 1 \Rightarrow \lim_{n\to\infty} \left|\frac{(x^2-1)^{n+1}}{2^{n+1}} \cdot \frac{2^n}{(x^2+1)^n}\right| < 1 \Rightarrow |x^2 - 1| < 2 \Rightarrow -\sqrt{3} < x < \sqrt{3}$; when $x = \pm\sqrt{3}$ we have $\sum_{n=0}^{\infty} 1^n$, a divergent series; the interval of convergence is $-\sqrt{3} < x < \sqrt{3}$; the series $\sum_{n=0}^{\infty} \left(\frac{x^2-1}{2}\right)^n$ is a convergent geometric series when $-\sqrt{3} < x < \sqrt{3}$ and its sum is $\frac{1}{1 - \left(\frac{x^2-1}{2}\right)} = \frac{1}{\left(\frac{2 - (x^2-1)}{2}\right)} = \frac{2}{3 - x^2}$

39. $\lim_{n\to\infty} \left|\frac{(x-3)^{n+1}}{2^{n+1}} \cdot \frac{2^n}{(x-3)^n}\right| < 1 \Rightarrow |x - 3| < 2 \Rightarrow 1 < x < 5$; when $x = 1$ we have $\sum_{n=1}^{\infty} (1)^n$ which diverges; when $x = 5$ we have $\sum_{n=1}^{\infty} (-1)^n$ which also diverges; the interval of convergence is $1 < x < 5$; the sum of this convergent geometric series is $\frac{1}{1 + \left(\frac{x-3}{2}\right)} = \frac{2}{x-1}$. If $f(x) = 1 - \frac{1}{2}(x-3) + \frac{1}{4}(x-3)^2 + \ldots + \left(-\frac{1}{2}\right)^n (x-3)^n + \ldots$ $= \frac{2}{x-1}$ then $f'(x) = -\frac{1}{2} + \frac{1}{2}(x-3) + \ldots + \left(-\frac{1}{2}\right)^n n(x-3)^{n-1} + \ldots$ is convergent when $1 < x < 5$, and diverges when $x = 1$ or 5. The sum for $f'(x)$ is $\frac{-2}{(x-1)^2}$, the derivative of $\frac{2}{x-1}$.

40. If $f(x) = 1 - \frac{1}{2}(x-3) + \frac{1}{4}(x-3)^2 + \ldots + \left(-\frac{1}{2}\right)^n (x-3)^n + \ldots = \frac{2}{x-1}$ then $\int f(x)\,dx$ $= x - \frac{(x-3)^2}{4} + \frac{(x-3)^3}{12} + \ldots + \left(-\frac{1}{2}\right)^n \frac{(x-3)^{n+1}}{n+1} + \ldots$. At $x = 1$ the series $\sum_{n=1}^{\infty} \frac{-2}{n+1}$ diverges; at $x = 5$ the series $\sum_{n=1}^{\infty} \frac{(-1)^n 2}{n+1}$ converges. Therefore the interval of convergence is $1 < x \le 5$ and the sum is $2 \ln|x - 1| + (3 - \ln 4)$, since $\int \frac{2}{x-1}\,dx = 2 \ln|x - 1| + C$, where $C = 3 - \ln 4$ when $x = 3$.

41. (a) Differentiate the series for sin x to get $\cos x = 1 - \frac{3x^2}{3!} + \frac{5x^4}{5!} - \frac{7x^6}{7!} + \frac{9x^8}{9!} - \frac{11x^{10}}{11!} + \ldots$ $= 1 - \frac{x^2}{2!} + \frac{x^4}{4!} - \frac{x^6}{6!} + \frac{x^8}{8!} - \frac{x^{10}}{10!} + \ldots$. The series converges for all values of x since

$\lim_{n\to\infty} \left|\frac{x^{n+1}}{(n+1)!}\cdot\frac{n!}{x^n}\right| = |x| \lim_{n\to\infty} \left(\frac{1}{n+1}\right) = 0 < 1$ for all x

(b) $\sin 2x = 2x - \frac{2^3x^3}{3!} + \frac{2^5x^5}{5!} - \frac{2^7x^7}{7!} + \frac{2^9x^9}{9!} - \frac{2^{11}x^{11}}{11!} + \ldots = 2x - \frac{8x^3}{3!} + \frac{32x^5}{5!} - \frac{128x^7}{7!} + \frac{512x^9}{9!} - \frac{2048x^{11}}{11!} + \ldots$

(c) $2\sin x\cos x = 2\Big[(0\cdot1) + (0\cdot0 + 1\cdot1)x + \left(0\cdot\frac{-1}{2} + 1\cdot0 + 0\cdot1\right)x^2 + \left(0\cdot0 - 1\cdot\frac{1}{2} + 0\cdot0 - 1\cdot\frac{1}{3!}\right)x^3$

$+\left(0\cdot\frac{1}{4!} + 1\cdot0 - 0\cdot\frac{1}{2} - 0\cdot\frac{1}{3!} + 0\cdot1\right)x^4 + \left(0\cdot0 + 1\cdot\frac{1}{4!} + 0\cdot0 + \frac{1}{2}\cdot\frac{1}{3!} + 0\cdot0 + 1\cdot\frac{1}{5!}\right)x^5$

$+\left(0\cdot\frac{1}{6!} + 1\cdot0 + 0\cdot\frac{1}{4!} + 0\cdot\frac{1}{3!} + 0\cdot\frac{1}{2} + 0\cdot\frac{1}{5!} + 0\cdot1\right)x^6 + \ldots\Big] = 2\left[x - \frac{4x^3}{3!} + \frac{16x^5}{5!} - \ldots\right]$

$= 2x - \frac{2^3x^3}{3!} + \frac{2^5x^5}{5!} - \frac{2^7x^7}{7!} + \frac{2^9x^9}{9!} - \frac{2^{11}x^{11}}{11!} + \ldots$

42. (a) $\frac{d}{x}(e^x) = 1 + \frac{2x}{2!} + \frac{3x^2}{3!} + \frac{4x^3}{4!} + \frac{5x^4}{5!} + \ldots = 1 + x + \frac{x^2}{2!} + \frac{x^3}{3!} + \frac{x^4}{4!} + \ldots = e^x$; thus the derivative of e^x is e^x itself

(b) $\int e^x\,dx = e^x + C = x + \frac{x^2}{2} + \frac{x^3}{3!} + \frac{x^4}{4!} + \frac{x^5}{5!} + \ldots + C$, which is the general antiderivative of e^x

(c) $e^{-x} = 1 - x + \frac{x^2}{2!} - \frac{x^3}{3!} + \frac{x^4}{4!} - \frac{x^5}{5!} + \ldots$; $e^{-x}\cdot e^x = 1\cdot1 + (1\cdot1 - 1\cdot1)x + \left(1\cdot\frac{1}{2!} - 1\cdot1 + \frac{1}{2!}\cdot1\right)x^2$

$+\left(1\cdot\frac{1}{3!} - 1\cdot\frac{1}{2!} + \frac{1}{2!}\cdot1 - \frac{1}{3!}\cdot1\right)x^3 + \left(1\cdot\frac{1}{4!} - 1\cdot\frac{1}{3!} + \frac{1}{2!}\cdot\frac{1}{2!} - \frac{1}{3!}\cdot1 + \frac{1}{4!}\cdot1\right)x^4$

$+\left(1\cdot\frac{1}{5!} - 1\cdot\frac{1}{4!} + \frac{1}{2!}\cdot\frac{1}{3!} - \frac{1}{3!}\cdot\frac{1}{2!} + \frac{1}{4!}\cdot1 - \frac{1}{5!}\cdot1\right)x^5 + \ldots = 1 + 0 + 0 + 0 + 0 + 0 + \ldots$

43. (a) $\ln|\sec x| + C = \int \tan x\,dx = \int\left(x + \frac{x^3}{3} + \frac{2x^5}{15} + \frac{17x^7}{315} + \frac{62x^9}{2835} + \ldots\right)dx$

$= \frac{x^2}{2} + \frac{x^4}{12} + \frac{x^6}{45} + \frac{17x^8}{2520} + \frac{31x^{10}}{14{,}175} + \ldots + C$; $x = 0 \Rightarrow C = 0 \Rightarrow \ln|\sec x| = \frac{x^2}{2} + \frac{x^4}{12} + \frac{x^6}{45} + \frac{17x^8}{2520} + \frac{31x^{10}}{14{,}175} + \ldots$,

converges when $-\frac{\pi}{2} < x < \frac{\pi}{2}$

(b) $\sec^2 x = \frac{d(\tan x)}{dx} = \frac{d}{dx}\left(x + \frac{x^3}{3} + \frac{2x^5}{15} + \frac{17x^7}{315} + \frac{62x^9}{2835} + \ldots\right) = 1 + x^2 + \frac{2x^4}{3} + \frac{17x^6}{45} + \frac{62x^8}{315} + \ldots$, converges

when $-\frac{\pi}{2} < x < \frac{\pi}{2}$

(c) $\sec^2 x = (\sec x)(\sec x) = \left(1 + \frac{x^2}{2} + \frac{5x^4}{24} + \frac{61x^6}{720} + \ldots\right)\left(1 + \frac{x^2}{2} + \frac{5x^4}{24} + \frac{61x^6}{720} + \ldots\right)$

$= 1 + \left(\frac{1}{2} + \frac{1}{2}\right)x^2 + \left(\frac{5}{24} + \frac{1}{4} + \frac{5}{24}\right)x^4 + \left(\frac{61}{720} + \frac{5}{48} + \frac{5}{48} + \frac{61}{720}\right)x^6 + \ldots$

$= 1 + x^2 + \frac{2x^4}{3} + \frac{17x^6}{45} + \frac{62x^8}{315} + \ldots,\ -\frac{\pi}{2} < x < \frac{\pi}{2}$

44. (a) $\ln|\sec x + \tan x| + C = \int \sec x\,dx = \int\left(1 + \frac{x^2}{2} + \frac{5x^4}{24} + \frac{61x^6}{720} + \ldots\right)dx$

$= x + \frac{x^3}{6} + \frac{x^5}{24} + \frac{61x^7}{5040} + \frac{277x^9}{72{,}576} + \ldots + C$; $x = 0 \Rightarrow C = 0 \Rightarrow \ln|\sec x + \tan x|$

$= x + \frac{x^3}{6} + \frac{x^5}{24} + \frac{61x^7}{5040} + \frac{277x^9}{72{,}576} + \ldots$, converges when $-\frac{\pi}{2} < x < \frac{\pi}{2}$

(b) $\sec x \tan x = \frac{d(\sec x)}{dx} = \frac{d}{dx}\left(1 + \frac{x^2}{2} + \frac{5x^4}{24} + \frac{61x^6}{720} + \ldots\right) = x + \frac{5x^3}{6} + \frac{61x^5}{120} + \frac{277x^7}{1008} + \ldots$, converges when $-\frac{\pi}{2} < x < \frac{\pi}{2}$

(c) $(\sec x)(\tan x) = \left(1 + \frac{x^2}{2} + \frac{5x^4}{24} + \frac{61x^6}{720} + \ldots\right)\left(x + \frac{x^3}{3} + \frac{2x^5}{15} + \frac{17x^7}{315} + \ldots\right)$

$= x + \left(\frac{1}{3} + \frac{1}{2}\right)x^3 + \left(\frac{2}{15} + \frac{1}{6} + \frac{5}{24}\right)x^5 + \left(\frac{17}{315} + \frac{1}{15} + \frac{5}{72} + \frac{61}{720}\right)x^7 + \ldots = x + \frac{5x^3}{6} + \frac{61x^5}{120} + \frac{277x^7}{1008} + \ldots$, $-\frac{\pi}{2} < x < \frac{\pi}{2}$

45. (a) If $f(x) = \sum_{n=0}^{\infty} a_n x^n$, then $f^{(k)}(x) = \sum_{n=k}^{\infty} n(n-1)(n-2)\cdots(n-(k-1))\, a_n x^{n-k}$ and $f^{(k)}(0) = k!a_k$ $\Rightarrow a_k = \frac{f^{(k)}(0)}{k!}$; likewise if $f(x) = \sum_{n=0}^{\infty} b_n x^n$, then $b_k = \frac{f^{(k)}(0)}{k!} \Rightarrow a_k = b_k$ for every nonnegative integer k

(b) If $f(x) = \sum_{n=0}^{\infty} a_n x^n = 0$ for all x, then $f^{(k)}(x) = 0$ for all $x \Rightarrow$ from part (a) that $a_k = 0$ for every nonnegative integer k

46. $\frac{1}{1-x} = 1 + x + x^2 + x^3 + x^4 + \ldots \Rightarrow x\left[\frac{1}{(1-x)^2}\right] = x(1 + 2x + 3x^2 + 4x^3 + \ldots) \Rightarrow \frac{x}{(1-x)^2}$

$= x + 2x^2 + 3x^3 + 4x^4 + \ldots \Rightarrow x\left[\frac{1+x}{(1-x)^3}\right] = x(1 + 4x + 9x^2 + 16x^3 + \ldots) \Rightarrow \frac{x+x^2}{(1-x)^3}$

$= x + 4x^2 + 9x^3 + 16x^4 + \ldots \Rightarrow \frac{\left(\frac{1}{2}+\frac{1}{4}\right)}{\left(\frac{1}{8}\right)} = \frac{1}{2} + \frac{4}{4} + \frac{9}{8} + \frac{16}{16} + \ldots \Rightarrow \sum_{n=1}^{\infty} \frac{n^2}{2^n} = 6$

47. The series $\sum_{n=1}^{\infty} \frac{x^n}{n}$ converges conditionally at the left-hand endpoint of its interval of convergence $[-1, 1]$; the series $\sum_{n=1}^{\infty} \frac{x^n}{(n^2)}$ converges absolutely at the left-hand endpoint of its interval of convergence $[-1, 1]$

48. Answers will vary. For instance:

(a) $\sum_{n=1}^{\infty} \left(\frac{x}{3}\right)^n$ (b) $\sum_{n=1}^{\infty} (x+1)^n$ (c) $\sum_{n=1}^{\infty} \left(\frac{x-3}{2}\right)^n$

8.9 TAYLOR AND MACLAURIN SERIES

1. $f(x) = \ln x$, $f'(x) = \frac{1}{x}$, $f''(x) = -\frac{1}{x^2}$, $f'''(x) = \frac{2}{x^3}$; $f(1) = \ln 1 = 0$, $f'(1) = 1$, $f''(1) = -1$, $f'''(1) = 2 \Rightarrow P_0(x) = 0$, $P_1(x) = (x-1)$, $P_2(x) = (x-1) - \frac{1}{2}(x-1)^2$, $P_3(x) = (x-1) - \frac{1}{2}(x-1)^2 + \frac{1}{3}(x-1)^3$

2. $f(x) = \ln(1+x)$, $f'(x) = \frac{1}{1+x} = (1+x)^{-1}$, $f''(x) = -(1+x)^{-2}$, $f'''(x) = 2(1+x)^{-3}$; $f(0) = \ln 1 = 0$, $f'(0) = \frac{1}{1} = 1$, $f''(0) = -(1)^{-2} = -1$, $f'''(0) = 2(1)^{-3} = 2 \Rightarrow P_0(x) = 0$, $P_1(x) = x$, $P_2(x) = x - \frac{x^2}{2}$, $P_3(x) = x - \frac{x^2}{2} + \frac{x^3}{3}$

3. $f(x) = \frac{1}{x} = x^{-1}$, $f'(x) = -x^{-2}$, $f''(x) = 2x^{-3}$, $f'''(x) = -6x^{-4}$; $f(2) = \frac{1}{2}$, $f'(2) = -\frac{1}{4}$, $f''(2) = \frac{1}{4}$, $f'''(x) = -\frac{3}{8}$ $\Rightarrow P_0(x) = \frac{1}{2}$, $P_1(x) = \frac{1}{2} - \frac{1}{4}(x-2)$, $P_2(x) = \frac{1}{2} - \frac{1}{4}(x-2) + \frac{1}{8}(x-2)^2$, $P_3(x) = \frac{1}{2} - \frac{1}{4}(x-2) + \frac{1}{8}(x-2)^2 - \frac{1}{16}(x-2)^3$

4. $f(x) = (x+2)^{-1}$, $f'(x) = -(x+2)^{-2}$, $f''(x) = 2(x+2)^{-3}$, $f'''(x) = -6(x+2)^{-4}$; $f(0) = (2)^{-1} = \frac{1}{2}$, $f'(0) = -(2)^{-2} = -\frac{1}{4}$, $f''(0) = 2(2)^{-3} = \frac{1}{4}$, $f'''(0) = -6(2)^{-4} = -\frac{3}{8} \Rightarrow P_0(x) = \frac{1}{2}$, $P_1(x) = \frac{1}{2} - \frac{x}{4}$, $P_2(x) = \frac{1}{2} - \frac{x}{4} + \frac{x^2}{8}$, $P_3(x) = \frac{1}{2} - \frac{x}{4} + \frac{x^2}{8} - \frac{x^3}{16}$

5. $f(x) = \sin x$, $f'(x) = \cos x$, $f''(x) = -\sin x$, $f'''(x) = -\cos x$; $f\left(\frac{\pi}{4}\right) = \sin\frac{\pi}{4} = \frac{\sqrt{2}}{2}$, $f'\left(\frac{\pi}{4}\right) = \cos\frac{\pi}{4} = \frac{\sqrt{2}}{2}$, $f''\left(\frac{\pi}{4}\right) = -\sin\frac{\pi}{4} = -\frac{\sqrt{2}}{2}$, $f'''\left(\frac{\pi}{4}\right) = -\cos\frac{\pi}{4} = -\frac{\sqrt{2}}{2} \Rightarrow P_0 = \frac{\sqrt{2}}{2}$, $P_1(x) = \frac{\sqrt{2}}{2} + \frac{\sqrt{2}}{2}\left(x - \frac{\pi}{4}\right)$, $P_2(x) = \frac{\sqrt{2}}{2} + \frac{\sqrt{2}}{2}\left(x - \frac{\pi}{4}\right) - \frac{\sqrt{2}}{4}\left(x - \frac{\pi}{4}\right)^2$, $P_3(x) = \frac{\sqrt{2}}{2} + \frac{\sqrt{2}}{2}\left(x - \frac{\pi}{4}\right) - \frac{\sqrt{2}}{4}\left(x - \frac{\pi}{4}\right)^2 - \frac{\sqrt{2}}{12}\left(x - \frac{\pi}{4}\right)^3$

6. $f(x) = \cos x$, $f'(x) = -\sin x$, $f''(x) = -\cos x$, $f'''(x) = \sin x$; $f\left(\frac{\pi}{4}\right) = \cos\frac{\pi}{4} = \frac{1}{\sqrt{2}}$, $f'\left(\frac{\pi}{4}\right) = -\sin\frac{\pi}{4} = -\frac{1}{\sqrt{2}}$, $f''\left(\frac{\pi}{4}\right) = -\cos\frac{\pi}{4} = -\frac{1}{\sqrt{2}}$, $f'''\left(\frac{\pi}{4}\right) = \sin\frac{\pi}{4} = \frac{1}{\sqrt{2}} \Rightarrow P_0(x) = \frac{1}{\sqrt{2}}$, $P_1(x) = \frac{1}{\sqrt{2}} - \frac{1}{\sqrt{2}}\left(x - \frac{\pi}{4}\right)$, $P_2(x) = \frac{1}{\sqrt{2}} - \frac{1}{\sqrt{2}}\left(x - \frac{\pi}{4}\right) - \frac{1}{2\sqrt{2}}\left(x - \frac{\pi}{4}\right)^2$, $P_3(x) = \frac{1}{\sqrt{2}} - \frac{1}{\sqrt{2}}\left(x - \frac{\pi}{4}\right) - \frac{1}{2\sqrt{2}}\left(x - \frac{\pi}{4}\right)^2 + \frac{1}{6\sqrt{2}}\left(x - \frac{\pi}{4}\right)^3$

7. $f(x) = \sqrt{x} = x^{1/2}$, $f'(x) = \left(\frac{1}{2}\right)x^{-1/2}$, $f''(x) = \left(-\frac{1}{4}\right)x^{-3/2}$, $f'''(x) = \left(\frac{3}{8}\right)x^{-5/2}$; $f(4) = \sqrt{4} = 2$, $f'(4) = \left(\frac{1}{2}\right)4^{-1/2} = \frac{1}{4}$, $f''(4) = \left(-\frac{1}{4}\right)4^{-3/2} = -\frac{1}{32}$, $f'''(4) = \left(\frac{3}{8}\right)4^{-5/2} = \frac{3}{256} \Rightarrow P_0(x) = 2$, $P_1(x) = 2 + \frac{1}{4}(x-4)$, $P_2(x) = 2 + \frac{1}{4}(x-4) - \frac{1}{64}(x-4)^2$, $P_3(x) = 2 + \frac{1}{4}(x-4) - \frac{1}{64}(x-4)^2 + \frac{1}{512}(x-4)^3$

8. $f(x) = (x+4)^{1/2}$, $f'(x) = \left(\frac{1}{2}\right)(x+4)^{-1/2}$, $f''(x) = \left(-\frac{1}{4}\right)(x+4)^{-3/2}$, $f'''(x) = \left(\frac{3}{8}\right)(x+4)^{-5/2}$; $f(0) = (4)^{1/2} = 2$, $f'(0) = \left(\frac{1}{2}\right)(4)^{-1/2} = \frac{1}{4}$, $f''(0) = \left(-\frac{1}{4}\right)(4)^{-3/2} = -\frac{1}{32}$, $f'''(0) = \left(\frac{3}{8}\right)(4)^{-5/2} = \frac{3}{256} \Rightarrow P_0(x) = 2$, $P_1(x) = 2 + \frac{1}{4}x$, $P_2(x) = 2 + \frac{1}{4}x - \frac{1}{64}x^2$, $P_3(x) = 2 + \frac{1}{4}x - \frac{1}{64}x^2 + \frac{1}{512}x^3$

9. $e^x = \sum_{n=0}^{\infty} \frac{x^n}{n!} \Rightarrow e^{-x} = \sum_{n=0}^{\infty} \frac{(-x)^n}{n!} = 1 - x + \frac{x^2}{2!} - \frac{x^3}{3!} + \frac{x^4}{4!} - \ldots$

10. $e^x = \sum_{n=0}^{\infty} \frac{x^n}{n!} \Rightarrow e^{x/2} = \sum_{n=0}^{\infty} \frac{\left(\frac{x}{2}\right)^n}{n!} = 1 + \frac{x}{2} + \frac{x^2}{4 \cdot 2!} + \frac{x^3}{2^3 \cdot 3!} + \frac{x^4}{2^4 \cdot 4!} + \ldots$

11. $f(x) = (1+x)^{-1} \Rightarrow f'(x) = -(1+x)^{-2}$, $f''(x) = 2(1+x)^{-3}$, $f'''(x) = -3!(1+x)^{-4} \Rightarrow \ldots f^{(k)}(x)$
$= (-1)^k k!(1+x)^{-k-1}$; $f(0) = 1$, $f'(0) = -1$, $f''(0) = 2$, $f'''(0) = -3!, \ldots, f^{(k)}(0) = (-1)^k k!$
$\Rightarrow \frac{1}{1+x} = 1 - x + x^2 - x^3 + \ldots = \sum_{n=0}^{\infty} (-x)^n = \sum_{n=0}^{\infty} (-1)^n x^n$

12. $f(x) = (1-x)^{-1} \Rightarrow f'(x) = (1-x)^{-2}$, $f''(x) = 2(1-x)^{-3}$, $f'''(x) = 3!(1-x)^{-4} \Rightarrow \ldots f^{(k)}(x)$
$= k!(1-x)^{-k-1}$; $f(0) = 1$, $f'(0) = 1$, $f''(0) = 2$, $f'''(0) = 3!, \ldots, f^{(k)}(0) = k!$
$\Rightarrow \frac{1}{1-x} = 1 + x + x^2 + x^3 + \ldots = \sum_{n=0}^{\infty} x^n$

13. $\sin x = \sum_{n=0}^{\infty} \frac{(-1)^n x^{2n+1}}{(2n+1)!} \Rightarrow \sin 3x = \sum_{n=0}^{\infty} \frac{(-1)^n (3x)^{2n+1}}{(2n+1)!} = \sum_{n=0}^{\infty} \frac{(-1)^n 3^{2n+1} x^{2n+1}}{(2n+1)!} = 3x - \frac{3^3 x^3}{3!} + \frac{3^5 x^5}{5!} - \ldots$

14. $\sin x = \sum_{n=0}^{\infty} \frac{(-1)^n x^{2n+1}}{(2n+1)!} \Rightarrow \sin \frac{x}{2} = \sum_{n=0}^{\infty} \frac{(-1)^n \left(\frac{x}{2}\right)^{2n+1}}{(2n+1)!} = \sum_{n=0}^{\infty} \frac{(-1)^n x^{2n+1}}{2^{2n+1}(2n+1)!} = \frac{x}{2} - \frac{x^3}{2^3 \cdot 3!} + \frac{x^5}{2^5 \cdot 5!} + \ldots$

15. $7 \cos(-x) = 7 \cos x = 7 \sum_{n=0}^{\infty} \frac{(-1)^n x^{2n}}{(2n)!} = 7 - \frac{7x^2}{2!} + \frac{7x^4}{4!} - \frac{7x^6}{6!} + \ldots$, since the cosine is an even function

16. $\cos x = \sum_{n=0}^{\infty} \frac{(-1)^n x^{2n}}{(2n)!} \Rightarrow 5 \cos \pi x = 5 \sum_{n=0}^{\infty} \frac{(-1)^n (\pi x)^{2n}}{(2n)!} = 5 - \frac{5\pi^2 x^2}{2!} + \frac{5\pi^4 x^4}{4!} - \frac{5\pi^6 x^6}{6!} + \ldots$

17. $\cosh x = \frac{e^x + e^{-x}}{2} = \frac{1}{2}\left[\left(1 + x^2 + \frac{x^2}{2!} + \frac{x^3}{3!} + \frac{x^4}{4!} + \ldots\right) + \left(1 - x + \frac{x^2}{2!} - \frac{x^3}{3!} + \frac{x^4}{4!} - \ldots\right)\right] = 1 + \frac{x^2}{2!} + \frac{x^4}{4!} + \frac{x^6}{6!} + \ldots$
$= \sum_{n=0}^{\infty} \frac{x^{2n}}{(2n)!}$

18. $\sinh x = \frac{e^x - e^{-x}}{2} = \frac{1}{2}\left[\left(1 + x + \frac{x^2}{2!} + \frac{x^3}{3!} + \frac{x^4}{4!} + \ldots\right) - \left(1 - x + \frac{x^2}{2!} - \frac{x^3}{3!} + \frac{x^4}{4!} - \ldots\right)\right] = x + \frac{x^3}{3!} + \frac{x^5}{5!} + \frac{x^6}{6!} + \ldots$
$= \sum_{n=0}^{\infty} \frac{x^{2n+1}}{(2n+1)!}$

19. $f(x) = x^4 - 2x^3 - 5x + 4 \Rightarrow f'(x) = 4x^3 - 6x^2 - 5$, $f''(x) = 12x^2 - 12x$, $f'''(x) = 24x - 12$, $f^{(4)}(x) = 24$
$\Rightarrow f^{(n)}(x) = 0$ if $n \geq 5$; $f(0) = 4$, $f'(0) = -5$, $f''(0) = 0$, $f'''(0) = -12$, $f^{(4)}(0) = 24$, $f^{(n)}(0) = 0$ if $n \geq 5$
$\Rightarrow x^4 - 2x^3 - 5x + 4 = 4 - 5x - \frac{12}{3!}x^3 + \frac{24}{4!}x^4 = x^4 - 2x^3 - 5x + 4$ itself

20. $f(x) = (x+1)^2 \Rightarrow f'(x) = 2(x+1);\ f''(x) = 2 \Rightarrow f^{(n)}(x) = 0$ if $n \ge 3$; $f(0) = 1,\ f'(0) = 2,\ f''(0) = 2,\ f^{(n)}(0) = 0$ if $n \ge 3 \Rightarrow (x+1)^2 = 1 + 2x + \frac{2}{2!}x^2 = 1 + 2x + x^2$

21. $f(x) = x^3 - 2x + 4 \Rightarrow f'(x) = 3x^2 - 2,\ f''(x) = 6x,\ f'''(x) = 6 \Rightarrow f^{(n)}(x) = 0$ if $n \ge 4$; $f(2) = 8,\ f'(2) = 10,$ $f''(2) = 12,\ f'''(2) = 6,\ f^{(n)}(2) = 0$ if $n \ge 4 \Rightarrow x^3 - 2x + 4 = 8 + 10(x-2) + \frac{12}{2!}(x-2)^2 + \frac{6}{3!}(x-2)^3$ $= 8 + 10(x-2) + 6(x-2)^2 + (x-2)^3$

22. $f(x) = 2x^3 + x^2 + 3x - 8 \Rightarrow f'(x) = 6x^2 + 2x + 3,\ f''(x) = 12x + 2,\ f'''(x) = 12 \Rightarrow f^{(n)}(x) = 0$ if $n \ge 4$; $f(1) = -2,$ $f'(1) = 11,\ f''(1) = 14,\ f'''(1) = 12,\ f^{(n)}(1) = 0$ if $n \ge 4 \Rightarrow 2x^3 + x^2 + 3x - 8$ $= -2 + 11(x-1) + \frac{14}{2!}(x-1)^2 + \frac{12}{3!}(x-1)^3 = -2 + 11(x-1) + 7(x-1)^2 + 2(x-1)^3$

23. $f(x) = x^4 + x^2 + 1 \Rightarrow f'(x) = 4x^3 + 2x,\ f''(x) = 12x^2 + 2,\ f'''(x) = 24x,\ f^{(4)}(x) = 24,\ f^{(n)}(x) = 0$ if $n \ge 5$; $f(-2) = 21,\ f'(-2) = -36,\ f''(-2) = 50,\ f'''(-2) = -48,\ f^{(4)}(-2) = 24,\ f^{(n)}(-2) = 0$ if $n \ge 5 \Rightarrow x^4 + x^2 + 1$ $= 21 - 36(x+2) + \frac{50}{2!}(x+2)^2 - \frac{48}{3!}(x+2)^3 + \frac{24}{4!}(x+2)^4 = 21 - 36(x+2) + 25(x+2)^2 - 8(x+2)^3 + (x+2)^4$

24. $f(x) = 3x^5 - x^4 + 2x^3 + x^2 - 2 \Rightarrow f'(x) = 15x^4 - 4x^3 + 6x^2 + 2x,\ f''(x) = 60x^3 - 12x^2 + 12x + 2,$ $f'''(x) = 180x^2 - 24x + 12,\ f^{(4)}(x) = 360x - 24,\ f^{(5)}(x) = 360,\ f^{(n)}(x) = 0$ if $n \ge 6$; $f(-1) = -7,$ $f'(-1) = 23,\ f''(-1) = -82,\ f'''(-1) = 216,\ f^{(4)}(-1) = -384,\ f^{(5)}(-1) = 360,\ f^{(n)}(-1) = 0$ if $n \ge 6$ $\Rightarrow 3x^5 - x^4 + 2x^3 + x^2 - 2 = -7 + 23(x+1) - \frac{82}{2!}(x+1)^2 + \frac{216}{3!}(x+1)^3 - \frac{384}{4!}(x+1)^4 + \frac{360}{5!}(x+1)^5$ $= -7 + 23(x+1) - 41(x+1)^2 + 36(x+1)^3 - 16(x+1)^4 + 3(x+1)^5$

25. $f(x) = x^{-2} \Rightarrow f'(x) = -2x^{-3},\ f''(x) = 3!\,x^{-4},\ f'''(x) = -4!\,x^{-5} \Rightarrow f^{(n)}(x) = (-1)^n(n+1)!\,x^{-n-2}$; $f(1) = 1,\ f'(1) = -2,\ f''(1) = 3!,\ f'''(1) = -4!,\ f^{(n)}(1) = (-1)^n(n+1)! \Rightarrow \frac{1}{x^2}$ $= 1 - 2(x-1) + 3(x-1)^2 - 4(x-1)^3 + \ldots = \sum_{n=0}^{\infty} (-1)^n(n+1)(x-1)^n$

26. $f(x) = \frac{x}{1-x} \Rightarrow f'(x) = (1-x)^{-2},\ f''(x) = 2(1-x)^{-3},\ f'''(x) = 3!\,(1-x)^{-4} \Rightarrow f^{(n)}(x) = n!\,(1-x)^{-n-1}$; $f(0) = 0,\ f'(0) = 1,\ f''(0) = 2,\ f'''(0) = 3! \Rightarrow \frac{x}{1-x} = x + x^2 + x^3 + \ldots = \sum_{n=0}^{\infty} x^{n+1}$

27. $f(x) = e^x \Rightarrow f'(x) = e^x,\ f''(x) = e^x \Rightarrow f^{(n)}(x) = e^x$; $f(2) = e^2,\ f'(2) = e^2,\ \ldots\ f^{(n)}(2) = e^2$ $\Rightarrow e^x = e^2 + e^2(x-2) + \frac{e^2}{2}(x-2)^2 + \frac{e^3}{3!}(x-2)^3 + \ldots = \sum_{n=0}^{\infty} \frac{e^2}{n!}(x-2)^n$

28. $f(x) = 2^x \Rightarrow f'(x) = 2^x \ln 2,\ f''(x) = 2^x(\ln 2)^2,\ f'''(x) = 2^x(\ln 2)^3 \Rightarrow f^{(n)}(x) = 2^x(\ln 2)^n$; $f(1) = 2,\ f'(1) = 2\ln 2,$ $f''(1) = 2(\ln 2)^2,\ f'''(1) = 2(\ln 2)^3,\ \ldots,\ f^{(n)}(1) = 2(\ln 2)^n$ $\Rightarrow 2^x = 2 + (2\ln 2)(x-1) + \frac{2(\ln 2)^2}{2}(x-1)^2 + \frac{2(\ln 2)^3}{3!}(x-1)^3 + \ldots = \sum_{n=0}^{\infty} \frac{2(\ln 2)^n(x-1)^n}{n!}$

29. If $e^x = \sum_{n=0}^{\infty} \frac{f^{(n)}(a)}{n!}(x-a)^n$ and $f(x) = e^x$, we have $f^{(n)}(a) = e^a$ f or all n = 0, 1, 2, 3, ...

$\Rightarrow e^x = e^a\left[\frac{(x-a)^0}{0!} + \frac{(x-a)^1}{1!} + \frac{(x-a)^2}{2!} + \ldots\right] = e^a\left[1 + (x-a) + \frac{(x-a)^2}{2!} + \ldots\right]$ at $x = a$

30. $f(x) = e^x \Rightarrow f^{(n)}(x) = e^x$ for all $n \Rightarrow f^{(n)}(1) = e$ for all n = 0, 1, 2, ...

$\Rightarrow e^x = e + e(x-1) + \frac{e}{2!}(x-1)^2 + \frac{e}{3!}(x-1)^3 + \ldots = e\left[1 + (x-1) + \frac{(x-1)^2}{2!} + \frac{(x-1)^3}{3!} + \ldots\right]$

31. $f(x) = f(a) + f'(a)(x-a) + \frac{f''(a)}{2}(x-a)^2 + \frac{f'''(a)}{3!}(x-a)^3 + \ldots \Rightarrow f'(x)$

$= f'(a) + f''(a)(x-a) + \frac{f'''(a)}{3!}3(x-a)^2 + \ldots \Rightarrow f''(x) = f''(a) + f'''(a)(x-a) + \frac{f^{(4)}(a)}{4!}4 \cdot 3(x-a)^2 + \ldots$

$\Rightarrow f^{(n)}(x) = f^{(n)}(a) + f^{(n+1)}(a)(x-a) + \frac{f^{(n+2)}(a)}{2}(x-a)^2 + \ldots$

$\Rightarrow f(a) = f(a) + 0,\ f'(a) = f'(a) + 0,\ \ldots,\ f^{(n)}(a) = f^{(n)}(a) + 0$

32. $E(x) = f(x) - b_0 - b_1(x-a) - b_2(x-a)^2 - b_3(x-a)^3 - \ldots - b_n(x-a)^n$

$\Rightarrow 0 = E(a) = f(a) - b_0 \Rightarrow b_0 = f(a)$; from condition (b),

$$\lim_{x\to a} \frac{f(x) - f(a) - b_1(x-a) - b_2(x-a)^2 - b_3(x-a)^3 - \ldots - b_n(x-a)^n}{(x-a)^n} = 0$$

$$\Rightarrow \lim_{x\to a} \frac{f'(x) - b_1 - 2b_2(x-a) - 3b_3(x-a)^2 - \ldots - nb_n(x-a)^{n-1}}{n(x-a)^{n-1}} = 0$$

$$\Rightarrow b_1 = f'(a) \Rightarrow \lim_{x\to a} \frac{f''(x) - 2b_2 - 3!\,b_3(x-a) - \ldots - n(n-1)b_n(x-a)^{n-2}}{n(n-1)(x-a)^{n-2}} = 0$$

$$\Rightarrow b_2 = \frac{1}{2}f''(a) \Rightarrow \lim_{x\to a} \frac{f'''(x) - 3!\,b_3 - \ldots - n(n-1)(n-2)b_n(x-a)^{n-3}}{n(n-1)(n-2)(x-a)^{n-3}} = 0$$

$= b_3 = \frac{1}{3!}f'''(a) \Rightarrow \lim_{x\to a} \frac{f^{(n)}(x) - n!\,b_n}{n!} = 0 \Rightarrow b_n = \frac{1}{n!}f^{(n)}(a)$; therefore,

$$g(x) = f(a) + f'(a)(x-a) + \frac{f''(a)}{2!}(x-a)^2 + \ldots + \frac{f^{(n)}(a)}{n!}(x-a)^n = P_n(x)$$

33. $f(x) = \ln(\cos x) \Rightarrow f'(x) = -\tan x$ and $f''(x) = -\sec^2 x$; $f(0) = 0$, $f'(0) = 0$, $f''(0) = -1$

$\Rightarrow L(x) = 0$ and $Q(x) = -\frac{x^2}{2}$

34. $f(x) = e^{\sin x} \Rightarrow f'(x) = (\cos x)e^{\sin x}$ and $f''(x) = (-\sin x)e^{\sin x} + (\cos x)^2 e^{\sin x}$; $f(0) = 1$, $f'(0) = 1$,
$f''(0) = 1 \Rightarrow L(x) = 1 + x$ and $Q(x) = 1 + x + \frac{x^2}{2}$

35. $f(x) = (1-x^2)^{-1/2} \Rightarrow f'(x) = x(1-x^2)^{-3/2}$ and $f''(x) = (1-x^2)^{-3/2} + 3x^2(1-x^2)^{-5/2}$; $f(0) = 1$,
$f'(0) = 0$, $f''(0) = 1 \Rightarrow L(x) = 1$ and $Q(x) = 1 = \frac{x^2}{2}$

36. $f(x) = \cosh x \Rightarrow f'(x) = \sinh x$ and $f''(x) = \cosh x$; $f(0) = 1$, $f'(0) = 0$, $f''(0) = 1 \Rightarrow L(x) = 1$ and $Q(x) = 1 + \frac{x^2}{2}$

37. $f(x) = \sin x \Rightarrow f'(x) = \cos x$ and $f''(x) = -\sin x$; $f(0) = 0$, $f'(0) = 1$, $f''(0) = 0 \Rightarrow L(x) = x$ and $Q(x) = x$

38. $f(x) = \tan x \Rightarrow f'(x) = \sec^2 x$ and $f''(x) = 2\sec^2 x \tan x$; $f(0) = 0$, $f'(0) = 1$, $f'' = 0 \Rightarrow L(x) = x$ and $Q(x) = x$

8.10 CONVERGENCE OF TAYLOR SERIES; ERROR ESTIMATES

1. $e^x = 1 + x + \frac{x^2}{2!} + \ldots = \sum_{n=0}^{\infty} \frac{x^n}{n!} \Rightarrow e^{-5x} = 1 + (-5x) + \frac{(-5x)^2}{2!} + \ldots = 1 - 5x + \frac{5^2x^2}{2!} - \frac{5^3x^3}{3!} + \ldots = \sum_{n=0}^{\infty} \frac{(-1)^n 5^n x^n}{n!}$

2. $e^x = 1 + x + \frac{x^2}{2!} + \ldots = \sum_{n=0}^{\infty} \frac{x^n}{n!} \Rightarrow e^{-x/2} = 1 + \left(\frac{-x}{2}\right) + \frac{\left(-\frac{x}{2}\right)^2}{2!} + \ldots = 1 - \frac{x}{2} + \frac{x^2}{2^2 2!} - \frac{x^3}{2^3 3!} + \ldots$
$= \sum_{n=0}^{\infty} \frac{(-1)^n x^n}{2^n n!}$

3. $\sin x = x - \frac{x^3}{3!} + \frac{x^5}{5!} - \ldots = \sum_{n=0}^{\infty} \frac{(-1)^n x^{2n+1}}{(2n+1)!} \Rightarrow 5\sin(-x) = 5\left[(-x) - \frac{(-x)^3}{3!} + \frac{(-x)^5}{5!} - \ldots\right]$
$= \sum_{n=0}^{\infty} \frac{5(-1)^{n+1} x^{2n+1}}{(2n+1)!}$

4. $\sin x = x - \frac{x^3}{3!} + \frac{x^5}{5!} - \ldots = \sum_{n=0}^{\infty} \frac{(-1)^n x^{2n+1}}{(2n+1)!} \Rightarrow \sin\frac{\pi x}{2} = \frac{\pi x}{2} - \frac{\left(\frac{\pi x}{2}\right)^3}{3!} + \frac{\left(\frac{\pi x}{2}\right)^5}{5!} - \frac{\left(\frac{\pi x}{2}\right)^7}{7!} + \ldots$
$= \sum_{n=0}^{\infty} \frac{(-1)^n \pi^{2n+1} x^{2n+1}}{2^{2n+1}(2n+1)!}$

5. $\cos x = \sum_{n=0}^{\infty} \frac{(-1)^n x^{2n}}{(2n)!} \Rightarrow \cos\sqrt{x} = \sum_{n=0}^{\infty} \frac{(-1)^n \left(x^{1/2}\right)^{2n}}{(2n)!} = \sum_{n=0}^{\infty} \frac{(-1)^n x^n}{(2n)!} = 1 - \frac{x}{2!} + \frac{x^2}{4!} - \frac{x^3}{6!} + \ldots$

6. $\cos x = \sum_{n=0}^{\infty} \frac{(-1)^n x^{2n}}{(2n)!} \Rightarrow \cos\left(\frac{x^{3/2}}{\sqrt{2}}\right) = \cos\left(\left(\frac{x^3}{2}\right)^{1/2}\right) = \sum_{n=0}^{\infty} \frac{(-1)^n \left(\left(\frac{x^3}{2}\right)^{1/2}\right)^{2n}}{(2n)!} = \sum_{n=0}^{\infty} \frac{(-1)^n x^{3n}}{2^n (2n)!}$
$= 1 - \frac{x^3}{2 \cdot 2!} + \frac{x^6}{2^2 \cdot 4!} - \frac{x^9}{2^3 \cdot 6!} + \ldots$

7. $e^x = \sum_{n=0}^{\infty} \frac{x^n}{n!} \Rightarrow xe^x = x\left(\sum_{n=0}^{\infty} \frac{x^n}{n!}\right) = \sum_{n=0}^{\infty} \frac{x^{n+1}}{n!} = x + x^2 + \frac{x^3}{2!} + \frac{x^4}{3!} + \frac{x^5}{4!} + \ldots$

8. $\sin x = \sum_{n=0}^{\infty} \frac{(-1)^n x^{2n+1}}{(2n+1)!} \Rightarrow x^2 \sin x = x^2\left(\sum_{n=0}^{\infty} \frac{(-1)^n x^{2n+1}}{(2n+1)!}\right) = \sum_{n=0}^{\infty} \frac{(-1)^n x^{2n+3}}{(2n+1)!} = x^3 - \frac{x^5}{3!} + \frac{x^7}{5!} - \frac{x^9}{7!} + \ldots$

9. $\cos x = \sum_{n=0}^{\infty} \frac{(-1)^n x^{2n}}{(2n)!} \Rightarrow \frac{x^2}{2} - 1 + \cos x = \frac{x^2}{2} - 1 + \sum_{n=0}^{\infty} \frac{(-1)^n x^{2n}}{(2n)!} = \frac{x^2}{2} - 1 + 1 - \frac{x^2}{2} + \frac{x^4}{4!} - \frac{x^6}{6!} + \frac{x^8}{8!} - \frac{x^{10}}{10!} + \ldots$
$= \frac{x^4}{4!} - \frac{x^6}{6!} + \frac{x^8}{8!} - \frac{x^{10}}{10!} + \ldots = \sum_{n=2}^{\infty} \frac{(-1)^n x^{2n}}{(2n)!}$

10. $\sin x = \sum_{n=0}^{\infty} \frac{(-1)^n x^{2n+1}}{(2n+1)!} \Rightarrow \sin x - x + \frac{x^3}{3!} = \left(\sum_{n=0}^{\infty} \frac{(-1)^n x^{2n+1}}{(2n+1)!} \right) - x + \frac{x^3}{3!}$
$= \left(x - \frac{x^3}{3!} + \frac{x^5}{5!} - \frac{x^7}{7!} + \frac{x^9}{9!} - \frac{x^{11}}{11!} + \ldots \right) - x + \frac{x^3}{3!} = \frac{x^5}{5!} - \frac{x^7}{7!} + \frac{x^9}{9!} - \frac{x^{11}}{11!} + \ldots = \sum_{n=2}^{\infty} \frac{(-1)^n x^{2n+1}}{(2n+1)!}$

11. $\cos x = \sum_{n=0}^{\infty} \frac{(-1)^n x^{2n}}{(2n)!} \Rightarrow x \cos \pi x = x \sum_{n=0}^{\infty} \frac{(-1)^n (\pi x)^{2n}}{(2n)!} = \sum_{n=0}^{\infty} \frac{(-1)^n \pi^{2n} x^{2n+1}}{(2n)!} = x - \frac{\pi^2 x^3}{2!} + \frac{\pi^4 x^5}{4!} - \frac{\pi^6 x^7}{6!} + \ldots$

12. $\cos x = \sum_{n=0}^{\infty} \frac{(-1)^n x^{2n}}{(2n)!} \Rightarrow x^2 \cos\left(x^2\right) = x^2 \sum_{n=0}^{\infty} \frac{(-1)^n \left(x^2\right)^{2n}}{(2n)!} = \sum_{n=0}^{\infty} \frac{(-1)^n x^{4n+2}}{(2n)!} = x^2 - \frac{x^6}{2!} + \frac{x^{10}}{4!} - \frac{x^{14}}{6!} + \ldots$

13. $\cos^2 x = \frac{1}{2} + \frac{\cos 2x}{2} = \frac{1}{2} + \frac{1}{2} \sum_{n=0}^{\infty} \frac{(-1)^n (2x)^{2n}}{(2n)!} = \frac{1}{2} + \frac{1}{2}\left[1 - \frac{(2x)^2}{2!} + \frac{(2x)^4}{4!} - \frac{(2x)^6}{6!} + \frac{(2x)^8}{8!} - \ldots \right]$
$= 1 - \frac{(2x)^2}{2 \cdot 2!} + \frac{(2x)^4}{2 \cdot 4!} - \frac{(2x)^6}{2 \cdot 6!} + \frac{(2x)^8}{2 \cdot 8!} - \ldots = 1 + \sum_{n=1}^{\infty} \frac{(-1)^n (2x)^{2n}}{2 \cdot (2n)!}$

14. $\sin^2 x = \left(\frac{1 - \cos 2x}{2} \right) = \frac{1}{2} - \frac{1}{2} \cos 2x = \frac{1}{2} - \frac{1}{2}\left(1 - \frac{(2x)^2}{2!} + \frac{(2x)^4}{4!} - \frac{(2x)^6}{6!} + \ldots \right) = \frac{(2x)^2}{2 \cdot 2!} - \frac{(2x)^4}{2 \cdot 4!} + \frac{(2x)^6}{2 \cdot 6!} - \ldots$
$= \sum_{n=1}^{\infty} \frac{(-1)^{n+1} (2x)^{2n}}{2 \cdot (2n)!}$

15. $\frac{x^2}{1-2x} = x^2 \left(\frac{1}{1-2x} \right) = x^2 \sum_{n=0}^{\infty} (2x)^n = \sum_{n=0}^{\infty} 2^n x^{n+2} = x^2 + 2x^3 + 2^2 x^4 + 2^3 x^5 + \ldots$

16. $x \ln(1+2x) = x \sum_{n=1}^{\infty} \frac{(-1)^{n-1} (2x)^n}{n} = \sum_{n=1}^{\infty} \frac{(-1)^{n-1} 2^n x^{n+1}}{n} = 2x^2 - \frac{2^2 x^3}{2} + \frac{2^3 x^4}{4} - \frac{2^4 x^5}{5} + \ldots$

17. $\frac{1}{1-x} = \sum_{n=0}^{\infty} x^n = 1 + x + x^2 + x^3 + \ldots \Rightarrow \frac{d}{dx}\left(\frac{1}{1-x} \right) = \frac{1}{(1-x)^2} = 1 + 2x + 3x^2 + \ldots = \sum_{n=1}^{\infty} n x^{n-1}$
$= \sum_{n=0}^{\infty} (n+1) x^n$

18. $\frac{2}{(1-x^3)} = \frac{d^2}{dx^2}\left(\frac{1}{1-x} \right) = \frac{d}{dx}\left(\frac{1}{(1-x)^2} \right) = \frac{d}{dx}(1 + 2x + 3x^2 + \ldots) = 2 + 6x + 12x^2 + \ldots = \sum_{n=2}^{\infty} n(n-1) x^{n-2}$
$= \sum_{n=0}^{\infty} (n+2)(n+1) x^n$

19. By the Alternating Series Estimation Theorem, the error is less than $\frac{|x|^5}{5!} \Rightarrow |x|^5 < (5!)(5 \times 10^{-4})$
$\Rightarrow |x|^5 < 600 \times 10^{-4} \Rightarrow |x| < \sqrt[5]{6 \times 10^{-2}} \approx 0.56968$

20. If $\cos x = 1 - \frac{x^2}{2}$ and $|x| < 0.5$, then the $|\text{error}| = |R_3(x)| = \left|\frac{\cos c}{4!}x^4\right| < \left|\frac{(.5)^4}{24}\right| = 0.0026$, where c is between 0 and x; since the next term in the series is positive, the approximation $1 - \frac{x^2}{2}$ is too small, by the Alternating Series Estimation Theorem

21. If $\sin x = x$ and $|x| < 10^{-3}$, then the $|\text{error}| = |R_2(x)| = \left|\frac{-\cos c}{3!}x^3\right| < \frac{(10^{-3})^3}{3!} \approx 1.67 \times 10^{-10}$, where c is between 0 and x. The Alternating Series Estimation Theorem says $R_2(x)$ has the same sign as $-\frac{x^3}{3!}$. Moreover, $x < \sin x \Rightarrow 0 < \sin x - x = R_2(x) \Rightarrow x < 0 \Rightarrow -10^{-3} < x < 0$.

22. $\sqrt{1+x} = 1 + \frac{x}{2} - \frac{x^2}{8} + \frac{x^3}{16} - \ldots$. By the Alternating Series Estimation Theorem the $|\text{error}| < \left|\frac{-x^2}{8}\right| < \frac{(0.01)^2}{8}$
$= 1.25 \times 10^{-5}$

23. $|R_2(x)| = \left|\frac{e^c x^3}{3!}\right| < \frac{3^{(0.1)}(0.1)^3}{3!} < 1.87 \times 10^{-5}$, where c is between 0 and x

24. $|R_2(x)| = \left|\frac{e^c x^3}{3!}\right| < \frac{(0.1)^3}{3!} = 1.67 \times 10^{-4}$, where c is between 0 and x

25. $|R_4(x)| < \left|\frac{\cosh c}{5!}x^5\right| = \left|\frac{e^c + e^{-c}}{2}\frac{x^5}{5!}\right| < \frac{1.65 + \frac{1}{1.65}}{2} \cdot \frac{(0.5)^5}{5!} = (1.3)\frac{(0.5)^5}{5!} \approx 0.000293653$

26. If we approximate e^h with $1 + h$ and $0 \le h \le 0.01$, then $|\text{error}| < \left|\frac{e^c h^2}{2}\right| \le \frac{e^{0.01}h \cdot h}{2} = \left(\frac{e^{0.01}(0.01)}{2}\right)h$
$= 0.005005h < 0.006h = (0.6\%)h$, where c is between 0 and h.

27. $|R_1| = \left|\frac{1}{(1+c)^2}\frac{x^2}{2!}\right| < \frac{x^2}{2} = \left|\frac{x}{2}\right||x| < .01\,|x| = (1\%)\,|x| \Rightarrow \left|\frac{x}{2}\right| < .01 \Rightarrow 0 < |x| < .02$

28. $\tan^{-1} x = x - \frac{x^3}{3} + \frac{x^5}{5} - \frac{x^7}{7} + \ldots \Rightarrow \frac{\pi}{4} = \tan^{-1} 1 = 1 - \frac{1}{3} + \frac{1}{5} - \frac{1}{7} + \ldots$; $|\text{error}| < \frac{1}{2n+1} < .01$
$\Rightarrow 2n + 1 > 100 \Rightarrow n > 49$

29. (a) $\sin x = x - \frac{x^3}{3!} + \frac{x^5}{5!} - \frac{x^7}{7!} + \ldots \Rightarrow \frac{\sin x}{x} = 1 - \frac{x^2}{3!} + \frac{x^4}{5!} - \frac{x^6}{7!} + \ldots$, $s_1 = 1$ and $s_2 = 1 - \frac{x^2}{6}$; if L is the sum of the series representing $\frac{\sin x}{x}$, then by the Alternating Series Estimation Theorem, $L - s_1 = \frac{\sin x}{x} - 1 < 0$ and $L - s_2 = \frac{\sin x}{x} - \left(1 - \frac{x^2}{6}\right) > 0$. Therefore $1 - \frac{x^2}{6} < \frac{\sin x}{x} < 1$

(b) The graph of $y = \frac{\sin x}{x}$, $x \neq 0$, is bounded below by the graph of $y = 1 - \frac{x^2}{6}$ and above by the graph of $y = 1$ as derived in part (a).

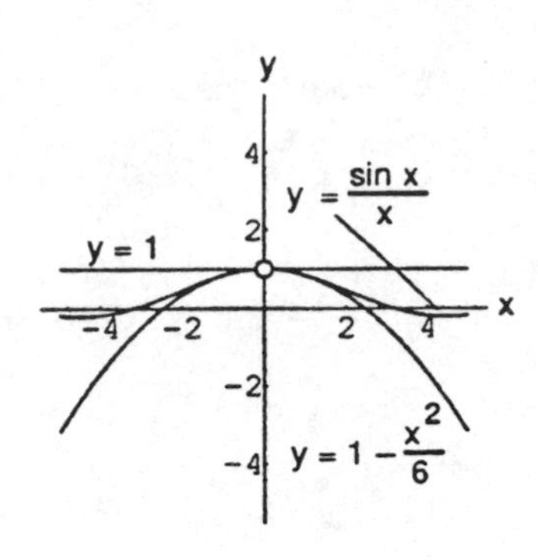

30. (a) $\cos x = 1 - \frac{x^2}{2!} + \frac{x^4}{4!} - \frac{x^6}{6!} + \ldots \Rightarrow 1 - \cos x = \frac{x^2}{2!} - \frac{x^4}{4!} + \frac{x^6}{6!} - \frac{x^8}{8!} + \ldots \Rightarrow \frac{1 - \cos x}{x^2} = \frac{1}{2} - \frac{x^2}{4!} + \frac{x^4}{6!} - \frac{x^6}{8!} + \ldots$; if L is the sum of the series representing $\frac{1 - \cos x}{x^2}$, then by the Alternating Series Estimation Theorem $L - s_1 = \frac{1 - \cos x}{x^2} - \frac{1}{2} < 0$ and $\frac{1 - \cos x}{x^2} - \left(\frac{1}{2} - \frac{x^2}{4!}\right) > 0$. Therefore $\frac{1}{2} - \frac{x^2}{24} < \frac{1 - \cos x}{x^2} < \frac{1}{2}$.

(b) The graph of $y = \frac{1 - \cos x}{x^2}$ is bounded below by the graph of $y = \frac{1}{2} - \frac{x^2}{24}$ and above by the graph of $y = \frac{1}{2}$ as indicated in part (a).

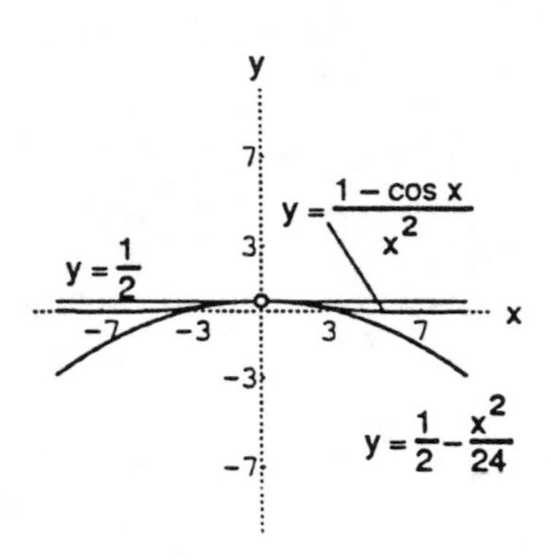

31. $\sin x$ when $x = 0.1$; the sum is $\sin(0.1) \approx 0.099833416$

32. $\cos x$ when $x = \frac{\pi}{4}$; the sum is $\cos\left(\frac{\pi}{4}\right) = \frac{1}{\sqrt{2}} \approx 0.707106781$

33. $\tan^{-1} x$ when $x = \frac{\pi}{3}$; the sum is $\tan^{-1}\left(\frac{\pi}{3}\right) = \sqrt{3} \approx 0.808448$

34. $\ln(1 + x)$ when $x = \pi$; the sum is $\ln(1 + \pi) \approx 1.421080$

35. $e^x \sin x = 0 + x + x^2 + x^3\left(-\frac{1}{3!} + \frac{1}{2!}\right) + x^4\left(-\frac{1}{3!} + \frac{1}{3!}\right) + x^5\left(\frac{1}{5!} - \frac{1}{2!}\frac{1}{3!} + \frac{1}{4!}\right) + x^6\left(\frac{1}{5!} - \frac{1}{3!}\frac{1}{3!} + \frac{1}{5!}\right) + \ldots$
$= x + x^2 + \frac{1}{3}x^3 = \frac{1}{30}x^5 - \frac{1}{90}x^6 + \ldots$

36. $e^x \cos x = 1 + x + x^2\left(-\frac{1}{2!} + \frac{1}{2!}\right) + x^3\left(-\frac{1}{2!} + \frac{1}{3!}\right) + x^4\left(\frac{1}{4!} - \frac{1}{2!}\frac{1}{2!} + \frac{1}{4!}\right) + x^5\left(\frac{1}{4!} - \frac{1}{2!}\frac{1}{3!} + \frac{1}{5!}\right) + \ldots$
$= 1 + x - \frac{1}{3}x^3 - \frac{1}{6}x^4 - \frac{1}{30}x^5 + \ldots$

37. $\sin^2 x = \left(\frac{1-\cos 2x}{2}\right) = \frac{1}{2} - \frac{1}{2}\cos 2x = \frac{1}{2} - \frac{1}{2}\left(1 - \frac{(2x)^2}{2!} + \frac{(2x)^4}{4!} - \frac{(2x)^6}{6!} + \ldots\right) = \frac{2x^2}{2!} - \frac{2^3x^4}{4!} + \frac{2^5x^6}{6!} - \ldots$

$\Rightarrow \frac{d}{dx}(\sin^2 x) = \frac{d}{dx}\left(\frac{2x^2}{2!} - \frac{2^3x^4}{4!} + \frac{2^5x^6}{6!} - \ldots\right) = 2x - \frac{(2x)^3}{3!} + \frac{(2x)^5}{5!} - \frac{(2x)^7}{7!} + \ldots \Rightarrow 2\sin x\cos x$

$= 2x - \frac{(2x)^3}{3!} + \frac{(2x)^5}{5!} - \frac{(2x)^7}{7!} + \ldots = \sin 2x$, which checks

38. $\cos^2 x = \cos 2x + \sin^2 x = \left(1 - \frac{(2x)^2}{2!} + \frac{(2x)^4}{4!} - \frac{(2x)^6}{6!} + \frac{(2x)^8}{8!} + \ldots\right) + \left(\frac{2x^2}{2!} - \frac{2^3x^4}{4!} + \frac{2^5x^6}{6!} - \frac{2^7x^8}{8!} + \ldots\right)$

$= 1 - \frac{2x^2}{2!} + \frac{2^3x^4}{4!} - \frac{2^5x^6}{6!} + \ldots = 1 - x^2 + \frac{1}{3}x^4 - \frac{2}{45}x^6 + \frac{1}{315}x^8 - \ldots$

39. A special case of Taylor's Formula is $f(x) = f(a) + f'(c)(x-a)$. Let $x = b$ and this becomes $f(b) - f(a) = f'(c)(b-a)$, the Mean Value Theorem

40. If $f(x)$ is twice differentiable and at $x = a$ there is a point of inflection, then $f''(a) = 0$. Therefore, $L(x) = Q(x) = f(a) + f'(a)(x-a)$.

41. (a) $f'' \le 0$, $f'(a) = 0$ and $x = a$ interior to the interval $I \Rightarrow f(x) - f(a) = \frac{f''(c_2)}{2}(x-a)^2 \le 0$ throughout I $\Rightarrow f(x) \le f(a)$ throughout I $\Rightarrow$ f has a local maximum at $x = a$

(b) similar reasoning gives $f(x) - f(a) = \frac{f''(c_2)}{2}(x-a)^2 \ge 0$ throughout I $\Rightarrow f(x) \ge f(a)$ throughout I $\Rightarrow$ f has a local minimum at $x = a$

42. (a) $f(x) = (1-x)^{-1} \Rightarrow f'(x) = (1-x)^{-2} \Rightarrow f''(x) = 2(1-x)^{-3} \Rightarrow f^{(3)}(x) = 6(1-x)^{-4}$ $\Rightarrow f^{(4)}(x) = 24(1-x)^{-5}$; therefore $\frac{1}{1-x} \approx 1 + x + x^2 + x^3$

(b) $|x| < 0.1 \Rightarrow \frac{10}{11} < \frac{1}{1-x} < \frac{10}{9} \Rightarrow \left|\frac{1}{(1-x)^5}\right| < \left(\frac{10}{9}\right)^5 \Rightarrow \left|\frac{x^4}{(1-x)^5}\right| < x^4\left(\frac{10}{9}\right)^5 \Rightarrow$ the error

$e_3 \le \left|\frac{\max f^{(4)}(x)\, x^4}{4!}\right| < (0.1)^4\left(\frac{10}{9}\right)^5 = 0.00016935 < 0.00017$, since $\left|\frac{f^{(4)}(x)}{4!}\right| = \left|\frac{1}{(1-x)^5}\right|$.

43. (a) $f(x) = (1+x)^k \Rightarrow f'(x) = k(1+x)^{k-1} \Rightarrow f''(x) = k(k-1)(1+x)^{k-2}$; $f(0) = 1$, $f'(0) = k$, and $f''(0) = k(k-1)$ $\Rightarrow Q(x) = 1 + kx + \frac{k(k-1)}{2}x^2$

(b) $|R_2(x)| = \left|\frac{3\cdot 2\cdot 1}{3!}x^3\right| < \frac{1}{100} \Rightarrow |x^3| < \frac{1}{100} \Rightarrow 0 < x < \frac{1}{100^{1/3}}$ or $0 < x < .21544$

44. Let $P = x + \pi \Rightarrow |x| = |P - \pi| < .5 \times 10^{-n}$ since P approximates π accurate to n decimals. Then, $P + \sin P = (\pi + x) + \sin(\pi + x) = (\pi + x) - \sin x = \pi + (x - \sin x) \Rightarrow |(P + \sin P) - \pi|$ $= |\sin x - x| \le \frac{|x|^3}{3!} < \frac{0.125}{3!} \times 10^{-3n} < .5 \times 10^{-3n} \Rightarrow P + \sin P$ gives an approximation to π correct to 3n decimals.

45. If $f(x) = \sum_{n=0}^{\infty} a_n x^n$, then $f^{(k)}(x) = \sum_{n=k}^{\infty} n(n-1)(n-2)\cdots(n-k+1)a_n x^{n-k}$ and $f^{(k)}(0) = k!\, a_k$ $\Rightarrow a_k = \frac{f^{(k)}(0)}{k!}$ for k a nonnegative integer. Therefore, the coefficients of f(x) are identical with the corresponding coefficients in the Maclaurin series of f(x) and the statement follows.

46. Note: f even $\Rightarrow f(-x) = f(x) \Rightarrow -f'(-x) = f'(x) \Rightarrow f'(-x) = -f'(x) \Rightarrow f'$ odd;
f odd $\Rightarrow f(-x) = -f(x) \Rightarrow -f'(-x) = -f'(x) \Rightarrow f'(-x) = f'(x) \Rightarrow f'$ even;
also, f odd $\Rightarrow f(-0) = f(0) \Rightarrow 2f(0) = 0 \Rightarrow f(0) = 0$
(a) If f(x) is even, then any odd-order derivative is odd and equal to 0 at $x = 0$. Therefore, $a_1 = a_3 = a_5 = \ldots = 0$; that is, the Maclaurin series for f contains only even powers.
(b) If f(x) is odd, then any even-order derivative is odd and equal to 0 at $x = 0$. Therefore, $a_0 = a_2 = a_4 = \ldots = 0$; that is, the Maclaurin series for f contains only odd powers.

47. (a) Suppose f(x) is a continuous periodic function with period p. Let x_0 be an arbitrary real number. Then f assumes a minimum m_1 and a maximum m_2 in the interval $[x_0, x_0 + p]$; i.e., $m_1 \le f(x) \le m_2$ for all x in $[x_0, x_0 + p]$. Since f is periodic it has exactly the same values on all other intervals $[x_0 + p, x_0 + 2p]$, $[x_0 + 2p, x_0 + 3p]$, ..., and $[x_0 - p, x_0]$, $[x_0 - 2p, x_0 - p]$, ..., and so forth. That is, for all real numbers $-\infty < x < \infty$ we have $m_1 \le f(x) \le m_2$. Now choose $M = \max\{|m_1|, |m_2|\}$. Then $-M \le -|m_1| \le m_1 \le f(x) \le m_2 \le |m_2| \le M \Rightarrow |f(x)| \le M$ for all x.
(b) The dominate term in the nth order Taylor polynomial generated by cos x about $x = a$ is $\frac{\sin(a)}{n!}(x-a)^n$ or $\frac{\cos(a)}{n!}(x-a)^n$. In both cases, as $|x|$ increases the absolute value of these dominate terms tends to ∞, causing the graph of $P_n(x)$ to move away from cos x.

48. (b) $\tan^{-1} x = x - \frac{x^3}{3} + \frac{x^5}{5} - \ldots \Rightarrow \frac{x - \tan^{-1} x}{x^3} = \frac{1}{3} - \frac{x^2}{5} + \ldots$; from the Alternating Series Estimation Theorem, $\frac{x - \tan^{-1} x}{x^3} - \frac{1}{3} < 0$ $\Rightarrow \frac{x - \tan^{-1} x}{x^3} - \left(\frac{1}{3} - \frac{x^2}{5}\right) > 0 \Rightarrow \frac{1}{3} < \frac{x - \tan^{-1} x}{x^3} < \frac{1}{3} - \frac{x^2}{5}$; therefore, the $\lim_{x \to 0} \frac{x - \tan^{-1} x}{x^3} = \frac{1}{3}$

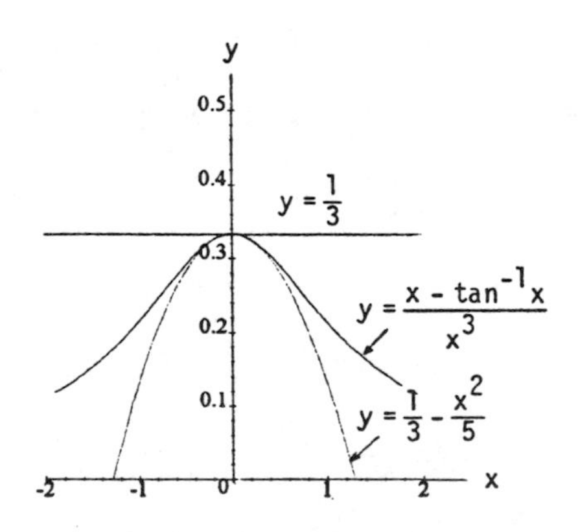

49. (a) $e^{-i\pi} = \cos(-\pi) + i \sin(-\pi) = -1 + i(0) = -1$
(b) $e^{i\pi/4} = \cos\left(\frac{\pi}{4}\right) + i \sin\left(\frac{\pi}{4}\right) = \frac{1}{\sqrt{2}} + \frac{i}{\sqrt{2}} = \left(\frac{1}{\sqrt{2}}\right)(1 + i)$
(c) $e^{-i\pi/2} = \cos\left(-\frac{\pi}{2}\right) + i \sin\left(-\frac{\pi}{2}\right) = 0 + i(-1) = -i$

50. $e^{i\theta} = \cos\theta + i\sin\theta \Rightarrow e^{-i\theta} = e^{i(-\theta)} = \cos(-\theta) + i\sin(-\theta) = \cos\theta - i\sin\theta$;

$e^{i\theta} + e^{-i\theta} = \cos\theta + i\sin\theta + \cos\theta - i\sin\theta = 2\cos\theta \Rightarrow \cos\theta = \frac{e^{i\theta}+e^{-i\theta}}{2}$;

$e^{i\theta} - e^{-i\theta} = \cos\theta + i\sin\theta - (\cos\theta - i\sin\theta) = 2i\sin\theta \Rightarrow \sin\theta = \frac{e^{i\theta}-e^{-i\theta}}{2i}$

51. $e^x = 1 + x + \frac{x^2}{2!} + \frac{x^3}{3!} + \frac{x^4}{4!} + \ldots \Rightarrow e^{i\theta} = 1 + i\theta + \frac{(i\theta)^2}{2!} + \frac{(i\theta)^3}{3!} + \frac{(i\theta)^4}{4!} + \ldots$ and

$e^{-i\theta} = 1 - i\theta + \frac{(-i\theta)^2}{2!} + \frac{(-i\theta)^3}{3!} + \frac{(-i\theta)^4}{4!} + \ldots = 1 - i\theta + \frac{(i\theta)^2}{2!} - \frac{(i\theta)^3}{3!} + \frac{(i\theta)^4}{4!} - \ldots$

$$\Rightarrow \frac{e^{i\theta}+e^{-i\theta}}{2} = \frac{\left(1 + i\theta + \frac{(i\theta)^2}{2!} + \frac{(i\theta)^3}{3!} + \frac{(i\theta)^4}{4!} + \ldots\right) + \left(1 - i\theta + \frac{(i\theta)^2}{2!} - \frac{(i\theta)^3}{3!} + \frac{(i\theta)^4}{4!} - \ldots\right)}{2}$$

$= 1 - \frac{\theta^2}{2!} + \frac{\theta^4}{4!} - \frac{\theta^6}{6!} + \ldots = \cos\theta$;

$$\frac{e^{i\theta}-e^{-i\theta}}{2} = \frac{\left(1 + i\theta + \frac{(i\theta)^2}{2!} + \frac{(i\theta)^3}{3!} + \frac{(i\theta)^4}{4!} + \ldots\right) - \left(1 - i\theta + \frac{(i\theta)^2}{2!} - \frac{(i\theta)^3}{3!} + \frac{(i\theta)^4}{4!} - \ldots\right)}{2}$$

$= \theta - \frac{\theta^3}{3!} + \frac{\theta^5}{5!} - \frac{\theta^7}{7!} + \ldots = \sin\theta$

52. $e^{i\theta} = \cos\theta + i\sin\theta \Rightarrow e^{-i\theta} = e^{i(-\theta)} = \cos(-\theta) + i\sin(-\theta) = \cos\theta - i\sin\theta$

(a) $e^{i\theta} + e^{-i\theta} = (\cos\theta + i\sin\theta) + (\cos\theta - i\sin\theta) = 2\cos\theta \Rightarrow \cos\theta = \frac{e^{i\theta}+e^{-i\theta}}{2} = \cosh i\theta$

(b) $e^{i\theta} - e^{-i\theta} = (\cos\theta + i\sin\theta) - (\cos\theta - i\sin\theta) = 2i\sin\theta \Rightarrow i\sin\theta = \frac{e^{i\theta}-e^{-i\theta}}{2} = \sinh i\theta$

53. $e^x \sin x = \left(1 + x + \frac{x^2}{2!} + \frac{x^3}{3!} + \frac{x^4}{4!} + \ldots\right)\left(x - \frac{x^3}{3!} + \frac{x^5}{5!} - \frac{x^7}{7!} + \ldots\right)$

$= (1)x + (1)x^2 + \left(-\frac{1}{6} + \frac{1}{2}\right)x^3 + \left(-\frac{1}{6} + \frac{1}{6}\right)x^4 + \left(\frac{1}{120} - \frac{1}{12} + \frac{1}{24}\right)x^5 + \ldots = x + x^2 + \frac{1}{3}x^3 - \frac{1}{30}x^5 + \ldots$;

$e^x \cdot e^{ix} = e^{(1+i)x} = e^x(\cos x + i\sin x) = e^x\cos x + i(e^x \sin x) \Rightarrow e^x \sin x$ is the series of the imaginary part of $e^{(1+i)x}$ which we calculate next; $e^{(1+i)x} = \sum_{n=0}^{\infty} \frac{(x+ix)^n}{n!} = 1 + (x+ix) + \frac{(x+ix)^2}{2!} + \frac{(x+ix)^3}{3!} + \frac{(x+ix)^4}{4!} + \ldots$

$= 1 + x + ix + \frac{1}{2!}(2ix^2) + \frac{1}{3!}(2ix^3 - 2x^3) + \frac{1}{4!}(-4x^4) + \frac{1}{5!}(-4x^5 - 4ix^5) + \frac{1}{6!}(-8ix^6) + \ldots \Rightarrow$ the imaginary part of $e^{(1+i)x}$ is $x + \frac{2}{2!}x^2 + \frac{2}{3!}x^3 - \frac{4}{5!}x^5 - \frac{8}{6!}x^6 + \ldots = x + x^2 + \frac{1}{3}x^3 - \frac{1}{30}x^5 - \frac{1}{90}x^6 + \ldots$ in agreement with our product calculation

54. $\frac{d}{dx}\left(e^{(a+ib)}\right) = \frac{d}{dx}\left[e^{ax}(\cos bx + i\sin bx)\right] = ae^{ax}(\cos bx + i\sin bx) + e^{ax}(-b\sin bx + bi\cos bx)$

$= ae^{ax}(\cos bx + i\sin bx) + bie^{ax}(\cos bx + i\sin bx) = ae^{(a+ib)x} + ibe^{(a+ib)x} = (a+ib)e^{(a+ib)x}$

55. (a) $e^{i\theta_1}e^{i\theta_2} = (\cos\theta_1 + i\sin\theta_1)(\cos\theta_2 + i\sin\theta_2) = (\cos\theta_1\cos\theta_2 - \sin\theta_1\sin\theta_2) + i(\sin\theta_1\cos\theta_2 + \sin\theta_2\cos\theta_1)$

$= \cos(\theta_1+\theta_2) + i\sin(\theta_1+\theta_2) = e^{i(\theta_1+\theta_2)}$

(b) $e^{-i\theta} = \cos(-\theta) + i\sin(-\theta) = \cos\theta - i\sin\theta = (\cos\theta - i\sin\theta)\left(\frac{\cos\theta + i\sin\theta}{\cos\theta + i\sin\theta}\right) = \frac{1}{\cos\theta + i\sin\theta} = \frac{1}{e^{i\theta}}$

56. $\frac{a-bi}{a^2+b^2}e^{(a+bi)x}+C_1+iC_2=\left(\frac{a-bi}{a^2+b^2}\right)e^{ax}(\cos bx+i\sin bx)+C_1+iC_2$

$=\frac{e^{ax}}{a^2+b^2}(a\cos bx+ia\sin bx-ib\cos bx+b\sin bx)+C_1+iC_2$

$=\frac{e^{ax}}{a^2+b^2}[(a\cos bx+b\sin bx)+(a\sin bx-b\cos bx)i]+C_1+iC_2$

$=\frac{e^{ax}(a\cos bx+b\sin bx)}{a^2+b^2}+C_1+\frac{ie^{ax}(a\sin bx-b\cos bx)}{a^2+b^2}+iC_2;$

$e^{(a+bi)x}=e^{ax}e^{ibx}=e^{ax}(\cos bx+i\sin bx)=e^{ax}\cos bx+ie^{ax}\sin bx$, so that given

$\int e^{(a+bi)x}\,dx=\frac{a-bi}{a^2+b^2}e^{(a+bi)x}+C_1+iC_2$ we conclude that $\int e^{ax}\cos bx\,dx=\frac{e^{ax}(a\cos bx+b\sin bx)}{a^2+b^2}+C_1$

and $\int e^{ax}\sin bx\,dx=\frac{e^{ax}(a\sin bx-b\cos bx)}{a^2+b^2}+C_2$

57-62. Example CAS commands:

Maple:

```
f:= x -> (1+x)^(3/2);
plot(f(x), x = -1..2);
mp:=proc(n):
convert(series(f(x),x=0,n),polynom) end:
p1:= mp(2); p2:= mp(3); p3:=mp(4);
der:=proc(n):
simplify(subs(x=z,diff(f(x),x$(n+1)))) end:
der(2); der(3); der(4);
plot(der(3),z=0..2, title = `3rd Derivative`);
Max:= 0.56: r:= (x,n) -> Max*x^(n+1)/(n+1)!;
r(x,2);
plot(r(x,2),x=0..2, title = `Maximum Remainder Term Using P2`);
plot({f(x),mp(3)}, x = -1..2, title = `Function and Taylor Polynomial P2`);
plot(f(x) - mp(3), x=-1..2, title = `Maximum Error Function `);
R:= (x,z,n) -> der(n)*x^(n+1)/(n+1)!;
R(x,z,3);
with(plots):
plot3d(R(x,z,3), x=-1..2, z=0..2);
```

Mathematica:

```
Clear[f,x,c]
f[x_] = (1+x) ^ (3/2)
{a,b} = {-1/2,2};
Plot[ f[x], {x,a,b} ]
p1[x_] = Series[ f[x], {x,0,1} ] // Normal
p2[x_] = Series[ f[x], {x,0,2} ] // Normal
p3[x_] = Series[ f[x], {x,0,3} ] // Normal
f''[c]
Plot[ f''[c], {c,a,b} ]
m1 = f''[a]
f'''[c]
Plot[ f'''[c], {c,a,b} ]
m2 = -f'''[a]
f''''[c]
```

```
Plot[ f''''[c], {c,a,b} ]
m3 = f''''[a]
r1 [x_] = m1 x^2/2!
Plot[ r1[x], {x,a,b} ]
r2 [x_] = m2 x^3/3!
Plot[ r2[x], {x,a,b} ]
r3 [x_] = m3 x^4/4!
Plot[ r3[x], {x,a,b} ]
```

Note: In estimating Rn from these graphs, consider only the portions where c is between 0 and x. (Mathematica has no simple way to plot only that portion.)

```
Plot3D[ f''[c] x^2/2!, {x,a,b}, {c,a,b}, PlotRange -> All ]
Plot3D[ f'''[c] x^3/3!, {x,a,b}, {c,a,b}, PlotRange -> All ]
Plot3D[ f''''[c] x^4/4!, {x,a,b}, {c,a,b}, PlotRange -> All ]
Plot[ {f[x],p1[x],p2[x],p3[x]}, {x,a,b} ]
```

8.11 APPLICATIONS OF POWER SERIES

1. $(1+x)^{1/2} = 1 + \frac{1}{2}x + \frac{\left(\frac{1}{2}\right)\left(-\frac{1}{2}\right)x^2}{2!} + \frac{\left(\frac{1}{2}\right)\left(-\frac{1}{2}\right)\left(-\frac{3}{2}\right)x^3}{3!} + \ldots = 1 + \frac{1}{2}x - \frac{1}{8}x^2 + \frac{1}{16}x^3 - \ldots$

2. $(1+x)^{1/3} = 1 + \frac{1}{3}x + \frac{\left(\frac{1}{3}\right)\left(-\frac{2}{3}\right)x^2}{2!} + \frac{\left(\frac{1}{3}\right)\left(-\frac{2}{3}\right)\left(-\frac{5}{3}\right)x^3}{3!} + \ldots = 1 + \frac{1}{3}x - \frac{1}{9}x^2 + \frac{5}{81}x^3 - \ldots$

3. $(1-x)^{-1/2} = 1 - \frac{1}{2}(-x) + \frac{\left(-\frac{1}{2}\right)\left(-\frac{3}{2}\right)(-x)^2}{2!} + \frac{\left(-\frac{1}{2}\right)\left(-\frac{3}{2}\right)\left(-\frac{5}{2}\right)(-x)^3}{3!} + \ldots = 1 + \frac{1}{2}x + \frac{3}{8}x^2 + \frac{5}{16}x^3 + \ldots$

4. $(1-2x)^{1/2} = 1 + \frac{1}{2}(-2x) + \frac{\left(\frac{1}{2}\right)\left(-\frac{1}{2}\right)(-2x)^2}{2!} + \frac{\left(\frac{1}{2}\right)\left(-\frac{1}{2}\right)\left(-\frac{3}{2}\right)(-2x)^3}{3!} + \ldots = 1 - x - \frac{1}{2}x^2 - \frac{1}{2}x^3 - \ldots$

5. $\left(1+\frac{x}{2}\right)^{-2} = 1 - 2\left(\frac{x}{2}\right) + \frac{(-2)(-3)\left(\frac{x}{2}\right)^2}{2!} + \frac{(-2)(-3)(-4)\left(\frac{x}{2}\right)^3}{3!} + \ldots = 1 - x + \frac{3}{4}x^2 - \frac{1}{2}x^3$

6. $\left(1-\frac{x}{2}\right)^{-2} = 1 - 2\left(-\frac{x}{2}\right) + \frac{(-2)(-3)\left(-\frac{x}{2}\right)^2}{2!} + \frac{(-2)(-3)(-4)\left(-\frac{x}{2}\right)^3}{3!} + \ldots = 1 + x + \frac{3}{4}x^2 + \frac{1}{2}x^3 + \ldots$

7. $\left(1+x^3\right)^{-1/2} = 1 - \frac{1}{2}x^3 + \frac{\left(-\frac{1}{2}\right)\left(-\frac{3}{2}\right)\left(x^3\right)^2}{2!} + \frac{\left(-\frac{1}{2}\right)\left(-\frac{3}{2}\right)\left(-\frac{5}{2}\right)\left(x^3\right)^3}{3!} + \ldots = 1 - \frac{1}{2}x^3 + \frac{3}{8}x^6 - \frac{5}{16}x^9 + \ldots$

8. $\left(1+x^2\right)^{-1/3} = 1 - \frac{1}{3}x^2 + \frac{\left(-\frac{1}{3}\right)\left(-\frac{4}{3}\right)\left(x^2\right)^2}{2!} + \frac{\left(-\frac{1}{3}\right)\left(-\frac{4}{3}\right)\left(-\frac{7}{3}\right)\left(x^2\right)^3}{3!} + \ldots = 1 - \frac{1}{3}x^2 + \frac{2}{9}x^4 - \frac{14}{81}x^6 + \ldots$

9. $\left(1+\frac{1}{x}\right)^{1/2} = 1+\frac{1}{2}\left(\frac{1}{x}\right)+\frac{\left(\frac{1}{2}\right)\left(-\frac{1}{2}\right)\left(\frac{1}{x}\right)^2}{2!}+\frac{\left(\frac{1}{2}\right)\left(-\frac{1}{2}\right)\left(-\frac{3}{2}\right)\left(\frac{1}{x}\right)^3}{3!}+\ldots = 1+\frac{1}{2x}-\frac{1}{8x^2}+\frac{1}{16x^3}$

10. $\left(1-\frac{2}{x}\right)^{1/3} = 1+\frac{1}{3}\left(-\frac{2}{x}\right)+\frac{\left(\frac{1}{3}\right)\left(-\frac{2}{3}\right)\left(-\frac{2}{x}\right)^2}{2!}+\frac{\left(\frac{1}{3}\right)\left(-\frac{2}{3}\right)\left(-\frac{5}{3}\right)\left(-\frac{2}{x}\right)^3}{3!}+\ldots = 1-\frac{2}{3x}-\frac{4}{9x^2}-\frac{40}{81x^3}-\ldots$

11. $(1+x)^4 = 1+4x+\frac{(4)(3)x^2}{2!}+\frac{(4)(3)(2)x^3}{3!}+\frac{(4)(3)(2)x^4}{4!} = 1+4x+6x^2+4x^3+x^4$

12. $\left(1+x^2\right)^3 = 1+3x^2+\frac{(3)(2)\left(x^2\right)^2}{2!}+\frac{(3)(2)(1)\left(x^2\right)^3}{3!} = 1+3x^2+3x^4+x^6$

13. $(1-2x)^3 = 1+3(-2x)+\frac{(3)(2)(-2x)^2}{2!}+\frac{(3)(2)(1)(-2x)^3}{3!} = 1-6x+12x^2-8x^3$

14. $\left(1-\frac{x}{2}\right)^4 = 1+4\left(-\frac{x}{2}\right)+\frac{(4)(3)\left(-\frac{x}{2}\right)^2}{2!}+\frac{(4)(3)(2)\left(-\frac{x}{2}\right)^3}{3!}+\frac{(4)(3)(2)(1)\left(-\frac{x}{2}\right)^4}{4!} = 1-2x+\frac{3}{2}x^2-\frac{1}{2}x^3+\frac{1}{16}x^4$

15. Assume the solution has the form $y = a_0+a_1x+a_2x^2+\ldots+a_{n-1}x^{n-1}+a_nx^n+\ldots$

$\Rightarrow \frac{dy}{dx} = a_1+2a_2x+\ldots+na_nx^{n-1}+\ldots$

$\Rightarrow \frac{dy}{dx}+y = (a_1+a_0)+(2a_2+a_1)x+(3a_3+a_2)x^2+\ldots+(na_n+a_{n-1})x^{n-1}+\ldots = 0$

$\Rightarrow a_1+a_0 = 0,\ 2a_2+a_1 = 0,\ 3a_3+a_2 = 0$ and in general $na_n+a_{n-1} = 0$. Since $y = 1$ when $x = 0$ we have $a_0 = 1$. Therefore $a_1 = -1,\ a_2 = \frac{-a_1}{2\cdot 1} = \frac{1}{2},\ a_3 = \frac{-a_2}{3} = -\frac{1}{3\cdot 2},\ \ldots,\ a_n = \frac{-a_{n-1}}{n} = \frac{(-1)^n}{n!}$

$\Rightarrow y = 1-x+\frac{1}{2}x^2-\frac{1}{3!}x^3+\ldots+\frac{(-1)^n}{n!}x^n+\ldots = \sum_{n=0}^{\infty}\frac{(-1)^nx^n}{n!} = e^{-x}$

16. Assume the solution has the form $y = a_0+a_1x+a_2x^2+\ldots+a_{n-1}x^{n-1}+a_nx^n+\ldots$

$\Rightarrow \frac{dy}{dx} = a_1+2a_2x+\ldots+na_nx^{n-1}+\ldots$

$\Rightarrow \frac{dy}{dx}-2y = (a_1-2a_0)+(2a_2-2a_1)x+(3a_3-2a_2)x^2+\ldots+(na_n-2a_{n-1})x^{n-1}+\ldots = 0$

$\Rightarrow a_1-2a_0 = 0,\ 2a_2-2a_1 = 0,\ 3a_3-2a_2 = 0$ and in general $na_n-2a_{n-1} = 0$. Since $y = 1$ when $x = 0$ we have $a_0 = 1$. Therefore $a_1 = 2a_0 = 2(1) = 2,\ a_2 = \frac{2}{2}a_1 = \frac{2}{2}(2) = \frac{2^2}{2},\ a_3 = \frac{2}{3}a_2 = \frac{2}{3}\left(\frac{2^2}{2}\right) = \frac{2^3}{3\cdot 2},\ \ldots,$

$a_n = \left(\frac{2}{n}\right)a_{n-1} = \left(\frac{2}{n}\right)\left(\frac{2^{n-1}}{n-1}\right)a_{n-2} = \frac{2^n}{n!} \Rightarrow y = 1+2x+\frac{2^2}{2}x^2+\frac{2^3}{3!}x^3+\ldots+\frac{2^n}{n!}x^n+\ldots$

$= 1+(2x)+\frac{(2x)^2}{2!}+\frac{(2x)^3}{3!}+\ldots+\frac{(2x)^n}{n!}+\ldots = \sum_{n=0}^{\infty}\frac{(2x)^n}{n!} = e^{2x}$

17. Assume the solution has the form $y = a_0+a_1x+a_2x^2+\ldots+a_{n-1}x^{n-1}+a_nx^n+\ldots$

$\Rightarrow \frac{dy}{dx} = a_1+2a_2x+\ldots+na_nx^{n-1}+\ldots$

$\Rightarrow \frac{dy}{dx} - y = (a_1 - a_0) + (2a_2 - a_1)x + (3a_3 - a_2)x^2 + \ldots + (na_n - a_{n-1})x^{n-1} + \ldots = 1$

$\Rightarrow a_1 - a_0 = 1,\ 2a_2 - a_1 = 0,\ 3a_3 - a_2 = 0$ and in general $na_n - a_{n-1} = 0$. Since $y = 0$ when $x = 0$ we have $a_0 = 0$. Therefore $a_1 = 1,\ a_2 = \frac{a_1}{2} = \frac{1}{2},\ a_3 = \frac{a_2}{3} = \frac{1}{3 \cdot 2},\ a_4 = \frac{a_3}{4} = \frac{1}{4 \cdot 3 \cdot 2}, \ldots, a_n = \frac{a_{n-1}}{n} = \frac{1}{n!}$

$\Rightarrow y = 0 + 1x + \frac{1}{2}x^2 + \frac{1}{3 \cdot 2}x^3 + \frac{1}{4 \cdot 3 \cdot 2}x^4 + \ldots + \frac{1}{n!}x^n + \ldots$

$= \left(1 + 1x + \frac{1}{2}x^2 + \frac{1}{3 \cdot 2}x^3 + \frac{1}{4 \cdot 3 \cdot 2}x^4 + \ldots + \frac{1}{n!}x^n + \ldots\right) - 1 = \sum_{n=0}^{\infty} \frac{x^n}{n!} - 1 = e^x - 1$

18. Assume the solution has the form $y = a_0 + a_1x + a_2x^2 + \ldots + a_{n-1}x^{n-1} + a_nx^n + \ldots$

$\Rightarrow \frac{dy}{dx} = a_1 + 2a_2x + \ldots + na_nx^{n-1} + \ldots$

$\Rightarrow \frac{dy}{dx} + y = (a_1 + a_0) + (2a_2 + a_1)x + (3a_3 + a_2)x^2 + \ldots + (na_n + a_{n-1})x^{n-1} + \ldots = 1$

$\Rightarrow a_1 + a_0 = 1,\ 2a_2 + a_1 = 0,\ 3a_3 + a_2 = 0$ and in general $na_n + a_{n-1} = 0$. Since $y = 2$ when $x = 0$ we have $a_0 = 2$. Therefore $a_1 = 1 - a_0 = -1,\ a_2 = \frac{-a_1}{2 \cdot 1} = \frac{1}{2},\ a_3 = \frac{-a_2}{3} = -\frac{1}{3 \cdot 2}, \ldots, a_n = \frac{-a_{n-1}}{n} = \frac{(-1)^n}{n!}$

$\Rightarrow y = 2 - x + \frac{1}{2}x^2 - \frac{1}{3 \cdot 2}x^3 + \ldots + \frac{(-1)^n}{n!}x^n + \ldots = 1 + \left(1 - x + \frac{1}{2}x^2 - \frac{1}{3 \cdot 2}x^3 + \ldots + \frac{(-1)^n}{n!}x^n + \ldots\right)$

$= 1 + \sum_{n=0}^{\infty} \frac{(-1)^n x^n}{n!} = 1 + e^{-x}$

19. Assume the solution has the form $y = a_0 + a_1x + a_2x^2 + \ldots + a_{n-1}x^{n-1} + a_nx^n + \ldots$

$\Rightarrow \frac{dy}{dx} = a_1 + 2a_2x + \ldots + na_nx^{n-1} + \ldots$

$\Rightarrow \frac{dy}{dx} - y = (a_1 - a_0) + (2a_2 - a_1)x + (3a_3 - a_2)x^2 + \ldots + (na_n - a_{n-1})x^{n-1} + \ldots = x$

$\Rightarrow a_1 - a_0 = 0,\ 2a_2 - a_1 = 1,\ 3a_3 - a_2 = 0$ and in general $na_n - a_{n-1} = 0$. Since $y = 0$ when $x = 0$ we have $a_0 = 0$. Therefore $a_1 = 0,\ a_2 = \frac{1 + a_1}{2} = \frac{1}{2},\ a_3 = \frac{a_2}{3} = \frac{1}{3 \cdot 2},\ a_4 = \frac{a_3}{4} = \frac{1}{4 \cdot 3 \cdot 2}, \ldots, a_n = \frac{a_{n-1}}{n} = \frac{1}{n!}$

$\Rightarrow y = 0 + 0x + \frac{1}{2}x^2 + \frac{1}{3 \cdot 2}x^3 + \frac{1}{4 \cdot 3 \cdot 2}x^4 + + \ldots + \frac{1}{n!}x^n + \ldots$

$= \left(1 + 1x + \frac{1}{2}x^2 + \frac{1}{3 \cdot 2}x^3 + \frac{1}{4 \cdot 3 \cdot 2}x^4 + \ldots + \frac{1}{n!}x^n + \ldots\right) - 1 - x = \sum_{n=0}^{\infty} \frac{x^n}{n!} - 1 - x = e^x - x - 1$

20. Assume the solution has the form $y = a_0 + a_1x + a_2x^2 + \ldots + a_{n-1}x^{n-1} + a_nx^n + \ldots$

$\Rightarrow \frac{dy}{dx} = a_1 + 2a_2x + \ldots + na_nx^{n-1} + \ldots$

$\Rightarrow \frac{dy}{dx} + y = (a_1 + a_0) + (2a_2 + a_1)x + (3a_3 + a_2)x^2 + \ldots + (na_n + a_{n-1})x^{n-1} + \ldots = 2x$

$\Rightarrow a_1 + a_0 = 0,\ 2a_2 + a_1 = 2,\ 3a_3 + a_2 = 0$ and in general $na_n + a_{n-1} = 0$. Since $y = -1$ when $x = 0$ we have $a_0 = -1$. Therefore $a_1 = 1,\ a_2 = \frac{2 - a_1}{2} = \frac{1}{2},\ a_3 = \frac{-a_2}{3} = -\frac{1}{3 \cdot 2}, \ldots, a_n = \frac{-a_{n-1}}{n} = \frac{(-1)^n}{n!}$

$\Rightarrow y = -1 + 1x + \frac{1}{2}x^2 - \frac{1}{3 \cdot 2}x^3 + \ldots + \frac{(-1)^n}{n!}x^n + \ldots$

$= \left(1 - 1x + \frac{1}{2}x^2 - \frac{1}{3 \cdot 2}x^3 + \ldots + \frac{(-1)^n}{n!}x^n + \ldots\right) - 2 + 2x = \sum_{n=0}^{\infty} \frac{(-1)^n x^n}{n!} - 2 + 2x = e^{-x} + 2x - 2$

21. $y' - xy = a_1 + (2a_2 - a_0)x + (3a_3 - a_1)x + \ldots + (na_n - a_{n-2})x^{n-1} + \ldots = 0 \Rightarrow a_1 = 0,\ 2a_2 - a_0 = 0,\ 3a_3 - a_1 = 0,$ $4a_4 - a_2 = 0$ and in general $na_n - a_{n-2} = 0$. Since $y = 1$ when $x = 0$, we have $a_0 = 1$. Therefore $a_2 = \frac{a_0}{2} = \frac{1}{2}$, $a_3 = \frac{a_1}{3} = 0,\ a_4 = \frac{a_2}{4} = \frac{1}{2 \cdot 4},\ a_5 = \frac{a_3}{5} = 0,\ \ldots,\ a_{2n} = \frac{1}{2 \cdot 4 \cdot 6 \cdots 2n}$ and $a_{2n+1} = 0$

$$\Rightarrow y = 1 + \frac{1}{2}x^2 + \frac{1}{2 \cdot 4}x^4 + \frac{1}{2 \cdot 4 \cdot 6}x^6 + \ldots + \frac{1}{2 \cdot 4 \cdot 6 \cdots 2n}x^{2n} + \ldots = \sum_{n=0}^{\infty} \frac{x^{2n}}{2^n n!} = \sum_{n=0}^{\infty} \frac{\left(\frac{x^2}{2}\right)^n}{n!} = e^{x^2/2}$$

22. $y' - x^2 y = a_1 + 2a_2 x + (3a_3 - a_0)x^2 + (4a_4 - a_1)x^3 + \ldots + (na_n - a_{n-3})x^{n-1} + \ldots = 0 \Rightarrow a_1 = 0,\ a_2 = 0,$ $3a_3 - a_0 = 0,\ 4a_4 - a_1 = 0$ and in general $na_n - a_{n-3} = 0$. Since $y = 1$ when $x = 0$, we have $a_0 = 1$. Therefore $a_3 = \frac{a_0}{3} = \frac{1}{3},\ a_4 = \frac{a_1}{4} = 0,\ a_5 = \frac{a_2}{5} = 0,\ a_6 = \frac{a_3}{6} = \frac{1}{3 \cdot 6},\ \ldots,\ a_{3n} = \frac{1}{3 \cdot 6 \cdot 9 \cdots 3n},\ a_{3n+1} = 0$ and $a_{3n+2} = 0$

$$\Rightarrow y = 1 + \frac{1}{3}x^3 + \frac{1}{3 \cdot 6}x^6 + \frac{1}{3 \cdot 6 \cdot 9}x^9 + \ldots + \frac{1}{3 \cdot 6 \cdot 9 \cdots 3n}x^{3n} + \ldots = \sum_{n=0}^{\infty} \frac{x^{3n}}{3^n n!} = \sum_{n=0}^{\infty} \frac{\left(\frac{x^3}{3}\right)^n}{n!} = e^{x^3/3}$$

23. $(1 - x)y' - y = (a_1 - a_0) + (2a_2 - a_1 - a_1)x + (3a_3 - 2a_2 - a_2)x^2 + (4a_4 - 3a_3 - a_3)x^3 + \ldots$ $+ (na_n - (n-1)a_{n-1} - a_{n-1})x^{n-1} + \ldots = 0 \Rightarrow a_1 - a_0 = 0,\ 2a_2 - 2a_1 = 0,\ 3a_3 - 3a_2 = 0$ and in general $(na_n - na_{n-1}) = 0$. Since $y = 2$ when $x = 0$, we have $a_0 = 2$. Therefore $a_1 = 2,\ a_2 = 2,\ \ldots,\ a_n = 2 \Rightarrow y = 2 + 2x + 2x^2 + \ldots = \sum_{n=0}^{\infty} 2x^n = \frac{2}{1 - x}$

24. $(1 + x^2)y' + 2xy = a_1 + (2a_2 + 2a_0)x + (3a_3 + 2a_1 + a_1)x^2 + (4a_4 + 2a_2 + 2a_2)x^3 + \ldots + (na_n + na_{n-2})x^{n-1} + \ldots$ $= 0 \Rightarrow a_1 = 0,\ 2a_2 + 2a_0 = 0,\ 3a_3 + 3a_1 = 0,\ 4a_4 + 4a_2 = 0$ and in general $na_n + na_{n-2} = 0$. Since $y = 3$ when $x = 0$, we have $a_0 = 3$. Therefore $a_2 = -3,\ a_3 = 0,\ a_4 = 3,\ \ldots,\ a_{2n+1} = 0,\ a_{2n} = (-1)^n 3$

$$\Rightarrow y = 3 - 3x^2 + 3x^4 - \ldots = \sum_{n=0}^{\infty} 3(-1)^n x^{2n} = \sum_{n=0}^{\infty} 3\left(-x^2\right)^n = \frac{3}{1 + x^2}$$

25. $y = a_0 + a_1 x + a_2 x^2 + \ldots + a_n x^n + \ldots \Rightarrow y'' = 2a_2 + 3 \cdot 2a_3 x + \ldots + n(n-1)a_n x^{n-2} + \ldots \Rightarrow y'' - y$ $= (2a_2 - a_0) + (3 \cdot 2a_3 - a_1)x + (4 \cdot 3a_4 - a_2)x^2 + \ldots + (n(n-1)a_n - a_{n-2})x^{n-2} + \ldots = 0 \Rightarrow 2a_2 - a_0 = 0,$ $3 \cdot 2a_3 - a_1 = 0,\ 4 \cdot 3a_4 - a_2 = 0$ and in general $n(n-1)a_n - a_{n-2} = 0$. Since $y' = 1$ and $y = 0$ when $x = 0$, we have $a_0 = 0$ and $a_1 = 1$. Therefore $a_2 = 0,\ a_3 = \frac{1}{3 \cdot 2},\ a_4 = 0,\ a_5 = \frac{1}{5 \cdot 4 \cdot 3 \cdot 2},\ \ldots,\ a_{2n+1} = \frac{1}{(2n+1)!}$ and $a_{2n} = 0 \Rightarrow y = x + \frac{1}{3!}x^3 + \frac{1}{5!}x^5 + \ldots = \sum_{n=0}^{\infty} \frac{x^{2n+1}}{(2n+1)!} = \sinh x$

26. $y = a_0 + a_1 x + a_2 x^2 + \ldots + a_n x^n + \ldots \Rightarrow y'' = 2a_2 + 3 \cdot 2a_3 x + \ldots + n(n-1)a_n x^{n-2} + \ldots \Rightarrow y'' + y$ $= (2a_2 + a_0) + (3 \cdot 2a_3 + a_1)x + (4 \cdot 3a_4 + a_2)x^2 + \ldots + (n(n-1)a_n + a_{n-2})x^{n-2} + \ldots = 0 \Rightarrow 2a_2 + a_0 = 0,$ $3 \cdot 2a_3 + a_1 = 0,\ 4 \cdot 3a_4 + a_2 = 0$ and in general $n(n-1)a_n + a_{n-2} = 0$. Since $y' = 0$ and $y = 1$ when $x = 0$, we have $a_0 = 1$ and $a_1 = 0$. Therefore $a_2 = -\frac{1}{2},\ a_3 = 0,\ a_4 = \frac{1}{4 \cdot 3 \cdot 2},\ a_5 = 0,\ \ldots,\ a_{2n+1} = 0$ and $a_{2n} = \frac{(-1)^n}{(2n)!}$

$$\Rightarrow y = 1 - \frac{1}{2}x^2 + \frac{1}{4!}x^4 - \ldots = \sum \frac{(-1)^n x^{2n}}{(2n)!} = \cos x$$

27. $y = a_0 + a_1x + a_2x^2 + \ldots + a_nx^n + \ldots \Rightarrow y'' = 2a_2 + 3\cdot 2a_3x + \ldots + n(n-1)a_nx^{n-2} + \ldots \Rightarrow y'' + y$
$= (2a_2 + a_0) + (3\cdot 2a_3 + a_1)x + (4\cdot 3a_4 + a_2)x^2 + \ldots + (n(n-1)a_n + a_{n-2})x^{n-2} + \ldots = x \Rightarrow 2a_2 + a_0 = 0,$
$3\cdot 2a_3 + a_1 = 1$, $4\cdot 3a_4 + a_2 = 0$ and in general $n(n-1)a_n + a_{n-2} = 0$. Since $y' = 1$ and $y = 2$ when $x = 0$,
we have $a_0 = 2$ and $a_1 = 1$. Therefore $a_2 = -1$, $a_3 = 0$, $a_4 = \frac{1}{4\cdot 3}$, $a_5 = 0, \ldots, a_{2n} = -2\cdot\frac{(-1)^{n+1}}{(2n)!}$ and
$a_{2n+1} = 0 \Rightarrow y = 2 + x - x^2 + 2\cdot\frac{x^4}{4!} + \ldots = 2 + x - 2\sum_{n=1}^{\infty}\frac{(-1)^{n+1}x^{2n}}{(2n)!}$

28. $y = a_0 + a_1x + a_2x^2 + \ldots + a_nx^n + \ldots \Rightarrow y'' = 2a_2 + 3\cdot 2a_3x + \ldots + n(n-1)a_nx^{n-2} + \ldots \Rightarrow y'' - y$
$= (2a_2 - a_0) + (3\cdot 2a_3 - a_1)x + (4\cdot 3a_4 - a_2)x^2 + \ldots + (n(n-1)a_n - a_{n-2})x^{n-2} + \ldots = x \Rightarrow 2a_2 - a_0 = 0,$
$3\cdot 2a_3 - a_1 = 1$, $4\cdot 3a_4 - a_2 = 0$ and in general $n(n-1)a_n - a_{n-2} = 0$. Since $y' = 2$ and $y = -1$ when $x = 0$,
we have $a_0 = -1$ and $a_1 = 2$. Therefore $a_2 = \frac{-1}{2}$, $a_3 = \frac{1}{2}$, $a_4 = \frac{-1}{2\cdot 3\cdot 4}$, $a_5 = \frac{1}{5\cdot 4\cdot 2} = \frac{3}{5!}, \ldots, a_{2n} = \frac{-1}{(2n)!}$
and $a_{2n+1} = \frac{3}{(2n+1)!} \Rightarrow y = -1 + 2x - \frac{1}{2}x^2 + \frac{3}{3!}x^3 - \ldots = -1 + 2x - \sum_{n=1}^{\infty}\frac{x^{2n}}{(2n)!} + \sum_{n=1}^{\infty}\frac{3x^{2n+1}}{(2n+1)!}$

29. $y = a_0 + a_1x + a_2x^2 + \ldots + a_nx^n + \ldots \Rightarrow y'' = 2a_2 + 3\cdot 2a_3x + \ldots + n(n-1)a_nx^{n-2} + \ldots \Rightarrow y'' - y$
$= (2a_2 - a_0) + (3\cdot 2a_3 - a_1)(x-2) + (4\cdot 3a_4 - a_2)(x-2)^2 + \ldots + (n(n-1)a_n - a_{n-2})(x-2)^{n-2} + \ldots = 0$
$\Rightarrow 2a_2 - a_0 = 0$, $3\cdot 2a_3 - a_1 = 0$, $4\cdot 3a_4 - a_2 = 0$ and in general $n(n-1)a_n - a_{n-2} = 0$. Since $y' = -2$ and
$y = 0$ when $x = 2$, we have $a_0 = 0$ and $a_1 = -2$. Therefore $a_2 = 0$, $a_3 = \frac{-2}{3\cdot 2}$, $a_4 = 0$, $a_5 = \frac{-2}{5!}, \ldots, a_{2n} = 0$, and
$a_{2n+1} = \frac{-2}{(2n+1)!} \Rightarrow y = -2(x-2) - \frac{2}{3!}(x-2)^3 - \ldots = \sum_{n=0}^{\infty}\frac{-2(x-2)^{2n+1}}{(2n+1)!}$

30. $y'' - x^2y = 2a_2 + 6a_3x + (4\cdot 3a_4 - a_0)x^2 + \ldots + (n(n-1)a_n - a_{n-4})x^{n-2} + \ldots = 0 \Rightarrow 2a_2 = 0$, $6a_3 = 0$,
$4\cdot 3a_4 - a_0 = 0$, $5\cdot 4a_5 - a_1 = 0$, and in general $n(n-1)a_n - a_{n-4} = 0$. Since $y' = b$ and $y = a$ when $x = 0$,
we have $a_0 = a$, $a_1 = b$, $a_2 = 0$, $a_3 = 0$, $a_4 = \frac{a}{3\cdot 4}$, $a_5 = \frac{b}{4\cdot 5}$, $a_6 = 0$, $a_7 = 0$, $a_8 = \frac{a}{3\cdot 4\cdot 7\cdot 8}$, $a_9 = \frac{b}{4\cdot 5\cdot 8\cdot 9}$
$\Rightarrow y = a + bx + \frac{a}{3\cdot 4}x^4 + \frac{b}{4\cdot 5}x^5 + \frac{a}{3\cdot 4\cdot 7\cdot 8}x^8 + \frac{b}{4\cdot 5\cdot 8\cdot 9}x^9 + \ldots$

31. $y'' + x^2y = 2a_2 + 6a_3x + (4\cdot 3a_4 + a_0)x^2 + \ldots + (n(n-1)a_n + a_{n-4})x^{n-2} + \ldots = x \Rightarrow 2a_2 = 0$, $6a_3 = 1$,
$4\cdot 3a_4 + a_0 = 0$, $5\cdot 4a_5 + a_1 = 0$, and in general $n(n-1)a_n + a_{n-4} = 0$. Since $y' = b$ and $y = a$ when $x = 0$,
we have $a_0 = a$ and $a_1 = b$. Therefore $a_2 = 0$, $a_3 = \frac{1}{2\cdot 3}$, $a_4 = -\frac{a}{3\cdot 4}$, $a_5 = -\frac{b}{4\cdot 5}$, $a_6 = 0$, $a_7 = \frac{1}{2\cdot 3\cdot 6\cdot 7}$
$\Rightarrow y = a + bx + \frac{1}{2\cdot 3}x^3 - \frac{a}{3\cdot 4}x^4 - \frac{b}{4\cdot 5}x^5 - \frac{1}{2\cdot 3\cdot 6\cdot 7}x^7 + \frac{ax^8}{3\cdot 4\cdot 7\cdot 8} + \frac{bx^9}{4\cdot 5\cdot 8\cdot 9} + \ldots$

32. $y'' - 2y' + y = (2a_2 - 2a_1 + a_0) + (2\cdot 3a_3 - 4a_2 + a_1)x + (3\cdot 4a_4 - 2\cdot 3a_3 + a_2)x^2 + \ldots$
$+ ((n-1)na_n - 2(n-1)a_{n-1} + a_{n-2})x^{n-2} + \ldots = 0 \Rightarrow 2a_2 - 2a_1 + a_0 = 0$, $2\cdot 3a_3 - 4a_2 + a_1 = 0$,
$3\cdot 4a_4 - 2\cdot 3a_3 + a_2 = 0$ and in general $(n-1)na_n - 2(n-1)a_{n-1} + a_{n-2} = 0$. Since $y' = 1$ and $y = 0$ when

when $x = 0$, we have $a_0 = 0$ and $a_1 = 1$. Therefore $a_2 = 1$, $a_3 = \frac{1}{2}$, $a_4 = \frac{1}{6}$, $a_5 = \frac{1}{24}$ and $a_n = \frac{1}{(n-1)!}$
$\Rightarrow y = x + x^2 + \frac{1}{2}x^3 + \frac{1}{6}x^4 + \frac{1}{24}x^5 + \ldots = \sum_{n=1}^{\infty} \frac{x^n}{(n-1)!} = \sum_{n=0}^{\infty} \frac{x^{n+1}}{n!} = x \sum_{n=0}^{\infty} \frac{x^n}{n!} = xe^x$

33. $\int_0^{0.2} \sin x^2 \, dx = \int_0^{0.2} \left(x^2 - \frac{x^6}{3!} + \frac{x^{10}}{5!} - \ldots\right) dx = \left[\frac{x^3}{3} - \frac{x^7}{7 \cdot 3!} + \ldots\right]_0^{0.2} \approx \left[\frac{x^3}{3}\right]_0^{0.2} \approx 0.00267$ with error

$|E| \le \frac{(.2)^7}{7 \cdot 3!} \approx 0.0000003$

34. $\int_0^{0.2} \frac{e^{-x} - 1}{x} \, dx = \int_0^{0.2} \frac{1}{x}\left(1 - x + \frac{x^2}{2!} - \frac{x^3}{3!} + \frac{x^4}{4!} - \ldots - 1\right) dx = \int_0^{0.2} \left(-1 + \frac{x}{2} - \frac{x^2}{6} + \frac{x^3}{24} - \ldots\right) dx$

$= \left[-x + \frac{x^2}{4} - \frac{x^3}{18} + \ldots\right]_0^{0.2} \approx -0.19044$ with error $|E| \le \frac{(0.2)^4}{96} \approx 0.00002$

35. $\int_0^{0.1} \frac{1}{\sqrt{1 + x^4}} \, dx = \int_0^{0.1} \left(1 - \frac{x^4}{2} + \frac{3x^8}{8} - \ldots\right) dx = \left[x - \frac{x^5}{10} + \ldots\right]_0^{0.1} \approx [x]_0^{0.1} \approx 0.1$ with error

$|E| \le \frac{(0.1)^5}{10} = 0.000001$

36. $\int_0^{0.25} \sqrt[3]{1 + x^2} \, dx = \int_0^{0.25} \left(1 + \frac{x^2}{3} - \frac{x^4}{9} + \ldots\right) dx = \left[x + \frac{x^3}{9} - \frac{x^5}{45} + \ldots\right]_0^{0.25} \approx \left[x + \frac{x^3}{9}\right]_0^{0.25} \approx 0.25174$ with error

$|E| \le \frac{(0.25)^5}{45} \approx 0.0000217$

37. $\int_0^{0.1} \frac{\sin x}{x} \, dx = \int_0^{0.1} \left(1 - \frac{x^2}{3!} + \frac{x^4}{5!} - \frac{x^6}{7!} + \ldots\right) dx = \left[x - \frac{x^3}{3 \cdot 3!} + \frac{x^5}{5 \cdot 5!} - \frac{x^7}{7 \cdot 7!} + \ldots\right]_0^{0.1} \approx \left[x - \frac{x^3}{3 \cdot 3!} + \frac{x^5}{5 \cdot 5!}\right]_0^{0.1}$

≈ 0.0999444611

38. $\int_0^{0.1} \exp(-x^2) \, dx = \int_0^{0.1} \left(1 - x^2 + \frac{x^4}{2!} - \frac{x^6}{3!} + \frac{x^8}{4!} - \ldots\right) dx = \left[x - \frac{x^3}{3} + \frac{x^5}{10} + \frac{x^7}{42} + \ldots\right]_0^{0.1} \approx \left[x - \frac{x^3}{3} + \frac{x^5}{10} - \frac{x^7}{42}\right]_0^{0.1}$

≈ 0.0996676643

39. $(1 + x^4)^{1/2} = (1)^{1/2} + \frac{\left(\frac{1}{2}\right)}{1}(1)^{-1/2}(x^4) + \frac{\left(\frac{1}{2}\right)\left(-\frac{1}{2}\right)}{2!}(1)^{-3/2}(x^4)^2 + \frac{\left(\frac{1}{2}\right)\left(-\frac{1}{2}\right)\left(-\frac{3}{2}\right)}{3!}(1)^{-5/2}(x^4)^3$

$+ \frac{\left(\frac{1}{2}\right)\left(-\frac{1}{2}\right)\left(-\frac{3}{2}\right)\left(-\frac{5}{2}\right)}{4!}(1)^{-7/2}(x^4)^4 + \ldots = 1 + \frac{x^4}{2} - \frac{x^8}{8} + \frac{x^{12}}{16} - \frac{5x^{16}}{128} + \ldots$

$\Rightarrow \int_0^{0.1} \left(1 + \frac{x^4}{2} - \frac{x^8}{8} + \frac{x^{12}}{16} - \frac{5x^{16}}{128} + \ldots\right) dx = \left[x + \frac{x^5}{10} - \frac{x^9}{72} + \frac{x^{13}}{208} - \frac{5x^{17}}{2176} + \ldots\right]_0^{0.1} \approx 0.100001$

40. $\int_0^1 \left(\frac{1-\cos x}{x^2}\right) dx = \int_0^1 \left(\frac{1}{2} - \frac{x^2}{4!} + \frac{x^4}{6!} - \frac{x^6}{8!} + \frac{x^8}{10!} - \ldots\right) dx = \left[\frac{x}{2} - \frac{x^3}{3 \cdot 4!} + \frac{x^5}{5 \cdot 6!} - \frac{x^7}{7 \cdot 8!} + \frac{x^9}{9 \cdot 10!} - \ldots\right]_0^1$

≈ 0.48638534764 since $\frac{1-\cos x}{x^2} = \frac{1}{2} - \frac{x^2}{4!} + \frac{x^4}{6!} - \frac{x^6}{8!} + \frac{x^8}{10!} - \ldots$

41. $\int_0^1 \cos t^2 \, dt = \int_0^1 \left(1 - \frac{t^4}{2} + \frac{t^8}{4!} - \frac{t^{12}}{6!} + \ldots\right) dt = \left[t - \frac{t^5}{10} + \frac{t^9}{9 \cdot 4!} - \frac{t^{13}}{13 \cdot 6!} + \ldots\right]_0^1 \Rightarrow |\text{error}| < \frac{1}{13 \cdot 6!} \approx .00011$

42. $\int_0^1 \cos \sqrt{t} \, dt = \int_0^1 \left(1 - \frac{t}{2} + \frac{t^2}{4!} - \frac{t^3}{6!} + \frac{t^4}{8!} - \ldots\right) dt = \left[t - \frac{t^2}{4} + \frac{t^3}{3 \cdot 4!} - \frac{t^4}{4 \cdot 6!} + \frac{t^5}{5 \cdot 8!} - \ldots\right]_0^1$

$\Rightarrow |\text{error}| < \frac{1}{5 \cdot 8!} \approx 0.000004960$

43. $F(x) = \int_0^x \left(t^2 - \frac{t^6}{3!} + \frac{t^{10}}{5!} - \frac{t^{14}}{7!} + \ldots\right) dt = \left[\frac{t^3}{3} - \frac{t^7}{7 \cdot 3!} + \frac{t^{11}}{11 \cdot 5!} - \frac{t^{15}}{15 \cdot 7!} + \ldots\right]_0^x \approx \frac{x^3}{3} - \frac{x^7}{7 \cdot 3!} + \frac{x^{11}}{11 \cdot 5!}$

$\Rightarrow |\text{error}| < \frac{1}{15 \cdot 7!} \approx 0.00002$

44. $F(x) = \int_0^x \left(t^2 - t^4 + \frac{t^6}{2!} - \frac{t^8}{3!} + \frac{t^{10}}{4!} - \frac{t^{12}}{5!} + \ldots\right) dt = \left[\frac{t^3}{3} - \frac{t^5}{5} + \frac{t^7}{7 \cdot 2!} - \frac{t^9}{9 \cdot 3!} + \frac{t^{11}}{11 \cdot 4!} - \frac{t^{13}}{13 \cdot 5!} + \ldots\right]_0^x$

$\approx \frac{x^3}{3} - \frac{x^5}{5} + \frac{x^7}{7 \cdot 2!} - \frac{x^9}{9 \cdot 3!} + \frac{x^{11}}{11 \cdot 4!} \Rightarrow |\text{error}| < \frac{1}{13 \cdot 5!} \approx 0.00064$

45. (a) $F(x) = \int_0^x \left(t - \frac{t^3}{3} + \frac{t^5}{5} - \frac{t^7}{7} + \ldots\right) dt = \left[\frac{t^2}{2} - \frac{t^4}{12} + \frac{t^6}{30} - \ldots\right]_0^x \approx \frac{x^2}{2} - \frac{x^4}{12} \Rightarrow |\text{error}| < \frac{(0.5)^6}{30} \approx .00052$

(b) $|\text{error}| < \frac{1}{33 \cdot 34} \approx .00089$ so $F(x) \approx \frac{x^2}{2} - \frac{x^4}{3 \cdot 4} + \frac{x^6}{5 \cdot 6} - \frac{x^8}{7 \cdot 8} + \ldots + (-1)^{15} \frac{x^{32}}{31 \cdot 32}$

46. (a) $F(x) = \int_0^x \left(1 - \frac{t}{2} + \frac{t^2}{3} - \frac{t^3}{4} + \ldots\right) dt = \left[t - \frac{t^2}{2 \cdot 2} + \frac{t^3}{3 \cdot 3} - \frac{t^4}{4 \cdot 4} + \frac{t^5}{5 \cdot 5} - \ldots\right]_0^x \approx x - \frac{x^2}{2^2} + \frac{x^3}{3^2} - \frac{x^4}{4^2} + \frac{x^5}{5^2}$

$\Rightarrow |\text{error}| < \frac{(0.5)^6}{6^2} \approx .00043$

(b) $|\text{error}| < \frac{1}{32^2} \approx .00097$ so $F(x) \approx x - \frac{x^2}{2^2} + \frac{x^3}{3^2} - \frac{x^4}{4^2} + \ldots + (-1)^{31} \frac{x^{31}}{31^2}$

47. $\frac{1}{x^2}(e^x - (1+x)) = \frac{1}{x^2}\left(\left(1 + x + \frac{x^2}{2} + \frac{x^3}{3!} + \ldots\right) - 1 - x\right) = \frac{1}{2} + \frac{x}{3!} + \frac{x^2}{4!} + \ldots \Rightarrow \lim_{x\to 0} \frac{e^x - (1+x)}{x^2}$

$= \lim_{x\to 0} \left(\frac{1}{2} + \frac{x}{3!} + \frac{x^2}{4!} + \ldots\right) = \frac{1}{2}$

48. $\frac{1}{x}\left(e^x - e^{-x}\right) = \frac{1}{x}\left[\left(1 + x + \frac{x^2}{2!} + \frac{x^3}{3!} + \frac{x^4}{4!} + \ldots\right) - \left(1 - x + \frac{x^2}{2!} - \frac{x^3}{3!} + \frac{x^4}{4!} - \ldots\right)\right] = \frac{1}{x}\left(2x + \frac{2x^3}{3!} + \frac{2x^5}{5!} + \frac{2x^7}{7!} + \ldots\right)$

$= 2 + \frac{2x^2}{3!} + \frac{2x^4}{5!} + \frac{2x^6}{7!} + \ldots \Rightarrow \lim_{x \to 0} \frac{e^x - e^{-x}}{x} = \lim_{x \to \infty} \left(2 + \frac{2x^2}{3!} + \frac{2x^4}{5!} + \frac{2x^6}{7!} + \ldots\right) = 2$

49. $\frac{1}{t^4}\left(1 - \cos t - \frac{t^2}{2}\right) = \frac{1}{t^4}\left[1 - \frac{t^2}{2} - \left(1 - \frac{t^2}{2} + \frac{t^4}{4!} - \frac{t^6}{6!} + \ldots\right)\right] = -\frac{1}{4!} + \frac{t^2}{6!} - \frac{t^4}{8!} + \ldots \Rightarrow \lim_{t \to 0} \frac{1 - \cos t - \left(\frac{t^2}{2}\right)}{t^4}$

$= \lim_{t \to 0} \left(-\frac{1}{4!} + \frac{t^2}{6!} - \frac{t^4}{8!} + \ldots\right) = -\frac{1}{24}$

50. $\frac{1}{\theta^5}\left(-\theta + \frac{\theta^3}{6} + \sin\theta\right) = \frac{1}{\theta^5}\left(-\theta + \frac{\theta^3}{6} + \theta - \frac{\theta^3}{3!} + \frac{\theta^5}{5!} - \ldots\right) = \frac{1}{5!} - \frac{\theta^2}{7!} + \frac{\theta^4}{9!} - \ldots \Rightarrow \lim_{\theta \to 0} \frac{\sin\theta - \theta + \left(\frac{\theta^3}{6}\right)}{\theta^5}$

$= \lim_{\theta \to 0} \left(\frac{1}{5!} - \frac{\theta^2}{7!} + \frac{\theta^4}{9!} - \ldots\right) = \frac{1}{120}$

51. $\frac{1}{y^3}\left(y - \tan^{-1} y\right) = \frac{1}{y^3}\left[y - \left(y - \frac{y^3}{3} + \frac{y^5}{5} - \ldots\right)\right] = \frac{1}{3} - \frac{y^2}{5} + \frac{y^4}{7} - \ldots \Rightarrow \lim_{y \to 0} \frac{y - \tan^{-1} y}{y^3} = \lim_{y \to 0} \left(\frac{1}{3} - \frac{y^2}{5} + \frac{y^4}{7} - \ldots\right)$

$= \frac{1}{3}$

52. $\frac{\tan^{-1} y - \sin y}{y^3 \cos y} = \frac{\left(y - \frac{y^3}{3} + \frac{y^5}{5} - \ldots\right) - \left(y - \frac{y^3}{3!} + \frac{y^5}{5!} - \ldots\right)}{y^3 \cos y} = \frac{\left(-\frac{y^3}{6} + \frac{23y^5}{5!} - \ldots\right)}{y^3 \cos y} = \frac{\left(-\frac{1}{6} + \frac{23y^2}{5!} - \ldots\right)}{\cos y}$

$\Rightarrow \lim_{y \to 0} \frac{\tan^{-1} y - \sin y}{y^3 \cos y} = \lim_{y \to 0} \frac{\left(-\frac{1}{6} + \frac{23y^2}{5!} - \ldots\right)}{\cos y} = -\frac{1}{6}$

53. $x^2\left(-1 + e^{-1/x^2}\right) = x^2\left(-1 + 1 - \frac{1}{x^2} + \frac{1}{2x^4} - \frac{1}{6x^6} + \ldots\right) = -1 + \frac{1}{2x^2} - \frac{1}{6x^4} + \ldots \Rightarrow \lim_{x \to \infty} x^2\left(e^{-1/x^2} - 1\right)$

$= \lim_{x \to \infty} \left(-1 + \frac{1}{2x^2} - \frac{1}{6x^4} + \ldots\right) = -1$

54. $(x+1)\sin\left(\frac{1}{x+1}\right) = (x+1)\left(\frac{1}{x+1} - \frac{1}{3!(x+1)^3} + \frac{1}{5!(x+1)^5} - \ldots\right) = 1 - \frac{1}{3!(x+1)^2} + \frac{1}{5!(x+1)^4} - \ldots$

$\Rightarrow \lim_{x \to \infty} (x+1)\sin\left(\frac{1}{x+1}\right) = \lim_{x \to \infty} \left(1 - \frac{1}{3!(x+1)^2} + \frac{1}{5!(x+1)^4} - \ldots\right) = 1$

55. $\frac{\ln(1+x^2)}{1 - \cos x} = \frac{\left(x^2 - \frac{x^4}{2} + \frac{x^6}{3} - \ldots\right)}{1 - \left(1 - \frac{x^2}{2!} + \frac{x^4}{4!} - \ldots\right)} = \frac{\left(1 - \frac{x^2}{2} + \frac{x^4}{3} - \ldots\right)}{\left(\frac{1}{2!} - \frac{x^2}{4!} + \ldots\right)} \Rightarrow \lim_{x \to 0} \frac{\ln(1+x^2)}{1 - \cos x} = \lim_{x \to 0} \frac{\left(1 - \frac{x^2}{2} + \frac{x^4}{3} - \ldots\right)}{\left(\frac{1}{2!} - \frac{x^2}{4!} + \ldots\right)} = 2!$

$= 2$

56. $\frac{x^2-4}{\ln(x-1)} = \frac{(x-2)(x+2)}{\left[(x-2)-\frac{(x-2)^2}{2}+\frac{(x-2)^3}{3}-\ldots\right]} = \frac{x+2}{\left[1-\frac{x-2}{2}+\frac{(x-2)^2}{3}-\ldots\right]} \Rightarrow \lim_{x\to 2} \frac{x^2-4}{\ln(x-1)}$

$= \lim_{x\to 2} \frac{x+2}{\left[1-\frac{x-2}{2}+\frac{(x-2)^2}{3}-\ldots\right]} = 4$

57. $\ln\left(\frac{1+x}{1-x}\right) = \ln(1+x) - \ln(1-x) = \left(x-\frac{x^2}{2}+\frac{x^3}{3}-\frac{x^4}{4}+\ldots\right)-\left(-x-\frac{x^2}{2}-\frac{x^3}{3}-\frac{x^4}{4}-\ldots\right) = 2\left(x+\frac{x^3}{3}+\frac{x^5}{5}+\ldots\right)$

58. $\ln(1+x) = x-\frac{x^2}{2}+\frac{x^3}{3}-\frac{x^4}{4}+\ldots+\frac{(-1)^{n-1}x^n}{n}+\ldots \Rightarrow |\text{error}| = \left|\frac{(-1)^{n-1}x^n}{n}\right| = \frac{1}{n10^n}$ when $x = 0.1$;

$\frac{1}{n10^n} < \frac{1}{10^8} \Rightarrow n10^n > 10^8$ when $n \geq 8 \Rightarrow 7$ terms

59. $\tan^{-1} x = x-\frac{x^3}{3}+\frac{x^5}{5}-\frac{x^7}{7}+\frac{x^9}{9}-\ldots+\frac{(-1)^{n-1}x^{2n-1}}{2n-1}+\ldots \Rightarrow |\text{error}| = \left|\frac{(-1)^{n-1}x^{2n-1}}{2n-1}\right| = \frac{1}{2n-1}$ when $x = 1$;

$\frac{1}{2n-1} < \frac{1}{10^3} \Rightarrow n > \frac{1001}{2} = 500.5 \Rightarrow$ the first term not used is the $501^{\text{st}} \Rightarrow$ we must use 500 terms

60. $\tan^{-1} x = x-\frac{x^3}{3}+\frac{x^5}{5}-\frac{x^7}{7}+\frac{x^9}{9}-\ldots+\frac{(-1)^{n-1}x^{2n-1}}{2n-1}+\ldots$ and $\lim_{n\to\infty}\left|\frac{x^{2n+1}}{2n+1}\cdot\frac{2n-1}{x^{2n-1}}\right| = x^2 \lim_{n\to\infty}\left|\frac{2n-1}{2n+1}\right| = x^2$

$\Rightarrow \tan^{-1} x$ converges for $|x| < 1$; when $x = -1$ we have $\sum_{n=1}^{\infty} \frac{(-1)^n}{2n-1}$ which is a convergent series; when $x = 1$ we have $\sum_{n=1}^{\infty} \frac{(-1)^{n+1}}{2n-1}$ which is a convergent series $\Rightarrow$ the series representing $\tan^{-1} x$ diverges for $|x| > 1$

61. $\tan^{-1} x = x-\frac{x^3}{3}+\frac{x^5}{5}-\frac{x^7}{7}+\frac{x^9}{9}-\ldots+\frac{(-1)^{n-1}x^{2n-1}}{2n-1}+\ldots$ and when the series representing $48\tan^{-1}\left(\frac{1}{18}\right)$ has an error of magnitude less than 10^{-6}, then the series representing the sum $48\tan^{-1}\left(\frac{1}{18}\right)+32\tan^{-1}\left(\frac{1}{57}\right)-20\tan^{-1}\left(\frac{1}{239}\right)$ also has an error of magnitude less than 10^{-6}; thus

$|\text{error}| = \frac{\left(\frac{1}{18}\right)^{2n-1}}{2n-1} < \frac{1}{10^6} \Rightarrow n \geq 3$ using a calculator $\Rightarrow 3$ terms

62. $\ln(\sec x) = \int_0^1 \tan t\, dt = \int_1^x \left(t+\frac{t^3}{3}+\frac{2t^5}{15}+\ldots\right) dt \approx \frac{x^2}{2}+\frac{x^4}{12}+\frac{x^6}{45}+\ldots$

63. (a) $\left(1-x^2\right)^{-1/2} \approx 1+\frac{x^2}{2}+\frac{3x^4}{8}+\frac{5x^6}{16} \Rightarrow \sin^{-1} x \approx x+\frac{x^3}{6}+\frac{3x^5}{40}+\frac{5x^7}{112}$;

$\lim_{n\to\infty}\left|\frac{1\cdot 3\cdot 5\cdots(2n-1)(2n+1)x^{2n+3}}{2\cdot 4\cdot 6\cdots(2n)(2n+2)(2n+3)}\cdot\frac{2\cdot 4\cdot 6\cdots(2n)(2n+1)}{1\cdot 3\cdot 5\cdots(2n-1)x^{2n+1}}\right| < 1 \Rightarrow x^2 \lim_{n\to\infty}\left|\frac{(2n+1)(2n+1)}{(2n+2)(2n+3)}\right| < 1$

$\Rightarrow |x| < 1 \Rightarrow$ the radius of convergence is 1

(b) $\frac{d}{dx}(\cos^{-1}x) = -(1-x^2)^{-1/2} \Rightarrow \cos^{-1}x = \frac{\pi}{2} - \sin^{-1}x \approx \frac{\pi}{2} - \left(x + \frac{x^3}{6} + \frac{3x^5}{40} + \frac{5x^7}{112}\right) \approx \frac{\pi}{2} - x - \frac{x^3}{6} - \frac{3x^5}{40} - \frac{5x^7}{112}$

64. (a) $(1+t^2)^{-1/2} \approx (1)^{-1/2} + \left(-\frac{1}{2}\right)(1)^{-3/2}(t^2) + \frac{\left(-\frac{1}{2}\right)\left(-\frac{3}{2}\right)(1)^{-5/2}(t^2)^2}{2!} + \frac{\left(-\frac{1}{2}\right)\left(-\frac{3}{2}\right)\left(-\frac{5}{2}\right)(1)^{-7/2}(t^2)^3}{3!}$

$= 1 - \frac{t^2}{2} + \frac{3t^4}{2^2 \cdot 2!} - \frac{3 \cdot 5t^6}{2^3 \cdot 3!} \Rightarrow \sinh^{-1}x \approx \int_0^x \left(1 - \frac{t^2}{2} + \frac{3t^4}{8} - \frac{5t^6}{16}\right) dt = x - \frac{x^3}{6} + \frac{3x^5}{40} - \frac{5x^7}{112}$

(b) $\sinh^{-1}\left(\frac{1}{4}\right) \approx \frac{1}{4} - \frac{1}{384} + \frac{3}{40{,}960} = 0.24746908$; the error is less than the absolute value of the first unused term, $\frac{5x^7}{112}$, evaluated at $t = \frac{1}{4}$ since the series is alternating $\Rightarrow |\text{error}| < \frac{5\left(\frac{1}{4}\right)^7}{112} \approx 2.725 \times 10^{-6}$

65. $\frac{-1}{1+x} = -\frac{1}{1-(-x)} = -1 + x - x^2 + x^3 - \ldots \Rightarrow \frac{d}{dx}\left(\frac{-1}{1+x}\right) = \frac{1}{1+x^2} = \frac{d}{dx}(-1 + x - x^2 + x^3 - \ldots)$
$= 1 - 2x + 3x^2 - 4x^3 + \ldots$

66. $\frac{1}{1-x^2} = 1 + x^2 + x^4 + x^6 + \ldots \Rightarrow \frac{d}{dx}\left(\frac{1}{1-x^2}\right) = \frac{2x}{(1-x^2)^2} = \frac{d}{dx}(1 + x^2 + x^4 + x^6 + \ldots) = 2x + 4x^3 + 6x^5 + \ldots$

67. Wallis' formula gives the approximation $\pi \approx 4\left[\frac{2 \cdot 4 \cdot 4 \cdot 6 \cdot 6 \cdot 8 \cdots (2n-2) \cdot (2n)}{3 \cdot 3 \cdot 5 \cdot 5 \cdot 7 \cdot 7 \cdots (2n-1) \cdot (2n-1)}\right]$ to produce the table

n	$\sim \pi$
10	3.221088998
20	3.181104886
30	3.167880758
80	3.151425420
90	3.150331383
93	3.150049112
94	3.149959030
95	3.149870848
100	3.149456425

At n = 1929 we obtain the first approximation accurate to 3 decimals: 3.141999845. At n = 30,000 we still do not obtain accuracy to 4 decimals: 3.141617732, so the convergence to π is very slow. Here is a Maple CAS procedure to produce these approximations:

```
pie :=
  proc(n)
  local i,j;
    a(2) := evalf(8/9);
    for i from 3 to n do a(i) := evalf(2*(2*i-2)*i/(2*i-1)^2*a(i-1)) od;
    [[j,4*a(j)] $ (j = n-5 .. n)]
  end
```

68. $\ln 1 = 0$; $\ln 2 = \ln \frac{1+\left(\frac{1}{3}\right)}{1-\left(\frac{1}{3}\right)} \approx 2\left(\frac{1}{3}+\frac{\left(\frac{1}{3}\right)^3}{3}+\frac{\left(\frac{1}{3}\right)^5}{5}+\frac{\left(\frac{1}{3}\right)^7}{7}\right) \approx 0.69314$; $\ln 3 = \ln 2 + \ln\left(\frac{3}{2}\right) = \ln 2 + \ln \frac{1+\left(\frac{1}{5}\right)}{1-\left(\frac{1}{5}\right)}$

$\approx \ln 2 + 2\left(\frac{1}{5}+\frac{\left(\frac{1}{5}\right)^3}{3}+\frac{\left(\frac{1}{5}\right)^5}{5}+\frac{\left(\frac{1}{5}\right)^7}{7}\right) \approx 1.09861$; $\ln 4 = 2\ln 2 \approx 1.38628$; $\ln 5 = \ln 4 + \ln\left(\frac{5}{4}\right) = \ln 4 + \ln \frac{1+\left(\frac{1}{9}\right)}{1-\left(\frac{1}{9}\right)}$

≈ 1.60943; $\ln 6 = \ln 2 + \ln 3 \approx 1.79175$; $\ln 7 = \ln 6 + \ln\left(\frac{7}{6}\right) = \ln 6 + \ln \frac{1+\left(\frac{1}{13}\right)}{1-\left(\frac{1}{13}\right)} \approx 1.94591$; $\ln 8 = 3\ln 2$

≈ 2.07944; $\ln 9 = 2\ln 3 \approx 2.19722$; $\ln 10 = \ln 2 + \ln 5 \approx 2.30258$

69. $(1-x^2)^{-1/2} = (1+(-x^2))^{-1/2} = (1)^{-1/2} + \left(-\frac{1}{2}\right)(1)^{-3/2}(-x^2) + \frac{\left(-\frac{1}{2}\right)\left(-\frac{3}{2}\right)(1)^{-5/2}(-x^2)^2}{2!}$

$+\frac{\left(-\frac{1}{2}\right)\left(-\frac{3}{2}\right)\left(-\frac{5}{2}\right)(1)^{-7/2}(-x^2)^3}{3!} + \ldots = 1 + \frac{x^2}{2} + \frac{1\cdot 3x^4}{2^2\cdot 2!} + \frac{1\cdot 3\cdot 5x^6}{2^3\cdot 3!} + \ldots = 1 + \sum_{n=1}^{\infty} \frac{1\cdot 3\cdot 5\cdots(2n-1)x^{2n}}{2^n\cdot n!}$

$\Rightarrow \sin^{-1} x = \int_0^x (1-t^2)^{-1/2}\,dt = \int_0^x \left(1+\sum_{n=1}^{\infty} \frac{1\cdot 3\cdot 5\cdots(2n-1)x^{2n}}{2^n\cdot n!}\right) dt = x + \sum_{n=1}^{\infty} \frac{1\cdot 3\cdot 5\cdots(2n-1)x^{2n+1}}{2\cdot 4\cdots(2n)(2n+1)}$,

where $|x| < 1$

70. $\left[\tan^{-1} t\right]_x^{\infty} = \frac{\pi}{2} - \tan^{-1} x = \int_x^{\infty} \frac{dt}{1+t^2} = \int_x^{\infty} \left[\frac{\left(\frac{1}{t^2}\right)}{1+\left(\frac{1}{t^2}\right)}\right] dt = \int_x^{\infty} \frac{1}{t^2}\left(1-\frac{1}{t^2}+\frac{1}{t^4}-\frac{1}{t^6}+\ldots\right) dt$

$= \int_x^{\infty} \left(\frac{1}{t^2}-\frac{1}{t^4}+\frac{1}{t^6}-\frac{1}{t^8}+\ldots\right) dt = \lim_{b\to\infty} \left[-\frac{1}{t}+\frac{1}{3t^3}-\frac{1}{5t^5}+\frac{1}{7t^7}-\ldots\right]_x^b = \frac{1}{x}-\frac{1}{3x^3}+\frac{1}{5x^5}-\frac{1}{7x^7}+\ldots$

$\Rightarrow \tan^{-1} x = \frac{\pi}{2} - \frac{1}{x} + \frac{1}{3x^3} - \frac{1}{5x^5} + \ldots$, $x > 1$; $\left[\tan^{-1} t\right]_{-\infty}^{x} = \tan^{-1} x + \frac{\pi}{2} = \int_{-\infty}^{x} \frac{dt}{1+t^2}$

$= \lim_{b\to-\infty} \left[-\frac{1}{t}+\frac{1}{3t^3}-\frac{1}{5t^5}+\frac{1}{7t^7}-\ldots\right]_b^x = -\frac{1}{x}+\frac{1}{3x^3}-\frac{1}{5x^5}+\frac{1}{7x^7}-\ldots \Rightarrow \tan^{-1} x = -\frac{\pi}{2}-\frac{1}{x}+\frac{1}{3x^3}-\frac{1}{5x^5}+\ldots$,

$x < -1$

71. (a) $\tan\left(\tan^{-1}(n+1) - \tan^{-1}(n-1)\right) = \frac{\tan\left(\tan^{-1}(n+1)\right) - \tan\left(\tan^{-1}(n-1)\right)}{1+\tan\left(\tan^{-1}(n+1)\right)\tan\left(\tan^{-1}(n-1)\right)} = \frac{(n+1)-(n-1)}{1+(n+1)(n-1)} = \frac{2}{n^2}$

(b) $\sum_{n=1}^{N} \tan^{-1}\left(\frac{2}{n^2}\right) = \sum_{n=1}^{N} \left[\tan^{-1}(n+1) - \tan^{-1}(n-1)\right] = \left(\tan^{-1} 2 - \tan^{-1} 0\right) + \left(\tan^{-1} 3 - \tan^{-1} 1\right)$

$+\left(\tan^{-1} 4 - \tan^{-1} 2\right) + \ldots + \left(\tan^{-1}(N+1) - \tan^{-1}(N-1)\right) = \tan^{-1}(N+1) + \tan^{-1} N - \frac{\pi}{4}$

(c) $\sum_{n=1}^{\infty} \tan^{-1}\left(\frac{2}{n^2}\right) = \lim_{n\to\infty} \left[\tan^{-1}(N+1) + \tan^{-1} N - \frac{\pi}{4}\right] = \frac{\pi}{2} + \frac{\pi}{2} - \frac{\pi}{4} = \frac{3\pi}{4}$

CHAPTER 8 PRACTICE EXERCISES

1. converges to 1, since $\lim_{n\to\infty} a_n = \lim_{n\to\infty} \left(1 + \frac{(-1)^n}{n}\right) = 1$

2. converges to 0, since $0 \le a_n \le \frac{2}{\sqrt{n}}$, $\lim_{n\to\infty} 0 = 0$, $\lim_{n\to\infty} \frac{2}{\sqrt{n}} = 0$ using the Sandwich Theorem for Sequences

3. converges to -1, since $\lim_{n\to\infty} a_n = \lim_{n\to\infty} \left(\frac{1-2^n}{2^n}\right) = \lim_{n\to\infty} \left(\frac{1}{2^n} - 1\right) = -1$

4. converges to 1, since $\lim_{n\to\infty} a_n = \lim_{n\to\infty} \left[1 + (0.9)^n\right] = 1 + 0 = 1$

5. diverges, since $\left\{\sin \frac{n\pi}{2}\right\} = \{0, 1, 0, -1, 0, 1, \ldots\}$

6. converges to 0, since $\{\sin n\pi\} = \{0, 0, 0, \ldots\}$

7. converges to 0, since $\lim_{n\to\infty} a_n = \lim_{n\to\infty} \frac{\ln n^2}{n} = 2 \lim_{n\to\infty} \frac{\left(\frac{1}{n}\right)}{1} = 0$

8. converges to 0, since $\lim_{n\to\infty} a_n = \lim_{n\to\infty} \frac{\ln(2n+1)}{n} = \lim_{n\to\infty} \frac{\left(\frac{2}{2n+1}\right)}{1} = 0$

9. converges to 1, since $\lim_{n\to\infty} a_n = \lim_{n\to\infty} \left(\frac{n + \ln n}{n}\right) = \lim_{n\to\infty} \frac{1 + \left(\frac{1}{n}\right)}{1} = 1$

10. converges to 0, since $\lim_{n\to\infty} a_n = \lim_{n\to\infty} \frac{\ln(2n^3+1)}{n} = \lim_{n\to\infty} \frac{\left(\frac{6n^2}{2n^3+1}\right)}{1} = \lim_{n\to\infty} \frac{12n}{6n^2} = \lim_{n\to\infty} \frac{2}{n} = 0$

11. converges to e^{-5}, since $\lim_{n\to\infty} a_n = \lim_{n\to\infty} \left(\frac{n-5}{n}\right)^n = \lim_{n\to\infty} \left(1 + \frac{(-5)}{n}\right)^n = e^{-5}$ by Table 8.1

12. converges to $\frac{1}{e}$, since $\lim_{n\to\infty} a_n = \lim_{n\to\infty} \left(1 + \frac{1}{n}\right)^{-n} = \lim_{n\to\infty} \frac{1}{\left(1 + \frac{1}{n}\right)^n} = \frac{1}{e}$ by Table 8.1

13. converges to 3, since $\lim_{n\to\infty} a_n = \lim_{n\to\infty} \left(\frac{3^n}{n}\right)^{1/n} = \lim_{n\to\infty} \frac{3}{n^{1/n}} = \frac{3}{1} = 3$ by Table 8.1

14. converges to 1, since $\lim_{n\to\infty} a_n = \lim_{n\to\infty} \left(\frac{3}{n}\right)^{1/n} = \lim_{n\to\infty} \frac{3^{1/n}}{n^{1/n}} = \frac{1}{1} = 1$ by Table 8.1

15. converges to $\ln 2$, since $\lim_{n\to\infty} a_n = \lim_{n\to\infty} n\left(2^{1/n} - 1\right) = \lim_{n\to\infty} \frac{2^{1/n} - 1}{\left(\frac{1}{n}\right)} = \lim_{n\to\infty} \frac{\left[\frac{\left(-2^{1/n} \ln 2\right)}{n^2}\right]}{\left(\frac{-1}{n^2}\right)} = \lim_{n\to\infty} 2^{1/n} \ln 2$

$= 2^0 \cdot \ln 2 = \ln 2$

16. converges to 1, since $\lim_{n\to\infty} a_n = \lim_{n\to\infty} \sqrt[n]{2n+1} = \lim_{n\to\infty} \exp\left(\frac{\ln(2n+1)}{n}\right) = \lim_{n\to\infty} \exp\left(\frac{\frac{2}{2n+1}}{1}\right) = e^0 = 1$

17. diverges, since $\lim_{n\to\infty} a_n = \lim_{n\to\infty} \frac{(n+1)!}{n!} = \lim_{n\to\infty} (n+1) = \infty$

18. converges to 0, since $\lim_{n\to\infty} a_n = \lim_{n\to\infty} \frac{(-4)^n}{n!} = 0$ by Table 8.1

19. $\frac{1}{(2n-3)(2n-1)} = \frac{\left(\frac{1}{2}\right)}{2n-3} - \frac{\left(\frac{1}{2}\right)}{2n-1} \Rightarrow s_n = \left[\frac{\left(\frac{1}{2}\right)}{3} - \frac{\left(\frac{1}{2}\right)}{5}\right] + \left[\frac{\left(\frac{1}{2}\right)}{5} - \frac{\left(\frac{1}{2}\right)}{7}\right] + \ldots + \left[\frac{\left(\frac{1}{2}\right)}{2n-3} - \frac{\left(\frac{1}{2}\right)}{2n-1}\right] = \frac{\left(\frac{1}{2}\right)}{3} - \frac{\left(\frac{1}{2}\right)}{2n-1}$

$\Rightarrow \lim_{n\to\infty} s_n = \lim_{n\to\infty} \left[\frac{1}{6} - \frac{\left(\frac{1}{2}\right)}{2n-1}\right] = \frac{1}{6}$

20. $\frac{-2}{n(n+1)} = \frac{-2}{n} + \frac{2}{n+1} \Rightarrow s_n = \left(\frac{-2}{2} + \frac{2}{3}\right) + \left(\frac{-2}{3} + \frac{2}{4}\right) + \ldots + \left(\frac{-2}{n} + \frac{2}{n+1}\right) = -\frac{2}{2} + \frac{2}{n+1} \Rightarrow \lim_{n\to\infty} s_n$

$= \lim_{n\to\infty} \left(-1 + \frac{2}{n+1}\right) = -1$

21. $\frac{9}{(3n-1)(3n+2)} = \frac{3}{3n-1} - \frac{3}{3n+2} \Rightarrow s_n = \left(\frac{3}{2} - \frac{3}{5}\right) + \left(\frac{3}{5} - \frac{3}{8}\right) + \left(\frac{3}{8} - \frac{3}{11}\right) + \ldots + \left(\frac{3}{3n-1} - \frac{3}{3n+2}\right)$

$= \frac{3}{2} - \frac{3}{3n+2} \Rightarrow \lim_{n\to\infty} s_n = \lim_{n\to\infty} \left(\frac{3}{2} - \frac{3}{3n+2}\right) = \frac{3}{2}$

22. $\frac{-8}{(4n-3)(4n+1)} = \frac{-2}{4n-3} + \frac{2}{4n+1} \Rightarrow s_n = \left(\frac{-2}{9} + \frac{2}{13}\right) + \left(\frac{-2}{13} + \frac{2}{17}\right) + \left(\frac{-2}{17} + \frac{2}{21}\right) + \ldots + \left(\frac{-2}{4n-3} + \frac{2}{4n+1}\right)$

$= -\frac{2}{9} + \frac{2}{4n+1} \Rightarrow \lim_{n\to\infty} s_n = \lim_{n\to\infty} \left(-\frac{2}{9} + \frac{2}{4n+1}\right) = -\frac{2}{9}$

23. $\sum_{n=0}^{\infty} e^{-n} = \sum_{n=0}^{\infty} \frac{1}{e^n}$, a convergent geometric series with $r = \frac{1}{e}$ and $a = 1 \Rightarrow$ the sum is $\frac{1}{1 - \left(\frac{1}{e}\right)} = \frac{e}{e-1}$

24. $\sum_{n=1}^{\infty} (-1)^n \frac{3}{4^n} = \sum_{n=0}^{\infty} \left(-\frac{3}{4}\right)\left(\frac{-1}{4}\right)^n$ a convergent geometric series with $r = -\frac{1}{4}$ and $a = \frac{-3}{4} \Rightarrow$ the sum is

$\frac{\left(-\frac{3}{4}\right)}{1 - \left(\frac{-1}{4}\right)} = -\frac{3}{5}$

25. diverges, a p-series with $p = \frac{1}{2}$

26. $\sum_{n=1}^{\infty} \frac{-5}{n} = -5 \sum_{n=1}^{\infty} \frac{1}{n}$, diverges since it is a nonzero multiple of the divergent harmonic series

27. Since $f(x) = \frac{1}{x^{1/2}} \Rightarrow f'(x) = -\frac{1}{2x^{3/2}} < 0 \Rightarrow f(x)$ is decreasing $\Rightarrow a_{n+1} < a_n$, and $\lim_{n\to\infty} a_n = \lim_{n\to\infty} \frac{(-1)}{\sqrt{n}} = 0$, the series $\sum_{n=1}^{\infty} \frac{(-1)^n}{\sqrt{n}}$ converges by the Alternating Series Test. Since $\sum_{n=1}^{\infty} \frac{1}{\sqrt{n}}$ diverges, the given series converges conditionally.

28. converges absolutely by the Direct Comparison Test since $\frac{1}{2n^3} < \frac{1}{n^3}$ for $n \geq 1$, which is the nth term of a convergent p-series

29. The given series does not converge absolutely by the Direct Comparison Test since $\frac{1}{\ln(n+1)} > \frac{1}{n+1}$, which is the nth term of a divergent series. Since $f(x) = \frac{1}{\ln(x+1)} \Rightarrow f'(x) = -\frac{1}{(\ln(x+1))^2(x+1)} < 0 \Rightarrow f(x)$ is decreasing $\Rightarrow a_{n+1} < a_n$, and $\lim_{n\to\infty} a_n = \lim_{n\to\infty} \frac{1}{\ln(n+1)} = 0$, the given series converges conditionally by the Alternating Series Test.

30. $\int_2^{\infty} \frac{1}{x(\ln x)^2}\, dx = \lim_{b\to\infty} \int_2^{1} \frac{1}{x(\ln x)^2}\, dx = \lim_{b\to\infty} \left[-(\ln x)^{-1}\right]_2^b = -\lim_{b\to\infty} \left(\frac{1}{\ln b} - \frac{1}{\ln 2}\right) = \frac{1}{\ln 2} \Rightarrow$ the series converges absolutely by the Integral Test

31. converges absolutely by the Direct Comparison Test since $\frac{\ln n}{n^3} < \frac{n}{n^3} = \frac{1}{n^2}$, the nth term of a convergent p-series

32. diverges by the Direct Comparison Test for $e^{n^n} > n \Rightarrow \ln\left(e^{n^n}\right) > \ln n \Rightarrow n^n > \ln n \Rightarrow \ln n^n > \ln(\ln n)$ $\Rightarrow n \ln n > \ln(\ln n) \Rightarrow \frac{\ln n}{\ln(\ln n)} > \frac{1}{n}$, the nth term of the divergent harmonic series

33. $\lim_{n\to\infty} \frac{\left(\frac{1}{n\sqrt{n^2+1}}\right)}{\left(\frac{1}{n^2}\right)} = \sqrt{\lim_{n\to\infty} \frac{n^2}{n^2+1}} = \sqrt{1} = 1 \Rightarrow$ converges absolutely by the Limit Comparison Test

34. Since $f(x) = \frac{3x^2}{x^3+1} \Rightarrow f'(x) = \frac{3x(2-x^3)}{(x^3+1)^2} < 0$ when $x \geq 2 \Rightarrow a_{n+1} < a_n$ for $n \geq 2$ and $\lim_{n\to\infty} \frac{3n^2}{n^3+1} = 0$, the series converges by the Alternating Series Test. The series does not converge absolutely: By the Limit Comparison Test, $\lim_{n\to\infty} \frac{\left(\frac{3n^2}{n^3+1}\right)}{\left(\frac{1}{n}\right)} = \lim_{n\to\infty} \frac{3n^3}{n^3+1} = 3$. Therefore the convergence is conditional.

35. converges absolutely by the Ratio Test since $\lim_{n\to\infty} \left[\frac{n+2}{(n+1)!} \cdot \frac{n!}{n+1}\right] = \lim_{n\to\infty} \frac{n+2}{(n+1)^2} = 0 < 1$

36. diverges since $\lim_{n\to\infty} a_n = \lim_{n\to\infty} \frac{(-1)^n(n^2+1)}{2n^2+n-1}$ does not exist

37. converges absolutely by the Ratio Test since $\lim\limits_{n\to\infty}\left[\frac{3^{n+1}}{(n+1)!}\cdot\frac{n!}{3^n}\right]=\lim\limits_{n\to\infty}\frac{3}{n+1}=0<1$

38. converges absolutely by the Root Test since $\lim\limits_{n\to\infty}\sqrt[n]{a_n}=\lim\limits_{n\to\infty}\sqrt[n]{\frac{2^n3^n}{n^n}}=\lim\limits_{n\to\infty}\frac{6}{n}=0<1$

39. converges absolutely by the Limit Comparison Test since $\lim\limits_{n\to\infty}\frac{\left(\frac{1}{n^{3/2}}\right)}{\left(\frac{1}{\sqrt{n(n+1)(n+2)}}\right)}=\sqrt{\lim\limits_{n\to\infty}\frac{n(n+1)(n+2)}{n^3}}$ $=1$

40. converges absolutely by the Limit Comparison Test since $\lim\limits_{n\to\infty}\frac{\left(\frac{1}{n^2}\right)}{\left(\frac{1}{n\sqrt{n^2-1}}\right)}=\sqrt{\lim\limits_{n\to\infty}\frac{n^2(n^2-1)}{n^4}}=1$

41. $\lim\limits_{n\to\infty}\left|\frac{u_{n+1}}{u_n}\right|<1\Rightarrow\lim\limits_{n\to\infty}\left|\frac{(x+4)^{n+1}}{(n+1)3^{n+1}}\cdot\frac{n3^n}{(x+4)^n}\right|<1\Rightarrow\frac{|x+4|}{3}\lim\limits_{n\to\infty}\left(\frac{n}{n+1}\right)<1\Rightarrow\frac{|x+4|}{3}<1$

$\Rightarrow|x+4|<3\Rightarrow-3<x+4<3\Rightarrow-7<x<-1$; at $x=-7$ we have $\sum\limits_{n=1}^{\infty}\frac{(-1)^n3^n}{n3^n}=\sum\limits_{n=1}^{\infty}\frac{(-1)^n}{n}$, the alternating harmonic series, which converges conditionally; at $x=-1$ we have $\sum\limits_{n=1}^{\infty}\frac{3^n}{n3^n}=\sum\limits_{n=1}^{\infty}\frac{1}{n}$, the divergent harmonic series

(a) the radius is 3; the interval of convergence is $-7\le x<-1$

(b) the interval of absolute convergence is $-7<x<-1$

(c) the series converges conditionally at $x=-7$

42. $\lim\limits_{n\to\infty}\left|\frac{u_{n+1}}{u_n}\right|<1\Rightarrow\lim\limits_{n\to\infty}\left|\frac{(x-1)^{2n}}{(2n+1)!}\cdot\frac{(2n-1)!}{(x-1)^{2n-2}}\right|<1\Rightarrow(x-1)^2\lim\limits_{n\to\infty}\frac{1}{(2n)(2n+1)}=0<1$, which holds for all x

(a) the radius is ∞; the series converges for all x

(b) the series converges absolutely for all x

(c) there are no values for which the series converges conditionally

43. $\lim\limits_{n\to\infty}\left|\frac{u_{n+1}}{u_n}\right|<1\Rightarrow\lim\limits_{n\to\infty}\left|\frac{(3x-1)^{n+1}}{(n+1)^2}\cdot\frac{n^2}{(3x-1)^n}\right|<1\Rightarrow|3x-1|\lim\limits_{n\to\infty}\frac{n^2}{(n+1)^2}<1\Rightarrow|3x-1|<1$

$\Rightarrow-1<3x-1<1\Rightarrow0<3x<2\Rightarrow0<x<\frac{2}{3}$; at $x=0$ we have $\sum\limits_{n=1}^{\infty}\frac{(-1)^{n-1}(-1)^n}{n^2}=\sum\limits_{n=1}^{\infty}\frac{(-1)^{2n-1}}{n^2}$ $=-\sum\limits_{n=1}^{\infty}\frac{1}{n^2}$, a nonzero constant multiple of a convergent p-series, which is absolutely convergent; at $x=\frac{2}{3}$ we have $\sum_{n=1}^{\infty} \frac{(-1)^{n-1}(1)^n}{n^2} = \sum_{n=1}^{\infty} \frac{(-1)^{n-1}}{n^2}$, which converges absolutely

(a) the radius is $\frac{1}{3}$; the interval of convergence is $0 \le x \le \frac{2}{3}$

(b) the interval of absolute convergence is $0 \le x \le \frac{2}{3}$

(c) there are no values for which the series converges conditionally

44. $\lim_{n\to\infty} \left|\frac{u_{n+1}}{u_n}\right| < 1 \Rightarrow \lim_{n\to\infty} \left|\frac{n+2}{2n+3} \cdot \frac{(2x+1)^{n+1}}{2^{n+1}} \cdot \frac{2n+1}{n+1} \cdot \frac{2^n}{(2x+1)^n}\right| < 1 \Rightarrow \frac{|2x+1|}{2} \lim_{n\to\infty} \left|\frac{n+2}{2n+3} \cdot \frac{2n+1}{n+1}\right| < 1$

$\Rightarrow \frac{|2x+1|}{2}(1) < 1 \Rightarrow |2x+1| < 2 \Rightarrow -2 < 2x+1 < 2 \Rightarrow -3 < 2x < 1 \Rightarrow -\frac{3}{2} < x < \frac{1}{2}$; at $x = -\frac{3}{2}$ we have

$\sum_{n=1}^{\infty} \frac{n+1}{2n+1} \cdot \frac{(-2)^n}{2^n} = \sum_{n=1}^{\infty} \frac{(-1)^n(n+1)}{2n+1}$ which diverges by the nth-Term Test for Divergence since

$\lim_{n\to\infty} \left(\frac{n+1}{2n+1}\right) = \frac{1}{2} \ne 0$; at $x = \frac{1}{2}$ we have $\sum_{n=1}^{\infty} \frac{n+1}{2n+1} \cdot \frac{2^n}{2^n} = \sum_{n=1}^{\infty} \frac{n+1}{2n+1}$, which diverges by the nth-Term Test

(a) the radius is 1; the interval of convergence is $-\frac{3}{2} < x < \frac{1}{2}$

(b) the interval of absolute convergence is $-\frac{3}{2} < x < \frac{1}{2}$

(c) there are no values for which the series converges conditionally

45. $\lim_{n\to\infty} \left|\frac{u_{n+1}}{u_n}\right| < 1 \Rightarrow \lim_{n\to\infty} \left|\frac{x^{n+1}}{(n+1)^{n+1}} \cdot \frac{n^n}{x^n}\right| < 1 \Rightarrow |x| \lim_{n\to\infty} \left|\left(\frac{n}{n+1}\right)^n \left(\frac{1}{n+1}\right)\right| < 1 \Rightarrow \frac{|x|}{e} \lim_{n\to\infty} \left(\frac{1}{n+1}\right) < 1$

$\Rightarrow \frac{|x|}{e} \cdot 0 < 1$, which holds for all x

(a) the radius is ∞; the series converges for all x

(b) the series converges absolutely for all x

(c) there are no values for which the series converges conditionally

46. $\lim_{n\to\infty} \left|\frac{u_{n+1}}{u_n}\right| < 1 \Rightarrow \lim_{n\to\infty} \left|\frac{x^{n+1}}{\sqrt{n+1}} \cdot \frac{\sqrt{n}}{x^n}\right| < 1 \Rightarrow |x| \lim_{n\to\infty} \sqrt{\frac{n}{n+1}} < 1 \Rightarrow |x| < 1$; when $x = -1$ we have

$\sum_{n=1}^{\infty} \frac{(-1)^n}{\sqrt{n}}$, which converges by the Alternating Series Test; when $x = 1$ we have $\sum_{n=1}^{\infty} \frac{1}{\sqrt{n}}$, a divergent p-series

(a) the radius is 1; the interval of convergence is $-1 \le x < 1$

(b) the interval of absolute convergence is $-1 < x < 1$

(c) the series converges conditionally at $x = -1$

47. $\lim_{n\to\infty} \left|\frac{u_{n+1}}{u_n}\right| < 1 \Rightarrow \lim_{n\to\infty} \left|\frac{(n+2)x^{2n+1}}{3^{n+1}} \cdot \frac{3^n}{(n+1)x^{2n-1}}\right| < 1 \Rightarrow \frac{x^2}{3} \lim_{n\to\infty} \left(\frac{n+2}{n+1}\right) < 1 \Rightarrow -\sqrt{3} < x < \sqrt{3}$;

the series $\sum_{n=1}^{\infty} -\frac{n+1}{\sqrt{3}}$ and $\sum_{n=1}^{\infty} \frac{n+1}{\sqrt{3}}$, obtained with $x = \pm\sqrt{3}$, both diverge

(a) the radius is $\sqrt{3}$; the interval of convergence is $-\sqrt{3} < x < \sqrt{3}$

(b) the interval of absolute convergence is $-\sqrt{3} < x < \sqrt{3}$

(c) there are no values for which the series converges conditionally

48. $\lim_{n\to\infty} \left|\frac{u_{n+1}}{u_n}\right| < 1 \Rightarrow \lim_{n\to\infty} \left|\frac{(x-1)x^{2n+3}}{2n+3} \cdot \frac{2n+1}{(x-1)^{2n+1}}\right| < 1 \Rightarrow (x-1)^2 \lim_{n\to\infty} \left(\frac{2n+1}{2n+3}\right) < 1 \Rightarrow (x-1)^2(1) < 1$

$\Rightarrow (x-1)^2 < 1 \Rightarrow |x-1| < 1 \Rightarrow -1 < x-1 < 1 \Rightarrow 0 < x < 2$; at $x = 0$ we have $\sum_{n=1}^{\infty} \frac{(-1)^n(-1)^{2n+1}}{2n+1}$ $= \sum_{n=1}^{\infty} \frac{(-1)^{3n+1}}{2n+1} = \sum_{n=1}^{\infty} \frac{(-1)^{n-1}}{2n+1}$ which converges conditionally by the Alternating Series Test and the fact that $\sum_{n=1}^{\infty} \frac{1}{2n+1}$ diverges; at $x = 2$ we have $\sum_{n=1}^{\infty} \frac{(-1)^n(1)^{2n+1}}{2n+1} = \sum_{n=1}^{\infty} \frac{(-1)^n}{2n+1}$, which also converges conditionally

(a) the radius is 1; the interval of convergence is $0 \le x \le 2$

(b) the interval of absolute convergence is $0 < x < 2$

(c) the series converges conditionally at $x = 0$ and $x = 2$

49. $\lim_{n\to\infty} \left|\frac{u_{n+1}}{u_n}\right| < 1 \Rightarrow \lim_{n\to\infty} \left|\frac{\operatorname{csch}(n+1)x^{n+1}}{\operatorname{csch}(n)x^n}\right| < 1 \Rightarrow |x| \lim_{n\to\infty} \left|\frac{\left(\frac{2}{e^{n+1}-e^{-n-1}}\right)}{\left(\frac{2}{e^n-e^{-n}}\right)}\right| < 1$

$\Rightarrow |x| \lim_{n\to\infty} \left|\frac{e^{-1}-e^{-2n-1}}{1-e^{-2n-2}}\right| < 1 \Rightarrow \frac{|x|}{e} < 1 \Rightarrow -e < x < e$; the series $\sum_{n=1}^{\infty} (\pm e)^n \operatorname{csch} n$, obtained with $x = \pm e$, both diverge since $\lim_{n\to\infty} (\pm e)^n \operatorname{csch} n \ne 0$

(a) the radius is e; the interval of convergence is $-e < x < e$

(b) the interval of absolute convergence is $-e < x < e$

(c) there are no values for which the series converges conditionally

50. $\lim_{n\to\infty} \left|\frac{u_{n+1}}{u_n}\right| < 1 \Rightarrow \lim_{n\to\infty} \left|\frac{x^{n+1}\coth(n+1)}{x^n \coth(n)}\right| < 1 \Rightarrow |x| \lim_{n\to\infty} \left|\frac{1+e^{-2n-2}}{1-e^{-2n-2}} \cdot \frac{1-e^{-2n}}{1+e^{-2n}}\right| < 1 \Rightarrow |x| < 1$

$\Rightarrow -1 < x < x$; the series $\sum_{n=1}^{\infty} (\pm 1)^n \coth n$, obtained with $x = \pm 1$, both diverge since $\lim_{n\to\infty} (\pm 1)^n \coth n \ne 0$

(a) the radius is 1; the interval of convergence is $-1 < x < 1$

(b) the interval of absolute convergence is $-1 < x < 1$

(c) there are no values for which the series converges conditionally

51. The given series has the form $1 - x + x^2 - x^3 + \ldots + (-x)^n + \ldots = \frac{1}{1+x}$, where $x = \frac{1}{4}$; the sum is $\frac{1}{1+\left(\frac{1}{4}\right)} = \frac{4}{5}$

52. The given series has the form $x - \frac{x^2}{2} + \frac{x^3}{3} - \ldots + (-1)^{n-1}\frac{x^n}{n} + \ldots = \ln(1+x)$, where $x = \frac{2}{3}$; the sum is $\ln\left(\frac{5}{3}\right) \approx 0.510825624$

53. The given series has the form $x - \frac{x^3}{3!} + \frac{x^5}{5!} - \ldots + (-1)^n \frac{x^{2n+1}}{(2n+1)!} + \ldots = \sin x$, where $x = \pi$; the sum is $\sin \pi = 0$

54. The given series has the form $1 - \frac{x^2}{2!} + \frac{x^4}{4!} - \ldots + (-1)^n \frac{x^{2n}}{(2n)!} + \ldots = \cos x$, where $x = \frac{\pi}{3}$; the sum is $\cos \frac{\pi}{3} = \frac{1}{2}$

55. The given series has the form $1 + x + \frac{x^2}{2!} + \frac{x^2}{3!} + \ldots + \frac{x^n}{n!} + \ldots = e^x$, where $x = \ln 2$; the sum is $e^{\ln(2)} = 2$

56. The given series has the form $x - \frac{x^3}{3} + \frac{x^5}{5} - \ldots + (-1)^n \frac{x^{2n-1}}{(2n-1)} + \ldots = \tan^{-1} x$, where $x = \frac{1}{\sqrt{3}}$; the sum is $\tan^{-1}\left(\frac{1}{\sqrt{3}}\right) = \frac{\pi}{6}$

57. Consider $\frac{1}{1-2x}$ as the sum of a convergent geometric series with $a = 1$ and $r = 2x \Rightarrow \frac{1}{1-2x}$ $= 1 + (2x) + (2x)^2 + (2x)^3 + \ldots = \sum_{n=0}^{\infty} (2x)^n = \sum_{n=0}^{\infty} 2^n x^n$ where $|2x| < 1 \Rightarrow |x| < \frac{1}{2}$

58. Consider $\frac{1}{1+x^3}$ as the sum of a convergent geometric series with $a = 1$ and $r = -x^3 \Rightarrow \frac{1}{1+x^3} = \frac{1}{1-(-x^3)}$ $= 1 + (-x^3) + (-x^3)^2 + (-x^3)^3 + \ldots = \sum_{n=0}^{\infty} (-1)^n x^{3n}$ where $|-x^3| < 1 \Rightarrow |x^3| < 1$

59. $\sin x = \sum_{n=0}^{\infty} \frac{(-1)^n x^{2n+1}}{(2n+1)!} \Rightarrow \sin \pi x = \sum_{n=0}^{\infty} \frac{(-1)^n (\pi x)^{2n+1}}{(2n+1)!} = \sum_{n=0}^{\infty} \frac{(-1)^n \pi^{2n+1} x^{2n+1}}{(2n+1)!}$

60. $\sin x = \sum_{n=0}^{\infty} \frac{(-1)^n x^{2n+1}}{(2n+1)!} \Rightarrow \sin \frac{2x}{3} = \sum_{n=0}^{\infty} \frac{(-1)^n \left(\frac{2x}{3}\right)^{2n+1}}{(2n+1)!} = \sum_{n=0}^{\infty} \frac{(-1)^n 2^{2n+1} x^{2n+1}}{3^{2n+1}(2n+1)!}$

61. $\cos x = \sum_{n=0}^{\infty} \frac{(-1)^n x^{2n}}{(2n)!} \Rightarrow \cos(x^{5/2}) = \sum_{n=0}^{\infty} \frac{(-1)^n (x^{5/2})^{2n}}{(2n)!} = \sum_{n=0}^{\infty} \frac{(-1)^n x^{5n}}{(2n)!}$

62. $\cos x = \sum_{n=0}^{\infty} \frac{(-1)^n x^{2n}}{(2n)!} \Rightarrow \cos \sqrt{5x} = \cos\left((5x)^{1/2}\right) = \sum_{n=0}^{\infty} \frac{(-1)^n \left((5x)^{1/2}\right)^{2n}}{(2n)!} = \sum_{n=0}^{\infty} \frac{(-1)^n 5^n x^n}{(2n)!}$

63. $e^x = \sum_{n=0}^{\infty} \frac{x^n}{n!} \Rightarrow e^{(\pi x/2)} = \sum_{n=0}^{\infty} \frac{\left(\frac{\pi x}{2}\right)^n}{n!} = \sum_{n=0}^{\infty} \frac{\pi^n x^n}{2^n n!}$

64. $e^x = \sum_{n=0}^{\infty} \frac{x^n}{n!} \Rightarrow e^{-x^2} = \sum_{n=0}^{\infty} \frac{(-x^2)^n}{n!} = \sum_{n=0}^{\infty} \frac{(-1)^n x^{2n}}{n!}$

65. $f(x) = \sqrt{3+x^2} = (3+x^2)^{1/2} \Rightarrow f'(x) = x(3+x^2)^{-1/2} \Rightarrow f''(x) = -x^2(3+x^2)^{-3/2} + (3+x^2)^{-1/2}$ $\Rightarrow f'''(x) = 3x^3(3+x^2)^{-5/2} - 3x(3+x^2)^{-3/2}$; $f(-1) = 2$, $f'(-1) = -\frac{1}{2}$, $f''(-1) = -\frac{1}{8} + \frac{1}{2} = \frac{3}{8}$,

$f'''(-1) = -\frac{3}{32} + \frac{3}{8} = \frac{9}{32} \Rightarrow \sqrt{3+x^2} = 2 - \frac{(x+1)}{2 \cdot 1!} + \frac{3(x+1)^2}{2^3 \cdot 2!} + \frac{9(x+1)^3}{2^5 \cdot 3!} + \ldots$

66. $f(x) = \frac{1}{1-x} = (1-x)^{-1} \Rightarrow f'(x) = (1-x)^{-2} \Rightarrow f''(x) = 2(1-x)^{-3} \Rightarrow f'''(x) = 6(1-x)^{-4}$; $f(2) = -1$, $f'(2) = 1$, $f''(2) = -2$, $f'''(2) = 6 \Rightarrow \frac{1}{1-x} = -1 + (x-2) - (x-2)^2 + (x-2)^3 - \ldots$

67. $f(x) = \frac{1}{x+1} = (x+1)^{-1} \Rightarrow f'(x) = -(x+1)^{-2} \Rightarrow f''(x) = 2(x+1)^{-3} \Rightarrow f'''(x) = -6(x+1)^{-4}$; $f(3) = \frac{1}{4}$, $f'(3) = -\frac{1}{4^2}$, $f''(3) = \frac{2}{4^3}$, $f'''(2) = \frac{-6}{4^4} \Rightarrow \frac{1}{x+1} = \frac{1}{4} - \frac{1}{4^2}(x-3) + \frac{1}{4^3}(x-3)^2 - \frac{1}{4^4}(x-3)^3 + \ldots$

68. $f(x) = \frac{1}{x} = x^{-1} \Rightarrow f'(x) = -x^{-2} \Rightarrow f''(x) = 2x^{-3} \Rightarrow f'''(x) = -6x^{-4}$; $f(a) = \frac{1}{a}$, $f'(a) = -\frac{1}{a^2}$, $f''(a) = \frac{2}{a^3}$, $f'''(a) = \frac{-6}{a^4} \Rightarrow \frac{1}{x} = \frac{1}{a} - \frac{1}{a^2}(x-a) + \frac{1}{a^3}(x-a)^2 - \frac{1}{a^4}(x-a)^3 + \ldots$

69. Assume the solution has the form $y = a_0 + a_1x + a_2x^2 + \ldots + a_{n-1}x^{n-1} + a_nx^n + \ldots$

$\Rightarrow \frac{dy}{dx} = a_1 + 2a_2x + \ldots + na_nx^{n-1} + \ldots \Rightarrow \frac{dy}{dx} + y$

$= (a_1 + a_0) + (2a_2 + a_1)x + (3a_3 + a_2)x^2 + \ldots + (na_n + a_{n-1})x^{n-1} + \ldots = 0 \Rightarrow a_1 + a_0 = 0$, $2a_2 + a_1 = 0$, $3a_3 + a_2 = 0$ and in general $na_n + a_{n-1} = 0$. Since $y = -1$ when $x = 0$ we have $a_0 = -1$. Therefore $a_1 = 1$, $a_2 = \frac{-a_1}{2 \cdot 1} = -\frac{1}{2}$, $a_3 = \frac{-a_2}{3} = \frac{1}{3 \cdot 2}$, $a_4 = \frac{-a_3}{4} = -\frac{1}{4 \cdot 3 \cdot 2}$, ..., $a_n = \frac{-a_{n-1}}{n} = \frac{-1}{n}\frac{(-1)^n}{(n-1)!} = \frac{(-1)^{n+1}}{n!}$

$\Rightarrow y = -1 + x - \frac{1}{2}x^2 + \frac{1}{3 \cdot 2}x^3 - \ldots + \frac{(-1)^{n+1}}{n!}x^n + \ldots = -\sum_{n=0}^{\infty} \frac{(-1)^n x^n}{n!} = -e^{-x}$

70. Assume the solution has the form $y = a_0 + a_1x + a_2x^2 + \ldots + a_{n-1}x^{n-1} + a_nx^n + \ldots$

$\Rightarrow \frac{dy}{dx} = a_1 + 2a_2x + \ldots + na_nx^{n-1} + \ldots \Rightarrow \frac{dy}{dx} - y$

$= (a_1 - a_0) + (2a_2 - a_1)x + (3a_3 - a_2)x^2 + \ldots + (na_n - a_{n-1})x^{n-1} + \ldots = 0 \Rightarrow a_1 - a_0 = 0$, $2a_2 - a_1 = 0$, $3a_3 - a_2 = 0$ and in general $na_n - a_{n-1} = 0$. Since $y = -3$ when $x = 0$ we have $a_0 = -3$. Therefore $a_1 = -3$, $a_2 = \frac{a_1}{2} = \frac{-3}{2}$, $a_3 = \frac{a_2}{3} = \frac{-3}{3 \cdot 2}$, $a_n = \frac{a_{n-1}}{n} = \frac{-3}{n!} \Rightarrow y = -3 - 3x - \frac{3}{2 \cdot 1}x^2 - \frac{3}{3 \cdot 2}x^3 - \ldots - \frac{-3}{n!}x^n + \ldots$

$= -3\left(1 + x + \frac{x^2}{2!} + \frac{x^3}{3!} + \ldots + \frac{x^n}{n!} + \ldots\right) = -3 \sum_{n=0}^{\infty} \frac{x^n}{n!} = -3e^x$

71. Assume the solution has the form $y = a_0 + a_1x + a_2x^2 + \ldots + a_{n-1}x^{n-1} + a_nx^n + \ldots$

$\Rightarrow \frac{dy}{dx} = a_1 + 2a_2x + \ldots + na_nx^{n-1} + \ldots \Rightarrow \frac{dy}{dx} + 2y$

$= (a_1 + 2a_0) + (2a_2 + 2a_1)x + (3a_3 + 2a_2)x^2 + \ldots + (na_n + 2a_{n-1})x^{n-1} + \ldots = 0$. Since $y = 3$ when $x = 0$ we have $a_0 = 3$. Therefore $a_1 = -2a_0 = -2(3) = -3(2)$, $a_2 = -\frac{2}{2}a_1 = -\frac{2}{2}(-2 \cdot 3) = 3\left(\frac{2^2}{2}\right)$, $a_3 = -\frac{2}{3}a_2$

$$= -\frac{2}{3}\left[3\left(\frac{2^2}{2}\right)\right] = -3\left(\frac{2^3}{3\cdot 2}\right), \ldots, a_n = \left(-\frac{2}{n}\right)a_{n-1} = \left(-\frac{2}{n}\right)\left(3\left(\frac{(-1)^{n-1}2^{n-1}}{(n-1)!}\right)\right) = 3\left(\frac{(-1)^n 2^n}{n!}\right)$$

$$\Rightarrow y = 3 - 3(2x) + 3\frac{(2)^2}{2}x^2 - 3\frac{(2)^3}{3\cdot 2}x^3 + \ldots + 3\frac{(-1)^n 2^n}{n!}x^n + \ldots$$

$$= 3\left[1 - (2x) + \frac{(2x)^2}{2!} - \frac{(2x)^3}{3!} + \ldots + \frac{(-1)^n(2x)^n}{n!} + \ldots\right] = 3\sum_{n=0}^{\infty}\frac{(-1)^n(2x)^n}{n!} = 3e^{-2x}$$

72. Assume the solution has the form $y = a_0 + a_1x + a_2x^2 + \ldots + a_{n-1}x^{n-1} + a_nx^n + \ldots$

$$\Rightarrow \frac{dy}{dx} = a_1 + 2a_2x + \ldots + na_nx^{n-1} + \ldots \Rightarrow \frac{dy}{dx} + y$$

$= (a_1 + a_0) + (2a_2 + a_1)x + (3a_3 + a_2)x^2 + \ldots + (na_n + a_{n-1})x^{n-1} + \ldots = 1 \Rightarrow a_1 + a_0 = 1,\ 2a_2 + a_1 = 0,$ $3a_3 + a_2 = 0$ and in general $na_n + a_{n-1} = 0$ for $n > 1$. Since $y = 0$ when $x = 0$ we have $a_0 = 0$. Therefore

$$a_1 = 1 - a_0 = 1,\ a_2 = \frac{-a_1}{2\cdot 1} = -\frac{1}{2},\ a_3 = \frac{-a_2}{3} = \frac{1}{3\cdot 2},\ a_4 = \frac{-a_3}{4} = -\frac{1}{4\cdot 3\cdot 2}, \ldots, a_n$$

$$= \frac{-a_{n-1}}{n} = \left(\frac{-1}{n}\right)\frac{(-1)^n}{(n-1)!} = \frac{(-1)^{n+1}}{n!} \Rightarrow y = 0 + x - \frac{1}{2}x^2 + \frac{1}{3\cdot 2}x^3 - \ldots + \frac{(-1)^{n+1}}{n!}x^n + \ldots$$

$$= -1\left[1 - x + \frac{1}{2}x^2 - \frac{1}{3\cdot 2}x^3 - \ldots + \frac{(-1)^n}{n!}x^n + \ldots\right] + 1 = -\sum_{n=0}^{\infty}\frac{(-1)^nx^n}{n!} + 1 = 1 - e^{-x}$$

73. Assume the solution has the form $y = a_0 + a_1x + a_2x^2 + \ldots + a_{n-1}x^{n-1} + a_nx^n + \ldots$

$$\Rightarrow \frac{dy}{dx} = a_1 + 2a_2x + \ldots + na_nx^{n-1} + \ldots \Rightarrow \frac{dy}{dx} - y$$

$= (a_1 - a_0) + (2a_2 - a_1)x + (3a_3 - a_2)x^2 + \ldots + (na_n - a_{n-1})x^{n-1} + \ldots = 3x \Rightarrow a_1 - a_0 = 0,\ 2a_2 - a_1 = 3,$ $3a_3 - a_2 = 0$ and in general $na_n - a_{n-1} = 0$ for $n > 2$. Since $y = -1$ when $x = 0$ we have $a_0 = -1$. Therefore

$$a_1 = -1,\ a_2 = \frac{3 + a_1}{2} = \frac{2}{2},\ a_3 = \frac{a_2}{3} = \frac{2}{3\cdot 2},\ a_4 = \frac{a_3}{4} = \frac{2}{4\cdot 3\cdot 2}, \ldots, a_n = \frac{a_{n-1}}{n} = \frac{2}{n!}$$

$$\Rightarrow y = -1 - x + \left(\frac{2}{2}\right)x^2 + \frac{3}{3\cdot 2}x^3 + \frac{2}{4\cdot 3\cdot 2}x^4 + \ldots + \frac{2}{n!}x^n + \ldots$$

$$= 2\left(1 + x + \frac{1}{2}x^2 + \frac{1}{3\cdot 2}x^3 + \frac{1}{4\cdot 3\cdot 2}x^4 + \ldots + \frac{1}{n!}x^n + \ldots\right) - 3 - 3x = 2\sum_{n=0}^{\infty}\frac{x^n}{n!} - 3 - 3x = 2e^x - 3x - 3$$

74. Assume the solution has the form $y = a_0 + a_1x + a_2x^2 + \ldots + a_{n-1}x^{n-1} + a_nx^n + \ldots$

$$\Rightarrow \frac{dy}{dx} = a_1 + 2a_2x + \ldots + na_nx^{n-1} + \ldots \Rightarrow \frac{dy}{dx} + y$$

$= (a_1 + a_0) + (2a_2 + a_1)x + (3a_3 + a_2)x^2 + \ldots + (na_n + a_{n-1})x^{n-1} + \ldots = x \Rightarrow a_1 + a_0 = 0,\ 2a_2 + a_1 = 1,$ $3a_3 + a_2 = 0$ and in general $na_n + a_{n-1} = 0$ for $n > 2$. Since $y = 0$ when $x = 0$ we have $a_0 = 0$. Therefore

$$a_1 = 0,\ a_2 = \frac{1 - a_1}{2} = \frac{1}{2},\ a_3 = \frac{-a_2}{3} = -\frac{1}{3\cdot 2}, \ldots, a_n = \frac{-a_{n-1}}{n} = \frac{(-1)^n}{n!}$$

$$\Rightarrow y = 0 - 0x + \frac{1}{2}x^2 - \frac{1}{3\cdot 2}x^3 + \ldots + \frac{(-1)^n}{n!}x^n + \ldots = \left(1 - x + \frac{1}{2}x^2 - \frac{1}{3\cdot 2}x^3 + \ldots + \frac{(-1)^n}{n!}x^n + \ldots\right) - 1 + x$$

$$= \sum_{n=0}^{\infty}\frac{(-1)^nx^n}{n!} - 1 + x = e^{-x} + x - 1$$

75. Assume the solution has the form $y = a_0 + a_1x + a_2x^2 + \ldots + a_{n-1}x^{n-1} + a_nx^n + \ldots$

$\Rightarrow \frac{dy}{dx} = a_1 + 2a_2x + \ldots + na_nx^{n-1} + \ldots \Rightarrow \frac{dy}{dx} - y$

$= (a_1 - a_0) + (2a_2 - a_1)x + (3a_3 - a_2)x^2 + \ldots + (na_n - a_{n-1})x^{n-1} + \ldots = x \Rightarrow a_1 - a_0 = 0,\ 2a_2 - a_1 = 1,$
$3a_3 - a_2 = 0$ and in general $na_n - a_{n-1} = 0$ for $n > 2$. Since $y = 1$ when $x = 0$ we have $a_0 = 1$. Therefore

$a_1 = 1,\ a_2 = \frac{1+a_1}{2} = \frac{2}{2},\ a_3 = \frac{a_2}{3} = \frac{2}{3\cdot 2},\ a_4 = \frac{a_3}{4} = \frac{2}{4\cdot 3\cdot 2}, \ldots, a_n = \frac{a_{n-1}}{n} = \frac{2}{n!}$

$\Rightarrow y = 1 + x + \left(\frac{2}{2}\right)x^2 + \frac{2}{3\cdot 2}x^3 + \frac{2}{4\cdot 2\cdot 2}x^4 + \ldots + \frac{2}{n!}x^n + \ldots$

$= 2\left(1 + x + \frac{1}{2}x^2 + \frac{1}{3\cdot 2}x^3 + \frac{1}{4\cdot 3\cdot 2}x^4 + \ldots + \frac{1}{n!}x^n + \ldots\right) - 1 - x = 2\sum_{n=0}^{\infty} \frac{x^n}{n!} - 1 - x = 2e^x - x - 1$

76. Assume the solution has the form $y = a_0 + a_1x + a_2x^2 + \ldots + a_{n-1}x^{n-1} + a_nx^n + \ldots$

$\Rightarrow \frac{dy}{dx} = a_1 + 2a_2x + \ldots + na_nx^{n-1} + \ldots \Rightarrow \frac{dy}{dx} - y$

$= (a_1 - a_0) + (2a_2 - a_1)x + (3a_3 - a_2)x^2 + \ldots + (na_n - a_{n-1})x^{n-1} + \ldots = -x \Rightarrow a_1 - a_0 = 0,\ 2a_2 - a_1 = -1,$
$3a_3 - a_2 = 0$ and in general $na_n - a_{n-1} = 0$ for $n > 2$. Since $y = 2$ when $x = 0$ we have $a_0 = 2$. Therefore

$a_1 = 2,\ a_2 = \frac{-1+a_1}{2} = \frac{1}{2},\ a_3 = \frac{a_2}{3} = \frac{1}{3\cdot 2},\ a_4 = \frac{a_3}{4} = \frac{1}{4\cdot 3\cdot 2}, \ldots, a_n = \frac{a_{n-1}}{n} = \frac{1}{n!}$

$\Rightarrow y = 2 + 2x + \frac{1}{2}x^2 + \frac{1}{3\cdot 2}x^3 + \frac{1}{4\cdot 3\cdot 2}x^4 + \ldots + \frac{1}{n!}x^n + \ldots$

$= \left(1 + x + \frac{1}{2}x^2 + \frac{1}{3\cdot 2}x^3 + \frac{1}{4\cdot 3\cdot 2}x^4 + \ldots + \frac{1}{n!}x^n + \ldots\right) + 1 + x = \sum_{n=0}^{\infty} \frac{x^n}{n!} + 1 + x = e^x + x + 1$

77. $\int_0^{1/2} \exp(-x^3)\,dx = \int_0^{1/2} \left(1 - x^3 + \frac{x^6}{2!} - \frac{x^9}{3!} + \frac{x^{12}}{4!} + \ldots\right) dx = \left[x - \frac{x^4}{4} + \frac{x^7}{7\cdot 2!} - \frac{x^{10}}{10\cdot 3!} + \frac{x^{13}}{13\cdot 4!} - \ldots\right]_0^{1/2}$

$\approx \frac{1}{2} - \frac{1}{2^4\cdot 4} + \frac{1}{2^7\cdot 7\cdot 2!} - \frac{1}{2^{10}\cdot 10\cdot 3!} + \frac{1}{2^{13}\cdot 13\cdot 4!} - \frac{1}{2^{16}\cdot 16\cdot 5!} \approx 0.484917143$

78. $\int_0^1 x\sin(x^3)\,dx = \int_0^1 x\left(x^3 - \frac{x^9}{3!} + \frac{x^{15}}{5!} - \frac{x^{21}}{7!} + \frac{x^{27}}{9!} + \ldots\right) dx = \int_0^1 \left(x^4 - \frac{x^{10}}{3!} + \frac{x^{16}}{5!} - \frac{x^{22}}{7!} + \frac{x^{28}}{9!} - \ldots\right) dx$

$= \left[\frac{x^5}{5} - \frac{x^{11}}{11\cdot 3!} + \frac{x^{17}}{17\cdot 5!} - \frac{x^{23}}{23\cdot 7!} + \frac{x^{29}}{29\cdot 9!} - \ldots\right]_0^1 \approx 0.185330149$

79. $\int_1^{1/2} \frac{\tan^{-1}x}{x}\,dx = \int_1^{1/2} \left(1 - \frac{x^2}{3} + \frac{x^4}{5} - \frac{x^6}{7} + \frac{x^8}{9} - \frac{x^{10}}{11} + \ldots\right) dx = \left[x - \frac{x^3}{9} + \frac{x^5}{25} - \frac{x^7}{49} + \frac{x^9}{81} - \frac{x^{11}}{121} + \ldots\right]_0^{1/2}$

$\approx \frac{1}{2} - \frac{1}{9\cdot 2^3} + \frac{1}{5^2\cdot 2^5} - \frac{1}{7^2\cdot 2^7} + \frac{1}{9^2\cdot 2^9} - \frac{1}{11^2\cdot 2^{11}} + \frac{1}{13^2\cdot 2^{13}} - \frac{1}{15^2\cdot 2^{15}} + \frac{1}{17^2\cdot 2^{17}} - \frac{1}{19^2\cdot 2^{19}} + \frac{1}{21^2\cdot 2^{21}}$

≈ 0.4872223583

80. $$\int_0^{1/64} \frac{\tan^{-1} x}{\sqrt{x}}\, dx = \int_0^{1/64} \frac{1}{\sqrt{x}}\left(x - \frac{x^3}{3} + \frac{x^5}{5} - \frac{x^7}{7} + \ldots\right) dx = \int_0^{1/64} \left(x^{1/2} - \frac{1}{3}x^{5/2} + \frac{1}{5}x^{9/2} - \frac{1}{7}x^{13/2} + \ldots\right) dx$$

$$= \left[\frac{2}{3}x^{3/2} - \frac{2}{21}x^{7/2} + \frac{2}{55}x^{11/2} - \frac{2}{105}x^{15/2} + \ldots\right]_0^{1/64} = \left(\frac{2}{3\cdot 8^3} - \frac{2}{21\cdot 8^7} + \frac{2}{55\cdot 8^{11}} - \frac{2}{105\cdot 8^{15}} + \ldots\right) \approx 0.0013020379$$

81. $$\lim_{x\to 0} \frac{7\sin x}{e^{2x} - 1} = \lim_{x\to 0} \frac{7\left(x - \frac{x^3}{3!} + \frac{x^5}{5!} - \ldots\right)}{\left(2x + \frac{2^2x^2}{2!} + \frac{2^3x^3}{3!} + \ldots\right)} = \lim_{x\to 0} \frac{7\left(1 - \frac{x^2}{3!} + \frac{x^4}{5!} - \ldots\right)}{\left(2 + \frac{2^2x}{2!} + \frac{2^3x^2}{3!} + \ldots\right)} = \frac{7}{2}$$

82. $$\lim_{\theta\to 0} \frac{e^\theta - e^{-\theta} - 2\theta}{\theta - \sin\theta} = \lim_{\theta\to 0} \frac{\left(1 + \theta + \frac{\theta^2}{2!} + \frac{\theta^3}{3!} + \ldots\right) - \left(1 - \theta + \frac{\theta^2}{2!} - \frac{\theta^3}{3!} + \ldots\right) - 2\theta}{\theta - \left(\theta - \frac{\theta^3}{3!} + \frac{\theta^5}{5!} - \ldots\right)} = \lim_{\theta\to 0} \frac{2\left(\frac{\theta^3}{3!} + \frac{\theta^5}{5!} + \ldots\right)}{\left(\frac{\theta^3}{3!} - \frac{\theta^5}{5!} + \ldots\right)}$$

$$= \lim_{\theta\to 0} \frac{2\left(\frac{1}{3!} + \frac{\theta^2}{5!} + \ldots\right)}{\left(\frac{1}{3!} - \frac{\theta^2}{5!} + \ldots\right)} = 2$$

83. $$\lim_{t\to 0} \left(\frac{1}{2 - 2\cos t} - \frac{1}{t^2}\right) = \lim_{t\to 0} \frac{t^2 - 2 + 2\cos t}{2t^2(1 - \cos t)} = \lim_{t\to 0} \frac{t^2 - 2 + 2\left(1 - \frac{t^2}{2} + \frac{t^4}{4!} - \ldots\right)}{2t^2\left(1 - 1 + \frac{t^2}{2} - \frac{t^4}{4!} + \ldots\right)} = \lim_{t\to 0} \frac{2\left(\frac{t^4}{4!} - \frac{t^6}{6!} + \ldots\right)}{\left(t^4 - \frac{2t^6}{4!} + \ldots\right)}$$

$$= \lim_{t\to 0} \frac{2\left(\frac{1}{4!} - \frac{t^2}{6!} + \ldots\right)}{\left(1 - \frac{2t^2}{4!} + \ldots\right)} = \frac{1}{12}$$

84. $$\lim_{h\to 0} \frac{\left(\frac{\sin h}{h}\right) - \cos h}{h^2} = \lim_{h\to 0} \frac{\left(1 - \frac{h^2}{3!} + \frac{h^4}{5!} - \ldots\right) - \left(1 - \frac{h^2}{2!} + \frac{h^4}{4!} - \ldots\right)}{h^2}$$

$$= \lim_{h\to 0} \frac{\left(\frac{h^2}{2!} - \frac{h^2}{3!} + \frac{h^4}{5!} - \frac{h^4}{4!} + \frac{h^6}{6!} - \frac{h^6}{7!} + \ldots\right)}{h^2} = \lim_{h\to 0} \left(\frac{1}{2!} - \frac{1}{3!} + \frac{h^2}{5!} - \frac{h^2}{4!} + \frac{h^4}{6!} - \frac{h^4}{7!} + \ldots\right) = \frac{1}{3}$$

85. $$\lim_{z\to 0} \frac{1 - \cos^2 z}{\ln(1 - z) + \sin z} = \lim_{z\to 0} \frac{1 - \left(1 - z^2 + \frac{z^4}{3} - \ldots\right)}{\left(-z - \frac{z^2}{2} - \frac{z^3}{3} - \ldots\right) + \left(z - \frac{z^3}{3!} + \frac{z^5}{5!} - \ldots\right)} = \lim_{z\to 0} \frac{\left(z^2 - \frac{z^4}{3} + \ldots\right)}{\left(-\frac{z^2}{2} - \frac{2z^3}{3} - \frac{z^4}{4} - \ldots\right)}$$

$$= \lim_{z\to 0} \frac{\left(1 - \frac{z^2}{3} + \ldots\right)}{\left(-\frac{1}{2} - \frac{2z}{3} - \frac{z^2}{4} - \ldots\right)} = -2$$

86. $\lim_{y\to 0} \frac{y^2}{\cos y - \cosh y} = \lim_{y\to 0} \frac{y^2}{\left(1-\frac{y^2}{2}+\frac{y^4}{4!}-\frac{y^6}{6!}+\ldots\right)-\left(1+\frac{y^2}{2!}+\frac{y^4}{4!}+\frac{y^6}{6!}+\ldots\right)} = \lim_{y\to 0} \frac{y^2}{\left(-\frac{2y^2}{2}-\frac{2y^6}{6!}-\ldots\right)}$

$= \lim_{y\to 0} \frac{1}{\left(-1-\frac{2y^4}{6!}-\ldots\right)} = -1$

87. $\lim_{x\to 0} \left(\frac{\sin 3x}{x^3}+\frac{r}{x^2}+s\right) = \lim_{x\to 0} \left[\frac{\left(3x-\frac{(3x)^3}{6}+\frac{(3x)^5}{120}-\ldots\right)}{x^3}+\frac{r}{x^2}+s\right] = \lim_{x\to 0} \left(\frac{3}{x^2}-\frac{9}{2}+\frac{81x^2}{40}+\ldots+\frac{r}{x^2}+s\right) = 0$

$\Rightarrow \frac{r}{x^2}+\frac{3}{x^2} = 0$ and $s-\frac{9}{2} = 0 \Rightarrow r = -3$ and $s = \frac{9}{2}$

88. (a) $\csc x \approx \frac{1}{x}+\frac{x}{6} \Rightarrow \csc x \approx \frac{6+x^2}{6x} \Rightarrow \sin x \approx \frac{6x}{6+x^2}$

(b) The approximation $\sin x \approx \frac{6x}{6+x^2}$ is better than $\sin x \approx x$.

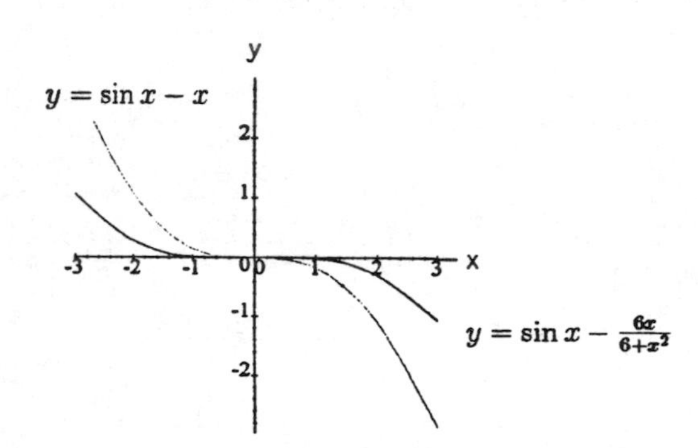

89. (a) $\sum_{n=1}^{\infty} \left(\sin\frac{1}{2n}-\sin\frac{1}{2n+1}\right) = \left(\sin\frac{1}{2}-\sin\frac{1}{3}\right)+\left(\sin\frac{1}{4}-\sin\frac{1}{5}\right)+\left(\sin\frac{1}{6}-\sin\frac{1}{7}\right)+\ldots+\left(\sin\frac{1}{2n}-\sin\frac{1}{2n+1}\right)$

$+\ldots = \sum_{n=2}^{\infty} (-1)^n \sin\frac{1}{n}$; $f(x) = \sin\frac{1}{x} \Rightarrow f'(x) = \frac{-\cos\left(\frac{1}{x}\right)}{x^2} < 0$ if $x \ge 2 \Rightarrow \sin\frac{1}{n+1} < \sin\frac{1}{n}$, and

$\lim_{n\to\infty} \sin\frac{1}{n} = 0 \Rightarrow \sum_{n=2}^{\infty} (-1)^n \sin\frac{1}{n}$ converges by the Alternating Series Test

(b) $|\text{error}| < \left|\sin\frac{1}{42}\right| \approx 0.02381$ and the sum is an underestimate because the remainder is positive

90. (a) $\sum_{n=1}^{\infty} \left(\tan\frac{1}{2n}-\tan\frac{1}{2n+1}\right) = \sum_{n=2}^{\infty} (-1)^n \tan\frac{1}{n}$ (see Exercise 89); $f(x) = \tan\frac{1}{x} \Rightarrow f'(x) = \frac{-\sec^2\left(\frac{1}{x}\right)}{x^2} < 0$

$\Rightarrow \tan\frac{1}{n+1} < \tan\frac{1}{n}$, and $\lim_{n\to\infty} \tan\frac{1}{n} = 0 \Rightarrow \sum_{n=2}^{\infty} (-1)^n \tan\frac{1}{n}$ converges by the Alternating Series Test

(b) $|\text{error}| < \left|\tan\frac{1}{42}\right| \approx 0.02382$ and the sum is an underestimate because the remainder is positive

91. $\lim\limits_{n\to\infty} \left|\frac{2\cdot5\cdot8\cdots(3n-1)(3n+2)x^{n+1}}{2\cdot4\cdot6\cdots(2n)(2n+2)}\cdot\frac{2\cdot4\cdot6\cdots(2n)}{2\cdot5\cdot8\cdots(3n-1)x^n}\right| < 1 \Rightarrow |x|\lim\limits_{n\to\infty}\left|\frac{3n+2}{2n+2}\right| < 1 \Rightarrow |x| < \frac{2}{3}$

$\Rightarrow$ the radius of convergence is $\frac{2}{3}$

92. $\lim\limits_{n\to\infty} \left|\frac{3\cdot5\cdot7\cdots(2n+1)(2n+3)(x-1)^{n+1}}{4\cdot9\cdot14\cdots(5n-1)(5n+4)}\cdot\frac{4\cdot9\cdot14\cdots(5n-1)}{3\cdot5\cdot7\cdots(2n+1)x^n}\right| < 1 \Rightarrow |x|\lim\limits_{n\to\infty}\left|\frac{2n+3}{5n+4}\right| < 1 \Rightarrow |x| < \frac{5}{2}$

$\Rightarrow$ the radius of convergence is $\frac{5}{2}$

93. $\sum\limits_{k=2}^{n} \ln\left(1-\frac{1}{k^2}\right) = \sum\limits_{k=2}^{n}\left[\ln\left(1+\frac{1}{k}\right)+\ln\left(1-\frac{1}{k}\right)\right] = \sum\limits_{k=2}^{n}\left[\ln(k+1)-\ln k+\ln(k-1)-\ln k\right]$

$= [\ln 3-\ln 2+\ln 1-\ln 2]+[\ln 4-\ln 3+\ln 2-\ln 3]+[\ln 5-\ln 4+\ln 3-\ln 4]+[\ln 6-\ln 5+\ln 4-\ln 5]$

$+\ldots+[\ln(n+1)-\ln n+\ln(n-1)-\ln n] = [\ln 1-\ln 2]+[\ln(n+1)-\ln n]$ after cancellation

$\Rightarrow \sum\limits_{k=2}^{n}\ln\left(1-\frac{1}{k^2}\right) = \ln\left(\frac{n+1}{2n}\right) \Rightarrow \sum\limits_{k=2}^{\infty}\ln\left(1-\frac{1}{k^2}\right) = \lim\limits_{n\to\infty}\ln\left(\frac{n+1}{2n}\right) = \ln\frac{1}{2}$ is the sum

94. $\sum\limits_{k=2}^{n}\frac{1}{k^2-1} = \frac{1}{2}\sum\limits_{k=2}^{n}\left(\frac{1}{k-1}-\frac{1}{k+1}\right) = \frac{1}{2}\left[\left(\frac{1}{1}-\frac{1}{3}\right)+\left(\frac{1}{2}-\frac{1}{4}\right)+\left(\frac{1}{3}-\frac{1}{5}\right)+\left(\frac{1}{4}-\frac{1}{6}\right)+\ldots+\left(\frac{1}{n-2}-\frac{1}{n}\right)\right.$

$\left.+\left(\frac{1}{n-1}-\frac{1}{n+1}\right)\right] = \frac{1}{2}\left(\frac{1}{1}+\frac{1}{2}-\frac{1}{n}-\frac{1}{n+1}\right) = \frac{1}{2}\left(\frac{3}{2}-\frac{1}{n}-\frac{1}{n+1}\right) = \frac{1}{2}\left[\frac{3n(n+1)-2(n+1)-2n}{2n(n+1)}\right] = \frac{3n^2-n-2}{4n(n+1)}$

$\Rightarrow \sum\limits_{k=2}^{\infty}\frac{1}{k^2-1} = \lim\limits_{n\to\infty}\left(\frac{3n^2-n-2}{4n^2+4n}\right) = \frac{3}{4}$

95. (a) $\lim\limits_{n\to\infty}\left|\frac{1\cdot4\cdot7\cdots(3n-2)(3n+1)x^{3n+3}}{(3n+3)!}\cdot\frac{(3n)!}{1\cdot4\cdot7\cdots(3n-2)x^{3n}}\right| < 1 \Rightarrow |x^3|\lim\limits_{n\to\infty}\frac{(3n+1)}{(3n+1)(3n+2)(3n+3)}$

$= |x^3|\cdot 0 < 1 \Rightarrow$ the radius of convergence is ∞

(b) $y = 1+\sum\limits_{n=1}^{\infty}\frac{1\cdot4\cdot7\cdots(3n-2)}{(3n)!}x^{3n} \Rightarrow \frac{dy}{dx} = \sum\limits_{n=1}^{\infty}\frac{1\cdot4\cdot7\cdots(3n-2)}{(3n-1)!}x^{3n-1}$

$\Rightarrow \frac{d^2y}{dx^2} = \sum\limits_{n=1}^{\infty}\frac{1\cdot4\cdot7\cdots(3n-2)}{(3n-2)!}x^{3n-2} = x+\sum\limits_{n=2}^{\infty}\frac{1\cdot4\cdot7\cdots(3n-5)}{(3n-3)!}x^{3n-2}$

$= x\left(1+\sum\limits_{n=1}^{\infty}\frac{1\cdot4\cdot7\cdots(3n-2)}{(3n)!}x^{3n}\right) = xy+0 \Rightarrow a = 1$ and $b = 0$

96. (a) $\frac{x^2}{1+x} = \frac{x^2}{1-(-x)} = x^2+x^2(-x)+x^2(-x)^2+x^2(-x)^3+\ldots = x^2-x^3+x^4-x^5+\ldots = \sum\limits_{n=2}^{\infty}(-1)^n x^n$ which converges absolutely for $|x| < 1$

(b) $x = 1 \Rightarrow \sum\limits_{n=2}^{\infty}(-1)^n x^n = \sum\limits_{n=2}^{\infty}(-1)^n$ which diverges

97. Yes, the series $\sum_{n=1}^{\infty} a_n b_n$ converges as we now show. Since $\sum_{n=1}^{\infty} a_n$ converges it follows that $a_n \to 0 \Rightarrow a_n < 1$ for $n >$ some index $N \Rightarrow a_n b_n < b_n$ for $n > N \Rightarrow \sum_{n=1}^{\infty} a_n b_n$ converges by the Direct Comparison Test with $\sum_{n=1}^{\infty} b_n$

98. No, the series $\sum_{n=1}^{\infty} a_n b_n$ might diverge (as it would if a_n and b_n both equaled n) or it might converge (as it would if a_n and b_n both equaled $\frac{1}{n}$).

99. $\sum_{n=1}^{\infty} (x_{n+1} - x_n) = \lim_{n\to\infty} \sum_{k=1}^{\infty} (x_{k+1} - x_k) = \lim_{n\to\infty} (x_{n+1} - x_1) = \lim_{n\to\infty} (x_{n+1}) - x_1 \Rightarrow$ both the series and sequence must either converge or diverge.

100. It converges by the Limit Comparison Test since $\lim_{n\to\infty} \frac{\left(\frac{a_n}{1+a_n}\right)}{a_n} = \lim_{n\to\infty} \frac{1}{1+a_n} = 1$ because $\sum_{n=1}^{\infty} a_n$ converges

101. Newton's method gives $x_{n+1} = x_n - \frac{(x_n-1)^{40}}{40(x_n-1)^{39}} = \frac{39}{40}x_n + \frac{1}{40}$, and if the sequence $\{x_n\}$ has the limit L, then $L = \frac{39}{40}L + \frac{1}{40} \Rightarrow L = 1$

102. (a) $\sum_{n=1}^{\infty} \frac{a_n}{n} = a_1 + \frac{a_2}{2} + \frac{a_3}{3} + \frac{a_4}{4} + \ldots \ge a_1 + \left(\frac{1}{2}\right)a_2 + \left(\frac{1}{3}+\frac{1}{4}\right)a_4 + \left(\frac{1}{5}+\frac{1}{6}+\frac{1}{7}+\frac{1}{8}\right)a_8$ $+\left(\frac{1}{9}+\frac{1}{10}+\frac{1}{11}+\ldots+\frac{1}{16}\right)a_{16} + \ldots \ge \frac{1}{2}(a_2 + a_4 + a_8 + a_{16} + \ldots)$ which is a divergent series

(b) $a_n = \frac{1}{\ln n}$ for $n \ge 2 \Rightarrow a_2 \ge a_3 \ge a_4 \ge \ldots$, and $\frac{1}{\ln 2} + \frac{1}{\ln 4} + \frac{1}{\ln 8} + \ldots = \frac{1}{\ln 2} + \frac{1}{2\ln 2} + \frac{1}{3\ln 2} + \ldots$ $= \frac{1}{\ln 2}\left(1 + \frac{1}{2} + \frac{1}{3} + \ldots\right)$ which diverges so that $1 + \sum_{n=2}^{\infty} \frac{1}{n \ln n}$ diverges by part (a)

103. (a) $T = \frac{\left(\frac{1}{2}\right)}{2}\left(0 + 2\left(\frac{1}{2}\right)^2 e^{1/2} + e\right) = \frac{1}{8}e^{1/2} + \frac{1}{4}e \approx 0.885660616$

(b) $x^2e^x = x^2\left(1 + x + \frac{x^2}{2} + \ldots\right) = x^2 + x^3 + \frac{x^4}{2} + \ldots \Rightarrow \int_0^1 \left(x^2 + x^3 + \frac{x^4}{2}\right) dx = \left[\frac{x^3}{3} + \frac{x^4}{4} + \frac{x^5}{10}\right]_0^1 = \frac{41}{60} = 0.68\overline{3}$

(c) If the second derivative is positive, the curve is concave upward and the polygonal line segments used in the trapezoidal rule lie above the curve. The trapezoidal approximation is therefore greater than the actual area under the graph.

(d) All terms in the Maclaurin series are positive. If we truncate the series, we are omitting positive terms and hence the estimate is too small.

(e) $\int_0^1 x^2e^x\,dx = \left[x^2e^x - 2xe^x + 2e^x\right]_0^1 = e - 2e + 2e - 2 = e - 2 \approx 0.7182818285$

CHAPTER 8 ADDITIONAL EXERCISES–THEORY, EXAMPLES, APPLICATIONS

1. converges since $\frac{1}{(3n-2)^{(2n+1)/2}} < \frac{1}{(3n-2)^{3/2}}$ and $\sum_{n=1}^{\infty} \frac{1}{(3n-2)^{3/2}}$ converges by the Limit Comparison Test:

$$\lim_{n\to\infty} \frac{\left(\frac{1}{n^{3/2}}\right)}{\left(\frac{1}{(3n-2)^{3/2}}\right)} = \lim_{n\to\infty} \left(\frac{3n-2}{n}\right)^{3/2} = 3^{3/2}$$

2. converges by the Integral Test: $\int_1^{\infty} \left(\tan^{-1} x\right)^2 \frac{dx}{x^2+1} = \lim_{b\to\infty} \left[\frac{\left(\tan^{-1} x\right)^3}{3}\right]_1^b = \lim_{b\to\infty} \left[\frac{\left(\tan^{-1} b\right)^3}{3} - \frac{\pi^3}{192}\right]$

$= \left(\frac{\pi^3}{24} - \frac{\pi^3}{192}\right) = \frac{7\pi^3}{192}$

3. diverges by the nth-Term Test since $\lim_{n\to\infty} a_n = \lim_{n\to\infty} (-1)^n \tanh n = \lim_{b\to\infty} (-1)^n \left(\frac{1-e^{-2n}}{1+e^{-2n}}\right) = \lim_{n\to\infty} (-1)^n$

does not exist

4. converges by the Direct Comparison Test: $n! < n^n \Rightarrow \ln(n!) < n \ln(n) \Rightarrow \frac{\ln(n!)}{\ln(n)} < n$

$\Rightarrow \log_n(n!) < n \Rightarrow \frac{\log_n(n!)}{n^3} < \frac{1}{n^2}$, which is the nth-term of a convergent p-series

5. converges by the Direct Comparison Test: $a_1 = 1 = \frac{12}{(1)(3)(2)^2}$, $a_2 = \frac{1\cdot 2}{3\cdot 4} = \frac{12}{(2)(4)(3)^2}$, $a_3 = \left(\frac{2\cdot 3}{4\cdot 5}\right)\left(\frac{1\cdot 2}{3\cdot 4}\right)$

$= \frac{12}{(3)(5)(4)^2}$, $a_4 = \left(\frac{3\cdot 4}{5\cdot 6}\right)\left(\frac{2\cdot 3}{4\cdot 5}\right)\left(\frac{1\cdot 2}{3\cdot 4}\right) = \frac{12}{(4)(6)(5)^2}, \ldots \Rightarrow 1 + \sum_{n=1}^{\infty} \frac{12}{(n+1)(n+3)(n+2)^2}$ represents the

given series and $\frac{12}{(n+1)(n+3)(n+2)^2} < \frac{12}{n^4}$, which is the nth-term of a convergent p-series

6. converges by the Ratio Test: $\lim_{n\to\infty} \frac{a_{n+1}}{a_n} = \lim_{n\to\infty} \frac{n}{(n-1)(n+1)} = 0 < 1$

7. diverges by the nth-Term Test since if $a_n \to L$ as $n \to \infty$, then $L = \frac{1}{1+L} \Rightarrow L^2 + L - 1 = 0 \Rightarrow L = \frac{-1 \pm \sqrt{5}}{2}$

$\neq 0$

8. Split the given series into $\sum_{n=1}^{\infty} \frac{1}{3^{2n+1}}$ and $\sum_{n=1}^{\infty} \frac{2n}{3^{2n}}$; the first subseries is a convergent geometric series and the

second converges by the Root Test: $\lim_{n\to\infty} \sqrt[n]{\frac{2n}{3^{2n}}} = \lim_{n\to\infty} \frac{\sqrt[n]{2}\,\sqrt[n]{n}}{9} = \frac{1\cdot 1}{9} = \frac{1}{9} < 1$

9. $f(x) = \cos x$ with $a = \frac{\pi}{3} \Rightarrow f\left(\frac{\pi}{3}\right) = 0.5$, $f'\left(\frac{\pi}{3}\right) = -\frac{\sqrt{3}}{2}$, $f''\left(\frac{\pi}{3}\right) = -0.5$, $f'''\left(\frac{\pi}{3}\right) = \frac{\sqrt{3}}{2}$, $f^{(4)}\left(\frac{\pi}{3}\right) = 0.5$;

$\cos x = \frac{1}{2} - \frac{\sqrt{3}}{2}\left(x - \frac{\pi}{3}\right) - \frac{1}{4}\left(x - \frac{\pi}{3}\right)^2 + \frac{\sqrt{3}}{12}\left(x - \frac{\pi}{3}\right)^3 + \ldots$

10. $f(x) = \sin x$ with $a = 2\pi \Rightarrow f(2\pi) = 0,\ f'(2\pi) = 1,\ f''(2\pi) = 0,\ f'''(2\pi) = -1,\ f^{(4)}(2\pi) = 0,\ f^{(5)}(2\pi) = 1,$
$f^{(6)}(2\pi) = 0,\ f^{(7)}(2\pi) = -1;\ \sin x = (x - 2\pi) - \frac{(x-2\pi)^3}{3!} + \frac{(x-2\pi)^5}{5!} - \frac{(x-2\pi)^7}{7!} + \ldots$

11. $e^x = 1 + x + \frac{x^2}{2!} + \frac{x^3}{2!} + \ldots$ with $a = 0$

12. $f(x) = \ln x$ with $a = 1 \Rightarrow f(1) = 0,\ f'(1) = 1,\ f''(1) = -1,\ f'''(1) = 2,\ f^{(4)}(1) = -6;$
$\ln x = (x-1) - \frac{(x-1)^2}{2} + \frac{(x-1)^3}{3} - \frac{(x-1)^4}{4} + \ldots$

13. $f(x) = \cos x$ with $a = 22\pi \Rightarrow f(22\pi) = 1,\ f'(22\pi) = 0,\ f''(22\pi) = -1,\ f'''(22\pi) = 0,\ f^{(4)}(22\pi) = 1,$
$f^{(5)}(22\pi) = 0,\ f^{(6)}(22\pi) = -1;\ \cos x = 1 - \frac{1}{2}(x - 22\pi)^2 + \frac{1}{4!}(x-22\pi)^4 - \frac{1}{6!}(x - 22\pi)^6 + \ldots$

14. $f(x) = \tan^{-1} x$ with $a = 1 \Rightarrow f(1) = \frac{\pi}{4},\ f'(1) = \frac{1}{2},\ f''(1) = -\frac{1}{2},\ f'''(1) = \frac{1}{2};$
$\tan^{-1} x = \frac{\pi}{4} + \frac{(x-1)}{2} - \frac{(x-1)^2}{4} + \frac{(x-1)^3}{12} + \ldots$

15. Yes, the sequence converges: $c_n = (a^n + b^n)^{1/n} \Rightarrow c_n = b\left(\left(\frac{a}{b}\right)^n + 1\right)^{1/n} \Rightarrow \lim_{n\to\infty} c_n = \lim_{n\to\infty} b\left(\left(\frac{a}{b}\right)^n + 1\right)^{1/n} = b$
since $0 < a < b$

16. $1 + \frac{2}{10} + \frac{3}{10^2} + \frac{7}{10^3} + \frac{2}{10^4} + \frac{3}{10^5} + \frac{7}{10^6} + \ldots = 1 + \sum_{n=1}^{\infty} \frac{2}{10^{3n-2}} + \sum_{n=1}^{\infty} \frac{3}{10^{3n-1}} + \sum_{n=1}^{\infty} \frac{7}{10^{3n}}$

$$= 1 + \sum_{n=0}^{\infty} \frac{2}{10^{3n+1}} + \sum_{n=0}^{\infty} \frac{3}{10^{3n+2}} + \sum_{n=0}^{\infty} \frac{7}{10^{3n+3}} = 1 + \frac{\left(\frac{2}{10}\right)}{1 - \left(\frac{1}{10}\right)^3} + \frac{\left(\frac{3}{10^2}\right)}{1 - \left(\frac{1}{10}\right)^3} + \frac{\left(\frac{7}{10^3}\right)}{1 - \left(\frac{1}{10}\right)^3}$$

$$= 1 + \frac{200}{999} + \frac{30}{999} + \frac{7}{999} = \frac{999 + 237}{999} = \frac{412}{333}$$

17. $s_n = \sum_{k=0}^{n-1} \int_k^{k+1} \frac{dx}{1+x^2} \Rightarrow s_n = \int_0^1 \frac{dx}{1+x^2} + \int_1^2 \frac{dx}{1+x^2} + \ldots + \int_{n-1}^{n} \frac{dx}{1+x^2} \Rightarrow s_n = \int_0^n \frac{dx}{1+x^2}$
$\Rightarrow \lim_{n\to\infty} s_n = \lim_{n\to\infty} \left(\tan^{-1} n - \tan^{-1} 0\right) = \frac{\pi}{2}$

18. $\lim_{n\to\infty} \left|\frac{u_{n+1}}{u_n}\right| = \lim_{n\to\infty} \left|\frac{(n+1)x^{n+1}}{(n+2)(2x+1)^{n+1}} \cdot \frac{(n+1)(2x+1)^n}{nx^n}\right| = \lim_{n\to\infty} \left|\frac{x}{2x+1} \cdot \frac{(n+1)^2}{n(n+2)}\right| = \left|\frac{x}{2x+1}\right| < 1$
$\Rightarrow |x| < |2x+1|$; if $x > 0$, $|x| < |2x+1| \Rightarrow x < 2x + 1 \Rightarrow x > -1$; if $-\frac{1}{2} < x < 0$, $|x| < |2x+1|$
$\Rightarrow -x < 2x + 1 \Rightarrow 3x > -1 \Rightarrow x > -\frac{1}{3}$; if $x < -\frac{1}{2}$, $|x| < |2x+1| \Rightarrow -x < -2x - 1 \Rightarrow x < -1$. Therefore,
the series converges absolutely for $x < -1$ and $x > -\frac{1}{3}$.

19. (a) Each A_{n+1} fits into the corresponding upper triangular region, whose vertices are:
$(n, f(n) - f(n+1))$, $(n+1, f(n+1))$ and $(n, f(n))$ along the line whose slope is $f(n+2) - f(n+1)$.
All the A_n's fit into the first upper triangular region whose area is $\frac{f(1) - f(2)}{2} \Rightarrow \sum_{n=1}^{\infty} A_n < \frac{f(1) - f(2)}{2}$

(b) If $A_k = \dfrac{f(k+1)+f(k)}{2} - \int_k^{k+1} f(x)\,dx$, then

$$\sum_{k=1}^{n-1} A_k = \frac{f(1)+f(2)+f(2)+f(3)+f(3)+\ldots+f(n-1)+f(n)}{2} - \int_1^2 f(x)\,dx - \int_2^3 f(x)\,dx - \ldots - \int_{n-1}^{n} f(x)\,dx$$

$$= \frac{f(1)+f(n)}{2} + \sum_{k=2}^{n-1} f(k) - \int_1^n f(x)\,dx \Rightarrow \sum_{k=1}^{n-1} A_k = \sum_{k=1}^{n} f(k) - \frac{f(1)+f(n)}{2} - \int_1^n f(x)\,dx < \frac{f(1)-f(2)}{2}, \text{ from}$$

part (a). The sequence $\left\{\sum_{k=1}^{n-1} A_k\right\}$ is bounded above and increasing, so it converges and the limit in question must exist.

(c) From part (b) we have $\sum_{k=1}^{\infty} f(k) - \int_1^n f(x)\,dx < f(1) - \dfrac{f(2)}{2} + \dfrac{f(n)}{2}$

$$\Rightarrow \lim_{n\to\infty}\left[\sum_{k=1}^{n} f(k) - \int_1^n f(x)\,dx\right] < \lim_{n\to\infty}\left[f(1) - \frac{f(2)}{2} + \frac{f(n)}{2}\right] = f(1) - \frac{f(2)}{2}.$$ The sequence

$\left\{\sum_{k=1}^{n} f(k) - \int_1^n f(x)\,dx\right\}$ is bounded and increasing, so it converges and the limit in question must exist.

20. The number of triangles removed at stage n is 3^{n-1}; the side length at stage n is $\frac{b}{2^{n-1}}$; the area of a triangle at stage n is $\frac{\sqrt{3}}{4}\left(\frac{b}{2^{n-1}}\right)^2$.

(a) $\frac{\sqrt{3}}{4}b^2 + 3\frac{\sqrt{3}}{4}\left(\frac{b^2}{2^2}\right) + 3^2\frac{\sqrt{3}}{4}\left(\frac{b^2}{2^4}\right) + 3^3\frac{\sqrt{3}}{4}\left(\frac{b^2}{2^6}\right) + \ldots = \frac{\sqrt{3}}{4}b^2 \sum_{n=0}^{\infty} \frac{3^n}{2^{2n}} = \frac{\sqrt{3}}{4}b^2 \sum_{n=0}^{\infty}\left(\frac{3}{4}\right)^n$

(b) a geometric series with sum $\dfrac{\left(\frac{\sqrt{3}}{4}b^2\right)}{1-\left(\frac{3}{4}\right)} = \sqrt{3}b^2$

(c) No; for instance,the three vertices of the original triangle are not removed. However the total area removed is $\sqrt{3}b^2$ which equals the area of the original triangle. Thus the set of points not removed has area 0.

21. (a) No, the limit does not appear to depend on the value of the constant a

(b) Yes, the limit depends on the value of b

(c) $s = \left(1 - \frac{\cos\left(\frac{a}{n}\right)}{n}\right)^n \Rightarrow \log s = \dfrac{\log\left(1 - \frac{\cos\left(\frac{a}{n}\right)}{n}\right)}{\left(\frac{1}{n}\right)} \Rightarrow \lim_{n\to\infty} \log s = \dfrac{\left(\dfrac{1}{1-\frac{\cos\left(\frac{a}{n}\right)}{n}}\right)\left(\dfrac{-\frac{a}{n}\sin\left(\frac{a}{n}\right)+\cos\left(\frac{a}{n}\right)}{n^2}\right)}{\left(-\frac{1}{n^2}\right)}$

$$= \lim_{n\to\infty} \frac{\frac{a}{n}\sin\left(\frac{a}{n}\right) - \cos\left(\frac{a}{n}\right)}{1 - \frac{\cos\left(\frac{a}{n}\right)}{n}} = \frac{0-1}{1-0} = -1 \Rightarrow \lim_{n\to\infty} s = e^{-1} \approx 0.3678794412; \text{ similarly,}$$

$$\lim_{n\to\infty}\left(1 - \frac{\cos\left(\frac{a}{n}\right)}{bn}\right)^n = e^{-1/b}$$

22. $\sum_{n=1}^{\infty} a_n$ converges $\Rightarrow \lim_{n\to\infty} a_n = 0$; $\lim_{n\to\infty} \left[\left(\frac{1+\sin a_n}{2}\right)^n\right]^{1/n} = \lim_{n\to\infty} \left(\frac{1+\sin a_n}{2}\right) = \frac{1+\sin(\lim_{n\to\infty} a_n)}{2} = \frac{1+\sin 0}{2}$

$= \frac{1}{2} \Rightarrow$ the series converges by the nth-Root Test

23. $\lim_{n\to\infty} \left|\frac{u_{n+1}}{u_n}\right| < 1 \Rightarrow \lim_{n\to\infty} \left|\frac{b^{n+1}x^{n+1}}{\ln(n+1)} \cdot \frac{\ln n}{b^n x^n}\right| < 1 \Rightarrow |bx| < 1 \Rightarrow -\frac{1}{b} < x < \frac{1}{b} = 5 \Rightarrow b = \pm\frac{1}{5}$

24. A polynomial has only a finite number of nonzero terms in its Taylor series, but the functions sin x, ln x and e^x have infinitely many nonzero terms in their Taylor expansions.

25. $\lim_{x\to 0} \frac{\sin(ax) - \sin x - x}{x^3} = \lim_{x\to 0} \frac{\left(ax - \frac{a^3x^3}{3!} + \dots\right) - \left(x - \frac{x^3}{3!} + \dots\right) - x}{x^3}$

$= \lim_{x\to 0} \left[\frac{a-2}{x^2} - \frac{a^3}{3!} + \frac{1}{3!} - \left(\frac{a^5}{5!} - \frac{1}{5!}\right)x^2 + \dots\right]$ is finite if $a - 2 = 0 \Rightarrow a = 2$;

$\lim_{x\to 0} \frac{\sin 2x - \sin x - x}{x^3} = -\frac{2^3}{3!} + \frac{1}{3!} = -\frac{7}{6}$

26. $\lim_{x\to 0} \frac{\cos ax - b}{2x^2} = -1 \Rightarrow \lim_{x\to 0} \frac{\left(1 - \frac{a^2x^2}{2} + \frac{a^4x^4}{4!} - \dots\right) - b}{2x^2} = -1 \Rightarrow \lim_{x\to 0} \left(\frac{1-b}{2x^2} - \frac{a^2}{4} + \frac{a^2x^2}{48} - \dots\right) = -1$

$\Rightarrow b = 1$ and $a = \pm 2$

27. (a) $\frac{u_n}{u_{n+1}} = \frac{(n+1)^2}{n^2} = 1 + \frac{2}{n} + \frac{1}{n^2} \Rightarrow C = 2 > 1$ and $\sum_{n=1}^{\infty} \frac{1}{n^2}$ converges

(b) $\frac{u_n}{u_{n+1}} = \frac{n+1}{n} = 1 + \frac{1}{n} + \frac{0}{n^2} \Rightarrow C = 1 \le 1$ and $\sum_{n=1}^{\infty} \frac{1}{n}$ diverges

28. $\frac{u_n}{u_{n+1}} = \frac{2n(2n+1)}{(2n-1)^2} = \frac{4n^2+2n}{4n^2-4n+1} = 1 + \frac{\left(\frac{6}{4}\right)}{n} + \frac{5}{4n^2-4n+1} = 1 + \frac{\left(\frac{3}{2}\right)}{n} + \frac{\left[\frac{5n^2}{(4n^2-4n+1)}\right]}{n^2}$ after long division

$\Rightarrow C = \frac{3}{2} > 1$ and $|f(n)| = \frac{5n^2}{4n^2-4n+1} = \frac{5}{\left(4 - \frac{4}{n} + \frac{1}{n^2}\right)} \le 5 \Rightarrow \sum_{n=1}^{\infty} u_n$ converges by Raabe's Test

29. (a) $\sum_{n=1}^{\infty} a_n = L \Rightarrow a_n^2 \le a_n \sum_{n=1}^{\infty} a_n = a_n L \Rightarrow \sum_{n=1}^{\infty} a_n^2$ converges by the Direct Comparison Test

(b) converges by the Limit Comparison Test: $\lim_{n\to\infty} \frac{\left(\frac{a_n}{1-a_n}\right)}{a_n} = \lim_{n\to\infty} \frac{1}{1-a_n} = 1$ since $\sum_{n=1}^{\infty} a_n$ converges and

therefore $\lim_{x\to\infty} a_n = 0$

30. If $0 < a_n < 1$ then $|\ln(1-a_n)| = -\ln(1-a_n) = a_n + \frac{a_n^2}{2} + \frac{a_n^3}{3} + \ldots < a_n + a_n^2 + a_n^3 + \ldots = \frac{a_n}{1-a_n}$, a positive term of a convergent series, by the Limit Comparison Test and Exercise 29b

31. $(1-x)^{-1} = 1 + \sum_{n=1}^{\infty} x^n$ where $|x| < 1 \Rightarrow \frac{1}{(1-x)^2} = \frac{d}{dx}(1-x)^{-1} = \sum_{n=1}^{\infty} nx^{n-1}$ and when $x = \frac{1}{2}$ we have $4 = 1 + 2\left(\frac{1}{2}\right) + 3\left(\frac{1}{2}\right)^2 + 4\left(\frac{1}{2}\right)^3 + \ldots + n\left(\frac{1}{2}\right)^{n-1} + \ldots$

32. (a) $\sum_{n=1}^{\infty} x^{n+1} = \frac{x^2}{1-x} \Rightarrow \sum_{n=1}^{\infty} (n+1)x^n = \frac{2x-x^2}{(1-x)^2} \Rightarrow \sum_{n=1}^{\infty} n(n+1)x^{n-1} = \frac{2}{(1-x)^3} \Rightarrow \sum_{n=1}^{\infty} n(n+1)x^n = \frac{2x}{(1-x)^3}$

$\Rightarrow \sum_{n=1}^{\infty} \frac{n(n+1)}{x^n} = \frac{\frac{2}{x}}{\left(1-\frac{1}{x}\right)^3} = \frac{2x^2}{(x-1)^3}$, $|x| > 1$

(b) $x = \sum_{n=1}^{\infty} \frac{n(n+1)}{x^n} \Rightarrow x = \frac{2x^2}{(x-1)^3} \Rightarrow x^3 - 3x^2 + x - 1 = 0 \Rightarrow x = 1 + \left(1 + \frac{\sqrt{57}}{9}\right)^{1/3} + \left(1 - \frac{\sqrt{57}}{9}\right)^{1/3}$ ≈ 2.769292, using a CAS or calculator

33. The sequence $\{x_n\}$ converges to $\frac{\pi}{2}$ from below so $\epsilon_n = \frac{\pi}{2} - x_n > 0$ for each n. By the Alternating Series Estimation Theorem $\epsilon_{n+1} \approx \frac{1}{3!}(\epsilon_n)^3$ with $|\text{error}| < \frac{1}{5!}(\epsilon_n)^5$, and since the remainder is negative this is an overestimate $\Rightarrow 0 < \epsilon_{n+1} < \frac{1}{6}(\epsilon_n)^3$.

34. Yes, the series $\sum_{n=1}^{\infty} \ln(1+a_n)$ converges by the Direct Comparison Test: $1 + a_n < 1 + a_n + \frac{a_n^2}{2!} + \frac{a_n^3}{3!} + \ldots$ $\Rightarrow 1 + a_n < e^{a_n} \Rightarrow \ln(1+a_n) < a_n$

35. (a) $\frac{1}{(1-x)^2} = \frac{d}{dx}\left(\frac{1}{1-x}\right) = \frac{d}{dx}(1 + x + x^2 + x^3 + \ldots) = 1 + 2x + 3x^2 + 4x^3 + \ldots = \sum_{n=1}^{\infty} nx^{n-1}$

(b) from part (a) we have $\sum_{n=1}^{\infty} n\left(\frac{5}{6}\right)^{n-1}\left(\frac{1}{6}\right) = \left(\frac{1}{6}\right)\left[\frac{1}{1-\left(\frac{5}{6}\right)}\right] = 6$

(c) from part (a) we have $\sum_{n=1}^{\infty} np^{n-1}q = \frac{q}{(1-p)^2} = \frac{q}{q^2} = \frac{1}{q}$

36. (a) $\sum_{k=1}^{\infty} p_k = \sum_{k=1}^{\infty} 2^{-k} = \frac{\left(\frac{1}{2}\right)}{1-\left(\frac{1}{2}\right)} = 1$ and $E(x) = \sum_{k=1}^{\infty} kp_k = \sum_{k=1}^{\infty} k2^{-k} = \frac{1}{2}\sum_{k=1}^{\infty} k2^{1-k} = \left(\frac{1}{2}\right)\frac{1}{\left[1-\left(\frac{1}{2}\right)\right]^2} = 2$ by Exercise 35(a)

(b) $\sum_{k=1}^{\infty} p_k = \sum_{k=1}^{\infty} \frac{5^{k-1}}{6^k} = \frac{1}{5}\sum_{k=1}^{\infty} \left(\frac{5}{6}\right)^k = \left(\frac{1}{5}\right)\left[\frac{\left(\frac{5}{6}\right)}{1-\left(\frac{5}{6}\right)}\right] = 1$ and $E(x) = \sum_{k=1}^{\infty} kp_k = \sum_{k=1}^{\infty} k\,\frac{5^{k-1}}{6^k} = \frac{1}{6}\sum_{k=1}^{\infty} k\left(\frac{5}{6}\right)^{k-1}$

$= \left(\frac{1}{6}\right)\frac{1}{\left[1-\left(\frac{5}{6}\right)\right]^2} = 6$

(c) $\sum_{k=1}^{\infty} p_k = \sum_{k=1}^{\infty} \frac{1}{k(k+1)} = \sum_{k=1}^{\infty} \left(\frac{1}{k} - \frac{1}{k+1}\right) = \lim_{k\to\infty} \left(1 - \frac{1}{k+1}\right) = 1$ and $E(x) = \sum_{k=1}^{\infty} kp_k = \sum_{k=1}^{\infty} k\left(\frac{1}{k(k+1)}\right)$

$= \sum_{k=1}^{\infty} \frac{1}{k+1}$, a divergent series so that E(x) does not exist

37. (a) $R_n = C_0e^{-kt_0} + C_0e^{-2kt_0} + \ldots + C_0e^{-nkt_0} = \frac{C_0e^{-kt_0}\left(1 - e^{-nkt_0}\right)}{1 - e^{-kt_0}} \Rightarrow R = \lim_{n\to\infty} R_n = \frac{C_0e^{-kt_0}}{1 - e^{-kt_0}} = \frac{C_0}{e^{kt_0} - 1}$

(b) $R_n = \frac{e^{-1}\left(1 - e^{-n}\right)}{1 - e^{-1}} \Rightarrow R_1 = e^{-1} \approx 0.36787944$ and $R_{10} = \frac{e^{-1}\left(1 - e^{-10}\right)}{1 - e^{-1}} \approx 0.58195028$;

$R = \frac{1}{e - 1} \approx 0.58197671$; $R - R_{10} \approx 0.00002643 \Rightarrow \frac{R - R_{10}}{R} < 0.0001$

(c) $R_n = \frac{e^{-.1}\left(1 - e^{-.1n}\right)}{1 - e^{-.1}}$, $\frac{R}{2} = \frac{1}{2}\left(\frac{1}{e^{.1} - 1}\right) \approx 4.7541659$; $R_n > \frac{R}{2} \Rightarrow \frac{1 - e^{-.1n}}{e^{.1} - 1} > \left(\frac{1}{2}\right)\left(\frac{1}{e^{.1} - 1}\right)$

$\Rightarrow 1 - e^{-n/10} > \frac{1}{2} \Rightarrow e^{-n/10} < \frac{1}{2} \Rightarrow -\frac{n}{10} < \ln\left(\frac{1}{2}\right) \Rightarrow \frac{n}{10} > -\ln\left(\frac{1}{2}\right) \Rightarrow n > 6.93 \Rightarrow n = 7$

38. (a) $R = \frac{C_0}{e^{kt_0} - 1} \Rightarrow Re^{kt_0} = R + C_0 = C_H \Rightarrow e^{kt_0} = \frac{C_H}{C_L} \Rightarrow t_0 = \frac{1}{k} \ln\left(\frac{C_H}{C_L}\right)$

(b) $t_0 = \frac{1}{0.05} \ln e = 20$ hrs

(c) Give an initial dose that produces a concentration of 2 mg/ml followed every $t_0 = \frac{1}{0.02} \ln\left(\frac{2}{0.5}\right) \approx 69.31$ hrs by a dose that raises the concentration by 1.5 mg/ml

(d) $t_0 = \frac{1}{0.2} \ln\left(\frac{0.1}{0.03}\right) = 5 \ln\left(\frac{10}{3}\right) \approx 6$ hrs

39. The convergence of $\sum_{n=1}^{\infty} |a_n|$ implies that $\lim_{n\to\infty} |a_n| = 0$. Let $N > 0$ be such that $|a_n| < \frac{1}{2} \Rightarrow 1 - |a_n| > \frac{1}{2}$

$\Rightarrow \frac{|a_n|}{1 - |a_n|} < 2|a_n|$ for all $n > N$. Now $\left|\ln(1 + a_n)\right| = \left|a_n - \frac{a_n^2}{2} + \frac{a_n^3}{3} - \frac{a_n^4}{4} + \ldots\right| \le |a_n| + \left|\frac{a_n^2}{2}\right| + \left|\frac{a_n^3}{3}\right| + \left|\frac{a_n^4}{4}\right| + \ldots$

$< |a_n| + |a_n|^2 + |a_n|^3 + |a_n|^4 + \ldots = \frac{|a_n|}{1 - |a_n|} < 2|a_n|$. Therefore $\sum_{n=1}^{\infty} \ln(1 + a_n)$ converges by the Direct Comparison Test since $\sum_{n=1}^{\infty} |a_n|$ converges.

40. $\sum_{n=3}^{\infty} \frac{1}{n \ln n (\ln (\ln n))^p}$ converges if $p > 1$ and diverges otherwise by the Integral Test: when $p = 1$ we have

$\lim_{b\to\infty} \int_3^b \frac{dx}{x \ln x (\ln (\ln x))} = \lim_{b\to\infty} \left[\ln (\ln (\ln x))\right]_3^b = \infty$; when $p \ne 1$ we have $\lim_{b\to\infty} \int_3^b \frac{dx}{x \ln x (\ln (\ln x))^p}$

$$= \lim_{b\to\infty} \left[\frac{(\ln (\ln x))^{-p+1}}{1 - p}\right]_3^b = \begin{cases} \frac{(\ln (\ln 3))^{-p+1}}{1 - p} & \text{if } p > 1 \\ \infty & \text{if } p < 1 \end{cases}$$

41. (a) $s_{2n+1} = \frac{c_1}{1} + \frac{c_2}{2} + \frac{c_3}{3} + \ldots + \frac{c_{2n+1}}{2n+1} = \frac{t_1}{1} + \frac{t_2 - t_1}{2} + \frac{t_3 - t_2}{3} + \ldots + \frac{t_{2n+1} - t_{2n}}{2n+1}$

$= t_1\left(1 - \frac{1}{2}\right) + t_2\left(\frac{1}{2} - \frac{1}{3}\right) + \ldots + t_{2n}\left(\frac{1}{2n} - \frac{1}{2n+1}\right) + \frac{t_{2n+1}}{2n+1} = \sum_{k=1}^{2n} \frac{t_k}{k(k+1)} + \frac{t_{2n+1}}{2n+1}.$

(b) $\{c_n\} = \{(-1)^n\} \Rightarrow \sum_{n=1}^{\infty} \frac{(-1)^n}{n}$ converges

(c) $\{c_n\} = \{1, -1, -1, 1, 1, -1, -1, 1, 1, \ldots\} \Rightarrow$ the series $1 - \frac{1}{2} - \frac{1}{3} + \frac{1}{4} + \frac{1}{5} - \frac{1}{6} - \frac{1}{7} + \ldots$ converges

42. (a) $(1 - t + t^2 - t^3 + \ldots + (-1)^n t^n)(1 + t) = 1 - t + t^2 - t^3 + \ldots + (-1)^n t^n + t - t^2 + t^3 - t^4 + \ldots + (-1)^n t^{n+1}$

$= 1 + (-1)^n t^{n+1} \Rightarrow 1 - t + t^2 - t^3 + \ldots + (-1)^n t^n - \frac{(-1)^n t^{n+1}}{1+t} = \frac{1}{1+t}$

(b) $\int_0^x \frac{1}{1+t}\,dt = \int_0^x \left[1 - t + t^2 + \ldots + (-1)^n t^n + \frac{(-1)^{n+1} t^{n+1}}{1+t}\right] dt \Rightarrow \left[\ln|1+t|\right]_0^x$

$= \left[t - \frac{t^2}{2} + \frac{t^3}{3} + \ldots + \frac{(-1)^n t^{n+1}}{n+1}\right]_0^x + \int_0^x \frac{(-1)^{n+1} t^{n+1}}{n+1}\,dt \Rightarrow \ln|1+x|$

$= x - \frac{x^2}{2} + \frac{x^3}{3} - \ldots + \frac{(-1)^n x^{n+1}}{n+1} + R_{n+1}$, where $R_{n+1} = \int_0^x \frac{(-1)^{n+1} t^{n+1}}{n+1}\,dt$

(c) $x > 0$ and $R_{n+1} = (-1)^{n+1} \int_0^x \frac{t^{n+1}}{1+t}\,dt \Rightarrow |R_{n+1}| = \int_0^x \frac{t^{n+1}}{1+t}\,dt \le \int_0^x t^{n+1}\,dt = \frac{x^{n+2}}{n+2}$

(d) $-1 < x < 0$ and $R_{n+1} = (-1)^{n+1} \int_0^x \frac{t^{n+1}}{1+t}\,dt \Rightarrow |R_{n+1}| = \left|\int_0^x \frac{t^{n+1}}{1+t}\,dt\right| \le \int_0^x \left|\frac{t^{n+1}}{1+t}\right| dt$

$\le \int_0^x \frac{|t|^{n+1}}{1-|x|}\,dx = \frac{|x|^{n+2}}{(1-|x|)(n+2)}$ since $|1+t| \ge 1 - |x|$

(e) From part (d) we have $|R_{n+1}| \le \frac{|x|^{n+2}}{(1-|x|)(n+2)} \Rightarrow$ the given series converges since

$\lim_{n\to\infty} \frac{|x|^{n+2}}{(1-|x|)(n+2)} = 0 \Rightarrow |R_{n+1}| \to 0$ when $|x| < 1$

NOTES:

CHAPTER 9 CONIC SECTIONS, PARAMETRIZED CURVES, AND POLAR COORDINATES

9.1 CONIC SECTIONS AND QUADRATIC EQUATIONS

1. $x = \frac{y^2}{8} \Rightarrow 4p = 8 \Rightarrow p = 2$; focus is $(2,0)$, directrix is $x = -2$

2. $x = -\frac{y^2}{4} \Rightarrow 4p = 4 \Rightarrow p = 1$; focus is $(-1,0)$, directrix is $x = 1$

3. $y = -\frac{x^2}{6} \Rightarrow 4p = 6 \Rightarrow p = \frac{3}{2}$; focus is $\left(0,-\frac{3}{2}\right)$, directrix is $y = \frac{3}{2}$

4. $y = \frac{x^2}{2} \Rightarrow 4p = 2 \Rightarrow p = \frac{1}{2}$; focus is $\left(0,\frac{1}{2}\right)$, directrix is $y = -\frac{1}{2}$

5. $\frac{x^2}{4} - \frac{y^2}{9} = 1 \Rightarrow c = \sqrt{4+9} = \sqrt{13} \Rightarrow$ foci are $(\pm\sqrt{13},0)$; vertices are $(\pm 2,0)$; asymptotes are $y = \pm\frac{3}{2}x$

6. $\frac{x^2}{4} + \frac{y^2}{9} = 1 \Rightarrow c = \sqrt{9-4} = \sqrt{5} \Rightarrow$ foci are $(0, \pm\sqrt{5})$; vertices are $(0, \pm 3)$

7. $\frac{x^2}{2} + y^2 = 1 \Rightarrow c = \sqrt{2-1} = 1 \Rightarrow$ foci are $(\pm 1,0)$; vertices are $(\pm\sqrt{2},0)$

8. $\frac{y^2}{4} - x^2 = 1 \Rightarrow c = \sqrt{4+1} = \sqrt{5} \Rightarrow$ foci are $(0, \pm\sqrt{5})$; vertices are $(0, \pm 2)$; asymptotes are $y = \pm 2x$

9. $y^2 = 12x \Rightarrow x = \frac{y^2}{12} \Rightarrow 4p = 12 \Rightarrow p = 3$;

   focus is $(3,0)$, directrix is $x = -3$

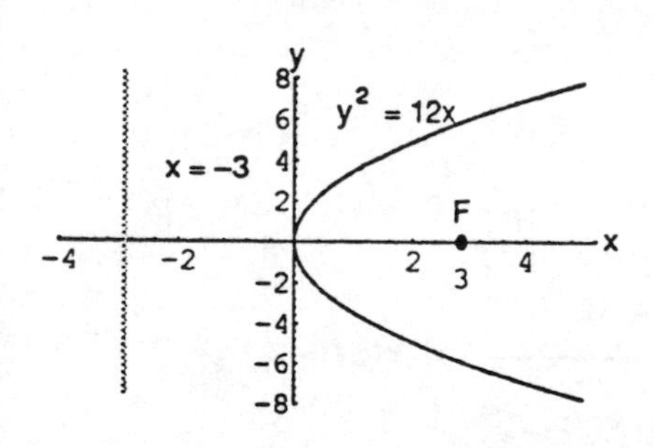

10. $x^2 = 6y \Rightarrow y = \frac{x^2}{6} \Rightarrow 4p = 6 \Rightarrow p = \frac{3}{2}$;

    focus is $\left(0,\frac{3}{2}\right)$, directrix is $y = -\frac{3}{2}$

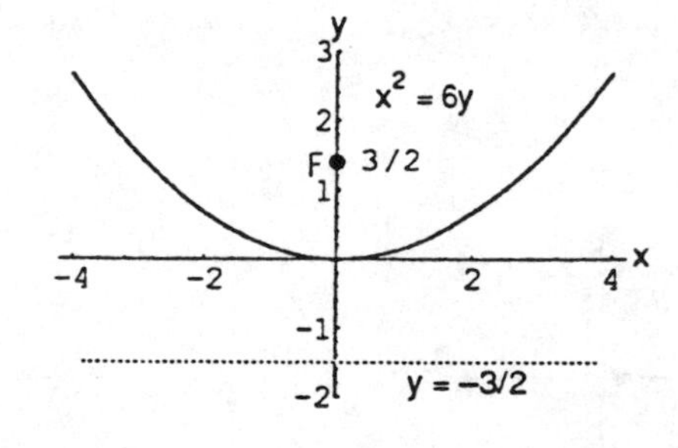

11. $x^2 = -8y \Rightarrow y = \frac{x^2}{-8} \Rightarrow 4p = 8 \Rightarrow p = 2;$

focus is $(0, -2)$, directrix is $y = 2$

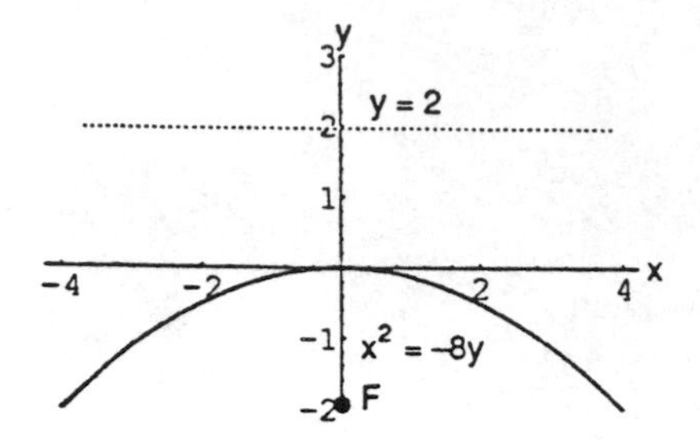

12. $y^2 = -2x \Rightarrow x = \frac{y^2}{-2} \Rightarrow 4p = 2 \Rightarrow p = \frac{1}{2};$

focus is $\left(-\frac{1}{2}, 0\right)$, directrix is $x = \frac{1}{2}$

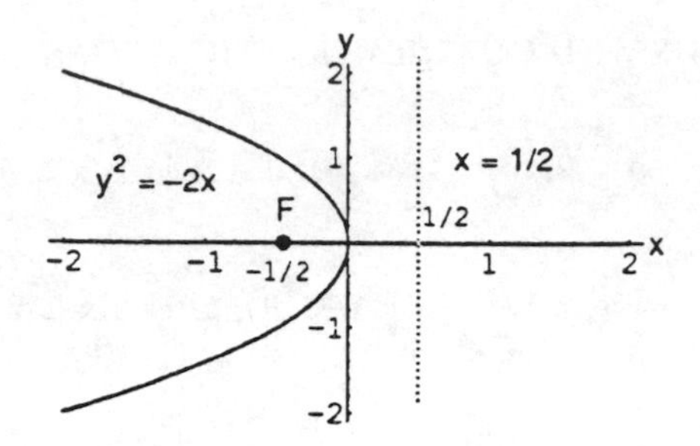

13. $y = 4x^2 \Rightarrow y = \frac{x^2}{\left(\frac{1}{4}\right)} \Rightarrow 4p = \frac{1}{4} \Rightarrow p = \frac{1}{16};$

focus is $\left(0, \frac{1}{16}\right)$, directrix is $y = -\frac{1}{16}$

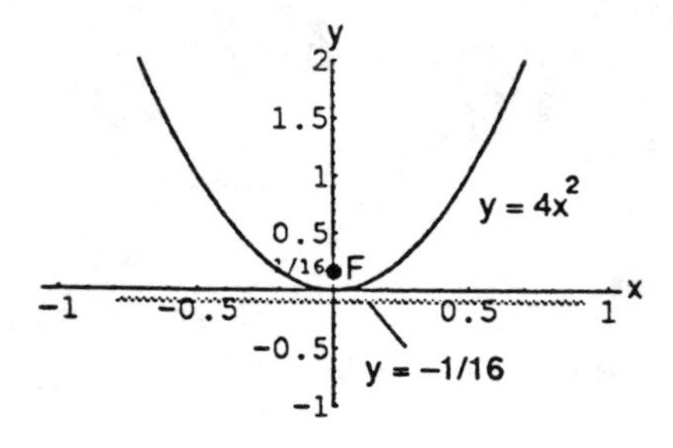

14. $y = -8x^2 \Rightarrow y = -\frac{x^2}{\left(\frac{1}{8}\right)} \Rightarrow 4p = \frac{1}{8} \Rightarrow p = \frac{1}{32};$

focus is $\left(0, -\frac{1}{32}\right)$, directrix is $y = \frac{1}{32}$

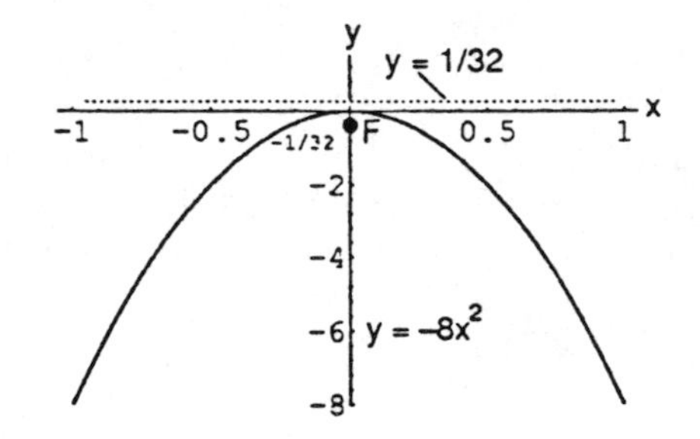

15. $x = -3y^2 \Rightarrow x = -\frac{y^2}{\left(\frac{1}{3}\right)} \Rightarrow 4p = \frac{1}{3} \Rightarrow p = \frac{1}{12};$

focus is $\left(-\frac{1}{12}, 0\right)$, directrix is $x = \frac{1}{12}$

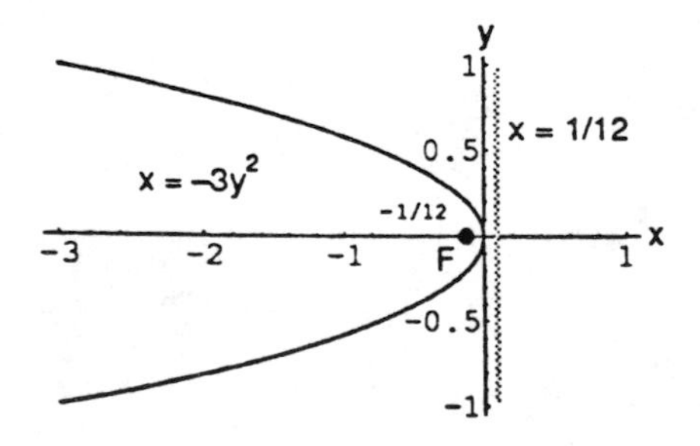

16. $x = 2y^2 \Rightarrow x = \frac{y^2}{\left(\frac{1}{2}\right)} \Rightarrow 4p = \frac{1}{2} \Rightarrow p = \frac{1}{8};$

focus is $\left(\frac{1}{8}, 0\right)$, directrix is $x = -\frac{1}{8}$

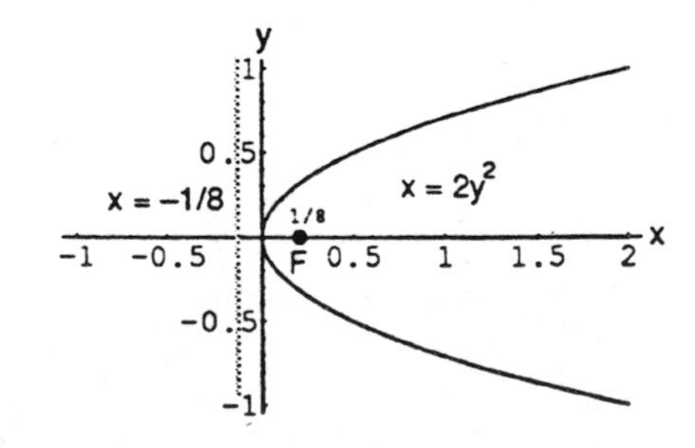

17. $16x^2 + 25y^2 = 400 \Rightarrow \frac{x^2}{25} + \frac{y^2}{16} = 1$

$\Rightarrow c = \sqrt{a^2 - b^2} = \sqrt{25 - 16} = 3$

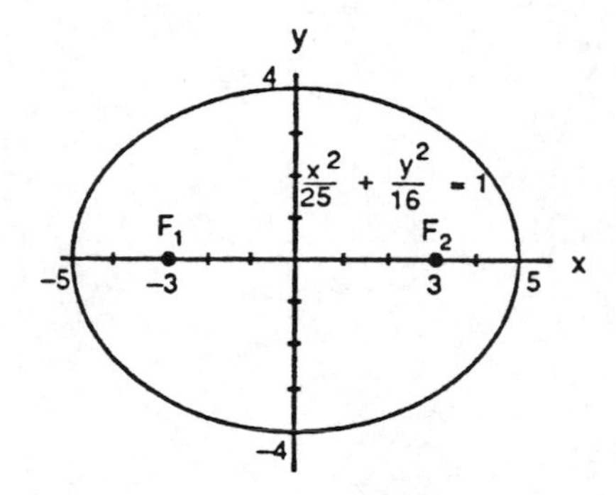

18. $7x^2 + 16y^2 = 112 \Rightarrow \frac{x^2}{16} + \frac{y^2}{7} = 1$

$\Rightarrow c = \sqrt{a^2 - b^2} = \sqrt{16 - 7} = 3$

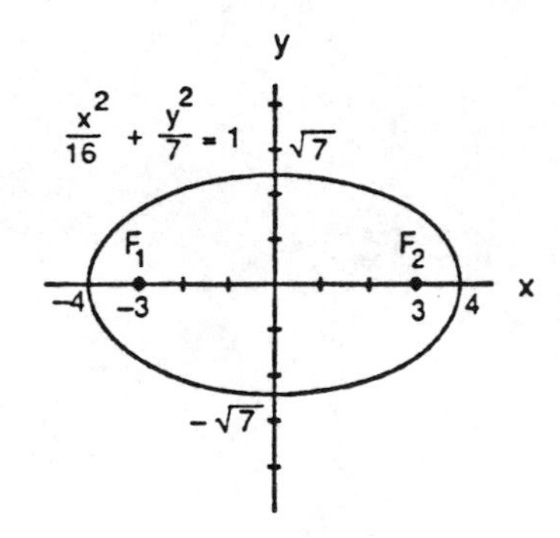

19. $2x^2 + y^2 = 2 \Rightarrow x^2 + \frac{y^2}{2} = 1$

$\Rightarrow c = \sqrt{a^2 - b^2} = \sqrt{2 - 1} = 1$

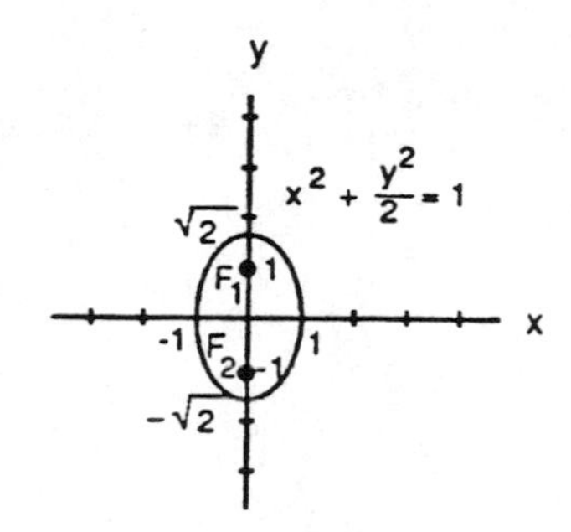

20. $2x^2 + y^2 = 4 \Rightarrow \frac{x^2}{2} + \frac{y^2}{4} = 1$

$\Rightarrow c = \sqrt{a^2 - b^2} = \sqrt{4 - 2} = \sqrt{2}$

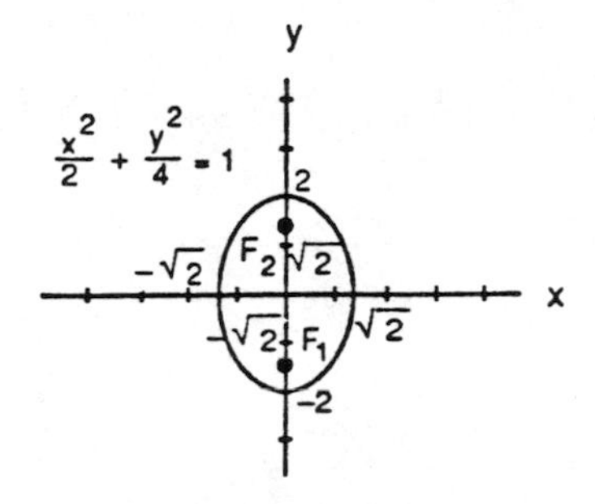

21. $3x^2 + 2y^2 = 6 \Rightarrow \frac{x^2}{2} + \frac{y^2}{3} = 1$

$\Rightarrow c = \sqrt{a^2 - b^2} = \sqrt{3 - 2} = 1$

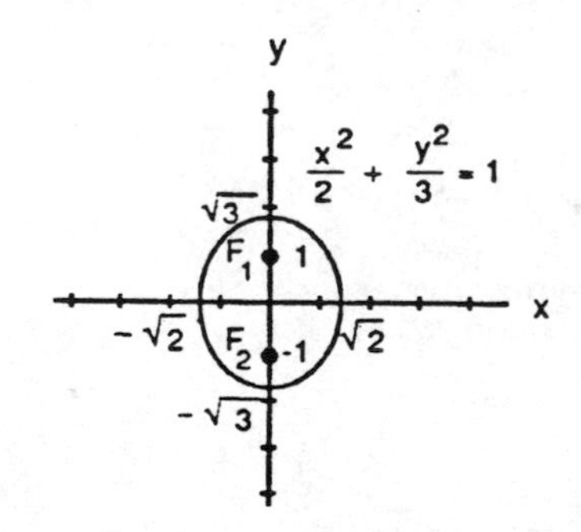

22. $9x^2 + 10y^2 = 90 \Rightarrow \frac{x^2}{10} + \frac{y^2}{9} = 1$

$\Rightarrow c = \sqrt{a^2 - b^2} = \sqrt{10 - 9} = 1$

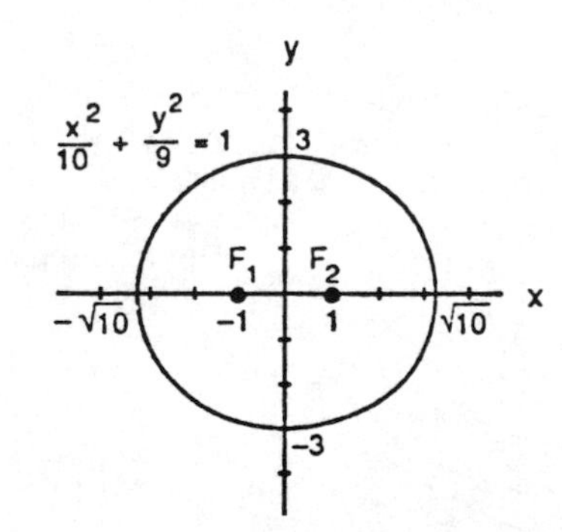

23. $6x^2 + 9y^2 = 54 \Rightarrow \frac{x^2}{9} + \frac{y^2}{6} = 1$

$\Rightarrow c = \sqrt{a^2 - b^2} = \sqrt{9-6} = \sqrt{3}$

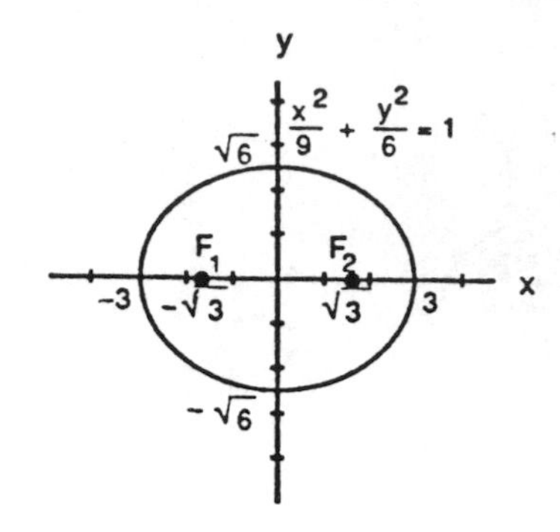

24. $169x^2 + 25y^2 = 4225 \Rightarrow \frac{x^2}{25} + \frac{y^2}{169} = 1$

$\Rightarrow c = \sqrt{a^2 - b^2} = \sqrt{169 - 25} = 12$

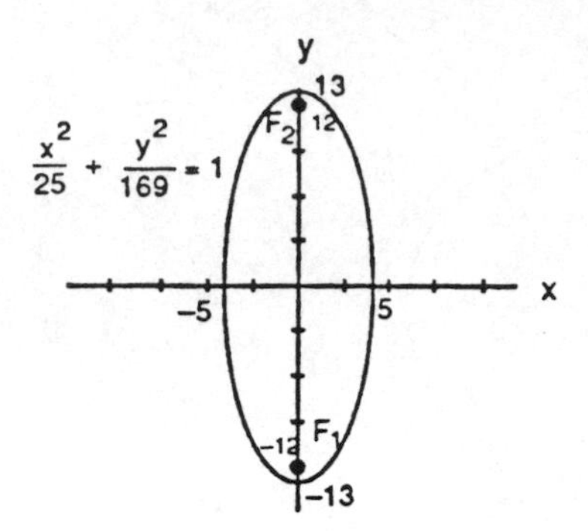

25. Foci: $(\pm\sqrt{2}, 0)$, Vertices: $(\pm 2, 0) \Rightarrow a = 2,\ c = \sqrt{2} \Rightarrow b^2 = a^2 - c^2 = 4 - (\sqrt{2})^2 = 2 \Rightarrow \frac{x^2}{4} + \frac{y^2}{2} = 1$

26. Foci: $(0, \pm 4)$, Vertices: $(0, \pm 5) \Rightarrow a = 5,\ c = 4 \Rightarrow b^2 = 25 - 16 = 9 \Rightarrow \frac{x^2}{9} + \frac{y^2}{25} = 1$

27. $x^2 - y^2 = 1 \Rightarrow c = \sqrt{a^2 + b^2} = \sqrt{1+1} = \sqrt{2}$;

asymptotes are $y = \pm x$

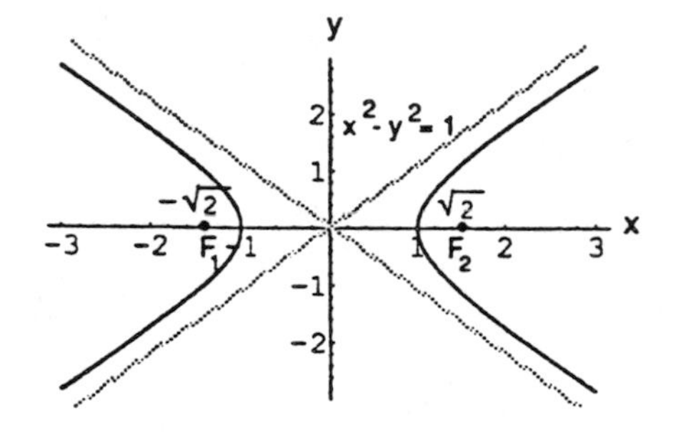

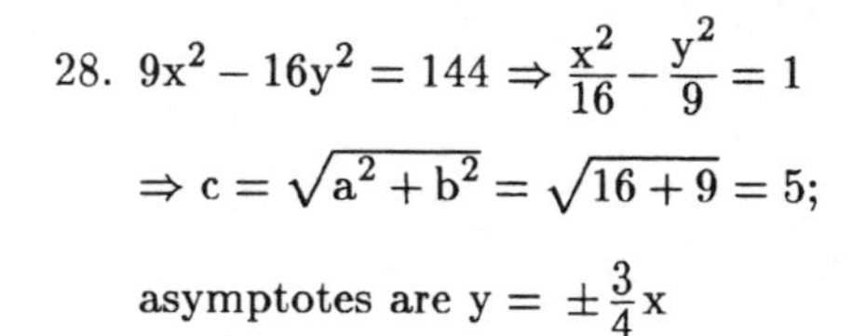
28. $9x^2 - 16y^2 = 144 \Rightarrow \frac{x^2}{16} - \frac{y^2}{9} = 1$

$\Rightarrow c = \sqrt{a^2 + b^2} = \sqrt{16 + 9} = 5$;

asymptotes are $y = \pm\frac{3}{4}x$

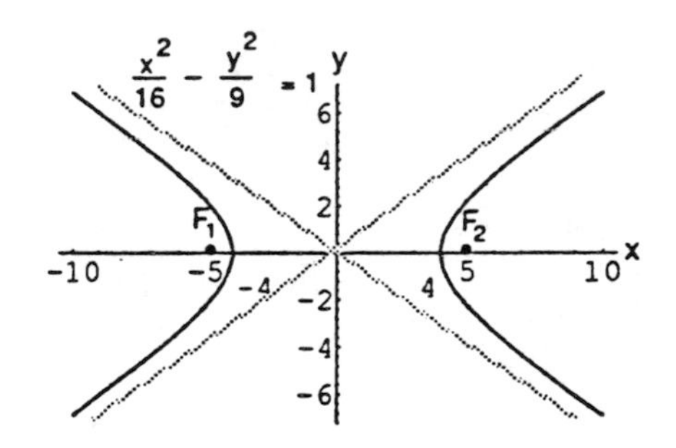

29. $y^2 - x^2 = 8 \Rightarrow \frac{y^2}{8} - \frac{x^2}{8} = 1 \Rightarrow c = \sqrt{a^2 + b^2}$

$= \sqrt{8+8} = 4$; asymptotes are $y = \pm x$

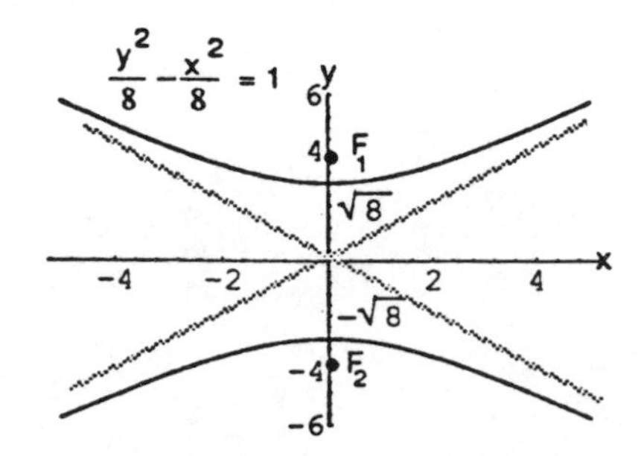

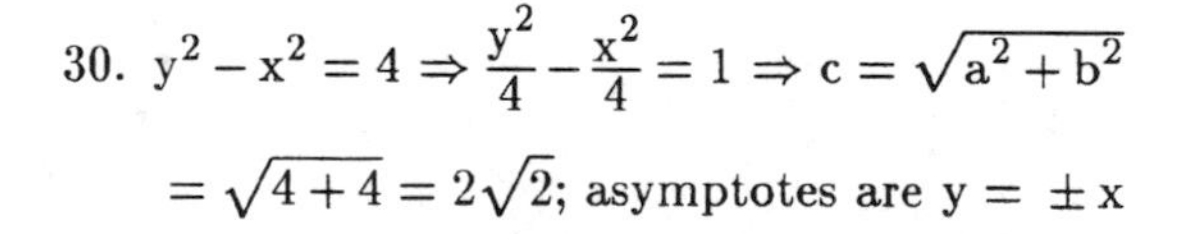
30. $y^2 - x^2 = 4 \Rightarrow \frac{y^2}{4} - \frac{x^2}{4} = 1 \Rightarrow c = \sqrt{a^2 + b^2}$

$= \sqrt{4+4} = 2\sqrt{2}$; asymptotes are $y = \pm x$

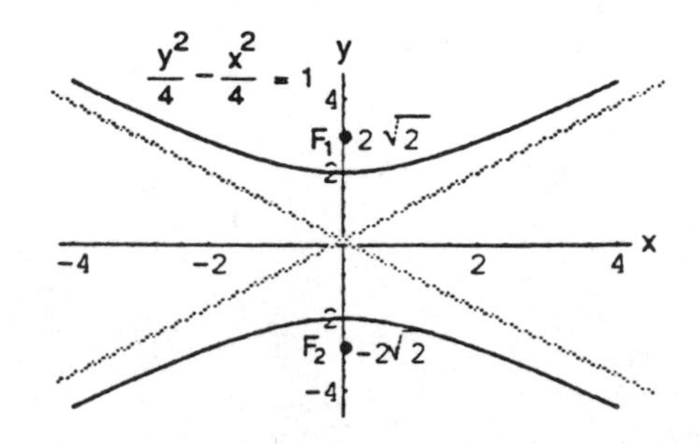

31. $8x^2 - 2y^2 = 16 \Rightarrow \frac{x^2}{2} - \frac{y^2}{8} = 1 \Rightarrow c = \sqrt{a^2 + b^2}$
$= \sqrt{2+8} = \sqrt{10}$; asymptotes are $y = \pm 2x$

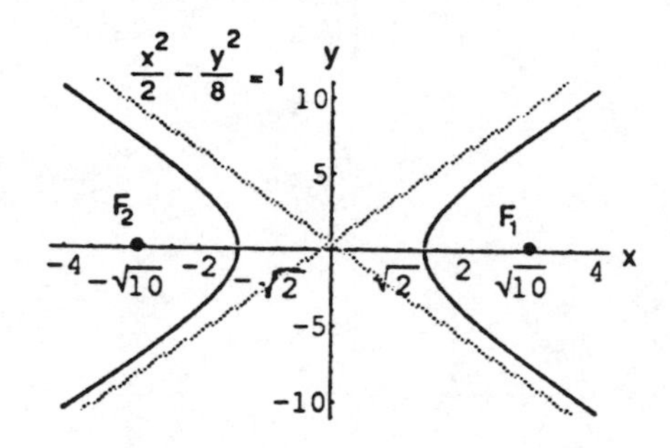

32. $y^2 - 3x^2 = 3 \Rightarrow \frac{y^2}{3} - x^2 = 1 \Rightarrow c = \sqrt{a^2 + b^2}$
$= \sqrt{3+1} = 2$; asymptotes are $y = \pm \sqrt{3}x$

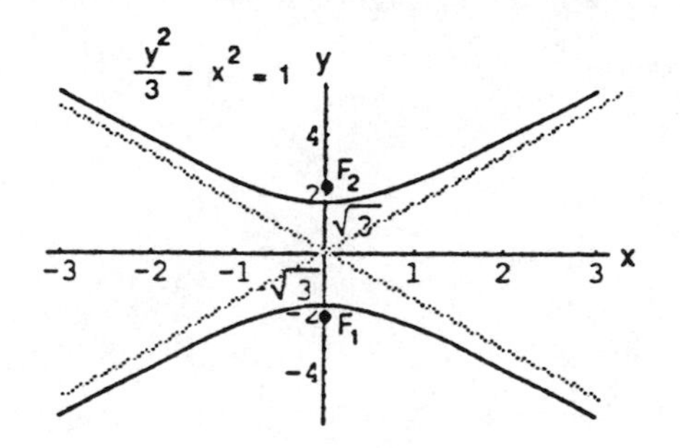

33. $8y^2 - 2x^2 = 16 \Rightarrow \frac{y^2}{2} - \frac{x^2}{8} = 1 \Rightarrow c = \sqrt{a^2 + b^2}$
$= \sqrt{2+8} = \sqrt{10}$; asymptotes are $y = \pm \frac{x}{2}$

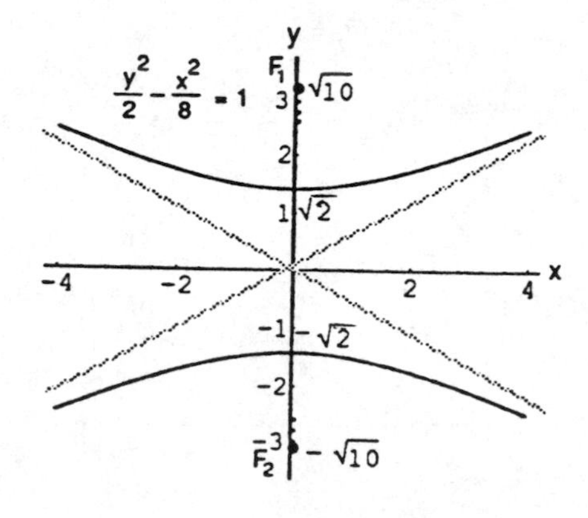

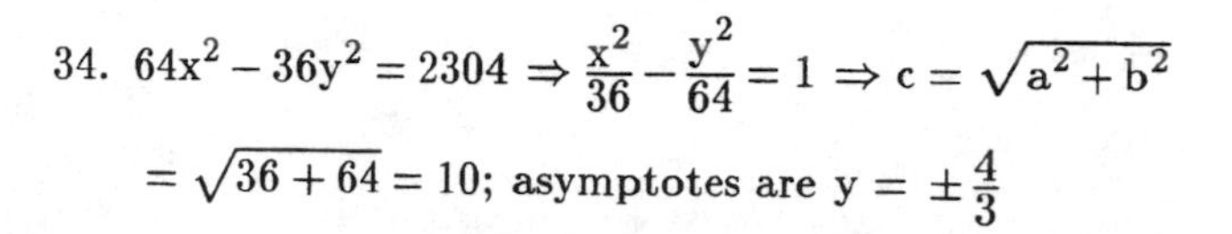

34. $64x^2 - 36y^2 = 2304 \Rightarrow \frac{x^2}{36} - \frac{y^2}{64} = 1 \Rightarrow c = \sqrt{a^2 + b^2}$
$= \sqrt{36+64} = 10$; asymptotes are $y = \pm \frac{4}{3}$

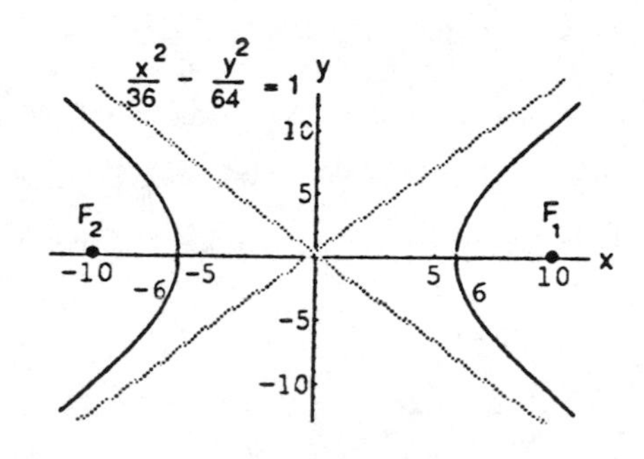

35. Foci: $(0, \pm\sqrt{2})$, Asymptotes: $y = \pm x \Rightarrow c = \sqrt{2}$ and $\frac{b}{a} = 1 \Rightarrow a = b \Rightarrow c^2 = a^2 + b^2 = 2a^2 \Rightarrow 2 = 2a^2$
$\Rightarrow a = 1 \Rightarrow b = 1 \Rightarrow y^2 - x^2 = 1$

36. Foci: $(\pm 2, 0)$, Asymptotes: $y = \pm \frac{1}{\sqrt{3}}x \Rightarrow c = 2$ and $\frac{b}{a} = \frac{1}{\sqrt{3}} \Rightarrow b = \frac{a}{\sqrt{3}} \Rightarrow c^2 = a^2 + b^2 = a^2 + \frac{a^2}{3} = \frac{4a^2}{3}$
$\Rightarrow 4 = \frac{4a^2}{3} \Rightarrow a^2 = 3 \Rightarrow a = \sqrt{3} \Rightarrow b = 1 \Rightarrow \frac{x^2}{3} - y^2 = 1$

37. Vertices: $(\pm 3, 0)$, Asymptotes: $y = \pm\frac{4}{3}x \Rightarrow a = 3$ and $\frac{b}{a} = \frac{4}{3} \Rightarrow b = \frac{4}{3}(3) = 4 \Rightarrow \frac{x^2}{9} - \frac{y^2}{16} = 1$

38. Vertices: $(0, \pm 2)$, Asymptotes: $y = \pm\frac{1}{2}x \Rightarrow a = 2$ and $\frac{a}{b} = \frac{1}{2} \Rightarrow b = 2(2) = 4 \Rightarrow \frac{y^2}{4} - \frac{x^2}{16} = 1$

39. (a) $y^2 = 8x \Rightarrow 4p = 8 \Rightarrow p = 2 \Rightarrow$ directrix is $x = -2$, focus is $(2,0)$, and vertex is $(0,0)$; therefore the new directrix is $x = -1$, the new focus is $(3,-2)$, and the new vertex is $(1,-2)$

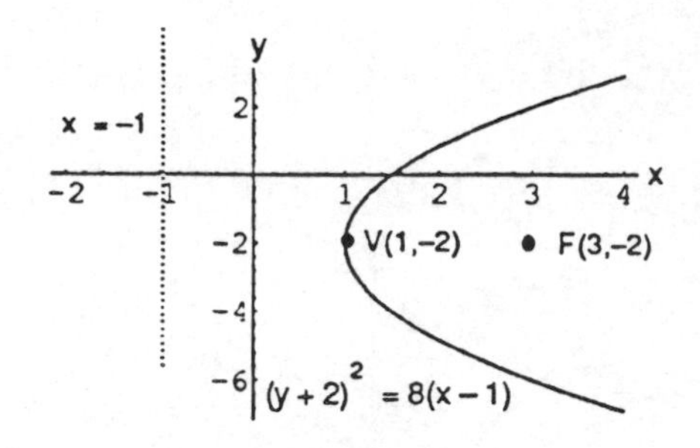

40. (a) $x^2 = -4y \Rightarrow 4p = 4 \Rightarrow p = 1 \Rightarrow$ directrix is $y = 1$, focus is $(0,-1)$, and vertex is $(0,0)$; therefore the new directrix is $y = 4$, the new focus is $(-1,2)$, and the new vertex is $(-1,3)$

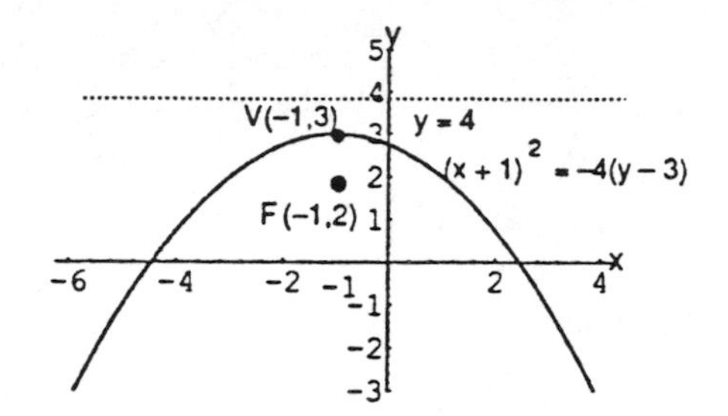

41. (a) $\frac{x^2}{16} + \frac{y^2}{9} = 1 \Rightarrow$ center is $(0,0)$, vertices are $(-4,0)$ and $(4,0)$; $c = \sqrt{a^2 - b^2} = \sqrt{7} \Rightarrow$ foci are $(\sqrt{7},0)$ and $(-\sqrt{7},0)$; therefore the new center is $(4,3)$, the new vertices are $(0,3)$ and $(8,3)$, and the new foci are $(4 \pm \sqrt{7},3)$

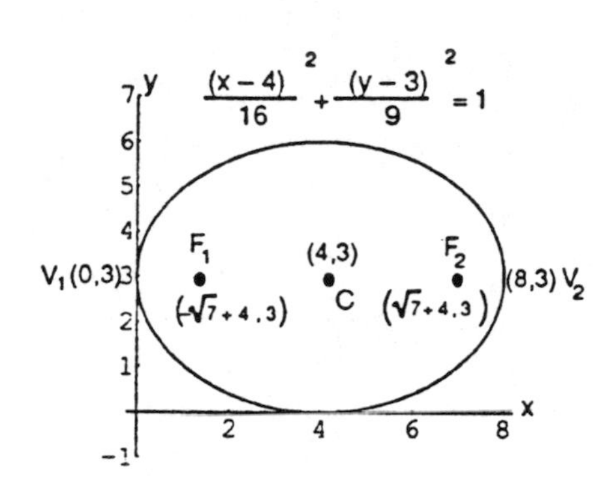

42. (a) $\frac{x^2}{9} + \frac{y^2}{25} = 1 \Rightarrow$ center is $(0,0)$, vertices are $(0,5)$ and $(0,-5)$; $c = \sqrt{a^2 - b^2} = \sqrt{16} = 4 \Rightarrow$ foci are $(0,4)$ and $(0,-4)$; therefore the new center is $(-3,-2)$, the new vertices are $(-3,3)$ and $(-3,-7)$, and the new foci are $(-3,2)$ and $(-3,-6)$

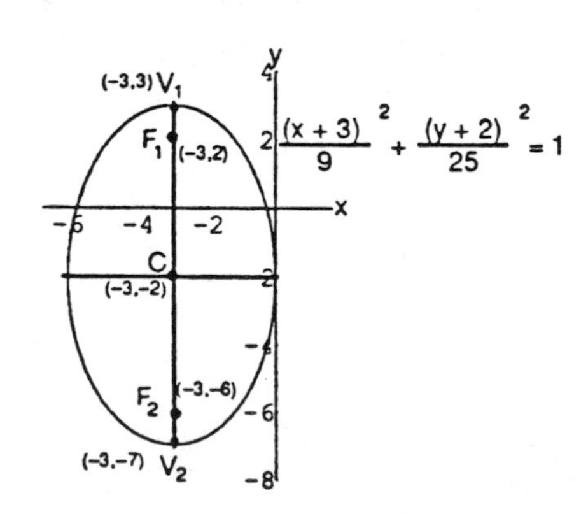

43. (a) $\frac{x^2}{16} - \frac{y^2}{9} = 1 \Rightarrow$ center is $(0,0)$, vertices are $(-4,0)$ and $(4,0)$, and the asymptotes are $\frac{x}{4} = \pm\frac{y}{3}$ or $y = \pm\frac{3x}{4}$; $c = \sqrt{a^2+b^2} = \sqrt{25} = 5 \Rightarrow$ foci are $(-5,0)$ and $(5,0)$; therefore the new center is $(2,0)$, the new vertices are $(-2,0)$ and $(6,0)$, the new foci are $(-3,0)$ and $(7,0)$, and the new asymptotes are $y = \pm\frac{3(x-2)}{4}$

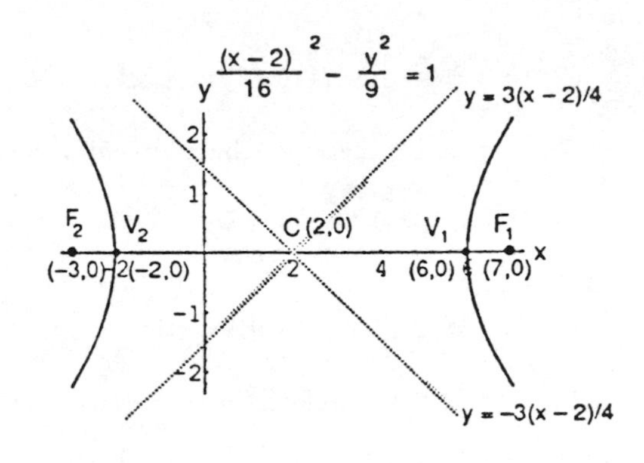

44. (a) $\frac{y^2}{4} - \frac{x^2}{5} = 1 \Rightarrow$ center is $(0,0)$, vertices are $(0,-2)$ and $(0,2)$, and the asymptotes are $\frac{y}{2} = \pm\frac{x}{\sqrt{5}}$ or $y = \pm\frac{2x}{\sqrt{5}}$; $c = \sqrt{a^2+b^2} = \sqrt{9} = 3 \Rightarrow$ foci are $(0,3)$ and $(0,-3)$; therefore the new center is $(0,-2)$, the new vertices are $(0,-4)$ and $(0,0)$, the new foci are $(0,1)$ and $(0,-5)$, and the new asymptotes are $y+2 = \pm\frac{2x}{\sqrt{5}}$

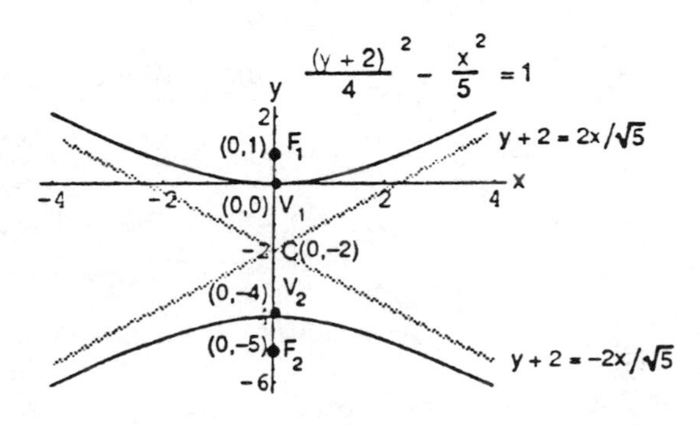

45. $y^2 = 4x \Rightarrow 4p = 4 \Rightarrow p = 1 \Rightarrow$ focus is $(1,0)$, directrix is $x = -1$, and vertex is $(0,0)$; therefore the new vertex is $(-2,-3)$, the new focus is $(-1,-3)$, and the new directrix is $x = -3$; the new equation is $(y+3)^2 = 4(x+2)$

46. $y^2 = -12x \Rightarrow 4p = 12 \Rightarrow p = 3 \Rightarrow$ focus is $(-3,0)$, directrix is $x = 3$, and vertex is $(0,0)$; therefore the new vertex is $(4,3)$, the new focus is $(1,3)$, and the new directrix is $x = 7$; the new equation is $(y-3)^2 = -12(x-4)$

47. $x^2 = 8y \Rightarrow 4p = 8 \Rightarrow p = 2 \Rightarrow$ focus is $(0,2)$, directrix is $y = -2$, and vertex is $(0,0)$; therefore the new vertex is $(1,-7)$, the new focus is $(1,-5)$, and the new directrix is $y = -9$; the new equation is $(x-1)^2 = 8(y+7)$

48. $x^2 = 6y \Rightarrow 4p = 6 \Rightarrow p = \frac{3}{2} \Rightarrow$ focus is $\left(0,\frac{3}{2}\right)$, directrix is $y = -\frac{3}{2}$, and vertex is $(0,0)$; therefore the new vertex is $(-3,-2)$, the new focus is $\left(-3,-\frac{1}{2}\right)$, and the new directrix is $y = -\frac{7}{2}$; the new equation is $(x+3)^2 = 6(y+2)$

49. $\frac{x^2}{6} + \frac{y^2}{9} = 1 \Rightarrow$ center is $(0,0)$, vertices are $(0,3)$ and $(0,-3)$; $c = \sqrt{a^2-b^2} = \sqrt{9-6} = \sqrt{3} \Rightarrow$ foci are $\left(0,\sqrt{3}\right)$ and $\left(0,-\sqrt{3}\right)$; therefore the new center is $(-2,-1)$, the new vertices are $(-2,2)$ and $(-2,-4)$, and the new foci are $\left(-2,-1\pm\sqrt{3}\right)$; the new equation is $\frac{(x+2)^2}{6} + \frac{(y+1)^2}{9} = 1$

50. $\frac{x^2}{2}+y^2=1 \Rightarrow$ center is $(0,0)$, vertices are $(\sqrt{2},0)$ and $(-\sqrt{2},0)$; $c=\sqrt{a^2-b^2}=\sqrt{2-1}=1 \Rightarrow$ foci are $(-1,0)$ and $(1,0)$; therefore the new center is $(3,4)$, the new vertices are $(3 \pm \sqrt{2},4)$, and the new foci are $(2,4)$ and $(4,4)$; the new equation is $\frac{(x-3)^2}{2}+(y-4)^2=1$

51. $\frac{x^2}{3}+\frac{y^2}{2}=1 \Rightarrow$ center is $(0,0)$, vertices are $(\sqrt{3},0)$ and $(-\sqrt{3},0)$; $c=\sqrt{a^2-b^2}=\sqrt{3-2}=1 \Rightarrow$ foci are $(-1,0)$ and $(1,0)$; therefore the new center is $(2,3)$, the new vertices are $(2 \pm \sqrt{3},3)$, and the new foci are $(1,3)$ and $(3,3)$; the new equation is $\frac{(x-2)^2}{3}+\frac{(y-3)^2}{2}=1$

52. $\frac{x^2}{16}+\frac{y^2}{25}=1 \Rightarrow$ center is $(0,0)$, vertices are $(0,5)$ and $(0,-5)$; $c=\sqrt{a^2-b^2}=\sqrt{25-16}=3 \Rightarrow$ foci are $(0,3)$ and $(0,-3)$; therefore the new center is $(-4,-5)$, the new vertices are $(-4,0)$ and $(-4,-10)$, and the new foci are $(-4,-2)$ and $(-4,-8)$; the new equation is $\frac{(x+4)^2}{16}+\frac{(y+5)^2}{25}=1$

53. $\frac{x^2}{4}-\frac{y^2}{5}=1 \Rightarrow$ center is $(0,0)$, vertices are $(2,0)$ and $(-2,0)$; $c=\sqrt{a^2+b^2}=\sqrt{4+5}=3 \Rightarrow$ foci are $(3,0)$ and $(-3,0)$; the asymptotes are $\pm\frac{x}{2}=\frac{y}{\sqrt{5}} \Rightarrow y=\pm\frac{\sqrt{5}x}{2}$; therefore the new center is $(2,2)$, the new vertices are $(4,2)$ and $(0,2)$, and the new foci are $(5,2)$ and $(-1,2)$; the new asymptotes are $y-2=\pm\frac{\sqrt{5}(x-2)}{2}$; the new equation is $\frac{(x-2)^2}{4}-\frac{(y-2)^2}{5}=1$

54. $\frac{x^2}{16}-\frac{y^2}{9}=1 \Rightarrow$ center is $(0,0)$, vertices are $(4,0)$ and $(-4,0)$; $c=\sqrt{a^2+b^2}=\sqrt{16+9}=5 \Rightarrow$ foci are $(-5,0)$ and $(5,0)$; the asymptotes are $\pm\frac{x}{4}=\frac{y}{3} \Rightarrow y=\pm\frac{3x}{4}$; therefore the new center is $(-5,-1)$, the new vertices are $(-1,-1)$ and $(-9,-1)$, and the new foci are $(-10,-1)$ and $(0,-1)$; the new asymptotes are $y+1=\pm\frac{3(x+5)}{4}$; the new equation is $\frac{(x+5)^2}{16}-\frac{(y+1)^2}{9}=1$

55. $y^2-x^2=1 \Rightarrow$ center is $(0,0)$, vertices are $(0,1)$ and $(0,-1)$; $c=\sqrt{a^2+b^2}=\sqrt{1+1}=\sqrt{2} \Rightarrow$ foci are $(0,\pm\sqrt{2})$; the asymptotes are $y=\pm x$; therefore the new center is $(-1,-1)$, the new vertices are $(-1,0)$ and $(-1,-2)$, and the new foci are $(-1,-1\pm\sqrt{2})$; the new asymptotes are $y+1=\pm(x+1)$; the new equation is $(y+1)^2-(x+1)^2=1$

56. $\frac{y^2}{3}-x^2=1 \Rightarrow$ center is $(0,0)$, vertices are $(0,\sqrt{3})$ and $(0,-\sqrt{3})$; $c=\sqrt{a^2+b^2}=\sqrt{3+1}=2 \Rightarrow$ foci are $(0,2)$ and $(0,-2)$; the asymptotes are $\pm x=\frac{y}{\sqrt{3}} \Rightarrow y=\pm\sqrt{3}x$; therefore the new center is $(1,3)$, the new vertices are $(1,3\pm\sqrt{3})$, and the new foci are $(1,5)$ and $(1,1)$; the new asymptotes are $y-3=\pm\sqrt{3}(x-1)$; the new equation is $\frac{(y-3)^2}{3}-(x-1)^2=1$

57. $x^2 + 4x + y^2 = 12 \Rightarrow x^2 + 4x + 4 + y^2 = 12 + 4 \Rightarrow (x+2)^2 + y^2 = 16$; this is a circle: center at $C(-2, 0)$, $a = 4$

58. $2x^2 + 2y^2 - 28x + 12y + 114 = 0 \Rightarrow x^2 - 14x + 49 + y^2 + 6y + 9 = -57 + 49 + 9 \Rightarrow (x-7)^2 + (y+3)^2 = 1$; this is a circle: center at $C(7, -3)$, $a = 1$

59. $x^2 + 2x + 4y - 3 = 0 \Rightarrow x^2 + 2x + 1 = -4y + 3 + 1 \Rightarrow (x+1)^2 = -4(y-1)$; this is a parabola: $V(-1, 1)$, $F(-1, 0)$

60. $y^2 - 4y - 8x - 12 = 0 \Rightarrow y^2 - 4y + 4 = 8x + 12 + 4 \Rightarrow (y-2)^2 = 8(x+2)$; this is a parabola: $V(-2, 2)$, $F(0, 2)$

61. $x^2 + 5y^2 + 4x = 1 \Rightarrow x^2 + 4x + 4 + 5y^2 = 5 \Rightarrow (x+2)^2 + 5y^2 = 5 \Rightarrow \frac{(x+2)^2}{5} + y^2 = 1$; this is an ellipse: the center is $(-2, 0)$, the vertices are $\left(-2 \pm \sqrt{5}, 0\right)$; $c = \sqrt{a^2 - b^2} = \sqrt{5-1} = 2 \Rightarrow$ the foci are $(-4, 0)$ and $(0, 0)$

62. $9x^2 + 6y^2 + 36y = 0 \Rightarrow 9x^2 + 6\left(y^2 + 6y + 9\right) = 54 \Rightarrow 9x^2 + 6(y+3)^2 = 54 \Rightarrow \frac{x^2}{6} + \frac{(y+3)^2}{9} = 1$; this is an ellipse: the center is $(0, -3)$, the vertices are $(0, 0)$ and $(0, -6)$; $c = \sqrt{a^2 - b^2} = \sqrt{9-6} = \sqrt{3} \Rightarrow$ the foci are $\left(0, -3 \pm \sqrt{3}\right)$

63. $x^2 + 2y^2 - 2x - 4y = -1 \Rightarrow x^2 - 2x + 1 + 2\left(y^2 - 2y + 1\right) = 2 \Rightarrow (x-1)^2 + 2(y-1)^2 = 2$ $\Rightarrow \frac{(x-1)^2}{2} + (y-1)^2 = 1$; this is an ellipse: the center is $(1, 1)$, the vertices are $\left(1 \pm \sqrt{2}, 1\right)$; $c = \sqrt{a^2 - b^2} = \sqrt{2-1} = 1 \Rightarrow$ the foci are $(2, 1)$ and $(0, 1)$

64. $4x^2 + y^2 + 8x - 2y = -1 \Rightarrow 4\left(x^2 + 2x + 1\right) + y^2 - 2y + 1 = 4 \Rightarrow 4(x+1)^2 + (y-1)^2 = 4$ $\Rightarrow (x+1)^2 + \frac{(y-1)^2}{4} = 1$; this is an ellipse: the center is $(-1, 1)$, the vertices are $(-1, 3)$ and $(-1, -1)$; $c = \sqrt{a^2 - b^2} = \sqrt{4-1} = \sqrt{3} \Rightarrow$ the foci are $\left(-1, 1 \pm \sqrt{3}\right)$

65. $x^2 - y^2 - 2x + 4y = 4 \Rightarrow x^2 - 2x + 1 - \left(y^2 - 4y + 4\right) = 1 \Rightarrow (x-1)^2 - (y-2)^2 = 1$; this is a hyperbola: the center is $(1, 2)$, the vertices are $(2, 2)$ and $(0, 2)$; $c = \sqrt{a^2 + b^2} = \sqrt{1+1} = \sqrt{2} \Rightarrow$ the foci are $\left(1 \pm \sqrt{2}, 2\right)$; the asymptotes are $y - 2 = \pm(x-1)$

66. $x^2 - y^2 + 4x - 6y = 6 \Rightarrow x^2 + 4x + 4 - \left(y^2 + 6y + 9\right) = 1 \Rightarrow (x+2)^2 - (y+3)^2 = 1$; this is a hyperbola: the center is $(-2, -3)$, the vertices are $(-1, -3)$ and $(-3, -3)$; $c = \sqrt{a^2 + b^2} = \sqrt{1+1} = \sqrt{2} \Rightarrow$ the foci are $\left(-2 \pm \sqrt{2}, -3\right)$; the asymptotes are $y + 3 = \pm(x+2)$

67. $2x^2 - y^2 + 6y = 3 \Rightarrow 2x^2 - \left(y^2 - 6y + 9\right) = -6 \Rightarrow \frac{(y-3)^2}{6} - \frac{x^2}{3} = 1$; this is a hyperbola: the center is $(0, 3)$, the vertices are $\left(0, 3 \pm \sqrt{6}\right)$; $c = \sqrt{a^2 + b^2} = \sqrt{6+3} = 3 \Rightarrow$ the foci are $(0, 6)$ and $(0, 0)$; the asymptotes are $\frac{y-3}{\sqrt{6}} = \pm\frac{x}{\sqrt{3}} \Rightarrow y = \pm\sqrt{2}x + 3$

68. $y^2 - 4x^2 + 16x = 24 \Rightarrow y^2 - 4(x^2 - 4x + 4) = 8 \Rightarrow \frac{y^2}{8} - \frac{(x-2)^2}{2} = 1$; this is a hyperbola: the center is $(2, 0)$, the vertices are $(2, \pm\sqrt{8})$; $c = \sqrt{a^2 + b^2} = \sqrt{8+2} = \sqrt{10} \Rightarrow$ the foci are $(2, \pm\sqrt{10})$; the asymptotes are $\frac{y}{\sqrt{8}} = \pm\frac{x-2}{\sqrt{2}} \Rightarrow y = \pm 2(x-2)$

69.

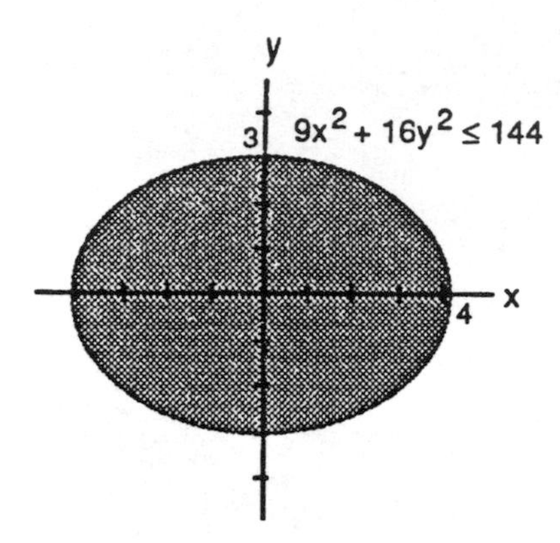

70.

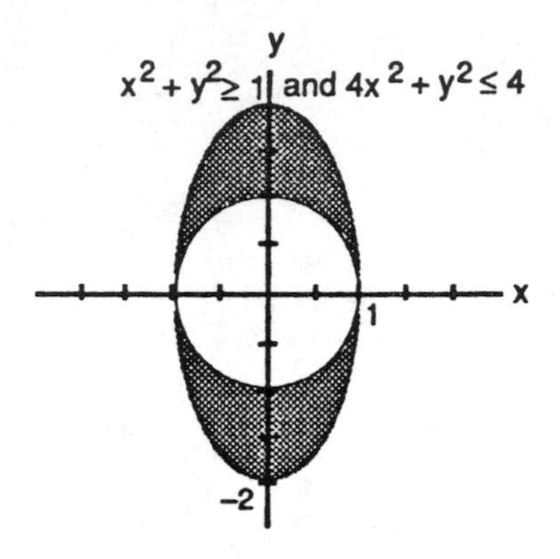

71.

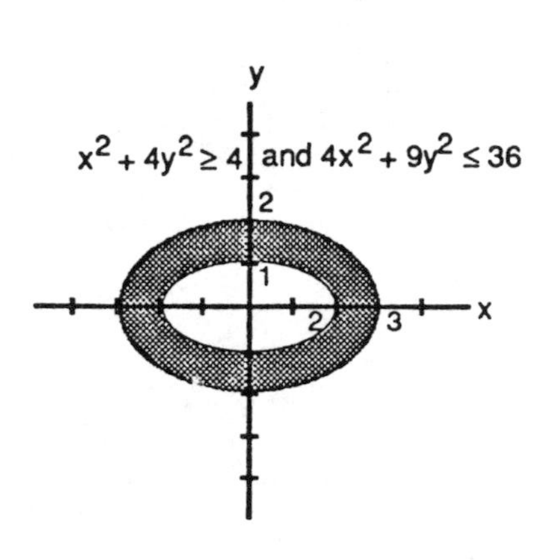

72.

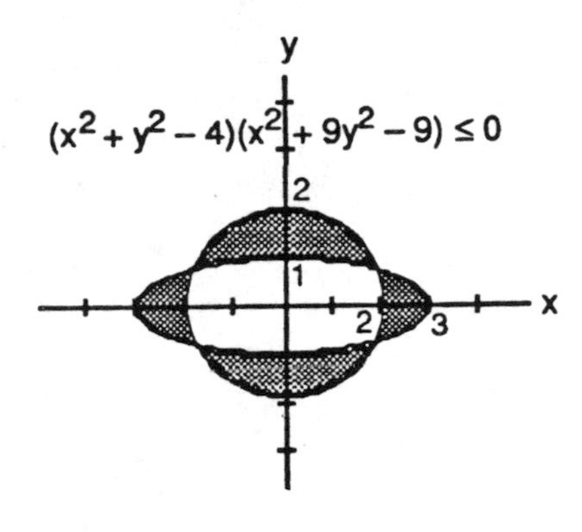

73.

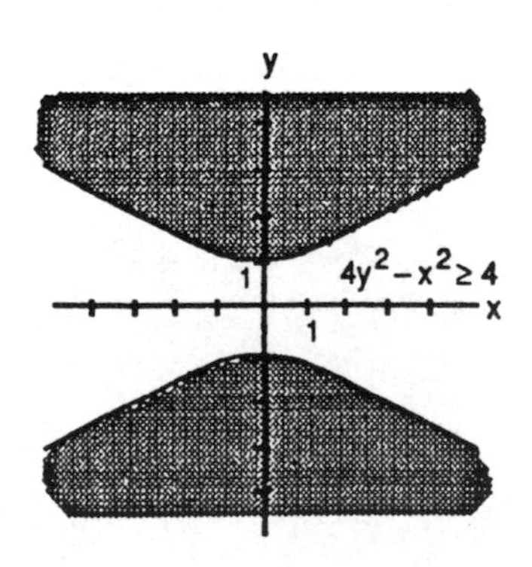

74. $|x^2 - y^2| \le 1 \Rightarrow -1 \le x^2 - y^2 \le 1 \Rightarrow -1 \le x^2 - y^2$ and $x^2 - y^2 \le 1 \Rightarrow 1 \ge y^2 - x^2$ and $x^2 - y^2 \le 1$

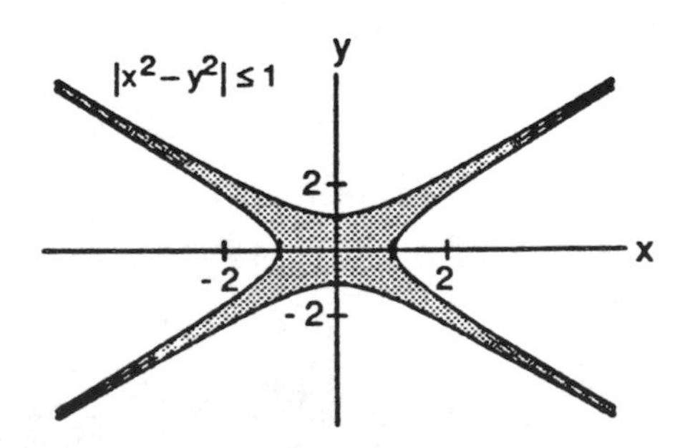

75. Volume of the Parabolic Solid: $V_1 = \int_0^{b/2} 2\pi x\left(h - \frac{4h}{b^2}x^2\right) dx = 2\pi h \int_0^{b/2} \left(x - \frac{4x^3}{b^2}\right) dx = 2\pi h\left[\frac{x^2}{2} - \frac{x^4}{b^2}\right]_0^{b/2}$

$= \frac{\pi hb^2}{8}$; Volume of the Cone: $V_2 = \frac{1}{3}\pi\left(\frac{b}{2}\right)^2 h = \frac{1}{3}\pi\left(\frac{b^2}{4}\right)h = \frac{\pi hb^2}{12}$; therefore $V_1 = \frac{3}{2}V_2$

76. $y = \int \frac{w}{H} x\, dx = \frac{w}{H}\left(\frac{x^2}{2}\right) + C = \frac{wx^2}{2H} + C$; $y = 0$ when $x = 0 \Rightarrow 0 = \frac{w(0)^2}{2H} + C \Rightarrow C = 0$; therefore $y = \frac{wx^2}{2H}$ is the equation of the cable's curve

77. A general equation of the circle is $x^2 + y^2 + ax + by + c = 0$, so we will substitute the three given points into this equation and solve the resulting system: $\left.\begin{array}{r} a \quad\quad + c = -1 \\ b + c = -1 \\ 2a + 2b + c = -8 \end{array}\right\} \Rightarrow c = \frac{4}{3}$ and $a = b = -\frac{7}{3}$; therefore $3x^2 + 3y^2 - 7x - 7y + 4 = 0$ represents the circle

78. A general equation of the circle is $x^2 + y^2 + ax + by + c = 0$, so we will substitute each of the three given points into this equation and solve the resulting system: $\left.\begin{array}{r} 2a + 3b + c = -13 \\ 3a + 2b + c = -13 \\ -4a + 3b + c = -25 \end{array}\right\} \Rightarrow a = 2,\ b = 2,$ and $c = -23$; therefore $x^2 + y^2 + 2x + 2y - 23 = 0$ represents the circle

79. $r^2 = (-2-1)^2 + (1-3)^2 = 13 \Rightarrow (x+2)^2 + (y-1)^2 = 13$ is an equation of the circle; the distance from the center to $(1.1, 2.8)$ is $\sqrt{(-2-1.1)^2 + (1-2.8)^2} = \sqrt{12.85} < \sqrt{13}$, the radius $\Rightarrow$ the point is inside the circle

80. $(x-2)^2 + (y-1)^2 = 5 \Rightarrow 2(x-2) + 2(y-1)\frac{dy}{dx} = 0 \Rightarrow \frac{dy}{dx} = -\frac{x-2}{y-1}$; $y = 0 \Rightarrow (x-2)^2 + (0-1)^2 = 5$
$\Rightarrow (x-2)^2 = 4 \Rightarrow x = 4$ or $x = 0 \Rightarrow$ the circle crosses the x-axis at $(4,0)$ and $(0,0)$; $x = 0$
$\Rightarrow (0-2)^2 + (y-1)^2 = 5 \Rightarrow (y-1)^2 = 1 \Rightarrow y = 2$ or $y = 0 \Rightarrow$ the circle crosses the y-axis at $(0,2)$ and $(0,0)$.
At $(4,0)$: $\frac{dy}{dx} = -\frac{4-2}{0-1} = 2 \Rightarrow$ the tangent line is $y = 2(x-4)$ or $y = 2x - 8$
At $(0,0)$: $\frac{dy}{dx} = -\frac{0-2}{0-1} = -2 \Rightarrow$ the tangent line is $y = -2x$
At $(0,2)$: $\frac{dy}{dx} = -\frac{0-2}{2-1} = 2 \Rightarrow$ the tangent line is $y - 2 = 2x$ or $y = 2x + 2$

81. (a) $y^2 = kx \Rightarrow x = \frac{y^2}{k}$; the volume of the solid formed by revolving R_1 about the y-axis is $V_1 = \int_0^{\sqrt{kx}} \pi\left(\frac{y^2}{k}\right)^2 dy$
$= \frac{\pi}{k^2}\int_0^{\sqrt{kx}} y^4\, dy = \frac{\pi x^2\sqrt{kx}}{5}$; the volume of the right circular cylinder formed by revolving PQ about the y-axis

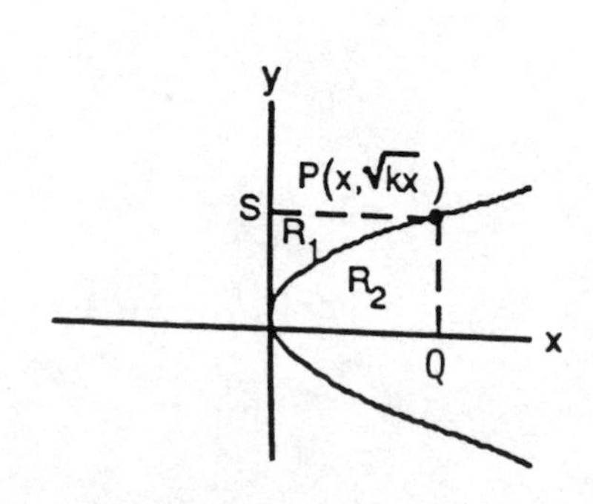

is $V_2 = \pi x^2\sqrt{kx} \Rightarrow$ the volume of the solid formed by revolving R_2 about the y-axis is $V_3 = V_2 - V_1$ $= \frac{4\pi x^2\sqrt{kx}}{5}$. Therefore we can see the ratio of V_3 to V_1 is 4:1.

(b) The volume of the solid formed by revolving R_2 about the x-axis is $V_1 = \int_0^x \pi\left(\sqrt{kt}\right)^2 dt = \pi k \int_0^x t\,dt$ $= \frac{\pi kx^2}{2}$. The volume of the right circular cylinder formed by revolving PS about the x-axis is $V_2 = \pi\left(\sqrt{kx}\right)^2 x = \pi kx^2 \Rightarrow$ the volume of the solid formed by revolving R_1 about the x-axis is $V_3 = V_2 - V_1 = \pi kx^2 - \frac{\pi kx^2}{2} = \frac{\pi kx^2}{2}$. Therefore the ratio of V_3 to V_1 is 1:1.

82. Let $P_1(-p, y_1)$ be any point on $x = -p$, and let $P(x, y)$ be a point where a tangent intersects $y^2 = 4px$. Now $y^2 = 4px \Rightarrow 2y\frac{dy}{dx} = 4p \Rightarrow \frac{dy}{dx} = \frac{2p}{y}$; then the slope of a tangent line from P_1 is $\frac{y - y_1}{x - (-p)} = \frac{dy}{dx} = \frac{2p}{y}$ $\Rightarrow y^2 - yy_1 = 2px + 2p^2$. Since $x = \frac{y^2}{4p}$, we have $y^2 - yy_1 = 2p\left(\frac{y^2}{4p}\right) + 2p^2 \Rightarrow y^2 - yy_1 = \frac{1}{2}y^2 + 2p^2$ $\Rightarrow \frac{1}{2}y^2 - yy_1 - 2p^2 = 0 \Rightarrow y = \frac{2y_1 \pm \sqrt{4y_1^2 + 16p^2}}{2} = y_1 \pm \sqrt{y_1^2 + 4p^2}$. Therefore the slopes of the two tangents from P_1 are $m_1 = \frac{2p}{y_1 + \sqrt{y_1^2 + 4p^2}}$ and $m_2 = \frac{2p}{y_1 - \sqrt{y_1^2 + 4p^2}} \Rightarrow m_1 m_2 = \frac{4p^2}{y_1^2 - \left(y_1^2 + 4p^2\right)} = -1$ $\Rightarrow$ the lines are perpendicular

83. Let $y = \sqrt{1 - \frac{x^2}{4}}$ on the interval $0 \le x \le 2$. The area of the inscribed rectangle is given by $A(x) = 2x\left(2\sqrt{1 - \frac{x^2}{4}}\right) = 4x\sqrt{1 - \frac{x^2}{4}}$ (since the length is 2x and the height is 2y) $\Rightarrow A'(x) = 4\sqrt{1 - \frac{x^2}{4}} - \frac{x^2}{\sqrt{1 - \frac{x^2}{4}}}$. Thus $A'(x) = 0 \Rightarrow 4\sqrt{1 - \frac{x^2}{4}} - \frac{x^2}{\sqrt{1 - \frac{x^2}{4}}} = 0 \Rightarrow 4\left(1 - \frac{x^2}{4}\right) - x^2 = 0 \Rightarrow x^2 = 2$ $\Rightarrow x = \sqrt{2}$ (only the positive square root lies in the interval). Since $A(0) = A(2) = 0$ we have that $A\left(\sqrt{2}\right) = 4$ is the maximum area when the length is $2\sqrt{2}$ and the height is $\sqrt{2}$.

84. (a) Around the x-axis: $9x^2 + 4y^2 = 36 \Rightarrow y^2 = 9 - \frac{9}{4}x^2 \Rightarrow y = \pm\sqrt{9 - \frac{9}{4}x^2}$ and we use the positive root $\Rightarrow V = 2\int_0^2 \pi\left(\sqrt{9 - \frac{9}{4}x^2}\right)^2 dx = 2\int_0^2 \pi\left(9 - \frac{9}{4}x^2\right) dx = 2\pi\left[9x - \frac{3}{4}x^3\right]_0^2 = 24\pi$

(b) Around the y-axis: $9x^2 + 4y^2 = 36 \Rightarrow x^2 = 4 - \frac{4}{9}y^2 \Rightarrow x = \pm\sqrt{4 - \frac{4}{9}y^2}$ and we use the positive root $\Rightarrow V = 2\int_0^3 \pi\left(\sqrt{4 - \frac{4}{9}y^2}\right)^2 dy = 2\int_0^3 \pi\left(4 - \frac{4}{9}y^2\right) dy = 2\pi\left[4y - \frac{4}{27}y^3\right]_0^3 = 16\pi$

85. $9x^2 - 4y^2 = 36 \Rightarrow y^2 = \frac{9x^2 - 36}{4} \Rightarrow y = \pm\frac{3}{2}\sqrt{x^2-4}$ on the interval $2 \le x \le 4 \Rightarrow V = \int_2^4 \pi\left(\frac{3}{2}\sqrt{x^2-4}\right)^2 dx$

$= \frac{9\pi}{4}\int_2^4 (x^2-4)\,dx = \frac{9\pi}{4}\left[\frac{x^3}{3} - 4x\right]_2^4 = \frac{9\pi}{4}\left[\left(\frac{64}{3} - 16\right) - \left(\frac{8}{3} - 8\right)\right] = \frac{9\pi}{4}\left(\frac{56}{3} - 8\right) = \frac{3\pi}{4}(56 - 24) = 24\pi$

86. $x^2 - y^2 = 1 \Rightarrow x = \pm\sqrt{1+y^2}$ on the interval $-3 \le y \le 3 \Rightarrow V = \int_{-3}^{3} \pi\left(\sqrt{1+y^2}\right)^2 dy = 2\int_0^3 \pi\left(\sqrt{1+y^2}\right)^2 dy$

$= 2\pi\int_0^3 (1+y^2)\,dy = 2\pi\left[y + \frac{y^3}{3}\right]_0^3 = 24\pi$

87. Let $y = \sqrt{16 - \frac{16}{9}x^2}$ on the interval $-3 \le x \le 3$. Since the plate is symmetric about the y-axis, $\overline{x} = 0$. For a vertical strip: $(\tilde{x}, \tilde{y}) = \left(x, \frac{\sqrt{16 - \frac{16}{9}x^2}}{2}\right)$, length $= \sqrt{16 - \frac{16}{9}x^2}$, width $= dx \Rightarrow$ area $= dA = \sqrt{16 - \frac{16}{9}x^2}\,dx$

$\Rightarrow$ mass $= dm = \delta\,dA = \delta\sqrt{16 - \frac{16}{9}x^2}\,dx$. Moment of the strip about the x-axis:

$\tilde{y}\,dm = \frac{\sqrt{16 - \frac{16}{9}x^2}}{2}\left(\delta\sqrt{16 - \frac{16}{9}x^2}\right)dx = \delta\left(8 - \frac{8}{9}x^2\right)dx$ so the moment of the plate about the x-axis is

$M_x = \int \tilde{y}\,dm = \int_{-3}^{3} \delta\left(8 - \frac{8}{9}x^2\right)dx = \delta\left[8x - \frac{8}{27}x^3\right]_{-3}^{3} = 32\delta$; also the mass of the plate is

$M = \int_{-3}^{3} \delta\sqrt{16 - \frac{16}{9}x^2}\,dx = \int_{-3}^{3} 4\delta\sqrt{1 - \left(\frac{1}{3}x\right)^2}\,dx = 4\delta\int_{-1}^{1} 3\sqrt{1-u^2}\,du$ where $u = \frac{x}{3} \Rightarrow 3\,du = dx$; $x = -3$

$\Rightarrow u = -1$ and $x = 3 \Rightarrow u = 1$. Hence, $4\delta\int_{-1}^{1} 3\sqrt{1-u^2}\,du = 12\delta\int_{-1}^{1}\sqrt{1-u^2}\,du$

$= 12\delta\left[\frac{1}{2}\left(u\sqrt{1-u^2} + \sin^{-1}u\right)\right]_{-1}^{1} = 6\pi\delta \Rightarrow \overline{y} = \frac{M_x}{M} = \frac{32\delta}{6\pi\delta} = \frac{16}{3\pi}$. Therefore the center of mass is $\left(0, \frac{16}{3\pi}\right)$.

88. $y = \sqrt{x^2+1} \Rightarrow \frac{dy}{dx} = \frac{1}{2}(x^2+1)^{-1/2}(2x) = \frac{x}{\sqrt{x^2+1}} \Rightarrow \left(\frac{dy}{dx}\right)^2 = \frac{x^2}{x^2+1} \Rightarrow \sqrt{1 + \left(\frac{dy}{dx}\right)^2} = \sqrt{1 + \frac{x^2}{x^2+1}}$

$= \sqrt{\frac{2x^2+1}{x^2+1}} \Rightarrow S = \int_0^{\sqrt{2}} 2\pi y\sqrt{1 + \left(\frac{dy}{dx}\right)^2}\,dx = \int_0^{\sqrt{2}} 2\pi\sqrt{x^2+1}\sqrt{\frac{2x^2+1}{x^2+1}}\,dx = \int_0^{\sqrt{2}} 2\pi\sqrt{2x^2+1}\,dx$;

$\left[\begin{array}{l} u = \sqrt{2}x \\ du = \sqrt{2}\,dx \end{array}\right] \rightarrow \frac{2\pi}{\sqrt{2}}\int_0^2 \sqrt{u^2+1}\,du = \frac{2\pi}{\sqrt{2}}\left[\frac{1}{2}\left(u\sqrt{u^2+1} + \ln\left(u + \sqrt{u^2+1}\right)\right)\right]_0^2 = \frac{\pi}{\sqrt{2}}\left[2\sqrt{5} + \ln\left(2+\sqrt{5}\right)\right]$

89. $\frac{dr_A}{dt} = \frac{dr_B}{dt} \Rightarrow \frac{d}{dt}(r_A - r_B) = 0 \Rightarrow r_A - r_B = C$, a constant $\Rightarrow$ the points $P(t)$ lie on a hyperbola with foci at A and B

90. (a) $\tan\beta = m_L \Rightarrow \tan\beta = f'(x_0)$ where $f(x) = \sqrt{4px}$;

$$f'(x) = \frac{1}{2}(4px)^{-1/2}(4p) = \frac{2p}{\sqrt{4px}} \Rightarrow f'(x_0) = \frac{2p}{\sqrt{4px_0}}$$

$$= \frac{2p}{y_0} \Rightarrow \tan\beta = \frac{2p}{y_0}.$$

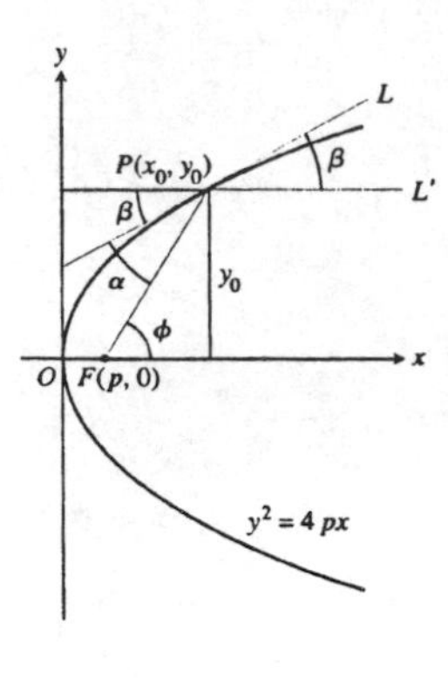

(b) $\tan\phi = m_{FP} = \frac{y_0 - 0}{x_0 - p} = \frac{y_0}{x_0 - p}$

(c) $$\tan\alpha = \frac{\tan\phi - \tan\beta}{1 + \tan\phi\tan\beta} = \frac{\left(\frac{y_0}{x_0 - p} - \frac{2p}{y_0}\right)}{1 + \left(\frac{y_0}{x_0 - p}\right)\left(\frac{2p}{y_0}\right)}$$

$$= \frac{y_0^2 - 2p(x_0 - p)}{y_0(x_0 - p + 2p)} = \frac{4px_0 - 2px_0 + 2p^2}{y_0(x_0 + p)} = \frac{2p(x_0 + p)}{y_0(x_0 + p)} = \frac{2p}{y_0}$$

91. PF will always equal PB because the string has constant length $AB = FP + PA = AP + PB$.

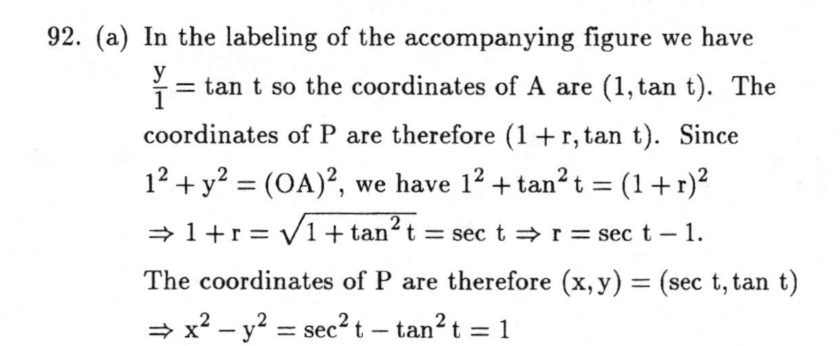

92. (a) In the labeling of the accompanying figure we have $\frac{y}{1} = \tan t$ so the coordinates of A are $(1, \tan t)$. The coordinates of P are therefore $(1 + r, \tan t)$. Since $1^2 + y^2 = (OA)^2$, we have $1^2 + \tan^2 t = (1 + r)^2$ $\Rightarrow 1 + r = \sqrt{1 + \tan^2 t} = \sec t \Rightarrow r = \sec t - 1$. The coordinates of P are therefore $(x, y) = (\sec t, \tan t)$ $\Rightarrow x^2 - y^2 = \sec^2 t - \tan^2 t = 1$

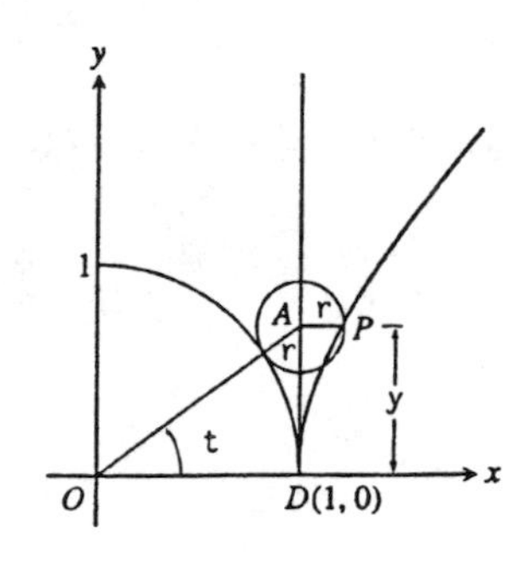

(b) In the labeling of the accompany figure the coordinates of A are $(\cos t, \sin t)$, the coordinates of C are $(1, \tan t)$, and the coordinates of P are $(1 + d, \tan t)$. By similar triangles, $\frac{d}{AB} = \frac{OC}{OA} \Rightarrow \frac{d}{1 - \cos t} = \frac{\sqrt{1 + \tan^2 t}}{1}$ $\Rightarrow d = (1 - \cos t)(\sec t) = \sec t - 1$. The coordinates of P are therefore $(\sec t, \tan t)$ and P moves on the hyperbola $x^2 - y^2 = 1$ as in part (a).

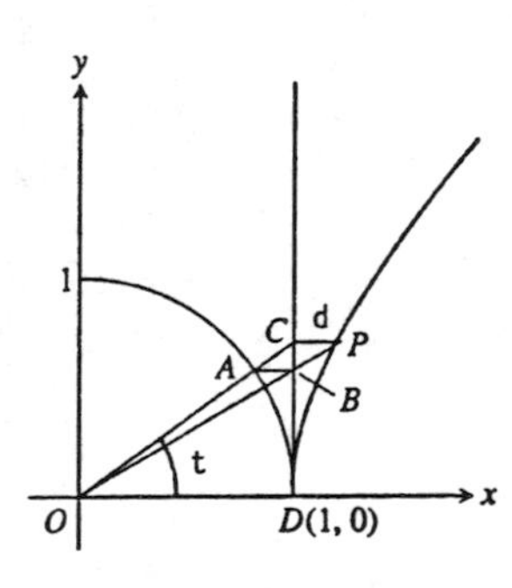

93. $x^2 = 4py$ and $y = p \Rightarrow x^2 = 4p^2 \Rightarrow x = \pm 2p$. Therefore the line $y = p$ cuts the parabola at points $(-2p, p)$ and $(2p, p)$, and these points are $\sqrt{[2p - (-2p)]^2 + (p - p)^2} = 4p$ units apart.

94. $$\lim_{x\to\infty}\left(\frac{b}{a}x - \frac{b}{a}\sqrt{x^2 - a^2}\right) = \frac{b}{a}\lim_{x\to\infty}\left(x - \sqrt{x^2 - a^2}\right) = \frac{b}{a}\lim_{x\to\infty}\left[\frac{\left(x - \sqrt{x^2 - a^2}\right)\left(x + \sqrt{x^2 - a^2}\right)}{x + \sqrt{x^2 - a^2}}\right]$$

$$= \frac{b}{a}\lim_{x\to\infty}\left[\frac{x^2 - (x^2 - a^2)}{x + \sqrt{x^2 - a^2}}\right] = \frac{b}{a}\lim_{x\to\infty}\left[\frac{a^2}{x + \sqrt{x^2 - a^2}}\right] = 0$$

9.2 CLASSIFYING CONIC SECTIONS BY ECCENTRICITY

1. $16x^2 + 25y^2 = 400 \Rightarrow \frac{x^2}{25} + \frac{y^2}{16} = 1 \Rightarrow c = \sqrt{a^2 - b^2}$
$= \sqrt{25 - 16} = 3 \Rightarrow e = \frac{c}{a} = \frac{3}{5}$; $F(\pm 3, 0)$;
directrices are $x = 0 \pm \frac{a}{e} = \pm \frac{5}{\left(\frac{3}{5}\right)} = \pm \frac{25}{3}$

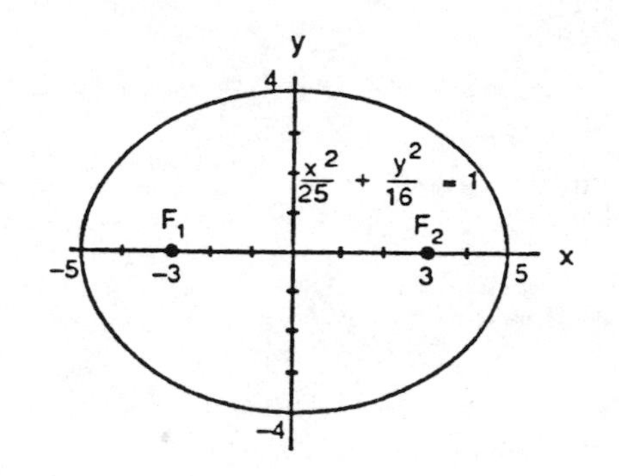

2. $7x^2 + 16y^2 = 112 \Rightarrow \frac{x^2}{16} + \frac{y^2}{7} = 1 \Rightarrow c = \sqrt{a^2 - b^2}$
$= \sqrt{16 - 7} = 3 \Rightarrow e = \frac{c}{a} = \frac{3}{4}$; $F(\pm 3, 0)$;
directrices are $x = 0 \pm \frac{a}{e} = \pm \frac{4}{\left(\frac{3}{4}\right)} = \pm \frac{16}{3}$

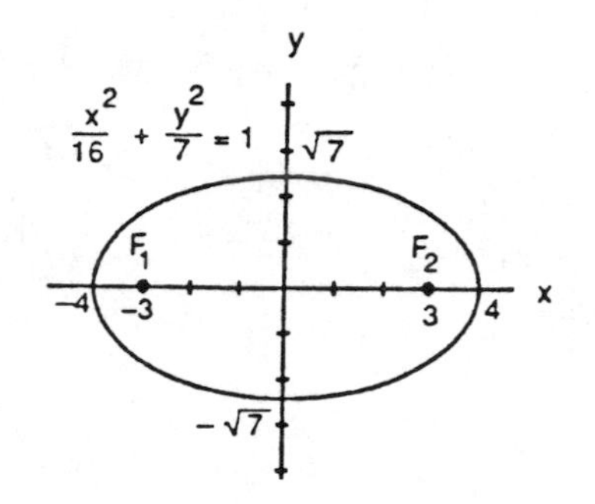

3. $2x^2 + y^2 = 2 \Rightarrow x^2 + \frac{y^2}{2} = 1 \Rightarrow c = \sqrt{a^2 - b^2}$
$= \sqrt{2 - 1} = 1 \Rightarrow e = \frac{c}{a} = \frac{1}{\sqrt{2}}$; $F(0, \pm 1)$;
directrices are $y = 0 \pm \frac{a}{e} = \pm \frac{\sqrt{2}}{\left(\frac{1}{\sqrt{2}}\right)} = \pm 2$

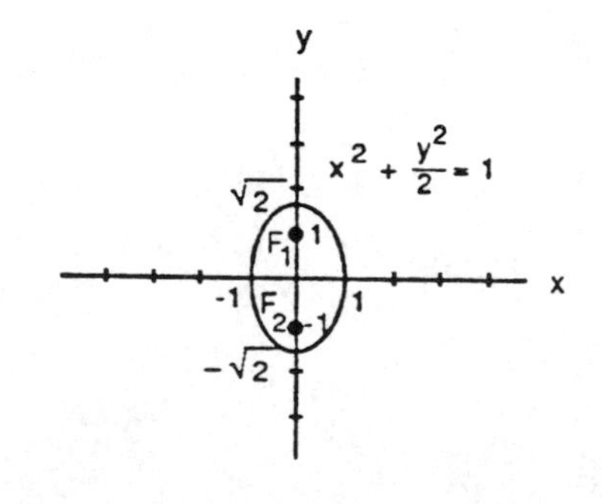

4. $2x^2 + y^2 = 4 \Rightarrow \frac{x^2}{2} + \frac{y^2}{4} = 1 \Rightarrow c = \sqrt{a^2 - b^2}$
$= \sqrt{4 - 2} = \sqrt{2} \Rightarrow e = \frac{c}{a} = \frac{\sqrt{2}}{2}$; $F(0, \pm\sqrt{2})$;
directrices are $y = 0 \pm \frac{a}{e} = \pm \frac{2}{\left(\frac{\sqrt{2}}{2}\right)} = \pm 2\sqrt{2}$

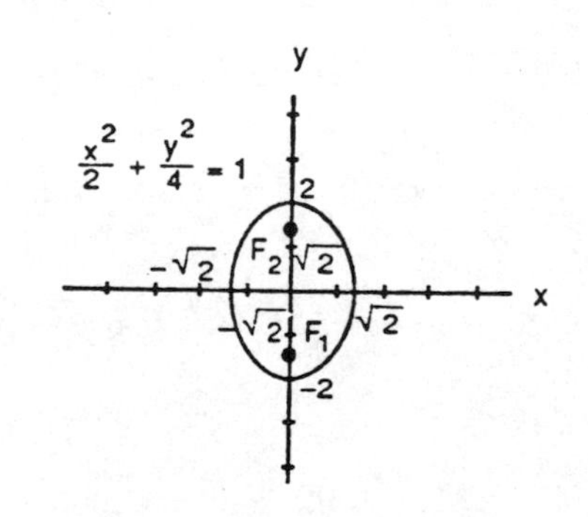

5. $3x^2 + 2y^2 = 6 \Rightarrow \frac{x^2}{2} + \frac{y^2}{3} = 1 \Rightarrow c = \sqrt{a^2 - b^2}$

$= \sqrt{3-2} = 1 \Rightarrow e = \frac{c}{a} = \frac{1}{\sqrt{3}}$; $F(0, \pm 1)$;

directrices are $y = 0 \pm \frac{a}{e} = \pm \frac{\sqrt{3}}{\left(\frac{1}{\sqrt{3}}\right)} = \pm 3$

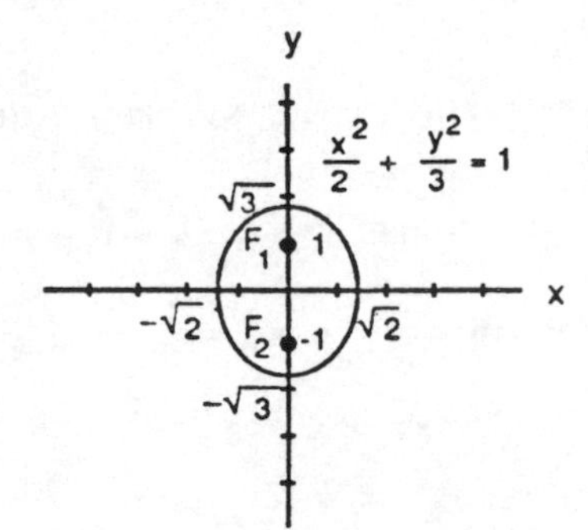

6. $9x^2 + 10y^2 = 90 \Rightarrow \frac{x^2}{10} + \frac{y^2}{9} = 1 \Rightarrow c = \sqrt{a^2 - b^2}$

$= \sqrt{10-9} = 1 \Rightarrow e = \frac{c}{a} = \frac{1}{\sqrt{10}}$; $F(\pm 1, 0)$;

directrices are $x = 0 \pm \frac{a}{e} = \pm \frac{\sqrt{10}}{\left(\frac{1}{\sqrt{10}}\right)} = \pm 10$

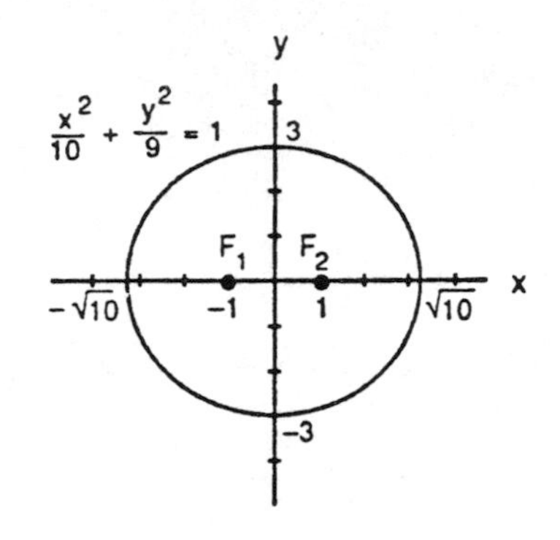

7. $6x^2 + 9y^2 = 54 \Rightarrow \frac{x^2}{9} + \frac{y^2}{6} = 1 \Rightarrow c = \sqrt{a^2 - b^2}$

$= \sqrt{9-6} = \sqrt{3} \Rightarrow e = \frac{c}{a} = \frac{\sqrt{3}}{3}$; $F(\pm \sqrt{3}, 0)$;

directrices are $x = 0 \pm \frac{a}{e} = \pm \frac{3}{\left(\frac{\sqrt{3}}{3}\right)} = \pm 3\sqrt{3}$

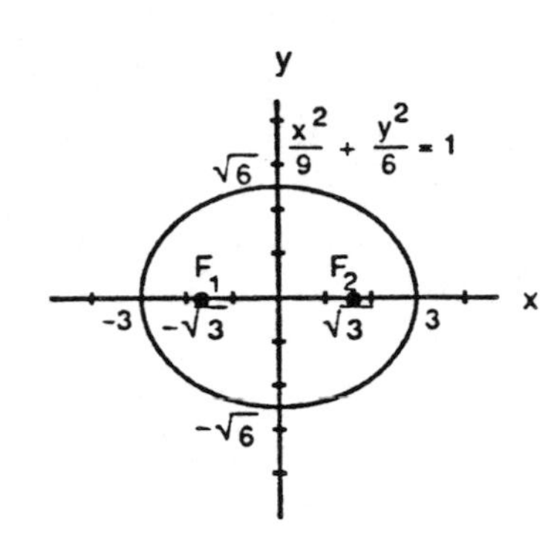

8. $169x^2 + 25y^2 = 4225 \Rightarrow \frac{x^2}{25} + \frac{y^2}{169} = 1 \Rightarrow c = \sqrt{a^2 - b^2}$

$= \sqrt{169-25} = 12 \Rightarrow e = \frac{c}{a} = \frac{12}{13}$; $F(0, \pm 12)$;

directrices are $y = 0 \pm \frac{a}{e} = \pm \frac{13}{\left(\frac{12}{13}\right)} = \pm \frac{169}{12}$

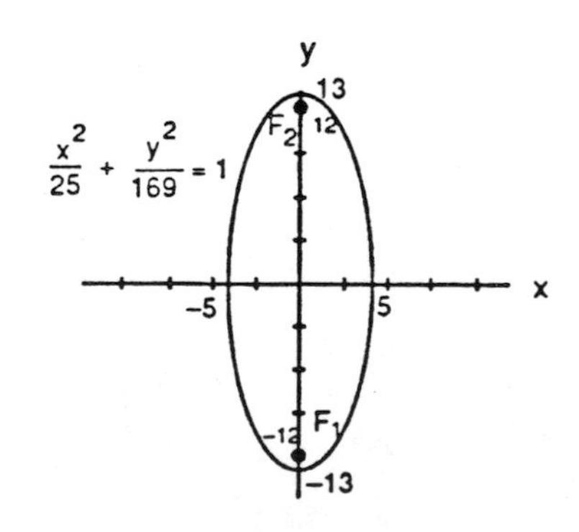

9. Foci: $(0, \pm 3)$, $e = 0.5 \Rightarrow c = 3$ and $a = \frac{c}{e} = \frac{3}{0.5} = 6 \Rightarrow b^2 = 36 - 9 = 27 \Rightarrow \frac{x^2}{27} + \frac{y^2}{36} = 1$

10. Foci: $(\pm 8, 0)$, $e = 0.2 \Rightarrow c = 8$ and $a = \frac{c}{e} = \frac{8}{0.2} = 40 \Rightarrow b^2 = 1600 - 64 = 1536 \Rightarrow \frac{x^2}{1600} + \frac{y^2}{1536} = 1$

11. Vertices: $(0, \pm 70)$, $e = 0.1 \Rightarrow a = 70$ and $c = ae = 70(0.1) = 7 \Rightarrow b^2 = 4900 - 49 = 4851 \Rightarrow \frac{x^2}{4851} + \frac{y^2}{4900} =$

12. Vertices: $(\pm 10, 0)$, $e = 0.24 \Rightarrow a = 10$ and $c = ae = 10(0.24) = 2.4 \Rightarrow b^2 = 100 - 5.76 = 94.24$
$\Rightarrow \frac{x^2}{100} + \frac{y^2}{94.24} = 1$

13. Focus: $(\sqrt{5}, 0)$, Directrix: $x = \frac{9}{\sqrt{5}} \Rightarrow c = ae = \sqrt{5}$ and $\frac{a}{e} = \frac{9}{\sqrt{5}} \Rightarrow \frac{ae}{e^2} = \frac{9}{\sqrt{5}} \Rightarrow \frac{\sqrt{5}}{e^2} = \frac{9}{\sqrt{5}} \Rightarrow e^2 = \frac{5}{9}$
$\Rightarrow e = \frac{\sqrt{5}}{3}$. Then $PF = \frac{\sqrt{5}}{3} PD \Rightarrow \sqrt{(x - \sqrt{5})^2 + (y - 0)^2} = \frac{\sqrt{5}}{3}\left|x - \frac{9}{\sqrt{5}}\right| \Rightarrow (x - \sqrt{5})^2 + y^2 = \frac{5}{9}\left(x - \frac{9}{\sqrt{5}}\right)^2$
$\Rightarrow x^2 - 2\sqrt{5}x + 5 + y^2 = \frac{5}{9}\left(x^2 - \frac{18}{\sqrt{5}}x + \frac{81}{5}\right) \Rightarrow \frac{4}{9}x^2 + y^2 = 4 \Rightarrow \frac{x^2}{9} + \frac{y^2}{4} = 1$

14. Focus: $(4, 0)$, Directrix: $x = \frac{16}{3} \Rightarrow c = ae = 4$ and $\frac{a}{e} = \frac{16}{3} \Rightarrow \frac{ae}{e^2} = \frac{16}{3} \Rightarrow \frac{4}{e^2} = \frac{16}{3} \Rightarrow e^2 = \frac{3}{4} \Rightarrow e = \frac{\sqrt{3}}{2}$. Then
$PF = \frac{\sqrt{3}}{2} PD \Rightarrow \sqrt{(x - 4)^2 + (y - 0)^2} = \frac{\sqrt{3}}{2}\left|x - \frac{16}{3}\right| \Rightarrow (x - 4)^2 + y^2 = \frac{3}{4}\left(x - \frac{16}{3}\right)^2 \Rightarrow x^2 - 8x + 16 + y^2$
$= \frac{3}{4}\left(x^2 - \frac{32}{3}x + \frac{256}{9}\right) \Rightarrow \frac{1}{4}x^2 + y^2 = \frac{16}{3} \Rightarrow \frac{x^2}{\left(\frac{64}{3}\right)} + \frac{y^2}{\left(\frac{16}{3}\right)} = 1$

15. Focus: $(-4, 0)$, Directrix: $x = -16 \Rightarrow c = ae = 4$ and $\frac{a}{e} = 16 \Rightarrow \frac{ae}{e^2} = 16 \Rightarrow \frac{4}{e^2} = 16 \Rightarrow e^2 = \frac{1}{4} \Rightarrow e = \frac{1}{2}$. Then
$PF = \frac{1}{2} PD \Rightarrow \sqrt{(x + 4)^2 + (y - 0)^2} = \frac{1}{2}|x + 16| \Rightarrow (x + 4)^2 + y^2 = \frac{1}{4}(x + 16)^2 \Rightarrow x^2 + 8x + 16 + y^2$
$= \frac{1}{4}(x^2 + 32x + 256) \Rightarrow \frac{3}{4}x^2 + y^2 = 48 \Rightarrow \frac{x^2}{64} + \frac{y^2}{48} = 1$

16. Focus: $(-\sqrt{2}, 0)$, Directrix: $x = -2\sqrt{2} \Rightarrow c = ae = \sqrt{2}$ and $\frac{a}{e} = 2\sqrt{2} \Rightarrow \frac{ae}{e^2} = 2\sqrt{2} \Rightarrow \frac{\sqrt{2}}{e^2} = 2\sqrt{2} \Rightarrow e^2 = \frac{1}{2}$
$\Rightarrow e = \frac{1}{\sqrt{2}}$. Then $PF = \frac{1}{\sqrt{2}} PD \Rightarrow \sqrt{(x + \sqrt{2})^2 + (y - 0)^2} = \frac{1}{\sqrt{2}}|x + 2\sqrt{2}| \Rightarrow (x + \sqrt{2})^2 + y^2$
$= \frac{1}{2}(x + 2\sqrt{2})^2 \Rightarrow x^2 + 2\sqrt{2}x + 2 + y^2 = \frac{1}{2}(x^2 + 4\sqrt{2}x + 8) \Rightarrow \frac{1}{2}x^2 + y^2 = 2 \Rightarrow \frac{x^2}{4} + \frac{y^2}{2} = 1$

17. $e = \frac{4}{5} \Rightarrow$ take $c = 4$ and $a = 5$; $c^2 = a^2 - b^2$
$\Rightarrow 16 = 25 - b^2 \Rightarrow b^2 = 9 \Rightarrow b = 3$; therefore
$\frac{x^2}{25} + \frac{y^2}{9} = 1$

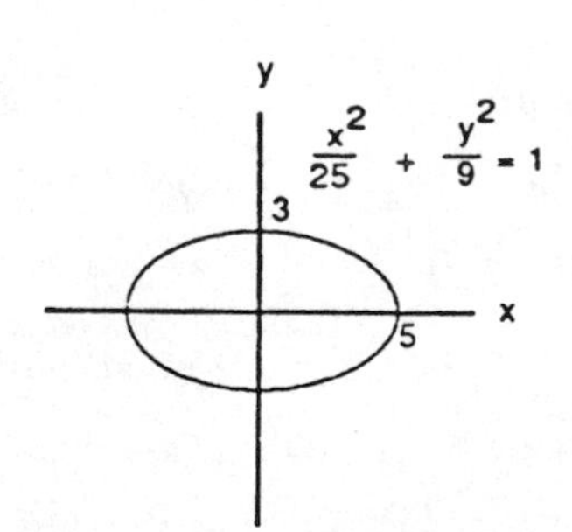

18. The eccentricity e for Pluto is 0.25 $\Rightarrow e = \frac{c}{a} = 0.25 = \frac{1}{4}$ $\Rightarrow$ take c = 1 and a = 4; $c^2 = a^2 - b^2 \Rightarrow 1 = 16 - b^2$ $\Rightarrow b^2 = 15 \Rightarrow b = \sqrt{15}$; therefore, $\frac{x^2}{16} + \frac{y^2}{15} = 1$ is a model of Pluto's orbit.

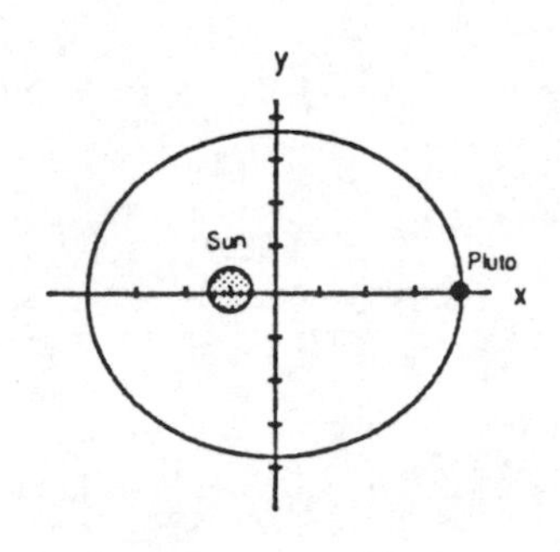

19. One axis is from A(1,1) to B(1,7) and is 6 units long; the other axis is from C(3,4) to D(−1,4) and is 4 units long. Therefore a = 3, b = 2 and the major axis is vertical. The center is the point C(1,4) and the ellipse is given by $\frac{(x-1)^2}{4} + \frac{(y-4)^2}{9} = 1$; $c^2 = a^2 - b^2 = 3^2 - 2^2 = 5$ $\Rightarrow c = \sqrt{5}$; therefore the foci are $F(1, 4 \pm \sqrt{5})$, the eccentricity is $e = \frac{c}{a} = \frac{\sqrt{5}}{3}$, and the directrices are $y = 4 \pm \frac{a}{e} = 4 \pm \frac{3}{\left(\frac{\sqrt{5}}{3}\right)} = 4 \pm \frac{9\sqrt{5}}{5}$.

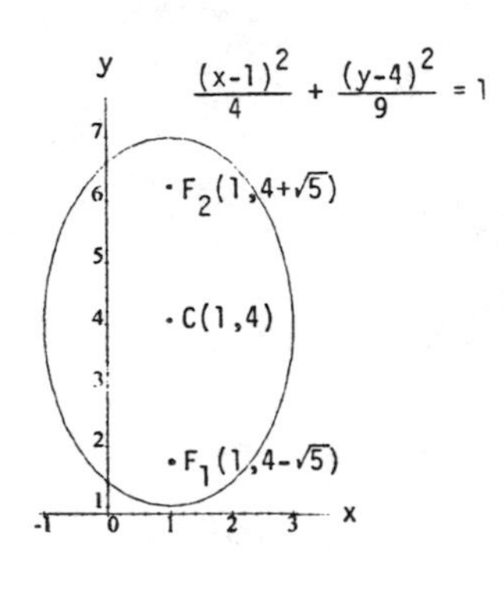

20. Using PF = e · PD, we have $\sqrt{(x-4)^2 + y^2} = \frac{2}{3}|x - 9| \Rightarrow (x-4)^2 + y^2 = \frac{4}{9}(x-9)^2 \Rightarrow x^2 - 8x + 16 + y^2$ $= \frac{4}{9}(x^2 - 18x + 81) \Rightarrow \frac{5}{9}x^2 + y^2 = 20 \Rightarrow 5x^2 + 9y^2 = 180$ or $\frac{x^2}{36} + \frac{y^2}{20} = 1$.

21. The ellipse must pass through (0,0) $\Rightarrow$ c = 0; the point (−1,2) lies on the ellipse $\Rightarrow -a + 2b = -8$. The ellipse is tangent to the x-axis $\Rightarrow$ its center is on the y-axis, so a = 0 and b = −4 $\Rightarrow$ the equation is $4x^2 + y^2 - 4y = 0$. Next, $4x^2 + y^2 - 4y + 16 = 16 \Rightarrow 4x^2 + (y-4)^2 = 16 \Rightarrow \frac{x^2}{4} + \frac{(y-4)^2}{16} = 1 \Rightarrow$ a = 4 and b = 2 (now using the standard symbols) $\Rightarrow c^2 = a^2 - b^2 = 16 - 4 = 12 \Rightarrow c = \sqrt{12} \Rightarrow e = \frac{c}{a} = \frac{\sqrt{12}}{4} = \frac{\sqrt{3}}{2}$.

22. We first prove a result which we will use: let m_1, and m_2 be two nonparallel, nonperpendicular lines. Let α be the acute angle between the lines. Then $\tan \alpha = \frac{m_1 - m_2}{1 + m_1 m_2}$. To see this result, let θ_1 be the angle of inclination of the line with slope m_1, and θ_2 be the angle of inclination of the line with slope m_2. Assume $m_1 > m_2$. Then $\theta_1 > \theta_2$ and we have $\alpha = \theta_1 - \theta_2$. Then $\tan \alpha = \tan(\theta_1 - \theta_2)$ $= \frac{\tan\theta_1 - \tan\theta_2}{1 + \tan\theta_1 \tan\theta_2} = \frac{m_1 - m_2}{1 + m_1 m_2}$, since $m_1 = \tan\theta_1$ and and $m_2 = \tan\theta_2$.

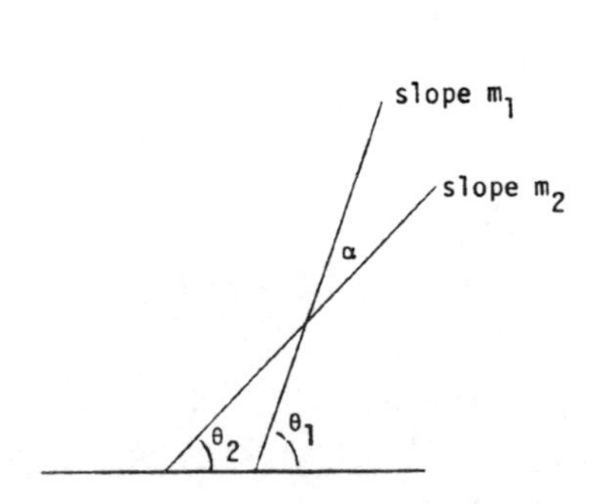

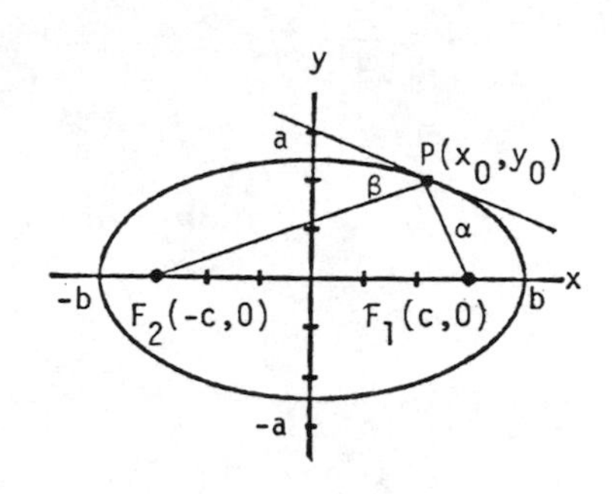

Now we prove the reflective property of ellipses (see the accompanying figure): If $\frac{x^2}{a^2}+\frac{y^2}{b^2}=1$, then $b^2x^2+a^2y^2=a^2b^2$ and $y=\frac{b}{a}\sqrt{a^2-x^2} \Rightarrow y'=\frac{-bx}{a\sqrt{a^2-x^2}}$. Let $P(x_0,y_0)$ be any point on the ellipse $\Rightarrow y'(x_0)=\frac{-bx_0}{a\sqrt{a^2-x_0^2}}=\frac{-b^2x_0}{a^2y_0}$. Let $F_1(c,0)$ and $F_2(-c,0)$ be the foci. Then $m_{PF_1}=\frac{y_0}{x_0-c}$ and $m_{PF_2}=\frac{y_0}{x_0+c}$. Let α and β be the angles between the tangent line and PF_1 and PF_2, respectively. Then

$$\tan\alpha=\frac{\left(-\frac{b^2x_0}{a^2y_0}-\frac{y_0}{x_0-c}\right)}{\left(1-\frac{b^2x_0y_0}{a^2y_0(x_0-c)}\right)}=\frac{-b^2x_0^2+b^2x_0c-a^2y_0^2}{a^2y_0x_0-a^2y_0c-b^2x_0y_0}=\frac{b^2x_0c-\left(b^2x_0^2+a^2y_0^2\right)}{-a^2y_0c+\left(a^2-b^2\right)x_0y_0}=\frac{b^2x_0c-a^2b^2}{-a^2y_0c+c^2x_0y_0}=\frac{b^2}{cy_0}.$$

Similarly, $\tan\beta=\frac{b^2}{cy_0}$. Since $\tan\alpha=\tan\beta$, and α and β are both less than 90°, we have $\alpha=\beta$.

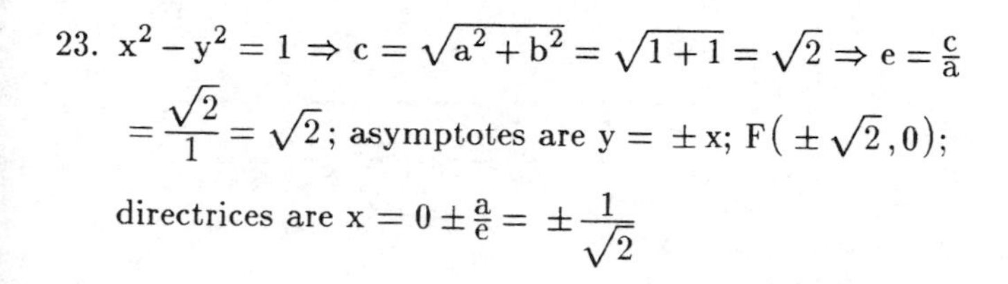

23. $x^2-y^2=1 \Rightarrow c=\sqrt{a^2+b^2}=\sqrt{1+1}=\sqrt{2} \Rightarrow e=\frac{c}{a}$
$=\frac{\sqrt{2}}{1}=\sqrt{2}$; asymptotes are $y=\pm x$; $F\left(\pm\sqrt{2},0\right)$;
directrices are $x=0\pm\frac{a}{e}=\pm\frac{1}{\sqrt{2}}$

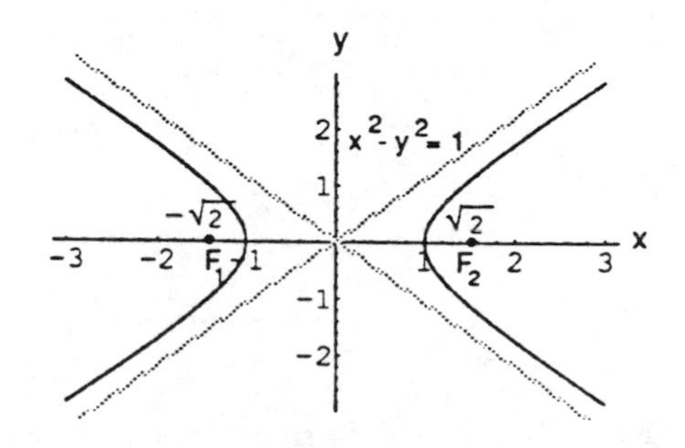

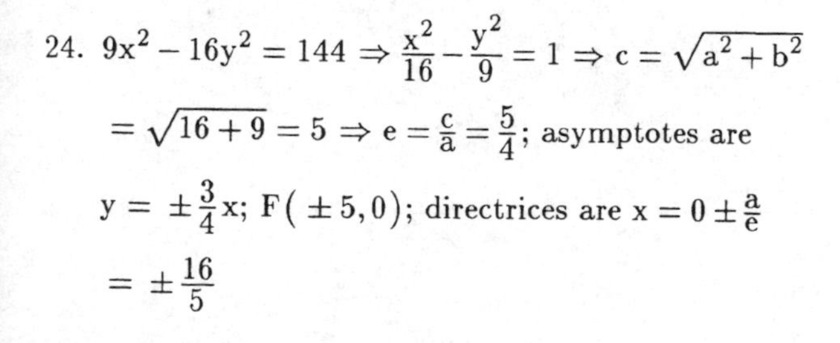

24. $9x^2-16y^2=144 \Rightarrow \frac{x^2}{16}-\frac{y^2}{9}=1 \Rightarrow c=\sqrt{a^2+b^2}$
$=\sqrt{16+9}=5 \Rightarrow e=\frac{c}{a}=\frac{5}{4}$; asymptotes are
$y=\pm\frac{3}{4}x$; $F\left(\pm 5,0\right)$; directrices are $x=0\pm\frac{a}{e}$
$=\pm\frac{16}{5}$

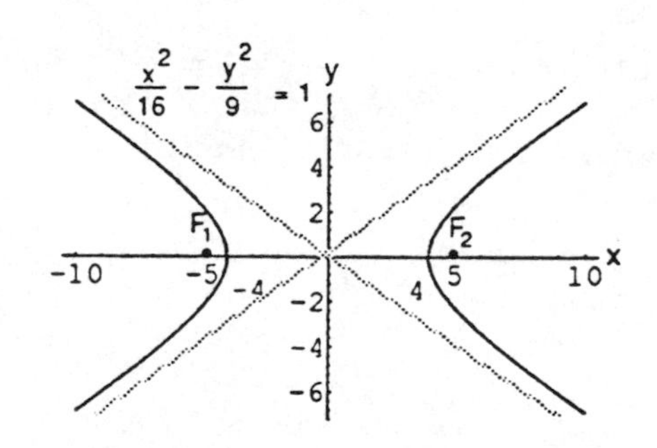

25. $y^2-x^2=8 \Rightarrow \frac{y^2}{8}-\frac{x^2}{8}=1 \Rightarrow c=\sqrt{a^2+b^2}$
$=\sqrt{8+8}=4 \Rightarrow e=\frac{c}{a}=\frac{4}{\sqrt{8}}=\sqrt{2}$; asymptotes are
$y=\pm x$; $F(0,\pm 4)$; directrices are $y=0\pm\frac{a}{e}$
$=\pm\frac{\sqrt{8}}{\sqrt{2}}=\pm 2$

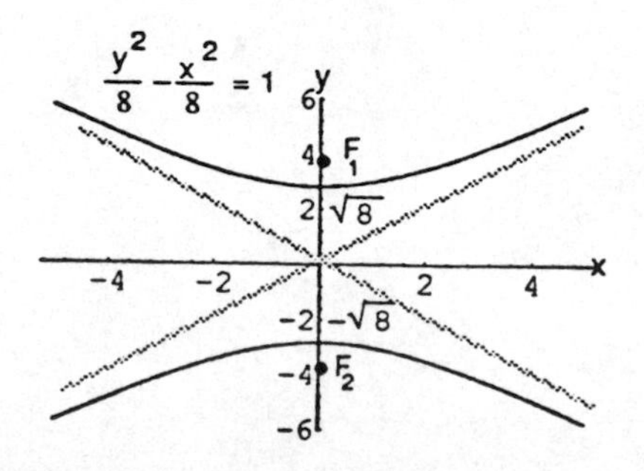

26. $y^2 - x^2 = 4 \Rightarrow \frac{y^2}{4} - \frac{x^2}{4} = 1 \Rightarrow c = \sqrt{a^2 + b^2}$
$= \sqrt{4+4} = 2\sqrt{2} \Rightarrow e = \frac{c}{a} = \frac{2\sqrt{2}}{2} = \sqrt{2}$; asymptotes are
$y = \pm x$; $F(0, \pm 2\sqrt{2})$; directrices are $y = 0 \pm \frac{a}{e}$
$= \pm \frac{2}{\sqrt{2}} = \pm \sqrt{2}$

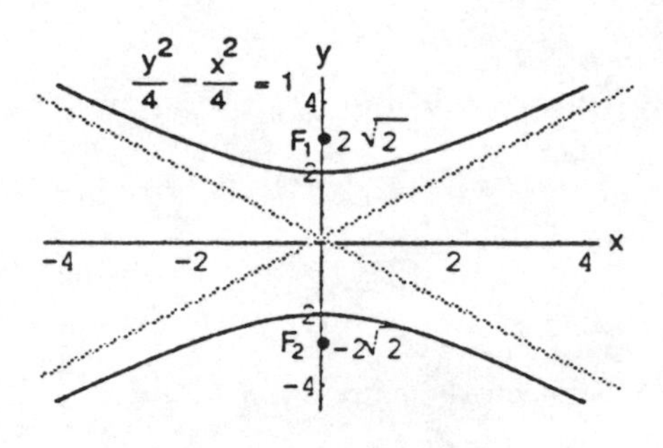

27. $8x^2 - 2y^2 = 16 \Rightarrow \frac{x^2}{2} - \frac{y^2}{8} = 1 \Rightarrow c = \sqrt{a^2 + b^2}$
$= \sqrt{2+8} = \sqrt{10} \Rightarrow e = \frac{c}{a} = \frac{\sqrt{10}}{\sqrt{2}} = \sqrt{5}$; asymptotes are
$y = \pm 2x$; $F(\pm\sqrt{10}, 0)$; directrices are $x = 0 \pm \frac{a}{e}$
$= \pm \frac{\sqrt{2}}{\sqrt{5}} = \pm \frac{2}{\sqrt{10}}$

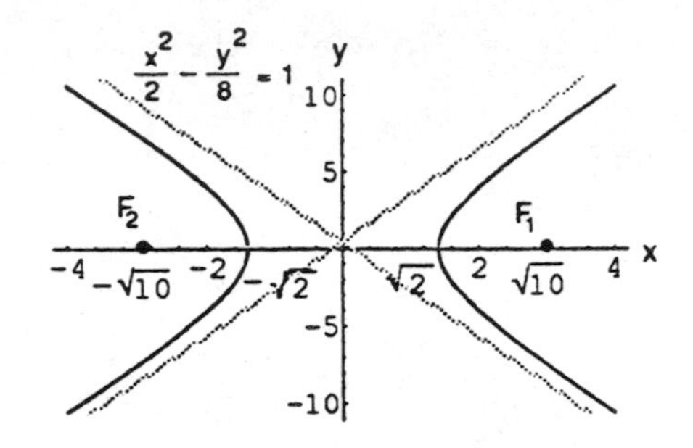

28. $y^2 - 3x^2 = 3 \Rightarrow \frac{y^2}{3} - x^2 = 1 \Rightarrow c = \sqrt{a^2 + b^2}$
$= \sqrt{3+1} = 2 \Rightarrow e = \frac{c}{a} = \frac{2}{\sqrt{3}}$; asymptotes are
$y = \pm\sqrt{3}x$; $F(0, \pm 2)$; directrices are $y = 0 \pm \frac{a}{e}$
$= \pm \frac{\sqrt{3}}{\left(\frac{2}{\sqrt{3}}\right)} = \pm \frac{3}{2}$

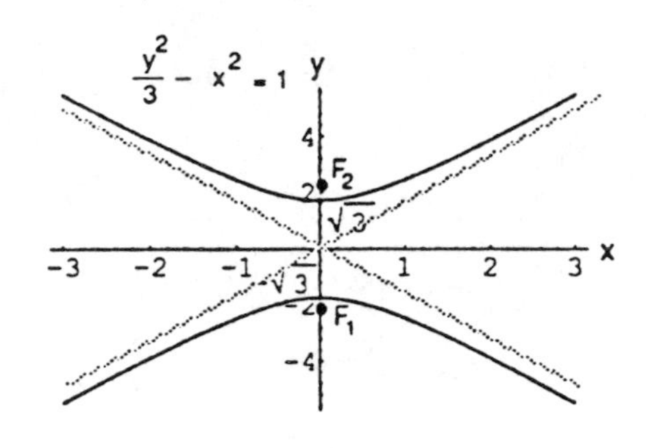

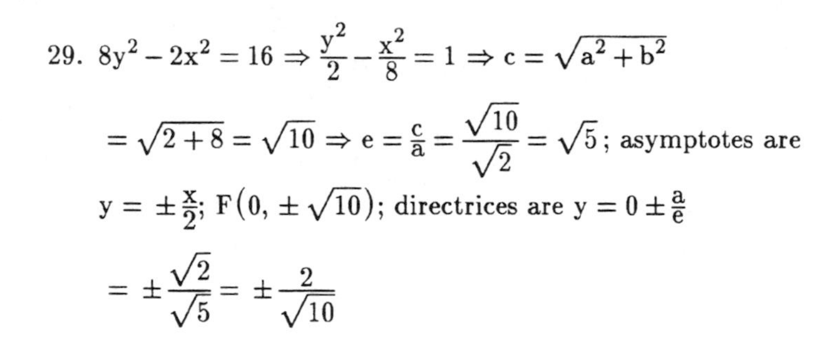

29. $8y^2 - 2x^2 = 16 \Rightarrow \frac{y^2}{2} - \frac{x^2}{8} = 1 \Rightarrow c = \sqrt{a^2 + b^2}$
$= \sqrt{2+8} = \sqrt{10} \Rightarrow e = \frac{c}{a} = \frac{\sqrt{10}}{\sqrt{2}} = \sqrt{5}$; asymptotes are
$y = \pm \frac{x}{2}$; $F(0, \pm\sqrt{10})$; directrices are $y = 0 \pm \frac{a}{e}$
$= \pm \frac{\sqrt{2}}{\sqrt{5}} = \pm \frac{2}{\sqrt{10}}$

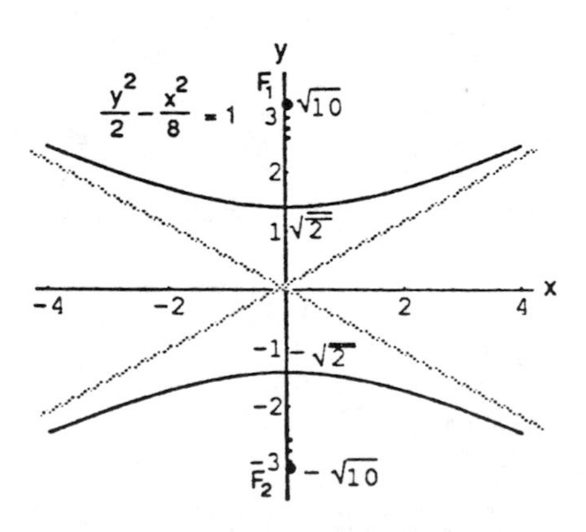

30. $64x^2 - 36y^2 = 2304 \Rightarrow \frac{x^2}{36} - \frac{y^2}{64} = 1 \Rightarrow c = \sqrt{a^2 + b^2}$
$= \sqrt{36 + 64} = 10 \Rightarrow e = \frac{c}{a} = \frac{10}{6} = \frac{5}{3}$; asymptotes are
$y = \pm\frac{4}{3}x$; $F(\pm 10, 0)$; directrices are $x = 0 \pm \frac{a}{e}$
$= \pm\frac{6}{\left(\frac{5}{3}\right)} = \pm\frac{18}{5}$

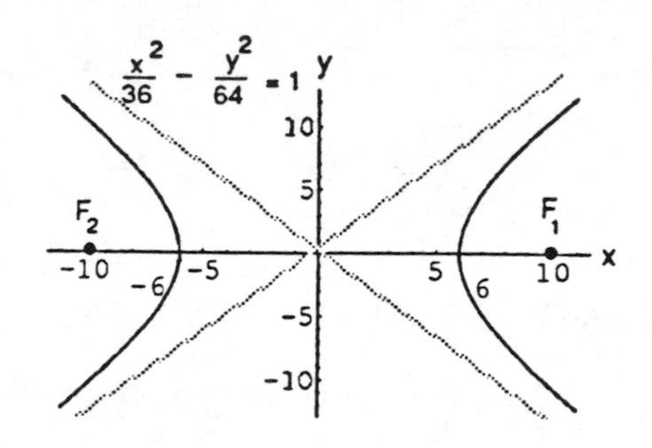

31. Vertices $(0, \pm 1)$ and $e = 3 \Rightarrow a = 1$ and $e = \frac{c}{a} = 3 \Rightarrow c = 3a = 3 \Rightarrow b^2 = c^2 - a^2 = 9 - 1 = 8 \Rightarrow y^2 - \frac{x^2}{8} = 1$

32. Vertices $(\pm 2, 0)$ and $e = 2 \Rightarrow a = 2$ and $e = \frac{c}{a} = 2 \Rightarrow c = 2a = 4 \Rightarrow b^2 = c^2 - a^2 = 16 - 4 = 12 \Rightarrow \frac{x^2}{4} - \frac{y^2}{12} = 1$

33. Foci $(\pm 3, 0)$ and $e = 3 \Rightarrow c = 3$ and $e = \frac{c}{a} = 3 \Rightarrow c = 3a \Rightarrow a = 1 \Rightarrow b^2 = c^2 - a^2 = 9 - 1 = 8 \Rightarrow x^2 - \frac{y^2}{8} = 1$

34. Foci $(0, \pm 5)$ and $e = 1.25 \Rightarrow c = 5$ and $e = \frac{c}{a} = 1.25 = \frac{5}{4} \Rightarrow c = \frac{5}{4}a \Rightarrow 5 = \frac{5}{4}a \Rightarrow a = 4 \Rightarrow b^2 = c^2 - a^2$
$= 25 - 16 = 9 \Rightarrow \frac{y^2}{16} - \frac{x^2}{9} = 1$

35. Focus $(4, 0)$ and Directrix $x = 2 \Rightarrow c = ae = 4$ and $\frac{a}{e} = 2 \Rightarrow \frac{ae}{e^2} = 2 \Rightarrow \frac{4}{e^2} = 2 \Rightarrow e^2 = 2 \Rightarrow e = \sqrt{2}$. Then
$PF = \sqrt{2}\,PD \Rightarrow \sqrt{(x-4)^2 + (y-0)^2} = \sqrt{2}\,|x - 2| \Rightarrow (x-4)^2 + y^2 = 2(x-2)^2 \Rightarrow x^2 - 8x + 16 + y^2$
$= 2(x^2 - 4x + 4) \Rightarrow -x^2 + y^2 = -8 \Rightarrow \frac{x^2}{8} - \frac{y^2}{8} = 1$

36. Focus $(\sqrt{10}, 0)$ and Directrix $x = \sqrt{2} \Rightarrow c = ae = \sqrt{10}$ and $\frac{a}{e} = \sqrt{2} \Rightarrow \frac{ae}{e^2} = \sqrt{2} \Rightarrow \frac{\sqrt{10}}{e^2} = \sqrt{2} \Rightarrow e^2 = \sqrt{5}$
$\Rightarrow e = \sqrt[4]{5}$. Then $PF = \sqrt[4]{5}\,PD \Rightarrow \sqrt{\left(x - \sqrt{10}\right)^2 + (y-0)^2} = \sqrt[4]{5}\left|x - \sqrt{2}\right| \Rightarrow \left(x - \sqrt{10}\right)^2 + y^2$
$= \sqrt{5}\left(x - \sqrt{2}\right)^2 \Rightarrow x^2 - 2\sqrt{10}\,x + 10 + y^2 = \sqrt{5}(x^2 - 2\sqrt{2}\,x + 2) \Rightarrow (1 - \sqrt{5})x^2 + y^2 = 2\sqrt{5} - 10$
$\Rightarrow \frac{(1-\sqrt{5})x^2}{2\sqrt{5} - 10} + \frac{y^2}{2\sqrt{5} - 10} = 1 \Rightarrow \frac{x^2}{2\sqrt{5}} - \frac{y^2}{10 - 2\sqrt{5}} = 1$

37. Focus $(-2, 0)$ and Directrix $x = -\frac{1}{2} \Rightarrow c = ae = 2$ and $\frac{a}{e} = \frac{1}{2} \Rightarrow \frac{ae}{e^2} = \frac{1}{2} \Rightarrow \frac{2}{e^2} = \frac{1}{2} \Rightarrow e^2 = 4 \Rightarrow e = 2$. Then
$PF = 2PD \Rightarrow \sqrt{(x+2)^2 + (y-0)^2} = 2\left|x + \frac{1}{2}\right| \Rightarrow (x+2)^2 + y^2 = 4\left(x + \frac{1}{2}\right)^2 \Rightarrow x^2 + 4x + 4 + y^2 = 4\left(x^2 + x + \frac{1}{4}\right)$
$\Rightarrow -3x^2 + y^2 = -3 \Rightarrow x^2 - \frac{y^2}{3} = 1$

38. Focus $(-6, 0)$ and Directrix $x = -2 \Rightarrow c = ae = 6$ and $\frac{a}{e} = 2 \Rightarrow \frac{ae}{e^2} = 2 \Rightarrow \frac{6}{e^2} = 2 \Rightarrow e^2 = 3 \Rightarrow e = \sqrt{3}$. Then
$PF = \sqrt{3}\,PD \Rightarrow \sqrt{(x+6)^2 + (y-0)^2} = \sqrt{3}\,|x + 2| \Rightarrow (x+6)^2 + y^2 = 3(x+2)^2 \Rightarrow x^2 + 12x + 36 + y^2$
$= 3(x^2 + 4x + 4) \Rightarrow -2x^2 + y^2 = -24 \Rightarrow \frac{x^2}{12} - \frac{y^2}{24} = 1$

39. $\sqrt{(x-1)^2+(y+3)^2} = \frac{3}{2}|y-2| \Rightarrow x^2-2x+1+y^2+6y+9 = \frac{9}{4}(y^2-4y+4) \Rightarrow 4x^2-5y^2-8x+60y+4=0$

$\Rightarrow 4(x^2-2x+1)-5(y^2-12y+36) = -4+4-180 \Rightarrow \frac{(y-6)^2}{36} - \frac{(x-1)^2}{45} = 1$

40. $c^2 = a^2+b^2 \Rightarrow b^2 = c^2-a^2$; $e = \frac{c}{a} \Rightarrow c = ea \Rightarrow c^2 = e^2a^2 \Rightarrow b^2 = e^2a^2 - a^2 = a^2(e^2-1)$; thus, $\frac{x^2}{a^2} - \frac{y^2}{b^2} = 1 \Rightarrow \frac{x^2}{a^2} - \frac{y^2}{a^2(e^2-1)} = 1$; the asymptotes of this hyperbola are $y = \pm(e^2-1) \Rightarrow$ as e increases, the slopes of the asymptotes increase and the hyperbola approaches a straight line.

41. To prove the reflective property for hyperbolas:

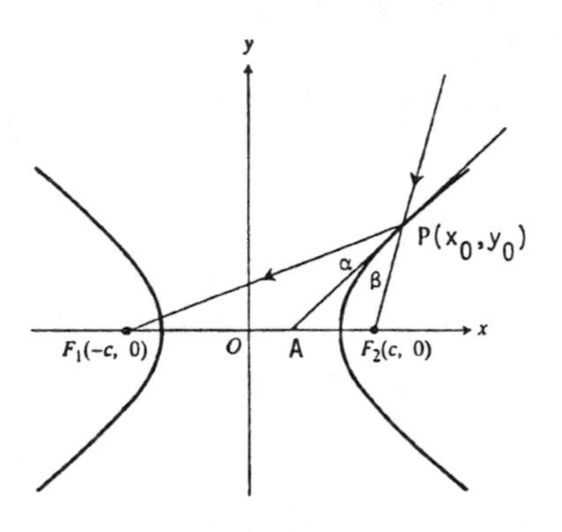

$\frac{x^2}{a^2} - \frac{y^2}{b^2} = 1 \Rightarrow a^2y^2 = b^2x^2 - a^2b^2$ and $\frac{dy}{dx} = \frac{xb^2}{ya^2}$.

Let $P(x_0, y_0)$ be a point of tangency (see the accompanying figure). The slope from P to F(−c, 0) is $\frac{y_0}{x_0+c}$ and from P to $F_2(c, 0)$ it is $\frac{y_0}{x_0-c}$. Let the tangent through P meet the x-axis in point A, and define the angles $\angle F_1PA = \alpha$ and $\angle F_2PA = \beta$. We will show that $\tan\alpha = \tan\beta$. From the preliminary result in Exercise 22,

$$\tan\alpha = \frac{\left(\frac{x_0b^2}{y_0a^2} - \frac{y_0}{x_0+c}\right)}{1+\left(\frac{x_0b^2}{y_0a^2}\right)\left(\frac{y_0}{x_0+c}\right)} = \frac{x_0^2b^2 + x_0b^2c - y_0^2a^2}{x_0y_0a^2 + y_0a^2c + x_0y_0b^2} = \frac{a^2b^2 + x_0b^2c}{x_0y_0c^2 + y_0a^2c} = \frac{b^2}{y_0c}.$$ In a similar manner,

$$\tan\beta = \frac{\left(\frac{y_0}{x_0-c} - \frac{x_0b^2}{y_0a^2}\right)}{1+\left(\frac{y_0}{x_0-c}\right)\left(\frac{x_0b^2}{y_0a^2}\right)} = \frac{b^2}{y_0c}.$$ Since $\tan\alpha = \tan\beta$, and α and β are acute angles, we have $\alpha = \beta$.

42. From the accompanying figure, a ray of light emanating from the focus A that met the parabola at P would be reflected from the hyperbola as if it came directly from B (Exercise 41). The same light ray would be reflected off the ellipse to pass through B. Thus BPC is a straight line.

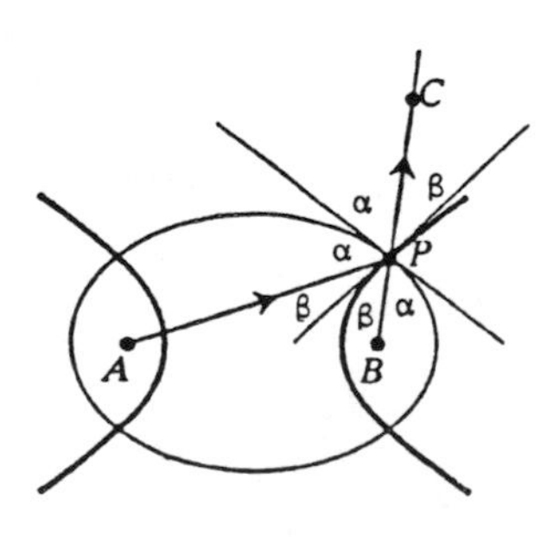

Let β be the angle of incidence of the light ray on the hyperbola. Let α be the angle of incidence of the light ray on the ellipse. Note that $\alpha+\beta$ is the angle between the tangent lines to the ellipse and hyperbola at P. Since BPC is a straight line, $2\alpha + 2\beta = 180°$. Thus $\alpha + \beta = 90°$.

9.3 QUADRATIC EQUATIONS AND ROTATIONS

1. $x^2 - 3xy + y^2 - x = 0 \Rightarrow B^2 - 4AC = (-3)^2 - 4(1)(1) = 5 > 0 \Rightarrow$ Hyperbola

2. $3x^2 - 18xy + 27y^2 - 5x + 7y = -4 \Rightarrow B^2 - 4AC = (-18)^2 - 4(3)(27) = 0 \Rightarrow$ Parabola

3. $3x^2 - 7xy + \sqrt{17}y^2 = 1 \Rightarrow B^2 - 4AC = (-7)^2 - 4(3)\sqrt{17} \approx -0.477 < 0 \Rightarrow$ Ellipse

4. $2x^2 - \sqrt{15}\,xy + 2y^2 + x + y = 0 \Rightarrow B^2 - 4AC = \left(-\sqrt{15}\right)^2 - 4(2)(2) = -1 < 0 \Rightarrow$ Ellipse

5. $x^2 + 2xy + y^2 + 2x - y + 2 = 0 \Rightarrow B^2 - 4AC = 2^2 - 4(1)(1) = 0 \Rightarrow$ Parabola

6. $2x^2 - y^2 + 4xy - 2x + 3y = 6 \Rightarrow B^2 - 4AC = 4^2 - 4(2)(-1) = 24 > 0 \Rightarrow$ Hyperbola

7. $x^2 + 4xy + 4y^2 - 3x = 6 \Rightarrow B^2 - 4AC = 4^2 - 4(1)(4) = 0 \Rightarrow$ Parabola

8. $x^2 + y^2 + 3x - 2y = 10 \Rightarrow B^2 - 4AC = 0^2 - 4(1)(1) = -4 < 0 \Rightarrow$ Ellipse (circle)

9. $xy + y^2 - 3x = 5 \Rightarrow B^2 - 4AC = 1^2 - 4(0)(1) = 1 > 0 \Rightarrow$ Hyperbola

10. $3x^2 + 6xy + 3y^2 - 4x + 5y = 12 \Rightarrow B^2 - 4AC = 6^2 - 4(3)(3) = 0 \Rightarrow$ Parabola

11. $3x^2 - 5xy + 2y^2 - 7x - 14y = -1 \Rightarrow B^2 - 4AC = (-5)^2 - 4(3)(2) = 1 > 0 \Rightarrow$ Hyperbola

12. $2x^2 - 4.9xy + 3y^2 - 4x = 7 \Rightarrow B^2 - 4AC = (-4.9)^2 - 4(2)(3) = 0.01 > 0 \Rightarrow$ Hyperbola

13. $x^2 - 3xy + 3y^2 + 6y = 7 \Rightarrow B^2 - 4AC = (-3)^2 - 4(1)(3) = -3 < 0 \Rightarrow$ Ellipse

14. $25x^2 + 21xy + 4y^2 - 350x = 0 \Rightarrow B^2 - 4AC = 21^2 - 4(25)(4) = 41 > 0 \Rightarrow$ Hyperbola

15. $6x^2 + 3xy + 2y^2 + 17y + 2 = 0 \Rightarrow B^2 - 4AC = 3^2 - 4(6)(2) = -39 < 0 \Rightarrow$ Ellipse

16. $3x^2 + 12xy + 12y^2 + 435x - 9y + 72 = 0 \Rightarrow B^2 - 4AC = 12^2 - 4(3)(12) = 0 \Rightarrow$ Parabola

17. $\cot 2\alpha = \frac{A - C}{B} = \frac{0}{1} = 0 \Rightarrow 2\alpha = \frac{\pi}{2} \Rightarrow \alpha = \frac{\pi}{4}$; therefore $x = x'\cos\alpha - y'\sin\alpha$,

$y = x'\sin\alpha + y'\cos\alpha \Rightarrow x = x'\frac{\sqrt{2}}{2} - y'\frac{\sqrt{2}}{2},\ y = x'\frac{\sqrt{2}}{2} + y'\frac{\sqrt{2}}{2}$

$\Rightarrow \left(\frac{\sqrt{2}}{2}x' - \frac{\sqrt{2}}{2}y'\right)\left(\frac{\sqrt{2}}{2}x' + \frac{\sqrt{2}}{2}y'\right) = 2 \Rightarrow \frac{1}{2}x'^2 - \frac{1}{2}y'^2 = 2 \Rightarrow x'^2 - y'^2 = 4 \Rightarrow$ Hyperbola

18. $\cot 2\alpha = \frac{A - C}{B} = \frac{1 - 1}{1} = 0 \Rightarrow 2\alpha = \frac{\pi}{2} \Rightarrow \alpha = \frac{\pi}{4}$; therefore $x = x'\cos\alpha - y'\sin\alpha$,

$y = x'\sin\alpha + y'\cos\alpha \Rightarrow x = x'\frac{\sqrt{2}}{2} - y'\frac{\sqrt{2}}{2},\ y = x'\frac{\sqrt{2}}{2} + y'\frac{\sqrt{2}}{2}$

$\Rightarrow \left(\frac{\sqrt{2}}{2}x' - \frac{\sqrt{2}}{2}y'\right)^2 + \left(\frac{\sqrt{2}}{2}x' + \frac{\sqrt{2}}{2}y'\right)\left(\frac{\sqrt{2}}{2}x' - \frac{\sqrt{2}}{2}y'\right) + \left(\frac{\sqrt{2}}{2}x' + \frac{\sqrt{2}}{2}y'\right)^2 = 1$

$\Rightarrow \frac{1}{2}x'^2 - x'y' + \frac{1}{2}y'^2 + \frac{1}{2}x'^2 - \frac{1}{2}y'^2 + \frac{1}{2}x'^2 + x'y' + \frac{1}{2}y'^2 = 1 \Rightarrow \frac{3}{2}x'^2 + \frac{1}{2}y'^2 = 1 \Rightarrow 3x'^2 + y'^2 = 2 \Rightarrow$ Ellipse

19. $\cot 2\alpha = \frac{A-C}{B} = \frac{3-1}{2\sqrt{3}} = \frac{1}{\sqrt{3}} \Rightarrow 2\alpha = \frac{\pi}{3} \Rightarrow \alpha = \frac{\pi}{6}$; therefore $x = x' \cos\alpha - y' \sin\alpha$,

$y = x' \sin\alpha + y' \cos\alpha \Rightarrow x = \frac{\sqrt{3}}{2}x' - \frac{1}{2}y'$, $y = \frac{1}{2}x' + \frac{\sqrt{3}}{2}y'$

$\Rightarrow 3\left(\frac{\sqrt{3}}{2}x' - \frac{1}{2}y'\right)^2 + 2\sqrt{3}\left(\frac{\sqrt{3}}{2}x' + \frac{1}{2}y'\right)\left(\frac{1}{2}x' + \frac{\sqrt{3}}{2}y'\right) + \left(\frac{1}{2}x' + \frac{\sqrt{3}}{2}y'\right)^2 - 8\left(\frac{\sqrt{3}}{2}x' - \frac{1}{2}y'\right)$

$+ 8\sqrt{3}\left(\frac{1}{2}x' + \frac{\sqrt{3}}{2}y'\right) = 0 \Rightarrow 4x'^2 + 16y' = 0 \Rightarrow$ Parabola

20. $\cot 2\alpha = \frac{A-C}{B} = \frac{1-2}{-\sqrt{3}} = \frac{1}{\sqrt{3}} \Rightarrow 2\alpha = \frac{\pi}{3} \Rightarrow \alpha = \frac{\pi}{6}$; therefore $x = x' \cos\alpha - y' \sin\alpha$,

$y = x' \sin\alpha + y' \cos\alpha \Rightarrow x = \frac{\sqrt{3}}{2}x' - \frac{1}{2}y'$, $y = \frac{1}{2}x' + \frac{\sqrt{3}}{2}y'$

$\Rightarrow \left(\frac{\sqrt{3}}{2}x' - \frac{1}{2}y'\right)^2 - \sqrt{3}\left(\frac{\sqrt{3}}{2}x' - \frac{1}{2}y'\right)\left(\frac{1}{2}x' + \frac{\sqrt{3}}{2}y'\right) + 2\left(\frac{1}{2}x' + \frac{\sqrt{3}}{2}y'\right)^2 = 1 \Rightarrow \frac{1}{2}x'^2 + \frac{5}{2}y'^2 = 1$

$\Rightarrow x'^2 + 5y'^2 = 2 \Rightarrow$ Ellipse

21. $\cot 2\alpha = \frac{A-C}{B} = \frac{1-1}{-2} = 0 \Rightarrow 2\alpha = \frac{\pi}{2} \Rightarrow \alpha = \frac{\pi}{4}$; therefore $x = x' \cos\alpha - y' \sin\alpha$,

$y = x' \sin\alpha + y' \cos\alpha \Rightarrow x = \frac{\sqrt{2}}{2}x' - \frac{\sqrt{2}}{2}y'$, $y = \frac{\sqrt{2}}{2}x' + \frac{\sqrt{2}}{2}y'$

$\Rightarrow \left(\frac{\sqrt{2}}{2}x' - \frac{\sqrt{2}}{2}y'\right)^2 - 2\left(\frac{\sqrt{2}}{2}x' - \frac{\sqrt{2}}{2}y'\right)\left(\frac{\sqrt{2}}{2}x' + \frac{\sqrt{2}}{2}y'\right) + \left(\frac{\sqrt{2}}{2}x' + \frac{\sqrt{2}}{2}y'\right)^2 = 2 \Rightarrow y'^2 = 1$

$\Rightarrow$ Parallel horizontal lines

22. $\cot 2\alpha = \frac{A-C}{B} = \frac{3-1}{-2\sqrt{3}} = -\frac{1}{\sqrt{3}} \Rightarrow 2\alpha = \frac{2\pi}{3} \Rightarrow \alpha = \frac{\pi}{3}$; therefore $x = x' \cos\alpha - y' \sin\alpha$,

$y = x' \sin\alpha + y' \cos\alpha \Rightarrow x = \frac{1}{2}x' - \frac{\sqrt{3}}{2}y'$, $y = \frac{\sqrt{3}}{2}x' + \frac{1}{2}y'$

$\Rightarrow 3\left(\frac{1}{2}x' - \frac{\sqrt{3}}{2}y'\right)^2 - 2\sqrt{3}\left(\frac{1}{2}x' - \frac{\sqrt{3}}{2}y'\right)\left(\frac{\sqrt{3}}{2}x' + \frac{1}{2}y'\right) + \left(\frac{\sqrt{3}}{2}x' + \frac{1}{2}y'\right)^2 = 1 \Rightarrow 4y'^2 = 1$

$\Rightarrow$ Parallel horizontal lines

23. $\cot 2\alpha = \frac{A-C}{B} = \frac{\sqrt{2}-\sqrt{2}}{2\sqrt{2}} = 0 \Rightarrow 2\alpha = \frac{\pi}{2} \Rightarrow \alpha = \frac{\pi}{4}$; therefore $x = x' \cos\alpha - y' \sin\alpha$,

$y = x' \sin\alpha + y' \cos\alpha \Rightarrow x = \frac{\sqrt{2}}{2}x' - \frac{\sqrt{2}}{2}y'$, $y = \frac{\sqrt{2}}{2}x' + \frac{\sqrt{2}}{2}y'$

$\Rightarrow \sqrt{2}\left(\frac{\sqrt{2}}{2}x' - \frac{\sqrt{2}}{2}y'\right)^2 + 2\sqrt{2}\left(\frac{\sqrt{2}}{2}x' - \frac{\sqrt{2}}{2}y'\right)\left(\frac{\sqrt{2}}{2}x' + \frac{\sqrt{2}}{2}y'\right) + \sqrt{2}\left(\frac{\sqrt{2}}{2}x' + \frac{\sqrt{2}}{2}y'\right)^2$

$- 8\left(\frac{\sqrt{2}}{2}x' - \frac{\sqrt{2}}{2}y'\right) + 8\left(\frac{\sqrt{2}}{2}x' + \frac{\sqrt{2}}{2}y'\right) = 0 \Rightarrow 2\sqrt{2}x'^2 + 8\sqrt{2}\,y' = 0 \Rightarrow$ Parabola

24. $\cot 2\alpha = \frac{A-C}{B} = \frac{0-0}{1} = 0 \Rightarrow 2\alpha = \frac{\pi}{2} \Rightarrow \alpha = \frac{\pi}{4}$; therefore $x = x' \cos\alpha - y' \sin\alpha$,

$y = x' \sin\alpha + y' \cos\alpha \Rightarrow x = \frac{\sqrt{2}}{2}x' - \frac{\sqrt{2}}{2}y'$, $y = \frac{\sqrt{2}}{2}x' + \frac{\sqrt{2}}{2}y'$

$\Rightarrow \left(\frac{\sqrt{2}}{2}x' - \frac{\sqrt{2}}{2}y'\right)\left(\frac{\sqrt{2}}{2}x' + \frac{\sqrt{2}}{2}y'\right) - \left(\frac{\sqrt{2}}{2}x' - \frac{\sqrt{2}}{2}y'\right) - \left(\frac{\sqrt{2}}{2}x' + \frac{\sqrt{2}}{2}y'\right) + 1 = 0 \Rightarrow x'^2 - y'^2 - 2\sqrt{2}x' + 2$
$= 0 \Rightarrow$ Hyperbola

25. $\cot 2\alpha = \frac{A-C}{B} = \frac{3-3}{2} = 0 \Rightarrow 2\alpha = \frac{\pi}{2} \Rightarrow \alpha = \frac{\pi}{4}$; therefore $x = x' \cos\alpha - y' \sin\alpha$,

$y = x' \sin\alpha + y' \cos\alpha \Rightarrow x = \frac{\sqrt{2}}{2}x' - \frac{\sqrt{2}}{2}y'$, $y = \frac{\sqrt{2}}{2}x' + \frac{\sqrt{2}}{2}y'$

$\Rightarrow 3\left(\frac{\sqrt{2}}{2}x' - \frac{\sqrt{2}}{2}y'\right)^2 + 2\left(\frac{\sqrt{2}}{2}x' - \frac{\sqrt{2}}{2}y'\right)\left(\frac{\sqrt{2}}{2}x' + \frac{\sqrt{2}}{2}y'\right) + 3\left(\frac{\sqrt{2}}{2}x' + \frac{\sqrt{2}}{2}y'\right)^2 = 19 \Rightarrow 4x'^2 + 2y'^2 = 19$

$\Rightarrow$ Ellipse

26. $\cot 2\alpha = \frac{A-C}{B} = \frac{3-(-1)}{4\sqrt{3}} = \frac{1}{\sqrt{3}} \Rightarrow 2\alpha = \frac{\pi}{3} \Rightarrow \alpha = \frac{\pi}{6}$; therefore $x = x' \cos\alpha - y' \sin\alpha$,

$y = x' \sin\alpha + y' \cos\alpha \Rightarrow x = \frac{\sqrt{3}}{2}x' - \frac{1}{2}y'$, $y = \frac{1}{2}x' + \frac{\sqrt{3}}{2}y'$

$\Rightarrow 3\left(\frac{\sqrt{3}}{2}x' - \frac{1}{2}y'\right)^2 + 4\sqrt{3}\left(\frac{\sqrt{3}}{2}x' - \frac{1}{2}y'\right)\left(\frac{1}{2}x' + \frac{\sqrt{3}}{2}y'\right) - \left(\frac{1}{2}x' + \frac{\sqrt{3}}{2}y'\right)^2 = 7 \Rightarrow 5x'^2 - 3y'^2 = 7$

$\Rightarrow$ Hyperbola

27. $\cot 2\alpha = \frac{14-2}{16} = \frac{3}{4} \Rightarrow \cos 2\alpha = \frac{3}{5}$ (if we choose 2α in Quadrant I); thus $\sin\alpha = \sqrt{\frac{1-\cos 2\alpha}{2}} = \sqrt{\frac{1-\left(\frac{3}{5}\right)}{2}} = \frac{1}{\sqrt{5}}$

and $\cos\alpha = \sqrt{\frac{1+\cos 2\alpha}{2}} = \sqrt{\frac{1+\left(\frac{3}{5}\right)}{2}} = \frac{2}{\sqrt{5}}$ (or $\sin\alpha = -\frac{2}{\sqrt{5}}$ and $\cos\alpha = \frac{1}{\sqrt{5}}$)

28. $\cot 2\alpha = \frac{A-C}{B} = \frac{4-1}{-4} = -\frac{3}{4} \Rightarrow \cos 2\alpha = -\frac{3}{5}$ (if we choose 2α in Quadrant II); thus $\sin\alpha = \sqrt{\frac{1-\cos 2\alpha}{2}}$

$= \sqrt{\frac{1-\left(\frac{3}{5}\right)}{2}} = \frac{2}{\sqrt{5}}$ and $\cos\alpha = \sqrt{\frac{1+\cos 2\alpha}{2}} = \sqrt{\frac{1-\left(\frac{3}{5}\right)}{2}}$

$= \frac{1}{\sqrt{5}}$ (or $\sin\alpha = -\frac{1}{\sqrt{5}}$ and $\cos\alpha = \frac{2}{\sqrt{5}}$)

29. $\tan 2\alpha = \frac{-1}{1-3} = \frac{1}{2} \Rightarrow 2\alpha \approx 26.57° \Rightarrow \alpha \approx 13.28° \Rightarrow \sin\alpha \approx 0.23$, $\cos\alpha \approx 0.97$; then $A' \approx 0.88$, $B' \approx 0.00$, $C' \approx 3.10$, $D' \approx 0.74$, $E' \approx -1.20$, and $F' = -3 \Rightarrow 0.88x'^2 + 3.10y'^2 + 0.74x' - 1.20y' - 3 = 0$, an ellipse

30. $\tan 2\alpha = \frac{-1}{2-(-3)} = \frac{1}{5} \Rightarrow 2\alpha \approx 11.31° \Rightarrow \alpha \approx 5.65° \Rightarrow \sin\alpha \approx 0.10$, $\cos\alpha \approx 0.99$; then $A' \approx 2.05$, $B' \approx 0.00$, $C' \approx -3.05$, $D' \approx 2.98$, $E' \approx -0.30$, and $F' = -7 \Rightarrow 2.05x'^2 - 3.05y'^2 + 2.99x' - 0.30y' - 7 = 0$, a hyperbola

31. $\tan 2\alpha = \frac{-4}{1-4} = \frac{4}{3} \Rightarrow 2\alpha \approx 53.13° \Rightarrow \alpha \approx 26.56° \Rightarrow \sin\alpha \approx 0.45$, $\cos\alpha \approx 0.89$; then $A' \approx 0.00$, $B' \approx 0.00$, $C' \approx 5.00$, $D' \approx 0$, $E' \approx 0$, and $F' = -5 \Rightarrow 5.00y'^2 - 5 = 0$ or $y' = \pm 1.00$, parallel lines

32. $\tan 2\alpha = \frac{-12}{2-18} = \frac{3}{4} \Rightarrow 2\alpha \approx 36.87^\circ \Rightarrow \alpha \approx 18.43^\circ \Rightarrow \sin\alpha \approx 0.32$, $\cos\alpha \approx 0.95$; then $A' \approx 0.00$, $B' \approx 0.00$, $C' \approx 20.00$, $D' \approx 0$, $E' \approx 0$, and $F' = -49 \Rightarrow 20.00y'^2 - 49 = 0$, parallel lines

33. $\tan 2\alpha = \frac{5}{3-2} = 5 \Rightarrow 2\alpha \approx 78.69^\circ \Rightarrow \alpha \approx 39.34^\circ \Rightarrow \sin\alpha \approx 0.63$, $\cos\alpha \approx 0.77$; then $A' \approx 5.05$, $B' \approx 0.00$, $C' \approx -0.05$, $D' \approx -5.07$, $E' \approx -6.18$, and $F' = -1 \Rightarrow 5.05x'^2 - 0.05y'^2 - 5.07x' - 6.18y' - 1 = 0$, a hyperbola

34. $\tan 2\alpha = \frac{7}{2-9} = -1 \Rightarrow 2\alpha \approx -45.00^\circ \Rightarrow \alpha \approx -22.5^\circ \Rightarrow \sin\alpha \approx 0.38$, $\cos\alpha \approx 0.92$; then $A' \approx 0.55$, $B' \approx 0.00$, $C' \approx 10.45$, $D' \approx 18.48$, $E' \approx -7.65$, and $F' = -86 \Rightarrow 0.55x'^2 + 10.45y'^2 + 18.48x' - 7.65y' - 86 = 0$, an ellipse

35. $\alpha = 90^\circ \Rightarrow x = x'\cos 90^\circ - y'\sin 90^\circ = -y'$ and $y = x'\sin 90^\circ + y'\cos 90^\circ = x'$

(a) $\frac{x'^2}{b^2} + \frac{y'^2}{a^2} = 1$ (b) $\frac{y'^2}{a^2} - \frac{x'^2}{b^2} = 1$ (c) $x'^2 + y'^2 = a^2$

(d) $y = mx \Rightarrow y - mx = 0 \Rightarrow D = -m$ and $E = 1$; $\alpha = 90^\circ \Rightarrow D' = 1$ and $E' = m \Rightarrow my' + x' = 0 \Rightarrow y' = -\frac{1}{m}x'$

(e) $y = mx + b \Rightarrow y - mx - b = 0 \Rightarrow D = -m$ and $E = 1$; $\alpha = 90^\circ \Rightarrow D' = 1$, $E' = m$ and $F' = -b$
$\Rightarrow my' + x' - b = 0 \Rightarrow y' = -\frac{1}{m}x' + \frac{b}{m}$

36. $\alpha = 180^\circ \Rightarrow x = x'\cos 180^\circ - y'\sin 180^\circ = -x'$ and $y = x'\sin 180^\circ + y'\cos 180^\circ = -y'$

(a) $\frac{x'^2}{a^2} + \frac{y'^2}{b^2} = 1$ (b) $\frac{x'^2}{a^2} - \frac{y'^2}{b^2} = 1$ (c) $x'^2 + y'^2 = a^2$

(d) $y = mx \Rightarrow y - mx = 0 \Rightarrow D = -m$ and $E = 1$; $\alpha = 180^\circ \Rightarrow D' = m$ and $E' = -1 \Rightarrow -y' + mx' = 0 \Rightarrow$
$y' = mx'$

(e) $y = mx + b \Rightarrow y - mx - b = 0 \Rightarrow D = -m$ and $E = 1$; $\alpha = 180^\circ \Rightarrow D' = m$, $E' = -1$ and $F' = -b$
$\Rightarrow -y' + mx' - b = 0 \Rightarrow y' = mx' - b$

37. (a) $A' = \cos 45^\circ \sin 45^\circ = \left(\frac{\sqrt{2}}{2}\right)\left(\frac{\sqrt{2}}{2}\right) = \frac{1}{2}$, $B' = 0$, $C' = -\cos 45^\circ \sin 45^\circ = -\frac{1}{2}$, $F' = -1$
$\Rightarrow \frac{1}{2}x'^2 - \frac{1}{2}y'^2 = 1 \Rightarrow x'^2 - y'^2 = 2$

(b) $A' = \frac{1}{2}$, $C' = -\frac{1}{2}$ (see part (a) above), $D' = E' = B' = 0$, $F' = -a \Rightarrow \frac{1}{2}x'^2 - \frac{1}{2}y'^2 = a \Rightarrow x'^2 - y'^2 = 2a$

38. $xy = 2 \Rightarrow x'^2 - y'^2 = 4 \Rightarrow \frac{x'^2}{4} - \frac{y'^2}{4} = 1$ (see Exercise 37(b)) $\Rightarrow a = 2$ and $b = 2 \Rightarrow c = \sqrt{4+4} = 2\sqrt{2}$
$\Rightarrow e = \frac{c}{a} = \frac{2\sqrt{2}}{2} = \sqrt{2}$

39. Yes, the graph is a hyperbola: with $AC < 0$ we have $-4AC > 0$ and $B^2 - 4AC > 0$.

40. The one curve that meets all three of the stated criteria is the ellipse $x^2 + 4xy + 5y^2 - 1 = 0$. The reasoning: The symmetry about the origin means that $(-x, -y)$ lies on the graph whenever (x, y) does. Adding $Ax^2 + Bxy + Cy^2 + Dx + Ey + F = 0$ and $A(-x)^2 + B(-x)(-y) + C(-y)^2 + D(-x) + E(-y) + F = 0$ and dividing the result by 2 produces the equivalent equation $Ax^2 + Bxy + Cy^2 + F = 0$. Substituting $x = 1$, $y = 0$ (because the point $(1,0)$ lies on the curve) shows further that $A = -F$. Then $-Fx^2 + Bxy + Cy^2 + F = 0$. By implicit

differentiation, $-2Fx + By + Bxy' + 2Cyy' = 0$, so substituting $x = -2$, $y = 1$, and $y' = 0$ (from Property 3) gives $4F + B = 0 \Rightarrow B = -4F \Rightarrow$ the conic is $-Fx^2 - 4Fxy + Cy^2 + F = 0$. Now substituting $x = -2$ and $y = 1$ again gives $-4F + 8F + C + F = 0 \Rightarrow C = -5F \Rightarrow$ the equation is now $-Fx^2 - 4Fxy - 5Fy^2 + F = 0$. Finally, dividing through by $-F$ gives the equation $x^2 + 4xy + 5y^2 - 1 = 0$.

41. Let α be any angle. Then $A' = \cos^2\alpha + \sin^2\alpha = 1$, $B' = 0$, $C' = \sin^2\alpha + \cos^2\alpha = 1$, $D' = E' = 0$ and $F' = -a^2$ $\Rightarrow x'^2 + y'^2 = a^2$.

42. If $A = C$, then $B' = B \cos 2\alpha + (C - A) \sin 2\alpha = B \cos 2\alpha$. Then $\alpha = \frac{\pi}{4} \Rightarrow 2\alpha = \frac{\pi}{2} \Rightarrow B' = B \cos \frac{\pi}{2} = 0$ so the xy-term is eliminated.

43. (a) $B^2 - 4AC = 4^2 - 4(1)(4) = 0$, so the discriminant indicates this conic is a parabola
    (b) The left-hand side of $x^2 + 4xy + 4y^2 + 6x + 12y + 9 = 0$ factors as a perfect square: $(x + 2y + 3)^2 = 0$ $\Rightarrow x + 2y + 3 = 0 \Rightarrow 2y = -x - 3$; thus the curve is a degenerate parabola (i.e., a straight line).

44. (a) $B^2 - 4AC = 6^2 - 4(9)(1) = 0$, so the discriminant indicates this conic is a parabola
    (b) The left-hand side of $9x^2 + 6xy + y^2 - 12x - 4y + 4 = 0$ factors as a perfect square: $(3x + y - 2)^2 = 0$ $\Rightarrow 3x + y - 2 = 0 \Rightarrow y = -3x + 2$; thus the curve is a degenerate parabola (i.e., a straight line).

45. (a) $B^2 - 4AC = 1 - 4(0)(0) = 1 \Rightarrow$ hyperbola
    (b) $xy + 2x - y = 0 \Rightarrow y(x - 1) = -2x \Rightarrow y = \frac{-2x}{x - 1}$
    (c) $y = \frac{-2x}{x-1} \Rightarrow \frac{dy}{dx} = \frac{2}{(x-1)^2}$ and we want $\frac{-1}{\left(\frac{dy}{dx}\right)} = -2$, the slope of $y = -2x \Rightarrow -2 = -\frac{(x-1)^2}{2} \Rightarrow (x - 1)^2 = 4$ $\Rightarrow x = 3$ or $x = -1$; $x = 3 \Rightarrow y = -3 \Rightarrow (3, -3)$ is a point on the hyperbola where the line with slope $m = -2$ is normal $\Rightarrow$ the line is $y + 3 = -2(x - 3)$ or

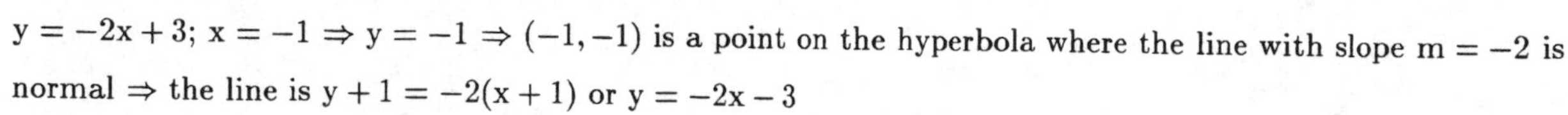

$y = -2x + 3$; $x = -1 \Rightarrow y = -1 \Rightarrow (-1, -1)$ is a point on the hyperbola where the line with slope $m = -2$ is normal $\Rightarrow$ the line is $y + 1 = -2(x + 1)$ or $y = -2x - 3$

46. (a) False: let $A = C = 1$, $B = 2 \Rightarrow B^2 - 4AC = 0 \Rightarrow$ parabola
    (b) False: see part (a) above
    (c) True: $AC < 0 \Rightarrow -4AC > 0 \Rightarrow B^2 - 4AC > 0 \Rightarrow$ hyperbola

47. Assume the ellipse has been rotated to eliminate the xy-term $\Rightarrow$ the new equation is $A'x'^2 + C'y'^2 = 1 \Rightarrow$ the semi-axes are $\sqrt{\frac{1}{A'}}$ and $\sqrt{\frac{1}{C'}} \Rightarrow$ the area is $\pi\left(\sqrt{\frac{1}{A'}}\right)\left(\sqrt{\frac{1}{C'}}\right) = \frac{\pi}{\sqrt{A'C'}} = \frac{2\pi}{\sqrt{4A'C'}}$. Since $B^2 - 4AC$ $= B'^2 - 4A'C' = -4A'C'$ (because $B' = 0$) we find that the area is $\frac{2\pi}{\sqrt{4AC - B^2}}$ as claimed.

48. (a) $A' + C' = (A\cos^2\alpha + B\cos\alpha\sin\alpha + C\sin^2\alpha) + (A\sin^2\alpha - B\cos\alpha\sin\alpha + C\sin^2\alpha)$
$= A(\cos^2\alpha + \sin^2\alpha) + C(\sin^2\alpha + \cos^2\alpha) = A + C$

(b) $D'^2 + E'^2 = (D\cos\alpha + E\sin\alpha)^2 + (-D\sin\alpha + E\cos\alpha)^2 = D^2\cos^2\alpha + 2DE\cos\alpha\sin\alpha + E^2\sin^2\alpha$
$+ D^2\sin^2\alpha - 2DE\sin\alpha\cos\alpha + E^2\cos^2\alpha = D^2(\cos^2\alpha + \sin^2\alpha) + E^2(\sin^2\alpha + \cos^2\alpha) = D^2 + E^2$

49. $B'^2 - 4A'C'$

$= (B\cos 2\alpha + (C - A)\sin 2\alpha)^2 - 4(A\cos^2\alpha + B\cos\alpha\sin\alpha + C\sin^2\alpha)(A\sin^2\alpha - B\cos\alpha\sin\alpha + C\cos^2\alpha)$

$= B^2\cos^2 2\alpha + 2B(C - A)\sin 2\alpha\cos 2\alpha + (C - A)^2\sin^2 2\alpha - 4A^2\cos^2\alpha\sin^2\alpha + 4AB\cos^3\alpha\sin\alpha$
$- 4AC\cos^4\alpha - 4AB\cos\alpha\sin^3\alpha + 4B^2\cos^2\alpha\sin^2\alpha - 4BC\cos^3\alpha\sin\alpha - 4AC\sin^4\alpha + 4BC\cos\alpha\sin^3\alpha$
$- 4C^2\cos^2\alpha\sin^2\alpha$

$= B^2\cos^2 2\alpha + 2BC\sin 2\alpha\cos 2\alpha - 2AB\sin 2\alpha\cos 2\alpha + C^2\sin^2 2\alpha - 2AC\sin^2 2\alpha + A^2\sin^2 2\alpha$
$- 4A^2\cos^2\alpha\sin^2\alpha + 4AB\cos^3\alpha\sin\alpha - 4AC\cos^4\alpha - 4AB\cos\alpha\sin^3\alpha + B^2\sin^2 2\alpha - 4BC\cos^3\alpha\sin\alpha$
$- 4AC\sin^4\alpha + 4BC\cos\alpha\sin^3\alpha - 4C^2\cos^2\alpha\sin^2\alpha$

$= B^2 + 2BC(2\sin\alpha\cos\alpha)(\cos^2\alpha - \sin^2\alpha) - 2AB(2\sin\alpha\cos\alpha)(\cos^2\alpha - \sin^2\alpha) + C^2(4\sin^2\alpha\cos^2\alpha)$
$- 2AC(4\sin^2\alpha\cos^2\alpha) + A^2(4\sin^2\alpha\cos^2\alpha) - 4A^2\cos^2\alpha\sin^2\alpha + 4AB\cos^3\alpha\sin\alpha - 4AC\cos^4\alpha$
$- 4AB\cos\alpha\sin^3\alpha - 4BC\cos^3\alpha\sin\alpha - 4AC\sin^4\alpha + 4BC\cos\alpha\sin^3\alpha - 4C^2\cos^2\alpha\sin^2\alpha$

$= B^2 - 8AC\sin^2\alpha\cos^2\alpha - 4AC\cos^4\alpha - 4AC\sin^4\alpha$

$= B^2 - 4AC(\cos^4\alpha + 2\sin^2\alpha\cos^2\alpha + \sin^4\alpha)$

$= B^2 - 4AC(\cos^2\alpha + \sin^2\alpha)^2$

$= B^2 - 4AC$

9.4 PARAMETRIZATIONS OF CURVES

1. $x = \cos t,\ y = \sin t,\ 0 \le t \le \pi$
$\Rightarrow \cos^2 t + \sin^2 t = 1 \Rightarrow x^2 + y^2 = 1$

2. $x = \cos 2t,\ y = \sin 2t,\ 0 \le t \le \pi$
$\Rightarrow \cos^2 2t + \sin^2 2t = 1 \Rightarrow x^2 + y^2 = 1$

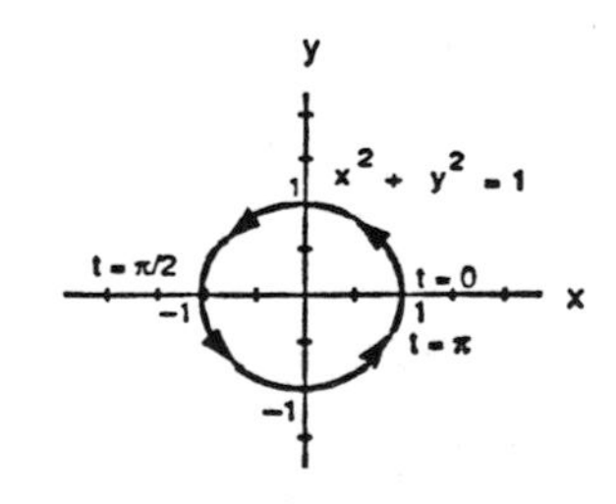

3. $x = \sin(2\pi(1-t)),\ y = \cos(2\pi(1-t)),\ 0 \le t \le 1$
$\Rightarrow \sin^2(2\pi(1-t)) + \cos^2(2\pi(1-t)) = 1$
$\Rightarrow x^2 + y^2 = 1$

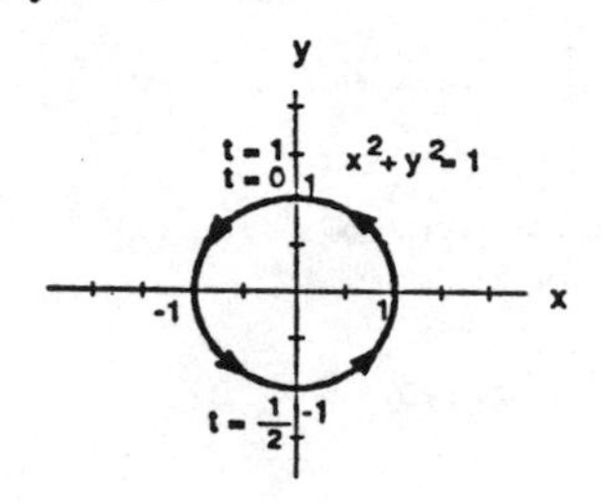

4. $x = \cos(\pi - t),\ y = \sin(\pi - t),\ 0 \le t \le \pi$
$\Rightarrow \cos^2(\pi - t) + \sin^2(\pi - t) = 1$
$\Rightarrow x^2 + y^2 = 1$

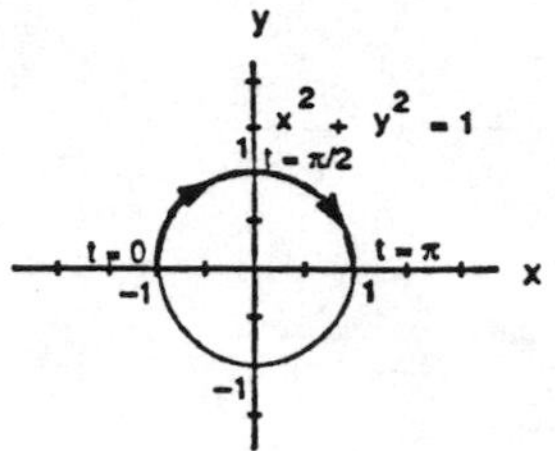

5. $x = 4\cos t,\ y = 2\sin t,\ 0 \le t \le 2\pi$
$\Rightarrow \frac{16\cos^2 t}{16} + \frac{4\sin^2 t}{4} = 1 \Rightarrow \frac{x^2}{16} + \frac{y^2}{4} = 1$

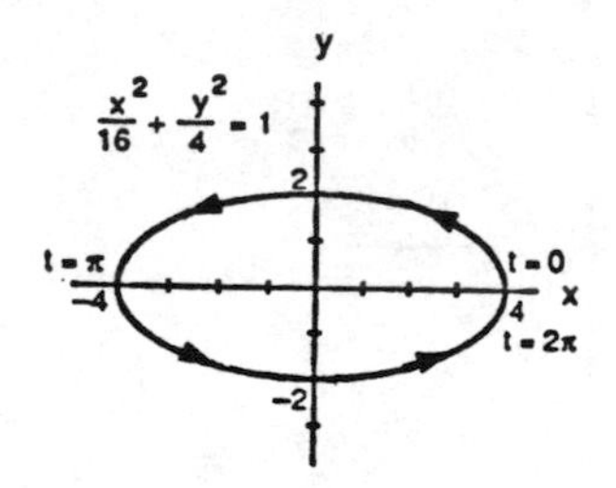

6. $x = 4\sin t,\ y = 2\cos t,\ 0 \le t \le \pi$
$\Rightarrow \frac{16\sin^2 t}{16} + \frac{4\cos^2 t}{4} = 1 \Rightarrow \frac{x^2}{16} + \frac{y^2}{4} = 1$

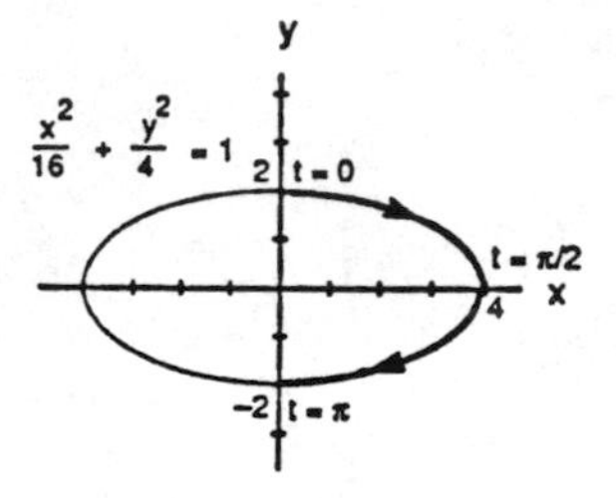

7. $x = 4\cos t,\ y = 5\sin t,\ 0 \le t \le \pi$
$\Rightarrow \frac{16\cos^2 t}{16} + \frac{25\sin^2 t}{25} = 1 \Rightarrow \frac{x^2}{16} + \frac{y^2}{25} = 1$

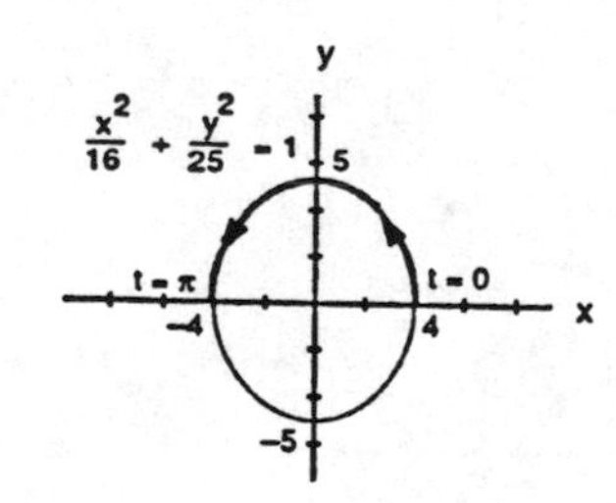

8. $x = 4\sin t,\ y = 5\cos t,\ 0 \le t \le 2\pi$
$\Rightarrow \frac{16\sin^2 t}{16} + \frac{25\cos^2 t}{25} = 1 \Rightarrow \frac{x^2}{16} + \frac{y^2}{25} = 1$

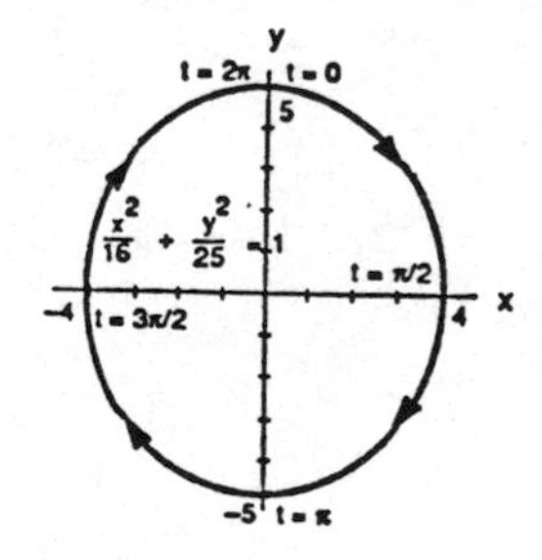

9. $x = 3t,\ y = 9t^2,\ -\infty < t < \infty \Rightarrow y = x^2$

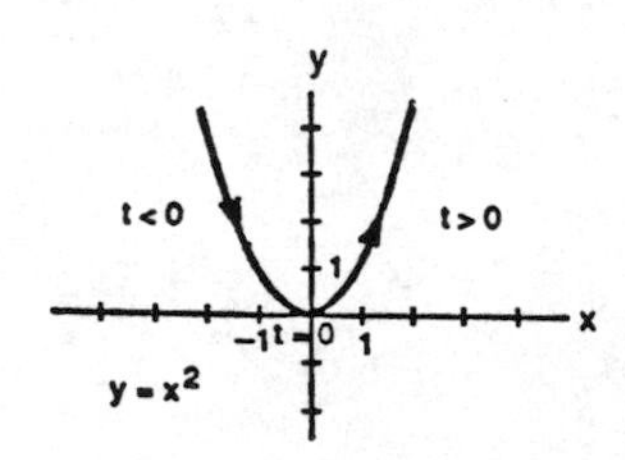

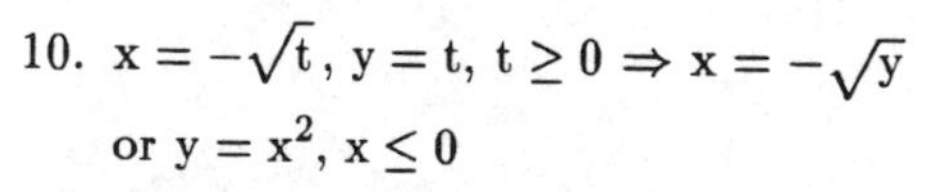
10. $x = -\sqrt{t},\ y = t,\ t \ge 0 \Rightarrow x = -\sqrt{y}$
or $y = x^2,\ x \le 0$

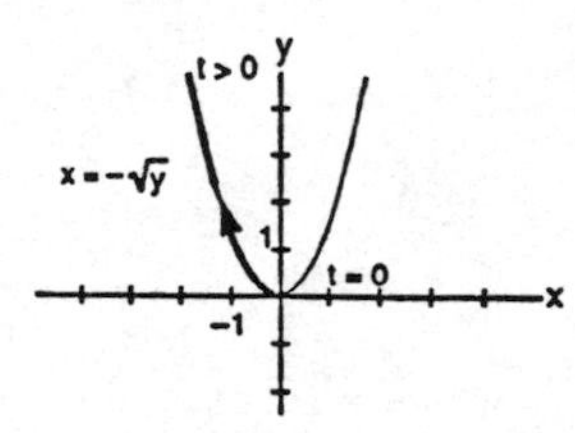

11. $x = t,\ y = \sqrt{t},\ t \geq 0 \Rightarrow y = \sqrt{x}$

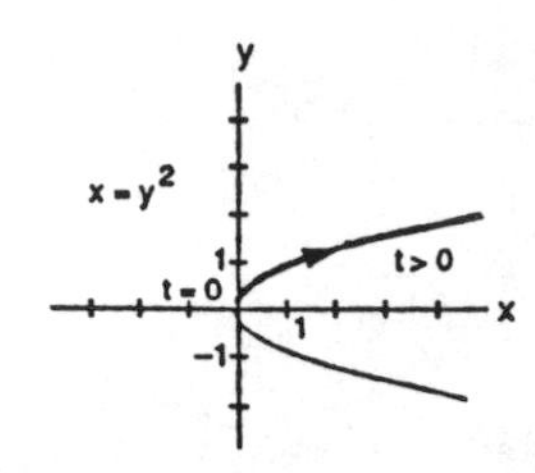

12. $x = \sec^2 t - 1,\ y = \tan t,\ -\frac{\pi}{2} < t < \frac{\pi}{2}$
$\Rightarrow \sec^2 t - 1 = \tan^2 t \Rightarrow x = y^2$

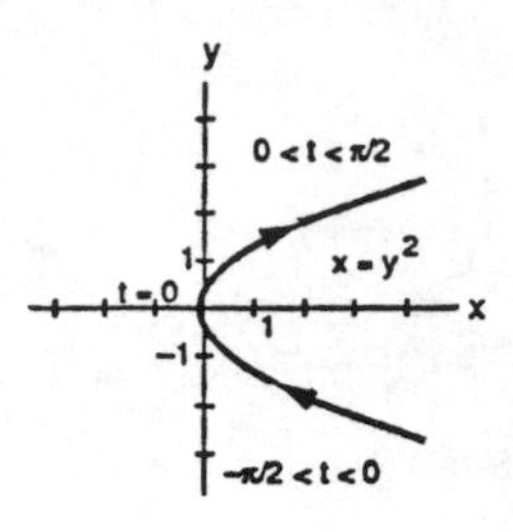

13. $x = -\sec t,\ y = \tan t,\ -\frac{\pi}{2} < t < \frac{\pi}{2}$
$\Rightarrow \sec^2 t - \tan^2 t = 1 \Rightarrow x^2 - y^2 = 1$

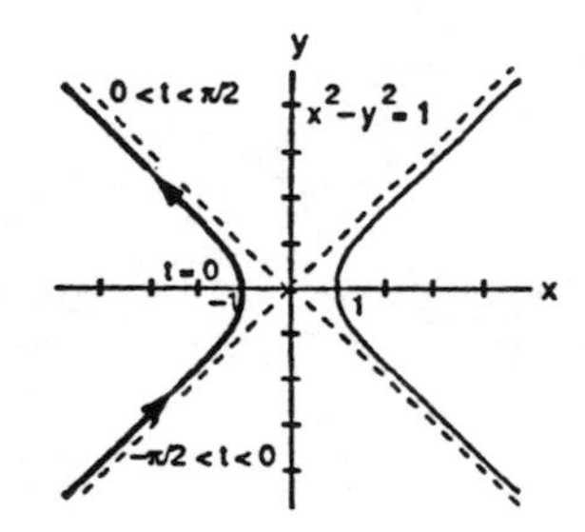

14. $x = \csc t,\ y = \cot t,\ 0 < t < \pi$
$\Rightarrow 1 + \cot^2 t = \csc^2 t \Rightarrow 1 + y^2 = x^2 \Rightarrow x^2 - y^2 = 1$

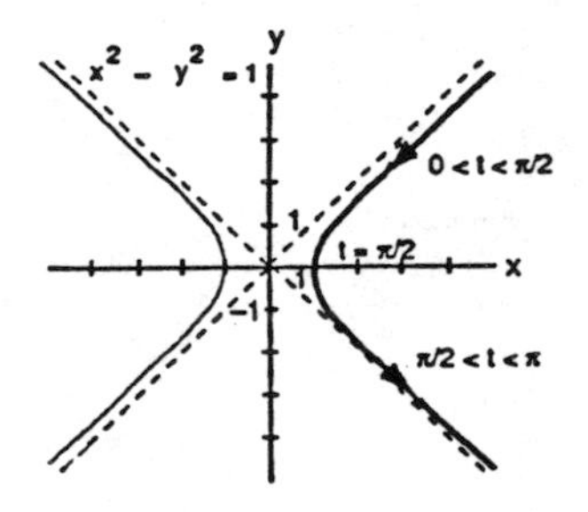

15. $x = 2t - 5,\ y = 4t - 7,\ -\infty < t < \infty$
$\Rightarrow x + 5 = 2t \Rightarrow 2(x + 5) = 4t$
$\Rightarrow y = 2(x + 5) - 7 \Rightarrow y = 2x + 3$

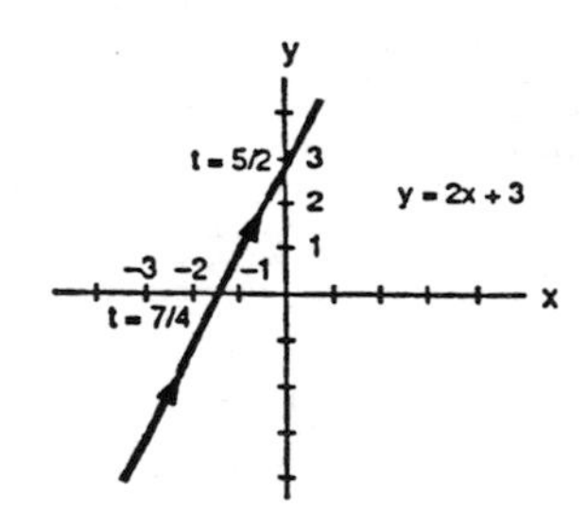

16. $x = 1 - t,\ y = 1 + t,\ -\infty < t < \infty$
$\Rightarrow 1 - x = t \Rightarrow y = 1 + (1 - x)$
$\Rightarrow y = -x + 2$

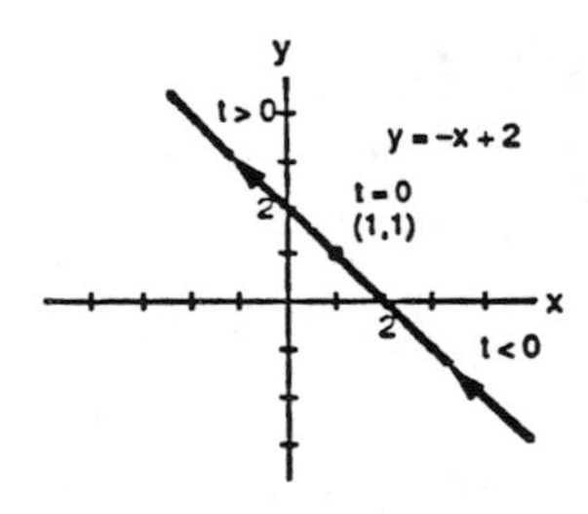

17. $x = t,\ y = 1 - t,\ 0 \leq t \leq 1$
$\Rightarrow y = 1 - x$

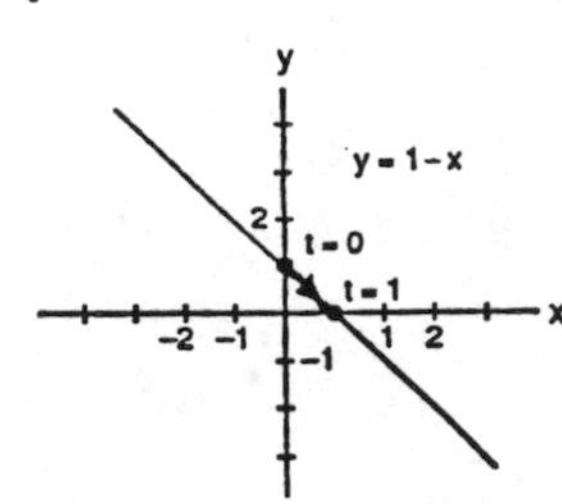

18. $x = 3 - 3t,\ y = 2t,\ 0 \leq t \leq 1 \Rightarrow \frac{y}{2} = t$
$\Rightarrow x = 3 - 3\left(\frac{y}{2}\right) \Rightarrow 2x = 6 - 3y \Rightarrow y = 2 - \frac{2}{3}x$

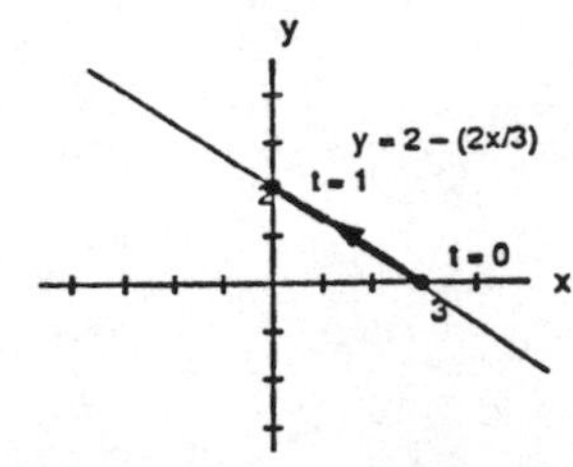

19. $x = t,\ y = \sqrt{1 - t^2},\ -1 \le t \le 0$
$\Rightarrow y = \sqrt{1 - x^2}$

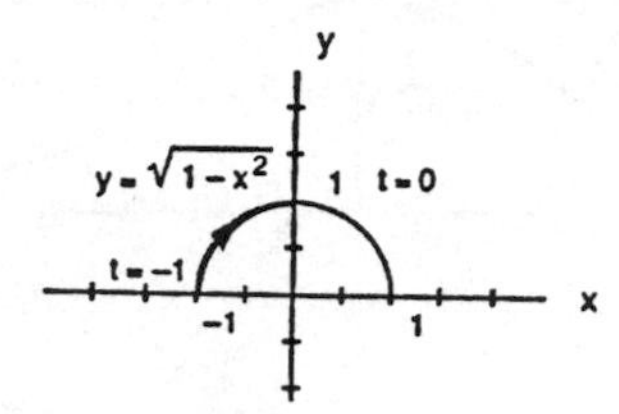

20. $x = t,\ y = \sqrt{4 - t^2},\ 0 \le t \le 2$
$\Rightarrow y = \sqrt{4 - x^2}$

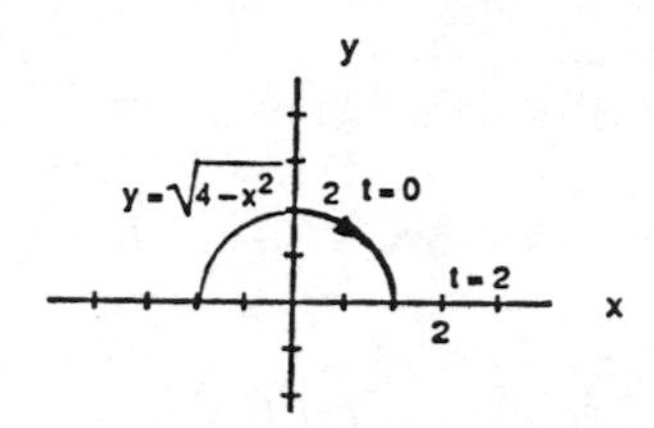

21. $x = t^2,\ y = \sqrt{t^4 + 1},\ t \ge 0$
$\Rightarrow y = \sqrt{x^2 + 1},\ x \ge 0$

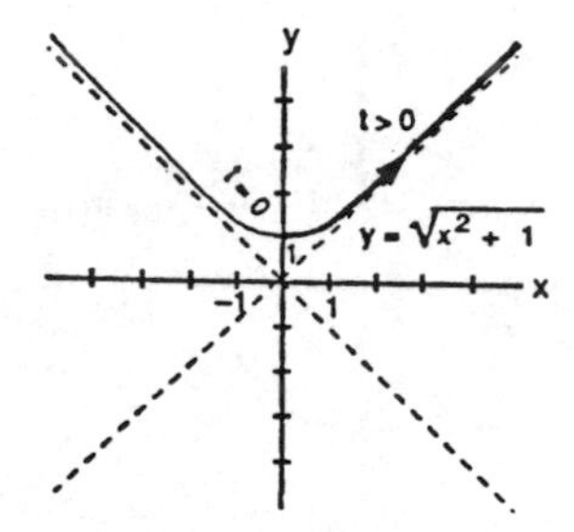

22. $x = \sqrt{t + 1},\ y = \sqrt{t},\ t \ge 0$
$\Rightarrow y^2 = t \Rightarrow x = \sqrt{y^2 + 1},\ y \ge 0$

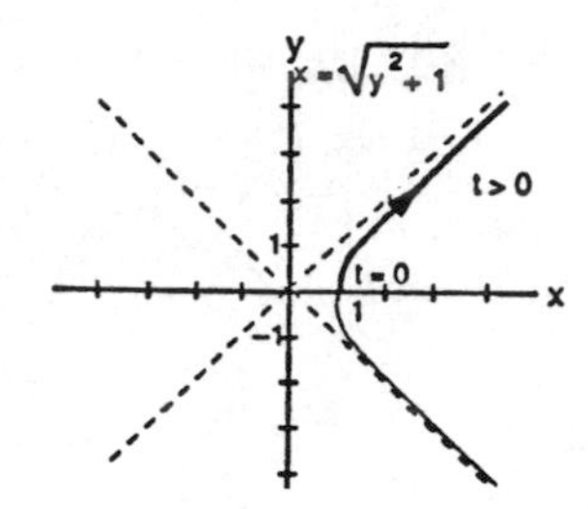

23. $x = -\cosh t,\ y = \sinh t,\ -\infty < 1 < \infty$
$\Rightarrow \cosh^2 t - \sinh^2 t = 1 \Rightarrow x^2 - y^2 = 1$

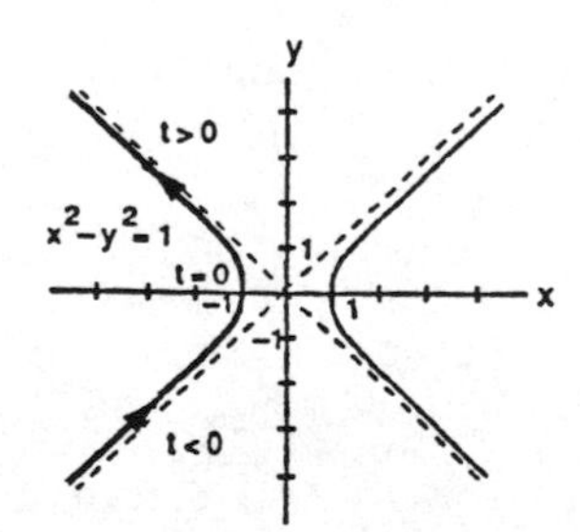

24. $x = 2 \sinh t,\ y = 2 \cosh t,\ -\infty < t < \infty$
$\Rightarrow 4 \cosh^2 t - 4 \sinh^2 t = 4 \Rightarrow y^2 - x^2 = 4$

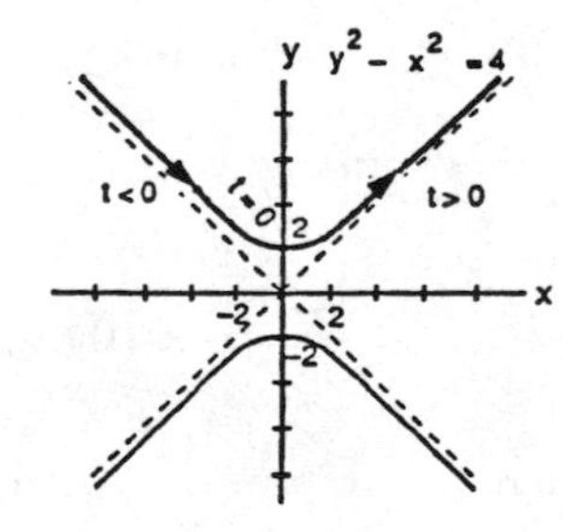

25. (a) $x = a \cos t,\ y = -a \sin t,\ 0 \le t \le 2\pi$
(b) $x = a \cos t,\ y = a \sin t,\ 0 \le t \le 2\pi$
(c) $x = a \cos t,\ y = -a \sin t,\ 0 \le t \le 4\pi$
(d) $x = a \cos t,\ y = a \sin t,\ 0 \le t \le 4\pi$

26. (a) $x = a \sin t,\ y = b \cos t,\ \frac{\pi}{2} \le t \le \frac{5\pi}{2}$
(b) $x = a \cos t,\ y = b \sin t,\ 0 \le t \le 2\pi$
(c) $x = a \sin t,\ y = b \cos t,\ \frac{\pi}{2} \le t \le \frac{9\pi}{2}$
(d) $x = a \cos t,\ y = b \sin t,\ 0 \le t \le 4\pi$

27. $x^2 + y^2 = a^2 \Rightarrow 2x + 2y \frac{dy}{dx} = 0 \Rightarrow \frac{dy}{dx} = -\frac{x}{y}$; let $t = \frac{dy}{dx} \Rightarrow -\frac{x}{y} = t \Rightarrow x = -yt$. Substitution yields $y^2t^2 + y^2 = a^2 \Rightarrow y = \frac{a}{\sqrt{1+t^2}}$ and $x = \frac{-at}{\sqrt{1+t}},\ -\infty < t < \infty$

28. In terms of θ, parametric equations for the circle are $x = a \cos \theta,\ y = a \sin \theta,\ 0 \le \theta < 2\pi$. Since $\theta = \frac{s}{a}$, the arc length parametrizations are: $x = a \cos \frac{s}{a},\ y = a \sin \frac{s}{a}$, and $0 \le \frac{s}{a} < 2\pi \Rightarrow 0 \le s \le 2\pi a$ is the interval for s.

29. Extend the vertical line through A to the x-axis and let C be the point of intersection. Then $OC = AQ = x$ and $\tan t = \frac{2}{OC} = \frac{2}{x} \Rightarrow x = \frac{2}{\tan t} = 2 \cot t$; $\sin t = \frac{2}{OA}$ $\Rightarrow OA = \frac{2}{\sin t}$; and $(AB)(OA) = (AQ)^2 \Rightarrow AB\left(\frac{2}{\sin t}\right) = x^2$ $\Rightarrow AB\left(\frac{2}{\sin t}\right) = \left(\frac{2}{\tan t}\right)^2 \Rightarrow AB = \frac{2 \sin t}{\tan^2 t}$. Next $y = 2 - AB \sin t \Rightarrow y = 2 - \left(\frac{2 \sin t}{\tan^2 t}\right) \sin t =$ $2 - \frac{2 \sin^2 t}{\tan^2 t} = 2 - 2 \cos^2 t = 2 \sin^2 t$. Therefore let $x = 2 \cot t$ and $y = 2 \sin^2 t$, $0 < t < \pi$.

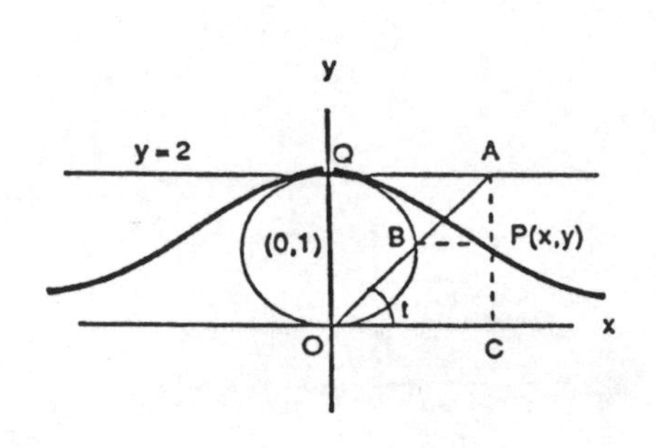

30. $\angle PQB = \angle QOB = t$ and $PQ = \text{arc}(AQ) = t$ since $PQ =$ length of the unwound string $=$ length of arc(AQ); thus $x = OB + BC = OB + DP = \cos t + t \sin t$, and $y = PC = QB - QD = \sin t - t \cos t$

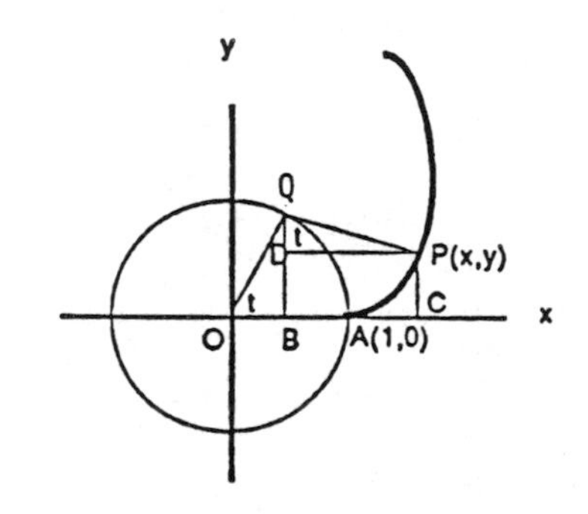

31. (a) $x = x_0 + (x_1 - x_0)t$ and $y = y_0 + (y_1 - y_0)t \Rightarrow t = \frac{x - x_0}{x_1 - x_0} \Rightarrow y = y_0 + (y_1 - y_0)\left(\frac{x - x_0}{x_1 - x_0}\right)$ $\Rightarrow y - y_0 = \left(\frac{y_1 - y_0}{x_1 - x_0}\right)(x - x_0)$ which is an equation of the line through the points (x_0, y_0) and (x_1, y_1)

(b) Let $x_0 = y_0 = 0$ in (a) $\Rightarrow x = x_1 t$, $y = y_1 t$ (the answer is not unique)

(c) Let $(x_0, y_0) = (-1, 0)$ and $(x_1, y_1) = (0, 1)$ or let $(x_0, y_0) = (0, 1)$ and $(x_1, y_1) = (-1, 0)$ in part (a) $\Rightarrow x = -1 + t$, $y = t$ or $x = -t$, $y = 1 - t$ (the answer is not unique)

32. (a) Let C be the point where the vertical line through P meets the x-axis. We have $AB = R$ and $AP = L$ $\Rightarrow PB = L - R$. Then $y = PB \sin \theta \Rightarrow y = (L - R) \sin \theta$, and $BC = PB \cos \theta \Rightarrow BC = (L - R) \cos \theta$. Also, in triangle OAB, $\cos \theta = \frac{OB}{AB} = \frac{OB}{R} \Rightarrow OB = R \cos \theta$. Therefore, $x = OB + BC = R \cos \theta + (L - R) \cos \theta$ $= L \cos \theta \Rightarrow x = L \cos \theta$ and $y = (L - R) \sin \theta$.

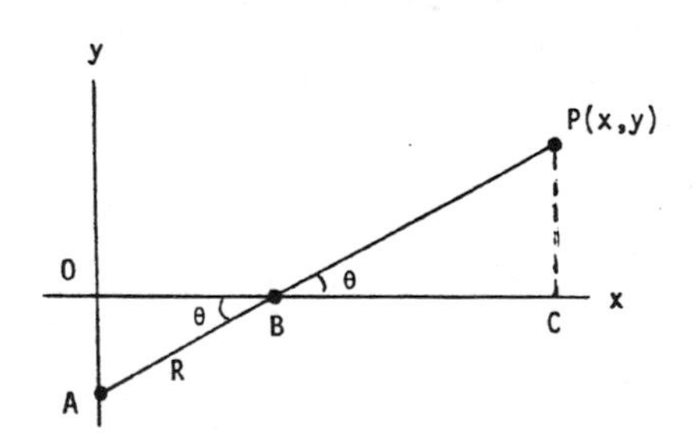

(b) Note that $\frac{x^2}{L^2} + \frac{y^2}{(L-R)^2} = \frac{L^2 \cos^2 \theta}{L^2} + \frac{(L-R)^2 \sin^2 \theta}{(L-R)^2} = 1$. Therefore, the points $P(x, y)$ satisfy the equation $\frac{x^2}{L^2} + \frac{y^2}{(L-R)^2} = 1$, which is an ellipse.

33. Arc PF = Arc AF since each is the distance rolled and $\frac{\text{Arc PF}}{b} = \angle FCP \Rightarrow \text{Arc PF} = b(\angle FCP)$; $\frac{\text{Arc AF}}{a} = \theta$ $\Rightarrow \text{Arc AF} = a\theta \Rightarrow a\theta = b(\angle FCP) \Rightarrow \angle FCP = \frac{a}{b}\theta$; $\angle OCG = \frac{\pi}{2} - \theta$; $\angle OCG = \angle OCP + \angle PCE$ $= \angle OCP + \left(\frac{\pi}{2} - \alpha\right)$. Now $\angle OCP = \pi - \angle FCP$ $= \pi - \frac{a}{b}\theta$. Thus $\angle OCG = \pi - \frac{a}{b}\theta + \frac{\pi}{2} - \alpha \Rightarrow \frac{\pi}{2} - \theta$ $= \pi - \frac{a}{b}\theta + \frac{\pi}{2} - \alpha \Rightarrow \alpha = \pi - \frac{a}{b}\theta + \theta = \pi - \left(\frac{a-b}{b}\theta\right)$.

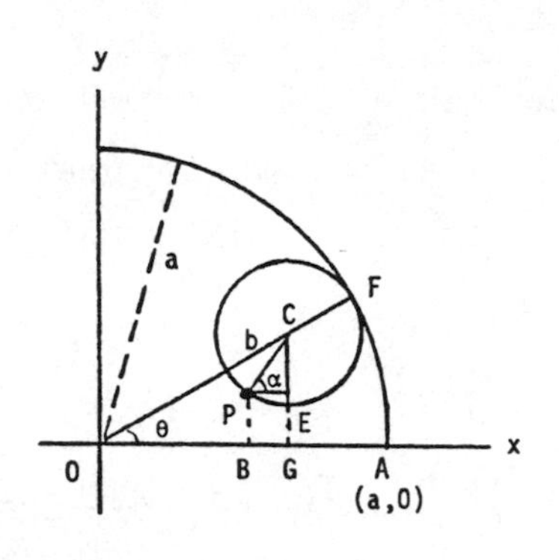

Then $x = OG - BG = OG - PE = (a-b)\cos\theta - b\cos\alpha = (a-b)\cos\theta - b\cos\left(\pi - \frac{a-b}{b}\theta\right)$ $= (a-b)\cos\theta + b\cos\left(\frac{a-b}{b}\theta\right)$. Also $y = EG = CG - CE = (a-b)\sin\theta - b\sin\alpha$ $= (a-b)\sin\theta - b\sin\left(\pi - \frac{a-b}{b}\theta\right) = (a-b)\sin\theta - b\sin\left(\frac{a-b}{b}\theta\right)$. Therefore $x = (a-b)\cos\theta + b\cos\left(\frac{a-b}{b}\theta\right)$ and $y = (a-b)\sin\theta - b\sin\left(\frac{a-b}{b}\theta\right)$.

If $b = \frac{a}{4}$, then $x = \left(a - \frac{a}{4}\right)\cos\theta + \frac{a}{4}\cos\left(\frac{a - \left(\frac{a}{4}\right)}{\left(\frac{a}{4}\right)}\theta\right)$

$= \frac{3a}{4}\cos\theta + \frac{a}{4}\cos 3\theta = \frac{3a}{4}\cos\theta + \frac{a}{4}(\cos\theta\cos 2\theta - \sin\theta\sin 2\theta)$

$= \frac{3a}{4}\cos\theta + \frac{a}{4}\left((\cos\theta)(\cos^2\theta - \sin^2\theta) - (\sin\theta)(2\sin\theta\cos\theta)\right)$

$= \frac{3a}{4}\cos\theta + \frac{a}{4}\cos^3\theta - \frac{a}{4}\cos\theta\sin^2\theta - \frac{2a}{4}\sin^2\theta\cos\theta$

$= \frac{3a}{4}\cos\theta + \frac{a}{4}\cos^3\theta - \frac{3a}{4}(\cos\theta)(1 - \cos^2\theta) = a\cos^3\theta$;

$y = \left(a - \frac{a}{4}\right)\sin\theta - \frac{a}{4}\sin\left(\frac{a - \left(\frac{a}{4}\right)}{\left(\frac{a}{4}\right)}\theta\right) = \frac{3a}{4}\sin\theta - \frac{a}{4}\sin 3\theta = \frac{3a}{4}\sin\theta - \frac{a}{4}(\sin\theta\cdot\cos 2\theta + \cos\theta\sin 2\theta)$

$= \frac{3a}{4}\sin\theta - \frac{a}{4}\left((\sin\theta)(\cos^2\theta - \sin^2\theta) + (\cos\theta)(2\sin\theta\cos\theta)\right)$

$= \frac{3a}{4}\sin\theta - \frac{a}{4}\sin\theta\cos^2\theta + \frac{a}{4}\sin^3\theta - \frac{2a}{4}\cos^2\theta\sin\theta$

$= \frac{3a}{4}\sin\theta - \frac{3a}{4}\sin\theta\cos^2\theta + \frac{a}{4}\sin^3\theta$

$= \frac{3a}{4}\sin\theta - \frac{3a}{4}(\sin\theta)(1 - \sin^2\theta) + \frac{a}{4}\sin^3\theta = a\sin^3\theta$.

34. P traces a hypocycloid where the larger radius is 2a and the smaller is a $\Rightarrow x = (2a - a)\cos\theta + a\cos\left(\frac{2a-a}{a}\theta\right)$ $= 2a\cos\theta$, $0 \le \theta \le 2\pi$, and $y = (2a-a)\sin\theta - a\sin\left(\frac{2a-a}{a}\theta\right) = a\sin\theta - a\sin\theta = 0$. Therefore P traces the diameter of the circle back and forth as θ goes from 0 to 2π.

35. Draw line AM in the figure and note that $\angle AMO$ is a right angle since it is an inscribed angle which spans the diameter of a circle. Then $AN^2 = MN^2 + AM^2$. Now, $OA = a$, $\frac{AN}{a} = \tan t$, and $\frac{AM}{a} = \sin t$. Next $MN = OP \Rightarrow OP^2 = AN^2 - AM^2$ $= a^2 \tan^2 t - a^2 \sin^2 t \Rightarrow OP = \sqrt{a^2 \tan^2 t - a^2 \sin^2 t}$ $= (a \sin t)\sqrt{\sec^2 t - 1} = \frac{a \sin^2 t}{\cos t}$. In triangle BPO, $x = OP \sin t = \frac{a \sin^3 t}{\cos t} = a \sin^2 t \tan t$ and $y = OP \cos t = a \sin^2 t \Rightarrow x = a \sin^2 t \tan t$ and $y = a \sin^2 t$.

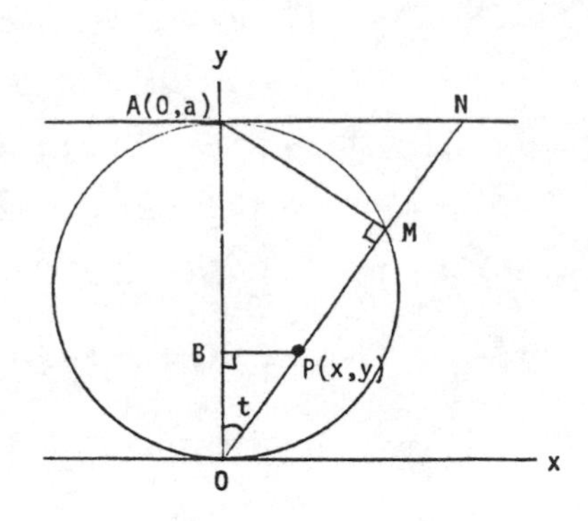

36. Let the x-axis be the line the wheel rolls along with the y-axis through a low point of the trochoid (see the accompanying figure).

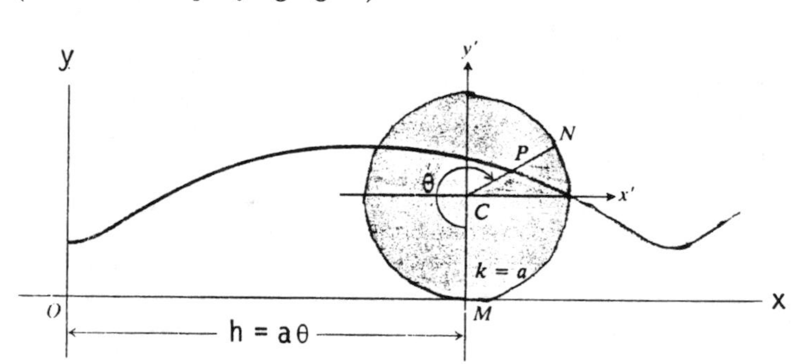

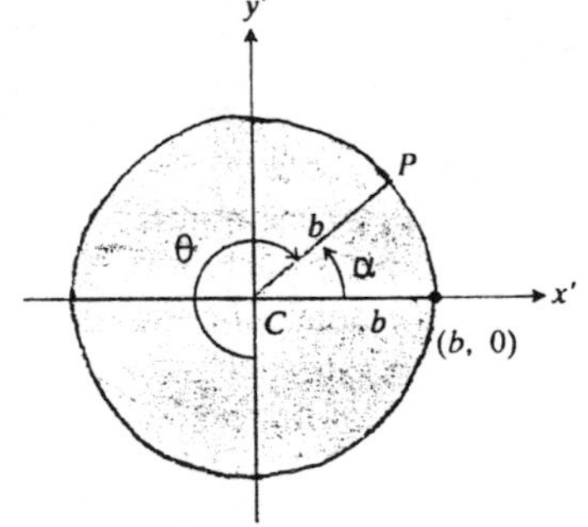

Let θ denote the angle through which the wheel turns. Then $h = a\theta$ and $k = a$. Next introduce $x'y'$-axes parallel to the xy-axes and having their origin at the center C of the wheel. Then $x' = b \cos \alpha$ and $y' = b \sin \alpha$, where $\alpha = \frac{3\pi}{2} - \theta$. It follows that $x' = b \cos\left(\frac{3\pi}{2} - \theta\right) = -b \sin \theta$ and $y' = b \sin\left(\frac{3\pi}{2} - \theta\right)$ $= -b \cos \theta \Rightarrow x = h + x' = a\theta - b \sin \theta$ and $y = k + y' = a - b \cos \theta$ are parametric equations of the trochoid.

37. $D = \sqrt{(x-2)^2 + \left(y - \frac{1}{2}\right)^2} \Rightarrow D^2 = (x-2)^2 + \left(y - \frac{1}{2}\right)^2 = (t-2)^2 + \left(t^2 - \frac{1}{2}\right)^2 \Rightarrow D^2 = t^4 - 4t + \frac{17}{4}$ $\Rightarrow \frac{d(D^2)}{dt} = 4t^3 - 4 = 0 \Rightarrow t = 1$. The second derivative is always positive for $t \neq 0 \Rightarrow t = 1$ gives a local minimum for D^2 (and hence D) which is an absolute minimum since it is the only extremum $\Rightarrow$ the closest point on the parabola is $(1, 1)$.

38. $D = \sqrt{\left(2 \cos t - \frac{3}{4}\right)^2 + (\sin t - 0)^2} \Rightarrow D^2 = \left(2 \cos t - \frac{3}{4}\right)^2 + \sin^2 t \Rightarrow \frac{d(D^2)}{dt}$ $= 2\left(2 \cos t - \frac{3}{4}\right)(-2 \sin t) + 2 \sin t \cos t = (-2 \sin t)\left(3 \cos t - \frac{3}{2}\right) = 0 \Rightarrow -2 \sin t = 0$ or $3 \cos t - \frac{3}{2} = 0$ $\Rightarrow t = 0, \pi$ or $t = \frac{\pi}{3}, \frac{5\pi}{3}$. Now $\frac{d^2(D^2)}{dt^2} = -6 \cos^2 t + 3 \cos t + 6 \sin^2 t$ so that $\frac{d^2(D^2)}{dt^2}(0) = -3 \Rightarrow$ relative maximum, $\frac{d^2(D^2)}{dt^2}(\pi) = -9 \Rightarrow$ relative maximum, $\frac{d^2(D^2)}{dt^2}\left(\frac{\pi}{3}\right) = \frac{9}{2} \Rightarrow$ relative minimum, and

$\frac{d^2(D^2)}{dt^2}\left(\frac{5\pi}{3}\right) = \frac{9}{2} \Rightarrow$ relative minimum. Therefore both $t = \frac{\pi}{3}$ and $t = \frac{5\pi}{3}$ give points on the ellipse closest to the point $\left(\frac{3}{4}, 0\right) \Rightarrow \left(1, \frac{\sqrt{3}}{2}\right)$ and $\left(1, -\frac{\sqrt{3}}{2}\right)$ are the desired points.

39. (a)

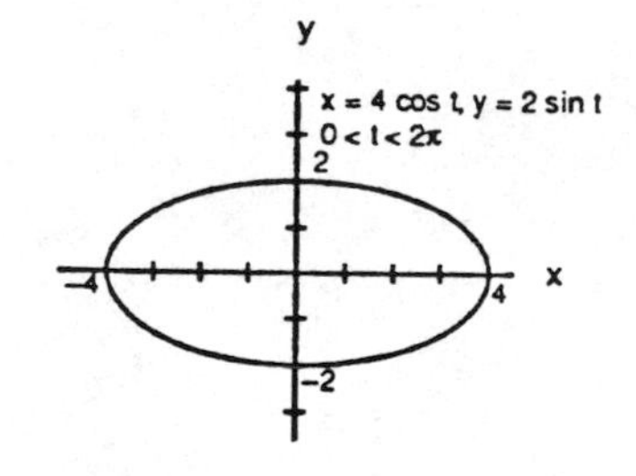

(b)

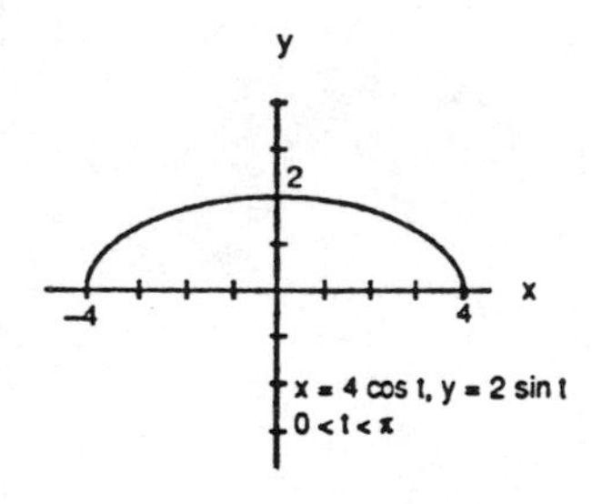

(c)

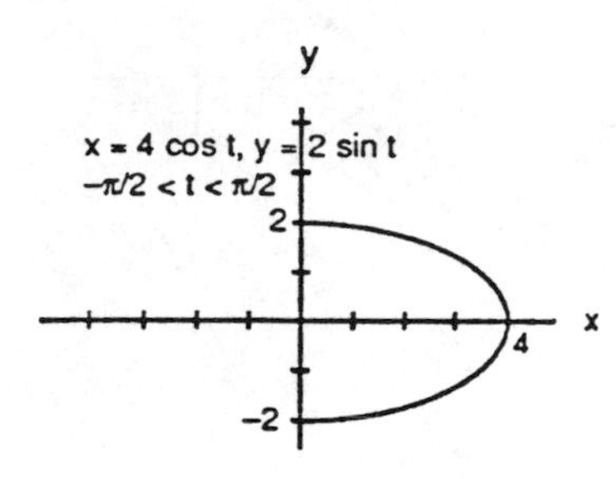

40. (a)

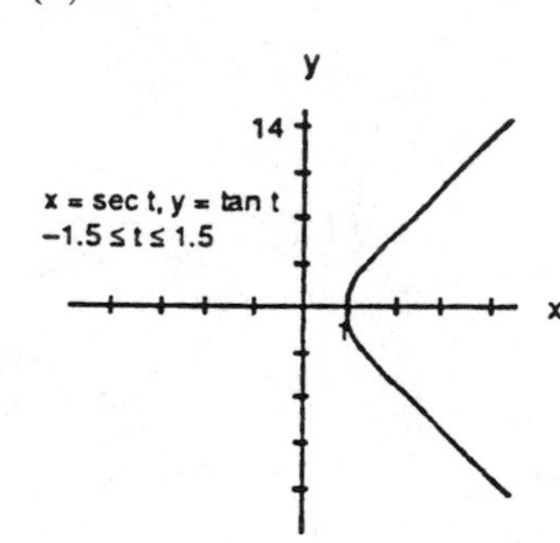

(b)

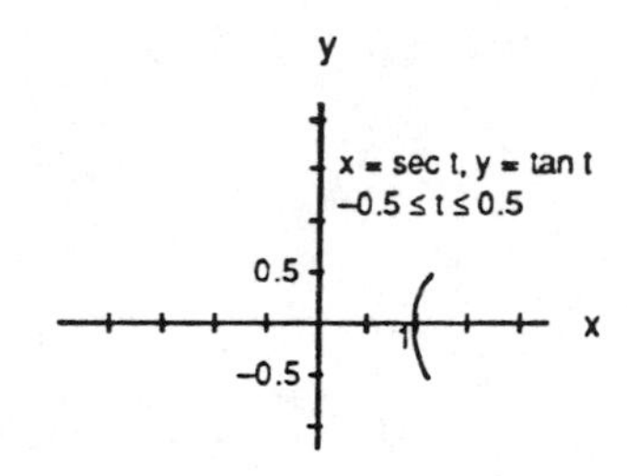

(c)

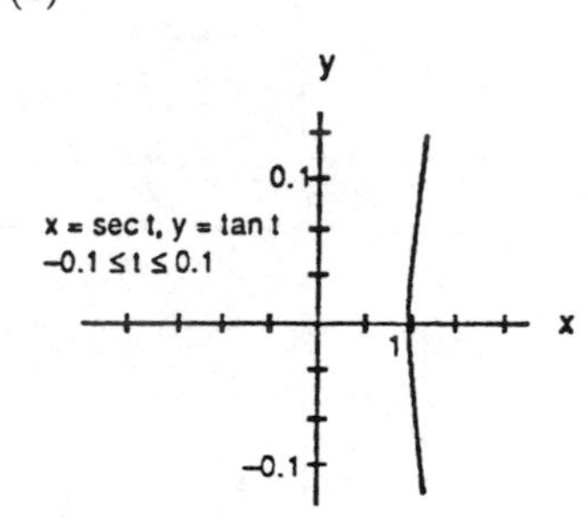

41.

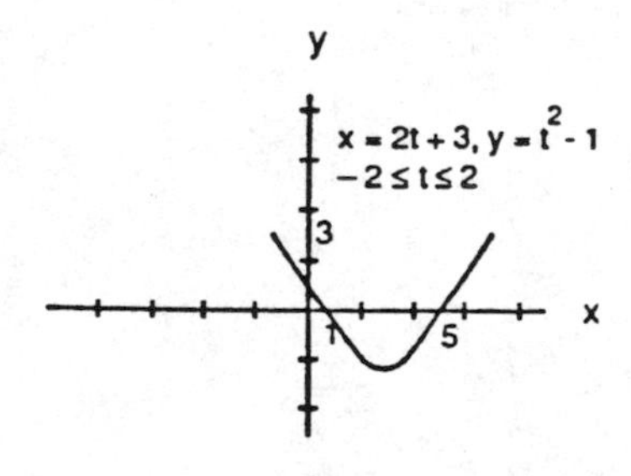

42. (a)

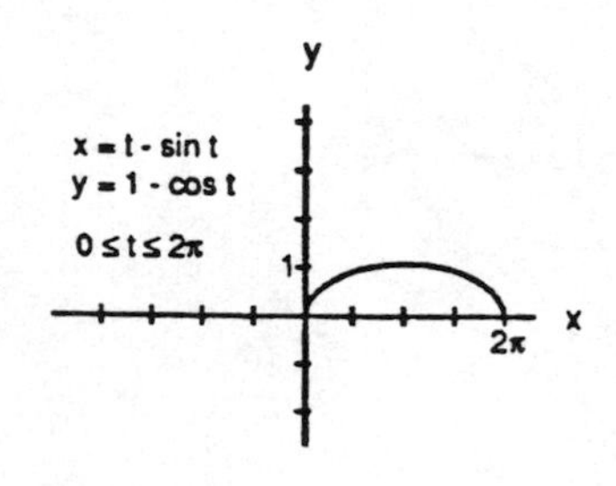

(b)

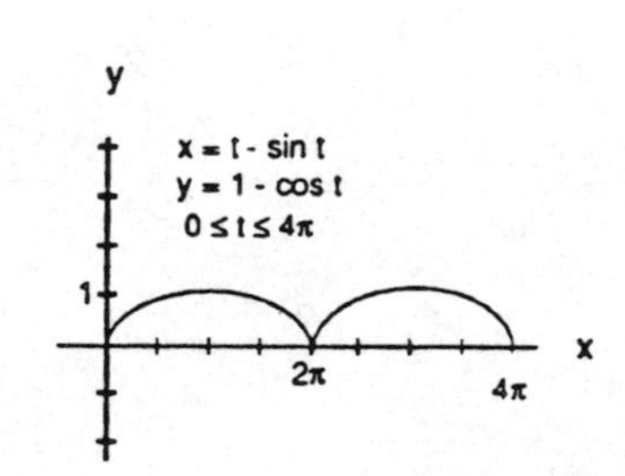

(c)

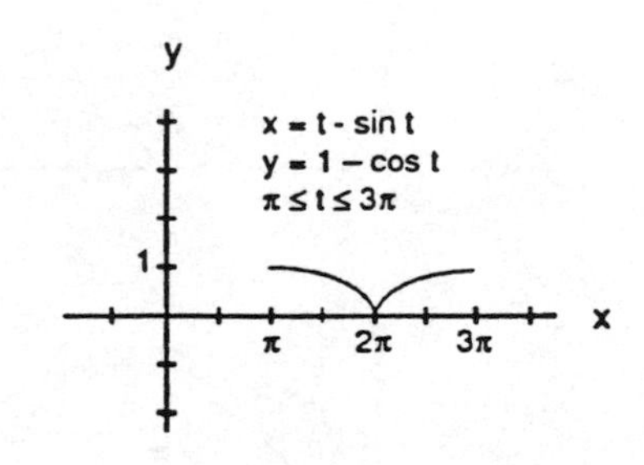

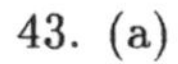

43. (a)

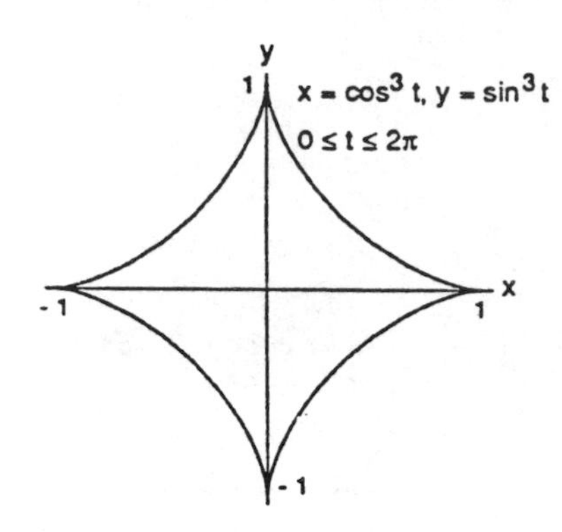

(b)

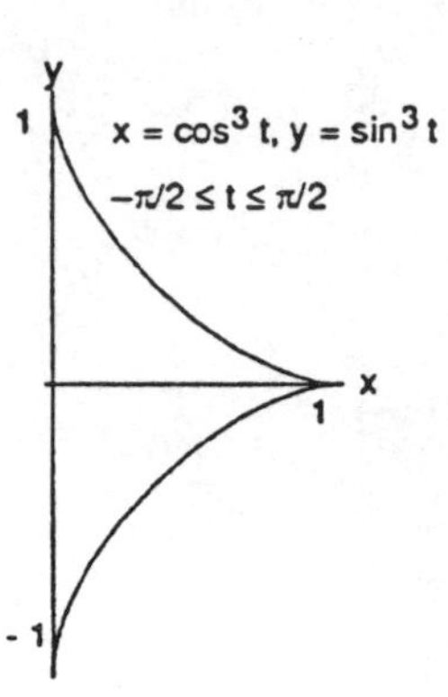

44. (a)

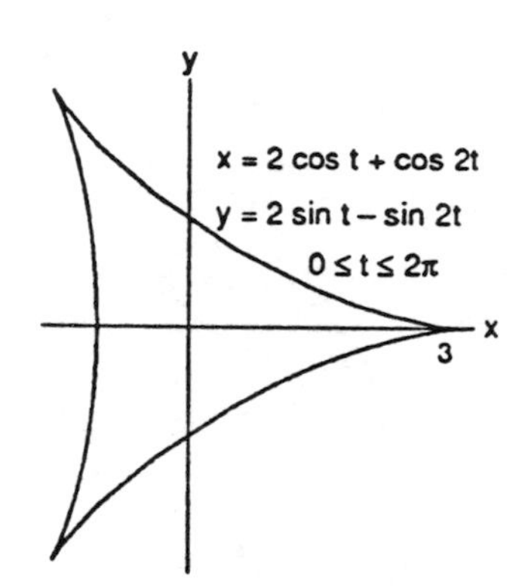

(b)

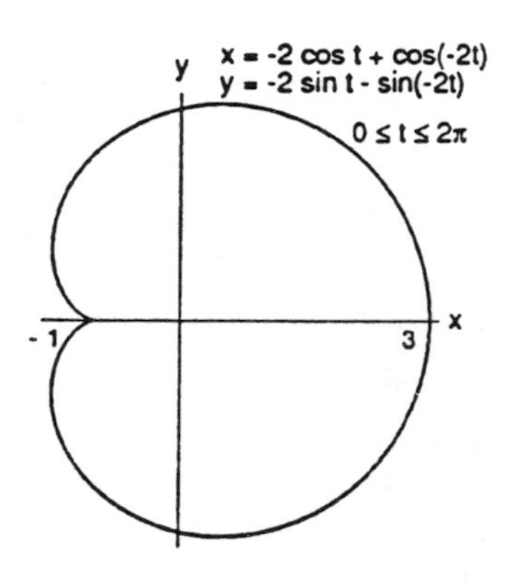

45. (a)

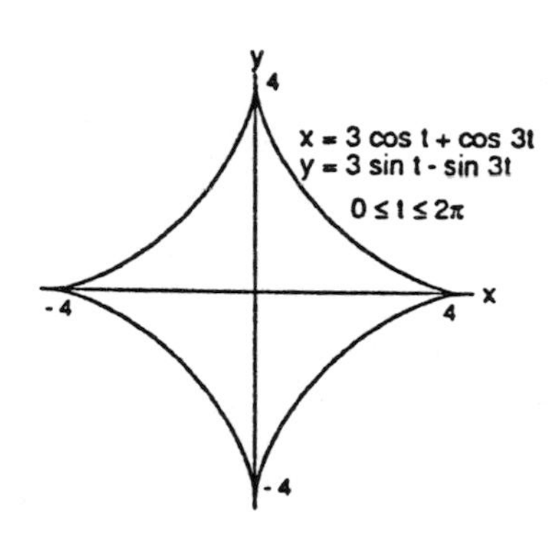

(b)

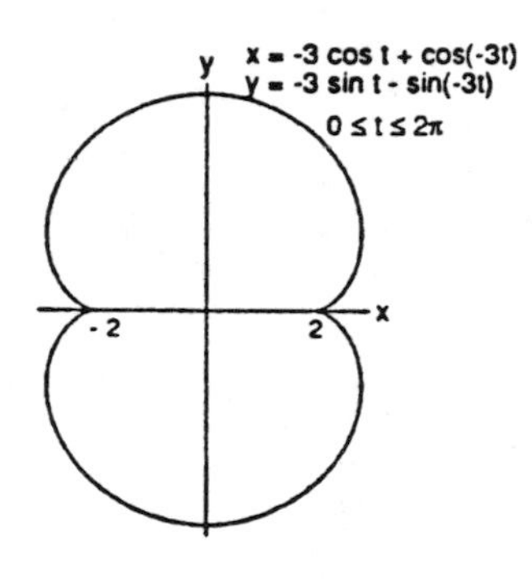

46. (a)

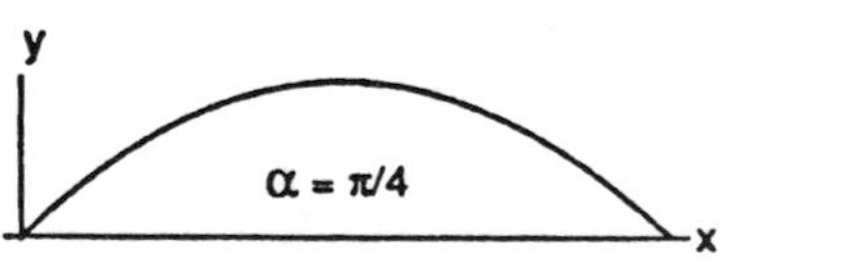

(b)

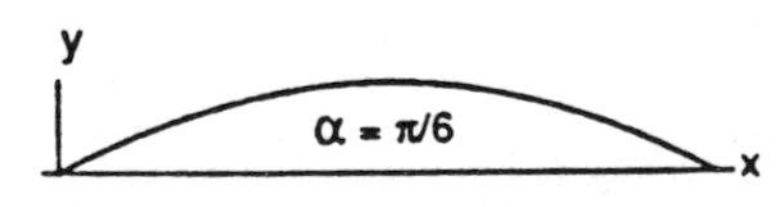

(c)

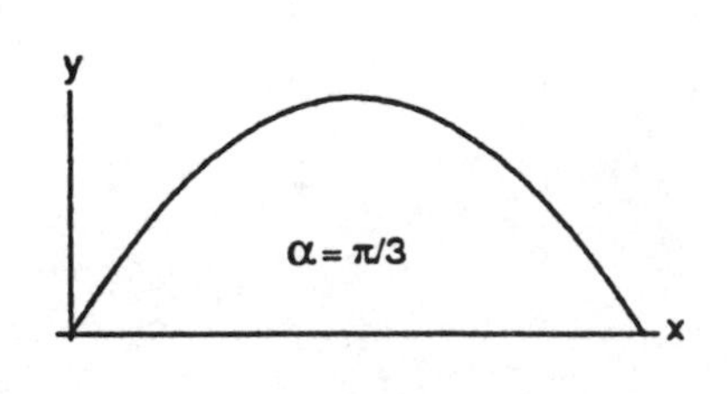

(d)

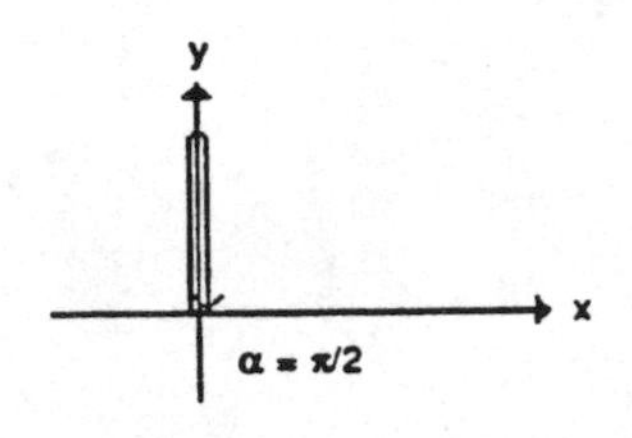

47. (a)

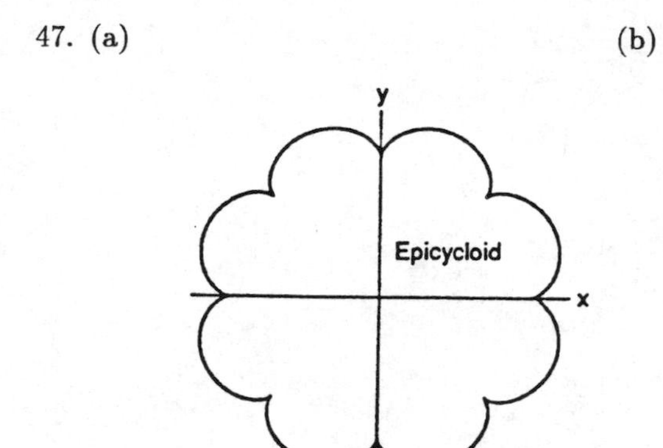

(b)

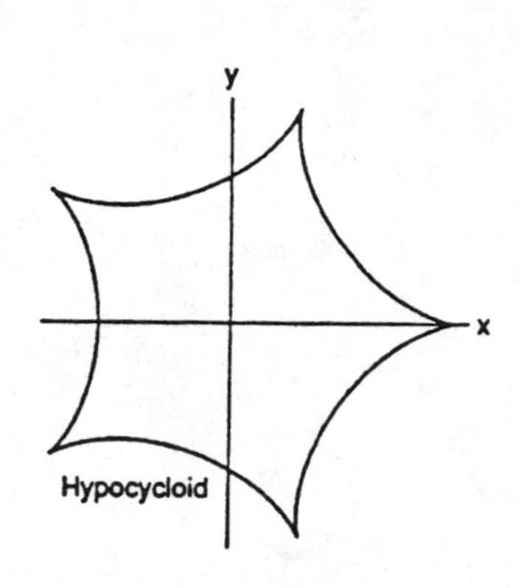

(c)

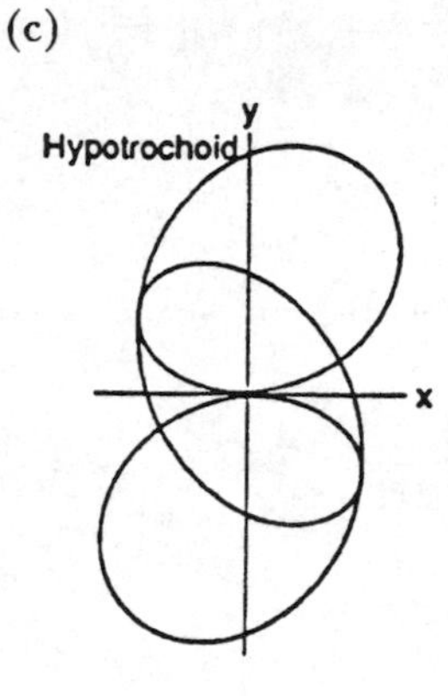

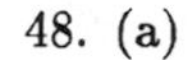
48. (a)

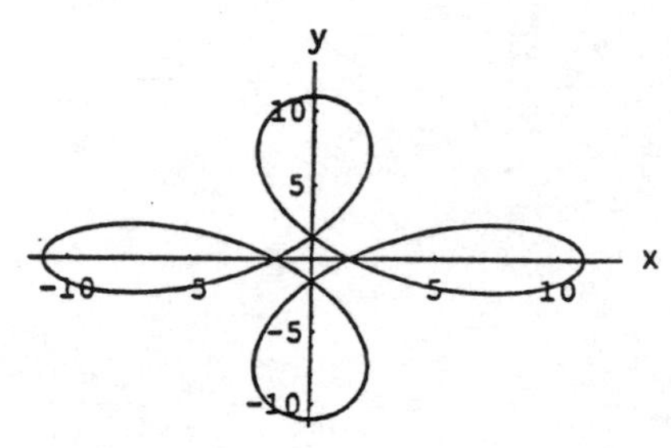

$x = 6\cos t + 5\cos 3t,\quad y = 6\sin t - 5\sin 3t,$
$0 \le t \le 2\pi$

(b)

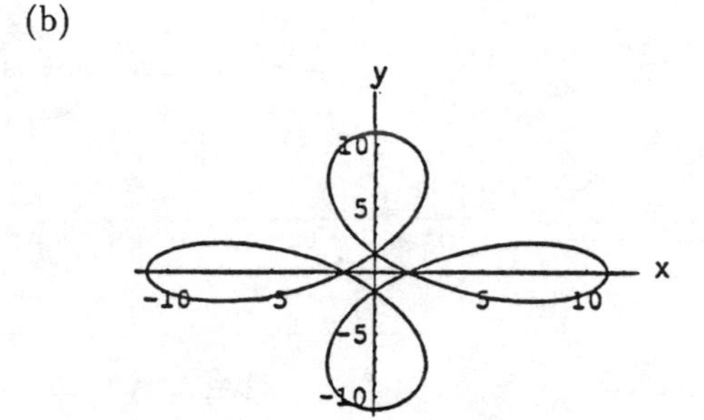

$x = 6\cos 2t + 5\cos 6t,\quad y = 6\sin 2t - 5\sin 6t,$
$0 \le t \le \pi$

(c)

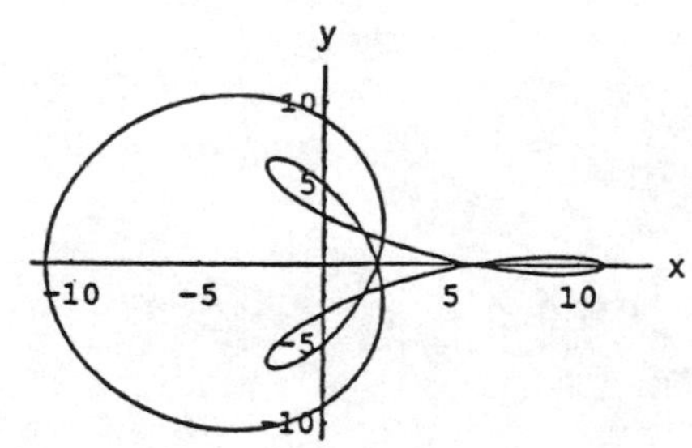

$x = 6\cos t + 5\cos 3t,\quad y = 6\sin 2t - 5\sin 3t,$
$0 \le t \le 2\pi$

(d)

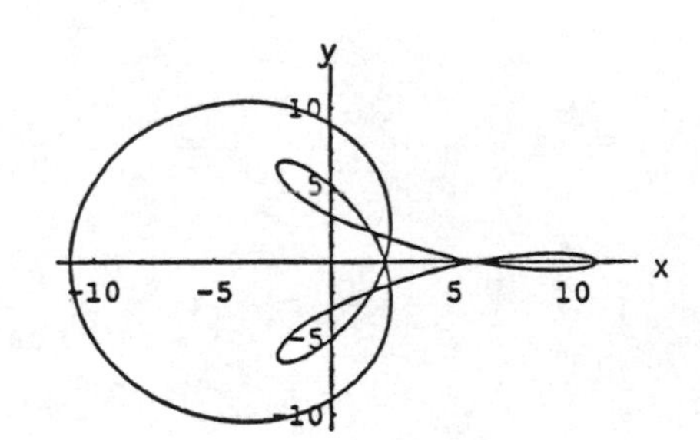

$x = 6\cos 2t + 5\cos 6t,\quad y = 6\sin 4t - 5\sin 6t,$
$0 \le t \le \pi$

9.5 CALCULUS WITH PARAMETRIZED CURVES

1. $t = \frac{\pi}{4} \Rightarrow x = 2\cos\frac{\pi}{4} = \sqrt{2},\ y = 2\sin\frac{\pi}{4} = \sqrt{2};\ \frac{dx}{dt} = -2\sin t,\ \frac{dy}{dt} = 2\cos t \Rightarrow \frac{dy}{dx} = \frac{dy/dt}{dx/dt} = \frac{2\cos t}{-2\sin t} = -\cot t$

$\Rightarrow \left.\frac{dy}{dx}\right|_{t=\frac{\pi}{4}} = -\cot\frac{\pi}{4} = -1$; tangent line is $y - \sqrt{2} = -1\left(x - \sqrt{2}\right)$ or $y = -x + 2\sqrt{2}$; $\frac{dy'}{dt} = \csc^2 t$

$\Rightarrow \frac{d^2y}{dx^2} = \frac{dy'/dt}{dx/dt} = \frac{\csc^2 t}{-2\sin t} = -\frac{1}{2\sin^3 t} \Rightarrow \left.\frac{d^2y}{dx^2}\right|_{t=\frac{\pi}{4}} = -\sqrt{2}$

2. $t=-\frac{1}{6} \Rightarrow x=\sin\left(2\pi\left(-\frac{1}{6}\right)\right)=\sin\left(-\frac{\pi}{3}\right)=-\frac{\sqrt{3}}{2}$, $y=\cos\left(2\pi\left(-\frac{1}{6}\right)\right)=\cos\left(-\frac{\pi}{3}\right)=\frac{1}{2}$; $\frac{dx}{dt}=2\pi\cos 2\pi t$, $\frac{dy}{dt}=-2\pi\sin 2\pi t \Rightarrow \frac{dy}{dx}=\frac{-2\pi\sin 2\pi t}{2\pi\cos 2\pi t}=-\tan 2\pi t \Rightarrow \left.\frac{dy}{dx}\right|_{t=-\frac{1}{6}}=-\tan\left(2\pi\left(-\frac{1}{6}\right)\right)=-\tan\left(-\frac{\pi}{3}\right)=\sqrt{3}$; tangent line is $y-\frac{1}{2}=\sqrt{3}\left[x-\left(-\frac{\sqrt{3}}{2}\right)\right]$ or $y=\sqrt{3}x+2$; $\frac{dy'}{dt}=-2\pi\sec^2 2\pi t \Rightarrow \frac{d^2y}{dx^2}=\frac{-2\pi\sec^2 2\pi t}{2\pi\cos 2\pi t}$ $=-\frac{1}{\cos^3 2\pi t} \Rightarrow \left.\frac{d^2y}{dx^2}\right|_{t=-\frac{1}{6}}=-8$

3. $t=\frac{\pi}{4} \Rightarrow x=4\sin\frac{\pi}{4}=2\sqrt{2}$, $y=2\cos\frac{\pi}{4}=\sqrt{2}$; $\frac{dx}{dt}=4\cos t$, $\frac{dy}{dt}=-2\sin t \Rightarrow \frac{dy}{dx}=\frac{dy/dt}{dx/dt}=\frac{-2\sin t}{4\cos t}$ $=-\frac{1}{2}\tan t \Rightarrow \left.\frac{dy}{dx}\right|_{t=\frac{\pi}{4}}=-\frac{1}{2}\tan\frac{\pi}{4}=-\frac{1}{2}$; tangent line is $y-\sqrt{2}=-\frac{1}{2}(x-2\sqrt{2})$ or $y=-\frac{1}{2}x+2\sqrt{2}$; $\frac{dy'}{dt}=-\frac{1}{2}\sec^2 t \Rightarrow \frac{d^2y}{dx^2}=\frac{dy'/dt}{dx/dt}=\frac{-\frac{1}{2}\sec^2 t}{4\cos t}=-\frac{1}{8\cos^3 t} \Rightarrow \left.\frac{d^2y}{dx^2}\right|_{t=\frac{\pi}{4}}=-\frac{\sqrt{2}}{4}$

4. $t=\frac{2\pi}{3} \Rightarrow x=\cos\frac{2\pi}{3}=-\frac{1}{2}$, $y=\sqrt{3}\cos\frac{2\pi}{3}=-\frac{\sqrt{3}}{2}$; $\frac{dx}{dt}=-\sin t$, $\frac{dy}{dt}=-\sqrt{3}\sin t \Rightarrow \frac{dy}{dx}=\frac{-\sqrt{3}\sin t}{-\sin t}=\sqrt{3}$ $\Rightarrow \left.\frac{dy}{dx}\right|_{t=\frac{2\pi}{3}}=\sqrt{3}$; tangent line is $y-\left(-\frac{\sqrt{3}}{2}\right)=\sqrt{3}\left[x-\left(-\frac{1}{2}\right)\right]$ or $y=\sqrt{3}x$; $\frac{dy'}{dt}=0 \Rightarrow \frac{d^2y}{dx^2}=\frac{0}{-\sin t}=0$ $\Rightarrow \left.\frac{d^2y}{dx^2}\right|_{t=\frac{2\pi}{3}}=0$

5. $t=\frac{1}{4} \Rightarrow x=\frac{1}{4}$, $y=\frac{1}{2}$; $\frac{dx}{dt}=1$, $\frac{dy}{dt}=\frac{1}{2\sqrt{t}} \Rightarrow \frac{dy}{dx}=\frac{dy/dt}{dx/dt}=\frac{1}{2\sqrt{t}} \Rightarrow \left.\frac{dy}{dx}\right|_{t=\frac{1}{4}}=\frac{1}{2\sqrt{\frac{1}{4}}}=1$; tangent line is $y-\frac{1}{2}=1\cdot\left(x-\frac{1}{4}\right)$ or $y=x+\frac{1}{4}$; $\frac{dy'}{dt}=-\frac{1}{4}t^{-3/2} \Rightarrow \frac{d^2y}{dx^2}=\frac{dy'/dt}{dx/dt}=-\frac{1}{4}t^{-3/2} \Rightarrow \left.\frac{d^2y}{dx^2}\right|_{t=\frac{1}{4}}=-2$

6. $t=-\frac{\pi}{4} \Rightarrow x=\sec^2\left(-\frac{\pi}{4}\right)-1=1$, $y=\tan\left(-\frac{\pi}{4}\right)=-1$; $\frac{dx}{dt}=2\sec^2 t\tan t$, $\frac{dy}{dt}=\sec^2 t$ $\Rightarrow \frac{dy}{dx}=\frac{\sec^2 t}{2\sec^2 t\tan t}=\frac{1}{2\tan t}=\frac{1}{2}\cot t \Rightarrow \left.\frac{dy}{dx}\right|_{t=-\frac{\pi}{4}}=\frac{1}{2}\cot\left(-\frac{\pi}{4}\right)=-\frac{1}{2}$; tangent line is $y-(-1)=-\frac{1}{2}(x-1)$ or $y=-\frac{1}{2}x-\frac{1}{2}$; $\frac{dy'}{dt}=-\frac{1}{2}\csc^2 t \Rightarrow \frac{d^2y}{dx^2}=\frac{-\frac{1}{2}\csc^2 t}{2\sec^2 t\tan t}=-\frac{1}{4}\cot^3 t$ $\Rightarrow \left.\frac{d^2y}{dx^2}\right|_{t=-\frac{\pi}{4}}=\frac{1}{4}$

7. $t = \frac{\pi}{6} \Rightarrow x = \sec\frac{\pi}{6} = \frac{2}{\sqrt{3}}$, $y = \tan\frac{\pi}{6} = \frac{1}{\sqrt{3}}$; $\frac{dx}{dt} = \sec t \tan t$, $\frac{dy}{dt} = \sec^2 t \Rightarrow \frac{dy}{dx} = \frac{dy/dt}{dx/dt}$ $= \frac{\sec^2 t}{\sec t \tan t} = \csc t \Rightarrow \left.\frac{dy}{dx}\right|_{t=\frac{\pi}{6}} = \csc\frac{\pi}{6} = 2$; tangent line is $y - \frac{1}{\sqrt{3}} = 2\left(x - \frac{2}{\sqrt{3}}\right)$ or $y = 2x - \sqrt{3}$;

$\frac{dy'}{dt} = -\csc t \cot t \Rightarrow \frac{d^2y}{dx^2} = \frac{dy'/dt}{dx/dt} = \frac{-\csc t \cot t}{\sec t \tan t} = -\cot^3 t \Rightarrow \left.\frac{d^2y}{dx^2}\right|_{t=\frac{\pi}{6}} = -3\sqrt{3}$

8. $t = 3 \Rightarrow x = -\sqrt{3+1} = -2$, $y = \sqrt{3(3)} = 3$; $\frac{dx}{dt} = -\frac{1}{2}(t+1)^{-1/2}$, $\frac{dy}{dt} = \frac{3}{2}(3t)^{-1/2} \Rightarrow \frac{dy}{dx} = \frac{\left(\frac{3}{2}\right)(3t)^{-1/2}}{\left(-\frac{1}{2}\right)(t+1)^{-1/2}}$ $= -\frac{3\sqrt{t+1}}{\sqrt{3t}} = \left.\frac{dy}{dx}\right|_{t=3} = \frac{-3\sqrt{3+1}}{\sqrt{3(3)}} = -2$; tangent line is $y - 3 = -2[x - (-2)]$ or $y = -2x - 1$;

$\frac{dy'}{dt} = \frac{\sqrt{3t}\left[-\frac{3}{2}(t+1)^{-1/2}\right] + 3\sqrt{t+1}\left[\frac{3}{2}(3t)^{-1/2}\right]}{3t} = \frac{3}{2t\sqrt{3t}\,\sqrt{t+1}} \Rightarrow \frac{d^2y}{dx^2} = \frac{\left(\frac{3}{2t\sqrt{3t}\,\sqrt{t+1}}\right)}{\left(\frac{-1}{2\sqrt{t+1}}\right)} = -\frac{3}{t\sqrt{3t}}$

$\Rightarrow \left.\frac{d^2y}{dx^2}\right|_{t=3} = -\frac{1}{3}$

9. $t = -1 \Rightarrow x = 5$, $y = 1$; $\frac{dx}{dt} = 4t$, $\frac{dy}{dt} = 4t^3 \Rightarrow \frac{dy}{dx} = \frac{dy/dt}{dx/dt} = \frac{4t^3}{4t} = t^2 \Rightarrow \left.\frac{dy}{dx}\right|_{t=-1} = (-1)^2 = 1$; tangent line is $y - 1 = 1\cdot(x-5)$ or $y = x - 4$; $\frac{dy'}{dt} = 2t \Rightarrow \frac{d^2y}{dx^2} = \frac{dy'/dt}{dx/dt} = \frac{2t}{4t} = \frac{1}{2} \Rightarrow \left.\frac{d^2y}{dx^2}\right|_{t=-1} = \frac{1}{2}$

10. $t = 1 \Rightarrow x = 1$, $y = -2$; $\frac{dx}{dt} = -\frac{1}{t^2}$, $\frac{dy}{dt} = \frac{1}{t} \Rightarrow \frac{dy}{dx} = \frac{\left(\frac{1}{t}\right)}{\left(-\frac{1}{t^2}\right)} = -t \Rightarrow \left.\frac{dy}{dx}\right|_{t=1} = -1$; tangent line is $y - (-2) = -1(x-1)$ or $y = -x - 1$; $\frac{dy'}{dt} = -1 \Rightarrow \frac{d^2y}{dx^2} = \frac{-1}{\left(-\frac{1}{t^2}\right)} = t^2 \Rightarrow \left.\frac{d^2y}{dx^2}\right|_{t=1} = 1$

11. $t = \frac{\pi}{3} \Rightarrow x = \frac{\pi}{3} - \sin\frac{\pi}{3} = \frac{\pi}{3} - \frac{\sqrt{3}}{2}$, $y = 1 - \cos\frac{\pi}{3} = 1 - \frac{1}{2} = \frac{1}{2}$; $\frac{dx}{dt} = 1 - \cos t$, $\frac{dy}{dt} = \sin t \Rightarrow \frac{dy}{dx} = \frac{dy/dt}{dx/dt}$ $= \frac{\sin t}{1 - \cos t} \Rightarrow \left.\frac{dy}{dx}\right|_{t=\frac{\pi}{3}} = \frac{\sin\left(\frac{\pi}{3}\right)}{1 - \cos\left(\frac{\pi}{3}\right)} = \frac{\left(\frac{\sqrt{3}}{2}\right)}{\left(\frac{1}{2}\right)} = \sqrt{3}$; tangent line is $y - \frac{1}{2} = \sqrt{3}\left(x - \frac{\pi}{3} + \frac{\sqrt{3}}{2}\right)$

$\Rightarrow y = \sqrt{3}x - \frac{\pi\sqrt{3}}{3} + 2$; $\frac{dy'}{dt} = \frac{(1-\cos t)(\cos t) - (\sin t)(\sin t)}{(1-\cos t)^2} = \frac{-1}{1-\cos t} \Rightarrow \frac{d^2y}{dx^2} = \frac{dy'/dt}{dx/dt} = \frac{\left(\frac{-1}{1-\cos t}\right)}{1-\cos t}$

$= \frac{-1}{(1-\cos t)^2} \Rightarrow \left.\frac{d^2y}{dx^2}\right|_{t=\frac{\pi}{3}} = -4$

12. $t = \frac{\pi}{2} \Rightarrow x = \cos \frac{\pi}{2} = 0,\ y = 1 + \sin \frac{\pi}{2} = 2;\ \frac{dx}{dt} = -\sin t,\ \frac{dy}{dt} = \cos t \Rightarrow \frac{dy}{dx} = \frac{\cos t}{-\sin t} = -\cot t$

$\Rightarrow \left.\frac{dy}{dx}\right|_{t=\frac{\pi}{2}} = -\cot \frac{\pi}{2} = 0$; tangent line is $y = 2$; $\frac{dy'}{dt} = \csc^2 t \Rightarrow \frac{d^2y}{dx^2} = \frac{\csc^2 t}{-\sin t} = -\csc^3 t \Rightarrow \left.\frac{d^2y}{dx^2}\right|_{t=\frac{\pi}{2}} = -1$

13. $x^2 - 2tx + 2t^2 = 4 \Rightarrow 2x\frac{dx}{dt} - 2x - 2t\frac{dx}{dt} + 4t = 0 \Rightarrow (2x - 2t)\frac{dx}{dt} = 2x - 4t \Rightarrow \frac{dx}{dt} = \frac{2x - 4t}{2x - 2t} = \frac{x - 2t}{x - t}$;

$2y^3 - 3t^2 = 4 \Rightarrow 6y^2\frac{dy}{dt} - 6t = 0 \Rightarrow \frac{dy}{dt} = \frac{6t}{6y^2} = \frac{t}{y^2}$; thus $\frac{dy}{dx} = \frac{dy/dt}{dx/dt} = \frac{\left(\frac{t}{y^2}\right)}{\left(\frac{x-2t}{x-t}\right)} = \frac{t(x-t)}{y^2(x-2t)}$; $t = 2$

$\Rightarrow x^2 - 2(2)x + 2(2)^2 = 4 \Rightarrow x^2 - 4x + 4 = 0 \Rightarrow (x-2)^2 = 0 \Rightarrow x = 2;\ t = 2 \Rightarrow 2y^3 - 3(2)^2 = 4$

$\Rightarrow 2y^3 = 16 \Rightarrow y^3 = 8 \Rightarrow y = 2$; therefore $\left.\frac{dy}{dx}\right|_{t=2} = \frac{2(2-2)}{(2)^2(2-2(2))} = 0$

14. $x = \sqrt{5 - \sqrt{t}} \Rightarrow \frac{dx}{dt} = \frac{1}{2}\left(5 - \sqrt{t}\right)^{-1/2}\left(-\frac{1}{2}t^{-1/2}\right) = \frac{1}{4\sqrt{t}\sqrt{5-\sqrt{t}}}$; $y(t-1) = \ln y \Rightarrow \frac{dy}{dt}(t-1) + y = \left(\frac{1}{y}\right)\frac{dy}{dt}$

$\Rightarrow \left(t - 1 - \frac{1}{y}\right)\frac{dy}{dt} = -y \Rightarrow \frac{dy}{dt} = \frac{-y}{\left(t - 1 - \frac{1}{y}\right)} = \frac{-y^2}{ty - y - 1}$; thus $\frac{dy}{dx} = \frac{\left(\frac{-y^2}{ty-y-1}\right)}{\left(\frac{-1}{4\sqrt{t}\sqrt{5-\sqrt{t}}}\right)} = \frac{4y^2\sqrt{t}\sqrt{5-\sqrt{t}}}{ty - y - 1}$;

$t = 1 \Rightarrow y(1-1) = \ln y \Rightarrow 0 = \ln y \Rightarrow y = 1$; therefore $\left.\frac{dy}{dx}\right|_{t=1} = \frac{4(1)^2\sqrt{1}\sqrt{5-\sqrt{1}}}{(1)(1) - 1 - 1} = -8$

15. $x + 2x^{3/2} = t^2 + t \Rightarrow \frac{dx}{dt} + 3x^{1/2}\frac{dx}{dt} = 2t + 1 \Rightarrow \left(1 + 3x^{1/2}\right)\frac{dx}{dt} = 2t + 1 \Rightarrow \frac{dx}{dt} = \frac{2t+1}{1+3x^{1/2}}$; $y\sqrt{t+1} + 2t\sqrt{y} = 4$

$\Rightarrow \frac{dy}{dt}\sqrt{t+1} + y\left(\frac{1}{2}\right)(t+1)^{-1/2} + 2\sqrt{y} + 2t\left(\frac{1}{2}y^{-1/2}\right)\frac{dy}{dt} = 0 \Rightarrow \frac{dy}{dt}\sqrt{t+1} + \frac{y}{2\sqrt{t+1}} + 2\sqrt{y} + \left(\frac{t}{\sqrt{y}}\right)\frac{dy}{dt} = 0$

$\Rightarrow \left(\sqrt{t+1} + \frac{t}{\sqrt{y}}\right)\frac{dy}{dt} = \frac{-y}{2\sqrt{t+1}} - 2\sqrt{y} \Rightarrow \frac{dy}{dt} = \frac{\left(\frac{-y}{2\sqrt{t+1}} - 2\sqrt{y}\right)}{\left(\sqrt{t+1} + \frac{t}{\sqrt{y}}\right)} = \frac{-y\sqrt{y} - 4y\sqrt{t+1}}{2\sqrt{y}(t+1) + 2t\sqrt{t+1}}$; thus

$\frac{dy}{dx} = \frac{dy/dt}{dx/dt} = \frac{\left(\frac{-y\sqrt{y} - 4y\sqrt{t+1}}{2\sqrt{y}(t+1) + 2t\sqrt{t+1}}\right)}{\left(\frac{2t+1}{1+3x^{1/2}}\right)}$; $t = 0 \Rightarrow x + 2x^{3/2} = 0 \Rightarrow x\left(1 + 2x^{1/2}\right) = 0 \Rightarrow x = 0;\ t = 0$

$\Rightarrow y\sqrt{0+1} + 2(0)\sqrt{y} = 4 \Rightarrow y = 4$; therefore $\left.\frac{dy}{dx}\right|_{t=0} = \frac{\left(\frac{-4\sqrt{4} - 4(4)\sqrt{0+1}}{2\sqrt{4}(0+1) + 2(0)\sqrt{0+1}}\right)}{\left(\frac{2(0)+1}{1+3(0)^{1/2}}\right)} = -6$

16. $x \sin t + 2x = t \Rightarrow \frac{dx}{dt} \sin t + x \cos t + 2\frac{dx}{dt} = 1 \Rightarrow (\sin t + 2)\frac{dx}{dt} = 1 - x \cos t \Rightarrow \frac{dx}{dt} = \frac{1 - x \cos t}{\sin t + 2}$;

$t \sin t - 2t = y \Rightarrow \sin t + t \cos t - 2 = \frac{dy}{dt}$; thus $\frac{dy}{dx} = \frac{\sin t + t \cos t - 2}{\left(\frac{1 - x \cos t}{\sin t + 2}\right)}$; $t = \pi \Rightarrow x \sin \pi + 2x = \pi$

$\Rightarrow x = \frac{\pi}{2}$; therefore $\left.\frac{dy}{dx}\right|_{t=\pi} = \frac{\sin \pi + \pi \cos \pi - 2}{\left[\frac{1 - \left(\frac{\pi}{2}\right)\cos \pi}{\sin \pi + 2}\right]} = \frac{-4\pi - 8}{2 + \pi} = -4$

17. $\frac{dx}{dt} = -\sin t$ and $\frac{dy}{dt} = 1 + \cos t \Rightarrow \sqrt{\left(\frac{dx}{dt}\right)^2 + \left(\frac{dy}{dt}\right)^2} = \sqrt{(-\sin t)^2 + (1 + \cos t)^2} = \sqrt{2 + 2\cos t}$

$\Rightarrow \text{Length} = \int_0^{\pi} \sqrt{2 + 2\cos t}\, dt = \sqrt{2}\int_0^{\pi} \sqrt{\left(\frac{1 - \cos t}{1 - \cos t}\right)(1 + \cos t)}\, dt = \sqrt{2}\int_0^{\pi} \sqrt{\frac{\sin^2 t}{1 - \cos t}}\, dt$

$= \sqrt{2}\int_0^{\pi} \frac{\sin t}{\sqrt{1 - \cos t}}\, dt$ (since $\sin t \ge 0$ on $[0, \pi]$); $[u = 1 - \cos t \Rightarrow du = \sin t\, dt;\ t = 0 \Rightarrow u = 0,$

$t = \pi \Rightarrow u = 2] \to \sqrt{2}\int_0^{2} u^{-1/2}\, du = \sqrt{2}\left[2u^{1/2}\right]_0^2 = 4$

18. $\frac{dx}{dt} = 3t^2$ and $\frac{dy}{dt} = 3t \Rightarrow \sqrt{\left(\frac{dx}{dt}\right)^2 + \left(\frac{dy}{dt}\right)^2} = \sqrt{(3t^2)^2 + (3t)^2} = \sqrt{9t^4 + 9t^2} = 3t\sqrt{t^2 + 1}$ (since $t \ge 0$ on $[0, \sqrt{3}]$)

$\Rightarrow \text{Length} = \int_0^{\sqrt{3}} 3t\sqrt{t^2 + 1}\, dt$; $\left[u = t^2 + 1 \Rightarrow \frac{3}{2}\, du = 3t\, dt;\ t = 0 \Rightarrow u = 1,\ t = \sqrt{3} \Rightarrow u = 4\right]$

$\to \int_1^4 \frac{3}{2}u^{1/2}\, du = \left[u^{3/2}\right]_1^4 = (8 - 1) = 7$

19. $\frac{dx}{dt} = t$ and $\frac{dy}{dt} = (2t + 1)^{1/2} \Rightarrow \sqrt{\left(\frac{dx}{dt}\right)^2 + \left(\frac{dy}{dt}\right)^2} = \sqrt{t^2 + (2t + 1)} = \sqrt{(t + 1)^2} = |t + 1| = t + 1$ since $0 \le t \le 4$

$\Rightarrow \text{Length} = \int_0^4 (t + 1)\, dt = \left[\frac{t^2}{2} + t\right]_0^4 = (8 + 4) = 12$

20. $\frac{dx}{dt} = (2t + 3)^{1/2}$ and $\frac{dy}{dt} = 1 + t \Rightarrow \sqrt{\left(\frac{dx}{dt}\right)^2 + \left(\frac{dy}{dt}\right)^2} = \sqrt{(2t + 3) + (1 + t)^2} = \sqrt{t^2 + 4t + 4} = |t + 2| = t + 2$

since $0 \le t \le 3 \Rightarrow \text{Length} = \int_0^3 (t + 2)\, dt = \left[\frac{t^2}{2} + 2t\right]_0^3 = \frac{21}{2}$

21. $\frac{dx}{dt} = 8t\cos t$ and $\frac{dy}{dt} = 8t\sin t \Rightarrow \sqrt{\left(\frac{dx}{dt}\right)^2 + \left(\frac{dy}{dt}\right)^2} = \sqrt{(8t\cos t)^2 + (8t\sin t)^2} = \sqrt{64t^2\cos^2 t + 64t^2\sin^2 t}$

$= |8t| = 8t$ since $0 \le t \le \frac{\pi}{2} \Rightarrow \text{Length} = \int_0^{\pi/2} 8t\,dt = \left[4t^2\right]_0^{\pi/2} = \pi^2$

22. $\frac{dx}{dt} = \left(\frac{1}{\sec t + \tan t}\right)(\sec t\tan t + \sec^2 t) - \cos t = \sec t - \cos t$ and $\frac{dy}{dt} = -\sin t \Rightarrow \sqrt{\left(\frac{dx}{dt}\right)^2 + \left(\frac{dy}{dt}\right)^2}$

$= \sqrt{(\sec t - \cos t)^2 + (-\sin t)^2} = \sqrt{\sec^2 t - 1} = \sqrt{\tan^2 t} = |\tan t| = \tan t$ since $0 \le t \le \frac{\pi}{3}$

$\Rightarrow \text{Length} = \int_0^{\pi/3} \tan t\,dt = \int_0^{\pi/3} \frac{\sin t}{\cos t}\,dt = \left[-\ln|\cos t|\right]_0^{\pi/3} = -\ln\frac{1}{2} + \ln 1 = \ln 2$

23. $\frac{dx}{dt} = -\sin t$ and $\frac{dy}{dt} = \cos t \Rightarrow \sqrt{\left(\frac{dx}{dt}\right)^2 + \left(\frac{dy}{dt}\right)^2} = \sqrt{(-\sin t)^2 + (\cos t)^2} = 1 \Rightarrow \text{Area} = \int 2\pi y\,ds$

$= \int_0^{2\pi} 2\pi(2 + \sin t)(1)\,dt = 2\pi[2t - \cos t]_0^{2\pi} = 2\pi[(4\pi - 1) - (0 - 1)] = 8\pi^2$

24. $\frac{dx}{dt} = t^{1/2}$ and $\frac{dy}{dt} = t^{-1/2} \Rightarrow \sqrt{\left(\frac{dx}{dt}\right)^2 + \left(\frac{dy}{dt}\right)^2} = \sqrt{t + t^{-1}} = \sqrt{\frac{t^2+1}{t}} \Rightarrow \text{Area} = \int 2\pi x\,ds$

$= \int_0^{\sqrt{3}} 2\pi\left(\frac{2}{3}t^{3/2}\right)\sqrt{\frac{t^2+1}{t}}\,dt = \frac{4\pi}{3}\int_0^{\sqrt{3}} t\sqrt{t^2+1}\,dt$; $\left[u = t^2 + 1 \Rightarrow du = 2t\,dt;\ t = 0 \Rightarrow u = 1,\right.$

$\left.t = \sqrt{3} \Rightarrow u = 4\right] \to \int_1^4 \frac{2\pi}{3}\sqrt{u}\,du = \left[\frac{4\pi}{9}u^{3/2}\right]_1^4 = \frac{28\pi}{9}$

Note: $\int_0^{\sqrt{3}} 2\pi\left(\frac{2}{3}t^{3/2}\right)\sqrt{\frac{t^2+1}{t}}\,dt$ is an improper integral but $\lim_{t\to 0^+} f(t)$ exists and is equal to 0, where

$f(t) = 2\pi\left(\frac{2}{3}t^{3/2}\right)\sqrt{\frac{t^2+1}{t}}$. Thus the discontinuity is removable: define $F(t) = f(t)$ for $t > 0$ and $F(0) = 0$

$\Rightarrow \int_0^{\sqrt{3}} F(t)\,dt = \frac{28\pi}{9}$.

25. $\frac{dx}{dt} = 1$ and $\frac{dy}{dt} = t + \sqrt{2} \Rightarrow \sqrt{\left(\frac{dx}{dt}\right)^2 + \left(\frac{dy}{dt}\right)^2} = \sqrt{1^2 + \left(t + \sqrt{2}\right)^2} = \sqrt{t^2 + 2\sqrt{2}t + 3} \Rightarrow \text{Area} = \int 2\pi x\,ds$

$= \int_{-\sqrt{2}}^{\sqrt{2}} 2\pi\left(t + \sqrt{2}\right)\sqrt{t^2 + 2\sqrt{2}t + 3}\,dt$; $\left[u = t^2 + 2\sqrt{2}t + 3 \Rightarrow du = \left(2t + 2\sqrt{2}\right)dt;\ t = -\sqrt{2} \Rightarrow u = 1,\right.$

$\left[t=\sqrt{2} \Rightarrow u=9\right] \rightarrow \int_1^9 \pi\sqrt{u}\,du=\left[\frac{2}{3}\pi u^{3/2}\right]_1^9=\frac{2\pi}{3}(27-1)=\frac{52\pi}{3}$

26. From Exercise 22, $\sqrt{\left(\frac{dx}{dt}\right)^2+\left(\frac{dy}{dt}\right)^2}=\tan t \Rightarrow \text{Area}=\int 2\pi y\,ds=\int_0^{\pi/3} 2\pi\cos t\tan t\,dt=2\pi\int_0^{\pi/3}\sin t\,dt$
$=2\pi[-\cos t]_0^{\pi/3}=2\pi\left[-\frac{1}{2}-(-1)\right]=\pi$

27. $\frac{dx}{dt}=2$ and $\frac{dy}{dt}=1 \Rightarrow \sqrt{\left(\frac{dx}{dt}\right)^2+\left(\frac{dy}{dt}\right)^2}=\sqrt{2^2+1^2}=\sqrt{5} \Rightarrow \text{Area}=\int 2\pi y\,ds=\int_0^1 2\pi(t+1)\sqrt{5}\,dt$
$=2\pi\sqrt{5}\left[\frac{t^2}{2}+t\right]_0^1=3\pi\sqrt{5}$. Check: slant height is $\sqrt{5} \Rightarrow$ Area is $\pi(1+2)\sqrt{5}=3\pi\sqrt{5}$.

28. $\frac{dx}{dt}=h$ and $\frac{dy}{dt}=r \Rightarrow \sqrt{\left(\frac{dx}{dt}\right)^2+\left(\frac{dy}{dt}\right)^2}=\sqrt{h^2+r^2} \Rightarrow \text{Area}=\int 2\pi y\,ds=\int_0^1 2\pi rt\sqrt{h^2+r^2}\,dt$
$=2\pi r\sqrt{h^2+r^2}\int_0^1 t\,dt=2\pi r\sqrt{h^2+r^2}\left[\frac{t^2}{2}\right]_0^1=\pi r\sqrt{h^2+r^2}$. Check: slant height is $\sqrt{h^2+r^2} \Rightarrow$ Area is $\pi r\sqrt{h^2+r^2}$.

29. (a) Let the density be $\delta=1$. Then $x=\cos t+t\sin t \Rightarrow \frac{dx}{dt}=t\cos t$, and $y=\sin t-t\cos t \Rightarrow \frac{dy}{dt}=t\sin t$
$\Rightarrow dm=1\cdot ds=\sqrt{\left(\frac{dx}{dt}\right)^2+\left(\frac{dy}{dt}\right)^2}\,dt=\sqrt{(t\cos t)^2+(t\sin t)^2}=|t|\,dt=t\,dt$ since $0\le t\le\frac{\pi}{2}$. The curve's mass is $M=\int dm=\int_0^{\pi/2} t\,dt=\frac{\pi^2}{8}$. Also $M_x=\int \tilde{y}\,dm=\int_0^{\pi/2}(\sin t-t\cos t)\,t\,dt$
$=\int_0^{\pi/2} t\sin t\,dt-\int_0^{\pi/2} t^2\cos t\,dt=[\sin t-t\cos t]_0^{\pi/2}-\left[t^2\sin t-2\sin t+2t\cos t\right]_0^{\pi/2}=3-\frac{\pi^2}{4}$, where we integrated by parts. Therefore, $\bar{y}=\frac{M_x}{M}=\frac{\left(3-\frac{\pi^2}{4}\right)}{\left(\frac{\pi^2}{8}\right)}=\frac{24}{\pi^2}-2$. Next, $M_y=\int \tilde{x}\,dm$
$=\int_0^{\pi/2}(\cos t+t\sin t)\,t\,dt=\int_0^{\pi/2} t\cos t\,dt+\int_0^{\pi/2} t^2\sin t\,dt$
$=\left[\cos t+t\sin t\right]_0^{\pi/2}+\left[-t^2\cos t+2\cos t+2t\sin t\right]_0^{\pi/2}=\frac{3\pi}{2}-3$, again integrating by parts.
Hence $\bar{x}=\frac{M_y}{M}=\frac{\left(\frac{3\pi}{2}-3\right)}{\left(\frac{\pi^2}{8}\right)}=\frac{12}{\pi}-\frac{24}{\pi^2}$. Therefore $(\bar{x},\bar{y})=\left(\frac{12}{\pi}-\frac{24}{\pi^2},\frac{24}{\pi^2}-2\right)$

(b) $(\overline{x},\overline{y}) \approx (1.4, 0.4)$

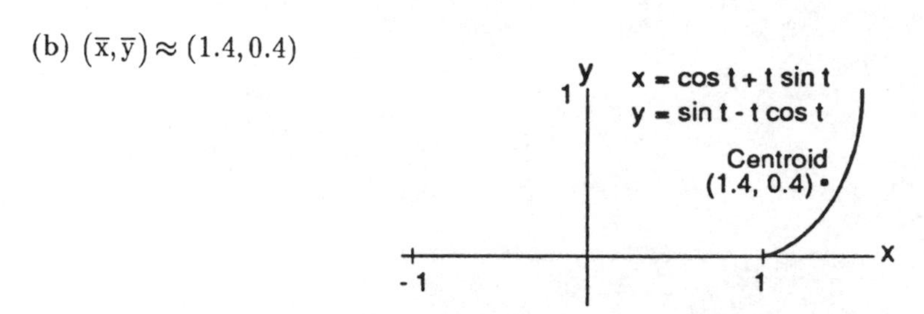

30. (a) Let the density be $\delta = 1$. Then $x = e^t \cos t \Rightarrow \frac{dx}{dt} = e^t \cos t - e^t \sin t$, and $y = e^t \sin t$

$\Rightarrow \frac{dy}{dt} = e^t \sin t + e^t \cos t \Rightarrow dm = 1 \cdot ds = \sqrt{\left(\frac{dx}{dt}\right)^2 + \left(\frac{dy}{dt}\right)^2}\, dt$

$= \sqrt{\left(e^t \cos t - e^t \sin t\right)^2 + \left(e^t \sin t + e^t \cos t\right)^2}\, dt = \sqrt{2e^{2t}}\, dt = \sqrt{2}\, e^t\, dt$. The curve's mass is

$M = \int dm = \int_0^{\pi} \sqrt{2}\, e^t\, dt = \sqrt{2}\, e^{\pi} - \sqrt{2}$. Also $M_x = \int \tilde{y}\, dm = \int_0^{\pi} \left(e^t \sin t\right)\left(\sqrt{2}\, e^t\right) dt$

$= \int_0^{\pi} \sqrt{2}\, e^{2t} \sin t\, dt = \sqrt{2}\left[\frac{e^{2t}}{5}(2 \sin t - \cos t)\right]_0^{\pi} = \sqrt{2}\left(\frac{e^{2\pi}}{5} + \frac{1}{5}\right) \Rightarrow \overline{y} = \frac{M_x}{M} = \frac{\sqrt{2}\left(\frac{e^{2\pi}}{5} + \frac{1}{5}\right)}{\sqrt{2}\, e^{\pi} - \sqrt{2}} = \frac{e^{2\pi} + 1}{5\left(e^{\pi} - 1\right)}$.

Next $M_y = \int \tilde{x}\, dm = \int_0^{\pi} \left(e^t \cos t\right)\left(\sqrt{2}\, e^t\right) dt = \int_0^{\pi} \sqrt{2}\, e^{2t} \cos t\, dt$

$= \sqrt{2}\left[\frac{e^{2t}}{5}(2 \cos t + \sin t)\right]_0^{\pi} = -\sqrt{2}\left(\frac{2e^{2\pi}}{5} + \frac{2}{5}\right) \Rightarrow \overline{x} = \frac{M_y}{M} = \frac{-\sqrt{2}\left(\frac{2e^{2\pi}}{5} + \frac{2}{5}\right)}{\sqrt{2}\, e^{\pi} - \sqrt{2}} = -\frac{2e^{2\pi} + 2}{5\left(e^{\pi} - 1\right)}$. Therefore

$(\overline{x},\overline{y}) = \left(-\frac{2e^{2\pi} + 2}{5\left(e^{\pi} - 1\right)}, \frac{e^{2\pi} + 1}{5\left(e^{\pi} - 1\right)}\right)$

(b) $(\overline{x},\overline{y}) \approx (-9.7, 4.8)$

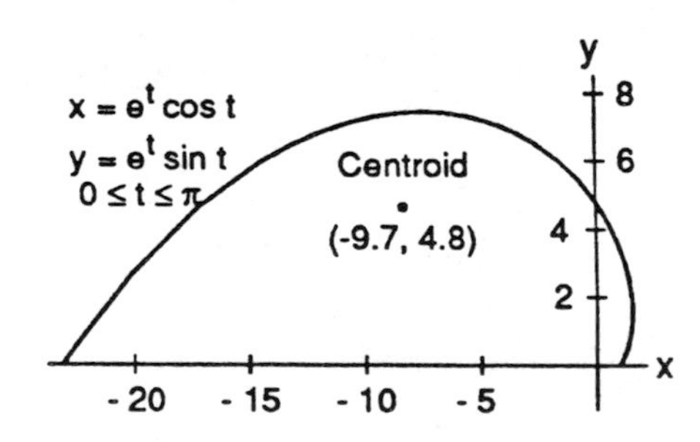

31. (a) Let the density be $\delta = 1$. Then $x = \cos t \Rightarrow \frac{dx}{dt} = -\sin t$, and $y = t + \sin t \Rightarrow \frac{dy}{dt} = 1 + \cos t$

$\Rightarrow dm = 1 \cdot ds = \sqrt{\left(\frac{dx}{dt}\right)^2 + \left(\frac{dy}{dt}\right)^2}\, dt = \sqrt{(-\sin t)^2 + (1 + \cos t)^2}\, dt = \sqrt{2 + 2 \cos t}\, dt$. The curve's mass

is $M = \int dm = \int_0^{\pi} \sqrt{2 + 2 \cos t}\, dt = \sqrt{2} \int_0^{\pi} \sqrt{1 + \cos t}\, dt = \sqrt{2} \int_0^{\pi} \sqrt{2 \cos^2\left(\frac{t}{2}\right)}\, dt = 2 \int_0^{\pi} \left|\cos\left(\frac{t}{2}\right)\right| dt$

$= 2\int_0^{\pi} \cos\left(\frac{t}{2}\right) dt \left(\text{since } 0 \le t \le \pi \Rightarrow 0 \le \frac{t}{2} \le \frac{\pi}{2}\right) = 2\left[2 \sin\left(\frac{t}{2}\right)\right]_0^{\pi} = 4.$ Also $M_x = \int \tilde{y}\, dm$

$= \int_0^{\pi} (t + \sin t)\left(2 \cos \frac{t}{2}\right) dt = \int_0^{\pi} 2t \cos\left(\frac{t}{2}\right) dt + \int_0^{\pi} 2 \sin t \cos\left(\frac{t}{2}\right) dt$

$= 2\left[4 \cos\left(\frac{t}{2}\right) + 2t \sin\left(\frac{t}{2}\right)\right]_0^{\pi} + 2\left[-\frac{1}{3} \cos\left(\frac{3}{2}t\right) - \cos\left(\frac{1}{2}t\right)\right]_0^{\pi} = 4\pi - \frac{16}{3} \Rightarrow \bar{y} = \frac{M_x}{M} = \frac{\left(4\pi - \frac{16}{3}\right)}{4} = \pi - \frac{4}{3}.$

Next $M_y = \int \tilde{x}\, dm = \int_0^{\pi} (\cos t)\left(2 \cos \frac{t}{2}\right) dt = 2\int_0^{\pi} \cos t \cos\left(\frac{t}{2}\right) dt = 2\left[\sin\left(\frac{t}{2}\right) + \frac{\sin\left(\frac{3}{2}t\right)}{3}\right]_0^{\pi} = 2 - \frac{2}{3}$

$= \frac{4}{3} \Rightarrow \bar{x} = \frac{M_y}{M} = \frac{\left(\frac{4}{3}\right)}{4} = \frac{1}{3}.$ Therefore $(\bar{x}, \bar{y}) = \left(\frac{1}{3}, \pi - \frac{4}{3}\right)$

(b) $(\bar{x}, \bar{y}) \approx (0.33, 1.81)$

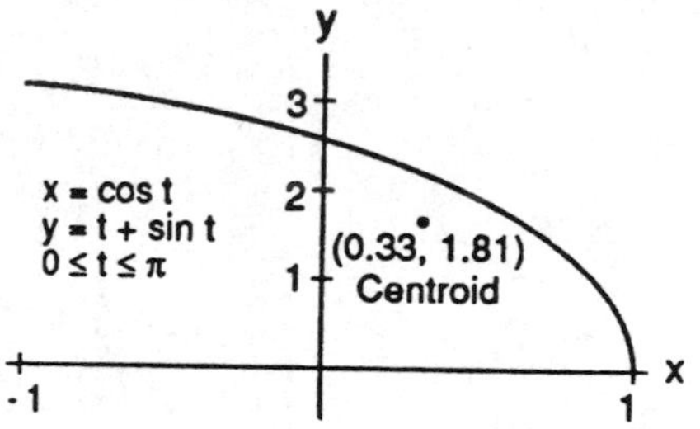

32. Let the density be $\delta = 1$. Then $x = t^3 \Rightarrow \frac{dx}{dt} = 3t^2$, and $y = \frac{3t^2}{2} \Rightarrow \frac{dy}{dt} = 3t \Rightarrow dm = 1 \cdot ds$

$= \sqrt{\left(\frac{dx}{dt}\right)^2 + \left(\frac{dy}{dt}\right)^2}\, dt = \sqrt{\left(3t^2\right)^2 + (3t)^2}\, dt = 3\,|t|\sqrt{t^2+1}\, dt$ since $0 \le t \le \sqrt{3}$. The curve's mass is

$M = \int dm = \int_0^{\sqrt{3}} 3t\sqrt{t^2+1}\, dt = \left[\left(t^2+1\right)^{3/2}\right]_0^{\sqrt{3}} = 7.$ Also $M_x = \int \tilde{y}\, dm = \int_0^{\sqrt{3}} \frac{3t^2}{2}\left(3t\sqrt{t^2+1}\right) dt$

$= \frac{9}{2}\int_0^{\sqrt{3}} t^3\sqrt{t^2+1}\, dt = \frac{87}{5} = 17.4$ (by computer) $\Rightarrow \bar{y} = \frac{M_x}{M} = \frac{17.4}{7} \approx 2.49.$ Next $M_y = \int \tilde{x}\, dm$

$= \int_0^{\sqrt{3}} t^3 \sqrt{3t\left(t^2+1\right)}\, dt = 3\int_0^{\sqrt{3}} t^4\sqrt{t^2+1}\, dt \approx 16.4849$ (by computer) $\Rightarrow \bar{x} = \frac{M_y}{M} = \frac{16.4849}{7} \approx 2.35.$

Therefore, $(\bar{x}, \bar{y}) \approx (2.35, 2.49)$

33. (a) $\frac{dx}{dt} = -2 \sin 2t$ and $\frac{dy}{dt} = 2 \cos 2t \Rightarrow \sqrt{\left(\frac{dx}{dt}\right)^2 + \left(\frac{dy}{dt}\right)^2} = \sqrt{(-2 \sin 2t)^2 + (2 \cos 2t)^2} = 2$

$\Rightarrow \text{Length} = \int_0^{\pi/2} 2\, dt = [2t]_0^{\pi/2} = \pi$

(b) $\frac{dx}{dt} = \pi \cos \pi t$ and $\frac{dy}{dt} = -\pi \sin \pi t = \sqrt{\left(\frac{dx}{dt}\right)^2 + \left(\frac{dy}{dt}\right)^2} = \sqrt{(\pi \cos \pi t)^2 + (-\pi \sin \pi t)^2} = \pi$

$\Rightarrow \text{Length} = \int_{-1/2}^{1/2} \pi \, dt = [\pi t]_{-1/2}^{1/2} = \pi$

34. (a) $a = 1, e = \frac{1}{2} \Rightarrow \text{Length} = 4 \int_0^{\pi/2} \sqrt{1 - \frac{1}{4} \cos^2 t} \, dt = 2 \int_0^{\pi/2} \sqrt{4 - \cos^2 t} \, dt = \int_0^{\pi/2} f(t) \, dt$; use the Trapezoid Rule with $n = 10 \Rightarrow h = \frac{b-a}{n} = \frac{\left(\frac{\pi}{2}\right) - 0}{10} = \frac{\pi}{20}$.

$\int_0^{\pi/2} \sqrt{4 - \cos^2 t} \, dt \approx \sum_{n=0}^{10} mf(x_n) = 37.3686183$

$\Rightarrow T = \frac{h}{2}(37.3686183) = \frac{\pi}{40}(37.3686183)$

$= 2.934924418 \Rightarrow \text{Length} = 2(2.934924418)$

≈ 5.870

	x_i	$f(x_i)$	m	$mf(x_i)$
x_0	0	1.732050808	1	1.732050808
x_1	$\pi/20$	1.739100843	2	3.478201686
x_2	$\pi/10$	1.759400893	2	3.518801786
x_3	$3\pi/20$	1.790560631	2	3.581121262
x_4	$\pi/5$	1.82906848	1	3.658136959
x_5	$\pi/4$	1.870828693	1	3.741657387
x_6	$3\pi/10$	1.911676881	2	3.823353762
x_7	$7\pi/20$	1.947791731	2	3.895583461
x_8	$2\pi/5$	1.975982919	2	3.951965839
x_9	$9\pi/20$	1.993872679	1	3.987745357
x_{10}	$\pi/2$	2	2	2

(b) $|f''(t)| < 1 \Rightarrow M = 1$

$\Rightarrow |E_T| \le \frac{b-a}{12}(h^2 M) \le \frac{\left(\frac{\pi}{2}\right) - 0}{12}\left(\frac{\pi}{20}\right)^2 1 \le 0.0032$

35. $x = x \Rightarrow \frac{dy}{dx} = 1$, and $y = f(x) \Rightarrow \frac{dy}{dx} = f'(x)$; then $\text{Length} = \int_a^b \sqrt{\left(\frac{dx}{dt}\right)^2 + \left(\frac{dy}{dt}\right)^2} \, dt = \int_a^b \sqrt{\left(\frac{dx}{dx}\right)^2 + \left(\frac{dy}{dx}\right)^2} \, dx$

$= \int_a^b \sqrt{1 + \left(\frac{dy}{dx}\right)^2} \, dx = \int_a^b \sqrt{1 + [f'(x)]^2} \, dx$

36. $x = g(y)$ has the parametrization $x = g(y)$ and $y = y$ for $c \le y \le d \Rightarrow \frac{dx}{dy} = g'(y)$ and $\frac{dy}{dy} = 1$; then

$\text{Length} = \int_c^d \sqrt{\left(\frac{dy}{dy}\right)^2 + \left(\frac{dx}{dy}\right)^2} \, dy = \int_c^d \sqrt{1 + \left(\frac{dy}{dx}\right)^2} \, dy = \int_c^d \sqrt{1 + [g'(y)]^2} \, dy$

37. For one arch of the cycloid we use the interval $0 \le \theta \le 2\pi$. Then, $A = \int_0^{2\pi} y(\theta) \, dx = \int_0^{2\pi} a(1 - \cos \theta)\left(\frac{dx}{d\theta}\right) d\theta$

and $\frac{dx}{d\theta} = a(1 - \cos \theta) \Rightarrow A = \int_0^{2\pi} a^2(1 - \cos \theta)^2 \, d\theta = a^2 \int_0^{2\pi} \left(1 - 2 \cos \theta + \cos^2 \theta\right) d\theta$

$= a^2\left[\int_0^{2\pi} d\theta - 2\int_0^{2\pi} \cos\theta\, d\theta + \int_0^{2\pi} \frac{1}{2}(1+\cos 2\theta)\, d\theta\right] = a^2\left([\theta]_0^{2\pi} - 2[\sin\theta]_0^{2\pi} + \frac{1}{2}\left[\theta + \frac{1}{2}\sin 2\theta\right]_0^{2\pi}\right)$

$= a^2\left[(2\pi - 0) - 2(0-0) + \frac{1}{2}(2\pi - 0)\right] = 3\pi a^2$

38. $x = a(\theta - \sin\theta) \Rightarrow \frac{dx}{d\theta} = a(1-\cos\theta) \Rightarrow \left(\frac{dx}{d\theta}\right)^2 = a^2(1 - 2\cos\theta + \cos^2\theta)$ and $y = a(1-\cos\theta)$

$\Rightarrow \frac{dy}{d\theta} = a\sin\theta \Rightarrow \left(\frac{dy}{d\theta}\right)^2 = a^2\sin^2\theta \Rightarrow \text{Length} = \int_0^{2\pi} \sqrt{\left(\frac{dx}{d\theta}\right)^2 + \left(\frac{dy}{d\theta}\right)^2}\, d\theta = \int_0^{2\pi} \sqrt{2a^2(1-\cos\theta)}\, d\theta$

$= a\sqrt{2}\int_0^{2\pi} \sqrt{2}\sqrt{\frac{1-\cos\theta}{2}}\, d\theta = 2a\int_0^{2\pi} \left|\sin\frac{\theta}{2}\right| d\theta = 2a\int_0^{2\pi} \sin\frac{\theta}{2}\, d\theta = -4a\left[\cos\frac{\theta}{2}\right]_0^{2\pi} = 8a$

39. $x = \theta - \sin\theta$ and $y = 1 - \cos\theta,\ 0 \le \theta \le 2\pi \Rightarrow ds = \sqrt{(1-\cos\theta)^2 + \sin^2\theta}\, d\theta = \sqrt{2 - 2\cos\theta}\, d\theta \Rightarrow S = \int 2\pi y\, ds$

$= 2\sqrt{2}\int_0^{2\pi} \pi(1-\cos\theta)^{3/2}\, d\theta = 2\sqrt{2}\int_0^{2\pi} \pi\left(\sqrt{2}\sin\frac{\theta}{2}\right)^3 d\theta = 8\pi\int_0^{2\pi} \left(1 - \cos^2\frac{\theta}{2}\right)\left(\sin\frac{\theta}{2}\right) d\theta$

$= 8\pi\left[-2\cos\frac{\theta}{2} + \frac{2}{3}\cos^3\frac{\theta}{2}\right]_0^{2\pi} = \frac{64\pi}{3}$

40. $V = \pi\int_0^{2\pi} y^2\, dx = \pi\int_0^{2\pi} y^2\left(\frac{dx}{d\theta}\right) d\theta = \pi\int_0^{2\pi} (1-\cos\theta)^2(1-\cos\theta)\, d\theta = \pi\int_0^{2\pi} (1 - 3\cos\theta + 3\cos^2\theta - \cos^3\theta)\, d\theta;$

evaluating each integral: $I_1 = \pi\int_0^{2\pi} d\theta = 2\pi^2;\ I_2 = \pi\int_0^{2\pi} (-3\cos\theta)\, d\theta = -3\pi[\sin\theta]_0^{2\pi} = 0;\ I_3 = \pi\int_0^{2\pi} 3\cos^2\theta\, d\theta$

$= 3\pi\left[\frac{1}{2}\theta + \frac{1}{4}\sin 2\theta\right]_0^{2\pi} = 3\pi^2;\ I_4 = \pi\int_0^{2\pi} \cos^3\theta\, d\theta = \pi\int_0^{2\pi} (1 - \sin^2\theta)(\cos\theta)\, d\theta = 0;$ therefore

$V = I_1 + I_2 + I_3 + I_4 = 5\pi^2$

41. $\frac{dx}{dt} = \cos t$ and $\frac{dy}{dt} = 2\cos 2t \Rightarrow \frac{dy}{dx} = \frac{dy/dt}{dx/dt} = \frac{2\cos 2t}{\cos t} = \frac{2(2\cos^2 t - 1)}{\cos t}$; then $\frac{dy}{dx} = 0 \Rightarrow \frac{2(2\cos^2 t - 1)}{\cos t} = 0$

$\Rightarrow 2\cos^2 t - 1 = 0 \Rightarrow \cos t = \pm\frac{1}{\sqrt{2}} \Rightarrow t = \frac{\pi}{4}, \frac{3\pi}{4}, \frac{5\pi}{4}, \frac{7\pi}{4}$. In the 1st quadrant: $t = \frac{\pi}{4} \Rightarrow x = \sin\frac{\pi}{4} = \frac{\sqrt{2}}{2}$ and

$y = \sin 2\left(\frac{\pi}{4}\right) = 1 \Rightarrow \left(\frac{\sqrt{2}}{2}, 1\right)$ is the point where the tangent line is horizontal. At the origin: $x = 0$ and $y = 0$

$\Rightarrow \sin t = 0 \Rightarrow t = 0$ or $t = \pi$ and $\sin 2t = 0 \Rightarrow t = 0, \frac{\pi}{2}, \pi, \frac{3\pi}{2}$; thus $t = 0$ and $t = \pi$ give the tangent lines at

the origin. Tangents at origin: $\left.\frac{dy}{dx}\right|_{t=0} = 2 \Rightarrow y = 2x$ and $\left.\frac{dy}{dx}\right|_{t=\pi} = -2 \Rightarrow y = -2x$

42. $\frac{dx}{dt} = 2\cos 2t$ and $\frac{dy}{dt} = 3\cos 3t \Rightarrow \frac{dy}{dx} = \frac{dy/dt}{dx/dt} = \frac{3\cos 3t}{2\cos 2t} = \frac{3(\cos 2t\cos t - \sin 2t\sin t)}{2(2\cos^2 t - 1)}$

$= \frac{3\left[(2\cos^2 t - 1)(\cos t) - 2\sin t\cos t\sin t\right]}{2(2\cos^2 t - 1)} = \frac{(3\cos t)(2\cos^2 t - 1 - 2\sin^2 t)}{2(2\cos^2 t - 1)} = \frac{(3\cos t)(4\cos^2 t - 3)}{2(2\cos^2 t - 1)}$; then

$\frac{dy}{dx} = 0 \Rightarrow \frac{(3\cos t)(4\cos^2 t - 3)}{2(2\cos^2 t - 1)} = 0 \Rightarrow 3\cos t = 0$ or $4\cos^2 t - 3 = 0$: $3\cos t = 0 \Rightarrow t = \frac{\pi}{2}, \frac{3\pi}{2}$ and

$4\cos^2 t - 3 = 0 \Rightarrow \cos t = \pm\frac{\sqrt{3}}{2} \Rightarrow t = \frac{\pi}{6}, \frac{5\pi}{6}, \frac{7\pi}{6}, \frac{11\pi}{6}$. In the 1st quadrant: $t = \frac{\pi}{6} \Rightarrow x = \sin 2\left(\frac{\pi}{6}\right) = \frac{\sqrt{3}}{2}$

and $y = \sin 3\left(\frac{\pi}{6}\right) = 1 \Rightarrow \left(\frac{\sqrt{3}}{2}, 1\right)$ is the point where the graph has a horizontal tangent. At the origin: $x = 0$

and $y = 0 \Rightarrow \sin 2t = 0$ and $\sin 3t = 0 \Rightarrow t = 0, \frac{\pi}{2}, \pi, \frac{3\pi}{2}$ and $t = 0, \frac{\pi}{3}, \frac{2\pi}{3}, \pi, \frac{4\pi}{3}, \frac{5\pi}{3} \Rightarrow t = 0$ and $t = \pi$ give

the tangent lines at the origin. Tangents at the origin: $\left.\frac{dy}{dx}\right|_{t=0} = \frac{3\cos 0}{2\cos 0} = \frac{3}{2} \Rightarrow y = \frac{3}{2}x$, and $\left.\frac{dy}{dx}\right|_{t=\pi}$

$= \frac{3\cos(3\pi)}{2\cos(2\pi)} = -\frac{3}{2} \Rightarrow y = -\frac{3}{2}x$

43.

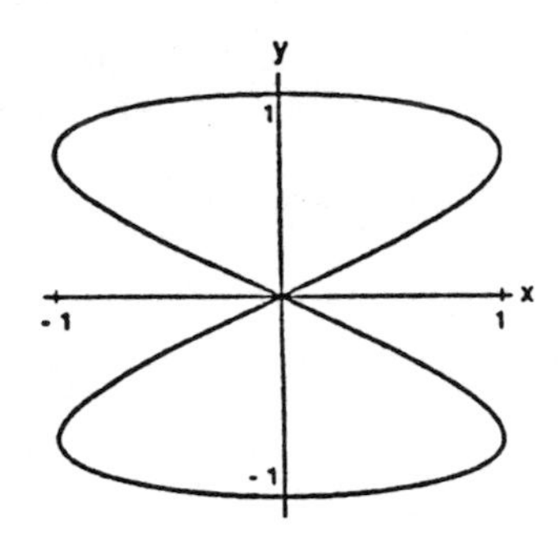

44.

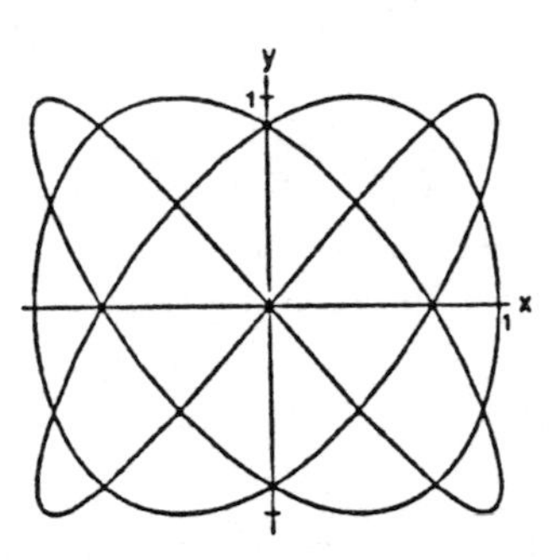

45.

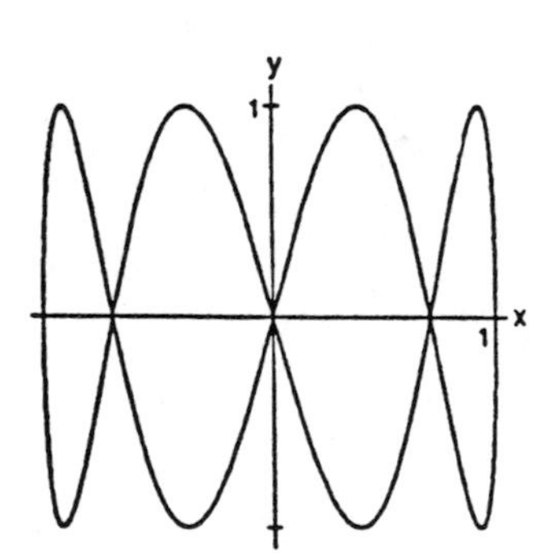

46.

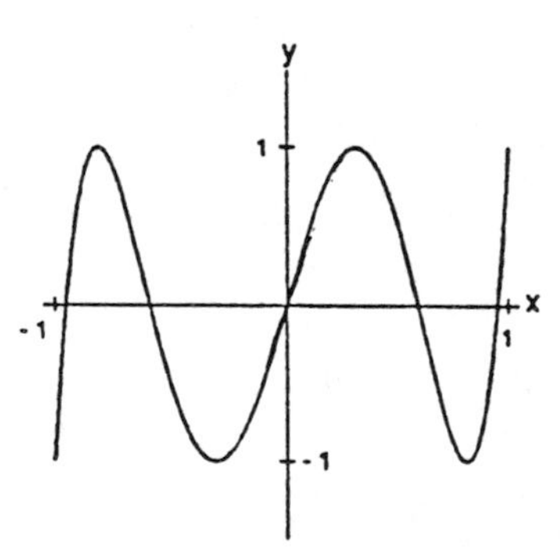

47.

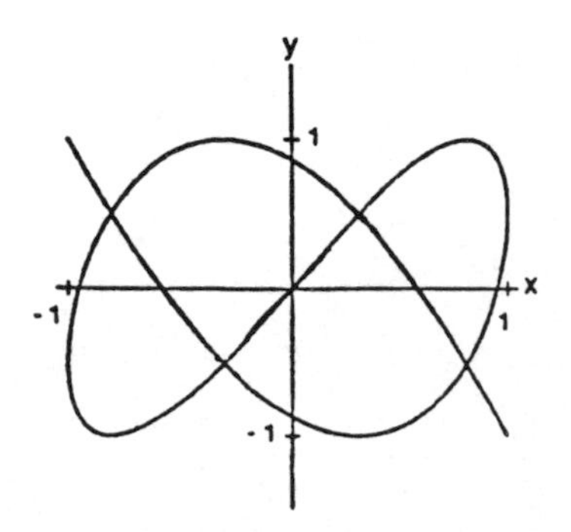

48.

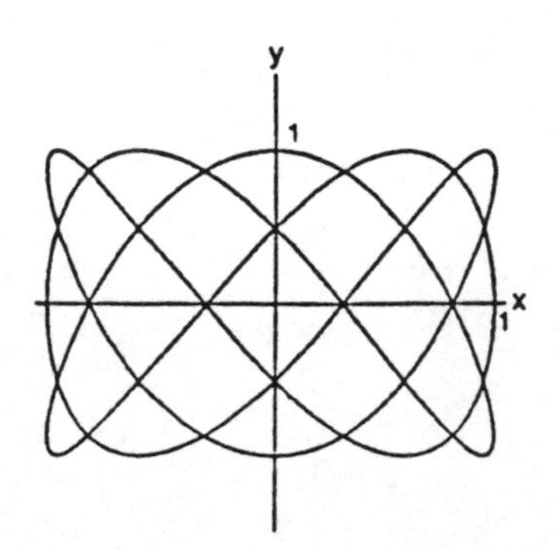

49.

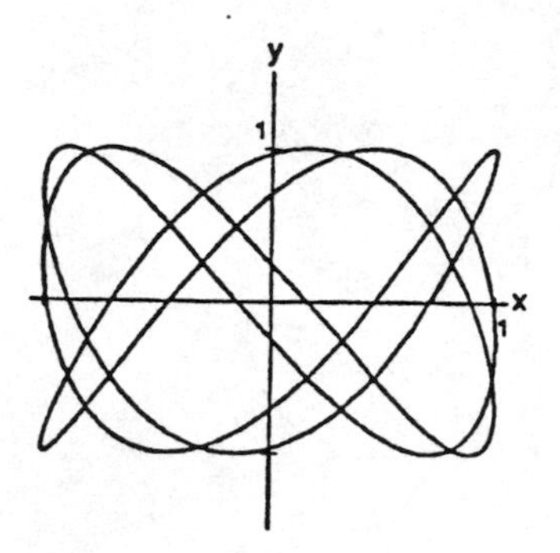

50-55. Example CAS commands:

Maple:

```
x:= t -> exp(t) – t^2;
y:= t -> t + exp(–t);
plot([x(t), y(t), t = –1..2]);
diff(x(t),t);
dx:= unapply(",t);
diff(y(t),t);
dy:= unapply(",t);
sqrt(dx(t)^2 + dy(t)^2);
simplify(");
ds:= unapply(",t);
arc:= int(abs(ds(t)), t =–1..2);
evalf(");
dy(t)/dx(t);
dydx:= unapply(",t);
diff(dydx(t),t);
simplify("): dy1:= unapply(",t);
dy1(t)/dx(t);
d2ydx2:= unapply(",t);
t0:=1: evalf(d2ydx2(t0));
tanline:= t -> y(t0) + (dy(t0)/dx(t0))*(t – x(t(0));
plot({[x(t), y(t), t = –1..2], [t, tanline(t), t=t0–1..t0+2]});
```

Mathematica:

```
Clear[x,y,t]
{a,b} = {–Pi,Pi}; t0 = Pi/4;
x[t_] = t – Cos[t]
y[t_] = 1 + Sin[t]
p1 = ParametricPlot[ {x[t],y[t]}, {t,a,b} ]
yp[t_] = y'[t]/x'[t]
ypp[t_] = yp'[t]/x'[t]
yp[t0] // N
ypp[t0] // N
tanline[x_] = y[t0] + yp[t0]*(x–x[t0])
p2 = Plot[ tanline[x], {x,0,0.2} ]
Show[ {p1,p2} ]
ds[t_] = Sqrt[ x'[t]^2 + y'[t]^2 ]
NIntegrate[ ds[t], {t,a,b} ]
```

56-57. Example CAS commands:

Maple:

```
with(plots):
eq1 := x^2*cos(t) + 2*x = t;
eq2 := t*sin(t) + 2*sqrt(y) = y;
solve(eq1,x);
x:= unapply(''[1],t);
solve(eq2,y);
y:= unapply(''[1],t);
t0:=-Pi/4: x0:= x(t0): y0:= y(t0):
plot([x(t), y(t), t = -2*Pi..2*Pi]);
diff(x(t),t);
simplify('');
dx:= unapply('',t);
diff(y(t),t);
dy:= unapply('',t);
dy(t)/dx(t);
dydx:= unapply('',t);
m:= dydx(t0);
tanline:= t -> y0 + m*(t - x0);
plot({[x(t), y(t), t = -2*Pi..2*Pi],[t, tanline(t), t = t0-1..t0 + 1]});
```

Mathematica:

```
{a,b} = {-3.436,2.073}; t0 = -Pi/4;
xeqn = x^2 Cos[t] + 2 x == t
yeqn = t Sin[t] + 2 Sqrt[y] == y
Solve[ xeqn, x ]
x[t_] = x /. %[[2]]
Solve[ yeqn, y ] // Simplify
y[t_] = y /. First[%]
yp[t_] = y'[t]/x'[t]
yp[t0] // N
tanline[x_] = y[t0] + yp[t0]*(x - x[t0])
p1 = ParametricPlot[ {x[t],y[t]}, {t,a,b} ]
p2 = Plot[ tanline[x], {x,-2,1} ]
Show[ {p1,p2} ]
```

9.6 POLAR COORDINATES

1. a, e; b, g; c, h; d, f

2. a, f; b, h; c, g; d, e

3. (a) $\left(2, \frac{\pi}{2} + 2n\pi\right)$ and $\left(-2, \frac{\pi}{2} + (2n+1)\pi\right)$, n an integer

(b) $(2, 2n\pi)$ and $(-2, (2n+1)\pi)$, n an integer

(c) $\left(2, \frac{3\pi}{2} + 2n\pi\right)$ and $\left(-2, \frac{3\pi}{2} + (2n+1)\pi\right)$, n an integer

(d) $(2, (2n+1)\pi)$ and $(-2, 2n\pi)$, n an integer

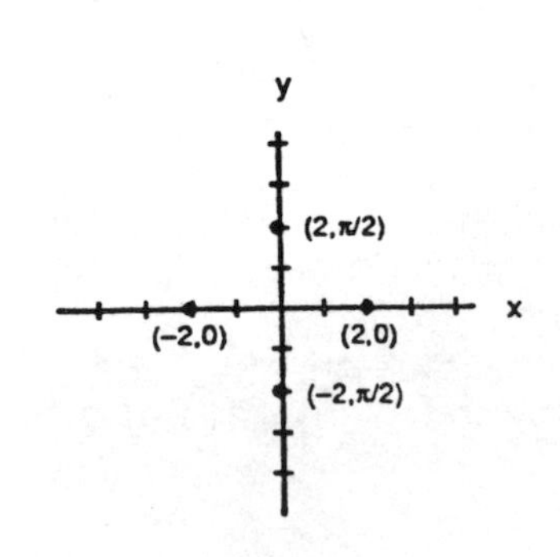

4. (a) $\left(3, \frac{\pi}{4} + 2n\pi\right)$ and $\left(-3, \frac{5\pi}{4} + 2n\pi\right)$, n an integer

(b) $\left(-3, \frac{\pi}{4} + 2n\pi\right)$ and $\left(3, \frac{5\pi}{4} + 2n\pi\right)$, n an integer

(c) $\left(3, -\frac{\pi}{4} + 2n\pi\right)$ and $\left(-3, \frac{3\pi}{4} + 2n\pi\right)$, n an integer

(d) $\left(-3, -\frac{\pi}{4} + 2n\pi\right)$ and $\left(3, \frac{3\pi}{4} + 2n\pi\right)$, n an integer

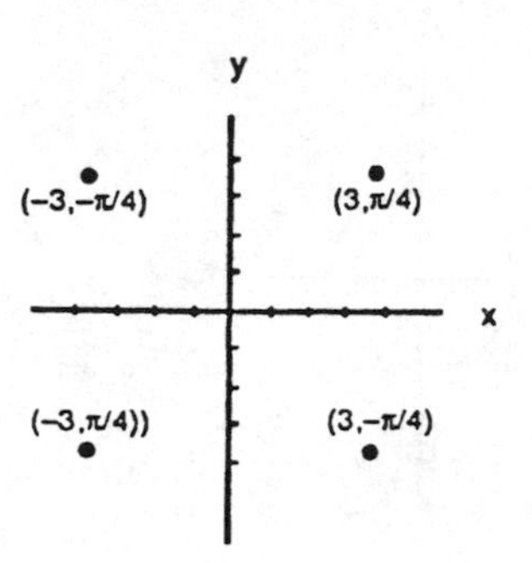

5. (a) $x = r\cos\theta = 3\cos 0 = 3$, $y = r\sin\theta = 3\sin 0 = 0 \Rightarrow$ Cartesian coordinates are $(3, 0)$

(b) $x = r\cos\theta = -3\cos 0 = -3$, $y = r\sin\theta = -3\sin 0 = 0 \Rightarrow$ Cartesian coordinates are $(-3, 0)$

(c) $x = r\cos\theta = 2\cos\frac{2\pi}{3} = -1$, $y = r\sin\theta = 2\sin\frac{2\pi}{3} = \sqrt{3} \Rightarrow$ Cartesian coordinates are $\left(-1, \sqrt{3}\right)$

(d) $x = r\cos\theta = 2\cos\frac{7\pi}{3} = 1$, $y = r\sin\theta = 2\sin\frac{7\pi}{3} = \sqrt{3} \Rightarrow$ Cartesian coordinates are $\left(1, \sqrt{3}\right)$

(e) $x = r\cos\theta = -3\cos\pi = 3$, $y = r\sin\theta = -3\sin\pi = 0 \Rightarrow$ Cartesian coordinates are $(3, 0)$

(f) $x = r\cos\theta = 2\cos\frac{\pi}{3} = 1$, $y = r\sin\theta = 2\sin\frac{\pi}{3} = \sqrt{3} \Rightarrow$ Cartesian coordinates are $\left(1, \sqrt{3}\right)$

(g) $x = r\cos\theta = -3\cos 2\pi = -3$, $y = r\sin\theta = -3\sin 2\pi = 0 \Rightarrow$ Cartesian coordinates are $(-3, 0)$

(h) $x = r\cos\theta = -2\cos\left(-\frac{\pi}{3}\right) = -1$, $y = r\sin\theta = -2\sin\left(-\frac{\pi}{3}\right) = \sqrt{3} \Rightarrow$ Cartesian coordinates are $\left(-1, \sqrt{3}\right)$

6. (a) $x = \sqrt{2}\cos\frac{\pi}{4} = 1$, $y = \sqrt{2}\sin\frac{\pi}{4} = 1 \Rightarrow$ Cartesian coordinates are $(1, 1)$

(b) $x = 1\cos 0 = 1$, $y = 1\sin 0 = 0 \Rightarrow$ Cartesian coordinates are $(1, 0)$

(c) $x = 0\cos\frac{\pi}{2} = 0$, $y = 0\sin\frac{\pi}{2} = 0 \Rightarrow$ Cartesian coordinates are $(0, 0)$

(d) $x = -\sqrt{2}\cos\left(-\frac{\pi}{4}\right) = -1$, $y = -\sqrt{2}\sin\left(-\frac{\pi}{4}\right) = 1 \Rightarrow$ Cartesian coordinates are $(-1, 1)$

(e) $x = -3\cos\frac{5\pi}{6} = \frac{3\sqrt{3}}{3}$, $y = -3\sin\frac{5\pi}{6} = -\frac{3}{2} \Rightarrow$ Cartesian coordinates are $\left(\frac{3\sqrt{3}}{2}, -\frac{3}{2}\right)$

(f) $x = 5\cos\left(\tan^{-1}\frac{4}{3}\right) = 3$, $y = 5\sin\left(\tan^{-1}\frac{4}{3}\right) = 4 \Rightarrow$ Cartesian coordinates are $(3, 4)$

(g) $x = -1\cos 7\pi = 1$, $y = -1\sin 7\pi = 0 \Rightarrow$ Cartesian coordinates are $(1, 0)$

(h) $x = 2\sqrt{3}\cos\frac{2\pi}{3} = -\sqrt{3}$, $y = 2\sqrt{3}\sin\frac{2\pi}{3} = 3 \Rightarrow$ Cartesian coordinates are $\left(-\sqrt{3}, 3\right)$

7.

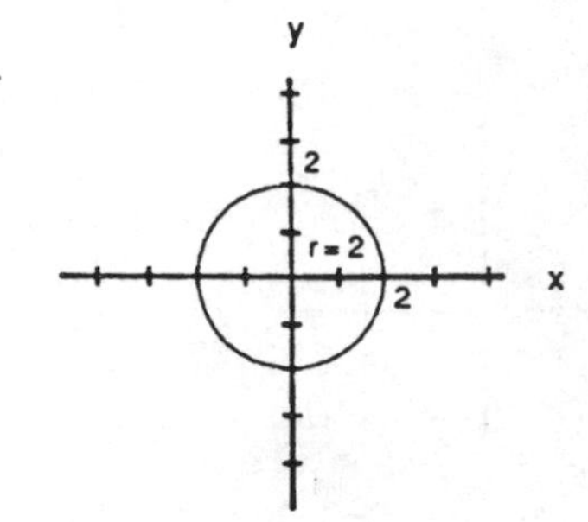

8.

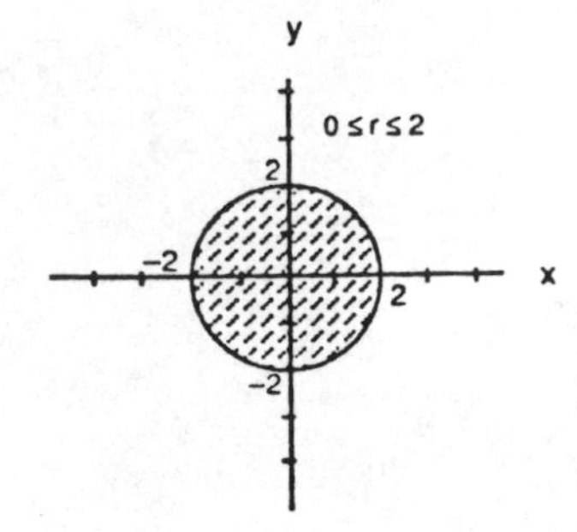

9.

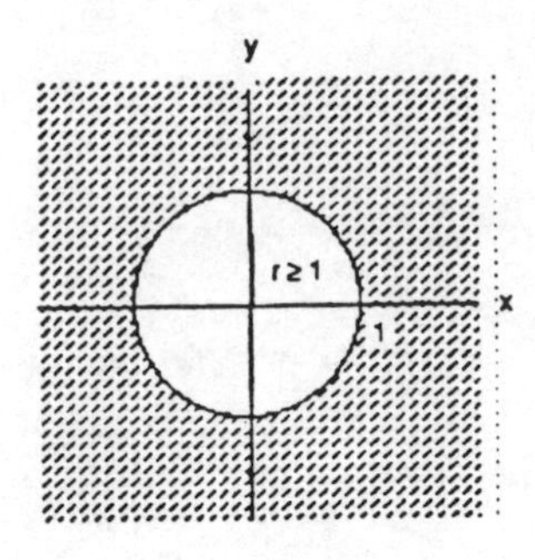

10.

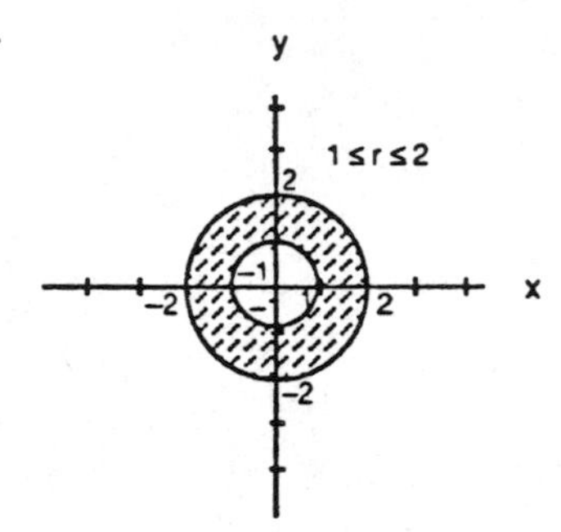

11.

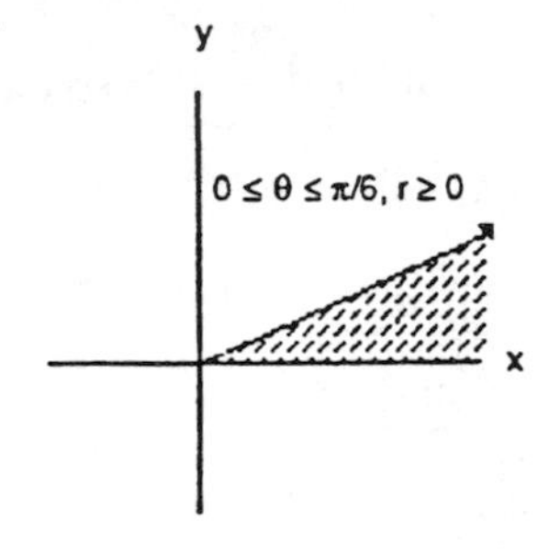

12.

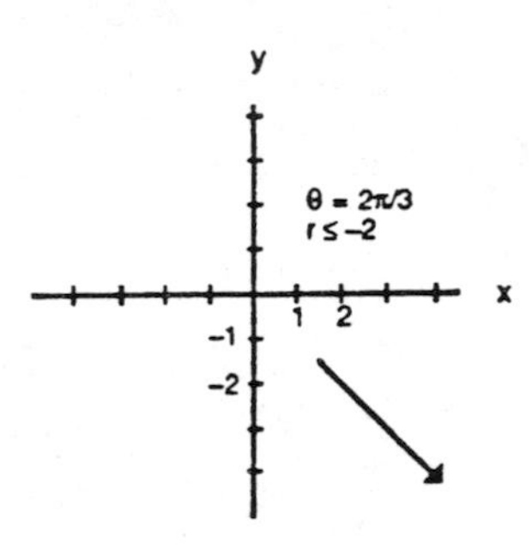

13.

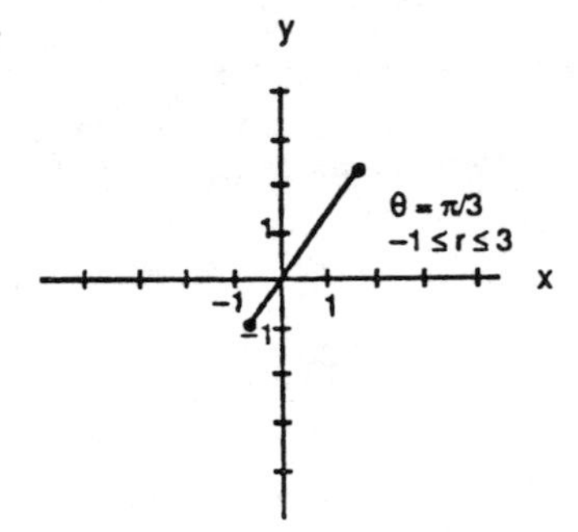

14.

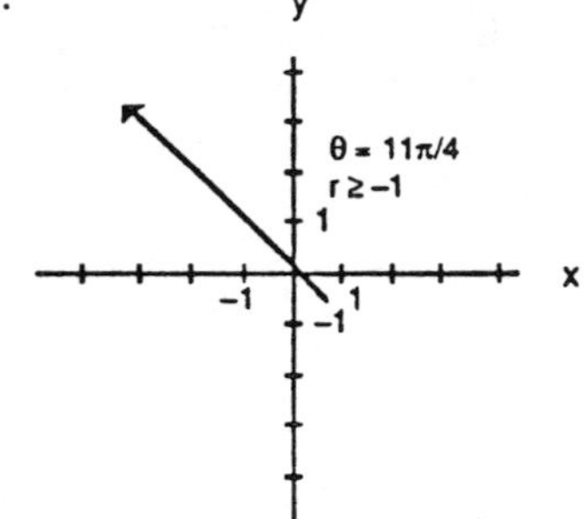

15.

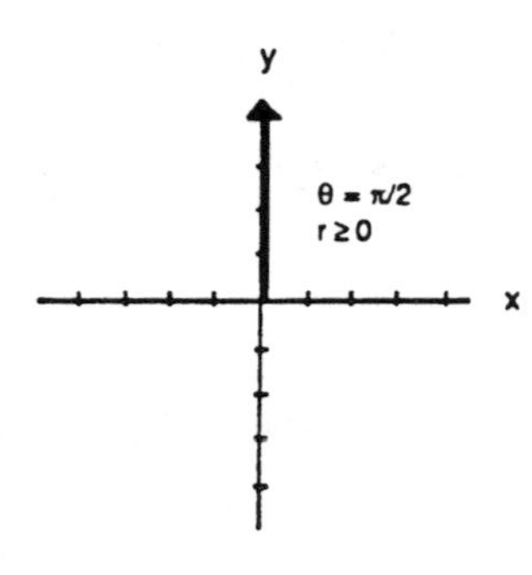

16.

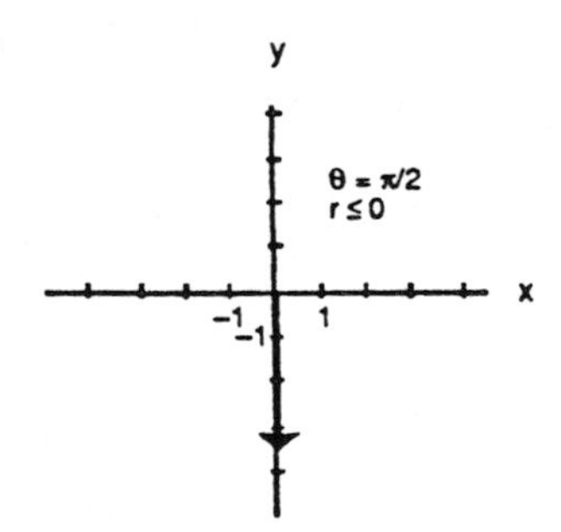

17.

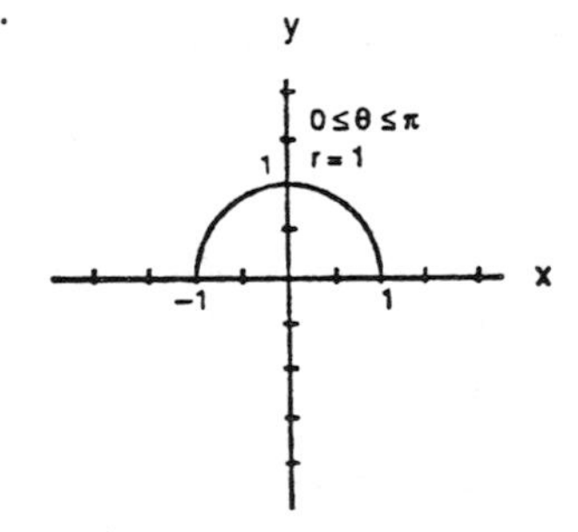

18.

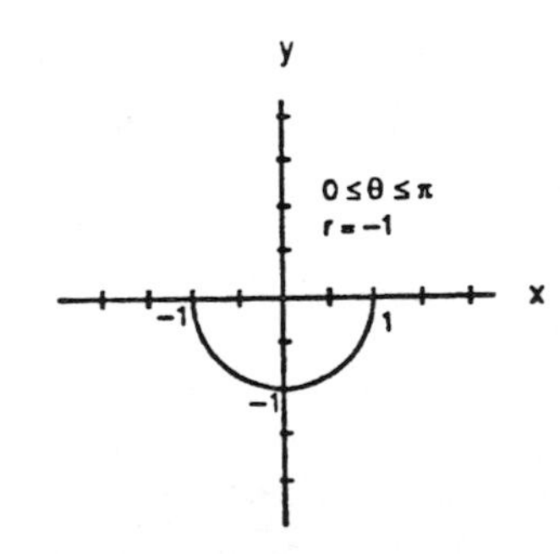

19.

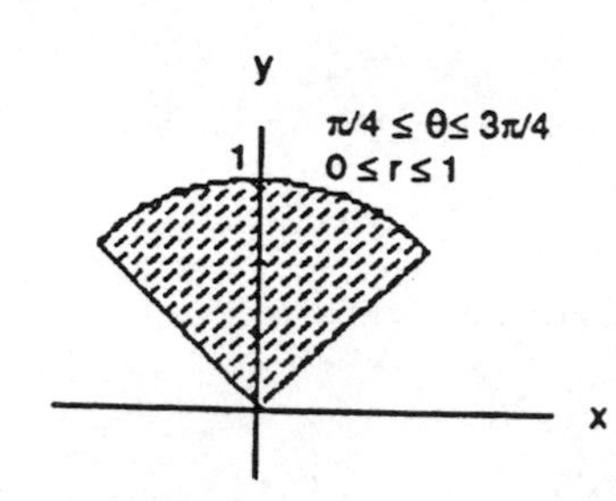

20.

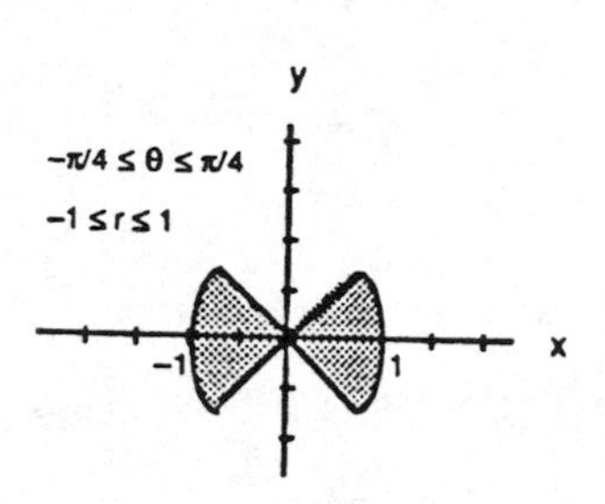

21.

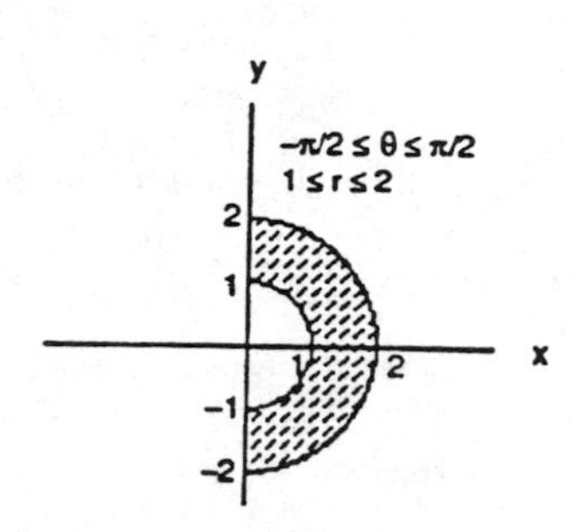

22.

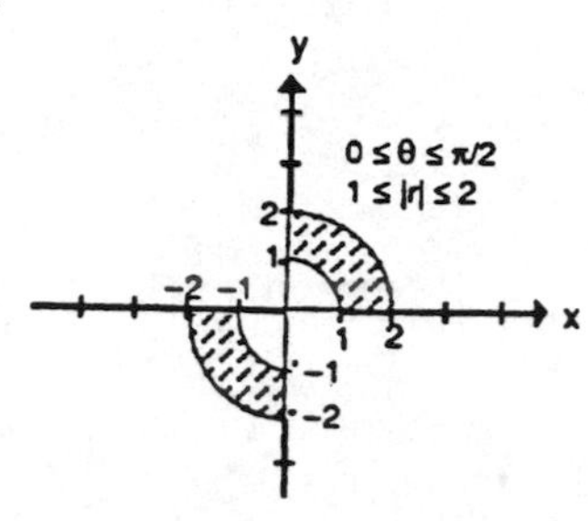

23. $r \cos \theta = 2 \Rightarrow x = 2$, vertical line through $(2, 0)$

24. $r \sin \theta = -1 \Rightarrow y = -1$, horizontal line through $(0, -1)$

25. $r \sin \theta = 0 \Rightarrow y = 0$, the x-axis

26. $r \cos \theta = 0 \Rightarrow x = 0$, the y-axis

27. $r = 4 \csc \theta \Rightarrow r = \frac{4}{\sin \theta} \Rightarrow r \sin \theta = 4 \Rightarrow y = 4$, a horizontal line through $(0, 4)$

28. $r = -3 \sec \theta \Rightarrow r = \frac{-3}{\cos \theta} \Rightarrow r \cos \theta = -3 \Rightarrow x = -3$, a vertical line through $(-3, 0)$

29. $r \cos \theta + r \sin \theta = 1 \Rightarrow x + y = 1$, line with slope $m = -1$ and intercept $b = 1$

30. $r \sin \theta = r \cos \theta \Rightarrow y = x$, line with slope $m = 1$ and intercept $b = 0$

31. $r^2 = 1 \Rightarrow x^2 + y^2 = 1$, circle with center $C = (0, 0)$ and radius 1

32. $r^2 = 4r \sin \theta \Rightarrow x^2 + y^2 = 4y \Rightarrow x^2 + y^2 - 4y + 4 = 4 \Rightarrow x^2 + (y - 2)^2 = 4$, circle with center $C = (0, 2)$ and radius 2

33. $r = \frac{5}{\sin \theta - 2 \cos \theta} \Rightarrow r \sin \theta - 2r \cos \theta = 5 \Rightarrow y - 2x = 5$, line with slope $m = 2$ and intercept $b = 5$

34. $r^2 \sin 2\theta = 2 \Rightarrow 2r^2 \sin \theta \cos \theta = 2 \Rightarrow (r \sin \theta)(r \cos \theta) = 1 \Rightarrow xy = 1$, hyperbola with focal axis $y = x$

35. $r = \cot \theta \csc \theta = \left(\frac{\cos \theta}{\sin \theta}\right)\left(\frac{1}{\sin \theta}\right) \Rightarrow r \sin^2 \theta = \cos \theta \Rightarrow r^2 \sin^2 \theta = r \cos \theta \Rightarrow y^2 = x$, parabola with vertex $(0, 0)$ which opens to the right

36. $r = 4 \tan \theta \sec \theta \Rightarrow r = 4\left(\frac{\sin \theta}{\cos^2 \theta}\right) \Rightarrow r \cos^2 \theta = 4 \sin \theta \Rightarrow r^2 \cos^2 \theta = 4r \sin \theta \Rightarrow x^2 = 4y$, parabola with vertex $= (0, 0)$ which opens upward

37. $r = (\csc \theta)\, e^{r \cos \theta} \Rightarrow r \sin \theta = e^{r \cos \theta} \Rightarrow y = e^x$, graph of the natural exponential function

38. $r \sin\theta = \ln r + \ln\cos\theta = \ln(r\cos\theta) \Rightarrow y = \ln x$, graph of the natural logarithm function

39. $r^2 + 2r^2\cos\theta\sin\theta = 1 \Rightarrow x^2 + y^2 + 2xy = 1 \Rightarrow x^2 + 2xy + y^2 = 1 \Rightarrow (x+y)^2 = 1 \Rightarrow x + y = \pm 1$, two parallel straight lines of slope -1 and y-intercepts $b = \pm 1$

40. $\cos^2\theta = \sin^2\theta \Rightarrow r^2\cos^2\theta = r^2\sin^2\theta \Rightarrow x^2 = y^2 \Rightarrow |x| = |y| \Rightarrow \pm x = y$, two perpendicular lines through the origin with slopes 1 and -1, respectively.

41. $r^2 = -4r\cos\theta \Rightarrow x^2 + y^2 = -4x \Rightarrow x^2 + 4x + y^2 = 0 \Rightarrow x^2 + 4x + 4 + y^2 = 4 \Rightarrow (x+2)^2 + y^2 = 4$, a circle with center $C(-2, 0)$ and radius 2

42. $r^2 = -6r\sin\theta \Rightarrow x^2 + y^2 = -6y \Rightarrow x^2 + y^2 + 6y = 0 \Rightarrow x^2 + y^2 + 6y + 9 = 9 \Rightarrow x^2 + (y+3)^2 = 9$, a circle with center $C(0, -3)$ and radius 3

43. $r = 8\sin\theta \Rightarrow r^2 = 8r\sin\theta \Rightarrow x^2 + y^2 = 8y \Rightarrow x^2 + y^2 - 8y = 0 \Rightarrow x^2 + y^2 - 8y + 16 = 16$
$\Rightarrow x^2 + (y-4)^2 = 16$, a circle with center $C(0, 4)$ and radius 4

44. $r = 3\cos\theta \Rightarrow r^2 = 3r\cos\theta \Rightarrow x^2 + y^2 = 3x \Rightarrow x^2 + y^2 - 3x = 0 \Rightarrow x^2 - 3x + \frac{9}{4} + y^2 = \frac{9}{4}$
$\Rightarrow \left(x - \frac{3}{2}\right)^2 + y^2 = \frac{9}{4}$, a circle with center $C\left(\frac{3}{2}, 0\right)$ and radius $\frac{3}{2}$

45. $r = 2\cos\theta + 2\sin\theta \Rightarrow r^2 = 2r\cos\theta + 2r\sin\theta \Rightarrow x^2 + y^2 = 2x + 2y \Rightarrow x^2 - 2x + y^2 - 2y = 0$
$\Rightarrow (x-1)^2 + (y-1)^2 = 2$, a circle with center $C(1, 1)$ and radius $\sqrt{2}$

46. $r = 2\cos\theta - \sin\theta \Rightarrow r^2 = 2r\cos\theta - r\sin\theta \Rightarrow x^2 + y^2 = 2x - y \Rightarrow x^2 - 2x + y^2 + y = 0$
$\Rightarrow (x-1)^2 + \left(y + \frac{1}{2}\right)^2 = \frac{5}{4}$, a circle with center $C\left(1, -\frac{1}{2}\right)$ and radius $\frac{\sqrt{5}}{2}$

47. $r\sin\left(\theta + \frac{\pi}{6}\right) = 2 \Rightarrow r\left(\sin\theta\cos\frac{\pi}{6} + \cos\theta\sin\frac{\pi}{6}\right) = 2 \Rightarrow \frac{\sqrt{3}}{2}r\sin\theta + \frac{1}{2}r\cos\theta = 2 \Rightarrow \frac{\sqrt{3}}{2}y + \frac{1}{2}x = 2$
$\Rightarrow \sqrt{3}\,y + x = 4$, line with slope $m = -\frac{1}{\sqrt{3}}$ and intercept $b = \frac{4}{\sqrt{3}}$

48. $r\sin\left(\frac{2\pi}{3} - \theta\right) = 5 \Rightarrow r\left(\sin\frac{2\pi}{3}\cos\theta - \cos\frac{2\pi}{3}\sin\theta\right) = 5 \Rightarrow \frac{\sqrt{3}}{2}r\cos\theta + \frac{1}{2}r\sin\theta = 5 \Rightarrow \frac{\sqrt{3}}{2}x + \frac{1}{2}y = 5$
$\Rightarrow \sqrt{3}\,x + y = 10$, line with slope $m = -\sqrt{3}$ and intercept $b = 10$

49. $x = 7 \Rightarrow r\cos\theta = 7$

50. $y = 1 \Rightarrow r\sin\theta = 1$

51. $x = y \Rightarrow r\cos\theta = r\sin\theta \Rightarrow \theta = \frac{\pi}{4}$

52. $x - y = 3 \Rightarrow r\cos\theta - r\sin\theta = 3$

53. $x^2 + y^2 = 4 \Rightarrow r^2 = 4 \Rightarrow r = 2$ or $r = -2$

54. $x^2 - y^2 = 1 \Rightarrow r^2\cos^2\theta - r^2\sin^2\theta = 1 \Rightarrow r^2\left(\cos^2\theta - \sin^2\theta\right) = 1 \Rightarrow r^2\cos 2\theta = 1$

55. $\frac{x^2}{9} + \frac{y^2}{4} = 1 \Rightarrow 4x^2 + 9y^2 = 36 \Rightarrow 4r^2\cos^2\theta + 9r^2\sin^2\theta = 36$

56. $xy = 2 \Rightarrow (r \cos \theta)(r \sin \theta) = 2 \Rightarrow r^2 \cos \theta \sin \theta = 2 \Rightarrow 2r^2 \cos \theta \sin \theta = 4 \Rightarrow r^2 \sin 2\theta = 4$

57. $y^2 = 4x \Rightarrow r^2 \sin^2 \theta = 4r \cos \theta \Rightarrow r \sin^2 \theta = 4 \cos \theta$

58. $x^2 + xy + y^2 = 1 \Rightarrow x^2 + y^2 + xy = 1 \Rightarrow r^2 + r^2 \sin \theta \cos \theta = 1 \Rightarrow r^2 (1 + \sin \theta \cos \theta) = 1$

59. $x^2 + (y-2)^2 = 4 \Rightarrow x^2 + y^2 - 4y + 4 = 4 \Rightarrow x^2 + y^2 = 4y \Rightarrow r^2 = 4r \sin \theta \Rightarrow r = 4 \sin \theta$

60. $(x-5)^2 + y^2 = 25 \Rightarrow x^2 - 10x + 25 + y^2 = 25 \Rightarrow x^2 + y^2 = 10x \Rightarrow r^2 = 10r \cos \theta \Rightarrow r = 10 \cos \theta$

61. $(x-3)^2 + (y+1)^2 = 4 \Rightarrow x^2 - 6x + 9 + y^2 + 2y + 1 = 4 \Rightarrow x^2 + y^2 = 6x - 2y - 6 \Rightarrow r^2 = 6r \cos \theta - 2r \sin \theta - 6$

62. $(x+2)^2 + (y-5)^2 = 16 \Rightarrow x^2 + 4x + 4 + y^2 - 10y + 25 = 16 \Rightarrow x^2 + y^2 = -4x + 10y - 13 \Rightarrow r^2$
$= -4r \cos \theta + 10r \sin \theta - 13$

63. $(0, \theta)$ where θ is any angle

64. (a) $x = a \Rightarrow r \cos \theta = a \Rightarrow r = \frac{a}{\cos \theta} \Rightarrow r = a \sec \theta$

(b) $y = b \Rightarrow r \sin \theta = b \Rightarrow r = \frac{b}{\sin \theta} \Rightarrow r = b \csc \theta$

9.7 GRAPHING IN POLAR COORDINATES

1. $1 + \cos(-\theta) = 1 + \cos \theta = r \Rightarrow$ symmetric about the x-axis; $1 + \cos(-\theta) \neq -r$ and $1 + \cos(\pi - \theta) = 1 - \cos \theta \neq r$ $\Rightarrow$ not symmetric about the y-axis; therefore not symmetric about the origin

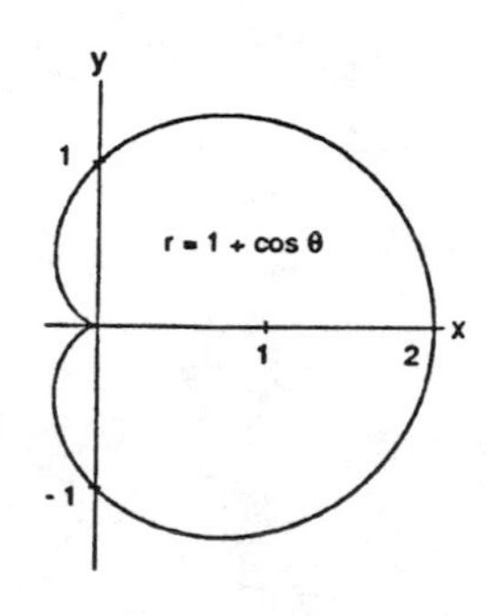

2. $2 - 2\cos(-\theta) = 2 - 2\cos \theta = r \Rightarrow$ symmetric about the x-axis; $2 - 2\cos(-\theta) \neq -r$ and $2 - 2\cos(\pi - \theta)$ $= 2 + 2\cos \theta \neq r \Rightarrow$ not symmetric about the y-axis; therefore not symmetric about the origin

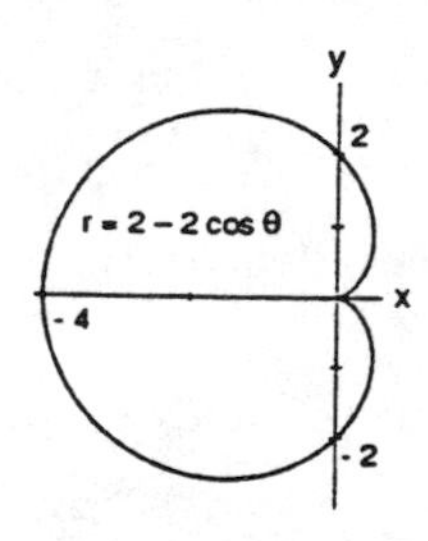

3. $1 - \sin(-\theta) = 1 + \sin \theta \neq r$ and $1 - \sin(\pi - \theta)$ $= 1 - \sin \theta \neq -r \Rightarrow$ not symmetric about the x-axis; $1 - \sin(\pi - \theta) = 1 - \sin \theta = r \Rightarrow$ symmetric about the y-axis; therefore not symmetric about the origin

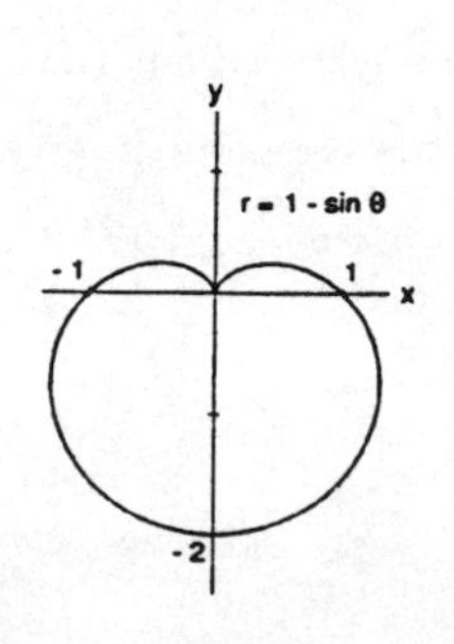

4. $1 + \sin(-\theta) = 1 - \sin\theta \neq r$ and $1 - \sin(\pi - \theta)$ $= 1 - \sin\theta \neq r \Rightarrow$ not symmetric about the x-axis; $1 + \sin(\pi - \theta) = 1 + \sin\theta = r \Rightarrow$ symmetric about the y-axis; therefore not symmetric about the origin

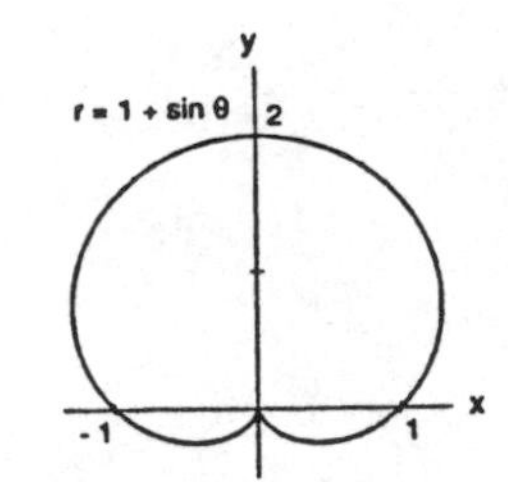

5. $2 + \sin(-\theta) = 2 - \sin\theta \neq r$ and $2 + \sin(\pi - \theta)$ $= 2 + \sin\theta \neq r \Rightarrow$ not symmetric about the x-axis; $2 + \sin(\pi - \theta) = 2 + \sin\theta = r \Rightarrow$ symmetric about the y-axis; therefore not symmetric about the origin

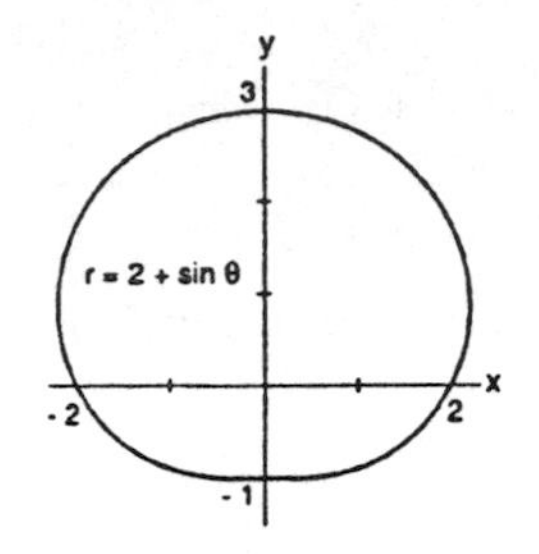

6. $1 + 2\sin(-\theta) = 1 - 2\sin\theta \neq r$ and $1 + 2\sin(\pi - \theta)$ $= 1 + 2\sin\theta \neq -r \Rightarrow$ not symmetric about the x-axis; $1 + 2\sin(\pi - \theta) = 1 + 2\sin\theta = r \Rightarrow$ symmetric about the y-axis; therefore not symmetric about the origin

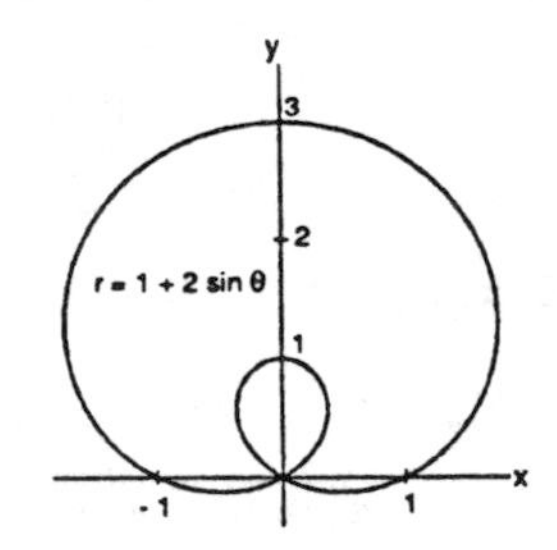

7. $\sin\left(-\frac{\theta}{2}\right) = -\sin\left(\frac{\theta}{2}\right) = -r \Rightarrow$ symmetric about the y-axis; $\sin\left(-\frac{\theta}{2}\right) = -\sin\left(\frac{\theta}{2}\right) = -r \neq r$ and $\sin\left(\frac{\pi - \theta}{2}\right) = \sin\left(\frac{\pi}{2} - \frac{\theta}{2}\right)$ $= \cos\left(\frac{\theta}{2}\right) \neq -r$, but clearly the graph <u>is</u> symmetric about the x-axis and the origin. The symmetry tests as stated do not necessarily tell when a graph is <u>not</u> symmetric. Note that $\sin\left(\frac{2\pi - \theta}{2}\right) = \sin\left(\frac{\theta}{2}\right)$, so the graph <u>is</u> symmetric about the x-axis, and hence the origin.

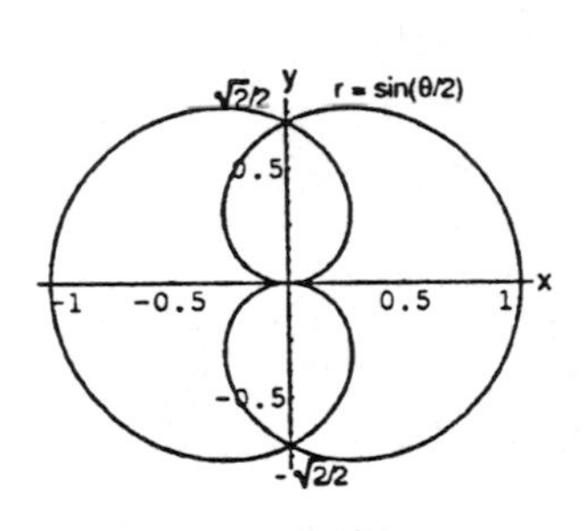

8. $\cos\left(-\frac{\theta}{2}\right) = \cos\left(\frac{\theta}{2}\right) = r \Rightarrow$ symmetric about the y-axis; $\cos\left(-\frac{\theta}{2}\right) \neq -r$ and $\cos\left(\frac{\pi - \theta}{2}\right) = \sin\left(\frac{\theta}{2}\right) \neq r$, but clearly the graph <u>is</u> symmetric about the y-axis and the origin. As in Exercise 7, the tests fail to give enough information. Note that $\cos\left(\frac{2\pi - \theta}{2}\right) = \cos\left(\frac{\theta}{2}\right)$, so the graph <u>is</u> symmetric about the y-axis.

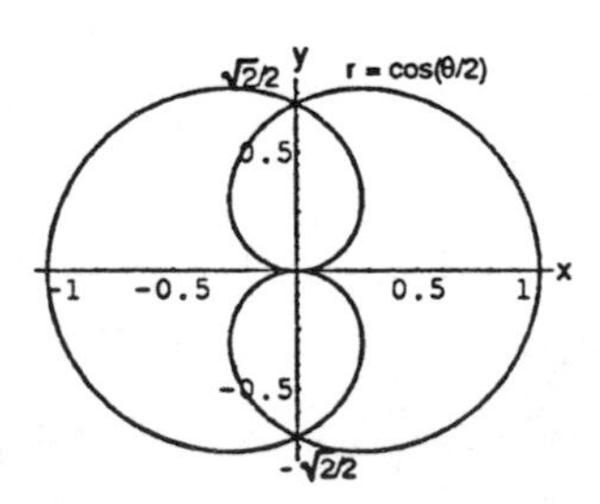

9. $\cos(-\theta) = \cos\theta = r^2 \Rightarrow (r, -\theta)$ and $(-r, -\theta)$ are on the graph when (r, θ) is on the graph $\Rightarrow$ symmetric about the x-axis and the y-axis; therefore symmetric about the origin

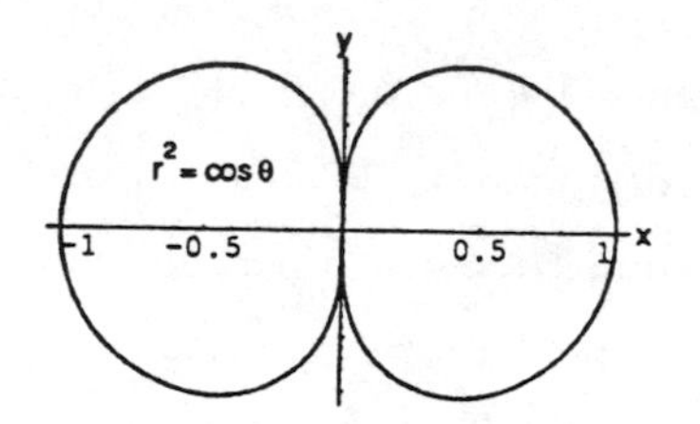

10. $\sin(\pi - \theta) = \sin\theta = r^2 \Rightarrow (r, \pi - \theta)$ and $(-r, \pi - \theta)$ are on the graph when (r, θ) is on the graph $\Rightarrow$ symmetric about the y-axis and the x-axis; therefore symmetric about the origin

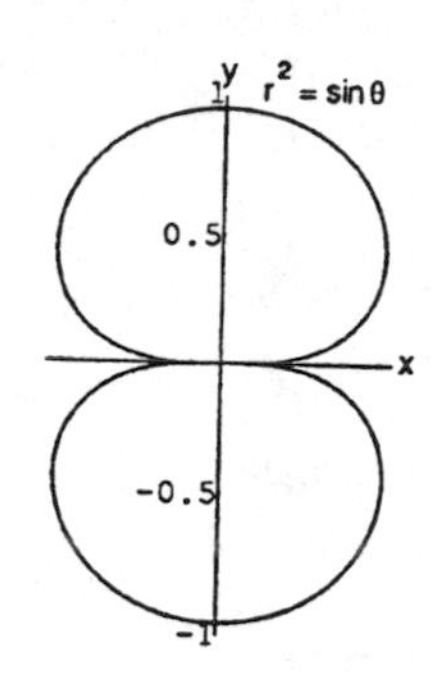

11. $-\sin(\pi - \theta) = -\sin\theta = r^2 \Rightarrow (r, \pi - \theta)$ and $(-r, \pi - \theta)$ are on the graph when (r, θ) is on the graph $\Rightarrow$ symmetric about the y-axis and the x-axis; therefore symmetric about the origin

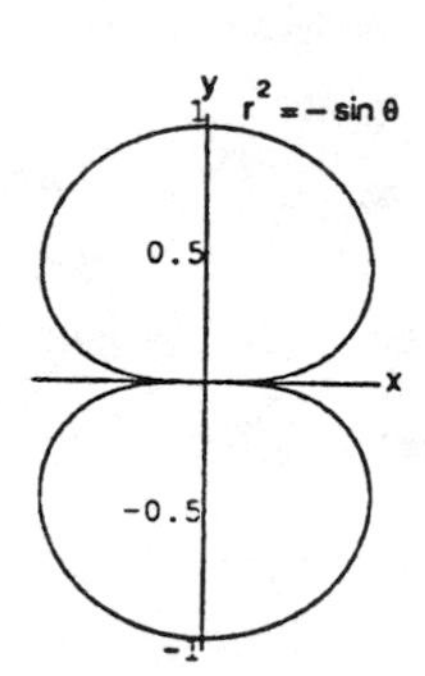

12. $-\cos(-\theta) = -\cos\theta = r^2 \Rightarrow (r, -\theta)$ and $(-r, -\theta)$ are on the graph when (r, θ) is on the graph $\Rightarrow$ symmetric about the x-axis and the y-axis; therefore symmetric about the origin

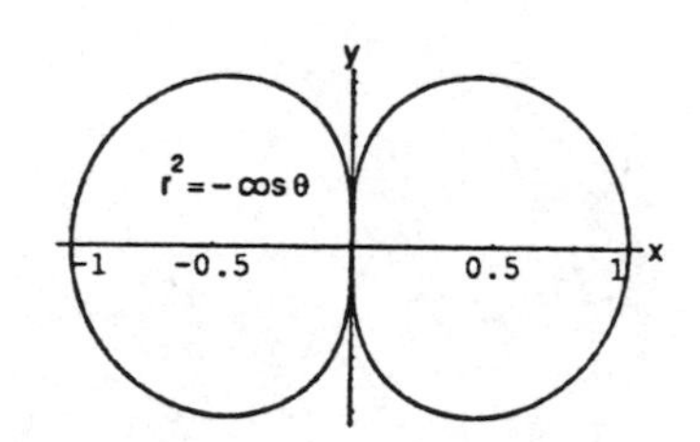

13. Since $(\pm r, -\theta)$ are on the graph when (r, θ) is on the graph $\left((\pm r)^2 = 4\cos 2(-\theta) \Rightarrow r^2 = 4\cos 2\theta\right)$, the graph is symmetric about the x-axis and the y-axis $\Rightarrow$ the graph is symmetric about the origin

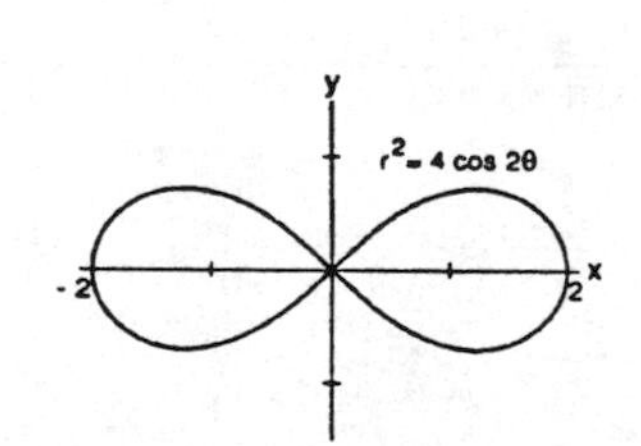

14. Since (r, θ) on the graph $\Rightarrow (-r, \theta)$ is on the graph $\left((\pm r)^2 = 4 \sin 2\theta \Rightarrow r^2 = 4 \sin 2\theta\right)$, the graph is symmetric about the origin. But $4 \sin 2(-\theta) = -4 \sin 2\theta \neq r^2$ and $4 \sin 2(\pi - \theta) = 4 \sin (2\pi - 2\theta) = 4 \sin (-2\theta) = -4 \sin 2\theta \neq r^2 \Rightarrow$ the graph is not symmetric about the x-axis; therefore the graph is not symmetric about the y-axis

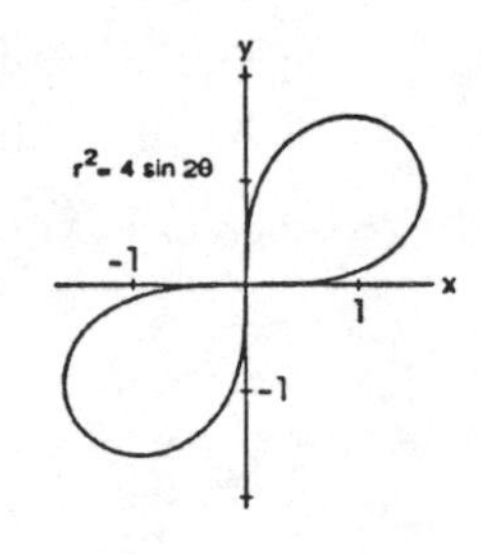

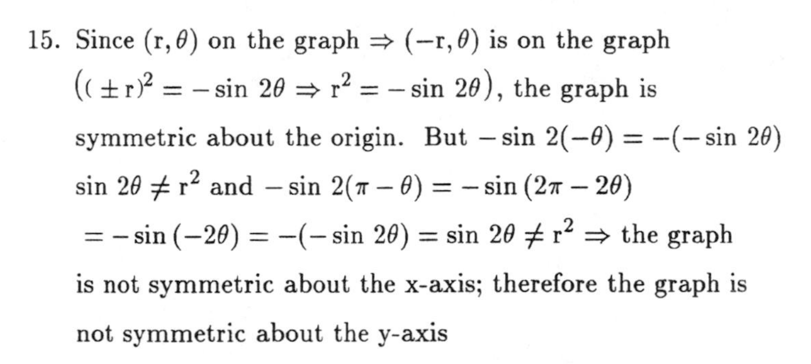

15. Since (r, θ) on the graph $\Rightarrow (-r, \theta)$ is on the graph $\left((\pm r)^2 = -\sin 2\theta \Rightarrow r^2 = -\sin 2\theta\right)$, the graph is symmetric about the origin. But $-\sin 2(-\theta) = -(-\sin 2\theta)$ $\sin 2\theta \neq r^2$ and $-\sin 2(\pi - \theta) = -\sin (2\pi - 2\theta) = -\sin (-2\theta) = -(-\sin 2\theta) = \sin 2\theta \neq r^2 \Rightarrow$ the graph is not symmetric about the x-axis; therefore the graph is not symmetric about the y-axis

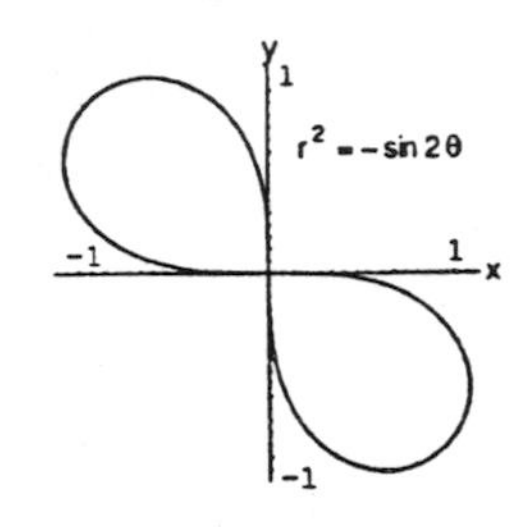

16. Since $(\pm r, -\theta)$ are on the graph when (r, θ) is on the graph $\left((\pm r)^2 = -\cos 2(-\theta) \Rightarrow r^2 = -\cos 2\theta\right)$, the graph is symmetric about the x-axis and the y-axis $\Rightarrow$ the graph is symmetric about the origin.

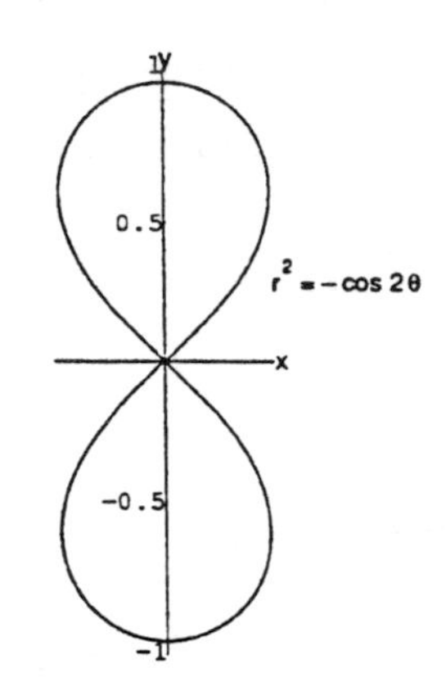

17. $\theta = \frac{\pi}{2} \Rightarrow r = -1 \Rightarrow \left(-1, \frac{\pi}{2}\right)$, and $\theta = -\frac{\pi}{2} \Rightarrow r = -1$ $\Rightarrow \left(-1, -\frac{\pi}{2}\right)$; $r' = \frac{dr}{d\theta} = -\sin \theta$; Slope $= \frac{r' \sin \theta + r \cos \theta}{r' \cos \theta - r \sin \theta}$ $= \frac{-\sin^2 \theta + r \cos \theta}{-\sin \theta \cos \theta - r \sin \theta} \Rightarrow$ Slope at $\left(-1, \frac{\pi}{2}\right)$ is $\frac{-\sin^2\left(\frac{\pi}{2}\right) + (-1) \cos \frac{\pi}{2}}{-\sin \frac{\pi}{2} \cos \frac{\pi}{2} - (-1) \sin \frac{\pi}{2}} = -1$; Slope at $\left(-1, -\frac{\pi}{2}\right)$ is

$$\frac{-\sin^2\left(-\frac{\pi}{2}\right) + (-1) \cos\left(-\frac{\pi}{2}\right)}{-\sin\left(-\frac{\pi}{2}\right) \cos\left(-\frac{\pi}{2}\right) - (-1) \sin\left(-\frac{\pi}{2}\right)} = 1$$

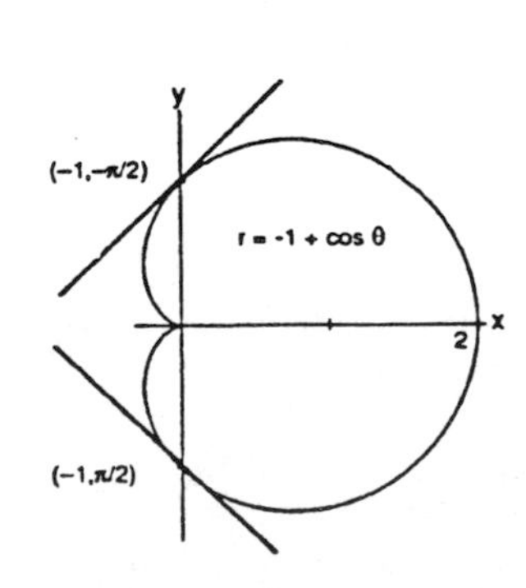

18. $\theta = 0 \Rightarrow r = -1 \Rightarrow (-1, 0)$, and $\theta = \pi \Rightarrow r = -1 \Rightarrow (-1, \pi)$;

$r' = \frac{dr}{d\theta} = \cos\theta$; Slope $= \frac{r' \sin\theta + r\cos\theta}{r'\cos\theta - r\sin\theta} = \frac{\cos\theta\sin\theta + r\cos\theta}{\cos\theta\cos\theta - r\sin\theta}$

$= \frac{\cos\theta\sin\theta + r\cos\theta}{\cos^2\theta - r\sin\theta} \Rightarrow$ Slope at $(-1, 0)$ is $\frac{\cos 0\sin 0 + (-1)\cos 0}{\cos^2 0 - (-1)\sin 0}$

$= -1$; Slope at $(-1, \pi)$ is $\frac{\cos\pi\sin\pi + (-1)\cos\pi}{\cos^2\pi - (-1)\sin\pi} = 1$

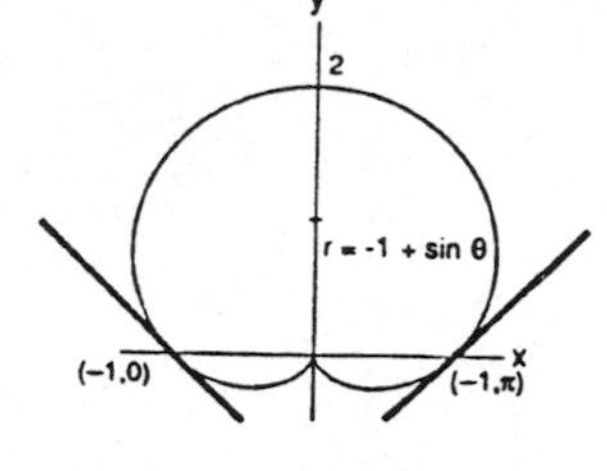

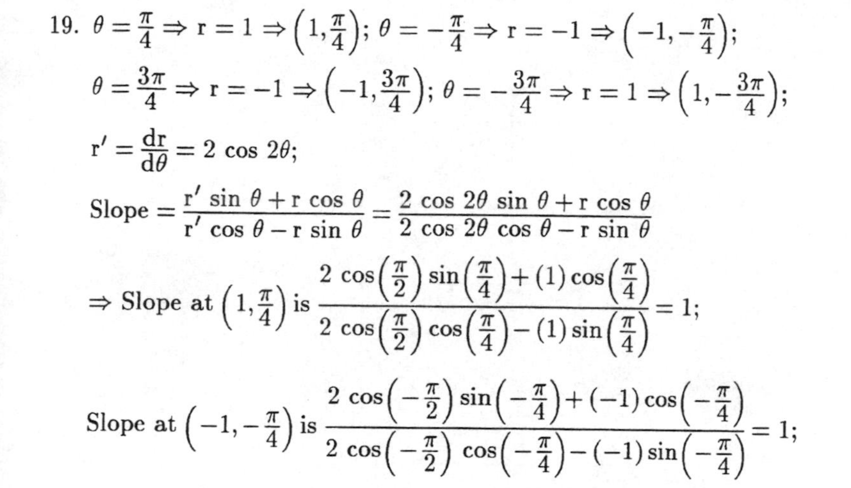

19. $\theta = \frac{\pi}{4} \Rightarrow r = 1 \Rightarrow \left(1, \frac{\pi}{4}\right)$; $\theta = -\frac{\pi}{4} \Rightarrow r = -1 \Rightarrow \left(-1, -\frac{\pi}{4}\right)$;

$\theta = \frac{3\pi}{4} \Rightarrow r = -1 \Rightarrow \left(-1, \frac{3\pi}{4}\right)$; $\theta = -\frac{3\pi}{4} \Rightarrow r = 1 \Rightarrow \left(1, -\frac{3\pi}{4}\right)$;

$r' = \frac{dr}{d\theta} = 2\cos 2\theta$;

Slope $= \frac{r'\sin\theta + r\cos\theta}{r'\cos\theta - r\sin\theta} = \frac{2\cos 2\theta\sin\theta + r\cos\theta}{2\cos 2\theta\cos\theta - r\sin\theta}$

$\Rightarrow$ Slope at $\left(1, \frac{\pi}{4}\right)$ is $\frac{2\cos\left(\frac{\pi}{2}\right)\sin\left(\frac{\pi}{4}\right) + (1)\cos\left(\frac{\pi}{4}\right)}{2\cos\left(\frac{\pi}{2}\right)\cos\left(\frac{\pi}{4}\right) - (1)\sin\left(\frac{\pi}{4}\right)} = 1$;

Slope at $\left(-1, -\frac{\pi}{4}\right)$ is $\frac{2\cos\left(-\frac{\pi}{2}\right)\sin\left(-\frac{\pi}{4}\right) + (-1)\cos\left(-\frac{\pi}{4}\right)}{2\cos\left(-\frac{\pi}{2}\right)\cos\left(-\frac{\pi}{4}\right) - (-1)\sin\left(-\frac{\pi}{4}\right)} = 1$;

Slope at $\left(-1, \frac{3\pi}{4}\right)$ is $\frac{2\cos\left(\frac{3\pi}{2}\right)\sin\left(\frac{3\pi}{4}\right) + (-1)\cos\left(\frac{3\pi}{4}\right)}{2\cos\left(\frac{3\pi}{2}\right)\cos\left(\frac{3\pi}{4}\right) - (-1)\sin\left(\frac{3\pi}{4}\right)} = 1$;

Slope at $\left(1, -\frac{3\pi}{4}\right)$ is $\frac{2\cos\left(-\frac{3\pi}{2}\right)\sin\left(-\frac{3\pi}{4}\right) + (1)\cos\left(-\frac{3\pi}{4}\right)}{2\cos\left(-\frac{3\pi}{2}\right)\cos\left(-\frac{3\pi}{4}\right) - (1)\sin\left(-\frac{3\pi}{4}\right)} = -1$

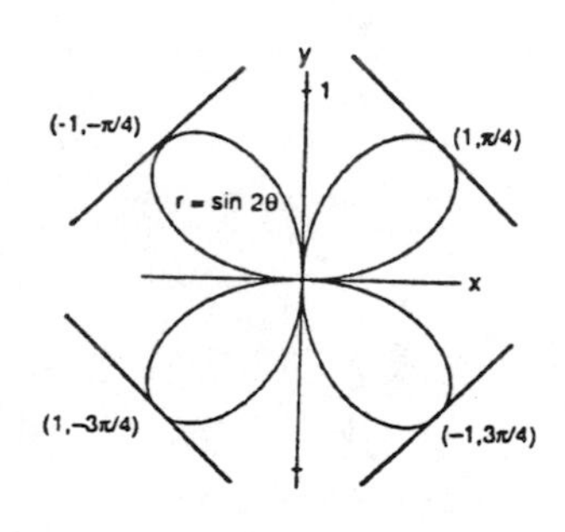

20. $\theta = 0 \Rightarrow r = 1 \Rightarrow (1, 0)$; $\theta = \frac{\pi}{2} \Rightarrow r = -1 \Rightarrow \left(-1, \frac{\pi}{2}\right)$; $\theta = -\frac{\pi}{2}$

$\Rightarrow r = -1 \Rightarrow \left(-1, -\frac{\pi}{2}\right)$; $\theta = \pi \Rightarrow r = 1 \Rightarrow (1, \pi)$; $r' = \frac{dr}{d\theta}$

$= -2\sin 2\theta$; Slope $= \frac{r'\sin\theta + r\cos\theta}{r'\cos\theta - r\sin\theta} = \frac{-2\sin 2\theta\sin\theta + r\cos\theta}{-2\sin 2\theta\cos\theta - r\sin\theta}$

$\Rightarrow$ Slope at $(1, 0)$ is $\frac{-2\sin 0\sin 0 + \cos 0}{-2\sin 0\cos 0 - \sin 0}$, which is undefined;

Slope at $\left(-1, \frac{\pi}{2}\right)$ is $\frac{-2\sin 2\left(\frac{\pi}{2}\right)\sin\left(\frac{\pi}{2}\right) + (-1)\cos\left(\frac{\pi}{2}\right)}{-2\sin 2\left(\frac{\pi}{2}\right)\cos\left(\frac{\pi}{2}\right) - (-1)\sin\left(\frac{\pi}{2}\right)} = 0$;

Slope at $\left(-1, -\frac{\pi}{2}\right)$ is $\frac{-2\cos 2\left(-\frac{\pi}{2}\right)\sin\left(-\frac{\pi}{2}\right) + (-1)\cos\left(-\frac{\pi}{2}\right)}{-2\sin 2\left(-\frac{\pi}{2}\right)\cos\left(-\frac{\pi}{2}\right) - (-1)\sin\left(-\frac{\pi}{2}\right)} = 0$;

Slope at $(1, \pi)$ is $\frac{-2\sin 2\pi\sin\pi + \cos\pi}{-2\sin 2\pi\cos\pi - \sin\pi}$, which is undefined

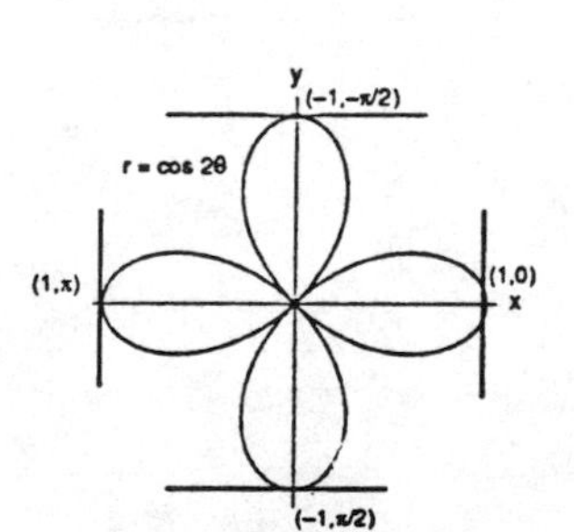

21. (a)

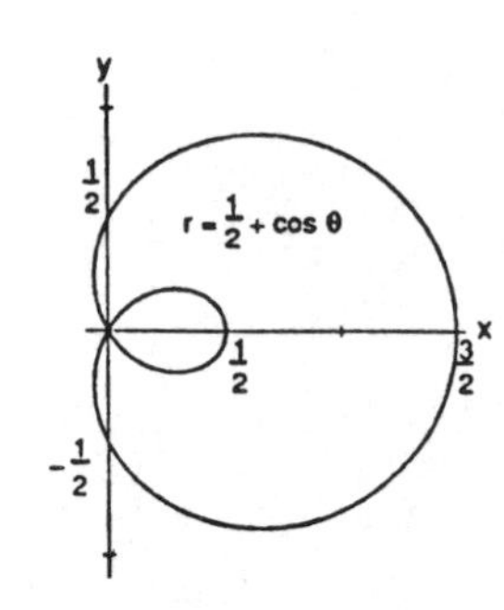

(b)

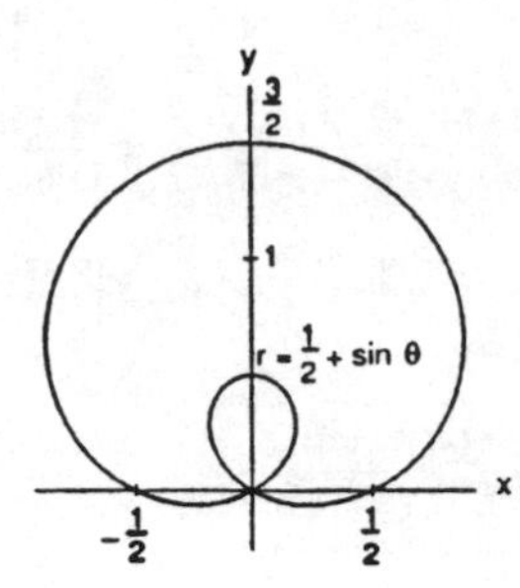

22. (a)

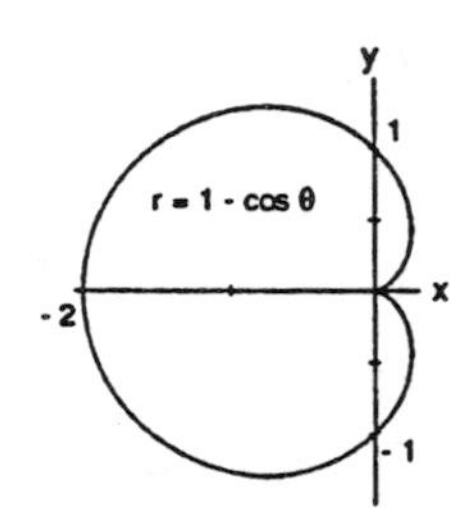

(b)

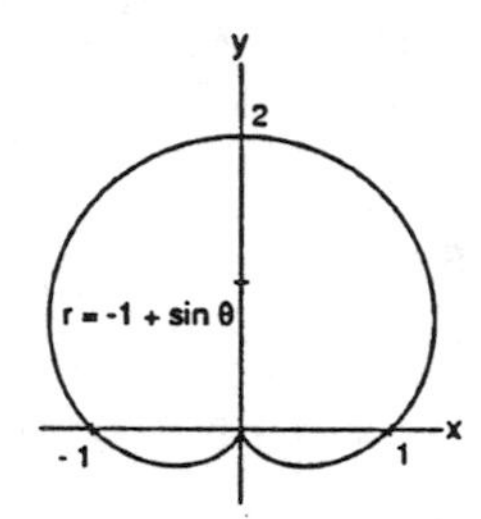

23. (a)

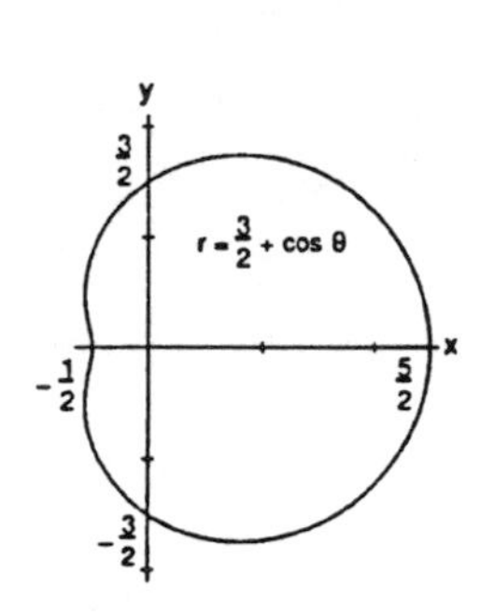

(b)

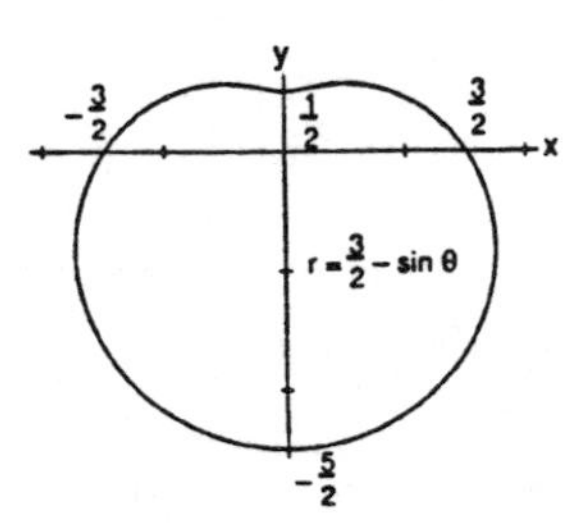

24. (a)

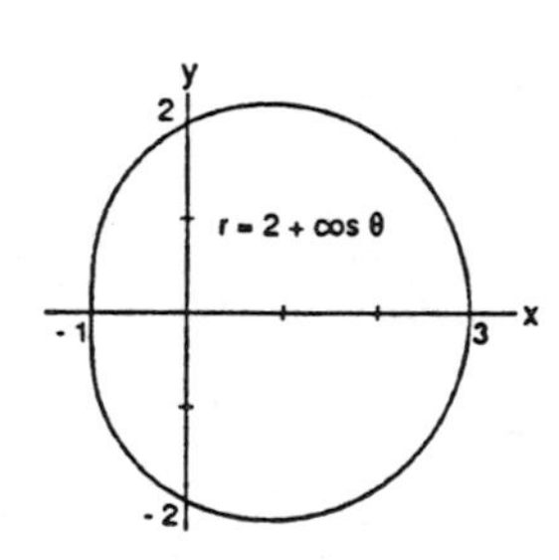

(b)

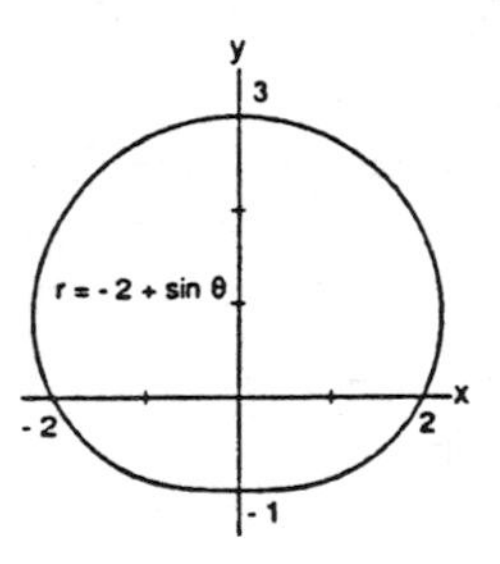

25.

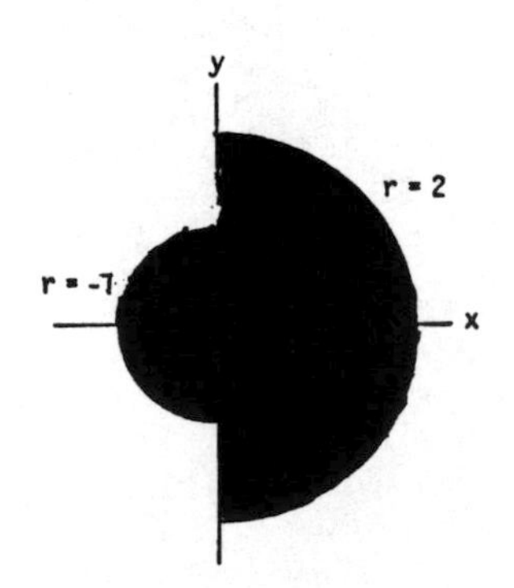

26. $r = 2 \sec \theta \Rightarrow r = \frac{2}{\cos \theta} \Rightarrow r \cos \theta = 2 \Rightarrow x = 2$

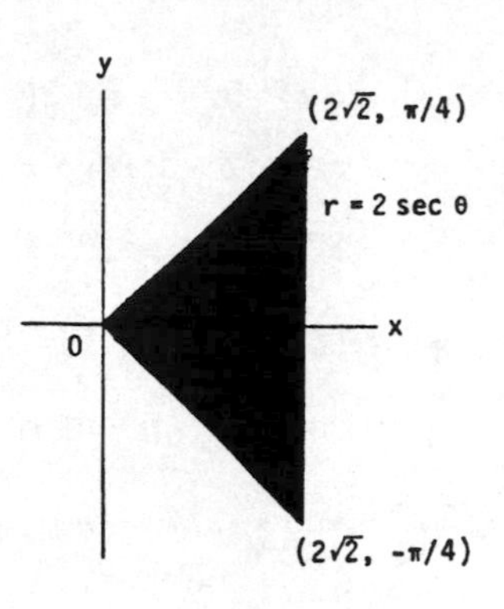

27.

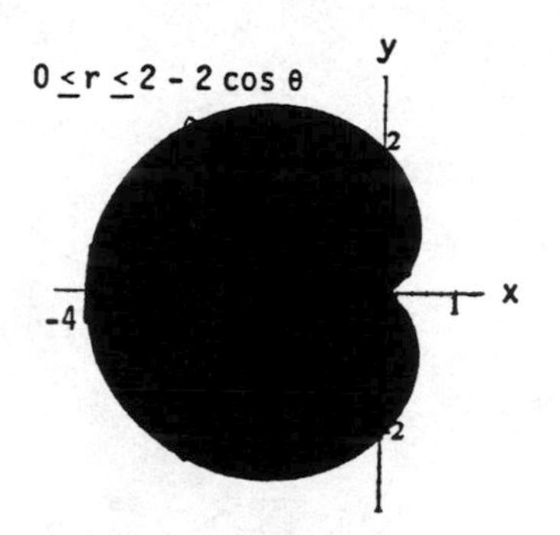

28.

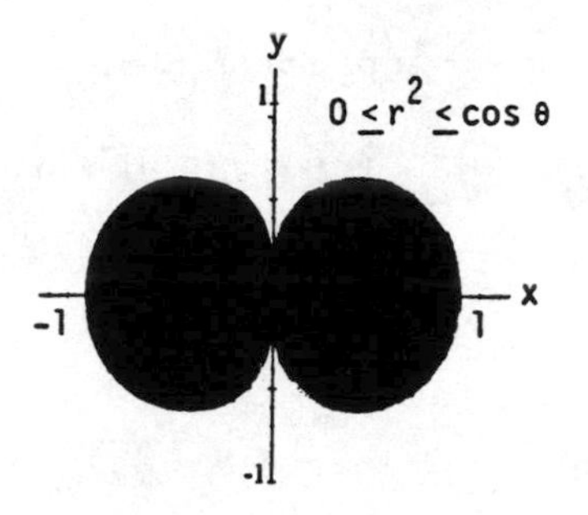

29. $\left(2, \frac{3\pi}{4}\right)$ is the same point as $\left(-2, -\frac{\pi}{4}\right)$; $r = 2 \sin 2\left(-\frac{\pi}{4}\right) = 2 \sin\left(-\frac{\pi}{2}\right) = -2 \Rightarrow \left(-2, -\frac{\pi}{4}\right)$ is on the graph $\Rightarrow \left(2, \frac{3\pi}{4}\right)$ is on the graph

30. $\left(\frac{1}{2}, \frac{3\pi}{2}\right)$ is the same point as $\left(-\frac{1}{2}, \frac{\pi}{2}\right)$; $r = -\sin\left(\frac{\left(\frac{\pi}{2}\right)}{3}\right) = -\sin \frac{\pi}{6} = -\frac{1}{2} \Rightarrow \left(-\frac{1}{2}, \frac{\pi}{2}\right)$ is on the graph $\Rightarrow \left(\frac{1}{2}, \frac{3\pi}{2}\right)$ is on the graph

31. $1 + \cos \theta = 1 - \cos \theta \Rightarrow \cos \theta = 0 \Rightarrow \theta = \frac{\pi}{2}, \frac{3\pi}{2} \Rightarrow r = 1$; points of intersection are $\left(1, \frac{\pi}{2}\right)$ and $\left(1, \frac{3\pi}{2}\right)$. The point of intersection $(0, 0)$ is found by graphing.

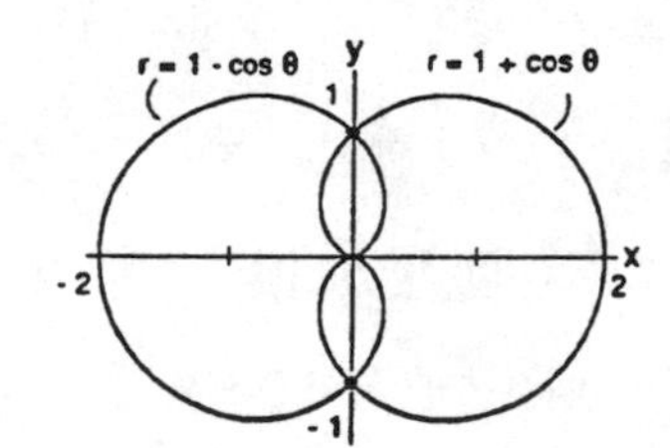

32. $1 + \sin \theta = 1 - \sin \theta \Rightarrow \sin \theta = 0 \Rightarrow \theta = 0, \pi \Rightarrow r = 1$; points of intersection are $(1, 0)$ and $(1, \pi)$. The point of intersection $(0, 0)$ is found by graphing.

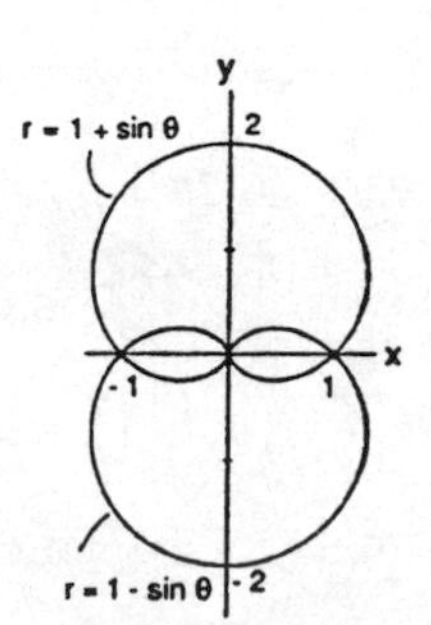

33. $2 \sin \theta = 2 \sin 2\theta \Rightarrow \sin \theta = \sin 2\theta \Rightarrow \sin \theta$
$= 2 \sin \theta \cos \theta \Rightarrow \sin \theta - 2 \sin \theta \cos \theta = 0$
$\Rightarrow (\sin \theta)(1 - 2 \cos \theta) = 0 \Rightarrow \sin \theta = 0$ or $\cos \theta = \frac{1}{2}$
$\Rightarrow \theta = 0, \frac{\pi}{3}$, or $-\frac{\pi}{3}$; $\theta = 0 \Rightarrow r = 0$, $\theta = \frac{\pi}{3} \Rightarrow r = \sqrt{3}$,
and $\theta = -\frac{\pi}{3} \Rightarrow r = -\sqrt{3}$; points of intersection are
$(0,0)$, $\left(\sqrt{3}, \frac{\pi}{3}\right)$, and $\left(-\sqrt{3}, -\frac{\pi}{3}\right)$

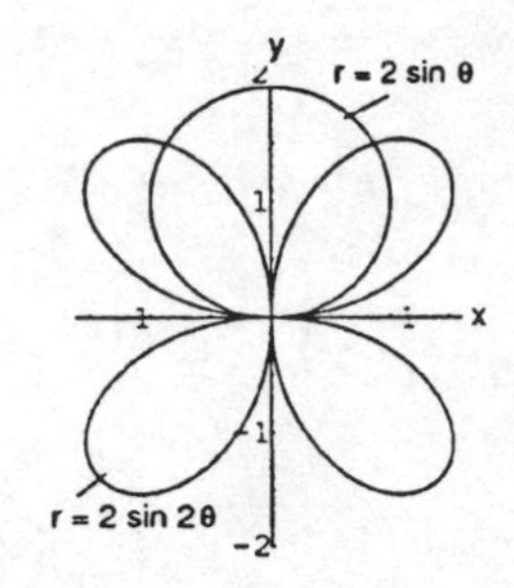

34. $\cos \theta = 1 - \cos \theta \Rightarrow 2 \cos \theta = 1 \Rightarrow \cos \theta = \frac{1}{2}$
$\Rightarrow \theta = \frac{\pi}{3}, -\frac{\pi}{3} \Rightarrow r = \frac{1}{2}$; points of intersection are
$\left(\frac{1}{2}, \frac{\pi}{3}\right)$ and $\left(\frac{1}{2}, -\frac{\pi}{3}\right)$. The point $(0,0)$ is found by
graphing.

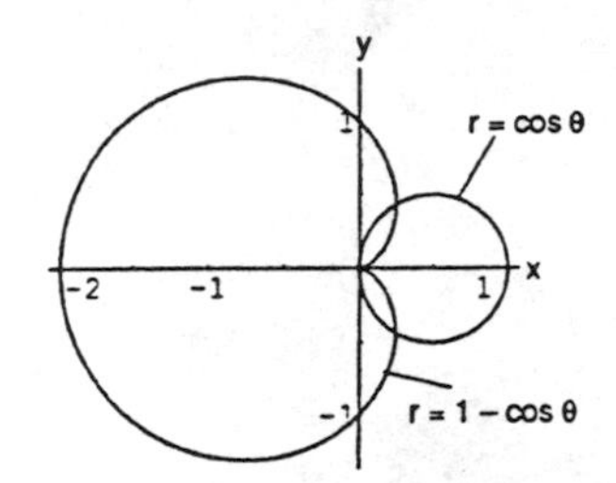

35. $\left(\sqrt{2}\right)^2 = 4 \sin \theta \Rightarrow \frac{1}{2} = \sin \theta \Rightarrow \theta = \frac{\pi}{6}, \frac{5\pi}{6}$; points
of intersection are $\left(\sqrt{2}, \frac{\pi}{6}\right)$ and $\left(\sqrt{2}, \frac{5\pi}{6}\right)$. The
points $\left(\sqrt{2}, -\frac{\pi}{6}\right)$ and $\left(\sqrt{2}, -\frac{5\pi}{6}\right)$ are found by
graphing.

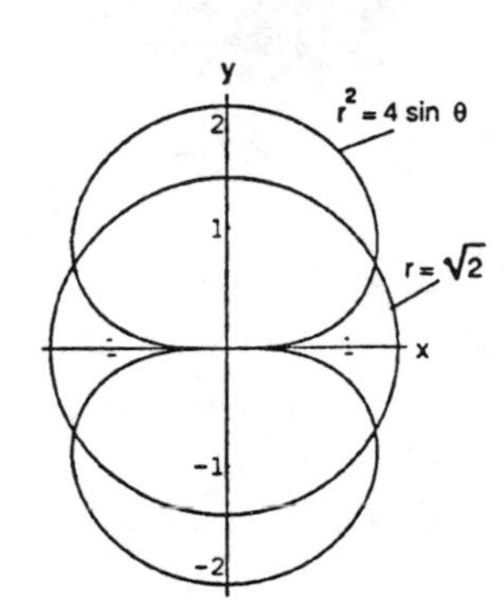

36. $\sqrt{2} \sin \theta = \sqrt{2} \cos \theta \Rightarrow \sin \theta = \cos \theta \Rightarrow \theta = \frac{\pi}{4}, \frac{5\pi}{4}$;
$\theta = \frac{\pi}{4} \Rightarrow r^2 = 1 \Rightarrow r = \pm 1$ and $\theta = \frac{5\pi}{4} \Rightarrow r^2 = -1$
$\Rightarrow$ no solution for r; points of intersection are $\left(\pm 1, \frac{\pi}{4}\right)$.
The points $(0,0)$ and $\left(\pm 1, \frac{3\pi}{4}\right)$ are found by graphing.

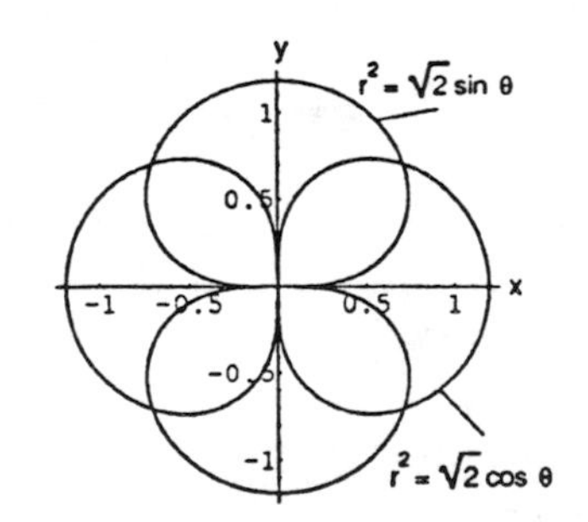

37. $1 = 2 \sin 2\theta \Rightarrow \sin 2\theta = \frac{1}{2} \Rightarrow 2\theta = \frac{\pi}{6}, \frac{5\pi}{6}, \frac{13\pi}{6}, \frac{17\pi}{6}$
$\Rightarrow \theta = \frac{\pi}{12}, \frac{5\pi}{12}, \frac{13\pi}{12}, \frac{17\pi}{12}$; points of intersection are
$\left(1, \frac{\pi}{12}\right)$, $\left(1, \frac{5\pi}{12}\right)$, $\left(1, \frac{13\pi}{12}\right)$, and $\left(1, \frac{17\pi}{12}\right)$. No other
points are found by graphing.

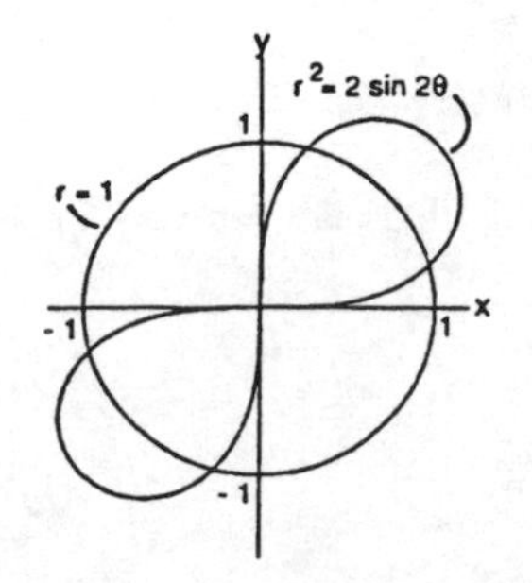

38. $\sqrt{2}\cos 2\theta = \sqrt{2}\sin 2\theta \Rightarrow \cos 2\theta = \sin 2\theta \Rightarrow 2\theta = \frac{\pi}{4}, \frac{5\pi}{4}, \frac{9\pi}{4}, \frac{13\pi}{4} \Rightarrow \theta = \frac{\pi}{8}, \frac{5\pi}{8}, \frac{9\pi}{8}, \frac{13\pi}{8}$; $\theta = \frac{\pi}{8}, \frac{9\pi}{8} \Rightarrow r^2 = 1$ $\Rightarrow r = \pm 1$; $\theta = \frac{5\pi}{8}, \frac{13\pi}{8} \Rightarrow r^2 = -1 \Rightarrow$ no solution for r; points of intersection are $\left(1, \frac{\pi}{8}\right)$ and $\left(1, \frac{9\pi}{8}\right)$. The point of intersection $(0,0)$ is found by graphing.

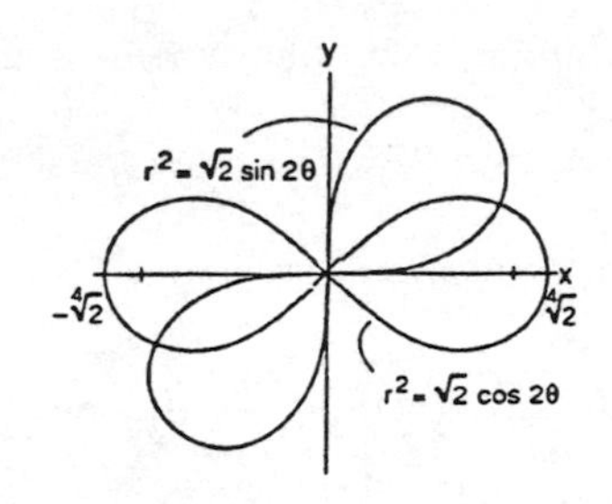

39. $r^2 = \sin 2\theta$ and $r^2 = \cos 2\theta$ are generated completely for $0 \le \theta \le \frac{\pi}{2}$. Then $\sin 2\theta = \cos 2\theta \Rightarrow 2\theta = \frac{\pi}{4}$ is the only solution on that interval $\Rightarrow \theta = \frac{\pi}{8} \Rightarrow r^2 = \sin 2\left(\frac{\pi}{8}\right) = \frac{1}{\sqrt{2}}$ $\Rightarrow r = \pm\frac{1}{\sqrt[4]{2}}$; points of intersection are $\left(\pm\frac{1}{\sqrt[4]{2}}, \frac{\pi}{8}\right)$. The point of intersection $(0,0)$ is found by graphing.

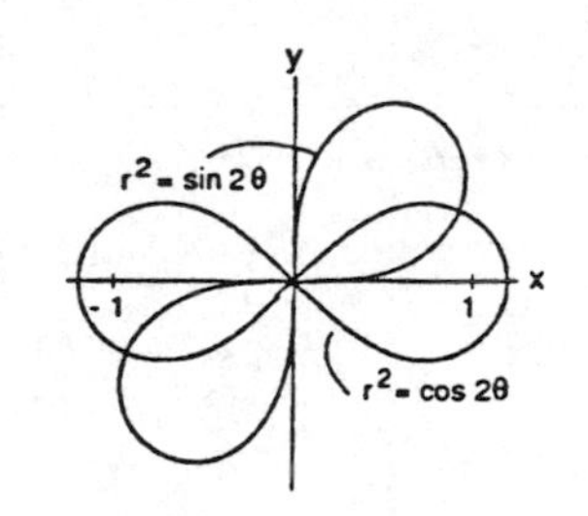

40. $1 - \sin\frac{\theta}{2} = 1 + \cos\frac{\theta}{2} \Rightarrow -\sin\frac{\theta}{2} = \cos\frac{\theta}{2} \Rightarrow \frac{\theta}{2} = \frac{3\pi}{4}, \frac{7\pi}{4}$ $\Rightarrow \theta = \frac{3\pi}{2}, \frac{7\pi}{2}$; $\theta = \frac{3\pi}{2} \Rightarrow r = 1 + \cos\frac{3\pi}{4} = 1 - \frac{\sqrt{2}}{2}$; $\theta = \frac{7\pi}{2} \Rightarrow r = 1 + \cos\frac{7\pi}{4} = 1 + \frac{\sqrt{2}}{2}$; points of intersection are $\left(1 - \frac{\sqrt{2}}{2}, \frac{3\pi}{2}\right)$ and $\left(1 + \frac{\sqrt{2}}{2}, \frac{7\pi}{2}\right)$. The three points of intersection $(0,0)$ and $\left(1 \pm \frac{\sqrt{2}}{2}, \frac{\pi}{2}\right)$ are found by graphing and symmetry.

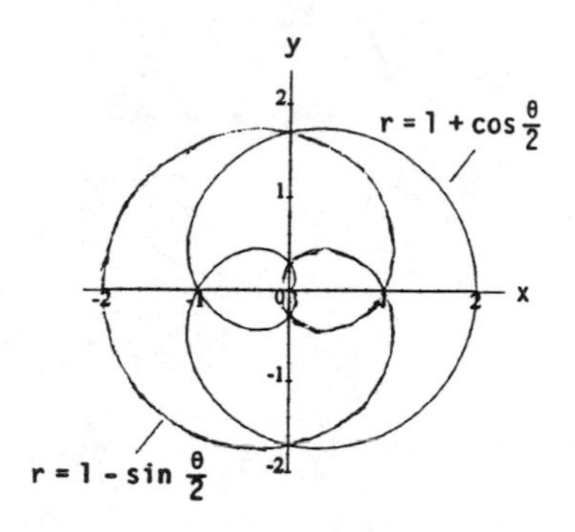

41. $1 = 2\sin 2\theta \Rightarrow \sin 2\theta = \frac{1}{2} \Rightarrow 2\theta = \frac{\pi}{6}, \frac{5\pi}{6}, \frac{13\pi}{6}, \frac{17\pi}{6}$ $\Rightarrow \theta = \frac{\pi}{12}, \frac{5\pi}{12}, \frac{13\pi}{12}, \frac{17\pi}{12}$; points of intersection are $\left(1, \frac{\pi}{12}\right), \left(1, \frac{5\pi}{12}\right), \left(1, \frac{13\pi}{12}\right)$, and $\left(1, \frac{17\pi}{12}\right)$. The points of intersection $\left(1, \frac{7\pi}{12}\right), \left(1, \frac{11\pi}{12}\right), \left(1, \frac{19\pi}{12}\right)$ and $\left(1, \frac{23\pi}{12}\right)$ are found by graphing and symmetry.

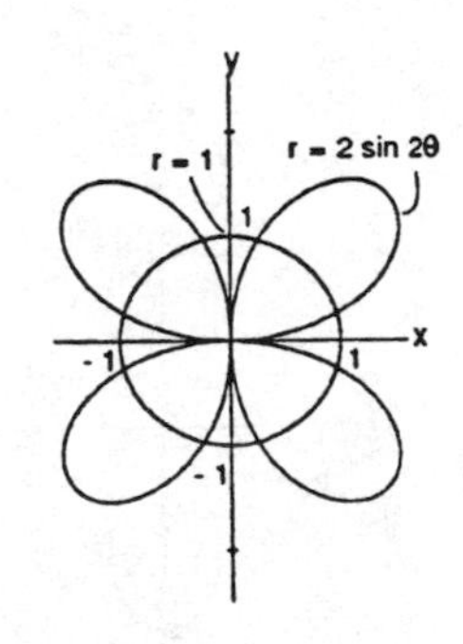

42. $r^2 = 2\sin 2\theta$ is completely generated on $0 \le \theta \le \frac{\pi}{2}$ so that $1 = 2\sin 2\theta \Rightarrow \sin 2\theta = \frac{1}{2} \Rightarrow 2\theta = \frac{\pi}{6}, \frac{5\pi}{6} \Rightarrow \theta = \frac{\pi}{12}, \frac{5\pi}{12}$; points of intersection are $\left(1, \frac{\pi}{12}\right)$ and $\left(1, \frac{5\pi}{12}\right)$. The points of intersection $\left(-1, \frac{\pi}{12}\right)$ and $\left(-1, \frac{5\pi}{12}\right)$ are found by graphing.

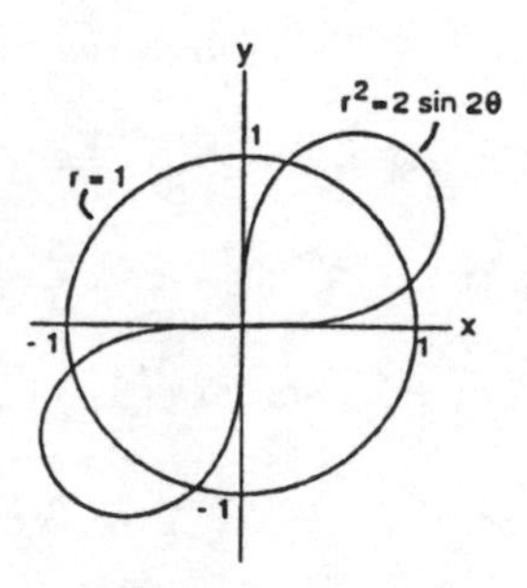

43. Note that (r, θ) and $(-r, \theta + \pi)$ describe the same point in the plane. Then $r = 1 - \cos\theta \Leftrightarrow -1 - \cos(\theta + \pi)$ $= -1 - (\cos\theta \cos\pi - \sin\theta \sin\pi) = -1 + \cos\theta = -(1 - \cos\theta) = -r$; therefore (r, θ) is on the graph of $r = 1 - \cos\theta \Leftrightarrow (-r, \theta + \pi)$ is on the graph of $r = -1 - \cos\theta \Rightarrow$ the answer is (a).

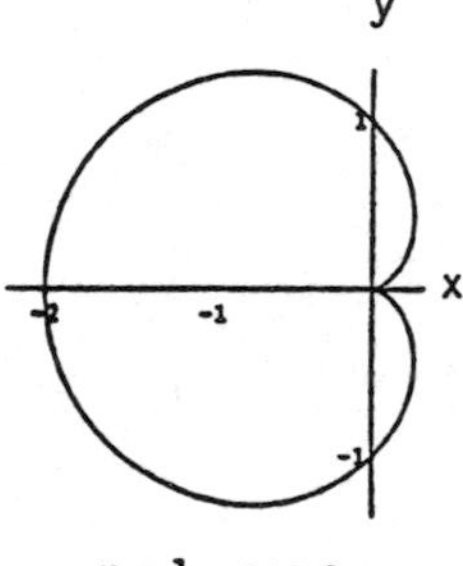

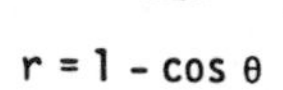
r = 1 - cos θ

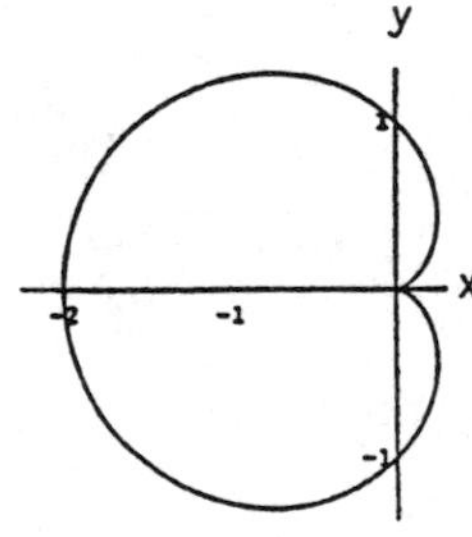
r = -1 - cos θ

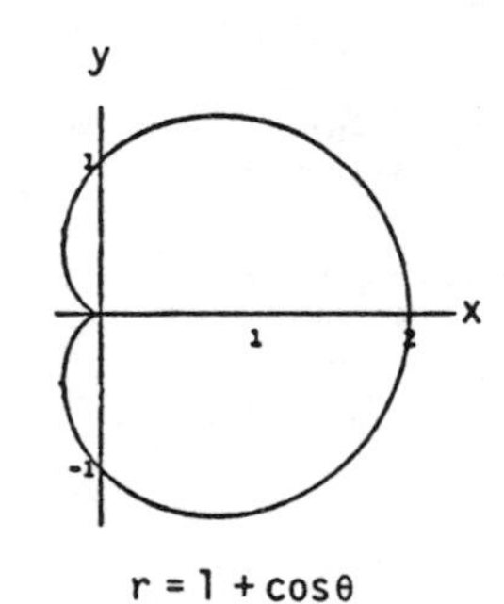
r = 1 + cos θ

44. Note that (r, θ) and $(-r, \theta + \pi)$ describe the same point in the plane. Then $r = \cos 2\theta \Leftrightarrow -\sin\left(2(\theta + \pi)) + \frac{\pi}{2}\right)$ $= -\sin\left(2\theta + \frac{5\pi}{2}\right) = -\sin(2\theta)\cos\left(\frac{5\pi}{2}\right) - \cos(2\theta)\sin\left(\frac{5\pi}{2}\right) = -\cos 2\theta = -r$; therefore (r, θ) is on the graph of $r = -\sin\left(2\theta + \frac{\pi}{2}\right) \Rightarrow$ the answer is (a).

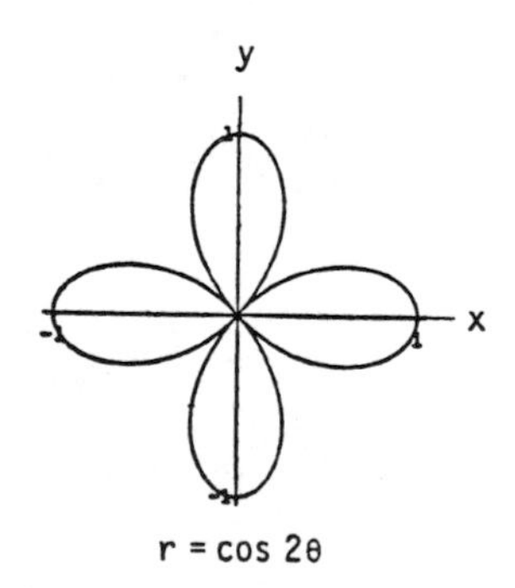
r = cos 2θ

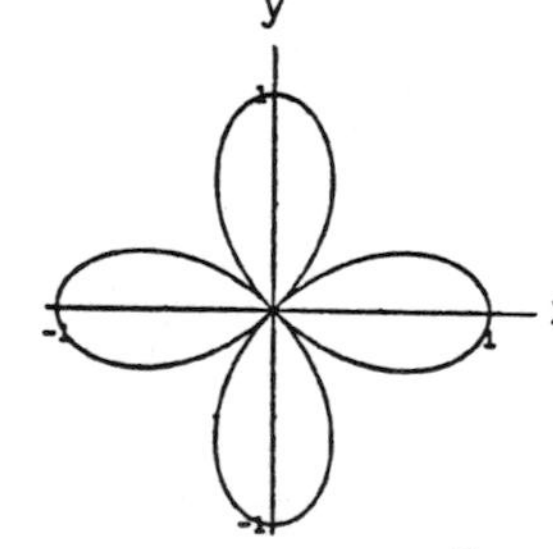
r = - sin (2θ + $\frac{\pi}{2}$)

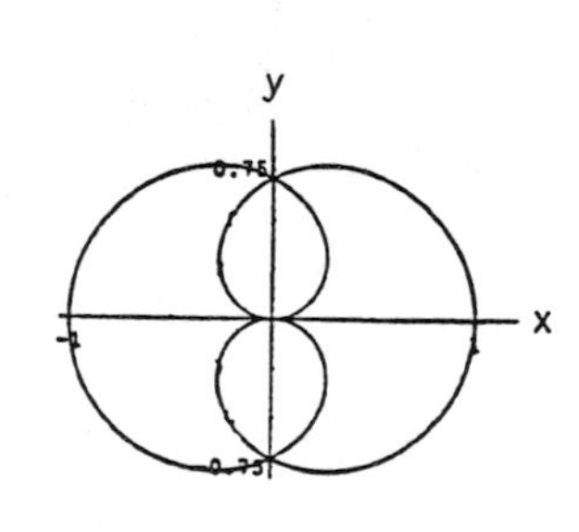
r = - cos $\frac{\theta}{2}$

45.

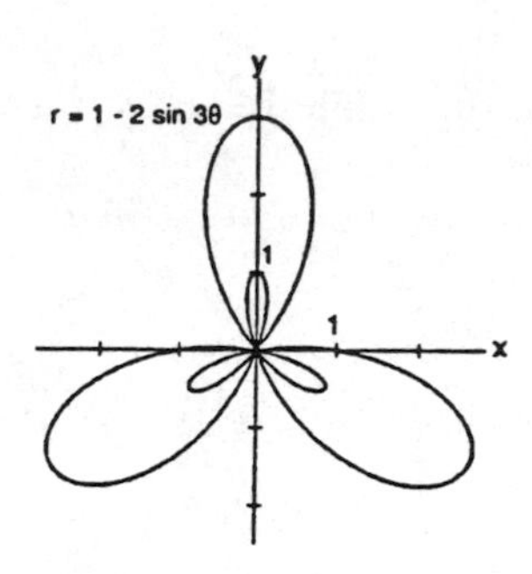

46.

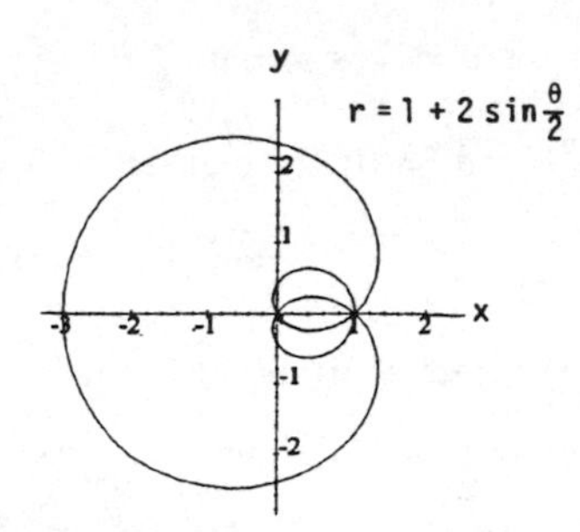

47. (a)

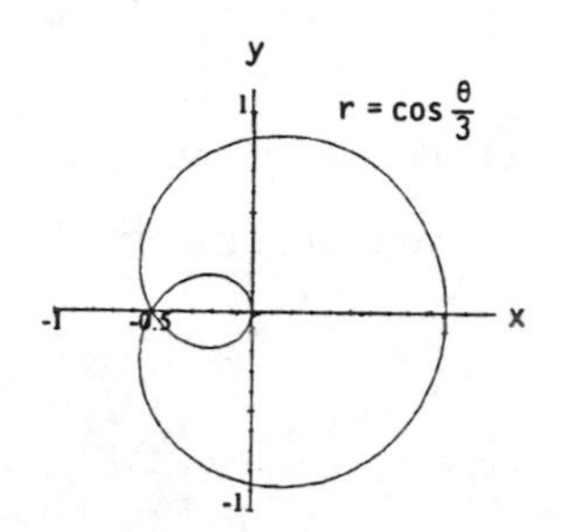

(b)

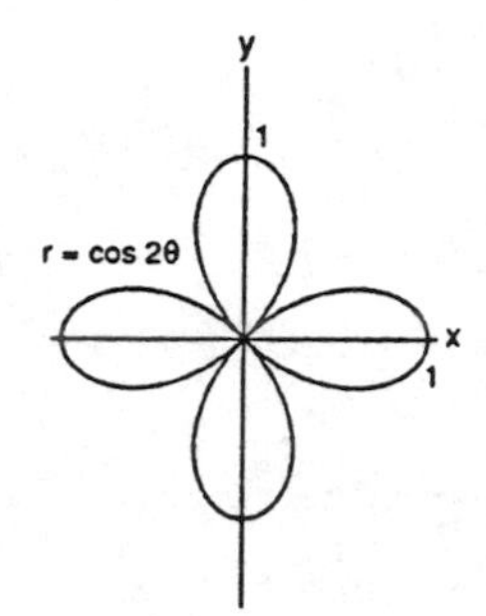

(c)

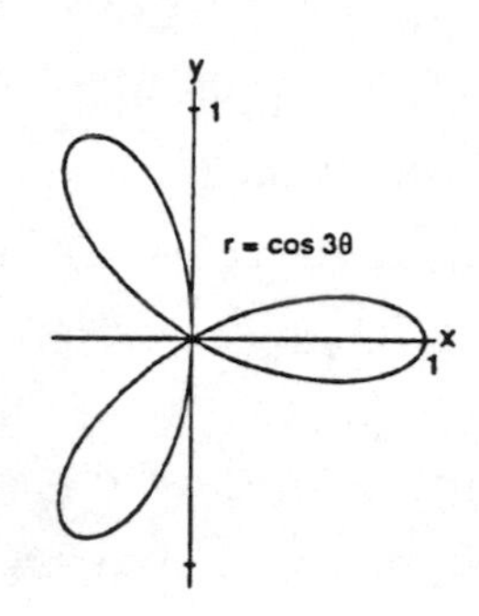

(d)

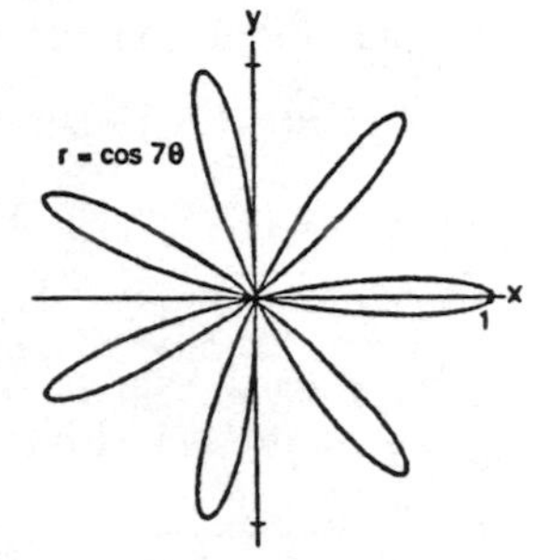

48. (a)

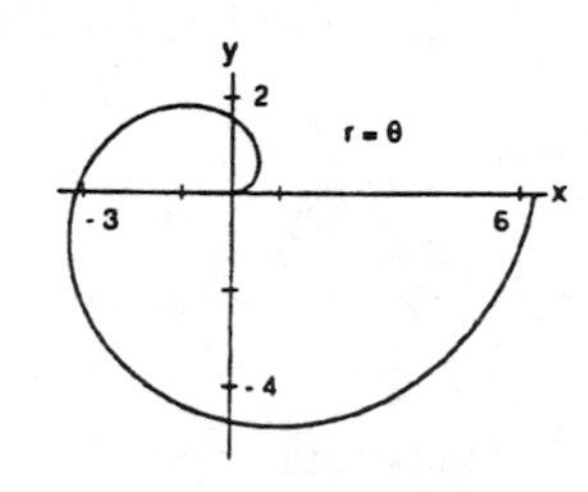

(b)

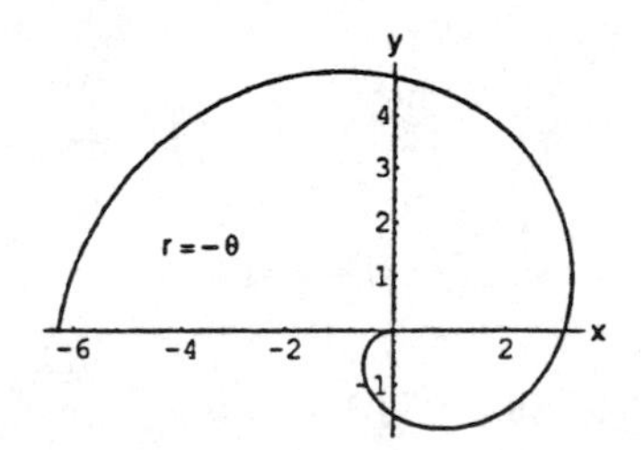

(c)

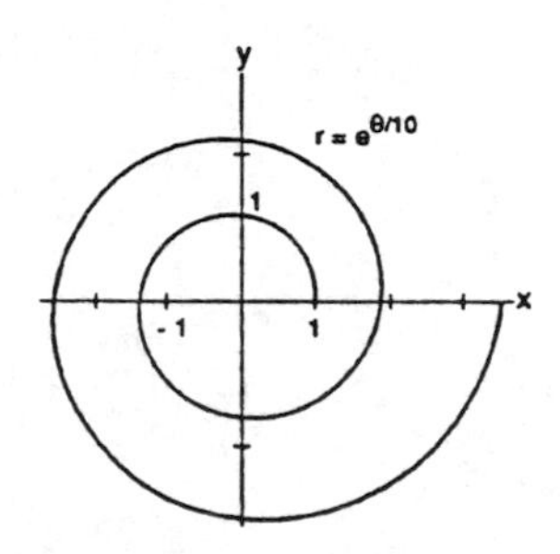

(d)

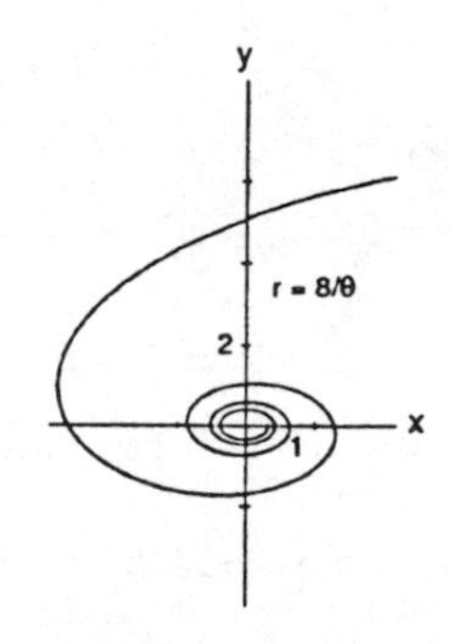

(e)

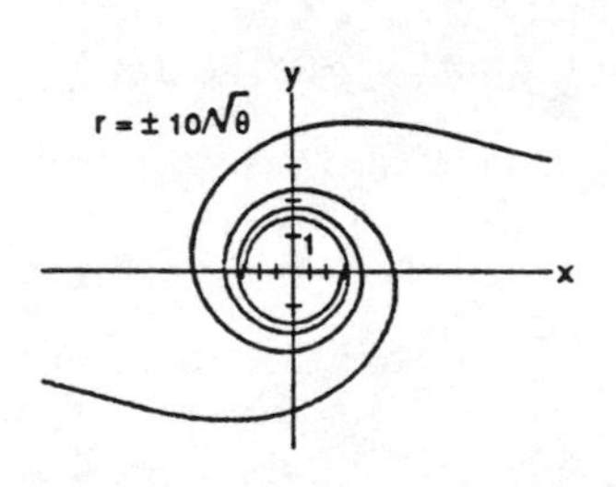

49. (a) $r^2 = -4\cos\theta \Rightarrow \cos\theta = -\frac{r^2}{4}$; $r = 1 - \cos\theta \Rightarrow r = 1 - \left(-\frac{r^2}{4}\right) \Rightarrow 0 = r^2 - 4r + 4 \Rightarrow (r-2)^2 = 0$

$\Rightarrow r = 2$; therefore $\cos\theta = -\frac{2^2}{4} = -1 \Rightarrow \theta = \pi \Rightarrow (2, \pi)$ is a point of intersection

(b) $r = 0 \Rightarrow 0^2 = 4 \cos \theta \Rightarrow \cos \theta = 0 \Rightarrow \theta = \frac{\pi}{2}, \frac{3\pi}{2} \Rightarrow \left(0, \frac{\pi}{2}\right)$ or $\left(0, \frac{3\pi}{2}\right)$ is on the graph; $r = 0 \Rightarrow 0 = 1 - \cos \theta$ $\Rightarrow \cos \theta = 1 \Rightarrow \theta = 0 \Rightarrow (0,0)$ is on the graph. Since $(0,0) = \left(0, \frac{\pi}{2}\right)$ for polar coordinates, the graphs intersect at the origin.

50. (a) Let $r = f(\theta)$ be symmetric about the x-axis and the y-axis. Then (r, θ) on the graph $\Rightarrow (r, -\theta)$ is on the graph because of symmetry about the x-axis. Then $(-r, -(-\theta)) = (-r, \theta)$ is on the graph because of symmetry about the y-axis. Therefore $r = f(\theta)$ is symmetric about the origin.

(b) Let $r = f(\theta)$ be symmetric about the x-axis and the origin. Then (r, θ) on the graph $\Rightarrow (r, -\theta)$ is on the graph because of symmetry about the x-axis. Then $(-r, -\theta)$ is on the graph because of symmetry about the origin. Therefore $r = f(\theta)$ is symmetric about the y-axis.

(c) Let $r = f(\theta)$ be symmetric about the y-axis and the origin. Then (r, θ) on the graph $\Rightarrow (-r, -\theta)$ is on the graph because of symmetry about the y-axis. Then $(-(-r), -\theta) = (r, -\theta)$ is on the graph because of symmetry about the origin. Therefore $r = f(\theta)$ is symmetric about the x-axis.

51. The maximum width of the petal of the rose which lies along the x-axis is twice the largest y value of the curve on the interval $0 \le \theta \le \frac{\pi}{4}$. So we wish to maximize $2y = 2r \sin \theta = 2 \cos 2\theta \sin \theta$ on $0 \le \theta \le \frac{\pi}{4}$. Let $f(\theta) = 2 \cos 2\theta \sin \theta = 2\left(1 - 2 \sin^2 \theta\right)(\sin \theta) = 2 \sin \theta - 4 \sin^3 \theta \Rightarrow f'(\theta) = 2 \cos \theta - 12 \sin^2 \theta \cos \theta$. Then $f'(\theta) = 0 \Rightarrow 2 \cos \theta - 12 \sin^2 \theta \cos \theta = 0 \Rightarrow (\cos \theta)\left(1 - 6 \sin^2 \theta\right) = 0 \Rightarrow \cos \theta = 0$ or $1 - 6 \sin^2 \theta = 0 \Rightarrow \theta = \frac{\pi}{2}$ or $\sin \theta = \frac{\pm 1}{\sqrt{6}}$. Since we want $0 \le \theta \le \frac{\pi}{4}$, we choose $\theta = \sin^{-1}\left(\frac{1}{\sqrt{6}}\right) \Rightarrow f(\theta) = 2 \sin \theta - 2 \sin^3 \theta$ $= 2\left(\frac{1}{\sqrt{6}}\right) - 4 \cdot \frac{1}{6\sqrt{6}} = \frac{2\sqrt{6}}{9}$. We can see from the graph of $r = \cos 2\theta$ that a maximum does occur in the interval $0 \le \theta \le \frac{\pi}{4}$. Therefore the maximum width occurs at $\theta = \sin^{-1}\left(\frac{1}{\sqrt{6}}\right)$, and the maximum width is $\frac{2\sqrt{6}}{9}$.

52. We wish to maximize $y = r \sin \theta = 2(1 + \cos \theta)(\sin \theta) = 2 \sin \theta + 2 \sin \theta \cos \theta$. Then $\frac{dy}{d\theta} = 2 \cos \theta + 2(\sin \theta)(-\sin \theta) + 2 \cos \theta \cos \theta = 2 \cos \theta - 2 \sin^2 \theta + 2 \cos^2 \theta = 2 \cos \theta + 4 \cos^2 \theta - 2$; thus $\frac{dy}{d\theta} = 0 \Rightarrow 4 \cos^2 \theta + 2 \cos \theta - 2 = 0 \Rightarrow 2 \cos^2 \theta + \cos \theta - 1 = 0 \Rightarrow (2 \cos \theta - 1)(\cos \theta + 1) = 0 \Rightarrow \cos \theta = \frac{1}{2}$ or $\cos \theta = -1 \Rightarrow \theta = \frac{\pi}{3}, \frac{5\pi}{3}, \pi$. From the graph, we can see that the maximum occurs in the first quadrant so we choose $\theta = \frac{\pi}{3}$. Then $y = 2 \sin \frac{\pi}{3} + 2 \sin \frac{\pi}{3} \cos \frac{\pi}{3} = \frac{3\sqrt{3}}{2}$. The x-coordinate of this point is $x = r \cos \frac{\pi}{3}$ $= 2\left(1 + \cos \frac{\pi}{3}\right)\left(\cos \frac{\pi}{3}\right) = \frac{3}{2}$. Thus the maximum height is $h = \frac{3\sqrt{3}}{2}$ occurring at $x = \frac{3}{2}$.

9.8 POLAR EQUATIONS OF CONIC SECTIONS

1. $r\cos\left(\theta - \frac{\pi}{6}\right) = 5 \Rightarrow r\left(\cos\theta\cos\frac{\pi}{6} + \sin\theta\sin\frac{\pi}{6}\right) = 5 \Rightarrow \frac{\sqrt{3}}{2}r\cos\theta + \frac{1}{2}r\sin\theta = 5 \Rightarrow \frac{\sqrt{3}}{2}x + \frac{1}{2}y = 5 \Rightarrow \sqrt{3}x + y$
$= 10 \Rightarrow y = -\sqrt{3}x + 10$

2. $r\cos\left(\theta - \frac{3\pi}{4}\right) = 2 \Rightarrow r\left(\cos\theta\cos\frac{3\pi}{4} + \sin\theta\sin\frac{3\pi}{4}\right) = 2 \Rightarrow -\frac{\sqrt{2}}{2}r\cos\theta + \frac{\sqrt{2}}{2}r\sin\theta = 2$
$\Rightarrow -\frac{\sqrt{2}}{2}x + \frac{\sqrt{2}}{2}y = 2 \Rightarrow -\sqrt{2}x + \sqrt{2}y = 4 \Rightarrow y = x + 2\sqrt{2}$

3. $r\cos\left(\theta - \frac{4\pi}{3}\right) = 3 \Rightarrow r\left(\cos\theta\cos\frac{4\pi}{3} + \sin\theta\sin\frac{4\pi}{3}\right) = 3 \Rightarrow -\frac{1}{2}r\cos\theta - \frac{\sqrt{3}}{2}r\sin\theta = 3$
$\Rightarrow -\frac{1}{2}x - \frac{\sqrt{3}}{2}y = 3 \Rightarrow x + \sqrt{3}y = -6 \Rightarrow y = -\frac{\sqrt{3}}{3}x - 2\sqrt{3}$

4. $r\cos\left(\theta - \left(-\frac{\pi}{4}\right)\right) = 4 \Rightarrow r\cos\left(\theta + \frac{\pi}{4}\right) = 4 \Rightarrow r\left(\cos\theta\cos\frac{\pi}{4} - \sin\theta\sin\frac{\theta}{4}\right) = 4$
$\Rightarrow \frac{\sqrt{2}}{2}r\cos\theta - \frac{\sqrt{2}}{2}r\sin\theta = 4 \Rightarrow \frac{\sqrt{2}}{2}x - \frac{\sqrt{2}}{2}y = r \Rightarrow \sqrt{2}x - \sqrt{2}y = 8 \Rightarrow y = x - 4\sqrt{2}$

5. $r\cos\left(\theta - \frac{\pi}{4}\right) = \sqrt{2} \Rightarrow r\left(\cos\theta\cos\frac{\pi}{4} + \sin\theta\sin\frac{\pi}{4}\right)$
$= \sqrt{2} \Rightarrow \frac{1}{\sqrt{2}}r\cos\theta + \frac{1}{\sqrt{2}}r\sin\theta = \sqrt{2} \Rightarrow \frac{1}{\sqrt{2}}x + \frac{1}{\sqrt{2}}y$
$= \sqrt{2} \Rightarrow x + y = 2 \Rightarrow y = 2 - x$

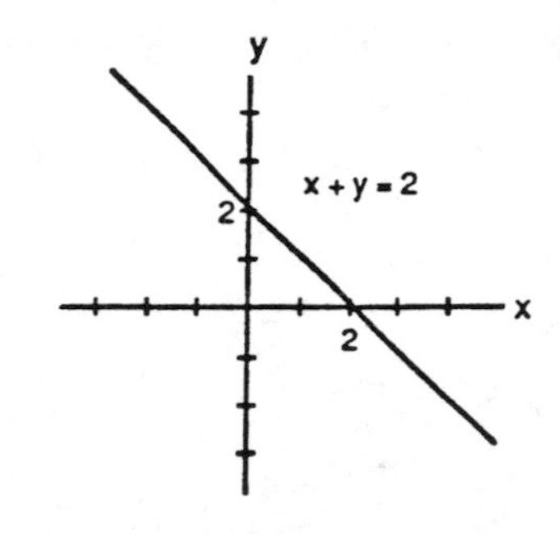

6. $r\cos\left(\theta + \frac{3\pi}{4}\right) = 1 \Rightarrow r\left(\cos\theta\cos\frac{3\pi}{4} - \sin\theta\sin\frac{3\pi}{4}\right) = 1$
$\Rightarrow -\frac{\sqrt{2}}{2}r\cos\theta - \frac{\sqrt{2}}{2}r\sin\theta = 1 \Rightarrow x + y = -\sqrt{2}$
$\Rightarrow y = -x - \sqrt{2}$

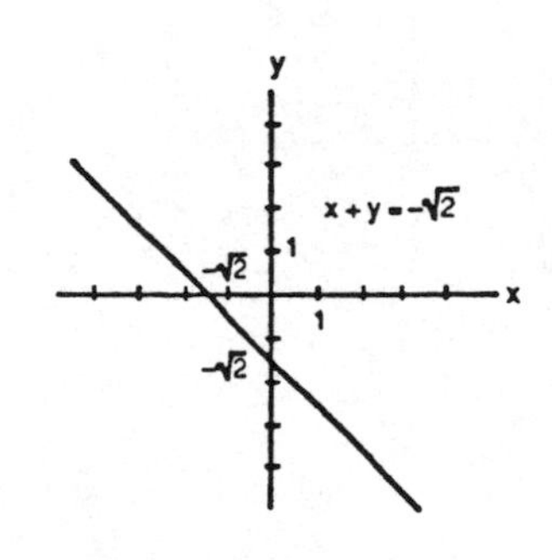

7. $r\cos\left(\theta - \frac{2\pi}{3}\right) = 3 \Rightarrow r\left(\cos\theta\cos\frac{2\pi}{3} + \sin\theta\sin\frac{2\pi}{3}\right) = 3$
$\Rightarrow -\frac{1}{2}r\cos\theta + \frac{\sqrt{3}}{2}r\sin\theta = 3 \Rightarrow -\frac{1}{2}x + \frac{\sqrt{3}}{2}y = 3$
$\Rightarrow -x + \sqrt{3}y = 6 \Rightarrow y = \frac{\sqrt{3}}{3}x + 2\sqrt{3}$

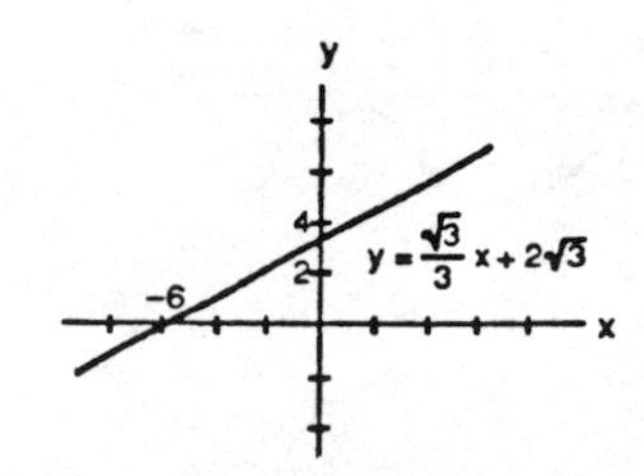

8. $r\cos\left(\theta+\frac{\pi}{3}\right)=2 \Rightarrow r\left(\cos\theta\cos\frac{\pi}{3}-\sin\theta\sin\frac{\pi}{3}\right)=2$

$\Rightarrow \frac{1}{2}r\cos\theta-\frac{\sqrt{3}}{2}r\sin\theta=2 \Rightarrow \frac{1}{2}x-\frac{\sqrt{3}}{2}y=2$

$\Rightarrow x-\sqrt{3}\,y=4 \Rightarrow y=\frac{\sqrt{3}}{3}x-\frac{4\sqrt{3}}{3}$

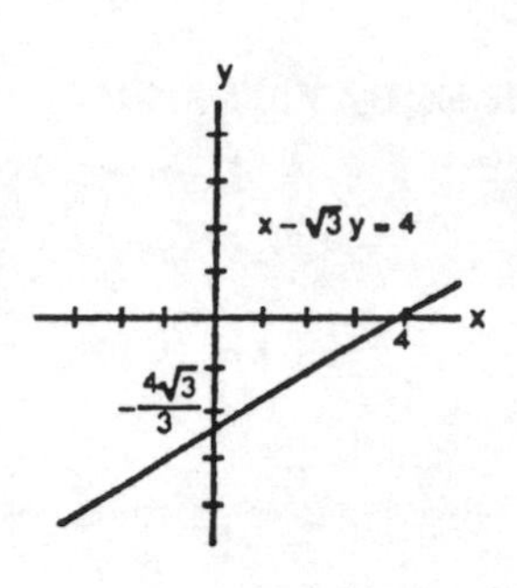

9. $\sqrt{2}x+\sqrt{2}y=6 \Rightarrow \sqrt{2}r\cos\theta+\sqrt{2}r\sin\theta=6 \Rightarrow r\left(\frac{\sqrt{2}}{2}\cos\theta+\frac{\sqrt{2}}{2}\sin\theta\right)=3 \Rightarrow r\left(\cos\frac{\pi}{4}\cos\theta+\sin\frac{\pi}{4}\sin\theta\right)$

$=3 \Rightarrow r\cos\left(\theta-\frac{\pi}{4}\right)=3$

10. $\sqrt{3}x-y=1 \Rightarrow \sqrt{3}r\cos\theta-r\sin\theta=1 \Rightarrow r\left(\frac{\sqrt{3}}{2}\cos\theta-\frac{1}{2}\sin\theta\right)=\frac{1}{2} \Rightarrow r\left(\cos\frac{\pi}{6}\cos\theta-\sin\frac{\pi}{6}\sin\theta\right)$

$=\frac{1}{2} \Rightarrow r\cos\left(\theta+\frac{\pi}{6}\right)=\frac{1}{2}$

11. $y=-5 \Rightarrow r\sin\theta=-5 \Rightarrow -r\sin\theta=5 \Rightarrow r\sin(-\theta)=5 \Rightarrow r\cos\left(\frac{\pi}{2}-(-\theta)\right)=5 \Rightarrow r\cos\left(\theta+\frac{\pi}{2}\right)=5$

12. $x=-4 \Rightarrow r\cos\theta=-4 \Rightarrow -r\cos\theta=4 \Rightarrow r\cos(\theta-\pi)=4$

13. $r=2(4)\cos\theta=8\cos\theta$

14. $r=-2(1)\sin\theta=-2\sin\theta$

15. $r=2\sqrt{2}\sin\theta$

16. $r=-2\left(\frac{1}{2}\right)\cos\theta=-\cos\theta$

17.

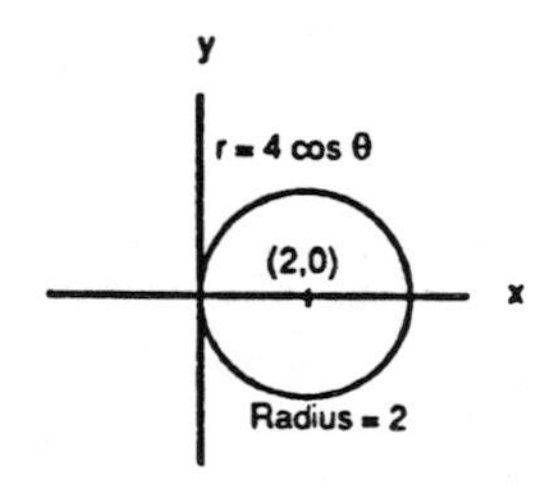

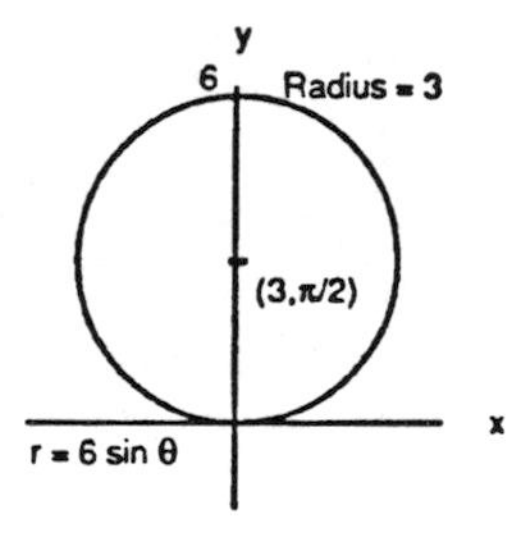

19.

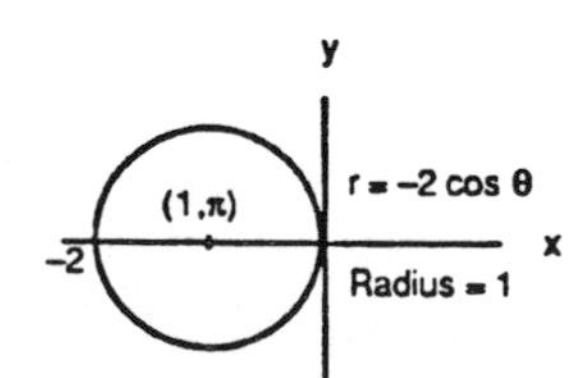

20.

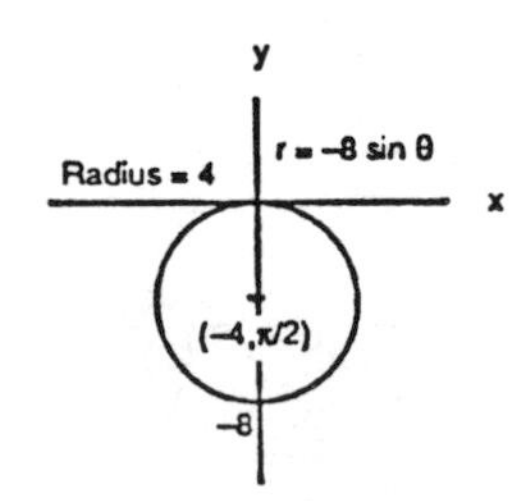

21. $(x-6)^2 + y^2 = 36 \Rightarrow C = (6,0)$, $a = 6$
$\Rightarrow r = 12 \cos\theta$ is the polar equation

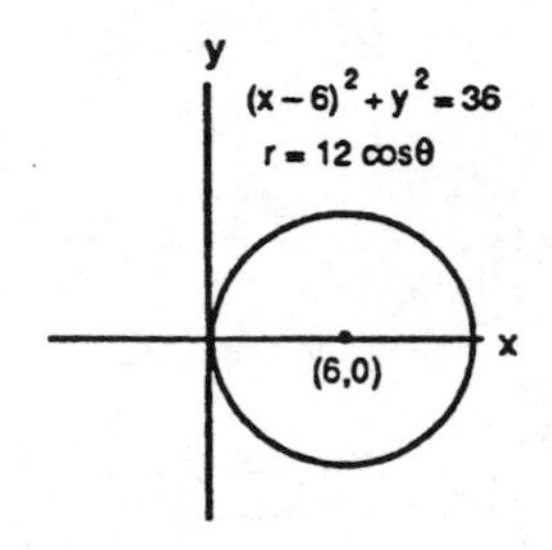

22. $(x+2)^2 + y^2 = 4 \Rightarrow C = (-2,0)$, $a = 2$
$\Rightarrow r = -4 \cos\theta$ is the polar equation

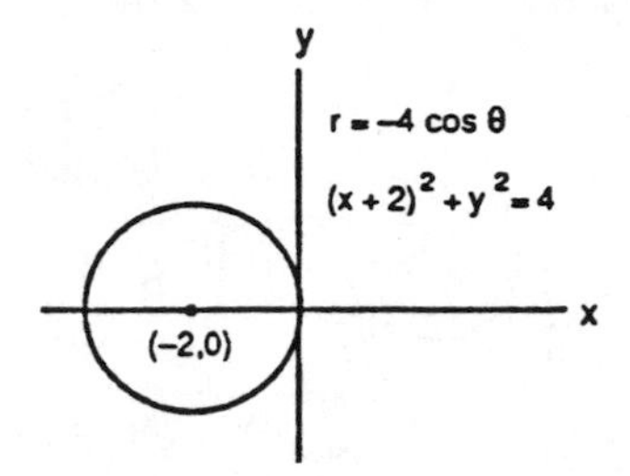

23. $x^2 + (y-5)^2 = 25 \Rightarrow C = (0,5)$, $a = 5$
$\Rightarrow r = 10 \sin\theta$ is the polar equation

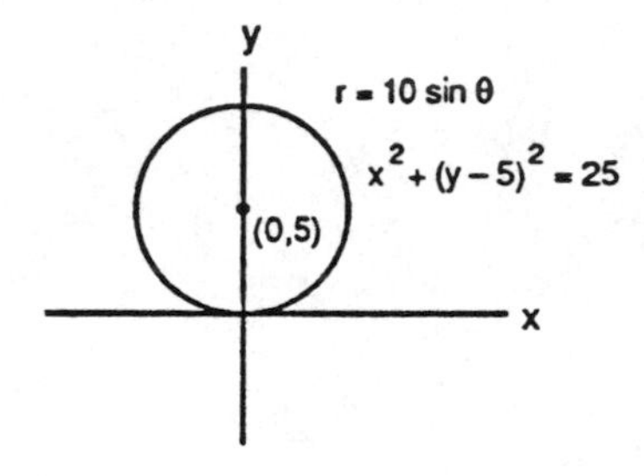

24. $x^2 + (y+7)^2 = 49 \Rightarrow C = (0,-7)$, $a = 7$
$\Rightarrow r = -14 \sin\theta$ is the polar equation

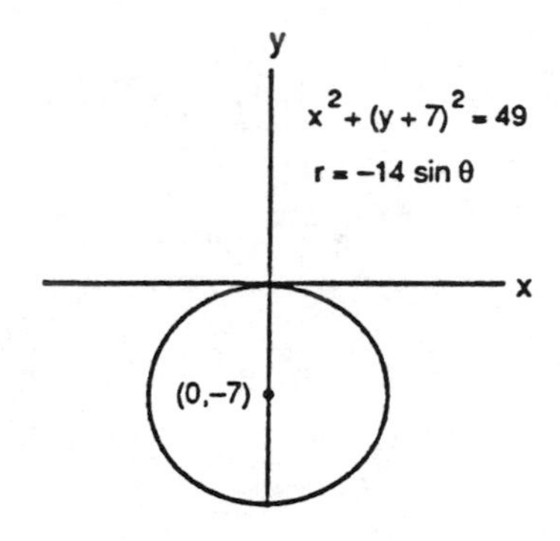

25. $x^2 + 2x + y^2 = 0 \Rightarrow (x+1)^2 + y^2 = 1$
$\Rightarrow C = (-1,0)$, $a = 1 \Rightarrow r = -2 \cos\theta$ is
the polar equation

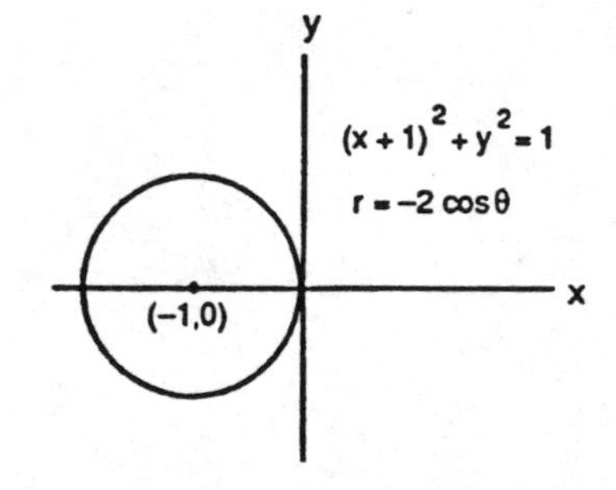

26. $x^2 - 16x + y^2 = 0 \Rightarrow (x-8)^2 + y^2 = 64$
$\Rightarrow C = (8,0)$, $a = 8 \Rightarrow r = 16 \cos\theta$ is the
polar equation

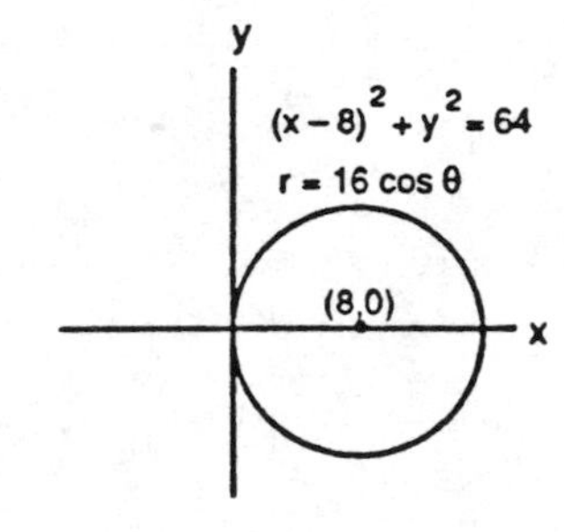

27. $x^2+y^2+y=0 \Rightarrow x^2+\left(y+\frac{1}{2}\right)^2=\frac{1}{4}$
$\Rightarrow C=\left(0,-\frac{1}{2}\right)$, $a=\frac{1}{2} \Rightarrow r=-\sin\theta$ is the polar equation

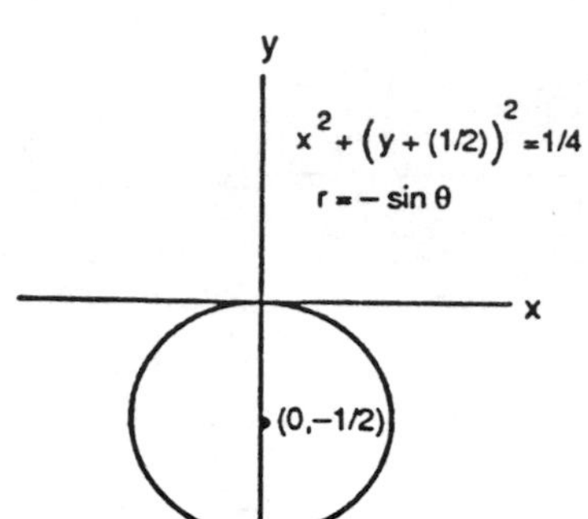

28. $x^2+y^2-\frac{4}{3}y=0 \Rightarrow x^2+\left(y-\frac{2}{3}\right)^2=\frac{4}{9}$
$\Rightarrow C=\left(0,\frac{2}{3}\right)$, $a=\frac{2}{3} \Rightarrow r=\frac{4}{3}\sin\theta$ is the polar equation

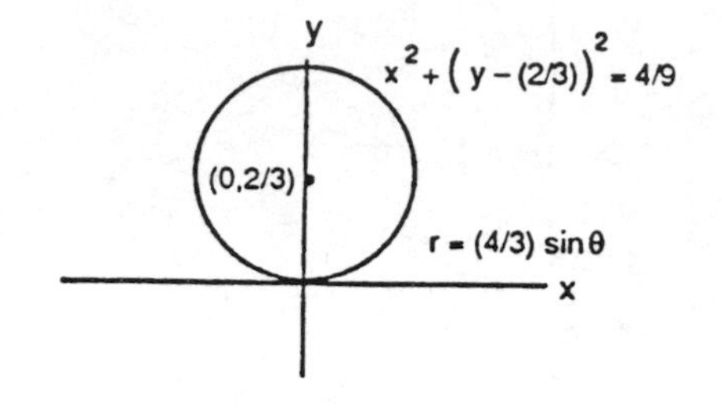

29. $e=1,\ x=2 \Rightarrow k=2 \Rightarrow r=\frac{2(1)}{1+(1)\cos\theta}=\frac{2}{1+\cos\theta}$

30. $e=1,\ y=2 \Rightarrow k=2 \Rightarrow r=\frac{2(1)}{1+(1)\sin\theta}=\frac{2}{1+\sin\theta}$

31. $e=5,\ y=-6 \Rightarrow k=6 \Rightarrow r=\frac{6(5)}{1-5\sin\theta}=\frac{30}{1-5\sin\theta}$

32. $e=2,\ x=4 \Rightarrow k=4 \Rightarrow r=\frac{4(2)}{1+2\cos\theta}=\frac{8}{1+2\cos\theta}$

33. $e=\frac{1}{2},\ x=1 \Rightarrow k=1 \Rightarrow r=\frac{\left(\frac{1}{2}\right)(1)}{1+\left(\frac{1}{2}\right)\cos\theta}=\frac{1}{2+\cos\theta}$

34. $e=\frac{1}{4},\ x=-2 \Rightarrow k=2 \Rightarrow r=\frac{\left(\frac{1}{4}\right)(2)}{1-\left(\frac{1}{4}\right)\cos\theta}=\frac{2}{4-\cos\theta}$

35. $e=\frac{1}{5},\ x=-10 \Rightarrow k=10 \Rightarrow r=\frac{\left(\frac{1}{5}\right)(10)}{1-\left(\frac{1}{5}\right)\sin\theta}=\frac{10}{5-\sin\theta}$

36. $e=\frac{1}{3},\ y=6 \Rightarrow k=6 \Rightarrow r=\frac{\left(\frac{1}{3}\right)(6)}{1+\left(\frac{1}{3}\right)\sin\theta}=\frac{6}{3+\sin\theta}$

37. $r=\frac{1}{1+\cos\theta} \Rightarrow e=1,\ k=1 \Rightarrow x=1$

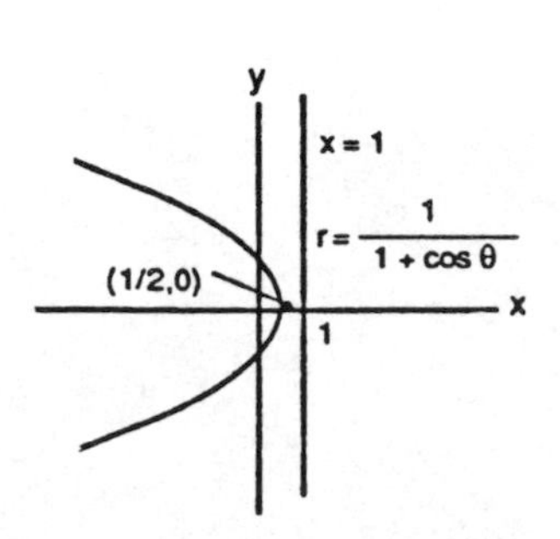

38. $r = \frac{6}{2 + \cos\theta} = \frac{3}{1 + \left(\frac{1}{2}\right)\cos\theta} \Rightarrow e = \frac{1}{2}$, $k = 6 \Rightarrow x = 6$;

$a(1 - e^2) = ke \Rightarrow a\left[1 - \left(\frac{1}{2}\right)^2\right] = 3 \Rightarrow \frac{3}{4}a = 3 \Rightarrow a = 4$

$\Rightarrow ea = 2$

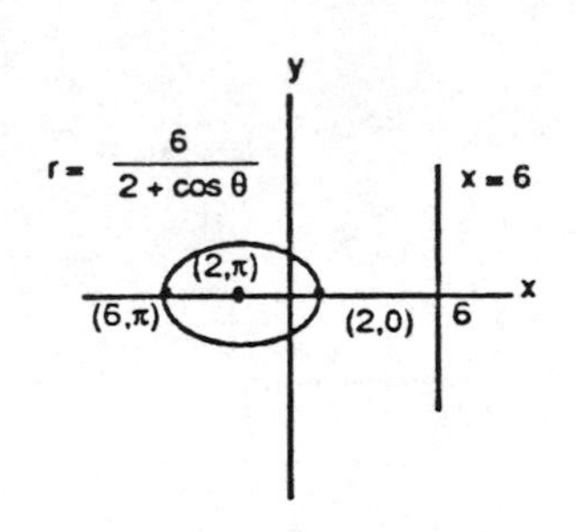

39. $r = \frac{25}{10 - 5\cos\theta} \Rightarrow r = \frac{\left(\frac{25}{10}\right)}{1 - \left(\frac{5}{10}\right)\cos\theta} = \frac{\left(\frac{5}{2}\right)}{1 - \left(\frac{1}{2}\right)\cos\theta}$

$\Rightarrow e = \frac{1}{2}$, $k = 5 \Rightarrow x = -5$; $a(1 - e^2) = ke \Rightarrow a\left[1 - \left(\frac{1}{2}\right)^2\right]$

$= \frac{5}{2} \Rightarrow \frac{3}{4}a = \frac{5}{2} \Rightarrow a = \frac{10}{3} \Rightarrow ea = \frac{5}{3}$

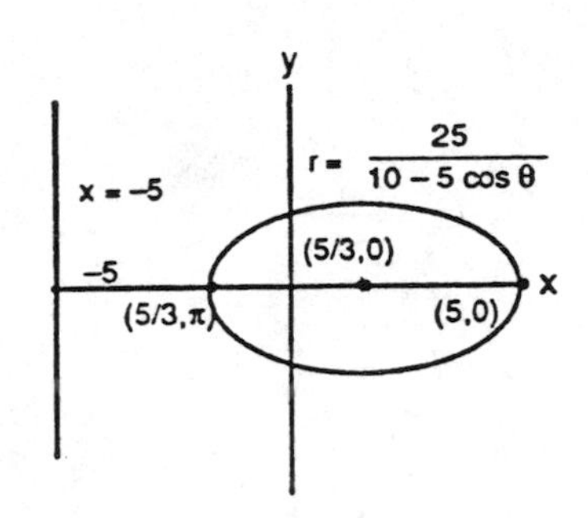

40. $r = \frac{4}{2 - 2\cos\theta} \Rightarrow r = \frac{2}{1 - \cos\theta} \Rightarrow e = 1$, $k = 2 \Rightarrow x = -2$

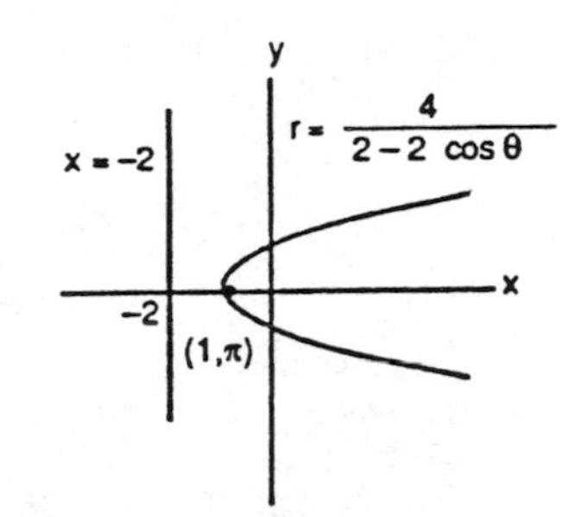

41. $r = \frac{400}{16 + 8\sin\theta} \Rightarrow r = \frac{\left(\frac{400}{16}\right)}{1 + \left(\frac{8}{16}\right)\sin\theta} \Rightarrow r = \frac{25}{1 + \left(\frac{1}{2}\right)\sin\theta}$

$e = \frac{1}{2}$, $k = 50 \Rightarrow y = 50$; $a(1 - e^2) = ke \Rightarrow a\left[1 - \left(\frac{1}{2}\right)^2\right]$

$= 25 \Rightarrow \frac{3}{4}a = 25 \Rightarrow a = \frac{100}{3} \Rightarrow ea = \frac{50}{3}$

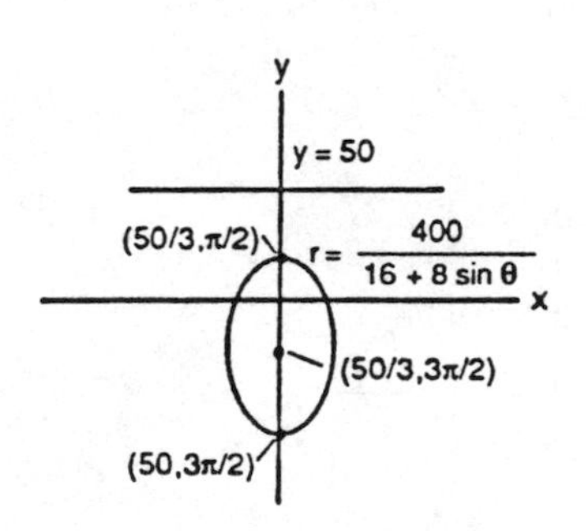

42. $r = \frac{12}{3 + 3\sin\theta} \Rightarrow r = \frac{4}{1 + \sin\theta} \Rightarrow e = 1,$

$k = 4 \Rightarrow y = 4$

43. $r = \frac{8}{2 - 2\sin\theta} \Rightarrow r = \frac{4}{1 - \sin\theta} \Rightarrow e = 1,$

$k = 4 \Rightarrow y = -4$

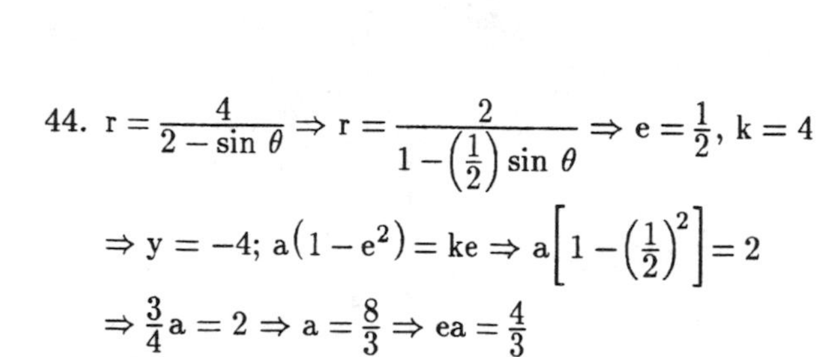

44. $r = \frac{4}{2 - \sin\theta} \Rightarrow r = \frac{2}{1 - \left(\frac{1}{2}\right)\sin\theta} \Rightarrow e = \frac{1}{2},\ k = 4$

$\Rightarrow y = -4;\ a\left(1 - e^2\right) = ke \Rightarrow a\left[1 - \left(\frac{1}{2}\right)^2\right] = 2$

$\Rightarrow \frac{3}{4}a = 2 \Rightarrow a = \frac{8}{3} \Rightarrow ea = \frac{4}{3}$

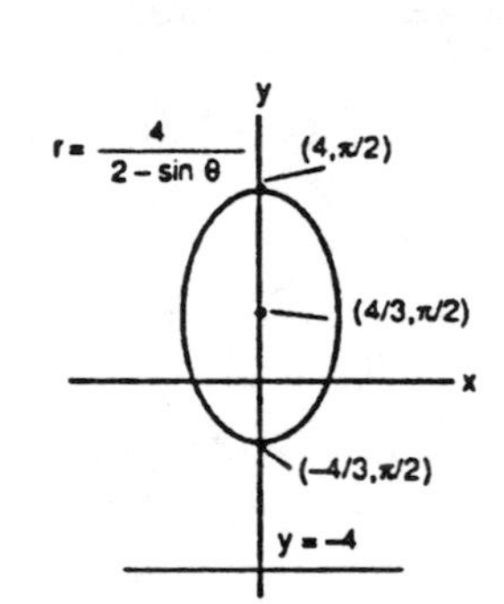

45.

46.

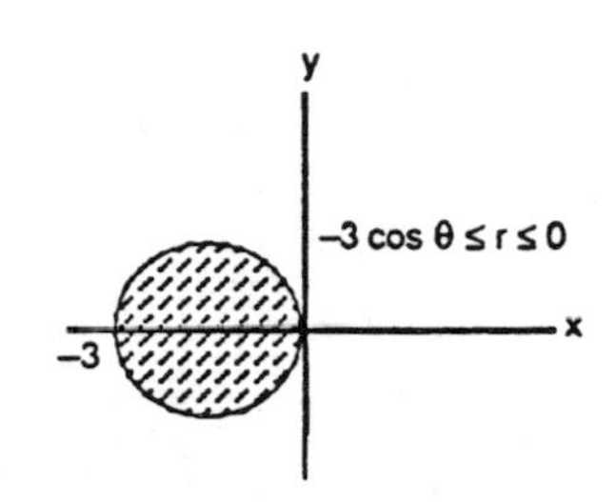

47.

48.

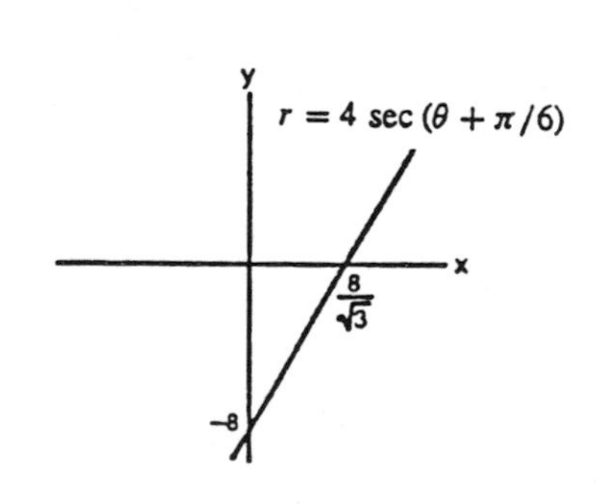

49.

50.

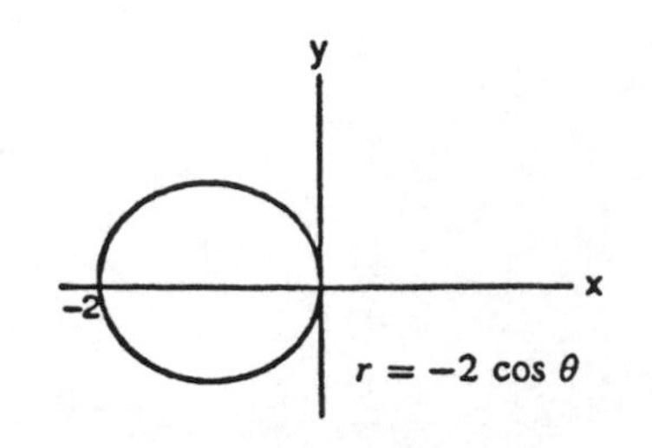

51.

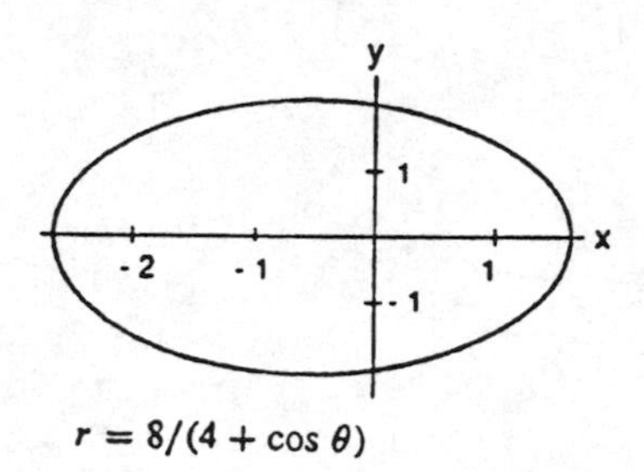

$r = 8/(4 + \cos\theta)$

52.

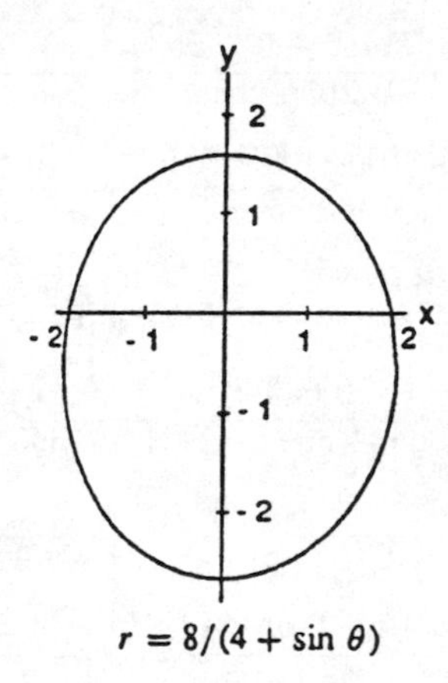

$r = 8/(4 + \sin\theta)$

53.

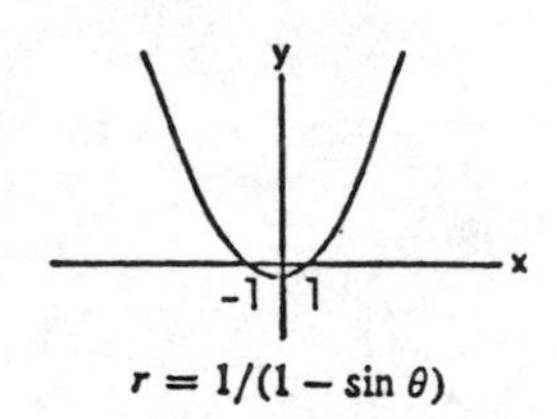

$r = 1/(1 - \sin\theta)$

54.

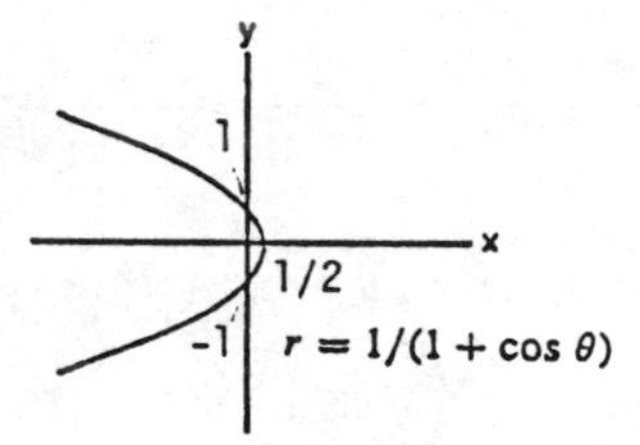

$r = 1/(1 + \cos\theta)$

55.

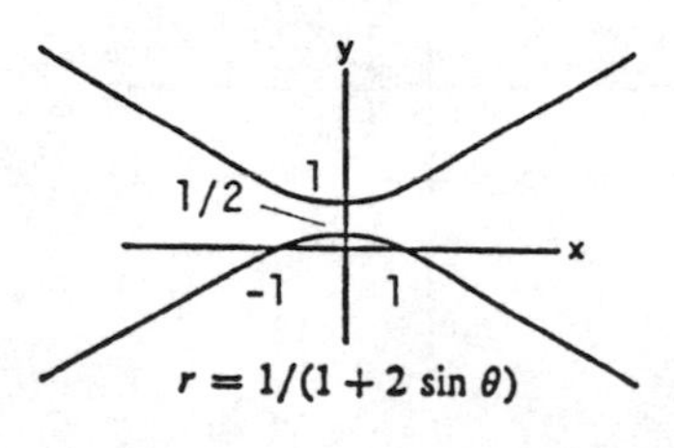

$r = 1/(1 + 2\sin\theta)$

56.

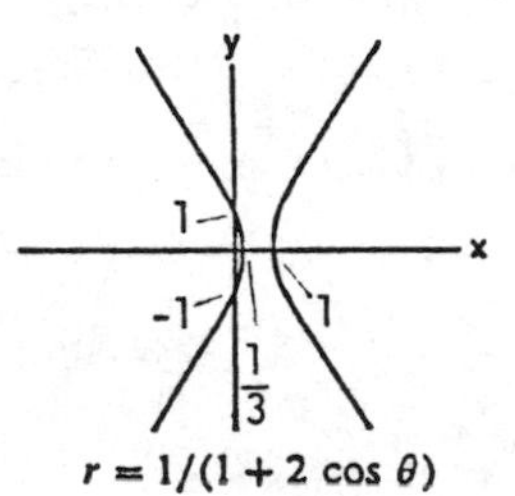

$r = 1/(1 + 2\cos\theta)$

57. (a) Perihelion $= a - ae = a(1 - e)$, Aphelion $= ea + a = a(1 + e)$

(b)

Planet	Perihelion	Aphelion
Mercury	0.3075 AU	0.4667 AU
Venus	0.7184 AU	0.7282 AU
Earth	0.9833 AU	1.0167 AU
Mars	1.3817 AU	1.6663 AU
Jupiter	4.9512 AU	5.4548 AU
Satrun	9.0210 AU	10.0570 AU
Uranus	18.2977 AU	20.0623 AU
Neptune	29.8135 AU	30.3065 AU
Pluto	29.6549 AU	49.2251 AU

58. Mercury: $r = \frac{(0.3871)(1-0.2056^2)}{1+0.2056 \cos\theta} = \frac{0.3707}{1+0.2056 \cos\theta}$

Venus: $r = \frac{(0.7233)(1-0.0068^2)}{1+0.0068 \cos\theta} = \frac{0.7233}{1+0.0068 \cos\theta}$

Earth: $r = \frac{1(1-0.0167^2)}{1+0.0167 \cos\theta} = \frac{0.9997}{1+0.0617 \cos\theta}$

Mars: $r = \frac{(1.524)(1-0.0934^2)}{1+0.0934 \cos\theta} = \frac{1.511}{1+0.0934 \cos\theta}$

Jupiter: $r = \frac{(5.203)(1-0.0484^2)}{1+0.0484 \cos\theta} = \frac{5.191}{1+0.0484 \cos\theta}$

Saturn: $r = \frac{(9.539)(1-0.0543^2)}{1+0.0543 \cos\theta} = \frac{9.511}{1+0.0543 \cos\theta}$

Uranus: $r = \frac{(19.18)(1-0.0460^2)}{1+0.0460 \cos\theta} = \frac{19.14}{1+0.0460 \cos\theta}$

Neptune: $r = \frac{(30.06)(1-0.0082^2)}{1+0.0082 \cos\theta} = \frac{30.06}{1+0.0082 \cos\theta}$

59. (a) $r = 4 \sin\theta \Rightarrow r^2 = 4r \sin\theta \Rightarrow x^2 + y^2 = 4y$;

$r = \sqrt{3} \sec\theta \Rightarrow r = \frac{\sqrt{3}}{\cos\theta} \Rightarrow r \cos\theta = \sqrt{3}$

$\Rightarrow x = \sqrt{3}$; $x = \sqrt{3} \Rightarrow (\sqrt{3})^2 + y^2 = 4y$

$\Rightarrow y^2 - 4y + 3 = 0 \Rightarrow (y-3)(y-1) = 0 \Rightarrow y = 3$

or $y = 1$. Therefore in Cartesian coordinates, the points of intersection are $(\sqrt{3}, 3)$ and $(\sqrt{3}, 1)$. In polar coordinates, $4 \sin\theta = \sqrt{3} \sec\theta \Rightarrow 4 \sin\theta \cos\theta = \sqrt{3}$

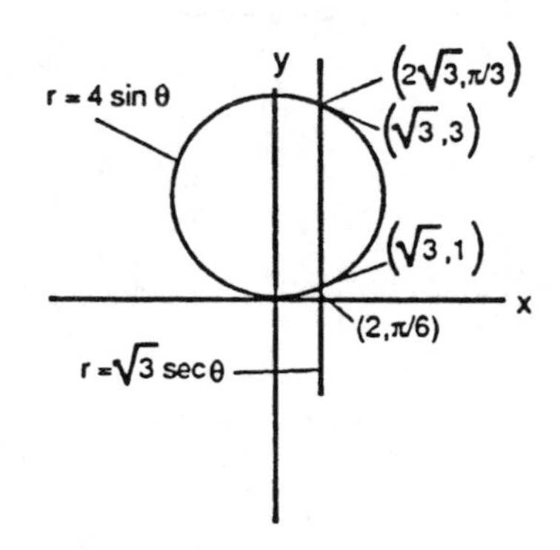

$\Rightarrow 2 \sin\theta \cos\theta = \frac{\sqrt{3}}{2} \Rightarrow \sin 2\theta = \frac{\sqrt{3}}{2} \Rightarrow 2\theta = \frac{\pi}{3}$ or

$\frac{2\pi}{3} \Rightarrow \theta = \frac{\pi}{6}$ or $\frac{\pi}{3}$; $\theta = \frac{\pi}{6} \Rightarrow r = 2$, and $\theta = \frac{\pi}{3} \Rightarrow r = 2\sqrt{3} \Rightarrow \left(2, \frac{\pi}{6}\right)$ and $\left(2\sqrt{3}, \frac{\pi}{3}\right)$ are the points of intersection in polar coordinates.

60. (a) $r = 8 \cos\theta \Rightarrow r^2 = 8r \cos\theta \Rightarrow x^2 + y^2 = 8x$

$\Rightarrow x^2 - 8x + y^2 = 0 \Rightarrow (x-4)^2 + y^2 = 16$;

$r = 2 \sec\theta \Rightarrow r = \frac{2}{\cos\theta} \Rightarrow r \cos\theta = 2$

$\Rightarrow x = 2$; $x = 2 \Rightarrow 2^2 - 8(2) + y^2 = 0$

$\Rightarrow y^2 = 12 \Rightarrow y = \pm 2\sqrt{3}$. Therefore $(2, \pm 2\sqrt{3})$ are the points of intersection in Cartesian coordinates.

In polar coordinates, $8 \cos\theta = 2 \sec\theta \Rightarrow 8 \cos^2\theta = 2$

$\Rightarrow \cos^2\theta = \frac{1}{4} \Rightarrow \cos\theta = \pm\frac{1}{2} \Rightarrow \theta = \frac{\pi}{3}, \frac{2\pi}{3}, \frac{4\pi}{3}$, or

$\frac{5\pi}{3}$; $\theta = \frac{\pi}{3}$ and $\frac{5\pi}{3} \Rightarrow r = 4$, and $\theta = \frac{2\pi}{3}$ and $\frac{4\pi}{3} \Rightarrow r = -4 \Rightarrow \left(4, \frac{\pi}{3}\right)$ and $\left(4, \frac{5\pi}{3}\right)$ are the points of intersection in polar coordinates. The points $\left(-4, \frac{2\pi}{3}\right)$ and $\left(-4, \frac{4\pi}{3}\right)$ are the same points.

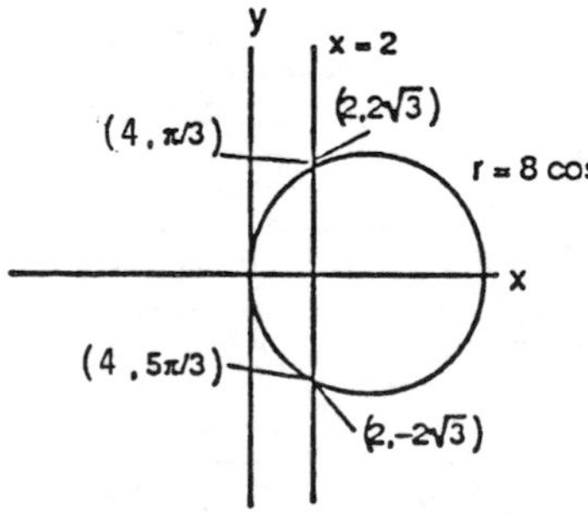

61. $r \cos\theta = 4 \Rightarrow x = 4 \Rightarrow k = 4$: parabola $\Rightarrow e = 1 \Rightarrow r = \frac{4}{1+\cos\theta}$

62. $r \cos\left(\theta - \frac{\pi}{2}\right) = 2 \Rightarrow r\left(\cos\theta \cos\frac{\pi}{2} + \sin\theta \sin\frac{\pi}{2}\right) = 2 \Rightarrow r \sin\theta = 2 \Rightarrow y = 2 \Rightarrow k = 2$: parabola $\Rightarrow e = 1$

$\Rightarrow r = \frac{2}{1+\sin\theta}$

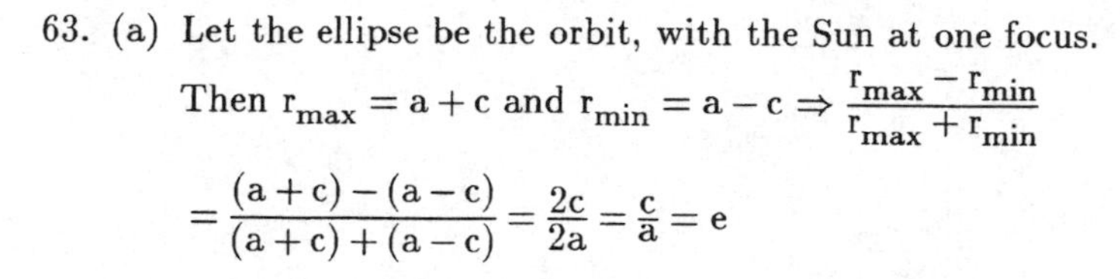

63. (a) Let the ellipse be the orbit, with the Sun at one focus.

Then $r_{max} = a + c$ and $r_{min} = a - c \Rightarrow \frac{r_{max} - r_{min}}{r_{max} + r_{min}}$

$= \frac{(a+c)-(a-c)}{(a+c)+(a-c)} = \frac{2c}{2a} = \frac{c}{a} = e$

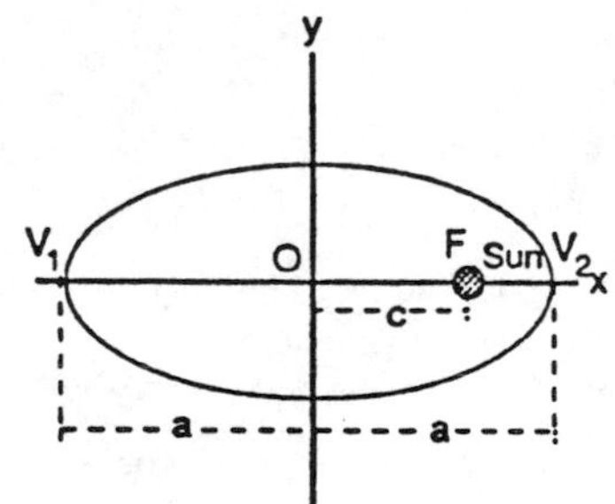

(b) Let F_1, F_2 be the foci. Then $PF_1 + PF_2 = 10$ where P is any point on the ellipse. If P is a vertex, then $PF_1 = a + c$ and $PF_2 = a - c \Rightarrow (a+c) + (a-c) = 10$ $\Rightarrow 2a = 10 \Rightarrow a = 5$. Since $e = \frac{c}{a}$ we have $0.2 = \frac{c}{5} \Rightarrow c = 1.0 \Rightarrow$ the pins should be 2 inches apart.

64. $e = 0.97$, Major axis $= 36.18$ AU $\Rightarrow a = 18.09$, Minor axis $= 9.12$ AU $\Rightarrow b = 4.56$ ($1 \text{ AU} \approx 1.49 \times 10^8$ km)

(a) $r = \frac{ke}{1+e\cos\theta} = \frac{a(1-e^2)}{1+e\cos\theta} = \frac{(18.09)[1-(0.97)^2]}{1+0.97\cos\theta} = \frac{1.07}{1+0.97\cos\theta}$ AU

(b) $\theta = 0 \Rightarrow r = \frac{1.07}{1+0.97} \approx 0.5427$ AU $\approx 8.08 \times 10^7$ km

(c) $\theta = \pi \Rightarrow r = \frac{1.07}{1-0.97} \approx 35.7$ AU $\approx 5.33 \times 10^9$ km

65. $x^2 + y^2 - 2ay = 0 \Rightarrow (r\cos\theta)^2 + (r\sin\theta)^2 - 2ar\sin\theta = 0$

$\Rightarrow r^2\cos^2\theta + r^2\sin^2\theta - 2ar\sin\theta = 0 \Rightarrow r^2 = 2ar\sin\theta$

$\Rightarrow r = 2a\sin\theta$

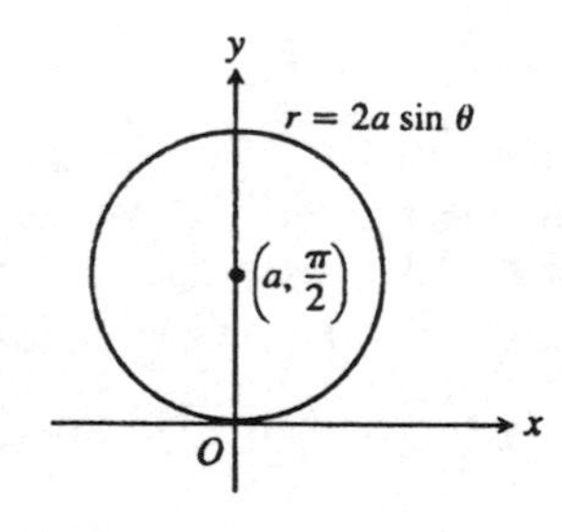

66. $y^2 = 4ax + 4a^2 \Rightarrow (r\sin\theta)^2 = 4ar\cos\theta + 4a^2 \Rightarrow r^2\sin^2\theta$

$= 4ar\cos\theta + 4a^2 \Rightarrow r^2(1-\cos^2\theta) = 4ar\cos\theta + 4a^2$

$\Rightarrow r^2 - r^2\cos^2\theta = 4ar\cos\theta + 4a^2 \Rightarrow r^2$

$= r^2\cos^2\theta + 4ar\cos\theta + 4a^2 \Rightarrow r^2 = (r\cos\theta + 2a)^2$

$\Rightarrow r = \pm(r\cos\theta + 2a) \Rightarrow r - r\cos\theta = 2a$ or

$r + r\cos\theta = -2a \Rightarrow r = \frac{2a}{1-\cos\theta}$ or $r = \frac{-2a}{1+\cos\theta}$;

the equations have the same graph, which is a parabola opening to the right

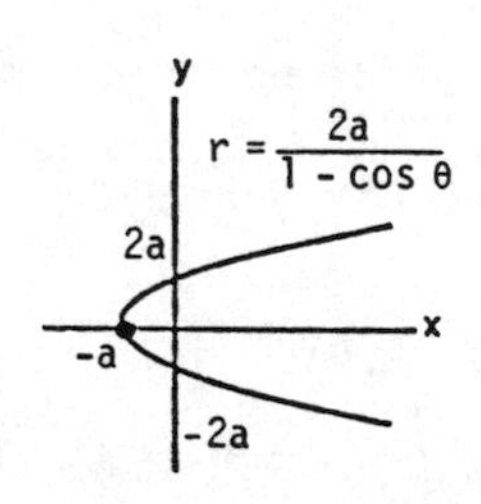

67. $x \cos \alpha + y \sin \alpha = p \Rightarrow r \cos \theta \cos \alpha + r \sin \theta \sin \alpha = p$

$\Rightarrow r(\cos \theta \cos \alpha + \sin \theta \sin \alpha) = p \Rightarrow r \cos(\theta - \alpha) = p$

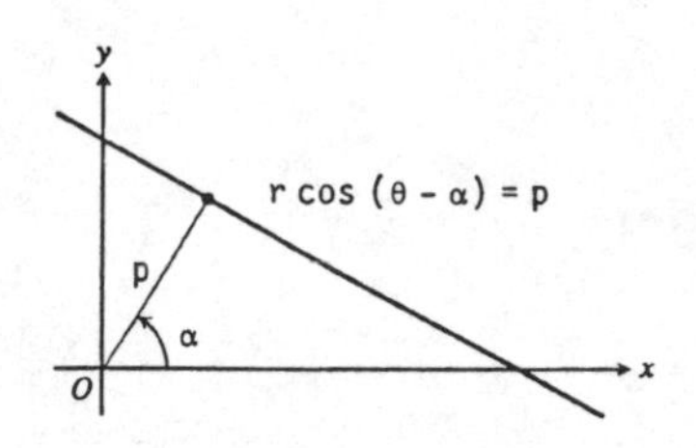

68. $(x^2+y^2)^2 + 2ax(x^2+y^2) - a^2y^2 = 0$

$\Rightarrow (r^2)^2 + 2a(r \cos \theta)(r^2) - a^2(r \sin \theta)^2 = 0$

$\Rightarrow r^4 + 2ar^3 \cos \theta - a^2r^2 \sin^2 \theta = 0$

$\Rightarrow r^2[r^2 + 2ar \cos \theta - a^2(1 - \cos^2 \theta)] = 0$ (assume $r \neq 0$)

$\Rightarrow r^2 + 2ar \cos \theta - a^2 + a^2 \cos^2 \theta = 0$

$\Rightarrow (r^2 + 2ar \cos \theta + a^2 \cos^2 \theta) - a^2 = 0$

$\Rightarrow (r + a \cos \theta)^2 = a^2 \Rightarrow r + a \cos \theta = \pm a$

$\Rightarrow r = a(1 - \cos \theta)$ or $r = -a(1 + \cos \theta)$;

the equations have the same graph, which is a cardioid

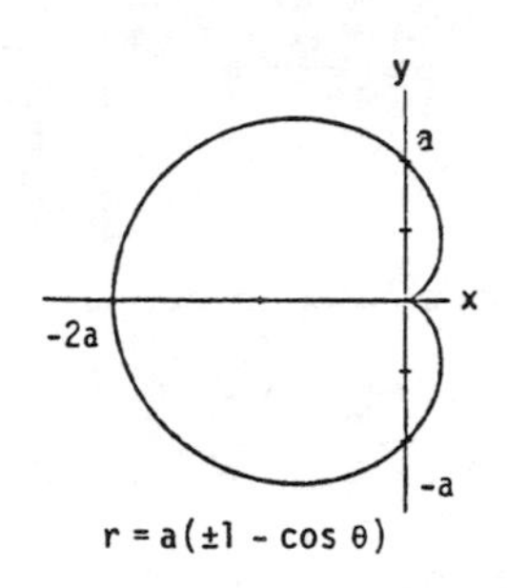

69. Example CAS commands:

Maple:

```
r:= t -> k*e/(1+e*cos(t));
k:=1; e:= 3/4;
plot(r(t),t=-Pi..Pi,coords=polar,view=[-4..1,-3..3]);
```

Mathematica:

```
Here we define a simple polar plotting function, for convenience:

polarplot[ r_, {ang_,a_,b_}, opts_ ] :=
ParametricPlot[{r Cos[ang], r Sin[ang]},
{ang,a,b}, opts, AspectRatio -> Automatic]
SetAttributes[ polarplot, HoldAll ]
Clear[e,k,t]
r[t_] = k e / (1 + e Cos[t])
k = -2;
e = 3/4; polarplot[ r[t], {t,-Pi,Pi} ]

Note: for the parabolas, a smaller range of theta gives a good picture:
e = 1; polarplot[ r[t], {t,-3,3} ]
e = 5/4; polarplot[ r[t], {t,-Pi,Pi} ]
```

70. Example CAS commands:

Maple:

```
r:= t -> a*(1-e^2)/(1+e*cos(t));
a:=2; e:= 1/3;
plot(r(t),t=-Pi..Pi,coords=polar,view=[-3..2,-3..3]);
```

Mathematica:

```
See Exercise 69 for the polar plotting function.

r[t_] = a (1 - e^2) / (1 + e Cos[t])
e = 9/10;
a = 1; polarplot[ r[t], {t,-Pi,Pi} ]
```

9.9 INTEGRATION IN POLAR COORDINATES

1. $A = \int_0^{2\pi} \frac{1}{2}(4 + 2\cos\theta)^2\, d\theta = \int_0^{2\pi} \frac{1}{2}\left(16 + 16\cos\theta + 4\cos^2\theta\right) d\theta = \int_0^{2\pi} \left[8 + 8\cos\theta + 2\left(\frac{1+\cos 2\theta}{2}\right)\right] d\theta$

$= \int_0^{2\pi} (9 + 8\cos\theta + \cos 2\theta)\, d\theta = \left[9\theta + 8\sin\theta + \frac{1}{2}\sin 2\theta\right]_0^{2\pi} = 18\pi$

2. $A = \int_0^{2\pi} \frac{1}{2}[a(1+\cos\theta)]^2\, d\theta = \int_0^{2\pi} \frac{1}{2}a^2\left(1 + 2\cos\theta + \cos^2\theta\right) d\theta = \frac{1}{2}a^2 \int_0^{2\pi} \left(1 + 2\cos\theta + \frac{1+\cos 2\theta}{2}\right) d\theta$

$= \frac{1}{2}a^2 \int_0^{2\pi} \left(\frac{3}{2} + 2\cos\theta + \frac{1}{2}\cos 2\theta\right) d\theta = \frac{1}{2}a^2\left[\frac{3}{2}\theta + 2\sin\theta + \frac{1}{4}\sin 2\theta\right]_0^{2\pi} = \frac{3}{2}\pi a^2$

3. $A = 2\int_0^{\pi/4} \frac{1}{2}\cos^2 2\theta\, d\theta = \int_0^{\pi/4} \frac{1+\cos 4\theta}{2}\, d\theta = \frac{1}{2}\left[\theta + \frac{\sin 4\theta}{4}\right]_0^{\pi/4} = \frac{\pi}{8}$

4. $A = 2\int_{-\pi/4}^{\pi/4} \frac{1}{2}\left(2a^2\cos 2\theta\right) d\theta = 2a^2 \int_{-\pi/4}^{\pi/4} \cos 2\theta\, d\theta = 2a^2\left[\frac{\sin 2\theta}{2}\right]_{-\pi/4}^{\pi/4} = 2a^2$

5. $A = 2\int_0^{\pi/2} \frac{1}{2}(4\sin 2\theta)\, d\theta = \int_0^{\pi/2} 2\sin 2\theta\, d\theta = [-\cos 2\theta]_0^{\pi/2} = 2$

6. $A = (6)(2)\int_0^{\pi/6} \frac{1}{2}(2\sin 3\theta)\, d\theta = 12\int_0^{\pi/6} \sin 3\theta\, d\theta = 12\left[-\frac{\cos 3\theta}{3}\right]_0^{\pi/6} = 4$

7. $r = 2\cos\theta$ and $r = 2\sin\theta \Rightarrow 2\cos\theta = 2\sin\theta$

$\Rightarrow \cos\theta = \sin\theta \Rightarrow \theta = \frac{\pi}{4}$; therefore

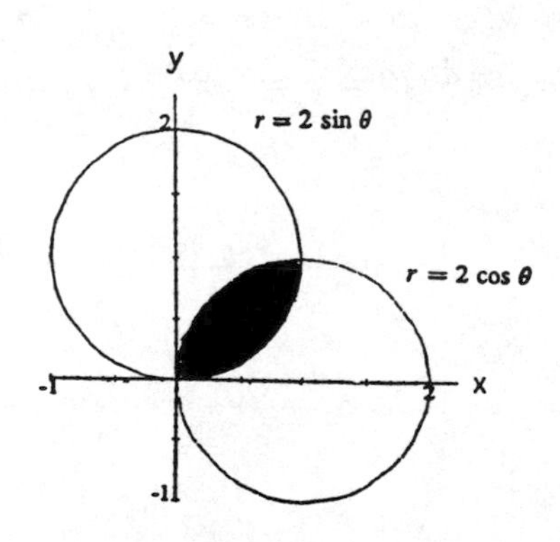

$A = 2\int_0^{\pi/4} \frac{1}{2}(2\sin\theta)^2\, d\theta = \int_0^{\pi/4} 4\sin^2\theta\, d\theta$

$= \int_0^{\pi/4} 4\left(\frac{1-\cos 2\theta}{2}\right) d\theta = \int_0^{\pi/4} (2 - 2\cos 2\theta)\, d\theta$

$= [2\theta - \sin 2\theta]_0^{\pi/4} = \frac{\pi}{2} - 1$

8. $r = 1$ and $r = 2\sin\theta \Rightarrow 2\sin\theta = 1 \Rightarrow \sin\theta = \frac{1}{2}$

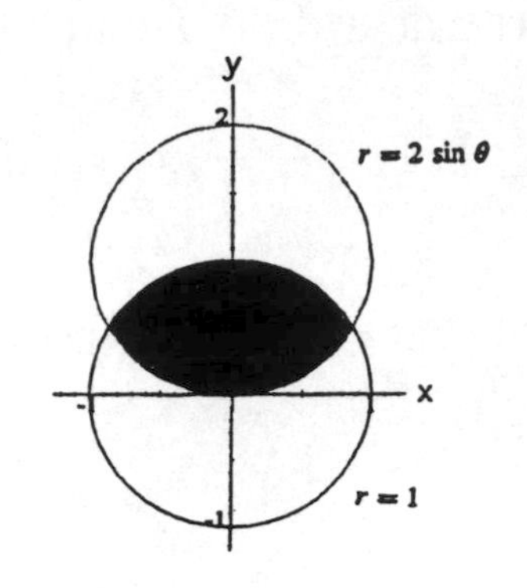

$\Rightarrow \theta = \frac{\pi}{6}$ or $\frac{5\pi}{6}$; therefore $A = \pi(1)^2 - \int_{\pi/6}^{5\pi/6} \frac{1}{2}\left[(2\sin\theta)^2 - 1^2\right] d\theta$

$= \pi - \int_{\pi/6}^{5\pi/6} \left(2\sin^2\theta - \frac{1}{2}\right) d\theta = \pi - \int_{\pi/6}^{5\pi/6} \left(1 - \cos 2\theta - \frac{1}{2}\right) d\theta$

$= \pi - \int_{\pi/6}^{5\pi/6} \left(\frac{1}{2} - \cos 2\theta\right) d\theta = \pi - \left[\frac{1}{2}\theta - \frac{\sin 2\theta}{2}\right]_{\pi/6}^{5\pi/6}$

$= \pi - \left(\frac{5\pi}{12} - \frac{1}{2}\sin\frac{5\pi}{3}\right) + \left(\frac{\pi}{12} - \frac{1}{2}\sin\frac{\pi}{3}\right) = \frac{4\pi - 3\sqrt{3}}{6}$

9. $r = 2$ and $r = 2(1 - \cos\theta) \Rightarrow 2 = 2(1 - \cos\theta) \Rightarrow \cos\theta = 0$

$\Rightarrow \theta = \pm\frac{\pi}{2}$; therefore $A = 2\int_0^{\pi/2} \frac{1}{2}[2(1 - \cos\theta)]^2\, d\theta$

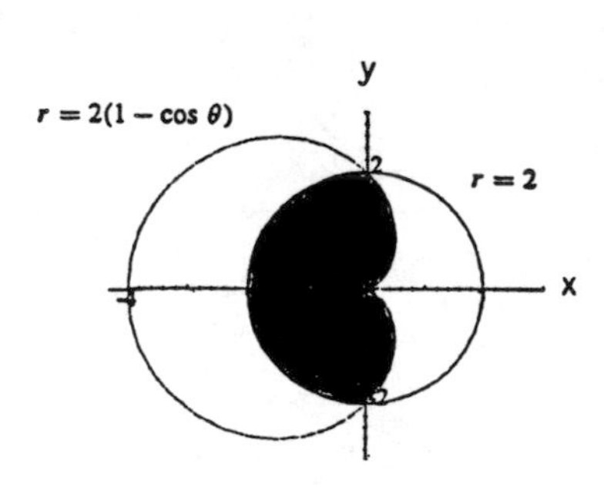

$+\frac{1}{2}$ area of the circle $= \int_0^{\pi/2} 4\left(1 - 2\cos\theta + \cos^2\theta\right) d\theta + \left(\frac{1}{2}\pi\right)(2)^2$

$= \int_0^{\pi/2} 4\left(1 - 2\cos\theta + \frac{1 + \cos 2\theta}{2}\right) d\theta + 2\pi$

$= \int_0^{\pi/2} (4 - 8\cos\theta + 2 + 2\cos 2\theta)\, d\theta + 2\pi$

$= \left[6\theta - 8\sin\theta + \sin 2\theta\right]_0^{\pi/2} + 2\pi = 5\pi - 8$

10. $r = 2(1 - \cos\theta)$ and $r = 2(1 + \cos\theta) \Rightarrow 1 - \cos\theta = 1 + \cos\theta$

$\Rightarrow \cos\theta = 0 \Rightarrow \theta = \frac{\pi}{2}$ or $\frac{3\pi}{2}$; the graph also gives the point of intersection $(0,0)$; therefore

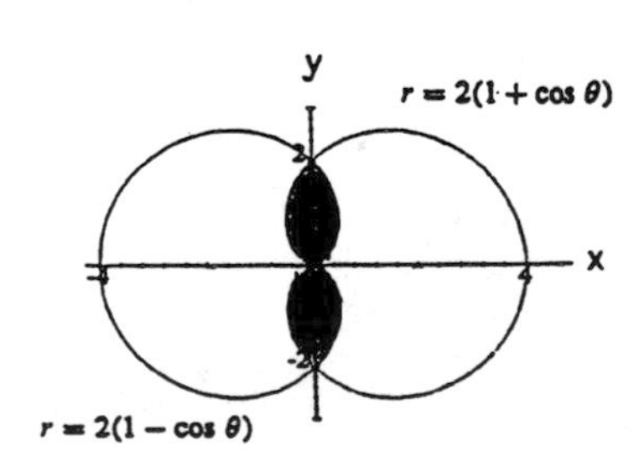

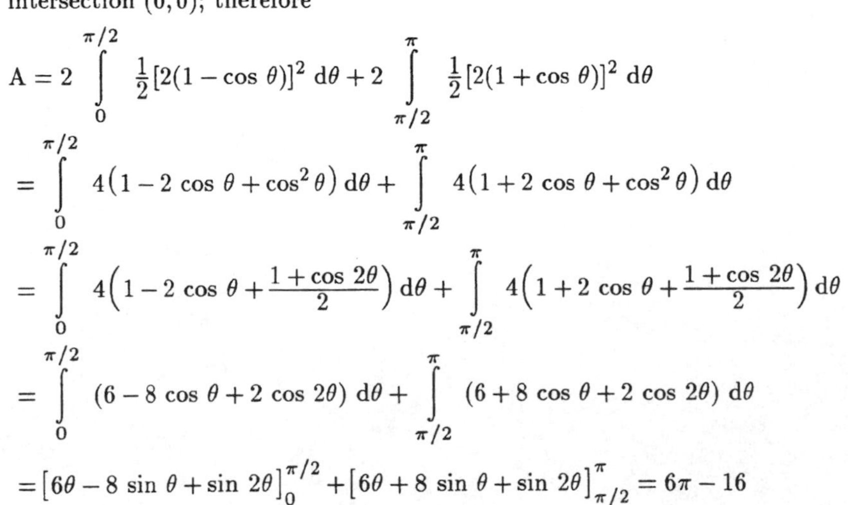

$A = 2\int_0^{\pi/2} \frac{1}{2}[2(1 - \cos\theta)]^2\, d\theta + 2\int_{\pi/2}^{\pi} \frac{1}{2}[2(1 + \cos\theta)]^2\, d\theta$

$= \int_0^{\pi/2} 4\left(1 - 2\cos\theta + \cos^2\theta\right) d\theta + \int_{\pi/2}^{\pi} 4\left(1 + 2\cos\theta + \cos^2\theta\right) d\theta$

$= \int_0^{\pi/2} 4\left(1 - 2\cos\theta + \frac{1 + \cos 2\theta}{2}\right) d\theta + \int_{\pi/2}^{\pi} 4\left(1 + 2\cos\theta + \frac{1 + \cos 2\theta}{2}\right) d\theta$

$= \int_0^{\pi/2} (6 - 8\cos\theta + 2\cos 2\theta)\, d\theta + \int_{\pi/2}^{\pi} (6 + 8\cos\theta + 2\cos 2\theta)\, d\theta$

$= \left[6\theta - 8\sin\theta + \sin 2\theta\right]_0^{\pi/2} + \left[6\theta + 8\sin\theta + \sin 2\theta\right]_{\pi/2}^{\pi} = 6\pi - 16$

11. $r = \sqrt{3}$ and $r^2 = 6 \cos 2\theta \Rightarrow 3 = 6 \cos 2\theta \Rightarrow \cos 2\theta = \frac{1}{2}$

$\Rightarrow \theta = \frac{\pi}{6}$ (in the 1st quadrant); we use symmetry of the

graph to find the area, so $A = 4 \int_0^{\pi/6} \left[\frac{1}{2}(6 \cos 2\theta) - \frac{1}{2}\left(\sqrt{3}\right)^2\right] d\theta$

$= 2 \int_0^{\pi/6} (6 \cos 2\theta - 3)\, d\theta = 2[3 \sin 2\theta - 3\theta]_0^{\pi/6} = 3\sqrt{3} - \pi$

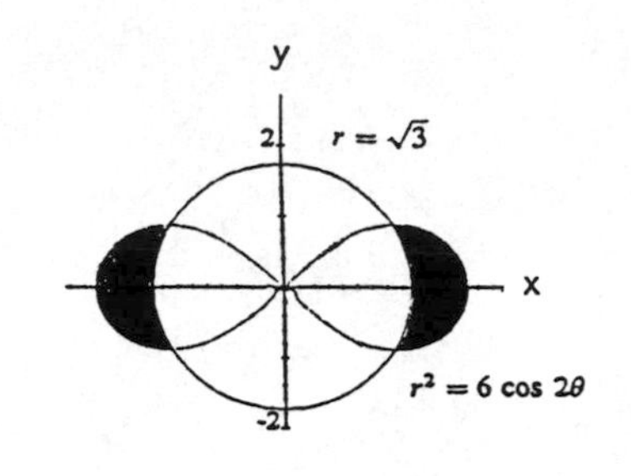

12. $r = 3a \cos \theta$ and $r = a(1 + \cos \theta) \Rightarrow 3a \cos \theta = a(1 + \cos \theta)$

$\Rightarrow 3 \cos \theta = 1 + \cos \theta \Rightarrow \cos \theta = \frac{1}{2} \Rightarrow \theta = \frac{\pi}{3}$ or $-\frac{\pi}{3}$; the

graph also gives the point of intersection $(0,0)$; therefore

$A = 2 \int_0^{\pi/3} \frac{1}{2}\left[(3a \cos \theta)^2 - a^2(1 + \cos \theta)^2\right] d\theta$

$= \int_0^{\pi/3} \left(9a^2 \cos^2\theta - a^2 - 2a^2 \cos \theta - a^2 \cos^2\theta\right) d\theta$

$= \int_0^{\pi/3} \left(8a^2 \cos^2\theta - 2a^2 \cos \theta - a^2\right) d\theta = \int_0^{\pi/3} \left[4a^2(1 + \cos 2\theta) - 2a^2 \cos \theta - a^2\right] d\theta$

$= \int_0^{\pi/3} \left(3a^2 + 4a^2 \cos 2\theta - 2a^2 \cos \theta\right) d\theta = \left[3a^2\theta + 2a^2 \sin 2\theta - 2a^2 \sin \theta\right]_0^{\pi/3} = \pi a^2 + 2a^2\left(\frac{1}{2}\right) - 2a^2\left(\frac{\sqrt{3}}{2}\right)$

$= a^2\left(\pi + 1 - \sqrt{3}\right)$

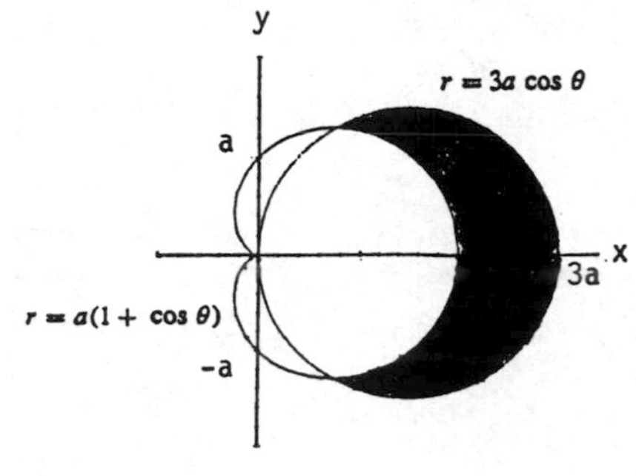

13. $r = 1$ and $r = -2 \cos \theta \Rightarrow 1 = -2 \cos \theta \Rightarrow \cos \theta = -\frac{1}{2}$

$\Rightarrow \theta = \frac{2\pi}{3}$ in quadrant II; therefore

$A = 2 \int_{2\pi/3}^{\pi} \frac{1}{2}\left[(-2 \cos \theta)^2 - 1^2\right] d\theta = \int_{2\pi/3}^{\pi} \left(4 \cos^2\theta - 1\right) d\theta$

$= \int_{2\pi/3}^{\pi} [2(1 + \cos 2\theta) - 1]\, d\theta = \int_{2\pi/3}^{\pi} (1 + 2 \cos 2\theta)\, d\theta$

$= \left[\theta + \sin 2\theta\right]_{2\pi/3}^{\pi} = \frac{\pi}{3} + \frac{\sqrt{3}}{2}$

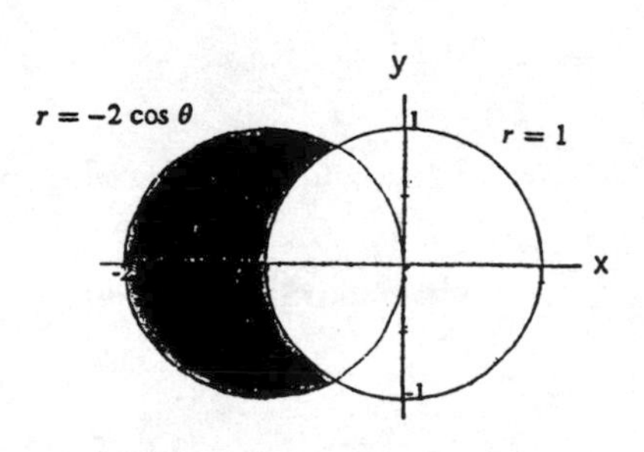

14. (a) $A = 2 \int_0^{2\pi/3} \frac{1}{2}(2 \cos \theta + 1)^2\, d\theta = \int_0^{2\pi/3} \left(4 \cos^2\theta + 4 \cos \theta + 1\right) d\theta = \int_0^{2\pi/3} [2(1 + \cos 2\theta) + 4 \cos \theta + 1]\, d\theta$

$= \int_0^{2\pi/3} (3 + 2 \cos 2\theta + 4 \cos \theta)\, d\theta = \left[3\theta + \sin 2\theta + 4 \sin \theta\right]_0^{2\pi/3} = 2\pi - \frac{\sqrt{3}}{2} + \frac{4\sqrt{3}}{2} = 2\pi + \frac{3\sqrt{3}}{2}$

(b) $A = \left(2\pi + \frac{3\sqrt{3}}{2}\right) - \left(\pi - \frac{3\sqrt{3}}{2}\right) = \pi + 3\sqrt{3}$ (from 14(a) above and Example 2 in the text)

15. $r = 6$ and $r = 3 \csc \theta \Rightarrow 6 \sin \theta = 3 \Rightarrow \sin \theta = \frac{1}{2} \Rightarrow \theta = \frac{\pi}{6}$

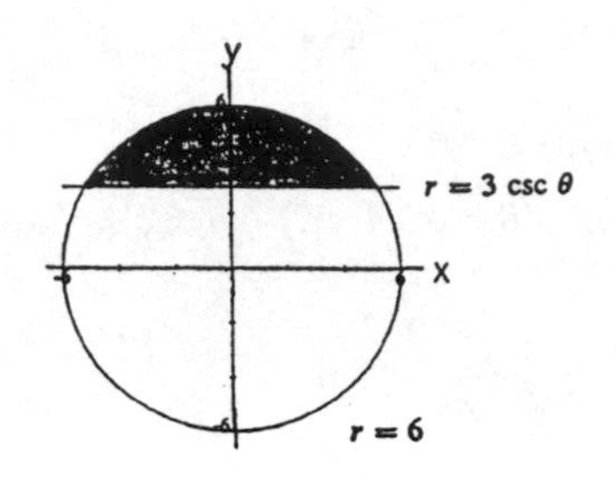

or $\frac{5\pi}{6}$; therefore $A = \int_{\pi/6}^{5\pi/6} \frac{1}{2}\left(6^2 - 9 \csc^2 \theta\right) d\theta$

$= \int_{\pi/6}^{5\pi/6} \left(18 - \frac{9}{2} \csc^2 \theta\right) d\theta = \left[18\theta + \frac{9}{2} \cot \theta\right]_{\pi/6}^{5\pi/6}$

$= \left(15\pi - \frac{9}{2}\sqrt{3}\right) - \left(3\pi + \frac{9}{2}\sqrt{3}\right) = 12\pi - 9\sqrt{3}$

16. $r^2 = 6 \cos 2\theta$ and $r = \frac{3}{2} \sec \theta \Rightarrow \frac{9}{4} \sec^2 \theta = 6 \cos 2\theta \Rightarrow \frac{9}{24} = \cos^2 \theta \cos 2\theta \Rightarrow \frac{3}{8} = \left(\cos^2 \theta\right)\left(2 \cos^2 \theta - 1\right)$

$\Rightarrow \frac{3}{8} = 2 \cos^4 \theta - \cos^2 \theta \Rightarrow 2 \cos^4 \theta - \cos^2 \theta - \frac{3}{8} = 0 \Rightarrow 16 \cos^4 \theta - 8 \cos^2 \theta - 3 = 0 \Rightarrow \left(4 \cos^2 \theta - 1\right)\left(4 \cos^2 \theta - 3\right)$

$= 0 \Rightarrow \cos^2 \theta = \frac{3}{4}$ or $\cos^2 \theta = -\frac{1}{4} \Rightarrow \cos \theta = \pm\frac{\sqrt{3}}{2}$ (the second equation has no real roots) $\Rightarrow \theta = \frac{\pi}{6}$ (in the first

quadrant); thus $A = 2 \int_0^{\pi/6} \frac{1}{2}\left(6 \cos 2\theta - \frac{9}{4} \sec^2 \theta\right) d\theta = \int_0^{\pi/6} \left(6 \cos 2\theta - \frac{9}{4} \sec^2 \theta\right) d\theta = \left[3 \sin 2\theta - \frac{9}{4} \tan \theta\right]_0^{\pi/6}$

$= 3\left(\frac{\sqrt{3}}{2}\right) - \frac{9}{4\sqrt{3}} = \frac{3\sqrt{3}}{2} - \frac{3\sqrt{3}}{4} = \frac{3\sqrt{3}}{4}$

17. (a) $r = \tan \theta$ and $r = \left(\frac{\sqrt{2}}{2}\right) \csc \theta \Rightarrow \tan \theta = \left(\frac{\sqrt{2}}{2}\right) \csc \theta$

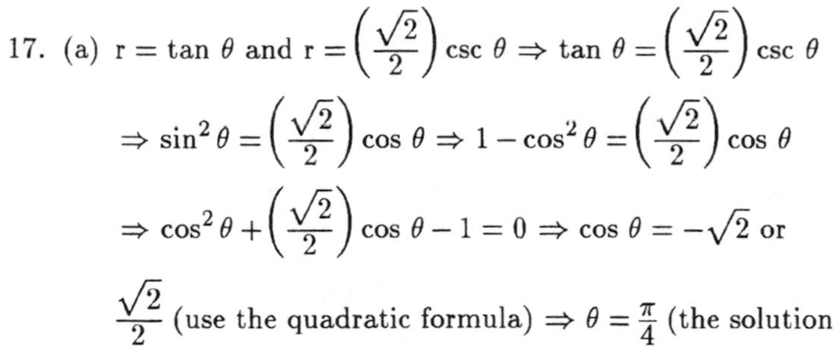

$\Rightarrow \sin^2 \theta = \left(\frac{\sqrt{2}}{2}\right) \cos \theta \Rightarrow 1 - \cos^2 \theta = \left(\frac{\sqrt{2}}{2}\right) \cos \theta$

$\Rightarrow \cos^2 \theta + \left(\frac{\sqrt{2}}{2}\right) \cos \theta - 1 = 0 \Rightarrow \cos \theta = -\sqrt{2}$ or

$\frac{\sqrt{2}}{2}$ (use the quadratic formula) $\Rightarrow \theta = \frac{\pi}{4}$ (the solution

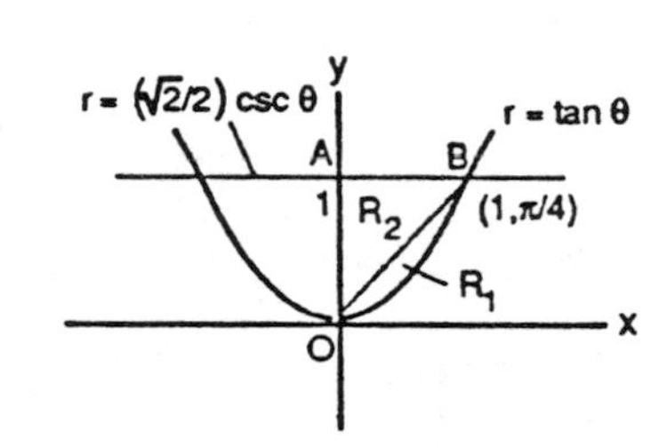

in the first quadrant); therefore the area of R_1 is $A_1 = \int_0^{\pi/4} \frac{1}{2} \tan^2 \theta \, d\theta = \frac{1}{2} \int_0^{\pi/4} \left(\sec^2 \theta - 1\right) d\theta$

$= \frac{1}{2}[\tan \theta - \theta]_0^{\pi/4} = \frac{1}{2}\left(\tan \frac{\pi}{4} - \frac{\pi}{4}\right) = \frac{1}{2} - \frac{\pi}{8}$; $AO = \left(\frac{\sqrt{2}}{2}\right) \csc \frac{\pi}{2} = \frac{\sqrt{2}}{2}$ and $OB = \left(\frac{\sqrt{2}}{2}\right) \csc \frac{\pi}{4} = 1$

$\Rightarrow AB = \sqrt{1^2 - \left(\frac{\sqrt{2}}{2}\right)^2} = \frac{\sqrt{2}}{2} \Rightarrow$ the area of R_2 is $A_2 = \frac{1}{2}\left(\frac{\sqrt{2}}{2}\right)\left(\frac{\sqrt{2}}{2}\right) = \frac{1}{4}$; therefore the area of the

region shaded in the text is $2\left(\frac{1}{2} - \frac{\pi}{8} + \frac{1}{4}\right) = \frac{3}{2} - \frac{\pi}{4}$. Note: The area must be found this way since no

common interval generates the region. For example, the interval $0 \le \theta \le \frac{\pi}{4}$ generates the arc OB of

$r = \tan \theta$ but does not generate the segment AB of the line $r = \frac{\sqrt{2}}{2} \csc \theta$. Instead the interval generates

the half-line from B to $+\infty$ on the line $r = \frac{\sqrt{2}}{2} \csc \theta$.

(b) $\lim_{\theta\to\pi/2^-} \tan\theta = \infty$ and the line $x = 1$ is $r = \sec\theta$ in polar coordinates; then $\lim_{\theta\to\pi/2^-} (\tan\theta - \sec\theta)$ $= \lim_{\theta\to\pi/2^-} \left(\frac{\sin\theta}{\cos\theta} - \frac{1}{\cos\theta}\right) = \lim_{\theta\to\pi/2^-} \left(\frac{\sin\theta - 1}{\cos\theta}\right) = \lim_{\theta\to\pi/2^-} \left(\frac{\cos\theta}{-\sin\theta}\right) = 0 \Rightarrow r = \tan\theta$ approaches $r = \sec\theta$ as $\theta \to \frac{\pi^-}{2} \Rightarrow r = \sec\theta$ (or $x = 1$) is a vertical asymptote of $r = \tan\theta$. Similarly, $r = -\sec\theta$ (or $x = -1$) is a vertical asymptote of $r = \tan\theta$.

18. It is not because the circle is generated twice from $\theta = 0$ to 2π. The area of the cardioid is

$$A = 2\int_0^\pi \tfrac{1}{2}(\cos\theta + 1)^2\, d\theta = \int_0^\pi (\cos^2\theta + 2\cos\theta + 1)\, d\theta = \int_0^\pi \left(\frac{1+\cos 2\theta}{2} + 2\cos\theta + 1\right) d\theta$$

$= \left[\frac{3\theta}{2} + \frac{\sin 2\theta}{4} + 2\sin\theta\right]_0^\pi = \frac{3\pi}{2}$. The area of the circle is $A = \pi\left(\frac{1}{2}\right)^2 = \frac{\pi}{4} \Rightarrow$ the area requested is actually $\frac{3\pi}{2} - \frac{\pi}{4} = \frac{5\pi}{4}$

19. $r = \theta^2,\ 0 \le \theta \le \sqrt{5} \Rightarrow \frac{dr}{d\theta} = 2\theta$; therefore Length $= \int_0^{\sqrt{5}} \sqrt{\left(\theta^2\right)^2 + (2\theta)^2}\, d\theta = \int_0^{\sqrt{5}} \sqrt{\theta^4 + 4\theta^2}\, d\theta$

$= \int_0^{\sqrt{5}} |\theta|\sqrt{\theta^2 + 4}\, d\theta =$ (since $\theta \ge 0$) $\int_0^{\sqrt{5}} \theta\sqrt{\theta^2+4}\, d\theta$; $\left[u = \theta^2 + 4 \Rightarrow \frac{1}{2}\, du = \theta\, d\theta;\ \theta = 0 \Rightarrow u = 4,\right.$ $\left.\theta = \sqrt{5} \Rightarrow u = 9\right] \to \int_4^9 \frac{1}{2}\sqrt{u}\, du = \frac{1}{2}\left[\frac{2}{3}u^{3/2}\right]_4^9 = \frac{19}{3}$

20. $r = \frac{e^\theta}{\sqrt{2}},\ 0 \le \theta \le \pi \Rightarrow \frac{dr}{d\theta} = \frac{e^\theta}{\sqrt{2}}$; therefore Length $= \int_0^\pi \sqrt{\left(\frac{e^\theta}{\sqrt{2}}\right)^2 + \left(\frac{e^\theta}{\sqrt{2}}\right)^2}\, d\theta = \int_0^\pi \sqrt{2\left(\frac{e^{2\theta}}{2}\right)}\, d\theta$

$= \int_0^\pi e^\theta\, d\theta = \left[e^\theta\right]_0^\pi = e^\pi - 1$

21. $r = 1 + \cos\theta \Rightarrow \frac{dr}{d\theta} = -\sin\theta$; therefore Length $= \int_0^{2\pi} \sqrt{(1+\cos\theta)^2 + (-\sin\theta)^2}\, d\theta$

$$= 2\int_0^\pi \sqrt{2 + 2\cos\theta}\, d\theta = 2\int_0^\pi \sqrt{\frac{4(1+\cos\theta)}{2}}\, d\theta = 4\int_0^\pi \sqrt{\frac{1+\cos\theta}{2}}\, d\theta = 4\int_0^\pi \cos\left(\frac{\theta}{2}\right) d\theta = 4\left[2\sin\frac{\theta}{2}\right]_0^\pi = 8$$

22. $r = a\sin^2\frac{\theta}{2},\ 0 \le \theta \le \pi,\ a > 0 \Rightarrow \frac{dr}{d\theta} = a\sin\frac{\theta}{2}\cos\frac{\theta}{2}$; therefore Length $= \int_0^\pi \sqrt{\left(a\sin^2\frac{\theta}{2}\right)^2 + \left(a\sin\frac{\theta}{2}\cos\frac{\theta}{2}\right)^2}\, d\theta$

$= \int_0^\pi \sqrt{a^2\sin^4\frac{\theta}{2} + a^2\sin^2\frac{\theta}{2}\cos^2\frac{\theta}{2}}\, d\theta = \int_0^\pi a\left|\sin\frac{\theta}{2}\right|\sqrt{\sin^2\frac{\theta}{2} + \cos^2\frac{\theta}{2}}\, d\theta =$ (since $0 \le \theta \le \pi$) $a\int_0^\pi \sin\left(\frac{\theta}{2}\right) d\theta$

$= \left[-2a\cos\frac{\theta}{2}\right]_0^\pi = 2a$

23. $r = \frac{6}{1+\cos\theta},\ 0 \le \theta \le \frac{\pi}{2} \Rightarrow \frac{dr}{d\theta} = \frac{6\sin\theta}{(1+\cos\theta)^2}$; therefore Length $= \int_0^{\pi/2} \sqrt{\left(\frac{6}{1+\cos\theta}\right)^2 + \left(\frac{6\sin\theta}{(1+\cos\theta)^2}\right)^2}\, d\theta$

$$= \int_0^{\pi/2} \sqrt{\frac{36}{(1+\cos\theta)^2} + \frac{36\sin^2\theta}{\left(1+\cos^2\theta\right)^4}}\, d\theta = 6 \int_0^{\pi/2} \left|\frac{1}{1+\cos\theta}\right| \sqrt{1 + \frac{\sin^2\theta}{(1+\cos\theta)^2}}\, d\theta$$

$$= \left(\text{since } \frac{1}{1+\cos\theta} > 0 \text{ on } 0 \le \theta \le \frac{\pi}{2}\right)\ 6 \int_0^{\pi/2} \left(\frac{1}{1+\cos\theta}\right) \sqrt{\frac{1+2\cos\theta+\cos^2\theta+\sin^2\theta}{(1+\cos\theta)^2}}\, d\theta$$

$$= 6\int_0^{\pi/2} \left(\frac{1}{1+\cos\theta}\right)\sqrt{\frac{2+2\cos\theta}{(1+\cos\theta)^2}}\, d\theta = 6\sqrt{2}\int_0^{\pi/2} \frac{d\theta}{(1+\cos\theta)^{3/2}} = 6\sqrt{2}\int_0^{\pi/2} \frac{d\theta}{\left(2\cos^2\frac{\theta}{2}\right)^{3/2}} = 6\int_0^{\pi/2} \left|\sec^3\frac{\theta}{2}\right| d\theta$$

$$= 6\int_0^{\pi/2} \sec^3\frac{\theta}{2}\, d\theta = 12\int_0^{\pi/4} \sec^3 u\, du = (\text{use tables})\ 6\left(\left[\frac{\sec u \tan u}{2}\right]_0^{\pi/4} + \frac{1}{2}\int_0^{\pi/4} \sec u\, du\right)$$

$$= 6\left(\frac{1}{\sqrt{2}} + \left[\frac{1}{2}\ln|\sec u + \tan u|\right]_0^{\pi/4}\right) = 3\left[\sqrt{2} + \ln\left(1+\sqrt{2}\right)\right]$$

24. $r = \frac{2}{1-\cos\theta},\ \frac{\pi}{2} \le \theta \le \pi \Rightarrow \frac{dr}{d\theta} = \frac{-2\sin\theta}{(1-\cos\theta)^2}$; therefore Length $= \int_{\pi/2}^{\pi} \sqrt{\left(\frac{2}{1-\cos\theta}\right)^2 + \left(\frac{-2\sin\theta}{(1-\cos\theta)^2}\right)^2}\, d\theta$

$$= \int_{\pi/2}^{\pi} \sqrt{\frac{4}{(1-\cos\theta)^2}\left(1 + \frac{\sin^2\theta}{\left(1-\cos^2\theta\right)^2}\right)}\, d\theta = 6\int_{\pi/2}^{\pi} \left|\frac{2}{1-\cos\theta}\right| \sqrt{\frac{(1-\cos\theta)^2+\sin^2\theta}{(1-\cos\theta)^2}}\, d\theta$$

$$= \left(\text{since } 1-\cos\theta \ge 0 \text{ on } \frac{\pi}{2} \le \theta \le \pi\right)\ 2\int_{\pi/2}^{\pi} \left(\frac{1}{1-\cos\theta}\right)\sqrt{\frac{1-2\cos\theta+\cos^2\theta+\sin^2\theta}{(1-\cos\theta)^2}}\, d\theta$$

$$= 2\int_{\pi/2}^{\pi} \left(\frac{1}{1-\cos\theta}\right)\sqrt{\frac{2-2\cos\theta}{(1-\cos\theta)^2}}\, d\theta = 2\sqrt{2}\int_{\pi/2}^{\pi} \frac{d\theta}{(1-\cos\theta)^{3/2}} = 2\sqrt{2}\int_{\pi/2}^{\pi} \frac{d\theta}{\left(2\sin^2\frac{\theta}{2}\right)^{3/2}} = \int_{\pi/2}^{\pi} \left|\csc^3\frac{\theta}{2}\right| d\theta$$

$$= 6\int_{\pi/2}^{\pi} \csc^3\left(\frac{\theta}{2}\right) d\theta = \left(\text{since } \csc\frac{\theta}{2} \ge 0 \text{ on } \frac{\pi}{2} \le \theta \le \pi\right)\ 2\int_{\pi/4}^{\pi/2} \csc^3 u\, du = (\text{use tables})$$

$$6\left(\left[-\frac{\csc u \cot u}{2}\right]_{\pi/4}^{\pi/2} + \frac{1}{2}\int_{\pi/4}^{\pi/2} \csc u\, du\right) = 2\left(\frac{1}{\sqrt{2}} - \left[\frac{1}{2}\ln|\csc u + \cot u|\right]_{\pi/4}^{\pi/2}\right) = 2\left[\frac{1}{\sqrt{2}} + \frac{1}{2}\ln\left(\sqrt{2}+1\right)\right]$$

$$= \sqrt{2} + \ln\left(1+\sqrt{2}\right)$$

25. $r = \cos^3 \frac{\theta}{3} \Rightarrow \frac{dr}{d\theta} = -\sin \frac{\theta}{3} \cos^2 \frac{\theta}{3}$; therefore Length $= \int_0^{\pi/4} \sqrt{\left(\cos^3 \frac{\theta}{3}\right)^2 + \left(-\sin \frac{\theta}{3} \cos^2 \frac{\theta}{3}\right)^2}\, d\theta$

$$= \int_0^{\pi/4} \sqrt{\cos^6\left(\frac{\theta}{3}\right) + \sin^2\left(\frac{\theta}{3}\right)\cos^4\left(\frac{\theta}{3}\right)}\, d\theta = \int_0^{\pi/4} \left(\cos^2 \frac{\theta}{3}\right)\sqrt{\cos^2\left(\frac{\theta}{3}\right) + \sin^2\left(\frac{\theta}{3}\right)}\, d\theta = \int_0^{\pi/4} \cos^2\left(\frac{\theta}{3}\right) d\theta$$

$$= \int_0^{\pi/4} \frac{1 + \cos\left(\frac{2\theta}{3}\right)}{2}\, d\theta = \frac{1}{2}\left[\theta + \frac{3}{2} \sin \frac{2\theta}{3}\right]_0^{\pi/4} = \frac{\pi}{8} + \frac{3}{8}$$

26. $r = \sqrt{1 + \sin 2\theta},\ 0 \le \theta \le \pi\sqrt{2} \Rightarrow \frac{dr}{d\theta} = \frac{1}{2}(1 + \sin 2\theta)^{-1/2}(2 \cos 2\theta) = (\cos 2\theta)(1 + \sin 2\theta)^{-1/2}$; therefore

$$\text{Length} = \int_0^{\pi\sqrt{2}} \sqrt{(1 + \sin 2\theta) + \frac{\cos^2 2\theta}{(1 + \sin 2\theta)}}\, d\theta = \int_0^{\pi\sqrt{2}} \sqrt{\frac{1 + 2 \sin 2\theta + \sin^2 2\theta + \cos^2 2\theta}{1 + \sin 2\theta}}\, d\theta$$

$$= \int_0^{\pi\sqrt{2}} \sqrt{\frac{2 + 2 \sin 2\theta}{1 + \sin 2\theta}}\, d\theta = \int_0^{\pi\sqrt{2}} \sqrt{2}\, d\theta = \left[\sqrt{2}\,\theta\right]_0^{\pi\sqrt{2}} = 2\pi$$

27. $r = \sqrt{1 + \cos 2\theta} \Rightarrow \frac{dr}{d\theta} = \frac{1}{2}(1 + \cos 2\theta)^{-1/2}(-2 \sin 2\theta)$; therefore Length $= \int_0^{\pi\sqrt{2}} \sqrt{(1 + \cos 2\theta) + \frac{\sin^2 2\theta}{(1 + \cos 2\theta)}}\, d\theta$

$$= \int_0^{\pi\sqrt{2}} \sqrt{\frac{1 + 2 \cos 2\theta + \cos^2 2\theta + \sin^2 2\theta}{1 + \cos 2\theta}}\, d\theta = \int_0^{\pi\sqrt{2}} \sqrt{\frac{2 + 2 \cos 2\theta}{1 + \cos 2\theta}}\, d\theta = \int_0^{\pi\sqrt{2}} \sqrt{2}\, d\theta = \left[\sqrt{2}\,\theta\right]_0^{\pi\sqrt{2}} = 2\pi$$

28. (a) $r = a \Rightarrow \frac{dr}{d\theta} = 0$; Length $= \int_0^{2\pi} \sqrt{a^2 + 0^2}\, d\theta = \int_0^{2\pi} |a|\, d\theta = [a\theta]_0^{2\pi} = 2\pi a$

(b) $r = a \cos \theta \Rightarrow \frac{dr}{d\theta} = -a \sin \theta$; Length $= \int_0^{\pi} \sqrt{(a \cos \theta)^2 + (-a \sin \theta)^2}\, d\theta = \int_0^{\pi} \sqrt{a^2(\cos^2 \theta + \sin^2 \theta)}\, d\theta$

$$= \int_0^{\pi} |a|\, d\theta = [a\theta]_0^{\pi} = \pi a$$

(c) $r = a \sin \theta \Rightarrow \frac{dr}{d\theta} = a \cos \theta$; Length $= \int_0^{\pi} \sqrt{(a \cos \theta)^2 + (a \sin \theta)^2}\, d\theta = \int_0^{\pi} \sqrt{a^2(\cos^2 \theta + \sin^2 \theta)}\, d\theta$

$$= \int_0^{\pi} |a|\, d\theta = [a\theta]_0^{\pi} = \pi a$$

29. $r = \sqrt{\cos 2\theta}$, $0 \le \theta \le \frac{\pi}{4} \Rightarrow \frac{dr}{d\theta} = \frac{1}{2}(\cos 2\theta)^{-1/2}(-\sin 2\theta)(2) = \frac{-\sin 2\theta}{\sqrt{\cos 2\theta}}$; therefore Surface Area

$$= \int_0^{\pi/4} (2\pi r \cos\theta)\sqrt{\left(\sqrt{\cos 2\theta}\right)^2 + \left(\frac{-\sin 2\theta}{\sqrt{\cos 2\theta}}\right)^2}\, d\theta = \int_0^{\pi/4} \left(2\pi\sqrt{\cos 2\theta}\right)(\cos\theta)\sqrt{\cos 2\theta + \frac{\sin^2 2\theta}{\cos 2\theta}}\, d\theta$$

$$= \int_0^{\pi/4} \left(2\pi\sqrt{\cos 2\theta}\right)(\cos\theta)\sqrt{\frac{1}{\cos 2\theta}}\, d\theta = \int_0^{\pi/4} 2\pi\cos\theta\, d\theta = [2\pi\sin\theta]_0^{\pi/4} = \pi\sqrt{2}$$

30. $r = \sqrt{2}e^{\theta/2}$, $0 \le \theta \le \frac{\pi}{2} \Rightarrow \frac{dr}{d\theta} = \sqrt{2}\left(\frac{1}{2}\right)e^{\theta/2} = \frac{\sqrt{2}}{2}e^{\theta/2}$; therefore Surface Area

$$= \int_0^{\pi/2} \left(2\pi\sqrt{2}\,e^{\theta/2}\right)(\sin\theta)\sqrt{\left(\sqrt{2}\,e^{\theta/2}\right)^2 + \left(\frac{\sqrt{2}}{2}e^{\theta/2}\right)^2}\, d\theta = \int_0^{\pi/2} \left(2\pi\sqrt{2}\,e^{\theta/2}\right)(\sin\theta)\sqrt{2e^\theta + \frac{1}{2}e^\theta}\, d\theta$$

$$= \int_0^{\pi/2} \left(2\pi\sqrt{2}\,e^{\theta/2}\right)(\sin\theta)\sqrt{\frac{5}{2}e^\theta}\, d\theta = \int_0^{\pi/2} \left(2\pi\sqrt{2}\,e^{\theta/2}\right)(\sin\theta)\left(\frac{\sqrt{5}}{\sqrt{2}}e^{\theta/2}\right) d\theta = 2\pi\sqrt{5}\int_0^{\pi/2} e^\theta \sin\theta\, d\theta$$

$$= 2\pi\sqrt{5}\left[\frac{e^\theta}{2}(\sin\theta - \cos\theta)\right]_0^{\pi/2} = \pi\sqrt{5}\left(e^{\pi/2} + 1\right) \text{ where we integrated by parts}$$

31. $r^2 = \cos 2\theta \Rightarrow r = \pm\sqrt{\cos 2\theta}$; use $r = \sqrt{\cos 2\theta}$ on $\left[0, \frac{\pi}{4}\right] \Rightarrow \frac{dr}{d\theta} = \frac{1}{2}(\cos 2\theta)^{-1/2}(-\sin 2\theta)(2) = \frac{-\sin 2\theta}{\sqrt{\cos 2\theta}}$;

therefore Surface Area $= 2\int_0^{\pi/4} \left(2\pi\sqrt{\cos 2\theta}\right)(\sin\theta)\sqrt{\cos 2\theta + \frac{\sin^2 2\theta}{\cos 2\theta}}\, d\theta = 4\pi\int_0^{\pi/4} \sqrt{\cos 2\theta}\,(\sin\theta)\sqrt{\frac{1}{\cos 2\theta}}\, d\theta$

$$= 4\pi\int_0^{\pi/4} \sin\theta\, d\theta = 4\pi[-\cos\theta]_0^{\pi/4} = 4\pi\left[-\frac{\sqrt{2}}{2} - (-1)\right] = 2\pi\left(2 - \sqrt{2}\right)$$

32. $r = 2a\cos 2\theta \Rightarrow \frac{dr}{d\theta} = -2a\sin\theta$; therefore Surface Area $= 2\int_0^{\pi} 2\pi(2a\cos\theta)(\cos\theta)\sqrt{(2a\cos\theta)^2 + (-2a\sin\theta)^2}\, d\theta$

$$= 4a\pi\int_0^{\pi} \left(\cos^2\theta\right)\sqrt{4a^2\left(\cos^2\theta + \sin^2\theta\right)}\, d\theta = 8a\pi\int_0^{\pi} \left(\cos^2\theta\right)|a|\, d\theta = 8a^2\pi\int_0^{\pi} \cos^2\theta\, d\theta$$

$$= 8a^2\pi\int_0^{\pi} \left(\frac{1 + \cos 2\theta}{2}\right) d\theta = 4a^2\pi\int_0^{\pi} (1 + \cos 2\theta)\, d\theta = 4a^2\pi\left[\theta + \frac{1}{2}\sin 2\theta\right]_0^{\pi} = 4a^2\pi^2$$

33. Let $r = f(\theta)$. Then $x = f(\theta)\cos\theta \Rightarrow \frac{dx}{d\theta} = f'(\theta)\cos\theta - f(\theta)\sin\theta \Rightarrow \left(\frac{dx}{d\theta}\right)^2 = \left[f'(\theta)\cos\theta - f(\theta)\sin\theta\right]^2$

$= \left[f'(\theta)\right]^2\cos^2\theta - 2f'(\theta)\,f(\theta)\sin\theta\cos\theta + [f(\theta)]^2\sin^2\theta$; $y = f(\theta)\sin\theta \Rightarrow \frac{dy}{d\theta} = f'(\theta)\sin\theta + f(\theta)\cos\theta$

$\Rightarrow \left(\frac{dy}{d\theta}\right)^2 = \left[f'(\theta)\sin\theta + f(\theta)\cos\theta\right]^2 = \left[f'(\theta)\right]^2\sin^2\theta + 2f'(\theta)f(\theta)\sin\theta\cos\theta + [f(\theta)]^2\cos^2\theta$. Therefore

$\left(\frac{dx}{d\theta}\right)^2 + \left(\frac{dy}{d\theta}\right)^2 = [f'(\theta)]^2(\cos^2\theta + \sin^2\theta) + [f(\theta)]^2(\cos^2\theta + \sin^2\theta) = [f'(\theta)]^2 + [f(\theta)]^2 = r^2 + \left(\frac{dr}{d\theta}\right)^2.$

Thus, $L = \int_\alpha^\beta \sqrt{\left(\frac{dx}{d\theta}\right)^2 + \left(\frac{dy}{d\theta}\right)^2}\, d\theta = \int_\alpha^\beta \sqrt{r^2 + \left(\frac{dr}{d\theta}\right)^2}\, d\theta.$

34. (a) $r_{av} = \frac{1}{2\pi - 0}\int_0^{2\pi} a(1 - \cos\theta)\, d\theta = \frac{a}{2\pi}[\theta - \sin\theta]_0^{2\pi} = a$

(b) $r_{av} = \frac{1}{2\pi - 0}\int_0^{2\pi} a\, d\theta = \frac{1}{2\pi}[a\theta]_0^{2\pi} = a$

(c) $r_{av} = \frac{1}{\left(\frac{\pi}{2}\right) - \left(-\frac{\pi}{2}\right)}\int_{-\pi/2}^{\pi/2} a\cos\theta\, d\theta = \frac{1}{\pi}[a\sin\theta]_{-\pi/2}^{\pi/2} = \frac{2a}{\pi}$

35. $r = 2f(\theta),\ \alpha \le \theta \le \beta \Rightarrow \frac{dr}{d\theta} = 2f'(\theta) \Rightarrow r^2 + \left(\frac{dr}{d\theta}\right)^2 = [2f(\theta)]^2 + [2f'(\theta)]^2 \Rightarrow \text{Length} = \int_\alpha^\beta \sqrt{4[f(\theta)]^2 + 4[f'(\theta)]^2}\, d\theta$

$= 2\int_\alpha^\beta \sqrt{[f(\theta)]^2 + [f'(\theta)]^2}\, d\theta$ which is twice the length of the curve $r = f(\theta)$ for $\alpha \le \theta \le \beta$.

36. Again $r = 2f(\theta) \Rightarrow r^2 + \left(\frac{dr}{d\theta}\right)^2 = [2f(\theta)] + [2f'(\theta)]^2 \Rightarrow \text{Surface Area} = \int_\alpha^\beta 2\pi[2f(\theta)\sin\theta]\sqrt{4[f(\theta)]^2 + 4[f'(\theta)]^2}\, d\theta$

$= 4\int_\alpha^\beta 2\pi[f(\theta)\sin\theta]\sqrt{[f(\theta)]^2 + [f'(\theta)]^2}\, d\theta$ which is four times the area of the surface generated by revolving $r = f(\theta)$ about the x-axis for $\alpha \le \theta \le \beta$.

37. $\bar{x} = \dfrac{\frac{2}{3}\int_0^{2\pi} r^3\cos\theta\, d\theta}{\int_0^{2\pi} r^2\, d\theta} = \dfrac{\frac{2}{3}\int_0^{2\pi} [a(1+\cos\theta)]^3(\cos\theta)\, d\theta}{\int_0^{2\pi} [a(1+\cos\theta)]^2\, d\theta} = \dfrac{\frac{2}{3}a^3\int_0^{2\pi} (1 + 3\cos\theta + 3\cos^2\theta + \cos^3\theta)(\cos\theta)\, d\theta}{a^2\int_0^{2\pi} (1 + 2\cos\theta + \cos^2\theta)\, d\theta}$

$= \dfrac{\frac{2}{3}a\int_0^{2\pi}\left[\cos\theta + 3\left(\frac{1+\cos 2\theta}{2}\right) + 3(1 - \sin^2\theta)(\cos\theta) + \left(\frac{1+\cos 2\theta}{2}\right)^2\right] d\theta}{\int_0^{2\pi}\left[1 + 2\cos\theta + \left(\frac{1+\cos 2\theta}{2}\right)\right] d\theta}$ = (After considerable algebra using the identity $\cos^2 A = \frac{1+\cos 2A}{2}\Big)$ $\dfrac{a\int_0^{2\pi}\left(\frac{15}{12}+\frac{8}{3}\cos\theta+\frac{4}{3}\cos 2\theta-2\cos\theta\sin^2\theta+\frac{1}{12}\cos 4\theta\right)d\theta}{\int_0^{2\pi}\left(\frac{3}{2}+2\cos\theta+\frac{1}{2}\cos 2\theta\right)d\theta}$

$$= \frac{a\left[\frac{15}{12}\theta+\frac{8}{3}\sin\theta+\frac{2}{3}\sin 2\theta-\frac{2}{3}\sin^3\theta+\frac{1}{48}\sin 4\theta\right]_0^{2\pi}}{\left[\frac{3}{2}\theta+2\sin\theta+\frac{1}{4}\sin 2\theta\right]_0^{2\pi}} = \frac{a\left(\frac{15}{6}\pi\right)}{3\pi} = \frac{5}{6}a;$$

$$\overline{y} = \frac{\frac{2}{3}\int_0^{2\pi} r^3\sin\theta\,d\theta}{\int_0^{2\pi} r^2\,d\theta} = \frac{\frac{2}{3}\int_0^{2\pi}[a(1+\cos\theta)]^3(\sin\theta)\,d\theta}{3\pi};\ \Big[u = a(1+\cos\theta) \Rightarrow -\tfrac{1}{a}\,du = \sin\theta\,d\theta;\ \theta = 0 \Rightarrow u = 2a;$$

$$\theta = 2\pi \Rightarrow u = 2a\Big] \to \frac{\frac{2}{3}\int_{2a}^{2a} -\frac{1}{a}u^3\,du}{3\pi} = \frac{0}{3\pi} = 0.$$ Therefore the centroid is $(\overline{x},\overline{y}) = \left(\frac{5}{6}a, 0\right)$

38. $$\int_0^{\pi} r^2\,d\theta = \int_0^{\pi} a^2\,d\theta = \left[a^2\theta\right]_0^{\pi} = a^2\pi;\ \overline{x} = \frac{\frac{2}{3}\int_0^{\pi} r^3\cos\theta\,d\theta}{\int_0^{\pi} r^2\,d\theta} = \frac{\frac{2}{3}\int_0^{\pi} a^3\cos\theta\,d\theta}{a^2\pi} = \frac{\frac{2}{3}a^3[-\sin\theta]_0^{\pi}}{a^2\pi} = \frac{0}{a^2\pi} = 0;$$

$$\overline{y} = \frac{\frac{2}{3}\int_0^{\pi} r^3\sin\theta\,d\theta}{\int_0^{\pi} r^2\,d\theta} = \frac{\frac{2}{3}\int_0^{\pi} a^3\sin\theta\,d\theta}{a^2\pi} = \frac{\frac{2}{3}a^3[-\cos\theta]_0^{\pi}}{a^2\pi} = \frac{\left(\frac{4}{3}\right)a^3}{a^2\pi} = \frac{4a}{3\pi}.$$ Therefore the centroid is $(\overline{x},\overline{y}) = \left(0, \frac{4a}{3\pi}\right)$.

CHAPTER 9 PRACTICE EXERCISES

1. $x^2 = -4y \Rightarrow y = -\frac{x^2}{4} \Rightarrow 4p = 4 \Rightarrow p = 1;$

therefore Focus is $(0,-1)$, Directrix is $y = 1$

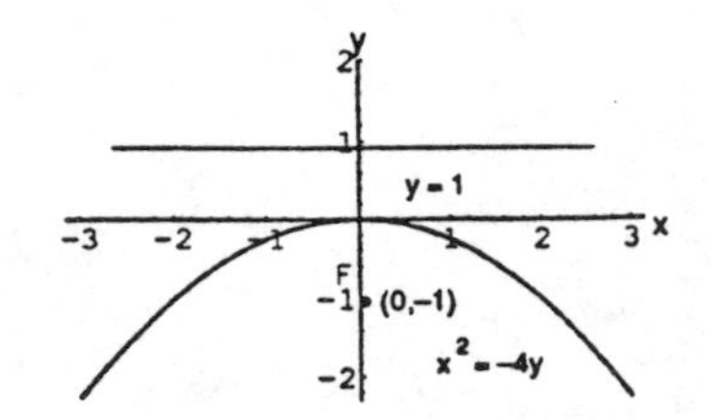

2. $x^2 = 2y \Rightarrow \frac{x^2}{2} = y \Rightarrow 4p = 2 \Rightarrow p = \frac{1}{2};$

therefore Focus is $\left(0,\frac{1}{2}\right)$; Directrix is $y = -\frac{1}{2}$

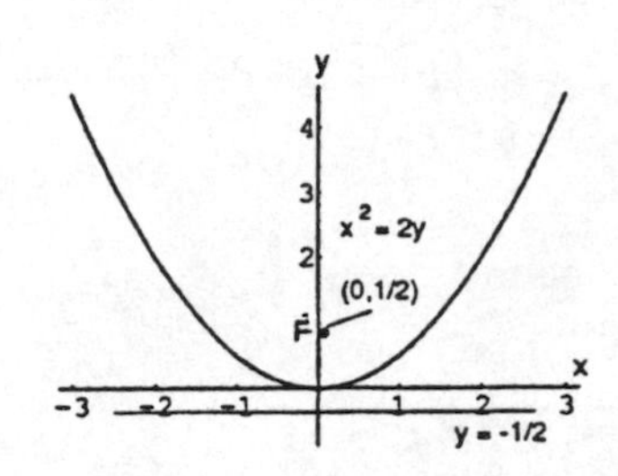

3. $y^2 = 3x \Rightarrow x = \frac{y^2}{3} \Rightarrow 4p = 3 \Rightarrow p = \frac{3}{4}$;

therefore Focus is $\left(\frac{3}{4}, 0\right)$, Directrix is $x = -\frac{3}{4}$

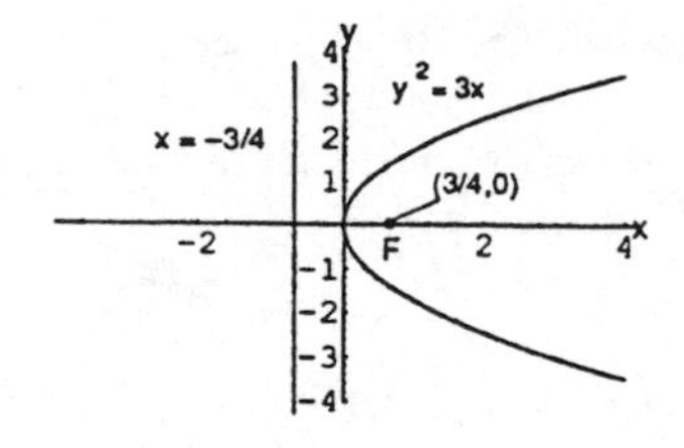

4. $y^2 = -\frac{8}{3}x \Rightarrow x = -\frac{y^2}{\left(\frac{8}{3}\right)} \Rightarrow 4p = \frac{8}{3} \Rightarrow p = \frac{2}{3}$;

therefore Focus is $\left(-\frac{2}{3}, 0\right)$, Directrix is $x = \frac{2}{3}$

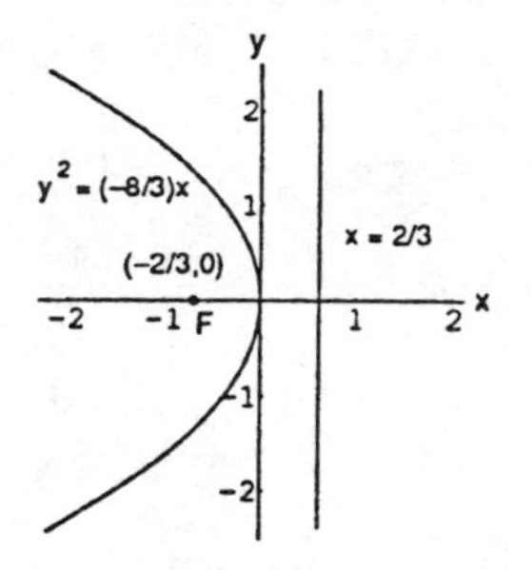

5. $16x^2 + 7y^2 = 112 \Rightarrow \frac{x^2}{7} + \frac{y^2}{16} = 1$

$\Rightarrow c^2 = 16 - 7 = 9 \Rightarrow c = 3$; $e = \frac{c}{a} = \frac{3}{4}$

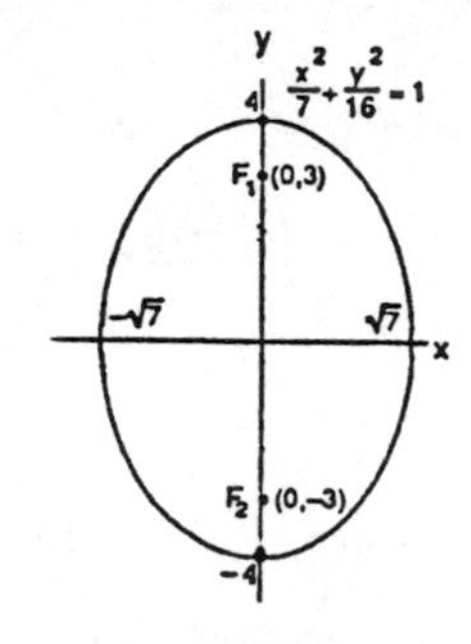

6. $x^2 + 2y^2 = 4 \Rightarrow \frac{x^2}{4} + \frac{y^2}{2} = 1 \Rightarrow c^2 = 4 - 2 = 2$

$\Rightarrow c = \sqrt{2}$; $e = \frac{c}{a} = \frac{\sqrt{2}}{2}$

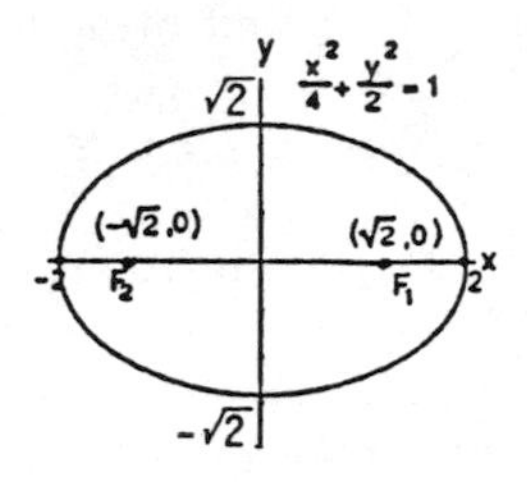

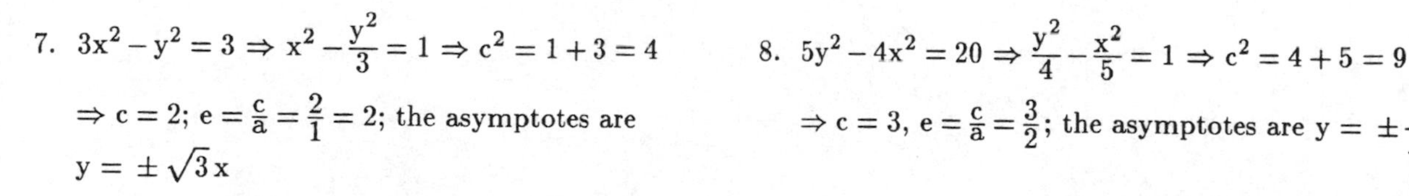

7. $3x^2 - y^2 = 3 \Rightarrow x^2 - \frac{y^2}{3} = 1 \Rightarrow c^2 = 1 + 3 = 4$

$\Rightarrow c = 2$; $e = \frac{c}{a} = \frac{2}{1} = 2$; the asymptotes are $y = \pm\sqrt{3}x$

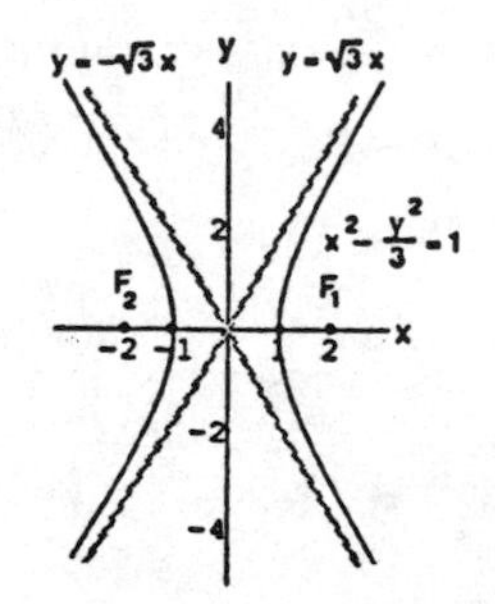

8. $5y^2 - 4x^2 = 20 \Rightarrow \frac{y^2}{4} - \frac{x^2}{5} = 1 \Rightarrow c^2 = 4 + 5 = 9$

$\Rightarrow c = 3$, $e = \frac{c}{a} = \frac{3}{2}$; the asymptotes are $y = \pm\frac{2}{\sqrt{5}}x$

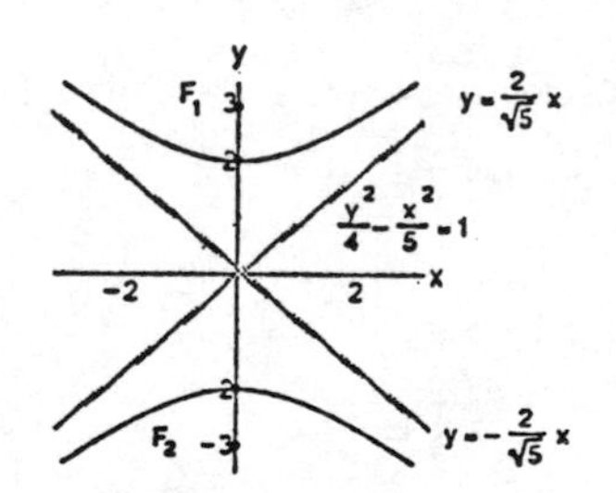

9. $x^2 = -12y \Rightarrow -\frac{x^2}{12} = y \Rightarrow 4p = 12 \Rightarrow p = 3 \Rightarrow$ focus is $(0, -3)$, directrix is $y = 3$, vertex is $(0, 0)$; therefore new vertex is $(2, 3)$, new focus is $(2, 0)$, new directrix is $y = 6$, and the new equation is $(x - 2)^2 = -12(y - 3)$

10. $y^2 = 10x \Rightarrow \frac{y^2}{10} = x \Rightarrow 4p = 10 \Rightarrow p = \frac{5}{2} \Rightarrow$ focus is $\left(\frac{5}{2}, 0\right)$, directrix is $x = -\frac{5}{2}$, vertex is $(0,0)$; therefore new vertex is $\left(-\frac{1}{2}, -1\right)$, new focus is $(2,-1)$, new directrix is $x = -3$, and the new equation is $(y+1)^2 = 10\left(x + \frac{1}{2}\right)$

11. $\frac{x^2}{9} + \frac{y^2}{25} = 1 \Rightarrow a = 5$ and $b = 3 \Rightarrow c = \sqrt{25 - 9} = 4 \Rightarrow$ foci are $(0, \pm 4)$, vertices are $(0, \pm 5)$, center is $(0,0)$; therefore the new center is $(-3,-5)$, new foci are $(-3,-1)$ and $(-3,-9)$, new vertices are $(-3,-10)$ and $(-3,0)$, and the new equation is $\frac{(x+3)^2}{9} + \frac{(y+5)^2}{25} = 1$

12. $\frac{x^2}{169} + \frac{y^2}{144} = 1 \Rightarrow a = 13$ and $b = 12 \Rightarrow c = \sqrt{169 - 144} = 5 \Rightarrow$ foci are $(\pm 5, 0)$, vertices are $(\pm 13, 0)$, center is $(0,0)$; therefore the new center is $(5,12)$, new foci are $(10,12)$ and $(0,12)$, new vertices are $(18,12)$ and $(-8,12)$, and the new equation is $\frac{(x-5)^2}{169} + \frac{(y-12)^2}{144} = 1$

13. $\frac{y^2}{8} - \frac{x^2}{2} = 1 \Rightarrow a = 2\sqrt{2}$ and $b = \sqrt{2} \Rightarrow c = \sqrt{8+2} = \sqrt{10} \Rightarrow$ foci are $(0, \pm\sqrt{10})$, vertices are $(0, \pm 2\sqrt{2})$, center is $(0,0)$, and the asymptotes are $y = \pm 2x$; therefore the new center is $(2, 2\sqrt{2})$, new foci are $(2, 2\sqrt{2} \pm \sqrt{10})$, new vertices are $(2, 4\sqrt{2})$ and $(2,0)$, the new asymptotes are $y = 2x - 4 + 2\sqrt{2}$ and $y = -2x + 4 + 2\sqrt{2}$; the new equation is $\frac{\left(y - 2\sqrt{2}\right)^2}{8} - \frac{(x-2)^2}{2} = 1$

14. $\frac{x^2}{36} - \frac{y^2}{64} = 1 \Rightarrow a = 6$ and $b = 8 \Rightarrow c = \sqrt{36 + 64} = 10 \Rightarrow$ foci are $(\pm 10, 0)$, vertices are $(\pm 6, 0)$, the center is $(0,0)$ and the asymptotes are $\frac{y}{8} = \pm \frac{x}{6}$ or $y = \pm \frac{4}{3}x$; therefore the new center is $(-10,-3)$, the new foci are $(-20,-3)$ and $(0,-3)$, the new vertices are $(-16,-3)$ and $(-4,-3)$, the new asymptotes are $y = \frac{4}{3}x + \frac{31}{3}$ and $y = -\frac{4}{3}x - \frac{49}{3}$; the new equation is $\frac{(x+10)^2}{36} - \frac{(y+3)^2}{64} = 1$

15. $x^2 - 4x - 4y^2 = 0 \Rightarrow x^2 - 4x + 4 - 4y^2 = 4 \Rightarrow (x-2)^2 - 4y^2 = 4 \Rightarrow \frac{(x-2)^2}{4} - y^2 = 1$, a hyperbola; $a = 2$ and $b = 1 \Rightarrow c = \sqrt{1+4} = \sqrt{5}$; the center is $(2,0)$, the vertices are $(0,0)$ and $(4,0)$; the foci are $(2 \pm \sqrt{5}, 0)$ and the asymptotes are $y = \pm \frac{x-2}{2}$

16. $4x^2 - y^2 + 4y = 8 \Rightarrow 4x^2 - y^2 + 4y - 4 = 4 \Rightarrow 4x^2 - (y-2)^2 = 4 \Rightarrow x^2 - \frac{(y-2)^2}{4} = 1$, a hyperbola; $a = 1$ and $b = 2 \Rightarrow c = \sqrt{1+4} = \sqrt{5}$; the center is $(0,2)$, the vertices are $(1,2)$ and $(-1,2)$, the foci are $(\pm\sqrt{5}, 2)$ and the asymptotes are $y = \pm 2x + 2$

17. $y^2 - 2y + 16x = -49 \Rightarrow y^2 - 2y + 1 = -16x - 48 \Rightarrow (y-1)^2 = -16(x+3)$, a parabola; the vertex is $(-3,1)$; $4p = 16 \Rightarrow p = 4 \Rightarrow$ the focus is $(-7,1)$ and the directrix is $x = 1$

18. $x^2 - 2x + 8y = -17 \Rightarrow x^2 - 2x + 1 = -8y - 16 \Rightarrow (x-1)^2 = -8(y+2)$, a parabola; the vertex is $(1,-2)$; $4p = 8 \Rightarrow p = 2 \Rightarrow$ the focus is $(1,-4)$ and the directrix is $y = 0$

19. $9x^2 + 16y^2 + 54x - 64y = -1 \Rightarrow 9(x^2 + 6x) + 16(y^2 - 4y) = -1 \Rightarrow 9(x^2 + 6x + 9) + 16(y^2 - 4y + 4) = 144$ $\Rightarrow 9(x+3)^2 + 16(y-2)^2 = 144 \Rightarrow \frac{(x+3)^2}{16} + \frac{(y-2)^2}{9} = 1$, an ellipse; the center is $(-3, 2)$; $a = 4$ and $b = 3$ $\Rightarrow c = \sqrt{16 - 9} = \sqrt{7}$; the foci are $(-3 \pm \sqrt{7}, 2)$; the vertices are $(1, 2)$ and $(-7, 2)$

20. $25x^2 + 9y^2 - 100x + 54y = 44 \Rightarrow 25(x^2 - 4x) + 9(y^2 + 6y) = 44 \Rightarrow 25(x^2 - 4x + 4) + 9(y^2 + 6y + 9) = 225$ $\Rightarrow \frac{(x-2)^2}{9} + \frac{(y+3)^2}{25} = 1$, an ellipse; the center is $(2, -3)$; $a = 5$ and $b = 3 \Rightarrow c = \sqrt{25 - 9} = 4$; the foci are $(2, 1)$ and $(2, -7)$; the vertices are $(2, 2)$ and $(2, -8)$

21. $x^2 + y^2 - 2x - 2y = 0 \Rightarrow x^2 - 2x + 1 + y^2 - 2y + 1 = 2 \Rightarrow (x-1)^2 + (y-1)^2 = 2$, a circle with center $(1, 1)$ and radius $= \sqrt{2}$

22. $x^2 + y^2 + 4x + 2y = 1 \Rightarrow x^2 + 4x + 4 + y^2 + 2y + 1 = 6 \Rightarrow (x+2)^2 + (y+1)^2 = 6$, a circle with center $(-2, -1)$ and radius $= \sqrt{6}$

23. $B^2 - 4AC = 1 - 4(1)(1) = -3 < 0 \Rightarrow$ ellipse

24. $B^2 - 4AC = 4^2 - 4(1)(4) = 0 \Rightarrow$ parabola

25. $B^2 - 4AC = 3^2 - 4(1)(2) = 1 > 0 \Rightarrow$ hyperbola

26. $B^2 - 4AC = 2^2 - 4(1)(-2) = 12 > 0 \Rightarrow$ hyperbola

27. $x^2 - 2xy + y^2 = 0 \Rightarrow (x-y)^2 = 0 \Rightarrow x - y = 0$ or $y = x$, a straight line

28. $B^2 - 4AC = (-3)^2 - 4(1)(4) = -7 < 0 \Rightarrow$ ellipse

29. $B^2 - 4AC = 1^2 - 4(2)(2) = -15 < 0 \Rightarrow$ ellipse; $\cot 2\alpha = \frac{A-C}{B} = 0 \Rightarrow 2\alpha = \frac{\pi}{2} \Rightarrow \alpha = \frac{\pi}{4}$; $x = \frac{\sqrt{2}}{2}x' - \frac{\sqrt{2}}{2}y'$ and $y = \frac{\sqrt{2}}{2}x' + \frac{\sqrt{2}}{2}y' \Rightarrow 2\left(\frac{\sqrt{2}}{2}x' - \frac{\sqrt{2}}{2}y'\right)^2 + \left(\frac{\sqrt{2}}{2}x' - \frac{\sqrt{2}}{2}y'\right)\left(\frac{\sqrt{2}}{2}x' + \frac{\sqrt{2}}{2}y'\right) + 2\left(\frac{\sqrt{2}}{2}x' + \frac{\sqrt{2}}{2}y'\right)^2 - 15 = 0$ $\Rightarrow 5x'^2 + 3y'^2 = 30$

30. $B^2 - 4AC = 2^2 - 4(3)(3) = -32 < 0 \Rightarrow$ ellipse; $\cot 2\alpha = \frac{A-C}{B} = 0 \Rightarrow 2\alpha = \frac{\pi}{2} \Rightarrow \alpha = \frac{\pi}{4}$; $x = \frac{\sqrt{2}}{2}x' - \frac{\sqrt{2}}{2}y'$ and $y = \frac{\sqrt{2}}{2}x' + \frac{\sqrt{2}}{2}y' \Rightarrow 3\left(\frac{\sqrt{2}}{2}x' - \frac{\sqrt{2}}{2}y'\right)^2 + 2\left(\frac{\sqrt{2}}{2}x' - \frac{\sqrt{2}}{2}y'\right)\left(\frac{\sqrt{2}}{2}x' + \frac{\sqrt{2}}{2}y'\right) + 3\left(\frac{\sqrt{2}}{2}x' + \frac{\sqrt{2}}{2}y'\right)^2 = 19$ $\Rightarrow 4x'^2 + 2y'^2 = 19$

31. $B^2 - 4AC = (2\sqrt{3})^2 - 4(1)(-1) = 16 \Rightarrow$ hyperbola; $\cot 2\alpha = \frac{A-C}{B} = \frac{1}{\sqrt{3}} \Rightarrow 2\alpha = \frac{\pi}{3} \Rightarrow \alpha = \frac{\pi}{6}$; $x = \frac{\sqrt{3}}{2}x' - \frac{1}{2}y'$ and $y = \frac{1}{2}x' + \frac{\sqrt{3}}{2}y' \Rightarrow \left(\frac{\sqrt{3}}{2}x' - \frac{1}{2}y'\right)^2 + 2\sqrt{3}\left(\frac{\sqrt{3}}{2}x' - \frac{1}{2}y'\right)\left(\frac{1}{2}x' + \frac{\sqrt{3}}{2}y'\right) - \left(\frac{1}{2}x' + \frac{\sqrt{3}}{2}y'\right)^2 = 4$ $\Rightarrow 2x'^2 - 2y'^2 = 4 \Rightarrow x'^2 - y'^2 = 2$

32. $B^2 - 4AC = (-3)^2 - 4(1)(1) = 5 > 0 \Rightarrow$ hyperbola; $\cot 2\alpha = \frac{A-C}{B} = 0 \Rightarrow 2\alpha = \frac{\pi}{2} \Rightarrow \alpha = \frac{\pi}{4}$; $x = \frac{\sqrt{2}}{2}x' - \frac{\sqrt{2}}{2}y'$ and $y = \frac{\sqrt{2}}{2}x' + \frac{\sqrt{2}}{2}y' \Rightarrow \left(\frac{\sqrt{2}}{2}x' - \frac{\sqrt{2}}{2}y'\right)^2 - 3\left(\frac{\sqrt{2}}{2}x' - \frac{\sqrt{2}}{2}y'\right)\left(\frac{\sqrt{2}}{2}x' + \frac{\sqrt{2}}{2}y'\right) + \left(\frac{\sqrt{2}}{2}x' + \frac{\sqrt{2}}{2}y'\right)^2 = 5$ $\Rightarrow \frac{5}{2}y'^2 - \frac{1}{2}x'^2 = 5$ or $5y'^2 - x'^2 = 10$

33. $x = \frac{t}{2}$ and $y = t + 1 \Rightarrow 2x = t \Rightarrow y = 2x + 1$

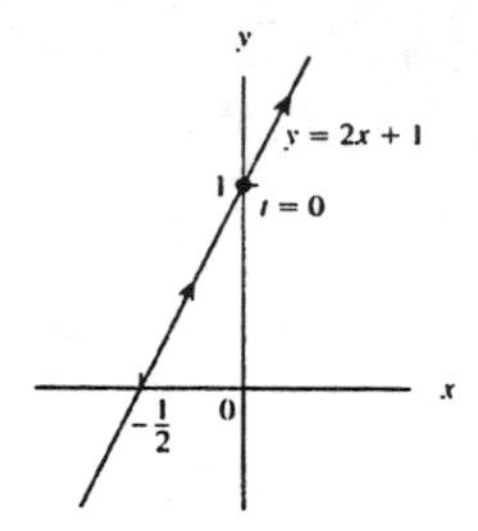

34. 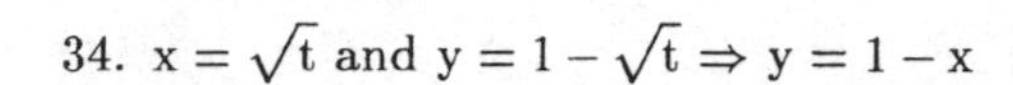 $x = \sqrt{t}$ and $y = 1 - \sqrt{t} \Rightarrow y = 1 - x$

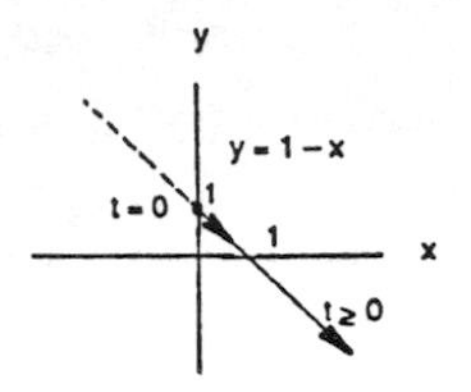

35. $x = \frac{1}{2}\tan t$ and $y = \frac{1}{2}\sec t \Rightarrow x^2 = \frac{1}{4}\tan^2 t$ and $y^2 = \frac{1}{4}\sec^2 t \Rightarrow 4x^2 = \tan^2 t$ and $4y^2 = \sec^2 t \Rightarrow 4x^2 + 1 = 4y^2 \Rightarrow 4y^2 - 4x^2 = 1$

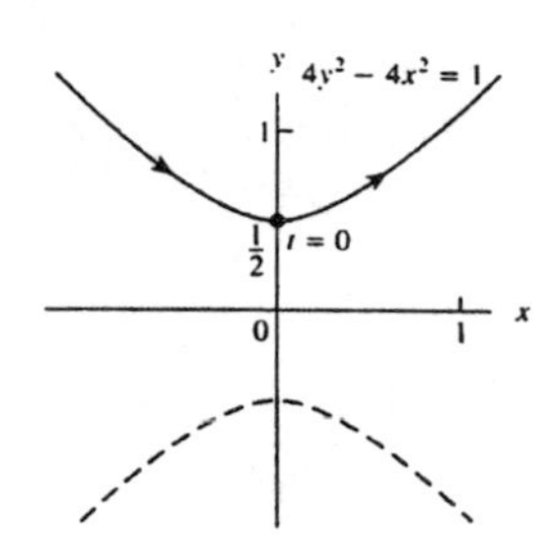

36. 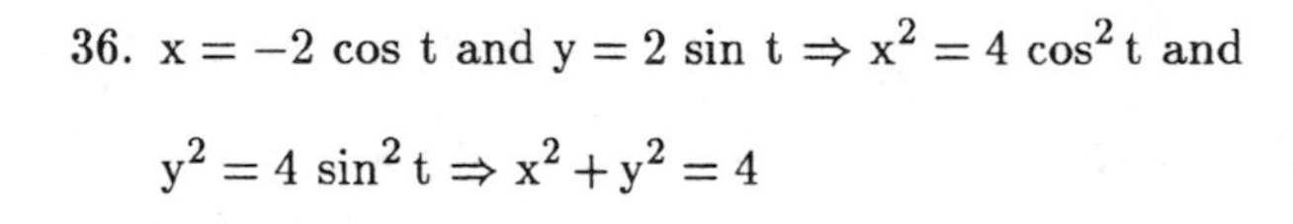 $x = -2\cos t$ and $y = 2\sin t \Rightarrow x^2 = 4\cos^2 t$ and $y^2 = 4\sin^2 t \Rightarrow x^2 + y^2 = 4$

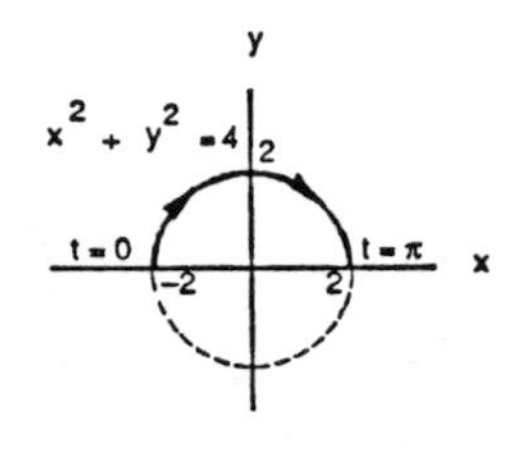

37. $x = -\cos t$ and $y = \cos^2 t \Rightarrow y = (-x)^2 = x^2$

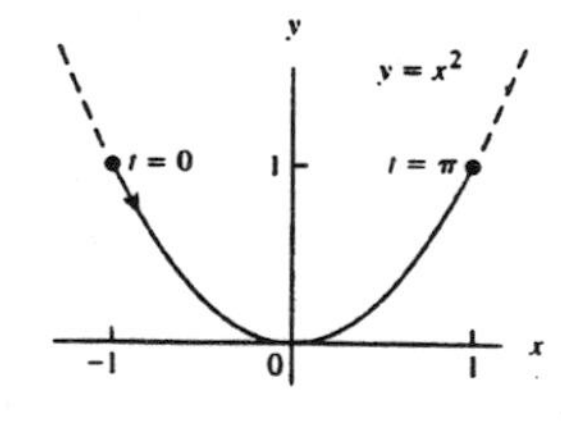

38. 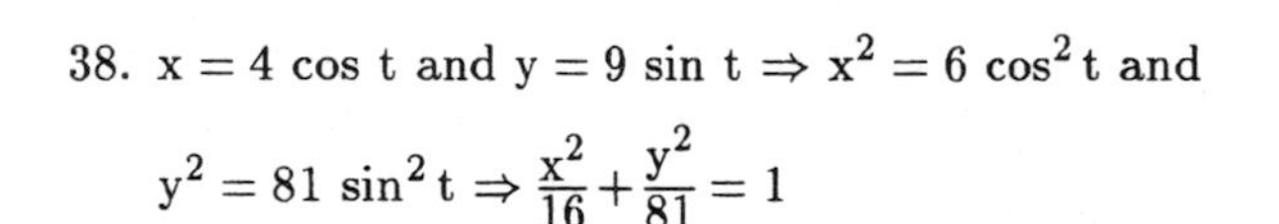 $x = 4\cos t$ and $y = 9\sin t \Rightarrow x^2 = 6\cos^2 t$ and $y^2 = 81\sin^2 t \Rightarrow \frac{x^2}{16} + \frac{y^2}{81} = 1$

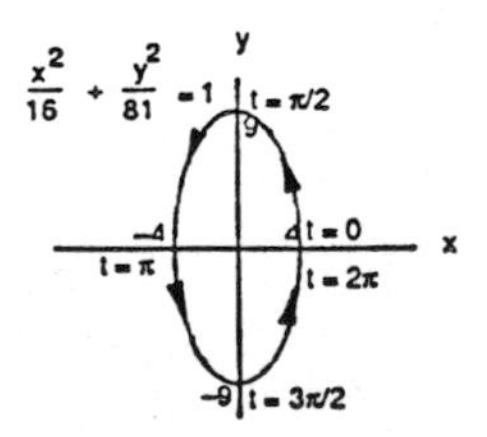

39. $16x^2 + 9y^2 = 144 \Rightarrow \frac{x^2}{9} + \frac{y^2}{16} = 1 \Rightarrow a = 3$ and $b = 4 \Rightarrow x = 3\cos t$ and $y = 4\sin t$, $0 \le t \le 2\pi$

40. $x^2 + y^2 = 4 \Rightarrow x = -2\cos t$ and $y = 2\sin t$, $0 \le t \le 6\pi$

41. $x = \frac{1}{2}\tan t,\ y = \frac{1}{2}\sec t \Rightarrow \frac{dy}{dx} = \frac{dy/dt}{dx/dt} = \frac{\frac{1}{2}\sec t\tan t}{\frac{1}{2}\sec^2 t} = \frac{\tan t}{\sec t} = \sin t \Rightarrow \left.\frac{dy}{dx}\right|_{t=\pi/3} = \sin\frac{\pi}{3} = \frac{\sqrt{3}}{2};\ t = \frac{\pi}{3}$
$\Rightarrow x = \frac{1}{2}\tan\frac{\pi}{3} = \frac{\sqrt{3}}{2}$ and $y = \frac{1}{2}\sec\frac{\pi}{3} = 1 \Rightarrow y = \frac{\sqrt{3}}{2}x + \frac{1}{4};\ \frac{d^2y}{dx^2} = \frac{dy'/dt}{dx/dt} = \frac{\cos t}{\frac{1}{2}\sec^2 t} = 2\cos^3 t \Rightarrow \left.\frac{d^2y}{dx^2}\right|_{t=\pi/3}$
$= 2\cos^3\left(\frac{\pi}{3}\right) = \frac{1}{4}$

42. $x = 1 + \frac{1}{t^2},\ y = 1 - \frac{3}{t} \Rightarrow \frac{dy}{dx} = \frac{dy/dt}{dx/dt} = \frac{\left(\frac{3}{t^2}\right)}{\left(-\frac{2}{t^3}\right)} = -\frac{3}{2}t \Rightarrow \left.\frac{dy}{dx}\right|_{t=2} = -\frac{3}{2}(2) = -3;\ t = 2 \Rightarrow x = 1 + \frac{1}{2^2} = \frac{5}{4}$ and
$y = 1 - \frac{3}{2} = -\frac{1}{2} \Rightarrow y = -3x + \frac{13}{4};\ \frac{d^2y}{dx^2} = \frac{dy'/dt}{dx/dt} = \frac{\left(-\frac{3}{2}\right)}{\left(-\frac{2}{t^3}\right)} = \frac{3}{4}t^3 \Rightarrow \left.\frac{d^2y}{dx^2}\right|_{t=2} = \frac{3}{4}(2)^3 = 6$

43. $x = e^{2t} - \frac{t}{8}$ and $y = e^t,\ 0 \le t \le \ln 2 \Rightarrow \frac{dx}{dt} = 2e^{2t} - \frac{1}{8}$ and $\frac{dy}{dt} = e^t \Rightarrow \text{Length} = \int_0^{\ln 2}\sqrt{\left(2e^{2t} - \frac{1}{8}\right)^2 + \left(e^t\right)^2}\,dt$
$= \int_0^{\ln 2}\sqrt{4e^{4t} + \frac{1}{2}e^{2t} + \frac{1}{64}}\,dt = \int_0^{\ln 2}\sqrt{\left(2e^{2t} + \frac{1}{8}\right)^2}\,dt = \int_0^{\ln 2}\left(2e^{2t} + \frac{1}{8}\right)dt = \left[e^{2t} + \frac{t}{8}\right]_0^{\ln 2} = 3 + \frac{\ln 2}{8}$

44. $x = t^2$ and $y = \frac{t^3}{3} - t,\ -\sqrt{3} \le t \le \sqrt{3} \Rightarrow \frac{dx}{dt} = 2t$ and $\frac{dy}{dt} = t^2 - 1 \Rightarrow \text{Length} = \int_{-\sqrt{3}}^{\sqrt{3}}\sqrt{(2t)^2 + \left(t^2 - 1\right)^2}\,dt$
$= \int_{-\sqrt{3}}^{\sqrt{3}}\sqrt{t^4 + 2t^2 + 1}\,dt = \int_{-\sqrt{3}}^{\sqrt{3}}\sqrt{\left(t^2 + 1\right)^2}\,dt = \int_{-\sqrt{3}}^{\sqrt{3}}\left(t^2 + 1\right)dt = \left[\frac{t^3}{3} + t\right]_{-\sqrt{3}}^{\sqrt{3}} = 4\sqrt{3}$

45. $x = \frac{t^2}{2}$ and $y = 2t,\ 0 \le t \le \sqrt{5} \Rightarrow \frac{dx}{dt} = t$ and $\frac{dy}{dt} = 2 \Rightarrow \text{Surface Area} = \int_0^{\sqrt{5}} 2\pi(2t)\sqrt{t^2 + 4}\,dt = \int_4^9 2\pi u^{1/2}\,du$
$= 2\pi\left[\frac{2}{3}u^{3/2}\right]_4^9 = \frac{76\pi}{3}$, where $u = t^2 + 4 \Rightarrow du = 2t\,dt;\ t = 0 \Rightarrow u = 4,\ t = \sqrt{5} \Rightarrow u = 9$

46. $x = t^2 + \frac{1}{2t}$ and $y = 4\sqrt{t},\ \frac{1}{\sqrt{2}} \le t \le 1 \Rightarrow \frac{dx}{dt} = 2t - \frac{1}{2t^2}$ and $\frac{dy}{dt} = \frac{2}{\sqrt{t}}$
$\Rightarrow \text{Surface Area} = \int_{1/\sqrt{2}}^{1} 2\pi\left(t^2 + \frac{1}{2t}\right)\sqrt{\left(2t - \frac{1}{2t^2}\right)^2 + \left(\frac{2}{\sqrt{t}}\right)^2}\,dt = 2\pi\int_{1/\sqrt{2}}^{1}\left(t^2 + \frac{1}{2t}\right)\sqrt{\left(2t + \frac{1}{2t^2}\right)^2}\,dt$
$= 2\pi\int_{1/\sqrt{2}}^{1}\left(t^2 + \frac{1}{2t}\right)\left(2t + \frac{1}{2t^2}\right)dt = 2\pi\int_{1/\sqrt{2}}^{1}\left(2t^3 + \frac{3}{2} + \frac{1}{4}t^{-3}\right)dt = 2\pi\left[\frac{1}{2}t^4 + \frac{3}{2}t - \frac{1}{8}t^{-2}\right]_{1/\sqrt{2}}^{1}$
$= 2\pi\left(2 - \frac{3\sqrt{2}}{4}\right)$

47.

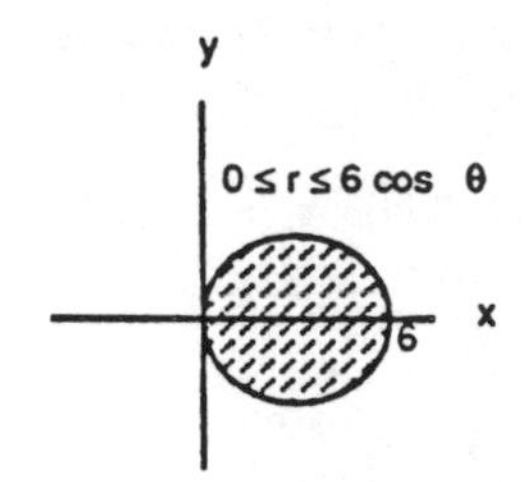

48.

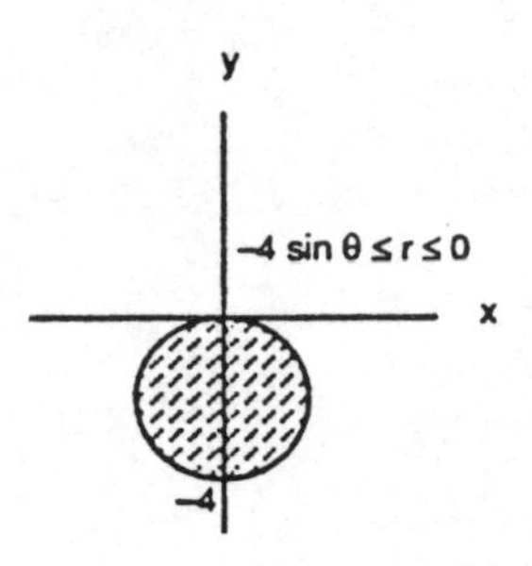

49. d
50. e
51. l
52. f
53. k
54. h
55. i
56. j

57. $r = \sin\theta$ and $r = 1 + \sin\theta \Rightarrow \sin\theta = 1 + \sin\theta \Rightarrow 0 = 1$ so no solutions exist. There are no points of intersection found by solving the system. The point of intersection $(0,0)$ is found by graphing.

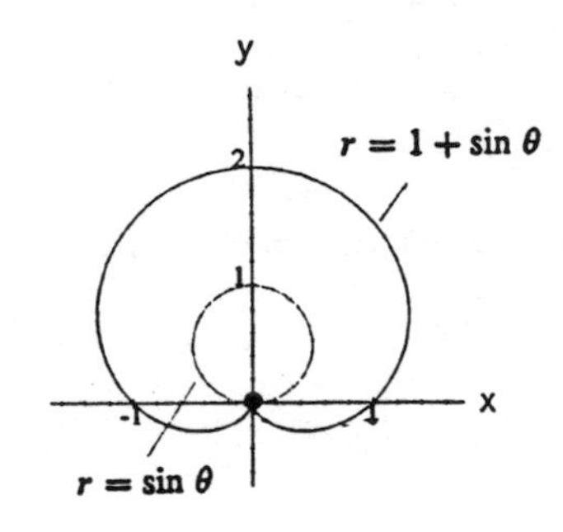

58. $r = \cos\theta$ and $r = 1 - \cos\theta \Rightarrow \cos\theta = 1 - \cos\theta$ $\Rightarrow \cos\theta = \frac{1}{2} \Rightarrow \theta = \frac{\pi}{3}, -\frac{\pi}{3};\ \theta = \frac{\pi}{3} \Rightarrow r = \frac{1}{2};\ \theta = -\frac{\pi}{3}$ $\Rightarrow r = \frac{1}{2}$. The points of intersection are $\left(\frac{1}{2}, \frac{\pi}{3}\right)$ and $\left(\frac{1}{2}, -\frac{\pi}{3}\right)$. The point of intersection $(0,0)$ is found by graphing.

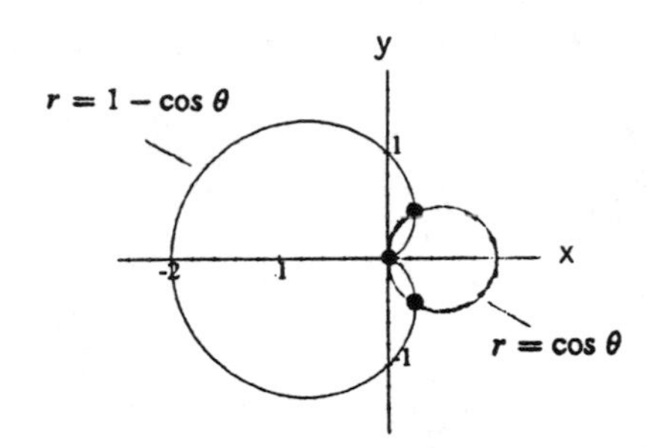

59. $r = 1 + \cos\theta$ and $r = 1 - \cos\theta \Rightarrow 1 + \cos\theta = 1 - \cos\theta$ $\Rightarrow 2\cos\theta = 0 \Rightarrow \cos\theta = 0 \Rightarrow \theta = \frac{\pi}{2}, \frac{3\pi}{2};\ \theta = \frac{\pi}{2}$ or $\frac{3\pi}{2}$ $\Rightarrow r = 1$. The points of intersection are $\left(1, \frac{\pi}{2}\right)$ and $\left(1, \frac{3\pi}{2}\right)$. The point of intersection $(0,0)$ is found by graphing.

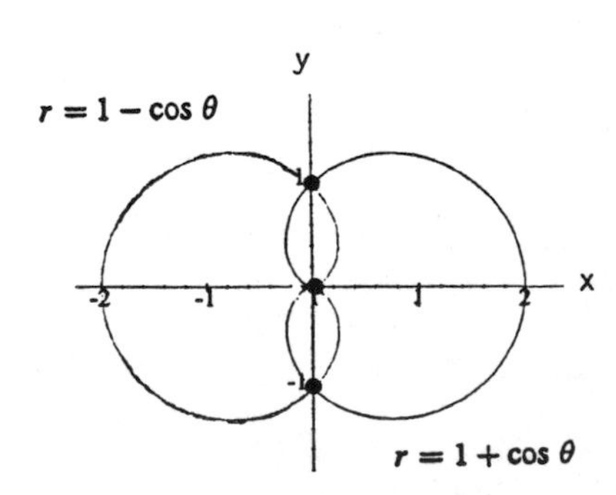

60. $r = 1 + \sin\theta$ and $r = 1 - \sin\theta \Rightarrow 1 + \sin\theta = 1 - \sin\theta$
$\Rightarrow 2\sin\theta = 0 \Rightarrow \sin\theta = 0 \Rightarrow \theta = 0,\ \pi;\ \theta = 0$ or π
$\Rightarrow r = 1$. The points of intersection are $(1, 0)$ and $(1, \pi)$.
The point of intersection $(0, 0)$ is found by graphing.

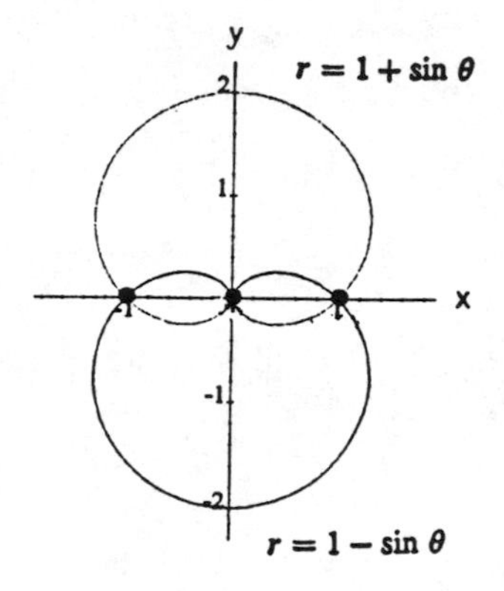

61. $r = 1 + \sin\theta$ and $r = -1 + \sin\theta$ intersect at all points of $r = 1 + \sin\theta$ because the graphs coincide. This can be seen by graphing them.

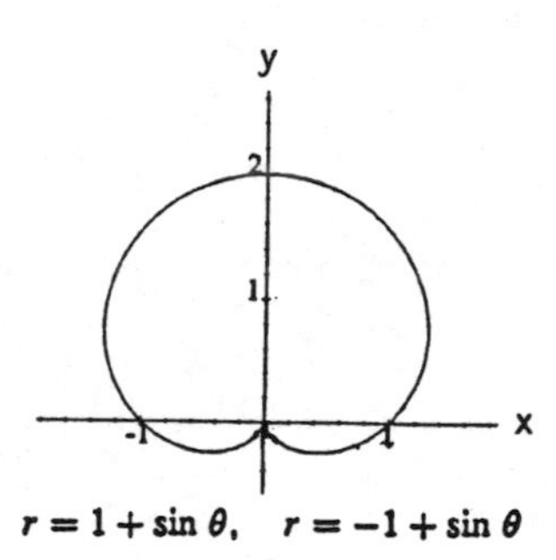

$r = 1 + \sin\theta,\quad r = -1 + \sin\theta$

62. $r = 1 + \cos\theta$ and $r = -1 + \cos\theta$ intersect at all points of $r = 1 + \cos\theta$ because the graphs coincide. This can be seen by graphing them.

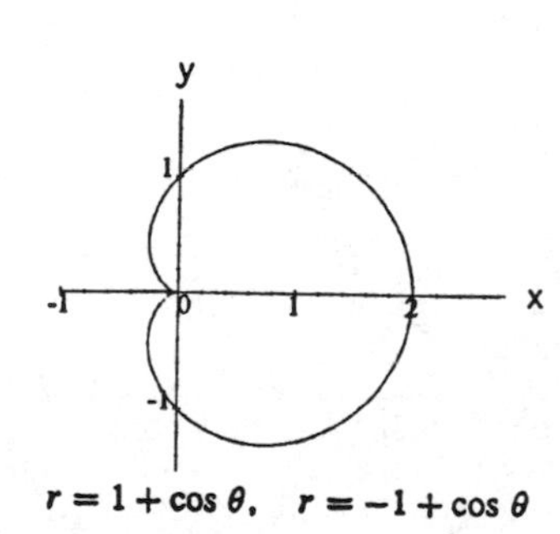

$r = 1 + \cos\theta,\quad r = -1 + \cos\theta$

63. $r = \sec\theta$ and $r = 2\sin\theta \Rightarrow \sec\theta = 2\sin\theta$
$\Rightarrow 1 = 2\sin\theta\cos\theta \Rightarrow 1 = \sin 2\theta \Rightarrow 2\theta = \frac{\pi}{2} \Rightarrow \theta = \frac{\pi}{4}$
$\Rightarrow r = 2\sin\frac{\pi}{4} = \sqrt{2} \Rightarrow$ the point of intersection is
$\left(\sqrt{2}, \frac{\pi}{4}\right)$. No other points of intersection exist.

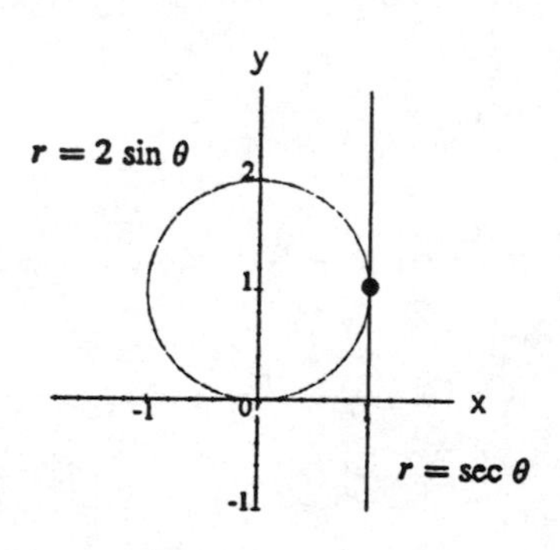

64. $r = -2 \csc \theta$ and $r = -4 \cos \theta \Rightarrow -2 \csc \theta = -4 \cos \theta$
$\Rightarrow 1 = 2 \sin \theta \cos \theta \Rightarrow 1 = \sin 2\theta \Rightarrow 2\theta = \frac{\pi}{2}, \frac{5\pi}{2}$
$\Rightarrow \theta = \frac{\pi}{4}, \frac{5\pi}{4}; \theta = \frac{\pi}{4} \Rightarrow r = -4 \cos \frac{\pi}{4} = -2\sqrt{2};$
$\theta = \frac{5\pi}{4} \Rightarrow r = -4 \cos \frac{5\pi}{4} = 2\sqrt{2}$. The point of intersection is $\left(2\sqrt{2}, \frac{5\pi}{4}\right)$ and the point $\left(-2\sqrt{2}, \frac{\pi}{4}\right)$ is the same point.

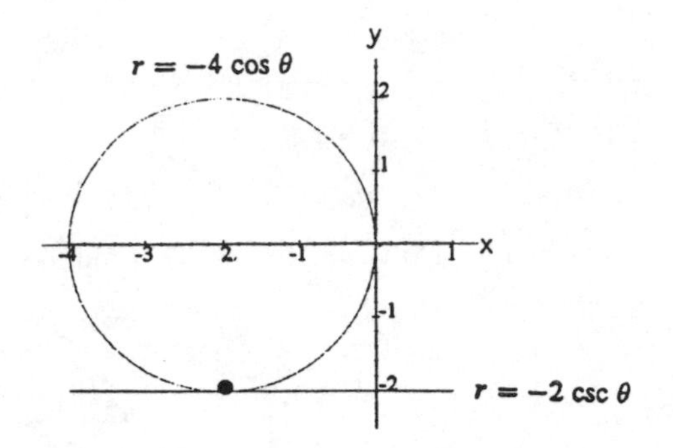

65. $r^2 = \cos 2\theta \Rightarrow r = 0$ when $\cos 2\theta = 0 \Rightarrow 2\theta = \frac{\pi}{2}, \frac{3\pi}{2} \Rightarrow \theta = \frac{\pi}{4}, \frac{3\pi}{4}; \theta_1 = \frac{\pi}{4} \Rightarrow m_1 = \tan \frac{\pi}{4} = 1 \Rightarrow y = x$ is one tangent line; $\theta_2 = \frac{3\pi}{4} \Rightarrow m_2 = \tan \frac{3\pi}{4} = -1 \Rightarrow y = -x$ is the other tangent line

66. $r^2 = 2 \cos \theta + 1 \Rightarrow r = 0$ when $2 \cos \theta + 1 = 0 \Rightarrow \cos \theta = -\frac{1}{2} \Rightarrow \theta = \frac{2\pi}{3}, \frac{4\pi}{3}; \theta_1 = \frac{2\pi}{3} \Rightarrow m_1 = \tan \frac{2\pi}{3} = -\sqrt{3}$
$\Rightarrow y = -\sqrt{3}x$ is one tangent line; $\theta_2 = \frac{4\pi}{3} \Rightarrow m_2 = \tan \frac{4\pi}{3} = \sqrt{3} \Rightarrow y = \sqrt{3}x$ is the other tangent line

67. The tips of the petals are at $\theta = \frac{\pi}{4}, \frac{3\pi}{4}, \frac{5\pi}{4}, \frac{7\pi}{4}$ and $r = 1$ at those values of θ. Then for $\theta = \frac{\pi}{4}$, the tangent line is $r \cos\left(\theta - \frac{\pi}{4}\right) = 1$; for $\theta = \frac{3\pi}{4}$, $r \cos\left(\theta - \frac{3\pi}{4}\right) = 1$; for $\theta = \frac{5\pi}{4}$, $r \cos\left(\theta - \frac{5\pi}{4}\right) = 1$; and for $\theta = \frac{7\pi}{4}$, $r \cos\left(\theta - \frac{7\pi}{4}\right) = 1$.

68. $r = 1 + \sin \theta$ crosses the x-axis at $(1,0)$, $(1,\pi)$, and $\left(0, \frac{3\pi}{2}\right)$, but at $\left(0, \frac{3\pi}{2}\right)$ there is no tangent line (see the graph in Exercise 61). Now, $\frac{dr}{d\theta} = \cos \theta \Rightarrow \text{Slope} = \frac{r' \sin \theta + r \cos \theta}{r' \cos \theta - r \sin \theta} = \frac{(\cos \theta)(\sin \theta) + (1 + \sin \theta)(\cos \theta)}{(\cos \theta)(\cos \theta) - (1 + \sin \theta)(\sin \theta)}$
$= \frac{(2 \cos \theta \sin \theta) + \cos \theta}{\cos^2 \theta - \sin^2 \theta - \sin \theta} \Rightarrow$ Slope at $(1,0)$ is $\frac{2 \cos 0 \sin 0 + \cos 0}{\cos^2 0 - \sin^2 0 - \sin 0} = 1 \Rightarrow$ the tangent line is $y = x - 1$
$\Rightarrow r \sin \theta = r \cos \theta - 1$; Slope at $(1,\pi)$ is $\frac{2 \cos \pi \sin \pi + \cos \pi}{\cos^2 \pi - \sin^2 \pi - \sin \pi} = -1 \Rightarrow$ the tangent line is $y = -x - 1$
$\Rightarrow r \sin \theta = -r \cos \theta - 1$

69. $r \cos\left(\theta + \frac{\pi}{3}\right) = 2\sqrt{3} \Rightarrow r\left(\cos \theta \cos \frac{\pi}{3} - \sin \theta \sin \frac{\pi}{3}\right)$
$= 2\sqrt{3} \Rightarrow \frac{1}{2} r \cos \theta - \frac{\sqrt{3}}{2} r \sin \theta = 2\sqrt{3}$
$\Rightarrow r \cos \theta - \sqrt{3} r \sin \theta = 4\sqrt{3} \Rightarrow x - \sqrt{3} y = 4\sqrt{3}$
$\Rightarrow y = \frac{\sqrt{3}}{3} x - 4$

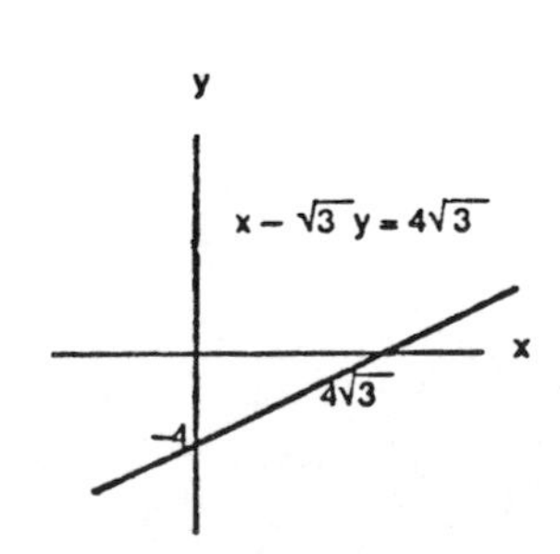

70. $r\cos\left(\theta-\frac{3\pi}{4}\right)=\frac{\sqrt{2}}{2}\Rightarrow r\left(\cos\theta\cos\frac{3\pi}{4}+\sin\theta\sin\frac{3\pi}{4}\right)$
$=\frac{\sqrt{2}}{2}\Rightarrow -\frac{\sqrt{2}}{2}r\cos\theta+\frac{\sqrt{2}}{2}r\sin\theta=\frac{\sqrt{2}}{2}\Rightarrow -x+y=1$
$\Rightarrow y=x+1$

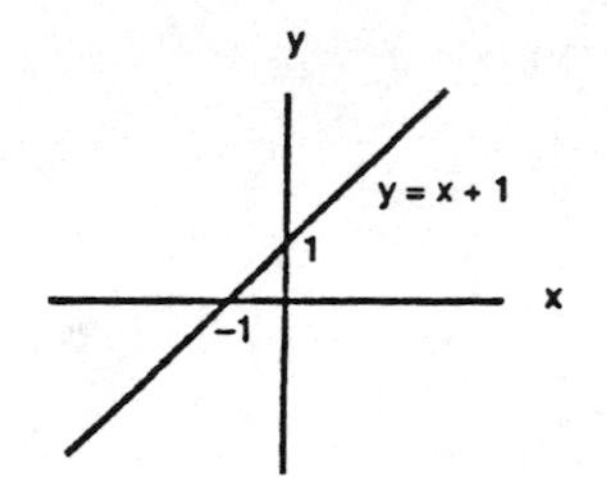

71. $r=2\sec\theta\Rightarrow r=\frac{2}{\cos\theta}\Rightarrow r\cos\theta=2\Rightarrow x=2$

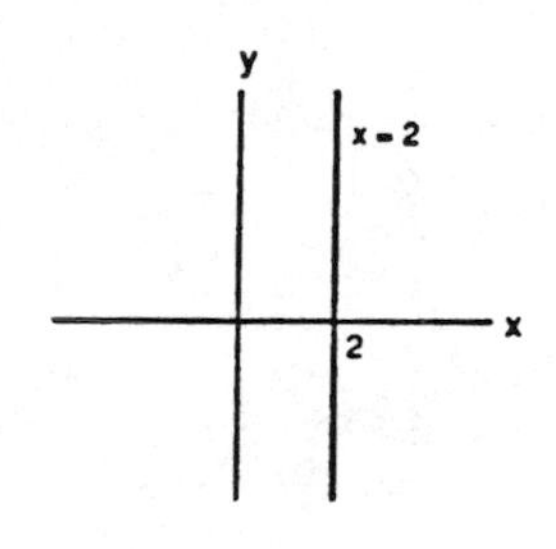

72. $r=-\sqrt{2}\sec\theta\Rightarrow r\cos\theta=-\sqrt{2}\Rightarrow x=-\sqrt{2}$

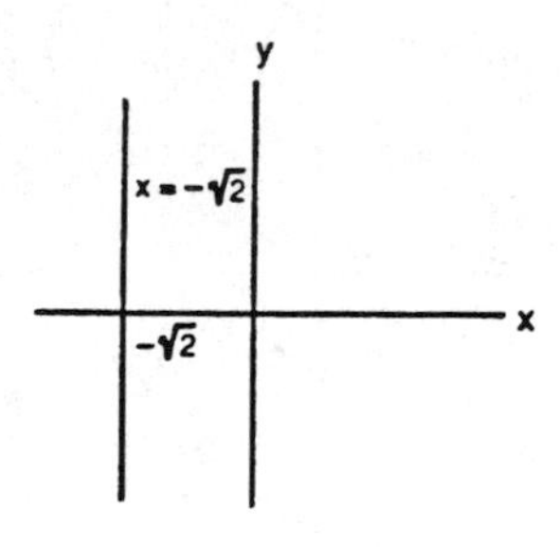

73. $r=-\frac{3}{2}\csc\theta\Rightarrow r\sin\theta=-\frac{3}{2}\Rightarrow y=-\frac{3}{2}$

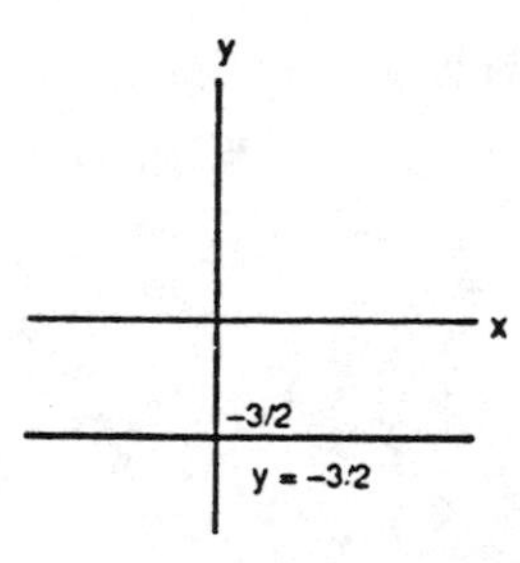

74. $r=3\sqrt{3}\csc\theta\Rightarrow r\sin\theta=3\sqrt{3}\Rightarrow y=3\sqrt{3}$

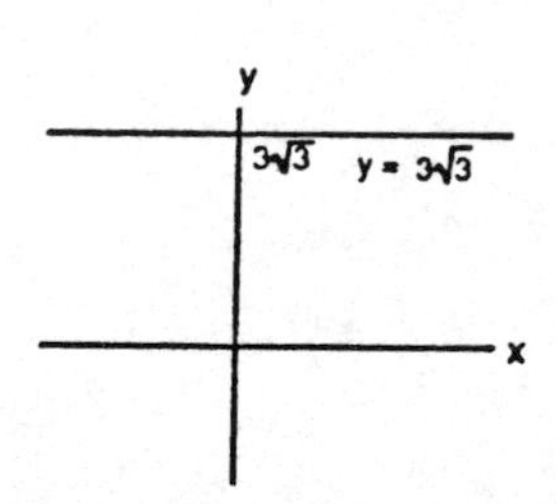

75. $r = -4 \sin \theta \Rightarrow r^2 = -4r \sin \theta \Rightarrow x^2 + y^2 + 4y = 0$
$\Rightarrow x^2 + (y+2)^2 = 4$; circle with center $(0, 2)$ and radius 2.

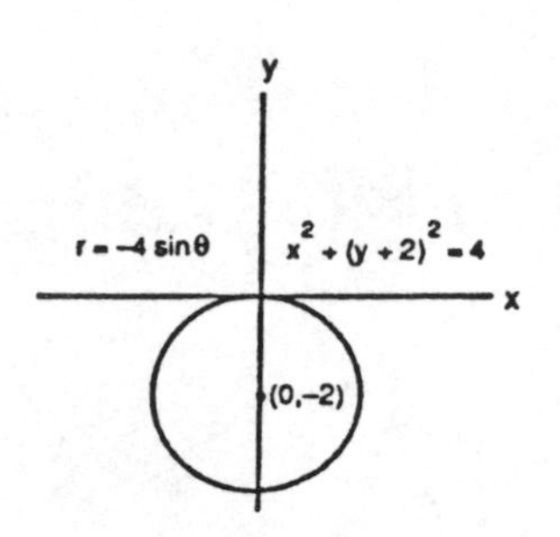

76. $r = 3\sqrt{3} \sin \theta \Rightarrow r^2 = 3\sqrt{3}\, r \sin \theta \Rightarrow x^2 + y^2 - 3\sqrt{3}\, y = 0$
$\Rightarrow x^2 + \left(y - \frac{3\sqrt{3}}{2}\right)^2 = \frac{27}{4}$; circle with center $\left(0, \frac{3\sqrt{3}}{2}\right)$
and radius $\frac{3\sqrt{3}}{2}$

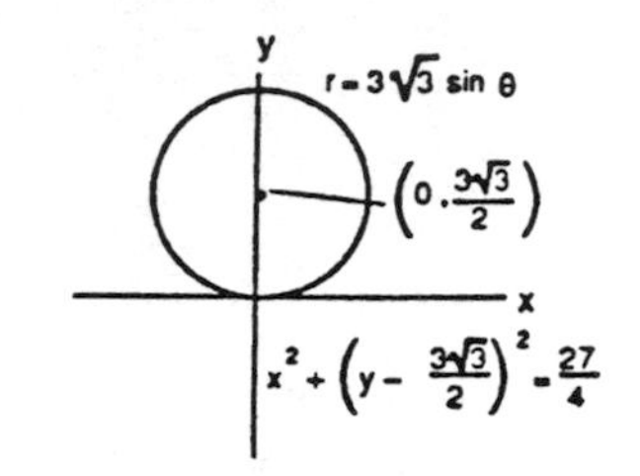

77. $r = 2\sqrt{2} \cos \theta \Rightarrow r^2 = 2\sqrt{2}\, r \cos \theta \Rightarrow x^2 + y^2 - 2\sqrt{2}\, x = 0$
$\Rightarrow \left(x - \sqrt{2}\right)^2 + y^2 = 2$; circle with center $\left(\sqrt{2}, 0\right)$ and
radius $\sqrt{2}$

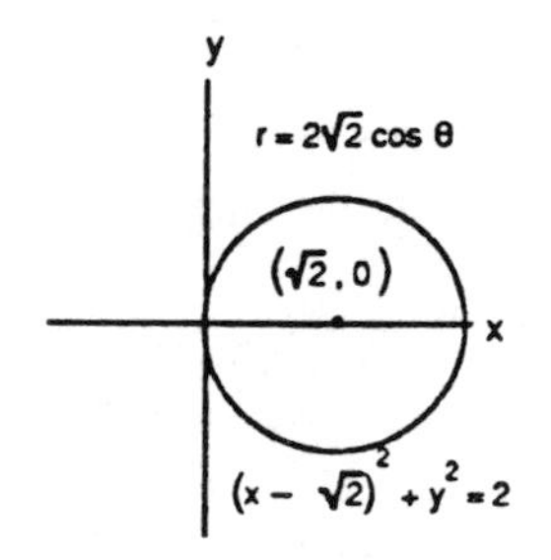

78. $r = -6 \cos \theta \Rightarrow r^2 = -6r \cos \theta \Rightarrow x^2 + y^2 + 6x = 0$
$\Rightarrow (x+3)^2 + y^2 = 9$; circle with center $(-3, 0)$ and
radius 3

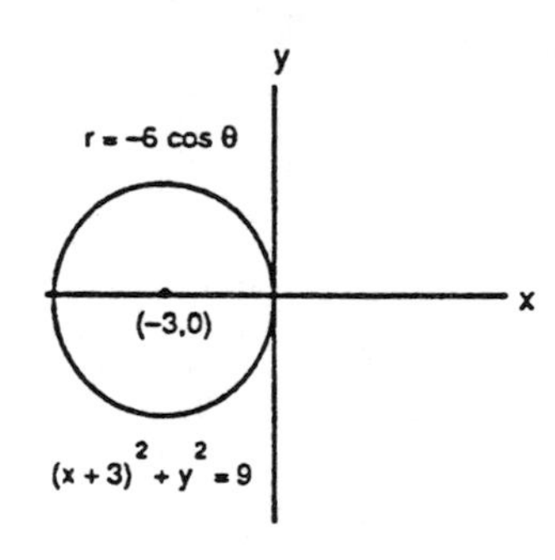

79. $x^2 + y^2 + 5y = 0 \Rightarrow x^2 + \left(y + \frac{5}{2}\right)^2 = \frac{25}{4} \Rightarrow C = \left(0, -\frac{5}{2}\right)$
and $a = \frac{5}{2}$; $r^2 + 5r \sin \theta = 0 \Rightarrow r = -5 \sin \theta$

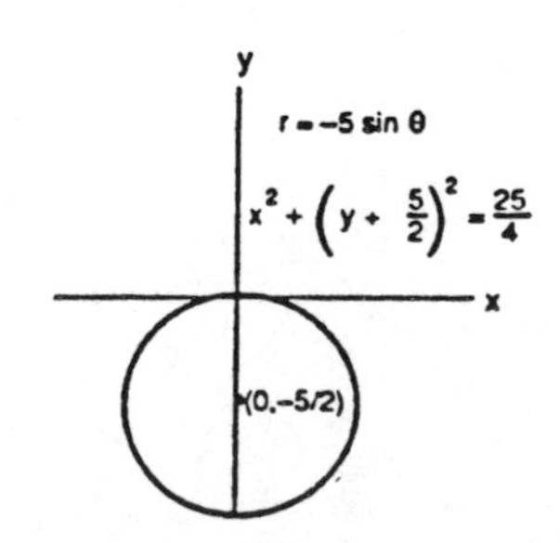

80. $x^2 + y^2 - 2y = 0 \Rightarrow x^2 + (y-1)^2 = 1 \Rightarrow C = (0,1)$ and $a = 1$; $r^2 - 2r \sin\theta = 0 \Rightarrow r = 2 \sin\theta$

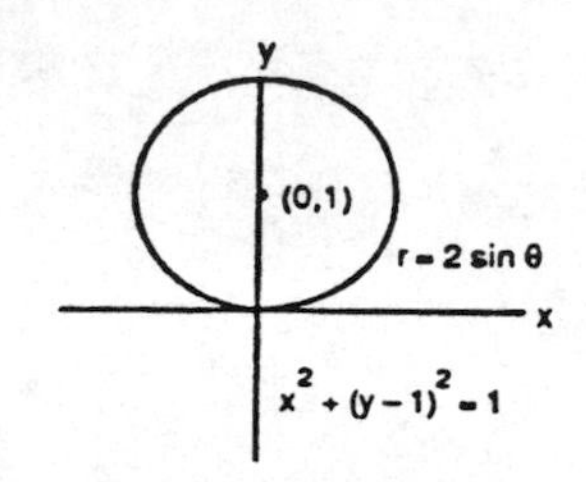

81. $x^2 + y^2 - 3x = 0 \Rightarrow \left(x - \frac{3}{2}\right)^2 + y^2 = \frac{9}{4} \Rightarrow C = \left(\frac{3}{2}, 0\right)$ and $a = \frac{3}{2}$; $r^2 - 3r \cos\theta = 0 \Rightarrow r = 3 \cos\theta$

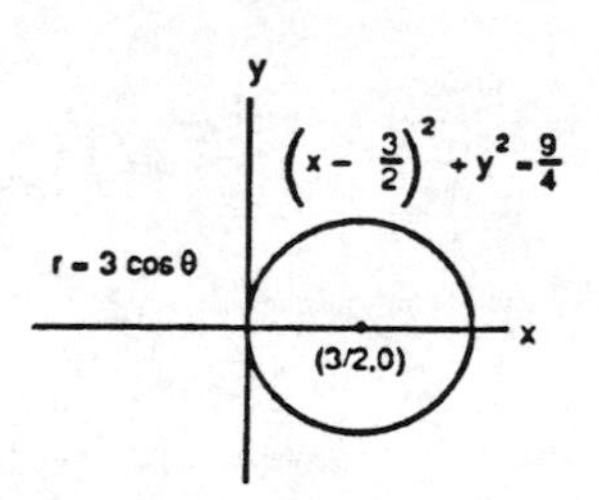

82. $x^2 + y^2 + 4x = 0 \Rightarrow (x+2)^2 + y^2 = 4 \Rightarrow C = (-2,0)$ and $a = 2$; $r^2 + 4r \cos\theta = 0 \Rightarrow r = -4 \cos\theta$

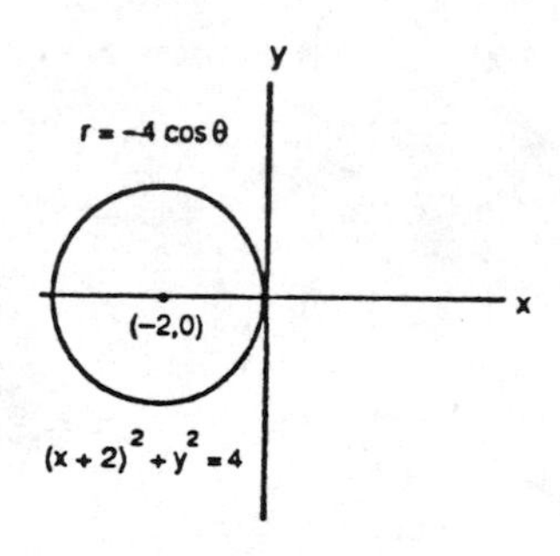

83. $r = \frac{2}{1 + \cos\theta} \Rightarrow e = 1 \Rightarrow$ parabola with vertex at $(1,0)$

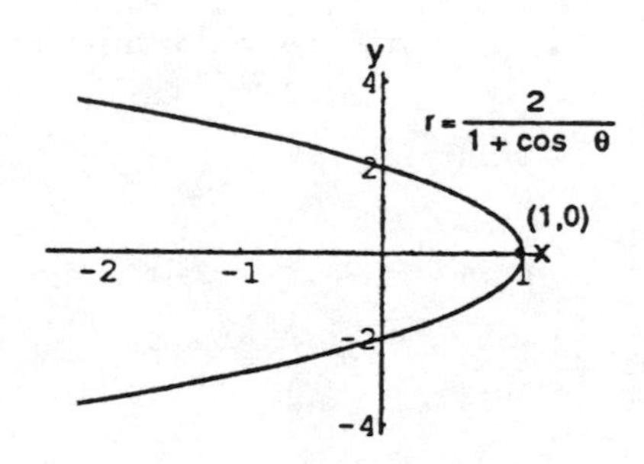

84. $r = \frac{8}{2 + \cos\theta} \Rightarrow r = \frac{4}{1 + \left(\frac{1}{2}\right) \cos\theta} \Rightarrow e = \frac{1}{2} \Rightarrow$ ellipse;

$ke = 4 \Rightarrow \frac{1}{2}k = 4 \Rightarrow k = 8$; $k = \frac{a}{e} - ea \Rightarrow 8 = \frac{a}{\left(\frac{1}{2}\right)} - \frac{1}{2}a$

$\Rightarrow a = \frac{16}{3} \Rightarrow ea = \left(\frac{1}{2}\right)\left(\frac{16}{3}\right) = \frac{8}{3}$; therefore the center is $\left(\frac{8}{3}, \pi\right)$; vertices are $(8, \pi)$ and $\left(\frac{8}{3}, 0\right)$

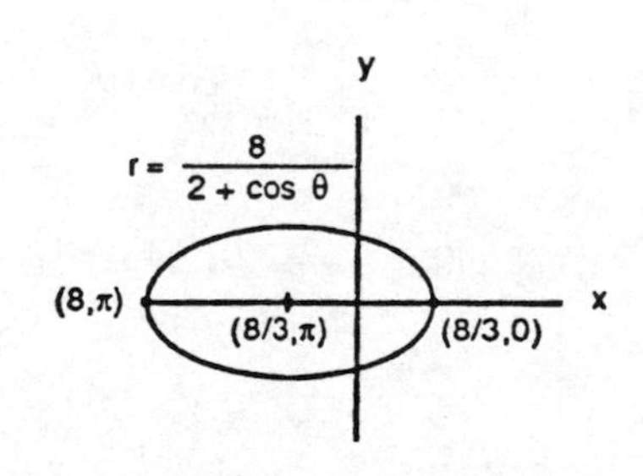

85. $r = \frac{6}{1 - 2\cos\theta} \Rightarrow e = 2 \Rightarrow$ hyperbola; $ke = 6 \Rightarrow 2k = 6$
$\Rightarrow k = 3 \Rightarrow$ vertices are $(2, \pi)$ and $(6, \pi)$

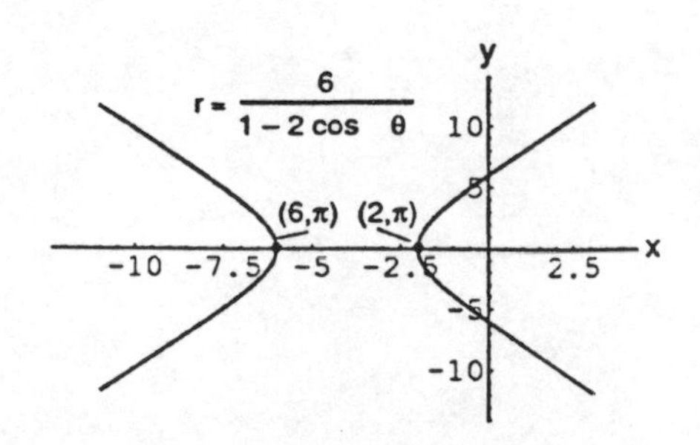

86. $r = \frac{12}{3 + \sin\theta} \Rightarrow r = \frac{4}{1 + \left(\frac{1}{3}\right)\sin\theta} \Rightarrow e = \frac{1}{3}$; $ke = 4$
$\Rightarrow \frac{1}{3}k = 4 \Rightarrow k = 12$; $a(1 - e^2) = 4 \Rightarrow a\left[1 - \left(\frac{1}{3}\right)^2\right]$
$= 4 \Rightarrow a = \frac{9}{2} \Rightarrow ea = \left(\frac{1}{3}\right)\left(\frac{9}{2}\right) = \frac{3}{2}$; therefore the
center is $\left(\frac{3}{2}, \frac{3\pi}{2}\right)$; vertices are $\left(3, \frac{\pi}{2}\right)$ and $\left(6, \frac{3\pi}{2}\right)$

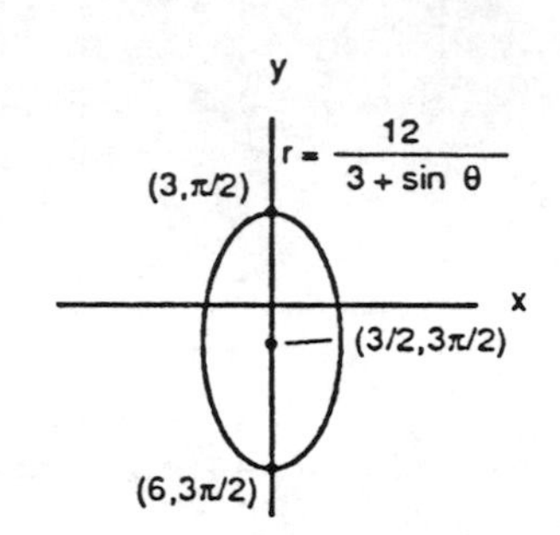

87. $e = 2$ and $r\cos\theta = 2 \Rightarrow x = 2$ is directrix $\Rightarrow k = 2$; the conic is a hyperbola; $r = \frac{ke}{1 + e\cos\theta} \Rightarrow r = \frac{(2)(2)}{1 + 2\cos\theta}$
$\Rightarrow r = \frac{4}{1 + 2\cos\theta}$

88. $e = 1$ and $r\cos\theta = -4 \Rightarrow x = -4$ is directrix $\Rightarrow k = 4$; the conic is a parabola; $r = \frac{ke}{1 - e\cos\theta} \Rightarrow r = \frac{(4)(1)}{1 - \cos\theta}$
$\Rightarrow r = \frac{4}{1 - \cos\theta}$

89. $e = \frac{1}{2}$ and $r\sin\theta = 2 \Rightarrow y = 2$ is directrix $\Rightarrow k = 2$; the conic is an ellipse; $r = \frac{ke}{1 + e\sin\theta} \Rightarrow r = \frac{(2)\left(\frac{1}{2}\right)}{1 + \left(\frac{1}{2}\right)\sin\theta}$
$\Rightarrow r = \frac{2}{2 + \sin\theta}$

90. $e = \frac{1}{3}$ and $r\sin\theta = -6 \Rightarrow y = -6$ is directrix $\Rightarrow k = 6$; the conic is an ellipse; $r = \frac{ke}{1 - e\sin\theta} \Rightarrow r = \frac{(6)\left(\frac{1}{3}\right)}{1 - \left(\frac{1}{3}\right)\sin\theta}$
$\Rightarrow r = \frac{6}{3 - \sin\theta}$

91. $A = 2\int_0^{\pi} \frac{1}{2}r^2\, d\theta = \int_0^{\pi} (2 - \cos\theta)^2\, d\theta = \int_0^{\pi} \left(4 - 2\cos\theta + \cos^2\theta\right) d\theta = \int_0^{\pi} \left(4 - 2\cos\theta + \frac{1 + \cos 2\theta}{2}\right) d\theta$
$= \int_0^{\pi} \left(\frac{9}{2} - 2\cos\theta + \frac{\cos 2\theta}{2}\right) d\theta = \left[\frac{9}{2}\theta - 2\sin\theta + \frac{\sin 2\theta}{4}\right]_0^{\pi} = \frac{9}{2}\pi$

92. $A = \int_0^{\pi/3} \frac{1}{2}\left(\sin^2 3\theta\right) d\theta = \int_0^{\pi/3} \frac{1}{2}\left(\frac{1 - \cos 6\theta}{2}\right) d\theta = \frac{1}{4}\left[\theta - \frac{1}{6}\sin 6\theta\right]_0^{\pi/3} = \frac{\pi}{12}$

93. $r = 1 + \cos 2\theta$ and $r = 1 \Rightarrow 1 = 1 + \cos 2\theta \Rightarrow 0 = \cos 2\theta \Rightarrow 2\theta = \frac{\pi}{2} \Rightarrow \theta = \frac{\pi}{4}$; therefore

$$A = 4\int_0^{\pi/4} \tfrac{1}{2}\left[(1+\cos 2\theta)^2 - 1^2\right] d\theta = 2\int_0^{\pi/4} \left(1 + 2\cos 2\theta + \cos^2 2\theta - 1\right) d\theta$$

$$= 2\int_0^{\pi/4} \left(2\cos 2\theta + \tfrac{1}{2} + \tfrac{\cos 4\theta}{2}\right) d\theta = 2\left[\sin 2\theta + \tfrac{1}{2}\theta + \tfrac{\sin 4\theta}{8}\right]_0^{\pi/4} = 2\left(1 + \tfrac{\pi}{8} + 0\right) = 2 + \tfrac{\pi}{4}$$

94. The circle lies interior to the cardiod (see the graphs in Exercises 61 and 63). Thus,

$$A = 2\int_{-\pi/2}^{\pi/2} \tfrac{1}{2}[2(1+\sin\theta)]^2\, d\theta - \pi \text{ (the integral is the area of the cardiod minus the area of the circle)}$$

$$= \int_{-\pi/2}^{\pi/2} 4\left(1 + 2\sin\theta + \sin^2\theta\right) d\theta - \pi = \int_{-\pi/2}^{\pi/2} (6 + 8\sin\theta - 2\cos 2\theta)\, d\theta - \pi = [6\theta - 8\cos\theta - \sin 2\theta]_{-\pi/2}^{\pi/2} - \pi$$

$$= [3\pi - (-3\pi)] - \pi = 5\pi$$

95. $r = -1 + \cos\theta \Rightarrow \frac{dr}{d\theta} = -\sin\theta$; Length $= \int_0^{2\pi} \sqrt{(-1+\cos\theta)^2 + (-\sin\theta)^2}\, d\theta = \int_0^{2\pi} \sqrt{2 - 2\cos\theta}\, d\theta$

$$= \int_0^{2\pi} \sqrt{\frac{4(1-\cos\theta)}{2}}\, d\theta = \int_0^{2\pi} 2\sin\frac{\theta}{2}\, d\theta = \left[-4\cos\frac{\theta}{2}\right]_0^{2\pi} = (-4)(-1) - (-4)(1) = 8$$

96. $r = 2\sin\theta + 2\cos\theta,\ 0 \le \theta \le \frac{\pi}{2} \Rightarrow \frac{dr}{d\theta} = 2\cos\theta - 2\sin\theta;\ r^2 + \left(\frac{dr}{d\theta}\right)^2 = (2\sin\theta + 2\cos\theta)^2 + (2\cos\theta - 2\sin\theta)^2$

$$= 8\left(\sin^2\theta + \cos^2\theta\right) = 8 \Rightarrow L = \int_0^{\pi/2} \sqrt{8}\, d\theta = \left[2\sqrt{2}\,\theta\right]_0^{\pi/2} = 2\sqrt{2}\left(\tfrac{\pi}{2}\right) = \pi\sqrt{2}$$

97. $r = 8\sin^3\left(\frac{\theta}{3}\right),\ 0 \le \theta \le \frac{\pi}{4} \Rightarrow \frac{dr}{d\theta} = 8\sin^2\left(\frac{\theta}{3}\right)\cos\left(\frac{\theta}{3}\right);\ r^2 + \left(\frac{dr}{d\theta}\right)^2 = \left[8\sin^3\left(\frac{\theta}{3}\right)\right]^2 + \left[8\sin^2\left(\frac{\theta}{3}\right)\cos\left(\frac{\theta}{3}\right)\right]^2$

$$= 64\sin^4\left(\tfrac{\theta}{3}\right) \Rightarrow L = \int_0^{\pi/4} \sqrt{64\sin^4\left(\tfrac{\theta}{3}\right)}\, d\theta = \int_0^{\pi/4} 8\sin^2\left(\tfrac{\theta}{3}\right) d\theta = \int_0^{\pi/4} 8\left[\frac{1-\cos\left(\frac{2\theta}{3}\right)}{2}\right] d\theta$$

$$= \int_0^{\pi/4} \left[4 - 4\cos\left(\tfrac{2\theta}{3}\right)\right] d\theta = \left[4\theta - 6\sin\left(\tfrac{2\theta}{3}\right)\right]_0^{\pi/4} = 4\left(\tfrac{\pi}{4}\right) - 6\sin\left(\tfrac{\pi}{6}\right) - 0 = \pi - 3$$

98. $r = \sqrt{1+\cos 2\theta} \Rightarrow \frac{dr}{d\theta} = \frac{1}{2}(1+\cos 2\theta)^{-1/2}(-2\sin 2\theta) = \frac{-\sin 2\theta}{\sqrt{1+\cos 2\theta}} \Rightarrow \left(\frac{dr}{d\theta}\right)^2 = \frac{\sin^2 2\theta}{1+\cos 2\theta}$

$$\Rightarrow r^2 + \left(\frac{dr}{d\theta}\right)^2 = 1 + \cos 2\theta + \frac{\sin^2 2\theta}{1+\cos 2\theta} = \frac{(1+\cos 2\theta)^2 + \sin^2 2\theta}{1+\cos 2\theta} = \frac{1 + 2\cos 2\theta + \cos^2 2\theta + \sin^2 2\theta}{1+\cos 2\theta}$$

$$= \frac{2 + 2\cos 2\theta}{1+\cos 2\theta} = 2 \Rightarrow L = \int_{-\pi/2}^{\pi/2} \sqrt{2}\, d\theta = \sqrt{2}\left[\tfrac{\pi}{2} - \left(-\tfrac{\pi}{2}\right)\right] = \sqrt{2}\,\pi$$

99. $r = \sqrt{\cos 2\theta} \Rightarrow \frac{dr}{d\theta} = \frac{-\sin 2\theta}{\sqrt{\cos 2\theta}}$; Surface Area $= \int_0^{\pi/4} 2\pi(r \sin\theta)\sqrt{r^2 + \left(\frac{dr}{d\theta}\right)^2}\, d\theta$

$= \int_0^{\pi/4} 2\pi\sqrt{\cos 2\theta}\,(\sin\theta)\sqrt{\cos 2\theta + \frac{\sin^2 2\theta}{\cos 2\theta}}\, d\theta = \int_0^{\pi/4} 2\pi\sqrt{\cos 2\theta}\,(\sin\theta)\sqrt{\frac{1}{\cos 2\theta}}\, d\theta = \int_0^{\pi/4} 2\pi \sin\theta\, d\theta$

$= \left[2\pi(-\cos\theta)\right]_0^{\pi/4} = 2\pi\left(1 - \frac{\sqrt{2}}{2}\right) = \left(2 - \sqrt{2}\right)\pi$

100. $r^2 = \sin 2\theta \Rightarrow 2r\frac{dr}{d\theta} = 2\cos 2\theta \Rightarrow r\frac{dr}{d\theta} = \cos 2\theta$; Surface Area $= 2\int_0^{\pi/2} 2\pi(r\cos\theta)\sqrt{r^2 + \left(\frac{dr}{d\theta}\right)^2}\, d\theta$

$= 2\int_0^{\pi/2} 2\pi(\cos\theta)\sqrt{r^4 + \left(r\frac{dr}{dt}\right)^2}\, d\theta = 2\int_0^{\pi/2} 2\pi(\cos\theta)\sqrt{(\sin 2\theta)^2 + (\cos 2\theta)^2}\, d\theta = 2\int_0^{\pi/2} 2\pi\cos\theta\, d\theta$

$= 2\left[2\pi\sin\theta\right]_0^{\pi/2} = 4\pi$

101. (a) Around the x-axis: $9x^2 + 4y^2 = 36 \Rightarrow y^2 = 9 - \frac{9}{4}x^2 \Rightarrow y = \pm\sqrt{9 - \frac{9}{4}x^2}$ and we use the positive root:

$$V = 2\int_0^2 \pi\left(\sqrt{9 - \frac{9}{4}x^2}\right)^2 dx = 2\int_0^2 \pi\left(9 - \frac{9}{4}x^2\right) dx = 2\pi\left[9x - \frac{3}{4}x^3\right]_0^2 = 24\pi$$

(b) Around the y-axis: $9x^2 + 4y^2 = 36 \Rightarrow x^2 = 4 - \frac{4}{9}y^2 \Rightarrow x = \pm\sqrt{4 - \frac{4}{9}y^2}$ and we use the positive root:

$$V = 2\int_0^3 \pi\left(\sqrt{4 - \frac{4}{9}y^2}\right)^2 dy = 2\int_0^3 \pi\left(4 - \frac{4}{9}y^2\right) dy = 2\pi\left[4y - \frac{4}{27}y^3\right]_0^3 = 16\pi$$

102. $9x^2 - 4y^2 = 36,\ x = 4 \Rightarrow y^2 = \frac{9x^2 - 36}{4} \Rightarrow y = \frac{3}{2}\sqrt{x^2 - 4};\ V = \int_2^4 \pi\left(\frac{3}{2}\sqrt{x^2 - 4}\right)^2 dx = \frac{9\pi}{4}\int_2^4 \left(x^2 - 4\right) dx$

$= \frac{9\pi}{4}\left[\frac{x^3}{3} - 4x\right]_2^4 = \frac{9\pi}{4}\left[\left(\frac{64}{3} - 16\right) - \left(\frac{8}{3} - 8\right)\right] = \frac{9\pi}{4}\left(\frac{56}{3} - \frac{24}{3}\right) = \frac{3\pi}{4}(32) = 24\pi$

103. Each portion of the wave front reflects to the other focus, and since the wave front travels at a constant speed as it expands, the different portions of the wave arrive at the second focus simultaneously, from all directions, causing a spurt at the second focus.

104. The velocity of the signals is $v = 980$ ft/ms. Let t_1 be the time it takes for the signal to go from A to S. Then $d_1 = 980t_1$ and $d_2 = 980(t_1 + 1400) \Rightarrow d_2 - d_1 = 980(1400) = 1.372 \times 10^6$ ft or 259.8 miles. The ship is 259.8 miles closer to A than to B. The difference of the distances is always constant (259.8 miles) so the ship is traveling along a branch of a hyperbola with foci at the two towers. The branch is the one having tower A as its focus.

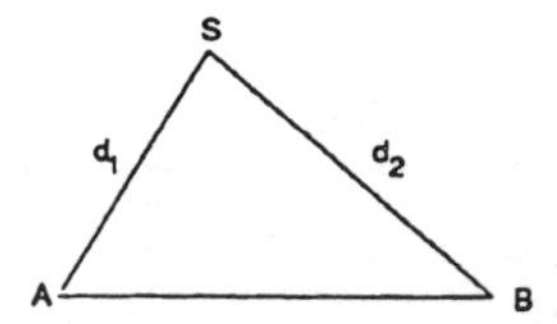

105. The time for the bullet to hit the target remains constant, say $t = t_0$. Let the time it takes for sound to travel from the target to the listener be t_2. Since the listener hears the sounds simultaneously, $t_1 = t_0 + t_2$ where t_1 is the time for the sound to travel from the rifle to the listener. If v is the velocity of sound, then $vt_1 = vt_0 + vt_2$ or $vt_1 - vt_2 = vt_0$. Now vt_1 is the distance from the rifle to the listener and vt_2 is the distance from the target to the listener. Therefore the difference of the distances is constant since vt_0 is constant so the listener is on a branch of a hyperbola with foci at the rifle and the target. The branch is the one with the target as focus.

106. Let (r_1, θ_1) be a point on the graph where $r_1 = a\theta_1$. Let (r_2, θ_2) be on the graph where $r_2 = a\theta_2$ and $\theta_2 = \theta_1 + 2\pi$. Then r_1 and r_2 lie on the same ray on consecutive turns of the spiral and the distance between the two points is $r_2 - r_1 = a\theta_2 - a\theta_1 = a(\theta_2 - \theta_1) = 2\pi a$, which is constant.

107. (a) $r = \frac{k}{1 + e\cos\theta} \Rightarrow r + er\cos\theta = k \Rightarrow \sqrt{x^2+y^2} + ex = k \Rightarrow \sqrt{x^2+y^2} = k - ex \Rightarrow x^2 + y^2$
$= k^2 - 2kex + e^2x^2 \Rightarrow x^2 - e^2x^2 + y^2 + 2kex - k^2 = 0 \Rightarrow (1-e^2)x^2 + y^2 + 2kex - k^2 = 0$

(b) $c = 0 \Rightarrow x^2 + y^2 - k^2 = 0 \Rightarrow x^2 + y^2 = k^2 \Rightarrow$ circle;
$0 < e < 1 \Rightarrow e^2 < 1 \Rightarrow e^2 - 1 < 0 \Rightarrow B^2 - 4AC = 0^2 - 4(1-e^2)(1) = 4(e^2 - 1) < 0 \Rightarrow$ ellipse;
$e = 1 \Rightarrow B^2 - 4AC = 0^2 - 4(0)(1) = 0 \Rightarrow$ parabola;
$e > 1 \Rightarrow e^2 > 1 \Rightarrow B^2 - 4AC = 0^2 - 4(1-e^2)(1) = 4e^2 - 4 > 0 \Rightarrow$ hyperbola

108. (a) The length of the major axis is 300 miles + 8000 miles + 1000 miles = 2a ⇒ a = 4650 miles. If the center of the earth is one focus and the distance from the center of the earth to the satellite's low point is 4300 miles (half the diameter plus the distance above the North Pole), then the distance from the center of the ellipse to the focus (center of the earth) is 4650 miles − 4300 miles = 350 miles = c. Therefore $e = \frac{c}{a} = \frac{350 \text{ miles}}{4650 \text{ miles}} = \frac{7}{93}$.

(b) $r = \frac{a(1-e^2)}{1 + e\cos\theta} \Rightarrow r = \frac{4650\left[1 - \left(\frac{7}{93}\right)^2\right]}{\left(1 + \frac{7}{93}\cos\theta\right)} = \frac{430{,}000}{93 + 7\cos\theta}$ miles

109. $\beta = \psi_2 - \psi_1 \Rightarrow \tan\beta = \tan(\psi_2 - \psi_1) = \frac{\tan\psi_2 - \tan\psi_1}{1 + \tan\psi_2 \tan\psi_1}$;

the curves will be orthogonal when $\tan\beta$ is undefined, or

when $\tan\psi_2 = \frac{-1}{\tan\psi_1} \Rightarrow \frac{r}{g'(\theta)} = \frac{-1}{\left[\frac{r}{f'(\theta)}\right]} \Rightarrow r^2 = -f'(\theta)\,g'(\theta)$

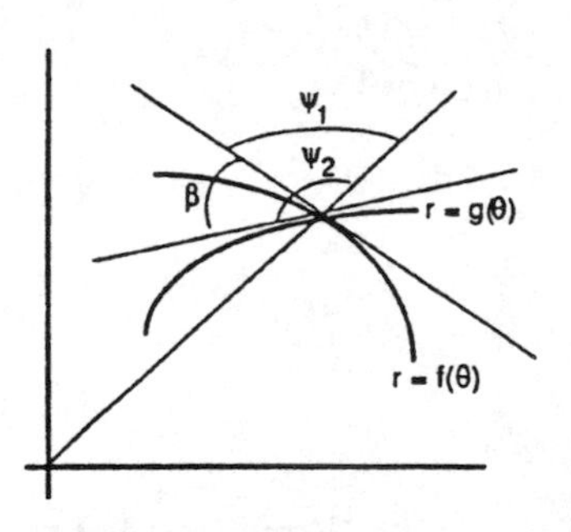

110. $r = \sin^4\left(\frac{\theta}{4}\right) \Rightarrow \frac{dr}{d\theta} = \sin^3\left(\frac{\theta}{4}\right)\cos\left(\frac{\theta}{4}\right) \Rightarrow \tan\psi = \frac{\sin^4\left(\frac{\theta}{4}\right)}{\sin^3\left(\frac{\theta}{4}\right)\cos\left(\frac{\theta}{4}\right)} = \tan\left(\frac{\theta}{4}\right)$

111. $r = 2a \sin 3\theta \Rightarrow \frac{dr}{d\theta} = 6a \cos 3\theta \Rightarrow \tan \psi = \frac{r}{\left(\frac{dr}{d\theta}\right)} = \frac{2a \sin 3\theta}{6a \cos 3\theta} = \frac{1}{3} \tan 3\theta$; when $\theta = \frac{\pi}{6}$, $\tan \psi = \frac{1}{3} \tan \frac{\pi}{2}$ $\Rightarrow \psi = \frac{\pi}{2}$

112. (b) $r\theta = 1 \Rightarrow r = \theta^{-1} \Rightarrow \frac{dr}{d\theta} = -\theta^{-2} \Rightarrow \tan \psi|_{\theta=1}$ $= \frac{\theta^{-1}}{-\theta^{-2}} = -\theta \Rightarrow \lim_{\theta\to\infty} \tan \psi = -\infty \Rightarrow \psi \to \frac{\pi}{2}$ from the right as the spiral winds in around the origin.

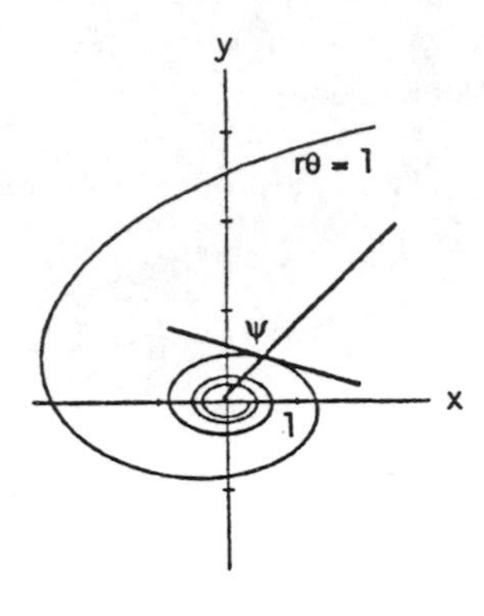

113. $\tan \psi_1 = \frac{\sqrt{3} \cos \theta}{-\sqrt{3} \sin \theta} = -\cot \theta$ is $-\frac{1}{\sqrt{3}}$ at $\theta = \frac{\pi}{3}$; $\tan \psi_2 = \frac{\sin \theta}{\cos \theta} = \tan \theta$ is $\sqrt{3}$ at $\theta = \frac{\pi}{3}$; since the product of these slopes is -1, the tangents are perpendicular

114. $a(1 + \cos \theta) = 3a \cos \theta \Rightarrow 1 = 2 \cos \theta \Rightarrow \cos \theta = \frac{1}{2}$ or $\theta = \frac{\pi}{3}$; $\tan \psi_2 = \frac{a(1 + \cos \theta)}{-a \sin \theta}$ is $-\sqrt{3}$ at $\theta = \frac{\pi}{3}$; $\tan \psi_1 = \frac{3a \cos \theta}{-3a \sin \theta}$ is $-\frac{1}{\sqrt{3}}$ at $\theta = \frac{\pi}{3}$. Then

$$\tan \beta = \frac{-\sqrt{3} - \left(-\frac{1}{\sqrt{3}}\right)}{1 + (-\sqrt{3})\left(-\frac{1}{\sqrt{3}}\right)} = \frac{-3 + \left(\frac{1}{\sqrt{3}}\right)}{2} = -\frac{1}{\sqrt{3}} \Rightarrow \beta = \frac{\pi}{6}$$

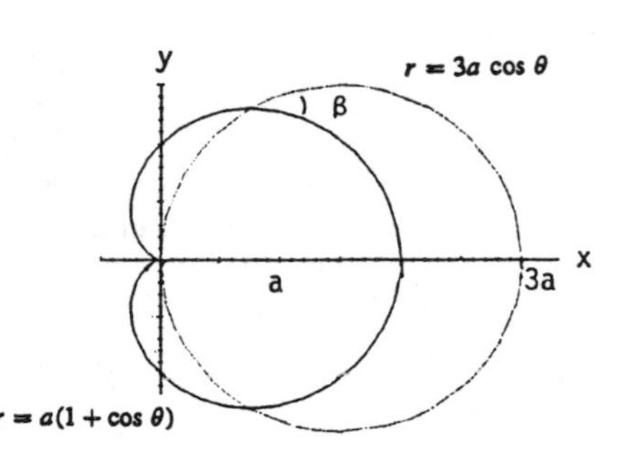

115. $r_1 = \frac{1}{1 - \cos \theta} \Rightarrow \frac{dr_1}{d\theta} = -\frac{\sin \theta}{(1 - \cos \theta)^2}$; $r_2 = \frac{3}{1 + \cos \theta} \Rightarrow \frac{dr_2}{d\theta} = \frac{3 \sin \theta}{(1 + \cos \theta)^2}$; $\frac{1}{1 - \cos \theta} = \frac{3}{1 + \cos \theta}$ $\Rightarrow 1 + \cos \theta = 3 - 3 \cos \theta \Rightarrow 4 \cos \theta = 2 \Rightarrow \cos \theta = \frac{1}{2} \Rightarrow \theta = \pm\frac{\pi}{3} \Rightarrow r_1 = r_2 = 2 \Rightarrow$ the curves intersect at the points $\left(2, \pm\frac{\pi}{3}\right)$; $\tan \psi_1 = \frac{\left(\frac{1}{1 - \cos \theta}\right)}{\left[\frac{-\sin \theta}{(1 - \cos \theta)^2}\right]} = -\frac{1 - \cos \theta}{\sin \theta}$ is $-\frac{1}{\sqrt{3}}$ at $\theta = \frac{\pi}{3}$; $\tan \psi_2 = \frac{\left(\frac{3}{1 + \cos \theta}\right)}{\left[\frac{3 \sin \theta}{(1 + \cos \theta)^2}\right]} = \frac{1 + \cos \theta}{\sin \theta}$ is $\sqrt{3}$ at $\theta = \frac{\pi}{3}$; therefore $\tan \beta$ is undefined at $\theta = \frac{\pi}{3}$ since $1 + \tan \psi_1 \tan \psi_2 = 1 + \left(-\frac{1}{\sqrt{3}}\right)(\sqrt{3}) = 0 \Rightarrow \beta = \frac{\pi}{2}$;

$$\tan \psi_1|_{\theta=-\pi/3} = -\frac{1 - \cos\left(-\frac{\pi}{3}\right)}{\sin\left(-\frac{\pi}{3}\right)} = \frac{1}{\sqrt{3}} \text{ and } \tan \psi_2|_{\theta=-\pi/3} = \frac{1 + \cos\left(-\frac{\pi}{3}\right)}{\sin\left(-\frac{\pi}{3}\right)} = -\sqrt{3} \Rightarrow \tan \beta \text{ is also undefined}$$

at $\theta = -\frac{\pi}{3} \Rightarrow \beta = \frac{\pi}{2}$

116. (a) We need $\psi + \theta = \pi$, so that $\tan \psi = \tan(\pi - \theta) = -\tan \theta$.

Now $\tan \psi = \frac{r}{\left(\frac{dr}{d\theta}\right)} = \frac{a(1+\cos\theta)}{-a\sin\theta} = -\tan\theta = -\frac{\sin\theta}{\cos\theta}$

$\Rightarrow \cos\theta + \cos^2\theta = \sin^2\theta \Rightarrow \cos\theta + \cos^2\theta = 1 - \cos^2\theta$

$\Rightarrow 2\cos^2\theta + \cos\theta - 1 = 0 \Rightarrow \cos\theta = \frac{1}{2}$ or $\cos\theta = -1$;

$\cos\theta = \frac{1}{2} \Rightarrow \theta = \pm\frac{\pi}{3} \Rightarrow r = \frac{3a}{2}$; $\cos\theta = -1 \Rightarrow \theta = \pi$

$\Rightarrow r = 0$. Therefore the points where the tangent line is horizontal are $\left(\frac{3a}{2}, \pm\frac{\pi}{3}\right)$ and $(0, \pi)$.

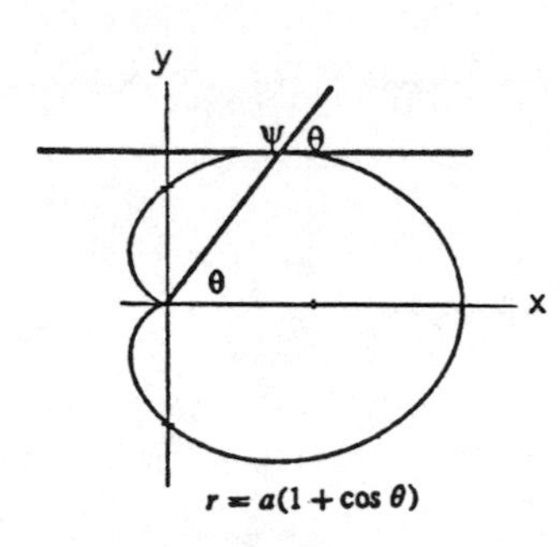

(b) We need $\psi + \theta = \frac{\pi}{2}$ so that $\tan\psi = \tan\left(\frac{\pi}{2} - \theta\right) = \cot\theta$. Thus $\tan\psi = \frac{r}{\left(\frac{dr}{d\theta}\right)} = \frac{a(1+\cos\theta)}{-a\sin\theta} = \cot\theta$

$= \frac{\cos\theta}{\sin\theta} \Rightarrow \sin\theta + \sin\theta\cos\theta = -\sin\theta\cos\theta \Rightarrow \cos\theta = -\frac{1}{2}$ or $\sin\theta = 0$; $\cos\theta = -\frac{1}{2} \Rightarrow \theta = \pm\frac{2\pi}{3}$

$\Rightarrow r = \frac{a}{2}$; $\sin\theta = 0 \Rightarrow \theta = 0$ (not π, see part (a)) $\Rightarrow r = 2a$. Therefore the points where the tangent line is vertical are $\left(\frac{a}{2}, \pm\frac{2\pi}{3}\right)$ and $(2a, 0)$.

117. $r_1 = \frac{a}{1+\cos\theta} \Rightarrow \frac{dr_1}{d\theta} = \frac{a\sin\theta}{(1+\cos\theta)^2}$ and $r_2 = \frac{b}{1-\cos\theta} \Rightarrow \frac{dr_2}{d\theta} = -\frac{b\sin\theta}{(1-\cos\theta)^2}$; then

$\tan\psi_1 = \frac{\left(\frac{a}{1+\cos\theta}\right)}{\left[\frac{a\sin\theta}{(1+\cos\theta)^2}\right]} = \frac{1+\cos\theta}{\sin\theta}$ and $\tan\psi_2 = \frac{\left(\frac{b}{1-\cos\theta}\right)}{\left[\frac{-b\sin\theta}{(1-\cos\theta)^2}\right]} = \frac{1-\cos\theta}{-\sin\theta} \Rightarrow 1 + \tan\psi_1\tan\psi_2$

$= 1 + \left(\frac{1+\cos\theta}{\sin\theta}\right)\left(\frac{1-\cos\theta}{-\sin\theta}\right) = 1 - \frac{1-\cos^2\theta}{\sin^2\theta} = 0 \Rightarrow \beta$ is undefined $\Rightarrow$ the parabolas are orthogonal at each point of intersection

118. $\tan\psi = \frac{r}{\left(\frac{dr}{d\theta}\right)} = \frac{a(1-\cos\theta)}{a\sin\theta}$ is 1 at $\theta = \frac{\pi}{2} \Rightarrow \psi = \frac{\pi}{4}$

119. $r = 3\sec\theta \Rightarrow r = \frac{3}{\cos\theta}$; $\frac{3}{\cos\theta} = 4 + 4\cos\theta \Rightarrow 3 = 4\cos\theta + 4\cos^2\theta \Rightarrow (2\cos\theta + 3)(2\cos\theta - 1) = 0$

$\Rightarrow \cos\theta = \frac{1}{2}$ or $\cos\theta = -\frac{3}{2} \Rightarrow \theta = \frac{\pi}{3}$ or $\frac{5\pi}{3}$ (the second equation has no solutions); $\tan\psi_2 = \frac{4(1+\cos\theta)}{-4\sin\theta}$

$= -\frac{1+\cos\theta}{\sin\theta}$ is $-\sqrt{3}$ at $\frac{\pi}{3}$ and $\tan\psi_1 = \frac{3\sec\theta}{3\sec\theta\tan\theta} = \cot\theta$ is $\frac{1}{\sqrt{3}}$ at $\frac{\pi}{3}$. Then $\tan\beta$ is undefined since

$1 + \tan\psi_1\tan\psi_2 = 1 + \left(\frac{1}{\sqrt{3}}\right)\left(-\sqrt{3}\right) = 0 \Rightarrow \beta = \frac{\pi}{2}$. Also, $\tan\psi_2|_{5\pi/3} = \sqrt{3}$ and $\tan\psi_1|_{5\pi/3} = -\frac{1}{\sqrt{3}}$

$\Rightarrow 1 + \tan\psi_1\tan\psi_2 = 1 + \left(-\frac{1}{\sqrt{3}}\right)\left(\sqrt{3}\right) = 0 \Rightarrow \tan\beta$ is also undefined $\Rightarrow \beta = \frac{\pi}{2}$.

120. $\tan\psi = \frac{a\tan\left(\frac{\theta}{2}\right)}{\frac{a}{2}\sec^2\left(\frac{\theta}{2}\right)} = 1$ at $\theta = \frac{\pi}{2} \Rightarrow \psi = \frac{\pi}{4}$; $m_{tan} = \tan(\theta + \psi) = \tan\frac{3\pi}{4} = -1$

121. $\frac{1}{1-\cos\theta} = \frac{1}{1-\sin\theta} \Rightarrow 1-\cos\theta = 1-\sin\theta \Rightarrow \cos\theta = \sin\theta \Rightarrow \theta = \frac{\pi}{4}$; $\tan\psi_1 = \frac{\left(\frac{1}{1-\cos\theta}\right)}{\left[\frac{-\sin\theta}{(1-\cos\theta)^2}\right]} = \frac{1-\cos\theta}{-\sin\theta}$;

$\tan\psi_2 = \frac{\left(\frac{1}{1-\sin\theta}\right)}{\left[\frac{\cos\theta}{(1-\sin\theta)^2}\right]} = \frac{1-\sin\theta}{\cos\theta}$. Thus at $\theta = \frac{\pi}{4}$, $\tan\psi_1 = \frac{1-\cos\left(\frac{\pi}{4}\right)}{-\sin\left(\frac{\pi}{4}\right)} = 1-\sqrt{2}$ and

$\tan\psi_2 = \frac{1-\sin\left(\frac{\pi}{4}\right)}{\cos\left(\frac{\pi}{4}\right)} = \sqrt{2}-1$. Then $\tan\beta = \frac{(\sqrt{2}-1)-(1-\sqrt{2})}{1+(\sqrt{2}-1)(1-\sqrt{2})} = \frac{2\sqrt{2}-2}{2\sqrt{2}-2} = 1 \Rightarrow \beta = \frac{\pi}{4}$

122. (b) $r^2 = 2\csc 2\theta = \frac{2}{\sin 2\theta} = \frac{2}{2\sin\theta\cos\theta}$

$\Rightarrow r^2\sin\theta\cos\theta = 1 \Rightarrow xy = 1$, a hyperbola

(c) At $\theta = \frac{\pi}{4}$, $x = y = 1 \Rightarrow \frac{dy}{dx} = -\frac{1}{x^2} = -1$

$= m_{\tan} \Rightarrow \phi = \frac{3\pi}{4} \Rightarrow \psi = \phi - \theta = \frac{3\pi}{4} - \frac{\pi}{4} = \frac{\pi}{2}$

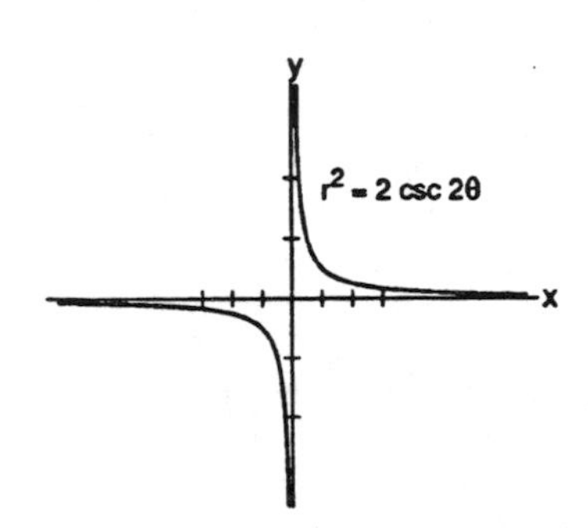

123. (a) $\tan\alpha = \frac{r}{\left(\frac{dr}{d\theta}\right)} \Rightarrow \frac{dr}{r} = \frac{d\theta}{\tan\alpha} \Rightarrow \ln r = \frac{\theta}{\tan\alpha} + C$ (by integration) $\Rightarrow r = Be^{\theta/(\tan\alpha)}$ for some constant B;

$A = \frac{1}{2}\int_{\theta_1}^{\theta_2} B^2e^{2\theta/(\tan\alpha)}\,d\theta = \left[\frac{B^2(\tan\alpha)e^{2\theta/(\tan\alpha)}}{4}\right]_{\theta_1}^{\theta_2} = \frac{\tan\alpha}{4}\left[B^2e^{2\theta_2/(\tan\alpha)} - B^2e^{2\theta_1/(\tan\alpha)}\right]$

$= \frac{\tan\alpha}{4}\left(r_2^2 - r_1^2\right)$ since $r_2^2 = B^2e^{2\theta_2/(\tan\alpha)}$ and $r_1^2 = B^2e^{2\theta_1/(\tan\alpha)}$; constant of proportionality $K = \frac{\tan\alpha}{4}$

(b) $\tan\alpha = \frac{r}{\left(\frac{dr}{d\theta}\right)} \Rightarrow \frac{dr}{d\theta} = \frac{r}{\tan\alpha} \Rightarrow \left(\frac{dr}{d\theta}\right)^2 = \frac{r^2}{\tan^2\alpha} \Rightarrow r^2 + \left(\frac{dr}{d\theta}\right)^2 = r^2 + \frac{r^2}{\tan^2\alpha} = r^2\left(\frac{\tan^2\alpha+1}{\tan^2\alpha}\right)$

$= r^2\left(\frac{\sec^2\alpha}{\tan^2\alpha}\right) \Rightarrow \text{Length} = \int_{\theta_1}^{\theta_2} r\left(\frac{\sec\alpha}{\tan\alpha}\right)d\theta = \int_{\theta_1}^{\theta_2} Be^{\theta/(\tan\alpha)}\cdot\frac{\sec\alpha}{\tan\alpha}\,d\theta = \left[B(\sec\alpha)e^{\theta/(\tan\alpha)}\right]_{\theta_1}^{\theta_2}$

$= (\sec\alpha)\left[Be^{\theta_2/(\tan\alpha)} - Be^{\theta_1(\tan\alpha)}\right] = K(r_2 - r_1)$ where $K = \sec\alpha$ is the constant of proportionality

124. $r^2\sin 2\theta = 2a^2 \Rightarrow r^2\sin\theta\cos\theta = a^2 \Rightarrow xy = a^2$ and $\frac{dy}{dx} = -\frac{a^2}{x^2}$. If $P(x_1, y_1)$ is a point on the curve, the tangent line is $y - y_1 = -\frac{a^2}{x_1^2}(x - x_1)$, so the tangent line crosses the x-axis when $y = 0 \Rightarrow -y_1 = -\frac{a^2}{x_1^2}(x - x_1)$

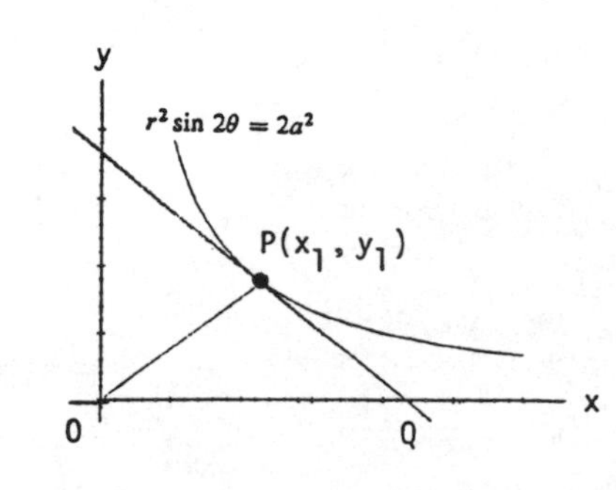

$\Rightarrow \frac{x_1^2 y_1}{a^2} = x - x_1 \Rightarrow x = \frac{x_1^2 y_1}{a^2} + x_1 = x_1 + x_1 = 2x_1$ since $\frac{x_1 y_1}{a^2} = 1$. Let Q be $(2x_1, 0)$. Then

$PQ = \sqrt{(2x_1 - x_1)^2 + (y_1 - 0)^2} = \sqrt{x_1^2 + y_1^2}$ and $OP = r = \sqrt{(x_1 - 0)^2 + (y_1 - 0)^2} = \sqrt{x_1^2 + y_1^2} \Rightarrow OP = PQ$ and

the triangle is isosceles.

CHAPTER 9 ADDITIONAL EXERCISES–THEORY, EXAMPLES, APPLICATIONS

1. Directrix $x = 3$ and focus $(4,0) \Rightarrow$ vertex is $\left(\frac{7}{2}, 0\right)$

   $\Rightarrow p = \frac{1}{2} \Rightarrow$ the equation is $x - \frac{7}{2} = \frac{y^2}{2}$

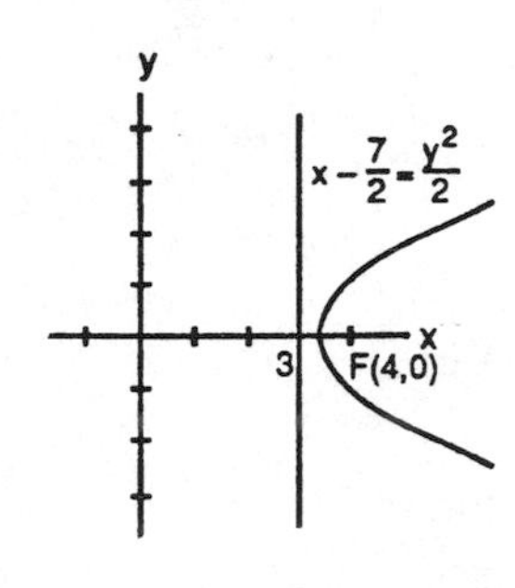

2. $x^2 - 6x - 12y + 9 = 0 \Rightarrow x^2 - 6x + 9 = 12y \Rightarrow \frac{(x-3)^2}{12} = y \Rightarrow$ vertex is $(3,0)$ and $p = 3 \Rightarrow$ focus is $(3,3)$ and the directrix is $y = -3$

3. $x^2 = 4y \Rightarrow$ vertex is $(0,0)$ and $p = 1 \Rightarrow$ focus is $(0,1)$; thus the distance from $P(x,y)$ to the vertex is $\sqrt{x^2 + y^2}$ and the distance from P to the focus is $\sqrt{x^2 + (y-1)^2} \Rightarrow \sqrt{x^2 + y^2} = 2\sqrt{x^2 + (y-1)^2}$

   $\Rightarrow x^2 + y^2 = 4[x^2 + (y-1)^2] \Rightarrow x^2 + y^2 = 4x^2 + 4y^2 - 8y + 4 \Rightarrow 3x^2 + 3y^2 - 8y + 4 = 0$, which is a circle

4. Let the segment $a + b$ intersect the y-axis in point A and intersect the x-axis in point B so that $PB = b$ and $PA = a$ (see figure). Draw the horizontal line through P and let it intersect the y-axis in point C. Let $\angle PBO = \theta \Rightarrow \angle APC = \theta$. Then $\sin\theta = \frac{y}{b}$ and $\cos\theta = \frac{x}{a} \Rightarrow \frac{x^2}{a^2} + \frac{y^2}{b^2} = \cos^2\theta + \sin^2\theta = 1$.

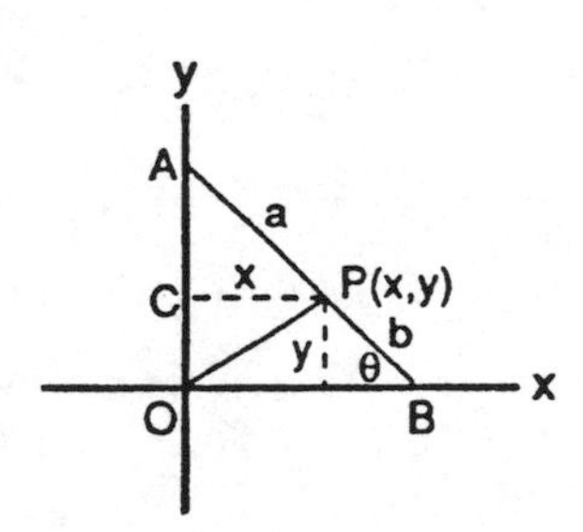

5. Vertices are $(0, \pm 2) \Rightarrow a = 2$; $e = \frac{c}{a} \Rightarrow 0.5 = \frac{c}{2} \Rightarrow c = 1 \Rightarrow$ foci are $(0, \pm 1)$

6. Let the center of the ellipse be $(x,0)$; directrix $x = 2$, focus $(4,0)$, and $e = \frac{2}{3} \Rightarrow \frac{a}{e} - c = 2 \Rightarrow \frac{a}{e} = 2 + c$

   $\Rightarrow a = \frac{2}{3}(2 + c)$. Also $c = ae = \frac{2}{3}a \Rightarrow a = \frac{2}{3}\left(2 + \frac{2}{3}a\right) \Rightarrow a = \frac{4}{3} + \frac{4}{9}a \Rightarrow \frac{5}{9}a = \frac{4}{3} \Rightarrow a = \frac{12}{5}$; $x - 2 = \frac{a}{e}$

   $\Rightarrow x - 2 = \left(\frac{12}{5}\right)\left(\frac{3}{2}\right) = \frac{18}{5} \Rightarrow x = \frac{28}{5} \Rightarrow$ the center is $\left(\frac{28}{5}, 0\right)$; $x - 4 = c \Rightarrow c = \frac{28}{5} - 4 = \frac{8}{5}$ so that $c^2 = a^2 - b^2$

   $= \left(\frac{12}{5}\right)^2 - \left(\frac{8}{5}\right)^2 = \frac{80}{25}$; therefore the equation is $\frac{\left(x - \frac{28}{5}\right)^2}{\left(\frac{144}{25}\right)} + \frac{y^2}{\left(\frac{80}{25}\right)} = 1$ or $\frac{25\left(x - \frac{28}{5}\right)^2}{144} + \frac{5y^2}{16} = 1$

7. Let the center of the hyperbola be $(0, y)$.

(a) Directrix $y = -1$, focus $(0, -7)$ and $e = 2 \Rightarrow c = -\frac{a}{e} = 6 \Rightarrow \frac{a}{e} = c - 6 \Rightarrow a = 2c - 12$. Also $c = ae = 2a$ $\Rightarrow a = 2(2a) - 12 \Rightarrow a = 4 \Rightarrow c = 8$; $y - (-1) = \frac{a}{e} = \frac{4}{2} = 2 \Rightarrow y = 1 \Rightarrow$ the center is $(0, 1)$; $c^2 = a^2 + b^2$ $\Rightarrow b^2 = c^2 - a^2 = 64 - 16 = 48$; therefore the equation is $\frac{(y-1)^2}{16} - \frac{x^2}{48} = 1$

(b) $e = 5 \Rightarrow c - \frac{a}{e} = 6 \Rightarrow \frac{a}{e} = c - 6 \Rightarrow a = 5c - 30$. Also, $c = ae = 5a \Rightarrow a = 5(5a) - 30 \Rightarrow 24a = 30 \Rightarrow a = \frac{5}{4}$ $\Rightarrow c = \frac{25}{4}$; $y - (-1) = \frac{a}{e} = \frac{\left(\frac{5}{4}\right)}{5} = \frac{1}{4} \Rightarrow y = -\frac{3}{4} \Rightarrow$ the center is $\left(0, -\frac{3}{4}\right)$; $c^2 = a^2 + b^2 \Rightarrow b^2 = c^2 - a^2$ $= \frac{625}{16} - \frac{24}{16} = \frac{75}{2}$; therefore the equation is $\frac{\left(y + \frac{3}{4}\right)^2}{\left(\frac{25}{16}\right)} - \frac{x^2}{\left(\frac{75}{2}\right)} = 1$ or $\frac{16\left(y + \frac{3}{4}\right)^2}{25} - \frac{2x^2}{75} = 1$

8. The center is $(0, 0)$ and $c = 2 \Rightarrow 4 = a^2 + b^2 \Rightarrow b^2 = 4 - a^2$. The equation is $\frac{y^2}{a^2} - \frac{x^2}{b^2} = 1 \Rightarrow \frac{49}{a^2} - \frac{144}{b^2} = 1$ $\Rightarrow \frac{49}{a^2} - \frac{144}{(4 - a^2)} = 1 \Rightarrow 49(4 - a^2) - 144a^2 = a^2(4 - a^2) \Rightarrow 196 - 49a^2 - 144a^2 = 4a^2 - a^4 \Rightarrow a^4 - 197a^2 + 196$ $= 0 \Rightarrow (a^2 - 196)(a^2 - 1) = 0 \Rightarrow a = 14$ or $a = 1$; $a = 14 \Rightarrow b^2 = 4 - (14)^2 < 0$ which is impossible; $a = 1$ $\Rightarrow b^2 = 4 - 1 = 3$; therefore the equation is $y^2 - \frac{x^2}{3} = 1$

9. $xy = 2 \Rightarrow x\frac{dy}{dx} + y = 0 \Rightarrow \frac{dy}{dx} = -\frac{y}{x}$; $x^2 - y^2 = 3$ $\Rightarrow 2x - 2y\frac{dy}{dx} = 0 \Rightarrow \frac{dy}{dx} = \frac{x}{y}$. If (x_0, y_0) is a point of intersection, then the product of the slopes is $\left(-\frac{y_0}{x_0}\right)\left(\frac{x_0}{y_0}\right) = -1 \Rightarrow$ the curves are orthogonal.

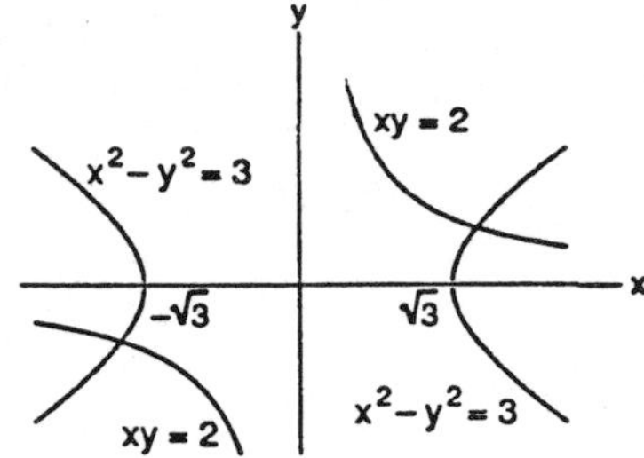

10. $y^2 = 4x + 4 \Rightarrow 2y\frac{dy}{dx} = 4 \Rightarrow \frac{dy}{dx} = \frac{2}{y}$; $y^2 = 64 - 16x$ $\Rightarrow 2y\frac{dy}{dx} = -16 \Rightarrow \frac{dy}{dx} = -\frac{8}{y}$. If (x_0, y_0) is a point of intersection, then the product of the slopes is $\left(\frac{2}{y_0}\right)\left(-\frac{8}{y_0}\right) = -\frac{16}{y_0^2}$; now $y^2 = 4x + 4$ and $y^2 = 64 - 16x$ $\Rightarrow 4x + 4 = 64 - 16x \Rightarrow 20x = 60 \Rightarrow x = 3 \Rightarrow y = \pm 4$ $\Rightarrow -\frac{16}{y_0^2} = -\frac{16}{(\pm 4)^2} = -1 \Rightarrow$ the curves are orthogonal at their points of intersection.

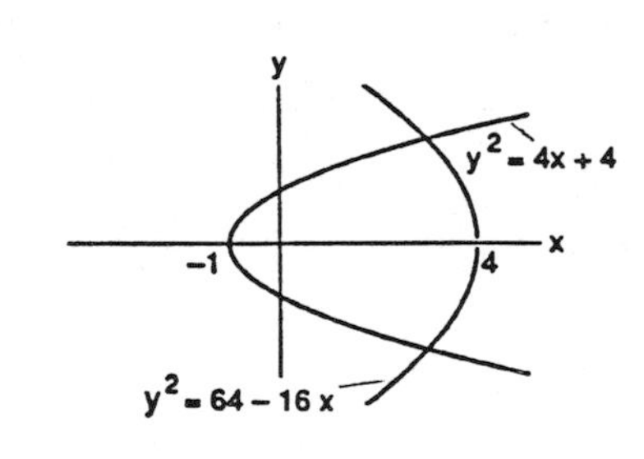

11. $2x^2 + 3y^2 = a^2 \Rightarrow 4x + 6y\frac{dy}{dx} = 0 \Rightarrow \frac{dy}{dx} = -\frac{2x}{3y}$; $ky^2 = x^3$ where k is a constant $\Rightarrow 2ky\frac{dy}{dx} = 3x^2 \Rightarrow \frac{dy}{dx} = \frac{3x^2}{2ky} = \frac{3x^2y}{2ky^2}$ $= \frac{3x^2y}{2x^3}$ (since $ky^2 = x^3$). If (x_0, y_0) is a point of intersection then the product of the slopes is $\left(-\frac{2x_0}{3y_0}\right)\left(\frac{3x_0^2y_0}{2x_0^3}\right) = -1$ $\Rightarrow$ the curves are orthogonal at their points of intersection.

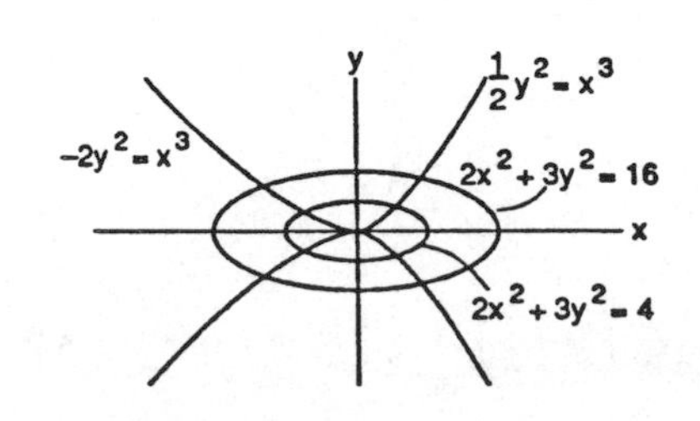

12. $y^2 = 4a(a-x) = -4a(x-a)$ has vertex $(a, 0)$ and $p = a \Rightarrow$ the focus is $(0,0)$; $y^2 = 4b(x+b)$ has vertex $(-b, 0)$ and $p = b \Rightarrow$ the focus is $(0,0)$; $4a(a-x) = 4b(x+b) \Rightarrow a^2 - b^2 = (a+b)x \Rightarrow x = a - b$ $\Rightarrow y^2 = 4a[a-(a-b)] = 4ab \Rightarrow y = \pm 2\sqrt{ab} \Rightarrow$ the points of intersection are $\left(a-b, \pm 2\sqrt{ab}\right)$; $y^2 = 4a(a-x)$ $\Rightarrow 2y\frac{dy}{dx} = -4a \Rightarrow \frac{dy}{dx} = -\frac{2a}{y}$ and $y^2 = 4b(x+b) \Rightarrow 2y\frac{dy}{dx} = 4b \Rightarrow \frac{dy}{dx} = \frac{2b}{y}$. Therefore at the points of intersection the product of the slopes is $\left(-\frac{2a}{y_0}\right)\left(\frac{2b}{y_0}\right) = -\frac{4ab}{y_0^2} = -\frac{4ab}{4ab} = -1 \Rightarrow$ the parabolas are orthogonal at their points of intersection.

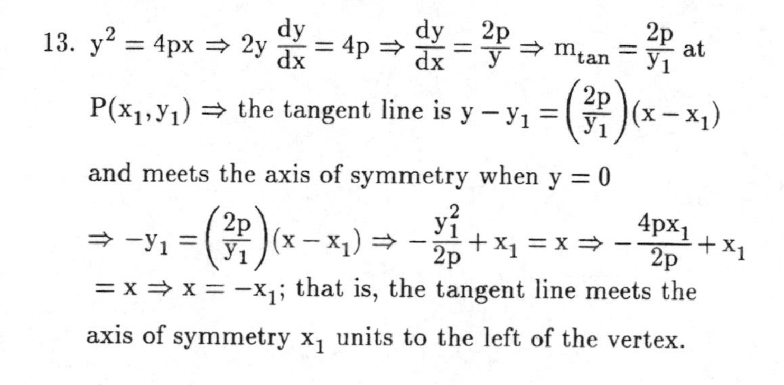

13. $y^2 = 4px \Rightarrow 2y\frac{dy}{dx} = 4p \Rightarrow \frac{dy}{dx} = \frac{2p}{y} \Rightarrow m_{tan} = \frac{2p}{y_1}$ at $P(x_1, y_1) \Rightarrow$ the tangent line is $y - y_1 = \left(\frac{2p}{y_1}\right)(x - x_1)$ and meets the axis of symmetry when $y = 0$ $\Rightarrow -y_1 = \left(\frac{2p}{y_1}\right)(x - x_1) \Rightarrow -\frac{y_1^2}{2p} + x_1 = x \Rightarrow -\frac{4px_1}{2p} + x_1$ $= x \Rightarrow x = -x_1$; that is, the tangent line meets the axis of symmetry x_1 units to the left of the vertex.

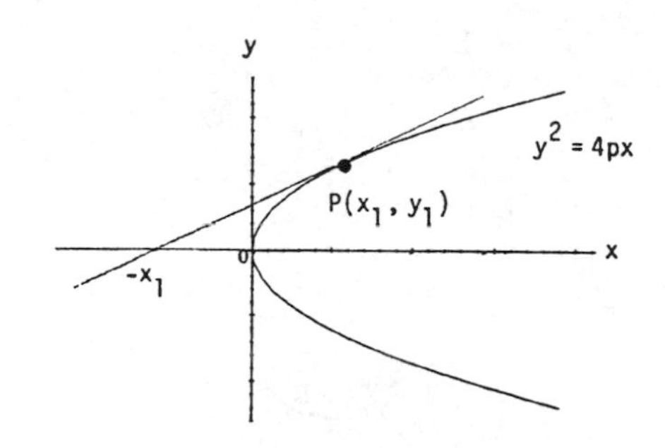

14. Let $P(x, y)$ be a point on the hyperbola. If the tangent line goes through the origin, then its slope is $\frac{y}{x}$. But this is the slope of an asymptote through the origin $\Rightarrow$ the tangent line is an asymptote. This is a contradiction.

15. $xy = a^2$ is a hyperbola whose asymptotes are the x and y axes; $xy = a^2 \Rightarrow y + x\frac{dy}{dx} = 0 \Rightarrow \frac{dy}{dx} = -\frac{y}{x}$. Let $P(x_1, y_1)$ be a point on the hyperbola $\Rightarrow m_{tan} = -\frac{y_1}{x_1} \Rightarrow$ the equation of the tangent line is $y - y_1 = \left(-\frac{y_1}{x_1}\right)(x - x_1)$. The tangent line intersects the coordinate axes to form the triangle (see figure). If $x = 0$, then $y - y_1 = \left(-\frac{y_1}{x_1}\right)(0 - x_1)$ $\Rightarrow y - y_1 = y_1 \Rightarrow y = 2y_1$; if $y = 0$, then $-y_1 = \left(-\frac{y_1}{x_1}\right)(x - x_1)$ $\Rightarrow x_1 = x - x_1 \Rightarrow x = 2x_1$. Therefore the area is $A = \frac{1}{2}(2x_1)(2y_1) = 2x_1y_1 = 2a^2$.

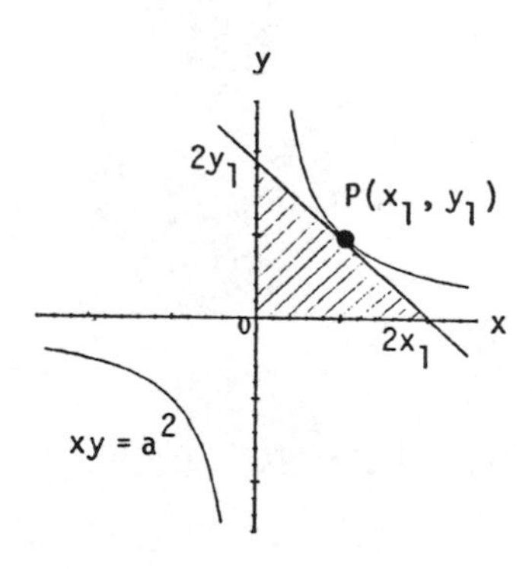

16. (a) $b^2x^2 + a^2y^2 = a^2b^2 \Rightarrow \frac{dy}{dx} = -\frac{b^2x}{a^2y}$; at (x_1, y_1) the tangent line is $y - y_1 = \left(-\frac{b^2x_1}{a^2y_1}\right)(x - x_1)$

$\Rightarrow a^2yy_1 + b^2xx_1 = b^2x_1^2 + a^2y_1^2 = a^2b^2 \Rightarrow b^2xx_1 + a^2yy_1 - a^2b^2 = 0$

(b) $b^2x^2 - a^2y^2 = a^2b^2 \Rightarrow \frac{dy}{dx} = \frac{b^2x}{a^2y}$; at (x_1, y_1) the tangent line is $y - y_1 = \left(\frac{b^2x_1}{a^2y_1}\right)(x - x_1)$

$\Rightarrow b^2xx_1 - a^2yy_1 = b^2x_1^2 - a^2y_1^2 = a^2b^2 \Rightarrow b^2xx_1 - a^2yy_1 - a^2b^2 = 0$

(c) $Ax^2 + Bxy + Cy^2 + Dx + Ey + F = 0$ has the derivative $\frac{dy}{dx} = \frac{-2Ax - By - D}{Bx + 2Cy + E}$; at (x_1, y_1) the tangent line is $y - y_1 = \left(\frac{-2Ax_1 - By_1 - D}{Bx_1 + 2Cy_1 + E}\right)(x - x_1) \Rightarrow Byx_1 + 2Cyy_1 + Ey - By_1x_1 - 2Cy_1^2 - Ey_1$ $= -2Axx_1 - Bxy_1 - Dx + 2Ax_1^2 + Bx_1y_1 + Dx_1 \Rightarrow 2Axx_1 + B(yx_1 + xy_1) + 2Cyy_1 + Dx - Dx_1 + Ey - Ey_1$ $= -2ax_1^2 + 2Bx_1y_1 + 2Cy_1^2$. Now add $2Dx_1 + 2Ey_1$ to both sides of this last equation, divide the result by 2, and represent the constant value on the right by $-F$ to get:

$$Axx_1 + B\left(\frac{yx_1 + xy_1}{2}\right) + Cyy_1 + D\left(\frac{x + x_1}{2}\right) + E\left(\frac{y + y_1}{2}\right) = -F$$

17.

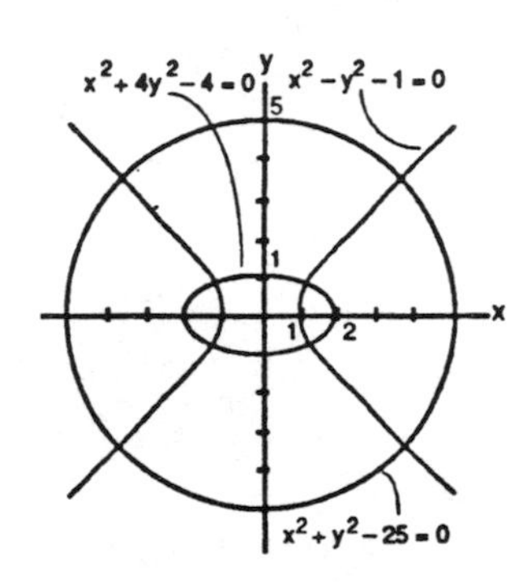

18.

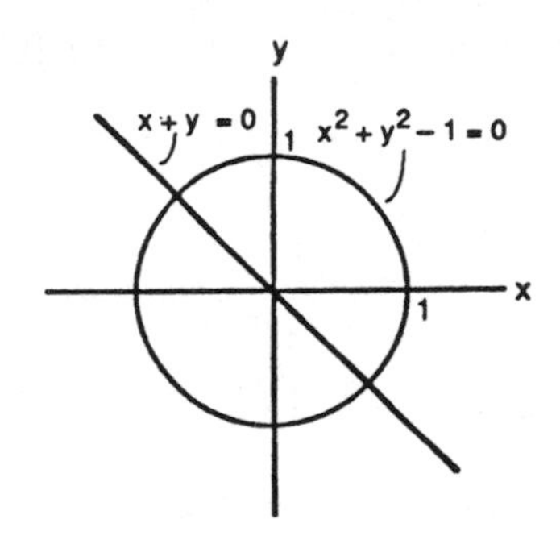

19.

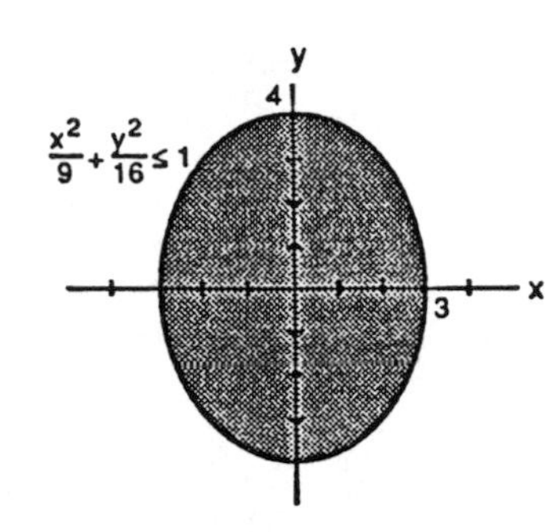

20.

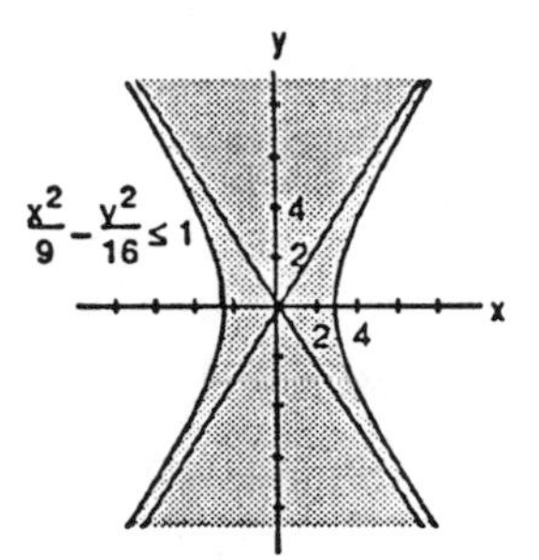

21. $(9x^2 + 4y^2 - 36)(4x^2 + 9y^2 - 16) \le 0 \Rightarrow 9x^2 + 4y^2 - 36 \le 0$ and $4x^2 + 9y^2 - 16 \ge 0$ or $9x^2 + 4y^2 - 36 \ge 0$ and $4x^2 + 9y^2 - 16 \le 0$

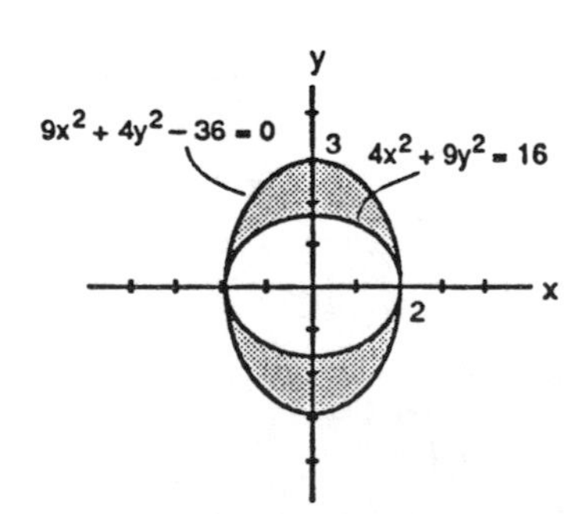

22. $(9x^2 + 4y^2 - 36)(4x^2 + 9y^2 - 16) > 0$, which is the complement of the set in Exercise 21

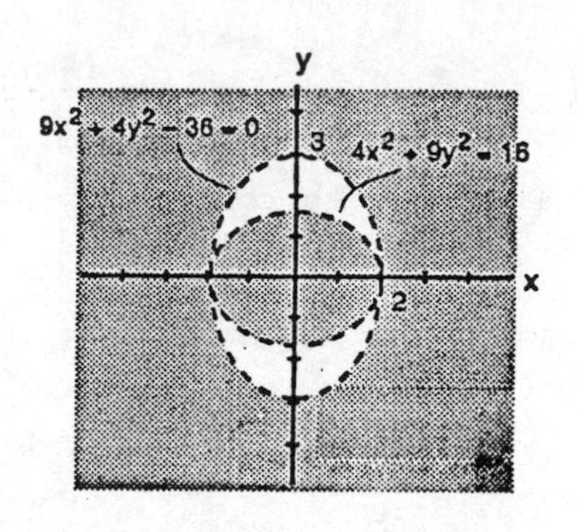

23. $x^4 - (y^2 - 9)^2 = 0 \Rightarrow x^2 - (y^2 - 9) = 0$ or
$x^2 + (y^2 - 9) = 0 \Rightarrow y^2 - x^2 = 9$ or $x^2 + y^2 = 9$

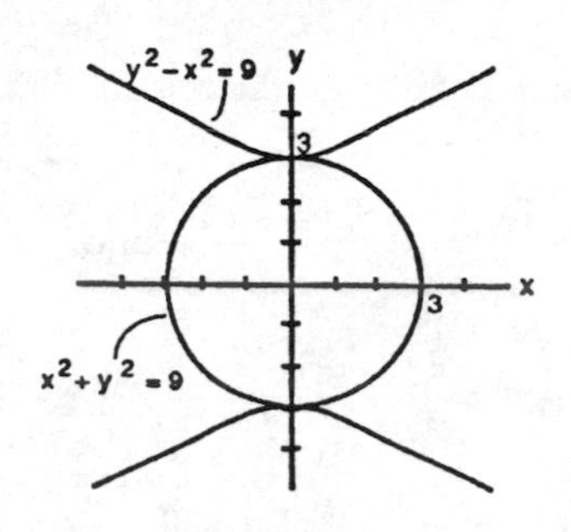

24. $x^2 + xy + y^2 < 3 \Rightarrow \tan 2\alpha = \frac{1}{1-1}$ which is undefined
$\Rightarrow 2\alpha = 90° \Rightarrow \alpha = 45° \Rightarrow A'$
$= \cos^2 45° + \cos 45° \sin 45° + \sin^2 45° = \frac{3}{2}$, $B' = 0$,
$C' = \sin^2 45° - \sin 45° \cos 45° + \cos^2 45° = \frac{1}{2}$
$\Rightarrow \frac{3}{2}x'^2 + \frac{1}{2}y'^2 < 3$ which is the interior of a rotated ellipse

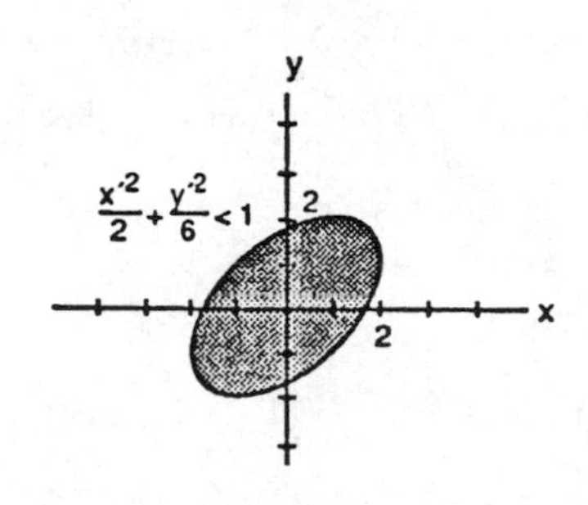

25. Arc PF = Arc AF since each is the distance rolled;
$\angle PCF = \frac{\text{Arc PF}}{b} \Rightarrow \text{Arc PF} = b(\angle PCF)$; $\theta = \frac{\text{Arc AF}}{a}$
$\Rightarrow \text{Arc AF} = a\theta \Rightarrow a\theta = b(\angle PCF) \Rightarrow \angle PCF = \left(\frac{a}{b}\right)\theta$;
$\angle OCB = \frac{\pi}{2} - \theta$ and $\angle OCB = \angle PCF - \angle PCE$
$= \angle PCF - \left(\frac{\pi}{2} - \alpha\right) = \left(\frac{a}{b}\right)\theta - \left(\frac{\pi}{2} - \alpha\right) \Rightarrow \frac{\pi}{2} - \theta$
$= \left(\frac{a}{b}\right)\theta - \left(\frac{\pi}{2} - \alpha\right) \Rightarrow \frac{\pi}{2} - \theta = \left(\frac{a}{b}\right)\theta - \frac{\pi}{2} + \alpha$
$\Rightarrow \alpha = \pi - \theta - \left(\frac{a}{b}\right)\theta \Rightarrow \alpha = \pi - \left(\frac{a+b}{b}\right)\theta$.

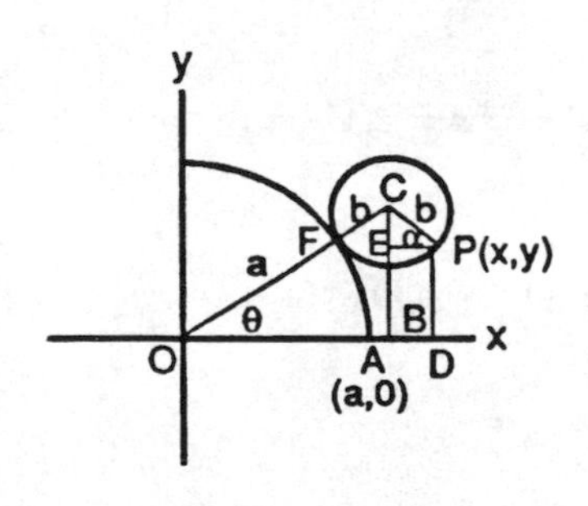

Now $x = OB + BD = OB + EP = (a+b)\cos\theta + b\cos\alpha = (a+b)\cos\theta + b\cos\left(\pi - \left(\frac{a+b}{b}\right)\theta\right)$
$= (a+b)\cos\theta + b\cos\pi\cos\left(\left(\frac{a+b}{b}\right)\theta\right) + b\sin\pi\sin\left(\left(\frac{a+b}{b}\right)\theta\right) = (a+b)\cos\theta - b\cos\left(\left(\frac{a+b}{b}\right)\theta\right)$ and

$y = PD = CB - CE = (a+b)\sin\theta - b\sin\alpha = (a+b)\sin\theta - b\sin\left(\pi - \left(\frac{a+b}{b}\right)\theta\right)$

$= (a+b)\sin\theta - b\sin\pi\cos\left(\left(\frac{a+b}{b}\right)\theta\right) + b\cos\pi\sin\left(\left(\frac{a+b}{b}\right)\theta\right) = (a+b)\sin\theta - b\sin\left(\left(\frac{a+b}{b}\right)\theta\right)$;

therefore $x = (a+b)\cos\theta - b\cos\left(\left(\frac{a+b}{b}\right)\theta\right)$ and $y = (a+b)\sin\theta - b\sin\left(\left(\frac{a+b}{b}\right)\theta\right)$

26. $\frac{dy}{dt} = \cos t \Rightarrow y = \int \cos t\, dt = \sin t + C$; $y = 0$ when $t = 0 \Rightarrow 0 + \sin 0 + C \Rightarrow C = 0 \Rightarrow y = \sin t$; then

$\frac{dx}{dt} = -2y = -2\sin t \Rightarrow x = \int -2\sin t\, dt = 2\cos t + C$; $x = 3$ when $t = 0 \Rightarrow 3 = 2\cos 0 + C \Rightarrow C = 1$

$\Rightarrow x = 2\cos t + 1$. Thus $\frac{x-1}{2} = \cos t \Rightarrow \left(\frac{x-1}{2}\right)^2 + y^2 = \cos^2 t + \sin^2 t = 1 \Rightarrow \frac{(x-1)^2}{4} + y^2 = 1$ is the

Cartesian equation, which is an ellipse.

27. $x = \frac{1-t^2}{1+t^2} \Rightarrow x^2 = \frac{(1-t^2)^2}{(1+t^2)^2}$ and $y = \frac{2t}{1+t^2} \Rightarrow y^2 = \frac{4t^2}{(1+t^2)^2}$

$\Rightarrow x^2 + y^2 = \frac{(1-t^2)^2 + 4t^2}{(1+t^2)^2} = \frac{t^4 + 2t^2 + 1}{(1+t^2)^2} = \frac{(t^2+1)^2}{(1+t^2)^2} = 1$;

$y = 0 \Rightarrow \frac{2t}{1+t^2} = 0 \Rightarrow t = 0 \Rightarrow x = 1 \Rightarrow (-1, 0)$ is not covered; $t = -1$ gives $(0, -1)$, $t = 0$ gives $(1, 0)$, and $t = 1$ gives $(0, 1)$. Note that as $t \to \pm\infty$, $x \to -1$ and $y \to 0$.

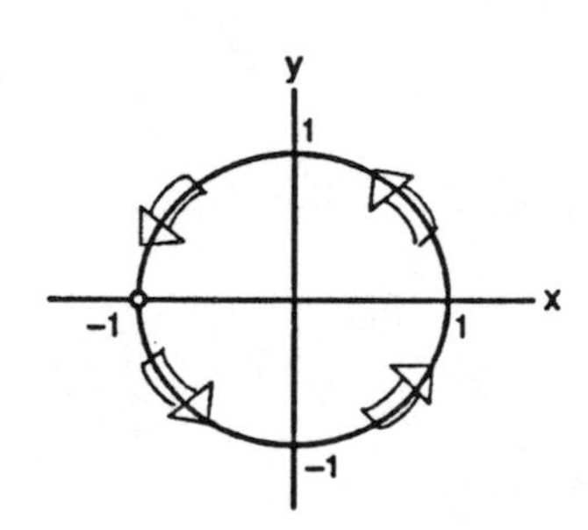

28. (a) $x = a(t - \sin t) \Rightarrow \frac{dx}{dt} = a(1 - \cos t)$ and let $\delta = 1 \Rightarrow dm = dA = y\, dx = y\left(\frac{dx}{dt}\right) dt$

$= a(1-\cos t)\, a(1-\cos t)\, dt = a^2(1-\cos t)^2\, dt$; then $A = \int_0^{2\pi} a^2(1-\cos t)^2\, dt$

$= a^2 \int_0^{2\pi} (1 - 2\cos t + \cos^2 t)\, dt = a^2 \int_0^{2\pi} \left(1 - 2\cos t + \frac{1}{2} + \frac{1}{2}\cos 2t\right) dt = a^2\left[\frac{3}{2}t - 2\sin t + \frac{\sin 2t}{4}\right]_0^{2\pi}$

$= 3\pi a^2$; $\tilde{x} = x = a(t - \sin t)$ and $\tilde{y} = \frac{1}{2}y = \frac{1}{2}a(1-\cos t) \Rightarrow M_x = \int \tilde{y}\, dm = \int \tilde{y}\, \delta\, dA$

$= \int_0^{2\pi} \frac{1}{2}a(1-\cos t)\, a^2(1-\cos t)^2\, dt = \frac{1}{2}a^3 \int_0^{2\pi} (1-\cos t)^3\, dt = \frac{a^3}{2} \int_0^{2\pi} (1 - 3\cos t + 3\cos^2 t - \cos^3 t)\, dt$

$= \frac{a^3}{2} \int_0^{2\pi} \left[1 - 3\cos t + \frac{3}{2} + \frac{3\cos 2t}{2} - (1 - \sin^2 t)(\cos t)\right] dt = \frac{a^3}{2}\left[\frac{5}{2}t - 3\sin t + \frac{3\sin 2t}{4} - \sin t + \frac{\sin^3 t}{3}\right]_0^{2\pi}$

$= \frac{5\pi a^3}{2}$. Therefore $\bar{y} = \frac{M_x}{M} = \frac{\left(\frac{5\pi a^3}{2}\right)}{3\pi a^2} = \frac{5}{6}a$. Also, $M_y = \int \tilde{x}\, dm = \int \tilde{x}\, \delta\, dA$

$= \int_0^{2\pi} a(t - \sin t)\, a^2(1-\cos t)^2\, dt = a^3 \int_0^{2\pi} (t - 2t\cos t + t\cos^2 t - \sin t + 2\sin t\cos t - \sin t\cos^2 t)\, dt$

$= a^3\left[\frac{t^2}{2} - 2\cos t - 2t\sin t + \frac{1}{4}t^2 + \frac{1}{8}\cos 2t + \frac{t}{4}\sin 2t + \cos t + \sin^2 t + \frac{\cos^3 t}{3}\right]_0^{2\pi} = 3\pi^2 a^3$. Thus $\overline{x} = \frac{M_y}{M} = \frac{3\pi^2 a^3}{3\pi a^2} = \pi a \Rightarrow \left(\pi a, \frac{5}{6}a\right)$ is the center of mass.

(b) $x = \frac{2}{3}t^{3/2} \Rightarrow \frac{dx}{dt} = t^{1/2}$ and $y = 2t^{1/2} \Rightarrow \frac{dy}{dt} = t^{-1/2}$; let $\delta = 1 \Rightarrow dm = dA = y\,dx = y\left(\frac{dx}{dt}\right)dt$

$\Rightarrow (2t^{1/2})(t^{1/2})\,dt = 2t\,dt$; $\tilde{x} = x = \frac{2}{3}t^{3/2}$ and $\tilde{y} = \frac{y}{2} = t^{1/2} \Rightarrow M_x = \int \tilde{y}\,dm = \int_0^{\sqrt{3}} t^{1/2}(2t\,dt)$

$= \int_0^{\sqrt{3}} 2t^{3/2}\,dt = \left[\frac{4}{5}t^{5/2}\right]_0^{\sqrt{3}} = \frac{36}{5}\sqrt{3}$. Also, $M_y = \int \tilde{x}\,dm = \int \tilde{x}\,dA = \int_0^{\sqrt{3}} \frac{2}{3}t^{3/2}(2t\,dt)$

$= \int_0^{\sqrt{3}} \frac{4}{3}t^{5/2}\,dt = \left[\frac{8}{21}t^{7/2}\right]_0^{\sqrt{3}} = \frac{72}{7}\sqrt{3}$.

29. (a) $x = e^{2t}\cos t$ and $y = e^{2t}\sin t \Rightarrow x^2 + y^2 = e^{4t}\cos^2 t + e^{4t}\sin^2 t = e^{4t}$. Also $\frac{y}{x} = \frac{e^{2t}\sin t}{e^{2t}\cos t} = \tan t$ $\Rightarrow t = \tan^{-1}\left(\frac{y}{x}\right) \Rightarrow x^2 + y^2 = e^{4\tan^{-1}(y/x)}$ is the Cartesian equation. Since $r^2 = x^2 + y^2$ and $\theta = \tan^{-1}\left(\frac{y}{x}\right)$, the polar equation is $r^2 = e^{4\theta}$ or $r = e^{2\theta}$ for $r > 0$

(b) $ds^2 = r^2\,d\theta^2 + dr^2$; $r = e^{2\theta} \Rightarrow dr = 2e^{2\theta}\,d\theta$

$\Rightarrow ds^2 = r^2\,d\theta^2 + \left(2e^{2\theta}\,d\theta\right)^2 = \left(e^{2\theta}\right)^2 d\theta^2 + 4e^{4\theta}\,d\theta^2$

$= 5e^{4\theta}\,d\theta^2 \Rightarrow ds = \sqrt{5}\,e^{2\theta}\,d\theta \Rightarrow L = \int_0^{2\pi} \sqrt{5}\,e^{2\theta}\,d\theta$

$= \left[\frac{\sqrt{5}\,e^{2\theta}}{2}\right]_0^{2\pi} = \frac{\sqrt{5}}{2}\left(e^{4\pi} - 1\right)$

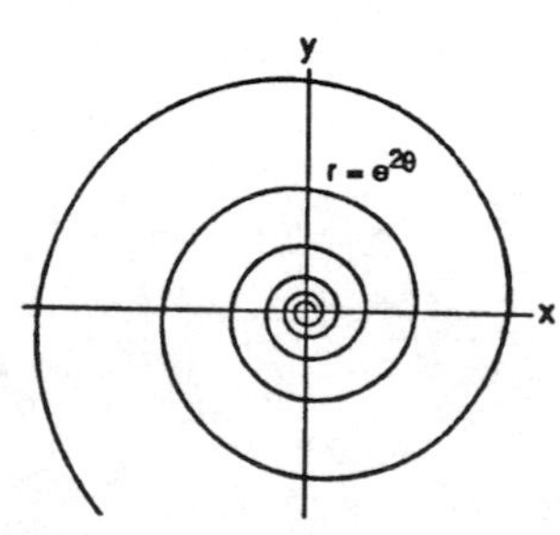

30. $r = 2\sin^3\left(\frac{\theta}{3}\right) \Rightarrow dr = 2\sin^2\left(\frac{\theta}{3}\right)\cos\left(\frac{\theta}{3}\right)d\theta \Rightarrow ds^2 = r^2\,d\theta^2 + dr^2 = \left[2\sin^3\left(\frac{\theta}{3}\right)\right]^2 d\theta^2 + \left[2\sin^2\left(\frac{\theta}{3}\right)\cos\left(\frac{\theta}{3}\right)d\theta\right]^2$

$= 4\sin^6\left(\frac{\theta}{3}\right)d\theta^2 + 4\sin^4\left(\frac{\theta}{3}\right)\cos^2\left(\frac{\theta}{3}\right)d\theta^2 = \left[4\sin^4\left(\frac{\theta}{3}\right)\right]\left[\sin^2\left(\frac{\theta}{3}\right) + \cos^2\left(\frac{\theta}{3}\right)\right]d\theta^2 = 4\sin^4\left(\frac{\theta}{3}\right)d\theta^2$

$\Rightarrow ds = 2\sin^2\left(\frac{\theta}{3}\right)d\theta$. Then $L = \int_0^{3\pi} 2\sin^2\left(\frac{\theta}{3}\right)d\theta = \int_0^{3\pi}\left[1 - \cos\left(\frac{2\theta}{3}\right)\right]d\theta = \left[\theta - \frac{3}{2}\sin\left(\frac{2\theta}{3}\right)\right]_0^{3\pi} = 3\pi$

31. $r = 1 + \cos\theta$ and $S = \int 2\pi\rho\,ds$, where $\rho = y = r\sin\theta$; $ds = \sqrt{r^2\,d\theta^2 + dr^2}$

$= \sqrt{(1+\cos\theta)^2\,d\theta^2 + \sin^2\theta\,d\theta^2}\ \sqrt{1 + 2\cos\theta + \cos^2\theta + \sin^2\theta}\,d\theta = \sqrt{2 + 2\cos\theta}\,d\theta = \sqrt{4\cos^2\left(\frac{\theta}{2}\right)}\,d\theta$

$= 2\cos\left(\frac{\theta}{2}\right)d\theta$ since $0 \le \theta \le \frac{\pi}{2}$. Then $S = \int_0^{\pi/2} 2\pi(r\sin\theta)\cdot 2\cos\left(\frac{\theta}{2}\right)d\theta = \int_0^{\pi/2} 4\pi(1+\cos\theta)\cdot\sin\theta\cos\left(\frac{\theta}{2}\right)d\theta$

$$= \int_0^{\pi/2} 4\pi\left[2\cos^2\left(\frac{\theta}{2}\right)\right]\left[2\sin\left(\frac{\theta}{2}\right)\cos\left(\frac{\theta}{2}\right)\cos\left(\frac{\theta}{2}\right)\right] d\theta = \int_0^{\pi/2} 16\pi\cos^4\left(\frac{\theta}{2}\right)\sin\left(\frac{\theta}{2}\right) d\theta = \left[\frac{-32\pi\cos^5\left(\frac{\theta}{2}\right)}{5}\right]_0^{\pi/2}$$

$$= \frac{(-32\pi)\left(\frac{\sqrt{2}}{2}\right)^5}{5} - \left(-\frac{32\pi}{5}\right) = \frac{32\pi - 4\pi\sqrt{2}}{5}$$

32. The region in question is the figure eight in the middle. The arc of $r = 2a\sin^2\left(\frac{\theta}{2}\right)$ in the first quadrant gives $\frac{1}{4}$ of that region. Therefore the area is $A = 4\int_0^{\pi/2} \frac{1}{2}r^2\, d\theta$

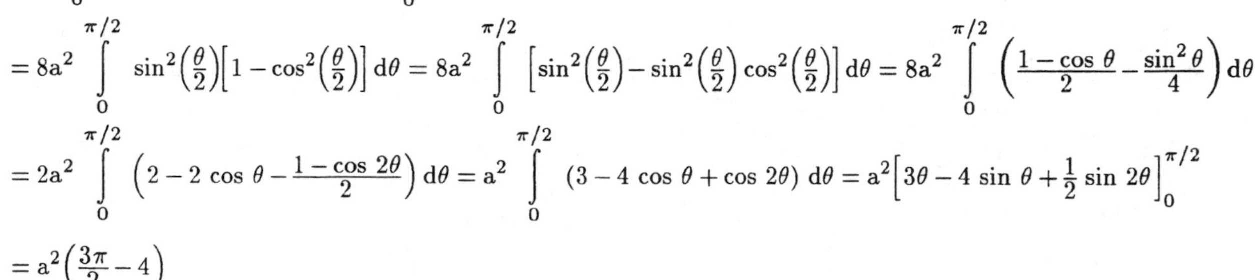

$$= 4\int_0^{\pi/2} \frac{1}{2}\left[2a\sin^2\left(\frac{\theta}{2}\right)\right]^2 d\theta = 8a^2\int_0^{\pi/2} \sin^4\left(\frac{\theta}{2}\right) d\theta$$

$$= 8a^2\int_0^{\pi/2} \sin^2\left(\frac{\theta}{2}\right)\left[1-\cos^2\left(\frac{\theta}{2}\right)\right] d\theta = 8a^2\int_0^{\pi/2}\left[\sin^2\left(\frac{\theta}{2}\right) - \sin^2\left(\frac{\theta}{2}\right)\cos^2\left(\frac{\theta}{2}\right)\right] d\theta = 8a^2\int_0^{\pi/2}\left(\frac{1-\cos\theta}{2} - \frac{\sin^2\theta}{4}\right) d\theta$$

$$= 2a^2\int_0^{\pi/2}\left(2 - 2\cos\theta - \frac{1-\cos 2\theta}{2}\right) d\theta = a^2\int_0^{\pi/2}(3 - 4\cos\theta + \cos 2\theta)\, d\theta = a^2\left[3\theta - 4\sin\theta + \frac{1}{2}\sin 2\theta\right]_0^{\pi/2}$$

$$= a^2\left(\frac{3\pi}{2} - 4\right)$$

33. $e = 2$ and $r\cos\theta = 2 \Rightarrow x = 2$ is the directrix $\Rightarrow k = 2$; the conic is a hyperbola with $r = \frac{ke}{1 + e\cos\theta}$

$$\Rightarrow r = \frac{(2)(2)}{1 + 2\cos\theta} = \frac{4}{1 + 2\cos\theta}$$

34. $e = 1$ and $r\cos\theta = -4 \Rightarrow x = -4$ is the directrix $\Rightarrow k = 4$; the conic is a parabola with $r = \frac{ke}{1 - e\cos\theta}$

$$\Rightarrow r = \frac{(4)(1)}{1 - \cos\theta} = \frac{4}{1 - \cos\theta}$$

35. $e = \frac{1}{2}$ and $r\sin\theta = 2 \Rightarrow y = 2$ is the directrix $\Rightarrow k = 2$; the conic is an ellipse with $r = \frac{ke}{1 + e\sin\theta}$

$$\Rightarrow r = \frac{2\left(\frac{1}{2}\right)}{1 + \left(\frac{1}{2}\right)\sin\theta} = \frac{2}{2 + \sin\theta}$$

36. $e = \frac{1}{3}$ and $r\sin\theta = -6 \Rightarrow y = -6$ is the directrix $\Rightarrow k = 6$; the conic is an ellipse with $r = \frac{ke}{1 - e\sin\theta}$

$$\Rightarrow r = \frac{6\left(\frac{1}{3}\right)}{1 - \left(\frac{1}{3}\right)\sin\theta} = \frac{6}{3 - \sin\theta}$$

37. The length of the rope is $L = 2x + 2c + y \geq 8C$.

(a) The angle A ($\angle BED$) occurs when the distance $CF = \ell$ is maximized. Now $\ell = \sqrt{x^2 - c^2} + y$
$\Rightarrow \ell = \sqrt{x^2 - c^2} + L - 2x - 2c$
$\Rightarrow \frac{d\ell}{dx} = \frac{1}{2}(x^2 - c^2)^{-1/2}(2x) - 2 = \frac{x}{\sqrt{x^2 - c^2}} - 2.$

Thus $\frac{d\ell}{dx} = 0 \Rightarrow \frac{x}{\sqrt{x^2 - c^2}} - 2 = 0 \Rightarrow x = 2\sqrt{x^2 - c^2}$

$\Rightarrow x^2 = 4x^2 - 4c^2 \Rightarrow 3x^2 = 4c^2 \Rightarrow \frac{c^2}{x^2} = \frac{3}{4} \Rightarrow \frac{c}{x} = \frac{\sqrt{3}}{2}.$

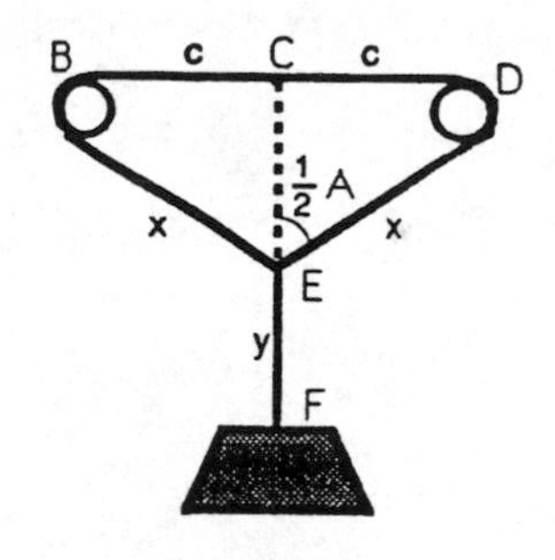

Since $\frac{c}{x} = \sin\frac{A}{2}$ we have $\sin\frac{A}{2} = \frac{\sqrt{3}}{2} \Rightarrow \frac{A}{2} = 60^\circ \Rightarrow A = 120^\circ$

(b) If the ring is fixed at E (i.e., y is held constant) and E is moved to the right, for example, the rope will slip around the pegs so that BE lengthens and DE becomes shorter $\Rightarrow$ BE + ED is always $2x = L - y - 2c$, which is constant $\Rightarrow$ the point E lies on an ellipse with the pegs as foci.

(c) Minimal potential energy occurs when the weight is at its lowest point $\Rightarrow$ E is at the intersection of the ellipse and its minor axis.

38. $\frac{d_1}{c} + \frac{d_2}{c} = \frac{30}{c} \Rightarrow d_1 + d_2 = 30$; $\frac{d_3}{c} + \frac{d_4}{c} = \frac{30}{c} \Rightarrow d_3 + d_4 = 30.$
Therefore P and Q lie on an ellipse with F_1 and F_2 as foci. Now $2a = d_1 + d_2 = 30 \Rightarrow a = 15$ and the focal distance is 10 $\Rightarrow b^2 = 15^2 - 10^2 = 125 \Rightarrow$ an equation of the ellipse is $\frac{x^2}{225} + \frac{y^2}{125} = 1$. Next $x_2 = x_1 + v_0 t = x_1 + v_0\left(\frac{10}{v_0}\right) = x_1 + 10.$

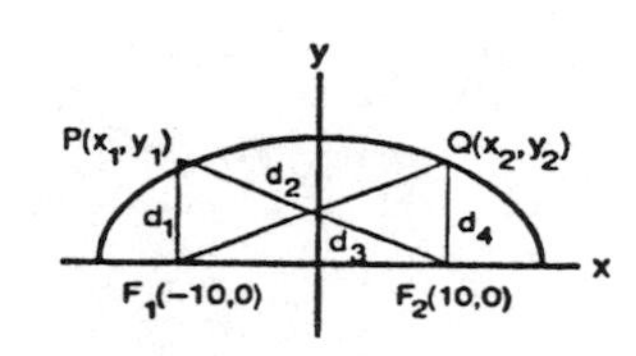

If the plane is flying level, then P and Q must be symmetric to the y-axis $\Rightarrow x_1 = -x_2 \Rightarrow x_2 = -x_2 + 10$
$\Rightarrow x_2 = 5 \Rightarrow \frac{5^2}{225} + \frac{y_2^2}{125} = 1 \Rightarrow y_2^2 = \frac{1000}{9} \Rightarrow y_2 = \frac{10\sqrt{10}}{3}$ since y_2 must be positive. Therefore the position of the plane is $\left(5, \frac{10\sqrt{10}}{3}\right)$ where the origin $(0,0)$ is located midway between the two stations.

39. If the vertex is $(0,0)$, then the focus is $(p,0)$. Let $P(x,y)$ be the present position of the comet. Then $\sqrt{(x-p)^2 + y^2} = 4 \times 10^7$. Since $y^2 = 4px$ we have $\sqrt{(x-p)^2 + 4px} = 4 \times 10^7 \Rightarrow (x-p)^2 + 4px = 16 \times 10^{14}$. Also, $x - p = 4 \times 10^7 \cos 60^\circ = 2 \times 10^7 \Rightarrow x = p + 2 \times 10^7$. Therefore $\left(2 \times 10^7\right)^2 + 4p\left(p + 2 \times 10^7\right) = 16 \times 10^{14}$
$\Rightarrow 4 \times 10^{14} + 4p^2 + 8p \times 10^7 = 16 \times 10^{14} \Rightarrow 4p^2 + 8p \times 10^7 - 12 \times 10^{14} = 0 \Rightarrow p^2 + 2p \times 10^7 - 3 \times 10^{14} = 0$
$\Rightarrow (p + 3 \times 10^7)(p - 10^7) = 0 \Rightarrow p = -3 \times 10^7$ or $p = 10^7$. Since p is positive we obtain $p = 10^7$ miles.

40. $x = 2t$ and $y = t^2 \Rightarrow y = \frac{x^2}{4}$; let $D = \sqrt{(x-0)^2 + \left(\frac{x^2}{4} - 3\right)^2} = \sqrt{x^2 + \frac{x^4}{16} - \frac{3}{2}x^2 + 9} = \sqrt{\frac{x^4}{16} - \frac{1}{2}x^2 + 9}$
$= \frac{1}{4}\sqrt{x^4 - 8x^2 + 144}$ be the distance from any point on the parabola to $(0,3)$. We want to minimize D. Then

$\frac{dD}{dx} = \frac{1}{8}(x^4 - 8x^2 + 144)^{-1/2}(4x^3 - 16x) = \frac{\left(\frac{1}{2}\right)x^3 - 2x}{\sqrt{x^4 - 8x^2 + 144}} = 0 \Rightarrow \frac{1}{2}x^3 - 2x = 0 \Rightarrow x^3 - 4x = 0 \Rightarrow x = 0$ or

$x = \pm 2$. Now $x = 0 \Rightarrow y = 0$ and $x = \pm 2 \Rightarrow y = 1$. The distance from $(0,0)$ to $(0,3)$ is $D = 3$. The distance from $(\pm 2, 1)$ to $(0,3)$ is $D = \sqrt{(\pm 2)^2 + (1-3)^2} = 2\sqrt{2}$ which is less than 3. Therefore the points closest to $(0,3)$ are $(\pm 2, 1)$.

41. $\cot\alpha = \frac{A-C}{B} = 0 \Rightarrow \alpha = 45^\circ$ is the angle of rotation $\Rightarrow A' = \cos^2 45^\circ + \cos 45^\circ \sin 45^\circ + \sin^2 45^\circ = \frac{3}{2}$, $B' = 0$, and $C' = \sin^2 45^\circ - \sin 45^\circ \cos 45^\circ + \cos^2 45^\circ = \frac{1}{2} \Rightarrow \frac{3}{2}x'^2 + \frac{1}{2}y'^2 = 1 \Rightarrow b = \sqrt{\frac{2}{3}}$ and $a = \sqrt{2} \Rightarrow c^2 = a^2 - b^2 = 2 - \frac{2}{3} = \frac{4}{3} \Rightarrow c = \frac{2}{\sqrt{3}}$. Therefore the eccentricity is $e = \frac{c}{a} = \frac{\left(\frac{2}{\sqrt{3}}\right)}{\sqrt{2}} = \sqrt{\frac{2}{3}} \approx 0.82$.

42. The angle of rotation is $\alpha = \frac{\pi}{4} \Rightarrow A' = \sin\frac{\pi}{4}\cos\frac{\pi}{4} = \frac{1}{2}$, $B' = 0$, and $C' = -\sin\frac{\pi}{4}\cos\frac{\pi}{4} = -\frac{1}{2} \Rightarrow \frac{x'^2}{2} - \frac{y'^2}{2} = 1$ $\Rightarrow a = \sqrt{2}$ and $b = \sqrt{2} \Rightarrow c^2 = a^2 + b^2 = 4 \Rightarrow c = 2$. Therefore the eccentricity is $e = \frac{c}{a} = \frac{2}{\sqrt{2}} = \sqrt{2}$.

43. $\sqrt{x} + \sqrt{y} = 1 \Rightarrow x + 2\sqrt{xy} + y = 1 \Rightarrow 2\sqrt{xy} = 1 - (x+y) \Rightarrow 4xy = 1 - 2(x+y) + (x+y)^2$
$\Rightarrow 4xy = x^2 + 2xy + y^2 - 2x - 2y + 1 \Rightarrow x^2 - 2xy + y^2 - 2x - 2y + 1 = 0 \Rightarrow B^2 - 4AC = (-2)^2 - 4(1)(1) = 0$
$\Rightarrow$ the curve is part of a parabola

44. $\alpha = \frac{\pi}{4} \Rightarrow A' = 2\sin\frac{\pi}{4}\cos\frac{\pi}{4} = 1$, $B' = 0$, $C' = -2\sin\frac{\pi}{4}\cos\frac{\pi}{4} = -1$, $D' = -\sqrt{2}\sin\frac{\pi}{4} = -1$, $E' = -\sqrt{2}\cos\frac{\pi}{4}$ $= -1$, $F' = 2 \Rightarrow x'^2 - y'^2 - x' - y' + 2 = 0 \Rightarrow (x'^2 - x') - (y'^2 + y') = -2 \Rightarrow \left(x'^2 - x' + \frac{1}{4}\right) - \left(y'^2 + y' + \frac{1}{4}\right)$ $= -2 \Rightarrow \frac{\left(y' + \frac{1}{2}\right)^2}{2} - \frac{\left(x' - \frac{1}{2}\right)^2}{2} = 1$. The center is $(x', y') = \left(\frac{1}{2}, -\frac{1}{2}\right) \Rightarrow x = \frac{1}{2}\cos\frac{\pi}{4} - \left(-\frac{1}{2}\right)\sin\frac{\pi}{4} = \frac{\sqrt{2}}{2}$ and $y = \frac{1}{2}\sin\frac{\pi}{4} - \frac{1}{2}\cos\frac{\pi}{4} = 0$ or the center is $(x, y) = \left(\frac{\sqrt{2}}{2}, 0\right)$. Next $a = \sqrt{2} \Rightarrow$ the vertices are $(x', y') = \left(\frac{1}{2}, \sqrt{2} - \frac{1}{2}\right)$ and $\left(\frac{1}{2}, -\sqrt{2} - \frac{1}{2}\right) \Rightarrow x = \frac{1}{2}\cos\frac{\pi}{4} - \left(\sqrt{2} - \frac{1}{2}\right)\sin\frac{\pi}{4} = \frac{\sqrt{2}}{2} - 1$ and $y = \frac{1}{2}\sin\frac{\pi}{4} + \left(\sqrt{2} - \frac{1}{2}\right)\cos\frac{\pi}{4} = 1$ or $(x, y) = \left(\frac{\sqrt{2}}{2} - 1, 1\right)$ is one vertex, and $x = \frac{1}{2}\cos\frac{\pi}{4} - \left(-\sqrt{2} - \frac{1}{2}\right)\sin\frac{\pi}{4}$ $= \frac{\sqrt{2}}{2} + 1$ and $y = \frac{1}{2}\sin\frac{\pi}{4} + \left(-\sqrt{2} - \frac{1}{2}\right)\sin\frac{\pi}{4} = -1$ or $(x, y) = \left(\frac{\sqrt{2}}{2} + 1, -1\right)$ is the other vertex. Also $c^2 = 2 + 2 = 4 \Rightarrow c = 2 \Rightarrow$ the foci are $(x', y') = \left(\frac{1}{2}, \frac{3}{2}\right)$ and $\left(\frac{1}{2}, -\frac{5}{2}\right) \Rightarrow x = \frac{1}{2}\cos\frac{\pi}{4} - \frac{3}{2}\sin\frac{\pi}{4} = -\frac{\sqrt{2}}{2}$ and $y = \frac{1}{2}\sin\frac{\pi}{4} + \frac{3}{2}\cos\frac{\pi}{4} = \sqrt{2}$ or $(x, y) = \left(-\frac{\sqrt{2}}{2}, \sqrt{2}\right)$ is one focus, and $x = \frac{1}{2}\cos\frac{\pi}{4} + \frac{5}{2}\sin\frac{\pi}{4} = \frac{3\sqrt{2}}{2}$ and $y = \frac{1}{2}\sin\frac{\pi}{4} - \frac{5}{2}\cos\frac{\pi}{4} = -\sqrt{2}$ or $(x, y) = \left(\frac{3\sqrt{2}}{2}, -\sqrt{2}\right)$ is the other focus. The asymptotes are $y' + \frac{1}{2} = \pm\left(x' - \frac{1}{2}\right)$ in the rotated system. Since $x = \frac{1}{\sqrt{2}}x' - \frac{1}{\sqrt{2}}y'$ and $y = \frac{1}{\sqrt{2}}x' + \frac{1}{\sqrt{2}}y' \Rightarrow x + y = \frac{2}{\sqrt{2}}x'$ $\Rightarrow \frac{\sqrt{2}}{2}x + \frac{\sqrt{2}}{2}y = x'$ and $x - y = -\frac{2}{\sqrt{2}}y' \Rightarrow -\frac{\sqrt{2}}{2}x + \frac{\sqrt{2}}{2}y = y'$; the asymptotes are

$-\frac{\sqrt{2}}{2}x+\frac{\sqrt{2}}{2}y+\frac{1}{2}=\pm\left(\frac{\sqrt{2}}{2}x+\frac{\sqrt{2}}{2}y-\frac{1}{2}\right)\Rightarrow$ the asymptotes are $-\sqrt{2}x+1=0$ or $x=\frac{1}{\sqrt{2}}$ and $\sqrt{2}y=0$ or $y=0$. Finally, the x′-axis is the line through $\left(\frac{\sqrt{2}}{2},0\right)$ with a slope of 1 $\left(\text{recall that } \alpha=\frac{\pi}{4}\right)\Rightarrow y=x-\frac{\sqrt{2}}{2}$. The y′-axis is the line through $\left(\frac{\sqrt{2}}{2},0\right)$ with a slope of $-1\Rightarrow y=-x+\frac{\sqrt{2}}{2}$.

45. (a) The equation of a parabola with focus (0,0) and vertex (a,0) is $r=\frac{2a}{1+\cos\theta}$ and rotating this parabola through $\alpha=45°$ gives $r=\frac{2a}{1+\cos\left(\theta-\frac{\pi}{4}\right)}$.

(b) Foci at (0,0) and (2,0) $\Rightarrow$ the center is (1,0) $\Rightarrow a=3$ and $c=1$ since one vertex is at (4,0). Then $e=\frac{c}{a}=\frac{1}{3}$. For ellipses with one focus at the origin and major axis along the x-axis we have $r=\frac{a(1-e^2)}{1-e\cos\theta}=\frac{3\left(1-\frac{1}{9}\right)}{1-\left(\frac{1}{3}\right)\cos\theta}=\frac{8}{3-\cos\theta}$.

(c) Center at $\left(2,\frac{\pi}{2}\right)$ and focus at (0,0) $\Rightarrow c=2$; center at $\left(2,\frac{\pi}{2}\right)$ and vertex at $\left(1,\frac{\pi}{2}\right)\Rightarrow a=1$. Then $e=\frac{c}{a}=\frac{2}{1}=2$. Also $k=ae-\frac{a}{e}=(1)(2)-\frac{1}{2}=\frac{3}{2}$. Therefore $r=\frac{ke}{1+e\sin\theta}=\frac{\left(\frac{3}{2}\right)(2)}{1+2\sin\theta}=\frac{3}{1+2\sin\theta}$.

46. Let (d_1,θ_1) and (d_2,θ_2) be the polar coordinates of P_1 and P_2, respectively. Then $\theta_2=\theta_1+\pi$, and we have $d_1=\frac{3}{2+\cos\theta_1}$ and $d_2=\frac{3}{2+\cos(\theta_1+\pi)}$. Therefore $\frac{1}{d_1}+\frac{1}{d_2}=\frac{2+\cos\theta_1}{3}+\frac{2+\cos(\theta_1+\pi)}{3}$
$=\frac{4+\cos\theta_1+\cos\theta_1\cos\pi-\sin\theta_1\sin\pi}{3}=\frac{4}{3}$.

47. Arc PT = Arc TO since each is the same distance rolled. Now Arc PT = a($\angle$TAP) and Arc TO = a($\angle$TBO) $\Rightarrow \angle TAP=\angle TBO$. Since AP = a = BO we have that ΔADP is congruent to ΔBCO $\Rightarrow$ CO = DP $\Rightarrow$ OP is parallel to AB $\Rightarrow \angle TBO=\angle TAP=\theta$. Then OPDC is a square $\Rightarrow r=CD=AB-AD-CB=AB-2CB$ $\Rightarrow r=2a-2a\cos\theta=2a(1-\cos\theta)$, which is the polar equation of a cardiod.

48. Note first that the point P traces out a circular arc as the door closes until the second door panel PQ is tangent to the circle. This happens when P is located at $\left(\frac{1}{\sqrt{2}},\frac{1}{\sqrt{2}}\right)$, since $\angle$OPQ is 90° at that time. Thus the curve is the circle $x^2+y^2=1$ for $0\le x\le\frac{1}{\sqrt{2}}$. When $x\ge\frac{1}{\sqrt{2}}$, the second door panel is tangent to the curve at P. Now let t represent $\angle$POQ so that as t runs from $\frac{\pi}{2}$ to 0, the door closes. The coordinates of P are given by (cos t, sin t), and the coordinates of Q by (2 cos t, 0) (since triangle POQ is isosceles). Therefore at a fixed instant of time t, the slope of the line

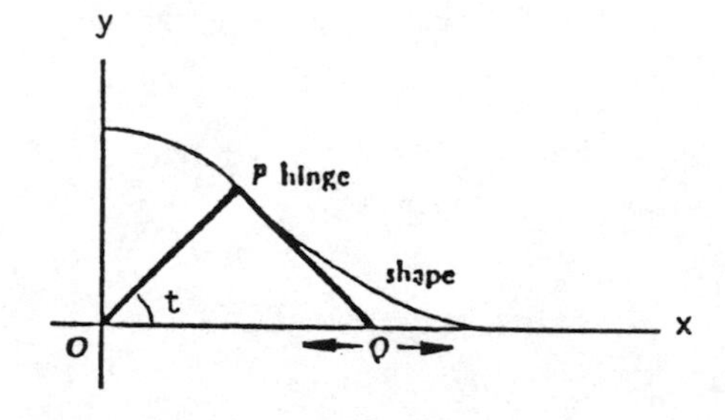

formed by the second panel PQ is $m = \frac{\Delta y}{\Delta x} = \frac{\sin t - 0}{\cos t - 2\cos t} = -\tan t \Rightarrow$ the tangent line PQ is $y - 0 = (-\tan t)(x - 2\cos t) \Rightarrow y = (-\tan t)x + 2\sin t$. Now, to find an equation of the curve for $\frac{1}{\sqrt{2}} \le x \le 1$, we want to find, for fixed x, the largest value of y as t ranges over the interval $0 \le t \le \frac{\pi}{4}$. We solve $\frac{dy}{dt} = 0 \Rightarrow (-\sec^2 t)x + 2\cos t = 0 \Rightarrow (-\sec^2 t)x = -2\cos t \Rightarrow x = 2\cos^3 t$. (Note that $\frac{d^2y}{dt^2} = (-2\sec^2 t \tan t)x - 2\sin t < 0$ on $0 \le t \le \frac{\pi}{2}$, so a maximum occurs for y.) Now $x = 2\cos^3 t \Rightarrow$ the corresponding y value is $y = (-\tan t)(2\cos^3 t) + 2\sin t = -2\sin t\cos^2 t + 2\sin t = (2\sin t)(-\cos^2 t + 1)$ $= 2\sin^3 t$. Therefore parametric equations for the path of the curve are given by $x = 2\cos^3 t$ and $y = 2\sin^3 t$ for $0 \le t \le \frac{\pi}{4}$. In Cartesian coordinates, we have the curve $x^{2/3} + y^{2/3} = (2\cos^3 t)^{2/3} + (2\sin^3 t)^{2/3}$ $= 2^{2/3}(\cos^2 t + \sin^2 t) = 2^{2/3} \Rightarrow$ the curve traced out by the door is given by

$$\left.\begin{array}{ll} x^2 + y^2 = 1 & \text{for } 0 \le x \le \frac{1}{\sqrt{2}} \\ x^{2/3} + y^{2/3} = 2^{2/3} & \text{for } \frac{1}{\sqrt{2}} \le x \le 1 \end{array}\right\}$$

49.

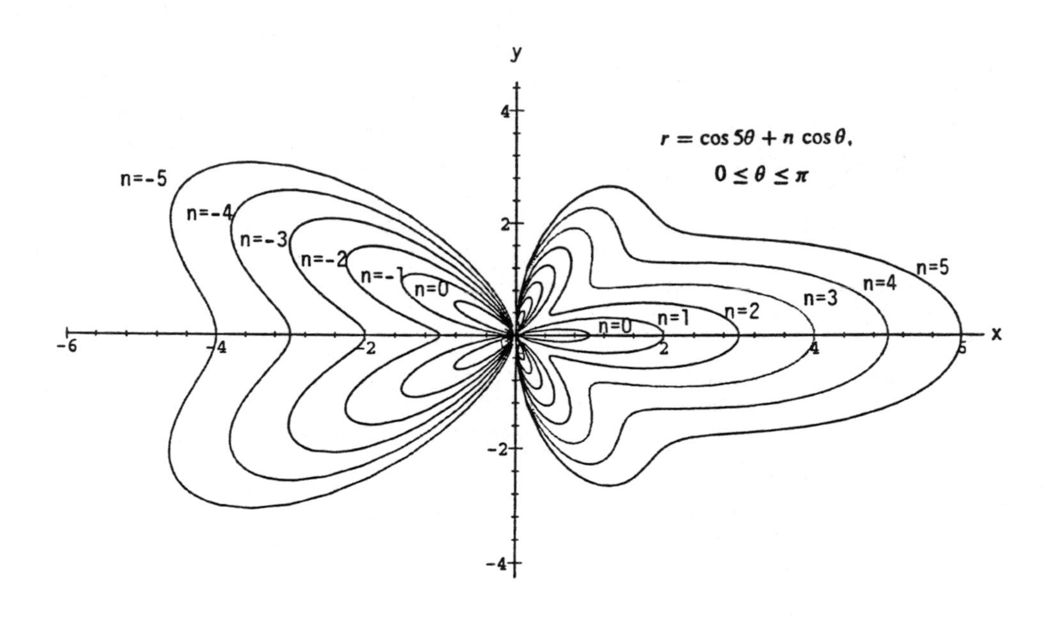

NOTES:

CHAPTER 10 VECTORS AND ANALYTIC GEOMETRY IN SPACE

10.1 VECTORS IN THE PLANE

1. (a)

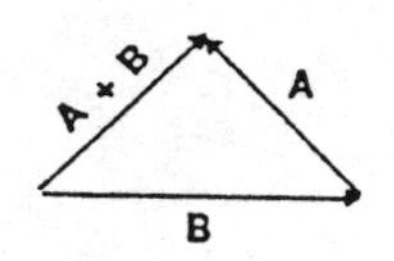

(b)

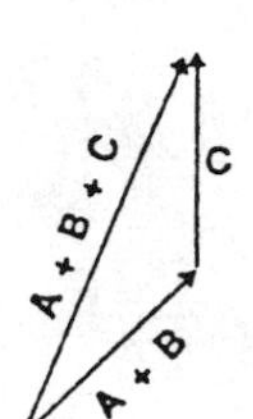

(c)

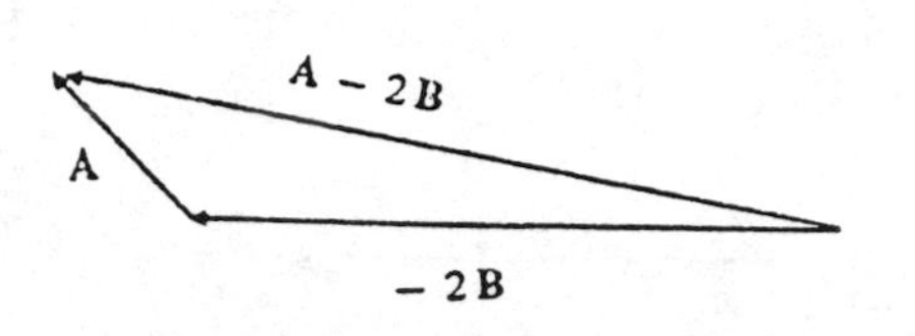

(d)

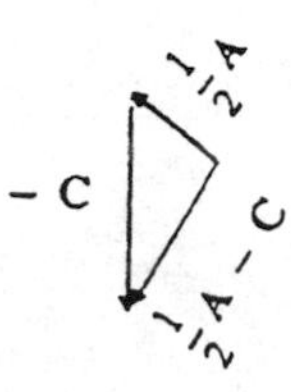

2. (a)

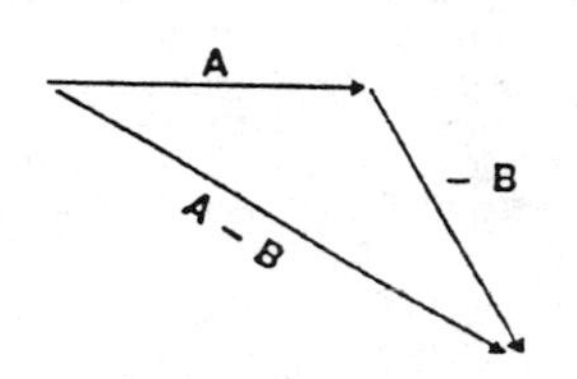

(b)

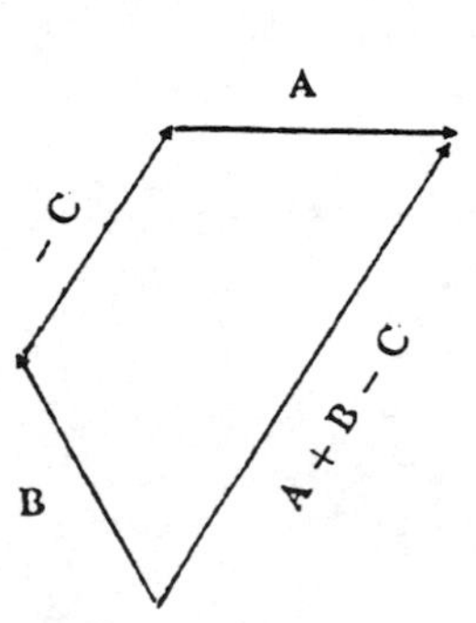

(c)

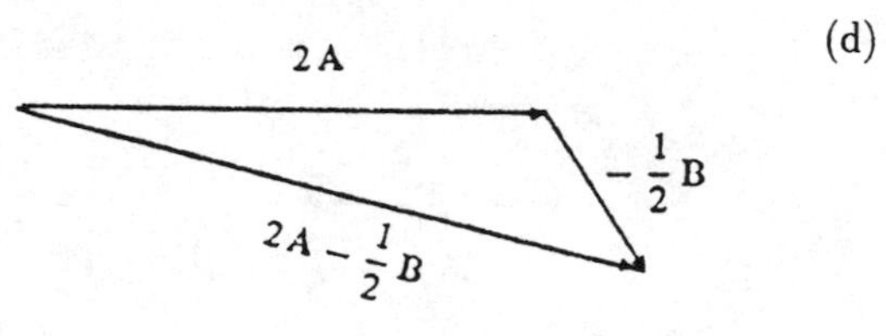

(d)

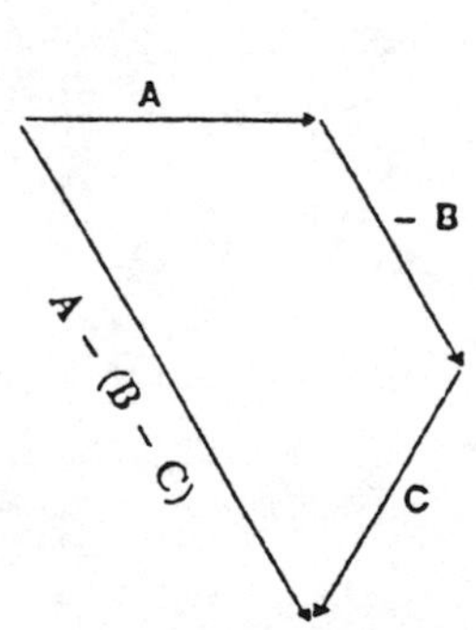

3. $\mathbf{A}+2\mathbf{B}=(2\mathbf{i}-7\mathbf{j})+2(\mathbf{i}+6\mathbf{j})=(2\mathbf{i}-7\mathbf{j})+(2\mathbf{i}+12\mathbf{j})=4\mathbf{i}+5\mathbf{j}$

4. $\mathbf{A}+\mathbf{B}-\mathbf{C}=(2\mathbf{i}-7\mathbf{j})+(\mathbf{i}+6\mathbf{j})-(\sqrt{3}\mathbf{i}-\pi\mathbf{j})=(3-\sqrt{3})\mathbf{i}+(\pi-1)\mathbf{j}$

5. $3\mathbf{A}-\frac{1}{\pi}\mathbf{C}=3(2\mathbf{i}-7\mathbf{j})-\frac{1}{\pi}(\sqrt{3}\mathbf{i}-\pi\mathbf{j})=(6\mathbf{i}-21\mathbf{j})-\left(\frac{\sqrt{3}}{\pi}\mathbf{i}-\mathbf{j}\right)=\left(6-\frac{\sqrt{3}}{\pi}\right)\mathbf{i}-20\mathbf{j}$

6. $2\mathbf{A}-3\mathbf{B}+32\mathbf{j}=2(2\mathbf{i}-7\mathbf{j})-3(\mathbf{i}+6\mathbf{j})+32\mathbf{j}=(4\mathbf{i}-14\mathbf{j})+(-3\mathbf{i}-18\mathbf{j})+32\mathbf{j}=\mathbf{i}+0\mathbf{j}=\mathbf{i}$

7. (a) $\mathbf{w}=\mathbf{u}+\mathbf{v}$ (b) $\mathbf{v}=\mathbf{w}+(-\mathbf{u})=\mathbf{w}-\mathbf{u}$

8. $\mathbf{a}=\mathbf{u}+\overrightarrow{BP}=\mathbf{u}+\frac{(\mathbf{w}-\mathbf{u})}{2}=\frac{2\mathbf{u}+\mathbf{w}-\mathbf{u}}{2}=\frac{\mathbf{u}+\mathbf{w}}{2}$

9. $\overrightarrow{P_1P_2}=(2-5)\mathbf{i}+(9-7)\mathbf{j}=-3\mathbf{i}+2\mathbf{j}$

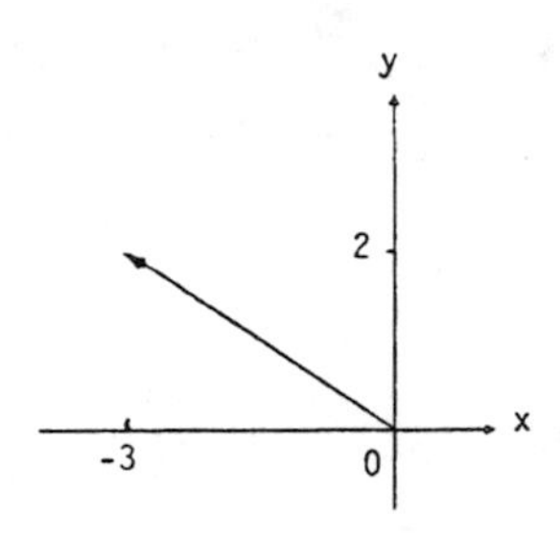

10. $\overrightarrow{P_1P_2}=(-3-1)\mathbf{i}+(5-2)\mathbf{j}=-4\mathbf{i}+3\mathbf{j}$

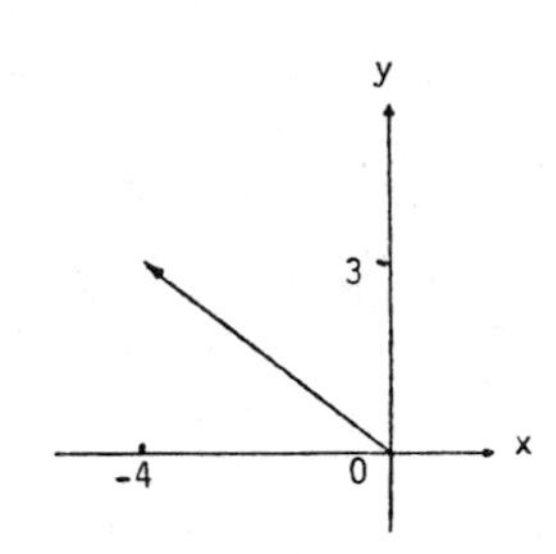

11. $\overrightarrow{AB}=(-10-(-5))\mathbf{i}+(8-3)\mathbf{j}=-5\mathbf{i}+5\mathbf{j}$

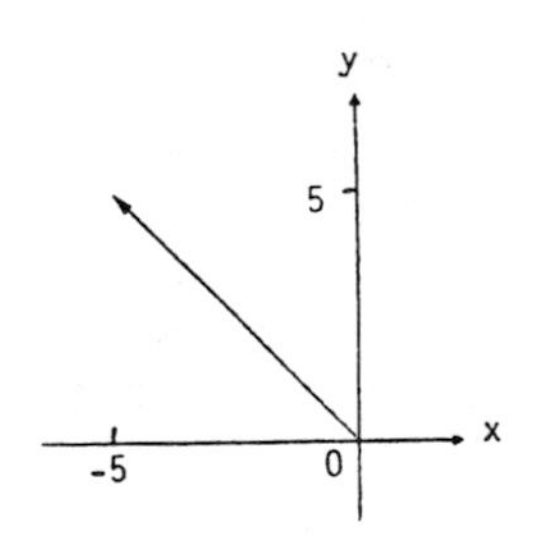

12. $\overrightarrow{AB}=(6-(-7))\mathbf{i}+(11-(-8))\mathbf{j}=13\mathbf{i}+19\mathbf{j}$

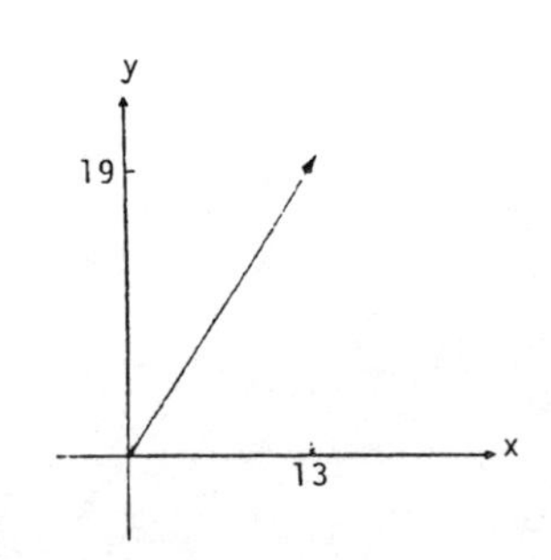

13. $\overrightarrow{P_1P_2} = (2-1)\mathbf{i} + (-1-3)\mathbf{j} = \mathbf{i} - 4\mathbf{j}$

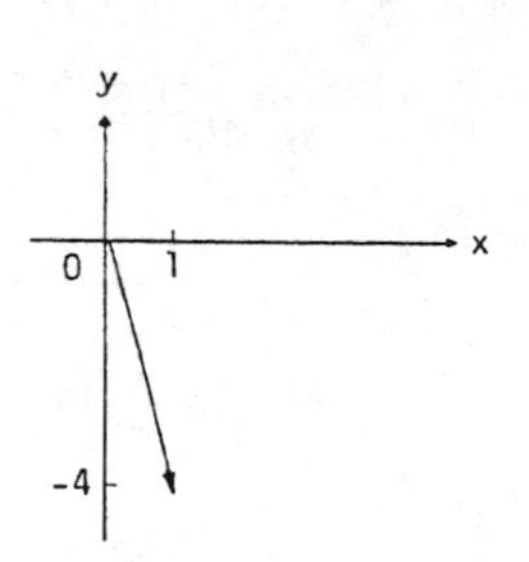

14. P_4 is $\left(\frac{2-4}{2}, \frac{-1+3}{2}\right) = (-1,1)$

$\Rightarrow \overrightarrow{P_3P_4} = (-1-1)\mathbf{i} + (1-3)\mathbf{j} = -2\mathbf{i} - 2\mathbf{j}$

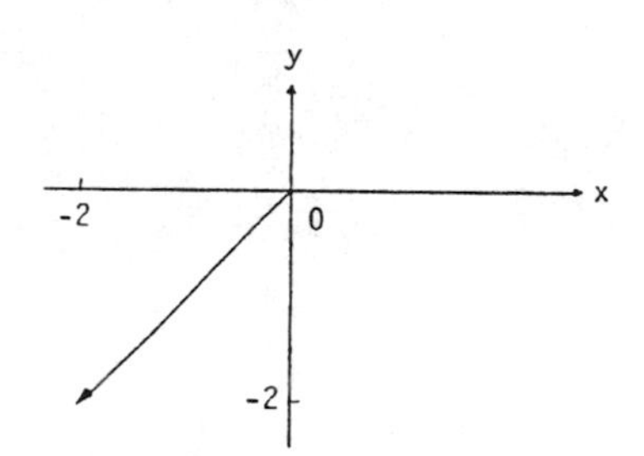

15. $\overrightarrow{AB} = (2-1)\mathbf{i} + (0-(-1))\mathbf{j} = \mathbf{i} + \mathbf{j}$ and

$\overrightarrow{CD} = (-2-(-1))\mathbf{i} + (2-3)\mathbf{j} = -\mathbf{i} - \mathbf{j}$

$\Rightarrow \overrightarrow{AB} + \overrightarrow{CD} = \mathbf{0}$

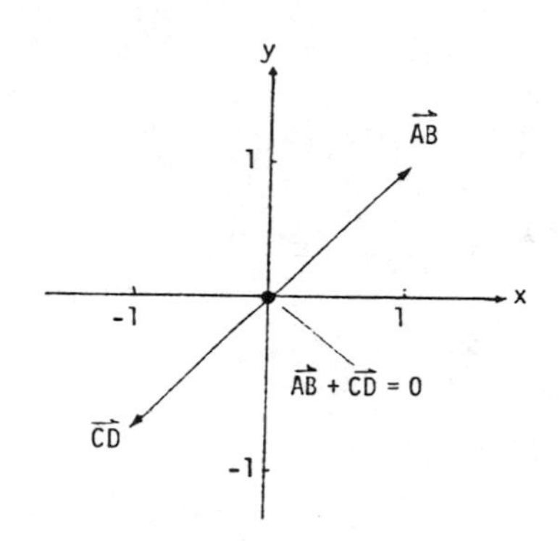

16. $\overrightarrow{AB} = (-2-a)\mathbf{i} + (5-b)\mathbf{j} = 4\mathbf{i} - 2\mathbf{j} \Rightarrow -2 - a = 4$ and $5 - b = -2$

$\Rightarrow a = -6$ and $b = 7 \Rightarrow \overrightarrow{AO} = -\overrightarrow{OA} = -(-6\mathbf{i} + 7\mathbf{j}) = 6\mathbf{i} - 7\mathbf{j}$

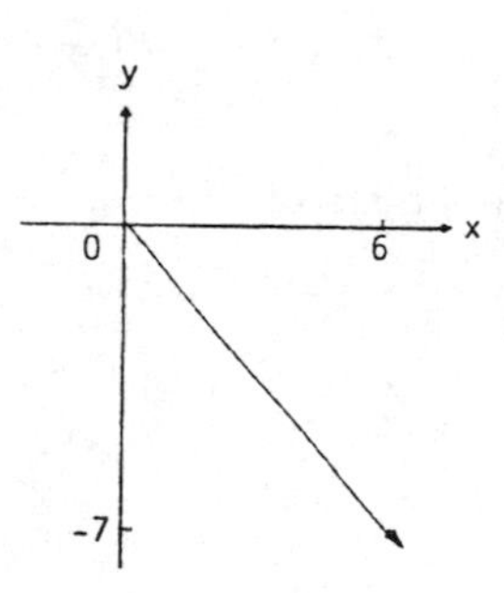

17. $\overrightarrow{AB} = (a-2)\mathbf{i} + (b-9)\mathbf{j} = 3\mathbf{i} - \mathbf{j} \Rightarrow a - 2 = 3$ and $b - 9 = -1 \Rightarrow a = 5$ and $b = 8 \Rightarrow$ B is the point $(5,8)$

18. $\overrightarrow{PQ} = (3-a)\mathbf{i} + (3-b)\mathbf{j} = -6\mathbf{i} - 4\mathbf{j} \Rightarrow 3 - a = -6$ and $3 - b = -4 \Rightarrow a = 9$ and $b = 7 \Rightarrow$ P is the point $(9,7)$

19. $\mathbf{u}=\left(\cos\frac{\pi}{6}\right)\mathbf{i}+\left(\sin\frac{\pi}{6}\right)\mathbf{j}=\frac{\sqrt{3}}{2}\mathbf{i}+\frac{1}{2}\mathbf{j}$;

$\mathbf{u}=\left(\cos\frac{2\pi}{3}\right)\mathbf{i}+\left(\sin\frac{2\pi}{3}\right)\mathbf{j}=-\frac{1}{2}\mathbf{i}+\frac{\sqrt{3}}{2}\mathbf{j}$

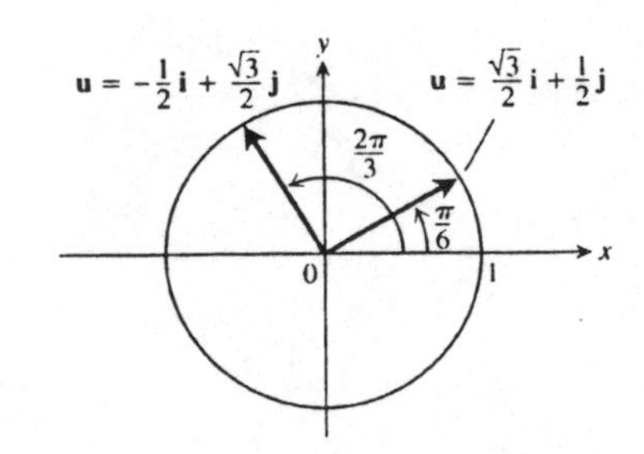

20. $\mathbf{u}=\left(\cos\left(-\frac{\pi}{4}\right)\right)\mathbf{i}+\left(\sin\left(-\frac{\pi}{4}\right)\right)\mathbf{j}=\frac{1}{\sqrt{2}}\mathbf{i}-\frac{1}{\sqrt{2}}\mathbf{j}$;

$\mathbf{u}=\left(\cos\left(-\frac{3\pi}{4}\right)\right)\mathbf{i}+\left(\sin\left(-\frac{3\pi}{4}\right)\right)\mathbf{j}=-\frac{1}{\sqrt{2}}\mathbf{i}-\frac{1}{\sqrt{2}}\mathbf{j}$

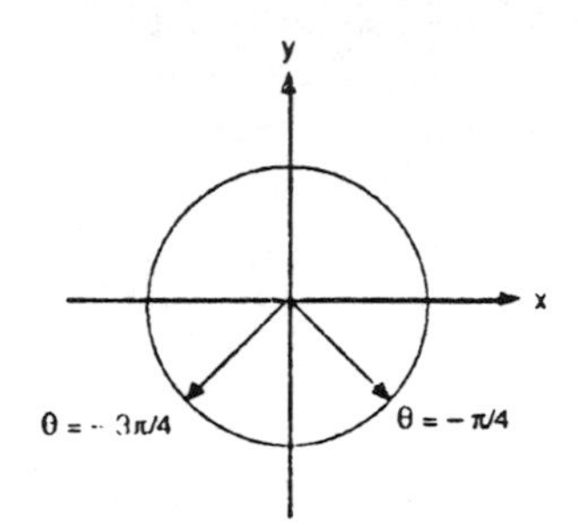

21. $\mathbf{u}=\left(\cos\left(\frac{\pi}{2}+\frac{3\pi}{4}\right)\right)\mathbf{i}+\left(\sin\left(\frac{\pi}{2}+\frac{3\pi}{4}\right)\right)\mathbf{j}$

$=\left(\cos\left(\frac{5\pi}{4}\right)\right)\mathbf{i}+\left(\sin\left(\frac{5\pi}{4}\right)\right)\mathbf{j}$

$=-\frac{\sqrt{2}}{2}\mathbf{i}-\frac{\sqrt{2}}{2}\mathbf{j}$

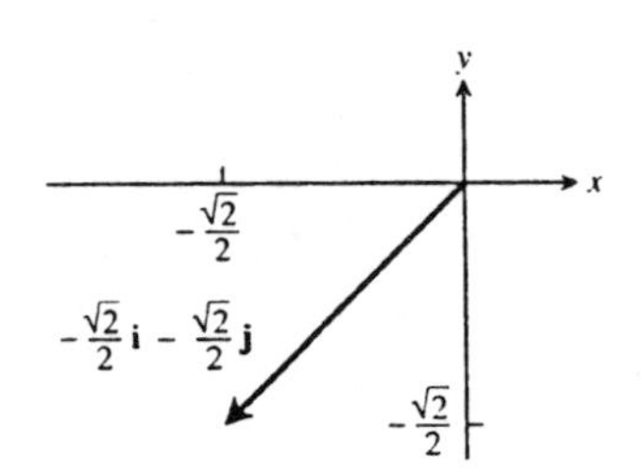

22. $\mathbf{u}=\left(\cos\left(\frac{\pi}{2}-\frac{2\pi}{3}\right)\right)\mathbf{i}+\left(\sin\left(\frac{\pi}{2}-\frac{2\pi}{3}\right)\right)\mathbf{j}$

$=\left(\cos\left(-\frac{\pi}{6}\right)\right)\mathbf{i}+\left(\sin\left(-\frac{\pi}{6}\right)\right)\mathbf{j}$

$=\frac{\sqrt{3}}{2}\mathbf{i}-\frac{1}{2}\mathbf{j}$

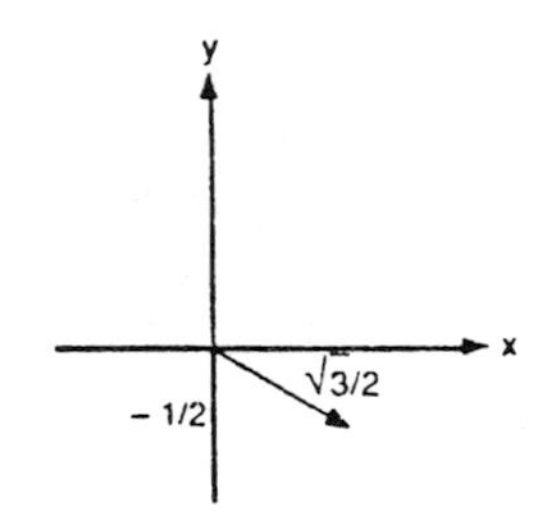

23. $|6\mathbf{i}-8\mathbf{j}|=\sqrt{36+64}=10 \Rightarrow \frac{\mathbf{v}}{|\mathbf{v}|}=\frac{6}{10}\mathbf{i}-\frac{8}{10}\mathbf{j}=\frac{3}{5}\mathbf{i}-\frac{4}{5}\mathbf{j}$

24. $|-\mathbf{i}+3\mathbf{j}|=\sqrt{1+9}=\sqrt{10} \Rightarrow \frac{\mathbf{v}}{|\mathbf{v}|}=-\frac{1}{\sqrt{10}}\mathbf{i}+\frac{3}{\sqrt{10}}\mathbf{j}$

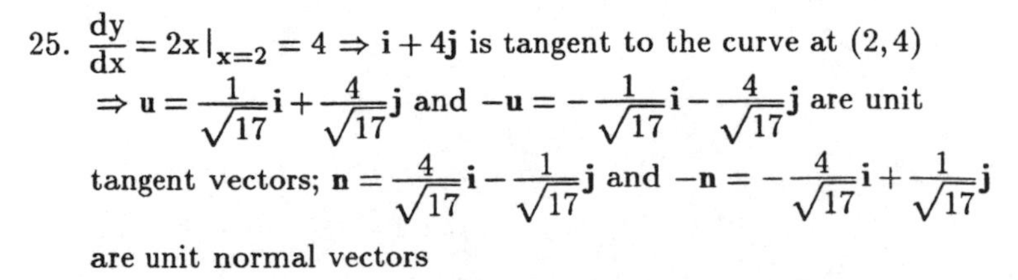

25. $\frac{dy}{dx}=2x\,|_{x=2}=4 \Rightarrow \mathbf{i}+4\mathbf{j}$ is tangent to the curve at $(2,4)$ $\Rightarrow \mathbf{u}=\frac{1}{\sqrt{17}}\mathbf{i}+\frac{4}{\sqrt{17}}\mathbf{j}$ and $-\mathbf{u}=-\frac{1}{\sqrt{17}}\mathbf{i}-\frac{4}{\sqrt{17}}\mathbf{j}$ are unit tangent vectors; $\mathbf{n}=\frac{4}{\sqrt{17}}\mathbf{i}-\frac{1}{\sqrt{17}}\mathbf{j}$ and $-\mathbf{n}=-\frac{4}{\sqrt{17}}\mathbf{i}+\frac{1}{\sqrt{17}}\mathbf{j}$ are unit normal vectors

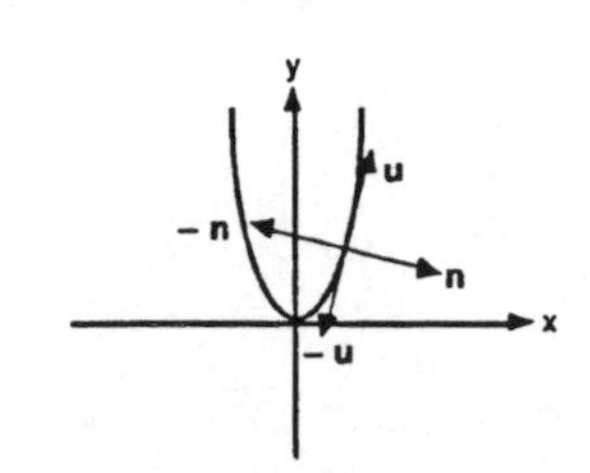

26. $2x + 4y\frac{dy}{dx} = 0 \Rightarrow \frac{dy}{dx} = -\frac{2x}{4y}\Big|_{(2,1)} = -1 \Rightarrow \mathbf{i} - \mathbf{j}$ is tangent to the curve at $(2,1) \Rightarrow \mathbf{u} = \frac{1}{\sqrt{2}}\mathbf{i} - \frac{1}{\sqrt{2}}\mathbf{j}$ and $-\mathbf{u} = -\frac{1}{\sqrt{2}}\mathbf{i} + \frac{1}{\sqrt{2}}\mathbf{j}$ are unit tangent vectors; $\mathbf{n} = \frac{1}{\sqrt{2}}\mathbf{i} + \frac{1}{\sqrt{2}}\mathbf{j}$ and $-\mathbf{n} = -\frac{1}{\sqrt{2}}\mathbf{i} - \frac{1}{\sqrt{2}}\mathbf{j}$ are unit normal vectors

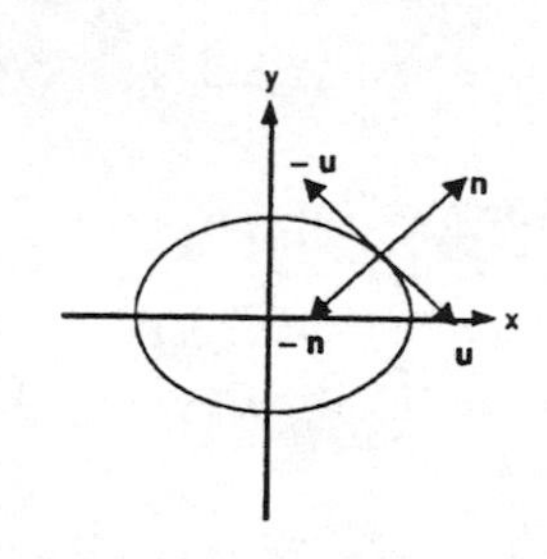

27. $\frac{dy}{dx} = \frac{1}{1+x^2}\Big|_{x=1} = \frac{1}{2} \Rightarrow \mathbf{i} + \frac{1}{2}\mathbf{j}$ is tangent to the curve at $(1,1) \Rightarrow 2\mathbf{i} + \mathbf{j}$ is tangent $\Rightarrow \mathbf{u} = \frac{2}{\sqrt{5}}\mathbf{i} + \frac{1}{\sqrt{5}}\mathbf{j}$ and $-\mathbf{u} = -\frac{2}{\sqrt{5}}\mathbf{i} - \frac{1}{\sqrt{5}}\mathbf{j}$ are unit tangent vectors; $\mathbf{n} = -\frac{1}{\sqrt{5}}\mathbf{i} + \frac{2}{\sqrt{5}}\mathbf{j}$ and $-\mathbf{n} = \frac{1}{\sqrt{5}}\mathbf{i} - \frac{2}{\sqrt{5}}\mathbf{j}$ are unit normal vectors

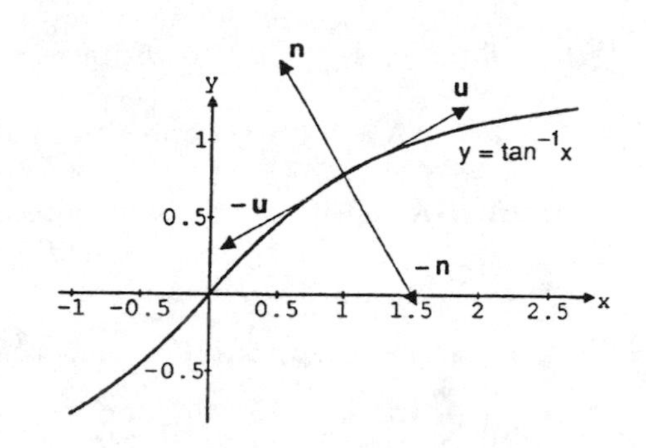

28. $\frac{dy}{dx} = \sum_{n=1}^{\infty} \frac{x^{n-1}}{(n-1)!} = \sum_{n=0}^{\infty} \frac{x^n}{n!} = e^x\big|_{(0,1)} = 1 \Rightarrow \mathbf{i} + \mathbf{j}$ is tangent to the curve at $(0,1) \Rightarrow \mathbf{u} = \frac{1}{\sqrt{2}}\mathbf{i} + \frac{1}{\sqrt{2}}\mathbf{j}$ and $-\mathbf{u} = -\frac{1}{\sqrt{2}}\mathbf{i} - \frac{1}{\sqrt{2}}\mathbf{j}$ are unit tangent vectors; $\mathbf{n} = \frac{1}{\sqrt{2}}\mathbf{i} - \frac{1}{\sqrt{2}}\mathbf{j}$ and $-\mathbf{n} = -\frac{1}{\sqrt{2}}\mathbf{i} + \frac{1}{\sqrt{2}}\mathbf{j}$ are unit normal vectors

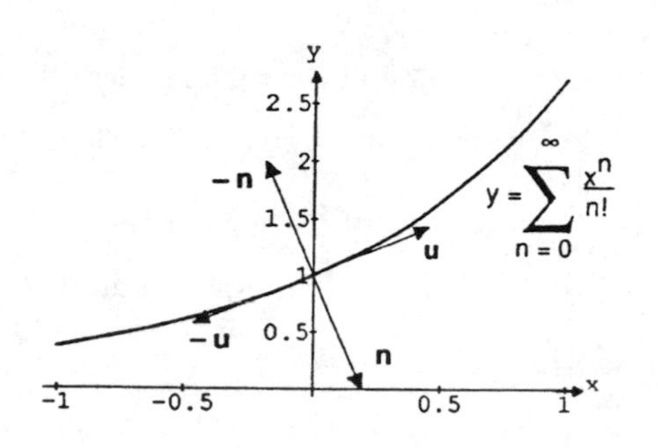

29. $6x + 8y + 8x\frac{dy}{dx} + 4y\frac{dy}{dx} = 0 \Rightarrow \frac{dy}{dx} = -\frac{3x+4y}{4x+2y}\Big|_{(1,0)} = -\frac{3}{4} \Rightarrow 4\mathbf{i} - 3\mathbf{j}$ is tangent to the curve at $(1,0)$ $\Rightarrow \mathbf{u} = \pm\frac{1}{5}(4\mathbf{i} - 3\mathbf{j})$ are unit tangent vectors and $\mathbf{v} = \pm\frac{1}{5}(3\mathbf{i} + 4\mathbf{j})$ are unit normal vectors

30. $2x - 6y - 6x\frac{dy}{dx} + 16y\frac{dy}{dx} - 2 = 0 \Rightarrow \frac{dy}{dx} = -\frac{x - 3y - 1}{8y - 3x}\Big|_{(1,1)} = \frac{3}{5} \Rightarrow 5\mathbf{i} + 3\mathbf{j}$ is tangent to the curve at $(1,1)$ $\Rightarrow \mathbf{u} = \pm\frac{1}{\sqrt{34}}(5\mathbf{i} + 3\mathbf{j})$ are unit tangent vectors and $\mathbf{v} = \pm\frac{1}{\sqrt{34}}(-3\mathbf{i} + 5\mathbf{j})$ are unit normal vectors

31. $\frac{dy}{dx} = \sqrt{3 + x^4}\big|_{(0,0)} = \sqrt{3} \Rightarrow \mathbf{i} + \sqrt{3}\mathbf{j}$ is tangent to the curve at $(0,0) \Rightarrow \mathbf{u} = \pm\frac{1}{2}(\mathbf{i} + \sqrt{3}\mathbf{j})$ are unit tangent vectors and $\mathbf{v} = \pm\frac{1}{2}(-\sqrt{3}\mathbf{i} + \mathbf{j})$ are unit normal vectors

32. $\frac{dy}{dx} = \ln(\ln x)\big|_{(e,0)} = \ln 1 = 0 \Rightarrow \mathbf{u} = \pm\mathbf{i}$ are unit tangent vectors and $\mathbf{v} = \pm\mathbf{j}$ are unit normal vectors

33. $\mathbf{v} = 5\mathbf{i} + 12\mathbf{j} \Rightarrow |\mathbf{v}| = \sqrt{25+144} = 13 \Rightarrow \mathbf{v} = |\mathbf{v}|\left(\frac{\mathbf{v}}{|\mathbf{v}|}\right) = 13\left(\frac{5}{13}\mathbf{i} + \frac{12}{13}\mathbf{j}\right)$

34. $\mathbf{v} = 2\mathbf{i} - 3\mathbf{j} \Rightarrow |\mathbf{v}| = \sqrt{4+9} = \sqrt{13} \Rightarrow \mathbf{v} = |\mathbf{v}|\left(\frac{\mathbf{v}}{|\mathbf{v}|}\right) = \sqrt{13}\left(\frac{2}{\sqrt{13}}\mathbf{i} - \frac{3}{\sqrt{13}}\mathbf{j}\right)$

35. $\mathbf{v} = 3\mathbf{i} - 4\mathbf{j} \Rightarrow |\mathbf{v}| = \sqrt{9+16} = 5 \Rightarrow \mathbf{u} = \pm\left(\frac{\mathbf{v}}{|\mathbf{v}|}\right) = \pm\frac{1}{5}(3\mathbf{i} - 4\mathbf{j})$

36. $\mathbf{A} = -\mathbf{i} + 2\mathbf{j} \Rightarrow |\mathbf{A}| = \sqrt{1+4} = \sqrt{5} \Rightarrow \mathbf{v} = -2\frac{\mathbf{A}}{|\mathbf{A}|} = -2\left(-\frac{1}{\sqrt{5}}\mathbf{i} + \frac{2}{\sqrt{5}}\mathbf{j}\right) = \frac{2}{\sqrt{5}}\mathbf{i} - \frac{4}{\sqrt{5}}\mathbf{j}$ is a vector of length 2 whose direction is opposite to $\mathbf{A}$; there is only one such vector

37. $\mathbf{A} = -3\mathbf{B} \Rightarrow \mathbf{A}$ and $\mathbf{B}$ have opposite directions

38. $\mathbf{A} = 6\mathbf{B} \Rightarrow \mathbf{A}$ and $\mathbf{B}$ have the same direction $\mathbf{v} = \frac{1}{\sqrt{5}}(\mathbf{i} + 2\mathbf{j})$

39. If $|\mathrm{x}|$ is the magnitude of the x-component, then $\cos 30^\circ = \frac{|\mathrm{x}|}{|\mathrm{F}|} \Rightarrow |\mathrm{x}| = |\mathrm{F}| \cos 30^\circ = (10)\left(\frac{\sqrt{3}}{2}\right) = 5\sqrt{3}$ lb $\Rightarrow \mathrm{x} = 5\sqrt{3}\,\mathbf{i}$;
if $|\mathrm{y}|$ is the magnitude of the y-component, then $\sin 30^\circ = \frac{|\mathrm{y}|}{|\mathrm{F}|} \Rightarrow |\mathrm{y}| = |\mathrm{F}| \sin 30^\circ = (10)\left(\frac{1}{2}\right) = 5$ lb $\Rightarrow \mathrm{y} = 5\mathbf{j}$.

40. If $|\mathrm{x}|$ is the magnitude of the x-component, then $\cos 45^\circ = \frac{|\mathrm{x}|}{|\mathrm{F}|} \Rightarrow |\mathrm{x}| = |\mathrm{F}| \cos 45^\circ = (12)\left(\frac{\sqrt{2}}{2}\right) = 6\sqrt{2}$ lb $\Rightarrow \mathrm{x} = -6\sqrt{2}\,\mathbf{i}$ (he negative sign is indicated by the diagram)
if $|\mathrm{y}|$ is the magnitude of the y-component, then $\sin 45^\circ = \frac{|\mathrm{y}|}{|\mathrm{F}|} \Rightarrow |\mathrm{y}| = |\mathrm{F}| \sin 45^\circ = (12)\left(\frac{\sqrt{2}}{2}\right) = 6\sqrt{2}$ lb $\Rightarrow \mathrm{y} = -6\sqrt{2}\,\mathbf{j}$ (the negative sign is indicated by the diagram)

41. $2\mathbf{i} + \mathbf{j} = \alpha(\mathbf{i} + \mathbf{j}) + \beta(\mathbf{i} - \mathbf{j}) = (\alpha + \beta)\mathbf{i} + (\alpha - \beta)\mathbf{j} \Rightarrow \alpha + \beta = 2$ and $\alpha - \beta = 1 \Rightarrow 2\alpha = 3 \Rightarrow \alpha = \frac{3}{2}$ and $\beta = \alpha - 1 = \frac{1}{2}$

42. $\mathbf{i} - 2\mathbf{j} = \alpha(2\mathbf{i} + 3\mathbf{j}) + \beta(\mathbf{i} + \mathbf{j}) = (2\alpha + \beta)\mathbf{i} + (3\alpha + \beta)\mathbf{j} \Rightarrow 2\alpha + \beta = 1$ and $3\alpha + \beta = -2 \Rightarrow \alpha = -3$ and $\beta = 1 - 2\alpha = 7 \Rightarrow \mathbf{A}_1 = \alpha(2\mathbf{i} + 3\mathbf{j}) = -6\mathbf{i} - 9\mathbf{j}$ and $\mathbf{A}_2 = \beta(\mathbf{i} + \mathbf{j}) = 7\mathbf{i} + 7\mathbf{j}$

43. (a) The tree is located at the tip of the vector $\overrightarrow{OP} = (5 \cos 60^\circ)\mathbf{i} + (5 \sin 60^\circ)\mathbf{j} = \frac{5}{2}\mathbf{i} + \frac{5\sqrt{3}}{2}\mathbf{j} \Rightarrow P = \left(\frac{5}{2}, \frac{5\sqrt{3}}{2}\right)$

(b) The telephone pole is located at the point Q, which is the tip of the vector $\overrightarrow{OP} + \overrightarrow{PQ}$
$= \left(\frac{5}{2}\mathbf{i} + \frac{5\sqrt{3}}{2}\mathbf{j}\right) + (10 \cos 315^\circ)\mathbf{i} + (10 \sin 315^\circ)\mathbf{j} = \left(\frac{5}{2} + \frac{\sqrt{2}}{2}\right)\mathbf{i} + \left(\frac{5\sqrt{3}}{2} - \frac{10\sqrt{2}}{2}\right)\mathbf{j}$
$\Rightarrow Q = \left(\frac{5+\sqrt{2}}{2}, \frac{5\sqrt{3} - 10\sqrt{2}}{2}\right)$

44. (a) The tree is located at the tip of the vector $\overrightarrow{OP} = (7 \cos 45^\circ)\mathbf{i} + (7 \sin 45^\circ)\mathbf{j} = \frac{7\sqrt{2}}{2}\mathbf{i} + \frac{7\sqrt{2}}{2}\mathbf{j}$
$\Rightarrow P = \left(\frac{7\sqrt{2}}{2}, \frac{7\sqrt{2}}{2}\right)$

(b) The telephone pole is located at the point Q which is the tip of the vector $\overrightarrow{OP} + \overrightarrow{PQ}$
$= \left(\frac{7\sqrt{2}}{2}\mathbf{i} + \frac{7\sqrt{2}}{2}\mathbf{j}\right) + (8 \cos 210^\circ)\mathbf{i} + (8 \sin 210^\circ)\mathbf{j} = \left(\frac{7\sqrt{2}}{2} - \frac{8\sqrt{3}}{2}\right)\mathbf{i} + \left(\frac{7\sqrt{2}}{2} - \frac{8}{2}\right)\mathbf{j}$

$\Rightarrow Q = \left(\frac{7\sqrt{2}}{2} - 4\sqrt{3}, \frac{7\sqrt{2}}{2} - 4\right)$

45. The slope of $-\mathbf{v} = -\alpha\mathbf{i} - b\mathbf{j}$ is $\frac{-b}{-a} = \frac{b}{a}$, which is the same as the slope of $\mathbf{v}$.

10.2 CARTESIAN (RECTANGULAR) COORDINATES AND VECTORS IN SPACE

1. The line through the point $(2,3,0)$ parallel to the z-axis
2. The line through the point $(-1,0,0)$ parallel to the y-axis
3. The x-axis
4. The line through the point $(1,0,0)$ parallel to the z-axis
5. The circle $x^2 + y^2 = 4$ in the xy-plane
6. The circle $x^2 + y^2 = 4$ in the plane $z = -2$
7. The circle $x^2 + z^2 = 4$ in the xz-plane
8. The circle $y^2 + z^2 = 1$ in the yz-plane
9. The circle $y^2 + z^2 = 1$ in the yz-plane
10. The circle $x^2 + z^2 = 9$ in the plane $y = -4$
11. The circle $x^2 + y^2 = 16$ in the xy-plane
12. The circle $x^2 + z^2 = 3$ in the xz-plane
13. (a) The first quadrant of the xy-plane (b) The fourth quadrant of the xy-plane
14. (a) The slab bounded by the planes $x = 0$ and $x = 1$
    (b) The square column bounded by the planes $x = 0$, $x = 1$, $y = 0$, $y = 1$
    (c) The unit cube in the first octant having one vertex at the origin
15. (a) The ball of radius 1 centered at the origin
    (b) All points at distance greater than 1 unit from the origin
16. (a) The circumference and interior of the circle $x^2 + y^2 = 1$ in the xy-plane
    (b) The circumference and interior of the circle $x^2 + y^2 = 1$ in the plane $z = 3$
    (c) A solid cylindrical column of radius 1 whose axis is the z-axis
17. (a) The upper hemisphere of radius 1 centered at the origin
    (b) The solid upper hemisphere of radius 1 centered at the origin
18. (a) The line $y = x$ in the xy-plane
    (b) The plane $y = x$ consisting of all points of the form (x,x,z)
19. (a) $x = 3$ (b) $y = -1$ (c) $z = -2$
20. (a) $x = 3$ (b) $y = -1$ (c) $z = 2$
21. (a) $z = 1$ (b) $x = 3$ (c) $y = -1$
22. (a) $x^2 + y^2 = 4,\ z = 0$ (b) $y^2 + z^2 = 4,\ x = 0$ (c) $x^2 + z^2 = 4,\ y = 0$
23. (a) $x^2 + (y-2)^2 = 4,\ z = 0$ (b) $(y-2)^2 + z^2 = 4,\ x = 0$ (c) $x^2 + z^2 = 4,\ y = 2$
24. (a) $(x+3)^2 + (y-4)^2 = 1,\ z = 1$ (b) $(y-4)^2 + (z-1)^2 = 1,\ x = -3$
    (c) $(x+3)^2 + (z-1)^2 = 1,\ y = 4$
25. (a) $y = 3,\ z = -1$ (b) $x = 1,\ z = -1$ (c) $x = 1,\ y = 3$

26. $\sqrt{x^2+y^2+z^2} = \sqrt{x^2+(y-2)^2+z^2} \Rightarrow x^2+y^2+z^2 = x^2+(y-2)^2+z^2 \Rightarrow y^2 = y^2-4y+4 \Rightarrow y=1$

27. $x^2+y^2+z^2 = 25,\ z=3$

28. $x^2+y^2+(z-1)^2 = 4$ and $x^2+y^2+(z+1)^2 = 4 \Rightarrow x^2+y^2+(z-1)^2 = x^2+y^2+(z+1)^2 \Rightarrow z=0,\ x^2+y^2=3$

29. $0 \le z \le 1$

30. $0 \le x \le 2,\ 0 \le y \le 2,\ 0 \le z \le 2$

31. $z \le 0$

32. $z = \sqrt{1-x^2-y^2}$

33. (a) $(x-1)^2+(y-1)^2+(z-1)^2 < 1$

(b) $(x-1)^2+(y-1)^2+(z-1)^2 > 1$

34. $1 \le x^2+y^2+z^2 \le 4$

35. length $= |2\mathbf{i}+\mathbf{j}-2\mathbf{k}| = \sqrt{2^2+1^2+(-2)^2} = 3$, the direction is $\frac{2}{3}\mathbf{i}+\frac{1}{3}\mathbf{j}-\frac{2}{3}\mathbf{k} \Rightarrow 2\mathbf{i}+\mathbf{j}-2\mathbf{k} = 3\left(\frac{2}{3}\mathbf{i}+\frac{1}{3}\mathbf{j}-\frac{2}{3}\mathbf{k}\right)$

36. length $= |3\mathbf{i}-6\mathbf{j}+2\mathbf{k}| = \sqrt{3^2+(-6)^2+2^2} = 7$, the direction is $\frac{3}{7}\mathbf{i}-\frac{6}{7}\mathbf{j}+\frac{2}{7}\mathbf{k} \Rightarrow 3\mathbf{i}-6\mathbf{j}+2\mathbf{k} = 7\left(\frac{3}{7}\mathbf{i}-\frac{6}{7}\mathbf{j}+\frac{2}{7}\mathbf{k}\right)$

37. length $= |\mathbf{i}+4\mathbf{j}-8\mathbf{k}| = \sqrt{1+16+64} = 9$, the direction is $\frac{1}{9}\mathbf{i}+\frac{4}{9}\mathbf{j}-\frac{8}{9}\mathbf{k} \Rightarrow \mathbf{i}+4\mathbf{j}-8\mathbf{k} = 9\left(\frac{1}{9}\mathbf{i}+\frac{4}{9}\mathbf{j}-\frac{8}{9}\mathbf{k}\right)$

38. length $= |9\mathbf{i}-2\mathbf{j}+6\mathbf{k}| = \sqrt{81+4+36} = 11$, the direction is $\frac{9}{11}\mathbf{i}-\frac{2}{11}\mathbf{j}+\frac{6}{11}\mathbf{k} \Rightarrow 9\mathbf{i}-2\mathbf{j}+6\mathbf{k}$
$= 11\left(\frac{9}{11}\mathbf{i}-\frac{2}{11}\mathbf{j}+\frac{6}{11}\mathbf{k}\right)$

39. length $= |5\mathbf{k}| = \sqrt{25} = 5$, the direction is $\mathbf{k} \Rightarrow 5\mathbf{k} = 5(\mathbf{k})$

40. length $= |-4\mathbf{j}| = \sqrt{(-4)^2} = \sqrt{16} = 4$, the direction is $-\mathbf{j} \Rightarrow -4\mathbf{j} = 4(-\mathbf{j})$

41. length $= \left|\frac{3}{5}\mathbf{i}+\frac{4}{5}\mathbf{k}\right| = \sqrt{\frac{9}{25}+\frac{16}{25}} = 1$, the direction is $\frac{3}{5}\mathbf{i}+\frac{4}{5}\mathbf{k} \Rightarrow \frac{3}{5}\mathbf{i}+\frac{4}{5}\mathbf{k} = 1\left(\frac{3}{5}\mathbf{i}+\frac{4}{5}\mathbf{k}\right)$

42. length $= \left|\frac{1}{\sqrt{2}}\mathbf{i}-\frac{1}{\sqrt{2}}\mathbf{k}\right| = \sqrt{\frac{1}{2}+\frac{1}{2}} = 1$, the direction is $\frac{1}{\sqrt{2}}\mathbf{i}-\frac{1}{\sqrt{2}}\mathbf{k} \Rightarrow \frac{1}{\sqrt{2}}\mathbf{i}-\frac{1}{\sqrt{2}}\mathbf{k} = 1\left(\frac{1}{\sqrt{2}}\mathbf{i}-\frac{1}{\sqrt{2}}\mathbf{k}\right)$

43. length $= \left|\frac{1}{\sqrt{6}}\mathbf{i}-\frac{1}{\sqrt{6}}\mathbf{j}-\frac{1}{\sqrt{6}}\mathbf{k}\right| = \sqrt{3\left(\frac{1}{\sqrt{6}}\right)^2} = \sqrt{\frac{1}{2}}$, the direction is $\frac{1}{\sqrt{3}}\mathbf{i}-\frac{1}{\sqrt{3}}\mathbf{j}-\frac{1}{\sqrt{3}}\mathbf{k}$
$\Rightarrow \frac{1}{\sqrt{6}}\mathbf{i}-\frac{1}{\sqrt{6}}\mathbf{j}-\frac{1}{\sqrt{6}}\mathbf{k} = \sqrt{\frac{1}{2}}\left(\frac{1}{\sqrt{3}}\mathbf{i}-\frac{1}{\sqrt{3}}\mathbf{j}-\frac{1}{\sqrt{3}}\mathbf{k}\right)$

44. length $= \left|\frac{1}{\sqrt{3}}\mathbf{i}+\frac{1}{\sqrt{3}}\mathbf{j}+\frac{1}{\sqrt{3}}\mathbf{k}\right| = \sqrt{3\left(\frac{1}{\sqrt{3}}\right)^2} = 1$, the direction is $\frac{1}{\sqrt{3}}\mathbf{i}+\frac{1}{\sqrt{3}}\mathbf{j}+\frac{1}{\sqrt{3}}\mathbf{k}$
$\Rightarrow \frac{1}{\sqrt{3}}\mathbf{i}+\frac{1}{\sqrt{3}}\mathbf{j}+\frac{1}{\sqrt{3}}\mathbf{k} = 1\left(\frac{1}{\sqrt{3}}\mathbf{i}+\frac{1}{\sqrt{3}}\mathbf{j}+\frac{1}{\sqrt{3}}\mathbf{k}\right)$

45. (a) $2\mathbf{i}$ (b) $-\sqrt{3}\mathbf{k}$ (c) $\frac{3}{10}\mathbf{j}+\frac{2}{5}\mathbf{k}$ (d) $6\mathbf{i}-2\mathbf{j}+3\mathbf{k}$

46. (a) $-7\mathbf{j}$ (b) $-\frac{3\sqrt{2}}{5}\mathbf{i}-\frac{4\sqrt{2}}{5}\mathbf{k}$ (c) $\frac{1}{4}\mathbf{i}-\frac{1}{3}\mathbf{j}-\mathbf{k}$ (d) $\frac{a}{\sqrt{2}}\mathbf{i}+\frac{a}{\sqrt{3}}\mathbf{j}-\frac{a}{\sqrt{6}}\mathbf{k}$

47. $|\mathbf{A}|=\sqrt{12^2+5^2}=\sqrt{169}=13$; $\frac{\mathbf{A}}{|\mathbf{A}|}=\frac{1}{13}\mathbf{A}=\frac{1}{13}(12\mathbf{i}-5\mathbf{k})\Rightarrow$ the desired vector is $\frac{7}{13}(12\mathbf{i}-5\mathbf{k})$

48. $|\mathbf{A}|=|\mathbf{i}+\mathbf{j}+\mathbf{k}|=\sqrt{3}$; $\frac{\mathbf{A}}{|\mathbf{A}|}=\frac{1}{\sqrt{3}}\mathbf{i}+\frac{1}{\sqrt{3}}\mathbf{j}+\frac{1}{\sqrt{3}}\mathbf{k}\Rightarrow$ the desired vector is $\sqrt{5}\left(\frac{1}{\sqrt{3}}\mathbf{i}+\frac{1}{\sqrt{3}}\mathbf{j}+\frac{1}{\sqrt{3}}\mathbf{k}\right)$

49. $|\mathbf{A}|=|2\mathbf{i}-3\mathbf{j}+6\mathbf{k}|=\sqrt{2^2+(-3)^2+6^2}=\sqrt{49}=7$; $\frac{\mathbf{A}}{|\mathbf{A}|}=\frac{2}{7}\mathbf{i}-\frac{3}{7}\mathbf{j}+\frac{6}{7}\mathbf{k}\Rightarrow$ the desired vector is $-5\left(\frac{2}{7}\mathbf{i}-\frac{3}{7}\mathbf{j}+\frac{6}{7}\mathbf{k}\right)=-\frac{10}{7}\mathbf{i}+\frac{15}{7}\mathbf{j}-\frac{30}{7}\mathbf{k}$

50. $|\mathbf{A}|=\sqrt{\frac{1}{4}+\frac{1}{4}+\frac{1}{4}}=\frac{\sqrt{3}}{2}$; $\frac{\mathbf{A}}{|\mathbf{A}|}=\frac{1}{\sqrt{3}}\mathbf{i}-\frac{1}{\sqrt{3}}\mathbf{j}-\frac{1}{\sqrt{3}}\mathbf{k}\Rightarrow$ the desired vector is $-3\left(\frac{1}{\sqrt{3}}\mathbf{i}-\frac{1}{\sqrt{3}}\mathbf{j}-\frac{1}{\sqrt{3}}\mathbf{k}\right)$
$=-\sqrt{3}\mathbf{i}+\sqrt{3}\mathbf{j}+\sqrt{3}\mathbf{k}$

51. (a) the distance = the length $=\left|\overrightarrow{P_1P_2}\right|=|2\mathbf{i}+2\mathbf{j}-\mathbf{k}|=\sqrt{2^2+2^2+(-1)^2}=3$
(b) $2\mathbf{i}+2\mathbf{j}-\mathbf{k}=3\left(\frac{2}{3}\mathbf{i}+\frac{2}{3}\mathbf{j}-\frac{1}{3}\mathbf{k}\right)\Rightarrow$ the direction is $\frac{2}{3}\mathbf{i}+\frac{2}{3}\mathbf{j}-\frac{1}{3}\mathbf{k}$
(c) the midpoint is $\left(2,2,\frac{1}{2}\right)$

52. (a) the distance = the length $=\left|\overrightarrow{P_1P_2}\right|=|3\mathbf{i}+4\mathbf{j}-5\mathbf{k}|=\sqrt{9+16+25}=5\sqrt{2}$
(b) $3\mathbf{i}+4\mathbf{j}-5\mathbf{k}=5\sqrt{2}\left(\frac{3}{5\sqrt{2}}\mathbf{i}+\frac{4}{5\sqrt{2}}\mathbf{j}-\frac{1}{\sqrt{2}}\mathbf{k}\right)\Rightarrow$ the direction is $\frac{3}{5\sqrt{2}}\mathbf{i}+\frac{4}{5\sqrt{2}}\mathbf{j}-\frac{1}{\sqrt{2}}\mathbf{k}$
(c) the midpoint is $\left(\frac{1}{2},3,\frac{5}{2}\right)$

53. (a) the distance = the length $=\left|\overrightarrow{P_1P_2}\right|=|3\mathbf{i}-6\mathbf{j}+2\mathbf{k}|=\sqrt{9+36+4}=7$
(b) $3\mathbf{i}-6\mathbf{j}+2\mathbf{k}=7\left(\frac{3}{7}\mathbf{i}-\frac{6}{7}\mathbf{j}+\frac{2}{7}\mathbf{k}\right)\Rightarrow$ the direction is $\frac{3}{7}\mathbf{i}-\frac{6}{7}\mathbf{j}+\frac{2}{7}\mathbf{k}$
(c) the midpoint is $\left(\frac{5}{2},1,6\right)$

54. (a) the distance = the length $=\left|\overrightarrow{P_1P_2}\right|=|-\mathbf{i}-\mathbf{j}-\mathbf{k}|=\sqrt{3}$
(b) $-\mathbf{i}-\mathbf{j}-\mathbf{k}=\sqrt{3}\left(-\frac{1}{\sqrt{3}}\mathbf{i}-\frac{1}{\sqrt{3}}\mathbf{j}-\frac{1}{\sqrt{3}}\mathbf{k}\right)\Rightarrow$ the direction is $-\frac{1}{\sqrt{3}}\mathbf{i}-\frac{1}{\sqrt{3}}\mathbf{j}-\frac{1}{\sqrt{3}}\mathbf{k}$
(c) the midpoint is $\left(\frac{5}{2},\frac{7}{2},\frac{9}{2}\right)$

55. (a) the distance = the length $=\left|\overrightarrow{P_1P_2}\right|=|2\mathbf{i}-2\mathbf{j}-2\mathbf{k}|=\sqrt{3\cdot 2^2}=2\sqrt{3}$
(b) $2\mathbf{i}-2\mathbf{j}-2\mathbf{k}=2\sqrt{3}\left(\frac{1}{\sqrt{3}}\mathbf{i}-\frac{1}{\sqrt{3}}\mathbf{j}-\frac{1}{\sqrt{3}}\mathbf{k}\right)\Rightarrow$ the direction is $\frac{1}{\sqrt{3}}\mathbf{i}-\frac{1}{\sqrt{3}}\mathbf{j}-\frac{1}{\sqrt{3}}\mathbf{k}$
(c) the midpoint is $(1,-1,-1)$

56. (a) the distance = the length $= \left|\overrightarrow{P_1P_2}\right| = |-5\mathbf{i} - 3\mathbf{j} + 2\mathbf{k}| = \sqrt{38}$

(b) $-5\mathbf{i} - 3\mathbf{j} + 2\mathbf{k} = \sqrt{38}\left(-\frac{5}{\sqrt{38}}\mathbf{i} - \frac{3}{\sqrt{38}}\mathbf{j} + \frac{2}{\sqrt{38}}\mathbf{k}\right) \Rightarrow$ the direction is $-\frac{5}{\sqrt{38}}\mathbf{i} - \frac{3}{\sqrt{38}}\mathbf{j} + \frac{2}{\sqrt{38}}\mathbf{k}$

(c) the midpoint is $\left(\frac{5}{2}, \frac{3}{2}, -1\right)$

57. $\overrightarrow{AB} = (5 - a)\mathbf{i} + (1 - b)\mathbf{j} + (3 - c)\mathbf{k} = \mathbf{i} + 4\mathbf{j} - 2\mathbf{k} \Rightarrow 5 - a = 1,\ 1 - b = 4$, and $3 - c = -2 \Rightarrow a = 4,\ b = -3$, and $c = 5 \Rightarrow$ A is the point $(4, -3, 5)$

58. $\overrightarrow{AB} = (a + 2)\mathbf{i} + (b + 3)\mathbf{j} + (c - 6)\mathbf{k} = -7\mathbf{i} + 3\mathbf{j} + 8\mathbf{k} \Rightarrow a + 2 = -7,\ b + 3 = 3$, and $c - 6 = 8 \Rightarrow a = -9,\ b = 0$, and $c = 14 \Rightarrow$ B is the point $(-9, 0, 14)$

59. center $(-2, 0, 2)$, radius $2\sqrt{2}$

60. center $\left(-\frac{1}{2}, -\frac{1}{2}, -\frac{1}{2}\right)$, radius $\frac{\sqrt{21}}{2}$

61. center $\left(\sqrt{2}, \sqrt{2}, -\sqrt{2}\right)$, radius $\sqrt{2}$

62. center $\left(0, -\frac{1}{3}, \frac{1}{3}\right)$, radius $\frac{\sqrt{29}}{3}$

63. $(x - 1)^2 + (y - 2)^2 + (z - 3)^2 = 14$

64. $x^2 + (y + 1)^2 + (z - 5)^2 = 4$

65. $(x + 2)^2 + y^2 + z^2 = 3$

66. $x^2 + (y + 7)^2 + z^2 = 49$

67. $x^2 + y^2 + z^2 + 4x - 4z = 0 \Rightarrow (x^2 + 4x + 4) + y^2 + (z^2 - 4z + 4) = 4 + 4 \Rightarrow (x + 2)^2 + (y - 0)^2 + (z - 2)^2 = \left(\sqrt{8}\right)^2$
$\Rightarrow$ the center is at $(-2, 0, 2)$ and the radius is $\sqrt{8}$

68. $x^2 + y^2 + z^2 - 6y + 8z = 0 \Rightarrow x^2 + (y^2 - 6y + 9) + (z^2 + 8z + 16) = 9 + 16 \Rightarrow (x - 0)^2 + (y - 3)^2 + (z + 4)^2 = 5^2$
$\Rightarrow$ the center is at $(0, 3, -4)$ and the radius is 5

69. $2x^2 + 2y^2 + 2z^2 + x + y + z = 9 \Rightarrow x^2 + \frac{1}{2}x + y^2 + \frac{1}{2}y + z^2 + \frac{1}{2}z = \frac{9}{2}$
$\Rightarrow \left(x^2 + \frac{1}{2}x + \frac{1}{16}\right) + \left(y^2 + \frac{1}{2}y + \frac{1}{16}\right) + \left(z^2 + \frac{1}{2}z + \frac{1}{16}\right) = \frac{9}{2} + \frac{3}{16} = \frac{75}{16} \Rightarrow \left(x + \frac{1}{4}\right)^2 + \left(y + \frac{1}{4}\right)^2 + \left(z + \frac{1}{4}\right)^2 = \left(\frac{5\sqrt{3}}{4}\right)^2$
$\Rightarrow$ the center is at $\left(-\frac{1}{4}, -\frac{1}{4}, -\frac{1}{4}\right)$ and the radius is $\frac{5\sqrt{3}}{4}$

70. $3x^2 + 3y^2 + 3z^2 + 2y - 2z = 9 \Rightarrow x^2 + y^2 + \frac{2}{3}y + z^2 - \frac{2}{3}z = 3 \Rightarrow x^2 + \left(y^2 + \frac{2}{3}y + \frac{1}{9}\right) + \left(z^2 - \frac{2}{3}z + \frac{1}{9}\right) = 3 + \frac{2}{9}$
$\Rightarrow (x - 0)^2 + \left(y + \frac{1}{3}\right)^2 + \left(z - \frac{1}{3}\right)^2 = \left(\frac{\sqrt{29}}{3}\right)^2 \Rightarrow$ the center is at $\left(0, -\frac{1}{3}, \frac{1}{3}\right)$ and the radius is $\frac{\sqrt{29}}{3}$

71. (a) the distance between (x, y, z) and $(x, 0, 0)$ is $\sqrt{y^2 + z^2}$

(b) the distance between (x, y, z) and $(0, y, 0)$ is $\sqrt{x^2 + z^2}$

(c) the distance between (x, y, z) and $(0, 0, z)$ is $\sqrt{x^2 + y^2}$

72. (a) the distance between (x, y, z) and $(x, y, 0)$ is z
(b) the distance between (x, y, z) and $(0, y, z)$ is x
(c) the distance between (x, y, z) and $(x, 0, z)$ is y

73. (a) the midpoint of AB is $M\left(\frac{5}{2}, \frac{5}{2}, 0\right)$ and $\overrightarrow{CM} = \left(\frac{5}{2} - 1\right)\mathbf{i} + \left(\frac{5}{2} - 1\right)\mathbf{j} + (0 - 3)\mathbf{k} = \frac{3}{2}\mathbf{i} + \frac{3}{2}\mathbf{j} - 3\mathbf{k}$

(b) the desired vector is $\left(\frac{2}{3}\right)\overrightarrow{CM} = \frac{2}{3}\left(\frac{3}{2}\mathbf{i} + \frac{3}{2}\mathbf{j} - 3\mathbf{k}\right) = \mathbf{i} + \mathbf{j} - 2\mathbf{k}$

(c) the vector whose sum is the vector from the origin to C and the result of part (b) will terminate at the center of mass $\Rightarrow$ the terminal point of $(\mathbf{i}+\mathbf{j}+3\mathbf{k})+(\mathbf{i}+\mathbf{j}-2\mathbf{k}) = 2\mathbf{i}+2\mathbf{j}+\mathbf{k}$ is the point $(2,2,1)$, which is the location of the center of mass

74. The midpoint of AB is $M\left(\frac{3}{2},0,\frac{5}{2}\right)$ and $\left(\frac{2}{3}\right)\overrightarrow{CM} = \frac{2}{3}\left[\left(\frac{3}{2}+1\right)\mathbf{i}+(0-2)\mathbf{j}+\left(\frac{5}{2}+1\right)\mathbf{k}\right] = \frac{2}{3}\left(\frac{5}{2}\mathbf{i}-2\mathbf{j}+\frac{7}{2}\mathbf{k}\right)$ $= \frac{5}{3}\mathbf{i}-\frac{4}{3}\mathbf{j}+\frac{7}{3}\mathbf{k}$. The terminal point of $\left(\frac{5}{3}\mathbf{i}-\frac{4}{3}\mathbf{j}+\frac{7}{3}\mathbf{k}\right)+\overrightarrow{OC} = \left(\frac{5}{3}\mathbf{i}-\frac{4}{3}\mathbf{j}+\frac{7}{3}\mathbf{k}\right)+(-\mathbf{i}+2\mathbf{j}-\mathbf{k})$ $= \frac{2}{3}\mathbf{i}+\frac{2}{3}\mathbf{j}+\frac{4}{3}\mathbf{k}$ is the point $\left(\frac{2}{3},\frac{2}{3},\frac{4}{3}\right)$ which is the location of the intersection of the medians.

75. Without loss of generality we identify the vertices of the quadrilateral such that $A(0,0,0)$, $B(x_b,0,0)$, $C(x_c,y_c,0)$ and $D(x_d,y_d,z_d)$ $\Rightarrow$ the midpoint of AB is $M_{AB}\left(\frac{x_b}{2},0,0\right)$, the midpoint of BC is $M_{BC}\left(\frac{x_b+x_c}{2},\frac{y_c}{2},0\right)$, the midpoint of CD is $M_{CD}\left(\frac{x_c+x_d}{2},\frac{y_c+y_d}{2},\frac{z_d}{2}\right)$ and the midpoint of AD is $M_{AD}\left(\frac{x_d}{2},\frac{y_d}{2},\frac{z_d}{2}\right)$ $\Rightarrow$ the midpoint of $M_{AB}M_{CD}$ is $\left(\frac{\frac{x_b}{2}+\frac{x_c+x_d}{2}}{2},\frac{y_c+y_d}{4},\frac{z_d}{4}\right)$ which is the same as the midpoint of $M_{AD}M_{BC} = \left(\frac{\frac{x_b+x_c}{2}+\frac{x_d}{2}}{2},\frac{y_c+y_d}{4},\frac{z_d}{4}\right)$.

76. Let $V_1, V_2, V_3, \ldots, V_n$ be the vertices of a regular n-sided polygon and $\mathbf{v}_i$ denote the vector from the center to V_i for $i = 1, 2, 3, \ldots, n$. If $\mathbf{S} = \sum_{i=1}^{n} \mathbf{v}_i$ and the polygon is rotated through an angle of $\frac{i(2\pi)}{n}$ where $i = 1, 2, 3, \ldots, n$, then $\mathbf{S}$ would remain the same. Since $\mathbf{S}$ does not change with these rotations we conclude that $\mathbf{S} = \mathbf{0}$.

77. Without loss of generality we can coordinatize the vertices of the triangle such that $A(0,0)$, $B(b,0)$ and $C(x_c,y_c) \Rightarrow$ a is located at $\left(\frac{b+x_c}{2},\frac{y_c}{2}\right)$, b is at $\left(\frac{x_c}{2},\frac{y_c}{2}\right)$ and c is at $\left(\frac{b}{2},0\right)$. Therefore, $\overrightarrow{Aa} = \left(\frac{b}{2}+\frac{x_c}{2}\right)\mathbf{i}+\left(\frac{y_c}{2}\right)\mathbf{j}$, $\overrightarrow{Bb} = \left(\frac{x_c}{2}-b\right)\mathbf{i}+\left(\frac{y_c}{2}\right)\mathbf{j}$, and $\overrightarrow{Cc} = \left(\frac{b}{2}-x_c\right)\mathbf{i}+(-y_c)\mathbf{j} \Rightarrow \overrightarrow{Aa}+\overrightarrow{Bb}+\overrightarrow{Cc} = \mathbf{0}$.

10.3 DOT PRODUCTS

NOTE: In Exercises 1-10 below we calculate $\text{proj}_{\mathbf{A}}\,\mathbf{B}$ as the vector $\left(\frac{|\mathbf{B}|\cos\theta}{|\mathbf{A}|}\right)\mathbf{A}$, so the scalar multiplier of $\mathbf{A}$ is the number in column 5 divided by the number in column 2.

	$\mathbf{A}\cdot\mathbf{B}$	$\lvert\mathbf{A}\rvert$	$\lvert\mathbf{B}\rvert$	$\cos\theta$	$\lvert\mathbf{B}\rvert\cos\theta$	$\text{proj}_{\mathbf{A}}\,\mathbf{B}$
1.	-25	5	5	-1	-5	$-2\mathbf{i}+4\mathbf{j}-\sqrt{5}\mathbf{k}$
2.	3	1	13	$\frac{3}{13}$	3	$3\left(\frac{3}{5}\mathbf{i}+\frac{4}{5}\mathbf{k}\right)$
3.	25	15	5	$\frac{1}{3}$	$\frac{5}{3}$	$\frac{1}{9}(10\mathbf{i}+11\mathbf{j}-2\mathbf{k})$
4.	13	15	3	$\frac{13}{45}$	$\frac{13}{15}$	$\frac{13}{225}(2\mathbf{i}+10\mathbf{j}-11\mathbf{k})$

	$\mathbf{A}\cdot\mathbf{B}$	$\lvert\mathbf{A}\rvert$	$\lvert\mathbf{B}\rvert$	$\cos\theta$	$\lvert\mathbf{B}\rvert\cos\theta$	$\text{proj}_{\mathbf{A}}\,\mathbf{B}$
5.	0	$\sqrt{53}$	1	0	0	$\mathbf{0}$
6.	0	1	$\sqrt{\frac{3}{2}}$	0	0	$\mathbf{0}$
7.	2	$\sqrt{34}$	$\sqrt{3}$	$\frac{2}{\sqrt{3}\sqrt{34}}$	$\frac{2}{\sqrt{34}}$	$\frac{1}{17}(5\mathbf{j}-3\mathbf{k})$
8.	2	$\sqrt{2}$	$\sqrt{3}$	$\sqrt{\frac{2}{3}}$	$\sqrt{2}$	$\mathbf{i}+\mathbf{k}$
9.	$\sqrt{3}-\sqrt{2}$	$\sqrt{2}$	3	$\frac{\sqrt{3}-\sqrt{2}}{3\sqrt{2}}$	$\frac{\sqrt{3}-\sqrt{2}}{\sqrt{2}}$	$\frac{\sqrt{3}-\sqrt{2}}{2}(-\mathbf{i}+\mathbf{j})$
10.	$-10+\sqrt{17}$	$\sqrt{26}$	11	$\frac{\sqrt{17}-10}{11\sqrt{26}}$	$\frac{\sqrt{17}-10}{\sqrt{26}}$	$\frac{\sqrt{17}-10}{26}(-5\mathbf{i}+\mathbf{j})$

11. $\mathbf{B}=\left(\frac{\mathbf{A}\cdot\mathbf{B}}{\mathbf{A}\cdot\mathbf{A}}\,\mathbf{A}\right)+\left(\mathbf{B}-\frac{\mathbf{A}\cdot\mathbf{B}}{\mathbf{A}\cdot\mathbf{A}}\,\mathbf{A}\right)=\frac{3}{2}(\mathbf{i}+\mathbf{j})+\left[(3\mathbf{j}+4\mathbf{k})-\frac{3}{2}(\mathbf{i}+\mathbf{j})\right]=\left(\frac{3}{2}\mathbf{i}+\frac{3}{2}\mathbf{j}\right)+\left(-\frac{3}{2}\mathbf{i}+\frac{3}{2}\mathbf{j}+4\mathbf{k}\right)$, where $\mathbf{A}\cdot\mathbf{B}=3$ and $\mathbf{A}\cdot\mathbf{A}=2$

12. $\mathbf{B}=\left(\frac{\mathbf{A}\cdot\mathbf{B}}{\mathbf{A}\cdot\mathbf{A}}\,\mathbf{A}\right)+\left(\mathbf{B}-\frac{\mathbf{A}\cdot\mathbf{B}}{\mathbf{A}\cdot\mathbf{A}}\,\mathbf{A}\right)=\frac{1}{2}\mathbf{A}+\left(\mathbf{B}-\frac{1}{2}\mathbf{A}\right)=\frac{1}{2}(\mathbf{i}+\mathbf{j})+\left[(\mathbf{j}+\mathbf{k})-\frac{1}{2}(\mathbf{i}+\mathbf{j})\right]=\left(\frac{1}{2}\mathbf{i}+\frac{1}{2}\mathbf{j}\right)+\left(-\frac{1}{2}\mathbf{i}+\frac{1}{2}\mathbf{j}+\mathbf{k}\right)$, where $\mathbf{A}\cdot\mathbf{B}=1$ and $\mathbf{A}\cdot\mathbf{A}=2$

13. $\mathbf{B}=\left(\frac{\mathbf{A}\cdot\mathbf{B}}{\mathbf{A}\cdot\mathbf{A}}\,\mathbf{A}\right)+\left(\mathbf{B}-\frac{\mathbf{A}\cdot\mathbf{B}}{\mathbf{A}\cdot\mathbf{A}}\,\mathbf{A}\right)=\frac{14}{3}(\mathbf{i}+2\mathbf{j}-\mathbf{k})+\left[(8\mathbf{i}+4\mathbf{j}-12\mathbf{k})-\left(\frac{14}{3}\mathbf{i}+\frac{28}{3}\mathbf{j}-\frac{14}{3}\mathbf{k}\right)\right]$
$=\left(\frac{14}{3}\mathbf{i}+\frac{28}{3}\mathbf{j}-\frac{14}{3}\mathbf{k}\right)+\left(\frac{10}{3}\mathbf{i}-\frac{16}{3}\mathbf{j}-\frac{22}{3}\mathbf{k}\right)$, where $\mathbf{A}\cdot\mathbf{B}=28$ and $\mathbf{A}\cdot\mathbf{A}=6$

14. $\mathbf{B}=\left(\frac{\mathbf{A}\cdot\mathbf{B}}{\mathbf{A}\cdot\mathbf{A}}\,\mathbf{A}\right)+\left(\mathbf{B}-\frac{\mathbf{A}\cdot\mathbf{B}}{\mathbf{A}\cdot\mathbf{A}}\,\mathbf{A}\right)=\frac{1}{1}(\mathbf{A})+\left[(\mathbf{i}+\mathbf{j}+\mathbf{k})-\left(\frac{1}{1}\right)\mathbf{A}\right]=(\mathbf{i})+(\mathbf{j}+\mathbf{k})$, where $\mathbf{A}\cdot\mathbf{B}=1$ and $\mathbf{A}\cdot\mathbf{A}=1$; yes

15. The sum of two vectors of equal length is *always* orthogonal to their difference, as we can see from the equation $(\mathbf{v}_1+\mathbf{v}_2)\cdot(\mathbf{v}_1-\mathbf{v}_2)=\mathbf{v}_1\cdot\mathbf{v}_1+\mathbf{v}_2\cdot\mathbf{v}_1-\mathbf{v}_1\cdot\mathbf{v}_2-\mathbf{v}_2\cdot\mathbf{v}_2=\lvert\mathbf{v}_1\rvert^2-\lvert\mathbf{v}_2\rvert^2=0$

16. $\overrightarrow{CA}\cdot\overrightarrow{CB}=(-\mathbf{v}+(-\mathbf{u}))\cdot(-\mathbf{v}+\mathbf{u})=\mathbf{v}\cdot\mathbf{v}-\mathbf{v}\cdot\mathbf{u}+\mathbf{u}\cdot\mathbf{v}-\mathbf{u}\cdot\mathbf{u}=\lvert\mathbf{v}\rvert^2-\lvert\mathbf{u}\rvert^2=0$ because $\lvert\mathbf{u}\rvert=\lvert\mathbf{v}\rvert$ since both equal the radius of the circle. Therefore, $\overrightarrow{CA}$ and $\overrightarrow{CB}$ are orthogonal.

17. Let $\mathbf{u}$ and $\mathbf{v}$ be the sides of a rhombus $\Rightarrow$ the diagonals are $\mathbf{d}_1=\mathbf{u}+\mathbf{v}$ and $\mathbf{d}_2=-\mathbf{u}+\mathbf{v}$ $\Rightarrow \mathbf{d}_1\cdot\mathbf{d}_2=(\mathbf{u}+\mathbf{v})\cdot(-\mathbf{u}+\mathbf{v})=-\mathbf{u}\cdot\mathbf{u}+\mathbf{u}\cdot\mathbf{v}-\mathbf{v}\cdot\mathbf{u}+\mathbf{v}\cdot\mathbf{v}=\lvert\mathbf{v}\rvert^2-\lvert\mathbf{u}\rvert^2=0$ because $\lvert\mathbf{u}\rvert=\lvert\mathbf{v}\rvert$, since a rhombus has equal sides.

18. Let $\mathbf{u}$ and $\mathbf{v}$ be the sides of a rectangle $\Rightarrow$ the diagonals are $\mathbf{d}_1=\mathbf{u}+\mathbf{v}$ and $\mathbf{d}_2=-\mathbf{u}+\mathbf{v}$. Since the diagonals are perpendicular we have $\mathbf{d}_1\cdot\mathbf{d}_2=0 \Leftrightarrow (\mathbf{u}+\mathbf{v})\cdot(-\mathbf{u}+\mathbf{v})=-\mathbf{u}\cdot\mathbf{u}+\mathbf{u}\cdot\mathbf{v}-\mathbf{v}\cdot\mathbf{u}+\mathbf{v}\cdot\mathbf{v}=0 \Leftrightarrow \lvert\mathbf{v}\rvert^2-\lvert\mathbf{u}\rvert^2=0$ $\Leftrightarrow (\lvert\mathbf{v}\rvert+\lvert\mathbf{u}\rvert)(\lvert\mathbf{v}\rvert-\lvert\mathbf{u}\rvert)=0 \Leftrightarrow (\lvert\mathbf{v}\rvert+\lvert\mathbf{u}\rvert)=0$ which is not possible, or $(\lvert\mathbf{v}\rvert-\lvert\mathbf{u}\rvert)=0$ which is equivalent to $\lvert\mathbf{v}\rvert=\lvert\mathbf{u}\rvert \Rightarrow$ the rectangle is a square.

19. Clearly the diagonals of a rectangle are equal in length. What is not as obvious is the statement that equal diagonals happen only in a rectangle. We show this is true by letting the opposite sides of a parallelogram be the vectors $(v_1\mathbf{i}+v_2\mathbf{j})$ and $(u_1\mathbf{i}+u_2\mathbf{j})$. The equal diagonals of the parallelogram are

$d_1 = (v_1\mathbf{i}+v_2\mathbf{j})+(u_1\mathbf{i}+u_2\mathbf{j})$ and $d_2 = (v_1\mathbf{i}+v_2\mathbf{j})-(u_1\mathbf{i}+u_2\mathbf{j})$. Hence $|d_1| = |d_2| = |(v_1\mathbf{i}+v_2\mathbf{j})+(u_1\mathbf{i}+u_2\mathbf{j})|$ $= |(v_1\mathbf{i}+v_2\mathbf{j})-(u_1\mathbf{i}+u_2\mathbf{j})| \Rightarrow |(v_1+u_1)\mathbf{i}+(v_2+u_2)\mathbf{j}| = |(v_1-u_1)\mathbf{i}+(v_2-u_2)\mathbf{j}|$

$\Rightarrow \sqrt{(v_1+u_1)^2+(v_2+u_2)^2} = \sqrt{(v_1-u_1)^2+(v_2-u_2)^2} \Rightarrow v_1^2+2v_1u_1+u_1^2+v_2^2+2v_2u_2+u_2^2$ $= v_1^2-2v_1u_1+u_1^2+v_2^2-2v_2u_2+u_2^2 \Rightarrow 2(v_1u_1+v_2u_2) = -2(v_1u_1+v_2u_2) \Rightarrow v_1u_1+v_2u_2 = 0$

$\Rightarrow (v_1\mathbf{i}+v_2\mathbf{j})\cdot(u_1\mathbf{i}+u_2\mathbf{j}) = 0 \Rightarrow$ the vectors $(v_1\mathbf{i}+v_2\mathbf{j})$ and $(u_1\mathbf{i}+u_2\mathbf{j})$ are perpendicular and the parallelogram must be a rectangle.

20. If $|\mathbf{u}| = |\mathbf{v}|$ and $\mathbf{u}+\mathbf{v}$ is the indicated diagonal, then $(\mathbf{u}+\mathbf{v})\cdot\mathbf{u} = \mathbf{u}\cdot\mathbf{u}+\mathbf{v}\cdot\mathbf{u} = |\mathbf{u}|^2+\mathbf{v}\cdot\mathbf{u} = \mathbf{u}\cdot\mathbf{v}+|\mathbf{v}|^2$ $= \mathbf{u}\cdot\mathbf{v}+\mathbf{v}\cdot\mathbf{v} = (\mathbf{u}+\mathbf{v})\cdot\mathbf{v} \Rightarrow$ the angle $\cos^{-1}\left(\frac{(\mathbf{u}+\mathbf{v})\cdot\mathbf{u}}{|\mathbf{u}+\mathbf{v}||\mathbf{u}|}\right)$ between the diagonal and $\mathbf{u}$ and the angle $\cos^{-1}\left(\frac{(\mathbf{u}+\mathbf{v})\cdot\mathbf{v}}{|\mathbf{u}+\mathbf{v}||\mathbf{v}|}\right)$ between the diagonal and $\mathbf{v}$ are equal because the inverse cosine function is one-to-one. Therefore, the diagonal bisects the angle between $\mathbf{u}$ and $\mathbf{v}$.

21. Let M be the midpoint of OB. By the Pythagorean Theorem $OB = \sqrt{1^2+1^2} = \sqrt{2}$ and $OM = \frac{\sqrt{2}}{2}$. Hence the angle θ between $\overrightarrow{OB}$ and $\overrightarrow{OD}$ has a tangent of $\frac{DM}{OB} = \frac{1}{\left(\frac{\sqrt{2}}{2}\right)} = \frac{2}{\sqrt{2}} = \sqrt{2}$. Therefore, $\tan\theta = \sqrt{2}$ $\Rightarrow \theta = \tan^{-1}\sqrt{2} \approx 54.7°$.

22. (a) $\cos\alpha = \frac{\mathbf{i}\cdot\mathbf{v}}{|\mathbf{i}||\mathbf{v}|} = \frac{a}{|\mathbf{v}|}$, $\cos\beta = \frac{\mathbf{j}\cdot\mathbf{v}}{|\mathbf{j}||\mathbf{v}|} = \frac{b}{|\mathbf{v}|}$, $\cos\gamma = \frac{\mathbf{k}\cdot\mathbf{v}}{|\mathbf{k}||\mathbf{v}|} = \frac{c}{|\mathbf{v}|}$ and

$$\cos^2\alpha+\cos^2\beta+\cos^2\gamma = \left(\frac{a}{|\mathbf{v}|}\right)^2+\left(\frac{b}{|\mathbf{v}|}\right)^2+\left(\frac{c}{|\mathbf{v}|}\right)^2 = \frac{a^2+b^2+c^2}{|\mathbf{v}||\mathbf{v}|} = \frac{|\mathbf{v}||\mathbf{v}|}{|\mathbf{v}||\mathbf{v}|} = 1$$

(b) $|\mathbf{v}| = 1 \Rightarrow \cos\alpha = \frac{a}{|\mathbf{v}|} = a$, $\cos\beta = \frac{b}{|\mathbf{v}|} = b$ and $\cos\gamma = \frac{c}{|\mathbf{v}|} = c$ are the direction cosines of $\mathbf{v}$

23. $\theta = \cos^{-1}\left(\frac{\mathbf{A}\cdot\mathbf{B}}{|\mathbf{A}||\mathbf{B}|}\right) = \cos^{-1}\left(\frac{(2)(1)+(1)(2)+(0)(-1)}{\sqrt{2^2+1^2+0^2}\sqrt{1^2+2^2+(-1)^2}}\right) = \cos^{-1}\left(\frac{4}{\sqrt{5}\sqrt{6}}\right) = \cos^{-1}\left(\frac{4}{\sqrt{30}}\right) \approx 0.75$ rad

24. $\theta = \cos^{-1}\left(\frac{\mathbf{A}\cdot\mathbf{B}}{|\mathbf{A}||\mathbf{B}|}\right) = \cos^{-1}\left(\frac{(2)(3)+(-2)(0)+(1)(4)}{\sqrt{2^2+(-2)^2+1^2}\sqrt{3^2+0^2+4^2}}\right) = \cos^{-1}\left(\frac{10}{\sqrt{9}\sqrt{25}}\right) = \cos^{-1}\left(\frac{2}{3}\right) \approx 0.84$ rad

25. $\theta = \cos^{-1}\left(\frac{\mathbf{A}\cdot\mathbf{B}}{|\mathbf{A}||\mathbf{B}|}\right) = \cos^{-1}\left(\frac{(\sqrt{3})(\sqrt{3})+(-7)(1)+(0)(-2)}{\sqrt{(\sqrt{3})^2+(-7)^2+0^2}\sqrt{(\sqrt{3})^2+(1)^2+(-2)^2}}\right) = \cos^{-1}\left(\frac{3-7}{\sqrt{52}\sqrt{8}}\right)$ $= \cos^{-1}\left(\frac{-1}{\sqrt{26}}\right) \approx 1.77$ rad

26. $\theta = \cos^{-1}\left(\frac{\mathbf{A}\cdot\mathbf{B}}{|\mathbf{A}||\mathbf{B}|}\right) = \cos^{-1}\left(\frac{(1)(-1)+(\sqrt{2})(1)+(-\sqrt{2})(1)}{\sqrt{(1)^2+(\sqrt{2})^2+(-\sqrt{2})^2}\sqrt{(-1)^2+(1)^2+(1)^2}}\right) = \cos^{-1}\left(\frac{-1}{\sqrt{5}\sqrt{3}}\right)$
$= \cos^{-1}\left(\frac{-1}{\sqrt{15}}\right) \approx 1.83$ rad

27. $\overrightarrow{AB} = 3\mathbf{i}+\mathbf{j}-3\mathbf{k}$, $\overrightarrow{AC} = 2\mathbf{i}-2\mathbf{j}$, $\overrightarrow{BA} = -3\mathbf{i}-\mathbf{j}+3\mathbf{k}$, $\overrightarrow{CA} = -2\mathbf{i}+2\mathbf{j}$, $\overrightarrow{CB} = \mathbf{i}+3\mathbf{j}-3\mathbf{k}$, $\overrightarrow{BC} = -\mathbf{i}-3\mathbf{j}+3\mathbf{k}$; thus $\angle A = \cos^{-1}\left(\frac{\overrightarrow{AB}\cdot\overrightarrow{AC}}{|\overrightarrow{AB}||\overrightarrow{AC}|}\right) = \cos^{-1}\left(\frac{4}{\sqrt{152}}\right) \approx 1.24 \text{ rad} \approx 71.07°$; $\angle B = \cos^{-1}\left(\frac{\overrightarrow{BA}\cdot\overrightarrow{BC}}{|\overrightarrow{BA}||\overrightarrow{BC}|}\right) = \cos^{-1}\left(\frac{15}{19}\right)$
$\approx 0.66 \text{ rad} \approx 37.86°$; $\angle C = \cos^{-1}\left(\frac{\overrightarrow{CA}\cdot\overrightarrow{CB}}{|\overrightarrow{CA}||\overrightarrow{CB}|}\right) = \cos^{-1}\left(\frac{4}{\sqrt{152}}\right) \approx 1.24 \text{ rad} \approx 71.07°$

28. $\theta = \cos^{-1}\left(\frac{\mathbf{A}\cdot\mathbf{B}}{|\mathbf{A}||\mathbf{B}|}\right) = \cos^{-1}\left(\frac{13}{45}\right) \approx 1.28 \text{ rad} \approx 73.21°$

29. Let $\mathbf{A} = \mathbf{i}+\mathbf{k}$ and $\mathbf{B} = \mathbf{i}+\mathbf{j}+\mathbf{k} \Rightarrow \theta = \cos^{-1}\left(\frac{\mathbf{A}\cdot\mathbf{B}}{|\mathbf{A}||\mathbf{B}|}\right) = \cos^{-1}\left(\frac{2}{\sqrt{2}\sqrt{3}}\right) \approx 0.62 \text{ rad} \approx 35.26°$

30. $\mathbf{A} = 10\mathbf{i}+2\mathbf{k}$ is parallel to the pipe in the north direction and $\mathbf{B} = 10\mathbf{j}+\mathbf{k}$ is parallel to the pipe in the east direction. The angle between the two pipes is $\theta = \cos^{-1}\left(\frac{\mathbf{A}\cdot\mathbf{B}}{|\mathbf{A}||\mathbf{B}|}\right) = \cos^{-1}\left(\frac{2}{\sqrt{104}\sqrt{101}}\right) \approx 1.55 \text{ rad} \approx 88.88°$.

31. (a) Since $|\cos\theta| \le 1$, we have $|\mathbf{u}\cdot\mathbf{v}| = |\mathbf{u}||\mathbf{v}||\cos\theta| \le |\mathbf{u}||\mathbf{v}|(1) = |\mathbf{u}||\mathbf{v}|$.
(b) We have equality precisely when $|\cos\theta| = 1$ or when one or both of $\mathbf{u}$ and $\mathbf{v}$ is $\mathbf{0}$. In the case of nonzero vectors, we have equality when $\theta = 0$ or π, i.e., when the vectors are parallel.

32. $(x\mathbf{i}+y\mathbf{j})\cdot\mathbf{v} = |x\mathbf{i}+y\mathbf{j}||\mathbf{v}|\cos\theta \le 0$ when $\frac{\pi}{2} \le \theta \le \pi$. This means (x, y) has to be a point whose position vector makes an angle with $\mathbf{v}$ that is a right angle or bigger.

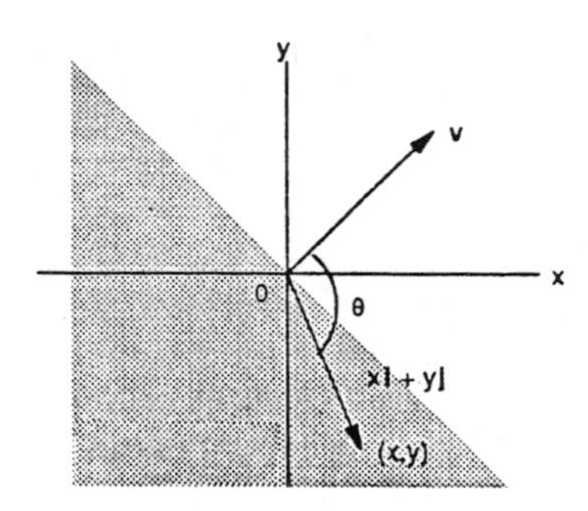

33. $\mathbf{v}\cdot\mathbf{u}_1 = (a\mathbf{u}_1 + b\mathbf{u}_2)\cdot\mathbf{u}_1 = a\mathbf{u}_1\cdot\mathbf{u}_1 + b\mathbf{u}_2\cdot\mathbf{u}_1 = a|\mathbf{u}_1|^2 + b(\mathbf{u}_2\cdot\mathbf{u}_1) = a(1)^2 + b(0) = a$

34. No, $\mathbf{B}_1$ need not equal $\mathbf{B}_2$. For example, $\mathbf{i}+\mathbf{j} \ne \mathbf{i}+2\mathbf{j}$ but $\mathbf{i}\cdot(\mathbf{i}+\mathbf{j}) = \mathbf{i}\cdot\mathbf{i}+\mathbf{i}\cdot\mathbf{j} = 1+0 = 1$ and $\mathbf{i}\cdot(\mathbf{i}+2\mathbf{j}) = \mathbf{i}\cdot\mathbf{i}+2\mathbf{i}\cdot\mathbf{j} = 1+2\cdot 0 = 1$.

35. (a) $|\mathbf{D}|^2 = \mathbf{D}\cdot\mathbf{D} = 25\mathbf{A}\cdot\mathbf{A} + 36\mathbf{B}\cdot\mathbf{B} + 9\mathbf{C}\cdot\mathbf{C} = 25+36+9 = 70 \Rightarrow |\mathbf{D}| = \sqrt{70}$
(b) $|\mathbf{D}|^2 = 25|\mathbf{A}|^2 + 36|\mathbf{B}|^2 + 9|\mathbf{C}|^2 = (25)(4)+(36)(9)+(9)(16) = 568 \Rightarrow |\mathbf{D}| = \sqrt{568}$

36. $\mathbf{D}\cdot\mathbf{A} = (\alpha\mathbf{A}+\beta\mathbf{B}+\gamma\mathbf{C})\cdot\mathbf{A} = \alpha\mathbf{A}\cdot\mathbf{A}+\beta\mathbf{B}\cdot\mathbf{A}+\gamma\mathbf{C}\cdot\mathbf{A} = (\alpha)(1)+(\beta)(0)+(\gamma)(0) = \alpha$; similarly $\mathbf{D}\cdot\mathbf{B} = \beta$ and $\mathbf{D}\cdot\mathbf{C} = \gamma$

37. $P(0,0,0)$, $Q(1,1,1)$ and $\mathbf{F} = 5\mathbf{k} \Rightarrow \overrightarrow{PQ} = \mathbf{i}+\mathbf{j}+\mathbf{k}$ and $\mathbf{W} = \mathbf{F}\cdot\overrightarrow{PQ} = (5\mathbf{k})\cdot(\mathbf{i}+\mathbf{j}+\mathbf{k}) = 5\ \text{N}\cdot\text{m} = 5\ \text{J}$

38. $\mathbf{W} = |\mathbf{F}|\,(\text{distance})\cos\theta = (602{,}148\ \text{N})(605\ \text{km})(\cos 0) = 364{,}299{,}540\ \text{N}\cdot\text{km} = (364{,}299{,}540)(1000)\ \text{N}\cdot\text{m}$
$= 3.6429954\times 10^{11}\ \text{J}$

39. $\mathbf{W} = |\mathbf{F}|\left|\overrightarrow{PQ}\right|\cos\theta = (200)(20)(\cos 30^\circ) = 2000\sqrt{3} = 3464.10\ \text{N}\cdot\text{m} = 3464.10\ \text{J}$

40. $\mathbf{W} = |\mathbf{F}|\left|\overrightarrow{PQ}\right|\cos\theta = (1000)(5280)(\cos 60^\circ) = 2{,}640{,}000\ \text{ft}\cdot\text{lb}$

41. $P(x_1,y_1) = P\left(x_1, \frac{c}{b} - \frac{a}{b}x_1\right)$ and $Q(x_2,y_2) = Q\left(x_2, \frac{c}{b} - \frac{a}{b}x_2\right)$ are any two points P and Q on the line with $b \neq 0$
$\Rightarrow \overrightarrow{PQ} = (x_2-x_1)\mathbf{i} + \frac{a}{b}(x_2-x_1)\mathbf{j} \Rightarrow \overrightarrow{PQ}\cdot\mathbf{v} = \left[(x_2-x_1)\mathbf{i} + \frac{a}{b}(x_2-x_1)\mathbf{j}\right]\cdot(a\mathbf{i}+b\mathbf{j}) = a(x_2-x_1) + b\left(\frac{a}{b}\right)(x_2-x_1)$
$= 0 \Rightarrow \mathbf{v}$ is perpendicular to $\overrightarrow{PQ}$ for $b \neq 0$. If $b = 0$, then $\mathbf{v} = a\mathbf{i}$ is perpendicular to the vertical line $ax = c$.

Alternatively, the slope of $\mathbf{v}$ is $\frac{b}{a}$ and the slope of the line $ax + by = c$ is $-\frac{a}{b}$, so the slopes are negative reciprocals $\Rightarrow$ the vector $\mathbf{v}$ and the line are perpendicular.

42. The slope of $\mathbf{v}$ is $\frac{b}{a}$ and the slope of $bx - ay = c$ is $\frac{b}{a}$, provided that $a \neq 0$. If $a = 0$, then $\mathbf{v} = b\mathbf{j}$ is parallel to the vertical line $bx = c$. In either case, the vector $\mathbf{v}$ is parallel to the line $ax - by = c$.

43. $\mathbf{v} = \mathbf{i} + 2\mathbf{j}$ is perpendicular to the line $x + 2y = c$;
$P(2,1)$ on the line $\Rightarrow 2 + 2 = c \Rightarrow x + 2y = 4$

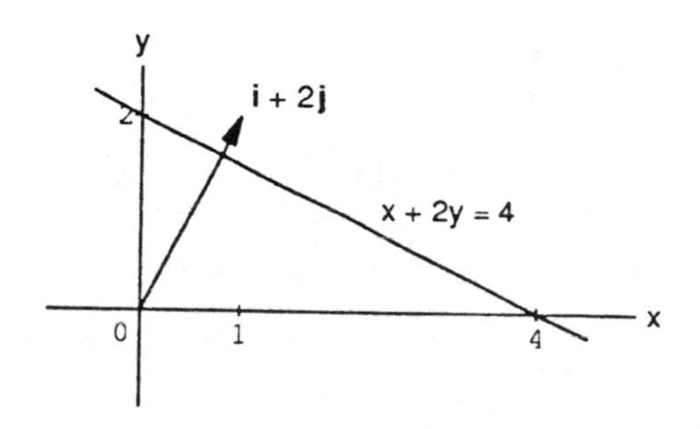

44. $\mathbf{v} = -2\mathbf{i} - \mathbf{j}$ is perpendicular to the line $-2x - y = c$;
$P(-1,2)$ on the line $\Rightarrow (-2)(-1) - 2 = c \Rightarrow -2x - y = 0$

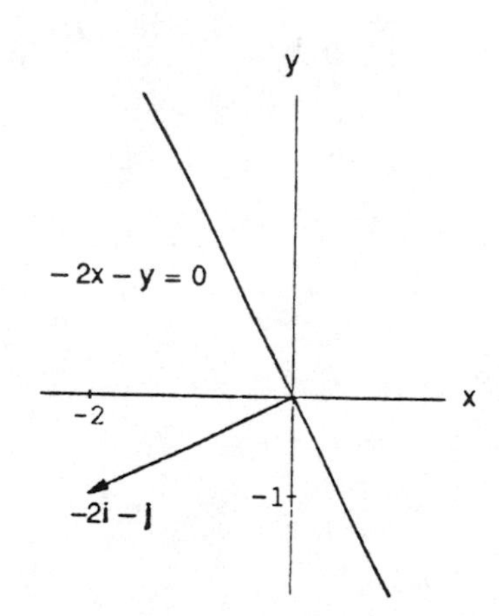

45. $\mathbf{v} = -2\mathbf{i} + \mathbf{j}$ is perpendicular to the line $-2x + y = c$;
P$(-2,-7)$ on the line $\Rightarrow (-2)(-2) - 7 = c \Rightarrow -2x + y = -3$

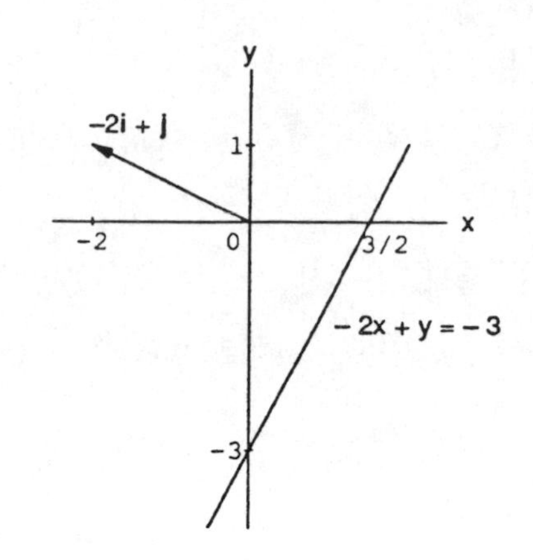

46. $\mathbf{v} = 2\mathbf{i} - 3\mathbf{j}$ is perpendicular to the line $2x - 3y = c$;
P$(11,10)$ on the line $\Rightarrow (2)(11) - (3)(10) = c$
$\Rightarrow 2x - 3y = -8$

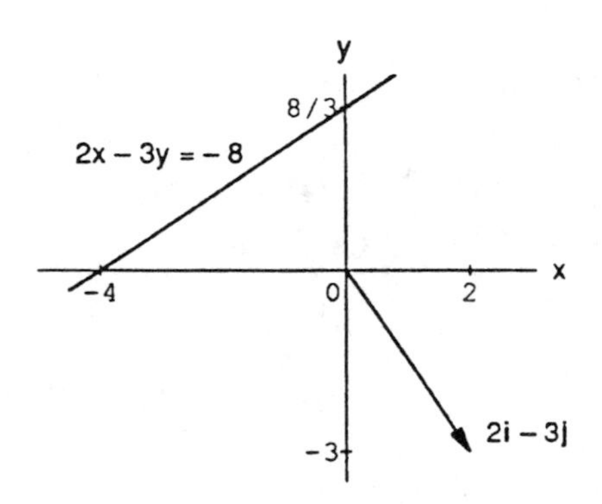

47. $\mathbf{v} = \mathbf{i} - \mathbf{j}$ is parallel to the line $x + y = c$;
P$(-2,1)$ on the line $\Rightarrow -2 + 1 = c \Rightarrow x + y = -1$

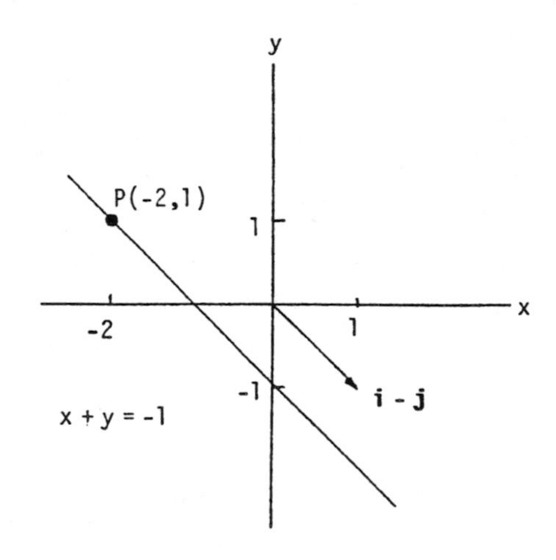

48. $\mathbf{v} = 2\mathbf{i} + 3\mathbf{j}$ is parallel to the line $3x - 2y = c$;
$P(0, -2)$ on the line $\Rightarrow 0 - 2(-2) = c \Rightarrow 3x - 2y = 4$

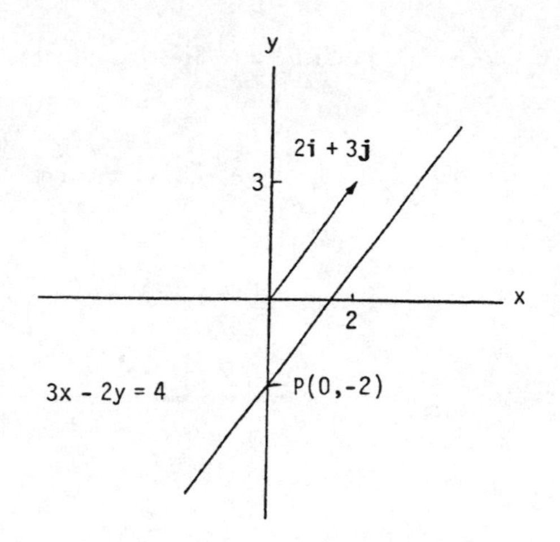

49. $\mathbf{v} = -\mathbf{i} - 2\mathbf{j}$ is parallel to the line $2x - y = c$;
$P(1, 2)$ on the line $\Rightarrow (2)(1) - 2 = c \Rightarrow 2x - y = 0$

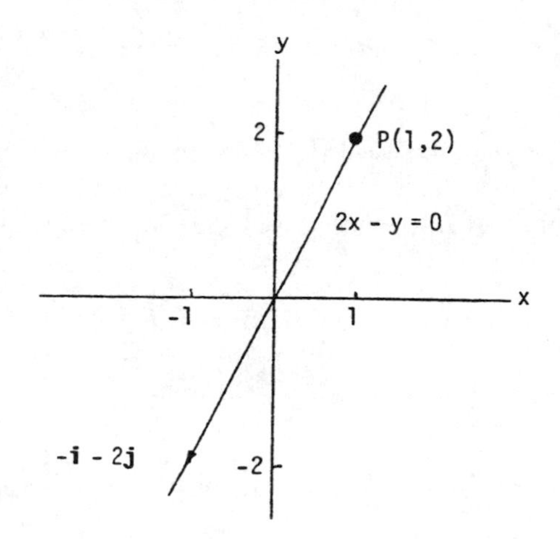

50. $\mathbf{v} = 3\mathbf{i} - 2\mathbf{j}$ is parallel to the line $2x + 3y = c$;
$P(1, 3)$ on the line $\Rightarrow (2)(1) + (3)(3) = c$
$\Rightarrow 2x + 3y = 11$

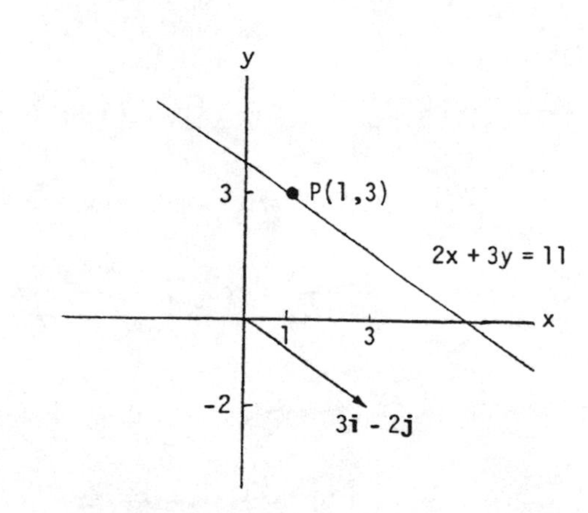

In Exercises 51-56 we use the fact that $\mathbf{n} = a\mathbf{i} + b\mathbf{j}$ is normal to the line $ax + by = c$.

51. $\mathbf{n}_1 = 3\mathbf{i} + \mathbf{j}$ and $\mathbf{n}_2 = 2\mathbf{i} - \mathbf{j} \Rightarrow \theta = \cos^{-1}\left(\frac{\mathbf{n}_1 \cdot \mathbf{n}_2}{|\mathbf{n}_1||\mathbf{n}_2|}\right) = \cos^{-1}\left(\frac{6-1}{\sqrt{10}\sqrt{5}}\right) = \cos^{-1}\left(\frac{1}{\sqrt{2}}\right) = \frac{\pi}{4}$

52. $\mathbf{n}_1 = -\sqrt{3}\mathbf{i}+\mathbf{j}$ and $\mathbf{n}_2 = \sqrt{3}\mathbf{i}+\mathbf{j} \Rightarrow \theta = \cos^{-1}\left(\frac{\mathbf{n}_1 \cdot \mathbf{n}_2}{|\mathbf{n}_1||\mathbf{n}_2|}\right) = \cos^{-1}\left(\frac{-3+1}{\sqrt{4}\sqrt{4}}\right) = \cos^{-1}\left(-\frac{1}{2}\right) = \frac{2\pi}{3}$

53. $\mathbf{n}_1 = \sqrt{3}\mathbf{i}-\mathbf{j}$ and $\mathbf{n}_2 = \mathbf{i}-\sqrt{3}\mathbf{j} \Rightarrow \theta = \cos^{-1}\left(\frac{\mathbf{n}_1 \cdot \mathbf{n}_2}{|\mathbf{n}_1||\mathbf{n}_2|}\right) = \cos^{-1}\left(\frac{\sqrt{3}+\sqrt{3}}{\sqrt{4}\sqrt{4}}\right) = \cos^{-1}\left(\frac{\sqrt{3}}{2}\right) = \frac{\pi}{6}$

54. $\mathbf{n}_1 = \mathbf{i}+\sqrt{3}\mathbf{j}$ and $\mathbf{n}_2 = (1-\sqrt{3})\mathbf{i}+(1+\sqrt{3})\mathbf{j} \Rightarrow \theta = \cos^{-1}\left(\frac{\mathbf{n}_1 \cdot \mathbf{n}_2}{|\mathbf{n}_1||\mathbf{n}_2|}\right)$

$= \cos^{-1}\left(\frac{1-\sqrt{3}+\sqrt{3}+3}{\sqrt{1+3}\sqrt{1-2\sqrt{3}+3+1+2\sqrt{3}+3}}\right) = \cos^{-1}\left(\frac{4}{2\sqrt{8}}\right) = \cos^{-1}\left(\frac{1}{\sqrt{2}}\right) = \frac{\pi}{4}$

55. $\mathbf{n}_1 = 3\mathbf{i}-4\mathbf{j}$ and $\mathbf{n}_2 = \mathbf{i}-\mathbf{j} \Rightarrow \theta = \cos^{-1}\left(\frac{\mathbf{n}_1 \cdot \mathbf{n}_2}{|\mathbf{n}_1||\mathbf{n}_2|}\right) = \cos^{-1}\left(\frac{3+4}{\sqrt{25}\sqrt{2}}\right) = \cos^{-1}\left(\frac{7}{5\sqrt{2}}\right) \approx 0.14$ rad

56. $\mathbf{n}_1 = 12\mathbf{i}+5\mathbf{j}$ and $\mathbf{n}_2 = 2\mathbf{i}-2\mathbf{j} \Rightarrow \theta = \cos^{-1}\left(\frac{\mathbf{n}_1 \cdot \mathbf{n}_2}{|\mathbf{n}_1||\mathbf{n}_2|}\right) = \cos^{-1}\left(\frac{24-10}{\sqrt{169}\sqrt{8}}\right) = \cos^{-1}\left(\frac{14}{26\sqrt{2}}\right) \approx 1.18$ rad

57. The angle between the corresponding normals is equal to the angle between the corresponding tangents. The points of intersection are $\left(-\frac{\sqrt{3}}{2},\frac{3}{4}\right)$ and $\left(\frac{\sqrt{3}}{2},\frac{3}{4}\right)$. At $\left(-\frac{\sqrt{3}}{2},\frac{3}{4}\right)$ the tangent line for $f(x) = x^2$ is $y-\frac{3}{4} = f'\left(-\frac{\sqrt{3}}{2}\right)\left(x-\left(-\frac{\sqrt{3}}{2}\right)\right) \Rightarrow y = -\sqrt{3}\left(x+\frac{\sqrt{3}}{2}\right)+\frac{3}{4} \Rightarrow y = -\sqrt{3}x-\frac{3}{4}$, and the tangent line for $f(x) = \left(\frac{3}{2}\right)-x^2$ is $y-\frac{3}{4} = f'\left(-\frac{\sqrt{3}}{2}\right)\left(x-\left(-\frac{\sqrt{3}}{2}\right)\right) \Rightarrow y = \sqrt{3}\left(x+\frac{\sqrt{3}}{2}\right)+\frac{3}{4} = \sqrt{3}x+\frac{9}{4}$. The corresponding normals are $\mathbf{n}_1 = \sqrt{3}\mathbf{i}+\mathbf{j}$ and $\mathbf{n}_2 = -\sqrt{3}\mathbf{i}+\mathbf{j}$. The angle at $\left(-\frac{\sqrt{3}}{2},\frac{3}{4}\right)$ is $\theta = \cos^{-1}\left(\frac{\mathbf{n}_1 \cdot \mathbf{n}_2}{|\mathbf{n}_1||\mathbf{n}_2|}\right)$ $= \cos^{-1}\left(\frac{-3+1}{\sqrt{4}\sqrt{4}}\right) = \cos^{-1}\left(-\frac{1}{2}\right) = \frac{2\pi}{3}$, the angle is $\frac{\pi}{3}$ and $\frac{2\pi}{3}$. At $\left(\frac{\sqrt{3}}{2},\frac{3}{4}\right)$ the tangent line for $f(x) = x^2$ is $y = \sqrt{3}\left(x+\frac{\sqrt{3}}{2}\right)+\frac{3}{4} = \sqrt{3}x+\frac{9}{4}$ and the tangent line for $f(x) = \frac{3}{2}-x^2$ is $y = -\sqrt{3}\left(x+\frac{\sqrt{3}}{2}\right)+\frac{3}{4}$ $= -\sqrt{3}x-\frac{3}{4}$. The corresponding normals are $\mathbf{n}_1 = -\sqrt{3}\mathbf{i}+\mathbf{j}$ and $\mathbf{n}_2 = \sqrt{3}\mathbf{i}+\mathbf{j}$. The angle at $\left(\frac{\sqrt{3}}{2},\frac{3}{4}\right)$ is $\theta = \cos^{-1}\left(\frac{\mathbf{n}_1 \cdot \mathbf{n}_2}{|\mathbf{n}_1||\mathbf{n}_2|}\right) = \cos^{-1}\left(\frac{-3+1}{\sqrt{4}\sqrt{4}}\right) = \cos^{-1}\left(-\frac{1}{2}\right) = \frac{2\pi}{3}$, the angle is $\frac{\pi}{3}$ and $\frac{2\pi}{3}$.

58. The points of intersection are $\left(0,\frac{\sqrt{3}}{2}\right)$ and $\left(0,-\frac{\sqrt{3}}{2}\right)$. The curve $x = \frac{3}{4}-y^2$ has derivative $\frac{dy}{dx} = -\frac{1}{2y} \Rightarrow$ the tangent line at $\left(0,\frac{\sqrt{3}}{2}\right)$ is $y-\frac{\sqrt{3}}{2} = -\frac{1}{\sqrt{3}}(x-0) \Rightarrow \mathbf{n}_1 = \frac{1}{\sqrt{3}}\mathbf{i}+\mathbf{j}$ is normal to the curve at that point. The curve $x = y^2-\frac{3}{4}$ has derivative $\frac{dy}{dx} = \frac{1}{2y} \Rightarrow$ the tangent line at $\left(0,\frac{\sqrt{3}}{2}\right)$ is $y-\frac{\sqrt{3}}{2} = \frac{1}{\sqrt{3}}(x-0)$ $\Rightarrow \mathbf{n}_2 = -\frac{1}{\sqrt{3}}\mathbf{i}+\mathbf{j}$ is normal to the curve. The angle between the curves is $\theta = \cos^{-1}\left(\frac{\mathbf{n}_1 \cdot \mathbf{n}_2}{|\mathbf{n}_1||\mathbf{n}_2|}\right)$

$= \cos^{-1}\left(\dfrac{-\frac{1}{3}+1}{\sqrt{\frac{1}{3}+1}\,\sqrt{\frac{1}{3}+1}}\right) = \cos^{-1}\left(\dfrac{\left(\frac{2}{3}\right)}{\left(\frac{4}{3}\right)}\right) = \cos^{-1}\left(\frac{1}{2}\right) = \frac{\pi}{3}$ and $\frac{2\pi}{3}$. Because of symmetry the angles between the curves at the two points of intersection are the same.

59. The curves intersect when $y = x^3 = \left(y^2\right)^3 = y^6 \Rightarrow y = 0$ or $y = 1$. The points of intersection are $(0,0)$ and $(1,1)$. Note that $y \ge 0$ since $y = y^6$. At $(0,0)$ the tangent line for $y = x^3$ is $y = 0$ and the tangent line for $y = \sqrt{x}$ is $x = 0$. Therefore, the angle of intersection at $(0,0)$ is $\frac{\pi}{2}$. At $(1,1)$ the tangent line for $y = x^3$ is $y = 3x - 2$ and the tangent line for $y = \sqrt{x}$ is $y = \frac{1}{2}x + \frac{1}{2}$. The corresponding normal vectors are $\mathbf{n}_1 = -3\mathbf{i} + \mathbf{j}$ and $\mathbf{n}_2 = -\frac{1}{2}\mathbf{i} + \mathbf{j} \Rightarrow \theta = \cos^{-1}\left(\frac{\mathbf{n}_1 \cdot \mathbf{n}_2}{|\mathbf{n}_1||\mathbf{n}_2|}\right) = \cos^{-1}\left(\frac{1}{\sqrt{2}}\right) = \frac{\pi}{4}$, the angle is $\frac{\pi}{4}$ and $\frac{3\pi}{4}$.

60. The points of intersection for the curves $y = -x^2$ and $y = \sqrt[3]{x}$ are $(0,0)$ and $(-1,-1)$. At $(0,0)$ the tangent line for $y = -x^2$ is $y = 0$ and the tangent line for $y = \sqrt[3]{x}$ is $x = 0$. Therefore, the angle of intersection at $(0,0)$ is $\frac{\pi}{2}$. At $(-1,-1)$ the tangent line for $y = -x^2$ is $y = 2x + 1$ and the tangent line for $y = \sqrt[3]{x}$ is $y = \frac{1}{3}x - \frac{2}{3}$. The corresponding normal vectors are $\mathbf{n}_1 = 2\mathbf{i} - \mathbf{j}$ and $\mathbf{n}_2 = \frac{1}{3}\mathbf{i} - \mathbf{j} \Rightarrow \theta = \cos^{-1}\left(\frac{\mathbf{n}_1 \cdot \mathbf{n}_2}{|\mathbf{n}_1||\mathbf{n}_2|}\right)$

$= \cos^{-1}\left(\dfrac{\frac{2}{3}+1}{\sqrt{5}\,\sqrt{\frac{1}{9}+1}}\right) = \cos^{-1}\left(\dfrac{\left(\frac{5}{3}\right)}{\frac{\sqrt{5}\,\sqrt{10}}{3}}\right) = \cos^{-1}\left(\frac{1}{\sqrt{2}}\right) = \frac{\pi}{4}$, the angle is $\frac{\pi}{4}$ and $\frac{3\pi}{4}$.

10.4 CROSS PRODUCTS

1. $\mathbf{A} \times \mathbf{B} = \begin{vmatrix} \mathbf{i} & \mathbf{j} & \mathbf{k} \\ 2 & -2 & -1 \\ 1 & 0 & -1 \end{vmatrix} = 3\left(\frac{2}{3}\mathbf{i} + \frac{1}{3}\mathbf{j} + \frac{2}{3}\mathbf{k}\right) \Rightarrow$ length $= 3$ and the direction is $\frac{2}{3}\mathbf{i} + \frac{1}{3}\mathbf{j} + \frac{2}{3}\mathbf{k}$;

$\mathbf{B} \times \mathbf{A} = -(\mathbf{A} \times \mathbf{B}) = -3\left(\frac{2}{3}\mathbf{i} + \frac{1}{3}\mathbf{j} + \frac{2}{3}\mathbf{k}\right) \Rightarrow$ length $= 3$ and the direction is $-\frac{2}{3}\mathbf{i} - \frac{1}{3}\mathbf{j} - \frac{2}{3}\mathbf{k}$

2. $\mathbf{A} \times \mathbf{B} = \begin{vmatrix} \mathbf{i} & \mathbf{j} & \mathbf{k} \\ 2 & 3 & 0 \\ -1 & 1 & 0 \end{vmatrix} = 5(\mathbf{k}) \Rightarrow$ length $= 5$ and the direction is $\mathbf{k}$

$\mathbf{B} \times \mathbf{A} = -(\mathbf{A} \times \mathbf{B}) = -5(\mathbf{k}) \Rightarrow$ length $= 5$ and the direction is $-\mathbf{k}$

3. $\mathbf{A} \times \mathbf{B} = \begin{vmatrix} \mathbf{i} & \mathbf{j} & \mathbf{k} \\ 2 & -2 & 4 \\ -1 & 1 & -2 \end{vmatrix} = \mathbf{0} \Rightarrow$ length $= 0$ and has no direction

$\mathbf{B} \times \mathbf{A} = -(\mathbf{A} \times \mathbf{B}) = \mathbf{0} \Rightarrow$ length $= 0$ and has no direction

4. $\mathbf{A}\times\mathbf{B} = \begin{vmatrix} \mathbf{i} & \mathbf{j} & \mathbf{k} \\ 1 & 1 & -1 \\ 0 & 0 & 0 \end{vmatrix} = \mathbf{0} \Rightarrow$ length $= 0$ and has no direction

$\mathbf{B}\times\mathbf{A} = -(\mathbf{A}\times\mathbf{B}) = \mathbf{0} \Rightarrow$ length $= 0$ and has no direction

5. $\mathbf{A}\times\mathbf{B} = \begin{vmatrix} \mathbf{i} & \mathbf{j} & \mathbf{k} \\ 2 & 0 & 0 \\ 0 & -3 & 0 \end{vmatrix} = -6(\mathbf{k}) \Rightarrow$ length $= 6$ and the direction is $-\mathbf{k}$

$\mathbf{B}\times\mathbf{A} = -(\mathbf{A}\times\mathbf{B}) = 6(\mathbf{k}) \Rightarrow$ length $= 6$ and the direction is $\mathbf{k}$

6. $\mathbf{A}\times\mathbf{B} = (\mathbf{i}\times\mathbf{j})\times(\mathbf{j}\times\mathbf{k}) = \mathbf{k}\times\mathbf{i} = \begin{vmatrix} \mathbf{i} & \mathbf{j} & \mathbf{k} \\ 0 & 0 & 1 \\ 1 & 0 & 0 \end{vmatrix} = \mathbf{j} \Rightarrow$ length $= 1$ and the direction is $\mathbf{j}$

$\mathbf{B}\times\mathbf{A} = -(\mathbf{A}\times\mathbf{B}) = -\mathbf{j} \Rightarrow$ length $= 1$ and the direction is $-\mathbf{j}$

7. $\mathbf{A}\times\mathbf{B} = \begin{vmatrix} \mathbf{i} & \mathbf{j} & \mathbf{k} \\ -8 & -2 & -4 \\ 2 & 2 & 1 \end{vmatrix} = 6\mathbf{i} - 12\mathbf{k} \Rightarrow$ length $= 6\sqrt{5}$ and the direction is $\frac{1}{\sqrt{5}}\mathbf{i} - \frac{2}{\sqrt{5}}\mathbf{k}$

$\mathbf{B}\times\mathbf{A} = -(\mathbf{A}\times\mathbf{B}) = -(6\mathbf{i} - 12\mathbf{k}) \Rightarrow$ length $= 6\sqrt{5}$ and the direction is $-\frac{1}{\sqrt{5}}\mathbf{i} + \frac{2}{\sqrt{5}}\mathbf{k}$

8. $\mathbf{A}\times\mathbf{B} = \begin{vmatrix} \mathbf{i} & \mathbf{j} & \mathbf{k} \\ \frac{3}{2} & -\frac{1}{2} & 1 \\ 1 & 1 & 2 \end{vmatrix} = -2\mathbf{i} - 2\mathbf{j} + 2\mathbf{k} \Rightarrow$ length $= 2\sqrt{3}$ and the direction is $-\frac{1}{\sqrt{3}}\mathbf{i} - \frac{1}{\sqrt{3}}\mathbf{j} + \frac{1}{\sqrt{3}}\mathbf{k}$

$\mathbf{B}\times\mathbf{A} = -(\mathbf{A}\times\mathbf{B}) = -(-2\mathbf{i} - 2\mathbf{j} + 2\mathbf{k}) \Rightarrow$ length $= 2\sqrt{3}$ and the direction is $\frac{1}{\sqrt{3}}\mathbf{i} + \frac{1}{\sqrt{3}}\mathbf{j} - \frac{1}{\sqrt{3}}\mathbf{k}$

9. $\mathbf{A}\times\mathbf{B} = \begin{vmatrix} \mathbf{i} & \mathbf{j} & \mathbf{k} \\ 1 & 0 & 0 \\ 0 & 1 & 0 \end{vmatrix} = \mathbf{k}$

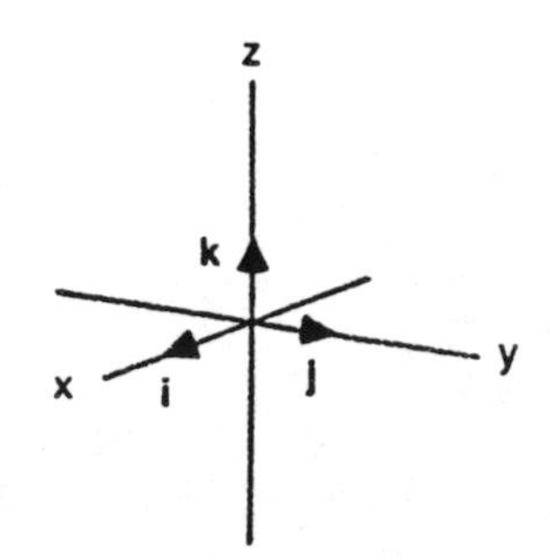

10. $\mathbf{A}\times\mathbf{B} = \begin{vmatrix} \mathbf{i} & \mathbf{j} & \mathbf{k} \\ 1 & 0 & -1 \\ 0 & 1 & 0 \end{vmatrix} = \mathbf{i} + \mathbf{k}$

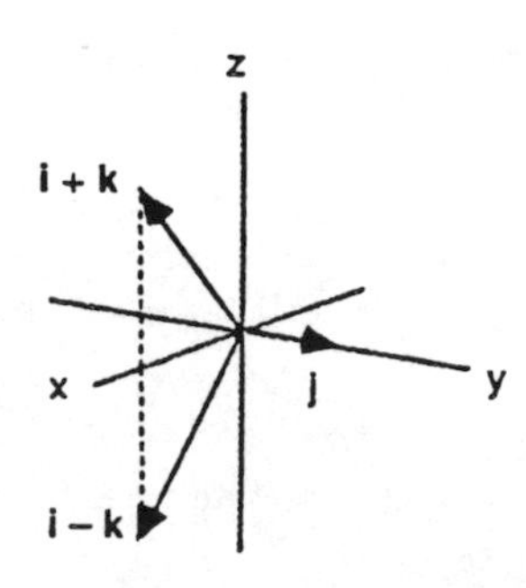

11. $\mathbf{A}\times\mathbf{B} = \begin{vmatrix} \mathbf{i} & \mathbf{j} & \mathbf{k} \\ 1 & 0 & -1 \\ 0 & 1 & 1 \end{vmatrix} = \mathbf{i}-\mathbf{j}+\mathbf{k}$

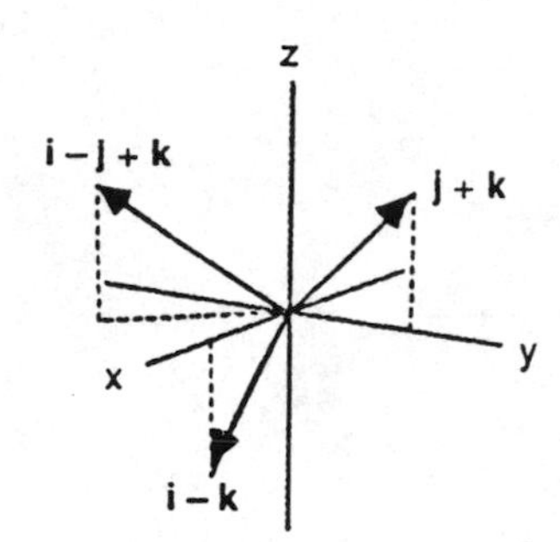

12. $\mathbf{A}\times\mathbf{B} = \begin{vmatrix} \mathbf{i} & \mathbf{j} & \mathbf{k} \\ 2 & -1 & 0 \\ 1 & 2 & 0 \end{vmatrix} = 5\mathbf{k}$

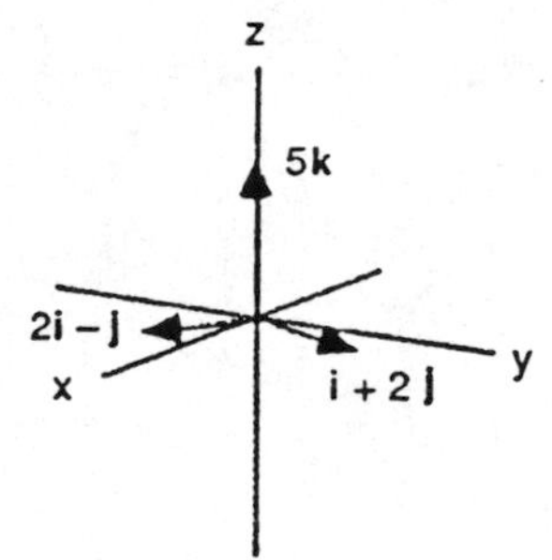

13. $\mathbf{A}\times\mathbf{B} = \begin{vmatrix} \mathbf{i} & \mathbf{j} & \mathbf{k} \\ 1 & 1 & 0 \\ 1 & -1 & 0 \end{vmatrix} = -2\mathbf{k}$

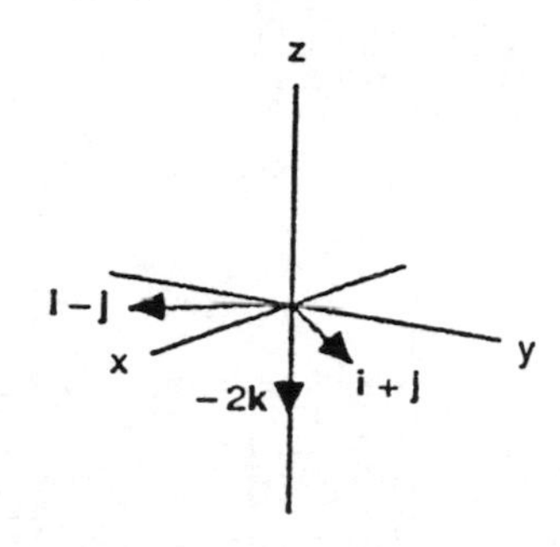

14. 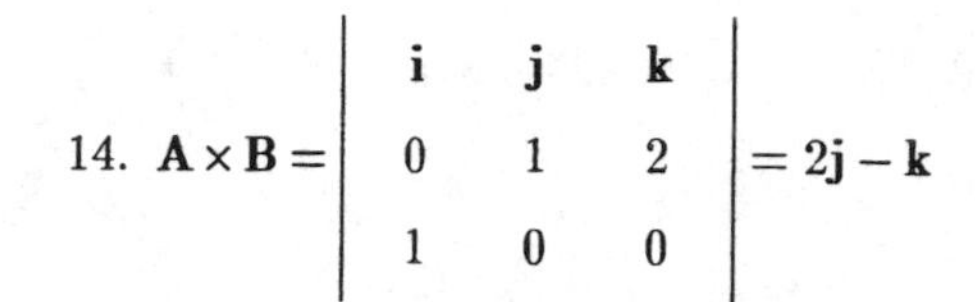
$\mathbf{A}\times\mathbf{B} = \begin{vmatrix} \mathbf{i} & \mathbf{j} & \mathbf{k} \\ 0 & 1 & 2 \\ 1 & 0 & 0 \end{vmatrix} = 2\mathbf{j}-\mathbf{k}$

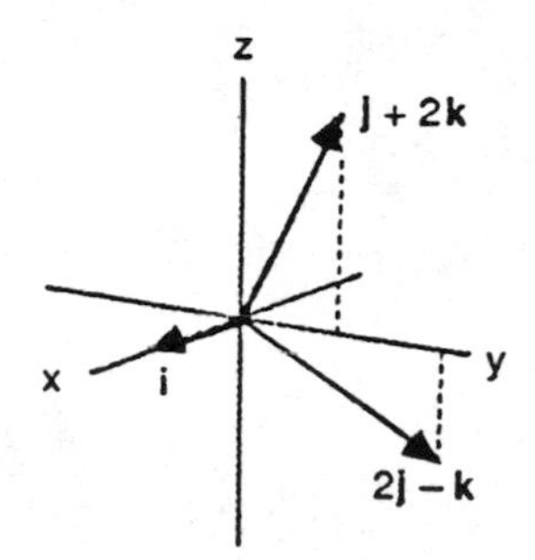

15. (a) $\overrightarrow{PQ}\times\overrightarrow{PR} = \begin{vmatrix} \mathbf{i} & \mathbf{j} & \mathbf{k} \\ 1 & 1 & -3 \\ -1 & 3 & -1 \end{vmatrix} = 8\mathbf{i}+4\mathbf{j}+4\mathbf{k} \Rightarrow \text{Area} = \frac{1}{2}\left|\overrightarrow{PQ}\times\overrightarrow{PR}\right| = \frac{1}{2}\sqrt{64+16+16} = 2\sqrt{6}$

(b) $\mathbf{u} = \pm\frac{\overrightarrow{PQ}\times\overrightarrow{PR}}{\left|\overrightarrow{PQ}\times\overrightarrow{PR}\right|} = \pm\frac{1}{\sqrt{6}}(2\mathbf{i}+\mathbf{j}+\mathbf{k})$

16. (a) $\overrightarrow{PQ}\times\overrightarrow{PR} = \begin{vmatrix} \mathbf{i} & \mathbf{j} & \mathbf{k} \\ 1 & 0 & 2 \\ 2 & -2 & 0 \end{vmatrix} = 4\mathbf{i}+4\mathbf{j}-2\mathbf{k} \Rightarrow \text{Area} = \frac{1}{2}\left|\overrightarrow{PQ}\times\overrightarrow{PR}\right| = \frac{1}{2}\sqrt{16+16+4} = 3$

(b) $\mathbf{u} = \pm\frac{\overrightarrow{PQ}\times\overrightarrow{PR}}{\left|\overrightarrow{PQ}\times\overrightarrow{PR}\right|} = \pm\frac{1}{3}(2\mathbf{i}+2\mathbf{j}-\mathbf{k})$

17. (a) $\overrightarrow{PQ} \times \overrightarrow{PR} = \begin{vmatrix} \mathbf{i} & \mathbf{j} & \mathbf{k} \\ 1 & 1 & 1 \\ 1 & 1 & 0 \end{vmatrix} = -\mathbf{i} + \mathbf{j} \Rightarrow \text{Area} = \frac{1}{2}\left|\overrightarrow{PQ} \times \overrightarrow{PR}\right| = \frac{1}{2}\sqrt{1+1} = \frac{\sqrt{2}}{2}$

(b) $\mathbf{u} = \pm\frac{\overrightarrow{PQ} \times \overrightarrow{PR}}{\left|\overrightarrow{PQ} \times \overrightarrow{PR}\right|} = \pm\frac{1}{\sqrt{2}}(-\mathbf{i}+\mathbf{j}) = \pm\frac{1}{\sqrt{2}}(\mathbf{i}-\mathbf{j})$

18. (a) $\overrightarrow{PQ} \times \overrightarrow{PR} = \begin{vmatrix} \mathbf{i} & \mathbf{j} & \mathbf{k} \\ 2 & -1 & -1 \\ 1 & 0 & -2 \end{vmatrix} = 2\mathbf{i} + 3\mathbf{j} + \mathbf{k} \Rightarrow \text{Area} = \frac{1}{2}\left|\overrightarrow{PQ} \times \overrightarrow{PR}\right| = \frac{1}{2}\sqrt{4+9+1} = \frac{\sqrt{14}}{2}$

(b) $\mathbf{u} = \pm\frac{\overrightarrow{PQ} \times \overrightarrow{PR}}{\left|\overrightarrow{PQ} \times \overrightarrow{PR}\right|} = \pm\frac{1}{\sqrt{14}}(2\mathbf{i}+3\mathbf{j}+\mathbf{k})$

19. (a) $\mathbf{A} \cdot \mathbf{B} = -6$, $\mathbf{A} \cdot \mathbf{C} = -81$, $\mathbf{B} \cdot \mathbf{C} = 18 \Rightarrow$ none

(b) $\mathbf{A} \times \mathbf{B} = \begin{vmatrix} \mathbf{i} & \mathbf{j} & \mathbf{k} \\ 5 & -1 & 1 \\ 0 & 1 & -5 \end{vmatrix} \neq \mathbf{0}$, $\mathbf{A} \times \mathbf{C} = \begin{vmatrix} \mathbf{i} & \mathbf{j} & \mathbf{k} \\ 5 & -1 & 1 \\ -15 & 3 & -3 \end{vmatrix} = \mathbf{0}$, $\mathbf{B} \times \mathbf{C} = \begin{vmatrix} \mathbf{i} & \mathbf{j} & \mathbf{k} \\ 0 & 1 & -5 \\ -15 & 3 & -3 \end{vmatrix} \neq \mathbf{0}$

$\Rightarrow$ **A** and **C** are parallel

20. (a) $\mathbf{A} \cdot \mathbf{B} = 0$, $\mathbf{A} \cdot \mathbf{C} = 0$, $\mathbf{A} \cdot \mathbf{D} = -3\pi$, $\mathbf{B} \cdot \mathbf{C} = 0$, $\mathbf{B} \cdot \mathbf{D} = 0$, $\mathbf{C} \cdot \mathbf{D} = 0 \Rightarrow \mathbf{A} \perp \mathbf{B}$, $\mathbf{A} \perp \mathbf{C}$, $\mathbf{B} \perp \mathbf{C}$, $\mathbf{B} \perp \mathbf{D}$ and $\mathbf{C} \perp \mathbf{D}$

(b) $\mathbf{A} \times \mathbf{B} = \begin{vmatrix} \mathbf{i} & \mathbf{j} & \mathbf{k} \\ 1 & 2 & -1 \\ -1 & 1 & 1 \end{vmatrix} \neq \mathbf{0}$, $\mathbf{A} \times \mathbf{C} = \begin{vmatrix} \mathbf{i} & \mathbf{j} & \mathbf{k} \\ 1 & 2 & -1 \\ 1 & 0 & 1 \end{vmatrix} \neq \mathbf{0}$, $\mathbf{A} \times \mathbf{D} = \begin{vmatrix} \mathbf{i} & \mathbf{j} & \mathbf{k} \\ 1 & 2 & -1 \\ -\frac{\pi}{2} & -\pi & \frac{\pi}{2} \end{vmatrix} = \mathbf{0}$

$\mathbf{B} \times \mathbf{C} = \begin{vmatrix} \mathbf{i} & \mathbf{j} & \mathbf{k} \\ -1 & 1 & 1 \\ 1 & 0 & 1 \end{vmatrix} \neq \mathbf{0}$, $\mathbf{B} \times \mathbf{D} = \begin{vmatrix} \mathbf{i} & \mathbf{j} & \mathbf{k} \\ -1 & 1 & 1 \\ -\frac{\pi}{2} & -\pi & \frac{\pi}{2} \end{vmatrix} \neq \mathbf{0}$, $\mathbf{C} \times \mathbf{D} = \begin{vmatrix} \mathbf{i} & \mathbf{j} & \mathbf{k} \\ 1 & 0 & 1 \\ -\frac{\pi}{2} & -\pi & \frac{\pi}{2} \end{vmatrix} \neq \mathbf{0}$

$\Rightarrow$ **A** and **D** are parallel

21. $\left|\overrightarrow{PQ} \times \mathbf{F}\right| = \left|\overrightarrow{PQ}\right|\left|\mathbf{F}\right| \sin(60°) = \frac{2}{3} \cdot 30 \cdot \frac{\sqrt{3}}{2}$ ft · lb $= 10\sqrt{3}$ ft · lb

22. $\left|\overrightarrow{PQ} \times \mathbf{F}\right| = \left|\overrightarrow{PQ}\right|\left|\mathbf{F}\right| \sin(135°) = \frac{2}{3} \cdot 30 \cdot \frac{\sqrt{2}}{2}$ ft · lb $= 10\sqrt{2}$ ft · lb

23. If $\mathbf{A} = a_1\mathbf{i} + a_2\mathbf{j} + a_3\mathbf{k}$, $\mathbf{B} = b_1\mathbf{i} + b_2\mathbf{j} + b_3\mathbf{k}$, and $\mathbf{C} = c_1\mathbf{i} + c_2\mathbf{j} + c_3\mathbf{k}$, then $\mathbf{A}\cdot(\mathbf{B}\times\mathbf{C}) = \begin{vmatrix} a_1 & a_2 & a_3 \\ b_1 & b_2 & b_3 \\ c_1 & c_2 & c_3 \end{vmatrix}$,

$\mathbf{B}\cdot(\mathbf{C}\times\mathbf{A}) = \begin{vmatrix} b_1 & b_2 & b_3 \\ c_1 & c_2 & c_3 \\ a_1 & a_2 & a_3 \end{vmatrix}$ and $\mathbf{C}\cdot(\mathbf{A}\times\mathbf{B}) = \begin{vmatrix} c_1 & c_2 & c_3 \\ a_1 & a_2 & a_3 \\ b_1 & b_2 & b_3 \end{vmatrix}$ which all have the same value, since the

interchanging of two pair of rows in a determinant does not change its value $\Rightarrow$ the volume is

$$|(\mathbf{A}\times\mathbf{B})\cdot\mathbf{C}| = \text{abs}\begin{vmatrix} 2 & 0 & 0 \\ 0 & 2 & 0 \\ 0 & 0 & 2 \end{vmatrix} = 8$$

24. $|(\mathbf{A}\times\mathbf{B})\cdot\mathbf{C}| = \text{abs}\begin{vmatrix} 1 & -1 & 1 \\ 2 & 1 & -2 \\ -1 & 2 & -1 \end{vmatrix} = 4$ (for details about verification, see Exercise 23)

25. $|(\mathbf{A}\times\mathbf{B})\cdot\mathbf{C}| = \text{abs}\begin{vmatrix} 2 & 1 & 0 \\ 2 & -1 & 1 \\ 1 & 0 & 2 \end{vmatrix} = |-7| = 7$ (for details about verification, see Exercise 23)

26. $|(\mathbf{A}\times\mathbf{B})\cdot\mathbf{C}| = \text{abs}\begin{vmatrix} 1 & 1 & -2 \\ -1 & 0 & -1 \\ 2 & 4 & -2 \end{vmatrix} = 8$ (for details about verification, see Exercise 23)

27. (a) true, $|\mathbf{A}| = \sqrt{a_1^2 + a_2^2 + a_3^2} = \sqrt{\mathbf{A}\cdot\mathbf{A}}$

(b) not always true, $\mathbf{A}\cdot\mathbf{A} = |\mathbf{A}|^2$

(c) true, $\mathbf{A}\times\mathbf{0} = \begin{vmatrix} \mathbf{i} & \mathbf{j} & \mathbf{k} \\ a_1 & a_2 & a_3 \\ 0 & 0 & 0 \end{vmatrix} = 0\mathbf{i} + 0\mathbf{j} + 0\mathbf{k} = \mathbf{0}$

(d) true, $\mathbf{A}\times(-\mathbf{A}) = \begin{vmatrix} \mathbf{i} & \mathbf{j} & \mathbf{k} \\ a_1 & a_2 & a_3 \\ -a_1 & -a_2 & -a_3 \end{vmatrix} = (-a_2a_3 + a_2a_3)\mathbf{i} + (-a_1a_3 + a_1a_3)\mathbf{j} + (-a_1a_2 + a_1a_2)\mathbf{k} = \mathbf{0}$

(e) not always true, $\mathbf{i}\times\mathbf{j} = \mathbf{k} \neq -\mathbf{k} = \mathbf{j}\times\mathbf{i}$ for example

(f) true, Eqn. (6)

(g) true, $(\mathbf{A}\times\mathbf{B})\cdot\mathbf{B} = \mathbf{A}\cdot(\mathbf{B}\times\mathbf{B}) = \mathbf{A}\cdot\mathbf{0} = 0$

(h) true, Eqn. (13)

28. (a) true, $\mathbf{A}\cdot\mathbf{B} = a_1b_1 + a_2b_2 + a_3b_3 = b_1a_1 + b_2a_2 + b_3a_3 = \mathbf{B}\cdot\mathbf{A}$

(b) true, $\mathbf{A}\times\mathbf{B} = \begin{vmatrix} \mathbf{i} & \mathbf{j} & \mathbf{k} \\ a_1 & a_2 & a_3 \\ b_1 & b_2 & b_3 \end{vmatrix} = -\begin{vmatrix} \mathbf{i} & \mathbf{j} & \mathbf{k} \\ b_1 & b_2 & b_3 \\ a_1 & a_2 & a_3 \end{vmatrix} = -(\mathbf{B}\times\mathbf{A})$

(c) true, $(-\mathbf{A})\times\mathbf{B} = \begin{vmatrix} \mathbf{i} & \mathbf{j} & \mathbf{k} \\ -a_1 & -a_2 & -a_3 \\ b_1 & b_2 & b_3 \end{vmatrix} = -\begin{vmatrix} \mathbf{i} & \mathbf{j} & \mathbf{k} \\ a_1 & a_2 & a_3 \\ b_1 & b_2 & b_3 \end{vmatrix} = -(\mathbf{A}\times\mathbf{B})$

(d) true, $(c\mathbf{A})\cdot\mathbf{B} = (ca_1)b_1 + (ca_2)b_2 + (ca_3)b_3 = a_1(cb_1) + a_2(cb_2) + a_3(cb_3) = \mathbf{A}\cdot(c\mathbf{B}) = c(a_1b_1 + a_2b_2 + a_3b_3)$
$= c(\mathbf{A}\cdot\mathbf{B})$

(e) true, $c(\mathbf{A}\times\mathbf{B}) = c\begin{vmatrix} \mathbf{i} & \mathbf{j} & \mathbf{k} \\ a_1 & a_2 & a_3 \\ b_1 & b_2 & b_3 \end{vmatrix} = \begin{vmatrix} \mathbf{i} & \mathbf{j} & \mathbf{k} \\ ca_1 & ca_2 & ca_3 \\ b_1 & b_2 & b_3 \end{vmatrix} = (c\mathbf{A})\times\mathbf{B} = \begin{vmatrix} \mathbf{i} & \mathbf{j} & \mathbf{k} \\ a_1 & a_2 & a_3 \\ cb_1 & cb_2 & cb_3 \end{vmatrix} = \mathbf{A}\times(c\mathbf{B})$

(f) true, $\mathbf{A}\cdot\mathbf{A} = a_1^2 + a_2^2 + a_3^2 = \left(\sqrt{a_1^2 + a_2^2 + a_3^2}\right)^2 = |\mathbf{A}|^2$

(g) true, $(\mathbf{A}\times\mathbf{A})\cdot\mathbf{A} = \mathbf{0}\cdot\mathbf{A} = 0$

(h) true, $\mathbf{A}\times\mathbf{B}\perp\mathbf{A}$ and $\mathbf{A}\times\mathbf{B}\perp\mathbf{B} \Rightarrow (\mathbf{A}\times\mathbf{B})\cdot\mathbf{A} = \mathbf{B}\cdot(\mathbf{A}\times\mathbf{B}) = 0$

29. (a) $\text{proj}_{\mathbf{B}}\,\mathbf{A} = \left(\frac{\mathbf{A}\cdot\mathbf{B}}{\mathbf{B}\cdot\mathbf{B}}\right)\mathbf{B}$ (b) $\pm(\mathbf{A}\times\mathbf{B})$ (c) $\pm(\mathbf{A}\times\mathbf{B})\times\mathbf{C}$ (d) $|(\mathbf{A}\times\mathbf{B})\cdot\mathbf{C}|$

30. (a) $(\mathbf{A}\times\mathbf{B})\times(\mathbf{A}\times\mathbf{C})$

(b) $(\mathbf{A}+\mathbf{B})\times(\mathbf{A}-\mathbf{B}) = (\mathbf{A}+\mathbf{B})\times\mathbf{A} - (\mathbf{A}+\mathbf{B})\times\mathbf{B} = \mathbf{A}\times\mathbf{A} + \mathbf{B}\times\mathbf{A} - \mathbf{A}\times\mathbf{B} - \mathbf{B}\times\mathbf{B}$
$= \mathbf{0} + \mathbf{B}\times\mathbf{A} - \mathbf{A}\times\mathbf{B} - \mathbf{0} = 2(\mathbf{B}\times\mathbf{A})$, or simply $\mathbf{A}\times\mathbf{B}$

(c) $|\mathbf{A}|\frac{\mathbf{B}}{|\mathbf{B}|}$

(d) $|\mathbf{A}\times\mathbf{C}|$

31. (a) yes, $\mathbf{A}\times\mathbf{B}$ and $\mathbf{C}$ are both vectors
(b) no, $\mathbf{A}$ is a vector but $\mathbf{B}\cdot\mathbf{C}$ is a scalar
(c) yes, $\mathbf{A}$ and $\mathbf{A}\times\mathbf{C}$ are both vectors
(d) no, $\mathbf{A}$ is a vector but $\mathbf{B}\cdot\mathbf{C}$ is a scalar

32. $(\mathbf{A}\times\mathbf{B})\times\mathbf{C}$ is perpendicular to $\mathbf{A}\times\mathbf{B}$, and $\mathbf{A}\times\mathbf{B}$ is perpendicular to both $\mathbf{A}$ and $\mathbf{B} \Rightarrow (\mathbf{A}\times\mathbf{B})\times\mathbf{C}$ is parallel to a vector in the plane of $\mathbf{A}$ and $\mathbf{B}$ which means it lies in the plane determined by $\mathbf{A}$ and $\mathbf{B}$. The situation is degenerate if $\mathbf{A}$ and $\mathbf{B}$ are parallel so $\mathbf{A}\times\mathbf{B} = \mathbf{0}$ and the vectors do not determine a plane. Similar reasoning shows that $\mathbf{A}\times(\mathbf{B}\times\mathbf{C})$ lies in the plane of $\mathbf{B}$ and $\mathbf{C}$ provided $\mathbf{B}$ and $\mathbf{C}$ are nonparallel.

33. No, $\mathbf{B}$ need not equal $\mathbf{C}$. For example, $\mathbf{i}+\mathbf{j} \neq -\mathbf{i}+\mathbf{j}$, but $\mathbf{i}\times(\mathbf{i}+\mathbf{j}) = \mathbf{i}\times\mathbf{i} + \mathbf{i}\times\mathbf{j} = \mathbf{0} + \mathbf{k} = \mathbf{k}$ and $\mathbf{i}\times(-\mathbf{i}+\mathbf{j}) = -\mathbf{i}\times\mathbf{i} + \mathbf{i}\times\mathbf{j} = \mathbf{0} + \mathbf{k} = \mathbf{k}$.

34. Yes. If $\mathbf{A} \times \mathbf{B} = \mathbf{A} \times \mathbf{C}$ and $\mathbf{A} \cdot \mathbf{B} = \mathbf{A} \cdot \mathbf{C}$, then $\mathbf{A} \times (\mathbf{B} - \mathbf{C}) = \mathbf{0}$ and $\mathbf{A} \cdot (\mathbf{B} - \mathbf{C}) = 0$. Suppose now that $\mathbf{B} \neq \mathbf{C}$. Then $\mathbf{A} \times (\mathbf{B} - \mathbf{C}) = \mathbf{0}$ implies that $\mathbf{B} - \mathbf{C} = \mathrm{k}\mathbf{A}$ for some real number $\mathrm{k} \neq 0$. This in turn implies that $\mathbf{A} \cdot (\mathbf{B} - \mathbf{C}) = \mathbf{A} \cdot (\mathrm{k}\mathbf{A}) = \mathrm{k}\,|\mathbf{A}|^2 = 0$, which implies that $\mathbf{A} = \mathbf{0}$. Since $\mathbf{A} \neq \mathbf{0}$, it cannot be true that $\mathbf{B} \neq \mathbf{C}$, so $\mathbf{B} = \mathbf{C}$.

35. $\overrightarrow{AB} = -\mathbf{i} + \mathbf{j}$ and $\overrightarrow{AD} = -\mathbf{i} - \mathbf{j} \Rightarrow \overrightarrow{AB} \times \overrightarrow{AD} = \begin{vmatrix} \mathbf{i} & \mathbf{j} & \mathbf{k} \\ -1 & 1 & 0 \\ -1 & -1 & 0 \end{vmatrix} = 2\mathbf{k} \Rightarrow \text{area} = \left|\overrightarrow{AB} \times \overrightarrow{AD}\right| = 2$

36. $\overrightarrow{AB} = 7\mathbf{i} + 3\mathbf{j}$ and $\overrightarrow{AD} = 2\mathbf{i} + 5\mathbf{j} \Rightarrow \overrightarrow{AB} \times \overrightarrow{AD} = \begin{vmatrix} \mathbf{i} & \mathbf{j} & \mathbf{k} \\ 7 & 3 & 0 \\ 2 & 5 & 0 \end{vmatrix} = 29\mathbf{k} \Rightarrow \text{area} = \left|\overrightarrow{AB} \times \overrightarrow{AD}\right| = 29$

37. $\overrightarrow{AB} = 3\mathbf{i} - 2\mathbf{j}$ and $\overrightarrow{AD} = 5\mathbf{i} + \mathbf{j} \Rightarrow \overrightarrow{AB} \times \overrightarrow{AD} = \begin{vmatrix} \mathbf{i} & \mathbf{j} & \mathbf{k} \\ 3 & -2 & 0 \\ 5 & 1 & 0 \end{vmatrix} = 13\mathbf{k} \Rightarrow \text{area} = \left|\overrightarrow{AB} \times \overrightarrow{AD}\right| = 13$

38. $\overrightarrow{AB} = 7\mathbf{i} - 4\mathbf{j}$ and $\overrightarrow{AD} = 2\mathbf{i} + 5\mathbf{j} \Rightarrow \overrightarrow{AB} \times \overrightarrow{AD} = \begin{vmatrix} \mathbf{i} & \mathbf{j} & \mathbf{k} \\ 7 & -4 & 0 \\ 2 & 5 & 0 \end{vmatrix} = 43\mathbf{k} \Rightarrow \text{area} = \left|\overrightarrow{AB} \times \overrightarrow{AD}\right| = 43$

39. $\overrightarrow{AB} = -2\mathbf{i} + 3\mathbf{j}$ and $\overrightarrow{AC} = 3\mathbf{i} + \mathbf{j} \Rightarrow \overrightarrow{AB} \times \overrightarrow{AC} = \begin{vmatrix} \mathbf{i} & \mathbf{j} & \mathbf{k} \\ -2 & 3 & 0 \\ 3 & 1 & 0 \end{vmatrix} = -11\mathbf{k} \Rightarrow \text{area} = \frac{1}{2}\left|\overrightarrow{AB} \times \overrightarrow{AC}\right| = \frac{11}{2}$

40. $\overrightarrow{AB} = 4\mathbf{i} + 4\mathbf{j}$ and $\overrightarrow{AC} = 3\mathbf{i} + 2\mathbf{j} \Rightarrow \overrightarrow{AB} \times \overrightarrow{AC} = \begin{vmatrix} \mathbf{i} & \mathbf{j} & \mathbf{k} \\ 4 & 4 & 0 \\ 3 & 2 & 0 \end{vmatrix} = -4\mathbf{k} \Rightarrow \text{area} = \frac{1}{2}\left|\overrightarrow{AB} \times \overrightarrow{AC}\right| = 2$

41. $\overrightarrow{AB} = 6\mathbf{i} - 5\mathbf{j}$ and $\overrightarrow{AC} = 11\mathbf{i} - 5\mathbf{j} \Rightarrow \overrightarrow{AB} \times \overrightarrow{AC} = \begin{vmatrix} \mathbf{i} & \mathbf{j} & \mathbf{k} \\ 6 & -5 & 0 \\ 11 & -5 & 0 \end{vmatrix} = 25\mathbf{k} \Rightarrow \text{area} = \frac{1}{2}\left|\overrightarrow{AB} \times \overrightarrow{AC}\right| = \frac{25}{2}$

42. $\overrightarrow{AB} = 16\mathbf{i} - 5\mathbf{j}$ and $\overrightarrow{AC} = 4\mathbf{i} + 4\mathbf{j} \Rightarrow \overrightarrow{AB} \times \overrightarrow{AC} = \begin{vmatrix} \mathbf{i} & \mathbf{j} & \mathbf{k} \\ 16 & -5 & 0 \\ 4 & 4 & 0 \end{vmatrix} = 84\mathbf{k} \Rightarrow \text{area} = \frac{1}{2}\left|\overrightarrow{AB} \times \overrightarrow{AC}\right| = 42$

43. If $\mathbf{A} = a_1\mathbf{i} + a_2\mathbf{j}$ and $\mathbf{B} = b_1\mathbf{i} + b_2\mathbf{j}$, then $\mathbf{A} \times \mathbf{B} = \begin{vmatrix} \mathbf{i} & \mathbf{j} & \mathbf{k} \\ a_1 & a_2 & 0 \\ b_1 & b_2 & 0 \end{vmatrix} = \begin{vmatrix} a_1 & a_2 \\ b_1 & b_2 \end{vmatrix} \mathbf{k}$ and the triangle's area is $\frac{1}{2}|\mathbf{A} \times \mathbf{B}| = \pm\frac{1}{2}\begin{vmatrix} a_1 & a_2 \\ b_1 & b_2 \end{vmatrix}$. The applicable sign is (+) if the acute angle from **A** to **B** runs counterclockwise in the xy-plane, and (−) if it runs clockwise, because the area must be a nonnegative number.

44. If $\mathbf{A} = a_1\mathbf{i} + a_2\mathbf{j}$, $\mathbf{B} = b_1\mathbf{i} + b_2\mathbf{j}$, and $\mathbf{C} = c_1\mathbf{i} + c_2\mathbf{j}$, then the area of the triangle is $\frac{1}{2}|\overrightarrow{AB} \times \overrightarrow{AC}|$. Now,

$$\overrightarrow{AB} \times \overrightarrow{AC} = \begin{vmatrix} \mathbf{i} & \mathbf{j} & \mathbf{k} \\ b_1 - a_1 & b_2 - a_2 & 0 \\ c_1 - a_1 & c_2 - a_2 & 0 \end{vmatrix} = \begin{vmatrix} b_1 - a_1 & b_2 - a_2 \\ c_1 - a_1 & c_2 - a_2 \end{vmatrix} \mathbf{k} \Rightarrow \frac{1}{2}|\overrightarrow{AB} \times \overrightarrow{AC}|$$

$$= \frac{1}{2}|(b_1 - a_1)(c_2 - a_2) - (c_1 - a_1)(b_2 - a_2)| = \frac{1}{2}|a_1(b_2 - c_2) + a_2(c_1 - b_1) + (b_1c_2 - c_1b_2)|$$

$$= \pm\frac{1}{2}\begin{vmatrix} a_1 & a_2 & 1 \\ b_1 & b_2 & 1 \\ c_1 & c_2 & 1 \end{vmatrix}.$$ The applicable sign ensures the area formula gives a nonnegative number.

10.5 LINES AND PLANES IN SPACE

1. The direction $\mathbf{i} + \mathbf{j} + \mathbf{k}$ and $P(3, -4, -1) \Rightarrow x = 3 + t,\ y = -4 + t,\ z = -1 + t$

2. The direction $\overrightarrow{PQ} = -2\mathbf{i} - 2\mathbf{j} + 2\mathbf{k}$ and $P(1, 2, -1) \Rightarrow x = 1 - 2t,\ y = 2 - 2t,\ z = -1 + 2t$

3. The direction $\overrightarrow{PQ} = 5\mathbf{i} + 5\mathbf{j} - 5\mathbf{k}$ and $P(-2, 0, 3) \Rightarrow x = -2 + 5t,\ y = 5t,\ z = 3 - 5t$

4. The direction $\overrightarrow{PQ} = -\mathbf{j} - \mathbf{k}$ and $P(1, 2, 0) \Rightarrow x = 1,\ y = 2 - t,\ z = -t$

5. The direction $2\mathbf{j} + \mathbf{k}$ and $P(0, 0, 0) \Rightarrow x = 0,\ y = 2t,\ z = t$

6. The direction $2\mathbf{i} - \mathbf{j} + 3\mathbf{k}$ and $P(3, -2, 1) \Rightarrow x = 3 + 2t,\ y = -2 - t,\ z = 1 + 3t$

7. The direction $\mathbf{k}$ and $P(1, 1, 1) \Rightarrow x = 1,\ y = 1,\ z = 1 + t$

8. The direction $3\mathbf{i} + 7\mathbf{j} - 5\mathbf{k}$ and $P(2, 4, 5) \Rightarrow x = 2 + 3t,\ y = 4 + 7t,\ z = 5 - 5t$

9. The direction $\mathbf{i} + 2\mathbf{j} + 2\mathbf{k}$ and $P(0, -7, 0) \Rightarrow x = t,\ y = -7 + 2t,\ z = 2t$

10. The direction is $\mathbf{A} \times \mathbf{B} = \begin{vmatrix} \mathbf{i} & \mathbf{j} & \mathbf{k} \\ 1 & 2 & 3 \\ 3 & 4 & 5 \end{vmatrix} = -2\mathbf{i} + 4\mathbf{j} - 2\mathbf{k}$ and $P(2, 3, 0) \Rightarrow x = 2 - 2t,\ y = 3 + 4t,\ z = -2t$

11. The direction $\mathbf{i}$ and $P(0, 0, 0) \Rightarrow x = t,\ y = 0,\ z = 0$

12. The direction $\mathbf{k}$ and $P(0,0,0) \Rightarrow x = 0,\ y = 0,\ z = t$

13. The direction $\overrightarrow{PQ} = \mathbf{i} + \mathbf{j} + \frac{3}{2}\mathbf{k}$ and $P(0,0,0) \Rightarrow x = t,\ y = t,\ z = \frac{3}{2}t$, where $0 \le t \le 1$

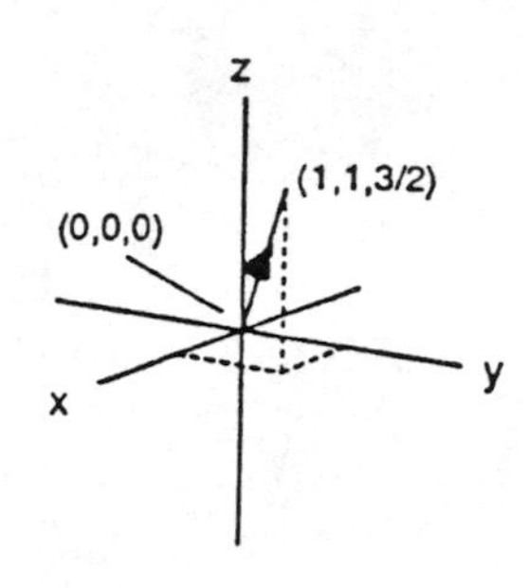

14. The direction $\overrightarrow{PQ} = \mathbf{i}$ and $P(0,0,0) \Rightarrow x = t,\ y = 0,\ z = 0$, where $0 \le t \le 1$

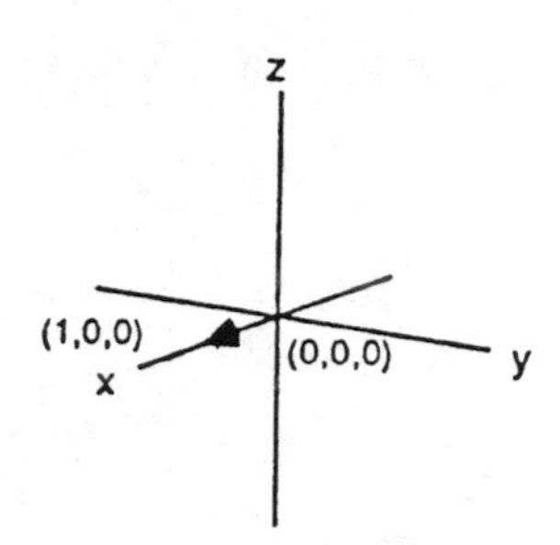

15. The direction $\overrightarrow{PQ} = \mathbf{j}$ and $P(1,1,0) \Rightarrow x = 1,\ y = 1 + t,\ z = 0$, where $-1 \le t \le 0$

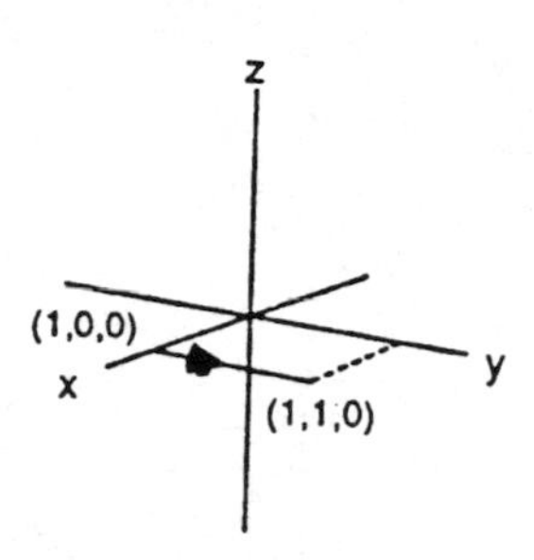

16. The direction $\overrightarrow{PQ} = \mathbf{k}$ and $P(1,1,0) \Rightarrow x = 1,\ y = 1,\ z = t$, where $0 \le t \le 1$

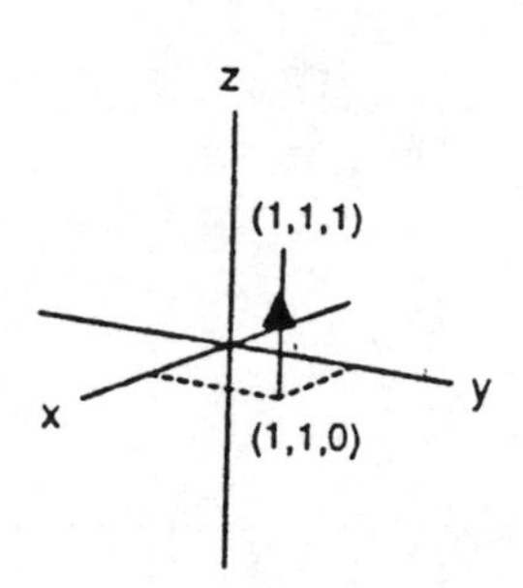

17. The direction $\overrightarrow{PQ} = -2\mathbf{j}$ and $P(0,1,1) \Rightarrow x = 0,\ y = 1 - 2t,\ z = 1$, where $0 \le t \le 1$

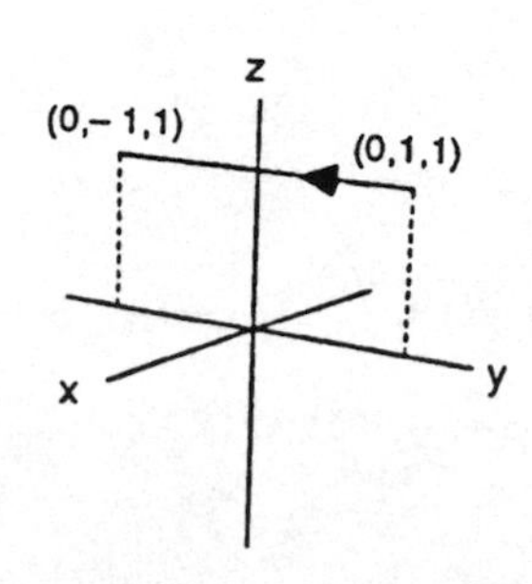

18. The direction $\overrightarrow{PQ} = 3\mathbf{i} - 2\mathbf{j}$ and $P(0,2,0) \Rightarrow x = 3t,\ y = 2 - 2t,$ $z = 0$, where $0 \le t \le 1$

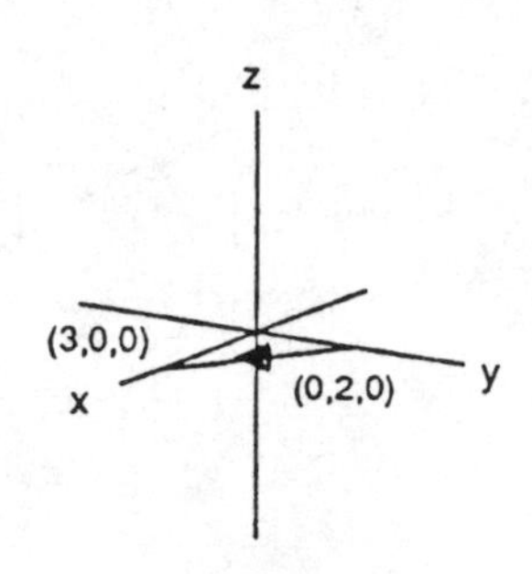

19. The direction $\overrightarrow{PQ} = -2\mathbf{i} + 2\mathbf{j} - 2\mathbf{k}$ and $P(2,0,2) \Rightarrow x = 2 - 2t,$ $y = 2t,\ z = 2 - 2t$, where $0 \le t \le 1$

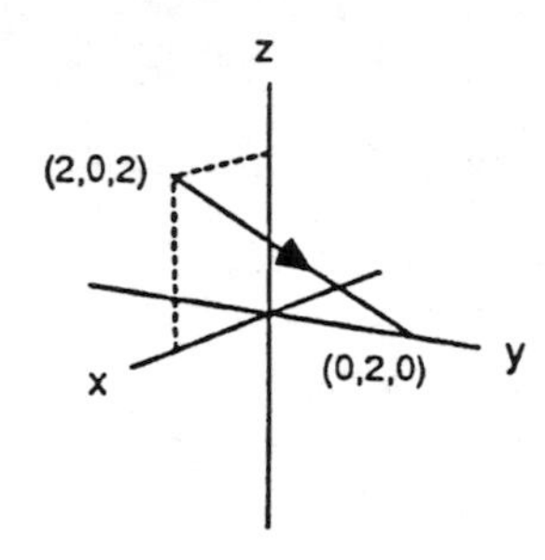

20. The direction $\overrightarrow{PQ} = -\mathbf{i} + 3\mathbf{j} + \mathbf{k}$ and $P(1,0,-1) \Rightarrow x = 1 - t,$ $y = 3t,\ z = -1 + t$, where $0 \le t \le 1$

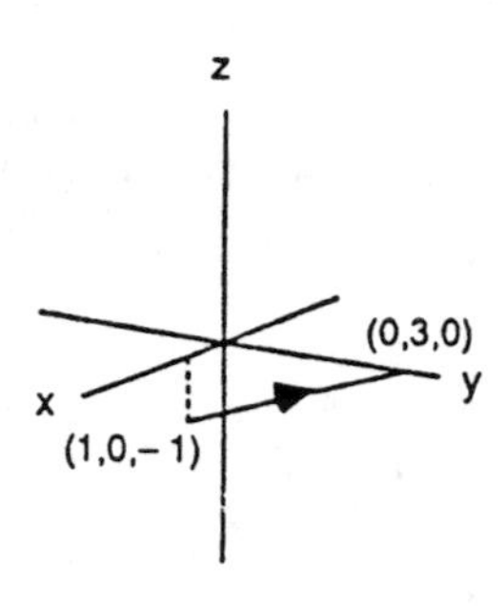

21. $3(x - 0) + (-2)(y - 2) + (-1)(z + 1) = 0 \Rightarrow 3x - 2y - z = -3$

22. $3(x - 1) + (1)(y + 1) + (1)(z - 3) = 0 \Rightarrow 3x + y + z = 5$

23. $\overrightarrow{PQ} = \mathbf{i} - \mathbf{j} + 3\mathbf{k},\ \overrightarrow{PS} = -\mathbf{i} - 3\mathbf{j} + 2\mathbf{k} \Rightarrow \overrightarrow{PQ} \times \overrightarrow{PS} = \begin{vmatrix} \mathbf{i} & \mathbf{j} & \mathbf{k} \\ 1 & -1 & 3 \\ -1 & -3 & 2 \end{vmatrix} = 7\mathbf{i} - 5\mathbf{j} - 4\mathbf{k}$ is normal to the plane

$\Rightarrow 7(x - 2) + (-5)(y - 0) + (-4)(z - 2) = 0 \Rightarrow 7x - 5y - 4z = 6$

24. $\overrightarrow{PQ} = -\mathbf{i} + \mathbf{j} + 2\mathbf{k},\ \overrightarrow{PS} = -3\mathbf{i} + 2\mathbf{j} + 3\mathbf{k} \Rightarrow \overrightarrow{PQ} \times \overrightarrow{PS} = \begin{vmatrix} \mathbf{i} & \mathbf{j} & \mathbf{k} \\ -1 & 1 & 2 \\ -3 & 2 & 3 \end{vmatrix} = -\mathbf{i} - 3\mathbf{j} + \mathbf{k}$ is normal to the plane

$\Rightarrow (-1)(x - 1) + (-3)(y - 5) + (1)(z - 7) = 0 \Rightarrow x + 3y - z = 9$

25. $\mathbf{n} = \mathbf{i} + 3\mathbf{j} + 4\mathbf{k},\ P(2,4,5) = (1)(x - 2) + (3)(y - 4) + (4)(z - 5) = 0 \Rightarrow x + 3y + 4z = 34$

26. $\mathbf{n} = \mathbf{i} - 2\mathbf{j} + \mathbf{k}$, $P(1,-2,1) = (1)(x-1) + (-2)(y+2) + (1)(z-1) = 0 \Rightarrow x - 2y + z = 6$

27. $\begin{cases} x = 2t+1 = s+2 \\ y = 3t+2 = 2s+4 \end{cases} \Rightarrow \begin{cases} 2t - s = 1 \\ 3t - 2s = 2 \end{cases} \Rightarrow \begin{cases} 4t - 2s = 2 \\ 3t - 2s = 2 \end{cases} \Rightarrow t = 0$ and $s = -1$; then $z = 4t + 3 = -4s - 1$

$\Rightarrow 4(0) + 3 = (-4)(-1) - 1$ is satisfied $\Rightarrow$ the lines do intersect when $t = 0$ and $s = -1 \Rightarrow$ the point of intersection is $x = 1$, $y = 2$, and $z = 3$ or $P(1,2,3)$. A vector normal to the plane determined by these lines is

$\mathbf{n}_1 \times \mathbf{n}_2 = \begin{vmatrix} \mathbf{i} & \mathbf{j} & \mathbf{k} \\ 2 & 3 & 4 \\ 1 & 2 & -4 \end{vmatrix} = -20\mathbf{i} + 12\mathbf{j} + \mathbf{k}$, where $\mathbf{n}_1$ and $\mathbf{n}_2$ are directions of the lines $\Rightarrow$ the plane

containing the lines is represented by$(-20)(x-1) + (12)(y-2) + (1)(z-3) = 0 \Rightarrow -20x + 12y + z = 7$.

28. $\begin{cases} x = t = 2s+2 \\ y = -t+2 = s+3 \end{cases} \Rightarrow \begin{cases} t - 2s = 2 \\ -t - s = 1 \end{cases} \Rightarrow s = -1$ and $t = 0$; then $z = t + 1 = 5s + 6 \Rightarrow 0 + 1 = 5(-1) + 6$

is satisfied $\Rightarrow$ the lines do intersect when $s = -1$ and $t = 0 \Rightarrow$ the point of intersection is $x = 0$, $y = 2$ and $z = 1$

or $P(0,2,1)$. A vector normal to the plane determined by these lines is $\mathbf{n}_1 \times \mathbf{n}_2 = \begin{vmatrix} \mathbf{i} & \mathbf{j} & \mathbf{k} \\ 1 & -1 & 1 \\ 2 & 1 & 5 \end{vmatrix}$

$= -6\mathbf{i} - 3\mathbf{j} + 3\mathbf{k}$, where $\mathbf{n}_1$ and $\mathbf{n}_2$ are directions of the lines $\Rightarrow$ the plane containing the lines is represented by $(-6)(x-0) + (-3)(y-2) + (3)(z-1) = 0 \Rightarrow 6x + 3y - 3z = 3$.

29. The cross product of $\mathbf{i} + \mathbf{j} - \mathbf{k}$ and $-4\mathbf{i} + 2\mathbf{j} - 2\mathbf{k}$ has the same direction as the normal to the plane

$\Rightarrow \mathbf{n} = \begin{vmatrix} \mathbf{i} & \mathbf{j} & \mathbf{k} \\ 1 & 1 & -1 \\ -4 & 2 & -2 \end{vmatrix} = 6\mathbf{j} + 6\mathbf{k}$. Select a point on either line, such as $P(-1,2,1)$. Since the lines are given

to intersect, the desired plane is $0(x+1) + 6(y-2) + 6(z-1) = 0 \Rightarrow 6y + 6z = 18 \Rightarrow y + z = 3$.

30. The cross product of $\mathbf{i} - 3\mathbf{j} - \mathbf{k}$ and $\mathbf{i} + \mathbf{j} + \mathbf{k}$ has the same direction as the normal to the plane

$\mathbf{n} = \begin{vmatrix} \mathbf{i} & \mathbf{j} & \mathbf{k} \\ 1 & -3 & -1 \\ 1 & 1 & 1 \end{vmatrix} = -2\mathbf{i} - 2\mathbf{j} + 4\mathbf{k}$. Select a point on either line, such as $P(0,3,-2)$. Since the lines are

given to intersect, the desired plane is $(-2)(x-0) + (-2)(y-3) + (4)(z+2) = 0 \Rightarrow -2x - 2y + 4z = -14$ $\Rightarrow x + y - 2z = 7$.

31. $\mathbf{n}_1 \times \mathbf{n}_2 = \begin{vmatrix} \mathbf{i} & \mathbf{j} & \mathbf{k} \\ 2 & 1 & -1 \\ 1 & 2 & 1 \end{vmatrix} = 3\mathbf{i} - 3\mathbf{j} + 3\mathbf{k}$ is a vector in the direction of the line of intersection of the planes $\Rightarrow 3(x-2) + (-3)(y-1) + 3(z+1) = 0 \Rightarrow 3x - 3y + 3z = 0 \Rightarrow x - y + z = 0$ is the desired plane containing $P_0(2, 1, -1)$

32. A vector normal to the desired plane is $\overrightarrow{P_1P_2} \times \mathbf{n} = \begin{vmatrix} \mathbf{i} & \mathbf{j} & \mathbf{k} \\ 2 & 0 & -2 \\ 4 & -1 & 2 \end{vmatrix} = -2\mathbf{i} - 12\mathbf{j} - 2\mathbf{k}$; choosing $P_1(1, 2, 3)$ as a point on the plane $\Rightarrow (-2)(x-1) + (-12)(y-2) + (-2)(z-3) = 0 \Rightarrow -2x - 12y - 2z = -32 \Rightarrow x + 6y + z = 16$ is the desired plane

33. $S(0, 0, 12)$, $P(0, 0, 0)$ and $\mathbf{v} = 4\mathbf{i} - 2\mathbf{j} + 2\mathbf{k} \Rightarrow \overrightarrow{PS} \times \mathbf{v} = \begin{vmatrix} \mathbf{i} & \mathbf{j} & \mathbf{k} \\ 0 & 0 & 12 \\ 4 & -2 & 2 \end{vmatrix} = 24\mathbf{i} + 48\mathbf{j} = 24(\mathbf{i} + 2\mathbf{j})$

$\Rightarrow d = \frac{|\overrightarrow{PS} \times \mathbf{v}|}{|\mathbf{v}|} = \frac{24\sqrt{1+4}}{\sqrt{16+4+4}} = \frac{24\sqrt{5}}{\sqrt{24}} = \sqrt{5 \cdot 24} = 2\sqrt{30}$ is the distance from S to the line

34. $S(0, 0, 0)$, $P(5, 5, -3)$ and $\mathbf{v} = 3\mathbf{i} + 4\mathbf{j} - 5\mathbf{k} \Rightarrow \overrightarrow{PS} \times \mathbf{v} = \begin{vmatrix} \mathbf{i} & \mathbf{j} & \mathbf{k} \\ -5 & -5 & 3 \\ 3 & 4 & -5 \end{vmatrix} = 13\mathbf{i} + 16\mathbf{j} - 5\mathbf{k}$

$\Rightarrow d = \frac{|\overrightarrow{PS} \times \mathbf{v}|}{|\mathbf{v}|} = \frac{\sqrt{169+256+25}}{\sqrt{9+16+25}} = \frac{\sqrt{450}}{\sqrt{50}} = \sqrt{9} = 3$ is the distance from S to the line

35. $S(2, 1, 3)$, $P(2, 1, 3)$ and $\mathbf{v} = 2\mathbf{i} + 6\mathbf{j} \Rightarrow \overrightarrow{PS} \times \mathbf{v} = \mathbf{0} \Rightarrow d = \frac{|\overrightarrow{PS} \times \mathbf{v}|}{|\mathbf{v}|} = \frac{0}{\sqrt{40}} = 0$ is the distance from S to the line (i.e., the point S lies on the line)

36. $S(2, 1, -1)$, $P(0, 1, 0)$ and $\mathbf{v} = 2\mathbf{i} + 2\mathbf{j} + 2\mathbf{k} \Rightarrow \overrightarrow{PS} \times \mathbf{v} = \begin{vmatrix} \mathbf{i} & \mathbf{j} & \mathbf{k} \\ 2 & 0 & -1 \\ 2 & 2 & 2 \end{vmatrix} = 2\mathbf{i} - 6\mathbf{j} + 4\mathbf{k}$

$\Rightarrow d = \frac{|\overrightarrow{PS} \times \mathbf{v}|}{|\mathbf{v}|} = \frac{\sqrt{4+36+16}}{\sqrt{4+4+4}} = \frac{\sqrt{56}}{\sqrt{12}} = \sqrt{\frac{14}{3}}$ is the distance from S to the line

37. $S(3, -1, 4)$, $P(4, 3, -5)$ and $\mathbf{v} = -\mathbf{i} + 2\mathbf{j} + 3\mathbf{k} \Rightarrow \overrightarrow{PS} \times \mathbf{v} = \begin{vmatrix} \mathbf{i} & \mathbf{j} & \mathbf{k} \\ -1 & -4 & 9 \\ -1 & 2 & 3 \end{vmatrix} = -30\mathbf{i} - 6\mathbf{j} - 6\mathbf{k}$

$\Rightarrow d = \frac{|\overrightarrow{PS} \times \mathbf{v}|}{|\mathbf{v}|} = \frac{\sqrt{900+36+36}}{\sqrt{1+4+9}} = \frac{\sqrt{972}}{\sqrt{14}} = \frac{\sqrt{486}}{\sqrt{7}} = \frac{\sqrt{81 \cdot 6}}{\sqrt{7}} = \frac{9\sqrt{42}}{7}$ is the distance from S to the line

38. S(−1, 4, 3), P(10, −3, 0) and $\mathbf{v} = 4\mathbf{i} + 4\mathbf{k} \Rightarrow \overrightarrow{PS} \times \mathbf{v} = \begin{vmatrix} \mathbf{i} & \mathbf{j} & \mathbf{k} \\ -11 & 7 & 3 \\ 4 & 0 & 4 \end{vmatrix} = 28\mathbf{i} + 56\mathbf{j} - 28\mathbf{k} = 28(\mathbf{i} + 2\mathbf{j} - \mathbf{k})$

$\Rightarrow d = \frac{|\overrightarrow{PS} \times \mathbf{v}|}{|\mathbf{v}|} = \frac{28\sqrt{1+4+1}}{4\sqrt{1+1}} = 7\sqrt{3}$ is the distance from S to the line

39. S(2, −3, 4), $x + 2y + 2z = 13$ and P(13, 0, 0) is on the plane $\Rightarrow \overrightarrow{PS} = -11\mathbf{i} - 3\mathbf{j} + 4\mathbf{k}$ and $\mathbf{n} = \mathbf{i} + 2\mathbf{j} + 2\mathbf{k}$

$\Rightarrow d = \left|\overrightarrow{PS} \cdot \frac{\mathbf{n}}{|\mathbf{n}|}\right| = \left|\frac{-11-6+8}{\sqrt{1+4+4}}\right| = \left|\frac{-9}{\sqrt{9}}\right| = 3$

40. S(0, 0, 0), $3x + 2y + 6z = 6$ and P(2, 0, 0) is on the plane $\Rightarrow \overrightarrow{PS} = -2\mathbf{i}$ and $\mathbf{n} = 3\mathbf{i} + 2\mathbf{j} + 6\mathbf{k}$

$\Rightarrow d = \left|\overrightarrow{PS} \cdot \frac{\mathbf{n}}{|\mathbf{n}|}\right| = \left|\frac{-6}{\sqrt{9+4+36}}\right| = \frac{6}{\sqrt{49}} = \frac{6}{7}$

41. S(0, 1, 1), $4y + 3z = -12$ and P(0, −3, 0) is on the plane $\Rightarrow \overrightarrow{PS} = 4\mathbf{j} + \mathbf{k}$ and $\mathbf{n} = 4\mathbf{j} + 3\mathbf{k}$

$\Rightarrow d = \left|\overrightarrow{PS} \cdot \frac{\mathbf{n}}{|\mathbf{n}|}\right| = \left|\frac{16+3}{\sqrt{16+9}}\right| = \frac{19}{5}$

42. S(2, 2, 3), $2x + y + 2z = 4$ and P(2, 0, 0) is on the plane $\Rightarrow \overrightarrow{PS} = 2\mathbf{j} + 3\mathbf{k}$ and $\mathbf{n} = 2\mathbf{i} + \mathbf{j} + 2\mathbf{k}$

$\Rightarrow d = \left|\overrightarrow{PS} \cdot \frac{\mathbf{n}}{|\mathbf{n}|}\right| = \left|\frac{2+6}{\sqrt{4+1+4}}\right| = \frac{8}{3}$

43. S(0, −1, 0), $2x + y + 2z = 4$ and P(2, 0, 0) is on the plane $\Rightarrow \overrightarrow{PS} = -2\mathbf{i} - \mathbf{j}$ and $\mathbf{n} = 2\mathbf{i} + \mathbf{j} + 2\mathbf{k}$

$\Rightarrow d = \left|\overrightarrow{PS} \cdot \frac{\mathbf{n}}{|\mathbf{n}|}\right| = \left|\frac{-4-1+0}{\sqrt{4+1+4}}\right| = \frac{5}{3}$

44. S(1, 0, −1), $-4x + y + z = 4$ and P(−1, 0, 0) is on the plane $\Rightarrow \overrightarrow{PS} = 2\mathbf{i} - \mathbf{k}$ and $\mathbf{n} = -4\mathbf{i} + \mathbf{j} + \mathbf{k}$

$\Rightarrow d = \left|\overrightarrow{PS} \cdot \frac{\mathbf{n}}{|\mathbf{n}|}\right| = \left|\frac{-8-1}{\sqrt{16+1+1}}\right| = \frac{9}{\sqrt{18}} = \frac{3\sqrt{2}}{2}$

45. The point P(1, 0, 0) is on the first plane and S(10, 0, 0) is a point on the second plane $\Rightarrow \overrightarrow{PS} = 9\mathbf{i}$, and $\mathbf{n} = \mathbf{i} + 2\mathbf{j} + 6\mathbf{k}$ is normal to the first plane $\Rightarrow$ the distance from S to the first plane is $d = \left|\overrightarrow{PS} \cdot \frac{\mathbf{n}}{|\mathbf{n}|}\right|$ $= \left|\frac{9}{\sqrt{1+4+36}}\right| = \frac{9}{\sqrt{41}}$, which is also the distance between the planes.

46. The line is parallel to the plane since $\mathbf{v} \cdot \mathbf{n} = \left(\mathbf{i} + \mathbf{j} - \frac{1}{2}\mathbf{k}\right) \cdot (\mathbf{i} + 2\mathbf{j} + 6\mathbf{k}) = 1 + 2 - 3 = 0$. Also the point S(1, 0, 0) when $t = -1$ lies on the line, and the point P(10, 0, 0) lies on the plane $\Rightarrow \overrightarrow{PS} = -9\mathbf{i}$. The distance from S to the plane is $d = \left|\overrightarrow{PS} \cdot \frac{\mathbf{n}}{|\mathbf{n}|}\right| = \left|\frac{-9}{\sqrt{1+4+36}}\right| = \frac{9}{\sqrt{41}}$, which is also the distance from the line to the plane.

47. $\mathbf{n}_1 = \mathbf{i}+\mathbf{j}$ and $\mathbf{n}_2 = 2\mathbf{i}+\mathbf{j}-2\mathbf{k} \Rightarrow \theta = \cos^{-1}\left(\frac{\mathbf{n}_1\cdot\mathbf{n}_2}{|\mathbf{n}_1||\mathbf{n}_2|}\right) = \cos^{-1}\left(\frac{2+1}{\sqrt{2}\,\sqrt{9}}\right) = \cos^{-1}\left(\frac{1}{\sqrt{2}}\right) = \frac{\pi}{4}$

48. $\mathbf{n}_1 = 5\mathbf{i}+\mathbf{j}-\mathbf{k}$ and $\mathbf{n}_2 = \mathbf{i}-2\mathbf{j}+3\mathbf{k} \Rightarrow \theta = \cos^{-1}\left(\frac{\mathbf{n}_1\cdot\mathbf{n}_2}{|\mathbf{n}_1||\mathbf{n}_2|}\right) = \cos^{-1}\left(\frac{5-2-3}{\sqrt{27}\,\sqrt{14}}\right) = \cos^{-1}(0) = \frac{\pi}{2}$

49. $\mathbf{n}_1 = 2\mathbf{i}+2\mathbf{j}+2\mathbf{k}$ and $\mathbf{n}_2 = 2\mathbf{i}-2\mathbf{j}-\mathbf{k} \Rightarrow \theta = \cos^{-1}\left(\frac{\mathbf{n}_1\cdot\mathbf{n}_2}{|\mathbf{n}_1||\mathbf{n}_2|}\right) = \cos^{-1}\left(\frac{4-4-2}{\sqrt{12}\,\sqrt{9}}\right) = \cos^{-1}\left(\frac{-1}{3\sqrt{3}}\right) \approx 1.76$ rad

50. $\mathbf{n}_1 = \mathbf{i}+\mathbf{j}+\mathbf{k}$ and $\mathbf{n}_2 = \mathbf{k} \Rightarrow \theta = \cos^{-1}\left(\frac{\mathbf{n}_1\cdot\mathbf{n}_2}{|\mathbf{n}_1||\mathbf{n}_2|}\right) = \cos^{-1}\left(\frac{1}{\sqrt{3}\,\sqrt{1}}\right) \approx 0.96$ rad

51. $\mathbf{n}_1 = 2\mathbf{i}+2\mathbf{j}-\mathbf{k}$ and $\mathbf{n}_2 = \mathbf{i}+2\mathbf{j}+\mathbf{k} \Rightarrow \theta = \cos^{-1}\left(\frac{\mathbf{n}_1\cdot\mathbf{n}_2}{|\mathbf{n}_1||\mathbf{n}_2|}\right) = \cos^{-1}\left(\frac{2+4-1}{\sqrt{9}\,\sqrt{6}}\right) = \cos^{-1}\left(\frac{5}{3\sqrt{6}}\right) \approx 0.82$ rad

52. $\mathbf{n}_1 = 4\mathbf{j}+3\mathbf{k}$ and $\mathbf{n}_2 = 3\mathbf{i}+2\mathbf{j}+6\mathbf{k} \Rightarrow \theta = \cos^{-1}\left(\frac{\mathbf{n}_1\cdot\mathbf{n}_2}{|\mathbf{n}_1||\mathbf{n}_2|}\right) = \cos^{-1}\left(\frac{8+18}{\sqrt{25}\,\sqrt{49}}\right) = \cos^{-1}\left(\frac{26}{35}\right) \approx 0.73$ rad

53. $2x - y + 3z = 6 \Rightarrow 2(1-t) - (3t) + 3(1+t) = 6 \Rightarrow -2t + 5 = 6 \Rightarrow t = -\frac{1}{2} \Rightarrow x = \frac{3}{2},\ y = -\frac{3}{2}$ and $z = \frac{1}{2}$
$\Rightarrow \left(\frac{3}{2}, -\frac{3}{2}, \frac{1}{2}\right)$ is the point

54. $6x + 3y - 4z = -12 \Rightarrow 6(2) + 3(3+2t) - 4(-2-2t) = -12 \Rightarrow 14t + 29 = -12 \Rightarrow t = -\frac{41}{14} \Rightarrow x = 2,\ y = 3 - \frac{41}{7}$,
and $z = -2 + \frac{41}{7} \Rightarrow \left(2, -\frac{20}{7}, \frac{27}{7}\right)$ is the point

55. $x + y + z = 2 \Rightarrow (1+2t) + (1+5t) + (3t) = 2 \Rightarrow 10t + 2 = 2 \Rightarrow t = 0 \Rightarrow x = 1,\ y = 1$ and $z = 0$
$\Rightarrow (1,1,0)$ is the point

56. $2x - 3z = 7 \Rightarrow 2(-1+3t) - 3(5t) = 7 \Rightarrow -9t - 2 = 7 \Rightarrow t = -1 \Rightarrow x = -1-3,\ y = -2$ and $z = -5$
$\Rightarrow (-4,-2,-5)$ is the point

57. $\mathbf{n}_1 = \mathbf{i}+\mathbf{j}+\mathbf{k}$ and $\mathbf{n}_2 = \mathbf{i}+\mathbf{j} \Rightarrow \mathbf{n}_1\times\mathbf{n}_2 = \begin{vmatrix} \mathbf{i} & \mathbf{j} & \mathbf{k} \\ 1 & 1 & 1 \\ 1 & 1 & 0 \end{vmatrix} = -\mathbf{i}+\mathbf{j}$, the direction of the desired line; $(1,1,-1)$
is on both planes $\Rightarrow$ the desired line is $x = 1 - t,\ y = 1 + t,\ z = -1$

58. $\mathbf{n}_1 = 3\mathbf{i}-6\mathbf{j}-2\mathbf{k}$ and $\mathbf{n}_2 = 2\mathbf{i}+\mathbf{j}-2\mathbf{k} \Rightarrow \mathbf{n}_1\times\mathbf{n}_2 = \begin{vmatrix} \mathbf{i} & \mathbf{j} & \mathbf{k} \\ 3 & -6 & -2 \\ 2 & 1 & -2 \end{vmatrix} = 14\mathbf{i}+2\mathbf{j}+15\mathbf{k}$, the direction of the
desired line; $(1,0,0)$ is on both planes $\Rightarrow$ the desired line is $x = 1 + 14t,\ y = 2t,\ z = 15t$

59. $\mathbf{n}_1 = \mathbf{i} - 2\mathbf{j} + 4\mathbf{k}$ and $\mathbf{n}_2 = \mathbf{i} + \mathbf{j} - 2\mathbf{k} \Rightarrow \mathbf{n}_1 \times \mathbf{n}_2 = \begin{vmatrix} \mathbf{i} & \mathbf{j} & \mathbf{k} \\ 1 & -2 & 4 \\ 1 & 1 & -2 \end{vmatrix} = 6\mathbf{j} + 3\mathbf{k}$, the direction of the desired line; $(4,3,1)$ is on both planes $\Rightarrow$ the desired line is $x = 4$, $y = 3 + 6t$, $z = 1 + 3t$

60. $\mathbf{n}_1 = 5\mathbf{i} - 2\mathbf{j}$ and $\mathbf{n}_2 = 4\mathbf{j} - 5\mathbf{k} \Rightarrow \mathbf{n}_1 \times \mathbf{n}_2 = \begin{vmatrix} \mathbf{i} & \mathbf{j} & \mathbf{k} \\ 5 & -2 & 0 \\ 0 & 4 & -5 \end{vmatrix} = 10\mathbf{i} + 25\mathbf{j} + 20\mathbf{k}$, the direction of the desired line; $(1,-3,1)$ is on both planes $\Rightarrow$ the desired line is $x = 1 + 10t$, $y = -3 + 25t$, $z = 1 + 20t$

61. <u>L1 & L2</u>: $x = 3 + 2t = 1 + 4s$ and $y = -1 + 4t = 1 + 2s \Rightarrow \begin{cases} 2t - 4s = -2 \\ 4t - 2s = 2 \end{cases} \Rightarrow \begin{cases} 2t - 4s = -2 \\ 2t - s = 1 \end{cases}$
$\Rightarrow -3s = -3 \Rightarrow s = 1$ and $t = 1 \Rightarrow$ on L1, $z = 1$ and on L2, $z = 1 \Rightarrow$ L1 and L2 intersect at $(5,3,1)$.
<u>L2 & L3</u>: The direction of L2 is $\frac{1}{6}(4\mathbf{i} + 2\mathbf{j} + 4\mathbf{k}) = \frac{1}{3}(2\mathbf{i} + \mathbf{j} + 2\mathbf{k})$ which is the same as the direction $\frac{1}{3}(2\mathbf{i} + \mathbf{j} + 2\mathbf{k})$ of L3; hence L2 and L3 are parallel.
<u>L1 & L3</u>: $x = 3 + 2t = 3 + 2r$ and $y = -1 + 4t = 2 + r \Rightarrow \begin{cases} 2t - 2r = 0 \\ 4t - r = 3 \end{cases} \Rightarrow \begin{cases} t - r = 0 \\ 4t - r = 3 \end{cases} \Rightarrow 3t = 3$
$\Rightarrow t = 1$ and $r = 1 \Rightarrow$ on L1, $z = 2$ while on L3, $z = 0 \Rightarrow$ L1 and L2 do not intersect. The direction of L1 is $\frac{1}{\sqrt{21}}(2\mathbf{i} + 4\mathbf{j} - \mathbf{k})$ while the direction of L3 is $\frac{1}{3}(2\mathbf{i} + \mathbf{j} + 2\mathbf{k})$ and neither is a multiple of the other; hence L1 and L3 are skew.

62. <u>L1 & L2</u>: $x = 1 + 2t = 2 - s$ and $y = -1 - t = 3s \Rightarrow \begin{cases} 2t + s = 1 \\ -t - 3s = 1 \end{cases} \Rightarrow -5s = 3 \Rightarrow s = -\frac{3}{5}$ and $t = \frac{4}{5} \Rightarrow$ on L1, $z = \frac{12}{5}$ while on L2, $z = 1 - \frac{3}{5} = \frac{2}{5} \Rightarrow$ L1 and L2 do not intersect. The direction of L1 is $\frac{1}{\sqrt{14}}(2\mathbf{i} - \mathbf{j} + 3\mathbf{k})$ while the direction of L2 is $\frac{1}{\sqrt{11}}(-\mathbf{i} + 3\mathbf{j} + \mathbf{k})$ and neither is a multiple of the other; hence, L1 and L2 are skew.
<u>L2 & L3</u>: $x = 2 - s = 5 + 2r$ and $y = 3s = 1 - r \Rightarrow \begin{cases} -s - 2r = 3 \\ 3s + r = 1 \end{cases} \Rightarrow 5s = 5 \Rightarrow s = 1$ and $r = -2 \Rightarrow$ on L2, $z = 2$ and on L3, $z = 2 \Rightarrow$ L2 and L3 intersect at $(1,3,2)$.
<u>L1 & L3</u>: L1 and L3 have the same direction $\frac{1}{\sqrt{14}}(2\mathbf{i} - \mathbf{j} + 3\mathbf{k})$; hence L1 and L3 are parallel.

63. $x = 2 + 2t$, $y = -4 - t$, $z = 7 + 3t$; $x = -2 - t$, $y = -2 + \frac{1}{2}t$, $z = 1 - \frac{3}{2}t$

64. $1(x - 4) - 2(y - 1) + 1(z - 5) = 0 \Rightarrow x - 4 - 2y + 2 + z - 5 = 0 \Rightarrow x - 2y + z = 7$;
$-\sqrt{2}(x - 3) + 2\sqrt{2}(y + 2) - \sqrt{2}(z - 0) = 0 \Rightarrow -\sqrt{2}x + 2\sqrt{2}y - \sqrt{2}z = -7\sqrt{2}$

65. $x=0 \Rightarrow t=-\frac{1}{2}$, $y=-\frac{1}{2}$, $z=-\frac{3}{2} \Rightarrow \left(0,-\frac{1}{2},-\frac{3}{2}\right)$; $y=0 \Rightarrow t=-1$, $x=-1$, $z=-3 \Rightarrow (-1,0,-3)$; $z=0$ $\Rightarrow t=0$, $x=1$, $y=-1 \Rightarrow (1,-1,0)$

66. The line contains $(0,0,3)$ and $\left(\sqrt{3},1,3\right)$ because the projection of the line onto the xy-plane contains the origin and intersects the positive x-axis at a 30° angle. The direction of the line is $\sqrt{3}\mathbf{i}+\mathbf{j}+0\mathbf{k} \Rightarrow$ the line in question is $x=\sqrt{3}t$, $y=t$, $z=3$.

67. With substitution of the line into the plane we have $2(1-2t)+(2+5t)-(-3t)=8 \Rightarrow 2-4t+2+5t+3t=8$ $\Rightarrow 4t+4=8 \Rightarrow t=1 \Rightarrow$ the point $(-1,7,-3)$ is contained in both the line and plane, so they are not parallel.

68. The planes are parallel when either vector $A_1\mathbf{i}+B_1\mathbf{j}+C_1\mathbf{k}$ or $A_2\mathbf{i}+B_2\mathbf{j}+C_2\mathbf{k}$ is a multiple of the other or when $\left|(A_1\mathbf{i}+B_1\mathbf{j}+C_1\mathbf{k})\times(A_2\mathbf{i}+B_2\mathbf{j}+C_2\mathbf{k}\right|=0$. The planes are perpendicular when their normals are perpendicular, or$(A_1\mathbf{i}+B_1\mathbf{j}+C_1\mathbf{k})\cdot(A_2\mathbf{i}+B_2\mathbf{j}+C_2\mathbf{k})=0$.

69. There are many possible answers. One is found as follows: eliminate t to get $t=x-1=2-y=\frac{z-3}{2}$ $\Rightarrow x-1=2-y$ and $2-y=\frac{z-3}{2} \Rightarrow x+y=3$ and $2y+z=7$ are two such planes.

70. Since the plane passes through the origin, its general equation is of the form $Ax+By+Cz=0$. Since it meets the plane M at a right angle, their normal vectors are perpendicular $\Rightarrow 2A+3B+C=0$. One choice satisfying this equation is $A=1$, $B=-1$ and $C=1 \Rightarrow x-y+z=0$. Any plane $Ax+By+Cz=0$ with $2A+3B+C=0$ will pass through the origin and be perpendicular to M.

71. The points $(a,0,0)$, $(0,b,0)$ and $(0,0,c)$ are the x, y, and z intercepts of the plane. Since a, b, and c are all nonzero, the plane must intersect all three coordinate axes and cannot pass through the origin. Thus, $\frac{x}{a}+\frac{y}{b}+\frac{z}{c}=1$ describes all planes <u>except</u> those through the origin or parallel to a coordinate axis.

72. Yes. If $\mathbf{v}_1$ and $\mathbf{v}_2$ are nonzero vectors parallel to the lines, then $\mathbf{v}_1\times\mathbf{v}_2\neq\mathbf{0}$ is perpendicular to the lines.

73. (a) $\overrightarrow{EP}=c\overrightarrow{EP_1} \Rightarrow -x_0\mathbf{i}+y\mathbf{j}+z\mathbf{k}=c\left[(x_1-x_0)\mathbf{i}+y_1\mathbf{j}+z_1\mathbf{k}\right] \Rightarrow -x_0=c(x_1-x_0)$, $y=cy_1$ and $z=cz_1$, where c is a positive real number

(b) At $x_1=0 \Rightarrow c=1 \Rightarrow y=y_1$ and $z=z_1$; at $x_1=x_0 \Rightarrow x_0=0$, $y=0$, $z=0$; $\lim\limits_{x_0\to\infty} c=\lim\limits_{x_0\to\infty}\frac{-x_0}{x_1-x_0}$ $=\lim\limits_{x_0\to\infty}\frac{-1}{-1}=1 \Rightarrow c\to 1$ so that $y\to y_1$ and $z\to z_1$

74. The plane which contains the triangular plane is $x+y+z=2$. The line containing the endpoints of the line segment is $x=1-t$, $y=2t$, $z=2t$. The plane and the line intersect at $\left(\frac{2}{3},\frac{2}{3},\frac{2}{3}\right)$. The visible section of the line segment is $\sqrt{\left(\frac{1}{3}\right)^2+\left(\frac{2}{3}\right)^2+\left(\frac{2}{3}\right)^2}=1$ unit in length. The length of the line segment is $\sqrt{1^2+2^2+2^2}=3 \Rightarrow \frac{2}{3}$ of the line segment is hidden from view.

10.6 CYLINDERS AND QUADRIC SURFACES

1. d, ellipsoid
2. i, hyperboloid
3. a, cylinder
4. g, cone
5. l, hyperbolic paraboloid
6. e, paraboloid

7. b, cylinder

8. j, hyperboloid

9. k, hyperbolic paraboloid

10. f, paraboloid

11. h, cone

12. c, ellipsoid

13. $x^2 + y^2 = 4$

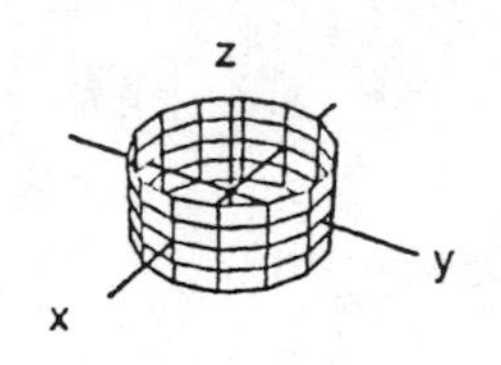

14. $x^2 + z^2 = 4$

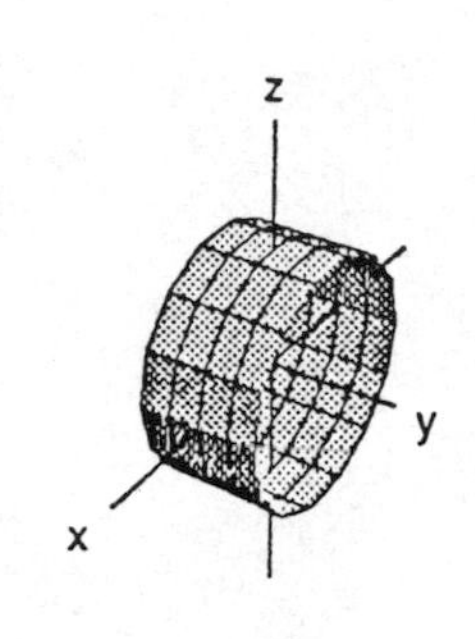

15. $z = y^2 - 1$

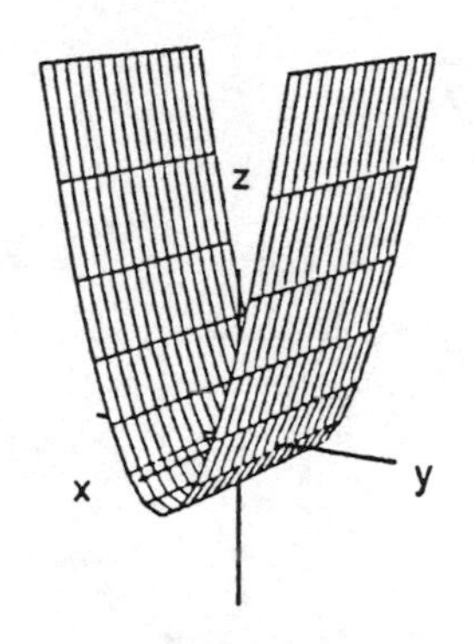

16. $x = y^2$

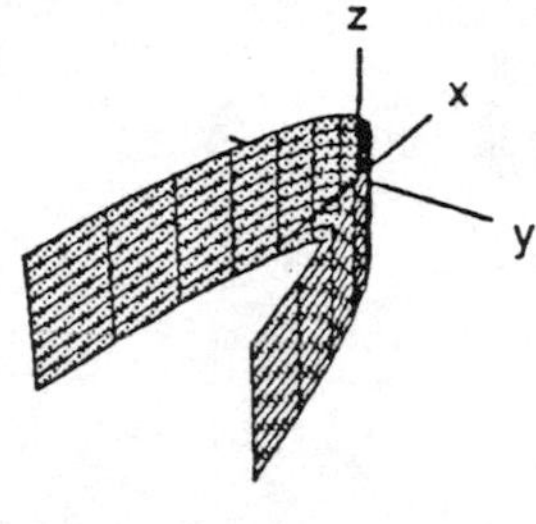

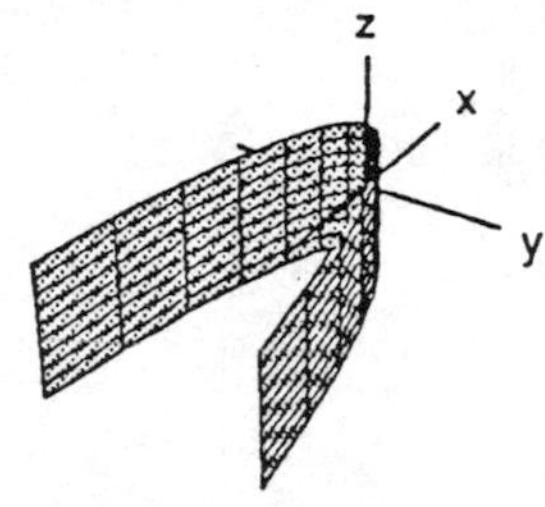

17. $x^2 + 4z^2 = 16$

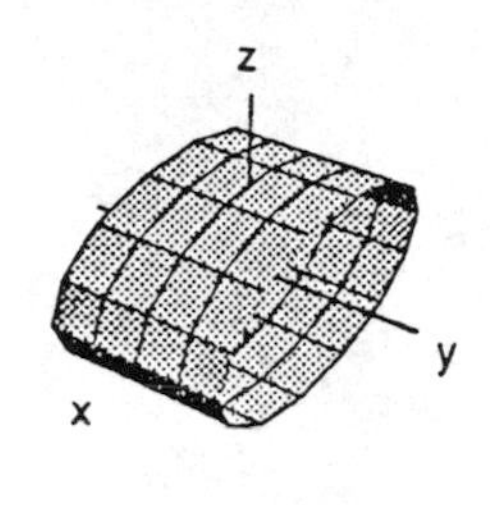

18. $4x^2 + y^2 = 36$

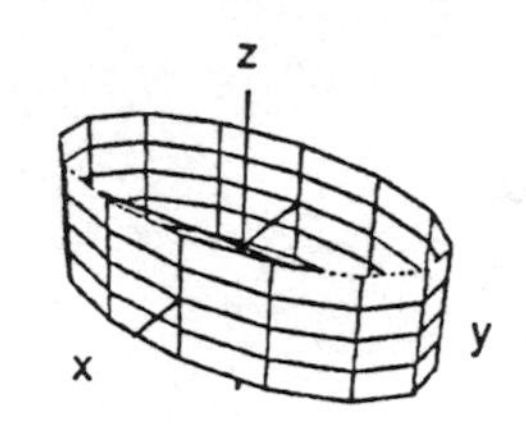

19. $z^2 - y^2 = 1$

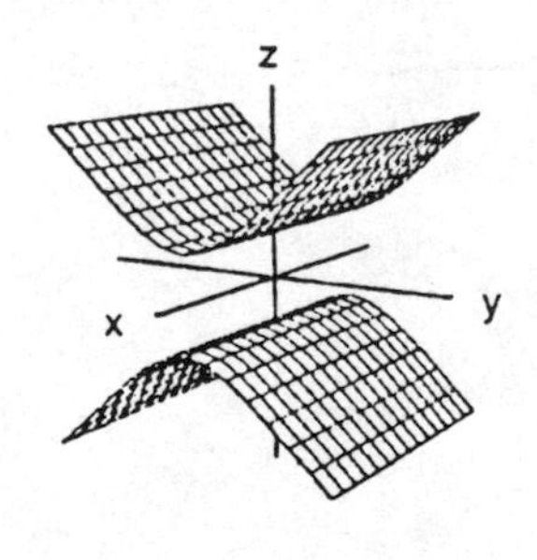

20. $yz = 1$

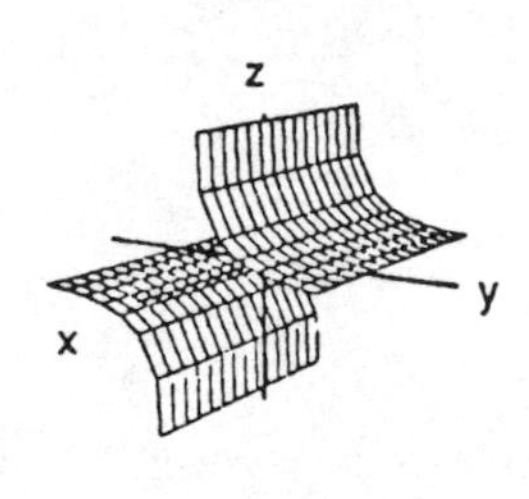

21. $9x^2 + y^2 + z^2 = 9$

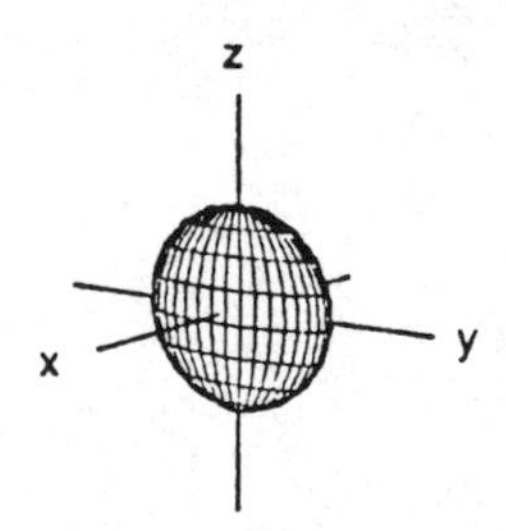

22. $4x^2 + 4y^2 + z^2 = 16$

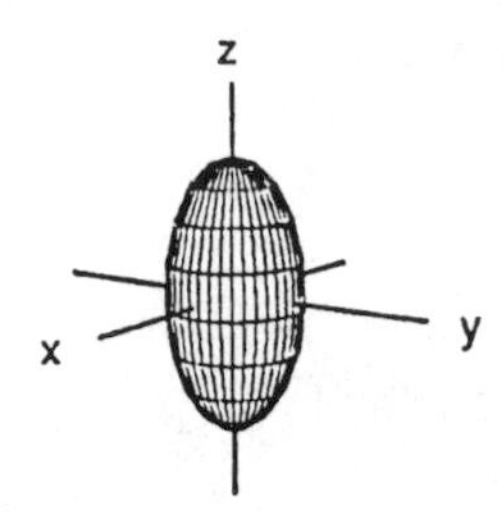

23. $4x^2 + 9y^2 + 4z^2 = 36$

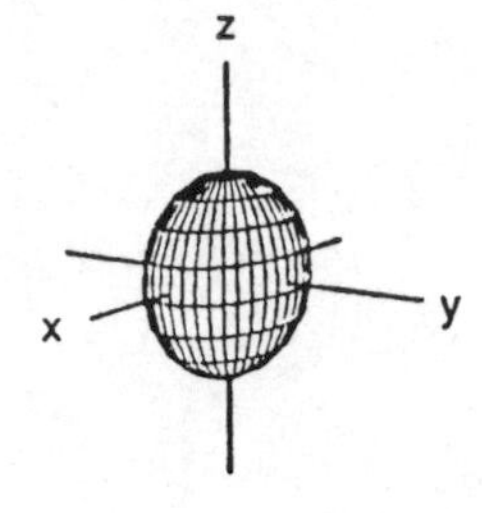

24. $9x^2 + 4y^2 + 36z^2 = 36$

25. $x^2 + 4y^2 = z$

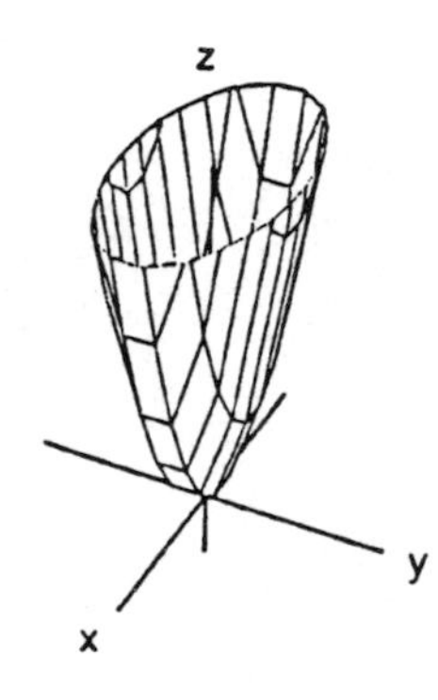

26. $z = x^2 + 9y^2$

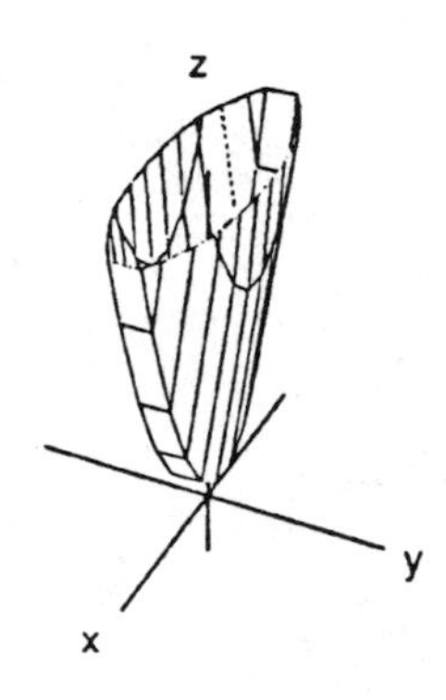

27. $z = 8 - x^2 - y^2$

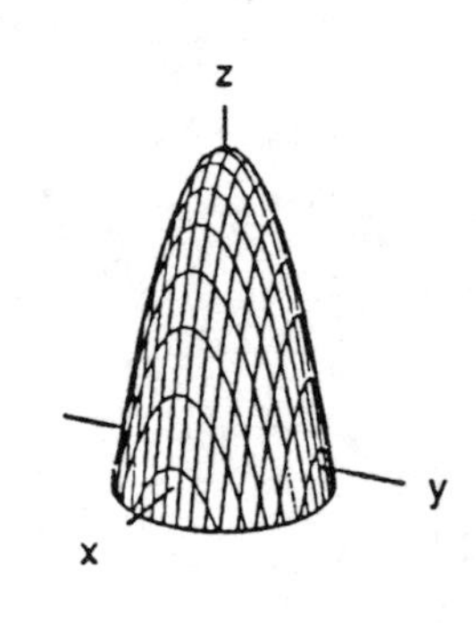

28. $z = 18 - x^2 - 9y^2$

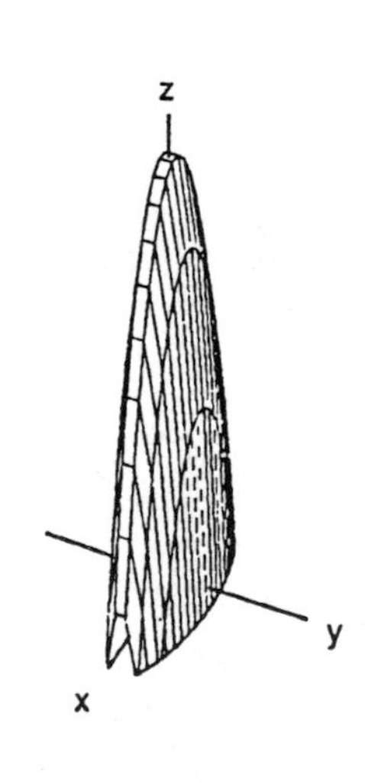

29. $x = 4 - 4y^2 - z^2$

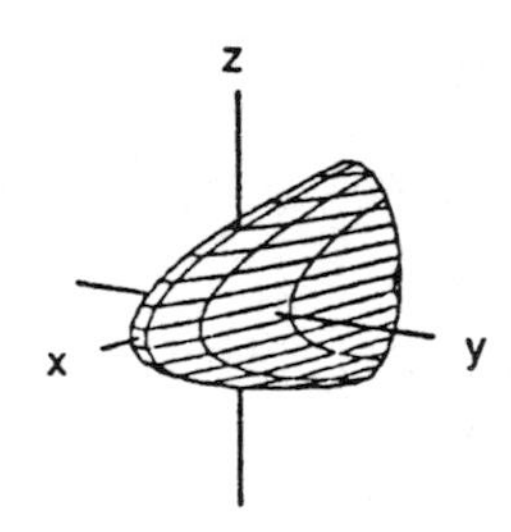

30. $y = 1 - x^2 - z^2$

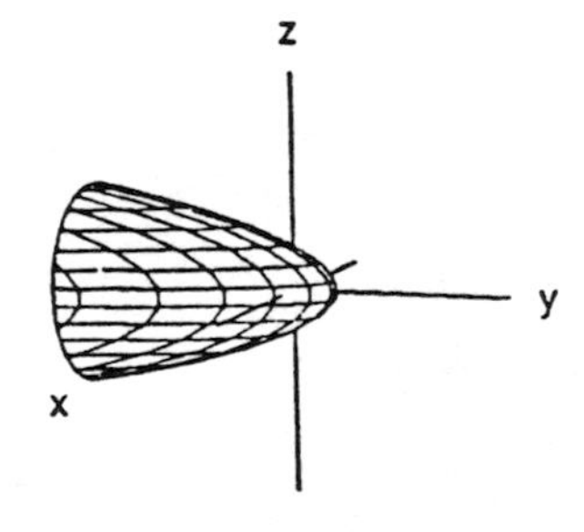

31. $x^2 + y^2 = z^2$

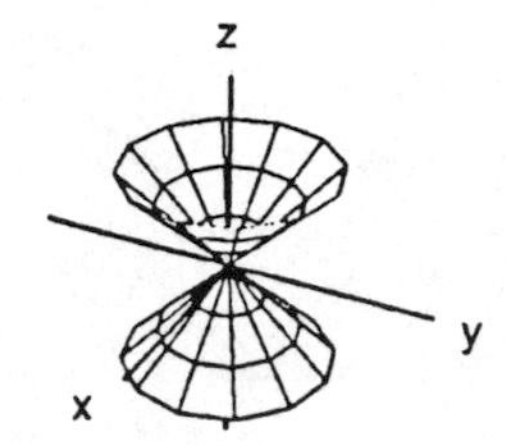

32. $y^2 + z^2 = x^2$

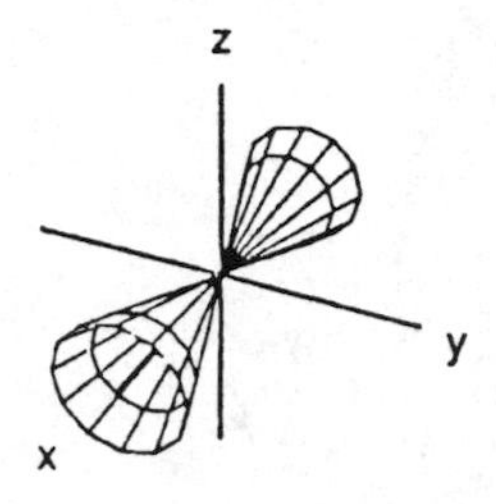

33. $4x^2 + 9z^2 = 9y^2$

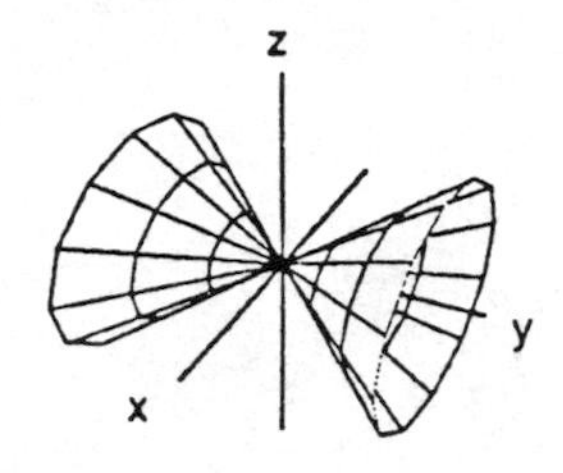

34. $9x^2 + 4y^2 = 36z^2$

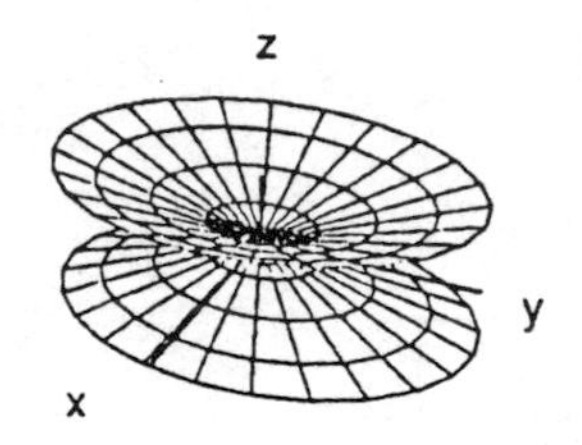

35. $x^2 + y^2 - z^2 = 1$

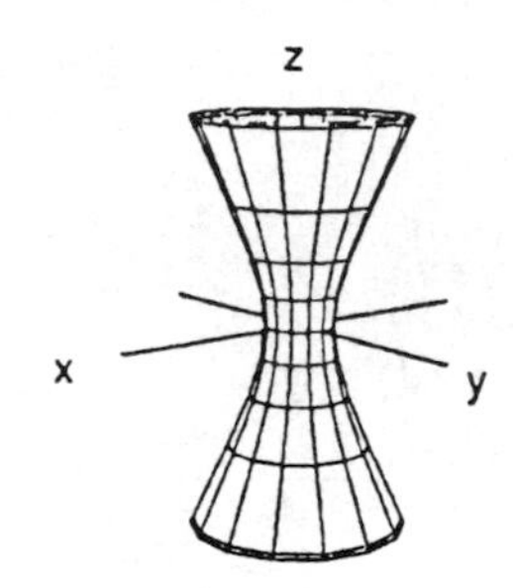

36. $y^2 + z^2 - x^2 = 1$

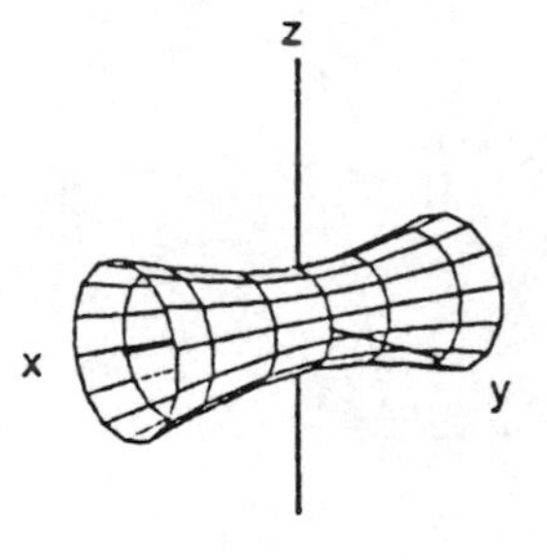

37. $\frac{y^2}{4} + \frac{z^2}{9} - \frac{x^2}{4} = 1$

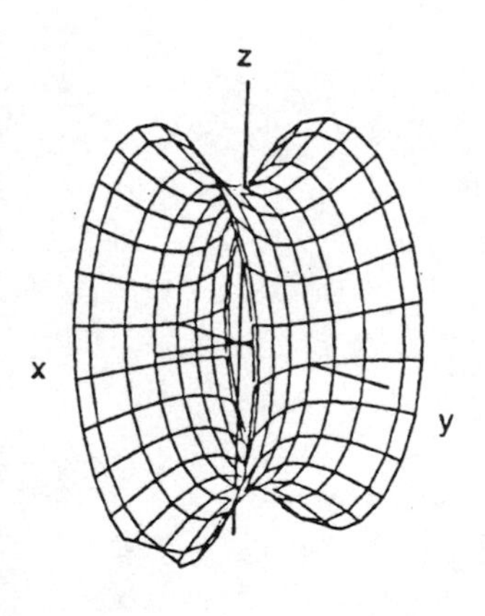

38. $\frac{x^2}{4} + \frac{y^2}{4} - \frac{z^2}{9} = 1$

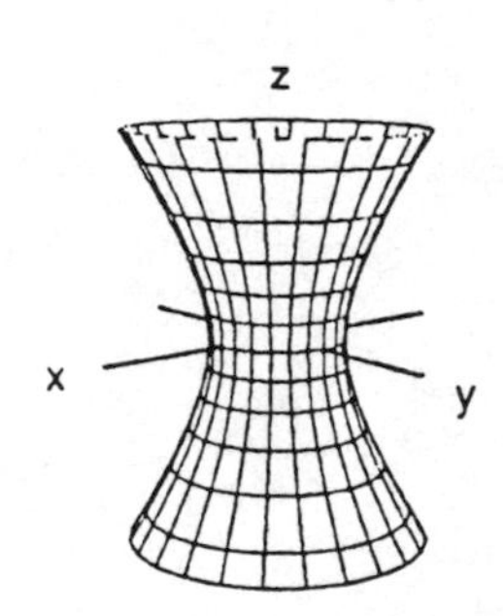

39. $z^2 - x^2 - y^2 = 1$

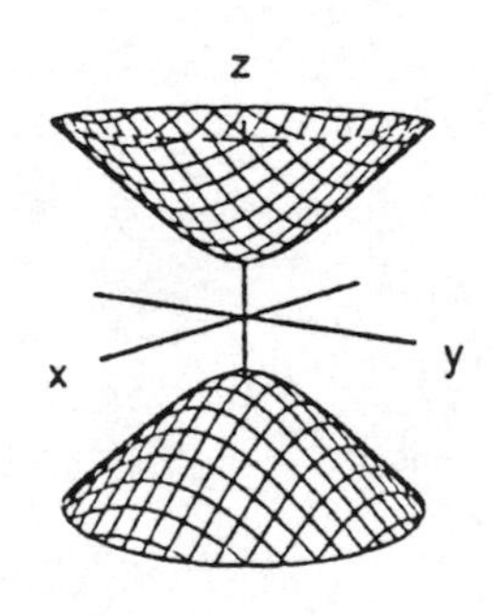

40. $\frac{y^2}{4} - \frac{x^2}{4} - z^2 = 1$

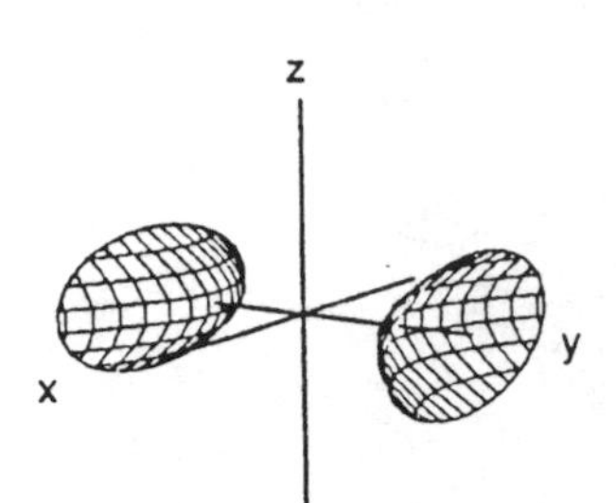

41. $x^2 - y^2 - \frac{z^2}{4} = 1$

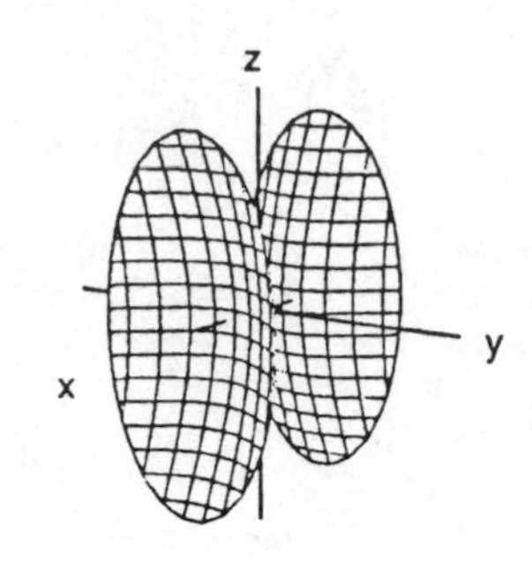

42. $\frac{x^2}{4} - y^2 - \frac{z^2}{4} = 1$

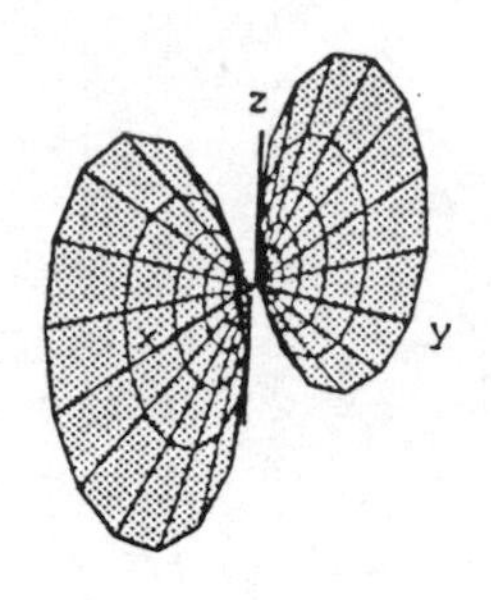

43. $y^2 - x^2 = z$

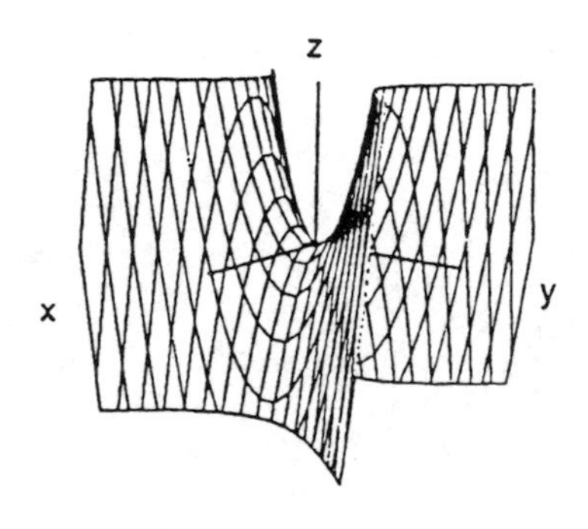

44. $x^2 - y^2 = z$

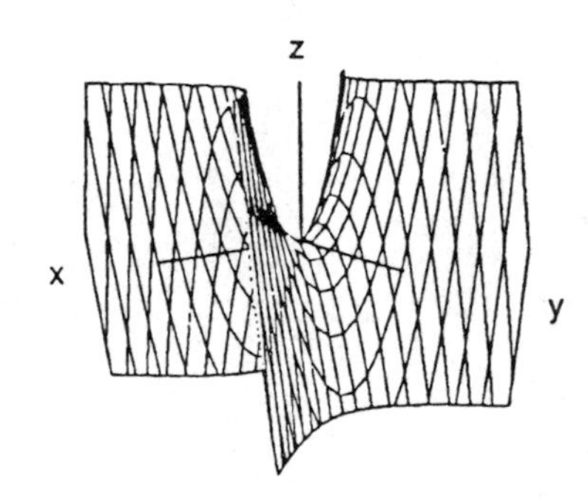

45. $x^2 + y^2 + z^2 = 4$

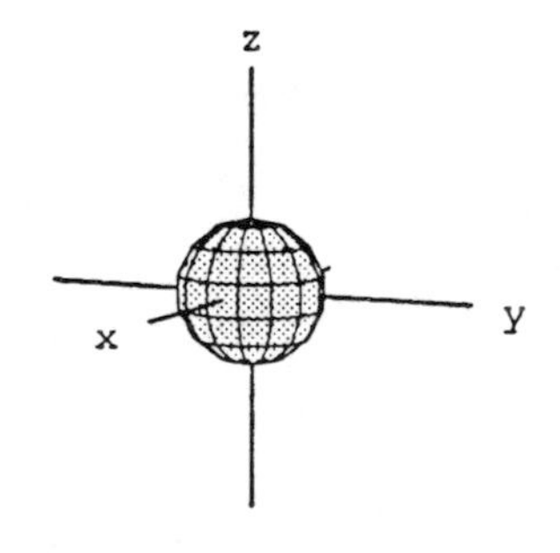

46. $4x^2 + 4y^2 = z^2$

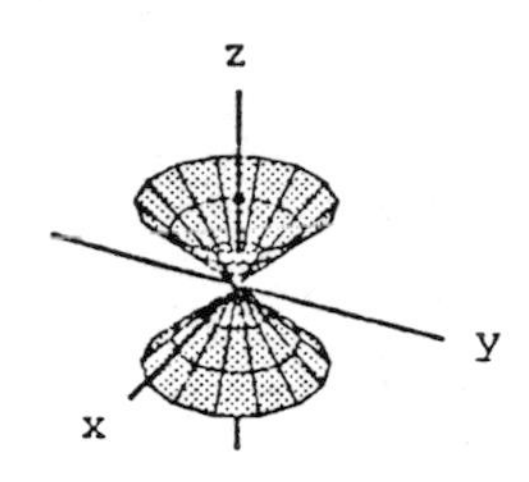

47. $z = 1 + y^2 - x^2$

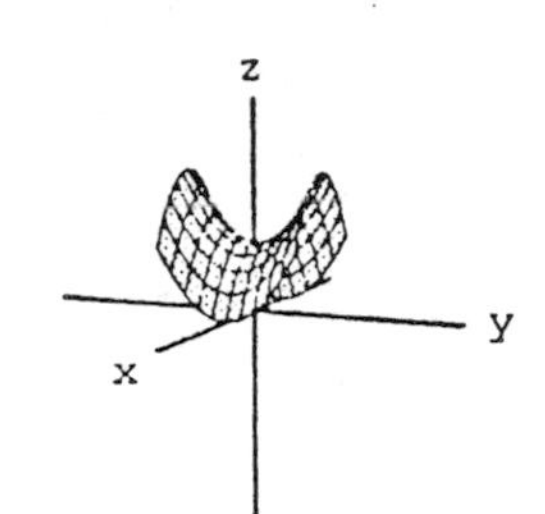

48. $y^2 - z^2 = 4$

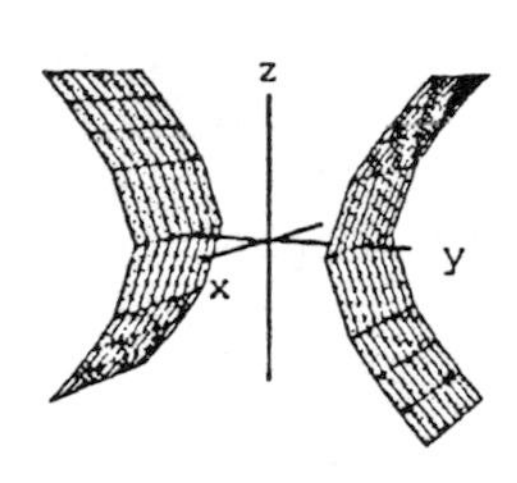

49. $y = -(x^2 + z^2)$

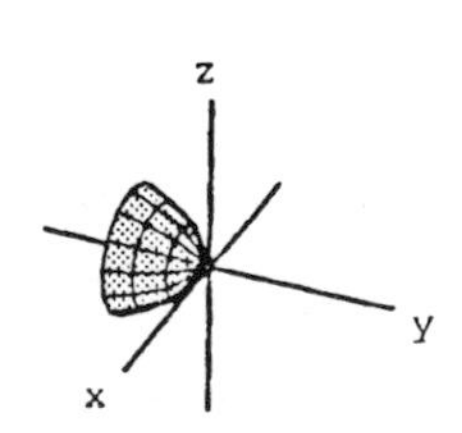

50. $z^2 - 4x^2 - 4y^2 = 4$

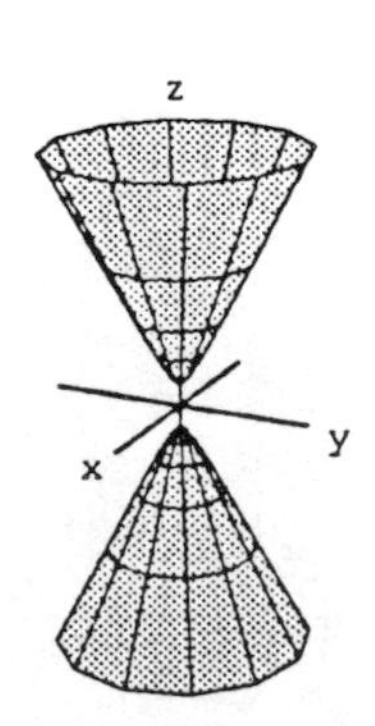

51. $16x^2 + 4y^2 = 1$

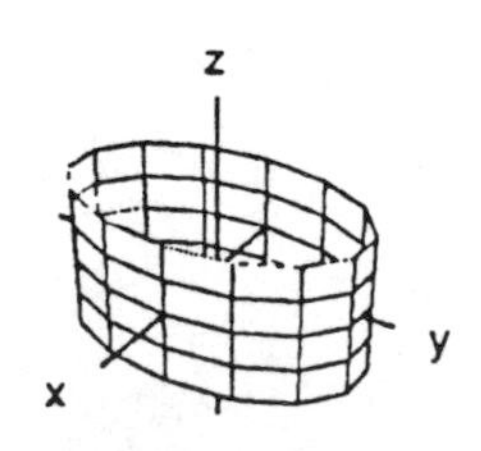

52. $z = x^2 + y^2 + 1$

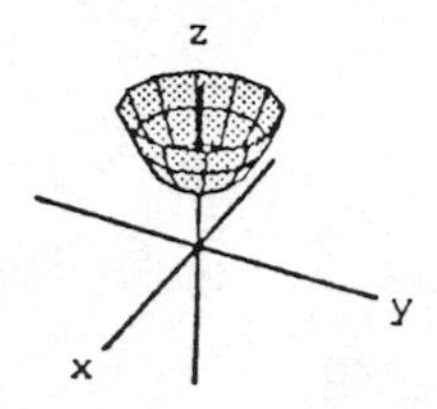

53. $x^2 + y^2 - z^2 = 4$

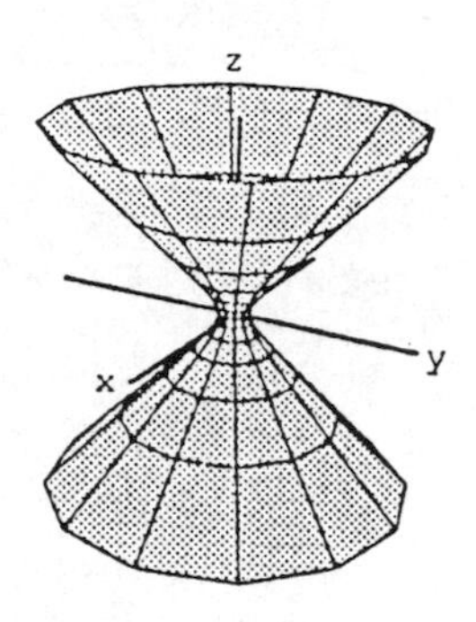

54. $x = 4 - y^2$

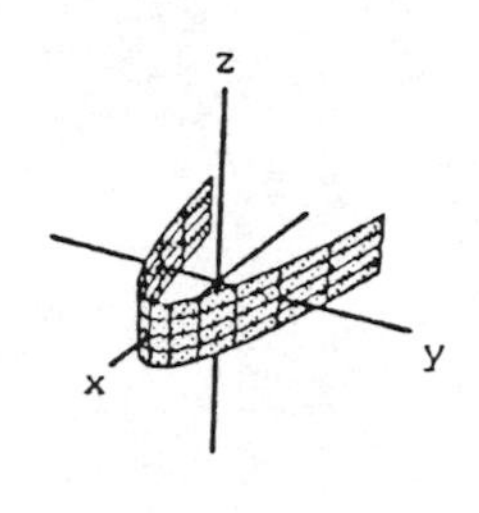

55. $x^2 + z^2 = y$

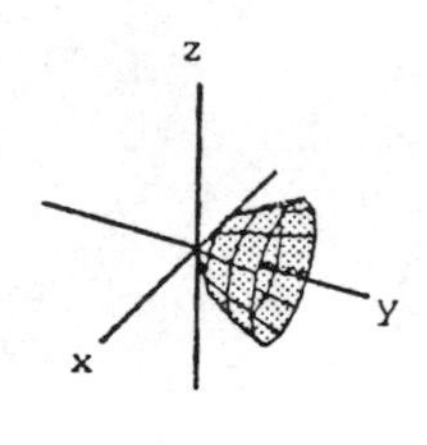

56. $z^2 - \frac{x^2}{4} - y^2 = 1$

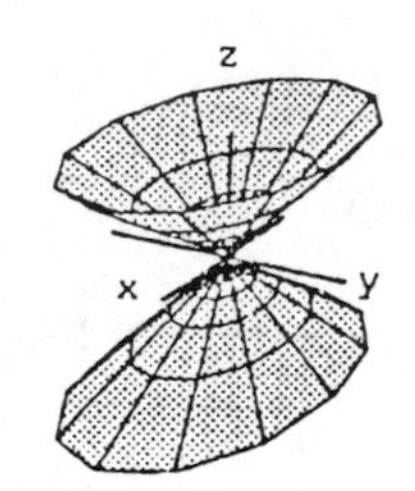

57. $x^2 + z^2 = 1$

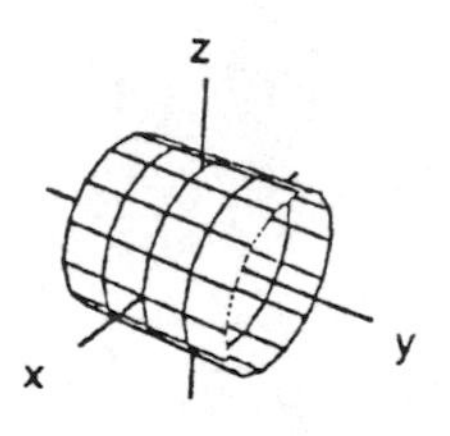

58. $4x^2 + 4y^2 + z^2 = 4$

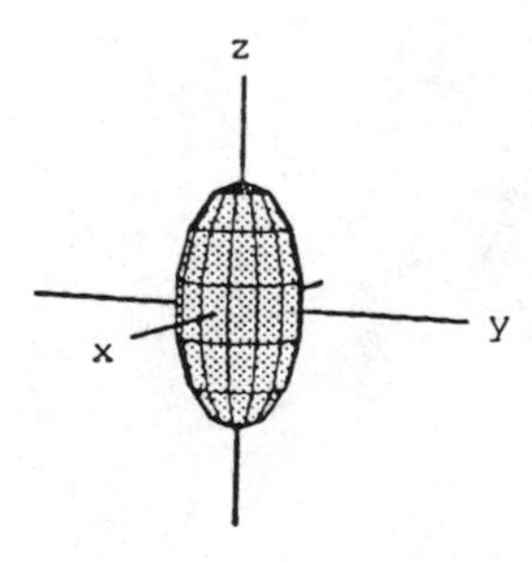

59. $16y^2 + 9z^2 = 4x^2$

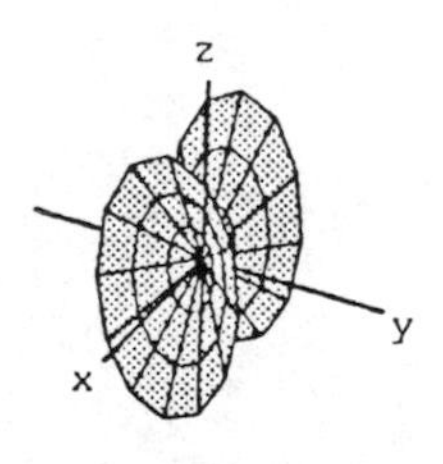

60. $z = x^2 - y^2 - 1$

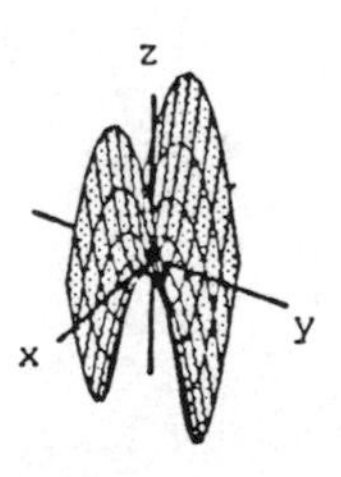

61. $9x^2 + 4y^2 + z^2 = 36$

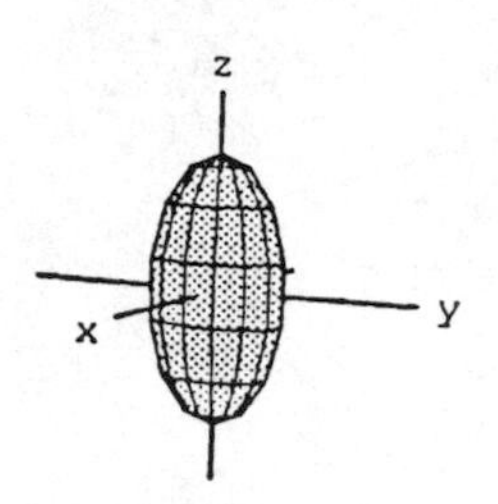

62. $4x^2 + 9z^2 = y^2$

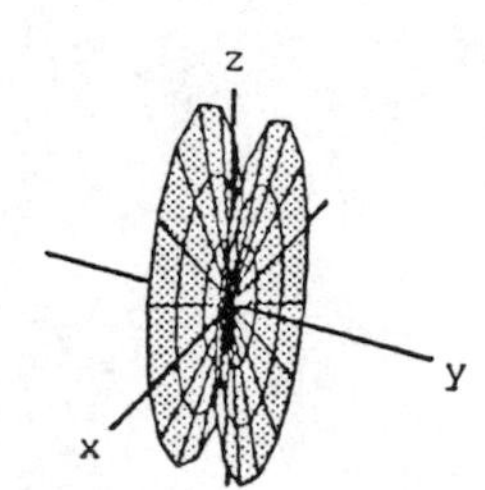

63. $x^2 + y^2 - 16z^2 = 16$

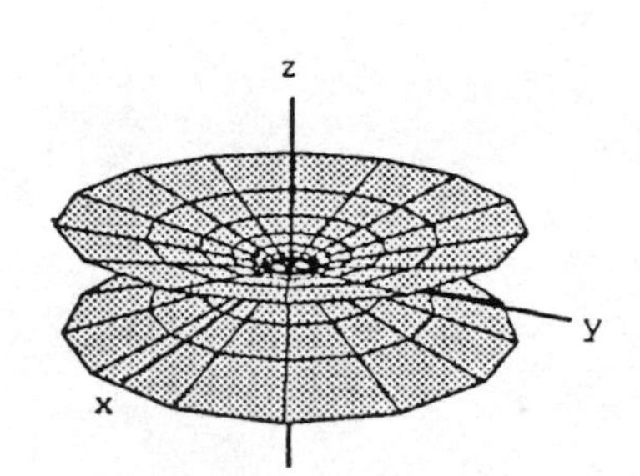

64. $z^2 + 4y^2 = 9$

65. $z = -(x^2 + y^2)$

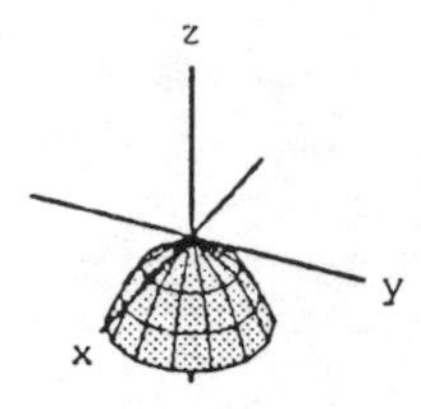

66. $y^2 - x^2 - z^2 = 1$

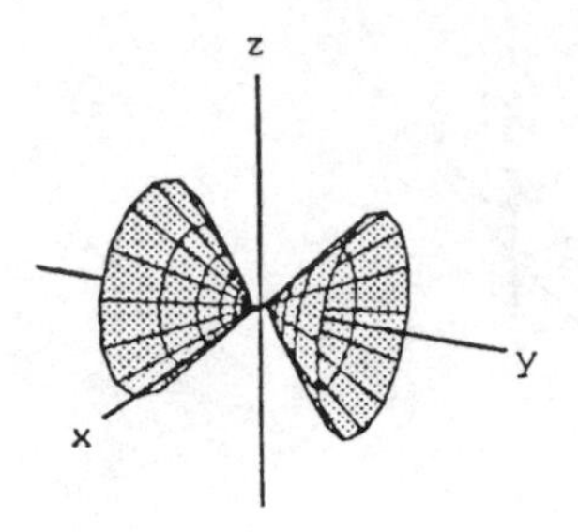

67. $x^2 - 4y^2 = 1$

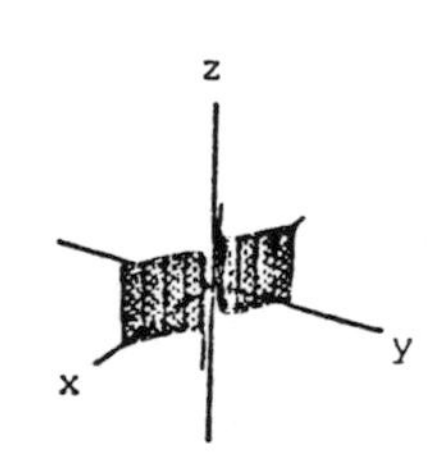

68. $z = 4x^2 + y^2 - 4$

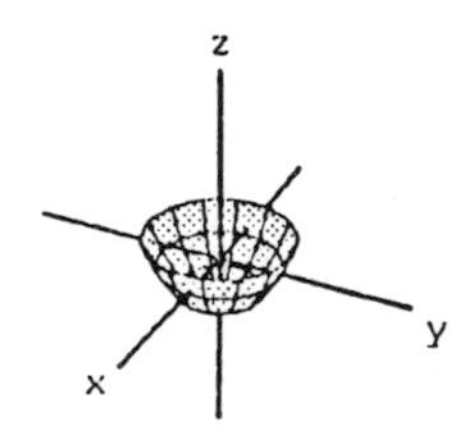

69. $4y^2 + z^2 - 4x^2 = 4$

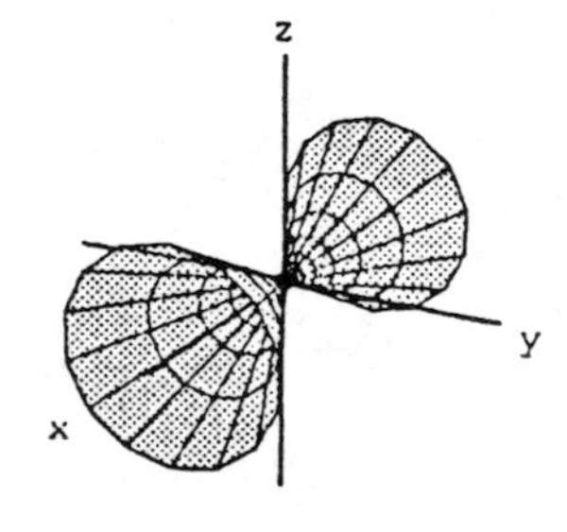

70. $z = 1 - x^2$

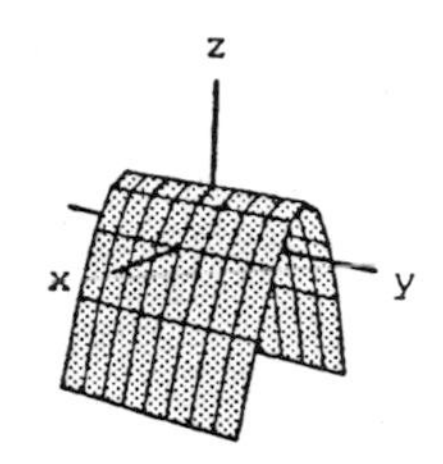

71. $x^2 + y^2 = z$

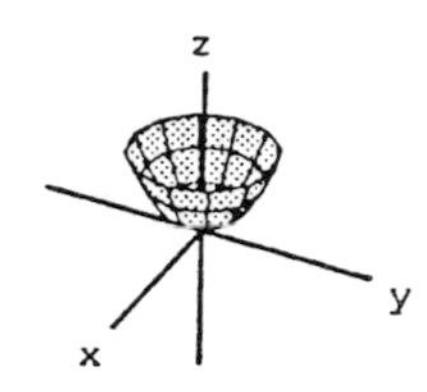

72. $\frac{x^2}{4} + y^2 - z^2 = 1$

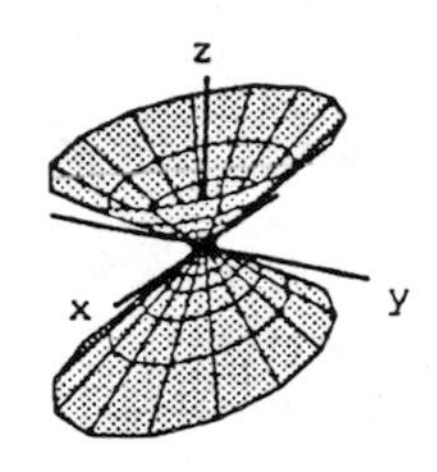

73. $yz = 1$

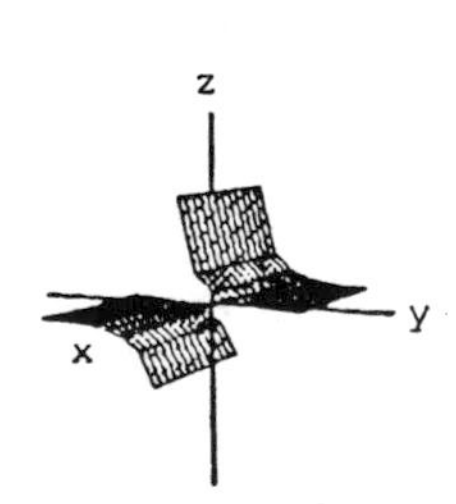

74. $36x^2 + 9y^2 + 4z^2 = 36$

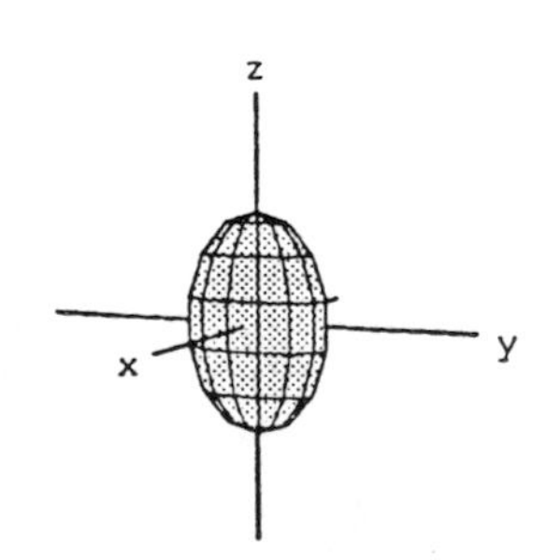

75. $9x^2 + 16y^2 = 4z^2$

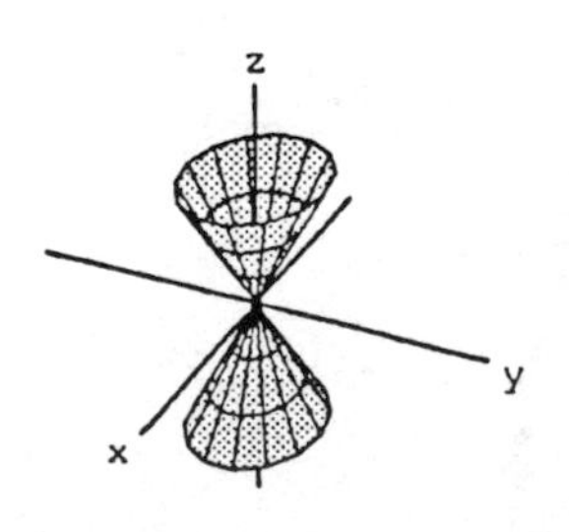

76. $4z^2 - x^2 - y^2 = 4$

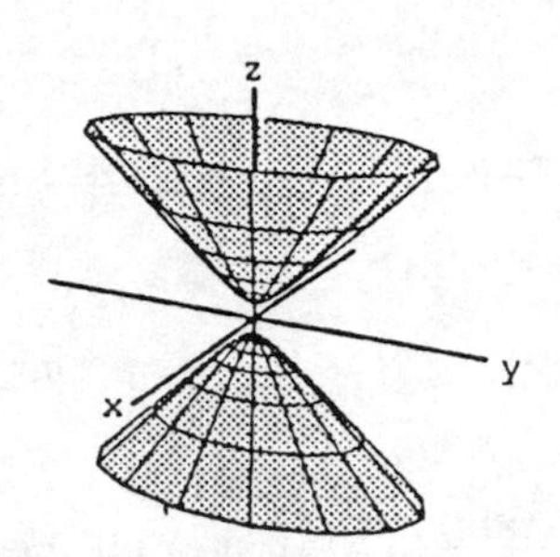

77. (a) If $x^2 + \frac{y^2}{4} + \frac{z^2}{9} = 1$ and $z = c$, then $x^2 + \frac{y^2}{4} = \frac{9 - c^2}{9} \Rightarrow \frac{x^2}{\left(\frac{9-c^2}{9}\right)} + \frac{y^2}{\left[\frac{4(9-c^2)}{9}\right]} = 1 \Rightarrow A = ab\pi$

$= \pi\left(\frac{\sqrt{9-c^2}}{3}\right)\left(\frac{2\sqrt{9-c^2}}{3}\right) = \frac{2\pi(9-c^2)}{9}$

(b) From part (a), each slice has the area $\frac{2\pi(9-z^2)}{9}$, where $-3 \le z \le 3$. Thus $V = 2\int_0^3 \frac{2\pi}{9}(9-z^2)\,dz$

$= \frac{4\pi}{9}\int_0^3 (9-z^2)\,dz = \frac{4\pi}{9}\left[9z - \frac{z^3}{3}\right]_0^3 = \frac{4\pi}{9}(27-9) = 8\pi$

(c) $\frac{x^2}{a^2} + \frac{y^2}{b^2} + \frac{z^2}{c^2} = 1 \Rightarrow \frac{x^2}{\left[\frac{a^2(c^2-z^2)}{c^2}\right]} + \frac{y^2}{\left[\frac{b^2(c^2-z^2)}{c^2}\right]} = 1 \Rightarrow A = \pi\left(\frac{a\sqrt{c^2-z^2}}{c}\right)\left(\frac{b\sqrt{c^2-z^2}}{c}\right)$

$\Rightarrow V = 2\int_0^c \frac{\pi ab}{c^2}(c^2-z^2)\,dz = \frac{2\pi ab}{c^2}\left[c^2z - \frac{z^3}{3}\right]_0^c = \frac{2\pi ab}{c^2}\left(\frac{2}{3}c^3\right) = \frac{4\pi abc}{3}$. Note that if $r = a = b = c$,

then $V = \frac{4\pi r^3}{3}$, which is the volume of a sphere.

78. The ellipsoid has the form $\frac{x^2}{R^2} + \frac{y^2}{R^2} + \frac{z^2}{c^2} = 1$. To determine c^2 we note that the point $(0, r, h)$ lies on the surface of the barrel. Thus, $\frac{r^2}{R^2} + \frac{h^2}{c^2} = 1 \Rightarrow c^2 = \frac{h^2R^2}{R^2 - r^2}$. We calculate the volume by the disk method:

$V = \pi\int_{-h}^{h} y^2\,dz$. Now, $\frac{y^2}{R^2} + \frac{z^2}{c^2} = 1 \Rightarrow y^2 = R^2\left(1 - \frac{z^2}{c^2}\right) = R^2\left[1 - \frac{z^2(R^2-r^2)}{h^2R^2}\right] = R^2 - \left(\frac{R^2-r^2}{h^2}\right)z^2$

$\Rightarrow V = \pi\int_{-h}^{h}\left[R^2 - \left(\frac{R^2-r^2}{h^2}\right)z^2\right]dz = \pi\left[R^2z - \frac{1}{3}\left(\frac{R^2-r^2}{h^2}\right)z^3\right]_{-h}^{h} = 2\pi\left[R^2h - \frac{1}{3}(R^2-r^2)h\right] = 2\pi\left(\frac{2R^2h}{3} + \frac{r^2h}{3}\right)$

$= \frac{4}{3}\pi R^2h + \frac{2}{3}\pi r^2h$, the volume of the barrel. If $r = R$, then $V = 2\pi R^2h$ which is the volume of a cylinder of radius R and height 2h. If $r = 0$ and $h = R$, then $V = \frac{4}{3}\pi R^3$ which is the volume of a sphere.

79. We calculate the volume by the slicing method, taking slices parallel to the xy-plane. For fixed z, $\frac{x^2}{a^2}+\frac{y^2}{b^2}=\frac{z}{c}$ gives the ellipse $\frac{x^2}{\left(\frac{za^2}{c}\right)}+\frac{y^2}{\left(\frac{zb^2}{c}\right)}=1$. The area of this ellipse is $\pi\left(a\sqrt{\frac{z}{c}}\right)\left(b\sqrt{\frac{z}{c}}\right)=\frac{\pi abz}{c}$ (see Exercise 77a). Hence the volume is given by $V=\int_0^h \frac{\pi abz}{c}\,dz=\left[\frac{\pi abz^2}{2c}\right]_0^h=\frac{\pi abh^2}{c}$. Now the area of the elliptic base when $z=h$ is $A=\frac{\pi abh}{c}$, as determined previously. Thus, $V=\frac{\pi abh^2}{c}=\frac{1}{2}\left(\frac{\pi abh}{c}\right)h=\frac{1}{2}(\text{base})(\text{altitude})$, as claimed.

80. (a) For each fixed value of z, the hyperboloid $\frac{x^2}{a^2}+\frac{y^2}{b^2}-\frac{z^2}{c^2}=1$ results in a cross-sectional ellipse $\frac{x^2}{\left[\frac{a^2(c^2+z^2)}{c^2}\right]}+\frac{y^2}{\left[\frac{b^2(c^2+z^2)}{c^2}\right]}=1$. The area of the cross-sectional ellipse (see Exercise 77a) is $A(z)=\pi\left(\frac{a}{c}\sqrt{c^2+z^2}\right)\left(\frac{b}{c}\sqrt{c^2+z^2}\right)=\frac{\pi ab}{c^2}(c^2+z^2)$. The volume of the solid by the method of slices is $V=\int_0^h A(z)\,dz=\int_0^h \frac{\pi ab}{c^2}(c^2+z^2)\,dz=\frac{\pi ab}{c^2}\left[c^2z+\frac{1}{3}z^3\right]_0^h=\frac{\pi ab}{c^2}\left(c^2h+\frac{1}{3}h^3\right)=\frac{\pi abh}{3c^2}(3c^2+h^2)$

(b) $A_0=A(0)=\pi ab$ and $A_h=A(h)=\frac{\pi ab}{c^2}(c^2+h^2)$, from part (a) $\Rightarrow V=\frac{\pi abh}{3c^2}(3c^2+h^2)$
$=\frac{\pi abh}{3}\left(2+1+\frac{h^2}{c^2}\right)=\frac{\pi abh}{3}\left(2+\frac{c^2+h^2}{c^2}\right)=\frac{h}{3}\left[2\pi ab+\frac{\pi ab}{c^2}(c^2+h^2)\right]=\frac{h}{3}(2A_0+A_h)$

(c) $A_m=A\left(\frac{h}{2}\right)=\frac{\pi ab}{c^2}+\left(c^2+\frac{h^2}{4}\right)=\frac{\pi ab}{4c^2}(4c^2+h^2)\Rightarrow \frac{h}{6}(A_0+4A_m+A_h)$
$=\frac{h}{6}\left[\pi ab+\frac{\pi ab}{c^2}(4c^2+h^2)+\frac{\pi ab}{c^2}(c^2+h^2)\right]=\frac{\pi abh}{6c^2}(c^2+4c^2+h^2+c^2+h^2)=\frac{\pi abh}{6c^2}(6c^2+2h^2)$
$=\frac{\pi abh}{3c^2}(3c^2+h^2)=V$ from part (a)

81. $y=y_1\Rightarrow\frac{z}{c}=\frac{y_1^2}{b^2}-\frac{x^2}{a^2}$, a parabola in the plane $y=y_1\Rightarrow$ vertex when $\frac{dz}{dx}=0$ or $c\,\frac{dz}{dx}=-\frac{2x}{a^2}=0\Rightarrow x=0$
$\Rightarrow$ Vertex$\left(0,y_1,\frac{cy_1^2}{b^2}\right)$; writing the parabola as $x^2=-\frac{a^2}{c}z+\frac{cy_1^2}{b^2}$ we see that $4p=-\frac{a^2}{c}\Rightarrow p=-\frac{a^2}{4c}$
$\Rightarrow$ Focus$\left(0,y_1,\frac{cy_1^2}{b^2}-\frac{a^2}{4c}\right)$

82. The curve has the general form $Ax^2+By^2+Dxy+Gx+Hy+K=0$ which is the same form as Eq. (1) in Section 9.3 for a conic section (including the degenerate cases) in the xy-plane.

83. No, it is not mere coincidence. A plane parallel to one of the coordinate planes will set one of the variables x, y, or z equal to a constant in the general equation $Ax^2+By^2+Cz^2+Dxy+Eyz+Fxz+Gx+Hy+Jz+K=0$ for a quadric surface. The resulting equation then has the general form for a conic in that parallel plane. For example, setting $y=y_1$ results in the equation $Ax^2+Cz^2+D'x+E'z+Fxz+Gx+Jz+K'=0$ where

$D' = Dy_1$, $E' = Ey_1$, and $K' = K + By_1^2 + Hy_1$, which is the general form of a conic section in the plane $y = y_1$ by Section 9.3.

84. The trace will be a conic section. To see why, solve the plane's equation $Ax + By + Cz = 0$ for one of the variables in terms of the other two and substitute into the equation $Ax^2 + By^2 + Cz^2 + \ldots + K = 0$. The result will be a second degree equation in the remaining two variables. By Section 9.3, this equation will represent a conic section. (See also the discussion in Exercises 82 and 83.)

85. $z = y^2$

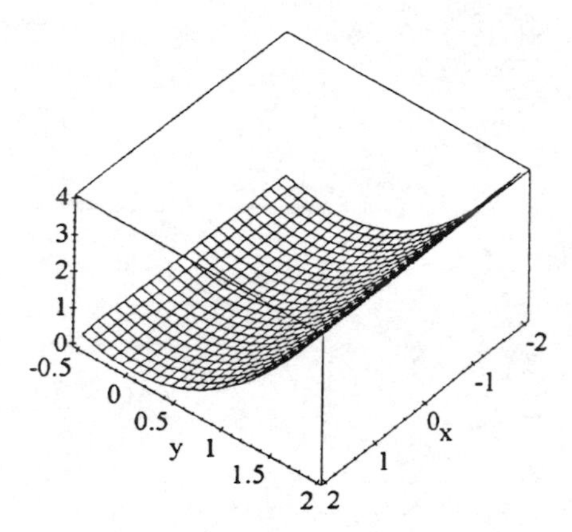

86. $z = 1 - y^2$

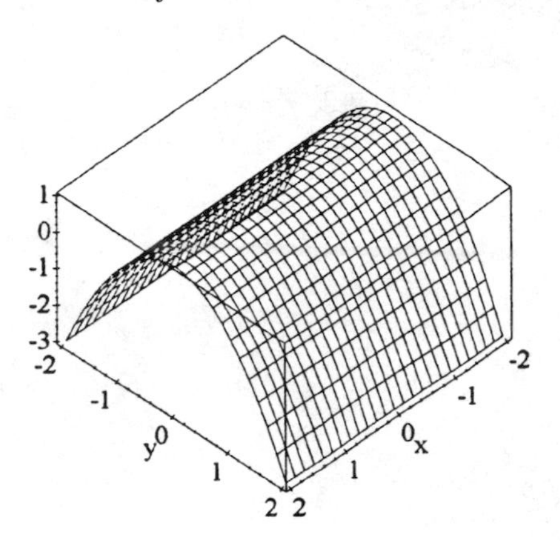

87. $z = x^2 + y^2$

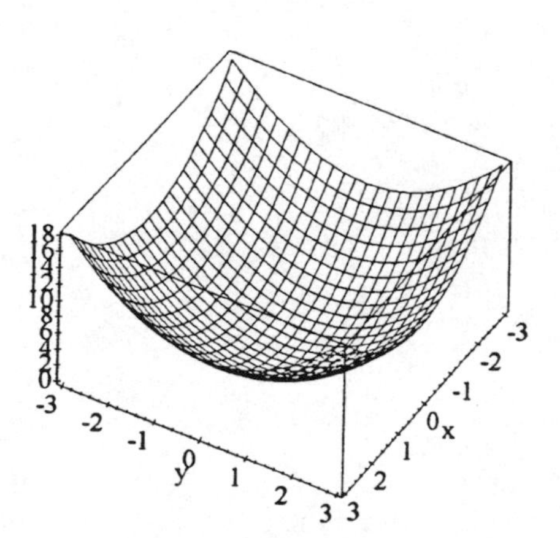

88. $z = x^2 + 2y^2$

(a)

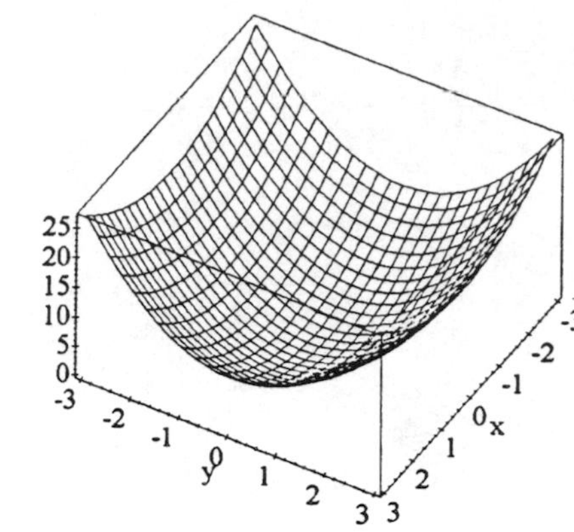

(b)

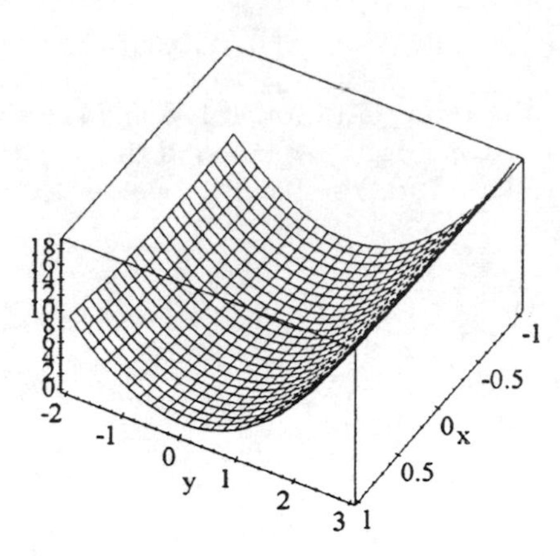

(c)

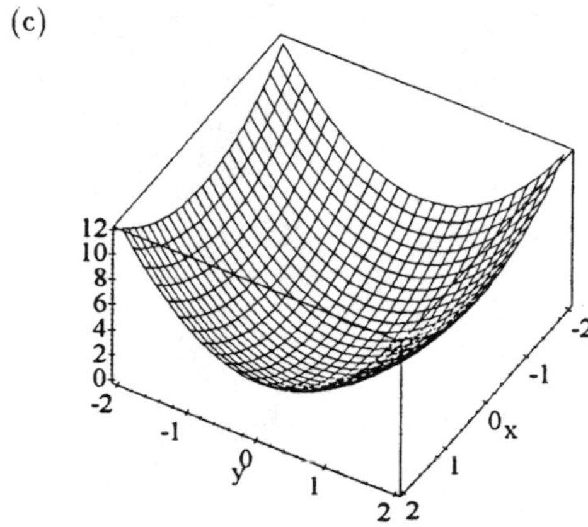

(d)

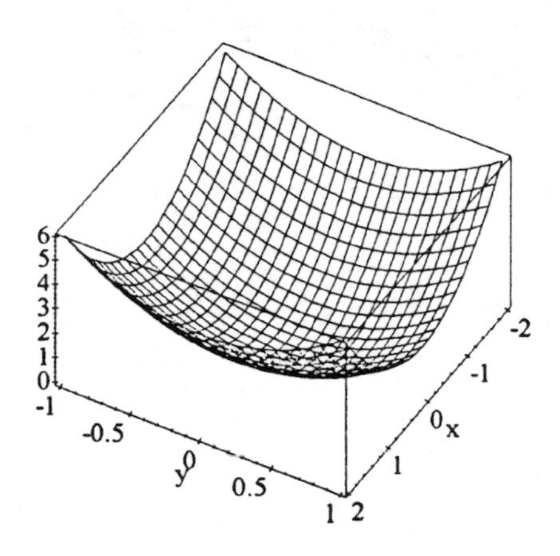

89-94. Example CAS commands:

<u>Maple</u>:

```
with(plots):
eq1:= x^2/9 - y^2/16 -z^2/2 = 1;
implicitplot3d(eq1, x = -15..15, y = -9..9, z = -7..7, title = `Hyperboloid of Two Sheets`);
```

<u>Mathematica</u>:

```
ContourPlot3D[ x^2/9 - y^2/16 - z^2/2 - 1,
{x,-9,9}, {y,-12,12}, {z,-5,5},
PlotLabel -> "Elliptic Hyperboloid of Two Sheets" ]
```

10.7 CYLINDRICAL AND SPHERICAL COORDINATES

	Rectangular	Cylindrical	Spherical
1.	$(0,0,0)$	$(0,0,0)$	$(0,0,0)$
2.	$(1,0,0)$	$(1,0,0)$	$\left(1,\frac{\pi}{2},0\right)$
3.	$(0,1,0)$	$\left(1,\frac{\pi}{2},0\right)$	$\left(1,\frac{\pi}{2},\frac{\pi}{2}\right)$
4.	$(0,0,1)$	$(0,0,1)$	$(1,0,0)$
5.	$(1,0,0)$	$(1,0,0)$	$\left(1,\frac{\pi}{2},0\right)$
6.	$(\sqrt{2},0,1)$	$(\sqrt{2},0,1)$	$\left(\sqrt{3},\cos\frac{1}{\sqrt{3}},0\right)$
7.	$(0,1,1)$	$\left(1,\frac{\pi}{2},1\right)$	$\left(\sqrt{2},\frac{\pi}{4},\frac{\pi}{2}\right)$
8.	$\left(0,-\frac{3}{2},\frac{\sqrt{3}}{2}\right)$	$\left(\frac{3}{2},\frac{3\pi}{2},\frac{\sqrt{3}}{2}\right)$	$\left(\sqrt{3},\frac{\pi}{3},-\frac{\pi}{2}\right)$
9.	$(0,-2\sqrt{2},0)$	$\left(2\sqrt{2},\frac{3\pi}{2},0\right)$	$\left(2\sqrt{2},\frac{\pi}{2},\frac{3\pi}{2}\right)$
10.	$(0,0,-\sqrt{2})$	$(0,0,-\sqrt{2})$	$\left(\sqrt{2},\pi,\frac{3\pi}{2}\right)$

11. $r = 0 \Rightarrow$ rectangular, $x^2 + y^2 = 0$; spherical, $\phi = 0$ or $\phi = \pi$; the z-axis

12. $x^2 + y^2 = 5 \Rightarrow$ cylindrical, $r = \sqrt{5}$; spherical, $\rho \sin\phi = \sqrt{5}$; a cylinder

13. $z = 0 \Rightarrow$ cylindrical, $z = 0$; spherical, $\phi = \frac{\pi}{2}$; the xy-plane

14. $z = -2 \Rightarrow$ cylindrical, $z = -2$; spherical, $\rho \cos\phi = -2$; the plane $z = -2$

15. $z = \sqrt{x^2+y^2}$, $z \le 1 \Rightarrow$ cylindrical, $z = r$, $0 \le r \le 1$; spherical, $\phi = \tan^{-1}\frac{\sqrt{x^2+y^2}}{z} = \tan^{-1} 1 = \frac{\pi}{4}$,
$\rho = \sqrt{x^2+y^2+z^2} = \sqrt{x^2+y^2+x^2+y^2} = \sqrt{2(x^2+y^2)} = \sqrt{2z^2} = \sqrt{2}\,|z| \le \sqrt{2} \Rightarrow 0 \le \rho \le \sqrt{2}$; a (finite) cone

16. $z = \sqrt{x^2+y^2}$, $1 \le z \le 2 \Rightarrow$ cylindrical, $z = r$, $1 \le r \le 2$; spherical, $\phi = \tan^{-1}\frac{\sqrt{x^2+y^2}}{z} = \tan^{-1} 1 = \frac{\pi}{4}$,
$1 \le z \le 2 \Rightarrow 1 \le \rho\cos\phi \le 2 \Rightarrow 1 \le \rho\cos\left(\frac{\pi}{4}\right) \le 2 \Rightarrow 1 \le \frac{\rho}{\sqrt{2}} \le 2 \Rightarrow \sqrt{2} \le \rho \le 2\sqrt{2}$, a frustum of a cone

17. $\rho \sin\phi \cos\theta = 0 \Rightarrow$ rectangular, $x = 0$; cylindrical $\theta = \frac{\pi}{2}$; the yz-plane

18. $\tan^2\phi = 1 \Rightarrow$ rectangular, $x^2 + y^2 = z^2$; cylindrical, $r^2 = z^2$; a circular cone symmetric about the z-axis

19. $x^2+y^2+z^2=4 \Rightarrow$ cylindrical, $r^2+z^2=4$; spherical, $\rho=2$; a sphere of radius 2 centered at the origin

20. $x^2+y^2+\left(z-\frac{1}{2}\right)^2=\frac{1}{4} \Rightarrow$ cylindrical, $r^2+\left(z-\frac{1}{2}\right)^2=\frac{1}{4}$; spherical, $(\rho \sin \phi)^2+\left(\rho \cos \phi-\frac{1}{2}\right)^2=\frac{1}{4}$
$\Rightarrow \rho^2 \sin^2 \phi+\rho^2 \cos^2 \phi-\rho \cos \phi+\frac{1}{4}=\frac{1}{4} \Rightarrow \rho^2\left(\sin^2 \phi+\cos^2 \phi\right)-\rho \sin \phi=0 \Rightarrow \rho^2-\rho \cos \phi=0$
$\Rightarrow \rho(\rho-\cos \phi)=0 \Rightarrow \rho=\cos \phi$ since $\rho \neq 0$, a sphere of radius $\frac{1}{2}$ centered at $\left(0,0,\frac{1}{2}\right)$ (rectangular)

21. $\rho=5 \cos \phi \Rightarrow$ rectangular, $\sqrt{x^2+y^2+z^2}=5 \cos\left(\cos^{-1}\left(\frac{z}{\sqrt{x^2+y^2+z^2}}\right)\right)=\sqrt{x^2+y^2+z^2}=\frac{5z}{\sqrt{x^2+y^2+z^2}}$
$\Rightarrow x^2+y^2+z^2=5z \Rightarrow x^2+y^2+z^2-5z+\frac{25}{4}=\frac{25}{4} \Rightarrow x^2+y^2+\left(z-\frac{5}{2}\right)^2=\frac{25}{4}$; cylindrical,
$r^2+\left(z-\frac{5}{2}\right)^2=\frac{25}{4} \Rightarrow r^2+z^2=5z$, a sphere of radius $\frac{5}{2}$ centered at $\left(0,0,\frac{5}{2}\right)$ (rectangular)

22. $\rho=-6 \cos \phi \Rightarrow$ rectangular, $\sqrt{x^2+y^2+z^2}=-6 \cos\left(\cos^{-1}\left(\frac{z}{\sqrt{x^2+y^2+z^2}}\right)\right)=\frac{-6z}{\sqrt{x^2+y^2+z^2}}$
$\Rightarrow x^2+y^2+z^2=-6z \Rightarrow x^2+y^2+z^2+6z=0 \Rightarrow x^2+y^2+z^2+6z+9=9 \Rightarrow x^2+y^2+(z+3)^2=9$;
cylindrical, $r^2+(z+3)^2=9 \Rightarrow r^2+z^2=-6z$, a sphere of radius 3 centered at $(0,0,-3)$ (rectangular)

23. $r=\csc \theta \Rightarrow$ rectangular, $r=\frac{r}{y} \Rightarrow y=1$ since $r \neq 0$; spherical, $\rho \sin \phi=\csc \theta \Rightarrow \rho \sin \phi \sin \theta=1$, the plane $y=1$

24. $r=-3 \sec \theta \Rightarrow r \cos \theta=-3 \Rightarrow$ rectangular, $x=-3$; spherical, $\rho \sin \phi \cos \theta=-3$, the plane $x=-3$

25. $\rho=\sqrt{2} \sec \phi \Rightarrow \rho=\frac{\sqrt{2}}{\cos \phi} \Rightarrow$ rectangular, $\sqrt{x^2+y^2+z^2}=\frac{\sqrt{2}}{\cos\left(\cos^{-1}\left(\frac{z}{\sqrt{x^2+y^2+z^2}}\right)\right)}$
$\Rightarrow \sqrt{x^2+y^2+z^2}=\frac{\sqrt{2}}{\left(\frac{z}{\sqrt{x^2+y^2+z^2}}\right)} \Rightarrow z\sqrt{x^2+y^2+z^2}=\sqrt{2}\sqrt{x^2+y^2+z^2} \Rightarrow z=\sqrt{2}$ since $x^2+y^2+z^2 \neq 0$;
cylindrical, $z=\sqrt{2}$, the plane $z=\sqrt{2}$

26. $\rho=9 \csc \phi \Rightarrow \rho \sin \phi=9 \Rightarrow$ cylindrical, $r=9$; rectangular $\sqrt{x^2+y^2}=9 \Rightarrow x^2+y^2=81$, a circular cylinder of radius 9 with axis about the z-axis

27. $x^2+y^2+(z-1)^2=1$, $z \leq 1 \Rightarrow$ cylindrical, $r^2+(z-1)^2=1 \Rightarrow r^2+z^2-2z+1=1 \Rightarrow r^2+z^2=2z$, $z \leq 1$;
spherical, $x^2+y^2+z^2-2z=0 \Rightarrow \rho^2-2\rho \cos \phi=0 \Rightarrow \rho(\rho-2 \cos \phi)=0 \Rightarrow \rho=2 \cos \phi$, $\frac{\pi}{4} \leq \phi \leq \frac{\pi}{2}$ since $\rho \neq 0$, the lower half (hemisphere) of the sphere of radius 1 centered at $(0,0,1)$ (rectangular)

28. $r^2+z^2=4$, $z \leq -\sqrt{2} \Rightarrow$ spherical, $\rho=2$ and $\frac{3\pi}{4} \leq \phi \leq \pi$; rectangular, $x^2+y^2+z^2=4$, the lower cap cut by the plane $z=-\sqrt{2}$ from the sphere of radius 2 centered at the origin

29. $\rho = 3,\ \frac{\pi}{3} \le \phi \le \frac{2\pi}{3} \Rightarrow$ rectangular, $\sqrt{x^2+y^2+z^2} = 3$ and $3\cos\left(\frac{\pi}{3}\right) \ge z \ge 3\cos\left(\frac{2\pi}{3}\right) \Rightarrow x^2+y^2+z^2 = 9$ and $-\frac{3}{2} \le z \le \frac{3}{2}$; cylindrical, $r^2+z^2 = 9$ and $-\frac{3}{2} \le z \le \frac{3}{2}$, the portion of the sphere of radius 3 centered at the origin between the planes $z = -\frac{3}{2}$ and $z = \frac{3}{2}$

30. $x^2+y^2+z^2 = 3,\ 0 \le z \le \frac{\sqrt{3}}{2} \Rightarrow$ cylindrical, $r^2+z^2 = 3$ and $0 \le z \le \frac{\sqrt{3}}{2}$; spherical, $\rho^2 = 3$ and $\cos^{-1}\left(\frac{\left(\frac{\sqrt{3}}{2}\right)}{\sqrt{3}}\right) \le \phi \le \cos^{-1}\left(\frac{0}{\sqrt{3}}\right) \Rightarrow \rho = \sqrt{3}$ and $\frac{\pi}{3} \le \phi \le \frac{\pi}{2}$, the portion of the sphere of radius $\sqrt{3}$ centered at the origin between the xy-plane and the plane $z = \frac{\sqrt{3}}{2}$

31. $z = 4 - 4r^2,\ 0 \le r \le 1 \Rightarrow$ spherical, $\rho\cos\phi = 4 - 4\rho^2\sin^2\phi$ and $0 \le \phi \le \frac{\pi}{2}$; rectangular $z = 4 - 4(x^2+y^2)$ and $0 \le z \le 4$, the upper portion cut from the paraboloid $z = 4 - 4(x^2+y^2)$ by the xy-plane

32. $z = 4 - r,\ 0 \le r \le 4 \Rightarrow$ spherical, $\rho\cos\phi = 4 - \rho\sin\phi$ and $0 \le \phi \le \frac{\pi}{2} \Rightarrow \rho(\cos\phi + \sin\phi) = 4$ and $0 \le \phi \le \frac{\pi}{2}$; rectangular, $z = 4 - \sqrt{x^2+y^2}$ and $0 \le z \le 4$, a cone with vertex at $(0,0,4)$ (rectangular) and base the circle $x^2+y^2 = 16$ in the xy-plane

33. $\phi = \frac{3\pi}{4},\ 0 \le \rho \le \sqrt{2} \Rightarrow$ rectangular, $\cos\frac{3\pi}{4} = \cos\phi = \frac{z}{\sqrt{x^2+y^2+z^2}} \Rightarrow -\frac{1}{\sqrt{2}} = \frac{z}{\sqrt{x^2+y^2+z^2}}$ $\Rightarrow \sqrt{x^2+y^2+z^2} = -\sqrt{2}z \Rightarrow x^2+y^2+z^2 = 2z^2 \Rightarrow x^2+y^2-z^2 = 0$ with $z \le 0 \Rightarrow z = -\sqrt{x^2+y^2}$ and $0 \ge z \ge \sqrt{2}\cos\frac{3\pi}{4} \Rightarrow z = -\sqrt{x^2+y^2}$ and $-1 \le z \le 0$; cylindrical $x^2+y^2-z^2 = 0 \Rightarrow r^2 - z^2 = 0$ $\Rightarrow r = -z$ or $r = z$, but $r \ge 0$ and $z \le 0 \Rightarrow r = -z$, a cone with vertex at the origin and base the circle $x^2+y^2 = 1$ in the plane $z = -1$

34. $\phi = \frac{\pi}{2},\ 0 \le \rho \le \sqrt{7} \Rightarrow$ cylindrical, $0 \le r \le \sqrt{7}$; rectangular, $x^2+y^2 \le 7$ and $z = 0$, the disk of radius $\sqrt{7}$ centered at the origin in the xy-plane

35. $z + r^2\cos 2\theta = 0 \Rightarrow z + r^2(\cos^2\theta - \sin^2\theta) = 0 \Rightarrow z + (r\cos\theta)^2 - (r\sin\theta)^2 = 0 \Rightarrow$ rectangular, $z + x^2 - y^2 = 0$ or $z = y^2 - x^2$; spherical, $z + r^2\cos 2\theta = 0 \Rightarrow \rho\cos\phi + (\rho\sin\phi)^2(\cos 2\theta) = 0 \Rightarrow \rho(\cos\phi + \rho\sin^2\phi\cos 2\theta) = 0$ $\Rightarrow \cos\phi + \rho\sin^2\phi\cos 2\theta = 0$ since $\rho \ne 0$, a hyperbolic paraboloid

36. $z^2 - r^2 = 1 \Rightarrow$ rectangular, $z^2 - (x^2+y^2) = 1$; spherical, $\rho^2\cos^2\phi - \rho^2\sin^2\phi = 1 \Rightarrow \rho^2 = \frac{1}{\cos 2\phi}$ $\Rightarrow \rho^2\cos 2\phi = 1$, hyperboloid of 2 sheets

37. $r^2+z^2 = 4r\cos\theta + 6r\sin\theta + 2z \Rightarrow x^2+y^2+z^2 = 4x + 6y + 2z \Rightarrow (x^2-4x+4)+(y^2-6y+9)+(z^2-2z+1)$ $= 14 \Rightarrow (x-2)^2+(y-3)^2+(z-1)^2 = 14 \Rightarrow$ the center is located at $(2,3,1)$ in rectangular coordinates

38. $\rho = (2 \sin \phi)(\cos \theta - 2 \sin \theta) \Rightarrow \rho = 2 (\sin \phi \cos \theta) - 4(\sin \phi \sin \theta) \Rightarrow \rho = 2\left(\frac{x}{\rho}\right) - 4\left(\frac{y}{\rho}\right) \Rightarrow \rho^2 = 2x - 4y$
$\Rightarrow x^2 + y^2 + z^2 = 2x - 4y \Rightarrow (x^2 - 2x + 1) + (y^2 + 4y + 4) + z^2 = 1 + 4 \Rightarrow (x-1)^2 + (y+2)^2 + (z-0)^2 = 5$
$\Rightarrow$ the center is located at $(1, -2, 0)$ in rectangular coordinates

39. Right circular cylinder parallel to the z-axis generated by the circle $r = -2 \sin \theta$ in the $r\theta$-plane

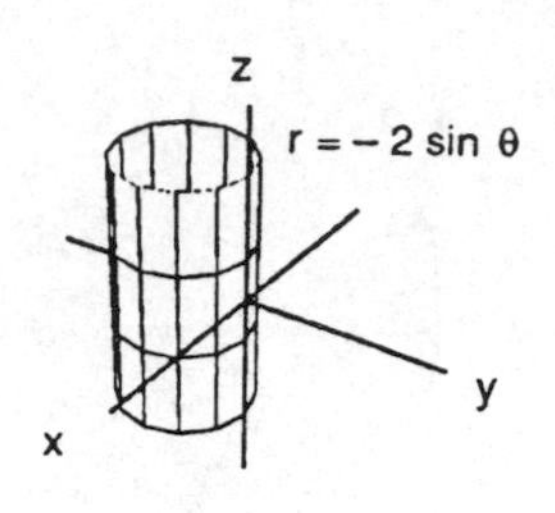

40. Right circular cylinder parallel to the z-axis generated by the circle $r = 2 \cos \theta$ in the $r\theta$-plane

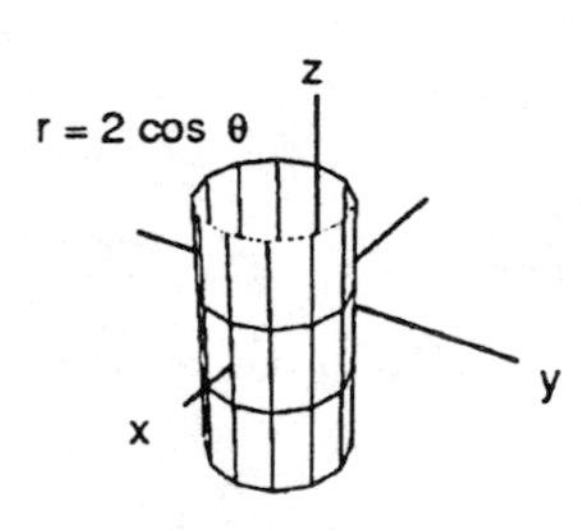

41. Cylinder of lines parallel to the z-axis generated by the cardioid $r = 1 - \cos \theta$ in the $r\theta$-plane

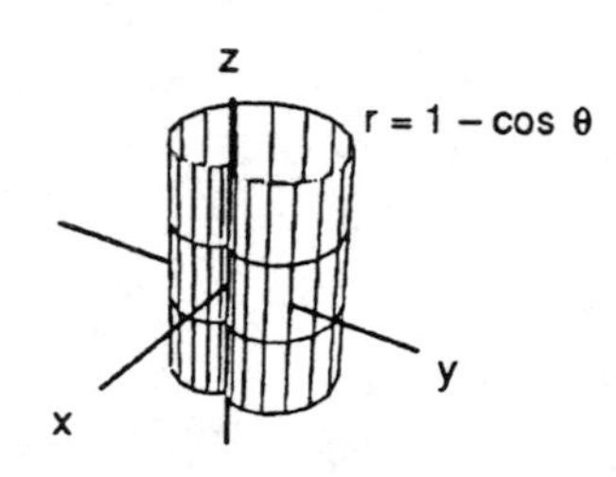

42. Cylinder of lines parallel to the z-axis generated by the cardioid $r = 1 + \sin \theta$ in the $r\theta$-plane

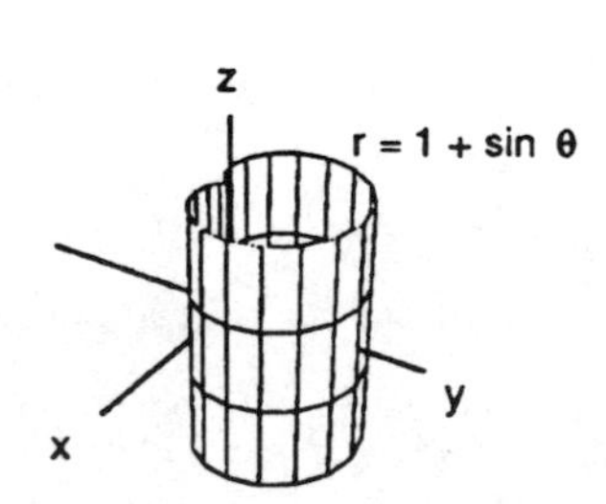

43. Cardioid of revolution symmetric about the y-axis, cusp at the origin pointing down

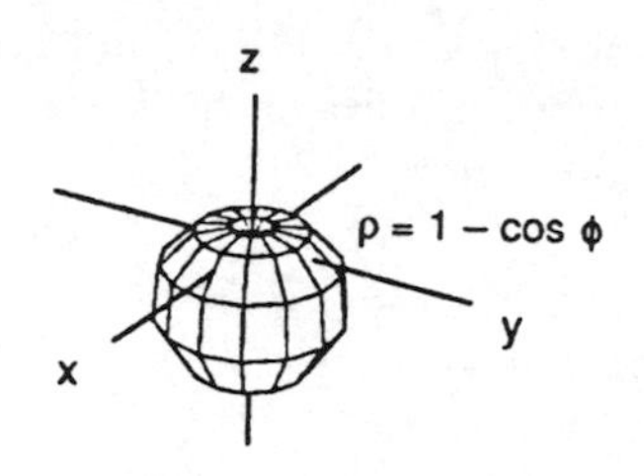

44. Cardioid of revolution symmetric about the y-axis, cusp at the origin pointing up

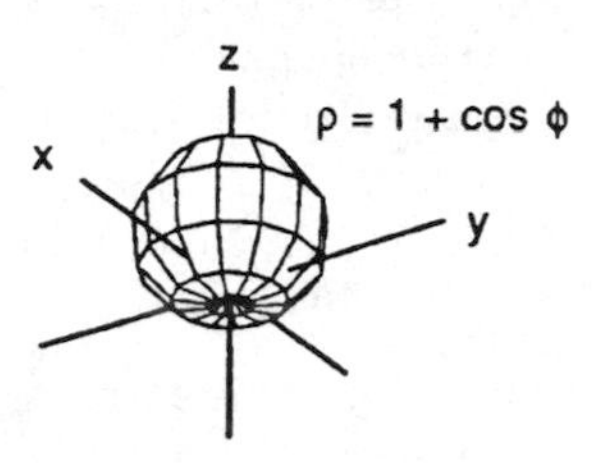

45. (a) $z = c \Rightarrow \rho \cos \phi = c \Rightarrow \rho = \frac{c}{\csc \phi} \Rightarrow \rho = c \sec \phi$

(b) The xy-plane is perpendicular to the z-axis $\Rightarrow \phi = \frac{\pi}{2}$

46. $x^2 + y^2 = a^2 \Rightarrow (\rho \sin \phi \cos \theta)^2 + (\rho \sin \phi \sin \theta)^2 = a^2 \Rightarrow (\rho^2 \sin^2 \phi)(\cos^2 \theta + \sin^2 \theta) = a^2 \Rightarrow \rho^2 \sin^2 \phi = a^2$
$\Rightarrow \rho \sin \phi = a$ or $\rho \sin \phi = -a \Rightarrow \rho \sin \phi = a$ or $\rho = a \csc \phi$, since $0 \le \phi \le \pi$ and $\rho \ge 0$

47. (a) A plane perpendicular to the x-axis has the form $x = a$ in rectangular coordinates $\Rightarrow r \cos \theta = a$
$\Rightarrow r = \frac{a}{\cos \theta} \Rightarrow r = a \sec \theta$, in cylindrical coordinates

(b) A plane perpendicular to the y-axis has the form $y = b$ in rectangular coordinates $\Rightarrow r \sin \theta = b$
$\Rightarrow r = \frac{b}{\sin \theta} \Rightarrow r = b \csc \theta$, in cylindrical coordinates

48. $ax + by = c \Rightarrow a(r \cos \theta) + b(r \sin \theta) = c \Rightarrow r(a \cos \theta + b \sin \theta) = c \Rightarrow r = \frac{c}{a \cos \theta + b \sin \theta}$

49. The equation $r = f(z)$ implies that the point (r, θ, z) $= (f(z), \theta, z)$ will lie on the surface for all θ. In particular $(f(z), \theta + \pi, z)$ lies on the surface whenever $(f(z), \theta, z)$ does $\Rightarrow$ the surface is symmetric with respect to the z-axis.

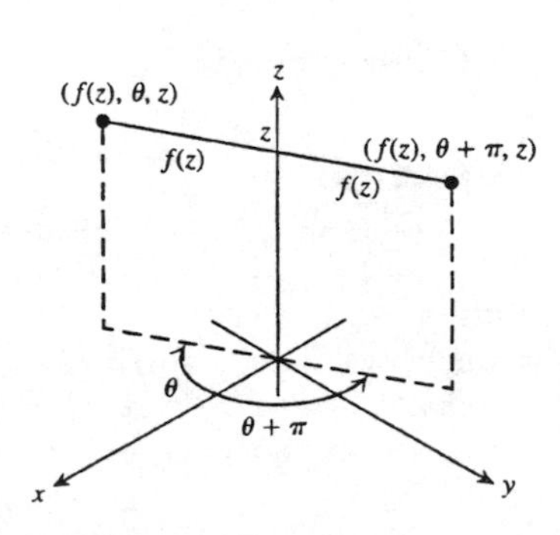

50. The equation $\rho = f(\phi)$ implies that the point $(\rho, \phi, \theta) = (f(\phi), \phi, \theta)$ lies on the surface for all θ. In particular, if $(f(\phi), \phi, \theta)$ lies on the surface, then $(f(\phi), \phi, \theta + \pi)$ lies on the surface, so the surface is symmetric with respect to the z-axis.

51. Example CAS commands:

Maple:

```
with(plots):
eq:= r^2 + z^2 = 2*r*(cos(theta) + sin(theta)) + 2;
subs(r = sqrt(x^2 + y^2), theta = arctan(y/x), eq);
simplify(",trig);
eq2: = x^2 + y^2 + z^2 - 2*x - 2*y = 0;
implicitplot3d(eq2, x=0..3, y=0..3, z=-2..2);
are:=solve(eq,r);
simplify(are[1],trig);
r:= unapply(",(theta,z));
cylinderplot(r, Pi/4..9*Pi/4, -2..2, grid = [100,100]);
```

Mathematica:

```
(We need the ParametricPlot3D package for SphericalPlot3D)
<< Graphics`ParametricPlot3D`

(ContourPlot3D allows implicit plotting in 3D)
<< Graphics`ContourPlot3D`

Clear[r,theta,x,y,z]
eqn = r^2 + z^2 == 2 r (Cos[theta]+Sin[theta]) + 2
Solve[ eqn, r ]
r[theta_,z_] = r /. %[[2]] // Simplify

Note: the CylindricalPlot3D function only handles plotting z(r,theta), not
r(theta,z), so we must use the more general ParametricPlot3D.

ParametricPlot3D[
{r[theta,z] Cos[theta], r[theta,z] Sin[theta], z},
{theta,Pi/4,9Pi/4}, {z,-2,2} ]
Map[Expand, eqn]
% /. {r Cos[theta] -> x, r Sin[theta] -> y,
r^2 -> x^2 + y^2}
eqn2 = Map[ (# -2x -2y +2)&, % ]
ContourPlot3D[ eqn2[[1]],
{x,-1,3}, {y,-1,3}, {z,-2,2},
Contours -> {eqn2[[2]]} ]
Clear[rho,theta,phi,x,y,z]
```

52. Example CAS commands:

Maple:

```
with(plots):
eq:= rho^2 = 2*rho*(cos(theta)*sin(phi) - cos(phi)) + 2;
solve(",rho);
simplify("[1],trig);
f:= unapply(",(theta,phi));
sphereplot(f,0..2*Pi,-Pi..Pi);
subs(theta=arctan(y/x), phi = arccos(z/sqrt(x^2 + y^2 + z^2)), rho = sqrt(x^2 + y^2
+ z^2), eq);
eq1:= simplify(",trig);
```

```
#It's now easy to simplify the equation by hand to obtain:
eq2:= x^2 + y^2 + z^2 = 2*x - 2*z + 2;
Implicitplot3d(eq2,x=-1..3, y=-2..2, z=-3..1);
```

Mathematica:

```
eqn = rho^2 == 2 rho (Cos[theta]Sin[phi] - Cos[phi]) + 2
Solve[ eqn, rho ]
rho[theta_,phi_] = rho /. %[[2]] // Simplify
Note: in the SphericalPlot3D function, the range for the polar angle (phi)
must come before that for the azimuthal angle (theta).
SphericalPlot3D[
rho[theta,phi], {phi,0,Pi}, {theta,0,2Pi} ]
Map[Expand,eqn]
% l. {
 rho Cos[theta] Sin[phi] -> x,
 rho Sin[theta] Sin[phi] -> y,
 rho Cos[phi] -> z,
 rho^2 -> x^2 + y^2 + z^2 }
eqn2 = Map[ (# -2x +2z +2)&, % ]
ContourPlot3D[ eqn2[[1]],
{x,-1,3}, {y,-2,2}, {z,-3,1},
Contours -> {eqn2[[2]]} ]
```

CHAPTER 10 PRACTICE EXERCISES

1. $\theta = 0 \Rightarrow \mathbf{u} = \mathbf{i}$; $\theta = \frac{\pi}{2} \Rightarrow \mathbf{u} = \mathbf{j}$; $\theta = \frac{2\pi}{3} \Rightarrow \mathbf{u} = -\frac{1}{2}\mathbf{i} + \frac{\sqrt{3}}{2}\mathbf{j}$;

   $\theta = \frac{5\pi}{4} \Rightarrow \mathbf{u} = -\frac{\sqrt{2}}{2}\mathbf{i} - \frac{\sqrt{2}}{2}\mathbf{j}$; $\theta = \frac{5\pi}{3} \Rightarrow \mathbf{u} = \frac{1}{2}\mathbf{i} - \frac{\sqrt{3}}{2}\mathbf{j}$

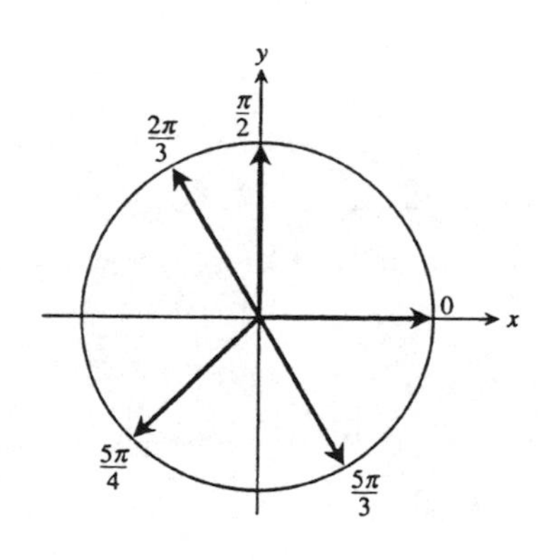

2. (a) Rotating $\mathbf{i}$ clockwise 45° yields $\frac{\sqrt{2}}{2}\mathbf{i} - \frac{\sqrt{2}}{2}\mathbf{j}$ (b) Rotating $\mathbf{j}$ counterclockwise 120° yields $-\frac{\sqrt{3}}{2}\mathbf{i} - \frac{1}{2}\mathbf{j}$

3. length $= |\sqrt{2}\mathbf{i} + \sqrt{2}\mathbf{j}| = \sqrt{2+2} = 2$, $\sqrt{2}\mathbf{i} + \sqrt{2}\mathbf{j} = 2\left(\frac{1}{\sqrt{2}}\mathbf{i} + \frac{1}{\sqrt{2}}\mathbf{j}\right) \Rightarrow$ the direction is $\frac{1}{\sqrt{2}}\mathbf{i} + \frac{1}{\sqrt{2}}\mathbf{j}$

4. length $= |-\mathbf{i} - \mathbf{j}| = \sqrt{1+1} = \sqrt{2}$, $-\mathbf{i} - \mathbf{j} = \sqrt{2}\left(-\frac{1}{\sqrt{2}}\mathbf{i} - \frac{1}{\sqrt{2}}\mathbf{j}\right) \Rightarrow$ the direction is $-\frac{1}{\sqrt{2}}\mathbf{i} - \frac{1}{\sqrt{2}}\mathbf{j}$

5. length $= |2\mathbf{i} - 3\mathbf{j} + 6\mathbf{k}| = \sqrt{4+9+36} = 7$, $2\mathbf{i} - 3\mathbf{j} + 6\mathbf{k} = 7\left(\frac{2}{7}\mathbf{i} - \frac{3}{7}\mathbf{j} + \frac{6}{7}\mathbf{k}\right) \Rightarrow$ the direction is $\frac{2}{7}\mathbf{i} - \frac{3}{7}\mathbf{j} + \frac{6}{7}\mathbf{k}$

6. length $= |\mathbf{i} + 2\mathbf{j} - \mathbf{k}| = \sqrt{1+4+1} = \sqrt{6}$, $\mathbf{i} + 2\mathbf{j} - \mathbf{k} = \sqrt{6}\left(\frac{1}{\sqrt{6}}\mathbf{i} + \frac{2}{\sqrt{6}}\mathbf{j} - \frac{1}{\sqrt{6}}\mathbf{k}\right) \Rightarrow$ the direction is $\frac{1}{\sqrt{6}}\mathbf{i} + \frac{2}{\sqrt{6}}\mathbf{j} - \frac{1}{\sqrt{6}}\mathbf{k}$

7. $2\frac{\mathbf{A}}{|\mathbf{A}|} = 2 \cdot \frac{4\mathbf{i} - \mathbf{j} + 4\mathbf{k}}{\sqrt{4^2 + (-1)^2 + 4^2}} = 2 \cdot \frac{4\mathbf{i} - \mathbf{j} + 4\mathbf{k}}{\sqrt{33}} = \frac{8}{\sqrt{33}}\mathbf{i} - \frac{2}{\sqrt{33}}\mathbf{j} + \frac{8}{\sqrt{33}}\mathbf{k}$

8. $-5\frac{\mathbf{A}}{|\mathbf{A}|} = -5 \cdot \frac{\left(\frac{3}{5}\right)\mathbf{i} + \left(\frac{4}{5}\right)\mathbf{j}}{\sqrt{\left(\frac{3}{5}\right)^2 + \left(\frac{4}{5}\right)^2}} = -5 \cdot \frac{\left(\frac{3}{5}\right)\mathbf{i} + \left(\frac{4}{5}\right)\mathbf{j}}{\sqrt{\frac{9}{25} + \frac{16}{25}}} = -3\mathbf{i} - 4\mathbf{j}$

9. (a) $\overrightarrow{BD} = \overrightarrow{AD} - \overrightarrow{AB}$

(b) $\overrightarrow{AP} = \overrightarrow{AB} + \frac{1}{2}\overrightarrow{BD} = \overrightarrow{AB} + \frac{1}{2}\left(\overrightarrow{AD} - \overrightarrow{AB}\right) = \overrightarrow{AB} + \frac{1}{2}\overrightarrow{AD} - \frac{1}{2}\overrightarrow{AB} = \frac{1}{2}\overrightarrow{AB} + \frac{1}{2}\overrightarrow{AD} = \frac{1}{2}\left(\overrightarrow{AB} + \overrightarrow{AD}\right)$

(c) $\overrightarrow{PC} = \overrightarrow{AC} - \overrightarrow{AP} = \left(\overrightarrow{AB} + \overrightarrow{AD}\right) - \frac{1}{2}\left(\overrightarrow{AB} + \overrightarrow{AD}\right) = \frac{1}{2}\left(\overrightarrow{AB} + \overrightarrow{AD}\right) = \overrightarrow{AP} \Rightarrow$ P is the midpoint of AC

10. $\overrightarrow{OD} = \overrightarrow{OC} + \overrightarrow{OB}$, $\overrightarrow{OE} = \overrightarrow{OD} + \overrightarrow{OA} = \overrightarrow{OC} + \overrightarrow{OB} + \overrightarrow{OA} = \overrightarrow{OC} + \overrightarrow{OA} + \overrightarrow{OB}$

11.

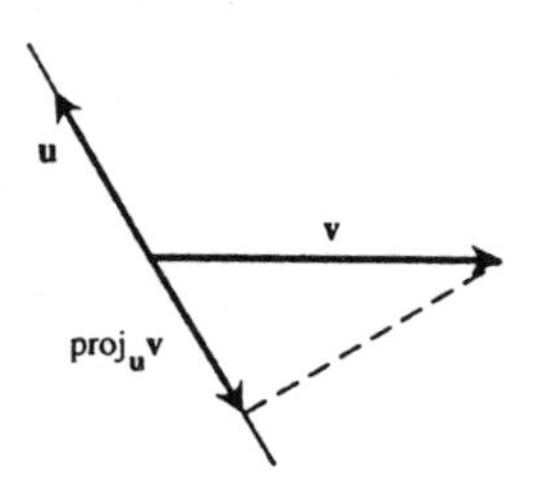

12. $\mathbf{a} = \text{proj}_{\mathbf{v}}\,\mathbf{u}$, $\mathbf{b} = \text{proj}_{\mathbf{u}}\,\mathbf{v}$, $\mathbf{c} = \mathbf{v} - \mathbf{b} = \mathbf{v} - \text{proj}_{\mathbf{u}}\,\mathbf{v}$

13. $|\mathbf{A}| = \sqrt{1+1} = \sqrt{2}$, $|\mathbf{B}| = \sqrt{4+1+4} = 3$, $\mathbf{A} \cdot \mathbf{B} = 3$, $\mathbf{B} \cdot \mathbf{A} = 3$, $\mathbf{A} \times \mathbf{B} = \begin{vmatrix} \mathbf{i} & \mathbf{j} & \mathbf{k} \\ 1 & 1 & 0 \\ 2 & 1 & -2 \end{vmatrix} = -2\mathbf{i} + 2\mathbf{j} - \mathbf{k}$,

$\mathbf{B} \times \mathbf{A} = -(\mathbf{A} \times \mathbf{B}) = 2\mathbf{i} - 2\mathbf{j} + \mathbf{k}$, $|\mathbf{A} \times \mathbf{B}| = \sqrt{4+4+1} = 3$, $\theta = \cos^{-1}\left(\frac{\mathbf{A} \cdot \mathbf{B}}{|\mathbf{A}||\mathbf{B}|}\right) = \cos^{-1}\left(\frac{1}{\sqrt{2}}\right) = \frac{\pi}{4}$,

$|\mathbf{B}| \cos\theta = \frac{3}{\sqrt{2}}$, $\text{proj}_{\mathbf{A}}\,\mathbf{B} = \left(\frac{\mathbf{A} \cdot \mathbf{B}}{\mathbf{A} \cdot \mathbf{A}}\right)\mathbf{A} = \frac{3}{2}(\mathbf{i} + \mathbf{j})$

14. $|\mathbf{A}| = \sqrt{1^2 + 1^2 + 2^2} = \sqrt{6}$, $|\mathbf{B}| = \sqrt{(-1)^2 + (-1)^2} = \sqrt{2}$, $\mathbf{A} \cdot \mathbf{B} = (1)(-1) + (1)(0) + (2)(-1) = -3$,

$\mathbf{B} \cdot \mathbf{A} = -3$, $\mathbf{A} \times \mathbf{B} = \begin{vmatrix} \mathbf{i} & \mathbf{j} & \mathbf{k} \\ 1 & 1 & 2 \\ -1 & 0 & -1 \end{vmatrix} = -\mathbf{i} - \mathbf{j} + \mathbf{k}$, $\mathbf{B} \times \mathbf{A} = -(\mathbf{A} \times \mathbf{B}) = \mathbf{i} + \mathbf{j} - \mathbf{k}$,

$|\mathbf{A}\times\mathbf{B}| = \sqrt{(-1)^2+(-1)^2+1^2} = \sqrt{3}$, $\theta = \cos^{-1}\left(\frac{\mathbf{A}\cdot\mathbf{B}}{|\mathbf{A}||\mathbf{B}|}\right) = \cos^{-1}\left(\frac{-3}{\sqrt{6}\sqrt{2}}\right) = \cos^{-1}\left(\frac{-3}{\sqrt{12}}\right)$

$= \cos^{-1}\left(-\frac{\sqrt{3}}{2}\right) = \frac{5\pi}{6}$, $|\mathbf{B}|\cos\theta = \frac{-3}{\sqrt{6}} = -\sqrt{\frac{9}{6}} = -\sqrt{\frac{3}{2}}$, $\text{proj}_{\mathbf{A}}\,\mathbf{B} = \left(\frac{\mathbf{A}\cdot\mathbf{B}}{\mathbf{A}\cdot\mathbf{A}}\right)\mathbf{A} = \frac{-3}{6}(\mathbf{i}+\mathbf{j}+2\mathbf{k}) = -\frac{1}{2}(\mathbf{i}+\mathbf{j}+\mathbf{k})$

15. $\mathbf{B} = \left(\frac{\mathbf{A}\cdot\mathbf{B}}{\mathbf{A}\cdot\mathbf{A}}\right)\mathbf{A} + \left[\mathbf{B} - \left(\frac{\mathbf{A}\cdot\mathbf{B}}{\mathbf{A}\cdot\mathbf{A}}\right)\mathbf{A}\right] = \frac{4}{3}(2\mathbf{i}+\mathbf{j}-\mathbf{k}) + \left[(\mathbf{i}+\mathbf{j}-5\mathbf{k}) - \frac{4}{3}(2\mathbf{i}+\mathbf{j}-\mathbf{k})\right] = \frac{4}{3}(2\mathbf{i}+\mathbf{j}-\mathbf{k}) - \frac{1}{3}(5\mathbf{i}+\mathbf{j}+11\mathbf{k})$, where $\mathbf{A}\cdot\mathbf{B} = 8$ and $\mathbf{A}\cdot\mathbf{A} = 6$

16. $\mathbf{B} = \left(\frac{\mathbf{A}\cdot\mathbf{B}}{\mathbf{A}\cdot\mathbf{A}}\right)\mathbf{A} + \left[\mathbf{B} - \left(\frac{\mathbf{A}\cdot\mathbf{B}}{\mathbf{A}\cdot\mathbf{A}}\right)\mathbf{A}\right] = -\frac{1}{5}(\mathbf{i}-2\mathbf{j}) + \left[(\mathbf{i}+\mathbf{j}+\mathbf{k}) - \left(\frac{-1}{5}\right)(\mathbf{i}-2\mathbf{j})\right] = -\frac{1}{5}(\mathbf{i}-2\mathbf{j}) + \left(\frac{6}{5}\mathbf{i}+\frac{3}{5}\mathbf{j}+\mathbf{k}\right)$, where $\mathbf{A}\cdot\mathbf{B} = -1$ and $\mathbf{A}\cdot\mathbf{A} = 5$

17. $\mathbf{A}\times\mathbf{B} = \begin{vmatrix} \mathbf{i} & \mathbf{j} & \mathbf{k} \\ 1 & 0 & 0 \\ 1 & 1 & 0 \end{vmatrix} = \mathbf{k}$

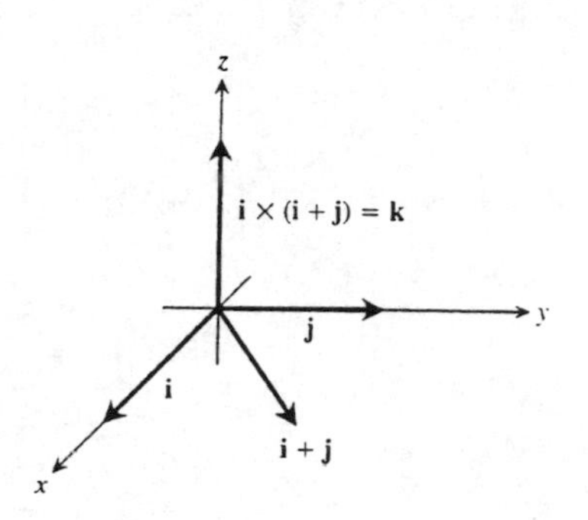

18. $\mathbf{A}\times\mathbf{B} = \begin{vmatrix} \mathbf{i} & \mathbf{j} & \mathbf{k} \\ 1 & -1 & 0 \\ 1 & 1 & 0 \end{vmatrix} = 2\mathbf{k}$

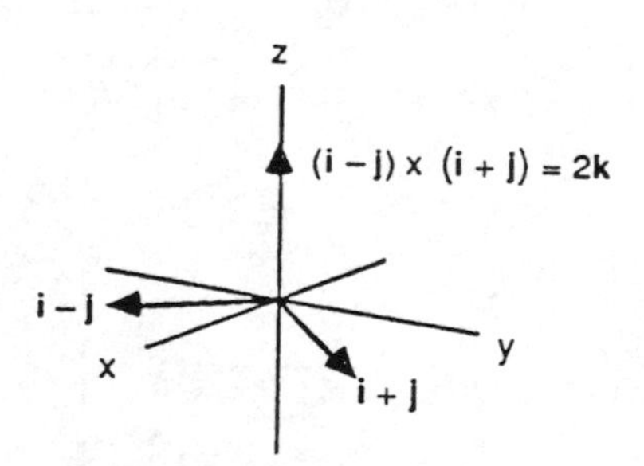

19. $y = \tan x \Rightarrow [y']_{\pi/4} = [\sec^2 x]_{\pi/4} = 2 = \frac{2}{1} \Rightarrow \mathbf{T} = \mathbf{i}+2\mathbf{j} \Rightarrow$ the unit tangents are $\pm\left(\frac{1}{\sqrt{5}}\mathbf{i}+\frac{2}{\sqrt{5}}\mathbf{j}\right)$ and the unit normals are $\pm\left(-\frac{2}{\sqrt{5}}\mathbf{i}+\frac{1}{\sqrt{5}}\mathbf{j}\right)$

20. $x^2+y^2 = 25 \Rightarrow [y']_{(3,4)} = \left[-\frac{x}{y}\right]_{(3,4)} = \frac{-3}{4} \Rightarrow \mathbf{T} = 4\mathbf{i}-3\mathbf{j} \Rightarrow$ the unit tangents are $\pm\frac{1}{5}(4\mathbf{i}-3\mathbf{j})$ and the unit normals are $\pm\frac{1}{5}(3\mathbf{i}+4\mathbf{j})$

21. Let $\mathbf{A} = a_1\mathbf{i} + a_2\mathbf{j} + a_3\mathbf{k}$ and $\mathbf{B} = b_1\mathbf{i} + b_2\mathbf{j} + b_3\mathbf{k}$. Then $|\mathbf{A} + \mathbf{B}|^2 + |\mathbf{A} - \mathbf{B}|^2$

$$= \left[(a_1 + b_1)^2 + (a_2 + b_2)^2 + (a_3 + b_3)^2\right] + \left[(a_1 - b_1)^2 + (a_2 - b_2)^2 + (a_3 - b_3)^2\right]$$

$$= \left(a_1^2 + 2a_1b_1 + b_1^2 + a_2^2 + 2a_2b_2 + b_2^2 + a_3^2 + 2a_3b_3 + b_3^2\right)$$

$$+ \left(a_1^2 - 2a_1b_1 + b_1^2 + a_2^2 - 2a_2b_2 + b_2^2 + a_3^2 - 2a_3b_3 + b_3^2\right)$$

$$= 2\left(a_1^2 + a_2^2 + a_3^2\right) + 2\left(b_1^2 + b_2^2 + b_3^2\right) = 2\,|\mathbf{A}|^2 + 2\,|\mathbf{B}|^2.$$

22. (a) area $= \frac{1}{2}|\mathbf{v}|\,h = \frac{1}{2}|\mathbf{v}||\mathbf{u}|\sin(\angle BAC) = \frac{1}{2}|\mathbf{u} \times \mathbf{v}|$ since $0 < \angle BAC < \frac{\pi}{2}$

(b) $h = |\mathbf{u}|\sin(\angle BAC) = \dfrac{|\mathbf{v}||\mathbf{u}|\sin(\angle BAC)}{|\mathbf{v}|} = \dfrac{|\mathbf{u} \times \mathbf{v}|}{|\mathbf{v}|}$

(c) area $= \frac{1}{2}\left|(\mathbf{i} - \mathbf{j} + \mathbf{k}) \times (2\mathbf{i} + \mathbf{k})\right| = \frac{1}{2}\begin{vmatrix} \mathbf{i} & \mathbf{j} & \mathbf{k} \\ 1 & -1 & 1 \\ 2 & 0 & 1 \end{vmatrix} = \frac{1}{2}\left|-\mathbf{i} + \mathbf{j} + 2\mathbf{k}\right| = \frac{1}{2}\sqrt{(-1)^2 + 1^2 + 2^2}$

$$= \frac{1}{2}\sqrt{6} = \frac{\sqrt{6}}{2} \text{ and } h = \frac{\left|(\mathbf{i} - \mathbf{j} + \mathbf{k}) \times (2\mathbf{i} + \mathbf{k})\right|}{|2\mathbf{i} + \mathbf{k}|} = \frac{|-\mathbf{i} + \mathbf{j} + 2\mathbf{k}|}{|2\mathbf{i} + \mathbf{k}|} = \frac{\sqrt{6}}{\sqrt{2^2 + 1^2}} = \frac{\sqrt{6}}{\sqrt{5}} = \sqrt{\frac{6}{5}}$$

23. Let $\mathbf{v} = v_1\mathbf{i} + v_2\mathbf{j} + v_3\mathbf{k}$ and $\mathbf{w} = w_1\mathbf{i} + w_2\mathbf{j} + w_3\mathbf{k}$. Then $|\mathbf{v} - 2\mathbf{w}|^2 = \left|(v_1\mathbf{i} + v_2\mathbf{j} + v_3\mathbf{k}) - 2(w_1\mathbf{i} + w_2\mathbf{j} + w_3\mathbf{k})\right|^2$

$$= \left|(v_1 - 2w_1)\mathbf{i} + (v_2 - 2w_2)\mathbf{j} + (v_3 - 2w_3)\mathbf{k}\right|^2 = \left(\sqrt{(v_1 - 2w_1)^2 + (v_2 - 2w_2)^2 + (v_3 - 2w_3)^2}\right)^2$$

$$= \left(v_1^2 + v_2^2 + v_3^2\right) - 4(v_1w_1 + v_2w_2 + v_3w_3) + 4\left(w_1^2 + w_2^2 + w_3^2\right) = |\mathbf{v}|^2 - 4\mathbf{v} \cdot \mathbf{w} + 4\,|\mathbf{w}|^2$$

$$= |\mathbf{v}|^2 - 4\,|\mathbf{u}||\mathbf{w}|\cos\theta + 4\,|\mathbf{w}|^2 = 4 - 4(2)(3)\left(\cos\frac{\pi}{3}\right) + 36 = 40 - 24\left(\frac{1}{2}\right) = 40 - 12 = 28 \Rightarrow |\mathbf{v} - 2\mathbf{w}| = \sqrt{28}$$

$$= 2\sqrt{7}$$

24. $\mathbf{u}$ and $\mathbf{v}$ are parallel when $\mathbf{u} \times \mathbf{v} = \mathbf{0} \Rightarrow \begin{vmatrix} \mathbf{i} & \mathbf{j} & \mathbf{k} \\ 2 & 4 & -5 \\ -4 & -8 & a \end{vmatrix} = \mathbf{0} \Rightarrow (4a - 40)\mathbf{i} + (20 - 2a)\mathbf{j} + (0)\mathbf{k} = \mathbf{0}$

$\Rightarrow 4a - 40 = 0$ and $20 - 2a = 0 \Rightarrow a = 10$

25. (a) area $= |\mathbf{A} \times \mathbf{B}| = \text{abs}\begin{vmatrix} \mathbf{i} & \mathbf{j} & \mathbf{k} \\ 1 & 1 & -1 \\ 2 & 1 & 1 \end{vmatrix} = |2\mathbf{i} - 3\mathbf{j} - \mathbf{k}| = \sqrt{4 + 9 + 1} = \sqrt{14}$

(b) volume $= \mathbf{A} \cdot (\mathbf{B} \times \mathbf{C}) = \begin{vmatrix} 1 & 1 & -1 \\ 2 & 1 & 1 \\ -1 & -2 & 3 \end{vmatrix} = 1(3 + 2) + 1(-1 - 6) - 1(-4 + 1) = 1$

26. (a) area $= |\mathbf{A}\times\mathbf{B}| = \text{abs}\begin{vmatrix} \mathbf{i} & \mathbf{j} & \mathbf{k} \\ 1 & 1 & 0 \\ 0 & 1 & 0 \end{vmatrix} = |\mathbf{k}| = 1$

(b) volume $= \mathbf{A}\cdot(\mathbf{B}\times\mathbf{C}) = \begin{vmatrix} 1 & 1 & 0 \\ 0 & 1 & 0 \\ 1 & 1 & 1 \end{vmatrix} = 1(1-0)+1(0-0)+0 = 1$

27. The desired vector is $\mathbf{n}\times\mathbf{v}$ or $\mathbf{v}\times\mathbf{n}$ since $\mathbf{n}\times\mathbf{v}$ is perpendicular to both $\mathbf{n}$ and $\mathbf{v}$ and, therefore, also parallel to the plane.

28. If $a = 0$ and $b \neq 0$, then the line $by = c$ and $\mathbf{i}$ are parallel. If $a \neq 0$ and $b = 0$, then the line $ax = c$ and $\mathbf{j}$ are parallel. If a and b are both $\neq 0$, then $ax + by = c$ contains the points $\left(\frac{c}{a}, 0\right)$ and $\left(0, \frac{c}{b}\right) \Rightarrow$ the vector $ab\left(\frac{c}{a}\mathbf{i} - \frac{c}{b}\mathbf{j}\right) = c(b\mathbf{i} - a\mathbf{j})$ and the line are parallel. Therefore, the vector $b\mathbf{i} - a\mathbf{j}$ is parallel to the line $ax + by = c$ in every case.

29. The line L passes through the point $P(0,0,-1)$ parallel to $\mathbf{v} = -\mathbf{i}+\mathbf{j}+\mathbf{k}$. With $\overrightarrow{PS} = 2\mathbf{i}+2\mathbf{j}+\mathbf{k}$ and

$\overrightarrow{PS}\times\mathbf{v} = \begin{vmatrix} \mathbf{i} & \mathbf{j} & \mathbf{k} \\ 2 & 2 & 1 \\ -1 & 1 & 1 \end{vmatrix} = (2-1)\mathbf{i} + (-1-2)\mathbf{j} + (2+2)\mathbf{k} = \mathbf{i} - 3\mathbf{j} + 4\mathbf{k}$, we find the distance

$$d = \frac{|\overrightarrow{PS}\times\mathbf{v}|}{|\mathbf{v}|} = \frac{\sqrt{1+9+16}}{\sqrt{1+1+1}} = \frac{\sqrt{26}}{\sqrt{3}} = \frac{\sqrt{78}}{3}.$$

30. The line L passes through the point $P(2,2,0)$ parallel to $\mathbf{v} = \mathbf{i}+\mathbf{j}+\mathbf{k}$. With $\overrightarrow{PS} = -2\mathbf{i}+2\mathbf{j}+\mathbf{k}$ and

$\overrightarrow{PS}\times\mathbf{v} = \begin{vmatrix} \mathbf{i} & \mathbf{j} & \mathbf{k} \\ -2 & 2 & 1 \\ 1 & 1 & 1 \end{vmatrix} = (2-1)\mathbf{i} + (1+2)\mathbf{j} + (-2-2)\mathbf{k} = \mathbf{i} + 3\mathbf{j} - 4\mathbf{k}$, we find the distance

$$d = \frac{|\overrightarrow{PS}\times\mathbf{v}|}{|\mathbf{v}|} = \frac{\sqrt{1+9+16}}{\sqrt{1+1+1}} = \frac{\sqrt{26}}{\sqrt{3}} = \frac{\sqrt{78}}{3}.$$

31. Parametric equations for the line are $x = 1 - 3t$, $y = 2$, $z = 3 + 7t$.

32. The line is parallel to $\overrightarrow{PQ} = 0\mathbf{i}+\mathbf{j}-\mathbf{k}$ and contains the point $P(1,2,0) \Rightarrow$ parametric equations are $x = 1$, $y = 2 + t$, $z = -t$ for $0 \le t \le 1$.

33. The point $P(4,0,0)$ lies on the plane $x - y = 4$, and $\overrightarrow{PS} = (6-4)\mathbf{i} + 0\mathbf{j} + (-6+0)\mathbf{k} = 2\mathbf{i} - 6\mathbf{k}$ with $\mathbf{n} = \mathbf{i} - \mathbf{j}$

$$\Rightarrow d = \frac{|\mathbf{n}\cdot\overrightarrow{PS}|}{|\mathbf{n}|} = \left|\frac{2+0+0}{\sqrt{1+1+0}}\right| = \frac{2}{\sqrt{2}} = \sqrt{2}.$$

34. The point $P(0,0,2)$ lies on the plane $2x+3y+z=2$, and $\overrightarrow{PS}=(3-0)\mathbf{i}+(0-0)\mathbf{j}+(10+2)\mathbf{k}=3\mathbf{i}+8\mathbf{k}$ with $\mathbf{n}=2\mathbf{i}+3\mathbf{j}+\mathbf{k} \Rightarrow d=\frac{\left|\mathbf{n}\cdot\overrightarrow{PS}\right|}{|\mathbf{n}|}=\left|\frac{6+0+8}{\sqrt{4+9+1}}\right|=\frac{14}{\sqrt{14}}=\sqrt{14}.$

35. $P(3,-2,1)$ and $\mathbf{n}=2\mathbf{i}+\mathbf{j}-\mathbf{k} \Rightarrow (2)(x-3)+(1)(y-(-2))+(-1)(z-1)=0 \Rightarrow 2x+y-z=3$

36. $P(-1,6,0)$ and $\mathbf{n}=\mathbf{i}-2\mathbf{j}+3\mathbf{k} \Rightarrow (1)(x-(-1))+(-2)(y-6)+(3)(z-0)=0 \Rightarrow x-2y+3z=-13$

37. $P(1,-1,2)$, $Q(2,1,3)$ and $R(-1,2,-1) \Rightarrow \overrightarrow{PQ}=\mathbf{i}+2\mathbf{j}+\mathbf{k}$, $\overrightarrow{PR}=-2\mathbf{i}+3\mathbf{j}-3\mathbf{k}$ and $\overrightarrow{PQ}\times\overrightarrow{PR}$

$$=\begin{vmatrix}\mathbf{i} & \mathbf{j} & \mathbf{k}\\ 1 & 2 & 1\\ -2 & 3 & -3\end{vmatrix}=-9\mathbf{i}+\mathbf{j}+7\mathbf{k} \text{ is normal to the plane} \Rightarrow (-9)(x-1)+(1)(y+1)+(7)(z-2)=0$$

$\Rightarrow -9x+y+7z=4$

38. $P(1,0,0)$, $Q(0,1,0)$ and $R(0,0,1) \Rightarrow \overrightarrow{PQ}=-\mathbf{i}+\mathbf{j}$, $\overrightarrow{PR}=-\mathbf{i}+\mathbf{k}$ and $\overrightarrow{PQ}\times\overrightarrow{PR}$

$$=\begin{vmatrix}\mathbf{i} & \mathbf{j} & \mathbf{k}\\ -1 & 1 & 0\\ -1 & 0 & 1\end{vmatrix}=\mathbf{i}+\mathbf{j}+\mathbf{k} \text{ is normal to the plane} \Rightarrow (1)(x-1)+(1)(y-0)+(1)(z-0)=0$$

$\Rightarrow x+y+z=1$

39. $\left(0,-\frac{1}{2},-\frac{3}{2}\right)$, since $t=-\frac{1}{2}$, $y=-\frac{1}{2}$ and $z=-\frac{3}{2}$ when $x=0$; $(-1,0,-3)$, since $t=-1$, $x=-1$ and $z=-3$ when $y=0$; $(1,-1,0)$, since $t=0$, $x=1$ and $y=-1$ when $z=0$

40. $x=2t$, $y=-t$, $z=-t$ represents a line containing the origin and perpendicular to the plane $2x-y-z=4$; this line intersects the plane $3x-5y+2z=6$ when t is the solution of $3(2t)-5(-t)+2(-t)=6$ $\Rightarrow t=\frac{2}{3} \Rightarrow \left(\frac{4}{3},-\frac{2}{3},-\frac{2}{3}\right)$ is the point of intersection

41. $\mathbf{n}_1=\mathbf{i}$ and $\mathbf{n}_2=\mathbf{i}+\mathbf{j}+\sqrt{2}\mathbf{k} \Rightarrow$ the desired angle is $\cos^{-1}\left(\frac{\mathbf{n}_1\cdot\mathbf{n}_2}{|\mathbf{n}_1||\mathbf{n}_2|}\right)=\cos^{-1}\left(\frac{1}{2}\right)=\frac{\pi}{3}$

42. $\mathbf{n}_1=\mathbf{i}+\mathbf{j}$ and $\mathbf{n}_2=\mathbf{j}+\mathbf{k} \Rightarrow$ the desired angle is $\cos^{-1}\left(\frac{\mathbf{n}_1\cdot\mathbf{n}_2}{|\mathbf{n}_1||\mathbf{n}_2|}\right)=\cos^{-1}\left(\frac{1}{2}\right)=\frac{\pi}{3}$

43. The direction of the line is $\mathbf{n}_1\times\mathbf{n}_2=\begin{vmatrix}\mathbf{i} & \mathbf{j} & \mathbf{k}\\ 1 & 2 & 1\\ 1 & -1 & 2\end{vmatrix}=5\mathbf{i}-\mathbf{j}-3\mathbf{k}$. Since the point $(-5,3,0)$ is on both planes, the desired line is $x=-5+5t$, $y=3-t$, $z=-3t$.

44. The direction of the intersection is $\mathbf{n}_1 \times \mathbf{n}_2 = \begin{vmatrix} \mathbf{i} & \mathbf{j} & \mathbf{k} \\ 1 & 2 & -2 \\ 5 & -2 & -1 \end{vmatrix} = -6\mathbf{i} - 9\mathbf{j} - 12\mathbf{k} = -3(2\mathbf{i} + 3\mathbf{j} + 4\mathbf{k})$ and is the same as the direction of the given line.

45. (a) The corresponding normals are $\mathbf{n}_1 = 3\mathbf{i} + 6\mathbf{k}$ and $\mathbf{n}_2 = 2\mathbf{i} + 2\mathbf{j} - \mathbf{k}$ and since $\mathbf{n}_1 \cdot \mathbf{n}_2 = (3)(2) + (0)(2) + (6)(-1) = 6 + 0 - 6 = 0$, we have that the planes are orthogonal

(b) The line of intersection is parallel to $\mathbf{n}_1 \times \mathbf{n}_2 = \begin{vmatrix} \mathbf{i} & \mathbf{j} & \mathbf{k} \\ 3 & 0 & 6 \\ 2 & 2 & -1 \end{vmatrix} = -12\mathbf{i} + 15\mathbf{j} + 6\mathbf{k}$. Now to find a point in the intersection, solve $\begin{cases} 3x + 6z = 1 \\ 2x + 2y - z = 3 \end{cases} \Rightarrow \begin{cases} 3x + 6z = 1 \\ 12x + 12y - 6z = 18 \end{cases} \Rightarrow 15x + 12y = 19 \Rightarrow x = 0$ and $y = \frac{19}{12}$ $\Rightarrow \left(0, \frac{19}{12}, \frac{1}{6}\right)$ is a point on the line we seek. Therefore, the line is $x = -12t$, $y = \frac{19}{12} + 15t$ and $z = \frac{1}{6} + 6t$.

46. A vector in the direction of the plane's normal is $\mathbf{n} = \mathbf{u} \times \mathbf{v} = \begin{vmatrix} \mathbf{i} & \mathbf{j} & \mathbf{k} \\ 2 & 3 & 1 \\ 1 & -1 & 2 \end{vmatrix} = 7\mathbf{i} - 3\mathbf{j} - 5\mathbf{k}$ and $P(1, 2, 3)$ on the plane $\Rightarrow 7(x - 1) - 3(y - 2) - 5(z - 3) = 0 \Rightarrow 7x - 3y - 5z = -14$.

47. Yes; $\mathbf{v} \cdot \mathbf{n} = (2\mathbf{i} - 4\mathbf{j} + \mathbf{k}) \cdot (2\mathbf{i} + \mathbf{j} + 0\mathbf{k}) = 2 \cdot 2 - 4 \cdot 1 + 1 \cdot 0 = 0 \Rightarrow$ the vector is orthogonal to the plane's normal $\Rightarrow \mathbf{v}$ is parallel to the plane

48. $\mathbf{n} \cdot \overrightarrow{PP_0} > 0$ represents the half-space of points lying on one side of the plane in the direction which the normal $\mathbf{n}$ points

49. A normal to the plane is $\mathbf{n} = \overrightarrow{AB} \times \overrightarrow{AC} = \begin{vmatrix} \mathbf{i} & \mathbf{j} & \mathbf{k} \\ 2 & 0 & -1 \\ 2 & -1 & 0 \end{vmatrix} = -\mathbf{i} - 2\mathbf{j} - 2\mathbf{k} \Rightarrow$ the distance is $d = \left|\frac{\overrightarrow{AP} \cdot \mathbf{n}}{\mathbf{n}}\right|$

$= \left|\frac{(\mathbf{i} + 4\mathbf{j}) \cdot (-\mathbf{i} - 2\mathbf{j} - 2\mathbf{k})}{\sqrt{1 + 4 + 4}}\right| = \left|\frac{-1 - 8 + 0}{3}\right| = 3$

50. $P(0, 0, 0)$ lies on the plane $2x + 3y + 5z = 0$, and $\overrightarrow{PS} = 2\mathbf{i} + 2\mathbf{j} + 3\mathbf{k}$ with $\mathbf{n} = 2\mathbf{i} + 3\mathbf{j} + 5\mathbf{k} \Rightarrow$

$d = \left|\frac{\mathbf{n} \cdot \overrightarrow{PS}}{|\mathbf{n}|}\right| = \left|\frac{4 + 6 + 15}{\sqrt{4 + 9 + 25}}\right| = \frac{25}{\sqrt{38}}$

51. $\mathbf{n} = 2\mathbf{i} - \mathbf{j} - \mathbf{k}$ is normal to the plane $\Rightarrow \mathbf{n} \times \mathbf{v} = \begin{vmatrix} \mathbf{i} & \mathbf{j} & \mathbf{k} \\ 2 & -1 & -1 \\ 1 & 1 & 1 \end{vmatrix} = 0\mathbf{i} - 3\mathbf{j} + 3\mathbf{k} = -3\mathbf{j} + 3\mathbf{k}$ is orthogonal to $\mathbf{v}$ and parallel to the plane

52. The vector $\mathbf{B}\times\mathbf{C}$ is normal to the plane of $\mathbf{B}$ and $\mathbf{C} \Rightarrow \mathbf{A}\times(\mathbf{B}\times\mathbf{C})$ is orthogonal to $\mathbf{A}$ and parallel to the plane:

$$\mathbf{B}\times\mathbf{C} = \begin{vmatrix} \mathbf{i} & \mathbf{j} & \mathbf{k} \\ 1 & 2 & 1 \\ 1 & 1 & -2 \end{vmatrix} = -5\mathbf{i}+3\mathbf{j}-\mathbf{k} \text{ and } \mathbf{A}\times(\mathbf{B}\times\mathbf{C}) = \begin{vmatrix} \mathbf{i} & \mathbf{j} & \mathbf{k} \\ 2 & -1 & 1 \\ -5 & 3 & -1 \end{vmatrix} = -2\mathbf{i}-3\mathbf{j}+\mathbf{k}$$

$\Rightarrow |\mathbf{A}\times(\mathbf{B}\times\mathbf{C})| = \sqrt{4+9+1} = \sqrt{14}$ and $\mathbf{u} = \frac{1}{\sqrt{14}}(-2\mathbf{i}-3\mathbf{j}+\mathbf{k})$ is the desired unit vector.

53. A vector parallel to the line of intersection is $\mathbf{v} = \mathbf{n}_1\times\mathbf{n}_2 = \begin{vmatrix} \mathbf{i} & \mathbf{j} & \mathbf{k} \\ 1 & 2 & 1 \\ 1 & -1 & 2 \end{vmatrix} = 5\mathbf{i}-\mathbf{j}-3\mathbf{k}$

$\Rightarrow |\mathbf{v}| = \sqrt{25+1+9} = \sqrt{35} \Rightarrow 2\left(\frac{\mathbf{v}}{|\mathbf{v}|}\right) = \frac{2}{\sqrt{35}}(5\mathbf{i}-\mathbf{j}-3\mathbf{k})$ is the desired vector.

54. The line containing $(0,0,0)$ normal to the plane is represented by $x = 2t$, $y = -t$, and $z = -t$. This line intersects the plane $3x-5y+2z = 6$ when $3(2t)-5(-t)+2(-t) = 6 \Rightarrow t = \frac{2}{3} \Rightarrow$ the point is $\left(\frac{4}{3}, -\frac{2}{3}, -\frac{2}{3}\right)$.

55. The line is represented by $x = 3+2t$, $y = 2-t$, and $z = 1+2t$. It meets the plane $2x-y+2z = -2$ when $2(3+2t)-(2-t)+2(1+2t) = -2 \Rightarrow t = -\frac{8}{9} \Rightarrow$ the point is $\left(\frac{11}{9}, \frac{26}{9}, -\frac{7}{9}\right)$.

56. The direction of the intersection is $\mathbf{v} = \mathbf{n}_1\times\mathbf{n}_2 = \begin{vmatrix} \mathbf{i} & \mathbf{j} & \mathbf{k} \\ 2 & 1 & -1 \\ 1 & 1 & 2 \end{vmatrix} = 3\mathbf{i}-5\mathbf{j}+\mathbf{k} \Rightarrow \theta = \cos^{-1}\left(\frac{\mathbf{v}\cdot\mathbf{i}}{|\mathbf{v}||\mathbf{i}|}\right)$

$= \cos^{-1}\left(\frac{3}{\sqrt{35}}\right) \approx 59.5^\circ$

57. The intersection occurs when $(3+2t)+3(2t)-t = -4 \Rightarrow t = -1 \Rightarrow$ the point is $(1,-2,-1)$. The required line must be perpendicular to both the given line and to the normal, and hence is parallel to $\begin{vmatrix} \mathbf{i} & \mathbf{j} & \mathbf{k} \\ 2 & 2 & 1 \\ 1 & 3 & -1 \end{vmatrix}$

$= -5\mathbf{i}+3\mathbf{j}+4\mathbf{k} \Rightarrow$ the line is represented by $x = 1-5t$, $y = -2+3t$, and $z = -1+4t$.

58. If $P(a,b,c)$ is a point on the line of intersection, then P lies in both planes $\Rightarrow a-2b+c+3 = 0$ and $2a-b-c+1 = 0 \Rightarrow (a-2b+c+3)+k(2a-b-c+1) = 0$.

59. The vector $\overrightarrow{AB}\times\overrightarrow{CD} = \begin{vmatrix} \mathbf{i} & \mathbf{j} & \mathbf{k} \\ 3 & -2 & 4 \\ \frac{26}{5} & 0 & -\frac{26}{5} \end{vmatrix} = \frac{26}{5}(2\mathbf{i}+7\mathbf{j}+2\mathbf{k})$ is normal to the plane and $A(-2,0,-3)$ lies on the plane $\Rightarrow 2(x+2)+7(y-0)+2(z-(-3)) = 0 \Rightarrow 2x+7y+2z+10 = 0$ is an equation of the plane.

60. Yes; the line's direction vector is $2\mathbf{i} + 3\mathbf{j} - 5\mathbf{k}$ which is parallel to the line and also parallel to the normal $-4\mathbf{i} - 6\mathbf{j} + 10\mathbf{k}$ to the plane $\Rightarrow$ the line is orthogonal to the plane.

61. The vector $\overrightarrow{PQ} \times \overrightarrow{PR} = \begin{vmatrix} \mathbf{i} & \mathbf{j} & \mathbf{k} \\ 2 & -1 & 3 \\ -3 & 0 & 1 \end{vmatrix} = -\mathbf{i} - 11\mathbf{j} - 3\mathbf{k}$ is normal to the plane.

(a) No, the plane is not orthogonal to $\overrightarrow{PQ} \times \overrightarrow{PR}$.
(b) No, these equations represent a line, not a plane.
(c) No, the plane $(x+2) + 11(y-1) - 3z = 0$ has normal $\mathbf{i} + 11\mathbf{j} - 3\mathbf{k}$ which is not parallel to $\overrightarrow{PQ} \times \overrightarrow{PR}$.
(d) No, this vector equation is equivalent to the equations $3y + 3z = 3$, $3x - 2z = -6$, and $3x + 2y = -4$ $\Rightarrow x = -\frac{4}{3} - \frac{2}{3}t$, $y = t$, $z = 1 - t$, which represents a line, not a plane.
(e) Yes, this is a plane containing the point $R(-2, 1, 0)$ with normal $\overrightarrow{PQ} \times \overrightarrow{PR}$.

62. (a) The line through A and B is $x = 1 + t$, $y = -t$, $z = -1 + 5t$; the line through C and D must be parallel and is L_1: $x = 1 + t$, $y = 2 - t$, $z = 3 + 5t$. The line through B and C is $x = 1$, $y = 2 + 2s$, $z = 3 + 4s$; the line through A and D must be parallel and is L_2: $x = 2$, $y = -1 + 2s$, $z = 4 + 4s$. The lines L_1 and L_2 intersect at $D(2, 1, 8)$ where $t = 1$ and $s = 1$.

(b) $\cos\theta = \frac{(2\mathbf{j} + 4\mathbf{k}) \cdot (\mathbf{i} - \mathbf{j} + 5\mathbf{k})}{\sqrt{20}\,\sqrt{27}} = \frac{3}{\sqrt{15}}$

(c) $\left(\frac{\overrightarrow{BA} \cdot \overrightarrow{BC}}{\overrightarrow{BC} \cdot \overrightarrow{BC}}\right)\overrightarrow{BC} = \frac{18}{20}\overrightarrow{BC} = \frac{9}{5}(\mathbf{j} + 2\mathbf{k})$ where $\overrightarrow{BA} = \mathbf{i} - \mathbf{j} + 5\mathbf{k}$ and $\overrightarrow{BC} = 2\mathbf{j} + 4\mathbf{k}$

(d) area $= |(2\mathbf{j} + 4\mathbf{k}) \times (\mathbf{i} - \mathbf{j} + 5\mathbf{k})| = |14\mathbf{i} + 4\mathbf{j} - 2\mathbf{k}| = 6\sqrt{6}$

(e) From part (d), $\mathbf{n} = 14\mathbf{i} + 4\mathbf{j} - 2\mathbf{k}$ is normal to the plane $\Rightarrow 14(x-1) + 4(y-0) - 2(z+1) = 0$ $\Rightarrow 7x + 2y - z = 8$.

(f) From part (d), $\mathbf{n} = 14\mathbf{i} + 4\mathbf{j} - 2\mathbf{k} \Rightarrow$ the area of the projection on the yz-plane is $|\mathbf{n} \cdot \mathbf{i}| = 14$; the area of the projection on the xy-plane is $|\mathbf{n} \cdot \mathbf{j}| = 4$; and the area of the projection on the xy-plane is $|\mathbf{n} \cdot \mathbf{k}| = 2$.

63. $\overrightarrow{AB} = -2\mathbf{i} + \mathbf{j} + \mathbf{k}$, $\overrightarrow{CD} = \mathbf{i} + 4\mathbf{j} - \mathbf{k}$, and $\overrightarrow{AC} = 2\mathbf{i} + \mathbf{j} \Rightarrow \mathbf{n} = \begin{vmatrix} \mathbf{i} & \mathbf{j} & \mathbf{k} \\ -2 & 1 & 1 \\ 1 & 4 & -1 \end{vmatrix} = -5\mathbf{i} - \mathbf{j} - 9\mathbf{k} \Rightarrow$ the distance is

$d = \left|\frac{(2\mathbf{i} + \mathbf{j}) \cdot (-5\mathbf{i} - \mathbf{j} - 9\mathbf{k})}{\sqrt{25 + 1 + 81}}\right| = \frac{11}{\sqrt{107}}$

64. $\overrightarrow{AB} = -2\mathbf{i} + 4\mathbf{j} - \mathbf{k}$, $\overrightarrow{CD} = \mathbf{i} - \mathbf{j} + 2\mathbf{k}$, and $\overrightarrow{AC} = -3\mathbf{i} + 3\mathbf{j} \Rightarrow \mathbf{n} = \begin{vmatrix} \mathbf{i} & \mathbf{j} & \mathbf{k} \\ -2 & 4 & -1 \\ 1 & -1 & 2 \end{vmatrix} = 7\mathbf{i} + 3\mathbf{j} - 2\mathbf{k} \Rightarrow$ the distance

is $d = \left|\frac{(-3\mathbf{i} + 3\mathbf{j}) \cdot (7\mathbf{i} + 3\mathbf{j} - 2\mathbf{k})}{\sqrt{49 + 9 + 4}}\right| = \frac{12}{\sqrt{62}}$

65. $x^2 + y^2 + z^2 = 4$

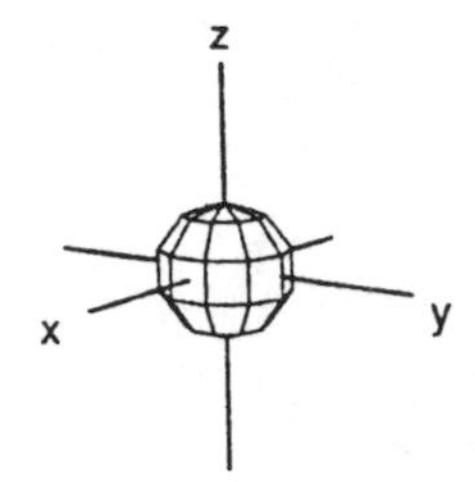

66. $x^2 + (y - 1)^2 + z^2 = 1$

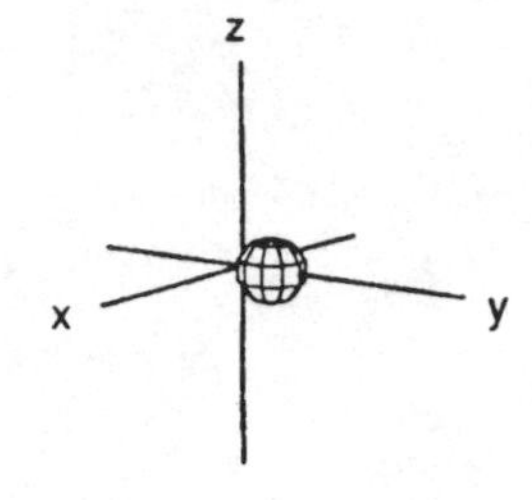

67. $4x^2 + 4y^2 + z^2 = 4$

68. $36x^2 + 9y^2 + 4z^2 = 36$

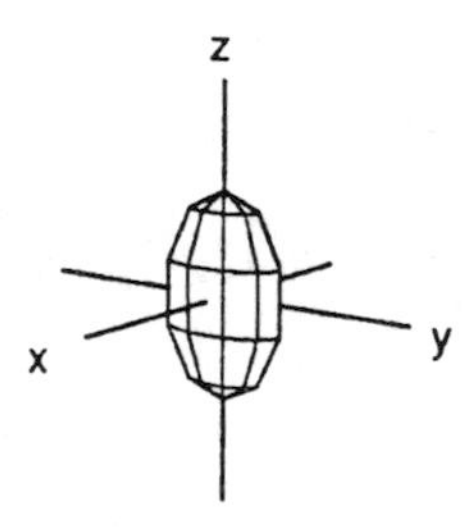

69. $z = -(x^2 + y^2)$

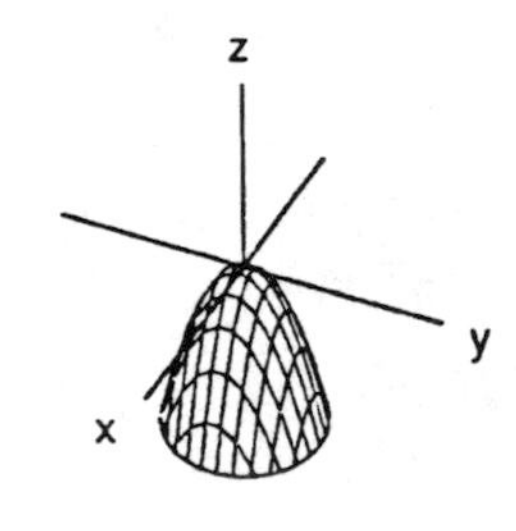

70. $y = -(x^2 + z^2)$

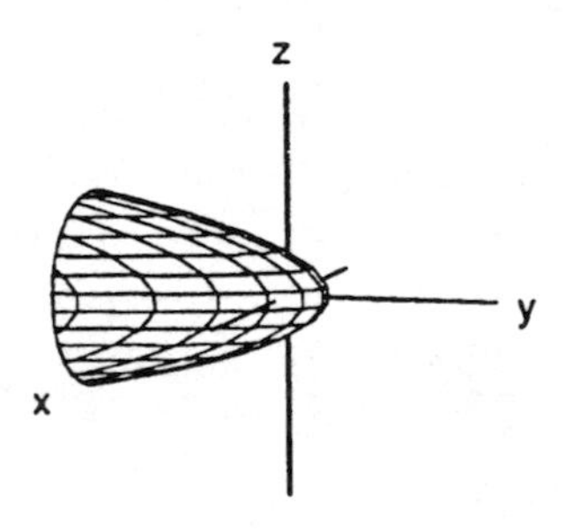

71. $x^2 + y^2 = z^2$

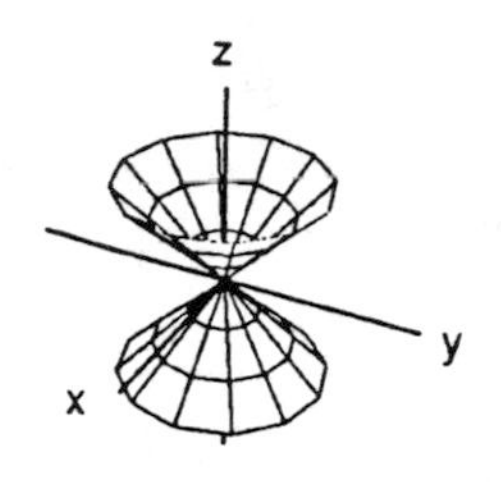

72. $x^2 + z^2 = y^2$

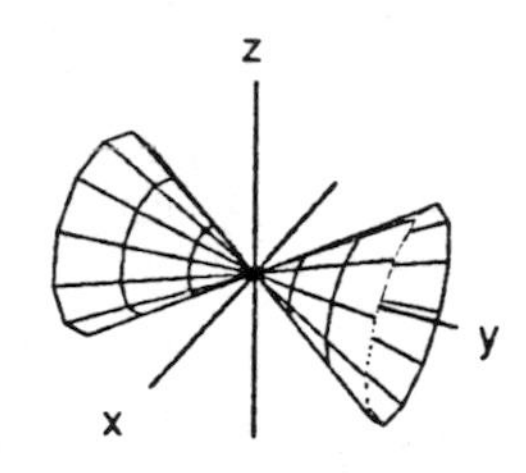

73. $x^2 + y^2 - z^2 = 4$

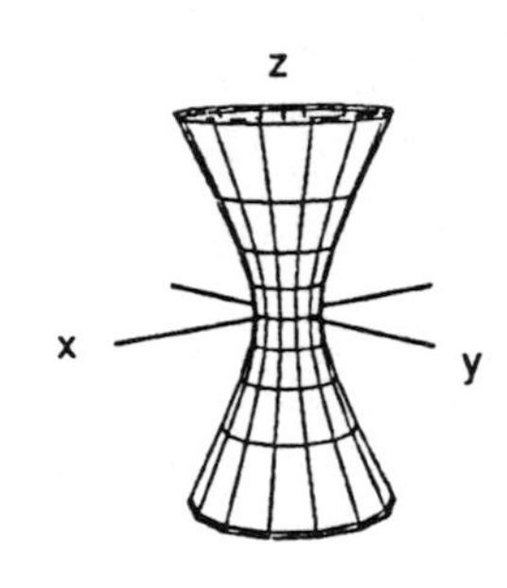

74. $4y^2 + z^2 - 4x^2 = 4$

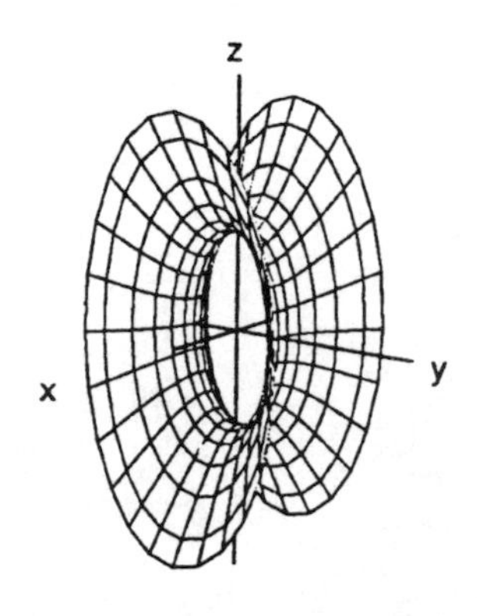

75. $y^2 - x^2 - z^2 = 1$

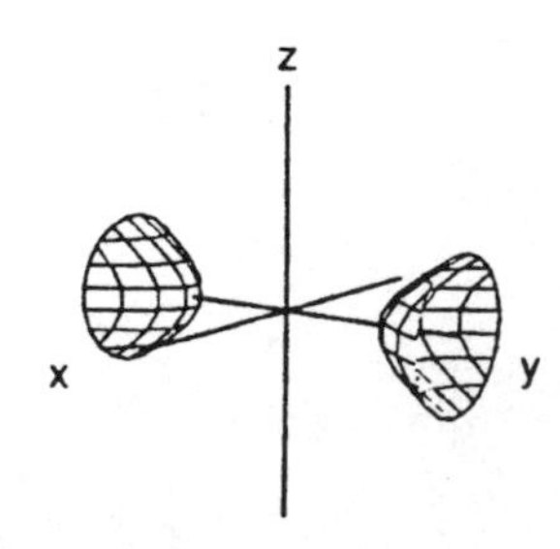

76. $z^2 - x^2 - y^2 = 1$

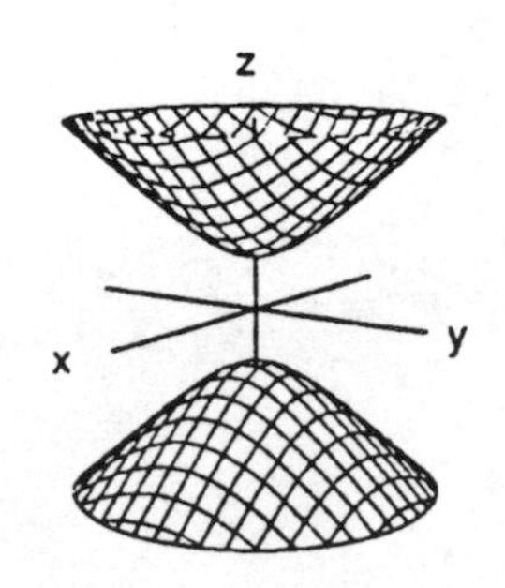

77. The y-axis in the xy-plane; the yz-plane in three dimensional space

78. The line $x + y = 1$ in the xy-plane; the plane $x + y = 1$ in three dimensional space

79. The circle centered at $(0,0)$ with radius 2 in the xy-plane; the cylinder parallel to the z-axis in three dimensional space with the circle as a generating curve

80. The ellipse $x^2 + 4y^2 = 16$ in the xy-plane; a cylinder parallel to the z-axis in three dimensional space with the ellipse as a generating curve

81. The parabola $x = y^2$ in the xy-plane; the cylinder parallel to the z-axis in three dimensional space with the parabola as a generating curve

82. The hyperbola $y^2 - x^2 = 1$ in the xy-plane; the cylinder parallel to the z-axis in three dimensional space with the hyperbola as a generating curve

83. A cardioid in the $r\theta$-plane; a cylinder parallel to the z-axis in three dimensional space with the cardioid as a generating curve

84. The circle centered at $\left(0,\frac{1}{2}\right)$ with radius $\frac{1}{2}$ in the xy-plane; the cylinder parallel to the z-axis in three dimensional space with the circle as a generating curve

85. A horizontal lemniscate of length $2\sqrt{2}$ in the $r\theta$-plane; the cylinder parallel to the z-axis in three dimensional space with the lemniscate as a generating curve

86. A rose in the $r\theta$-plane; the cylinder parallel to the z-axis in three dimensional space with the rose as a generating curve

87. The sphere of radius 2 centered at the origin

88. The plane that intersects the xy-plane at a right angle along the line $y = x$

89. The upper nappe of a cone having its vertex at the origin and making an angle of $\frac{\pi}{6}$ rad with the z-axis

90. The circle $x^2 + y^2 = 1$ in the xy-plane

91. The upper hemisphere of the sphere of radius 1 centered at the origin

92. The spheres centered at the origin having radii of 1 and 2 and all points between the spheres

Rectangular	Cylindrical	Spherical
93. $(1,0,0)$	$(1,0,0)$	$\left(1,\frac{\pi}{2},0\right)$
94. $(0,1,0)$	$\left(1,\frac{\pi}{2},0\right)$	$\left(1,\frac{\pi}{2},\frac{\pi}{2}\right)$
95. $(0,1,1)$	$\left(1,\frac{\pi}{2},1\right)$	$\left(\sqrt{2},\frac{\pi}{4},\frac{\pi}{2}\right)$
96. $(1,0,-\sqrt{3})$	$(1,0,-\sqrt{3})$	$\left(2,\frac{5\pi}{6},0\right)$

Rectangular	Cylindrical	Spherical
97. $(-1,0,-1)$	$(1,\pi,-1)$	$\left(\sqrt{2},\frac{3\pi}{4},\pi\right)$
98. $(0,-1,1)$	$\left(1,\frac{3\pi}{2},1\right)$	$\left(\sqrt{2},\frac{\pi}{4},\frac{3\pi}{2}\right)$

99. $z = 2 \Rightarrow$ cylindrical, $z = 2$; spherical, $\rho \cos \phi = 2$; a plane parallel to the xy-plane

100. $z = \sqrt{3x^2 + 3y^2} \Rightarrow$ cylindrical, $z = \sqrt{3r^2} \Rightarrow z = \sqrt{3}r$, $r \ge 0$; spherical, $\rho \cos \phi = \sqrt{3}\rho \sin \phi$ $\Rightarrow \rho\left(\cos \phi - \sqrt{3} \sin \phi\right) = 0 \Rightarrow \tan \phi = \frac{1}{\sqrt{3}}$ or $\phi = \frac{\pi}{6}$ (since $\rho = 0$ is only the origin), the upper nappe of a cone making an angle of $\frac{\pi}{6}$ with the positive z-axis and having vertex at the origin

101. $x^2 + y^2 + (z+1)^2 = 1 \Rightarrow$ cylindrical, $r^2 + (z+1)^2 = 1 \Rightarrow r^2 + z^2 + 2z + 1 = 1 \Rightarrow r^2 + z^2 = -2z$; spherical, $x^2 + y^2 + z^2 + 2z = 0 \Rightarrow \rho^2 + 2\rho \cos \phi = 0 \Rightarrow \rho(\rho + 2 \cos \phi) = 0 \Rightarrow \rho = -2 \cos \phi$ (since $\rho \ne 0$), a sphere of radius 1 centered at $(0,0,-1)$ (rectangular)

102. $x^2 + y^2 + (z-3)^2 = 9 \Rightarrow$ cylindrical, $r^2 + (z-3)^2 = 9 \Rightarrow r^2 + z^2 - 6z + 9 = 9 \Rightarrow r^2 + z^2 = 6z$; spherical, $x^2 + y^2 + z^2 - 6z = 0 \Rightarrow \rho^2 - 6\rho \cos \phi = 0 \Rightarrow \rho = 6 \cos \phi$ (since $\rho \ne 0$), a sphere of radius 3 centered at $(0,0,3)$ (rectangular)

103. $z = r^2 \Rightarrow$ rectangular, $z = x^2 + y^2$; spherical, $\rho \cos \phi = \rho^2 \sin^2 \phi \Rightarrow \rho^2 \sin^2 \phi - \rho \cos \phi = 0$ $\Rightarrow \rho\left(\rho \sin^2 \phi - \cos \phi\right) = 0 \Rightarrow \rho = \frac{\cos \phi}{\sin^2 \phi}$, $0 < \phi \le \frac{\pi}{2}$ (since $\rho \ne 0$ unless $\phi = \pi$), a circular paraboloid symmetric to the z-axis opening upward with vertex at the origin

104. $z = |r| \Rightarrow$ rectangular, $z = |r| = \sqrt{r^2} = \sqrt{x^2 + y^2} \Rightarrow z = \sqrt{x^2 + y^2}$; spherical, $\rho \cos \phi = |\rho \sin \phi|$ $\Rightarrow \rho \cos \phi = \rho \sin \phi \Rightarrow \tan \phi = 1 \Rightarrow \phi = \frac{\pi}{4}$, the upper nappe of a cone making an angle of $\frac{\pi}{4}$ with the positive z-axis with vertex at the origin

105. $r = 7 \sin \theta \Rightarrow$ rectangular, $r = 7 \sin \theta \Rightarrow r = 7\left(\frac{y}{r}\right) \Rightarrow r^2 = 7y \Rightarrow x^2 + y^2 - 7y = 0 \Rightarrow x^2 + y^2 - 7y + \frac{49}{4}$ $= \frac{49}{4} \Rightarrow x^2 + \left(y - \frac{7}{2}\right)^2 = \frac{49}{4}$; spherical, $r = 7 \sin \theta \Rightarrow \rho \sin \phi = 7 \sin \theta$, a circular cylinder parallel to the z-axis generated by the circle

106. $r = 4 \cos \theta \Rightarrow$ rectangular, $r = 4 \cos \theta \Rightarrow r = 4\left(\frac{x}{r}\right) \Rightarrow r^2 = 4x \Rightarrow x^2 - 4x + y^2 = 0 \Rightarrow x^2 - 4x + 4 + y^2 = 4$ $\Rightarrow (x-2)^2 + y^2 = 2^2$; spherical, $r = \cos \theta \Rightarrow \rho \sin \phi = 4 \cos \theta$, a circular cylinder parallel to the z-axis generated by the circle

107. $\rho = 4 \Rightarrow$ rectangular, $\sqrt{x^2 + y^2 + z^2} = 4 \Rightarrow x^2 + y^2 + z^2 = 16$; cylindrical, $r^2 + z^2 = 16$, a sphere of radius 4 centered at the origin

108. $\rho = \sqrt{3} \sec \phi \Rightarrow$ rectangular, $\rho = \sqrt{3} \sec \phi \Rightarrow \rho \cos \phi = \sqrt{3} \Rightarrow z = \sqrt{3}$; cylindrical $z = \sqrt{3}$, the plane $z = \sqrt{3}$

109. $\phi = \frac{3\pi}{4} \Rightarrow$ cylindrical, $\tan^{-1}\left(\frac{r}{z}\right) = \frac{3\pi}{4} \Rightarrow \frac{r}{z} = -1 \Rightarrow z = -r$, $r \geq 0$; rectangular, $z = -\sqrt{x^2+y^2}$, the lower nappe of a cone making an angle of $\frac{3\pi}{4}$ with the positive z-axis and having vertex at the origin

110. $\rho \cos \phi + \rho^2 \sin^2 \phi = 1 \Rightarrow$ cylindrical, $z + r^2 = 1 \Rightarrow z = 1 - r^2$; rectangular, $z = 1 - (x^2+y^2) \Rightarrow z + x^2 + y^2 = 1$, a circular paraboloid opening downward from the point $(0,0,1)$ (rectangular), axis along the z-axis

CHAPTER 10 ADDITIONAL EXERCISES–THEORY, EXAMPLES, APPLICATIONS

1. Information from ship A indicates the submarine is now on the line L_1: $x = 4 + 2t$, $y = 3t$, $z = -\frac{1}{3}t$; information from ship B indicates the submarine is now on the line L_2: $x = 18s$, $y = 5 - 6s$, $z = -s$. The current position of the sub is $\left(6, 3, -\frac{1}{3}\right)$ and occurs when the lines intersect at $t = 1$ and $s = \frac{1}{3}$. The straight line path of the submarine contains both points $P\left(2, -1, -\frac{1}{3}\right)$ and $Q\left(6, 3, -\frac{1}{3}\right)$; the line representing this path is L: $x = 2 + 4t$, $y = -1 + 4t$, $z = -\frac{1}{3}$. The submarine traveled the distance between P and Q in 4 minutes $\Rightarrow$ a speed of $\frac{\left|\overrightarrow{PQ}\right|}{4} = \frac{\sqrt{32}}{4} = \sqrt{2}$ thousand ft/min. In 20 minutes the submarine will move $20\sqrt{2}$ thousand ft from Q along the line L $\Rightarrow 20\sqrt{2} = \sqrt{(2+4t-6)^2 + (-1+4t-3)^2 + 0^2} \Rightarrow 800 = 16(t-1)^2 + 16(t-1)^2 = 32(t-1)^2$ $\Rightarrow (t-1)^2 = \frac{800}{32} = 25 \Rightarrow t = 6 \Rightarrow$ the submarine will be located at $\left(26, 23, -\frac{1}{3}\right)$ in 20 minutes.

2. H_2 stops its flight when $6 + 110t = 446 \Rightarrow t = 4$ hours. After 6 hours, H_1 is at $P(246, 57, 9)$ while H_2 is at $(446, 13, 0)$. The distance between P and Q is $\sqrt{(246-446)^2 + (57-13)^2 + (9-0)^2} \approx 204.98$ miles. At 150 mph, it would take about 1.37 hours for H_1 to reach H_2.

3. $\text{Work} = \mathbf{F} \cdot \overrightarrow{PQ} = |\mathbf{F}|\left|\overrightarrow{PQ}\right| \cos \theta = |160||250| \cos \frac{\pi}{6} = (40{,}000)\left(\frac{\sqrt{3}}{2}\right) \approx 34{,}641$ J

4. $\text{Torque} = \left|\overrightarrow{PQ} \times \mathbf{F}\right| \Rightarrow 15 \text{ ft-lb} = \left|\overrightarrow{PQ}\right||\mathbf{F}| \sin \frac{\pi}{2} = \frac{3}{4} \text{ ft} \cdot |\mathbf{F}| \Rightarrow |\mathbf{F}| = 20$ lb

5. $|\mathbf{A}+\mathbf{B}|^2 = (\mathbf{A}+\mathbf{B}) \cdot (\mathbf{A}+\mathbf{B}) = \mathbf{A} \cdot \mathbf{A} + 2\mathbf{A} \cdot \mathbf{B} + \mathbf{B} \cdot \mathbf{B} \leq |\mathbf{A}|^2 + 2|\mathbf{A}||\mathbf{B}| + |\mathbf{B}|^2 = (|\mathbf{A}| + |\mathbf{B}|)^2 \Rightarrow |\mathbf{A}+\mathbf{B}| \leq |\mathbf{A}| + |\mathbf{B}|$

6. $\mathbf{C} = \text{proj}_{\mathbf{B}} \mathbf{A} = \left(\frac{\mathbf{A} \cdot \mathbf{B}}{\mathbf{B} \cdot \mathbf{B}}\right)\mathbf{B}$ and $\mathbf{D} = \mathbf{A} - \mathbf{C} = \mathbf{A} - \left(\frac{\mathbf{A} \cdot \mathbf{B}}{\mathbf{B} \cdot \mathbf{B}}\right)\mathbf{B}$

7. Let α denote the angle between **C** and **A**, and β the angle between **C** and **B**. Let $a = |\mathbf{A}|$ and $b = |\mathbf{B}|$. Then $\cos \alpha = \frac{\mathbf{C} \cdot \mathbf{A}}{|\mathbf{C}||\mathbf{A}|} = \frac{(a\mathbf{B} + b\mathbf{A}) \cdot \mathbf{A}}{|\mathbf{C}||\mathbf{A}|} = \frac{(a\mathbf{B} \cdot \mathbf{A} + b\mathbf{A} \cdot \mathbf{A})}{|\mathbf{C}||\mathbf{A}|} = \frac{(a\mathbf{B} \cdot \mathbf{A} + b\mathbf{A} \cdot \mathbf{A})}{|\mathbf{C}||\mathbf{A}|} = \frac{(a\mathbf{B} \cdot \mathbf{A} + ba^2)}{|\mathbf{C}|a} = \frac{\mathbf{B} \cdot \mathbf{A} + ba}{|\mathbf{C}|}$, and likewise, $\cos \beta = \frac{\mathbf{A} \cdot \mathbf{B} + ba}{|\mathbf{C}|}$. Since the angle between A and B is always $\leq \frac{\pi}{2}$ and $\cos \alpha = \cos \beta$, we have that $\alpha = \beta \Rightarrow$ **C** bisects the angle between **A** and **B**.

8. $(\mathbf{aB}+\mathbf{bA})\cdot(\mathbf{bA}-\mathbf{aB}) = \mathbf{aB}\cdot\mathbf{bA}+\mathbf{bA}\cdot\mathbf{bA}-\mathbf{aB}\cdot\mathbf{aB}-\mathbf{bA}\cdot\mathbf{aB} = \mathbf{bA}\cdot\mathbf{aB}+b^2\mathbf{A}\cdot\mathbf{A}-a^2\mathbf{B}\cdot\mathbf{B}-\mathbf{bA}\cdot\mathbf{aB}$
$= b^2a^2 - a^2b^2 = 0$, where $a = |\mathbf{A}|$ and $b = |\mathbf{B}|$

9. If $\mathbf{A} = a\mathbf{i}+b\mathbf{j}+c\mathbf{k}$, then $\mathbf{A}\cdot\mathbf{A} = a^2+b^2+c^2 \geq 0$ and $\mathbf{A}\cdot\mathbf{A} = 0$ iff $a = b = c = 0$.

10. If $\mathbf{A} = (\cos\alpha)\mathbf{i}+(\sin\alpha)\mathbf{j}$ and $\mathbf{B} = (\cos\beta)\mathbf{i}+(\sin\beta)\mathbf{j}$, where $\beta > \alpha$, then $\mathbf{A}\times\mathbf{B} = [|\mathbf{A}||\mathbf{B}|\sin(\beta-\alpha)]\mathbf{k}$

$$= \begin{vmatrix} \mathbf{i} & \mathbf{j} & \mathbf{k} \\ \cos\alpha & \sin\alpha & 0 \\ \cos\beta & \sin\beta & 0 \end{vmatrix} = (\cos\alpha\sin\beta - \sin\alpha\cos\beta)\mathbf{k} \Rightarrow \sin(\beta-\alpha) = \cos\alpha\sin\beta - \sin\alpha\cos\beta, \text{ since}$$

$|\mathbf{A}| = 1$ and $|\mathbf{B}| = 1$.

11. If $\mathbf{A} = a\mathbf{i}+b\mathbf{j}$ and $\mathbf{B} = c\mathbf{i}+d\mathbf{j}$, then $\mathbf{A}\cdot\mathbf{B} = |\mathbf{A}||\mathbf{B}|\cos\theta \Rightarrow ac+bd = \sqrt{a^2+b^2}\,\sqrt{c^2+d^2}\cos\theta$
$\Rightarrow (ac+bd)^2 = (a^2+b^2)(c^2+d^2)\cos^2\theta \Rightarrow (ac+bd)^2 \leq (a^2+b^2)(c^2+d^2)$, since $\cos^2\theta \leq 1$.

12. Extend $\overrightarrow{CD}$ to $\overrightarrow{CG}$ so that $\overrightarrow{CD} = \overrightarrow{DG}$. Then $\overrightarrow{CG} = t\,\overrightarrow{CF} = \overrightarrow{CB}+\overrightarrow{BG}$ and $t\,\overrightarrow{CF} = 3\,\overrightarrow{CE}+\overrightarrow{CA}$, since ACBG is a parallelogram. If $t\,\overrightarrow{CF} - 3\,\overrightarrow{CE} - \overrightarrow{CA} = \mathbf{0}$, then $t-3-1 = 0 \Rightarrow t = 4$, since F, E, and A are collinear. Therefore, $\overrightarrow{CG} = 4\,\overrightarrow{CF} \Rightarrow \overrightarrow{CD} = 2\,\overrightarrow{CF} \Rightarrow$ F is the midpoint of $\overline{CG}$.

13. (a) If $P(x,y,z)$ is a point in the plane determined by the three points $P_1(x_1,y_1,z_1)$, $P_2(x_2,y_2,z_2)$ and $P_3(x_3,y_3,z_3)$, then the vectors $\overrightarrow{PP_1}$, $\overrightarrow{PP_2}$ and $\overrightarrow{PP_3}$ all lie in the plane. Thus $\overrightarrow{PP_1}\cdot(\overrightarrow{PP_2}\times\overrightarrow{PP_3}) = 0$

$$\Rightarrow \begin{vmatrix} x_1-x & y_1-y & z_1-z \\ x_2-x & y_2-y & z_2-z \\ x_3-x & y_3-y & z_3-z \end{vmatrix} = 0 \text{ by the determinant formula for the triple scalar product in Section 10.4.}$$

(b) Subtract row 1 from rows 2, 3, and 4 and evaluate the resulting determinant (which has the same value as the given determinant) by cofactor expansion about column 4. This expansion is exactly the determinant in part (a) so we have all points $P(x,y,z)$ in the plane determined by $P_1(x_1,y_1,z_1)$, $P_2(x_2,y_2,z_2)$, and $P_3(x_3,y_3,z_3)$.

14. Let L_1: $x = a_1s+b_1$, $y = a_2s+b_2$, $z = a_3s+b_3$ and L_2: $x = c_1t+d_1$, $y = c_2t+d_2$, $z = c_3t+d_3$. If $L_1 \parallel L_2$,

then for some k, $a_i = kc_i$, $i = 1, 2, 3$ and the determinant $\begin{vmatrix} a_1 & c_1 & b_1-d_1 \\ a_2 & c_2 & b_2-d_2 \\ a_3 & c_3 & b_3-d_3 \end{vmatrix} = \begin{vmatrix} kc_1 & c_1 & b_1-d_1 \\ kc_2 & c_2 & b_2-d_2 \\ kc_3 & c_3 & b_3-d_3 \end{vmatrix} = 0$,

since the first column is a multiple of the second column. The lines L_1 and L_2 intersect if and only if the

system $\begin{cases} a_1s - c_1t + (b_1-d_1) = 0 \\ a_2s - c_2t + (b_2-d_2) = 0 \\ a_3s - c_3t + (b_3-d_3) = 0 \end{cases}$ has a nontrivial solution $\Leftrightarrow$ the determinant of the coefficients is zero.

15. If $Q(x,y)$ is a point on the line $ax+by=c$, then $\overrightarrow{P_1Q}=(x-x_1)\mathbf{i}+(y-y_1)\mathbf{j}$, and $\mathbf{n}=a\mathbf{i}+b\mathbf{j}$ is normal to the line. The distance is $\left|\text{proj}_{\mathbf{n}}\, \overrightarrow{P_1Q}\right| = \left|\frac{[(x-x_1)\mathbf{i}+(y-y_1)\mathbf{j}]\cdot(a\mathbf{i}+b\mathbf{j})}{\sqrt{a^2+b^2}}\right| = \frac{|a(x-x_1)+b(y-y_1)|}{\sqrt{a^2+b^2}}$ $=\frac{|ax_1+by_1-c|}{\sqrt{a^2+b^2}}$, since $c=ax+by$.

16. (a) Let (x,y,z) be any point on $Ax+By+Cz-D=0$. Let $\overrightarrow{QP_1}=(x-x_1)\mathbf{i}+(y-y_1)\mathbf{j}+(z-z_1)\mathbf{k}$, and $\mathbf{n}=\frac{A\mathbf{i}+B\mathbf{j}+C\mathbf{k}}{\sqrt{A^2+B^2+C^2}}$. The distance is $\left|\text{proj}_{\mathbf{n}}\, \overrightarrow{QP_1}\right| = \left|((x-x_1)\mathbf{i}+(y-y_1)\mathbf{j}+(z-z_1)\mathbf{k})\cdot\left(\frac{A\mathbf{i}+B\mathbf{j}+C\mathbf{k}}{\sqrt{A^2+B^2+C^2}}\right)\right|$ $=\frac{|Ax_1+By_1+Cz_1-(Ax+By+Cz)|}{\sqrt{a^2+b^2+c^2}}=\frac{|Ax_1+By_1+Cz_1-D|}{\sqrt{A^2+B^2+C^2}}$.

(b) Since both tangent planes are parallel, one-half of the distance between them is equal to the radius of the sphere, i.e., $r=\frac{1}{2}\frac{|3-9|}{\sqrt{1+1+1}}=\sqrt{3}$ (see also Exercise 17a). Clearly, the points $(1,2,3)$ and $(-1,-2,-3)$ are on the line containing the sphere's center. Hence, the line containing the center is $x=1+2t$, $y=2+4t$, $z=3+6t$. The distance from the plane $x+y+z-3=0$ to the center is $\sqrt{3}$ $\Rightarrow \frac{|(1+2t)+(2+4t)+(3+6t)-3|}{\sqrt{1+1+1}}=\sqrt{3}$ from part (a) $\Rightarrow t=0 \Rightarrow$ the center is at $(1,2,3)$. Therefore an equation of the sphere is $(x-1)^2+(y-2)^2+(z-3)^2=3$.

17. (a) If (x_1,y_1,z_1) is on the plane $Ax+By+Cz=D_1$, then the distance d between the planes is $d=\frac{|Ax_1+By_1+Cz_1-D_2|}{\sqrt{A^2+B^2+C^2}}=\frac{|D_1-D_2|}{|A\mathbf{i}+B\mathbf{j}+C\mathbf{k}|}$, since $Ax_1+By_1+Cz_1=D_1$, by Exercise 16(a).

(b) $d=\frac{|12-6|}{\sqrt{4+9+1}}=\frac{6}{\sqrt{14}}$

(c) $\frac{|2(3)+(-1)(2)+2(-1)+4|}{\sqrt{14}}=\frac{|2(3)+(-1)(2)+2(-1)+D|}{\sqrt{14}} \Rightarrow D=-8$ or $4 \Rightarrow$ the desired plane is $2x-y+2x=8$

(d) Choose the point $(2,0,1)$ on the plane. Then $\frac{|3-D|}{\sqrt{6}}=5 \Rightarrow D=3\pm5\sqrt{6} \Rightarrow$ the desired planes are $x-2y+z=3+5\sqrt{6}$ and $x-2y+z=3-5\sqrt{6}$.

18. Let $\mathbf{n}=\overrightarrow{AB}\times\overrightarrow{BC}$ and $P(x,y,x)$ be any point in the plane determined by A, B and C. Then the point D lies in this plane if and only if $\overrightarrow{AD}\cdot\mathbf{n}=0 \Leftrightarrow \overrightarrow{AD}\cdot(\overrightarrow{AB}\times\overrightarrow{BC})=0$.

19. (a) $\mathbf{A}\times\mathbf{B}=4\mathbf{i}\times\mathbf{j}=4\mathbf{k} \Rightarrow (\mathbf{A}\times\mathbf{B})\times\mathbf{C}=\mathbf{0}$; $(\mathbf{A}\cdot\mathbf{C})\mathbf{B}-(\mathbf{B}\cdot\mathbf{C})\mathbf{A}=0\mathbf{B}-0\mathbf{A}=\mathbf{0}$; $\mathbf{B}\times\mathbf{C}=4\mathbf{i} \Rightarrow \mathbf{A}\times(\mathbf{B}\times\mathbf{C})=\mathbf{0}$; $(\mathbf{A}\cdot\mathbf{C})\mathbf{B}-(\mathbf{A}\cdot\mathbf{B})\mathbf{C}=0\mathbf{B}-0\mathbf{C}=\mathbf{0}$

(b) $\mathbf{A}\times\mathbf{B}=\begin{vmatrix}\mathbf{i} & \mathbf{j} & \mathbf{k}\\ 1 & -1 & 1\\ 2 & 1 & -2\end{vmatrix}=\mathbf{i}+4\mathbf{j}+3\mathbf{k} \Rightarrow (\mathbf{A}\times\mathbf{B})\times\mathbf{C}=\begin{vmatrix}\mathbf{i} & \mathbf{j} & \mathbf{k}\\ 1 & 4 & 3\\ -1 & 2 & -1\end{vmatrix}=-10\mathbf{i}-2\mathbf{j}+6\mathbf{k}$;

$(\mathbf{A}\cdot\mathbf{C})\mathbf{B}-(\mathbf{B}\cdot\mathbf{C})\mathbf{A}=-4(2\mathbf{i}+\mathbf{j}-2\mathbf{k})-2(\mathbf{i}-\mathbf{j}+\mathbf{k})=-10\mathbf{i}-2\mathbf{j}+6\mathbf{k}$;

$$\mathbf{B}\times\mathbf{C}=\begin{vmatrix}\mathbf{i}&\mathbf{j}&\mathbf{k}\\2&1&-2\\-1&2&-1\end{vmatrix}=3\mathbf{i}+4\mathbf{j}+5\mathbf{k}\Rightarrow\mathbf{A}\times(\mathbf{B}\times\mathbf{C})=\begin{vmatrix}\mathbf{i}&\mathbf{j}&\mathbf{k}\\1&-1&1\\3&4&5\end{vmatrix}=-9\mathbf{i}-2\mathbf{j}+7\mathbf{k};$$

$(\mathbf{A}\cdot\mathbf{C})\mathbf{B}-(\mathbf{A}\cdot\mathbf{B})\mathbf{C}=-4(2\mathbf{i}+\mathbf{j}-2\mathbf{k})-(-1)(-\mathbf{i}+2\mathbf{j}-\mathbf{k})=-9\mathbf{i}-2\mathbf{j}+7\mathbf{k}$

(c) $$\mathbf{A}\times\mathbf{B}=\begin{vmatrix}\mathbf{i}&\mathbf{j}&\mathbf{k}\\2&1&0\\2&-1&1\end{vmatrix}=\mathbf{i}-2\mathbf{j}-4\mathbf{k}\Rightarrow(\mathbf{A}\times\mathbf{B})\times\mathbf{C}=\begin{vmatrix}\mathbf{i}&\mathbf{j}&\mathbf{k}\\1&-2&-4\\1&0&2\end{vmatrix}=-4\mathbf{i}-6\mathbf{j}+2\mathbf{k};$$

$(\mathbf{A}\cdot\mathbf{C})\mathbf{B}-(\mathbf{B}\cdot\mathbf{C})\mathbf{A}=2(2\mathbf{i}-\mathbf{j}+\mathbf{k})-4(2\mathbf{i}+\mathbf{j})=-4\mathbf{i}-6\mathbf{j}+2\mathbf{k}$;

$$\mathbf{B}\times\mathbf{C}=\begin{vmatrix}\mathbf{i}&\mathbf{j}&\mathbf{k}\\2&-1&1\\1&0&2\end{vmatrix}=-2\mathbf{i}-3\mathbf{j}+\mathbf{k}\Rightarrow\mathbf{A}\times(\mathbf{B}\times\mathbf{C})=\begin{vmatrix}\mathbf{i}&\mathbf{j}&\mathbf{k}\\2&1&0\\-2&-3&1\end{vmatrix}=\mathbf{i}-2\mathbf{j}-4\mathbf{k};$$

$(\mathbf{A}\cdot\mathbf{C})\mathbf{B}-(\mathbf{A}\cdot\mathbf{B})\mathbf{C}=2(2\mathbf{i}-\mathbf{j}+\mathbf{k})-3(\mathbf{i}+2\mathbf{k})=\mathbf{i}-2\mathbf{j}-4\mathbf{k}$

(d) $$\mathbf{A}\times\mathbf{B}=\begin{vmatrix}\mathbf{i}&\mathbf{j}&\mathbf{k}\\1&1&-2\\-1&0&-1\end{vmatrix}=-\mathbf{i}+3\mathbf{j}+\mathbf{k}\Rightarrow(\mathbf{A}\times\mathbf{B})\times\mathbf{C}=\begin{vmatrix}\mathbf{i}&\mathbf{j}&\mathbf{k}\\-1&3&1\\2&4&-2\end{vmatrix}=-10\mathbf{i}-10\mathbf{k};$$

$(\mathbf{A}\cdot\mathbf{C})\mathbf{B}-(\mathbf{B}\cdot\mathbf{C})\mathbf{A}=10(-\mathbf{i}-\mathbf{k})-0(\mathbf{i}+\mathbf{j}-2\mathbf{k})=-10\mathbf{i}-10\mathbf{k}$;

$$\mathbf{B}\times\mathbf{C}=\begin{vmatrix}\mathbf{i}&\mathbf{j}&\mathbf{k}\\-1&0&-1\\2&4&-2\end{vmatrix}=4\mathbf{i}-4\mathbf{j}-4\mathbf{k}\Rightarrow\mathbf{A}\times(\mathbf{B}\times\mathbf{C})=\begin{vmatrix}\mathbf{i}&\mathbf{j}&\mathbf{k}\\1&1&-2\\4&-4&-4\end{vmatrix}=-12\mathbf{i}-4\mathbf{j}-8\mathbf{k};$$

$(\mathbf{A}\cdot\mathbf{C})\mathbf{B}-(\mathbf{A}\cdot\mathbf{B})\mathbf{C}=10(-\mathbf{i}-\mathbf{k})-1(2\mathbf{i}+4\mathbf{j}-2\mathbf{k})=-12\mathbf{i}-4\mathbf{j}-8\mathbf{k}$

20. (a) $\mathbf{A}\times(\mathbf{B}\times\mathbf{C})+\mathbf{B}\times(\mathbf{C}\times\mathbf{A})+\mathbf{C}\times(\mathbf{A}\times\mathbf{B})=(\mathbf{A}\cdot\mathbf{C})\mathbf{B}-(\mathbf{A}\cdot\mathbf{B})\mathbf{C}+(\mathbf{B}\cdot\mathbf{A})\mathbf{C}-(\mathbf{B}\cdot\mathbf{C})\mathbf{A}+(\mathbf{C}\cdot\mathbf{B})\mathbf{A}-(\mathbf{C}\cdot\mathbf{A})\mathbf{B}=\mathbf{0}$

(b) $[\mathbf{A}\cdot(\mathbf{B}\times\mathbf{i})]\mathbf{i}+[(\mathbf{A}\cdot(\mathbf{B}\times\mathbf{j})]\mathbf{j}+[(\mathbf{A}\cdot(\mathbf{B}\times\mathbf{k})]\mathbf{k}=[(\mathbf{A}\times\mathbf{B})\cdot\mathbf{i}]\mathbf{i}+[(\mathbf{A}\times\mathbf{B})\cdot\mathbf{j}]\mathbf{j}+[(\mathbf{A}\times\mathbf{B})\cdot\mathbf{k}]\mathbf{k}=\mathbf{A}\times\mathbf{B}$

(c) $(\mathbf{A}\times\mathbf{B})\cdot(\mathbf{C}\times\mathbf{D})=\mathbf{A}\cdot[\mathbf{B}\times(\mathbf{C}\times\mathbf{D})]=\mathbf{A}\cdot[(\mathbf{B}\cdot\mathbf{D})\mathbf{C}-(\mathbf{B}\cdot\mathbf{C})\mathbf{D}]=(\mathbf{A}\cdot\mathbf{C})(\mathbf{B}\cdot\mathbf{D})-(\mathbf{A}\cdot\mathbf{D})(\mathbf{B}\cdot\mathbf{C})$

$$=\begin{vmatrix}\mathbf{A}\cdot\mathbf{C}&\mathbf{B}\cdot\mathbf{C}\\\mathbf{A}\cdot\mathbf{D}&\mathbf{B}\cdot\mathbf{D}\end{vmatrix}$$

21. The formula is always true; $\mathbf{A}\times[\mathbf{A}\times(\mathbf{A}\times\mathbf{B})]\cdot\mathbf{C}=\mathbf{A}\times[(\mathbf{A}\cdot\mathbf{B})\mathbf{A}-(\mathbf{A}\cdot\mathbf{A})\mathbf{B}]\cdot\mathbf{C}$
$=[(\mathbf{A}\cdot\mathbf{B})\mathbf{A}\times\mathbf{A}-(\mathbf{A}\cdot\mathbf{A})\mathbf{A}\times\mathbf{B}]\cdot\mathbf{C}=-|\mathbf{A}|^2\mathbf{A}\times\mathbf{B}\cdot\mathbf{C}=-|\mathbf{A}|^2\mathbf{A}\cdot\mathbf{B}\times\mathbf{C}$

22. $\mathbf{n} = \mathbf{i} + 2\mathbf{j} + 6\mathbf{k}$ is normal to the plane $x + 2y + 6z = 6$; $\mathbf{v} \times \mathbf{n} = \begin{vmatrix} \mathbf{i} & \mathbf{j} & \mathbf{k} \\ 1 & 1 & 1 \\ 1 & 2 & 6 \end{vmatrix} = 4\mathbf{i} - 5\mathbf{j} + \mathbf{k}$ is parallel to the plane and perpendicular to the plane of $\mathbf{v}$ and $\mathbf{n} \Rightarrow \mathbf{w} = \mathbf{n} \times (\mathbf{v} \times \mathbf{n}) = \begin{vmatrix} \mathbf{i} & \mathbf{j} & \mathbf{k} \\ 1 & 2 & 6 \\ 4 & -5 & 1 \end{vmatrix} = 32\mathbf{i} + 23\mathbf{j} - 13\mathbf{k}$ is a vector parallel to the plane $x + 2y + 6z = 6$ in the direction of the projection vector $\text{proj}_P\, \mathbf{v}$. Therefore,

$$\text{proj}_P\, \mathbf{v} = \text{proj}_{\mathbf{w}}\, \mathbf{v} = \left(\mathbf{v} \cdot \frac{\mathbf{w}}{|\mathbf{w}|}\right)\frac{\mathbf{w}}{|\mathbf{w}|} = \left(\frac{\mathbf{v}\cdot\mathbf{w}}{|\mathbf{w}|^2}\right)\mathbf{w} = \left(\frac{32 + 23 - 13}{32^2 + 23^2 + 13^2}\right)\mathbf{w} = \frac{42}{1722}\mathbf{w} = \frac{1}{41}\mathbf{w} = \frac{32}{41}\mathbf{i} + \frac{23}{41}\mathbf{j} - \frac{13}{41}\mathbf{k}$$

23. $\text{proj}_{\mathbf{z}}\, \mathbf{w} = -\text{proj}_{\mathbf{z}}\, \mathbf{v}$ and $\mathbf{w} - \text{proj}_{\mathbf{z}}\, \mathbf{w} = \mathbf{v} - \text{proj}_{\mathbf{z}}\, \mathbf{v}$ lies along the line $L \Rightarrow \mathbf{w} = (\mathbf{w} - \text{proj}_{\mathbf{z}}\, \mathbf{w}) + \text{proj}_{\mathbf{z}}\, \mathbf{w}$ $= (\mathbf{v} - \text{proj}_{\mathbf{z}}\, \mathbf{v}) + \text{proj}_{\mathbf{z}}\, \mathbf{w} = \mathbf{v} - 2\,\text{proj}_{\mathbf{z}}\, \mathbf{v} = \mathbf{v} - 2\left(\frac{\mathbf{v}\cdot\mathbf{z}}{|\mathbf{z}|^2}\right)\mathbf{z}$

24. (a) Let (x, y, z) be the Cartesian coordinates of P. By (i) we have $x^2 + y^2 = \alpha\beta$. From (iii), $x = r\cos\theta = \sqrt{\alpha\beta}\cos\gamma$ and $y = r\sin\theta = \sqrt{\alpha\beta}\sin\gamma$. Finally, from (ii), $|\alpha - \beta| = 2|z|$ $\Rightarrow z = \frac{\alpha - \beta}{2}$ since P lies above the xy-plane if $\alpha - \beta > 0$ and below if $\alpha - \beta < 0$. Therefore, $(x, y, z) = \left(\sqrt{\alpha\beta}\cos\gamma,\ \sqrt{\alpha\beta}\sin\gamma,\ \frac{\alpha - \beta}{2}\right)$.

(b) Fix α. From part (a), $\beta = \alpha - 2z \Rightarrow x^2 + y^2 = \alpha\beta = \alpha(\alpha - 2z) \Rightarrow z = -\frac{1}{2\alpha}(x^2 + y^2) + \frac{\alpha^2}{2}$ which is an equation of a paraboloid for fixed α. Similarly the graph is a paraboloid for fixed β.

25. (a) The vector from $(0, d)$ to $(kd, 0)$ is $\mathbf{r}_k = kd\mathbf{i} - d\mathbf{j} \Rightarrow |\mathbf{r}_k|^3 = \frac{1}{d^3(k^2+1)^{3/2}} \Rightarrow \frac{\mathbf{r}_k}{|\mathbf{r}_k|^3} = \frac{k\mathbf{i} - \mathbf{j}}{d^2(k^2+1)^{3/2}}$. The total force on the mass $(0, d)$ due to the masses Q_k for $k = -n, -n+1, \ldots, n-1, n$ is

$$\mathbf{F} = \frac{GMm}{d^2}(-\mathbf{j}) + \frac{GMm}{2d^2}\left(\frac{\mathbf{i}-\mathbf{j}}{\sqrt{2}}\right) + \frac{GMm}{5d^2}\left(\frac{2\mathbf{i}-\mathbf{j}}{\sqrt{5}}\right) + \ldots + \frac{GMm}{(n^2+1)d^2}\left(\frac{n\mathbf{i}-\mathbf{j}}{\sqrt{n^2+1}}\right) + \frac{GMm}{2d^2}\left(\frac{-\mathbf{i}-\mathbf{j}}{\sqrt{2}}\right)$$
$$+ \frac{GMm}{5d^2}\left(\frac{-2\mathbf{i}-\mathbf{j}}{\sqrt{5}}\right) + \ldots + \frac{GMm}{(n^2+1)d^2}\left(\frac{-n\mathbf{i}-\mathbf{j}}{\sqrt{n^2+1}}\right)$$

The $\mathbf{i}$ components cancel, giving

$$\mathbf{F} = \frac{GMm}{d^2}\left(-1 - \frac{2}{2\sqrt{2}} - \frac{2}{5\sqrt{5}} - \ldots - \frac{2}{(n^2+1)(n^2+1)^{1/2}}\right)\mathbf{j} \Rightarrow \text{the magnitude of the force is}$$

$$|\mathbf{F}| = \frac{GMm}{d^2}\left(1 + \sum_{i=1}^{n} \frac{2}{(i^2+1)^{3/2}}\right).$$

(b) Yes, it is finite: $\lim_{n\to\infty} |\mathbf{F}| = \frac{GMm}{d^2}\left(1 + \sum_{i=1}^{\infty} \frac{2}{(i^2+1)^{3/2}}\right)$ is finite since $\sum_{i=1}^{\infty} \frac{2}{(i^2+1)^{3/2}}$ converges.

26. (a) If $\vec{x}\cdot\vec{y}=0$, then $\vec{x}\times(\vec{x}\times\vec{y})=(\vec{x}\cdot\vec{y})\vec{x}-(\vec{x}\cdot\vec{x})\vec{y}=-(\vec{x}\cdot\vec{x})\vec{y}$. This means that

$$\vec{x}\oplus\vec{y}=\vec{x}+\vec{y}+\frac{1}{c^2}\cdot\frac{1}{1+\sqrt{1-\frac{\vec{x}\cdot\vec{x}}{c^2}}}(-(\vec{x}\cdot\vec{x}))\vec{y}=\vec{x}+\left(1-\frac{|\vec{x}|^2}{c^2+\sqrt{c^4-c^2|\vec{x}|^2}}\right)\vec{y}.$$ Since $\vec{x}$ and $\vec{y}$ are orthogonal, then $|\vec{x}\oplus\vec{y}|^2=|\vec{x}|^2+\left(1-\frac{|\vec{x}|^2}{c^2+\sqrt{c^4-c^2|\vec{x}|^2}}\right)^2|\vec{y}|^2$. A calculation will show that

$|\vec{x}|^2+\left(1-\frac{|\vec{x}|^2}{c^2+\sqrt{c^4-c^2|\vec{x}|^2}}\right)^2c^2=c^2$. Since $|\vec{y}|<c$, then $|\vec{y}|^2<c^2$ so

$\left(1-\frac{|\vec{x}|^2}{c^2+\sqrt{c^4-c^2|\vec{x}|^2}}\right)^2|\vec{y}|^2<\left(1-\frac{|\vec{x}|^2}{c^2+\sqrt{c^4-c^2|\vec{x}|^2}}\right)c^2$. This means that

$$|\vec{x}\oplus\vec{y}|^2=|\vec{x}|^2+\left(1-\frac{|\vec{x}|^2}{c^2+\sqrt{c^4-c^2|\vec{x}|^2}}\right)^2|\vec{y}|^2<|\vec{x}|^2+\left(1-\frac{|\vec{x}|^2}{c^2+\sqrt{c^4-c^2|\vec{x}|^2}}\right)^2c^2=c^2.$$

We now have $|\vec{x}\oplus\vec{y}|^2<c^2$, so $|\vec{x}\oplus\vec{y}|<c$.

(b) If $\vec{x}$ and $\vec{y}$ are parallel, then $\vec{x}\times(\vec{x}\times\vec{y})=\vec{0}$. This gives $\vec{x}\oplus\vec{y}=\dfrac{\vec{x}+\vec{y}}{1+\frac{\vec{x}\cdot\vec{y}}{c^2}}$.

(i) If $\vec{x}$ and $\vec{y}$ have the same direction, then $\vec{x}\oplus\vec{y}=\dfrac{\vec{x}+\vec{y}}{1+\frac{|\vec{x}|}{c}\cdot\frac{|\vec{y}|}{c}}$ and $|\vec{x}\oplus\vec{y}|=\dfrac{|\vec{x}|+|\vec{y}|}{1+\frac{|\vec{x}|}{c}\cdot\frac{|\vec{y}|}{c}}$.

Since $|\vec{y}|<c$, $|\vec{x}|<c$, we have $|\vec{y}|\left(1-\frac{|\vec{x}|}{c}\right)<c\left(1-\frac{|\vec{x}|}{c}\right)\Rightarrow|\vec{y}|-\frac{|\vec{y}||\vec{x}|}{c}<c-|\vec{x}|$

$\Rightarrow|\vec{x}|+|\vec{y}|<c+\frac{|\vec{x}||\vec{y}|}{c}=c\left(1+\frac{|\vec{x}|}{c}\cdot\frac{|\vec{y}|}{c}\right)\Rightarrow\dfrac{|\vec{x}|+|\vec{y}|}{1+\frac{|\vec{x}|}{c}\cdot\frac{|\vec{y}|}{c}}<c$. This means that $|\vec{x}\oplus\vec{y}|<c$.

(ii) If $\vec{x}$ and $\vec{y}$ have opposite directions, then $\vec{x}\cdot\vec{y}=-|\vec{x}||\vec{y}|$ and $\vec{x}\oplus\vec{y}=\dfrac{\vec{x}+\vec{y}}{1-\frac{|\vec{x}||\vec{y}|}{c^2}}$.

Assume $|\vec{x}|\geq|\vec{y}|$, then $|\vec{x}\oplus\vec{y}|=\dfrac{|\vec{x}|-|\vec{y}|}{1-\frac{|\vec{x}||\vec{y}|}{c^2}}$. Since $|\vec{x}|<c$, we have $|\vec{x}|\left(1+\frac{|\vec{y}|}{c}\right)<c\left(1+\frac{|\vec{y}|}{c}\right)$

$\Rightarrow|\vec{x}|+\frac{|\vec{x}||\vec{y}|}{c}<c+|\vec{y}|\Rightarrow|\vec{x}|-|\vec{y}|<c-\frac{|\vec{x}||\vec{y}|}{c}=c\left(1-\frac{|\vec{x}||\vec{y}|}{c^2}\right)\Rightarrow\dfrac{|\vec{x}|-|\vec{y}|}{1-\frac{|\vec{x}||\vec{y}|}{c^2}}<c$.

This means that $|\vec{x}\oplus\vec{y}|<c$. A similar argument holds if $|\vec{x}|>|\vec{y}|$.

(c) $\lim_{c\to\infty}\vec{x}\oplus\vec{y}=\vec{x}+\vec{y}$.

CHAPTER 11 VECTOR-VALUED FUNCTIONS AND MOTION IN SPACE

11.1 VECTOR-VALUED FUNCTIONS AND SPACE CURVES

1. $x = t + 1$ and $y = t^2 - 1 \Rightarrow y = (x-1)^2 - 1 = x^2 - 2x$; $\mathbf{v} = \frac{d\mathbf{r}}{dt} = \mathbf{i} + 2t\mathbf{j} \Rightarrow \mathbf{a} = \frac{d\mathbf{v}}{dt} = 2\mathbf{j} \Rightarrow \mathbf{v} = \mathbf{i} + 2\mathbf{j}$ and $\mathbf{a} = 2\mathbf{j}$ at $t = 1$

2. $x = t^2 + 1$ and $y = 2t - 1 \Rightarrow x = \left(\frac{y+1}{2}\right)^2 + 1 \Rightarrow x = \frac{1}{4}(y+1)^2 + 1$; $\mathbf{v} = \frac{d\mathbf{r}}{dt} = 2t\mathbf{i} + 2\mathbf{j} \Rightarrow \mathbf{a} = \frac{d\mathbf{v}}{dt} = 2\mathbf{i}$ $\Rightarrow \mathbf{v} = \mathbf{i} + 2\mathbf{j}$ and $\mathbf{a} = 2\mathbf{i}$ at $t = \frac{1}{2}$

3. $x = e^t$ and $y = \frac{2}{9}e^{2t} \Rightarrow y = \frac{2}{9}x^2$; $\mathbf{v} = \frac{d\mathbf{r}}{dt} = e^t\mathbf{i} + \frac{4}{9}e^{2t}\mathbf{j} \Rightarrow \mathbf{a} = e^t\mathbf{i} + \frac{8}{9}e^{2t}\mathbf{j} \Rightarrow \mathbf{v} = 3\mathbf{i} + 4\mathbf{j}$ and $\mathbf{a} = 3\mathbf{i} + 8\mathbf{j}$ at $t = \ln 3$

4. $x = \cos 2t$ and $y = 3 \sin 2t \Rightarrow x^2 + \frac{1}{9}y^2 = 1$; $\mathbf{v} = \frac{d\mathbf{r}}{dt} = (-2 \sin 2t)\mathbf{i} + (6 \cos 2t)\mathbf{j} \Rightarrow \mathbf{a} = \frac{d\mathbf{v}}{dt}$ $= (-4 \cos 2t)\mathbf{i} + (12 \sin 2t)\mathbf{j} \Rightarrow \mathbf{v} = 6\mathbf{j}$ and $\mathbf{a} = -4\mathbf{i}$ at $t = 0$

5. $\mathbf{v} = \frac{d\mathbf{r}}{dt} = (\cos t)\mathbf{i} - (\sin t)\mathbf{j}$ and $\mathbf{a} = \frac{d\mathbf{v}}{dt} = -(\sin t)\mathbf{i} - (\cos t)\mathbf{j}$
$\Rightarrow$ for $t = \frac{\pi}{4}$, $\mathbf{v}\left(\frac{\pi}{4}\right) = \frac{\sqrt{2}}{2}\mathbf{i} - \frac{\sqrt{2}}{2}\mathbf{j}$ and $\mathbf{a}\left(\frac{\pi}{4}\right) = -\frac{\sqrt{2}}{2}\mathbf{i} - \frac{\sqrt{2}}{2}\mathbf{j}$;
for $t = \frac{\pi}{2}$, $\mathbf{v}\left(\frac{\pi}{2}\right) = -\mathbf{j}$ and $\mathbf{a}\left(\frac{\pi}{2}\right) = -\mathbf{i}$

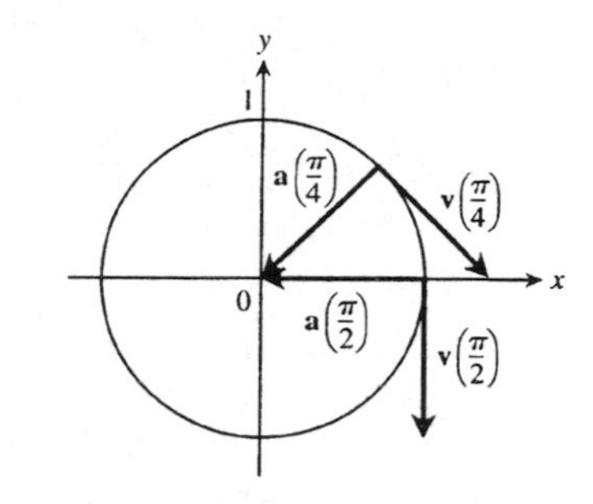

6. $\mathbf{v} = \frac{d\mathbf{r}}{dt} = \left(-2 \sin \frac{t}{2}\right)\mathbf{i} + \left(2 \cos \frac{t}{2}\right)\mathbf{j}$ and $\mathbf{a} = \frac{d\mathbf{v}}{dt} = \left(-\cos \frac{t}{2}\right)\mathbf{i} + \left(-\sin \frac{t}{2}\right)\mathbf{j}$
$\Rightarrow$ for $t = \pi$, $\mathbf{v}(\pi) = -2\mathbf{i}$ and $\mathbf{a}(\pi) = -\mathbf{j}$; for $t = \frac{3\pi}{2}$,
$\mathbf{v}\left(\frac{3\pi}{2}\right) = -\sqrt{2}\,\mathbf{i} - \sqrt{2}\,\mathbf{j}$ and $\mathbf{a}\left(\frac{3\pi}{2}\right) = \frac{\sqrt{2}}{2}\mathbf{i} - \frac{\sqrt{2}}{2}\mathbf{j}$

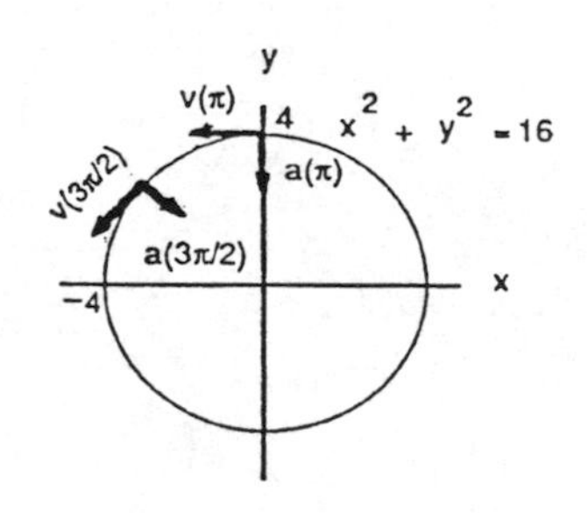

7. . $\mathbf{v} = \frac{d\mathbf{r}}{dt} = (1 - \cos t)\mathbf{i} + (\sin t)\mathbf{j}$ and $\mathbf{a} = \frac{d\mathbf{v}}{dt} = (\sin t)\mathbf{i} + (\cos t)\mathbf{j}$
$\Rightarrow$ for $t = \pi$, $\mathbf{v}(\pi) = 2\mathbf{i}$ and $\mathbf{a}(\pi) = -\mathbf{j}$; for $t = \frac{3\pi}{2}$,
$\mathbf{v}\left(\frac{3\pi}{2}\right) = \mathbf{i} - \mathbf{j}$ and $\mathbf{a}\left(\frac{3\pi}{2}\right) = -\mathbf{i}$

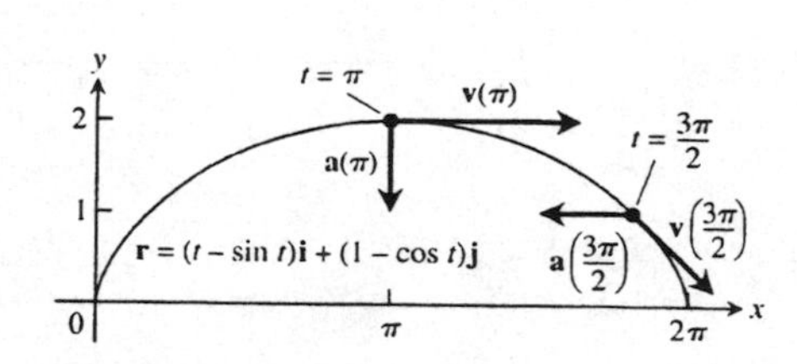

8. $\mathbf{v} = \frac{d\mathbf{r}}{dt} = \mathbf{i} + 2t\mathbf{j}$ and $\mathbf{a} = \frac{d\mathbf{v}}{dt} = 2\mathbf{j} \Rightarrow$ for $t = -1$, $\mathbf{v}(-1) = \mathbf{i} - 2\mathbf{j}$ and $\mathbf{a}(-1) = 2\mathbf{j}$; for $t = 0$, $\mathbf{v}(0) = \mathbf{i}$ and $\mathbf{a}(0) = 2\mathbf{j}$; for $t = 1$, $\mathbf{v}(1) = \mathbf{i} + 2\mathbf{j}$ and $\mathbf{a}(1) = 2\mathbf{j}$

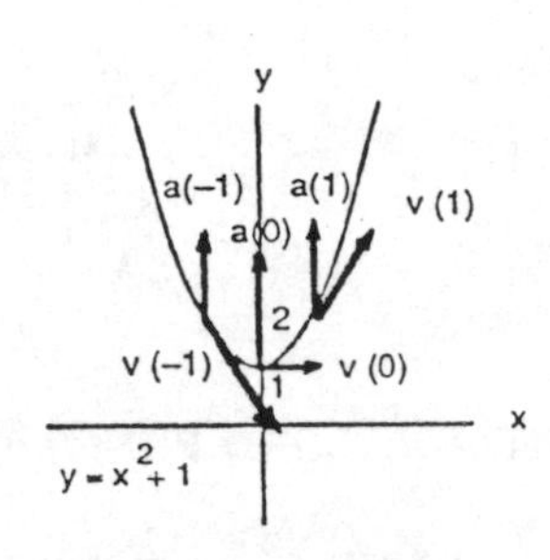

9. $\mathbf{r} = (t+1)\mathbf{i} + (t^2-1)\mathbf{j} + 2t\mathbf{k} \Rightarrow \mathbf{v} = \frac{d\mathbf{r}}{dt} = \mathbf{i} + 2t\mathbf{j} + 2\mathbf{k} \Rightarrow \mathbf{a} = \frac{d^2\mathbf{r}}{dt^2} = 2\mathbf{j}$; Speed: $|\mathbf{v}(1)| = \sqrt{1^2 + (2(1))^2 + 2^2} = 3$; Direction: $\frac{\mathbf{v}(1)}{|\mathbf{v}(1)|} = \frac{\mathbf{i} + 2(1)\mathbf{j} + 2\mathbf{k}}{3} = \frac{1}{3}\mathbf{i} + \frac{2}{3}\mathbf{j} + \frac{2}{3}\mathbf{k} \Rightarrow \mathbf{v}(1) = 3\left(\frac{1}{3}\mathbf{i} + \frac{2}{3}\mathbf{j} + \frac{2}{3}\mathbf{k}\right)$

10. $\mathbf{r} = (1+t)\mathbf{i} + \frac{t^2}{\sqrt{2}}\mathbf{j} + \frac{t^3}{3}\mathbf{k} \Rightarrow \mathbf{v} = \frac{d\mathbf{r}}{dt} = \mathbf{i} + \frac{2t}{\sqrt{2}}\mathbf{j} + t^2\mathbf{k} \Rightarrow \mathbf{a} = \frac{d^2\mathbf{r}}{dt^2} = \frac{2}{\sqrt{2}}\mathbf{j} + 2t\mathbf{k}$; Speed: $|\mathbf{v}(1)|$ $= \sqrt{1^2 + \left(\frac{2(1)}{\sqrt{2}}\right)^2 + (1^2)^2} = 2$; Direction: $\frac{\mathbf{v}(1)}{|\mathbf{v}(1)|} = \frac{\mathbf{i} + \frac{2(1)}{\sqrt{2}}\mathbf{j} + (1^2)\mathbf{k}}{2} = \frac{1}{2}\mathbf{i} + \frac{1}{\sqrt{2}}\mathbf{j} + \frac{1}{2}\mathbf{k} \Rightarrow \mathbf{v}(1)$ $= 2\left(\frac{1}{2}\mathbf{i} + \frac{1}{\sqrt{2}}\mathbf{j} + \frac{1}{2}\mathbf{k}\right)$

11. $\mathbf{r} = (2\cos t)\mathbf{i} + (3\sin t)\mathbf{j} + 4t\mathbf{k} \Rightarrow \mathbf{v} = \frac{d\mathbf{r}}{dt} = (-2\sin t)\mathbf{i} + (3\cos t)\mathbf{j} + 4\mathbf{k} \Rightarrow \mathbf{a} = \frac{d^2\mathbf{r}}{dt^2} = (-2\cos t)\mathbf{i} - (3\sin t)\mathbf{j}$; Speed: $\left|\mathbf{v}\left(\frac{\pi}{2}\right)\right| = \sqrt{\left(-2\sin\frac{\pi}{2}\right)^2 + \left(3\cos\frac{\pi}{2}\right)^2 + 4^2} = 2\sqrt{5}$; Direction: $\frac{\mathbf{v}\left(\frac{\pi}{2}\right)}{\left|\mathbf{v}\left(\frac{\pi}{2}\right)\right|}$ $= \left(-\frac{2}{2\sqrt{5}}\sin\frac{\pi}{2}\right)\mathbf{i} + \left(\frac{3}{2\sqrt{5}}\cos\frac{\pi}{2}\right)\mathbf{j} + \frac{4}{2\sqrt{5}}\mathbf{k} = -\frac{1}{\sqrt{5}}\mathbf{i} + \frac{2}{\sqrt{5}}\mathbf{k} \Rightarrow \mathbf{v}\left(\frac{\pi}{2}\right) = 2\sqrt{5}\left(-\frac{1}{\sqrt{5}}\mathbf{i} + \frac{2}{\sqrt{5}}\mathbf{k}\right)$

12. $\mathbf{r} = (\sec t)\mathbf{i} + (\tan t)\mathbf{j} + \frac{4}{3}t\mathbf{k} \Rightarrow \mathbf{v} = \frac{d\mathbf{r}}{dt} = (\sec t\tan t)\mathbf{i} + (\sec^2 t)\mathbf{j} + \frac{4}{3}\mathbf{k} \Rightarrow \mathbf{a} = \frac{d^2\mathbf{r}}{dt^2}$ $= (\sec t\tan^2 t + \sec^3 t)\mathbf{i} + (2\sec^2 t\tan t)\mathbf{j}$; Speed: $\left|\mathbf{v}\left(\frac{\pi}{6}\right)\right| = \sqrt{\left(\sec\frac{\pi}{6}\tan\frac{\pi}{6}\right)^2 + \left(\sec^2\frac{\pi}{6}\right)^2 + \left(\frac{4}{3}\right)^2} = 2$; Direction: $\frac{\mathbf{v}\left(\frac{\pi}{6}\right)}{\left|\mathbf{v}\left(\frac{\pi}{6}\right)\right|} = \frac{\left(\sec\frac{\pi}{6}\tan\frac{\pi}{6}\right)\mathbf{i} + \left(\sec^2\frac{\pi}{6}\right)\mathbf{j} + \frac{4}{3}\mathbf{k}}{2} = \frac{1}{3}\mathbf{i} + \frac{2}{3}\mathbf{j} + \frac{2}{3}\mathbf{k} \Rightarrow \mathbf{v}\left(\frac{\pi}{6}\right) = 2\left(\frac{1}{3}\mathbf{i} + \frac{2}{3}\mathbf{j} + \frac{2}{3}\mathbf{k}\right)$

13. $\mathbf{r} = (2\ln(t+1))\mathbf{i} + t^2\mathbf{j} + \frac{t^2}{2}\mathbf{k} \Rightarrow \mathbf{v} = \frac{d\mathbf{r}}{dt} = \left(\frac{2}{t+1}\right)\mathbf{i} + 2t\mathbf{j} + t\mathbf{k} \Rightarrow \mathbf{a} = \frac{d^2\mathbf{r}}{dt^2} = \left[\frac{-2}{(t+1)^2}\right]\mathbf{i} + 2\mathbf{j} + \mathbf{k}$; Speed: $|\mathbf{v}(1)| = \sqrt{\left(\frac{2}{1+1}\right)^2 + (2(1))^2 + 1^2} = \sqrt{6}$; Direction: $\frac{\mathbf{v}(1)}{|\mathbf{v}(1)|} = \frac{\left(\frac{2}{1+1}\right)\mathbf{i} + 2(1)\mathbf{j} + (1)\mathbf{k}}{\sqrt{6}}$

$= \frac{1}{\sqrt{6}}\mathbf{i} + \frac{2}{\sqrt{6}}\mathbf{j} + \frac{1}{\sqrt{6}}\mathbf{k} \Rightarrow \mathbf{v}(1) = \sqrt{6}\left(\frac{1}{\sqrt{6}}\mathbf{i} + \frac{2}{\sqrt{6}}\mathbf{j} + \frac{1}{\sqrt{6}}\mathbf{k}\right)$

14. $\mathbf{r} = (e^{-t})\mathbf{i} + (2\cos 3t)\mathbf{j} + (2\sin 3t)\mathbf{k} \Rightarrow \mathbf{v} = \frac{d\mathbf{r}}{dt} = (-e^{-t})\mathbf{i} - (6\sin 3t)\mathbf{j} + (6\cos 3t)\mathbf{k} \Rightarrow \mathbf{a} = \frac{d^2\mathbf{r}}{dt^2}$

$= (e^{-t})\mathbf{i} - (18\cos 3t)\mathbf{j} - (18\sin 3t)\mathbf{k}$; Speed: $|\mathbf{v}(0)| = \sqrt{(-e^0)^2 + [-6\sin 3(0)]^2 + [6\cos 3(0)]^2} = \sqrt{37}$;

Direction: $\frac{\mathbf{v}(0)}{|\mathbf{v}(0)|} = \frac{(-e^0)\mathbf{i} - 6\sin 3(0)\mathbf{j} + 6\cos 3(0)\mathbf{k}}{\sqrt{37}} = -\frac{1}{\sqrt{37}}\mathbf{i} + \frac{6}{\sqrt{37}}\mathbf{k} \Rightarrow \mathbf{v}(0) = \sqrt{37}\left(-\frac{1}{\sqrt{37}}\mathbf{i} + \frac{6}{\sqrt{37}}\mathbf{k}\right)$

15. $\mathbf{v} = 3\mathbf{i} + \sqrt{3}\mathbf{j} + 2t\mathbf{k}$ and $\mathbf{a} = 2\mathbf{k} \Rightarrow \mathbf{v}(0) = 3\mathbf{i} + \sqrt{3}\mathbf{j}$ and $\mathbf{a}(0) = 2\mathbf{k} \Rightarrow |\mathbf{v}(0)| = \sqrt{3^2 + (\sqrt{3})^2 + 0^2} = \sqrt{12}$ and $|\mathbf{a}(0)| = \sqrt{2^2} = 2$; $\mathbf{v}(0)\cdot\mathbf{a}(0) = 0 \Rightarrow \cos\theta = 0 \Rightarrow \theta = \frac{\pi}{2}$

16. $\mathbf{v} = \frac{\sqrt{2}}{2}\mathbf{i} + \left(\frac{\sqrt{2}}{2} - 32t\right)\mathbf{j}$ and $\mathbf{a} = -32\mathbf{j} \Rightarrow \mathbf{v}(0) = \frac{\sqrt{2}}{2}\mathbf{i} + \frac{\sqrt{2}}{2}\mathbf{j}$ and $\mathbf{a}(0) = -32\mathbf{j} \Rightarrow |\mathbf{v}(0)| = \sqrt{\left(\frac{\sqrt{2}}{2}\right)^2 + \left(\frac{\sqrt{2}}{2}\right)^2}$

$= 1$ and $|\mathbf{a}(0)| = \sqrt{(-32)^2} = 32$; $\mathbf{v}(0)\cdot\mathbf{a}(0) = \left(\frac{\sqrt{2}}{2}\right)(-32) = -16\sqrt{2} \Rightarrow \cos\theta = \frac{-16\sqrt{2}}{1(32)} = -\frac{\sqrt{2}}{2} \Rightarrow \theta = \frac{3\pi}{4}$

17. $\mathbf{v} = \left(\frac{2t}{t^2+1}\right)\mathbf{i} + \left(\frac{1}{t^2+1}\right)\mathbf{j} + t(t^2+1)^{-1/2}\mathbf{k}$ and $\mathbf{a} = \left[\frac{-2t^2+2}{(t^2+1)^2}\right]\mathbf{i} - \left[\frac{2t}{(t^2+1)^2}\right]\mathbf{j} + \left[\frac{1}{(t^2+1)^{3/2}}\right]\mathbf{k} \Rightarrow \mathbf{v}(0) = \mathbf{j}$ and $\mathbf{a}(0) = 2\mathbf{i} + \mathbf{k} \Rightarrow |\mathbf{v}(0)| = 1$ and $|\mathbf{a}(0)| = \sqrt{2^2 + 1^2} = \sqrt{5}$; $\mathbf{v}(0)\cdot\mathbf{a}(0) = 0 \Rightarrow \cos\theta = 0 \Rightarrow \theta = \frac{\pi}{2}$

18. $\mathbf{v} = \frac{2}{3}(1+t)^{1/2}\mathbf{i} - \frac{2}{3}(1-t)^{1/2}\mathbf{j} + \frac{1}{3}\mathbf{k}$ and $\mathbf{a} = \frac{1}{3}(1+t)^{-1/2}\mathbf{i} + \frac{1}{3}(1-t)^{-1/2}\mathbf{j} \Rightarrow \mathbf{v}(0) = \frac{2}{3}\mathbf{i} - \frac{2}{3}\mathbf{j} + \frac{1}{3}\mathbf{k}$ and

$\mathbf{a}(0) = \frac{1}{3}\mathbf{i} + \frac{1}{3}\mathbf{j} \Rightarrow |\mathbf{v}(0)| = \sqrt{\left(\frac{2}{3}\right)^2 + \left(-\frac{2}{3}\right)^2 + \left(\frac{1}{3}\right)^2} = 1$ and $|\mathbf{a}(0)| = \sqrt{\left(\frac{1}{3}\right)^2 + \left(\frac{1}{3}\right)^2} = \frac{\sqrt{2}}{3}$; $\mathbf{v}(0)\cdot\mathbf{a}(0) = \frac{2}{9} - \frac{2}{9}$

$= 0 \Rightarrow \cos\theta = 0 \Rightarrow \theta = \frac{\pi}{2}$

19. $\mathbf{v} = (1 - \cos t)\mathbf{i} + (\sin t)\mathbf{j}$ and $\mathbf{a} = (\sin t)\mathbf{i} + (\cos t)\mathbf{j} \Rightarrow \mathbf{v}\cdot\mathbf{a} = (\sin t)(1 - \cos t) + (\sin t)(\cos t) = \sin t$. Thus, $\mathbf{v}\cdot\mathbf{a} = 0 \Rightarrow \sin t = 0 \Rightarrow t = 0, \pi$, or 2π

20. $\mathbf{v} = (\cos t)\mathbf{i} + \mathbf{j} - (\sin t)\mathbf{k}$ and $\mathbf{a} = (-\sin t)\mathbf{i} - (\cos t)\mathbf{k} \Rightarrow \mathbf{v}\cdot\mathbf{a} = -\sin t\cos t + \sin t\cos t = 0$ for all $t \ge 0$

21. $\int_0^1 [t^3\mathbf{i} + 7\mathbf{j} + (t+1)\mathbf{k}]\,dt = \left[\frac{t^4}{4}\right]_0^1\mathbf{i} + [7t]_0^1\mathbf{j} + \left[\frac{t^2}{2} + t\right]_0^1\mathbf{k} = \frac{1}{4}\mathbf{i} + 7\mathbf{j} + \frac{3}{2}\mathbf{k}$

22. $\int_1^2 \left[(6-6t)\mathbf{i} + 3\sqrt{t}\,\mathbf{j} + \left(\frac{4}{t^2}\right)\mathbf{k}\right]dt = \left[6t - 3t^2\right]_1^2\mathbf{i} + \left[2t^{3/2}\right]_1^2\mathbf{j} + \left[-4t^{-1}\right]_1^2\mathbf{k} = -3\mathbf{i} + (4\sqrt{2} - 2)\mathbf{j} + 2\mathbf{k}$

23. $\int_{-\pi/4}^{\pi/4} \left[(\sin t)\mathbf{i} + (1 + \cos t)\mathbf{j} + (\sec^2 t)\mathbf{k}\right] dt = [-\cos t]_{-\pi/4}^{\pi/4}\mathbf{i} + [t + \sin t]_{-\pi/4}^{\pi/4}\mathbf{j} + [\tan t]_{-\pi/4}^{\pi/4}\mathbf{k}$

$= \left(\frac{\pi + 2\sqrt{2}}{2}\right)\mathbf{j} + 2\mathbf{k}$

24. $\int_0^{\pi/3} [(\sec t \tan t)\mathbf{i} + (\tan t)\mathbf{j} + (2 \sin t \cos t)\mathbf{k}]\, dt = \int_0^{\pi/3} [(\sec t \tan t)\mathbf{i} + (\tan t)\mathbf{j} + (\sin 2t)\mathbf{k}]\, dt$

$= [\sec t]_0^{\pi/3}\mathbf{i} + [-\ln(\cos t)]_0^{\pi/3}\mathbf{j} + \left[-\frac{1}{2}\cos 2t\right]_0^{\pi/3}\mathbf{k} = \mathbf{i} + (\ln 2)\mathbf{j} + \frac{3}{4}\mathbf{k}$

25. $\int_1^4 \left(\frac{1}{t}\mathbf{i} + \frac{1}{5-t}\mathbf{j} + \frac{1}{2t}\mathbf{k}\right) dt = = [\ln t]_1^4\mathbf{i} + [-\ln(5-t)]_1^4\mathbf{j} + \left[\frac{1}{2}\ln t\right]_1^4\mathbf{k} = (\ln 4)\mathbf{i} + (\ln 4)\mathbf{j} + (\ln 2)\mathbf{k}$

26. $\int_0^1 \left(\frac{2}{\sqrt{1-t^2}}\mathbf{j} + \frac{\sqrt{3}}{1+t^2}\mathbf{k}\right) dt = [2\sin^{-1} t]_0^1\mathbf{i} + [\sqrt{3}\tan^{-1} t]_0^1\mathbf{k} = \pi\mathbf{i} + \frac{\pi\sqrt{3}}{4}\mathbf{k}$

27. $\mathbf{r} = \int (-t\mathbf{i} - t\mathbf{j} - t\mathbf{k})\, dt = -\frac{t^2}{2}\mathbf{i} - \frac{t^2}{2}\mathbf{j} - \frac{t^2}{2}\mathbf{k} + \mathbf{C}$; $\mathbf{r}(0) = 0\mathbf{i} - 0\mathbf{j} - 0\mathbf{k} + \mathbf{C} = \mathbf{i} + 2\mathbf{j} + 3\mathbf{k} \Rightarrow \mathbf{C} = \mathbf{i} + 2\mathbf{j} + 3\mathbf{k}$

$\Rightarrow \mathbf{r} = \left(-\frac{t^2}{2} + 1\right)\mathbf{i} + \left(-\frac{t^2}{2} + 2\right)\mathbf{j} + \left(-\frac{t^2}{2} + 3\right)\mathbf{k}$

28. $\mathbf{r} = \int \left[(180t)\mathbf{i} + (180t - 16t^2)\mathbf{j}\right] dt = 90t^2\mathbf{i} + \left(90t^2 - \frac{16}{3}t^3\right)\mathbf{j} + \mathbf{C}$; $\mathbf{r}(0) = 90(0)^2\mathbf{i} + \left[90(0)^2 - \frac{16}{3}(0)^3\right]\mathbf{j} + \mathbf{C}$

$= 100\mathbf{j} \Rightarrow \mathbf{C} = 100\mathbf{j} \Rightarrow \mathbf{r} = 90t^2\mathbf{i} + \left(90t^2 - \frac{16}{3}t^3 + 100\right)\mathbf{j}$

29. $\mathbf{r} = \int \left[\left(\frac{3}{2}(t+1)^{1/2}\right)\mathbf{i} + e^{-t}\mathbf{j} + \left(\frac{1}{t+1}\right)\mathbf{k}\right] dt = (t+1)^{3/2}\mathbf{i} - e^{-t}\mathbf{j} + \ln(t+1)\mathbf{k} + \mathbf{C}$;

$\mathbf{r}(0) = (0+1)^{3/2}\mathbf{i} - e^{-0}\mathbf{j} + \ln(0+1)\mathbf{k} + \mathbf{C} = \mathbf{k} \Rightarrow \mathbf{C} = -\mathbf{i} + \mathbf{j} + \mathbf{k}$

$\Rightarrow \mathbf{r} = \left[(t+1)^{3/2} - 1\right]\mathbf{i} + (1 - e^{-t})\mathbf{j} + [1 + \ln(t+1)]\mathbf{k}$

30. $\mathbf{r} = \int [(t^3 + 4t)\mathbf{i} + t\mathbf{j} + 2t^2\mathbf{k}]\, dt = \left(\frac{t^4}{4} + 2t^2\right)\mathbf{i} + \frac{t^2}{2}\mathbf{j} + \frac{2t^3}{3}\mathbf{k} + \mathbf{C}$; $\mathbf{r}(0) = \left[\frac{0^4}{4} + 2(0)^2\right]\mathbf{i} + \frac{0^2}{2}\mathbf{j} + \frac{2(0)^3}{3}\mathbf{k} + \mathbf{C}$

$= \mathbf{i} + \mathbf{j} \Rightarrow \mathbf{C} = \mathbf{i} + \mathbf{j} \Rightarrow \mathbf{r} = \left(\frac{t^4}{4} + 2t^2 + 1\right)\mathbf{i} + \left(\frac{t^2}{2} + 1\right)\mathbf{j} + \frac{2t^3}{3}\mathbf{k}$

31. $\frac{d\mathbf{r}}{dt} = \int (-32\mathbf{k})\ dt = -32t\mathbf{k} + \mathbf{C}_1;\ \frac{d\mathbf{r}}{dt}(0) = 8\mathbf{i} + 8\mathbf{j} \Rightarrow -32(0)\mathbf{k} + \mathbf{C}_1 = 8\mathbf{i} + 8\mathbf{j} \Rightarrow \mathbf{C}_1 = 8\mathbf{i} + 8\mathbf{j}$

$\Rightarrow \frac{d\mathbf{r}}{dt} = 8\mathbf{i} + 8\mathbf{j} - 32t\mathbf{k};\ \mathbf{r} = \int (8\mathbf{i} + 8\mathbf{j} - 32t\mathbf{k})\ dt = 8t\mathbf{i} + 8t\mathbf{j} - 16t^2\mathbf{k} + \mathbf{C}_2;\ \mathbf{r}(0) = 100\mathbf{k}$

$\Rightarrow 8(0)\mathbf{i} + 8(0)\mathbf{j} - 16(0)^2\mathbf{k} + \mathbf{C}_2 = 100\mathbf{k} \Rightarrow \mathbf{C}_2 = 100\mathbf{k} \Rightarrow \mathbf{r} = 8t\mathbf{i} + 8t\mathbf{j} + \left(100 - 16t^2\right)\mathbf{k}$

32. $\frac{d\mathbf{r}}{dt} = \int -(\mathbf{i} + \mathbf{j} + \mathbf{k})\ dt = -(t\mathbf{i} + t\mathbf{j} + t\mathbf{k}) + \mathbf{C}_1;\ \frac{d\mathbf{r}}{dt}(0) = \mathbf{0} \Rightarrow -(0\mathbf{i} + 0\mathbf{j} + 0\mathbf{k}) + \mathbf{C}_1 = \mathbf{0} \Rightarrow \mathbf{C}_1 = \mathbf{0}$

$\Rightarrow \frac{d\mathbf{r}}{dt} = -(t\mathbf{i} + t\mathbf{j} + t\mathbf{k});\ \mathbf{r} = \int -(t\mathbf{i} + t\mathbf{j} + t\mathbf{k})\ dt = -\left(\frac{t^2}{2}\mathbf{i} + \frac{t^2}{2}\mathbf{j} + \frac{t^2}{2}\mathbf{k}\right) + \mathbf{C}_2;\ \mathbf{r}(0) = 10\mathbf{i} + 10\mathbf{j} + 10\mathbf{k}$

$\Rightarrow -\left(\frac{0^2}{2}\mathbf{i} + \frac{0^2}{2}\mathbf{j} + \frac{0^2}{2}\mathbf{k}\right) + \mathbf{C}_2 = 10\mathbf{i} + 10\mathbf{j} + 10\mathbf{k} \Rightarrow \mathbf{C}_2 = 10\mathbf{i} + 10\mathbf{j} + 10\mathbf{k}$

$\Rightarrow \mathbf{r} = \left(-\frac{t^2}{2} + 10\right)\mathbf{i} + \left(-\frac{t^2}{2} + 10\right)\mathbf{j} + \left(-\frac{t^2}{2} + 10\right)\mathbf{k}$

33. $\mathbf{r}(t) = (\sin t)\mathbf{i} + \left(t^2 - \cos t\right)\mathbf{j} + e^t\mathbf{k} \Rightarrow \mathbf{v}(t) = (\cos t)\mathbf{i} + (2t + \sin t)\mathbf{j} + e^t\mathbf{k};\ t_0 = 0 \Rightarrow \mathbf{v}(0) = \mathbf{i} + \mathbf{k}$ and $\mathbf{r}(0) = P_0 = (0, -1, 1) \Rightarrow x = 0 + t = t,\ y = -1$, and $z = 1 + t$ are parametric equations of the tangent line

34. $\mathbf{r}(t) = (2 \sin t)\mathbf{i} + (2 \cos t)\mathbf{j} + 5t\mathbf{k} \Rightarrow \mathbf{v}(t) = (2 \cos t)\mathbf{i} - (2 \sin t)\mathbf{j} + 5\mathbf{k};\ t_0 = 4\pi \Rightarrow \mathbf{v}(0) = 2\mathbf{i} + 5\mathbf{k}$ and $\mathbf{r}(0) = P_0 = (0, 2, 20\pi) \Rightarrow x = 0 + 2t = 2t,\ y = 2$, and $z = 20\pi + 5t$ are parametric equations of the tangent line

35. $\mathbf{r}(t) = (a \sin t)\mathbf{i} + (a \cos t)\mathbf{j} + bt\mathbf{k} \Rightarrow \mathbf{v}(t) = (a \cos t)\mathbf{i} - (a \sin t)\mathbf{j} + b\mathbf{k};\ t_0 = 2\pi \Rightarrow \mathbf{v}(0) = a\mathbf{i} + b\mathbf{k}$ and $\mathbf{r}(0) = P_0 = (0, a, 2b\pi) \Rightarrow x = 0 + at = at,\ y = a$, and $z = 2\pi b + bt$ are parametric equations of the tangent line

36. $\mathbf{r}(t) = (\cos t)\mathbf{i} + (\sin t)\mathbf{j} + (\sin 2t)\mathbf{k} \Rightarrow \mathbf{v}(t) = (-\sin t)\mathbf{i} + (\cos t)\mathbf{j} + (2 \cos 2t)\mathbf{k};\ t_0 = \frac{\pi}{2} \Rightarrow \mathbf{v}(0) = -\mathbf{i} - 2\mathbf{k}$ and $\mathbf{r}(0) = P_0 = (0, 1, 0) \Rightarrow x = 0 - t = -t,\ y = 1$, and $z = 0 - 2t = -2t$ are parametric equations of the tangent line

37. (a) $\mathbf{v}(t) = -(\sin t)\mathbf{i} + (\cos t)\mathbf{j} \Rightarrow \mathbf{a}(t) = -(\cos t)\mathbf{i} - (\sin t)\mathbf{j}$;

(i) $|\mathbf{v}(t)| = \sqrt{(-\sin t)^2 + (\cos t)^2} = 1 \Rightarrow$ constant speed;

(ii) $\mathbf{v} \cdot \mathbf{a} = (\sin t)(\cos t) - (\cos t)(\sin t) = 0 \Rightarrow$ yes, orthogonal;

(iii) counterclockwise movement;

(iv) yes, $\mathbf{r}(0) = \mathbf{i} + 0\mathbf{j}$

(b) $\mathbf{v}(t) = -(2 \sin 2t)\mathbf{i} + (2 \cos 2t)\mathbf{j} \Rightarrow \mathbf{a}(t) = -(4 \cos 2t)\mathbf{i} - (4 \sin 2t)\mathbf{j}$;

(i) $|\mathbf{v}(t)| = \sqrt{4 \sin^2 2t + 4 \cos^2 2t} = 2 \Rightarrow$ constant speed;

(ii) $\mathbf{v} \cdot \mathbf{a} = 8 \sin 2t \cos 2t - 8 \cos 2t \sin 2t = 0 \Rightarrow$ yes, orthogonal;

(iii) counterclockwise movement;

(iv) yes, $\mathbf{r}(0) = \mathbf{i} + 0\mathbf{j}$

(c) $\mathbf{v}(t) = -\sin\left(t - \frac{\pi}{2}\right)\mathbf{i} + \cos\left(t - \frac{\pi}{2}\right)\mathbf{j} \Rightarrow \mathbf{a}(t) = -\cos\left(t - \frac{\pi}{2}\right)\mathbf{i} - \sin\left(t - \frac{\pi}{2}\right)\mathbf{j}$;

(i) $|\mathbf{v}(t)| = \sqrt{\sin^2\left(t - \frac{\pi}{2}\right) + \cos^2\left(t - \frac{\pi}{2}\right)} = 1 \Rightarrow$ constant speed;

(ii) $\mathbf{v}\cdot\mathbf{a}=\sin\left(t-\frac{\pi}{2}\right)\cos\left(t-\frac{\pi}{2}\right)-\cos\left(t-\frac{\pi}{2}\right)\sin\left(t-\frac{\pi}{2}\right)=0\Rightarrow$ yes, orthogonal;

(iii) counterclockwise movement;

(iv) no, $\mathbf{r}(0)=0\mathbf{i}-\mathbf{j}$ instead of $\mathbf{i}+0\mathbf{j}$

(d) $\mathbf{v}(t)=-(\sin t)\mathbf{i}-(\cos t)\mathbf{j}\Rightarrow\mathbf{a}(t)=-(\cos t)\mathbf{i}+(\sin t)\mathbf{j}$;

(i) $|\mathbf{v}(t)|=\sqrt{(-\sin t)^2+(-\cos t)^2}=1\Rightarrow$ constant speed;

(ii) $\mathbf{v}\cdot\mathbf{a}=(\sin t)(\cos t)-(\cos t)(\sin t)=0\Rightarrow$ yes, orthogonal;

(iii) clockwise movement;

(iv) yes, $\mathbf{r}(0)=\mathbf{i}-0\mathbf{j}$

(e) $\mathbf{v}(t)=-(2t\sin t)\mathbf{i}+(2t\cos t)\mathbf{j}\Rightarrow\mathbf{a}(t)=-(2\sin t+2t\cos t)\mathbf{i}+(2\cos t-2t\sin t)\mathbf{j}$;

(i) $|\mathbf{v}(t)|=2\sqrt{(\sin t+t\cos t)^2+(\cos t-t\sin t)^2}$

$=2\sqrt{\sin^2 t+\cos^2 t+2t\sin t\cos t-2t\sin t\cos t+t^2\cos^2 t+t^2\sin^2 t}$

$=2\sqrt{1+t^2}\Rightarrow$ variable speed;

(ii) $\mathbf{v}\cdot\mathbf{a}=4\left(t\sin^2 t+t^2\sin t\cos t\right)+4\left(t\cos^2 t-t^2\cos t\sin t\right)=4t\neq 0$ in general $\Rightarrow$ not orthogonal in general;

(iii) counterclockwise movement;

(iv) yes, $\mathbf{r}(0)=\mathbf{i}+0\mathbf{j}$

38. Let $\mathbf{p}=2\mathbf{i}+2\mathbf{j}+\mathbf{k}$ denote the position vector of the point, $\mathbf{u}=\frac{1}{\sqrt{2}}\mathbf{i}-\frac{1}{\sqrt{2}}\mathbf{j}$ and $\mathbf{v}=\frac{1}{\sqrt{3}}\mathbf{i}+\frac{1}{\sqrt{3}}\mathbf{j}+\frac{1}{\sqrt{3}}\mathbf{k}$. Then $\mathbf{r}(t)=\mathbf{p}+(\cos t)\mathbf{u}+(\sin t)\mathbf{v}$. Note that $(2,2,1)$ is a point on the plane and $\mathbf{n}=\mathbf{i}+\mathbf{j}-2\mathbf{k}$ is normal to the plane. Moreover, $\mathbf{u}$ and $\mathbf{v}$ are orthogonal unit vectors with $\mathbf{u}\cdot\mathbf{n}=\mathbf{v}\cdot\mathbf{n}=0\Rightarrow\mathbf{u}$ and $\mathbf{v}$ are parallel to the plane. Therefore, $\mathbf{r}(t)$ identifies a point that lies in the plane for each t. Also, for each t, $(\cos t)\mathbf{u}+(\sin t)\mathbf{v}$ is a unit vector. Starting at the point $(2,2,1)$ the vector $(\cos t)\mathbf{u}+(\sin t)\mathbf{v}$ traces out a circle of radius 1 and center $(2,2,1)$ in the plane $x+y-2z=2$.

39. $\frac{d\mathbf{v}}{dt}=\mathbf{a}=3\mathbf{i}-\mathbf{j}+\mathbf{k}\Rightarrow\mathbf{v}(t)=3t\mathbf{i}-t\mathbf{j}+t\mathbf{k}+\mathbf{C}_1$; the particle travels in the direction of the vector $(4-1)\mathbf{i}+(1-2)\mathbf{j}+(4-3)\mathbf{k}=3\mathbf{i}-\mathbf{j}+\mathbf{k}$ (since it travels in a straight line), and at time $t=0$ it has speed $2\Rightarrow\mathbf{v}(0)=\frac{2}{\sqrt{9+1+1}}(3\mathbf{i}-\mathbf{j}+\mathbf{k})=\mathbf{C}_1\Rightarrow\frac{d\mathbf{r}}{dt}=\mathbf{v}(t)=\left(3t+\frac{6}{\sqrt{11}}\right)\mathbf{i}-\left(t+\frac{2}{\sqrt{11}}\right)\mathbf{j}+\left(t+\frac{2}{\sqrt{11}}\right)\mathbf{k}$

$\Rightarrow\mathbf{r}(t)=\left(\frac{3}{2}t^2+\frac{6}{\sqrt{11}}t\right)\mathbf{i}-\left(\frac{1}{2}t^2+\frac{2}{\sqrt{11}}t\right)\mathbf{j}+\left(\frac{1}{2}t^2+\frac{2}{\sqrt{22}}t\right)\mathbf{k}+\mathbf{C}_2$; $\mathbf{r}(0)=\mathbf{i}+2\mathbf{j}+3\mathbf{k}=\mathbf{C}_2$

$\Rightarrow\mathbf{r}(t)=\left(\frac{3}{2}t^2+\frac{6}{\sqrt{11}}t+1\right)\mathbf{i}-\left(\frac{1}{2}t^2+\frac{2}{\sqrt{11}}t-2\right)\mathbf{j}+\left(\frac{1}{2}t^2+\frac{2}{\sqrt{11}}t+3\right)\mathbf{k}$

$=\left(\frac{1}{2}t^2+\frac{2}{\sqrt{11}}t\right)(3\mathbf{i}-\mathbf{j}+\mathbf{k})+(\mathbf{i}+2\mathbf{j}+3\mathbf{k})$

40. $\frac{d\mathbf{v}}{dt}=\mathbf{a}=2\mathbf{i}+\mathbf{j}+\mathbf{k}\Rightarrow\mathbf{v}(t)=2t\mathbf{i}+t\mathbf{j}+t\mathbf{k}+\mathbf{C}_1$; the particle travels in the direction of the vector

$(3-1)\mathbf{i}+(0-(-1))\mathbf{j}+(3-2)\mathbf{k}=2\mathbf{i}+\mathbf{j}+\mathbf{k}$ (since it travels in a straight line), and at time $t=0$ it has speed 2

$\Rightarrow \mathbf{v}(0) = \frac{2}{\sqrt{4+1+1}}(2\mathbf{i}+\mathbf{j}+\mathbf{k}) = \mathbf{C}_1 \Rightarrow \frac{d\mathbf{r}}{dt} = \mathbf{v}(t) = \left(2t + \frac{4}{\sqrt{6}}\right)\mathbf{i} + \left(t + \frac{2}{\sqrt{6}}\right)\mathbf{j} + \left(t + \frac{2}{\sqrt{6}}\right)\mathbf{k}$

$\Rightarrow \mathbf{r}(t) = \left(t^2 + \frac{4}{\sqrt{6}}t\right)\mathbf{i} + \left(\frac{1}{2}t^2 + \frac{2}{\sqrt{6}}t\right)\mathbf{j} + \left(\frac{1}{2}t^2 + \frac{2}{\sqrt{6}}t\right)\mathbf{k} + \mathbf{C}_2$; $\mathbf{r}(0) = \mathbf{i} - \mathbf{j} + 2\mathbf{k} = \mathbf{C}_2$

$\Rightarrow \mathbf{r}(t) = \left(t^2 + \frac{4}{\sqrt{6}}t + 1\right)\mathbf{i} + \left(\frac{1}{2}t^2 + \frac{2}{\sqrt{6}}t - 1\right)\mathbf{j} + \left(\frac{1}{2}t^2 + \frac{2}{\sqrt{6}}t + 2\right)\mathbf{k} = \left(\frac{1}{2}t^2 + \frac{2}{\sqrt{6}}t\right)(2\mathbf{i}+\mathbf{j}+\mathbf{k}) + (\mathbf{i}-\mathbf{j}+2\mathbf{k})$

41. The velocity vector is tangent to the graph of $y^2 = 2x$ at the point $(2,2)$, has length 5, and a positive $\mathbf{i}$ component. Now, $y^2 = 2x \Rightarrow 2y\frac{dy}{dx} = 2 \Rightarrow \left.\frac{dy}{dx}\right|_{(2,2)} = \frac{2}{2\cdot 2} = \frac{1}{2} \Rightarrow$ the tangent vector lies in the direction of the vector $\mathbf{i} + \frac{1}{2}\mathbf{j} \Rightarrow$ the velocity vector is $\mathbf{v} = \frac{5}{\sqrt{1+\frac{1}{4}}}\left(\mathbf{i} + \frac{1}{2}\mathbf{j}\right) = \frac{5}{\left(\frac{\sqrt{5}}{2}\right)}\left(\mathbf{i} + \frac{1}{2}\mathbf{j}\right) = 2\sqrt{5}\mathbf{i} + \sqrt{5}\mathbf{j}$

42. $\mathbf{v} = (1 - \cos t)\mathbf{i} + (\sin t)\mathbf{j}$ and $\mathbf{a} = (\sin t)\mathbf{i} + (\cos t)\mathbf{j}$; $|\mathbf{v}|^2 = (1-\cos t)^2 + \sin^2 t = 2 - 2\cos t \Rightarrow |\mathbf{v}|^2$ is at a max when $\cos t = -1 \Rightarrow t = \pi, 3\pi, 5\pi$, etc., and at these values of t, $|\mathbf{v}|^2 = 4 \Rightarrow \max|\mathbf{v}| = \sqrt{4} = 2$; $|\mathbf{v}|^2$ is at a min when $\cos t = 1 \Rightarrow t = 0, 2\pi, 4\pi$, etc., and at these values of t, $|\mathbf{v}|^2 = 0 \Rightarrow \min|\mathbf{v}| = 0$; $|\mathbf{a}|^2 = \sin^2 t + \cos^2 t = 1$ for every $t \Rightarrow \max|\mathbf{a}| = \min|\mathbf{a}| = \sqrt{1} = 1$

43. $\mathbf{v} = (-3\sin t)\mathbf{j} + (2\cos t)\mathbf{k}$ and $\mathbf{a} = (-3\cos t)\mathbf{j} - (2\sin t)\mathbf{k}$; $|\mathbf{v}|^2 = 9\sin^2 t + 4\cos^2 t \Rightarrow \frac{d}{dt}(|\mathbf{v}|^2)$ $= 18\sin t\cos t - 8\cos t\sin t = 10\sin t\cos t$; $\frac{d}{dt}(|\mathbf{v}|^2) = 0 \Rightarrow 10\sin t\cos t = 0 \Rightarrow \sin t = 0$ or $\cos t = 0$ $\Rightarrow t = 0, \pi$ or $t = \frac{\pi}{2}, \frac{3\pi}{2}$. When $t = 0, \pi$, $|\mathbf{v}|^2 = 4 \Rightarrow |\mathbf{v}| = \sqrt{4} = 2$; when $t = \frac{\pi}{2}, \frac{3\pi}{2}$, $|\mathbf{v}| = \sqrt{9} = 3$. Therefore $\max|\mathbf{v}|$ is 3 when $t = \frac{\pi}{2}, \frac{3\pi}{2}$, and $\min|\mathbf{v}| = 2$ when $t = 0, \pi$. Next, $|\mathbf{a}|^2 = 9\cos^2 t + 4\sin^2 t$ $\Rightarrow \frac{d}{dt}(|\mathbf{a}|^2) = -18\cos t\sin t + 8\sin t\cos t = -10\sin t\cos t$; $\frac{d}{dt}(|\mathbf{a}|^2) = 0 \Rightarrow -10\sin t\cos t = 0 \Rightarrow \sin t = 0$ or $\cos t = 0 \Rightarrow t = 0, \pi$ or $t = \frac{\pi}{2}, \frac{3\pi}{2}$. When $t = 0, \pi$, $|\mathbf{a}|^2 = 9 \Rightarrow |\mathbf{a}| = 3$; when $t = \frac{\pi}{2}, \frac{3\pi}{2}$, $|\mathbf{a}|^2 = 4 \Rightarrow |\mathbf{a}| = 2$. Therefore, $\max|\mathbf{a}| = 3$ when $t = 0, \pi$, and $\min|\mathbf{a}| = 2$ when $t = \frac{\pi}{2}, \frac{3\pi}{2}$.

44. (a) $\mathbf{r}(t) = (r_0\cos\theta)\mathbf{i} + (r_0\sin\theta)\mathbf{j}$, and the distance traveled along the circle in time t is vt (rate times time) which equals the circular arc length $r_0\theta \Rightarrow \theta = \frac{vt}{r_0} \Rightarrow \mathbf{r}(t) = \left(r_0\cos\frac{vt}{r_0}\right)\mathbf{i} + \left(r_0\sin\frac{vt}{r_0}\right)\mathbf{j}$

(b) $\mathbf{v}(t) = \frac{d\mathbf{r}}{dt} = \left(-v\sin\frac{vt}{r_0}\right)\mathbf{i} + \left(v\cos\frac{vt}{r_0}\right)\mathbf{j} \Rightarrow \mathbf{a}(t) = \frac{d\mathbf{v}}{dt} = \left(-\frac{v^2}{r_0}\cos\frac{vt}{r_0}\right)\mathbf{i} + \left(-\frac{v^2}{r_0}\sin\frac{vt}{r_0}\right)\mathbf{j}$

$= -\frac{v^2}{r_0^2}\left[\left(r_0\cos\frac{vt}{r_0}\right)\mathbf{i} + \left(r_0\sin\frac{vt}{r_0}\right)\mathbf{j}\right] = -\frac{v^2}{r_0^2}\mathbf{r}(t)$

(c) $\mathbf{F} = m\mathbf{a} \Rightarrow \left(-\frac{GmM}{r_0^2}\right)\frac{\mathbf{r}}{r_0} = m\left(-\frac{v^2}{r_0^2}\right)\mathbf{r} \Rightarrow -\frac{GmM}{r_0^3} = -\frac{mv^2}{r_0^2} \Rightarrow v^2 = \frac{GM}{r_0}$

(d) T is the time for the satellite to complete one full orbit $\Rightarrow vT =$ circumference of circle $\Rightarrow vT = 2\pi r_0$

(e) Substitute $v = \frac{2\pi r_0}{T}$ into $v^2 = \frac{GM}{r_0} \Rightarrow \frac{4\pi^2 r_0^2}{T^2} = \frac{GM}{r_0} \Rightarrow T^2 = \frac{4\pi^2 r_0^3}{GM} \Rightarrow T^2$ is proportional to r_0^3 since $\frac{4\pi^2}{GM}$ is a constant

45. $\frac{d}{dt}(\mathbf{v}\cdot\mathbf{v}) = \mathbf{v}\cdot\frac{d\mathbf{v}}{dt} + \frac{d\mathbf{v}}{dt}\cdot\mathbf{v} = 2\mathbf{v}\cdot\frac{d\mathbf{v}}{dt} = 2\cdot 0 = 0 \Rightarrow \mathbf{v}\cdot\mathbf{v}$ is a constant $\Rightarrow |\mathbf{v}| = \sqrt{\mathbf{v}\cdot\mathbf{v}}$ is constant

46. (a) $\frac{d}{dt}(\mathbf{u}\cdot\mathbf{v}\times\mathbf{w}) = \frac{d\mathbf{u}}{dt}\cdot(\mathbf{v}\times\mathbf{w}) + \mathbf{u}\cdot\frac{d}{dt}(\mathbf{v}\times\mathbf{w}) = \frac{d\mathbf{u}}{dt}\cdot(\mathbf{v}\times\mathbf{w}) + \mathbf{u}\cdot\left(\frac{d\mathbf{v}}{dt}\times\mathbf{w} + \mathbf{v}\times\frac{d\mathbf{w}}{dt}\right)$
$= \frac{d\mathbf{u}}{dt}\cdot(\mathbf{v}\times\mathbf{w}) + \mathbf{u}\cdot\frac{d\mathbf{v}}{dt}\times\mathbf{w} + \mathbf{u}\cdot\mathbf{v}\times\frac{d\mathbf{w}}{dt}$

(b) Each of the determinants is equivalent to each expression in Eq. (6) in part (a) because of Eq. (13) in Section 10.4 expressing the triple scalar product as a determinant.

47. $\frac{d}{dt}\left[\mathbf{r}\cdot\left(\frac{d\mathbf{r}}{dt}\times\frac{d^2\mathbf{r}}{dt^2}\right)\right] = \frac{d\mathbf{r}}{dt}\cdot\left(\frac{d\mathbf{r}}{dt}\times\frac{d^2\mathbf{r}}{dt^2}\right) + \mathbf{r}\cdot\left(\frac{d^2\mathbf{r}}{dt^2}\times\frac{d^2\mathbf{r}}{dt^2}\right) + \mathbf{r}\cdot\left(\frac{d\mathbf{r}}{dt}\times\frac{d^3\mathbf{r}}{dt^3}\right) = \mathbf{r}\cdot\left(\frac{d\mathbf{r}}{dt}\times\frac{d^3\mathbf{r}}{dt^3}\right)$, since $\mathbf{A}\cdot(\mathbf{A}\times\mathbf{B}) = 0$ and $\mathbf{A}\cdot(\mathbf{B}\times\mathbf{B}) = 0$ for any vectors $\mathbf{A}$ and $\mathbf{B}$

48. $\mathbf{u} = \mathbf{C} = a\mathbf{i} + b\mathbf{j} + c\mathbf{k}$ with a, b, c real constants $\Rightarrow \frac{d\mathbf{u}}{dt} = \frac{da}{dt}\mathbf{i} + \frac{db}{dt}\mathbf{j} + \frac{dc}{dt}\mathbf{k} = 0\mathbf{i} + 0\mathbf{j} + 0\mathbf{k} = \mathbf{0}$

49. (a) $\mathbf{u} = f(t)\mathbf{i} + g(t)\mathbf{j} + h(t)\mathbf{k} \Rightarrow c\mathbf{u} = cf(t)\mathbf{i} + cg(t)\mathbf{j} + ch(t)\mathbf{k} \Rightarrow \frac{d}{dt}(c\mathbf{u}) = c\frac{df}{dt}\mathbf{i} + c\frac{dg}{dt}\mathbf{j} + c\frac{dh}{dt}\mathbf{k}$
$= c\left(\frac{df}{dt}\mathbf{i} + \frac{dg}{dt}\mathbf{j} + \frac{dh}{dt}\mathbf{k}\right) = c\frac{d\mathbf{u}}{dt}$

(b) $f\mathbf{u} = ff(t)\mathbf{i} + fg(t)\mathbf{j} + fh(t)\mathbf{k} \Rightarrow \frac{d}{dt}(f\mathbf{u}) = \left[\frac{df}{dt}f(t) + f\frac{df}{dt}\right]\mathbf{i} + \left[\frac{df}{dt}g(t) + f\frac{dg}{dt}\right]\mathbf{j} + \left[\frac{df}{dt}h(t) + f\frac{dh}{dt}\right]\mathbf{k}$
$= \frac{df}{dt}[f(t)\mathbf{i} + g(t)\mathbf{j} + h(t)\mathbf{k}] + f\left[\frac{df}{dt}\mathbf{i} + \frac{dg}{dt}\mathbf{j} + \frac{dh}{dt}\mathbf{k}\right] = \frac{df}{dt}\mathbf{u} + f\frac{d\mathbf{u}}{dt}$

50. Let $\mathbf{u} = f_1(t)\mathbf{i} + f_2(t)\mathbf{j} + f_3(t)\mathbf{k}$ and $\mathbf{v} = g_1(t)\mathbf{i} + g_2(t)\mathbf{j} + g_3(t)\mathbf{k}$. Then
$\mathbf{u} + \mathbf{v} = [f_1(t) + g_1(t)]\mathbf{i} + [f_2(t) + g_2(t)]\mathbf{j} + [f_3(t) + g_3(t)]\mathbf{k}$
$\Rightarrow \frac{d}{dt}(\mathbf{u} + \mathbf{v}) = [f_1'(t) + g_1'(t)]\mathbf{i} + [f_2'(t) + g_2'(t)]\mathbf{j} + [f_3'(t) + g_3'(t)]\mathbf{k}$
$= [f_1'(t)\mathbf{i} + f_2'(t)\mathbf{j} + f_3'(t)\mathbf{k}] + [g_1'(t)\mathbf{i} + g_2'(t)\mathbf{j} + g_3'(t)\mathbf{k}] = \frac{d\mathbf{u}}{dt} + \frac{d\mathbf{v}}{dt}$;
$\mathbf{u} - \mathbf{v} = [f_1(t) - g_1(t)]\mathbf{i} + [f_2(t) - g_2(t)]\mathbf{j} + [f_3(t) - g_3(t)]\mathbf{k}$
$\Rightarrow \frac{d}{dt}(\mathbf{u} - \mathbf{v}) = [f_1'(t) - g_1'(t)]\mathbf{i} + [f_2'(t) - g_2'(t)]\mathbf{j} + [f_3'(t) - g_3'(t)]\mathbf{k}$
$= [f_1'(t)\mathbf{i} + f_2'(t)\mathbf{j} + f_3'(t)\mathbf{k}] - [g_1'(t)\mathbf{i} + g_2'(t)\mathbf{j} + g_3'(t)\mathbf{k}] = \frac{d\mathbf{u}}{dt} - \frac{d\mathbf{v}}{dt}$

51. Suppose $\mathbf{r}$ is continuous at $t = t_0$. Then $\lim_{t\to t_0} \mathbf{r}(t) = \mathbf{r}(t_0) \Leftrightarrow \lim_{t\to t_0} [f(t)\mathbf{i} + g(t)\mathbf{j} + h(t)\mathbf{k}]$
$= f(t_0)\mathbf{i} + g(t_0)\mathbf{j} + h(t_0)\mathbf{k} \Leftrightarrow \lim_{t\to t_0} f(t) = f(t_0)$, $\lim_{t\to t_0} g(t) = g(t_0)$, and $\lim_{t\to t_0} h(t) = h(t_0) \Leftrightarrow$ f, g, and h are continuous at $t = t_0$.

52. $\lim_{t\to t_0} [\mathbf{r}_1(t)\times\mathbf{r}_2(t)] = \lim_{t\to t_0} \begin{vmatrix} \mathbf{i} & \mathbf{j} & \mathbf{k} \\ f_1(t) & f_2(t) & f_3(t) \\ g_1(t) & g_2(t) & g_3(t) \end{vmatrix} = \begin{vmatrix} \mathbf{i} & \mathbf{j} & \mathbf{k} \\ \lim_{t\to t_0} f_1(t) & \lim_{t\to t_0} f_2(t) & \lim_{t\to t_0} f_3(t) \\ \lim_{t\to t_0} g_1(t) & \lim_{t\to t_0} g_2(t) & \lim_{t\to t_0} g_3(t) \end{vmatrix}$
$= \lim_{t\to t_0} \mathbf{r}_1(t) \times \lim_{t\to t_0} \mathbf{r}_2(t) = \mathbf{A}\times\mathbf{B}$

53. $\mathbf{r}'(t_0)$ exists $\Rightarrow f'(t_0)\mathbf{i} + g'(t_0)\mathbf{j} + h'(t_0)\mathbf{k}$ exists $\Rightarrow f'(t_0), g'(t_0), h'(t_0)$ all exist $\Rightarrow$ f, g, and h are continuous at $t = t_0 \Rightarrow \mathbf{r}(t)$ is continuous at $t = t_0$

54. (a) $\int_a^b k\mathbf{r}(t)\,dt = \int_a^b [kf(t)\mathbf{i} + kg(t)\mathbf{j} + kh(t)\mathbf{k}]\,dt = \int_a^b [kf(t)]\,dt\,\mathbf{i} + \int_a^b [kg(t)]\,dt\,\mathbf{j} + \int_a^b [kh(t)]\,dt\,\mathbf{k}$

$$= k\left(\int_a^b f(t)\,dt\,\mathbf{i} + \int_a^b g(t)\,dt\,\mathbf{j} + \int_a^b h(t)\,dt\,\mathbf{k}\right) = k\int_a^b \mathbf{r}(t)\,dt$$

(b) $\int_a^b [\mathbf{r}_1(t) \pm \mathbf{r}_2(t)]\,dt = \int_a^b ([f_1(t)\mathbf{i} + g_1(t)\mathbf{j} + h_1(t)\mathbf{k}] \pm [f_2(t)\mathbf{i} + g_2(t)\mathbf{j} + h_2(t)\mathbf{k}])\,dt$

$$= \int_a^b ([f_1(t) \pm f_2(t)]\mathbf{i} + [g_1(t) \pm g_2(t)]\mathbf{j} + [h_1(t) \pm h_2(t)]\mathbf{k})\,dt$$

$$= \int_a^b [f_1(t) \pm f_2(t)]\,dt\,\mathbf{i} + \int_a^b [g_1(t) \pm g_2(t)]\,dt\,\mathbf{j} + \int_a^b [h_1(t) \pm h_2(t)]\,dt\,\mathbf{k}$$

$$= \left[\int_a^b f_1(t)\,dt\,\mathbf{i} \pm \int_a^b f_2(t)\,dt\,\mathbf{i}\right] + \left[\int_a^b g_1(t)\,dt\,\mathbf{j} \pm \int_a^b g_2(t)\,dt\,\mathbf{j}\right] + \left[\int_a^b h_1(t)\,dt\,\mathbf{k} \pm \int_a^b h_2(t)\,dt\,\mathbf{k}\right]$$

$$= \int_a^b \mathbf{r}_1(t)\,dt \pm \int_a^b \mathbf{r}_2(t)\,dt$$

(c) Let $\mathbf{C} = c_1\mathbf{i} + c_2\mathbf{j} + c_3\mathbf{k}$. Then $\int_a^b \mathbf{C}\cdot\mathbf{r}(t)\,dt = \int_a^b [c_1f(t) + c_2g(t) + c_3h(t)]\,dt$

$$= c_1\int_a^b f(t)\,dt + c_2\int_a^b g(t)\,dt + c_3\int_a^b h(t)\,dt = \mathbf{C}\cdot\int_a^b \mathbf{r}(t)\,dt;$$

$$\int_a^b \mathbf{C}\times\mathbf{r}(t)\,dt = \int_a^b [c_2h(t) - c_3f(t)]\mathbf{i} + [c_3(t) - c_1h(t)]\mathbf{j} + [c_1g(t) - c_2f(t)]\mathbf{k}\,dt$$

$$= \left[c_2\int_a^b h(t)\,dt - c_3\int_a^b f(t)\,dt\right]\mathbf{i} + \left[c_3\int_a^b f(t)\,dt - c_1\int_a^b h(t)\,dt\right]\mathbf{j} + \left[c_1\int_a^b g(t)\,dt - c_2\int_a^b f(t)\,dt\right]\mathbf{k}$$

$$= \mathbf{C}\times\int_a^b \mathbf{r}(t)\,dt$$

55. (a) Let u and $\mathbf{r}$ be continuous on $[a, b]$. Then $\lim_{t\to t_0} u(t)\mathbf{r}(t) = \lim_{t\to t_0} [u(t)f(t)\mathbf{i} + u(t)g(t)\mathbf{j} + u(t)h(t)\mathbf{k}]$

$= u(t_0)f(t_0)\mathbf{i} + u(t_0)g(t_0)\mathbf{j} + u(t_0)h(t_0)\mathbf{k} = u(t_0)\mathbf{r}(t_0) \Rightarrow u\mathbf{r}$ is continuous for every t_0 in $[a, b]$.

(b) Let u and $\mathbf{r}$ be differentiable. Then $\frac{d}{dt}(u\mathbf{r}) = \frac{d}{dt}[u(t)f(t)\mathbf{i} + u(t)g(t)\mathbf{j} + u(t)h(t)\mathbf{k}]$

$$= \left(\frac{du}{dt}f(t) + u(t)\frac{df}{dt}\right)\mathbf{i} + \left(\frac{du}{dt}g(t) + u(t)\frac{dg}{dt}\right)\mathbf{j} + \left(\frac{du}{dt}h(t) + u(t)\frac{dh}{dt}\right)\mathbf{k}$$

$= [f(t)\mathbf{i} + g(t)\mathbf{j} + h(t)\mathbf{k}]\frac{du}{dt} + u(t)\left(\frac{df}{dt}\mathbf{i} + \frac{dg}{dt}\mathbf{j} + \frac{dh}{dt}\mathbf{k}\right) = \mathbf{r}\frac{du}{dt} + u\frac{d\mathbf{r}}{dt}$

56. (a) If $\mathbf{R}_1(t)$ and $\mathbf{R}_2(t)$ have identical derivatives on $\mathbf{i}$, then $\frac{d\mathbf{R}_1}{dt} = \frac{df_1}{dt}\mathbf{i} + \frac{dg_1}{dt}\mathbf{j} + \frac{dh_1}{dt}\mathbf{k} = \frac{df_2}{dt}\mathbf{i} + \frac{dg_2}{dt}\mathbf{j} + \frac{dh_2}{dt}\mathbf{k}$ $= \frac{d\mathbf{R}_2}{dt} \Rightarrow \frac{df_1}{dt} = \frac{df_2}{dt}, \frac{dg_1}{dt} = \frac{dg_2}{dt}, \frac{dh_1}{dt} = \frac{dh_2}{dt} \Rightarrow f_1(t) = f_2(t) + c_1,\ g_1(t) = g_2(t) + c_2,\ h_1(t) = h_2(t) + c_3$ $\Rightarrow f_1(t)\mathbf{i} + g_1(t)\mathbf{j} + h_1(t)\mathbf{k} = [f_2(t) + c_1]\mathbf{i} + [g_2(t) + c_2]\mathbf{j} + [h_2(t) + c_3]\mathbf{k} \Rightarrow \mathbf{R}_1(t) = \mathbf{R}_2(t) + \mathbf{C}$, where $\mathbf{C} = c_1\mathbf{i} + c_2\mathbf{j} + c_3\mathbf{k}$.

(b) Let $\mathbf{R}(t)$ be an antiderivative of $\mathbf{r}(t)$ on $\mathbf{i}$. Then $\mathbf{R}'(t) = \mathbf{r}(t)$. If $\mathbf{U}(t)$ is an antiderivative of $\mathbf{r}(t)$ on $\mathbf{i}$, then $\mathbf{U}'(t) = \mathbf{r}(t)$. Thus $\mathbf{U}'(t) = \mathbf{R}'(t)$ on $\mathbf{i} \Rightarrow \mathbf{U}(t) = \mathbf{R}(t) + \mathbf{C}$.

57. $\frac{d}{dt}\int_a^t \mathbf{r}(\tau)\,d\tau = \frac{d}{dt}\int_a^t [f(\tau)\mathbf{i} + g(\tau)\mathbf{j} + h(\tau)\mathbf{k}]\,d\tau = \frac{d}{dt}\int_a^t f(\tau)\,d\tau\,\mathbf{i} + \frac{d}{dt}\int_a^t g(\tau)\,d\tau\,\mathbf{j} + \frac{d}{dt}\int_a^t h(\tau)\,d\tau\,\mathbf{k}$

$= f(t)\mathbf{i} + g(t)\mathbf{j} + h(t)\mathbf{k} = \mathbf{r}(t)$. Since $\frac{d}{dt}\int_a^t \mathbf{r}(\tau)\,d\tau = \mathbf{r}(t)$, we have that $\int_a^t \mathbf{r}(\tau)\,d\tau$ is an antiderivative of $\mathbf{r}$. If $\mathbf{R}$ is any antiderivative of $\mathbf{r}$, then $\mathbf{R}(t) = \int_a^t \mathbf{r}(\tau)\,d\tau + \mathbf{C}$ by Exercise 56(b). Then $\mathbf{R}(a) = \int_a^a \mathbf{r}(\tau)\,d\tau + \mathbf{C}$ $= \mathbf{0} + \mathbf{C} \Rightarrow \mathbf{C} = \mathbf{R}(a) \Rightarrow \int_a^t \mathbf{r}(\tau)\,d\tau = \mathbf{R}(t) - \mathbf{C} = \mathbf{R}(t) - \mathbf{R}(a) \Rightarrow \int_a^b \mathbf{r}(\tau)\,d\tau = \mathbf{R}(b) - \mathbf{R}(a)$.

58-61. Example CAS commands:

Maple:

```
with(plots):
x:= t -> sin(t) - t*cos(t);
y:= t -> cos(t) + t*sin(t);
z:= t -> t^2;
s1:= spacecurve([x(t),y(t),z(t)], t=0..6*Pi, numpoints = 120, axes=NORMAL):
dx:= t -> D(x)(t);
dy:= t -> D(y)(t);
dz:= t -> D(z)(t);
t0:= 3*Pi/2:
s2:=spacecurve([x(t0)+t*dx(t0),y(t0)+t*dy(t0),z(t0)+t*dz(t0),t=-2..2]):
display({s1,s2},title = `Space Curve and Tangent Line at t0=3 Pi/2`);
```

Mathematica:

```
Clear[x,y,z,t]
r[t_] = {x[t],y[t],z[t]}
x[t_] = Sin[t] - t Cos[t]
y[t_] = Cos[t] + t Sin[t]
z[t_] = t^2
{a,b} = {0, 6 Pi};
t0 = 3/2 Pi;
p1 = ParametricPlot3D[ {x[t],y[t],z[t]}, {t,a,b} ]
v[t_] = r'[t]
v0 = v[t0]
```

```
line[t_] = r[t0] + t v0
p2 = ParametricPlot3D[ Evaluate[ line[t] ], {t,-2,2} ]
Show[ p1, p2 ]
```

62-63. Example CAS commands:

Maple:

```
with(plots):
x:= t -> cos(a*t):
y:= t -> sin(a*t):
z:= t -> b*t: a:=2: b:= 1:
s1:=spacecurve([x(t),y(t),z(t)], t=0..4*Pi, numpoints = 400, axes=NORMAL):
dx:= t -> D(x)(t);
dy:= t -> D(y)(t);
dz:= t -> D(z)(t);
t0:= 3*Pi/2:
s2:=spacecurve([x(t0)+t*dx(t0),y(t0)+t*dy(t0),z(t0)+t*dz(t0),t=-2..2]):
display({s1,s2},title = ` Helix With a = 2 and b = 1`);
```

Mathematica:

```
Clear[a,b]
x[t_] = Cos[a t]
y[t_] = Sin[a t]
z[t_] = b t
t0 = 3/2 Pi;
v[t_] = r'[t]
v0 = v[t0]
line[t_] = r[t0] + t v0
b = 1
a = 2
p1 = ParametricPlot3D[ {x[t],y[t],z[t]}, {t,0,4Pi} ]
p2 = ParametricPlot3D[ Evaluate[ line[t] ], {t,-2,2} ]
Show[ p1, p2 ]
```

11.2 MODELING PROJECTILE MOTION

1. $x = (v_0 \cos \alpha)t \Rightarrow (21 \text{ km})\left(\frac{1000 \text{ m}}{1 \text{ km}}\right) = (840 \text{ m/s})(\cos 60°)t \Rightarrow t = \frac{21{,}000 \text{ m}}{(840 \text{ m/s})(\cos 60°)} = 50$ seconds

2. $R = \frac{v_0^2}{g} \sin 2\alpha$ and maximum R occurs when $\alpha = 45° \Rightarrow 24.5 \text{ km} = \left(\frac{v_0^2}{9.8 \text{ m/s}^2}\right)(\sin 90°)$

   $\Rightarrow v_0 = \sqrt{(9.8)(24{,}500) \text{ m}^2/\text{s}^2} = 490 \text{ m/s}$

3. (a) $t = \frac{2v_0 \sin \alpha}{g} = \frac{2(500 \text{ m/s})(\sin 45°)}{9.8 \text{ m/s}^2} = 72.2$ seconds; $R = \frac{v_0^2}{g} \sin 2\alpha = \frac{(500 \text{ m/s})^2}{9.8 \text{ m/s}^2}(\sin 90°) = 25{,}510.2$ m

   (b) $x = (v_0 \cos \alpha)t \Rightarrow 5000 \text{ m} = (500 \text{ m/s})(\cos 45°)t \Rightarrow t = \frac{5000 \text{ m}}{(500 \text{ m/s})(\cos 45°)} \approx 14.14$ s; thus,

   $y = (v_0 \sin \alpha)t - \frac{1}{2}gt^2 \Rightarrow y \approx (500 \text{ m/s})(\sin 45°)(14.14 \text{ s}) - \frac{1}{2}(9.8 \text{ m/s}^2)(14.14 \text{ s})^2 \approx 4020$ m

(c) $y_{max} = \frac{(v_0 \sin \alpha)^2}{2g} = \frac{((500 \text{ m/s})(\sin 45°))^2}{2(9.8 \text{ m/s}^2)} = 6378 \text{ m}$

4. $y = y_0 + (v_0 \sin \alpha)t - \frac{1}{2}gt^2 \Rightarrow y = 32 \text{ ft} + (32 \text{ ft/sec})(\sin 30°)t - \frac{1}{2}(32 \text{ ft/sec}^2)t^2 \Rightarrow y = 32 + 16t - 16t^2$; the ball hits the ground when $y = 0 \Rightarrow 0 = 32 + 16t - 16t^2 \Rightarrow t = -1$ or $t = 2 \Rightarrow t = 2$ sec since $t > 0$; thus, $x = (v_0 \cos \alpha)t \Rightarrow x = (32 \text{ ft/sec})(\cos 30°)t = 32\left(\frac{\sqrt{3}}{2}\right)(2) \approx 55.4 \text{ ft}$

5. $x = x_0 + (v_0 \cos \alpha)t = 0 + (44 \cos 45°)t = 22\sqrt{2}t$ and $y = y_0 + (v_0 \sin \alpha)t - \frac{1}{2}gt^2 = 6.5 + (44 \sin 45°)t - 16t^2$ $= 6.5 + 22\sqrt{2}t - 16t^2$; the shot lands when $y = 0 \Rightarrow t = \frac{22\sqrt{2} \pm \sqrt{968 + 416}}{32} \approx 2.135$ sec since $t > 0$; thus $x = 22\sqrt{2}t \approx (22\sqrt{2})(2.134839) \approx 66.42 \text{ ft}$

6. $x = 0 + (44 \cos 40°)t = 33.706t$ and $y = 6.5 + (44 \sin 40°)t - 16t^2 \approx 6.5 + 28.283t - 16t^2$; $y = 0$ $\Rightarrow t \approx \frac{28.283 + \sqrt{(28.283)^2 + 416}}{32} \approx 1.9735$ sec since $t > 0$; thus $x = (33.706)(1.9735) \approx 66.51 \text{ ft} \Rightarrow$ the difference in distances is about $66.51 - 66.42 = 0.09$ ft or about 1 inch

7. $R = \frac{v_0^2}{g} \sin 2\alpha \Rightarrow 10 \text{ m} = \left(\frac{v_0^2}{9.8 \text{ m/s}^2}\right)(\sin 90°) \Rightarrow v_0^2 = 98 \text{ m}^2\text{s}^2 \Rightarrow v_0 \approx 9.9 \text{ m/s}$;

$6\text{m} \approx \frac{(9.9 \text{ m/s})^2}{9.8 \text{ m/s}^2}(\sin 2\alpha) \Rightarrow \sin 2\alpha \approx 0.59999 \Rightarrow 2\alpha \approx 36.87°$ or $143.12° \Rightarrow \alpha \approx 18.4°$ or $71.6°$

8. $v_0 = 5 \times 10^6$ m/s and $x = 40$ cm $= 0.4$ m; thus $x = (v_0 \cos \alpha)t \Rightarrow 0.4\text{m} = (5 \times 10^6 \text{ m/s})(\cos 0°)t$ $\Rightarrow t = 0.08 \times 10^{-6} \text{ s} = 8 \times 10^{-8}$ s; also, $y = y_0 + (v_0 \sin \alpha)t - \frac{1}{2}gt^2$ $\Rightarrow y = (5 \times 10^6 \text{ m/s})(\sin 0°)(8 \times 10^{-8} \text{ s}) - \frac{1}{2}(9.8 \text{ m/s}^2)(8 \times 10^{-8} \text{ s})^2 = -3.136 \times 10^{-14}$ m or -3.136×10^{-12} cm. Therefore, it drops 3.136×10^{-12} cm.

9. $R = \frac{v_0^2}{g} \sin 2\alpha \Rightarrow 3(248.8) \text{ ft} = \left(\frac{v_0^2}{32 \text{ ft/sec}^2}\right)(\sin 18°) \Rightarrow v_0^2 \approx 77{,}292.84 \text{ ft}^2/\text{sec}^2 \Rightarrow v_0 \approx 278.01 \text{ ft/sec} \approx 190 \text{ mph}$

10. $v_0 = \frac{80\sqrt{10}}{3}$ ft/sec and $R = 200 \text{ ft} \Rightarrow 200 = \frac{\left(\frac{80\sqrt{10}}{3}\right)^2}{32}(\sin 2\alpha) \Rightarrow \sin 2\alpha = 0.9 \Rightarrow 2\alpha \approx 64.2° \Rightarrow \alpha \approx 32.1°$;

$y_{max} = \frac{\left[\left(\frac{80\sqrt{10}}{3}\right)(\sin 32.1°)\right]^2}{2(32)} \approx 31.4$ ft. In order to reach the cushion, the angle of elevation will need to be about 32.1°. At this angle, the circus performer will go 31.4 ft into the air at maximum height and will not strike the 75 ft high ceiling.

11. $x = (v_0 \cos \alpha)t \Rightarrow 135 \text{ ft} = (90 \text{ ft/sec})(\cos 30°)t \Rightarrow t \approx 1.732$ sec; $y = (v_0 \sin \alpha)t - \frac{1}{2}gt^2$ $\Rightarrow y \approx (90 \text{ ft/sec})(\sin 30°)(1.732 \text{ sec}) - \frac{1}{2}(32 \text{ ft/sec}^2)(1.732 \text{ sec})^2 \Rightarrow y \approx 29.94 \text{ ft} \Rightarrow$ the golf ball will clip the leaves at the top

12. $v_0 = 116$ ft/sec, $\alpha = 45°$, and $x = (v_0 \cos \alpha)t$
$\Rightarrow 369 = (116 \cos 45°)t \Rightarrow t \approx 4.50$ sec;
also $y = (v_0 \sin \alpha)t - \frac{1}{2}gt^2$
$\Rightarrow y = (116 \sin 45°)(4.50) - \frac{1}{2}(32)(4.50)^2$
≈ 45.11 ft. It will take the ball 4.50 sec to travel 369 ft. At that time the ball will be 45.11 ft in the air and will hit the green just past the pin.

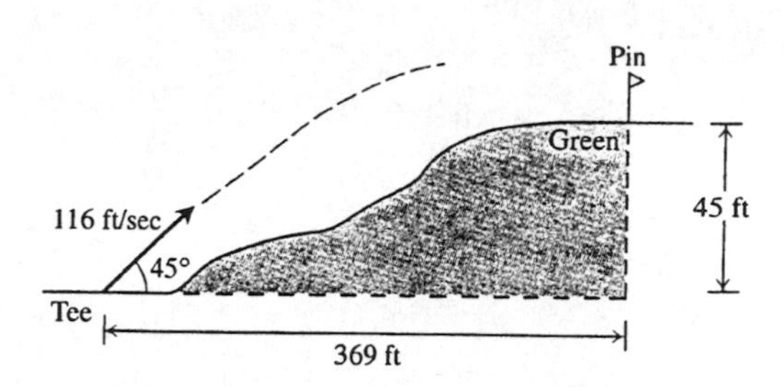

13. $x = x_0 + (v_0 \cos \alpha)t = 0 + (v_0 \cos 40°)t \approx 0.766\, v_0 t$ and $y = y_0 + (v_0 \sin \alpha)t - \frac{1}{2}gt^2 = 6.5 + (v_0 \sin 40°)t - 16t^2$ $\approx 6.5 + 0.643\, v_0 t - 16t^2$; now the shot went 73.833 ft $\Rightarrow 73.833 = 0.766\, v_0 t \Rightarrow t \approx \frac{96.383}{v_0}$ sec; the shot lands when $y = 0 \Rightarrow 0 = 6.5 + (0.643)(96.383) - 16\left(\frac{96.383}{v_0}\right)^2 \Rightarrow 0 \approx 68.474 - \frac{148{,}634}{v_0^2} \Rightarrow v_0 \approx \sqrt{\frac{148{,}634}{68.474}}$ ≈ 46.6 ft/sec, the shot's initial speed

14. $x = (v_0 \cos \alpha)t \Rightarrow 315 \text{ ft} = (v_0 \cos 20°)t \Rightarrow v_0 = \frac{315}{t \cos 20°}$; also $y = (v_0 \sin \alpha)t - \frac{1}{2}gt^2$
$\Rightarrow 34 \text{ ft} = \left(\frac{315}{t \cos 20°}\right)(t \sin 20°) - \frac{1}{2}(32)t^2 \Rightarrow 34 = 315 \tan 20° - 16t^2 \Rightarrow t^2 \approx 5.04 \text{ sec}^2 \Rightarrow t \approx 2.25$ sec
$\Rightarrow v_0 = \frac{315}{(2.25)(\cos 20°)} \approx 149$ ft/sec

15. $R = \frac{v_0^2}{g} \sin 2\alpha = \frac{v_0^2}{g}(2 \sin \alpha \cos \alpha) = \frac{v_0^2}{g}[2 \cos(90° - \alpha) \sin(90° - \alpha)] = \frac{v_0^2}{g}[\sin 2(90° - \alpha)]$

16. $R = \frac{v_0^2}{g} \sin 2\alpha \Rightarrow 16{,}000 \text{ m} = \frac{(400 \text{ m/s})^2}{9.8 \text{ m/s}^2} \sin 2\alpha \Rightarrow \sin 2\alpha = 0.98 \Rightarrow 2\alpha \approx 78.5°$ or $2\alpha \approx 101.5° \Rightarrow \alpha \approx 39.3°$ or $50.7°$

17. $R = \frac{(2v_0)^2}{g} \sin 2\alpha = \frac{4v_0^2}{g} \sin 2\alpha = 4\left(\frac{v_0^2}{g} \sin \alpha\right)$ or 4 times the original range. Now, let the initial range be $R = \frac{v_0^2}{g} \sin 2\alpha$. Then we want the factor p so that pv_0 will double the range $\Rightarrow \frac{(pv_0)^2}{g} \sin 2\alpha = 2\left(\frac{v_0^2}{g} \sin 2\alpha\right)$ $\Rightarrow p^2 = 2 \Rightarrow p = \sqrt{2}$ or about 141%. The same percentage will approximately double the height.

18. $y_{max} = \frac{(v_0 \sin \alpha)^2}{2g} \Rightarrow \frac{3}{4} y_{max} = \frac{3(v_0 \sin \alpha)^2}{8g}$ and $y = (v_0 \sin \alpha)t - \frac{1}{2}gt^2 \Rightarrow \frac{3(v_0 \sin \alpha)^2}{8g} = (v_0 \sin \alpha)t - \frac{1}{2}gt^2$ $\Rightarrow 3(v_0 \sin \alpha)^2 = (8gv_0 \sin \alpha)t - 4g^2t^2 \Rightarrow 4g^2t^2 - (8gv_0 \sin \alpha)t + 3(v_0 \sin \alpha)^2 = 0 \Rightarrow 2gt - 3v_0 \sin \alpha = 0$ or $2gt - v_0 \sin \alpha = 0 \Rightarrow t = \frac{3v_0 \sin \alpha}{2g}$ or $t = \frac{v_0 \sin \alpha}{2g}$. Since the time it takes to reach y_{max} is $t_{max} = \frac{v_0 \sin \alpha}{g}$, then the time it takes the projectile to reach $\frac{3}{4}$ of y_{max} is the shorter time $t = \frac{v_0 \sin \alpha}{2g}$ or half the time it takes to reach the maximum height.

19. $\frac{d\mathbf{r}}{dt} = \int (-g\mathbf{j})\, dt = -gt\mathbf{j} + \mathbf{C}_1$ and $\frac{d\mathbf{r}}{dt}(0) = (v_0 \cos \alpha)\mathbf{i} + (v_0 \sin \alpha)\mathbf{j} \Rightarrow -g(0)\mathbf{j} + \mathbf{C}_1 = (v_0 \cos \alpha)\mathbf{i} + (v_0 \sin \alpha)\mathbf{j}$

$\Rightarrow \mathbf{C}_1 = (v_0 \cos \alpha)\mathbf{i} + (v_0 \sin \alpha)\mathbf{j} \Rightarrow \frac{d\mathbf{r}}{dt} = (v_0 \cos \alpha)\mathbf{i} + (v_0 \sin \alpha - gt)\mathbf{j}$; $\mathbf{r} = \int [(v_0 \cos \alpha)\mathbf{i} + (v_0 \sin \alpha - gt)\mathbf{j}]\, dt$

$= (v_0 t \cos \alpha)\mathbf{i} + \left(v_0 t \sin \alpha - \frac{1}{2}gt^2\right)\mathbf{j} + \mathbf{C}_2$ and $\mathbf{r}(0) = x_0\mathbf{i} + y_0\mathbf{j} \Rightarrow [v_0(0) \cos \alpha]\mathbf{i} + \left[v_0(0) \sin \alpha - \frac{1}{2}g(0)^2\right]\mathbf{j} + \mathbf{C}_2$

$= x_0\mathbf{i} + y_0\mathbf{j} \Rightarrow \mathbf{C}_2 = x_0\mathbf{i} + y_0\mathbf{j} \Rightarrow \mathbf{r} = (x_0 + v_0 t \cos \alpha)\mathbf{i} + \left(y_0 + v_0 t \sin \alpha - \frac{1}{2}gt^2\right)\mathbf{j} \Rightarrow x = x_0 + v_0 t \cos \alpha$ and

$y = y_0 + v_0 t \sin \alpha - \frac{1}{2}gt^2$

20. From Example 3(b) in the text, $v_0 \sin \alpha = \sqrt{(68)(64)} \Rightarrow v_0 \sin 57° \approx 65.97 \Rightarrow v_0 \approx 79$ ft/sec

21. The horizontal distance from Rebollo to the center of the cauldron is 90 ft $\Rightarrow$ the horizontal distance to the nearest rim is $x = 90 - \frac{1}{2}(12) = 84 \Rightarrow 84 = x_0 + (v_0 \cos \alpha)t \approx 0 + \left(\frac{90g}{v_0 \sin \alpha}\right)t \Rightarrow 84 = \frac{(90)(32)}{\sqrt{(68)(64)}}t$

$\Rightarrow t = 1.92$ sec. The vertical distance at this time is $y = y_0 + (v_0 \sin \alpha)t - \frac{1}{2}gt^2$

$\approx 6 + \sqrt{(68)(64)}\,(1.92) - 16(1.92)^2 \approx 73.7$ ft $\Rightarrow$ the arrow clears the rim by 3.7 ft

22. Flight time = 1 sec and the measure of the angle of elevation is about 64° (using a protractor) so that

$t = \frac{2v_0 \sin \alpha}{g} \Rightarrow 1 = \frac{2v_0 \sin 64°}{32} \Rightarrow v_0 \approx 17.80$ ft/sec. Then $y_{max} = \frac{(17.80 \sin 64°)^2}{2(32)} \approx 4.00$ ft and

$R = \frac{v_0^2}{g} \sin 2\alpha \Rightarrow R = \frac{(17.80)^2}{32} \sin 128° \approx 7.80$ ft $\Rightarrow$ the engine traveled about 7.80 ft in 1 sec $\Rightarrow$ the engine velocity was about 7.80 ft/sec

23. When marble A is located R units downrange, we have $x = (v_0 \cos \alpha)t \Rightarrow R = (v_0 \cos \alpha)t \Rightarrow t = \frac{R}{v_0 \cos \alpha}$. At that time the height of marble A is $y = y_0 + (v_0 \sin \alpha)t - \frac{1}{2}gt^2 = (v_0 \sin \alpha)\left(\frac{R}{v_0 \cos \alpha}\right) - \frac{1}{2}g\left(\frac{R}{v_0 \cos \alpha}\right)^2$

$\Rightarrow y = R \tan \alpha - \frac{1}{2}g\left(\frac{R^2}{v_0^2 \cos^2 \alpha}\right)$. The height of marble B at the same time $t = \frac{R}{v_0 \cos \alpha}$ seconds is

$h = R \tan \alpha - \frac{1}{2}gt^2 = R \tan \alpha - \frac{1}{2}g\left(\frac{R^2}{v_0^2 \cos^2 \alpha}\right)$. Since the heights are the same, the marbles collide regardless of the initial velocity v_0.

24. (a) At the time t when the projectile hits the line OR we have $\tan \beta = \frac{y}{x}$;

$x = [v_0 \cos(\alpha - \beta)]t$ and $y = [v_0 \sin(\alpha - \beta)]t - \frac{1}{2}gt^2 < 0$ since R is below level ground. Therefore let $|y| = \frac{1}{2}gt^2 - [v_0 \sin(\alpha - \beta)]t > 0$

so that $\tan \beta = \frac{\left[\frac{1}{2}gt^2 (v_0 \sin(\alpha - \beta))t\right]}{[v_0 \cos(\alpha - \beta)]t} = \frac{\left[\frac{1}{2}gt - v_0 \sin(\alpha - \beta)\right]}{v_0 \cos(\alpha - \beta)}$

$\Rightarrow v_0 \cos(\alpha - \beta) \tan \beta = \frac{1}{2}gt - v_0 \sin(\alpha - \beta)$

$\Rightarrow t = \frac{2v_0 \sin(\alpha - \beta) + 2v_0 \cos(\alpha - \beta) \tan \beta}{g}$, which is the time when the projectile hits the downhill slope. Therefore,

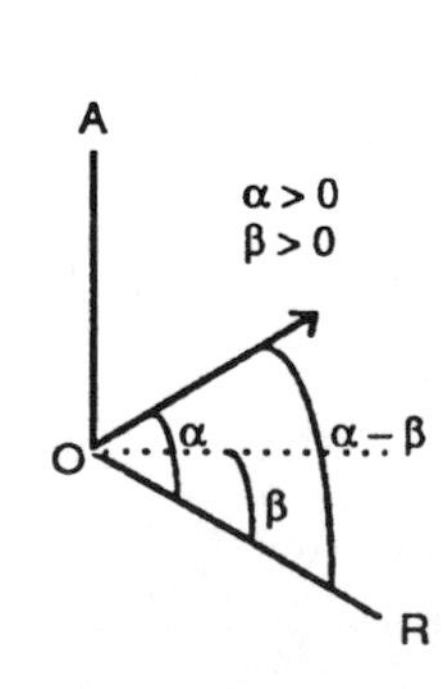

$x = [v_0 \cos(\alpha - \beta)]\left[\frac{2v_0 \sin(\alpha - \beta) + 2v_0 \cos(\alpha - \beta) \tan \beta}{g}\right]$

$= \frac{2v_0^2}{g}\left[\cos^2(\alpha - \beta) \tan \beta + \sin(\alpha - \beta) \cos(\alpha - \beta)\right]$. If x is maximized, then OR is maximized:

$\frac{dx}{d\alpha} = \frac{2v_0^2}{g}[-\sin 2(\alpha - \beta) \tan \beta + \cos 2(\alpha - \beta)] = 0 \Rightarrow -\sin 2(\alpha - \beta) \tan \beta + \cos 2(\alpha - \beta) = 0$

$\Rightarrow \tan \beta = \cot 2(\alpha - \beta) \Rightarrow 2(\alpha - \beta) = 90° - \beta \Rightarrow \alpha - \beta = \frac{1}{2}(90° - \beta) \Rightarrow \alpha = \frac{1}{2}(90° + \beta) = \frac{1}{2}$ of $\angle AOR$.

(b) At the time t when the projectile hits OR we have $\tan \beta = \frac{y}{x}$;

$x = [v_0 \cos(\alpha + \beta)]t$ and $y = [v_0 \sin(\alpha + \beta)]t - \frac{1}{2}gt^2$

$\Rightarrow \tan \beta = \frac{[v_0 \sin(\alpha + \beta)]t - \frac{1}{2}gt^2}{[v_0 \cos(\alpha + \beta)]t} = \frac{\left[v_0 \sin(\alpha + \beta) - \frac{1}{2}gt\right]}{v_0 \cos(\alpha + \beta)}$

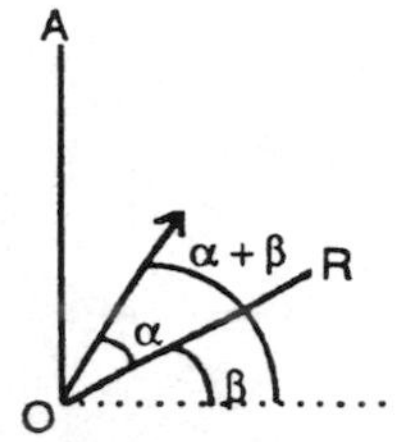

$\Rightarrow v_0 \cos(\alpha + \beta) \tan \beta = v_0 \sin(\alpha + \beta) - \frac{1}{2}gt$

$\Rightarrow t = \frac{2v_0 \sin(\alpha + \beta) - 2v_0 \cos(\alpha + \beta) \tan \beta}{g}$, which is the time when the projectile hits the uphill slope. Therefore,

A
α + β
R
α
β
O

$x = [v_0 \cos(\alpha + \beta)]\left[\frac{2v_0 \sin(\alpha + \beta) - 2v_0 \cos(\alpha + \beta) \tan \beta}{g}\right]$

$= \frac{2v_0^2}{g}\left[\sin(\alpha + \beta) \cos(\alpha + \beta) - \cos^2(\alpha + \beta) \tan \beta\right]$. If x is maximized, then OR is maximized:

$\frac{dx}{d\alpha} = \frac{2v_0^2}{g}[\cos 2(\alpha + \beta) + \sin 2(\alpha + \beta) \tan \beta] = 0 \Rightarrow \cos 2(\alpha + \beta) + \sin 2(\alpha + \beta) \tan \beta = 0$

$\Rightarrow \cot 2(\alpha + \beta) + \tan \beta = 0 \Rightarrow \cot 2(\alpha + \beta) = -\tan \beta = \tan(-\beta) \Rightarrow 2(\alpha + \beta) = 90° - (-\beta)$

$= 90° + \beta \Rightarrow \alpha = \frac{1}{2}(90° - \beta) = \frac{1}{2}$ of $\angle AOR$. Therefore v_0 would bisect $\angle AOR$ for maximum range uphill.

25. $\mathbf{a}(t) = -g\mathbf{k} \Rightarrow \mathbf{v}(t) = -gt\mathbf{k} + \mathbf{C}$; $\mathbf{v}(0) = \mathbf{v}_0 \Rightarrow \mathbf{C} = \mathbf{v}_0 \Rightarrow \mathbf{v}(t) = -gt\mathbf{k} + \mathbf{v}_0 \Rightarrow \mathbf{r}(t) = -\frac{1}{2}gt^2\mathbf{k} + \mathbf{v}_0 t + \mathbf{C}_1$;

$\mathbf{r}(0) = \mathbf{0} \Rightarrow \mathbf{C}_1 = \mathbf{0} \Rightarrow \mathbf{r}(t) = -\frac{1}{2}gt^2\mathbf{k} + \mathbf{v}_0 t$

26. $m\frac{d^2\mathbf{r}}{dt^2} = -mg\mathbf{j} - k\frac{d\mathbf{r}}{dt} \Rightarrow m\frac{d\mathbf{r}}{dt} = -mgt\mathbf{j} - k\mathbf{r} + \mathbf{C} \Rightarrow \frac{d\mathbf{r}}{dt} = -gt\mathbf{j} - \frac{k}{m}\mathbf{r} + \mathbf{C}_1$; $\mathbf{r}(0) = \mathbf{0} \Rightarrow \mathbf{C}_1 = \mathbf{v}_0 \Rightarrow \frac{d\mathbf{r}}{dt}$

$= -gt\mathbf{j} - \frac{k}{m}\mathbf{r} + \mathbf{v}_0 \Rightarrow \frac{d\mathbf{r}}{dt} + \frac{k}{m}\mathbf{r} = \mathbf{v}_0 - gt\mathbf{j} \Rightarrow e^{(k/m)t}\left[\frac{d\mathbf{r}}{dt} + \frac{k}{m}\mathbf{r}\right] = e^{(k/m)t}[\mathbf{v}_0 - gt\mathbf{j}] \Rightarrow e^{(k/m)t}\frac{d\mathbf{r}}{dt} + e^{(k/m)t}\left(\frac{k}{m}\mathbf{r}\right)$

$= \mathbf{v}_0 e^{(k/m)t} - gte^{(k/m)t}\mathbf{j}$. Integrating we get $e^{(k/m)t}\mathbf{r} = \frac{m}{k}\mathbf{v}_0 e^{(k/m)t} - g\left[\frac{m}{k}te^{(k/m)t} - \frac{m^2}{k^2}e^{(k/m)t}\right]\mathbf{j} + \mathbf{C}$

$\Rightarrow \mathbf{r} = \frac{m}{k}\mathbf{v}_0 - g\left(\frac{m}{k}t - \frac{m^2}{k^2}\right)\mathbf{j} + \mathbf{C}e^{-(k/m)t}$

27. From Eq. (7) in the text, the maximum height is $y = \frac{(v_0 \sin \alpha)^2}{2g}$ and this occurs for $x = \frac{v_0^2}{2g}\sin 2\alpha$ $= \frac{v_0^2 \sin \alpha \cos \alpha}{g}$. These equations describe parametrically the points on a curve in the xy-plane associated with the maximum heights on the parabolic trajectories in terms of the parameter (launch angle) α.

Eliminating the parameter α, we have $x^2 = \frac{v_0^4 \sin^2\alpha \cos^2\alpha}{g^2} = \frac{\left(v_0^4 \sin^2\alpha\right)\left(1 - \sin^2\alpha\right)}{g^2} = \frac{v_0^4 \sin^2\alpha}{g^2} - \frac{v_0^4 \sin^4\alpha}{g^2}$

$= \frac{v_0^2}{g}(2y) - (2y)^2 \Rightarrow x^2 + 4y^2 - \left(\frac{2v_0^2}{g}\right)y = 0 \Rightarrow x^2 + 4\left[y^2 - \left(\frac{v_0^2}{2g}\right)y + \frac{v_0^4}{16g^2}\right] = \frac{v_0^4}{16g^2} \Rightarrow x^2 + 4\left(y - \frac{v_0^2}{4g}\right)^2 = \frac{v_0^4}{16g^2}$,

where $x \geq 0$.

28. (a)

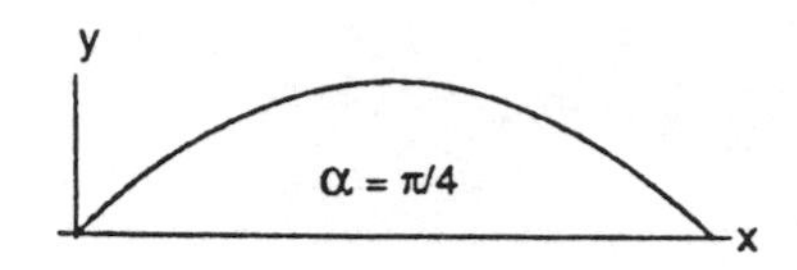

(b)

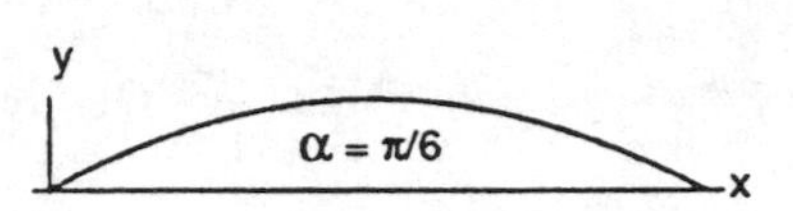

(c)

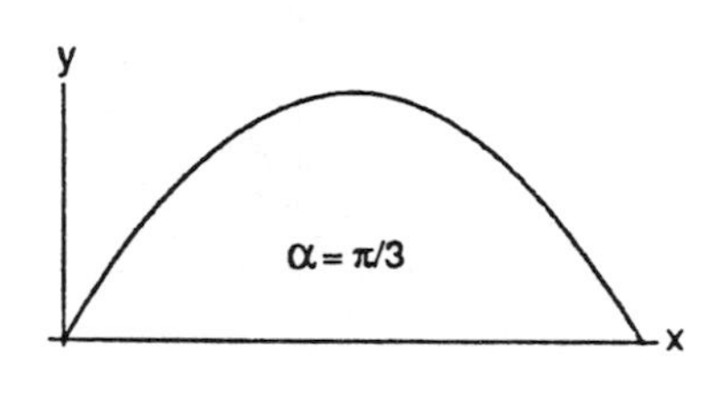

(d)

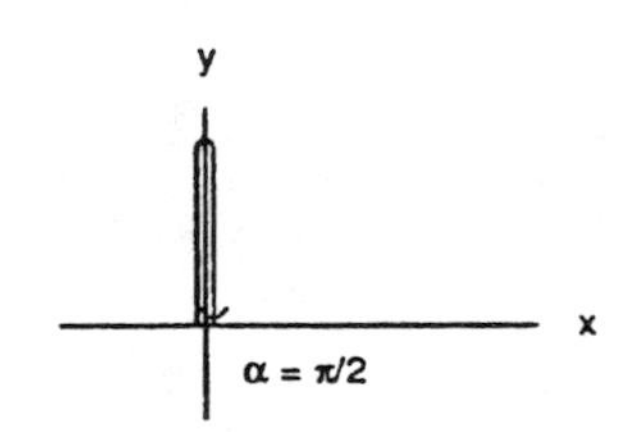

11.3 ARC LENGTH AND THE UNIT TANGENT VECTOR T

1. $\mathbf{r} = (2\cos t)\mathbf{i} + (2\sin t)\mathbf{j} + \sqrt{5}t\mathbf{k} \Rightarrow \mathbf{v} = (-2\sin t)\mathbf{i} + (2\cos t)\mathbf{j} + \sqrt{5}\mathbf{k}$

$\Rightarrow |\mathbf{v}| = \sqrt{(-2\sin t)^2 + (2\cos t)^2 + \left(\sqrt{5}\right)^2} = \sqrt{4\sin^2 t + 4\cos^2 t + 5} = 3;\ \mathbf{T} = \frac{\mathbf{v}}{|\mathbf{v}|}$

$= \left(-\frac{2}{3}\sin t\right)\mathbf{i} + \left(\frac{2}{3}\cos t\right)\mathbf{j} + \frac{\sqrt{5}}{3}\mathbf{k}$ and Length $= \int_0^{\pi} |\mathbf{v}|\,dt = \int_0^{\pi} 3\,dt = [3t]_0^{\pi} = 3\pi$

2. $\mathbf{r} = (6\sin 2t)\mathbf{i} + (6\cos 2t)\mathbf{j} + 5t\mathbf{k} \Rightarrow \mathbf{v} = (12\cos 2t)\mathbf{i} + (-12\sin 2t)\mathbf{j} + 5\mathbf{k}$

$\Rightarrow |\mathbf{v}| = \sqrt{(12\cos 2t)^2 + (-12\sin 2t)^2 + 5^2} = \sqrt{144\cos^2 2t + 144\sin^2 2t + 25} = 13;\ \mathbf{T} = \frac{\mathbf{v}}{|\mathbf{v}|}$

$= \left(\frac{12}{13}\cos 2t\right)\mathbf{i} - \left(\frac{12}{13}\sin 2t\right)\mathbf{j} + \frac{5}{13}\mathbf{k}$ and Length $= \int_0^{\pi} |\mathbf{v}|\,dt = \int_0^{\pi} 13\,dt = [13t]_0^{\pi} = 13\pi$

3. $\mathbf{r} = t\mathbf{i} + \frac{2}{3}t^{3/2}\mathbf{k} \Rightarrow \mathbf{v} = \mathbf{i} + t^{1/2}\mathbf{k} \Rightarrow |\mathbf{v}| = \sqrt{1^2 + \left(t^{1/2}\right)^2} = \sqrt{1+t};\ \mathbf{T} = \frac{\mathbf{v}}{|\mathbf{v}|} = \frac{1}{\sqrt{1+t}}\mathbf{i} + \frac{\sqrt{t}}{\sqrt{1+t}}\mathbf{k}$

and Length $= \int_0^{8} \sqrt{1+t}\,dt = \left[\frac{2}{3}(1+t)^{3/2}\right]_0^{8} = \frac{52}{3}$

4. $\mathbf{r}=(2+t)\mathbf{i}-(t+1)\mathbf{j}+t\mathbf{k} \Rightarrow \mathbf{v}=\mathbf{i}-\mathbf{j}+\mathbf{k} \Rightarrow |\mathbf{v}|=\sqrt{1^2+(-1)^2+1^2}=\sqrt{3}$; $\mathbf{T}=\frac{\mathbf{v}}{|\mathbf{v}|}=\frac{1}{\sqrt{3}}\mathbf{i}-\frac{1}{\sqrt{3}}\mathbf{j}+\frac{1}{\sqrt{3}}\mathbf{k}$

and Length $=\int_0^3 \sqrt{3}\,dt=\left[\sqrt{3}t\right]_0^3=3\sqrt{3}$

5. $\mathbf{r}=(\cos^3 t)\mathbf{j}+(\sin^3 t)\mathbf{k} \Rightarrow \mathbf{v}=(-3\cos^2 t\sin t)\mathbf{j}+(3\sin^2 t\cos t)\mathbf{k} \Rightarrow |\mathbf{v}|$

$=\sqrt{(-3\cos^2 t\sin t)^2+(3\sin^2 t\cos t)^2}=\sqrt{(9\cos^2 t\sin^2 t)(\cos^2 t+\sin^2 t)}=3|\cos t\sin t|$;

$\mathbf{T}=\frac{\mathbf{v}}{|\mathbf{v}|}=\frac{-3\cos^2 t\sin t}{3|\cos t\sin t|}\mathbf{j}+\frac{3\sin^2 t\cos t}{3|\cos t\sin t|}\mathbf{k}=(-\cos t)\mathbf{j}+(\sin t)\mathbf{k}$, if $0\le t\le\frac{\pi}{2}$, and

Length $=\int_0^{\pi/2} 3|\cos t\sin t|\,dt=\int_0^{\pi/2} 3\cos t\sin t\,dt=\int_0^{\pi/2}\frac{3}{2}\sin 2t\,dt=\left[-\frac{3}{4}\cos 2t\right]_0^{\pi/2}=\frac{3}{2}$

6. $\mathbf{r}=6t^3\mathbf{i}-2t^3\mathbf{j}-3t^3\mathbf{k} \Rightarrow \mathbf{v}=18t^2\mathbf{i}-6t^2\mathbf{j}-9t^2\mathbf{k} \Rightarrow |\mathbf{v}|=\sqrt{(18t^2)^2+(-6t^2)^2+(-9t^2)^2}=\sqrt{441t^4}=21t^2$;

$\mathbf{T}=\frac{\mathbf{v}}{|\mathbf{v}|}=\frac{18t^2}{21t^2}\mathbf{i}-\frac{6t^2}{21t^2}\mathbf{j}-\frac{9t^2}{21t^2}\mathbf{k}=\frac{6}{7}\mathbf{i}-\frac{2}{7}\mathbf{j}-\frac{3}{7}\mathbf{k}$ and Length $=\int_1^2 21t^2\,dt=\left[7t^3\right]_1^2=49$

7. $\mathbf{r}=(t\cos t)\mathbf{i}+(t\sin t)\mathbf{j}+\frac{2\sqrt{2}}{3}t^{3/2}\mathbf{k} \Rightarrow \mathbf{v}=(\cos t-t\sin t)\mathbf{i}+(\sin t+t\cos t)\mathbf{j}+\left(\sqrt{2}\,t^{1/2}\right)\mathbf{k}$

$\Rightarrow |\mathbf{v}|=\sqrt{(\cos t-t\sin t)^2+(\sin t+t\cos t)^2+(\sqrt{2t})^2}=\sqrt{1+t^2+2t}=\sqrt{(t+1)^2}=|t+1|=t+1$, if $t\ge 0$;

$\mathbf{T}=\frac{\mathbf{v}}{|\mathbf{v}|}=\left(\frac{\cos t-t\sin t}{t+1}\right)\mathbf{i}+\left(\frac{\sin t+t\cos t}{t+1}\right)\mathbf{j}+\left(\frac{\sqrt{2}t^{1/2}}{t+1}\right)\mathbf{k}$ and Length $=\int_0^{\pi}(t+1)\,dt=\left[\frac{t^2}{2}+t\right]_0^{\pi}=\frac{\pi^2}{2}+\pi$

8. $\mathbf{r}=(t\sin t+\cos t)\mathbf{i}+(t\cos t-\sin t)\mathbf{j} \Rightarrow \mathbf{v}=(\sin t+t\cos t-\sin t)\mathbf{i}+(\cos t-t\sin t-\cos t)\mathbf{j}$

$=(t\cos t)\mathbf{i}-(t\sin t)\mathbf{j} \Rightarrow |\mathbf{v}|=\sqrt{(t\cos t)^2+(-t\sin t)^2}=\sqrt{t^2}=|t|=t$ if $\sqrt{2}\le t\le 2$; $\mathbf{T}=\frac{\mathbf{v}}{|\mathbf{v}|}$

$=\left(\frac{t\cos t}{t}\right)\mathbf{i}-\left(\frac{t\sin t}{t}\right)\mathbf{j}=(\cos t)\mathbf{i}-(\sin t)\mathbf{j}$ and Length $=\int_{\sqrt{2}}^{2} t\,dt=\left[\frac{t^2}{2}\right]_{\sqrt{2}}^{2}=1$

9. Let $P(t_0)$ denote the point. Then $\mathbf{v}=(5\cos t)\mathbf{i}-(5\sin t)\mathbf{j}+12\mathbf{k}$ and $26\pi=\int_0^{t_0}\sqrt{25\cos^2 t+25\sin^2 t+144}\,dt$

$=\int_0^{t_0} 13\,dt=13t_0 \Rightarrow t_0=2\pi$, and the point is $P(2\pi)=(5\sin 2\pi, 5\cos 2\pi, 24\pi)=(5,0,24\pi)$

10. Let $P(t_0)$ denote the point. Then $\mathbf{v} = (12 \cos t)\mathbf{i} + (12 \sin t)\mathbf{j} + 5\mathbf{k}$ and

$$-13\pi = \int_0^{t_0} \sqrt{144 \cos^2 t + 144 \sin^2 t + 25}\, dt = \int_0^{t_0} 13\, dt = 13t_0 \Rightarrow t_0 = -\pi, \text{ and the point is}$$

$P(-\pi) = (12 \sin(-\pi), -12 \cos(-\pi), -5\pi) = (0, 12, -5\pi)$

11. $\mathbf{r} = (4 \cos t)\mathbf{i} + (4 \sin t)\mathbf{j} + 3t\mathbf{k} \Rightarrow \mathbf{v} = (-4 \sin t)\mathbf{i} + (4 \cos t)\mathbf{j} + 3\mathbf{k} \Rightarrow |\mathbf{v}| = \sqrt{(-4 \sin t)^2 + (4 \cos t)^2 + 3^2}$

$$= \sqrt{25} = 5 \Rightarrow s(t) = \int_0^t 5\, d\tau = 5t \Rightarrow \text{Length} = s\left(\frac{\pi}{2}\right) = \frac{5\pi}{2}$$

12. $\mathbf{r} = (\cos t + t \sin t)\mathbf{i} + (\sin t - t \cos t)\mathbf{j} \Rightarrow \mathbf{v} = (-\sin t + \sin t + t \cos t)\mathbf{i} + (\cos t - \cos t + t \sin t)\mathbf{j}$

$$= (t \cos t)\mathbf{i} + (t \sin t)\mathbf{j} \Rightarrow |\mathbf{v}| = \sqrt{(t \cos t)^2 + (t \cos t)^2} = = \sqrt{t^2} = t, \text{ since } \frac{\pi}{2} \le t \le \pi \Rightarrow s(t) = \int_0^t \tau\, d\tau = \frac{t^2}{2}$$

$$\Rightarrow \text{Length} = s(\pi) - s\left(\frac{\pi}{2}\right) = \frac{\pi^2}{2} - \frac{\left(\frac{\pi}{2}\right)^2}{2} = \frac{3\pi^2}{8}$$

13. $\mathbf{r} = (e^t \cos t)\mathbf{i} + (e^t \sin t)\mathbf{j} + e^t\mathbf{k} \Rightarrow \mathbf{v} = (e^t \cos t - e^t \sin t)\mathbf{i} + (e^t \sin t + e^t \cos t)\mathbf{j} + e^t\mathbf{k}$

$$\Rightarrow |\mathbf{v}| = \sqrt{(e^t \cos t - e^t \sin t)^2 + (e^t \sin t + e^t \cos t)^2 + (e^t)^2} = = \sqrt{3e^{2t}} = \sqrt{3}\, e^t \Rightarrow s(t) = \int_0^t \sqrt{3}\, e^\tau\, d\tau$$

$$= \sqrt{3}\, e^t - \sqrt{3} \Rightarrow \text{Length} = s(0) - s(-\ln 4) = 0 - \left(\sqrt{3}\, e^{-\ln 4} - \sqrt{3}\right) = \frac{3\sqrt{3}}{4}$$

14. $\mathbf{r} = (1 + 2t)\mathbf{i} + (1 + 3t)\mathbf{j} + (6 - 6t)\mathbf{k} \Rightarrow \mathbf{v} = 2\mathbf{i} + 3\mathbf{j} - 6\mathbf{k} \Rightarrow |\mathbf{v}| = \sqrt{2^2 + 3^2 + (-6)^2} = 7 \Rightarrow s(t) = \int_0^t 7\, d\tau = 7t$

$$\Rightarrow \text{Length} = s(0) - s(-1) = 0 - (-7) = 7$$

15. $\mathbf{r} = (\sqrt{2}t)\mathbf{i} + (\sqrt{2}t)\mathbf{j} + (1 - t^2)\mathbf{k} \Rightarrow \mathbf{v} = \sqrt{2}\mathbf{i} + \sqrt{2}\mathbf{j} - 2t\mathbf{k} \Rightarrow |\mathbf{v}| = \sqrt{(\sqrt{2})^2 + (\sqrt{2})^2 + (-2t)^2} = \sqrt{4 + 4t^2}$

$$= 2\sqrt{1 + t^2} \Rightarrow \text{Length} = \int_0^1 2\sqrt{1 + t^2}\, dt = \left[2\left(\frac{t}{2}\sqrt{1 + t^2} + \frac{1}{2} \ln\left(t + \sqrt{1 + t^2}\right)\right)\right]_0^1 = \sqrt{2} + \ln\left(1 + \sqrt{2}\right)$$

16. Let the helix make one complete turn from $t = 0$ to $t = 2\pi$. Note that the radius of the cylinder is $1 \Rightarrow$ the circumference of the base is 2π. When $t = 2\pi$, the point P is $(\cos 2\pi, \sin 2\pi, 2\pi) = (1, 0, 2\pi) \Rightarrow$ the cylinder is 2π units high. Cut the cylinder along PQ and flatten. The resulting rectangle has a width equal to the circumference of the cylinder $= 2\pi$ and a height equal to 2π, the height of the cylinder. Therefore, the rectangle is a square and the portion of the helix from $t = 0$ to $t = 2\pi$ is its diagonal.

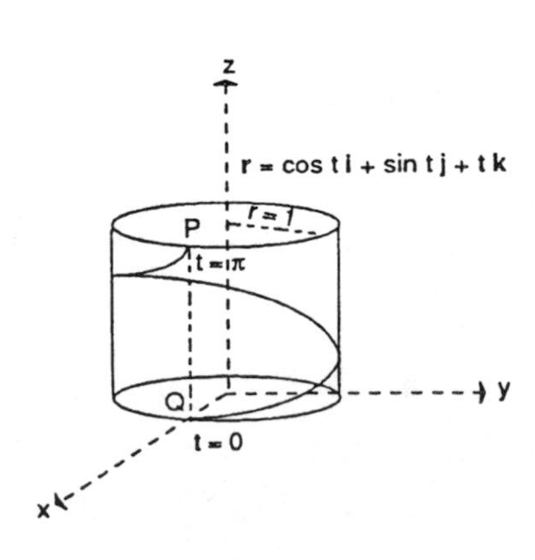

17. (a) $\mathbf{r} = (\cos t)\mathbf{i} + (\sin t)\mathbf{j} + (1 - \cos t)\mathbf{k}$, $0 \le t \le 2\pi \Rightarrow x = \cos t$, $y = \sin t$, $z = 1 - \cos t \Rightarrow x^2 + y^2$ $= \cos^2 t + \sin^2 t = 1$, a right circular cylinder with the z-axis as the axis and radius $= 1$. Therefore $P(\cos t, \sin t, 1 - \cos t)$ lies on the cylinder $x^2 + y^2 = 1$; $t = 0 \Rightarrow P(1,0,0)$ is on the curve; $t = \frac{\pi}{2} \Rightarrow Q(0,1,1)$ is on the curve; $t = \pi \Rightarrow R(-1,0,2)$ is on the curve. Then $\overrightarrow{PQ} = -\mathbf{i} + \mathbf{j} + \mathbf{k}$ and $\overrightarrow{PR} = -2\mathbf{i} + 2\mathbf{k}$

$$\Rightarrow \overrightarrow{PQ} \times \overrightarrow{PR} = \begin{bmatrix} \mathbf{i} & \mathbf{j} & \mathbf{k} \\ -1 & 1 & 1 \\ -2 & 0 & 2 \end{bmatrix} = 2\mathbf{i} + 2\mathbf{k}$$ is a vector normal to the plane of P, Q, and R. Then the plane containing P, Q, and R has an equation $2x + 2z = 2(1) + 2(0)$ or $x + z = 1$. Any point on the curve will satisfy this equation since $x + z = \cos t + (1 - \cos t) = 1$. Therefore, any point on the curve lies on the intersection of the cylinder $x^2 + y^2 = 1$ and the plane $x + z = 1 \Rightarrow$ the curve is an ellipse.

(b) $\mathbf{v} = (-\sin t)\mathbf{i} + (\cos t)\mathbf{j} + (\sin t)\mathbf{k} \Rightarrow |\mathbf{v}| = \sqrt{\sin^2 t + \cos^2 t + \sin^2 t} = \sqrt{1 + \sin^2 t} \Rightarrow \mathbf{T} = \frac{\mathbf{v}}{|\mathbf{v}|}$

$$= \frac{(-\sin t)\mathbf{i} + (\cos t)\mathbf{j} + (\sin t)\mathbf{k}}{\sqrt{1 + \sin^2 t}} \Rightarrow \mathbf{T}(0) = \mathbf{j},\ \mathbf{T}\left(\frac{\pi}{2}\right) = \frac{-\mathbf{i} + \mathbf{k}}{\sqrt{2}},\ \mathbf{T}(\pi) = -\mathbf{j},\ \mathbf{T}\left(\frac{3\pi}{2}\right) = \frac{\mathbf{i} - \mathbf{k}}{\sqrt{2}}$$

(c) $\mathbf{a} = (-\cos t)\mathbf{i} - (\sin t)\mathbf{j} + (\cos t)\mathbf{k}$; $\mathbf{n} = \mathbf{i} + \mathbf{k}$ is normal to the plane $x + z = 1 \Rightarrow \mathbf{n} \cdot \mathbf{a} = -\cos t + \cos t$ $= 0 \Rightarrow \mathbf{a}$ is orthogonal to $\mathbf{n} \Rightarrow \mathbf{a}$ is parallel to the plane; $\mathbf{a}(0) = -\mathbf{i} + \mathbf{k}$, $\mathbf{a}\left(\frac{\pi}{2}\right) = -\mathbf{j}$, $\mathbf{a}\left(\frac{3\pi}{2}\right) = \mathbf{j}$

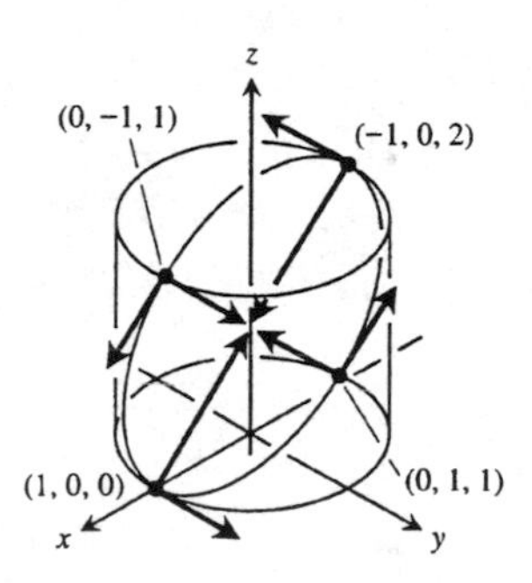

(d) $|\mathbf{v}| = \sqrt{1 + \sin^2 t}$ (See part (b) $\Rightarrow L = \int_0^{2\pi} \sqrt{1 + \sin^2 t}\, dt$

(e) $L \approx 7.64$ (by *Mathematica*)

18. (a) $\mathbf{r} = (\cos 4t)\mathbf{i} + (\sin 4t)\mathbf{j} + 4t\mathbf{k} \Rightarrow \mathbf{v} = (-4 \sin 4t)\mathbf{i} + (4 \cos 4t)\mathbf{j} + 4\mathbf{k} \Rightarrow |\mathbf{v}| = \sqrt{(-4 \sin 4t)^2 + (4 \cos 4t)^2 + 4^2}$ $= \sqrt{32} = 4\sqrt{2} \Rightarrow \text{Length} = \int_0^{\pi/2} 4\sqrt{2}\, dt = \left[4\sqrt{2}\, t\right]_0^{\pi/2} = 2\pi\sqrt{2}$

(b) $\mathbf{r} = \left(\cos \frac{t}{2}\right)\mathbf{i} + \left(\sin \frac{t}{2}\right)\mathbf{j} + \frac{t}{2}\mathbf{k} \Rightarrow \mathbf{v} = \left(-\frac{1}{2} \sin \frac{t}{2}\right)\mathbf{i} + \left(\frac{1}{2} \cos \frac{t}{2}\right)\mathbf{j} + \frac{1}{2}\mathbf{k}$

$$\Rightarrow |\mathbf{v}| = \sqrt{\left(-\frac{1}{2} \sin \frac{t}{2}\right)^2 + \left(\frac{1}{2} \cos \frac{t}{2}\right)^2 + \left(\frac{1}{2}\right)^2} = \sqrt{\frac{1}{4} + \frac{1}{4}} = \frac{\sqrt{2}}{2} \Rightarrow \text{Length} = \int_0^{4\pi} \frac{\sqrt{2}}{2}\, dt = \left[\frac{\sqrt{2}}{2} t\right]_0^{4\pi} = 2\pi\sqrt{2}$$

(c) $\mathbf{r} = (\cos t)\mathbf{i} - (\sin t)\mathbf{j} - t\mathbf{k} \Rightarrow \mathbf{v} = (-\sin t)\mathbf{i} - (\cos t)\mathbf{j} - \mathbf{k} \Rightarrow |\mathbf{v}| = \sqrt{(-\sin t)^2 + (-\cos t)^2 + (-1)^2} = \sqrt{1+1}$

$= \sqrt{2} \Rightarrow \text{Length} = \int_{-2\pi}^{0} \sqrt{2}\, dt = \left[\sqrt{2}\, t\right]_{-2\pi}^{0} = 2\pi\sqrt{2}$

11.4 CURVATURE, TORSION, AND THE TNB FRAME

1. $\mathbf{r} = t\mathbf{i} + \ln(\cos t)\mathbf{j} \Rightarrow \mathbf{v} = \mathbf{i} + \left(\frac{-\sin t}{\cos t}\right)\mathbf{j} = \mathbf{i} - (\tan t)\mathbf{j} \Rightarrow |\mathbf{v}| = \sqrt{1^2 + (-\tan t)^2} = \sqrt{\sec^2 t} = |\sec t| = \sec t$, since

$-\frac{\pi}{2} < t < \frac{\pi}{2} \Rightarrow \mathbf{T} = \frac{\mathbf{v}}{|\mathbf{v}|} = \left(\frac{1}{\sec t}\right)\mathbf{i} - \left(\frac{\tan t}{\sec t}\right)\mathbf{j} = (\cos t)\mathbf{i} - (\sin t)\mathbf{j}$; $\frac{d\mathbf{T}}{dt} = (-\sin t)\mathbf{i} - (\cos t)\mathbf{j}$

$\Rightarrow \left|\frac{d\mathbf{T}}{dt}\right| = \sqrt{(-\sin t)^2 + (-\cos t)^2} = 1 \Rightarrow \mathbf{N} = \frac{\left(\frac{d\mathbf{T}}{dt}\right)}{\left|\frac{d\mathbf{T}}{dt}\right|} = (-\sin t)\mathbf{i} - (\cos t)\mathbf{j}$; $\mathbf{a} = \left(-\sec^2 t\right)\mathbf{j}$

$\Rightarrow \mathbf{v} \times \mathbf{a} = \begin{vmatrix} \mathbf{i} & \mathbf{j} & \mathbf{k} \\ 1 & -\tan t & 0 \\ 0 & -\sec^2 t & 0 \end{vmatrix} = \left(-\sec^2 t\right)\mathbf{k} \Rightarrow |\mathbf{v} \times \mathbf{a}| = \sqrt{\left(-\sec^2 t\right)^2} = \sec^2 t \Rightarrow \kappa = \frac{|\mathbf{v} \times \mathbf{a}|}{|\mathbf{v}|^3} = \frac{\sec^2 t}{\sec^3 t} = \cos t$

2. $\mathbf{r} = \ln(\sec t)\mathbf{i} + t\mathbf{j} \Rightarrow \mathbf{v} = \left(\frac{\sec t \tan t}{\sec t}\right)\mathbf{i} + \mathbf{j} = (\tan t)\mathbf{i} + \mathbf{j} \Rightarrow |\mathbf{v}| = \sqrt{(\tan t)^2 + 1^2} = \sqrt{\sec^2 t} = |\sec t| = \sec t$,

since $-\frac{\pi}{2} < t < \frac{\pi}{2} \Rightarrow \mathbf{T} = \frac{\mathbf{v}}{|\mathbf{v}|} = \left(\frac{\tan t}{\sec t}\right)\mathbf{i} - \left(\frac{1}{\sec t}\right)\mathbf{j} = (\sin t)\mathbf{i} + (\cos t)\mathbf{j}$; $\frac{d\mathbf{T}}{dt} = (\cos t)\mathbf{i} - (\sin t)\mathbf{j}$

$\Rightarrow \left|\frac{d\mathbf{T}}{dt}\right| = \sqrt{(\cos t)^2 + (-\sin t)^2} = 1 \Rightarrow \mathbf{N} = \frac{\left(\frac{d\mathbf{T}}{dt}\right)}{\left|\frac{d\mathbf{T}}{dt}\right|} = (\cos t)\mathbf{i} - (\sin t)\mathbf{j}$; $\mathbf{a} = \left(\sec^2 t\right)\mathbf{i}$

$\Rightarrow \mathbf{v} \times \mathbf{a} = \begin{vmatrix} \mathbf{i} & \mathbf{j} & \mathbf{k} \\ \tan t & 1 & 0 \\ \sec^2 t & 0 & 0 \end{vmatrix} = \left(-\sec^2 t\right)\mathbf{k} \Rightarrow |\mathbf{v} \times \mathbf{a}| = \sqrt{\left(-\sec^2 t\right)^2} = \sec^2 t \Rightarrow \kappa = \frac{|\mathbf{v} \times \mathbf{a}|}{|\mathbf{v}|^3} = \frac{\sec^2 t}{\sec^3 t} = \cos t$

3. $\mathbf{r} = (2t+3)\mathbf{i} + \left(5 - t^2\right)\mathbf{j} \Rightarrow \mathbf{v} = 2\mathbf{i} - 2t\mathbf{j} \Rightarrow |\mathbf{v}| = \sqrt{2^2 + (-2t)^2} = 2\sqrt{1+t^2} \Rightarrow \mathbf{T} = \frac{\mathbf{v}}{|\mathbf{v}|} = \frac{2}{2\sqrt{1+t^2}}\mathbf{i} + \frac{-2t}{2\sqrt{1+t^2}}\mathbf{j}$

$= \frac{1}{\sqrt{1+t^2}}\mathbf{i} - \frac{t}{\sqrt{1+t^2}}\mathbf{j}$; $\frac{d\mathbf{T}}{dt} = \frac{-t}{\left(\sqrt{1+t^2}\right)^3} - \frac{1}{\left(\sqrt{1+t^2}\right)^3}\mathbf{j} \Rightarrow \left|\frac{d\mathbf{T}}{dt}\right| = \sqrt{\left(\frac{-t}{\left(\sqrt{1+t^2}\right)^3}\right)^2 + \left(-\frac{1}{\left(\sqrt{1+t^2}\right)^3}\right)^2}$

$= \sqrt{\frac{1}{\left(1+t^2\right)^2}} = \frac{1}{1+t^2} \Rightarrow \mathbf{N} = \frac{\left(\frac{d\mathbf{T}}{dt}\right)}{\left|\frac{d\mathbf{T}}{dt}\right|} = \frac{-t}{\sqrt{1+t^2}}\mathbf{i} - \frac{1}{\sqrt{1+t^2}}\mathbf{j}$; $\mathbf{a} = -2\mathbf{j} \Rightarrow \mathbf{v} \times \mathbf{a} = \begin{vmatrix} \mathbf{i} & \mathbf{j} & \mathbf{k} \\ 2 & -2t & 0 \\ 0 & -2 & 0 \end{vmatrix} = -4\mathbf{k}$

$\Rightarrow |\mathbf{v} \times \mathbf{a}| = \sqrt{(-4)^2} = 4 \Rightarrow \kappa = \frac{|\mathbf{v} \times \mathbf{a}|}{|\mathbf{v}|^3} = \frac{4}{\left(2\sqrt{1+t^2}\right)^3} = \frac{1}{2\left(\sqrt{1+t^2}\right)^3}$

4. $\mathbf{r}=(\cos t+t\sin t)\mathbf{i}+(\sin t-t\cos t)\mathbf{j} \Rightarrow \mathbf{v}=(t\cos t)\mathbf{i}+(t\sin t)\mathbf{j} \Rightarrow |\mathbf{v}|=\sqrt{(t\cos t)^2+(t\sin t)^2}=\sqrt{t^2}=|t|$ $=t$, since $t>0 \Rightarrow \mathbf{T}=\frac{\mathbf{v}}{|\mathbf{v}|}=\frac{(t\cos t)\mathbf{i}+(t\sin t)\mathbf{j}}{t}=(\cos t)\mathbf{i}+(\sin t)\mathbf{j}$; $\frac{d\mathbf{T}}{dt}=(-\sin t)\mathbf{i}+(\cos t)\mathbf{j}$ $\Rightarrow \left|\frac{d\mathbf{T}}{dt}\right|=\sqrt{(-\sin t)^2+(\cos t)^2}=1 \Rightarrow \mathbf{N}=\frac{\left(\frac{d\mathbf{T}}{dt}\right)}{\left|\frac{d\mathbf{T}}{dt}\right|}=(-\sin t)\mathbf{i}+(\cos t)\mathbf{j}$; $\mathbf{a}=(\cos t-t\sin t)\mathbf{i}+(\sin t+t\cos t)\mathbf{j}$

$$\Rightarrow \mathbf{v}\times\mathbf{a}=\begin{vmatrix}\mathbf{i} & \mathbf{j} & \mathbf{k}\\ t\cos t & t\sin t & 0\\ \cos t-t\sin t & \sin t+t\cos t & 0\end{vmatrix}$$

$=[(t\cos t)(\sin t+t\cos t)-(t\sin t)(\cos t-t\sin t)]\mathbf{k}=t^2\mathbf{k} \Rightarrow |\mathbf{v}\times\mathbf{a}|=\sqrt{(t^2)^2}=t^2 \Rightarrow \kappa=\frac{|\mathbf{v}\times\mathbf{a}|}{|\mathbf{v}|^3}=\frac{t^2}{t^3}=\frac{1}{t}$

5. $\mathbf{r}=(2t+3)\mathbf{i}+(t^2-1)\mathbf{j} \Rightarrow \mathbf{v}=2\mathbf{i}+2t\mathbf{j} \Rightarrow |\mathbf{v}|=\sqrt{2^2+(2t)^2}=2\sqrt{1+t^2} \Rightarrow a_T=\frac{d}{dt}|\mathbf{v}|=2\left(\frac{1}{2}\right)(1+t^2)^{-1/2}(2t)$ $=\frac{2t}{\sqrt{1+t^2}}$; $\mathbf{a}=2\mathbf{j} \Rightarrow |\mathbf{a}|=2 \Rightarrow a_N=\sqrt{|\mathbf{a}|^2-a_T^2}=\sqrt{2^2-\left(\frac{2t}{\sqrt{1+t^2}}\right)^2}=\frac{2}{\sqrt{1+t^2}}$; $\mathbf{a}=\frac{2t}{\sqrt{1+t^2}}\mathbf{T}+\frac{2}{\sqrt{1+t^2}}\mathbf{N}$

6. $\mathbf{r}=\ln(t^2+1)\mathbf{i}+(t-2\tan^{-1}t)\mathbf{j} \Rightarrow \mathbf{v}=\left(\frac{2t}{t^2+1}\right)\mathbf{i}+\left(1-\frac{2}{t^2+1}\right)\mathbf{j}=\left(\frac{2t}{t^2+1}\right)\mathbf{i}+\left(\frac{t^2-1}{t^2+1}\right)\mathbf{j}$ $\Rightarrow |\mathbf{v}|=\sqrt{\left(\frac{2t}{t^2+1}\right)^2+\left(\frac{t^2-1}{t^2+1}\right)^2}=1 \Rightarrow a_T=0$; $\mathbf{a}=\frac{2-2t^2}{(t^2+1)^2}\mathbf{i}+\frac{4t}{(t^2+1)^2}\mathbf{j}$ $\Rightarrow |\mathbf{a}|=\sqrt{\left[\frac{2-2t^2}{(t^2+1)^2}\right]^2+\left[\frac{4t}{(t^2+1)^2}\right]^2}=\frac{2}{(t^2+1)} \Rightarrow a_N=\sqrt{|\mathbf{a}|^2-a_T^2}=\sqrt{|\mathbf{a}|^2-0^2}=|\mathbf{a}|=\frac{2}{(t^2+1)}$; $\mathbf{a}=(0)\mathbf{T}+\left(\frac{2}{t^2+1}\right)\mathbf{N}=\left(\frac{2}{t^2+1}\right)\mathbf{N}$

7. (a) $\mathbf{r}=x\mathbf{i}+f(x)\mathbf{j} \Rightarrow \mathbf{v}=\mathbf{i}+f'(x)\mathbf{j} \Rightarrow \mathbf{a}=f''(x)\mathbf{j} \Rightarrow \mathbf{v}\times\mathbf{a}=\begin{vmatrix}\mathbf{i} & \mathbf{j} & \mathbf{k}\\ 1 & f'(x) & 0\\ 0 & f''(x) & 0\end{vmatrix}=f''(x)\mathbf{k}$

$\Rightarrow |\mathbf{v}\times\mathbf{a}|=\sqrt{(f''(x))^2}=|f''(x)|$ and $|\mathbf{v}|=\sqrt{1^2+[f'(x)]^2}=\sqrt{1+[f'(x)]^2} \Rightarrow \kappa=\frac{|\mathbf{v}\times\mathbf{a}|}{|\mathbf{v}|^3}$ $=\frac{|f''(x)|}{\left[1+(f'(x))^2\right]^{3/2}}$

(b) $y=\ln(\cos x) \Rightarrow \frac{dy}{dx}=\left(\frac{1}{\cos x}\right)(-\sin x)=-\tan x \Rightarrow \frac{d^2y}{dx^2}=-\sec^2 x \Rightarrow \kappa=\frac{|-\sec^2 x|}{[1+(-\tan x)^2]^{3/2}}=\frac{\sec^2 x}{|\sec^3 x|}$ $=\frac{1}{\sec x}=\cos x$, since $-\frac{\pi}{2}<x<\frac{\pi}{2}$

(c) $x = x_0$ gives a point of inflection $\Rightarrow f''(x_0) = 0$ (since f is twice differentiable) $\Rightarrow \kappa = 0$

8. (a) $\mathbf{r} = f(t)\mathbf{i} + g(t)\mathbf{j} = x\mathbf{i} + y\mathbf{j} \Rightarrow \mathbf{v} = \dot{x}\mathbf{i} + \dot{y}\mathbf{j} \Rightarrow \mathbf{a} = \ddot{x}\mathbf{i} + \ddot{y}\mathbf{j} \Rightarrow \mathbf{v}\times\mathbf{a} = \begin{vmatrix} \mathbf{i} & \mathbf{j} & \mathbf{k} \\ \dot{x} & \dot{y} & 0 \\ \ddot{x} & \ddot{y} & 0 \end{vmatrix} = (\dot{x}\ddot{y} - \dot{y}\ddot{x})\mathbf{k}$

$\Rightarrow |\mathbf{v}\times\mathbf{a}| = |\dot{x}\ddot{y} - \dot{y}\ddot{x}|$ and $|\mathbf{v}| = \sqrt{\dot{x}^2 + \dot{y}^2} \Rightarrow \kappa = \frac{|\mathbf{v}\times\mathbf{a}|}{|\mathbf{v}|^3} = \frac{|\dot{x}\ddot{y} - \dot{y}\ddot{x}|}{(\dot{x}^2 + \dot{y}^2)^{3/2}}$

(b) $\mathbf{r}(t) = t\mathbf{i} + \ln(\sin t)\mathbf{j},\ 0 < t < \pi \Rightarrow x = t$ and $y = \ln(\sin t) \Rightarrow \dot{x} = 1,\ \ddot{x} = 0;\ \dot{y} = \frac{\cos t}{\sin t} = \cot t,\ \ddot{y} = -\csc^2 t$

$\Rightarrow \kappa = \frac{|-\csc^2 t - 0|}{(1 + \cot^2 t)^{3/2}} = \frac{\csc^2 t}{\csc^3 t} = \sin t$

(c) $\mathbf{r}(t) = \tan^{-1}(\sinh t)\mathbf{i} + \ln(\cosh t)\mathbf{j} \Rightarrow x = \tan^{-1}(\sinh t)$ and $y = \ln(\cosh t) \Rightarrow \dot{x} = \frac{\cosh t}{1 + \sinh^2 t} = \frac{1}{\cosh t}$

$= \operatorname{sech} t,\ \ddot{x} = -\operatorname{sech} t \tanh t;\ \dot{y} = \frac{\sinh t}{\cosh t} = \tanh t,\ \ddot{y} = \operatorname{sech}^2 t \Rightarrow \kappa = \frac{|\operatorname{sech}^3 t + \operatorname{sech} t \tanh^2 t|}{(\operatorname{sech}^2 t + \tanh^2 t)} = |\operatorname{sech} t|$

$= \operatorname{sech} t$

9. (a) $\mathbf{r}(t) = f(t)\mathbf{i} + g(t)\mathbf{j} \Rightarrow \mathbf{v} = f'(t)\mathbf{i} + g'(t)\mathbf{j}$ is tangent to the curve at the point $(f(t), g(t))$;

$\mathbf{n}\cdot\mathbf{v} = [-g'(t)\mathbf{i} + f'(t)\mathbf{j}]\cdot[f'(t)\mathbf{i} + g'(t)\mathbf{j}] = -g'(t)f'(t) + f'(t)g'(t) = 0;\ -\mathbf{n}\cdot\mathbf{v} = -(\mathbf{n}\cdot\mathbf{v}) = 0$; thus,

$\mathbf{n}$ and $-\mathbf{n}$ are both normal to the curve at the point

(b) $\mathbf{r}(t) = t\mathbf{i} + e^{2t}\mathbf{j} \Rightarrow \mathbf{v} = \mathbf{i} + 2e^{2t}\mathbf{j} \Rightarrow \mathbf{n} = -2e^{2t}\mathbf{i} + \mathbf{j}$ points toward the concave side of the curve; $\mathbf{N} = \frac{\mathbf{n}}{|\mathbf{n}|}$ and

$|\mathbf{n}| = \sqrt{4e^{4t} + 1} \Rightarrow \mathbf{N} = \frac{-2e^{2t}}{\sqrt{1 + 4e^{4t}}}\mathbf{i} + \frac{1}{\sqrt{1 + 4e^{4t}}}\mathbf{j}$

(c) $\mathbf{r}(t) = \sqrt{4 - t^2}\,\mathbf{i} + t\mathbf{j} \Rightarrow \mathbf{v} = \frac{-t}{\sqrt{4 - t^2}}\mathbf{i} + \mathbf{j} \Rightarrow \mathbf{n} = -\mathbf{i} - \frac{t}{\sqrt{4 - t^2}}\mathbf{j}$ points toward the concave side of the curve;

$\mathbf{N} = \frac{\mathbf{n}}{|\mathbf{n}|}$ and $|\mathbf{n}| = \sqrt{1 + \frac{t^2}{4 - t^2}} = \frac{2}{\sqrt{4 - t^2}} \Rightarrow \mathbf{N} = -\frac{1}{2}\left(\sqrt{4 - t^2}\,\mathbf{i} + t\mathbf{j}\right)$

10. (a) $\mathbf{r}(t) = t\mathbf{i} + \frac{1}{3}t^3\mathbf{j} \Rightarrow \mathbf{v} = \mathbf{i} + t^2\mathbf{j} \Rightarrow \mathbf{n} = t^2\mathbf{i} - \mathbf{j}$ points toward the concave side of the curve when $t < 0$ and

$-\mathbf{n} = -t^2\mathbf{i} + \mathbf{j}$ points toward the concave side when $t > 0 \Rightarrow \mathbf{N} = \frac{1}{\sqrt{1 + t^4}}(t^2\mathbf{i} - \mathbf{j})$ for $t < 0$ and

$\mathbf{N} = \frac{1}{\sqrt{1 + t^4}}(-t^2\mathbf{i} + \mathbf{j})$ for $t > 0$

(b) From part (a), $|\mathbf{v}| = \sqrt{1 + t^4} \Rightarrow \mathbf{T} = \frac{1}{\sqrt{1 + t^4}}(\mathbf{i} + t^2\mathbf{j}) \Rightarrow \frac{d\mathbf{T}}{dt} = \frac{-2t^3}{(1 + t^4)^{3/2}}(\mathbf{i} + t^2\mathbf{j}) + \frac{1}{\sqrt{1 + t^4}}(2t\mathbf{j})$

$= \frac{-2t^3}{(1 + t^4)^{3/2}}\left[\mathbf{i} + t^2\mathbf{j} - \left(\frac{1 + t^4}{t^2}\right)\mathbf{j}\right] = \frac{-2t^3}{(1 + t^4)^{3/2}}\left(\mathbf{i} - \frac{1}{t^2}\mathbf{j}\right) \Rightarrow \left|\frac{d\mathbf{T}}{dt}\right| = \frac{2|t|^3}{(1 + t^4)^{3/2}}\sqrt{1 + \frac{1}{t^4}} = \frac{2|t|}{(1 + t^4)^{3/2}}\sqrt{1 + t^4}$

$= \frac{2|t|}{1 + t^4};\ \mathbf{N} = \frac{\left(\frac{d\mathbf{T}}{dt}\right)}{\left|\frac{d\mathbf{T}}{dt}\right|} = \left(\frac{1 + t^4}{2|t|}\right)\left[\frac{2t^3}{(1 + t^4)^{3/2}}\right]\left(-\mathbf{i} + \frac{1}{t^2}\mathbf{j}\right) = \frac{t}{|t|}\left(-\frac{t^2}{\sqrt{1 + t^4}}\mathbf{i} + \frac{1}{\sqrt{1 + t^4}}\mathbf{j}\right),\ t \neq 0$. The normal

$\mathbf{N}$ does not exist at $t = 0$, where the curve has a point of inflection; $\left.\frac{d\mathbf{T}}{dt}\right|_{t=0} = 0$ so the curvature $\kappa = \left|\frac{d\mathbf{T}}{ds}\right|$ $= \left|\frac{d\mathbf{T}}{dt} \cdot \frac{dt}{ds}\right| = 0$ at $t = 0 \Rightarrow \mathbf{N} = \frac{1}{\kappa} \frac{d\mathbf{T}}{ds}$ is undefined. Since $x = t$ and $y = \frac{1}{3}t^3 \Rightarrow y = \frac{1}{3}x^3$, the curve is the cubic power curve which is concave down for $x = t < 0$ and concave up for $x = t > 0$.

11. $\mathbf{r} = (3 \sin t)\mathbf{i} + (3 \cos t)\mathbf{j} + 4t\mathbf{k} \Rightarrow \mathbf{v} = (3 \cos t)\mathbf{i} + (-3 \sin t)\mathbf{j} + 4\mathbf{k} \Rightarrow |\mathbf{v}| = \sqrt{(3 \cos t)^2 + (-3 \sin t)^2 + 4^2}$ $= \sqrt{25} = 5 \Rightarrow \mathbf{T} = \frac{\mathbf{v}}{|\mathbf{v}|} = \left(\frac{3}{5} \cos t\right)\mathbf{i} - \left(\frac{3}{5} \sin t\right)\mathbf{j} + \frac{4}{5}\mathbf{k} \Rightarrow \frac{d\mathbf{T}}{dt} = \left(-\frac{3}{5} \sin t\right)\mathbf{i} - \left(\frac{3}{5} \cos t\right)\mathbf{j}$

$\Rightarrow \left|\frac{d\mathbf{T}}{dt}\right| = \sqrt{\left(-\frac{3}{5} \sin t\right)^2 + \left(-\frac{3}{5} \cos t\right)^2} = \frac{3}{5} \Rightarrow \mathbf{N} = \frac{\left(\frac{d\mathbf{T}}{dt}\right)}{\left|\frac{d\mathbf{T}}{dt}\right|} = (-\sin t)\mathbf{i} - (\cos t)\mathbf{j}$; $\mathbf{a} = (-3 \sin t)\mathbf{i} + (-3 \cos t)\mathbf{j}$

$$\Rightarrow \mathbf{v} \times \mathbf{a} = \begin{vmatrix} \mathbf{i} & \mathbf{j} & \mathbf{k} \\ 3 \cos t & -3 \sin t & 4 \\ -3 \sin t & -3 \cos t & 0 \end{vmatrix} = (12 \cos t)\mathbf{i} - (12 \sin t)\mathbf{j} - 9\mathbf{k} \Rightarrow |\mathbf{v} \times \mathbf{a}|$$

$= \sqrt{(12 \cos t)^2 + (-12 \sin t)^2 + (-9)^2} = \sqrt{225} = 15 \Rightarrow \kappa = \frac{|\mathbf{v} \times \mathbf{a}|}{|\mathbf{v}|^3} = \frac{15}{5^3} = \frac{3}{25}$; $\mathbf{B} = \mathbf{T} \times \mathbf{N}$

$$= \begin{vmatrix} \mathbf{i} & \mathbf{j} & \mathbf{k} \\ \frac{3}{5} \cos t & -\frac{3}{5} \sin t & \frac{4}{5} \\ -\sin t & -\cos t & 0 \end{vmatrix} = \left(\frac{4}{5} \cos t\right)\mathbf{i} - \left(\frac{4}{5} \sin t\right)\mathbf{j} - \frac{3}{5}\mathbf{k}; \ \frac{d\mathbf{a}}{dt} = (-3 \cos t)\mathbf{i} + (3 \sin t)\mathbf{j}$$

$$\Rightarrow \tau = \frac{\begin{vmatrix} 3 \cos t & -3 \sin t & 4 \\ -3 \sin t & -3 \sin t & 0 \\ -3 \cos t & 3 \sin t & 0 \end{vmatrix}}{|\mathbf{v} \times \mathbf{a}|^2} = \frac{-36 \sin^2 t - 36 \cos^2 t}{15^2} = -\frac{4}{25}$$

12. $\mathbf{r} = (\cos t + t \sin t)\mathbf{i} + (\sin t - t \cos t)\mathbf{j} + 3\mathbf{k} \Rightarrow \mathbf{v} = (t \cos t)\mathbf{i} + (t \sin t)\mathbf{j} \Rightarrow |\mathbf{v}| = \sqrt{(t \cos t)^2 + (t \sin t)^2} = \sqrt{t^2}$ $= |t| = t$, if $t > 0 \Rightarrow \mathbf{T} = \frac{\mathbf{v}}{|\mathbf{v}|} = (\cos t)\mathbf{i} - (\sin t)\mathbf{j}$, $t > 0 \Rightarrow \frac{d\mathbf{T}}{dt} = (-\sin t)\mathbf{i} + (\cos t)\mathbf{j}$

$\Rightarrow \left|\frac{d\mathbf{T}}{dt}\right| = \sqrt{(-\sin t)^2 + (\cos t)^2} = 1 \Rightarrow \mathbf{N} = \frac{\left(\frac{d\mathbf{T}}{dt}\right)}{\left|\frac{d\mathbf{T}}{dt}\right|} = (-\sin t)\mathbf{i} + (\cos t)\mathbf{j}$; $\mathbf{a} = (\cos t - t \sin t)\mathbf{i} + (\sin t + t \cos t)\mathbf{j}$

$$\Rightarrow \mathbf{v} \times \mathbf{a} = \begin{vmatrix} \mathbf{i} & \mathbf{j} & \mathbf{k} \\ t \cos t & t \sin t & 0 \\ \cos t - t \sin t & \sin t + t \cos t & 0 \end{vmatrix}$$

$= [(t \cos t)(\sin t + t \cos t) - (t \sin t)(\cos t - t \sin t)]\mathbf{k} = t^2\mathbf{k} \Rightarrow |\mathbf{v} \times \mathbf{a}| = \sqrt{\left(t^2\right)^2} = t^2$

$$\Rightarrow \kappa = \frac{|\mathbf{v}\times\mathbf{a}|}{|\mathbf{v}|^3} = \frac{t^2}{t^3} = \frac{1}{t};\ \mathbf{B} = \mathbf{T}\times\mathbf{N} = \begin{vmatrix} \mathbf{i} & \mathbf{j} & \mathbf{k} \\ \cos t & \sin t & 0 \\ -\sin t & \cos t & 0 \end{vmatrix} = (\cos^2 t + \sin^2 t)\mathbf{k} = \mathbf{k};$$

$$\frac{d\mathbf{a}}{dt} = (-2\sin t - t\cos t)\mathbf{i} + (2\cos t - t\sin t)\mathbf{j} \Rightarrow \tau = \frac{\begin{vmatrix} t\cos t & t\sin t & 0 \\ \cos t - t\sin t & \sin t + t\cos t & 0 \\ -2\sin t - t\cos t & 2\cos t - t\sin t & 0 \end{vmatrix}}{|\mathbf{v}\times\mathbf{a}|^2}$$

$$= \frac{0}{|\mathbf{v}\times\mathbf{a}|^2} = 0$$

13. $\mathbf{r} = (e^t\cos t)\mathbf{i} + (e^t\sin t)\mathbf{j} + 2\mathbf{k} \Rightarrow \mathbf{v} = (e^t\cos t - e^t\sin t)\mathbf{i} + (e^t\sin t + e^t\cos t)\mathbf{j} \Rightarrow$

$$|\mathbf{v}| = \sqrt{(e^t\cos t - e^t\sin t)^2 + (e^t\sin t + e^t\cos t)^2} = \sqrt{2e^{2t}} = e^t\sqrt{2};$$

$$\mathbf{T} = \frac{\mathbf{v}}{|\mathbf{v}|} = \left(\frac{\cos t - \sin t}{\sqrt{2}}\right)\mathbf{i} + \left(\frac{\sin t + \cos t}{\sqrt{2}}\right)\mathbf{j} \Rightarrow \frac{d\mathbf{T}}{dt} = \left(\frac{-\sin t - \cos t}{\sqrt{2}}\right)\mathbf{i} + \left(\frac{\cos t - \sin t}{\sqrt{2}}\right)\mathbf{j}$$

$$\Rightarrow \left|\frac{d\mathbf{T}}{dt}\right| = \sqrt{\left(\frac{-\sin t - \cos t}{\sqrt{2}}\right)^2 + \left(\frac{\cos t - \sin t}{\sqrt{2}}\right)^2} = 1 \Rightarrow \mathbf{N} = \frac{\left(\frac{d\mathbf{T}}{dt}\right)}{\left|\frac{d\mathbf{T}}{dt}\right|} = \left(\frac{-\cos t - \sin t}{\sqrt{2}}\right)\mathbf{i} + \left(\frac{-\sin t + \cos t}{\sqrt{2}}\right)\mathbf{j};$$

$$\mathbf{a} = (-2e^t\sin t)\mathbf{i} + (2e^t\cos t)\mathbf{j} \Rightarrow \mathbf{v}\times\mathbf{a} = \begin{vmatrix} \mathbf{i} & \mathbf{j} & \mathbf{k} \\ e^t\cos t - e^t\sin t & e^t\sin t + e^t\cos t & 0 \\ -2e^t\sin t & 2e^t\cos t & 0 \end{vmatrix} = 2e^{2t}\mathbf{k}$$

$$\Rightarrow |\mathbf{v}\times\mathbf{a}| = \sqrt{(2e^{2t})^2} = 2e^{2t} \Rightarrow \kappa = \frac{|\mathbf{v}\times\mathbf{a}|}{|\mathbf{v}|^3} = \frac{2e^{2t}}{(e^t\sqrt{2})^3} = \frac{1}{e^t\sqrt{2}};$$

$$\mathbf{B} = \mathbf{T}\times\mathbf{N} = \begin{vmatrix} \mathbf{i} & \mathbf{j} & \mathbf{k} \\ \frac{\cos t - \sin t}{\sqrt{2}} & \frac{\sin t + \cos t}{\sqrt{2}} & 0 \\ \frac{-\cos t - \sin t}{\sqrt{2}} & \frac{-\sin t + \cos t}{\sqrt{2}} & 0 \end{vmatrix}$$

$$= \left[\frac{1}{2}(\cos t - \sin t)(-\sin t + \cos t) - \frac{1}{2}(-\cos t - \sin t)(\sin t + \cos t)\right]\mathbf{k}$$

$$= \left[\frac{1}{2}(\cos^2 t - 2\cos t\sin t + \sin^2 t) + \frac{1}{2}(\cos^2 t + 2\sin t\cos t + \sin^2 t)\right]\mathbf{k} = \mathbf{k};$$

$\frac{d\mathbf{a}}{dt} = \left(-2e^t \sin t - 2e^t \cos t\right)\mathbf{i} + \left(2e^t \cos t - 2e^t \sin t\right)\mathbf{j}$

$$\Rightarrow \tau = \frac{\begin{vmatrix} e^t \cos t - e^t \sin t & e^t \sin t + e^t \cos t & 0 \\ -2e^t \sin t & 2e^t \cos t & 0 \\ -2e^t \sin t - 2e^t \cos t & 2e^t \cos t - 2e^t \sin t & 0 \end{vmatrix}}{|\mathbf{v}\times\mathbf{a}|^2} = 0$$

14. $\mathbf{r} = (6 \sin 2t)\mathbf{i} + (6 \cos 2t)\mathbf{j} + 5t\mathbf{k} \Rightarrow \mathbf{v} = (12 \cos 2t)\mathbf{i} - (12 \sin 2t)\mathbf{j} + 5\mathbf{k}$

$\Rightarrow |\mathbf{v}| = \sqrt{(12 \cos 2t)^2 + (-12 \sin 2t)^2 + 5^2} = \sqrt{169} = 13 \Rightarrow \mathbf{T} = \frac{\mathbf{v}}{|\mathbf{v}|}$

$= \left(\frac{12}{13} \cos 2t\right)\mathbf{i} - \left(\frac{12}{13} \sin 2t\right)\mathbf{j} + \frac{5}{13}\mathbf{k} \Rightarrow \frac{d\mathbf{T}}{dt} = \left(-\frac{24}{13} \sin 2t\right)\mathbf{i} - \left(\frac{24}{13} \cos 2t\right)\mathbf{j}$

$\Rightarrow \left|\frac{d\mathbf{T}}{dt}\right| = \sqrt{\left(-\frac{24}{13} \sin 2t\right)^2 + \left(-\frac{24}{13} \cos 2t\right)^2} = \frac{24}{13} \Rightarrow \mathbf{N} = \frac{\left(\frac{d\mathbf{T}}{dt}\right)}{\left|\frac{d\mathbf{T}}{dt}\right|} = (-\sin 2t)\mathbf{i} - (\cos 2t)\mathbf{j};$

$$\mathbf{a} = (-24 \sin 2t)\mathbf{i} - (24 \cos 2t)\mathbf{j} \Rightarrow \mathbf{v}\times\mathbf{a} = \begin{vmatrix} \mathbf{i} & \mathbf{j} & \mathbf{k} \\ 12 \cos 2t & -12 \sin 2t & 5 \\ -24 \sin 2t & -24 \cos 2t & 0 \end{vmatrix}$$

$= (120 \cos 2t)\mathbf{i} - (120 \sin 2t)\mathbf{j} - 288\mathbf{k} \Rightarrow |\mathbf{v}\times\mathbf{a}| = \sqrt{(120 \cos 2t)^2 + (-120 \sin 2t)^2 + (-288)^2} = 312$

$$\Rightarrow \kappa = \frac{|\mathbf{v}\times\mathbf{a}|}{|\mathbf{v}|^3} = \frac{312}{13^3} = \frac{24}{169};\ \mathbf{B} = \mathbf{T}\times\mathbf{N} = \begin{vmatrix} \mathbf{i} & \mathbf{j} & \mathbf{k} \\ \frac{12}{13} \cos 2t & -\frac{12}{13} \sin 2t & \frac{5}{13} \\ -\sin 2t & -\cos 2t & 0 \end{vmatrix}$$

$= \left(\frac{5}{13} \cos 2t\right)\mathbf{i} - \left(\frac{5}{13} \sin 2t\right)\mathbf{j} - \frac{12}{13}\mathbf{k};\ \frac{d\mathbf{a}}{dt} = (-48 \cos 2t)\mathbf{i} + (48 \sin 2t)\mathbf{j}$

$$\Rightarrow \tau = \frac{\begin{vmatrix} 12 \cos 2t & -12 \sin 2t & 5 \\ -24 \sin 2t & -24 \cos 2t & 0 \\ -48 \cos 2t & 48 \sin 2t & 0 \end{vmatrix}}{|\mathbf{v}\times\mathbf{a}|^2} = -\frac{(5)(24)(48)}{(312)^2} = -\frac{5\cdot 1\cdot 2}{13\cdot 13} = -\frac{10}{169}$$

15. $\mathbf{r} = \left(\frac{t^3}{3}\right)\mathbf{i} + \left(\frac{t^2}{2}\right)\mathbf{j},\ t > 0 \Rightarrow \mathbf{v} = t^2\mathbf{i} + t\mathbf{j} \Rightarrow |\mathbf{v}| = \sqrt{t^4 + t^2} = t\sqrt{t^2+1}$, since $t > 0 \Rightarrow \mathbf{T} = \frac{\mathbf{v}}{|\mathbf{v}|}$

$= \frac{t}{\sqrt{t^2+t}}\mathbf{i} + \frac{1}{\sqrt{t^2+1}}\mathbf{j} \Rightarrow \frac{d\mathbf{T}}{dt} = \frac{1}{\left(t^2+1\right)^{3/2}}\mathbf{i} - \frac{t}{\left(t^2+1\right)^{3/2}}\mathbf{j} \Rightarrow \left|\frac{d\mathbf{T}}{dt}\right| = \sqrt{\left(\frac{1}{\left(t^2+1\right)^{3/2}}\right)^2 + \left(\frac{-t}{\left(t^2+1\right)^{3/2}}\right)^2}$

$$= \sqrt{\frac{1+t^2}{\left(t^2+1\right)^3}} = \frac{1}{t^2+1} \Rightarrow \mathbf{N} = \frac{\left(\frac{d\mathbf{T}}{dt}\right)}{\left|\frac{d\mathbf{T}}{dt}\right|} = \frac{1}{\sqrt{t^2+1}}\mathbf{i} - \frac{t}{\sqrt{t^2+1}}\mathbf{j};\ \mathbf{a} = 2t\mathbf{i} + \mathbf{j} \Rightarrow \mathbf{v}\times\mathbf{a} = \begin{vmatrix} \mathbf{i} & \mathbf{j} & \mathbf{k} \\ t^2 & t & 0 \\ 2t & 1 & 0 \end{vmatrix} = -t^2\mathbf{k}$$

$\Rightarrow |\mathbf{v}\times\mathbf{a}| = \sqrt{(-t^2)^2} = t^2 \Rightarrow \kappa = \frac{|\mathbf{v}\times\mathbf{a}|}{|\mathbf{v}|^3} = \frac{t^2}{\left(t\sqrt{t^2+1}\right)^3} = \frac{1}{t\left(t^2+1\right)^{3/2}}$;

$$\mathbf{B} = \mathbf{T}\times\mathbf{N} = \begin{vmatrix} \mathbf{i} & \mathbf{j} & \mathbf{k} \\ \frac{t}{\sqrt{t^2+1}} & \frac{1}{\sqrt{t^2+1}} & 0 \\ \frac{1}{\sqrt{t^2+1}} & \frac{-t}{\sqrt{t^2+1}} & 0 \end{vmatrix} = -\mathbf{k};\ \frac{d\mathbf{a}}{dt} = 2\mathbf{i} \Rightarrow \tau = \frac{\begin{vmatrix} t^2 & t & 0 \\ 2t & 1 & 0 \\ 2 & 0 & 0 \end{vmatrix}}{|\mathbf{v}\times\mathbf{a}|^2} = 0$$

16. $\mathbf{r} = (\cos^3 t)\mathbf{i} + (\sin^3 t)\mathbf{j},\ 0 < t < \frac{\pi}{2} \Rightarrow \mathbf{v} = (-3\cos^2 t\sin t)\mathbf{i} + (3\sin^2 t\cos t)\mathbf{j}$

$\Rightarrow |\mathbf{v}| = \sqrt{(-3\cos^2 t\sin t)^2 + (3\sin^2 t\cos t)^2} = \sqrt{9\cos^4 t\sin^2 t + 9\sin^4 t\cos^2 t} = 3\cos t\sin t$, since $0 < t < \frac{\pi}{2}$

$\Rightarrow \mathbf{T} = \frac{\mathbf{v}}{|\mathbf{v}|} = (-\cos t)\mathbf{i} + (\sin t)\mathbf{j} \Rightarrow \frac{d\mathbf{T}}{dt} = (\sin t)\mathbf{i} + (\cos t)\mathbf{j} \Rightarrow \left|\frac{d\mathbf{T}}{dt}\right| = \sqrt{\sin^2 t + \cos^2 t} = 1 \Rightarrow \mathbf{N} = \frac{\left(\frac{d\mathbf{T}}{dt}\right)}{\left|\frac{d\mathbf{T}}{dt}\right|}$

$= (\sin t)\mathbf{i} + (\cos t)\mathbf{j} \Rightarrow \mathbf{a} = (6\cos t\sin^2 t - 3\cos^3 t)\mathbf{i} + (6\sin t\cos^2 t - 3\sin^3 t)\mathbf{j}$

$$\Rightarrow \mathbf{v}\times\mathbf{a} = \begin{vmatrix} \mathbf{i} & \mathbf{j} & \mathbf{k} \\ -3\cos^2 t\sin t & 3\sin^2 t\cos t & 0 \\ 6\cos t\sin^2 t - 3\cos^3 t & 6\sin t\cos^2 t - 3\sin^3 t & 0 \end{vmatrix}$$

$= (-18\sin^2 t\cos^4 t + 9\cos^2 t\sin^4 t - 18\sin^4 t\cos^2 t + 9\sin^2 t\cos^4 t)\mathbf{k} = (-9\sin^2 t\cos^4 t - 9\cos^2 t\sin^4 t)\mathbf{k}$

$= (-9\sin^2 t\cos^2 t)(\cos^2 t + \sin^2 t)\mathbf{k} = (-9\sin^2 t\cos^2 t)\mathbf{k} \Rightarrow |\mathbf{v}\times\mathbf{a}| = 9\sin^2 t\cos^2 t \Rightarrow \kappa = \frac{|\mathbf{v}\times\mathbf{a}|}{|\mathbf{v}|^3}$

$$= \frac{9\cos^2 t\sin^2 t}{(3\cos t\sin t)^3} = \frac{1}{3\cos t\sin t};\ \mathbf{B} = \mathbf{T}\times\mathbf{N} = \begin{vmatrix} \mathbf{i} & \mathbf{j} & \mathbf{k} \\ -\cos t & \sin t & 0 \\ \sin t & \cos t & 0 \end{vmatrix} = -\mathbf{k};\ \frac{d\mathbf{a}}{dt} = f(t)\mathbf{i} + g(t)\mathbf{j} \text{ where}$$

$f(t) = \frac{d}{dt}(6\cos t\sin^2 t - 3\cos^3 t)$ and $g(t) = \frac{d}{dt}(6\sin t\cos^2 t - 3\sin^3 t)$

$$\Rightarrow \tau = \frac{\begin{vmatrix} -3\cos^2 t\sin t & 3\sin^2 t\cos t & 0 \\ 6\cos t\sin^2 t - 3\cos^3 t & 6\sin t\cos^2 t - 3\sin^3 t & 0 \\ f(t) & g(t) & 0 \end{vmatrix}}{|\mathbf{v}\times\mathbf{a}|^2} = 0$$

17. $\mathbf{r} = t\mathbf{i} + \left(a\cosh\frac{t}{a}\right)\mathbf{j},\ a > 0 \Rightarrow \mathbf{v} = \mathbf{i} + \left(\sinh\frac{t}{a}\right)\mathbf{j} \Rightarrow |\mathbf{v}| = \sqrt{1 + \sinh^2\left(\frac{t}{a}\right)} = \sqrt{\cosh^2\left(\frac{t}{a}\right)} = \cosh\frac{t}{a}$

$\Rightarrow \mathbf{T} = \frac{\mathbf{v}}{|\mathbf{v}|} = \left(\operatorname{sech}\frac{t}{a}\right)\mathbf{i} + \left(\tanh\frac{t}{a}\right)\mathbf{j} \Rightarrow \frac{d\mathbf{T}}{dt} = \left(-\frac{1}{a}\operatorname{sech}\frac{t}{a}\tanh\frac{t}{a}\right)\mathbf{i} + \left(\frac{1}{a}\operatorname{sech}^2\frac{t}{a}\right)\mathbf{j}$

$\Rightarrow \left|\frac{d\mathbf{T}}{dt}\right| = \sqrt{\frac{1}{a^2}\operatorname{sech}^2\left(\frac{t}{a}\right)\tanh^2\left(\frac{t}{a}\right) + \frac{1}{a^2}\operatorname{sech}^4\left(\frac{t}{a}\right)} = \frac{1}{a}\operatorname{sech}\left(\frac{t}{a}\right) \Rightarrow \mathbf{N} = \frac{\left(\frac{d\mathbf{T}}{dt}\right)}{\left|\frac{d\mathbf{T}}{dt}\right|} = \left(-\tanh\frac{t}{a}\right)\mathbf{i} + \left(\operatorname{sech}\frac{t}{a}\right)\mathbf{j}$;

$$\mathbf{a} = \left(\tfrac{1}{a}\cosh\tfrac{t}{a}\right)\mathbf{j} \Rightarrow \mathbf{v}\times\mathbf{a} = \begin{vmatrix} \mathbf{i} & \mathbf{j} & \mathbf{k} \\ 1 & \sinh\left(\tfrac{t}{a}\right) & 0 \\ 0 & \tfrac{1}{a}\cosh\left(\tfrac{t}{a}\right) & 0 \end{vmatrix} = \left(\tfrac{1}{a}\cosh\tfrac{t}{a}\right)\mathbf{k} \Rightarrow |\mathbf{v}\times\mathbf{a}| = \tfrac{1}{a}\cosh\left(\tfrac{t}{a}\right) \Rightarrow \kappa = \frac{|\mathbf{v}\times\mathbf{a}|}{|\mathbf{v}|^3}$$

$$= \frac{\frac{1}{a}\cosh\left(\frac{t}{a}\right)}{\cosh^3\left(\frac{t}{a}\right)} = \tfrac{1}{a}\operatorname{sech}^2\left(\tfrac{t}{a}\right);\ \mathbf{B} = \mathbf{T}\times\mathbf{N} = \begin{vmatrix} \mathbf{i} & \mathbf{j} & \mathbf{k} \\ \operatorname{sech}\left(\tfrac{t}{a}\right) & \tanh\left(\tfrac{t}{a}\right) & 0 \\ -\tanh\left(\tfrac{t}{a}\right) & \operatorname{sech}\left(\tfrac{t}{a}\right) & 0 \end{vmatrix} = \mathbf{k};\ \frac{d\mathbf{a}}{dt} = \frac{1}{a^2}\sinh\left(\tfrac{t}{a}\right)\mathbf{j}$$

$$\tau = \frac{\begin{vmatrix} 1 & \sinh\left(\tfrac{t}{a}\right) & 0 \\ 0 & \tfrac{1}{a}\cosh\left(\tfrac{t}{a}\right) & 0 \\ 0 & \tfrac{1}{a^2}\sinh\left(\tfrac{t}{a}\right) & 0 \end{vmatrix}}{|\mathbf{v}\times\mathbf{a}|^2} = 0$$

18. $\mathbf{r} = (\cosh t)\mathbf{i} + (\sinh t)\mathbf{j} + t\mathbf{k} \Rightarrow \mathbf{v} = (\sinh t)\mathbf{i} - (\cosh t)\mathbf{j} + \mathbf{k} \Rightarrow |\mathbf{v}| = \sqrt{\sinh^2 t + (-\cosh t)^2 + 1} = \sqrt{2}\cosh t$

$$\Rightarrow \mathbf{T} = \frac{\mathbf{v}}{|\mathbf{v}|} = \left(\frac{1}{\sqrt{2}}\tanh t\right)\mathbf{i} - \frac{1}{\sqrt{2}}\mathbf{j} + \left(\frac{1}{\sqrt{2}}\operatorname{sech} t\right)\mathbf{k} \Rightarrow \frac{d\mathbf{T}}{dt} = \left(\frac{1}{\sqrt{2}}\operatorname{sech}^2 t\right)\mathbf{i} - \left(\frac{1}{\sqrt{2}}\operatorname{sech} t\tanh t\right)\mathbf{k}$$

$$\Rightarrow \left|\frac{d\mathbf{T}}{dt}\right| = \sqrt{\tfrac{1}{2}\operatorname{sech}^4 t + \tfrac{1}{2}\operatorname{sech}^2 t\tanh^2 t} = \frac{1}{\sqrt{2}}\operatorname{sech} t \Rightarrow \mathbf{N} = \frac{\left(\frac{d\mathbf{T}}{dt}\right)}{\left|\frac{d\mathbf{T}}{dt}\right|} = (\operatorname{sech} t)\mathbf{i} - (\tanh t)\mathbf{k};$$

$$\mathbf{a} = (\cosh t)\mathbf{i} - (\sinh t)\mathbf{j} \Rightarrow \mathbf{v}\times\mathbf{a} = \begin{vmatrix} \mathbf{i} & \mathbf{j} & \mathbf{k} \\ \sinh t & -\cosh t & 1 \\ \cosh t & -\sinh t & 0 \end{vmatrix} = (\sinh t)\mathbf{i} + (\cosh t)\mathbf{j} - \mathbf{k}$$

$$\Rightarrow |\mathbf{v}\times\mathbf{a}| = \sqrt{\sinh^2 t + \cosh^2 t + 1} = \sqrt{2}\cosh t \Rightarrow \kappa = \frac{|\mathbf{v}\times\mathbf{a}|}{|\mathbf{v}|^3} = \frac{\sqrt{2}\cosh t}{\left(\sqrt{2}\right)^3\cosh^3 t} = \frac{1}{2}\operatorname{sech}^2 t;$$

$$\mathbf{B} = \mathbf{T}\times\mathbf{N} = \begin{vmatrix} \mathbf{i} & \mathbf{j} & \mathbf{k} \\ \frac{1}{\sqrt{2}}\tanh t & \frac{-1}{\sqrt{2}} & \frac{1}{\sqrt{2}}\operatorname{sech} t \\ \operatorname{sech} t & 0 & -\tanh t \end{vmatrix} = \left(\frac{1}{\sqrt{2}}\tanh t\right)\mathbf{i} + \frac{1}{\sqrt{2}}\mathbf{j} + \left(\frac{1}{\sqrt{2}}\operatorname{sech} t\right)\mathbf{k};$$

$\frac{d\mathbf{a}}{dt} = (\sinh t)\mathbf{i} - (\cosh t)\mathbf{j} \Rightarrow \tau = \dfrac{\begin{vmatrix} \sinh t & -\cosh t & 1 \\ \cosh t & -\sinh t & 0 \\ \sinh t & -\cosh t & 0 \end{vmatrix}}{|\mathbf{v} \times \mathbf{a}|^2} = \frac{-1}{2\cosh^2 t} = -\frac{1}{2}\operatorname{sech}^2 t$

19. $\mathbf{r} = (a \cos t)\mathbf{i} + (a \sin t)\mathbf{j} + bt\mathbf{k} \Rightarrow \mathbf{v} = (-a \sin t)\mathbf{i} + (a \cos t)\mathbf{j} + b\mathbf{k} \Rightarrow |\mathbf{v}| = \sqrt{(-a \sin t)^2 + (a \cos t)^2 + b^2}$
$= \sqrt{a^2 + b^2} \Rightarrow a_T = \frac{d}{dt}|\mathbf{v}| = 0$; $\mathbf{a} = (-a \cos t)\mathbf{i} + (-a \sin t)\mathbf{j} \Rightarrow |\mathbf{a}| = \sqrt{(-a \cos t)^2 + (-a \sin t)^2} = \sqrt{a^2} = |a|$
$\Rightarrow a_N = \sqrt{|\mathbf{a}|^2 - a_T^2} = \sqrt{|\mathbf{a}|^2 - 0^2} = |\mathbf{a}| = |a| \Rightarrow \mathbf{a} = (0)\mathbf{T} + |a|\mathbf{N} = |a|\mathbf{N}$

20. $\mathbf{r} = (1 + 3t)\mathbf{i} + (t - 2)\mathbf{j} - 3t\mathbf{k} \Rightarrow \mathbf{v} = 3\mathbf{i} + \mathbf{j} - 3\mathbf{k} \Rightarrow |\mathbf{v}| = \sqrt{3^2 + 1^2 + (-3)^2} = \sqrt{19} \Rightarrow a_T = \frac{d}{dt}|\mathbf{v}| = 0$; $\mathbf{a} = \mathbf{0}$
$\Rightarrow a_N = \sqrt{|\mathbf{a}|^2 - a_T^2} = 0 \Rightarrow \mathbf{a} = (0)\mathbf{T} + (0)\mathbf{N} = \mathbf{0}$

21. $\mathbf{r} = (t + 1)\mathbf{i} + 2t\mathbf{j} + t^2\mathbf{k} \Rightarrow \mathbf{v} = \mathbf{i} + 2\mathbf{j} + 2t\mathbf{k} \Rightarrow |\mathbf{v}| = \sqrt{1^2 + 2^2 + (2t)^2} = \sqrt{5 + 4t^2} \Rightarrow a_T = \frac{1}{2}(5 + 4t^2)^{-1/2}(8t)$
$= 4t(5 + 4t^2)^{-1/2} \Rightarrow a_T(1) = \frac{4}{\sqrt{9}} = \frac{4}{3}$; $\mathbf{a} = 2\mathbf{k} \Rightarrow \mathbf{a}(1) = 2\mathbf{k} \Rightarrow |\mathbf{a}(1)| = 2 \Rightarrow a_N = \sqrt{|\mathbf{a}|^2 - a_T^2} = \sqrt{2^2 - \left(\frac{4}{3}\right)^2}$
$= \sqrt{\frac{20}{9}} = \frac{2\sqrt{5}}{3} \Rightarrow \mathbf{a}(1) = \frac{4}{3}\mathbf{T} + \frac{2\sqrt{5}}{3}\mathbf{N}$

22. $\mathbf{r} = (t \cos t)\mathbf{i} + (t \sin t)\mathbf{j} + t^2\mathbf{k} \Rightarrow \mathbf{v} = (\cos t - t \sin t)\mathbf{i} + (\sin t + t \cos t)\mathbf{j} + 2t\mathbf{k}$
$\Rightarrow |\mathbf{v}| = \sqrt{(\cos t - t \sin t)^2 + (\sin t + t \cos t)^2 + (2t)^2} = \sqrt{5t^2 + 1} \Rightarrow a_T = \frac{1}{2}(5t^2 + 1)^{-1/2}(10t)$
$= \frac{5t}{\sqrt{5t^2 + 1}} \Rightarrow a_T(0) = 0$; $\mathbf{a} = (-2 \sin t - t \cos t)\mathbf{i} + (2 \cos t - t \sin t)\mathbf{j} + 2\mathbf{k} \Rightarrow \mathbf{a}(0) = 2\mathbf{j} + 2\mathbf{k} \Rightarrow |\mathbf{a}(0)|$
$= \sqrt{2^2 + 2^2} = 2\sqrt{2} \Rightarrow a_N = \sqrt{|\mathbf{a}|^2 - a_T^2} = \sqrt{(2\sqrt{2})^2 - 0^2} = 2\sqrt{2} \Rightarrow \mathbf{a}(0) = (0)\mathbf{T} + 2\sqrt{2}\mathbf{N} = 2\sqrt{2}\mathbf{N}$

23. $\mathbf{r} = t^2\mathbf{i} + \left(t + \frac{1}{3}t^3\right)\mathbf{j} + \left(t - \frac{1}{3}t^3\right)\mathbf{k} \Rightarrow \mathbf{v} = 2t\mathbf{i} + (1 + t^2)\mathbf{j} + (1 - t^2)\mathbf{k} \Rightarrow |\mathbf{v}| = \sqrt{(2t)^2 + (1 + t^2)^2 + (1 - t^2)^2}$
$= \sqrt{2(t^4 + 2t^2 + 1)} = \sqrt{2}(1 + t^2) \Rightarrow a_T = 2t\sqrt{2} \Rightarrow a_T(0) = 0$; $\mathbf{a} = 2\mathbf{i} + 2t\mathbf{j} - 2t\mathbf{k} \Rightarrow \mathbf{a}(0) = 2\mathbf{i} \Rightarrow |\mathbf{a}(0)| = 2$
$\Rightarrow a_N = \sqrt{|\mathbf{a}|^2 - a_T^2} = \sqrt{2^2 - 0^2} = 2 \Rightarrow \mathbf{a}(0) = (0)\mathbf{T} + 2\mathbf{N} = 2\mathbf{N}$

24. $\mathbf{r} = (e^t \cos t)\mathbf{i} + (e^t \sin t)\mathbf{j} + \sqrt{2}e^t\mathbf{k} \Rightarrow \mathbf{v} = (e^t \cos t - e^t \sin t)\mathbf{i} + (e^t \sin t + e^t \cos t)\mathbf{j} + \sqrt{2}e^t\mathbf{k}$
$\Rightarrow |\mathbf{v}| = \sqrt{(e^t \cos t - e^t \sin t)^2 + (e^t \sin t + e^t \cos t)^2 + (\sqrt{2}e^t)^2} = \sqrt{4e^{2t}} = 2e^t \Rightarrow a_T = 2e^t \Rightarrow a_T(0) = 2$;
$\mathbf{a} = (e^t \cos t - e^t \sin t - e^t \sin t - e^t \cos t)\mathbf{i} + (e^t \sin t + e^t \cos t + e^t \cos t - e^t \sin t)\mathbf{j} + \sqrt{2}e^t\mathbf{k}$
$= (-2e^t \sin t)\mathbf{i} + (2e^t \cos t)\mathbf{j} + \sqrt{2}e^t\mathbf{k} \Rightarrow \mathbf{a}(0) = 2\mathbf{j} + \sqrt{2}\mathbf{k} \Rightarrow |\mathbf{a}(0)| = \sqrt{2^2 + (\sqrt{2})^2} = \sqrt{6}$
$\Rightarrow a_N = \sqrt{|\mathbf{a}|^2 - a_T^2} = \sqrt{(\sqrt{6})^2 - 2^2} = \sqrt{2} \Rightarrow \mathbf{a}(0) = 2\mathbf{T} + \sqrt{2}\mathbf{N}$

25. $\mathbf{r} = (\cos t)\mathbf{i} + (\sin t)\mathbf{j} - \mathbf{k} \Rightarrow \mathbf{v} = (-\sin t)\mathbf{i} + (\cos t)\mathbf{j} \Rightarrow |\mathbf{v}| = \sqrt{(-\sin t)^2 + (\cos t)^2} = 1 \Rightarrow \mathbf{T} = \frac{\mathbf{v}}{|\mathbf{v}|}$ $= (-\sin t)\mathbf{i} + (\cos t)\mathbf{j} \Rightarrow \mathbf{T}\left(\frac{\pi}{4}\right) = -\frac{\sqrt{2}}{2}\mathbf{i} + \frac{\sqrt{2}}{2}\mathbf{j}$; $\frac{d\mathbf{T}}{dt} = (-\cos t)\mathbf{i} - (\sin t)\mathbf{j} \Rightarrow \left|\frac{d\mathbf{T}}{dt}\right| = \sqrt{(-\cos t)^2 + (-\sin t)^2}$ $= 1 \Rightarrow \mathbf{N} = \frac{\left(\frac{d\mathbf{T}}{dt}\right)}{\left|\frac{d\mathbf{T}}{dt}\right|} = (-\cos t)\mathbf{i} - (\sin t)\mathbf{j} \Rightarrow \mathbf{N}\left(\frac{\pi}{4}\right) = -\frac{\sqrt{2}}{2}\mathbf{i} - \frac{\sqrt{2}}{2}\mathbf{j}$; $\mathbf{B} = \mathbf{T} \times \mathbf{N} = \begin{vmatrix} \mathbf{i} & \mathbf{j} & \mathbf{k} \\ -\sin t & \cos t & 0 \\ -\cos t & -\sin t & 0 \end{vmatrix} = \mathbf{k}$ $\Rightarrow \mathbf{B}\left(\frac{\pi}{4}\right) = \mathbf{k}$, the normal to the osculating plane; $\mathbf{r}\left(\frac{\pi}{4}\right) = \frac{\sqrt{2}}{2}\mathbf{i} + \frac{\sqrt{2}}{2}\mathbf{j} - \mathbf{k} \Rightarrow P = \left(\frac{\sqrt{2}}{2}, \frac{\sqrt{2}}{2}, -1\right)$ lies on the osculating plane $\Rightarrow 0\left(x - \frac{\sqrt{2}}{2}\right) + 0\left(y - \frac{\sqrt{2}}{2}\right) + (z - (-1)) = 0 \Rightarrow z = -1$ is the osculating plane; $\mathbf{T}$ is normal to the normal plane $\Rightarrow \left(-\frac{\sqrt{2}}{2}\right)\left(x - \frac{\sqrt{2}}{2}\right) + \left(\frac{\sqrt{2}}{2}\right)\left(y - \frac{\sqrt{2}}{2}\right) + 0(z - (-1)) = 0 \Rightarrow -\frac{\sqrt{2}}{2}x + \frac{\sqrt{2}}{2}y = 0$ $\Rightarrow -x + y = 0$ is the normal plane; $\mathbf{N}$ is normal to the rectifying plane $\Rightarrow \left(-\frac{\sqrt{2}}{2}\right)\left(x - \frac{\sqrt{2}}{2}\right) + \left(-\frac{\sqrt{2}}{2}\right)\left(y - \frac{\sqrt{2}}{2}\right) + 0(z - (-1)) = 0 \Rightarrow -\frac{\sqrt{2}}{2}x - \frac{\sqrt{2}}{2}y = -1 \Rightarrow x + y = \sqrt{2}$ is the rectifying plane

26. $\mathbf{r} = (\cos t)\mathbf{i} + (\sin t)\mathbf{j} + t\mathbf{k} \Rightarrow \mathbf{v} = (-\sin t)\mathbf{i} + (\cos t)\mathbf{j} + \mathbf{k} \Rightarrow |\mathbf{v}| = \sqrt{\sin^2 t + \cos^2 t + 1} = \sqrt{2} \Rightarrow \mathbf{T} = \frac{\mathbf{v}}{|\mathbf{v}|}$ $= \left(-\frac{1}{\sqrt{2}}\sin t\right)\mathbf{i} + \left(\frac{1}{\sqrt{2}}\cos t\right)\mathbf{j} + \frac{1}{\sqrt{2}}\mathbf{k} \Rightarrow \frac{d\mathbf{T}}{dt} = \left(-\frac{1}{\sqrt{2}}\cos t\right)\mathbf{i} + \left(-\frac{1}{\sqrt{2}}\sin t\right)\mathbf{j} \Rightarrow \left|\frac{d\mathbf{T}}{dt}\right|$ $= \sqrt{\frac{1}{2}\cos^2 t + \frac{1}{2}\sin^2 t} = \frac{1}{\sqrt{2}} \Rightarrow \mathbf{N} = \frac{\left(\frac{d\mathbf{T}}{dt}\right)}{\left|\frac{d\mathbf{T}}{dt}\right|} = (-\cos t)\mathbf{i} - (\sin t)\mathbf{j}$; thus $\mathbf{T}(0) = \frac{1}{\sqrt{2}}\mathbf{j} + \frac{1}{\sqrt{2}}\mathbf{k}$ and $\mathbf{N}(0) = -\mathbf{i}$ $\Rightarrow \mathbf{B}(0) = \begin{vmatrix} \mathbf{i} & \mathbf{j} & \mathbf{k} \\ 0 & \frac{1}{\sqrt{2}} & \frac{1}{\sqrt{2}} \\ -1 & 0 & 0 \end{vmatrix} = -\frac{1}{\sqrt{2}}\mathbf{j} + \frac{1}{\sqrt{2}}\mathbf{k}$, the normal to the osculating plane; $\mathbf{r}(0) = \mathbf{i} \Rightarrow P(1,0,0)$ lies on the osculating plane $\Rightarrow 0(x-1) - \frac{1}{\sqrt{2}}(y-0) + \frac{1}{\sqrt{2}}(z-0) = 0 \Rightarrow y - z = 0$ is the osculating plane; $\mathbf{T}$ is normal to the normal plane $\Rightarrow 0(x-1) + \frac{1}{\sqrt{2}}(y-0) + \frac{1}{\sqrt{2}}(z-0) = 0 \Rightarrow y + z = 0$ is the normal plane; $\mathbf{N}$ is normal to the rectifying plane $\Rightarrow -1(x-1) + 0(y-0) + 0(z-0) = 0 \Rightarrow x = 1$ is the rectifying plane

27. Yes. If the car is moving along a curved path, then $\kappa \neq 0$ and $a_N = \kappa|\mathbf{v}|^2 \neq 0 \Rightarrow \mathbf{a} = a_T\mathbf{T} + a_N\mathbf{N} \neq \mathbf{0}$.

28. $|\mathbf{v}|$ constant $\Rightarrow a_T = \frac{d}{dt}|\mathbf{v}| = 0 \Rightarrow \mathbf{a} = a_N\mathbf{N}$ is orthogonal to $\mathbf{T} \Rightarrow$ the acceleration is normal to the path

29. $\mathbf{a} \perp \mathbf{v} \Rightarrow \mathbf{a} \perp \mathbf{T} \Rightarrow a_T = 0 \Rightarrow \frac{d}{dt}|\mathbf{v}| = 0 \Rightarrow |\mathbf{v}|$ is constant

30. $\mathbf{a}(t) = a_T\mathbf{T} + a_N\mathbf{N}$, where $a_T = \frac{d}{dt}|\mathbf{v}| = \frac{d}{dt}(10) = 0$ and $a_N = \kappa|\mathbf{v}|^2 = 100\kappa \Rightarrow \mathbf{a} = 0\mathbf{T} + 100\kappa\mathbf{N}$. Now, from Exercise 7(a), we find for $y = f(x) = x^2$ that $\kappa = \frac{|f''(x)|}{\left[1+(f'(x))^2\right]^{3/2}} = \frac{2}{\left[1+(2x)^2\right]^{3/2}} = \frac{2}{\left(1+4x^2\right)^{3/2}}$; also, $\mathbf{r}(t) = t\mathbf{i} + t^2\mathbf{j}$ is the position vector of the moving mass $\Rightarrow \mathbf{v} = \mathbf{i} + 2t\mathbf{j} \Rightarrow |\mathbf{v}| = \sqrt{1+4t^2}$ $\Rightarrow \mathbf{T} = \frac{1}{\sqrt{1+4t^2}}(\mathbf{i}+2t\mathbf{j})$. At $(0,0)$: $\mathbf{T}(0) = \mathbf{i}$, $\mathbf{N}(0) = \mathbf{j}$ and $\kappa(0) = 2 \Rightarrow \mathbf{F} = m\mathbf{a} = m(100\kappa)\mathbf{N} = 200m\,\mathbf{j}$;
At $(\sqrt{2},2)$: $\mathbf{T}(\sqrt{2}) = \frac{1}{3}(\mathbf{i}+2\sqrt{2}\mathbf{j}) = \frac{1}{3}\mathbf{i} + \frac{2\sqrt{2}}{3}\mathbf{j}$, $\mathbf{N}(\sqrt{2}) = -\frac{2\sqrt{2}}{3}\mathbf{i} + \frac{1}{3}\mathbf{j}$, and $\kappa(\sqrt{2}) = \frac{2}{27} \Rightarrow \mathbf{F} = m\mathbf{a}$ $= m(100\kappa)\mathbf{N} = \left(\frac{200}{27}m\right)\left(-\frac{2\sqrt{2}}{3}\mathbf{i} + \frac{1}{3}\mathbf{j}\right) = -\frac{400\sqrt{2}}{81}m\,\mathbf{i} + \frac{200}{81}m\,\mathbf{j}$

31. $\mathbf{a} = a_T\mathbf{T} + a_N\mathbf{N}$, where $a_T = \frac{d}{dt}|\mathbf{v}| = \frac{d}{dt}(\text{constant}) = 0$ and $a_N = \kappa|\mathbf{v}|^2 \Rightarrow \mathbf{F} = m\mathbf{a} = m\kappa|\mathbf{v}|^2\mathbf{N} \Rightarrow |\mathbf{F}| = m\kappa|\mathbf{v}|^2$ $= \left(m|\mathbf{v}|^2\right)\kappa$, a constant multiple of the curvature κ of the trajectory

32. $a_N = 0 \Rightarrow \kappa|\mathbf{v}|^2 = 0 \Rightarrow \kappa = 0$ (since the particle is moving, we cannot have zero speed) $\Rightarrow$ the curvature is zero so the particle is moving along a straight line

33. $y = ax^2 \Rightarrow y' = 2ax \Rightarrow y'' = 2a$; from Exercise 7(a), $\kappa(x) = \frac{|2a|}{\left(1+4a^2x^2\right)^{3/2}} = |2a|\left(1+4a^2x^2\right)^{-3/2}$ $\Rightarrow \kappa'(x) = -\frac{3}{2}|2a|\left(1+4a^2x^2\right)^{-5/2}\left(8a^2x\right)$; thus, $\kappa'(x) = 0 \Rightarrow x = 0$. Now, $\kappa'(x) > 0$ for $x < 0$ and $\kappa'(x) < 0$ for $x > 0$ so that $\kappa(x)$ has an absolute maximum at $x = 0$ which is the vertex of the parabola. Since $x = 0$ is the only critical point for $\kappa(x)$, the curvature has no minimum value.

34. $\mathbf{r} = (a\cos t)\mathbf{i} + (b\sin t)\mathbf{j} \Rightarrow \mathbf{v} = (-a\sin t)\mathbf{i} + (b\cos t)\mathbf{j} \Rightarrow \mathbf{a} = (-a\cos t)\mathbf{i} - (b\sin t)\mathbf{j} \Rightarrow \mathbf{v}\times\mathbf{a}$

$$= \begin{vmatrix} \mathbf{i} & \mathbf{j} & \mathbf{k} \\ -a\sin t & b\cos t & 0 \\ -a\cos t & -b\sin t & 0 \end{vmatrix} = ab\mathbf{k} \Rightarrow |\mathbf{v}\times\mathbf{a}| = |ab| = ab, \text{ since } a > b > 0;\ \kappa(t) = \frac{|\mathbf{v}\times\mathbf{a}|}{|\mathbf{v}|^3}$$

$= ab\left(a^2\sin^2 t + b^2\cos^2 t\right)^{-3/2}$; $\kappa'(t) = -\frac{3}{2}(ab)\left(a^2\sin^2 t + b^2\cos^2 t\right)^{-5/2}\left(2a^2\sin t\cos t - 2b^2\sin t\cos t\right)$ $= -\frac{3}{2}(ab)\left(a^2-b^2\right)(\sin 2t)\left(a^2\sin^2 t + b^2\cos^2 t\right)^{-5/2}$; thus, $\kappa'(t) = 0 \Rightarrow \sin 2t = 0 \Rightarrow t = 0, \pi$ identifying points on the major axis, or $t = \frac{\pi}{2}, \frac{3\pi}{2}$ identifying points on the minor axis. Furthermore, $\kappa'(t) < 0$ for $0 < t < \frac{\pi}{2}$ and for $\pi < t < \frac{3\pi}{2}$; $\kappa'(t) > 0$ for $\frac{\pi}{2} < t < \pi$ and $\frac{3\pi}{2} < t < 2\pi$. Therefore, the points associated with $t = 0$ and $t = \pi$ on the major axis give absolute maximum curvature and the points associated with $t = \frac{\pi}{2}$ and $t = \frac{3\pi}{2}$ on the minor axis give absolute minimum curvature.

35. $\kappa = \frac{a}{a^2+b^2} \Rightarrow \frac{d\kappa}{da} = \frac{-a^2+b^2}{(a^2+b^2)^2}$; $\frac{d\kappa}{da} = 0 \Rightarrow a^2+b^2=0 \Rightarrow a = \pm b \Rightarrow a = b$ since a, $b > 0$. Now, $\frac{d\kappa}{da} > 0$ if $a < b$ and $\frac{d\kappa}{da} < 0$ if $a > b \Rightarrow \kappa$ is at a maximum for $a = b$ and $\kappa(b) = \frac{b}{b^2+b^2} = \frac{1}{2b}$ is the maximum value

36. From Example 6, $|\mathbf{v}| = t$ and $a_N = t$ so that $a_N = \kappa|\mathbf{v}|^2 \Rightarrow \kappa = \frac{a_N}{|\mathbf{v}|^2} = \frac{t}{t^2} = \frac{1}{t}$, $t \neq 0 \Rightarrow \rho = \frac{1}{\kappa} = t$

37. $\mathbf{r} = (x_0 + At)\mathbf{i} + (y_0 + Bt)\mathbf{j} + (z_0 + Ct)\mathbf{k} \Rightarrow \mathbf{v} = A\mathbf{i} + B\mathbf{j} + C\mathbf{k} \Rightarrow \mathbf{a} = \mathbf{0} \Rightarrow \mathbf{v} \times \mathbf{a} = \mathbf{0} \Rightarrow \kappa = 0$. Since the curve is a plane curve, $\tau = 0$.

38. From Example 7, the curvature of the helix $\mathbf{r}(t) = (a \cos t)\mathbf{i} + (a \sin t)\mathbf{j} + bt\mathbf{k}$, a, $b \geq 0$ is $\kappa = \frac{a}{a^2+b^2}$; also $|\mathbf{v}| = \sqrt{a^2+b^2}$. For the helix $\mathbf{r}(t) = (3 \cos t)\mathbf{i} + (3 \sin t)\mathbf{j} + t\mathbf{k}$, $0 \leq t \leq 4\pi$, $a = 3$ and $b = 1 \Rightarrow \kappa = \frac{3}{3^2+1^2} = \frac{3}{10}$ and $|\mathbf{v}| = \sqrt{10} \Rightarrow K = \int_0^{4\pi} \frac{3}{10}\sqrt{10}\,dt = \left[\frac{3}{\sqrt{10}}t\right]_0^{4\pi} = \frac{12\pi}{\sqrt{10}}$

39. (a) From Exercise 36, $\kappa = \frac{1}{t}$ and $|\mathbf{v}| = t \Rightarrow K = \int_a^b \left(\frac{1}{t}\right)(t)\,dt = b - a$

(b) $y = x^2 \Rightarrow x = t$ and $y = t^2$, $-\infty < t < \infty \Rightarrow \mathbf{r}(t) = t\mathbf{i} + t^2\mathbf{j} \Rightarrow \mathbf{v} = \mathbf{i} + 2t\mathbf{j} \Rightarrow |\mathbf{v}| = \sqrt{1+4t^2}$; also $\mathbf{a} = 2\mathbf{j}$

$$\Rightarrow \mathbf{v} \times \mathbf{a} = \begin{vmatrix} \mathbf{i} & \mathbf{j} & \mathbf{k} \\ 1 & 2t & 0 \\ 0 & 2 & 0 \end{vmatrix} = 2\mathbf{k} \Rightarrow |\mathbf{v} \times \mathbf{a}| = 2 \Rightarrow \kappa = \frac{|\mathbf{v} \times \mathbf{a}|}{|\mathbf{v}|^3} = \frac{2}{\left(\sqrt{1+4t^2}\right)^3}.$$ Then

$$K = \int_{-\infty}^{\infty} \frac{2}{\left(\sqrt{1+4t^2}\right)^3}\left(\sqrt{1+4t^2}\right)dt = \int_{-\infty}^{\infty} \frac{2}{1+4t^2}\,dt = \lim_{a \to -\infty} \int_a^0 \frac{2}{1+4t^2}\,dt + \lim_{b \to \infty} \int_0^b \frac{2}{1+4t^2}\,dt$$

$$= \lim_{a \to -\infty} \left[\tan^{-1} 2t\right]_a^0 + \lim_{b \to \infty} \left[\tan^{-1} 2t\right]_0^b = \lim_{a \to -\infty} \left(-\tan^{-1} 2a\right) + \lim_{b \to \infty} \left(\tan^{-1} 2b\right) = \frac{\pi}{2} + \frac{\pi}{2} = \pi$$

40. (a) $\mathbf{r} = t\mathbf{i} + (\sin t)\mathbf{j} \Rightarrow \mathbf{v} = \mathbf{i} + (\cos t)\mathbf{j} \Rightarrow |\mathbf{v}| = \sqrt{1^2 + (\cos t)^2} = \sqrt{1 + \cos^2 t} \Rightarrow \left|\mathbf{v}\left(\frac{\pi}{2}\right)\right| = \sqrt{1 + \cos^2\left(\frac{\pi}{2}\right)} = 1$;

$$\mathbf{a} = (-\sin t)\mathbf{j} \Rightarrow \mathbf{v} \times \mathbf{a} = \begin{vmatrix} \mathbf{i} & \mathbf{j} & \mathbf{k} \\ 1 & \cos t & 0 \\ 0 & -\sin t & 0 \end{vmatrix} = (-\sin t)\mathbf{k} \Rightarrow |\mathbf{v} \times \mathbf{a}| = \sqrt{(-\sin t)^2} = |\sin t| \Rightarrow |\mathbf{v} \times \mathbf{a}|\left(\frac{\pi}{2}\right)$$

$$= \left|\sin\left(\frac{\pi}{2}\right)\right| = 1 \Rightarrow \kappa\left(\frac{\pi}{2}\right) = \frac{|\mathbf{v} \times \mathbf{a}|}{|\mathbf{v}|^3} = \frac{1}{1^3} = 1 \Rightarrow \rho = \frac{1}{1} = 1$$ and the center is $\left(\frac{\pi}{2}, 0\right) \Rightarrow \left(x - \frac{\pi}{2}\right)^2 + y^2 = 1$

(b) $\mathbf{r} = (2 \ln t)\mathbf{i} - \left(t + \frac{1}{t}\right)\mathbf{j} \Rightarrow \mathbf{v} = \left(\frac{2}{t}\right)\mathbf{i} - \left(1 - \frac{1}{t^2}\right)\mathbf{j} \Rightarrow \mathbf{v}(1) = 2\mathbf{i}$ and $|\mathbf{v}(1)| = 2$; $\mathbf{a} = \left(-\frac{2}{t^2}\right)\mathbf{i} - \left(\frac{2}{t^3}\right)\mathbf{j}$

$$\Rightarrow \mathbf{a}(1) = -2\mathbf{i} - 2\mathbf{j} \Rightarrow (\mathbf{v} \times \mathbf{a})(1) = \begin{vmatrix} \mathbf{i} & \mathbf{j} & \mathbf{k} \\ 2 & 0 & 0 \\ -2 & -2 & 0 \end{vmatrix} = -4\mathbf{k} \Rightarrow |\mathbf{v} \times \mathbf{a}|(1) = 4 \Rightarrow \kappa(1) = \frac{|\mathbf{v} \times \mathbf{a}|}{|\mathbf{v}|^3} = \frac{4}{2^3} = \frac{1}{2}$$

$\Rightarrow \rho = \frac{1}{\kappa} = 2$. The circle of curvature is tangent to the curve at $P(0, -2) \Rightarrow$ circle has same tangent as the curve $\Rightarrow \mathbf{v}(1) = 2\mathbf{i}$ is tangent to the circle $\Rightarrow$ the center lies on the y-axis. If $t \neq 1$ $(t > 0)$, then $(t-1)^2 > 0$ $\Rightarrow t^2 - 2t + 1 > 0 \Rightarrow t^2 + 1 > 2t \Rightarrow \frac{t^2+1}{t} > 2$ since $t > 0 \Rightarrow t + \frac{1}{t} > 2 \Rightarrow -\left(t + \frac{1}{t}\right) < -2 \Rightarrow y < -2$ on both sides of $(0, -2) \Rightarrow$ the curve is concave down $\Rightarrow$ center of circle of curvature is $(0, -4) \Rightarrow x^2 + (y+4)^2 = 4$ is an equation of the circle of curvature

41. If a plane curve is sufficiently differentiable the torsion is zero as the following argument shows:
$\mathbf{r} = f(t)\mathbf{i} + g(t)\mathbf{j} \Rightarrow \mathbf{v} = f'(t)\mathbf{i} + g'(t)\mathbf{j} \Rightarrow \mathbf{a} = f''(t)\mathbf{i} + g''(t)\mathbf{j} \Rightarrow \frac{d\mathbf{a}}{dt} = f'''(t)\mathbf{i} + g'''(t)\mathbf{j}$

$$\Rightarrow \tau = \frac{\begin{vmatrix} f'(t) & g'(t) & 0 \\ f''(t) & g''(t) & 0 \\ f'''(t) & g'''(t) & 0 \end{vmatrix}}{|\mathbf{v} \times \mathbf{a}|^2} = 0$$

42. From Example 7, $\tau = \frac{b}{a^2 + b^2} \Rightarrow \tau'(b) = \frac{a^2 - b^2}{\left(a^2 + b^2\right)^2}$; $\tau'(b) = 0 \Rightarrow \frac{a^2 - b^2}{\left(a^2 + b^2\right)^2} = 0 \Rightarrow a^2 - b^2 = 0 \Rightarrow b = \pm a$ $\Rightarrow b = a$ since $a, b > 0$. Also $b < a \Rightarrow \tau' > 0$ and $b > a \Rightarrow \tau' < 0$ so τ_{max} occurs when $b = a \Rightarrow \tau_{max} = \frac{a}{a^2 + a^2}$ $= \frac{1}{2a}$

43. $\mathbf{r}(t) = f(t)\mathbf{i} + g(t)\mathbf{j} + h(t)\mathbf{k} \Rightarrow \mathbf{v} = f'(t)\mathbf{i} + g'(t)\mathbf{j} + h'(t)\mathbf{k}$; $\mathbf{v} \cdot \mathbf{k} = 0 \Rightarrow h'(t) = 0 \Rightarrow h(t) = C$ $\Rightarrow \mathbf{r}(t) = f(t)\mathbf{i} + g(t)\mathbf{j} + C\mathbf{k}$ and $\mathbf{r}(a) = f(a)\mathbf{i} + g(a)\mathbf{j} + C\mathbf{k} = \mathbf{0} \Rightarrow f(a) = 0$, $g(a) = 0$ and $C = 0 \Rightarrow h(t) = 0$.

44. From Example 7, $\mathbf{v} = -(a \sin t)\mathbf{i} + (a \cos t)\mathbf{j} + b\mathbf{k} \Rightarrow |\mathbf{v}| = \sqrt{a^2 + b^2} \Rightarrow \mathbf{T} = \frac{\mathbf{v}}{|\mathbf{v}|}$

$$= \frac{1}{\sqrt{a^2+b^2}}[-(a \sin t)\mathbf{i} + (a \cos t)\mathbf{j} + b\mathbf{k}]; \frac{d\mathbf{T}}{dt} = \frac{1}{\sqrt{a^2+b^2}}[-(a \cos t)\mathbf{i} - (a \sin t)\mathbf{j}] \Rightarrow \mathbf{N} = \frac{\left(\frac{d\mathbf{T}}{dt}\right)}{\left|\frac{d\mathbf{T}}{dt}\right|}$$

$$= -(\cos t)\mathbf{i} - (\sin t)\mathbf{j}; \mathbf{B} = \mathbf{T} \times \mathbf{N} = \begin{vmatrix} \mathbf{i} & \mathbf{j} & \mathbf{k} \\ -\frac{a \sin t}{\sqrt{a^2+b^2}} & \frac{a \cos t}{\sqrt{a^2+b^2}} & \frac{b}{\sqrt{a^2+b^2}} \\ -\cos t & -\sin t & 0 \end{vmatrix}$$

$$= \frac{b \sin t}{\sqrt{a^2+b^2}}\mathbf{i} - \frac{b \cos t}{\sqrt{a^2+b^2}}\mathbf{j} + \frac{a}{\sqrt{a^2+b^2}}\mathbf{k} \Rightarrow \frac{d\mathbf{B}}{dt} = \frac{1}{\sqrt{a^2+b^2}}[(b \cos t)\mathbf{i} + (b \sin t)\mathbf{j}] \Rightarrow \frac{d\mathbf{B}}{dt} \cdot \mathbf{N} = -\frac{b}{\sqrt{a^2+b^2}}$$

$\Rightarrow \tau = -\frac{1}{|\mathbf{v}|}\left(\frac{d\mathbf{B}}{dt}\cdot \mathbf{N}\right) = \left(-\frac{1}{\sqrt{a^2+b^2}}\right)\left(-\frac{b}{\sqrt{a^2+b^2}}\right) = \frac{b}{a^2+b^2}$, which is consistent with the result in Example 7.

45. $y = x^2 \Rightarrow f'(x) = 2x$ and $f''(x) = 2$

$$\Rightarrow \kappa = \frac{|2|}{\left(1+(2x)^2\right)^{3/2}} = \frac{2}{\left(1+4x^2\right)^{3/2}}$$

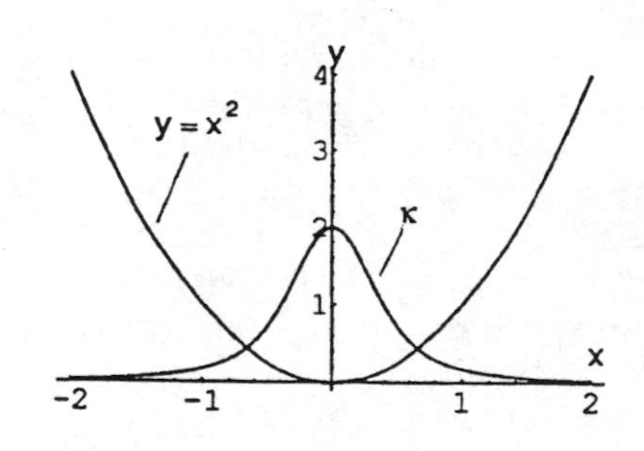

46. $y = \frac{x^4}{4} \Rightarrow f'(x) = x^3$ and $f''(x) = 3x^2$

$$\Rightarrow \kappa = \frac{|3x^2|}{\left(1+\left(x^3\right)^2\right)^{3/2}} = \frac{3x^2}{\left(1+x^6\right)^{3/2}}$$

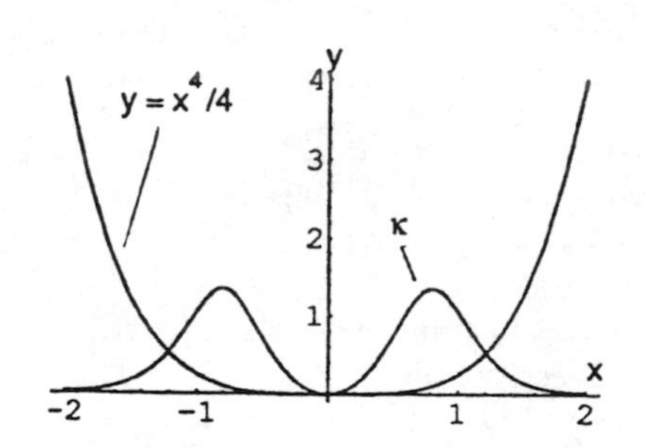

47. $y = \sin x \Rightarrow f'(x) = \cos x$ and $f''(x) = -\sin x$

$$\Rightarrow \kappa = \frac{|-\sin x|}{\left(1+\cos^2 x\right)^{3/2}} = \frac{|\sin x|}{\left(1+\cos^2 x\right)^{3/2}}$$

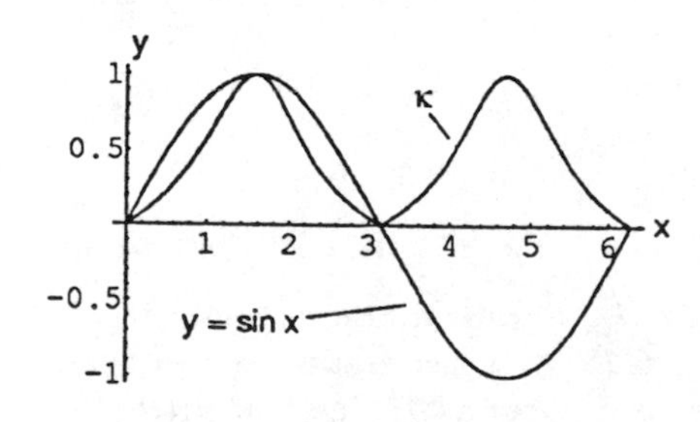

48. $y = e^x \Rightarrow f'(x) = e^x$ and $f''(x) = e^x$

$$\Rightarrow \kappa = \frac{|e^x|}{\left(1+\left(e^x\right)^2\right)^{3/2}} = \frac{e^x}{\left(1+e^{2x}\right)^{3/2}}$$

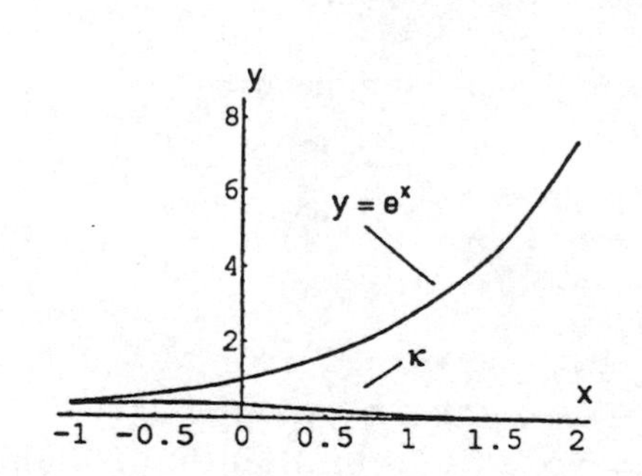

49-56. Example CAS commands:

Maple:

```
with(plots):
x:= t -> t^3 - 2*t^2 - t;
y:= t -> 3*t/sqrt(1 + t^2);
```

```
dx:= t -> D(x)(t);
dy:= t -> D(y)(t);
ds:= t -> sqrt((dx^2)(t) + (dy^2)(t));
d2x:= t -> D(dx)(t);
d2y:= t -> D(dy)(t);
kap:= t -> abs(dx(t)*d2y(t) - dy(t)*d2x(t))/((ds)(t))^3;
a:= t -> x0 - (1/kap(t))*(dy(t)/ds(t));
b:= t -> x0 + (1/kap(t))*(dx(t)/ds(t));
s1:= plot([x(t),y(t), t = -2..5], -15..5, -10..4, scaling=CONSTRAINED):
display(s1);
t0:=1: x0:= x(t0): y0:= y(t0):
circle:= ((x-a(t0))^2 + (y-b(t0))^2 = (1/kap(t0))^2);
s2:=implicitplot(circle, x=-15..6,y=-10..5,scaling=CONSTRAINED):
s3:=plot([a(t0),b(t0),x0,y0]):
display({s1,s2,s3});
```

Mathematica:

```
Clear[x,y,t]
r[t_] = {x[t],y[t]}
x[t_] = t^3 - 2 t^2 - t
y[t_] = 3 t / Sqrt[1+t^2]
{a,b} = {-2,5};
t0 = 1;
p1 = ParametricPlot[ {x[t],y[t]}, {t,a,b},
 AspectRatio -> Automatic ]
v0 = r'[t0]
s0 = Sqrt[ v0 . v0 ]
k0 = Abs[ x' [t0] y''[t0] - y' [t0] x'' [t0] ]/s0^3
N[%]
n0 = { -y' [t0], x' [t0] } / s0
r0 = r[t0]
c0 = r0 + 1/k0 n0

Note: Plot the circle parametrically rather than implicitly:

circ = ParametricPlot[ Evaluate[c0 + 1/k0 {Cos[t],Sin[t]}],
 {t,0,2Pi}, AspectRatio -> Automatic ]
line = Graphics[{Line[{c0,r0}]}]
Show[ p1, circ, line ]
```

57-60. Example CAS commands:

Maple:

```
with(linalg):
x:= t -> t*cos(t);
y:= t -> t*sin(t);
z:= t -> t;
t0:=sqrt(3);
r:= vector([x(t),y(t),z(t)]);
v:= vector([D(x)(t0),D(y)(t0),D(z)(t0)]):evalf('',5);
a:= vector([D@@2)(x)(t0),(D@@2)(y)(t0),(D@@2)(z)(t0)]):evalf('',5);
j:= vector([D@@3)(x)(t0),(D@@3)(y)(t0),(D@@3)(z)(t0)]):evalf('',5);
s:= sqrt(dotprod(v,v)):evalf('',5);
tvec:= scalarmul(v,1/s):evalf('',5);
at:= dotprod(a,tvec):evalf('',5);
n1vec:= add(a, -scalarmul(tvec,at)):evalf('',5);
```

```
an:= sqrt(dotprod(n1vec,n1vec)):evalf('',5);
nvec:= scalarmul(n1vec,1/an):evalf('',5);
bvec:= crossprod(tvec,nvec):evalf('',5);
k:= an/s^2:evalf('',5);
vca:= crossprod(v,a):evalf('',5);
tau:= (dotprod(vca,j))/(dotprod(vca,vca)):evalf('',5);
```

Mathematica:

```
x[t_] = t Cos[t]
y[t_] = t Sin[t]
z[t_] = t
t0 = Sqrt[3];
v = N[ r' [t0] ]
a = N[ r''[t0] ]
j = N[ r'''[t0] ]
s = Sqrt[ v . v ]
tvec = v / s
at = a . tvec
nvec = a - at tvec
an = Sqrt[ nvec . nvec ]
nvec = nvec / an
bvec = Cross[ tvec, nvec ]
k = an / s^2
vca = Cross[ v, a ]
tau = (vca . j)/(vca . vca)
```

11.5 PLANETARY MOTION AND SATELLITES

1. $\frac{T^2}{a^3} = \frac{4\pi^2}{GM} \Rightarrow T^2 = \frac{4\pi^2}{GM} a^3 \Rightarrow T^2 = \frac{4\pi^2}{(6.6720 \times 10^{-11}\ Nm^2kg^{-2})(5.975 \times 10^{24}\ kg)}(6{,}808{,}000\ m)^3$

   $\approx 3.125 \times 10^7\ sec^2 \Rightarrow T \approx \sqrt{3125 \times 10^4\ sec^2} \approx 55.90 \times 10^2\ sec \approx 93.2\ min$

2. $e = 0.0167$ and perihelion distance $= 149{,}577{,}000$ km and $e = \frac{r_0 v_0^2}{GM} - 1$

   $\Rightarrow 0.0167 = \frac{(149{,}577{,}000{,}000\ m)v_0^2}{(6.6720 \times 10^{-11}\ Nm^2kg^{-2})(1.99 \times 10^{30}\ kg)} - 1 \Rightarrow v_0^2 \approx 9.02 \times 10^8\ m^2/sec^2$

   $\Rightarrow v_0 \approx \sqrt{9.02 \times 10^8\ m^2/sec^2} \approx 3.00 \times 10^4\ m/sec$

3. $92.25\ min = 5535\ sec$ and $\frac{T^2}{a^3} = \frac{4\pi^2}{GM} \Rightarrow a^3 = \frac{GM}{4\pi^2} T^2$

   $\Rightarrow a^3 = \frac{(6.6720 \times 10^{-11}\ Nm^2kg^{-2})(5.975 \times 10^{24}\ kg)}{4\pi^2}(5535\ sec)^2 = 3.094 \times 10^{20}\ m^3 \Rightarrow a \approx \sqrt[3]{3.094 \times 10^{20}\ m^3}$

   $= 6.763 \times 10^6\ m \approx 6763\ km$; the mean distance from center of the Earth $= \frac{12{,}757\ km + 183\ km + 589\ km}{2}$

   $= 6765\ km$

4. (a) $T = 1639$ min $= 98{,}340$ sec and mass of Mars $= 6.418 \times 10^{23}$ kg $\Rightarrow a^3 = \frac{GM}{4\pi^2} T^2$

$= \frac{(6.6720 \times 10^{-11}\ \text{Nm}^2\text{kg}^{-2})(6.418 \times 10^{23}\ \text{kg})(98{,}340\ \text{sec})^2}{4\pi^2} \approx 1.049 \times 10^{22}\ \text{m}^3 \Rightarrow a \approx \sqrt[3]{1.049 \times 10^{22}\ \text{m}^3}$

$= 2.19 \times 10^7\ \text{m} = 21{,}900$ km

(b) $2a =$ diameter of Mars + perigee height + apogee height $= D + 1499$ km $+ 35{,}800$ km

$\Rightarrow 2(21{,}900)\ \text{km} = D + 37{,}800\ \text{km} \Rightarrow D = 6501$ km

5. $a = 22{,}030$ km $= 2.203 \times 10^7$ m and $T^2 = \frac{4\pi^2}{GM} a^3$

$\Rightarrow T^2 = \frac{4\pi^2}{(6.6720 \times 10^{-11}\ \text{Nm}^2\text{kg}^{-2})(6.418 \times 10^{23}\ \text{kg})}(2.203 \times 10^7\ \text{sec})^3 \approx 9.857 \times 10^9\ \text{sec}^2$

$\Rightarrow T \approx \sqrt{9.857 \times 10^8\ \text{sec}^2} \approx 9.928 \times 10^4\ \text{sec} \approx 1655$ min

6. (a) Period of the satellite = rotational period of the Earth $\Rightarrow$ period of the satellite $= 1436.1$ min

$= 86{,}166$ sec; $a^3 = \frac{GMT^2}{4\pi^2} \Rightarrow a^3 = \frac{(6.6720 \times 10^{-11}\ \text{Nm}^2\text{kg}^{-2})(5.975 \times 10^{24}\ \text{kg})(86{,}166\ \text{sec})^2}{4\pi^2}$

$\approx 7.4973 \times 10^{22}\ \text{m}^3 \Rightarrow a \approx \sqrt[3]{74.973 \times 10^{21}\ \text{m}^3} \approx 4.2167 \times 10^7\ \text{m} = 42{,}167$ km

(b) The radius of the Earth is approximately 6379 km $\Rightarrow$ the height of the orbit is $42{,}167 - 6379 = 35{,}788$ km

(c) Symcom 3, GOES 4, and Intelsat 5

7. $T = 1477.4$ min $= 88{,}644$ sec $\Rightarrow a^3 = \frac{GMT^2}{4\pi^2}$

$= \frac{(6.6720 \times 10^{-11}\ \text{Nm}^2\text{kg}^{-2})(6.418 \times 10^{23}\ \text{kg})(88{,}644\ \text{sec})^2}{4\pi^2} = 8.523 \times 10^{21}\ \text{m}^3 \Rightarrow a \approx \sqrt[3]{8.523 \times 10^{21}\ \text{m}^3}$

$\approx 2.043 \times 10^7\ \text{m} = 20{,}430$ km

8. Period of the Moon $= 2.36055 \times 10^6$ sec $\Rightarrow a^3 = \frac{GMT^2}{4\pi^2}$

$= \frac{(6.6720 \times 10^{-11}\ \text{Nm}^2\text{kg}^{-2})(5.975 \times 10^{24}\ \text{kg})(2.36055 \times 10^6\ \text{sec})^2}{4\pi^2} \approx 5.627 \times 10^{25}\ \text{m}^3 \Rightarrow a \approx \sqrt[3]{5.627 \times 10^{25}\ \text{m}^3}$

$\approx 3.832 \times 10^8\ \text{m} = 383{,}200$ km from the center of the Earth, or about 376,821 km from the surface

9. $r = \frac{GM}{v^2} \Rightarrow v^2 = \frac{GM}{r} \Rightarrow |v| = \sqrt{\frac{GM}{r}} = \sqrt{\frac{(6.6720 \times 10^{-11}\ \text{Nm}^2\text{kg}^{-2})(5.975 \times 10^{24}\ \text{kg})}{r}} \approx 1.9966 \times 10^7 r^{-1/2}$ m/sec

10. Solar System: $\frac{T^2}{a^3} = \frac{4\pi^2}{(6.6720 \times 10^{-11}\ \text{Nm}^2\text{kg}^{-2})(1.99 \times 10^{30}\ \text{kg})} \approx 2.97 \times 10^{-19}\ \text{sec}^2/\text{m}^3$;

Earth: $\frac{T^2}{a^3} = \frac{4\pi^2}{(6.6720 \times 10^{-11}\ \text{Nm}^2\text{kg}^{-2})(5.975 \times 10^{24}\ \text{kg})} \approx 9.903 \times 10^{-14}\ \text{sec}^2/\text{m}^3$;

Moon: $\frac{T^2}{a^3} = \frac{4\pi^2}{(6.6720 \times 10^{-11}\ \text{Nm}^2\text{kg}^{-2})(7.354 \times 10^{22}\ \text{kg})} \approx 8.046 \times 10^{-12}\ \text{sec}^2/\text{m}^3$;

11. $e = \frac{r_0 v_0^2}{GM} - 1 \Rightarrow v_0^2 = \frac{GM(e+1)}{r_0} \Rightarrow v_0 = \sqrt{\frac{GM(e+1)}{r_0}}$;

Circle: $e = 0 \Rightarrow v_0 = \sqrt{\frac{GM}{r_0}}$

Ellipse: $0 < e < 1 \Rightarrow \sqrt{\frac{GM}{r_0}} < v_0 < \sqrt{\frac{2GM}{r_0}}$

Parabola: $e = 1 \Rightarrow v_0 = \sqrt{\frac{2GM}{r_0}}$

Hyperbola: $e > 1 \Rightarrow v_0 > \sqrt{\frac{2GM}{r_0}}$

12. $r = \frac{GM}{v^2} \Rightarrow v^2 = \frac{GM}{r} \Rightarrow v = \sqrt{\frac{GM}{r}}$ which is constant since G, M, and r (the radius of orbit) are constant

13. $\Delta A = \frac{1}{2}\left|\mathbf{r}(t+\Delta t) \times \mathbf{r}(t)\right| \Rightarrow \frac{\Delta A}{\Delta t} = \frac{1}{2}\left|\frac{\mathbf{r}(t+\Delta t)}{\Delta t} \times \mathbf{r}(t)\right| = \frac{1}{2}\left|\frac{\mathbf{r}(t+\Delta t) - \mathbf{r}(t) + \mathbf{r}(t)}{\Delta t} \times \mathbf{r}(t)\right|$

$= \frac{1}{2}\left|\frac{\mathbf{r}(t+\Delta t) - \mathbf{r}(t)}{\Delta t} \times \mathbf{r}(t) + \frac{1}{\Delta t}\mathbf{r}(t) \times \mathbf{r}(t)\right| = \frac{1}{2}\left|\frac{\mathbf{r}(t+\Delta t) - \mathbf{r}(t)}{\Delta t} \times \mathbf{r}(t)\right| \Rightarrow \frac{dA}{dt} = \lim_{\Delta t \to 0} \frac{1}{2}\left|\frac{\mathbf{r}(t+\Delta t) - \mathbf{r}(t)}{\Delta t} \times \mathbf{r}(t)\right|$

$= \frac{1}{2}\left|\frac{d\mathbf{r}}{dt} \times \mathbf{r}(t)\right| = \frac{1}{2}\left|\mathbf{r}(t) \times \frac{d\mathbf{r}}{dt}\right| = \frac{1}{2}\left|\mathbf{r} \times \dot{\mathbf{r}}\right|$

14. $T = \left(\frac{2\pi a^2}{r_0 v_0}\right)\sqrt{1-e^2} \Rightarrow T^2 = \left(\frac{4\pi^2 a^4}{r_0^2 v_0^2}\right)(1-e^2) = \left(\frac{4\pi^2 a^4}{r_0^2 v_0^2}\right)\left[1 - \left(\frac{r_0 v_0^2}{GM} - 1\right)^2\right]$ (from Equation 32)

$= \left(\frac{4\pi^2 a^4}{r_0^2 v_0^2}\right)\left[-\frac{r_0^2 v_0^4}{G^2M^2} + 2\left(\frac{r_0 v_0^2}{GM}\right)\right] = \left(\frac{4\pi^2 a^4}{r_0^2 v_0^2}\right)\left[\frac{2GMr_0v_0^2 - r_0^2 v_0^4}{G^2M^2}\right] = \frac{(4\pi^2 a^4)(2GM - r_0 v_0^2)}{r_0 G^2 M^2}$

$= (4\pi^2 a^4)\left(\frac{2GM - r_0 v_0^2}{2r_0 GM}\right)\left(\frac{2}{GM}\right) = (4\pi^2 a^4)\left(\frac{1}{2a}\right)\left(\frac{2}{GM}\right)$ (from Equation 35) $\Rightarrow T^2 = \frac{4\pi^2 a^3}{GM} \Rightarrow \frac{T^2}{a^3} = \frac{4\pi^2}{GM}$

15. (a) Let $\mathbf{r}_{AB}(t)$ denote the vector from planet A to planet B at time t. Then $\mathbf{r}_{AB}(t) = \mathbf{r}_B(t) - \mathbf{r}_A(t)$

$= [3\cos(\pi t) - 2\cos(2\pi t)]\mathbf{i} + [3\sin(\pi t) - 2\sin(2\pi t)]\mathbf{j}$

$= [3\cos(\pi t) - 2(\cos^2(\pi t) - \sin^2(\pi t))]\mathbf{i} + [3\sin(\pi t) - 4\sin(\pi t)\cos(\pi t)]\mathbf{j}$

$= [3\cos(\pi t) - 4\cos^2(\pi t) + 2]\mathbf{i} + [(3 - 4\cos(\pi t))\sin(\pi t)]\mathbf{j} \Rightarrow$ parametric equations for the path are $x(t) = 2 + [3 - 4\cos(\pi t)]\cos(\pi t)$ and $y(t) = [3 - 4\cos(\pi t)]\sin(\pi t)$

(b) Setting $\theta = \pi t$ and $r = 3 - 4\cos\theta$, we see that $x - 2 = r\cos\theta$ and $y = r\sin\theta \Rightarrow$ the graph of the path of planet B is the limacon $r = 3 - 4\cos\theta$ shown at the right. The planet A is located at $x = -2$.

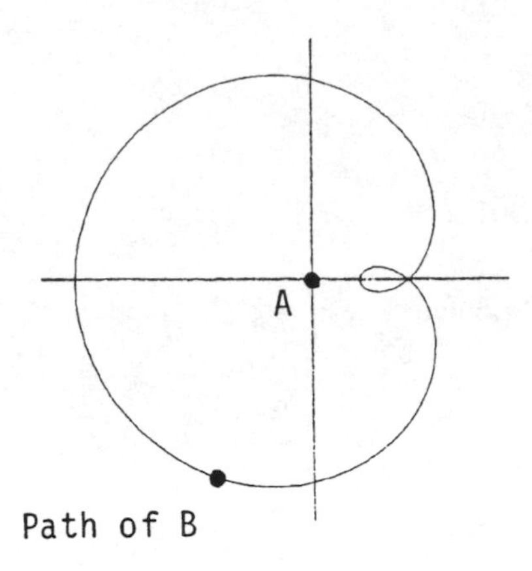

16. (i) Perihelion is the time t such that $|\mathbf{r}(t)|$ is a minimum.
(ii) Aphelion is the time t such that $|\mathbf{r}(t)|$ is a maximum.
(iii) Equinox is the time t such that $\mathbf{r}(t)\cdot\mathbf{w} = 0$.
(iv) Summer solstice is the time t such that the angle between $\mathbf{r}(t)$ and $\mathbf{w}$ is a maximum.
(v) Winter solstice is the time t such that the angle between $\mathbf{r}(t)$ and $\mathbf{w}$ is a minimum.

CHAPTER 11 PRACTICE EXERCISES

1. $\mathbf{r}(t) = (4\cos t)\mathbf{i} + (\sqrt{2}\sin t)\mathbf{j} \Rightarrow x = \cos t$ and $y = \sqrt{2}\sin t \Rightarrow \frac{x^2}{16} + \frac{y^2}{2} = 1$; $\mathbf{v} = (-4\sin t)\mathbf{i} + (\sqrt{2}\cos t)\mathbf{j}$ and $\mathbf{a} = (-4\cos t)\mathbf{i} - (\sqrt{2}\sin t)\mathbf{j}$; $\mathbf{r}(0) = \mathbf{i}$, $\mathbf{v}(0) = \sqrt{2}\mathbf{j}$, $\mathbf{a}(0) = -4\mathbf{i}$; $\mathbf{r}\left(\frac{\pi}{4}\right) = 2\sqrt{2}\mathbf{i} + \mathbf{j}$, $\mathbf{v}\left(\frac{\pi}{4}\right) = -2\sqrt{2}\mathbf{i} + \mathbf{j}$, $\mathbf{a}\left(\frac{\pi}{4}\right) = -2\sqrt{2}\mathbf{i} - \mathbf{j}$; $|\mathbf{v}| = \sqrt{16\sin^2 t + 2\cos^2 t}$

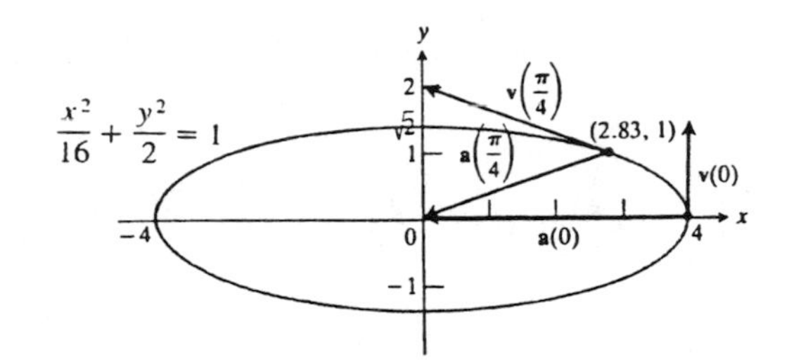

$\Rightarrow a_T = \frac{d}{dt}|\mathbf{v}| = \frac{14\sin t\cos t}{\sqrt{16\sin^2 t + 2\cos^2 t}}$; at $t = 0$: $a_T = 0$, $a_N = \sqrt{|\mathbf{a}|^2 - 0} = 4$, $\kappa = \frac{a_N}{|\mathbf{v}|^2} = \frac{4}{2} = 2$;

at $t = \frac{\pi}{4}$: $a_T = \frac{7}{\sqrt{8+1}} = \frac{7}{3}$, $a_N = \sqrt{9 - \frac{49}{9}} = \frac{4\sqrt{2}}{3}$, $\kappa = \frac{a_N}{|\mathbf{v}|^2} = \frac{4\sqrt{2}}{27}$

2. $\mathbf{r}(t) = (\sqrt{3}\sec t)\mathbf{i} + (\sqrt{3}\tan t)\mathbf{j} \Rightarrow x = \sqrt{3}\sec t$ and $y = \sqrt{3}\tan t \Rightarrow \frac{x^2}{3} - \frac{y^2}{3} = \sec^2 t - \tan^2 t = 1$; $\Rightarrow x^2 - y^2 = 3$; $\mathbf{v} = (\sqrt{3}\sec t\tan t)\mathbf{i} + (\sqrt{3}\sec^2 t)\mathbf{j}$ and $\mathbf{a} = (\sqrt{3}\sec t\tan^2 t + \sqrt{3}\sec^3 t)\mathbf{i} - (2\sqrt{3}\sec^2 t\tan t)\mathbf{j}$; $\mathbf{r}(0) = \sqrt{3}\mathbf{i}$, $\mathbf{v}(0) = \sqrt{3}\mathbf{j}$, $\mathbf{a}(0) = \sqrt{3}\mathbf{i}$;

$|\mathbf{v}| = \sqrt{3\sec^2 t\tan^2 t + 3\sec^4 t}$

$\Rightarrow a_T = \frac{d}{dt}|\mathbf{v}| = \frac{6\sec^2 t\tan^3 t + 12\sec^4 t\tan t}{2\sqrt{3\sec^2 t\tan^2 t + 3\sec^4 t}}$;

at $t = 0$: $a_T = 0$, $a_N = \sqrt{|\mathbf{a}|^2 - 0} = \sqrt{3}$,

$\kappa = \frac{a_N}{|\mathbf{v}|^2} = \frac{\sqrt{3}}{3} = \frac{1}{\sqrt{3}}$

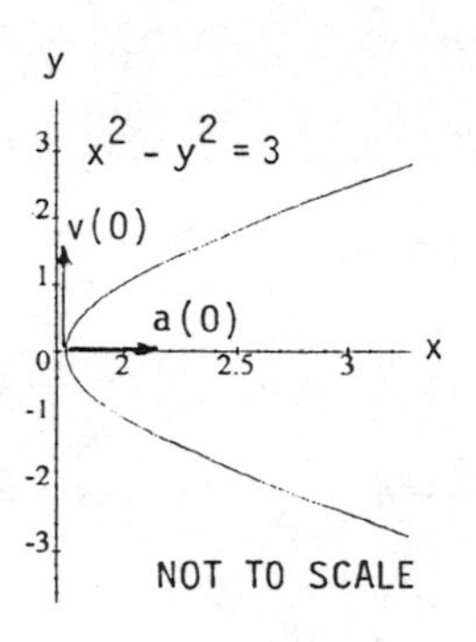

3. $\mathbf{r} = \frac{1}{\sqrt{1+t^2}}\mathbf{i} + \frac{t}{\sqrt{1+t^2}}\mathbf{j} \Rightarrow \mathbf{v} = -t\left(1+t^2\right)^{-3/2}\mathbf{i} + \left(1+t^2\right)^{-3/2}\mathbf{j}$

$\Rightarrow |\mathbf{v}| = \sqrt{\left[-t\left(1+t^2\right)^{-3/2}\right]^2 + \left[\left(1+t^2\right)^{-3/2}\right]^2} = \frac{1}{1+t^2}$. We want to maximize $|\mathbf{v}|$: $\frac{d|\mathbf{v}|}{dt} = \frac{-2t}{\left(1+t^2\right)^2}$ and $\frac{d|\mathbf{v}|}{dt} = 0 \Rightarrow \frac{-2t}{\left(1+t^2\right)^2} = 0 \Rightarrow t = 0$. For $t < 0$, $\frac{-2t}{\left(1+t^2\right)^2} > 0$; for $t > 0$, $\frac{-2t}{\left(1+t^2\right)^2} < 0 \Rightarrow |\mathbf{v}|_{max}$ occurs when $t = 0 \Rightarrow |\mathbf{v}|_{max} = 1$

4. $\mathbf{r} = \left(e^t\cos t\right)\mathbf{i} + \left(e^t\sin t\right)\mathbf{j} \Rightarrow \mathbf{v} = \left(e^t\cos t - e^t\sin t\right)\mathbf{i} + \left(e^t\sin t + e^t\cos t\right)\mathbf{j}$
$\Rightarrow \mathbf{a} = \left(e^t\cos t - e^t\sin t - e^t\sin t - e^t\cos t\right)\mathbf{i} + \left(e^t\sin t + e^t\cos t + e^t\cos t - e^t\sin t\right)\mathbf{j}$
$= \left(-2e^t\sin t\right)\mathbf{i} + \left(2e^t\cos t\right)\mathbf{j}$. Let θ be the angle between $\mathbf{r}$ and $\mathbf{a}$. Then $\theta = \cos^{-1}\left(\frac{\mathbf{r}\cdot\mathbf{a}}{|\mathbf{r}||\mathbf{a}|}\right)$
$= \cos^{-1}\left(\frac{-2e^{2t}\sin t\cos t + 2e^{2t}\sin t\cos t}{\sqrt{\left(e^t\cos t\right)^2 + \left(e^t\sin t\right)^2}\sqrt{\left(-2e^t\sin t\right)^2 + \left(2e^t\cos t\right)^2}}\right) = \cos^{-1}\left(\frac{0}{2e^{2t}}\right) = \cos^{-1} 0 = \frac{\pi}{2}$ for all t

5. $\mathbf{v} = 3\mathbf{i} + 4\mathbf{j}$ and $\mathbf{a} = 5\mathbf{i} + 15\mathbf{j} \Rightarrow \mathbf{v}\times\mathbf{a} = \begin{vmatrix} \mathbf{i} & \mathbf{j} & \mathbf{k} \\ 3 & 4 & 0 \\ 5 & 15 & 0 \end{vmatrix} = 25\mathbf{k} \Rightarrow |\mathbf{v}\times\mathbf{a}| = 25$; $|\mathbf{v}| = \sqrt{3^2 + 4^2} = 5$

$\Rightarrow \kappa = \frac{|\mathbf{v}\times\mathbf{a}|}{|\mathbf{v}|^3} = \frac{25}{5^3} = \frac{1}{5}$

6. $\kappa = \frac{|y''|}{\left[1 + (y')^2\right]^{3/2}} = e^x\left(1 + e^{2x}\right)^{-3/2} \Rightarrow \frac{d\kappa}{dx} = e^x\left(1 + e^{2x}\right)^{-3/2} + e^x\left[-\frac{3}{2}\left(1 + e^{2x}\right)^{-5/2}\left(2e^{2x}\right)\right]$
$= e^x\left(1 + e^{2x}\right)^{-3/2} - 3e^{3x}\left(1 + e^{2x}\right)^{-5/2} = e^x\left(1 + e^{2x}\right)^{-5/2}\left[\left(1 + e^{2x}\right) - 3e^{2x}\right] = e^x\left(1 + e^{2x}\right)^{-5/2}\left(1 - 2e^{2x}\right)$;
$\frac{d\kappa}{dx} = 0 \Rightarrow \left(1 - 2e^{2x}\right) = 0 \Rightarrow e^{2x} = \frac{1}{2} \Rightarrow 2x = -\ln 2 \Rightarrow x = -\frac{1}{2}\ln 2 = -\ln\sqrt{2} \Rightarrow y = \frac{1}{\sqrt{2}}$; therefore κ is at a maximum at the point $\left(-\ln\sqrt{2}, \frac{1}{\sqrt{2}}\right)$

7. $\mathbf{r} = x\mathbf{i} + y\mathbf{j} \Rightarrow \mathbf{v} = \frac{dx}{dt}\mathbf{i} + \frac{dy}{dt}\mathbf{j}$ and $\mathbf{v} \cdot \mathbf{i} = y \Rightarrow \frac{dx}{dt} = y$. Since the particle moves around the unit circle $x^2 + y^2 = 1$, $2x\frac{dx}{dt} + 2y\frac{dy}{dt} = 0 \Rightarrow \frac{dy}{dt} = -\frac{x}{y}\frac{dx}{dt} \Rightarrow \frac{dy}{dt} = -\frac{x}{y}(y) = -x$. Since $\frac{dx}{dt} = y$ and $\frac{dy}{dt} = -x$, we have $\mathbf{v} = y\mathbf{i} - x\mathbf{j} \Rightarrow$ at $(1,0)$, $\mathbf{v} = -\mathbf{j}$ and the motion is clockwise.

8. $9y = x^3 \Rightarrow 9\frac{dy}{dt} = 3x^2\frac{dx}{dt} \Rightarrow \frac{dy}{dt} = \frac{1}{3}x^2\frac{dx}{dt}$. If $\mathbf{r} = x\mathbf{i} + y\mathbf{j}$, where x and y are differentiable functions of t, then $\mathbf{v} = \frac{dx}{dt}\mathbf{i} + \frac{dy}{dt}\mathbf{j}$. Hence $\mathbf{v} \cdot \mathbf{i} = 4 \Rightarrow \frac{dx}{dt} = 4$ and $\mathbf{v} \cdot \mathbf{j} = \frac{dy}{dt} = \frac{1}{3}x^2\frac{dx}{dt} = \frac{1}{3}(3)^2(4) = 12$ at $(3,3)$. Also, $\mathbf{a} = \frac{d^2x}{dt^2}\mathbf{i} + \frac{d^2y}{dt^2}\mathbf{j}$ and $\frac{d^2y}{dt^2} = \left(\frac{2}{3}x\right)\left(\frac{dx}{dt}\right)^2 + \left(\frac{1}{3}x^2\right)\frac{d^2x}{dt^2}$. Hence $\mathbf{a} \cdot \mathbf{i} = -2 \Rightarrow \frac{d^2x}{dt^2} = -2$ and $\mathbf{a} \cdot \mathbf{j} = \frac{d^2y}{dt^2} = \frac{2}{3}(3)(4)^2 + \frac{1}{3}(3)^2(-2) = 26$ at the point $(x,y) = (3,3)$.

9. $\frac{d\mathbf{r}}{dt}$ orthogonal to $\mathbf{r} \Rightarrow 0 = \frac{d\mathbf{r}}{dt} \cdot \mathbf{r} = \frac{1}{2}\frac{d\mathbf{r}}{dt} \cdot \mathbf{r} + \frac{1}{2}\mathbf{r} \cdot \frac{d\mathbf{r}}{dt} = \frac{1}{2}\frac{d}{dt}(\mathbf{r} \cdot \mathbf{r}) \Rightarrow \mathbf{r} \cdot \mathbf{r} = K$, a constant. If $\mathbf{r} = x\mathbf{i} + y\mathbf{j}$, where x and y are differentiable functions of t, then $\mathbf{r} \cdot \mathbf{r} = x^2 + y^2 \Rightarrow x^2 + y^2 = K$, which is the equation of a circle centered at the origin.

10. (b) $\mathbf{v} = (\pi - \pi\cos\pi t)\mathbf{i} + (\pi\sin\pi t)\mathbf{j}$
$\Rightarrow \mathbf{a} = (\pi^2\sin\pi t)\mathbf{i} + (\pi^2\cos\pi t)\mathbf{j}$;
$\mathbf{v}(0) = \mathbf{0}$ and $\mathbf{a}(0) = \pi^2\mathbf{j}$;
$\mathbf{v}(1) = 2\pi\mathbf{i}$ and $\mathbf{a}(1) = -\pi^2\mathbf{j}$;
$\mathbf{v}(2) = \mathbf{0}$ and $\mathbf{a}(2) = \pi^2\mathbf{j}$;
$\mathbf{v}(3) = 2\pi\mathbf{i}$ and $\mathbf{a}(3) = -\pi^2\mathbf{j}$

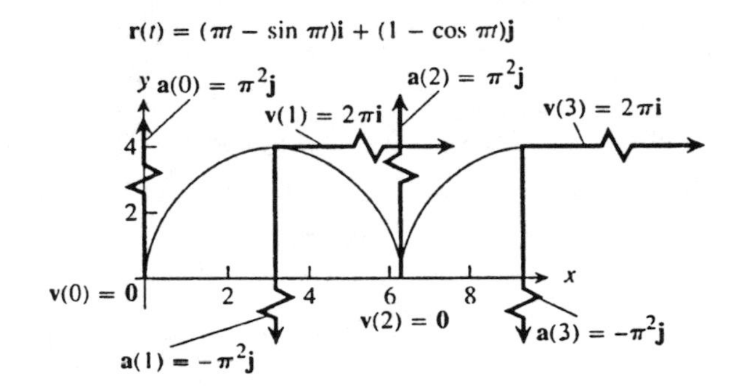

(c) Forward speed at the topmost point is $|\mathbf{v}(1)| = |\mathbf{v}(3)| = 2\pi$ ft/sec; since the circle makes $\frac{1}{2}$ revolution per second, the center moves π ft parallel to the x-axis each second $\Rightarrow$ the forward speed of C is π ft/sec.

11. $y = y_0 + (v_0\sin\alpha)t - \frac{1}{2}gt^2 \Rightarrow y = 6.5 + (44 \text{ ft/sec})(\sin 45°)(3 \text{ sec}) - \frac{1}{2}(32 \text{ ft/sec}^2)(3 \text{ sec})^2 = 6.5 + 66\sqrt{2} - 144 \approx -41.36$ ft $\Rightarrow$ the shot put is on the ground. Now, $y = 0 \Rightarrow 6.5 + 22\sqrt{2}t - 16t^2 = 0 \Rightarrow t \approx 2.13$ sec (the positive root) $\Rightarrow x \approx (44 \text{ ft/sec})(\cos 45°)(2.13 \text{ sec}) \approx 66.42$ ft or about 66 ft, 5 in. from the stopboard

12. $y_{max} = y_0 + \frac{(v_0\sin\alpha)^2}{2g} = 7 \text{ ft} + \frac{[(80 \text{ ft/sec})(\sin 45°)]^2}{(2)(32 \text{ ft/sec}^2)} \approx 57$ ft

13. $x = (v_0\cos\alpha)t$ and $y = (v_0\sin\alpha)t - \frac{1}{2}gt^2 \Rightarrow \tan\phi = \frac{y}{x} = \frac{(v_0\sin\alpha)t - \frac{1}{2}gt^2}{(v_0\cos\alpha)t} = \frac{(v_0\sin\alpha) - \frac{1}{2}gt}{v_0\cos\alpha}$
$\Rightarrow v_0\cos\alpha\tan\phi = v_0\sin\alpha - \frac{1}{2}gt \Rightarrow t = \frac{2v_0\sin\alpha - 2v_0\cos\alpha\tan\phi}{g}$, which is the time when the golf ball

hits the upward slope. At this time

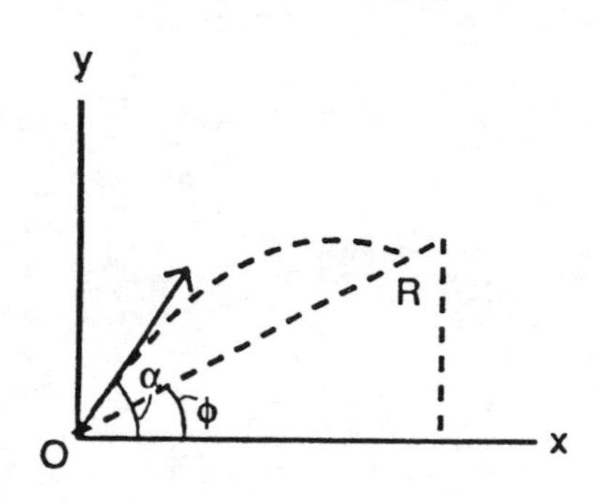

$x = (v_0 \cos\alpha)\left(\frac{2v_0 \sin\alpha - 2v_0 \cos\alpha \tan\phi}{g}\right)$

$= \left(\frac{2}{g}\right)\left(v_0^2 \sin\alpha \cos\alpha - v_0^2 \cos^2\alpha \tan\phi\right)$. Now

$OR = \frac{x}{\cos\phi} \Rightarrow OR = \left(\frac{2}{g}\right)\left(\frac{v_0^2 \sin\alpha \cos\alpha - v_0^2 \cos^2\alpha \tan\phi}{\cos\phi}\right)$

$= \left(\frac{2v_0^2 \cos\alpha}{g}\right)\left(\frac{\sin\alpha}{\cos\phi} - \frac{\cos\alpha \tan\phi}{\cos\phi}\right)$

$= \left(\frac{2v_0^2 \cos\alpha}{g}\right)\left(\frac{\sin\alpha \cos\phi - \cos\alpha \sin\phi}{\cos^2\phi}\right)$

$= \left(\frac{2v_0^2 \cos\alpha}{g \cos^2\phi}\right)[\sin(\alpha - \phi)]$. The distance OR is maximized when x is maximized:

$\frac{dx}{d\alpha} = \left(\frac{2v_0^2}{g}\right)(\cos 2\alpha + \sin 2\alpha \tan\phi) = 0 \Rightarrow (\cos 2\alpha + \sin 2\alpha \tan\phi) = 0 \Rightarrow \cot 2\alpha + \tan\phi = 0$

$\Rightarrow \cot 2\alpha = \tan(-\phi) \Rightarrow 2\alpha = \frac{\pi}{2} + \phi \Rightarrow \alpha = \frac{\phi}{2} + \frac{\pi}{4}$

14. $R = \frac{v_0^2}{g} \sin 2\alpha \Rightarrow v_0 = \sqrt{\frac{Rg}{\sin 2\alpha}}$; for 4325 yards: 4325 yards = 12,975 ft $\Rightarrow v_0 = \sqrt{\frac{(12{,}975 \text{ ft})(32 \text{ ft/sec}^2)}{(\sin 90^\circ)}}$

≈ 644 ft/sec; for 4752 yards: 4752 yards = 14,256 ft $\Rightarrow v_0 = \sqrt{\frac{(14{,}256 \text{ ft})(32 \text{ ft/sec}^2)}{(\sin 90^\circ)}} \approx 675$ ft/sec

15. (a) $R = \frac{v_0^2}{g} \sin 2\alpha \Rightarrow 109.5 \text{ ft} = \left(\frac{v_0^2}{32 \text{ ft/sec}^2}\right)(\sin 90^\circ) \Rightarrow v_0^2 = 3504 \text{ ft}^2/\text{sec}^2 \Rightarrow v_0 = \sqrt{3504 \text{ ft}^2/\text{sec}^2}$

≈ 59.19 ft/sec

(b) $x = (v_0 \cos\alpha)t$ and $y = 4 + (v_0 \sin\alpha)t - \frac{1}{2}gt^2$; when the cork hits the ground, $x = 177.75$ ft and $y = 0$

$\Rightarrow 177.75 = \left(v_0 \frac{1}{\sqrt{2}}\right)t$ and $0 = 4 + \left(v_0 \frac{1}{\sqrt{2}}\right)t - 16t^2 \Rightarrow 16t^2 = 4 + 177.75 \Rightarrow t = \frac{\sqrt{181.75}}{4}$

$\Rightarrow v_0 = \frac{(177.75)\sqrt{2}}{t} = \frac{4(177.75)\sqrt{2}}{\sqrt{181.75}} \approx 74.58$ ft/sec

16. (a) $x = v_0(\cos 40^\circ)t$ and $y = 6.5 + v_0(\sin 40^\circ)t - \frac{1}{2}gt^2 = 6.5 + v_0(\sin 40^\circ)t - 16t^2$; $x = 262\frac{5}{12}$ ft and $y = 0$ ft

$\Rightarrow 262\frac{5}{12} = v_0(\cos 40^\circ)t$ or $v_0 = \frac{262.4167}{(\cos 40^\circ)t}$ and $0 = 6.5 + \left[\frac{262.4167}{(\cos 40^\circ)t}\right](\sin 40^\circ)t - 16t^2 \Rightarrow t^2 = 14.1684$

$\Rightarrow t \approx 3.764$ sec. Therefore, $262.4167 \approx v_0(\cos 40^\circ)(3.764 \text{ sec}) \Rightarrow v_0 \approx \frac{262.4167}{(\cos 40^\circ)(3.764 \text{ sec})} \Rightarrow v_0 \approx 91$ ft/sec

(b) $y_{max} = y_0 + \frac{(v_0 \sin\alpha)^2}{2g} \approx 6.5 + \frac{(91)(\sin 40^\circ)}{(2)(32)} \approx 60$ ft

17. $x^2 = \left(v_0^2 \cos^2\alpha\right)t^2$ and $\left(y + \frac{1}{2}gt^2\right)^2 = \left(v_0^2 \sin^2\alpha\right)t^2 \Rightarrow x^2 + \left(y + \frac{1}{2}gt^2\right) = v_0^2 t^2$

18. $\ddot{s} = \frac{d}{dt}\sqrt{\dot{x}^2+\dot{y}^2} = \frac{\dot{x}\ddot{x}+\dot{y}\ddot{y}}{\sqrt{\dot{x}^2+\dot{y}^2}} \Rightarrow \ddot{x}^2+\ddot{y}^2-\ddot{s}^2 = \ddot{x}^2+\ddot{y}^2-\frac{(\dot{x}\ddot{x}+\dot{y}\ddot{y})^2}{\dot{x}^2+\dot{y}^2}$

$= \frac{(\ddot{x}^2+\ddot{y}^2)(\dot{x}^2+\dot{y}^2)-(\dot{x}^2\ddot{x}^2+2\dot{x}\ddot{x}\dot{y}\ddot{y}+\dot{y}^2\ddot{y}^2)}{\dot{x}^2+\dot{y}^2} = \frac{\dot{x}^2\ddot{y}^2+\dot{y}^2\ddot{x}^2-2\dot{x}\ddot{x}\dot{y}\ddot{y}}{\dot{x}^2+\dot{y}^2} = \frac{(\dot{x}\ddot{y}-\dot{y}\ddot{x})^2}{\dot{x}^2+\dot{y}^2}$

$\Rightarrow \sqrt{\ddot{x}^2+\ddot{y}^2-\ddot{s}^2} = \frac{|\dot{x}\ddot{y}-\dot{y}\ddot{x}|}{\sqrt{\dot{x}^2+\dot{y}^2}} \Rightarrow \frac{\dot{x}^2+\dot{y}^2}{\sqrt{\ddot{x}^2+\ddot{y}^2-\ddot{s}^2}} = \frac{(\dot{x}^2+\dot{y}^2)^{3/2}}{|\dot{x}\ddot{y}-\dot{y}\ddot{x}|} = \frac{1}{\kappa} = \rho$

19. $\mathbf{r}(t) = \left[\int_0^t \cos\left(\frac{1}{2}\pi\theta^2\right) d\theta\right]\mathbf{i} + \left[\int_0^t \sin\left(\frac{1}{2}\pi\theta^2\right) d\theta\right]\mathbf{j} \Rightarrow \mathbf{v}(t) = \cos\left(\frac{\pi t^2}{2}\right)\mathbf{i} + \sin\left(\frac{\pi t^2}{2}\right)\mathbf{j} \Rightarrow |\mathbf{v}| = 1;$

$\mathbf{a}(t) = -\pi t \sin\left(\frac{\pi t^2}{2}\right)\mathbf{i} + \pi t \cos\left(\frac{\pi t^2}{2}\right)\mathbf{j} \Rightarrow \mathbf{v}\times\mathbf{a} = \begin{vmatrix} \mathbf{i} & \mathbf{j} & \mathbf{k} \\ \cos\left(\frac{\pi t^2}{2}\right) & \sin\left(\frac{\pi t^2}{2}\right) & 0 \\ -\pi t \sin\left(\frac{\pi t^2}{2}\right) & \pi t \cos\left(\frac{\pi t^2}{2}\right) & 0 \end{vmatrix}$

$= \pi t\mathbf{k} \Rightarrow \kappa = \frac{|\mathbf{v}\times\mathbf{a}|}{|\mathbf{v}|^3} = \pi t;\ |\mathbf{v}(t)| = \frac{ds}{dt} = 1 \Rightarrow s = t + C;\ \mathbf{r}(0) = \mathbf{0} \Rightarrow s(0) = 0 \Rightarrow C = 0 \Rightarrow \kappa = \pi s$

20. $s = a\theta \Rightarrow \theta = \frac{s}{a} \Rightarrow \phi = \frac{s}{a} + \frac{\pi}{2} \Rightarrow \frac{d\phi}{ds} = \frac{1}{a} \Rightarrow \kappa = \left|\frac{1}{a}\right| = \frac{1}{a}$ since $a > 0$

21. $\mathbf{r} = (2\cos t)\mathbf{i} + (2\sin t)\mathbf{j} + t^2\mathbf{k} \Rightarrow \mathbf{v} = (-2\sin t)\mathbf{i} + (2\cos t)\mathbf{j} + 2t\mathbf{k} \Rightarrow |\mathbf{v}| = \sqrt{(-2\sin t)^2 + (2\cos t)^2 + (2t)^2}$

$= 2\sqrt{1+t^2} \Rightarrow \text{Length} = \int_0^{\pi/4} 2\sqrt{1+t^2}\,dt = \left[t\sqrt{1+t^2} + \ln\left|t+\sqrt{1+t^2}\right|\right]_0^{\pi/4} = \frac{\pi}{4}\sqrt{1+\frac{\pi^2}{16}} + \ln\left(\frac{\pi}{4} + \sqrt{1+\frac{\pi^2}{16}}\right)$

22. $\mathbf{r} = (3\cos t)\mathbf{i} + (3\sin t)\mathbf{j} + 2t^{3/2}\mathbf{k} \Rightarrow \mathbf{v} = (-3\sin t)\mathbf{i} + (3\cos t)\mathbf{j} + 3t^{1/2}\mathbf{k}$

$\Rightarrow |\mathbf{v}| = \sqrt{(-3\sin t)^2 + (3\cos t)^2 + \left(3t^{1/2}\right)^2} = \sqrt{9+9t} = 3\sqrt{1+t} \Rightarrow \text{Length} = \int_0^3 3\sqrt{1+t}\,dt = \left[2(1+t)^{3/2}\right]_0^3$

$= 14$

23. $\mathbf{r} = \frac{4}{9}(1+t)^{3/2}\mathbf{i} + \frac{4}{9}(1-t)^{3/2}\mathbf{j} + \frac{1}{3}t\mathbf{k} \Rightarrow \mathbf{v} = \frac{2}{3}(1+t)^{1/2}\mathbf{i} - \frac{2}{3}(1-t)^{1/2}\mathbf{j} + \frac{1}{3}\mathbf{k}$

$\Rightarrow |\mathbf{v}| = \sqrt{\left[\frac{2}{3}(1+t)^{1/2}\right]^2 + \left[-\frac{2}{3}(1-t)^{1/2}\right]^2 + \left(\frac{1}{3}\right)^2} = 1 \Rightarrow \mathbf{T} = \frac{2}{3}(1+t)^{1/2}\mathbf{i} - \frac{2}{3}(1-t)^{1/2}\mathbf{j} + \frac{1}{3}\mathbf{k}$

$\Rightarrow \mathbf{T}(0) = \frac{2}{3}\mathbf{i} - \frac{2}{3}\mathbf{j} + \frac{1}{3}\mathbf{k};\ \frac{d\mathbf{T}}{dt} = \frac{1}{3}(1+t)^{-1/2}\mathbf{i} + \frac{1}{3}(1-t)^{-1/2}\mathbf{j} \Rightarrow \frac{d\mathbf{T}}{dt}(0) = \frac{1}{3}\mathbf{i} + \frac{1}{3}\mathbf{j} \Rightarrow \left|\frac{d\mathbf{T}}{dt}(0)\right| = \frac{\sqrt{2}}{3}$

$\Rightarrow \mathbf{N}(0) = \frac{1}{\sqrt{2}}\mathbf{i} + \frac{1}{\sqrt{2}}\mathbf{j}$; $\mathbf{B}(0) = \mathbf{T}(0) \times \mathbf{N}(0) = \begin{vmatrix} \mathbf{i} & \mathbf{j} & \mathbf{k} \\ \frac{2}{3} & -\frac{2}{3} & \frac{1}{3} \\ \frac{1}{\sqrt{2}} & \frac{1}{\sqrt{2}} & 0 \end{vmatrix} = -\frac{1}{3\sqrt{2}}\mathbf{i} + \frac{1}{3\sqrt{2}}\mathbf{j} + \frac{4}{3\sqrt{2}}\mathbf{k}$;

$\mathbf{a} = \frac{1}{3}(1+t)^{-1/2}\mathbf{i} + \frac{1}{3}(1-t)^{-1/2}\mathbf{j} \Rightarrow \mathbf{a}(0) = \frac{1}{3}\mathbf{i} + \frac{1}{3}\mathbf{j}$ and $\mathbf{v}(0) = \frac{2}{3}\mathbf{i} - \frac{2}{3}\mathbf{j} + \frac{1}{3}\mathbf{k} \Rightarrow \mathbf{v}(0) \times \mathbf{a}(0)$

$= \begin{vmatrix} \mathbf{i} & \mathbf{j} & \mathbf{k} \\ \frac{2}{3} & -\frac{2}{3} & \frac{1}{3} \\ \frac{1}{3} & \frac{1}{3} & 0 \end{vmatrix} = -\frac{1}{9}\mathbf{i} + \frac{1}{9}\mathbf{j} + \frac{4}{9}\mathbf{k} \Rightarrow |\mathbf{v} \times \mathbf{a}| = \frac{\sqrt{2}}{3} \Rightarrow \kappa(0) = \frac{|\mathbf{v} \times \mathbf{a}|}{|\mathbf{v}|^3} = \frac{\left(\frac{\sqrt{2}}{3}\right)}{1^3} = \frac{\sqrt{2}}{3}$;

$\dot{\mathbf{a}} = -\frac{1}{6}(1+t)^{-3/2}\mathbf{i} + \frac{1}{6}(1-t)^{-3/2}\mathbf{j} \Rightarrow \dot{\mathbf{a}}(0) = -\frac{1}{6}\mathbf{i} + \frac{1}{6}\mathbf{j} \Rightarrow \tau(0) = \frac{\begin{vmatrix} \frac{2}{3} & -\frac{2}{3} & \frac{1}{3} \\ \frac{1}{3} & \frac{1}{3} & 0 \\ -\frac{1}{6} & \frac{1}{6} & 0 \end{vmatrix}}{|\mathbf{v} \times \mathbf{a}|^2} = \frac{\left(\frac{1}{3}\right)\left(\frac{2}{18}\right)}{\left(\frac{\sqrt{2}}{3}\right)^2} = \frac{1}{6}$;

$t = 0 \Rightarrow \left(\frac{4}{9}, \frac{4}{9}, 0\right)$ is the point on the curve

24. $\mathbf{r} = (e^t \sin 2t)\mathbf{i} + (e^t \cos 2t)\mathbf{j} + 2e^t\mathbf{k} \Rightarrow \mathbf{v} = (e^t \sin 2t + 2e^t \cos 2t)\mathbf{i} + (e^t \cos 2t - 2e^t \sin 2t)\mathbf{j} + 2e^t\mathbf{k}$

$\Rightarrow |\mathbf{v}| = \sqrt{(e^t \sin 2t + 2e^t \cos 2t)^2 + (e^t \cos 2t - 2e^t \sin 2t)^2 + (2e^t)^2} = 3e^t \Rightarrow \mathbf{T} = \frac{\mathbf{v}}{|\mathbf{v}|}$

$= \left(\frac{1}{3} \sin 2t + \frac{2}{3} \cos 2t\right)\mathbf{i} + \left(\frac{1}{3} \cos 2t - \frac{2}{3} \sin 2t\right)\mathbf{j} + \frac{2}{3}\mathbf{k} \Rightarrow \mathbf{T}(0) = \frac{2}{3}\mathbf{i} + \frac{1}{3}\mathbf{j} + \frac{2}{3}\mathbf{k}$;

$\frac{d\mathbf{T}}{dt} = \left(\frac{2}{3} \cos 2t - \frac{4}{3} \sin 2t\right)\mathbf{i} + \left(-\frac{2}{3} \sin 2t - \frac{4}{3} \cos 2t\right)\mathbf{j} \Rightarrow \frac{d\mathbf{T}}{dt}(0) = \frac{2}{3}\mathbf{i} - \frac{4}{3}\mathbf{j} \Rightarrow \left|\frac{d\mathbf{T}}{dt}(0)\right| = \frac{2}{3}\sqrt{5}$

$\Rightarrow \mathbf{N}(0) = \frac{\left(\frac{2}{3}\mathbf{i} - \frac{4}{3}\mathbf{j}\right)}{\left(\frac{2\sqrt{5}}{3}\right)} = \frac{1}{\sqrt{5}}\mathbf{i} - \frac{2}{\sqrt{5}}\mathbf{j}$; $\mathbf{B}(0) = \mathbf{T}(0) \times \mathbf{N}(0) = \begin{vmatrix} \mathbf{i} & \mathbf{j} & \mathbf{k} \\ \frac{2}{3} & \frac{1}{3} & \frac{2}{3} \\ \frac{1}{\sqrt{5}} & -\frac{2}{\sqrt{5}} & 0 \end{vmatrix} = \frac{4}{3\sqrt{5}}\mathbf{i} + \frac{2}{3\sqrt{5}}\mathbf{j} - \frac{5}{3\sqrt{5}}\mathbf{k}$;

$\mathbf{a} = (4e^t \cos 2t - 3e^t \sin 2t)\mathbf{i} + (-3e^t \cos 2t - 4e^t \sin 2t)\mathbf{j} + 2e^t\mathbf{k} \Rightarrow \mathbf{a}(0) = 4\mathbf{i} - 3\mathbf{j} + 2\mathbf{k}$ and $\mathbf{v}(0) = 2\mathbf{i} + \mathbf{j} + 2\mathbf{k}$

$\Rightarrow \mathbf{v}(0) \times \mathbf{a}(0) = \begin{vmatrix} \mathbf{i} & \mathbf{j} & \mathbf{k} \\ 2 & 1 & 2 \\ 4 & -3 & 2 \end{vmatrix} = 8\mathbf{i} + 4\mathbf{j} - 10\mathbf{k} \Rightarrow |\mathbf{v} \times \mathbf{a}| = \sqrt{64 + 16 + 100} = 6\sqrt{5}$ and $|\mathbf{v}(0)| = 3$

$\Rightarrow \kappa(0) = \frac{6\sqrt{5}}{3^3} = \frac{2\sqrt{5}}{9}$;

$\dot{\mathbf{a}} = (4e^t \cos 2t - 8e^t \sin 2t - 3e^t \sin 2t - 6e^t \cos 2t)\mathbf{i} + (-3e^t \cos 2t + 6e^t \sin 2t - 4e^t \sin 2t - 8e^t \cos 2t)\mathbf{j} + 2e^t\mathbf{k}$
$= (-2e^t \cos 2t - 11e^t \sin 2t)\mathbf{i} + (-11\, e^t \cos 2t + 2e^t \sin 2t)\mathbf{j} + 2e^t\mathbf{k} \Rightarrow \dot{\mathbf{a}}(0) = -2\mathbf{i} - 11\mathbf{j} + 2\mathbf{k}$

$$\Rightarrow \tau(0) = \frac{\begin{vmatrix} 2 & 1 & 2 \\ 4 & -3 & 2 \\ -2 & -11 & 2 \end{vmatrix}}{|\mathbf{v}\times\mathbf{a}|^2} = \frac{-80}{180} = -\frac{4}{9};\ t = 0 \Rightarrow (0,1,2) \text{ is on the curve}$$

25. $\mathbf{r} = t\mathbf{i} + \frac{1}{2}e^{2t}\mathbf{j} \Rightarrow \mathbf{v} = \mathbf{i} + e^{2t}\mathbf{j} \Rightarrow |\mathbf{v}| = \sqrt{1+e^{4t}} \Rightarrow \mathbf{T} = \frac{1}{\sqrt{1+e^{4t}}}\mathbf{i} + \frac{e^{2t}}{\sqrt{1+e^{4t}}}\mathbf{j} \Rightarrow \mathbf{T}(\ln 2) = \frac{1}{\sqrt{17}}\mathbf{i} + \frac{4}{\sqrt{17}}\mathbf{j};$

$\frac{d\mathbf{T}}{dt} = \frac{-2e^{4t}}{(1+e^{4t})^{3/2}}\mathbf{i} + \frac{2e^{2t}}{(1+e^{4t})^{3/2}}\mathbf{j} \Rightarrow \frac{d\mathbf{T}}{dt}(\ln 2) = \frac{-32}{17\sqrt{17}}\mathbf{i} + \frac{8}{17\sqrt{17}}\mathbf{j} \Rightarrow \mathbf{N}(\ln 2) = -\frac{4}{\sqrt{17}}\mathbf{i} + \frac{1}{\sqrt{17}}\mathbf{j};$

$$\mathbf{B}(\ln 2) = \mathbf{T}(\ln 2) \times \mathbf{N}(\ln 2) = \begin{vmatrix} \mathbf{i} & \mathbf{j} & \mathbf{k} \\ \frac{1}{\sqrt{17}} & \frac{4}{\sqrt{17}} & 0 \\ -\frac{4}{\sqrt{17}} & \frac{1}{\sqrt{17}} & 0 \end{vmatrix} = \mathbf{k};\ \mathbf{a} = 2e^{2t}\mathbf{j} \Rightarrow \mathbf{a}(\ln 2) = 8\mathbf{j} \text{ and } \mathbf{v}(\ln 2) = \mathbf{i} + 4\mathbf{j}$$

$$\Rightarrow \mathbf{v}(\ln 2) \times \mathbf{a}(\ln 2) = \begin{vmatrix} \mathbf{i} & \mathbf{j} & \mathbf{k} \\ 1 & 4 & 0 \\ 0 & 8 & 0 \end{vmatrix} = 8\mathbf{k} \Rightarrow |\mathbf{v}\times\mathbf{a}| = 8 \text{ and } |\mathbf{v}(\ln 2)| = \sqrt{17} \Rightarrow \kappa(\ln 2) = \frac{8}{17\sqrt{17}};\ \dot{\mathbf{a}} = 4e^{2t}\mathbf{j}$$

$$\Rightarrow \dot{\mathbf{a}}(\ln 2) = 16\mathbf{j} \Rightarrow \tau(\ln 2) = \frac{\begin{vmatrix} 1 & 4 & 0 \\ 0 & 8 & 0 \\ 0 & 16 & 0 \end{vmatrix}}{|\mathbf{v}\times\mathbf{a}|^2} = 0;\ t = \ln 2 \Rightarrow (\ln 2, 2, 0) \text{ is on the curve}$$

26. $\mathbf{r} = (3\cosh 2t)\mathbf{i} + (3\sinh 2t)\mathbf{j} + 6t\mathbf{k} \Rightarrow \mathbf{v} = (6\sinh 2t)\mathbf{i} + (6\cosh 2t)\mathbf{j} + 6\mathbf{k}$

$\Rightarrow |\mathbf{v}| = \sqrt{36\sinh^2 2t + 36\cosh^2 2t + 36} = 6\sqrt{2}\cosh 2t \Rightarrow \mathbf{T} = \frac{\mathbf{v}}{|\mathbf{v}|} = \left(\frac{1}{\sqrt{2}}\tanh 2t\right)\mathbf{i} + \frac{1}{\sqrt{2}}\mathbf{j} + \left(\frac{1}{\sqrt{2}}\operatorname{sech} 2t\right)\mathbf{k}$

$\Rightarrow \mathbf{T}(\ln 2) = \frac{15}{17\sqrt{2}}\mathbf{i} + \frac{1}{\sqrt{2}}\mathbf{j} + \frac{8}{17\sqrt{2}}\mathbf{k};\ \frac{d\mathbf{T}}{dt} = \left(\frac{2}{\sqrt{2}}\operatorname{sech}^2 2t\right)\mathbf{i} - \left(\frac{2}{\sqrt{2}}\operatorname{sech} 2t \tanh 2t\right)\mathbf{k} \Rightarrow \frac{d\mathbf{T}}{dt}(\ln 2)$

$= \left(\frac{2}{\sqrt{2}}\right)\left(\frac{8}{17}\right)^2\mathbf{i} - \left(\frac{2}{\sqrt{2}}\right)\left(\frac{8}{17}\right)\left(\frac{15}{17}\right)\mathbf{k} = \frac{128}{289\sqrt{2}}\mathbf{i} - \frac{240}{289\sqrt{2}}\mathbf{k} \Rightarrow \left|\frac{d\mathbf{T}}{dt}(\ln 2)\right| = \sqrt{\left(\frac{128}{289\sqrt{2}}\right)^2 + \left(-\frac{240}{289\sqrt{2}}\right)^2} = \frac{8\sqrt{2}}{17}$

$$\Rightarrow \mathbf{N}(\ln 2) = \frac{8}{17}\mathbf{i} - \frac{15}{17}\mathbf{k};\ \mathbf{B}(\ln 2) = \mathbf{T}(\ln 2)\times\mathbf{N}(\ln 2) = \begin{vmatrix} \mathbf{i} & \mathbf{j} & \mathbf{k} \\ \frac{15}{17\sqrt{2}} & \frac{1}{\sqrt{2}} & \frac{8}{17\sqrt{2}} \\ \frac{8}{17} & 0 & -\frac{15}{17} \end{vmatrix} = -\frac{15}{17\sqrt{2}}\mathbf{i} + \frac{1}{\sqrt{2}}\mathbf{j} - \frac{8}{17\sqrt{2}}\mathbf{k};$$

$\mathbf{a} = (12 \cosh 2t)\mathbf{i} + (12 \sinh 2t)\mathbf{j} \Rightarrow \mathbf{a}(\ln 2) = 12\left(\frac{17}{8}\right)\mathbf{i} + 12\left(\frac{15}{8}\right)\mathbf{j} = \frac{51}{2}\mathbf{i} + \frac{45}{2}\mathbf{j}$ and

$$\mathbf{v}(\ln 2) = 6\left(\frac{15}{8}\right)\mathbf{i} + 6\left(\frac{17}{8}\right)\mathbf{j} + 6\mathbf{k} = \frac{45}{4}\mathbf{i} + \frac{51}{4}\mathbf{j} + 6\mathbf{k} \Rightarrow \mathbf{v}(\ln 2) \times \mathbf{a}(\ln 2) = \begin{vmatrix} \mathbf{i} & \mathbf{j} & \mathbf{k} \\ \frac{45}{4} & \frac{51}{4} & 6 \\ \frac{51}{2} & \frac{45}{2} & 0 \end{vmatrix}$$

$$= -135\mathbf{i} + 153\mathbf{j} - 72\mathbf{k} \Rightarrow |\mathbf{v} \times \mathbf{a}| = 153\sqrt{2} \text{ and } |\mathbf{v}(\ln 2)| = \frac{51}{4}\sqrt{2} \Rightarrow \kappa(\ln 2) = \frac{153\sqrt{2}}{\left(\frac{51}{4}\sqrt{2}\right)^3} = \frac{32}{867};$$

$$\dot{\mathbf{a}} = (24 \sinh 2t)\mathbf{i} + (24 \cosh 2t)\mathbf{j} \Rightarrow \dot{\mathbf{a}}(\ln 2) = 45\mathbf{i} + 51\mathbf{j} \Rightarrow \tau(\ln 2) = \frac{\begin{vmatrix} \frac{45}{4} & \frac{51}{4} & 6 \\ \frac{51}{2} & \frac{45}{2} & 0 \\ 45 & 51 & 0 \end{vmatrix}}{|\mathbf{v} \times \mathbf{a}|^2} = \frac{32}{867};$$

$t = \ln 2 \Rightarrow \left(\frac{51}{8}, \frac{45}{8}, 6 \ln 2\right)$ is on the curve

27. $\mathbf{r} = (2 + 3t + 3t^2)\mathbf{i} + (4t + 4t^2)\mathbf{j} - (6 \cos t)\mathbf{k} \Rightarrow \mathbf{v} = (3 + 6t)\mathbf{i} + (4 + 8t)\mathbf{j} + (6 \sin t)\mathbf{k}$

$\Rightarrow |\mathbf{v}| = \sqrt{(3 + 6t)^2 + (4 + 8t)^2 + (6 \sin t)^2} = \sqrt{25 + 100t + 100t^2 + 36 \sin^2 t}$

$\Rightarrow \frac{d|\mathbf{v}|}{dt} = \frac{1}{2}(25 + 100t + 100t^2 + 36 \sin^2 t)^{-1/2}(100 + 200t + 72 \sin t \cos t) \Rightarrow a_T(0) = \frac{d|\mathbf{v}|}{dt}(0) = 10;$

$\mathbf{a} = 6\mathbf{i} + 8\mathbf{j} + (t \cos t)\mathbf{k} \Rightarrow |\mathbf{a}| = \sqrt{6^2 + 8^2 + (6 \cos t)^2} = \sqrt{100 + 36 \cos^2 t} \Rightarrow |\mathbf{a}(0)| = \sqrt{136}$

$\Rightarrow a_N = \sqrt{|\mathbf{a}|^2 - a_T^2} = \sqrt{136 - 10^2} = \sqrt{36} = 6 \Rightarrow \mathbf{a}(0) = 10\mathbf{T} + 6\mathbf{N}$

28. $\mathbf{r} = (2 + t)\mathbf{i} + (t + 2t^2)\mathbf{j} + (1 + t^2)\mathbf{k} \Rightarrow \mathbf{v} = \mathbf{i} + (1 + 4t)\mathbf{j} + 2t\mathbf{k} \Rightarrow |\mathbf{v}| = \sqrt{1^2 + (1 + 4t)^2 + (2t)^2}$

$= \sqrt{2 + 8t + 20t^2} \Rightarrow \frac{d|\mathbf{v}|}{dt} = \frac{1}{2}(2 + 8t + 20t^2)^{-1/2}(8 + 40t) \Rightarrow a_T = \frac{d|\mathbf{v}|}{dt}(0) = 2\sqrt{2};\ \mathbf{a} = 4\mathbf{j} + 2\mathbf{k}$

$\Rightarrow |\mathbf{a}| = \sqrt{4^2 + 2^2} = \sqrt{20} \Rightarrow a_N = \sqrt{|\mathbf{a}|^2 - a_T^2} = \sqrt{20 - (2\sqrt{2})^2} = \sqrt{12} = 2\sqrt{3} \Rightarrow \mathbf{a}(0) = 2\sqrt{2}\mathbf{T} + 2\sqrt{3}\mathbf{N}$

29. $\mathbf{r} = (\sin t)\mathbf{i} + (\sqrt{2} \cos t)\mathbf{j} + (\sin t)\mathbf{k} \Rightarrow \mathbf{v} = (\cos t)\mathbf{i} - (\sqrt{2} \sin t)\mathbf{j} + (\cos t)\mathbf{k}$

$\Rightarrow |\mathbf{v}| = \sqrt{(\cos t)^2 + (-\sqrt{2} \sin t)^2 + (\cos t)^2} = \sqrt{2} \Rightarrow \mathbf{T} = \frac{\mathbf{v}}{|\mathbf{v}|} = \left(\frac{1}{\sqrt{2}} \cos t\right)\mathbf{i} - (\sin t)\mathbf{j} + \left(\frac{1}{\sqrt{2}} \cos t\right)\mathbf{k};$

$\frac{d\mathbf{T}}{dt} = \left(-\frac{1}{\sqrt{2}} \sin t\right)\mathbf{i} - (\cos t)\mathbf{j} - \left(\frac{1}{\sqrt{2}} \sin t\right)\mathbf{k} \Rightarrow \left|\frac{d\mathbf{T}}{dt}\right| = \sqrt{\left(-\frac{1}{\sqrt{2}} \sin t\right)^2 + (-\cos t)^2 + \left(-\frac{1}{\sqrt{2}} \sin t\right)^2} = 1$

$$\Rightarrow \mathbf{N} = \frac{\left(\frac{d\mathbf{T}}{dt}\right)}{\left|\frac{d\mathbf{T}}{dt}\right|} = \left(-\frac{1}{\sqrt{2}}\sin t\right)\mathbf{i} - (\cos t)\mathbf{j} - \left(\frac{1}{\sqrt{2}}\sin t\right)\mathbf{k};\ \mathbf{B} = \mathbf{T}\times\mathbf{N} = \begin{vmatrix} \mathbf{i} & \mathbf{j} & \mathbf{k} \\ \frac{1}{\sqrt{2}}\cos t & -\sin t & \frac{1}{\sqrt{2}}\cos t \\ -\frac{1}{\sqrt{2}}\sin t & -\cos t & -\frac{1}{\sqrt{2}}\sin t \end{vmatrix}$$

$$= \frac{1}{\sqrt{2}}\mathbf{i} - \frac{1}{\sqrt{2}}\mathbf{k};\ \mathbf{a} = (-\sin t)\mathbf{i} - \left(\sqrt{2}\cos t\right)\mathbf{j} - (\sin t)\mathbf{k} \Rightarrow \mathbf{v}\times\mathbf{a} = \begin{vmatrix} \mathbf{i} & \mathbf{j} & \mathbf{k} \\ \cos t & -\sqrt{2}\sin t & \cos t \\ -\sin t & -\sqrt{2}\cos t & -\sin t \end{vmatrix}$$

$$= \sqrt{2}\,\mathbf{i} - \sqrt{2}\,\mathbf{k} \Rightarrow |\mathbf{v}\times\mathbf{a}| = \sqrt{4} = 2 \Rightarrow \kappa = \frac{|\mathbf{v}\times\mathbf{a}|}{|\mathbf{v}|^3} = \frac{2}{\left(\sqrt{2}\right)^3} = \frac{1}{\sqrt{2}};\ \dot{\mathbf{a}} = (-\cos t)\mathbf{i} + \left(\sqrt{2}\sin t\right)\mathbf{j} - (\cos t)\mathbf{k}$$

$$\Rightarrow \tau = \frac{\begin{vmatrix} \cos t & -\sqrt{2}\sin t & \cos t \\ -\sin t & -\sqrt{2}\cos t & -\sin t \\ -\cos t & \sqrt{2}\sin t & -\cos t \end{vmatrix}}{|\mathbf{v}\times\mathbf{a}|^2} = \frac{(\cos t)\left(\sqrt{2}\right) - \left(\sqrt{2}\sin t\right)(0) + (\cos t)\left(-\sqrt{2}\right)}{4} = 0$$

30. $\mathbf{r} = \mathbf{i} + (5\cos t)\mathbf{j} + (3\sin t)\mathbf{k} \Rightarrow \mathbf{v} = (-5\sin t)\mathbf{j} + (3\cos t)\mathbf{k} \Rightarrow \mathbf{a} = (-5\cos t)\mathbf{j} - (3\sin t)\mathbf{k}$

$\Rightarrow \mathbf{v}\cdot\mathbf{a} = 25\sin t\cos t - 9\sin t\cos t = 16\sin t\cos t$; $\mathbf{v}\cdot\mathbf{a} = 0 \Rightarrow 16\sin t\cos t = 0 \Rightarrow \sin t = 0$ or $\cos t = 0$

$\Rightarrow t = 0, \frac{\pi}{2}$ or π

31. $\mathbf{r} = 2\mathbf{i} + \left(4\sin\frac{t}{2}\right)\mathbf{j} + \left(3 - \frac{t}{\pi}\right)\mathbf{k} \Rightarrow 0 = \mathbf{r}\cdot(\mathbf{i}-\mathbf{j}) = 2(1) + \left(4\sin\frac{t}{2}\right)(-1) \Rightarrow 0 = 2 - 4\sin\frac{t}{2} \Rightarrow \sin\frac{t}{2} = \frac{1}{2} \Rightarrow \frac{t}{2} = \frac{\pi}{6}$

$\Rightarrow t = \frac{\pi}{3}$ (for the first time)

32. $\mathbf{r}(t) = t\mathbf{i} + t^2\mathbf{j} + t^3\mathbf{k} \Rightarrow \mathbf{v} = \mathbf{i} + 2t\mathbf{j} + 3t^2\mathbf{k} \Rightarrow |\mathbf{v}| = \sqrt{1+4t^2+9t^4} \Rightarrow |\mathbf{v}(1)| = \sqrt{14}$

$\Rightarrow \mathbf{T}(1) = \frac{1}{\sqrt{14}}\mathbf{i} + \frac{2}{\sqrt{14}}\mathbf{j} + \frac{3}{\sqrt{14}}\mathbf{k}$, which is normal to the normal plane

$\Rightarrow \frac{1}{\sqrt{14}}(x-1) + \frac{2}{\sqrt{14}}(y-1) + \frac{3}{\sqrt{14}}(z-1) = 0$ or $x + 2y + 3z = 6$ is an equation of the normal plane. Next we calculate $\mathbf{N}(1)$ which is normal to the rectifying plane. Now, $\mathbf{a} = 2\mathbf{j} + 6t\mathbf{k} \Rightarrow \mathbf{a}(1) = 2\mathbf{j} + 6\mathbf{k} \Rightarrow \mathbf{v}(1)\times\mathbf{a}(1)$

$$= \begin{vmatrix} \mathbf{i} & \mathbf{j} & \mathbf{k} \\ 1 & 2 & 3 \\ 0 & 2 & 6 \end{vmatrix} = 6\mathbf{i} - 6\mathbf{j} + 2\mathbf{k} \Rightarrow |\mathbf{v}(1)\times\mathbf{a}(1)| = \sqrt{76} \Rightarrow \kappa(1) = \frac{\sqrt{76}}{\left(\sqrt{14}\right)^3} = \frac{\sqrt{19}}{7\sqrt{14}};\ \frac{ds}{dt} = |\mathbf{v}(t)| \Rightarrow \left.\frac{d^2s}{dt^2}\right|_{t=1}$$

$$= \frac{1}{2}\left(1 + 4t^2 + 9t^4\right)^{-1/2}\left(8t + 36t^3\right)\Big|_{t=1} = \frac{22}{\sqrt{14}},\ \text{so } \mathbf{a} = \frac{d^2s}{dt^2}\mathbf{T} + \kappa\left(\frac{ds}{dt}\right)^2\mathbf{N} \Rightarrow 2\mathbf{j} + 6\mathbf{k}$$

$= \frac{22}{\sqrt{14}}\left(\frac{\mathbf{i}+2\mathbf{j}+3\mathbf{k}}{\sqrt{14}}\right) + \frac{\sqrt{19}}{7\sqrt{14}}\left(\sqrt{14}\right)^2 \mathbf{N} \Rightarrow \mathbf{N} = \frac{\sqrt{14}}{2\sqrt{19}}\left(-\frac{11}{7}\mathbf{i} - \frac{8}{7}\mathbf{j} + \frac{9}{7}\mathbf{k}\right) \Rightarrow -\frac{11}{7}(x-1) - \frac{8}{7}(y-1) + \frac{9}{7}(z-1)$ $= 0$ or $11x + 8y - 9z = 10$ is an equation of the rectifying plane. Finally, $\mathbf{B}(1) = \mathbf{T}(1) \times \mathbf{N}(1)$

$$= \left(\frac{\sqrt{14}}{2\sqrt{19}}\right)\left(\frac{1}{\sqrt{14}}\right)\left(\frac{1}{7}\right)\begin{vmatrix} \mathbf{i} & \mathbf{j} & \mathbf{k} \\ 1 & 2 & 3 \\ -11 & -8 & 9 \end{vmatrix} = \frac{1}{\sqrt{19}}(3\mathbf{i} - 3\mathbf{j} + \mathbf{k}) \Rightarrow 3(x-1) - 3(y-1) + (z-1) = 0 \text{ or } 3x - 3y + z$$

$= 1$ is an equation of the osculating plane.

33. $\mathbf{r} = e^t\mathbf{i} + (\sin t)\mathbf{j} + \ln(1-t)\mathbf{k} \Rightarrow \mathbf{v} = e^t\mathbf{i} + (\cos t)\mathbf{j} - \left(\frac{1}{1-t}\right)\mathbf{k} \Rightarrow \mathbf{v}(0) = \mathbf{i} + \mathbf{j} - \mathbf{k}$; $\mathbf{r}(0) = \mathbf{i} \Rightarrow (1,0,0)$ is on the line $\Rightarrow x = 1 + t$, $y = t$, and $z = -t$ are parametric equations of the line

34. $\mathbf{r} = \left(\sqrt{2}\cos t\right)\mathbf{i} + \left(\sqrt{2}\sin t\right)\mathbf{j} + t\mathbf{k} \Rightarrow \mathbf{v} = \left(-\sqrt{2}\sin t\right)\mathbf{i} + \left(\sqrt{2}\cos t\right)\mathbf{j} + \mathbf{k} \Rightarrow \mathbf{v}\left(\frac{\pi}{4}\right)$ $= \left(-\sqrt{2}\sin\frac{\pi}{4}\right)\mathbf{i} + \left(\sqrt{2}\cos\frac{\pi}{4}\right)\mathbf{j} + \mathbf{k} = -\mathbf{i} + \mathbf{j} + \mathbf{k}$ is a vector tangent to the helix when $t = \frac{\pi}{4} \Rightarrow$ the tangent line is parallel to $\mathbf{v}\left(\frac{\pi}{4}\right)$; also $\mathbf{r}\left(\frac{\pi}{4}\right) = \left(\sqrt{2}\cos\frac{\pi}{4}\right)\mathbf{i} + \left(\sqrt{2}\sin\frac{\pi}{4}\right)\mathbf{j} + \frac{\pi}{4}\mathbf{k} \Rightarrow$ the point $\left(1, 1, \frac{\pi}{4}\right)$ is on the line $\Rightarrow x = 1 - t$, $y = 1 + t$, and $z = \frac{\pi}{4} + t$ are parametric equations of the line

35. $\Delta SOT \approx \Delta TOD \Rightarrow \frac{DO}{OT} = \frac{OT}{SO} \Rightarrow \frac{y_0}{6380} = \frac{6380}{6380 + 437}$ $\Rightarrow y_0 = \frac{6380^2}{6817} \Rightarrow y_0 \approx 5971$ km;

$$VA = \int_{5971}^{6380} 2\pi x \sqrt{1 + \left(\frac{dx}{dy}\right)^2}\, dy$$

$$= 2\pi \int_{5971}^{6817} \sqrt{6380^2 - y^2}\left(\frac{6380}{\sqrt{6380^2 - y^2}}\right) dy$$

$$= 2\pi \int_{5971}^{6817} 6380\, dy = 2\pi\left[6380y\right]_{5971}^{6817}$$

$= 16{,}395{,}469$ km$^2 \approx 1.639 \times 10^7$ km^2;

percentage visible $\approx \frac{16{,}395{,}469 \text{ km}^2}{4\pi(6380 \text{ km})^2} \approx 3.21\%$

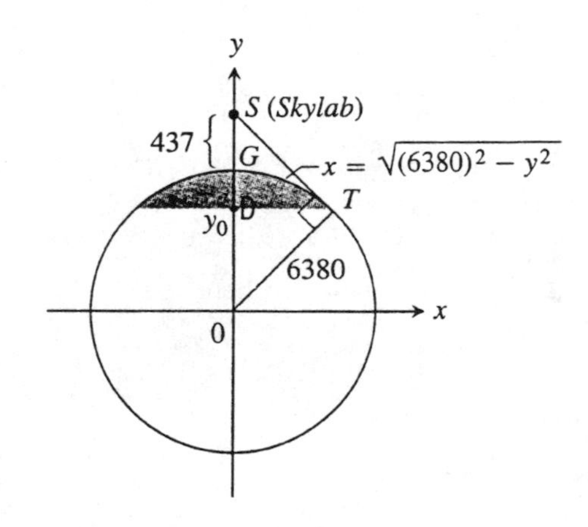

CHAPTER 11 ADDITIONAL EXERCISES–THEORY, EXAMPLES, APPLICATIONS

1. (a) The velocity of the boat at (x, y) relative to land is the sum of the velocity due to the rower and the velocity of the river, or $\mathbf{v} = \left[-\frac{1}{250}(y-50)^2 + 10\right]\mathbf{i} - 20\mathbf{j}$. Now, $\frac{dy}{dt} = -20 \Rightarrow y = -20t + c$; $y(0) = 100$ $\Rightarrow c = 100 \Rightarrow y = -20t + 100 \Rightarrow \mathbf{v} = \left[-\frac{1}{250}(-20t+50)^2 + 10\right]\mathbf{i} - 20\mathbf{j} = \left(-\frac{8}{5}t^2 + 8t\right)\mathbf{i} - 20\mathbf{j}$

$\Rightarrow \mathbf{r}(t) = \left(-\frac{8}{15}t^3 + 4t^2\right)\mathbf{i} - 20t\mathbf{j} + \mathbf{C}_1$; $\mathbf{r}(0) = 0\mathbf{i} + 100\mathbf{j} \Rightarrow 100\mathbf{j} = \mathbf{C}_1 \Rightarrow \mathbf{r}(t)$
$= \left(-\frac{8}{15}t^3 + 4t^2\right)\mathbf{i} + (100 - 20t)\mathbf{j}$

(b) The boat reaches the shore when $y = 0 \Rightarrow 0 = -20t + 100$ from part (a) $\Rightarrow t = 5$
$\Rightarrow \mathbf{r}(5) = \left(-\frac{8}{15} \cdot 125 + 4 \cdot 25\right)\mathbf{i} + (100 - 20 \cdot 5)\mathbf{j} = \left(-\frac{200}{3} + 100\right)\mathbf{i} = \frac{100}{3}\mathbf{i}$; the distance downstream is therefore $\frac{100}{3}$ m

2. (a) Let $a\mathbf{i} + b\mathbf{j}$ be the velocity of the boat. The velocity of the boat relative to an observer on the bank of the river is $\mathbf{v} = a\mathbf{i} + \left[b - \frac{3x(20-x)}{100}\right]\mathbf{j}$. The distance x of the boat as it crosses the river is related to time by $x = at \Rightarrow \mathbf{v} = a\mathbf{i} + \left[b - \frac{3at(20-at)}{100}\right]\mathbf{j} = a\mathbf{i} + \left(b + \frac{3a^2t^2 - 60at}{100}\right)\mathbf{j} \Rightarrow \mathbf{r}(t) = at\mathbf{i} + \left(bt + \frac{a^2t^3}{100} - \frac{30at^2}{100}\right)\mathbf{j} + \mathbf{C}$;
$\mathbf{r}(0) = 0\mathbf{i} + 0\mathbf{j} \Rightarrow \mathbf{C} = 0 \Rightarrow \mathbf{r}(t) = at\mathbf{i} + \left(bt + \frac{a^2t^3 - 30at^2}{100}\right)\mathbf{j}$. The boat reaches the shore when $x = 20$
$\Rightarrow 20 = at \Rightarrow t = \frac{20}{a}$ and $y = 0 \Rightarrow 0 = b\left(\frac{20}{a}\right) + \frac{a^2\left(\frac{20}{a}\right)^3 - 30a\left(\frac{20}{a}\right)^2}{100} = \frac{20b}{a} + \frac{(20)^3 - 30(20)^2}{100a}$
$= \frac{2000b + 8000 - 12{,}000}{100a} \Rightarrow b = 2$; the speed of the boat is $\sqrt{20} = |\mathbf{v}| = \sqrt{a^2 + b^2} = \sqrt{a^2 + 4} \Rightarrow a^2 = 16$
$\Rightarrow a = 4$; thus, $\mathbf{v} = 4\mathbf{i} + 2\mathbf{j}$ is the velocity of the boat

(b) $\mathbf{r}(t) = at\mathbf{i} + \left(bt + \frac{a^2t^3 - 30at^2}{100}\right)\mathbf{j} = 4t\mathbf{i} + \left(2t + \frac{16t^3}{100} - \frac{120t^2}{100}\right)\mathbf{j}$ by part (a), where $0 \le t \le 5$

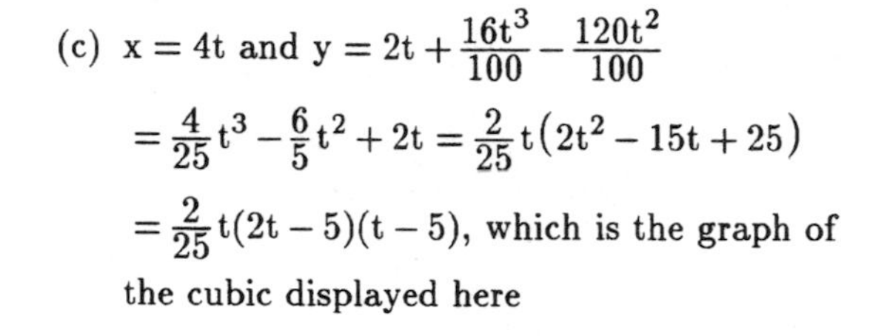

(c) $x = 4t$ and $y = 2t + \frac{16t^3}{100} - \frac{120t^2}{100}$
$= \frac{4}{25}t^3 - \frac{6}{5}t^2 + 2t = \frac{2}{25}t(2t^2 - 15t + 25)$
$= \frac{2}{25}t(2t-5)(t-5)$, which is the graph of the cubic displayed here

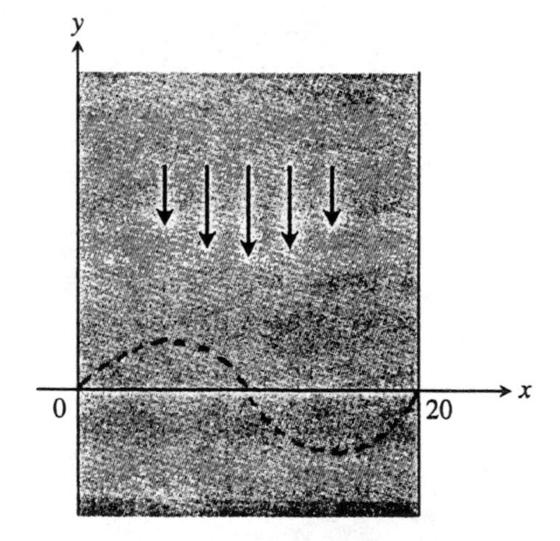

3. Let $\mathbf{a} = \mathbf{i} + \mathbf{j} + \mathbf{k}$ be the vector from O to A and $\mathbf{b} = \mathbf{i} + 3\mathbf{j} + 2\mathbf{k}$ be the vector from O to B. The vector $\mathbf{v}$ orthogonal to $\mathbf{a}$ and $\mathbf{b} \Rightarrow \mathbf{v}$ is parallel to $\mathbf{b} \times \mathbf{a}$ (since the rotation is clockwise). Now $\mathbf{b} \times \mathbf{a} = \mathbf{i} + \mathbf{j} - 2\mathbf{k}$; $\text{proj}_{\mathbf{a}}\, \mathbf{b} = \left(\frac{\mathbf{a} \cdot \mathbf{b}}{\mathbf{a} \cdot \mathbf{a}}\right)\mathbf{a} = 2\mathbf{i} + 2\mathbf{j} + 2\mathbf{k} \Rightarrow (2, 2, 2)$ is the center of the circular path $(1, 3, 2)$ takes $\Rightarrow$ radius $= \sqrt{1^2 + (-1)^2 + 0^2} = \sqrt{2} \Rightarrow$ arc length per second covered by the point is $\frac{3}{2}\sqrt{2}$ units/sec $= |\mathbf{v}|$ (velocity is

constant). A unit vector in the direction of $\mathbf{v}$ is $\frac{\mathbf{b}\times\mathbf{a}}{|\mathbf{b}\times\mathbf{a}|} = \frac{1}{\sqrt{6}}\mathbf{i} + \frac{1}{\sqrt{6}}\mathbf{j} - \frac{2}{\sqrt{6}}\mathbf{k} \Rightarrow \mathbf{v} = |\mathbf{v}|\left(\frac{\mathbf{b}\times\mathbf{a}}{|\mathbf{b}\times\mathbf{a}|}\right)$
$= \frac{3}{2}\sqrt{2}\left(\frac{1}{\sqrt{6}}\mathbf{i} + \frac{1}{\sqrt{6}}\mathbf{j} - \frac{2}{\sqrt{6}}\mathbf{k}\right) = \frac{\sqrt{3}}{2}\mathbf{i} + \frac{\sqrt{3}}{2}\mathbf{j} - \sqrt{3}\mathbf{k}$

4. $\mathbf{r} = (2\sqrt{t}\cos t)\mathbf{i} + (3\sqrt{t}\sin t)\mathbf{j} + \sqrt{1-t}\mathbf{k}$, $0 \le t \le 1 \Rightarrow x = 2\sqrt{t}\cos t$, $y = 3\sqrt{t}\sin t$, and $z = \sqrt{1-t}$
$\Rightarrow x^2 = 4t\cos^2 t$ and $y^2 = 9t\sin^2 t \Rightarrow \frac{x^2}{4} + \frac{y^2}{9} = t\cos^2 t + t\sin^2 t = t$; $z^2 = 1 - t \Rightarrow t = 1 - z^2 \Rightarrow \frac{x^2}{4} + \frac{y^2}{9}$
$= 1 - z^2 \Rightarrow \frac{x^2}{4} + \frac{y^2}{9} + z^2 = 1$, an ellipsoid

5. (a) $\mathbf{r}(\theta) = (a\cos\theta)\mathbf{i} + (a\sin\theta)\mathbf{j} + b\theta\mathbf{k} \Rightarrow \frac{d\mathbf{r}}{dt} = [(-a\sin\theta)\mathbf{i} + (a\cos\theta)\mathbf{j} + b\mathbf{k}]\frac{d\theta}{dt}$; $|\mathbf{v}| = \sqrt{2gz} = \left|\frac{d\mathbf{r}}{dt}\right|$
$= \sqrt{a^2+b^2}\,\frac{d\theta}{dt} \Rightarrow \frac{d\theta}{dt} = \sqrt{\frac{2gz}{a^2+b^2}} = \sqrt{\frac{2gb\theta}{a^2+b^2}} \Rightarrow \left.\frac{d\theta}{dt}\right|_{\theta=2\pi} = \sqrt{\frac{4\pi gb}{a^2+b^2}} = 2\sqrt{\frac{\pi gb}{a^2+b^2}}$

(b) $\frac{d\theta}{dt} = \sqrt{\frac{2gb\theta}{a^2+b^2}} \Rightarrow \frac{d\theta}{\sqrt{\theta}} = \sqrt{\frac{2gb}{a^2+b^2}}\,dt \Rightarrow 2\theta^{1/2} = \sqrt{\frac{2gb}{a^2+b^2}}\,t + C$; $t = 0 \Rightarrow \theta = 0 \Rightarrow C = 0$
$\Rightarrow 2\theta^{1/2} = \sqrt{\frac{2gb}{a^2+b^2}}\,t \Rightarrow \theta = \frac{gbt^2}{2(a^2+b^2)}$; $z = b\theta \Rightarrow z = \frac{gb^2t^2}{2(a^2+b^2)}$

(c) $\mathbf{v}(t) = \frac{d\mathbf{r}}{dt} = [(-a\sin\theta)\mathbf{i} + (a\cos\theta)\mathbf{j} + b\mathbf{k}]\frac{d\theta}{dt} = [(-a\sin\theta)\mathbf{i} + (a\cos\theta)\mathbf{j} + b\mathbf{k}]\left(\frac{gbt}{a^2+b^2}\right)$, from part (b)
$\Rightarrow \mathbf{v}(t) = \left[\frac{(-a\sin\theta)\mathbf{i} + (a\cos\theta)\mathbf{j} + b\mathbf{k}}{\sqrt{a^2+b^2}}\right]\left(\frac{gbt}{\sqrt{a^2+b^2}}\right) = \frac{gbt}{\sqrt{a^2+b^2}}\mathbf{T}$;
$\frac{d^2\mathbf{r}}{dt^2} = [(-a\cos\theta)\mathbf{i} - (a\sin\theta)\mathbf{j}]\left(\frac{d\theta}{dt}\right)^2 + [(-a\sin\theta)\mathbf{i} + (a\cos\theta)\mathbf{j} + b\mathbf{k}]\frac{d^2\theta}{dt^2}$
$= \left(\frac{gbt}{a^2+b^2}\right)^2[(-a\cos\theta)\mathbf{i} - (a\sin\theta)\mathbf{j}] + [(-a\sin\theta)\mathbf{i} + (a\cos\theta)\mathbf{j} + b\mathbf{k}]\left(\frac{gb}{a^2+b^2}\right)$
$= \left[\frac{(-a\sin\theta)\mathbf{i} + (a\cos\theta)\mathbf{j} + b\mathbf{k}}{\sqrt{a^2+b^2}}\right]\left(\frac{gb}{\sqrt{a^2+b^2}}\right) + a\left(\frac{gbt}{a^2+b^2}\right)^2[(-\cos\theta)\mathbf{i} - (\sin\theta)\mathbf{j}]$
$= \frac{gb}{\sqrt{a^2+b^2}}\mathbf{T} + a\left(\frac{gbt}{a^2+b^2}\right)^2\mathbf{N}$ (there is no component in the direction of $\mathbf{B}$).

6. (a) $\mathbf{r}(\theta) = (a\theta\cos\theta)\mathbf{i} + (a\theta\sin\theta)\mathbf{j} + b\theta\mathbf{k} \Rightarrow \frac{d\mathbf{r}}{dt} = [(a\cos\theta - a\theta\sin\theta)\mathbf{i} + (a\sin\theta + a\theta\cos\theta)\mathbf{j} + b\mathbf{k}]\frac{d\theta}{dt}$;
$|\mathbf{v}| = \sqrt{2gz} = \left|\frac{d\mathbf{r}}{dt}\right| = \left(a^2 + a^2\theta^2 + b^2\right)^{1/2}\left(\frac{d\theta}{dt}\right) \Rightarrow \frac{d\theta}{dt} = \frac{\sqrt{2gb\theta}}{\sqrt{a^2 + a^2\theta^2 + b^2}}$

(b) $s = \int_0^t |\mathbf{v}|\,dt = \int_0^t (a^2 + a^2\theta^2 + b^2)\frac{d\theta}{dt}\,dt = \int_0^t (a^2 + a^2\theta^2 + b^2)\,d\theta = \int_0^\theta \left(a^2 + a^2u^2 + b^2\right)^{1/2}du$
$= \int_0^\theta a\sqrt{\frac{a^2+b^2}{a^2} + u^2}\,du = a\int_0^\theta \sqrt{c^2 + u^2}\,du$, where $c = \frac{a^2+b^2}{a^2}$

$$\Rightarrow s = a\left[\frac{u}{2}\sqrt{c^2+u^2} + \frac{c^2}{2}\ln\left|u+\sqrt{c^2+u^2}\right|\right]_0^\theta = \frac{a}{2}\left(\theta\sqrt{c^2+\theta^2} + c^2\ln\left|\theta+\sqrt{c^2+\theta^2}\right| - c^2\ln c\right)$$

7. $r = \frac{(1+e)r_0}{1+e\cos\theta} \Rightarrow \frac{dr}{d\theta} = \frac{(1+e)r_0(e\sin\theta)}{(1+e\cos\theta)^2}$; $\frac{dr}{d\theta} = 0 \Rightarrow \frac{(1+e)r_0(e\sin\theta)}{(1+e\cos\theta)^2} = 0 \Rightarrow (1+e)r_0(e\sin\theta) = 0$

$\Rightarrow \sin\theta = 0 \Rightarrow \theta = 0$ or π. Note that $\frac{dr}{d\theta} > 0$ when $\sin\theta > 0$ and $\frac{dr}{d\theta} < 0$ when $\sin\theta < 0$. Since $\sin\theta < 0$ on $-\pi < \theta < 0$ and $\sin\theta > 0$ on $0 < \theta < \pi$, r is a minimum when $\theta = 0$ and $r(0) = \frac{(1+e)r_0}{1+e\cos 0} = r_0$

8. (a) $f(x) = x - 1 - \frac{1}{2}\sin x = 0 \Rightarrow f(0) = -1$ and $f(2) = 2 - 1 - \frac{1}{2}\sin 2 \ge \frac{1}{2}$ since $|\sin 2| \le 1$; since f is continuous on $[0,2]$, the Intermediate Value Theorem implies there is a root between 0 and 2

(b) Root ≈ 1.4987011335179

9. (a) $\mathbf{v} = \frac{dx}{dt}\mathbf{i} + \frac{dy}{dt}\mathbf{j}$ and $\mathbf{v} = \frac{dr}{dt}\mathbf{u}_r + r\frac{d\theta}{dt}\mathbf{u}_\theta = \left(\frac{dr}{dt}\right)[(\cos\theta)\mathbf{i} + (\sin\theta)\mathbf{j}] + \left(r\frac{d\theta}{dt}\right)[(-\sin\theta)\mathbf{i} + (\cos\theta)\mathbf{j}] \Rightarrow \mathbf{v}\cdot\mathbf{i}\ \frac{dx}{dt}$ and $\mathbf{v}\cdot\mathbf{i} = \frac{dr}{dt}\cos\theta - r\frac{d\theta}{dt}\sin\theta \Rightarrow \frac{dx}{dt} = \frac{dr}{dt}\cos\theta - r\frac{d\theta}{dt}\sin\theta$; $\mathbf{v}\cdot\mathbf{j} = \frac{dy}{dt}$ and $\mathbf{v}\cdot\mathbf{j} = \frac{dr}{dt}\sin\theta + r\frac{d\theta}{dt}\cos\theta$

$\Rightarrow \frac{dy}{dt} = \frac{dr}{dt}\sin\theta + r\frac{d\theta}{dt}\cos\theta$

(b) $\mathbf{u}_r = (\cos\theta)\mathbf{i} + (\sin\theta)\mathbf{j} \Rightarrow \mathbf{v}\cdot\mathbf{u}_r = \frac{dx}{dt}\cos\theta + \frac{dy}{dt}\sin\theta$

$= \left(\frac{dr}{dt}\cos\theta - r\frac{d\theta}{dt}\sin\theta\right)(\cos\theta) + \left(\frac{dr}{dt}\sin\theta + r\frac{d\theta}{dt}\cos\theta\right)(\sin\theta)$ by part (a),

$\Rightarrow \mathbf{v}\cdot\mathbf{u}_r = \frac{dr}{dt}$; therefore, $\frac{dr}{dt} = \frac{dx}{dt}\cos\theta + \frac{dy}{dt}\sin\theta$;

$\mathbf{u}_\theta = -(\sin\theta)\mathbf{i} + (\cos\theta)\mathbf{j} \Rightarrow \mathbf{v}\cdot\mathbf{u}_\theta = -\frac{dx}{dt}\sin\theta + \frac{dy}{dt}\cos\theta$

$= \left(\frac{dr}{dt}\cos\theta - r\frac{d\theta}{dt}\sin\theta\right)(-\sin\theta) + \left(\frac{dr}{dt}\sin\theta + r\frac{d\theta}{dt}\cos\theta\right)(\cos\theta)$ by part (a) $\Rightarrow \mathbf{v}\cdot\mathbf{u}_\theta = r\frac{d\theta}{dt}$;

therefore, $r\frac{d\theta}{dt} = -\frac{dx}{dt}\sin\theta + \frac{dy}{dt}\cos\theta$

10. $\mathbf{r} = f(\theta) \Rightarrow \frac{dr}{dt} = f'(\theta)\frac{d\theta}{dt} \Rightarrow \frac{d^2r}{dt^2} = f''(\theta)\left(\frac{d\theta}{dt}\right)^2 + f'(\theta)\frac{d^2\theta}{dt^2}$; $\mathbf{v} = \frac{dr}{dt}\mathbf{u}_r + r\frac{d\theta}{dt}\mathbf{u}_\theta$

$= \left(\cos\theta\frac{dr}{dt} - r\sin\theta\frac{d\theta}{dt}\right)\mathbf{i} + \left(\sin\theta\frac{dr}{dt} + r\cos\theta\frac{d\theta}{dt}\right)\mathbf{j} \Rightarrow |\mathbf{v}| = \left[\left(\frac{dr}{dt}\right)^2 + r^2\left(\frac{d\theta}{dt}\right)^2\right]^{1/2} = \left[(f')^2 + f^2\right]^{1/2}\left(\frac{d\theta}{dt}\right)$;

$|\mathbf{v}\times\mathbf{a}| = |\dot{x}\ddot{y} - \dot{y}\ddot{x}|$, where $x = r\cos\theta$ and $y = r\sin\theta$. Then $\frac{dx}{dt} = (-r\sin\theta)\frac{d\theta}{dt} + (\cos\theta)\frac{dr}{dt}$

$\Rightarrow \frac{d^2x}{dt^2} = (-2\sin\theta)\frac{d\theta}{dt}\frac{dr}{dt} - (r\cos\theta)\left(\frac{d\theta}{dt}\right)^2 - (r\sin\theta)\frac{d^2\theta}{dt^2} + (\cos\theta)\frac{d^2r}{dt^2}$; $\frac{dy}{dt} = (r\cos\theta)\frac{d\theta}{dt} + (\sin\theta)\frac{dr}{dt}$

$\Rightarrow \frac{d^2y}{dt^2} = (2\cos\theta)\frac{d\theta}{dt}\frac{dr}{dt} - (r\sin\theta)\left(\frac{d\theta}{dt}\right)^2 + (r\cos\theta)\frac{d^2\theta}{dt^2} + (\sin\theta)\frac{d^2r}{dt^2}$. Then $|\mathbf{v}\times\mathbf{a}|$

$= $ (after much algebra) $r^2\left(\frac{d\theta}{dt}\right)^3 + r\frac{d^2\theta}{dt^2}\frac{dr}{dt} - r\frac{d\theta}{dt}\left(\frac{dr}{dt}\right)^2 \Rightarrow \kappa = \frac{r^2\left(\frac{d\theta}{dt}\right)^3 + r\left(\frac{d^2\theta}{dt^2}\right)\left(\frac{dr}{dt}\right) - r\left(\frac{d\theta}{dt}\right)\left(\frac{dr}{dt}\right)^2}{\left[(f')^2 + f^2\right]^{3/2}}$

$$= \frac{r^2 + r\left(\frac{d^2\theta}{dt^2}\right)(f')\left(\frac{dt}{d\theta}\right)^2 - rf'' - rf'\left(\frac{d^2\theta}{dt^2}\right)\left(\frac{dt}{d\theta}\right)^2 + 2(f')^2}{\left[(f')^2 + f^2\right]^{3/2}} = \frac{f^2 - ff'' + 2(f')^2}{\left[(f')^2 + f^2\right]^{3/2}}$$

11. (a) Let $r = 2 - t$ and $\theta = 3t \Rightarrow \frac{dr}{dt} = -1$ and $\frac{d\theta}{dt} = 3 \Rightarrow \frac{d^2r}{dt^2} = \frac{d^2\theta}{dt^2} = 0$. The halfway point is $(1,3) \Rightarrow t = 1$;

$\mathbf{v} = \frac{dr}{dt}\mathbf{u}_r + r\frac{d\theta}{dt}\mathbf{u}_\theta \Rightarrow \mathbf{v}(1) = -\mathbf{u}_r + 3\mathbf{u}_\theta$; $\mathbf{a} = \left[\frac{d^2r}{dt^2} - r\left(\frac{d\theta}{dt}\right)^2\right]\mathbf{u}_r + \left[r\frac{d^2\theta}{dt^2} + 2\frac{dr}{dt}\frac{d\theta}{dt}\right]\mathbf{u}_\theta \Rightarrow \mathbf{a}(1) = -9\mathbf{u}_r - 6\mathbf{u}_\theta$

(b) It takes the beetle 2 min to crawl to the origin $\Rightarrow$ the rod has revolved 6 radians

$$\Rightarrow L = \int_0^6 \sqrt{[f(\theta)]^2 + [f'(\theta)]^2}\, d\theta = \int_0^6 \sqrt{\left(2 - \frac{\theta}{3}\right)^2 + \left(-\frac{1}{3}\right)^2}\, d\theta = \int_0^6 \sqrt{4 - \frac{4\theta}{3} + \frac{\theta^2}{9} + \frac{1}{9}}\, d\theta$$

$$= \int_0^6 \sqrt{\frac{37 - 12\theta + \theta^2}{9}}\, d\theta = \frac{1}{3}\int_0^6 \sqrt{(\theta - 6)^2 + 1}\, d\theta = \frac{1}{3}\left[\frac{(\theta - 6)}{2}\sqrt{(\theta - 6)^2 + 1} + \frac{1}{2}\ln\left|\theta - 6 + \sqrt{(\theta - 6)^2 + 1}\right|\right]_0^6$$

$$= \sqrt{37} - \frac{1}{6}\ln\left(\sqrt{37} - 6\right) \approx 6.5 \text{ in.}$$

12. $\mathbf{L}(t) = \mathbf{r}(t) + m\mathbf{v}(t) \Rightarrow \frac{d\mathbf{L}}{dt} = \left(\frac{d\mathbf{r}}{dt} \times m\mathbf{v}\right) + \left(\mathbf{r} + m\frac{d^2\mathbf{r}}{dt^2}\right) \Rightarrow \frac{d\mathbf{L}}{dt} = (\mathbf{v} \times m\mathbf{v}) + (\mathbf{r} \times m\mathbf{a}) = \mathbf{r} \times m\mathbf{a}$; $\mathbf{F} = m\mathbf{a} \Rightarrow -\frac{c}{|\mathbf{r}|^3}\mathbf{r}$

$= m\mathbf{a} \Rightarrow \frac{d\mathbf{L}}{dt} = \mathbf{r} \times m\mathbf{a} = \mathbf{r} \times \left(-\frac{c}{|\mathbf{r}|^3}\mathbf{r}\right) = -\frac{c}{|\mathbf{r}|^3}(\mathbf{r} \times \mathbf{r}) = \mathbf{0} \Rightarrow \mathbf{L} =$ constant vector

13. (a) $\mathbf{u}_r \times \mathbf{u}_\theta = \begin{vmatrix} \mathbf{i} & \mathbf{j} & \mathbf{k} \\ \cos\theta & \sin\theta & 0 \\ -\sin\theta & \cos\theta & 0 \end{vmatrix} = \mathbf{k} \Rightarrow$ a right-handed frame of unit vectors

(b) $\frac{d\mathbf{u}_r}{d\theta} = (-\sin\theta)\mathbf{i} + (\cos\theta)\mathbf{j} = \mathbf{u}_\theta$ and $\frac{d\mathbf{u}_\theta}{d\theta} = (-\cos\theta)\mathbf{i} - (\sin\theta)\mathbf{j} = -\mathbf{u}_r$

(c) From Eq. (7), $\mathbf{v} = \dot{r}\mathbf{u}_r + r\dot{\theta}\mathbf{u}_\theta + \dot{z}\mathbf{k} \Rightarrow \mathbf{a} = \dot{\mathbf{v}} = \left(\ddot{r}\mathbf{u}_r + \dot{r}\,\dot{\mathbf{u}}_r\right) + \left(\dot{r}\,\dot{\theta}\mathbf{u}_\theta + r\ddot{\theta}\mathbf{u}_\theta + r\dot{\theta}\,\dot{\mathbf{u}}_\theta\right) + \ddot{z}\,\mathbf{k}$

$= \left(\ddot{r} - r\dot{\theta}^2\right)\mathbf{u}_r + \left(r\ddot{\theta} + 2\dot{r}\,\dot{\theta}\right)\mathbf{u}_\theta + \ddot{z}\,\mathbf{k}$

14. (a) $x = r\cos\theta \Rightarrow dx = \cos\theta\, dr - r\sin\theta\, d\theta$; $y = r\sin\theta \Rightarrow dy = \sin\theta\, dr + r\cos\theta\, d\theta$; thus

$dx^2 = \cos^2\theta\, dr^2 - 2r\sin\theta\cos\theta\, dr\, d\theta + r^2\sin^2\theta\, d\theta^2$ and

$dy^2 = \sin^2\theta\, dr^2 + 2r\sin\theta\cos\theta\, dr\, d\theta + r^2\cos^2\theta\, d\theta^2 \Rightarrow dx^2 + dy^2 + dz^2 = dr^2 + r^2\, d\theta^2 + dz^2$

(c) $r = e^\theta \Rightarrow dr = e^\theta\, d\theta$

$$\Rightarrow L = \int_0^{\ln 8} \sqrt{dr^2 + r^2\, d\theta^2 + dz^2}$$

$$= \int_0^{\ln 8} \sqrt{e^{2\theta} + e^{2\theta} + e^{2\theta}}\, d\theta$$

$$= \int_0^{\ln 8} \sqrt{3}e^\theta\, d\theta = \left[\sqrt{3}\,e^\theta\right]_0^{\ln 8}$$

$$= 8\sqrt{3} - \sqrt{3} = 7\sqrt{3}$$

(b)

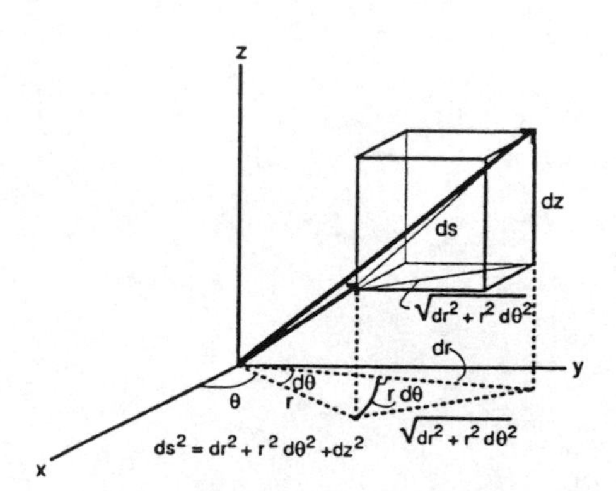

15. (a) $\mathbf{u}_\rho = (\sin\phi\cos\theta)\mathbf{i} + (\sin\phi\sin\theta)\mathbf{j} + (\cos\phi)\mathbf{k}$, $\mathbf{u}_\phi = (\cos\phi\cos\theta)\mathbf{i} + (\cos\phi\sin\theta)\mathbf{j} - (\sin\phi)\mathbf{k}$, and $\mathbf{u}_\theta = \mathbf{u}_\rho \times \mathbf{u}_\phi = (-\sin\theta)\mathbf{i} + (\cos\theta)\mathbf{j}$ (we choose $\rho = 1$ for unit vectors)

(b) $\mathbf{u}_\rho \cdot \mathbf{u}_\phi = (\sin\phi\cos\theta)(\cos\phi\cos\theta) + (\sin\phi\sin\theta)(\cos\phi\sin\theta) + (\cos\phi)(-\sin\phi)$
$= (\sin\phi\cos\phi)(\cos^2\theta) + (\sin\phi\cos\phi)(\sin^2\theta) - \cos\phi\sin\phi = (\sin\phi\cos\phi) - (\cos\phi\sin\phi) = 0$

(c) $\mathbf{u}_\rho \times \mathbf{u}_\phi = \begin{vmatrix} \mathbf{i} & \mathbf{j} & \mathbf{k} \\ \sin\phi\cos\theta & \sin\phi\sin\theta & \cos\phi \\ \cos\phi\cos\theta & \cos\phi\sin\theta & -\sin\phi \end{vmatrix} = (-\sin\theta)\mathbf{i} + (\cos\theta)\mathbf{j} = \mathbf{u}_\theta$

(d) $\mathbf{u}_\rho \times \mathbf{u}_\phi = \mathbf{u}_\theta \Rightarrow$ right-handed frame

16. (a) $x = \rho\sin\phi\cos\theta \Rightarrow dx = \sin\phi\cos\theta\, d\rho + \rho\cos\phi\cos\theta\, d\phi - \rho\sin\phi\sin\theta\, d\theta$;
$y = \rho\sin\phi\sin\theta \Rightarrow dy = \sin\phi\sin\theta\, d\rho + \rho\cos\phi\sin\theta\, d\phi + \rho\sin\phi\cos\theta\, d\theta$;
$z = \rho\cos\phi \Rightarrow dz = \cos\phi\, d\rho - \rho\sin\phi\, d\phi$; thus $dx^2 + dy^2 + dz^2 = d\rho^2 + \rho^2\, d\phi^2 + \rho^2\sin^2\phi\, d\theta^2$
(The above concluding statement involves much algebra and several trig identities.)

(c) $\rho = 2e^\theta \Rightarrow d\rho = 2e^\theta\, d\theta$; $d\phi = 0$; thus
$ds^2 = d\rho^2 + \rho^2\, d\phi^2 + \rho^2\sin^2\phi\, d\theta^2$
$= (4e^{2\theta})d\theta^2 + (4e^{2\theta})(0)^2 + (4e^{2\theta})\left(\frac{1}{2}\right)^2 d\theta^2$
$= 5e^{2\theta}\, d\theta^2$;
$L = \int_0^{\ln 8} \left(\frac{ds}{d\theta}\right) d\theta = \int_0^{\ln 8} \sqrt{5e^{2\theta}}\, d\theta$
$= \int_0^{\ln 8} \sqrt{5}\, e^\theta\, d\theta = \left[\sqrt{5}\, e^\theta\right]_0^{\ln 8}$
$= 8\sqrt{5} - \sqrt{5} = 7\sqrt{5}$

(b)

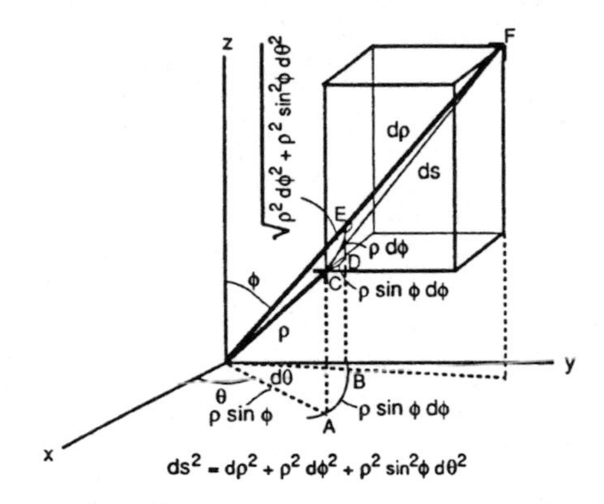

CHAPTER 12 MULTIVARIABLE FUNCTIONS AND PARTIAL DERIVATIVES

12.1 FUNCTIONS OF SEVERAL VARIABLES

1. (a) Domain: all points in the xy-plane
   (b) Range: all real numbers
   (c) level curves are straight lines $y - x = c$ parallel to the line $y = x$
   (d) no boundary points
   (e) both open and closed
   (f) unbounded

2. (a) Domain: set of all (x, y) so that $y - x \geq 0 \Rightarrow y \geq x$
   (b) Range: $z \geq 0$
   (c) level curves are straight lines of the form $y - x = c$ where $c \geq 0$
   (d) boundary is $\sqrt{y - x} = 0 \Rightarrow y = x$, a straight line
   (e) closed
   (f) unbounded

3. (a) Domain: all points in the xy-plane
   (b) Range: $z \geq 0$
   (c) level curves: for $f(x, y) = 0$, the origin; for $f(x, y) = c > 0$, ellipses with center $(0, 0)$ and major and minor axes along the x- and y-axes, respectively
   (d) no boundary points
   (e) both open and closed
   (f) unbounded

4. (a) Domain: all points in the xy-plane
   (b) Range: all real numbers
   (c) level curves: for $f(x, y) = 0$, the union of the lines $y = \pm x$; for $f(x, y) = c \neq 0$, hyperbolas centered at $(0, 0)$ with foci on the x-axis if $c > 0$ and on the y-axis if $c < 0$
   (d) no boundary points
   (e) both open and closed
   (f) unbounded

5. (a) Domain: all points in the xy-plane
   (b) Range: all real numbers
   (c) level curves are hyperbolas with the x- and y-axes as asymptotes when $f(x, y) \neq 0$, and the x- and y-axes when $f(x, y) = 0$
   (d) no boundary points
   (e) both open and closed
   (f) unbounded

6. (a) Domain: all $(x, y) \neq (0, y)$
   (b) Range: all real numbers
   (c) level curves: for $f(x, y) = 0$, the x-axis minus the origin; for $f(x, y) = c \neq 0$, the parabolas $y = cx^2$ minus the origin
   (d) boundary is the line $x = 0$

(e) open
(f) unbounded

7. (a) Domain: all (x, y) satisfying $x^2 + y^2 < 16$
   (b) Range: $z \geq \frac{1}{4}$
   (c) level curves are circles centered at the origin with radii $r < 4$
   (d) boundary is the circle $x^2 + y^2 = 16$
   (e) open
   (f) bounded

8. (a) Domain: all (x, y) satisfying $x^2 + y^2 \leq 9$
   (b) Range: $0 \leq z \leq 3$
   (c) level curves are circles centered at the origin with radii $r \leq 3$
   (d) boundary is the circle $x^2 + y^2 = 9$
   (e) closed
   (f) bounded

9. (a) Domain: $(x, y) \neq (0, 0)$
   (b) Range: all real numbers
   (c) level curves are circles with center $(0, 0)$ and radii $r > 0$
   (d) boundary is the single point $(0, 0)$
   (e) open
   (f) unbounded

10. (a) Domain: all points in the xy-plane
    (b) Range: $0 < z \leq 1$
    (c) level curves are the origin itself and the circles with center $(0, 0)$ and radii $r > 0$
    (d) no boundary points
    (e) both open and closed
    (f) unbounded

11. (a) Domain: all (x, y) satisfying $-1 \leq y - x \leq 1$
    (b) Range: $-\frac{\pi}{2} \leq z \leq \frac{\pi}{2}$
    (c) level curves are straight lines of the form $y - x = c$ where $-1 \leq c \leq 1$
    (d) boundary is the two straight lines $y = 1 + x$ and $y = -1 + x$
    (e) closed
    (f) unbounded

12. (a) Domain: all (x, y), $x \neq 0$
    (b) Range: $-\frac{\pi}{2} < z < \frac{\pi}{2}$
    (c) level curves are the straight lines of the form $y = cx$, c any real number and $x \neq 0$
    (d) boundary is the line $x = 0$
    (e) open
    (f) unbounded

13. f

14. e

15. a

16. c

17. d

18. b

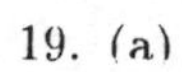
19. (a)

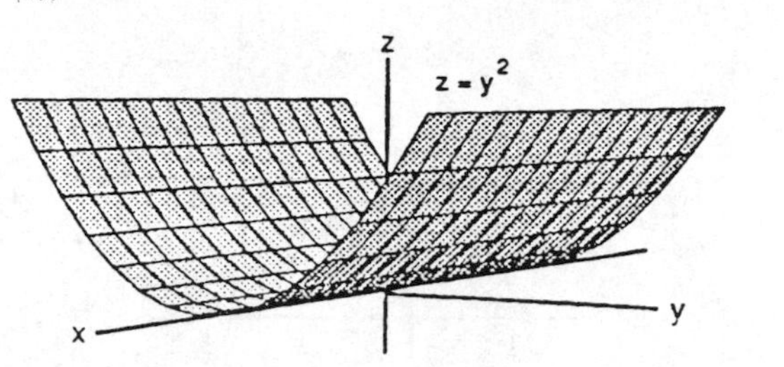

(b)

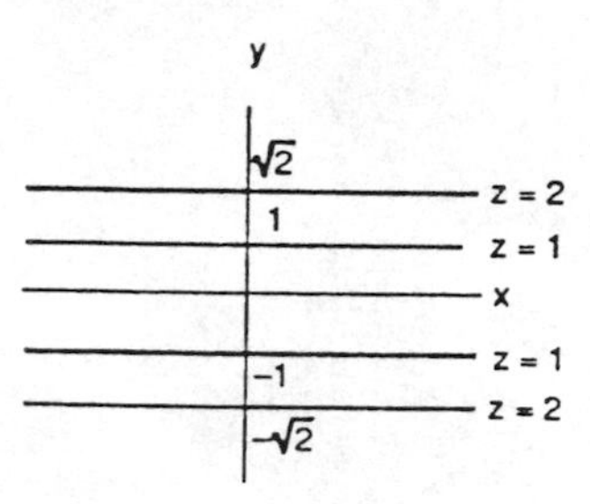

20. (a)

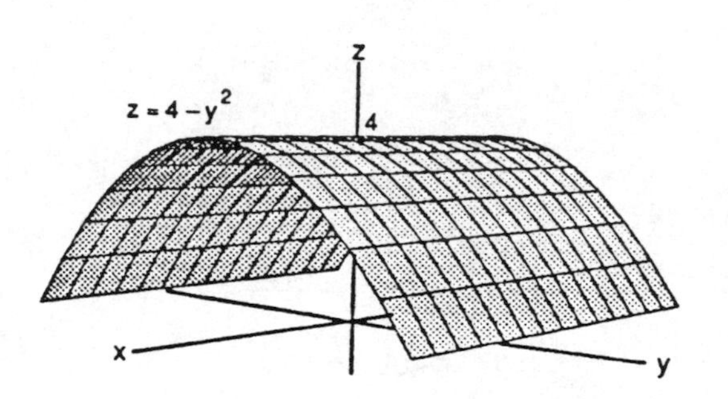

(b)

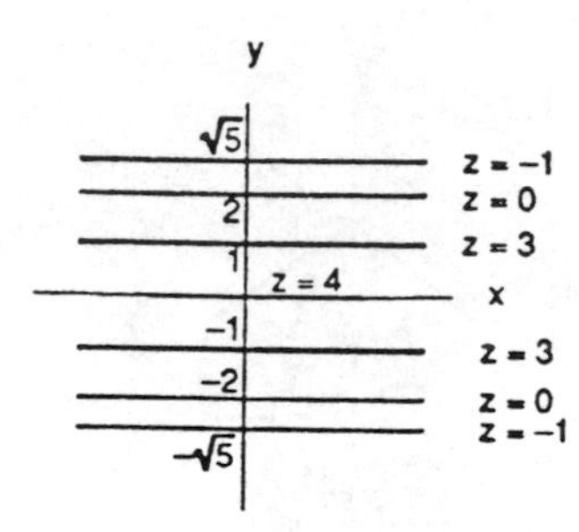

21. (a)

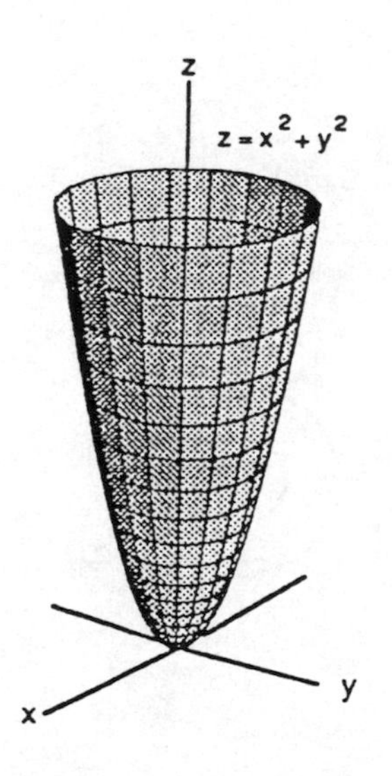

(b)

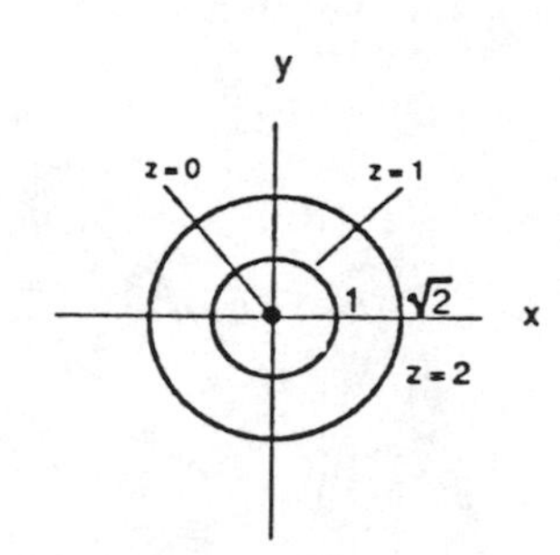

22. (a) (b)

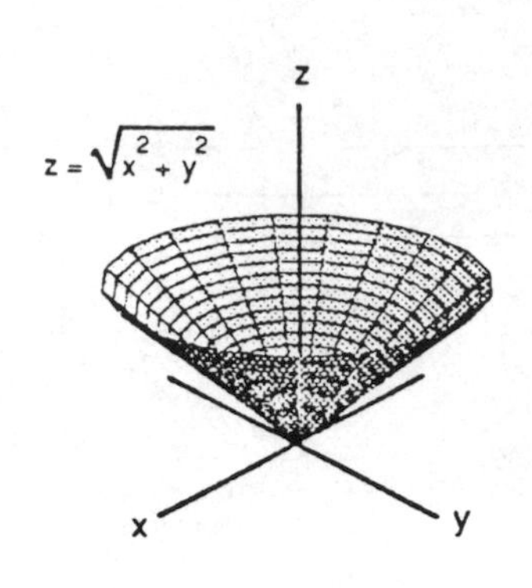

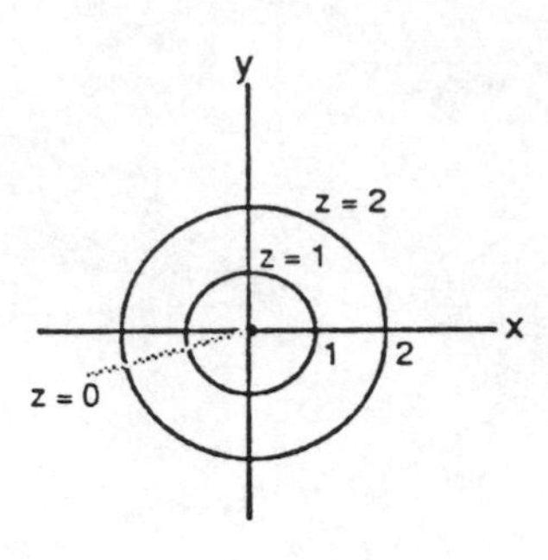

23. (a) (b)

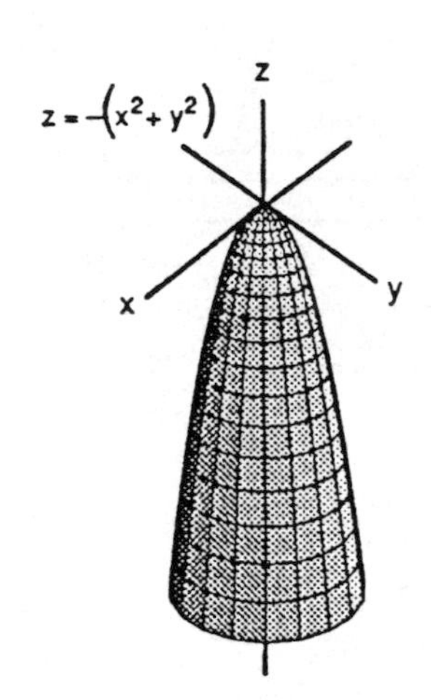

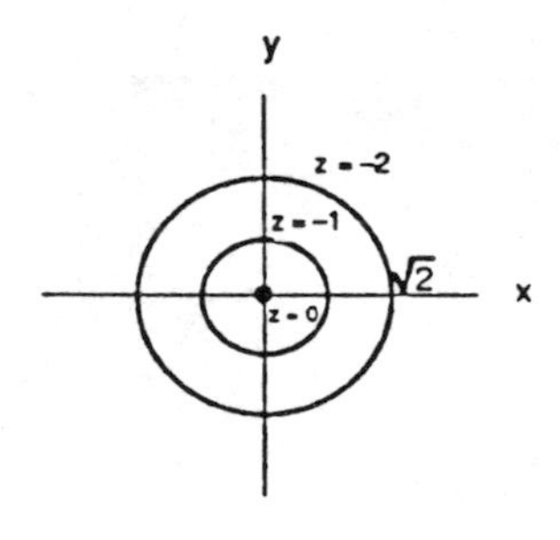

24. (a) (b)

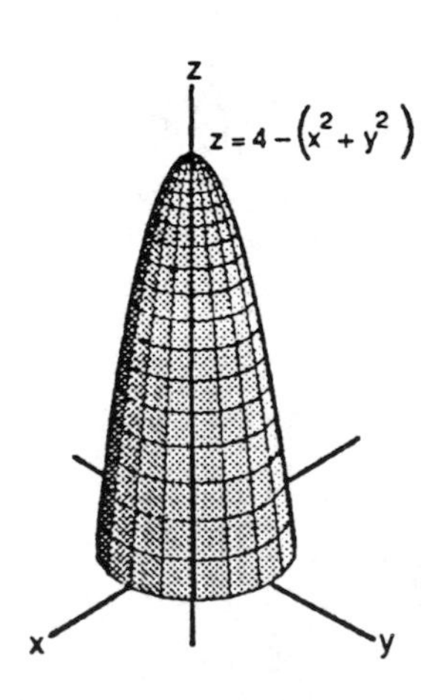

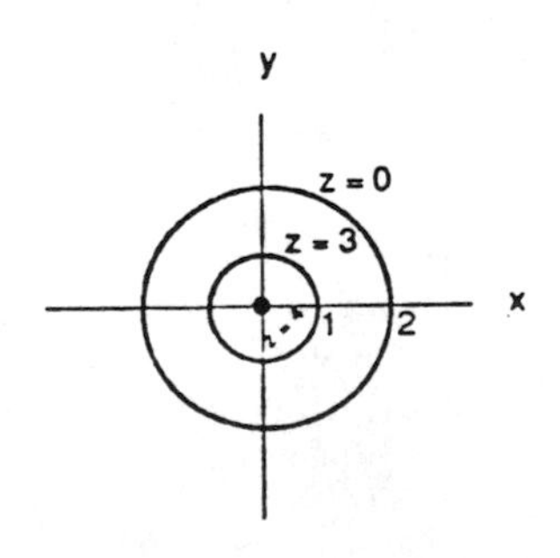

25. (a)

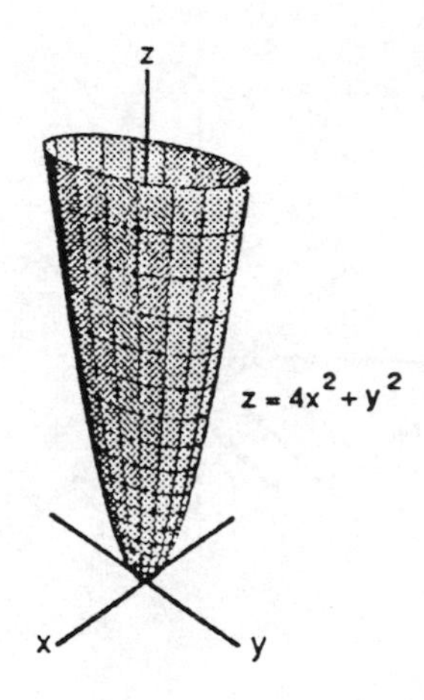

(b)

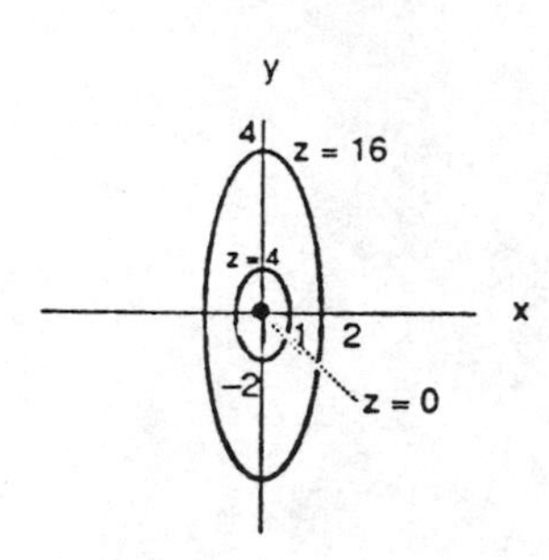

26. (a)

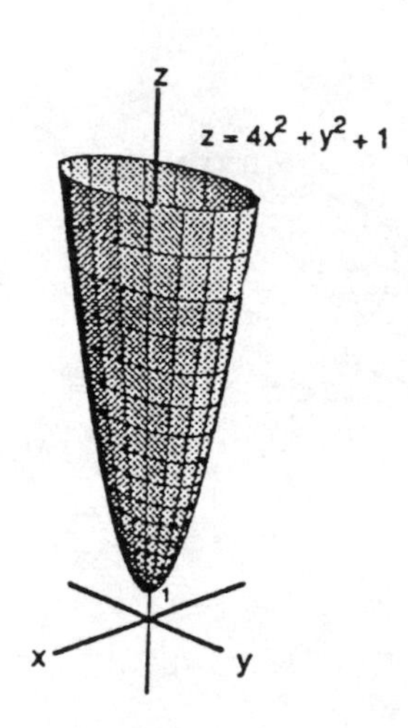

(b)

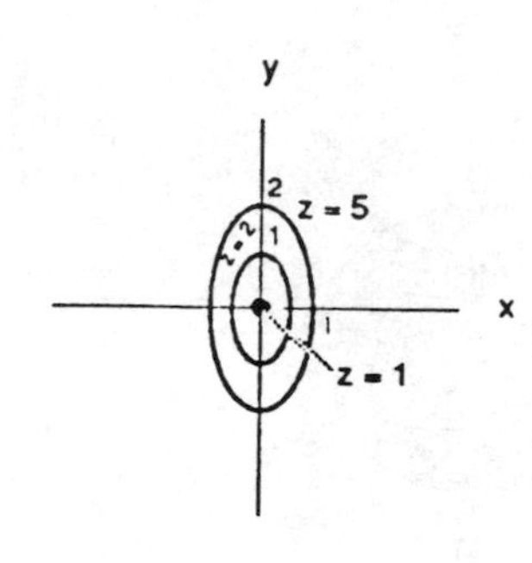

27. (a)

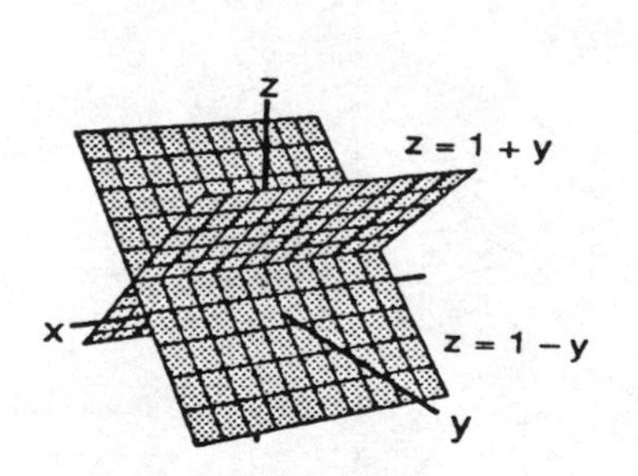

(b)

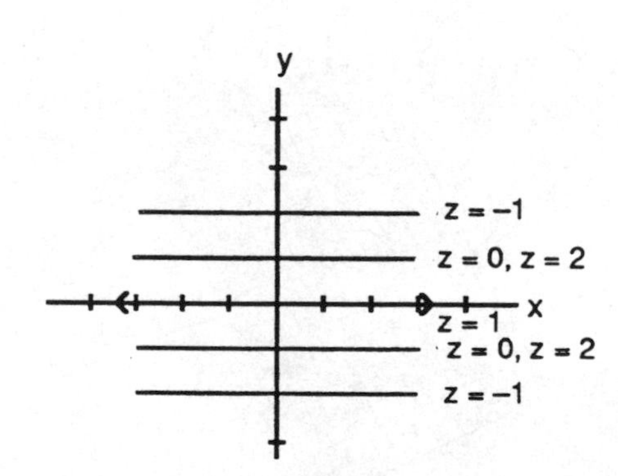

28. (a)

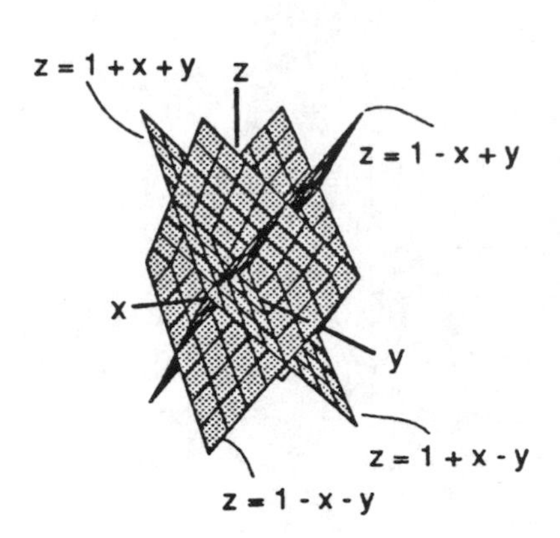

(b)

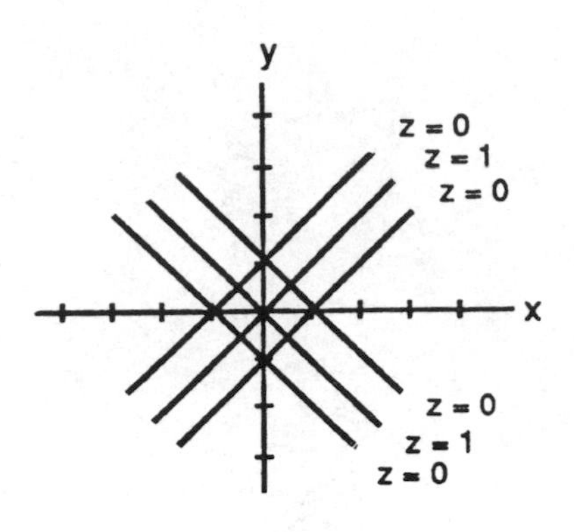

29.

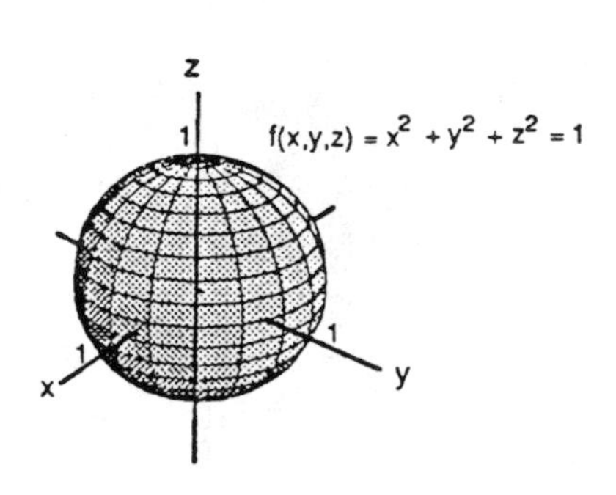

30.

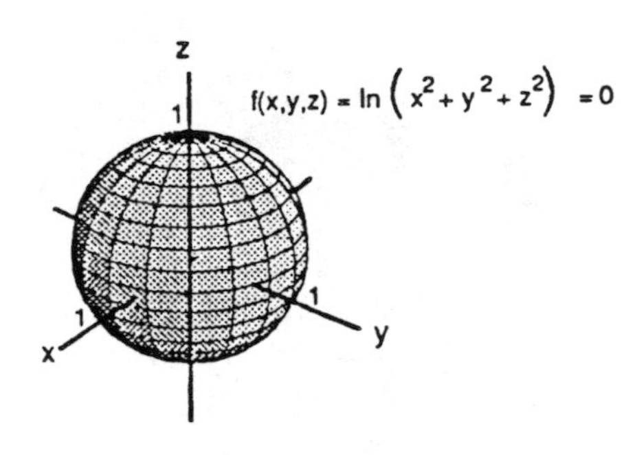

31.

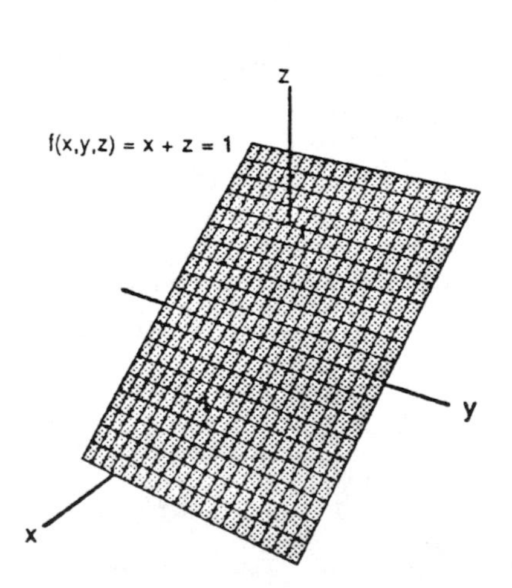

32.

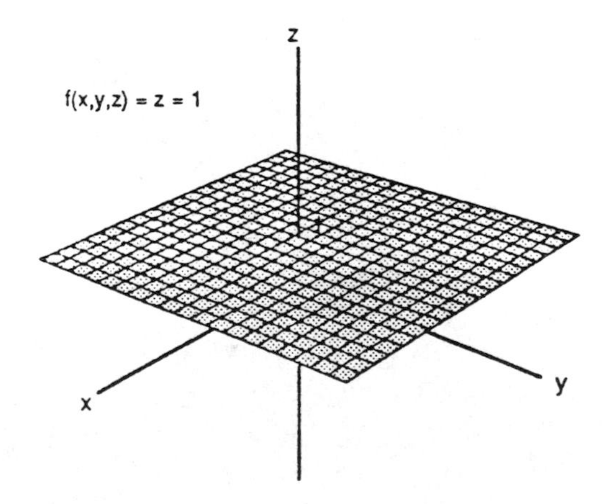

33.

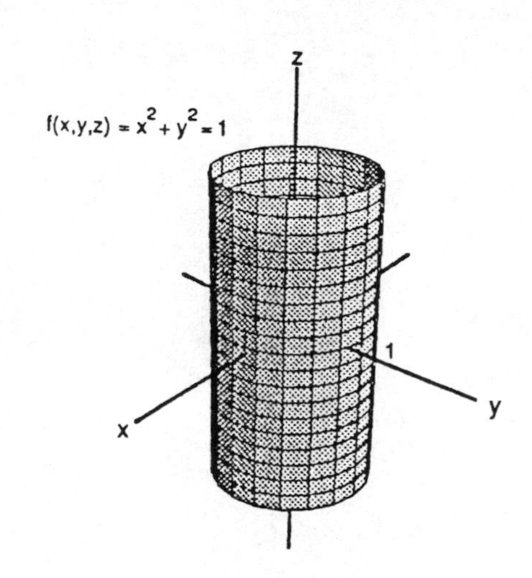

34.

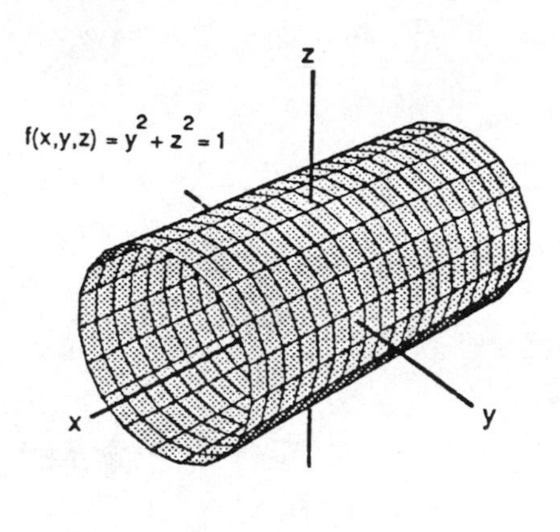

35.

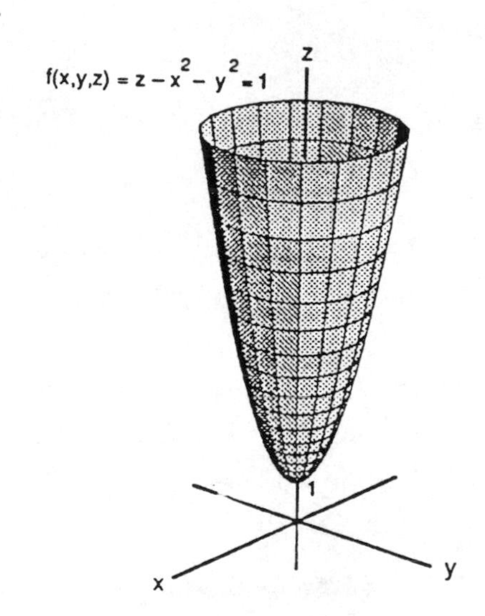

36.

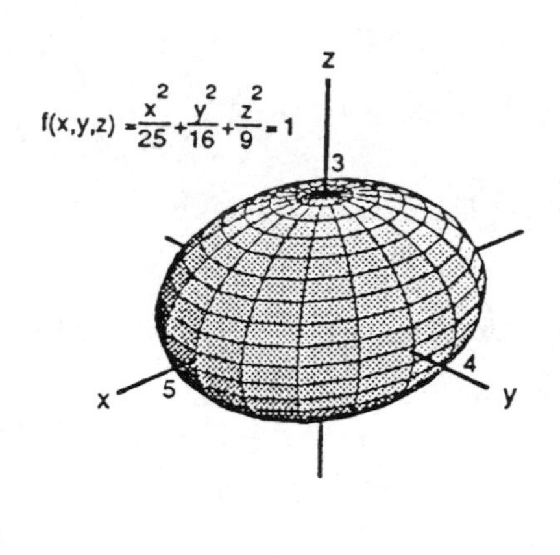

37. $f(x,y) = 16 - x^2 - y^2$ and $(2\sqrt{2}, \sqrt{2}) \Rightarrow z = 16 - (2\sqrt{2})^2 - (\sqrt{2})^2 = 6 \Rightarrow 6 = 16 - x^2 - y^2 \Rightarrow x^2 + y^2 = 10$

38. $f(x,y) = \sqrt{x^2 - 1}$ and $(1,0) \Rightarrow z = \sqrt{1^2 - 1} = 0 \Rightarrow x^2 - 1 = 0 \Rightarrow x = 1$ or $x = -1$

39. $f(x,y) = \int_x^y \frac{1}{1+t^2}\,dt$ at $(-\sqrt{2}, \sqrt{2}) \Rightarrow z = \tan^{-1} y - \tan^{-1} x$; at $(-\sqrt{2}, \sqrt{2}) \Rightarrow z = \tan^{-1}\sqrt{2} - \tan^{-1}(-\sqrt{2})$ $= 2\tan^{-1}\sqrt{2} \Rightarrow \tan^{-1} y - \tan^{-1} x = 2\tan^{-1}\sqrt{2}$

40. $f(x,y) = \sum_{n=0}^{\infty} \left(\frac{x}{y}\right)^n$ at $(1,2) \Rightarrow z = \frac{1}{1-\left(\frac{x}{y}\right)} = \frac{y}{y-x}$; at $(1,2) \Rightarrow z = \frac{2}{2-1} = 2 \Rightarrow 2 = \frac{y}{y-x} \Rightarrow 2y - 2x = y$ $\Rightarrow y = 2x$

41. $f(x,y,z) = \sqrt{x-y} - \ln z$ at $(3,-1,1) \Rightarrow w = \sqrt{x-y} - \ln z$; at $(3,-1,1) \Rightarrow w = \sqrt{3-(-1)} - \ln 1 = 2$
$\Rightarrow \sqrt{x-y} - \ln z = 2$

42. $f(x,y,z) = \ln(x^2+y+z^2)$ at $(-1,2,1) \Rightarrow w = \ln(x^2+y+z^2)$; at $(-1,2,1) \Rightarrow w = \ln(1+2+1) = \ln 4$
$\Rightarrow \ln 4 = \ln(x^2+y+z^2) \Rightarrow x^2+y+z^2 = 4$

43. $g(x,y,z) = \sum_{n=0}^{\infty} \frac{(x+y)^n}{n!\,z^n}$ at $(\ln 2, \ln 4, 3) \Rightarrow w = \sum_{n=0}^{\infty} \frac{(x+y)^n}{n!\,z^n} = e^{(x+y)/z}$; at $(\ln 2, \ln 4, 3) \Rightarrow w = e^{(\ln 2+\ln 4)/3}$
$= e^{(\ln 8)/3} = e^{\ln 2} = 2 \Rightarrow 2 = e^{(x+y)/z} \Rightarrow \frac{x+y}{z} = \ln 2$

44. $g(x,y,z) = \int_x^y \frac{d\theta}{\sqrt{1-\theta^2}} + \int_{\sqrt{2}}^z \frac{dt}{t\sqrt{t^2-1}}$ at $\left(0,\frac{1}{2},2\right) \Rightarrow w = \left[\sin^{-1}\theta\right]_x^y + \left[\sec^{-1} t\right]_{\sqrt{2}}^z$
$= \sin^{-1} y - \sin^{-1} x + \sec^{-1} z - \sec^{-1}\left(\sqrt{2}\right) \Rightarrow w = \sin^{-1} y - \sin^{-1} x + \sec^{-1} z - \frac{\pi}{4}$; at $\left(0,\frac{1}{2},2\right)$
$\Rightarrow w = \sin^{-1}\frac{1}{2} - \sin^{-1} 0 + \sec^{-1} 2 - \frac{\pi}{4} = \frac{\pi}{4} \Rightarrow \frac{\pi}{2} = \sin^{-1} y - \sin^{-1} x + \sec^{-1} z$

45. $f(x,y,z) = xyz$ and $x = 20-t$, $y = t$, $z = 20 \Rightarrow w = (20-t)(t)(20)$ along the line $\Rightarrow w = 400t - 20t^2$
$\Rightarrow \frac{dw}{dt} = 400 - 40t$; $\frac{dw}{dt} = 0 \Rightarrow 400 - 40t = 0 \Rightarrow t = 10$ and $\frac{d^2w}{dt^2} = -40$ for all $t \Rightarrow$ yes, maximum at $t = 10$
$\Rightarrow x = 20 - 10 = 10$, $y = 10$, $z = 20 \Rightarrow$ maximum of f along the line is $f(10,10,20) = (10)(10)(20) = 2000$

46. $f(x,y,z) = xy - z$ and $x = t-1$, $y = t-2$, $z = t+7 \Rightarrow w = (t-1)(t-2) - (t+7) = t^2 - 4t - 5$ along the line
$\Rightarrow \frac{dw}{dt} = 2t - 4$; $\frac{dw}{dt} = 0 \Rightarrow 2t - 4 = 0 \Rightarrow t = 2$ and $\frac{d^2w}{dt^2} = 2$ for all $t \Rightarrow$ yes, minimum at $t = 2 \Rightarrow x = 2 - 1 = 1$,
$y = 2 - 2 = 0$, and $z = 2 + 7 = 9 \Rightarrow$ minimum of f along the line is $f(1,0,9) = (1)(0) - 9 = -9$

47. $w = 4\left(\frac{Th}{d}\right)^{1/2} = 4\left[\frac{(290\text{ K})(16.8\text{ km})}{5\text{ K/km}}\right]^{1/2} \approx 124.86$ km $\Rightarrow$ must be $\frac{1}{2}(124.86) \approx 63$ km south of Nantucket

48. The graph of $f(x_1,x_2,x_3,x_4)$ is a set in a five-dimensional space. It is the set of points $(x_1,x_2,x_3,x_4,f(x_1,x_2,x_3,x_4))$ for (x_1,x_2,x_3,x_4) in the domain of f. The graph of $f(x_1,x_2,x_3,\ldots,x_n)$ is a set in an $(n+1)$-dimensional space. It is the set of points $(x_1,x_2,x_3,\ldots,x_n,f(x_1,x_2,x_3,\ldots,x_n))$ for $(x_1,x_2,x_3,\ldots,x_n)$ in the domain of f.

49-52. Example CAS commands:

Maple:

```
with(plots):
f:= (x,y) -> x*sin(y/2) + y*sin(2*x):
plot3d(f(x,y), x = 0..3*Pi, y=0..3*Pi, axes=FRAMED, title = `x sin y/2 + y sin 2x`);
contourplot(f(x,y), x=0..5*Pi, y=0..5*Pi);
eq:= f(x,y) = f(3*Pi,3*Pi);
implicitplot(eq, x=0..3*Pi, y=0..10*Pi);
```

Mathematica:

```
Clear[x,y]
<< Graphics`ImplicitPlot`
SetOptions[Plot3D, PlotPoints -> 25];
SetOptions[ContourPlot, PlotPoints -> 25,
   ContourShading -> False];
f[x_,y_] = x Sin[y/2] + y Sin[2x]
{xa,xb} = {0, 5 Pi};
{ya,yb} = {0, 5 Pi};
{x0,y0} = {3Pi, 3Pi};
Plot3D[ f[x,y], {x,xa,xb}, {y,ya,yb} ]
ContourPlot[ f[x,y], {x,xa,xb}, {y,ya,yb} ]
ImplicitPlot[ f[x,y] == f[x0,y0], {x,xa,xb}, {y,ya,yb} ]
```

53-56. Example CAS commands:

Maple:

```
with(plots):
eq:= ln(x^2 + y^2 + z^2) = 0.25;
implicitplot3d(eq, x=-1..1, y=-1..1,z=-1..1, axes=BOXED,scaling=CONSTRAINED);
```

Mathematica:

```
ContourPlot3D[ 4 Log[x^2+y^2+z^2],
   {x,-1.1,1.2}, {y,-1.1,1.2}, {z,-1.1,1.2},
   Contours->{1.} ]
```

57-60. Example CAS commands:

Maple:

```
with(plots):
x:= (u,v) -> u*cos(v);
y:= (u,v) -> u*sin(v);
z:= (u,v) -> u;
plot3d([x(u,v), y(u,v), z(u,v)], u = 0..2, v = 0..2*Pi, axes=FRAMED);
contourplot([x(u,v),y(u,v),z(u,v)],u=0..2, v=0..2*Pi);
```

Mathematica:

```
Note: While in Maple it is trivial to get contours from any 3D surface,
in Mathematica it is not obvious for parametric surfaces. In these examples,
z only depends on one parameter, so we can solve for that parameter in terms
of z, and substitute to get x & y in terms of z and the other parameter, then
parametrically plot level curves for several equally spaced values of z (using
"Table").
```

```
ParametricPlot3D[ {u Cos[v], u Sin[v], u},
   {u,0,2}, {v,0,2Pi} ]
ParametricPlot[ Evaluate[Table[
   {z Cos[v], z Sin[v]}, {z,0,2,1/3} ]],
   {v,0,2Pi}, AspectRatio -> Automatic ]
```

12.2 LIMITS AND CONTINUITY

1. $\lim_{(x,y)\to(0,0)} \frac{3x^2 - y^2 + 5}{x^2 + y^2 + 2} = \frac{3(0)^2 - 0^2 + 5}{0^2 + 0^2 + 2} = \frac{5}{2}$

2. $\lim_{(x,y)\to(0,4)} \frac{x}{\sqrt{y}} = \frac{0}{\sqrt{4}} = 0$

3. $\lim_{(x,y)\to(3,4)} \sqrt{x^2 + y^2 - 1} = \sqrt{3^2 + 4^2 - 1} = \sqrt{24} = 2\sqrt{6}$

4. $\lim_{(x,y)\to(2,-3)} \left(\frac{1}{x} + \frac{1}{y}\right)^2 = \left[\frac{1}{2} + \left(\frac{1}{-3}\right)\right]^2 = \left(\frac{1}{6}\right)^2 = \frac{1}{36}$

5. $\lim_{(x,y)\to\left(0,\frac{\pi}{4}\right)} \sec x \tan y = (\sec 0)\left(\tan \frac{\pi}{4}\right) = (1)(1) = 1$

6. $\lim_{(x,y)\to(0,0)} \cos\left(\frac{x^2 + y^3}{x + y + 1}\right) = \cos\left(\frac{0^2 + 0^3}{0 + 0 + 1}\right) = \cos 0 = 1$

7. $\lim_{(x,y)\to(0,\ln 2)} e^{x-y} = e^{0 - \ln 2} = e^{\ln\left(\frac{1}{2}\right)} = \frac{1}{2}$

8. $\lim_{(x,y)\to(1,1)} \ln\left|1 + x^2y^2\right| = \ln\left|1 + (1)^2(1)^2\right| = \ln 2$

9. $\lim_{(x,y)\to(0,0)} \frac{e^y \sin x}{x} = \lim_{(x,y)\to(0,0)} (e^y)\left(\frac{\sin x}{x}\right) = e^0 \cdot \lim_{x\to 0} \left(\frac{\sin x}{x}\right) = 1 \cdot 1 = 1$

10. $\lim_{(x,y)\to(1,1)} \cos\left(\sqrt[3]{|xy| - 1}\right) = \cos\left(\sqrt[3]{(1)(1) - 1}\right) = \cos 0 = 1$

11. $\lim_{(x,y)\to(1,0)} \frac{x \sin y}{x^2 + 1} = \frac{1 \cdot \sin 0}{1^2 + 1} = \frac{0}{2} = 0$

12. $\lim_{(x,y)\to\left(\frac{\pi}{2},0\right)} \frac{\cos y + 1}{y - \sin x} = \frac{(\cos 0) + 1}{0 - \sin\left(\frac{\pi}{2}\right)} = \frac{1 + 1}{-1} = -2$

13. $\lim_{\substack{(x,y)\to(1,1)\\ x\neq y}} \frac{x^2 - 2xy + y^2}{x - y} = \lim_{(x,y)\to(1,1)} \frac{(x - y)^2}{x - y} = \lim_{(x,y)\to(1,2)} (x - y) = (1 - 1) = 0$

14. $\lim_{\substack{(x,y)\to(1,1)\\ x\neq y}} \frac{x^2 - y^2}{x - y} = \lim_{(x,y)\to(1,1)} \frac{(x + y)(x - y)}{x - y} = \lim_{(x,y)\to(1,1)} (x + y) = (1 + 1) = 2$

15. $\lim_{\substack{(x,y)\to(1,1)\\ x\neq 1}} \frac{xy - y - 2x + 2}{x - 1} = \lim_{\substack{(x,y)\to(1,1)\\ x\neq 1}} \frac{(x - 1)(y - 2)}{x - 1} = \lim_{(x,y)\to(1,1)} (y - 2) = (1 - 2) = -1$

16. $\lim\limits_{\substack{(x,y)\to(2,-4)\\ y\neq -4,\ x\neq x^2}} \frac{y+4}{x^2y-xy+4x^2-4x} = \lim\limits_{\substack{(x,y)\to(2,-4)\\ y\neq -4,\ x\neq x^2}} \frac{y+4}{x(x-1)(y+4)} = \lim\limits_{\substack{(x,y)\to(2,-4)\\ x\neq x^2}} \frac{1}{x(x-1)} = \frac{1}{2(2-1)} = \frac{1}{2}$

17. $\lim\limits_{\substack{(x,y)\to(0,0)\\ x\neq y}} \frac{x-y+2\sqrt{x}-2\sqrt{y}}{\sqrt{x}-\sqrt{y}} = \lim\limits_{\substack{(x,y)\to(0,0)\\ x\neq y}} \frac{\left(\sqrt{x}-\sqrt{y}\right)\left(\sqrt{x}+\sqrt{y}+2\right)}{\sqrt{x}-\sqrt{y}} = \lim\limits_{(x,y)\to(0,0)} \left(\sqrt{x}+\sqrt{y}+2\right)$

$= \left(\sqrt{0}+\sqrt{0}+2\right) = 2$

Note: (x,y) must approach $(0,0)$ through the first quadrant only with $x \neq y$.

18. $\lim\limits_{\substack{(x,y)\to(2,2)\\ x+y\neq 4}} \frac{x+y-4}{\sqrt{x+y}-2} = \lim\limits_{\substack{(x,y)\to(2,2)\\ x+y\neq 4}} \frac{\left(\sqrt{x+y}+2\right)\left(\sqrt{x+y}-2\right)}{\sqrt{x+y}-2} = \lim\limits_{\substack{(x,y)\to(2,2)\\ x+y\neq 4}} \left(\sqrt{x+y}+2\right)$

$= \left(\sqrt{2+2}+2\right) = 2+2 = 4$

19. $\lim\limits_{\substack{(x,y)\to(2,0)\\ 2x-y\neq 4}} \frac{\sqrt{2x-y}-2}{2x-y-4} = \lim\limits_{\substack{(x,y)\to(2,0)\\ 2x-y\neq 4}} \frac{\sqrt{2x-y}-2}{\left(\sqrt{2x-y}+2\right)\left(\sqrt{2x-y}-2\right)} = \lim\limits_{(x,y)\to(2,0)} \frac{1}{\sqrt{2x-y}+2}$

$= \frac{1}{\sqrt{(2)(2)-0}+2} = \frac{1}{2+2} = \frac{1}{4}$

20. $\lim\limits_{\substack{(x,y)\to(4,3)\\ x-y\neq 1}} \frac{\sqrt{x}-\sqrt{y+1}}{x-y-1} = \lim\limits_{\substack{(x,y)\to(4,3)\\ x-y\neq 1}} \frac{\sqrt{x}-\sqrt{y+1}}{\left(\sqrt{x}+\sqrt{y+1}\right)\left(\sqrt{x}-\sqrt{y+1}\right)} = \lim\limits_{(x,y)\to(4,3)} \frac{1}{\sqrt{x}+\sqrt{y+1}}$

$= \frac{1}{\sqrt{4}+\sqrt{3+1}} = \frac{1}{2+2} = \frac{1}{4}$

21. $\lim\limits_{P\to(1,3,4)} \left(\frac{1}{x}+\frac{1}{y}+\frac{1}{z}\right) = \frac{1}{1}+\frac{1}{3}+\frac{1}{4} = \frac{12+4+3}{12} = \frac{19}{12}$

22. $\lim\limits_{P\to(1,-1,-1)} \frac{2xy+yz}{x^2+z^2} = \frac{2(1)(-1)+(-1)(-1)}{1^2+(-1)^2} = \frac{-2+1}{1+1} = -\frac{1}{2}$

23. $\lim\limits_{P\to(3,3,0)} \left(\sin^2 x+\cos^2 y+\sec^2 z\right) = \left(\sin^2 3+\cos^2 3\right)+\sec^2 0 = 1+1^2 = 2$

24. $\lim\limits_{P\to\left(-\frac{1}{4},\frac{\pi}{2},2\right)} \tan^{-1}(xyz) = \tan^{-1}\left(-\frac{1}{4}\cdot\frac{\pi}{2}\cdot 2\right) = \tan^{-1}\left(-\frac{\pi}{4}\right)$

25. $\lim\limits_{P\to(\pi,0,3)} ze^{-2y}\cos 2x = 3e^{-2(0)}\cos 2\pi = (3)(1)(1) = 3$

26. $\lim\limits_{P\to(0,-2,0)} \ln\sqrt{x^2+y^2+z^2} = \ln\sqrt{0^2+(-2)^2+0^2} = \ln\sqrt{4} = \ln 2$

27. (a) All (x,y)
(b) All (x,y) except $(0,0)$

28. (a) All (x, y) so that $x \neq y$
(b) All (x, y)

29. (a) All (x, y) except where $x = 0$ or $y = 0$
(b) All (x, y)

30. (a) All (x, y) so that $x^2 - 3x + 2 \neq 0 \Rightarrow (x-2)(x-1) \neq 0 \Rightarrow x \neq 2$ and $x \neq 1$
(b) All (x, y) so that $y \neq x^2$

31. (a) All (x, y, z)
(b) All (x, y, z) except the interior of the cylinder $x^2 + y^2 = 1$

32. (a) All (x, y, z) so that $xyz > 0$
(b) All (x, y, z)

33. (a) All (x, y, z) with $z \neq 0$
(b) All (x, y, z) with $x^2 + z^2 \neq 1$

34. (a) All (x, y, z) except (x, 0, 0)
(b) All (x, y, z) except (0, y, 0) or (x, 0, 0)

35. $\lim\limits_{\substack{(x,y)\to(0,0)\\ \text{along } y=x\\ x>0}} -\frac{x}{\sqrt{x^2+y^2}} = \lim\limits_{x\to 0} -\frac{x}{\sqrt{x^2+x^2}} = \lim\limits_{x\to 0} -\frac{x}{\sqrt{2}\,|x|} = \lim\limits_{x\to 0} -\frac{x}{\sqrt{2}x} = \lim\limits_{x\to 0} -\frac{1}{\sqrt{2}} = -\frac{1}{\sqrt{2}}$;

$\lim\limits_{\substack{(x,y)\to(0,0)\\ \text{along } y=x\\ x<0}} -\frac{x}{\sqrt{x^2+y^2}} = \lim\limits_{x\to 0} -\frac{x}{\sqrt{2}\,|x|} = \lim\limits_{x\to 0} -\frac{x}{\sqrt{2}(-x)} = \lim\limits_{x\to 0} \frac{1}{\sqrt{2}} = \frac{1}{\sqrt{2}}$

36. $\lim\limits_{\substack{(x,y)\to(0,0)\\ \text{along } y=0}} \frac{x^4}{x^4+y^2} = \lim\limits_{x\to 0} \frac{x^4}{x^4+0^2} = 1$; $\lim\limits_{\substack{(x,y)\to(0,0)\\ \text{along } y=x^2}} \frac{x^4}{x^4+y^2} = \lim\limits_{x\to 0} \frac{x^4}{x^4+(x^2)^2} = \lim\limits_{x\to 0} \frac{x^4}{2x^4} = \frac{1}{2}$

37. $\lim\limits_{\substack{(x,y)\to(0,0)\\ \text{along } y=kx^2}} \frac{x^4-y^2}{x^4+y^2} = \lim\limits_{x\to 0} \frac{x^4-(kx^2)^2}{x^4+(kx^2)^2} = \lim\limits_{x\to 0} \frac{x^4-k^2x^4}{x^4+k^2x^4} = \frac{1-k^2}{1+k^2} \Rightarrow$ different limits for different values of k

38. $\lim\limits_{\substack{(x,y)\to(0,0)\\ \text{along } y=kx\\ k\neq 0}} \frac{xy}{|xy|} = \lim\limits_{x\to 0} \frac{x(kx)}{|x(kx)|} = \lim\limits_{x\to 0} \frac{kx^2}{|kx^2|} = \lim\limits_{x\to 0} \frac{k}{|k|}$; if $k > 0$, the limit is 1; but if $k < 0$, the limit is -1

39. $\lim\limits_{\substack{(x,y)\to(0,0)\\ \text{along } y=kx\\ k\neq -1}} \frac{x-y}{x+y} = \lim\limits_{x\to 0} \frac{x-kx}{x+kx} = \frac{1-k}{1+k} \Rightarrow$ different limits for different values of k, $k \neq -1$

40. $\lim\limits_{\substack{(x,y)\to(0,0)\\ \text{along } y=kx\\ k\neq 1}} \frac{x+y}{x-y} = \lim\limits_{x\to 0} \frac{x+kx}{x-kx} = \frac{1+k}{1-k} \Rightarrow$ different limits for different values of k, $k \neq 1$

41. $\lim\limits_{\substack{(x,y)\to(0,0)\\ \text{along } y=kx^2\\ k\neq 0}} \frac{x^2+y}{y} = \lim\limits_{x\to 0} \frac{x^2+kx^2}{kx^2} = \frac{1+k}{k} \Rightarrow$ different limits for different values of k, $k \neq 0$

42. $\lim\limits_{\substack{(x,y)\to(0,0)\\ \text{along } y=kx^2\\ k\neq 1}} \frac{x^2}{x^2-y} = \lim\limits_{x\to 0} \frac{x^2}{x^2-kx^2} = \frac{1}{1-k} \Rightarrow$ different limits for different values of k, $k \neq 1$

43. No, the limit depends only on the values $f(x,y)$ has when $(x,y) \neq (x_0,y_0)$

44. If f is continuous at (x_0,y_0), then $\lim\limits_{(x,y)\to(x_0,y_0)} f(x,y)$ must equal $f(x_0,y_0) = 3$. If f is not continuous at (x_0,y_0), the limit could have any value different from 3, and need not even exist.

45. $\lim\limits_{(x,y)\to(0,0)} \left(1-\frac{x^2y^2}{3}\right) = 1$ and $\lim\limits_{(x,y)\to(0,0)} 1 = 1 \Rightarrow \lim\limits_{(x,y)\to(0,0)} \frac{\tan^{-1} xy}{xy} = 1$, by the Sandwich Theorem

46. If $xy > 0$, $\lim\limits_{(x,y)\to(0,0)} \frac{2\,|xy| - \left(\frac{x^2y^2}{6}\right)}{|xy|} = \lim\limits_{(x,y)\to(0,0)} \frac{2xy - \left(\frac{x^2y^2}{6}\right)}{xy} = \lim\limits_{(x,y)\to(0,0)} \left(2-\frac{xy}{6}\right) = 2$ and $\lim\limits_{(x,y)\to(0,0)} \frac{2\,|xy|}{|xy|} = \lim\limits_{(x,y)\to(0,0)} 2 = 2$; if $xy < 0$, $\lim\limits_{(x,y)\to(0,0)} \frac{2\,|xy| - \left(\frac{x^2y^2}{6}\right)}{|xy|} = \lim\limits_{(x,y)\to(0,0)} \frac{-2xy - \left(\frac{x^2y^2}{6}\right)}{-xy}$ $= \lim\limits_{(x,y)\to(0,0)} \left(2+\frac{xy}{6}\right) = 2$ and $\lim\limits_{(x,y)\to(0,0)} \frac{2\,|xy|}{|xy|} = 2 \Rightarrow \lim\limits_{(x,y)\to(0,0)} \frac{4 - 4\cos\sqrt{|xy|}}{|xy|} = 2$, by the Sandwich Theorem

47. The limit is 0 since $\left|\sin\left(\frac{1}{x}\right)\right| \leq 1 \Rightarrow -1 \leq \sin\left(\frac{1}{x}\right) \leq 1 \Rightarrow -y \leq y \sin\left(\frac{1}{x}\right) \leq y$ for $y \geq 0$, and $-y \geq y \sin\left(\frac{1}{x}\right) \geq y$ for $y \leq 0$. Thus as $(x,y) \to (0,0)$, both $-y$ and y approach $0 \Rightarrow y \sin\left(\frac{1}{x}\right) \to 0$, by the Sandwich Theorem.

48. The limit is 0 since $\left|\cos\left(\frac{1}{y}\right)\right| \leq 1 \Rightarrow -1 \leq \cos\left(\frac{1}{y}\right) \leq 1 \Rightarrow -x \leq x \cos\left(\frac{1}{y}\right) \leq x$ for $x \geq 0$, and $-x \geq x \cos\left(\frac{1}{y}\right) \geq x$ for $x \leq 0$. Thus as $(x,y) \to (0,0)$, both $-x$ and x approach $0 \Rightarrow x \cos\left(\frac{1}{y}\right) \to 0$, by the Sandwich Theorem.

49. (a) $f(x,y)\big|_{y=mx} = \frac{2m}{1+m^2} = \frac{2\tan\theta}{1+\tan^2\theta} = \sin 2\theta$. The value of $f(x,y) = \sin 2\theta$ varies with θ, which is the line's angle of inclination.

(b) Since $f(x,y)\big|_{y=mx} = \sin 2\theta$ and since $-1 \leq \sin 2\theta \leq 1$ for every θ, $\lim\limits_{(x,y)\to(0,0)} f(x,y)$ varies from -1 to 1 along $y = mx$.

50. $\left|xy\,(x^2-y^2)\right| = |xy|\,|x^2-y^2| \leq |x|\,|y|\,|x^2+y^2| = \sqrt{x^2}\,\sqrt{y^2}\,|x^2+y^2| \leq \sqrt{x^2+y^2}\,\sqrt{x^2+y^2}\,|x^2+y^2|$ $= (x^2+y^2)^2 \Rightarrow \left|\frac{xy\,(x^2-y^2)}{x^2+y^2}\right| \leq \frac{(x^2+y^2)^2}{x^2+y^2} = x^2+y^2 \Rightarrow -(x^2+y^2) \leq \frac{xy\,(x^2-y^2)}{x^2+y^2} \leq (x^2+y^2)$

$\Rightarrow \lim_{(x,y)\to(0,0)} xy\,\frac{x^2-y^2}{x^2+y^2} = 0$ by the Sandwich Theorem, since $\lim_{(x,y)\to(0,0)} \pm(x^2+y^2) = 0$; thus, define $f(0,0) = 0$

51. $\lim_{(x,y)\to(0,0)} \frac{x^3-xy^2}{x^2+y^2} = \lim_{r\to 0} \frac{r^3\cos^3\theta - (r\cos\theta)(r^2\sin^2\theta)}{r^2\cos^2\theta + r^2\sin^2\theta} = \lim_{r\to 0} \frac{r(\cos^3\theta - \cos\theta\sin^2\theta)}{1} = 0$

52. $\lim_{(x,y)\to(0,0)} \cos\left(\frac{x^3-y^3}{x^2+y^2}\right) = \lim_{r\to 0} \cos\left(\frac{r^3\cos^3\theta - r^3\sin^3\theta}{r^2\cos^2\theta + r^2\sin^2\theta}\right) = \lim_{r\to 0} \cos\left[\frac{r(\cos^3\theta - \sin^3\theta)}{1}\right] = \cos 0 = 1$

53. $\lim_{(x,y)\to(0,0)} \frac{y^2}{x^2+y^2} = \lim_{r\to 0} \frac{r^2\sin^2\theta}{r^2} = \lim_{r\to 0} (\sin^2\theta) = \sin^2\theta$; the limit does not exist since $\sin^2\theta$ is between 0 and 1 depending on θ

54. $\lim_{(x,y)\to(0,0)} \frac{2x}{x^2+x+y^2} = \lim_{r\to 0} \frac{2r\cos\theta}{r^2+r\cos\theta} = \lim_{r\to 0} \frac{2\cos\theta}{r+\cos\theta} = \frac{2\cos\theta}{\cos\theta}$; the limit does not exist for $\cos\theta = 0$

55. $\lim_{(x,y)\to(0,0)} \tan^{-1}\left[\frac{|x|+|y|}{x^2+y^2}\right] = \lim_{r\to 0} \tan^{-1}\left[\frac{|r\cos\theta|+|r\sin\theta|}{r^2}\right] = \lim_{r\to 0} \tan^{-1}\left[\frac{|r|(|\cos\theta|+|\sin\theta|)}{r^2}\right]$;

if $r\to 0^+$, then $\lim_{r\to 0^+} \tan^{-1}\left[\frac{|r|(|\cos\theta|+|\sin\theta|)}{r^2}\right] = \lim_{r\to 0^+} \tan^{-1}\left[\frac{|\cos\theta|+|\sin\theta|}{r}\right] = \frac{\pi}{2}$; if $r\to 0^-$, then

$\lim_{r\to 0^-} \tan^{-1}\left[\frac{|r|(|\cos\theta|+|\sin\theta|)}{r^2}\right] = \lim_{r\to 0^-} \tan^{-1}\left(\frac{|\cos\theta|+|\sin\theta|}{-r}\right) = \frac{\pi}{2} \Rightarrow$ the limit is $\frac{\pi}{2}$

56. $\lim_{(x,y)\to(0,0)} \frac{x^2-y^2}{x^2+y^2} = \lim_{r\to 0} \frac{r^2\cos^2\theta - r^2\sin^2\theta}{r^2} = \lim_{r\to 0} (\cos^2\theta - \sin^2\theta) = \lim_{r\to 0} (\cos 2\theta)$ which ranges between -1 and 1 depending on $\theta \Rightarrow$ the lmit does not exist

57. $\lim_{(x,y)\to(0,0)} \ln\left(\frac{3x^2-x^2y^2+3y^2}{x^2+y^2}\right) = \lim_{r\to 0} \ln\left(\frac{3r^2\cos^2\theta - r^4\cos^2\theta\sin^2\theta + 3r^2\sin^2\theta}{r^2}\right)$

$= \lim_{r\to 0} \ln(3 - r^2\cos^2\theta\sin^2\theta) = \ln 3 \Rightarrow$ define $f(0,0) = \ln 3$

58. $\lim_{(x,y)\to(0,0)} \frac{2xy^2}{x^2+y^2} = \lim_{r\to 0} \frac{(2r\cos\theta)(r^2\sin^2\theta)}{r^2} = \lim_{r\to 0} 2r\cos\theta\sin^2\theta = 0 \Rightarrow$ define $f(0,0) = 0$

59. In Eq. (1), if the point (x,y) lies within a disk centered at (x_0,y_0) and radius less than δ, then $|f(x,y)-L|<\epsilon$; in Eq. (2), if the point (x,y) lies within a square centered at (x_0,y_0) with the side length less than 2δ, then $|f(x,y)-L|<\epsilon$. Since every circle of radius δ is circumscribed by a square of side length 2δ, $\sqrt{(x-x_0)^2+(y-y_0)^2}<\delta \Rightarrow |x-x_0|<\delta$ and $|y-y_0|<\delta$; likewise, every square of side length $\frac{2\delta}{\sqrt{2}}$ is circumscribed by a circle of radius δ so that $|x-x_0|<\frac{\delta}{\sqrt{2}}$ and $|y-y_0|<\frac{\delta}{\sqrt{2}}$ $\Rightarrow \sqrt{(x-x_0)^2+(y-y_0)^2}<\delta$. Thus the requirements are equivalent: small circles give small inscribed squares, and small squares give small inscribed circles.

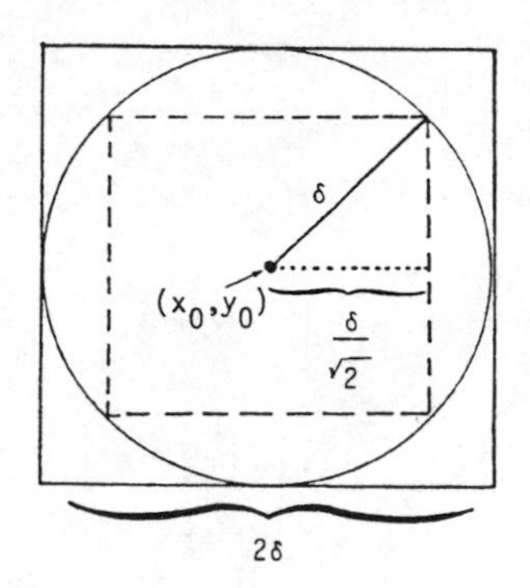

60. $\lim_{(x,y,z)\to(x_0,y_0,z_0)} g(x,y,z)=L$ if, for every number $\epsilon>0$, there exists a corresponding $\delta>0$ such that for all (x,y,z) in the domain of g, $0<\sqrt{(x-x_0)^2+(y-y_0)^2+(z-z_0)^2}<\delta \Rightarrow |g(x,y,z)-L|<\epsilon$. With four independent variables and $P=(x,y,z,t)$, $\lim_{P\to(x_0,y_0,z_0,t_0)} h(x,y,z,t)=L$ if, for every number $\epsilon>0$, there exists a correspoding $\delta>0$ such that for all P in the domain of h, $0<\sqrt{(x-x_0)^2+(y-y_0)^2+(z-z_0)^2+(t-t_0)^2}$ $<\delta \Rightarrow |h(x,y,x,t)-L|<\epsilon$.

61. Let $\delta=0.1$. Then $\sqrt{x^2+y^2}<\delta \Rightarrow \sqrt{x^2+y^2}<0.1 \Rightarrow x^2+y^2<0.01 \Rightarrow |x^2+y^2-0|<0.01 \Rightarrow |f(x,y)-f(0,0)|$ $<0.01=\epsilon$.

62. Let $\delta=0.05$. Then $|x|<\delta$ and $|y|<\delta \Rightarrow |f(x,y)-f(0,0)|=\left|\frac{y}{x^2+1}-0\right|=\left|\frac{y}{x^2+1}\right|\le|y|<0.05=\epsilon$.

63. Let $\delta=0.005$. Then $|x|<\delta$ and $|y|<\delta \Rightarrow |f(x,y)-f(0,0)|=\left|\frac{x+y}{x^2+1}-0\right|=\left|\frac{x+y}{x^2+1}\right|\le|x+y|<|x|+|y|$ $<0.005+0.005=0.01=\epsilon$.

64. Let $\delta=0.01$. Since $-1\le\cos x\le1 \Rightarrow 1\le 2+\cos x\le 3 \Rightarrow \frac{1}{3}\le\frac{1}{2+\cos x}\le 1 \Rightarrow \frac{|x+y|}{3}\le\left|\frac{x+y}{2+\cos x}\right|\le|x+y|$ $\le|x|+|y|$. Then $|x|<\delta$ and $|y|<\delta \Rightarrow |f(x,y)-f(0,0)|=\left|\frac{x+y}{2+\cos x}-0\right|=\left|\frac{x+y}{2+\cos x}\right|\le|x|+|y|<0.01+0.01$ $=0.02=\epsilon$.

65. Let $\delta=\sqrt{0.015}$. Then $\sqrt{x^2+y^2+z^2}<\delta \Rightarrow |f(x,y,z)-f(0,0,0)|=|x^2+y^2+z^2-0|=|x^2+y^2+z^2|$ $=\left(\sqrt{x^2+t^2+x^2}\right)^2<\left(\sqrt{0.015}\right)^2=0.015=\epsilon$.

66. Let $\delta = 0.2$. Then $|x| < \delta$, $|y| < \delta$, and $|z| < \delta \Rightarrow |f(x,y,z) - f(0,0,0)| = |xyz - 0| = |xyz| = |x||y||z| < (0.2)^3 = 0.008 = \epsilon$.

67. Let $\delta = 0.005$. Then $|x| < \delta$, $|y| < \delta$, and $|z| < \delta \Rightarrow |f(x,y,z) - f(0,0,0)| = \left|\frac{x+y+z}{x^2+y^2+z^2+1} - 0\right|$ $= \left|\frac{x+y+z}{x^2+y^2+z^2+1}\right| \le |x+y+z| \le |x|+|y|+|z| < 0.005 + 0.005 + 0.005 = 0.015 = \epsilon$.

68. Let $\delta = \tan^{-1}(0.1)$. Then $|x| < \delta$, $|y| < \delta$, and $|z| < \delta \Rightarrow |f(x,y,z) - f(0,0,0)| = |\tan^2 x + \tan^2 y + \tan^2 z|$ $\le |\tan^2 x| + |\tan^2 y| + |\tan^2 z| = \tan^2 x + \tan^2 y + \tan^2 z < \tan^2 \delta + \tan^2 \delta + \tan^2 \delta = 0.01 + 0.01 + 0.01 = 0.03$ $= \epsilon$.

69. $\lim\limits_{(x,y,z)\to(x_0,y_0,z_0)} f(x,y,z) = \lim\limits_{(x,y,z)\to(x_0,y_0,z_0)} (x+y+z) = x_0 + y_0 + z_0 = f(x_0,y_0,z_0) \Rightarrow$ f is continuous at every (x_0,y_0,z_0)

70. $\lim\limits_{(x,y,z)\to(x_0,y_0,z_0)} f(x,y,z) = \lim\limits_{(x,y,z)\to(x_0,y_0,z_0)} (x^2+y^2+z^2) = x_0^2 + y_0^2 + z_0^2 = f(x_0,y_0,z_0) \Rightarrow$ f is continuous at every point (x_0,y_0,z_0)

12.3 PARTIAL DERIVATIVES

1. $\frac{\partial f}{\partial x} = 4x$, $\frac{\partial f}{\partial y} = -3$

2. $\frac{\partial f}{\partial x} = 2x - y$, $\frac{\partial f}{\partial y} = -x + 2y$

3. $\frac{\partial f}{\partial x} = 2x(y+2)$, $\frac{\partial f}{\partial y} = x^2 - 1$

4. $\frac{\partial f}{\partial x} = 5y - 14x + 3$, $\frac{\partial f}{\partial y} = 5x - 2y - 6$

5. $\frac{\partial f}{\partial x} = 2y(xy-1)$, $\frac{\partial f}{\partial y} = 2x(xy-1)$

6. $\frac{\partial f}{\partial x} = 6(2x-3y)^2$, $\frac{\partial f}{\partial y} = -9(2x-3y)^2$

7. $\frac{\partial f}{\partial x} = \frac{x}{\sqrt{x^2+y^2}}$, $\frac{\partial f}{\partial y} = \frac{y}{\sqrt{x^2+y^2}}$

8. $\frac{\partial f}{\partial x} = \frac{2x^2}{\sqrt[3]{x^3+\left(\frac{y}{2}\right)}}$, $\frac{\partial f}{\partial y} = \frac{1}{3\sqrt[3]{x^3+\left(\frac{y}{2}\right)}}$

9. $\frac{\partial f}{\partial x} = -\frac{1}{(x+y)^2} \cdot \frac{\partial}{\partial x}(x+y) = -\frac{1}{(x+y)^2}$, $\frac{\partial f}{\partial y} = -\frac{1}{(x+y)^2} \cdot \frac{\partial}{\partial y}(x+y) = -\frac{1}{(x+y)^2}$

10. $\frac{\partial f}{\partial x} = \frac{(x^2+y^2)(1) - x(2x)}{(x^2+y^2)^2} = \frac{y^2-x^2}{(x^2+y^2)^2}$, $\frac{\partial f}{\partial y} = \frac{(x^2+y^2)(0) - x(2y)}{(x^2+y^2)^2} = -\frac{2xy}{(x^2+y^2)^2}$

11. $\frac{\partial f}{\partial x} = = \frac{(xy-1)(1) - (x+y)(y)}{(xy-1)^2} = \frac{-y^2-1}{(xy-1)^2}$, $\frac{\partial f}{\partial y} = \frac{(xy-1)(1) - (x+y)(x)}{(xy-1)^2} = \frac{-x^2-1}{(xy-1)^2}$

12. $\frac{\partial f}{\partial x} = \frac{1}{1+\left(\frac{y}{x}\right)^2} \cdot \frac{\partial}{\partial x}\left(\frac{y}{x}\right) = -\frac{y}{x^2\left[1+\left(\frac{y}{x}\right)^2\right]} = -\frac{y}{x^2+y^2}$, $\frac{\partial f}{\partial y} = \frac{1}{1+\left(\frac{y}{x}\right)^2} \cdot \frac{\partial}{\partial y}\left(\frac{y}{x}\right) = \frac{1}{x\left[1+\left(\frac{y}{x}\right)^2\right]} = \frac{x}{x^2+y^2}$

13. $\frac{\partial f}{\partial x} = e^{(x+y+1)} \cdot \frac{\partial}{\partial x}(x+y+1) = e^{(x+y+1)}$, $\frac{\partial f}{\partial y} = e^{(x+y+1)} \cdot \frac{\partial}{\partial y}(x+y+1) = e^{(x+y+1)}$

14. $\frac{\partial f}{\partial x} = -e^{-x}\sin(x+y) + e^{-x}\cos(x+y)$, $\frac{\partial f}{\partial y} = e^{-x}\cos(x+y)$

15. $\frac{\partial f}{\partial x} = \frac{1}{x+y} \cdot \frac{\partial}{\partial x}(x+y) = \frac{1}{x+y}$, $\frac{\partial f}{\partial y} = \frac{1}{x+y} \cdot \frac{\partial}{\partial y}(x+y) = \frac{1}{x+y}$

16. $\frac{\partial f}{\partial x} = e^{xy} \cdot \frac{\partial}{\partial x}(xy) \cdot \ln y = ye^{xy}\ln y$, $\frac{\partial f}{\partial y} = e^{xy} \cdot \frac{\partial}{\partial y}(xy) \cdot \ln y + e^{xy} \cdot \frac{1}{y} = xe^{xy}\ln y + \frac{e^{xy}}{y}$

17. $\frac{\partial f}{\partial x} = 2\sin(x-3y) \cdot \frac{\partial}{\partial x}\sin(x-3y) = 2\sin(x-3y)\cos(x-3y) \cdot \frac{\partial}{\partial x}(x-3y) = 2\sin(x-3y)\cos(x-3y)$,

$\frac{\partial f}{\partial y} = 2\sin(x-3y) \cdot \frac{\partial}{\partial y}\sin(x-3y) = 2\sin(x-3y)\cos(x-3y) \cdot \frac{\partial}{\partial y}(x-3y) = -6\sin(x-3y)\cos(x-3y)$

18. $\frac{\partial f}{\partial x} = 2\cos(3x-y^2) \cdot \frac{\partial}{\partial x}\cos(3x-y^2) = -2\cos(3x-y^2)\sin(3x-y^2) \cdot \frac{\partial}{\partial x}(3x-y^2)$

$= -6\cos(3x-y^2)\sin(3x-y^2)$,

$\frac{\partial f}{\partial y} = 2\cos(3x-y^2) \cdot \frac{\partial}{\partial y}\cos(3x-y^2) = -2\cos(3x-y^2)\sin(3x-y^2) \cdot \frac{\partial}{\partial y}(3x-y^2)$

$= 4y\cos(3x-y^2)\sin(3x-y^2)$

19. $\frac{\partial f}{\partial x} = yx^{y-1}$, $\frac{\partial f}{\partial y} = x^y \ln x$

20. $f(x,y) = \frac{\ln x}{\ln y} \Rightarrow \frac{\partial f}{\partial x} = \frac{1}{x \ln y}$ and $\frac{\partial f}{\partial y} = \frac{-\ln x}{y(\ln y)^2}$

21. $\frac{\partial f}{\partial x} = -g(x)$, $\frac{\partial f}{\partial y} = g(y)$

22. $f(x,y) = \sum_{n=0}^{\infty}(xy)^n$, $|xy| < 1 \Rightarrow f(x,y) = \frac{1}{1-xy} \Rightarrow \frac{\partial f}{\partial x} = -\frac{1}{(1-xy)^2} \cdot \frac{\partial}{\partial x}(1-xy) = \frac{y}{(1-xy)^2}$ and

$\frac{\partial f}{\partial y} = -\frac{1}{(1-xy)^2} \cdot \frac{\partial}{\partial y}(1-xy) = \frac{x}{(1-xy)^2}$

23. $f_x = 1 + y^2$, $f_y = 2xy$, $f_z = -4z$

24. $f_x = y + z$, $f_y = x + z$, $f_z = y + x$

25. $f_x = 1$, $f_y = -\frac{y}{\sqrt{y^2+z^2}}$, $f_z = -\frac{z}{\sqrt{y^2+z^2}}$

26. $f_x = -x(x^2+y^2+z^2)^{-3/2}$, $f_y = -y(x^2+y^2+z^2)^{-3/2}$, $f_z = -z(x^2+y^2+z^2)^{-3/2}$

27. $f_x = \frac{yz}{\sqrt{1-x^2y^2z^2}}$, $f_y = \frac{xz}{\sqrt{1-x^2y^2z^2}}$, $f_z = \frac{xy}{\sqrt{1-x^2y^2z^2}}$

28. $f_x = \frac{1}{|x+yz|\sqrt{(x+yz)^2-1}}$, $f_y = \frac{z}{|x+yz|\sqrt{(x+yz)^2-1}}$, $f_z = \frac{y}{|x+yz|\sqrt{(x+yz)^2-1}}$

29. $f_x = \frac{1}{x+2y+3z}$, $f_y = \frac{2}{x+2y+3z}$, $f_z = \frac{3}{x+2y+3z}$

30. $f_x = yz \cdot \frac{1}{xy} \cdot \frac{\partial}{\partial x}(xy) = \frac{(yz)(y)}{xy} = \frac{yz}{x}$, $f_y = z \ln(xy) + yz \cdot \frac{\partial}{\partial y} \ln(xy) = z \ln(xy) + \frac{yz}{xy} \cdot \frac{\partial}{\partial y}(xy) = z \ln(xy) + z$, $f_z = y \ln(xy) + yz \cdot \frac{\partial}{\partial z} \ln(xy) = y \ln(xy)$

31. $f_x = -2xe^{-(x^2+y^2+z^2)}$, $f_y = -2ye^{-(x^2+y^2+z^2)}$, $f_z = -2ze^{-(x^2+y^2+z^2)}$

32. $f_x = -yze^{-xyz}$, $f_y = -xze^{-xyz}$, $f_z = -xye^{-xyz}$

33. $f_x = \operatorname{sech}^2(x + 2y + 3z)$, $f_y = 2 \operatorname{sech}^2(x + 2y + 3z)$, $f_z = 3 \operatorname{sech}^2(x + 2y + 3z)$

34. $f_x = y \cosh(xy - z^2)$, $f_y = x \cosh(xy - z^2)$, $f_z = -2z \cosh(xy - z^2)$

35. $\frac{\partial f}{\partial t} = -2\pi \sin(2\pi t - \alpha)$, $\frac{\partial f}{\partial \alpha} = \sin(2\pi t - \alpha)$

36. $\frac{\partial g}{\partial u} = v^2 e^{(2u/v)} \cdot \frac{\partial}{\partial u}\left(\frac{2u}{v}\right) = 2ve^{(2u/v)}$, $\frac{\partial g}{\partial v} = 2ve^{(2u/v)} + v^2 e^{(2u/v)} \cdot \frac{\partial}{\partial v}\left(\frac{2u}{v}\right) = 2ve^{(2u/v)} - 2ue^{(2u/v)}$

37. $\frac{\partial h}{\partial \rho} = \sin\phi \cos\theta$, $\frac{\partial h}{\partial \phi} = \rho \cos\phi \cos\theta$, $\frac{\partial h}{\partial \theta} = -\rho \sin\phi \sin\theta$

38. $\frac{\partial g}{\partial r} = 1 - \cos\theta$, $\frac{\partial g}{\partial \theta} = r \sin\theta$, $\frac{\partial g}{\partial z} = -1$

39. $W_P = V$, $W_v = P + \frac{\delta v^2}{2g}$, $W_\delta = \frac{Vv^2}{2g}$, $W_v = \frac{2V\delta v}{2g} = \frac{V\delta v}{g}$, $W_g = -\frac{V\delta v^2}{2g^2}$

40. $\frac{\partial A}{\partial c} = m$, $\frac{\partial A}{\partial h} = \frac{q}{2}$, $\frac{\partial A}{\partial k} = \frac{m}{q}$, $\frac{\partial A}{\partial m} = \frac{k}{q} + c$, $\frac{\partial A}{\partial q} = -\frac{km}{q^2} + \frac{h}{2}$

41. $\frac{\partial f}{\partial x} = 1 + y$, $\frac{\partial f}{\partial y} = 1 + x$, $\frac{\partial^2 f}{\partial x^2} = 0$, $\frac{\partial^2 f}{\partial y^2} = 0$, $\frac{\partial^2 f}{\partial y \partial x} = \frac{\partial^2 f}{\partial x \partial y} = 1$

42. $\frac{\partial f}{\partial x} = y \cos xy$, $\frac{\partial f}{\partial y} = x \cos xy$, $\frac{\partial^2 f}{\partial x^2} = -y^2 \sin xy$, $\frac{\partial^2 f}{\partial y^2} = -x^2 \sin xy$, $\frac{\partial^2 f}{\partial y \partial x} = \frac{\partial^2 f}{\partial x \partial y} = \cos xy - xy \sin xy$

43. $\frac{\partial g}{\partial x} = 2xy + y \cos x$, $\frac{\partial g}{\partial y} = x^2 - \sin y + \sin x$, $\frac{\partial^2 g}{\partial x^2} = 2y - y \sin x$, $\frac{\partial^2 g}{\partial y^2} = -\cos y$, $\frac{\partial^2 g}{\partial y \partial x} = \frac{\partial^2 g}{\partial x \partial y} = 2x + \cos x$

44. $\frac{\partial h}{\partial x} = e^y$, $\frac{\partial h}{\partial y} = xe^y + 1$, $\frac{\partial^2 h}{\partial x^2} = 0$, $\frac{\partial^2 h}{\partial y^2} = xe^y$, $\frac{\partial^2 h}{\partial y \partial x} = \frac{\partial^2 h}{\partial x \partial y} = e^y$

45. $\frac{\partial r}{\partial x} = \frac{1}{x+y}$, $\frac{\partial r}{\partial y} = \frac{1}{x+y}$, $\frac{\partial^2 r}{\partial x^2} = \frac{-1}{(x+y)^2}$, $\frac{\partial^2 r}{\partial y^2} = \frac{-1}{(x+y)^2}$, $\frac{\partial^2 r}{\partial y \partial x} = \frac{\partial^2 r}{\partial x \partial y} = \frac{-1}{(x+y)^2}$

46. $\frac{\partial s}{\partial x} = \left[\frac{1}{1+\left(\frac{y}{x}\right)^2}\right] \cdot \frac{\partial}{\partial x}\left(\frac{y}{x}\right) = \left(-\frac{y}{x^2}\right)\left[\frac{1}{1+\left(\frac{y}{x}\right)^2}\right] = \frac{-y}{x^2+y^2}$, $\frac{\partial s}{\partial y} = \left[\frac{1}{1+\left(\frac{y}{x}\right)^2}\right] \cdot \frac{\partial}{\partial y}\left(\frac{y}{x}\right) = \left(\frac{1}{x}\right)\left[\frac{1}{1+\left(\frac{y}{x}\right)^2}\right] = \frac{x}{x^2+y^2}$,

$$\frac{\partial^2 s}{\partial x^2} = \frac{y(2x)}{(x^2+y^2)^2} = \frac{2xy}{(x^2+y^2)^2}, \quad \frac{\partial^2 s}{\partial y^2} = \frac{-x(2y)}{(x^2+y^2)^2} = -\frac{2xy}{(x^2+y^2)^2},$$

$$\frac{\partial^2 s}{\partial y \partial x} = \frac{\partial^2 s}{\partial x \partial y} = \frac{(x^2+y^2)(-1) + y(2y)}{(x^2+y^2)^2} = \frac{y^2 - x^2}{(x^2+y^2)^2}$$

47. $\frac{\partial w}{\partial x} = \frac{2}{2x+3y}$, $\frac{\partial w}{\partial y} = \frac{3}{2x+3y}$, $\frac{\partial^2 w}{\partial y \partial x} = \frac{-6}{(2x+3y)^2}$, and $\frac{\partial^2 w}{\partial x \partial y} = \frac{-6}{(2x+3y)^2}$

48. $\frac{\partial w}{\partial x} = e^x + \ln y + \frac{y}{x}$, $\frac{\partial w}{\partial y} = \frac{x}{y} + \ln x$, $\frac{\partial^2 w}{\partial y \partial x} = = \frac{1}{y} + \frac{1}{x}$, and $\frac{\partial^2 w}{\partial x \partial y} = \frac{1}{y} + \frac{1}{x}$

49. $\frac{\partial w}{\partial x} = y^2 + 2xy^3 + 3x^2y^4$, $\frac{\partial w}{\partial y} = 2xy + 3x^2y^2 + 4x^3y^3$, $\frac{\partial^2 w}{\partial y \partial x} = 2y + 6xy^2 + 12x^2y^3$, and $\frac{\partial^2 w}{\partial x \partial y} = 2y + 6xy^2 + 12x^2y^3$

50. $\frac{\partial w}{\partial x} = \sin y + y \cos x + y$, $\frac{\partial w}{\partial y} = x \cos y + \sin x + x$, $\frac{\partial^2 w}{\partial y \partial x} = \cos y + \cos x + 1$, and $\frac{\partial^2 w}{\partial x \partial y} = \cos y + \cos x + 1$

51. (a) x first (b) y first (c) x first (d) x first (e) y first (f) y first

52. (a) y first three times (b) y first three times (c) y first twice (d) x first twice

53. $f_x(1,2) = \lim_{h \to 0} \frac{f(1+h,2) - f(1,2)}{h} = \lim_{h \to 0} \frac{[1-(1+h)+2-6(1+h)^2]-(2-6)}{h} = \lim_{h \to 0} \frac{-h-6(1+2h+h^2)+6}{h}$
$= \lim_{h \to 0} \frac{-13h-6h^2}{h} = \lim_{h \to 0} (-13-6h) = -13$,
$f_y(1,2) = \lim_{h \to 0} \frac{f(1,2+h) - f(1,2)}{h} = \lim_{h \to 0} \frac{[1-1+(2+h)-3(2+h)]-(2-6)}{h} = \lim_{h \to 0} \frac{(2-6-2h)-(2-6)}{h}$
$= \lim_{h \to 0} (-2) = -2$

54. $f_x(-2,1) = \lim_{h \to 0} \frac{f(-2+h,1) - f(-2,1)}{h} = \lim_{h \to 0} \frac{[4+2(-2+h)-3-(-2+h)]-(-3+2)}{h}$
$= \lim_{h \to 0} \frac{(2h-1-h)+1}{h} = \lim_{h \to 0} 1 = 1$,
$f_y(-2,1) = \lim_{h \to 0} \frac{f(-2,1+h) - f(-2,1)}{h} = \lim_{h \to 0} \frac{[4-4-3(1+h)+2(1+h^2)]-(-3+2)}{h}$
$= \lim_{h \to 0} \frac{(-3-3h+2+4h+2h^2)+1}{h} = \lim_{h \to 0} \frac{h+2h^2}{h} = \lim_{h \to 0} (1+2h) = 1$

55. $f_z(x_0,y_0,z_0) = \lim_{h \to 0} \frac{f(x_0,y_0,z_0+h) - f(x_0,y_0,z_0)}{h}$;
$f_z(1,2,3) = \lim_{h \to 0} \frac{f(1,2,3+h) - f(1,2,3)}{h} = \lim_{h \to 0} \frac{2(3+h)^2 - 2(9)}{h} = \lim_{h \to 0} \frac{12h+2h^2}{h} = \lim_{h \to 0} (12+2h) = 12$

56. $f_y(x_0,y_0,z_0) = \lim_{h \to 0} \frac{f(x_0,y_0+h,z_0) - f(x_0,y_0,z_0)}{h}$;
$f_y(-1,0,3) = \lim_{h \to 0} \frac{f(-1,h,3) - f(-1,0,3)}{h} = \lim_{h \to 0} \frac{(2h^2+9h)-0}{h} = \lim_{h \to 0} (2h+9) = 9$

57. $y+\left(3z^2\frac{\partial z}{\partial x}\right)x+z^3-2y\frac{\partial z}{\partial x}=0 \Rightarrow \left(3xz^2-2y\right)\frac{\partial z}{\partial x}=-y-z^3 \Rightarrow$ at $(1,1,1)$ we have $(3-2)\frac{\partial z}{\partial x}=-1-1$ or $\frac{\partial z}{\partial x}=-2$

58. $\left(\frac{\partial x}{\partial z}\right)z+x+\left(\frac{y}{x}\right)\frac{\partial x}{\partial z}-2x\frac{\partial x}{\partial z}=0 \Rightarrow \left(z+\frac{y}{x}-2x\right)\frac{\partial x}{\partial z}=-x \Rightarrow$ at $(1,-1,-3)$ we have $(-3-1-2)\frac{\partial x}{\partial z}=-1$ or $\frac{\partial x}{\partial z}=\frac{1}{6}$

59. $a^2=b^2+c^2-2bc\cos A \Rightarrow 2a=(2bc\sin A)\frac{\partial A}{\partial a} \Rightarrow \frac{\partial A}{\partial a}=\frac{a}{bc\sin A}$; also $0=2b-2c\cos A+(2bc\sin A)\frac{\partial A}{\partial b}$ $\Rightarrow 2c\cos A-2b=(2bc\sin A)\frac{\partial A}{\partial b} \Rightarrow \frac{\partial A}{\partial b}=\frac{c\cos A-b}{bc\sin A}$

60. $\frac{a}{\sin A}=\frac{b}{\sin B} \Rightarrow \frac{(\sin A)\frac{\partial a}{\partial A}-a\cos A}{\sin^2 A}=0 \Rightarrow (\sin A)\frac{\partial a}{\partial x}-a\cos A=0 \Rightarrow \frac{\partial a}{\partial A}=\frac{a\cos A}{\sin A}$; also $\left(\frac{1}{\sin A}\right)\frac{\partial a}{\partial B}=b(-\csc B\cot B) \Rightarrow \frac{\partial a}{\partial B}=-b\csc B\cot B\sin A$

61. Differentiating each equation implicitly gives $1=v_x\ln u+\left(\frac{v}{u}\right)u_x$ and $0=u_x\ln v+\left(\frac{u}{v}\right)v_x$ or

$$\left.\begin{aligned}(\ln u)\,v_x \quad +\left(\tfrac{v}{u}\right)u_x&=1\\ \left(\tfrac{u}{v}\right)v_x+(\ln v)\,u_x&=0\end{aligned}\right\} \Rightarrow v_x=\frac{\begin{vmatrix}1 & \frac{v}{u}\\ 0 & \ln v\end{vmatrix}}{\begin{vmatrix}\ln u & \frac{v}{u}\\ \frac{u}{v} & \ln v\end{vmatrix}}=\frac{\ln v}{(\ln u)(\ln v)-1}$$

62. Differentiating each equation implicitly gives $1=(2x)x_u-(2y)y_u$ and $0=(2x)x_u-y_u$ or

$$\left.\begin{aligned}(2x)x_u-(2y)y_u&=1\\ (2x)x_u- \qquad y_u&=0\end{aligned}\right\} \Rightarrow x_u=\frac{\begin{vmatrix}1 & -2y\\ 0 & -1\end{vmatrix}}{\begin{vmatrix}2x & -2y\\ 2x & -1\end{vmatrix}}=\frac{-1}{-2x+4xy}=\frac{1}{2x-4xy} \text{ and}$$

$$y_u=\frac{\begin{vmatrix}2x & 1\\ 2x & 0\end{vmatrix}}{-2x+4xy}=\frac{-2x}{-2x+4xy}=\frac{2x}{2x-4xy}=\frac{1}{1-2y}; \text{ next } s=x^2+y^2 \Rightarrow \frac{\partial s}{\partial u}=2x\frac{\partial x}{\partial u}+2y\frac{\partial y}{\partial u}$$

$$=2x\left(\frac{1}{2x-4xy}\right)+2y\left(\frac{1}{1-2y}\right)=\frac{1}{1-2y}+\frac{2y}{1-2y}=\frac{1+2y}{1-2y}$$

63. $\frac{\partial f}{\partial x}=2x,\ \frac{\partial f}{\partial y}=2y,\ \frac{\partial f}{\partial z}=-4z \Rightarrow \frac{\partial^2 f}{\partial x^2}=2,\ \frac{\partial^2 f}{\partial y^2}=2,\ \frac{\partial^2 f}{\partial z^2}=-4 \Rightarrow \frac{\partial^2 f}{\partial x^2}+\frac{\partial^2 f}{\partial y^2}+\frac{\partial^2 f}{\partial z^2}=2+2+(-4)=0$

64. $\frac{\partial f}{\partial x} = -6xz,\ \frac{\partial f}{\partial y} = -6yz,\ \frac{\partial f}{\partial z} = 6z^2 - 3(x^2+y^2),\ \frac{\partial^2 f}{\partial x^2} = -6z,\ \frac{\partial^2 f}{\partial y^2} = -6z,\ \frac{\partial^2 f}{\partial z^2} = 12z \Rightarrow \frac{\partial^2 f}{\partial x^2} + \frac{\partial^2 f}{\partial y^2} + \frac{\partial^2 f}{\partial z^2}$
$= -6z - 6z + 12z = 0$

65. $\frac{\partial f}{\partial x} = -2e^{-2y} \sin 2x,\ \frac{\partial f}{\partial y} = -e^{-2y} \cos 2x,\ \frac{\partial^2 f}{\partial x^2} = -4e^{-2y} \cos 2x,\ \frac{\partial^2 f}{\partial y^2} = 4e^{-2y} \cos 2x \Rightarrow \frac{\partial^2 f}{\partial x^2} + \frac{\partial^2 f}{\partial y^2}$
$= -4e^{-2y} \cos 2x + 4e^{-2y} \cos 2x = 0$

66. $\frac{\partial f}{\partial x} = \frac{x}{x^2+y^2},\ \frac{\partial f}{\partial y} = \frac{y}{x^2+y^2},\ \frac{\partial^2 f}{\partial x^2} = \frac{y^2-x^2}{(x^2+y^2)^2},\ \frac{\partial^2 f}{\partial y^2} = \frac{x^2-y^2}{(x^2+y^2)^2} \Rightarrow \frac{\partial^2 f}{\partial x^2} + \frac{\partial^2 f}{\partial y^2} = \frac{y^2-x^2}{(x^2+y^2)^2} + \frac{x^2-y^2}{(x^2+y^2)^2} = 0$

67. $\frac{\partial f}{\partial x} = -\frac{1}{2}(x^2+y^2+z^2)^{-3/2}(2x) = -x(x^2+y^2+z^2)^{-3/2},\ \frac{\partial f}{\partial y} = -\frac{1}{2}(x^2+y^2+z^2)^{-3/2}(2y)$
$= -y(x^2+y^2+z^2)^{-3/2},\ \frac{\partial f}{\partial z} = -\frac{1}{2}(x^2+y^2+z^2)^{-3/2}(2z) = -z(x^2+y^2+z^2)^{-3/2};$

$\frac{\partial^2 f}{\partial x^2} = -(x^2+y^2+z^2)^{-3/2} + 3x^2(x^2+y^2+z^2)^{-5/2},\ \frac{\partial^2 f}{\partial y^2} = -(x^2+y^2+z^2)^{-3/2} + 3y^2(x^2+y^2+z^2)^{-5/2},$
$\frac{\partial^2 f}{\partial z^2} = -(x^2+y^2+z^2)^{-3/2} + 3z^2(x^2+y^2+z^2)^{-5/2} \Rightarrow \frac{\partial^2 f}{\partial x^2} + \frac{\partial^2 f}{\partial y^2} + \frac{\partial^2 f}{\partial z^2}$
$= \left[-(x^2+y^2+z^2)^{-3/2} + 3x^2(x^2+y^2+z^2)^{-5/2}\right] + \left[-(x^2+y^2+z^2)^{-3/2} + 3y^2(x^2+y^2+z^2)^{-5/2}\right]$
$+ \left[-(x^2+y^2+z^2)^{-3/2} + 3z^2(x^2+y^2+z^2)^{-5/2}\right] = -3(x^2+y^2+z^2)^{-3/2} + (3x^2+3y^2+3z^2)(x^2+y^2+z^2)^{-5/2}$
$= 0$

68. $\frac{\partial f}{\partial x} = 3e^{3x+4y} \cos 5z,\ \frac{\partial f}{\partial y} = 4e^{3x+4y} \cos 5z,\ \frac{\partial f}{\partial z} = -5e^{3x+4y} \sin 5z;\ \frac{\partial^2 f}{\partial x^2} = 9e^{3x+4y} \cos 5z,\ \frac{\partial^2 f}{\partial y^2} = 16e^{3x+4y} \cos 5z,$
$\frac{\partial^2 f}{\partial z^2} = -25e^{3x+4y} \cos 5z \Rightarrow \frac{\partial^2 f}{\partial x^2} + \frac{\partial^2 f}{\partial y^2} + \frac{\partial^2 f}{\partial z^2} = 9e^{3x+4y} \cos 5z + 16e^{3x+4y} \cos 5z - 25e^{3x+4y} \cos 5z = 0$

69. $\frac{\partial w}{\partial x} = \cos(x+ct),\ \frac{\partial w}{\partial t} = c\cos(x+ct);\ \frac{\partial^2 w}{\partial x^2} = -\sin(x+ct),\ \frac{\partial^2 w}{\partial t^2} = -c^2 \sin(x+ct) \Rightarrow \frac{\partial^2 w}{\partial t^2} = c^2[-\sin(x+ct)]$
$= c^2 \frac{\partial^2 w}{\partial x^2}$

70. $\frac{\partial w}{\partial x} = -2\sin(2x+2ct),\ \frac{\partial w}{\partial t} = -2c\sin(2x+2ct);\ \frac{\partial^2 w}{\partial x^2} = -4\cos(2x+2ct),\ \frac{\partial^2 w}{\partial t^2} = -4c^2 \cos(2x+2ct)$
$\Rightarrow \frac{\partial^2 w}{\partial t^2} = c^2[-4\cos(2x+2ct)] = c^2 \frac{\partial^2 w}{\partial x^2}$

71. $\frac{\partial w}{\partial x} = \cos(x+ct) - 2\sin(2x+2ct),\ \frac{\partial w}{\partial t} = c\cos(x+ct) - 2c\sin(2x+2ct);$
$\frac{\partial^2 w}{\partial x^2} = -\sin(x+ct) - 4\cos(2x+2ct),\ \frac{\partial^2 w}{\partial t^2} = -c^2\sin(x+ct) - 4c^2\cos(2x+2ct)$
$\Rightarrow \frac{\partial^2 w}{\partial t^2} = c^2[-\sin(x+ct) - 4\cos(2x+2ct)] = c^2 \frac{\partial^2 w}{\partial x^2}$

72. $\frac{\partial w}{\partial x} = \frac{1}{x+ct},\ \frac{\partial w}{\partial t} = \frac{c}{x+ct};\ \frac{\partial^2 w}{\partial x^2} = \frac{-1}{(x+ct)^2},\ \frac{\partial^2 w}{\partial t^2} = \frac{-c^2}{(x+ct)^2} \Rightarrow \frac{\partial^2 w}{\partial t^2} = c^2\left[\frac{-1}{(x+ct)^2}\right] = c^2 \frac{\partial^2 w}{\partial x^2}$

73. $\frac{\partial w}{\partial x} = 2 \sec^2(2x - 2ct)$, $\frac{\partial w}{\partial t} = -2c \sec^2(2x - 2ct)$; $\frac{\partial^2 w}{\partial x^2} = 8 \sec^2(2x - 2ct) \tan(2x - 2ct)$,

$\frac{\partial^2 w}{\partial t^2} = 8c^2 \sec^2(2x - 2ct) \tan(2x - 2ct) \Rightarrow \frac{\partial^2 w}{\partial t^2} = c^2[8 \sec^2(2x - 2ct) \tan(2x - 2ct)] = c^2 \frac{\partial^2 w}{\partial x^2}$

74. $\frac{\partial w}{\partial x} = -15 \sin(3x + 3ct) + e^{x+ct}$, $\frac{\partial w}{\partial t} = -15c \sin(3x + 3ct) + ce^{x+ct}$; $\frac{\partial^2 w}{\partial x^2} = -45 \cos(3x + 3ct) + e^{x+ct}$,

$\frac{\partial^2 w}{\partial t^2} = -45c^2 \cos(3x + 3ct) + c^2 e^{x+ct} \Rightarrow \frac{\partial^2 w}{\partial t^2} = c^2[-45 \cos(3x + 3ct) + e^{x+ct}] = c^2 \frac{\partial^2 w}{\partial x^2}$

75. $\frac{\partial w}{\partial t} = \frac{\partial f}{\partial u}\frac{\partial u}{\partial t} = \frac{\partial f}{\partial u}(ac) \Rightarrow \frac{\partial^2 w}{\partial t^2} = (ac)\left(\frac{\partial^2 f}{\partial u^2}\right)(ac) = a^2c^2 \frac{\partial^2 f}{\partial u^2}$; $\frac{\partial w}{\partial x} = \frac{\partial f}{\partial u}\frac{\partial u}{\partial x} = \frac{\partial f}{\partial u} \cdot a \Rightarrow \frac{\partial^2 w}{\partial x^2} = \left(a \frac{\partial^2 f}{\partial u^2}\right) \cdot a$

$= a^2 \frac{\partial^2 f}{\partial u^2} \Rightarrow \frac{\partial^2 w}{\partial t^2} = a^2c^2 \frac{\partial^2 f}{\partial u^2} = c^2\left(a^2 \frac{\partial^2 f}{\partial u^2}\right) = c^2 \frac{\partial^2 w}{\partial x^2}$

12.4 DIFFERENTIABILITY, LINEARIZATION, AND DIFFERENTIALS

1. (a) $f(0,0) = 1$, $f_x(x,y) = 2x \Rightarrow f_x(0,0) = 0$, $f_y(x,y) = 2y \Rightarrow f_y(0,0) = 0 \Rightarrow L(x,y) = 1 + 0(x - 0) + 0(y - 0) = 1$

   (b) $f(1,1) = 3$, $f_x(1,1) = 2$, $f_y(1,1) = 2 \Rightarrow L(x,y) = 3 + 2(x - 1) + 2(y - 1) = 2x + 2y - 1$

2. (a) $f(0,0) = 4$, $f_x(x,y) = 2(x + y + 2) \Rightarrow f_x(0,0) = 4$, $f_y(x,y) = 2(x + y + 2) \Rightarrow f_y(0,0) = 4$

   $\Rightarrow L(x,y) = 4 + 4(x - 0) + 4(y - 0) = 4x + 4y + 4$

   (b) $f(1,2) = 25$, $f_x(1,2) = 10$, $f_y(1,2) = 10 \Rightarrow L(x,y) = 25 + 10(x - 1) + 10(y - 2) = 10x + 10y - 5$

3. (a) $f(0,0) = 5$, $f_x(x,y) = 3$ for all (x,y), $f_y(x,y) = -4$ for all $(x,y) \Rightarrow L(x,y) = 5 + 3(x - 0) - 4(y - 0)$

   $= 3x - 4y + 5$

   (b) $f(1,1) = 4$, $f_x(1,1) = 3$, $f_y(1,1) = -4 \Rightarrow L(x,y) = 4 + 3(x - 1) - 4(y - 1) = 3x - 4y + 5$

4. (a) $f(1,1) = 1$, $f_x(x,y) = 3x^2y^4 \Rightarrow f_x(1,1) = 3$, $f_y(x,y) = 4x^3y^3 \Rightarrow f_y(1,1) = 4$

   $\Rightarrow L(x,y) = 1 + 3(x - 1) + 4(y - 1) = 3x + 4y - 6$

   (b) $f(0,0) = 0$, $f_x(0,0) = 0$, $f_y(0,0) = 0 \Rightarrow L(x,y) = 0$

5. (a) $f(0,0) = 1$, $f_x(x,y) = e^x \cos y \Rightarrow f_x(0,0) = 1$, $f_y(x,y) = -e^x \sin y \Rightarrow f_y(0,0) = 0$

   $\Rightarrow L(x,y) = 1 + 1(x - 0) + 0(y - 0) = x + 1$

   (b) $f\left(0,\frac{\pi}{2}\right) = 0$, $f_x\left(0,\frac{\pi}{2}\right) = 0$, $f_y\left(0,\frac{\pi}{2}\right) = -1 \Rightarrow L(x,y) = 0 + 0(x - 0) - 1\left(y - \frac{\pi}{2}\right) = -y + \frac{\pi}{2}$

6. (a) $f(0,0) = 1$, $f_x(x,y) = -e^{2y-x} \Rightarrow f_x(0,0) = -1$, $f_y(x,y) = 2e^{2y-x} \Rightarrow f_y(0,0) = 2$

   $\Rightarrow L(x,y) = 1 - 1(x - 0) + 2(y - 0) = -x + 2y + 1$

   (b) $f(1,2) = e^3$, $f_x(1,2) = -e^3$, $f_y(1,2) = 2e^3 \Rightarrow L(x,y) = e^3 - e^3(x - 1) + 2e^3(y - 2)$

   $= -e^3x + 2e^3y - 2e^3$

7. $f(2,1) = 3$, $f_x(x,y) = 2x - 3y \Rightarrow f_x(2,1) = 1$, $f_y(x,y) = -3x \Rightarrow f_y(2,1) = -6 \Rightarrow L(x,y) = 3 + 1(x-2) - 6(y-1)$ $= 7 + x - 6y$; $f_{xx}(x,y) = 2$, $f_{yy}(x,y) = 0$, $f_{xy}(x,y) = -3 \Rightarrow M = 3$; thus $|E(x,y)| \le \left(\frac{1}{2}\right)(3)\left(|x-2|+|y-1|\right)^2$ $\le \left(\frac{3}{2}\right)(0.1+0.1)^2 = 0.06$

8. $f(2,2) = 11$, $f_x(x,y) = x + y + 3 \Rightarrow f_x(2,2) = 7$, $f_y(x,y) = x + \frac{y}{2} - 3 \Rightarrow f_y(2,2) = 0$

$\Rightarrow L(x,y) = 11 + 7(x-2) + 0(y-2) = 7x - 3$; $f_{xx}(x,y) = 1$, $f_{yy}(x,y) = \frac{1}{2}$, $f_{xy}(x,y) = 1$

$\Rightarrow M = 1$; thus $|E(x,y)| \le \left(\frac{1}{2}\right)(1)\left(|x-2|+|y-2|\right)^2 \le \left(\frac{1}{2}\right)(0.1+0.1)^2 = 0.02$

9. $f(0,0) = 1$, $f_x(x,y) = \cos y \Rightarrow f_x(0,0) = 1$, $f_y(x,y) = 1 - x \sin y \Rightarrow f_y(0,0) = 1$

$\Rightarrow L(x,y) = 1 + 1(x-0) + 1(y-0) = x + y + 1$; $f_{xx}(x,y) = 0$, $f_{yy}(x,y) = -x \cos y$, $f_{xy}(x,y) = -\sin y \Rightarrow M = 1$; thus $|E(x,y)| \le \left(\frac{1}{2}\right)(1)\left(|x|+|y|\right)^2 \le \left(\frac{1}{2}\right)(0.2+0.2)^2 = 0.08$

10. $f(1,2) = 6$, $f_x(x,y) = y^2 - y \sin(x-1) \Rightarrow f_x(1,2) = 4$, $f_y(x,y) = 2xy + \cos(x-1) \Rightarrow f_y(1,2) = 5$

$\Rightarrow L(x,y) = 6 + 4(x-1) + 5(y-2) = 4x + 5y - 8$; $f_{xx}(x,y) = -y \cos(x-1)$, $f_{yy}(x,y) = 2x$,

$f_{xy}(x,y) = 2y - \sin(x-1)$; $|x-1| \le 0.1 \Rightarrow 0.9 \le x \le 1.1$ and $|y-2| \le 0.1 \Rightarrow 1.9 \le y \le 2.1$; thus the max of $|f_{xx}(x,y)|$ on R is 2.1, the max of $|f_{yy}(x,y)|$ on R is 2.2, and the max of $|f_{xy}(x,y)|$ on R is $2(2.1) - \sin(0.9-1)$

$\le 4.3 \Rightarrow M = 4.3$; thus $|E(x,y)| \le \left(\frac{1}{2}\right)(4.3)\left(|x-1|+|y-2|\right)^2 \le (2.15)(0.1+0.1)^2 = 0.086$

11. $f(0,0) = 1$, $f_x(x,y) = e^x \cos y \Rightarrow f_x(0,0) = 1$, $f_y(x,y) = -e^x \sin y \Rightarrow f_y(0,0) = 0$

$\Rightarrow L(x,y) = 1 + 1(x-0) + 0(y-0) = 1 + x$; $f_{xx}(x,y) = e^x \cos y$, $f_{yy}(x,y) = -e^x \cos y$, $f_{xy}(x,y) = -e^x \sin y$;

$|x| \le 0.1 \Rightarrow -0.1 \le x \le 0.1$ and $|y| \le 0.1 \Rightarrow -0.1 \le y \le 0.1$; thus the max of $|f_{xx}(x,y)|$ on R is $e^{0.1} \cos(0.1)$ ≤ 1.11, the max of $|f_{yy}(x,y)|$ on R is $e^{0.1} \cos(0.1) \le 1.11$, and the max of $|f_{xy}(x,y)|$ on R is $e^{0.1} \sin(0.1)$ $\le 0.002 \Rightarrow M = 1.11$; thus $|E(x,y)| \le \left(\frac{1}{2}\right)(1.11)\left(|x|+|y|\right)^2 \le (0.555)(0.1+0.1)^2 = 0.0222$

12. $f(1,1) = 0$, $f_x(x,y) = \frac{1}{x} \Rightarrow f_x(1,1) = 1$, $f_y(x,y) = \frac{1}{y} \Rightarrow f_y(1,1) = 1 \Rightarrow L(x,y) = 0 + 1(x-1) + 1(y-1)$ $= x + y - 2$; $f_{xx}(x,y) = -\frac{1}{x^2}$, $f_{yy}(x,y) = -\frac{1}{y^2}$, $f_{xy}(x,y) = 0$; $|x-1| \le 0.2 \Rightarrow 0.98 \le x \le 1.2$ so the max of $|f_{xx}(x,y)|$ on R is $\frac{1}{(0.98)^2} \le 1.04$; $|y-1| \le 0.2 \Rightarrow 0.98 \le y \le 1.2$ so the max of $|f_{yy}(x,y)|$ on R is $\frac{1}{(0.98)^2} \le 1.04 \Rightarrow M = 1.04$; thus $|E(x,y)| \le \left(\frac{1}{2}\right)(1.04)\left(|x-1|+|y-1|\right)^2 \le (0.52)(0.2+0.2)^2 = 0.0832$

13. $A = xy \Rightarrow dA = x\,dy + y\,dx$; if $x > y$ then a 1-unit change in y gives a greater change in dA than a 1-unit change in x. Thus, pay more attention to y which is the smaller of the two dimensions.

14. (a) $f_x(x,y) = 2x(y+1) \Rightarrow f_x(1,0) = 2$ and $f_y(x,y) = x^2 \Rightarrow f_y(1,0) = 1 \Rightarrow df = 2\,dx + 1\,dy \Rightarrow$ df is more sensitive to changes in x

(b) $df = 0 \Rightarrow 2\,dx + dy = 0 \Rightarrow 2\frac{dx}{dy} + 1 = 0 \Rightarrow \frac{dx}{dy} = -\frac{1}{2}$

15. $T_x(x,y) = e^y + e^{-y}$ and $T_y(x,y) = x(e^y - e^{-y}) \Rightarrow dT = T_x(x,y)\ dx + T_y(x,y)\ dy$
$= (e^y + e^{-y})dx + x(e^y - e^{-y})\ dy \Rightarrow dT|_{(2,\ln 2)} = 2.5\ dx + 3.0\ dy$. If $|dx| \le 0.1$ and $|dy| \le 0.02$, then the maximum possible error in the computed value of T is $(2.5)(0.1) + (3.0)(0.02) = 0.31$ in magnitude.

16. $V_r = 2\pi rh$ and $V_h = \pi r^2 \Rightarrow dV = V_r\ dr + V_h\ dh \Rightarrow \frac{dV}{V} = \frac{2\pi rh\ dr + \pi r^2\ dh}{\pi r^2 h} = \frac{2}{r}\ dr + \frac{1}{h}\ dh$; now $\left|\frac{dr}{r} \cdot 100\right| \le 1$ and $\left|\frac{dh}{h} \cdot 100\right| \le 1 \Rightarrow \left|\frac{dV}{V} \cdot 100\right| \le \left|\left(2\,\frac{dr}{r}\right)(100) + \left(\frac{dh}{h}\right)(100)\right| \le 2\left|\frac{dr}{r} \cdot 100\right| + \left|\frac{dh}{h} \cdot 100\right| \le 2(1) + 1 = 3 \Rightarrow 3\%$

17. $V_r = 2\pi rh$ and $V_h = \pi r^2 \Rightarrow dV = V_r\ dr + V_h\ dh \Rightarrow dV = 2\pi rh\ dr + \pi r^2\ dh \Rightarrow dV|_{(5,12)} = 120\pi\ dr + 25\pi\ dh$; $|dr| \le 0.1$ cm and $|dh| \le 0.1$ cm $\Rightarrow dV \le (120\pi)(0.1) + (25\pi)(0.1) = 14.5\pi\ \text{cm}^3$; $V(5,12) = 300\pi\ \text{cm}^3$
$\Rightarrow$ maximum percentage error is $\pm\frac{14.5\pi}{300\pi} \times 100 = \pm 4.83\%$

18. $V_r = 2\pi rh$ and $V_h = \pi r^2 \Rightarrow dV = V_r\ dr + V_h\ dh \Rightarrow dV = 2\pi rh\ dr = \pi r^2\ dh$; assuming $dr = dh$
$\Rightarrow dV = 2\pi rh\ dr + \pi r^2\ dr = (2\pi rh + \pi r^2)\ dr$; $dV \le 0.1\ \text{m}^3$ when $r = 2$ m and $h = 3$ m $\Rightarrow \left[2\pi(2)(3) + \pi(2)^2\right] dr \le 0.1 \Rightarrow dr \le \frac{0.1}{16\pi} \approx 0.001$ m (rounded down). Thus, the absolute value of the error in measuring r and h should be less than or equal to 0.002 m.

19. $df = f_x(x,y)\ dx + f_y(x,y)\ dy = 3x^2y^4\ dx + 4x^3y^3\ dy \Rightarrow df|_{(1,1)} = 3\ dx + 4\ dy$; for a square, $dx = dy$
$\Rightarrow df = 7\ dx$ so that $|df| \le 0.1 \Rightarrow 7|dx| \le 0.1 \Rightarrow |dx| \le \frac{0.1}{7} \approx 0.014 \Rightarrow$ for the square, $|x - 1| \le 0.014$ and $|y - 1| \le 0.014$

20. (a) $\frac{1}{R} = \frac{1}{R_1} + \frac{1}{R_2} \Rightarrow -\frac{1}{R^2}\ dR = -\frac{1}{R_1^2}\ dR_1 - \frac{1}{R_2^2}\ dR_2 \Rightarrow dR = \left(\frac{R}{R_1}\right)^2 dR_1 + \left(\frac{R}{R_2}\right)^2 dR_2$

(b) $dR = R^2\left[\left(\frac{1}{R_1^2}\right) dR_1 + \left(\frac{1}{R_2^2}\right) dR_2\right] \Rightarrow dR|_{(100,400)} = R^2\left[\frac{1}{(100)^2}\ dR_1 + \frac{1}{(400)^2}\ dR_2\right] \Rightarrow$ R will be more sensitive to a variation in R_1 since $\frac{1}{(100)^2} > \frac{1}{(400)^2}$

21. From Exercise 20, $dR = \left(\frac{R}{R_1}\right)^2 dR_1 + \left(\frac{R}{R_2}\right)^2 dR_2$ so that R_1 changing from 20 to 20.1 ohms $\Rightarrow dR_1 = 0.1$ ohm and R_2 changing from 25 to 24.9 ohms $\Rightarrow dR_2 = -0.1$ ohms; $\frac{1}{R} = \frac{1}{R_1} + \frac{1}{R_2} \Rightarrow R = \frac{100}{9}$ ohms
$\Rightarrow dR|_{(20,25)} = \frac{\left(\frac{100}{9}\right)^2}{(20)^2}(0.1) + \frac{\left(\frac{100}{9}\right)^2}{(25)^2}(-0.1) \approx 0.011$ ohms $\Rightarrow$ percentage change is $\left.\frac{dR}{R}\right|_{(20,25)} \times 100$
$= \frac{0.011}{\left(\frac{100}{9}\right)} \times 100 \approx 0.1\%$

22. (a) $r^2 = x^2 + y^2 \Rightarrow 2r\ dr = 2x\ dx + 2y\ dy \Rightarrow dr = \frac{x}{r}\ dx + \frac{y}{r}\ dy \Rightarrow dr|_{(3,4)} = \left(\frac{3}{5}\right)(\pm 0.01) + \left(\frac{4}{5}\right)(\pm 0.01)$
$= \pm\frac{0.07}{5} = \pm 0.014 \Rightarrow \left|\frac{dr}{r} \times 100\right| = \left|\pm\frac{0.014}{5} \times 100\right| = 0.28\%$; $d\theta = \frac{\left(-\frac{y}{x^2}\right)}{\left(\frac{y}{x}\right)^2 + 1}\ dx + \frac{\left(\frac{1}{x}\right)}{\left(\frac{y}{x}\right)^2 + 1}\ dy$

$= \frac{-y}{y^2+x^2}\,dx + \frac{x}{y^2+x^2}\,dy \Rightarrow d\theta|_{(3,4)} = \left(\frac{-4}{25}\right)(\pm 0.01) + \left(\frac{3}{25}\right)(\pm 0.01) = \frac{\mp 0.04}{25} + \frac{\pm 0.03}{25}$

$\Rightarrow$ maximum change in $d\theta$ occurs when dx and dy have opposite signs ($dx = 0.01$ and $dy = -0.01$ or vice versa) $\Rightarrow d\theta = \frac{\pm 0.07}{25} \approx \pm 0.0028$; $\theta = \tan^{-1}\left(\frac{4}{3}\right) \approx 0.927255218 \Rightarrow \left|\frac{d\theta}{\theta} \times 100\right| = \left|\frac{\pm 0.0028}{0.927255218} \times 100\right|$ $\approx 0.30\%$

(b) the radius r is more sensitive to changes in y, and the angle θ is more sensitive to changes in x

23. (a) $f(1,1,1) = 3$, $f_x(1,1,1) = y+z|_{(1,1,1)} = 2$, $f_y(1,1,1) = x+z|_{(1,1,1)} = 2$, $f_z(1,1,1) = y+x|_{(1,1,1)} = 2$ $\Rightarrow L(x,y,z) = 3 + 2(x-1) + 2(y-1) + 2(z-1) = 2x + 2y + 2z - 3$

(b) $f(1,0,0) = 0$, $f_x(1,0,0) = 0$, $f_y(1,0,0) = 1$, $f_z(1,0,0) = 1 \Rightarrow L(x,y,z) = 0 + 0(x-1) + (y-0) + (z-0)$ $= y + z$

(c) $f(0,0,0) = 0$, $f_x(0,0,0) = 0$, $f_y(0,0,0) = 0$, $f_z(0,0,0) = 0 \Rightarrow L(x,y,z) = 0$

24. (a) $f(1,1,1) = 3$, $f_x(1,1,1) = 2x|_{(1,1,1)} = 2$, $f_y(1,1,1) = 2y|_{(1,1,1)} = 2$, $f_z(1,1,1) = 2z|_{(1,1,1)} = 2$ $\Rightarrow L(x,y,z) = 3 + 2(x-1) + 2(y-1) + 2(z-1) = 2x + 2y + 2z - 3$

(b) $f(0,1,0) = 1$, $f_x(0,1,0) = 0$, $f_y(0,1,0) = 2$, $f_z(0,1,0) = 0 \Rightarrow L(x,y,z) = 1 + 0(x-0) + 2(y-1) + 0(z-0)$ $= 2y - 1$

(c) $f(1,0,0) = 1$, $f_x(1,0,0) = 2$, $f_y(1,0,0) = 0$, $f_z(1,0,0) = 0 \Rightarrow L(x,y,z) = 1 + 2(x-1) + 0(y-0) + 0(z-0)$ $= 2x - 1$

25. (a) $f(1,0,0) = 1$, $f_x(1,0,0) = \frac{x}{\sqrt{x^2+y^2+z^2}}\Big|_{(1,0,0)} = 1$, $f_y(1,0,0) = \frac{y}{\sqrt{x^2+y^2+z^2}}\Big|_{(1,0,0)} = 0$, $f_z(1,0,0) = \frac{z}{\sqrt{x^2+y^2+z^2}}\Big|_{(1,0,0)} = 0 \Rightarrow L(x,y,z) = 1 + 1(x-1) + 0(y-0) + 0(z-0) = x$

(b) $f(1,1,0) = \sqrt{2}$, $f_x(1,1,0) = \frac{1}{\sqrt{2}}$, $f_y(1,1,0) = \frac{1}{\sqrt{2}}$, $f_z(1,1,0) = 0$ $\Rightarrow L(x,y,z) = \sqrt{2} + \frac{1}{\sqrt{2}}(x-1) + \frac{1}{\sqrt{2}}(y-1) + 0(z-0) = \frac{1}{\sqrt{2}}x + \frac{1}{\sqrt{2}}y$

(c) $f(1,2,2) = 3$, $f_x(1,2,2) = \frac{1}{3}$, $f_y(1,2,2) = \frac{2}{3}$, $f_z(1,2,2) = \frac{2}{3} \Rightarrow L(x,y,z) = 3 + \frac{1}{3}(x-1) + \frac{2}{3}(y-2) + \frac{2}{3}(z-2)$ $= \frac{1}{3}x + \frac{2}{3}y + \frac{2}{3}z$

26. (a) $f\left(\frac{\pi}{2},1,1\right) = 1$, $f_x\left(\frac{\pi}{2},1,1\right) = \frac{y\cos xy}{z}\Big|_{\left(\frac{\pi}{2},1,1\right)} = 0$, $f_y\left(\frac{\pi}{2},1,1\right) = \frac{x\cos xy}{z}\Big|_{\left(\frac{\pi}{2},1,1\right)} = 0$, $f_z\left(\frac{\pi}{2},1,1\right) = \frac{-\sin xy}{z^2}\Big|_{\left(\frac{\pi}{2},1,1\right)} = -1 \Rightarrow L(x,y,z) = 1 + 0\left(x - \frac{\pi}{2}\right) + 0(y-1) - 1(z-1) = 2 - z$

(b) $f(2,0,1) = 0$, $f_x(2,0,1) = 0$, $f_y(2,0,1) = 2$, $f_z(2,0,1) = 0 \Rightarrow L(x,y,z) = 0 + 0(x-2) + 2(y-0) + 0(z-1) = 2y$

27. (a) $f(0,0,0) = 2$, $f_x(0,0,0) = e^x|_{(0,0,0)} = 1$, $f_y(0,0,0) = -\sin(y+z)|_{(0,0,0)} = 0$, $f_z(0,0,0) = -\sin(y+z)|_{(0,0,0)} = 0 \Rightarrow L(x,y,z) = 2 + 1(x-0) + 0(y-0) + 0(z-0) = 2 + x$

(b) $f\left(0,\frac{\pi}{2},0\right)=1$, $f_x\left(0,\frac{\pi}{2},0\right)=1$, $f_y\left(0,\frac{\pi}{2},0\right)=-1$, $f_z\left(0,\frac{\pi}{2},0\right)=-1 \Rightarrow L(x,y,z)$
$=1+1(x-0)-1\left(y-\frac{\pi}{2}\right)-1(z-0)=x-y-z+\frac{\pi}{2}+1$

(c) $f\left(0,\frac{\pi}{4},\frac{\pi}{4}\right)=1$, $f_x\left(0,\frac{\pi}{4},\frac{\pi}{4}\right)=1$, $f_y\left(0,\frac{\pi}{4},\frac{\pi}{4}\right)=-1$, $f_z\left(0,\frac{\pi}{4},\frac{\pi}{4}\right)=-1 \Rightarrow L(x,y,z)$
$=1+1(x-0)-1\left(y-\frac{\pi}{4}\right)-1\left(z-\frac{\pi}{4}\right)=x-y-z+\frac{\pi}{2}+1$

28. (a) $f(1,0,0)=0$, $f_x(1,0,0)=\left.\frac{yz}{(xyz)^2+1}\right|_{(1,0,0)}=0$, $f_y(1,0,0)=\left.\frac{xz}{(xyz)^2+1}\right|_{(1,0,0)}=0$,

$f_z(1,0,0)=\left.\frac{xy}{(xyz)^2+1}\right|_{(1,0,0)}=0 \Rightarrow L(x,y,z)=0$

(b) $f(1,1,0)=0$, $f_x(1,1,0)=0$, $f_y(1,1,0)=0$, $f_z(1,1,0)=1 \Rightarrow L(x,y,z)=0+0(x-1)+0(y-1)+1(z-0)=z$

(c) $f(1,1,1)=\frac{\pi}{4}$, $f_x(1,1,1)=\frac{1}{2}$, $f_y(1,1,1)=\frac{1}{2}$, $f_z(1,1,1)=\frac{1}{2} \Rightarrow L(x,y,z)=\frac{\pi}{4}+\frac{1}{2}(x-1)+\frac{1}{2}(y-1)+\frac{1}{2}(z-1)$
$=\frac{1}{2}x+\frac{1}{2}y+\frac{1}{2}z+\frac{\pi}{4}-\frac{3}{2}$

29. $f(x,y,z)=xz-3yz+2$ at $P_0(1,1,2) \Rightarrow f(1,1,2)=-2$; $f_x=z$, $f_y=-3z$, $f_z=x-3y \Rightarrow L(x,y,z)$
$=-2+2(x-1)-6(y-1)-2(z-2)=2x-6y-2z+6$; $f_{xx}=0$, $f_{yy}=0$, $f_{zz}=0$, $f_{xy}=0$, $f_{yz}=-3$
$\Rightarrow M=3$; thus, $|E(x,y,z)|\le\left(\frac{1}{2}\right)(3)(0.01+0.01+0.02)^2=0.0024$

30. $f(x,y,z)=x^2+xy+yz+\frac{1}{4}z^2$ at $P_0(1,1,2) \Rightarrow f(1,1,2)=5$; $f_x=2x+y$, $f_y=x+z$, $f_z=y+\frac{1}{2}z$
$\Rightarrow L(x,y,z)=5+3(x-1)+3(y-1)+2(z-2)=3x+3y+2z-5$; $f_{xx}=2$, $f_{yy}=0$, $f_{zz}=\frac{1}{2}$, $f_{xy}=1$, $f_{xz}=0$,
$f_{yz}=1 \Rightarrow M=2$; thus $|E(x,y,z)|\le\left(\frac{1}{2}\right)(2)(0.01+0.01+0.08)^2=0.01$

31. $f(x,y,z)=xy+2yz-3xz$ at $P_0(1,1,0) \Rightarrow f(1,1,0)=1$; $f_x=y-3z$, $f_y=x+2z$, $f_z=2y-3x$
$\Rightarrow L(x,y,z)=1+(x-1)+(y-1)-(z-0)=x+y-z-1$; $f_{xx}=0$, $f_{yy}=0$, $f_{zz}=0$, $f_{xy}=1$, $f_{xz}=-3$,
$f_{yz}=2 \Rightarrow M=3$; thus $|E(x,y,z)|\le\left(\frac{1}{2}\right)(3)(0.01+0.01+0.01)^2=0.00135$

32. $f(x,y,z)=\sqrt{2}\cos x\sin(y+z)$ at $P_0\left(x,y,\frac{\pi}{4}\right) \Rightarrow f\left(0,0,\frac{\pi}{4}\right)=1$; $f_x=-\sqrt{2}\sin x\sin(y+z)$,
$f_y=\sqrt{2}\cos x\cos(y+z)$, $f_z=\sqrt{2}\cos x\cos(y+z) \Rightarrow L(x,y,z)=1-0(x-0)+(y-0)+\left(z-\frac{\pi}{4}\right)$
$=y+z-\frac{\pi}{4}+1$; $f_{xx}=-\sqrt{2}\cos x\sin(y+z)$, $f_{yy}=-\sqrt{2}\cos x\sin(y+z)$, $f_{zz}=-\sqrt{2}\cos x\sin(y+z)$,
$f_{xy}=-\sqrt{2}\sin x\cos(y+z)$, $f_{xz}=-\sqrt{2}\sin x\cos(y+z)$, $f_{yz}=-\sqrt{2}\cos x\sin(y+z)$. The absolute value of each of these second partial derivatives is bounded above by $\sqrt{2} \Rightarrow M=\sqrt{2}$; thus $|E(x,y,z)|$
$\le\left(\frac{1}{2}\right)\left(\sqrt{2}\right)(0.01+0.01+0.01)^2=0.000636$.

33. (a) $dS=S_p\,dp+S_x\,dx+S_w\,dw+S_h\,dh=C\left(\frac{x^4}{wh^3}\,dp+\frac{4px^3}{wh^3}\,dx-\frac{px^4}{w^2h^3}\,dw-\frac{3px^4}{wh^4}\,dh\right)$
$=C\left(\frac{px^4}{wh^3}\right)\left(\frac{1}{p}\,dp+\frac{4}{x}\,dx-\frac{1}{w}\,dw-\frac{3}{h}\,dh\right)=S_0\left(\frac{1}{p_0}\,dp+\frac{4}{x_0}\,dx-\frac{1}{w_0}\,dw-\frac{3}{h_0}\,dh\right)$

$= S_0\left(\frac{1}{100}\,dp + dx - 5\,dw - 30\,dh\right)$, where $p_0 = 100$ N/m, $x_0 = 4$ m, $w_0 = 0.2$ m, $h_0 = 0.1$ m

(b) More sensitive to a change in height

34. (a) $V = \pi r^2 h \Rightarrow dV = 2\pi rh\,dr + \pi r^2\,dh \Rightarrow$ at $r = 1$ and $h = 5$ we have $dV = 10\pi\,dr + \pi\,dh \Rightarrow$ the volume is about 10 times more sensitive to a change in r

(b) $dV = 0 \Rightarrow 0 = 2\pi rh\,dr + \pi r^2\,dh = 2h\,dr + r\,dh = 10\,dr + dh \Rightarrow dr = -\frac{1}{10}\,dh$; choose $dh = 1.5$ $\Rightarrow dr = -0.15 \Rightarrow h = 6.5$ in. and $r = 0.85$ in. is one solution for $\Delta V \approx dV = 0$

35. $f(a,b,c,d) = \begin{vmatrix} a & b \\ c & d \end{vmatrix} = ad - bc \Rightarrow f_a = d,\ f_b = -c,\ f_c = -b,\ f_d = a \Rightarrow df = d\,da - c\,db - b\,dc + a\,dd$; since $|a|$ is much greater than $|b|$, $|c|$, and $|d|$, the function f is most sensitive to a change in d.

36. $p(a,b,c) = abc \Rightarrow p_a = bc,\ p_b = ac,\ p_c = ab \Rightarrow dp = bc\,da + ac\,db + ab\,dc \Rightarrow \frac{dp}{p} = \frac{bc\,da + ac\,db + ab\,dc}{abc}$ $= \frac{da}{a} + \frac{db}{b} + \frac{dc}{c}$. Now $\left|\frac{da}{a}\cdot 100\right| = 2$, $\left|\frac{db}{b}\cdot 100\right| = 2$, and $\left|\frac{dc}{c}\cdot 100\right| = 2 \Rightarrow \left|\frac{dp}{p}\cdot 100\right|$ $= \left|\frac{da}{a}\cdot 100 + \frac{db}{b}\cdot 100 + \frac{dc}{c}\cdot 100\right| \le \left|\frac{da}{a}\cdot 100\right| + \left|\frac{db}{b}\cdot 100\right| + \left|\frac{dc}{c}\cdot 100\right| = 2 + 2 + 2 = 6$ or 6%

37. $V = lwh \Rightarrow V_l = wh,\ V_w = lh,\ V_h = lw \Rightarrow dV = wh\,dl + lh\,dw + lw\,dh \Rightarrow dV|_{(5,3,2)} = 6\,dl + 10\,dw + 15\,dh$; $dl = 1$ in. $= \frac{1}{12}$ ft, $dw = 1$ in. $= \frac{1}{12}$ ft, $dh = \frac{1}{2}$ in. $= \frac{1}{24}$ ft $\Rightarrow dV = 6\left(\frac{1}{12}\right) + 10\left(\frac{1}{12}\right) + 15\left(\frac{1}{24}\right) = \frac{47}{24}$ ft^3

38. $A = \frac{1}{2}ab\sin C \Rightarrow A_a = \frac{1}{2}b\sin C,\ A_b = \frac{1}{2}a\sin C,\ A_c = \frac{1}{2}ab\cos C$ $\Rightarrow dA = \left(\frac{1}{2}b\sin C\right)da + \left(\frac{1}{2}a\sin C\right)db + \left(\frac{1}{2}ab\cos C\right)dC$; $dC = |2°| = |0.0349|$ radians, $da = |0.5|$ ft, $db = |0.5|$ ft; at $a = 150$ ft, $b = 200$ ft, and $C = 60°$, we see that the change is approximately $dA = \frac{1}{2}(200)(\sin 60°)\,|0.5| + \frac{1}{2}(150)(\sin 60°)\,|0.5| + \frac{1}{2}(200)(150)(\cos 60°)\,|0.0349| = \pm 338$ ft^2

39. $u_x = e^y,\ u_y = xe^y + \sin z,\ u_z = y\cos z \Rightarrow du = e^y\,dx + (xe^y + \sin z)\,dy + (y\cos z)\,dz$ $\Rightarrow du|_{(2,\ln 3,\frac{\pi}{2})} = 3\,dx + 7\,dy + 0\,dz = 3\,dx + 7\,dy \Rightarrow$ magnitude of the maximum possible error $\le 3(0.2) + 7(0.6) = 4.8$

40. $Q_K = \frac{1}{2}\left(\frac{2KM}{h}\right)^{-1/2}\left(\frac{2M}{h}\right)$, $Q_M = \frac{1}{2}\left(\frac{2KM}{h}\right)^{-1/2}\left(\frac{2K}{h}\right)$, and $Q_h = \frac{1}{2}\left(\frac{2KM}{h}\right)^{-1/2}\left(\frac{-2KM}{h^2}\right)$

$\Rightarrow dQ = \frac{1}{2}\left(\frac{2KM}{h}\right)^{-1/2}\left(\frac{2M}{h}\right)dK + \frac{1}{2}\left(\frac{2KM}{h}\right)^{-1/2}\left(\frac{2K}{h}\right)dM + \frac{1}{2}\left(\frac{2KM}{h}\right)^{-1/2}\left(\frac{-2KM}{h^2}\right)dh$

$= \frac{1}{2}\left(\frac{2KM}{h}\right)^{-1/2}\left[\frac{2M}{h}\,dK + \frac{2K}{h}\,dM - \frac{2KM}{h^2}\,dh\right] \Rightarrow dQ|_{(2,20,0.0.05)}$

$= \frac{1}{2}\left[\frac{(2)(2)(20)}{0.05}\right]^{-1/2}\left[\frac{(2)(20)}{0.05}\,dK + \frac{(2)(2)}{0.05}\,dM - \frac{(2)(2)(20)}{(0.05)^2}\,dh\right] = (0.0125)(800\,dK + 80\,dM - 32{,}000\,dh)$

$\Rightarrow$ Q is most sensitive to changes in h

41. If the first partial derivatives are continuous throughout an open region R, then by Eq. (3) in this section of the text, $f(x,y) = f(x_0,y_0) + f_x(x_0,y_0)\,\Delta x + f_y(x_0,y_0)\,\Delta y + \epsilon_1 \Delta x + \epsilon_2 \Delta y$, where $\epsilon_1, \epsilon_2 \to 0$ as $\Delta x, \Delta y \to 0$. Then as $(x,y) \to (x_0,y_0)$, $\Delta x \to 0$ and $\Delta y \to 0 \Rightarrow \lim_{(x,y)\to(x_0,y_0)} f(x,y) = f(x_0,y_0) \Rightarrow$ f is continuous at every point (x_0,y_0) in R.

42. Yes, since f_{xx}, f_{yy}, f_{xy}, and f_{yx} are all continuous on R, use the same reasoning as in Exercise 41 with $f_x(x,y) = f_x(x_0,y_0) + f_{xx}(x_0,y_0)\,\Delta x + f_{xy}(x_0,y_0)\,\Delta y + \epsilon_1 \Delta x + \epsilon_2 \Delta y$ and $f_y(x,y) = f_y(x_0,y_0) + f_{yx}(x_0,y_0)\,\Delta x + f_{yy}(x_0,y_0)\,\Delta y + \hat{\epsilon}_1 \Delta x + \hat{\epsilon}_2 \Delta y$. Then $\lim_{(x,y)\to(x_0,y_0)} f_x(x,y) = f_x(x_0,y_0)$ and $\lim_{(x,y)\to(x_0,y_0)} f_y(x,y) = f_y(x_0,y_0)$.

12.5 THE CHAIN RULE

1. (a) $\frac{\partial w}{\partial x} = 2x,\ \frac{\partial w}{\partial y} = 2y,\ \frac{dx}{dt} = -\sin t,\ \frac{dy}{dt} = \cos t \Rightarrow \frac{dw}{dt} = -2x \sin t + 2y \cos t = -2 \cos t \sin t + 2 \sin t \cos t$
$= 0;\ w = x^2 + y^2 = \cos^2 t + \sin^2 t = 1 \Rightarrow \frac{dw}{dt} = 0$

   (b) $\frac{dw}{dt}(\pi) = 0$

2. (a) $\frac{\partial w}{\partial x} = 2x,\ \frac{\partial w}{\partial y} = 2y,\ \frac{dx}{dt} = -\sin t + \cos t,\ \frac{dy}{dt} = -\sin t - \cos t \Rightarrow \frac{dw}{dt}$
$= (2x)(-\sin t + \cos t) + (2y)(-\sin t - \cos t)$
$= 2(\cos t + \sin t)(\cos t - \sin t) - 2(\cos t - \sin t)(\sin t + \cos t) = (2\cos^2 t - 2\sin^2 t) - (2\cos^2 t - 2\sin^2 t)$
$= 0;\ w = x^2 + y^2 = (\cos t + \sin t)^2 + (\cos t - \sin t)^2 = 2\cos^2 t + 2\sin^2 t = 2 \Rightarrow \frac{dw}{dt} = 0$

   (b) $\frac{dw}{dt}(0) = 0$

3. (a) $\frac{\partial w}{\partial x} = \frac{1}{z},\ \frac{\partial w}{\partial y} = \frac{1}{z},\ \frac{\partial w}{\partial z} = \frac{-(x+y)}{z^2},\ \frac{dx}{dt} = -2\cos t \sin t,\ \frac{dy}{dt} = 2 \sin t \cos t,\ \frac{dz}{dt} = -\frac{1}{t^2}$
$\Rightarrow \frac{dw}{dt} = -\frac{2}{z}\cos t \sin t + \frac{2}{z}\sin t \cos t + \frac{x+y}{z^2 t^2} = \frac{\cos^2 t + \sin^2 t}{\left(\frac{1}{t^2}\right)(t^2)} = 1;\ w = \frac{x}{z} + \frac{y}{z} = \frac{\cos^2 t}{\left(\frac{1}{t}\right)} + \frac{\sin^2 t}{\left(\frac{1}{t}\right)} = t \Rightarrow \frac{dw}{dt} = 1$

   (b) $\frac{dw}{dt}(3) = 1$

4. (a) $\frac{\partial w}{\partial x} = \frac{2x}{x^2+y^2+z^2},\ \frac{\partial w}{\partial y} = \frac{2y}{x^2+y^2+z^2},\ \frac{\partial w}{\partial z} = \frac{2z}{x^2+y^2+z^2},\ \frac{dx}{dt} = -\sin t,\ \frac{dy}{dt} = \cos t,\ \frac{dz}{dt} = 2t^{-1/2}$
$\Rightarrow \frac{dw}{dt} = \frac{-2x \sin t}{x^2+y^2+z^2} + \frac{2y \cos t}{x^2+y^2+z^2} + \frac{4zt^{-1/2}}{x^2+y^2+z^2} = \frac{-2\cos t \sin t + 2 \sin t \cos t + 4\left(4t^{-1/2}\right)t^{-1/2}}{\cos^2 t + \sin^2 t + 16t}$
$= \frac{16}{1+16t};\ w = \ln\left(x^2+y^2+z^2\right) = \ln\left(\cos^2 t + \sin^2 t + 16t\right) = \ln(1+16t) \Rightarrow \frac{dw}{dt} = \frac{16}{1+16t}$

   (b) $\frac{dw}{dt}(3) = \frac{16}{49}$

5. (a) $\frac{\partial w}{\partial x} = 2ye^x,\ \frac{\partial w}{\partial y} = 2e^x,\ \frac{\partial w}{\partial z} = -\frac{1}{z},\ \frac{dx}{dt} = \frac{2t}{t^2+1},\ \frac{dy}{dt} = \frac{1}{t^2+1},\ \frac{dz}{dt} = e^t \Rightarrow \frac{dw}{dt} = \frac{4yte^x}{t^2+1} + \frac{2e^x}{t^2+1} - \frac{e^t}{z}$
$= \frac{(4t)\left(\tan^{-1} t\right)\left(t^2+1\right)}{t^2+1} + \frac{2\left(t^2+1\right)}{t^2+1} - \frac{e^t}{e^t} = 4t \tan^{-1} t + 1;\ w = 2ye^x - \ln z = \left(2\tan^{-1} t\right)\left(t^2+1\right) - t$

$\Rightarrow \frac{dw}{dt} = \left(\frac{2}{t^2+1}\right)(t^2+1) + (2\tan^{-1}t)(2t) - 1 = 4t\tan^{-1}t + 1$

(b) $\frac{dw}{dt}(1) = (4)(1)\left(\frac{\pi}{4}\right) + 1 = \pi + 1$

6. (a) $\frac{\partial w}{\partial x} = -y\cos xy,\ \frac{\partial w}{\partial y} = -x\cos xy,\ \frac{\partial w}{\partial z} = 1,\ \frac{dx}{dt} = 1,\ \frac{dy}{dt} = \frac{1}{t},\ \frac{dz}{dt} = e^{t-1} \Rightarrow \frac{dw}{dt} = -y\cos xy - \frac{x\cos xy}{t} + e^{t-1}$

$= -(\ln t)[\cos(t\ln t)] - \frac{t\cos(t\ln t)}{t} + e^{t-1} = -(\ln t)[\cos(t\ln t)] - \cos(t\ln t) + e^{t-1};\ w = z - \sin xy$

$= e^{t-1} - \sin(t\ln t) \Rightarrow \frac{dw}{dt} = e^{t-1} - [\cos(t\ln t)]\left[\ln t + t\left(\frac{1}{t}\right)\right] = e^{t-1} - (1+\ln t)\cos(t\ln t)$

(b) $\frac{dw}{dt}(1) = 1 - (1+0)(1) = 0$

7. (a) $\frac{\partial z}{\partial r} = \frac{\partial z}{\partial x}\frac{\partial x}{\partial r} + \frac{\partial z}{\partial y}\frac{\partial y}{\partial r} = (4e^x\ln y)\left(\frac{\cos\theta}{r\cos\theta}\right) + \left(\frac{4e^x}{y}\right)(\sin\theta) = \frac{4e^x\ln y}{r} + \frac{4e^x\sin\theta}{y}$

$= \frac{4(r\cos\theta)\ln(r\sin\theta)}{r} + \frac{4(r\cos\theta)(\sin\theta)}{r\sin\theta} = (4\cos\theta)\ln(r\sin\theta) + 4\cos\theta;$

$\frac{\partial z}{\partial \theta} = \frac{\partial z}{\partial x}\frac{\partial x}{\partial \theta} + \frac{\partial z}{\partial y}\frac{\partial y}{\partial \theta} = (4e^x\ln y)\left(\frac{-r\sin\theta}{r\cos\theta}\right) + \left(\frac{4e^x}{y}\right)(r\cos\theta) = -(4e^x\ln y)(\tan\theta) + \frac{4e^x r\cos\theta}{y}$

$= [-4(r\cos\theta)\ln(r\sin\theta)](\tan\theta) + \frac{4(r\cos\theta)(r\cos\theta)}{r\sin\theta} = (-4r\sin\theta)\ln(r\sin\theta) + \frac{4r\cos^2\theta}{\sin\theta};$

$z = 4e^x\ln y = 4(r\cos\theta)\ln(r\sin\theta) \Rightarrow \frac{\partial z}{\partial r} = (4\cos\theta)\ln(r\sin\theta) + 4(r\cos\theta)\left(\frac{\sin\theta}{r\sin\theta}\right)$

$= (4\cos\theta)\ln(r\sin\theta) + 4\cos\theta$; also $\frac{\partial z}{\partial \theta} = (-4r\sin\theta)\ln(r\sin\theta) + 4(r\cos\theta)\left(\frac{r\cos\theta}{r\sin\theta}\right)$

$= (-4r\sin\theta)\ln(r\sin\theta) + \frac{4r\cos^2\theta}{\sin\theta}$

(b) At $\left(2, \frac{\pi}{4}\right)$: $\frac{\partial z}{\partial r} = 4\cos\frac{\pi}{4}\ln\left(2\sin\frac{\pi}{4}\right) + 4\cos\frac{\pi}{4} = 2\sqrt{2}\ln\sqrt{2} + 2\sqrt{2} = \sqrt{2}(\ln 2 + 2);$

$\frac{\partial z}{\partial \theta} = (-4)(2)\sin\frac{\pi}{4}\ln\left(2\sin\frac{\pi}{4}\right) + \frac{(4)(2)\left(\cos^2\frac{\pi}{4}\right)}{\left(\sin\frac{\pi}{4}\right)} = -4\sqrt{2}\ln\sqrt{2} + 4\sqrt{2} = -2\sqrt{2}\ln 2 + 4\sqrt{2}$

8. (a) $\frac{\partial z}{\partial r} = \left[\frac{\left(\frac{1}{y}\right)}{\left(\frac{x}{y}\right)^2 + 1}\right]\cos\theta + \left[\frac{\left(\frac{-x}{y^2}\right)}{\left(\frac{x}{y}\right)^2 + 1}\right]\sin\theta = \frac{y\cos\theta}{x^2+y^2} - \frac{x\sin\theta}{x^2+y^2} = \frac{(r\sin\theta)(\cos\theta) - (r\cos\theta)(\sin\theta)}{r^2} = 0;$

$\frac{\partial z}{\partial \theta} = \left[\frac{\left(\frac{1}{y}\right)}{\left(\frac{x}{y}\right)^2 + 1}\right](-r\sin\theta) + \left[\frac{\left(\frac{-x}{y^2}\right)}{\left(\frac{x}{y}\right)^2 + 1}\right]r\cos\theta = -\frac{yr\sin\theta}{x^2+y^2} - \frac{xr\cos\theta}{x^2+y^2} = \frac{-(r\sin\theta)(r\sin\theta) - (r\cos\theta)(r\cos\theta)}{r^2}$

$= -\sin^2\theta - \cos^2\theta = -1;\ z = \tan^{-1}\left(\frac{x}{y}\right) = \tan^{-1}(\cot\theta) \Rightarrow \frac{\partial z}{\partial r} = 0$ and $\frac{\partial z}{\partial \theta} = \left(\frac{1}{1+\cot^2\theta}\right)(-\csc^2\theta)$

$= \frac{-1}{\sin^2\theta + \cos^2\theta} = -1$

(b) At $\left(1.3, \frac{\pi}{6}\right)$: $\frac{\partial z}{\partial r} = 0$ and $\frac{\partial z}{\partial \theta} = -1$

9. (a) $\frac{\partial w}{\partial u} = \frac{\partial w}{\partial x}\frac{\partial x}{\partial u} + \frac{\partial w}{\partial y}\frac{\partial y}{\partial u} + \frac{\partial w}{\partial z}\frac{\partial z}{\partial u} = (y+z)(1) + (x+z)(1) + (y+x)(v) = x + y + 2z + v(y+x)$

$= (u+v) + (u-v) + 2uv + v(2u) = 2u + 4uv;\ \frac{\partial w}{\partial v} = \frac{\partial w}{\partial x}\frac{\partial x}{\partial v} + \frac{\partial w}{\partial y}\frac{\partial y}{\partial v} + \frac{\partial w}{\partial z}\frac{\partial z}{\partial v}$

$= (y+z)(1) + (x+z)(-1) + (y+x)(u) = y - x + (y+x)u = -2v + (2u)u = -2v + 2u^2$;
$w = xy + yz + xz = (u^2 - v^2) + (u^2v - uv^2) + (u^2v + uv^2) = u^2 - v^2 + 2u^2v \Rightarrow \frac{\partial w}{\partial u} = 2u + 4uv$ and
$\frac{\partial w}{\partial v} = -2v + 2u^2$

(b) At $\left(\frac{1}{2}, 1\right)$: $\frac{\partial w}{\partial u} = 2\left(\frac{1}{2}\right) + 4\left(\frac{1}{2}\right)(1) = 3$ and $\frac{\partial w}{\partial v} = -2(1) + 2\left(\frac{1}{2}\right)^2 = -\frac{3}{2}$

10. (a) $\frac{\partial w}{\partial u} = \left(\frac{2x}{x^2+y^2+z^2}\right)(e^v \sin u + ue^v \cos u) + \left(\frac{2y}{x^2+y^2+z^2}\right)(e^v \cos u - ue^v \sin u) + \left(\frac{2z}{x^2+y^2+z^2}\right)(e^v)$

$$= \left(\frac{2ue^v \sin u}{u^2e^{2v}\sin^2 u + u^2e^{2v}\cos^2 u + u^2e^{2v}}\right)(e^v \sin u + ue^v \cos u)$$

$$+ \left(\frac{2ue^v \cos u}{u^2e^{2v}\sin^2 u + u^2e^{2v}\cos^2 u + u^2e^{2v}}\right)(e^v \cos u - ue^v \sin u)$$

$$+ \left(\frac{2ue^v}{u^2e^{2v}\sin^2 u + u^2e^{2v}\cos^2 u + u^2e^{2v}}\right)(e^v) = \frac{2}{u};$$

$$\frac{\partial w}{\partial v} = \left(\frac{2x}{x^2+y^2+z^2}\right)(ue^v \sin u) + \left(\frac{2y}{x^2+y^2+z^2}\right)(ue^v \cos u) + \left(\frac{2z}{x^2+y^2+z^2}\right)(ue^v)$$

$$= \left(\frac{2ue^v \sin u}{u^2e^{2v}\sin^2 u + u^2e^{2v}\cos^2 u + u^2e^{2v}}\right)(ue^v \sin u)$$

$$+ \left(\frac{2ue^v \cos u}{u^2e^{2v}\sin^2 u + u^2e^{2v}\cos^2 u + u^2e^{2v}}\right)(ue^v \cos u)$$

$+ \left(\frac{2ue^v}{u^2e^{2v}\sin^2 u + u^2e^{2v}\cos^2 u + u^2e^{2v}}\right)(ue^v) = 2$; $w = \ln\left(u^2e^{2v}\sin^2 u + u^2e^{2v}\cos^2 u + u^2e^{2v}\right) = \ln\left(2u^2e^{2v}\right)$

$= \ln 2 + 2\ln u + 2v \Rightarrow \frac{\partial w}{\partial u} = \frac{2}{u}$ and $\frac{\partial w}{\partial v} = 2$

(b) At $(-2, 0)$: $\frac{\partial w}{\partial u} = \frac{2}{-2} = -1$ and $\frac{\partial w}{\partial v} = 2$

11. (a) $\frac{\partial u}{\partial x} = \frac{\partial u}{\partial p}\frac{\partial p}{\partial x} + \frac{\partial u}{\partial q}\frac{\partial q}{\partial x} + \frac{\partial u}{\partial r}\frac{\partial r}{\partial x} = \frac{1}{q-r} + \frac{r-p}{(q-r)^2} + \frac{p-q}{(q-r)^2} = \frac{q-r+r-p+p-q}{(q-r)^2} = 0$;

$\frac{\partial u}{\partial y} = \frac{\partial u}{\partial p}\frac{\partial p}{\partial y} + \frac{\partial u}{\partial q}\frac{\partial q}{\partial y} + \frac{\partial u}{\partial r}\frac{\partial r}{\partial y} = \frac{1}{q-r} - \frac{r-p}{(q-r)^2} + \frac{p-q}{(q-r)^2} = \frac{q-r-r+p+p-q}{(q-r)^2} = \frac{2p-2r}{(q-r)^2}$

$= \frac{(2x+2y+2z)-(2x+2y-2z)}{(2z-2y)^2} = \frac{z}{(z-y)^2}$; $\frac{\partial u}{\partial z} = \frac{\partial u}{\partial p}\frac{\partial p}{\partial z} + \frac{\partial u}{\partial q}\frac{\partial q}{\partial z} + \frac{\partial u}{\partial r}\frac{\partial r}{\partial z}$

$= \frac{1}{q-r} + \frac{r-p}{(q-r)^2} - \frac{p-q}{(q-r)^2} = \frac{q-r+r-p-p+q}{(q-r)^2} = \frac{2q-2p}{(q-r)^2} = \frac{-4y}{(2z-2y)^2} = -\frac{y}{(z-y)^2}$;

$u = \frac{p-q}{q-r} = \frac{2y}{2z-2y} = \frac{y}{z-y} \Rightarrow \frac{\partial u}{\partial x} = 0$, $\frac{\partial u}{\partial y} = \frac{(z-y)-y(-1)}{(z-y)^2} = \frac{z}{(z-y)^2}$, and $\frac{\partial u}{\partial z} = \frac{(z-y)(0)-y(1)}{(z-y)^2}$

$= -\frac{y}{(z-y)^2}$

(b) At $\left(\sqrt{3}, 2, 1\right)$: $\frac{\partial u}{\partial x} = 0$, $\frac{\partial u}{\partial y} = \frac{1}{(1-2)^2} = 1$, and $\frac{\partial u}{\partial z} = \frac{-2}{(1-2)^2} = -2$

12. (a) $\frac{\partial u}{\partial x} = \frac{e^{qr}}{\sqrt{1-p^2}}(\cos x) + (re^{qr}\sin^{-1}p)(0) + (qe^{qr}\sin^{-1}p)(0) = \frac{e^{qr}\cos x}{\sqrt{1-p^2}} = \frac{e^{z\ln y}\cos x}{\sqrt{1-\sin^2 x}} = y^z$ if $-\frac{\pi}{2} < x < \frac{\pi}{2}$;

$\frac{\partial u}{\partial y} = \frac{e^{qr}}{\sqrt{1-p^2}}(0) + (re^{qr}\sin^{-1}p)\left(\frac{z^2}{y}\right) + (qe^{qr}\sin^{-1}p)(0) = \frac{z^2\, re^{qr}\sin^{-1}p}{y} = \frac{z^2\left(\frac{1}{z}\right)y^z x}{y} = xzy^{z-1}$;

$\frac{\partial u}{\partial z} = \frac{e^{qr}}{\sqrt{1-p^2}}(0) + (re^{qr}\sin^{-1}p)(2z\ln y) + (qe^{qr}\sin^{-1}p)\left(-\frac{1}{z^2}\right) = (2zre^{qr}\sin^{-1}p)(\ln y) - \frac{qe^{qr}\sin^{-1}p}{z^2}$

$= (2z)\left(\frac{1}{z}\right)(y^z x\ln y) - \frac{(z^2\ln y)(y^z)x}{z^2} = xy^z\ln y$; $u = e^{z\ln y}\sin^{-1}(\sin x) = xy^z$ if $-\frac{\pi}{2} \le x \le \frac{\pi}{2} \Rightarrow \frac{\partial u}{\partial x} = y^z$,

$\frac{\partial u}{\partial y} = xzy^{z-1}$, and $\frac{\partial u}{\partial z} = \; = xy^z\ln y$ from direct calculations

(b) At $\left(\frac{\pi}{4}, \frac{1}{2}, -\frac{1}{2}\right)$: $\frac{\partial u}{\partial x} = \left(\frac{1}{2}\right)^{-1/2} = \sqrt{2}$, $\frac{\partial u}{\partial y} = \left(\frac{\pi}{4}\right)\left(-\frac{1}{2}\right)\left(\frac{1}{2}\right)^{(-1/2)-1} = -\frac{\pi\sqrt{2}}{4}$, $\frac{\partial u}{\partial z} = \left(\frac{\pi}{4}\right)\left(\frac{1}{2}\right)^{-1/2}\ln\left(\frac{1}{2}\right)$

$= -\frac{\pi\sqrt{2}\ln 2}{4}$

13. $\frac{dz}{dt} = \frac{\partial z}{\partial x}\frac{dx}{dt} + \frac{\partial z}{\partial y}\frac{dy}{dt}$

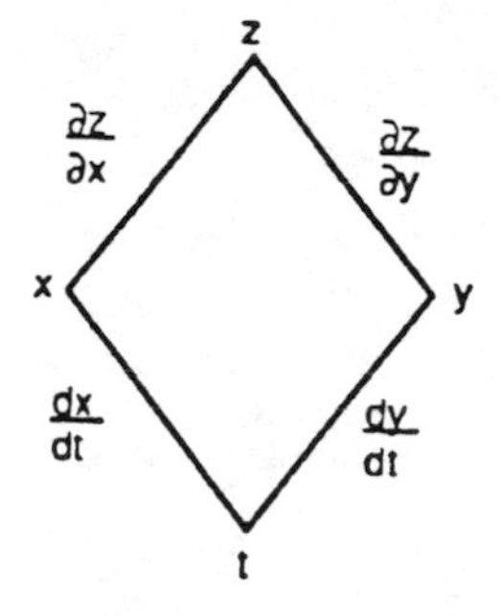

14. $\frac{dz}{dt} = \frac{\partial z}{\partial u}\frac{du}{dt} + \frac{\partial z}{\partial v}\frac{dv}{dt} + \frac{\partial x}{\partial w}\frac{dw}{dt}$

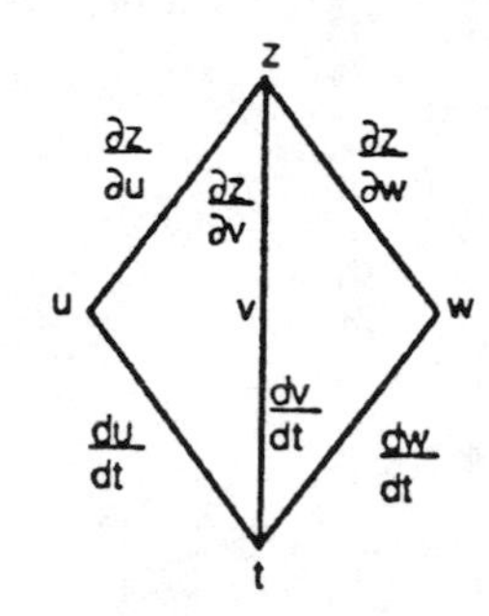

15. $\frac{\partial w}{\partial u} = \frac{\partial w}{\partial x}\frac{\partial x}{\partial u} + \frac{\partial w}{\partial y}\frac{\partial y}{\partial u} + \frac{\partial w}{\partial z}\frac{\partial z}{\partial u}$

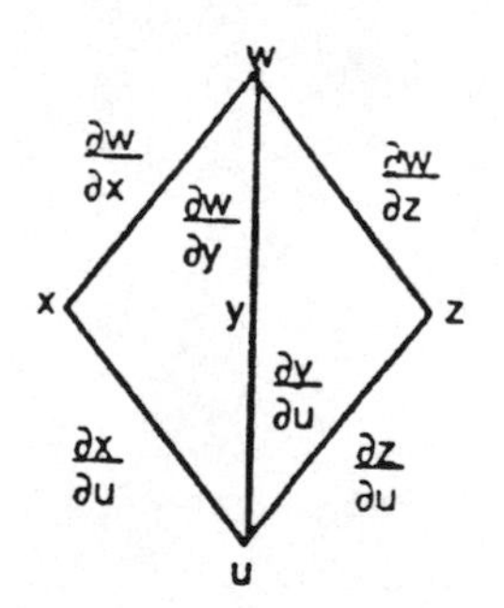

$\frac{\partial w}{\partial v} = \frac{\partial w}{\partial x}\frac{\partial x}{\partial v} + \frac{\partial w}{\partial y}\frac{\partial y}{\partial v} + \frac{\partial w}{\partial z}\frac{\partial z}{\partial v}$

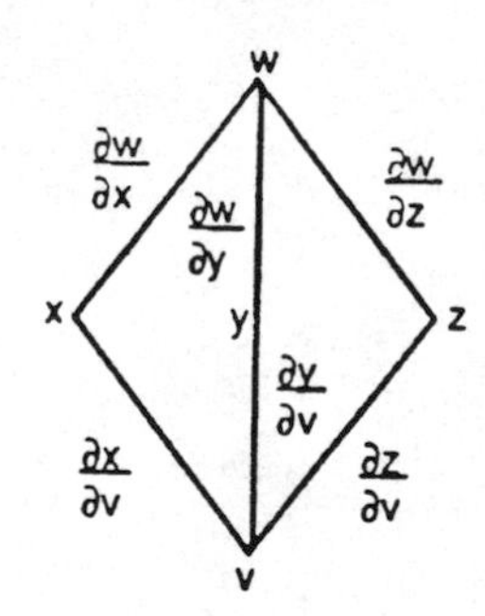

16. $\frac{\partial w}{\partial x} = \frac{\partial w}{\partial r}\frac{\partial r}{\partial x} + \frac{\partial w}{\partial s}\frac{\partial s}{\partial x} + \frac{\partial w}{\partial t}\frac{\partial t}{\partial x}$

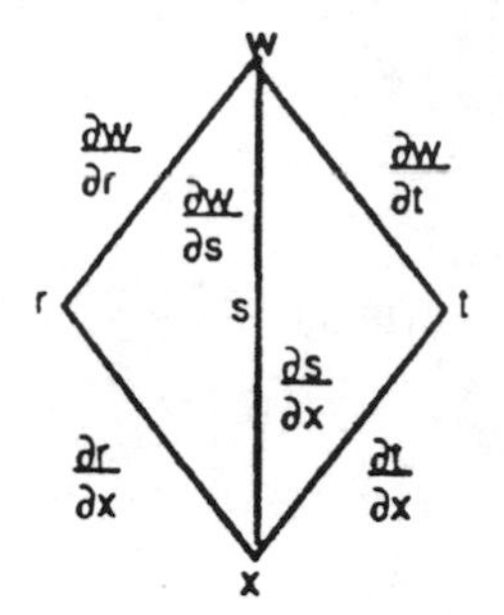

$\frac{\partial w}{\partial y} = \frac{\partial w}{\partial r}\frac{\partial r}{\partial y} + \frac{\partial w}{\partial s}\frac{\partial s}{\partial y} + \frac{\partial w}{\partial t}\frac{\partial t}{\partial y}$

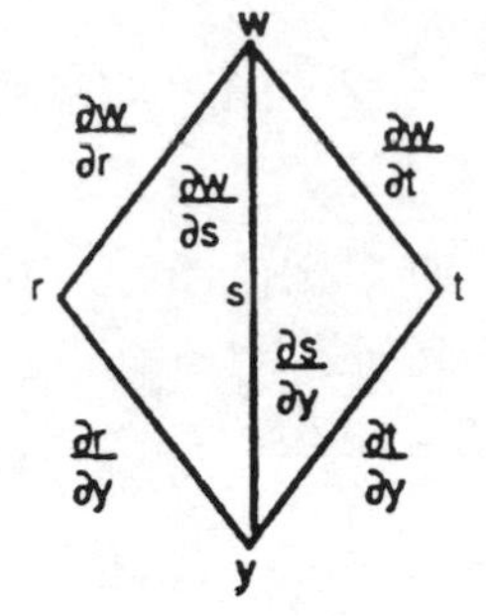

17. $\frac{\partial w}{\partial u} = \frac{\partial w}{\partial x}\frac{\partial x}{\partial u} + \frac{\partial w}{\partial y}\frac{\partial y}{\partial u}$

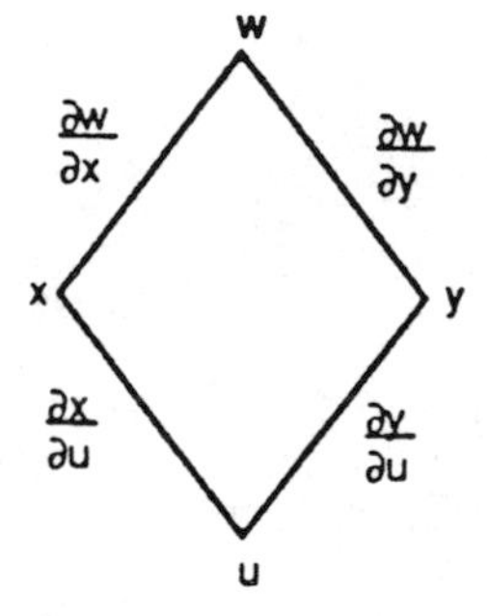

$\frac{\partial w}{\partial v} = \frac{\partial w}{\partial x}\frac{\partial x}{\partial v} + \frac{\partial w}{\partial y}\frac{\partial y}{\partial v}$

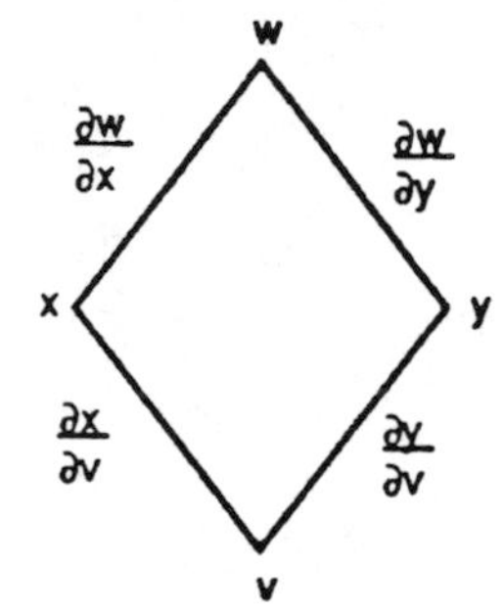

18. $\frac{\partial w}{\partial x} = \frac{\partial w}{\partial u}\frac{\partial u}{\partial x} + \frac{\partial w}{\partial v}\frac{\partial v}{\partial x}$

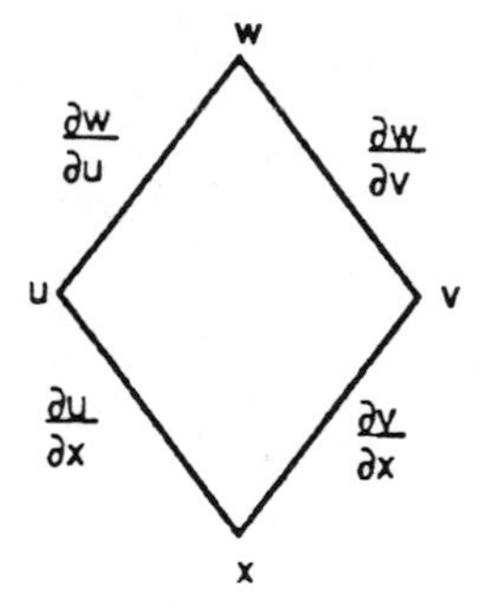

$\frac{\partial w}{\partial y} = \frac{\partial w}{\partial u}\frac{\partial u}{\partial y} + \frac{\partial w}{\partial v}\frac{\partial v}{\partial y}$

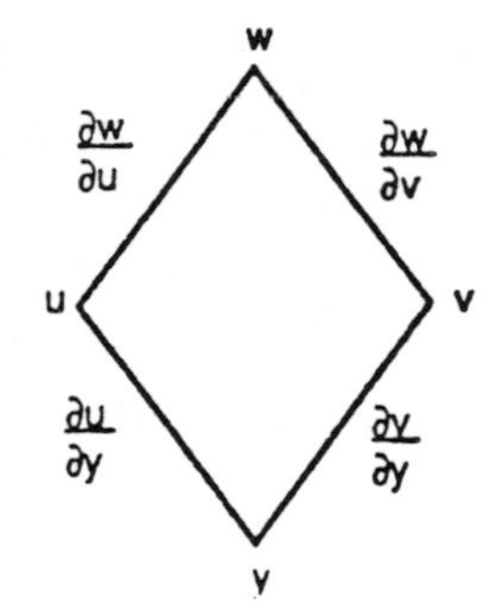

19. $\frac{\partial z}{\partial t} = \frac{\partial z}{\partial x}\frac{\partial x}{\partial t} + \frac{\partial z}{\partial y}\frac{\partial y}{\partial t}$

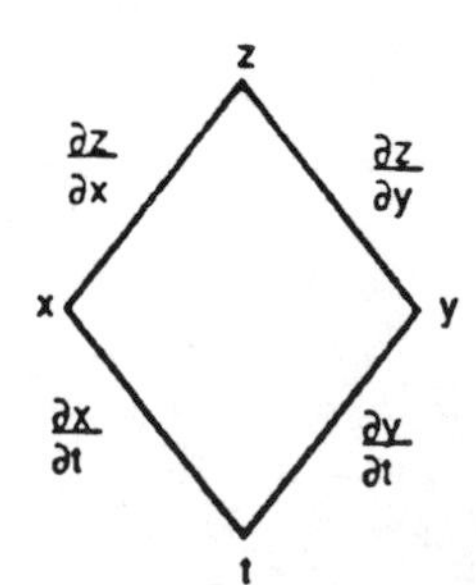

$\frac{\partial z}{\partial s} = \frac{\partial z}{\partial x}\frac{\partial x}{\partial s} + \frac{\partial z}{\partial y}\frac{\partial y}{\partial s}$

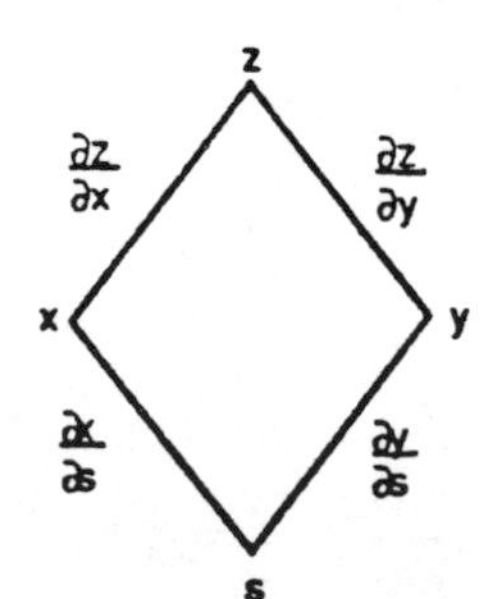

20. $\frac{\partial y}{\partial r} = \frac{dy}{du}\frac{\partial u}{\partial r}$

21. $\frac{\partial w}{\partial s} = \frac{dw}{du}\frac{\partial u}{\partial s}$

$\frac{\partial w}{\partial t} = \frac{dw}{du}\frac{\partial u}{\partial t}$

22. $\frac{\partial w}{\partial p} = \frac{\partial w}{\partial x}\frac{\partial x}{\partial p} + \frac{\partial w}{\partial y}\frac{\partial y}{\partial p} + \frac{\partial w}{\partial z}\frac{\partial z}{\partial p} + \frac{\partial w}{\partial v}\frac{\partial v}{\partial p}$

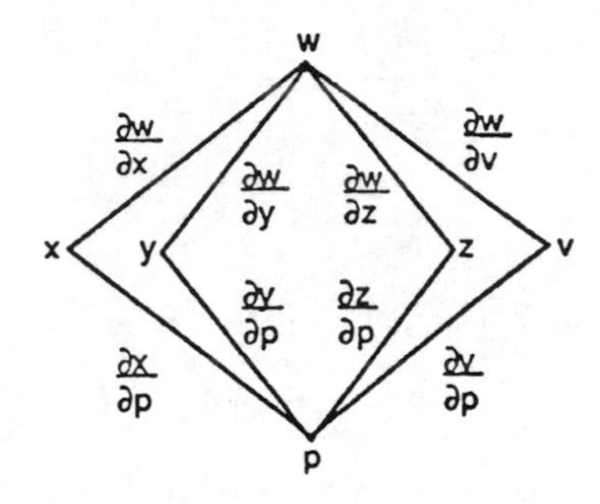

23. $\frac{\partial w}{\partial r} = \frac{\partial w}{\partial x}\frac{dx}{dr} + \frac{\partial w}{\partial y}\frac{dy}{dr} = \frac{\partial w}{\partial x}\frac{dx}{dr}$ since $\frac{dy}{dr} = 0$

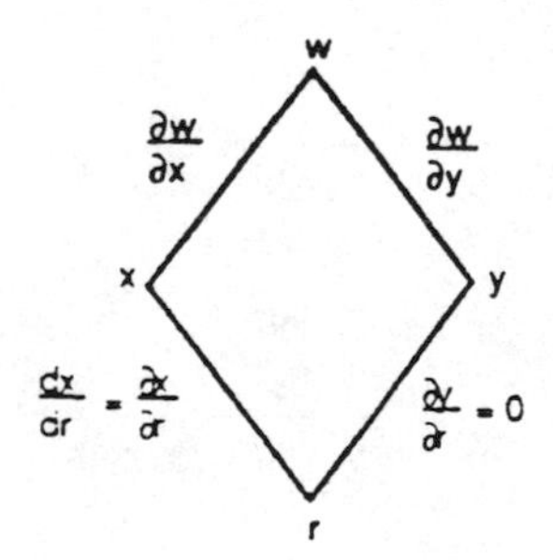

$\frac{\partial w}{\partial s} = \frac{\partial w}{\partial x}\frac{dx}{ds} + \frac{\partial w}{\partial y}\frac{dy}{ds} = \frac{\partial w}{\partial y}\frac{dy}{ds}$ since $\frac{dx}{ds} = 0$

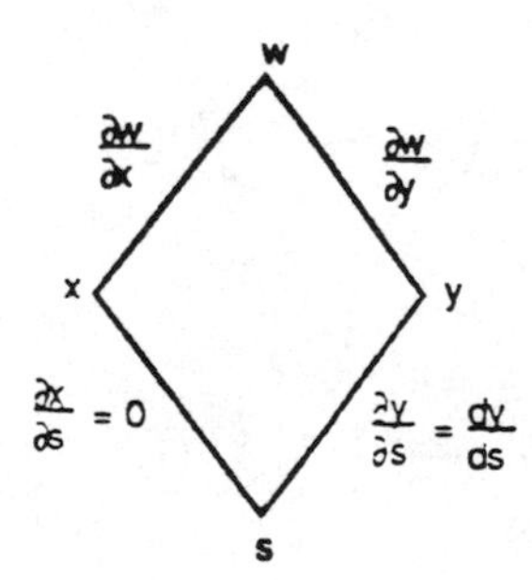

24. $\frac{\partial w}{\partial s} = \frac{\partial w}{\partial x}\frac{\partial x}{\partial s} + \frac{\partial w}{\partial y}\frac{\partial y}{\partial s}$

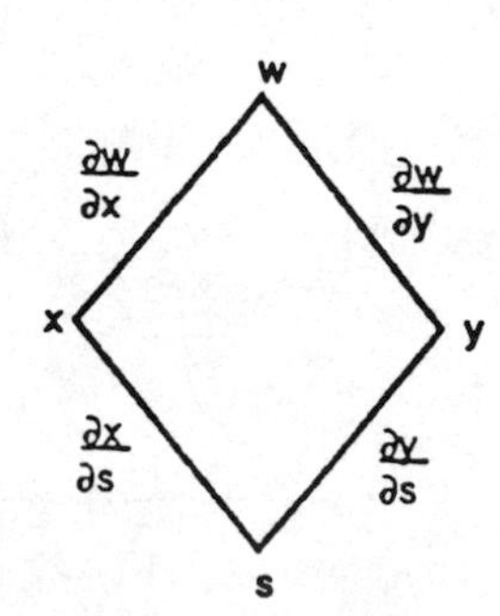

25. Let $F(x,y) = x^3 - 2y^2 + xy = 0 \Rightarrow F_x(x,y) = 3x^2 + y$ and $F_y(x,y) = -4y + x \Rightarrow \frac{dy}{dx} = -\frac{F_x}{F_y} = -\frac{3x^2+y}{(-4y+x)}$ $\Rightarrow \frac{dy}{dx}(1,1) = \frac{4}{3}$

26. Let $F(x,y) = xy + y^2 - 3x - 3 = 0 \Rightarrow F_x(x,y) = y - 3$ and $F_y(x,y) = x + 2y \Rightarrow \frac{dy}{dx} = -\frac{F_x}{F_y} = -\frac{y-3}{x+2y}$ $\Rightarrow \frac{dy}{dx}(-1,1) = 2$

27. Let $F(x,y) = x^2 + xy + y^2 - 7 = 0 \Rightarrow F_x(x,y) = 2x + y$ and $F_y(x,y) = x + 2y \Rightarrow \frac{dy}{dx} = -\frac{F_x}{F_y} = -\frac{2x+y}{x+2y}$ $\Rightarrow \frac{dy}{dx}(1,2) = -\frac{4}{5}$

28. Let $F(x,y) = xe^y + \sin xy + y - \ln 2 = 0 \Rightarrow F_x(x,y) = e^y + y \cos xy$ and $F_y(x,y) = xe^y + x \sin xy + 1$ $\Rightarrow \frac{dy}{dx} = -\frac{F_x}{F_y} = -\frac{e^y + y\cos xy}{xe^y + x\sin xy + 1} \Rightarrow \frac{dy}{dx}(0, \ln 2) = -(2 + \ln 2)$

29. Let $F(x,y,z) = z^3 - xy + yz + y^3 - 2 = 0 \Rightarrow F_x(x,y,z) = -y$, $F_y(x,y,z) = -x + z + 3y^2$, $F_z(x,y,z) = 3z^2 + y$ $\Rightarrow \frac{\partial z}{\partial x} = -\frac{F_x}{F_y} = -\frac{-y}{3x^2+y} = \frac{y}{3z^2+y} \Rightarrow \frac{\partial z}{\partial x}(1,1,1) = \frac{1}{4}$; $\frac{\partial z}{\partial y} = -\frac{F_y}{F_z} = -\frac{-x+z+3y^2}{3z^2+y} = \frac{x-z-3y^2}{3z^2+y}$ $\Rightarrow \frac{\partial z}{\partial y}(1,1,1) = -\frac{3}{4}$

30. Let $F(x,y,z) = \frac{1}{x} + \frac{1}{y} + \frac{1}{z} - 1 = 0 \Rightarrow F_x(x,y,z) = -\frac{1}{x^2}$, $F_y(x,y,z) = -\frac{1}{y^2}$, $F_z(x,y,z) = -\frac{1}{z^2}$ $\Rightarrow \frac{\partial z}{\partial x} = -\frac{F_x}{F_z} = -\frac{\left(-\frac{1}{x^2}\right)}{\left(-\frac{1}{z^2}\right)} = -\frac{z^2}{x^2} \Rightarrow \frac{\partial z}{\partial x}(2,3,6) = -9$; $\frac{\partial z}{\partial y} = -\frac{F_y}{F_z} = -\frac{\left(-\frac{1}{y^2}\right)}{\left(-\frac{1}{z^2}\right)} = -\frac{z^2}{y^2} \Rightarrow \frac{\partial z}{\partial y}(2,3,6) = -4$

31. Let $F(x,y,z) = \sin(x+y) + \sin(y+z) + \sin(z+z) = 0 \Rightarrow F_x(x,y,z) = \cos(x+y) + \cos(x+z)$, $F_y(x,y,z) = \cos(x+y) + \cos(y+z)$, $F_z(x,y,z) = \cos(y+z) + \cos(x+z) \Rightarrow \frac{\partial z}{\partial x} = -\frac{F_x}{F_z}$ $= -\frac{\cos(x+y) + \cos(x+z)}{\cos(y+z) + \cos(x+z)} \Rightarrow \frac{\partial z}{\partial x}(\pi,\pi,\pi) = -1$; $\frac{\partial z}{\partial y} = -\frac{F_y}{F_z} = -\frac{\cos(x+y)+\cos(y+z)}{\cos(y+z)+\cos(x+z)} \Rightarrow \frac{\partial z}{\partial y}(\pi,\pi,\pi) = -1$

32. Let $F(x,y,z) = xe^y + ye^z + 2\ln x - 2 - 3\ln 2 = 0 \Rightarrow F_x(x,y,z) = e^y + \frac{2}{x}$, $F_y(x,y,z) = xe^y + e^z$, $F_z(x,y,z) = ye^z$ $\Rightarrow \frac{\partial z}{\partial x} = -\frac{F_x}{F_z} = -\frac{\left(e^y + \frac{2}{x}\right)}{ye^z} \Rightarrow \frac{\partial z}{\partial x}(1, \ln 2, \ln 3) = -\frac{4}{3\ln 2}$; $\frac{\partial z}{\partial y} = -\frac{F_y}{F_z} = -\frac{xe^y + e^z}{ye^z} \Rightarrow \frac{\partial z}{\partial y}(1, \ln 2, \ln 3) = -\frac{5}{3\ln 2}$

33. $\frac{\partial w}{\partial r} = \frac{\partial w}{\partial x}\frac{\partial x}{\partial r} + \frac{\partial w}{\partial y}\frac{\partial y}{\partial r} + \frac{\partial w}{\partial z}\frac{\partial z}{\partial r} = 2(x+y+z)(1) + 2(x+y+z)[-\sin(r+s)] + 2(x+y+z)[\cos(r+s)]$ $= 2(x+y+z)[1 - \sin(r+s) + \cos(r+s)] = 2[r - s + \cos(r+s) + \sin(r+s)][1 - \sin(r+s) + \cos(r+s)]$ $\Rightarrow \left.\frac{\partial w}{\partial r}\right|_{r=1,s=-1} = 2(3)(2) = 12$

34. $\frac{\partial w}{\partial v} = \frac{\partial w}{\partial x}\frac{\partial x}{\partial v} + \frac{\partial w}{\partial y}\frac{\partial y}{\partial v} + \frac{\partial w}{\partial z}\frac{\partial z}{\partial v} = y\left(\frac{2v}{u}\right) + x(1) + \left(\frac{1}{z}\right)(0) = (u+v)\left(\frac{2v}{u}\right) + \frac{v^2}{u} \Rightarrow \left.\frac{\partial w}{\partial v}\right|_{u=-1,v=2} = (1)\left(\frac{4}{-1}\right) + \left(\frac{4}{-1}\right)$ $= -8$

35. $\frac{\partial w}{\partial v} = \frac{\partial w}{\partial x}\frac{\partial x}{\partial v} + \frac{\partial w}{\partial y}\frac{\partial y}{\partial v} = \left(2x - \frac{y}{x^2}\right)(-2) + \left(\frac{1}{x}\right)(1) = \left[2(u-2v+1) - \frac{2u+v-2}{(u-2v+1)^2}\right](-2) + \frac{1}{u-2v+1}$ $\Rightarrow \left.\frac{\partial w}{\partial v}\right|_{u=0,v=0} = -7$

36. $\frac{\partial z}{\partial u} = \frac{\partial z}{\partial x}\frac{\partial x}{\partial u} + \frac{\partial z}{\partial y}\frac{\partial y}{\partial u} = (y \cos xy + \sin y)(2u) + (x \cos xy + x \cos y)(v)$

$= [uv \cos(u^3v + uv^3) + \sin uv](2u) + [(u^2+v^2)\cos(u^3v + uv^3) + (u^2+v^2)\cos uv](v)$

$\Rightarrow \left.\frac{\partial z}{\partial u}\right|_{u=0,\, v=1} = 0 + (\cos 0 + \cos 0)(1) = 2$

37. $\frac{\partial z}{\partial u} = \frac{dz}{dx}\frac{\partial x}{\partial u} = \left(\frac{5}{1+x^2}\right)e^u = \left[\frac{5}{1+(e^u + \ln v)^2}\right]e^u \Rightarrow \left.\frac{\partial z}{\partial u}\right|_{u=\ln 2,\, v=1} = \left[\frac{5}{1+(2)^2}\right](2) = 2;$

$\frac{\partial z}{\partial v} = \frac{dz}{dx}\frac{\partial x}{\partial v} = \left(\frac{5}{1+x^2}\right)\left(\frac{1}{v}\right) = \left[\frac{5}{1+(e^u + \ln v)^2}\right]\left(\frac{1}{v}\right) \Rightarrow \left.\frac{\partial z}{\partial v}\right|_{u=\ln 2,\, v=1} = \left[\frac{5}{1+(2)^2}\right](1) = 1$

38. $\frac{\partial z}{\partial u} = \frac{dz}{dq}\frac{\partial q}{\partial u} = \left(\frac{1}{q}\right)\left(\frac{\sqrt{v+3}}{1+u^2}\right) = \left(\frac{1}{\sqrt{v+3}\tan^{-1}u}\right)\left(\frac{\sqrt{v+3}}{1+u^2}\right) = \frac{1}{(\tan^{-1}u)(1+u^2)}$

$\Rightarrow \left.\frac{\partial z}{\partial u}\right|_{u=1,\, v=-2} = \frac{1}{(\tan^{-1}1)(1+1^2)} = \frac{2}{\pi};\ \frac{\partial z}{\partial v} = \frac{dz}{dq}\frac{\partial q}{\partial v} = \left(\frac{1}{q}\right)\left(\frac{\tan^{-1}u}{2\sqrt{v+3}}\right)$

$= \left(\frac{1}{\sqrt{v+3}\tan^{-1}u}\right)\left(\frac{\tan^{-1}u}{2\sqrt{v+3}}\right) = \frac{1}{2(v+3)} \Rightarrow \left.\frac{\partial z}{\partial v}\right|_{u=1,\, v=-2} = \frac{1}{2}$

39. $V = IR \Rightarrow \frac{\partial V}{\partial I} = R$ and $\frac{\partial V}{\partial R} = I;\ \frac{dV}{dt} = \frac{\partial V}{\partial I}\frac{dI}{dt} + \frac{\partial V}{\partial R}\frac{dR}{dt} = R\frac{dI}{dt} + I\frac{dR}{dt} \Rightarrow -0.01$ volts/sec

$= (600 \text{ ohms})\frac{dI}{dt} + (0.04 \text{ amps})(0.5 \text{ ohms/sec}) \Rightarrow \frac{dI}{dt} = -0.00005$ amps/sec

40. $V = abc \Rightarrow \frac{dV}{dt} = \frac{\partial V}{\partial a}\frac{da}{dt} + \frac{\partial V}{\partial b}\frac{db}{dt} + \frac{\partial V}{\partial c}\frac{dc}{dt} = (bc)\frac{da}{dt} + (ac)\frac{db}{dt} + (ab)\frac{dc}{dt}$

$\Rightarrow \left.\frac{dV}{dt}\right|_{a=1,\, b=2,\, c=3} = (2\text{ m})(3\text{ m})(1\text{ m/sec}) + (1\text{ m})(3\text{ m})(1\text{ m/sec}) + (1\text{ m})(2\text{ m})(-3\text{ m/sec}) = 3\text{ m}^3/\text{sec}$

and the volume is increasing; $S = 2ab + 2ac + 2bc \Rightarrow \frac{dS}{dt} = \frac{\partial S}{\partial a}\frac{da}{dt} + \frac{\partial S}{\partial b}\frac{db}{dt} + \frac{\partial S}{\partial c}\frac{dc}{dt}$

$= 2(b+c)\frac{da}{dt} + 2(a+c)\frac{db}{dt} + 2(a+b)\frac{dc}{dt} \Rightarrow \left.\frac{dS}{dt}\right|_{a=1,\, b=2,\, c=3}$

$= 2(5\text{ m})(1\text{ m/sec}) + 2(4\text{ m})(1\text{ m/sec}) + 2(3\text{ m})(-3\text{ m/sec}) = 0\text{ m}^2/\text{sec}$ and the surface area is not changing;

$D = \sqrt{a^2+b^2+c^2} \Rightarrow \frac{dD}{dt} = \frac{\partial D}{\partial a}\frac{da}{dt} + \frac{\partial D}{\partial b}\frac{db}{dt} + \frac{\partial D}{\partial c}\frac{dc}{dt} = \frac{1}{\sqrt{a^2+b^2+c^2}}\left(a\frac{da}{dt} + b\frac{db}{dt} + c\frac{dc}{dt}\right) \Rightarrow \left.\frac{dD}{dt}\right|_{a=1,\, b=2,\, c=3}$

$= \left(\frac{1}{\sqrt{14}\text{ m}}\right)[(1\text{ m})(1\text{ m/sec}) + (2\text{ m})(1\text{ m/sec}) + (3\text{ m})(-3\text{ m/sec})] = -\frac{6}{\sqrt{14}}\text{ m/sec} < 0 \Rightarrow$ the diagonals are decreasing in length

41. $\frac{\partial f}{\partial x} = \frac{\partial f}{\partial u}\frac{\partial u}{\partial x} + \frac{\partial f}{\partial v}\frac{\partial v}{\partial x} + \frac{\partial f}{\partial w}\frac{\partial w}{\partial x} = \frac{\partial f}{\partial u}(1) + \frac{\partial f}{\partial v}(0) + \frac{\partial f}{\partial w}(-1) = \frac{\partial f}{\partial u} - \frac{\partial f}{\partial w},$

$\frac{\partial f}{\partial y} = \frac{\partial f}{\partial u}\frac{\partial u}{\partial y} + \frac{\partial f}{\partial v}\frac{\partial v}{\partial y} + \frac{\partial f}{\partial w}\frac{\partial w}{\partial y} = \frac{\partial f}{\partial u}(-1) + \frac{\partial f}{\partial v}(1) + \frac{\partial f}{\partial w}(0) = -\frac{\partial f}{\partial u} + \frac{\partial f}{\partial v}$, and

$\frac{\partial f}{\partial z} = \frac{\partial f}{\partial u}\frac{\partial u}{\partial z} + \frac{\partial f}{\partial v}\frac{\partial v}{\partial z} + \frac{\partial f}{\partial w}\frac{\partial w}{\partial z} = \frac{\partial f}{\partial u}(0) + \frac{\partial f}{\partial v}(-1) + \frac{\partial f}{\partial w}(1) = -\frac{\partial f}{\partial v} + \frac{\partial f}{\partial w} \Rightarrow \frac{\partial f}{\partial x} + \frac{\partial f}{\partial y} + \frac{\partial f}{\partial z} = 0$

42. (a) $\frac{\partial w}{\partial r} = f_x \frac{\partial x}{\partial r} + f_y \frac{\partial y}{\partial r} = f_x \cos\theta + f_y \sin\theta$ and $\frac{\partial w}{\partial \theta} = f_x(-r\sin\theta) + f_y(r\cos\theta) \Rightarrow \frac{1}{r}\frac{\partial w}{\partial \theta} = -f_x \sin\theta + f_y \cos\theta$

(b) $\frac{\partial w}{\partial r}\sin\theta = f_x \sin\theta\cos\theta + f_y \sin^2\theta$ and $\left(\frac{\cos\theta}{r}\right)\frac{\partial w}{\partial \theta} = -f_x \sin\theta\cos\theta + f_y\cos^2\theta$

$\Rightarrow f_y = (\sin\theta)\frac{\partial w}{\partial r} + \left(\frac{\cos\theta}{r}\right)\frac{\partial w}{\partial \theta}$; then $\frac{\partial w}{\partial r} = f_x\cos\theta + \left[(\sin\theta)\frac{\partial w}{\partial r} + \left(\frac{\cos\theta}{r}\right)\frac{\partial w}{\partial \theta}\right](\sin\theta) \Rightarrow f_x\cos\theta$

$= \frac{\partial w}{\partial r} - (\sin^2\theta)\frac{\partial w}{\partial r} - \left(\frac{\sin\theta\cos\theta}{r}\right)\frac{\partial w}{\partial \theta} = (1-\sin^2\theta)\frac{\partial w}{\partial r} - \left(\frac{\sin\theta\cos\theta}{r}\right)\frac{\partial w}{\partial \theta} \Rightarrow f_x = (\cos\theta)\frac{\partial w}{\partial r} - \left(\frac{\sin\theta}{r}\right)\frac{\partial w}{\partial \theta}$

(c) $(f_x)^2 = (\cos^2\theta)\left(\frac{\partial w}{\partial r}\right)^2 - \left(\frac{2\sin\theta\cos\theta}{r}\right)\left(\frac{\partial w}{\partial r}\frac{\partial w}{\partial \theta}\right) + \left(\frac{\sin^2\theta}{r^2}\right)\left(\frac{\partial w}{\partial \theta}\right)^2$ and

$(f_y)^2 = (\sin^2\theta)\left(\frac{\partial w}{\partial r}\right)^2 + \left(\frac{2\sin\theta\cos\theta}{r}\right)\left(\frac{\partial w}{\partial r}\frac{\partial w}{\partial \theta}\right) + \left(\frac{\cos^2\theta}{r^2}\right)\left(\frac{\partial w}{\partial \theta}\right)^2 \Rightarrow (f_x)^2 + (f_y)^2 = \left(\frac{\partial w}{\partial r}\right)^2 + \frac{1}{r^2}\left(\frac{\partial w}{\partial \theta}\right)^2$

43. $w_x = \frac{\partial w}{\partial x} = \frac{\partial w}{\partial u}\frac{\partial u}{\partial x} + \frac{\partial w}{\partial v}\frac{\partial v}{\partial x} = x\frac{\partial w}{\partial u} + y\frac{\partial w}{\partial v} \Rightarrow w_{xx} = \frac{\partial w}{\partial u} + x\frac{\partial}{\partial x}\left(\frac{\partial w}{\partial u}\right) + y\frac{\partial}{\partial x}\left(\frac{\partial w}{\partial v}\right)$

$= \frac{\partial w}{\partial u} + x\left(\frac{\partial^2 w}{\partial u^2}\frac{\partial u}{\partial x} + \frac{\partial^2 w}{\partial v\partial u}\frac{\partial v}{\partial x}\right) + y\left(\frac{\partial^2 w}{\partial u\partial v}\frac{\partial u}{\partial x} + \frac{\partial^2 w}{\partial v^2}\frac{\partial v}{\partial x}\right) = \frac{\partial w}{\partial u} + x\left(x\frac{\partial^2 w}{\partial u^2} + y\frac{\partial^2 w}{\partial v\partial u}\right) + y\left(x\frac{\partial^2 w}{\partial u\partial v} + y\frac{\partial^2 w}{\partial v^2}\right)$

$= \frac{\partial w}{\partial u} + x^2\frac{\partial^2 w}{\partial u^2} + 2xy\frac{\partial^2 w}{\partial v\partial u} + y^2\frac{\partial^2 w}{\partial v^2}$; $w_y = \frac{\partial w}{\partial y} = \frac{\partial w}{\partial u}\frac{\partial u}{\partial y} + \frac{\partial w}{\partial v}\frac{\partial v}{\partial y} = -y\frac{\partial w}{\partial u} + x\frac{\partial w}{\partial v}$

$\Rightarrow w_{yy} = -\frac{\partial w}{\partial u} - y\left(\frac{\partial^2 w}{\partial u^2}\frac{\partial u}{\partial y} + \frac{\partial^2 w}{\partial v\partial u}\frac{\partial v}{\partial y}\right) + x\left(\frac{\partial^2 w}{\partial u\partial v}\frac{\partial u}{\partial y} + \frac{\partial^2 w}{\partial v^2}\frac{\partial v}{\partial y}\right)$

$= -\frac{\partial w}{\partial u} - y\left(-y\frac{\partial^2 w}{\partial u^2} + x\frac{\partial^2 w}{\partial v\partial u}\right) + x\left(-y\frac{\partial^2 w}{\partial u\partial v} + x\frac{\partial^2 w}{\partial v^2}\right) = -\frac{\partial w}{\partial u} + y^2\frac{\partial^2 w}{\partial u^2} - 2xy\frac{\partial^2 w}{\partial v\partial u} + x^2\frac{\partial^2 w}{\partial v^2}$; thus

$w_{xx} + w_{yy} = (x^2+y^2)\frac{\partial^2 w}{\partial u^2} + (x^2+y^2)\frac{\partial^2 w}{\partial v^2} = (x^2+y^2)(w_{uu}+w_{vv}) = 0$, since $w_{uu} + w_{vv} = 0$

44. $\frac{\partial w}{\partial x} = f'(u)(1) + g'(v)(1) = f'(u) + g'(v) \Rightarrow w_{xx} = f''(u)(1) + g''(v)(1) = f''(u) + g''(v)$;

$\frac{\partial w}{\partial y} = f'(u)(i) + g'(v)(-i) \Rightarrow w_{yy} = f''(u)(i^2) + g''(v)(i^2) = -f''(u) - g''(v) \Rightarrow w_{xx} + w_{yy} = 0$

45. $f_x(x,y,z) = \cos t$, $f_y(x,y,z) = \sin t$, and $f_z(x,y,z) = t^2 + t - 2 \Rightarrow \frac{df}{dt} = \frac{\partial f}{\partial x}\frac{dx}{dt} + \frac{\partial f}{\partial y}\frac{dy}{dt} + \frac{\partial f}{\partial z}\frac{dz}{dt}$

$= (\cos t)(-\sin t) + (\sin t)(\cos t) + (t^2+t-2)(1) = t^2+t-2$; $\frac{df}{dt} = 0 \Rightarrow t^2+t-2 = 0 \Rightarrow t = -2$ or $t = 1$; $t = -2 \Rightarrow x = \cos(-2)$, $y = \sin(-2)$, $z = -2$ for the point $(\cos(-2), \sin(-2), -2)$; $t = 1 \Rightarrow x = \cos 1$, $y = \sin 1$, $z = 1$ for the point $(\cos 1, \sin 1, 1)$

46. $\frac{dw}{dt} = \frac{\partial w}{\partial x}\frac{dx}{dt} + \frac{\partial w}{\partial y}\frac{dy}{dt} + \frac{\partial w}{\partial z}\frac{dz}{dt} = (2xe^{2y}\cos 3z)(-\sin t) + (2x^2e^{2y}\cos 3z)\left(\frac{1}{t+2}\right) + (-3x^2e^{2y}\sin 3z)(1)$

$= -2xe^{2y}\cos 3z\sin t + \frac{2x^2e^{2y}\cos 3z}{t+2} - 3x^2e^{2y}\sin 3z$; at the point on the curve $z = 0 \Rightarrow t = z = 0$

$\Rightarrow \left.\frac{dw}{dt}\right|_{(1,\ln 2,0)} = 0 + \frac{2(1)^2(4)(1)}{2} - 0 = 4$

47. (a) $\frac{\partial T}{\partial x} = 8x - 4y$ and $\frac{\partial T}{\partial y} = 8y - 4x \Rightarrow \frac{dT}{dt} = \frac{\partial T}{\partial x}\frac{dx}{dt} + \frac{\partial T}{\partial y}\frac{dy}{dt} = (8x - 4y)(-\sin t) + (8y - 4x)(\cos t)$

$= (8 \cos t - 4 \sin t)(-\sin t) + (8 \sin t - 4 \cos t)(\cos t) = 4 \sin^2 t - 4 \cos^2 t \Rightarrow \frac{d^2T}{dt^2} = 16 \sin t \cos t;$

$\frac{dT}{dt} = 0 \Rightarrow 4 \sin^2 t - 4 \cos^2 t = 0 \Rightarrow \sin^2 t = \cos^2 t \Rightarrow \sin t = \cos t$ or $\sin t = -\cos t \Rightarrow t = \frac{\pi}{4}, \frac{5\pi}{4}, \frac{3\pi}{4}, \frac{7\pi}{4}$ on the interval $0 \le t \le 2\pi$;

$\left.\frac{d^2T}{dt^2}\right|_{t=\frac{\pi}{4}} = 16 \sin \frac{\pi}{4} \cos \frac{\pi}{4} > 0 \Rightarrow$ T has a minimum at $(x, y) = \left(\frac{\sqrt{2}}{2}, \frac{\sqrt{2}}{2}\right)$;

$\left.\frac{d^2T}{dt^2}\right|_{t=\frac{3\pi}{4}} = 16 \sin \frac{3\pi}{4} \cos \frac{3\pi}{4} < 0 \Rightarrow$ T has a maximum at $(x, y) = \left(-\frac{\sqrt{2}}{2}, \frac{\sqrt{2}}{2}\right)$;

$\left.\frac{d^2T}{dt^2}\right|_{t=\frac{5\pi}{4}} = 16 \sin \frac{5\pi}{4} \cos \frac{5\pi}{4} > 0 \Rightarrow$ T has a minimum at $(x, y) = \left(-\frac{\sqrt{2}}{2}, -\frac{\sqrt{2}}{2}\right)$;

$\left.\frac{d^2T}{dt^2}\right|_{t=\frac{7\pi}{4}} = 16 \sin \frac{7\pi}{4} \cos \frac{7\pi}{4} < 0 \Rightarrow$ T has a maximum at $(x, y) = \left(\frac{\sqrt{2}}{2}, -\frac{\sqrt{2}}{2}\right)$

(b) $T = 4x^2 - 4xy + 4y^2 \Rightarrow \frac{\partial T}{\partial x} = 8x - 4y$, and $\frac{\partial T}{\partial y} = 8y - 4x$ so the extreme values occur at the four points found in part (a): $T\left(-\frac{\sqrt{2}}{2}, \frac{\sqrt{2}}{2}\right) = T\left(\frac{\sqrt{2}}{2}, -\frac{\sqrt{2}}{2}\right) = 4\left(\frac{1}{2}\right) - 4\left(-\frac{1}{2}\right) + 4\left(\frac{1}{2}\right) = 6$, the maximum and $T\left(\frac{\sqrt{2}}{2}, \frac{\sqrt{2}}{2}\right) = T\left(-\frac{\sqrt{2}}{2}, -\frac{\sqrt{2}}{2}\right) = 4\left(\frac{1}{2}\right) - 4\left(\frac{1}{2}\right) + 4\left(\frac{1}{2}\right) = 2$, the minimum

48. (a) $\frac{\partial T}{\partial x} = y$ and $\frac{\partial T}{\partial y} = x \Rightarrow \frac{dT}{dt} = \frac{\partial T}{\partial x}\frac{dx}{dt} + \frac{\partial T}{\partial y}\frac{dy}{dt} = y(-2\sqrt{2} \sin t) + x(\sqrt{2} \cos t)$

$= (\sqrt{2} \sin t)(-2\sqrt{2} \sin t) + (2\sqrt{2} \cos t)(\sqrt{2} \cos t) = -4 \sin^2 t + 4 \cos^2 t = -4 \sin^2 t + 4(1 - \sin^2 t)$

$= 4 - 8 \sin^2 t \Rightarrow \frac{d^2T}{dt^2} = -16 \sin t \text{ cost } t$; $\frac{dT}{dt} = 0 \Rightarrow 4 \cdot 8 \sin^2 t = 0 \Rightarrow \sin^2 t = \frac{1}{2} \Rightarrow \sin t = \pm\frac{1}{\sqrt{2}} \Rightarrow t = \frac{\pi}{4}, \frac{3\pi}{4}, \frac{5\pi}{4}, \frac{7\pi}{4}$ on the interval $0 \le t \le 2\pi$;

$\left.\frac{d^2T}{dt^2}\right|_{t=\frac{\pi}{4}} = -8 \sin 2\left(\frac{\pi}{4}\right) = -8 \Rightarrow$ T has a maximum at $(x, y) = (2, 1)$;

$\left.\frac{d^2T}{dt^2}\right|_{t=\frac{3\pi}{4}} = -8 \sin 2\left(\frac{3\pi}{4}\right) = 8 \Rightarrow$ T has a minimum at $(x, y) = (-2, 1)$;

$\left.\frac{d^2T}{dt^2}\right|_{t=\frac{5\pi}{4}} = -8 \sin 2\left(\frac{5\pi}{4}\right) = -8 \Rightarrow$ T has a maximum at $(x, y) = (-2, -1)$;

$\left.\frac{d^2T}{dt^2}\right|_{t=\frac{7\pi}{4}} = -8 \sin 2\left(\frac{7\pi}{4}\right) = 8 \Rightarrow$ T has a minimum at $(x, y) = (2, -1)$

(b) $T = xy - 2 \Rightarrow \frac{\partial T}{\partial x} = y$ and $\frac{\partial T}{\partial y} = x$ so the extreme values occur at the four points found in part (a): $T(2, 1) = T(-2, -1) = 0$, the maximum and $T(-2, 1) = T(2, -1) = -4$, the minimum

49. $G(u,x) = \int_a^u g(t,x)\,dt$ where $u = f(x) \Rightarrow \frac{dG}{dx} = \frac{\partial G}{\partial u}\frac{du}{dx} + \frac{\partial G}{\partial x}\frac{dx}{dx} = g(u,x)f'(x) + \int_a^u g_x(t,x)\,dt$; thus

$$F(x) = \int_0^{x^2} \sqrt{t^4 + x^3}\,dt \Rightarrow F'(x) = \sqrt{(x^2)^4 + x^3}\,(2x) + \int_0^{x^2} \frac{\partial}{\partial x}\sqrt{t^4 + x^3}\,dt = 2x\sqrt{x^8 + x^3} + \int_0^{x^2} \frac{3x^2}{2\sqrt{t^4 + x^3}}\,dt$$

50. Using the result in Exercise 49, $F(x) = \int_{x^2}^{1} \sqrt{t^3 + x^2}\,dt = -\int_1^{x^2} \sqrt{t^3 + x^2}\,dt \Rightarrow F'(x)$

$$= -\sqrt{(x^2)^3 + x^2}\,(2x) - \int_1^{x^2} \frac{\partial}{\partial x}\sqrt{t^3 + x^2}\,dt = \int_{x^2}^{1} \frac{x}{\sqrt{t^3 + x^2}}\,dt - 2x\sqrt{x^6 + x^2}$$

12.6 PARTIAL DERIVATIVES WITH CONSTRAINED VARIABLES

1. $w = x^2 + y^2 + z^2$ and $z = x^2 + y^2$:

(a) $$\begin{pmatrix} y \\ z \end{pmatrix} \to \begin{pmatrix} x = x(y,z) \\ y = y \\ z = z \end{pmatrix} \to w \Rightarrow \left(\frac{\partial w}{\partial y}\right)_z = \frac{\partial w}{\partial x}\frac{\partial x}{\partial y} + \frac{\partial w}{\partial y}\frac{\partial y}{\partial y} + \frac{\partial w}{\partial z}\frac{\partial z}{\partial y};\ \frac{\partial z}{\partial y} = 0 \text{ and } \frac{\partial z}{\partial y} = 2x\frac{\partial x}{\partial y} + 2y\frac{\partial y}{\partial y}$$

$$= 2x\frac{\partial x}{\partial y} + 2y \Rightarrow 0 = 2x\frac{\partial x}{\partial y} + 2y \Rightarrow \frac{\partial x}{\partial y} = -\frac{y}{x} \Rightarrow \left(\frac{\partial w}{\partial y}\right)_z = (2x)\left(-\frac{y}{x}\right) + (2y)(1) + (2z)(0) = -2y + 2y = 0$$

(b) $$\begin{pmatrix} x \\ z \end{pmatrix} \to \begin{pmatrix} x = x \\ y = y(x,z) \\ z = z \end{pmatrix} \to w \Rightarrow \left(\frac{\partial w}{\partial z}\right)_x = \frac{\partial w}{\partial x}\frac{\partial x}{\partial z} + \frac{\partial w}{\partial y}\frac{\partial y}{\partial z} + \frac{\partial w}{\partial z}\frac{\partial z}{\partial z};\ \frac{\partial x}{\partial z} = 0 \text{ and } \frac{\partial z}{\partial z} = 2x\frac{\partial x}{\partial z} + 2y\frac{\partial y}{\partial z}$$

$$\Rightarrow 1 = 2y\frac{\partial y}{\partial z} \Rightarrow \frac{\partial y}{\partial z} = \frac{1}{2y} \Rightarrow \left(\frac{\partial w}{\partial z}\right)_x = (2x)(0) + (2y)\left(\frac{1}{2y}\right) + (2z)(1) = 1 + 2z$$

(c) $$\begin{pmatrix} y \\ z \end{pmatrix} \to \begin{pmatrix} x = x(y,z) \\ y = y \\ z = z \end{pmatrix} \to w \Rightarrow \left(\frac{\partial w}{\partial z}\right)_y = \frac{\partial w}{\partial x}\frac{\partial x}{\partial z} + \frac{\partial w}{\partial y}\frac{\partial y}{\partial z} + \frac{\partial w}{\partial z}\frac{\partial z}{\partial z};\ \frac{\partial y}{\partial z} = 0 \text{ and } \frac{\partial z}{\partial z} = 2x\frac{\partial x}{\partial z} + 2y\frac{\partial y}{\partial z}$$

$$\Rightarrow 1 = 2x\frac{\partial x}{\partial z} \Rightarrow \frac{\partial x}{\partial z} = \frac{1}{2x} \Rightarrow \left(\frac{\partial w}{\partial z}\right)_y = (2x)\left(\frac{1}{2x}\right) + (2y)(0) + (2z)(1) = 1 + 2z$$

2. $w = x^2 + y - z + \sin t$ and $x + y = t$:

(a) $$\begin{pmatrix} x \\ y \\ z \end{pmatrix} \to \begin{pmatrix} x = x \\ y = y \\ z = z \\ t = x + y \end{pmatrix} \to w \Rightarrow \left(\frac{\partial w}{\partial y}\right)_{x,z} = \frac{\partial w}{\partial x}\frac{\partial x}{\partial y} + \frac{\partial w}{\partial y}\frac{\partial y}{\partial y} + \frac{\partial w}{\partial z}\frac{\partial z}{\partial y} + \frac{\partial w}{\partial t}\frac{\partial t}{\partial y};\ \frac{\partial x}{\partial y} = 0,\ \frac{\partial z}{\partial y} = 0, \text{ and}$$

$$\frac{\partial t}{\partial y} = 1 \Rightarrow \left(\frac{\partial w}{\partial y}\right)_{x,t} = (2x)(0) + (1)(1) + (-1)(0) + (\cos t)(1) = 1 + \cos t = 1 + \cos(x + y)$$

(b) $\begin{pmatrix} y \\ z \\ t \end{pmatrix} \to \begin{pmatrix} x = t - y \\ y = y \\ z = z \\ t = t \end{pmatrix} \to w \Rightarrow \left(\frac{\partial w}{\partial y}\right)_{z,t} = \frac{\partial w}{\partial x}\frac{\partial x}{\partial y} + \frac{\partial w}{\partial y}\frac{\partial y}{\partial y} + \frac{\partial w}{\partial z}\frac{\partial z}{\partial y} + \frac{\partial w}{\partial t}\frac{\partial t}{\partial y}$; $\frac{\partial z}{\partial y} = 0$ and $\frac{\partial t}{\partial y} = 0$

$\Rightarrow \frac{\partial x}{\partial y} = \frac{\partial t}{\partial y} - \frac{\partial y}{\partial y} = -1 \Rightarrow \left(\frac{\partial w}{\partial y}\right)_{z,t} = (2x)(-1) + (1)(1) + (-1)(0) + (\cos t)(0) = 1 - 2x$

(c) $\begin{pmatrix} x \\ y \\ z \end{pmatrix} \to \begin{pmatrix} x = x \\ y = y \\ z = z \\ t = x + y \end{pmatrix} \to w \Rightarrow \left(\frac{\partial w}{\partial z}\right)_{x,y} = \frac{\partial w}{\partial x}\frac{\partial x}{\partial z} + \frac{\partial w}{\partial y}\frac{\partial y}{\partial z} + \frac{\partial w}{\partial z}\frac{\partial z}{\partial z} + \frac{\partial w}{\partial t}\frac{\partial t}{\partial z}$; $\frac{\partial x}{\partial z} = 0$ and $\frac{\partial y}{\partial z} = 0$

$\Rightarrow \left(\frac{\partial w}{\partial z}\right)_{x,y} = (2x)(0) + (1)(0) + (-1)(1) + (\cos t)(0) = -1$

(d) $\begin{pmatrix} y \\ z \\ t \end{pmatrix} \to \begin{pmatrix} x = t - y \\ y = y \\ z = z \\ t = t \end{pmatrix} \to w \Rightarrow \left(\frac{\partial w}{\partial z}\right)_{y,t} = \frac{\partial w}{\partial x}\frac{\partial x}{\partial z} + \frac{\partial w}{\partial y}\frac{\partial y}{\partial z} + \frac{\partial w}{\partial z}\frac{\partial z}{\partial z} + \frac{\partial w}{\partial t}\frac{\partial t}{\partial z}$; $\frac{\partial y}{\partial z} = 0$ and $\frac{\partial t}{\partial z} = 0$

$\Rightarrow \left(\frac{\partial w}{\partial z}\right)_{y,t} = (2x)(0) + (1)(0) + (-1)(1) + (\cos t)(0) = -1$

(e) $\begin{pmatrix} x \\ z \\ t \end{pmatrix} \to \begin{pmatrix} x = x \\ y = t - x \\ z = z \\ t = t \end{pmatrix} \to w \Rightarrow \left(\frac{\partial w}{\partial t}\right)_{x,z} = \frac{\partial w}{\partial x}\frac{\partial x}{\partial t} + \frac{\partial w}{\partial y}\frac{\partial y}{\partial t} + \frac{\partial w}{\partial z}\frac{\partial z}{\partial t} + \frac{\partial w}{\partial t}\frac{\partial t}{\partial t}$; $\frac{\partial x}{\partial t} = 0$ and $\frac{\partial z}{\partial t} = 0$

$\Rightarrow \left(\frac{\partial w}{\partial t}\right)_{x,z} = (2x)(0) + (1)(1) + (-1)(0) + (\cos t)(1) = 1 + \cos t$

(f) $\begin{pmatrix} y \\ z \\ t \end{pmatrix} \to \begin{pmatrix} x = t - y \\ y = y \\ z = z \\ t = t \end{pmatrix} \to w \Rightarrow \left(\frac{\partial w}{\partial t}\right)_{y,z} = \frac{\partial w}{\partial x}\frac{\partial x}{\partial t} + \frac{\partial w}{\partial y}\frac{\partial y}{\partial t} + \frac{\partial w}{\partial z}\frac{\partial z}{\partial t} + \frac{\partial w}{\partial t}\frac{\partial t}{\partial t}$; $\frac{\partial y}{\partial t} = 0$ and $\frac{\partial z}{\partial t} = 0$

$\Rightarrow \left(\frac{\partial w}{\partial t}\right)_{y,z} = (2x)(1) + (1)(0) + (-1)(0) + (\cos t)(1) = \cos t + 2x = \cos t + 2(t - y)$

3. $U = f(P, V, T)$ and $PV = nRT$

(a) $\begin{pmatrix} P \\ V \end{pmatrix} \to \begin{pmatrix} P = P \\ V = V \\ T = \frac{PV}{nR} \end{pmatrix} \to U \Rightarrow \left(\frac{\partial U}{\partial P}\right)_V = \frac{\partial U}{\partial P}\frac{\partial P}{\partial P} + \frac{\partial U}{\partial V}\frac{\partial V}{\partial P} + \frac{\partial U}{\partial T}\frac{\partial T}{\partial P} = \frac{\partial U}{\partial P} + \left(\frac{\partial U}{\partial V}\right)(0) + \left(\frac{\partial U}{\partial T}\right)\left(\frac{V}{nR}\right)$

$= \frac{\partial U}{\partial P} + \left(\frac{\partial U}{\partial T}\right)\left(\frac{V}{nR}\right)$

(b) $\begin{pmatrix} V \\ T \end{pmatrix} \to \begin{pmatrix} P = \frac{nRT}{V} \\ V = V \\ T = T \end{pmatrix} \to U \Rightarrow \left(\frac{\partial U}{\partial T}\right)_V = \frac{\partial U}{\partial P}\frac{\partial P}{\partial T} + \frac{\partial U}{\partial V}\frac{\partial V}{\partial T} + \frac{\partial U}{\partial T}\frac{\partial T}{\partial T} = \left(\frac{\partial U}{\partial P}\right)\left(\frac{nR}{V}\right) + \left(\frac{\partial U}{\partial V}\right)(0) + \frac{\partial U}{\partial T}$

$= \left(\frac{\partial U}{\partial P}\right)\left(\frac{nR}{V}\right) + \frac{\partial U}{\partial T}$

4. $w = x^2 + y^2 + z^2$ and $y \sin z + z \sin x = 1$

(a) $\begin{pmatrix} x \\ y \end{pmatrix} \to \begin{pmatrix} x = x \\ y = y \\ z = z(x, y) \end{pmatrix} \to w \Rightarrow \left(\frac{\partial w}{\partial x}\right)_y = \frac{\partial w}{\partial x}\frac{\partial x}{\partial x} + \frac{\partial w}{\partial y}\frac{\partial y}{\partial x} + \frac{\partial w}{\partial z}\frac{\partial z}{\partial x};\ \frac{\partial y}{\partial x} = 0$ and

$(y \cos z)\frac{\partial z}{\partial x} + (\sin x)\frac{\partial z}{\partial x} + z \cos x = 0 \Rightarrow \frac{\partial z}{\partial x} = \frac{-z \cos x}{y \cos z + \sin x}$. At $(0, 1, \pi)$, $\frac{\partial z}{\partial x} = \frac{-\pi}{-1} = \pi$

$\Rightarrow \left(\frac{\partial w}{\partial x}\right)_y \bigg|_{(0,1,\pi)} = (2x)(1) + (2y)(0) + (2z)(\pi)\big|_{(0,1,\pi)} = 2\pi^2$

(b) $\begin{pmatrix} y \\ z \end{pmatrix} \to \begin{pmatrix} x = x(y, z) \\ y = y \\ z = z \end{pmatrix} \to w \Rightarrow \left(\frac{\partial w}{\partial z}\right)_y = \frac{\partial w}{\partial x}\frac{\partial x}{\partial z} + \frac{\partial w}{\partial y}\frac{\partial y}{\partial z} + \frac{\partial w}{\partial z}\frac{\partial z}{\partial z} = (2x)\frac{\partial x}{\partial z} + (2y)(0) + (2z)(1)$

$= (2x)\frac{\partial x}{\partial z} + 2z$. Now $(\sin z)\frac{\partial y}{\partial z} + y \cos z + \sin x + (z \cos x)\frac{\partial x}{\partial z} = 0$ and $\frac{\partial y}{\partial z} = 0$

$\Rightarrow y \cos z + \sin x + (z \cos x)\frac{\partial x}{\partial z} = 0 \Rightarrow \frac{\partial x}{\partial z} = \frac{-y \cos z - \sin x}{z \cos x}$. At $(0, 1, \pi)$, $\frac{\partial x}{\partial z} = \frac{1 - 0}{(\pi)(1)} = \frac{1}{\pi}$

$\Rightarrow \left(\frac{\partial w}{\partial z}\right)_y \bigg|_{(0,1,\pi)} = 2(0)\left(\frac{1}{\pi}\right) + 2\pi = 2\pi$

5. $w = x^2y^2 + yz - z^3$ and $x^2 + y^2 + z^2 = 6$

(a) $\begin{pmatrix} x \\ y \end{pmatrix} \to \begin{pmatrix} x = x \\ y = y \\ z = z(x, y) \end{pmatrix} \to w \Rightarrow \left(\frac{\partial w}{\partial y}\right)_x = \frac{\partial w}{\partial x}\frac{\partial x}{\partial y} + \frac{\partial w}{\partial y}\frac{\partial y}{\partial y} + \frac{\partial w}{\partial z}\frac{\partial z}{\partial y}$

$= (2xy^2)(0) + (2x^2y + z)(1) + (y - 3z^2)\frac{\partial z}{\partial y} = 2x^2y + z + (y - 3z^2)\frac{\partial z}{\partial y}$. Now $(2x)\frac{\partial x}{\partial y} + 2y + (2z)\frac{\partial z}{\partial y} = 0$ and

$\frac{\partial x}{\partial y} = 0 \Rightarrow 2y + (2z)\frac{\partial z}{\partial y} = 0 \Rightarrow \frac{\partial z}{\partial y} = -\frac{y}{z}$. At $(w,x,y,z) = (4,2,1,-1)$, $\frac{\partial z}{\partial y} = -\frac{1}{-1} = 1 \Rightarrow \left(\frac{\partial w}{\partial y}\right)_x\Big|_{(4,2,1,-1)}$
$= [(2)(2)^2(1) + (-1)] + [1 - 3(-1)^2](1) = 5$

(b) $\begin{pmatrix} y \\ z \end{pmatrix} \to \begin{pmatrix} x = x(y,z) \\ y = y \\ z = z \end{pmatrix} \to w \Rightarrow \left(\frac{\partial w}{\partial y}\right)_z = \frac{\partial w}{\partial x}\frac{\partial x}{\partial y} + \frac{\partial w}{\partial y}\frac{\partial y}{\partial y} + \frac{\partial w}{\partial z}\frac{\partial z}{\partial y}$

$= (2xy^2)\frac{\partial x}{\partial y} + (2x^2y + z)(1) + (y - 3z^2)(0) = (2x^2y)\frac{\partial x}{\partial y} + 2x^2y + z$. Now $(2x)\frac{\partial x}{\partial y} + 2y + (2z)\frac{\partial z}{\partial y} = 0$ and

$\frac{\partial z}{\partial y} = 0 \Rightarrow (2x)\frac{\partial x}{\partial y} + 2y = 0 \Rightarrow \frac{\partial x}{\partial y} = -\frac{y}{x}$. At $(w,x,y,z) = (4,2,1,-1)$, $\frac{\partial x}{\partial y} = -\frac{1}{2} \Rightarrow \left(\frac{\partial w}{\partial y}\right)_z\Big|_{(4,2,1,-1)}$
$= (2)(2)(1)^2\left(-\frac{1}{2}\right) + (2)(2)^2(1) + (-1) = 5$

6. $y = uv \Rightarrow 1 = v\frac{\partial u}{\partial y} + u\frac{\partial v}{\partial y}$; $x = u^2 + v^2$ and $\frac{\partial x}{\partial y} = 0 \Rightarrow 0 = 2u\frac{\partial u}{\partial y} + 2v\frac{\partial v}{\partial y} \Rightarrow \frac{\partial v}{\partial y} = \left(-\frac{u}{v}\right)\frac{\partial u}{\partial y} \Rightarrow 1$
$= v\frac{\partial u}{\partial y} + u\left(-\frac{u}{v}\frac{\partial u}{\partial y}\right) = \left(\frac{v^2 - u^2}{v}\right)\frac{\partial u}{\partial y} \Rightarrow \frac{\partial u}{\partial y} = \frac{v}{v^2 - u^2}$. At $(u,v) = (\sqrt{2}, 1)$, $\frac{\partial u}{\partial y} = \frac{1}{1^2 - (\sqrt{2})^2} = -1$
$\Rightarrow \left(\frac{\partial u}{\partial y}\right)_x = -1$

7. (a) $\begin{pmatrix} r \\ \theta \end{pmatrix} \to \begin{pmatrix} x = r\cos\theta \\ y = r\sin\theta \end{pmatrix} \Rightarrow \left(\frac{\partial x}{\partial r}\right)_\theta = \cos\theta$

(b) $x^2 + y^2 = r^2 \Rightarrow 2x + 2y\frac{\partial y}{\partial x} = 2r\frac{\partial r}{\partial x}$ and $\frac{\partial y}{\partial x} = 0 \Rightarrow 2x = 2r\frac{\partial r}{\partial x} \Rightarrow \frac{\partial r}{\partial x} = \frac{x}{r} \Rightarrow \left(\frac{\partial r}{\partial x}\right)_y = \frac{x}{\sqrt{x^2 + y^2}}$

8. If x, y, and z are independent, then $\left(\frac{\partial w}{\partial x}\right)_{y,z} = \frac{\partial w}{\partial x}\frac{\partial x}{\partial x} + \frac{\partial w}{\partial y}\frac{\partial y}{\partial x} + \frac{\partial w}{\partial z}\frac{\partial z}{\partial x} + \frac{\partial w}{\partial t}\frac{\partial t}{\partial x}$
$= (2x)(1) + (-2y)(0) + (4)(0) + (1)\left(\frac{\partial t}{\partial x}\right) = 2x + \frac{\partial t}{\partial x}$. Thus $x + 2z + t = 25 \Rightarrow 1 + 0 + \frac{\partial t}{\partial x} = 0 \Rightarrow \frac{\partial t}{\partial x} = -1$
$\Rightarrow \left(\frac{\partial w}{\partial x}\right)_{y,z} = 2x - 1$. On the other hand, if x, y, and t are independent, then $\left(\frac{\partial w}{\partial x}\right)_{y,t}$
$= \frac{\partial w}{\partial x}\frac{\partial x}{\partial x} + \frac{\partial w}{\partial y}\frac{\partial y}{\partial x} + \frac{\partial w}{\partial z}\frac{\partial z}{\partial x} + \frac{\partial w}{\partial t}\frac{\partial t}{\partial x} = (2x)(1) + (-2y)(0) + 4\frac{\partial z}{\partial x} + (1)(0) = 2x + 4\frac{\partial z}{\partial x}$. Thus, $x + 2z + t = 25$
$\Rightarrow 1 + 2\frac{\partial z}{\partial x} + 0 = 0 \Rightarrow \frac{\partial z}{\partial x} = -\frac{1}{2} \Rightarrow \left(\frac{\partial w}{\partial x}\right)_{y,t} = 2x + 4\left(-\frac{1}{2}\right) = 2x - 2$.

9. If x is a differentiable function of y and z, then $f(x,y,z) = 0 \Rightarrow \frac{\partial f}{\partial x}\frac{\partial x}{\partial x} + \frac{\partial f}{\partial y}\frac{\partial y}{\partial x} + \frac{\partial f}{\partial z}\frac{\partial z}{\partial x} = 0 \Rightarrow \frac{\partial f}{\partial x} + \frac{\partial f}{\partial y}\frac{\partial y}{\partial x} = 0$
$\Rightarrow \left(\frac{\partial x}{\partial y}\right)_z = -\frac{\partial f/\partial y}{\partial f/\partial z}$. Similarly, if y is a differentiable function of x and z, $\left(\frac{\partial y}{\partial z}\right)_x = -\frac{\partial f/\partial z}{\partial f/\partial x}$ and if z is a differentiable function of x and y, $\left(\frac{\partial z}{\partial x}\right)_y = -\frac{\partial f/\partial x}{\partial f/\partial y}$. Then $\left(\frac{\partial x}{\partial y}\right)_z\left(\frac{\partial y}{\partial z}\right)_x\left(\frac{\partial z}{\partial x}\right)_y$
$= \left(-\frac{\partial f/\partial y}{\partial f/\partial z}\right)\left(-\frac{\partial f/\partial z}{\partial f/\partial x}\right)\left(-\frac{\partial f/\partial x}{\partial f/\partial y}\right) = -1$.

10. $z = z + f(u)$ and $u = xy \Rightarrow \frac{\partial z}{\partial x} = 1 + \frac{df}{du}\frac{\partial u}{\partial x} = 1 + y\,\frac{df}{du}$; also $\frac{\partial z}{\partial y} = 0 + \frac{df}{du}\frac{\partial u}{\partial y} = x\,\frac{df}{du}$ so that $x\,\frac{\partial z}{\partial x} - y\,\frac{\partial z}{\partial y}$
$= x\left(1 + y\,\frac{df}{du}\right) - y\left(x\,\frac{df}{du}\right) = x$

11. If x and y are independent, then $g(x,y,z) = 0 \Rightarrow \frac{\partial g}{\partial x}\frac{\partial x}{\partial y} + \frac{\partial g}{\partial y}\frac{\partial y}{\partial y} + \frac{\partial g}{\partial z}\frac{\partial z}{\partial y} = 0$ and $\frac{\partial x}{\partial y} = 0 \Rightarrow \frac{\partial g}{\partial y} + \frac{\partial g}{\partial z}\frac{\partial z}{\partial y} = 0$
$\Rightarrow \left(\frac{\partial z}{\partial y}\right)_x = -\frac{\partial g/\partial y}{\partial g/\partial z}$, as claimed.

12. Let x and y be independent. Then $f(x,y,z,w) = 0$, $g(x,y,z,w) = 0$ and $\frac{\partial y}{\partial x} = 0$
$\Rightarrow \frac{\partial f}{\partial x}\frac{\partial x}{\partial x} + \frac{\partial f}{\partial y}\frac{\partial y}{\partial x} + \frac{\partial f}{\partial z}\frac{\partial z}{\partial x} + \frac{\partial f}{\partial w}\frac{\partial w}{\partial x} = \frac{\partial f}{\partial x} + \frac{\partial f}{\partial z}\frac{\partial z}{\partial x} + \frac{\partial f}{\partial w}\frac{\partial w}{\partial x} = 0$ and
$\frac{\partial g}{\partial x}\frac{\partial x}{\partial x} + \frac{\partial g}{\partial y}\frac{\partial y}{\partial x} + \frac{\partial g}{\partial z}\frac{\partial z}{\partial x} + \frac{\partial g}{\partial w}\frac{\partial w}{\partial x} = \frac{\partial g}{\partial x} + \frac{\partial g}{\partial z}\frac{\partial z}{\partial x} + \frac{\partial g}{\partial w}\frac{\partial w}{\partial x} = 0$ imply

$$\begin{cases} \frac{\partial f}{\partial z}\frac{\partial z}{\partial x} + \frac{\partial f}{\partial w}\frac{\partial w}{\partial x} = -\frac{\partial f}{\partial x} \\ \frac{\partial g}{\partial z}\frac{\partial z}{\partial x} + \frac{\partial g}{\partial w}\frac{\partial w}{\partial x} = -\frac{\partial g}{\partial x} \end{cases} \Rightarrow \left(\frac{\partial z}{\partial x}\right)_y = \frac{\begin{vmatrix} -\frac{\partial f}{\partial x} & \frac{\partial f}{\partial w} \\ -\frac{\partial g}{\partial x} & \frac{\partial g}{\partial w} \end{vmatrix}}{\begin{vmatrix} \frac{\partial f}{\partial z} & \frac{\partial f}{\partial w} \\ \frac{\partial g}{\partial z} & \frac{\partial g}{\partial w} \end{vmatrix}} = \frac{-\frac{\partial f}{\partial x}\frac{\partial g}{\partial w} + \frac{\partial g}{\partial x}\frac{\partial f}{\partial w}}{\frac{\partial f}{\partial z}\frac{\partial g}{\partial w} - \frac{\partial g}{\partial z}\frac{\partial f}{\partial w}} = -\frac{\frac{\partial f}{\partial x}\frac{\partial g}{\partial w} - \frac{\partial f}{\partial w}\frac{\partial g}{\partial x}}{\frac{\partial f}{\partial z}\frac{\partial g}{\partial w} - \frac{\partial f}{\partial w}\frac{\partial g}{\partial z}},$$ as claimed.

Likewise, $f(x,y,z,w) = 0$, $g(x,y,z,w) = 0$ and $\frac{\partial x}{\partial y} = 0 \Rightarrow \frac{\partial f}{\partial x}\frac{\partial x}{\partial y} + \frac{\partial f}{\partial y}\frac{\partial y}{\partial y} + \frac{\partial f}{\partial z}\frac{\partial z}{\partial y} + \frac{\partial f}{\partial w}\frac{\partial w}{\partial y}$
$= \frac{\partial f}{\partial y} + \frac{\partial f}{\partial z}\frac{\partial z}{\partial y} + \frac{\partial f}{\partial w}\frac{\partial w}{\partial y} = 0$ and (similarly) $\frac{\partial g}{\partial y} + \frac{\partial g}{\partial z}\frac{\partial z}{\partial y} + \frac{\partial g}{\partial w}\frac{\partial w}{\partial y} = 0$ imply

$$\begin{cases} \frac{\partial f}{\partial z}\frac{\partial z}{\partial y} + \frac{\partial f}{\partial w}\frac{\partial w}{\partial y} = -\frac{\partial f}{\partial y} \\ \frac{\partial g}{\partial z}\frac{\partial z}{\partial y} + \frac{\partial g}{\partial w}\frac{\partial w}{\partial y} = -\frac{\partial g}{\partial y} \end{cases} \Rightarrow \left(\frac{\partial w}{\partial y}\right)_x = \frac{\begin{vmatrix} \frac{\partial f}{\partial z} & -\frac{\partial f}{\partial y} \\ \frac{\partial g}{\partial z} & -\frac{\partial g}{\partial y} \end{vmatrix}}{\begin{vmatrix} \frac{\partial f}{\partial z} & \frac{\partial f}{\partial w} \\ \frac{\partial g}{\partial z} & \frac{\partial g}{\partial w} \end{vmatrix}} = \frac{-\frac{\partial f}{\partial z}\frac{\partial g}{\partial y} + \frac{\partial g}{\partial z}\frac{\partial f}{\partial y}}{\frac{\partial f}{\partial z}\frac{\partial g}{\partial w} - \frac{\partial g}{\partial z}\frac{\partial f}{\partial w}} = -\frac{\frac{\partial f}{\partial z}\frac{\partial g}{\partial y} - \frac{\partial f}{\partial y}\frac{\partial g}{\partial z}}{\frac{\partial f}{\partial z}\frac{\partial g}{\partial w} - \frac{\partial f}{\partial w}\frac{\partial g}{\partial z}},$$ as claimed.

12.7 DIRECTIONAL DERIVATIVES, GRADIENT VECTORS, AND TANGENT PLANES

1. $\frac{\partial f}{\partial x} = -1$, $\frac{\partial f}{\partial y} = 1 \Rightarrow \nabla f = -\mathbf{i} + \mathbf{j}$; $f(2,1) = -1$
$\Rightarrow -1 = y - x$ is the level curve

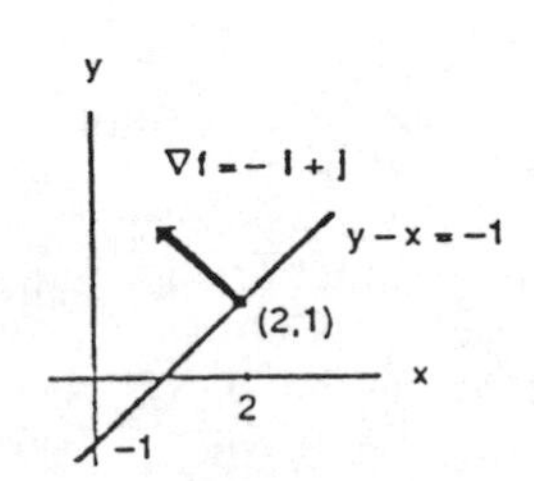

2. $\frac{\partial f}{\partial x} = \frac{2x}{x^2+y^2} \Rightarrow \frac{\partial f}{\partial x}(1,1) = 1$; $\frac{\partial f}{\partial y} = \frac{2y}{x^2+y^2}$
$\Rightarrow \frac{\partial f}{\partial y}(1,1) = 1 \Rightarrow \nabla f = \mathbf{i}+\mathbf{j}$; $f(1,1) = \ln 2 \Rightarrow \ln 2$
$= \ln(x^2+y^2) \Rightarrow 2 = x^2+y^2$ is the level curve

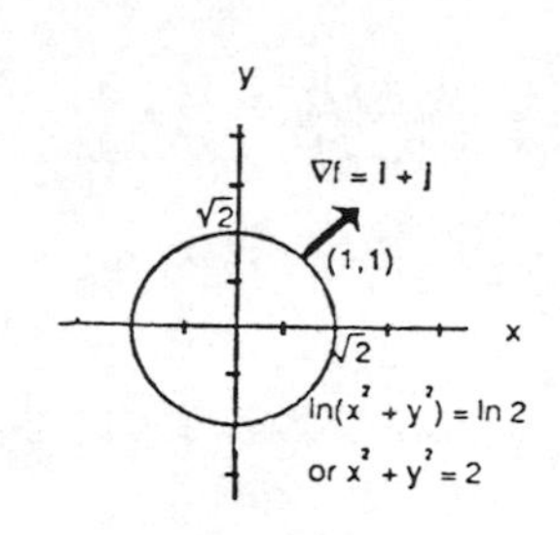

3. $\frac{\partial g}{\partial x} = -2x \Rightarrow \frac{\partial g}{\partial x}(-1,0) = 2$; $\frac{\partial g}{\partial y} = 1$

$\Rightarrow \nabla g = 2\mathbf{i}+\mathbf{j}$; $g(-1,0) = -1$

$\Rightarrow -1 = y - x^2$ is the level curve

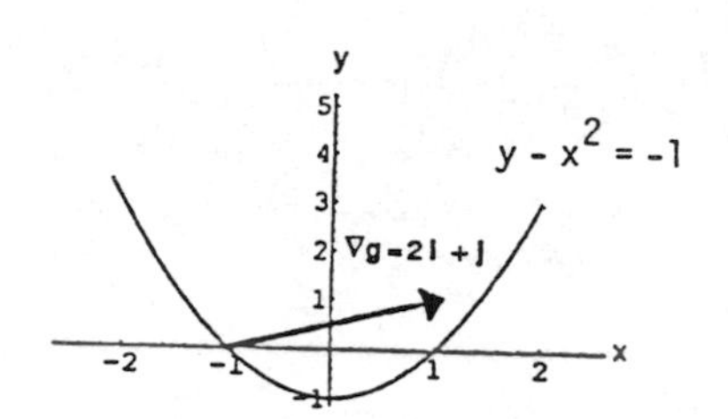

4. $\frac{\partial g}{\partial x} = x \Rightarrow \frac{\partial g}{\partial x}(\sqrt{2},1) = \sqrt{2}$; $\frac{\partial g}{\partial y} = -y$
$\Rightarrow \frac{\partial g}{\partial y}(\sqrt{2},1) = -1 \Rightarrow \nabla g = \sqrt{2}\mathbf{i} - \mathbf{j}$; $g(\sqrt{2},1) = \frac{1}{2}$
$\Rightarrow \frac{1}{2} = \frac{x^2}{2} - \frac{y^2}{2}$ or $1 = x^2 - y^2$ is the level curve

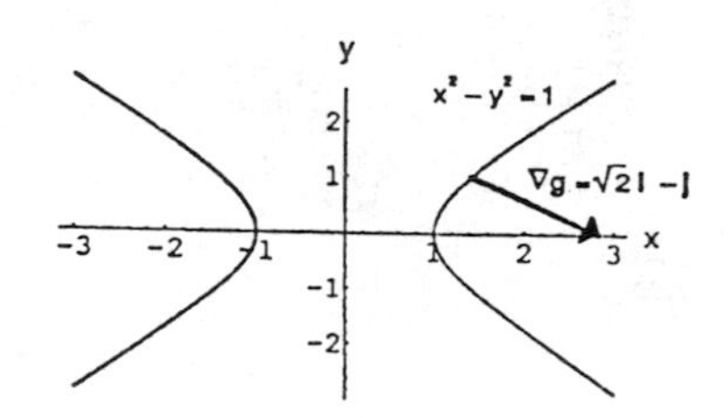

5. $\frac{\partial f}{\partial x} = 2x + \frac{z}{x} \Rightarrow \frac{\partial f}{\partial x}(1,1,1) = 3$; $\frac{\partial f}{\partial y} = 2y \Rightarrow \frac{\partial f}{\partial y}(1,1,1) = 2$; $\frac{\partial f}{\partial z} = -4z + \ln x \Rightarrow \frac{\partial f}{\partial z}(1,1,1) = -4$;
thus $\nabla f = 3\mathbf{i} + 2\mathbf{j} - 4\mathbf{k}$

6. $\frac{\partial f}{\partial x} = -6xz + \frac{z}{x^2z^2+1} \Rightarrow \frac{\partial f}{\partial x}(1,1,1) = -\frac{11}{2}$; $\frac{\partial f}{\partial y} = -6yz \Rightarrow \frac{\partial f}{\partial y}(1,1,1) = -6$; $\frac{\partial f}{\partial z} = 6z^2 - 3(x^2+y^2) + \frac{x}{x^2z^2+1}$
$\Rightarrow \frac{\partial f}{\partial z}(1,1,1) = \frac{1}{2}$; thus $\nabla f = -\frac{11}{2}\mathbf{i} - 6\mathbf{j} + \frac{1}{2}\mathbf{k}$

7. $\frac{\partial f}{\partial x} = -\frac{x}{(x^2+y^2+z^2)^{3/2}} + \frac{1}{x} \Rightarrow \frac{\partial f}{\partial x}(-1,2,-2) = -\frac{26}{27}$; $\frac{\partial f}{\partial y} = -\frac{y}{(x^2+y^2+z^2)^{3/2}} + \frac{1}{y} \Rightarrow \frac{\partial f}{\partial y}(-1,2,-2) = \frac{23}{54}$;
$\frac{\partial f}{\partial z} = -\frac{z}{(x^2+y^2+z^2)^{3/2}} + \frac{1}{z} \Rightarrow \frac{\partial f}{\partial z}(-1,2,-2) = -\frac{23}{54}$; thus $\nabla f = -\frac{26}{27}\mathbf{i} + \frac{23}{54}\mathbf{j} - \frac{23}{54}\mathbf{k}$

8. $\frac{\partial f}{\partial x} = e^{x+y}\cos z + \frac{y+1}{\sqrt{1-x^2}} \Rightarrow \frac{\partial f}{\partial x}\left(0,0,\frac{\pi}{6}\right) = \frac{\sqrt{3}}{2} + 1$; $\frac{\partial f}{\partial y} = e^{x+y}\cos z + \sin^{-1} x \Rightarrow \frac{\partial f}{\partial y}\left(0,0,\frac{\pi}{6}\right) = \frac{\sqrt{3}}{2}$;
$\frac{\partial f}{\partial z} = -e^{x+y}\sin z \Rightarrow \frac{\partial f}{\partial z}\left(0,0,\frac{\pi}{6}\right) = -\frac{1}{2}$; thus $\nabla f = \left(\frac{\sqrt{3}+2}{2}\right)\mathbf{i} + \frac{\sqrt{3}}{2}\mathbf{j} - \frac{1}{2}\mathbf{k}$

9. $\mathbf{u} = \frac{\mathbf{A}}{|\mathbf{A}|} = \frac{4\mathbf{i} + 3\mathbf{j}}{\sqrt{4^2 + 3^2}} = \frac{4}{5}\mathbf{i} + \frac{3}{5}\mathbf{j}$; $f_x(x, y) = 2y \Rightarrow f_x(5, 5) = 10$; $f_y(x, y) = 2x - 6y \Rightarrow f_y(5, 5) = -20$

$\Rightarrow \nabla f = 10\mathbf{i} - 20\mathbf{j} \Rightarrow (D_{\mathbf{u}}f)_{P_0} = \nabla f \cdot \mathbf{u} = 10\left(\frac{4}{5}\right) - 20\left(\frac{3}{5}\right) = -4$

10. $\mathbf{u} = \frac{\mathbf{A}}{|\mathbf{A}|} = \frac{3\mathbf{i} - 4\mathbf{j}}{\sqrt{3^2 + (-4)^2}} = \frac{3}{5}\mathbf{i} - \frac{4}{5}\mathbf{j}$; $f_x(x, y) = 4x \Rightarrow f_x(-1, 1) = -4$; $f_y(x, y) = 2y \Rightarrow f_y(-1, 1) = 2$

$\Rightarrow \nabla f = -4\mathbf{i} + 2\mathbf{j} \Rightarrow (D_{\mathbf{u}}f)_{P_0} = \nabla f \cdot \mathbf{u} = -\frac{12}{5} - \frac{8}{5} = -4$

11. $\mathbf{u} = \frac{\mathbf{A}}{|\mathbf{A}|} = \frac{12\mathbf{i} + 5\mathbf{j}}{\sqrt{12^2 + 5^2}} = \frac{12}{13}\mathbf{i} + \frac{5}{13}\mathbf{j}$; $g_x(x, y) = 1 + \frac{y^2}{x^2} + \frac{2y\sqrt{3}}{2xy\sqrt{4x^2y^2 - 1}} \Rightarrow g_x(1, 1) = 3$; $g_y(x, y)$

$= -\frac{2y}{x} + \frac{2x\sqrt{3}}{2xy\sqrt{4x^2y^2 - 1}} \Rightarrow g_y(1, 1) = -1 \Rightarrow \nabla g = 3\mathbf{i} - \mathbf{j} \Rightarrow (D_{\mathbf{u}}g)_{P_0} = \nabla g \cdot \mathbf{u} = \frac{36}{13} - \frac{5}{13} = \frac{31}{13}$

12. $\mathbf{u} = \frac{\mathbf{A}}{|\mathbf{A}|} = \frac{3\mathbf{i} - 2\mathbf{j}}{\sqrt{3^2 + (-2)^2}} = \frac{3}{\sqrt{13}}\mathbf{i} - \frac{2}{\sqrt{13}}\mathbf{j}$; $h_x(x, y) = \frac{\left(\frac{-y}{x^2}\right)}{\left(\frac{y}{x}\right)^2 + 1} + \frac{\left(\frac{y}{2}\right)\sqrt{3}}{\sqrt{1 - \left(\frac{x^2y^2}{4}\right)}} \Rightarrow h_x(1, 1) = \frac{1}{2}$;

$h_y(x, y) = \frac{\left(\frac{1}{x}\right)}{\left(\frac{y}{x}\right)^2 + 1} + \frac{\left(\frac{x}{2}\right)\sqrt{3}}{\sqrt{1 - \left(\frac{x^2y^2}{4}\right)}} \Rightarrow h_y(1, 1) = \frac{3}{2} \Rightarrow \nabla h = \frac{1}{2}\mathbf{i} + \frac{3}{2}\mathbf{j} \Rightarrow (D_{\mathbf{u}}h)_{P_0} = \nabla h \cdot \mathbf{u} = \frac{3}{2\sqrt{13}} - \frac{6}{2\sqrt{13}}$

$= -\frac{3}{2\sqrt{13}}$

13. $\mathbf{u} = \frac{\mathbf{A}}{|\mathbf{A}|} = \frac{3\mathbf{i} + 6\mathbf{j} - 2\mathbf{k}}{\sqrt{3^2 + 6^2 + (-2)^2}} = \frac{3}{7}\mathbf{i} + \frac{6}{7}\mathbf{j} - \frac{2}{7}\mathbf{k}$; $f_x(x, y, z) = y + z \Rightarrow f_x(1, -1, 2) = 1$; $f_y(x, y, z) = x + z$

$\Rightarrow f_y(1, -1, 2) = 3$; $f_z(x, y, z) = y + x \Rightarrow f_z(1, -1, 2) = 0 \Rightarrow \nabla f = \mathbf{i} + 3\mathbf{j} \Rightarrow (D_{\mathbf{u}}f)_{P_0} = \nabla f \cdot \mathbf{u} = \frac{3}{7} + \frac{18}{7} = 3$

14. $\mathbf{u} = \frac{\mathbf{A}}{|\mathbf{A}|} = \frac{\mathbf{i} + \mathbf{j} + \mathbf{k}}{\sqrt{1^2 + 1^2 + 1^2}} = \frac{1}{\sqrt{3}}\mathbf{i} + \frac{1}{\sqrt{3}}\mathbf{j} + \frac{1}{\sqrt{3}}\mathbf{k}$; $f_x(x, y, z) = 2x \Rightarrow f_x(1, 1, 1) = 2$; $f_y(x, y, z) = 4y$

$\Rightarrow f_y(1, 1, 1) = 4$; $f_z(x, y, z) = -6z \Rightarrow f_z(1, 1, 1) = -6 \Rightarrow \nabla f = 2\mathbf{i} + 4\mathbf{j} - 6\mathbf{k} \Rightarrow (D_{\mathbf{u}}f)_{P_0} = \nabla f \cdot \mathbf{u}$

$= 2\left(\frac{1}{\sqrt{3}}\right) + 4\left(\frac{1}{\sqrt{3}}\right) - 6\left(\frac{1}{\sqrt{3}}\right) = 0$

15. $\mathbf{u} = \frac{\mathbf{A}}{|\mathbf{A}|} = \frac{2\mathbf{i} + \mathbf{j} - 2\mathbf{k}}{\sqrt{2^2 + 1^2 + (-2)^2}} = \frac{2}{3}\mathbf{i} + \frac{1}{3}\mathbf{j} - \frac{2}{3}\mathbf{k}$; $g_x(x, y, z) = 3e^x \cos yz \Rightarrow g_x(0, 0, 0) = 3$; $g_y(x, y, z) = -3ze^x \sin yz$

$\Rightarrow g_y(0, 0, 0) = 0$; $g_z(x, y, z) = -3ye^x \sin yz \Rightarrow g_z(0, 0, 0) = 0 \Rightarrow \nabla g = 3\mathbf{i} \Rightarrow (D_{\mathbf{u}}g)_{P_0} = \nabla g \cdot \mathbf{u} = 2$

16. $\mathbf{u} = \frac{\mathbf{A}}{|\mathbf{A}|} = \frac{\mathbf{i} + 2\mathbf{j} + 2\mathbf{k}}{\sqrt{1^2 + 2^2 + 2^2}} = \frac{1}{3}\mathbf{i} + \frac{2}{3}\mathbf{j} + \frac{2}{3}\mathbf{k}$; $h_x(x, y, z) = -y \sin xy + \frac{1}{x} \Rightarrow h_x\left(1, 0, \frac{1}{2}\right) = 1$;

$h_y(x, y, z) = -x \sin xy + ze^{yz} \Rightarrow h_y\left(1, 0, \frac{1}{2}\right) = \frac{1}{2}$; $h_z(x, y, z) = ye^{yz} + \frac{1}{z} \Rightarrow h_z\left(1, 0, \frac{1}{2}\right) = 2 \Rightarrow \nabla h = \mathbf{i} + \frac{1}{2}\mathbf{j} + 2\mathbf{k}$

$\Rightarrow (D_{\mathbf{u}}h)_{P_0} = \nabla h \cdot \mathbf{u} = \frac{1}{3} + \frac{1}{3} + \frac{4}{3} = 2$

17. $\nabla f = (2x+y)\mathbf{i} + (x+2y)\mathbf{j} \Rightarrow \nabla f(-1,1) = -\mathbf{i}+\mathbf{j} \Rightarrow \mathbf{u} = \frac{\nabla f}{|\nabla f|} = \frac{-\mathbf{i}+\mathbf{j}}{\sqrt{(-1)^2+1^2}} = -\frac{1}{\sqrt{2}}\mathbf{i} + \frac{1}{\sqrt{2}}\mathbf{j}$; f increases most rapidly in the direction $\mathbf{u} = -\frac{1}{\sqrt{2}}\mathbf{i} + \frac{1}{\sqrt{2}}\mathbf{j}$ and decreases most rapidly in the direction $-\mathbf{u} = \frac{1}{\sqrt{2}}\mathbf{i} - \frac{1}{\sqrt{2}}\mathbf{j}$; $(D_{\mathbf{u}}f)_{P_0} = \nabla f \cdot \mathbf{u} = |\nabla f| = \sqrt{2}$ and $(D_{-\mathbf{u}}f)_{P_0} = -\sqrt{2}$

18. $\nabla f = (2xy + ye^{xy}\sin y)\mathbf{i} + (x^2 + xe^{xy}\sin y + e^{xy}\cos y)\mathbf{j} \Rightarrow \nabla f(1,0) = 2\mathbf{j} \Rightarrow \mathbf{u} = \frac{\nabla f}{|\nabla f|} = \mathbf{j}$; f increases most rapidly in the direction $\mathbf{u} = \mathbf{j}$ and decreases most rapidly in the direction $-\mathbf{u} = -\mathbf{j}$; $(D_{\mathbf{u}}f)_{P_0} = \nabla f \cdot \mathbf{u} = |\nabla f| = 2$ and $(D_{-\mathbf{u}}f)_{P_0} = -2$

19. $\nabla f = \frac{1}{y}\mathbf{i} - \left(\frac{x}{y^2} + z\right)\mathbf{j} - y\mathbf{k} \Rightarrow \nabla f(4,1,1) = \mathbf{i} - 5\mathbf{j} - \mathbf{k} \Rightarrow \mathbf{u} = \frac{\nabla f}{|\nabla f|} = \frac{\mathbf{i}-5\mathbf{j}-\mathbf{k}}{\sqrt{1^2+(-5)^2+(-1)^2}}$ $= \frac{1}{3\sqrt{3}}\mathbf{i} - \frac{5}{3\sqrt{3}}\mathbf{j} - \frac{1}{3\sqrt{3}}\mathbf{k}$; f increases most rapidly in the direction of $\mathbf{u} = \frac{1}{3\sqrt{3}}\mathbf{i} - \frac{5}{3\sqrt{3}}\mathbf{j} - \frac{1}{3\sqrt{3}}\mathbf{k}$ and decreases most rapidly in the direction $-\mathbf{u} = -\frac{1}{3\sqrt{3}}\mathbf{i} + \frac{5}{3\sqrt{3}}\mathbf{j} + \frac{1}{3\sqrt{3}}\mathbf{k}$; $(D_{\mathbf{u}}f)_{P_0} = \nabla f \cdot \mathbf{u} = |\nabla f| = 3\sqrt{3}$ and $(D_{-\mathbf{u}}f)_{P_0} = -3\sqrt{3}$

20. $\nabla g = e^y\mathbf{i} + xe^y\mathbf{j} + 2z\mathbf{k} \Rightarrow \nabla g\left(1, \ln 2, \frac{1}{2}\right) = 2\mathbf{i} + 2\mathbf{j} + \mathbf{k} \Rightarrow \mathbf{u} = \frac{\nabla g}{|\nabla g|} = \frac{2\mathbf{i}+2\mathbf{j}+\mathbf{k}}{\sqrt{2^2+2^2+1^2}} = \frac{2}{3}\mathbf{i} + \frac{2}{3}\mathbf{j} + \frac{1}{3}\mathbf{k}$; g increases most rapidly in the direction $\mathbf{u} = \frac{2}{3}\mathbf{i} + \frac{2}{3}\mathbf{j} + \frac{1}{3}\mathbf{k}$ and decreases most rapidly in the direction $-\mathbf{u} = -\frac{2}{3}\mathbf{i} - \frac{2}{3}\mathbf{j} - \frac{1}{3}\mathbf{k}$; $(D_{\mathbf{u}}g)_{P_0} = \nabla g \cdot \mathbf{u} = |\nabla g| = 3$ and $(D_{-\mathbf{u}}g)_{P_0} = -3$

21. $\nabla f = \left(\frac{1}{x} + \frac{1}{x}\right)\mathbf{i} + \left(\frac{1}{y} + \frac{1}{y}\right)\mathbf{j} + \left(\frac{1}{z} + \frac{1}{z}\right)\mathbf{k} \Rightarrow \nabla f(1,1,1) = 2\mathbf{i} + 2\mathbf{j} + 2\mathbf{k} \Rightarrow \mathbf{u} = \frac{\nabla f}{|\nabla f|} = \frac{1}{\sqrt{3}}\mathbf{i} + \frac{1}{\sqrt{3}}\mathbf{j} + \frac{1}{\sqrt{3}}\mathbf{k}$; f increases most rapidly in the direction $\mathbf{u} = \frac{1}{\sqrt{3}}\mathbf{i} + \frac{1}{\sqrt{3}}\mathbf{j} + \frac{1}{\sqrt{3}}\mathbf{k}$; $(D_{\mathbf{u}}f)_{P_0} = \nabla f \cdot \mathbf{u} = |\nabla f| = 2\sqrt{3}$ and $(D_{-\mathbf{u}}f)_{P_0} = -2\sqrt{3}$

22. $\nabla h = \left(\frac{2x}{x^2+y^2-1}\right)\mathbf{i} + \left(\frac{2y}{x^2+y^2-1}\right)\mathbf{j} + 6\mathbf{k} \Rightarrow \nabla h(1,1,0) = 2\mathbf{i} + 3\mathbf{j} + 6\mathbf{k} \Rightarrow \mathbf{u} = \frac{\nabla h}{|\nabla h|} = \frac{2\mathbf{i}+3\mathbf{j}+6\mathbf{k}}{\sqrt{2^2+3^2+6^2}}$ $= \frac{2}{7}\mathbf{i} + \frac{3}{7}\mathbf{j} + \frac{6}{7}\mathbf{k}$; h increases most rapidly in the direction $\mathbf{u} = \frac{2}{7}\mathbf{i} + \frac{3}{7}\mathbf{j} + \frac{6}{7}\mathbf{k}$ and decreases most rapidly in the direction $-\mathbf{u} = -\frac{2}{7}\mathbf{i} - \frac{3}{7}\mathbf{j} - \frac{6}{7}\mathbf{k}$; $(D_{\mathbf{u}}h)_{P_0} = \nabla h \cdot \mathbf{u} = |\nabla h| = 7$ and $(D_{-\mathbf{u}}h)_{P_0} = -7$

23. $\nabla f = \left(\frac{x}{x^2+y^2+z^2}\right)\mathbf{i} + \left(\frac{y}{x^2+y^2+z^2}\right)\mathbf{j} + \left(\frac{z}{x^2+y^2+z^2}\right)\mathbf{k} \Rightarrow \nabla f(3,4,12) = \frac{3}{169}\mathbf{i} + \frac{4}{169}\mathbf{j} + \frac{12}{169}\mathbf{k}$; $\mathbf{u} = \frac{\mathbf{A}}{|\mathbf{A}|} = \frac{3\mathbf{i}+6\mathbf{j}-2\mathbf{k}}{\sqrt{3^2+6^2+(-2)^2}} = \frac{3}{7}\mathbf{i} + \frac{6}{7}\mathbf{j} - \frac{2}{7}\mathbf{k} \Rightarrow \nabla f \cdot \mathbf{u} = \frac{9}{1183}$ and $df = (\nabla f \cdot \mathbf{u})\,ds = \left(\frac{9}{1183}\right)(0.1) \approx 0.000760$

24. $\nabla f = (e^x \cos yz)\mathbf{i} - (ze^x \sin yz)\mathbf{j} - (ye^x \sin yz)\mathbf{k} \Rightarrow \nabla f(0,0,0) = \mathbf{i}$; $\mathbf{u} = \frac{\mathbf{A}}{|\mathbf{A}|} = \frac{2\mathbf{i}+2\mathbf{j}-2\mathbf{k}}{\sqrt{2^2+2^2+(-2)^2}}$
$= \frac{1}{\sqrt{3}}\mathbf{i} + \frac{1}{\sqrt{3}}\mathbf{j} - \frac{1}{\sqrt{3}}\mathbf{k} \Rightarrow \nabla f \cdot \mathbf{u} = \frac{1}{\sqrt{3}}$ and $df = (\nabla f \cdot \mathbf{u})\, ds = \frac{1}{\sqrt{3}}(0.1) \approx 0.0577$

25. $\nabla g = (1+\cos z)\mathbf{i} + (1-\sin z)\mathbf{j} + (-x \sin z - y \cos z)\mathbf{k} \Rightarrow \nabla g(2,-1,0) = 2\mathbf{i} + \mathbf{j} + \mathbf{k}$; $\mathbf{A} = \overrightarrow{P_0P_1} = -2\mathbf{i} + 2\mathbf{j} + 2\mathbf{k}$
$\Rightarrow \mathbf{u} = \frac{\mathbf{A}}{|\mathbf{A}|} = \frac{-2\mathbf{i}+2\mathbf{j}+2\mathbf{k}}{\sqrt{(-2)^2+2^2+2^2}} = -\frac{1}{\sqrt{3}}\mathbf{i} + \frac{1}{\sqrt{3}}\mathbf{j} + \frac{1}{\sqrt{3}}\mathbf{k} \Rightarrow \nabla g \cdot \mathbf{u} = 0$ and $dg = (\nabla g \cdot \mathbf{u})\, ds = (0)(0.2) = 0$

26. $\nabla h = [-\pi y \sin(\pi xy) + z^2]\mathbf{i} - [\pi x \sin(\pi xy)]\mathbf{j} + 2xz\mathbf{k} \Rightarrow \nabla h(-1,-1,-1) = (\pi \sin \pi + 1)\mathbf{i} + (\pi \sin \pi)\mathbf{j} + 2\mathbf{k}$
$= \mathbf{i} + 2\mathbf{k}$; $\mathbf{A} = \overrightarrow{P_0P_1} = \mathbf{i} + \mathbf{j} + \mathbf{k}$ where $P_1 = (0,0,0) \Rightarrow \mathbf{u} = \frac{\mathbf{A}}{|\mathbf{A}|} = \frac{\mathbf{i}+\mathbf{j}+\mathbf{k}}{\sqrt{1^2+1^2+1^2}} = \frac{1}{\sqrt{3}}\mathbf{i} + \frac{1}{\sqrt{3}}\mathbf{j} + \frac{1}{\sqrt{3}}\mathbf{k}$
$\Rightarrow \nabla h \cdot \mathbf{u} = \frac{3}{\sqrt{3}} = \sqrt{3}$ and $dh = (\nabla h \cdot \mathbf{u})\, ds = \sqrt{3}(0.1) \approx 0.1732$

27. $\nabla f = 2x\mathbf{i} + 2y\mathbf{j} + 2z\mathbf{k} \Rightarrow \nabla f(1,1,1) = 2\mathbf{i} + 2\mathbf{j} + 2\mathbf{k} \Rightarrow$ Tangent plane: $2(x-1) + 2(y-1) + 2(z-1) = 0$
$\Rightarrow x + y + z = 3$; Normal line: $x = 1 + 2t,\ y = 1 + 2t,\ z = 1 + 2t$

28. $\nabla f = 2x\mathbf{i} + 2y\mathbf{j} - 2z\mathbf{k} \Rightarrow \nabla f(3,5,-4) = 6\mathbf{i} + 10\mathbf{j} + 8\mathbf{k} \Rightarrow$ Tangent plane: $6(x-3) + 10(y-5) + 8(z+4) = 0$
$\Rightarrow 3x + 5y + 4z = 18$; Normal line: $x = 3 + 6t,\ y = 5 + 10t,\ z = -4 + 8t$

29. $\nabla f = -2x\mathbf{i} + 2\mathbf{k} \Rightarrow \nabla f(2,0,2) = -4\mathbf{i} + 2\mathbf{k} \Rightarrow$ Tangent plane: $-4(x-2) + 2(z-2) = 0 \Rightarrow -4x + 2z + 4 = 0$
$\Rightarrow -2x + z + 2 = 0$; Normal line: $x = 2 - 4t,\ y = 0,\ z = 2 + 2t$

30. $\nabla f = (2x+2y)\mathbf{i} + (2x-2y)\mathbf{j} + 2z\mathbf{k} \Rightarrow \nabla f(1,-1,3) = 4\mathbf{j} + 6\mathbf{k} \Rightarrow$ Tangent plane: $4(y+1) + 6(z-3) = 0$
$\Rightarrow 2y + 3z = 7$; Normal line: $x = 1,\ y = -1 + 4t,\ z = 3 + 6t$

31. $\nabla f = (-\pi \sin \pi x - 2xy + ze^{xz})\mathbf{i} + (-x^2 + z)\mathbf{j} + (xe^{xz} + y)\mathbf{k} \Rightarrow \nabla f(0,1,2) = 2\mathbf{i} + 2\mathbf{j} + \mathbf{k} \Rightarrow$ Tangent plane:
$2(x-0) + 2(y-1) + 1(z-2) = 0 \Rightarrow 2x + 2y + z - 4 = 0$; Normal line: $x = 2t,\ y = 1 + 2t,\ z = 2 + t$

32. $\nabla f = (2x-y)\mathbf{i} - (x+2y)\mathbf{j} - \mathbf{k} \Rightarrow \nabla f(1,1,-1) = \mathbf{i} - 3\mathbf{j} - \mathbf{k} \Rightarrow$ Tangent plane: $1(x-1) - 3(y-1) - 1(z+1) = 0$
$\Rightarrow x - 3y - z = -1$; Normal line: $x = 1 + t,\ y = 1 - 3t,\ z = -1 - t$

33. $\nabla f = \mathbf{i} + \mathbf{j} + \mathbf{k}$ for all points $\Rightarrow \nabla f(0,1,0) = \mathbf{i} + \mathbf{j} + \mathbf{k} \Rightarrow$ Tangent plane: $1(x-0) + 1(y-1) + 1(z-0) = 0$
$\Rightarrow x + y + z - 1 = 0$; Normal line: $x = t,\ y = 1 + t,\ z = t$

34. $\nabla f = (2x-2y-1)\mathbf{i} + (2y-2x+3)\mathbf{j} - \mathbf{k} \Rightarrow \nabla f(2,-3,18) = 9\mathbf{i} - 7\mathbf{j} - \mathbf{k} \Rightarrow$ Tangent plane:
$9(x-2) - 7(y+3) - 1(z-18) = 0 \Rightarrow 9x - 7y - z = 21$; Normal line: $x = 2 + 9t,\ y = -3 - 7t,\ z = 18 - t$

35. $z = f(x,y) = \ln(x^2+y^2) \Rightarrow f_x(x,y) = \frac{2x}{x^2+y^2}$ and $f_y(x,y) = \frac{2y}{x^2+y^2} \Rightarrow f_x(1,0) = 2$ and $f_y(1,0) = 0 \Rightarrow$ from Eq. (10) the tangent plane at $(1,0,0)$ is $2(x-1) - z = 0$ or $2x - z - 2 = 0$

36. $z = f(x,y) = e^{-(x^2+y^2)} \Rightarrow f_x(x,y) = -2xe^{-(x^2+y^2)}$ and $f_y(x,y) = -2ye^{-(x^2+y^2)} \Rightarrow f_x(0,0) = 0$ and $f_y(0,0) = 0$
$\Rightarrow$ from Eq. (10) the tangent plane at $(0,0,1)$ is $z - 1 = 0$ or $z = 1$

37. $z = f(x,y) = \sqrt{y-x} \Rightarrow f_x(x,y) = -\frac{1}{2}(y-x)^{-1/2}$ and $f_y(x,y) = \frac{1}{2}(y-x)^{-1/2} \Rightarrow f_x(1,2) = -\frac{1}{2}$ and $f_y(1,2) = \frac{1}{2}$ $\Rightarrow$ from Eq. (10) the tangent plane at $(1,2,1)$ is $-\frac{1}{2}(x-1)+\frac{1}{2}(y-2)-(z-1)=0 \Rightarrow x-y+2z-1=0$

38. $z = f(x,y) = 4x^2+y^2 \Rightarrow f_x(x,y) = 8x$ and $f_y(x,y) = 2y \Rightarrow f_x(1,1) = 8$ and $f_y(1,1) = 2 \Rightarrow$ from Eq. (10) the tangent plane at $(1,1,5)$ is $8(x-1)+2(y-1)-(z-5)=0$ or $8x+2y-z-5=0$

39. $\nabla f = 2x\mathbf{i}+2y\mathbf{j} \Rightarrow \nabla f(\sqrt{2},\sqrt{2}) = 2\sqrt{2}\,\mathbf{i}+2\sqrt{2}\,\mathbf{j}$
$\Rightarrow$ Tangent line: $2\sqrt{2}(x-\sqrt{2})+2\sqrt{2}(y-\sqrt{2})=0$
$\Rightarrow \sqrt{2}x+\sqrt{2}y=4$

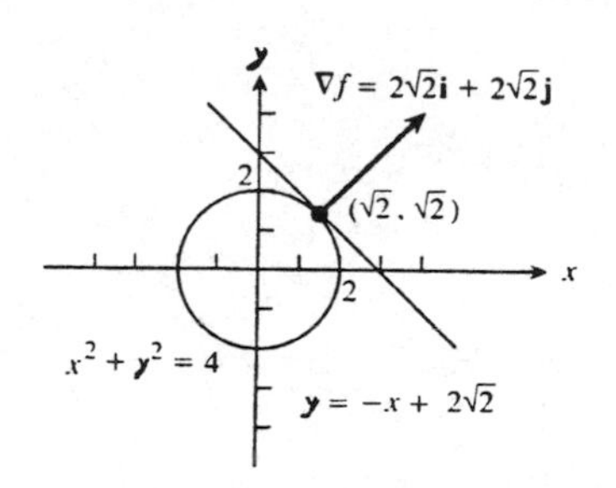

40. $\nabla f = 2x\mathbf{i}-\mathbf{j} \Rightarrow \nabla f(\sqrt{2},1) = 2\sqrt{2}\,\mathbf{i}-\mathbf{j}$
$\Rightarrow$ Tangent line: $2\sqrt{2}(x-\sqrt{2})-(y-1)=0$
$\Rightarrow y = 2\sqrt{2}x-3$

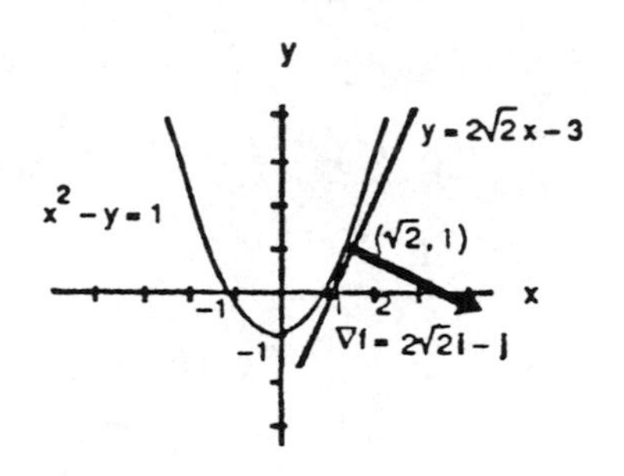

41. $\nabla f = y\mathbf{i}+x\mathbf{j} \Rightarrow \nabla f(2,-2) = -2\mathbf{i}+2\mathbf{j}$
$\Rightarrow$ Tangent line: $-2(x-2)+2(y+2)=0$
$\Rightarrow y = x-4$

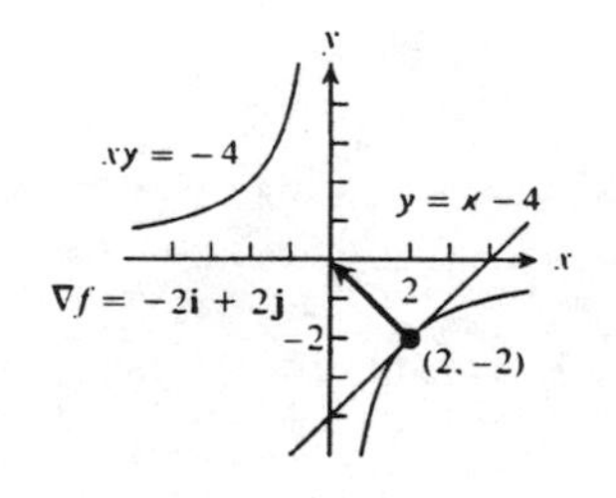

42. $\nabla f = (2x-y)\mathbf{i}+(2y-x)\mathbf{j} \Rightarrow \nabla f(-1,2) = -4\mathbf{i}+5\mathbf{j}$
$\Rightarrow$ Tangent line: $-4(x+1)+5(y-2)=0$
$\Rightarrow -4x+5y-14=0$

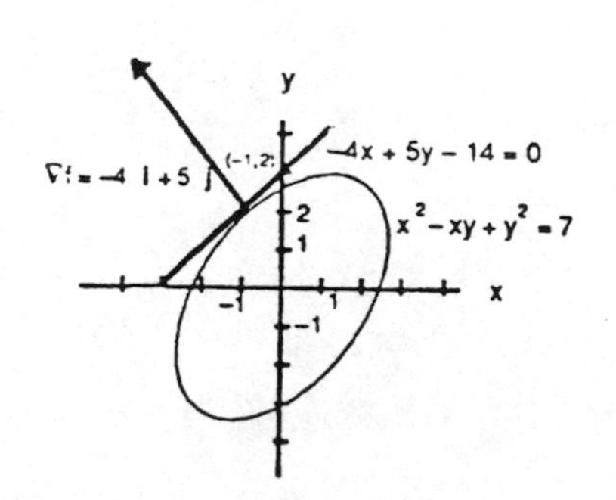

43. $\nabla f = \mathbf{i} + 2y\mathbf{j} + 2\mathbf{k} \Rightarrow \nabla f(1,1,1) = \mathbf{i} + 2\mathbf{j} + 2\mathbf{k}$ and $\nabla g = \mathbf{i}$ for all points; $\mathbf{v} = \nabla f \times \nabla g$

$$\Rightarrow \mathbf{v} = \begin{vmatrix} \mathbf{i} & \mathbf{j} & \mathbf{k} \\ 1 & 2 & 2 \\ 1 & 0 & 0 \end{vmatrix} = 2\mathbf{j} - 2\mathbf{k} \Rightarrow \text{Tangent line: } x = 1,\ y = 1 + 2t,\ z = 1 - 2t$$

44. $\nabla f = yz\mathbf{i} + xz\mathbf{j} + xy\mathbf{k} \Rightarrow \nabla f(1,1,1) = \mathbf{i} + \mathbf{j} + \mathbf{k}$; $\nabla g = 2x\mathbf{i} + 4y\mathbf{j} + 6z\mathbf{k} \Rightarrow \nabla g(1,1,1) = 2\mathbf{i} + 4\mathbf{j} + 6\mathbf{k}$;

$$\Rightarrow \mathbf{v} = \nabla f \times \nabla g \Rightarrow \begin{vmatrix} \mathbf{i} & \mathbf{j} & \mathbf{k} \\ 1 & 1 & 1 \\ 2 & 4 & 6 \end{vmatrix} = 2\mathbf{i} - 4\mathbf{j} + 2\mathbf{k} \Rightarrow \text{Tangent line: } x = 1 + 2t,\ y = 1 - 4t,\ z = 1 + 2t$$

45. $\nabla f = 2x\mathbf{i} + 2\mathbf{j} + 2\mathbf{k} \Rightarrow \nabla f\left(1, 1, \frac{1}{2}\right) = 2\mathbf{i} + 2\mathbf{j} + 2\mathbf{k}$ and $\nabla g = \mathbf{j}$ for all points; $\mathbf{v} = \nabla f \times \nabla g$

$$\Rightarrow \mathbf{v} = \begin{vmatrix} \mathbf{i} & \mathbf{j} & \mathbf{k} \\ 2 & 2 & 2 \\ 0 & 1 & 0 \end{vmatrix} = -2\mathbf{i} + 2\mathbf{k} \Rightarrow \text{Tangent line: } x = 1 - 2t,\ y = 1,\ z = \frac{1}{2} + 2t$$

46. $\nabla f = \mathbf{i} + 2y\mathbf{j} + \mathbf{k} \Rightarrow \nabla f\left(\frac{1}{2}, 1, \frac{1}{2}\right) = \mathbf{i} + 2\mathbf{j} + \mathbf{k}$ and $\nabla g = \mathbf{j}$ for all points; $\mathbf{v} = \nabla f \times \nabla g$

$$\Rightarrow \mathbf{v} = \begin{vmatrix} \mathbf{i} & \mathbf{j} & \mathbf{k} \\ 1 & 2 & 1 \\ 0 & 1 & 0 \end{vmatrix} = -\mathbf{i} + \mathbf{k} \Rightarrow \text{Tangent line: } x = \frac{1}{2} - t,\ y = 1,\ z = \frac{1}{2} + t$$

47. $\nabla f = \left(3x^2 + 6xy^2 + 4y\right)\mathbf{i} + \left(6x^2y + 3y^2 + 4x\right)\mathbf{j} - 2z\mathbf{k} \Rightarrow \nabla f(1,1,3) = 13\mathbf{i} + 13\mathbf{j} - 6\mathbf{k}$; $\nabla g = 2x\mathbf{i} + 2y\mathbf{j} + 2z\mathbf{k}$

$$\Rightarrow \nabla g(1,1,3) = 2\mathbf{i} + 2\mathbf{j} + 6\mathbf{k};\ \mathbf{v} = \nabla f \times \nabla g \Rightarrow \mathbf{v} = \begin{vmatrix} \mathbf{i} & \mathbf{j} & \mathbf{k} \\ 13 & 13 & -6 \\ 2 & 2 & 6 \end{vmatrix} = 90\mathbf{i} - 90\mathbf{j} \Rightarrow \text{Tangent line:}$$

$x = 1 + 90t,\ y = 1 - 90t,\ z = 3$

48. $\nabla f = 2x\mathbf{i} + 2y\mathbf{j} \Rightarrow \nabla f(\sqrt{2}, \sqrt{2}, 4) = 2\sqrt{2}\mathbf{i} + 2\sqrt{2}\mathbf{j}$; $\nabla g = 2x\mathbf{i} + 2y\mathbf{j} - \mathbf{k} \Rightarrow \nabla g(\sqrt{2}, \sqrt{2}, 4)$

$$= 2\sqrt{2}\mathbf{i} + 2\sqrt{2}\mathbf{j} - \mathbf{k};\ \mathbf{v} = \nabla f \times \nabla g \Rightarrow \mathbf{v} = \begin{vmatrix} \mathbf{i} & \mathbf{j} & \mathbf{k} \\ 2\sqrt{2} & 2\sqrt{2} & 0 \\ 2\sqrt{2} & 2\sqrt{2} & -1 \end{vmatrix} = -2\sqrt{2}\mathbf{i} + 2\sqrt{2}\mathbf{j} \Rightarrow \text{Tangent line:}$$

$x = \sqrt{2} - 2\sqrt{2}t,\ y = \sqrt{2} + 2\sqrt{2}t,\ z = 4$

49. $\nabla f = y\mathbf{i} + (x + 2y)\mathbf{j} \Rightarrow \nabla f(3,2) = 2\mathbf{i} + 7\mathbf{j}$; a vector orthogonal to ∇f is $\mathbf{A} = 7\mathbf{i} - 2\mathbf{j} \Rightarrow \mathbf{u} = \frac{\mathbf{A}}{|\mathbf{A}|} = \frac{7\mathbf{i} - 2\mathbf{j}}{\sqrt{7^2 + (-2)^2}}$

$= \frac{7}{\sqrt{53}}\mathbf{i} - \frac{2}{\sqrt{53}}\mathbf{j}$ and $-\mathbf{u} = -\frac{7}{\sqrt{53}}\mathbf{i} + \frac{2}{\sqrt{53}}\mathbf{j}$ are the directions where the derivative is zero

50. $\nabla f = \frac{4xy^2}{(x^2+y^2)^2}\mathbf{i} - \frac{4x^2y}{(x^2+y^2)^2}\mathbf{j} \Rightarrow \nabla f(1,1) = \mathbf{i} - \mathbf{j}$; a vector orthogonal to ∇f is $\mathbf{A} = \mathbf{i} + \mathbf{j}$
$\Rightarrow \mathbf{u} = \frac{\mathbf{A}}{|\mathbf{A}|} = \frac{\mathbf{i}+\mathbf{j}}{\sqrt{1^2+1^2}} = \frac{1}{\sqrt{2}}\mathbf{i} + \frac{1}{\sqrt{2}}\mathbf{j}$ and $-\mathbf{u} = -\frac{1}{\sqrt{2}}\mathbf{i} - \frac{1}{\sqrt{2}}\mathbf{j}$ are the directions where the derivative is zero

51. $\nabla f = (2x-3y)\mathbf{i} + (-3x+8y)\mathbf{j} \Rightarrow \nabla f(1,2) = -4\mathbf{i} + 13\mathbf{j} \Rightarrow |\nabla f(1,2)| = \sqrt{(-4)^2+(13)^2} = \sqrt{185}$; no, the maximum rate of change is $\sqrt{185} < 14$

52. $\nabla T = 2y\mathbf{i} + (2x-z)\mathbf{j} - y\mathbf{k} \Rightarrow \nabla T(1,-1,1) = -2\mathbf{i} + \mathbf{j} + \mathbf{k} \Rightarrow |\nabla T(1,-1,1)| = \sqrt{(-2)^2+1^2+1^2} = \sqrt{6}$; no, the minimum rate of change is $-\sqrt{6} > -3$

53. $\nabla f = f_x(1,2)\mathbf{i} + f_y(1,2)\mathbf{j}$ and $\mathbf{u}_1 = \frac{\mathbf{i}+\mathbf{j}}{\sqrt{1^2+1^2}} = \frac{1}{\sqrt{2}}\mathbf{i} + \frac{1}{\sqrt{2}}\mathbf{j} \Rightarrow (D_{\mathbf{u}_1}f)(1,2) = f_x(1,2)\left(\frac{1}{\sqrt{2}}\right) + f_y(1,2)\left(\frac{1}{\sqrt{2}}\right)$
$= 2\sqrt{2} \Rightarrow f_x(1,2) + f_y(1,2) = 4$; $\mathbf{u}_2 = -\mathbf{j} \Rightarrow (D_{\mathbf{u}_2}f)(1,2) = f_x(1,2)(0) + f_y(1,2)(-1) = -3 \Rightarrow -f_y(1,2) = -3$
$\Rightarrow f_y(1,2) = 3$; then $f_x(1,2) + 3 = 4 \Rightarrow f_x(1,2) = 1$; thus $\nabla f(1,2) = \mathbf{i} + 3\mathbf{j}$ and $\mathbf{u} = \frac{\mathbf{A}}{|\mathbf{A}|} = \frac{-\mathbf{i}-2\mathbf{j}}{\sqrt{(-1)^2+(-2)^2}}$
$= -\frac{1}{\sqrt{5}}\mathbf{i} - \frac{2}{\sqrt{5}}\mathbf{j} \Rightarrow (D_{\mathbf{u}}f)_{P_0} = \nabla f \cdot \mathbf{u} = -\frac{1}{\sqrt{5}} - \frac{6}{\sqrt{5}} = -\frac{7}{\sqrt{5}}$

54. (a) $(D_{\mathbf{u}}f)_P = 2\sqrt{3} \Rightarrow |\nabla f| = 2\sqrt{3}$; $\mathbf{u} = \frac{\mathbf{A}}{|\mathbf{A}|} = \frac{\mathbf{i}+\mathbf{j}-\mathbf{k}}{\sqrt{1^2+1^2+(-1)^2}} = \frac{1}{\sqrt{3}}\mathbf{i} + \frac{1}{\sqrt{3}}\mathbf{j} - \frac{1}{\sqrt{3}}\mathbf{k}$; thus $\mathbf{u} = \frac{\nabla f}{|\nabla f|}$
$\Rightarrow \nabla f = |\nabla f|\mathbf{u} \Rightarrow \nabla f = 2\sqrt{3}\left(\frac{1}{\sqrt{3}}\mathbf{i} + \frac{1}{\sqrt{3}}\mathbf{j} - \frac{1}{\sqrt{3}}\mathbf{k}\right) = 2\mathbf{i} + 2\mathbf{j} - 2\mathbf{k}$

(b) $\mathbf{A} = \mathbf{i} + \mathbf{j} \Rightarrow \mathbf{u} = \frac{\mathbf{A}}{|\mathbf{A}|} = \frac{\mathbf{i}+\mathbf{j}}{\sqrt{1^2+1^2}} = \frac{1}{\sqrt{2}}\mathbf{i} + \frac{1}{\sqrt{2}}\mathbf{j} \Rightarrow (D_{\mathbf{u}}f)_{P_0} = \nabla f \cdot \mathbf{u} = 2\left(\frac{1}{\sqrt{2}}\right) + 2\left(\frac{1}{\sqrt{2}}\right) - 2(0) = 2\sqrt{2}$

55. (a) The unit tangent vector at $\left(\frac{1}{2}, \frac{\sqrt{3}}{2}\right)$ in the direction of motion is $\mathbf{u} = \frac{\sqrt{3}}{2}\mathbf{i} - \frac{1}{2}\mathbf{j}$;
$\nabla T = (\sin 2y)\mathbf{i} + (2x \cos 2y)\mathbf{j} \Rightarrow \nabla T\left(\frac{1}{2}, \frac{\sqrt{3}}{2}\right) = (\sin \sqrt{3})\mathbf{i} + (\cos \sqrt{3})\mathbf{j} \Rightarrow D_{\mathbf{u}}T\left(\frac{1}{2}, \frac{\sqrt{3}}{2}\right) = \nabla T \cdot \mathbf{u}$
$= \frac{\sqrt{3}}{2} \sin \sqrt{3} - \frac{1}{2} \cos \sqrt{3} \approx 0.935°$ C/ft

(b) $\mathbf{r}(t) = (\sin 2t)\mathbf{i} + (\cos 2t)\mathbf{j} \Rightarrow \mathbf{v}(t) = (2 \cos 2t)\mathbf{i} - (2 \sin 2t)\mathbf{j}$ and $|\mathbf{v}| = 2$; $\frac{dT}{dt} = \frac{\partial T}{\partial x}\frac{dx}{dt} + \frac{\partial T}{\partial y}\frac{dy}{dt}$
$= \nabla T \cdot \mathbf{v} = \left(\nabla T \cdot \frac{\mathbf{v}}{|\mathbf{v}|}\right)|\mathbf{v}| = (D_{\mathbf{u}}T)\,|\mathbf{v}|$, where $\mathbf{u} = \frac{\mathbf{v}}{|\mathbf{v}|}$; at $\left(\frac{1}{2}, \frac{\sqrt{3}}{2}\right)$ we have $\mathbf{u} = \frac{\sqrt{3}}{2}\mathbf{i} - \frac{1}{2}\mathbf{j}$ from part (a)
$\Rightarrow \frac{dT}{dt} = \left(\frac{\sqrt{3}}{2} \sin \sqrt{3} - \frac{1}{2} \cos \sqrt{3}\right) \cdot 2 = \sqrt{3} \sin \sqrt{3} - \cos \sqrt{3} \approx 1.87°$ C/sec

56. $\nabla f = 2x\mathbf{i} + 2y\mathbf{j} = 2(\cos t + t \sin t)\mathbf{i} + 2(\sin t - t \cos t)\mathbf{j}$ and $\mathbf{v} = (t \cos t)\mathbf{i} + (t \sin t)\mathbf{j} \Rightarrow \mathbf{u} = \frac{\mathbf{v}}{|\mathbf{v}|}$
$= \frac{(t \cos t)\mathbf{i} + (t \sin t)\mathbf{j}}{\sqrt{(t \cos t)^2 + (t \sin t)^2}} = (\cos t)\mathbf{i} + (\sin t)\mathbf{j}$ since $t > 0 \Rightarrow (D_{\mathbf{u}}f)_{P_0} = \nabla f \cdot \mathbf{u}$
$= 2(\cos t + t \sin t)(\cos t) + 2(\sin t - t \cos t)(\sin t) = 2$

57. $\nabla f = 2x\mathbf{i} + 2y\mathbf{j} + 2z\mathbf{k} = (2\cos t)\mathbf{i} + (2\sin t)\mathbf{j} + 2t\mathbf{k}$ and $\mathbf{v} = (-\sin t)\mathbf{i} + (\cos t)\mathbf{j} + \mathbf{k} \Rightarrow \mathbf{u} = \frac{\mathbf{v}}{|\mathbf{v}|}$

$= \frac{(-\sin t)\mathbf{i} + (\cos t)\mathbf{j} + \mathbf{k}}{\sqrt{(\sin t)^2 + (\cos t)^2 + 1^2}} = \left(\frac{-\sin t}{\sqrt{2}}\right)\mathbf{i} + \left(\frac{\cos t}{\sqrt{2}}\right)\mathbf{j} + \frac{1}{\sqrt{2}}\mathbf{k} \Rightarrow (D_{\mathbf{u}}f)_{P_0} = \nabla f \cdot \mathbf{u}$

$= (2\cos t)\left(\frac{-\sin t}{\sqrt{2}}\right) + (2\sin t)\left(\frac{\cos t}{\sqrt{2}}\right) + (2t)\left(\frac{1}{\sqrt{2}}\right) = \frac{2t}{\sqrt{2}} \Rightarrow (D_{\mathbf{u}}f)\left(\frac{-\pi}{4}\right) = \frac{-\pi}{2\sqrt{2}}$, $(D_{\mathbf{u}}f)(0) = 0$ and

$(D_{\mathbf{u}}f)\left(\frac{\pi}{4}\right) = \frac{\pi}{2\sqrt{2}}$

58. (a) $\nabla T = (4x - yz)\mathbf{i} - xz\mathbf{j} - xy\mathbf{k} \Rightarrow \nabla T(8,6,-4) = 56\mathbf{i} + 32\mathbf{j} - 48\mathbf{k}$; $\mathbf{r}(t) = 2t^2\mathbf{i} + 3t\mathbf{j} - t^2\mathbf{k} \Rightarrow$ the particle is at the point $P(8,6,-4)$ when $t = 2$; $\mathbf{v}(t) = 4t\mathbf{i} + 3\mathbf{j} - 2t\mathbf{k} \Rightarrow \mathbf{v}(2) = 8\mathbf{i} + 3\mathbf{j} - 4\mathbf{k} \Rightarrow \mathbf{u} = \frac{\mathbf{v}}{|\mathbf{v}|}$

$= \frac{8}{\sqrt{89}}\mathbf{i} + \frac{3}{\sqrt{89}}\mathbf{j} - \frac{4}{\sqrt{89}}\mathbf{k} \Rightarrow D_{\mathbf{u}}T(8,6,-4) = \nabla T \cdot \mathbf{u} = \frac{1}{\sqrt{89}}[56 \cdot 8 + 32 \cdot 3 - 48 \cdot (-4)] = \frac{736}{\sqrt{89}}\,^\circ\text{C/m}$

(b) $\frac{dT}{dt} = \frac{\partial T}{\partial x}\frac{dx}{dt} + \frac{\partial T}{\partial y}\frac{dy}{dt} = \nabla T \cdot \mathbf{v} = (\nabla T \cdot \mathbf{u})\,|\mathbf{v}| \Rightarrow$ at $t = 2$, $\frac{dT}{dt} = D_{\mathbf{u}}T\Big|_{t=2} \mathbf{v}(2) = \left(\frac{736}{\sqrt{89}}\right)\sqrt{89} = 736^\circ\text{C/sec}$

59. If (x,y) is a point on the line, then $\mathbf{T}(x,y) = (x - x_0)\mathbf{i} + (y - y_0)\mathbf{j}$ is a vector parallel to the line $\Rightarrow \mathbf{T} \cdot \mathbf{N} = 0$ $\Rightarrow A(x - x_0) + B(y - y_0) = 0$, as claimed.

60. (a) $\mathbf{r} = \sqrt{t}\mathbf{i} + \sqrt{t}\mathbf{j} - \frac{1}{4}(t+3)\mathbf{k} \Rightarrow \mathbf{v} = \frac{1}{2}t^{-1/2}\mathbf{i} + \frac{1}{2}t^{-1/2}\mathbf{j} - \frac{1}{4}\mathbf{k}$; $t = 1 \Rightarrow x = 1$, $y = 1$, $z = -1 \Rightarrow P_0 = (1,1,-1)$ and $\mathbf{v}(1) = \frac{1}{2}\mathbf{i} + \frac{1}{2}\mathbf{j} - \frac{1}{4}\mathbf{k}$; $f(x,y,z) = x^2 + y^2 - z - 3 = 0 \Rightarrow \nabla f = 2x\mathbf{i} + 2y\mathbf{j} - \mathbf{k} \Rightarrow \nabla f(1,1,-1) = 2\mathbf{i} + 2\mathbf{j} - \mathbf{k}$; therefore $\mathbf{v} = \frac{1}{4}(\nabla f) \Rightarrow$ the curve is normal to the surface

(b) $\mathbf{r} = \sqrt{t}\mathbf{i} + \sqrt{t}\mathbf{j} + (2t - 1)\mathbf{k} \Rightarrow \mathbf{v} = \frac{1}{2}t^{-1/2}\mathbf{i} + \frac{1}{2}t^{-1/2}\mathbf{j} + 2\mathbf{k}$; $t = 1 \Rightarrow x = 1$, $y = 1$, $z = 1 \Rightarrow P_0 = (1,1,1)$ and $\mathbf{v}(1) = \frac{1}{2}\mathbf{i} + \frac{1}{2}\mathbf{j} + 2\mathbf{k}$; $f(x,y,z) = x^2 + y^2 - z - 1 = 0 \Rightarrow \nabla f = 2x\mathbf{i} + 2y\mathbf{j} - \mathbf{k} \Rightarrow \nabla f(1,1,1) = 2\mathbf{i} + 2\mathbf{j} - \mathbf{k}$; therefore $\mathbf{v} \cdot \nabla f = \frac{1}{2}(2) + \frac{1}{2}(2) + 2(-1) = 0 \Rightarrow$ the curve is tangent to the surface when $t = 1$

61. $x = g(t)$ and $y = h(t) \Rightarrow \mathbf{r} = g(t)\mathbf{i} + h(t)\mathbf{j} \Rightarrow \mathbf{v} = g'(t)\mathbf{i} + h'(t)\mathbf{j} \Rightarrow \mathbf{T} = \frac{\mathbf{v}}{|\mathbf{v}|} = \frac{g'(t)\mathbf{i} + h'(t)\mathbf{j}}{\sqrt{[g'(t)]^2 + [h'(t)]^2}}$;

$z = f(x,y) \Rightarrow \frac{df}{dt} = \frac{\partial f}{\partial x}\frac{dx}{dt} + \frac{\partial f}{\partial y}\frac{dy}{dt} = \frac{\partial f}{\partial x}g'(t) + \frac{\partial f}{\partial y}h'(t) = \nabla f \cdot \mathbf{T}$. If $f(g(t),h(t)) = c$, then $\frac{df}{dt} = 0$ $\Rightarrow \frac{\partial f}{\partial x}g'(t) + \frac{\partial f}{\partial y}h'(t) = 0 \Rightarrow \nabla f \cdot \mathbf{T} = 0 \Rightarrow \nabla f$ is normal to $\mathbf{T}$

62. $z = f(x,y) \Rightarrow g(x,y,z) = f(x,y) - z = 0 \Rightarrow g_x(x,y,z) = f_x(x,y)$, $g_y(x,y,z) = f_y(x,y)$ and $g_z(x,y,z) = -1$ $\Rightarrow g_x(x_0,y_0,f(x_0,y_0)) = f_x(x_0,y_0)$, $g_y(x_0,y_0,f(x_0,y_0)) = f_y(x_0,y_0)$ and $g_z(x_0,y_0,f(x_0,y_0)) = -1 \Rightarrow$ the tangent plane at the point P_0 is $f_x(x_0,y_0)(x - x_0) + f_y(x_0,y_0)(y - y_0) - [z - f(x_0,y_0)] = 0$ or $z = f_x(x_0,y_0)(x - x_0) + f_y(x_0,y_0)(y - y_0) + f(x_0,y_0)$

63. The directional derivative is the scalar component. With ∇f evaluated at P_0, the scalar component of ∇f in the direction of $\mathbf{u}$ is $\nabla f \cdot \mathbf{u} = (D_{\mathbf{u}}f)_{P_0}$.

64. $D_{\mathbf{i}}f = \nabla f \cdot \mathbf{i} = (f_x\mathbf{i} + f_y\mathbf{j} + f_z\mathbf{k}) \cdot \mathbf{i} = f_x$; similarly, $D_{\mathbf{j}}f = \nabla f \cdot \mathbf{j} = f_y$ and $D_{\mathbf{k}}f = \nabla f \cdot \mathbf{k} = f_z$

65. (a) $\nabla(kf) = \frac{\partial(kf)}{\partial x}\mathbf{i} + \frac{\partial(kf)}{\partial y}\mathbf{j} + \frac{\partial(kf)}{\partial z}\mathbf{k} = k\left(\frac{\partial f}{\partial x}\right)\mathbf{i} + k\left(\frac{\partial f}{\partial y}\right)\mathbf{j} + k\left(\frac{\partial f}{\partial z}\right)\mathbf{k} = k\left(\frac{\partial f}{\partial x}\mathbf{i} + \frac{\partial f}{\partial y}\mathbf{j} + \frac{\partial f}{\partial z}\mathbf{k}\right) = k\,\nabla f$

(b) $\nabla(f+g) = \frac{\partial(f+g)}{\partial x}\mathbf{i} + \frac{\partial(f+g)}{\partial y}\mathbf{j} + \frac{\partial(f+g)}{\partial z}\mathbf{k} = \left(\frac{\partial f}{\partial x} + \frac{\partial g}{\partial x}\right)\mathbf{i} + \left(\frac{\partial f}{\partial y} + \frac{\partial g}{\partial y}\right)\mathbf{j} + \left(\frac{\partial f}{\partial z} + \frac{\partial g}{\partial z}\right)\mathbf{k}$

$= \frac{\partial f}{\partial x}\mathbf{i} + \frac{\partial g}{\partial x}\mathbf{i} + \frac{\partial f}{\partial y}\mathbf{j} + \frac{\partial g}{\partial y}\mathbf{j} + \frac{\partial f}{\partial z}\mathbf{k} + \frac{\partial g}{\partial z}\mathbf{k} = \left(\frac{\partial f}{\partial x}\mathbf{i} + \frac{\partial f}{\partial y}\mathbf{j} + \frac{\partial f}{\partial z}\mathbf{k}\right) + \left(\frac{\partial g}{\partial x}\mathbf{i} + \frac{\partial g}{\partial y}\mathbf{j} + \frac{\partial g}{\partial z}\mathbf{k}\right) = \nabla f + \nabla g$

(c) $\nabla(f-g) = \nabla f - \nabla g$ (Substitute $-g$ for g in part (b) above)

(d) $\nabla(fg) = \frac{\partial(fg)}{\partial x}\mathbf{i} + \frac{\partial(fg)}{\partial y}\mathbf{j} + \frac{\partial(fg)}{\partial z}\mathbf{k} = \left(\frac{\partial f}{\partial x}g + \frac{\partial g}{\partial x}f\right)\mathbf{i} + \left(\frac{\partial f}{\partial y}g + \frac{\partial g}{\partial y}f\right)\mathbf{j} + \left(\frac{\partial f}{\partial z}g + \frac{\partial g}{\partial z}f\right)\mathbf{k}$

$= \left(\frac{\partial f}{\partial x}g\right)\mathbf{i} + \left(\frac{\partial g}{\partial x}f\right)\mathbf{i} + \left(\frac{\partial f}{\partial y}g\right)\mathbf{j} + \left(\frac{\partial g}{\partial y}f\right)\mathbf{j} + \left(\frac{\partial f}{\partial z}g\right)\mathbf{k} + \left(\frac{\partial g}{\partial z}f\right)\mathbf{k}$

$= f\left(\frac{\partial g}{\partial x}\mathbf{i} + \frac{\partial g}{\partial y}\mathbf{j} + \frac{\partial g}{\partial z}\mathbf{k}\right) + g\left(\frac{\partial f}{\partial x}\mathbf{i} + \frac{\partial f}{\partial y}\mathbf{j} + \frac{\partial f}{\partial z}\mathbf{k}\right) = f\,\nabla g + g\,\nabla f$

(e) $\nabla\left(\frac{f}{g}\right) = \frac{\partial\left(\frac{f}{g}\right)}{\partial x}\mathbf{i} + \frac{\partial\left(\frac{f}{g}\right)}{\partial y}\mathbf{j} + \frac{\partial\left(\frac{f}{g}\right)}{\partial z}\mathbf{k} = \left(\frac{g\frac{\partial f}{\partial x} - f\frac{\partial g}{\partial x}}{g^2}\right)\mathbf{i} + \left(\frac{g\frac{\partial f}{\partial y} - f\frac{\partial g}{\partial y}}{g^2}\right)\mathbf{j} + \left(\frac{g\frac{\partial f}{\partial z} - f\frac{\partial g}{\partial z}}{g^2}\right)\mathbf{k}$

$= \left(\frac{g\frac{\partial f}{\partial x}\mathbf{i} + g\frac{\partial f}{\partial y}\mathbf{j} + g\frac{\partial f}{\partial z}\mathbf{k}}{g^2}\right) - \left(\frac{f\frac{\partial g}{\partial x}\mathbf{i} + f\frac{\partial g}{\partial y}\mathbf{j} + f\frac{\partial g}{\partial z}\mathbf{k}}{g^2}\right) = \frac{g\left(\frac{\partial f}{\partial x}\mathbf{i} + \frac{\partial f}{\partial y}\mathbf{j} + \frac{\partial f}{\partial z}\mathbf{k}\right)}{g^2} - \frac{f\left(\frac{\partial g}{\partial x}\mathbf{i} + \frac{\partial g}{\partial y}\mathbf{j} + \frac{\partial g}{\partial z}\mathbf{k}\right)}{g^2}$

$= \frac{g\,\nabla f}{g^2} - \frac{f\,\nabla g}{g^2} = \frac{g\,\nabla f - f\,\nabla g}{g^2}$

12.8 EXTREME VALUES AND SADDLE POINTS

1. $f_x(x,y) = 2x + y + 3 = 0$ and $f_y(x,y) = x + 2y - 3 = 0 \Rightarrow x = -3$ and $y = 3 \Rightarrow$ critical point is $(-3,3)$; $f_{xx}(-3,3) = 2$, $f_{yy}(-3,3) = 2$, $f_{xy}(-3,3) = 1 \Rightarrow f_{xx}f_{yy} - f_{xy}^2 = 3 > 0$ and $f_{xx} > 0 \Rightarrow$ local minimum of $f(-3,3) = -5$

2. $f_x(x,y) = 2x + 3y - 6 = 0$ and $f_y(x,y) = 3x + 6y + 3 = 0 \Rightarrow x = 15$ and $y = -8 \Rightarrow$ critical point is $(15,-8)$; $f_{xx}(15,-8) = 2$, $f_{yy}(15,-8) = 6$, $f_{xy}(15,-8) = 3 \Rightarrow f_{xx}f_{yy} - f_{xy}^2 = 3 > 0$ and $f_{xx} > 0 \Rightarrow$ local minimum of $f(15,-8) = -63$

3. $f_x(x,y) = 2y - 10x + 4 = 0$ and $f_y(x,y) = 2x - 4y + 4 = 0 \Rightarrow x = \frac{2}{3}$ and $y = \frac{4}{3} \Rightarrow$ critical point is $\left(\frac{2}{3},\frac{4}{3}\right)$; $f_{xx}\left(\frac{2}{3},\frac{4}{3}\right) = -10$, $f_{yy}\left(\frac{2}{3},\frac{4}{3}\right) = -4$, $f_{xy}\left(\frac{2}{3},\frac{4}{3}\right) = 2 \Rightarrow f_{xx}f_{yy} - f_{xy}^2 = 36 > 0$ and $f_{xx} < 0 \Rightarrow$ local maximum of $f\left(\frac{2}{3},\frac{4}{3}\right) = 0$

4. $f_x(x,y) = 2y - 10x + 4 = 0$ and $f_y(x,y) = 2x - 4y = 0 \Rightarrow x = \frac{4}{9}$ and $y = \frac{2}{9} \Rightarrow$ critical point is $\left(\frac{4}{9},\frac{2}{9}\right)$; $f_{xx}\left(\frac{4}{9},\frac{2}{9}\right) = -10$, $f_{yy}\left(\frac{4}{9},\frac{2}{9}\right) = -4$, $f_{xy}\left(\frac{4}{9},\frac{2}{9}\right) = 2 \Rightarrow f_{xx}f_{yy} - f_{xy}^2 = 36 > 0$ and $f_{xx} < 0 \Rightarrow$ local maximum of $f\left(\frac{4}{9},\frac{2}{9}\right) = -\frac{28}{9}$

5. $f_x(x,y) = 2x + y + 3 = 0$ and $f_y(x,y) = x + 2 = 0 \Rightarrow x = -2$ and $y = 1 \Rightarrow$ critical point is $(-2,1)$; $f_{xx}(-2,1) = 2$, $f_{yy}(-2,1) = 0$, $f_{xy}(-2,1) = 1 \Rightarrow f_{xx}f_{yy} - f_{xy}^2 = -1 < 0 \Rightarrow$ saddle point

6. $f_x(x,y) = y - 2 = 0$ and $f_y(x,y) = 2y + x - 2 = 0 \Rightarrow x = -2$ and $y = 2 \Rightarrow$ critical point is $(-2,2)$; $f_{xx}(-2,2) = 0$, $f_{yy}(-2,2) = 2$, $f_{xy}(-2,2) = 1 \Rightarrow f_{xx}f_{yy} - f_{xy}^2 = -1 < 0 \Rightarrow$ saddle point

7. $f_x(x,y) = 5y - 14x + 3 = 0$ and $f_y(x,y) = 5x - 6 = 0 \Rightarrow x = \frac{6}{5}$ and $y = \frac{69}{25} \Rightarrow$ critical point is $\left(\frac{6}{5}, \frac{69}{25}\right)$; $f_{xx}\left(\frac{6}{5}, \frac{69}{25}\right) = -14$, $f_{yy}\left(\frac{6}{5}, \frac{69}{25}\right) = 0$, $f_{xy}\left(\frac{6}{5}, \frac{69}{25}\right) = 5 \Rightarrow f_{xx}f_{yy} - f_{xy}^2 = -25 < 0 \Rightarrow$ saddle point

8. $f_x(x,y) = 2y - 2x + 3 = 0$ and $f_y(x,y) = 2x - 4y = 0 \Rightarrow x = 3$ and $y = \frac{3}{2} \Rightarrow$ critical point is $\left(3, \frac{3}{2}\right)$; $f_{xx}\left(3, \frac{3}{2}\right) = -2$, $f_{yy}\left(3, \frac{3}{2}\right) = -4$, $f_{xy}\left(3, \frac{3}{2}\right) = 2 \Rightarrow f_{xx}f_{yy} - f_{xy}^2 = 4 > 0$ and $f_{xx} < 0 \Rightarrow$ local maximum of $f\left(3, \frac{3}{2}\right) = \frac{17}{2}$

9. $f_x(x,y) = 2x - 4y = 0$ and $f_y(x,y) = -4x + 2y + 6 = 0 \Rightarrow x = 2$ and $y = 1 \Rightarrow$ critical point is $(2,1)$; $f_{xx}(2,1) = 2$, $f_{yy}(2,1) = 2$, $f_{xy}(2,1) = -4 \Rightarrow f_{xx}f_{yy} - f_{xy}^2 = -12 < 0 \Rightarrow$ saddle point

10. $f_x(x,y) = 6x + 6y - 2 = 0$ and $f_y(x,y) = 6x + 14y + 4 = 0 \Rightarrow x = \frac{13}{12}$ and $y = -\frac{3}{4} \Rightarrow$ critical point is $\left(\frac{13}{12}, -\frac{3}{4}\right)$; $f_{xx}\left(\frac{13}{12}, -\frac{3}{4}\right) = 6$, $f_{yy}\left(\frac{13}{12}, -\frac{3}{4}\right) = 14$, $f_{xy}\left(\frac{13}{12}, -\frac{3}{4}\right) = 6 \Rightarrow f_{xx}f_{yy} - f_{xy}^2 = 48 > 0$ and $f_{xx} > 0 \Rightarrow$ local minimum of $f\left(\frac{13}{12}, -\frac{3}{4}\right) = -\frac{31}{12}$

11. $f_x(x,y) = 4x + 3y - 5 = 0$ and $f_y(x,y) = 3x + 8y + 2 = 0 \Rightarrow x = 2$ and $y = -1 \Rightarrow$ critical point is $(2,-1)$; $f_{xx}(2,-1) = 4$, $f_{yy}(2,-1) = 8$, $f_{xy}(2,-1) = 3 \Rightarrow f_{xx}f_{yy} - f_{xy}^2 = 23 > 0$ and $f_{xx} > 0 \Rightarrow$ local minimum of $f(2,-1) = -6$

12. $f_x(x,y) = 8x - 6y - 20 = 0$ and $f_y(x,y) = -6x + 10y + 26 = 0 \Rightarrow x = 1$ and $y = -2 \Rightarrow$ critical point is $(1,-2)$; $f_{xx}(1,-2) = 8$, $f_{yy}(1,-2) = 10$, $f_{xy}(1,-2) = -6 \Rightarrow f_{xx}f_{yy} - f_{xy}^2 = 44 > 0$ and $f_{xx} > 0 \Rightarrow$ local minimum of $f(1,-2) = -36$

13. $f_x(x,y) = 2x - 2 = 0$ and $f_y(x,y) = -2y + 4 = 0 \Rightarrow x = 1$ and $y = 2 \Rightarrow$ critical point is $(1,2)$; $f_{xx}(1,2) = 2$, $f_{yy}(1,2) = -2$, $f_{xy}(1,2) = 0 \Rightarrow f_{xx}f_{yy} - f_{xy}^2 = -4 < 0 \Rightarrow$ saddle point

14. $f_x(x,y) = 2x - 2y - 2 = 0$ and $f_y(x,y) = -2x + 4y + 2 = 0 \Rightarrow x = 1$ and $y = 0 \Rightarrow$ critical point is $(1,0)$; $f_{xx}(1,0) = 2$, $f_{yy}(1,0) = 4$, $f_{xy}(1,0) = -2 \Rightarrow f_{xx}f_{yy} - f_{xy}^2 = 4 > 0$ and $f_{xx} > 0 \Rightarrow$ local minimum of $f(1,0) = 0$

15. $f_x(x,y) = 2x + 2y = 0$ and $f_y(x,y) = 2x = 0 \Rightarrow x = 0$ and $y = 0 \Rightarrow$ critical point is $(0,0)$; $f_{xx}(0,0) = 2$, $f_{yy}(0,0) = 0$, $f_{xy}(0,0) = 2 \Rightarrow f_{xx}f_{yy} - f_{xy}^2 = -4 < 0 \Rightarrow$ saddle point

16. $f_x(x,y) = 2 - 4x - 2y = 0$ and $f_y(x,y) = 2 - 2x - 2y = 0 \Rightarrow x = 0$ and $y = 1 \Rightarrow$ critical point is $(0,1)$; $f_{xx}(0,1) = -4$, $f_{yy}(0,1) = -2$, $f_{xy}(0,1) = -2 \Rightarrow f_{xx}f_{yy} - f_{xy}^2 = 4 > 0$ and $f_{xx} < 0 \Rightarrow$ local maximum of $f(0,1) = 4$

17. $f_x(x,y) = 3x^2 - 2y = 0$ and $f_y(x,y) = -3y^2 - 2x = 0 \Rightarrow x = 0$ and $y = 0$, or $x = -\frac{2}{3}$ and $y = \frac{2}{3} \Rightarrow$ critical points are $(0,0)$ and $\left(-\frac{2}{3}, \frac{2}{3}\right)$; for $(0,0)$: $f_{xx}(0,0) = 6x\big|_{(0,0)} = 0$, $f_{yy}(0,0) = -6y\big|_{(0,0)} = 0$, $f_{xy}(0,0) = -2$ $\Rightarrow f_{xx}f_{yy} - f_{xy}^2 = -4 < 0 \Rightarrow$ saddle point; for $\left(-\frac{2}{3}, \frac{2}{3}\right)$: $f_{xx}\left(-\frac{2}{3}, \frac{2}{3}\right) = -4$, $f_{yy}\left(-\frac{2}{3}, \frac{2}{3}\right) = -4$, $f_{xy}\left(-\frac{2}{3}, \frac{2}{3}\right) = -2$ $\Rightarrow f_{xx}f_{yy} - f_{xy}^2 = 12 > 0$ and $f_{xx} < 0 \Rightarrow$ local maximum of $f\left(-\frac{2}{3}, \frac{2}{3}\right) = \frac{170}{27}$

18. $f_x(x,y) = 3x^2 + 3y = 0$ and $f_y(x,y) = 3x + 3y^2 = 0 \Rightarrow x = 0$ and $y = 0$, or $x = -1$ and $y = -1 \Rightarrow$ critical points are $(0,0)$ and $(-1,-1)$; for $(0,0)$: $f_{xx}(0,0) = 6x\big|_{(0,0)} = 0$, $f_{yy}(0,0) = 6y\big|_{(0,0)} = 0$, $f_{xy}(0,0) = 3 \Rightarrow f_{xx}f_{yy} - f_{xy}^2$ $= -9 < 0 \Rightarrow$ saddle point; for $(-1,-1)$: $f_{xx}(-1,-1) = -6$, $f_{yy}(-1,-1) = -6$, $f_{xy}(-1,-1) = 3 \Rightarrow f_{xx}f_{yy} - f_{xy}^2$ $= 27 > 0$ and $f_{xx} < 0 \Rightarrow$ local maximum of $f(-1,-1) = 1$

19. $f_x(x,y) = 12x - 6x^2 + 6y = 0$ and $f_y(x,y) = 6y + 6x = 0 \Rightarrow x = 0$ and $y = 0$, or $x = 1$ and $y = -1 \Rightarrow$ critical points are $(0,0)$ and $(1,-1)$; for $(0,0)$: $f_{xx}(0,0) = 12 - 12x\big|_{(0,0)} = 12$, $f_{yy}(0,0) = 6$, $f_{xy}(0,0) = 6 \Rightarrow f_{xx}f_{yy} - f_{xy}^2$ $= 36 > 0$ and $f_{xx} > 0 \Rightarrow$ local minimum of $f(0,0) = 0$; for $(1,-1)$: $f_{xx}(1,-1) = 0$, $f_{yy}(1,-1) = 6$, $f_{xy}(1,-1) = 6 \Rightarrow f_{xx}f_{yy} - f_{xy}^2 = -36 < 0 \Rightarrow$ saddle point

20. $f_x(x,y) = -6x + 6y = 0 \Rightarrow x = y$; $f_y(x,y) = 6y - 6y^2 + 6x = 0 \Rightarrow 12y - 6y^2 = 0 \Rightarrow 6y(2-y) = 0 \Rightarrow y = 0$ or $y = 2 \Rightarrow (0,0)$ and $(2,2)$ are the critical points; $f_{xx}(x,y) = -6$, $f_{yy}(x,y) = 6 - 12y$, $f_{xy}(x,y) = 6$; for $(0,0)$: $f_{xx}(0,0) = -6$, $f_{yy}(0,0) = 6$, $f_{xy}(0,0) = 6 \Rightarrow f_{xx}f_{yy} - f_{xy}^2 = -72 < 0 \Rightarrow$ saddle point; for $(2,2)$: $f_{xx}(2,2) = -6$, $f_{yy}(2,2) = -18$, $f_{xy}(2,2) = 6 \Rightarrow f_{xx}f_{yy} - f_{xy}^2 = 72 > 0$ and $f_{xx} < 0 \Rightarrow$ local maximum of $f(2,2) = 8$

21. $f_x(x,y) = 27x^2 - 4y = 0$ and $f_y(x,y) = y^2 - 4x = 0 \Rightarrow x = 0$ and $y = 0$, or $x = \frac{4}{9}$ and $y = \frac{4}{3} \Rightarrow$ critical points are $(0,0)$ and $\left(\frac{4}{9}, \frac{4}{3}\right)$; for $(0,0)$: $f_{xx}(0,0) = 54x\big|_{(0,0)} = 0$, $f_{yy}(0,0) = 2y\big|_{(0,0)} = 0$, $f_{xy}(0,0) = -4 \Rightarrow f_{xx}f_{yy} - f_{xy}^2$ $= -16 < 0 \Rightarrow$ saddle point; for $\left(\frac{4}{9}, \frac{4}{3}\right)$: $f_{xx}\left(\frac{4}{9}, \frac{4}{3}\right) = 24$, $f_{yy}\left(\frac{4}{9}, \frac{4}{3}\right) = \frac{8}{3}$, $f_{xy}\left(\frac{4}{9}, \frac{4}{3}\right) = -4 \Rightarrow f_{xx}f_{yy} - f_{xy}^2 = 48 > 0$ and $f_{xx} > 0 \Rightarrow$ local minimum of $f\left(\frac{4}{9}, \frac{4}{3}\right) = -\frac{64}{81}$

22. $f_x(x,y) = 24x^2 + 6y = 0 \Rightarrow y = -4x^2$; $f_y(x,y) = 3y^2 + 6x = 0 \Rightarrow 3\left(-4x^2\right)^2 + 6x = 0 \Rightarrow 16x^4 + 2x = 0$ $\Rightarrow 2x\left(8x^3 + 1\right) = 0 \Rightarrow x = 0$ or $x = -\frac{1}{2} \Rightarrow (0,0)$ and $\left(-\frac{1}{2}, -1\right)$ are the critical points; $f_{xx}(x,y) = 48x$, $f_{yy}(x,y) = 6y$, and $f_{xy}(x,y) = 6$; for $(0,0)$: $f_{xx}(0,0) = 0$, $f_{yy}(0,0) = 0$, $f_{xy}(0,0) = 6 \Rightarrow f_{xx}f_{yy} - f_{xy}^2 = -36 < 0$ $\Rightarrow$ saddle point; for $\left(-\frac{1}{2}, -1\right)$: $f_{xx}\left(-\frac{1}{2}, -1\right) = -24$, $f_{yy}\left(-\frac{1}{2}, -1\right) = -6$, $f_{xy}\left(-\frac{1}{2}, -1\right) = 6$ $\Rightarrow f_{xx}f_{yy} - f_{xy}^2 = 108 > 0$ and $f_{xx} < 0 \Rightarrow$ local maximum of $f\left(-\frac{1}{2}, -1\right) = 1$

23. $f_x(x,y) = 3x^2 + 6x = 0 \Rightarrow x = 0$ or $x = -2$; $f_y(x,y) = 3y^2 - 6y = 0 \Rightarrow y = 0$ or $y = 2 \Rightarrow$ the critical points are $(0,0)$, $(0,2)$, $(-2,0)$, and $(-2,2)$; for $(0,0)$: $f_{xx}(0,0) = 6x + 6\big|_{(0,0)} = 6$, $f_{yy}(0,0) = 6y - 6\big|_{(0,0)} = -6$, $f_{xy}(0,0) = 0 \Rightarrow f_{xx}f_{yy} - f_{xy}^2 = -36 < 0 \Rightarrow$ saddle point; for $(0,2)$: $f_{xx}(0,2) = 6$, $f_{yy}(0,2) = 6$, $f_{xy}(0,2) = 0$ $\Rightarrow f_{xx}f_{yy} - f_{xy}^2 = 36 > 0$ and $f_{xx} > 0 \Rightarrow$ local minimum of $f(0,2) = -12$; for $(-2,0)$: $f_{xx}(-2,0) = -6$, $f_{yy}(-2,0) = -6$, $f_{xy}(-2,0) = 0 \Rightarrow f_{xx}f_{yy} - f_{xy}^2 = 36 > 0$ and $f_{xx} < 0 \Rightarrow$ local maximum of $f(-2,0) = -4$; for $(-2,2)$: $f_{xx}(-2,2) = -6$, $f_{yy}(-2,2) = 6$, $f_{xy}(-2,2) = 0 \Rightarrow f_{xx}f_{yy} - f_{xy}^2 = -36 < 0 \Rightarrow$ saddle point

24. $f_x(x,y) = 6x^2 - 18x = 0 \Rightarrow 6x(x-3) = 0 \Rightarrow x = 0$ or $x = 3$; $f_y(x,y) = 6y^2 + 6y - 12 = 0 \Rightarrow 6(y+2)(y-1) = 0$ $\Rightarrow y = -2$ or $y = 1 \Rightarrow$ the critical points are $(0,-2)$, $(0,1)$, $(3,-2)$, and $(3,1)$; $f_{xx}(x,y) = 12x - 18$, $f_{yy}(x,y) = 12y + 6$, and $f_{xy}(x,y) = 0$; for $(0,-2)$: $f_{xx}(0,-2) = -18$, $f_{yy}(0,-2) = -18$, $f_{xy}(0,-2) = 0$ $\Rightarrow f_{xx}f_{yy} - f_{xy}^2 = 324 > 0$ and $f_{xx} < 0 \Rightarrow$ local maximum of $f(0,-2) = 20$; for $(0,1)$: $f_{xx}(0,1) = -18$, $f_{yy}(0,1) = 18$, $f_{xy}(0,1) = 0 \Rightarrow f_{xx}f_{yy} - f_{xy}^2 = -324 < 0 \Rightarrow$ saddle point; for $(3,-2)$: $f_{xx}(3,-2) = 18$, $f_{yy}(3,-2) = -18$, $f_{xy}(3,-2) = 0 \Rightarrow f_{xx}f_{yy} - f_{xy}^2 = -324 < 0 \Rightarrow$ saddle point; for $(3,1)$: $f_{xx}(3,1) = 18$, $f_{yy}(3,1) = 18$, $f_{xy}(3,1) = 0 \Rightarrow f_{xx}f_{yy} - f_{xy}^2 = 324 > 0$ and $f_{xx} > 0 \Rightarrow$ local minimum of $f(3,1) = -34$

25. $f_x(x,y) = 4y - 4x^3 = 0$ and $f_y(x,y) = 4x - 4y^3 = 0 \Rightarrow x = y \Rightarrow x(1-x^2) = 0 \Rightarrow x = 0,\ 1,\ -1 \Rightarrow$ the critical points are $(0,0)$, $(1,1)$, and $(-1,-1)$; for $(0,0)$: $f_{xx}(0,0) = -12x^2\big|_{(0,0)} = 0$, $f_{yy}(0,0) = -12y^2\big|_{(0,0)} = 0$, $f_{xy}(0,0) = 4 \Rightarrow f_{xx}f_{yy} - f_{xy}^2 = -16 < 0 \Rightarrow$ saddle point; for $(1,1)$: $f_{xx}(1,1) = -12$, $f_{yy}(1,1) = -12$, $f_{xy}(1,1) = 4$ $\Rightarrow f_{xx}f_{yy} - f_{xy}^2 = 128 > 0$ and $f_{xx} < 0 \Rightarrow$ local maximum of $f(1,1) = 2$; for $(-1,-1)$: $f_{xx}(-1,-1) = -12$, $f_{yy}(-1,-1) = -12$, $f_{xy}(-1,-1) = 4 \Rightarrow f_{xx}f_{yy} - f_{xy}^2 = 128 > 0$ and $f_{xx} < 0 \Rightarrow$ local maximum of $f(-1,-1) = 2$

26. $f_x(x,y) = 4x^3 + 4y = 0$ and $f_y(x,y) = 4y^3 + 4x = 0 \Rightarrow x = -y \Rightarrow -x^3 + x = 0 \Rightarrow x(1-x^2) = 0 \Rightarrow x = 0,\ 1,\ -1$ $\Rightarrow$ the critical points are $(0,0)$, $(1,-1)$, and $(-1,1)$; $f_{xx}(x,y) = 12x^2$, $f_{yy}(x,y) = 12y^2$, and $f_{xy}(x,y) = 4$; for $(0,0)$: $f_{xx}(0,0) = 0$, $f_{yy}(0,0) = 0$, $f_{xy}(0,0) = 4 \Rightarrow f_{xx}f_{yy} - f_{xy}^2 = -16 < 0 \Rightarrow$ saddle point; for $(1,-1)$: $f_{xx}(1,-1) = 12$, $f_{yy}(1,-1) = 12$, $f_{xy}(1,-1) = 4 \Rightarrow f_{xx}f_{yy} - f_{xy}^2 = 128 > 0$ and $f_{xx} > 0 \Rightarrow$ local minimum of $f(1,-1) = -2$; for $(-1,1)$: $f_{xx}(-1,1) = 12$, $f_{yy}(-1,1) = 12$, $f_{xy}(-1,1) = 4 \Rightarrow f_{xx}f_{yy} - f_{xy}^2 = 128 > 0$ and $f_{xx} > 0 \Rightarrow$ local minimum of $f(-1,1) = -2$

27. $f_x(x,y) = \dfrac{-2x}{(x^2+y^2-1)^2} = 0$ and $f_y(x,y) = \dfrac{-2y}{(x^2+y^2-1)^2} = 0 \Rightarrow x = 0$ and $y = 0 \Rightarrow$ the critical point is $(0,0)$;

$f_{xx} = \dfrac{4x^2 - 2y^2 + 2}{(x^2+y^2-1)^3}$, $f_{yy} = \dfrac{-2x^2 + 4y^2 + 2}{(x^2+y^2-1)^3}$, $f_{xy} = \dfrac{6xy}{(x^2+y^2-1)^3}$; $f_{xx}(0,0) = -2$, $f_{yy}(0,0) = -2$, $f_{xy}(0,0) = 0$ $\Rightarrow f_{xx}f_{yy} - f_{xy}^2 = 4 > 0$ and $f_{xx} < 0 \Rightarrow$ local maximum of $f(0,0) = -1$

28. $f_x(x,y) = -\frac{1}{x^2} + y = 0$ and $f_y(x,y) = x - \frac{1}{y^2} = 0 \Rightarrow x = 1$ and $y = 1 \Rightarrow$ the critical point is $(1,1)$; $f_{xx} = \frac{2}{x^3}$, $f_{yy} = \frac{2}{y^3}$, $f_{xy} = 1$; $f_{xx}(1,1) = 2$, $f_{yy}(1,1) = 2$, $f_{xy}(1,1) = 1 \Rightarrow f_{xx}f_{yy} - f_{xy}^2 = 3 > 0$ and $f_{xx} > 2 \Rightarrow$ local

minimum of $f(1,1) = 3$

29. $f_x(x,y) = y\cos x = 0$ and $f_y(x,y) = \sin x = 0 \Rightarrow x = n\pi$, n an integer, and $y = 0 \Rightarrow$ the critical points are $(n\pi, 0)$, n an integer (Note: cos x and sin x cannot both be 0 for the same x, so sin x must be 0 and $y = 0$); $f_{xx} = -y\sin x$, $f_{yy} = 0$, $f_{xy} = \cos x$; $f_{xx}(n\pi,0) = 0$, $f_{yy}(n\pi,0) = 0$, $f_{xy}(n\pi,0) = 1$ if n is even and $f_{xy}(n\pi,0) = -1$ if n is odd $\Rightarrow f_{xx}f_{yy} - f_{xy}^2 = -1 < 0 \Rightarrow$ saddle point; $f(n\pi,0) = 0$ for every n

30. $f_x(x,y) = 2e^{2x}\cos y = 0$ and $f_y(x,y) = -e^{2x}\sin y = 0 \Rightarrow$ no solution since $e^{2x} \neq 0$ for any x and the functions cos y and sin y cannot equal 0 for the same y $\Rightarrow$ no critical points $\Rightarrow$ no extrema and no saddle points

31. (i) On OA, $f(x,y) = f(0,y) = y^2 - 4y + 1$ on $0 \le y \le 2$; $f'(0,y) = 2y - 4 = 0 \Rightarrow y = 2$; $f(0,0) = 1$ and $f(0,2) = -3$

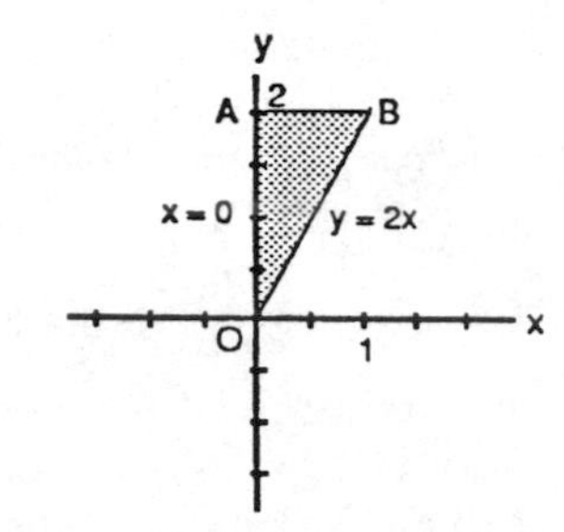

(ii) On AB, $f(x,y) = f(x,2) = 2x^2 - 4x - 3$ on $0 \le x \le 1$; $f'(x,2) = 4x - 4 = 0 \Rightarrow x = 1$; $f(0,2) = -3$ and $f(1,2) = -5$

(iii) On OB, $f(x,y) = f(x,2x) = 6x^2 - 12x + 1$ on $0 \le x \le 1$; endpoint values have been found above; $f'(x,2x) = 12x - 12 = 0 \Rightarrow x = 1$ and $y = 2$, but $(1,2)$ is not an interior point of OB

(iv) For interior points of the triangular region, $f_x(x,y) = 4x - 4 = 0$ and $f_y(x,y) = 2y - 4 = 0$ $\Rightarrow x = 1$ and $y = 2$, but $(1,2)$ is not an interior point of the region. Therefore, the absolute maximum is 1 at $(0,0)$ and the absolute minimum is -5 at $(1,2)$.

32. (i) On OA, $D(x,y) = D(0,y) = y^2 + 1$ on $0 \le y \le 4$; $D'(0,y) = 2y = 0 \Rightarrow y = 0$; $D(0,0) = 1$ and $D(0,4) = 17$

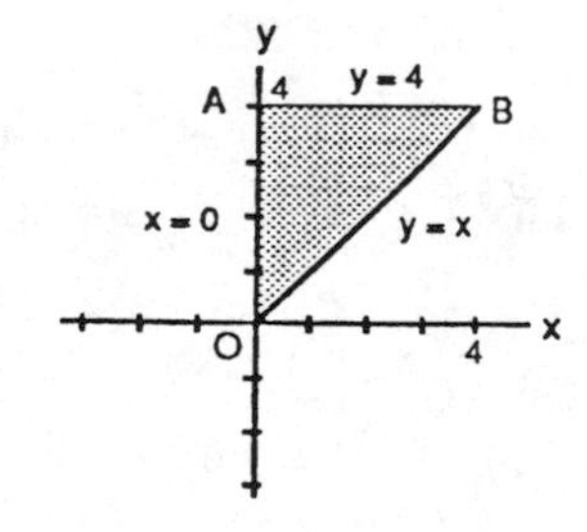

(ii) On AB, $D(x,y) = D(x,4) = x^2 - 4x + 17$ on $0 \le x \le 4$; $D'(x,4) = 2x - 4 = 0 \Rightarrow x = 2$ and $(2,4)$ is an interior point of OA; $D(2,4) = 13$ and $D(4,4) = D(0,4) = 17$

(iii) On OB, $D(x,y) = D(x,x) = x^2 + 1$ on $0 \le x \le 4$; $D'(x,x) = 2x = 0 \Rightarrow x = 0$ and $y = 0$, which is not an interior point of OB; endpoint values have been found above

(iv) For interior points of the triangular region, $f_x(x,y) = 2x - y = 0$ and $f_y(x,y) = -x + 2y = 0 \Rightarrow x = 0$ and $y = 0$, which is not an interior point of the region. Therefore, the absolute maximum is 17 at $(0,4)$ and $(4,4)$, and the absolute minimum is 1 at $(0,0)$.

33. (i) On OA, $f(x,y) = f(0,y) = y^2$ on $0 \le y \le 2$; $f'(0,y) = 2y = 0$ $\Rightarrow y = 0$ and $x = 0$; $f(0,0) = 0$ and $f(0,2) = 4$

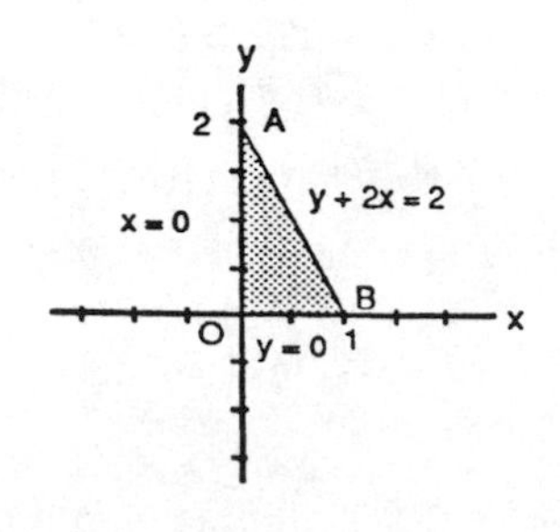

(ii) On OB, $f(x,y) = f(x,0) = x^2$ on $0 \le x \le 1$; $f'(x,0) = 2x = 0$ $\Rightarrow x = 0$ and $y = 0$; $f(0,0) = 0$ and $f(1,0) = 1$

(iii) On AB, $f(x,y) = f(x,-2x+2) = 5x^2 - 8x + 4$ on $0 \le x \le 1$; $f'(x,-2x+2) = 10x - 8 = 0 \Rightarrow x = \frac{4}{5}$ and $y = \frac{2}{5}$; $f\left(\frac{4}{5},\frac{2}{5}\right) = \frac{4}{5}$ and $f(0,2) = 4$

(iv) For interior points of the triangular region, $f_x(x,y) = 2x = 0$ and $f_y(x,y) = 2y = 0$ $\Rightarrow x = 0$ and $y = 0$, but $(0,0)$ is not an interior point of the region. Therefore the absolute maximum is 4 at $(0,2)$ and the absolute minimum is 0 at $(0,0)$.

34. (i) 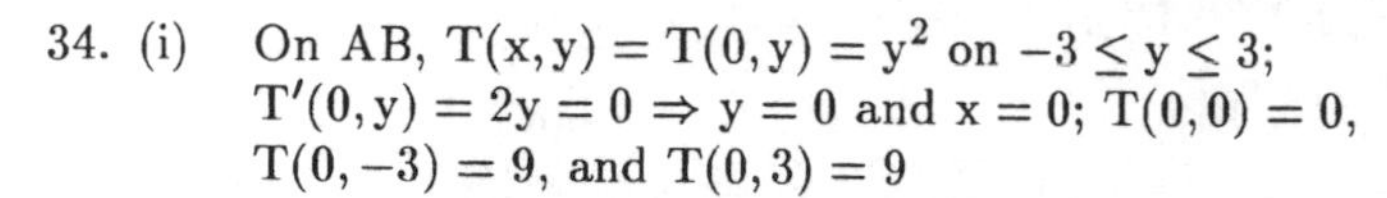
On AB, $T(x,y) = T(0,y) = y^2$ on $-3 \le y \le 3$; $T'(0,y) = 2y = 0 \Rightarrow y = 0$ and $x = 0$; $T(0,0) = 0$, $T(0,-3) = 9$, and $T(0,3) = 9$

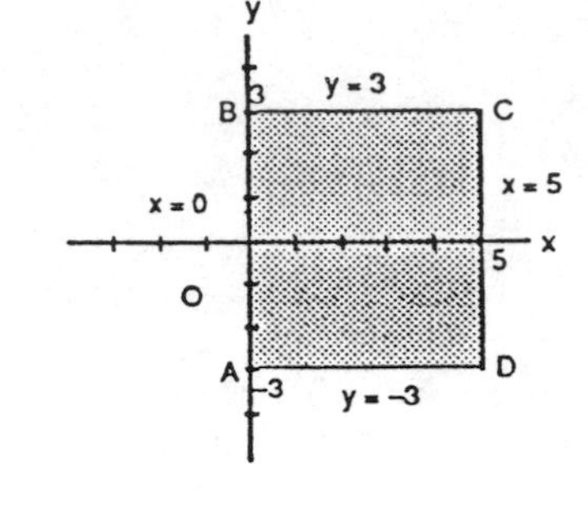

(ii) 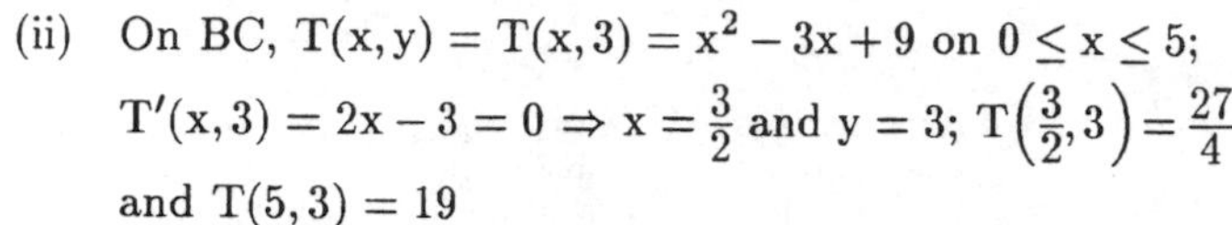
On BC, $T(x,y) = T(x,3) = x^2 - 3x + 9$ on $0 \le x \le 5$; $T'(x,3) = 2x - 3 = 0 \Rightarrow x = \frac{3}{2}$ and $y = 3$; $T\left(\frac{3}{2}, 3\right) = \frac{27}{4}$ and $T(5,3) = 19$

(iii) 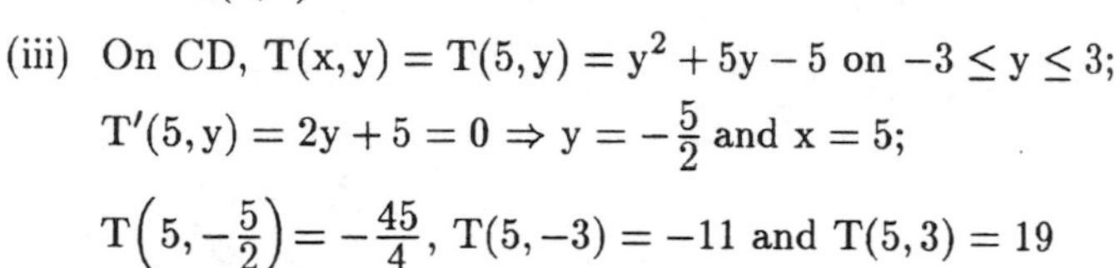
On CD, $T(x,y) = T(5,y) = y^2 + 5y - 5$ on $-3 \le y \le 3$; $T'(5,y) = 2y + 5 = 0 \Rightarrow y = -\frac{5}{2}$ and $x = 5$; $T\left(5, -\frac{5}{2}\right) = -\frac{45}{4}$, $T(5,-3) = -11$ and $T(5,3) = 19$

(iv) On AD, $T(x,y) = T(x,-3) = x^2 - 9x + 9$ on $0 \le x \le 5$; $T'(x,-3) = 2x - 9 = 0 \Rightarrow x = \frac{9}{2}$ and $y = -3$; $T\left(\frac{9}{2}, -3\right) = -\frac{45}{4}$, $T(0,-3) = 9$ and $T(5,-3) = -11$

(v) For interior points of the rectangular region, $T_x(x,y) = 2x + y - 6 = 0$ and $T_y(x,y) = x + 2y = 0 \Rightarrow x = 4$ and $y = -2 \Rightarrow (4,-2)$ is an interior critical point with $T(4,-2) = -12$. Therefore the absolute maximum is 19 at $(5,3)$ and the absolute minimum is -12 at $(4,-2)$.

35. (i) On OC, $T(x,y) = T(x,0) = x^2 - 6x + 2$ on $0 \le x \le 5$; $T'(x,0) = 2x - 6 = 0 \Rightarrow x = 3$ and $y = 0$; $T(3,0) = -7$, $T(0,0) = 2$, and $T(5,0) = -3$

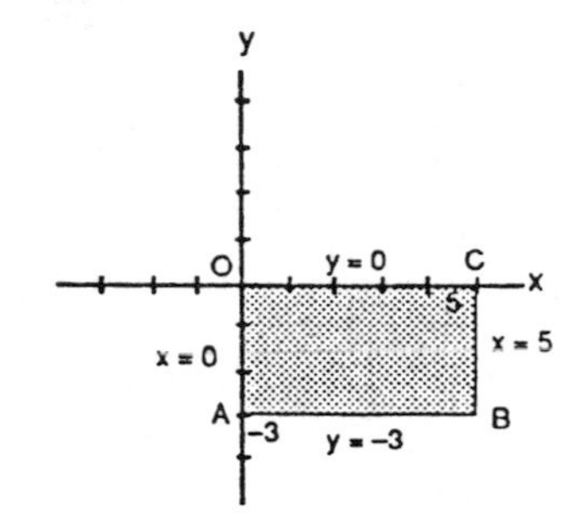

(ii) 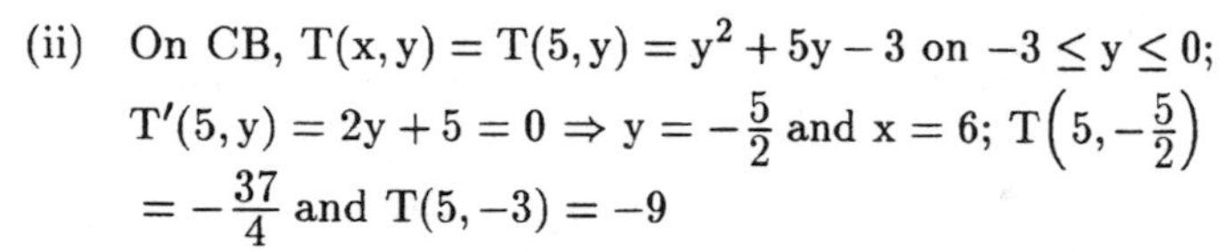
On CB, $T(x,y) = T(5,y) = y^2 + 5y - 3$ on $-3 \le y \le 0$; $T'(5,y) = 2y + 5 = 0 \Rightarrow y = -\frac{5}{2}$ and $x = 6$; $T\left(5, -\frac{5}{2}\right) = -\frac{37}{4}$ and $T(5,-3) = -9$

(iii) 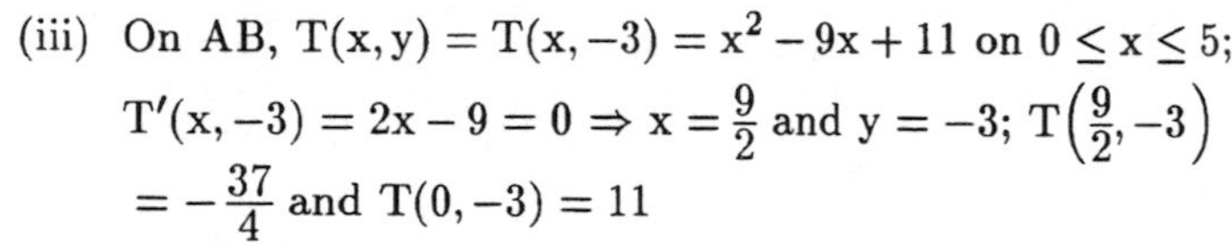
On AB, $T(x,y) = T(x,-3) = x^2 - 9x + 11$ on $0 \le x \le 5$; $T'(x,-3) = 2x - 9 = 0 \Rightarrow x = \frac{9}{2}$ and $y = -3$; $T\left(\frac{9}{2}, -3\right) = -\frac{37}{4}$ and $T(0,-3) = 11$

(iv) On AO, $T(x,y) = T(0,y) = y^2 + 2$ on $-3 \le y \le 0$; $T'(0,y) = 2y = 0 \Rightarrow y = 0$ and $x = 0$, but $(0,0)$ is not an interior point of AO

(v) For interior points of the rectangular region, $T_x(x,y) = 2x + y - 6 = 0$ and $T_y(x,y) = x + 2y = 0 \Rightarrow x = 4$ and $y = -2$, an interior critical point with $T(4,-2) = -10$. Therefore the absolute maximum is 11 at $(0,-3)$ and the absolute minimum is -10 at $(4,-2)$.

36. (i) On OA, $f(x,y) = f(0,y) = -24y^2$ on $0 \le y \le 1$; $f'(0,y) = -48y = 0 \Rightarrow y = 0$ and $x = 0$, but $(0,0)$ is not an interior point of OA; $f(0,0) = 0$ and $f(0,1) = -24$

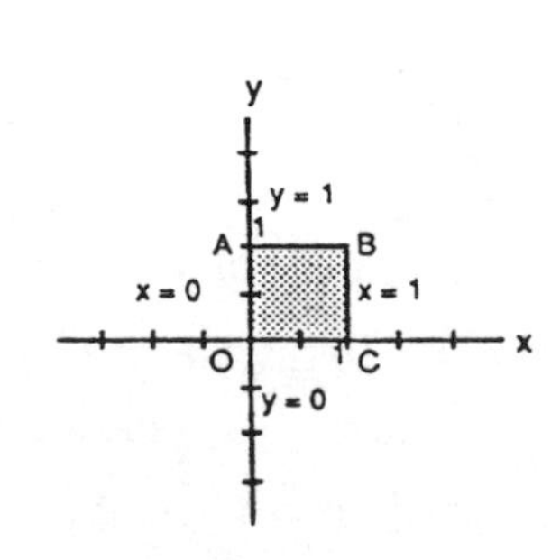

(ii) 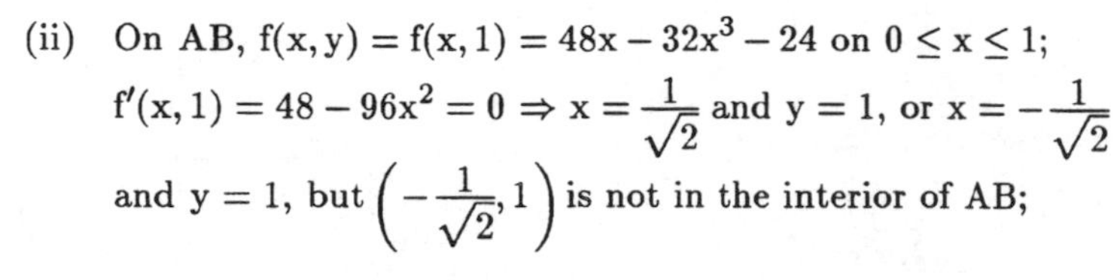
On AB, $f(x,y) = f(x,1) = 48x - 32x^3 - 24$ on $0 \le x \le 1$; $f'(x,1) = 48 - 96x^2 = 0 \Rightarrow x = \frac{1}{\sqrt{2}}$ and $y = 1$, or $x = -\frac{1}{\sqrt{2}}$ and $y = 1$, but $\left(-\frac{1}{\sqrt{2}}, 1\right)$ is not in the interior of AB;

$f\left(\frac{1}{\sqrt{2}}, 1\right) = 16\sqrt{2} - 24$ and $f(1,1) = -8$

(iii) On BC, $f(x,y) = f(1,y) = 48y - 32 - 24y^2$ on $0 \le y \le 1$; $f'(1,y) = 48 - 48y = 0 \Rightarrow y = 1$ and $x = 1$, but $(1,1)$ is not an interior point of BC; $f(1,0) = -32$ and $f(1,1) = -8$

(iv) On OC, $f(x,y) = f(x,0) = -32x^3$ on $0 \le x \le 1$; $f'(x,0) = -96x^2 = 0 \Rightarrow x = 0$ and $y = 0$, but $(0,0)$ is not an interior point of OC; $f(0,0) = 0$ and $f(1,0) = -32$

(v) For interior points of the rectangular region, $f_x(x,y) = 48y - 96x^2 = 0$ and $f_y(x,y) = 48x - 48y = 0$ $\Rightarrow x = 0$ and $y = 0$, or $x = \frac{1}{2}$ and $y = \frac{1}{2}$, but $(0,0)$ is not an interior point of the region; $f\left(\frac{1}{2}, \frac{1}{2}\right) = 2$.

Therefore the absolute maximum is 2 at $\left(\frac{1}{2}, \frac{1}{2}\right)$ and the absolute minimum is -32 at $(1,0)$.

37. (i) On AB, $f(x,y) = f(1,y) = 3 \cos y$ on $-\frac{\pi}{4} \le y \le \frac{\pi}{4}$;
$f'(1,y) = -3 \sin y = 0 \Rightarrow y = 0$ and $x = 1$; $f(1,0) = 3$,
$f\left(1, -\frac{\pi}{4}\right) = \frac{3\sqrt{2}}{2}$, and $f\left(1, \frac{\pi}{4}\right) = \frac{3\sqrt{2}}{2}$

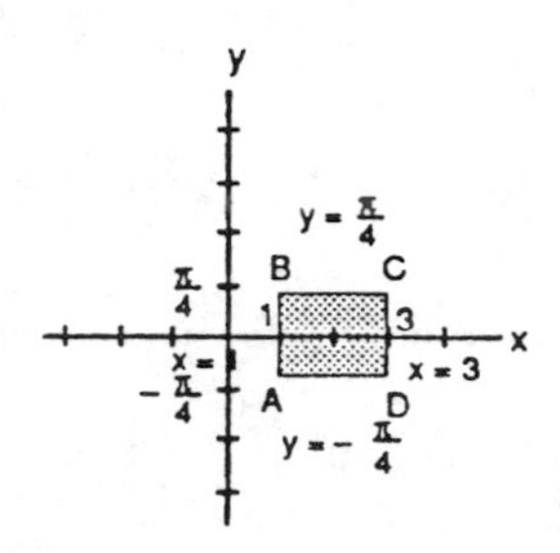

(ii) On CD, $f(x,y) = f(3,y) = 3 \cos y$ on $-\frac{\pi}{4} \le y \le \frac{\pi}{4}$;
$f'(3,y) = -3 \sin y = 0 \Rightarrow y = 0$ and $x = 3$; $f(3,0) = 3$,
$f\left(3, -\frac{\pi}{4}\right) = \frac{3\sqrt{2}}{2}$ and $f\left(3, \frac{\pi}{4}\right) = \frac{3\sqrt{2}}{2}$

(iii) On BC, $f(x,y) = f\left(x, \frac{\pi}{4}\right) = \frac{\sqrt{2}}{2}(4x - x^2)$ on $1 \le x \le 3$;
$f'\left(x, \frac{\pi}{4}\right) = \sqrt{2}(2 - x) = 0 \Rightarrow x = 2$ and $y = \frac{\pi}{4}$; $f\left(2, \frac{\pi}{4}\right) = 2\sqrt{2}$, $f\left(1, \frac{\pi}{4}\right) = \frac{3\sqrt{2}}{2}$, and $f\left(3, \frac{\pi}{4}\right) = \frac{3\sqrt{2}}{2}$

(iv) On AD, $f(x,y) = f\left(x, -\frac{\pi}{4}\right) = \frac{\sqrt{2}}{2}(4x - x^2)$ on $1 \le x \le 3$; $f'\left(x, -\frac{\pi}{4}\right) = \sqrt{2}(2 - x) = 0 \Rightarrow x = 2$ and $y = -\frac{\pi}{4}$;
$f\left(2, -\frac{\pi}{4}\right) = 2\sqrt{2}$, $f\left(1, -\frac{\pi}{4}\right) = \frac{3\sqrt{2}}{2}$, and $f\left(3, -\frac{\pi}{4}\right) = \frac{3\sqrt{2}}{2}$

(v) For interior points of the region, $f_x(x,y) = (4 - 2x) \cos y = 0$ and $f_y(x,y) = -(4x - x^2)\sin y = 0 \Rightarrow x = 2$ and $y = 0$, which is an interior critical point with $f(2,0) = 4$. Therefore the absolute maximum is 4 at $(2,0)$ and the absolute minimum is $\frac{3\sqrt{2}}{2}$ at $\left(3, -\frac{\pi}{4}\right)$, $\left(3, \frac{\pi}{4}\right)$, $\left(1, -\frac{\pi}{4}\right)$, and $\left(1, \frac{\pi}{4}\right)$.

38. (i) On OA, $f(x,y) = f(0,y) = 2y + 1$ on $0 \le y \le 1$;
$f'(0,y) = 2 \Rightarrow$ no interior critical points; $f(0,0) = 1$ and $f(0,1) = 3$

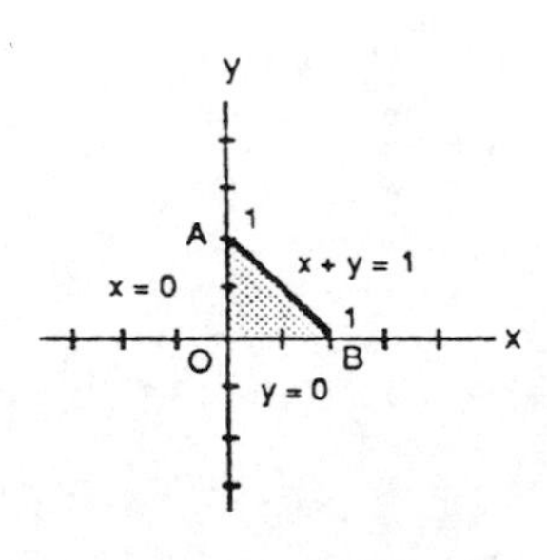

(ii) On OB, $f(x,y) = f(x,0) = 4x + 1$ on $0 \le x \le 1$;
$f'(x,0) = 4 \Rightarrow$ no interior critical points; $f(1,0) = 5$

(iii) On AB, $f(x,y) = f(x, -x + 1) = 8x^2 - 6x + 3$ on $0 \le x \le 1$;
$f'(x, -x + 1) = 16x - 6 = 0 \Rightarrow x = \frac{3}{8}$ and $y = \frac{5}{8}$;
$f\left(\frac{3}{8}, \frac{5}{8}\right) = \frac{15}{8}$, $f(0,1) = 3$, and $f(1,0) = 5$

(iv) For interior points of the triangular region,
$f_x(x,y) = 4 - 8y = 0$ and $f_y(x,y) = -8x + 2 = 0$
$\Rightarrow y = \frac{1}{2}$ and $x = \frac{1}{4}$ which is an interior critical

point with $f\left(\frac{1}{4},\frac{1}{2}\right) = 2$. Therefore the absolute maximum is 5 at $(1,0)$ and the absolute minimum is 1 at $(0,0)$.

39. Let $F(a,b) = \int_a^b (6 - x - x^2)\,dx$ where $a \le b$. The boundary of the domain of F is the line $a = b$ in the ab-plane, and $F(a,a) = 0$, so F is identically 0 on the boundary of its domain. For interior critical points we have: $\frac{\partial F}{\partial a} = -(6 - a - a^2) = 0 \Rightarrow a = -3,\ 2$ and $\frac{\partial F}{\partial b} = (6 - b - b^2) = 0 \Rightarrow b = -3,\ 2$. Since $a \le b$, there is only one interior critical point $(-3,2)$ and $F(-3,2) = \int_{-3}^{2} (6 - x - x^2)\,dx$ gives the area under the parabola $y = 6 - x - x^2$ that is above the x-axis. Therefore, $a = -3$ and $b = 2$.

40. Let $F(a,b) = \int_a^b (24 - 2x - x^2)^{1/3}\,dx$ where $a \le b$. The boundary of the domain of F is the line $a = b$ and on this line F is identically 0. For interior critical points we have: $\frac{\partial F}{\partial a} = -(24 - 2a - a^2)^{1/3} = 0 \Rightarrow a = -4,\ 6$ and $\frac{\partial F}{\partial b} = (24 - 2b - b^2)^{1/3} = 0 \Rightarrow b = -4,\ 6$. Since $a \le b$, there is only one critical point $(-4,6)$ and $F(-4,6) = \int_{-4}^{6} (24 - 2x - x^2)\,dx$ gives the area under the curve $y = (24 - 2x - x^2)^{1/3}$ that is above the x-axis. Therefore, $a = -4$ and $b = 6$.

41. $T_x(x,y) = 2x - 1 = 0$ and $T_y(x,y) = 4y = 0 \Rightarrow x = \frac{1}{2}$ and $y = 0$ with $T\left(\frac{1}{2},0\right) = -\frac{1}{4}$; on the boundary $x^2 + y^2 = 1$: $T(x,y) = -x^2 - x + 2$ for $-1 \le x \le 1 \Rightarrow T'(x,y) = -2x - 1 = 0 \Rightarrow x = -\frac{1}{2}$ and $y = \pm\frac{\sqrt{3}}{2}$; $T\left(-\frac{1}{2},\frac{\sqrt{3}}{2}\right) = \frac{9}{4}$, $T\left(-\frac{1}{2},-\frac{\sqrt{3}}{2}\right) = \frac{9}{4}$, $T(-1,0) = 2$, and $T(1,0) = 0 \Rightarrow$ the hottest is $2\frac{1}{4}^\circ$ at $\left(-\frac{1}{2},\frac{\sqrt{3}}{2}\right)$ and $\left(-\frac{1}{2},-\frac{\sqrt{3}}{2}\right)$; the coldest is $-\frac{1}{4}^\circ$ at $\left(\frac{1}{2},0\right)$.

42. $f_x(x,y) = y + 2 - \frac{2}{x} = 0$ and $f_y(x,y) = x - \frac{1}{y} = 0 \Rightarrow x = \frac{1}{2}$ and $y = 2$; $f_{xx}\left(\frac{1}{2},2\right) = \left.\frac{2}{x^2}\right|_{\left(\frac{1}{2},2\right)} = 8$, $f_{yy}\left(\frac{1}{2},2\right) = \left.\frac{1}{y^2}\right|_{\left(\frac{1}{2},2\right)} = \frac{1}{4}$, $f_{xy}\left(\frac{1}{2},2\right) = 1 \Rightarrow f_{xx}f_{yy} - f_{xy}^2 = 1 > 0$ and $f_{xx} > 0 \Rightarrow$ a local minimum of $f\left(\frac{1}{2},2\right) = 2 - \ln\frac{1}{2} = 2 + \ln 2$

43. (a) $f_x(x,y) = 2x - 4y = 0$ and $f_y(x,y) = 2y - 4x = 0 \Rightarrow x = 0$ and $y = 0$; $f_{xx}(0,0) = 2$, $f_{yy}(0,0) = 2$, $f_{xy}(0,0) = -4 \Rightarrow f_{xx}f_{yy} - f_{xy}^2 = -12 < 0 \Rightarrow$ saddle point at $(0,0)$

(b) $f_x(x,y) = 2x - 2 = 0$ and $f_y(x,y) = 2y - 4 = 0 \Rightarrow x = 1$ and $y = 2$; $f_{xx}(1,2) = 2$, $f_{yy}(1,2) = 2$, $f_{xy}(1,2) = 0 \Rightarrow f_{xx}f_{yy} - f_{xy}^2 = 4 > 0$ and $f_{xx} > 0 \Rightarrow$ local minimum at $(1,2)$

(c) $f_x(x,y) = 9x^2 - 9 = 0$ and $f_y(x,y) = 2y + 4 = 0 \Rightarrow x = \pm 1$ and $y = -2$; $f_{xx}(1,-2) = 18x\big|_{(1,-2)} = 18$, $f_{yy}(1,-2) = 2$, $f_{xy}(1,-2) = 0 \Rightarrow f_{xx}f_{yy} - f_{xy}^2 = 36 > 0$ and $f_{xx} > 0 \Rightarrow$ local minimum at $(1,-2)$; $f_{xx}(-1,-2) = -18$, $f_{yy}(-1,-2) = 2$, $f_{xy}(-1,-2) = 0 \Rightarrow f_{xx}f_{yy} - f_{xy}^2 = -36 < 0 \Rightarrow$ saddle point at $(-1,-2)$

44. (a) Minimum at $(0,0)$ since $f(x,y) > 0$ for all other (x,y)
(b) Maximum of 1 at $(0,0)$ since $f(x,y) < 1$ for all other (x,y)
(c) Neither since $f(x,y) < 0$ for $x < 0$ and $f(x,y) > 0$ for $x > 0$
(d) Neither since $f(x,y) < 0$ for $x < 0$ and $f(x,y) > 0$ for $x > 0$
(e) Neither since $f(x,y) < 0$ for $x < 0$ and $y > 0$, but $f(x,y) > 0$ for $x > 0$ and $y > 0$
(f) Minimum at $(0,0)$ since $f(x,y) > 0$ for all other (x,y)

45. If $k = 0$, then $f(x,y) = x^2 + y^2 \Rightarrow f_x(x,y) = 2x = 0$ and $f_y(x,y) = 2y = 0 \Rightarrow x = 0$ and $y = 0 \Rightarrow (0,0)$ is the only critical point. If $k \neq 0$, $f_x(x,y) = 2x + ky = 0 \Rightarrow y = -\frac{2}{k}x$; $f_y(x,y) = kx + 2y = 0 \Rightarrow kx + 2\left(-\frac{2}{k}x\right) = 0$ $\Rightarrow kx - \frac{4x}{k} = 0 \Rightarrow \left(k - \frac{4}{k}\right)x = 0 \Rightarrow x = 0$ or $k = \pm 2 \Rightarrow y = \left(-\frac{2}{k}\right)(0) = 0$ or $y = -x$; in any case $(0,0)$ is a critical point.

46. (See Exercise 45 above): $f_{xx}(x,y) = 2$, $f_{yy}(x,y) = 2$, and $f_{xy}(x,y) = k \Rightarrow f_{xx}f_{yy} - f_{xy}^2 = 4 - k^2$; f will have a saddle point at $(0,0)$ if $4 - k^2 < 0 \Rightarrow k > 2$ or $k < -2$; f will have a local minimum at $(0,0)$ if $4 - k^2 > 0$ $\Rightarrow -2 < k < 2$; the test is inconclusive if $4 - k^2 = 0 \Rightarrow k = \pm 2$.

47. (a) No; for example $f(x,y) = xy$ has a saddle point at $(a,b) = (0,0)$ where $f_x = f_y = 0$.
(b) If $f_{xx}(a,b)$ and $f_{yy}(a,b)$ differ in sign, then $f_{xx}(a,b)\, f_{yy}(a,b) < 0$ so $f_{xx}f_{yy} - f_{xy}^2 < 0$. The surface must therefore have a saddle point at (a,b) by the second derivative test.

48. Suppose that f has a local minimum value at an interior point (a,b) of its domain. Then
(i) $x = a$ is an interior point of the domain of the curve $z = f(x,b)$ in which the plane $y = b$ cuts the surface $z = f(x,y)$.
(ii) The function $z = f(x,b)$ is a differentiable function of x at $x = a$ (the derivative is $f_x(a,b)$).
(iii) The function $z = f(x,b)$ has a local minimum value at $x = a$.
(iv) The value of the derivative of $z = f(x,b)$ at $x = a$ is therefore zero (Theorem 2, Section 3.1). Since this derivative is $f_x(a,b)$, we conclude that $f_x(a,b) = 0$.

A similar argument with the function $z = f(a,y)$ shows that $f_y(a,b) = 0$.

49. We want the point on $z = 10 - x^2 - y^2$ where the tangent plane is parallel to the plane $x + 2y + 3z = 0$. To find a normal vector to $z = 10 - x^2 - y^2$ let $w = z + x^2 + y^2 - 10$. Then $\nabla w = 2x\mathbf{i} + 2y\mathbf{j} + \mathbf{k}$ is normal to $z = 10 - x^2 - y^2$ at (x,y). The vector ∇w is parallel to $\mathbf{i} + 2\mathbf{j} + 3\mathbf{k}$ which is normal to the plane $x + 2y + 3z = 0$ if $6x\mathbf{i} + 6y\mathbf{j} + 3\mathbf{k} = \mathbf{i} + 2\mathbf{j} + 3\mathbf{k}$ or $x = \frac{1}{6}$ and $y = \frac{1}{3}$. Thus the point is $\left(\frac{1}{6}, \frac{1}{3}, 10 - \frac{1}{36} - \frac{1}{9}\right)$ or $\left(\frac{1}{6}, \frac{1}{3}, \frac{355}{36}\right)$.

50. We want the point on $z = x^2 + y^2 + 10$ where the tangent plane is parallel to the plane $x + 2y - z = 0$. Let $w = z - x^2 - y^2 - 10$, then $\nabla w = -2x\mathbf{i} - 2y\mathbf{j} + \mathbf{k}$ is normal to $z = x^2 + y^2 + 10$ at (x,y). The vector ∇w is parallel to $\mathbf{i} + 2\mathbf{j} - \mathbf{k}$ which is normal to the plane if $x = \frac{1}{2}$ and $y = 1$. Thus the point $\left(\frac{1}{2}, 1, \frac{1}{4} + 1 + 10\right)$ or $\left(\frac{1}{2}, 1, \frac{45}{4}\right)$ is the point on the surface $z = x^2 + y^2 + 10$ nearest the plane $x + 2y - z = 0$.

51. No, because the domain $x \geq 0$ and $y \geq 0$ is unbounded since x and y can be as large as we please. Absolute extrema are guaranteed for continuous functions defined over closed and bounded domains in the plane. Since the domain is unbounded, the continuous function $f(x,y) = x + y$ need not have an absolute maximum (although, in this case, it does have an absolute minimum value of $f(0,0) = 0$).

52. (a) (i) On $x = 0$, $f(x,y) = f(0,y) = y^2 - y + 1$ for $0 \leq y \leq 1$; $f'(0,y) = 2y - 1 = 0 \Rightarrow y = \frac{1}{2}$ and $x = 0$; $f\left(0, \frac{1}{2}\right) = \frac{3}{4}$, $f(0,0) = 1$, and $f(0,1) = 1$

(ii) On $y = 1$, $f(x,y) = f(x,1) = x^2 + x + 1$ for $0 \leq x \leq 1$; $f'(x,1) = 2x + 1 = 0 \Rightarrow x = -\frac{1}{2}$ and $y = 1$, but $\left(-\frac{1}{2}, 1\right)$ is outside the domain; $f(0,1) = 1$ and $f(1,1) = 3$

(iii) On $x = 1$, $f(x,y) = f(1,y) = y^2 + y + 1$ for $0 \leq y \leq 1$; $f'(1,y) = 2y + 1 = 0 \Rightarrow y = -\frac{1}{2}$ and $x = 1$, but $\left(1, -\frac{1}{2}\right)$ is outside the domain; $f(1,0) = 1$ and $f(1,1) = 3$

(iv) On $y = 0$, $f(x,y) = f(x,0) = x^2 - x + 1$ for $0 \leq x \leq 1$; $f'(x,0) = 2x - 1 = 0 \Rightarrow x = \frac{1}{2}$ and $y = 0$; $f\left(\frac{1}{2}, 0\right) = \frac{3}{4}$; $f(0,0) = 1$, and $f(1,0) = 1$

(v) On the interior of the square, $f_x(x,y) = 2x + 2y - 1 = 0$ and $f_y(x,y) = 2y + 2x - 1 = 0 \Rightarrow 2x + 2y = 1$ $\Rightarrow (x + y) = \frac{1}{2}$. Then $f(x,y) = x^2 + y^2 + 2xy - x - y + 1 = (x+y)^2 - (x+y) + 1 = \frac{3}{4}$ is the absolute minimum value when $2x + 2y = 1$.

(b) The absolute maximum is $f(1,1) = 3$ and the absolute minimum is $\frac{3}{4}$ along the line $x + y = \frac{1}{2}$ in the square $0 \leq x \leq 1$ and $0 \leq y \leq 1$, as found in part (a).

53. (a) $\frac{df}{dt} = \frac{\partial f}{\partial x}\frac{dx}{dt} + \frac{\partial f}{\partial y}\frac{dy}{dt} = \frac{dx}{dt} + \frac{dy}{dt} = -2 \sin t + 2 \cos t = 0 \Rightarrow \cos t = \sin t \Rightarrow x = y$

(i) On the semicircle $x^2 + y^2 = 4$, $y \geq 0$, we have $t = \frac{\pi}{4}$ and $x = y = \sqrt{2} \Rightarrow f\left(\sqrt{2}, \sqrt{2}\right) = 2\sqrt{2}$. At the endpoints, $f(-2,0) = -2$ and $f(2,0) = 2$. Therefore the absolute minimum is $f(-2,0) = -2$ when $t = \pi$; the absolute maximum is $f\left(\sqrt{2}, \sqrt{2}\right) = 2\sqrt{2}$ when $t = \frac{\pi}{4}$.

(ii) On the quartercircle $x^2 + y^2 = 4$, $x \geq 0$ and $y \geq 0$, the endpoints give $f(0,2) = 2$ and $f(2,0) = 2$. Therefore the absolute minimum is $f(2,0) = 2$ and $f(0,2) = 2$ when $t = 0, \frac{\pi}{2}$ respectively; the absolute maximum is $f\left(\sqrt{2}, \sqrt{2}\right) = 2\sqrt{2}$ when $t = \frac{\pi}{4}$.

(b) $\frac{dg}{dt} = \frac{\partial g}{\partial x}\frac{dx}{dt} + \frac{\partial g}{\partial y}\frac{dy}{dt} = y\frac{dx}{dt} + x\frac{dy}{dt} = -4 \sin^2 t + 4 \cos^2 t = 0 \Rightarrow \cos t = \pm \sin t \Rightarrow x = \pm y$.

(i) On the semicircle $x^2 + y^2 = 4$, $y \geq 0$, we obtain $x = y = \sqrt{2}$ at $t = \frac{\pi}{4}$ and $x = -\sqrt{2}$, $y = \sqrt{2}$ at $t = \frac{3\pi}{4}$. Then $g\left(\sqrt{2}, \sqrt{2}\right) = 2$ and $g\left(-\sqrt{2}, \sqrt{2}\right) = -2$. At the endpoints, $g(-2,0) = g(2,0) = 0$. Therefore the absolute minimum is $g\left(-\sqrt{2}, \sqrt{2}\right) = -2$ when $t = \frac{3\pi}{4}$; the absolute maximum is $g\left(\sqrt{2}, \sqrt{2}\right) = 2$ when $t = \frac{\pi}{4}$.

(ii) On the quartercircle $x^2 + y^2 = 4$, $x \geq 0$ and $y \geq 0$, the endpoints give $g(0,2) = 0$ and $g(2,0) = 0$. Therefore the absolute minimum is $g(2,0) = 0$ and $g(0,2) = 0$ when $t = 0, \frac{\pi}{2}$ respectively; the absolute maximum is $g\left(\sqrt{2}, \sqrt{2}\right) = 2$ when $t = \frac{\pi}{4}$.

(c) $\frac{dh}{dt} = \frac{\partial h}{\partial x}\frac{dx}{dt} + \frac{\partial h}{\partial y}\frac{dy}{dt} = 4x\frac{dx}{dt} + 2y\frac{dy}{dt} = (8 \cos t)(-2 \sin t) + (4 \sin t)(2 \cos t) = -8 \cos t \sin t = 0$

$\Rightarrow t = 0, \frac{\pi}{2}, \pi$ yielding the points $(2,0)$, $(0,2)$, and $(-2,0)$, respectively.

(i) On the semicircle $x^2 + y^2 = 4$, $y \ge 0$ we have $h(2,0) = 8$, $h(0,2) = 4$, and $h(-2,0) = 8$. Therefore, the absolute minimum is $h(0,2) = 4$ when $t = \frac{\pi}{2}$; the absolute maximum is $h(2,0) = 8$ and $h(-2,0) = 8$ when $t = 0, \pi$ respectively.

(ii) On the quartercircle $x^2 + y^2 = 4$, $x \ge 0$ and $y \ge 0$ the absolute minimum is $h(0,2) = 4$ when $t = \frac{\pi}{2}$; the absolute maximum is $h(2,0) = 8$ when $t = 0$.

54. (a) $\frac{df}{dt} = \frac{\partial f}{\partial x}\frac{dx}{dt} + \frac{\partial f}{\partial y}\frac{dy}{dt} = 2\frac{dx}{dt} + 3\frac{dy}{dt} = -6 \sin t + 6 \cos t = 0 \Rightarrow \sin t = \cos t \Rightarrow t = \frac{\pi}{4}$ for $0 \le t \le \pi$.

(i) On the semi-ellipse, $\frac{x^2}{9} + \frac{y^2}{4} = 1$, $y \ge 0$, $f(x,y) = 2x + 3y = 6 \cos t + 6 \sin t = 6\left(\frac{\sqrt{2}}{2}\right) + 6\left(\frac{\sqrt{2}}{2}\right) = 6\sqrt{2}$ at $t = \frac{\pi}{4}$. At the endpoints, $f(-3,0) = -6$ and $f(3,0) = 6$. The absolute minimum is $f(-3,0) = -6$ when $t = \pi$; the absolute maximum is $f\left(\frac{3\sqrt{2}}{2}, \sqrt{2}\right) = 6\sqrt{2}$ when $t = \frac{\pi}{4}$.

(ii) On the quarter ellipse, at the endpoints $f(0,2) = 6$ and $f(3,0) = 6$. The absolute minimum is $f(3,0) = 6$ and $f(0,2) = 6$ when $t = 0, \frac{\pi}{2}$ respectively; the absolute maximum is $f\left(\frac{3\sqrt{2}}{2}, \sqrt{2}\right) = 6\sqrt{2}$ when $t = \frac{\pi}{4}$.

(b) $\frac{dg}{dt} = \frac{\partial g}{\partial x}\frac{dx}{dt} + \frac{\partial g}{\partial y}\frac{dy}{dt} = y\frac{dx}{dt} + x\frac{dy}{dt} = (2 \sin t)(-3 \sin t) + (3 \cos t)(2 \cos t) = 6\left(\cos^2 t - \sin^2 t\right) = 6 \cos 2t = 0$

$\Rightarrow t = \frac{\pi}{4}, \frac{3\pi}{4}$ for $0 \le t \le \pi$.

(i) On the semi-ellipse, $g(x,y) = xy = 6 \sin t \cos t$. Then $g\left(\frac{3\sqrt{2}}{2}, \sqrt{2}\right) = 3$ when $t = \frac{\pi}{4}$, and $g\left(-\frac{3\sqrt{2}}{2}, \sqrt{2}\right) = -3$ when $t = \frac{3\pi}{4}$. At the endpoints, $g(-3,0) = g(3,0) = 0$. The absolute minimum is $g\left(-\frac{3\sqrt{2}}{2}, \sqrt{2}\right) = -3$ when $t = \frac{3\pi}{4}$; the absolute maximum is $g\left(\frac{3\sqrt{2}}{2}, \sqrt{2}\right) = 3$ when $t = \frac{\pi}{4}$.

(ii) On the quarter ellipse, at the endpoints $g(0,2) = 0$ and $g(3,0) = 0$. The absolute minimum is $g(3,0) = 0$ and $g(0,2) = 0$ at $t = 0, \frac{\pi}{2}$ respectively; the absolute maximum is $g\left(\frac{3\sqrt{2}}{2}, \sqrt{2}\right) = 3$ when $t = \frac{\pi}{4}$.

(c) $\frac{dh}{dt} = \frac{\partial h}{\partial x}\frac{dx}{dt} + \frac{\partial h}{\partial y}\frac{dy}{dt} = 2x\frac{dx}{dt} + 6y\frac{dy}{dt} = (6 \cos t)(-3 \sin t) + (12 \sin t)(2 \cos t) = 6 \sin t \cos t = 0$

$\Rightarrow t = 0, \frac{\pi}{2}, \pi$ for $0 \le t \le \pi$, yielding the points $(3,0)$, $(0,2)$, and $(-3,0)$.

(i) On the semi-ellipse, $y \ge 0$ so that $h(3,0) = 9$, $h(0,2) = 12$, and $h(-3,0) = 9$. The absolute minimum is $h(3,0) = 9$ and $h(-3,0) = 9$ when $t = 0, \pi$ respectively; the absolute maximum is $h(0,2) = 12$ when $t = \frac{\pi}{2}$.

(ii) On the quarter ellipse, the absolute minimum is $h(3,0) = 9$ when $t = 0$; the absolute maximum is $h(0,2) = 12$ when $t = \frac{\pi}{2}$.

55. $\frac{df}{dt} = \frac{\partial f}{\partial x}\frac{dx}{dt} + \frac{\partial f}{\partial y}\frac{dy}{dt} = y\frac{dx}{dt} + x\frac{dy}{dt}$

(i) $x = 2t$ and $y = t + 1 \Rightarrow \frac{df}{dt} = (t+1)(2) + (2t)(1) = 4t + 2 = 0 \Rightarrow t = -\frac{1}{2} \Rightarrow x = -1$ and $y = \frac{1}{2}$ with $f\left(-1, \frac{1}{2}\right) = -\frac{1}{2}$. The absolute minimum is $f\left(-1, \frac{1}{2}\right) = -\frac{1}{2}$ when $t = -\frac{1}{2}$; there is no absolute maximum.

(ii) For the endpoints: $t = -1 \Rightarrow x = -2$ and $y = 0$ with $f(-2,0) = 0$; $t = 0 \Rightarrow x = 0$ and $y = 1$ with $f(0,1) = 0$. The absolute minimum is $f\left(-1, \frac{1}{2}\right) = -\frac{1}{2}$ when $t = -\frac{1}{2}$; the absolute maximum is $f(0,1) = 0$ and $f(-2,0) = 0$ when $t = -1, 0$ respectively.

(iii) There are no interior critical points. for the endpoints: $t = 0 \Rightarrow x = 0$ and $y = 1$ with $f(0,1) = 0$; $t = 1 \Rightarrow x = 2$ and $y = 2$ with $f(2,2) = 4$. The absolute minimum is $f(0,1) = 0$ when $t = 0$; the absolute maximum is $f(2,2) = 4$ when $t = 1$.

56. (a) $\frac{df}{dt} = \frac{\partial f}{\partial x}\frac{dx}{dt} + \frac{\partial f}{\partial y}\frac{dy}{dt} = 2x\frac{dx}{dt} + 2y\frac{dy}{dt}$

(i) $x = t$ and $y = 2 - 2t \Rightarrow \frac{df}{dt} = (2t)(1) + 2(2-2t)(-2) = 10t - 8 = 0 \Rightarrow t = \frac{4}{5} \Rightarrow x = \frac{4}{5}$ and $y = \frac{2}{5}$ with $f\left(\frac{4}{5}, \frac{2}{5}\right) = \frac{16}{25} + \frac{4}{25} = \frac{4}{5}$. The absolute minimum is $f\left(\frac{4}{5}, \frac{2}{5}\right) = \frac{4}{5}$ when $t = \frac{4}{5}$; there is no absolute maximum along the line.

(ii) For the endpoints: $t = 0 \Rightarrow x = 0$ and $y = 2$ with $f(0,2) = 4$; $t = 1 \Rightarrow x = 1$ and $y = 0$ with $f(1,0) = 1$. The absolute minimum is $f\left(\frac{4}{5}, \frac{2}{5}\right) = \frac{4}{5}$ at the interior critical point when $t = \frac{4}{5}$; the absolute maximum is $f(0,2) = 4$ at the endpoint when $t = 0$.

(b) $\frac{dg}{dt} = \frac{\partial g}{\partial x}\frac{dx}{dt} + \frac{\partial g}{\partial y}\frac{dy}{dt} = \left[\frac{-2x}{(x^2+y^2)^2}\right]\frac{dx}{dt} + \left[\frac{-2y}{(x^2+y^2)^2}\right]\frac{dy}{dt}$

(i) $x = t$ and $y = 2 - 2t \Rightarrow x^2 + y^2 = 5t^2 - 8t + 4 \Rightarrow \frac{dg}{dt} = -\left(5t^2 - 8t + 4\right)^{-2}[(-2t)(1) + (-2)(2-2t)(-2)]$ $= -\left(5t^2 - 8t + 4\right)^{-2}(-10t + 8) = 0 \Rightarrow t = \frac{4}{5} \Rightarrow x = \frac{4}{5}$ and $y = \frac{2}{5}$ with $g\left(\frac{4}{5}, \frac{2}{5}\right) = \frac{1}{\left(\frac{4}{5}\right)} = \frac{5}{4}$. The absolute maximum is $g\left(\frac{4}{5}, \frac{2}{5}\right) = \frac{5}{4}$ when $t = \frac{4}{5}$; there is no absolute minimum along the line since x and y can be as large as we please.

(ii) For the endpoints: $t = 0 \Rightarrow x = 0$ and $y = 2$ with $g(0,2) = \frac{1}{4}$; $t = 1 \Rightarrow x = 1$ and $y = 0$ with $g(1,0) = 1$. The absolute minimum is $g(0,2) = \frac{1}{4}$ when $t = 0$; the absolute maximum is $g\left(\frac{4}{5}, \frac{2}{5}\right) = \frac{5}{4}$ when $t = \frac{4}{5}$.

57. $m = \frac{(2)(-1) - 3(-14)}{(2)^2 - 3(10)} = -\frac{20}{13}$ and

$b = \frac{1}{3}\left[-1 - \left(-\frac{20}{13}\right)(2)\right] = \frac{9}{13}$

$\Rightarrow y = -\frac{20}{13}x + \frac{9}{13};\ y\Big|_{x=4} = -\frac{71}{13}$

k	x_k	y_k	x_k^2	$x_k y_k$
1	−1	2	1	−2
2	0	1	0	0
3	3	−4	9	−12
Σ	2	−1	10	−14

58. $m = \frac{(0)(5) - 3(6)}{(0)^2 - 3(8)} = \frac{3}{4}$ and

$b = \frac{1}{3}\left[5 - \frac{3}{4}(0)\right] = \frac{5}{3}$

$\Rightarrow y = \frac{3}{4}x + \frac{5}{3};\ y\Big|_{x=4} = \frac{14}{3}$

k	x_k	y_k	x_k^2	$x_k y_k$
1	−2	0	4	0
2	0	2	0	0
3	2	3	4	6
Σ	0	5	8	6

59. $m = \frac{(3)(5) - 3(8)}{(3)^2 - 3(5)} = \frac{3}{2}$ and

$b = \frac{1}{3}\left[5 - \frac{3}{2}(3)\right] = \frac{1}{6}$

$\Rightarrow y = \frac{3}{2}x + \frac{1}{6};\ y\Big|_{x=4} = \frac{37}{6}$

k	x_k	y_k	x_k^2	$x_k y_k$
1	0	0	0	0
2	1	2	1	2
3	2	3	4	6
Σ	3	5	5	8

60. $m = \frac{(5)(5) - 3(10)}{(5)^2 - 3(13)} = \frac{5}{14}$ and

$b = \frac{1}{3}\left[5 - \frac{5}{14}(5)\right] = \frac{15}{14}$

$\Rightarrow y = \frac{5}{14}x + \frac{15}{14};\ y\Big|_{x=4} = \frac{35}{14} = \frac{5}{2}$

k	x_k	y_k	x_k^2	$x_k y_k$
1	0	1	0	0
2	2	2	4	4
3	3	2	9	6
Σ	5	5	13	10

61. $m = \frac{(162)(41.32) - 6(1192.8)}{(162)^2 - 6(5004)} \approx 0.122$ and

$b = \frac{1}{6}[41.32 - (0.122)(162)] \approx 3.58$

$\Rightarrow y = 0.122x + 3.58$

k	x_k	y_k	x_k^2	$x_k y_k$
1	12	5.27	144	63.24
2	18	5.68	324	102.24
3	24	6.25	576	150
4	30	7.21	900	216.3
5	36	8.20	1296	295.2
6	42	8.71	1764	365.82
Σ	162	41.32	5004	1192.8

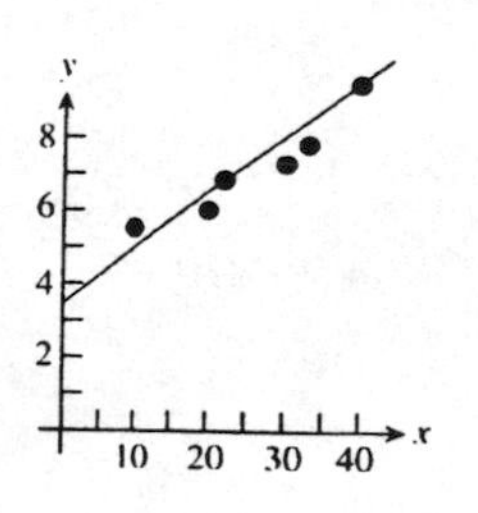

62. $m = \dfrac{(0.001863)(91) - 4(0.065852)}{(0.001863)^2 - 4(0.000001323)} \approx 51{,}545$

and $b = \frac{1}{4}(91 - 51{,}545(0.001863)) \approx -1.26$

$\Rightarrow F = 51{,}545\,\frac{1}{D^2} - 1.26$

k	$\left(\frac{1}{D^2}\right)_k$	F_k	$\left(\frac{1}{D^2}\right)_k^2$	$\left(\frac{1}{D^2}\right)_k F_k$
1	0.001	51	0.000001	0.051
2	0.0005	22	0.00000025	0.011
3	0.00024	14	0.0000000576	0.00336
4	0.000123	4	0.0000000153	0.000492
Σ	0.001863	91	0.000001323	0.065852

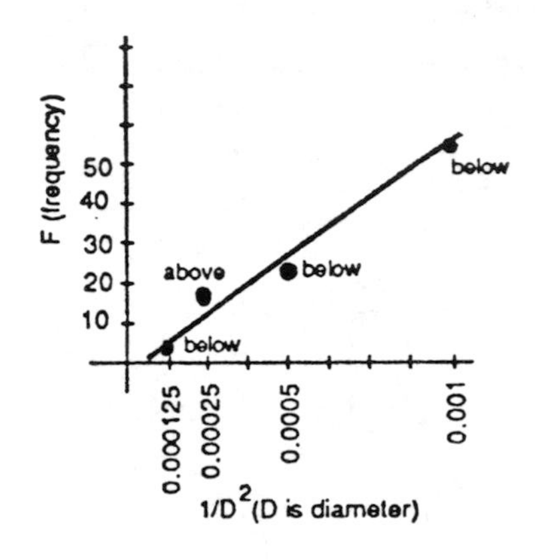

63. (b) $m = \dfrac{(3201)(17{,}785) - 10(5{,}710{,}292)}{(3201)^2 - 10(1{,}430{,}389)}$

≈ 0.0427 and $b = \frac{1}{10}[17{,}785 - (0.0427)(3201)]$

$\approx 1764.8 \Rightarrow y = 0.0427K + 1764.8$

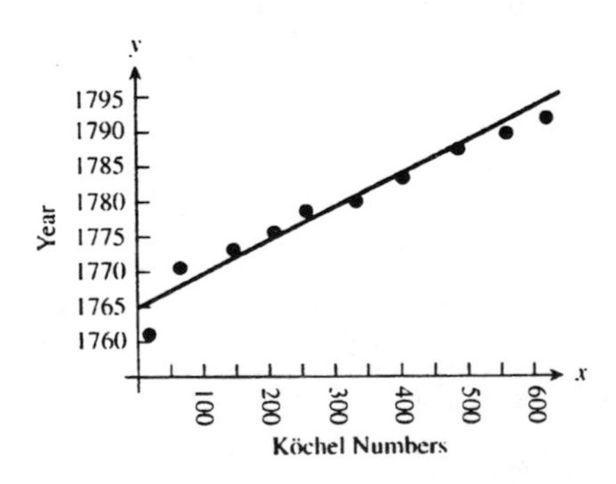

(c) $K = 364 \Rightarrow y = (0.0427)(364)$

$\Rightarrow y = (0.0427)(364) + 1764.8$

≈ 1780

k	K_k	y_k	K_k^2	$K_k y_k$
1	1	1761	1	1761
2	75	1771	5625	132,825
3	155	1772	24,025	274,660
4	219	1775	47,961	388,725
5	271	1777	73,441	481,567
6	351	1780	123,201	624,780
7	425	1783	180,625	757,775
8	503	1786	253,009	898,358
9	575	1789	330,625	1,028,675
10	626	1791	391,876	1,121,166
Σ	3201	17,785	1,430,389	5,710,292

64. $m = \frac{(123)(140) - 16(1431)}{(123)^2 - 16(1267)} \approx 1.04$ and

$b = \frac{1}{16}[140 - (1.04)(123)] \approx 0.763$

$\Rightarrow y = 1.10x + 0.763$

k	x_k	y_k	x_k^2	$x_k y_k$
1	3	3	9	9
2	2	2	4	4
3	4	6	16	24
4	2	3	4	6
5	5	4	25	20
6	5	3	25	15
7	9	11	81	99
8	12	9	144	108
9	8	10	64	80
10	13	16	169	208
11	14	13	196	182
12	3	5	9	15
13	4	6	16	24
14	13	19	169	247
15	10	15	100	150
16	16	15	256	240
Σ	123	140	1287	1431

65-70. Example CAS commands:

Maple:

```
with(plots):
f:= (x,y) -> 2*x^4 + y^4 - 2*x^2 - 2*y^2 + 3;
```

```
plot3d(f(x,y), x=-1..1, y=-1..1, axes=BOXED);
contourplot(f(x,y), x=-3/2..3/2, y=-3/2..3/2, axes=NORMAL);
exp1:= diff(f(x,y),x) = 0;
exp2:= diff(f(x,y),y) = 0;
critical:= evalf(solve({exp1,exp2}, {x,y}));
diff(diff(f(x,y),x),x): fxx:= unapply('',(x,y));
diff(diff(f(x,y),x),y): fxy:= unapply('',(x,y));
diff(diff(f(x,y),y),y): fyy:= unapply('',(x,y));
fxx(x,y)*fyy(x,y) - (fxy(x,y))^2: disc:= unapply('',(x,y));
subs(critical[1],[fxx(x,y),disc(x,y)]);
subs(critical[2],[fxx(x,y),disc(x,y)]);
subs(critical[3],[fxx(x,y),disc(x,y)]);
subs(critical[4],[fxx(x,y),disc(x,y)]);
subs(critical[5],[fxx(x,y),disc(x,y)]);
subs(critical[6],[fxx(x,y),disc(x,y)]);
```

Mathematica:

```
Clear[x,y]
SetOptions[ContourPlot, PlotPoints -> 25,
 Contours -> 20, ContourShading -> False];
f[x_,y_] = 2 x^4 + y^4 - 2 x^2 - 2 y^2 + 3
{xa,xb} = {-3/2,3/2};
{ya,yb} = {-3/2,3/2};
Plot3D[ f[x,y], {x,xa,xb}, {y,ya,yb} ]
ContourPlot[ f[x,y], {x,xa,xb}, {y,ya,yb} ]
fx = D[f[x,y],x]
fy = D[f[x,y],y]
crit = Solve[{fx==0,fy==0}]
critpts = {x,y} /. crit
fxx = D[fx,x]
fxy = D[fx,y]
fyy = D[fy,y]
disc = fxx fyy - fxy^2
{{x,y},disc,fxx} /. crit
```

12.9 LAGRANGE MULTIPLIERS

1. $\nabla f = y\mathbf{i} + x\mathbf{j}$ and $\nabla g = 2x\mathbf{i} + 4y\mathbf{j}$ so that $\nabla f = \lambda \nabla g \Rightarrow y\mathbf{i} + x\mathbf{j} = \lambda(2x\mathbf{i} + 4y\mathbf{j}) \Rightarrow y = 2x\lambda$ and $x = 4y\lambda$ $\Rightarrow x = 8x\lambda^2 \Rightarrow \lambda = \pm\frac{\sqrt{2}}{4}$ or $x = 0$.

   CASE 1: If $x = 0$, then $y = 0$. But $(0,0)$ is not on the ellipse so $x \neq 0$.

   CASE 2: $x \neq 0 \Rightarrow \lambda = \pm\frac{\sqrt{2}}{4} \Rightarrow x = \pm\sqrt{2}y \Rightarrow \left(\pm\sqrt{2}y\right)^2 + 2y^2 = 1 \Rightarrow y = \pm\frac{1}{2}$.

   Therefore f takes on its extreme values at $\left(\pm\sqrt{2}, \frac{1}{2}\right)$ and $\left(\pm\sqrt{2}, -\frac{1}{2}\right)$. The extreme values of f on the ellipse are $\pm\frac{\sqrt{2}}{2}$.

2. $\nabla f = y\mathbf{i} + x\mathbf{j}$ and $\nabla g = 2x\mathbf{i} + 2y\mathbf{j}$ so that $\nabla f = \lambda \nabla g \Rightarrow y\mathbf{i} + x\mathbf{j} = \lambda(2x\mathbf{i} + 2y\mathbf{j}) \Rightarrow y = 2x\lambda$ and $x = 2y\lambda$ $\Rightarrow x = 4x\lambda^2 \Rightarrow x = 0$ or $\lambda = \pm\frac{1}{2}$.

CASE 1: If $x = 0$, then $y = 0$. But $(0,0)$ is not on the circle $x^2 + y^2 - 10 = 0$ so $x \neq 0$.

CASE 2: $x \neq 0 \Rightarrow \lambda = \pm\frac{1}{2} \Rightarrow y = 2x\left(\pm\frac{1}{2}\right) = \pm x \Rightarrow x^2 + (\pm x)^2 - 10 = 0 \Rightarrow x = \pm\sqrt{5} \Rightarrow y = \pm\sqrt{5}$.

Therefore f takes on its extreme values at $(\pm\sqrt{5}, \sqrt{5})$ and $(\pm\sqrt{5}, -\sqrt{5})$. The extreme values of f on the circle are 5 and -5.

3. $\nabla f = -2x\mathbf{i} - 2y\mathbf{j}$ and $\nabla g = \mathbf{i} + 3\mathbf{j}$ so that $\nabla f = \lambda \nabla g \Rightarrow -2x\mathbf{i} - 2y\mathbf{j} = \lambda(\mathbf{i} + 3\mathbf{j}) \Rightarrow x = -\frac{\lambda}{2}$ and $y = -\frac{3\lambda}{2}$ $\Rightarrow \left(-\frac{\lambda}{2}\right) + 3\left(-\frac{3\lambda}{2}\right) = 10 \Rightarrow \lambda = -2 \Rightarrow x = 1$ and $y = 3 \Rightarrow$ f takes on its extreme value at $(1,3)$ on the line. The extreme value is $f(1,3) = 49 - 1 - 9 = 39$.

4. $\nabla f = 2xy\mathbf{i} + x^2\mathbf{j}$ and $\nabla g = \mathbf{i} + \mathbf{j}$ so that $\nabla f = \lambda \nabla g \Rightarrow 2xy\mathbf{i} + x^2\mathbf{j} = \lambda(\mathbf{i} + \mathbf{j}) \Rightarrow 2xy = \lambda$ and $x^2 = \lambda$ $\Rightarrow 2xy = x^2 \Rightarrow x = 0$ or $2y = x$.

   CASE 1: If $x = 0$, then $x + y = 3 \Rightarrow y = 3$.

   CASE 2: If $x \neq 0$, then $2y = x$ so that $x + y = 3 \Rightarrow 2y + y = 3 \Rightarrow y = 1 \Rightarrow x = 2$.

   Therefore f takes on its extreme values at $(0,3)$ and $(2,1)$. The extreme values of f are $f(0,3) = 0$ and $f(2,1) = 4$.

5. We optimize $f(x,y) = x^2 + y^2$, the square of the distance to the origin, subject to the constraint $g(x,y) = xy^2 - 54 = 0$. Thus $\nabla f = 2x\mathbf{i} + 2y\mathbf{j}$ and $\nabla g = y^2\mathbf{i} + 2xy\mathbf{j}$ so that $\nabla f = \lambda \nabla g \Rightarrow 2x\mathbf{i} + 2y\mathbf{j}$ $= \lambda(y^2\mathbf{i} + 2xy\mathbf{j}) \Rightarrow 2x = \lambda y^2$ and $2y = 2\lambda xy$.

   CASE 1: If $y = 0$, then $x = 0$. But $(0,0)$ does not satisfy the constraint $xy^2 = 54$ so $y \neq 0$.

   CASE 2: If $y \neq 0$, then $2 = 2\lambda x \Rightarrow x = \frac{1}{\lambda} \Rightarrow 2\left(\frac{1}{\lambda}\right) = \lambda y^2 \Rightarrow y^2 = \frac{2}{\lambda^2}$. Then $xy^2 = 54 \Rightarrow \left(\frac{1}{\lambda}\right)\left(\frac{2}{\lambda^2}\right) = 54$

   $\Rightarrow \lambda^3 = \frac{1}{27} \Rightarrow \lambda = \pm\frac{1}{3} \Rightarrow x = \pm 3$. Since $xy^2 = 54$ we cannot have $x = -3$, so $x = 3$ and $y^2 = 18$

   $\Rightarrow x = 3$ and $y = \pm 3\sqrt{2}$.

   Therefore $(3, \pm 3\sqrt{2})$ are the points on the curve $xy^2 = 54$ nearest the origin (since $xy^2 = 54$ has points increasingly far away as y gets close to 0, no points are farthest away).

6. We optimize $f(x,y) = x^2 + y^2$, the square of the distance to the origin subject to the constraint $g(x,y)$ $= x^2y - 2 = 0$. Thus $\nabla f = 2x\mathbf{i} + 2y\mathbf{j}$ and $\nabla g = 2xy\mathbf{i} + x^2\mathbf{j}$ so that $\nabla f = \lambda \nabla g \Rightarrow 2x = 2xy\lambda$ and $2y = x^2\lambda$ $\Rightarrow \lambda = \frac{2y}{x^2}$, since $x = 0 \Rightarrow y = 0$ (but $g(0,0) \neq 0$). Thus $x \neq 0$ and $2x = 2xy\left(\frac{2y}{x^2}\right) \Rightarrow x^2 = 2y^2$ $\Rightarrow (2y^2)y - 2 = 0 \Rightarrow y = 1$ (since $y > 0$) $\Rightarrow x = \pm\sqrt{2}$. Therefore $(\pm\sqrt{2}, 1)$ are the points on the curve $x^2y = 2$ nearest the origin (since $x^2y = 2$ has points increasingly far away as x gets close to 0, no points are farthest away).

7. (a) $\nabla f = \mathbf{i} + \mathbf{j}$ and $\nabla g = y\mathbf{i} + x\mathbf{j}$ so that $\nabla f = \lambda \nabla g \Rightarrow \mathbf{i} + \mathbf{j} = \lambda(y\mathbf{i} + x\mathbf{j}) \Rightarrow 1 = \lambda y$ and $1 = \lambda x \Rightarrow y = \frac{1}{\lambda}$ and $x = \frac{1}{\lambda} \Rightarrow \frac{1}{\lambda^2} = 16 \Rightarrow \lambda = \pm\frac{1}{4}$. Use $\lambda = \frac{1}{4}$ since $x > 0$ and $y > 0$. Then $x = 4$ and $y = 4 \Rightarrow$ the minimum value is 8 at the point $(4,4)$. Now, $xy = 16$, $x > 0$, $y > 0$ is a branch of a hyperbola in the first quadrant with the x-and y-axes as asymptotes. The equations $x + y = c$ give a family of parallel lines with $m = -1$.

As these lines move away from the origin, the number c increases. Thus the minimum value of c occurs where $x + y = c$ is tangent to the hyperbola's branch.

(b) $\nabla f = y\mathbf{i} + x\mathbf{j}$ and $\nabla g = \mathbf{i} + \mathbf{j}$ so that $\nabla f = \lambda \nabla g \Rightarrow y\mathbf{i} + x\mathbf{j} = \lambda(\mathbf{i} + \mathbf{j}) \Rightarrow y = \lambda = x \Rightarrow y + y = 16 \Rightarrow y = 8$ $\Rightarrow x = 8 \Rightarrow f(8,8) = 64$ is the maximum value. The equations $xy = c$ ($x > 0$ and $y > 0$ or $x < 0$ and $y < 0$ to get a maximum value) give a family of hyperbolas in the first and third quadrants with the x- and y-axes as asymptotes. The maximum value of c occurs where the hyperbola $xy = c$ is tangent to the line $x + y = 16$.

8. Let $f(x,y) = x^2 + y^2$ be the square of the distance from the origin. Then $\nabla f = 2x\mathbf{i} + 2y\mathbf{j}$ and $\nabla g = (2x + y)\mathbf{i} + (2y + x)\mathbf{j}$ so that $\nabla f = \lambda \nabla g \Rightarrow 2x = \lambda(2x + y)$ and $2y = \lambda(2y + x) \Rightarrow \frac{2y}{2y + x} = \lambda$

$\Rightarrow 2x = \left(\frac{2y}{2y + x}\right)(2x + y) \Rightarrow x(2y + x) = y(2x + y) \Rightarrow x^2 = y^2 \Rightarrow y = \pm x.$

CASE 1: $y = x \Rightarrow x^2 + x(x) + x^2 - 1 = 0 \Rightarrow x = \pm\frac{1}{\sqrt{3}}$ and $y = x$.

CASE 2: $y = -x \Rightarrow x^2 + x(-x) + (-x)^2 - 1 = 0 \Rightarrow x = \pm 1$ and $y = -x$. Thus $f\left(\frac{1}{\sqrt{3}}, \frac{1}{\sqrt{3}}\right) = \frac{2}{3}$

$= f\left(-\frac{1}{\sqrt{3}}, -\frac{1}{\sqrt{3}}\right)$ and $f(1,-1) = 2 = f(-1,1)$.

Therefore the points $(1,-1)$ and $(-1,1)$ are the farthest away; $\left(\frac{1}{\sqrt{3}}, \frac{1}{\sqrt{3}}\right)$ and $\left(-\frac{1}{\sqrt{3}}, -\frac{1}{\sqrt{3}}\right)$ are the closest points to the origin.

9. $V = \pi r^2 h \Rightarrow 16\pi = \pi r^2 h \Rightarrow 16 = r^2 h \Rightarrow g(r,h) = r^2 h - 16$; $S = 2\pi rh + 2\pi r^2 \Rightarrow \nabla S = (2\pi h + 4\pi r)\mathbf{i} + 2\pi r\mathbf{j}$ and $\nabla g = 2rh\mathbf{i} + r^2\mathbf{j}$ so that $\nabla S = \lambda \nabla g \Rightarrow (2\pi rh + 4\pi r)\mathbf{i} + 2\pi r\mathbf{j} = \lambda(2rh\mathbf{i} + r^2\mathbf{j}) \Rightarrow 2\pi rh + 4\pi r = 2rh\lambda$ and $2\pi r = \lambda r^2 \Rightarrow r = 0$ or $\lambda = \frac{2\pi}{r}$. But $r = 0$ gives no physical can, so $r \neq 0 \Rightarrow \lambda = \frac{2\pi}{r} \Rightarrow 2\pi h + 4\pi r$ $= 2rh\left(\frac{2\pi}{r}\right) \Rightarrow 2r = h \Rightarrow 16 = r^2(2r) \Rightarrow r = 2 \Rightarrow h = 4$; thus $r = 2$ cm and $h = 4$ cm give the only extreme surface area of 24π cm^2. Since $r = 4$ cm and $h = 1$ cm $\Rightarrow V = 16\pi$ cm^3 and $S = 40\pi$ cm^2, which is a larger surface area, then 24π cm^2 must be the minimum surface area.

10. For a cylinder of radius r and height h we want to maximize the surface area $S = 2\pi rh$ subject to the constraint $g(r,h) = r^2 + \left(\frac{h}{2}\right)^2 - a^2 = 0$. Thus $\nabla S = 2\pi h\mathbf{i} + 2\pi r\mathbf{j}$ and $\nabla g = 2r\mathbf{i} + \frac{h}{2}\mathbf{j}$ so that $\nabla S = \lambda \nabla g \Rightarrow 2\pi h = 2\lambda r$ and $2\pi r = \frac{\lambda h}{2} \Rightarrow \frac{\pi h}{r} = \lambda$ and $2\pi r = \left(\frac{\pi h}{r}\right)\left(\frac{h}{2}\right) \Rightarrow 4r^2 = h^2 \Rightarrow h = 2r \Rightarrow r^2 + \frac{4r^2}{4} = a^2 \Rightarrow 2r^2 = a^2 \Rightarrow r = \frac{a}{\sqrt{2}}$ $\Rightarrow h = a\sqrt{2} \Rightarrow S = 2\pi\left(\frac{a}{\sqrt{2}}\right)\left(a\sqrt{2}\right) = 2\pi a^2$.

11. $A = (2x)(2y) = 4xy$ subject to $g(x,y) = \frac{x^2}{16} + \frac{y^2}{9} - 1 = 0$; $\nabla A = 4y\mathbf{i} + 4x\mathbf{j}$ and $\nabla g = \frac{x}{8}\mathbf{i} + \frac{2y}{9}\mathbf{j}$ so that ∇A $= \lambda \nabla g \Rightarrow 4y\mathbf{i} + 4x\mathbf{j} = \lambda\left(\frac{x}{8}\mathbf{i} + \frac{2y}{9}\mathbf{j}\right) \Rightarrow 4y = \left(\frac{x}{8}\right)\lambda$ and $4x = \left(\frac{2y}{9}\right)\lambda \Rightarrow \lambda = \frac{32y}{x}$ and $4x = \left(\frac{2y}{9}\right)\left(\frac{32y}{x}\right)$ $\Rightarrow y = \pm\frac{3}{4}x \Rightarrow \frac{x^2}{16} + \frac{\left(\pm\frac{3}{4}x\right)^2}{9} = 1 \Rightarrow x^2 = 8 \Rightarrow x = \pm 2\sqrt{2}$. We use $x = 2\sqrt{2}$ since x represents distance. Then $y = \frac{3}{4}(2\sqrt{2}) = \frac{3\sqrt{2}}{2}$, so the length is $2x = 4\sqrt{2}$ and the width is $2y = 3\sqrt{2}$.

12. $P = 4x + 4y$ subject to $g(x,y) = \frac{x^2}{a^2} + \frac{y^2}{b^2} - 1 = 0$; $\nabla P = 4\mathbf{i} + 4\mathbf{j}$ and $\nabla g = \frac{2x}{a^2}\mathbf{i} + \frac{2y}{b^2}\mathbf{j}$ so that $\nabla P = \lambda \nabla g$

$\Rightarrow 4 = \left(\frac{2x}{a^2}\right)\lambda$ and $4 = \left(\frac{2y}{b^2}\right)\lambda \Rightarrow \lambda = \frac{2a^2}{x}$ and $4 = \left(\frac{2y}{b^2}\right)\left(\frac{2a^2}{x}\right) \Rightarrow y = \left(\frac{b^2}{a^2}\right)x \Rightarrow \frac{x^2}{a^2} + \frac{\left(\frac{b^2}{a^2}\right)^2 x^2}{b^2} = 1 \Rightarrow \frac{x^2}{a^2} + \frac{b^2x^2}{a^4}$

$= 1 \Rightarrow (a^2 + b^2)x^2 = a^4 \Rightarrow x = \frac{a^2}{\sqrt{a^2+b^2}}$, since $x > 0 \Rightarrow y = \left(\frac{b^2}{a^2}\right)x = \frac{b^2}{\sqrt{a^2+b^2}} \Rightarrow$ width $= 2x = \frac{2a^2}{\sqrt{a^2+b^2}}$

and height $= 2y = \frac{2b^2}{\sqrt{a^2+b^2}} \Rightarrow$ perimeter is $P = 4x + 4y = \frac{4a^2 + 4b^2}{\sqrt{a^2+b^2}} = 4\sqrt{a^2+b^2}$

13. $\nabla f = 2x\mathbf{i} + 2y\mathbf{j}$ and $\nabla g = (2x-2)\mathbf{i} + (2y-4)\mathbf{j}$ so that $\nabla f = \lambda \nabla g = 2x\mathbf{i} + 2y\mathbf{j} = \lambda[(2x-2)\mathbf{i} + (2y-4)\mathbf{j}]$
$\Rightarrow 2x = \lambda(2x-2)$ and $2y = \lambda(2y-4) \Rightarrow x = \frac{\lambda}{\lambda - 1}$ and $y = \frac{2\lambda}{\lambda-1}$, $\lambda \neq 1 \Rightarrow y = 2x \Rightarrow x^2 - 2x + (2x)^2 - 4(2x)$
$= 0 \Rightarrow x = 0$ and $y = 0$, or $x = 2$ and $y = 4$. Therefore $f(0,0) = 0$ is the minimum value and $f(2,4) = 20$ is the maximum value. (Note that $\lambda = 1$ gives $2x = 2x - 2$ or $0 = -2$, which is impossible.)

14. $\nabla f = 3\mathbf{i} - \mathbf{j}$ and $\nabla g = 2x\mathbf{i} + 2y\mathbf{j}$ so that $\nabla f = \lambda \nabla g \Rightarrow 3 = 2\lambda x$ and $-1 = 2\lambda y \Rightarrow \lambda = \frac{3}{2x}$ and $-1 = 2\left(\frac{3}{2x}\right)y$
$\Rightarrow y = -\frac{x}{3} \Rightarrow x^2 + \left(-\frac{x}{3}\right)^2 = 4 \Rightarrow 10x^2 = 36 \Rightarrow x = \pm\frac{6}{\sqrt{10}} \Rightarrow x = \frac{6}{\sqrt{10}}$ and $y = -\frac{2}{\sqrt{10}}$, or $x = -\frac{6}{\sqrt{10}}$ and
$y = \frac{2}{\sqrt{10}}$. Therefore $f\left(\frac{6}{\sqrt{10}}, -\frac{2}{\sqrt{10}}\right) = \frac{20}{\sqrt{10}} + 6 = 2\sqrt{10} + 6 \approx 12.325$ is the maximum value, and
$f\left(-\frac{6}{\sqrt{10}}, \frac{2}{\sqrt{10}}\right) = -2\sqrt{10} + 6 \approx -0.325$ is the minimum value.

15. $\nabla T = (8x - 4y)\mathbf{i} + (-4x + 2y)\mathbf{j}$ and $g(x,y) = x^2 + y^2 - 25 = 0 \Rightarrow \nabla g = 2x\mathbf{i} + 2y\mathbf{j}$ so that $\nabla T = \lambda \nabla g$
$\Rightarrow (8x - 4y)\mathbf{i} + (-4x + 2y)\mathbf{j} = \lambda(2x\mathbf{i} + 2y\mathbf{j}) \Rightarrow 8x - 4y = 2\lambda x$ and $-4x + 2y = 2\lambda y \Rightarrow y = \frac{-2x}{\lambda - 1}$, $\lambda \neq 1$
$\Rightarrow 8x - 4\left(\frac{-2x}{\lambda-1}\right) = 2\lambda x \Rightarrow x = 0$, or $\lambda = 0$, or $\lambda = 5$.
CASE 1: $x = 0 \Rightarrow y = 0$; but $(0,0)$ is not on $x^2 + y^2 = 25$ so $x \neq 0$.
CASE 2: $\lambda = 0 \Rightarrow y = 2x \Rightarrow x^2 + (2x)^2 = 25 \Rightarrow x = \pm\sqrt{5}$ and $y = 2x$.
CASE 3: $\lambda = 5 \Rightarrow y = \frac{-2x}{4} = -\frac{x}{2} \Rightarrow x^2 + \left(-\frac{x}{2}\right)^2 = 25 \Rightarrow x = \pm 2\sqrt{5} \Rightarrow x = 2\sqrt{5}$ and $y = -\sqrt{5}$, or $x = -2\sqrt{5}$ and $y = \sqrt{5}$.
Therefore $T(\sqrt{5}, 2\sqrt{5}) = 0° = T(-\sqrt{5}, -2\sqrt{5})$ is the minimum value and $T(2\sqrt{5}, -\sqrt{5}) = 125°$ $= T(-2\sqrt{5}, \sqrt{5})$ is the maximum value. (Note: $\lambda = 1 \Rightarrow x = 0$ from the equation $-4x + 2y = 2\lambda y$; but we found $x \neq 0$ in CASE 1.)

16. The surface area is given by $S = 4\pi r^2 + 2\pi rh$ subject to the constraint $V(r,h) = \frac{4}{3}\pi r^3 + \pi r^2 h = 8000$. Thus $\nabla S = (8\pi r + 2\pi h)\mathbf{i} + 2\pi r\mathbf{j}$ and $\nabla V = (4\pi r^2 + 2\pi rh)\mathbf{i} + \pi r^2\mathbf{j}$ so that $\nabla S = \lambda \nabla V = (8\pi r + 2\pi h)\mathbf{i} + 2\pi r\mathbf{j}$ $= \lambda[(4\pi r^2 + 2\pi rh)\mathbf{i} + \pi r^2\mathbf{j}] \Rightarrow 8\pi r + 2\pi h = \lambda(4\pi r^2 + 2\pi rh)$ and $2\pi r = \lambda\pi r^2 \Rightarrow r = 0$ or $2 = r\lambda$. But $r \neq 0$ so $2 = r\lambda \Rightarrow \lambda = \frac{2}{r} \Rightarrow 4r + h = \frac{2}{r}(2r^2 + rh) \Rightarrow h = 0 \Rightarrow$ the tank is a sphere (there is no cylindrical part) and $\frac{4}{3}\pi r^3 = 8000 \Rightarrow r = 10\left(\frac{6}{\pi}\right)^{1/3}$.

17. Let $f(x,y,z) = (x-1)^2 + (y-1)^2 + (z-1)^2$ be the square of the distance from $(1,1,1)$. Then $\nabla f = 2(x-1)\mathbf{i} + 2(y-1)\mathbf{j} + 2(z-1)\mathbf{k}$ and $\nabla g = \mathbf{i} + 2\mathbf{j} + 3\mathbf{k}$ so that $\nabla f = \lambda \nabla g$ $\Rightarrow 2(x-1)\mathbf{i} + 2(y-1)\mathbf{j} + 2(z-1)\mathbf{k} = \lambda(\mathbf{i} + 2\mathbf{j} + 3\mathbf{k}) \Rightarrow 2(x-1) = \lambda,\ 2(y-1) = 2\lambda,\ 2(z-1) = 3\lambda$ $\Rightarrow 2(y-1) = 2[2(x-1)]$ and $2(z-1) = 3[2(x-1)] \Rightarrow x = \frac{y+1}{2} \Rightarrow z + 2 = 3\left(\frac{y+1}{2}\right)$ or $z = \frac{3y-1}{2}$; thus $\frac{y+1}{2} + 2y + 3\left(\frac{3y-1}{2}\right) - 13 = 0 \Rightarrow y = 2 \Rightarrow x = \frac{3}{2}$ and $z = \frac{5}{2}$. Therefore the point $\left(\frac{3}{2}, 2, \frac{5}{2}\right)$ is closest (since no point on the plane is farthest from the point $(1,1,1)$).

18. Let $f(x,y,z) = (x-1)^2 + (y+1)^2 + (z-1)^2$ be the square of the distance from $(1,-1,1)$. Then $\nabla f = 2(x-1)\mathbf{i} + 2(y+1)\mathbf{j} + 2(z-1)\mathbf{k}$ and $\nabla g = 2x\mathbf{i} + 2y\mathbf{j} + 2z\mathbf{k}$ so that $\nabla f = \lambda \nabla g \Rightarrow x - 1 = \lambda x,\ y + 1 = \lambda y$ and $z - 1 = \lambda z \Rightarrow x = \frac{1}{1-\lambda},\ y = -\frac{1}{1-\lambda}$, and $z = \frac{1}{1-\lambda}$ for $\lambda \neq 1 \Rightarrow \left(\frac{1}{1-\lambda}\right)^2 + \left(\frac{-1}{1-\lambda}\right)^2 + \left(\frac{1}{1-\lambda}\right)^2 = 4$ $\Rightarrow \frac{1}{1-\lambda} = \pm\frac{2}{\sqrt{3}} \Rightarrow x = \frac{2}{\sqrt{3}},\ y = -\frac{2}{\sqrt{3}},\ z = \frac{2}{\sqrt{3}}$ or $x = -\frac{2}{\sqrt{3}},\ y = \frac{2}{\sqrt{3}},\ z = -\frac{2}{\sqrt{3}}$. The largest value of f occurs where $x < 0$, $y > 0$, and $z < 0$ or at the point $\left(-\frac{2}{\sqrt{3}}, \frac{2}{\sqrt{3}}, -\frac{2}{\sqrt{3}}\right)$ on the sphere.

19. Let $f(x,y,z) = x^2 + y^2 + z^2$ be the square of the distance from the origin. Then $\nabla f = 2x\mathbf{i} + 2y\mathbf{j} + 2z\mathbf{k}$ and $\nabla g = 2x\mathbf{i} - 2y\mathbf{j} - 2z\mathbf{k}$ so that $\nabla f = \lambda \nabla g \Rightarrow 2x\mathbf{i} + 2y\mathbf{j} + 2z\mathbf{k} = \lambda(2x\mathbf{i} - 2y\mathbf{j} - 2z\mathbf{k}) \Rightarrow 2x = 2x\lambda,\ 2y = -2y\lambda$, and $2z = -2z\lambda \Rightarrow x = 0$ or $\lambda = 1$.

CASE 1: $\lambda = 1 \Rightarrow 2y = -2y \Rightarrow y = 0;\ 2z = -2z \Rightarrow z = 0 \Rightarrow x^2 - 1 = 0 \Rightarrow x = \pm 1$ and $y = z = 0$.

CASE 2: $x = 0 \Rightarrow -y^2 - z^2 = 1$, which has no solution.

Therefore the points $(\pm 1, 0, 0)$ are closest to the origin $\Rightarrow$ the minimum distance from the surface to the origin is 1 (since there is no maximum distance from the surface to the origin).

20. Let $f(x,y,z) = x^2 + y^2 + z^2$ be the square of the distance to the origin. Then $\nabla f = 2x\mathbf{i} + 2y\mathbf{j} + 2z\mathbf{k}$ and $\nabla g = y\mathbf{i} + x\mathbf{j} - \mathbf{k}$ so that $\nabla f = \lambda \nabla g \Rightarrow 2x\mathbf{i} + 2y\mathbf{j} + 2z\mathbf{k} = \lambda(y\mathbf{i} + x\mathbf{j} - \mathbf{k}) \Rightarrow 2x = \lambda y,\ 2y = \lambda x$, and $2z = -\lambda$ $\Rightarrow x = \frac{\lambda y}{2} \Rightarrow 2y = \lambda\left(\frac{\lambda y}{2}\right) \Rightarrow y = 0$ or $\lambda = \pm 2$.

CASE 1: $y = 0 \Rightarrow x = 0 \Rightarrow -z + 1 = 0 \Rightarrow z = 1$.

CASE 2: $\lambda = 2 \Rightarrow x = y$ and $z = -1 \Rightarrow x^2 - (-1) + 1 = 0 \Rightarrow x^2 + 2 = 0$, so no solution.

CASE 3: $\lambda = -2 \Rightarrow x = -y$ and $z = 1 \Rightarrow (-y)y - 1 + 1 = 0 \Rightarrow y = 0$, again.

Therefore $(0,0,1)$ is the point on the surface closest to the origin since this point gives the only extreme value and there is no maximum distance from the surface to the origin.

21. Let $f(x,y,z) = x^2 + y^2 + z^2$ be the square of the distance to the origin. Then $\nabla f = 2x\mathbf{i} + 2y\mathbf{j} + 2z\mathbf{k}$ and $\nabla g = -y\mathbf{i} - x\mathbf{j} + 2z\mathbf{k}$ so that $\nabla f = \lambda \nabla g \Rightarrow 2x\mathbf{i} + 2y\mathbf{j} + 2z\mathbf{k} = \lambda(-y\mathbf{i} - x\mathbf{j} + 2z\mathbf{k}) \Rightarrow 2x = -y\lambda,\ 2y = -x\lambda$, and $2z = 2z\lambda \Rightarrow \lambda = 1$ or $z = 0$.

CASE 1: $\lambda = 1 \Rightarrow 2x = -y$ and $2y = -x \Rightarrow y = 0$ and $x = 0 \Rightarrow z^2 - 4 = 0 \Rightarrow z = \pm 2$ and $x = y = 0$.

CASE 2: $z = 0 \Rightarrow -xy - 4 = 0 \Rightarrow y = -\frac{4}{x}$. Then $2x = \frac{4}{x}\lambda \Rightarrow \lambda = \frac{x^2}{2}$, and $-\frac{8}{x} = -x\lambda \Rightarrow -\frac{8}{x} = -x\left(\frac{x^2}{2}\right)$ $\Rightarrow x^4 = 16 \Rightarrow x = \pm 2$. Thus, $x = 2$ and $y = -2$, or $x = -2$ and $y = 2$.

Therefore we get four points: $(2,-2,0)$, $(-2,2,0)$, $(0,0,2)$ and $(0,0,-2)$. But the points $(0,0,2)$ and $(0,0,-2)$ are closest to the origin since they are 2 units away and the others are $2\sqrt{2}$ units away.

22. Let $f(x,y,z) = x^2+y^2+z^2$ be the square of the distance to the origin. Then $\nabla f = 2x\mathbf{i}+2y\mathbf{j}+2z\mathbf{k}$ and $\nabla g = yz\mathbf{i}+xz\mathbf{j}+xy\mathbf{k}$ so that $\nabla f = \lambda \nabla g \Rightarrow 2x = \lambda yz$, $2y = \lambda yz$, and $2z = \lambda xy \Rightarrow 2x^2 = \lambda xyz$ and $2y^2 = \lambda yxz$ $\Rightarrow x^2 = y^2 \Rightarrow y = \pm x \Rightarrow z = \pm x \Rightarrow x(\pm x)(\pm x) = 1 \Rightarrow x = \pm 1 \Rightarrow$ the points are $(1,1,1)$, $(1,-1,-1)$, $(-1,-1,1)$, and $(-1,1,-1)$.

23. $\nabla f = \mathbf{i}-2\mathbf{j}+5\mathbf{k}$ and $\nabla g = 2x\mathbf{i}+2y\mathbf{j}+2z\mathbf{k}$ so that $\nabla f = \lambda \nabla g \Rightarrow \mathbf{i}-2\mathbf{j}+5\mathbf{k} = \lambda(2x\mathbf{i}+2y\mathbf{j}+2z\mathbf{k}) \Rightarrow 1 = 2x\lambda$, $-2 = 2y\lambda$, and $5 = 2z\lambda \Rightarrow x = \frac{1}{2\lambda}$, $y = -\frac{1}{\lambda} = -2x$, and $z = \frac{5}{2\lambda} = 5x \Rightarrow x^2+(-2x)^2+(5x)^2 = 30 \Rightarrow x = \pm 1$. Thus, $x = 1$, $y = -2$, $z = 5$ or $x = -1$, $y = 2$, $z = -5$. Therefore $f(1,-2,5) = 30$ is the maximum value and $f(-1,2,-5) = -30$ is the minimum value.

24. $\nabla f = \mathbf{i}+2\mathbf{j}+3\mathbf{k}$ and $\nabla g = 2x\mathbf{i}+2y\mathbf{j}+2z\mathbf{k}$ so that $\nabla f = \lambda \nabla g \Rightarrow \mathbf{i}+2\mathbf{j}+3\mathbf{k} = \lambda(2x\mathbf{i}+2y\mathbf{j}+2z\mathbf{k}) \Rightarrow 1 = 2x\lambda$, $2 = 2y\lambda$, and $3 = 2z\lambda \Rightarrow x = \frac{1}{2\lambda}$, $y = \frac{1}{\lambda} = 2x$, and $z = \frac{3}{2\lambda} = 3x \Rightarrow x^2+(2x)^2+(3x)^2 = 25 \Rightarrow x = \pm\frac{5}{\sqrt{14}}$. Thus, $x = \frac{5}{\sqrt{14}}$, $y = \frac{10}{\sqrt{14}}$, $z = \frac{15}{\sqrt{14}}$ or $x = -\frac{5}{\sqrt{14}}$, $y = -\frac{10}{\sqrt{14}}$, $z = -\frac{15}{\sqrt{14}}$. Therefore $f\left(\frac{5}{\sqrt{14}},\frac{10}{\sqrt{14}},\frac{15}{\sqrt{14}}\right)$ $= 5\sqrt{14}$ is the maximum value and $f\left(-\frac{5}{\sqrt{14}},-\frac{10}{\sqrt{14}},-\frac{15}{\sqrt{14}}\right) = -5\sqrt{14}$ is the minimum value.

25. $f(x,y,z) = x^2+y^2+z^2$ and $g(x,y,z) = x+y+z-9 = 0 \Rightarrow \nabla f = 2x\mathbf{i}+2y\mathbf{j}+2z\mathbf{k}$ and $\nabla g = \mathbf{i}+\mathbf{j}+\mathbf{k}$ so that $\nabla f = \lambda \nabla g \Rightarrow 2x\mathbf{i}+2y\mathbf{j}+2z\mathbf{k} = \lambda(\mathbf{i}+\mathbf{j}+\mathbf{k}) \Rightarrow 2x = \lambda$, $2y = \lambda$, and $2z = \lambda \Rightarrow x = y = z \Rightarrow x+x+x-9 = 0$ $\Rightarrow x = 3$, $y = 3$, and $z = 3$.

26. $f(x,y,z) = xyz$ and $g(x,y,z) = x+y+z^2-16 = 0 \Rightarrow \nabla f = yz\mathbf{i}+xz\mathbf{j}+xy\mathbf{k}$ and $\nabla g = \mathbf{i}+\mathbf{j}+2z\mathbf{k}$ so that $\nabla f = \lambda \nabla g \Rightarrow yz\mathbf{i}+xz\mathbf{j}+xy\mathbf{k} = \lambda(\mathbf{i}+\mathbf{j}+2z\mathbf{k}) \Rightarrow yz = \lambda$, $xz = \lambda$, and $xy = 2z\lambda \Rightarrow yz = xz \Rightarrow z = 0$ or $y = x$. But $z > 0$ so that $y = x \Rightarrow x^2 = 2z\lambda$ and $xz = \lambda$. Then $x^2 = 2z(xz) \Rightarrow x = 0$ or $x = 2z^2$. But $x > 0$ so that $x = 2z^2 \Rightarrow y = 2z^2 \Rightarrow 2z^2+2z^2+z^2 = 16 \Rightarrow z = \pm\frac{4}{\sqrt{5}}$. We use $z = \frac{4}{\sqrt{5}}$ since $z > 0$. Then $x = \frac{32}{5}$ and $y = \frac{32}{5}$ which yields $f\left(\frac{32}{5},\frac{32}{5},\frac{4}{\sqrt{5}}\right) = \frac{4096}{25\sqrt{5}}$.

27. $V = 6xyz$ and $g(x,y,z) = x^2+y^2+z^2-1 = 0 \Rightarrow \nabla V = 6yz\mathbf{i}+6xz\mathbf{j}+6xy\mathbf{k}$ and $\nabla g = 2x\mathbf{i}+2y\mathbf{j}+2z\mathbf{k}$ so that $\nabla f = \lambda \nabla g \Rightarrow 3yz = \lambda x$, $3xz = \lambda y$, and $3xy = \lambda z \Rightarrow 3xyz = \lambda x^2$ and $3xyz = \lambda y^2 \Rightarrow y = \pm x \Rightarrow z = \pm x$ $\Rightarrow x^2+x^2+x^2 = 1 \Rightarrow x = \frac{1}{\sqrt{3}}$ since $x > 0 \Rightarrow$ the dimensions of the box are $\frac{2}{\sqrt{3}}$ by $\frac{2}{\sqrt{3}}$ by $\frac{2}{\sqrt{3}}$ for maximum volume. (Note that there is no minimum volume since the box could be made arbitrarily thin.)

28. $V = xyz$ with x,y,z all positive and $\frac{x}{a}+\frac{y}{b}+\frac{z}{c} = 1$; thus $V = xyz$ and $g(x,y,z) = bcx+acy+abz-abc = 0$ $\Rightarrow \nabla V = yz\mathbf{i}+xz\mathbf{j}+xy\mathbf{k}$ and $\nabla g = bc\mathbf{i}+ac\mathbf{j}+ab\mathbf{k}$ so that $\nabla V = \lambda \nabla g \Rightarrow yz = \lambda bc$, $xz = \lambda ac$, and $xy = \lambda ab$ $\Rightarrow xyz = \lambda bcx$, $xyz = \lambda acy$, and $xyz = \lambda abz \Rightarrow \lambda \neq 0$. Also, $\lambda bcx = \lambda acy = \lambda abz \Rightarrow bx = ay$, $cy = bz$, and $cx = ax \Rightarrow y = \frac{b}{a}x$ and $z = \frac{c}{a}x$. Then $\frac{x}{a}+\frac{y}{b}+\frac{c}{z} = 1 \Rightarrow \frac{x}{a}+\frac{1}{b}\left(\frac{b}{a}x\right)+\frac{1}{c}\left(\frac{c}{a}x\right) = 1 \Rightarrow \frac{3x}{a} = 1 \Rightarrow x = \frac{a}{3}$ $\Rightarrow y = \left(\frac{b}{a}\right)\left(\frac{a}{3}\right) = \frac{b}{3}$ and $z = \left(\frac{c}{a}\right)\left(\frac{a}{3}\right) = \frac{c}{3} \Rightarrow V = xyz = \left(\frac{a}{3}\right)\left(\frac{b}{3}\right)\left(\frac{c}{3}\right) = \frac{abc}{27}$ is the maximum volume. (Note that there is no minimum volume since the box could be made arbitrarily thin.)

29. $\nabla T = 16x\mathbf{i} + 4z\mathbf{j} + (4y - 16)\mathbf{k}$ and $\nabla g = 8x\mathbf{i} + 2y\mathbf{j} + 8z\mathbf{k}$ so that $\nabla T = \lambda \nabla g \Rightarrow 16x\mathbf{i} + 4z\mathbf{j} + (4y - 16)\mathbf{k}$ $= \lambda(8x\mathbf{i} + 2y\mathbf{j} + 8z\mathbf{k}) \Rightarrow 16x = 8x\lambda$, $4z = 2y\lambda$, and $4y - 16 = 8z\lambda \Rightarrow \lambda = 2$ or $x = 0$.

CASE 1: $\lambda = 2 \Rightarrow 4z = 2y(2) \Rightarrow z = y$. Then $4z - 16 = 16z \Rightarrow z = -\frac{4}{3} \Rightarrow y = -\frac{4}{3}$. Then $4x^2 + \left(-\frac{4}{3}\right)^2 + 4\left(-\frac{4}{3}\right)^2 = 16 \Rightarrow x = \pm\frac{4}{3}$.

CASE 2: $x = 0 \Rightarrow \lambda = \frac{2z}{y} \Rightarrow 4y - 16 = 8z\left(\frac{2z}{y}\right) \Rightarrow y^2 - 4y = 4z^2 \Rightarrow 4(0)^2 + y^2 + \left(y^2 - 4y\right) - 16 = 0$ $\Rightarrow y^2 - 2y - 8 = 0 \Rightarrow (y - 4)(y + 2) = 0 \Rightarrow y = 4$ or $y = -2$. Now $y = 4 \Rightarrow 4z^2 = 4^2 - 4(4)$ $\Rightarrow z = 0$ and $y = -2 \Rightarrow 4z^2 = (-2)^2 - 4(-2) \Rightarrow z = \pm\sqrt{3}$.

The temperatures are $T\left(\pm\frac{4}{3}, -\frac{4}{3}, -\frac{4}{3}\right) = 642\frac{2}{3}^{\circ}$, $T(0,4,0) = 600^{\circ}$, $T\left(0,-2,\sqrt{3}\right) = \left(600 - 24\sqrt{3}\right)^{\circ}$, and $T\left(0,-2,-\sqrt{3}\right) = \left(600 + 24\sqrt{3}\right)^{\circ} \approx 641.6^{\circ}$. Therefore $\left(\pm\frac{4}{3}, -\frac{4}{3}, -\frac{4}{3}\right)$ are the hottest points on the space probe.

30. $\nabla T = 400yz^2\mathbf{i} + 400xz^2\mathbf{j} + 800xyz\mathbf{k}$ and $\nabla g = 2x\mathbf{i} + 2y\mathbf{j} + 2z\mathbf{k}$ so that $\nabla T = \lambda \nabla g$ $\Rightarrow 400yz^2\mathbf{i} + 400xz^2\mathbf{j} + 800xyz\mathbf{k} = \lambda(2x\mathbf{i} + 2y\mathbf{j} + 2z\mathbf{k}) \Rightarrow 400yz^2 = 2x\lambda$, $400xz^2 = 2y\lambda$, and $800xyz = 2z\lambda$. Solving this system yields the points $(0, \pm 1, 0)$, $(\pm 1, 0, 0)$, and $\left(\pm\frac{1}{2}, \pm\frac{1}{2}, \pm\frac{\sqrt{2}}{2}\right)$. The corresponding temperatures are $T(0, \pm 1, 0) = 0$, $T(\pm 1, 0, 0) = 0$, and $T\left(\pm\frac{1}{2}, \pm\frac{1}{2}, \pm\frac{\sqrt{2}}{2}\right) = \pm 50$. Therefore 50 is the maximum temperature at $\left(\frac{1}{2}, \frac{1}{2}, \pm\frac{\sqrt{2}}{2}\right)$ and $\left(-\frac{1}{2}, -\frac{1}{2}, \pm\frac{\sqrt{2}}{2}\right)$; -50 is the minimum temperature at $\left(\frac{1}{2}, -\frac{1}{2}, \pm\frac{\sqrt{2}}{2}\right)$ and $\left(-\frac{1}{2}, \frac{1}{2}, \pm\frac{\sqrt{2}}{2}\right)$.

31. $\nabla U = (y + 2)\mathbf{i} + x\mathbf{j}$ and $\nabla g = 2\mathbf{i} + \mathbf{j}$ so that $\nabla U = \lambda \nabla g \Rightarrow (y + 2)\mathbf{i} + x\mathbf{j} = \lambda(2\mathbf{i} + \mathbf{j}) \Rightarrow y + 2 = 2\lambda$ and $x = \lambda \Rightarrow y + 2 = 2x \Rightarrow y = 2x - 2 \Rightarrow 2x + (2x - 2) = 30 \Rightarrow x = 8$ and $y = 14$. Therefore $U(8, 14) = \$128$ is the maximum value of U under the constraint.

32. $\nabla M = (6 + z)\mathbf{i} - 2y\mathbf{j} + x\mathbf{k}$ and $\nabla g = 2x\mathbf{i} + 2y\mathbf{j} + 2z\mathbf{k}$ so that $\nabla M = \lambda \nabla g \Rightarrow (6 + z)\mathbf{i} - 2y\mathbf{j} + x\mathbf{k}$ $= \lambda(2x\mathbf{i} + 2y\mathbf{j} + 2z\mathbf{k}) \Rightarrow 6 + z = 2x\lambda$, $-2y = 2y\lambda$, $x = 2z\lambda \Rightarrow \lambda = -1$ or $y = 0$.

CASE 1: $\lambda = -1 \Rightarrow 6 + z = -2x$ and $x = -2z \Rightarrow 6 + z = -2(-2z) \Rightarrow z = 2$ and $x = -4$. Then $(-4)^2 + y^2 + 2^2 - 36 = 0 \Rightarrow y = \pm 4$.

CASE 2: $y = 0$, $6 + z = 2x\lambda$, and $x = 2z\lambda \Rightarrow \lambda = \frac{x}{2z} \Rightarrow 6 + z = 2x\left(\frac{x}{2z}\right) \Rightarrow 6z + z^2 = x^2$ $\Rightarrow \left(6z + z^2\right) + 0^2 + z^2 = 36 \Rightarrow z = -6$ or $z = 3$. Now $z = -6 \Rightarrow x^2 = 0 \Rightarrow x = 0$; $z = 3$ $\Rightarrow x^2 + 27 \Rightarrow x = \pm 3\sqrt{3}$.

Therefore we have the points $\left(\pm 3\sqrt{3}, 0, 3\right)$, $(0, 0, -6)$, and $(-4, 2, \pm 4)$. Then $M\left(3\sqrt{3}, 0, 3\right)$ $= 27\sqrt{3} + 60 \approx 106.8$, $M\left(-3\sqrt{3}, 0, 3\right) = 60 - 27\sqrt{3} \approx 13.2$, $M(0, 0, -6) = 60$, and $M(-4, 4, 2) = 12$ $= M(-4, -4, 2)$. Therefore, the weakest field is at $(-4, \pm 4, 2)$.

33. Let $g_1(x, y, z) = 2x - y = 0$ and $g_2(x, y, z) = y + z = 0 \Rightarrow \nabla g_1 = 2\mathbf{i} - \mathbf{j}$, $\nabla g_2 = \mathbf{j} + \mathbf{k}$, and $\nabla f = 2x\mathbf{i} + 2\mathbf{j} - 2z\mathbf{k}$ so that $\nabla f = \lambda \nabla g_1 + \mu \nabla g_2 \Rightarrow 2x\mathbf{i} + 2\mathbf{j} - 2z\mathbf{k} = \lambda(2\mathbf{i} - \mathbf{j}) + \mu(\mathbf{j} + \mathbf{k}) \Rightarrow 2x\mathbf{i} + 2\mathbf{j} - 2z\mathbf{k} = 2\lambda\mathbf{i} + (\mu - \lambda)\mathbf{j} + \mu\mathbf{k}$ $\Rightarrow 2x = 2\lambda$, $2 = \mu - \lambda$, and $-2z = \mu \Rightarrow x = \lambda$. Then $2 = -2z - x \Rightarrow x = -2z - 2$ so that $2x - y = 0$ $\Rightarrow 2(-2z - 2) - y = 0 \Rightarrow -4z - 4 - y = 0$. This equation coupled with $y + z = 0$ implies $z = -\frac{4}{3}$ and $y = \frac{4}{3}$.

Then $x = \frac{2}{3}$ so that $\left(\frac{2}{3}, \frac{4}{3}, -\frac{4}{3}\right)$ is the point that gives the maximum value $f\left(\frac{2}{3}, \frac{4}{3}, -\frac{4}{3}\right) = \left(\frac{2}{3}\right)^2 + 2\left(\frac{4}{3}\right) - \left(-\frac{4}{3}\right)^2$ $= \frac{4}{3}$.

34. Let $g_1(x,y,z) = x + 2y + 3z - 6 = 0$ and $g_2(x,y,z) = x + 3y + 9z - 9 = 0 \Rightarrow \nabla g_1 = \mathbf{i} + 2\mathbf{j} + 3\mathbf{k}$, $\nabla g_2 = \mathbf{i} + 3\mathbf{j} + 9\mathbf{k}$, and $\nabla f = 2x\mathbf{i} + 2y\mathbf{j} + 2z\mathbf{k}$ so that $\nabla f = \lambda \nabla g_1 + \mu \nabla g_2 \Rightarrow 2x\mathbf{i} + 2y\mathbf{j} + 2z\mathbf{k}$ $= \lambda(\mathbf{i} + 2\mathbf{j} + 3\mathbf{k}) + \mu(\mathbf{i} + 3\mathbf{j} + 9\mathbf{k}) \Rightarrow 2x = \lambda + \mu$, $2y = 2\lambda + 3\mu$, and $2z = 3\lambda + 9\mu$. Then $0 = x + 2y + 3z - 6$ $= \frac{1}{2}(\lambda + \mu) + (2\lambda + 3\mu) + \left(\frac{9}{2}\lambda + \frac{27}{2}\mu\right) - 6 \Rightarrow 7\lambda + 17\mu = 6$; $0 = x + 3y + 9z - 9$ $\Rightarrow \frac{1}{2}(\lambda + \mu) + \left(3\lambda + \frac{9}{2}\mu\right) + \left(\frac{27}{2}\lambda + \frac{81}{2}\mu\right) - 9 \Rightarrow 34\lambda + 91\mu = 18$. Solving these two equations for λ and μ gives $\lambda = \frac{240}{59}$ and $\mu = -\frac{78}{59} \Rightarrow x = \frac{\lambda + \mu}{2} = \frac{81}{59}$, $y = \frac{2\lambda + 3\mu}{2} = \frac{123}{59}$, and $z = \frac{3\lambda + 9\mu}{2} = \frac{9}{59}$. The minimum value is $f\left(\frac{81}{59}, \frac{123}{59}, \frac{9}{59}\right) = \frac{21{,}771}{59^2} = \frac{369}{59}$. (Note that there is no maximum value of f subject to the constraints because at least one of the variables x, y, or z can be made arbitrary and assume a value as large as we please.)

35. Let $f(x,y,z) = x^2 + y^2 + z^2$ be the square of the distance from the origin. We want to minimize $f(x,y,z)$ subject to the constraints $g_1(x,y,z) = y + 2z - 12 = 0$ and $g_2(x,y,z) = x + y - 6 = 0$. Thus $\nabla f = 2x\mathbf{i} + 2y\mathbf{j} + 2z\mathbf{k}$, $\nabla g_1 = \mathbf{j} + 2\mathbf{k}$, and $\nabla g_2 = \mathbf{i} + \mathbf{j}$ so that $\nabla f = \lambda \nabla g_1 + \mu \nabla g_2 \Rightarrow 2x = \mu$, $2y = \lambda + \mu$, and $2z = 2\lambda$. Then $0 = y + 2z - 12 = \left(\frac{\lambda}{2} + \frac{\mu}{2}\right) + 2\lambda - 12 \Rightarrow \frac{5}{2}\lambda + \frac{1}{2}\mu = 12 \Rightarrow 5\lambda + \mu = 24$; $0 = x + y - 6 = \frac{\mu}{2} + \left(\frac{\lambda}{2} + \frac{\mu}{2}\right) - 6$ $\Rightarrow \frac{1}{2}\lambda + \mu = 6 \Rightarrow \lambda + 2\mu = 12$. Solving these two equations for λ and μ gives $\lambda = 4$ and $\mu = 4 \Rightarrow x = \frac{\mu}{2} = 2$, $y = \frac{\lambda + \mu}{2} = 4$, and $z = \lambda = 4$. The point $(2,4,4)$ on the line of intersection is closest to the origin. (There is no maximum distance from the origin since points on the line can be arbitrarily far away.)

36. The maximum value is $f\left(\frac{2}{3}, \frac{4}{3}, -\frac{4}{3}\right) = \frac{4}{3}$ from Exercise 33 above.

37. Let $g_1(x,y,z) = z - 1 = 0$ and $g_2(x,y,z) = x^2 + y^2 + z^2 - 10 = 0 \Rightarrow \nabla g_1 = \mathbf{k}$, $\nabla g_2 = 2x\mathbf{i} + 2y\mathbf{j} + 2z\mathbf{k}$, and $\nabla f = 2xyz\mathbf{i} + x^2z\mathbf{j} + x^2y\mathbf{k}$ so that $\nabla f = \lambda \nabla g_1 + \mu \nabla g_2 \Rightarrow 2xyz\mathbf{i} + x^2z\mathbf{j} + x^2y\mathbf{k} = \lambda(\mathbf{k}) + \mu(2x\mathbf{i} + 2y\mathbf{j} + 2z\mathbf{k})$ $\Rightarrow 2xyz = 2x\mu$, $x^2z = 2y\mu$, and $x^2y = 2z\mu + \lambda \Rightarrow xyz = x\mu \Rightarrow x = 0$ or $yz = \mu \Rightarrow \mu = y$ since $z = 1$.

CASE 1: $x = 0$ and $z = 1 \Rightarrow y^2 - 9 = 0$ (from g_2) $\Rightarrow y = \pm 3$ yielding the points $(0, \pm 3, 1)$.

CASE 2: $\mu = y \Rightarrow x^2z = 2y^2 \Rightarrow x^2 = 2y^2$ (since $z = 1$) $\Rightarrow 2y^2 + y^2 + 1 - 10 = 0$ (from g_2) $\Rightarrow 3y^2 - 9 = 0$ $\Rightarrow y = \pm\sqrt{3} \Rightarrow x^2 = 2\left(\pm\sqrt{3}\right)^2 \Rightarrow x = \pm\sqrt{6}$ yielding the points $\left(\pm\sqrt{6}, \pm\sqrt{3}, 1\right)$.

Now $f(0, \pm 3, 1) = 1$ and $f\left(\pm\sqrt{6}, \pm\sqrt{3}, 1\right) = 6\left(\pm\sqrt{3}\right) + 1 = 1 \pm 6\sqrt{3}$. Therefore the maximum of f is $1 + 6\sqrt{3}$ at $\left(\pm\sqrt{6}, \sqrt{3}, 1\right)$, and the minimum of f is $1 - 6\sqrt{3}$ at $\left(\pm\sqrt{6}, -\sqrt{3}, 1\right)$.

38. (a) Let $g_1(x,y,z) = x + y + z - 40 = 0$ and $g_2(x,y,z) = x + y - z = 0 \Rightarrow \nabla g_1 = \mathbf{i} + \mathbf{j} + \mathbf{k}$, $\nabla g_2 = \mathbf{i} + \mathbf{j} - \mathbf{k}$, and $\nabla w = yz\mathbf{i} + xz\mathbf{j} + xy\mathbf{k}$ so that $\nabla w = \lambda \nabla g_1 + \mu \nabla g_2 \Rightarrow yz\mathbf{i} + xz\mathbf{j} + xy\mathbf{k} = \lambda(\mathbf{i} + \mathbf{j} + \mathbf{k}) + \mu(\mathbf{i} + \mathbf{j} - \mathbf{k})$ $\Rightarrow yz = \lambda + \mu$, $xz = \lambda + \mu$, and $xy = \lambda - \mu \Rightarrow yz = xz \Rightarrow z = 0$ or $y = x$.

CASE 1: $z = 0 \Rightarrow x + y = 40$ and $x + y = 0 \Rightarrow$ no solution.

CASE 2: $x = y \Rightarrow 2x + z - 40 = 0$ and $2x - z = 0 \Rightarrow z = 20 \Rightarrow x = 10$ and $y = 10 \Rightarrow w = (10)(10)(20)$ $= 2000$

(b) $\mathbf{n} = \begin{vmatrix} \mathbf{i} & \mathbf{j} & \mathbf{k} \\ 1 & 1 & 1 \\ 1 & 1 & -1 \end{vmatrix} = -2\mathbf{i} + 2\mathbf{j}$ is parallel to the line of intersection $\Rightarrow$ the line is $x = -2t + 10$, $y = 2t + 10$, $z = 20$. Since $z = 20$, we see that $w = xyz = (-2t + 10)(2t + 10)(20) = (-4t^2 + 100)(20)$ which has its maximum when $t = 0 \Rightarrow x = 10$, $y = 10$, and $z = 20$.

39. Let $g_1(x,y,z) = y - x = 0$ and $g_2(x,y,z) = x^2 + y^2 + z^2 - 4 = 0$. Then $\nabla f = y\mathbf{i} + x\mathbf{j} + 2z\mathbf{k}$, $\nabla g_1 = -\mathbf{i} + \mathbf{j}$, and $\nabla g_2 = 2x\mathbf{i} + 2y\mathbf{j} + 2z\mathbf{k}$ so that $\nabla f = \lambda \nabla g_1 + \mu \nabla g_2 \Rightarrow y\mathbf{i} + x\mathbf{j} + 2z\mathbf{k} = \lambda(-\mathbf{i} + \mathbf{j}) + \mu(2x\mathbf{i} + 2y\mathbf{j} + 2z\mathbf{k})$ $\Rightarrow y = -\lambda + 2x\mu$, $x = \lambda + 2y\mu$, and $2z = 2z\mu \Rightarrow z = 0$ or $\mu = 1$.

CASE 1: $z = 0 \Rightarrow x^2 + y^2 - 4 = 0 \Rightarrow 2x^2 - 4 = 0$ (since $x = y$) $\Rightarrow x = \pm\sqrt{2}$ and $y = \pm\sqrt{2}$ yielding the points $(\pm\sqrt{2}, \pm\sqrt{2}, 0)$.

CASE 2: $\mu = 1 \Rightarrow y = -\lambda + 2x$ and $x = \lambda + 2y \Rightarrow x + y = 2(x + y) \Rightarrow 2x = 2(2x)$ since $x = y \Rightarrow x = 0 \Rightarrow y = 0$ $\Rightarrow z^2 - 4 = 0 \Rightarrow z = \pm 2$ yielding the points $(0, 0, \pm 2)$.

Now, $f(0, 0, \pm 2) = 4$ and $f(\pm\sqrt{2}, \pm\sqrt{2}, 0) = 2$. Therefore the maximum value of f is 4 at $(0, 0, \pm 2)$ and the minimum value of f is 2 at $(\pm\sqrt{2}, \pm\sqrt{2}, 0)$.

40. Let $f(x,y,z) = x^2 + y^2 + z^2$ be the square of the distance from the origin. We want to minimize $f(x,y,z)$ subject to the constraints $g_1(x,y,z) = 2y + 4z - 5 = 0$ and $g_2(x,y,z) = 4x^2 + 4y^2 - z^2 = 0$. Thus $\nabla f = 2x\mathbf{i} + 2y\mathbf{j} + 2z\mathbf{k}$, $\nabla g_1 = 2\mathbf{j} + 4\mathbf{k}$, and $\nabla g_2 = 8x\mathbf{i} + 8y\mathbf{j} - 2z\mathbf{k}$ so that $\nabla f = \lambda \nabla g_1 + \mu \nabla g_2 \Rightarrow 2x\mathbf{i} + 2y\mathbf{j} + 2z\mathbf{k}$ $= \lambda(2\mathbf{j} + 4\mathbf{k}) + \mu(8x\mathbf{i} + 8y\mathbf{j} - 2z\mathbf{k}) \Rightarrow 2x = 8x\mu$, $2y = 2\lambda + 8y\mu$, and $2z = 4\lambda - 2z\mu \Rightarrow x = 0$ or $\mu = \frac{1}{4}$.

CASE 1: $x = 0 \Rightarrow 4(0)^2 + 4y^2 - z^2 = 0 \Rightarrow z = \pm 2y \Rightarrow 2y + 4(2y) - 5 = 0 \Rightarrow y = \frac{1}{2}$, or $2y + 4(-2y) - 5 = 0$ $\Rightarrow y = -\frac{5}{6}$ yielding the points $\left(0, \frac{1}{2}, 1\right)$ and $\left(0, -\frac{5}{6}, \frac{5}{3}\right)$.

CASE 2: $\mu = \frac{1}{4} \Rightarrow y = \lambda + y \Rightarrow \lambda = 0 \Rightarrow 2z = 4(0) - 2z\left(\frac{1}{4}\right) \Rightarrow z = 0 \Rightarrow 2y + 4(0) = 5 \Rightarrow y = \frac{5}{2}$ and $(0)^2 = 4x^2 + 4\left(\frac{5}{2}\right)^2 \Rightarrow$ no solution.

Then $f\left(0, \frac{1}{2}, 1\right) = \frac{5}{4}$ and $f\left(0, -\frac{5}{6}, \frac{5}{3}\right) = 25\left(\frac{1}{36} + \frac{1}{9}\right) = \frac{125}{36} \Rightarrow$ the point $\left(0, \frac{1}{2}, 1\right)$ is closest to the origin.

41. $\nabla f = \mathbf{i} + \mathbf{j}$ and $\nabla g = y\mathbf{i} + x\mathbf{j}$ so that $\nabla f = \lambda \nabla g \Rightarrow \mathbf{i} + \mathbf{j} = \lambda(y\mathbf{i} + x\mathbf{j}) \Rightarrow 1 = y\lambda$ and $1 = x\lambda \Rightarrow y = x$ $\Rightarrow y^2 = 16 \Rightarrow y = \pm 4 \Rightarrow (4,4)$ and $(-4,-4)$ are candidates for the location of extreme values. But as $x \to \infty$, $y \to \infty$ and $f(x,y) \to \infty$; as $x \to -\infty$, $y \to 0$ and $f(x,y) \to -\infty$. Therefore no maximum or minimum value exists subject to the constraint.

42. Let $f(A,B,C) = \sum_{k=1}^{4} (Ax_k + By_k + C - z_k)^2 = C^2 + (B + C - 1)^2 + (A + B + C - 1)^2 + (A + C - 1)^2$. We want to minimize f. Then $f_A(A,B,C) = 4A + 2B + 4C$, $f_B(A,B,C) = 2A + 4B + 4C - 4$, and $f_C(A,B,C) = 4A + 4B + 8C - 2$. Set each partial derivative equal to 0 and solve the system to get $A = -\frac{1}{2}$, $B = \frac{3}{2}$, and $C = -\frac{1}{4}$ or the critical point of f is $\left(-\frac{1}{2}, \frac{3}{2}, -\frac{1}{4}\right)$.

43. (a) Maximize $f(a,b,c) = a^2b^2c^2$ subject to $a^2 + b^2 + c^2 = r^2$. Thus $\nabla f = 2ab^2c^2\mathbf{i} + 2a^2bc^2\mathbf{j} + 2a^2b^2c\mathbf{k}$ and $\nabla g = 2a\mathbf{i} + 2b\mathbf{j} + 2c\mathbf{k}$ so that $\nabla f = \lambda \nabla g \Rightarrow 2ab^2c^2 = 2a\lambda$, $2a^2bc^2 = 2b\lambda$, and $2a^2b^2c = 2c\lambda$

$\Rightarrow 2a^2b^2c^2 = 2a^2\lambda = 2b^2\lambda = 2c^2\lambda \Rightarrow \lambda = 0$ or $a^2 = b^2 = c^2$.

CASE 1: $\lambda = 0 \Rightarrow a^2b^2c^2 = 0$.

CASE 2: $a^2 = b^2 = c^2 \Rightarrow f(a,b,c) = a^2a^2a^2$ and $3a^2 = r^2 \Rightarrow f(a,b,c) = \left(\frac{r^2}{3}\right)^3$ is the maximum value.

(b) The point $\left(\sqrt{a}, \sqrt{b}, \sqrt{c}\right)$ is on the sphere if $a + b + c = r^2$. Moreover, by part (a), $abc = f\left(\sqrt{a}, \sqrt{b}, \sqrt{c}\right) \le \left(\frac{r^2}{3}\right)^3 \Rightarrow (abc)^{1/3} \le \frac{r^2}{3} = \frac{a+b+c}{3}$, as claimed.

44. Let $f(x_1, x_2, \ldots, x_n) = \sum_{i=1}^{n} a_i x_i = a_1x_1 + a_2x_2 + \ldots + a_nx_n$ and $g(x_1, x_2, \ldots, x_n) = x_1^2 + x_2^2 + \ldots + x_n^2 - 1$. Then we want $\nabla f = \lambda \nabla g \Rightarrow a_1 = \lambda(2x_1),\ a_2 = \lambda(2x_2),\ \ldots,\ a_n = \lambda(2x_n),\ \lambda \ne 0 \Rightarrow x_i = \frac{a_i}{2\lambda} \Rightarrow \frac{a_1^2}{4\lambda^2} + \frac{a_2^2}{4\lambda^2} + \ldots + \frac{a_n^2}{4\lambda^2} = 1$ $\Rightarrow 4\lambda^2 = \sum_{i=1}^{n} a_i^2 \Rightarrow 2\lambda = \left(\sum_{i=1}^{n} a_i^2\right)^{1/2} \Rightarrow f(x_1, x_2, \ldots, x_n) = \sum_{i=1}^{n} a_i x_i = \sum_{i=1}^{n} a_i \left(\frac{a_i}{2\lambda}\right) = \frac{1}{2\lambda} \sum_{i=1}^{n} a_i^2 = \left(\sum_{i=1}^{n} a_i^2\right)^{1/2}$ is the maximum value.

45-50. Example CAS commands:

Maple:

```
f:= (x,y,z) -> x*y + y*z;
g1:= (x,y,z) -> x^2 + y^2 -2;
g2:= (x,y,z) -> x^2 + z^2 -2;
lambda1:= `lambda1`: lambda2:= `lambda2`:
h:= (x,y,z) -> f(x,y,z) - lambda1*g1(x,y,z) -lambda2*g2(x,y,z);
expn1:= diff(h(x,y,z),x) = 0;
expn2:= diff(h(x,y,z),y) = 0;
expn3:= diff(h(x,y,z),z) = 0;
expn4:= diff(h(x,y,z), lambda1) = 0;
expn5:= diff(h(x,y,z), lambda2) = 0;
s:= evalf(solve({expn1,expn2,expn3,expn4,expn5}, {x,y,z,lambda1,lambda2}));
subs({x=-1.306562965,y=.541196100,z=.5411961001},f(x,y,z));
subs({x=1.306562965,y=-.541196100,z=-.5411961001},f(x,y,z));
subs({x=.541196100,y=-1.306562965,z=1.306562965},f(x,y,z));
subs({x=-.541196100,y=1.306562965,z=-1.306562965},f(x,y,z));
```

Mathematica:

```
Clear[w,x,y,z,l1,l2]
f[x_,y_,z_] = x y + y z
g1[x_,y_,z_] = x^2 + y^2 -2
g2[x_,y_,z_] = x^2 + z^2 -2
h = f[x,y,z] - l1 g1[x,y,z] - l2 g2[x,y,z]
hx = D[h,x]
hy = D[h,y]
hz = D[h,z]
hl1 = D[h,l1]
hl2 = D[h,l2]
crit = NSolve[{hx==0,hy==0,hz==0,hl1==0,hl2==0}]
{{x,y,z},f[x,y,z]} /. crit
```

12.10 TAYLOR'S FORMULA

1. $f(x,y) = xe^y \Rightarrow f_x = e^y,\ f_y = xe^y,\ f_{xx} = 0,\ f_{xy} = e^y,\ f_{yy} = xe^y$

   $\Rightarrow f(x,y) \approx f(0,0) + xf_x(0,0) + yf_y(0,0) + \frac{1}{2}\left[x^2f_{xx}(0,0) + 2xyf_{xy}(0,0) + y^2f_{yy}(0,0)\right]$

   $= 0 + x\cdot 1 + y\cdot 0 + \frac{1}{2}\left(x^2\cdot 0 + 2xy\cdot 1 + y^2\cdot 0\right) = x + xy$ quadratic approximation;

   $f_{xxx} = 0,\ f_{xxy} = 0,\ f_{xyy} = e^y,\ f_{yyy} = xe^y$

   $\Rightarrow f(x,y) \approx \text{quadratic} + \frac{1}{6}\left[x^3f_{xxx}(0,0) + 3x^2yf_{xxy}(0,0) + 3xy^2f_{xyy}(0,0) + y^3f_{yyy}(0,0)\right]$

   $= x + xy + \frac{1}{6}\left(x^3\cdot 0 + 3x^2y\cdot 0 + 3xy^2\cdot 1 + y^3\cdot 0\right) = x + xy + \frac{1}{2}xy^2$, cubic approximation

2. $f(x,y) = e^x \cos y \Rightarrow f_x = e^x \cos y,\ f_y = -e^x \sin y,\ f_{xx} = e^x \cos y,\ f_{xy} = -e^x \sin y,\ f_{yy} = -e^x \cos y$

   $\Rightarrow f(x,y) \approx f(0,0) + xf_x(0,0) + yf_y(0,0) + \frac{1}{2}\left[x^2f_{xx}(0,0) + 2xyf_{xy}(0,0) + y^2f_{yy}(0,0)\right]$

   $= 1 + x\cdot 1 + y\cdot 0 + \frac{1}{2}[x^2\cdot 1 + 2xy\cdot 0 + y^2\cdot(-1)] = 1 + x + \frac{1}{2}(x^2 - y^2)$, quadratic approximation;

   $f_{xxx} = e^x \cos y,\ f_{xxy} = -e^x \sin y,\ f_{xyy} = -e^x \cos y,\ f_{yyy} = e^x \sin y$

   $\Rightarrow f(x,y) \approx \text{quadratic} + \frac{1}{6}\left[x^3f_{xxx}(0,0) + 3x^2yf_{xxy}(0,0) + 3xy^2f_{xyy}(0,0) + y^3f_{yyy}(0,0)\right]$

   $= 1 + x + \frac{1}{2}(x^2 - y^2) + \frac{1}{6}[x^3\cdot 1 + 3x^2y\cdot 0 + 3xy^2\cdot(-1) + y^3\cdot 0]$

   $= 1 + x + \frac{1}{2}(x^2 - y^2) + \frac{1}{6}(x^3 - 3xy^2)$, cubic approximation

3. $f(x,y) = y \sin x \Rightarrow f_x = y \cos x,\ f_y = \sin x,\ f_{xx} = -y \sin x,\ f_{xy} = \cos x,\ f_{yy} = 0$

   $\Rightarrow f(x,y) \approx f(0,0) + xf_x(0,0) + yf_y(0,0) + \frac{1}{2}\left[x^2f_{xx}(0,0) + 2xyf_{xy}(0,0) + y^2f_{yy}(0,0)\right]$

   $= 0 + x\cdot 0 + y\cdot 0 + \frac{1}{2}\left(x^2\cdot 0 + 2xy\cdot 1 + y^2\cdot 0\right) = xy$, quadratic approximation;

   $f_{xxx} = -y \cos x,\ f_{xxy} = -\sin x,\ f_{xyy} = 0,\ f_{yyy} = 0$

   $\Rightarrow f(x,y) \approx \text{quadratic} + \frac{1}{6}\left[x^3f_{xxx}(0,0) + 3x^2yf_{xxy}(0,0) + 3xy^2f_{xyy}(0,0) + y^3f_{yyy}(0,0)\right]$

   $= xy + \frac{1}{6}\left(x^3\cdot 0 + 3x^2y\cdot 0 + 3xy^2\cdot 0 + y^3\cdot 0\right) = xy$, cubic approximation

4. $f(x,y) = \sin x \cos y \Rightarrow f_x = \cos x \cos y,\ f_y = -\sin x \sin y,\ f_{xx} = -\sin x \cos y,\ f_{xy} = -\cos x \sin y,$

   $f_{yy} = -\sin x \cos y \Rightarrow f(x,y) \approx f(0,0) + xf_x(0,0) + yf_y(0,0) + \frac{1}{2}\left[x^2f_{xx}(0,0) + 2xyf_{xy}(0,0) + y^2f_{yy}(0,0)\right]$

   $= 0 + x\cdot 1 + y\cdot 0 + \frac{1}{2}\left(x^2\cdot 0 + 2xy\cdot 0 + y^2\cdot 0\right) = x$, quadratic approximation;

   $f_{xxx} = -\cos x \cos y,\ f_{xxy} = \sin x \sin y,\ f_{xyy} = -\cos x \cos y,\ f_{yyy} = \sin x \sin y$

   $\Rightarrow f(x,y) \approx \text{quadratic} + \frac{1}{6}\left[x^3f_{xxx}(0,0) + 3x^2yf_{xxy}(0,0) + 3xy^2f_{xyy}(0,0) + y^3f_{yyy}(0,0)\right]$

   $= x + \frac{1}{6}[x^3\cdot(-1) + 3x^2y\cdot 0 + 3xy^2\cdot(-1) + y^3\cdot 0] = x - \frac{1}{6}(x^3 + 3xy^2)$, cubic approximation

5. $f(x,y) = e^x \ln(1+y) \Rightarrow f_x = e^x \ln(1+y),\ f_y = \frac{e^x}{1+y},\ f_{xx} = e^x \ln(1+y),\ f_{xy} = \frac{e^x}{1+y},\ f_{yy} = -\frac{e^x}{(1+y)^2}$

$\Rightarrow f(x,y) \approx f(0,0) + xf_x(0,0) + yf_y(0,0) + \frac{1}{2}\left[x^2 f_{xx}(0,0) + 2xyf_{xy}(0,0) + y^2 f_{yy}(0,0)\right]$

$= 0 + x \cdot 0 + y \cdot 1 + \frac{1}{2}[x^2 \cdot 0 + 2xy \cdot 1 + y^2 \cdot (-1)] = y + \frac{1}{2}(2xy - y^2)$, quadratic approximation;

$f_{xxx} = e^x \ln(1+y),\ f_{xxy} = \frac{e^x}{1+y},\ f_{xyy} = -\frac{e^x}{(1+y)^2},\ f_{yyy} = \frac{2e^x}{(1+y)^3}$

$\Rightarrow f(x,y) \approx \text{quadratic} + \frac{1}{6}\left[x^3 f_{xxx}(0,0) + 3x^2 y f_{xxy}(0,0) + 3xy^2 f_{xyy}(0,0) + y^3 f_{yyy}(0,0)\right]$

$= y + \frac{1}{2}(2xy - y^2) + \frac{1}{6}[x^3 \cdot 0 + 3x^2 y \cdot 1 + 3xy^2 \cdot (-1) + y^3 \cdot 2]$

$= y + \frac{1}{2}(2xy - y^2) + \frac{1}{6}(3x^2 y - 3xy^2 + 2y^3)$, cubic approximation

6. $f(x,y) = \ln(2x+y+1) \Rightarrow f_x = \frac{2}{2x+y+1},\ f_y = \frac{1}{2x+y+1},\ f_{xx} = \frac{-4}{(2x+y+1)^2},\ f_{xy} = \frac{-2}{(2x+y+1)^2},$

$f_{yy} = \frac{-1}{(2x+y+1)^2} \Rightarrow f(x,y) \approx f(0,0) + xf_x(0,0) + yf_y(0,0) + \frac{1}{2}\left[x^2 f_{xx}(0,0) + 2xyf_{xy}(0,0) + y^2 f_{yy}(0,0)\right]$

$= 0 + x \cdot 2 + y \cdot 1 + \frac{1}{2}[x^2 \cdot (-4) + 2xy \cdot (-2) + y^2 \cdot (-1)] = 2x + y + \frac{1}{2}(-4x^2 - 4xy - y^2)$

$= (2x+y) - \frac{1}{2}\,(2x+y)^2$, quadratic approximation;

$f_{xxx} = \frac{16}{(2x+y+1)^3},\ f_{xxy} = \frac{8}{(2x+y+1)^3},\ f_{xyy} = \frac{4}{(2x+y+1)^3},\ f_{yyy} = \frac{2}{(2x+y+1)^3}$

$\Rightarrow f(x,y) \approx \text{quadratic} + \frac{1}{6}\left[x^3 f_{xxx}(0,0) + 3x^2 y f_{xxy}(0,0) + 3xy^2 f_{xyy}(0,0) + y^3 f_{yyy}(0,0)\right]$

$= (2x+y) - \frac{1}{2}(2x+y)^2 + \frac{1}{6}(x^3 \cdot 16 + 3x^2 y \cdot 8 + 3xy^2 \cdot 4 + y^3 \cdot 2)$

$= (2x+y) - \frac{1}{2}(2x+y)^2 + \frac{1}{3}(8x^3 + 12x^2 y + 6xy^2 + y^2)$

$= (2x+y) - \frac{1}{2}(2x+y)^2 + \frac{1}{3}(2x+y)^3$, cubic approximation

7. $f(x,y) = \sin(x^2+y^2) \Rightarrow f_x = 2x\cos(x^2+y^2),\ f_y = 2y\cos(x^2+y^2),\ f_{xx} = 2\cos(x^2+y^2) - 4x^2\sin(x^2+y^2),$
$f_{xy} = -4xy\sin(x^2+y^2),\ f_{yy} = 2\cos(x^2+y^2) - 4y^2\sin(x^2+y^2)$

$\Rightarrow f(x,y) \approx f(0,0) + xf_x(0,0) + yf_y(0,0) + \frac{1}{2}\left[x^2 f_{xx}(0,0) + 2xyf_{xy}(0,0) + y^2 f_{yy}(0,0)\right]$

$= 0 + x \cdot 0 + y \cdot 0 + \frac{1}{2}(x^2 \cdot 2 + 2xy \cdot 0 + y^2 \cdot 2) = x^2 + y^2$, quadratic approximation;

$f_{xxx} = -12x\sin(x^2+y^2) - 8x^3\cos(x^2+y^2),\ f_{xxy} = -4y\sin(x^2+y^2) - 8x^2 y\cos(x^2+y^2),$
$f_{xyy} = -4x\sin(x^2+y^2) - 8xy^2\cos(x^2+y^2),\ f_{yyy} = -12y\sin(x^2+y^2) - 8y^3\cos(x^2+y^2)$

$\Rightarrow f(x,y) \approx \text{quadratic} + \frac{1}{6}\left[x^3 f_{xxx}(0,0) + 3x^2 y f_{xxy}(0,0) + 3xy^2 f_{xyy}(0,0) + y^3 f_{yyy}(0,0)\right]$

$= x^2 + y^2 + \frac{1}{6}(x^3 \cdot 0 + 3x^2 y \cdot 0 + 3xy^2 \cdot 0 + y^3 \cdot 0) = x^2 + y^2$, cubic approximation

8. $f(x,y) = \cos(x^2+y^2) \Rightarrow f_x = -2x \sin(x^2+y^2)$, $f_y = -2y \sin(x^2+y^2)$,
$f_{xx} = -2\sin(x^2+y^2) - 4x^2\cos(x^2+y^2)$, $f_{xy} = -4xy\cos(x^2+y^2)$, $f_{yy} = -2\sin(x^2+y^2) - 4y^2\cos(x^2+y^2)$

$\Rightarrow f(x,y) \approx f(0,0) + xf_x(0,0) + yf_y(0,0) + \frac{1}{2}\left[x^2 f_{xx}(0,0) + 2xyf_{xy}(0,0) + y^2 f_{yy}(0,0)\right]$

$= 1 + x\cdot 0 + y \cdot 0 + \frac{1}{2}[x^2\cdot 0 + 2xy\cdot 0 + y^2 \cdot 0] = 1$, quadratic approximation;

$f_{xxx} = -12x\cos(x^2+y^2) + 8x^3\sin(x^2+y^2)$, $f_{xxy} = -4y\cos(x^2+y^2) + 8x^2y\sin(x^2+y^2)$,
$f_{xyy} = -4x\cos(x^2+y^2) + 8xy^2\sin(x^2+y^2)$, $f_{yyy} = -12y\cos(x^2+y^2) + 8y^3\sin(x^2+y^2)$

$\Rightarrow f(x,y) \approx \text{quadratic} + \frac{1}{6}\left[x^3 f_{xxx}(0,0) + 3x^2 y f_{xxy}(0,0) + 3xy^2 f_{xyy}(0,0) + y^3 f_{yyy}(0,0)\right]$

$= 1 + \frac{1}{6}(x^3\cdot 0 + 3x^2y\cdot 0 + 3xy^2\cdot 0 + y^3\cdot 0) = 1$, cubic approximation

9. $f(x,y) = \frac{1}{1-x-y} \Rightarrow f_x = \frac{1}{(1-x-y)^2} = f_y$, $f_{xx} = \frac{2}{(1-x-y)^3} = f_{xy} = f_{yy}$

$\Rightarrow f(x,y) \approx f(0,0) + xf_x(0,0) + yf_y(0,0) + \frac{1}{2}\left[x^2 f_{xx}(0,0) + 2xyf_{xy}(0,0) + y^2 f_{yy}(0,0)\right]$

$= 1 + x\cdot 1 + y\cdot 1 + \frac{1}{2}(x^2\cdot 2 + 2xy\cdot 2 + y^2\cdot 2) = 1 + (x+y) + (x^2+2xy+y^2)$

$= 1 + (x+y) + (x+y)^2$, quadratic approximation; $f_{xxx} = \frac{6}{(1-x-y)^4} = f_{xxy} = f_{xyy} = f_{yyy}$

$\Rightarrow f(x,y) \approx \text{quadratic} + \frac{1}{6}\left[x^3 f_{xxx}(0,0) + 3x^2 y f_{xxy}(0,0) + 3xy^2 f_{xyy}(0,0) + y^3 f_{yyy}(0,0)\right]$

$= 1 + (x+y) + (x+y)^2 + \frac{1}{6}(x^3\cdot 6 + 3x^2y\cdot 6 + 3xy^2\cdot 6 + y^3\cdot 6)$

$= 1 + (x+y) + (x+y)^2 + (x^3 + 3x^2y + 3xy^2 + y^3) = 1 + (x+y) + (x+y)^2 + (x+y)^3$, cubic approximation

10. $f(x,y) = \frac{1}{1-x-y+xy} \Rightarrow f_x = \frac{1-y}{(1-x-y+xy)^2}$, $f_y = \frac{1-x}{(1-x-y+xy)^2}$, $f_{xx} = \frac{2(1-y)^2}{(1-x-y+xy)^3}$,

$f_{xy} = \frac{1-3x-3y+3xy}{(1-x-y+xy)^3}$, $f_{yy} = \frac{2(1-x)^2}{(1-x-y+xy)^3}$

$\Rightarrow f(x,y) \approx f(0,0) + xf_x(0,0) + yf_y(0,0) + \frac{1}{2}\left[x^2 f_{xx}(0,0) + 2xyf_{xy}(0,0) + y^2 f_{yy}(0,0)\right]$

$= 1 + x\cdot 1 + y\cdot 1 + \frac{1}{2}(x^2\cdot 2 + 2xy\cdot 1 + y^2\cdot 2) = 1 + x + y + x^2 + xy + y^2$, quadratic approximation;

$f_{xxx} = \frac{6(1-y)^3}{(1-x-y+xy)^4}$, $f_{xxy} = \frac{[-4(1-x-y+xy) + 6(1-y)(1-x)](1-y)}{(1-x-y+xy)^4}$,

$f_{xyy} = \frac{[-4(1-x-y+xy) + 6(1-x)(1-y)](1-x)}{(1-x-y+xy)^4}$, $f_{yyy} = \frac{6(1-x)^3}{(1-x-y+xy)^4}$

$\Rightarrow f(x,y) \approx \text{quadratic} + \frac{1}{6}\left[x^3 f_{xxx}(0,0) + 3x^2 y f_{xxy}(0,0) + 3xy^2 f_{xyy}(0,0) + y^3 f_{yyy}(0,0)\right]$

$= 1 + x + y + x^2 + xy + y^2 + \frac{1}{6}(x^3\cdot 6 + 3x^2y\cdot 2 + 3xy^2\cdot 2 + y^3\cdot 6)$

$= 1 + x + y + x^2 + xy + y^2 + x^3 + x^2y + xy^2 + y^3$, cubic approximation

11. $f(x,y) = \cos x \cos y \Rightarrow f_x = -\sin x \cos y$, $f_y = -\cos x \sin y$, $f_{xx} = -\cos x \cos y$, $f_{xy} = \sin x \sin y$,

$f_{yy} = -\cos x \cos y \Rightarrow f(x,y) \approx f(0,0) + xf_x(0,0) + yf_y(0,0) + \frac{1}{2}\left[x^2 f_{xx}(0,0) + 2xyf_{xy}(0,0) + y^2 f_{yy}(0,0)\right]$

$= 1 + x \cdot 0 + y \cdot 0 + \frac{1}{2}[x^2 \cdot (-1) + 2xy \cdot 0 + y^2 \cdot (-1)] = 1 - \frac{x^2}{2} - \frac{y^2}{2}$, quadratic approximation. Since all partial derivatives of f are products of sines and cosines, the absolute value of these derivatives is less than or equal to $1 \Rightarrow E(x,y) \le \frac{1}{6}[(0.1)^3 + 3(0.1)^3 + 3(0.1)^3 + 0.1)^3] \le 0.00134$.

12. $f(x,y) = e^x \sin y \Rightarrow f_x = e^x \sin y$, $f_y = e^x \cos y$, $f_{xx} = e^x \sin y$, $f_{xy} = e^x \cos y$, $f_{yy} = -e^x \sin y$

$\Rightarrow f(x,y) \approx f(0,0) + xf_x(0,0) + yf_y(0,0) + \frac{1}{2}\left[x^2 f_{xx}(0,0) + 2xyf_{xy}(0,0) + y^2 f_{yy}(0,0)\right]$

$= 0 + x \cdot 0 + y \cdot 1 + \frac{1}{2}(x^2 \cdot 0 + 2xy \cdot 1 + y^2 \cdot 0) = y + xy$, quadratic approximation. Now, $f_{xxx} = e^x \sin y$,

$f_{xxy} = e^x \cos y$, $f_{xyy} = -e^x \sin y$, and $f_{yyy} = -e^x \cos y$. Since $|x| \le 0.1$, $|e^x \sin y| \le |e^{0.1} \sin 0.1| \approx 0.11$ and $|e^x \cos y| \le |e^{0.1} \cos 0.1| \approx 1.11$. Therefore,

$E(x,y) \le \frac{1}{6}[(0.11)(0.1)^3 + 3(1.11)(0.1)^3 + 3(0.11)(0.1)^3 + (1.11)(0.1)^3] \le 0.000814$.

CHAPTER 12 PRACTICE EXERCISES

1. Domain: All points in the xy-plane

Range: $z \ge 0$

Level curves are ellipses with major axis along the y-axis and minor axis along the x-axis.

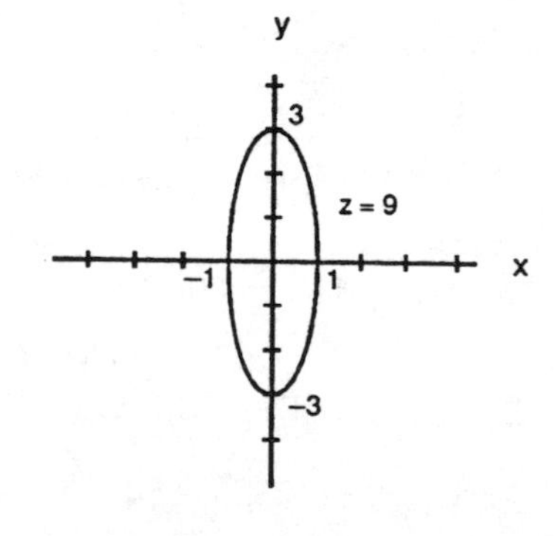

2. Domain: All points in the xy-plane

Range: $0 < z < \infty$

Level curves are the straight lines $x + y = \ln z$ with slope -1, and $z > 0$.

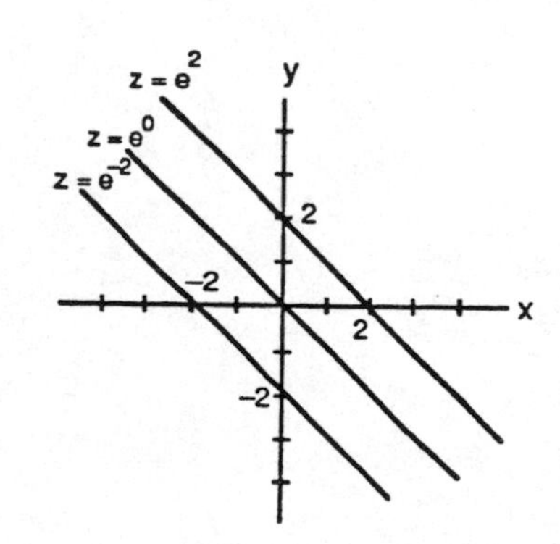

3. Domain: All (x,y) such that $x \neq 0$ and $y \neq 0$

   Range: $z \neq 0$

   Level curves are hyperbolas with the x- and y-axes as asymptotes.

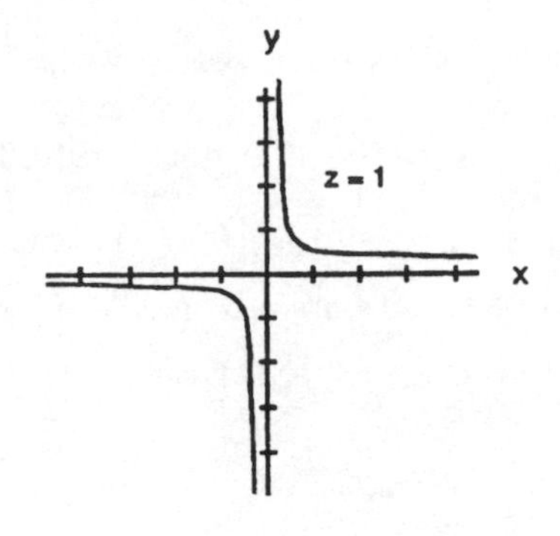

4. Domain: All (x,y) so that $x^2 - y \geq 0$

   Range: $z \geq 0$

   Level curves are the parabolas $y = x^2 - c$, $c \geq 0$.

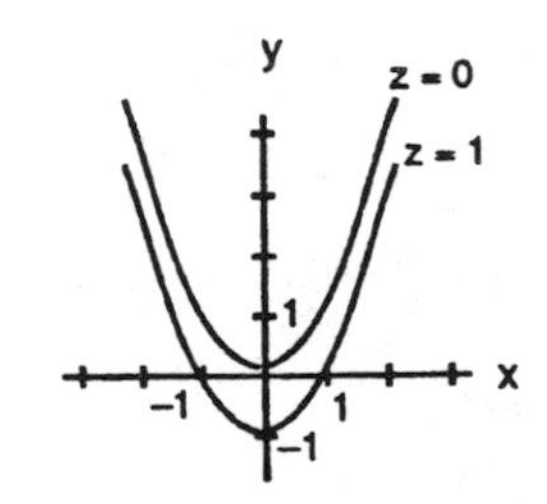

5. Domain: All (x,y,z) such that $(x,y,z) \neq (0,0,0)$

   Range: All real numbers

   Level surfaces are paraboloids of revolution with the z-axis as axis.

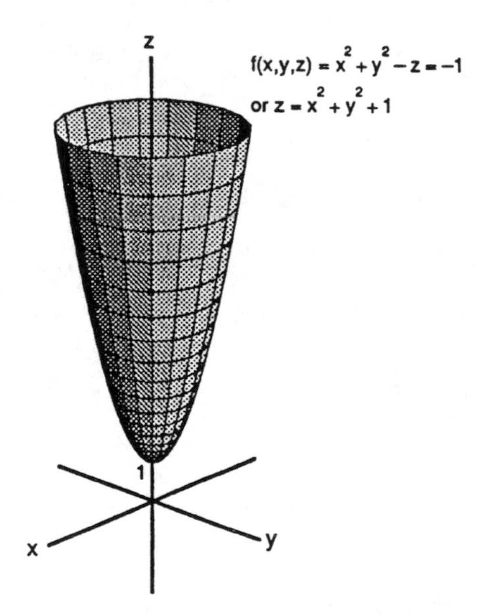

6. Domain: All points (x, y, z) in space

   Range: Nonnegative real numbers

   Level surfaces are ellipsoids with center $(0, 0, 0)$.

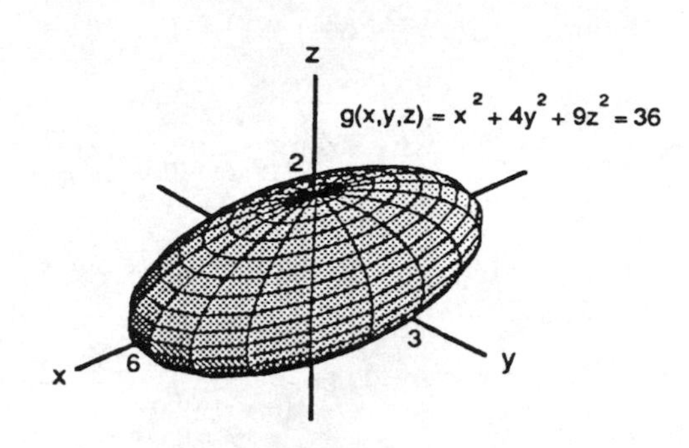

7. Domain: All (x, y, z) such that $(x, y, z) \neq (0, 0, 0)$

   Range: Positive real numbers

   Level surfaces are spheres with center $(0, 0, 0)$ and radius $r > 0$.

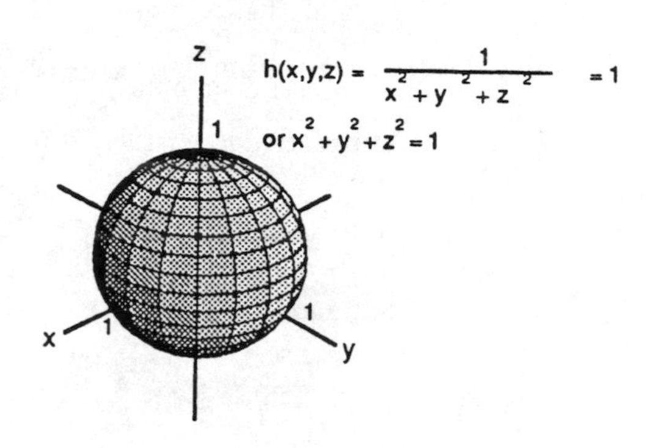

8. Domain: All points (x, y, z) in space

   Range: $(0, 1]$

   Level surfaces are spheres with center $(0, 0, 0)$ and radius $r > 0$.

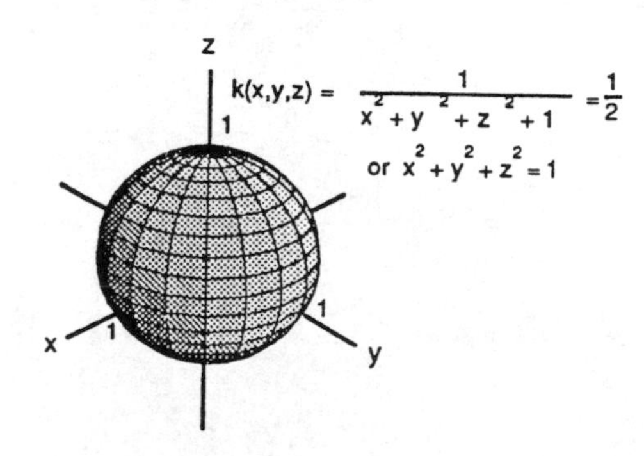

9. $\lim\limits_{(x,y)\to(\pi,\ln 2)} e^y \cos x = e^{\ln 2} \cos \pi = (2)(-1) = -2$

10. $\lim\limits_{(x,y)\to(0,0)} \dfrac{2+y}{x+\cos y} = \dfrac{2+0}{0+\cos 0} = 2$

11. $\lim\limits_{\substack{(x,y)\to(1,1)\\ x\neq \pm y}} \dfrac{x-y}{x^2-y^2} = \lim\limits_{\substack{(x,y)\to(1,1)\\ x\neq \pm y}} \dfrac{x-y}{(x-y)(x+y)} = \lim\limits_{(x,y)\to(1,1)} \dfrac{1}{x+y} = \dfrac{1}{1+1} = \dfrac{1}{2}$

12. $\lim\limits_{(x,y)\to(1,1)} \dfrac{x^3y^3-1}{xy-1} = \lim\limits_{(x,y)\to(1,1)} \dfrac{(xy-1)(x^2y^2+xy+1)}{xy-1} = \lim\limits_{(x,y)\to(1,1)} (x^2y^2+xy+1) = 1^2\cdot 1^2 + 1\cdot 1 + 1 = 3$

13. $\lim\limits_{P\to(1,-1,e)} \ln|x+y+z| = \ln|1+(-1)+e| = \ln e = 1$

14. $\lim\limits_{P\to(1,-1,-1)} \tan^{-1}(x+y+z) = \tan^{-1}(1+(-1)+(-1)) = \tan^{-1}(-1) = -\frac{\pi}{4}$

15. Let $y = kx^2$, $k \neq 1$. Then $\lim\limits_{\substack{(x,y)\to(0,0) \\ y\neq x^2}} \frac{y}{x^2-y} = \lim\limits_{(x,kx^2)\to(0,0)} \frac{kx^2}{x^2-kx^2} = \frac{k}{1-k^2}$ which gives different limits for different values of k $\Rightarrow$ the limit does not exist.

16. Let $y = kx$, $k \neq 0$. Then $\lim\limits_{\substack{(x,y)\to(0,0) \\ xy\neq 0}} \frac{x^2+y^2}{xy} = \lim\limits_{(x,kx)\to(0,0)} \frac{x^2+(kx)^2}{x(kx)} = \frac{1+k^2}{k}$ which gives different limits for different values of k $\Rightarrow$ the limit does not exist.

17. (a) Let $y = kx$. Then $\lim\limits_{(x,y)\to(0,0)} \frac{x^2-y^2}{x^2+y^2} = \frac{x^2-k^2x^2}{x^2+k^2x^2} = \frac{1-k^2}{1+k^2}$ which gives different limits for different values of k $\Rightarrow$ the limit does not exist so $f(0,0)$ cannot be defined in a way that makes f continuous at the origin.

(b) Along the x-axis, $y = 0$ and $\lim\limits_{(x,y)\to(0,0)} \frac{\sin(x-y)}{|x+y|} = \lim\limits_{x\to 0} \frac{\sin x}{|x|} = \begin{cases} 1, & x>0 \\ -1, & x<0 \end{cases}$, so the limit fails to exist $\Rightarrow$ f is not continuous at $(0,0)$.

18. (a) $\lim\limits_{r\to 0} \frac{\sin 6r}{6r} = \lim\limits_{t\to 0} \frac{\sin t}{t} = 1$, where $t = 6r$

(b) $f_r(0,0) = \lim\limits_{h\to 0} \frac{f(0+h,0)-f(0,0)}{h} = \lim\limits_{h\to 0} \frac{\left(\frac{\sin 6h}{6h}\right)-1}{h} = \lim\limits_{h\to 0} \frac{\sin 6h - 6h}{6h^2} = \lim\limits_{h\to 0} \frac{6\cos 6h - 6}{12h}$

$= \lim\limits_{h\to 0} \frac{-36\sin 6h}{12} = 0$ (applying l'Hôpital's rule twice)

(c) $f_\theta(r,\theta) = \lim\limits_{h\to 0} \frac{f(r,\theta+h)-f(r,\theta)}{h} = \lim\limits_{h\to 0} \frac{\left(\frac{\sin 6r}{6r}\right)-\left(\frac{\sin 6r}{6r}\right)}{h} = \lim\limits_{h\to 0} \frac{0}{h} = 0$

19. $\frac{\partial g}{\partial r} = \cos\theta + \sin\theta$, $\frac{\partial g}{\partial \theta} = -r\sin\theta + r\cos\theta$

20. $\frac{\partial f}{\partial x} = \frac{1}{2}\left(\frac{2x}{x^2+y^2}\right) + \frac{\left(-\frac{y}{x^2}\right)}{1+\left(\frac{y}{x}\right)^2} = \frac{x}{x^2+y^2} - \frac{y}{x^2+y^2} = \frac{x-y}{x^2+y^2}$,

$\frac{\partial f}{\partial y} = \frac{1}{2}\left(\frac{2y}{x^2+y^2}\right) + \frac{\left(\frac{1}{x}\right)}{1+\left(\frac{y}{x}\right)^2} = \frac{y}{x^2+y^2} + \frac{x}{x^2+y^2} = \frac{x+y}{x^2+y^2}$

21. $\frac{\partial f}{\partial R_1} = -\frac{1}{R_1^2}$, $\frac{\partial f}{\partial R_2} = -\frac{1}{R_2^2}$, $\frac{\partial f}{\partial R_3} = -\frac{1}{R_3^2}$

22. $h_x(x,y,z) = 2\pi\cos(2\pi x + y - 3z)$, $h_y(x,y,z) = \cos(2\pi x + y - 3z)$, $h_z(x,y,z) = -3\cos(2\pi x + y - 3z)$

23. $\frac{\partial P}{\partial n} = \frac{RT}{V}$, $\frac{\partial P}{\partial R} = \frac{nT}{V}$, $\frac{\partial P}{\partial T} = \frac{nR}{V}$, $\frac{\partial P}{\partial V} = -\frac{nRT}{V^2}$

24. $f_r(r,\ell,T,w) = -\frac{1}{2r^2\ell}\sqrt{\frac{T}{\pi w}}$, $f_\ell(r,\ell,T,w) = -\frac{1}{2r\ell^2}\sqrt{\frac{T}{\pi w}}$, $f_T(r,\ell,T,w) = \left(\frac{1}{2r\ell}\right)\left(\frac{1}{\sqrt{\pi w}}\right)\left(\frac{1}{2\sqrt{T}}\right)$

$= \frac{1}{4r\ell}\sqrt{\frac{1}{T\pi w}} = \frac{1}{4r\ell T}\sqrt{\frac{T}{\pi w}}$, $f_w(r,\ell,T,w) = \left(\frac{1}{2r\ell}\right)\sqrt{\frac{T}{\pi}}\left(-\frac{1}{2}w^{-3/2}\right) = -\frac{1}{4r\ell w}\sqrt{\frac{T}{\pi w}}$

25. $\frac{\partial g}{\partial x} = \frac{1}{y}$, $\frac{\partial g}{\partial y} = 1 - \frac{x}{y^2} \Rightarrow \frac{\partial^2 g}{\partial x^2} = 0$, $\frac{\partial^2 g}{\partial y^2} = \frac{2x}{y^3}$, $\frac{\partial^2 g}{\partial y \partial x} = \frac{\partial^2 g}{\partial x \partial y} = -\frac{1}{y^2}$

26. $g_x(x,y) = e^x + y\cos x$, $g_y(x,y) = \sin x \Rightarrow g_{xx}(x,y) = e^x - y\sin x$, $g_{yy}(x,y) = 0$, $g_{xy}(x,y) = g_{yx}(x,y) = \cos x$

27. $\frac{\partial f}{\partial x} = 1 + y - 15x^2 + \frac{2x}{x^2+1}$, $\frac{\partial f}{\partial y} = x \Rightarrow \frac{\partial^2 f}{\partial x^2} = -30x + \frac{2-2x^2}{(x^2+1)^2}$, $\frac{\partial^2 f}{\partial y^2} = 0$, $\frac{\partial^2 f}{\partial y \partial x} = \frac{\partial^2 f}{\partial x \partial y} = 1$

28. $f_x(x,y) = -3y$, $f_y(x,y) = 2y - 3x - \sin y + 7e^y \Rightarrow f_{xx}(x,y) = 0$, $f_{yy}(x,y) = 2 - \cos y + 7e^y$, $f_{xy}(x,y) = f_{yx}(x,y)$
$= -3$

29. $f\left(\frac{\pi}{4},\frac{\pi}{4}\right) = \frac{1}{2}$, $f_x\left(\frac{\pi}{4},\frac{\pi}{4}\right) = \cos x \cos y\big|_{(\pi/4,\pi/4)} = \frac{1}{2}$, $f_y\left(\frac{\pi}{4},\frac{\pi}{4}\right) = -\sin x \sin y\big|_{(\pi/4,\pi/4)} = -\frac{1}{2}$
$\Rightarrow L(x,y) = \frac{1}{2} + \frac{1}{2}\left(x - \frac{\pi}{4}\right) - \frac{1}{2}\left(y - \frac{\pi}{4}\right) = \frac{1}{2} + \frac{1}{2}x - \frac{1}{2}y$; $f_{xx}(x,y) = -\sin x \cos y$, $f_{yy}(x,y) = -\sin x \cos y$, and
$f_{xy}(x,y) = -\cos x \sin y$. Thus an upper bound for E depends on the bound M used for $|f_{xx}|$, $|f_{xy}|$, and $|f_{yy}|$.
With $M = \frac{\sqrt{2}}{2}$ we have $|E(x,y)| \le \frac{1}{2}\left(\frac{\sqrt{2}}{2}\right)\left(\left|x - \frac{\pi}{4}\right| + \left|y - \frac{\pi}{4}\right|\right)^2 \le \frac{\sqrt{2}}{4}(0.2)^2 \le 0.0142$;

with $M = 1$, $|E(x,y)| \le \frac{1}{2}(1)\left(\left|x - \frac{\pi}{4}\right| + \left|y - \frac{\pi}{4}\right|\right)^2 = \frac{1}{2}(0.2)^2 = 0.02$.

30. $f(1,1) = 0$, $f_x(1,1) = y\big|_{(1,1)} = 1$, $f_y(1,1) = x - 6y\big|_{(1,1)} = -5 \Rightarrow L(x,y) = (x-1) - 5(y-1) = x - 5y + 4$;
$f_{xx}(x,y) = 0$, $f_{yy}(x,y) = -6$, and $f_{xy}(x,y) = 1 \Rightarrow$ maximum of $|f_{xx}|$, $|f_{yy}|$, and $|f_{xy}|$ is $6 \Rightarrow M = 6$
$\Rightarrow |E(x,y)| \le \frac{1}{2}(6)(|x-1| + |y-1|)^2 = \frac{1}{2}(6)(0.1 + 0.2)^2 = 0.27$

31. $f(1,0,0) = 0$, $f_x(1,0,0) = y - 3z\big|_{(1,0,0)} = 0$, $f_y(1,0,0) = x + 2z\big|_{(1,0,0)} = 1$, $f_z(1,0,0) = 2y - 3x\big|_{(1,0,0)} = -3$
$\Rightarrow L(x,y,z) = 0(x-1) + (y-0) - 3(z-0) = y - 3z$; $f(1,1,0) = 1$, $f_x(1,1,0) = 1$, $f_y(1,1,0) = 1$, $f_z(1,1,0) = -1$
$\Rightarrow L(x,y,z) = 1 + (x-1) + (y-1) - 1(z-0) = x + y - z - 1$

32. $f\left(0,0,\frac{\pi}{4}\right) = 1$, $f_x\left(0,0,\frac{\pi}{4}\right) = -\sqrt{2}\sin x \sin(y+z)\big|_{\left(0,0,\frac{\pi}{4}\right)} = 0$, $f_y\left(0,0,\frac{\pi}{4}\right) = \sqrt{2}\cos x \cos(y+z)\big|_{\left(0,0,\frac{\pi}{4}\right)} = 1$,
$f_z\left(0,0,\frac{\pi}{4}\right) = \sqrt{2}\cos x \cos(y+z)\big|_{\left(0,0,\frac{\pi}{4}\right)} = 1 \Rightarrow L(x,y,z) = 1 + 1(y-0) + 1\left(z - \frac{\pi}{4}\right) = 1 + y + z - \frac{\pi}{4}$;

$f\left(\frac{\pi}{4},\frac{\pi}{4},0\right) = \frac{\sqrt{2}}{2}$, $f_x\left(\frac{\pi}{4},\frac{\pi}{4},0\right) = -\frac{\sqrt{2}}{2}$, $f_y\left(\frac{\pi}{4},\frac{\pi}{4},0\right) = \frac{\sqrt{2}}{2}$, $f_z\left(\frac{\pi}{4},\frac{\pi}{4},0\right) = \frac{\sqrt{2}}{2}$
$\Rightarrow L(x,y,z) = \frac{\sqrt{2}}{2} - \frac{\sqrt{2}}{2}\left(x - \frac{\pi}{4}\right) + \frac{\sqrt{2}}{2}\left(y - \frac{\pi}{4}\right) + \frac{\sqrt{2}}{2}(z-0) = \frac{\sqrt{2}}{2} - \frac{\sqrt{2}}{2}x + \frac{\sqrt{2}}{2}y + \frac{\sqrt{2}}{2}z$

33. $V = \pi r^2 h \Rightarrow dV = 2\pi rh\ dr + \pi r^2\ dh \Rightarrow dV|_{(1.5,\,5280)} = 2\pi(1.5)(5280)\ dr + \pi(1.5)^2\ dh = 15{,}840\pi\ dr + 2.25\pi\ dh.$ You should be more careful with the diameter since it has a greater effect on dV.

34. $df = (2x - y)\ dx + (-x + 2y)\ dy \Rightarrow df|_{(1,2)} = 3\ dy \Rightarrow$ f is more sensitive to changes in y; in fact, near the point (1, 2) a change in x does not change f.

35. $dI = \frac{1}{R} dV - \frac{V}{R^2} dR \Rightarrow dI\Big|_{(24,\,100)} = \frac{1}{100} dV - \frac{24}{100^2} dR \Rightarrow dI\Big|_{dV = -1,\, dR = -20} = -0.01 + (480)(.0001) = 0.038,$ or increases by 0.038 amps; % change in $V = (100)\left(-\frac{1}{24}\right) \approx -4.17\%$; % change in $R = \left(-\frac{20}{100}\right)(100) = -20\%$; $I = \frac{24}{100} = 0.24 \Rightarrow$ estimated % change in $I = \frac{dI}{I} \times 100 = \frac{0.038}{0.24} \times 100 \approx 15.83\%$

36. $A = \pi ab \Rightarrow dA = \pi b\ da + \pi a\ db \Rightarrow dA|_{(10,16)} = 16\pi\ da + 10\pi\ db$; $da = \pm 0.1$ and $db = \pm 0.1$ $\Rightarrow dA = \pm 26\pi(0.1) = \pm 2.6\pi$ and $A = \pi(10)(16) = 160\pi \Rightarrow \left|\frac{dA}{A} \times 100\right| = \left|\frac{2.6\pi}{160\pi} \times 100\right| \approx 1.625\%$

37. (a) $y = uv \Rightarrow dy = v\ du + u\ dv$; percentage change in $u \le 2\% \Rightarrow |du| \le 0.02$, and percentage change in $v \le 3\%$ $\Rightarrow |dv| \le 0.03$; $\frac{dy}{y} = \frac{v\ du + u\ dv}{uv} = \frac{du}{u} + \frac{dv}{v} \Rightarrow \left|\frac{dy}{y} \times 100\right| = \left|\frac{du}{u} \times 100 + \frac{dv}{v} \times 100\right| \le \left|\frac{du}{u} \times 100\right| + \left|\frac{dv}{v} \times 100\right|$ $\le 2\% + 3\% = 5\%$

(b) $z = u + v \Rightarrow \frac{dz}{z} = \frac{du + dv}{u + v} = \frac{du}{u + v} + \frac{dv}{u + v} \le \frac{du}{u} + \frac{dv}{v}$ (since $u > 0$, $v > 0$) $\Rightarrow \left|\frac{dz}{z} \times 100\right| \le \left|\frac{du}{u} \times 100 + \frac{dv}{v} \times 100\right| = \left|\frac{dy}{y} \times 100\right|$

38. $C = \frac{7}{71.84 w^{0.425} h^{0.725}} \Rightarrow C_w = \frac{(-0.425)(7)}{71.84 w^{1.425} h^{0.725}}$ and $C_h = \frac{(-0.725)(7)}{71.84 w^{0.425} h^{1.725}}$ $\Rightarrow dC = \frac{-2.975}{71.84 w^{1.425} h^{0.725}}\ dw + \frac{-5.075}{71.84 w^{0.425} h^{1.725}}\ dh$; thus when $w = 70$ and $h = 180$ we have $dC|_{(70,\,180)} \approx -(0.00000225)\ dw - (0.0000149)\ dh \Rightarrow$ 1 cm error in height has more effect

39. $\frac{\partial w}{\partial x} = y \cos(xy + \pi),\ \frac{\partial w}{\partial y} = x \cos(xy + \pi),\ \frac{dx}{dt} = e^t,\ \frac{dy}{dt} = \frac{1}{t+1}$

$\Rightarrow \frac{dw}{dt} = [y \cos(xy + \pi)]e^t + [x \cos(xy + \pi)]\left(\frac{1}{t+1}\right)$; $t = 0 \Rightarrow x = 1$ and $y = 0$

$\Rightarrow \frac{dw}{dt}\Big|_{t=0} = 0 \cdot 1 + [1 \cdot (-1)]\left(\frac{1}{0+1}\right) = -1$

40. $\frac{\partial w}{\partial x} = e^y,\ \frac{\partial w}{\partial y} = xe^y + \sin z,\ \frac{\partial w}{\partial z} = y \cos z + \sin z,\ \frac{dx}{dt} = t^{-1/2},\ \frac{dy}{dt} = 1 + \frac{1}{t},\ \frac{dz}{dt} = \pi$

$\Rightarrow \frac{dw}{dt} = e^y t^{-1/2} + (xe^y + \sin z)\left(1 + \frac{1}{t}\right) + (y \cos z + \sin z)\pi$; $t = 1 \Rightarrow x = 2$, $y = 0$, and $z = \pi$

$\Rightarrow \frac{dw}{dt}\Big|_{t=1} = 1 \cdot 1 + (2 \cdot 1 - 0)(2) + (0 + 0)\pi = 5$

41. $\frac{\partial w}{\partial x} = 2\cos(2x-y),\ \frac{\partial w}{\partial y} = -\cos(2x-y),\ \frac{\partial x}{\partial r} = 1,\ \frac{\partial x}{\partial s} = \cos s,\ \frac{\partial y}{\partial r} = s,\ \frac{\partial y}{\partial s} = r$

$\Rightarrow \frac{\partial w}{\partial r} = [2\cos(2x-y)](1) + [-\cos(2x-y)](s)$; $r = \pi$ and $s = 0 \Rightarrow x = \pi$ and $y = 0$

$\Rightarrow \left.\frac{\partial w}{\partial r}\right|_{(\pi,0)} = (2\cos 2\pi) - (\cos 2\pi)(0) = 2$; $\frac{\partial w}{\partial s} = [2\cos(2x-y)](\cos s) + [-\cos(2x-y)](r)$

$\Rightarrow \left.\frac{\partial w}{\partial s}\right|_{(\pi,0)} = (2\cos 2\pi)(\cos 0) - (\cos 2\pi)(\pi) = 2 - \pi$

42. $\frac{\partial w}{\partial u} = \frac{dw}{dx}\frac{\partial x}{\partial u} = \left(\frac{x}{1+x^2} - \frac{1}{x^2+1}\right)(2e^u \cos v)$; $u = v = 0 \Rightarrow x = 2 \Rightarrow \left.\frac{\partial w}{\partial u}\right|_{(0,0)} = \left(\frac{2}{5} - \frac{1}{5}\right)(2) = \frac{2}{5}$;

$\frac{\partial w}{\partial v} = \frac{dw}{dx}\frac{\partial x}{\partial v} = \left(\frac{x}{1+x^2} - \frac{1}{x^2+1}\right)(-2e^u \sin v) \Rightarrow \left.\frac{\partial w}{\partial v}\right|_{(0,0)} = \left(\frac{2}{5} - \frac{1}{5}\right)(0) = 0$

43. $\frac{\partial f}{\partial x} = y+z,\ \frac{\partial f}{\partial y} = x+z,\ \frac{\partial f}{\partial z} = y+x,\ \frac{dx}{dt} = -\sin t,\ \frac{dy}{dt} = \cos t,\ \frac{dz}{dt} = -2\sin 2t$

$\Rightarrow \frac{df}{dt} = -(y+z)(\sin t) + (x+z)(\cos t) - 2(y+x)(\sin 2t)$; $t = 1 \Rightarrow x = \cos 1$, $y = \sin 1$, and $z = \cos 2$

$\Rightarrow \left.\frac{df}{dt}\right|_{t=1} = -(\sin 1 + \cos 2)(\sin 1) + (\cos 1 + \cos 2)(\cos 1) - 2(\sin 1 + \cos 1)(\sin 2)$

44. $\frac{\partial w}{\partial x} = \frac{dw}{ds}\frac{\partial s}{\partial x} = (5)\frac{dw}{ds}$ and $\frac{\partial w}{\partial y} = \frac{dw}{ds}\frac{\partial s}{\partial y} = (1)\frac{dw}{ds} = \frac{dw}{ds} \Rightarrow \frac{\partial w}{\partial x} - 5\frac{\partial w}{\partial y} = 5\frac{dw}{ds} - 5\frac{dw}{ds} = 0$

45. $F(x,y) = 1 - x - y^2 - \sin xy \Rightarrow F_x = -1 - y\cos xy$ and $F_y = -2y - x\cos xy \Rightarrow \frac{dy}{dx} = -\frac{F_x}{F_y} = \frac{-1 - y\cos xy}{-2y - x\cos xy}$

$= \frac{1 + y\cos xy}{-2y - x\cos xy} \Rightarrow$ at $(x,y) = (0,1)$ we have $\left.\frac{dy}{dx}\right|_{(0,1)} = \frac{1+1}{-2} = -1$

46. $F(x,y) = 2xy + e^{x+y} - 2 \Rightarrow F_x = 2y + e^{x+y}$ and $F_y = 2x + e^{x+y} \Rightarrow \frac{dy}{dx} = -\frac{F_x}{F_y} = -\frac{2y + e^{x+y}}{2x + e^{x+y}}$

$\Rightarrow$ at $(x,y) = (0, \ln 2)$ we have $\left.\frac{dy}{dx}\right|_{(0,\ln 2)} = -\frac{2\ln 2 + 2}{0+2} = -(\ln 2 + 1)$

47. (a) y, z are independent with $w = x^2 e^{yz}$ and $z = x^2 - y^2 \Rightarrow \frac{\partial w}{\partial y} = \frac{\partial w}{\partial x}\frac{\partial x}{\partial y} + \frac{\partial w}{\partial y}\frac{\partial y}{\partial y} + \frac{\partial w}{\partial z}\frac{\partial z}{\partial y}$

$= (2xe^{yz})\frac{\partial x}{\partial y} + (zx^2 e^{yz})(1) + (yx^2 e^{yz})(0)$; $z = x^2 - y^2 \Rightarrow 0 = 2x\frac{\partial x}{\partial y} - 2y \Rightarrow \frac{\partial x}{\partial y} = \frac{y}{x}$; therefore,

$\left(\frac{\partial w}{\partial y}\right)_z = (2xe^{yz})\left(\frac{y}{x}\right) + zx^2 e^{yz} = (2y + zx^2)e^{yz}$

(b) z, x are independent with $w = x^2 e^{yz}$ and $z = x^2 - y^2 \Rightarrow \frac{\partial w}{\partial z} = \frac{\partial w}{\partial x}\frac{\partial x}{\partial z} + \frac{\partial w}{\partial y}\frac{\partial y}{\partial z} + \frac{\partial w}{\partial z}\frac{\partial z}{\partial z}$

$= (2xe^{yz})(0) + (zx^2 e^{yz})\frac{\partial y}{\partial z} + (yx^2 e^{yz})(1)$; $z = x^2 - y^2 \Rightarrow 1 = 0 - 2y\frac{\partial y}{\partial z} \Rightarrow \frac{\partial y}{\partial z} = -\frac{1}{2y}$; therefore,

$\left(\frac{\partial w}{\partial z}\right)_x = (zx^2 e^{yz})\left(-\frac{1}{2y}\right) + yx^2 e^{yz} = x^2 e^{yz}\left(y - \frac{z}{2y}\right)$

(c) z, y are independent with $w = x^2e^{yz}$ and $z = x^2 - y^2 \Rightarrow \frac{\partial w}{\partial z} = \frac{\partial w}{\partial x}\frac{\partial x}{\partial z} + \frac{\partial w}{\partial y}\frac{\partial y}{\partial z} + \frac{\partial w}{\partial z}\frac{\partial z}{\partial z}$

$= (2xe^{yz})\frac{\partial x}{\partial z} + (zx^2e^{yz})(0) + (yx^2e^{yz})(1)$; $z = x^2 - y^2 \Rightarrow 1 = 2x\frac{\partial x}{\partial z} - 0 \Rightarrow \frac{\partial x}{\partial z} = \frac{1}{2x}$; therefore,

$\left(\frac{\partial w}{\partial z}\right)_y = (2xe^{yz})\left(\frac{1}{2x}\right) + yx^2e^{yz} = (1 + x^2y)e^{yz}$

48. (a) T, P are independent with $U = f(P, V, T)$ and $PV = nRT \Rightarrow \frac{\partial U}{\partial T} = \frac{\partial U}{\partial P}\frac{\partial P}{\partial T} + \frac{\partial U}{\partial V}\frac{\partial V}{\partial T} + \frac{\partial U}{\partial T}\frac{\partial T}{\partial T}$

$= \left(\frac{\partial U}{\partial P}\right)(0) + \left(\frac{\partial U}{\partial V}\right)\left(\frac{\partial V}{\partial T}\right) + \left(\frac{\partial U}{\partial T}\right)(1)$; $PV = nRT \Rightarrow P\frac{\partial V}{\partial T} = nR \Rightarrow \frac{\partial V}{\partial T} = \frac{nR}{P}$; therefore,

$\left(\frac{\partial U}{\partial T}\right)_P = \left(\frac{\partial U}{\partial V}\right)\left(\frac{nR}{P}\right) + \frac{\partial U}{\partial T}$

(b) V, T are independent with $U = f(P, V, T)$ and $PV = nRT \Rightarrow \frac{\partial U}{\partial V} = \frac{\partial U}{\partial P}\frac{\partial P}{\partial V} + \frac{\partial U}{\partial V}\frac{\partial V}{\partial V} + \frac{\partial U}{\partial T}\frac{\partial T}{\partial V}$

$= \left(\frac{\partial U}{\partial P}\right)\left(\frac{\partial P}{\partial V}\right) + \left(\frac{\partial U}{\partial V}\right)(1) + \left(\frac{\partial U}{\partial T}\right)(0)$; $PV = nRT \Rightarrow V\frac{\partial P}{\partial V} + P = (nR)\left(\frac{\partial T}{\partial V}\right) = 0 \Rightarrow \frac{\partial P}{\partial V} = -\frac{P}{V}$; therefore,

$\left(\frac{\partial U}{\partial V}\right)_T = \left(\frac{\partial U}{\partial P}\right)\left(-\frac{P}{V}\right) + \frac{\partial U}{\partial V}$

49. $\nabla f = (-\sin x \cos y)\mathbf{i} - (\cos x \sin y)\mathbf{j} \Rightarrow \nabla f|_{\left(\frac{\pi}{4}, \frac{\pi}{4}\right)} = -\frac{1}{2}\mathbf{i} - \frac{1}{2}\mathbf{j} \Rightarrow |\nabla f| = \sqrt{\left(-\frac{1}{2}\right)^2 + \left(-\frac{1}{2}\right)^2} = \frac{1}{\sqrt{2}} = \frac{\sqrt{2}}{2}$;

$\mathbf{u} = \frac{\nabla f}{|\nabla f|} = -\frac{\sqrt{2}}{2}\mathbf{i} - \frac{\sqrt{2}}{2}\mathbf{j} \Rightarrow$ f increases most rapidly in the direction $\mathbf{u} = -\frac{\sqrt{2}}{2}\mathbf{i} - \frac{\sqrt{2}}{2}\mathbf{j}$ and decreases most

rapidly in the direction $-\mathbf{u} = \frac{\sqrt{2}}{2}\mathbf{i} + \frac{\sqrt{2}}{2}\mathbf{j}$; $(D_{\mathbf{u}}f)_{P_0} = |\nabla f| = \frac{\sqrt{2}}{2}$ and $(D_{-\mathbf{u}}f)_{P_0} = -\frac{\sqrt{2}}{2}$;

$\mathbf{u}_1 = \frac{\mathbf{A}}{|\mathbf{A}|} = \frac{3\mathbf{i} + 4\mathbf{j}}{\sqrt{3^2 + 4^2}} = \frac{3}{5}\mathbf{i} + \frac{4}{5}\mathbf{j} \Rightarrow (D_{\mathbf{u}_1}f)_{P_0} = \nabla f \cdot \mathbf{u}_1 = \left(-\frac{1}{2}\right)\left(\frac{3}{5}\right) + \left(-\frac{1}{2}\right)\left(\frac{4}{5}\right) = -\frac{7}{10}$

50. $\nabla f = 2xe^{-2y}\mathbf{i} - 2x^2e^{-2y}\mathbf{j} \Rightarrow \nabla f|_{(1,0)} = 2\mathbf{i} - 2\mathbf{j} \Rightarrow |\nabla f| = \sqrt{2^2 + (-2)^2} = 2\sqrt{2}$; $\mathbf{u} = \frac{\nabla f}{|\nabla f|} = \frac{1}{\sqrt{2}}\mathbf{i} - \frac{1}{\sqrt{2}}\mathbf{j}$

$\Rightarrow$ f increases most rapidly in the direction $\mathbf{u} = \frac{1}{\sqrt{2}}\mathbf{i} - \frac{1}{\sqrt{2}}\mathbf{j}$ and decreases most rapidly in the direction

$-\mathbf{u} = -\frac{1}{\sqrt{2}}\mathbf{i} + \frac{1}{\sqrt{2}}\mathbf{j}$; $(D_{\mathbf{u}}f)_{P_0} = |\nabla f| = 2\sqrt{2}$ and $(D_{-\mathbf{u}}f)_{P_0} = -2\sqrt{2}$; $\mathbf{u}_1 = \frac{\mathbf{A}}{|\mathbf{A}|} = \frac{\mathbf{i} + \mathbf{j}}{\sqrt{1^2 + 1^2}} = \frac{1}{\sqrt{2}}\mathbf{i} + \frac{1}{\sqrt{2}}\mathbf{j}$

$\Rightarrow (D_{\mathbf{u}_1}f)_{P_0} = \nabla f \cdot \mathbf{u}_1 = (2)\left(\frac{1}{\sqrt{2}}\right) + (-2)\left(\frac{1}{\sqrt{2}}\right) = 0$

51. $\nabla f = \left(\frac{2}{2x + 3y + 6z}\right)\mathbf{i} + \left(\frac{3}{2x + 3y + 6z}\right)\mathbf{j} + \left(\frac{6}{2x + 3y + 6z}\right)\mathbf{k} \Rightarrow \nabla f|_{(-1,-1,1)} = 2\mathbf{i} + 3\mathbf{j} + 6\mathbf{k}$;

$\mathbf{u} = \frac{\nabla f}{|\nabla f|} = \frac{2\mathbf{i} + 3\mathbf{j} + 6\mathbf{k}}{\sqrt{2^2 + 3^2 + 6^2}} = \frac{2}{7}\mathbf{i} + \frac{3}{7}\mathbf{j} + \frac{6}{7}\mathbf{k} \Rightarrow$ f increases most rapidly in the direction $\mathbf{u} = \frac{2}{7}\mathbf{i} + \frac{3}{7}\mathbf{j} + \frac{6}{7}\mathbf{k}$ and

decreases most rapidly in the direction $-\mathbf{u} = -\frac{2}{7}\mathbf{i} - \frac{3}{7}\mathbf{j} - \frac{6}{7}\mathbf{k}$; $(D_{\mathbf{u}}f)_{P_0} = |\nabla f| = 7$, $(D_{-\mathbf{u}}f)_{P_0} = -7$;

$\mathbf{u}_1 = \frac{\mathbf{A}}{|\mathbf{A}|} = \frac{2}{7}\mathbf{i} + \frac{3}{7}\mathbf{j} + \frac{6}{7}\mathbf{k} \Rightarrow (D_{\mathbf{u}_1}f)_{P_0} = (D_{\mathbf{u}}f)_{P_0} = 7$

52. $\nabla f = (2x + 3y)\mathbf{i} + (3x + 2)\mathbf{j} + (1 - 2z)\mathbf{k} \Rightarrow \nabla f|_{(0,0,0)} = 2\mathbf{j} + \mathbf{k}$; $\mathbf{u} = \frac{\nabla f}{|\nabla f|} = \frac{2}{\sqrt{5}}\mathbf{j} + \frac{1}{\sqrt{5}}\mathbf{k} \Rightarrow$ f increases most

rapidly in the direction $\mathbf{u} = \frac{2}{\sqrt{5}}\mathbf{j} + \frac{1}{\sqrt{5}}\mathbf{k}$ and decreases most rapidly in the direction $-\mathbf{u} = -\frac{2}{\sqrt{5}}\mathbf{j} - \frac{1}{\sqrt{5}}\mathbf{k}$;

$(D_{\mathbf{u}}f)_{P_0} = |\nabla f| = \sqrt{5}$ and $(D_{-\mathbf{u}}f)_{P_0} = -\sqrt{5}$; $\mathbf{u}_1 = \frac{\mathbf{A}}{|\mathbf{A}|} = \frac{\mathbf{i}+\mathbf{j}+\mathbf{k}}{\sqrt{1^2+1^2+1^2}} = \frac{1}{\sqrt{3}}\mathbf{i} + \frac{1}{\sqrt{3}}\mathbf{j} + \frac{1}{\sqrt{3}}\mathbf{k}$

$\Rightarrow (D_{\mathbf{u}_1}f)_{P_0} = \nabla f \cdot \mathbf{u}_1 = (0)\left(\frac{1}{\sqrt{3}}\right) + (2)\left(\frac{1}{\sqrt{3}}\right) + (1)\left(\frac{1}{\sqrt{3}}\right) = \frac{3}{\sqrt{3}} = \sqrt{3}$

53. $\mathbf{r} = (\cos 3t)\mathbf{i} + (\sin 3t)\mathbf{j} + 3t\mathbf{k} \Rightarrow \mathbf{v}(t) = (-3 \sin 3t)\mathbf{i} + (3 \cos 3t)\mathbf{j} + 3\mathbf{k} \Rightarrow \mathbf{v}\left(\frac{\pi}{3}\right) = -3\mathbf{j} + 3\mathbf{k}$

$\Rightarrow \mathbf{u} = -\frac{1}{\sqrt{2}}\mathbf{j} + \frac{1}{\sqrt{2}}\mathbf{k}$; $f(x,y,z) = xyz \Rightarrow \nabla f = yz\mathbf{i} + xz\mathbf{j} + xy\mathbf{k}$; $t = \frac{\pi}{3}$ yields the point on the helix $(-1, 0, \pi)$

$\Rightarrow \nabla f|_{(1,0,\pi)} = -\pi\mathbf{j} \Rightarrow \nabla f \cdot \mathbf{u} = (-\pi\mathbf{j}) \cdot \left(-\frac{1}{\sqrt{2}}\mathbf{j} + \frac{1}{\sqrt{2}}\mathbf{k}\right) = \frac{\pi}{\sqrt{2}}$

54. $f(x,y,z) = xyz \Rightarrow \nabla f = yz\mathbf{i} + xz\mathbf{j} + xy\mathbf{k}$; at $(1,1,1)$ we get $\nabla f = \mathbf{i} + \mathbf{j} + \mathbf{k} \Rightarrow$ the maximum value of $D_{\mathbf{u}}f|_{(1,1,1)} = |\nabla f| = \sqrt{3}$

55. (a) Let $\nabla f = a\mathbf{i} + b\mathbf{j}$ at $(1,2)$. The direction toward $(2,2)$ is determined by $\mathbf{v}_1 = (2-1)\mathbf{i} + (2-2)\mathbf{j} = \mathbf{i} = \mathbf{u}$ so that $\nabla f \cdot \mathbf{u} = 2 \Rightarrow a = 2$. The direction toward $(1,1)$ is determined by $\mathbf{v}_2 + (1-1)\mathbf{i} + (1-2)\mathbf{j} = -\mathbf{j} = \mathbf{u}$ so that $\nabla f \cdot \mathbf{u} = -2 \Rightarrow -b = -2 \Rightarrow b = 2$. Therefore $\nabla f = 2\mathbf{i} + 2\mathbf{j}$.

(b) The direction toward $(4,6)$ is determined by $\mathbf{v}_3 = (4-1)\mathbf{i} + (6-2)\mathbf{j} = 3\mathbf{i} + 4\mathbf{j} \Rightarrow \mathbf{u} = \frac{3}{5}\mathbf{i} + \frac{4}{5}\mathbf{j}$

$\Rightarrow \nabla f \cdot \mathbf{u} = \frac{14}{5}$.

56. (a) True (b) False (c) True (d) True

57. $\nabla f = 2x\mathbf{i} + \mathbf{j} + 2z\mathbf{k} \Rightarrow$

$\nabla f|_{(0,-1,-1)} = \mathbf{j} - 2\mathbf{k}$,

$\nabla f|_{(0,0,0)} = \mathbf{j}$,

$\nabla f|_{(0,-1,1)} = \mathbf{j} + 2\mathbf{k}$

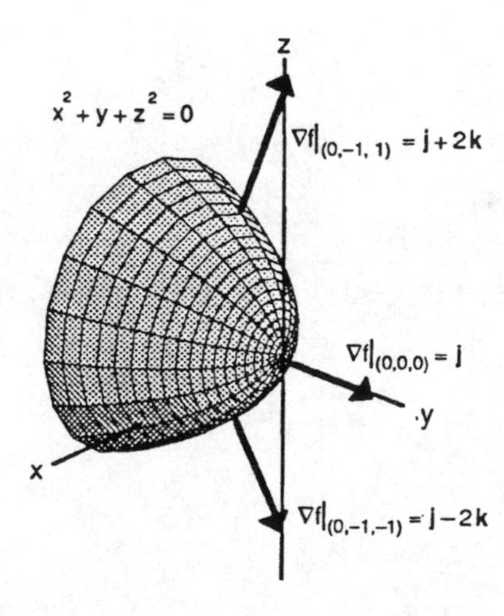

58. $\nabla f = 2y\mathbf{j} + 2z\mathbf{k} \Rightarrow$
$\nabla f|_{(2,2,0)} = 4\mathbf{j}$,
$\nabla f|_{(2,-2,0)} = -4\mathbf{j}$,
$\nabla f|_{(2,0,2)} = 4\mathbf{k}$,
$\nabla f|_{(2,0,-2)} = -4\mathbf{k}$

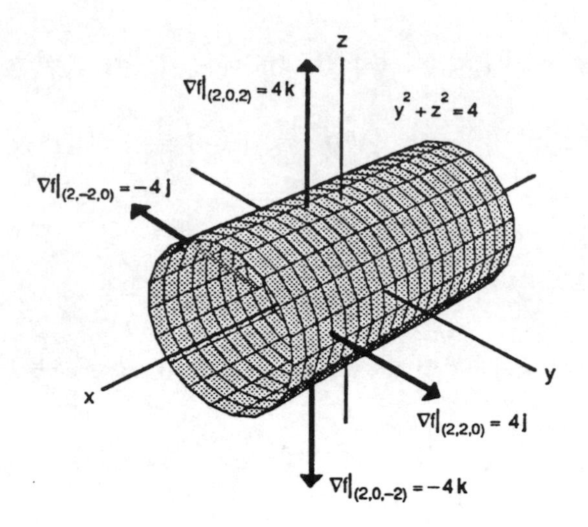

59. $\nabla f = 2x\mathbf{i} - \mathbf{j} - 5\mathbf{k} \Rightarrow \nabla f|_{(2,-1,1)} = 4\mathbf{i} - \mathbf{j} - 5\mathbf{k} \Rightarrow$ Tangent Plane: $4(x-2) - (y+1) - 5(z-1) = 0$ $\Rightarrow 4x - y - 5z = 4$; Normal Line: $x = 2 + 4t,\ y = -1 - t,\ z = 1 - 5t$

60. $\nabla f = 2x\mathbf{i} + 2y\mathbf{j} + \mathbf{k} \Rightarrow \nabla f|_{(1,1,2)} = 2\mathbf{i} + 2\mathbf{j} + \mathbf{k} \Rightarrow$ Tangent Plane: $2(x-1) + 2(y-1) + (z-2) = 0$ $\Rightarrow 2x + 2y + z - 6 = 0$; Normal Line: $x = 1 + 2t,\ y = 1 + 2t,\ z = 2 + t$

61. $\frac{\partial z}{\partial x} = \frac{2x}{x^2+y^2} \Rightarrow \left.\frac{\partial z}{\partial x}\right|_{(0,1,0)} = 0$ and $\frac{\partial z}{\partial y} = \frac{2y}{x^2+y^2} \Rightarrow \left.\frac{\partial z}{\partial y}\right|_{(0,1,0)} = 2$; thus the tangent plane is $2(y-1) - (z-0) = 0$ or $2y - z - 2 = 0$

62. $\frac{\partial z}{\partial x} = -2x\left(x^2+y^2\right)^{-2} \Rightarrow \left.\frac{\partial z}{\partial x}\right|_{\left(1,1,\frac{1}{2}\right)} = -\frac{1}{2}$ and $\frac{\partial z}{\partial y} = -2y\left(x^2+y^2\right)^{-2} \Rightarrow \left.\frac{\partial z}{\partial y}\right|_{\left(1,1,\frac{1}{2}\right)} = -\frac{1}{2}$; thus the tangent plane is $-\frac{1}{2}(x-1) - \frac{1}{2}(y-1) - \left(z - \frac{1}{2}\right) = 0$ or $x + y + 2z - 3 = 0$

63. $\nabla f = (-\cos x)\mathbf{i} + \mathbf{j} \Rightarrow \nabla f|_{(\pi,1)} = \mathbf{i} + \mathbf{j} \Rightarrow$ the tangent line is $(x - \pi) + (y - 1) = 0 \Rightarrow x + y = \pi + 1$; the normal line is $y - 1 = 1(x - \pi) \Rightarrow y = x - \pi + 1$

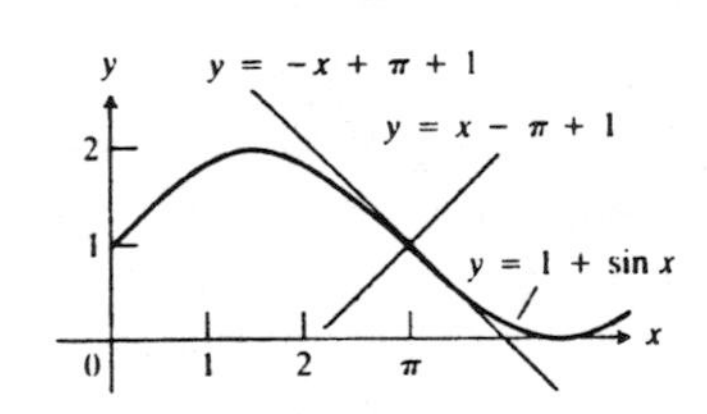

64. $\nabla f = -x\mathbf{i} + y\mathbf{j} \Rightarrow \nabla f|_{(1,2)} = -\mathbf{i} + 2\mathbf{j} \Rightarrow$ the tangent line is $-(x-1) + 2(y-2) = 0 = \frac{1}{2}x + \frac{3}{2}$; the normal line is $y - 2 = -2(x-1) \Rightarrow y = -2x + 4$

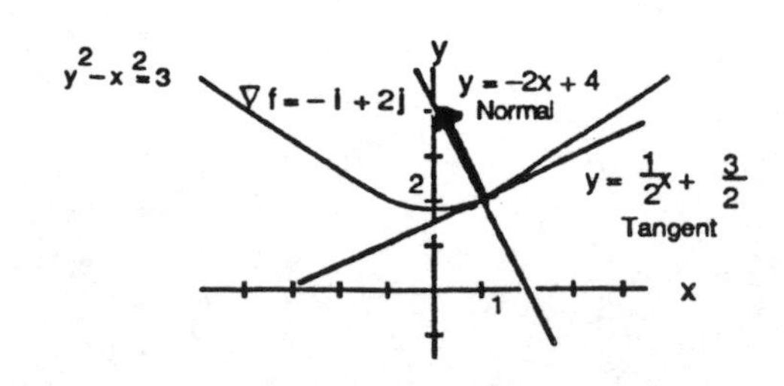

65. Let $f(x,y,z) = x^2 + 2y + 2z - 4$ and $g(x,y,z) = y - 1$. Then $\nabla f = 2x\mathbf{i} + 2\mathbf{j} + 2\mathbf{k}\big|_{\left(1,1,\frac{1}{2}\right)} = 2\mathbf{i} + 2\mathbf{j} + 2\mathbf{k}$

and $\nabla g = \mathbf{j} \Rightarrow \nabla f \times \nabla g = \begin{vmatrix} \mathbf{i} & \mathbf{j} & \mathbf{k} \\ 2 & 2 & 2 \\ 0 & 1 & 0 \end{vmatrix} = -2\mathbf{i} + 2\mathbf{k} \Rightarrow$ the line is $x = 1 - 2t$, $y = 1$, $z = \frac{1}{2} + 2t$

66. Let $f(x,y,z) = x + y^2 + z - 2$ and $g(x,y,z) = y - 1$. Then $\nabla f = \mathbf{i} + 2y\mathbf{j} + \mathbf{k}\big|_{\left(\frac{1}{2},1,\frac{1}{2}\right)} = \mathbf{i} + 2\mathbf{j} + \mathbf{k}$ and

$\nabla g = \mathbf{j} \Rightarrow \nabla f \times \nabla g = \begin{vmatrix} \mathbf{i} & \mathbf{j} & \mathbf{k} \\ 1 & 2 & 1 \\ 0 & 1 & 0 \end{vmatrix} = -\mathbf{i} + \mathbf{k} \Rightarrow$ the line is $x = \frac{1}{2} - t$, $y = 1$, $z = \frac{1}{2} + t$

67. $f_x(x,y) = 2x - y + 2 = 0$ and $f_y(x,y) = -x + 2y + 2 = 0 \Rightarrow x = -2$ and $y = -2 \Rightarrow (-2,-2)$ is the critical point; $f_{xx}(-2,-2) = 2$, $f_{yy}(-2,-2) = 2$, $f_{xy}(-2,-2) = -1 \Rightarrow f_{xx}f_{yy} - f_{xy}^2 = 3 > 0$ and $f_{xx} > 0 \Rightarrow$ local minimum value of $f(-2,-2) = -8$

68. $f_x(x,y) = 10x + 4y + 4 = 0$ and $f_y(x,y) = 4x - 4y - 4 = 0 \Rightarrow x = 0$ and $y = -1 \Rightarrow (0,-1)$ is the critical point; $f_{xx}(0,-1) = 10$, $f_{yy}(0,-1) = -4$, $f_{xy}(0,-1) = 4 \Rightarrow f_{xx}f_{yy} - f_{xy}^2 = -56 < 0 \Rightarrow$ saddle point with $f(0,-1) = 2$

69. $f_x(x,y) = 6x^2 + 3y = 0$ and $f_y(x,y) = 3x + 6y^2 = 0 \Rightarrow y = -2x^2$ and $3x + 6\left(4x^4\right) = 0 \Rightarrow x\left(1 + 8x^3\right) = 0$ $\Rightarrow x = 0$ and $y = 0$, or $x = -\frac{1}{2}$ and $y = -\frac{1}{2} \Rightarrow$ the critical points are $(0,0)$ and $\left(-\frac{1}{2},-\frac{1}{2}\right)$. For $(0,0)$: $f_{xx}(0,0) = 12x\big|_{(0,0)} = 0$, $f_{yy}(0,0) = 12y\big|_{(0,0)} = 0$, $f_{xy}(0,0) = 3 \Rightarrow f_{xx}f_{yy} - f_{xy}^2 = -9 < 0 \Rightarrow$ saddle point with $f(0,0) = 0$. For $\left(-\frac{1}{2},-\frac{1}{2}\right)$: $f_{xx} = -6$, $f_{yy} = -6$, $f_{xy} = 3 \Rightarrow f_{xx}f_{yy} - f_{xy}^2 = 27 > 0$ and $f_{xx} < 0 \Rightarrow$ local maximum value of $f\left(-\frac{1}{2},-\frac{1}{2}\right) = \frac{1}{4}$

70. $f_x(x,y) = 3x^2 - 3y = 0$ and $f_y(x,y) = 3y^2 - 3x = 0 \Rightarrow y = x^2$ and $x^4 - x = 0 \Rightarrow x\left(x^3 - 1\right) = 0 \Rightarrow$ the critical points are $(0,0)$ and $(1,1)$. For $(0,0)$: $f_{xx}(0,0) = 6x\big|_{(0,0)} = 0$, $f_{yy}(0,0) = 6y\big|_{(0,0)} = 0$, $f_{xy}(0,0) = -3$ $\Rightarrow f_{xx}f_{yy} - f_{xy}^2 = -9 < 0 \Rightarrow$ saddle point with $f(0,0) = 15$. For $(1,1)$: $f_{xx}(1,1) = 6$, $f_{yy}(1,1) = 6$, $f_{xy}(1,1) = -3$ $\Rightarrow f_{xx}f_{yy} - f_{xy}^2 = 27 > 0$ and $f_{xx} > 0 \Rightarrow$ local minimum value of $f(1,1) = 14$

71. $f_x(x,y) = 3x^2 + 6x = 0$ and $f_y(x,y) = 3y^2 - 6y = 0 \Rightarrow x(x+2) = 0$ and $y(y-2) = 0 \Rightarrow x = 0$ or $x = -2$ and $y = 0$ or $y = 2 \Rightarrow$ the critical points are $(0,0)$, $(0,2)$, $(-2,0)$, and $(-2,2)$. For $(0,0)$: $f_{xx}(0,0) = 6x + 6\big|_{(0,0)}$ $= 6$, $f_{yy}(0,0) = 6y - 6\big|_{(0,0)} = -6$, $f_{xy}(0,0) = 0 \Rightarrow f_{xx}f_{yy} - f_{xy}^2 = -36 < 0 \Rightarrow$ saddle point with $f(0,0) = 0$. For $(0,2)$: $f_{xx}(0,2) = 6$, $f_{yy}(0,2) = 6$, $f_{xy}(0,2) = 0 \Rightarrow f_{xx}f_{yy} - f_{xy}^2 = 36 > 0$ and $f_{xx} > 0 \Rightarrow$ local minimum value of $f(0,2) = -4$. For $(-2,0)$: $f_{xx}(-2,0) = -6$, $f_{yy}(-2,0) = -6$, $f_{xy}(-2,0) = 0 \Rightarrow f_{xx}f_{yy} - f_{xy}^2 = 36 > 0$ and $f_{xx} < 0$ $\Rightarrow$ local maximum value of $f(-2,0) = 4$. For $(-2,2)$: $f_{xx}(-2,2) = -6$, $f_{yy}(-2,2) = 6$, $f_{xy}(-2,2) = 0$ $\Rightarrow f_{xx}f_{yy} - f_{xy}^2 = -36 < 0 \Rightarrow$ saddle point with $f(-2,2) = 0$.

72. $f_x(x,y) = 4x^3 - 16x = 0 \Rightarrow 4x\left(x^2 - 4\right) = 0 \Rightarrow x = 0,\ 2,\ -2;\ f_y(x,y) = 6y - 6 = 0 \Rightarrow y = 1$. Therefore the critical points are $(0,1)$, $(2,1)$, and $(-2,1)$. For $(0,1)$: $f_{xx}(0,1) = 12x^2 - 16\big|_{(0,1)} = -16$, $f_{yy}(0,1) = 6$, $f_{xy}(0,1) = 0$ $\Rightarrow f_{xx}f_{yy} - f_{xy}^2 = -96 < 0 \Rightarrow$ saddle point with $f(0,1) = -3$. For $(2,1)$: $f_{xx}(2,1) = 32$, $f_{yy}(2,1) = 6$, $f_{xy}(2,1) = 0 \Rightarrow f_{xx}f_{yy} - f_{xy}^2 = 192 > 0$ and $f_{xx} > 0 \Rightarrow$ local minimum value of $f(2,1) = -19$. For $(-2,1)$: $f_{xx}(-2,1) = 32$, $f_{yy}(-2,1) = 6$, $f_{xy}(-2,1) = 0 \Rightarrow f_{xx}f_{yy} - f_{xy}^2 = 192 > 0$ and $f_{xx} > 0 \Rightarrow$ local minimum value of $f(-2,1) = -19$.

73. (i) On OA, $f(x,y) = f(0,y) = y^2 + 3y$ for $0 \le y \le 4$ $\Rightarrow f'(0,y) = 2y + 3 = 0 \Rightarrow y = -\frac{3}{2}$. But $\left(0, -\frac{3}{2}\right)$ is not in the region.
Endpoints: $f(0,0) = 0$ and $f(0,4) = 28$.

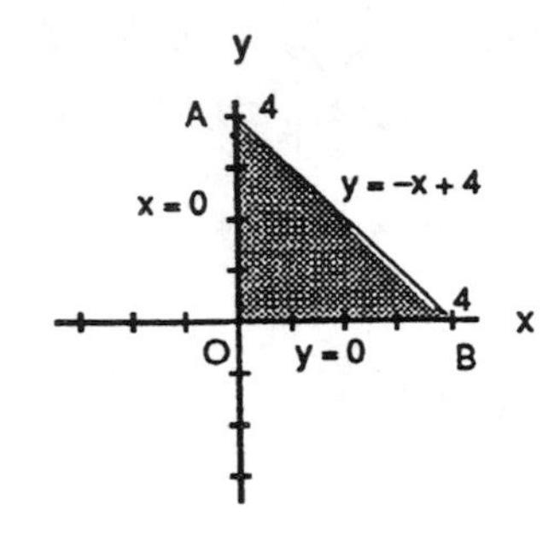

(ii) On AB, $f(x,y) = f(x,-x+4) = x^2 - 10x + 28$ for $0 \le x \le 4 \Rightarrow f'(x,-x+4) = 2x - 10 = 0$ $\Rightarrow x = 5,\ y = -1$. But $(5,-1)$ is not in the region.
Endpoints: $f(4,0) = 4$ and $f(0,4) = 28$.

(iii) On OB, $f(x,y) = f(x,0) = x^2 - 3x$ for $0 \le x \le 4 \Rightarrow f'(x,0) = 2x - 3 \Rightarrow x = \frac{3}{2}$ and $y = 0 \Rightarrow \left(\frac{3}{2}, 0\right)$ is a critical point with $f\left(\frac{3}{2}, 0\right) = -\frac{9}{4}$. Endpoints: $f(0,0) = 0$ and $f(4,0) = 4$.

(iv) For the interior of the triangular region, $f_x(x,y) = 2x + y - 3 = 0$ and $f_y(x,y) = x + 2y + 3 = 0 \Rightarrow x = 3$ and $y = -3$. But $(3,-3)$ is not in the region. Therefore the absolute maximum is 28 at $(0,4)$ and the absolute minimum is $-\frac{9}{4}$ at $\left(\frac{3}{2}, 0\right)$.

74. (i) On OA, $f(x,y) = f(0,y) = -y^2 + 4y + 1$ for $0 \le y \le 2$ $\Rightarrow f'(0,y) = -2y + 4 = 0 \Rightarrow y = 2$ and $x = 0$. But $(0,2)$ is not in the interior of OA.
Endpoints: $f(0,0) = 1$ and $f(0,2) = 5$.

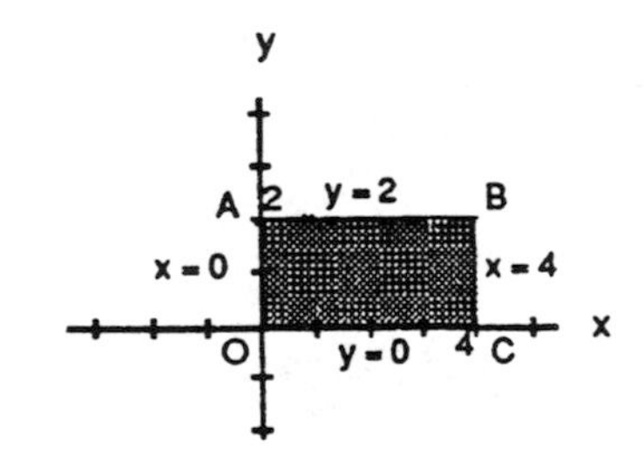

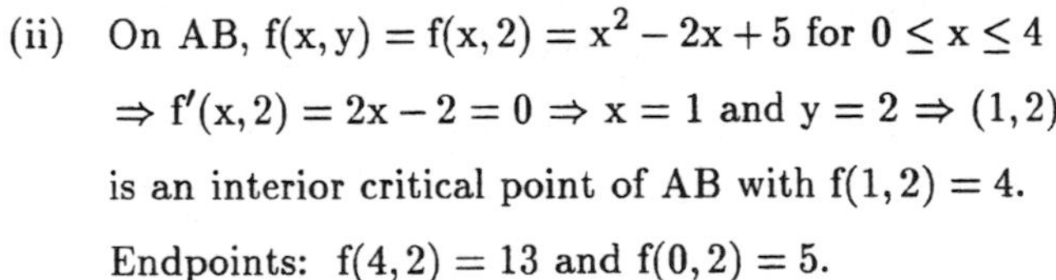

(ii) On AB, $f(x,y) = f(x,2) = x^2 - 2x + 5$ for $0 \le x \le 4$ $\Rightarrow f'(x,2) = 2x - 2 = 0 \Rightarrow x = 1$ and $y = 2 \Rightarrow (1,2)$ is an interior critical point of AB with $f(1,2) = 4$.
Endpoints: $f(4,2) = 13$ and $f(0,2) = 5$.

(iii) On BC, $f(x,y) = f(4,y) = -y^2 + 4y + 9$ for $0 \le y \le 2 \Rightarrow f'(4,y) = -2y + 4 = 0 \Rightarrow y = 2$ and $x = 4$. But $(4,2)$ is not in the interior of BC. Endpoints: $f(4,0) = 9$ and $f(4,2) = 13$.

(iv) On OC, $f(x,y) = f(x,0) = x^2 - 2x + 1$ for $0 \le x \le 4 \Rightarrow f'(x,0) = 2x - 2 = 0 \Rightarrow x = 1$ and $y = 0 \Rightarrow (1,0)$ is an interior critical point of OC with $f(1,0) = 0$. Endpoints: $f(0,0) = 1$ and $f(4,0) = 9$.

(v) For the interior of the rectangular region, $f_x(x,y) = 2x - 2 = 0$ and $f_y(x,y) = -2y + 4 = 0 \Rightarrow x = 1$ and $y = 2$. But $(1,2)$ is not in the interior of the region. Therefore the absolute maximum is 13 at $(4,2)$ and the absolute minimum is 0 at $(1,0)$.

75. (i) On AB, $f(x,y) = f(-2,y) = y^2 - y - 4$ for $-2 \le y \le 2$
$\Rightarrow f'(-2,y) = 2y - 1 \Rightarrow y = \frac{1}{2}$ and $x = -2 \Rightarrow \left(-2, \frac{1}{2}\right)$
is an interior critical point in AB with $f\left(-2, \frac{1}{2}\right)$
$= -\frac{17}{4}$.
Endpoints: $f(-2,-2) = 2$ and $f(2,2) = -2$.

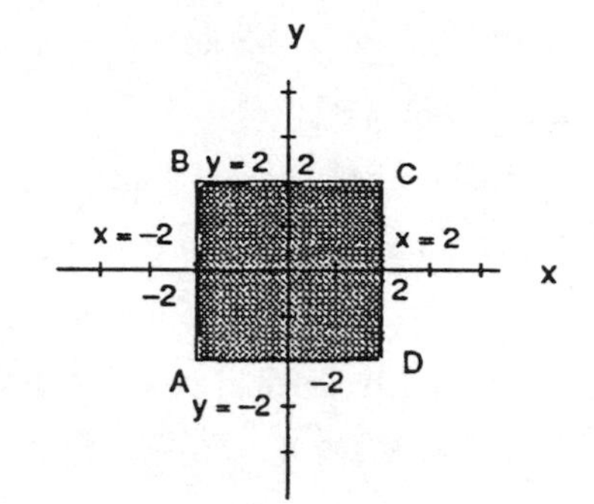

(ii) On BC, $f(x,y) = f(x,2) = -2$ for $-2 \le x \le 2$
$\Rightarrow f'(x,2) = 0 \Rightarrow$ no critical points in the interior of BC.
Endpoints: $f(-2,2) = -2$ and $f(2,2) = -2$.

(iii) On CD, $f(x,y) = f(2,y) = y^2 - 5y + 4$ for $-2 \le y \le 2$
$\Rightarrow f'(2,y) = 2y - 5 = 0 \Rightarrow y = \frac{5}{2}$ and $x = 2$. But $\left(2, \frac{5}{2}\right)$ is not in the region.
Endpoints: $f(2,-2) = 18$ and $f(2,2) = -2$.

(iv) On AD, $f(x,y) = f(x,-2) = 4x + 10$ for $-2 \le x \le 2 \Rightarrow f'(x,-2) = 4 \Rightarrow$ no critical points in the interior of AD. Endpoints: $f(-2,-2) = 2$ and $f(2,-2) = 18$.

(v) For the interior of the square, $f_x(x,y) = -y + 2 = 0$ and $f_y(x,y) = 2y - x - 3 = 0 \Rightarrow y = 2$ and $x = 1$ $\Rightarrow (1,2)$ is an interior critical point of the square with $f(1,2) = -2$. Therefore the absolute maximum is 18 at $(2,-2)$ and the absolute minimum is $-\frac{17}{4}$ at $\left(-2, \frac{1}{2}\right)$.

76. (i) On OA, $f(x,y) = f(0,y) = 2y - y^2$ for $0 \le y \le 2$
$\Rightarrow f'(0,y) = 2 - 2y = 0 \Rightarrow y = 1$ and $x = 0 \Rightarrow (0,1)$
is an interior critical point of OA with $f(0,1) = 1$.
Endpoints: $f(0,0) = 0$ and $f(0,2) = 0$.

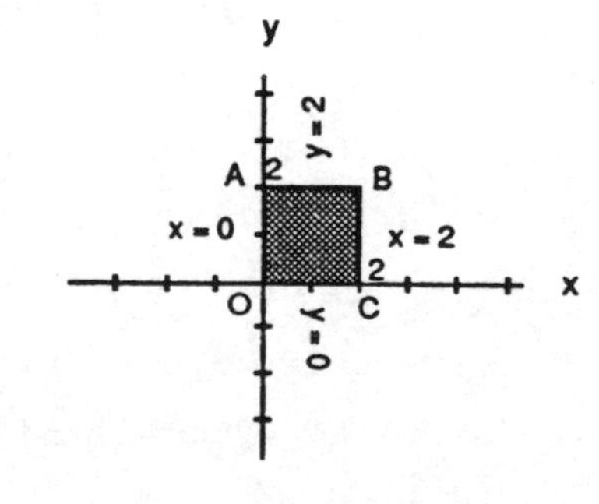

(ii) On AB, $f(x,y) = f(x,2) = 2x - x^2$ for $0 \le x \le 2$
$\Rightarrow f'(x,2) = 2 - 2x = 0 \Rightarrow x = 1$ and $y = 2 \Rightarrow (1,2)$
is an interior critical point of AB with $f(1,2) = 1$.
Endpoints: $f(0,2) = 0$ and $f(2,2) = 0$.

(iii) On BC, $f(x,y) = f(2,y) = 2y - y^2$ for $0 \le y \le 2 \Rightarrow f'(2,y) = 2 - 2y = 0 \Rightarrow y = 1$ and $x = 2$ $\Rightarrow (2,1)$ is an interior critical point of BC with $f(2,1) = 1$. Endpoints: $f(2,0) = 0$ and $f(2,2) = 0$.

(iv) On OC, $f(x,y) = f(x,0) = 2x - x^2$ for $0 \le x \le 2 \Rightarrow f'(x,0) = 2 - 2x = 0 \Rightarrow x = 1$ and $y = 0 \Rightarrow (1,0)$ is an interior critical point of OC with $f(1,0) = 1$. Endpoints: $f(0,0) = 0$ and $f(0,2) = 0$.

(v) For the interior of the rectangular region, $f_x(x,y) = 2 - 2x = 0$ and $f_y(x,y) = 2 - 2y = 0 \Rightarrow x = 1$ and $y = 1 \Rightarrow (1,1)$ is an interior critical point of the square with $f(1,1) = 2$. Therefore the absolute maximum is 2 at $(1,1)$ and the absolute minimum is 0 at the four corners $(0,0)$, $(0,2)$, $(2,2)$, and $(2,0)$.

77. (i) On AB, $f(x,y) = f(x,x+2) = -2x+4$ for $-2 \le x \le 2$ $\Rightarrow f'(x,x+2) = -2 = 0 \Rightarrow$ no critical points in the interior of AB.
Endpoints: $f(-2,0) = 8$ and $f(2,4) = 0$.

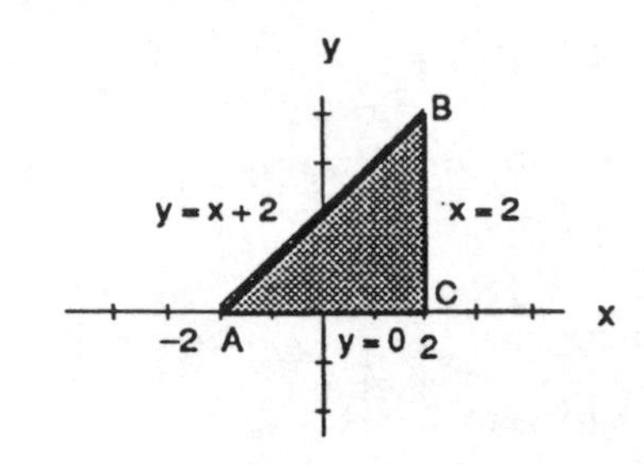

(ii) On BC, $f(x,y) = f(2,y) = -y^2 + 4y$ for $0 \le y \le 4$ $\Rightarrow f'(2,y) = -2y + 4 = 0 \Rightarrow y = 2$ and $x = 2 \Rightarrow (2,2)$ is an interior critical point of BC with $f(2,2) = 4$.
Endpoints: $f(2,0) = 0$ and $f(2,4) = 0$.

(iii) On AC, $f(x,y) = f(x,0) = x^2 - 2x$ for $-2 \le x \le 2$
$\Rightarrow f'(x,0) = 2x - 2 \Rightarrow x = 1$ and $y = 0 \Rightarrow (1,0)$ is an interior critical point of AC with $f(1,0) = -1$.
Endpoints: $f(-2,0) = 8$ and $f(2,0) = 0$.

(iv) For the interior of the triangular region, $f_x(x,y) = 2x - 2 = 0$ and $f_y(x,y) = -2y + 4 = 0 \Rightarrow x = 1$ and $y = 2 \Rightarrow (1,2)$ is an interior critical point of the region with $f(1,2) = 3$. Therefore the absolute maximum is 8 at $(-2,0)$ and the absolute minimum is -1 at $(1,0)$.

78. (i) On AB, $f(x,y) = f(x,x) = 4x^2 - 2x^4 + 16$ for $-2 \le x \le 2$ $\Rightarrow f'(x,x) = 8x - 8x^3 = 0 \Rightarrow x = 0$ and $y = 0$, or $x = 1$ and $y = 1$, or $x = -1$ and $y = -1 \Rightarrow (0,0)$, $(1,1)$, $(-1,-1)$ are all interior points of AB with $f(0,0) = 16$, $f(1,1) = 18$, and $f(-1,-1) = 18$.
Endpoints: $f(-2,-2) = 0$ and $f(2,2) = 0$.

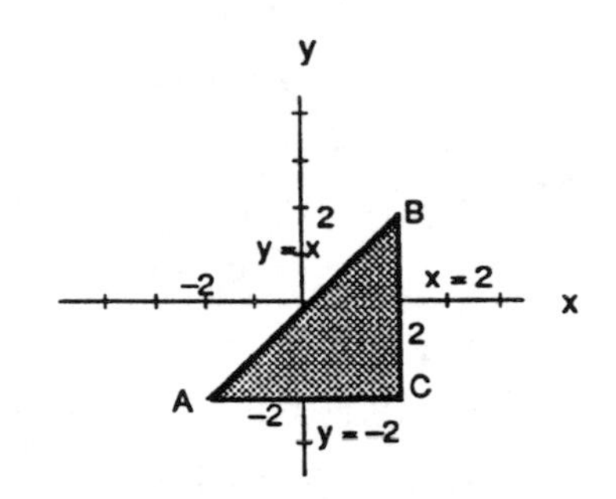

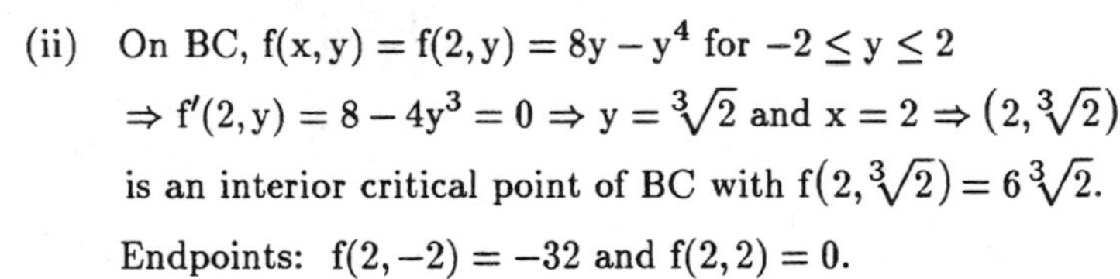

(ii) On BC, $f(x,y) = f(2,y) = 8y - y^4$ for $-2 \le y \le 2$ $\Rightarrow f'(2,y) = 8 - 4y^3 = 0 \Rightarrow y = \sqrt[3]{2}$ and $x = 2 \Rightarrow (2, \sqrt[3]{2})$ is an interior critical point of BC with $f(2,\sqrt[3]{2}) = 6\sqrt[3]{2}$.
Endpoints: $f(2,-2) = -32$ and $f(2,2) = 0$.

(iii) On AC, $f(x,y) = f(x,-2) = -8x - x^4$ for $-2 \le x \le 2 \Rightarrow f'(x,-2) = -8 - 4x^3 = 0 \Rightarrow x = \sqrt[3]{-2}$ and $y = -2$ $\Rightarrow (\sqrt[3]{-2}, -2)$ is an interior critical point of AC with $f(\sqrt[3]{-2}, -2) = 6\sqrt[3]{2}$. Endpoints: $f(-2,-2) = 0$ and $f(2,-2) = -32$.

(iv) For the interior of the triangular region, $f_x(x,y) = 4y - 4x^3 = 0$ and $f_y(x,y) = 4x - 4y^3 = 0 \Rightarrow x = 0$ and $y = 0$, or $x = 1$ and $y = 1$. But neither of the points $(0,0)$ and $(1,1)$ are interior to the region. Therefore the absolute maximum is 18 at $(1,1)$ and $(-1,-1)$, and the absolute minimum is -32 at $(2,-2)$.

79. (i) On AB, $f(x,y) = f(-1,y) = y^3 - 3y^2 + 2$ for $-1 \le y \le 1$ $\Rightarrow f'(-1,y) = 3y^2 - 6y = 0 \Rightarrow y = 0$ and $x = -1$, or $y = 2$ and $x = -1 \Rightarrow (-1,0)$ is an interior critical point of AB with $f(-1,0) = 2$; $(-1,2)$ is outside the boundary.
Endpoints: $f(-1,-1) = -2$ and $f(-1,1) = 0$.

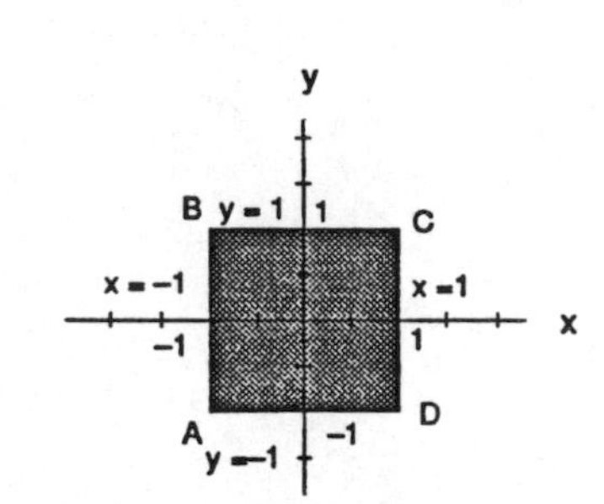

(ii) On BC, $f(x,y) = f(x,1) = x^3 + 3x^2 - 2$ for $-1 \le x \le 1 \Rightarrow f'(x,1) = 3x^2 + 6x = 0 \Rightarrow x = 0$ and $y = 1$, or $x = 2$ and $y = 1 \Rightarrow (0,1)$ is an interior critical point of BC with $f(0,1) = -2$; $(2,1)$ is outside the boundary. Endpoints: $f(-1,1) = 0$ and $f(1,1) = 2$.

(iii) On CD, $f(x,y) = f(1,y) = y^3 - 3y^2 + 4$ for $-1 \le y \le 1 \Rightarrow f'(1,y) = 3y^2 - 6y = 0 \Rightarrow y = 0$ and $x = 1$, or $y = 2$ and $x = 1 \Rightarrow (1,0)$ is an interior critical point of CD with $f(1,0) = 4$; $(1,2)$ is outside the boundary. Endpoints: $f(1,1) = 2$ and $f(1,-1) = 0$.

(iv) On AD, $f(x,y) = f(x,-1) = x^3 + 3x^2 - 4$ for $-1 \le x \le 1 \Rightarrow f'(x,-1) = 3x^2 + 6x = 0 \Rightarrow x = 0$ and $y = -1$, or $x = -2$ and $y = -1 \Rightarrow (0,-1)$ is an interior point of AD with $f(0,-1) = -4$; $(-2,-1)$ is outside the boundary. Endpoints: $f(-1,-1) = -2$ and $f(1,-1) = 0$.

(v) For the interior of the square, $f_x(x,y) = 3x^2 + 6x = 0$ and $f_y(x,y) = 3y^2 - 6y = 0 \Rightarrow x = 0$ or $x = -2$, and $y = 0$ or $y = 2 \Rightarrow (0,0)$ is an interior critical point of the square region with $f(0,0) = 0$; the points $(0,2)$, $(-2,0)$, and $(-2,2)$ are outside the region. Therefore the absolute maximum is 4 at $(1,0)$ and the absolute minimum is -4 at $(0,-1)$.

80. (i) On AB, $f(x,y) = f(-1,y) = y^3 - 3y$ for $-1 \le y \le 1$ $\Rightarrow f'(-1,y) = 3y^2 - 3 = 0 \Rightarrow y = \pm 1$ and $x = -1$ yielding the corner points $(-1,-1)$ and $(-1,1)$ with $f(-1,-1) = 2$ and $f(-1,1) = -2$.

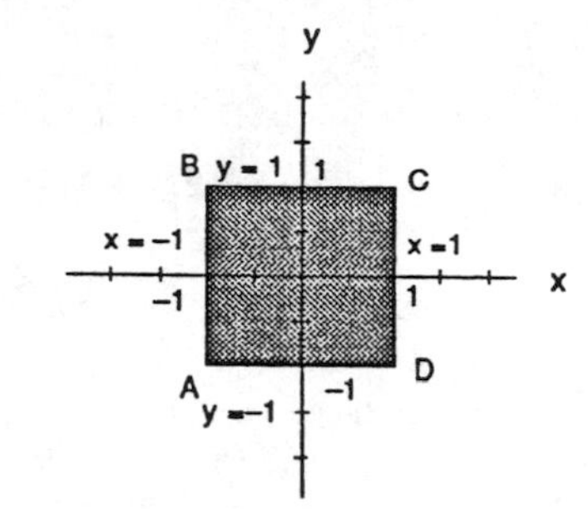

(ii) On BC, $f(x,y) = f(x,1) = x^3 + 3x + 2$ for $-1 \le x \le 1$ $\Rightarrow f'(x,1) = 3x^2 + 3 = 0 \Rightarrow$ no solution. Endpoints: $f(-1,1) = -2$ and $f(1,1) = 6$.

(iii) On CD, $f(x,y) = f(1,y) = y^3 + 3y + 2$ for $-1 \le y \le 1$ $\Rightarrow f'(1,y) = 3y^2 + 3 = 0 \Rightarrow$ no solution. Endpoints: $f(1,1) = 6$ and $f(1,-1) = -2$.

(iv) On AD, $f(x,y) = f(x,-1) = x^3 - 3x$ for $-1 \le x \le 1 \Rightarrow f'(x,-1) = 3x^2 - 3 = 0 \Rightarrow x = \pm 1$ and $y = -1$ yielding the corner points $(-1,-1)$ and $(1,-1)$ with $f(-1,-1) = 2$ and $f(1,-1) = -2$

(v) For the interior of the square, $f_x(x,y) = 3x^2 + 3y = 0$ and $f_y(x,y) = 3y^2 + 3x = 0 \Rightarrow y = -x^2$ and $x^4 + x = 0 \Rightarrow x = 0$ or $x = -1 \Rightarrow y = 0$ or $y = -1 \Rightarrow (0,0)$ is an interior critical point of the square region with $f(0,0) = 1$; $(-1,-1)$ is on the boundary. Therefore the absolute maximum is 6 at $(1,1)$ and the absolute minimum is -2 at $(1,-1)$ and $(-1,1)$.

81. $\nabla f = 3x^2\mathbf{i} + 2y\mathbf{j}$ and $\nabla g = 2x\mathbf{i} + 2y\mathbf{j}$ so that $\nabla f = \lambda \nabla g \Rightarrow 3x^2\mathbf{i} + 2y\mathbf{j} = \lambda(2x\mathbf{i} + 2y\mathbf{j}) \Rightarrow 3x^2 = 2x\lambda$ and $2y = 2y\lambda \Rightarrow \lambda = 1$ or $y = 0$.

CASE 1: $\lambda = 1 \Rightarrow 3x^2 = 2x \Rightarrow x = 0$ or $x = \frac{2}{3}$; $x = 0 \Rightarrow y = \pm 1$ yielding the points $(0,1)$ and $(0,-1)$; $x = \frac{2}{3}$ $\Rightarrow y = \pm\frac{\sqrt{5}}{3}$ yielding the points $\left(\frac{2}{3}, \frac{\sqrt{5}}{3}\right)$ and $\left(\frac{2}{3}, -\frac{\sqrt{5}}{3}\right)$.

CASE 2: $y = 0 \Rightarrow x^2 - 1 = 0 \Rightarrow x = \pm 1$ yielding the points $(1,0)$ and $(-1,0)$.

Evaluations give $f(0, \pm 1) = 1$, $f\left(\frac{2}{3}, \pm\frac{\sqrt{5}}{3}\right) = \frac{23}{27}$, $f(1,0) = 1$, and $f(-1,0) = -1$. Therefore the absolute maximum is 1 at $(0, \pm 1)$ and $(1,0)$, and the absolute minimum is -1 at $(-1,0)$.

82. $\nabla f = y\mathbf{i} + x\mathbf{j}$ and $\nabla g = 2x\mathbf{i} + 2y\mathbf{j}$ so that $\nabla f = \lambda \nabla g \Rightarrow y\mathbf{i} + x\mathbf{j} = \lambda(2x\mathbf{i} + 2y\mathbf{j}) \Rightarrow y = 2\lambda x$ and $xy = 2\lambda y \Rightarrow x = 2\lambda(2\lambda x) = 4\lambda^2 x \Rightarrow x = 0$ or $4\lambda^2 = 1$.

CASE 1: $x = 0 \Rightarrow y = 0$ but $(0,0)$ does not lie on the circle, so no solution.

CASE 2: $4\lambda^2 = 1 \Rightarrow \lambda = \frac{1}{2}$ or $\lambda = -\frac{1}{2}$. For $\lambda = \frac{1}{2}$, $y = x \Rightarrow 1 = x^2 + y^2 = 2x^2 \Rightarrow x = y = \pm\frac{1}{\sqrt{2}}$ yielding the points $\left(\frac{1}{\sqrt{2}}, \frac{1}{\sqrt{2}}\right)$ and $\left(-\frac{1}{\sqrt{2}}, -\frac{1}{\sqrt{2}}\right)$. For $\lambda = -\frac{1}{2}$, $y = -x \Rightarrow 1 = x^2 + y^2 = 2x^2 \Rightarrow x = \pm\frac{1}{\sqrt{2}}$ and $y = -x$ yielding the points $\left(-\frac{1}{\sqrt{2}}, \frac{1}{\sqrt{2}}\right)$ and $\left(\frac{1}{\sqrt{2}}, -\frac{1}{\sqrt{2}}\right)$.

Evaluations give the absolute maximum value $f\left(\frac{1}{\sqrt{2}}, \frac{1}{\sqrt{2}}\right) = f\left(-\frac{1}{\sqrt{2}}, -\frac{1}{\sqrt{2}}\right) = \frac{1}{2}$ and the absolute minimum value $f\left(-\frac{1}{\sqrt{2}}, \frac{1}{\sqrt{2}}\right) = f\left(\frac{1}{\sqrt{2}}, -\frac{1}{\sqrt{2}}\right) = -\frac{1}{2}$.

83. (i) $f(x,y) = x^2 + 3y^2 + 2y$ on $x^2 + y^2 = 1 \Rightarrow \nabla f = 2x\mathbf{i} + (6y+2)\mathbf{j}$ and $\nabla g = 2x\mathbf{i} + 2y\mathbf{j}$ so that $\nabla f = \lambda \nabla g$ $\Rightarrow 2x\mathbf{i} + (6y+2)\mathbf{j} = \lambda(2x\mathbf{i} + 2y\mathbf{j}) \Rightarrow 2x = 2x\lambda$ and $6y + 2 = 2y\lambda \Rightarrow \lambda = 1$ or $x = 0$.

CASE 1: $\lambda = 1 \Rightarrow 6y + 2 = 2y \Rightarrow y = -\frac{1}{2}$ and $x = \pm\frac{\sqrt{3}}{2}$ yielding the points $\left(\pm\frac{\sqrt{3}}{2}, -\frac{1}{2}\right)$.

CASE 2: $x = 0 \Rightarrow y^2 = 1 \Rightarrow y = \pm 1$ yielding the points $(0, \pm 1)$.

Evaluations give $f\left(\pm\frac{\sqrt{3}}{2}, -\frac{1}{2}\right) = \frac{1}{2}$, $f(0,1) = 5$, and $f(0,-1) = 1$. Therefore $\frac{1}{2}$ and 5 are the extreme values on the boundary of the disk.

(ii) For the interior of the disk, $f_x(x,y) = 2x = 0$ and $f_y(x,y) = 6y + 2 = 0 \Rightarrow x = 0$ and $y = -\frac{1}{3}$ $\Rightarrow \left(0, -\frac{1}{3}\right)$ is an interior critical point with $f\left(0, -\frac{1}{3}\right) = -\frac{1}{3}$. Therefore the absolute maximum of f on the disk is 5 at $(0,1)$ and the absolute minimum of f on the disk is $-\frac{1}{3}$ at $\left(0, -\frac{1}{3}\right)$.

84. (i) $f(x,y) = x^2 + y^2 - 3x - xy$ on $x^2 + y^2 = 9 \Rightarrow \nabla f = (2x - 3 - y)\mathbf{i} + (2y - x)\mathbf{j}$ and $\nabla g = 2x\mathbf{i} + 2y\mathbf{j}$ so that $\nabla f = \lambda \nabla g \Rightarrow (2x - 3 - y)\mathbf{i} + (2y - x)\mathbf{j} = \lambda(2x\mathbf{i} + 2y\mathbf{j}) \Rightarrow 2x - 3 - y = 2x\lambda$ and $2y - x = 2y\lambda$ $\Rightarrow 2x(1-\lambda) - y = 3$ and $-x + 2y(1-\lambda) = 0 \Rightarrow 1 - \lambda = \frac{x}{2y}$ and $(2x)\left(\frac{x}{2y}\right) - y = 3 \Rightarrow x^2 - y^2 = 3y$ $\Rightarrow x^2 = y^2 + 3y$. Thus, $9 = x^2 + y^2 = y^2 + 3y + y^2 \Rightarrow 2y^2 + 3y - 9 = 0 \Rightarrow (2y-3)(y+3) = 0$ $\Rightarrow y = -3, \frac{3}{2}$. For $y = -3$, $x^2 + y^2 = 9 \Rightarrow x = 0$ yielding the point $(0,-3)$. For $y = \frac{3}{2}$, $x^2 + y^2 = 9$ $\Rightarrow x^2 + \frac{9}{4} = 9 \Rightarrow x^2 = \frac{27}{4} \Rightarrow x = \pm\frac{3\sqrt{3}}{2}$. Evaluations give $f(0,-3) = 9$, $f\left(-\frac{3\sqrt{3}}{2}, \frac{3}{2}\right) = 9 + \frac{27\sqrt{3}}{4}$ ≈ 20.691, and $f\left(\frac{3\sqrt{3}}{2}, \frac{3}{2}\right) = 9 - \frac{27\sqrt{3}}{4} \approx -2.691$.

(ii) For the interior of the disk, $f_x(x,y) = 2x - 3 - y = 0$ and $f_y(x,y) = 2y - x = 0 \Rightarrow x = 2$ and $y = 1$ $\Rightarrow (2,1)$ is an interior critical point of the disk with $f(2,1) = -3$. Therefore, the absolute maximum of f on the disk is $9 + \frac{27\sqrt{3}}{4}$ at $\left(-\frac{3\sqrt{3}}{2}, \frac{3}{2}\right)$ and the absolute minimum of f on the disk is -3 at $(2,1)$.

85. $\nabla f = \mathbf{i} - \mathbf{j} + \mathbf{k}$ and $\nabla g = 2x\mathbf{i} + 2y\mathbf{j} + 2z\mathbf{k}$ so that $\nabla f = \lambda \nabla g \Rightarrow \mathbf{i} - \mathbf{j} + \mathbf{k} = \lambda(2x\mathbf{i} + 2y\mathbf{j} + 2z\mathbf{k}) \Rightarrow 1 = 2x\lambda$, $-1 = 2y\lambda$, $1 = 2z\lambda \Rightarrow x = -y = z = \frac{1}{\lambda}$. Thus $x^2 + y^2 + z^2 = 1 \Rightarrow 3x^2 = 1 \Rightarrow x = \pm\frac{1}{\sqrt{3}}$ yielding the points $\left(\frac{1}{\sqrt{3}}, -\frac{1}{\sqrt{3}}, \frac{1}{\sqrt{3}}\right)$ and $\left(-\frac{1}{\sqrt{3}}, \frac{1}{\sqrt{3}}, -\frac{1}{\sqrt{3}}\right)$. Evaluations give the absolute maximum value of $f\left(\frac{1}{\sqrt{3}}, -\frac{1}{\sqrt{3}}, \frac{1}{\sqrt{3}}\right) = \frac{3}{\sqrt{3}} = \sqrt{3}$ and the absolute minimum value of $f\left(-\frac{1}{\sqrt{3}}, \frac{1}{\sqrt{3}}, -\frac{1}{\sqrt{3}}\right) = -\sqrt{3}$.

86. Let $f(x,y,z) = x^2 + y^2 + z^2$ be the square of the distance to the origin and $g(x,y,z) = z^2 - xy - 4$. Then $\nabla f = 2x\mathbf{i} + 2y\mathbf{j} + 2z\mathbf{k}$ and $\nabla g = -y\mathbf{i} - x\mathbf{j} + 2z\mathbf{k}$ so that $\nabla f = \lambda \nabla g \Rightarrow 2x = -\lambda y$, $2y = -\lambda x$, and $2z = 2\lambda z$ $\Rightarrow z = 0$ or $\lambda = 1$.

CASE 1: $z = 0 \Rightarrow xy = -4 \Rightarrow x = -\frac{4}{y}$ and $y = -\frac{4}{x} \Rightarrow 2\left(-\frac{4}{y}\right) = -\lambda y$ and $2\left(-\frac{4}{x}\right) = -\lambda x \Rightarrow \frac{8}{\lambda} = y^2$ and $\frac{8}{\lambda} = x^2$ $\Rightarrow y^2 = x^2 \Rightarrow y = \pm x$. But $y = x \Rightarrow x^2 = -4$ leads to no solution, so $y = -x \Rightarrow x^2 = 4 \Rightarrow x = \pm 2$ yielding the points $(-2,2,0)$ and $(2,-2,0)$.

CASE 2: $\lambda = 1 \Rightarrow 2x = -y$ and $2y = -x \Rightarrow 2y = -\left(-\frac{y}{2}\right) \Rightarrow 4y = y \Rightarrow y = 0 \Rightarrow x = 0 \Rightarrow z^2 - 4 = 0 \Rightarrow z = \pm 2$ yielding the points $(0,0,-2)$ and $(0,0,2)$.

Evaluations give $f(-2,2,0) = f(2,-2,0) = 8$ and $f(0,0,-2) = f(0,0,2) = 4$. Thus the points $(0,0,-2)$ and $(0,0,2)$ on the surface are closest to the origin.

87. The cost is $f(x,y,z) = 2axy + 2bxz + 2cyz$ subject to the constraint $xyz = V$. Then $\nabla f = \lambda \nabla g$ $\Rightarrow 2ay + 2bz = \lambda yz$, $2ax + 2cz = \lambda xz$, and $2bx + 2cy = \lambda xy \Rightarrow 2axy + 2bxz = \lambda xyz$, $2axy + 2cyz = \lambda xyz$, and $2bxz + 2cyz = \lambda xyz \Rightarrow 2axy + 2bxz = 2axy + 2cyz \Rightarrow y = \left(\frac{b}{c}\right)x$. Also $2axy + 2bxz = 2bxz + 2cyz \Rightarrow z = \left(\frac{a}{c}\right)x$. Then $x\left(\frac{b}{c}x\right)\left(\frac{a}{c}x\right) = V \Rightarrow x^3 = \frac{c^2V}{ab} \Rightarrow \text{width} = x = \left(\frac{c^2V}{ab}\right)^{1/3}$, $\text{Depth} = y = \left(\frac{b}{c}\right)\left(\frac{c^2V}{ab}\right)^{1/3} = \left(\frac{b^2V}{ac}\right)^{1/3}$, and $\text{Height} = z = \left(\frac{a}{c}\right)\left(\frac{c^2V}{ab}\right)^{1/3} = \left(\frac{a^2V}{bc}\right)^{1/3}$.

88. The volume of the pyramid in the first octant formed by the plane is $V(a,b,c) = \frac{1}{3}\left(\frac{1}{2}ab\right)c = \frac{1}{6}abc$. The point $(2,1,2)$ on the plane $\Rightarrow \frac{2}{a} + \frac{1}{b} + \frac{2}{c} = 1$. We want to maximize V subject to the constraint $2bc + ac + 2ab = abc$. Thus, $\nabla V = \frac{bc}{6}\mathbf{i} + \frac{ac}{6}\mathbf{j} + \frac{ab}{6}\mathbf{k}$ and $\nabla g = (c + 2b - bc)\mathbf{i} + (2c + 2a - ac)\mathbf{j} + (2b + a - ab)\mathbf{k}$ so that $\nabla V = \lambda \nabla g$ $\Rightarrow \frac{bc}{6} = \lambda(c + 2b - bc)$, $\frac{ac}{6} = \lambda(2c + 2a - ac)$, and $\frac{ab}{6} = \lambda(2b + a - ab) \Rightarrow \frac{abc}{6} = \lambda(ac + 2ab - abc)$, $\frac{abc}{6} = \lambda(2bc + 2ab - abc)$, and $\frac{abc}{6} = \lambda(2bc + ac - abc) \Rightarrow \lambda ac = 2\lambda bc$ and $2\lambda ab = 2\lambda bc$. Now $\lambda \neq 0$ since $a \neq 0$, $b \neq 0$, and $c \neq 0 \Rightarrow ac = 2bc$ and $ab = bc \Rightarrow a = 2b = c$. Substituting into the constraint equation gives $\frac{2}{a} + \frac{2}{a} + \frac{2}{a} = 1 \Rightarrow a = 6 \Rightarrow b = 3$ and $c = 6$. Therefore the desired plane is $\frac{x}{6} + \frac{y}{3} + \frac{z}{6} = 1$ or $x + 2y + z = 6$.

89. $\nabla f = (y+z)\mathbf{i} + x\mathbf{j} + x\mathbf{k}$, $\nabla g = 2x\mathbf{i} + 2y\mathbf{j}$, and $\nabla h = z\mathbf{i} + x\mathbf{k}$ so that $\nabla f = \lambda \nabla g + \mu \nabla h$ $\Rightarrow (y+z)\mathbf{i} + x\mathbf{j} + x\mathbf{k} = \lambda(2x\mathbf{i} + 2y\mathbf{j}) + \mu(z\mathbf{i} + x\mathbf{k}) \Rightarrow y + z = 2\lambda x + \mu z$, $x = 2\lambda y$, $x = \mu x \Rightarrow x = 0$ or $\mu = 1$.

CASE 1: $x = 0$ which is impossible since $xz = 1$.

CASE 2: $\mu = 1 \Rightarrow y + z = 2\lambda x + z \Rightarrow y = 2\lambda x$ and $x = 2\lambda y \Rightarrow y = (2\lambda)(2\lambda y) \Rightarrow y = 0$ or $4\lambda^2 = 1$. If $y = 0$, then $x^2 = 1 \Rightarrow x = \pm 1$ so with $xz = 1$ we obtain the points $(1,0,1)$ and $(-1,0,-1)$. If $4\lambda^2 = 1$, then $\lambda = \pm\frac{1}{2}$. For $\lambda = -\frac{1}{2}$, $y = -x$ so $x^2 + y^2 = 1 \Rightarrow x^2 = \frac{1}{2}$ $\Rightarrow x = \pm\frac{1}{\sqrt{2}}$ with $xz = 1 \Rightarrow z = \pm\sqrt{2}$, and we obtain the points $\left(\frac{1}{\sqrt{2}}, -\frac{1}{\sqrt{2}}, \sqrt{2}\right)$ and $\left(-\frac{1}{\sqrt{2}}, \frac{1}{\sqrt{2}}, -\sqrt{2}\right)$. For $\lambda = \frac{1}{2}$, $y = x \Rightarrow x^2 = \frac{1}{2} \Rightarrow x = \pm\frac{1}{\sqrt{2}}$ with $xz = 1 \Rightarrow z = \pm\sqrt{2}$, and we obtain the points $\left(\frac{1}{\sqrt{2}}, \frac{1}{\sqrt{2}}, \sqrt{2}\right)$ and $\left(-\frac{1}{\sqrt{2}}, -\frac{1}{\sqrt{2}}, -\sqrt{2}\right)$.

Evaluations give $f(1,0,1) = 1$, $f(-1,0,-1) = 1$, $f\left(\frac{1}{\sqrt{2}}, -\frac{1}{\sqrt{2}}, \sqrt{2}\right) = \frac{1}{2}$, $f\left(-\frac{1}{\sqrt{2}}, \frac{1}{\sqrt{2}}, -\sqrt{2}\right) = \frac{1}{2}$, $f\left(\frac{1}{\sqrt{2}}, \frac{1}{\sqrt{2}}, \sqrt{2}\right) = \frac{3}{2}$, and $f\left(-\frac{1}{\sqrt{2}}, -\frac{1}{\sqrt{2}}, -\sqrt{2}\right) = \frac{3}{2}$. Therefore the absolute maximum is $\frac{3}{2}$ at $\left(\frac{1}{\sqrt{2}}, \frac{1}{\sqrt{2}}, \sqrt{2}\right)$ and $\left(-\frac{1}{\sqrt{2}}, -\frac{1}{\sqrt{2}}, -\sqrt{2}\right)$, and the absolute minimum is $\frac{1}{2}$ at $\left(-\frac{1}{\sqrt{2}}, \frac{1}{\sqrt{2}}, -\sqrt{2}\right)$ and $\left(\frac{1}{\sqrt{2}}, -\frac{1}{\sqrt{2}}, \sqrt{2}\right)$.

90. Let $f(x,y,z) = x^2 + y^2 + z^2$ be the square of the distance to the origin. Then $\nabla f = 2x\mathbf{i} + 2y\mathbf{j} + 2z\mathbf{k}$, $\nabla g = \mathbf{i} + \mathbf{j} + \mathbf{k}$, and $\nabla h = 4x\mathbf{i} + 4y\mathbf{j} - 2z\mathbf{k}$ so that $\nabla f = \lambda \nabla g + \mu \nabla h \Rightarrow 2x = \lambda + 4x\mu$, $2y = \lambda + 4y\mu$, and $2z = \lambda - 2z\mu \Rightarrow \lambda = 2x(1 - 2\mu) = 2y(1 - 2\mu) = 2z(1 + 2\mu) \Rightarrow x = y$ or $\mu = \frac{1}{2}$.

CASE 1: $x = y \Rightarrow z^2 = 4x^2 \Rightarrow z = \pm 2x$ so that $x + y + z = 1 \Rightarrow x + x + 2x = 1$ or $x + x - 2x = 1$ (impossible) $\Rightarrow x = \frac{1}{4} \Rightarrow y = \frac{1}{4}$ and $z = \frac{1}{2}$ yielding the point $\left(\frac{1}{4}, \frac{1}{4}, \frac{1}{2}\right)$.

CASE 2: $\mu = \frac{1}{2} \Rightarrow \lambda = 0 \Rightarrow 0 = 2z(1+1) \Rightarrow z = 0$ so that $2x^2 + 2y^2 = 0 \Rightarrow x = y = 0$. But the origin $(0,0,0)$ fails to satisfy the first constraint $x + y + z = 1$.

Therefore, the point $\left(\frac{1}{4}, \frac{1}{4}, \frac{1}{2}\right)$ on the curve of intersection is closest to the origin.

91. Note that $x = r\cos\theta$ and $y = r\sin\theta \Rightarrow r = \sqrt{x^2 + y^2}$ and $\theta = \tan^{-1}\left(\frac{y}{x}\right)$. Thus,

$$\frac{\partial w}{\partial x} = \frac{\partial w}{\partial r}\frac{\partial r}{\partial x} + \frac{\partial w}{\partial \theta}\frac{\partial \theta}{\partial x} = \left(\frac{\partial w}{\partial r}\right)\left(\frac{x}{\sqrt{x^2+y^2}}\right) + \left(\frac{\partial w}{\partial \theta}\right)\left(\frac{-y}{x^2+y^2}\right) = (\cos\theta)\frac{\partial w}{\partial r} - \left(\frac{\sin\theta}{r}\right)\frac{\partial w}{\partial \theta};$$

$$\frac{\partial w}{\partial y} = \frac{\partial w}{\partial r}\frac{\partial r}{\partial y} + \frac{\partial w}{\partial \theta}\frac{\partial \theta}{\partial y} = \left(\frac{\partial w}{\partial r}\right)\left(\frac{y}{\sqrt{x^2+y^2}}\right) + \left(\frac{\partial w}{\partial \theta}\right)\left(\frac{x}{x^2+y^2}\right) = (\sin\theta)\frac{\partial w}{\partial r} + \left(\frac{\cos\theta}{r}\right)\frac{\partial w}{\partial \theta}$$

92. $z_x = f_u\frac{\partial u}{\partial x} + f_v\frac{\partial v}{\partial x} = af_u + af_v$, and $z_y = f_u\frac{\partial u}{\partial y} + f_v\frac{\partial v}{\partial y} = bf_u - bf_v$

93. $\frac{\partial u}{\partial y} = b$ and $\frac{\partial u}{\partial x} = a \Rightarrow \frac{\partial w}{\partial x} = \frac{dw}{du}\frac{\partial u}{\partial x} = a\frac{dw}{du}$ and $\frac{\partial w}{\partial y} = \frac{dw}{du}\frac{\partial u}{\partial y} = b\frac{dw}{du} \Rightarrow \frac{1}{a}\frac{\partial w}{\partial x} = \frac{dw}{du}$ and $\frac{1}{b}\frac{\partial w}{\partial y} = \frac{dw}{du}$
$\Rightarrow \frac{1}{a}\frac{\partial w}{\partial x} = \frac{1}{b}\frac{\partial w}{\partial y} \Rightarrow b\frac{\partial w}{\partial x} = a\frac{\partial w}{\partial y}$

94. $\frac{\partial w}{\partial x} = \frac{2x}{x^2+y^2+2z} = \frac{2(r+s)}{(r+s)^2+(r-s)^2+4rs} = \frac{2(r+s)}{2(r^2+2rs+s^2)} = \frac{1}{r+s}$, $\frac{\partial w}{\partial y} = \frac{2y}{x^2+y^2+2z} = \frac{2(r-s)}{2(r+s)^2} = \frac{r-s}{(r+s)^2}$,
and $\frac{\partial w}{\partial z} = \frac{2}{x^2+y^2+2z} = \frac{1}{(r+s)^2} \Rightarrow \frac{\partial w}{\partial r} = \frac{\partial w}{\partial x}\frac{\partial x}{\partial r} + \frac{\partial w}{\partial y}\frac{\partial y}{\partial r} + \frac{\partial w}{\partial z}\frac{\partial z}{\partial r} = \frac{1}{r+s} + \frac{r-s}{(r+s)^2} + \left[\frac{1}{(r+s)^2}\right](2s) = \frac{2r+2s}{(r+s)^2}$
$= \frac{2}{r+s}$ and $\frac{\partial w}{\partial s} = \frac{\partial w}{\partial x}\frac{\partial x}{\partial s} + \frac{\partial w}{\partial y}\frac{\partial y}{\partial s} + \frac{\partial w}{\partial z}\frac{\partial z}{\partial s} = \frac{1}{r+s} - \frac{r-s}{(r+s)^2} + \left[\frac{1}{(r+s)^2}\right](2r) = \frac{2}{r+s}$

95. $e^u \cos v - x = 0 \Rightarrow (e^u \cos v)\frac{\partial u}{\partial x} - (e^u \sin v)\frac{\partial v}{\partial x} = 1$; $e^u \sin v - y = 0 \Rightarrow (e^u \sin v)\frac{\partial u}{\partial x} + (e^u \cos v)\frac{\partial v}{\partial x} = 0$.
Solving this system yields $\frac{\partial u}{\partial x} = e^{-u}\cos v$ and $\frac{\partial v}{\partial x} = -e^{-u}\sin v$. Similarly, $e^u \cos v - x = 0$
$\Rightarrow (e^u \cos v)\frac{\partial u}{\partial y} - (e^u \sin v)\frac{\partial v}{\partial y} = 0$ and $e^u \sin v - y = 0 \Rightarrow (e^u \sin v)\frac{\partial u}{\partial y} + (e^u \cos v)\frac{\partial v}{\partial y} = 1$. Solving this second system yields $\frac{\partial u}{\partial y} = e^{-u}\sin v$ and $\frac{\partial v}{\partial y} = e^{-u}\cos v$. Therefore $\left(\frac{\partial u}{\partial x}\mathbf{i} + \frac{\partial u}{\partial y}\mathbf{j}\right)\cdot\left(\frac{\partial v}{\partial x}\mathbf{i} + \frac{\partial v}{\partial y}\mathbf{j}\right)$
$= [(e^{-u}\cos v)\mathbf{i} + (e^{-u}\sin v)\mathbf{j}]\cdot[(-e^{-u}\sin v)\mathbf{i} + (e^{-u}\cos v)\mathbf{j}] = 0 \Rightarrow$ the vectors are orthogonal $\Rightarrow$ the angle between the vectors is the constant $\frac{\pi}{2}$.

96. $\frac{\partial g}{\partial \theta} = \frac{\partial f}{\partial x}\frac{\partial x}{\partial \theta} + \frac{\partial f}{\partial y}\frac{\partial y}{\partial \theta} = (-r\sin\theta)\frac{\partial f}{\partial x} + (r\cos\theta)\frac{\partial f}{\partial y}$
$\Rightarrow \frac{\partial^2 g}{\partial \theta^2} = (-r\sin\theta)\left(\frac{\partial^2 f}{\partial x^2}\frac{\partial x}{\partial \theta} + \frac{\partial^2 f}{\partial y\partial x}\frac{\partial y}{\partial \theta}\right) - (r\cos\theta)\frac{\partial f}{\partial x} + (r\cos\theta)\left(\frac{\partial^2 f}{\partial x\partial y}\frac{\partial x}{\partial \theta} + \frac{\partial^2 f}{\partial y^2}\frac{\partial y}{\partial \theta}\right) - (r\sin\theta)\frac{\partial f}{\partial y}$
$= (-r\sin\theta)\left(\frac{\partial x}{\partial \theta} + \frac{\partial y}{\partial \theta}\right) - (r\cos\theta) + (r\cos\theta)\left(\frac{\partial x}{\partial \theta} + \frac{\partial y}{\partial \theta}\right) - (r\sin\theta)$
$= (-r\sin\theta + r\cos\theta)(-r\sin\theta + r\cos\theta) - (r\cos\theta + r\sin\theta) = (-2)(-2) - (0+2) = 4 - 2 = 2$ at $(r,\theta) = \left(2, \frac{\pi}{2}\right)$.

97. $(y+z)^2 + (z-x)^2 = 16 \Rightarrow \nabla f = -2(z-x)\mathbf{i} + 2(y+z)\mathbf{j} + 2(y+2z-x)\mathbf{k}$; if the normal line is parallel to the yz-plane, then x is constant $\Rightarrow \frac{\partial f}{\partial x} = 0 \Rightarrow -2(z-x) = 0 \Rightarrow z = x \Rightarrow (y+z)^2 + (z-z)^2 = 16 \Rightarrow y + z = \pm 4$.
Let $x = t \Rightarrow z = t \Rightarrow y = -t \pm 4$. Therefore the points are $(t, -t \pm 4, t)$, t a real number.

98. Let $f(x,y,z) = xy + yz + zx - x - z^2 = 0$. If the tangent plane is to be parallel to the xy-plane, then ∇f is perpendicular to the xy-plane $\Rightarrow \nabla f \cdot \mathbf{i} = 0$ and $\nabla f \cdot \mathbf{j} = 0$. Now $\nabla f = (y+z-1)\mathbf{i} + (x+z)\mathbf{j} + (y+x-2z)\mathbf{k}$ so that $\nabla f \cdot \mathbf{i} = y + z - 1 = 0 \Rightarrow y + z = 1 \Rightarrow y = 1 - z$, and $\nabla f \cdot \mathbf{j} = x + z = 0 \Rightarrow x = -z$. Then
$-z(1-z) + (1-z)z + z(-z) - (-z) - z^2 = 0 \Rightarrow z - 2z^2 = 0 \Rightarrow z = \frac{1}{2}$ or $z = 0$. Now $z = \frac{1}{2} \Rightarrow x = -\frac{1}{2}$ and $y = \frac{1}{2}$
$\Rightarrow \left(-\frac{1}{2}, \frac{1}{2}, \frac{1}{2}\right)$ is one desired point; $z = 0 \Rightarrow x = 0$ and $y = 1 \Rightarrow (0,1,0)$ is a second desired point.

99. $\nabla f = \lambda(x\mathbf{i} + y\mathbf{j} + z\mathbf{k}) \Rightarrow \frac{\partial f}{\partial x} = \lambda x \Rightarrow f(x,y,z) = \frac{1}{2}\lambda x^2 + g(y,z)$ for some function $g \Rightarrow \lambda y = \frac{\partial f}{\partial y} = \frac{\partial g}{\partial y}$ $\Rightarrow g(y,z) = \frac{1}{2}\lambda y^2 + h(z)$ for some function $h \Rightarrow yz = \frac{\partial f}{\partial z} = \frac{\partial g}{\partial z} = h'(z) \Rightarrow h(z) = \frac{1}{2}\lambda z^2 + C$ for some arbitrary constant $C \Rightarrow g(y,z) = \frac{1}{2}\lambda y^2 + \left(\frac{1}{2}\lambda z^2 + C\right) \Rightarrow f(x,y,z) = \frac{1}{2}\lambda x^2 + \frac{1}{2}\lambda y^2 + \frac{1}{2}\lambda z^2 + C \Rightarrow f(0,0,a) = \frac{1}{2}\lambda a^2 + C$ and $f(0,0,-a) = \frac{1}{2}\lambda(-a)^2 + C \Rightarrow f(0,0,a) = f(0,0,-a)$ for any constant a, as claimed.

100. $$\left(\frac{df}{ds}\right)_{\mathbf{u},(0,0,0)} = \lim_{s\to 0} \frac{f(0+su_1, 0+su_2, 0+su_3) - f(0,0,0)}{s},\ s>0$$
$$= \lim_{s\to 0} \frac{\sqrt{s^2u_1^2 + s^2u_2^2 + s^2u_3^2} - 0}{s},\ s>0$$
$$= \lim_{s\to 0} \frac{s\sqrt{u_1^2+u_2^2+u_3^2}}{s} = \lim_{s\to 0} |\mathbf{u}| = 1;$$
however, $\nabla f = \frac{x}{\sqrt{x^2+y^2+z^2}}\mathbf{i} + \frac{y}{\sqrt{x^2+y^2+z^2}}\mathbf{j} + \frac{z}{\sqrt{x^2+y^2+z^2}}\mathbf{k}$ fails to exist at the origin $(0,0,0)$

101. Let $f(x,y,z) = xy + z - 2 \Rightarrow \nabla f = y\mathbf{i} + z\mathbf{j} + \mathbf{k}$. At $(1,1,1)$, we have $\nabla f = \mathbf{i} + \mathbf{j} + \mathbf{k} \Rightarrow$ the normal line is $x = 1 + t,\ y = 1 + t,\ z = 1 + t$, so at $t = -1 \Rightarrow x = 0,\ y = 0,\ z = 0$ and the normal line passes through the origin.

102. (b) $f(x,y,z) = x^2 - y^2 + z^2 = 4$
$\Rightarrow \nabla f = 2x\mathbf{i} - 2y\mathbf{j} + 2z\mathbf{k} \Rightarrow$ at $(2,-3,3)$ the gradient is $\nabla f = 4\mathbf{i} + 6\mathbf{j} + 6\mathbf{k}$ which is normal to the surface

(c) Tangent plane: $4x + 6y + 6z = 8$ or $2x + 3y + 3z = 4$
Normal line: $x = 2 + 4t,\ y = -3 + 6t,\ z = 3 + 6t$

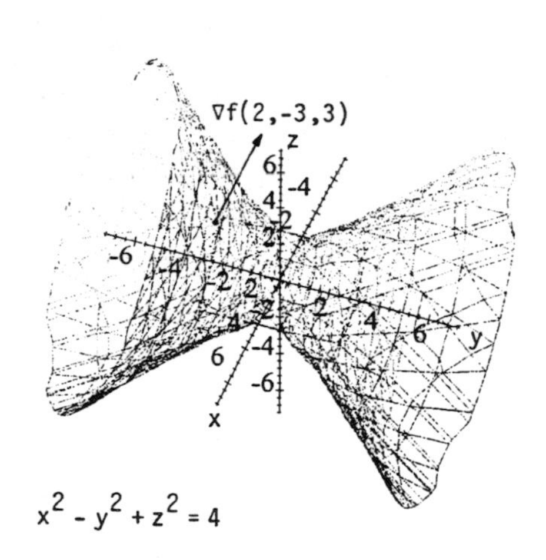

CHAPTER 12 ADDITIONAL EXERCISES–THEORY, EXAMPLES, APPLICATIONS

1. By definition, $f_{xy}(0,0) = \lim_{h\to 0} \frac{f_x(0,h) - f_x(0,0)}{h}$ so we need to calculate the first partial derivatives in the numerator. For $(x,y) \neq (0,0)$ we calculate $f_x(x,y)$ by applying the differentiation rules to the formula for $f(x,y)$: $f_x(x,y) = \frac{x^2y - y^3}{x^2+y^2} + (xy)\frac{(x^2+y^2)(2x) - (x^2-y^2)(2x)}{(x^2+y^2)^2} = \frac{x^2y - y^3}{x^2+y^2} + \frac{4x^2y^3}{(x^2+y^2)^2} \Rightarrow f_x(0,h) = -\frac{h^3}{h^2} = -h.$

For $(x,y)=(0,0)$ we apply the definition: $f_x(0,0)=\lim_{h\to 0}\frac{f(h,0)-f(0,0)}{h}=\lim_{h\to 0}\frac{0-0}{h}=0$. Then by definition $f_{xy}(0,0)=\lim_{h\to 0}\frac{-h-0}{h}=-1$. Similarly, $f_{yx}(0,0)=\lim_{h\to 0}\frac{f_y(h,0)-f_y(0,0)}{h}$, so for $(x,y)\neq(0,0)$ we have $f_y(x,y)=\frac{x^3-xy^2}{x^2+y^2}-\frac{4x^3y^2}{(x^2+y^2)^2}\Rightarrow f_y(h,0)=\frac{h^3}{h^2}=h$; for $(x,y)=(0,0)$ we obtain $f_y(0,0)=\lim_{h\to 0}\frac{f(0,h)-f(0,0)}{h}$ $=\lim_{h\to 0}\frac{0-0}{h}=0$. Then by definition $f_{yx}(0,0)=\lim_{h\to 0}\frac{h-0}{h}=1$. Note that $f_{xy}(0,0)\neq f_{yx}(0,0)$ in this case.

2. $\frac{\partial w}{\partial x}=1+e^x\cos y\Rightarrow w=x+e^x\cos y+g(y)$; $\frac{\partial w}{\partial y}=-e^x\sin y+g'(y)=2y-e^x\sin y\Rightarrow g'(y)=2y$ $\Rightarrow g(y)=y^2+C$; $w=\ln 2$ when $x=\ln 2$ and $y=0\Rightarrow \ln 2=\ln 2+e^{\ln 2}\cos 0+0^2+C\Rightarrow 0=2+C$ $\Rightarrow C=-2$. Thus, $w=x+e^x\cos y+g(y)=x+e^x\cos y+y^2-2$.

3. Substitution of $u+u(x)$ and $v=v(x)$ in $g(u,v)$ gives $g(u(x),v(x))$ which is a function of the independent variable x. Then, $g(u,v)=\int_u^v f(t)\,dt\Rightarrow\frac{dg}{dx}=\frac{\partial g}{\partial u}\frac{du}{dx}+\frac{\partial g}{\partial v}\frac{dv}{dx}=\left(\frac{\partial}{\partial u}\int_u^v f(t)\,dt\right)\frac{du}{dx}+\left(\frac{\partial}{\partial v}\int_u^v f(t)\,dt\right)\frac{dv}{dx}$ $=\left(-\frac{\partial}{\partial u}\int_v^u f(t)\,dt\right)\frac{du}{dx}+\left(\frac{\partial}{\partial v}\int_u^v f(t)\,dt\right)\frac{dv}{dx}=-f(u(x))\frac{du}{dx}+f(v(x))\frac{dv}{dx}=f(v(x))\frac{dv}{dx}-f(u(x))\frac{du}{dx}$

4. Applying the chain rules, $f_x=\frac{df}{dr}\frac{\partial r}{\partial x}\Rightarrow f_{xx}=\left(\frac{d^2f}{dr^2}\right)\left(\frac{\partial r}{\partial x}\right)^2+\frac{df}{dr}\frac{\partial^2 r}{\partial x^2}$. Similarly, $f_{yy}=\left(\frac{d^2f}{dr^2}\right)\left(\frac{\partial r}{\partial y}\right)^2+\frac{df}{dr}\frac{\partial^2 r}{\partial y^2}$ and $f_{zz}=\left(\frac{d^2f}{dr^2}\right)\left(\frac{\partial r}{\partial z}\right)^2+\frac{df}{dr}\frac{\partial^2 r}{\partial z^2}$. Moreover, $\frac{\partial r}{\partial x}=\frac{x}{\sqrt{x^2+y^2+z^2}}\Rightarrow\frac{\partial^2 r}{\partial x^2}=\frac{y^2+z^2}{\left(\sqrt{x^2+y^2+z^2}\right)^2}$; $\frac{\partial r}{\partial y}=\frac{y}{\sqrt{x^2+y^2+z^2}}$ $\Rightarrow\frac{\partial^2 r}{\partial y^2}=\frac{x^2+z^2}{\left(\sqrt{x^2+y^2+z^2}\right)^2}$; and $\frac{\partial r}{\partial z}=\frac{z}{\sqrt{x^2+y^2+z^2}}\Rightarrow\frac{\partial^2 r}{\partial z^2}=\frac{x^2+y^2}{\left(\sqrt{x^2+y^2+z^2}\right)^2}$. Next, $f_{xx}+f_{yy}+f_{zz}=0$

$$\Rightarrow\left(\frac{d^2f}{dr^2}\right)\left(\frac{x^2}{x^2+y^2+z^2}\right)+\left(\frac{df}{dr}\right)\left(\frac{y^2+z^2}{\left(\sqrt{x^2+y^2+z^2}\right)^2}\right)+\left(\frac{d^2f}{dr^2}\right)\left(\frac{y^2}{x^2+y^2+z^2}\right)+\left(\frac{df}{dr}\right)\left(\frac{x^2+z^2}{\left(\sqrt{x^2+y^2+z^2}\right)^2}\right)$$

$$+\left(\frac{d^2f}{dr^2}\right)\left(\frac{z^2}{x^2+y^2+z^2}\right)+\left(\frac{df}{dr}\right)\left(\frac{x^2+y^2}{\left(\sqrt{x^2+y^2+z^2}\right)^2}\right)=0\Rightarrow\frac{d^2f}{dr^2}+\left(\frac{2}{\sqrt{x^2+y^2+z^2}}\right)\frac{df}{dr}=0\Rightarrow\frac{d^2f}{dr^2}+\frac{2}{r}\frac{df}{dr}=0$$

$\Rightarrow\frac{d}{dr}(f')=\left(-\frac{2}{r}\right)f'$, where $f'=\frac{df}{dr}\Rightarrow\frac{df'}{f'}=-\frac{2\,dr}{r}\Rightarrow\ln f'=-2\ln r+\ln C\Rightarrow f'=Cr^{-2}$, or $\frac{df}{dr}=Cr^{-2}\Rightarrow f(r)=-\frac{C}{r}+b=\frac{a}{r}+b$ for some constants a and b (setting $a=-C$)

5. (a) Let $u=tx$, $v=ty$, and $w=f(u,v)=f(u(t,x),v(t,y))=f(tx,ty)=t^nf(x,y)$, where t, x, and y are independent variables. Then $nt^{n-1}f(x,y)=\frac{\partial w}{\partial t}=\frac{\partial w}{\partial u}\frac{\partial u}{\partial t}+\frac{\partial w}{\partial v}\frac{\partial v}{\partial t}=x\frac{\partial w}{\partial u}+y\frac{\partial w}{\partial v}$. Now, $\frac{\partial w}{\partial x}=\frac{\partial w}{\partial u}\frac{\partial u}{\partial x}+\frac{\partial w}{\partial v}\frac{\partial v}{\partial x}=\left(\frac{\partial w}{\partial u}\right)(t)+\left(\frac{\partial w}{\partial v}\right)(0)=t\frac{\partial w}{\partial u}\Rightarrow\frac{\partial w}{\partial u}=\left(\frac{1}{t}\right)\left(\frac{\partial w}{\partial x}\right)$. Likewise, $\frac{\partial w}{\partial y}=\frac{\partial w}{\partial u}\frac{\partial u}{\partial y}+\frac{\partial w}{\partial v}\frac{\partial v}{\partial y}=\left(\frac{\partial w}{\partial u}\right)(0)+\left(\frac{\partial w}{\partial v}\right)(t)\Rightarrow\frac{\partial w}{\partial v}=\left(\frac{1}{t}\right)\left(\frac{\partial w}{\partial y}\right)$. Therefore,

$nt^{n-1}f(x,y) = x\,\frac{\partial w}{\partial u} + y\,\frac{\partial w}{\partial v} = \left(\frac{x}{t}\right)\left(\frac{\partial w}{\partial x}\right) + \left(\frac{y}{t}\right)\left(\frac{\partial w}{\partial y}\right) = u\,\frac{\partial w}{\partial x} + v\,\frac{\partial w}{\partial y}$. When $t = 1$, $u = x$, $v = y$, and $w = f(x,y) \Rightarrow \frac{\partial w}{\partial x} = \frac{\partial f}{\partial x}$ and $\frac{\partial w}{\partial y} = \frac{\partial f}{\partial x} \Rightarrow nf(x,y) = x\,\frac{\partial f}{\partial x} + y\,\frac{\partial f}{\partial y}$, as claimed.

(b) From part (a), $nt^{n-1}f(x,y) = x\,\frac{\partial w}{\partial u} + y\,\frac{\partial w}{\partial v}$. Differentiating with respect to t again we obtain

$n(n-1)t^{n-2}f(x,y) = x\,\frac{\partial^2 w}{\partial u^2}\,\frac{\partial u}{\partial t} + x\,\frac{\partial^2 w}{\partial v \partial w}\,\frac{\partial v}{\partial t} + y\,\frac{\partial^2 w}{\partial u \partial v}\,\frac{\partial u}{\partial t} + y\,\frac{\partial^2 w}{\partial v^2}\,\frac{\partial v}{\partial t} = x^2\,\frac{\partial^2 w}{\partial u^2} + 2xy\,\frac{\partial^2 w}{\partial u \partial v} + y^2\,\frac{\partial^2 w}{\partial v^2}$.

Also from part (a), $\frac{\partial^2 w}{\partial x^2} = \frac{\partial}{\partial x}\left(\frac{\partial w}{\partial x}\right) = \frac{\partial}{\partial x}\left(t\,\frac{\partial w}{\partial u}\right) = t\,\frac{\partial^2 w}{\partial u^2}\,\frac{\partial u}{\partial x} + t\,\frac{\partial^2 w}{\partial v \partial u}\,\frac{\partial v}{\partial x} = t^2\,\frac{\partial^2 w}{\partial u^2}$, $\frac{\partial^2 w}{\partial y^2} = \frac{\partial}{\partial y}\left(\frac{\partial w}{\partial y}\right)$

$= \frac{\partial}{\partial y}\left(t\,\frac{\partial w}{\partial v}\right) = t\,\frac{\partial^2 w}{\partial u \partial v}\,\frac{\partial u}{\partial y} + t\,\frac{\partial^2 w}{\partial v^2}\,\frac{\partial v}{\partial y} = t^2\,\frac{\partial^2 w}{\partial v^2}$, and $\frac{\partial^2 w}{\partial y \partial x} = \frac{\partial}{\partial y}\left(\frac{\partial w}{\partial x}\right) = \frac{\partial}{\partial y}\left(t\,\frac{\partial w}{\partial u}\right) = t\,\frac{\partial^2 w}{\partial u^2}\,\frac{\partial u}{\partial y} + t\,\frac{\partial^2 w}{\partial v \partial u}\,\frac{\partial v}{\partial y}$

$= t^2\,\frac{\partial^2 w}{\partial v \partial u} \Rightarrow \left(\frac{1}{t^2}\right)\frac{\partial^2 w}{\partial x^2} = \frac{\partial^2 w}{\partial u^2}$, $\left(\frac{1}{t^2}\right)\frac{\partial^2 w}{\partial y^2} = \frac{\partial^2 w}{\partial v^2}$, and $\left(\frac{1}{t^2}\right)\frac{\partial^2 w}{\partial y \partial x} = \frac{\partial^2 w}{\partial v \partial u}$

$\Rightarrow n(n-1)t^{n-2}f(x,y) = \left(\frac{x^2}{t^2}\right)\left(\frac{\partial^2 w}{\partial x^2}\right) + \left(\frac{2xy}{t^2}\right)\left(\frac{\partial^2 w}{\partial y \partial x}\right) + \left(\frac{y^2}{t^2}\right)\left(\frac{\partial^2 w}{\partial y^2}\right)$ for $t \neq 0$. When $t = 1$, $w = f(x,y)$ and

we have $n(n-1)f(x,y) = x^2\left(\frac{\partial^2 f}{\partial x^2}\right) + 2xy\left(\frac{\partial^2 f}{\partial x \partial y}\right) + y^2\left(\frac{\partial^2 f}{\partial y^2}\right)$ as claimed.

6. $x = \rho \sin\phi \cos\theta$, $y = \rho \sin\phi \sin\theta$, and $z = \rho \cos\phi \Rightarrow \mathbf{r} = (\rho \sin\phi \cos\theta)\mathbf{i} + (\rho \sin\phi \sin\theta)\mathbf{j} + (\rho \cos\phi)\mathbf{k}$ so that $\frac{\partial \mathbf{r}}{\partial \rho} = (\sin\phi \cos\theta)\mathbf{i} + (\sin\phi \sin\theta)\mathbf{j} + (\cos\phi)\mathbf{k} \Rightarrow \frac{\partial \mathbf{r}}{\partial \rho} = \mathbf{u}_\rho$;
$\frac{\partial \mathbf{r}}{\partial \phi} = (\rho \cos\phi \cos\theta)\mathbf{i} + (\rho \cos\phi \sin\theta)\mathbf{j} - (\rho \sin\phi)\mathbf{k} \Rightarrow \frac{\partial \mathbf{r}}{\partial \phi} = \rho \mathbf{u}_\phi$; $\frac{\partial \mathbf{r}}{\partial \theta} = (-\rho \sin\phi \sin\theta)\mathbf{i} + (\rho \sin\phi \cos\theta)\mathbf{j}$
$\Rightarrow \frac{\partial \mathbf{r}}{\partial \theta} = (\rho \sin\phi)\,\mathbf{u}_\theta$

7. (a) $\mathbf{r} = x\mathbf{i} + y\mathbf{j} + z\mathbf{k} \Rightarrow r = |\mathbf{r}| = \sqrt{x^2+y^2+z^2}$ and $\nabla r = \frac{x}{\sqrt{x^2+y^2+z^2}}\mathbf{i} + \frac{y}{\sqrt{x^2+y^2+z^2}}\mathbf{j} + \frac{z}{\sqrt{x^2+y^2+z^2}}\mathbf{k}$
$= \frac{\mathbf{r}}{r}$

(b) $r^n = \left(\sqrt{x^2+y^2+z^2}\right)^n$
$\Rightarrow \nabla(r^n) = nx\left(x^2+y^2+z^2\right)^{(n/2)-1}\mathbf{i} + ny\left(x^2+y^2+z^2\right)^{(n/2)-1}\mathbf{j} + nz\left(x^2+y^2+z^2\right)^{(n/2)-1}\mathbf{k}$
$= nr^{n-2}\mathbf{r}$

(c) Let $n = 2$ in part (b). Then $\frac{1}{2}\nabla(r^2) = \mathbf{r} \Rightarrow \nabla\left(\frac{1}{2}r^2\right) = \mathbf{r} \Rightarrow \frac{r^2}{2} = \frac{1}{2}(x^2+y^2+z^2)$ is the function.

(d) $d\mathbf{r} = dx\mathbf{i} + dy\mathbf{j} + dz\mathbf{k} \Rightarrow \mathbf{r} \cdot d\mathbf{r} = x\,dx + y\,dy + z\,dz$, and $dr = r_x\,dx + r_y\,dy + r_z\,dz = \frac{x}{r}\,dx + \frac{y}{r}\,dy + \frac{z}{r}\,dz$
$\Rightarrow r\,dr = x\,dx + y\,dy + z\,dz = \mathbf{r} \cdot d\mathbf{r}$

(e) $\mathbf{A} = a\mathbf{i} + b\mathbf{j} + c\mathbf{k} \Rightarrow \mathbf{A} \cdot \mathbf{r} = ax + by + cz \Rightarrow \nabla(\mathbf{A} \cdot \mathbf{r}) = a\mathbf{i} + b\mathbf{j} + c\mathbf{k} = \mathbf{A}$

8. $f(g(t),h(t)) = c \Rightarrow 0 = \frac{df}{dt} = \frac{\partial f}{\partial x}\,\frac{dx}{dt} + \frac{\partial f}{\partial y}\,\frac{dy}{dt} = \left(\frac{\partial f}{\partial x}\mathbf{i} + \frac{\partial f}{\partial y}\mathbf{j}\right) \cdot \left(\frac{dx}{dt}\mathbf{i} + \frac{dy}{dt}\mathbf{j}\right)$, where $\frac{dx}{dt}\mathbf{i} + \frac{dy}{dt}\mathbf{j}$ is the tangent vector
$\Rightarrow \nabla f$ is orthogonal to the tangent vector

9. $f(x,y,z) = xz^2 - yz + \cos xy - 1 \Rightarrow \nabla f = (z^2 - y \sin xy)\mathbf{i} + (-z - x \sin xy)\mathbf{j} + (2xz - y)\mathbf{k} \Rightarrow \nabla f(0,0,1) = \mathbf{i} - \mathbf{j}$
$\Rightarrow$ the tangent plane is $x - y = 0$; $\mathbf{r} = (\ln t)\mathbf{i} + (t \ln t)\mathbf{j} + t\mathbf{k} \Rightarrow \mathbf{r}' = \left(\frac{1}{t}\right)\mathbf{i} + (\ln t + 1)\mathbf{j} + \mathbf{k}$; $x = y = 0$, $z = 1$

$\Rightarrow t = 1 \Rightarrow \mathbf{r}'(1) = \mathbf{i} + \mathbf{j} + \mathbf{k}$. Since $(\mathbf{i}+\mathbf{j}+\mathbf{k}) \cdot (\mathbf{i}-\mathbf{j}) = \mathbf{r}'(1) \cdot \nabla f = 0$, $\mathbf{r}$ is parallel to the plane, and $\mathbf{r}(1) = 0\mathbf{i} + 0\mathbf{j} + \mathbf{k} \Rightarrow \mathbf{r}$ is contained in the plane.

10. Let $f(x,y,z) = x^3 + y^3 + z^3 - xyz \Rightarrow \nabla f = (3x^2 - yz)\mathbf{i} + (3y^2 - xz)\mathbf{j} + (3z^2 - xy)\mathbf{k} \Rightarrow \nabla f(0,-1,1) = \mathbf{i} + 3\mathbf{j} + 3\mathbf{k}$ $\Rightarrow$ the tangent plane is $x + 3y + 3z = 0$; $\mathbf{r} = \left(\frac{t^3}{4} - 2\right)\mathbf{i} + \left(\frac{4}{t} - 3\right)\mathbf{j} + (\cos(t-2))\,\mathbf{k}$ $\Rightarrow \mathbf{r}' = \left(\frac{3t^2}{4}\right)\mathbf{i} - \left(\frac{4}{t^2}\right)\mathbf{j} - (\sin(t-2))\,\mathbf{k}$; $x = 0,\ y = -1,\ z = 1 \Rightarrow t = 2 \Rightarrow \mathbf{r}'(2) = 3\mathbf{i} - \mathbf{j}$. Since $\mathbf{r}'(2) \cdot \nabla f = 0 \Rightarrow \mathbf{r}$ is parallel to the plane, and $\mathbf{r}(2) = -\mathbf{i} + \mathbf{k} \Rightarrow \mathbf{r}$ is contained in the plane.

11. $\frac{\partial w}{\partial r} = \frac{\partial w}{\partial x}\frac{\partial x}{\partial r} + \frac{\partial w}{\partial y}\frac{\partial y}{\partial r} = \frac{\partial w}{\partial x}(\cos\theta) + \frac{\partial w}{\partial y}(\sin\theta)$; $\frac{\partial w}{\partial \theta} = \frac{\partial w}{\partial x}\frac{\partial x}{\partial \theta} + \frac{\partial w}{\partial y}\frac{\partial y}{\partial \theta} = \frac{\partial w}{\partial x}(-r\sin\theta) + \frac{\partial w}{\partial y}(r\cos\theta)$;

$\frac{\partial w}{\partial r}\mathbf{u}_r = \left[\frac{\partial w}{\partial x}(\cos\theta) + \frac{\partial w}{\partial y}(\sin\theta)\right][(\cos\theta)\mathbf{i} + (\sin\theta)\mathbf{j}]$ and

$\frac{1}{r}\frac{\partial w}{\partial \theta}\mathbf{u}_\theta = \left[\frac{\partial w}{\partial x}(-\sin\theta) + \frac{\partial w}{\partial y}(\cos\theta)\right][(-\sin\theta)\mathbf{i} + (\cos\theta)\mathbf{j}] \Rightarrow \frac{\partial w}{\partial r}\mathbf{u}_r + \frac{1}{r}\frac{\partial w}{\partial \theta}\mathbf{u}_\theta + \frac{\partial w}{\partial z}\mathbf{k}$

$= \frac{\partial w}{\partial x}(\cos^2\theta + \sin^2\theta)\mathbf{i} + \frac{\partial w}{\partial y}(\cos^2\theta + \sin^2\theta)\mathbf{j} + \frac{\partial w}{\partial z}\mathbf{k} = \frac{\partial w}{\partial x}\mathbf{i} + \frac{\partial w}{\partial y}\mathbf{j} + \frac{\partial w}{\partial z}\mathbf{k} = \nabla w$

12. $\frac{\partial w}{\partial \rho} = \frac{\partial w}{\partial x}\frac{\partial x}{\partial \rho} + \frac{\partial w}{\partial y}\frac{\partial y}{\partial \rho} + \frac{\partial w}{\partial z}\frac{\partial z}{\partial \rho} = \frac{\partial w}{\partial x}(\sin\phi\cos\theta) + \frac{\partial w}{\partial y}(\sin\phi\sin\theta) + \frac{\partial w}{\partial z}(\cos\phi)$; $\frac{\partial w}{\partial \phi} = \frac{\partial w}{\partial x}\frac{\partial x}{\partial \phi} + \frac{\partial w}{\partial y}\frac{\partial y}{\partial \phi} + \frac{\partial w}{\partial z}\frac{\partial z}{\partial \phi}$

$= \rho\left[\frac{\partial w}{\partial x}(\cos\phi\cos\theta) + \frac{\partial w}{\partial y}(\cos\phi\sin\theta) - \frac{\partial w}{\partial z}(\sin\phi)\right]$; $\frac{\partial w}{\partial \theta} = \frac{\partial w}{\partial x}\frac{\partial x}{\partial \theta} + \frac{\partial w}{\partial y}\frac{\partial y}{\partial \theta} + \frac{\partial w}{\partial z}\frac{\partial z}{\partial \theta}$

$= \rho\left[\frac{\partial w}{\partial x}(-\sin\phi\sin\theta) + \frac{\partial w}{\partial y}(\sin\phi\cos\theta)\right] \Rightarrow \frac{\partial w}{\partial \rho}\mathbf{u}_\rho + \frac{1}{\rho}\frac{\partial w}{\partial \phi}\mathbf{u}_\phi + \frac{1}{\rho\sin\phi}\frac{\partial w}{\partial \theta}\mathbf{u}_\theta$

$= \left(\frac{\partial w}{\partial x}\sin\phi\cos\theta + \frac{\partial w}{\partial y}\sin\phi\sin\theta + \frac{\partial w}{\partial z}\cos\phi\right)[(\sin\phi\cos\theta)\mathbf{i} + (\sin\phi\sin\theta)\mathbf{j} + (\cos\phi)\,\mathbf{k}]$

$+\left(\frac{\partial w}{\partial x}\cos\phi\cos\theta + \frac{\partial w}{\partial y}\cos\phi\sin\theta - \frac{\partial w}{\partial z}\sin\phi\right)[(\cos\phi\cos\theta)\mathbf{i} + (\cos\phi\sin\theta)\mathbf{j} - (\sin\phi)\mathbf{k}]$

$+\left(-\frac{\partial w}{\partial x}\sin\theta + \frac{\partial w}{\partial y}\cos\theta\right)[(-\sin\theta)\mathbf{i} + (\cos\theta)\mathbf{j}]$

$= \frac{\partial w}{\partial x}[(\sin^2\phi + \cos^2\phi)\cos^2\theta + \sin^2\theta]\mathbf{i} + \frac{\partial w}{\partial y}[(\sin^2\phi\ \cos^2\phi)\sin^2\theta + \cos^2\theta]\mathbf{j} + \frac{\partial w}{\partial z}(\cos^2\phi + \sin^2\phi)\mathbf{k}$

$= \frac{\partial w}{\partial x}\mathbf{i} + \frac{\partial w}{\partial y}\mathbf{j} + \frac{\partial w}{\partial z}\mathbf{k} = \nabla w$

13. $\frac{\partial z}{\partial x} = 3x^2 - 9y = 0$ and $\frac{\partial z}{\partial y} = 3y^2 - 9x = 0 \Rightarrow y = \frac{1}{3}x^2$ and $3\left(\frac{1}{3}x^2\right)^2 - 9x = 0 \Rightarrow \frac{1}{3}x^4 - 9x = 0$ $\Rightarrow x(x^3 - 27) = 0 \Rightarrow x = 0$ or $x = 3$. Now $x = 0 \Rightarrow y = 0$ or $(0,0)$ and $x = 3 \Rightarrow y = 3$ or $(3,3)$. Next $\frac{\partial^2 z}{\partial x^2} = 6x$, $\frac{\partial^2 z}{\partial y^2} = 6y$, and $\frac{\partial^2 z}{\partial x \partial y} = -9$. For $(0,0)$, $\frac{\partial^2 z}{\partial x^2}\frac{\partial^2 z}{\partial y^2} - \left(\frac{\partial^2 z}{\partial x \partial y}\right)^2 = -81 \Rightarrow$ no extremum (a saddle point), and for $(3,3)$, $\frac{\partial^2 z}{\partial x^2}\frac{\partial^2 z}{\partial y^2} - \left(\frac{\partial^2 z}{\partial x \partial y}\right)^2 = 243 > 0$ and $\frac{\partial^2 z}{\partial x^2} = 18 > 0 \Rightarrow$ a local minimum.

14. $f(x,y) = 6xye^{-(2x+3y)} \Rightarrow f_x(x,y) = 6y(1-2x)e^{-(2x+3y)} = 0$ and $f_y(x,y) = 6x(1-3y)e^{-(2x+3y)} = 0 \Rightarrow x = 0$ and $y = 0$, or $x = \frac{1}{2}$ and $y = \frac{1}{3}$. The value $f(0,0) = 0$ is on the boundary, and $f\left(\frac{1}{2},\frac{1}{3}\right) = \frac{1}{e^2}$. On the positive y-axis, $f(0,y) = 0$, and on the positive x-axis, $f(x,0) = 0$. As $x \to \infty$ or $y \to \infty$ we see that $f(x,y) \to 0$. Thus the absolute maximum of f in the closed first quadrant is $\frac{1}{e^2}$ at the point $\left(\frac{1}{2},\frac{1}{3}\right)$.

15. Let $f(x,y,z) = \frac{x^2}{a^2}+\frac{y^2}{b^2}+\frac{z^2}{c^2}-1 \Rightarrow \nabla f = \frac{2x}{a^2}\mathbf{i}+\frac{2y}{b^2}\mathbf{j}+\frac{2z}{c^2}\mathbf{k} \Rightarrow$ an equation of the plane tangent at the point $P_0(x_0,y_0,y_0)$ is $\left(\frac{2x_0}{a^2}\right)x+\left(\frac{2y_0}{b^2}\right)y+\left(\frac{2z_0}{c^2}\right)z = \frac{2x_0^2}{a^2}+\frac{2y_0^2}{b^2}+\frac{2z_0^2}{c^2} = 2$ or $\left(\frac{x_0}{a^2}\right)x+\left(\frac{y_0}{b^2}\right)y+\left(\frac{z_0}{c^2}\right)z = 1$.

The intercepts of the plane are $\left(\frac{a^2}{x_0},0,0\right)$, $\left(0,\frac{b^2}{y_0},0\right)$ and $\left(0,0,\frac{c^2}{z_0}\right)$. The volume of the tetrahedron formed by the plane and the coordinate planes is $V = \left(\frac{1}{3}\right)\left(\frac{1}{2}\right)\left(\frac{a^2}{x_0}\right)\left(\frac{b^2}{y_0}\right)\left(\frac{c^2}{z_0}\right) \Rightarrow$ we need to maximize $V(x,y,z) = \frac{(abc)^2}{2}(xyz)^{-1}$ subject to the constraint $f(x,y,z) = \frac{x^2}{a^2}+\frac{y^2}{b^2}+\frac{z^2}{c^2} = 1$. Thus,

$\left[-\frac{(abc)^2}{6}\right]\left(\frac{1}{x^2yz}\right) = \frac{2x}{a^2}\lambda$, $\left[-\frac{(abc)^2}{6}\right]\left(\frac{1}{xy^2z}\right) = \frac{2y}{b^2}\lambda$, and $\left[-\frac{(abc)^2}{6}\right]\left(\frac{1}{xyz^2}\right) = \frac{2z}{c^2}\lambda$. Multiply the first equation by a^2yz, the second by b^2xz, and the third by c^2xy. Then equate the first and second $\Rightarrow a^2y^2 = b^2x^2$ $\Rightarrow y = \frac{b}{a}x$, $x > 0$; equate the first and third $\Rightarrow a^2z^2 = c^2x^2 \Rightarrow z = \frac{c}{a}x$, $x > 0$; substitute into $f(x,y,z) = 0$ $\Rightarrow x = \sqrt{\frac{a}{3}} \Rightarrow y = \sqrt{\frac{b}{3}} \Rightarrow z = \sqrt{\frac{c}{3}} \Rightarrow V = \frac{\sqrt{3}}{2}abc$.

16. $2(x-u) = -\lambda$, $2(y-v) = \lambda$, $-2(x-u) = \mu$, and $-2(y-v) = 2\mu v \Rightarrow x-u = v-y$, $x-u = \frac{\mu}{2}$, and $y-v = -\mu v$ $\Rightarrow x-\mu = \mu v = \frac{\mu}{2} \Rightarrow v = \frac{1}{2}$ or $\mu = 0$.

CASE 1: $\mu = 0 \Rightarrow x = u$, $y = v$, and $\lambda = 0$; then $y = x+1 \Rightarrow v = u+1$ and $v^2 = u \Rightarrow v = v^2+1$ $\Rightarrow v^2-v+1 = 0 \Rightarrow v = \frac{1\pm\sqrt{1-4}}{2} \Rightarrow$ no real solution.

CASE 2: $v = \frac{1}{2}$ and $u = v^2 \Rightarrow u = \frac{1}{4}$; $x-\frac{1}{4} = \frac{1}{2}-y$ and $y = x+1 \Rightarrow x-\frac{1}{4} = -x-\frac{1}{2} \Rightarrow 2x = -\frac{1}{4}$ $\Rightarrow x = -\frac{1}{8} \Rightarrow y = \frac{7}{8}$. Then $f\left(-\frac{1}{8},\frac{7}{8},\frac{1}{4},\frac{1}{2}\right) = \left(-\frac{1}{8}-\frac{1}{4}\right)^2+\left(\frac{7}{8}-\frac{1}{2}\right)^2 = 2\left(\frac{3}{8}\right)^2 \Rightarrow$ the minimum distance is $\frac{3}{8}\sqrt{2}$. (Notice that f has no maximum value.)

17. Let (x_0,y_0) be any point in R. We must show $\lim_{(x,y)\to(x_0,y_0)} f(x,y) = f(x_0,y_0)$ or, equivalently that $\lim_{(h,k)\to(0,0)} |f(x_0+h,y_0+k)-f(x_0,y_0)| = 0$. Consider $f(x_0+h,y_0+k)-f(x_0,y_0)$ $= [f(x_0+h,y_0+k)-f(x_0,y_0+k)]+[f(x_0,y_0+k)-f(x_0,y_0)]$. Let $F(x) = f(x,y_0+k)$ and apply the Mean Value Theorem: there exists ξ with $x_0 < \xi < x_0+h$ such that $F'(\xi)h = F(x_0+h)-F(x_0) \Rightarrow hf_x(\xi,y_0+k)$ $= f(x_0+h,y_0+k)-f(x_0,y_0+k)$. Similarly, $k\,f_y(x_0,\eta) = f(x_0,y_0+k)-f(x_0,y_0)$ for some η with

$y_0 < \eta < y_0 + k$. Then $|f(x_0+h, y_0+k) - f(x_0, y_0)| \le |hf_x(\xi, y_0+k)| + |kf_y(x_0, \eta)|$. If M, N are positive real numbers such that $|f_x| \le M$ and $|f_y| \le N$ for all (x, y) in the xy-plane, then $|f(x_0+h, y_0+k) - f(x_0, y_0)|$ $\le M|h| + N|k|$. As $(h,k) \to 0$, $|f(x_0+h, y_0+k) - f(x_0, y_0)| \to 0 \Rightarrow \lim_{(h,k)\to(0,0)} |f(x_0+h, y_0+k) - f(x_0, y_0)|$ $= 0 \Rightarrow$ f is continuous at (x_0, y_0).

18. At extreme values, ∇f and $\mathbf{v} = \frac{d\mathbf{r}}{dt}$ are orthogonal because $\frac{df}{dt} = \nabla f \cdot \frac{d\mathbf{r}}{dt} = 0$ by the First Derivative Theorem for Local Extreme Values.

19. $\frac{\partial f}{\partial x} = 0 \Rightarrow f(x,y) = h(y)$ is a function of y only. Also, $\frac{\partial g}{\partial y} = \frac{\partial f}{\partial x} = 0 \Rightarrow g(x,y) = k(x)$ is a function of x only. Moreover, $\frac{\partial f}{\partial y} = \frac{\partial g}{\partial x} \Rightarrow h'(y) = k'(x)$ for all x and y. This can happen only if $h'(y) = k'(x) = c$ is a constant. Integration gives $h(y) = cy + c_1$ and $k(x) = cx + c_2$, where c_1 and c_2 are constants. Therefore $f(x,y) = cy + c_1$ and $g(x,y) = cx + c_2$. Then $f(1,2) = g(1,2) = 5 \Rightarrow 5 = 2c + c_1 = c + c_2$, and $f(0,0) = 4 \Rightarrow c_1 = 4 \Rightarrow c = \frac{1}{2}$ $\Rightarrow c_2 = \frac{9}{2}$. Thus, $f(x,y) = \frac{1}{2}y + 4$ and $g(x,y) = \frac{1}{2}x + \frac{9}{2}$.

20. Let $g(x,y) = D_{\mathbf{u}}f(x,y) = f_x(x,y)a + f_y(x,y)b$. Then $D_{\mathbf{u}}g(x,y) = g_x(x,y)a + g_y(x,y)b$
$= f_{xx}(x,y)a^2 + f_{yx}(x,y)ab + f_{xy}(x,y)ba + f_{yy}(x,y)b^2 = f_{xx}(x,y)a^2 + 2f_{xy}(x,y)ab + f_{yy}(x,y)b^2$.

21. Since the particle is heat-seeking, at each point (x, y) it moves in the direction of maximal temperature increase, that is in the direction of $\nabla T(x,y) = (e^{-2y}\sin x)\mathbf{i} + (2e^{-2y}\cos x)\mathbf{j}$. Since $\nabla T(x,y)$ is parallel to the particle's velocity vector, it is tangent to the path $y = f(x)$ of the particle $\Rightarrow f'(x) = \frac{2e^{-2y}\cos x}{e^{-2y}\sin x} = 2\cot x$. Integration gives $f(x) = 2\ln|\sin x| + C$ and $f\left(\frac{\pi}{4}\right) = 0 \Rightarrow 0 = 2\ln\left|\sin\frac{\pi}{4}\right| + C \Rightarrow C = -2\ln\frac{\sqrt{2}}{2} = \ln\left(\frac{2}{\sqrt{2}}\right)^2$ $= \ln 2$. Therefore, the path of the particle is the graph of $y = 2\ln|\sin x| + \ln 2$.

22. The line of travel is $x = t$, $y = t$, $z = 30 - 5t$, and the bullet hits the surface $z = 2x^2 + 3y^2$ when $30 - 5t = 2t^2 + 3t^2 \Rightarrow t^2 + t - 6 = 0 \Rightarrow (t+3)(t-2) = 0 \Rightarrow t = 2$ (since $t > 0$). Thus the bullet hits the surface at the point (2, 2, 20). Now, the vector $4x\mathbf{i} + 6y\mathbf{j} - \mathbf{k}$ is normal to the surface at any (x, y, z), so that $\mathbf{n} = 8\mathbf{i} + 12\mathbf{j} - \mathbf{k}$ is normal to the surface at (2, 2, 20). If $\mathbf{v} = \mathbf{i} + \mathbf{j} - 5\mathbf{k}$, then the velocity of the particle after the ricochet is $\mathbf{w} = \mathbf{v} - 2\,\text{proj}_{\mathbf{n}}\,\mathbf{v} = \mathbf{v} - \left(\frac{2\mathbf{v}\cdot\mathbf{n}}{|\mathbf{n}|^2}\right)\mathbf{n} = \mathbf{v} - \left(\frac{2\cdot 25}{209}\right)\mathbf{n} = (\mathbf{i} + \mathbf{j} - 5\mathbf{k}) - \left(\frac{400}{209}\mathbf{i} + \frac{600}{209}\mathbf{j} - \frac{50}{209}\mathbf{k}\right)$ $= -\frac{191}{209}\mathbf{i} - \frac{391}{209}\mathbf{j} - \frac{995}{209}\mathbf{k}$.

23. (a) $\mathbf{k}$ is a vector normal to $z = 10 - x^2 - y^2$ at the point (0, 0, 10). So directions tangential to S at (0, 0, 10) will be unit vectors $\mathbf{u} = a\mathbf{i} + b\mathbf{j}$. Also, $\nabla T(x,y,z) = (2xy + 4)\mathbf{i} + (x^2 + 2yz + 14)\mathbf{j} + (y^2 + 1)\mathbf{k}$ $\Rightarrow \nabla T(0,0,10) = 4\mathbf{i} + 14\mathbf{j} + \mathbf{k}$. We seek the unit vector $\mathbf{u} = a\mathbf{i} + b\mathbf{j}$ such that $D_{\mathbf{u}}T(0,0,10)$ $= (4\mathbf{i} + 14\mathbf{j} + \mathbf{k}) \cdot (a\mathbf{i} + b\mathbf{j}) = (4\mathbf{i} + 14\mathbf{j}) \cdot (a\mathbf{i} + b\mathbf{j})$ is a maximum. The maximum will occur when $a\mathbf{i} + b\mathbf{j}$ has the same direction as $4\mathbf{i} + 14\mathbf{j}$, or $\mathbf{u} = \frac{1}{\sqrt{53}}(2\mathbf{i} + 7\mathbf{j})$.

(b) A vector normal to S at $(1,1,8)$ is $\mathbf{n} = 2\mathbf{i} + 2\mathbf{j} + \mathbf{k}$. Now, $\nabla T(1,1,8) = 6\mathbf{i} + 31\mathbf{j} + 2\mathbf{k}$ and we seek the unit vector $\mathbf{u}$ such that $D_{\mathbf{u}}T(1,1,8) = \nabla T \cdot \mathbf{u}$ has its largest value. Now write $\nabla T = \mathbf{v} + \mathbf{w}$, where $\mathbf{v}$ is parallel to ∇T and $\mathbf{w}$ is orthogonal to ∇T. Then $D_{\mathbf{u}}T = \nabla T \cdot \mathbf{u} = (\mathbf{v} + \mathbf{w}) \cdot \mathbf{u} = \mathbf{v} \cdot \mathbf{u} + \mathbf{w} \cdot \mathbf{u} = \mathbf{w} \cdot \mathbf{u}$. Thus $D_{\mathbf{u}}T(1,1,8)$ is a maximum when $\mathbf{u}$ has the same direction as $\mathbf{w}$. Now, $\mathbf{w} = \nabla T - \left(\frac{\nabla T \cdot \mathbf{n}}{|\mathbf{n}|^2}\right)\mathbf{n}$

$= (6\mathbf{i} + 31\mathbf{j} + 2\mathbf{k}) - \left(\frac{12 + 62 + 2}{4 + 4 + 1}\right)(2\mathbf{i} + 2\mathbf{j} + \mathbf{k}) = \left(6 - \frac{152}{9}\right)\mathbf{i} + \left(31 - \frac{152}{9}\right)\mathbf{j} + \left(2 - \frac{76}{9}\right)\mathbf{k}$

$= -\frac{98}{9}\mathbf{i} + \frac{127}{9}\mathbf{j} - \frac{58}{9}\mathbf{k} \Rightarrow \mathbf{u} = \frac{\mathbf{w}}{|\mathbf{w}|} = -\frac{1}{\sqrt{29{,}097}}(98\mathbf{i} - 127\mathbf{j} + 58\mathbf{k})$.

24. Suppose the surface (boundary) of the mineral deposit is the graph of $z = f(x,y)$ (where the z-axis points up into the air). Then $-\frac{\partial f}{\partial x}\mathbf{i} - \frac{\partial f}{\partial y}\mathbf{j} + \mathbf{k}$ is an outer normal to the mineral deposit at (x,y) and $\frac{\partial f}{\partial x}\mathbf{i} + \frac{\partial f}{\partial y}\mathbf{j}$ points in the direction of steepest ascent of the mineral deposit. This is in the direction of the vector $\frac{\partial f}{\partial x}\mathbf{i} + \frac{\partial f}{\partial y}\mathbf{j}$ at $(0,0)$ (the location of the 1st borehole) that the geologists should drill their fourth borehole. To approximate this vector we use the fact that $(0,0,-1000)$, $(0,100,-950)$, and $(100,0,-1025)$ lie on the graph of $z = f(x,y)$. The plane containing these three points is a good approximation to the tangent plane to $z = f(x,y)$ at the point $(0,0,0)$. A normal to this plane is $\begin{vmatrix} \mathbf{i} & \mathbf{j} & \mathbf{k} \\ 0 & 100 & 50 \\ 100 & 0 & -25 \end{vmatrix} = -2500\mathbf{i} + 5000\mathbf{j} - 10{,}000\mathbf{k}$, or $-\mathbf{i} + 2\mathbf{j} - 4\mathbf{k}$. So at $(0,0)$ the vector $\frac{\partial f}{\partial x}\mathbf{i} + \frac{\partial f}{\partial y}\mathbf{j}$ is approximately $-\mathbf{i} + 2\mathbf{j}$. Thus the geologists should drill their fourth borehole in the direction of $\frac{1}{\sqrt{5}}(-\mathbf{i} + 2\mathbf{j})$ from the first borehole.

25. $w = e^{rt} \sin \pi x \Rightarrow w_t = re^{rt} \sin \pi x$ and $w_x = \pi e^{rt} \cos \pi x \Rightarrow w_{xx} = -\pi^2 e^{rt} \sin \pi x$; $w_{xx} = \frac{1}{c^2} w_t$, where c^2 is the positive constant determined by the material of the rod $\Rightarrow -\pi^2 e^{rt} \sin \pi x = \frac{1}{c^2}\left(re^{rt} \sin \pi x\right)$

$\Rightarrow \left(r + c^2\pi^2\right)e^{rt} \sin \pi x = 0 \Rightarrow r = -c^2\pi^2 \Rightarrow w = e^{-c^2\pi^2 t} \sin \pi x$

26. $w = e^{rt} \sin kx \Rightarrow w_t = re^{rt} \sin kx$ and $w_x = ke^{rt} \cos kx \Rightarrow w_{xx} = -k^2 e^{rt} \sin kx$; $w_{xx} = \frac{1}{c^2} w_t$

$\Rightarrow -k^2 e^{rt} \sin kx = \frac{1}{c^2}\left(re^{rt} \sin kx\right) \Rightarrow \left(r + c^2k^2\right)e^{rt} \sin kx = 0 \Rightarrow r = -c^2k^2 \Rightarrow w = e^{-c^2k^2 t} \sin kx$.

Now, $w(L,t) = 0 \Rightarrow e^{-c^2k^2 t} \sin kL = 0 \Rightarrow kL = n\pi$ for n an integer $\Rightarrow k = \frac{n\pi}{L} \Rightarrow w = e^{-c^2n^2\pi^2 t/L^2} \sin\left(\frac{n\pi}{L}x\right)$.

As $t \to \infty$, $w \to 0$ since $\left|\sin\left(\frac{n\pi}{L}x\right)\right| \le 1$ and $e^{-c^2n^2\pi^2 t/L^2} \to 0$.

CHAPTER 13 MULTIPLE INTEGRALS

13.1 DOUBLE INTEGRALS

1. $\int_0^3 \int_0^2 (4 - y^2)\,dy\,dx = \int_0^3 \left[4y - \frac{y^3}{3}\right]_0^2 dx = \frac{16}{3}\int_0^3 dx = 16$

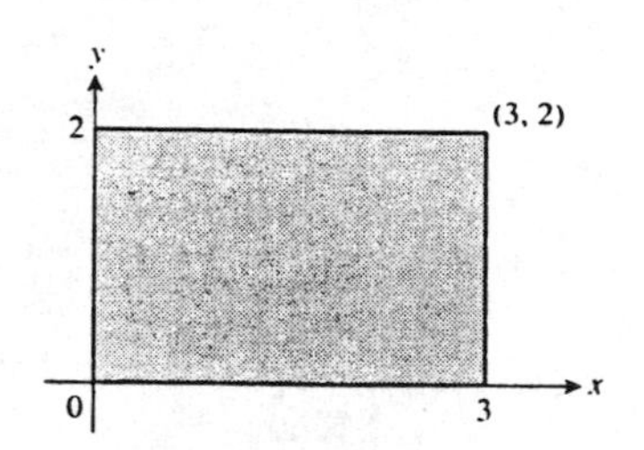

2. $\int_0^3 \int_{-2}^0 ((x^2 y - 2xy)\,dy\,dx = \int_0^3 \left[\frac{x^2y^2}{2} - xy^2\right]_{-2}^0 dx$

$= \int_0^3 (4x - 2x^2)\,dx = \left[2x^2 - \frac{2x^3}{3}\right]_0^3 = 0$

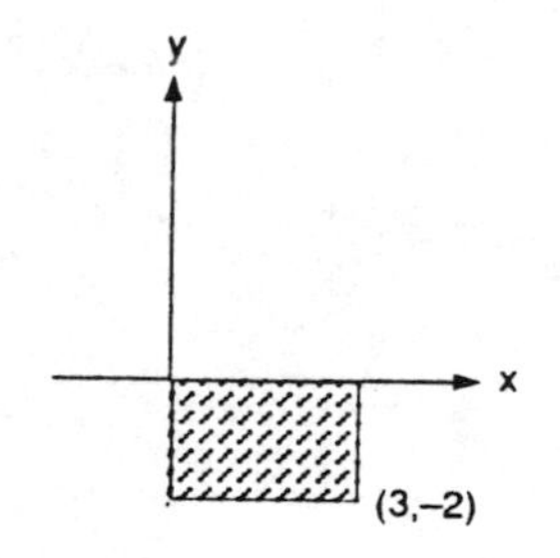

3. $\int_{-1}^0 \int_{-1}^1 (x + y + 1)\,dx\,dy = \int_{-1}^0 \left[\frac{x^2}{2} + yx + x\right]_{-1}^1 dy$

$= \int_{-1}^0 (2y + 2)\,dy = \left[y^2 + 2y\right]_{-1}^0 = 1$

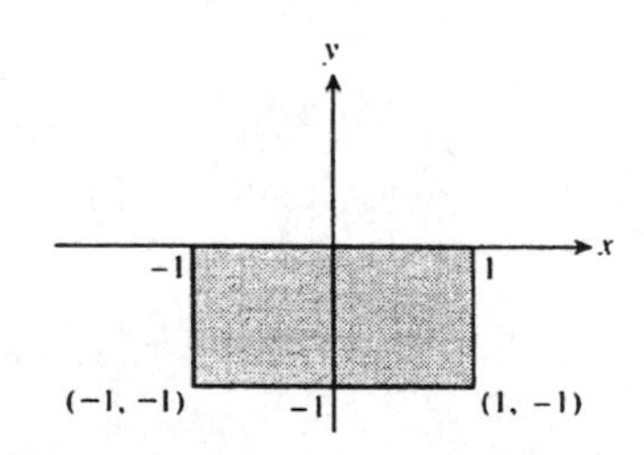

4. $\int_\pi^{2\pi} \int_0^\pi (\sin x + \cos y)\,dx\,dy = \int_\pi^{2\pi} \left[(-\cos x) + (\cos y)x\right]_0^\pi dy$

$= \int_\pi^{2\pi} (\pi \cos y + 2)\,dy = \left[\pi \sin y + 2y\right]_\pi^{2\pi} = 2\pi$

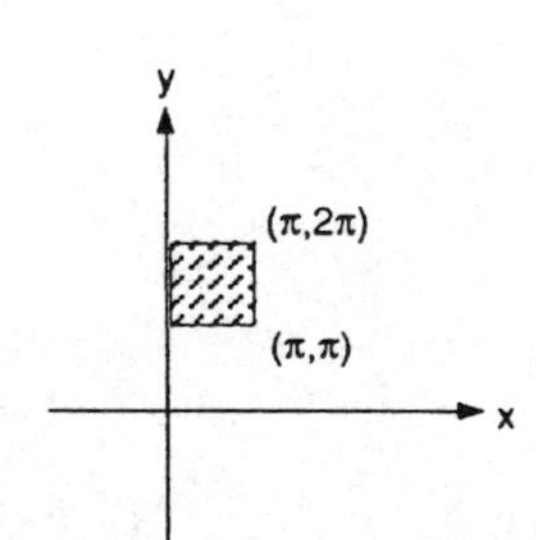

5. $\int_0^{\pi}\int_0^{x} (x \sin y)\, dy\, dx = \int_0^{\pi} [-x \cos y]_0^x\, dx$

$= \int_0^{\pi} (x - x \cos x)\, dx = \left[\frac{x^2}{2} - (\cos x + x \sin x)\right]_0^{\pi} = \frac{\pi^2}{2} + 2$

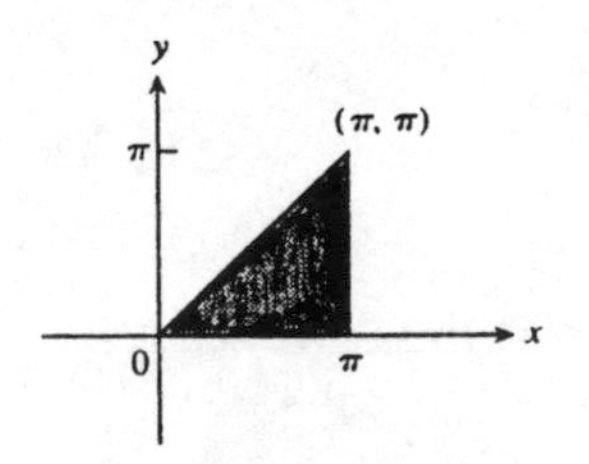

6. $\int_0^{\pi}\int_0^{\sin x} y\, dy\, dx = \int_0^{\pi} \left[\frac{y^2}{2}\right]_0^{\sin x} dx = \int_0^{\pi} \frac{1}{2} \sin^2 x\, dx$

$= \frac{1}{4}\int_0^{\pi} (1 - \cos 2x)\, dx = \frac{1}{4}\left[x - \frac{1}{2} \sin 2x\right]_0^{\pi} = \frac{\pi}{4}$

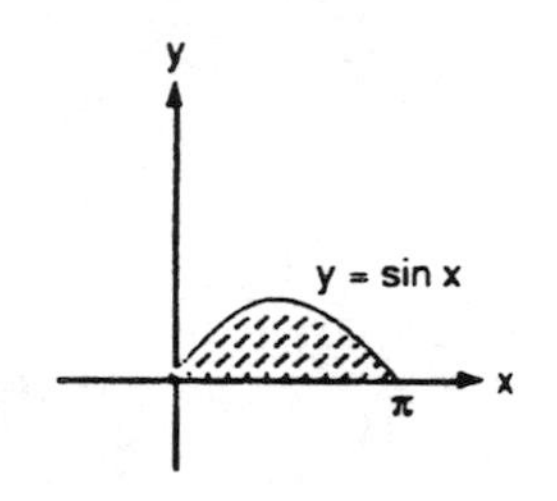

7. $\int_1^{\ln 8}\int_0^{\ln y} e^{x+y}\, dx\, dy = \int_1^{\ln 8} \left[e^{x+y}\right]_0^{\ln y} dy = \int_1^{\ln 8} \left(ye^y - e^y\right) dy$

$= \left[(y-1)e^y - e^y\right]_1^{\ln 8} = 8(\ln 8 - 1) - 8 + e = 8 \ln 8 - 16 + e$

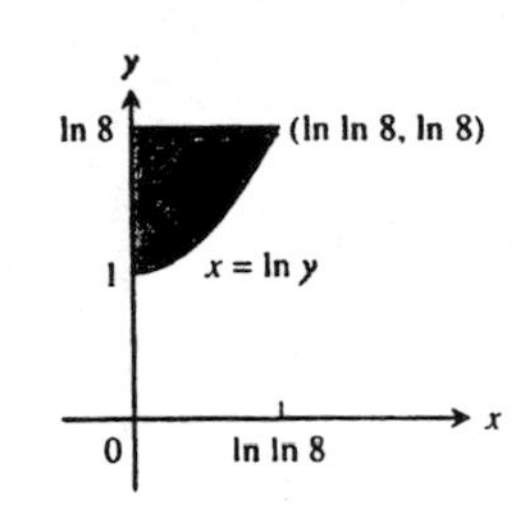

8. $\int_1^{2}\int_y^{y^2} dx\, dy = \int_1^{2} \left(y^2 - y\right) dy = \left[\frac{y^3}{3} - \frac{y^2}{2}\right]_1^2$

$= \left(\frac{8}{3} - 2\right) - \left(\frac{1}{3} - \frac{1}{2}\right) = \frac{7}{3} - \frac{3}{2} = \frac{5}{6}$

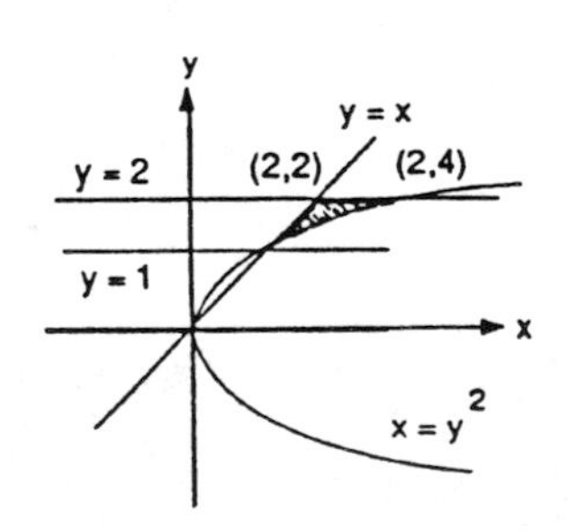

9. $\int_0^1 \int_0^{y^2} 3y^3 e^{xy}\, dx\, dy = \int_0^1 \left[3y^2 e^{xy}\right]_0^{y^2} dy$

$= \int_0^1 \left(3y^2 e^{y^3} - 3y^2\right) dy = \left[e^{y^3} - y^3\right]_0^1 = e - 2$

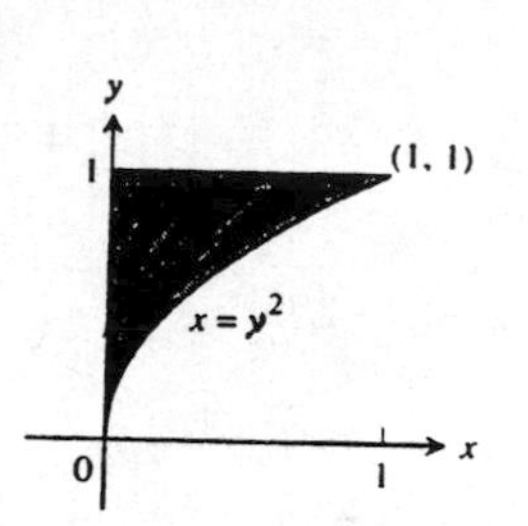

10. $\int_1^4 \int_0^{\sqrt{x}} \frac{3}{2} e^{y/\sqrt{x}}\, dy\, dx = \int_1^4 \left[\frac{3}{2}\sqrt{x}\, e^{y/\sqrt{x}}\right]_0^{\sqrt{x}} dx$

$= \frac{3}{2}(e-1) \int_1^4 \sqrt{x}\, dx = \left[\frac{3}{2}(e-1)\left(\frac{2}{3}\right)x^{3/2}\right]_1^4 = 7(e-1)$

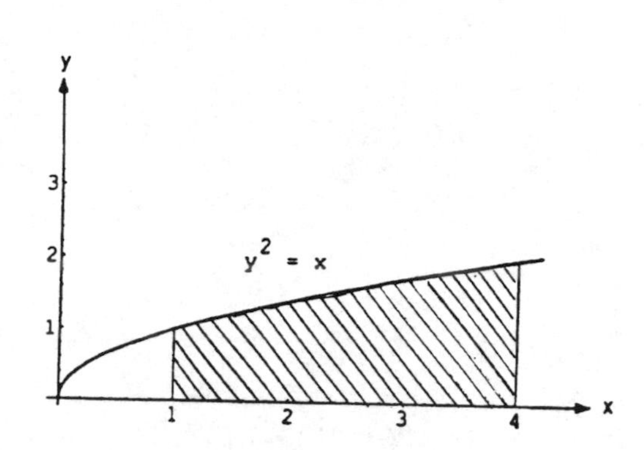

11. $\int_1^2 \int_x^{2x} \frac{x}{y}\, dy\, dx = \int_1^2 \left[x \ln y\right]_x^{2x} dx = (\ln 2) \int_1^2 x\, dx = \frac{3}{2} \ln 2$

12. $\int_1^2 \int_1^2 \frac{1}{xy}\, dy\, dx = \int_1^2 \frac{1}{x}(\ln 2 - \ln 1)\, dx = (\ln 2) \int_1^2 \frac{1}{x}\, dx = (\ln 2)^2$

13. $\int_0^1 \int_0^{1-x} (x^2 + y^2)\, dy\, dx = \int_0^1 \left[x^2 y + \frac{y^3}{3}\right]_0^{1-x} dx = \int_0^1 \left[x^2(1-x) + \frac{(1-x)^3}{3}\right] dx = \int_0^1 \left[x^2 - x^3 + \frac{(1-x)^3}{3}\right] dx$

$= \left[\frac{x^3}{3} - \frac{x^4}{4} - \frac{(1-x)^4}{12}\right]_0^1 = \left(\frac{1}{3} - \frac{1}{4} - 0\right) - \left(0 - 0 - \frac{1}{12}\right) = \frac{1}{6}$

14. $\int_0^1 \int_0^{\pi} y \cos xy\, dx\, dy = \int_0^1 \left[\sin xy\right]_0^{\pi} dy = \int_0^1 \sin \pi y\, dy = \left[-\frac{1}{\pi} \cos \pi y\right]_0^1 = -\frac{1}{\pi}(-1-1) = \frac{2}{\pi}$

15. $\int_0^1 \int_0^{1-u} (v - \sqrt{u})\, dv\, du = \int_0^1 \left[\frac{v^2}{2} - v\sqrt{u}\right]_0^{1-u} du = \int_0^1 \left[\frac{1 - 2u + u^2}{2} - \sqrt{u}(1-u)\right] du$

$= \int_0^1 \left(\frac{1}{2} - u + \frac{u^2}{2} - u^{1/2} + u^{3/2}\right) du = \left[\frac{u}{2} - \frac{u^2}{2} + \frac{u^3}{6} - \frac{2}{3}u^{3/2} + \frac{2}{5}u^{5/2}\right]_0^1 = \frac{1}{2} - \frac{1}{2} + \frac{1}{6} - \frac{2}{3} + \frac{2}{5} = -\frac{1}{2} + \frac{2}{5} = -\frac{1}{10}$

16. $\int_1^2 \int_0^{\ln t} e^s \ln t \, ds\, dt = \int_1^2 \left[e^s \ln t\right]_0^{\ln t} dt = \int_1^2 (t \ln t - \ln t)\, dt = \left[\frac{t^2}{2} \ln t - \frac{t^2}{4} - t \ln t + t\right]_1^2$

$= (2 \ln 2 - 1 - 2 \ln 2 + 2) - \left(-\frac{1}{4} + 1\right) = \frac{1}{4}$

17. $\int_{-2}^{0} \int_{v}^{-v} 2\, dp\, dv = \int_{-2}^{0} [p]_v^{-v}\, dv = 2\int_{-2}^{0} -2v\, dv$

$= -2\left[v^2\right]_{-2}^{0} = 8$

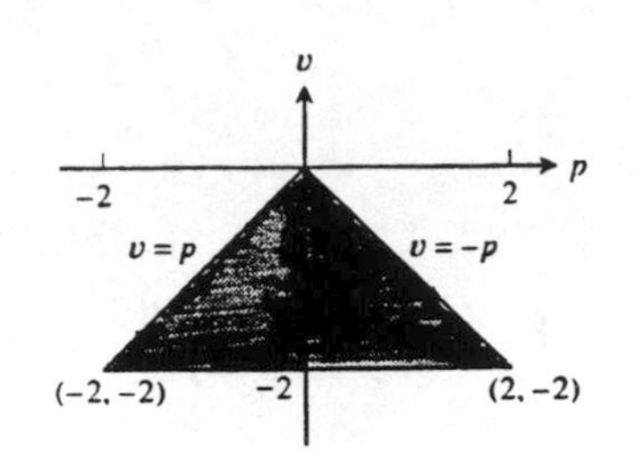

18. $\int_0^1 \int_0^{\sqrt{1-s^2}} 8t\, dt\, ds = \int_0^1 \left[4t^2\right]_0^{\sqrt{1-s^2}} ds$

$= \int_0^1 4\left(1 - s^2\right) ds = 4\left[s - \frac{s^3}{3}\right]_0^1 = \frac{8}{3}$

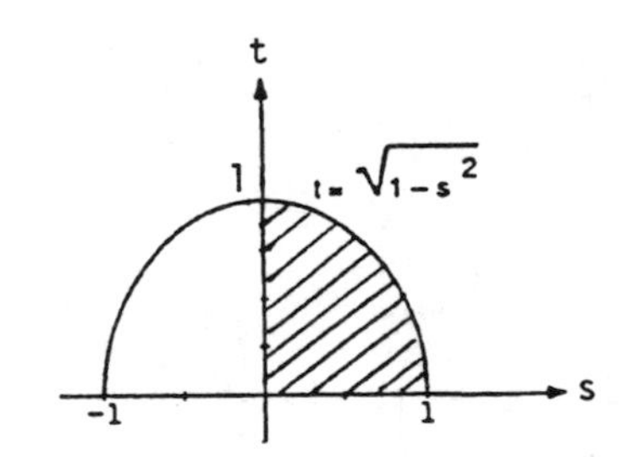

19. $\int_{-\pi/3}^{\pi/3} \int_0^{\sec t} 3 \cos t\, du\, dt = \int_{-\pi/3}^{\pi/3} \left[(3 \cos t)u\right]_0^{\sec t}$

$= \int_{-\pi/3}^{\pi/3} 3\, dt = 2\pi$

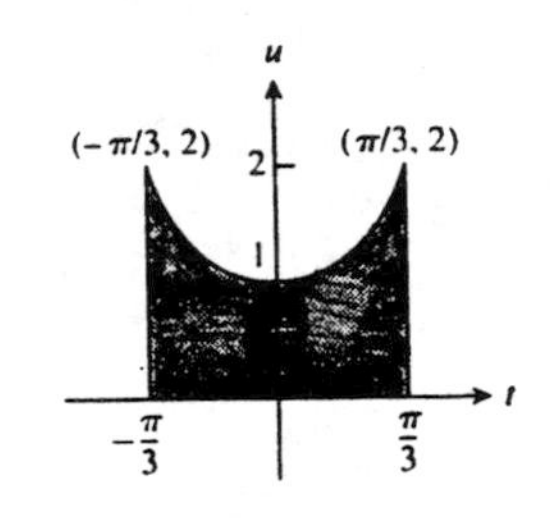

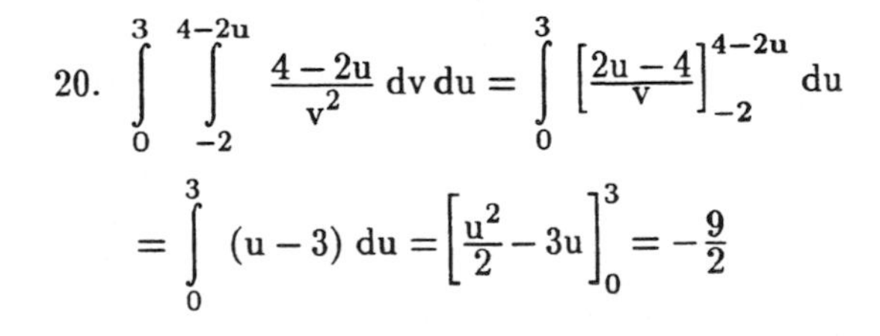

20. $\int_0^3 \int_{-2}^{4-2u} \frac{4 - 2u}{v^2}\, dv\, du = \int_0^3 \left[\frac{2u - 4}{v}\right]_{-2}^{4-2u} du$

$= \int_0^3 (u - 3)\, du = \left[\frac{u^2}{2} - 3u\right]_0^3 = -\frac{9}{2}$

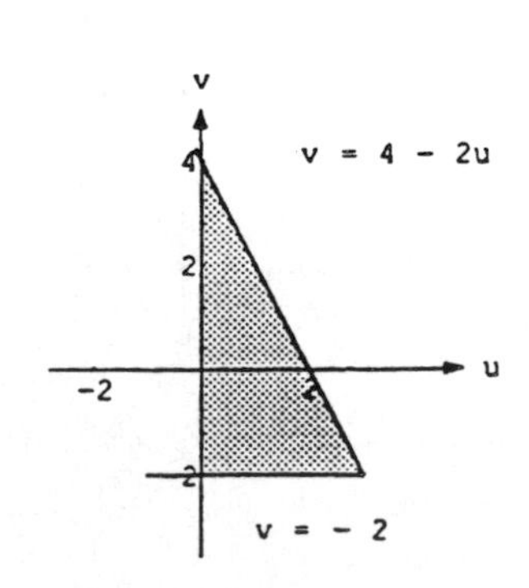

21. $\int_2^4 \int_0^{(4-y)/2} dx\,dy$

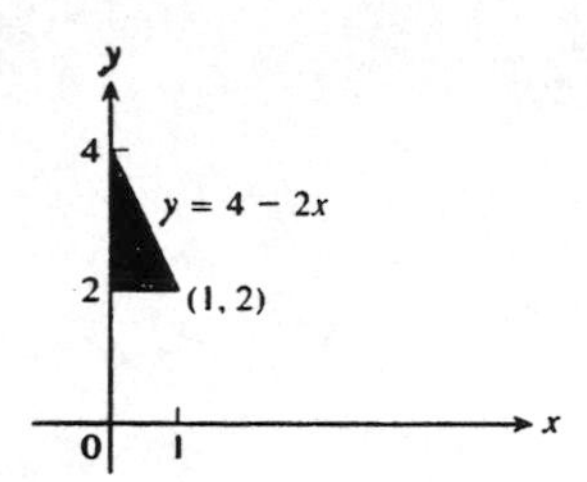

22. $\int_{-2}^0 \int_0^{x+2} dy\,dx$

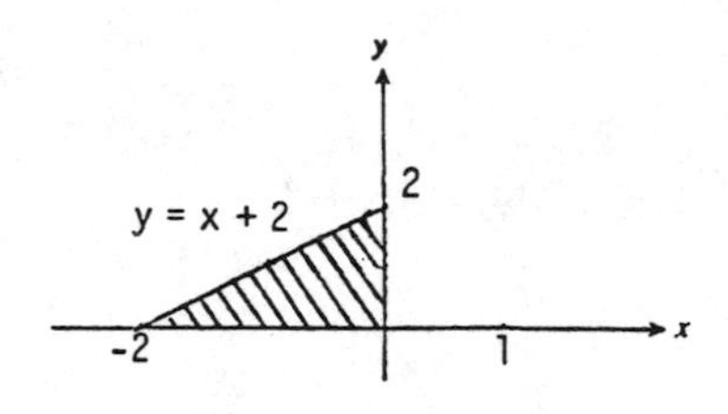

23. $\int_0^1 \int_{x^2}^{x} dy\,dx$

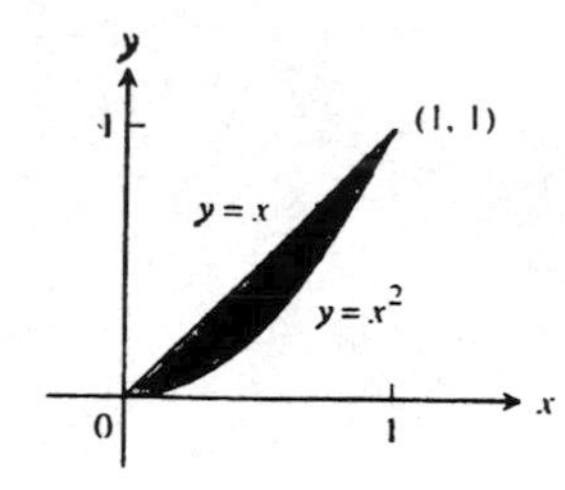

24. $\int_0^1 \int_{1-y}^{\sqrt{1-y}} dx\,dy$

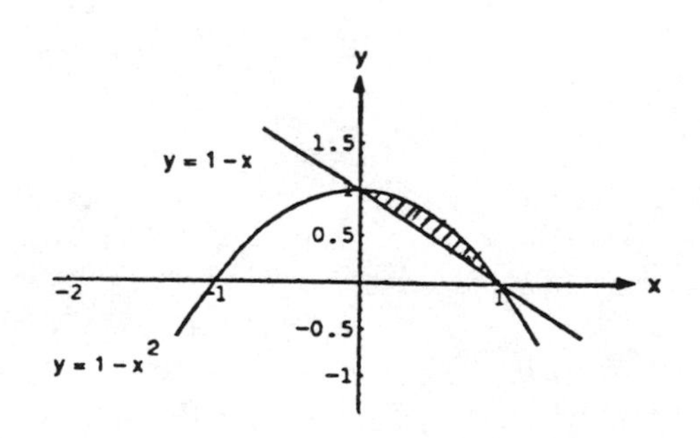

25. $\int_1^e \int_{\ln y}^{1} dx\,dy$

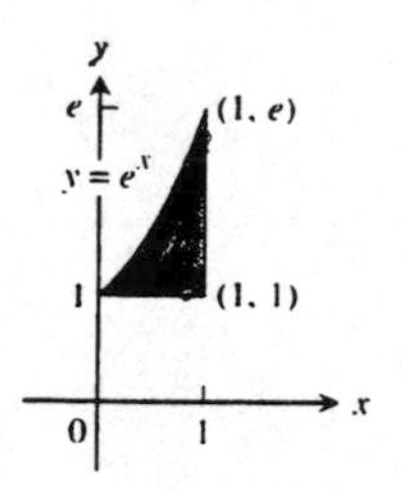

26. $\int_{1}^{2}\int_{0}^{\ln x} dy\, dx$

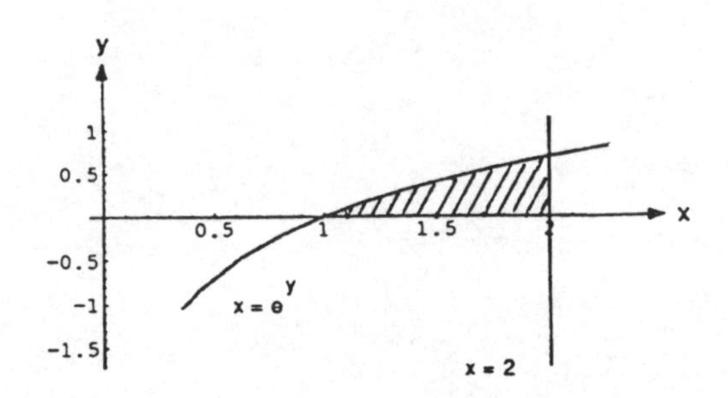

27. $\int_{0}^{9}\int_{0}^{\frac{1}{2}\sqrt{9-y}} 16x\, dx\, dy$

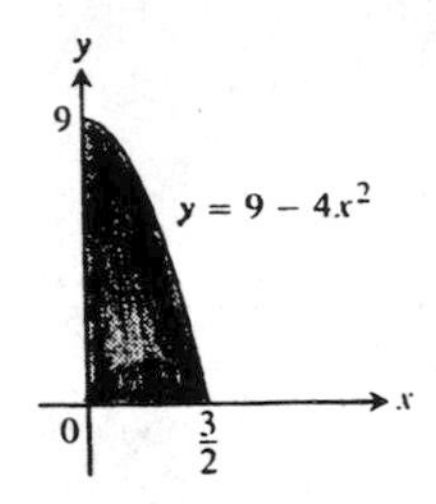

28. $\int_{0}^{4}\int_{0}^{\sqrt{4-x}} y\, dy\, dx$

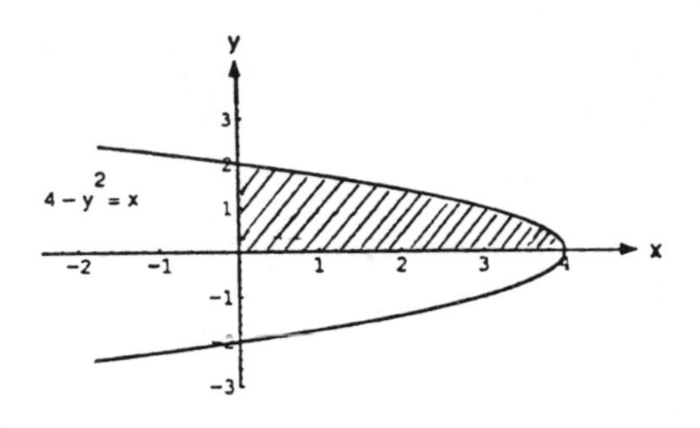

29. $\int_{-1}^{1}\int_{0}^{\sqrt{1-x^2}} 3y\, dy\, dx$

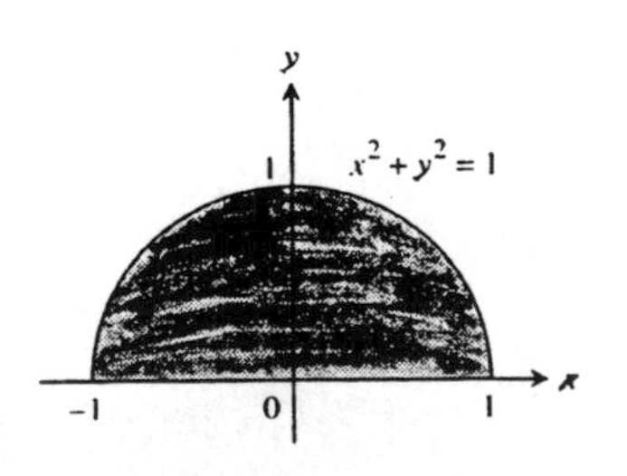

30. $\int_{-2}^{2}\int_{0}^{\sqrt{4-y^2}} 6x\,dx\,dy$

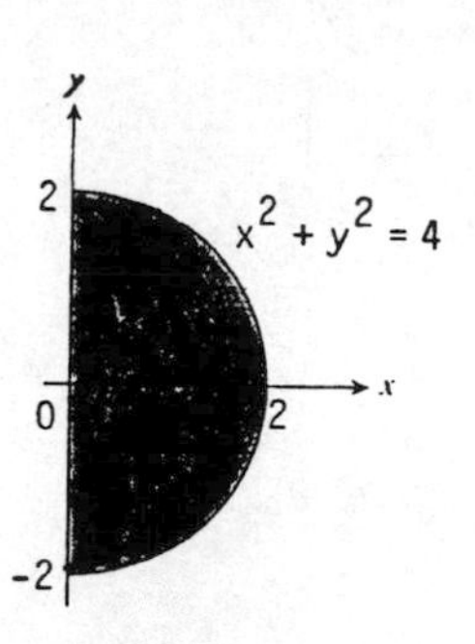

31. $\int_{0}^{\pi}\int_{x}^{\pi} \frac{\sin y}{y}\,dy\,dx = \int_{0}^{\pi}\int_{0}^{y} \frac{\sin y}{y}\,dx\,dy = \int_{0}^{\pi} \sin y\,dy = 2$

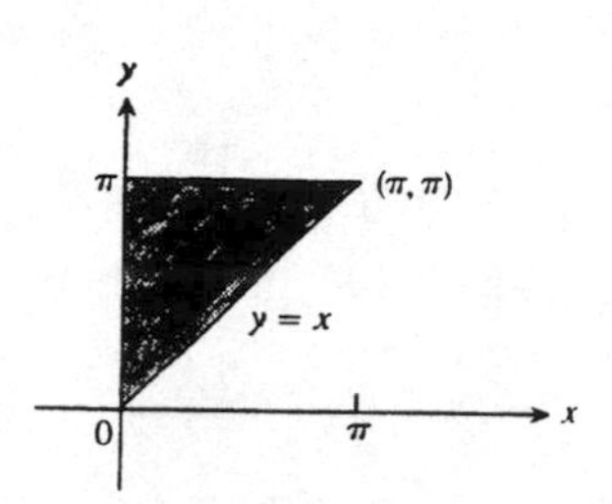

32. $\int_{0}^{2}\int_{x}^{2} 2y^2 \sin xy\,dy\,dx = \int_{0}^{2}\int_{0}^{y} 2y^2 \sin xy\,dx\,dy$

$= \int_{0}^{2} \left[-2y \cos xy\right]_0^y dy = \int_{0}^{2} \left(-2y \cos y^2 + 2y\right) dy$

$= \left[-\sin y^2 + y^2\right]_0^2 = 4 - \sin 4$

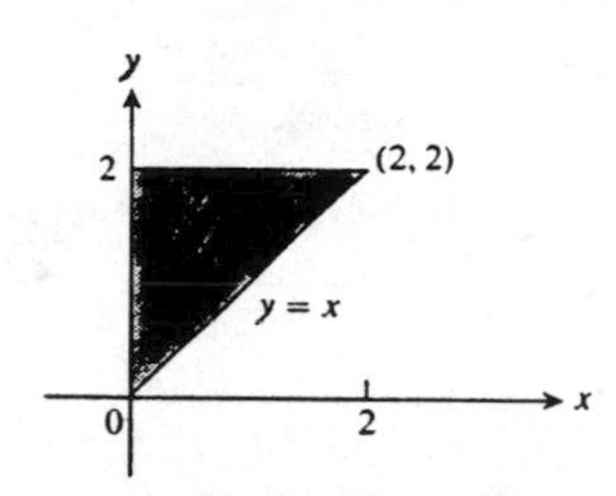

33. $\int_{0}^{1}\int_{y}^{1} x^2 e^{xy}\,dx\,dy = \int_{0}^{1}\int_{0}^{x} x^2 e^{xy}\,dy\,dx = \int_{0}^{1} \left[xe^{xy}\right]_0^x dx$

$= \int_{0}^{1} \left(xe^{x^2} - x\right) dx = \left[\frac{1}{2}e^{x^2} - \frac{x^2}{2}\right]_0^1 = \frac{e-2}{2}$

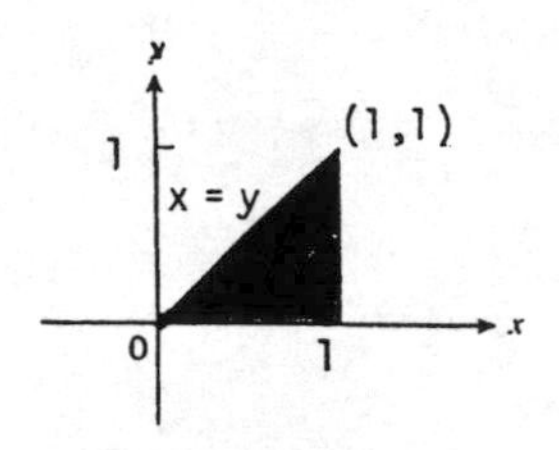

34. $\int_{0}^{2}\int_{0}^{4-x^2} \frac{xe^{2y}}{4-y}\,dy\,dx = \int_{0}^{4}\int_{0}^{\sqrt{4-y}} \frac{xe^{2y}}{4-y}\,dx\,dy$

$= \int_{0}^{4} \left[\frac{x^2 e^{2y}}{2(4-y)}\right]_0^{\sqrt{4-y}} dy = \int_{0}^{4} \frac{e^{2y}}{2}\,dy = \left[\frac{e^{2y}}{4}\right]_0^4 = \frac{e^8 - 1}{4}$

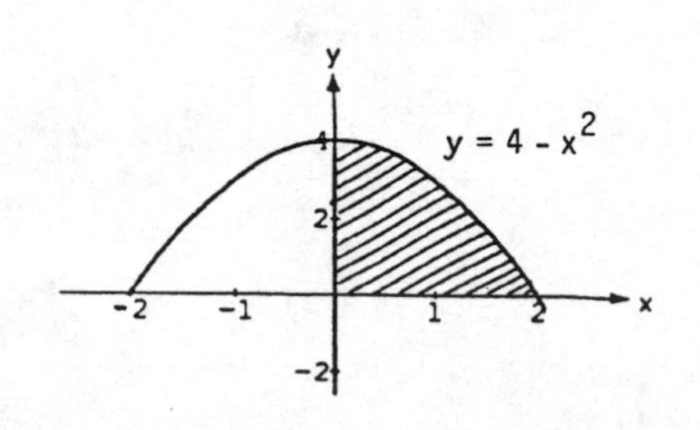

35. $$\int_0^{2\sqrt{\ln 3}} \int_{y/2}^{\sqrt{\ln 3}} e^{x^2}\,dx\,dy = \int_0^{\sqrt{\ln 3}} \int_0^{2x} e^{x^2}\,dy\,dx$$

$$= \int_0^{\sqrt{\ln 3}} 2xe^{x^2}\,dx = \left[e^{x^2}\right]_0^{\sqrt{\ln 3}} = e^{\ln 3} - 1 = 2$$

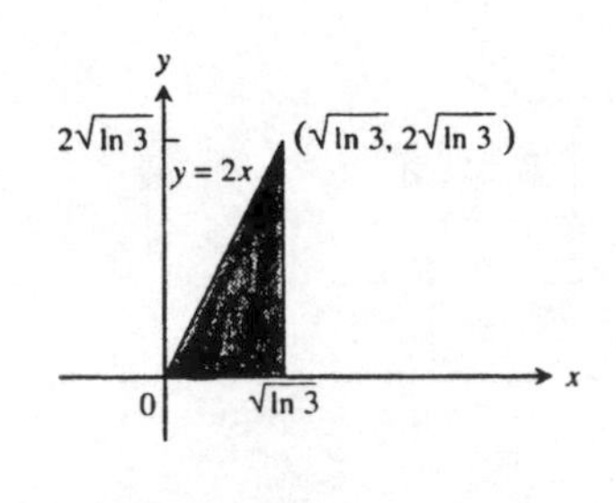

36. $$\int_0^3 \int_{\sqrt{x/3}}^1 e^{y^3}\,dy\,dx = \int_0^1 \int_0^{3y^2} e^{y^3}\,dx\,dy$$

$$= \int_0^1 3y^2e^{y^3}\,dy = \left[e^{y^3}\right]_0^1 = e - 1$$

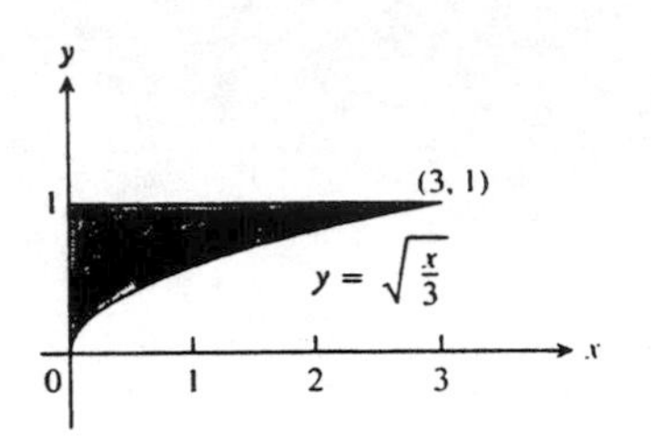

37. $$\int_0^{1/16} \int_{y^{1/4}}^{1/2} \cos(16\pi x^5)\,dx\,dy = \int_0^{1/2} \int_0^{x^4} \cos(16\pi x^5)\,dy\,dx$$

$$= \int_0^{1/2} x^4 \cos(16\pi x^5)\,dx = \left[\frac{\sin(16\pi x^5)}{80\pi}\right]_0^{1/2} = \frac{1}{80\pi}$$

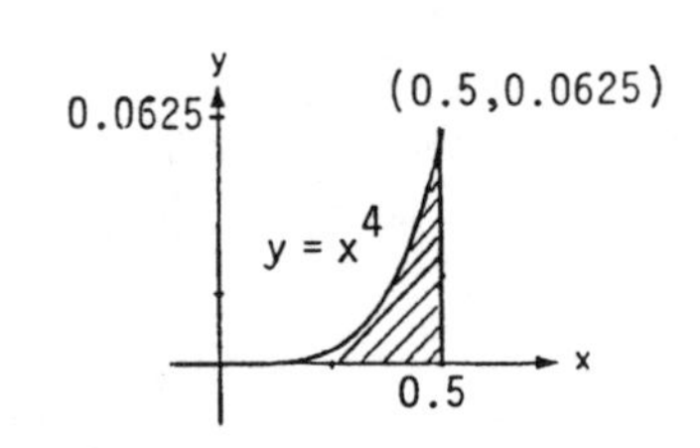

38. 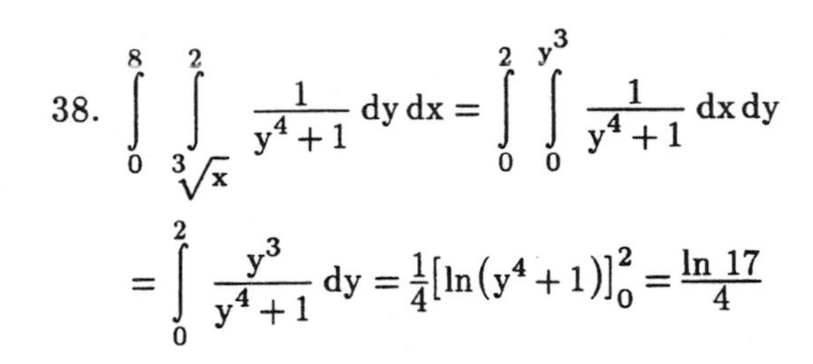

$$\int_0^8 \int_{\sqrt[3]{x}}^2 \frac{1}{y^4+1}\,dy\,dx = \int_0^2 \int_0^{y^3} \frac{1}{y^4+1}\,dx\,dy$$

$$= \int_0^2 \frac{y^3}{y^4+1}\,dy = \tfrac{1}{4}\left[\ln(y^4+1)\right]_0^2 = \frac{\ln 17}{4}$$

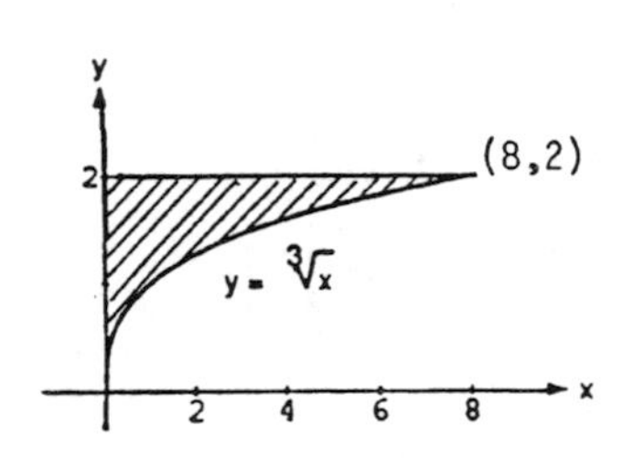

39. $$\iint_R (y - 2x^2)\,dA = \int_{-1}^0 \int_{-x-1}^{x+1} (y - 2x^2)\,dy\,dx + \int_0^1 \int_{x-1}^{1-x} (y - 2x^2)\,dy\,dx$$

$$= \int_{-1}^0 \left[\tfrac{1}{2}y^2 - 2x^2y\right]_{-x-1}^{x+1} dx + \int_0^1 \left[\tfrac{1}{2}y^2 - 2x^2y\right]_{x-1}^{1-x} dx$$

$$= \int_{-1}^0 \left[\tfrac{1}{2}(x+1)^2 - 2x^2(x+1) - \tfrac{1}{2}(-x-1)^2 + 2x^2(-x-1)\right] dx$$

$$+ \int_0^1 \left[\tfrac{1}{2}(1-x)^2 - 2x^2(1-x) - \tfrac{1}{2}(x-1)^2 + 2x^2(x-1)\right] dx$$

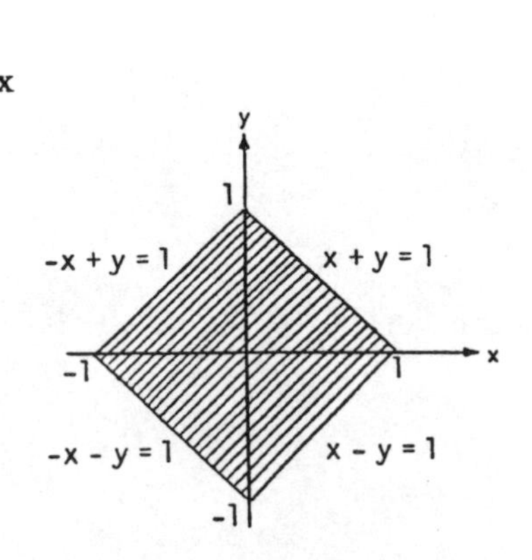

$$= -4\int_{-1}^{0}(x^3+x^2)\,dx + 4\int_{0}^{1}(x^3-x^2)\,dx = -4\left[\frac{x^4}{4}+\frac{x^3}{3}\right]_{-1}^{0} + 4\left[\frac{x^4}{4}-\frac{x^3}{3}\right]_{0}^{1}$$

$$= 4\left[\frac{(-1)^4}{4}+\frac{(-1)^3}{3}\right] + 4\left(\frac{1}{4}-\frac{1}{3}\right) = 8\left(\frac{3}{12}-\frac{4}{12}\right) = -\frac{8}{12} = -\frac{2}{3}$$

40. $$\iint_R xy\,dA = \int_0^{2/3}\int_0^{2x} xy\,dy\,dx + \int_{2/3}^{1}\int_x^{2-x} xy\,dy\,dx$$

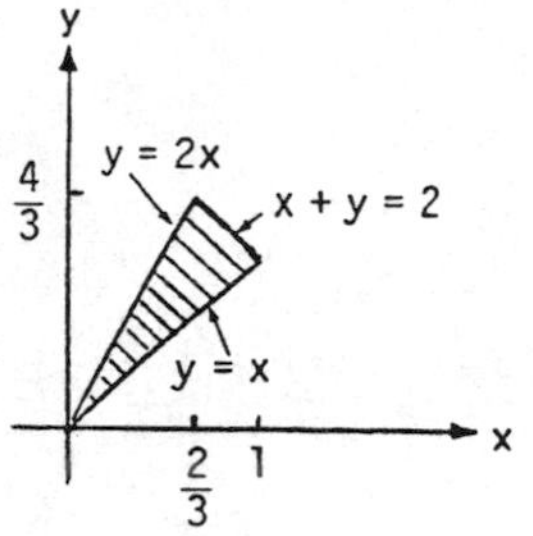

$$= \int_0^{2/3}\left[\frac{1}{2}xy^2\right]_x^{2x} dx + \int_{2/3}^{1}\left[\frac{1}{2}xy^2\right]_x^{2-x} dx$$

$$= \int_0^{2/3}\left(2x^3-\frac{1}{2}x^3\right)dx + \int_{2/3}^{1}\left[\frac{1}{2}x(2-x)^2-\frac{1}{2}x^3\right]dx$$

$$= \int_0^{2/3}\frac{3}{2}x^3\,dx + \int_{2/3}^{1}(2x-x^2)\,dx$$

$$= \left[\frac{3}{8}x^4\right]_0^{2/3} + \left[x^2-\frac{2}{3}x^3\right]_{2/3}^{1} = \left(\frac{3}{8}\right)\left(\frac{16}{81}\right)+\left(1-\frac{2}{3}\right)-\left[\frac{4}{9}-\left(\frac{2}{3}\right)\left(\frac{8}{27}\right)\right] = \frac{6}{81}+\frac{27}{81}-\left(\frac{36}{81}-\frac{16}{81}\right) = \frac{13}{81}$$

41. $$V = \int_0^1\int_x^{2-x}(x^2+y^2)\,dy\,dx = \int_0^1\left[x^2y+\frac{y^3}{3}\right]_x^{2-x} dx = \int_0^1\left[2x^2-\frac{7x^3}{3}+\frac{(2-x)^3}{3}\right]dx = \left[\frac{2x^3}{3}-\frac{7x^4}{12}-\frac{(2-x)^4}{12}\right]_0^1$$

$$= \left(\frac{2}{3}-\frac{7}{12}-\frac{1}{12}\right)-\left(0-0-\frac{16}{12}\right) = \frac{4}{3}$$

42. $$V = \int_{-2}^{1}\int_x^{2-x^2} x^2\,dy\,dx = \int_{-2}^{1}\left[x^2y\right]_x^{2-x^2} dx = \int_{-2}^{1}(2x^2-x^4-x^3)\,dx = \left[\frac{2}{3}x^3-\frac{1}{5}x^5-\frac{1}{4}x^4\right]_{-2}^{1}$$

$$= \left(\frac{2}{3}-\frac{1}{5}-\frac{1}{4}\right)-\left(-\frac{16}{3}+\frac{32}{5}-\frac{16}{4}\right) = \left(\frac{40}{60}-\frac{12}{60}-\frac{15}{60}\right)-\left(-\frac{320}{60}+\frac{384}{60}-\frac{240}{60}\right) = \frac{189}{60} = \frac{63}{20}$$

43. $$V = \int_{-4}^{1}\int_{3x}^{4-x^2}(x+4)\,dy\,dx = \int_{-4}^{1}\left[xy+4y\right]_{3x}^{4-x^2} dx = \int_{-4}^{1}\left[x(4-x^2)+4(4-x^2)-3x^2-12x\right]dx$$

$$= \int_{-4}^{1}(-x^3-7x^2-8x+16)\,dx = \left[-\frac{1}{4}x^4-\frac{7}{3}x^3-4x^2+16x\right]_{-4}^{1} = \left(-\frac{1}{4}-\frac{7}{3}+12\right)-\left(\frac{64}{3}-64\right)$$

$$= \frac{157}{3}-\frac{1}{4} = \frac{625}{12}$$

44. $V = \int_0^2 \int_0^{\sqrt{4-x^2}} (3-y)\,dy\,dx = \int_0^2 \left[3y - \frac{y^2}{2}\right]_0^{\sqrt{4-x^2}} dx = \int_0^2 \left[3\sqrt{4-x^2} - \left(\frac{4-x^2}{2}\right)\right] dx$

$= \left[\frac{3}{2}x\sqrt{4-x^2} + 6\sin^{-1}\left(\frac{x}{2}\right) - 2x + \frac{x^3}{6}\right]_0^2 = 6\left(\frac{\pi}{2}\right) - 4 + \frac{8}{6} = 3\pi - \frac{16}{6} = \frac{9\pi - 8}{3}$

45. $V = \int_0^2 \int_0^3 (4-y^2)\,dx\,dy = \int_0^2 \left[4x - y^2x\right]_0^3 dy = \int_0^2 (12 - 3y^2)\,dy = \left[12y - y^3\right]_0^2 = 24 - 8 = 16$

46. $V = \int_0^2 \int_0^{4-x^2} (4 - x^2 - y)\,dy\,dx = \int_0^2 \left[(4-x^2)y - \frac{y^2}{2}\right]_0^{4-x^2} dx = \int_0^2 \frac{1}{2}(4-x^2)^2\,dx = \int_0^2 \left(8 - 4x^2 + \frac{x^4}{2}\right) dx$

$= \left[8x - \frac{4}{3}x^3 + \frac{1}{10}x^5\right]_0^2 = 16 - \frac{32}{3} + \frac{32}{10} = \frac{480 - 320 + 96}{30} = \frac{128}{15}$

47. $V = \int_0^2 \int_0^{2-x} (12 - 3y^2)\,dy\,dx = \int_0^2 \left[12y - y^3\right]_0^{2-x} dx = \int_0^2 \left[24 - 12x - (2-x)^3\right] dx$

$= \left[24x - 6x^2 + \frac{(2-x)^4}{4}\right]_0^2 = 20$

48. $V = \int_{-1}^0 \int_{-x-1}^{x+1} (3 - 3x)\,dy\,dx + \int_0^1 \int_{x-1}^{1-x} (3-3x)\,dy\,dx = 6\int_{-1}^0 (1-x^2)\,dx + 6\int_0^1 (1-x)^2\,dx = 2 + 4 = 6$

49. $V = \int_1^2 \int_{-1/x}^{1/x} (x+1)\,dy\,dx = \int_1^2 \left[xy + y\right]_{-1/x}^{1/x} dx = \int_1^2 \left[1 + \frac{1}{x} - \left(-1 - \frac{1}{x}\right)\right] = 2\int_1^2 \left(1 + \frac{1}{x}\right) dx$

$= 2\left[x + \ln x\right]_1^2 = 2(1 + \ln 2)$

50. $V = 4\int_0^{\pi/3} \int_0^{\sec x} (1 + y^2)\,dy\,dx = 4\int_0^{\pi/3} \left[y + \frac{y^3}{3}\right]_0^{\sec x} dx = 4\int_0^{\pi/3} \left(\sec x + \frac{\sec^3 x}{3}\right) dx$

$= \frac{2}{3}\left[7\ln|\sec x + \tan x| + \sec x \tan x\right]_0^{\pi/3} = \frac{2}{3}\left[7\ln(2 + \sqrt{3}) + 2\sqrt{3}\right]$

51. $\int_1^\infty \int_{e^{-x}}^1 \frac{1}{x^3 y}\,dy\,dx = \int_1^\infty \left[\frac{\ln y}{x^3}\right]_{e^{-x}}^1 dx = \int_1^\infty -\left(\frac{-x}{x^3}\right) dx = -\lim_{b\to\infty} \left[\frac{1}{x}\right]_1^b = -\lim_{b\to\infty} \left(\frac{1}{b} - 1\right) = 1$

52. $\int_{-1}^{1}\int_{-1/\sqrt{1-x^2}}^{1/\sqrt{1-x^2}} (2y+1)\,dy\,dx = \int_{-1}^{1}\left[y^2+y\right]_{-1/(1-x^2)^{1/2}}^{1/(1-x^2)^{1/2}} dx = \int_{-1}^{1}\frac{2}{\sqrt{1-x^2}}\,dx = 4\lim_{b\to 1^-}\left[\sin^{-1}x\right]_0^b$

$= 4\lim_{b\to 1^-}\left[\sin^{-1}b - 0\right] = 2\pi$

53. $\int_{-\infty}^{\infty}\int_{-\infty}^{\infty}\frac{1}{(x^2+1)(y^2+1)}\,dx\,dy = 2\int_0^{\infty}\left(\frac{2}{y^2+1}\right)\left(\lim_{b\to\infty}\tan^{-1}b - \tan^{-1}0\right)dy = 2\pi\lim_{b\to\infty}\int_0^b\frac{1}{y^2+1}\,dy$

$= 2\pi\left(\lim_{b\to\infty}\tan^{-1}b - \tan^{-1}0\right) = (2\pi)\left(\frac{\pi}{2}\right) = \pi^2$

54. $\int_0^{\infty}\int_0^{\infty} xe^{-(x+2y)}\,dx\,dy = \int_0^{\infty} e^{-2y}\lim_{b\to\infty}\left[-xe^{-x}-e^{-x}\right]_0^b dy = \int_0^{\infty} e^{-2y}\lim_{b\to\infty}\left(-be^{-b}-e^{-b}+1\right)dy$

$= \int_0^{\infty} e^{-2y}\,dy = \frac{1}{2}\lim_{b\to\infty}\left(-e^{-2b}+1\right) = \frac{1}{2}$

55. $\iint_R f(x,y)\,dA \approx \frac{1}{4}f\left(-\frac{1}{2},0\right)+\frac{1}{8}f(0,0)+\frac{1}{8}f\left(\frac{1}{4},0\right)+\frac{1}{4}f\left(\frac{1}{2},0\right)+\frac{1}{4}f\left(-\frac{1}{2},\frac{1}{2}\right)+\frac{1}{8}f\left(0,\frac{1}{2}\right)+\frac{1}{8}f\left(\frac{1}{4},\frac{1}{2}\right)$

$= \frac{1}{4}\left(-\frac{1}{2}+\frac{1}{2}+0\right)+\frac{1}{8}\left(0+\frac{1}{4}+\frac{1}{2}+\frac{3}{4}\right) = \frac{3}{16}$

56. $\iint_R f(x,y)\,dA \approx \frac{1}{4}\left[f\left(\frac{7}{4},\frac{9}{4}\right)+f\left(\frac{9}{4},\frac{9}{4}\right)+f\left(\frac{5}{4},\frac{11}{4}\right)+f\left(\frac{7}{4},\frac{11}{4}\right)+f\left(\frac{9}{4},\frac{11}{4}\right)+f\left(\frac{11}{4},\frac{11}{4}\right)+f\left(\frac{5}{4},\frac{13}{4}\right)+f\left(\frac{7}{4},\frac{13}{4}\right)\right.$

$\left.+f\left(\frac{9}{4},\frac{13}{4}\right)+f\left(\frac{11}{4},\frac{13}{4}\right)+f\left(\frac{7}{4},\frac{15}{4}\right)+f\left(\frac{9}{4},\frac{15}{4}\right)\right]$

$= \frac{1}{16}(25+27+27+29+31+33+31+33+35+37+37+39) = \frac{384}{16} = 24$

57. The ray $\theta = \frac{\pi}{6}$ meets the circle $x^2+y^2=4$ at the point $\left(\sqrt{3},1\right) \Rightarrow$ the ray is represented by the line $y = \frac{x}{\sqrt{3}}$.

Thus, $\iint_R f(x,y)\,dA = \int_0^{\sqrt{3}}\int_{x/\sqrt{3}}^{\sqrt{4-x^2}}\sqrt{4-x^2}\,dy\,dx = \int_0^{\sqrt{3}}\left[(4-x^2)-\frac{x}{3}\sqrt{4-x^2}\right]dx = \left[4x-\frac{x^3}{3}+\frac{(4-x^2)^{3/2}}{3\sqrt{3}}\right]_0^{\sqrt{3}}$

$= \frac{20\sqrt{3}}{9}$

58. $\int_2^{\infty}\int_0^2\frac{1}{(x^2-x)(y-1)^{2/3}}\,dy\,dx = \int_2^{\infty}\left[\frac{3(y-1)^{1/3}}{(x^2-x)}\right]_0^2 dx = \int_2^{\infty}\left(\frac{3}{x^2-x}+\frac{3}{x^2-x}\right)dx = 6\int_2^{\infty}\frac{dx}{x(x-1)}$

$= 6\lim_{b\to\infty}\int_2^b\left(\frac{1}{x-1}-\frac{1}{x}\right)dx = 6\lim_{b\to\infty}\left[\ln(x-1)-\ln x\right]_2^b = 6\lim_{b\to\infty}\left[\ln(b-1)-\ln b-\ln 1+\ln 2\right]$

$$= 6\left[\lim_{b\to\infty} \ln\left(1-\tfrac{1}{b}\right)+\ln 2\right] = 6 \ln 2$$

59. $V = \int_0^1 \int_x^{2-x} (x^2+y^2)\, dy\, dx = \int_0^1 \left[x^2 y + \frac{y^3}{3}\right]_x^{2-x} dx$

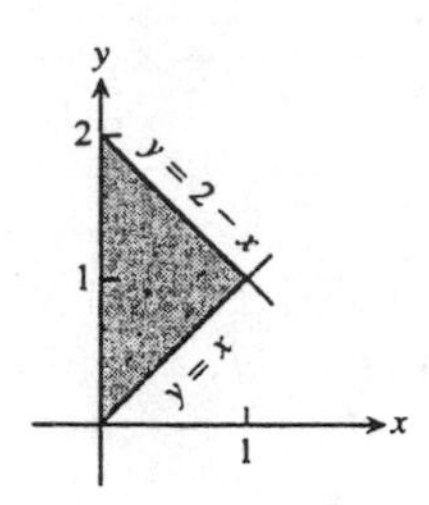

$$= \int_0^1 \left[2x^2 - \frac{7x^3}{3} + \frac{(2-x)^3}{3}\right] dx = \left[\frac{2x^3}{3} - \frac{7x^4}{12} - \frac{(2-x)^4}{12}\right]_0^1$$

$$= \left(\tfrac{2}{3} - \tfrac{7}{12} - \tfrac{1}{12}\right) - \left(0 - 0 - \tfrac{16}{12}\right) = \left(\tfrac{2}{3} + \tfrac{8}{12}\right) = \tfrac{4}{3}$$

60. $\int_0^2 \left(\tan^{-1} \pi x - \tan^{-1} x\right) dx = \int_0^2 \int_x^{\pi x} \frac{1}{1+y^2}\, dy\, dx = \int_0^2 \int_{y/\pi}^{y} \frac{1}{1+y^2}\, dx\, dy + \int_2^{2\pi} \int_{y/\pi}^{2} \frac{1}{1+y^2}\, dx\, dy$

$$= \int_0^2 \frac{\left(1-\frac{1}{\pi}\right)y}{1+y^2}\, dy + \int_2^{2\pi} \frac{\left(2-\frac{y}{\pi}\right)}{1+y^2}\, dy = \left(\frac{\pi-1}{2\pi}\right)\left[\ln(1+y^2)\right]_0^2 + \left[2 \tan^{-1} y - \frac{1}{2\pi} \ln(1+y^2)\right]_2^{2\pi}$$

$$= \left(\frac{\pi-1}{2\pi}\right) \ln 5 + 2 \tan^{-1} 2\pi - \frac{1}{2\pi} \ln(1+4\pi^2) - 2 \tan^{-1} 2 - \frac{1}{2\pi} \ln 5$$

$$= 2 \tan^{-1} 2\pi - 2 \tan^{-1} 2 - \frac{1}{2\pi} \ln(1+4\pi^2) + \frac{\ln 5}{2}$$

61. To maximize the integral, we want the domain to include all points where the integrand is positive and to exclude all points where the integrand is negative. These criteria are met by the points (x, y) such that $4 - x^2 - 2y^2 \geq 0$ or $x^2 + 2y^2 \leq 4$, which is the ellipse $x^2 + 2y^2 = 4$ together with its interior.

62. To minimize the integral, we want the domain to include all points where the integrand is negative and to exclude all points where the integrand is positive. These criteria are met by the points (x, y) such that $x^2 + y^2 - 9 \leq 0$ or $x^2 + y^2 \leq 9$, which is the closed disk of radius 3 centered at the origin.

63. No, it is not all right. By Fubini's theorem, the two orders of integration must give the same result.

64. One way would be to partition R into two triangles with the line $y = 1$. The integral of f over R could then be written as a sum of integrals that could be evaluated by integrating first with respect to x and then with respect to y:

$$\iint_R f(x,y)\, dA = \int_0^1 \int_{2-2y}^{2-(y/2)} f(x,y)\, dx\, dy + \int_1^2 \int_{y-1}^{2-(y/2)} f(x,y)\, dx\, dy.$$

Partitioning R with the line $x = 1$ would let us write the integral of f over R as a sum of iterated integrals with order dy dx.

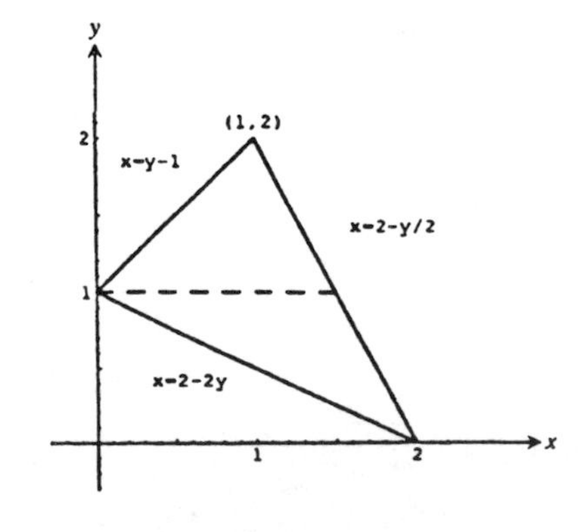

65. $\int_{-b}^{b}\int_{-b}^{b} e^{-x^2-y^2}\,dx\,dy = \int_{-b}^{b}\int_{-b}^{b} e^{-y^2}e^{-x^2}\,dx\,dy = \int_{-b}^{b} e^{-y^2}\left(\int_{-b}^{b} e^{-x^2}\,dx\right)dy = \left(\int_{-b}^{b} e^{-x^2}\,dx\right)\left(\int_{-b}^{b} e^{-y^2}\,dy\right)$

$= \left(\int_{-b}^{b} e^{-x^2}\,dx\right)^2 = \left(2\int_{0}^{b} e^{-x^2}\,dx\right)^2 = 4\left(\int_{0}^{b} e^{-x^2}\,dx\right)^2$; taking limits as $b \to \infty$ gives the stated result.

66. $\int_{0}^{1}\int_{0}^{3} \frac{x^2}{(y-1)^{2/3}}\,dy\,dx = \int_{0}^{3}\int_{0}^{1} \frac{x^2}{(y-1)^{2/3}}\,dx\,dy = \int_{0}^{3} \frac{1}{(y-1)^{2/3}}\left[\frac{x^3}{3}\right]_0^1 dy = \frac{1}{3}\int_{0}^{3} \frac{dy}{(y-1)^{2/3}}$

$= \frac{1}{3}\lim_{b\to 1^-}\int_{0}^{b} \frac{dy}{(y-1)^{2/3}} + \frac{1}{3}\lim_{b\to 1^+}\int_{b}^{3} \frac{dy}{(y-1)^{2/3}} = \lim_{b\to 1^-}\left[(y-1)^{1/3}\right]_0^b + \lim_{b\to 1^+}\left[(y-1)^{1/3}\right]_b^3$

$= \left[\lim_{b\to 1^-}(b-1)^{1/3} - (-1)^{1/3}\right] + \left[\lim_{b\to 1^+}(b-1)^{1/3} - (2)^{1/3}\right] = (0+1) - (0 - \sqrt[3]{2}) = 1 + \sqrt[3]{2}$

67. $\int_{1}^{3}\int_{1}^{x} \frac{1}{xy}\,dy\,dx \approx 0.603$

68. $\int_{0}^{1}\int_{0}^{1} e^{-(x^2+y^2)}\,dy\,dx \approx 0.558$

69. $\int_{0}^{1}\int_{0}^{1} \tan^{-1} xy\,dy\,dx \approx 0.233$

70. $\int_{-1}^{1}\int_{0}^{\sqrt{1-x^2}} 3\sqrt{1-x^2-y^2}\,dy\,dx \approx 3.142$

13.2 AREAS, MOMENTS, AND CENTERS OF MASS

1. $\int_{0}^{2}\int_{0}^{2-x} dy\,dx = \int_{0}^{2} (2-x)\,dx = \left[2x - \frac{x^2}{2}\right]_0^2 = 2,$

or $\int_{0}^{2}\int_{0}^{2-y} dx\,dy = \int_{0}^{2} (2-y)\,dy = 2$

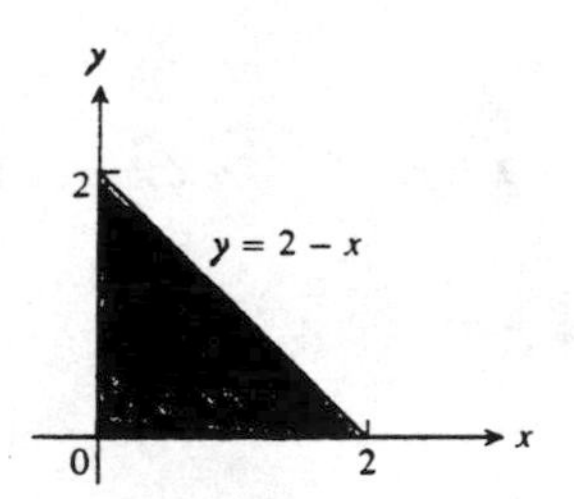

2. $\int_{0}^{2}\int_{2x}^{4} dy\,dx = \int_{0}^{2} (4-2x)\,dx = \left[4x - x^2\right]_0^2 = 4,$

or $\int_{0}^{4}\int_{0}^{y/2} dx\,dy = \int_{0}^{4} \frac{y}{2}\,dy = 4$

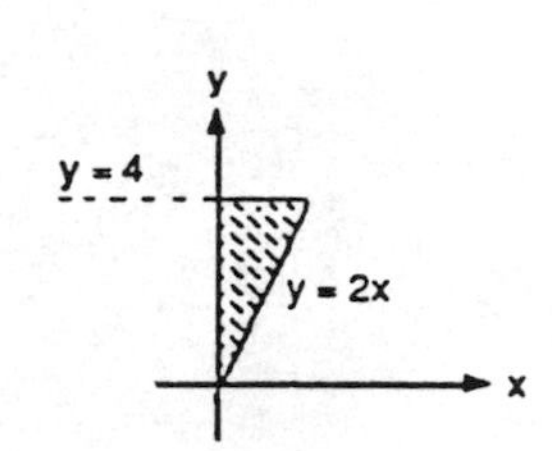

3. $\int_{-2}^{1}\int_{y-2}^{-y^2} dx\,dy = \int_{-2}^{1}(-y^2-y+2)\,dy = \left[-\frac{y^3}{3}-\frac{y^2}{2}+2y\right]_{-2}^{1}$
$= \left(-\frac{1}{3}-\frac{1}{2}+2\right)-\left(\frac{8}{3}-2-4\right)=\frac{9}{2}$

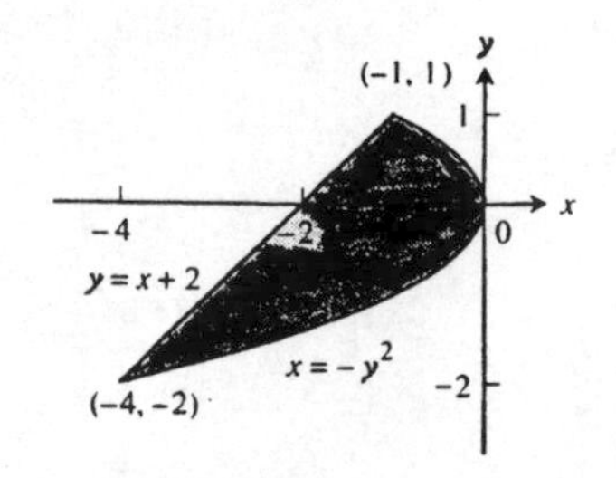

4. $\int_{0}^{2}\int_{-y}^{y-y^2} dx\,dy = \int_{0}^{2}(2y-y^2)\,dy = \left[y^2-\frac{y^3}{3}\right]_0^2 = 4-\frac{8}{3}=\frac{4}{3}$

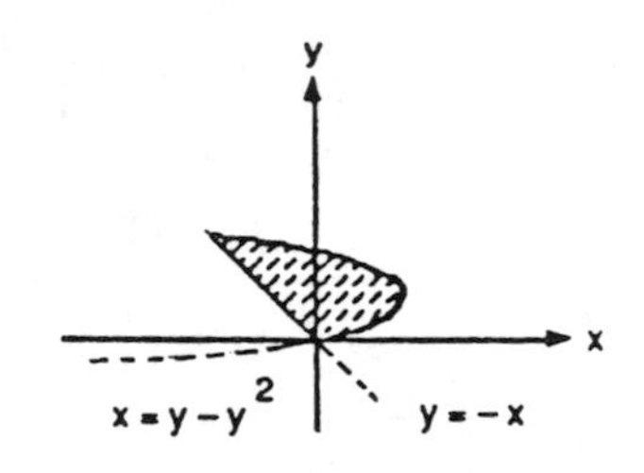

5. $\int_{0}^{\ln 2}\int_{0}^{e^x} dy\,dx = \int_{0}^{\ln 2} e^x\,dx = \left[e^x\right]_0^{\ln 2} = 2-1=1$

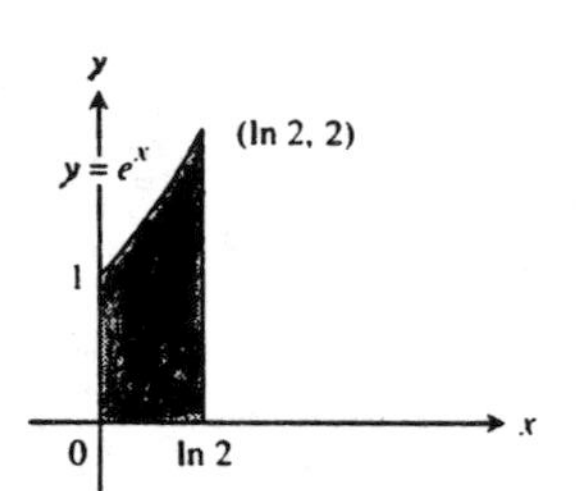

6. $\int_{1}^{e}\int_{\ln x}^{2\ln x} dy\,dx = \int_{1}^{e}\ln x\,dx = [x\ln x - x]_1^e$
$= (e-e)-(0-1)=1$

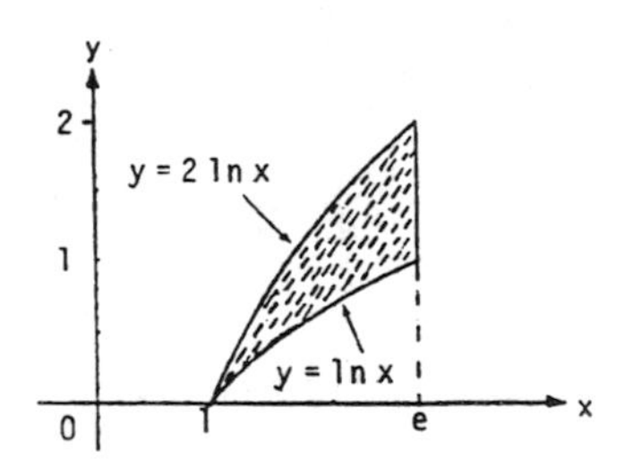

7. $\int_{0}^{1}\int_{y^2}^{2y-y^2} dx\,dy = \int_{0}^{1}(2y-2y^2)\,dy = \left[y^2-\frac{2}{3}y^3\right]_0^1 = \frac{1}{3}$

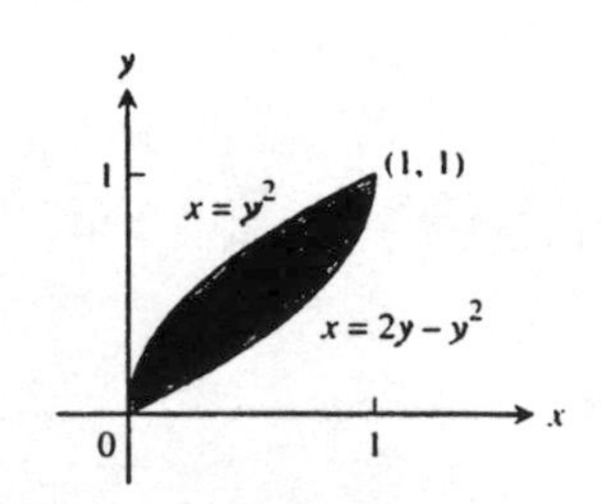

8. $\int_{-1}^{1}\int_{2y^2-2}^{y^2-1} dx\,dy = \int_{-1}^{1}\left(y^2-1-2y^2+2\right)dy$

$= \int_{-1}^{1}\left(1-y^2\right)dy = \left[y-\frac{y^3}{3}\right]_{-1}^{1} = \frac{4}{3}$

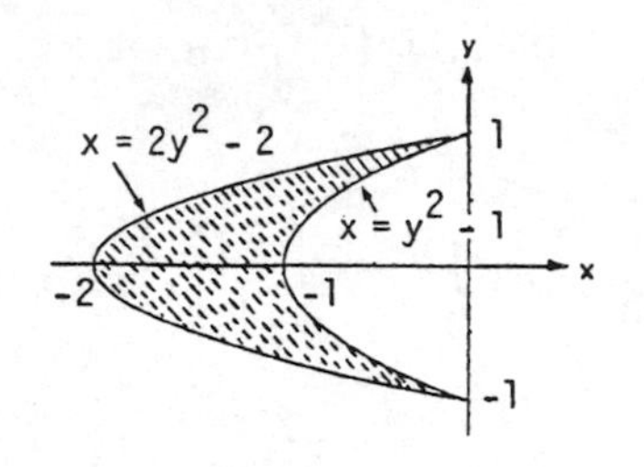

9. $\int_{0}^{6}\int_{y^2/3}^{2y} dx\,dy = \int_{0}^{6}\left(2y-\frac{y^2}{3}\right)dy = \left[y^2-\frac{y^3}{9}\right]_{0}^{6}$

$= 36 - \frac{216}{9} = 12$

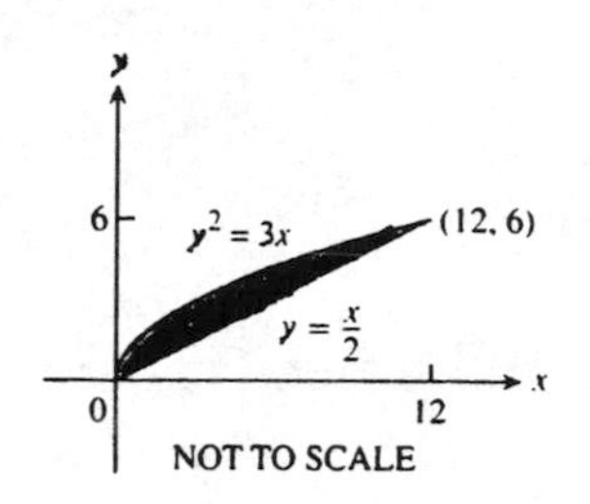

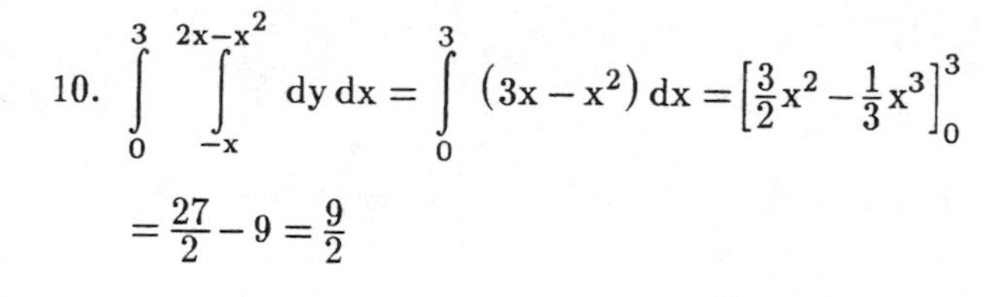

10. $\int_{0}^{3}\int_{-x}^{2x-x^2} dy\,dx = \int_{0}^{3}\left(3x-x^2\right)dx = \left[\frac{3}{2}x^2-\frac{1}{3}x^3\right]_{0}^{3}$

$= \frac{27}{2} - 9 = \frac{9}{2}$

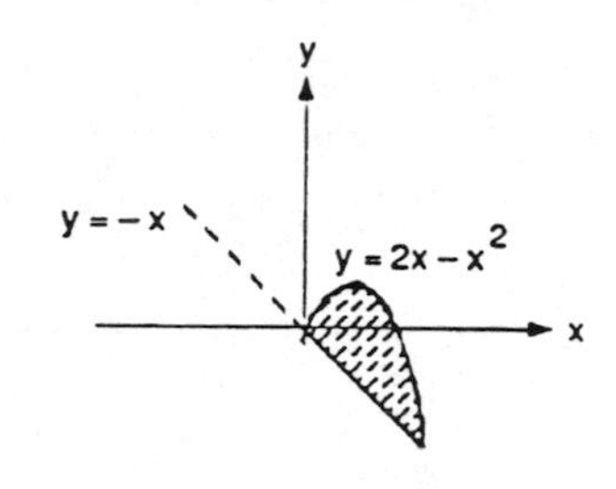

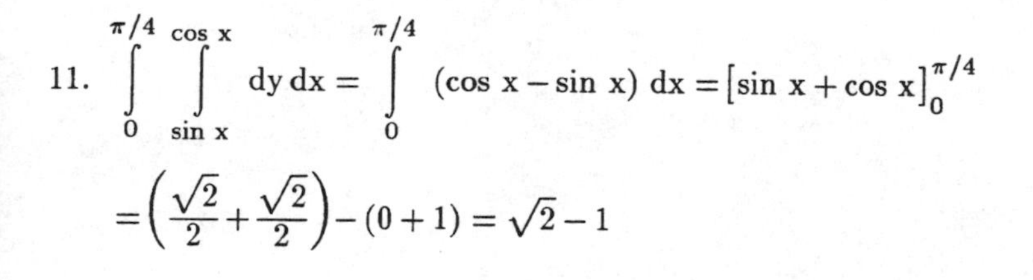

11. $\int_{0}^{\pi/4}\int_{\sin x}^{\cos x} dy\,dx = \int_{0}^{\pi/4}\left(\cos x - \sin x\right)dx = \left[\sin x + \cos x\right]_{0}^{\pi/4}$

$= \left(\frac{\sqrt{2}}{2}+\frac{\sqrt{2}}{2}\right) - (0+1) = \sqrt{2}-1$

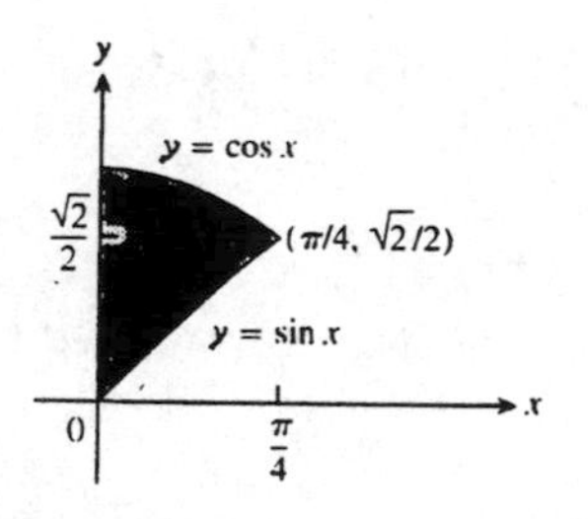

12. $\int_{-1}^{2}\int_{y^2}^{y+2} dx\,dy = \int_{-1}^{2}\left(y+2-y^2\right)dy = \left[\frac{y^2}{2}+2y-\frac{y^3}{3}\right]_{-1}^{2}$

$= \left(2+4-\frac{8}{3}\right) - \left(\frac{1}{2}-2+\frac{1}{3}\right) = 5-\frac{1}{2} = \frac{9}{2}$

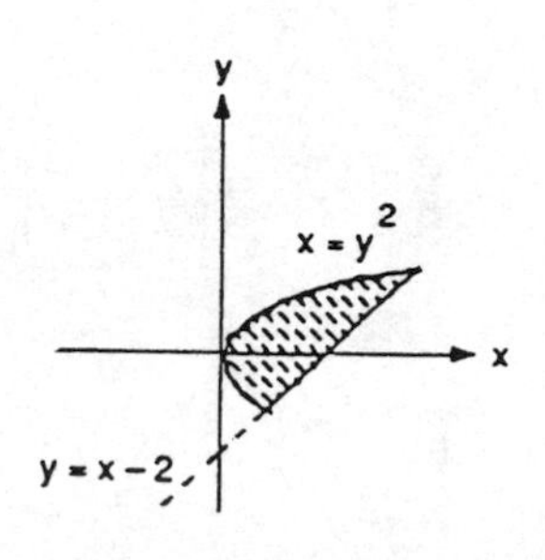

13. $\int_{-1}^{0}\int_{-2x}^{1-x} dy\,dx + \int_{0}^{2}\int_{-x/2}^{1-x} dy\,dx$

$= \int_{-1}^{0} (1+x)\,dx + \int_{0}^{2} \left(1-\frac{x}{2}\right) dx$

$= \left[x+\frac{x^2}{2}\right]_{-1}^{0} + \left[x-\frac{x^2}{4}\right]_{0}^{2} = -\left(-1+\frac{1}{2}\right)+(2-1) = \frac{3}{2}$

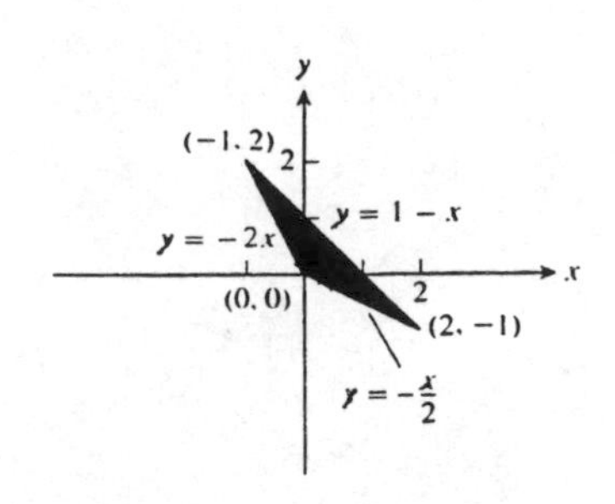

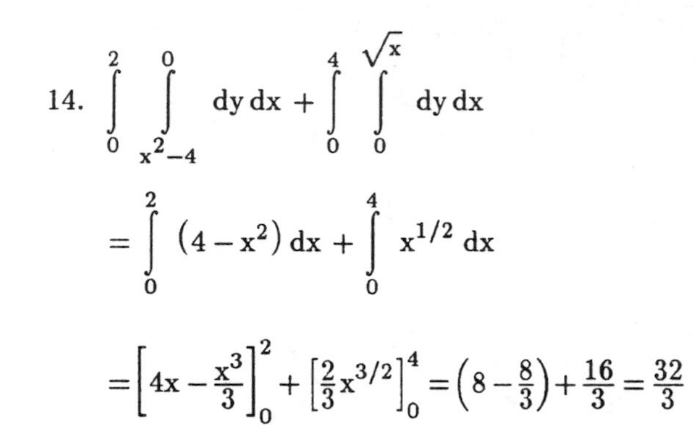

14. $\int_{0}^{2}\int_{x^2-4}^{0} dy\,dx + \int_{0}^{4}\int_{0}^{\sqrt{x}} dy\,dx$

$= \int_{0}^{2} \left(4-x^2\right) dx + \int_{0}^{4} x^{1/2}\,dx$

$= \left[4x-\frac{x^3}{3}\right]_{0}^{2} + \left[\frac{2}{3}x^{3/2}\right]_{0}^{4} = \left(8-\frac{8}{3}\right)+\frac{16}{3} = \frac{32}{3}$

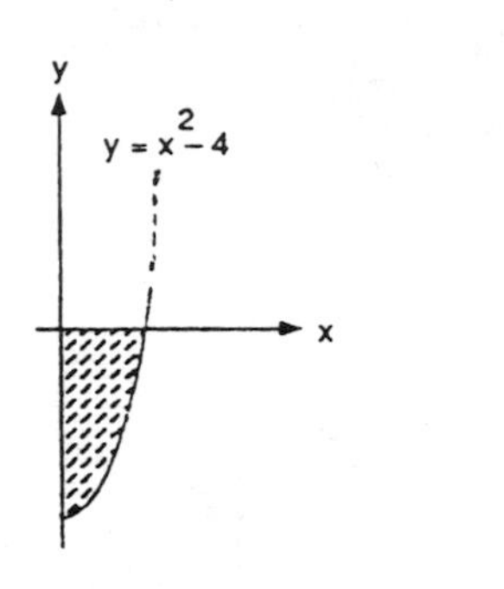

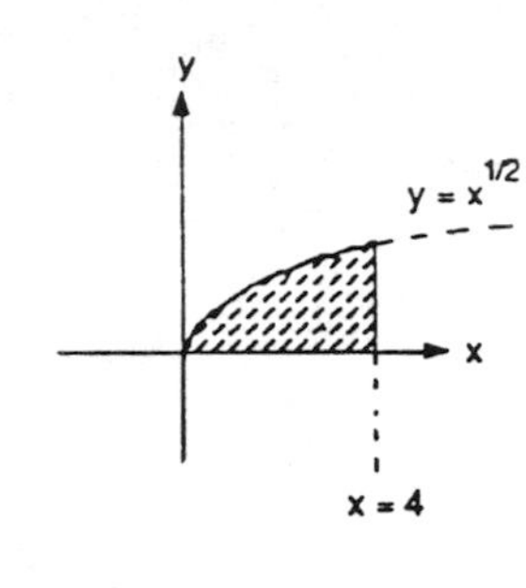

15. (a) average $= \frac{1}{\pi^2}\int_{0}^{\pi}\int_{0}^{\pi} \sin(x+y)\,dy\,dx = \frac{1}{\pi^2}\int_{0}^{\pi} \left[-\cos(x+y)\right]_{0}^{\pi} dx = \frac{1}{\pi^2}\int_{0}^{\pi} \left[-\cos(x+\pi)+\cos x\right] dx$

$= \frac{1}{\pi^2}\left[-\sin(x+\pi)+\sin x\right]_{0}^{\pi} = \frac{1}{\pi^2}\left[(-\sin 0+\sin\pi)-(-\sin\pi+\sin 0)\right] = 0$

(b) average $= \frac{1}{\left(\frac{\pi^2}{2}\right)}\int_{0}^{\pi}\int_{0}^{\pi/2} \sin(x+y)\,dy\,dx = \frac{2}{\pi^2}\int_{0}^{\pi} \left[-\cos(x+y)\right]_{0}^{\pi/2} dx = \frac{2}{\pi^2}\int_{0}^{\pi} \left[-\cos\left(x+\frac{\pi}{2}\right)+\cos x\right] dx$

$= \frac{2}{\pi^2}\left[-\sin\left(x+\frac{\pi}{2}\right)+\sin x\right]_{0}^{\pi} = \frac{2}{\pi^2}\left[\left(-\sin\frac{3\pi}{2}+\sin\pi\right)-\left(-\sin\frac{\pi}{2}+\sin 0\right)\right] = \frac{4}{\pi^2}$

16. average value over the square $= \int_{0}^{1}\int_{0}^{1} xy\,dy\,dx = \int_{0}^{1} \left[\frac{xy^2}{2}\right]_{0}^{1} dx = \int_{0}^{1} \frac{x}{2}\,dx = \frac{1}{4} = 0.25;$

average value over the quarter circle $= \frac{1}{\left(\frac{\pi}{4}\right)}\int_{0}^{1}\int_{0}^{\sqrt{1-x^2}} xy\,dy\,dx = \frac{4}{\pi}\int_{0}^{1} \left[\frac{xy^2}{2}\right]_{0}^{\sqrt{1-x^2}} dx$

$= \frac{2}{\pi}\int_{0}^{1} \left(x-x^3\right) dx = \frac{2}{\pi}\left[\frac{x^2}{2}-\frac{x^2}{4}\right]_{0}^{1} = \frac{1}{2\pi} \approx 0.159$

17. average height $= \frac{1}{4}\int_{0}^{2}\int_{0}^{2} \left(x^2+y^2\right) dy\,dx = \frac{1}{4}\int_{0}^{2} \left[x^2y+\frac{y^3}{3}\right]_{0}^{2} dx = \frac{1}{4}\int_{0}^{2} \left(2x^2+\frac{8}{3}\right) dx = \frac{1}{2}\left[\frac{x^3}{3}+\frac{4x}{3}\right]_{0}^{2} = \frac{8}{3}$

18. average $= \frac{1}{(\ln 2)^2} \int_{\ln 2}^{2\ln 2} \int_{\ln 2}^{2\ln 2} \frac{1}{xy}\, dy\, dx = \frac{1}{(\ln 2)^2} \int_{\ln 2}^{2\ln 2} \left[\frac{\ln y}{x}\right]_{\ln 2}^{2\ln 2} dx$

$= \frac{1}{(\ln 2)^2} \int_{\ln 2}^{2\ln 2} \frac{1}{x}(\ln 2 + \ln \ln 2 - \ln \ln 2)\, dx = \left(\frac{1}{\ln 2}\right) \int_{\ln 2}^{2\ln 2} \frac{dx}{x} = \left(\frac{1}{\ln 2}\right) [\ln x]_{\ln 2}^{2\ln 2}$

$= \left(\frac{1}{\ln 2}\right)(\ln 2 + \ln \ln 2 - \ln \ln 2) = 1$

19. $M = \int_0^1 \int_x^{2-x^2} 3\, dy\, dx = 3 \int_0^1 (2 - x^2 - x)\, dx = \frac{7}{2}$; $M_y = \int_0^1 \int_x^{2-x^2} 3x\, dy\, dx = 3 \int_0^1 [xy]_x^{2-x^2}\, dx$

$= 3 \int_0^1 (2x - x^3 - x^2)\, dx = \frac{5}{4}$; $M_x = \int_0^1 \int_x^{2-x^2} 3y\, dy\, dx = \frac{3}{2} \int_0^1 [y^2]_x^{2-x^2}\, dx = \frac{3}{2} \int_0^1 (4 - 5x^2 + x^4)\, dx = \frac{19}{5}$

$\Rightarrow \bar{x} = \frac{5}{14}$ and $\bar{y} = \frac{38}{35}$

20. $M = \delta \int_0^3 \int_0^3 dy\, dx = \delta \int_0^3 3\, dx = 9\delta$; $I_x = \delta \int_0^3 \int_0^3 y^2\, dy\, dx = \delta \int_0^3 \left[\frac{y^3}{3}\right]_0^3 dx = 27\delta$; $R_x = \sqrt{\frac{I_x}{M}} = \sqrt{3}$;

$I_y = \delta \int_0^3 \int_0^3 x^2\, dy\, dx = \delta \int_0^3 [x^2 y]_0^3\, dx = \delta \int_0^3 3x^2\, dx = 27\delta$; $R_y = \sqrt{\frac{I_y}{M}} = \sqrt{3}$

21. $M = \int_0^2 \int_{y^2/2}^{4-y} dx\, dy = \int_0^2 \left(4 - y - \frac{y^2}{2}\right) dy = \frac{14}{3}$; $M_y = \int_0^2 \int_{y^2/2}^{4-y} x\, dx\, dy = \frac{1}{2} \int_0^2 [x^2]_{y^2/2}^{4-y}\, dy$

$= \frac{1}{2} \int_0^2 \left(16 - 8y + y^2 - \frac{y^4}{4}\right) dy = \frac{128}{15}$; $M_x = \int_0^2 \int_{y^2/2}^{4-y} y\, dx\, dy = \int_0^2 \left(4y - y^2 - \frac{y^3}{2}\right) dy = \frac{10}{3}$

$\Rightarrow \bar{x} = \frac{64}{35}$ and $\bar{y} = \frac{5}{7}$

22. $M = \int_0^3 \int_0^{3-x} dy\, dx = \int_0^3 (3 - x)\, dx = \frac{9}{2}$; $M_y = \int_0^3 \int_0^{3-x} x\, dy\, dx = \int_0^3 [xy]_0^{3-x}\, dx = \int_0^3 (3x - x^2)\, dx = \frac{9}{2}$

$\Rightarrow \bar{x} = 1$ and $\bar{y} = 1$, by symmetry

23. $M = 2 \int_0^1 \int_0^{\sqrt{1-x^2}} dy\, dx = 2 \int_0^1 \sqrt{1-x^2}\, dx = 2\left(\frac{\pi}{4}\right) = \frac{\pi}{2}$; $M_x = 2 \int_0^1 \int_0^{\sqrt{1-x^2}} y\, dy\, dx = \int_0^1 [y^2]_0^{\sqrt{1-x^2}}\, dx$

$= \int_0^1 (1 - x^2)\, dx = \left[x - \frac{x^3}{3}\right]_0^1 = \frac{2}{3} \Rightarrow \bar{y} = \frac{4}{3\pi}$ and $\bar{x} = 0$, by symmetry

24. $M = \frac{125\delta}{6}$; $M_y = \delta \int_0^5 \int_x^{6x-x^2} x\,dy\,dx = \delta \int_0^5 [xy]_x^{6x-x^2}\,dx = \delta \int_0^5 (5x^2 - x^3)\,dx = \frac{625\delta}{12}$;

$M_x = \delta \int_0^5 \int_x^{6x-x^2} y\,dy\,dx = \frac{\delta}{2} \int_0^5 [y^2]_x^{6x-x^2}\,dx = \frac{\delta}{2} \int_0^5 (35x^2 - 12x^3 + x^4)\,dx = \frac{625\delta}{6} \Rightarrow \overline{x} = \frac{5}{2}$ and $\overline{y} = 5$

25. $M = \int_0^a \int_0^{\sqrt{a^2-x^2}} dy\,dx = \frac{\pi a^2}{4}$; $M_y = \int_0^a \int_0^{\sqrt{a^2-x^2}} x\,dy\,dx = \int_0^a [xy]_0^{\sqrt{a^2-x^2}}\,dx = \int_0^a x\sqrt{a^2-x^2}\,dx = \frac{a^3}{3}$

$\Rightarrow \overline{x} = \overline{y} = \frac{4a}{3\pi}$, by symmetry

26. $I_x = \int_{-2}^2 \int_{-\sqrt{4-x^2}}^{\sqrt{4-x^2}} y^2\,dy\,dx = \int_{-2}^2 \left[\frac{y^3}{3}\right]_{-\sqrt{4-x^2}}^{\sqrt{4-x^2}} dx = \frac{2}{3} \int_{-2}^2 (4-x^2)^{3/2}\,dx = 4\pi$; $I_y = 4\pi$, by symmetry;

$I_o = I_x + I_y = 8\pi$

27. $M = \int_0^\pi \int_0^{\sin x} dy\,dx = \int_0^\pi \sin x\,dx = 2$; $M_x = \int_0^\pi \int_0^{\sin x} y\,dy\,dx = \frac{1}{2} \int_0^\pi [y^2]_0^{\sin x}\,dx = \frac{1}{2} \int_0^\pi \sin^2 x\,dx$

$= \frac{1}{4} \int_0^\pi (1 - \cos 2x)\,dx = \frac{\pi}{4} \Rightarrow \overline{x} = \frac{\pi}{2}$ and $\overline{y} = \frac{\pi}{8}$

28. $I_y = \int_\pi^{2\pi} \int_0^{(\sin^2 x)/x^2} x^2\,dy\,dx = \int_0^{2\pi} (\sin^2 x - 0)\,dx = \frac{1}{2} \int_\pi^{2\pi} (1 - \cos 2x)\,dx = \frac{\pi}{2}$

29. $M = \int_{-\infty}^0 \int_0^{e^x} dy\,dx = \int_{-\infty}^0 e^x\,dx = \lim_{b\to-\infty} \int_b^0 e^x\,dx = 1 - \lim_{b\to-\infty} e^b = 1$; $M_y = \int_{-\infty}^0 \int_0^{e^x} x\,dy\,dx = \int_{-\infty}^0 xe^x\,dx$

$= \lim_{b\to-\infty} \int_b^0 xe^x\,dx = \lim_{b\to-\infty} [xe^x - e^x]_b^0 = -1 - \lim_{b\to-\infty} (be^b - e^b) = -1$; $M_x = \int_{-\infty}^0 \int_0^{e^x} y\,dy\,dx$

$= \frac{1}{2} \int_{-\infty}^0 e^{2x}\,dx = \frac{1}{2} \lim_{b\to-\infty} \int_b^0 e^{2x}\,dx = \frac{1}{4} \Rightarrow \overline{x} = -1$ and $\overline{y} = \frac{1}{4}$

30. $M_y = \int_0^\infty \int_0^{e^{-x^2/2}} x\,dy\,dx = \lim_{b\to\infty} \int_0^b xe^{-x^2/2}\,dx = -\lim_{b\to\infty} \left[\frac{1}{e^{x^2/2}} - 1\right]_0^b = 1$

31. $M = \int_0^2 \int_{-y}^{y-y^2} (x+y)\, dx\, dy = \int_0^2 \left[\frac{x^2}{2} + xy\right]_{-y}^{y-y^2} dy = \int_0^2 \left(\frac{y^4}{2} - 2y^3 + 2y^2\right) dy = \left[\frac{y^5}{10} - \frac{y^4}{2} + \frac{2y^3}{3}\right]_0^2 = \frac{8}{15}$;

$I_x = \int_0^2 \int_{-y}^{y-y^2} y^2(x+y)\, dx\, dy = \int_0^2 \left[\frac{x^2y^2}{2} + xy^3\right]_{-y}^{y-y^2} dy = \int_0^2 \left(\frac{y^6}{2} - 2y^5 + 2y^4\right) dy = \frac{64}{105}$;

$R_x = \sqrt{\frac{I_x}{M}} = \sqrt{\frac{8}{7}} = 2\sqrt{\frac{2}{7}}$

32. $M = \int_{-\sqrt{3}/2}^{\sqrt{3}/2} \int_{4y^2}^{\sqrt{12-4y^2}} 5x\, dx\, dy = 5 \int_{-\sqrt{3}/2}^{\sqrt{3}/2} \left[\frac{x^2}{2}\right]_{4y^2}^{\sqrt{12-4y^2}} dy = \frac{5}{2} \int_{-\sqrt{3}/2}^{\sqrt{3}/2} (12 - 4y^2 - 16y^4)\, dy = 23\sqrt{3}$

33. $M = \int_0^1 \int_x^{2-x} (6x + 3y + 3)\, dy\, dx = \int_0^1 \left[6xy + \frac{3}{2}y^2 + 3y\right]_x^{2-x} dx = \int_0^1 (12 - 12x^2)\, dx = 8$;

$M_y = \int_0^1 \int_x^{2-x} x(6x + 3y + 3)\, dy\, dx = \int_0^1 (12x - 12x^3)\, dx = 3$; $M_x = \int_0^1 \int_x^{2-x} y(6x + 3y + 3)\, dy\, dx$

$= \int_0^1 (14 - 6x - 6x^2 - 2x^3)\, dx = \frac{17}{2} \Rightarrow \bar{x} = \frac{3}{8}$ and $\bar{y} = \frac{17}{16}$

34. $M = \int_0^1 \int_{y^2}^{2y-y^2} (y+1)\, dx\, dy = \int_0^1 (2y - 2y^3)\, dy = \frac{1}{2}$; $M_x = \int_0^1 \int_{y^2}^{2y-y^2} y(y+1)\, dx\, dy = \int_0^1 (2y^2 - 2y^4)\, dy = \frac{4}{15}$;

$M_y = \int_0^1 \int_{y^2}^{2y-y^2} x(y+1)\, dx\, dy = \int_0^1 (2y^2 - 2y^4)\, dy = \frac{4}{15} \Rightarrow \bar{x} = \frac{8}{15}$ and $\bar{y} = \frac{8}{15}$; $I_x = \int_0^1 \int_{y^2}^{2y-y^2} y^2(y+1)\, dx\, dy$

$= 2\int_0^1 (y^3 - y^5)\, dy = \frac{1}{6}$

35. $M = \int_0^1 \int_0^6 (x + y + 1)\, dx\, dy = \int_0^1 (6y + 24)\, dy = 27$; $M_x = \int_0^1 \int_0^6 y(x + y + 1)\, dx\, dy = \int_0^1 y(6y + 24)\, dy = 14$;

$M_y = \int_0^1 \int_0^6 x(x + y + 1)\, dx\, dy = \int_0^1 (18y + 90)\, dy = 99 \Rightarrow \bar{x} = \frac{11}{3}$ and $\bar{y} = \frac{14}{27}$; $I_y = \int_0^1 \int_0^6 x^2(x + y + 1)\, dx\, dy$

$= 216 \int_0^1 \left(\frac{y}{3} + \frac{11}{6}\right) dy = 432$; $R_y = \sqrt{\frac{I_y}{M}} = 4$

36. $M = \int_{-1}^{1}\int_{x^2}^{1} (y+1)\,dy\,dx = -\int_{-1}^{1} \left(\frac{x^4}{2}+x^2-\frac{3}{2}\right) dx = \frac{32}{15}$; $M_x = \int_{-1}^{1}\int_{x^2}^{1} y(y+1)\,dy\,dx = \int_{-1}^{1} \left(\frac{5}{6}-\frac{x^6}{3}-\frac{x^4}{2}\right) dx$ $= \frac{48}{35}$; $M_y = \int_{-1}^{1}\int_{x^2}^{1} x(y+1)\,dy\,dx = \int_{-1}^{1} \left(\frac{3x}{2}-\frac{x^5}{2}-x^3\right) dx = 0 \Rightarrow \bar{x} = 0$ and $\bar{y} = \frac{9}{14}$; $I_y = \int_{-1}^{1}\int_{x^2}^{1} x^2(y+1)\,dy\,dx$ $= \int_{-1}^{1} \left(\frac{3x^2}{2}-\frac{x^6}{2}-x^4\right) dx = \frac{16}{35}$; $R_y = \sqrt{\frac{I_y}{M}} = \sqrt{\frac{3}{14}}$

37. $M = \int_{-1}^{1}\int_{0}^{x^2} (7y+1)\,dy\,dx = \int_{-1}^{1} \left(\frac{7x^4}{2}+x^2\right) dx = \frac{31}{15}$; $M_x = \int_{-1}^{1}\int_{0}^{x^2} y(7y+1)\,dy\,dx = \int_{-1}^{1} \left(\frac{7x^6}{3}+\frac{x^4}{2}\right) dx = \frac{13}{15}$; $M_y = \int_{-1}^{1}\int_{0}^{x^2} x(7y+1)\,dy\,dx = \int_{-1}^{1} \left(\frac{7x^5}{2}+x^3\right) dx = 0 \Rightarrow \bar{x} = 0$ and $\bar{y} = \frac{13}{31}$; $I_y = \int_{-1}^{1}\int_{0}^{x^2} x^2(7y+1)\,dy\,dx$ $= \int_{-1}^{1} \left(\frac{7x^6}{2}+x^4\right) dx = \frac{7}{5}$; $R_y = \sqrt{\frac{I_y}{M}} = \sqrt{\frac{21}{31}}$

38. $M = \int_{0}^{20}\int_{-1}^{1} \left(1+\frac{x}{20}\right) dy\,dx = \int_{0}^{20} \left(2+\frac{x}{10}\right) dx = 60$; $M_x = \int_{0}^{20}\int_{-1}^{1} y\left(1+\frac{x}{20}\right) dy\,dx = \int_{0}^{20} \left[\left(1+\frac{x}{20}\right)\left(\frac{y^2}{2}\right)\right]_{-1}^{1} dx = 0$; $M_y = \int_{0}^{20}\int_{-1}^{1} x\left(1+\frac{x}{20}\right) dy\,dx = \int_{0}^{20} \left(2x+\frac{x^2}{10}\right) dx = \frac{2000}{3} \Rightarrow \bar{x} = \frac{100}{9}$ and $\bar{y} = 0$; $I_x = \int_{0}^{20}\int_{-1}^{1} y^2\left(1+\frac{x}{20}\right) dy\,dx$ $= \frac{2}{3}\int_{0}^{20} \left(1+\frac{x}{20}\right) dx = 20$; $R_x = \sqrt{\frac{I_x}{M}} = \sqrt{\frac{1}{3}}$

39. $M = \int_{0}^{1}\int_{-y}^{y} (y+1)\,dx\,dy = \int_{0}^{1} (2y^2+2y)\,dy = \frac{5}{3}$; $M_x = \int_{0}^{1}\int_{-y}^{y} y(y+1)\,dx\,dy = 2\int_{0}^{1} (y^3+y^2)\,dy = \frac{7}{6}$; $M_y = \int_{0}^{1}\int_{-y}^{y} x(y+1)\,dx\,dy = \int_{0}^{1} 0\,dy = 0 \Rightarrow \bar{x} = 0$ and $\bar{y} = \frac{7}{10}$; $I_x = \int_{0}^{1}\int_{-y}^{y} y^2(y+1)\,dx\,dy = \int_{0}^{1} (2y^4+2y^3)\,dy$ $= \frac{9}{10} \Rightarrow R_x = \sqrt{\frac{I_x}{M}} = \frac{3\sqrt{6}}{10}$; $I_y = \int_{0}^{1}\int_{-y}^{y} x^2(y+1)\,dx\,dy = \frac{1}{3}\int_{0}^{1} (2y^4+2y^3)\,dy = \frac{3}{10} \Rightarrow R_y = \sqrt{\frac{I_y}{M}} = \frac{3\sqrt{2}}{10}$; $I_o = I_x + I_y = \frac{6}{5} \Rightarrow R_0 = \sqrt{\frac{I_o}{M}} = \frac{3\sqrt{2}}{5}$

40. $M = \int_0^1 \int_{-y}^{y} (3x^2+1)\,dx\,dy = \int_0^1 (2y^3+2y)\,dy = \frac{3}{2}$; $M_x = \int_0^1 \int_{-y}^{y} y(3x^2+1)\,dx\,dy = \int_0^1 (2y^4+2y^2)\,dy = \frac{16}{15}$;

$M_y = \int_0^1 \int_{-y}^{y} x(3x^2+1)\,dx\,dy = 0 \Rightarrow \overline{x} = 0$ and $\overline{y} = \frac{32}{45}$; $I_x = \int_0^1 \int_{-y}^{y} y^2(3x^2+1)\,dx\,dy = \int_0^1 (2y^5+2y^3)\,dy = \frac{5}{6}$

$\Rightarrow R_x = \sqrt{\frac{I_x}{M}} = \frac{\sqrt{5}}{3}$; $I_y = \int_0^1 \int_{-y}^{y} x^2(3x^2+1)\,dx\,dy = 2\int_0^1 \left(\frac{3}{5}y^5 + \frac{1}{3}y^3\right) dy = \frac{11}{30} \Rightarrow R_y = \sqrt{\frac{I_y}{M}} = \sqrt{\frac{11}{45}}$;

$I_o = I_x + I_y = \frac{6}{5} \Rightarrow R_o = \sqrt{\frac{I_o}{M}} = \frac{2}{\sqrt{5}}$

41. $\int_{-5}^{5} \int_{-2}^{0} \frac{10{,}000e^y}{1+\frac{|x|}{2}}\,dy\,dx = 10{,}000\left(1-e^{-2}\right) \int_{-5}^{5} \frac{dx}{1+\frac{|x|}{2}} = 10{,}000\left(1-e^{-2}\right)\left[\int_{-5}^{0} \frac{dx}{1-\frac{x}{2}} + \int_0^5 \frac{dx}{1+\frac{x}{2}}\right]$

$= 10{,}000\left(1-e^{-2}\right)\left[-2\ln\left(1-\frac{x}{2}\right)\right]_{-5}^{0} + 10{,}000\left(1-e^{-2}\right)\left[2\ln\left(1+\frac{x}{2}\right)\right]_0^5$

$= 10{,}000\left(1-e^{-2}\right)\left[2\ln\left(1+\frac{5}{2}\right)\right] + 10{,}000\left(1-e^{-2}\right)\left[2\ln\left(1+\frac{5}{2}\right)\right] = 40{,}000\left(1-e^{-2}\right)\ln\left(\frac{7}{2}\right) \approx 43{,}329$

42. $\int_0^1 \int_{y^2}^{2y-y^2} 100(y+1)\,dx\,dy = \int_0^1 \left[100(y+1)x\right]_{y^2}^{2y-y^2} dy = \int_0^1 100(y+1)\left(2y-2y^2\right) dy = 200\int_0^1 \left(y-y^3\right) dy$

$= 200\left[\frac{y^2}{2} - \frac{y^4}{4}\right]_0^1 = (200)\left(\frac{1}{4}\right) = 50$

43. $M = \int_{-1}^{1} \int_0^{a(1-x^2)} dy\,dx = 2a\int_0^1 \left(1-x^2\right) dx = 2a\left[x - \frac{x^3}{3}\right]_0^1 = \frac{4a}{3}$; $M_x = \int_{-1}^{1} \int_0^{a(1-x^2)} y\,dy\,dx$

$= \frac{2a^2}{2}\int_0^1 \left(1-2x^2+x^4\right) dx = a^2\left[x - \frac{2x^3}{3} + \frac{x^5}{5}\right]_0^1 = \frac{8a^2}{15} \Rightarrow \overline{y} = \frac{M_x}{M} = \frac{\left(\frac{8a^2}{15}\right)}{\left(\frac{4a}{3}\right)} = \frac{2a}{5}$. The angle θ between the x-axis and the line segment from the fulcrum to the center of mass on the y-axis plus 45° must be no more than 90° if the center of mass is to lie on the left side of the line $x = 1 \Rightarrow \theta + \frac{\pi}{4} \le \frac{\pi}{2} \Rightarrow \tan^{-1}\left(\frac{2a}{5}\right) \le \frac{\pi}{4} \Rightarrow a \le \frac{5}{2}$. Thus, if $0 < a \le \frac{5}{2}$, then the appliance will have to be tipped more than 45° to fall over.

44. $f(a) = I_a = \int_0^4 \int_0^2 (y-a)^2\,dy\,dx = \int_0^4 \left[\frac{(2-a)^3}{3} + \frac{a^3}{3}\right] dx = \frac{4}{3}\left[(2-a)^3 + a^3\right]$; thus $f'(a) = 0 \Rightarrow -4(2-a)^2 + 4a^2 = 0 \Rightarrow a^2 - (2-a)^2 = 0 \Rightarrow -4 + 4a = 0 \Rightarrow a = 1$. Since $f''(a) = 8(2-a) + 8a = 16 > 0$, $a = 1$ gives a minimum value of I_a.

45. $M = \int_0^1 \int_{-1/\sqrt{1-x^2}}^{1/\sqrt{1-x^2}} dy\,dx = \int_0^1 \frac{2}{\sqrt{1-x^2}}\,dx = \left[2\sin^{-1}x\right]_0^1 = 2\left(\frac{\pi}{2}-0\right) = \pi$; $M_y = \int_0^1 \int_{-1/\sqrt{1-x^2}}^{1/\sqrt{1-x^2}} x\,dy\,dx$

$= \int_0^1 \frac{2x}{\sqrt{1-x^2}}\,dx = \left[-2\left(1-x^2\right)^{1/2}\right]_0^1 = 2 \Rightarrow \bar{x} = \frac{2}{\pi}$ and $\bar{y} = 0$ by symmetry

46. (a) $I = \int_{-L/2}^{L/2} \delta x^2\,dx = \frac{\delta L^3}{12} \Rightarrow R = \sqrt{\frac{\delta L^3}{12}\cdot\frac{1}{\delta L}} = \frac{L}{2\sqrt{3}}$

(b) $I = \int_0^L \delta x^2\,dx = \frac{\delta L^3}{3} \Rightarrow R = \sqrt{\frac{\delta L^3}{3}\cdot\frac{1}{\delta L}} = \frac{L}{\sqrt{3}}$

47. (a) $\frac{1}{2} = M = \int_0^1 \int_{y^2}^{2y-y^2} \delta\,dx\,dy = 2\delta \int_0^1 \left(y-y^2\right)dy = 2\delta\left[\frac{y^2}{2}-\frac{y^3}{3}\right]_0^1 = 2\delta\left(\frac{1}{6}\right) = \frac{\delta}{3} \Rightarrow \delta = \frac{3}{2}$

(b) average value $= \dfrac{\int_0^1 \int_{y^2}^{2y-y^2} (y+1)\,dx\,dy}{\int_0^1 \int_{y^2}^{2y-y^2} dx\,dy} = \dfrac{\left(\frac{1}{2}\right)}{\left(\frac{1}{3}\right)} = \frac{3}{2} = \delta$, so the values are the same

48. Let (x_i, y_i) be the location of the weather station in county i for $i = 1, \ldots, 254$. The average temperature in Texas at time t_0 is approximately $\dfrac{\sum_{i=1}^{254} T(x_i, y_i)\,\Delta_i A}{A}$, where $T(x_i, y_i)$ is the temperature at time t_0 at the weather station in county i, $\Delta_i A$ is the area of county i, and A is the area of Texas.

49. (a) $\bar{x} = \frac{M_y}{M} = 0 \Rightarrow M_y = \iint_R x\delta(x,y)\,dy\,dx = 0$

(b) $I_L = \iint_R (x-h)^2\,\delta(x,y)\,dA = \iint_R x^2\,\delta(x,y)\,dA - \iint_R 2hx\,\delta(x,y)\,dA + \iint_R h^2\,\delta(x,y)\,dA$

$= I_y - 0 + h^2 \iint_R \delta(x,y)\,dA = I_{c.m.} + mh^2$

50. (a) $I_{c.m.} = I_L - mh^2 \Rightarrow I_{x=5/7} = I_y - mh^2 = \frac{39}{5} - 14\left(\frac{5}{7}\right)^2 = \frac{23}{35}$; $I_{y=11/14} = I_x - mh^2 = 12 - 14\left(\frac{11}{14}\right)^2 = \frac{47}{14}$

(b) $I_{x=1} = I_{x=5/7} + mh^2 = \frac{23}{35} + 14\left(\frac{2}{7}\right)^2 = \frac{9}{5}$; $I_{y=2} = I_{y=11/14} + mh^2 = \frac{47}{14} + 14\left(\frac{17}{14}\right)^2 = 24$

51. $M_{x_{P_1 \cup P_2}} = \iint\limits_{R_1} y\,dA_1 + \iint\limits_{R_2} y\,dA_2 = M_{x_1} + M_{x_2} \Rightarrow \bar{x} = \frac{M_{x_1} + M_{x_2}}{m_1 + m_2}$; likewise, $\bar{y} = \frac{M_{y_1} + M_{y_2}}{m_1 + m_2}$;

thus $\mathbf{c} = \bar{x}\mathbf{i} + \bar{y}\mathbf{j} = \frac{1}{m_1 + m_2}\left[\left(M_{x_1} + M_{x_2}\right)\mathbf{i} + \left(M_{y_1} + M_{y_2}\right)\mathbf{j}\right] = \frac{1}{m_1 + m_2}[(m_1\bar{x}_1 + m_2\bar{x}_2)\mathbf{i} + (m_1\bar{y}_1 + m_2\bar{y}_2)\mathbf{j}]$

$= \frac{1}{m_1 + m_2}[m_1(\bar{x}_1\mathbf{i} + \bar{y}_1\mathbf{j}) + m_2(\bar{x}_2\mathbf{i} + \bar{y}_2\mathbf{j})] = \frac{m_1\mathbf{c}_1 + m_2\mathbf{c}_2}{m_1 + m_2}$

52. From Exercise 51 we have that Pappus's formula is true for $n = 2$. Assume that Pappus's formula is true for $n = k - 1$, i.e., that $\mathbf{c}(k-1) = \frac{\sum_{i=1}^{k-1} m_i\mathbf{c}_i}{\sum_{i=1}^{k-1} m_i}$. The first moment about x of k nonoverlapping plates is

$\sum_{i=1}^{k-1}\left(\iint\limits_{R_i} y\,dA_i\right) + \iint\limits_{R_k} y\,dA_k = M_{x_{c(k-1)}} + M_{x_k} \Rightarrow \bar{x} = \frac{M_{x_{c(k-1)}} + M_{x_k}}{\left(\sum_{i=1}^{k-1} m_i\right) + m_k}$; similarly, $\bar{y} = \frac{M_{y_{c(k-1)}} + M_{y_k}}{\left(\sum_{i=1}^{k-1} m_i\right) + m_k}$;

thus $\mathbf{c}(k) = \bar{x}\mathbf{i} + \bar{y}\mathbf{j} = \frac{1}{\sum_{i=1}^{k} m_i}\left[\left(M_{x_{c(k-1)}} + M_{x_k}\right)\mathbf{i} + \left(M_{y_{c(k-1)}} + M_{y_k}\right)\mathbf{j}\right]$

$= \frac{1}{\sum_{i=1}^{k} m_i}\left[\left(\left(\sum_{i=1}^{k-1} m_i\right)\bar{x}_c + m_k\bar{x}_k\right)\mathbf{i} + \left(\left(\sum_{i=1}^{k-1} m_i\right)\bar{y}_c + m_k\bar{y}_k\right)\mathbf{j}\right]$

$= \frac{1}{\sum_{i=1}^{k} m_i}\left[\left(\sum_{i=1}^{k-1} m_i\right)(\bar{x}_c\mathbf{i} + \bar{y}_c\mathbf{j}) + m_k(\bar{x}_k\mathbf{i} + \bar{y}_k\mathbf{j})\right] = \frac{\left(\sum_{i=1}^{k-1} m_i\right)\mathbf{c}(k-1) + m_k\mathbf{c}_k}{\sum_{i=1}^{k} m_i}$

$= \frac{m_1\mathbf{c}_1 + m_2\mathbf{c}_2 + \ldots + m_{k-1}\mathbf{c}_{k-1} + m_k\mathbf{c}_k}{m_1 + m_2 + \ldots + m_{k-1} + m_k}$, and by mathematical induction the statement follows.

53. (a) $\mathbf{c} = \frac{8(\mathbf{i} + 3\mathbf{j}) + 2(3\mathbf{i} + 3.5\mathbf{j})}{8 + 2} = \frac{14\mathbf{j} + 31\mathbf{k}}{10} \Rightarrow \bar{x} = \frac{7}{5}$ and $\bar{y} = \frac{31}{10}$

(b) $\mathbf{c} = \frac{8(\mathbf{i} + 3\mathbf{j}) + 6(5\mathbf{i} + 2\mathbf{j})}{14} = \frac{38\mathbf{i} + 36\mathbf{j}}{14} \Rightarrow \bar{x} = \frac{19}{7}$ and $\bar{y} = \frac{18}{7}$

(c) $\mathbf{c} = \frac{2(3\mathbf{i} + 3.5\mathbf{j}) + 6(5\mathbf{i} + 2\mathbf{j})}{8} = \frac{36\mathbf{i} + 19\mathbf{j}}{8} \Rightarrow \bar{x} = \frac{9}{2}$ and $\bar{y} = \frac{19}{8}$

(d) $\mathbf{c} = \frac{8(\mathbf{i} + 3\mathbf{j}) + 2(3\mathbf{i} + 3.5\mathbf{j}) + 6(5\mathbf{i} + 2\mathbf{j})}{16} = \frac{44\mathbf{i} + 43\mathbf{j}}{16} \Rightarrow \bar{x} = \frac{11}{4}$ and $\bar{y} = \frac{43}{16}$

54. $\mathbf{c} = \frac{15\left(\frac{3}{4}\mathbf{i} + 7\mathbf{j}\right) + 48(12\mathbf{i} + \mathbf{j})}{15 + 48} = \frac{15(3\mathbf{i} + 28\mathbf{j}) + 48(48\mathbf{i} + 4\mathbf{j})}{4 \cdot 63} = \frac{2349\mathbf{i} + 612\mathbf{j}}{4 \cdot 63} = \frac{261\mathbf{i} + 68\mathbf{j}}{4 \cdot 7}$

$\Rightarrow \bar{x} = \frac{261}{28}$ and $\bar{y} = \frac{17}{7}$

55. Place the midpoint of the triangle's base at the origin and above the semicircle. Then the center of mass of the triangle is $\left(0, \frac{h}{3}\right)$, and the center of mass of the disk is $\left(0, -\frac{4a}{3\pi}\right)$ from Exercise 25. From

Pappus's formula, $\mathbf{c} = \frac{(ah)\left(\frac{h}{3}\mathbf{j}\right)+\left(\frac{\pi a^2}{2}\right)\left(-\frac{4a}{3\pi}\mathbf{j}\right)}{\left(ah+\frac{\pi a^2}{2}\right)} = \frac{\left(\frac{ah^2-2a^3}{3}\right)\mathbf{j}}{\left(ah+\frac{\pi a^2}{2}\right)}$, so the centroid is on the boundary if $ah^2 - 2a^3 = 0 \Rightarrow h^2 = 2a^2 \Rightarrow h = a\sqrt{2}$. In order for the center of mass to be inside T we must have $ah^2 - 2a^3 > 0$ or $h > a\sqrt{2}$.

56. Place the midpoint of the triangle's base at the origin and above the square. From Pappus's formula, $\mathbf{c} = \frac{\left(\frac{sh}{2}\right)\left(\frac{h}{3}\mathbf{j}\right)+s^2\left(-\frac{s}{2}\mathbf{j}\right)}{\left(\frac{sh}{2}+s^2\right)}$, so the centroid is on the boundary if $\frac{sh^2}{6} - \frac{s^3}{2} = 0 \Rightarrow h^2 - 3s^2 = 0 \Rightarrow h = s\sqrt{3}$.

13.3 DOUBLE INTEGRALS IN POLAR FORM

1. $$\int_{-1}^{1}\int_{0}^{\sqrt{1-x^2}} dy\,dx = \int_{0}^{\pi}\int_{0}^{1} r\,dr\,d\theta = \frac{1}{2}\int_{0}^{\pi} d\theta = \frac{\pi}{2}$$

2. $$\int_{-1}^{1}\int_{-\sqrt{1-x^2}}^{\sqrt{1-x^2}} dy\,dx = \int_{0}^{2\pi}\int_{0}^{1} r\,dr\,d\theta = \frac{1}{2}\int_{0}^{2\pi} d\theta = \pi$$

3. $$\int_{0}^{1}\int_{0}^{\sqrt{1-y^2}} (x^2+y^2)\,dx\,dy = \int_{0}^{\pi/2}\int_{0}^{1} r^3\,dr\,d\theta = \frac{1}{4}\int_{0}^{\pi/2} d\theta = \frac{\pi}{8}$$

4. $$\int_{-1}^{1}\int_{-\sqrt{1-y^2}}^{\sqrt{1-y^2}} (x^2+y^2)\,dx\,dy = \int_{0}^{2\pi}\int_{0}^{1} r^3\,dr\,d\theta = \frac{1}{4}\int_{0}^{2\pi} d\theta = \frac{\pi}{2}$$

5. $$\int_{-a}^{a}\int_{-\sqrt{a^2-x^2}}^{\sqrt{a^2-x^2}} dy\,dx = \int_{0}^{2\pi}\int_{0}^{a} r\,dr\,d\theta = \frac{a^2}{2}\int_{0}^{2\pi} d\theta = \pi a^2$$

6. $$\int_{0}^{2}\int_{0}^{\sqrt{4-y^2}} (x^2+y^2)\,dx\,dy = \int_{0}^{\pi/2}\int_{0}^{2} r^3\,dr\,d\theta = 4\int_{0}^{\pi/2} d\theta = 2\pi$$

7. $$\int_{0}^{6}\int_{0}^{y} x\,dx\,dy = \int_{\pi/4}^{\pi/2}\int_{0}^{6\csc\theta} r^2\cos\theta\,dr\,d\theta = 72\int_{\pi/4}^{\pi/2} \cot\theta\csc^2\theta\,d\theta = -36\left[\cot^2\theta\right]_{\pi/4}^{\pi/2} = 36$$

8. $\int_0^2 \int_0^x y\, dy\, dx = \int_0^{\pi/4} \int_0^{2 \sec \theta} r^2 \sin \theta\, dr\, d\theta = \frac{8}{3} \int_0^{\pi/4} \tan \theta \sec^2 \theta\, d\theta = \frac{4}{3}$

9. $\int_{-1}^0 \int_{-\sqrt{1-x^2}}^0 \frac{2}{1+\sqrt{x^2+y^2}}\, dy\, dx = \int_{\pi}^{3\pi/2} \int_0^1 \frac{2r}{1+r}\, dr\, d\theta = 2 \int_{\pi}^{3\pi/2} \int_0^1 \left(1 - \frac{1}{1+r}\right) dr\, d\theta = 2 \int_{\pi}^{3\pi/2} (1 - \ln 2)\, d\theta$
$= (1 - \ln 2)\pi$

10. $\int_{-1}^1 \int_{-\sqrt{1-y^2}}^0 \frac{4\sqrt{x^2+y^2}}{1+x^2+y^2}\, dx\, dy = \int_{\pi/2}^{3\pi/2} \int_0^1 \frac{4r^2}{1+r^2}\, dr\, d\theta = 4 \int_{\pi/2}^{3\pi/2} \int_0^1 \left(1 - \frac{1}{1+r^2}\right) dr\, d\theta = 4 \int_{\pi/2}^{3\pi/2} \left(1 - \frac{\pi}{4}\right) d\theta$
$= 4\pi - \pi^2$

11. $\int_0^{\ln 2} \int_0^{\sqrt{(\ln 2)^2 - y^2}} e^{\sqrt{x^2+y^2}}\, dx\, dy = \int_0^{\pi/2} \int_0^{\ln 2} re^r\, dr\, d\theta = \int_0^{\pi/2} (2 \ln 2 - 1)\, d\theta = \frac{\pi}{2}(2 \ln 2 - 1)$

12. $\int_0^1 \int_0^{\sqrt{1-x^2}} e^{-(x^2+y^2)}\, dy\, dx = \int_0^{\pi/2} \int_0^1 re^{-r^2}\, dr\, d\theta = -\frac{1}{2} \int_0^{\pi/2} \left(\frac{1}{e} - 1\right) d\theta = \frac{\pi(e-1)}{4e}$

13. $\int_0^2 \int_0^{\sqrt{1-(x-1)^2}} \frac{x+y}{x^2+y^2}\, dy\, dx = \int_0^{\pi/2} \int_0^{2 \cos \theta} \frac{r(\cos \theta + \sin \theta)}{r^2} r\, dr\, d\theta = \int_0^{\pi/2} (2 \cos^2 \theta + 2 \sin \theta \cos \theta)\, d\theta$
$= \left[\theta + \frac{\sin 2\theta}{2} + \sin^2 \theta\right]_0^{\pi/2} = \frac{\pi + 2}{2} = \frac{\pi}{2} + 1$

14. $\int_0^2 \int_{-\sqrt{1-(y-1)^2}}^0 xy^2\, dx\, dy = \int_{\pi/2}^{\pi} \int_0^{2 \sin \theta} \sin^2 \theta \cos \theta\, r^4\, dr\, d\theta = \frac{32}{5} \int_{\pi/2}^{\pi} \sin^7 \theta \cos \theta\, d\theta = \frac{4}{5}\left[\sin^8 \theta\right]_{\pi/2}^{\pi} = -\frac{4}{5}$

15. $\int_{-1}^1 \int_{-\sqrt{1-y^2}}^{\sqrt{1-y^2}} \ln(x^2+y^2+1)\, dx\, dy = 4 \int_0^{\pi/2} \int_0^1 \ln(r^2+1)\, r\, dr\, d\theta = 2 \int_0^{\pi/2} (\ln 4 - 1)\, d\theta = \pi(\ln 4 - 1)$

16. $\int_{-1}^1 \int_{-\sqrt{1-x^2}}^{\sqrt{1-x^2}} \frac{2}{(1+x^2+y^2)^2}\, dy\, dx = 4 \int_0^{\pi/2} \int_0^1 \frac{2r}{(1+r^2)^2}\, dr\, d\theta = 4 \int_0^{\pi/2} \left[-\frac{1}{1+r^2}\right]_0^1 d\theta = 2 \int_0^{\pi/2} d\theta = \pi$

17. $\int_0^{\pi/2} \int_0^{2\sqrt{2 - \sin 2\theta}} r\, dr\, d\theta = 2 \int_0^{\pi/2} (2 - \sin 2\theta)\, d\theta = 2(\pi - 1)$

18. $A = 2\int_0^{\pi/2}\int_1^{1+\cos\theta} r\,dr\,d\theta = \int_0^{\pi/2}\left(2\cos\theta + \cos^2\theta\right)d\theta = \frac{8+\pi}{4}$

19. $A = 2\int_0^{\pi/6}\int_0^{12\cos 3\theta} r\,dr\,d\theta = 144\int_0^{\pi/6}\cos^2 3\theta\,d\theta = 12\pi$

20. $A = \int_0^{2\pi}\int_0^{4\theta/3} r\,dr\,d\theta = \frac{8}{9}\int_0^{2\pi}\theta^2\,d\theta = \frac{64\pi^3}{27}$

21. $A = \int_0^{\pi/2}\int_0^{1+\sin\theta} r\,dr\,d\theta = \frac{1}{2}\int_0^{\pi/2}\left(\frac{3}{2} + 2\sin\theta - \frac{\cos 2\theta}{2}\right)d\theta = \frac{3\pi}{8} + 1$

22. $A = 4\int_0^{\pi/2}\int_0^{1-\cos\theta} r\,dr\,d\theta = 2\int_0^{\pi/2}\left(\frac{3}{2} - 2\cos\theta + \frac{\cos 2\theta}{2}\right)d\theta = \frac{3\pi}{2} - 4$

23. $M_x = \int_0^{\pi}\int_0^{1-\cos\theta} 3r^2\sin\theta\,dr\,d\theta = 2\int_0^{\pi}(1-\cos\theta)^3\sin\theta\,d\theta = 4$

24. $I_x = \int_{-a}^{a}\int_{-\sqrt{a^2-x^2}}^{\sqrt{a^2-x^2}} y^2\left[k\left(x^2+y^2\right)\right]dy\,dx = k\int_0^{2\pi}\int_0^{a} r^5\sin^2\theta\,dr\,d\theta = \frac{ka^6}{6}\int_0^{2\pi}\frac{1-\cos 2\theta}{2}\,d\theta = \frac{ka^6\pi}{6}$;

$I_o = \int_{-a}^{a}\int_{-\sqrt{a^2-x^2}}^{\sqrt{a^2-x^2}} k\left(x^2+y^2\right)^2 dy\,dx = k\int_0^{2\pi}\int_0^{a} r^5\,dr\,d\theta = \frac{ka^6}{6}\int_0^{2\pi}d\theta = \frac{ka^6\pi}{3}$

25. $M = 2\int_{\pi/6}^{\pi/2}\int_3^{6\sin\theta} dr\,d\theta = 2\int_{\pi/6}^{\pi/2}(6\sin\theta - 3)\,d\theta = 6\left[-2\cos\theta - \theta\right]_{\pi/6}^{\pi/2} = 6\sqrt{3} - 2\pi$

26. $I_o = \int_{\pi/2}^{3\pi/2}\int_1^{1-\cos\theta} r\,dr\,d\theta = \frac{1}{2}\int_{\pi/2}^{3\pi/2}\left(\cos^2\theta - 2\cos\theta\right)d\theta = \frac{1}{2}\left[\frac{\sin 2\theta}{4} + \frac{\theta}{2} - 2\sin\theta\right]_{\pi/2}^{3\pi/2} = 2 + \frac{\pi}{4}$

27. $M = 2\int_0^{\pi}\int_0^{1+\cos\theta} r\,dr\,d\theta = \int_0^{\pi}(1+\cos\theta)^2\,d\theta = \frac{3\pi}{2}$; $M_y = 2\int_0^{2\pi}\int_0^{1+\cos\theta} r^2\cos\theta\,dr\,d\theta$

$= \int_0^{2\pi}\left(\frac{4\cos\theta}{3} + \frac{15}{24} + \cos 2\theta - \sin^2\theta\cos\theta + \frac{\cos 4\theta}{4}\right)d\theta = \frac{5\pi}{4} \Rightarrow \bar{x} = \frac{5}{6}$ and $\bar{y} = 0$, by symmetry

28. $I_o = \int_0^{2\pi}\int_0^{1+\cos\theta} r^3\,dr\,d\theta = \frac{1}{4}\int_0^{2\pi}(1+\cos\theta)^4\,d\theta = \frac{35\pi}{16}$

29. $\text{average} = \frac{4}{\pi a^2}\int_0^{\pi/2}\int_0^{a} r\sqrt{a^2-r^2}\,dr\,d\theta = \frac{4}{3\pi a^2}\int_0^{\pi/2} a^3\,d\theta = \frac{2a}{3}$

30. $\text{average} = \frac{4}{\pi a^2}\int_0^{\pi/2}\int_0^{a} r^2\,dr\,d\theta = \frac{4}{3\pi a^2}\int_0^{\pi/2} a^3\,d\theta = \frac{2a}{3}$

31. $\text{average} = \frac{1}{\pi a^2}\int_{-a}^{a}\int_{-\sqrt{a^2-x^2}}^{\sqrt{a^2-x^2}} \sqrt{x^2+y^2}\,dy\,dx = \frac{1}{\pi a^2}\int_0^{2\pi}\int_0^{a} r^2\,dr\,d\theta = \frac{a}{3\pi}\int_0^{2\pi} d\theta = \frac{2a}{3}$

32. $\text{average} = \frac{1}{\pi}\iint_R \left[(1-x)^2+y^2\right]dy\,dx = \frac{1}{\pi}\int_0^{2\pi}\int_0^{1}\left[(1-r\cos\theta)^2+r^2\sin^2\theta\right]r\,dr\,d\theta$

$= \frac{1}{\pi}\int_0^{2\pi}\int_0^{1}\left(r^3-2r^2\cos\theta+r\right)dr\,d\theta = \frac{1}{\pi}\int_0^{2\pi}\left(\frac{3}{4}-\frac{2\cos\theta}{3}\right)d\theta = \frac{1}{\pi}\left[\frac{3}{4}\theta-\frac{2\sin\theta}{3}\right]_0^{2\pi} = \frac{3}{2}$

33. $\int_0^{2\pi}\int_1^{\sqrt{e}}\left(\frac{\ln r^2}{r}\right)r\,dr\,d\theta = \int_0^{2\pi}\int_1^{\sqrt{e}} 2\ln r\,dr\,d\theta = 2\int_0^{2\pi}\left[r\ln r - r\right]_1^{e^{1/2}}d\theta = 2\int_0^{2\pi}\sqrt{e}\left[\left(\frac{1}{2}-1\right)+1\right]d\theta = 2\pi\sqrt{e}$

34. $\int_0^{2\pi}\int_1^{e}\left(\frac{\ln r^2}{r}\right)dr\,d\theta = \int_0^{2\pi}\int_1^{e}\left(\frac{2\ln r}{r}\right)dr\,d\theta = \int_0^{2\pi}\left[(\ln r)^2\right]_1^{e}d\theta = \int_0^{2\pi}d\theta = 2\pi$

35. $V = 2\int_0^{\pi/2}\int_1^{1+\cos\theta} r^2\cos\theta\,dr\,d\theta = \frac{2}{3}\int_0^{\pi/2}\left(3\cos^2\theta+3\cos^3\theta+\cos^4\theta\right)d\theta$

$= \frac{2}{3}\left[\frac{15\theta}{8}+\sin 2\theta+3\sin\theta-\sin^3\theta+\frac{\sin 4\theta}{32}\right]_0^{\pi/2} = \frac{4}{3}+\frac{5\pi}{8}$

36. $V = 4\int_0^{\pi/4}\int_0^{\sqrt{2\cos 2\theta}} r\sqrt{2-r^2}\,dr\,d\theta = -\frac{4}{3}\int_0^{\pi/4}\left[(2-2\cos 2\theta)^{3/2}-2^{3/2}\right]d\theta$

$= \frac{2\pi\sqrt{2}}{3}-\frac{32}{3}\int_0^{\pi/4}\left(1-\cos^2\theta\right)\sin\theta\,d\theta = \frac{2\pi\sqrt{2}}{3}-\frac{32}{3}\left[\frac{\cos^3\theta}{3}-\cos\theta\right]_0^{\pi/4} = \frac{6\pi\sqrt{2}+40\sqrt{2}-64}{9}$

37. (a) $I^2 = \int_0^\infty \int_0^\infty e^{-(x^2+y^2)}\, dx\, dy = \int_0^{\pi/2} \int_0^\infty \left(e^{-r^2}\right) r\, dr\, d\theta = \int_0^{\pi/2} \left[\lim_{b\to\infty} \int_0^b re^{-r^2}\, dr \right] d\theta$

$= -\frac{1}{2} \int_0^{\pi/2} \lim_{b\to\infty} \left(e^{-b^2} - 1\right) d\theta = \frac{1}{2} \int_0^{\pi/2} d\theta = \frac{\pi}{4} \Rightarrow I = \frac{\sqrt{\pi}}{2}$

(b) $\lim_{x\to\infty} \int_0^x \frac{2e^{-t^2}}{\sqrt{\pi}}\, dt = \frac{2}{\sqrt{\pi}} \int_0^\infty e^{-t^2}\, dt = \left(\frac{2}{\sqrt{\pi}}\right)\left(\frac{\sqrt{\pi}}{2}\right) = 1$, from part (a)

38. $\int_0^\infty \int_0^\infty \frac{1}{\left(1+x^2+y^2\right)^2}\, dx\, dy = \int_0^{\pi/2} \int_0^\infty \frac{r}{\left(1+r^2\right)^2}\, dr\, d\theta = \frac{\pi}{2} \lim_{b\to\infty} \int_0^b \frac{r}{\left(1+r^2\right)^2}\, dr = \frac{\pi}{4} \lim_{b\to\infty} \left[-\frac{1}{1+r^2}\right]_0^b$

$= \frac{\pi}{4} \lim_{b\to\infty} \left(1 - \frac{1}{1+b^2}\right) = \frac{\pi}{4}$

39. Over the disk $x^2+y^2 \le \frac{3}{4}$: $\iint_R \frac{1}{1-x^2-y^2}\, dA = \int_0^{2\pi} \int_0^{\sqrt{3}/2} \frac{r}{1-r^2}\, dr\, d\theta = \int_0^{2\pi} \left[-\frac{1}{2} \ln\left(1-r^2\right)\right]_0^{\sqrt{3}/2} d\theta$

$= \int_0^{2\pi} \left(-\frac{1}{2} \ln \frac{1}{4}\right) d\theta = (\ln 2) \int_0^{2\pi} d\theta = \pi \ln 4$

Over the disk $x^2+y^2 \le 1$: $\iint_R \frac{1}{1-x^2-y^2}\, dA = \int_0^{2\pi} \int_0^1 \frac{r}{1-r^2}\, dr\, d\theta = \int_0^{2\pi} \left[\lim_{a\to 1^-} \int_0^a \frac{r}{1-r^2}\, dr \right] d\theta$

$= \int_0^{2\pi} \lim_{a\to 1^-} \left[-\frac{1}{2} \ln\left(1-a^2\right)\right] d\theta = 2\pi \cdot \lim_{a\to 1^-} \left[-\frac{1}{2} \ln\left(1-a^2\right)\right] = 2\pi \cdot \infty$, so the integral does not exist over $x^2+y^2 \le 1$

40. The area in polar coordinates is given by $A = \int_\alpha^\beta \int_0^{f(\theta)} r\, dr\, d\theta = \int_\alpha^\beta \left[\frac{r^2}{2}\right]_0^{f(\theta)} d\theta = \frac{1}{2} \int_\alpha^\beta f^2(\theta)\, d\theta = \int_\alpha^\beta \frac{1}{2} r^2\, d\theta$, where $r = f(\theta)$

41. $\text{average} = \frac{1}{\pi a^2} \int_0^{2\pi} \int_0^a \left[(r \cos\theta - h)^2 + r^2 \sin^2\theta\right] r\, dr\, d\theta = \frac{1}{\pi a^2} \int_0^{2\pi} \int_0^a \left(r^3 - 2r^2 h \cos\theta + rh^2\right) dr\, d\theta$

$= \frac{1}{\pi a^2} \int_0^{2\pi} \left(\frac{a^4}{4} - \frac{2a^3 h \cos\theta}{3} + \frac{a^2 h^2}{2}\right) d\theta = \frac{1}{\pi} \int_0^{2\pi} \left(\frac{a^2}{4} - \frac{2ah \cos\theta}{3} + \frac{h^2}{2}\right) d\theta = \frac{1}{\pi}\left[\frac{a^2\theta}{4} - \frac{2ah \sin\theta}{3} + \frac{h^2\theta}{2}\right]_0^{2\pi}$

$= \frac{1}{2}\left(a^2 + 2h^2\right)$

42. (a) $A = \int_{\pi/4}^{3\pi/4} \int_{\csc\theta}^{2\sin\theta} r\,dr\,d\theta = \frac{1}{2} \int_{\pi/4}^{3\pi/4} \left(4\sin^2\theta - \csc^2\theta\right) d\theta$

$= \frac{1}{2}\left[2\theta - \sin 2\theta + \cot\theta\right]_{\pi/4}^{3\pi/4} = \frac{\pi}{2}$

(b) $V = 2\pi\bar{y}A = 2\pi\left(\frac{3\pi+4}{3\pi}\right)\left(\frac{\pi}{2}\right) = \pi^2 + \frac{4\pi}{3}$

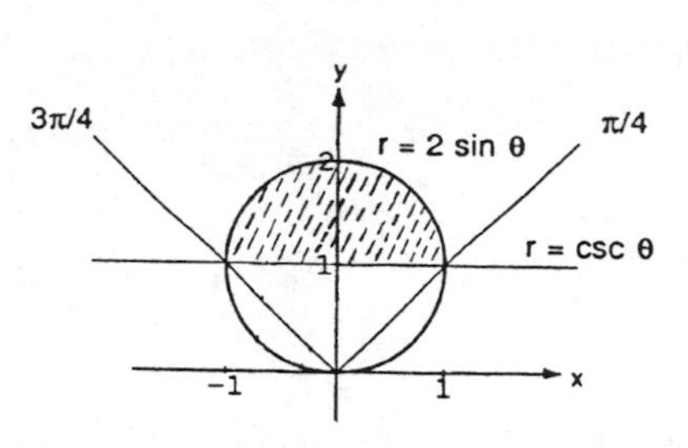

43-46. Example CAS commands:

Maple:

```
with(plots): y:='y'; x:='x';
bdy1:= y = 0; bdy2:= y = 2 - x; bdy3:= y = x;
implicitplot({bdy1, bdy2, bdy3}, x=0..2,y=0..1, scaling=CONSTRAINED,title=`ORIGINAL PLOT`);
X:= r*cos(theta): Y:= r*sin(theta):
r1:= solve(Y=0,r); theta1:= evalf(solve(Y=0,theta));
r2:=solve(Y=2-X,r);theta2:=solve(Y=2-X,theta);
r3:=solve(Y=X,r); theta3:=solve(Y=X,theta);
trbdy1:= theta=theta1; trbyd2:= r = r2; trbdy3:= theta=theta3;
implicitplot({trbdy1,trbdy2,trbdy3}, theta=0..1, r=0..2, title=`TRANSFORMED PLOT`);
f:= (x,y) -> sqrt(x+y);
subs(x=X, y=Y, f(x,y));
g:= unapply('',(r,theta));
int(int(g(r,theta), r=0..r2),theta=0..theta3);
evalf('');
```

Mathematica:

```
Clear[x,y,r,t]
topolar = {x -> r Cos[t], y -> r Sin[t]}
<< Graphics`ImplicitPlot`
f = Sqrt[x+y]
bdy1 = x == y
bdy2 = x == 2-y
ImplicitPlot[{bdy1,bdy2},{x,0,2},{y,0,1}]
bdy3 = y == 0
bdy1 /. topolar
```

Note: Mathematica cannot solve this directly, so we need to help by dividing the equation by the right-hand side:

```
%[[1]]/%[[2]] == 1
Solve[ %, t ]
t1 = t /. First[%]
bdy2 /. topolar
Solve[ %, r ]
r2 = r /. First[%]
bdy3 /. topolar
Solve[ %, t ]
t2 = t /. First[%]
r1 = 0
ImplicitPlot[{r==r1,r==r2,t==t1,t==t2},{t,0,1},{r,0,2}]
f /. topolar
f = Simplify[%]
Integrate[ f r, {t,t2,t1}, {r,r1,r2} ]
N[%]
```

13.4 TRIPLE INTEGRALS IN RECTANGULAR COORDINATES

1. $\int_0^1 \int_0^{1-z} \int_0^2 dx\,dy\,dz = 2\int_0^1 \int_0^{1-z} dy\,dz = 2\int_0^1 (1-z)\,dz = 2\left[z - \frac{z^2}{2}\right]_0^1 = 2\left(1 - \frac{1}{2}\right) = 1$

2. $\int_0^1 \int_0^2 \int_0^3 dz\,dy\,dx = \int_0^1 \int_0^2 3\,dy\,dx = \int_0^1 6\,dx = 6,\ \int_0^2 \int_0^1 \int_0^3 dz\,dx\,dy,\ \int_0^3 \int_0^2 \int_0^1 dx\,dy\,dz,\ \int_0^2 \int_0^3 \int_0^1 dx\,dz\,dy,$

$\int_0^3 \int_0^1 \int_0^2 dy\,dx\,dz,\ \int_0^1 \int_0^3 \int_0^2 dy\,dz\,dx$

3. $\int_0^1 \int_0^{2-2x} \int_0^{3-3x-3y/2} dz\,dy\,dx = \int_0^1 \int_0^{2-2x} \left(3 - 3x - \frac{3}{2}y\right) dy\,dx$

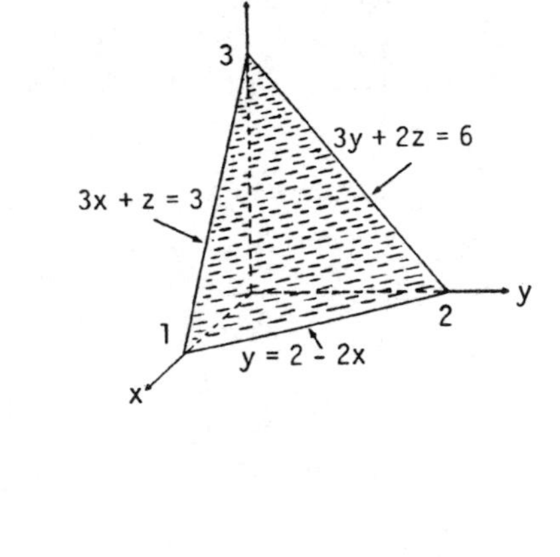

$= \int_0^1 \left[3(1-x)\cdot 2(1-x) - \frac{3}{4}\cdot 4(1-x)^2\right] dx$

$= 3\int_0^1 (1-x)^2\,dx = \left[-(1-x)^3\right]_0^1 = 1,$

$\int_0^2 \int_0^{1-y/2} \int_0^{3-3x-3y/2} dz\,dx\,dy,\ \int_0^1 \int_0^{3-3x} \int_0^{2-2x-2z/3} dy\,dz\,dx,$

$\int_0^3 \int_0^{1-z/3} \int_0^{2-2x-2z/3} dy\,dx\,dz,\ \int_0^2 \int_0^{3-3y/2} \int_0^{1-y/2-z/3} dx\,dz\,dy,\ \int_0^3 \int_0^{2-2z/3} \int_0^{1-y/2-z/3} dx\,dy\,dz$

4. $\int_0^2 \int_0^3 \int_0^{\sqrt{4-x^2}} dz\,dy\,dx = \int_0^2 \int_0^3 \sqrt{4-x^2}\,dy\,dx = \int_0^2 3\sqrt{4-x^2}\,dx = \frac{3}{2}\left[x\sqrt{4-x^2} + 4\sin^{-1}\frac{x}{2}\right]_0^2 = 6\sin^{-1}1 = 3\pi,$

$\int_0^3 \int_0^2 \int_0^{\sqrt{4-x^2}} dz\,dx\,dy,\ \int_0^2 \int_0^{\sqrt{4-x^2}} \int_0^3 dy\,dz\,dx,\ \int_0^2 \int_0^{\sqrt{4-z^2}} \int_0^3 dy\,dx\,dz,\ \int_0^2 \int_0^3 \int_0^{\sqrt{4-z^2}} dx\,dy\,dz,$

$\int_0^3 \int_0^2 \int_0^{\sqrt{4-z^2}} dx\,dz\,dy$

5. $\int_{-2}^{2}\int_{-\sqrt{4-x^2}}^{\sqrt{4-x^2}}\int_{x^2+y^2}^{8-x^2-y^2} dz\,dy\,dx = 4\int_{0}^{2}\int_{0}^{\sqrt{4-x^2}}\int_{x^2+y^2}^{8-x^2-y^2} dz\,dy\,dx$

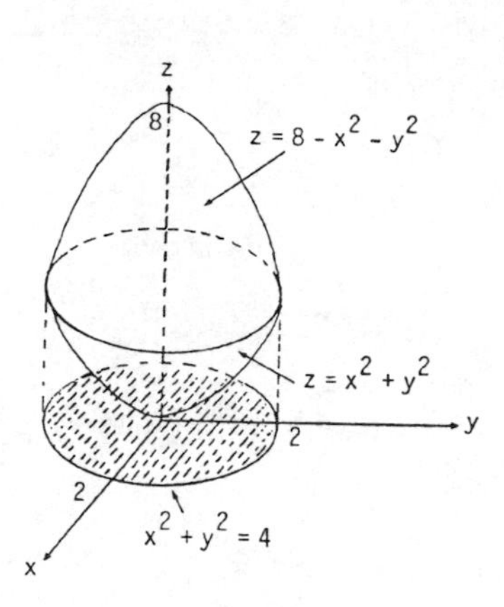

$= 4\int_{0}^{2}\int_{0}^{\sqrt{4-x^2}} \left[8-2(x^2+y^2)\right] dy\,dx$

$= 8\int_{0}^{2}\int_{0}^{\sqrt{4-x^2}} (4-x^2-y^2)\,dy\,dx = 8\int_{0}^{\pi/2}\int_{0}^{2} (4-r^2)\,r\,dr\,d\theta$

$= 8\int_{0}^{\pi/2}\left[2r^2-\frac{r^4}{4}\right]_0^2 d\theta = 32\int_{0}^{\pi/2} d\theta = 32\left(\frac{\pi}{2}\right) = 16\pi,$

$\int_{-2}^{2}\int_{-\sqrt{4-y^2}}^{\sqrt{4-y^2}}\int_{x^2+y^2}^{8-x^2-y^2} dz\,dx\,dy, \quad \int_{-2}^{2}\int_{y^2}^{4}\int_{-\sqrt{z-y^2}}^{\sqrt{z-y^2}} dx\,dz\,dy + \int_{-2}^{2}\int_{4}^{8-y^2}\int_{-\sqrt{8-z-y^2}}^{\sqrt{8-z-y^2}} dx\,dz\,dy,$

$\int_{0}^{4}\int_{-\sqrt{z}}^{\sqrt{z}}\int_{-\sqrt{z-y^2}}^{\sqrt{z-y^2}} dx\,dy\,dz + \int_{4}^{8}\int_{-\sqrt{8-z}}^{\sqrt{8-z}}\int_{-\sqrt{8-z-y^2}}^{\sqrt{8-z-y^2}} dx\,dy\,dz,$

$\int_{-2}^{2}\int_{x^2}^{4}\int_{-\sqrt{z-x^2}}^{\sqrt{z-x^2}} dy\,dz\,dx + \int_{-2}^{2}\int_{4}^{8-x^2}\int_{-\sqrt{8-z-x^2}}^{\sqrt{8-z-x^2}} dy\,dz\,dx,$

$\int_{0}^{4}\int_{-\sqrt{z}}^{\sqrt{z}}\int_{-\sqrt{z-x^2}}^{\sqrt{z-x^2}} dy\,dx\,dz + \int_{4}^{8}\int_{-\sqrt{8-z}}^{\sqrt{8-z}}\int_{-\sqrt{8-z-x^2}}^{\sqrt{8-z-x^2}} dy\,dx\,dz$

6. The projection of D onto the xy-plane has the boundary $x^2+y^2=2y \Rightarrow x^2+(y-1)^2=1$, which is a circle. Therefore the two integrals are:

$\int_{0}^{2}\int_{-\sqrt{2y-y^2}}^{\sqrt{2y-y^2}}\int_{x^2+y^2}^{2y} dz\,dx\,dy$ and $\int_{-1}^{1}\int_{1-\sqrt{1-x^2}}^{1+\sqrt{1-x^2}}\int_{x^2+y^2}^{2y} dz\,dy\,dx$

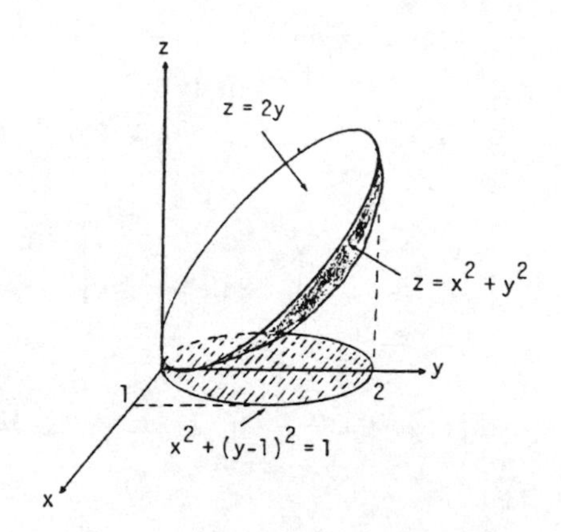

7. $\int_0^1\int_0^1\int_0^1 (x^2+y^2+z^2)\,dz\,dy\,dx = \int_0^1\int_0^1 \left(x^2+y^2+\frac{1}{3}\right)dy\,dx = \int_0^1 \left(x^2+\frac{2}{3}\right)dx = 1$

8. $\int_0^{\sqrt{2}}\int_0^{3y}\int_{x^2+3y^2}^{8-x^2-y^2} dz\,dx\,dy = \int_0^{\sqrt{2}}\int_0^{3y} (8-2x^2-4y^2)\,dx\,dy = \int_0^{\sqrt{2}} \left[8x-\frac{2}{3}x^3-4xy^2\right]_0^{3y} dy$

$= \int_0^{\sqrt{2}} (24y-18y^3-12y^3)\,dy = \left[12y^2-\frac{15}{2}y^4\right]_0^{\sqrt{2}} = 24-30 = -6$

9. $\int_1^e\int_1^e\int_1^e \frac{1}{xyz}\,dx\,dy\,dz = \int_1^e\int_1^e \left[\frac{\ln x}{yz}\right]_1^e dy\,dz = \int_1^e\int_1^e \frac{1}{yz}\,dy\,dz = \int_1^e \left[\frac{\ln y}{z}\right]_1^e dz = \int_1^e \frac{1}{z}\,dz = 1$

10. $\int_0^1\int_0^{3-3x}\int_0^{3-3x-y} dz\,dy\,dx = \int_0^1\int_0^{3-3x} (3-3x-y)\,dy\,dx = \int_0^1 \left[(3-3x)^2-\frac{1}{2}(3-3x)^2\right]dx = \frac{9}{2}\int_0^1 (1-x)^2\,dx$

$= -\frac{3}{2}\left[(1-x)^3\right]_0^1 = \frac{3}{2}$

11. $\int_0^1\int_0^{\pi}\int_0^{\pi} y\sin z\,dx\,dy\,dz = \int_0^1\int_0^{\pi} \pi y\sin z\,dy\,dz = \frac{\pi^3}{2}\int_0^1 \sin z\,dz = \frac{\pi^3}{2}(1-\cos 1)$

12. $\int_{-1}^1\int_{-1}^1\int_{-1}^1 (x+y+z)\,dy\,dx\,dz = \int_{-1}^1\int_{-1}^1 \left[xy+\frac{1}{2}y^2+zy\right]_{-1}^1 dx\,dz = \int_{-1}^1\int_{-1}^1 (2x+2z)\,dx\,dz = \int_{-1}^1 \left[x^2+2zx\right]_{-1}^1 dz$

$= \int_{-1}^1 4z\,dz = 0$

13. $\int_0^3\int_0^{\sqrt{9-x^2}}\int_0^{\sqrt{9-x^2}} dz\,dy\,dx = \int_0^3\int_0^{\sqrt{9-x^2}} \sqrt{9-x^2}\,dy\,dx = \int_0^3 (9-x^2)\,dx = \left[9x-\frac{x^3}{3}\right]_0^3 = 18$

14. $\int_0^2\int_{-\sqrt{4-y^2}}^{\sqrt{4-y^2}}\int_0^{2x+y} dz\,dx\,dy = \int_0^2\int_{-\sqrt{4-y^2}}^{\sqrt{4-y^2}} (2x+y)\,dx\,dy = \int_0^2 \left[x^2+xy\right]_{-\sqrt{4-y^2}}^{\sqrt{4-y^2}} dy = \int_0^2 (4-y^2)^{1/2}(2y)\,dy$

$= \left[-\frac{2}{3}(4-y^2)^{3/2}\right]_0^2 = \frac{2}{3}(4)^{3/2} = \frac{16}{3}$

15. $\int_0^1 \int_0^{2-x} \int_0^{2-x-y} dz\,dy\,dx = \int_0^1 \int_0^{2-x} (2-x-y)\,dy\,dx = \int_0^1 \left[(2-x)^2 - \frac{1}{2}(2-x)^2\right] dx = \frac{1}{2} \int_0^1 (2-x)^2\,dx$
$= \left[-\frac{1}{6}(2-x)^3\right]_0^1 = -\frac{1}{6} + \frac{8}{6} = \frac{7}{6}$

16. $\int_0^1 \int_0^{1-x^2} \int_3^{4-x^2-y} x\,dz\,dy\,dx = \int_0^1 \int_0^{1-x^2} x\left(1-x^2-y\right) dy\,dx = \int_0^1 x\left[\left(1-x^2\right)^2 - \frac{1}{2}\left(1-x^2\right)\right] dx = \int_0^1 \frac{1}{2}x\left(1-x^2\right)^2 dx$
$= \left[-\frac{1}{12}\left(1-x^2\right)^3\right]_0^1 = \frac{1}{12}$

17. $\int_0^\pi \int_0^\pi \int_0^\pi \cos(u+v+w)\,du\,dv\,dw = \int_0^\pi \int_0^\pi [\sin(w+v+\pi) - \sin(w+v)]\,dv\,dw$
$= \int_0^\pi [(-\cos(w+2\pi) + \cos(w+\pi)) + (\cos(w+\pi) - \cos w)]\,dw$
$= [-\sin(w+2\pi) + \sin(w+\pi) - \sin w + \sin(w+\pi)]_0^\pi = 0$

18. $\int_1^e \int_1^e \int_1^e \ln r \ln s \ln t\,dt\,dr\,ds = \int_1^e \int_1^e (\ln r \ln s)[t \ln t - t]_1^e\,dr\,ds = \int_1^e (\ln s)[r \ln r - r]_1^e\,ds = [s \ln s - s]_1^e = 1$

19. $\int_0^{\pi/4} \int_0^{\ln \sec v} \int_{-\infty}^{2t} e^x\,dx\,dt\,dv = \int_0^{\pi/4} \int_0^{\ln \sec v} \lim_{b \to -\infty} \left(e^{2t} - e^b\right) dt\,dv = \int_0^{\pi/4} \int_0^{\ln \sec v} e^{2t}\,dt\,dv = \int_0^{\pi/4} \frac{1}{2} e^{2 \ln \sec v}\,dv$
$= \int_0^{\pi/4} \frac{\sec^2 v}{2}\,dv = \left[\frac{\tan v}{2}\right]_0^{\pi/4} = \frac{1}{2}$

20. $\int_0^7 \int_0^2 \int_0^{\sqrt{4-q^2}} \frac{q}{r+1}\,dp\,dq\,dr = \int_0^7 \int_0^2 \frac{q\sqrt{4-q^2}}{r+1}\,dq\,dr = \int_0^7 \frac{1}{3(r+1)}\left[-\left(4-q^2\right)^{3/2}\right]_0^2 dr = \frac{8}{3} \int_0^7 \frac{1}{r+1}\,dr$
$= \frac{8 \ln 8}{3} = 8 \ln 2$

21. (a) $\int_{-1}^1 \int_0^{1-x^2} \int_{x^2}^{1-z} dy\,dz\,dx$ (b) $\int_0^1 \int_{-\sqrt{1-z}}^{\sqrt{1-z}} \int_{x^2}^{1-z} dy\,dx\,dz$ (c) $\int_0^1 \int_0^{1-z} \int_{-\sqrt{y}}^{\sqrt{y}} dx\,dy\,dz$

(d) $\int_0^1 \int_0^{1-y} \int_{-\sqrt{y}}^{\sqrt{y}} dx\,dz\,dy$ (e) $\int_0^1 \int_{-\sqrt{y}}^{\sqrt{y}} \int_0^{1-y} dz\,dx\,dy$

22. (a) $\int_0^1 \int_0^1 \int_{-1}^{-\sqrt{z}} dy\, dz\, dx$ (b) $\int_0^1 \int_0^1 \int_{-1}^{-\sqrt{z}} dy\, dx\, dz$ (c) $\int_0^1 \int_{-1}^{-\sqrt{z}} \int_0^1 dx\, dy\, dz$

(d) $\int_{-1}^0 \int_0^{y^2} \int_0^1 dx\, dz\, dy$ (e) $\int_{-1}^0 \int_0^1 \int_0^{y^2} dz\, dx\, dy$

23. $V = \int_0^1 \int_{-1}^1 \int_0^{y^2} dz\, dy\, dx = \int_0^1 \int_{-1}^1 y^2\, dy\, dx = \frac{2}{3} \int_0^1 dx = \frac{2}{3}$

24. $V = \int_0^1 \int_0^{1-x} \int_0^{2-2z} dy\, dz\, dx = \int_0^1 \int_0^{1-x} (2-2z)\, dz\, dx = \int_0^1 \left[2z - z^2\right]_0^{1-x} dx = \int_0^1 \left(1-x^2\right) dx = \left[x - \frac{x^3}{3}\right]_0^1 = \frac{2}{3}$

25. $V = \int_0^4 \int_0^{\sqrt{4-x}} \int_0^{2-y} dz\, dy\, dx = \int_0^4 \int_0^{\sqrt{4-x}} (2-y)\, dy\, dx = \int_0^4 \left[2\sqrt{4-x} - \left(\frac{4-x}{2}\right)\right] dx$

$= \left[-\frac{4}{3}(4-x)^{3/2} + \frac{1}{4}(4-x)^2\right]_0^4 = \frac{4}{3}(4)^{3/2} - \frac{1}{4}(16) = \frac{32}{3} - 4 = \frac{20}{3}$

26. $V = 2 \int_0^1 \int_{-\sqrt{1-x^2}}^0 \int_0^{-y} dz\, dy\, dx = -2 \int_0^1 \int_{-\sqrt{1-x^2}}^0 y\, dy\, dx = \int_0^1 \left(1-x^2\right) dx = \frac{2}{3}$

27. $V = \int_0^1 \int_0^{2-2x} \int_0^{3-3x-3y/2} dz\, dy\, dx = \int_0^1 \int_0^{2-2x} \left(3 - 3x - \frac{3}{2}y\right) dy\, dx = \int_0^1 \left[6(1-x)^2 - \frac{3}{4} \cdot 4(1-x)^2\right] dx$

$= \int_0^1 3(1-x)^2\, dx = \left[-(1-x)^3\right]_0^1 = 1$

28. $V = \int_0^1 \int_0^{1-x} \int_0^{\cos(\pi x/2)} dz\, dy\, dx = \int_0^1 \int_0^{1-x} \cos\left(\frac{\pi x}{2}\right) dy\, dx = \int_0^1 \left(\cos \frac{\pi x}{2}\right)(1-x)\, dx$

$= \int_0^1 \cos\left(\frac{\pi x}{2}\right) dx - \int_0^1 x \cos\left(\frac{\pi x}{2}\right) dx = \left[\frac{2}{\pi} \sin \frac{\pi x}{2}\right]_0^1 - \frac{4}{\pi^2} \int_0^{\pi/2} u \cos u\, du = \frac{2}{\pi} - \frac{4}{\pi^2}\left[\cos u + u \sin u\right]_0^{\pi/2}$

$= \frac{2}{\pi} - \frac{4}{\pi^2}\left(\frac{\pi}{2} - 1\right) = \frac{4}{\pi^2}$

29. $V = 8 \int_0^1 \int_0^{\sqrt{1-x^2}} \int_0^{\sqrt{1-x^2}} dz\, dy\, dx = 8 \int_0^1 \int_0^{\sqrt{1-x^2}} \sqrt{1-x^2}\, dy\, dx = 8 \int_0^1 \left(1-x^2\right) dx = \frac{16}{3}$

30. $V = \int_0^2 \int_0^{4-x^2} \int_0^{4-x^2-y} dz\,dy\,dx = \int_0^2 \int_0^{4-x^2} (4-x^2-y)\,dy\,dx = \int_0^2 \left[(4-x^2)^2 - \frac{1}{2}(4-x^2)^2\right] dx$

$= \frac{1}{2}\int_0^2 (4-x^2)^2\,dx = \int_0^2 \left(8 - 4x^2 + \frac{x^4}{2}\right) dx = \frac{128}{15}$

31. $V = \int_0^4 \int_0^{\left(\sqrt{16-y^2}\right)/2} \int_0^{4-y} dx\,dz\,dy = \int_0^4 \int_0^{\left(\sqrt{16-y^2}\right)/2} (4-y)\,dz\,dy = \int_0^4 \frac{\sqrt{16-y^2}}{2}(4-y)\,dy$

$= \int_0^4 2\sqrt{16-y^2}\,dy - \frac{1}{2}\int_0^4 y\sqrt{16-y^2}\,dy = \left[y\sqrt{16-y^2} + 16\sin^{-1}\frac{y}{4}\right]_0^4 + \left[\frac{1}{6}(16-y^2)^{3/2}\right]_0^4$

$= 16\left(\frac{\pi}{2}\right) - \frac{1}{6}(16)^{3/2} = 8\pi - \frac{32}{3}$

32. $V = \int_{-2}^2 \int_{-\sqrt{4-x^2}}^{\sqrt{4-x^2}} \int_0^{3-x} dz\,dy\,dx = \int_{-2}^2 \int_{-\sqrt{4-x^2}}^{\sqrt{4-x^2}} (3-x)\,dy\,dx = 2\int_{-2}^2 (3-x)\sqrt{4-x^2}\,dx$

$= 3\int_{-2}^2 2\sqrt{4-x^2}\,dx - 2\int_{-2}^2 x\sqrt{4-x^2}\,dx = 3\left[x\sqrt{4-x^2} + 4\sin^{-1}\frac{x}{2}\right]_{-2}^2 + \left[\frac{2}{3}(4-x^2)^{3/2}\right]_{-2}^2$

$= 12\sin^{-1}1 - 12\sin^{-1}(-1) = 12\left(\frac{\pi}{2}\right) - 12\left(-\frac{\pi}{2}\right) = 12\pi$

33. $\int_0^2 \int_0^{2-x} \int_{(2-x-y)/2}^{4-2x-2y} dz\,dy\,dx = \int_0^2 \int_0^{2-x} \left(3 - \frac{3x}{2} - \frac{3y}{2}\right) dy\,dx$

$= \int_0^2 \left[3\left(1-\frac{x}{2}\right)(2-x) - \frac{3}{4}(2-x)^2\right] dx$

$= \int_0^2 \left[6 - 6x + \frac{3x^2}{2} - \frac{3(2-x)^2}{4}\right] dx$

$= \left[6x - 3x^2 + \frac{x^3}{2} + \frac{(2-x)^3}{4}\right]_0^2 = (12 - 12 + 4 + 0) - \frac{2^3}{4} = 2$

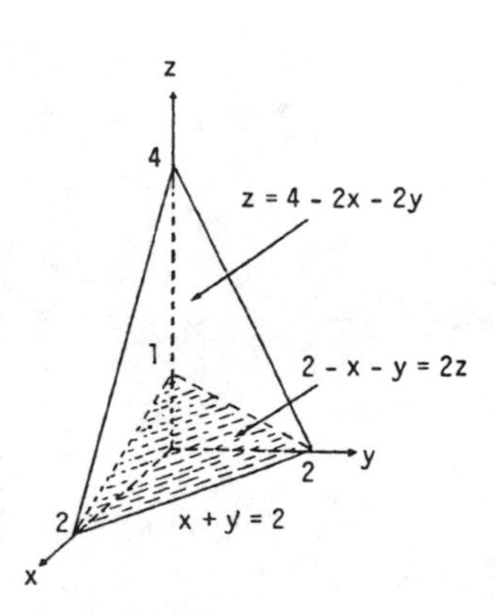

34. $V = \int_0^4 \int_z^8 \int_z^{8-z} dx\,dy\,dz = \int_0^4 \int_z^8 (8-2z)\,dy\,dz = \int_0^4 (8-2z)(8-z)\,dz = \int_0^4 (64 - 24z + 2z^2)\,dz$

$= \left[64z - 12z^2 + \frac{2}{3}z^3\right]_0^4 = \frac{320}{3}$

35. $V = 2\int_{-2}^{2}\int_{0}^{\sqrt{4-x^2}/2}\int_{0}^{x+2} dz\,dy\,dx = 2\int_{-2}^{2}\int_{0}^{\sqrt{4-x^2}/2} (x+2)\,dy\,dx = \int_{-2}^{2} (x+2)\sqrt{4-x^2}\,dx$

$= \int_{-2}^{2} 2\sqrt{4-x^2}\,dx + \int_{-2}^{2} x\sqrt{4-x^2}\,dx = \left[x\sqrt{4-x^2} + 4\sin^{-1}\frac{x}{2}\right]_{-2}^{2} + \left[-\frac{1}{3}(4-x^2)^{3/2}\right]_{-2}^{2}$

$= 4\left(\frac{\pi}{2}\right) - 4\left(-\frac{\pi}{2}\right) = 4\pi$

36. $V = 2\int_{0}^{1}\int_{0}^{1-y^2}\int_{0}^{x^2+y^2} dz\,dx\,dy = 2\int_{0}^{1}\int_{0}^{1-y^2} (x^2+y^2)\,dx\,dy = 2\int_{0}^{1}\left[\frac{x^3}{3} + xy^2\right]_{0}^{1-y^2} dy$

$= 2\int_{0}^{1} (1-y^2)\left[\frac{1}{3}(1-y^2)^2 + y^2\right] dy = 2\int_{0}^{1} (1-y^2)\left(\frac{1}{3} + \frac{1}{3}y^2 + \frac{1}{3}y^4\right) dy = \frac{2}{3}\int_{0}^{1} (1-y^6)\,dy$

$= \frac{2}{3}\left[y - \frac{y^7}{7}\right]_{0}^{1} = \left(\frac{2}{3}\right)\left(\frac{6}{7}\right) = \frac{4}{7}$

37. $\text{average} = \frac{1}{8}\int_{0}^{2}\int_{0}^{2}\int_{0}^{2} (x^2+9)\,dz\,dy\,dx = \frac{1}{8}\int_{0}^{2}\int_{0}^{2} (2x^2+18)\,dy\,dx = \frac{1}{8}\int_{0}^{2} (4x^2+36)\,dx = \frac{31}{3}$

38. $\text{average} = \frac{1}{2}\int_{0}^{1}\int_{0}^{1}\int_{0}^{2} (x+y-z)\,dz\,dy\,dx = \frac{1}{2}\int_{0}^{1}\int_{0}^{1} (2x+2y-2)\,dy\,dx = \frac{1}{2}\int_{0}^{1} (2x-1)\,dx = 0$

39. $\text{average} = \int_{0}^{1}\int_{0}^{1}\int_{0}^{1} (x^2+y^2+z^2)\,dz\,dy\,dx = \int_{0}^{1}\int_{0}^{1} \left(x^2+y^2+\frac{1}{3}\right) dy\,dx = \int_{0}^{1} \left(x^2+\frac{2}{3}\right) dx = 1$

40. $\text{average} = \frac{1}{8}\int_{0}^{2}\int_{0}^{2}\int_{0}^{2} xyz\,dz\,dy\,dx = \frac{1}{4}\int_{0}^{2}\int_{0}^{2} xy\,dy\,dx = \frac{1}{2}\int_{0}^{2} x\,dx = 1$

41. $\int_{0}^{4}\int_{0}^{1}\int_{2y}^{2} \frac{4\cos(x^2)}{2\sqrt{z}}\,dx\,dy\,dz = \int_{0}^{4}\int_{0}^{2}\int_{0}^{x/2} \frac{4\cos(x^2)}{2\sqrt{z}}\,dy\,dx\,dz = \int_{0}^{4}\int_{0}^{2} \frac{x\cos(x^2)}{2\sqrt{z}}\,dx\,dz = \int_{0}^{4} \left(\frac{\sin 4}{2}\right) z^{-1/2}\,dz$

$= \left[(\sin 4)z^{1/2}\right]_{0}^{4} = 2\sin 4$

42. $\int_{0}^{1}\int_{0}^{1}\int_{x^2}^{1} 12xz\,e^{zy^2}\,dy\,dx\,dz = \int_{0}^{1}\int_{0}^{1}\int_{0}^{\sqrt{y}} 12xz\,e^{zy^2}\,dx\,dy\,dz = \int_{0}^{1}\int_{0}^{1} 6yz\,e^{zy^2}\,dy\,dz = \int_{0}^{1} \left[3e^{zy^2}\right]_{0}^{1} dz$

$= 3\int_{0}^{1} (e^z - 1)\,dz = 3\left[e^z - 1\right]_{0}^{1} = 3e - 6$

43. $\int_0^1 \int_{\sqrt[3]{z}}^1 \int_0^{\ln 3} \frac{\pi e^{2x} \sin(\pi y^2)}{y^2}\, dx\, dy\, dz = \int_0^1 \int_{\sqrt[3]{z}}^1 \frac{4\pi \sin(\pi y^2)}{y^2}\, dy\, dz = \int_0^1 \int_0^{y^3} \frac{4\pi \sin(\pi y^2)}{y^2}\, dz\, dy$

$= \int_0^1 4\pi y \sin(\pi y^2)\, dy = \left[-2\cos(\pi y^2)\right]_0^1 = -2(-1) + 2(1) = 4$

44. $\int_0^2 \int_0^{4-x^2} \int_0^x \frac{\sin 2z}{4-z}\, dy\, dz\, dx = \int_0^2 \int_0^{4-x^2} \frac{x \sin 2z}{4-z}\, dz\, dx = \int_0^4 \int_0^{\sqrt{4-z}} \left(\frac{\sin 2z}{4-z}\right) x\, dx\, dz = \int_0^4 \left(\frac{\sin 2z}{4-z}\right) \frac{1}{2}(4-z)\, dz$

$= \left[-\frac{1}{4}\cos 2z\right]_0^4 = \left[-\frac{1}{4} + \frac{1}{2}\sin^2 z\right]_0^4 = \frac{\sin^2 4}{2}$

45. $\int_0^1 \int_0^{4-a-x^2} \int_a^{4-x^2-y} dz\, dy\, dx = \frac{4}{15} \Rightarrow \int_0^1 \int_0^{4-a-x^2} (4 - x^2 - y - a)\, dy\, dx = \frac{4}{15}$

$\Rightarrow \int_0^1 \left[(4-a-x^2)^2 - \frac{1}{2}(4-a-x^2)^2\right] dx = \frac{4}{15} \Rightarrow \frac{1}{2}\int_0^1 (4-a-x^2)^2\, dx = \frac{4}{15} \Rightarrow \int_0^1 \left[(4-a)^2 - 2x^2(4-a) + x^4\right] dx$

$= \frac{8}{15} \Rightarrow \left[(4-a)^2 x - \frac{2}{3}x^3(4-a) + \frac{x^5}{5}\right]_0^1 = \frac{8}{15} \Rightarrow (4-a)^2 - \frac{2}{3}(4-a) + \frac{1}{5} = \frac{8}{15} \Rightarrow 15(4-a)^2 - 10(4-a) - 5 = 0$

$\Rightarrow 3(4-a)^2 - 2(4-a) - 1 = 0 \Rightarrow [3(4-a)+1][(4-a)-1] = 0 \Rightarrow 4 - a = -\frac{1}{3}$ or $4 - a = 1 \Rightarrow a = \frac{13}{3}$ or $a = 3$

46. The volume of the ellipsoid $\frac{x^2}{a^2} + \frac{y^2}{b^2} + \frac{z^2}{c^2} = 1$ is $\frac{4abc\pi}{3}$ so that $\frac{4(1)(2)(c)\pi}{3} = 8\pi \Rightarrow c = 3$.

47. To minimize the integral, we want the domain to include all points where the integrand is negative and to exclude all points where it is positive. These criteria are met by the points (x, y, z) such that $4x^2 + 4y^2 + z^2 - 4 \le 0$ or $4x^2 + 4y^2 + z^2 \le 4$, which is a solid ellipsoid centered at the origin.

48. To maximize the integral, we want the domain to include all points where the integrand is positive and to exclude all points where it is negative. These criteria are met by the points (x, y, z) such that $1 - x^2 - y^2 - z^2 \ge 0$ or $x^2 + y^2 + z^2 \le 1$, which is a solid sphere of radius 1 centered at the origin.

49-52. Example CAS commands:

Maple:

```
int(int(int(z/(x^2+y^2+z^2)^(3/2), z=sqrt(x^2+y^2)..1), y=-sqrt(1-x^2)..sqrt(1-x^2)), x =-1..1);
evalf('');
```

Mathematica:

```
Integrate[ z/(x^2+y^2+z^2)^(3/2),
  {x,-1,1}, {y,-Sqrt[1-x^2], Sqrt[1-x^2]},
  {z,Sqrt [x^2+y^2],1} ]
N[%]
```

13.5 MASSES AND MOMENTS IN THREE DIMENSIONS

1. $I_x = \int_{-c/2}^{c/2} \int_{-b/2}^{b/2} \int_{-a/2}^{a/2} (y^2+z^2)\,dx\,dy\,dz = a \int_{-c/2}^{c/2} \int_{-b/2}^{b/2} (y^2+z^2)\,dy\,dz = a \int_{-c/2}^{c/2} \left[\frac{y^3}{3}+yz^2\right]_{-b/2}^{b/2} dz$

$= a \int_{-c/2}^{c/2} \left(\frac{b^3}{12}+bz^2\right) dz = ab\left[\frac{b^2}{12}z+\frac{z^3}{3}\right]_{-c/2}^{c/2} = ab\left(\frac{b^2c}{12}+\frac{c^3}{12}\right) = \frac{abc}{12}(b^2+c^2) = \frac{M}{12}(b^2+c^2);$

$R_x = \sqrt{\frac{b^2+c^2}{12}}$; likewise $R_y = \sqrt{\frac{a^2+c^2}{12}}$ and $R_z = \sqrt{\frac{a^2+b^2}{12}}$, by symmetry

2. The plane $z = \frac{4-2y}{3}$ is the top of the wedge $\Rightarrow I_x = \int_{-3}^{3} \int_{-2}^{4} \int_{-4/3}^{(4-2y)/3} (y^2+z^2)\,dz\,dy\,dx$

$= \int_{-3}^{3} \int_{-2}^{4} \left[\frac{8y^2}{3}-\frac{2y^3}{3}+\frac{8(2-y)^3}{81}+\frac{64}{81}\right] dy\,dx = \int_{-3}^{3} \frac{104}{3}\,dx = 208;\ I_y = \int_{-3}^{3} \int_{-2}^{4} \int_{-4/3}^{(4-2y)/3} (x^2+z^2)\,dz\,dy\,dx$

$= \int_{-3}^{3} \int_{-2}^{4} \left[\frac{(4-2y)^3}{81}+\frac{x^2(4-2y)}{3}+\frac{4x^2}{3}+\frac{64}{81}\right] dy\,dx = \int_{-3}^{3} \left(12x^2+\frac{32}{3}\right) dx = 280;$

$I_z = \int_{-3}^{3} \int_{-2}^{4} \int_{-4/3}^{(4-2y)/3} (x^2+y^2)\,dz\,dy\,dx = \int_{-3}^{3} \int_{-2}^{4} (x^2+y^2)\left(\frac{8}{3}-\frac{2y}{3}\right) dy\,dx = 12 \int_{-3}^{3} (x^2+2)\,dx = 360$

3. $I_x = \int_0^a \int_0^b \int_0^c (y^2+z^2)\,dz\,dy\,dx = \int_0^a \int_0^b \left(cy^2+\frac{c^3}{3}\right) dy\,dx = \int_0^a \left(\frac{cb^3}{3}+\frac{c^3b}{3}\right) dx = \frac{abc(b^2+c^2)}{3}$

$= \frac{M}{3}(b^2+c^2)$ where $M = abc$; $I_y = \frac{M}{3}(a^2+c^2)$ and $I_z = \frac{M}{3}(a^2+b^2)$, by symmetry

4. (a) $M = \int_0^1 \int_0^{1-x} \int_0^{1-x-y} dz\,dy\,dx = \int_0^1 \int_0^{1-x} (1-x-y)\,dy\,dx = \int_0^1 \left(\frac{x^2}{2}-x+\frac{1}{2}\right) dx = \frac{1}{6};$

$M_{yz} = \int_0^1 \int_0^{1-x} \int_0^{1-x-y} x\,dz\,dy\,dx = \int_0^1 \int_0^{1-x} x(1-x-y)\,dy\,dx = \frac{1}{2} \int_0^1 (x^3-2x^2+x)\,dx = \frac{1}{24}$

$\Rightarrow \bar{x} = \bar{y} = \bar{z} = \frac{1}{4}$, by symmetry; $I_z = \int_0^1 \int_0^{1-x} \int_0^{1-x-y} (y^2+z^2)\,dz\,dy\,dx$

$= \int_0^1 \int_0^{1-x} \left[y^2-xy^2-y^3+\frac{(1-x-y)^3}{3}\right] dy\,dx = \frac{1}{6} \int_0^1 (1-x)^4\,dx = \frac{1}{30} \Rightarrow I_y = I_x = \frac{1}{30}$, by symmetry

(b) $R_x = \sqrt{\frac{I_x}{M}} = \sqrt{\frac{1}{5}} = \frac{\sqrt{5}}{5} \approx 0.4472$; the distance from the centroid to the x-axis is $\sqrt{0^2 + \frac{1}{16} + \frac{1}{16}} = \sqrt{\frac{1}{8}} = \frac{\sqrt{2}}{4}$ ≈ 0.3536

5. $M = 4\int_0^1\int_0^1\int_{4y^2}^4 dz\,dy\,dx = 4\int_0^1\int_0^1 (4-4y^2)\,dy\,dx = 16\int_0^1 \frac{2}{3}\,dx = \frac{32}{3}$; $M_{xy} = 4\int_0^1\int_0^1\int_{4y^2}^4 z\,dz\,dy\,dx$

$= 2\int_0^1\int_0^1 (16-16y^4)\,dy\,dx = \frac{128}{5}\int_0^1 dx = \frac{128}{5} \Rightarrow \bar{z} = \frac{12}{5}$, and $\bar{x} = \bar{y} = 0$, by symmetry;

$I_x = 4\int_0^1\int_0^1\int_{4y^2}^4 (y^2+z^2)\,dz\,dy\,dx = 4\int_0^1\int_0^1 \left[\left(4y^2+\frac{64}{3}\right)-\left(4y^4+\frac{64y^6}{3}\right)\right] dy\,dx = 4\int_0^1 \frac{1976}{105}\,dx = \frac{7904}{105}$;

$I_y = 4\int_0^1\int_0^1\int_{4y^2}^4 (x^2+z^2)\,dz\,dy\,dx = 4\int_0^1\int_0^1 \left[\left(4x^2+\frac{64}{3}\right)-\left(4x^2y^2+\frac{64y^6}{3}\right)\right] dy\,dx = 4\int_0^1 \left(\frac{8}{3}x^2+\frac{128}{7}\right) dx$

$= \frac{4832}{63}$; $I_z = 4\int_0^1\int_0^1\int_{4y^2}^4 (x^2+y^2)\,dz\,dy\,dx = 16\int_0^1\int_0^1 (x^2 - x^2y^2 + y^2 - y^4)\,dy\,dx$

$= 16\int_0^1 \left(\frac{2x^2}{3}+\frac{2}{15}\right) dx = \frac{256}{45}$

6. (a) $M = \int_{-2}^{2}\int_{\left(-\sqrt{4-x^2}\right)/2}^{\left(\sqrt{4-x^2}\right)/2}\int_0^{2-x} dz\,dy\,dx = \int_{-2}^{2}\int_{\left(-\sqrt{4-x^2}\right)/2}^{\left(\sqrt{4-x^2}\right)/2} (2-x)\,dy\,dx = \int_{-2}^{2} (2-x)\left(\sqrt{4-x^2}\right) dx = 4\pi$;

$M_{yz} = \int_{-2}^{2}\int_{\left(-\sqrt{4-x^2}\right)/2}^{\left(\sqrt{4-x^2}\right)/2}\int_0^{2-x} x\,dz\,dy\,dx = \int_{-2}^{2}\int_{\left(-\sqrt{4-x^2}\right)/2}^{\left(\sqrt{4-x^2}\right)/2} x(2-x)\,dy\,dx = \int_{-2}^{2} x(2-x)\left(\sqrt{4-x^2}\right) dx = -2\pi$;

$M_{xz} = \int_{-2}^{2}\int_{\left(-\sqrt{4-x^2}\right)/2}^{\left(\sqrt{4-x^2}\right)/2}\int_0^{2-x} y\,dz\,dy\,dx = \int_{-2}^{2}\int_{\left(-\sqrt{4-x^2}\right)/2}^{\left(\sqrt{4-x^2}\right)/2} y(2-x)\,dy\,dx$

$= \frac{1}{2}\int_{-2}^{2} (2-x)\left[\frac{4-x^2}{4} - \frac{4-x^2}{4}\right] dx = 0 \Rightarrow \bar{x} = -\frac{1}{2}$ and $\bar{y} = 0$

(b) $M_{xy} = \int_{-2}^{2} \int_{(-\sqrt{4-x^2})/2}^{(\sqrt{4-x^2})/2} \int_{0}^{2-x} z\,dz\,dy\,dx = \frac{1}{2}\int_{-2}^{2} \int_{(-\sqrt{4-x^2})/2}^{(\sqrt{4-x^2})/2} (2-x)^2\,dy\,dx = \frac{1}{2}\int_{-2}^{2} (2-x)^2\left(\sqrt{4-x^2}\right)dx$

$= 5\pi \Rightarrow \bar{z} = \frac{5}{4}$

7. (a) $M = 4\int_0^2 \int_0^{\sqrt{4-x^2}} \int_{x^2+y^2}^{4} dz\,dy\,dx = 4\int_0^{\pi/2} \int_0^2 \int_{r^2}^{4} r\,dz\,dr\,d\theta = 4\int_0^{\pi/2} \int_0^2 \left(4r - r^3\right)dr\,d\theta = 4\int_0^{\pi/2} 4\,d\theta = 8\pi;$

$M_{xy} = \int_0^{2\pi} \int_0^2 \int_{r^2}^{4} zr\,dz\,dr\,d\theta = \int_0^{2\pi} \int_0^2 \frac{r}{2}\left(16 - r^4\right)dr\,d\theta = \frac{32}{3}\int_0^{2\pi} d\theta = \frac{64\pi}{3} \Rightarrow \bar{z} = \frac{8}{3}$, and $\bar{x} = \bar{y} = 0$,

by symmetry

(b) $M = 8\pi \Rightarrow 4\pi = \int_0^{2\pi} \int_0^{\sqrt{c}} \int_{r^2}^{c} r\,dz\,dr\,d\theta = \int_0^{2\pi} \int_0^{\sqrt{c}} \left(cr - r^3\right)dr\,d\theta = \int_0^{2\pi} \frac{c^2}{4}\,d\theta = \frac{c^2\pi}{2} \Rightarrow c^2 = 8 \Rightarrow c = 2\sqrt{2}$,

since $c > 0$

8. $M = 8$; $M_{xy} = \int_{-1}^{1} \int_3^5 \int_{-1}^{1} z\,dz\,dy\,dx = \int_{-1}^{1} \int_3^5 \left[\frac{z^2}{2}\right]_{-1}^{1} dy\,dx = 0$; $M_{yz} = \int_{-1}^{1} \int_3^5 \int_{-1}^{1} x\,dz\,dy\,dx$

$= 2\int_{-1}^{1} \int_3^5 x\,dy\,dx = 4\int_{-1}^{1} x^2\,dx = 0$; $M_{xz} = \int_{-1}^{1} \int_3^5 \int_{-1}^{1} y\,dz\,dy\,dx = 2\int_{-1}^{1} \int_3^5 y\,dy\,dx = 16\int_{-1}^{1} dx = 32$

$\Rightarrow \bar{x} = 0,\ \bar{y} = 4,\ \bar{z} = 0$; $I_x = \int_{-1}^{1} \int_3^5 \int_{-1}^{1} \left(y^2 + z^2\right)dz\,dy\,dx = \int_{-1}^{1} \int_3^5 \left(2y^2 + \frac{2}{3}\right)dy\,dx = \frac{2}{3}\int_{-1}^{1} 100\,dx = \frac{400}{3}$;

$I_y = \int_{-1}^{1} \int_3^5 \int_{-1}^{1} \left(x^2 + z^2\right)dz\,dy\,dx = \int_{-1}^{1} \int_3^5 \left(2x^2 + \frac{2}{3}\right)dy\,dx = \frac{4}{3}\int_{-1}^{1} \left(3x^2 + 1\right)dx = \frac{16}{3}$;

$I_z = \int_{-1}^{1} \int_3^5 \int_{-1}^{1} \left(x^2 + y^2\right)dz\,dy\,dx = 2\int_{-1}^{1} \int_3^5 \left(x^2 + y^2\right)dy\,dx = 2\int_{-1}^{1} \left(2x^2 + \frac{98}{3}\right)dx = \frac{400}{3} \Rightarrow R_x = R_z = \sqrt{\frac{50}{3}}$

and $R_y = \sqrt{\frac{2}{3}}$

9. The plane $y + 2z = 2$ is the top of the wedge $\Rightarrow I_L = \int_{-2}^{2} \int_{-2}^{4} \int_{-1}^{(2-y)/2} \left[(y-6)^2 + z^2\right]dz\,dy\,dx$

$= \int_{-2}^{2} \int_{-2}^{4} \left[\frac{(y-6)^2(4-y)}{2} + \frac{(2-y)^3}{24} + \frac{1}{3}\right]dy\,dx$; let $t = 2 - y \Rightarrow I_L = 4\int_{-2}^{4} \left(\frac{13t^3}{24} + 5t^2 + 16t + \frac{49}{3}\right)dt = 1386$;

$M = \frac{1}{2}(3)(6)(4) = 36 \Rightarrow R_L = \sqrt{\frac{I_L}{M}} = \sqrt{\frac{77}{2}}$

10. The plane $y + 2z = 2$ is the top of the wedge $\Rightarrow I_L = \int_{-2}^{2}\int_{-2}^{4}\int_{-1}^{(2-y)/2} \left[(x-4)^2 + y^2\right] dz\,dy\,dx$

$= \frac{1}{2}\int_{-2}^{2}\int_{-2}^{4} (x^2 - 8x + 16 + y^2)(4-y)\,dy\,dx = \int_{-2}^{2} (9x^2 - 72x + 162)\,dx = 696;\ M = \frac{1}{2}(3)(6)(4) = 36$

$\Rightarrow R_L = \sqrt{\frac{I_L}{M}} = \sqrt{\frac{58}{3}}$

11. $M = 8;\ I_L = \int_0^4\int_0^2\int_0^1 \left[z^2 + (y-2)^2\right] dz\,dy\,dx = \int_0^4\int_0^2 \left(y^2 - 4y + \frac{13}{3}\right) dy\,dx = \frac{10}{3}\int_0^4 dx = \frac{40}{3}$

$\Rightarrow R_L = \sqrt{\frac{I_L}{M}} = \sqrt{\frac{5}{3}}$

12. $M = 8;\ I_L = \int_0^4\int_0^2\int_0^1 \left[(x-4)^2 + y^2\right] dz\,dy\,dx = \int_0^4\int_0^2 \left[(x-4)^2 + y^2\right] dy\,dx = \int_0^4 \left[2(x-4)^2 + \frac{8}{3}\right] dx = \frac{160}{3}$

$\Rightarrow R_L = \sqrt{\frac{I_L}{M}} = \sqrt{\frac{20}{3}}$

13. (a) $M = \int_0^2\int_0^{2-x}\int_0^{2-x-y} 2x\,dz\,dy\,dx = \int_0^2\int_0^{2-x} \left(4x - 2x^2 - 2xy\right) dy\,dx = \int_0^2 \left(x^3 - 4x^2 + 4x\right) dx = \frac{4}{3}$

(b) $M_{xy} = \int_0^2\int_0^{2-x}\int_0^{2-x-y} 2xz\,dz\,dy\,dx = \int_0^2\int_0^{2-x} x(2-x-y)^2\,dy\,dx = \int_0^2 \frac{x(2-x)^3}{3}\,dx = \frac{8}{15};\ M_{xz} = \frac{8}{15}$ by

symmetry; $M_{yz} = \int_0^2\int_0^{2-x}\int_0^{2-x-y} 2x^2\,dz\,dy\,dx = \int_0^2\int_0^{2-x} 2x^2(2-x-y)\,dy\,dx = \int_0^2 \left(2x - x^2\right)^2 dx = \frac{16}{15}$

$\Rightarrow \bar{x} = \frac{4}{5}$, and $\bar{y} = \bar{z} = \frac{2}{5}$

14. (a) $M = \int_0^2\int_0^{\sqrt{x}}\int_0^{4-x^2} kxy\,dz\,dy\,dx = k\int_0^2\int_0^{\sqrt{x}} xy\left(4 - x^2\right) dy\,dx = \frac{k}{2}\int_0^2 \left(4x^2 - x^4\right) dx = \frac{32k}{15}$

(b) $M_{yz} = \int_0^2\int_0^{\sqrt{x}}\int_0^{4-x^2} kx^2y\,dz\,dy\,dx = k\int_0^2\int_0^{\sqrt{x}} x^2y\left(4 - x^2\right) dy\,dx = \frac{k}{2}\int_0^2 \left(4x^3 - x^5\right) dx = \frac{8k}{3}$

$\Rightarrow \bar{x} = \frac{5}{4};\ M_{xz} = \int_0^2\int_0^{\sqrt{x}}\int_0^{4-x^2} kxy^2\,dz\,dy\,dx = k\int_0^2\int_0^{\sqrt{x}} xy^2\left(4 - x^2\right) dy\,dx = \frac{k}{3}\int_0^2 \left(4x^{5/2} - x^{9/2}\right) dx$

$= \frac{256\sqrt{2}k}{231} \Rightarrow \bar{y} = \frac{40\sqrt{2}}{77};\ M_{xy} = \int_0^2\int_0^{\sqrt{x}}\int_0^{4-x^2} kxyz\,dz\,dy\,dx = \frac{k}{2}\int_0^2\int_0^{\sqrt{x}} xy\left(4 - x^2\right)^2 dy\,dx$

$$= \frac{k}{4} \int_0^2 \left(16x^2 - 8x^4 + x^6\right) dx = \frac{256k}{105} \Rightarrow \bar{z} = \frac{8}{7}$$

15. (a) $M = \int_0^1 \int_0^1 \int_0^1 (x + y + z + 1)\, dz\, dy\, dx = \int_0^1 \int_0^1 \left(x + y + \frac{3}{2}\right) dy\, dx = \int_0^1 (x + 2)\, dx = \frac{5}{2}$

(b) $M_{xy} = \int_0^1 \int_0^1 \int_0^1 z(x + y + z + 1)\, dz\, dy\, dx = \frac{1}{2} \int_0^1 \int_0^1 \left(x + y + \frac{5}{3}\right) dy\, dx = \frac{1}{2} \int_0^1 \left(x + \frac{13}{6}\right) dx = \frac{4}{3}$

$\Rightarrow M_{xy} = M_{yz} = M_{xz} = \frac{4}{3}$, by symmetry $\Rightarrow \bar{x} = \bar{y} = \bar{z} = \frac{8}{15}$

(c) $I_z = \int_0^1 \int_0^1 \int_0^1 (x^2 + y^2)(x + y + z + 1)\, dz\, dy\, dx = \int_0^1 \int_0^1 (x^2 + y^2)\left(x + y + \frac{3}{2}\right) dy\, dx$

$= \int_0^1 \left(x^3 + 2x^2 + \frac{1}{3}x + \frac{3}{4}\right) dx = \frac{11}{6} \Rightarrow I_x = I_y = I_z = \frac{11}{6}$, by symmetry

(d) $R_x = R_y = R_z = \sqrt{\frac{I_z}{M}} = \sqrt{\frac{11}{15}}$

16. The plane $y + 2z = 2$ is the top of the wedge.

(a) $M = \int_{-1}^{1} \int_{-2}^{4} \int_{-1}^{(2-y)/2} (x + 1)\, dz\, dy\, dx = \int_{-1}^{1} \int_{-2}^{4} (x + 1)\left(2 - \frac{y}{2}\right) dy\, dx = 18$

(b) $M_{yz} = \int_{-1}^{1} \int_{-2}^{4} \int_{-1}^{(2-y)/2} x(x + 1)\, dz\, dy\, dx = \int_{-1}^{1} \int_{-2}^{4} x(x + 1)\left(2 - \frac{y}{2}\right) dy\, dx = 6;$

$M_{xz} = \int_{-1}^{1} \int_{-2}^{4} \int_{-1}^{(2-y)/2} y(x + 1)\, dz\, dy\, dx = \int_{-1}^{1} \int_{-2}^{4} y(x + 1)\left(2 - \frac{y}{2}\right) dy\, dx = 0;$

$M_{xy} = \int_{-1}^{1} \int_{-2}^{4} \int_{-1}^{(2-y)/2} z(x + 1)\, dz\, dy\, dx = \frac{1}{2} \int_{-1}^{1} \int_{-2}^{4} (x + 1)\left(\frac{y^2}{4} - y\right) dy\, dx = 0 \Rightarrow \bar{x} = \frac{4}{3}$, and $\bar{y} = \bar{z} = 0$

(c) $I_x = \int_{-1}^{1} \int_{-2}^{4} \int_{-1}^{(2-y)/2} (x + 1)(y^2 + z^2)\, dz\, dy\, dx = \int_{-1}^{1} \int_{-2}^{4} (x + 1)\left[2y^2 + \frac{1}{3} - \frac{y^3}{2} + \frac{1}{3}\left(1 - \frac{y}{2}\right)^3\right] dy\, dx = 45;$

$I_y = \int_{-1}^{1} \int_{-2}^{4} \int_{-1}^{(2-y)/2} (x + 1)(x^2 + z^2)\, dz\, dy\, dx = \int_{-1}^{1} \int_{-2}^{4} (x + 1)\left[2x^2 + \frac{1}{3} - \frac{x^2 y}{2} + \frac{1}{3}\left(1 - \frac{y}{2}\right)^3\right] dy\, dx = 15;$

$I_z = \int_{-1}^{1} \int_{-2}^{4} \int_{-1}^{(2-y)/2} (x + 1)(x^2 + y^2)\, dz\, dy\, dx = \int_{-1}^{1} \int_{-2}^{4} (x + 1)\left(2 - \frac{y}{2}\right)(x^2 + y^2)\, dy\, dx = 42$

(d) $R_x = \sqrt{\frac{I_x}{M}} = \sqrt{\frac{5}{2}}$, $R_y = \sqrt{\frac{I_y}{M}} = \sqrt{\frac{5}{6}}$, and $R_z = \sqrt{\frac{I_z}{M}} = \sqrt{\frac{7}{3}}$

17. $M = \int_0^1 \int_{z-1}^{1-z} \int_0^{\sqrt{z}} (2y+5)\,dy\,dx\,dz = \int_0^1 \int_{z-1}^{1-z} (z+5\sqrt{z})\,dx\,dz = \int_0^1 2(z+5\sqrt{z})(1-z)\,dz$

$= 2\int_0^1 \left(5z^{1/2}+z-5z^{3/2}-z^2\right)dz = 2\left[\frac{10}{3}z^{3/2}+\frac{1}{2}z^2-2z^{5/2}-\frac{1}{3}z^3\right]_0^1 = 2\left(\frac{9}{3}-\frac{3}{2}\right) = 3$

18. $M = \int_{-2}^{2} \int_{-\sqrt{4-x^2}}^{\sqrt{4-x^2}} \int_{2(x^2+y^2)}^{16-2(x^2+y^2)} \sqrt{x^2+y^2}\,dz\,dy\,dx = \int_{-2}^{2} \int_{-\sqrt{4-x^2}}^{\sqrt{4-x^2}} \sqrt{x^2+y^2}\left[16-4\left(x^2+y^2\right)\right]dy\,dx$

$= 4\int_0^{2\pi} \int_0^2 r\left(4-r^2\right) r\,dr\,d\theta = 4\int_0^{2\pi} \left[\frac{4r^3}{3}-\frac{r^5}{5}\right]_0^2 d\theta = 4\int_0^{2\pi} \frac{64}{15}\,d\theta = \frac{512\pi}{15}$

19. (a) Let ΔV_i be the volume of the ith piece, and let (x_i, y_i, z_i) be a point in the ith piece. Then the work done by gravity in moving the ith piece to the xy-plane is approximately $W_i = m_i g z_i = (x_i+y_i+z_i+1)g\,\Delta V_i\, z_i$ $\Rightarrow$ the total work done is the triple integral $W = \int_0^1 \int_0^1 \int_0^1 (x+y+z+1)gz\,dz\,dy\,dx$

$= g\int_0^1 \int_0^1 \left[\frac{1}{2}xz^2+\frac{1}{2}yz^2+\frac{1}{3}z^3+\frac{1}{2}z^2\right]_0^1 dy\,dx = g\int_0^1 \int_0^1 \left(\frac{1}{2}x+\frac{1}{2}y+\frac{5}{6}\right)dy\,dx = g\int_0^1 \left[\frac{1}{2}xy+\frac{1}{4}y^2+\frac{5}{6}y\right]_0^1 dx$

$= g\int_0^1 \left(\frac{1}{2}x+\frac{13}{12}\right)dx = g\left[\frac{x^2}{4}+\frac{13}{12}x\right]_0^1 = g\left(\frac{16}{12}\right) = \frac{4}{3}g$

(b) From Exercise 15 the center of mass is $\left(\frac{8}{15}, \frac{8}{15}, \frac{8}{15}\right)$ and the mass of the liquid is $\frac{5}{2}$ $\Rightarrow$ the work done by gravity in moving the center of mass to the xy-plane is $W = mgd = \left(\frac{5}{2}\right)(g)\left(\frac{8}{15}\right) = \frac{4}{3}g$, which is the same as the work done in part (a).

20. (a) From Exercise 19(a) we see that the work done is $W = g\int_0^2 \int_0^{\sqrt{x}} \int_0^{4-x^2} kxyz\,dz\,dy\,dx$

$= kg\int_0^2 \int_0^{\sqrt{x}} \frac{1}{2}xy\left(4-x^2\right)^2 dy\,dx = \frac{kg}{4}\int_0^2 x^2\left(4-x^2\right)^2 dx = \frac{kg}{4}\int_0^2 \left(16x^2-8x^4+x^6\right)dx$

$= \frac{kg}{4}\left[\frac{16}{3}x^3-\frac{8}{5}x^5+\frac{1}{7}x^7\right]_0^2 = \frac{256kg}{105}$

(b) From Exercise 14 the center of mass is $\left(\frac{5}{4}, \frac{40\sqrt{2}}{77}, \frac{8}{7}\right)$ and the mass of the liquid is $\frac{32k}{15}$ $\Rightarrow$ the work done by gravity in moving the center of mass to the xy-plane is $W = mgd = \left(\frac{32k}{15}\right)(g)\left(\frac{8}{7}\right) = \frac{256kg}{105}$

21. (a) $\bar{x} = \frac{M_{yz}}{M} = 0 \Rightarrow \iiint\limits_R x\delta(x,y,z)\,dx\,dy\,dz = 0 \Rightarrow M_{yz} = 0$

(b) $I_L = \iiint\limits_R |\mathbf{v} - h\mathbf{i}|^2\,dm = \iiint\limits_R |(x-h)\,\mathbf{i} + y\mathbf{j}|^2\,dm = \iiint\limits_R (x^2 - 2xh + h^2 + y^2)\,dm$

$= \iiint\limits_R (x^2+y^2)\,dm - 2h \iiint\limits_R x\,dm + h^2 \iiint\limits_R dm = I_x - 0 + h^2m = I_{c.m.} + h^2m$

22. $I_L = I_{c.m.} + mh^2 = \frac{2}{5}ma^2 + ma^2 = \frac{7}{5}ma^2$

23. (a) $(\bar{x},\bar{y},\bar{z}) = \left(\frac{a}{2},\frac{b}{2},\frac{c}{2}\right) \Rightarrow I_z = I_{c.m.} + abc\left(\sqrt{\frac{a^2}{4}+\frac{b^2}{4}}\right)^2 \Rightarrow I_{c.m.} = I_z - \frac{abc(a^2+b^2)}{4}$

$= \frac{abc(a^2+b^2)}{3} - \frac{abc(a^2+b^2)}{4} = \frac{abc(a^2+b^2)}{12}$; $R_{c.m.} = \sqrt{\frac{I_{c.m.}}{M}} = \sqrt{\frac{a^2+b^2}{12}}$

(b) $I_L = I_{c.m.} + abc\left(\sqrt{\frac{a^2}{4}+\left(\frac{b}{2}-2b\right)^2}\right)^2 = \frac{abc(a^2+b^2)}{12} + \frac{abc(a^2+9b^2)}{4} = \frac{abc(4a^2+28b^2)}{12}$

$= \frac{abc(a^2+7b^2)}{3}$; $R_L = \sqrt{\frac{I_L}{M}} = \sqrt{\frac{a^2+7b^2}{3}}$

24. $M = \int_{-3}^{3}\int_{-2}^{4}\int_{-4/3}^{(4-2y)/3} dz\,dy\,dx = \int_{-3}^{3}\int_{-2}^{4} \frac{2}{3}(4-y)\,dy\,dx = \int_{-3}^{3} \frac{2}{3}\left[4y - \frac{y^2}{2}\right]_{-2}^{4} dx = 12\int_{-3}^{3} dx = 72$;

$\bar{x} = \bar{y} = \bar{z} = 0$ from Exercise 2 $\Rightarrow I_x = I_{c.m.} + 72\left(\sqrt{0^2+0^2}\right)^2 = I_{c.m.} \Rightarrow I_L = I_{c.m.} + 72\left(\sqrt{16+\frac{16}{9}}\right)^2$

$= 208 + 72\left(\frac{160}{9}\right) = 1488$

25. $M_{yz_{B_1 \cup B_2}} = \iiint\limits_{B_1} x\,dV_1 + \iiint\limits_{B_2} x\,dV_2 = M_{(yz)_1} + M_{(yz)_2} \Rightarrow \bar{x} = \frac{M_{(yz)_1} + M_{(yz)_2}}{m_1+m_2}$; similarly,

$\bar{y} = \frac{M_{(xz)_1} + M_{(xz)_2}}{m_1+m_2}$ and $\bar{z} = \frac{M_{(xy)_1} + M_{(xy)_2}}{m_1+m_2} \Rightarrow \mathbf{c} = \bar{x}\mathbf{i} + \bar{y}\mathbf{j} + \bar{z}\mathbf{k}$

$= \frac{1}{m_1+m_2}\left[\left(M_{(yz)_1} + M_{(yz)_2}\right)\mathbf{i} + \left(M_{(xz)_1} + M_{(xz)_2}\right)\mathbf{j} + \left(M_{(xy)_1} + M_{(xy)_2}\right)\mathbf{k}\right]$

$= \frac{1}{m_1+m_2}[(m_1\bar{x}_1 + m_2\bar{x}_2)\mathbf{i} + (m_1\bar{y}_1 + m_2\bar{y}_2)\mathbf{j} + (m_1\bar{z}_1 + m_2\bar{z}_2)\mathbf{k}]$

$= \frac{1}{m_1+m_2}[m_1(\bar{x}_1\mathbf{i} + \bar{y}_1\mathbf{j} + \bar{z}_1\mathbf{k}) + m_2(\bar{x}_2\mathbf{i} + \bar{y}_2\mathbf{j} + \bar{z}_2\mathbf{k})] = \frac{m_1\mathbf{c}_1 + m_2\mathbf{c}_2}{m_1+m_2}$

26. (a) $\mathbf{c} = \frac{12\left(\mathbf{i}+\frac{3}{2}\mathbf{j}+\mathbf{k}\right)+2\left(\frac{1}{2}\mathbf{i}+4\mathbf{j}+\frac{1}{2}\mathbf{k}\right)}{12+2} = \frac{\frac{13}{2}\mathbf{i}+13\mathbf{j}+\frac{13}{2}\mathbf{k}}{7} \Rightarrow \overline{x} = \frac{13}{14},\ \overline{y} = \frac{13}{7},\ \overline{z} = \frac{13}{14}$

(b) $\mathbf{c} = \frac{12\left(\mathbf{i}+\frac{3}{2}\mathbf{j}+\mathbf{k}\right)+12\left(\mathbf{i}+\frac{11}{2}\mathbf{j}-\frac{1}{2}\mathbf{k}\right)}{12+12} = \frac{2\mathbf{i}+7\mathbf{j}+\frac{1}{2}\mathbf{k}}{2} \Rightarrow \overline{x} = 1,\ \overline{y} = \frac{7}{2},\ \overline{z} = \frac{1}{4}$

(c) $\mathbf{c} = \frac{2\left(\frac{1}{2}\mathbf{i}+4\mathbf{j}+\frac{1}{2}\mathbf{k}\right)+12\left(\mathbf{i}+\frac{11}{2}\mathbf{j}-\frac{1}{2}\mathbf{k}\right)}{2+12} = \frac{13\mathbf{i}+74\mathbf{j}-5\mathbf{k}}{14} \Rightarrow \overline{x} = \frac{13}{14},\ \overline{y} = \frac{37}{7},\ \overline{z} = -\frac{5}{14}$

(d) $\mathbf{c} = \frac{12\left(\mathbf{i}+\frac{3}{2}\mathbf{j}+\mathbf{k}\right)+2\left(\frac{1}{2}\mathbf{i}+4\mathbf{j}+\frac{1}{2}\mathbf{k}\right)+12\left(\mathbf{i}+\frac{11}{2}\mathbf{j}-\frac{1}{2}\mathbf{k}\right)}{12+2+12} = \frac{25\mathbf{i}+92\mathbf{j}+7\mathbf{k}}{26} \Rightarrow \overline{x} = \frac{25}{26},\ \overline{y} = \frac{46}{13},\ \overline{z} = \frac{7}{26}$

27. (a) $\mathbf{c} = \frac{\left(\frac{\pi a^2 h}{3}\right)\left(\frac{h}{4}\mathbf{k}\right)+\left(\frac{2\pi a^3}{3}\right)\left(-\frac{3a}{8}\mathbf{k}\right)}{m_1+m_2} = \frac{\left(\frac{a^2\pi}{3}\right)\left(\frac{h^2-3a^2}{4}\mathbf{k}\right)}{m_1+m_2}$, where $m_1 = \frac{\pi a^2 h}{3}$ and $m_2 = \frac{2\pi a^3}{3}$; if $\frac{h^2-3a^2}{4} = 0$, or $h = a\sqrt{3}$, then the centroid is on the common base

(b) See the solution to Exercise 55, Section 13.2, to see that $h = a\sqrt{2}$.

28. $\mathbf{c} = \frac{\left(\frac{s^2 h}{3}\right)\left(\frac{h}{4}\mathbf{k}\right)+s^3\left(-\frac{s}{2}\mathbf{k}\right)}{m_1+m_2} = \frac{\left(\frac{s^2}{12}\right)[(h^2-6s^2)\mathbf{k}]}{m_1+m_2}$, where $m_1 = \frac{s^2 h}{3}$ and $m_2 = s^3$; if $h^2 - 6s^2 < 0$, or $h < \sqrt{6}s$, then the centroid is in the base of the pyramid.

13.6 TRIPLE INTEGRALS IN CYLINDRICAL AND SPHERICAL COORDINATES

1. $\int_0^{2\pi}\int_0^1\int_r^{\sqrt{2-r^2}} dz\, r\, dr\, d\theta = \int_0^{2\pi}\int_0^1 \left[r\left(2-r^2\right)^{1/2} - r^2\right] dr\, d\theta = \int_0^{2\pi}\left[-\frac{1}{3}\left(2-r^2\right)^{3/2} - \frac{r^3}{3}\right]_0^1 d\theta$
$= \int_0^{2\pi}\left(\frac{2^{3/2}}{3} - \frac{2}{3}\right) d\theta = \frac{4\pi\left(\sqrt{2}-1\right)}{3}$

2. $\int_0^{2\pi}\int_0^3\int_{r^2/3}^{\sqrt{18-r^2}} dz\, r\, dr\, d\theta = \int_0^{2\pi}\int_0^3\left[r\left(18-r^2\right)^{1/2} - \frac{r^3}{3}\right] dr\, d\theta = \int_0^{2\pi}\left[-\frac{1}{3}\left(18-r^2\right)^{3/2} - \frac{r^4}{12}\right]_0^3 d\theta$
$= \frac{9\pi\left(8\sqrt{2}-7\right)}{2}$

3. $\int_0^{2\pi}\int_0^{\theta/2\pi}\int_0^{3+24r^2} dz\, r\, dr\, d\theta = \int_0^{2\pi}\int_0^{\theta/2\pi}\left(3r+24r^3\right) dr\, d\theta = \int_0^{2\pi}\left[\frac{3}{2}r^2+6r^4\right]_0^{\theta/2\pi} d\theta = \frac{3}{2}\int_0^{2\pi}\left(\frac{\theta^2}{4\pi^2}+\frac{4\theta^4}{16\pi^4}\right) d\theta$
$= \frac{3}{2}\left[\frac{\theta^3}{12\pi^2}+\frac{\theta^5}{5\pi^4}\right]_0^{2\pi} = \frac{17\pi}{5}$

4. $$\int_0^{\pi}\int_0^{\theta/\pi}\int_{-\sqrt{4-r^2}}^{3\sqrt{4-r^2}} z\ dz\ r\ dr\, d\theta = \int_0^{\pi}\int_0^{\theta/\pi} \tfrac{1}{2}[9(4-r^2)-(4-r^2)]r\ dr\, d\theta = 4\int_0^{\pi}\int_0^{\theta/\pi} (4r-r^3)\, dr\, d\theta$$

$$= 4\int_0^{\pi}\left[2r^2-\frac{r^4}{4}\right]_0^{\theta/\pi} = 4\int_0^{\pi}\left(\frac{2\theta^2}{\pi^2}-\frac{\theta^4}{4\pi^4}\right) d\theta = \frac{37\pi}{15}$$

5. $$\int_0^{2\pi}\int_0^{1}\int_r^{(2-r^2)^{-1/2}} 3\ dz\ r\ dr\, d\theta = 3\int_0^{2\pi}\int_0^{1}\left[r(2-r^2)^{-1/2}-r^2\right] dr\, d\theta = 3\int_0^{2\pi}\left[-(2-r^2)^{1/2}-\frac{r^3}{3}\right]_0^1 d\theta$$

$$= 3\int_0^{2\pi}\left(\sqrt{2}-\tfrac{4}{3}\right) d\theta = \pi(6\sqrt{2}-8)$$

6. $$\int_0^{2\pi}\int_0^{1}\int_{-1/2}^{1/2} (r^2\sin^2\theta+z^2)\, dz\ r\ dr\, d\theta = \int_0^{2\pi}\int_0^{1}\left(r^3\sin^2\theta+\tfrac{r}{12}\right) dr\, d\theta = \int_0^{2\pi}\left(\frac{\sin^2\theta}{4}+\frac{1}{24}\right) d\theta = \frac{\pi}{3}$$

7. $$\int_0^{2\pi}\int_0^{3}\int_0^{z/3} r^3\, dr\, dz\, d\theta = \int_0^{2\pi}\int_0^{3} \frac{z^4}{324}\, dz\, d\theta = \int_0^{2\pi} \frac{3}{20}\, d\theta = \frac{3\pi}{10}$$

8. $$\int_{-1}^{1}\int_0^{2\pi}\int_0^{1+\cos\theta} 4r\ dr\, d\theta\, dz = \int_{-1}^{1}\int_0^{2\pi} 2(1+\cos\theta)^2\, d\theta\, dz = \int_{-1}^{1} 6\pi\ d\theta = 12\pi$$

9. $$\int_0^{1}\int_0^{\sqrt{z}}\int_0^{2\pi} (r^2\cos^2\theta+z^2)\, r\ d\theta\, dr\, dz = \int_0^{1}\int_0^{\sqrt{z}}\left[\frac{r^2\theta}{2}+\frac{r^2\sin 2\theta}{4}+z^2\theta\right]_0^{2\pi} r\ dr\, dz = \int_0^{1}\int_0^{\sqrt{z}} (\pi r^3+2\pi r z^2)\, dr\, dz$$

$$= \int_0^{1}\left[\frac{\pi r^4}{4}+\pi r^2 z^2\right]_0^{\sqrt{z}} dz = \int_0^{1}\left(\frac{\pi z^2}{4}+\pi z^3\right) dz = \left[\frac{\pi z^3}{12}+\frac{\pi z^4}{4}\right]_0^1 = \frac{\pi}{3}$$

10. $$\int_0^{2}\int_{r-2}^{\sqrt{4-r^2}}\int_0^{2\pi} (r\sin\theta+1)\, r\ d\theta\, dz\, dr = \int_0^{2}\int_{r-2}^{\sqrt{4-r^2}} 2\pi r\ dz\, dr = 2\pi\int_0^{2}\left[r(4-r^2)^{1/2}-r^2+2r\right] dr$$

$$= 2\pi\left[-\tfrac{1}{3}(4-r^2)^{3/2}-\frac{r^3}{3}+r^2\right]_0^2 = 2\pi\left[-\tfrac{8}{3}+4+\tfrac{1}{3}(4)^{3/2}\right] = 8\pi$$

11. (a) $\int_0^{2\pi}\int_0^1\int_0^{\sqrt{4-r^2}} dz\ r\ dr\, d\theta$

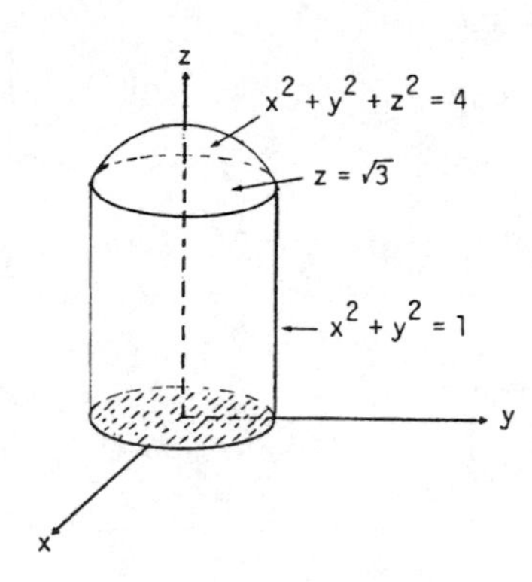

(b) $\int_0^{2\pi}\int_0^{\sqrt{3}}\int_0^1 r\ dr\, dz\, d\theta + \int_0^{2\pi}\int_{\sqrt{3}}^2\int_0^{\sqrt{4-z^2}} r\ dr\, dz\, d\theta$

(c) $\int_0^1\int_0^{\sqrt{4-r^2}}\int_0^{2\pi} r\ d\theta\, dz\, dr$

12. (a) $\int_0^{2\pi}\int_0^1\int_r^{2-r^2} dz\ r\ dr\, d\theta$

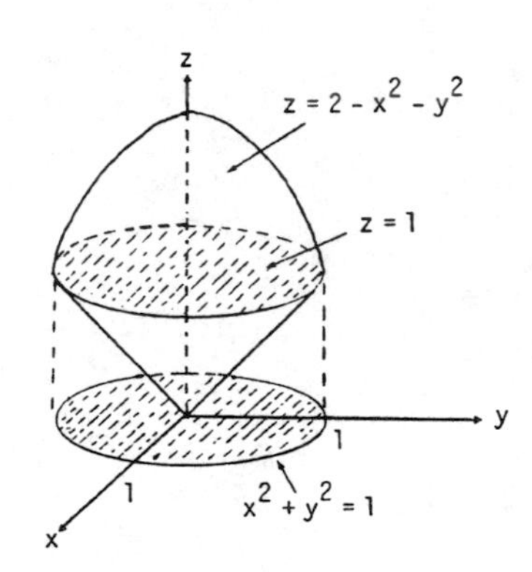

(b) $\int_0^{2\pi}\int_0^1\int_0^z r\ dr\, dz\, d\theta + \int_0^{2\pi}\int_1^2\int_0^{\sqrt{2-z}} r\ dr\, dz\, d\theta$

(c) $\int_0^1\int_r^{2-r^2}\int_0^{2\pi} r\ d\theta\, dz\, dr$

13. $\int_{-\pi/2}^{\pi/2}\int_0^{\cos\theta}\int_0^{3r^2} f(r,\theta,z)\ dz\ r\ dr\, d\theta$

14. $\int_{-\pi/2}^{\pi/2}\int_0^1\int_0^{r\cos\theta} r^3\ dz\, dr\, d\theta = \int_{-\pi/2}^{\pi/2}\int_0^1 r^4\cos\theta\ dr\, d\theta = \frac{1}{5}\int_{-\pi/2}^{\pi/2}\cos\theta\ d\theta = \frac{2}{5}$

15. $\int_0^{\pi}\int_0^{2\sin\theta}\int_0^{4-r\sin\theta} f(r,\theta,z)\ dz\ r\ dr\, d\theta$

16. $\int_{-\pi/2}^{\pi/2}\int_0^{3\cos\theta}\int_0^{5-r\cos\theta} f(r,\theta,z)\ dz\ r\ dr\, d\theta$

17. $\int_{-\pi/2}^{\pi/2}\int_1^{1+\cos\theta}\int_0^{4} f(r,\theta,z)\ dz\ r\ dr\, d\theta$

18. $\int_{-\pi/2}^{\pi/2}\int_{\cos\theta}^{2\cos\theta}\int_0^{3-r\sin\theta} f(r,\theta,z)\ dz\ r\ dr\, d\theta$

19. $\int_0^{\pi/4}\int_0^{\sec\theta}\int_0^{2-r\sin\theta} f(r,\theta,z)\ dz\ r\ dr\, d\theta$

20. $\int_{\pi/4}^{\pi/2}\int_0^{\csc\theta}\int_0^{2-r\sin\theta} f(r,\theta,z)\ dz\ r\ dr\, d\theta$

21. $\int_0^{\pi}\int_0^{\pi}\int_0^{2\sin\phi} \rho^2 \sin\phi \, d\rho \, d\phi \, d\theta = \frac{8}{3}\int_0^{\pi}\int_0^{\pi} \sin^4\phi \, d\phi \, d\theta = \frac{8}{3}\int_0^{\pi}\left(\left[-\frac{\sin^3\phi\cos\phi}{4}\right]_0^{\pi} + \frac{3}{4}\int_0^{\pi} \sin^2\phi \, d\phi\right) d\theta$

$= 2\int_0^{\pi}\int_0^{\pi} \sin^2\phi \, d\phi \, d\theta = \int_0^{\pi}\left[\theta - \frac{\sin 2\theta}{2}\right]_0^{\pi} d\theta = \int_0^{\pi} \pi \, d\theta = \pi^2$

22. $\int_0^{2\pi}\int_0^{\pi/4}\int_0^{2} (\rho\cos\phi)\,\rho^2 \sin\phi \, d\rho \, d\phi \, d\theta = \int_0^{2\pi}\int_0^{\pi/4} 4\cos\phi\sin\phi \, d\phi \, d\theta = \int_0^{2\pi}\left[2\sin^2\phi\right]_0^{\pi/4} d\theta = \int_0^{2\pi} d\theta = 2\pi$

23. $\int_0^{2\pi}\int_0^{\pi}\int_0^{(1-\cos\phi)/2} \rho^2 \sin\phi \, d\rho \, d\phi \, d\theta = \frac{1}{24}\int_0^{2\pi}\int_0^{\pi} (1-\cos\phi)^3 \sin\phi \, d\phi \, d\theta = \frac{1}{96}\int_0^{2\pi}\left[(1-\cos\phi)^4\right]_0^{\pi} d\theta$

$= \frac{1}{96}\int_0^{2\pi} (2^4 - 0) \, d\theta = \frac{16}{96}\int_0^{2\pi} d\theta = \frac{1}{6}(2\pi) = \frac{\pi}{3}$

24. $\int_0^{3\pi/2}\int_0^{\pi}\int_0^{1} 5\rho^3 \sin^3\phi \, d\rho \, d\phi \, d\theta = \frac{5}{4}\int_0^{3\pi/2}\int_0^{\pi} \sin^3\phi \, d\phi \, d\theta = \frac{5}{4}\int_0^{3\pi/2}\left(\left[-\frac{\sin^2\phi\cos\phi}{3}\right]_0^{\pi} + \frac{2}{3}\int_0^{\pi} \sin\phi \, d\phi\right) d\theta$

$= \frac{5}{6}\int_0^{3\pi/2}\left[-\cos\phi\right]_0^{\pi} d\theta = \frac{5}{3}\int_0^{3\pi/2} d\theta = \frac{5\pi}{2}$

25. $\int_0^{2\pi}\int_0^{\pi/3}\int_{\sec\phi}^{2} 3\rho^2 \sin\phi \, d\rho \, d\phi \, d\theta = \int_0^{2\pi}\int_0^{\pi/3} (8 - \sec^3\phi)\sin\phi \, d\phi \, d\theta = \int_0^{2\pi}\left[-8\cos\phi - \frac{1}{2}\sec^2\phi\right]_0^{\pi/3} d\theta$

$= \int_0^{2\pi}\left[(-4-2) - \left(-8 - \frac{1}{2}\right)\right] d\theta = \frac{5}{2}\int_0^{2\pi} d\theta = 5\pi$

26. $\int_0^{2\pi}\int_0^{\pi/4}\int_0^{\sec\phi} \rho^3 \sin\phi\cos\phi \, d\rho \, d\phi \, d\theta = \frac{1}{4}\int_0^{2\pi}\int_0^{\pi/4} \tan\phi\sec^2\phi \, d\phi \, d\theta = \frac{1}{4}\int_0^{2\pi}\left[\frac{1}{2}\tan^2\phi\right]_0^{\pi/4} d\theta$

$= \frac{1}{8}\int_0^{2\pi} d\theta = \frac{\pi}{4}$

27. $\int_0^{2}\int_{-\pi}^{0}\int_{\pi/4}^{\pi/2} \rho^3 \sin 2\phi \, d\phi \, d\theta \, d\rho = \int_0^{2}\int_{-\pi}^{0} \rho^3\left[-\frac{\cos 2\phi}{2}\right]_{\pi/4}^{\pi/2} d\theta \, d\rho = \int_0^{2}\int_{-\pi}^{0} \frac{\rho^3}{2} \, d\theta \, d\rho = \int_0^{2} \frac{\rho^3\pi}{2} \, d\rho$

$= \left[\frac{\pi\rho^4}{8}\right]_0^2 = 2\pi$

28. $\int_{\pi/6}^{\pi/3}\int_{\csc\phi}^{2\csc\phi}\int_0^{2\pi}\rho^2\sin\phi\,d\theta\,d\rho\,d\phi = 2\pi\int_{\pi/6}^{\pi/3}\int_{\csc\phi}^{2\csc\phi}\rho^2\sin\phi\,d\rho\,d\phi = \frac{2\pi}{3}\int_{\pi/6}^{\pi/3}\left[\rho^3\sin\phi\right]_{\csc\phi}^{2\csc\phi}\,d\phi$

$= \frac{14\pi}{3}\int_{\pi/6}^{\pi/3}\csc^2\phi\,d\phi = \frac{28\pi}{3\sqrt{3}}$

29. $\int_0^1\int_0^{\pi}\int_0^{\pi/4} 12\rho\sin^3\phi\,d\phi\,d\theta\,d\rho = \int_0^1\int_0^{\pi}\left(12\rho\left[\frac{-\sin^2\phi\cos\phi}{3}\right]_0^{\pi/4} + 8\rho\int_0^{\pi/4}\sin\phi\,d\phi\right)d\theta\,d\rho$

$= \int_0^1\int_0^{\pi}\left(-\frac{2\rho}{\sqrt{2}} - 8\rho\left[\cos\phi\right]_0^{\pi/4}\right)d\theta\,d\rho = \int_0^1\int_0^{\pi}\left(8\rho - \frac{10\rho}{\sqrt{2}}\right)d\theta\,d\rho = \pi\int_0^1\left(8\rho - \frac{10\rho}{\sqrt{2}}\right)d\rho = \pi\left[4\rho^2 - \frac{5\rho^2}{\sqrt{2}}\right]_0^1$

$= \frac{(4\sqrt{2}-5)\,\pi}{\sqrt{2}}$

30. $\int_{\pi/6}^{\pi/2}\int_{-\pi/2}^{\pi/2}\int_{\csc\phi}^{2} 5\rho^4\sin^3\phi\,d\rho\,d\theta\,d\phi = \int_{\pi/6}^{\pi/2}\int_{-\pi/2}^{\pi/2}(32-\csc^5\phi)\sin^3\phi\,d\theta\,d\phi = \int_{\pi/6}^{\pi/2}\int_{-\pi/2}^{\pi/2}(32\sin^3\phi - \csc^2\phi)\,d\theta\,d\phi$

$= \pi\int_{\pi/6}^{\pi/2}(32\sin^3\phi - \csc^2\phi)\,d\phi = \pi\left[-\frac{32\sin^2\phi\cos\phi}{3}\right]_{\pi/6}^{\pi/2} + \frac{64\pi}{3}\int_{\pi/6}^{\pi/2}\sin\phi\,d\phi + \pi\left[\cot\phi\right]_{\pi/6}^{\pi/2}$

$= \pi\left(\frac{32\sqrt{3}}{24}\right) - \frac{64\pi}{3}\left[\cos\phi\right]_{\pi/6}^{\pi/2} + \pi\left(\sqrt{3}\right) = \frac{\sqrt{3}}{3}\pi + \left(\frac{64\pi}{3}\right)\left(\frac{\sqrt{3}}{2}\right) = \frac{33\pi\sqrt{3}}{3} = 11\pi\sqrt{3}$

31. (a) $x^2+y^2=1 \Rightarrow \rho^2\sin^2\phi = 1$, and $\rho\sin\phi = 1 \Rightarrow \rho = \csc\phi$; thus

$$\int_0^{2\pi}\int_0^{\pi/6}\int_0^{2}\rho^2\sin\phi\,d\rho\,d\phi\,d\theta + \int_0^{2\pi}\int_{\pi/6}^{\pi/2}\int_0^{\csc\phi}\rho^2\sin\phi\,d\rho\,d\phi\,d\theta$$

(b) $\int_0^{2\pi}\int_1^{2}\int_{\pi/6}^{\sin^{-1}(1/\rho)}\rho^2\sin\phi\,d\phi\,d\rho\,d\theta + \int_0^{2\pi}\int_0^{2}\int_0^{\pi/6}\rho^2\sin\phi\,d\phi\,d\rho\,d\theta$

32. (a) $\int_0^{2\pi}\int_0^{\pi/4}\int_0^{\sec\phi}\rho^2\sin\phi\,d\rho\,d\phi\,d\theta$

(b) $\int_0^{2\pi}\int_0^1\int_0^{\pi/4} \rho^2 \sin\phi \, d\phi \, d\rho \, d\theta$

$+\int_0^{2\pi}\int_1^{\sqrt{2}}\int_{\cos^{-1}(1/\rho)}^{\pi/4} \rho^2 \sin\phi \, d\phi \, d\rho \, d\theta$

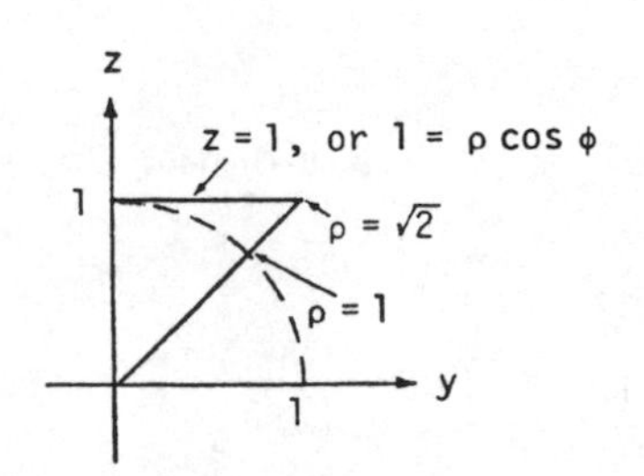

33. $V = \int_0^{2\pi}\int_0^{\pi/2}\int_{\cos\phi}^{2} \rho^2 \sin\phi \, d\rho \, d\phi \, d\theta = \frac{1}{3}\int_0^{2\pi}\int_0^{\pi/2} \left(8 - \cos^3\phi\right) \sin\phi \, d\phi \, d\theta$

$= \frac{1}{3}\int_0^{2\pi} \left[-8\cos\phi + \frac{\cos^4\phi}{4}\right]_0^{\pi/2} d\theta = \frac{1}{3}\int_0^{2\pi} \left(8 - \frac{1}{4}\right) d\theta = \left(\frac{31}{12}\right)(2\pi) = \frac{31\pi}{6}$

34. $V = \int_0^{2\pi}\int_0^{\pi/2}\int_1^{1+\cos\phi} \rho^2 \sin\phi \, d\rho \, d\phi \, d\theta = \frac{1}{3}\int_0^{2\pi}\int_0^{\pi/2} \left(3\cos\phi + 3\cos^2\phi + \cos^3\phi\right) \sin\phi \, d\phi \, d\theta$

$= \frac{1}{3}\int_0^{2\pi} \left[-\frac{3}{2}\cos^2\phi - \cos^3\phi - \frac{1}{4}\cos^4\phi\right]_0^{\pi/2} d\theta = \frac{1}{3}\int_0^{2\pi} \left(\frac{3}{2} + 1 + \frac{1}{4}\right) d\theta = \frac{11}{12}\int_0^{2\pi} d\theta = \left(\frac{11}{12}\right)(2\pi) = \frac{11\pi}{6}$

35. $V = \int_0^{2\pi}\int_0^{\pi}\int_0^{1-\cos\phi} \rho^2 \sin\phi \, d\rho \, d\phi \, d\theta = \frac{1}{3}\int_0^{2\pi}\int_0^{\pi} (1-\cos\phi)^3 \sin\phi \, d\phi \, d\theta = \frac{1}{3}\int_0^{2\pi} \left[\frac{(1-\cos\phi)^4}{4}\right]_0^{\pi} d\theta$

$= \frac{1}{12}(2)^4 \int_0^{2\pi} d\theta = \frac{4}{3}(2\pi) = \frac{8\pi}{3}$

36. $V = \int_0^{2\pi}\int_0^{\pi/2}\int_0^{1-\cos\phi} \rho^2 \sin\phi \, d\rho \, d\phi \, d\theta = \frac{1}{3}\int_0^{2\pi}\int_0^{\pi/2} (1-\cos\phi)^3 \sin\phi \, d\phi \, d\theta = \frac{1}{3}\int_0^{2\pi} \left[\frac{(1-\cos\phi)^4}{4}\right]_0^{\pi/2} d\theta$

$= \frac{1}{12}\int_0^{2\pi} d\theta = \frac{1}{12}(2\pi) = \frac{\pi}{6}$

37. $V = \int_0^{2\pi}\int_{\pi/4}^{\pi/2}\int_0^{2\cos\phi} \rho^2 \sin\phi \, d\rho \, d\phi \, d\theta = \frac{8}{3}\int_0^{2\pi}\int_{\pi/4}^{\pi/2} \cos^3\phi \sin\phi \, d\phi \, d\theta = \frac{8}{3}\int_0^{2\pi} \left[-\frac{\cos^4\phi}{4}\right]_{\pi/4}^{\pi/2} d\theta$

$= \left(\frac{8}{3}\right)\left(\frac{1}{16}\right)\int_0^{2\pi} d\theta = \frac{1}{6}(2\pi) = \frac{\pi}{3}$

38. $V = \int_0^{2\pi}\int_{\pi/3}^{\pi/2}\int_0^{2} \rho^2 \sin\phi \, d\rho \, d\phi \, d\theta = \frac{8}{3}\int_0^{2\pi}\int_{\pi/3}^{\pi/2} \sin\phi \, d\phi \, d\theta = \frac{8}{3}\int_0^{2\pi} \left[-\cos\phi\right]_{\pi/3}^{\pi/2} d\theta = \frac{4}{3}\int_0^{2\pi} d\theta = \frac{8\pi}{3}$

39. (a) $8\int_0^{\pi/2}\int_0^{\pi/2}\int_0^2 \rho^2 \sin\phi \, d\rho\, d\phi\, d\theta$ (b) $8\int_0^{\pi/2}\int_0^2\int_0^{\sqrt{4-r^2}} dz\, r\, dr\, d\theta$

(c) $8\int_0^2\int_0^{\sqrt{4-x^2}}\int_0^{\sqrt{4-x^2-y^2}} dz\, dy\, dx$

40. (a) $\int_0^{\pi/2}\int_0^{3/\sqrt{2}}\int_r^{\sqrt{9-r^2}} dz\, r\, dr\, d\theta$ (b) $\int_0^{\pi/2}\int_0^{\pi/4}\int_0^3 \rho^2 \sin\phi\, d\rho\, d\phi\, d\theta$

(c) $\int_0^{\pi/2}\int_0^{\pi/4}\int_0^3 \rho^2 \sin\phi\, d\rho\, d\phi\, d\theta = 9\int_0^{\pi/2}\int_0^{\pi/4} \sin\phi\, d\phi\, d\theta = -9\int_0^{\pi/2}\left(\frac{1}{\sqrt{2}}-1\right) d\theta = \frac{9\pi\left(2-\sqrt{2}\right)}{4}$

41. (a) $V = \int_0^{2\pi}\int_0^{\pi/3}\int_{\sec\phi}^{2} \rho^2 \sin\phi\, d\rho\, d\phi\, d\theta$ (b) $V = \int_0^{2\pi}\int_0^{\sqrt{3}}\int_1^{\sqrt{4-r^2}} dz\, r\, dr\, d\theta$

(c) $V = \int_{-\sqrt{3}}^{\sqrt{3}}\int_{-\sqrt{3-x^2}}^{\sqrt{3-x^2}}\int_1^{\sqrt{4-x^2-y^2}} dz\, dy\, dx$

(d) $V = \int_0^{2\pi}\int_0^{\sqrt{3}}\left[r\left(4-r^2\right)^{1/2}-r\right] dr\, d\theta = \int_0^{2\pi}\left[-\frac{\left(4-r^2\right)^{3/2}}{3}-\frac{r^2}{2}\right]_0^{\sqrt{3}} d\theta = \int_0^{2\pi}\left(-\frac{1}{3}-\frac{3}{2}+\frac{4^{3/2}}{3}\right) d\theta$

$= \frac{5}{6}\int_0^{2\pi} d\theta = \frac{5\pi}{3}$

42. (a) $I_z = \int_0^{2\pi}\int_0^1\int_0^{\sqrt{1-r^2}} r^2\, dz\, r\, dr\, d\theta$

(b) $I_z = \int_0^{2\pi}\int_0^{\pi/2}\int_0^1 \left(\rho^2 \sin^2\phi\right)\left(\rho^2 \sin\phi\right) d\rho\, d\phi\, d\theta$, since $r^2 = x^2+y^2 = \rho^2 \sin^2\phi \cos^2\theta + \rho^2 \sin^2\phi \sin^2\theta$

$= \rho^2 \sin^2\phi$

(c) $I_z = \int_0^{2\pi}\int_0^{\pi/2} \frac{1}{5}\sin^3\phi\, d\phi\, d\theta = \frac{1}{5}\int_0^{2\pi}\left(\left[-\frac{\sin^2\phi\cos\phi}{3}\right]_0^{\pi/2} + \frac{2}{3}\int_0^{\pi/2} \sin\phi\, d\phi\right) d\theta = \frac{2}{15}\int_0^{2\pi}\left[-\cos\phi\right]_0^{\pi/2} d\theta$

$= \frac{2}{15}(2\pi) = \frac{4\pi}{15}$

43. $V = 4 \int_0^{\pi/2} \int_0^1 \int_{r^4-1}^{4-4r^2} dz\, r\, dr\, d\theta = 4 \int_0^{\pi/2} \int_0^1 \left(5r - 4r^3 - r^5\right) dr\, d\theta = 4 \int_0^{\pi/2} \left(\frac{5}{2} - 1 - \frac{1}{6}\right) d\theta$

$= 4 \int_0^{\pi/2} \frac{8}{6}\, d\theta = \frac{8\pi}{3}$

44. $V = 4 \int_0^{\pi/2} \int_0^1 \int_{-\sqrt{1-r^2}}^{1-r} dz\, r\, dr\, d\theta = 4 \int_0^{\pi/2} \int_0^1 \left(r - r^2 + r\sqrt{1-r^2}\right) dr\, d\theta = 4 \int_0^{\pi/2} \left[\frac{r^2}{2} - \frac{r^3}{3} - \frac{1}{3}\left(1 - r^2\right)^{3/2}\right]_0^1 d\theta$

$= 4 \int_0^{\pi/2} \left(\frac{1}{2} - \frac{1}{3} + \frac{1}{3}\right) d\theta = 2 \int_0^{\pi/2} d\theta = 2\left(\frac{\pi}{2}\right) = \pi$

45. $V = \int_{3\pi/2}^{2\pi} \int_0^{3\cos\theta} \int_0^{-r\sin\theta} dz\, r\, dr\, d\theta = \int_{3\pi/2}^{2\pi} \int_0^{3\cos\theta} -r^2 \sin\theta\, dr\, d\theta = \int_{3\pi/2}^{2\pi} \left(-9\cos^3\theta\right)(\sin\theta)\, d\theta$

$= \left[\frac{9}{4}\cos^4\theta\right]_{3\pi/2}^{2\pi} = \frac{9}{4} - 0 = \frac{9}{4}$

46. $V = 2 \int_{\pi/2}^{\pi} \int_0^{-3\cos\theta} \int_0^{r} dz\, r\, dr\, d\theta = 2 \int_{\pi/2}^{\pi} \int_0^{-3\cos\theta} r^2\, dr\, d\theta = \frac{2}{3} \int_{\pi/2}^{\pi} -27\cos^3\theta\, d\theta$

$= -18\left(\left[\frac{\cos^2\theta\sin\theta}{3}\right]_{\pi/2}^{\pi} + \frac{2}{3}\int_{\pi/2}^{\pi} \cos\theta\, d\theta\right) = -12[\sin\theta]_{\pi/2}^{\pi} = 12$

47. $V = \int_0^{\pi/2} \int_0^{\sin\theta} \int_0^{\sqrt{1-r^2}} dz\, r\, dr\, d\theta = \int_0^{\pi/2} \int_0^{\sin\theta} r\sqrt{1-r^2}\, dr\, d\theta = \int_0^{\pi/2} \left[-\frac{1}{3}\left(1 - r^2\right)^{3/2}\right]_0^{\sin\theta} d\theta$

$= -\frac{1}{3} \int_0^{\pi/2} \left[\left(1 - \sin^2\theta\right)^{3/2} - 1\right] d\theta = -\frac{1}{3} \int_0^{\pi/2} \left(\cos^3\theta - 1\right) d\theta = -\frac{1}{3}\left(\left[\frac{\cos^2\theta\sin\theta}{3}\right]_0^{\pi/2} + \frac{2}{3}\int_0^{\pi/2} \cos\theta\, d\theta\right) + \left[\frac{\theta}{3}\right]_0^{\pi/2}$

$= -\frac{2}{9}[\sin\theta]_0^{\pi/2} + \frac{\pi}{6} = \frac{-4 + 3\pi}{18}$

48. $V = \int_0^{\pi/2} \int_0^{\cos\theta} \int_0^{3\sqrt{1-r^2}} dz\, r\, dr\, d\theta = \int_0^{\pi/2} \int_0^{\cos\theta} 3r\sqrt{1-r^2}\, dr\, d\theta = \int_0^{\pi/2} \left[-\left(1 - r^2\right)^{3/2}\right]_0^{\cos\theta} d\theta$

$= \int_0^{\pi/2} \left[-\left(1 - \cos^2\theta\right)^{3/2} + 1\right] d\theta = \int_0^{\pi/2} \left(1 - \sin^3\theta\right) d\theta = \left[\theta + \frac{\sin^2\theta\cos\theta}{3}\right]_0^{\pi/2} - \frac{2}{3} \int_0^{\pi/2} \sin\theta\, d\theta$

$$= \frac{\pi}{2} + \frac{2}{3}[\cos\theta]_0^{\pi/2} = \frac{\pi}{2} - \frac{2}{3} = \frac{3\pi - 4}{6}$$

49. $V = \int_0^{2\pi}\int_{\pi/3}^{2\pi/3}\int_0^{a} \rho^2 \sin\phi \, d\rho\, d\phi\, d\theta = \int_0^{2\pi}\int_{\pi/3}^{2\pi/3} \frac{a^3}{3}\sin\phi\, d\phi\, d\theta = \frac{a^3}{3}\int_0^{2\pi}[-\cos\phi]_{\pi/3}^{2\pi/3}\, d\theta = \frac{a^3}{3}\int_0^{2\pi}\left(\frac{1}{2}+\frac{1}{2}\right)d\theta = \frac{2\pi a^3}{3}$

50. $V = \int_0^{\pi/6}\int_0^{\pi/2}\int_0^{a} \rho^2 \sin\phi\, d\rho\, d\phi\, d\theta = \frac{a^3}{3}\int_0^{\pi/6}\int_0^{\pi/2} \sin\phi\, d\phi\, d\theta = \frac{a^3}{3}\int_0^{\pi/6} d\theta = \frac{a^3\pi}{18}$

51. $V = \int_0^{2\pi}\int_0^{\pi/3}\int_{\sec\phi}^{2} \rho^2 \sin\phi\, d\rho\, d\phi\, d\theta$

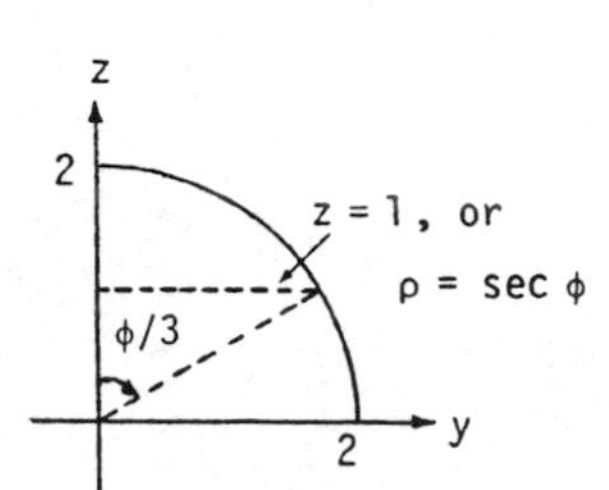

$$= \frac{1}{3}\int_0^{2\pi}\int_0^{\pi/3} \left(8\sin\phi - \tan\phi\sec^2\phi\right) d\phi\, d\theta$$

$$= \frac{1}{3}\int_0^{2\pi}\left[-8\cos\phi - \frac{1}{2}\tan^2\phi\right]_0^{\pi/3} d\theta$$

$$= \frac{1}{3}\int_0^{2\pi}\left[-4 - \frac{1}{2}(3) + 8\right] d\theta = \frac{1}{3}\int_0^{2\pi}\frac{5}{2}\, d\theta = \frac{5}{6}(2\pi) = \frac{5\pi}{3}$$

52. $V = 4\int_0^{\pi/2}\int_0^{\pi/4}\int_{\sec\phi}^{2\sec\phi} \rho^2\sin\phi\, d\rho\, d\phi\, d\theta = \frac{4}{3}\int_0^{\pi/2}\int_0^{\pi/4}\left(8\sec^3\phi - \sec^3\phi\right)\sin\phi\, d\phi\, d\theta$

$$= \frac{28}{3}\int_0^{\pi/2}\int_0^{\pi/4}\sec^3\phi\sin\phi\, d\phi\, d\theta = \frac{28}{3}\int_0^{\pi/2}\int_0^{\pi/4}\tan\phi\sec^2\phi\, d\phi\, d\theta = \frac{28}{3}\int_0^{\pi/2}\left[\frac{1}{2}\tan^2\phi\right]_0^{\pi/4} d\theta$$

$$= \frac{14}{3}\int_0^{\pi/2} d\theta = \frac{7\pi}{3}$$

53. $V = 4\int_0^{\pi/2}\int_0^{1}\int_0^{r^2} dz\, r\, dr\, d\theta = 4\int_0^{\pi/2}\int_0^{1} r^3\, dr\, d\theta = \int_0^{\pi/2} d\theta = \frac{\pi}{2}$

54. $V = 4\int_0^{\pi/2}\int_0^{1}\int_{r^2}^{r^2+1} dz\, r\, dr\, d\theta = 4\int_0^{\pi/2}\int_0^{1} r\, dr\, d\theta = 2\int_0^{\pi/2} d\theta = \pi$

55. $V = 8\int_0^{\pi/2}\int_1^{\sqrt{2}}\int_0^{r} dz\, r\, dr\, d\theta = 8\int_0^{\pi/2}\int_1^{\sqrt{2}} r^2\, dr\, d\theta = 8\left(\frac{2\sqrt{2}-1}{3}\right)\int_0^{\pi/2} d\theta = \frac{4\pi(2\sqrt{2}-1)}{3}$

56. $V = 8\int_0^{\pi/2}\int_1^{\sqrt{2}}\int_0^{\sqrt{2-r^2}} dz\ r\ dr\ d\theta = 8\int_0^{\pi/2}\int_1^{\sqrt{2}} r\sqrt{2-r^2}\ dr\ d\theta = 8\int_0^{\pi/2}\left[-\frac{1}{3}\left(2-r^2\right)^{3/2}\right]_1^{\sqrt{2}} d\theta$

$= \frac{8}{3}\int_0^{\pi/2} d\theta = \frac{4\pi}{3}$

57. $V = \int_0^{2\pi}\int_0^{2}\int_0^{4-r\sin\theta} dz\ r\ dr\ d\theta = \int_0^{2\pi}\int_0^{2}\left(4r - r^2\sin\theta\right) dr\ d\theta = 8\int_0^{2\pi}\left(1-\frac{\sin\theta}{3}\right) d\theta = 16\pi$

58. $V = \int_0^{2\pi}\int_0^{2}\int_0^{4-r\cos\theta-r\sin\theta} dz\ r\ dr\ d\theta = \int_0^{2\pi}\int_0^{2}\left[4r - r^2(\cos\theta+\sin\theta)\right] dr\ d\theta = \frac{8}{3}\int_0^{2\pi}(3-\cos\theta-\sin\theta)\ d\theta = 16\pi$

59. The paraboloids intersect when $4x^2+4y^2 = 5-x^2-y^2 \Rightarrow x^2+y^2 = 1$ and $z = 4$

$\Rightarrow V = 4\int_0^{\pi/2}\int_0^{1}\int_{4r^2}^{5-r^2} dz\ r\ dr\ d\theta = 4\int_0^{\pi/2}\int_0^{1}\left(5r-5r^3\right) dr\ d\theta = 20\int_0^{\pi/2}\left[\frac{r^2}{2}-\frac{r^4}{4}\right]_0^1 d\theta = 5\int_0^{\pi/2} d\theta = \frac{5\pi}{2}$

60. The paraboloid intersects the xy-plane when $9-x^2-y^2 = 0 \Rightarrow x^2+y^2 = 9 \Rightarrow$

$V = 4\int_0^{\pi/2}\int_1^{3}\int_0^{9-r^2} dz\ r\ dr\ d\theta = 4\int_0^{\pi/2}\int_1^{3}\left(9r-r^3\right) dr\ d\theta = 4\int_0^{\pi/2}\left[\frac{9r^2}{2}-\frac{r^4}{4}\right]_1^3 d\theta = 4\int_0^{\pi/2}\left(\frac{81}{4}-\frac{17}{4}\right) d\theta$

$= 64\int_0^{\pi/2} d\theta = 32\pi$

61. $V = 8\int_0^{\pi/2}\int_0^{1}\int_0^{\sqrt{4-r^2}} dz\ r\ dr\ d\theta = 8\int_0^{\pi/2}\int_0^{1} r\left(4-r^2\right)^{1/2} dr\ d\theta = 8\int_0^{\pi/2}\left[-\frac{1}{3}\left(4-r^2\right)^{3/2}\right]_0^1 d\theta$

$= -\frac{8}{3}\int_0^{\pi/2}\left(3^{3/2}-8\right) d\theta = \frac{4\pi\left(8-3\sqrt{3}\right)}{3}$

62. The sphere and paraboloid intersect when $x^2+y^2+z^2 = 2$ and $z = x^2+y^2 \Rightarrow z^2+z-2 = 0$ $\Rightarrow (z+2)(z-1) = 0 \Rightarrow z = 1$ or $z = -2 \Rightarrow z = 1$ since $z \ge 0$. Thus, $x^2+y^2 = 1$ and the volume is given by the triple integral $V = 4\int_0^{\pi/2}\int_0^{1}\int_{r^2}^{\sqrt{2-r^2}} dz\ r\ dr\ d\theta = 4\int_0^{\pi/2}\int_0^{1}\left[r\left(2-r^2\right)^{1/2} - r^3\right] dr\ d\theta$

$= 4\int_0^{\pi/2}\left[-\frac{1}{3}\left(2-r^2\right)^{3/2} - \frac{r^4}{4}\right]_0^1 d\theta = 4\int_0^{\pi/2}\left(\frac{2\sqrt{2}}{3}-\frac{7}{12}\right) d\theta = \frac{\pi\left(8\sqrt{2}-7\right)}{6}$

63. average $= \frac{1}{2\pi}\int_0^{2\pi}\int_0^1\int_{-1}^1 r^2\,dz\,dr\,d\theta = \frac{1}{2\pi}\int_0^{2\pi}\int_0^1 2r^2\,dr\,d\theta = \frac{1}{3\pi}\int_0^{2\pi} d\theta = \frac{2}{3}$

64. average $= \frac{1}{\left(\frac{4\pi}{3}\right)}\int_0^{2\pi}\int_0^1\int_{-\sqrt{1-r^2}}^{\sqrt{1-r^2}} r^2\,dz\,dr\,d\theta = \frac{3}{4\pi}\int_0^{2\pi}\int_0^1 2r^2\sqrt{1-r^2}\,dr\,d\theta$

$= \frac{3}{2\pi}\int_0^{2\pi}\left[\frac{1}{8}\sin^{-1}r - \frac{1}{8}r\sqrt{1-r^2}\left(1-r^2\right)\right]_0^1 d\theta = \frac{3}{16\pi}\int_0^{2\pi}\left(\frac{\pi}{2}+0\right)d\theta = \frac{3}{32}\int_0^{2\pi} d\theta = \left(\frac{3}{32}\right)(2\pi) = \frac{3\pi}{16}$

65. average $= \frac{1}{\left(\frac{4\pi}{3}\right)}\int_0^{2\pi}\int_0^{\pi}\int_0^1 \rho^3\sin\phi\,d\rho\,d\phi\,d\theta = \frac{3}{16\pi}\int_0^{2\pi}\int_0^{\pi}\sin\phi\,d\phi\,d\theta = \frac{3}{8\pi}\int_0^{2\pi} d\theta = \frac{3}{4}$

66. average $= \frac{1}{\left(\frac{2\pi}{3}\right)}\int_0^{2\pi}\int_0^{\pi/2}\int_0^1 \rho^3\cos\phi\sin\phi\,d\rho\,d\phi\,d\theta = \frac{3}{8\pi}\int_0^{2\pi}\int_0^{\pi/2}\cos\phi\sin\phi\,d\phi\,d\theta = \frac{3}{8\pi}\int_0^{2\pi}\left[\frac{\sin^2\phi}{2}\right]_0^{\pi/2} d\theta$

$= \frac{3}{16\pi}\int_0^{2\pi} d\theta = \left(\frac{3}{16\pi}\right)(2\pi) = \frac{3}{8}$

67. $M = 4\int_0^{\pi/2}\int_0^1\int_0^r dz\,r\,dr\,d\theta = 4\int_0^{\pi/2}\int_0^1 r^2\,dr\,d\theta = \frac{4}{3}\int_0^{\pi/2} d\theta = \frac{2\pi}{3}$; $M_{xy} = \int_0^{2\pi}\int_0^1\int_0^r z\,dz\,r\,dr\,d\theta$

$= \frac{1}{2}\int_0^{2\pi}\int_0^1 r^3\,dr\,d\theta = \frac{1}{8}\int_0^{2\pi} d\theta = \frac{\pi}{4} \Rightarrow \bar{z} = \frac{M_{xy}}{M} = \left(\frac{\pi}{4}\right)\left(\frac{3}{2\pi}\right) = \frac{3}{8}$, and $\bar{x} = \bar{y} = 0$, by symmetry

68. $M = \int_0^{\pi/2}\int_0^2\int_0^r dz\,r\,dr\,d\theta = \int_0^{\pi/2}\int_0^2 r^2\,dr\,d\theta = \frac{8}{3}\int_0^{\pi/2} d\theta = \frac{4\pi}{3}$; $M_{yz} = \int_0^{\pi/2}\int_0^2\int_0^r x\,dz\,r\,dr\,d\theta$

$= \int_0^{\pi/2}\int_0^2 r^3\cos\theta\,dr\,d\theta = 4\int_0^{\pi/2}\cos\theta\,d\theta = 4$; $M_{xz} = \int_0^{\pi/2}\int_0^2\int_0^r y\,dz\,r\,dr\,d\theta = \int_0^{\pi/2}\int_0^2 r^3\sin\theta\,dr\,d\theta$

$= 4\int_0^{\pi/2}\sin\theta\,d\theta = 4$; $M_{xy} = \int_0^{\pi/2}\int_0^2\int_0^r z\,dz\,r\,dr\,d\theta = \frac{1}{2}\int_0^{\pi/2}\int_0^2 r^3\,dr\,d\theta = 2\int_0^{\pi/2} d\theta = \pi \Rightarrow \bar{x} = \frac{M_{yz}}{M} = \frac{3}{\pi}$,

$\bar{y} = \frac{M_{xz}}{M} = \frac{3}{\pi}$, and $\bar{z} = \frac{M_{xy}}{M} = \frac{3}{4}$

69. $M = \frac{8\pi}{3}$; $M_{xy} = \int_0^{2\pi}\int_{\pi/3}^{\pi/2}\int_0^2 z\rho^2 \sin\phi \, d\rho \, d\phi \, d\theta = \int_0^{2\pi}\int_{\pi/3}^{\pi/2}\int_0^2 \rho^3 \cos\phi \sin\phi \, d\rho \, d\phi \, d\theta = 4\int_0^{2\pi}\int_{\pi/3}^{\pi/2} \cos\phi \sin\phi \, d\phi \, d\theta$

$= 4\int_0^{2\pi}\left[\frac{\sin^2\phi}{2}\right]_{\pi/3}^{\pi/2} d\theta = 4\int_0^{2\pi}\left(\frac{1}{2}-\frac{3}{8}\right) d\theta = \frac{1}{2}\int_0^{2\pi} d\theta = \pi \Rightarrow \bar{z} = \frac{M_{xy}}{M} = (\pi)\left(\frac{3}{8\pi}\right) = \frac{3}{8}$, and $\bar{x} = \bar{y} = 0$, by symmetry

70. $M = \int_0^{2\pi}\int_0^{\pi/4}\int_0^a \rho^2 \sin\phi \, d\rho \, d\phi \, d\theta = \frac{a^3}{3}\int_0^{2\pi}\int_0^{\pi/4} \sin\phi \, d\phi \, d\theta = \frac{a^3}{3}\int_0^{2\pi} \frac{2-\sqrt{2}}{2} \, d\theta = \frac{\pi a^3(2-\sqrt{2})}{3}$;

$M_{xy} = \int_0^{2\pi}\int_0^{\pi/4}\int_0^a \rho^3 \sin\phi \cos\phi \, d\rho \, d\phi \, d\theta = \frac{a^3}{4}\int_0^{2\pi}\int_0^{\pi/4} \sin\phi \cos\phi \, d\phi \, d\theta = \frac{a^4}{16}\int_0^{2\pi} d\theta = \frac{\pi a^4}{8}$

$\Rightarrow \bar{z} = \frac{M_{xy}}{M} = \left(\frac{\pi a^4}{8}\right)\left[\frac{3}{\pi a^3(2-\sqrt{2})}\right] = \left(\frac{3a}{8}\right)\left(\frac{2+\sqrt{2}}{2}\right) = \frac{3(2+\sqrt{2})a}{16}$, and $\bar{x} = \bar{y} = 0$, by symmetry

71. $M = \int_0^{2\pi}\int_0^4\int_0^{\sqrt{r}} dz \, r \, dr \, d\theta = \int_0^{2\pi}\int_0^4 r^{3/2} \, dr \, d\theta = \frac{64}{5}\int_0^{2\pi} d\theta = \frac{128\pi}{5}$; $M_{xy} = \int_0^{2\pi}\int_0^4\int_0^{\sqrt{r}} z \, dz \, r \, dr \, d\theta$

$= \frac{1}{2}\int_0^{2\pi}\int_0^4 r^2 \, dr \, d\theta = \frac{32}{3}\int_0^{2\pi} d\theta = \frac{64\pi}{3} \Rightarrow \bar{z} = \frac{M_{xy}}{M} = \frac{5}{6}$, and $\bar{x} = \bar{y} = 0$, by symmetry

72. $M = \int_{-\pi/3}^{\pi/3}\int_0^1\int_{-\sqrt{1-r^2}}^{\sqrt{1-r^2}} dz \, r \, dr \, d\theta = \int_{-\pi/3}^{\pi/3}\int_0^1 2r\sqrt{1-r^2} \, dr \, d\theta = \int_{-\pi/3}^{\pi/3}\left[-\frac{2}{3}(1-r^2)^{3/2}\right]_0^1 d\theta$

$= \frac{2}{3}\int_{-\pi/3}^{\pi/3} d\theta = \left(\frac{2}{3}\right)\left(\frac{2\pi}{3}\right) = \frac{4\pi}{9}$; $M_{yz} = \int_{-\pi/3}^{\pi/3}\int_0^1\int_{-\sqrt{1-r^2}}^{\sqrt{1-r^2}} r^2 \cos\theta \, dz \, dr \, d\theta = 2\int_{-\pi/3}^{\pi/3}\int_0^1 r^2\sqrt{1-r^2} \cos\theta \, dr \, d\theta$

$= 2\int_{-\pi/3}^{\pi/3}\left[\frac{1}{8}\sin^{-1} r - \frac{1}{8} r\sqrt{1-r^2}(1-2r^2)\right]_0^1 \cos\theta \, d\theta = \frac{\pi}{8}\int_{-\pi/3}^{\pi/3} \cos\theta \, d\theta = \frac{\pi}{8}[\sin\theta]_{-\pi/3}^{\pi/3} = \left(\frac{\pi}{8}\right)\left(2\cdot\frac{\sqrt{3}}{2}\right) = \frac{\pi\sqrt{3}}{8}$

$\Rightarrow \bar{x} = \frac{M_{yz}}{M} = \frac{9\sqrt{3}}{32}$, and $\bar{y} = \bar{z} = 0$, by symmetry

73. $I_z = \int_0^{2\pi}\int_1^2\int_0^4 (x^2+y^2)\,dz\,r\,dr\,d\theta = 4\int_0^{2\pi}\int_r^2 r^3\,dr\,d\theta = \int_0^{2\pi} 15\,d\theta = 30\pi$; $M = \int_0^{2\pi}\int_1^2\int_0^4 dz\,r\,dr\,d\theta$

$= \int_0^{2\pi}\int_1^2 4r\,dr\,d\theta = \int_0^{2\pi} 6\,d\theta = 12\pi \Rightarrow R_z = \sqrt{\frac{I_z}{M}} = \sqrt{\frac{5}{2}}$

74. (a) $I_z = \int_0^{2\pi}\int_0^1\int_{-1}^1 r^3\,dz\,dr\,d\theta = 2\int_0^{2\pi}\int_0^1 r^3\,dr\,d\theta = \frac{1}{2}\int_0^{2\pi} d\theta = \pi$

(b) $I_x = \int_0^{2\pi}\int_0^1\int_{-1}^1 (r^2\sin^2\theta + z^2)\,dz\,r\,dr\,d\theta = 2\int_0^{2\pi}\int_0^1 \left(2r^3\sin^2\theta + \frac{2r}{3}\right)dr\,d\theta = \int_0^{2\pi}\left(\frac{\sin^2\theta}{2} + \frac{1}{3}\right)d\theta$

$= \left[\frac{\theta}{4} - \frac{\sin 2\theta}{8} + \frac{\theta}{3}\right]_0^{2\pi} = \frac{\pi}{2} + \frac{2\pi}{3} = \frac{7\pi}{6}$

75. We orient the cone with its vertex at the origin and axis along the z-axis $\Rightarrow \phi = \frac{\pi}{4}$. We use the the x-axis which is through the vertex and parallel to the base of the cone $\Rightarrow I_x = \int_0^{2\pi}\int_0^1\int_r^1 (r^2\sin^2\theta + z^2)\,dz\,r\,dr\,d\theta$

$= \int_0^{2\pi}\int_0^1 \left(r^3\sin^2\theta - r^4\sin^2\theta + \frac{r}{3} - \frac{r^4}{4}\right)dr\,d\theta = \int_0^{2\pi}\left(\frac{\sin^2\theta}{20} + \frac{1}{10}\right)d\theta = \left[\frac{\theta}{40} - \frac{\sin 2\theta}{80} + \frac{\theta}{10}\right]_0^{2\pi} = \frac{\pi}{20} + \frac{\pi}{5} = \frac{\pi}{4}$

76. $I_z = \int_0^{2\pi}\int_0^a\int_{-\sqrt{a^2-r^2}}^{\sqrt{a^2-r^2}} r^3\,dz\,dr\,d\theta = \int_0^{2\pi}\int_0^a 2r^3\sqrt{a^2-r^2}\,dr\,d\theta = 2\int_0^{2\pi}\left[\left(-\frac{r^2}{5} - \frac{2a^2}{15}\right)(a^2-r^2)^{3/2}\right]_0^a d\theta$

$= 2\int_0^{2\pi} \frac{2}{15}a^5\,d\theta = \frac{8\pi a^5}{15}$

77. $I_z = \int_0^{2\pi}\int_0^a\int_{\left(\frac{h}{a}\right)r}^{h} (x^2+y^2)\,dz\,r\,dr\,d\theta = \int_0^{2\pi}\int_0^a\int_{\frac{hr}{a}}^{h} r^3\,dz\,dr\,d\theta = \int_0^{2\pi}\int_0^a \left(hr^3 - \frac{hr^4}{a}\right)dr\,d\theta$

$= \int_0^{2\pi} h\left[\frac{r^4}{4} - \frac{r^5}{5a}\right]_0^a d\theta = \int_0^{2\pi} h\left(\frac{a^4}{4} - \frac{a^5}{5a}\right)d\theta = \frac{ha^4}{20}\int_0^{2\pi} d\theta = \frac{\pi h a^4}{10}$

78. (a) $M = \int_0^{2\pi}\int_0^1\int_0^{r^2} z\,dz\,r\,dr\,d\theta = \int_0^{2\pi}\int_0^1 \frac{1}{2}r^5\,dr\,d\theta = \frac{1}{12}\int_0^{2\pi} d\theta = \frac{\pi}{6}$; $M_{xy} = \int_0^{2\pi}\int_0^1\int_0^{r^2} z^2\,dz\,r\,dr\,d\theta$

$= \frac{1}{3}\int_0^{2\pi}\int_0^1 r^7\,dr\,d\theta = \frac{1}{24}\int_0^{2\pi} d\theta = \frac{\pi}{12} \Rightarrow \bar{z} = \frac{1}{2}$, and $\bar{x} = \bar{y} = 0$, by symmetry;

$$I_z = \int_0^{2\pi}\int_0^1\int_0^{r^2} zr^3\,dz\,dr\,d\theta = \frac{1}{2}\int_0^{2\pi}\int_0^1 r^7\,dr\,d\theta = \frac{1}{16}\int_0^{2\pi} d\theta = \frac{\pi}{8} \Rightarrow R_z = \sqrt{\frac{I_z}{M}} = \frac{\sqrt{3}}{2}$$

(b) $M = \int_0^{2\pi}\int_0^1\int_0^{r^2} r^2\,dz\,dr\,d\theta = \int_0^{2\pi}\int_0^1 r^4\,dr\,d\theta = \frac{1}{5}\int_0^{2\pi} d\theta = \frac{2\pi}{5}$; $M_{xy} = \int_0^{2\pi}\int_0^1\int_0^{r^2} zr^2\,dz\,dr\,d\theta$

$= \frac{1}{2}\int_0^{2\pi}\int_0^1 r^6\,dr\,d\theta = \frac{1}{14}\int_0^{2\pi} d\theta = \frac{\pi}{7} \Rightarrow \bar{z} = \frac{5}{14}$, and $\bar{x} = \bar{y} = 0$, by symmetry; $I_z = \int_0^{2\pi}\int_0^1\int_0^{r^2} r^4\,dz\,dr\,d\theta$

$$= \int_0^{2\pi}\int_0^1 r^6\,dr\,d\theta = \frac{1}{7}\int_0^{2\pi} d\theta = \frac{2\pi}{7} \Rightarrow R_z = \sqrt{\frac{I_z}{M}} = \sqrt{\frac{5}{7}}$$

79. (a) $M = \int_0^{2\pi}\int_0^1\int_r^1 z\,dz\,r\,dr\,d\theta = \frac{1}{2}\int_0^{2\pi}\int_0^1 (r - r^3)\,dr\,d\theta = \frac{1}{8}\int_0^{2\pi} d\theta = \frac{\pi}{4}$; $M_{xy} = \int_0^{2\pi}\int_0^1\int_r^1 z^2\,dz\,r\,dr\,d\theta$

$= \frac{1}{3}\int_0^{2\pi}\int_0^1 (r - r^4)\,dr\,d\theta = \frac{1}{10}\int_0^{2\pi} d\theta = \frac{\pi}{5} \Rightarrow \bar{z} = \frac{4}{5}$, and $\bar{x} = \bar{y} = 0$, by symmetry; $I_z = \int_0^{2\pi}\int_0^1\int_r^1 zr^3\,dz\,dr\,d\theta$

$$= \frac{1}{2}\int_0^{2\pi}\int_0^1 (r^3 - r^5)\,dr\,d\theta = \frac{1}{24}\int_0^{2\pi} d\theta = \frac{\pi}{12} \Rightarrow R_z = \sqrt{\frac{I_z}{M}} = \sqrt{\frac{1}{3}}$$

(b) $M = \int_0^{2\pi}\int_0^1\int_r^1 z^2\,dz\,r\,dr\,d\theta = \frac{\pi}{5}$ from part (a); $M_{xy} = \int_0^{2\pi}\int_0^1\int_r^1 z^3\,dz\,r\,dr\,d\theta = \frac{1}{4}\int_0^{2\pi}\int_0^1 (r - r^5)\,dr\,d\theta$

$= \frac{1}{12}\int_0^{2\pi} d\theta = \frac{\pi}{6} \Rightarrow \bar{z} = \frac{5}{6}$, and $\bar{x} = \bar{y} = 0$, by symmetry; $I_z = \int_0^{2\pi}\int_0^1\int_r^1 z^2r^3\,dz\,dr\,d\theta = \frac{1}{3}\int_0^{2\pi}\int_0^1 (r^3 - r^6)\,dr\,d\theta$

$$= \frac{1}{28}\int_0^{2\pi} d\theta = \frac{\pi}{14} \Rightarrow R_z = \sqrt{\frac{I_z}{M}} = \sqrt{\frac{5}{14}}$$

80. (a) $M = \int_0^{2\pi}\int_0^{\pi}\int_0^a \rho^4 \sin\phi\,d\rho\,d\phi\,d\theta = \frac{a^5}{5}\int_0^{2\pi}\int_0^{\pi} \sin\phi\,d\phi\,d\theta = \frac{2a^5}{5}\int_0^{2\pi} d\theta = \frac{4\pi a^5}{5}$;

$I_z = \int_0^{2\pi}\int_0^{\pi}\int_0^a \rho^6 \sin^3\phi\,d\rho\,d\phi\,d\theta = \frac{a^7}{7}\int_0^{2\pi}\int_0^{\pi} (1 - \cos^2\phi)\sin\phi\,d\phi\,d\theta = \frac{a^7}{7}\int_0^{2\pi} \left[-\cos\phi + \frac{\cos^3\phi}{3}\right]_0^{\pi} d\theta$

$$= \frac{4a^7}{21}\int_0^{2\pi} d\theta = \frac{8a^7\pi}{21} \Rightarrow R_z = \sqrt{\frac{I_z}{M}} = \sqrt{\frac{10}{21}}\,a$$

(b) $M = \int_0^{2\pi}\int_0^{\pi}\int_0^{a} \rho^3 \sin^2\phi \, d\rho \, d\phi \, d\theta = \frac{a^4}{4}\int_0^{2\pi}\int_0^{\pi} \frac{(1-\cos 2\phi)}{2} \, d\phi \, d\theta = \frac{\pi a^4}{8}\int_0^{2\pi} d\theta = \frac{\pi^2 a^4}{4}$;

$$I_z = \int_0^{2\pi}\int_0^{\pi}\int_0^{a} \rho^5 \sin^4\phi \, d\rho \, d\phi \, d\theta = \frac{a^6}{6}\int_0^{2\pi}\int_0^{\pi} \sin^4\phi \, d\phi \, d\theta$$

$$= \frac{a^6}{6}\int_0^{2\pi}\left(\left[\frac{-\sin^3\phi\cos\phi}{4}\right]_0^{\pi} + \frac{3}{4}\int_0^{\pi}\sin^2\phi \, d\phi\right) d\theta = \frac{a^6}{8}\int_0^{2\pi}\left[\frac{\phi}{2} - \frac{\sin 2\phi}{4}\right]_0^{\pi} d\theta = \frac{\pi a^6}{16}\int_0^{2\pi} d\theta$$

$$= \frac{a^6\pi^2}{8} \Rightarrow R_z = \sqrt{\frac{I_z}{M}} = \frac{a}{\sqrt{2}}$$

81. $M = \int_0^{2\pi}\int_0^{a}\int_0^{\frac{h}{a}\sqrt{a^2-r^2}} dz \, r \, dr \, d\theta = \int_0^{2\pi}\int_0^{a} \frac{h}{a} r\sqrt{a^2-r^2} \, dr \, d\theta = \frac{h}{a}\int_0^{2\pi}\left[-\frac{1}{3}\left(a^2-r^2\right)^{3/2}\right]_0^{a} d\theta$

$$= \frac{h}{a}\int_0^{2\pi}\frac{a^3}{3} \, d\theta = \frac{2ha^2\pi}{3};\ M_{xy} = \int_0^{2\pi}\int_0^{a}\int_0^{\frac{h}{a}\sqrt{a^2-r^2}} z \, dz \, r \, dr \, d\theta = \frac{h^2}{2a^2}\int_0^{2\pi}\left(a^2 r - r^3\right) dr \, d\theta$$

$$= \frac{h^2}{2a^2}\int_0^{2\pi}\left(\frac{a^4}{2} - \frac{a^4}{4}\right) d\theta = \frac{a^2h^2\pi}{4} \Rightarrow \bar{z} = \left(\frac{\pi a^2 h^2}{4}\right)\left(\frac{3}{2ha^2\pi}\right) = \frac{3}{8}h, \text{ and } \bar{x} = \bar{y} = 0, \text{ by symmetry}$$

82. Let the base radius of the cone be a and the height h, and place the cone's axis of symmetry along the z-axis with the vertex at the origin. Then $M = \frac{\pi r^2 h}{3}$ and $M_{xy} = \int_0^{2\pi}\int_0^{a}\int_{\left(\frac{h}{a}\right)r}^{h} z \, dz \, r \, dr \, d\theta = \frac{1}{2}\int_0^{2\pi}\int_0^{a}\left(h^2 r - \frac{h^2}{a^2} r^3\right) dr \, d\theta$

$$= \frac{h^2}{2}\int_0^{2\pi}\left[\frac{r^2}{2} - \frac{r^4}{4a^2}\right]_0^{a} d\theta = \frac{h^2}{2}\int_0^{2\pi}\left(\frac{a^2}{2} - \frac{a^2}{4}\right) d\theta = \frac{h^2a^2}{8}\int_0^{2\pi} d\theta = \frac{h^2a^2\pi}{4} \Rightarrow \bar{z} = \frac{M_{xy}}{M} = \left(\frac{h^2a^2\pi}{4}\right)\left(\frac{3}{\pi a^2 h}\right) = \frac{3}{4}h, \text{ and}$$

$\bar{x} = \bar{y} = 0$, by symmetry $\Rightarrow$ the centroid is one fourth of the way from the base to the vertex

83. $M = \int_0^{2\pi}\int_0^{a}\int_0^{h} (z+1) \, dz \, r \, dr \, d\theta = \int_0^{2\pi}\int_0^{a}\left(\frac{h^2}{2} + h\right) r \, dr \, d\theta = \frac{a^2\left(h^2+2h\right)}{4}\int_0^{2\pi} d\theta = \frac{\pi a^2\left(h^2+2h\right)}{2}$;

$$M_{xy} = \int_0^{2\pi}\int_0^{a}\int_0^{h} \left(z^2+z\right) dz \, r \, dr \, d\theta = \int_0^{2\pi}\int_0^{a}\left(\frac{h^3}{3} + \frac{h^2}{2}\right) r \, dr \, d\theta = \frac{a^2\left(2h^3+3h^2\right)}{12}\int_0^{2\pi} d\theta = \frac{\pi a^2\left(2h^3+3h^2\right)}{6}$$

$$\Rightarrow \bar{z} = \left[\frac{\pi a^2\left(2h^3+3h^2\right)}{6}\right]\left[\frac{2}{\pi a^2\left(h^2+2h\right)}\right] = \frac{2h^2+3h}{3h+6}, \text{ and } \bar{x} = \bar{y} = 0, \text{ by symmetry;}$$

$$I_z = \int_0^{2\pi}\int_0^a\int_0^h (z+1)r^3\, dz\, dr\, d\theta = \left(\frac{h^2+2h}{2}\right)\int_0^{2\pi}\int_0^a r^3\, dr\, d\theta = \left(\frac{h^2+2h}{2}\right)\left(\frac{a^4}{4}\right)\int_0^{2\pi} d\theta = \frac{\pi a^4(h^2+2h)}{4};$$

$$R_z = \sqrt{\frac{I_z}{M}} = \sqrt{\frac{\pi a^4(h^2+2h)}{4}\cdot\frac{2}{\pi a^2(h^2+2h)}} = \frac{a}{\sqrt{2}}$$

84. The mass of the plant's atmosphere to an altitude h above the surface of the planet is the triple integral

$$M(h) = \int_0^{2\pi}\int_0^{\pi}\int_R^h \mu_0 e^{-c(\rho-R)}\rho^2 \sin\phi\, d\rho\, d\phi\, d\theta = \int_R^h\int_0^{2\pi}\int_0^{\pi} \mu_0 e^{-c(\rho-R)}\rho^2 \sin\phi\, d\phi\, d\theta\, d\rho$$

$$= \int_R^h\int_0^{2\pi}\left[\mu_0 e^{-c(\rho-R)}\rho^2(-\cos\phi)\right]_0^{\pi} d\theta\, d\rho = 2\int_R^h\int_0^{2\pi} \mu_0 e^{cR} e^{-c\rho}\rho^2\, d\theta\, d\rho = 4\pi\mu_0 e^{cR}\int_R^h e^{-c\rho}\rho^2\, d\rho$$

$$= 4\pi\mu_0 e^{cR}\left[-\frac{\rho^2 e^{-c\rho}}{c} - \frac{2\rho e^{-c\rho}}{c^2} - \frac{2e^{-c\rho}}{c^3}\right]_R^h \quad \text{(by parts)}$$

$$= 4\pi\mu_0 e^{cR}\left(-\frac{h^2 e^{-ch}}{c} - \frac{2he^{-ch}}{c^2} - \frac{2e^{-ch}}{c^3} + \frac{R^2 e^{-cR}}{c} + \frac{2Re^{-cR}}{c^2} + \frac{2e^{-cR}}{c^3}\right).$$

The mass of the planet's atmosphere is therefore $M = \lim_{h\to\infty} M(h) = 4\pi\mu_0\left(\frac{R^2}{c} + \frac{2R}{c^2} + \frac{2}{c^3}\right)$.

85. The density distribution function is linear so it has the form $\delta(\rho) = k\rho + C$, where ρ is the distance from the center of the planet. Now, $\delta(R) = 0 \Rightarrow kR + C = 0$, and $\delta(\rho) = k\rho - kR$. It remains to determine the constant k: $M = \int_0^{2\pi}\int_0^{\pi}\int_0^R (k\rho - kR)\,\rho^2 \sin\phi\, d\rho\, d\phi\, d\theta = \int_0^{2\pi}\int_0^{\pi}\left[k\frac{\rho^4}{4} - kR\frac{\rho^3}{3}\right]_0^R \sin\phi\, d\phi\, d\theta$

$$= \int_0^{2\pi}\int_0^{\pi} k\left(\frac{R^4}{4} - \frac{R^4}{3}\right)\sin\phi\, d\phi\, d\theta = \int_0^{2\pi} -\frac{k}{12}R^4\left[-\cos\phi\right]_0^{\pi} d\theta = \int_0^{2\pi} -\frac{k}{6}R^4\, d\theta = -\frac{k\pi R^4}{3} \Rightarrow k = -\frac{3M}{\pi R^4}$$

$\Rightarrow \delta(\rho) = -\frac{3M}{\pi R^4}\rho + \frac{3M}{\pi R^4}R$. At the center of the planet $\rho = 0 \Rightarrow \delta(0) = \left(\frac{3M}{\pi R^4}\right)R = \frac{3M}{\pi R^3}$.

13.7 SUBSTITUTIONS IN MULTIPLE INTEGRALS

1. (a) $x - y = u$ and $2x + y = v \Rightarrow 3x = u + v$ and $y = x - u \Rightarrow x = \frac{1}{3}(u+v)$ and $y = \frac{1}{3}(-2u+v)$;

$$\frac{\partial(x,y)}{\partial(u,v)} = \begin{vmatrix} \frac{1}{3} & \frac{1}{3} \\ -\frac{2}{3} & \frac{1}{3} \end{vmatrix} = \frac{1}{9} + \frac{2}{9} = \frac{1}{3}$$

(b) The line segment $y = x$ from $(0,0)$ to $(1,1)$ is $x - y = 0$ $\Rightarrow u = 0$; the line segment $y = -2x$ from $(0,0)$ to $(1,-2)$ is $2x + y = 0 \Rightarrow v = 0$; the line segment $x = 1$ from $(1,1)$ to $(1,-2)$ is $(x - y) + (2x + y) = 3 \Rightarrow u + v = 3$. The transformed region is sketched at the right.

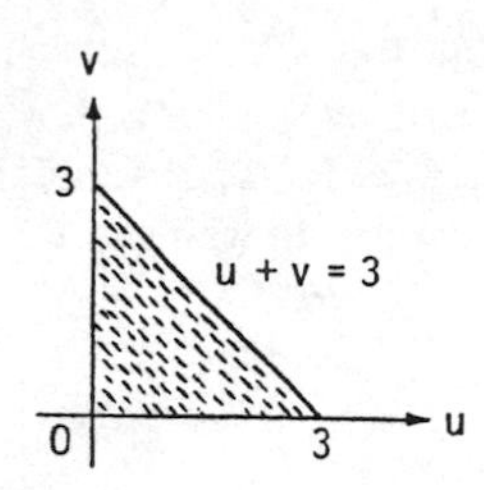

2. (a) $x + 2y = u$ and $x - y = v \Rightarrow 3y = u - v$ and $x = v + y \Rightarrow y = \frac{1}{3}(u - v)$ and $x = \frac{1}{3}(u + 2v)$;

$$\frac{\partial(x,y)}{\partial(u,v)} = \begin{vmatrix} \frac{1}{3} & \frac{2}{3} \\ \frac{1}{3} & -\frac{1}{3} \end{vmatrix} = -\frac{1}{9} - \frac{2}{9} = -\frac{1}{3}$$

(b) The triangular region in the xy-plane has vertices $(0,0)$, $(2,0)$, and $\left(\frac{2}{3}, \frac{2}{3}\right)$. The line segment $y = x$ from $(0,0)$ to $\left(\frac{2}{3}, \frac{2}{3}\right)$ is $x - y = 0 \Rightarrow v = 0$; the line segment $y = 0$ from $(0,0)$ to $(2,0) \Rightarrow u = v$; the line segment $x + 2y = 2$ from $\left(\frac{2}{3}, \frac{2}{3}\right)$ to $(2,0) \Rightarrow u = 2$. The transformed region is sketched at the right.

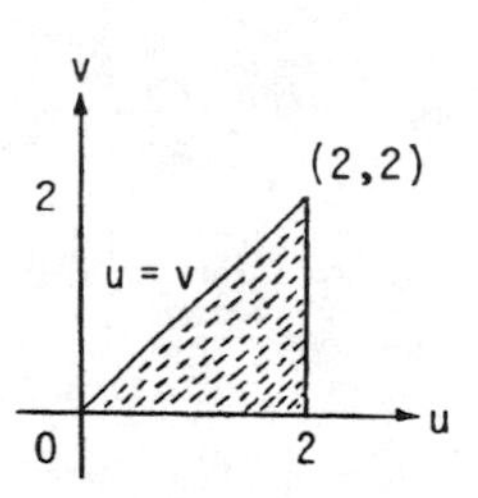

3. (a) $3x + 2y = u$ and $x + 4y = v \Rightarrow -5x = -2u + v$ and $y = \frac{1}{2}(u - 3x) \Rightarrow x = \frac{1}{5}(2u - v)$ and $y = \frac{1}{10}(3v - u)$;

$$\frac{\partial(x,y)}{\partial(u,v)} = \begin{vmatrix} \frac{2}{5} & -\frac{1}{5} \\ -\frac{1}{10} & \frac{3}{10} \end{vmatrix} = \frac{6}{50} - \frac{1}{50} = \frac{1}{10}$$

(b) The x-axis $y = 0 \Rightarrow u = 3v$; the y-axis $x = 0 \Rightarrow v = 2u$; the line $x + y = 1 \Rightarrow \frac{1}{5}(2u - v) + \frac{1}{10}(3v - u) = 1$ $\Rightarrow 2(2u - v) + (3v - u) = 10 \Rightarrow 3u + v = 10$. The transformed region is sketched at the right.

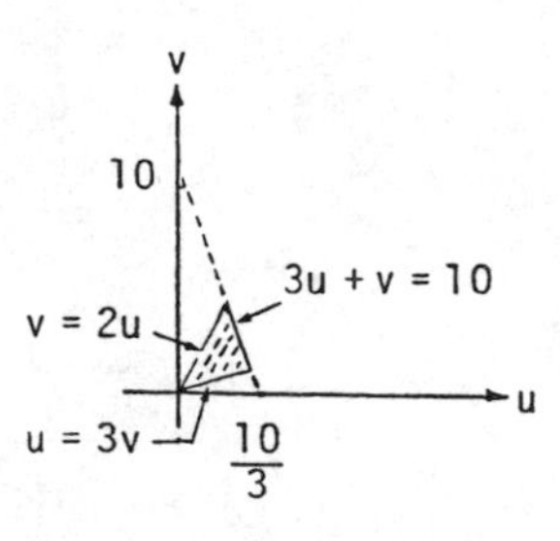

4. (a) $2x - 3y = u$ and $-x + y = v \Rightarrow -x = u + 3v$ and $y = v + x \Rightarrow x = -u - 3v$ and $y = -u - 2v$;

$$\frac{\partial(x,y)}{\partial(u,v)} = \begin{vmatrix} -1 & -3 \\ -1 & -2 \end{vmatrix} = 2 - 3 = -1$$

(b) The line $x = -3 \Rightarrow -u - 3v = -3$ or $u + 3v = 3$; $x = 0 \Rightarrow u + 3v = 0$; $y = x \Rightarrow v = 0$; $y = x + 1$ $\Rightarrow v = 1$. The transformed region is the parallelogram sketched at the right.

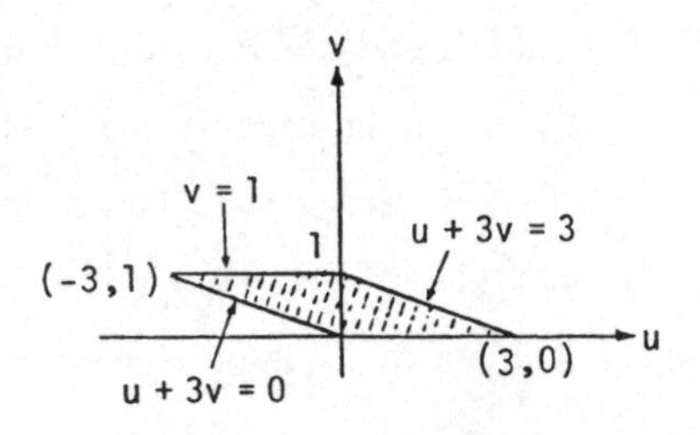

5. (a) $x = u \cos v$ and $y = u \sin v \Rightarrow \frac{\partial(x,y)}{\partial(u,v)} = \begin{vmatrix} \cos v & -u \sin v \\ \sin v & u \cos v \end{vmatrix} = u \cos^2 v + u \sin^2 v = u$

(b) $x = u \sin v$ and $y = u \cos v \Rightarrow \frac{\partial(x,y)}{\partial(u,v)} = \begin{vmatrix} \sin v & u \cos v \\ \cos v & -u \sin v \end{vmatrix} = -u \sin^2 v - u \cos^2 v = -u$

6. (a) $x = u \cos v$, $y = u \sin v$, $z = w \Rightarrow \frac{\partial(x,y,z)}{\partial(u,v,w)} = \begin{vmatrix} \cos v & -u \sin v & 0 \\ \sin v & u \cos v & 0 \\ 0 & 0 & 1 \end{vmatrix} = u \cos^2 v + u \sin^2 v = u$

(b) $x = 2u - 1$, $y = 3v - 4$, $z = \frac{1}{2}(w - 4) \Rightarrow \frac{\partial(x,y,z)}{\partial(u,v,w)} = \begin{vmatrix} 2 & 0 & 0 \\ 0 & 3 & 0 \\ 0 & 0 & \frac{1}{2} \end{vmatrix} = (2)(3)\left(\frac{1}{2}\right) = 3$

7. $\int_0^4 \int_{y/2}^{(y/2)+1} \left(x - \frac{y}{2}\right) dx\, dy = \int_0^4 \left[\frac{x^2}{2} - \frac{xy}{2}\right]_{\frac{y}{2}}^{\frac{y}{2}+1} dy = \frac{1}{2}\int_0^4 \left[\left(\frac{y}{2}+1\right)^2 - \left(\frac{y}{2}\right)^2 - \left(\frac{y}{2}+1\right)y + \left(\frac{y}{2}\right)y\right] dy$

$= \frac{1}{2}\int_0^4 (y + 1 - y)\, dy = \frac{1}{2}\int_0^4 dy = \frac{1}{2}(4) = 2$

8. $\iint_R (2x^2 - xy - y^2)\, dx\, dy = \iint_R (x - y)(2x + y)\, dx\, dy$

$= \iint_G uv \left|\frac{\partial(x,y)}{\partial(u,v)}\right| du\, dv = \frac{1}{3}\iint_G uv\, du\, dv$;

We find the boundaries of G from the boundaries of R, shown in the accompanying figure:

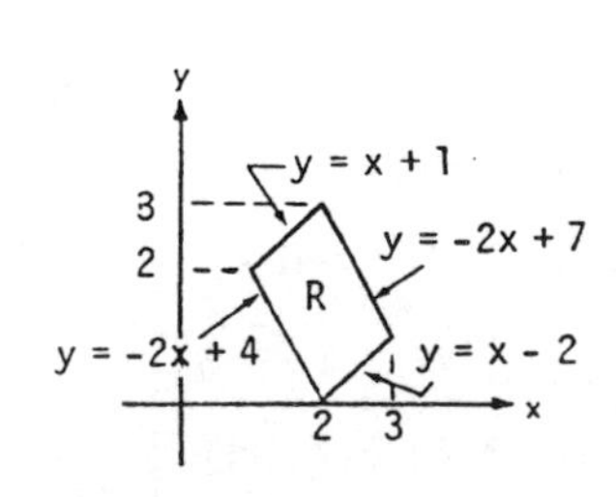

xy-equations for the boundary of R	Corresponding uv-equations for the boundary of G	Simplified uv-equations
$y = -2x + 4$	$\frac{1}{3}(-2u + v) = -\frac{2}{3}(u + v) + 4$	$v = 4$
$y = -2x + 7$	$\frac{1}{3}(-2u + v) = -\frac{2}{3}(u + v) + 7$	$v = 7$
$y = x - 2$	$\frac{1}{3}(-2u + v) = \frac{1}{3}(u + v) - 2$	$u = 2$
$y = x + 1$	$\frac{1}{3}(-2u + v) = \frac{1}{3}(u + v) + 1$	$u = -1$

$$\Rightarrow \frac{1}{3}\iint_G uv\,du\,dv = \frac{1}{3}\int_{-1}^{2}\int_{4}^{7} uv\,dv\,du = \frac{1}{3}\int_{-1}^{2} u\left[\frac{v^2}{2}\right]_4^7 du = \frac{11}{2}\int_{-1}^{2} u\,du = \left(\frac{11}{2}\right)\left[\frac{u^2}{2}\right]_{-1}^{2} = \left(\frac{11}{4}\right)(4-1) = \frac{33}{4}$$

9. $\iint_R (3x^2 + 14xy + 8y^2)\,dx\,dy = \iint_R (3x + 2y)(x + 4y)\,dx\,dy$

$$= \iint_G uv\left|\frac{\partial(x,y)}{\partial(u,v)}\right| du\,dv = \frac{1}{10}\iint_G uv\,du\,dv;$$

We find the boundaries of G from the boundaries of R, shown in the accompanying figure:

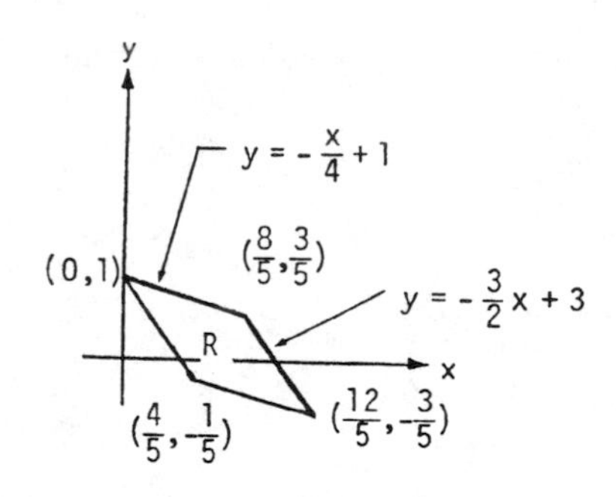

xy-equations for the boundary of R	Corresponding uv-equations for the boundary of G	Simplified uv-equations
$y = -\frac{3}{2}x + 1$	$\frac{1}{10}(3v - u) = -\frac{3}{10}(2u - v) + 1$	$u = 2$
$y = -\frac{3}{2}x + 3$	$\frac{1}{10}(3v - u) = -\frac{3}{10}(2u - v) + 3$	$u = 6$
$y = -\frac{1}{4}x$	$\frac{1}{10}(3v - u) = -\frac{1}{20}(2u - v)$	$v = 0$
$y = -\frac{1}{4}x + 1$	$\frac{1}{10}(3v - u) = -\frac{1}{20}(2u - v) + 1$	$v = 4$

$$\Rightarrow \frac{1}{10}\iint_G uv\,du\,dv = \frac{1}{10}\int_{2}^{6}\int_{0}^{4} uv\,dv\,du = \frac{1}{10}\int_{2}^{6} u\left[\frac{v^2}{2}\right]_0^4 du = \frac{4}{5}\int_{2}^{6} u\,du = \left(\frac{4}{5}\right)\left[\frac{u^2}{2}\right]_2^6 = \left(\frac{4}{5}\right)(18-2) = \frac{64}{5}$$

10. $\iint_R 2(x - y)\,dx\,dy = \iint_G -2v\left|\frac{\partial(x,y)}{\partial(u,v)}\right| du\,dv = \iint_G -2v\,du\,dv$; the region G is sketched in Exercise 4

$$\Rightarrow \iint_G -2v\,du\,dv = \int_0^1\int_{-3v}^{3-3v} -2v\,du\,dv = \int_0^1 -2v(3 - 3v + 3v)\,dv = \int_0^1 -6v\,dv = \left[-3v^2\right]_0^1 = -3$$

11. $x = \frac{u}{v}$ and $y = uv \Rightarrow \frac{y}{x} = v^2$ and $xy = u^2$; $\frac{\partial(x,y)}{\partial(u,v)} = J(u,v) = \begin{vmatrix} v^{-1} & -uv^{-2} \\ v & u \end{vmatrix} = v^{-1}u + v^{-1}u = \frac{2u}{v}$;

$y = x \Rightarrow uv = \frac{u}{v} \Rightarrow v = 1$, and $y = 4x \Rightarrow v = 2$; $xy = 1 \Rightarrow u = 1$, and $xy = 9 \Rightarrow u = 3$; thus

$$\iint_R \left(\sqrt{\frac{y}{x}} + \sqrt{xy}\right) dx\,dy = \int_1^3 \int_1^2 (v+u)\left(\frac{2u}{v}\right) dv\,du = \int_1^3 \int_1^2 \left(2u + \frac{2u^2}{v}\right) dv\,du = \int_1^3 \left[2uv + 2u^2 \ln v\right]_1^2 du$$

$$= \int_1^3 \left(2u + 2u^2 \ln 2\right) du = \left[u^2 + \frac{2}{3}u^2 \ln 2\right]_1^3 = 8 + \frac{2}{3}(26)(\ln 2) = 8 + \frac{52}{3}(\ln 2)$$

12. (a) $\frac{\partial(x,y)}{\partial(u,v)} = J(u,v) = \begin{vmatrix} 1 & 0 \\ v & u \end{vmatrix} = u$, and

the region G is sketched at the right

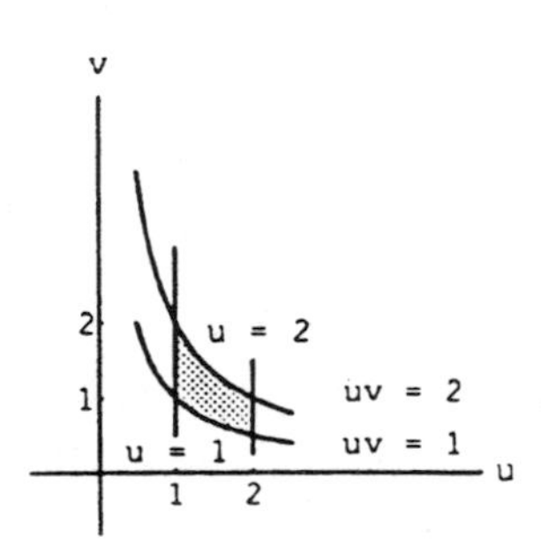

(b) $x = 1 \Rightarrow u = 1$, and $x = 2 \Rightarrow u = 2$; $y = 1 \Rightarrow uv = 1 \Rightarrow v = \frac{1}{u}$, and $y = 2 \Rightarrow uv = 2 \Rightarrow v = \frac{2}{u}$; thus,

$$\int_1^2 \int_1^2 \frac{y}{x}\,dy\,dx = \int_1^2 \int_{1/u}^{2/u} \left(\frac{uv}{u}\right) u\,dv\,du = \int_1^2 \int_{1/u}^{2/u} uv\,dv\,du = \int_1^2 u\left[\frac{v^2}{2}\right]_{1/u}^{2/u} du = \int_1^2 u\left(\frac{2}{u^2} - \frac{1}{2u^2}\right) du$$

$$= \frac{3}{2}\int_1^2 u\left(\frac{1}{u^2}\right) du = \frac{3}{2}\left[\ln u\right]_1^2 = \frac{3}{2} \ln 2$$

13. $x = ar\cos\theta$ and $y = ar\sin\theta \Rightarrow \frac{\partial(x,y)}{\partial(r,\theta)} = J(r,\theta) = \begin{vmatrix} a\cos\theta & -ar\sin\theta \\ b\sin\theta & br\cos\theta \end{vmatrix} = abr\cos^2\theta + abr\sin^2\theta = abr$;

$$I_0 = \iint_R \left(x^2+y^2\right) dA = \int_0^{2\pi} \int_0^1 r^2\left(a^2\cos^2\theta + b^2\sin^2\theta\right) |J(r,\theta)|\,dr\,d\theta = \int_0^{2\pi} \int_0^1 abr^3\left(a^2\cos^2\theta + b^2\sin^2\theta\right) dr\,d\theta$$

$$= \frac{ab}{4}\int_0^{2\pi} \left(a^2\cos^2\theta + b^2\sin^2\theta\right) d\theta = \frac{ab}{4}\left[\frac{a^2\theta}{2} + \frac{a^2\sin 2\theta}{4} + \frac{b^2\theta}{2} - \frac{b^2\sin 2\theta}{4}\right]_0^{2\pi} = \frac{ab\pi\left(a^2+b^2\right)}{4}$$

14. $\frac{\partial(x,y)}{\partial(u,v)} = J(u,v) = \begin{vmatrix} a & 0 \\ 0 & b \end{vmatrix} = ab$; $A = \iint_R dy\,dx = \iint_G ab\,du\,dv = \int_{-1}^{1} \int_{-\sqrt{1-u^2}}^{\sqrt{1-u^2}} ab\,dv\,du$

$$= 2ab\int_{-1}^{1} \sqrt{1-u^2}\,du = 2ab\left[\frac{u}{2}\sqrt{1-u^2} + \frac{1}{2}\sin^{-1}u\right]_{-1}^{1} = ab\left[\sin^{-1}1 - \sin^{-1}(-1)\right] = ab\left[\frac{\pi}{2} - \left(-\frac{\pi}{2}\right)\right] = ab\pi$$

15. The region of integration R in the xy-plane is sketched in the figure at the right. The boundaries of the image G are obtained as follows, with G sketched at the right:

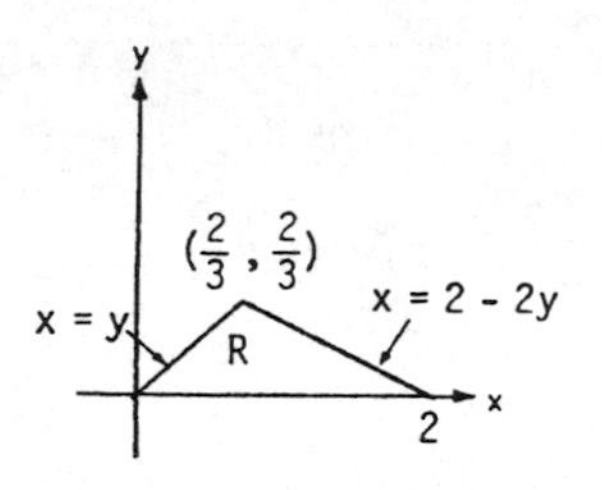

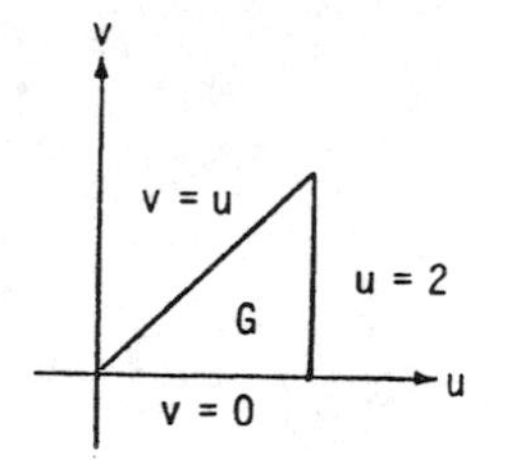

xy-equations for the boundary of R	Corresponding uv-equations for the boundary of G	Simplified uv-equations
$x = y$	$\frac{1}{3}(u + 2v) = \frac{1}{3}(u - v)$	$v = 0$
$x = 2 - 2y$	$\frac{1}{3}(u + 2v) = 2 - \frac{2}{3}(u - v)$	$u = 2$
$y = 0$	$0 = \frac{1}{3}(u - v)$	$v = u$

Also, from Exercise 2, $\frac{\partial(x,y)}{\partial(u,v)} = J(u,v) = -\frac{1}{3} \Rightarrow \int_0^{2/3}\int_y^{2-2y} (x+2y)\,e^{(y-x)}\,dx\,dy = \int_0^2\int_0^u ue^{-v}\left|-\frac{1}{3}\right| dv\,du$

$= \frac{1}{3}\int_0^2 u\left[-e^{-v}\right]_0^u du = \frac{1}{3}\int_0^2 u\left(1 - e^{-u}\right) du = \frac{1}{3}\left[u\left(u + e^{-u}\right) - \frac{u^2}{2} + e^{-u}\right]_0^2 = \frac{1}{3}\left[2\left(2 + e^{-2}\right) - 2 + e^{-2} - 1\right]$

$= \frac{1}{3}\left(3e^{-2} + 1\right) \approx 0.4687$

16. $x = u + \frac{v}{2}$ and $y = v \Rightarrow 2x - y = (2u + v) - v = 2u$ and

$\frac{\partial(x,y)}{\partial(u,v)} = J(u,v) = \begin{vmatrix} 1 & \frac{1}{2} \\ 0 & 1 \end{vmatrix} = 1$; next, $u = x - \frac{v}{2}$

$= x - \frac{y}{2}$ and $v = y$, so the boundaries of the region of integration R in the xy-plane are transformed to the boundaries of G:

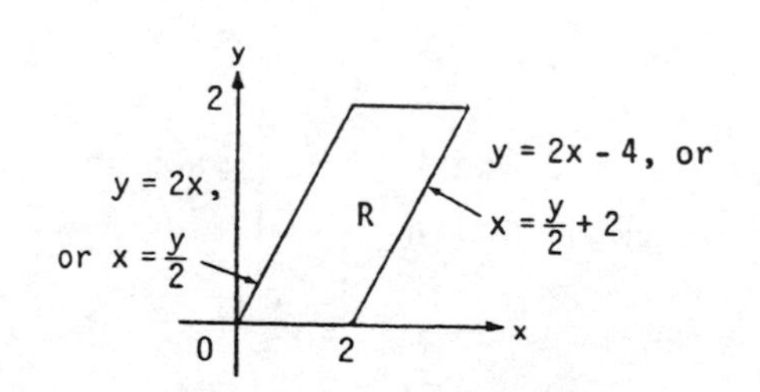

xy-equations for the boundary of R	Corresponding uv-equations for the boundary of G	Simplified uv-equations
$x = \frac{y}{2}$	$u + \frac{v}{2} = \frac{v}{2}$	$u = 0$
$x = \frac{y}{2} + 2$	$u + \frac{v}{2} = \frac{v}{2} + 2$	$u = 2$
$y = 0$	$v = 0$	$v = 0$
$y = 2$	$v = 2$	$v = 2$

$$\Rightarrow \int_0^2 \int_{y/2}^{(y/2)+2} y^3(2x - y)\, e^{(2x-y)^2}\, dx\, dy = \int_0^2 \int_0^2 v^3(2u)\, e^{4u^2}\, du\, dv = \int_0^2 v^3\left[\frac{1}{4}e^{4u^2}\right]_0^2 dv = \frac{1}{4}\int_0^2 v^3\left(e^{16} - 1\right) dv$$

$$= \frac{1}{4}\left(e^{16} - 1\right)\left[\frac{v^4}{4}\right]_0^2 = e^{16} - 1$$

17. $$\begin{vmatrix} \sin\phi\cos\theta & \rho\cos\phi\cos\theta & -\rho\sin\phi\sin\theta \\ \sin\phi\sin\theta & \rho\cos\phi\sin\theta & \rho\sin\phi\cos\theta \\ \cos\phi & -\rho\sin\phi & 0 \end{vmatrix}$$

$$= (\cos\phi)\begin{vmatrix} \rho\cos\phi\cos\theta & -\rho\sin\phi\sin\theta \\ \rho\cos\phi\sin\theta & \rho\sin\phi\cos\theta \end{vmatrix} + (\rho\sin\phi)\begin{vmatrix} \sin\phi\cos\theta & -\rho\sin\phi\sin\theta \\ \sin\phi\sin\theta & \rho\sin\phi\cos\theta \end{vmatrix}$$

$$= \left(\rho^2\cos\phi\right)\left(\sin\phi\cos\phi\cos^2\theta + \sin\phi\cos\phi\sin^2\theta\right) + \left(\rho^2\sin\phi\right)\left(\sin^2\phi\cos^2\theta + \sin^2\phi\sin^2\theta\right)$$
$$= \rho^2\sin\phi\cos^2\phi + \rho^2\sin^3\phi = \left(\rho^2\sin\phi\right)\left(\cos^2\phi + \sin^2\phi\right) = \rho^2\sin\phi$$

18. $$\int_0^3 \int_0^4 \int_{y/2}^{1+(y/2)} \left(\frac{2x - y}{2} + \frac{z}{3}\right) dx\, dy\, dz = \int_0^3 \int_0^4 \left[\frac{x^2}{2} - \frac{xy}{2} + \frac{xz}{3}\right]_{y/2}^{1+(y/2)} dy\, dz = \int_0^3 \int_0^4 \left[\frac{1}{2}(y + 1) - \frac{y}{2} + \frac{z}{3}\right] dy\, dz$$

$$= \int_0^3 \left[\frac{(y + 1)^2}{4} - \frac{y^2}{4} + \frac{yz}{3}\right]_0^4 dz = \int_0^3 \left(\frac{9}{4} + \frac{4z}{3} - \frac{1}{4}\right) dz = \int_0^3 \left(2 + \frac{4z}{3}\right) dz = \left[2z + \frac{2z^2}{3}\right]_0^3 = 12$$

19. $J(u, v, w) = \begin{vmatrix} a & 0 & 0 \\ 0 & b & 0 \\ 0 & 0 & c \end{vmatrix} = abc$; the transformation takes the ellipsoid region $\frac{x^2}{a^2} + \frac{y^2}{b^2} + \frac{z^2}{c^2} \le 1$ in xyz-space

into the spherical region $u^2 + v^2 + w^2 \le 1$ in uvw-space $\left(\text{which has volume } V = \frac{4}{3}\pi\right)$

$$\Rightarrow V = \iiint_R dx\, dy\, dz = \iiint_G abc\, du\, dv\, dw = \frac{4\pi abc}{3}$$

20. $J(u,v,w) = \begin{vmatrix} a & 0 & 0 \\ 0 & b & 0 \\ 0 & 0 & c \end{vmatrix} = abc$; for R and G as in Exercise 19, $\iiint\limits_R |xyz|\, dx\, dy\, dz$

$= \iiint\limits_G a^2b^2c^2uvw\, dw\, dv\, du = 8a^2b^2c^2 \int_0^{\pi/2}\int_0^{\pi/2}\int_0^1 (\rho \sin\phi \cos\theta)(\rho \sin\phi \sin\theta)(\rho\cos\phi)(\rho^2 \sin\phi)\, d\rho\, d\phi\, d\theta$

$= \frac{4a^2b^2c^2}{3}\int_0^{\pi/2}\int_0^{\pi/2} \sin\theta \cos\theta \sin^3\phi \cos\phi\, d\phi\, d\theta = \frac{a^2b^2c^2}{3}\int_0^{\pi/2} \sin\theta\cos\theta\, d\theta = \frac{a^2b^2c^2}{6}$

21. $u = x$, $v = xy$, and $w = 3z \Rightarrow x = u$, $y = \frac{v}{u}$, and $z = \frac{1}{3}w \Rightarrow J(u,v,w) = \begin{vmatrix} 1 & 0 & 0 \\ -\frac{v}{u^2} & \frac{1}{u} & 0 \\ 0 & 0 & \frac{1}{3} \end{vmatrix} = \frac{1}{3u}$;

$\iiint\limits_R (x^2y + 3xyz)\, dx\, dy\, dz = \iiint\limits_G \left[u^2\left(\frac{v}{u}\right) + 3u\left(\frac{v}{u}\right)\left(\frac{w}{3}\right)\right] |J(u,v,w)|\, du\, dv\, dw = \frac{1}{3}\int_0^3\int_0^2\int_1^2 \left(v + \frac{vw}{u}\right) du\, dv\, dw$

$= \frac{1}{3}\int_0^3\int_0^2 (v + vw \ln 2)\, dv\, dw = \frac{1}{3}\int_0^3 (1 + w\ln 2)\left[\frac{v^2}{2}\right]_0^2 dw = \frac{2}{3}\int_0^3 (1 + w\ln 2)\, dw = \frac{2}{3}\left[w + \frac{w^2}{2}\ln 2\right]_0^3$

$= \frac{2}{3}\left(3 + \frac{9}{2}\ln 2\right) = 2 + 3\ln 2 = 2 + \ln 8$

22. The first moment about the xy-coordinate plane for the semi-ellipsoid, $\frac{x^2}{a^2} + \frac{y^2}{b^2} + \frac{z^2}{c^2} = 1$ using the

transformation in Exercise 21 is, $M_{xy} = \iiint\limits_D z\, dz\, dy\, dx = \iiint\limits_G cw\, |J(u,v,w)|\, du\, dv\, dw$

$= abc^2 \iiint\limits_G w\, du\, dv\, dw = (abc^2)\cdot\left(M_{xy} \text{ of the hemisphere } x^2 + y^2 + z^2 = 1,\ z \ge 0\right) = \frac{abc^2\pi}{4}$;

the mass of the semi-ellipsoid is $\frac{2abc\pi}{3} \Rightarrow \bar{z} = \left(\frac{abc^2\pi}{4}\right)\left(\frac{3}{2abc\pi}\right) = \frac{3}{8}c$

23. Let $u = g(x) \Rightarrow J(x) = \frac{du}{dx} = g'(x) \Rightarrow \int_a^b f(u)\, du = \int_{g(a)}^{g(b)} f(g(x))g'(x)\, dx$ in accordance with formula (1) in Section 4.8. Note that $g'(x)$ represents the Jacobian of the transformation $u = g(x)$ or $x = g^{-1}(u)$. Several examples are presented in Section 4.8.

CHAPTER 13 PRACTICE EXERCISES

1. $$\int_1^{10}\int_0^{1/y} ye^{xy}\,dx\,dy = \int_1^{10} \left[e^{xy}\right]_0^{1/y} dy$$

$$= \int_1^{10} (e-1)\,dy = 9e - 9$$

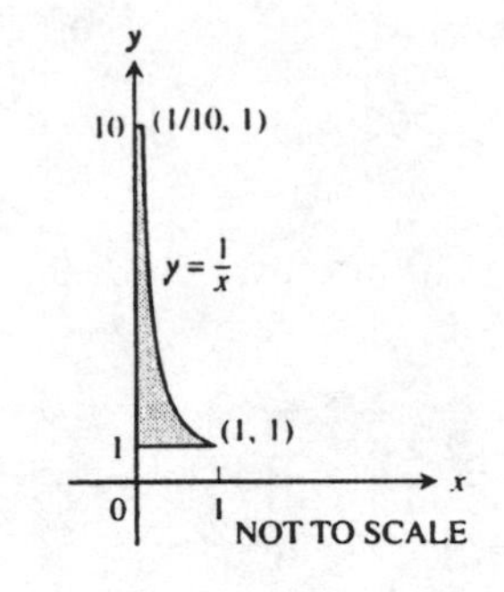

2. $$\int_0^1\int_0^{x^3} e^{y/x}\,dy\,dx = \int_1^1 x\left[e^{y/x}\right]_0^{x^3} dx$$

$$= \int_0^1 \left(xe^{x^2} - x\right) dx = \left[\frac{1}{2}e^{x^2} - \frac{x^2}{2}\right]_0^1 = \frac{e-2}{2}$$

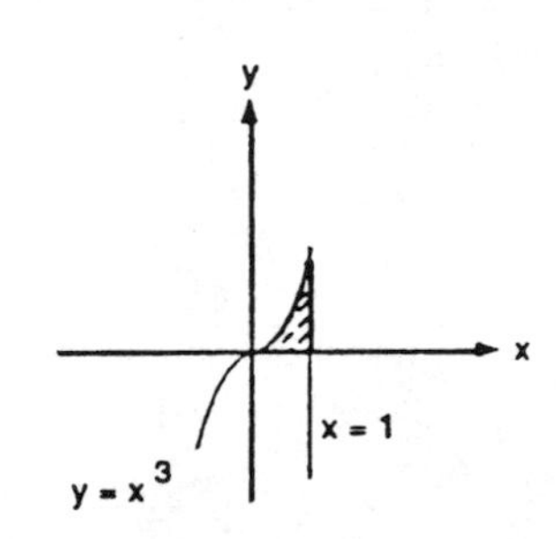

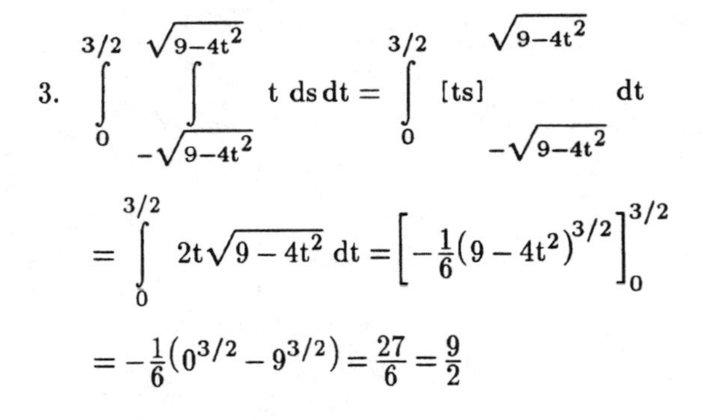

3. $$\int_0^{3/2}\int_{-\sqrt{9-4t^2}}^{\sqrt{9-4t^2}} t\,ds\,dt = \int_0^{3/2} [ts]_{-\sqrt{9-4t^2}}^{\sqrt{9-4t^2}}\,dt$$

$$= \int_0^{3/2} 2t\sqrt{9-4t^2}\,dt = \left[-\frac{1}{6}\left(9-4t^2\right)^{3/2}\right]_0^{3/2}$$

$$= -\frac{1}{6}\left(0^{3/2} - 9^{3/2}\right) = \frac{27}{6} = \frac{9}{2}$$

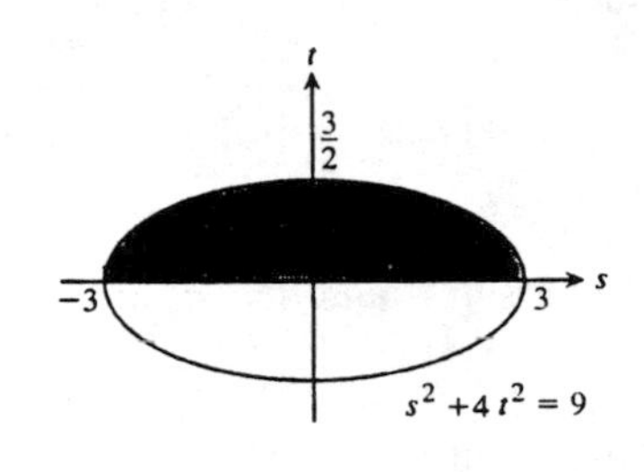

4. $$\int_0^3\int_{\sqrt{y}}^{2-\sqrt{y}} xy\,dx\,dy = \int_0^3 y\left[\frac{x^2}{2}\right]_{\sqrt{y}}^{2-\sqrt{y}} dy$$

$$= \frac{1}{2}\int_0^3 y\left(4 - 4\sqrt{y} + y - y\right) dy$$

$$= \int_0^3 \left(2y - 2y^{3/2}\right) dy = \left[y^2 - \frac{4y^{5/2}}{5}\right]_0^1 = \frac{1}{5}$$

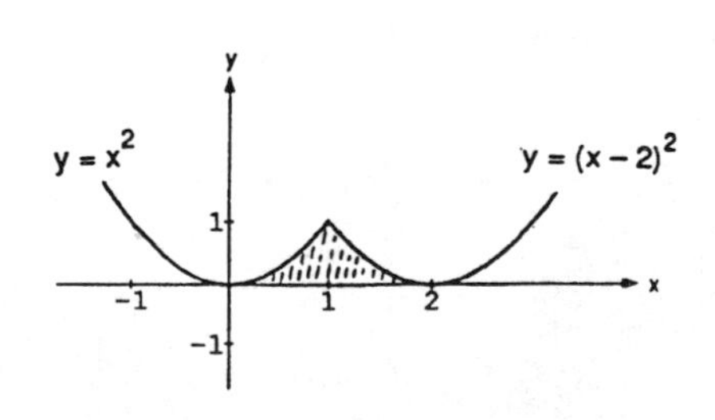

5. $\int_{-2}^{0}\int_{2x+4}^{4-x^2} dy\,dx = \int_{-2}^{0} \left(-x^2-2x\right) dx$

$= \left[-\frac{x^3}{3} - x^2\right]_{-2}^{0} = -\left(\frac{8}{3} - 4\right) = \frac{4}{3}$

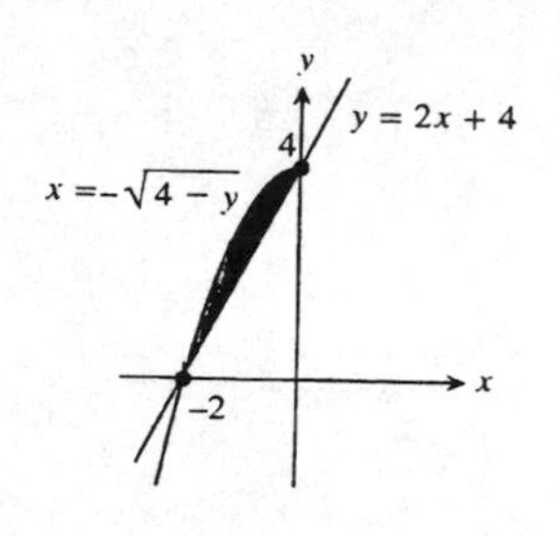

6. $\int_{0}^{1}\int_{y}^{\sqrt{y}} \sqrt{x}\,dx\,dy = \int_{0}^{1} \left[\frac{2}{3}x^{3/2}\right]_{y}^{\sqrt{y}} dy$

$= \frac{2}{3}\int_{0}^{1} \left(y^{3/4} - y^{3/2}\right) dy = \frac{2}{3}\left[\frac{4}{7}y^{7/4} - \frac{2}{5}y^{5/2}\right]_{0}^{1}$

$= \frac{2}{3}\left(\frac{4}{7} - \frac{2}{5}\right) = \frac{4}{35}$

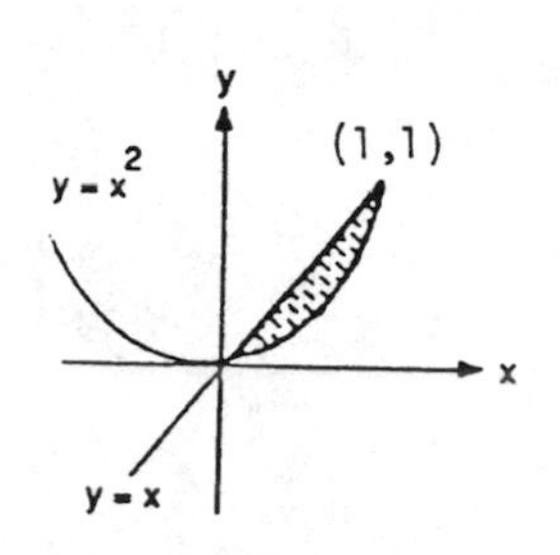

7. $\int_{-3}^{3}\int_{0}^{(1/2)\sqrt{9-x^2}} y\,dy\,dx = \int_{-3}^{3} \left[\frac{y^2}{2}\right]_{0}^{(1/2)\sqrt{9-x^2}} dx$

$= \int_{-3}^{3} \frac{1}{8}\left(9 - x^2\right) dx = \left[\frac{9x}{8} - \frac{x^3}{24}\right]_{-3}^{3}$

$= \left(\frac{27}{8} - \frac{27}{24}\right) - \left(-\frac{27}{8} + \frac{27}{24}\right) = \frac{27}{6} = \frac{9}{2}$

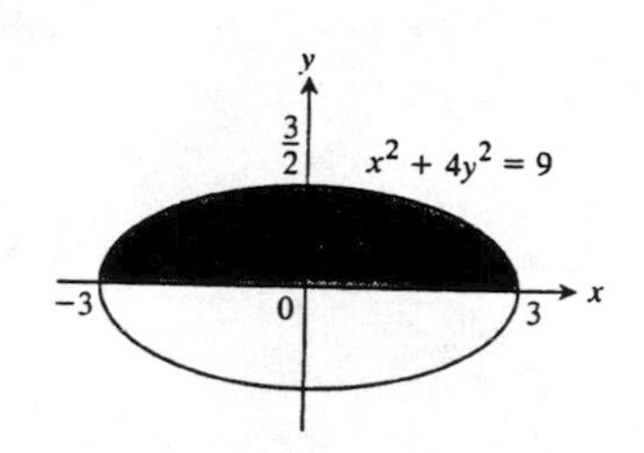

8. $\int_{0}^{4}\int_{0}^{\sqrt{4-y}} 2x\,dx\,dy = \int_{0}^{4} \left[x^2\right]_{0}^{\sqrt{4-y}} dy$

$= \int_{0}^{4} (4-y)\,dy = \left[4y - \frac{y^2}{2}\right]_{0}^{4} = 8$

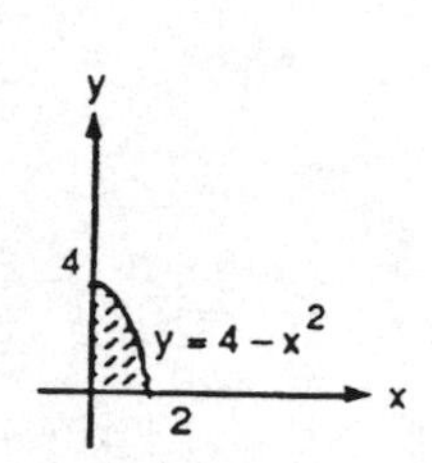

9. $\int_{0}^{1}\int_{2y}^{2} 4\cos\left(x^2\right) dx\,dy = \int_{0}^{2}\int_{0}^{x/2} 4\cos\left(x^2\right) dy\,dx = \int_{0}^{2} 2x\cos\left(x^2\right) dx = \left[\sin\left(x^2\right)\right]_{0}^{2} = \sin 4$

10. $\int_{0}^{2}\int_{y/2}^{1} e^{x^2}\,dx\,dy = \int_{0}^{1}\int_{0}^{2x} e^{x^2}\,dy\,dx = \int_{0}^{1} 2xe^{x^2}\,dx = \left[e^{x^2}\right]_{0}^{1} = e - 1$

11. $\int_0^8 \int_{\sqrt[3]{x}}^2 \frac{1}{y^4+1}\, dy\, dx = \int_0^2 \int_0^{y^3} \frac{1}{y^4+1}\, dx\, dy = \frac{1}{4}\int_0^2 \frac{4y^3}{y^4+1}\, dy = \frac{\ln 17}{4}$

12. $\int_0^1 \int_{\sqrt[3]{y}}^1 \frac{2\pi \sin(\pi x^2)}{x^2}\, dx\, dy = \int_0^1 \int_0^{x^3} \frac{2\pi \sin(\pi x^2)}{x^2}\, dy\, dx = \int_0^1 2\pi x \sin(\pi x^2)\, dx = \left[-\cos(\pi x^2)\right]_0^1 = -(-1)-(-1) = 2$

13. $A = \int_{-2}^0 \int_{2x+4}^{4-x^2} dy\, dx = \int_{-2}^0 (-x^2-2x)\, dx = \frac{4}{3}$

14. $A = \int_1^4 \int_{2-y}^{\sqrt{y}} dx\, dy = \int_1^4 (\sqrt{y}-2+y)\, dy = \frac{37}{6}$

15. $V = \int_0^1 \int_x^{2-x} (x^2+y^2)\, dy\, dx = \int_0^1 \left[x^2 y + \frac{y^3}{3}\right]_x^{2-x} dx = \int_0^1 \left[2x^2 + \frac{(2-x)^3}{3} - \frac{7x^3}{3}\right] dx = \left[\frac{2x^3}{3} - \frac{(2-x)^4}{12} - \frac{7x^4}{12}\right]_0^1$
$= \left(\frac{2}{3} - \frac{1}{12} - \frac{7}{12}\right) + \frac{2^4}{12} = \frac{4}{3}$

16. $V = \int_{-3}^2 \int_x^{6-x^2} x^2\, dy\, dx = \int_{-3}^2 \left[x^2 y\right]_x^{6-x^2} dx = \int_{-3}^2 (6x^2 - x^4 - x^3)\, dx = \frac{125}{4}$

17. average value $= \int_0^1 \int_0^1 xy\, dy\, dx = \int_0^1 \left[\frac{xy^2}{2}\right]_0^1 dx = \int_0^1 \frac{x}{2}\, dx = \frac{1}{4}$

18. average value $= \frac{1}{\left(\frac{\pi}{4}\right)} \int_0^1 \int_0^{\sqrt{1-x^2}} xy\, dy\, dx = \frac{4}{\pi} \int_0^1 \left[\frac{xy^2}{2}\right]_0^{\sqrt{1-x^2}} dx = \frac{2}{\pi} \int_0^1 (x - x^3)\, dx = \frac{1}{2\pi}$

19. $M = \int_1^2 \int_{2/x}^2 dy\, dx = \int_1^2 \left(2 - \frac{2}{x}\right) dx = 2 - \ln 4$; $M_y = \int_1^2 \int_{2/x}^2 x\, dy\, dx = \int_1^2 x\left(2 - \frac{2}{x}\right) dx = 1$;
$M_x = \int_1^2 \int_{2/x}^2 y\, dy\, dx = \int_1^2 \left(2 - \frac{2}{x^2}\right) dx = 1 \Rightarrow \bar{x} = \bar{y} = \frac{1}{2 - \ln 4}$

20. $M = \int_0^4 \int_{-2y}^{2y-y^2} dx\, dy = \int_0^4 (4y - y^2)\, dy = \frac{32}{3}$; $M_x = \int_0^4 \int_{-2y}^{2y-y^2} y\, dx\, dy = \int_0^4 (4y^2 - y^3)\, dy = \left[\frac{4y^3}{3} - \frac{y^4}{4}\right]_0^4 = \frac{64}{3}$;
$M_y = \int_0^4 \int_{-2y}^{2y-y^2} x\, dx\, dy = \int_0^4 \left[\frac{(2y-y^2)^2}{2} - 2y^2\right] dy = \left[\frac{y^5}{10} - \frac{y^4}{2}\right]_0^4 = -\frac{128}{5} \Rightarrow \bar{x} = \frac{M_y}{M} = -\frac{12}{5}$ and $\bar{y} = \frac{M_x}{M} = 2$

21. $I_o = \int_0^2 \int_{2x}^4 (x^2+y^2)(3)\,dy\,dx = 3\int_0^2 \left(4x^2 + \frac{64}{3} - \frac{14x^3}{3}\right) dx = 104$

22. (a) $I_o = \int_{-2}^2 \int_{-1}^1 (x^2+y^2)\,dy\,dx = \int_{-2}^2 \left(2x^2+\frac{2}{3}\right) dx = \frac{40}{3}$

(b) $I_x = \int_{-a}^a \int_{-b}^b y^2\,dy\,dx = \int_{-a}^a \frac{2b^3}{3}\,dx = \frac{4ab^3}{3}$; $I_y = \int_{-b}^b \int_{-a}^a x^2\,dx\,dy = \int_{-b}^b \frac{2a^3}{3}\,dy = \frac{4a^3b}{3} \Rightarrow I_o = I_x + I_y$
$= \frac{4ab^3}{3} + \frac{4a^3b}{3} = \frac{4ab(b^2+a^2)}{3}$

23. $M = \delta \int_0^3 \int_0^{2x/3} dy\,dx = \delta \int_0^3 \frac{2x}{3}\,dx = 3\delta$; $I_x = \delta \int_0^3 \int_0^{2x/3} y^2\,dy\,dx = \frac{8\delta}{81}\int_0^3 x^3\,dx = \left(\frac{8\delta}{81}\right)\left(\frac{3^4}{4}\right) = 2\delta \Rightarrow R_x = \sqrt{\frac{2}{3}}$

24. $M = \int_0^1 \int_{x^2}^x (x+1)\,dy\,dx = \int_0^1 (x-x^3)\,dx = \frac{1}{4}$; $M_x = \int_0^1 \int_{x^2}^x y(x+1)\,dy\,dx = \frac{1}{2}\int_0^1 (x^3 - x^5 + x^2 - x^4)\,dx = \frac{13}{120}$;
$M_y = \int_0^1 \int_{x^2}^x x(x+1)\,dy\,dx = \int_0^1 (x^2 - x^4)\,dx = \frac{2}{15} \Rightarrow \overline{x} = \frac{8}{15}$ and $\overline{y} = \frac{13}{30}$; $I_x = \int_0^1 \int_{x^2}^x y^2(x+1)\,dy\,dx$
$= \frac{1}{3}\int_0^1 (x^4 - x^7 + x^3 - x^6)\,dx = \frac{17}{280} \Rightarrow R_x = \sqrt{\frac{I_x}{M}} = \sqrt{\frac{17}{70}}$; $I_y = \int_0^1 \int_{x^2}^x x^2(x+1)\,dy\,dx = \int_0^1 (x^3 - x^5)\,dx$
$= \frac{1}{12} \Rightarrow R_y = \sqrt{\frac{I_y}{M}} = \sqrt{\frac{1}{3}}$

25. $M = \int_{-1}^1 \int_{-1}^1 \left(x^2+y^2+\frac{1}{3}\right) dy\,dx = \int_{-1}^1 \left(2x^2+\frac{4}{3}\right) dx = 4$; $M_x = \int_{-1}^1 \int_{-1}^1 y\left(x^2+y^2+\frac{1}{3}\right) dy\,dx = \int_{-1}^1 0\,dx = 0$;
$M_y = \int_{-1}^1 \int_{-1}^1 x\left(x^2+y^2+\frac{1}{3}\right) dy\,dx = \int_{-1}^1 \left(2x^3+\frac{4}{3}x\right) dx = 0$

26. Place the ΔABC with its vertices at $A(0,0)$, $B(b,0)$ and $C(a,h)$. The line through the points A and C is $y = \frac{h}{a}x$; the line through the points C and B is $y = \frac{h}{a-b}(x-b)$. Thus, $M = \int_0^h \int_{ay/h}^{(a-b)y/h+b} \delta\,dx\,dy$
$= b\delta \int_0^h \left(1 - \frac{y}{h}\right) dy = \frac{\delta bh}{2}$; $I_x = \int_0^h \int_{ay/h}^{(a-b)y/h+b} y^2\delta\,dx\,dy = b\delta \int_0^h \left(y^2 - \frac{y^3}{h}\right) dy = \frac{\delta bh^3}{12}$; $R_x = \sqrt{\frac{I_x}{M}} = \frac{h}{\sqrt{6}}$

27. $\int_{-1}^{1}\int_{-\sqrt{1-x^2}}^{\sqrt{1-x^2}} \frac{2}{(1+x^2+y^2)^2}\,dy\,dx = \int_0^{2\pi}\int_0^1 \frac{2r}{(1+r^2)^2}\,dr\,d\theta = \int_0^{2\pi}\left[-\frac{1}{1+r^2}\right]_0^1 d\theta = \frac{1}{2}\int_0^{2\pi} d\theta = \pi$

28. $\int_{-1}^{1}\int_{-\sqrt{1-y^2}}^{\sqrt{1-y^2}} \ln(x^2+y^2+1)\,dx\,dy = \int_0^{2\pi}\int_0^1 r\ln(r^2+1)\,dr\,d\theta = \int_0^{2\pi}\int_1^2 \frac{1}{2}\ln u\,du\,d\theta = \frac{1}{2}\int_0^{2\pi} [u\ln u - u]_1^2\,d\theta$

$= \frac{1}{2}\int_0^{2\pi} (2\ln 2 - 1)\,d\theta = [\ln(4) - 1]\,\pi$

29. $M = \int_{-\pi/3}^{\pi/3}\int_0^3 r\,dr\,d\theta = \frac{9}{2}\int_{-\pi/3}^{\pi/3} d\theta = 3\pi$; $M_y = \int_{-\pi/3}^{\pi/3}\int_0^3 r^2\cos\theta\,dr\,d\theta = 9\int_{-\pi/3}^{\pi/3}\cos\theta\,d\theta = 9\sqrt{3} \Rightarrow \bar{x} = \frac{3\sqrt{3}}{\pi}$,

and $\bar{y} = 0$ by symmetry

30. $M = \int_0^{\pi/2}\int_1^3 r\,dr\,d\theta = 4\int_0^{\pi/2} d\theta = 2\pi$; $M_y = \int_0^{\pi/2}\int_1^3 r^2\cos\theta\,dr\,d\theta = \frac{26}{3}\int_0^{\pi/2}\cos\theta\,d\theta = \frac{26}{3} \Rightarrow \bar{x} = \frac{13}{3\pi}$, and

$\bar{y} = \frac{13}{3\pi}$ by symmetry

31. (a) $M = 2\int_0^{\pi/2}\int_1^{1+\cos\theta} r\,dr\,d\theta$

$= \int_0^{\pi/2}\left(2\cos\theta + \frac{1+\cos 2\theta}{2}\right)d\theta = \frac{8+\pi}{4}$;

$M_y = \int_{-\pi/2}^{\pi/2}\int_1^{1+\cos\theta} (r\cos\theta)\,r\,dr\,d\theta$

$= \int_{-\pi/2}^{\pi/2}\left(\cos^2\theta + \cos^3\theta + \frac{\cos^4\theta}{3}\right)d\theta$

$= \frac{32+15\pi}{24} \Rightarrow \bar{x} = \frac{15\pi+32}{6\pi+48}$, and

$\bar{y} = 0$ by symmetry

(b)

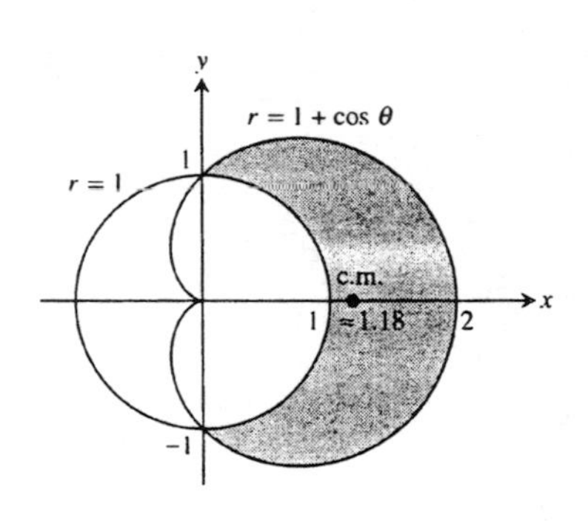

32. (a) $M = \int_{-\alpha}^{\alpha} \int_0^a r\, dr\, d\theta = \int_{-\alpha}^{\alpha} \frac{a^2}{2}\, d\theta = a^2\alpha$; $M_y = \int_{-\alpha}^{\alpha} \int_0^a (r \cos \theta)\, r\, dr\, d\theta = \int_{-\alpha}^{\alpha} \frac{a^3 \cos \theta}{3}\, d\theta = \frac{2a^3 \sin \alpha}{3}$

$\Rightarrow \overline{x} = \frac{2a \sin \alpha}{3\alpha}$, and $\overline{y} = 0$ by symmetry; $\lim\limits_{\alpha \to \pi^-} \overline{x} = \lim\limits_{\alpha \to \pi^-} \frac{2a \sin \alpha}{3\alpha} = 0$

(b) $\overline{x} = \frac{2a}{5\pi}$ and $\overline{y} = 0$

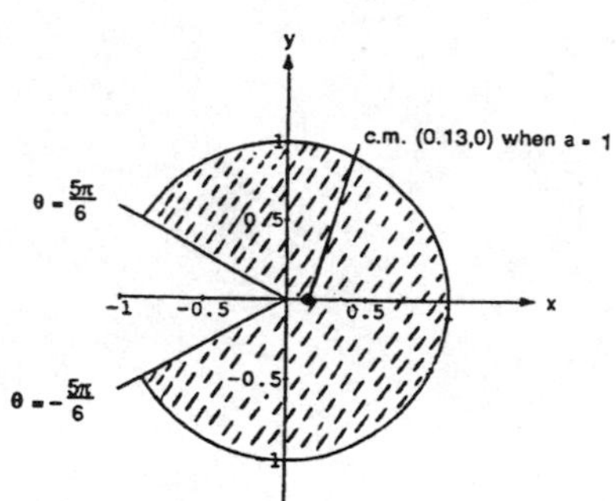

33. $(x^2+y^2)^2 - (x^2-y^2) = 0 \Rightarrow r^4 - r^2 \cos 2\theta = 0 \Rightarrow r^2 = \cos 2\theta$ so the integral is $\int_{-\pi/4}^{\pi/4} \int_0^{\sqrt{\cos 2\theta}} \frac{r}{(1+r^2)^2}\, dr\, d\theta$

$= \int_{-\pi/4}^{\pi/4} \left[-\frac{1}{2(1+r^2)}\right]_0^{\sqrt{\cos 2\theta}} d\theta = \frac{1}{2}\int_{-\pi/4}^{\pi/4} \left(1 - \frac{1}{1+\cos 2\theta}\right) d\theta = \frac{1}{2}\int_{-\pi/4}^{\pi/4} \left(1 - \frac{1}{2\cos^2\theta}\right) d\theta$

$= \frac{1}{2}\int_{-\pi/4}^{\pi/4} \left(1 - \frac{\sec^2\theta}{2}\right) d\theta = \frac{1}{2}\left[\theta - \frac{\tan\theta}{2}\right]_{-\pi/4}^{\pi/4} = \frac{\pi - 2}{4}$

34. (a) $\iint_R \frac{1}{(1+x^2+y^2)^2}\, dx\, dy = \int_0^{\pi/3} \int_0^{\sec\theta} \frac{r}{(1+r^2)^2}\, dr\, d\theta = \int_0^{\pi/3} \left[-\frac{1}{2(1+r^2)}\right]_0^{\sec\theta} d\theta$

$= \int_0^{\pi/3} \left[\frac{1}{2} - \frac{1}{2(1+\sec^2\theta)}\right] d\theta = \frac{1}{2}\int_0^{\pi/3} \frac{\sec^2\theta}{1+\sec^2\theta}\, d\theta$; $\begin{bmatrix} u = \tan\theta \\ du = \sec^2\theta\, d\theta \end{bmatrix} \to \frac{1}{2}\int_0^{\sqrt{3}} \frac{du}{2+u^2}$

$= \frac{1}{2}\left[\frac{1}{\sqrt{2}} \tan^{-1} \frac{u}{\sqrt{2}}\right]_0^{\sqrt{3}} = \frac{\sqrt{2}}{4} \tan^{-1} \sqrt{\frac{3}{2}}$

(b) $\iint_R \frac{1}{(1+x^2+y^2)^2}\, dx\, dy = \int_0^{\pi/2} \int_0^{\infty} \frac{r}{(1+r^2)^2}\, dr\, d\theta = \int_0^{\pi/2} \lim\limits_{b\to\infty} \left[-\frac{1}{2(1+r^2)}\right] d\theta$

$= \int_0^{\pi/2} \lim\limits_{b\to\infty} \left[\frac{1}{2} - \frac{1}{2(1+b^2)}\right] d\theta = \frac{1}{2}\int_0^{\pi/2} d\theta = \frac{\pi}{4}$

35. $\int_0^{\pi}\int_0^{\pi}\int_0^{\pi} \cos(x+y+z)\,dx\,dy\,dz = \int_0^{\pi}\int_0^{\pi} [\sin(z+y+\pi) - \sin(z+y)]\,dy\,dz$

$= \int_0^{\pi} [-\cos(z+2\pi) + \cos(z+\pi) + \cos z - \cos(z+\pi)]\,dz = 0$

36. $\int_{\ln 6}^{\ln 7}\int_0^{\ln 2}\int_{\ln 4}^{\ln 5} e^{(x+y+z)}\,dz\,dy\,dx = \int_{\ln 6}^{\ln 7}\int_0^{\ln 2} e^{(x+y)}\,dy\,dx = \int_{\ln 6}^{\ln 7} e^x\,dx = 1$

37. $\int_0^1\int_0^{x^2}\int_0^{x+y} (2x-y-z)\,dz\,dy\,dx = \int_0^1\int_0^{x^2} \left(\frac{3x^2}{2} - \frac{3y^2}{2}\right)dy\,dx = \int_0^1 \left(\frac{3x^4}{2} - \frac{x^6}{2}\right)dx = \frac{8}{35}$

38. $\int_1^e\int_1^x\int_0^z \frac{2y}{z^3}\,dy\,dz\,dx = \int_1^e\int_1^x \frac{1}{z}\,dz\,dx = \int_1^e \ln x\,dx = [x\ln x - x]_1^e = 1$

39. $V = 2\int_0^{\pi/2}\int_{-\cos y}^0\int_0^{-2x} dz\,dx\,dy = 2\int_0^{\pi/2}\int_{-\cos y}^0 -2x\,dx\,dy = 2\int_0^{\pi/2} \cos^2 y\,dy = 2\left[\frac{y}{2} + \frac{\sin 2y}{4}\right]_0^{\pi/2} = \frac{\pi}{2}$

40. $V = 4\int_0^2\int_0^{\sqrt{4-x^2}}\int_0^{4-x^2} dz\,dy\,dx = 4\int_0^2\int_0^{\sqrt{4-x^2}} (4-x^2)\,dy\,dx = 4\int_0^2 (4-x^2)^{3/2}\,dx$

$= \left[x(4-x^2)^{3/2} + 6x\sqrt{4-x^2} + 24\sin^{-1}\frac{x}{2}\right]_0^2 = 24\sin^{-1} 1 = 12\pi$

41. $\text{average} = \frac{1}{3}\int_0^1\int_0^3\int_0^1 30xz\sqrt{x^2+y}\,dz\,dy\,dx = \frac{1}{3}\int_0^1\int_0^3 15x\sqrt{x^2+y}\,dy\,dx = \frac{1}{3}\int_0^3\int_0^1 15x\sqrt{x^2+y}\,dx\,dy$

$= \frac{1}{3}\int_0^3 \left[5(x^2+y)^{3/2}\right]_0^1 dy = \frac{1}{3}\int_0^3 \left[5(1+y)^{3/2} - 5y^{3/2}\right]dy = \frac{1}{3}\left[2(1+y)^{5/2} - 2y^{5/2}\right]_0^3 = \frac{1}{3}\left[2(4)^{5/2} - 2(3)^{5/2} - 2\right]$

$= \frac{1}{3}\left[2\left(31 - 3^{5/2}\right)\right]$

42. $\text{average} = \frac{3}{4\pi a^3}\int_0^{2\pi}\int_0^{\pi}\int_0^a \rho^3 \sin\phi\,d\rho\,d\phi\,d\theta = \frac{3a}{16\pi}\int_0^{2\pi}\int_0^{\pi} \sin\phi\,d\phi\,d\theta = \frac{3a}{8\pi}\int_0^{2\pi} d\theta = \frac{3a}{4}$

43. (a) $\int_{-\sqrt{2}}^{\sqrt{2}}\int_{-\sqrt{2-y^2}}^{\sqrt{2-y^2}}\int_{\sqrt{x^2+y^2}}^{\sqrt{4-x^2-y^2}} 3\,dz\,dx\,dy$

(b) $\int_0^{2\pi}\int_0^{\pi/4}\int_0^2 3\rho^2 \sin\phi \, d\rho \, d\phi \, d\theta$

(c) $\int_0^{2\pi}\int_0^{\sqrt{2}}\int_r^{\sqrt{4-r^2}} 3 \, dz \, r \, dr \, d\theta = 3\int_0^{2\pi}\int_0^{\sqrt{2}} \left[r\left(4-r^2\right)^{1/2} - r^2\right] dr \, d\theta = 3\int_0^{2\pi}\left[-\frac{1}{3}\left(4-r^2\right)^{3/2} - \frac{r^3}{3}\right]_0^{\sqrt{2}} d\theta$

$= \int_0^{2\pi}\left(-2^{3/2} - 2^{3/2} + 4^{3/2}\right) d\theta = \left(8-4\sqrt{2}\right)\int_0^{2\pi} d\theta = 2\pi\left(8-4\sqrt{2}\right)$

44. (a) $\int_{-\pi/2}^{\pi/2}\int_0^1\int_{-r^2}^{r^2} 21(r\cos\theta)(r\sin\theta)^2 \, dz \, r \, dr \, d\theta = \int_{-\pi/2}^{\pi/2}\int_0^1\int_{-r^2}^{r^2} 21r^3\cos\theta\sin^2\theta \, dz \, r \, dr \, d\theta$

(b) $\int_{-\pi/2}^{\pi/2}\int_0^1\int_{-r^2}^{r^2} 21r^3\cos\theta\sin^2\theta \, dz \, r \, dr \, d\theta = 84\int_0^{\pi/2}\int_0^1 r^6\sin^2\theta\cos\theta \, dr \, d\theta = 12\int_0^{\pi/2}\sin^2\theta\cos\theta \, d\theta = 4$

45. (a) $\int_0^{2\pi}\int_0^{\pi/4}\int_0^{\sec\phi} \rho^2\sin\phi \, d\rho \, d\phi \, d\theta$

(b) $\int_0^{2\pi}\int_0^{\pi/4}\int_0^{\sec\phi} \rho^2\sin\phi \, d\rho \, d\phi \, d\theta = \frac{1}{3}\int_0^{2\pi}\int_0^{\pi/4}(\sec\phi)(\sec\phi\tan\phi) \, d\phi \, d\theta = \frac{1}{3}\int_0^{2\pi}\left[\frac{1}{2}\tan^2\phi\right]_0^{\pi/4} d\theta = \frac{1}{6}\int_0^{2\pi} d\theta = \frac{\pi}{3}$

46. (a) $\int_0^1\int_0^{\sqrt{1-x^2}}\int_0^{\sqrt{x^2+y^2}} (6+4y) \, dz \, dy \, dx$ (b) $\int_0^{\pi/2}\int_0^1\int_0^r (6+4r\sin\theta) \, dz \, r \, dr \, d\theta$

(c) $\int_0^{\pi/2}\int_{\pi/4}^{\pi/2}\int_0^{\csc\phi} (6+4\rho\sin\phi\sin\theta)\left(\rho^2\sin\phi\right) d\rho \, d\phi \, d\theta$

(d) $\int_0^{\pi/2}\int_0^1\int_0^r (6+4r\sin\theta) \, dz \, r \, dr \, d\theta = \int_0^{\pi/2}\int_0^1 \left(6r^2+4r^3\sin\theta\right) dr \, d\theta = \int_0^{\pi/2}\left[2r^3+r^4\sin\theta\right]_0^1 d\theta$

$= \int_0^{\pi/2}(2+\sin\theta) \, d\theta = [2\theta - \cos\theta]_0^{\pi/2} = \pi + 1$

47. $\int_0^1\int_{\sqrt{1-x^2}}^{\sqrt{3-x^2}}\int_1^{\sqrt{4-x^2-y^2}} z^2yx \, dz \, dy \, dx + \int_1^{\sqrt{3}}\int_0^{\sqrt{3-x^2}}\int_1^{\sqrt{4-x^2-y^2}} z^2yx \, dz \, dy \, dx$

48. (a) Bounded on the top and bottom by the sphere $x^2 + y^2 + z^2 = 4$, on the right by the right circular cylinder $(x-1)^2 + y^2 = 1$, on the left by the plane $y = 0$

(b) $\int_0^{\pi/2} \int_0^{2\cos\theta} \int_{-\sqrt{4-r^2}}^{\sqrt{4-r^2}} dz\, r\, dr\, d\theta$

49. (a) $\int_{-\sqrt{3}}^{\sqrt{3}} \int_{-\sqrt{3-x^2}}^{\sqrt{3-x^2}} \int_{1}^{\sqrt{4-x^2-y^2}} dz\, dy\, dx$ (b) $\int_0^{2\pi} \int_0^{\sqrt{3}} \int_1^{\sqrt{4-r^2}} dz\, r\, dr\, d\theta$

(c) $\int_0^{2\pi} \int_0^{\pi/3} \int_{\sec\phi}^{2} \rho^2 \sin\phi\, d\rho\, d\phi\, d\theta$

50. (a) $I_z = \int_{-1}^{1} \int_{-\sqrt{1-x^2}}^{\sqrt{1-x^2}} \int_0^{\sqrt{1-x^2-y^2}} (x^2 + y^2)\, dz\, dy\, dx$

(b) $I_z = \int_0^{2\pi} \int_0^1 \int_0^{\sqrt{1-r^2}} r^3\, dz\, dr\, d\theta$

(c) $I_z = \int_0^{2\pi} \int_0^{\pi/2} \int_0^1 \rho^4 \sin^3\phi\, d\rho\, d\phi\, d\theta$

51. (a) $V = \int_0^{2\pi} \int_0^2 \int_2^{\sqrt{8-r^2}} dz\, r\, dr\, d\theta = \int_0^{2\pi} \int_0^2 \left(r\sqrt{8-r^2} - 2r\right) dr\, d\theta = \int_0^{2\pi} \left[-\frac{1}{3}(8-r^2)^{3/2} - r^2\right]_0^2 d\theta$

$= \int_0^{2\pi} \left[-\frac{1}{3}(4)^{3/2} - 4 + \frac{1}{3}(8)^{3/2}\right] d\theta = \int_0^{2\pi} \frac{4}{3}\left(-2 - 3 + 2\sqrt{8}\right) d\theta = \frac{4}{3}\left(4\sqrt{2} - 5\right) \int_0^{2\pi} d\theta = \frac{8\pi\left(4\sqrt{2}-5\right)}{3}$

(b) $V = \int_0^{2\pi} \int_0^{\pi/4} \int_{2\sec\phi}^{\sqrt{8}} \rho^2 \sin\phi\, d\rho\, d\phi\, d\theta = \frac{8}{3} \int_0^{2\pi} \int_0^{\pi/4} \left(2\sqrt{2}\sin\phi - \sec^3\phi \sin\phi\right) d\phi\, d\theta$

$= \frac{8}{3} \int_0^{2\pi} \int_0^{\pi/4} \left(2\sqrt{2}\sin\phi - \tan\phi \sec^2\phi\right) d\phi\, d\theta = \frac{8}{3} \int_0^{2\pi} \left[-2\sqrt{2}\cos\phi - \frac{1}{2}\tan^2\phi\right]_0^{\pi/4} d\theta$

$= \frac{8}{3} \int_0^{2\pi} \left(-2 - \frac{1}{2} + 2\sqrt{2}\right) d\theta = \frac{8}{3} \int_0^{2\pi} \left(\frac{-5 + 4\sqrt{2}}{2}\right) d\theta = \frac{8\pi\left(4\sqrt{2}-5\right)}{3}$

52. $I_z = \int_0^{2\pi}\int_0^{\pi/3}\int_0^2 (\rho \sin\phi)^2(\rho^2 \sin\phi)\, d\rho\, d\phi\, d\theta = \int_0^{2\pi}\int_0^{\pi/3}\int_0^2 \rho^4 \sin^3\phi\, d\rho\, d\phi\, d\theta$

$= \frac{32}{5}\int_0^{2\pi}\int_0^{\pi/3} (\sin\phi - \cos^2\phi \sin\phi)\, d\phi\, d\theta = \frac{32}{5}\int_0^{2\pi}\left[-\cos\phi + \frac{\cos^3\phi}{3}\right]_0^{\pi/3} d\theta = \frac{8\pi}{3}$

53. With the centers of the spheres at the origin, $I_z = \int_0^{2\pi}\int_0^{\pi}\int_a^b \delta(\rho \sin\phi)^2(\rho^2 \sin\phi)\, d\rho\, d\phi\, d\theta$

$= \frac{\delta(b^5 - a^5)}{5}\int_0^{2\pi}\int_0^{\pi} \sin^3\phi\, d\phi\, d\theta = \frac{\delta(b^5 - a^5)}{5}\int_0^{2\pi}\int_0^{\pi} (\sin\phi - \cos^2\phi \sin\phi)\, d\phi\, d\theta$

$= \frac{\delta(b^5 - a^5)}{5}\int_0^{2\pi}\left[-\cos\phi + \frac{\cos^3\phi}{3}\right]_0^{\pi} d\theta = \frac{4\delta(b^5 - a^5)}{15}\int_0^{2\pi} d\theta = \frac{8\pi\delta(b^5 - a^5)}{15}$

54. $I_z = \int_0^{2\pi}\int_0^{\pi}\int_0^{1-\cos\theta} (\rho \sin\phi)^2(\rho^2 \sin\phi)\, d\rho\, d\phi\, d\theta = \int_0^{2\pi}\int_0^{\pi}\int_0^{1-\cos\theta} \rho^4 \sin^3\phi\, d\rho\, d\phi\, d\theta$

$= \frac{1}{5}\int_0^{2\pi}\int_0^{\pi} (1 - \cos\phi)^5 \sin^3\phi\, d\phi\, d\theta = \int_0^{2\pi}\int_0^{\pi} (1 - \cos\phi)^6(1 + \cos\phi) \sin\phi\, d\phi\, d\theta;$

$\begin{bmatrix} u = 1 - \cos\phi \\ du = \sin\phi\, d\phi \end{bmatrix} \to \frac{1}{5}\int_0^{2\pi}\int_0^{2} u^6(2 - u)\, du\, d\theta = \frac{1}{5}\int_0^{2\pi}\left[\frac{2u^7}{7} - \frac{u^8}{8}\right]_0^2 d\theta = \frac{1}{5}\int_0^{2\pi}\left(\frac{1}{7} - \frac{1}{8}\right)2^8\, d\theta$

$= \frac{1}{5}\int_0^{2\pi} \frac{2^3 \cdot 2^5}{56}\, d\theta = \frac{32}{35}\int_0^{2\pi} d\theta = \frac{64\pi}{35}$

55. $x = u + y$ and $y = v \Rightarrow x = u + v$ and $y = v$

$\Rightarrow J(u, v) = \begin{vmatrix} 1 & 1 \\ 0 & 1 \end{vmatrix} = 1$; the boundary of the image G is obtained from the boundary of R as follows:

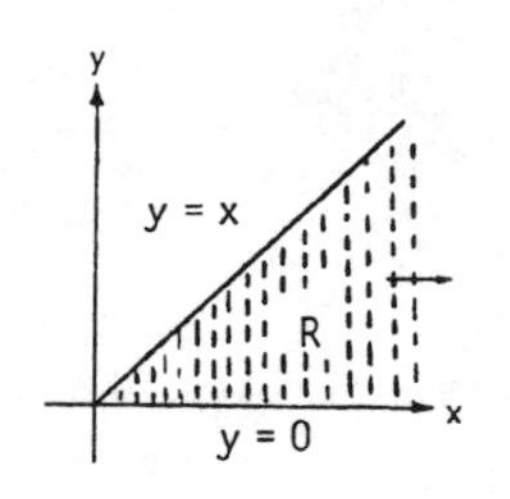

xy-equations for the boundary of R	Corresponding uv-equations for the boundary of G	Simplified uv-equations
$y = x$	$v = u + v$	$u = 0$
$y = 0$	$v = 0$	$v = 0$

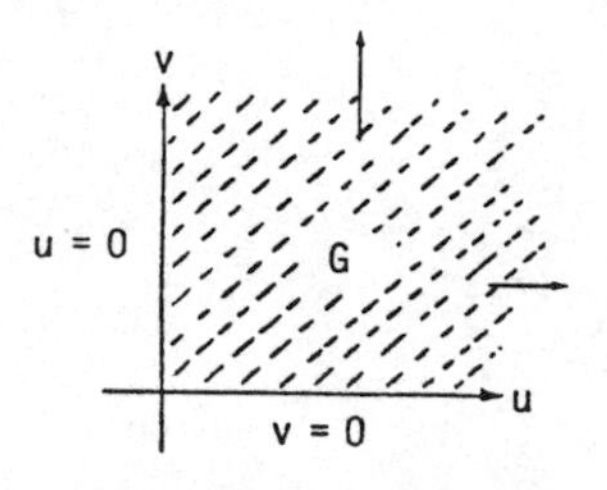

$$\Rightarrow \int_0^{\infty}\int_0^{x} e^{-sx} f(x-y,y)\,dy\,dx = \int_0^{\infty}\int_0^{\infty} e^{-s(u+v)} f(u,v)\,du\,dv$$

56. If $s = \alpha x + \beta y$ and $t = \gamma x + \delta y$ where $(\alpha\delta - \beta\gamma)^2 = ac - b^2$, then $x = \frac{\delta s - \beta t}{\alpha\delta - \beta\gamma}$, $y = \frac{-\gamma s + \alpha t}{\alpha\delta - \beta\gamma}$,

and $J(s,t) = \frac{1}{(\alpha\delta-\beta\gamma)^2}\begin{vmatrix} \delta & -\beta \\ -\gamma & \alpha \end{vmatrix} = \frac{1}{\alpha\delta - \beta\gamma} \Rightarrow \int_{-\infty}^{\infty}\int_{-\infty}^{\infty} e^{-\left(s^2+t^2\right)} \frac{1}{\sqrt{ac-b^2}}\,ds\,dt$

$= \frac{1}{\sqrt{ac-b^2}}\int_0^{2\pi}\int_0^{\infty} re^{-r^2}\,dr\,d\theta = \frac{1}{2\sqrt{ac-b^2}}\int_0^{2\pi} d\theta = \frac{\pi}{\sqrt{ac-b^2}}$. Therefore, $\frac{\pi}{\sqrt{ac-b^2}} = 1 \Rightarrow ac - b^2 = \pi^2$.

CHAPTER 13 ADDITIONAL EXERCISES–THEORY, EXAMPLES, APPLICATIONS

1. (a) $V = \int_{-3}^{2}\int_{x}^{6-x^2} x^2\,dy\,dx$ (b) $V = \int_{-3}^{2}\int_{x}^{6-x^2}\int_0^{x^2} dz\,dy\,dx$

(c) $V = \int_{-3}^{2}\int_{x}^{6-x^2} x^2\,dy\,dx = \int_{-3}^{2}\left(6x^2 - x^4 - x^3\right)dx = \left[2x^3 - \frac{x^5}{5} - \frac{x^4}{4}\right]_{-3}^{2} = \frac{125}{4}$

2. Place the sphere's center at the origin with the surface of the water at $z = -3$. Then $9 = 25 - x^2 - y^2 \Rightarrow x^2 + y^2 = 16$ is the projection of the volume of water onto the xy-plane

$\Rightarrow V = \int_0^{2\pi}\int_0^{4}\int_{-\sqrt{25-r^2}}^{-3} dz\ r\,dr\,d\theta = \int_0^{2\pi}\int_0^{4}\left(r\sqrt{25-r^2} - 3r\right)dr\,d\theta = \int_0^{2\pi}\left[-\frac{1}{3}\left(25-r^2\right)^{3/2} - \frac{3}{2}r^2\right]_0^4 d\theta$

$= \int_0^{2\pi}\left[-\frac{1}{3}(9)^{3/2} - 24 + \frac{1}{3}(25)^{3/2}\right]d\theta = \int_0^{2\pi}\frac{26}{3}\,d\theta = \frac{52\pi}{3}$

3. Using cylindrical coordinates, $V = \int_0^{2\pi}\int_0^{1}\int_0^{2-r(\cos\theta+\sin\theta)} dz\ r\,dr\,d\theta = \int_0^{2\pi}\int_0^{1}\left(2r - r^2\cos\theta - r^2\sin\theta\right)dr\,d\theta$

$= \int_0^{2\pi}\left(1 - \frac{1}{3}\cos\theta - \frac{1}{3}\sin\theta\right)d\theta = \left[\theta + \frac{1}{3}\sin\theta - \frac{1}{3}\cos\theta\right]_0^{2\pi} = 2\pi$

4. $V = 4\int_0^{\pi/2}\int_0^1\int_{r^2}^{\sqrt{2-r^2}} dz\, r\, dr\, d\theta = 4\int_0^{\pi/2}\int_0^1 \left(r\sqrt{2-r^2}-r^3\right) dr\, d\theta = 4\int_0^{\pi/2}\left[-\frac{1}{3}\left(2-r^2\right)^{3/2}-\frac{r^4}{4}\right]_0^1 d\theta$

$= 4\int_0^{\pi/2}\left(-\frac{1}{3}-\frac{1}{4}+\frac{2\sqrt{2}}{3}\right) d\theta = \left(\frac{8\sqrt{2}-7}{3}\right)\int_0^{\pi/2} d\theta = \frac{\pi\left(8\sqrt{2}-7\right)}{6}$

5. The surfaces intersect when $3-x^2-y^2 = 2x^2+2y^2 \Rightarrow x^2+y^2 = 1$. Thus the volume is

$V = 4\int_0^1\int_0^{\sqrt{1-x^2}}\int_{2x^2+2y^2}^{3-x^2-y^2} dz\, dy\, dx = 4\int_0^{\pi/2}\int_0^1\int_{2r^2}^{3-r^2} dz\, r\, dr\, d\theta = 4\int_0^{\pi/2}\int_0^1 \left(3r-3r^3\right) dr\, d\theta = 3\int_0^{\pi/2} d\theta = \frac{3\pi}{2}$

6. $V = 8\int_0^{\pi/2}\int_0^{\pi/2}\int_0^{2\sin\phi} \rho^2 \sin\phi\, d\rho\, d\phi\, d\theta = \frac{64}{3}\int_0^{\pi/2}\int_0^{\pi/2} \sin^4\phi\, d\phi\, d\theta$

$= \frac{64}{3}\int_0^{\pi/2}\left[-\frac{\sin^3\phi\cos\phi}{4}\Big|_0^{\pi/2}+\frac{3}{4}\int_0^{\pi/2}\sin^2\phi\, d\phi\right] d\theta = 16\int_0^{\pi/2}\left[\frac{\phi}{2}-\frac{\sin 2\phi}{4}\right]_0^{\pi/2} d\theta = 4\pi\int_0^{\pi/2} d\theta = 2\pi^2$

7. (a) The radius of the hole is 1, and the radius of the sphere is 2.

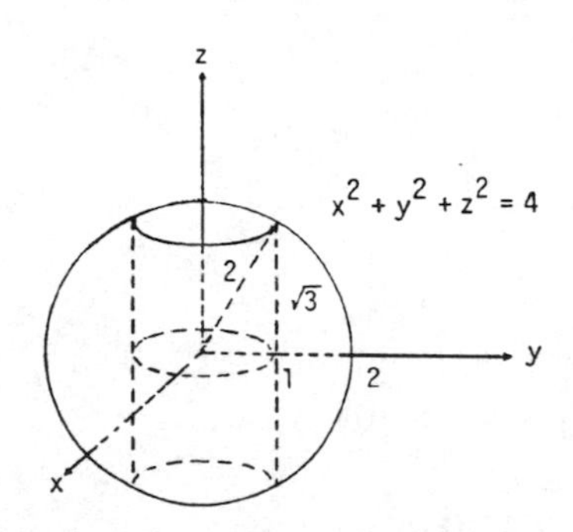

(b) $V = 2\int_0^{2\pi}\int_0^{\sqrt{3}}\int_1^{\sqrt{4-z^2}} r\, dr\, dz\, d\theta = \int_0^{2\pi}\int_0^{\sqrt{3}} \left(3-z^2\right) dz\, d\theta = 2\sqrt{3}\int_0^{2\pi} d\theta = 4\sqrt{3}\pi$

8. $V = \int_0^{\pi}\int_0^{3\sin\theta}\int_0^{\sqrt{9-r^2}} dz\, r\, dr\, d\theta = \int_0^{\pi}\int_0^{3\sin\theta} r\sqrt{9-r^2}\, dr\, d\theta = \int_0^{\pi}\left[-\frac{1}{3}\left(9-r^2\right)^{3/2}\right]_0^{3\sin\theta} d\theta$

$= \int_0^{\pi}\left[-\frac{1}{3}\left(9-9\sin^2\theta\right)^{3/2}+\frac{1}{3}(9)^{3/2}\right] d\theta = 9\int_0^{\pi}\left[1-\left(1-\sin^2\theta\right)^{3/2}\right] d\theta = 9\int_0^{\pi}\left(1-\cos^3\theta\right) d\theta$

$= \int_0^{\pi}\left(1-\cos\theta+\sin^2\theta\cos\theta\right) d\theta = 9\left[\theta+\sin\theta+\frac{\sin^3\theta}{3}\right]_0^{\pi} = 9\pi$

9. The surfaces intersect when $x^2+y^2=\frac{x^2+y^2+1}{2} \Rightarrow x^2+y^2=1$. Thus the volume in cylindrical coordinates is $V=4\int_0^{\pi/2}\int_0^1\int_{r^2}^{(r^2+1)/2} dz\, r\, dr\, d\theta = 4\int_0^{\pi/2}\int_0^1 \left(\frac{r}{2}-\frac{r^3}{2}\right) dr\, d\theta = 4\int_0^{\pi/2}\left[\frac{r^2}{4}-\frac{r^4}{8}\right]_0^1 d\theta$

$=\frac{1}{2}\int_0^{\pi/2} d\theta = \frac{\pi}{4}$

10. $V=\int_0^{\pi/2}\int_1^2\int_0^{r^2\sin\theta\cos\theta} dz\, r\, dr\, d\theta = \int_0^{\pi/2}\int_1^2 r^3 \sin\theta\cos\theta\, dr\, d\theta = \int_0^{\pi/2}\left[\frac{r^4}{4}\right]_1^2 \sin\theta\cos\theta\, d\theta$

$=\frac{15}{4}\int_0^{\pi/2}\sin\theta\cos\theta\, d\theta = \frac{15}{4}\left[\frac{\sin^2\theta}{2}\right]_0^{\pi/2} = \frac{15}{8}$

11. $\int_0^1\int_y^{\sqrt{y}} f(x,y)\, dx\, dy$

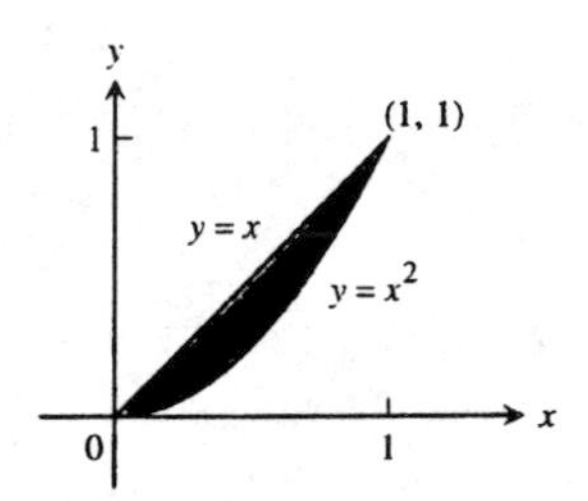

12. $\int_0^4\int_{x^2/4}^{x} f(x,y)\, dy\, dx$

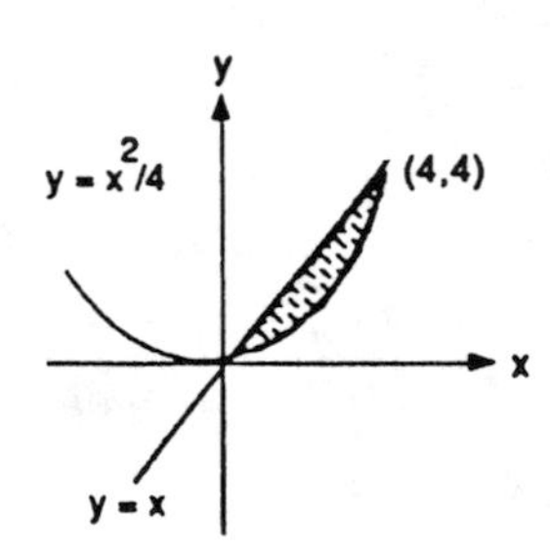

13. $\int_0^\infty \frac{e^{-ax}-e^{-bx}}{x}\, dx = \int_0^\infty\int_a^b e^{-xy}\, dy\, dx = \int_a^b\int_0^\infty e^{-xy}\, dx\, dy = \int_a^b \left(\lim_{t\to\infty}\int_0^t e^{-xy}\, dx\right) dy$

$=\int_a^b \lim_{t\to\infty}\left[-\frac{e^{-xy}}{y}\right]_0^t dy = \int_a^b \lim_{t\to\infty}\left(\frac{1}{y}-\frac{e^{-yt}}{y}\right) dy = \int_a^b \frac{1}{y}\, dy = [\ln y]_a^b = \ln\left(\frac{b}{a}\right)$

14. (a) The region of integration is sketched at the right

$$\Rightarrow \int_0^{a\sin\beta}\int_{y\cot\beta}^{\sqrt{a^2-y^2}} \ln(x^2+y^2)\,dx\,dy = \int_0^{\beta}\int_0^{a} r\ln(r^2)\,dr\,d\theta;$$

$$\begin{bmatrix} u = r^2 \\ du = 2r\,dr \end{bmatrix} \to \frac{1}{2}\int_0^{\beta}\int_0^{a^2} \ln u\,du\,d\theta = \frac{1}{2}\int_0^{\beta} \left[u\ln u - u\right]_0^{a^2} d\theta$$

$$= \frac{1}{2}\int_0^{\beta}\left[2a^2\ln a - a^2 - \lim_{t\to 0} t\ln t\right] d\theta = \frac{a^2}{2}\int_0^{\beta}(2\ln a - 1)\,d\theta = a^2\beta\left(\ln a - \frac{1}{2}\right)$$

(b) $\displaystyle\int_0^{a\cos\beta}\int_0^{(\tan\beta)x} \ln(x^2+y^2)\,dy\,dx + \int_{a\cos\beta}^{a}\int_0^{\sqrt{a^2-x^2}} \ln(x^2+y^2)\,dy\,dx$

15. $\displaystyle\int_0^x\int_0^u e^{m(x-t)}f(t)\,dt\,du = \int_0^x\int_1^x e^{m(x-t)}f(t)\,du\,dt = \int_0^x (x-t)e^{m(x-t)}f(t)\,dt$; also

$$\int_0^x\int_0^v\int_0^u e^{m(x-t)}f(t)\,dt\,du\,dv = \int_0^x\int_t^x\int_t^v e^{m(x-t)}f(t)\,du\,dv\,dt = \int_0^x\int_t^x (v-t)e^{m(x-t)}f(t)\,dv\,dt$$

$$= \int_0^x \left[\frac{1}{2}(v-t)^2 e^{m(x-t)}f(t)\right]_t^x dt = \int_0^x \frac{(x-t)^2}{2}e^{m(x-t)}f(t)\,dt$$

16. $\displaystyle\int_0^1 f(x)\left(\int_0^x g(x-y)f(y)\,dy\right)dx = \int_0^1\int_0^x g(x-y)f(x)f(y)\,dy\,dx$

$$= \int_0^1\int_y^1 g(x-y)f(x)f(y)\,dx\,dy = \int_0^1 f(y)\left(\int_y^1 g(x-y)f(x)\,dx\right)dy;$$

$$\int_0^1\int_0^1 g(|x-y|)f(x)f(y)\,dx\,dy = \int_0^1\int_0^x g(x-y)f(x)f(y)\,dy\,dx + \int_0^1\int_x^1 g(y-x)f(x)f(y)\,dy\,dx$$

$$= \int_0^1\int_y^1 g(x-y)f(x)f(y)\,dx\,dy + \int_0^1\int_x^1 g(y-x)f(x)f(y)\,dy\,dx$$

$$= \int_0^1\int_y^1 g(x-y)f(x)f(y)\,dx\,dy + \underbrace{\int_0^1\int_y^1 g(x-y)f(y)f(x)\,dx\,dy}_{\text{simply interchange x and y variable names}}$$

$$= 2\int_0^1\int_y^1 g(x-y)f(x)f(y)\,dx\,dy, \text{ and the statement now follows.}$$

17. $I_o(a) = \int_0^a \int_0^{x/a^2} (x^2+y^2)\,dy\,dx = \int_0^a \left[x^2y + \frac{y^3}{3}\right]_0^{x/a^2} dx = \int_0^a \left(\frac{x^3}{a^2} + \frac{x^3}{3a^6}\right) dx = \left[\frac{x^4}{4a^2} + \frac{x^4}{12a^6}\right]_0^a$

$= \frac{a^2}{4} + \frac{1}{12}a^{-2}$; $I_o'(a) = \frac{1}{2}a - \frac{1}{6}a^{-3} = 0 \Rightarrow a^4 = \frac{1}{3} \Rightarrow a = \sqrt[4]{\frac{1}{3}} = \frac{1}{\sqrt[4]{3}}$. Since $I_o''(a) = \frac{1}{2} + \frac{1}{2}a^{-4} > 0$, the value of a does provide a minimum for the polar moment of inertia $I_o(a)$.

18. $I_o = \int_0^2 \int_{2x}^4 (x^2+y^2)(3)\,dy\,dx = 3\int_0^2 \left(4x^2 - \frac{14x^3}{3} + \frac{64}{3}\right) dx = 104$

19. $M = 2\int_0^{\pi/2} \int_1^{1+\cos\theta} r\,dr\,d\theta = \int_0^{\pi/2} \left(2\cos\theta + \frac{1+\cos 2\theta}{2}\right) d\theta = \frac{8+\pi}{4}$;

$M_y = \int_{-\pi/2}^{\pi/2} \int_1^{1+\cos\theta} r^2\cos\theta\,dr\,d\theta = \int_{-\pi/2}^{\pi/2} \left(\cos^2\theta + \cos^3\theta + \frac{\cos^4\theta}{3}\right) d\theta$

$= \left[\frac{\theta}{2} + \frac{\sin 2\theta}{4} + \frac{\cos^2\theta\sin\theta}{3} + \frac{\cos^3\theta\sin\theta}{12}\right]_{-\pi/2}^{\pi/2} + \int_{-\pi/2}^{\pi/2} \left(\frac{2}{3}\cos\theta + \frac{1}{4}\cos^2\theta\right) d\theta$

$= \frac{\pi}{2} + \left[\frac{2}{3}\sin\theta + \frac{1}{4}\left(\frac{\theta}{2} + \frac{\sin 2\theta}{4}\right)\right]_{-\pi/2}^{\pi/2} = \frac{\pi}{2} + \frac{4}{3} + \frac{\pi}{8} = \frac{5\pi}{8} + \frac{4}{3} = \frac{32+15\pi}{24} \Rightarrow \bar{x} = \frac{15\pi+32}{6\pi+48}$, and $\bar{y} = 0$ by symmetry

20. $M = \int_{-2}^{2} \int_{1-(y^2/4)}^{2-(y^2/2)} dx\,dy = \int_{-2}^{2} \left(1 - \frac{y^2}{4}\right) dy = \left[y - \frac{y^3}{12}\right]_{-2}^{2} = \frac{8}{3}$; $M_y = \int_{-2}^{2} \int_{1-(y^2/4)}^{2-(y^2/2)} x\,dx\,dy$

$= \int_{-2}^{2} \left[\frac{x^2}{2}\right]_{1-(y^2/4)}^{2-(y^2/2)} dy = \int_{-2}^{2} \frac{3}{32}(4-y^2)\,dy = \frac{3}{32}\int_{-2}^{2} (16 - 8y^2 + y^4)\,dy = \frac{3}{16}\left[16y - \frac{8y^3}{3} + \frac{y^5}{5}\right]_0^2$

$= \frac{3}{16}\left(32 - \frac{64}{3} + \frac{32}{5}\right) = \left(\frac{3}{16}\right)\left(\frac{32\cdot 8}{15}\right) = \frac{48}{15} \Rightarrow \bar{x} = \frac{M_y}{M} = \left(\frac{48}{15}\right)\left(\frac{3}{8}\right) = \frac{6}{5}$, and $\bar{y} = 0$ by symmetry

21. $M = \int_{-\theta}^{\theta} \int_{b\sec\theta}^{a} r\,dr\,d\theta = \int_{-\theta}^{\theta} \left(\frac{a^2}{2} - \frac{b^2}{2}\sec^2\theta\right) d\theta$

$= a^2\theta - b^2\tan\theta = a^2\cos^{-1}\left(\frac{b}{a}\right) - b^2\left(\frac{\sqrt{a^2-b^2}}{b}\right)$

$= a^2\cos^{-1}\left(\frac{b}{a}\right) - b\sqrt{a^2-b^2}$; $I_o = \int_{-\theta}^{\theta} \int_{b\sec\theta}^{a} r^3\,dr\,d\theta$

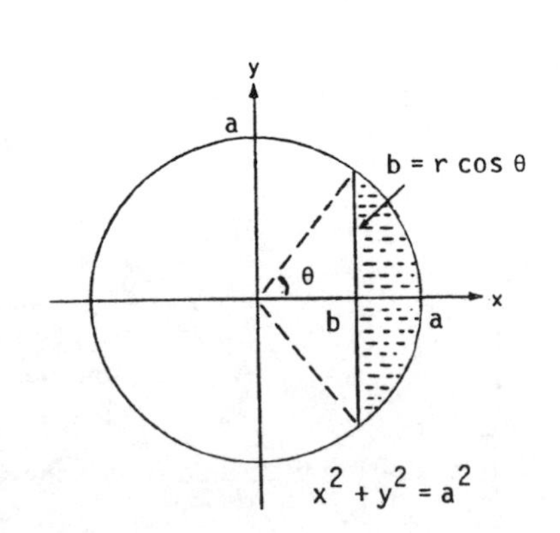

$$=\frac{1}{4}\int_{-\theta}^{\theta}\left(a^4+b^4\sec^4\theta\right)d\theta=\frac{1}{4}\int_{-\theta}^{\theta}\left[a^4+b^4\left(1+\tan^2\theta\right)\left(\sec^2\theta\right)\right]d\theta=\frac{1}{4}\left[a^4\theta-b^4\tan\theta-\frac{b^4\tan^3\theta}{3}\right]_{-\theta}^{\theta}$$

$$=\frac{a^4\theta}{2}-\frac{b^4\tan\theta}{2}-\frac{b^4\tan^3\theta}{6}=\frac{1}{2}a^4\cos^{-1}\left(\frac{b}{a}\right)-\frac{1}{2}b^3\sqrt{a^2-b^2}-\frac{1}{6}b^3\left(a^2-b^2\right)^{3/2}$$

22. $$M=\int_{-\pi/4}^{\pi/4}\int_{0}^{a\sqrt{2\cos 2\theta}} r\,dr\,d\theta=\int_{-\pi/4}^{\pi/4}\left(a^2\cos 2\theta\right)d\theta=a^2\left[\frac{\sin 2\theta}{2}\right]_{-\pi/4}^{\pi/4}=a^2;$$

$$I_x=\int_{-\pi/4}^{\pi/4}\int_{0}^{a\sqrt{2\cos 2\theta}}(r\sin\theta)^2\,r\,dr\,d\theta=\int_{-\pi/4}^{\pi/4}\left(a^4\cos^2 2\theta\sin^2\theta\right)d\theta=\frac{(3\pi-8)a^4}{24};$$

$$I_y=\int_{-\pi/4}^{\pi/4}\int_{0}^{a\sqrt{2\cos 2\theta}}(r\cos\theta)^2\,r\,dr\,d\theta=\int_{-\pi/4}^{\pi/4}\left(a^4\cos^2 2\theta\cos^2\theta\right)d\theta=\frac{(8+3\pi)a^4}{24}$$

$$\Rightarrow R_x=\sqrt{\frac{I_x}{M}}=\frac{a\sqrt{3\pi-8}}{2\sqrt{6}}\text{ and }R_y=\sqrt{\frac{I_y}{M}}=\frac{a\sqrt{3\pi+8}}{2\sqrt{6}}$$

23. (a) $$M=\int_0^{2\pi}\int_0^1\int_0^{r^2} z\,dz\,r\,dr\,d\theta=\int_0^{2\pi}\int_0^1\frac{1}{2}r^5\,dr\,d\theta=\frac{1}{12}\int_0^{2\pi}d\theta=\frac{\pi}{6};\ M_{xy}=\int_0^{2\pi}\int_0^1\int_0^{r^2} z^2\,dz\,r\,dr\,d\theta$$

$$=\frac{1}{3}\int_0^{2\pi}\int_0^1 r^7\,dr\,d\theta=\frac{1}{24}\int_0^{2\pi}d\theta=\frac{\pi}{12}\Rightarrow\bar{z}=\frac{1}{2},\text{ and }\bar{x}=\bar{y}=0\text{ by symmetry; }I_z=\int_0^{2\pi}\int_0^1\int_0^{r^2} zr^3\,dz\,dr\,d\theta$$

$$=\frac{1}{2}\int_0^{2\pi}\int_0^1 r^7\,dr\,d\theta=\frac{1}{16}\int_0^{2\pi}d\theta=\frac{\pi}{8}\Rightarrow R_z=\sqrt{\frac{I_z}{M}}=\frac{\sqrt{3}}{2}$$

(b) $$M=\int_0^{2\pi}\int_0^1\int_0^{r^2} r^2\,dz\,dr\,d\theta=\int_0^{2\pi}\int_0^1 r^4\,dr\,d\theta=\frac{1}{5}\int_0^{2\pi}d\theta=\frac{2\pi}{5};\ M_{xy}=\int_0^{2\pi}\int_0^1\int_0^{r^2} zr^2\,dz\,dr\,d\theta$$

$$=\frac{1}{2}\int_0^{2\pi}\int_0^1 r^6\,dr\,d\theta=\frac{1}{14}\int_0^{2\pi}d\theta=\frac{\pi}{7}\Rightarrow\bar{z}=\frac{5}{14},\text{ and }\bar{x}=\bar{y}=0\text{ by symmetry; }I_z=\int_0^{2\pi}\int_0^1\int_0^{r^2} r^4\,dz\,dr\,d\theta$$

$$=\int_0^{2\pi}\int_0^1 r^6\,dr\,d\theta=\frac{1}{7}\int_0^{2\pi}d\theta=\frac{2\pi}{7}\Rightarrow R_z=\sqrt{\frac{I_z}{M}}=\sqrt{\frac{5}{7}}$$

24. (a) $M = \int_0^{2\pi}\int_0^1\int_r^1 z\, dz\, r\, dr\, d\theta = \frac{1}{2}\int_0^{2\pi}\int_0^1 \left(r - r^3\right) dr\, d\theta = \frac{1}{8}\int_0^{2\pi} d\theta = \frac{\pi}{4}$; $M_{xy} = \int_0^{2\pi}\int_0^1\int_r^1 z^2\, dz\, r\, dr\, d\theta$

$= \frac{1}{3}\int_0^{2\pi}\int_0^1 \left(r - r^4\right) dr\, d\theta = \frac{1}{10}\int_0^{2\pi} d\theta = \frac{\pi}{5} \Rightarrow \bar{z} = \frac{4}{5}$, and $\bar{x} = \bar{y} = 0$ by symmetry; $I_z = \int_0^{2\pi}\int_0^1\int_r^1 zr^3\, dz\, dr\, d\theta$

$= \frac{1}{2}\int_0^{2\pi}\int_0^1 \left(r^3 - r^5\right) dr\, d\theta = \frac{1}{24}\int_0^{2\pi} d\theta = \frac{\pi}{12} \Rightarrow R_z = \sqrt{\frac{I_z}{M}} = \sqrt{\frac{1}{3}}$

(b) $M = \int_0^{2\pi}\int_0^1\int_r^1 z^2\, dz\, r\, dr\, d\theta = \frac{\pi}{5}$ from part (a); $M_{xy} = \int_0^{2\pi}\int_0^1\int_r^1 z^3\, dz\, r\, dr\, d\theta = \frac{1}{4}\int_0^{2\pi}\int_0^1 \left(r - r^5\right) dr\, d\theta$

$= \frac{1}{12}\int_0^{2\pi} d\theta = \frac{\pi}{6} \Rightarrow \bar{z} = \frac{5}{6}$, and $\bar{x} = \bar{y} = 0$ by symmetry; $I_z = \int_0^{2\pi}\int_0^1\int_r^1 z^2 r^3\, dz\, dr\, d\theta = \frac{1}{3}\int_0^{2\pi}\int_0^1 \left(r^3 - r^6\right) dr\, d\theta$

$= \frac{1}{28}\int_0^{2\pi} d\theta = \frac{\pi}{14} \Rightarrow R_z = \sqrt{\frac{I_z}{M}} = \sqrt{\frac{5}{14}}$

25. $M = \frac{2}{3}\pi a^3$; $M_{xy} = \int_0^{2\pi}\int_0^{\pi/2}\int_0^a \left(\rho^3 \sin\phi \cos\phi\right) d\rho\, d\phi\, d\theta = \frac{a^4}{4}\int_0^{2\pi}\int_0^{\pi/2} \sin\phi \cos\phi\, d\phi\, d\theta = \frac{a^4}{8}\int_0^{2\pi} d\theta = \frac{a^4\pi}{4}$

$\Rightarrow \bar{z} = \frac{3a}{8}$, and $\bar{x} = \bar{y} = 0$, by symmetry

26. $M = \frac{4}{3}\pi a^3$; $I_z = \int_0^{2\pi}\int_0^{\pi}\int_0^a \left(\rho^4 \sin^3\phi\right) d\rho\, d\phi\, d\theta = \frac{a^5}{5}\int_0^{2\pi}\int_0^{\pi} \sin^3\phi\, d\phi\, d\theta = \frac{a^5}{5}\int_0^{2\pi} \left[-\frac{\sin^2\phi \cos\phi}{3} - \frac{2}{3}\cos\phi\right]_0^{\pi} d\theta$

$= \frac{4a^5}{15}\int_0^{2\pi} d\theta = \frac{8a^5\pi}{15} \Rightarrow R_z = \sqrt{\frac{I_z}{M}} = a\sqrt{\frac{2}{5}}$

27. $\int_0^a\int_0^b e^{\max\left(b^2x^2,\, a^2y^2\right)}\, dy\, dx = \int_0^a\int_0^{bx/a} e^{b^2x^2}\, dy\, dx + \int_0^b\int_0^{ay/b} e^{a^2y^2}\, dx\, dy$

$= \int_0^a \left(\frac{b}{a}x\right)e^{b^2x^2}\, dx + \int_0^b \left(\frac{a}{b}y\right)e^{a^2y^2}\, dy = \left[\frac{1}{2ab}e^{b^2x^2}\right]_0^a + \left[\frac{1}{2ba}e^{a^2y^2}\right]_0^b = \frac{1}{2ab}\left(e^{b^2a^2} - 1\right) + \frac{1}{2ab}\left(e^{a^2b^2} - 1\right)$

$= \frac{1}{ab}\left(e^{a^2b^2} - 1\right)$

28. $\int_{y_0}^{y_1}\int_{x_0}^{x_1} \frac{\partial^2 F(x,y)}{\partial x\,\partial y}\,dx\,dy = \int_{y_0}^{y_1}\left[\frac{\partial F(x,y)}{\partial y}\right]_{x_0}^{x_1} dy = \int_{y_0}^{y_1}\left[\frac{\partial F(x_1,y)}{\partial y} - \frac{\partial F(x_0,y)}{\partial y}\right] dx = \left[F(x_1,y) - F(x_0,y)\right]_{y_0}^{y_1}$

$= F(x_1,y_1) - F(x_0,y_1) - F(x_1,y_0) + F(x_0,y_0)$

29. (a) (i) Fubini's Theorem
(ii) Treating G(y) as a constant
(iii) Algebraic rearrangement
(iv) The definite integral is a constant number

(b) $\int_0^{\ln 2}\int_0^{\pi/2} e^x \cos y\,dy\,dx = \left(\int_0^{\ln 2} e^x\,dx\right)\left(\int_0^{\pi/2} \cos y\,dy\right) = \left(e^{\ln 2} - e^0\right)\left(\sin\frac{\pi}{2} - \sin 0\right) = (1)(1) = 1$

(c) $\int_1^2\int_{-1}^1 \frac{x}{y^2}\,dx\,dy = \left(\int_1^2 \frac{1}{y^2}\,dy\right)\left(\int_{-1}^1 x\,dx\right) = \left[-\frac{1}{y}\right]_1^2\left[\frac{x^2}{2}\right]_{-1}^1 = \left(-\frac{1}{2}+1\right)\left(\frac{1}{2}-\frac{1}{2}\right) = 0$

30. (a) $\nabla f = x\mathbf{i} + y\mathbf{j} \Rightarrow D_u f = u_1 x + u_2 y$; the area of the region of integration is $\frac{1}{2}$

$\Rightarrow \text{average} = 2\int_0^1\int_0^{1-x} (u_1 x + u_2 y)\,dy\,dx = 2\int_0^1 \left[u_1 x(1-x) + \frac{1}{2}u_2(1-x)^2\right] dx$

$= 2\left[u_1\left(\frac{x^2}{2} - \frac{x^3}{3}\right) - \left(\frac{1}{2}u_2\right)\frac{(1-x)^3}{3}\right]_0^1 = 2\left(\frac{1}{6}u_1 + \frac{1}{6}u_2\right) = \frac{1}{3}(u_1 + u_2)$

(b) $\text{average} = \frac{1}{\text{area}}\iint_R (u_1 x + u_2 y)\,dA = \frac{u_1}{\text{area}}\iint_R x\,dA + \frac{u_2}{\text{area}}\iint_R y\,dA = u_1\left(\frac{M_y}{M}\right) + u_2\left(\frac{M_x}{M}\right) = u_1\bar{x} + u_2\bar{y}$

31. (a) $I^2 = \int_0^\infty\int_0^\infty e^{-(x^2+y^2)}\,dx\,dy = \int_0^{\pi/2}\int_0^\infty \left(e^{-r^2}\right) r\,dr\,d\theta = \int_0^{\pi/2}\left[\lim_{b\to\infty}\int_0^b re^{-r^2}\,dr\right] d\theta$

$= -\frac{1}{2}\int_0^{\pi/2} \lim_{b\to\infty}\left(e^{-b^2} - 1\right) d\theta = \frac{1}{2}\int_0^{\pi/2} d\theta = \frac{\pi}{4} \Rightarrow I = \frac{\sqrt{\pi}}{2}$

(b) $\Gamma\left(\frac{1}{2}\right) = \int_0^\infty t^{-1/2}e^{-t}\,dt = \int_0^\infty \left(y^2\right)^{-1/2} e^{-y^2}(2y)\,dy = 2\int_0^\infty e^{-y^2}\,dy = 2\left(\frac{\sqrt{\pi}}{2}\right) = \sqrt{\pi}$, where $y = \sqrt{t}$

32. $Q = \int_0^{2\pi}\int_0^R kr^2(1 - \sin\theta)\,dr\,d\theta = \frac{kR^3}{3}\int_0^{2\pi}(1 - \sin\theta)\,d\theta = \frac{kR^3}{3}\left[\theta + \cos\theta\right]_0^{2\pi} = \frac{2\pi kR^3}{3}$

33. For a height h in the bowl the volume of water is $V = \int_{-\sqrt{h}}^{\sqrt{h}} \int_{-\sqrt{h-x^2}}^{\sqrt{h-x^2}} \int_{x^2+y^2}^{h} dz\,dy\,dx$

$$= \int_{-\sqrt{h}}^{\sqrt{h}} \int_{-\sqrt{h-x^2}}^{\sqrt{h-x^2}} (h - x^2 - y^2)\,dy\,dx = \int_0^{2\pi} \int_0^{\sqrt{h}} (h - r^2)\,r\,dr\,d\theta = \int_0^{2\pi} \left[\frac{hr^2}{2} - \frac{r^4}{4}\right]_0^{\sqrt{h}} d\theta = \int_0^{2\pi} \frac{h^2}{4}\,d\theta = \frac{h^2\pi}{2}.$$

Since the top of the bowl has area 10π, then we calibrate the bowl by comparing it to a right circular cylinder whose cross sectional area is 10π from $z = 0$ to $z = 10$. If such a cylinder contains $\frac{h^2\pi}{2}$ cubic inches of water to a depth w then we have $10\pi w = \frac{h^2\pi}{2} \Rightarrow w = \frac{h^2}{20}$. So for 1 inch of rain, $w = 1$ and $h = \sqrt{20}$; for 3 inches of rain, $w = 3$ and $h = \sqrt{60}$.

34. (a) An equation for the satellite dish in standard position is $z = \frac{1}{2}x^2 + \frac{1}{2}y^2$. Since the axis is tilted 30°, a unit vector $\mathbf{v} = 0\mathbf{i} + a\mathbf{j} + b\mathbf{k}$ normal to the plane of the water level satisfies $b = \mathbf{v} \cdot \mathbf{k} = \cos\left(\frac{\pi}{6}\right) = \frac{\sqrt{3}}{2}$

$\Rightarrow a = -\sqrt{1 - b^2} = -\frac{1}{2} \Rightarrow \mathbf{v} = -\frac{1}{2}\mathbf{j} + \frac{\sqrt{3}}{2}\mathbf{k}$

$\Rightarrow -\frac{1}{2}(y - 1) + \frac{\sqrt{3}}{2}\left(z - \frac{1}{2}\right) = 0 \Rightarrow z = \frac{1}{\sqrt{3}}y + \left(\frac{1}{2} - \frac{1}{\sqrt{3}}\right)$

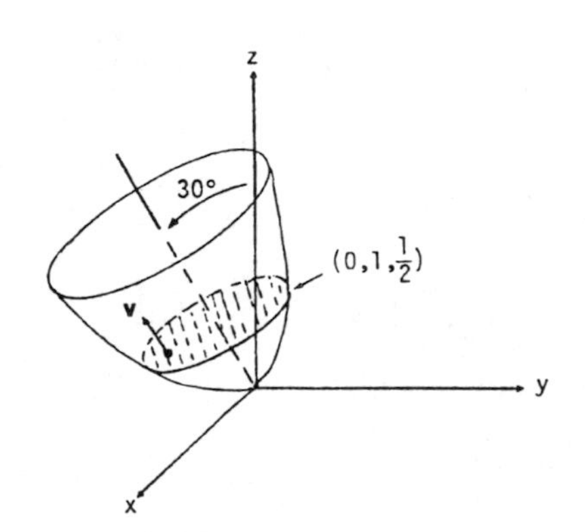

is an equation of the plane of the water level. Therefore the volume of water is $V = \iint_R \int_{\frac{1}{2}x^2+\frac{1}{2}y^2}^{\frac{1}{\sqrt{3}}y+\frac{1}{2}-\frac{1}{\sqrt{3}}} dz\,dy\,dx$, where R is the interior of the ellipse

$x^2 + y^2 - \frac{2}{3}y - 1 + \frac{2}{\sqrt{3}} = 0$. When $x = 0$, then $y = \alpha$ or $y = \beta$, where $\alpha = \frac{\frac{2}{3} + \sqrt{\frac{4}{9} - 4\left(\frac{2}{\sqrt{3}} - 1\right)}}{2}$

and $\beta = \frac{\frac{2}{3} - \sqrt{\frac{4}{9} - 4\left(\frac{2}{\sqrt{3}} - 1\right)}}{2} \Rightarrow V = \int_\alpha^\beta \int_{-\left(\frac{2}{3}y+1-\frac{2}{\sqrt{3}}-y^2\right)^{1/2}}^{\left(\frac{2}{3}y+1-\frac{2}{\sqrt{3}}-y^2\right)^{1/2}} \int_{\frac{1}{2}x^2+\frac{1}{2}y^2}^{\frac{1}{\sqrt{3}}y+\frac{1}{2}-\frac{1}{\sqrt{3}}} 1\,dz\,dx\,dy$

(b) $x = 0 \Rightarrow z = \frac{1}{2}y^2$ and $\frac{dz}{dy} = y$; $y = 1 \Rightarrow \frac{dz}{dy} = 1 \Rightarrow$ the tangent line has slope 1 or a 45° slant $\Rightarrow$ at 45° and thereafter, the dish will not hold water.

35. Using cylindrical coordinates with $x = r\cos\theta$, $y = y$, and

$z = r\sin\theta$, we have $V = \int_a^b \int_0^{2\pi} \int_0^{f(r)} r\, dy\, d\theta\, dr$

$$= \int_a^b \int_0^{2\pi} [yr]_0^{f(r)}\, r\, d\theta\, dr = \int_a^b \int_0^{2\pi} r f(r)\, d\theta\, dr$$

$$= \int_a^b 2\pi r f(r)\, dr = 2\pi \int_a^b x f(x)\, dx.$$

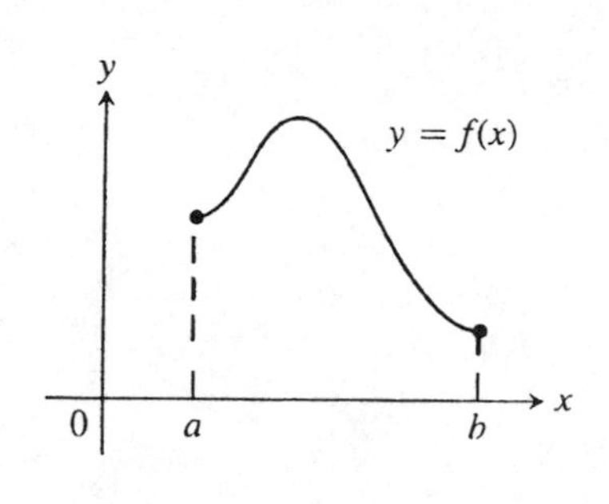

36. The cylinder is given by $x^2 + y^2 = 1$ from $z = 1$ to $\infty \Rightarrow \iiint_D z\left(r^2 + z^2\right)^{-5/2} dV$

$$= \int_0^{2\pi} \int_0^1 \int_1^{\infty} \frac{z}{\left(r^2 + z^2\right)^{5/2}}\, dz\, r\, dr\, d\theta = \lim_{a\to\infty} \int_0^{2\pi} \int_0^1 \int_1^a \frac{rz}{\left(r^2 + z^2\right)^{5/2}}\, dz\, dr\, d\theta$$

$$= \lim_{a\to\infty} \int_0^{2\pi} \int_0^1 \left[\left(-\frac{1}{3}\right)\frac{r}{\left(r^2 + z^2\right)^{3/2}}\right]_1^a dr\, d\theta = \lim_{a\to\infty} \int_0^{2\pi} \int_0^1 \left[\left(-\frac{1}{3}\right)\frac{r}{\left(r^2 + a^2\right)^{3/2}} + \left(\frac{1}{3}\right)\frac{r}{\left(r^2 + 1\right)^{3/2}}\right] dr\, d\theta$$

$$= \lim_{a\to\infty} \int_0^{2\pi} \left[\frac{1}{3}\left(r^2 + a^2\right)^{-1/2} - \frac{1}{3}\left(r^2 + 1\right)^{-1/2}\right]_0^1 d\theta = \lim_{a\to\infty} \int_0^{2\pi} \left[\frac{1}{3}\left(1 + a^2\right)^{-1/2} - \frac{1}{3}\left(2^{-1/2}\right) - \frac{1}{3}\left(a^2\right)^{-1/2} + \frac{1}{3}\right] d\theta$$

$$= \lim_{a\to\infty} 2\pi\left[\frac{1}{3}\left(1 + a^2\right)^{-1/2} - \frac{1}{3}\left(\frac{\sqrt{2}}{2}\right) - \frac{1}{3}\left(\frac{1}{a}\right) + \frac{1}{3}\right] = 2\pi\left[\frac{1}{3} - \left(\frac{1}{3}\right)\frac{\sqrt{2}}{2}\right].$$

In analogy with the unit 3-sphere, the volume inside the unit 4-sphere is

$V = \int_{-1}^{1} \int_{-\sqrt{1-w^2}}^{\sqrt{1-w^2}} \int_{-\sqrt{1-w^2-x^2}}^{\sqrt{1-w^2-x^2}} \int_{-\sqrt{1-w^2-x^2-y^2}}^{\sqrt{1-w^2-x^2-y^2}} 1\, dz\, dy\, dx\, dw$. The inner triple integral gives the volume inside the sphere $x^2 + y^2 + z^2 = \left(\sqrt{1 - w^2}\right)^2$, which is $\frac{4\pi}{3}\left(\sqrt{1 - w^2}\right)^3 \Rightarrow V = \int_{-1}^{1} \frac{4\pi}{3}\left(\sqrt{1 - w^2}\right)^3 dw$.

$$= \int_{-1}^{1} \frac{4\pi}{3}\left(1 - w^2\right)^{3/2} dw = \left(\frac{4\pi}{3}\right)\left(\frac{1}{4}\right)\left[w\left(1 - w^2\right)^{3/2} + \frac{3w}{2}\left(1 - w^2\right)^{1/2} + \frac{3}{2}\sin^{-1} w\right]_{-1}^{1}$$

$$= \left(\frac{\pi}{3}\right)\left[\frac{3}{2}\sin^{-1} 1 - \frac{3}{2}\sin^{-1}(-1)\right] = \left(\frac{\pi}{3}\right)\left(\frac{3\pi}{2}\right) = \frac{1}{2}\pi^2.$$

NOTES:

CHAPTER 14 INTEGRATION IN VECTOR FIELDS

14.1 LINE INTEGRALS

1. $\mathbf{r} = t\mathbf{i} + (1-t)\mathbf{j} \Rightarrow x = t$ and $y = 1 - t \Rightarrow y = 1 - x \Rightarrow$ (c)

2. $\mathbf{r} = \mathbf{i} + \mathbf{j} + t\mathbf{k} \Rightarrow x = 1,\ y = 1,$ and $z = t \Rightarrow$ (e)

3. $\mathbf{r} = (2\cos t)\mathbf{i} + (2\sin t)\mathbf{j} \Rightarrow x = 2\cos t$ and $y = 2\sin t \Rightarrow x^2 + y^2 = 4 \Rightarrow$ (g)

4. $\mathbf{r} = t\mathbf{i} \Rightarrow x = t,\ y = 0,$ and $z = 0 \Rightarrow$ (a)

5. $\mathbf{r} = t\mathbf{i} + t\mathbf{j} + t\mathbf{k} \Rightarrow x = t,\ y = t,$ and $z = t \Rightarrow$ (d)

6. $\mathbf{r} = t\mathbf{j} + (2-2t)\mathbf{k} \Rightarrow y = t$ and $z = 2 - 2t \Rightarrow z = 2 - 2y \Rightarrow$ (b)

7. $\mathbf{r} = \left(t^2 - 1\right)\mathbf{j} + 2t\mathbf{k} \Rightarrow y = t^2 - 1$ and $z = 2t \Rightarrow y = \frac{z^2}{4} - 1 \Rightarrow$ (f)

8. $\mathbf{r} = (2\cos t)\mathbf{i} + (2\sin t)\mathbf{k} \Rightarrow x = 2\cos t$ and $z = 2\sin t \Rightarrow x^2 + z^2 = 4 \Rightarrow$ (h)

9. $\mathbf{r}(t) = t\mathbf{i} + (1-t)\mathbf{j},\ 0 \le t \le 1 \Rightarrow \frac{d\mathbf{r}}{dt} = \mathbf{i} - \mathbf{j} \Rightarrow \left|\frac{d\mathbf{r}}{dt}\right| = \sqrt{2}\,\mathbf{j};\ x = t$ and $y = 1 - t \Rightarrow x + y = t + (1-t) = 1$

$\Rightarrow \int_C f(x,y,z)\,ds = \int_0^1 f(t, 1-t, 0)\,\frac{d\mathbf{r}}{dt}\,dt = \int_0^1 (1)\left(\sqrt{2}\right)dt = \left[\sqrt{2}\,t\right]_0^1 = \sqrt{2}$

0. $\mathbf{r}(t) = t\mathbf{i} + (1-t)\mathbf{j} + \mathbf{k},\ 0 \le t \le 1 \Rightarrow \frac{d\mathbf{r}}{dt} = \mathbf{i} - \mathbf{j} \Rightarrow \left|\frac{d\mathbf{r}}{dt}\right| = \sqrt{2};\ x = t,\ y = 1 - t,$ and $z = 1 \Rightarrow x - y + z - 2$

$= t - (1-t) + 1 - 2 = 2t - 2 \Rightarrow \int_C f(x,y,z)\,ds = \int_0^1 (2t-2)\sqrt{2}\,dt = \sqrt{2}\left[t^2 - 2t\right]_0^1 = -\sqrt{2}$

1. $\mathbf{r}(t) = 2t\mathbf{i} + t\mathbf{j} + (2-2t)\mathbf{k},\ 0 \le t \le 1 \Rightarrow \frac{d\mathbf{r}}{dt} = 2\mathbf{i} + \mathbf{j} - 2\mathbf{k} \Rightarrow \left|\frac{d\mathbf{r}}{dt}\right| = \sqrt{4+1+4} = 3;\ xy + y + z$

$= (2t)t + t + (2-2t) \Rightarrow \int_C f(x,y,z)\,ds = \int_0^1 \left(2t^2 - t + 2\right)3\,dt = 3\left[\frac{2}{3}t^3 - \frac{1}{2}t^2 + 2t\right]_0^1 = 3\left(\frac{2}{3} - \frac{1}{2} + 2\right) = \frac{13}{2}$

. $\mathbf{r}(t) = (4\cos t)\mathbf{i} + (4\sin t)\mathbf{j} + 3t\mathbf{k},\ -2\pi \le t \le 2\pi \Rightarrow \frac{d\mathbf{r}}{dt} = (-4\sin t)\mathbf{i} + (4\cos t)\mathbf{j} + 3\mathbf{k}$

$\Rightarrow \left|\frac{d\mathbf{r}}{dt}\right| = \sqrt{16\sin^2 t + 16\cos^2 t + 9} = 5;\ \sqrt{x^2 + y^2} = \sqrt{16\cos^2 t + 16\sin^2 t} = 4 \Rightarrow \int_C f(x,y,z)\,ds = \int_{-2\pi}^{2\pi} (4)(5)\,dt$

$= [20t]_{-2\pi}^{2\pi} = 80\pi$

13. $\mathbf{r}(t) = (\mathbf{i}+2\mathbf{j}+3\mathbf{k}) + t(-\mathbf{i}-3\mathbf{j}-2\mathbf{k}) = (1-t)\mathbf{i} + (2-3t)\mathbf{j} + (3-2t)\mathbf{k},\ 0 \le t \le 1 \Rightarrow \frac{d\mathbf{r}}{dt} = -\mathbf{i}-3\mathbf{j}-2\mathbf{k}$

$\Rightarrow \left|\frac{d\mathbf{r}}{dt}\right| = \sqrt{1+9+4} = \sqrt{14}$; $x+y+z = (1-t)+(2-3t)+(3-2t) = 6-6t \Rightarrow \int_C f(x,y,z)\,ds$

$= \int_0^1 (6-6t)\sqrt{14}\,dt = 6\sqrt{14}\left[1-\frac{t^2}{2}\right]_0^1 = \left(6\sqrt{14}\right)\left(\frac{1}{2}\right) = 3\sqrt{14}$

14. $\mathbf{r}(t) = t\mathbf{i}+t\mathbf{j}+t\mathbf{k},\ 1 \le t \le \infty \Rightarrow \frac{d\mathbf{r}}{dt} = \mathbf{i}+\mathbf{j}+\mathbf{k} \Rightarrow \left|\frac{d\mathbf{r}}{dt}\right| = \sqrt{3}$; $\frac{\sqrt{3}}{x^2+y^2+z^2} = \frac{\sqrt{3}}{t^2+t^2+t^2} = \frac{\sqrt{3}}{3t^2}$

$\Rightarrow \int_C f(x,y,z)\,ds = \int_1^\infty \left(\frac{\sqrt{3}}{3t^2}\right)\sqrt{3}\,dt = \left[-\frac{1}{t}\right]_1^\infty = \lim_{b\to\infty}\left(-\frac{1}{b}+1\right) = 1$

15. C_1: $\mathbf{r}(t) = t\mathbf{i}+t^2\mathbf{j},\ 0 \le t \le 1 \Rightarrow \frac{d\mathbf{r}}{dt} = \mathbf{i}+2t\mathbf{j} \Rightarrow \left|\frac{d\mathbf{r}}{dt}\right| = \sqrt{1+4t^2}$; $x+\sqrt{y}-z^2 = t+\sqrt{t^2}-0 = t+|t| = 2t$

$\Rightarrow \int_{C_1} f(x,y,z)\,ds = \int_0^1 2t\sqrt{1+4t^2}\,dt = \left[\frac{1}{6}\left(1+4t^2\right)^{3/2}\right]_0^1 = \frac{1}{6}(5)^{3/2}-\frac{1}{6} = \frac{1}{6}\left(5\sqrt{5}-1\right)$;

C_2: $\mathbf{r}(t) = \mathbf{i}+\mathbf{j}+t\mathbf{k},\ 0 \le t \le 1 \Rightarrow \frac{d\mathbf{r}}{dt} = \mathbf{k} \Rightarrow \left|\frac{d\mathbf{r}}{dt}\right| = 1$; $x+\sqrt{y}-z^2 = 1+\sqrt{1}-t^2 = 2-t^2$

$\Rightarrow \int_{C_2} f(x,y,z)\,ds = \int_0^1 \left(2-t^2\right)(1)\,dt = \left[2t-\frac{1}{3}t^3\right]_0^1 = 2-\frac{1}{3} = \frac{5}{3}$; therefore $\int_C f(x,y,z)\,ds$

$= \int_{C_1} f(x,y,z)\,ds + \int_{C_2} f(x,y,z)\,ds = \frac{5}{6}\sqrt{5}+\frac{3}{2}$

16. C_1: $\mathbf{r}(t) = t\mathbf{k},\ 0 \le t \le 1 \Rightarrow \frac{d\mathbf{r}}{dt} = \mathbf{k} \Rightarrow \left|\frac{d\mathbf{r}}{dt}\right| = 1$; $x+\sqrt{y}-z^2 = 0+\sqrt{0}-t^2 = -t^2$

$\Rightarrow \int_{C_1} f(x,y,z)\,ds = \int_0^1 \left(-t^2\right)(1)\,dt = \left[-\frac{t^3}{3}\right]_0^1 = -\frac{1}{3}$;

C_2: $\mathbf{r}(t) = t\mathbf{j}+\mathbf{k},\ 0 \le t \le 1 \Rightarrow \frac{d\mathbf{r}}{dt} = \mathbf{j} \Rightarrow \left|\frac{d\mathbf{r}}{dt}\right| = 1$; $x+\sqrt{y}-z^2 = 0+\sqrt{t}-1 = \sqrt{t}-1$

$\Rightarrow \int_{C_2} f(x,y,z)\,ds = \int_0^1 \left(\sqrt{t}-1\right)(1)\,dt = \left[\frac{2}{3}t^{3/2}-t\right]_0^1 = \frac{2}{3}-1 = -\frac{1}{3}$;

C_3: $\mathbf{r}(t) = t\mathbf{i}+\mathbf{j}+\mathbf{k},\ 0 \le t \le 1 \Rightarrow \frac{d\mathbf{r}}{dt} = \mathbf{i} \Rightarrow \left|\frac{d\mathbf{r}}{dt}\right| = 1$; $x+\sqrt{y}-z^2 = t+\sqrt{1}-1 = t$

$\Rightarrow \int_{C_3} f(x,y,z)\,ds = \int_0^1 (t)(1)\,dt = \left[\frac{t^2}{2}\right]_0^1 = \frac{1}{2} \Rightarrow \int_C f(x,y,z)\,ds = \int_{C_1} f\,ds + \int_{C_2} f\,ds + \int_{C_3} f\,ds = -\frac{1}{3}+\left(-\frac{1}{3}\right)+\frac{1}{2}$

$= -\frac{1}{6}$

17. $\mathbf{r}(t) = t\mathbf{i} + t\mathbf{j} + t\mathbf{k},\ 0 < a \le t \le b \Rightarrow \frac{d\mathbf{r}}{dt} = \mathbf{i} + \mathbf{j} + \mathbf{k} \Rightarrow \left|\frac{d\mathbf{r}}{dt}\right| = \sqrt{3};\ \frac{x+y+z}{x^2+y^2+z^2} = \frac{t+t+t}{t^2+t^2+t^2} = \frac{1}{t}$

$\Rightarrow \int_C f(x,y,z)\, ds = \int_a^b \left(\frac{1}{t}\right)\sqrt{3}\, dt = \left[\sqrt{3}\ \ln|t|\right]_a^b = \sqrt{3}\ \ln\left(\frac{b}{a}\right)$, since $0 < a \le b$

18. $\mathbf{r}(t) = (a\cos t)\mathbf{j} + (a\sin t)\mathbf{k},\ 0 \le t \le 2\pi \Rightarrow \frac{d\mathbf{r}}{dt} = (-a\sin t)\mathbf{j} + (a\cos t)\mathbf{k} \Rightarrow \left|\frac{d\mathbf{r}}{dt}\right| = \sqrt{a^2\sin^2 t + a^2\cos^2 t} = |a|;$

$-\sqrt{x^2+z^2} = -\sqrt{0 + a^2\sin^2 t} = \begin{cases} -|a|\sin t, & 0 \le t \le \pi \\ |a|\sin t, & \pi \le t \le 2\pi \end{cases} \Rightarrow \int_C f(x,y,z)\, ds = \int_0^{\pi} -|a|^2 \sin t\, dt + \int_{\pi}^{2\pi} |a|^2 \sin t\, dt$

$= \left[a^2\cos t\right]_0^{\pi} - \left[a^2\cos t\right]_{\pi}^{2\pi} = \left[a^2(-1) - a^2\right] - \left[a^2 - a^2(-1)\right] = -4a^2$

19. $\mathbf{r}(x) = x\mathbf{i} + y\mathbf{j} = x\mathbf{i} + \frac{x^2}{2}\mathbf{j},\ 0 \le x \le 2 \Rightarrow \frac{d\mathbf{r}}{dx} = \mathbf{i} + x\mathbf{j} \Rightarrow \left|\frac{d\mathbf{r}}{dx}\right| = \sqrt{1+x^2};\ f(x,y) = f\left(x, \frac{x^2}{2}\right) = \frac{x^3}{\left(\frac{x^2}{2}\right)} = 2x \Rightarrow \int_C f\, ds$

$= \int_0^2 (2x)\sqrt{1+x^2}\, dx = \left[\frac{2}{3}\left(1+x^2\right)^{3/2}\right]_0^2 = \frac{2}{3}\left(5^{3/2} - 1\right) = \frac{10\sqrt{5}-2}{3}$

20. $\mathbf{r}(x) = x\mathbf{i} + \frac{x^2}{2}\mathbf{j} \Rightarrow \frac{d\mathbf{r}}{dx} = \mathbf{i} + x\mathbf{j},\ 0 \le x \le 1 \Rightarrow \left|\frac{d\mathbf{r}}{dx}\right| = \sqrt{1+x^2};\ f(x,y) = f\left(x, \frac{x^2}{2}\right) = \frac{x + \left(\frac{x^4}{4}\right)}{\sqrt{1+x^2}} = \frac{4x + x^4}{4\sqrt{1+x^2}}$

$\Rightarrow \int_C f\, ds = \int_0^1 \left(\frac{4x+x^4}{4\sqrt{1+x^2}}\right)\sqrt{1+x^2}\, dx = \int_0^1 \left(x + \frac{x^4}{4}\right) dx = \left[\frac{x^2}{2} + \frac{x^5}{20}\right]_0^1 = \frac{1}{2} + \frac{1}{20} = \frac{11}{20}$

21. $\mathbf{r}(t) = (2\cos t)\mathbf{i} + (2\sin t)\mathbf{j},\ 0 \le t \le \frac{\pi}{2} \Rightarrow \frac{d\mathbf{r}}{dt} = (-2\sin t)\mathbf{i} + (2\cos t)\mathbf{j} \Rightarrow \left|\frac{d\mathbf{r}}{dt}\right| = 2;\ f(x,y) = f(2\cos t, 2\sin t)$

$= 2\cos t + 2\sin t \Rightarrow \int_C f\, ds = \int_0^{\pi/2} (2\cos t + 2\sin t)(2)\, dt = \left[4\sin t - 4\cos t\right]_0^{\pi/2} = 4 - (-4) = 8$

22. $\mathbf{r}(t) = (2\cos t)\mathbf{i} + (2\sin t)\mathbf{j},\ \frac{\pi}{2} \ge t \ge \frac{\pi}{4} \Rightarrow \frac{d\mathbf{r}}{dt} = (-2\sin t)\mathbf{i} + (2\cos t)\mathbf{j} \Rightarrow \left|\frac{d\mathbf{r}}{dt}\right| = 2;\ f(x,y) = f(2\cos t, 2\sin t)$

$= 4\cos^2 t - 2\sin t \Rightarrow \int_C f\, ds = \int_{\pi/2}^{\pi/4} \left(4\cos^2 t - 2\sin t\right)(2)\, dt = \left[4t + 2\sin 2t + 4\cos t\right]_{\pi/2}^{\pi/4}$

$= \left(\pi + 2 + 2\sqrt{2}\right) - (2\pi + 0 + 0) = 2\left(1 + \sqrt{2}\right) - \pi$

23. $\mathbf{r}(t) = \left(t^2 - 1\right)\mathbf{j} + 2t\mathbf{k},\ 0 \le t \le 1 \Rightarrow \frac{d\mathbf{r}}{dt} = 2t\mathbf{j} + 2\mathbf{k} \Rightarrow \left|\frac{d\mathbf{r}}{dt}\right| = 2\sqrt{t^2+1};\ M = \int_C \delta(x,y,z)\, ds = \int_0^1 \delta(t)\left(2\sqrt{t^2+1}\right) dt$

$= \int_0^1 \left(\frac{3}{2}t\right)\left(2\sqrt{t^2+1}\right) dt = \left[\left(t^2+1\right)\right]_0^1 = 2^{3/2} - 1 = 2\sqrt{2} - 1$

24. $\mathbf{r}(t) = (t^2-1)\mathbf{j} + 2t\mathbf{k},\ -1 \le t \le 1 \Rightarrow \frac{d\mathbf{r}}{dt} = 2t\mathbf{j} + 2\mathbf{k}$

$\Rightarrow \left|\frac{d\mathbf{r}}{dt}\right| = 2\sqrt{t^2+1};\ M = \int_C \delta(x,y,z)\,ds$

$= \int_{-1}^{1} \left(15\sqrt{(t^2-1)+2}\right)\left(2\sqrt{t^2+1}\right) dt$

$= \int_{-1}^{1} 30(t^2+1)\,dt = \left[30\left(\frac{t^3}{3}+t\right)\right]_{-1}^{1} = 60\left(\frac{1}{3}+1\right) = 80;$

$M_{xz} = \int_C y\delta(x,y,z)\,ds = \int_{-1}^{1} (t^2-1)[30(t^2+1)]\,dt = \int_{-1}^{1} 30(t^4-1)\,dt = \left[30\left(\frac{t^5}{5}-t\right)\right]_{-1}^{1} = 60\left(\frac{1}{5}-1\right) = -48$

$\Rightarrow \overline{y} = \frac{M_{xz}}{M} = -\frac{48}{80} = -\frac{3}{5};\ M_{yz} = \int_C x\delta(x,y,z)\,ds = \int_C 0\,\delta\,ds = 0 \Rightarrow \overline{x} = 0;\ \overline{z} = 0$ by symmetry (since δ is independent of z) $\Rightarrow (\overline{x},\overline{y},\overline{z}) = \left(0,-\frac{3}{5},0\right)$

25. $\mathbf{r}(t) = \sqrt{2}t\,\mathbf{i} + \sqrt{2}t\,\mathbf{j} + (4-t^2)\mathbf{k},\ 0 \le t \le 1 \Rightarrow \frac{d\mathbf{r}}{dt} = \sqrt{2}\mathbf{i} + \sqrt{2}\mathbf{j} - 2t\mathbf{k} \Rightarrow \left|\frac{d\mathbf{r}}{dt}\right| = \sqrt{2+2+4t^2} = 2\sqrt{1+t^2};$

(a) $M = \int_C \delta\,ds = \int_0^1 (3t)\left(2\sqrt{1+t^2}\right) dt = \left[2(1+t^2)^{3/2}\right]_0^1 = 2\left(2^{3/2}-1\right) = 4\sqrt{2}-2$

(b) $M = \int_C \delta\,ds = \int_0^1 (1)\left(2\sqrt{1+t^2}\right) dt = \left[t\sqrt{1+t^2} + \ln\left(t+\sqrt{1+t^2}\right)\right]_0^1 = \left[\sqrt{2} + \ln\left(1+\sqrt{2}\right)\right] - (0 + \ln 1)$
$= \sqrt{2} + \ln\left(1+\sqrt{2}\right)$

26. $\mathbf{r}(t) = t\mathbf{i} + 2t\mathbf{j} + \frac{2}{3}t^{3/2}\mathbf{k},\ 0 \le t \le 2 \Rightarrow \frac{d\mathbf{r}}{dt} = \mathbf{i} + 2\mathbf{j} + t^{1/2}\mathbf{k} \Rightarrow \left|\frac{d\mathbf{r}}{dt}\right| = \sqrt{1+4+t} = \sqrt{5+t};$

$M = \int_C \delta\,ds = \int_0^2 \left(3\sqrt{5+t}\right)\left(\sqrt{5+t}\right) dt = \int_0^2 3(5+t)\,dt = \left[\frac{3}{2}(5+t)^2\right]_0^2 = \frac{3}{2}(7^2-5^2) = \frac{3}{2}(24) = 36;$

$M_{yz} = \int_C x\delta\,ds = \int_0^2 t[3(5+t)]\,dt = \int_0^2 (15t+3t^2)\,dt = \left[\frac{15}{2}t^2 + t^3\right]_0^2 = 30 + 8 = 38;$

$M_{xz} = \int_C y\delta\,ds = \int_0^2 2t[3(5+t)]\,dt = 2\int_0^2 (15t+3t^2)\,dt = 76;\ M_{xy} = \int_C z\delta\,ds = \int_0^2 \frac{2}{3}t^{3/2}[3(5+t)]\,dt$

$= \int_0^2 \left(10t^{3/2} + 2t^{5/2}\right) dt = \left[4t^{5/2} + \frac{4}{7}t^{7/2}\right]_0^2 = 4(2)^{5/2} + \frac{4}{7}(2)^{7/2} = 16\sqrt{2} + \frac{32}{7}\sqrt{2} = \frac{144}{7}\sqrt{2} \Rightarrow \overline{x} = \frac{M_{yz}}{M}$

$= \frac{38}{36} = \frac{19}{18},\ \overline{y} = \frac{M_{xz}}{M} = \frac{76}{36} = \frac{19}{9}$, and $\overline{z} = \frac{M_{xy}}{M} = \frac{144\sqrt{2}}{7 \cdot 36} = \frac{4}{7}\sqrt{2}$

27. Let $x = a \cos t$ and $y = a \sin t$, $0 \le t \le 2\pi$. Then $\frac{dx}{dt} = -a \sin t$, $\frac{dy}{dt} = a \cos t$, $\frac{dz}{dt} = 0$

$$\Rightarrow \sqrt{\left(\frac{dx}{dt}\right)^2 + \left(\frac{dy}{dt}\right)^2 + \left(\frac{dz}{dt}\right)^2}\, dt = a\, dt;\ I_z = \int_C (x^2 + y^2)\delta\, ds = \int_0^{2\pi} (a^2 \sin^2 t + a^2 \cos^2 t)a\delta\, dt$$

$$= \int_0^{2\pi} a^3\delta\, dt = 2\pi\delta a^3;\ M = \int_C \delta(x,y,z)\, ds = \int_0^{2\pi} \delta a\, dt = 2\pi\delta a \Rightarrow R_z = \sqrt{\frac{I_z}{M}} = \sqrt{\frac{2\pi a^3\delta}{2\pi a\delta}} = a.$$

28. $\mathbf{r}(t) = t\mathbf{j} + (2 - 2t)\mathbf{k}$, $0 \le t \le 1 \Rightarrow \frac{d\mathbf{r}}{dt} = \mathbf{j} - 2\mathbf{k} \Rightarrow \left|\frac{d\mathbf{r}}{dt}\right| = \sqrt{5}$; $M = \int_C \delta\, ds = \int_0^1 \delta\sqrt{5}\, dt = \delta\sqrt{5}$;

$$I_x = \int_C (y^2 + z^2)\delta\, ds = \int_0^1 [t^2 + (2 - 2t)^2]\delta\sqrt{5}\, dt = \int_0^1 (5t^2 - 8t + 4)\delta\sqrt{5}\, dt = \delta\sqrt{5}\left[\frac{5}{3}t^3 - 4t^2 + 4t\right]_0^1 = \frac{5}{3}\delta\sqrt{5};$$

$$I_y = \int_C (x^2 + z^2)\delta\, ds = \int_0^1 [0^2 + (2 - 2t)^2]\delta\sqrt{5}\, dt = \int_0^1 (4t^2 - 8t + 4)\delta\sqrt{5}\, dt = \delta\sqrt{5}\left[\frac{4}{3}t^3 - 4t^2 + 4t\right]_0^1 = \frac{4}{3}\delta\sqrt{5};$$

$$I_z = \int_C (x^2 + y^2)\delta\, ds = \int_0^1 (0^2 + t^2)\delta\sqrt{5}\, dt = \delta\sqrt{5}\left[\frac{t^3}{3}\right]_0^1 = \frac{1}{3}\delta\sqrt{5} \Rightarrow R_x = \sqrt{\frac{I_x}{M}} = \sqrt{\frac{5}{3}},\ R_y = \sqrt{\frac{I_y}{M}} = \sqrt{\frac{4}{3}} = \frac{2}{\sqrt{3}},$$

and $R_z = \sqrt{\frac{I_z}{M}} = \frac{1}{\sqrt{3}}$

29. $\mathbf{r}(t) = (\cos t)\mathbf{i} + (\sin t)\mathbf{j} + t\mathbf{k}$, $0 \le t \le 2\pi \Rightarrow \frac{d\mathbf{r}}{dt} = (-\sin t)\mathbf{i} + (\cos t)\mathbf{j} + \mathbf{k} \Rightarrow \left|\frac{d\mathbf{r}}{dt}\right| = \sqrt{\sin^2 t + \cos^2 t + 1} = \sqrt{2}$;

(a) $M = \int_C \delta\, ds = \int_0^{2\pi} \delta\sqrt{2}\, dt = 2\pi\delta\sqrt{2}$; $I_z = \int_C (x^2 + y^2)\delta\, ds = \int_0^{2\pi} (\cos^2 t + \sin^2 t)\delta\sqrt{2}\, dt = 2\pi\delta\sqrt{2}$

$\Rightarrow R_z = \sqrt{\frac{I_z}{M}} = 1$

(b) $M = \int_C \delta(x,y,z)\, ds = \int_0^{4\pi} \delta\sqrt{2}\, dt = 4\pi\delta\sqrt{2}$ and $I_z = \int_C (x^2 + y^2)\delta\, ds = \int_0^{4\pi} \delta\sqrt{2}\, dt = 4\pi\delta\sqrt{2}$

$\Rightarrow R_z = \sqrt{\frac{I_z}{M}} = 1$

30. $\mathbf{r}(t) = (t \cos t)\mathbf{i} + (t \sin t)\mathbf{j} + \frac{2\sqrt{2}}{3}t^{3/2}\mathbf{k}$, $0 \le t \le 1 \Rightarrow \frac{d\mathbf{r}}{dt} = (\cos t - t \sin t)\mathbf{i} + (\sin t + t \cos t)\mathbf{j} + \sqrt{2}\mathbf{k}$

$$\Rightarrow \left|\frac{d\mathbf{r}}{dt}\right| = \sqrt{(t+1)^2} = t + 1 \text{ for } 0 \le t \le 1;\ M = \int_C \delta\, ds = \int_0^1 (t+1)\, dt = \left[\frac{1}{2}(t+1)^2\right]_0^1 = \frac{1}{2}(2^2 - 1^2) = \frac{3}{2};$$

$$M_{xy} = \int_C z\delta\, ds = \int_0^1 \left(\frac{2\sqrt{2}}{3}t^{3/2}\right)(t+1)\, dt = \frac{2\sqrt{2}}{3}\int_0^1 (t^{5/2} + t^{3/2})\, dt = \frac{2\sqrt{2}}{3}\left[\frac{2}{7}t^{7/2} + \frac{2}{5}t^{5/2}\right]_0^1$$

$= \frac{2\sqrt{2}}{3}\left(\frac{2}{7}+\frac{2}{5}\right) = \frac{2\sqrt{2}}{3}\left(\frac{24}{35}\right) = \frac{16\sqrt{2}}{35} \Rightarrow \bar{z} = \frac{M_{xy}}{M} = \left(\frac{16\sqrt{2}}{35}\right)\left(\frac{2}{3}\right) = \frac{32\sqrt{2}}{105}$; $I_z = \int_C (x^2+y^2)\delta\, ds$

$= \int_0^1 (t^2\cos^2 t + t^2 \sin^2 t)(t+1)\, dt = \int_0^1 (t^3+t^2)\, dt = \left[\frac{t^4}{4}+\frac{t^3}{3}\right]_0^1 = \frac{1}{4}+\frac{1}{3} = \frac{7}{12} \Rightarrow R_z = \sqrt{\frac{I_z}{M}} = \sqrt{\frac{7}{18}}$

31. $\delta(x,y,z) = 2 - z$ and $\mathbf{r}(t) = (\cos t)\mathbf{j} + (\sin t)\mathbf{k}$, $0 \le t \le \pi \Rightarrow M = 2\pi - 2$ as found in Example 4 of the text;

also $\left|\frac{d\mathbf{r}}{dt}\right| = 1$; $I_x = \int_C (y^2+z^2)\delta\, ds = \int_0^\pi (\cos^2 t + \sin^2 t)(2-\sin t)\, dt = \int_0^\pi (2-\sin t)\, dt = 2\pi - 2 \Rightarrow R_x = \sqrt{\frac{I_x}{M}}$

$= 1$

32. $\mathbf{r}(t) = t\mathbf{i} + \frac{2\sqrt{2}}{3}t^{3/2}\mathbf{j} + \frac{t^2}{2}\mathbf{k}$, $0 \le t \le 2 \Rightarrow \frac{d\mathbf{r}}{dt} = \mathbf{i} + \sqrt{2}\,t^{1/2}\mathbf{j} + t\mathbf{k} \Rightarrow \left|\frac{d\mathbf{r}}{dt}\right| = \sqrt{1+2t+t^2} = \sqrt{(1+t)^2} = 1+t$ for

$0 \le t \le 2$; $M = \int_C \delta\, ds = \int_0^2 \left(\frac{1}{t+1}\right)(1+t)\, dt = \int_0^2 dt = 2$; $M_{yz} = \int_C x\delta\, ds = \int_0^2 t\left(\frac{1}{t+1}\right)(1+t)\, dt = \left[\frac{t^2}{2}\right]_0^2 = 2$;

$M_{xz} = \int_C y\delta\, ds = \int_0^2 \frac{2\sqrt{2}}{3}t^{3/2}\, dt = \left[\frac{4\sqrt{2}}{15}t^{5/2}\right]_0^2 = \frac{32}{15}$; $M_{xy} = \int_C z\delta\, ds = \int_0^2 \frac{t^2}{2}\, dt = \left[\frac{t^3}{6}\right]_0^2 = \frac{8}{3} \Rightarrow \bar{x} = \frac{M_{yz}}{M} = 1$,

$\bar{y} = \frac{M_{xz}}{M} = \frac{16}{15}$, and $\bar{z} = \frac{M_{xy}}{M} = \frac{4}{3}$; $I_x = \int_C (y^2+z^2)\delta\, ds = \int_0^2 \left(\frac{8}{9}t^3 + \frac{1}{4}t^4\right) dt = \left[\frac{2}{9}t^4 + \frac{t^5}{20}\right]_0^2 = \frac{32}{9} + \frac{32}{20} = \frac{232}{45}$;

$I_y = \int_C (x^2+z^2)\delta\, ds = \int_0^2 \left(t^2 + \frac{1}{4}t^4\right) dt = \left[\frac{t^3}{3} + \frac{t^5}{20}\right]_0^2 = \frac{8}{3} + \frac{32}{20} = \frac{64}{15}$; $I_z = \int_C (x^2+y^2)\delta\, ds$

$= \int_0^2 \left(t^2 + \frac{8}{9}t^3\right) dt = \left[\frac{t^3}{3} + \frac{2}{9}t^4\right]_0^2 = \frac{8}{3} + \frac{32}{9} = \frac{56}{9} \Rightarrow R_x = \sqrt{\frac{I_x}{M}} = \frac{2}{3}\sqrt{\frac{29}{5}}$, $R_y = \sqrt{\frac{I_y}{M}} = \sqrt{\frac{32}{15}}$, and

$R_z = \sqrt{\frac{I_z}{M}} = \frac{2}{3}\sqrt{7}$

33-36. Example CAS commands:

Maple:

```
x:= t -> cos(2*t); y:= t -> sin(2*t);
z:= t -> t^(5/2);
f:= (x,y,z) -> (1 + (9/4)*z^(1/3))^(1/4);
sqrt(D(x)(t)^2 + D(y)(t)^2 + D(z)(t)^2: absvee := unapply('',t);
a:= 0: b:= 2*Pi:
integrand:= simplify(f(x(t),y(t),z(t))*absvee(t));
int(integrand,t=a..b);
evalf('');
```

Mathematica:

```
Clear[x,y,z,t]
r[t_] = {x[t],y[t],z[t]}
f[x_,y_,z_] = (1 +  9/4  z^(1/3))^(1/4)
x[t_]  =  Cos[2 t]
y[t_]  =  Sin[2 t]
```

```
z[t_] = t^(5/2)
{a,b} = {0,2Pi};
v[t_] = r'[t]
s[t_] = Sqrt[ v[t] . v[t] ]
integrand = f[x[t],y[t],z[t]] s[t]
NIntegrate[ integrand, {t,a,b} ]
```

14.2 VECTOR FIELDS, WORK, CIRCULATION, AND FLUX

1. $f(x,y,z) = (x^2+y^2+z^2)^{-1/2} \Rightarrow \frac{\partial f}{\partial x} = -\frac{1}{2}(x^2+y^2+z^2)^{-3/2}(2x) = -x(x^2+y^2+z^2)^{-3/2}$; similarly, $\frac{\partial f}{\partial y} = -y(x^2+y^2+z^2)^{-3/2}$ and $\frac{\partial f}{\partial z} = -z(x^2+y^2+z^2)^{-3/2} \Rightarrow \nabla f = \frac{-x\mathbf{i} - y\mathbf{j} - z\mathbf{k}}{(x^2+y^2+z^2)^{3/2}}$

2. $f(x,y,z) = \ln\sqrt{x^2+y^2+z^2} = \frac{1}{2}\ln(x^2+y^2+z^2) \Rightarrow \frac{\partial f}{\partial x} = \frac{1}{2}\left(\frac{1}{x^2+y^2+z^2}\right)(2x) = \frac{x}{x^2+y^2+z^2}$;

   similarly, $\frac{\partial f}{\partial y} = \frac{y}{x^2+y^2+z^2}$ and $\frac{\partial f}{\partial z} = \frac{z}{x^2+y^2+z^2} \Rightarrow \nabla f = \frac{x\mathbf{i}+y\mathbf{j}+z\mathbf{k}}{x^2+y^2+z^2}$

3. $g(x,y,z) = e^x - \ln(x^2+y^2) \Rightarrow \frac{\partial g}{\partial x} = -\frac{2x}{x^2+y^2}$, $\frac{\partial g}{\partial y} = -\frac{2y}{x^2+y^2}$ and $\frac{\partial g}{\partial z} = e^z$

   $\Rightarrow \nabla g = \left(\frac{-2x}{x^2+y^2}\right)\mathbf{i} - \left(\frac{2y}{x^2+y^2}\right)\mathbf{j} + e^z\mathbf{k}$

4. $g(x,y,z) = xy + yz + xz \Rightarrow \frac{\partial g}{\partial x} = y + z$, $\frac{\partial g}{\partial y} = x + z$, and $\frac{\partial g}{\partial z} = y + x \Rightarrow \nabla g = (y+z)\mathbf{i} + (x+z)\mathbf{j} + (x+y)\mathbf{k}$

5. $|\mathbf{F}|$ inversely proportional to the square of the distance from (x,y) to the origin $\Rightarrow \sqrt{(M(x,y))^2 + (N(x,y))^2}$ $= \frac{k}{x^2+y^2}$, $k > 0$; $\mathbf{F}$ points toward the origin $\Rightarrow \mathbf{F}$ is in the direction of $\mathbf{n} = \frac{-x}{\sqrt{x^2+y^2}}\mathbf{i} - \frac{y}{\sqrt{x^2+y^2}}\mathbf{j}$

   $\Rightarrow \mathbf{F} = a\mathbf{n}$, for some constant $a > 0$. Then $M(x,y) = \frac{-ax}{\sqrt{x^2+y^2}}$ and $N(x,y) = \frac{-ay}{\sqrt{x^2+y^2}}$

   $\Rightarrow \sqrt{(M(x,y))^2 + (N(x,y))^2} = a \Rightarrow a = \frac{k}{x^2+y^2} \Rightarrow \mathbf{F} = \frac{-kx}{(x^2+y^2)^{3/2}}\mathbf{i} - \frac{ky}{(x^2+y^2)^{3/2}}\mathbf{j}$, for any constant $k > 0$

6. Given $x^2 + y^2 = a^2 + b^2$, let $x = \sqrt{a^2+b^2}\cos t$ and $y = -\sqrt{a^2+b^2}\sin t$. Then $\mathbf{r} = \left(\sqrt{a^2+b^2}\cos t\right)\mathbf{i} - \left(\sqrt{a^2+b^2}\sin t\right)\mathbf{j}$ traces the circle in a clockwise direction as t goes from 0 to 2π $\Rightarrow \mathbf{v} = \left(-\sqrt{a^2+b^2}\sin t\right)\mathbf{i} - \left(\sqrt{a^2+b^2}\cos t\right)\mathbf{j}$ is tangent to the circle in a clockwise direction. Thus, let $\mathbf{F} = \mathbf{v} \Rightarrow \mathbf{F} = y\mathbf{i} - x\mathbf{j}$ and $\mathbf{F}(0,0) = \mathbf{0}$.

7. Substitute the parametric representations for $\mathbf{r}(t) = x(t)\mathbf{i} + y(t)\mathbf{j} + z(t)\mathbf{k}$ representing each path into the vector field $\mathbf{F}$, and calculate the work $W = \int_C \mathbf{F} \cdot \frac{d\mathbf{r}}{dt}$.

(a) $\mathbf{F} = 3t\mathbf{i} + 2t\mathbf{j} + 4t\mathbf{k}$ and $\frac{d\mathbf{r}}{dt} = \mathbf{i} + \mathbf{j} + \mathbf{k} \Rightarrow \mathbf{F} \cdot \frac{d\mathbf{r}}{dt} = 9t \Rightarrow W = \int_0^1 9t\,dt = \frac{9}{2}$

(b) $\mathbf{F} = 3t^2\mathbf{i} + 2t\mathbf{j} + 4t^4\mathbf{k}$ and $\frac{d\mathbf{r}}{dt} = \mathbf{i} + 2t\mathbf{j} + 4t^3\mathbf{k} \Rightarrow \mathbf{F} \cdot \frac{d\mathbf{r}}{dt} = 7t^2 + 16t^7 \Rightarrow W = \int_0^1 (7t^2 + 16t^7)\,dt = \left[\frac{7}{3}t^3 + 2t^8\right]_0^1$
$= \frac{7}{3} + 2 = \frac{13}{3}$

(c) $\mathbf{r}_1 = t\mathbf{i} + t\mathbf{j}$ and $\mathbf{r}_2 = \mathbf{i} + \mathbf{j} + t\mathbf{k}$; $\mathbf{F}_1 = 3t\mathbf{i} + 2t\mathbf{j}$ and $\frac{d\mathbf{r}_1}{dt} = \mathbf{i} + \mathbf{j} \Rightarrow \mathbf{F}_1 \cdot \frac{d\mathbf{r}_1}{dt} = 5t \Rightarrow W_1 = \int_0^1 5t\,dt = \frac{5}{2}$;

$\mathbf{F}_2 = 3\mathbf{i} + 2\mathbf{j} + 4t\mathbf{k}$ and $\frac{d\mathbf{r}_2}{dt} = \mathbf{k} \Rightarrow \mathbf{F}_2 \cdot \frac{d\mathbf{r}_2}{dt} = 4t \Rightarrow W_2 = \int_0^1 4t\,dt = 2 \Rightarrow W = W_1 + W_2 = \frac{9}{2}$

8. Substitute the parametric representation for $\mathbf{r}(t) = x(t)\mathbf{i} + y(t)\mathbf{j} + z(t)\mathbf{k}$ representing each path into the vector field $\mathbf{F}$, and calculate the work $W = \int_C \mathbf{F} \cdot \frac{d\mathbf{r}}{dt}$.

(a) $\mathbf{F} = \left(\frac{1}{t^2+1}\right)\mathbf{j}$ and $\frac{d\mathbf{r}}{dt} = \mathbf{i} + \mathbf{j} + \mathbf{k} \Rightarrow \mathbf{F} \cdot \frac{d\mathbf{r}}{dt} = \frac{1}{t^2+1} \Rightarrow W = \int_0^1 \frac{1}{t^2+1}\,dt = \left[\tan^{-1} t\right]_0^1 = \frac{\pi}{4}$

(b) $\mathbf{F} = \left(\frac{1}{t^2+1}\right)\mathbf{j}$ and $\frac{d\mathbf{r}}{dt} = \mathbf{i} + 2t\mathbf{j} + 4t^3\mathbf{k} \Rightarrow \mathbf{F} \cdot \frac{d\mathbf{r}}{dt} = \frac{2t}{t^2+1} \Rightarrow W = \int_0^1 \frac{2t}{t^2+1}\,dt = \left[\ln(t^2+1)\right]_0^1 = \ln 2$

(c) $\mathbf{r}_1 = t\mathbf{i} + t\mathbf{j}$ and $\mathbf{r}_2 = \mathbf{i} + \mathbf{j} + t\mathbf{k}$; $\mathbf{F}_1 = \left(\frac{1}{t^2+1}\right)\mathbf{j}$ and $\frac{d\mathbf{r}_1}{dt} = \mathbf{i} + \mathbf{j} \Rightarrow \mathbf{F}_1 \cdot \frac{d\mathbf{r}_1}{dt} = \frac{1}{t^2+1}$; $\mathbf{F}_2 = \frac{1}{2}\mathbf{j}$ and $\frac{d\mathbf{r}_2}{dt} = \mathbf{k}$

$\Rightarrow \mathbf{F}_2 \cdot \frac{d\mathbf{r}_2}{dt} = 0 \Rightarrow W = \int_0^1 \frac{1}{t^2+1}\,dt = \frac{\pi}{4}$

9. Substitute the parametric representation for $\mathbf{r}(t) = x(t)\mathbf{i} + y(t)\mathbf{j} + z(t)\mathbf{k}$ representing each path into the vector field $\mathbf{F}$, and calculate the work $W = \int_C \mathbf{F} \cdot \frac{d\mathbf{r}}{dt}$.

(a) $\mathbf{F} = \sqrt{t}\mathbf{i} - 2t\mathbf{j} + \sqrt{t}\mathbf{k}$ and $\frac{d\mathbf{r}}{dt} = \mathbf{i} + \mathbf{j} + \mathbf{k} \Rightarrow \mathbf{F} \cdot \frac{d\mathbf{r}}{dt} = 2\sqrt{t} - 2t \Rightarrow W = \int_0^1 (2\sqrt{t} - 2t)\,dt = \left[\frac{4}{3}t^{3/2} - t^2\right]_0^1 = \frac{1}{3}$

(b) $\mathbf{F} = t^2\mathbf{i} - 2t\mathbf{j} + t\mathbf{k}$ and $\frac{d\mathbf{r}}{dt} = \mathbf{i} + 2t\mathbf{j} + 4t^3\mathbf{k} \Rightarrow \mathbf{F} \cdot \frac{d\mathbf{r}}{dt} = 4t^4 - 3t^2 \Rightarrow W = \int_0^1 (4t^4 - 3t^2)\,dt = \left[\frac{4}{5}t^5 - t^3\right]_0^1 = -\frac{1}{5}$

(c) $\mathbf{r}_1 = t\mathbf{i} + t\mathbf{j}$ and $\mathbf{r}_2 = \mathbf{i} + \mathbf{j} + t\mathbf{k}$; $\mathbf{F}_1 = -2t\mathbf{j} + \sqrt{t}\,\mathbf{k}$ and $\frac{d\mathbf{r}_1}{dt} = \mathbf{i} + \mathbf{j} \Rightarrow \mathbf{F}_1 \cdot \frac{d\mathbf{r}_1}{dt} = -2t \Rightarrow W_1 = \int_0^1 -2t\,dt$

$= -1$; $\mathbf{F}_2 = \sqrt{t}\mathbf{i} - 2\mathbf{j} + \mathbf{k}$ and $\frac{d\mathbf{r}_2}{dt} = \mathbf{k} \Rightarrow \mathbf{F}_2 \cdot \frac{d\mathbf{r}_2}{dt} = 1 \Rightarrow W_2 = \int_0^1 dt = 1 \Rightarrow W = W_1 + W_2 = 0$

10. Substitute the parametric representation for $\mathbf{r}(t) = x(t)\mathbf{i} + y(t)\mathbf{j} + z(t)\mathbf{k}$ representing each path into the vector field $\mathbf{F}$, and calculate the work $W = \int_C \mathbf{F} \cdot \frac{d\mathbf{r}}{dt}$.

(a) $\mathbf{F} = t^2\mathbf{i} + t^2\mathbf{j} + t^2\mathbf{k}$ and $\frac{d\mathbf{r}}{dt} = \mathbf{i} + \mathbf{j} + \mathbf{k} \Rightarrow \mathbf{F} \cdot \frac{d\mathbf{r}}{dt} = 3t^2 \Rightarrow W = \int_0^1 3t^2\,dt = 1$

(b) $\mathbf{F} = t^3\mathbf{i} - t^6\mathbf{j} + t^5\mathbf{k}$ and $\frac{d\mathbf{r}}{dt} = \mathbf{i} + 2t\mathbf{j} + 4t^3\mathbf{k} \Rightarrow \mathbf{F} \cdot \frac{d\mathbf{r}}{dt} = t^3 + 2t^7 + 4t^8 \Rightarrow W = \int_0^1 \left(t^3 + 2t^7 + 4t^8\right) dt$

$= \left[\frac{t^4}{4} + \frac{t^8}{4} + \frac{4}{9}t^9\right]_0^1 = \frac{17}{18}$

(c) $\mathbf{r}_1 = t\mathbf{i} + t\mathbf{j}$ and $\mathbf{r}_2 = \mathbf{i} + \mathbf{j} + t\mathbf{k}$; $\mathbf{F}_1 = t^2\mathbf{i}$ and $\frac{d\mathbf{r}_1}{dt} = \mathbf{i} + \mathbf{j} \Rightarrow \mathbf{F}_1 \cdot \frac{d\mathbf{r}_1}{dt} = t^2 \Rightarrow W_1 = \int_0^1 t^2\,dt = \frac{1}{3}$;

$\mathbf{F}_2 = \mathbf{i} + t\mathbf{j} + t\mathbf{k}$ and $\frac{d\mathbf{r}_2}{dt} = \mathbf{k} \Rightarrow \mathbf{F}_2 \cdot \frac{d\mathbf{r}_2}{dt} = t \Rightarrow W_2 = \int_0^1 t\,dt = \frac{1}{2} \Rightarrow W = W_1 + W_2 = \frac{5}{6}$

11. Substitute the parametric representation for $\mathbf{r}(t) = x(t)\mathbf{i} + y(t)\mathbf{j} + z(t)\mathbf{k}$ representing each path into the vector field $\mathbf{F}$, and calculate the work $W = \int_C \mathbf{F} \cdot \frac{d\mathbf{r}}{dt}$.

(a) $\mathbf{F} = \left(3t^2 - 3t\right)\mathbf{i} + 3t\mathbf{j} + \mathbf{k}$ and $\frac{d\mathbf{r}}{dt} = \mathbf{i} + \mathbf{j} + \mathbf{k} \Rightarrow \mathbf{F} \cdot \frac{d\mathbf{r}}{dt} = 3t^2 + 1 \Rightarrow W = \int_0^1 \left(3t^2 + 1\right) dt = \left[t^3 + t\right]_0^1 = 2$

(b) $\mathbf{F} = \left(3t^2 - 3t\right)\mathbf{i} + 3t^4\mathbf{j} + \mathbf{k}$ and $\frac{d\mathbf{r}}{dt} = \mathbf{i} + 2t\mathbf{j} + 4t^3\mathbf{k} \Rightarrow \mathbf{F} \cdot \frac{d\mathbf{r}}{dt} = 6t^5 + 4t^3 + 3t^2 - 3t$

$\Rightarrow W = \int_0^1 \left(6t^5 + 4t^3 + 3t^2 - 3t\right) dt = \left[t^6 + t^4 + t^3 - \frac{3}{2}t^2\right]_0^1 = \frac{3}{2}$

(c) $\mathbf{r}_1 = t\mathbf{i} + t\mathbf{j}$ and $\mathbf{r}_2 = \mathbf{i} + \mathbf{j} + t\mathbf{k}$; $\mathbf{F}_1 = \left(3t^2 - 3t\right)\mathbf{i} + \mathbf{k}$ and $\frac{d\mathbf{r}_1}{dt} = \mathbf{i} + \mathbf{j} \Rightarrow \mathbf{F}_1 \cdot \frac{d\mathbf{r}_1}{dt} = 3t^2 - 3t$

$\Rightarrow W_1 = \int_0^1 \left(3t^2 - 3t\right) dt = \left[t^3 - \frac{3}{2}t^2\right]_0^1 = -\frac{1}{2}$; $\mathbf{F}_2 = 3t\mathbf{j} + \mathbf{k}$ and $\frac{d\mathbf{r}_2}{dt} = \mathbf{k} \Rightarrow \mathbf{F}_2 \cdot \frac{d\mathbf{r}_2}{dt} = 1 \Rightarrow W_2 = \int_0^1 dt = 1$

$\Rightarrow W = W_1 + W_2 = \frac{1}{2}$

12. Substitute the parametric representation for $\mathbf{r}(t) = x(t)\mathbf{i} + y(t)\mathbf{j} + z(t)\mathbf{k}$ representing each path into the vector field $\mathbf{F}$, and calculate the work $W = \int_C \mathbf{F} \cdot \frac{d\mathbf{r}}{dt}$.

(a) $\mathbf{F} = 2t\mathbf{i} + 2t\mathbf{j} + 2t\mathbf{k}$ and $\frac{d\mathbf{r}}{dt} = \mathbf{i} + \mathbf{j} + \mathbf{k} \Rightarrow \mathbf{F} \cdot \frac{d\mathbf{r}}{dt} = 6t \Rightarrow W = \int_0^1 6t\, dt = \left[3t^2\right]_0^1 = 3$

(b) $\mathbf{F} = (t^2 + t^4)\mathbf{i} + (t^4 + t)\mathbf{j} + (t + t^2)\mathbf{k}$ and $\frac{d\mathbf{r}}{dt} = \mathbf{i} + 2t\mathbf{j} + 4t^3\mathbf{k} \Rightarrow \mathbf{F} \cdot \frac{d\mathbf{r}}{dt} = 6t^5 + 5t^4 + 3t^2$

$\Rightarrow W = \int_0^1 (6t^5 + 5t^4 + 3t^2)\, dt = \left[t^6 + t^5 + t^3\right]_0^1 = 3$

(c) $\mathbf{r}_1 = t\mathbf{i} + t\mathbf{j}$ and $\mathbf{r}_2 = \mathbf{i} + \mathbf{j} + t\mathbf{k}$; $\mathbf{F}_1 = t\mathbf{i} + t\mathbf{j} + 2t\mathbf{k}$ and $\frac{d\mathbf{r}_1}{dt} = \mathbf{i} + \mathbf{j} \Rightarrow \mathbf{F}_1 \cdot \frac{d\mathbf{r}_1}{dt} = 2t$

$\Rightarrow W_1 = \int_0^1 2t\, dt = 1$; $\mathbf{F}_2 = (1 + t)\mathbf{i} + (t + 1)\mathbf{j} + 2\mathbf{k}$ and $\frac{d\mathbf{r}_2}{dt} = \mathbf{k} \Rightarrow \mathbf{F}_2 \cdot \frac{d\mathbf{r}_2}{dt} = 2 \Rightarrow W_2 = \int_0^1 2\, dt = 2$

$\Rightarrow W = W_1 + W_2 = 3$

13. $\mathbf{r} = t\mathbf{i} + t^2\mathbf{j} + t\mathbf{k}$, $0 \le t \le 1$, and $\mathbf{F} = xy\mathbf{i} + y\mathbf{j} - yz\mathbf{k} \Rightarrow \mathbf{F} = t^3\mathbf{i} + t^2\mathbf{j} - t^3\mathbf{k}$ and $\frac{d\mathbf{r}}{dt} = \mathbf{i} + 2t\mathbf{j} + \mathbf{k}$

$\Rightarrow \mathbf{F} \cdot \frac{d\mathbf{r}}{dt} = 2t^3 \Rightarrow \text{work} = \int_0^1 2t^3\, dt = \frac{1}{2}$

14. $\mathbf{r} = (\cos t)\mathbf{i} + (\sin t)\mathbf{j} + \frac{t}{6}\mathbf{k}$, $0 \le t \le 2\pi$, and $\mathbf{F} = 2y\mathbf{i} + 3x\mathbf{j} + (x + y)\mathbf{k}$
$\Rightarrow \mathbf{F} = (2 \sin t)\mathbf{i} + (3 \cos t)\mathbf{j} + (\cos t + \sin t)\mathbf{k}$ and $\frac{d\mathbf{r}}{dt} = (-\sin t)\mathbf{i} + (\cos t)\mathbf{j} + \frac{1}{6}\mathbf{k} \Rightarrow \mathbf{F} \cdot \frac{d\mathbf{r}}{dt}$

$= 5 \cos^2 t - 2 + \frac{1}{6} \cos t + \frac{1}{6} \sin t \Rightarrow \text{work} = \int_0^{2\pi} \left(5 \cos^2 t - 2 + \frac{1}{6} \cos t + \frac{1}{6} \sin t\right) dt$

$= \left[\frac{5}{2}t + \frac{5}{4} \sin 2t - 2t + \frac{1}{6} \sin t - \frac{1}{6} \cos t\right]_0^{2\pi} = 5\pi - 4\pi = \pi$

15. $\mathbf{r} = (\sin t)\mathbf{i} + (\cos t)\mathbf{j} + t\mathbf{k}$, $0 \le t \le 2\pi$, and $\mathbf{F} = z\mathbf{i} + x\mathbf{j} + y\mathbf{k} \Rightarrow \mathbf{F} = t\mathbf{i} + (\sin t)\mathbf{j} + (\cos t)\mathbf{k}$ and

$\frac{d\mathbf{r}}{dt} = (\cos t)\mathbf{i} - (\sin t)\mathbf{j} + \mathbf{k} \Rightarrow \mathbf{F} \cdot \frac{d\mathbf{r}}{dt} = t \cos t - \sin^2 t + \cos t \Rightarrow \text{work} = \int_0^{2\pi} \left(t \cos t - \sin^2 t + \cos t\right) dt$

$= \left[\cos t + t \sin t - \frac{t}{2} + \frac{\sin 2t}{4} + \sin t\right]_0^{2\pi} = -\pi$

16. $\mathbf{r} = (\sin t)\mathbf{i} + (\cos t)\mathbf{j} + \frac{t}{6}\mathbf{k}$, $0 \le t \le 2\pi$, and $\mathbf{F} = 6z\mathbf{i} + y^2\mathbf{j} + 12x\mathbf{k} \Rightarrow \mathbf{F} = t\mathbf{i} + (\cos^2 t)\mathbf{j} + (12 \sin t)\mathbf{k}$ and

$\frac{d\mathbf{r}}{dt} = (\cos t)\mathbf{i} - (\sin t)\mathbf{j} + \frac{1}{6}\mathbf{k} \Rightarrow \mathbf{F} \cdot \frac{d\mathbf{r}}{dt} = t \cos t - \sin t \cos^2 t + 2 \sin t$

$\Rightarrow$ work $= \int_0^{2\pi} (t \cos t - \sin t \cos^2 t + 2 \sin t)\, dt = \left[\cos t + t \sin t + \frac{1}{3}\cos^3 t - 2 \cos t\right]_0^{2\pi} = 0$

17. $x = t$ and $y = x^2 = t^2 \Rightarrow \mathbf{r} = t\mathbf{i} + t^2\mathbf{j}$, $-1 \le t \le 2$, and $\mathbf{F} = xy\mathbf{i} + (x+y)\mathbf{j} \Rightarrow \mathbf{F} = t^3\mathbf{i} + (t+t^2)\mathbf{j}$ and

$\frac{d\mathbf{r}}{dt} = \mathbf{i} + 2t\mathbf{j} \Rightarrow \mathbf{F}\cdot\frac{d\mathbf{r}}{dt} = t^3 + (2t^2 + 2t^3) = 3t^3 + 2t^2 \Rightarrow \int_C xy\, dx + (x+y)\, dy = \int_C \mathbf{F}\cdot\frac{d\mathbf{r}}{dt}\, dt = \int_{-1}^{2} (3t^3 + 2t^2)\, dt$

$= \left[\frac{3}{4}t^4 + \frac{2}{3}t^3\right]_{-1}^{2} = \left(12 + \frac{16}{3}\right) - \left(\frac{3}{4} - \frac{2}{3}\right) = \frac{45}{4} + \frac{18}{3} = \frac{207}{12}$

18. Along $(0,0)$ to $(1,0)$: $\mathbf{r} = t\mathbf{i}$, $0 \le t \le 1$, and $\mathbf{F} = (x-y)\mathbf{i} + (x+y)\mathbf{j} \Rightarrow \mathbf{F} = t\mathbf{i} + t\mathbf{j}$ and $\frac{d\mathbf{r}}{dt} = \mathbf{i} \Rightarrow \mathbf{F}\cdot\frac{d\mathbf{r}}{dt} = t$;

Along $(1,0)$ to $(0,1)$: $\mathbf{r} = (1-t)\mathbf{i} + t\mathbf{j}$, $0 \le t \le 1$, and $\mathbf{F} = (x-y)\mathbf{i} + (x+y)\mathbf{j} \Rightarrow \mathbf{F} = (1-2t)\mathbf{i} + \mathbf{j}$ and

$\frac{d\mathbf{r}}{dt} = -\mathbf{i} + \mathbf{j} \Rightarrow \mathbf{F}\cdot\frac{d\mathbf{r}}{dt} = 2t$;

Along $(0,1)$ to $(0,0)$: $\mathbf{r} = (1-t)\mathbf{j}$, $0 \le t \le 1$, and $\mathbf{F} = (x-y)\mathbf{i} + (x+y)\mathbf{j} \Rightarrow \mathbf{F} = (t-1)\mathbf{i} + (1-t)\mathbf{j}$ and

$\frac{d\mathbf{r}}{dt} = -\mathbf{j} \Rightarrow \mathbf{F}\cdot\frac{d\mathbf{r}}{dt} = t - 1 \Rightarrow \int_C (x-y)\, dx + (x+y)\, dy = \int_0^1 t\, dt + \int_0^1 2t\, dt + \int_0^1 (t-1)\, dt = \int_0^1 (4t-1)\, dt$

$= \left[2t^2 - t\right]_0^1 = 2 - 1 = 1$

19. $\mathbf{r} = x\mathbf{i} + y\mathbf{j} = y^2\mathbf{i} + y\mathbf{j}$, $2 \ge y \ge -1$, and $\mathbf{F} = x^2\mathbf{i} - y\mathbf{j} = y^4\mathbf{i} - y\mathbf{j} \Rightarrow \frac{d\mathbf{r}}{dy} = 2y\mathbf{i} + \mathbf{j}$ and $\mathbf{F}\cdot\frac{d\mathbf{r}}{dy} = 2y^5 - y$

$\Rightarrow \int_C \mathbf{F}\cdot\mathbf{T}\, ds = \int_2^{-1} \mathbf{F}\cdot\frac{d\mathbf{r}}{dy}\, dy = \int_2^{-1} (2y^5 - y)\, dy = \left[\frac{1}{3}y^6 - \frac{1}{2}y^2\right]_2^{-1} = \left(\frac{1}{3} - \frac{1}{2}\right) - \left(\frac{64}{3} - \frac{4}{2}\right) = \frac{3}{2} - \frac{63}{3} = -\frac{39}{2}$

20. $\mathbf{r} = (\cos t)\mathbf{i} + (\sin t)\mathbf{j}$, $0 \le t \le \frac{\pi}{2}$, and $\mathbf{F} = y\mathbf{i} - x\mathbf{j} \Rightarrow \mathbf{F} = (\sin t)\mathbf{i} - (\cos t)\mathbf{j}$ and $\frac{d\mathbf{r}}{dt} = (-\sin t)\mathbf{i} + (\cos t)\mathbf{j}$

$\Rightarrow \mathbf{F}\cdot\frac{d\mathbf{r}}{dt} = -\sin^2 t - \cos^2 t = -1 \Rightarrow \int_C \mathbf{F}\cdot d\mathbf{r} = \int_0^{\pi/2} (-1)\, dt = -\frac{\pi}{2}$

21. $\mathbf{r} = (\mathbf{i} + \mathbf{j}) + t(\mathbf{i} + 2\mathbf{j}) = (1+t)\mathbf{i} + (1+2t)\mathbf{j}$, $0 \le t \le 1$, and $\mathbf{F} = xy\mathbf{i} + (y-x)\mathbf{j} \Rightarrow \mathbf{F} = (1 + 3t + 2t^2)\mathbf{i} + t\mathbf{j}$ and

$\frac{d\mathbf{r}}{dt} = \mathbf{i} + 2\mathbf{j} \Rightarrow \mathbf{F}\cdot\frac{d\mathbf{r}}{dt} = 1 + 5t + 2t^2 \Rightarrow$ work $= \int_C \mathbf{F}\cdot\frac{d\mathbf{r}}{dt}\, dt = \int_0^1 (1 + 5t + 2t^2)\, dt = \left[t + \frac{5}{2}t^2 + \frac{2}{3}t^3\right]_0^1 = \frac{25}{6}$

22. $\mathbf{r} = (2\cos t)\mathbf{i} + (2\sin t)\mathbf{j}$, $0 \le t \le 2\pi$, and $\mathbf{F} = \nabla f = 2(x+y)\mathbf{i} + 2(x+y)\mathbf{j}$

$\Rightarrow \mathbf{F} = 4(\cos t + \sin t)\mathbf{i} + 4(\cos t + \sin t)\mathbf{j}$ and $\frac{d\mathbf{r}}{dt} = (-2\sin t)\mathbf{i} + (2\cos t)\mathbf{j} \Rightarrow \mathbf{F}\cdot\frac{d\mathbf{r}}{dt}$

$= -8(\sin t \cos t + \sin^2 t) + 8(\cos^2 t + \cos t \sin t) = 8(\cos^2 t - \sin^2 t) = 8 \cos 2t \Rightarrow$ work $= \int_C \nabla f\cdot d\mathbf{r}$

$= \int_C \mathbf{F}\cdot\frac{d\mathbf{r}}{dt}\, dt = \int_0^{2\pi} 8 \cos 2t\, dt = [4 \sin 2t]_0^{2\pi} = 0$

23. (a) $\mathbf{r} = (\cos t)\mathbf{i} + (\sin t)\mathbf{j}$, $0 \le t \le 2\pi$, $\mathbf{F}_1 = x\mathbf{i} + y\mathbf{j}$, and $\mathbf{F}_2 = -y\mathbf{i} + x\mathbf{j} \Rightarrow \frac{d\mathbf{r}}{dt} = (-\sin t)\mathbf{i} + (\cos t)\mathbf{j}$,

$\mathbf{F}_1 = (\cos t)\mathbf{i} + (\sin t)\mathbf{j}$, and $\mathbf{F}_2 = (-\sin t)\mathbf{i} + (\cos t)\mathbf{j} \Rightarrow \mathbf{F}_1 \cdot \frac{d\mathbf{r}}{dt} = 0$ and $\mathbf{F}_2 \cdot \frac{d\mathbf{r}}{dt} = \sin^2 t + \cos^2 t = 1$

$\Rightarrow \text{Circ}_1 = \int_0^{2\pi} 0\, dt = 0$ and $\text{Circ}_2 = \int_0^{2\pi} dt = 2\pi$; $\mathbf{n} = (\cos t)\mathbf{i} + (\sin t)\mathbf{j} \Rightarrow \mathbf{F}_1 \cdot \mathbf{n} = \cos^2 t + \sin^2 t = 1$ and

$\mathbf{F}_2 \cdot \mathbf{n} = 0 \Rightarrow \text{Flux}_1 = \int_0^{2\pi} dt = 2\pi$ and $\text{Flux}_2 = \int_0^{2\pi} 0\, dt = 0$

(b) $\mathbf{r} = (\cos t)\mathbf{i} + (4 \sin t)\mathbf{j}$, $0 \le t \le 2\pi \Rightarrow \frac{d\mathbf{r}}{dt} = (-\sin t)\mathbf{i} + (4 \cos t)\mathbf{j}$, $\mathbf{F}_1 = (\cos t)\mathbf{i} + (4 \sin t)\mathbf{j}$, and

$\mathbf{F}_2 = (-4 \sin t)\mathbf{i} + (\cos t)\mathbf{j} \Rightarrow \mathbf{F}_1 \cdot \frac{d\mathbf{r}}{dt} = 15 \sin t \cos t$ and $\mathbf{F}_2 \cdot \frac{d\mathbf{r}}{dt} = 4 \Rightarrow \text{Circ}_1 = \int_0^{2\pi} 15 \sin t \cos t\, dt$

$= \left[\frac{15}{2} \sin^2 t\right]_0^{2\pi} = 0$ and $\text{Circ}_2 = \int_0^{2\pi} 4\, dt = 8\pi$; $\mathbf{n} = \left(\frac{4}{\sqrt{17}} \cos t\right)\mathbf{i} + \left(\frac{1}{\sqrt{17}} \sin t\right)\mathbf{j} \Rightarrow \mathbf{F}_1 \cdot \mathbf{n}$

$= \frac{4}{\sqrt{17}} \cos^2 t + \frac{4}{\sqrt{17}} \sin^2 t$ and $\mathbf{F}_2 \cdot \mathbf{n} = -\frac{15}{\sqrt{17}} \sin t \cos t \Rightarrow \text{Flux}_1 = \int_0^{2\pi} (\mathbf{F}_1 \cdot \mathbf{n})\, |\mathbf{v}|\, dt = \int_0^{2\pi} \left(\frac{4}{\sqrt{17}}\right)\sqrt{17}\, dt$

$= 8\pi$ and $\text{Flux}_2 = \int_0^{2\pi} (\mathbf{F}_2 \cdot \mathbf{n})\, |\mathbf{v}|\, dt = \int_0^{2\pi} \left(-\frac{15}{\sqrt{17}} \sin t \cos t\right)\sqrt{17}\, dt = \left[-\frac{15}{2} \sin^2 t\right]_0^{2\pi} = 0$

24. $\mathbf{r} = (a \cos t)\mathbf{i} + (a \sin t)\mathbf{j}$, $0 \le t \le 2\pi$, $\mathbf{F}_1 = 2x\mathbf{i} - 3y\mathbf{j}$, and $\mathbf{F}_2 = 2x\mathbf{i} + (x - y)\mathbf{j} \Rightarrow \frac{d\mathbf{r}}{dt} = (-a \sin t)\mathbf{i} + (a \cos t)\mathbf{j}$,
$\mathbf{F}_1 = (2a \cos t)\mathbf{i} - (3a \sin t)\mathbf{j}$, and $\mathbf{F}_2 = (2a \cos t)\mathbf{i} + (a \cos t - a \sin t)\mathbf{j} \Rightarrow \mathbf{n}\, |\mathbf{v}| = (a \cos t)\mathbf{i} + (a \sin t)\mathbf{j}$,
$\mathbf{F}_1 \cdot \mathbf{n}\, |\mathbf{v}| = 2a^2 \cos^2 t - 3a^2 \sin^2 t$, and $\mathbf{F}_2 \cdot \mathbf{n}\, |\mathbf{v}| = 2a^2 \cos^2 t + a^2 \sin t \cos t - a^2 \sin^2 t$

$\Rightarrow \text{Flux}_1 = \int_0^{2\pi} \left(2a^2 \cos^2 t + 3a^2 \sin^2 t\right) dt = 2a^2\left[\frac{t}{2} + \frac{\sin 2t}{4}\right]_0^{2\pi} - 3a^2\left[\frac{t}{2} - \frac{\sin 2t}{4}\right]_0^{2\pi} = -\pi a^2$, and

$\text{Flux}_2 = \int_0^{2\pi} \left(2a^2 \cos^2 t + a^2 \sin t \cos t - a^2 \sin^2 t\right) dt = 2a^2\left[\frac{t}{2} + \frac{\sin 2t}{4}\right]_0^{2\pi} + \frac{a^2}{2}\left[\sin^2 t\right]_0^{2\pi} - a^2\left[\frac{t}{2} - \frac{\sin 2t}{4}\right]_0^{2\pi} = \pi a^2$

25. $\mathbf{F}_1 = (a \cos t)\mathbf{i} + (a \sin t)\mathbf{j}$, $\frac{d\mathbf{r}_1}{dt} = (-a \sin t)\mathbf{i} + (a \cos t)\mathbf{j} \Rightarrow \mathbf{F}_1 \cdot \frac{d\mathbf{r}_1}{dt} = 0 \Rightarrow \text{Circ}_1 = 0$; $M_1 = a \cos t$,

$N_1 = a \sin t$, $dx = -a \sin t\, dt$, $dy = a \cos t\, dt \Rightarrow \text{Flux}_1 = \int_C M_1\, dy - N_1\, dx = \int_0^{\pi} \left(a^2 \cos^2 t + a^2 \sin^2 t\right) dt$

$= \int_0^{\pi} a^2\, dt = a^2 \pi$;

$\mathbf{F}_2 = t\mathbf{i}$, $\frac{d\mathbf{r}_2}{dt} = \mathbf{i} \Rightarrow \mathbf{F}_2 \cdot \frac{d\mathbf{r}_2}{dt} = t \Rightarrow \text{Circ}_2 = \int_{-a}^{a} t\,dt = 0$; $M_2 = t$, $N_2 = 0$, $dx = dt$, $dy = 0 \Rightarrow \text{Flux}_2$

$= \int_C M_2\,dy - N_2\,dx = \int_{-a}^{a} 0\,dt = 0$; therefore, $\text{Circ} = \text{Circ}_1 + \text{Circ}_2 = 0$ and $\text{Flux} = \text{Flux}_1 + \text{Flux}_2 = a^2\pi$

26. $\mathbf{F}_1 = (a^2 \cos^2 t)\mathbf{i} + (a^2 \sin^2 t)\mathbf{j}$, $\frac{d\mathbf{r}_1}{dt} = (-a \sin t)\mathbf{i} + (a \cos t)\mathbf{j} \Rightarrow \mathbf{F}_1 \cdot \frac{d\mathbf{r}_1}{dt} = -a^3 \sin t \cos^2 t + a^3 \cos t \sin^2 t$

$\Rightarrow \text{Circ}_1 = \int_0^{\pi} (-a^3 \sin t \cos^2 t + a^3 \cos t \sin^2 t)\,dt = -\frac{2a^3}{3}$; $M_1 = a^2 \cos^2 t$, $N_1 = a^2 \sin^2 t$, $dy = a \cos t\,dt$,

$dx = -a \sin t\,dt \Rightarrow \text{Flux}_1 = \int_C M_1\,dy - N_1\,dx = \int_0^{\pi} (a^3 \cos^3 t + a^3 \sin^3 t)\,dt = \frac{4}{3}a^3$;

$\mathbf{F}_2 = t^2\mathbf{i}$, $\frac{d\mathbf{r}_2}{dt} = \mathbf{i} \Rightarrow \mathbf{F}_2 \cdot \frac{d\mathbf{r}_2}{dt} = t^2 \Rightarrow \text{Circ}_2 = \int_{-a}^{a} t^2\,dt = \frac{2a^3}{3}$; $M_2 = t^2$, $N_2 = 0$, $dy = 0$, $dx = dt$

$\Rightarrow \text{Flux}_2 = \int_C M_2\,dy - N_2\,dx = 0$; therefore, $\text{Circ} = \text{Circ}_1 + \text{Circ}_2 = 0$ and $\text{Flux} = \text{Flux}_1 + \text{Flux}_2 = \frac{4}{3}a^3$

27. $\mathbf{F}_1 = (-a \sin t)\mathbf{i} + (a \cos t)\mathbf{j}$, $\frac{d\mathbf{r}_1}{dt} = (-a \sin t)\mathbf{i} + (a \cos t)\mathbf{j} \Rightarrow \mathbf{F}_1 \cdot \frac{d\mathbf{r}_1}{dt} = a^2 \sin^2 t + a^2 \cos^2 t = a^2$

$\Rightarrow \text{Circ}_1 = \int_0^{\pi} a^2\,dt = a^2\pi$; $M_1 = -a \sin t$, $N_1 = a \cos t$, $dx = -a \sin t\,dt$, $dy = a \cos t\,dt$

$\Rightarrow \text{Flux}_1 = \int_C M_1\,dy - N_1\,dx = \int_0^{\pi} (-a^2 \sin t \cos t + a^2 \sin t \cos t)\,dt = 0$; $\mathbf{F}_2 = t\mathbf{j}$, $\frac{d\mathbf{r}_2}{dt} = \mathbf{i} \Rightarrow \mathbf{F}_2 \cdot \frac{d\mathbf{r}_2}{dt} = 0$

$\Rightarrow \text{Circ}_2 = 0$; $M_2 = 0$, $N_2 = t$, $dx = dt$, $dy = 0 \Rightarrow \text{Flux}_2 = \int_C M_2\,dy - N_2\,dx = \int_{-a}^{a} -t\,dt = 0$; therefore,

$\text{Circ} = \text{Circ}_1 + \text{Circ}_2 = a^2\pi$ and $\text{Flux} = \text{Flux}_1 + \text{Flux}_2 = 0$

28. $\mathbf{F}_1 = (-a^2 \sin^2 t)\mathbf{i} + (a^2 \cos^2 t)\mathbf{j}$, $\frac{d\mathbf{r}_1}{dt} = (-a \sin t)\mathbf{i} + (a \cos t)\mathbf{j} \Rightarrow \mathbf{F}_1 \cdot \frac{d\mathbf{r}_1}{dt} = a^3 \sin^3 t + a^3 \cos^3 t$

$\Rightarrow \text{Circ}_1 = \int_0^{\pi} (a^2 \sin^3 t + a^3 \cos^3 t)\,dt = \frac{4}{3}a^3$; $M_1 = -a^2 \sin^2 t$, $N_1 = a^2 \cos^2 t$, $dy = a \cos t\,dt$, $dx = -a \sin t\,dt$

$\Rightarrow \text{Flux}_1 = \int_C M_1\,dy - N_1\,dx = \int_0^{\pi} (-a^3 \cos t \sin^2 t + a^3 \sin t \cos^2 t)\,dt = \frac{2}{3}a^3$; $\mathbf{F}_2 = t^2\mathbf{j}$, $\frac{d\mathbf{r}_2}{dt} = \mathbf{i} \Rightarrow \mathbf{F}_2 \cdot \frac{d\mathbf{r}_2}{dt} = 0$

$\Rightarrow \text{Circ}_2 = 0$; $M_2 = 0$, $N_2 = t^2$, $dy = 0$, $dx = dt \Rightarrow \text{Flux}_2 = \int_C M_2\,dy - N_2\,dx = \int_{-a}^{a} -t^2\,dt = -\frac{2}{3}a^3$; therefore,

$\text{Circ} = \text{Circ}_1 + \text{Circ}_2 = \frac{4}{3}a^3$ and $\text{Flux} = \text{Flux}_1 + \text{Flux}_2 = 0$

29. (a) $\mathbf{r} = (\cos t)\mathbf{i} + (\sin t)\mathbf{j}$, $0 \le t \le \pi$, and $\mathbf{F} = (x+y)\mathbf{i} - (x^2+y^2)\mathbf{j} \Rightarrow \frac{d\mathbf{r}}{dt} = (-\sin t)\mathbf{i} + (\cos t)\mathbf{j}$ and

$\mathbf{F} = (\cos t + \sin t)\mathbf{i} - (\cos^2 t + \sin^2 t)\mathbf{j} \Rightarrow \mathbf{F} \cdot \frac{d\mathbf{r}}{dt} = -\sin t \cos t - \sin^2 t - \cos t \Rightarrow \int_C \mathbf{F} \cdot \mathbf{T}\, ds$

$= \int_0^\pi (-\sin t \cos t - \sin^2 t - \cos t)\, dt = \left[-\frac{1}{2}\sin^2 t - \frac{t}{2} + \frac{\sin 2t}{4} - \sin t\right]_0^\pi = -\frac{\pi}{2}$

(b) $\mathbf{r} = (1-2t)\mathbf{i}$, $0 \le t \le 1$, and $\mathbf{F} = (x+y)\mathbf{i} - (x^2+y^2)\mathbf{j} \Rightarrow \frac{d\mathbf{r}}{dt} = -\mathbf{i}$ and $\mathbf{F} = (1-2t)\mathbf{i} - (1-2t)^2\mathbf{j} \Rightarrow$

$\mathbf{F} \cdot \frac{d\mathbf{r}}{dt} = 2t - 1 \Rightarrow \int_C \mathbf{F} \cdot \mathbf{T}\, ds = \int_0^1 (2t-1)\, dt = \left[t^2 - t\right]_0^1 = 0$

(c) $\mathbf{r}_1 = (1-t)\mathbf{i} - t\mathbf{j}$, $0 \le t \le 1$, and $\mathbf{F} = (x+y)\mathbf{i} - (x^2+y^2)\mathbf{j} \Rightarrow \frac{d\mathbf{r}_1}{dt} = -\mathbf{i} - \mathbf{j}$ and $\mathbf{F} = (1-2t)\mathbf{i} - (1-2t+2t^2)\mathbf{j}$

$\Rightarrow \mathbf{F} \cdot \frac{d\mathbf{r}_1}{dt} = (2t-1) + (1-2t+2t^2) = 2t^2 \Rightarrow \text{Flow}_1 = \int_{C_1} \mathbf{F} \cdot \frac{d\mathbf{r}_1}{dt} = \int_0^1 2t^2\, dt = \frac{2}{3}$; $\mathbf{r}_2 = -t\mathbf{i} + (t-1)\mathbf{j}$,

$0 \le t \le 1$, and $\mathbf{F} = (x+y)\mathbf{i} - (x^2+y^2)\mathbf{j} \Rightarrow \frac{d\mathbf{r}_2}{dt} = -\mathbf{i} + \mathbf{j}$ and $\mathbf{F} = -\mathbf{i} - (t^2 + t^2 - 2t + 1)\mathbf{j}$

$= -\mathbf{i} - (2t^2 - 2t + 1)\mathbf{j} \Rightarrow \mathbf{F} \cdot \frac{d\mathbf{r}_2}{dt} = 1 - (2t^2 - 2t + 1) = 2t - 2t^2 \Rightarrow \text{Flow}_2 = \int_{C_2} \mathbf{F} \cdot \frac{d\mathbf{r}_2}{dt} = \int_0^1 (2t - 2t^2)\, dt$

$= \left[t^2 - \frac{2}{3}t^3\right]_0^1 = \frac{1}{3} \Rightarrow \text{Flow} = \text{Flow}_1 + \text{Flow}_2 = \frac{2}{3} + \frac{1}{3} = 1$

30. From $(1,0)$ to $(0,1)$: $\mathbf{r}_1 = (1-t)\mathbf{i} + t\mathbf{j}$, $0 \le t \le 1$, and $\mathbf{F} = (x+y)\mathbf{i} - (x^2+y^2)\mathbf{j} \Rightarrow \frac{d\mathbf{r}_1}{dt} = -\mathbf{i} + \mathbf{j}$,

$\mathbf{F} = \mathbf{i} - (1-2t+2t^2)\mathbf{j}$, and $\mathbf{n}_1 |\mathbf{v}_1| = \mathbf{i} + \mathbf{j} \Rightarrow \mathbf{F} \cdot \mathbf{n}_1 |\mathbf{v}_1| = 2t - 2t^2 \Rightarrow \text{Flux}_1 = \int_0^1 (2t - 2t^2)\, dt$

$= \left[t^2 - \frac{2}{3}t^3\right]_0^1 = \frac{1}{3}$;

From $(0,1)$ to $(-1,0)$: $\mathbf{r}_2 = -t\mathbf{i} + (1-t)\mathbf{j}$, $0 \le t \le 1$, and $\mathbf{F} = (x+y)\mathbf{i} - (x^2+y^2)\mathbf{j} \Rightarrow \frac{d\mathbf{r}_2}{dt} = -\mathbf{i} - \mathbf{j}$,

$\mathbf{F} = (1-2t)\mathbf{i} - (1-2t+2t^2)\mathbf{j}$, and $\mathbf{n}_2 |\mathbf{v}_2| = -\mathbf{i} + \mathbf{j} \Rightarrow \mathbf{F} \cdot \mathbf{n}_2 |\mathbf{v}_2| = (2t-1) + (-1+2t-2t^2) = -2 + 4t - 2t^2$

$\Rightarrow \text{Flux}_2 = \int_0^1 (-2 + 4t - 2t^2)\, dt = \left[-2t + 2t^2 - \frac{2}{3}t^3\right]_0^1 = -\frac{2}{3}$;

From $(-1,0)$ to $(1,0)$: $\mathbf{r}_3 = (-1+2t)\mathbf{i}$, $0 \le t \le 1$, and $\mathbf{F} = (x+y)\mathbf{i} - (x^2+y^2)\mathbf{j} \Rightarrow \frac{d\mathbf{r}_3}{dt} = 2\mathbf{i}$,

$\mathbf{F} = (-1+2t)\mathbf{i} - (1-4t+4t^2)\mathbf{j}$, and $\mathbf{n}_3 |\mathbf{v}_3| = -2\mathbf{j} \Rightarrow \mathbf{F} \cdot \mathbf{n}_3 |\mathbf{v}_3| = 2(1-4t+4t^2)$

$\Rightarrow \text{Flux}_3 = 2\int_0^1 (1-4t+4t^2)\, dt = 2\left[t - 2t^2 + \frac{4}{3}t^3\right]_0^1 = \frac{2}{3} \Rightarrow \text{Flux} = \text{Flux}_1 + \text{Flux}_2 + \text{Flux}_3 = \frac{1}{3} - \frac{2}{3} + \frac{2}{3} = \frac{1}{3}$

31. $\mathbf{F} = -\frac{y}{\sqrt{x^2+y^2}}\mathbf{i} + \frac{x}{\sqrt{x^2+y^2}}\mathbf{j}$ on $x^2+y^2=4$;

at $(2,0)$, $\mathbf{F}=\mathbf{j}$; at $(0,2)$, $\mathbf{F}=-\mathbf{i}$; at $(-2,0)$,

$\mathbf{F}=-\mathbf{j}$; at $(0,-2)$, $\mathbf{F}=\mathbf{i}$; at $(1,\sqrt{3})$, $\mathbf{F}=-\frac{\sqrt{3}}{2}\mathbf{i}+\frac{1}{2}\mathbf{j}$;

at $(1,-\sqrt{3})$, $\mathbf{F}=\frac{\sqrt{3}}{2}\mathbf{i}+\frac{1}{2}\mathbf{j}$; at $(-1,\sqrt{3})$,

$\mathbf{F}=-\frac{\sqrt{3}}{2}\mathbf{i}-\frac{1}{2}\mathbf{j}$; at $(-1,-\sqrt{3})$, $\mathbf{F}=\frac{\sqrt{3}}{2}\mathbf{i}-\frac{1}{2}\mathbf{j}$

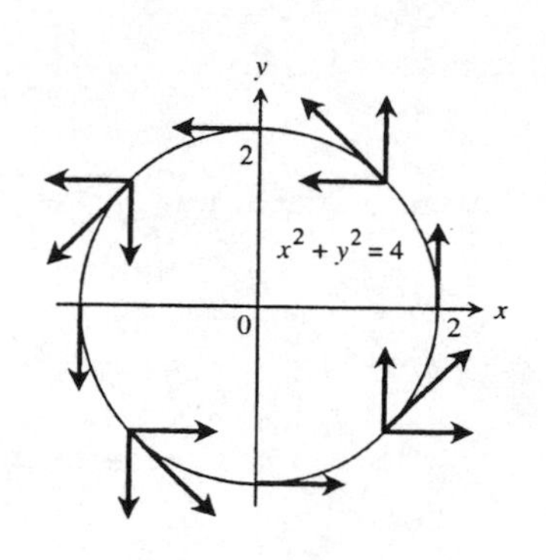

32. $\mathbf{F}=x\mathbf{i}+y\mathbf{j}$ on $x^2+y^2=1$; at $(1,0)$, $\mathbf{F}=\mathbf{i}$;

at $(-1,0)$, $\mathbf{F}=-\mathbf{i}$; at $(0,1)$, $\mathbf{F}=\mathbf{j}$; at $(0,-1)$,

$\mathbf{F}=-\mathbf{j}$; at $\left(\frac{1}{2},\frac{\sqrt{3}}{2}\right)$, $\mathbf{F}=\frac{1}{2}\mathbf{i}+\frac{\sqrt{3}}{2}\mathbf{j}$;

at $\left(-\frac{1}{2},\frac{\sqrt{3}}{2}\right)$, $\mathbf{F}=-\frac{1}{2}\mathbf{i}+\frac{\sqrt{3}}{2}\mathbf{j}$;

at $\left(\frac{1}{2},-\frac{\sqrt{3}}{2}\right)$, $\mathbf{F}=\frac{1}{2}\mathbf{i}-\frac{\sqrt{3}}{2}\mathbf{j}$; at $\left(-\frac{1}{2},-\frac{\sqrt{3}}{2}\right)$,

$\mathbf{F}=-\frac{1}{2}\mathbf{i}-\frac{\sqrt{3}}{2}\mathbf{j}$.

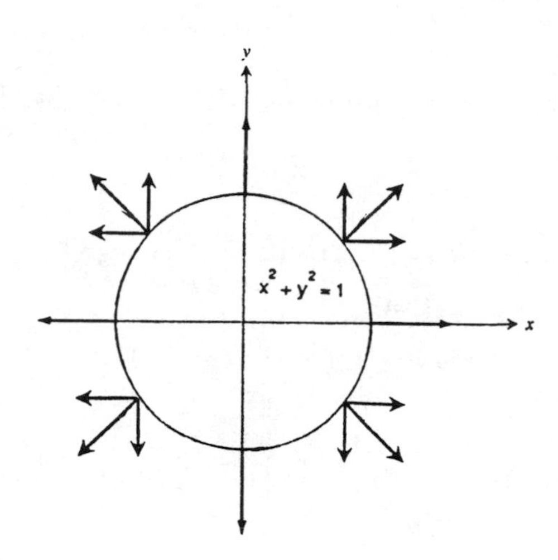

33. (a) $\mathbf{G}=P(x,y)\mathbf{i}+Q(x,y)\mathbf{j}$ is to have a magnitude $\sqrt{a^2+b^2}$ and to be tangent to $x^2+y^2=a^2+b^2$ in a counterclockwise direction. Thus $x^2+y^2=a^2+b^2 \Rightarrow 2x+2yy'=0 \Rightarrow y'=-\frac{x}{y}$ is the slope of the tangent line at any point on the circle $\Rightarrow y'=-\frac{a}{b}$ at (a,b). Let $\mathbf{v}=-b\mathbf{i}+a\mathbf{j} \Rightarrow |\mathbf{v}|=\sqrt{a^2+b^2}$, with $\mathbf{v}$ in a counterclockwise direction and tangent to the circle. Then let $P(x,y)=-y$ and $Q(x,y)=x$ $\Rightarrow \mathbf{G}=-y\mathbf{i}+x\mathbf{j} \Rightarrow$ for (a,b) on $x^2+y^2=a^2+b^2$ we have $\mathbf{G}=-b\mathbf{i}+a\mathbf{j}$ and $|\mathbf{G}|=\sqrt{a^2+b^2}$.

(b) $\mathbf{G}=\left(\sqrt{x^2+y^2}\right)\mathbf{F}=\left(\sqrt{a^2+b^2}\right)\mathbf{F}$, since $x^2+y^2=a^2+b^2$

34. (a) From Exercise 33, part a, $-y\mathbf{i}+x\mathbf{j}$ is a vector tangent to the circle and pointing in a counterclockwise direction $\Rightarrow y\mathbf{i}-x\mathbf{j}$ is a vector tangent to the circle pointing in a clockwise direction $\Rightarrow \mathbf{G}=\frac{y\mathbf{i}-x\mathbf{j}}{\sqrt{x^2+y^2}}$ is a unit vector tangent to the circle and pointing in a clockwise direction.

(b) $\mathbf{G}=-\mathbf{F}$

35. The slope of the line through (x,y) and the origin is $\frac{y}{x} \Rightarrow \mathbf{v}=x\mathbf{i}+y\mathbf{j}$ is a vector parallel to that line and pointing away from the origin $\Rightarrow \mathbf{F}=-\frac{x\mathbf{i}+y\mathbf{j}}{\sqrt{x^2+y^2}}$ is the unit vector pointing toward the origin.

36. (a) From Exercise 35, $-\frac{x\mathbf{i}+y\mathbf{j}}{\sqrt{x^2+y^2}}$ is a unit vector through (x, y) pointing toward the origin and we want

$|\mathbf{F}|$ to have magnitude $\sqrt{x^2+y^2} \Rightarrow \mathbf{F} = \sqrt{x^2+y^2}\left(-\frac{x\mathbf{i}+y\mathbf{j}}{\sqrt{x^2+y^2}}\right) = -x\mathbf{i} - y\mathbf{j}$.

(b) We want $|\mathbf{F}| = \frac{C}{\sqrt{x^2+y^2}} \Rightarrow \mathbf{F} = \frac{C}{\sqrt{x^2+y^2}}\left(-\frac{x\mathbf{i}+y\mathbf{j}}{\sqrt{x^2+y^2}}\right) = -C\left(\frac{x\mathbf{i}+y\mathbf{j}}{x^2+y^2}\right)$, $C \neq 0$, and constant

37. $\mathbf{F} = -4t^3\mathbf{i} + 8t^2\mathbf{j} + 2\mathbf{k}$ and $\frac{d\mathbf{r}}{dt} = \mathbf{i} + 2t\mathbf{j} \Rightarrow \mathbf{F} \cdot \frac{d\mathbf{r}}{dt} = 12t^3 \Rightarrow \text{Flow} = \int_0^2 12t^3\,dt = \left[3t^4\right]_0^2 = 48$

38. $\mathbf{F} = 12t^2\mathbf{j} + 9t^2\mathbf{k}$ and $\frac{d\mathbf{r}}{dt} = 3\mathbf{j} + 4\mathbf{k} \Rightarrow \mathbf{F} \cdot \frac{d\mathbf{r}}{dt} = 72t^2 \Rightarrow \text{Flow} = \int_0^1 72t^2\,dt = \left[24t^3\right]_0^1 = 24$

39. $\mathbf{F} = (\cos t - \sin t)\mathbf{i} + (\cos t)\mathbf{k}$ and $\frac{d\mathbf{r}}{dt} = (-\sin t)\mathbf{i} + (\cos t)\mathbf{k} \Rightarrow \mathbf{F} \cdot \frac{d\mathbf{r}}{dt} = -\sin t \cos t + 1$

$\Rightarrow \text{Flow} = \int_0^{\pi} (-\sin t \cos t + 1)\,dt = \left[\frac{1}{2}\cos^2 t + t\right]_0^{\pi} = \left(\frac{1}{2} + \pi\right) - \left(\frac{1}{2} + 0\right) = \pi$

40. $\mathbf{F} = (-2\sin t)\mathbf{i} - (2\cos t)\mathbf{j} + 2\mathbf{k}$ and $\frac{d\mathbf{r}}{dt} = (2\sin t)\mathbf{i} + (2\cos t)\mathbf{j} + 2\mathbf{k} \Rightarrow \mathbf{F} \cdot \frac{d\mathbf{r}}{dt} = -4\sin^2 t - 4\cos^2 t + 4 = 0$
$\Rightarrow \text{Flow} = 0$

41. C_1: $\mathbf{r} = (\cos t)\mathbf{i} + (\sin t)\mathbf{j} + t\mathbf{k}$, $0 \le t \le \frac{\pi}{2} \Rightarrow \mathbf{F} = (2\cos t)\mathbf{i} + 2t\mathbf{j} + (2\sin t)\mathbf{k}$ and $\frac{d\mathbf{r}}{dt} = (-\sin t)\mathbf{i} + (\cos t)\mathbf{j} + \mathbf{k}$

$\Rightarrow \mathbf{F} \cdot \frac{d\mathbf{r}}{dt} = -2\cos t \sin t + 2t\cos t + 2\sin t = -\sin 2t + 2t\cos t + 2\sin t$

$\Rightarrow \text{Flow}_1 = \int_0^{\pi/2} (-\sin 2t + 2t\cos t + 2\sin t)\,dt = \left[\frac{1}{2}\cos 2t + 2t\sin t + 2\cos t - 2\cos t\right]_0^{\pi/2} = -1 + \pi;$

C_2: $\mathbf{r} = \mathbf{j} + \frac{\pi}{2}(1-t)\mathbf{k}$, $0 \le t \le 1 \Rightarrow \mathbf{F} = \pi(1-t)\mathbf{j} + 2\mathbf{k}$ and $\frac{d\mathbf{r}}{dt} = -\frac{\pi}{2}\mathbf{k} \Rightarrow \mathbf{F} \cdot \frac{d\mathbf{r}}{dt} = -\pi$

$\Rightarrow \text{Flow}_2 = \int_0^1 -\pi\,dt = [-\pi t]_0^1 = -\pi;$

C_3: $\mathbf{r} = t\mathbf{i} + (1-t)\mathbf{j}$, $0 \le t \le 1 \Rightarrow \mathbf{F} = 2t\mathbf{i} + 2(1-t)\mathbf{j}$ and $\frac{d\mathbf{r}}{dt} = \mathbf{i} - \mathbf{j} \Rightarrow \mathbf{F} \cdot \frac{d\mathbf{r}}{dt} = 2t$

$\Rightarrow \text{Flow}_3 = \int_0^1 2t\,dt = \left[t^2\right]_0^1 = 1 \Rightarrow \text{Circulation} = (-1 + \pi) - \pi + 1 = 0$

42. $\mathbf{F}\cdot\frac{d\mathbf{r}}{dt} = x\frac{dx}{dt} + y\frac{dy}{dt} + z\frac{dz}{dt} = \frac{\partial f}{\partial x}\frac{dx}{dt} + \frac{\partial f}{\partial y}\frac{dy}{dt} + \frac{\partial f}{\partial z}\frac{dz}{dt}$, where $f(x,y,z) = \frac{1}{2}(x^2+y^2+x^2) \Rightarrow \mathbf{F}\cdot\frac{d\mathbf{r}}{dt} = \frac{df}{dt}$,

by the chain rule $\Rightarrow$ Circulation $= \int_C \mathbf{F}\cdot\frac{d\mathbf{r}}{dt}\,dt = \int_C \frac{df}{dt}\,dt = f(B) - f(A)$ for any two points A and B on the path

$\Rightarrow$ Circulation $= 0$ since $A = B$ all the way around the curve of intersection.

43. Let $x = t$ be the parameter $\Rightarrow y = x^2 = t^2$ and $z = x = t \Rightarrow \mathbf{r} = t\mathbf{i} + t^2\mathbf{j} + t\mathbf{k}$, $0 \le t \le 1$ from $(0,0,0)$ to $(1,1,1)$

$\Rightarrow \frac{d\mathbf{r}}{dt} = \mathbf{i} + 2t\mathbf{j} + \mathbf{k}$ and $\mathbf{F} = xy\mathbf{i} + y\mathbf{j} - yz\mathbf{k} = t^3\mathbf{i} + t^2\mathbf{j} - t^3\mathbf{k} \Rightarrow \mathbf{F}\cdot\frac{d\mathbf{r}}{dt} = t^3 + 2t^3 - t^3 = 2t^3 \Rightarrow$ Flow $= \int_0^1 2t^3\,dt$

$= \frac{1}{2}$

44. (a) $\mathbf{F} = \nabla(xy^2z^3) \Rightarrow \mathbf{F}\cdot\frac{d\mathbf{r}}{dt} = \frac{\partial f}{\partial x}\frac{dx}{dt} + \frac{\partial f}{\partial y}\frac{dy}{dt} + \frac{\partial z}{\partial z}\frac{dz}{dt} = \frac{df}{dt}$, where $f(x,y,z) = xy^2z^3 \Rightarrow \oint_C \mathbf{F}\cdot\frac{d\mathbf{r}}{dt}\,dt = \oint_C \frac{df}{dt}\,dt$

$= f(A) - f(A) = 0$ for any point A on the curve C

(b) $\int_C \mathbf{F}\cdot\frac{d\mathbf{r}}{dt} = \int_{(1,1,1)}^{(2,1,-1)} \frac{d}{dt}(xy^2z^3)\,dt = \left[xy^2z^3\right]_{(1,1,1)}^{(2,1,-1)} = (2)(1)^2(-1)^3 - (1)(1)^2(1)^3 = -2 - 1 = -3$

45. Yes. The work and area have the same numerical value because work $= \int_C \mathbf{F}\cdot d\mathbf{r} = \int_C y\mathbf{i}\cdot d\mathbf{r}$

$= \int_b^a [f(t)\mathbf{i}]\cdot\left[\mathbf{i} + \frac{df}{dt}\mathbf{j}\right]dt$ [On the path, y equals f(t)]

$= \int_a^b f(t)\,dt =$ Area under the curve [because $f(t) > 0$]

46. $\mathbf{r} = x\mathbf{i} + y\mathbf{j} = x\mathbf{i} + f(x)\mathbf{j} \Rightarrow \frac{d\mathbf{r}}{dx} = \mathbf{i} + f'(x)\mathbf{j}$; $\mathbf{F} = \frac{k}{\sqrt{x^2+y^2}}(x\mathbf{i} + y\mathbf{j})$ has constant magnitude k and points away

from the origin $\Rightarrow \mathbf{F}\cdot\frac{d\mathbf{r}}{dx} = \frac{kx}{\sqrt{x^2+y^2}} + \frac{kyf'(x)}{\sqrt{x^2+y^2}} = \frac{kx + kf(x)f'(x)}{\sqrt{x^2+[f(x)]^2}} = k\frac{d}{dx}\sqrt{x^2+[f(x)]^2}$, by the chain rule

$\Rightarrow \int_C \mathbf{F}\cdot\mathbf{T}\,ds = \int_C \mathbf{F}\cdot\frac{d\mathbf{r}}{dx}\,dx = \int_a^b k\frac{d}{dx}\sqrt{x^2+[f(x)]^2}\,dx = k\left[\sqrt{x^2+[f(x)]^2}\right]_a^b$

$= k\left(\sqrt{b^2+[f(b)]^2} - \sqrt{a^2+[f(a)]^2}\right)$, as claimed.

47-52. Example CAS commands:

Maple:

```
with(linalg):
x:= t -> cos(t);
y:= t -> sin(t);
z:= t -> 0;
```

```
a:=0; b:= Pi;
M:= (x,y,z) ->3/(1 + x^2);
N:= (x,y,z) ->2/(1 + y^2);
P:= (x,y,z) -> 0;
F:= t -> vector([M(x(t),y(t),z(t)), N(x(t),y(t),z(t)), P(x(t),y(t),z(t))]);
dr:= t -> vector([D(x)(t), D(y)(t), D(z)(t)]);
integrand:= dotprod(F(t), dr(t), orthogonal);
int(integrand, t=a..b);
evalf('');
```

Mathematica:

```
Clear[x,y,z,t]
r[t_] = {x[t],y[t],z[t]}
f[x_,y_] = {
  3/(1+x^2) ,
  2/(1+y^2) }
x[t_] = Cos[t]
y[t_] = Sin[t]
z[t_] = 0
{a,b} = {0,Pi};
v[t_] = r'[t]
integrand = f[x[t],y[t],z[t]] . v[t]
integrand = Simplify[ integrand ]
```

14.3 PATH INDEPENDENCE, POTENTIAL FUNCTIONS, AND CONSERVATIVE FIELDS

1. $\frac{\partial P}{\partial y} = x = \frac{\partial N}{\partial z}, \frac{\partial M}{\partial z} = y = \frac{\partial P}{\partial x}, \frac{\partial N}{\partial x} = z = \frac{\partial M}{\partial y} \Rightarrow$ Conservative

2. $\frac{\partial P}{\partial y} = x \cos z = \frac{\partial N}{\partial z}, \frac{\partial M}{\partial z} = y \cos z = \frac{\partial P}{\partial x}, \frac{\partial N}{\partial x} = \sin z = \frac{\partial M}{\partial y} \Rightarrow$ Conservative

3. $\frac{\partial P}{\partial y} = -1 \neq 1 = \frac{\partial N}{\partial z} \Rightarrow$ Not Conservative

4. $\frac{\partial N}{\partial x} = 1 \neq -1 = \frac{\partial M}{\partial y} \Rightarrow$ Not Conservative

5. $\frac{\partial N}{\partial x} = 0 \neq 1 = \frac{\partial M}{\partial y} \Rightarrow$ Not Conservative

6. $\frac{\partial P}{\partial y} = 0 = \frac{\partial N}{\partial z}, \frac{\partial M}{\partial z} = 0 = \frac{\partial P}{\partial x}, \frac{\partial N}{\partial x} = -e^x \sin y = \frac{\partial M}{\partial y} \Rightarrow$ Conservative

7. $\frac{\partial f}{\partial x} = 2x \Rightarrow f(x,y,z) = x^2 + g(y,z) \Rightarrow \frac{\partial f}{\partial y} = \frac{\partial g}{\partial y} = 3y \Rightarrow g(y,z) = \frac{3y^2}{2} + h(z) \Rightarrow f(x,y,z) = x^2 + \frac{3y^2}{2} + h(z)$
$\Rightarrow \frac{\partial f}{\partial z} = h'(z) = 4z \Rightarrow h(z) = 2z^2 + C \Rightarrow f(x,y,z) = x^2 + \frac{3y^2}{2} + 2z^2 + C$

8. $\frac{\partial f}{\partial x} = y + z \Rightarrow f(x,y,z) = (y+z)x + g(y,z) \Rightarrow \frac{\partial f}{\partial y} = x + \frac{\partial g}{\partial y} = x + z \Rightarrow \frac{\partial g}{\partial y} = z \Rightarrow g(y,z) = zy + h(z)$
$\Rightarrow f(x,y,z) = (y+z)x + zy + h(z) \Rightarrow \frac{\partial f}{\partial z} = x + y + h'(z) = x + y \Rightarrow h'(z) = 0 \Rightarrow h(z) = C \Rightarrow f(x,y,z)$
$= (y+z)x + zy + C$

9. $\frac{\partial f}{\partial x} = e^{y+2z} \Rightarrow f(x,y,z) = xe^{y+2z} + g(y,z) \Rightarrow \frac{\partial f}{\partial y} = xe^{y+2z} + \frac{\partial g}{\partial y} = xe^{y+2z} \Rightarrow \frac{\partial g}{\partial y} = 0 \Rightarrow f(x,y,z)$
$= xe^{y+2z} + h(z) \Rightarrow \frac{\partial f}{\partial z} = 2xe^{y+2z} + h'(z) = 2xe^{y+2z} \Rightarrow h'(z) = 0 \Rightarrow h(z) = C \Rightarrow f(x,y,z) = xe^{y+2z} + C$

10. $\frac{\partial f}{\partial x} = y \sin z \Rightarrow f(x,y,z) = xy \sin z + g(y,z) \Rightarrow \frac{\partial f}{\partial y} = x \sin z + \frac{\partial g}{\partial y} = x \sin z \Rightarrow \frac{\partial g}{\partial y} = 0 \Rightarrow g(y,z) = h(z)$
$\Rightarrow f(x,y,z) = xy \sin z + h(z) \Rightarrow \frac{\partial f}{\partial z} = xy \cos z + h'(z) = xy \cos z \Rightarrow h'(z) = 0 \Rightarrow h(z) + C \Rightarrow f(x,y,z)$
$= xy \sin z + C$

11. $\frac{\partial f}{\partial z} = \frac{z}{y^2+z^2} \Rightarrow f(x,y,z) = \frac{1}{2}\ln(y^2+z^2) + g(x,y) \Rightarrow \frac{\partial f}{\partial x} = \frac{\partial g}{\partial x} = \ln x + \sec^2(x+y) \Rightarrow g(x,y)$
$= (x \ln x - x) + \tan(x+y) + h(y) \Rightarrow f(x,y,z) = \frac{1}{2}\ln(y^2+z^2) + (x \ln x - x) + \tan(x+y) + h(y)$
$\Rightarrow \frac{\partial f}{\partial y} = \frac{y}{y^2+z^2} + \sec^2(x+y) + h'(y) = \sec^2(x+y) + \frac{y}{y^2+z^2} \Rightarrow h'(y) = 0 \Rightarrow h(y) = C \Rightarrow f(x,y,z)$
$= \frac{1}{2}\ln(y^2+z^2) + (x \ln x - x) + \tan(x+y) + C$

12. $\frac{\partial f}{\partial x} = \frac{y}{1+x^2y^2} \Rightarrow f(x,y,z) = \tan^{-1}(xy) + g(y,z) \Rightarrow \frac{\partial f}{\partial y} = \frac{x}{1+x^2y^2} + \frac{\partial g}{\partial y} = \frac{x}{1+x^2y^2} + \frac{z}{\sqrt{1-y^2z^2}}$
$\Rightarrow \frac{\partial g}{\partial y} = \frac{z}{\sqrt{1-y^2z^2}} \Rightarrow g(y,z) = \sin^{-1}(yz) + h(z) \Rightarrow f(x,y,z) = \tan^{-1}(xy) + \sin^{-1}(yz) + h(z)$
$\Rightarrow \frac{\partial f}{\partial z} = \frac{y}{\sqrt{1-y^2z^2}} + h'(z) = \frac{y}{\sqrt{1-y^2z^2}} + \frac{1}{z} \Rightarrow h'(z) = \frac{1}{z} \Rightarrow h(z) = \ln|z| + C$
$\Rightarrow f(x,y,z) = \tan^{-1}(xy) + \sin^{-1}(yz) + \ln|z| + C$

13. Let $\mathbf{F}(x,y,z) = 2x\mathbf{i} + 2y\mathbf{j} + 2z\mathbf{k} \Rightarrow \frac{\partial P}{\partial y} = 0 = \frac{\partial N}{\partial z}, \frac{\partial M}{\partial z} = 0 = \frac{\partial P}{\partial x}, \frac{\partial N}{\partial x} = 0 = \frac{\partial M}{\partial y} \Rightarrow M\,dx + N\,dy + P\,dz$ is exact; $\frac{\partial f}{\partial x} = 2x \Rightarrow f(x,y,z) = x^2 + g(y,z) \Rightarrow \frac{\partial f}{\partial y} = \frac{\partial g}{\partial y} = 2y \Rightarrow g(y,z) = y^2 + h(z) \Rightarrow f(x,y,z) = x^2 + y^2 = h(z)$
$\Rightarrow \frac{\partial f}{\partial z} = h'(z) = 2z \Rightarrow h(z) = z^2 + C \Rightarrow f(x,y,z) = x^2 + y^2 + z^2 + C \Rightarrow \int_{(0,0,0)}^{(2,3,-6)} 2x\,dx + 2y\,dy + 2z\,dz$
$= f(2,3,-6) - f(0,0,0) = 2^2 + 3^2 + (-6)^2 = 49$

14. Let $\mathbf{F}(x,y,z) = yz\mathbf{i} + xz\mathbf{j} + xy\mathbf{k} \Rightarrow \frac{\partial P}{\partial y} = x = \frac{\partial N}{\partial z}, \frac{\partial M}{\partial z} = y = \frac{\partial P}{\partial x}, \frac{\partial N}{\partial x} = z = \frac{\partial M}{\partial y} \Rightarrow M\,dx + N\,dy + P\,dz$ is exact; $\frac{\partial f}{\partial x} = yz \Rightarrow f(x,y,z) = xyz + g(y,z) \Rightarrow \frac{\partial f}{\partial y} = xz + \frac{\partial g}{\partial y} = xz \Rightarrow \frac{\partial g}{\partial y} = 0 \Rightarrow g(y,z) = h(z) \Rightarrow f(x,y,z)$
$= xyz + h(z) \Rightarrow \frac{\partial f}{\partial z} = xy + h'(z) = xy \Rightarrow h'(z) = 0 \Rightarrow h(z) = C \Rightarrow f(x,y,z) = xyz + C$
$\Rightarrow \int_{(1,1,2)}^{(3,5,0)} yz\,dx + xz\,dy + xy\,dz = f(3,5,0) - f(1,1,2) = 0 - 2 = -2$

15. Let $\mathbf{F}(x,y,z) = 2xy\mathbf{i} + (x^2 - z^2)\mathbf{j} - 2yz\mathbf{k} \Rightarrow \frac{\partial P}{\partial y} = -2z = \frac{\partial N}{\partial z}, \frac{\partial M}{\partial z} = 0 = \frac{\partial P}{\partial x}, \frac{\partial N}{\partial x} = 2x = \frac{\partial M}{\partial y}$
$\Rightarrow M\,dx + N\,dy + P\,dz$ is exact; $\frac{\partial f}{\partial x} = 2xy \Rightarrow f(x,y,z) = x^2y + g(y,z) \Rightarrow \frac{\partial f}{\partial y} = x^2 + \frac{\partial g}{\partial y} = x^2 - z^2 \Rightarrow \frac{\partial g}{\partial y} = -z^2$
$\Rightarrow g(y,z) = -yz^2 + h(z) \Rightarrow f(x,y,z) = x^2y - yz^2 + h(z) \Rightarrow \frac{\partial f}{\partial z} = -2yz + h'(z) = -2yz \Rightarrow h'(z) = 0 \Rightarrow h(z) = C$

$$\Rightarrow f(x,y,z) = x^2y - yz^2 + C \Rightarrow \int_{(0,0,0)}^{(1,2,3)} 2xy\,dx + (x^2 - z^2)\,dy - 2yz\,dz = f(1,2,3) - f(0,0,0) = 2 - 2(3)^2 = -16$$

16. Let $\mathbf{F}(x,y,z) = 2x\mathbf{i} - y^2\mathbf{j} - \left(\frac{4}{1+z^2}\right)\mathbf{k} \Rightarrow \frac{\partial P}{\partial y} = 0 = \frac{\partial N}{\partial z}, \frac{\partial M}{\partial z} = 0 = \frac{\partial P}{\partial x}, \frac{\partial N}{\partial x} = 0 = \frac{\partial M}{\partial y}$

$\Rightarrow M\,dx + N\,dy + P\,dz$ is exact; $\frac{\partial f}{\partial x} = 2x \Rightarrow f(x,y,z) = x^2 + g(y,z) \Rightarrow \frac{\partial f}{\partial y} = \frac{\partial g}{\partial y} = -y^2 \Rightarrow g(y,z) = -\frac{y^3}{3} + h(z)$

$\Rightarrow f(x,y,z) = x^2 - \frac{y^3}{3} + h(z) \Rightarrow \frac{\partial f}{\partial z} = h'(z) = -\frac{4}{1+z^2} \Rightarrow h(z) = -4\tan^{-1} z + C \Rightarrow f(x,y,z)$

$$= x^2 - \frac{y^3}{3} - 4\tan^{-1} z + C \Rightarrow \int_{(0,0,0)}^{(3,3,1)} 2x\,dx - y^2\,dy - \frac{4}{1-z^2}\,dz = f(3,3,1) - f(0,0,0)$$

$$= \left(9 - \frac{27}{3} - 4\cdot\frac{\pi}{4} + C\right) - (0 - 0 - 0 + C) = -\pi$$

17. Let $\mathbf{F}(x,y,z) = (\sin y \cos x)\mathbf{i} + (\cos y \sin x)\mathbf{j} + \mathbf{k} \Rightarrow \frac{\partial P}{\partial y} = 0 = \frac{\partial N}{\partial z}, \frac{\partial M}{\partial z} = 0 = \frac{\partial P}{\partial x}, \frac{\partial N}{\partial x} = \cos y \cos x = \frac{\partial M}{\partial y}$

$\Rightarrow M\,dx + N\,dy + P\,dz$ is exact; $\frac{\partial f}{\partial x} = \sin y \cos x \Rightarrow f(x,y,z) = \sin y \sin x + g(y,z) \Rightarrow \frac{\partial f}{\partial y} = \cos y \sin x + \frac{\partial g}{\partial y}$

$= \cos y \sin x \Rightarrow \frac{\partial g}{\partial y} = 0 \Rightarrow g(y,z) = h(z) \Rightarrow f(x,y,z) = \sin y \sin x + h(z) \Rightarrow \frac{\partial f}{\partial z} = h'(z) = 1 \Rightarrow h(z) = z + C$

$$\Rightarrow f(x,y,z) = \sin y \sin x + z + C \Rightarrow \int_{(1,0,0)}^{(0,1,1)} \sin y \cos x\,dx + \cos y \sin x\,dy + dz = f(0,1,1) - f(1,0,0)$$

$= (0 + 1 + C) - (0 + 0 + C) = 1$

18. Let $\mathbf{F}(x,y,z) = (2\cos y)\mathbf{i} + \left(\frac{1}{y} - 2x\sin y\right)\mathbf{j} + \left(\frac{1}{z}\right)\mathbf{k} \Rightarrow \frac{\partial P}{\partial y} = 0 = \frac{\partial N}{\partial z}, \frac{\partial M}{\partial z} = 0 = \frac{\partial P}{\partial x}, \frac{\partial N}{\partial x} = -2\sin y = \frac{\partial M}{\partial y}$

$\Rightarrow M\,dx + N\,dy + P\,dz$ is exact; $\frac{\partial f}{\partial x} = 2\cos y \Rightarrow f(x,y,z) = 2x\cos y + g(y,z) \Rightarrow \frac{\partial f}{\partial y} = -2x\sin y + \frac{\partial g}{\partial y}$

$= \frac{1}{y} - 2x\sin y \Rightarrow \frac{\partial g}{\partial y} = \frac{1}{y} \Rightarrow g(y,z) = \ln|y| + h(z) \Rightarrow f(x,y,z) = 2x\cos y + \ln|y| + h(z) \Rightarrow \frac{\partial f}{\partial z} = h'(z) = \frac{1}{z}$

$\Rightarrow h(z) = \ln|z| + C \Rightarrow f(x,y,z) = 2x\cos y + \ln|y| + \ln|z| + C$

$$\Rightarrow \int_{(0,2,1)}^{(1,\pi/2,2)} 2\cos y\,dx + \left(\frac{1}{y} - 2x\sin y\right)dy + \frac{1}{z}\,dz = f\left(1,\frac{\pi}{2},2\right) - f(0,2,1)$$

$$= \left(2\cdot 0 + \ln\frac{\pi}{2} + \ln 2 + C\right) - (0\cdot\cos 2 + \ln 2 + \ln 1 + C) = \ln\frac{\pi}{2}$$

19. Let $\mathbf{F}(x,y,z) = 3x^2\mathbf{i} + \left(\frac{z^2}{y}\right)\mathbf{j} + (2z\ln y)\mathbf{k} \Rightarrow \frac{\partial P}{\partial y} = \frac{2z}{y} = \frac{\partial N}{\partial z}, \frac{\partial M}{\partial z} = 0 = \frac{\partial P}{\partial x}, \frac{\partial N}{\partial x} = 0 = \frac{\partial M}{\partial y}$

$\Rightarrow M\,dx + N\,dy + P\,dz$ is exact; $\frac{\partial f}{\partial x} = 3x^2 \Rightarrow f(x,y,z) = x^3 + g(y,z) \Rightarrow \frac{\partial f}{\partial y} = \frac{\partial g}{\partial y} = \frac{z^2}{y} \Rightarrow g(y,z) = z^2\ln y + h(z)$

$\Rightarrow f(x,y,z) = x^3 + z^2\ln y + h(z) \Rightarrow \frac{\partial f}{\partial z} = 2z\ln y + h'(z) = 2z\ln y \Rightarrow h'(z) = 0 \Rightarrow h(z) = C \Rightarrow f(x,y,z)$

$= x^3 + z^2 \ln y + C \Rightarrow \int_{(1,1,1)}^{(1,2,3)} 3x^2\, dx + \frac{z^2}{y}\, dy + 2z \ln y\, dz = f(1,2,3) - f(1,1,1)$

$= (1 + 9 \ln 2 + C) - (1 + 0 + C) = 9 \ln 2$

20. Let $\mathbf{F}(x,y,z) = (2x \ln y - yz)\mathbf{i} + \left(\frac{x^2}{y} - xz\right)\mathbf{j} - (xy)\mathbf{k} \Rightarrow \frac{\partial P}{\partial y} = -x = \frac{\partial N}{\partial z}, \frac{\partial M}{\partial z} = -y = \frac{\partial P}{\partial x}, \frac{\partial N}{\partial x} = \frac{2x}{y} - z = \frac{\partial M}{\partial y}$

$\Rightarrow M\, dx + N\, dy + P\, dz$ is exact; $\frac{\partial f}{\partial x} = 2x \ln y - yz \Rightarrow f(x,y,z) = x^2 \ln y - xyz + g(y,z) \Rightarrow \frac{\partial f}{\partial y} = \frac{x^2}{y} - xz + \frac{\partial g}{\partial y}$

$= \frac{x^2}{y} - xz \Rightarrow \frac{\partial g}{\partial y} = 0 \Rightarrow g(y,z) = h(z) \Rightarrow f(x,y,z) = x^2 \ln y - xyz + h(z) \Rightarrow \frac{\partial f}{\partial z} = -xy + h'(z) = -xy \Rightarrow h'(z) = 0$

$\Rightarrow h(z) = C \Rightarrow f(x,y,z) = x^2 \ln y - xyz + C \Rightarrow \int_{(1,2,1)}^{(2,1,1)} (2x \ln y - yz)\, dx + \left(\frac{x^2}{y} - xz\right) dy - xy\, dz$

$= f(2,1,1) - f(1,2,1) = (4 \ln 1 - 2 + C) - (\ln 2 - 2 + C) = -\ln 2$

21. Let $\mathbf{F}(x,y,z) = \left(\frac{1}{y}\right)\mathbf{i} + \left(\frac{1}{z} - \frac{x}{y^2}\right)\mathbf{j} - \left(\frac{y}{z^2}\right)\mathbf{k} \Rightarrow \frac{\partial P}{\partial y} = -\frac{1}{z^2} = \frac{\partial N}{\partial z}, \frac{\partial M}{\partial z} = 0 = \frac{\partial P}{\partial x}, \frac{\partial N}{\partial x} = -\frac{1}{y^2} = \frac{\partial M}{\partial y}$

$\Rightarrow M\, dx + N\, dy + P\, dz$ is exact; $\frac{\partial f}{\partial x} = \frac{1}{y} \Rightarrow f(x,y,z) = \frac{x}{y} + g(y,z) \Rightarrow \frac{\partial f}{\partial y} = -\frac{x}{y^2} + \frac{\partial g}{\partial y} = \frac{1}{z} - \frac{x}{y^2}$

$\Rightarrow \frac{\partial g}{\partial y} = \frac{1}{z} \Rightarrow g(y,z) = \frac{y}{z} + h(z) \Rightarrow f(x,y,z) = \frac{x}{y} + \frac{y}{z} + h(z) \Rightarrow \frac{\partial f}{\partial z} = -\frac{y}{z^2} + h'(z) = -\frac{y}{z^2} \Rightarrow h'(z) = 0 \Rightarrow h(z) = C$

$\Rightarrow f(x,y,z) = \frac{x}{y} + \frac{y}{z} + C \Rightarrow \int_{(1,1,1)}^{(2,2,2)} \frac{1}{y}\, dx + \left(\frac{1}{z} - \frac{x}{y^2}\right) dy - \frac{y}{z^2}\, dz = f(2,2,2) - f(1,1,1) = \left(\frac{2}{2} + \frac{2}{2} + C\right) - \left(\frac{1}{1} + \frac{1}{1} + C\right)$

$= 0$

22. Let $\mathbf{F}(x,y,z) = \frac{2x\mathbf{i} + 2y\mathbf{j} + 2z\mathbf{k}}{x^2 + y^2 + z^2}$ $\left(\text{and let } \rho^2 = x^2 + y^2 + z^2 \Rightarrow \frac{\partial \rho}{\partial x} = \frac{x}{\rho}, \frac{\partial \rho}{\partial y} = \frac{y}{\rho}, \frac{\partial \rho}{\partial z} = \frac{z}{\rho}\right)$

$\Rightarrow \frac{\partial P}{\partial y} = -\frac{4yz}{\rho^4} = \frac{\partial N}{\partial z}, \frac{\partial M}{\partial z} = -\frac{4xz}{\rho^4} = \frac{\partial P}{\partial x}, \frac{\partial N}{\partial x} = -\frac{4xy}{\rho^4} = \frac{\partial M}{\partial y} \Rightarrow M\, dx + N\, dy + P\, dz$ is exact;

$\frac{\partial f}{\partial x} = \frac{2x}{x^2 + y^2 + z^2} \Rightarrow f(x,y,z) = \ln(x^2 + y^2 + z^2) + g(y,z) \Rightarrow \frac{\partial f}{\partial y} = \frac{2y}{x^2 + y^2 + z^2} + \frac{\partial g}{\partial y} = \frac{2y}{x^2 + y^2 + z^2}$

$\Rightarrow \frac{\partial g}{\partial y} = 0 \Rightarrow g(y,z) = h(z) \Rightarrow f(x,y,z) = \ln(x^2 + y^2 + z^2) + h(z) \Rightarrow \frac{\partial f}{\partial z} = \frac{2z}{x^2 + y^2 + z^2} + h'(z)$

$= \frac{2z}{x^2 + y^2 + z^2} \Rightarrow h'(z) = 0 \Rightarrow h(z) = C \Rightarrow f(x,y,z) = \ln(x^2 + y^2 + z^2) + C$

$\Rightarrow \int_{(-1,-1,-1)}^{(2,2,2)} \frac{2x\, dx + 2y\, dy + 2z\, dz}{x^2 + y^2 + z^2} = f(2,2,2) - f(-1,-1,-1) = \ln 12 - \ln 3 = \ln 4$

23. $\mathbf{r} = (\mathbf{i} + \mathbf{j} + \mathbf{k}) + t(\mathbf{i} + 2\mathbf{j} - 2\mathbf{k}) = (1 + t)\mathbf{i} + (1 + 2t)\mathbf{j} + (1 - 2t)\mathbf{k} \Rightarrow dx = dt,\ dy = 2\, dt,\ dz = -2\, dt$

$\Rightarrow \int_{(1,1,1)}^{(2,3,-1)} y\, dx + x\, dy + 4\, dz = \int_0^1 (2t + 1)\, dt + (t + 1)(2\, dt) + 4(-2)\, dt = \int_0^1 (4t - 5)\, dt = \left[2t^2 - 5t\right]_0^1 = -3$

24. $\mathbf{r} = t(3\mathbf{j} + 4\mathbf{k}) \Rightarrow dx = 0,\ dy = 3\ dt,\ dz = 4\ dt \Rightarrow \int_{(0,0,0)}^{(0,3,4)} x^2\ dx + yz\ dy + \left(\frac{y^2}{2}\right) dz$

$= \int_0^1 (12t^2)(3\ dt) + \left(\frac{9t^2}{2}\right)(4\ dt) = \int_0^1 54t^2\ dt = \left[18t^2\right]_0^1 = 18$

25. $\frac{\partial P}{\partial y} = 0 = \frac{\partial N}{\partial z},\ \frac{\partial M}{\partial z} = 2z = \frac{\partial P}{\partial x},\ \frac{\partial N}{\partial x} = 0 = \frac{\partial M}{\partial y} \Rightarrow M\ dx + N\ dy + P\ dz$ is exact $\Rightarrow$ **F** is conservative

$\Rightarrow$ path independence

26. $\frac{\partial P}{\partial y} = -\frac{yz}{\left(\sqrt{x^2+y^2+z^2}\right)^3} = \frac{\partial N}{\partial z},\ \frac{\partial M}{\partial z} = -\frac{xz}{\left(\sqrt{x^2+y^2+z^2}\right)^3} = \frac{\partial P}{\partial x},\ \frac{\partial N}{\partial x} = -\frac{xy}{\left(\sqrt{x^2+y^2+z^2}\right)^3} = \frac{\partial M}{\partial y}$

$\Rightarrow M\ dx + N\ dy + P\ dz$ is exact $\Rightarrow$ **F** is conservative $\Rightarrow$ path independence

27. $\frac{\partial P}{\partial y} = 0 = \frac{\partial N}{\partial z},\ \frac{\partial M}{\partial z} = 2z = \frac{\partial P}{\partial x},\ \frac{\partial N}{\partial x} = -\frac{2x}{y^2} = \frac{\partial M}{\partial y} \Rightarrow$ **F** is conservative $\Rightarrow$ there exists an f so that $\mathbf{F} = \nabla f$;

$\frac{\partial f}{\partial x} = \frac{2x}{y} \Rightarrow f(x,y) = \frac{x^2}{y} + g(y) \Rightarrow \frac{\partial f}{\partial y} = -\frac{x^2}{y^2} + g'(y) = \frac{1-x^2}{y^2} \Rightarrow g'(y) = \frac{1}{y^2} \Rightarrow g(y) = -\frac{1}{y} + C$

$\Rightarrow f(x,y) = \frac{x^2}{y} - \frac{1}{y} + C \Rightarrow \mathbf{F} = \nabla\left(\frac{x^2-1}{y}\right)$

28. $\frac{\partial P}{\partial y} = \cos z = \frac{\partial N}{\partial z},\ \frac{\partial M}{\partial z} = 0 = \frac{\partial P}{\partial x},\ \frac{\partial N}{\partial x} = \frac{e^x}{y} = \frac{\partial M}{\partial y} \Rightarrow$ **F** is conservative $\Rightarrow$ there exists an f so that $\mathbf{F} = \nabla f$;

$\frac{\partial f}{\partial x} = e^x \ln y \Rightarrow f(x,y,z) = e^x \ln y + g(y,z) \Rightarrow \frac{\partial f}{\partial y} = \frac{e^x}{y} + \frac{\partial g}{\partial y} = \frac{e^x}{y} + \sin z \Rightarrow \frac{\partial g}{\partial y} = \sin z \Rightarrow g(y,z)$

$= y \sin z + h(z) \Rightarrow f(x,y,z) = e^x \ln y + y \sin z + h(z) \Rightarrow \frac{\partial f}{\partial z} = y \cos z + h'(z) = y \cos z \Rightarrow h'(z) = 0$

$\Rightarrow h(z) = C \Rightarrow f(x,y,z) = e^x \ln y + y \sin z + C \Rightarrow \mathbf{F} = \nabla(e^x \ln y + y \sin z)$

29. $\frac{\partial P}{\partial y} = 0 = \frac{\partial N}{\partial z},\ \frac{\partial M}{\partial z} = 0 = \frac{\partial P}{\partial x},\ \frac{\partial N}{\partial x} = 1 = \frac{\partial M}{\partial y} \Rightarrow$ **F** is conservative $\Rightarrow$ there exists an f so that $\mathbf{F} = \nabla f$;

$\frac{\partial f}{\partial x} = x^2 + y \Rightarrow f(x,y,z) = \frac{1}{3}x^3 + xy + g(y,z) \Rightarrow \frac{\partial f}{\partial y} = x + \frac{\partial g}{\partial y} = y^2 + x \Rightarrow \frac{\partial g}{\partial y} = y^2 \Rightarrow g(y,z) = \frac{1}{3}y^3 + h(z)$

$\Rightarrow f(x,y,z) = \frac{1}{3}x^3 + xy + \frac{1}{3}y^3 + h(z) \Rightarrow \frac{\partial f}{\partial z} = h'(z) = ze^z \Rightarrow h(z) = ze^z - e^z + C \Rightarrow f(x,y,z)$

$= \frac{1}{3}x^3 + xy + \frac{1}{3}y^3 + ze^z - e^z + C \Rightarrow \mathbf{F} = \nabla\left(\frac{1}{3}x^3 + xy + \frac{1}{3}y^3 + ze^z - e^z\right)$

(a) work $= \int_A^B \mathbf{F} \cdot \frac{d\mathbf{r}}{dt}\ dt = \int_A^B \mathbf{F} \cdot d\mathbf{r} = \left[\frac{1}{3}x^3 + xy + \frac{1}{3}y^3 + ze^z - e^z\right]_{(1,0,0)}^{(1,0,1)} = \left(\frac{1}{3} + 0 + 0 + e - e\right) - \left(\frac{1}{3} + 0 + 0 - 1\right)$

$= 1$

(b) work $= \int_A^B \mathbf{F} \cdot d\mathbf{r} = \left[\frac{1}{3}x^3 + xy + \frac{1}{3}y^3 + ze^z - e^z\right]_{(1,0,0)}^{(1,0,1)} = 1$

(c) work $= \int_A^B \mathbf{F} \cdot d\mathbf{r} = \left[\frac{1}{3}x^3 + xy + \frac{1}{3}y^3 + ze^z - e^z\right]_{(1,0,0)}^{(1,0,1)} = 1$

Note: Since $\mathbf{F}$ is conservative, $\int_A^B \mathbf{F}\cdot d\mathbf{r}$ is independent of the path from $(1,0,0)$ to $(1,0,1)$.

30. $\frac{\partial P}{\partial y} = xe^{yz} + xyze^{yz} + \cos y = \frac{\partial N}{\partial z}$, $\frac{\partial M}{\partial z} = ye^{yz} = \frac{\partial P}{\partial x}$, $\frac{\partial N}{\partial x} = ze^{yz} = \frac{\partial M}{\partial y} \Rightarrow \mathbf{F}$ is conservative $\Rightarrow$ there exists an f so that $\mathbf{F} = \nabla f$; $\frac{\partial f}{\partial x} = e^{yz} \Rightarrow f(x,y,z) = xe^{yz} + g(y,z) \Rightarrow \frac{\partial f}{\partial y} = xze^{yz} + \frac{\partial g}{\partial y} = xze^{yz} + z\cos y \Rightarrow \frac{\partial g}{\partial y} = z\cos y$ $\Rightarrow g(y,z) = z\sin y + h(z) \Rightarrow f(x,y,z) = xe^{yz} + z\sin y + h(z) \Rightarrow \frac{\partial f}{\partial z} = xye^{yz} + \sin y + h'(z) = xye^{yz} + \sin y$ $\Rightarrow h'(z) = 0 \Rightarrow h(z) = C \Rightarrow f(x,y,z) = xe^{yz} + z\sin y + C \Rightarrow \mathbf{F} = \nabla(xe^{yz} + z\sin y)$

(a) work $= \int_A^B \mathbf{F}\cdot d\mathbf{r} = \left[xe^{yz} + z\sin y\right]_{(1,0,1)}^{(1,\pi/2,0)} = (1+0) - (1+0) = 0$

(b) work $= \int_A^B \mathbf{F}\cdot d\mathbf{r} = \left[xe^{yz} + z\sin y\right]_{(1,0,1)}^{(1,\pi/2,0)} = 0$

(c) work $= \int_A^B \mathbf{F}\cdot d\mathbf{r} = \left[xe^{yz} + z\sin y\right]_{(1,0,1)}^{(1,\pi/2,0)} = 0$

Note: Since $\mathbf{F}$ is conservative, $\int_A^B \mathbf{F}\cdot d\mathbf{r}$ is independent of the path from $(1,0,1)$ to $\left(1,\frac{\pi}{2},0\right)$.

31. (a) $\mathbf{F} = \nabla(x^3y^2) \Rightarrow \mathbf{F} = 3x^2y^2\mathbf{i} + 2x^3y\mathbf{j}$; let C_1 be the path from $(-1,1)$ to $(0,0) \Rightarrow x = t-1$ and $y = -t+1$, $0 \le t \le 1 \Rightarrow \mathbf{F} = 3(t-1)^2(-t+1)^2\mathbf{i} + 2(t-1)^3(-t+1)\mathbf{j} = 3(t-1)^4\mathbf{i} - 2(t-1)^4\mathbf{j}$ and $\mathbf{r}_1 = (t-1)\mathbf{i} + (-t+1)\mathbf{j} \Rightarrow d\mathbf{r}_1 = dt\,\mathbf{i} - dt\,\mathbf{j} \Rightarrow \int_{C_1} \mathbf{F}\cdot d\mathbf{r}_1 = \int_0^1 \left[3(t-1)^4 + 2(t-1)^4\right] dt$ $= \int_0^1 5(t-1)^4\,dt = \left[(t-1)^5\right]_0^1 = 1$; let C_2 be the path from $(0,0)$ to $(1,1) \Rightarrow x = t$ and $y = t$, $0 \le t \le 1 \Rightarrow \mathbf{F} = 3t^4\mathbf{i} + 2t^4\mathbf{j}$ and $\mathbf{r}_2 = t\mathbf{i} + t\mathbf{j} \Rightarrow d\mathbf{r}_2 = dt\,\mathbf{i} + dt\,\mathbf{j} \Rightarrow \int_{C_2} \mathbf{F}\cdot d\mathbf{r}_2 = \int_0^1 (3t^4 + 2t^4)\,dt$ $= \int_0^1 5t^4\,dt = 1 \Rightarrow \int_C \mathbf{F}\cdot d\mathbf{r} = \int_{C_1} \mathbf{F}\cdot d\mathbf{r}_1 + \int_{C_2} \mathbf{F}\cdot d\mathbf{r}_2 = 2$

(b) Since $f(x,y) = x^3y^2$ is a potential function for $\mathbf{F}$, $\int_{(-1,1)}^{(1,1)} \mathbf{F}\cdot d\mathbf{r} = f(1,1) - f(-1,1) = 2$

32. $\frac{\partial P}{\partial y} = 0 = \frac{\partial N}{\partial z}$, $\frac{\partial M}{\partial z} = 0 = \frac{\partial P}{\partial x}$, $\frac{\partial N}{\partial x} = -2x\sin y = \frac{\partial M}{\partial y} \Rightarrow \mathbf{F}$ is conservative $\Rightarrow$ there exists an f so that $\mathbf{F} = \nabla f$; $\frac{\partial f}{\partial x} = 2x\cos y \Rightarrow f(x,y,z) = x^2\cos y + g(y,z) \Rightarrow \frac{\partial f}{\partial y} = -x^2\sin y + \frac{\partial g}{\partial y} = -x^2\sin y \Rightarrow \frac{\partial g}{\partial y} = 0 \Rightarrow g(y,z) = h(z)$

$\Rightarrow f(x,y,z) = x^2 \cos y + h(z) \Rightarrow \frac{\partial f}{\partial z} = h'(z) = 0 \Rightarrow h(z) = C \Rightarrow f(x,y,z) = x^2 \cos y + C \Rightarrow \mathbf{F} = \nabla(x^2 \cos y)$

(a) $\int_C 2x \cos y\, dx - x^2 \sin y\, dy = \left[x^2 \cos y\right]_{(1,0)}^{(0,1)} = 0 - 1 = -1$

(b) $\int_C 2x \cos y\, dx - x^2 \sin y\, dy = \left[x^2 \cos y\right]_{(-1,\pi)}^{(1,0)} = 1 - (-1) = 2$

(c) $\int_C 2x \cos y\, dx - x^2 \sin y\, dy = \left[x^2 \cos y\right]_{(-1,0)}^{(1,0)} = 1 - 1 = 0$

(d) $\int_C 2x \cos y\, dx - x^2 \sin y\, dy = \left[x^2 \cos y\right]_{(1,0)}^{(1,0)} = 1 - 1 = 0$

33. Let $-GmM = C \Rightarrow \mathbf{F} = C\left[\frac{x}{(x^2+y^2+z^2)^{3/2}}\mathbf{i} + \frac{y}{(x^2+y^2+z^2)^{3/2}}\mathbf{j} + \frac{z}{(x^2+y^2+z^2)^{3/2}}\mathbf{k}\right]$

$\Rightarrow \frac{\partial P}{\partial y} = \frac{-3yzC}{(x^2+y^2+z^2)^{5/2}} = \frac{\partial N}{\partial z}, \frac{\partial M}{\partial z} = \frac{-3xzC}{(x^2+y^2+z^2)^{5/2}} = \frac{\partial P}{\partial x}, \frac{\partial N}{\partial x} = \frac{-3xyC}{(x^2+y^2+z^2)^{5/2}} = \frac{\partial M}{\partial y} \Rightarrow \mathbf{F} = \nabla f$ for some f; $\frac{\partial f}{\partial x} = \frac{xC}{(x^2+y^2+z^2)^{3/2}} \Rightarrow f(x,y,z) = -\frac{C}{(x^2+y^2+z^2)^{1/2}} + g(y,z) \Rightarrow \frac{\partial f}{\partial y} = \frac{yC}{(x^2+y^2+z^2)^{3/2}} + \frac{\partial g}{\partial y}$
$= \frac{yC}{(x^2+y^2+z^2)^{3/2}} \Rightarrow \frac{\partial g}{\partial y} = 0 \Rightarrow g(y,z) = h(z) \Rightarrow \frac{\partial f}{\partial z} = \frac{zC}{(x^2+y^2+z^2)^{3/2}} + h'(z) = \frac{zC}{(x^2+y^2+z^2)^{3/2}}$
$\Rightarrow h(z) = C_1 \Rightarrow f(x,y,z) = -\frac{C}{(x^2+y^2+z^2)^{1/2}} + C_1$. Let $C_1 = 0 \Rightarrow f(x,y,z) = \frac{GmM}{(x^2+y^2+z^2)^{1/2}}$ is a potential function for **F**.

34. If s is the distance of (x,y,z) from the origin, then $s = \sqrt{x^2+y^2+z^2}$. The work done by the gravitational field **F** is work $= \int_{P_1}^{P_2} \mathbf{F} \cdot d\mathbf{r} = \left[\frac{GmM}{\sqrt{x^2+y^2+z^2}}\right]_{P_1}^{P_2} = \frac{GmM}{s_2} - \frac{GmM}{s_1} = GmM\left(\frac{1}{s_2} - \frac{1}{s_1}\right)$, as claimed.

35. (a) If the differential form is exact, then $\frac{\partial P}{\partial y} = \frac{\partial N}{\partial z} \Rightarrow 2ay = cy$ for all $y \Rightarrow 2a = c$, $\frac{\partial M}{\partial z} = \frac{\partial P}{\partial x} \Rightarrow 2cx = 2cx$ for all x, and $\frac{\partial N}{\partial x} = \frac{\partial M}{\partial y} \Rightarrow by = 2ay$ for all $y \Rightarrow b = 2a$ and $c = 2a$

(b) $\mathbf{F} = \nabla f \Rightarrow$ the differential form with $a = 1$ in part (a) is exact $\Rightarrow b = 2$ and $c = 2$

36. $\mathbf{F} = \nabla f \Rightarrow g(x,y,z) = \int_{(0,0,0)}^{(x,y,z)} \mathbf{F} \cdot d\mathbf{r} = \int_{(0,0,0)}^{(x,y,z)} \nabla f \cdot d\mathbf{r} = f(x,y,z) - f(0,0,0) \Rightarrow \frac{\partial g}{\partial x} = \frac{\partial f}{\partial x} - 0, \frac{\partial g}{\partial y} = \frac{\partial f}{\partial y} - 0$, and $\frac{\partial g}{\partial z} = \frac{\partial f}{\partial z} - 0 \Rightarrow \nabla g = \nabla f = \mathbf{F}$, as claimed

37. The path will not matter; the work along any path will be the same because the field is conservative.

38. The field is not conservative, for otherwise the work would be the same along C_1 and C_2.

14.4 GREEN'S THEOREM IN THE PLANE

1. $M = -y = -a \sin t$, $N = x = a \cos t$, $dx = -a \sin t \, dt$, $dy = a \cos t \, dt \Rightarrow \frac{\partial M}{\partial x} = 0$, $\frac{\partial M}{\partial y} = -1$, $\frac{\partial N}{\partial x} = 1$, and $\frac{\partial N}{\partial y} = 0$;

Equation (11): $\oint_C M \, dy - N \, dx = \int_0^{2\pi} [(-a \sin t)(a \cos t) - (a \cos t)(-a \sin t)] \, dt = \int_0^{2\pi} 0 \, dt = 0$;

$\iint_R \left(\frac{\partial M}{\partial x} + \frac{\partial N}{\partial y}\right) dx \, dy = \iint_R 0 \, dx \, dy = 0$, Flux

Equation (12): $\oint_C M \, dx + N \, dy = \int_0^{2\pi} [(-a \sin t)(-a \sin t) - (a \cos t)(a \cos t)] \, dt = \int_0^{2\pi} a^2 \, dt = 2\pi a^2$;

$\iint_R \left(\frac{\partial N}{\partial x} - \frac{\partial M}{\partial y}\right) dx \, dy = \int_{-a}^{a} \int_{-\sqrt{a^2-x^2}}^{\sqrt{a^2-x^2}} 2 \, dy \, dx = \int_{-a}^{a} 4\sqrt{a^2 - x^2} \, dx = 4\left[\frac{x}{2}\sqrt{a^2 - x^2} + \frac{a^2}{2} \sin^{-1} \frac{x}{a}\right]_{-a}^{a}$

$= 2a^2\left(\frac{\pi}{2} + \frac{\pi}{2}\right) = 2a^2\pi$, Circulation

2. $M = y = a \sin t$, $N = 0$, $dx = -a \sin t \, dt$, $dy = a \cos t \, dt \Rightarrow \frac{\partial M}{\partial x} = 0$, $\frac{\partial M}{\partial y} = 1$, $\frac{\partial N}{\partial x} = 0$, and $\frac{\partial N}{\partial y} = 0$;

Equation (11): $\oint_C M \, dy - N \, dx = \int_0^{2\pi} a^2 \sin t \cos t \, dt = a^2\left[\frac{1}{2} \sin^2 t\right]_0^{2\pi} = 0$; $\iint_R 0 \, dx \, dy = 0$, Flux

Equation (12): $\oint_C M \, dx + N \, dy = \int_0^{2\pi} \left(-a^2 \sin^2 t\right) dt = -a^2\left[\frac{t}{2} - \frac{\sin 2t}{4}\right]_0^{2\pi} = -\pi a^2$; $\iint_R \left(\frac{\partial N}{\partial x} - \frac{\partial M}{\partial y}\right) dx \, dy$

$= \iint_R -1 \, dx \, dy = \int_0^{2\pi} \int_0^{a} -r \, dr \, d\theta = \int_0^{2\pi} -\frac{a^2}{2} \, d\theta = -\pi a^2$, Circulation

3. $M = 2x = 2a \cos t$, $N = -3y = -3a \sin t$, $dx = -a \sin t \, dt$, $dy = a \cos t \, dt \Rightarrow \frac{\partial M}{\partial x} = 2$, $\frac{\partial M}{\partial y} = 0$, $\frac{\partial N}{\partial x} = 0$, and $\frac{\partial N}{\partial y} = -3$;

Equation (11): $\oint_C M \, dy - N \, dx = \int_0^{2\pi} [(2a \cos t)(a \cos t) + (3a \sin t)(-a \sin t)] \, dt$

$$= \int_0^{2\pi} \left(2a^2\cos^2 t - 3a^2\sin^2 t\right) dt = 2a^2\left[\frac{t}{2} + \frac{\sin 2t}{4}\right]_0^{2\pi} - 3a^2\left[\frac{t}{2} - \frac{\sin 2t}{4}\right]_0^{2\pi} = 2\pi a^2 - 3\pi a^2 = -\pi a^2;$$

$$\iint_R \left(\frac{\partial M}{\partial x} + \frac{\partial N}{\partial y}\right) = \iint_R -1\, dx\, dy = \int_0^{2\pi}\int_0^a -r\, dr\, d\theta = \int_0^{2\pi} -\frac{a^2}{2}\, d\theta = -\pi a^2, \text{ Flux}$$

Equation (12): $\oint_C M\, dx + N\, dy = \int_0^{2\pi} [(2a\cos t)(-a\sin t) + (-3a\sin t)(a\cos t)]\, dt$

$$= \int_0^{2\pi} \left(-2a^2\sin t\cos t - 3a^2\sin t\cos t\right) dt = -5a^2\left[\frac{1}{2}\sin^2 t\right]_0^{2\pi} = 0; \quad \iint_R 0\, dx\, dy = 0, \text{ Circulation}$$

4. $M = -x^2y = -a^3\cos^2 t$, $N = xy^2 = a^3\cos t\sin^2 t$, $dx = -a\sin t\, dt$, $dy = a\cos t\, dt \Rightarrow \frac{\partial M}{\partial x} = -2a^2\cos t\sin t$, $\frac{\partial M}{\partial y} = -a^2\cos^2 t$, $\frac{\partial N}{\partial x} = a^2\sin^2 t$, and $\frac{\partial N}{\partial y} = 2a^2\cos t\sin t$;

Equation (11): $\oint_C M\, dy - N\, dx = \int_0^{2\pi} \left(-a^4\cos^3 t\sin t + a^4\cos t\sin^3 t\right) = \left[\frac{a^4}{4}\cos^4 t + \frac{a^4}{4}\sin^4 t\right]_0^{2\pi} = 0;$

$$\iint_R \left(\frac{\partial M}{\partial x} + \frac{\partial N}{\partial y}\right) dx\, dy = \iint_R (-2xy + 2xy)\, dx\, dy = 0, \text{ Flux}$$

Equation (12): $\oint_C M\, dx + N\, dy = \int_0^{2\pi} \left(a^4\cos^2 t\sin^2 t + a^4\cos^2 t\sin^2 t\right) dt = \int_0^{2\pi} \left(2a^4\cos^2 t\sin^2 t\right) dt$

$$= \int_0^{2\pi} \frac{1}{2}a^4\sin^2 2t\, dt = \frac{a^4}{4}\int_0^{4\pi} \sin^2 u\, du = \frac{a^4}{4}\left[\frac{u}{2} - \frac{\sin 2u}{4}\right]_0^{4\pi} = \frac{\pi a^4}{2}; \quad \iint_R \left(\frac{\partial N}{\partial x} - \frac{\partial M}{\partial y}\right) dx\, dy = \iint_R (y^2 + x^2)\, dx\, dy$$

$$= \int_0^{2\pi}\int_0^a r^2 \cdot r\, dr\, d\theta = \int_0^{2\pi} \frac{a^4}{4}\, d\theta = \frac{\pi a^4}{2}, \text{ Circulation}$$

5. $M = x - y$, $N = y - x \Rightarrow \frac{\partial M}{\partial x} = 1$, $\frac{\partial M}{\partial y} = -1$, $\frac{\partial N}{\partial x} = -1$, $\frac{\partial N}{\partial y} = 1 \Rightarrow \text{Flux} = \iint_R 2\, dx\, dy = \int_0^1\int_0^1 2\, dx\, dy = 2;$

$$\text{Circ} = \iint_R [-1 - (-1)]\, dx\, dy = 0$$

6. $M = x^2 + 4y$, $N = x + y^2 \Rightarrow \frac{\partial M}{\partial x} = 2x$, $\frac{\partial M}{\partial y} = 4$, $\frac{\partial N}{\partial x} = 1$, $\frac{\partial N}{\partial y} = 2y \Rightarrow \text{Flux} = \iint_R (2x + 2y)\, dx\, dy$

$$= \int_0^1\int_0^1 (2x + 2y)\, dx\, dy = \int_0^1 \left[x^2 + 2xy\right]_0^1 dy = \int_0^1 (1 + 2y)\, dy = \left[y + y^2\right]_0^1 = 2; \text{ Circ} = \iint_R (1 - 4)\, dx\, dy$$

$$= \int_0^1\int_0^1 -3\, dx\, dy = -3$$

7. $M = y^2 - x^2,\ N = x^2 + y^2 \Rightarrow \frac{\partial M}{\partial x} = -2x,\ \frac{\partial M}{\partial y} = 2y,\ \frac{\partial N}{\partial x} = 2x,\ \frac{\partial N}{\partial y} = 2y \Rightarrow \text{Flux} = \iint_R (-2x + 2y)\, dx\, dy$

$= \int_0^3 \int_0^x (-2x + 2y)\, dy\, dx = \int_0^3 (-2x^2 + x^2)\, dx = \left[-\frac{1}{3}x^3\right]_0^3 = -9;\ \text{Circ} = \iint_R (2x - 2y)\, dx\, dy$

$= \int_0^3 \int_0^x (2x - 2y)\, dy\, dx = \int_0^3 x^2\, dx = 9$

8. $M = x + y,\ N = -(x^2 + y^2) \Rightarrow \frac{\partial M}{\partial x} = 1,\ \frac{\partial M}{\partial y} = 1,\ \frac{\partial N}{\partial x} = -2x,\ \frac{\partial N}{\partial y} = -2y \Rightarrow \text{Flux} = \iint_R (1 - 2y)\, dx\, dy$

$= \int_0^1 \int_0^x (1 - 2y)\, dy\, dx = \int_0^1 (x - x^2)\, dx = \frac{1}{6};\ \text{Circ} = \iint_R (-2x - 1)\, dx\, dy = \int_0^1 \int_0^x (-2x - 1)\, dy\, dx$

$= \int_0^1 (-2x^2 - x)\, dx = -\frac{7}{6}$

9. $M = x + e^x \sin y,\ N = x + e^x \cos y \Rightarrow \frac{\partial M}{\partial x} = 1 + e^x \sin y,\ \frac{\partial M}{\partial y} = e^x \cos y,\ \frac{\partial N}{\partial x} = 1 + e^x \cos y,\ \frac{\partial N}{\partial y} = -e^x \sin y$

$\Rightarrow \text{Flux} = \iint_R dx\, dy = \int_{-\pi/4}^{\pi/4} \int_0^{\sqrt{\cos 2\theta}} r\, dr\, d\theta = \int_{-\pi/4}^{\pi/4} \left(\frac{1}{2} \cos 2\theta\right) d\theta = \left[\frac{1}{4} \sin 2\theta\right]_{-\pi/4}^{\pi/4} = \frac{1}{2};$

$\text{Circ} = \iint_R (1 + e^x \cos y - e^x \cos y)\, dx\, dy = \iint_R dx\, dy = \int_{-\pi/4}^{\pi/4} \int_0^{\sqrt{\cos 2\theta}} r\, dr\, d\theta = \int_{-\pi/4}^{\pi/4} \left(\frac{1}{2} \cos 2\theta\right) d\theta = \frac{1}{2}$

10. $M = \tan^{-1} \frac{y}{x},\ N = \ln(x^2 + y^2) \Rightarrow \frac{\partial M}{\partial x} = \frac{-y}{x^2 + y^2},\ \frac{\partial M}{\partial y} = \frac{x}{x^2 + y^2},\ \frac{\partial N}{\partial x} = \frac{2x}{x^2 + y^2},\ \frac{\partial N}{\partial y} = \frac{2y}{x^2 + y^2}$

$\Rightarrow \text{Flux} = \iint_R \left(\frac{-y}{x^2 + y^2} + \frac{2y}{x^2 + y^2}\right) dx\, dy = \int_0^\pi \int_1^2 \left(\frac{r \sin \theta}{r^2}\right) r\, dr\, d\theta = \int_0^\pi \sin \theta\, d\theta = 2;$

$\text{Circ} = \iint_R \left(\frac{2x}{x^2 + y^2} - \frac{x}{x^2 + y^2}\right) dx\, dy = \int_0^\pi \int_1^2 \left(\frac{r \cos \theta}{r^2}\right) r\, dr\, d\theta = \int_0^\pi \cos \theta\, d\theta = 0$

11. $M = xy,\ N = y^2 \Rightarrow \frac{\partial M}{\partial x} = y,\ \frac{\partial M}{\partial y} = x,\ \frac{\partial N}{\partial x} = 0,\ \frac{\partial N}{\partial y} = 2y \Rightarrow \text{Flux} = \iint_R (y + 2y)\, dy\, dx = \int_0^1 \int_{x^2}^x 3y\, dy\, dx$

$= \int_0^1 \left(\frac{3x^2}{2} - \frac{3x^4}{2}\right) dx = \frac{1}{5};\ \text{Circ} = \iint_R -x\, dy\, dx = \int_0^1 \int_{x^2}^x -x\, dy\, dx = \int_0^1 (-x^2 + x^3)\, dx = -\frac{1}{12}$

12. $M = -\sin y,\ N = x\cos y \Rightarrow \frac{\partial M}{\partial x} = 0,\ \frac{\partial M}{\partial y} = -\cos y,\ \frac{\partial N}{\partial x} = \cos y,\ \frac{\partial N}{\partial y} = -x\sin y$

$$\Rightarrow \text{Flux} = \iint_R (-x\sin y)\,dx\,dy = \int_0^{\pi/2}\int_0^{\pi/2} (-x\sin y)\,dx\,dy = \int_0^{\pi/2} \left(-\frac{\pi^2}{8}\sin y\right) dy = -\frac{\pi^2}{8};$$

$$\text{Circ} = \iint_R [\cos y - (-\cos y)]\,dx\,dy = \int_0^{\pi/2}\int_0^{\pi/2} 2\cos y\,dx\,dy = \int_0^{\pi/2} \pi\cos y\,dy = [\pi\sin y]_0^{\pi/2} = \pi$$

13. $M = 3xy - \frac{x}{1+y^2},\ N = e^x + \tan^{-1} y \Rightarrow \frac{\partial M}{\partial x} = 3y - \frac{1}{1+y^2},\ \frac{\partial N}{\partial y} = \frac{1}{1+y^2}$

$$\Rightarrow \text{Flux} = \iint_R \left(3y - \frac{1}{1+y^2} + \frac{1}{1+y^2}\right) dx\,dy = \iint_R 3y\,dx\,dy = \int_0^{2\pi}\int_0^{a(1+\cos\theta)} (3r\sin\theta)\,r\,dr\,d\theta$$

$$= \int_0^{2\pi} a^3(1+\cos\theta)^3(\sin\theta)\,d\theta = \left[-\frac{a^3}{4}(1+\cos\theta)^4\right]_0^{2\pi} = -4a^3 - (-4a^3) = 0$$

14. $M = y + e^x \ln y,\ N = \frac{e^x}{y} \Rightarrow \frac{\partial M}{\partial y} = 1 + \frac{e^x}{y},\ \frac{\partial N}{\partial x} = \frac{e^x}{y} \Rightarrow \text{Circ} = \iint_R \left[\frac{e^x}{y} - \left(1 + \frac{e^x}{y}\right)\right] dx\,dy = \iint_R (-1)\,dx\,dy$

$$= \int_{-1}^{1}\int_{x^4+1}^{3-x^2} -\,dy\,dx = -\int_{-1}^{1} \left[(3-x^2) - (x^4+1)\right] dx = \int_{-1}^{1} (x^4 + x^2 - 2)\,dx = -\frac{23}{15}$$

15. $M = 2xy^3,\ N = 4x^2y^2 \Rightarrow \frac{\partial M}{\partial y} = 6xy^2,\ \frac{\partial N}{\partial x} = 8xy^2 \Rightarrow \text{work} = \oint_C 2xy^3\,dx + 4x^2y^2\,dy = \iint_R (8xy^2 - 6xy^2)\,dx\,dy$

$$= \int_0^1\int_0^{x^3} 2xy^2\,dy\,dx = \int_0^1 \frac{2}{3}x^{10}\,dx = \frac{2}{33}$$

16. $M = 4x - 2y,\ N = 2x - 4y \Rightarrow \frac{\partial M}{\partial y} = -2,\ \frac{\partial N}{\partial x} = 2 \Rightarrow \text{work} = \oint_C (4x - 2y)\,dx + (2x - 4y)\,dy$

$$= \iint_R [2 - (-2)]\,dx\,dy = 4\iint_R dx\,dy = 4(\text{Area of the circle}) = 4(\pi \cdot 4) = 16\pi$$

17. $M = y^2,\ N = x^2 \Rightarrow \frac{\partial M}{\partial y} = 2y,\ \frac{\partial N}{\partial x} = 2x \Rightarrow \oint_C y^2\,dx + x^2\,dy = \iint_R (2x - 2y)\,dy\,dx$

$$= \int_0^1\int_0^{1-x} (2x - 2y)\,dy\,dx = \int_0^1 (-3x^2 + 4x - 1)\,dx = [-x^3 + 2x^2 - x]_0^1 = -1 + 2 - 1 = 0$$

18. $M = 3y,\ N = 2x \Rightarrow \frac{\partial M}{\partial y} = 3,\ \frac{\partial N}{\partial x} = 2 \Rightarrow \oint_C 3y\,dx + 2x\,dy = \iint_R (2-3)\,dx\,dy = \int_0^{\pi} \int_0^{\sin x} -1\,dy\,dx$
$= -\int_0^{\pi} \sin x\,dx = -2$

19. $M = 6y + x,\ N = y + 2x \Rightarrow \frac{\partial M}{\partial y} = 6,\ \frac{\partial N}{\partial x} = 2 \Rightarrow \oint_C (6y+x)\,dx + (y+2x)\,dy = \iint_R (2-6)\,dy\,dx$
$= -4(\text{Area of the circle}) = -16\pi$

20. $M = 2x + y^2,\ N = 2xy + 3y \Rightarrow \frac{\partial M}{\partial y} = 2y,\ \frac{\partial N}{\partial x} = 2y \Rightarrow \oint_C (2x+y^2)\,dx + (2xy+3y)\,dy = \iint_R (2y-2y)\,dx\,dy = 0$

21. $M = x = a\cos t,\ N = y = a\sin t \Rightarrow dx = -a\sin t\,dt,\ dy = a\cos t\,dt \Rightarrow \text{Area} = \frac{1}{2}\oint_C x\,dy - y\,dx$
$= \frac{1}{2}\int_0^{2\pi} (a^2\cos^2 t + a^2\sin^2 t)\,dt = \frac{1}{2}\int_0^{2\pi} a^2\,dt = \pi a^2$

22. $M = x = a\cos t,\ N = y = b\sin t \Rightarrow dx = -a\sin t\,dt,\ dy = b\cos t\,dt \Rightarrow \text{Area} = \frac{1}{2}\oint_C x\,dy - y\,dx$
$= \frac{1}{2}\int_0^{2\pi} (ab\cos^2 t + ab\sin^2 t)\,dt = \frac{1}{2}\int_0^{2\pi} ab\,dt = \pi ab$

23. $M = x = a\cos^3 t,\ N = y = \sin^3 t \Rightarrow dx = -3\cos^2 t\sin t\,dt,\ dy = 3\sin^2 t\cos t\,dt \Rightarrow \text{Area} = \frac{1}{2}\oint_C x\,dy - y\,dx$
$= \frac{1}{2}\int_0^{2\pi} (3\sin^2 t\cos^2 t)(\cos^2 t + \sin^2 t)\,dt = \frac{1}{2}\int_0^{2\pi} (3\sin^2 t\cos^2 t)\,dt = \frac{3}{8}\int_0^{2\pi} \sin^2 2t\,dt = \frac{3}{16}\int_0^{4\pi} \sin^2 u\,du$
$= \frac{3}{16}\left[\frac{u}{2} - \frac{\sin 2u}{4}\right]_0^{4\pi} = \frac{3}{8}\pi$

24. $M = x = t^2,\ N = y = \frac{t^3}{3} - t \Rightarrow dx = 2t\,dt,\ dy = (t^2-1)\,dt \Rightarrow \text{Area} = \frac{1}{2}\oint_C x\,dy - y\,dx$
$= \frac{1}{2}\int_{-\sqrt{3}}^{\sqrt{3}} \left[t^2(t^2-1) - \left(\frac{t^3}{3} - t\right)(2t)\right]dt = \frac{1}{2}\int_{-\sqrt{3}}^{\sqrt{3}} \left(\frac{1}{3}t^4 + t^2\right)dt = \frac{1}{2}\left[\frac{1}{15}t^5 + -\frac{1}{3}t^3\right]_{-\sqrt{3}}^{\sqrt{3}} = \frac{1}{15}(9\sqrt{3} + 15\sqrt{3})$
$= \frac{8}{5}\sqrt{3}$

25. (a) $M = f(x),\ N = g(y) \Rightarrow \frac{\partial M}{\partial y} = 0,\ \frac{\partial N}{\partial x} = 0 \Rightarrow \oint_C f(x)\,dx + g(y)\,dy = \iint_R \left(\frac{\partial N}{\partial x} - \frac{\partial M}{\partial y}\right) dx\,dy$

$= \iint_R 0\,dx\,dy = 0$

(b) $M = ky,\ N = hx \Rightarrow \frac{\partial M}{\partial y} = k,\ \frac{\partial N}{\partial x} = h \Rightarrow \oint_C ky\,dx + hx\,dy = \iint_R \left(\frac{\partial N}{\partial x} - \frac{\partial M}{\partial y}\right) dx\,dy$

$= \iint_R (h - k)\,dx\,dy = (h - k)(\text{Area of the region})$

26. $M = xy^2,\ N = x^2y + 2x \Rightarrow \frac{\partial M}{\partial y} = 2xy,\ \frac{\partial N}{\partial x} = 2xy + 2 \Rightarrow \oint_C xy^2\,dx + (x^2y + 2x)\,dy = \iint_R \left(\frac{\partial N}{\partial x} - \frac{\partial M}{\partial y}\right) dx\,dy$

$= \iint_R (2xy + 2 - 2xy)\,dx\,dy = 2\iint_R dx\,dy = 2$ times the area of the square

27. The integral is 0 for any simple closed plane curve C. The reasoning: By the tangential form of Green's Theorem, with $M = 4x^3y$ and $N = x^4$, $\oint_C 4x^3y\ x + x^4\,dy = \iint_R \left[\frac{\partial}{\partial x}(x^4) - \frac{\partial}{\partial y}(4x^3y)\right] dx\,dy$

$= \iint_R \underbrace{(4x^3 - 4x^3)}_{0}\,dx\,dy = 0.$

28. The integral is 0 for any simple closed curve C. The reasoning: By the normal form of Green's theorem, with $M = -y^3$ and $N = x^3$, $\oint_C -y^3\,dx + x^3\,dy = \iint_R \left[\underbrace{\frac{\partial}{\partial x}(-y^3)}_{0} + \underbrace{\frac{\partial}{\partial y}(x^3)}_{0}\right] dx\,dy = 0.$

29. Let $M = x$ and $N = 0 \Rightarrow \frac{\partial M}{\partial x} = 1$ and $\frac{\partial N}{\partial y} = 0 \Rightarrow \oint M\,dy - N\,dx = \iint_R \left(\frac{\partial M}{\partial x} + \frac{\partial N}{\partial y}\right) dx\,dy \Rightarrow \oint_C x\,dy$

$= \iint_R (1 + 0)\,dx\,dy \Rightarrow \text{Area of R} = \iint_R dx\,dy = \oint_C x\,dy$; similarly, $M = y$ and $N = 0 \Rightarrow \frac{\partial M}{\partial y} = 1$ and

$\frac{\partial N}{\partial x} = 0 \Rightarrow \oint_C M\,dx + N\,dy = \iint_R \left(\frac{\partial N}{\partial x} + \frac{\partial M}{\partial y}\right) dy\,dx \Rightarrow \oint_C y\,dx = \iint_R (0 - 1)\,dy\,dx \Rightarrow -\oint_C y\,dx$

$= \iint_R dx\,dy = \text{Area of R}$

30. $\int_a^b f(x)\,dx = \text{Area of R} = -\oint_C y\,dx$, from Exercise 29

31. Let $\delta(x,y) = 1 \Rightarrow \bar{x} = \frac{M_y}{M} = \frac{\iint_R x\,\delta(x,y)\,dA}{\iint_R \delta(x,y)\,dA} = \frac{\iint_R x\,dA}{\iint_R dA} = \frac{\iint_R x\,dA}{A} \Rightarrow A\bar{x} = \iint_R x\,dA = \iint_R (x+0)\,dx\,dy$

$= \oint_C \frac{x^2}{2}\,dy$, $A\bar{x} = \iint_R x\,dA = \iint_R (0+x)\,dx\,dy = -\oint_C xy\,dx$, and $A\bar{x} = \iint_R x\,dA = \iint_R \left(\frac{2}{3}x + \frac{1}{3}x\right) dx\,dy$

$= \oint_C \frac{1}{3}x^2\,dy - \frac{1}{3}xy\,dx \Rightarrow \frac{1}{2}\oint_C x^2\,dy = -\oint_C xy\,dx = \frac{1}{3}\oint_C x^2\,dy - xy\,dx = A\bar{x}$

32. If $\delta(x,y) = 1$, then $I_y = \iint_R x^2\,\delta(x,y)\,dA = \iint_R x^2\,dA = \iint_R (x^2+0)\,dy\,dx = \frac{1}{3}\oint_C x^3\,dy$,

$\iint_R x^2\,dA = \iint_R (0+x^2)\,dy\,dx = -\oint_C x^2y\,dx$, and $\iint_R x^2\,dA = \iint_R \left(\frac{3}{4}x^2 + \frac{1}{4}x^2\right) dy\,dx$

$= \oint_C \frac{1}{4}x^3\,dy - \frac{1}{4}x^2y\,dx = \frac{1}{4}\oint_C x^3\,dy - x^2y\,dx \Rightarrow \frac{1}{3}\oint_C x^3\,dy = -\oint_C x^2y\,dx = \frac{1}{4}\oint_C x^3\,dy - x^2y\,dx = I_y$

33. $M = \frac{\partial f}{\partial y}$, $N = -\frac{\partial f}{\partial x} \Rightarrow \frac{\partial M}{\partial y} = \frac{\partial^2 f}{\partial y^2}$, $\frac{\partial N}{\partial x} = -\frac{\partial^2 f}{\partial x^2} \Rightarrow \oint_C \frac{\partial f}{\partial y}\,dx - \frac{\partial f}{\partial x}\,dy = \iint_R \left(-\frac{\partial^2 f}{\partial x^2} - \frac{\partial^2 f}{\partial y^2}\right) dx\,dy = 0$ for such curves C

34. $M = \frac{1}{4}x^2y + \frac{1}{3}y^3$, $N = x \Rightarrow \frac{\partial M}{\partial y} = \frac{1}{4}x^2 + y^2$, $\frac{\partial N}{\partial x} = 1 \Rightarrow \text{Curl} = \frac{\partial N}{\partial x} - \frac{\partial M}{\partial y} = 1 - \left(\frac{1}{4}x^2 + y^2\right) > 0$ in the interior of the ellipse $\frac{1}{4}x^2 + y^2 = 1 \Rightarrow$ work $= \int_C \mathbf{F}\cdot d\mathbf{r} = \iint_R \left(1 - \frac{1}{4}x^2 - y^2\right) dx\,dy$ will be maximized on the region $R = \{(x,y)\,|\,\text{curl } \mathbf{F}\} \geq 0$ or over the region enclosed by $1 = \frac{1}{4}x^2 + y^2$

35. (a) $\nabla f = \left(\frac{2x}{x^2+y^2}\right)\mathbf{i} + \left(\frac{2y}{x^2+y^2}\right)\mathbf{j} \Rightarrow M = \frac{2x}{x^2+y^2}$, $N = \frac{2y}{x^2+y^2}$; since M, N are discontinuous at (0,0), we cannot apply Green's Theorem over C. Thus, let C_h be the circle $x = h\cos\theta$, $y = h\sin\theta$, $0 < h \leq a$ and let C_1 be the circle $x = a\cos t$, $y = a\sin t$, $a > 0$. Then $\oint_C \mathbf{F}\cdot\mathbf{n}\,ds = \oint_{C_1} M\,dy - N\,dx + \oint_{C_h} M\,dy - N\,dx$

$= \oint_{C_1} \frac{2x}{x^2+y^2}\,dx - \frac{2y}{x^2+y^2}\,dx + \oint_{C_h} \frac{2x}{x^2+y^2}\,dy - \frac{2y}{x^2+y^2}\,dx$. In the first integral, let $x = a\cos t$, $y = a\sin t$ $\Rightarrow dx = -a\sin t\,dt$, $dy = a\cos t\,dt$, $M = 2a\cos t$, $N = 2a\sin t$, $0 \leq \ \leq 2\pi$. In the second integral, let $x = h\cos\theta$, $y = h\sin\theta \Rightarrow dx = -h\sin\theta\,d\theta$, $dy = h\cos\theta\,d\theta$, $M = 2h\cos\theta$, $N = 2h\sin\theta$, $0 \leq \theta \leq 2\pi$.

Then $\oint_C \mathbf{F}\cdot\mathbf{n}\,ds = \oint_{C_1} \frac{2x}{x^2+y^2}\,dy - \frac{2y}{x^2+y^2}\,dx + \oint_{C_h} \frac{2x}{x^2+y^2}\,dy - \frac{2y}{x^2+y^2}\,dx$

$$= \oint_{C_1} \frac{(2a \cos t)(a \cos t)\, dt}{a^2} - \frac{(2a \sin t)(-a \sin t)\, dt}{a^2} + \oint_{C_h} \frac{(2h \cos \theta)(h \cos \theta)\, d\theta}{h^2} - \frac{(2h \sin \theta)(-h \sin \theta)\, d\theta}{h^2}$$

$$= \int_0^{2\pi} 2\, dt + \int_{2\pi}^{0} 2\, d\theta = 0 \text{ for every h}$$

(b) If K is any simple closed curve surrounding C_h (K contains $(0,0)$), then $\oint_C \mathbf{F} \cdot \mathbf{n}\, ds$

$= \oint_{C_1} M\, dy - N\, dx + \oint_{C_h} M\, dy - N\, dx$, and in polar coordinates, $\nabla f \cdot \mathbf{n} = M\, dy - N\, dx$

$= \left(\frac{2r \cos \theta}{r^2}\right)(r \cos \theta\, d\theta + \sin \theta\, dr) - \left(\frac{2r \sin \theta}{r^2}\right)(-r \sin \theta\, d\theta + \cos \theta\, dr) = \frac{2r^2}{r^2}\, d\theta = 2\, d\theta$. Now,

2θ increases by 4π as K is traversed once counterclockwise from $\theta = 0$ to $\theta = 2\pi \Rightarrow \oint_C \mathbf{F} \cdot \mathbf{n}\, ds = 0$

(since $\oint_{C_h} M\, dy - N\, dx = -4\pi$) when $(0,0)$ is in the region, but $\oint_K \mathbf{F} \cdot \mathbf{n}\, ds = 4\pi$ when $(0,0)$ is not in the region.

36. Assume a particle has a closed trajectory in R and let C_1 be the path $\Rightarrow C_1$ encloses a simply connected region $R_1 \Rightarrow C_1$ is a simple closed curve. Then the flux over R_1 is $\oint_{C_1} \mathbf{F} \cdot \mathbf{n}\, ds = 0$, since the velocity vectors $\mathbf{F}$ are tangent to C_1. But $0 = \oint_{C_1} \mathbf{F} \cdot \mathbf{n}\, ds = \oint_{C_1} M\, dy - N\, dx = \iint_{R_1} \left(\frac{\partial M}{\partial x} + \frac{\partial N}{\partial y}\right) dx\, dy \Rightarrow M_x + N_y = 0$, which is a contradiction. Therefore, C_1 cannot be a closed trajectory.

37. $\int_{g_1(y)}^{g_2(y)} \frac{\partial N}{\partial x}\, dx\, dy = N(g_2(y), y) - N(g_1(y), y) \Rightarrow \int_c^d \int_{g_1(y)}^{g_2(y)} \left(\frac{\partial N}{\partial x}\, dx\right) dy = \int_c^d [N(g_2(y), y) - N(g_1(y), y)]\, dy$

$= \int_c^d N(g_2(y), y)\, dy - \int_c^d N(g_1(y), y)\, dy = \int_c^d N(g_2(y), y)\, dy + \int_d^c N(g_1(y), y)\, dy = \int_{C_2} N\, dy + \int_{C_1} N\, dy$

$= \oint_C dy \Rightarrow \oint_C N\, dy = \iint_R \frac{\partial N}{\partial x}\, dx\, dy$

38. $\int_a^b \int_c^d \frac{\partial M}{\partial y}\, dy\, dx = \int_a^b [M(x,d) - M(x,c)]\, dx = \int_a^b M(x,d)\, dx + \int_b^a M(x,c)\, dx = -\int_{C_2} M\, dx - \int_{C_1} M\, dx.$

Because x is constant along C_2 and C_4, $\int_{C_2} M\, dx = \int_{C_4} M\, dx = 0$

$\Rightarrow -\left(\int_{C_1} M\, dx + \int_{C_2} M\, dx + \int_{C_3} M\, dx + \int_{C_4} M\, dx\right) = -\oint_C M\, dx \Rightarrow \int_a^b \int_c^d \frac{\partial M}{\partial y}\, dy\, dx = -\oint_C M\, dx.$

39. The curl of a conservative two-dimensional field is zero. The reasoning: A two-dimensional field $\mathbf{F} = M\mathbf{i} + N\mathbf{j}$ can be considered to be the restriction to the xy-plane of a three-dimensional field whose k component is zero, and whose **i** and **j** components are independent of z. For such a field to be conservative, we must have $\frac{\partial N}{\partial x} = \frac{\partial M}{\partial y}$ by the component test in Section 14.3 $\Rightarrow$ curl $\mathbf{F} = \frac{\partial N}{\partial x} - \frac{\partial M}{\partial y} = 0$.

40. Green's theorem tells us that the circulation of a conservative two-dimensional field around any simple closed curve in the xy-plane is zero. The reasoning: For a conservative field $\mathbf{F} = M\mathbf{i} + N\mathbf{j}$, we have $\frac{\partial N}{\partial x} = \frac{\partial M}{\partial y}$ (component test for conservative fields, Section 14.3, Eq. (3)), so curl $\mathbf{F} = \frac{\partial N}{\partial x} - \frac{\partial M}{\partial y} = 0$. By Green's theorem, the counterclockwise circulation around a simple closed plane curve C must equal the integral of curl $\mathbf{F}$ over the region R enclosed by C. Since curl $\mathbf{F} = 0$, the latter integral is zero and, therefore, so is the circulation.

The circulation $\oint_C \mathbf{F} \cdot \mathbf{T}\, ds$ is the same as the work $\oint_C \mathbf{F} \cdot d\mathbf{r}$ done by $\mathbf{F}$ around C, so our observation that circulation of a conservative two-dimensional field is zero agrees with the fact that the work done by a conservative field around a closed curve is always 0.

41-44. Example CAS commands:

Maple:

```
with(plots):
implicitplot({y=4 - 2*x}, x = 0..3, y = 0..5, scaling=CONSTRAINED);
M:= (x,y) -> x*exp(y);
N:= (x,y) -> 4*x^2*ln(y);
My:= diff(M(x,y),y);
Nx:= diff(N9x,y),x);
int(int(Nx - My, y = 0..4 - 2*x), x= 0..2);
evalf(");
```

Mathematica:

```
<< Graphics`ImplicitPlot`
SetOptions[ ImplicitPlot, AspectRatio -> Automatic ];
Clear[x,y]
f[x_,y_] = {
  x E^y ,
  4 x^2 Log[y] }
y1 = 0
y2 = -2 x + 4
Plot[ {y1,y2}, {x,0,2}, AspectRatio -> Automatic ]
integrand = D[f[x,y] [[2]],x] - D[f[x,y][[1]],y]
Solve[ c, y ]
{y1,y2} = y /. %
Integrate[ integrand, {x,0,2}, {y,y1,y2} ]
Simplify[%]
N[%]
```

14.5 SURFACE AREA AND SURFACE INTEGRALS

1. $\mathbf{p}=\mathbf{k},\ \nabla f = 2x\mathbf{i}+2y\mathbf{j}-\mathbf{k} \Rightarrow |\nabla f| = \sqrt{(2x)^2+(2y)^2+(-1)^2} = \sqrt{4x^2+4y^2+1}$ and $|\nabla f \cdot \mathbf{p}| = 1$;

   $z=2 \Rightarrow x^2+y^2=2$; thus $S = \iint\limits_R \frac{|\nabla f|}{|\nabla f\cdot \mathbf{p}|}\,dA = \iint\limits_R \sqrt{4x^2+4y^2+1}\,dx\,dy$

   $= \iint\limits_R \sqrt{4r^2\cos^2\theta + 4r^2\sin^2\theta+1}\ r\,dr\,d\theta = \int_0^{2\pi}\int_0^{\sqrt{2}} \sqrt{4r^2+1}\ r\,dr\,d\theta = \int_0^{2\pi}\left[\frac{1}{12}\left(4r^2+1\right)^{3/2}\right]_0^{\sqrt{2}} d\theta$

   $= \int_0^{2\pi} \frac{13}{6}\,d\theta = \frac{13}{3}\pi$

2. $\mathbf{p}=\mathbf{k},\ \nabla f = 2x\mathbf{i}+2y\mathbf{j}-\mathbf{k} \Rightarrow |\nabla f| = \sqrt{4x^2+4y^2+1}$ and $|\nabla f\cdot\mathbf{p}|=1$; $2\le x^2+y^2\le 6$

   $\Rightarrow S = \iint\limits_R \frac{|\nabla f|}{|\nabla f\cdot \mathbf{p}|}\,dA = \iint\limits_R \sqrt{4x^2+4y^2+1}\,dx\,dy = \iint\limits_R \sqrt{4r^2+1}\ r\,dr\,d\theta = \int_0^{2\pi}\int_{\sqrt{2}}^{\sqrt{6}} \sqrt{4r^2+1}\ r\,dr\,d\theta$

   $= \int_0^{2\pi}\left[\frac{1}{12}\left(4r^2+1\right)^{3/2}\right]_{\sqrt{2}}^{\sqrt{6}} d\theta = \int_0^{2\pi}\frac{49}{6}\,d\theta = \frac{49}{3}\pi$

3. $\mathbf{p}=\mathbf{k},\ \nabla f = \mathbf{i}+2\mathbf{j}+2\mathbf{k} \Rightarrow |\nabla f| = 3$ and $|\nabla f\cdot\mathbf{p}|=2$; $x=y^2$ and $x=2-y^2$ intersect at $(1,1)$ and $(1,-1)$

   $\Rightarrow S = \iint\limits_R \frac{|\nabla f|}{|\nabla f\cdot \mathbf{p}|}\,dA = \iint\limits_R \frac{3}{2}\,dx\,dy = \int_{-1}^{1}\int_{y^2}^{2-y^2}\frac{3}{2}\,dx\,dy = \int_{-1}^{1}\left(3-3y^2\right)dy = 4$

4. $\mathbf{p}=\mathbf{k},\ \nabla f = 2x\mathbf{i}-2\mathbf{k} \Rightarrow |\nabla f| = \sqrt{4x^2+4} = 2\sqrt{x^2+1}$ and $|\nabla f\cdot\mathbf{p}| = 2 \Rightarrow S = \iint\limits_R \frac{|\nabla f|}{|\nabla f\cdot \mathbf{p}|}\,dA$

   $= \iint\limits_R \frac{2\sqrt{x^2+1}}{2}\,dx\,dy = \int_0^{\sqrt{3}}\int_0^{x}\sqrt{x^2+1}\,dy\,dx = \int_0^{\sqrt{3}} x\sqrt{x^2+1}\,dx = \left[\frac{1}{3}\left(x^2+1\right)^{3/2}\right]_0^{\sqrt{3}} = \frac{1}{3}(4)^{3/2}-\frac{1}{3} = \frac{7}{3}$

5. $\mathbf{p}=\mathbf{k},\ \nabla f = 2x\mathbf{i}-2\mathbf{j}-2\mathbf{k} \Rightarrow |\nabla f| = \sqrt{(2x)^2+(-2)^2+(-2)^2} = \sqrt{4x^2+8} = 2\sqrt{x^2+2}$ and $|\nabla f\cdot\mathbf{p}| = 2$

   $\Rightarrow S = \iint\limits_R \frac{|\nabla f|}{|\nabla f\cdot \mathbf{p}|}\,dA = \iint\limits_R \frac{2\sqrt{x^2+2}}{2}\,dx\,dy = \int_0^{2}\int_0^{3x}\sqrt{x^2+2}\,dy\,dx = \int_0^{2} 3x\sqrt{x^2+2}\,dx = \left[\left(x^2+2\right)^{3/2}\right]_0^{2}$

   $= 6\sqrt{6}-2\sqrt{2}$

6. $\mathbf{p}=\mathbf{k},\ \nabla f=2x\mathbf{i}+2y\mathbf{j}+2z\mathbf{k} \Rightarrow |\nabla f|=\sqrt{4x^2+4y^2+4z^2}=\sqrt{8}=2\sqrt{2}$ and $|\nabla f\cdot\mathbf{p}|=2z$; $x^2+y^2+z^2=2$ and $z=\sqrt{x^2+y^2} \Rightarrow x^2+y^2=1$; thus, $S=\iint_R \frac{|\nabla f|}{|\nabla f\cdot\mathbf{p}|}\,dA=\iint_R \frac{2\sqrt{2}}{2z}\,dA=\sqrt{2}\iint_R \frac{1}{z}\,dA$

$$=\sqrt{2}\iint_R \frac{1}{\sqrt{2-(x^2+y^2)}}\,dA=\sqrt{2}\int_0^{2\pi}\int_0^1 \frac{r\,dr\,d\theta}{\sqrt{2-r^2}}=\sqrt{2}\int_0^{2\pi}(-1+\sqrt{2})\,d\theta=2\pi(2-\sqrt{2})$$

7. $\mathbf{p}=\mathbf{k},\ \nabla f=c\mathbf{i}-\mathbf{k} \Rightarrow |\nabla f|=\sqrt{c^2+1}$ and $|\nabla f\cdot\mathbf{p}|=1 \Rightarrow S=\iint_R \frac{|\nabla f|}{|\nabla f\cdot\mathbf{p}|}\,dA=\iint_R \sqrt{c^2+1}\,dx\,dy$

$$=\int_0^{2\pi}\int_0^1 \sqrt{c^2+1}\,r\,dr\,d\theta=\int_0^{2\pi}\frac{\sqrt{c^2+1}}{2}\,d\theta=\pi\sqrt{c^2+1}$$

8. $\mathbf{p}=\mathbf{k},\ \nabla f=2x\mathbf{i}+2z\mathbf{j} \Rightarrow |\nabla f|=\sqrt{(2x)^2+(2z)^2}=2$ and $|\nabla f\cdot\mathbf{p}|=2z$ for the upper surface, $z\ge 0$

$$\Rightarrow S=\iint_R \frac{|\nabla f|}{|\nabla f\cdot\mathbf{p}|}\,dA=\iint_R \frac{2}{2z}\,dA=\iint_R \frac{1}{\sqrt{1-x^2}}\,dy\,dx=2\int_{-1/2}^{1/2}\int_0^{1/2}\frac{1}{\sqrt{1-x^2}}\,dy\,dx=\int_{-1/2}^{1/2}\frac{1}{\sqrt{1-x^2}}\,dx$$

$$=\left[\sin^{-1}x\right]_{-1/2}^{1/2}=\frac{\pi}{6}-\left(-\frac{\pi}{6}\right)=\frac{\pi}{3}$$

9. $\mathbf{p}=\mathbf{i},\ \nabla f=\mathbf{i}+2y\mathbf{j}+2z\mathbf{k} \Rightarrow |\nabla f|=\sqrt{1^2+(2y)^2+(2z)^2}=\sqrt{1+4y^2+4z^2}$ and $|\nabla f\cdot\mathbf{p}|=1$; $1\le y^2+z^2\le 4$

$$\Rightarrow S=\iint_R \frac{|\nabla f|}{|\nabla f\cdot\mathbf{p}|}\,dA=\iint_R \sqrt{1+4y^2+4z^2}\,dy\,dz=\int_0^{2\pi}\int_1^2 \sqrt{1+4r^2\cos^2\theta+4r^2\sin^2\theta}\,r\,dr\,d\theta$$

$$=\int_0^{2\pi}\int_1^2 \sqrt{1+4r^2}\,r\,dr\,d\theta=\int_0^{2\pi}\left[\frac{1}{12}(1+4r^2)^{3/2}\right]_1^2 d\theta=\int_0^{2\pi}\frac{1}{12}(17\sqrt{17}-5\sqrt{5})\,d\theta=\frac{\pi}{6}(17\sqrt{17}-5\sqrt{5})$$

10. $\mathbf{p}=\mathbf{j},\ \nabla f=2x\mathbf{i}+\mathbf{j}+2z\mathbf{k} \Rightarrow |\nabla f|=\sqrt{4x^2+4z^2+1}$ and $|\nabla f\cdot\mathbf{p}|=1$; $y=0$ and $x^2+y+z^2=2 \Rightarrow x^2+z^2=2$;

$$\text{thus, } S=\iint_R \frac{|\nabla f|}{|\nabla f\cdot\mathbf{p}|}\,dA=\iint_R \sqrt{4x^2+4z^2+1}\,dx\,dz=\int_0^{2\pi}\int_0^{\sqrt{2}}\sqrt{4r^2+1}\,r\,dr\,d\theta=\int_0^{2\pi}\frac{13}{6}\,d\theta=\frac{13}{3}\pi$$

11. $\mathbf{p}=\mathbf{k},\ \nabla f=\left(2x-\frac{2}{x}\right)\mathbf{i}+\sqrt{15}\,\mathbf{j}-\mathbf{k} \Rightarrow |\nabla f|=\sqrt{\left(2x-\frac{2}{x}\right)^2+\left(\sqrt{15}\right)^2+(-1)^2}=\sqrt{4x^2+8+\frac{4}{x^2}}=\sqrt{\left(2x+\frac{2}{x}\right)^2}$

$=2x+\frac{2}{x}$, on $1\le x\le 2$ and $|\nabla f\cdot\mathbf{p}|=1 \Rightarrow S=\iint_R \frac{|\nabla f|}{|\nabla f\cdot\mathbf{p}|}\,dA=\iint_R (2x+2x^{-1})\,dx\,dy$

$$=\int_0^1\int_1^2 (2x+2x^{-1})\,dx\,dy=\int_0^1 \left[x^2+2\ln x\right]_1^2 dy=\int_0^1 (3+2\ln 2)\,dy=3+2\ln 2$$

12. $\mathbf{p}=\mathbf{k}$, $\nabla f = 3\sqrt{x}\,\mathbf{i}+3\sqrt{y}\,\mathbf{j}-3\mathbf{k} \Rightarrow |\nabla f| = \sqrt{9x+9y+9} = 3\sqrt{x+y+1}$ and $|\nabla f\cdot\mathbf{p}| = 3$

$$\Rightarrow S = \iint_R \frac{|\nabla f|}{|\nabla f\cdot \mathbf{p}|}\,dA = \iint_R \sqrt{x+y+1}\,dx\,dy = \int_0^1\int_0^1 \sqrt{x+y+1}\,dx\,dy = \int_0^1 \left[\frac{2}{3}(x+y+1)^{3/2}\right]_0^1 dy$$

$$= \int_0^1 \left[\frac{2}{3}(y+2)^{3/2} - \frac{2}{3}(y+1)^{3/2}\right] dy = \left[\frac{4}{15}(y+2)^{5/2} - \frac{4}{15}(y+1)^{5/2}\right]_0^1 = \frac{4}{15}\left[(3)^{5/2} - (2)^{5/2} - (2)^{5/2} + 1\right]$$

$$= \frac{4}{15}\left(9\sqrt{3} - 8\sqrt{2} + 1\right)$$

13. The bottom face S of the cube is in the xy-plane $\Rightarrow z = 0 \Rightarrow g(x,y,0) = x+y$ and $f(x,y,z) = z = 0 \Rightarrow \mathbf{p}=\mathbf{k}$ and $\nabla f = \mathbf{k} \Rightarrow |\nabla f| = 1$ and $|\nabla f\cdot\mathbf{p}| = 1 \Rightarrow d\sigma = dx\,dy \Rightarrow \iint_S g\,d\sigma = \iint_R (x+y)\,dx\,dy$

$= \int_0^a\int_0^a (x+y)\,dx\,dy = \int_0^a \left(\frac{a^2}{2}+ay\right) dy = a^3$. Because of symmetry, we also get a^3 over the face of the cube in the xz-plane and a^3 over the face of the cube in the yz-plane. Next, on the top of the cube, $g(x,y,z) = g(x,y,a) = x+y+a$ and $f(x,y,z) = z = a \Rightarrow \mathbf{p}=\mathbf{k}$ and $\nabla f = \mathbf{k} \Rightarrow |\nabla f| = 1$ and $|\nabla f\cdot\mathbf{p}| = 1 \Rightarrow d\sigma = dx\,dy$

$$\iint_S g\,d\sigma = \iint_R (x+y+a)\,dx\,dy = \int_0^a\int_0^a (x+y+a)\,dx\,dy = \int_0^a\int_0^a (x+y)\,dx\,dy + \int_0^a\int_0^a a\,dx\,dy = 2a^3.$$

Because of symmetry, the integral is also $2a^3$ over each of the other two faces. Therefore,

$$\iint_{\text{cube}} (x+y+z)\,d\sigma = 3\left(a^3+2a^3\right) = 9a^3.$$

14. On the face S in the xz-plane, we have $y = 0 \Rightarrow f(x,y,z) = y = 0$ and $g(x,y,z) = g(x,0,z) = z \Rightarrow \mathbf{p}=\mathbf{j}$ and $\nabla f = \mathbf{j} \Rightarrow |\nabla f| = 1$ and $|\nabla f\cdot\mathbf{p}| = 1 \Rightarrow d\sigma = dx\,dz \Rightarrow \iint_S g\,d\sigma = \iint_S (y+z)\,d\sigma = \int_0^1\int_0^2 z\,dx\,dz = \int_0^1 2z\,dz$

$= 1$.

On the face in the xy-plane, we have $z = 0 \Rightarrow f(x,y,z) = z = 0$ and $g(x,y,z) = g(x,y,0) = y \Rightarrow \mathbf{p}=\mathbf{k}$ and

$\nabla f = \mathbf{k} \Rightarrow |\nabla f| = 1$ and $|\nabla f\cdot\mathbf{p}| = 1 \Rightarrow d\sigma = dx\,dy \Rightarrow \iint_S g\,d\sigma = \iint_S y\,d\sigma = \int_0^1\int_0^2 y\,dx\,dy = 1.$

On the triangular face in the plane $x = 2$ we have $f(x,y,z) = x = 2$ and $g(x,y,z) = g(2,y,z) = y+z \Rightarrow \mathbf{p}=\mathbf{i}$ and

$\nabla f = \mathbf{i} \Rightarrow |\nabla f| = 1$ and $|\nabla f\cdot\mathbf{p}| = 1 \Rightarrow d\sigma = dz\,dy \Rightarrow \iint_S g\,d\sigma = \iint_S (y+z)\,d\sigma = \int_0^1\int_0^{1-y} (y+z)\,dz\,dy$

$= \int_0^1 \frac{1}{2}\left(1-y^2\right) dy = \frac{1}{3}.$

On the triangular face in the yz-plane, we have $x = 0 \Rightarrow f(x,y,z) = x = 0$ and $g(x,y,z) = g(0,y,z) = y + z$

$\Rightarrow \mathbf{p} = \mathbf{i}$ and $\nabla f = \mathbf{i} \Rightarrow |\nabla f| = 1$ and $|\nabla f \cdot \mathbf{p}| = 1 \Rightarrow d\sigma = dz\,dy \Rightarrow \iint_S g\,d\sigma = \iint_S (y+z)\,d\sigma$

$= \int_0^1 \int_0^{1-y} (y+z)\,dz\,dy = \frac{1}{3}.$

Finally, on the sloped face, we have $y + z = 1 \Rightarrow f(x,y,z) = y + z = 1$ and $g(x,y,z) = y + z = 1 \Rightarrow \mathbf{p} = \mathbf{k}$ and

$\nabla f = \mathbf{j} + \mathbf{k} \Rightarrow |\nabla f| = \sqrt{2}$ and $|\nabla f \cdot \mathbf{p}| = 1 \Rightarrow d\sigma = \sqrt{2}\,dx\,dy \Rightarrow \iint_S g\,d\sigma = \iint_S (y+z)\,d\sigma$

$= \int_0^1 \int_0^2 \sqrt{2}\,dx\,dy = 2\sqrt{2}.$ Therefore, $\iint_{\text{wedge}} g(x,y,z)\,d\sigma = 1 + 1 + \frac{1}{3} + \frac{1}{3} + 2\sqrt{2} = \frac{8}{3} + 2\sqrt{2}$

15. On the faces in the coordinate planes, $g(x,y,z) = 0 \Rightarrow$ the integral over these faces is 0.
On the face $x = a$, we have $f(x,y,z) = x = a$ and $g(x,y,z) = g(a,y,z) = ayz \Rightarrow \mathbf{p} = \mathbf{i}$ and $\nabla f = \mathbf{i} \Rightarrow |\nabla f| = 1$

and $|\nabla f \cdot \mathbf{p}| = 1 \Rightarrow d\sigma = dy\,dz \Rightarrow \iint_S g\,d\sigma = \iint_S ayz\,d\sigma = \int_0^c \int_0^b ayz\,dy\,dz = \frac{ab^2c^2}{4}.$

On the face $y = b$, we have $f(x,y,z) = y = b$ and $g(x,y,z) = g(x,b,z) = bxz \Rightarrow \mathbf{p} = \mathbf{j}$ and $\nabla f = \mathbf{j} \Rightarrow |\nabla f| = 1$

and $|\nabla f \cdot \mathbf{p}| = 1 \Rightarrow d\sigma = dx\,dz \Rightarrow \iint_S g\,d\sigma = \iint_S bxz\,d\sigma = \int_0^c \int_0^a bxz\,dz\,dx = \frac{a^2bc^2}{4}.$

On the face $z = c$, we have $f(x,y,z) = z = c$ and $g(x,y,z) = g(x,y,c) = cxy \Rightarrow \mathbf{p} = \mathbf{k}$ and $\nabla f = \mathbf{k} \Rightarrow |\nabla f| = 1$

and $|\nabla f \cdot \mathbf{p}| = 1 \Rightarrow d\sigma = dy\,dx \Rightarrow \iint_S g\,d\sigma = \iint_S cxy\,d\sigma = \int_0^b \int_0^a cxy\,dx\,dy = \frac{a^2b^2c}{4}.$ Therefore,

$\iint_S g(x,y,z)\,d\sigma = \frac{abc(ab+ac+bc)}{4}.$

16. On the face $x = a$, we have $f(x,y,z) = x = a$ and $g(x,y,z) = g(a,y,z) = ayz \Rightarrow \mathbf{p} = \mathbf{i}$ and $\nabla f = \mathbf{i} \Rightarrow |\nabla f| = 1$

and $|\nabla f \cdot \mathbf{p}| = 1 \Rightarrow d\sigma = dz\,dy \Rightarrow \iint_S g\,d\sigma = \iint_S ayz\,d\sigma = \int_{-b}^b \int_{-c}^c ayz\,dz\,dy = 0.$ Because of the symmetry

of g on all the other faces, all the integrals are 0, and $\iint_S g(x,y,z)\,d\sigma = 0.$

17. $f(x,y,z) = 2x + 2y + z = 2 \Rightarrow \nabla f = 2\mathbf{i} + 2\mathbf{j} + \mathbf{k}$ and $g(x,y,z) = x + y + (2 - 2x - 2y) = 2 - x - y \Rightarrow \mathbf{p} = \mathbf{k}$,

$|\nabla f| = 3$ and $|\nabla f \cdot \mathbf{p}| = 1 \Rightarrow d\sigma = 3\,dy\,dx$; $z = 0 \Rightarrow 2x + 2y = 2 \Rightarrow y = 1 - x \Rightarrow \iint_S g\,d\sigma = \iint_S (2 - x - y)\,d\sigma$

$$= 3\int_0^1\int_0^{1-x}(2-x-y)\,dy\,dx = 3\int_0^1\left[(2-x)(1-x)-\tfrac{1}{2}(1-x)^2\right]dx = 3\int_0^1\left(\tfrac{3}{2}-2x+\tfrac{x^2}{2}\right)dx = 2$$

18. $f(x,y,z) = y^2+4z = 16 \Rightarrow \nabla f = 2y\mathbf{j}+4\mathbf{k} \Rightarrow |\nabla f| = \sqrt{4y^2+16} = 2\sqrt{y^2+4}$ and $\mathbf{p} = \mathbf{k} \Rightarrow |\nabla f\cdot\mathbf{p}| = 4$

$$\Rightarrow d\sigma = \frac{2\sqrt{y^2+4}}{4}\,dx\,dy \Rightarrow \iint_S g\,d\sigma = \int_{-4}^{4}\int_0^1\left(x\sqrt{y^2+4}\right)\left(\frac{\sqrt{y^2+4}}{2}\right)dx\,dy = \int_{-4}^{4}\int_0^1 \frac{x(y^2+4)}{2}\,dx\,dy$$

$$= \int_{-4}^{4}\tfrac{1}{4}(y^2+4)\,dy = \tfrac{1}{2}\left[\frac{y^3}{3}+4y\right]_0^4 = \tfrac{1}{2}\left(\tfrac{64}{3}+16\right) = \tfrac{56}{3}$$

19. $g(x,y,z) = z,\ \mathbf{p} = \mathbf{k} \Rightarrow \nabla g = \mathbf{k} \Rightarrow |\nabla g| = 1$ and $|\nabla g\cdot\mathbf{p}| = 1 \Rightarrow \text{Flux} = \iint_S \mathbf{F}\cdot\mathbf{n}\,d\sigma = \iint_R (\mathbf{F}\cdot\mathbf{k})\,dA$

$$= \int_0^2\int_0^3 3\,dy\,dx = 18$$

20. $g(x,y,z) = y,\ \mathbf{p} = -\mathbf{j} \Rightarrow \nabla g = \mathbf{j} \Rightarrow |\nabla g| = 1$ and $|\nabla g\cdot\mathbf{p}| = 1 \Rightarrow \text{Flux} = \iint_S \mathbf{F}\cdot\mathbf{n}\,d\sigma = \iint_R (\mathbf{F}\cdot -\mathbf{j})\,dA$

$$= \int_{-1}^{2}\int_2^7 2\,dz\,dx = \int_{-1}^{2} 2(7-2)\,dx = 10(2+1) = 30$$

21. $\nabla g = 2x\mathbf{i}+2y\mathbf{j}+2z\mathbf{k} \Rightarrow |\nabla g| = \sqrt{4x^2+4y^2+4z^2} = 2a;\ \mathbf{n} = \dfrac{2x\mathbf{i}+2y\mathbf{j}+2z\mathbf{k}}{2\sqrt{x^2+y^2+z^2}} = \dfrac{x\mathbf{i}+y\mathbf{j}+z\mathbf{k}}{a} \Rightarrow \mathbf{F}\cdot\mathbf{n} = \frac{z^2}{a};$

$$|\nabla g\cdot\mathbf{k}| = 2z \Rightarrow d\sigma = \tfrac{2a}{2z}\,dA \Rightarrow \text{Flux} = \iint_R \left(\tfrac{z^2}{a}\right)\left(\tfrac{a}{z}\right)dA = \iint_R z\,dA = \iint_R \sqrt{a^2-(x^2+y^2)}\,dx\,dy$$

$$= \int_0^{\pi/2}\int_0^a \sqrt{a^2-r^2}\,r\,dr\,d\theta = \frac{\pi a^3}{6}$$

22. $\nabla g = 2x\mathbf{i}+2y\mathbf{j}+2z\mathbf{k} \Rightarrow |\nabla g| = \sqrt{4x^2+4y^2+4z^2} = 2a;\ \mathbf{n} = \dfrac{2x\mathbf{i}+2y\mathbf{j}+2z\mathbf{k}}{2\sqrt{x^2+y^2+z^2}} = \dfrac{x\mathbf{i}+y\mathbf{j}+z\mathbf{k}}{a} \Rightarrow \mathbf{F}\cdot\mathbf{n} = \frac{-xy}{a}+\frac{xy}{a}$

$$= 0;\ |\nabla g\cdot\mathbf{k}| = 2z \Rightarrow d\sigma = \tfrac{2a}{2z}\,dA \Rightarrow \text{Flux} = \iint_S \mathbf{F}\cdot\mathbf{n}\,d\sigma = \iint_S 0\,d\sigma = 0$$

23. From Exercise 21, $\mathbf{n} = \dfrac{x\mathbf{i}+y\mathbf{j}+z\mathbf{k}}{a}$ and $d\sigma = \frac{a}{z}\,dA \Rightarrow \mathbf{F}\cdot\mathbf{n} = \frac{xy}{a}-\frac{xy}{a}+\frac{z}{a} = \frac{z}{a} \Rightarrow \text{Flux} = \iint_R \left(\tfrac{z}{a}\right)\left(\tfrac{a}{z}\right)dA$

$$= \iint_R 1\,dA = \frac{\pi a^2}{4}$$

24. From Exercise 21, $\mathbf{n} = \frac{x\mathbf{i}+y\mathbf{j}+z\mathbf{k}}{a}$ and $d\sigma = \frac{a}{z}\,dA \Rightarrow \mathbf{F}\cdot\mathbf{n} = \frac{zx^2}{a} + \frac{zy^2}{a} + \frac{z^3}{a} = z\left(\frac{x^2+y^2+z^2}{a}\right) = az$

$\Rightarrow \text{Flux} = \iint\limits_R (za)\left(\frac{a}{z}\right) dx\,dy = \iint\limits_R a^2\,dx\,dy = a^2(\text{Area of R}) = \frac{1}{4}\pi a^4$

25. From Exercise 21, $\mathbf{n} = \frac{x\mathbf{i}+y\mathbf{j}+z\mathbf{k}}{a}$ and $d\sigma = \frac{a}{z}\,dA \Rightarrow \mathbf{F}\cdot\mathbf{n} = \frac{x^2}{a} + \frac{y^2}{a} + \frac{z^2}{a} = a \Rightarrow \text{Flux}$

$= \iint\limits_R a\left(\frac{a}{z}\right) dA = \iint\limits_R \frac{a^2}{z}\,dA = \iint\limits_R \frac{a^2}{\sqrt{a^2-(x^2+y^2)}}\,dA = \int_0^{\pi/2}\int_0^a \frac{a^2}{\sqrt{a^2-r^2}}\,r\,dr\,d\theta$

$= \int_0^{\pi/2} a^2\left[-\sqrt{a^2-r^2}\right]_0^a d\theta = \frac{\pi a^3}{2}$

26. From Exercise 21, $\mathbf{n} = \frac{x\mathbf{i}+y\mathbf{j}+z\mathbf{k}}{a}$ and $d\sigma = \frac{a}{z}\,dA \Rightarrow \mathbf{F}\cdot\mathbf{n} = \frac{\left(\frac{x^2}{a}\right)+\left(\frac{y^2}{a}\right)+\left(\frac{z^2}{a}\right)}{\sqrt{x^2+y^2+z^2}} = \frac{\left(\frac{a^2}{a}\right)}{a} = 1$

$\Rightarrow \text{Flux} = \iint\limits_R \frac{a}{z}\,dx\,dy = \iint\limits_R \frac{a}{\sqrt{a^2-(x^2+y^2)}}\,dx\,dy = \int_0^{\pi/2}\int_0^a \frac{a}{\sqrt{a^2-r^2}}\,r\,dr\,d\theta = \frac{\pi a^2}{2}$

27. $g(x,y,z) = y^2+z = 4 \Rightarrow \nabla g = 2y\mathbf{j}+\mathbf{k} \Rightarrow |\nabla g| = \sqrt{4y^2+1} \Rightarrow \mathbf{n} = \frac{2y\mathbf{j}+\mathbf{k}}{\sqrt{4y^2+1}}$

$\Rightarrow \mathbf{F}\cdot\mathbf{n} = \frac{2xy-3z}{\sqrt{4y^2+1}}$; $\mathbf{p} = \mathbf{k} \Rightarrow |\nabla g\cdot\mathbf{p}| = 1 \Rightarrow d\sigma = \sqrt{4y^2+1}\,dA \Rightarrow \text{Flux}$

$= \iint\limits_R \left(\frac{2xy-3z}{\sqrt{4y^2+1}}\right)\sqrt{4y^2+1}\,dA = \iint\limits_R (2xy-3z)\,dA$; $z = 0$ and $z = 4-y^2 \Rightarrow y^2 = 4$

$\Rightarrow \text{Flux} = \iint\limits_R \left[2xy-3\left(4-y^2\right)\right] dA = \int_0^1\int_{-2}^2 \left(2xy-12+3y^2\right) dy\,dx = \int_0^1 \left[xy^2-12y+y^3\right]_{-2}^2 dx$

$= \int_0^1 -32\,dx = -32$

28. $g(x,y,z) = x^2+y^2-z = 0 \Rightarrow \nabla g = 2x\mathbf{i}+2y\mathbf{j}-\mathbf{k} \Rightarrow |\nabla g| = \sqrt{4x^2+4y^2+1} = \sqrt{4\left(x^2+y^2\right)+1}$

$\Rightarrow \mathbf{n} = \frac{2x\mathbf{i}+2y\mathbf{j}-\mathbf{k}}{\sqrt{4\left(x^2+y^2\right)+1}} \Rightarrow \mathbf{F}\cdot\mathbf{n} = \frac{8x^2+8y^2-2}{\sqrt{4\left(x^2+y^2\right)+1}}$; $\mathbf{p} = \mathbf{k} \Rightarrow |\nabla g\cdot\mathbf{p}| = 1 \Rightarrow d\sigma = \sqrt{4\left(x^2+y^2\right)+1}\,dA$

$\Rightarrow \text{Flux} = \iint\limits_R \left(\frac{8x^2+8y^2-2}{\sqrt{4\left(x^2+y^2\right)+1}}\right)\sqrt{4\left(x^2+y^2\right)+1}\,dA = \iint\limits_R \left(8x^2+8y^2-2\right) dA$; $z = 1$ and $x^2+y^2 = z$

$\Rightarrow x^2+y^2=1 \Rightarrow \text{Flux} = \int_0^{2\pi}\int_0^1 (8r^2-2)\, r\, dr\, d\theta = 2\pi$

29. $g(x,y,z) = y - e^x = 0 \Rightarrow \nabla g = -e^x\mathbf{i}+\mathbf{j} \Rightarrow |\nabla g| = \sqrt{e^{2x}+1} \Rightarrow \mathbf{n} = \frac{e^x\mathbf{i}-\mathbf{j}}{\sqrt{e^{2x}+1}} \Rightarrow \mathbf{F}\cdot\mathbf{n} = \frac{-2e^x-2y}{\sqrt{e^{2x}+1}}$; $\mathbf{p}=\mathbf{i}$

$\Rightarrow |\nabla g\cdot\mathbf{p}| = e^x \Rightarrow d\sigma = \frac{\sqrt{e^{2x}+1}}{e^x}\, dA \Rightarrow \text{Flux} = \iint_R \left(\frac{-2e^x-2y}{\sqrt{e^{2x}+1}}\right)\left(\frac{\sqrt{e^{2x}+1}}{e^x}\right) dA = \iint_R \frac{-2e^x-2e^x}{e^x}\, dA$

$= \iint_R -4\, dA = \int_0^1\int_1^2 -4\, dy\, dz = -4$

30. $g(x,y,z) = y - \ln x = 0 \Rightarrow \nabla g = -\frac{1}{x}\mathbf{i}+\mathbf{j} \Rightarrow |\nabla g| = \sqrt{\frac{1}{x^2}+1} = \frac{\sqrt{1+x^2}}{x}$ since $1 \le x \le e$

$\Rightarrow \mathbf{n} = \frac{\left(-\frac{1}{x}\mathbf{i}+\mathbf{j}\right)}{\left(\frac{\sqrt{1+x^2}}{x}\right)} = \frac{-\mathbf{i}+x\mathbf{j}}{\sqrt{1+x^2}} \Rightarrow \mathbf{F}\cdot\mathbf{n} = \frac{2xy}{\sqrt{1+x^2}}$; $\mathbf{p}=\mathbf{j} \Rightarrow |\nabla g\cdot\mathbf{p}| = 1 \Rightarrow d\sigma = \frac{\sqrt{1+x^2}}{x}\, dA$

$\Rightarrow \text{Flux} = \iint_R \left(\frac{2xy}{\sqrt{1+x^2}}\right)\left(\frac{\sqrt{1+x^2}}{x}\right) dA = \int_0^1\int_1^e 2y\, dx\, dz = \int_1^e\int_0^1 2\ln x\, dz\, dx = \int_1^e 2\ln x\, dx$

$= 2[x\ln x - x]_1^e = 2(e-e) - 2(0-1) = 2$

31. On the face $z=a$: $g(x,y,z) = z \Rightarrow \nabla g = \mathbf{k} \Rightarrow |\nabla g| = 1$; $\mathbf{n}=\mathbf{k} \Rightarrow \mathbf{F}\cdot\mathbf{n} = 2xz = 2ax$ since $z=a$;

$d\sigma = dx\, dy \Rightarrow \text{Flux} = \iint_R 2ax\, dx\, dy = \int_0^a\int_0^a 2ax\, dx\, dy = a^4.$

On the face $z=0$: $g(x,y,z) = z \Rightarrow \nabla g = \mathbf{k} \Rightarrow |\nabla g| = 1$; $\mathbf{n}=-\mathbf{k} \Rightarrow \mathbf{F}\cdot\mathbf{n} = -2xz = 0$ since $z=0$;

$d\sigma = dx\, dy \Rightarrow \text{Flux} = \iint_R 0\, dx\, dy = 0.$

On the face $x=a$: $g(x,y,z) = x \Rightarrow \nabla g = \mathbf{i} \Rightarrow |\nabla g| = 1$; $\mathbf{n}=\mathbf{i} \Rightarrow \mathbf{F}\cdot\mathbf{n} = 2xy = 2ay$ since $x=a$;

$d\sigma = dy\, dz \Rightarrow \text{Flux} = \int_0^a\int_0^a 2ay\, dy\, dz = a^4.$

On the face $x=0$: $g(x,y,z) = x \Rightarrow \nabla g = \mathbf{i} \Rightarrow |\nabla g| = 1$; $\mathbf{n}=-\mathbf{i} \Rightarrow \mathbf{F}\cdot\mathbf{n} = -2xy = 0$ since $x=0$

$\Rightarrow \text{Flux} = 0.$

On the face $y=a$: $g(x,y,z) = y \Rightarrow \nabla g = \mathbf{j} \Rightarrow |\nabla g| = 1$; $\mathbf{n}=\mathbf{j} \Rightarrow \mathbf{F}\cdot\mathbf{n} = 2yz = 2az$ since $y=a$;

$d\sigma = dz\, dx \Rightarrow \text{Flux} = \int_0^a\int_0^a 2az\, dz\, dx = a^4.$

On the face $y=0$: $g(x,y,z) = y \Rightarrow \nabla g = \mathbf{j} \Rightarrow |\nabla g| = 1$; $\mathbf{n}=-\mathbf{j} \Rightarrow \mathbf{F}\cdot\mathbf{n} = -2yz = 0$ since $y=0$

$\Rightarrow \text{Flux} = 0.$ Therefore, Total Flux $= 3a^4$.

32. Across the cap: $g(x,y,z) = x^2 + y^2 + z^2 = 25 \Rightarrow \nabla g = 2x\mathbf{i} + 2y\mathbf{j} + 2z\mathbf{k} \Rightarrow |\nabla g| = \sqrt{4x^2 + 4y^2 + 4z^2} = 10$

$\Rightarrow \mathbf{n} = \frac{\nabla g}{|\nabla g|} = \frac{x\mathbf{i} + y\mathbf{j} + z\mathbf{k}}{5} \Rightarrow \mathbf{F}\cdot\mathbf{n} = \frac{x^2z}{5} + \frac{y^2z}{5} + \frac{z}{5}$; $\mathbf{p} = \mathbf{k} \Rightarrow |\nabla g \cdot \mathbf{p}| = 2z$ since $z \geq 0 \Rightarrow d\sigma = \frac{10}{2z}\,dA$

$\Rightarrow \text{Flux}_{\text{cap}} = \iint\limits_{\text{cap}} \mathbf{F}\cdot\mathbf{n}\,d\sigma = \iint\limits_R \left(\frac{x^2z}{5} + \frac{y^2z}{5} + \frac{z}{5}\right)\left(\frac{5}{z}\right)dA = \iint\limits_R (x^2 + y^2 + 1)\,dx\,dy = \int_0^{2\pi}\int_0^4 (r^2 + 1)\,r\,dr\,d\theta$

$= \int_0^{2\pi} 72\,d\theta = 144\pi.$

Across the bottom: $g(x,y,z) = z = 3 \Rightarrow \nabla g = \mathbf{k} \Rightarrow |\nabla g| = 1 \Rightarrow \mathbf{n} = -\mathbf{k} \Rightarrow \mathbf{F}\cdot\mathbf{n} = -1$; $\mathbf{p} = \mathbf{k} \Rightarrow |\nabla g \cdot \mathbf{p}| = 1$

$\Rightarrow d\sigma = dA \Rightarrow \text{Flux}_{\text{bottom}} = \iint\limits_{\text{bottom}} \mathbf{F}\cdot\mathbf{n}\,d\sigma = \iint\limits_R -1\,d\sigma = \iint\limits_R -1\,dA = -1$(Area of the circular region)

$= -16\pi$. Therefore, $\text{Flux} = \text{Flux}_{\text{cap}} + \text{Flux}_{\text{bottom}} = 128\pi$

33. $\nabla f = 2x\mathbf{i} + 2y\mathbf{j} + 2z\mathbf{k} \Rightarrow |\nabla f| = \sqrt{4x^2 + 4y^2 + 4z^2} = 2a$; $\mathbf{p} = \mathbf{k} \Rightarrow |\nabla f \cdot \mathbf{p}| = 2z$ since $z \geq 0 \Rightarrow d\sigma = \frac{2a}{2z}\,dA$

$= \frac{a}{z}\,dA$; $M = \iint\limits_S \delta\,d\sigma = \frac{\delta}{8}$ (surface area of sphere) $= \frac{\delta\pi a^2}{2}$; $M_{xy} = \iint\limits_S z\delta\,d\sigma = \delta \iint\limits_R z\left(\frac{a}{z}\right)dA$

$= a\delta \iint\limits_R dA = a\delta \int_0^{\pi/2}\int_0^a r\,dr\,d\theta = \frac{\delta\pi a^3}{4} \Rightarrow \bar{z} = \frac{M_{xy}}{M} = \left(\frac{\delta\pi a^3}{4}\right)\left(\frac{2}{\delta\pi a^2}\right) = \frac{a}{2}$. Because of symmetry, $\bar{x} = \bar{y}$

$= \frac{a}{2} \Rightarrow$ the centroid is $\left(\frac{a}{2}, \frac{a}{2}, \frac{a}{2}\right)$.

34. $\nabla f = 2y\mathbf{i} + 2z\mathbf{k} \Rightarrow |\nabla f| = \sqrt{4y^2 + 4z^2} = \sqrt{4(y^2 + z^2)} = 6$; $\mathbf{p} = \mathbf{k} \Rightarrow |\nabla f \cdot \mathbf{k}| = 2z$ since $z \geq 0 \Rightarrow d\sigma = \frac{6}{2z}\,dA$

$= \frac{3}{z}\,dA$; $M = \iint\limits_S 1\,d\sigma = \int_{-3}^{3}\int_0^3 \frac{3}{z}\,dx\,dy = \int_{-3}^{3}\int_0^3 \frac{3}{\sqrt{9 - y^2}}\,dx\,dy = 9\pi$; $M_{xy} = \iint\limits_S z\,d\sigma$

$= \int_{-3}^{3}\int_0^3 z\left(\frac{3}{z}\right)dx\,dy = 54$; $M_{xz} = \iint\limits_S y\,d\sigma = \int_{-3}^{3}\int_0^3 y\left(\frac{3}{z}\right)dx\,dy = \int_{-3}^{3}\int_0^3 \frac{3y}{\sqrt{9 - y^2}}\,dx\,dy = 0$;

$M_{yz} = \iint\limits_S x\,d\sigma = \int_{-3}^{3}\int_0^3 \frac{3x}{\sqrt{9 - y^2}}\,dx\,dy = \frac{27}{2}\pi$. Therefore, $\bar{x} = \frac{\left(\frac{27}{2}\pi\right)}{9\pi} = \frac{3}{2}$, $\bar{y} = 0$, and $\bar{z} = \frac{54}{9\pi} = \frac{6}{\pi}$.

35. Because of symmetry, $\bar{x} = \bar{y} = 0$; $M = \iint\limits_S \delta\,d\sigma = \delta \iint\limits_S d\sigma = (\text{Area of } S)\delta = 3\pi\sqrt{2}\,\delta$; $\nabla f = 2x\mathbf{i} + 2y\mathbf{j} - 2z\mathbf{k}$

$\Rightarrow |\nabla f| = \sqrt{4x^2 + 4y^2 + 4z^2} = 2\sqrt{x^2 + y^2 + z^2}$; $\mathbf{p} = \mathbf{k} \Rightarrow |\nabla f \cdot \mathbf{p}| = 2z \Rightarrow d\sigma = \frac{2\sqrt{x^2 + y^2 + z^2}}{2z}\,dA$

$$= \frac{\sqrt{x^2+y^2+(x^2+y^2)}}{z}\, dA = \frac{\sqrt{2}\sqrt{x^2+y^2}}{z}\, dA \Rightarrow M_{xy} = \delta \iint_S z\left(\frac{\sqrt{2}\sqrt{x^2+y^2}}{z}\right) dA$$

$$= \delta \iint_S \sqrt{2}\sqrt{x^2+y^2}\, dA = \delta \int_0^{2\pi}\int_1^2 \sqrt{2}\, r^2\, dr\, d\theta = \frac{14\pi\sqrt{2}}{3}\delta \Rightarrow \overline{z} = \frac{\left(\frac{14\pi\sqrt{2}}{3}\delta\right)}{3\pi\sqrt{2}\,\delta} = \frac{14}{9}$$

$$\Rightarrow (\overline{x},\overline{y},\overline{z}) = \left(0,0,\tfrac{14}{9}\right). \text{ Next, } I_z = \iint_S (x^2+y^2)\delta\, d\sigma = \iint_S (x^2+y^2)\left(\frac{\sqrt{2}\sqrt{x^2+y^2}}{z}\right)\delta\, dA$$

$$= \delta\sqrt{2}\iint_S (x^2+y^2)\, dA = \delta\sqrt{2}\int_0^{2\pi}\int_1^2 r^3\, dr\, d\theta = \frac{15\pi\sqrt{2}}{2}\delta \Rightarrow R_z = \sqrt{\frac{I_z}{M}} = \frac{\sqrt{10}}{2}$$

36. $f(x,y,z) = 4x^2+4y^2-z^2 = 0 \Rightarrow \nabla f = 8x\mathbf{i}+8y\mathbf{j}-2z\mathbf{k} \Rightarrow |\nabla f| = \sqrt{64x^2+64y^2+4z^2}$

$$= 2\sqrt{16x^2+16y^2+z^2} = 2\sqrt{4z^2+z^2} = 2\sqrt{5}\,z \text{ since } z \ge 0;\ \mathbf{p} = \mathbf{k} \Rightarrow |\nabla f\cdot\mathbf{p}| = 2z \Rightarrow d\sigma = \frac{2\sqrt{5}\,z}{2z}\, dA = \sqrt{5}\, dA$$

$$\Rightarrow I_z = \iint_S (x^2+y^2)\,\delta\, d\sigma = \delta\sqrt{5}\iint_R (x^2+y^2)\, dx\, dy = \delta\sqrt{5}\int_{-\pi/2}^{\pi/2}\int_0^{2\cos\theta} r^3\, dr\, d\theta = \frac{3\sqrt{5}\pi\delta}{2}$$

37. (a) Let the diameter lie on the z-axis and let $f(x,y,z) = x^2+y^2+z^2 = a^2$, $z \ge 0$ be the upper hemisphere

$$\Rightarrow \nabla f = 2x\mathbf{i}+2y\mathbf{j}+2z\mathbf{k} \Rightarrow |\nabla f| = \sqrt{4x^2+4y^2+4z^2} = 2a,\ a>0;\ \mathbf{p} = \mathbf{k} \Rightarrow |\nabla f\cdot\mathbf{p}| = 2z \text{ since } z \ge 0$$

$$\Rightarrow d\sigma = \tfrac{a}{z}\, dA \Rightarrow I_z = \iint_S \delta(x^2+y^2)\left(\tfrac{a}{z}\right) d\sigma = a\delta\iint_R \frac{x^2+y^2}{\sqrt{a^2-(x^2+y^2)}}\, dA = a\delta\int_0^{2\pi}\int_0^a \frac{r^2}{\sqrt{a^2-r^2}}\, r\, dr\, d\theta$$

$$= a\delta\int_0^{2\pi}\left[-r^2\sqrt{a^2-r^2}-\tfrac{2}{3}\left(a^2-r^2\right)^{3/2}\right]_0^a d\theta = a\delta\int_0^{2\pi}\tfrac{2}{3}a^3\, d\theta = \tfrac{4\pi}{3}a^4\delta \Rightarrow \text{the moment of inertia is } \tfrac{8\pi}{3}a^4\delta \text{ for}$$

the whole sphere

(b) $I_L = I_{c.m.} + mh^2$, where m is the mass of the body and h is the distance between the parallel lines; now,

$$I_{c.m.} = \tfrac{8\pi}{3}a^4\delta \text{ (from part a) and } \tfrac{m}{2} = \iint_S \delta\, d\sigma = \delta\iint_R \left(\tfrac{a}{z}\right) dA = a\delta\iint_R \frac{1}{\sqrt{a^2-(x^2+y^2)}}\, dy\, dx$$

$$= a\delta\int_0^{2\pi}\int_0^a \frac{1}{\sqrt{a^2-r^2}}\, r\, dr\, d\theta = a\delta\int_0^{2\pi}\left[-\sqrt{a^2-r^2}\right]_0^a d\theta = a\delta\int_0^{2\pi} a\, d\theta = 2\pi a^2\delta \text{ and } h = a$$

$$\Rightarrow I_L = \tfrac{8\pi}{3}a^4\delta + 4\pi a^2\delta a^2 = \tfrac{20\pi}{3}a^4\delta$$

38. (a) Let $z = \frac{h}{a}\sqrt{x^2+y^2}$ be the cone from $z = 0$ to $z = h$, $h > 0$. Because of symmetry, $\overline{x} = 0$ and $\overline{y} = 0$;

$$z = \tfrac{h}{a}\sqrt{x^2+y^2} \Rightarrow f(x,y,z) = \tfrac{h^2}{a^2}(x^2+y^2)-z^2 = 0 \Rightarrow \nabla f = \frac{2xh^2}{a^2}\mathbf{i}+\frac{2yh^2}{a^2}\mathbf{j}-2z\mathbf{k}$$

$\Rightarrow |\nabla f| = \sqrt{\frac{4x^2h^4}{a^4} + \frac{4y^2h^4}{a^4} + 4z^2} = 2\sqrt{\frac{h^4}{a^4}(x^2+y^2) + \frac{h^2}{a^2}(x^2+y^2)} = 2\sqrt{\left(\frac{h^2}{a^2}\right)(x^2+y^2)\left(\frac{h^2}{a^2}+1\right)}$

$= 2\sqrt{z^2\left(\frac{h^2+a^2}{a^2}\right)} = \left(\frac{2z}{a}\right)\sqrt{h^2+a^2}$ since $z \ge 0$; $\mathbf{p} = \mathbf{k} \Rightarrow |\nabla f \cdot \mathbf{p}| = 2z \Rightarrow d\sigma = \frac{\left(\frac{2z}{a}\right)\sqrt{h^2+a^2}}{2z}\,dA$

$= \frac{\sqrt{h^2+a^2}}{a}\,dA$; $M = \iint\limits_S d\sigma = \iint\limits_R \frac{\sqrt{h^2+a^2}}{a}\,dA = \frac{\sqrt{h^2+a^2}}{a}(\pi a^2) = \pi a\sqrt{h^2+a^2}$;

$M_{xy} = \iint\limits_S z\delta\,d\sigma = \iint\limits_R z\left(\frac{\sqrt{h^2+a^2}}{a}\right)dA = \frac{\sqrt{h^2+a^2}}{a}\iint\limits_R \frac{h}{a}\sqrt{x^2+y^2}\,dx\,dy = \frac{h\sqrt{h^2+a^2}}{a^2}\int_0^{2\pi}\int_0^a r^2\,dr\,d\theta$

$= \frac{2\pi ah\sqrt{h^2+a^2}}{3} \Rightarrow \bar{z} = \frac{M_{xy}}{M} = \frac{2h}{3} \Rightarrow$ the centroid is $\left(0,0,\frac{2h}{3}\right)$

(b) The base is a circle of radius a and center at $(0,0,h) \Rightarrow (0,0,h)$ is the centroid of the base and the mass is $M = \iint\limits_S d\sigma = \pi a^2$. In Pappus' formula, let $\mathbf{c}_1 = \frac{2h}{3}\mathbf{k}$, $\mathbf{c}_2 = h\mathbf{k}$, $m_1 = \pi a\sqrt{h^2+a^2}$, and $m_2 = \pi a^2$

$\Rightarrow \mathbf{c} = \frac{\pi a\sqrt{h^2+a^2}\left(\frac{2h}{3}\right)\mathbf{k} + \pi a^2 h\mathbf{k}}{\pi a\sqrt{h^2+a^2} + \pi a^2} = \frac{2h\sqrt{h^2+a^2}+3ah}{3\left(\sqrt{h^2+a^2}+a\right)}\mathbf{k} \Rightarrow$ the centroid is $\left(0,0,\frac{2h\sqrt{h^2+a^2}+3ah}{3\left(\sqrt{h^2+a^2}+a\right)}\right)$

(c) If the hemisphere is sitting so its base is in the plane $z = h$, then its centroid is $\left(0,0,h+\frac{a}{2}\right)$ and its mass is $2\pi a^2$. In Pappus' formula, let $\mathbf{c}_1 = \frac{2h}{3}\mathbf{k}$, $\mathbf{c}_2 = \left(h+\frac{a}{2}\right)\mathbf{k}$, $m_1 = \pi a\sqrt{h^2+a^2}$, and $m_2 = 2\pi a^2$

$\Rightarrow \mathbf{c} = \frac{\pi a\sqrt{h^2+a^2}\left(\frac{2h}{3}\right)\mathbf{k} + 2\pi a^2\left(h+\frac{a}{2}\right)\mathbf{k}}{\pi a\sqrt{h^2+a^2}+2\pi a^2} = \frac{2h\sqrt{h^2+a^2}+6ah+3a^2}{3\left(\sqrt{h^2+a^2}+2a\right)}\mathbf{k} \Rightarrow$ the centroid is

$\left(0,0,\frac{2h\sqrt{h^2+a^2}+6ah+3a^2}{3\left(\sqrt{h^2+a^2}+2a\right)}\right)$. Thus, for the centroid to be in the plane of the bases we must have $z = h$

$\Rightarrow \frac{2h\sqrt{h^2+a^2}+6ah+3a^2}{3\left(\sqrt{h^2+a^2}+2a\right)} = h \Rightarrow 2h\sqrt{h^2+a^2}+6ah+3a^2 = 3h\sqrt{h^2+a^2}+6ah \Rightarrow 3a^2 = h\sqrt{h^2+a^2}$

$\Rightarrow 9a^4 = h^2(h^2+a^2) \Rightarrow h^4 + a^2h^2 - 9a^4 = 0 \Rightarrow h^2 = \frac{(\sqrt{37}-1)a^2}{2}$ (the positive root) $\Rightarrow h = \frac{\sqrt{2\sqrt{37}-2}}{2}\,a$

39. $f_x(x,y) = 2x$, $f_y(x,y) = 2y \Rightarrow \sqrt{f_x^2+f_y^2+1} = \sqrt{4x^2+4y^2+1} \Rightarrow \text{Area} = \iint\limits_R \sqrt{4x^2+4y^2+1}\,dx\,dy$

$= \int_0^{2\pi}\int_0^{\sqrt{3}} \sqrt{4r^2+1}\,r\,dr\,d\theta = \frac{\pi}{6}\left(13\sqrt{13}-1\right)$

40. $f_y(y,z) = -2y,\ f_z(y,z) = -2z \Rightarrow \sqrt{f_y^2 + f_z^2 + 1} = \sqrt{4y^2 + 4z^2 + 1} \Rightarrow \text{Area} = \iint_R \sqrt{4y^2 + 4z^2 + 1}\, dy\, dz$

$= \int_0^{2\pi} \int_0^1 \sqrt{4r^2 + 1}\ r\, dr\, d\theta = \frac{\pi}{6}\left(5\sqrt{5} - 1\right)$

41. $f_x(x,y) = \frac{x}{\sqrt{x^2 + y^2}},\ f_y(x,y) = \frac{y}{\sqrt{x^2 + y^2}} \Rightarrow \sqrt{f_x^2 + f_y^2 + 1} = \sqrt{\frac{x^2}{x^2 + y^2} + \frac{y^2}{x^2 + y^2} + 1} = \sqrt{2}$

$\Rightarrow \text{Area} = \iint_{R_{xy}} \sqrt{2}\, dx\, dy = \sqrt{2}(\text{Area between the ellipse and the circle}) = \sqrt{2}(6\pi - \pi) = 5\pi\sqrt{2}$

42. Over R_{xy}: $z = 2 - \frac{2}{3}x - 2y \Rightarrow f_x(x,y) = -\frac{2}{3},\ f_y(x,y) = -2 \Rightarrow \sqrt{f_x^2 + f_y^2 + 1} = \sqrt{\frac{4}{9} + 4 + 1} = \frac{7}{3}$

$\Rightarrow \text{Area} = \iint_{R_{xy}} \frac{7}{3}\, dA = \frac{7}{3}(\text{Area of the shadow triangle in the xy-plane}) = \left(\frac{7}{3}\right)\left(\frac{1}{2}\right) = \frac{7}{2}.$

Over R_{xz}: $y = 1 - \frac{1}{3}x - \frac{1}{2}z \Rightarrow f_x(x,z) = -\frac{1}{3},\ f_z(x,z) = -\frac{1}{2} \Rightarrow \sqrt{f_x^2 + f_z^2 + 1} = \sqrt{\frac{1}{9} + \frac{1}{4} + 1} = \frac{7}{6}$

$\Rightarrow \text{Area} = \iint_{R_{xz}} \frac{7}{6}\, dA = \frac{7}{6}(\text{Area of the shadow triangle in the xz-plane}) = \left(\frac{7}{6}\right)(3) = \frac{7}{2}.$

Over R_{yz}: $x = 3 - 3y - \frac{3}{2}z \Rightarrow f_y(y,z) = -3,\ f_z(y,z) = -\frac{3}{2} \Rightarrow \sqrt{f_y^2 + f_z^2 + 1} = \sqrt{9 + \frac{9}{4} + 1} = \frac{7}{2}$

$\Rightarrow \text{Area} = \iint_{R_{yz}} \frac{7}{2}\, dA = \frac{7}{2}(\text{Area of the shadow triangle in the yz-plane}) = \left(\frac{7}{2}\right)(1) = \frac{7}{2}.$

43. $y = \frac{2}{3}z^{3/2} \Rightarrow f_x(x,z) = 0,\ f_z(x,z) = z^{1/2} \Rightarrow \sqrt{f_x^2 + f_z^2 + 1} = \sqrt{z + 1};\ y = \frac{16}{3} \Rightarrow \frac{16}{3} = \frac{2}{3}z^{3/2} \Rightarrow z = 4$

$\Rightarrow \text{Area} = \int_0^4 \int_0^1 \sqrt{z + 1}\, dx\, dz = \int_0^4 \sqrt{z + 1}\, dz = \frac{2}{3}\left(5\sqrt{5} - 1\right)$

44. $y = 4 - z \Rightarrow f_x(x,z) = 0,\ f_z(x,z) = -1 \Rightarrow \sqrt{f_x^2 + f_z^2 + 1} = \sqrt{2} \Rightarrow \text{Area} = \iint_{R_{xz}} \sqrt{2}\, dA = \int_0^2 \int_0^{4-z^2} \sqrt{2}\, dx\, dz$

$= \sqrt{2} \int_0^2 \left(4 - z^2\right) dz = \frac{16\sqrt{2}}{3}$

14.6 PARAMETRIZED SURFACES

1. In cylindrical coordinates, let $x = r\cos\theta$, $y = r\sin\theta$, $z = \left(\sqrt{x^2 + y^2}\right)^2 = r^2$. Then $\mathbf{r}(r,\theta) = (r\cos\theta)\mathbf{i} + (r\sin\theta)\mathbf{j} + r^2\mathbf{k}$, $0 \le r \le 2$, $0 \le \theta \le 2\pi$.

2. In cylindrical coordinates, let $x = r \cos \theta$, $y = r \sin \theta$, $z = 9 - x^2 - y^2 = 9 - r^2$. Then $\mathbf{r}(r,\theta) = (r \cos \theta)\mathbf{i} + (r \sin \theta)\mathbf{j} + (9 - r^2)\mathbf{k}$; $z \ge 0 \Rightarrow 9 - r^2 \ge 0 \Rightarrow r^2 \le 9 \Rightarrow -3 \le r \le 3$, $0 \le \theta \le 2\pi$. But $-3 \le r \le 0$ gives the same points as $0 \le r \le 3$, so let $0 \le r \le 3$.

3. In cylindrical coordinates, let $x = r \cos \theta$, $y = r \sin \theta$, $z = \frac{\sqrt{x^2+y^2}}{2} \Rightarrow z = \frac{r}{2}$. Then $\mathbf{r}(r,\theta) = (r \cos \theta)\mathbf{i} + (r \sin \theta)\mathbf{j} + \left(\frac{r}{2}\right)\mathbf{k}$. For $0 \le z \le 3$, $0 \le \frac{r}{2} \le 3 \Rightarrow 0 \le r \le 6$; to get only the first octant, let $0 \le \theta \le \frac{\pi}{2}$.

4. In cylindrical coordinates, let $x = r \cos \theta$, $y = r \sin \theta$, $z = 2\sqrt{x^2+y^2} \Rightarrow z = 2r$. Then $\mathbf{r}(r,\theta) = (r \cos \theta)\mathbf{i} + (r \sin \theta)\mathbf{j} + 2r\mathbf{k}$. For $2 \le z \le 4$, $2 \le 2r \le 4 \Rightarrow 1 \le r \le 2$, and let $0 \le \theta \le 2\pi$.

5. In cylindrical coordinates, let $x = r \cos \theta$, $y = r \sin \theta$ since $x^2 + y^2 = 9 \Rightarrow z^2 = 9 - (x^2+y^2) = 9 - r^2$ $\Rightarrow z = \sqrt{9 - r^2}$, $z \ge 0$. Then $\mathbf{r}(r,\theta) = (r \cos \theta)\mathbf{i} + (r \sin \theta)\mathbf{j} + \sqrt{9 - r^2}\mathbf{k}$. Let $0 \le \theta \le 2\pi$. For the domain of r: $z = \sqrt{x^2+y^2}$ and $x^2 + y^2 + z^2 = 9 \Rightarrow x^2 + y^2 + \left(\sqrt{x^2+y^2}\right)^2 = 9 \Rightarrow 2(x^2+y^2) = 9 \Rightarrow 2r^2 = 9$ $\Rightarrow r = \frac{3}{\sqrt{2}} \Rightarrow 0 \le r \le \frac{3}{\sqrt{2}}$.

6. In cylindrical coordinates, $\mathbf{r}(r,\theta) = (r \cos \theta)\mathbf{i} + (r \sin \theta)\mathbf{j} + \sqrt{4 - r^2}\,\mathbf{k}$ (see Exercise 5 above with $x^2 + y^2 + z^2 = 4$, instead of $x^2 + y^2 + z^2 = 9$). For the first octant, let $0 \le \theta \le \frac{\pi}{2}$. For the domain of r: $z = \sqrt{x^2+y^2}$ and $x^2 + y^2 + z^2 = 4 \Rightarrow x^2 + y^2 + \left(\sqrt{x^2+y^2}\right)^2 = 4 \Rightarrow 2(x^2+y^2) = 4 \Rightarrow 2r^2 = 4 \Rightarrow r = \sqrt{2}$. Thus, let $\sqrt{2} \le r \le 2$ (to get the portion of the sphere between the cone and the xy-plane).

7. In spherical coordinates, $x = \rho \sin \phi \cos \theta$, $y = \rho \sin \phi \sin \theta$, $\rho = \sqrt{x^2+y^2+z^2} \Rightarrow \rho^2 = 3 \Rightarrow \rho = \sqrt{3}$ $\Rightarrow z = \sqrt{3} \cos \phi$ for the sphere; $z = \frac{\sqrt{3}}{2} = \sqrt{3} \cos \phi \Rightarrow \cos \phi = \frac{1}{2} \Rightarrow \phi = \frac{\pi}{3}$; $z = -\frac{\sqrt{3}}{2} \Rightarrow -\frac{\sqrt{3}}{2} = \sqrt{3} \cos \phi$ $\Rightarrow \cos \phi = -\frac{1}{2} \Rightarrow \phi = \frac{2\pi}{3}$. Then $\mathbf{r}(r,\theta) = \left(\sqrt{3} \sin \phi \cos \theta\right)\mathbf{i} + \left(\sqrt{3} \sin \phi \sin \theta\right)\mathbf{j} + \left(\sqrt{3} \cos \phi\right)\mathbf{k}$, $\frac{\pi}{3} \le \phi \le \frac{2\pi}{3}$ and $0 \le \theta \le 2\pi$.

8. In spherical coordinates, $x = \rho \sin \phi \cos \theta$, $y = \rho \sin \phi \sin \theta$, $\rho = \sqrt{x^2+y^2+z^2} \Rightarrow \rho^2 = 8 \Rightarrow \rho = \sqrt{8} = 2\sqrt{2}$ $\Rightarrow x = 2\sqrt{2} \sin \phi \cos \theta$, $y = 2\sqrt{2} \sin \phi \sin \theta$, and $z = 2\sqrt{2} \cos \phi$. Thus let $\mathbf{r}(r,\theta) = \left(2\sqrt{2} \sin \phi \cos \theta\right)\mathbf{i} + \left(2\sqrt{2} \sin \phi \sin \theta\right)\mathbf{j} + \left(2\sqrt{2} \cos \phi\right)\mathbf{k}$; $z = -2 \Rightarrow -2 = 2\sqrt{2} \cos \phi$ $\Rightarrow \cos \phi = -\frac{1}{\sqrt{2}} \Rightarrow \phi = \frac{3\pi}{4}$; $z = 2\sqrt{2} \Rightarrow 2\sqrt{2} = 2\sqrt{2} \cos \phi \Rightarrow \cos \phi = 1 \Rightarrow \phi = 0$. Thus $0 \le \phi \le \frac{3\pi}{4}$ and $0 \le \theta \le 2\pi$.

9. Since $z = 4 - y^2$, we can let $\mathbf{r}$ be a function of x and y $\Rightarrow \mathbf{r}(x,y) = x\mathbf{i} + y\mathbf{j} + (4 - y^2)\mathbf{k}$. Then $z = 0$ $\Rightarrow 0 = 4 - y^2 \Rightarrow y = \pm 2$. Thus, let $-2 \le y \le 2$ and $0 \le x \le 2$.

10. Since $y = x^2$, we can let $\mathbf{r}$ be a function of x and z $\Rightarrow \mathbf{r}(x,z) = x\mathbf{i} + x^2\mathbf{j} + z\mathbf{k}$. Then $y = 2$ $\Rightarrow x^2 = 2 \Rightarrow x = \pm\sqrt{2}$. Thus, let $-\sqrt{2} \le x \le \sqrt{2}$ and $0 \le z \le 3$.

11. When $x = 0$, let $y^2 + z^2 = 9$ be the circular section in the yz-plane. Use polar coordinates in the yz-plane $\Rightarrow y = 3\cos\theta$ and $z = 3\sin\theta$. Thus let $x = u$ and $\theta = v \Rightarrow \mathbf{r}(u,v) = x\mathbf{i} + (3\cos v)\mathbf{j} + (3\sin v)\mathbf{k}$ where $0 \le u \le 3$, and $0 \le v \le 2\pi$.

12. When $y = 0$, let $x^2 + z^2 = 4$ be the circular section in the xz-plane. Use polar coordinates in the xz-plane $\Rightarrow x = 2\cos\theta$ and $z = 2\sin\theta$. Thus let $y = u$ and $\theta = v \Rightarrow \mathbf{r}(u,v) = (2\cos v)\mathbf{i} + u\mathbf{j} + (3\sin v)\mathbf{k}$ where $-2 \le u \le 2$, and $0 \le v \le \pi$ (since we want the portion above the xy-plane).

13. (a) $x + y + z = 1 \Rightarrow z = 1 - x - y$. In cylindrical coordinates, let $x = r\cos\theta$ and $y = r\sin\theta$ $\Rightarrow z = 1 - r\cos\theta - r\sin\theta \Rightarrow \mathbf{r}(r,\theta) = (r\cos\theta)\mathbf{i} + (r\sin\theta)\mathbf{j} + (1 - r\cos\theta - r\sin\theta)\mathbf{k}$, $0 \le \theta \le 2\pi$ and $0 \le r \le 3$.

    (b) In a fashion similar to cylindrical coordinates, but working in the yz-plane instead of the xy-plane, let $y = u\cos v$, $z = u\sin v$ where $u = \sqrt{y^2 + z^2}$ and v is the angle formed by (x,y,z), $(x,0,0)$, and $(x,y,0)$ with $(x,0,0)$ as vertex. Since $x + y + z = 1 \Rightarrow x = 1 - y - z \Rightarrow x = 1 - u\cos v - u\sin v$, then $\mathbf{r}$ is a function of u and v $\Rightarrow \mathbf{r}(u,v) = (1 - u\cos v - u\sin v)\mathbf{i} + (u\cos v)\mathbf{j} + (u\sin v)\mathbf{k}$, $0 \le u \le 3$ and $0 \le v \le 2\pi$.

14. (a) In a fashion similar to cylindrical coordinates, but working in the xz-plane instead of the xy-plane, let $x = u\cos v$, $z = u\sin v$ where $u = \sqrt{x^2 + z^2}$ and v is the angle formed by (x,y,z), $(y,0,0)$, and $(x,y,0)$ with vertex $(y,0,0)$. Since $x - y + 2z = 2 \Rightarrow y = x + 2z - 2$, then $\mathbf{r}(u,v)$ $= (u\cos v)\mathbf{i} + (u\cos v + 2u\sin v - 2)\mathbf{j} + (u\sin v)\mathbf{k}$, $0 \le u \le \sqrt{3}$ and $0 \le v \le 2\pi$.

    (b) In a fashion similar to cylindrical coordinates, but working in the yz-plane instead of the xy-plane, let $y = u\cos v$, $z = u\sin v$ where $u = \sqrt{y^2 + z^2}$ and v is the angle formed by (x,y,z), $(x,0,0)$, and $(x,y,0)$ with vertex $(x,0,0)$. Since $x - y + 2z = 2 \Rightarrow x = y - 2z + 2$, then $\mathbf{r}(u,v)$ $= (u\cos v - 2u\sin v + 2)\mathbf{i} + (u\cos v)\mathbf{j} + (u\sin v)\mathbf{k}$, $0 \le u \le \sqrt{2}$ and $0 \le v \le 2\pi$.

15. Let $x = w\cos v$ and $z = w\sin v$. Then $(x-2)^2 + z^2 = 4 \Rightarrow x^2 - 4x + z^2 = 0 \Rightarrow w^2\cos^2 v - 4w\cos v + w^2\sin^2 v$ $= 0 \Rightarrow w^2 - 4w\cos v = 0 \Rightarrow w = 0$ or $w - 4\cos v = 0 \Rightarrow w = 0$ or $w = 4\cos v$. Now $w = 0 \Rightarrow x = 0$ and $y = 0$, which is a line not a cylinder. Therefore, let $w = 4\cos v \Rightarrow x = (4\cos v)(\cos v) = 4\cos^2 v$ and $z = 4\cos v\sin v$. Finally, let $y = u$. Then $\mathbf{r}(u,v) = (4\cos^2 v)\mathbf{i} + u\mathbf{j} + (4\cos v\sin v)\mathbf{k}$, $-\frac{\pi}{2} \le v \le \frac{\pi}{2}$ and $0 \le u \le 3$.

16. Let $y = w\cos v$ and $z = w\sin v$. Then $y^2 + (z-5)^2 = 25 \Rightarrow y^2 + z^2 - 10z = 0$ $\Rightarrow w^2\cos^2 v + w^2\sin^2 v - 10w\sin v = 0 \Rightarrow w^2 - 10w\sin v = 0 \Rightarrow w(w - 10\sin v) = 0 \Rightarrow w = 0$ or $w = 10\sin v$. Now $w = 0 \Rightarrow y = 0$ and $z = 0$, which is a line not a cylinder. Therefore, let $w = 10\sin v$

$\Rightarrow y = 10 \sin v \cos v$ and $z = 10 \sin^2 v$. Finally, let $x = u$. Then $\mathbf{r}(u,v) = u\mathbf{i} + (10 \sin v \cos v)\mathbf{j} + (10 \sin^2 v)\mathbf{k}$, $0 \le u \le 10$ and $0 \le v \le \pi$.

17. Let $x = r \cos \theta$ and $y = r \sin \theta$. Then $\mathbf{r}(r,\theta) = (r \cos \theta)\mathbf{i} + (r \sin \theta)\mathbf{j} + \left(\frac{2 - r \sin \theta}{2}\right)\mathbf{k}$, $0 \le r \le 1$ and $0 \le \theta \le 2\pi$
$\Rightarrow \mathbf{r}_r = (\cos \theta)\mathbf{i} + (\sin \theta)\mathbf{j} - \left(\frac{\sin \theta}{2}\right)\mathbf{k}$ and $\mathbf{r}_\theta = (-r \sin \theta)\mathbf{i} + (r \cos \theta)\mathbf{j} - \left(\frac{r \cos \theta}{2}\right)\mathbf{k}$

$$\Rightarrow \mathbf{r}_r \times \mathbf{r}_\theta = \begin{vmatrix} \mathbf{i} & \mathbf{j} & \mathbf{k} \\ \cos \theta & \sin \theta & -\frac{\sin \theta}{2} \\ -r \sin \theta & r \cos \theta & -\frac{r \cos \theta}{2} \end{vmatrix}$$

$$= \left(\frac{-r \sin \theta \cos \theta}{2} + \frac{(\sin \theta)(r \cos \theta)}{2}\right)\mathbf{i} + \left(\frac{r \sin^2 \theta}{2} + \frac{r \cos^2 \theta}{2}\right)\mathbf{j} + (r \cos^2 \theta + r \sin^2 \theta)\mathbf{k} = \frac{r}{2}\mathbf{j} + r\mathbf{k}$$

$$\Rightarrow |\mathbf{r}_r \times \mathbf{r}_\theta| = \sqrt{\frac{r^2}{4} + r^2} = \frac{\sqrt{5}\,r}{2} \Rightarrow A = \int_0^{2\pi}\int_0^1 \frac{\sqrt{5}\,r}{2}\,dr\,d\theta = \int_0^{2\pi}\left[\frac{\sqrt{5}\,r^2}{4}\right]_0^1 d\theta = \int_0^{2\pi}\frac{\sqrt{5}}{4}\,d\theta = \frac{\pi\sqrt{5}}{2}$$

18. Let $x = r \cos \theta$ and $y = r \sin \theta \Rightarrow z = -x = -r \cos \theta$, $0 \le r \le 2$ and $0 \le \theta \le 2\pi$. Then $\mathbf{r}(r,\theta) = (r \cos \theta)\mathbf{i} + (r \sin \theta)\mathbf{j} - (r \cos \theta)\mathbf{k} \Rightarrow \mathbf{r}_r = (\cos \theta)\mathbf{i} + (\sin \theta)\mathbf{j} - (\cos \theta)\mathbf{k}$ and $\mathbf{r}_\theta = (-r \sin \theta)\mathbf{i} + (r \cos \theta)\mathbf{j} + (r \sin \theta)\mathbf{k}$

$$\Rightarrow \mathbf{r}_r \times \mathbf{r}_\theta = \begin{vmatrix} \mathbf{i} & \mathbf{j} & \mathbf{k} \\ \cos \theta & \sin \theta & -\cos \theta \\ -r \sin \theta & r \cos \theta & r \sin \theta \end{vmatrix}$$

$$= (r \sin^2 \theta + r \cos^2 \theta)\mathbf{i} + (r \sin \theta \cos \theta - r \sin \theta \cos \theta)\mathbf{j} + (r \cos^2 \theta + r \sin^2 \theta)\mathbf{k} = r\mathbf{i} + r\mathbf{k}$$

$$\Rightarrow |\mathbf{r}_r \times \mathbf{r}_\theta| = \sqrt{r^2 + r^2} = r\sqrt{2} \Rightarrow A = \int_0^{2\pi}\int_0^2 r\sqrt{2}\,dr\,d\theta = \int_0^{2\pi}\left[\frac{r^2\sqrt{2}}{2}\right]_0^2 d\theta = \int_0^{2\pi} 2\sqrt{2}\,d\theta = 4\pi\sqrt{2}$$

19. Let $x = r \cos \theta$ and $y = r \sin \theta \Rightarrow z = 2\sqrt{x^2 + y^2} = 2r$, $1 \le r \le 3$ and $0 \le \theta \le 2\pi$. Then $\mathbf{r}(r,\theta) = (r \cos \theta)\mathbf{i} + (r \sin \theta)\mathbf{j} + 2r\mathbf{k} \Rightarrow \mathbf{r}_r = (\cos \theta)\mathbf{i} + (\sin \theta)\mathbf{j} + 2\mathbf{k}$ and $\mathbf{r}_\theta = (-r \sin \theta)\mathbf{i} + (r \cos \theta)\mathbf{j}$

$$\Rightarrow \mathbf{r}_r \times \mathbf{r}_\theta = \begin{vmatrix} \mathbf{i} & \mathbf{j} & \mathbf{k} \\ \cos \theta & \sin \theta & 2 \\ -r \sin \theta & r \cos \theta & 0 \end{vmatrix} = (-2r \cos \theta)\mathbf{i} - (2r \sin \theta)\mathbf{j} + (r \cos^2 \theta + r \sin^2 \theta)\mathbf{k}$$

$$= (-2r \cos \theta)\mathbf{i} - (2r \sin \theta)\mathbf{j} + r\mathbf{k} \Rightarrow |\mathbf{r}_r \times \mathbf{r}_\theta| = \sqrt{4r^2 \cos^2 \theta + 4r^2 \sin^2 \theta + r^2} = \sqrt{5r^2} = r\sqrt{5}$$

$$\Rightarrow A = \int_0^{2\pi}\int_1^3 r\sqrt{5}\,dr\,d\theta = \int_0^{2\pi}\left[\frac{r^2\sqrt{5}}{2}\right]_1^3 d\theta = \int_0^{2\pi} 4\sqrt{5}\,d\theta = 8\pi\sqrt{5}$$

20. Let $x = r\cos\theta$ and $y = r\sin\theta \Rightarrow z = \frac{\sqrt{x^2+y^2}}{3} = \frac{r}{3}$, $3 \le r \le 4$ and $0 \le \theta \le 2\pi$. Then

$\mathbf{r}(r,\theta) = (r\cos\theta)\mathbf{i} + (r\sin\theta)\mathbf{j} + \left(\frac{r}{3}\right)\mathbf{k} \Rightarrow \mathbf{r}_r = (\cos\theta)\mathbf{i} + (\sin\theta)\mathbf{j} + \left(\frac{1}{3}\right)\mathbf{k}$ and $\mathbf{r}_\theta = (-r\sin\theta)\mathbf{i} + (r\cos\theta)\mathbf{j}$

$$\Rightarrow \mathbf{r}_r \times \mathbf{r}_\theta = \begin{vmatrix} \mathbf{i} & \mathbf{j} & \mathbf{k} \\ \cos\theta & \sin\theta & \frac{1}{3} \\ -r\sin\theta & r\cos\theta & 0 \end{vmatrix} = \left(-\frac{1}{3}r\cos\theta\right)\mathbf{i} - \left(\frac{1}{3}r\sin\theta\right)\mathbf{j} + \left(r\cos^2\theta + r\sin^2\theta\right)\mathbf{k}$$

$$= \left(-\frac{1}{3}r\cos\theta\right)\mathbf{i} - \left(\frac{1}{3}r\sin\theta\right)\mathbf{j} + r\mathbf{k} \Rightarrow |\mathbf{r}_r \times \mathbf{r}_\theta| = \sqrt{\frac{1}{9}r^2\cos^2\theta + \frac{1}{9}r^2\sin^2\theta + r^2} = \sqrt{\frac{10r^2}{9}} = \frac{r\sqrt{10}}{3}$$

$$\Rightarrow A = \int_0^{2\pi}\int_3^4 \frac{r\sqrt{10}}{3}\,dr\,d\theta = \int_0^{2\pi}\left[\frac{r^2\sqrt{10}}{6}\right]_3^4 d\theta = \int_0^{2\pi} \frac{7\sqrt{10}}{6}\,d\theta = \frac{7\pi\sqrt{10}}{3}$$

21. Let $x = r\cos\theta$ and $y = r\sin\theta \Rightarrow r^2 = x^2 + y^2 = 1$, $1 \le z \le 4$ and $0 \le \theta \le 2\pi$. Then

$\mathbf{r}(z,\theta) = (\cos\theta)\mathbf{i} + (\sin\theta)\mathbf{j} + z\mathbf{k} \Rightarrow \mathbf{r}_z = \mathbf{k}$ and $\mathbf{r}_\theta = (-\sin\theta)\mathbf{i} + (\cos\theta)\mathbf{j}$

$$\Rightarrow \mathbf{r}_\theta \times \mathbf{r}_z = \begin{vmatrix} \mathbf{i} & \mathbf{j} & \mathbf{k} \\ -\sin\theta & \cos\theta & 0 \\ 0 & 0 & 1 \end{vmatrix} = (\cos\theta)\mathbf{i} + (\sin\theta)\,\mathbf{j} = |\mathbf{r}_\theta \times \mathbf{r}_z| = \sqrt{\cos^2\theta + \sin^2\theta} = 1$$

$$\Rightarrow A = \int_0^{2\pi}\int_1^4 1\,dr\,d\theta = \int_0^{2\pi} 3\,d\theta = 6\pi$$

22. Let $x = u\cos v$ and $z = u\sin v \Rightarrow u^2 = x^2 + y^2 = 10$, $-1 \le y \le 1$, $0 \le v \le 2\pi$. Then

$\mathbf{r}(y,v) = (u\cos v)\mathbf{i} + y\mathbf{j} + (u\sin v)\mathbf{k} = \left(\sqrt{10}\cos v\right)\mathbf{i} + y\mathbf{j} + \left(\sqrt{10}\sin v\right)\mathbf{k}$

$$\Rightarrow \mathbf{r}_v = \left(-\sqrt{10}\sin v\right)\mathbf{i} + \left(\sqrt{10}\cos v\right)\mathbf{k} \text{ and } \mathbf{r}_y = \mathbf{j} \Rightarrow \mathbf{r}_v \times \mathbf{r}_y = \begin{vmatrix} \mathbf{i} & \mathbf{j} & \mathbf{k} \\ -\sqrt{10}\sin v & 0 & \sqrt{10}\cos v \\ 0 & 1 & 0 \end{vmatrix}$$

$$= \left(-\sqrt{10}\cos v\right)\mathbf{i} - \left(\sqrt{10}\sin v\right)\mathbf{k} = |\mathbf{r}_v \times \mathbf{r}_y| = \sqrt{10} \Rightarrow A = \int_0^{2\pi}\int_{-1}^{1} \sqrt{10}\,du\,dv = \int_0^{2\pi} \left[\sqrt{10}u\right]_{-1}^{1} dv$$

$$= \int_0^{2\pi} 2\sqrt{10}\,dv = 4\pi\sqrt{10}$$

23. $z = 2 - x^2 - y^2$ and $z = \sqrt{x^2+y^2} \Rightarrow z = 2 - z^2 \Rightarrow z^2 + z - 2 = 0 \Rightarrow z = -2$ or $z = 1$. Since $z = \sqrt{x^2+y^2} \ge 0$, we get $z = 1$ where the cone intersects the paraboloid. When $x = 0$ and $y = 0$, $z = 2 \Rightarrow$ the vertex of the paraboloid is $(0,0,2)$. Therefore, z ranges from 1 to 2 on the "cap" $\Rightarrow$ r ranges from 1 (when $x^2 + y^2 = 1$) to 0 (when $x = 0$ and $y = 0$ at the vertex). Let $x = r\cos\theta$, $y = r\sin\theta$, and $z = 2 - r^2$. Then $\mathbf{r}(r,\theta) = (r\cos\theta)\mathbf{i} + (r\sin\theta)\mathbf{j} + \left(2 - r^2\right)\mathbf{k}$, $0 \le r \le 1$, $0 \le \theta \le 2\pi \Rightarrow \mathbf{r}_r = (\cos\theta)\mathbf{i} + (\sin\theta)\mathbf{j} - 2r\mathbf{k}$ and

$\mathbf{r}_\theta = (-r \sin\theta)\mathbf{i} + (r\cos\theta)\mathbf{j} \Rightarrow \mathbf{r}_r \times \mathbf{r}_\theta = \begin{vmatrix} \mathbf{i} & \mathbf{j} & \mathbf{k} \\ \cos\theta & \sin\theta & -2r \\ -r\sin\theta & r\cos\theta & 0 \end{vmatrix}$

$= (2r^2\cos\theta)\mathbf{i} + (2r^2\sin\theta)\mathbf{j} + r\mathbf{k} \Rightarrow |\mathbf{r}_r \times \mathbf{r}_\theta| = \sqrt{4r^4\cos^2\theta + 4r^4\sin^2\theta + r^2} = r\sqrt{4r^2+1}$

$\Rightarrow A = \int_0^{2\pi}\int_0^1 r\sqrt{4r^2+1}\,dr\,d\theta = \int_0^{2\pi}\left[\frac{1}{12}(4r^2+1)^{3/2}\right]_0^1 d\theta = \int_0^{2\pi}\left(\frac{5\sqrt{5}-1}{12}\right)d\theta = \frac{\pi}{6}(5\sqrt{5}-1)$

24. Let $x = r\cos\theta$, $y = r\sin\theta$ and $z = x^2+y^2 = r^2$. Then $\mathbf{r}(r,\theta) = (r\cos\theta)\mathbf{i} + (r\sin\theta)\mathbf{j} + r^2\mathbf{k}$, $1 \le r \le 2$, $0 \le \theta \le 2\pi \Rightarrow \mathbf{r}_r = (\cos\theta)\mathbf{i} + (\sin\theta)\mathbf{j} + 2r\mathbf{k}$ and $\mathbf{r}_\theta = (-r\sin\theta)\mathbf{i} + (r\cos\theta)\mathbf{j}$

$\Rightarrow \mathbf{r}_r \times \mathbf{r}_\theta = \begin{vmatrix} \mathbf{i} & \mathbf{j} & \mathbf{k} \\ \cos\theta & \sin\theta & 2r \\ -r\sin\theta & r\cos\theta & 0 \end{vmatrix} = (-2r^2\cos\theta)\mathbf{i} - (2r^2\sin\theta)\mathbf{j} + r\mathbf{k} \Rightarrow |\mathbf{r}_r \times \mathbf{r}_\theta|$

$= \sqrt{4r^4\cos^2\theta + 4r^4\sin^2\theta + r^2} = r\sqrt{4r^2+1} \Rightarrow A = \int_0^{2\pi}\int_1^2 r\sqrt{4r^2+1}\,dr\,d\theta = \int_0^{2\pi}\left[\frac{1}{12}(4r^2+1)^{3/2}\right]_1^2 d\theta$

$= \int_0^{2\pi}\left(\frac{17\sqrt{17}-5\sqrt{5}}{12}\right)d\theta = \frac{\pi}{6}(17\sqrt{17}-5\sqrt{5})$

25. Let $x = \rho\sin\phi\cos\theta$, $y = \rho\sin\phi\sin\theta$, and $z = \rho\cos\phi \Rightarrow \rho = \sqrt{x^2+y^2+z^2} = \sqrt{2}$ on the sphere. Next, $x^2+y^2+z^2 = 2$ and $z = \sqrt{x^2+y^2} \Rightarrow z^2+z^2 = 2 \Rightarrow z^2 = 1 \Rightarrow z = 1$ since $z \ge 0 \Rightarrow \phi = \frac{\pi}{4}$. For the lower portion of the sphere cut by the cone, we get $\phi = \pi$. Then

$\mathbf{r}(\phi,\theta) = (\sqrt{2}\sin\phi\cos\theta)\mathbf{i} + (\sqrt{2}\sin\phi\sin\theta)\mathbf{j} + (\sqrt{2}\cos\phi)\mathbf{k}$, $\frac{\pi}{4} \le \phi \le \pi$, $0 \le \theta \le 2\pi$

$\Rightarrow \mathbf{r}_\phi = (\sqrt{2}\cos\phi\cos\theta)\mathbf{i} + (\sqrt{2}\cos\phi\sin\theta)\mathbf{j} - (\sqrt{2}\sin\phi)\mathbf{k}$ and $\mathbf{r}_\theta = (-\sqrt{2}\sin\phi\sin\theta)\mathbf{i} + (\sqrt{2}\sin\phi\cos\theta)\mathbf{j}$

$\Rightarrow \mathbf{r}_\phi \times \mathbf{r}_\theta = \begin{vmatrix} \mathbf{i} & \mathbf{j} & \mathbf{k} \\ \sqrt{2}\cos\phi\cos\theta & \sqrt{2}\cos\phi\sin\theta & -\sqrt{2}\sin\phi \\ -\sqrt{2}\sin\phi\sin\theta & \sqrt{2}\sin\phi\cos\theta & 0 \end{vmatrix}$

$= (2\sin^2\phi\cos\theta)\mathbf{i} + (2\sin^2\phi\sin\theta)\mathbf{j} + (2\sin\phi\cos\phi)\mathbf{k}$

$\Rightarrow |\mathbf{r}_\phi \times \mathbf{r}_\theta| = \sqrt{4\sin^4\phi\cos^2\theta + 4\sin^4\phi\sin^2\theta + 4\sin^2\phi\cos^2\phi} = \sqrt{4\sin^2\phi} = 2|\sin\phi| = 2\sin\phi$

$\Rightarrow A = \int_0^{2\pi}\int_{\pi/4}^{\pi} 2\sin\phi\,d\phi\,d\theta = \int_0^{2\pi}(2+\sqrt{2})\,d\theta = (4+2\sqrt{2})\pi$

26. Let $x = \rho \sin\phi \cos\theta$, $y = \rho \sin\phi \sin\theta$, and $z = \rho \cos\phi \Rightarrow \rho = \sqrt{x^2+y^2+z^2} = 2$ on the sphere. Next, $z = -1 \Rightarrow -1 = 2\cos\phi \Rightarrow \cos\phi = -\frac{1}{2} \Rightarrow \phi = \frac{2\pi}{3}$; $z = \sqrt{3} \Rightarrow \sqrt{3} = 2\cos\phi \Rightarrow \cos\phi = \frac{\sqrt{3}}{2} \Rightarrow \phi = \frac{\pi}{6}$. Then $\mathbf{r}(\phi,\theta) = (2\sin\phi\cos\theta)\mathbf{i} + (2\sin\phi\sin\theta)\mathbf{j} + (2\cos\phi)\mathbf{k}$, $\frac{\pi}{6} \le \phi \le \frac{2\pi}{3}$, $0 \le \theta \le 2\pi$

$\Rightarrow \mathbf{r}_\phi = (2\cos\phi\cos\theta)\mathbf{i} + (2\cos\phi\sin\theta)\mathbf{j} - (2\sin\phi)\mathbf{k}$ and

$\mathbf{r}_\theta = (-2\sin\phi\sin\theta)\mathbf{i} + (2\sin\phi\cos\theta)\mathbf{j}$

$$\Rightarrow \mathbf{r}_\phi \times \mathbf{r}_\theta = \begin{vmatrix} \mathbf{i} & \mathbf{j} & \mathbf{k} \\ 2\cos\phi\cos\theta & 2\cos\phi\sin\theta & -2\sin\phi \\ -2\sin\phi\sin\theta & 2\sin\phi\cos\theta & 0 \end{vmatrix}$$

$$= (4\sin^2\phi\cos\theta)\mathbf{i} + (4\sin^2\phi\sin\theta)\mathbf{j} + (4\sin\phi\cos\phi)\mathbf{k}$$

$$\Rightarrow |\mathbf{r}_\phi \times \mathbf{r}_\theta| = \sqrt{16\sin^4\phi\cos^2\theta + 16\sin^4\phi\sin^2\theta + 16\sin^2\phi\cos^2\phi} = \sqrt{16\sin^2\phi} = 4|\sin\phi| = 4\sin\phi$$

$$\Rightarrow A = \int_0^{2\pi}\int_{\pi/6}^{2\pi/3} 4\sin\phi\, d\phi\, d\theta = \int_0^{2\pi} \left(2+2\sqrt{3}\right) d\theta = \left(4+4\sqrt{3}\right)\pi$$

27. Let the parametrization be $\mathbf{r}(x,z) = x\mathbf{i} + x^2\mathbf{j} + z\mathbf{k} \Rightarrow \mathbf{r}_x = \mathbf{i} + 2x\mathbf{j}$ and $\mathbf{r}_z = \mathbf{k} \Rightarrow \mathbf{r}_x \times \mathbf{r}_z = \begin{vmatrix} \mathbf{i} & \mathbf{j} & \mathbf{k} \\ 1 & 2x & 0 \\ 0 & 0 & 1 \end{vmatrix}$

$$= 2x\mathbf{i} + \mathbf{j} \Rightarrow |\mathbf{r}_x \times \mathbf{r}_z| = \sqrt{4x^2+1} \Rightarrow \iint_S G(x,y,z)\, d\sigma = \int_0^3\int_0^2 x\sqrt{4x^2+1}\, dx\, dz = \int_0^3 \left[\tfrac{1}{12}\left(4x^2+1\right)^{3/2}\right]_0^2 dz$$

$$= \int_0^3 \tfrac{1}{12}\left(17\sqrt{17}-1\right) dz = \frac{17\sqrt{17}-1}{4}$$

28. Let the parametrization be $\mathbf{r}(x,y) = x\mathbf{i} + y\mathbf{j} + \sqrt{4-y^2}\,\mathbf{k}$, $-2 \le y \le 2 \Rightarrow \mathbf{r}_x = \mathbf{i}$ and $\mathbf{r}_y = \mathbf{j} - \frac{y}{\sqrt{4-y^2}}\mathbf{k}$

$$\Rightarrow \mathbf{r}_x \times \mathbf{r}_y = \begin{vmatrix} \mathbf{i} & \mathbf{j} & \mathbf{k} \\ 1 & 0 & 0 \\ 0 & 1 & -\frac{y}{\sqrt{4-y^2}} \end{vmatrix} = \frac{y}{\sqrt{4-y^2}}\mathbf{j} + \mathbf{k} \Rightarrow |\mathbf{r}_x \times \mathbf{r}_y| = \sqrt{\frac{y^2}{4-y^2}+1} = \frac{2}{\sqrt{4-y^2}}$$

$$\Rightarrow \iint_S G(x,y,z)\, d\sigma = \int_1^4\int_{-2}^2 \sqrt{4-y^2}\left(\frac{2}{\sqrt{4-y^2}}\right) dy\, dx = 24$$

29. Let the parametrization be $\mathbf{r}(\phi,\theta) = (\sin\phi\cos\theta)\mathbf{i} + (\sin\phi\sin\theta)\mathbf{j} + (\cos\phi)\mathbf{k}$ (spherical coordinates with $\rho = 1$ on the sphere), $0 \le \phi \le \pi$, $0 \le \theta \le 2\pi \Rightarrow \mathbf{r}_\phi = (\cos\phi\cos\theta)\mathbf{i} + (\cos\phi\sin\theta)\mathbf{j} - (\sin\phi)\mathbf{k}$ and

$\mathbf{r}_\theta = (-\sin\phi\sin\theta)\mathbf{i} + (\sin\phi\cos\theta)\mathbf{j} \Rightarrow \mathbf{r}_\phi \times \mathbf{r}_\theta = \begin{vmatrix} \mathbf{i} & \mathbf{j} & \mathbf{k} \\ \cos\phi\cos\theta & \cos\phi\sin\theta & -\sin\phi \\ -\sin\phi\sin\theta & \sin\phi\cos\theta & 0 \end{vmatrix}$

$= (\sin^2\phi\cos\theta)\mathbf{i} + (\sin^2\phi\sin\theta)\mathbf{j} + (\sin\phi\cos\phi)\mathbf{k} \Rightarrow |\mathbf{r}_\phi \times \mathbf{r}_\theta| = \sqrt{\sin^4\phi\cos^2\theta + \sin^4\phi\sin^2\theta + \sin^2\phi\cos^2\phi}$

$= \sin\phi$; $x = \sin\phi\cos\theta \Rightarrow G(x,y,z) = \cos^2\theta\sin^2\phi \Rightarrow \iint_S G(x,y,z)\,d\sigma = \int_0^{2\pi}\int_0^{\pi} (\cos^2\theta\sin^2\phi)(\sin\phi)\,d\phi\,d\theta$

$= \int_0^{2\pi}\int_0^{\pi} (\cos^2\theta)(1-\cos^2\phi)(\sin\phi)\,d\phi\,d\theta; \begin{bmatrix} u = \cos\phi \\ du = -\sin\phi\,d\phi \end{bmatrix} \to \int_0^{2\pi}\int_{-1}^{1} (\cos^2\theta)(u^2-1)\,du\,d\theta$

$= \int_0^{2\pi} (\cos^2\theta)\left[\frac{u^3}{3} - u\right]_1^{-1} d\theta = \frac{4}{3}\int_0^{2\pi} \cos^2\theta\,d\theta = \frac{4}{3}\left[\frac{\theta}{2} + \frac{\sin 2\theta}{4}\right]_0^{2\pi} = \frac{4\pi}{3}$

30. Let the parametrization be $\mathbf{r}(\phi,\theta) = (a\sin\phi\cos\theta)\mathbf{i} + (a\sin\phi\sin\theta)\mathbf{j} + (a\cos\phi)\mathbf{k}$ (spherical coordinates with $\rho = a$, $a \ge 0$, on the sphere), $0 \le \phi \le \frac{\pi}{2}$ (since $z \ge 0$), $0 \le \theta \le 2\pi$
$\Rightarrow \mathbf{r}_\phi = (a\cos\phi\cos\theta)\mathbf{i} + (a\cos\phi\sin\theta)\mathbf{j} - (a\sin\phi)\mathbf{k}$ and

$\mathbf{r}_\theta = (-a\sin\phi\sin\theta)\mathbf{i} + (a\sin\phi\cos\theta)\mathbf{j} \Rightarrow \mathbf{r}_\phi \times \mathbf{r}_\theta = \begin{vmatrix} \mathbf{i} & \mathbf{j} & \mathbf{k} \\ a\cos\phi\cos\theta & a\cos\phi\sin\theta & -a\sin\phi \\ -a\sin\phi\sin\theta & a\sin\phi\cos\theta & 0 \end{vmatrix}$

$= (a^2\sin^2\phi\cos\theta)\mathbf{i} + (a^2\sin^2\phi\sin\theta)\mathbf{j} + (a^2\sin\phi\cos\phi)\mathbf{k}$

$\Rightarrow |\mathbf{r}_\phi \times \mathbf{r}_\theta| = \sqrt{a^4\sin^4\phi\cos^2\theta + a^4\sin^4\phi\sin^2\theta + a^4\sin^2\phi\cos^2\phi} = a^2\sin\phi$; $z = a\cos\phi$

$\Rightarrow G(x,y,z) = a^2\cos^2\phi \Rightarrow \iint_S G(x,y,z)\,d\sigma = \int_0^{2\pi}\int_0^{\pi/2} (a^2\cos^2\phi)(a^2\sin\phi)\,d\phi\,d\theta = \frac{2}{3}\pi a^4$

31. Let the parametrization be $\mathbf{r}(x,y) = x\mathbf{i} + y\mathbf{j} + (4-x-y)\mathbf{k} \Rightarrow \mathbf{r}_x = \mathbf{i} - \mathbf{k}$ and $\mathbf{r}_y = \mathbf{j} - \mathbf{k}$

$\Rightarrow \mathbf{r}_x \times \mathbf{r}_y = \begin{vmatrix} \mathbf{i} & \mathbf{j} & \mathbf{k} \\ 1 & 0 & -1 \\ 0 & 1 & -1 \end{vmatrix} = \mathbf{i} + \mathbf{j} + \mathbf{k} \Rightarrow |\mathbf{r}_x \times \mathbf{r}_y| = \sqrt{3} \Rightarrow \iint_S F(x,y,z)\,d\sigma = \int_0^1\int_0^1 (4-x-y)\sqrt{3}\,dy\,dx$

$= \int_0^1 \sqrt{3}\left[4y - xy - \frac{y^2}{2}\right]_0^1 dx = \int_0^1 \sqrt{3}\left(\frac{7}{2} - x\right) dx = \sqrt{3}\left[\frac{7}{2}x - \frac{x^2}{2}\right]_0^1 = 3\sqrt{3}$

32. Let the parametrization be $\mathbf{r}(r,\theta) = (r\cos\theta)\mathbf{i} + (r\sin\theta)\mathbf{j} + r\mathbf{k}$, $0 \le r \le 1$ (since $0 \le z \le 1$) and $0 \le \theta \le 2\pi$

$$\Rightarrow \mathbf{r}_r = (\cos\theta)\mathbf{i} + (\sin\theta)\mathbf{j} + \mathbf{k} \text{ and } \mathbf{r}_\theta = (-r\sin\theta)\mathbf{i} + (r\cos\theta)\mathbf{j} \Rightarrow \mathbf{r}_r \times \mathbf{r}_\theta = \begin{vmatrix} \mathbf{i} & \mathbf{j} & \mathbf{k} \\ \cos\theta & \sin\theta & 1 \\ -r\sin\theta & r\cos\theta & 0 \end{vmatrix}$$

$= (-r\cos\theta)\mathbf{i} - (r\sin\theta)\mathbf{j} + r\mathbf{k} \Rightarrow |\mathbf{r}_r \times \mathbf{r}_\theta| = \sqrt{(-r\cos\theta)^2 + (-r\sin\theta)^2 + r^2} = r\sqrt{2}$; $z = r$ and $x = r\cos\theta$

$$\Rightarrow F(x,y,z) = r - r\cos\theta \Rightarrow \iint_S F(x,y,z)\,d\sigma = \int_0^{2\pi}\int_0^1 (r - r\cos\theta)\left(r\sqrt{2}\right)dr\,d\theta = \sqrt{2}\int_0^{2\pi}\int_0^1 (1-\cos\theta)\,r^2\,dr\,d\theta$$

$$= \frac{2\pi\sqrt{2}}{3}$$

33. Let the parametrization be $\mathbf{r}(r,\theta) = (r\cos\theta)\mathbf{i} + (r\sin\theta)\mathbf{j} + \left(1-r^2\right)\mathbf{k}$, $0 \le r \le 1$ (since $0 \le z \le 1$) and $0 \le \theta \le 2\pi$

$$\Rightarrow \mathbf{r}_r = (\cos\theta)\mathbf{i} + (\sin\theta)\mathbf{j} - 2r\mathbf{k} \text{ and } \mathbf{r}_\theta = (-r\sin\theta)\mathbf{i} + (r\cos\theta)\mathbf{j} \Rightarrow \mathbf{r}_r \times \mathbf{r}_\theta = \begin{vmatrix} \mathbf{i} & \mathbf{j} & \mathbf{k} \\ \cos\theta & \sin\theta & -2r \\ -r\sin\theta & r\cos\theta & 0 \end{vmatrix}$$

$= \left(2r^2\cos\theta\right)\mathbf{i} + \left(2r^2\sin\theta\right)\mathbf{j} + r\mathbf{k} \Rightarrow |\mathbf{r}_r \times \mathbf{r}_\theta| = \sqrt{\left(2r^2\cos\theta\right)^2 + \left(2r^2\sin\theta\right) + r^2} = r\sqrt{1+4r^2}$; $z = 1 - r^2$ and

$x = r\cos\theta \Rightarrow H(x,y,z) = \left(r^2\cos^2\theta\right)\sqrt{1+4r^2} \Rightarrow \iint_S H(x,y,z)\,d\sigma$

$$= \int_0^{2\pi}\int_0^1 \left(r^2\cos^2\theta\right)\left(\sqrt{1+4r^2}\right)\left(r\sqrt{1+4r^2}\right)dr\,d\theta = \int_0^{2\pi}\int_0^1 r^3\left(1+4r^2\right)\cos^2\theta\,dr\,d\theta = \frac{11\pi}{12}$$

34. Let the parametrization be $\mathbf{r}(\phi,\theta) = (2\sin\phi\cos\theta)\mathbf{i} + (2\sin\phi\sin\theta)\mathbf{j} + (2\cos\phi)\mathbf{k}$ (spherical coordinates with $\rho = 2$ on the sphere), $0 \le \phi \le \frac{\pi}{4}$; $x^2 + y^2 = z^2 = 4$ and $z = \sqrt{x^2+y^2} \Rightarrow z^2 + z^2 = 4 \Rightarrow z^2 = 2 \Rightarrow z = \sqrt{2}$ (since $z \ge 0) \Rightarrow 2\cos\phi = \sqrt{2} \Rightarrow \cos\phi = \frac{\sqrt{2}}{2} \Rightarrow \phi = \frac{\pi}{4}$, $0 \le \theta \le 2\pi$; $\mathbf{r}_\phi = (2\cos\phi\cos\theta)\mathbf{i} + (2\cos\phi\sin\theta)\mathbf{j} - (2\sin\phi)\mathbf{k}$

$$\text{and } \mathbf{r}_\theta = (-2\sin\phi\sin\theta)\mathbf{i} + (2\sin\phi\cos\theta)\mathbf{j} \Rightarrow \mathbf{r}_\phi \times \mathbf{r}_\theta = \begin{vmatrix} \mathbf{i} & \mathbf{j} & \mathbf{k} \\ 2\cos\phi\cos\theta & 2\cos\phi\sin\theta & -2\sin\phi \\ -2\sin\phi\sin\theta & 2\sin\phi\cos\theta & 0 \end{vmatrix}$$

$= \left(4\sin^2\phi\cos\theta\right)\mathbf{i} + \left(4\sin^2\phi\sin\theta\right)\mathbf{j} + (4\sin\phi\cos\phi)\mathbf{k}$

$\Rightarrow |\mathbf{r}_\phi \times \mathbf{r}_\theta| = \sqrt{16\sin^4\phi\cos^2\theta + 16\sin^4\phi\sin^2\theta + 16\sin^2\phi\cos^2\phi} = 4\sin\phi$; $y = 2\sin\phi\sin\theta$ and

$$z = 2\cos\phi \Rightarrow H(x,y,z) = 4\cos\phi\sin\phi\cos\theta \Rightarrow \iint_S H(x,y,z)\,d\sigma = \int_0^{2\pi}\int_0^{\pi/4} (4\cos\phi\sin\phi\sin\theta)(4\sin\phi)\,d\phi\,d\theta$$

$$= \int_0^{2\pi}\int_0^{\pi/4} 16\sin^2\phi\cos\phi\sin\theta\,d\sigma\,d\theta = 0$$

35. Let the parametrization be $\mathbf{r}(x,y) = x\mathbf{i} + y\mathbf{j} + (4-y^2)\mathbf{k}$, $0 \le x \le 1$, $-2 \le y \le 2$; $z = 0 \Rightarrow 0 = 4 - y^2$

$$\Rightarrow y = \pm 2;\ \mathbf{r}_x = \mathbf{i} \text{ and } \mathbf{r}_y = \mathbf{j} - 2y\mathbf{k} \Rightarrow \mathbf{r}_x \times \mathbf{r}_y = \begin{vmatrix} \mathbf{i} & \mathbf{j} & \mathbf{k} \\ 1 & 0 & 0 \\ 0 & 1 & -2y \end{vmatrix} = 2y\mathbf{j} + \mathbf{k} \Rightarrow \mathbf{F} \cdot \mathbf{n}\, d\sigma$$

$$= \mathbf{F} \cdot \frac{\mathbf{r}_x \times \mathbf{r}_y}{|\mathbf{r}_x \times \mathbf{r}_y|} |\mathbf{r}_x \times \mathbf{r}_y|\, dy\, dx = (2xy - 3z)\, dy\, dx = [2xy - 3(4-y^2)]\, dy\, dx \Rightarrow \iint_S \mathbf{F} \cdot \mathbf{n}\, d\sigma$$

$$= \int_0^1 \int_{-2}^2 (2xy + 3y^2 - 12)\, dy\, dx = \int_0^1 \left[xy^2 + y^3 - 12y\right]_{-2}^2 dx = \int_0^1 -32\, dx = -32$$

36. Let the parametrization be $\mathbf{r}(x,y) = x\mathbf{i} + x^2\mathbf{j} + z\mathbf{k}$, $-1 \le x \le 1$, $0 \le z \le 2 \Rightarrow \mathbf{r}_x = \mathbf{i} + 2x\mathbf{j}$ and $\mathbf{r}_z = \mathbf{k}$

$$\Rightarrow \mathbf{r}_x \times \mathbf{r}_z = \begin{vmatrix} \mathbf{i} & \mathbf{j} & \mathbf{k} \\ 1 & 2x & 0 \\ 0 & 0 & 1 \end{vmatrix} = 2x\mathbf{i} - \mathbf{j} \Rightarrow \mathbf{F} \cdot \mathbf{n}\, d\sigma = \mathbf{F} \cdot \frac{\mathbf{r}_x \times \mathbf{r}_z}{|\mathbf{r}_x \times \mathbf{r}_z|} |\mathbf{r}_x \times \mathbf{r}_z|\, dz\, dx = -x^2\, dz\, dx$$

$$\Rightarrow \iint_S \mathbf{F} \cdot \mathbf{n}\, d\sigma = \int_{-1}^1 \int_0^2 -x^2\, dz\, dx = -\frac{4}{3}$$

37. Let the parametrization be $\mathbf{r}(\phi,\theta) = (a \sin\phi \cos\theta)\mathbf{i} + (a \sin\phi \sin\theta)\mathbf{j} + (a \cos\phi)\mathbf{k}$ (spherical coordinates with $\rho = a$, $a \ge 0$, on the sphere), $0 \le \phi \le \frac{\pi}{2}$ (for the first octant)

$\Rightarrow \mathbf{r}_\phi = (a \cos\phi \cos\theta)\mathbf{i} + (a \cos\phi \sin\theta)\mathbf{j} - (a \sin\phi)\mathbf{k}$ and $\mathbf{r}_\theta = (-a \sin\phi \sin\theta)\mathbf{i} + (a \sin\phi \cos\theta)\mathbf{j}$

$$\Rightarrow \mathbf{r}_\phi \times \mathbf{r}_\theta = \begin{vmatrix} \mathbf{i} & \mathbf{j} & \mathbf{k} \\ a\cos\phi\cos\theta & a\cos\phi\sin\theta & -a\sin\phi \\ -a\sin\phi\sin\theta & a\sin\phi\cos\theta & 0 \end{vmatrix}$$

$$= (a^2 \sin^2\phi \cos\theta)\mathbf{i} + (a^2 \sin^2\phi \sin\theta)\mathbf{j} + (a^2 \sin\phi \cos\phi)\mathbf{k} \Rightarrow \mathbf{F} \cdot \mathbf{n}\, d\sigma = \mathbf{F} \cdot \frac{\mathbf{r}_\phi \times \mathbf{r}_\theta}{|\mathbf{r}_\phi \times \mathbf{r}_\theta|} |\mathbf{r}_\phi \times \mathbf{r}_\theta|\, d\theta\, d\phi$$

$$= a^3 \cos^2\phi \sin\phi\, d\theta\, d\phi \text{ since } \mathbf{F} = z\mathbf{k} = (a \cos\phi)\mathbf{k} \Rightarrow \iint_S \mathbf{F} \cdot \mathbf{n}\, d\sigma = \int_0^{\pi/2} \int_0^{\pi/2} a^3 \cos^2\phi \sin\phi\, d\phi\, d\theta = \frac{\pi a^3}{6}$$

38. Let the parametrization be $\mathbf{r}(\phi,\theta) = (a \sin\phi \cos\theta)\mathbf{i} + (a \sin\phi \sin\theta)\mathbf{j} + (a \cos\phi)\mathbf{k}$ (spherical coordinates with $\rho = a$, $a \ge 0$, on the sphere), $0 \le \phi \le \pi$, $0 \le \theta \le 2\pi$

$\Rightarrow \mathbf{r}_\phi = (a \cos\phi \cos\theta)\mathbf{i} + (a \cos\phi \sin\theta)\mathbf{j} - (a \sin\phi)\mathbf{k}$ and $\mathbf{r}_\theta = (-a \sin\phi \sin\theta)\mathbf{i} + (a \sin\phi \cos\theta)\mathbf{j}$

$$\Rightarrow \mathbf{r}_\phi \times \mathbf{r}_\theta = \begin{vmatrix} \mathbf{i} & \mathbf{j} & \mathbf{k} \\ a\cos\phi\cos\theta & a\cos\phi\sin\theta & -a\sin\phi \\ -a\sin\phi\sin\theta & a\sin\phi\cos\theta & 0 \end{vmatrix}$$

$= (a^2 \sin^2\phi \cos\theta)\mathbf{i} + (a^2 \sin^2\phi \sin\theta)\mathbf{j} + (a^2 \sin\phi \cos\phi)\mathbf{k} \Rightarrow \mathbf{F}\cdot\mathbf{n}\, d\sigma = \mathbf{F}\cdot\frac{\mathbf{r}_\phi\times\mathbf{r}_\theta}{|\mathbf{r}_\phi\times\mathbf{r}_\theta|}|\mathbf{r}_\phi\times\mathbf{r}_\theta|\, d\theta\, d\phi$

$= (a^3 \sin^3\phi \cos^2\phi + a^3 \sin^3\phi \sin^2\theta + a^3 \sin\phi \cos^2\phi)\, d\theta\, d\phi = a^3 \sin\phi\, d\theta\, d\phi$ since $\mathbf{F} = x\mathbf{i} + y\mathbf{j} + z\mathbf{k}$

$= (a \sin\phi \cos\theta)\mathbf{i} + (a \sin\phi \sin\theta)\mathbf{j} + (a \cos\phi)\mathbf{k} \Rightarrow \iint_S \mathbf{F}\cdot\mathbf{n}\, d\sigma = \int_0^{2\pi}\int_0^{\pi} a^3 \sin\phi\, d\phi\, d\theta = 4\pi a^3$

39. Let the parametrization be $\mathbf{r}(x,y) = x\mathbf{i} + y\mathbf{j} + (2a - x - y)\mathbf{k}$, $0 \le x \le a$, $0 \le y \le a \Rightarrow \mathbf{r}_x = \mathbf{i} - \mathbf{k}$ and $\mathbf{r}_y = \mathbf{j} - \mathbf{k}$

$$\Rightarrow \mathbf{r}_x\times\mathbf{r}_y = \begin{vmatrix} \mathbf{i} & \mathbf{j} & \mathbf{k} \\ 1 & 0 & -1 \\ 0 & 1 & -1 \end{vmatrix} = \mathbf{i} + \mathbf{j} + \mathbf{k} \Rightarrow \mathbf{F}\cdot\mathbf{n}\, d\sigma = \mathbf{F}\cdot\frac{\mathbf{r}_x\times\mathbf{r}_y}{|\mathbf{r}_x\times\mathbf{r}_y|}|\mathbf{r}_x\times\mathbf{r}_y|\, dy\, dx$$

$= [2xy + 2y(2a - x - y) + 2x(2a - x - y)]\, dy\, dx$ since $\mathbf{F} = 2xy\mathbf{i} + 2yz\mathbf{j} + 2xz\mathbf{k}$

$= 2xy\mathbf{i} + 2y(2a - x - y)\mathbf{j} + 2x(2a - x - y)\mathbf{k} \Rightarrow \iint_S \mathbf{F}\cdot\mathbf{n}\, d\sigma$

$$= \int_0^a\int_0^a [2xy + 2y(2a - x - y) + 2x(2a - x - y)]\, dy\, dx = \int_0^a\int_0^a \left(4ay - 2y^2 + 4ax - 2x^2 - 2xy\right) dy\, dx$$

$$= \int_0^a \left(\tfrac{4}{3}a^3 + 3a^2x - 2ax^2\right) dx = \left(\tfrac{4}{3} + \tfrac{3}{2} - \tfrac{2}{3}\right)a^4 = \tfrac{13a^4}{6}$$

40. Let the parametrization be $\mathbf{r}(\theta,z) = (\cos\theta)\mathbf{i} + (\sin\theta)\mathbf{j} + z\mathbf{k}$, $0 \le z \le a$, $0 \le \theta \le 2\pi$ (where $r = \sqrt{x^2 + y^2} = 1$ on

the cylinder) $\Rightarrow \mathbf{r}_\theta = (-\sin\theta)\mathbf{i} + (\cos\theta)\mathbf{j}$ and $\mathbf{r}_z = \mathbf{k} \Rightarrow \mathbf{r}_\theta\times\mathbf{r}_z = \begin{vmatrix} \mathbf{i} & \mathbf{j} & \mathbf{k} \\ -\sin\theta & \cos\theta & 0 \\ 0 & 0 & 1 \end{vmatrix} = (\cos\theta)\mathbf{i} + (\sin\theta)\mathbf{j}$

$\Rightarrow \mathbf{F}\cdot\mathbf{n}\, d\sigma = \mathbf{F}\cdot\frac{\mathbf{r}_\theta\times\mathbf{r}_z}{|\mathbf{r}_\theta\times\mathbf{r}_z|}|\mathbf{r}_\theta\times\mathbf{r}_z|\, dz\, d\theta = \left(\cos^2\theta + \sin^2\theta\right) dz\, d\theta = dz\, d\theta$, since $\mathbf{F} = (\cos\theta)\mathbf{i} + (\sin\theta)\mathbf{j} + z\mathbf{k}$

$$\Rightarrow \iint_S \mathbf{F}\cdot\mathbf{n}\, d\sigma = \int_0^{2\pi}\int_0^a 1\, dz\, d\theta = 2\pi a$$

41. Let the parametrization be $\mathbf{r}(r,\theta) = (r\cos\theta)\mathbf{i} + (r\sin\theta)\mathbf{j} + r\mathbf{k}$, $0 \le r \le 1$ (since $0 \le z \le 1$) and $0 \le \theta \le 2\pi$

$$\Rightarrow \mathbf{r}_r = (\cos\theta)\mathbf{i} + (\sin\theta)\mathbf{j} + \mathbf{k} \text{ and } \mathbf{r}_\theta = (-r\sin\theta)\mathbf{i} + (r\cos\theta)\mathbf{j} \Rightarrow \mathbf{r}_\theta\times\mathbf{r}_r = \begin{vmatrix} \mathbf{i} & \mathbf{j} & \mathbf{k} \\ -r\sin\theta & r\cos\theta & 0 \\ \cos\theta & \sin\theta & 1 \end{vmatrix}$$

$= (r\cos\theta)\mathbf{i} + (r\sin\theta)\mathbf{j} - r\mathbf{k} \Rightarrow \mathbf{F}\cdot\mathbf{n}\, d\sigma = \mathbf{F}\cdot\frac{\mathbf{r}_\theta\times\mathbf{r}_r}{|\mathbf{r}_\theta\times\mathbf{r}_r|}|\mathbf{r}_\theta\times\mathbf{r}_r|\, d\theta\, dr = \left(r^3 \sin\theta \cos^2\theta + r^2\right) d\theta\, dr$ since

$\mathbf{F} = (r^2 \sin\theta \cos\theta)\mathbf{i} - r\mathbf{k} \Rightarrow \iint_S \mathbf{F}\cdot\mathbf{n}\, d\sigma = \int_0^{2\pi}\int_0^1 (r^3 \sin\theta \cos^2\theta + r^2)\, dr\, d\theta = \int_0^{2\pi} \left(\frac{1}{4}\sin\theta\cos^2\theta + \frac{1}{3}\right) d\theta$

$= \left[-\frac{1}{12}\cos^3\theta + \frac{\theta}{3}\right]_0^{2\pi} = \frac{2\pi}{3}$

42. Let the parametrization be $\mathbf{r}(r,\theta) = (r\cos\theta)\mathbf{i} + (r\sin\theta)\mathbf{j} + 2r\mathbf{k}$, $0 \le r \le 2$ (since $0 \le z \le 2$) and $0 \le \theta \le 2\pi$

$\Rightarrow \mathbf{r}_r = (\cos\theta)\mathbf{i} + (\sin\theta)\mathbf{j} + 2\mathbf{k}$ and $\mathbf{r}_\theta = (-r\sin\theta)\mathbf{i} + (r\cos\theta)\mathbf{j} \Rightarrow \mathbf{r}_\theta \times \mathbf{r}_r = \begin{vmatrix} \mathbf{i} & \mathbf{j} & \mathbf{k} \\ -r\sin\theta & r\cos\theta & 0 \\ \cos\theta & \sin\theta & 2 \end{vmatrix}$

$= (2r\cos\theta)\mathbf{i} + (2r\sin\theta)\mathbf{j} - r\mathbf{k} \Rightarrow \mathbf{F}\cdot\mathbf{n}\, d\sigma = \mathbf{F}\cdot\frac{\mathbf{r}_\theta \times \mathbf{r}_r}{|\mathbf{r}_\theta \times \mathbf{r}_r|}|\mathbf{r}_\theta \times \mathbf{r}_r|\, d\theta\, dr$

$= (2r^3 \sin^2\theta \cos\theta + 4r^3 \cos\theta \sin\theta + r)\, d\theta\, dr$ since

$\mathbf{F} = (r^2 \sin^2\theta)\mathbf{i} + (2r^2\cos\theta)\mathbf{j} - \mathbf{k} \Rightarrow \iint_S \mathbf{F}\cdot\mathbf{n}\, d\sigma = \int_0^{2\pi}\int_0^1 (2r^3\sin^2\theta\cos\theta + 4r^3\cos\theta\sin\theta + r)\, dr\, d\theta$

$= \int_0^{2\pi} \left(\frac{1}{2}\sin^2\theta\cos\theta + \cos\theta\sin\theta + \frac{1}{2}\right) d\theta = \left[\frac{1}{6}\sin^3\theta + \frac{1}{2}\sin^2\theta + \frac{1}{2}\theta\right]_0^{2\pi} = \pi$

43. Let the parametrization be $\mathbf{r}(r,\theta) = (r\cos\theta)\mathbf{i} + (r\sin\theta)\mathbf{j} + r\mathbf{k}$, $1 \le r \le 2$ (since $1 \le z \le 2$) and $0 \le \theta \le 2\pi$

$\Rightarrow \mathbf{r}_r = (\cos\theta)\mathbf{i} + (\sin\theta)\mathbf{j} + \mathbf{k}$ and $\mathbf{r}_\theta = (-r\sin\theta)\mathbf{i} + (r\cos\theta)\mathbf{j} \Rightarrow \mathbf{r}_\theta \times \mathbf{r}_r = \begin{vmatrix} \mathbf{i} & \mathbf{j} & \mathbf{k} \\ -r\sin\theta & r\cos\theta & 0 \\ \cos\theta & \sin\theta & 1 \end{vmatrix}$

$= (r\cos\theta)\mathbf{i} + (r\sin\theta)\mathbf{j} - r\mathbf{k} \Rightarrow \mathbf{F}\cdot\mathbf{n}\, d\sigma = \mathbf{F}\cdot\frac{\mathbf{r}_\theta \times \mathbf{r}_r}{|\mathbf{r}_\theta \times \mathbf{r}_r|}|\mathbf{r}_\theta \times \mathbf{r}_r|\, d\theta\, dr = (-r^2\cos^2\theta - r^2\sin^2\theta - r^3)\, d\theta\, dr$

$= (-r^2 - r^3)\, d\theta\, dr$ since $\mathbf{F} = (-r\cos\theta)\mathbf{i} - (r\sin\theta)\mathbf{j} + r^2\mathbf{k} \Rightarrow \iint_S \mathbf{F}\cdot\mathbf{n}\, d\sigma = \int_0^{2\pi}\int_1^2 (-r^2 - r^3)\, dr\, d\theta = -\frac{73\pi}{6}$

44. Let the parametrization be $\mathbf{r}(r,\theta) = (r\cos\theta)\mathbf{i} + (r\sin\theta)\mathbf{j} + r^2\mathbf{k}$, $0 \le r \le 1$ (since $0 \le z \le 1$) and $0 \le \theta \le 2\pi$

$\Rightarrow \mathbf{r}_r = (\cos\theta)\mathbf{i} + (\sin\theta)\mathbf{j} + 2r\mathbf{k}$ and $\mathbf{r}_\theta = (-r\sin\theta)\mathbf{i} + (r\cos\theta)\mathbf{j} \Rightarrow \mathbf{r}_\theta \times \mathbf{r}_r = \begin{vmatrix} \mathbf{i} & \mathbf{j} & \mathbf{k} \\ -r\sin\theta & r\cos\theta & 0 \\ \cos\theta & \sin\theta & 2r \end{vmatrix}$

$= (2r^2\cos\theta)\mathbf{i} + (2r^2\sin\theta)\mathbf{j} - r\mathbf{k} \Rightarrow \mathbf{F}\cdot\mathbf{n}\, d\sigma = \mathbf{F}\cdot\frac{\mathbf{r}_\theta \times \mathbf{r}_r}{|\mathbf{r}_\theta \times \mathbf{r}_r|}|\mathbf{r}_\theta \times \mathbf{r}_r|\, d\theta\, dr = (8r^3\cos^2\theta + 8r^3\sin^2\theta - 2r)\, d\theta\, dr$

$= (8r^3 - 2r)\, d\theta\, dr$ since $\mathbf{F} = (4r\cos\theta)\mathbf{i} + (4r\sin\theta)\mathbf{j} + 2\mathbf{k} \Rightarrow \iint_S \mathbf{F}\cdot\mathbf{n}\, d\sigma = \int_0^{2\pi}\int_0^1 (8r^3 - 2r)\, dr\, d\theta = 2\pi$

45. Let the parametrization be $\mathbf{r}(\phi,\theta)=(a \sin\phi\cos\theta)\mathbf{i}+(a\sin\phi\sin\theta)\mathbf{j}+(a\cos\phi)\mathbf{k}$, $0\le\phi\le\frac{\pi}{2}$, $0\le\theta\le\frac{\pi}{2}$
$\Rightarrow \mathbf{r}_\phi=(a\cos\phi\cos\theta)\mathbf{i}+(a\cos\phi\sin\theta)\mathbf{j}-(a\sin\phi)\mathbf{k}$ and $\mathbf{r}_\theta=(-a\sin\phi\sin\theta)\mathbf{i}+(a\sin\phi\cos\theta)\mathbf{j}$

$$\Rightarrow \mathbf{r}_\phi\times\mathbf{r}_\theta=\begin{vmatrix}\mathbf{i} & \mathbf{j} & \mathbf{k}\\ a\cos\phi\cos\theta & a\cos\phi\sin\theta & -a\sin\phi\\ -a\sin\phi\sin\theta & a\sin\phi\cos\theta & 0\end{vmatrix}$$

$$=\left(a^2\sin^2\phi\cos\theta\right)\mathbf{i}+\left(a^2\sin^2\phi\sin\theta\right)\mathbf{j}+\left(a^2\sin\phi\cos\phi\right)\mathbf{k}$$

$\Rightarrow |\mathbf{r}_\phi\times\mathbf{r}_\theta|=\sqrt{a^4\sin^4\phi\cos^2\theta+a^4\sin^4\phi\sin^2\theta+a^4\sin^2\phi\cos^2\phi}=\sqrt{a^4\sin^2\phi}=a^2\sin\phi$. The mass is

$M=\iint_S d\sigma=\int_0^{\pi/2}\int_0^{\pi/2}\left(a^2\sin\phi\right)d\phi\,d\theta=\frac{a^2\pi}{2}$; the first moment is $M_{yz}=\iint_S x\,d\sigma$

$=\int_0^{\pi/2}\int_0^{\pi/2}(a\sin\phi\cos\theta)\left(a^2\sin\phi\right)d\phi\,d\theta=\frac{a^3\pi}{4}\Rightarrow \bar{x}=\frac{\left(\frac{a^3\pi}{4}\right)}{\left(\frac{a^2\pi}{2}\right)}=\frac{a}{2}\Rightarrow$ the centroid is located at $\left(\frac{a}{2},\frac{a}{2},\frac{a}{2}\right)$ by symmetry

46. Let the parametrization be $\mathbf{r}(r,\theta)=(r\cos\theta)\mathbf{i}+(r\sin\theta)\mathbf{j}+r\mathbf{k}$, $1\le r\le 2$ (since $1\le z\le 2$) and $0\le\theta\le 2\pi$

$\Rightarrow \mathbf{r}_r=(\cos\theta)\mathbf{i}+(\sin\theta)\mathbf{j}+\mathbf{k}$ and $\mathbf{r}_\theta=(-r\sin\theta)\mathbf{i}+(r\cos\theta)\mathbf{j}\Rightarrow \mathbf{r}_\theta\times\mathbf{r}_r=\begin{vmatrix}\mathbf{i} & \mathbf{j} & \mathbf{k}\\ -r\sin\theta & r\cos\theta & 0\\ \cos\theta & \sin\theta & 1\end{vmatrix}$

$=(r\cos\theta)\mathbf{i}+(r\sin\theta)\mathbf{j}-r\mathbf{k}\Rightarrow|\mathbf{r}_\theta\times\mathbf{r}_r|=\sqrt{r^2\cos^2\theta+r^2\sin^2\theta+r^2}=r\sqrt{2}$. The mass is

$M=\iint_S \delta\,d\sigma=\int_0^{2\pi}\int_1^2 \delta r\sqrt{2}\,dr\,d\theta=\left(3\sqrt{2}\right)\pi\delta$; the first moment is $M_{yz}=\iint_S \delta x\,d\sigma=\int_0^{2\pi}\int_1^2 \delta r\left(r\sqrt{2}\right)dr\,d\theta$

$=\frac{\left(14\sqrt{2}\right)\pi\delta}{3}\Rightarrow\bar{x}=\frac{\left(\frac{\left(14\sqrt{2}\right)\pi\delta}{3}\right)}{\left(3\sqrt{2}\right)\pi\delta}=\frac{14}{9}\Rightarrow$ the center of mass is located at $\left(0,0,\frac{14}{9}\right)$ by symmetry. The

moment of inertia is $I_z=\iint_S \delta\left(x^2+y^2\right)d\sigma=\int_0^{2\pi}\int_1^2 \delta r^2\left(r\sqrt{2}\right)dr\,d\theta=\frac{\left(15\sqrt{2}\right)\pi\delta}{2}\Rightarrow$ the radius of gyration is

$R_z=\sqrt{\frac{I_z}{M}}=\sqrt{\frac{5}{2}}$

47. Let the parametrization be $\mathbf{r}(\phi,\theta)=(a\sin\phi\cos\theta)\mathbf{i}+(a\sin\phi\sin\theta)\mathbf{j}+(a\cos\phi)\mathbf{k}$, $0\le\phi\le\pi$, $0\le\theta\le 2\pi$
$\Rightarrow \mathbf{r}_\phi=(a\cos\phi\cos\theta)\mathbf{i}+(a\cos\phi\sin\theta)\mathbf{j}-(a\sin\phi)\mathbf{k}$ and $\mathbf{r}_\theta=(-a\sin\phi\sin\theta)\mathbf{i}+(a\sin\phi\cos\theta)\mathbf{j}$

$$\Rightarrow \mathbf{r}_\phi\times\mathbf{r}_\theta=\begin{vmatrix}\mathbf{i} & \mathbf{j} & \mathbf{k}\\ a\cos\phi\cos\theta & a\cos\phi\sin\theta & -a\sin\phi\\ -a\sin\phi\sin\theta & a\sin\phi\cos\theta & 0\end{vmatrix}$$

$= (a^2 \sin^2 \phi \cos \theta)\,\mathbf{i} + (a^2 \sin^2 \phi \sin \theta)\mathbf{j} + (a^2 \sin \phi \cos \phi)\mathbf{k}$

$\Rightarrow |\mathbf{r}_\phi \times \mathbf{r}_\theta| = \sqrt{a^4 \sin^4 \phi \cos^2 \theta + a^4 \sin^4 \phi \sin^2 \theta + a^4 \sin^2 \phi \cos^2 \phi} = \sqrt{a^4 \sin^2 \phi} = a^2 \sin \phi$. The moment of inertia is $I_z = \iint_S \delta(x^2+y^2)\,d\sigma = \int_0^{2\pi}\int_0^{\pi} \delta[(a \sin \phi \cos \theta)^2 + (a \sin \phi \sin \theta)^2](a^2 \sin \phi)\,d\phi\,d\theta$

$= \int_0^{2\pi}\int_0^{\pi} \delta(a^2 \sin^2 \phi)(a^2 \sin \phi)\,d\phi\,d\theta = \int_0^{2\pi}\int_0^{\pi} \delta a^4 \sin^3 \phi\,d\phi\,d\theta = \int_0^{2\pi} \delta a^4\left[\left(-\frac{1}{3}\cos \phi\right)(\sin^2 \phi + 2)\right]_0^{\pi} d\theta = \frac{8\delta\pi a^4}{3}$

48. Let the parametrization be $\mathbf{r}(r,\theta) = (r \cos \theta)\mathbf{i} + (r \sin \theta)\mathbf{j} + r\mathbf{k}$, $0 \le r \le 1$ (since $0 \le z \le 1$) and $0 \le \theta \le 2\pi$

$\Rightarrow \mathbf{r}_r = (\cos \theta)\mathbf{i} + (\sin \theta)\mathbf{j} + \mathbf{k}$ and $\mathbf{r}_\theta = (-r \sin \theta)\mathbf{i} + (r \cos \theta)\mathbf{j} \Rightarrow \mathbf{r}_\theta \times \mathbf{r}_r = \begin{vmatrix} \mathbf{i} & \mathbf{j} & \mathbf{k} \\ -r \sin \theta & r \cos \theta & 0 \\ \cos \theta & \sin \theta & 1 \end{vmatrix}$

$= (r \cos \theta)\mathbf{i} + (r \sin \theta)\mathbf{j} - r\mathbf{k} \Rightarrow |\mathbf{r}_\theta \times \mathbf{r}_r| = \sqrt{r^2 \cos^2 \theta + r^2 \sin^2 \theta + r^2} = r\sqrt{2}$. The moment of inertia is

$I_z = \iint_S \delta(x^2+y^2)\,d\sigma = \int_0^{2\pi}\int_0^{1} \delta r^2(r\sqrt{2})\,dr\,d\theta = \frac{\pi\delta\sqrt{2}}{2}$

49. The parametrization $\mathbf{r}(r,\theta) = (r \cos \theta)\mathbf{i} + (r \sin \theta)\mathbf{j} + r\mathbf{k}$ at $P_0 = (\sqrt{2}, \sqrt{2}, 2) \Rightarrow \theta = \frac{\pi}{4}$, $r = 2$,

$\mathbf{r}_r = (\cos \theta)\mathbf{i} + (\sin \theta)\mathbf{j} + \mathbf{k} = \frac{\sqrt{2}}{2}\mathbf{i} + \frac{\sqrt{2}}{2}\mathbf{j} + \mathbf{k}$ and

$\mathbf{r}_\theta = (-r \sin \theta)\mathbf{i} + (r \cos \theta)\mathbf{j} = -\sqrt{2}\mathbf{i} + \sqrt{2}\mathbf{j}$

$\Rightarrow \mathbf{r}_r \times \mathbf{r}_\theta = \begin{vmatrix} \mathbf{i} & \mathbf{j} & \mathbf{k} \\ \sqrt{2}/2 & \sqrt{2}/2 & 1 \\ -\sqrt{2} & \sqrt{2} & 0 \end{vmatrix}$

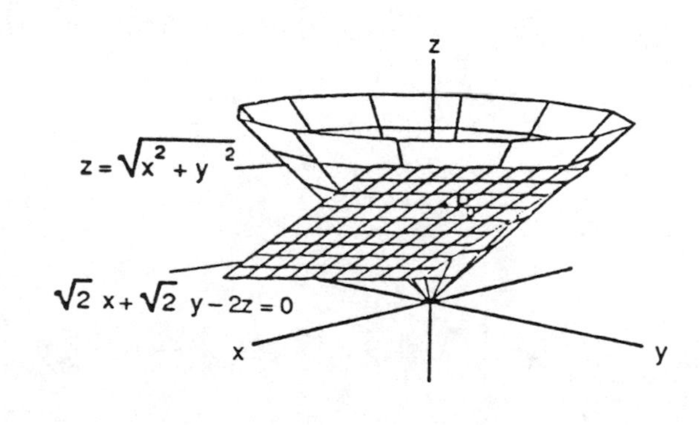

$= -\sqrt{2}\mathbf{i} - \sqrt{2}\mathbf{j} + 2\mathbf{k} \Rightarrow$ the tangent plane is

$(-\sqrt{2}\mathbf{i} - \sqrt{2}\mathbf{j} + 2\mathbf{k}) \cdot \left[(x - \sqrt{2})\mathbf{i} + (y - \sqrt{2})\mathbf{j} + (z - 2)\mathbf{k}\right] = \sqrt{2}x + \sqrt{2}y - 2z = 0$, or $x + y - \sqrt{2}z = 0$. The parametrization $\mathbf{r}(r,\theta) \Rightarrow x = r \cos \theta$, $y = r \sin \theta$ and $z = r \Rightarrow x^2 + y^2 = r^2 = z^2 \Rightarrow$ the surface is $z = \sqrt{x^2+y^2}$.

50. The parametrization $\mathbf{r}(\phi,\theta)$
$= (4 \sin \phi \cos \theta)\mathbf{i} + (4 \sin \phi \sin \theta)\mathbf{j} + (4 \cos \phi)\mathbf{k}$
at $P_0 = (\sqrt{2}, \sqrt{2}, 2\sqrt{3}) \Rightarrow \rho = 4$ and $z = 2\sqrt{3}$
$= 4 \cos \phi \Rightarrow\ = \frac{\pi}{6}$; also $x = \sqrt{2}$ and $y = \sqrt{2}$
$\Rightarrow \theta = \frac{\pi}{4}$. Then $\mathbf{r}_\phi$
$= (4 \cos \phi \cos \theta)\mathbf{i} + (r \cos \phi \sin \theta)\mathbf{j} - (4 \sin \phi)\mathbf{k}$
$= \sqrt{6}\mathbf{i} + \sqrt{6}\mathbf{j} - 2\mathbf{k}$ and
$\mathbf{r}_\theta = (-4 \sin \phi \sin \theta)\mathbf{i} + (4 \sin \phi \cos \theta)\mathbf{j}$
$= -\sqrt{2}\mathbf{i} + \sqrt{2}\mathbf{j}$ at P_0

$$\Rightarrow \mathbf{r}_\phi \times \mathbf{r}_\theta = \begin{vmatrix} \mathbf{i} & \mathbf{j} & \mathbf{k} \\ \sqrt{6} & \sqrt{6} & -2 \\ -\sqrt{2} & \sqrt{2} & 0 \end{vmatrix} = 2\sqrt{2}\mathbf{i} + 2\sqrt{2}\mathbf{j} + 4\sqrt{3}\mathbf{k} \Rightarrow \text{the tangent plane is}$$

$(2\sqrt{2}\mathbf{i} + 2\sqrt{2}\mathbf{j} + 4\sqrt{3}\mathbf{k}) \cdot [(x - \sqrt{2})\mathbf{i} + (y - \sqrt{2})\mathbf{j} + (z - 2\sqrt{3})\mathbf{k}] = 0 \Rightarrow \sqrt{2}x + \sqrt{2}y + 2\sqrt{3}z = 16$,
or $x + y + \sqrt{6}z = 8\sqrt{2}$. The parametrization $\Rightarrow x = 4 \sin \phi \cos \theta$, $y = 4 \sin \phi \sin \theta$, $z = 4 \cos \phi$
$\Rightarrow$ the surface is $x^2 + y^2 + z^2 = 16$, $z \ge 0$.

51. The parametrization $\mathbf{r}(\theta, z) = (3 \sin 2\theta)\mathbf{i} + (6 \sin^2 \theta)\mathbf{j} + z\mathbf{k}$
at $P_0 = \left(\frac{3\sqrt{3}}{2}, \frac{9}{2}, 0\right) \Rightarrow \theta = \frac{\pi}{3}$ and $z = 0$. Then
$\mathbf{r}_\theta = (6 \cos 2\theta)\mathbf{i} + (12 \sin \theta \cos \theta)\mathbf{j}$
$= -3\mathbf{i} + 3\sqrt{3}\mathbf{j}$ and $\mathbf{r}_z = \mathbf{k}$ at P_0

$$\Rightarrow \mathbf{r}_\theta \times \mathbf{r}_z = \begin{vmatrix} \mathbf{i} & \mathbf{j} & \mathbf{k} \\ -3 & 3\sqrt{3} & 0 \\ 0 & 0 & 1 \end{vmatrix} = 3\sqrt{3}\mathbf{i} + 3\mathbf{j} \Rightarrow \text{the tangent}$$

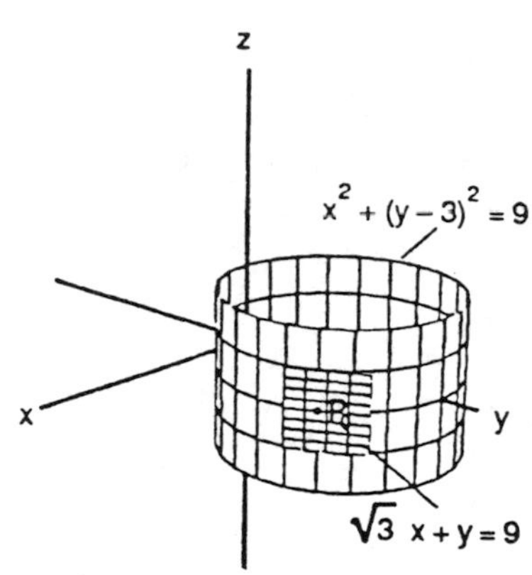

plane is $(3\sqrt{3}\mathbf{i} + 3\mathbf{j}) \cdot \left[\left(x - \frac{3\sqrt{3}}{2}\right)\mathbf{i} + \left(y - \frac{9}{2}\right)\mathbf{j} + (z - 0)\mathbf{k}\right] = 0$
$\Rightarrow \sqrt{3}x + y = 9$. The parametrization $\Rightarrow x = 3 \sin 2\theta$ and $y = 6 \sin^2 \theta \Rightarrow x^2 + y^2 = 9 \sin^2 2\theta + (6 \sin^2 \theta)^2$
$= 9(4 \sin^2 \theta \cos^2 \theta) + 36 \sin^4 \theta = 6(6 \sin^2 \theta) = 6y \Rightarrow x^2 + y^2 - 6y + 9 = 9 \Rightarrow x^2 + (y-3)^2 = 9$

52. The parametrization $\mathbf{r}(x,y) = x\mathbf{i} + y\mathbf{j} - x^2\mathbf{k}$ at $P_0 = (1,2,-1)$ $\Rightarrow \mathbf{r}_x = \mathbf{i} - 2x\mathbf{k} = \mathbf{i} - 2\mathbf{k}$ and $\mathbf{r}_y = \mathbf{j}$ at $P_0 \Rightarrow \mathbf{r}_x \times \mathbf{r}_y$ $= \begin{vmatrix} \mathbf{i} & \mathbf{j} & \mathbf{k} \\ 1 & 0 & -2 \\ 0 & 1 & 0 \end{vmatrix} = 2\mathbf{i} + \mathbf{k} \Rightarrow$ the tangent plane is $(2\mathbf{i} + \mathbf{k}) \cdot [(x-1)\mathbf{i} + (y-2)\mathbf{j} + (z+1)\mathbf{k}] = 0 \Rightarrow 2x + z = 1$. The parametrization $\Rightarrow x = x$, $y = y$ and $z = -x^2 \Rightarrow$ the surface is $z = -x^2$

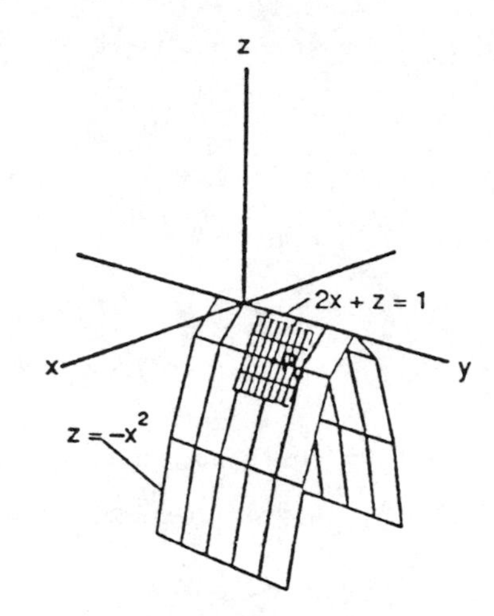

53. (a) An arbitrary point on the circle C is $(x,z) = (R + r\cos u, r\sin u) \Rightarrow (x,y,z)$ is on the torus with $x = (R + r\cos u)\cos v$, $y = (R + r\cos u)\sin v$, and $z = r\sin u$, $0 \le u \le 2\pi$, $0 \le v \le 2\pi$

(b) $\mathbf{r}_u = (-r\sin u\cos v)\mathbf{i} - (r\sin u\sin v)\mathbf{j} + (r\cos u)\mathbf{k}$ and $\mathbf{r}_v = (-(R + r\cos u)\sin v)\mathbf{i} + ((R + r\cos u)\cos v)\mathbf{j}$

$$\Rightarrow \mathbf{r}_u \times \mathbf{r}_v = \begin{vmatrix} \mathbf{i} & \mathbf{j} & \mathbf{k} \\ -r\sin u\cos v & -r\sin u\sin v & r\cos u \\ -(R + r\cos u)\sin v & (R + r\cos u)\cos v & 0 \end{vmatrix}$$

$$= -(R + r\cos u)(r\cos v\cos u)\mathbf{i} - (R + r\cos u)(r\sin v\cos u)\mathbf{j} + (-r\sin u)(R + r\cos u)\mathbf{k}$$

$$\Rightarrow |\mathbf{r}_u \times \mathbf{r}_v|^2 = (R + r\cos u)^2\left(r^2\cos^2 v\cos^2 u + r^2\sin^2 v\cos^2 u + r^2\sin^2 u\right) \Rightarrow |\mathbf{r}_u \times \mathbf{r}_v| = r(R + r\cos u)$$

$$\Rightarrow A = \int_0^{2\pi}\int_0^{2\pi} \left(rR + r^2\cos u\right) du\,dv = \int_0^{2\pi} 2\pi rR\,dv = 4\pi^2 rR$$

54. (a) The point (x,y,z) is on the surface for fixed $x = f(u)$ when $y = g(u)\sin\left(\frac{\pi}{2} - v\right)$ and $z = g(u)\cos\left(\frac{\pi}{2} - v\right)$ $\Rightarrow x = f(u)$, $y = g(u)\cos v$, and $z = g(u)\sin v \Rightarrow \mathbf{r}(u,v) = f(u)\mathbf{i} + (g(u)\cos v)\mathbf{j} + (g(u)\sin v)\mathbf{k}$, $0 \le v \le 2\pi$, $a \le u \le b$

(b) Let $u = y$ and $x = u^2 \Rightarrow f(u) = u^2$ and $g(u) = u \Rightarrow \mathbf{r}(u,v) = u^2\mathbf{i} + (u\cos v)\mathbf{j} + (u\sin v)\mathbf{k}$, $0 \le v \le 2\pi$, $0 \le u$

55. (a) Let $w^2 + \frac{z^2}{c^2} = 1$ where $w = \cos\phi$ and $\frac{z}{c} = \sin\phi \Rightarrow \frac{x^2}{a^2} + \frac{y^2}{b^2} = \cos^2\phi \Rightarrow \frac{x}{a} = \cos\phi\cos\theta$ and $\frac{y}{b} = \cos\phi\sin\theta$ $\Rightarrow x = a\cos\theta\cos\phi$, $y = b\sin\theta\cos\phi$, and $z = c\sin\phi$ $\Rightarrow \mathbf{r}(\theta,\phi) = (a\cos\theta\cos\phi)\mathbf{i} + (b\sin\theta\cos\phi)\mathbf{j} + (c\sin\phi)\mathbf{k}$

(b) $\mathbf{r}_\theta = (-a \sin\theta \cos\phi)\mathbf{i} + (b \cos\theta \cos\phi)\mathbf{j}$ and $\mathbf{r}_\phi = (-a \cos\theta \sin\phi)\mathbf{i} - (b \sin\theta \sin\phi)\mathbf{j} + (c \cos\phi)\mathbf{k}$

$$\Rightarrow \mathbf{r}_\theta \times \mathbf{r}_\phi = \begin{vmatrix} \mathbf{i} & \mathbf{j} & \mathbf{k} \\ -a \sin\theta \cos\phi & b \cos\theta \cos\phi & 0 \\ -a \cos\theta \sin\phi & -b \sin\theta \sin\phi & c \cos\phi \end{vmatrix}$$

$= (bc \cos\theta \cos^2\phi)\mathbf{i} + (ac \sin\theta \cos^2\phi)\mathbf{j} + (ab \sin\phi \cos\phi)\mathbf{k}$

$\Rightarrow |\mathbf{r}_\theta \times \mathbf{r}_\phi|^2 = b^2c^2 \cos^2\theta \cos^4\phi + a^2c^2 \sin^2\theta \cos^4\phi + a^2b^2 \sin^2\phi \cos^2\phi$, and the result follows.

56. (a) $\mathbf{r}(\theta, u) = (\cosh u \cos\theta)\mathbf{i} + (\cosh u \sin\theta)\mathbf{j} + (\sinh u)\mathbf{k}$

(b) $\mathbf{r}(\theta, u) = (a \cosh u \cos\theta)\mathbf{i} + (b \cosh u \sin\theta)\mathbf{j} + (c \sinh u)\mathbf{k}$

57. $\mathbf{r}(\theta, u) = (5 \cosh u \cos\theta)\mathbf{i} + (5 \cosh u \sin\theta)\mathbf{j} + (5 \sinh u)\mathbf{k} \Rightarrow \mathbf{r}_\theta = (-5 \cosh u \sin\theta)\mathbf{i} + (5 \cosh u \cos\theta)\mathbf{j}$ and $\mathbf{r}_u = (5 \sinh u \cos\theta)\mathbf{i} + (5 \sinh u \sin\theta)\mathbf{j} + (5 \cosh u)\mathbf{k}$

$$\Rightarrow \mathbf{r}_\theta \times \mathbf{r}_u = \begin{vmatrix} \mathbf{i} & \mathbf{j} & \mathbf{k} \\ -5 \cosh u \sin\theta & 5 \cosh u \cos\theta & 0 \\ 5 \sinh u \cos\theta & 5 \sinh u \sin\theta & 5 \cosh u \end{vmatrix}$$

$= (25 \cosh^2 u \cos\theta)\mathbf{i} + (25 \cosh^2 u \sin\theta)\mathbf{j} - (25 \cosh u \sinh u)\mathbf{k}$. At the point $(x_0, y_0, 0)$, where $x_0^2 + y_0^2 = 25$ we have $5 \sinh u = 0 \Rightarrow u = 0$ and $x_0 = 25 \cos\theta$, $y_0 = 25 \sin\theta \Rightarrow$ the tangent plane is $(x_0\mathbf{i} + y_0\mathbf{j}) \cdot [(x - x_0)\mathbf{i} + (y - y_0)\mathbf{j} + z\mathbf{k}] = 0 \Rightarrow x_0x - x_0^2 + y_0y - y_0^2 = 0 \Rightarrow x_0x + y_0y = 25$

58. Let $\frac{z^2}{c^2} - w^2 = 1$ where $\frac{z}{c} = \cosh u$ and $w = \sinh u \Rightarrow w^2 = \frac{x^2}{a^2} + \frac{y^2}{b^2} \Rightarrow \frac{x}{a} = w \cos\theta$ and $\frac{y}{b} = w \sin\theta$

$\Rightarrow x = a \sinh u \cos\theta$, $y = b \sinh u \sin\theta$, and $z = c \cosh u$

$\Rightarrow \mathbf{r}(\theta, u) = (a \sinh u \cos\theta)\mathbf{i} + (b \sinh u \sin\theta)\mathbf{j} + (c \cosh u)\mathbf{k}$, $0 \le \theta \le 2\pi$, $-\infty < u < \infty$

14.7 STOKES'S THEOREM

1. $$\text{curl } \mathbf{F} = \nabla \times \mathbf{F} = \begin{vmatrix} \mathbf{i} & \mathbf{j} & \mathbf{k} \\ \frac{\partial}{\partial x} & \frac{\partial}{\partial y} & \frac{\partial}{\partial z} \\ x^2 & 2x & z^2 \end{vmatrix} = 0\mathbf{i} + 0\mathbf{j} + (2 - 0)\mathbf{k} = 2\mathbf{k} \text{ and } \mathbf{n} = \mathbf{k} \Rightarrow \text{curl } \mathbf{F} \cdot \mathbf{n} = 2 \Rightarrow d\sigma = dx\,dy$$

$$\Rightarrow \oint_C \mathbf{F} \cdot d\mathbf{r} = \iint_R 2\, dA = 2(\text{Area of the ellipse}) = 4\pi$$

2. $\text{curl } \mathbf{F} = \nabla \times \mathbf{F} = \begin{vmatrix} \mathbf{i} & \mathbf{j} & \mathbf{k} \\ \frac{\partial}{\partial x} & \frac{\partial}{\partial y} & \frac{\partial}{\partial z} \\ 2y & 3x & -z^2 \end{vmatrix} = 0\mathbf{i} + 0\mathbf{j} + (3-2)\mathbf{k} = \mathbf{k}$ and $\mathbf{n} = \mathbf{k} \Rightarrow \text{curl } \mathbf{F} \cdot \mathbf{n} = 1 \Rightarrow d\sigma = dx\,dy$

$\Rightarrow \oint_C \mathbf{F} \cdot d\mathbf{r} = \iint_R dx\,dy = \text{Area of circle} = 9\pi$

3. $\text{curl } \mathbf{F} = \nabla \times \mathbf{F} = \begin{vmatrix} \mathbf{i} & \mathbf{j} & \mathbf{k} \\ \frac{\partial}{\partial x} & \frac{\partial}{\partial y} & \frac{\partial}{\partial z} \\ y & xz & x^2 \end{vmatrix} = -x\mathbf{i} - 2x\mathbf{j} + (z-1)\mathbf{k}$ and $\mathbf{n} = \frac{\mathbf{i}+\mathbf{j}+\mathbf{k}}{\sqrt{3}} \Rightarrow \text{curl } \mathbf{F} \cdot \mathbf{n}$

$= \frac{1}{\sqrt{3}}(-3x + z - 1) \Rightarrow d\sigma = \frac{\sqrt{3}}{1}\,dA \Rightarrow \oint_C \mathbf{F} \cdot d\mathbf{r} = \iint_R \frac{1}{\sqrt{3}}(-3x + z - 1)\sqrt{3}\,dA$

$= \int_0^1 \int_0^{1-x} [-3x + (1 - x - y) - 1]\,dy\,dx = \int_0^1 \int_0^{1-x} (-4x - y)\,dy\,dx = \int_0^1 -\left[4x(1-x) + \frac{1}{2}(1-x)^2\right] dx$

$= -\int_0^1 \left(\frac{1}{2} + 3x - \frac{7}{2}x^2\right) dx = -\frac{5}{6}$

4. $\text{curl } \mathbf{F} = \nabla \times \mathbf{F} = \begin{vmatrix} \mathbf{i} & \mathbf{j} & \mathbf{k} \\ \frac{\partial}{\partial x} & \frac{\partial}{\partial y} & \frac{\partial}{\partial z} \\ y^2 + z^2 & x^2 + z^2 & x^2 + y^2 \end{vmatrix} = (2y - 2z)\mathbf{i} + (2z - 2x)\mathbf{j} + (2x - 2y)\mathbf{k}$ and $\mathbf{n} = \frac{\mathbf{i}+\mathbf{j}+\mathbf{k}}{\sqrt{3}}$

$\Rightarrow \text{curl } \mathbf{F} \cdot \mathbf{n} = \frac{1}{\sqrt{3}}(2y - 2z + 2z - 2x + 2x - 2y) = 0 \Rightarrow \oint_C \mathbf{F} \cdot d\mathbf{r} = \iint_S d\sigma = 0$

5. $\text{curl } \mathbf{F} = \nabla \times \mathbf{F} = \begin{vmatrix} \mathbf{i} & \mathbf{j} & \mathbf{k} \\ \frac{\partial}{\partial x} & \frac{\partial}{\partial y} & \frac{\partial}{\partial z} \\ y^2 + z^2 & x^2 + y^2 & x^2 + y^2 \end{vmatrix} = 2y\mathbf{i} + (2z - 2x)\mathbf{j} + (2x - 2y)\mathbf{k}$ and $\mathbf{n} = \mathbf{k}$

$\Rightarrow \text{curl } \mathbf{F} \cdot \mathbf{n} = 2x - 2y \Rightarrow d\sigma = dx\,dy \Rightarrow \oint_C \mathbf{F} \cdot d\mathbf{r} = \int_{-1}^{1} \int_{-1}^{1} (2x - 2y)\,dx\,dy = \int_{-1}^{1} \left[x^2 - 2xy\right]_{-1}^{1} dy$

$= \int_{-1}^{1} -4y\,dy = 0$

6. $\text{curl } \mathbf{F} = \nabla \times \mathbf{F} = \begin{vmatrix} \mathbf{i} & \mathbf{j} & \mathbf{k} \\ \frac{\partial}{\partial x} & \frac{\partial}{\partial y} & \frac{\partial}{\partial z} \\ x^2y^3 & 1 & z \end{vmatrix} = 0\mathbf{i} + 0\mathbf{j} - 3x^2y^2\mathbf{k}$ and $\mathbf{n} = \frac{2x\mathbf{i} + 2y\mathbf{j} + 2z\mathbf{k}}{2\sqrt{x^2+y^2+z^2}} = \frac{x\mathbf{i} + y\mathbf{j} + z\mathbf{k}}{4}$

$\Rightarrow \text{curl } \mathbf{F} \cdot \mathbf{n} = -\frac{3}{4}x^2y^2z$; $d\sigma = \frac{4}{z}\, dA$ (Section 14.5, Example 5, with $a = 4$) $\Rightarrow \oint_C \mathbf{F} \cdot d\mathbf{r}$

$= \iint_R \left(-\frac{3}{4}x^2y^2z\right)\left(\frac{4}{z}\right) dA = -3\int_0^{2\pi}\int_0^2 (r^2\cos^2\theta)(r^2\sin^2\theta)\, r\, dr\, d\theta = -3\int_0^{2\pi}\left[\frac{r^6}{6}\right]_0^2 (\cos\theta\sin\theta)^2\, d\theta$

$= -32\int_0^{2\pi} \frac{1}{4}\sin^2 2\theta\, d\theta = -4\int_0^{4\pi} \sin^2 u\, du = -4\left[\frac{u}{2} - \frac{\sin 2u}{4}\right]_0^{4\pi} = -8\pi$

7. $x = 3\cos t$ and $y = 2\sin t \Rightarrow \mathbf{F} = (2\sin t)\mathbf{i} + (9\cos^2 t)\mathbf{j} + (9\cos^2 t + 16\sin^4 t)\sin e^{\sqrt{(6\sin t\cos t)(0)}}\mathbf{k}$ at the base of the shell; $\mathbf{r} = (3\cos t)\mathbf{i} + (2\sin t)\mathbf{j} \Rightarrow d\mathbf{r} = (-3\sin t)\mathbf{i} + (2\cos t)\mathbf{j} \Rightarrow \mathbf{F}\cdot\frac{d\mathbf{r}}{dt} = -6\sin^2 t + 18\cos^3 t$

$\Rightarrow \iint_S \nabla \times \mathbf{F}\cdot\mathbf{n}\, d\sigma = \int_0^{2\pi} (-6\sin^2 t + 18\cos^3 t)\, dt = \left[-3t + \frac{3}{2}\sin 2t + 6(\sin t)(\cos^2 t + 2)\right]_0^{2\pi} = -6\pi$

8. $\text{curl } \mathbf{F} = \nabla \times \mathbf{F} = \begin{vmatrix} \mathbf{i} & \mathbf{j} & \mathbf{k} \\ \frac{\partial}{\partial x} & \frac{\partial}{\partial y} & \frac{\partial}{\partial z} \\ -z + \frac{1}{2+x} & \tan^{-1} y & x + \frac{1}{4+z} \end{vmatrix} = -2\mathbf{j}$; $f(x,y,z) = 4x^2 + y + z^2 \Rightarrow \nabla f = 8x\mathbf{i} + \mathbf{j} + 2z\mathbf{k}$

$\Rightarrow \mathbf{n} = \frac{\nabla f}{|\nabla f|}$ and $\mathbf{p} = \mathbf{j} \Rightarrow |\nabla f\cdot\mathbf{p}| = 1 \Rightarrow d\sigma = \frac{|\nabla f|}{|\nabla f\cdot\mathbf{p}|}\, dA = |\nabla f|\, dA$; $\nabla \times \mathbf{F}\cdot\mathbf{n} = \frac{1}{|\nabla f|}(-2\mathbf{j}\cdot\nabla f) = \frac{-2}{|\nabla f|}$

$\Rightarrow \nabla \times \mathbf{F}\cdot\mathbf{n}\, d\sigma = -2\, dA \Rightarrow \iint_S \nabla \times \mathbf{F}\cdot\mathbf{n}\, d\sigma = \iint_R -2\, dA = -2(\text{Area of R}) = -2(\pi\cdot 1\cdot 2) = -4\pi$, where R is the elliptic region in the xz-plane enclosed by $4x^2 + z^2 = 4$.

9. Flux of $\nabla \times \mathbf{F} = \iint_S \nabla \times \mathbf{F}\cdot\mathbf{n}\, d\sigma = \oint_C \mathbf{F}\cdot d\mathbf{r}$, so let C be parametrized by $\mathbf{r} = (a\cos t)\mathbf{i} + (a\sin t)\mathbf{j}$, $0 \le t \le 2\pi \Rightarrow \frac{d\mathbf{r}}{dt} = (-a\sin t)\mathbf{i} + (a\cos t)\mathbf{j} \Rightarrow \mathbf{F}\cdot\frac{d\mathbf{r}}{dt} = ay\sin t + ax\cos t = a^2\sin^2 t + a^2\cos^2 t = a^2$

$\Rightarrow$ Flux of $\nabla \times \mathbf{F} = \oint_C \mathbf{F}\cdot d\mathbf{r} = \int_0^{2\pi} a^2\, dt = 2\pi a^2$

10. $\nabla \times (y\mathbf{i}) = \begin{vmatrix} \mathbf{i} & \mathbf{j} & \mathbf{k} \\ \frac{\partial}{\partial x} & \frac{\partial}{\partial y} & \frac{\partial}{\partial z} \\ y & 0 & 0 \end{vmatrix} = -\mathbf{k}$; $\mathbf{n} = \frac{\nabla f}{|\nabla f|} = \frac{2x\mathbf{i} + 2y\mathbf{j} + 2z\mathbf{k}}{2\sqrt{x^2+y^2+z^2}} = x\mathbf{i} + y\mathbf{j} + z\mathbf{k}$

$\Rightarrow \nabla \times (y\mathbf{i}) \cdot \mathbf{n} = -z$; $d\sigma = \frac{1}{z}\,dA$ (Section 14.5, Example 5, with $a = 1$) $\Rightarrow \iint_S \nabla \times (y\mathbf{i}) \cdot \mathbf{n}\, d\sigma$

$= \iint_R (-z)\left(\frac{1}{z}\,dA\right) = -\iint_R dA = -\pi$, where R is the circle $x^2 + y^2 = 1$ in the xy-plane.

11. Let S_1 and S_2 be oriented surfaces that span C and that induce the same positive direction on C. Then

$$\iint_{S_1} \nabla \times \mathbf{F} \cdot \mathbf{n}_1\, d\sigma_1 = \oint_C \mathbf{F} \cdot d\mathbf{r} = \iint_{S_2} \nabla \times \mathbf{F} \cdot \mathbf{n}_2\, d\sigma_2$$

12. $\iint_S \nabla \times \mathbf{F} \cdot \mathbf{n}\, d\sigma = \iint_{S_1} \nabla \times \mathbf{F} \cdot \mathbf{n}\, d\sigma + \iint_{S_2} \nabla \times \mathbf{F} \cdot \mathbf{n}\, d\sigma$, and since S_1 and S_2 are joined by the simple closed curve C, each of the above integrals will be equal to a circulation integral on C. But for one surface the circulation will be counterclockwise, and for the other surface the circulation will be clockwise. Since the integrands are the same, the sum will be $0 \Rightarrow \iint_S \nabla \times \mathbf{F} \cdot \mathbf{n}\, d\sigma = 0$.

13. $\nabla \times \mathbf{F} = \begin{vmatrix} \mathbf{i} & \mathbf{j} & \mathbf{k} \\ \frac{\partial}{\partial x} & \frac{\partial}{\partial y} & \frac{\partial}{\partial z} \\ 2z & 3x & 5y \end{vmatrix} = 5\mathbf{i} + 2\mathbf{j} + 3\mathbf{k}$; $\mathbf{r}_r = (\cos\theta)\mathbf{i} + (\sin\theta)\mathbf{j} - 2r\mathbf{k}$ and $\mathbf{r}_\theta = (-r\sin\theta)\mathbf{i} + (r\cos\theta)\mathbf{j}$

$\Rightarrow \mathbf{r}_r \times \mathbf{r}_\theta = \begin{vmatrix} \mathbf{i} & \mathbf{j} & \mathbf{k} \\ \cos\theta & \sin\theta & -2r \\ -r\sin\theta & r\cos\theta & 0 \end{vmatrix} = (2r^2\cos\theta)\mathbf{i} + (2r^2\sin\theta)\mathbf{j} + r\mathbf{k}$; $\mathbf{n} = \frac{\mathbf{r}_r \times \mathbf{r}_\theta}{|\mathbf{r}_r \times \mathbf{r}_\theta|}$ and $d\sigma = |\mathbf{r}_r \times \mathbf{r}_\theta|\, dr\, d\theta$

$\Rightarrow \nabla \times \mathbf{F} \cdot \mathbf{n}\, d\sigma = (\nabla \times \mathbf{F}) \cdot (\mathbf{r}_r \times \mathbf{r}_\theta)\, dr\, d\theta = \left(10r^2\cos\theta + 4r^2\sin\theta + 3r\right) dr\, d\theta \Rightarrow \iint_S \nabla \times \mathbf{F} \cdot \mathbf{n}\, d\sigma$

$= \int_0^{2\pi}\int_0^2 \left(10r^2\cos\theta + 4r^2\sin\theta + 3r\right) dr\, d\theta = \int_0^{2\pi} \left[\frac{10}{3}r^3\cos\theta + \frac{4}{3}r^3\sin\theta + \frac{3}{2}r^2\right]_0^2 d\theta$

$= \int_0^{2\pi} \left(\frac{80}{3}\cos\theta + \frac{32}{3}\sin\theta + 6\right) d\theta = 6(2\pi) = 12\pi$

14. $\nabla \times \mathbf{F} = \begin{vmatrix} \mathbf{i} & \mathbf{j} & \mathbf{k} \\ \frac{\partial}{\partial x} & \frac{\partial}{\partial y} & \frac{\partial}{\partial z} \\ y-z & z-x & x+z \end{vmatrix} = \mathbf{i} - \mathbf{j} - 2\mathbf{k}$; $\mathbf{r}_r \times \mathbf{r}_\theta = (2r^2 \cos\theta)\mathbf{i} + (2r^2 \sin\theta)\mathbf{j} + r\mathbf{k}$ and

$\nabla \times \mathbf{F} \cdot \mathbf{n}\, d\sigma = (\nabla \times \mathbf{F}) \cdot (\mathbf{r}_r \times \mathbf{r}_\theta)\, dr\, d\theta$ (see Exercise 13 above) $\Rightarrow \iint_S \nabla \times \mathbf{F} \cdot \mathbf{n}\, d\sigma$

$= \int_0^{2\pi} \int_0^3 (2r^2 \cos\theta - 2r^2 \sin\theta - 2r)\, dr\, d\theta = \int_0^{2\pi} \left[\frac{2}{3} r^3 \cos\theta - \frac{2}{3} r^3 \sin\theta - r^2\right]_0^3 d\theta$

$= \int_0^{2\pi} \left(\frac{54}{3} \cos\theta - \frac{54}{3} \sin\theta - 9\right) d\theta = -9(2\pi) = -18\pi$

15. $\nabla \times \mathbf{F} = \begin{vmatrix} \mathbf{i} & \mathbf{j} & \mathbf{k} \\ \frac{\partial}{\partial x} & \frac{\partial}{\partial y} & \frac{\partial}{\partial z} \\ x^2 y & 2y^3 z & 3z \end{vmatrix} = -2y^3\mathbf{i} + 0\mathbf{j} - x^2\mathbf{k}$; $\mathbf{r}_r \times \mathbf{r}_\theta = \begin{vmatrix} \mathbf{i} & \mathbf{j} & \mathbf{k} \\ \cos\theta & \sin\theta & 1 \\ -r\sin\theta & r\cos\theta & 0 \end{vmatrix}$

$= (-r\cos\theta)\mathbf{i} - (r\sin\theta)\mathbf{j} + r\mathbf{k}$ and $\nabla \times \mathbf{F} \cdot \mathbf{n}\, d\sigma = (\nabla \times \mathbf{F}) \cdot (\mathbf{r}_r \times \mathbf{r}_\theta)\, dr\, d\theta$ (see Exercise 13 above)

$\Rightarrow \iint_S \nabla \times \mathbf{F} \cdot \mathbf{n}\, d\sigma = \iint_R (2ry^3 \cos\theta - rx^2)\, dr\, d\theta = \int_0^{2\pi} \int_0^1 (2r^4 \sin\theta \cos\theta - r^3 \cos^2\theta)\, dr\, d\theta$

$= \int_0^{2\pi} \left(\frac{2}{5} \sin\theta \cos\theta - \frac{1}{4} \cos^2\theta\right) d\theta = \left[\frac{1}{5} \sin^2\theta - \frac{1}{4}\left(\frac{\theta}{2} + \frac{\sin 2\theta}{4}\right)\right]_0^{2\pi} = -\frac{\pi}{4}$

16. $\nabla \times \mathbf{F} = \begin{vmatrix} \mathbf{i} & \mathbf{j} & \mathbf{k} \\ \frac{\partial}{\partial x} & \frac{\partial}{\partial y} & \frac{\partial}{\partial z} \\ x-y & y-z & z-x \end{vmatrix} = \mathbf{i} + \mathbf{j} + \mathbf{k}$; $\mathbf{r}_r \times \mathbf{r}_\theta = \begin{vmatrix} \mathbf{i} & \mathbf{j} & \mathbf{k} \\ \cos\theta & \sin\theta & -1 \\ -r\sin\theta & r\cos\theta & 0 \end{vmatrix}$

$= (r\cos\theta)\mathbf{i} + (r\sin\theta)\mathbf{j} + r\mathbf{k}$ and $\nabla \times \mathbf{F} \cdot \mathbf{n}\, d\sigma = (\nabla \times \mathbf{F}) \cdot (\mathbf{r}_r \times \mathbf{r}_\theta)\, dr\, d\theta$ (see Exercise 13 above)

$\Rightarrow \iint_S \nabla \times \mathbf{F} \cdot \mathbf{n}\, d\sigma = \int_0^{2\pi} \int_0^5 (r\cos\theta + r\sin\theta + r)\, dr\, d\theta = \int_0^{2\pi} \left[(\cos\theta + \sin\theta + 1)\frac{r^2}{2}\right]_0^5 d\theta = \left(\frac{25}{2}\right)(2\pi) = 25\pi$

17. $\nabla \times \mathbf{F} = \begin{vmatrix} \mathbf{i} & \mathbf{j} & \mathbf{k} \\ \frac{\partial}{\partial x} & \frac{\partial}{\partial y} & \frac{\partial}{\partial z} \\ 3y & 5-2x & z^2-2 \end{vmatrix} = 0\mathbf{i} + 0\mathbf{j} - 5\mathbf{k}$;

$\mathbf{r}_\phi \times \mathbf{r}_\theta = \begin{vmatrix} \mathbf{i} & \mathbf{j} & \mathbf{k} \\ \sqrt{3}\cos\phi\cos\theta & \sqrt{3}\cos\phi\sin\theta & -\sqrt{3}\sin\phi \\ -\sqrt{3}\sin\phi\sin\theta & \sqrt{3}\sin\phi\cos\theta & 0 \end{vmatrix}$

$= (3\sin^2\phi\cos\theta)\mathbf{i} + (3\sin^2\phi\sin\theta)\mathbf{j} + (3\sin\phi\cos\phi)\mathbf{k}$; $\nabla\times\mathbf{F}\cdot\mathbf{n}\,d\sigma = (\nabla\times\mathbf{F})\cdot(\mathbf{r}_\phi\times\mathbf{r}_\theta)\,d\phi\,d\theta$ (see Exercise 13 above) $\Rightarrow \iint_S \nabla\times\mathbf{F}\cdot\mathbf{n}\,d\sigma = \int_0^{2\pi}\int_0^{\pi/2} -15\cos\phi\sin\phi\,d\phi\,d\theta = \int_0^{2\pi}\left[\frac{15}{2}\cos^2\phi\right]_0^{\pi/2} d\theta = \int_0^{2\pi} -\frac{15}{2}\,d\theta = -15\pi$

18. $\nabla\times\mathbf{F} = \begin{vmatrix} \mathbf{i} & \mathbf{j} & \mathbf{k} \\ \frac{\partial}{\partial x} & \frac{\partial}{\partial y} & \frac{\partial}{\partial z} \\ y^2 & z^2 & x \end{vmatrix} = -2z\mathbf{i} - \mathbf{j} - 2y\mathbf{k}$;

$$\mathbf{r}_\phi\times\mathbf{r}_\theta = \begin{vmatrix} \mathbf{i} & \mathbf{j} & \mathbf{k} \\ 2\cos\phi\cos\theta & 2\cos\phi\sin\theta & -2\sin\phi \\ -2\sin\phi\sin\theta & 2\sin\phi\cos\theta & 0 \end{vmatrix}$$

$= (4\sin^2\phi\cos\theta)\mathbf{i} + (4\sin^2\phi\sin\theta)\mathbf{j} + (4\sin\phi\cos\phi)\mathbf{k}$; $\nabla\times\mathbf{F}\cdot\mathbf{n}\,d\sigma = (\nabla\times\mathbf{F})\cdot(\mathbf{r}_\phi\times\mathbf{r}_\theta)\,d\phi\,d\theta$ (see Exercise 13 above) $\Rightarrow \iint_S \nabla\times\mathbf{F}\cdot\mathbf{n}\,d\sigma = \iint_R \left(-8z\sin^2\phi\cos\theta - 4\sin^2\phi\sin\theta - 8y\sin\phi\cos\theta\right)d\phi\,d\theta$

$$= \int_0^{2\pi}\int_0^{\pi/2} \left(-16\sin^2\phi\cos\phi\cos\theta - 4\sin^2\phi\sin\theta - 16\sin^2\phi\sin\theta\cos\theta\right)d\phi\,d\theta$$

$$= \int_0^{2\pi}\left[-\frac{16}{3}\sin^3\phi\cos\theta - 4\left(\frac{\phi}{2} - \frac{\sin 2\phi}{4}\right)(\sin\theta) - 16\left(\frac{\phi}{2} - \frac{\sin 2\phi}{4}\right)(\sin\theta\cos\theta)\right]_0^{\pi/2} d\theta$$

$$= \int_0^{2\pi}\left(-\frac{16}{3}\cos\theta - \pi\sin\theta - 4\pi\sin\theta\cos\theta\right)d\theta = \left[-\frac{16}{3}\sin\theta + \pi\cos\theta - 2\pi\sin^2\theta\right]_0^{2\pi} = 0$$

19. (a) $\mathbf{F} = 2x\mathbf{i} + 2y\mathbf{j} + 2z\mathbf{k} \Rightarrow \text{curl }\mathbf{F} = \mathbf{0} \Rightarrow \oint_C \mathbf{F}\cdot d\mathbf{r} = \iint_S \nabla\times\mathbf{F}\cdot\mathbf{n}\,d\sigma = \iint_S d\sigma = 0$

(b) Let $f(x,y,z) = x^2y^2z^3 \Rightarrow \nabla\times\mathbf{F} = \nabla\times\nabla f = \mathbf{0} \Rightarrow \text{curl }\mathbf{F} = \mathbf{0} \Rightarrow \oint_C \mathbf{F}\cdot d\mathbf{r} = \iint_S \nabla\times\mathbf{F}\cdot\mathbf{n}\,d\sigma$
$= \iint_S 0\,d\sigma = 0$

(c) $\mathbf{F} = \nabla\times(x\mathbf{i} + y\mathbf{j} + z\mathbf{k}) = \mathbf{0} \Rightarrow \nabla\times\mathbf{F} = \mathbf{0} \Rightarrow \oint_C \mathbf{F}\cdot d\mathbf{r} = \iint_S \nabla\times\mathbf{F}\cdot\mathbf{n}\,d\sigma = \iint_S 0\,d\sigma = 0$

(d) $\mathbf{F} = \nabla f \Rightarrow \nabla\times\mathbf{F} = \nabla\times\nabla f = \mathbf{0} \Rightarrow \oint_C \mathbf{F}\cdot d\mathbf{r} = \iint_S \nabla\times\mathbf{F}\cdot\mathbf{n}\,d\sigma = \iint_S 0\,d\sigma = 0$

20. $\mathbf{F} = \nabla f = -\frac{1}{2}(x^2+y^2+z^2)^{-3/2}(2x)\mathbf{i} - \frac{1}{2}(x^2+y^2+z^2)^{-3/2}(2y)\mathbf{j} - \frac{1}{2}(x^2+y^2+z^2)^{-3/2}(2z)\mathbf{k}$
$= -x(x^2+y^2+z^2)^{-3/2}\mathbf{i} - y(x^2+y^2+z^2)^{-3/2}\mathbf{j} - z(x^2+y^2+z^2)^{-3/2}\mathbf{k}$

(a) $\mathbf{r} = (a\cos t)\mathbf{i} + (a\sin t)\mathbf{j},\ 0 \le t \le 2\pi \Rightarrow \frac{d\mathbf{r}}{dt} = (-a\sin t)\mathbf{i} + (a\cos t)\mathbf{j}$

$\Rightarrow \mathbf{F}\cdot\frac{d\mathbf{r}}{dt} = -x(x^2+y^2+z^2)^{-3/2}(-a\sin t) - y(x^2+y^2+z^2)^{-3/2}(a\cos t)$

$= \left(-\frac{a\cos t}{a^3}\right)(-a\sin t) - \left(\frac{a\sin t}{a^3}\right)(a\cos t) = 0 \Rightarrow \oint_C \mathbf{F}\cdot d\mathbf{r} = 0$

(b) $\oint_C \mathbf{F}\cdot d\mathbf{r} = \iint_S \nabla\times\mathbf{F}\cdot\mathbf{n}\,d\sigma = \iint_S \nabla\times\nabla f\cdot\mathbf{n}\,d\sigma = \iint_S \mathbf{0}\cdot\mathbf{n}\,d\sigma = \iint_S 0\,d\sigma = 0$

21. Let $\mathbf{F} = 2y\mathbf{i} + 3z\mathbf{j} - x\mathbf{k} \Rightarrow \nabla\times\mathbf{F} = \begin{vmatrix} \mathbf{i} & \mathbf{j} & \mathbf{k} \\ \frac{\partial}{\partial x} & \frac{\partial}{\partial y} & \frac{\partial}{\partial z} \\ 2y & 3z & -x \end{vmatrix} = -3\mathbf{i} + \mathbf{j} - 2\mathbf{k};\ \mathbf{n} = \frac{2\mathbf{i}+2\mathbf{j}+\mathbf{k}}{3}$

$\Rightarrow \nabla\times\mathbf{F}\cdot\mathbf{n} = -2 \Rightarrow \oint_C 2y\,dx + 3z\,dy - x\,dz = \oint_C \mathbf{F}\cdot d\mathbf{r} = \iint_S \nabla\times\mathbf{F}\cdot\mathbf{n}\,d\sigma = \iint_S -2\,d\sigma$

$= -2\iint_S d\sigma$, where $\iint_S d\sigma$ is the area of the region enclosed by C on the plane S: $2x + 2y + z = 2$

22. $\nabla\times\mathbf{F} = \begin{vmatrix} \mathbf{i} & \mathbf{j} & \mathbf{k} \\ \frac{\partial}{\partial x} & \frac{\partial}{\partial y} & \frac{\partial}{\partial z} \\ x & y & z \end{vmatrix} = 0$

23. Suppose $\mathbf{F} = M\mathbf{i} + N\mathbf{j} + P\mathbf{k}$ exists such that $\nabla\times\mathbf{F} = \left(\frac{\partial P}{\partial y} - \frac{\partial N}{\partial z}\right)\mathbf{i} + \left(\frac{\partial M}{\partial z} - \frac{\partial P}{\partial x}\right)\mathbf{j} + \left(\frac{\partial N}{\partial x} - \frac{\partial M}{\partial y}\right)\mathbf{k}$

$= x\mathbf{i} + y\mathbf{j} + z\mathbf{k}$ Then $\frac{\partial}{\partial x}\left(\frac{\partial P}{\partial y} - \frac{\partial N}{\partial z}\right) = \frac{\partial}{\partial x}(x) \Rightarrow \frac{\partial^2 P}{\partial x\partial y} - \frac{\partial^2 N}{\partial x\partial z} = 1$. Likewise, $\frac{\partial}{\partial y}\left(\frac{\partial M}{\partial z} - \frac{\partial P}{\partial x}\right) = \frac{\partial}{\partial y}(y)$

$\Rightarrow \frac{\partial^2 M}{\partial y\partial z} - \frac{\partial^2 P}{\partial y\partial x} = 1$ and $\frac{\partial}{\partial z}\left(\frac{\partial N}{\partial x} - \frac{\partial M}{\partial y}\right) = \frac{\partial}{\partial z}(z) \Rightarrow \frac{\partial^2 N}{\partial z\partial x} - \frac{\partial^2 M}{\partial z\partial y} = 1$. Summing the calculated equations

$\Rightarrow \left(\frac{\partial^2 P}{\partial x\partial y} - \frac{\partial^2 P}{\partial y\partial x}\right) + \left(\frac{\partial^2 N}{\partial z\partial x} - \frac{\partial^2 N}{\partial y\partial z}\right) + \left(\frac{\partial^2 M}{\partial y\partial z} - \frac{\partial^2 M}{\partial z\partial y}\right) = 3$ or $0 = 3$ (assuming the second mixed partials are equal). This result is a contradiction, so there is no field $\mathbf{F}$ such that curl $\mathbf{F} = x\mathbf{i} + y\mathbf{j} + z\mathbf{k}$.

24. Yes: If $\nabla\times\mathbf{F} = \mathbf{0}$, then the circulation of $\mathbf{F}$ around the boundary C of any oriented surface S in the domain of $\mathbf{F}$ is zero. The reason is this: By Stokes's theorem, circulation $= \oint_C \mathbf{F}\cdot d\mathbf{r} = \iint_S \nabla\times\mathbf{F}\cdot\mathbf{n}\,d\sigma = \iint_S \mathbf{0}\cdot\mathbf{n}\,d\sigma$ $= 0$.

25. $r = \sqrt{x^2+y^2} \Rightarrow r^4 = (x^2+y^2)^2 \Rightarrow \mathbf{F} = \nabla(r^4) = 4x(x^2+y^2)\mathbf{i} + 4y(x^2+y^2)\mathbf{j} = M\mathbf{i} + N\mathbf{j}$

$\Rightarrow \oint_C \nabla(r^4)\cdot\mathbf{n}\,ds = \oint_C \mathbf{F}\cdot\mathbf{n}\,ds = \oint_C M\,dy - N\,dx = \iint_R \left(\frac{\partial M}{\partial x} + \frac{\partial N}{\partial y}\right)dx\,dy$

$= \iint\limits_R [4(x^2+y^2)+8x^2+4(x^2+y^2)+8y^2]\, dA = \iint\limits_R 16(x^2+y^2)\, dA = 16 \iint\limits_R x^2\, dA + 16 \iint\limits_R y^2\, dA$

$= 16I_y + 16I_x.$

26. $\frac{\partial P}{\partial y} = 0,\ \frac{\partial N}{\partial z} = 0,\ \frac{\partial M}{\partial z} = 0,\ \frac{\partial P}{\partial x} = 0,\ \frac{\partial N}{\partial x} = \frac{y^2-x^2}{(x^2+y^2)^2},\ \frac{\partial M}{\partial y} = \frac{y^2-x^2}{(x^2+y^2)^2} \Rightarrow \text{curl } \mathbf{F} = \left[\frac{y^2-x^2}{(x^2+y^2)^2} - \frac{y^2-x^2}{(x^2+y^2)^2}\right]\mathbf{k} = \mathbf{0}.$

However, $x^2+y^2 = 1 \Rightarrow \mathbf{r} = (a \cos t)\mathbf{i} + (a \sin t)\mathbf{j} \Rightarrow \frac{d\mathbf{r}}{dt} = (-a \sin t)\mathbf{i} + (a \cos t)\mathbf{j}$

$\Rightarrow \mathbf{F} = \left(\frac{-a \sin t}{a^2}\right)\mathbf{i} + \left(\frac{a \cos t}{a^2}\right)\mathbf{j} + 2\mathbf{k} \Rightarrow \mathbf{F}\cdot\frac{d\mathbf{r}}{dt} = \frac{a^2 \sin^2 t}{a^2} + \frac{a^2 \cos^2 t}{a^2} = 1 \Rightarrow \oint\limits_C \mathbf{F}\cdot d\mathbf{r} = \oint_0^{2\pi} 1\, dt = 2\pi$ which is not zero.

14.8 THE DIVERGENCE THEOREM AND A UNIFIED THEORY

1. $\mathbf{F} = \frac{-y\mathbf{i} + x\mathbf{j}}{\sqrt{x^2+y^2}} \Rightarrow \text{div } \mathbf{F} = \frac{xy - xy}{(x^2+y^2)^{3/2}} = 0$

2. $\mathbf{F} = x\mathbf{i} + y\mathbf{j} \Rightarrow \text{div } \mathbf{F} = 1 + 1 = 2$

3. $\mathbf{F} = -\frac{GM(x\mathbf{i}+y\mathbf{j}+z\mathbf{k})}{(x^2+y^2+z^2)^{3/2}} \Rightarrow \text{div } \mathbf{F} = -GM\left[\frac{(x^2+y^2+z^2)^{3/2} - 3x^2(x^2+y^2+z^2)^{1/2}}{(x^2+y^2+z^2)^3}\right]$

$-GM\left[\frac{(x^2+y^2+z^2)^{3/2} - 3y^2(x^2+y^2+z^2)^{1/2}}{(x^2+y^2+z^2)^3}\right] - GM\left[\frac{(x^2+y^2+z^2)^{3/2} - 3z^2(x^2+y^2+z^2)^{1/2}}{(x^2+y^2+z^2)^3}\right]$

$= -GM\left[\frac{3(x^2+y^2+z^2)^2 - 3(x^2+y^2+z^2)(x^2+y^2+z^2)}{(x^2+y^2+z^2)^{7/2}}\right] = 0$

4. $z = a^2 - r^2$ in cylindrical coordinates $\Rightarrow z = a^2 - (x^2+y^2) \Rightarrow \mathbf{v} = (a^2 - x^2 - y^2)\mathbf{k} \Rightarrow \text{div } \mathbf{v} = 0$

5. $\frac{\partial}{\partial x}(y-x) = -1,\ \frac{\partial}{\partial y}(z-y) = -1,\ \frac{\partial}{\partial z}(y-x) = 0 \Rightarrow \nabla\cdot\mathbf{F} = -2 \Rightarrow \text{Flux} = \int_{-1}^{1}\int_{-1}^{1}\int_{-1}^{1} -2\, dx\, dy\, dz = -2(2^3)$

$= -16$

6. $\frac{\partial}{\partial x}(x^2) = 2x,\ \frac{\partial}{\partial y}(y^2) = 2y,\ \frac{\partial}{\partial x}(z^2) = 2z \Rightarrow \nabla\cdot\mathbf{F} = 2x + 2y + 2z$

(a) $\text{Flux} = \int_0^1\int_0^1\int_0^1 (2x+2y+2z)\, dx\, dy\, dz = \int_0^1\int_0^1 \left[x^2 + 2x(y+z)\right]_0^1 dy\, dz = \int_0^1\int_0^1 (1+2y+2z)\, dy\, dz$

$= \int_0^1 \left[y(1+2z) + y^2\right]_0^1 dz = \int_0^1 (2+2z)\, dz = \left[2z + z^2\right]_0^1 = 3$

(b) $\text{Flux} = \int_{-1}^{1}\int_{-1}^{1}\int_{-1}^{1} (2x + 2y + 2z)\, dx\, dy\, dz = \int_{-1}^{1}\int_{-1}^{1} \left[x^2 + 2x(y + z)\right]_{-1}^{1} dy\, dz = \int_{-1}^{1}\int_{-1}^{1} (4y + 4z)\, dy\, dz$

$= \int_{-1}^{1} \left[2y^2 + 4yz\right]_{-1}^{1} dz = \int_{-1}^{1} 8z\, dz = \left[4z^2\right]_{-1}^{1} = 0$

(c) In cylindrical coordinates, $\text{Flux} = \iiint_D (2x + 2y + 2z)\, dx\, dy\, dz$

$= \int_0^1\int_0^{2\pi}\int_0^2 (2r\cos\theta + 2r\sin\theta + 2z)\, r\, dr\, d\theta\, dz = \int_0^1\int_0^{2\pi} \left[\frac{2}{3}r^3\cos\theta + \frac{2}{3}r^3\sin\theta + zr^2\right]_0^2 d\theta\, dz$

$= \int_0^1\int_0^{2\pi} \left(\frac{16}{3}\cos\theta + \frac{16}{3}\sin\theta + 4z\right) d\theta\, dz = \int_0^1 \left[\frac{16}{3}\sin\theta - \frac{16}{3}\cos\theta + 4z\theta\right]_0^{2\pi} dz = \int_0^1 8\pi z\, dz = \left[4\pi z^2\right]_0^1 = 4\pi$

7. $\frac{\partial}{\partial x}(y) = 0,\ \frac{\partial}{\partial y}(xy) = x,\ \frac{\partial}{\partial z}(-z) = -1 \Rightarrow \nabla\cdot\mathbf{F} = x - 1;\ z = x^2 + y^2 \Rightarrow z = r^2$ in cylindrical coordinates

$\Rightarrow \text{Flux} = \iiint_D (x - 1)\, dz\, dy\, dx = \int_0^{2\pi}\int_0^2\int_0^{r^2} (r\cos\theta - 1)\, dz\, r\, dr\, d\theta = \int_0^{2\pi}\int_0^2 \left(r^3\cos\theta - r^2\right) r\, dr\, d\theta$

$= \int_0^{2\pi} \left[\frac{r^5}{5}\cos\theta - \frac{r^4}{4}\right]_0^2 d\theta = \int_0^{2\pi} \left(\frac{32}{5}\cos\theta - 4\right) d\theta = \left[\frac{32}{5}\sin\theta - 4\theta\right]_0^{2\pi} = -8\pi$

8. $\frac{\partial}{\partial x}(x^2) = 2x,\ \frac{\partial}{\partial y}(xz) = 0,\ \frac{\partial}{\partial z}(3z) = 3 \Rightarrow \nabla\cdot\mathbf{F} = 2x + 3 \Rightarrow \text{Flux} = \iiint_D (2x + 3)\, dV$

$= \int_0^{2\pi}\int_0^{\pi}\int_0^2 (2\rho\sin\phi\cos\theta + 3)\left(\rho^2\sin\phi\right) d\rho\, d\phi\, d\theta = \int_0^{2\pi}\int_0^{\pi} \left[\frac{\rho^4}{2}\sin\phi\cos\theta + \rho^3\right]_0^2 \sin\phi\, d\phi\, d\theta$

$= \int_0^{2\pi}\int_0^{\pi} (8\sin\phi\cos\theta + 8)\sin\phi\, d\phi\, d\theta = \int_0^{2\pi} \left[8\left(\frac{\phi}{2} - \frac{\sin 2\phi}{4}\right)\cos\theta - 8\cos\phi\right]_0^{\pi} d\theta = \int_0^{2\pi} (4\pi\cos\theta + 16)\, d\theta$

$= 32\pi$

9. $\frac{\partial}{\partial x}(x^2) = 2x,\ \frac{\partial}{\partial y}(-2xy) = -2x,\ \frac{\partial}{\partial z}(3xz) = 3x \Rightarrow \text{Flux} = \iiint_D 3x\, dx\, dy\, dz$

$= \int_0^{\pi/2}\int_0^{\pi/2}\int_0^2 (3\rho\sin\phi\cos\theta)\left(\rho^2\sin\phi\right) d\rho\, d\phi\, d\theta = \int_0^{\pi/2}\int_0^{\pi/2} 12\sin^2\phi\cos\theta\, d\phi\, d\theta = \int_0^{\pi/2} 3\pi\cos\theta\, d\theta = 3\pi$

10. $\frac{\partial}{\partial x}(6x^2+2xy) = 12x+2y$, $\frac{\partial}{\partial y}(2y+x^2z) = 2$, $\frac{\partial}{\partial z}(4x^2y^3) = 0 \Rightarrow \nabla \cdot \mathbf{F} = 12x+2y+2$

$$\Rightarrow \text{Flux} = \iiint_D (12x+2y+2)\, dV = \int_0^3 \int_0^{\pi/2} \int_0^2 (12r \cos\theta + 2r \sin\theta + 2)\, r\, dr\, d\theta\, dz$$

$$= \int_0^3 \int_0^{\pi/2} \left(32\cos\theta + \frac{16}{3}\sin\theta + 4\right) d\theta\, dz = \int_0^3 \left(32 + 2\pi + \frac{16}{3}\right) dz = 112 + 6\pi$$

11. $\frac{\partial}{\partial x}(2xz) = 2z$, $\frac{\partial}{\partial y}(-xy) = -x$, $\frac{\partial}{\partial z}(-z^2) = -2z \Rightarrow \nabla \cdot \mathbf{F} = -x \Rightarrow \text{Flux} = \iiint_D -x\, dV$

$$= \int_0^2 \int_0^{\sqrt{16-4x^2}} \int_0^{4-y} -x\, dz\, dy\, dx = \int_0^2 \int_0^{\sqrt{16-4x^2}} (xy - 4x)\, dy\, dx = \int_0^2 \left[\frac{1}{2}x(16-4x^2) - 4x\sqrt{16-4x^2}\right] dx$$

$$= \left[4x^2 - \frac{1}{2}x^4 + \frac{1}{3}(16-4x^2)^{3/2}\right]_0^2 = -\frac{40}{3}$$

12. $\frac{\partial}{\partial x}(x^3) = 3x^2$, $\frac{\partial}{\partial y}(y^3) = 3y^2$, $\frac{\partial}{\partial z}(z^3) = 3z^2 \Rightarrow \nabla \cdot \mathbf{F} = 3x^2+3y^2+3z^2 \Rightarrow \text{Flux} = \iiint_D 3(x^2+y^2+z^2)\, dV$

$$= 3\int_0^{2\pi} \int_0^{\pi} \int_0^{a} \rho^2(\rho^2 \sin\phi)\, d\rho\, d\phi\, d\theta = 3\int_0^{2\pi} \int_0^{\pi} \frac{a^5}{5}\sin\phi\, d\phi\, d\theta = 3\int_0^{2\pi} \frac{2a^5}{5}\, d\theta = \frac{12\pi a^5}{5}$$

13. Let $\rho = \sqrt{x^2+y^2+z^2}$. Then $\frac{\partial \rho}{\partial x} = \frac{x}{\rho}$, $\frac{\partial \rho}{\partial y} = \frac{y}{\rho}$, $\frac{\partial \rho}{\partial z} = \frac{z}{\rho} \Rightarrow \frac{\partial}{\partial x}(\rho x) = \left(\frac{\partial \rho}{\partial x}\right)x + \rho = \frac{x^2}{\rho} + \rho$, $\frac{\partial}{\partial y}(\rho y) = \left(\frac{\partial \rho}{\partial y}\right)y + \rho$

$= \frac{y^2}{\rho} + \rho$, $\frac{\partial}{\partial z}(\rho z) = \left(\frac{\partial \rho}{\partial z}\right)z + \rho = \frac{z^2}{\rho} + \rho \Rightarrow \nabla \cdot \mathbf{F} = \frac{x^2+y^2+z^2}{\rho} + 3\rho = 4\rho$, since $\rho = \sqrt{x^2+y^2+z^2}$

$$\Rightarrow \text{Flux} = \iiint_D 4\rho\, dV = \int_0^{2\pi} \int_0^{\pi} \int_1^{\sqrt{2}} (4\rho)(\rho^2 \sin\phi)\, d\rho\, d\phi\, d\theta = \int_0^{2\pi} \int_0^{\pi} 3\sin\phi\, d\phi\, d\theta = \int_0^{2\pi} 6\, d\theta = 12\pi$$

14. Let $\rho = \sqrt{x^2+y^2+z^2}$. Then $\frac{\partial \rho}{\partial x} = \frac{x}{\rho}$, $\frac{\partial \rho}{\partial y} = \frac{y}{\rho}$, $\frac{\partial \rho}{\partial z} = \frac{z}{\rho} \Rightarrow \frac{\partial}{\partial x}\left(\frac{x}{\rho}\right) = \frac{1}{\rho} - \left(\frac{x}{\rho^2}\right)\frac{\partial \rho}{\partial x} = \frac{1}{\rho} - \frac{x^2}{\rho^3}$. Similarly,

$\frac{\partial}{\partial y}\left(\frac{y}{\rho}\right) = \frac{1}{\rho} - \frac{y^2}{\rho^3}$ and $\frac{\partial}{\partial z}\left(\frac{z}{\rho}\right) = \frac{1}{\rho} - \frac{z^2}{\rho^3} \Rightarrow \nabla \cdot \mathbf{F} = \frac{3}{\rho} - \frac{x^2+y^2+z^2}{\rho^3} = \frac{2}{\rho}$

$$\Rightarrow \text{Flux} = \iiint_D \frac{2}{\rho}\, dV = \int_0^{2\pi} \int_0^{\pi} \int_1^{2} \left(\frac{2}{\rho}\right)(\rho^2 \sin\phi)\, d\rho\, d\phi\, d\theta = \int_0^{2\pi} \int_0^{\pi} 15\sin\phi\, d\phi\, d\theta = \int_0^{2\pi} 30\, d\theta = 60\pi$$

15. $\frac{\partial}{\partial x}(5x^3+12xy^2)=15x^2+12y^2$, $\frac{\partial}{\partial y}(y^3+e^y \sin z)=3y^2+e^y \sin z$, $\frac{\partial}{\partial z}(5z^3+e^y \cos z)=15z^2-e^y \sin z$

$$\Rightarrow \nabla \cdot \mathbf{F}=15x^2+15y^2+15z^2=15\rho^2 \Rightarrow \text{Flux}=\iiint_D 15\rho^2\, dV=\int_0^{2\pi}\int_0^{\pi}\int_1^{\sqrt{2}}(15\rho^2)(\rho^2 \sin \phi)\, d\rho\, d\phi\, d\theta$$

$$=\int_0^{2\pi}\int_0^{\pi}(12\sqrt{2}-3)\sin \phi\, d\phi\, d\theta=\int_0^{2\pi}(24\sqrt{2}-6)\, d\theta=(48\sqrt{2}-12)\pi$$

16. $\frac{\partial}{\partial x}[\ln(x^2+y^2)]=\frac{2x}{x^2+y^2}$, $\frac{\partial}{\partial y}\left(-\frac{2z}{x}\tan^{-1}\frac{y}{x}\right)=\left(-\frac{2z}{x}\right)\left[\frac{\left(\frac{1}{x}\right)}{1+\left(\frac{y}{x}\right)^2}\right]=-\frac{2z}{x^2+y^2}$, $\frac{\partial}{\partial z}\left(z\sqrt{x^2+y^2}\right)=\sqrt{x^2+y^2}$

$$\Rightarrow \nabla \cdot \mathbf{F}=\frac{2x}{x^2+y^2}-\frac{2z}{x^2+y^2}+\sqrt{x^2+y^2} \Rightarrow \text{Flux}=\iiint_D\left(\frac{2x}{x^2+y^2}-\frac{2z}{x^2+y^2}+\sqrt{x^2+y^2}\right)dz\, dy\, dx$$

$$=\int_0^{2\pi}\int_1^{\sqrt{2}}\int_{-1}^{2}\left(\frac{2r\cos\theta}{r^2}-\frac{2z}{r^2}+r\right)dz\, r\, dr\, d\theta=\int_0^{2\pi}\int_1^{\sqrt{2}}\left(6\cos\theta-\frac{3}{r}+3r^2\right)dr\, d\theta$$

$$=\int_0^{2\pi}\left[6(\sqrt{2}-1)\cos\theta-3\ln\sqrt{2}+2\sqrt{2}-1\right]d\theta=2\pi\left(-\frac{3}{2}\ln 2+2\sqrt{2}-1\right)$$

17. (a) $\mathbf{G}=M\mathbf{i}+N\mathbf{j}+P\mathbf{k} \Rightarrow \nabla \times \mathbf{G}=\text{curl } \mathbf{G}=\left(\frac{\partial P}{\partial y}-\frac{\partial N}{\partial z}\right)\mathbf{i}+\left(\frac{\partial M}{\partial z}-\frac{\partial P}{\partial x}\right)\mathbf{k}+\left(\frac{\partial N}{\partial x}-\frac{\partial M}{\partial y}\right)\mathbf{k} \Rightarrow \nabla \cdot \nabla \times \mathbf{G}$

$$=\text{div}(\text{curl } \mathbf{G})=\frac{\partial}{\partial x}\left(\frac{\partial P}{\partial y}-\frac{\partial N}{\partial z}\right)+\frac{\partial}{\partial y}\left(\frac{\partial M}{\partial z}-\frac{\partial P}{\partial x}\right)+\frac{\partial}{\partial z}\left(\frac{\partial N}{\partial x}-\frac{\partial M}{\partial y}\right)$$

$$=\frac{\partial^2 P}{\partial x\partial y}-\frac{\partial^2 N}{\partial x\partial z}+\frac{\partial^2 M}{\partial y\partial z}-\frac{\partial^2 P}{\partial y\partial x}+\frac{\partial^2 N}{\partial z\partial x}-\frac{\partial^2 M}{\partial z\partial y}=0$$ if all first and second partial derivatives are continuous

(b) By the Divergence Theorem, the outward flux of $\nabla \times \mathbf{G}$ across a closed surface is zero because

outward flux of $\nabla \times \mathbf{G}=\iint_S(\nabla \times \mathbf{G})\cdot \mathbf{n}\, d\sigma$

$$=\iiint_D \nabla \cdot \nabla \times \mathbf{G}\, dV \qquad [\text{Divergence Theorem with } \mathbf{F}=\nabla \times \mathbf{G}]$$

$$=\iiint_D (0)\, dV=0 \qquad [\text{by part (a)}]$$

18. (a) Let $\mathbf{F}_1=M_1\mathbf{i}+N_1\mathbf{j}+P_1\mathbf{k}$ and $\mathbf{F}_2=M_2\mathbf{i}+N_2\mathbf{j}+P_2\mathbf{k} \Rightarrow a\mathbf{F}_1+b\mathbf{F}_2$

$$=(aM_1+bM_2)\mathbf{i}+(aN_1+bN_2)\mathbf{j}+(aP_1+bP_2)\mathbf{k} \Rightarrow \nabla \cdot (a\mathbf{F}_1+b\mathbf{F}_2)$$

$$=\left(a\frac{\partial M_1}{\partial x}+b\frac{\partial M_2}{\partial x}\right)+\left(a\frac{\partial N_1}{\partial y}+b\frac{\partial N_2}{\partial y}\right)+\left(a\frac{\partial P_1}{\partial z}+b\frac{\partial P_2}{\partial z}\right)$$

$$=a\left(\frac{\partial M_1}{\partial x}+\frac{\partial N_1}{\partial y}+\frac{\partial P_1}{\partial z}\right)+b\left(\frac{\partial M_2}{\partial x}+\frac{\partial N_2}{\partial y}+\frac{\partial P_2}{\partial z}\right)=a(\nabla \cdot \mathbf{F}_1)+b(\nabla \cdot \mathbf{F}_2)$$

(b) Define $\mathbf{F}_1$ and $\mathbf{F}_2$ as in part a $\Rightarrow \nabla \times (a\mathbf{F}_1 + b\mathbf{F}_2)$

$$= \left[\left(a\frac{\partial P_1}{\partial y} + b\frac{\partial P_2}{\partial y}\right) - \left(a\frac{\partial N_1}{\partial z} + b\frac{\partial N_2}{\partial z}\right)\right]\mathbf{i} + \left[\left(a\frac{\partial M_1}{\partial z} + b\frac{\partial M_2}{\partial z}\right) - \left(a\frac{\partial P_1}{\partial x} + b\frac{\partial P_2}{\partial x}\right)\right]\mathbf{j}$$

$$+ \left[\left(a\frac{\partial N_1}{\partial x} + b\frac{\partial N_2}{\partial x}\right) - \left(a\frac{\partial M_1}{\partial y} + b\frac{\partial M_2}{\partial y}\right)\right]\mathbf{k} = a\left[\left(\frac{\partial P_1}{\partial y} - \frac{\partial N_1}{\partial z}\right)\mathbf{i} + \left(\frac{\partial M_1}{\partial z} - \frac{\partial P_1}{\partial x}\right)\mathbf{j} + \left(\frac{\partial N_1}{\partial x} - \frac{\partial M_1}{\partial y}\right)\mathbf{k}\right]$$

$$+ b\left[\left(\frac{\partial P_2}{\partial y} - \frac{\partial N_2}{\partial z}\right)\mathbf{i} + \left(\frac{\partial M_2}{\partial z} - \frac{\partial P_2}{\partial x}\right)\mathbf{j} + \left(\frac{\partial N_2}{\partial x} - \frac{\partial M_2}{\partial y}\right)\mathbf{k}\right] = a\nabla \times \mathbf{F}_1 + b\nabla \times \mathbf{F}_2$$

(c) $$\mathbf{F}_1 \times \mathbf{F}_2 = \begin{vmatrix} \mathbf{i} & \mathbf{j} & \mathbf{k} \\ M_1 & N_1 & P_1 \\ M_2 & N_2 & P_2 \end{vmatrix} = (N_1P_2 - P_1N_2)\mathbf{i} - (M_1P_2 - P_1M_2)\mathbf{j} + (M_1N_2 - N_1M_2)\mathbf{k} \Rightarrow \nabla \cdot (\mathbf{F}_1 \times \mathbf{F}_2)$$

$$= \nabla \cdot [(N_1P_2 - P_1N_2)\mathbf{i} - (M_1P_2 - P_1M_2)\mathbf{j} + (M_1N_2 - N_1M_2)\mathbf{k}]$$

$$= \frac{\partial}{\partial x}(N_1P_2 - P_1N_2) - \frac{\partial}{\partial y}(M_1P_2 - P_1M_2) + \frac{\partial}{\partial z}(M_1N_2 - N_1M_2) = \left(P_2\frac{\partial N_1}{\partial x} + N_1\frac{\partial P_2}{\partial x} - N_2\frac{\partial P_1}{\partial x} - P_1\frac{\partial N_2}{\partial x}\right)$$

$$- \left(M_1\frac{\partial P_2}{\partial y} + P_2\frac{\partial M_1}{\partial y} - P_1\frac{\partial M_2}{\partial y} - M_2\frac{\partial P_1}{\partial y}\right) + \left(M_1\frac{\partial N_2}{\partial z} + N_2\frac{\partial M_1}{\partial z} - N_1\frac{\partial M_2}{\partial z} - M_2\frac{\partial N_1}{\partial z}\right)$$

$$= M_2\left(\frac{\partial P_1}{\partial y} - \frac{\partial N_1}{\partial z}\right) + N_2\left(\frac{\partial M_1}{\partial z} - \frac{\partial P_1}{\partial x}\right) + P_2\left(\frac{\partial N_1}{\partial x} - \frac{\partial M_1}{\partial y}\right) + M_1\left(\frac{\partial N_2}{\partial z} - \frac{\partial P_2}{\partial y}\right) + N_1\left(\frac{\partial P_2}{\partial x} - \frac{\partial M_2}{\partial z}\right)$$

$$+ P_1\left(\frac{\partial M_2}{\partial y} - \frac{\partial N_2}{\partial x}\right) = \mathbf{F}_2 \cdot \nabla \times \mathbf{F}_1 - \mathbf{F}_1 \cdot \nabla \times \mathbf{F}_2$$

19. (a) $$\operatorname{div}(g\mathbf{F}) = \nabla \cdot g\mathbf{F} = \frac{\partial}{\partial x}(gM) + \frac{\partial}{\partial y}(gN) + \frac{\partial}{\partial z}(gP) = \left(g\frac{\partial M}{\partial x} + M\frac{\partial g}{\partial x}\right) + \left(g\frac{\partial N}{\partial y} + N\frac{\partial g}{\partial y}\right) + \left(g\frac{\partial P}{\partial z} + P\frac{\partial g}{\partial x}\right)$$

$$= \left(M\frac{\partial g}{\partial x} + N\frac{\partial g}{\partial y} + P\frac{\partial g}{\partial z}\right) + g\left(\frac{\partial M}{\partial x} + \frac{\partial N}{\partial y} + \frac{\partial P}{\partial z}\right) = g\nabla \cdot \mathbf{F} + \nabla g \cdot \mathbf{F}$$

(b) $$\nabla \times (g\mathbf{F}) = \left[\frac{\partial}{\partial y}(gP) - \frac{\partial}{\partial z}(gN)\right]\mathbf{i} + \left[\frac{\partial}{\partial z}(gM) - \frac{\partial}{\partial x}(gP)\right]\mathbf{j} + \left[\frac{\partial}{\partial x}(gN) - \frac{\partial}{\partial y}(gM)\right]\mathbf{k}$$

$$= \left(P\frac{\partial g}{\partial y} + g\frac{\partial P}{\partial y} - N\frac{\partial g}{\partial z} - g\frac{\partial N}{\partial z}\right)\mathbf{i} + \left(M\frac{\partial g}{\partial z} + g\frac{\partial M}{\partial z} - P\frac{\partial g}{\partial x} - g\frac{\partial P}{\partial x}\right)\mathbf{j} + \left(N\frac{\partial g}{\partial x} + g\frac{\partial N}{\partial x} - M\frac{\partial g}{\partial y} - g\frac{\partial M}{\partial y}\right)\mathbf{k}$$

$$= \left(P\frac{\partial g}{\partial y} - N\frac{\partial g}{\partial z}\right)\mathbf{i} + \left(g\frac{\partial P}{\partial y} - g\frac{\partial N}{\partial z}\right)\mathbf{i} + \left(M\frac{\partial g}{\partial z} - P\frac{\partial g}{\partial x}\right)\mathbf{j} + \left(g\frac{\partial M}{\partial z} - g\frac{\partial P}{\partial x}\right)\mathbf{j} + \left(N\frac{\partial g}{\partial x} - M\frac{\partial g}{\partial y}\right)\mathbf{k}$$

$$+ \left(g\frac{\partial N}{\partial x} - g\frac{\partial M}{\partial y}\right)\mathbf{k} = g\nabla \times \mathbf{F} + \nabla g \times \mathbf{F}$$

20. Let $\mathbf{F}_1 = M_1\mathbf{i} + N_1\mathbf{j} + P_1\mathbf{k}$ and $\mathbf{F}_2 = M_2\mathbf{i} + N_2\mathbf{j} + P_2\mathbf{k}$.

(a) $$\mathbf{F}_1 \times \mathbf{F}_2 = (N_1P_2 - P_1N_2)\mathbf{i} + (P_1M_2 - M_1P_2)\mathbf{j} + (M_1N_2 - N_1M_2)\mathbf{k} \Rightarrow \nabla \times (\mathbf{F}_1 \times \mathbf{F}_2)$$

$$= \left[\frac{\partial}{\partial y}(M_1N_2 - N_1M_2) - \frac{\partial}{\partial z}(P_1M_2 - M_1P_2)\right]\mathbf{i} + \left[\frac{\partial}{\partial z}(N_1P_2 - P_1N_2) - \frac{\partial}{\partial x}(M_1N_2 - N_1M_2)\right]\mathbf{j}$$

$+\left[\frac{\partial}{\partial x}(P_1M_2 - M_1P_2) - \frac{\partial}{\partial y}(N_1P_2 - P_1N_2)\right]\mathbf{k}$

and consider the i-component only: $\frac{\partial}{\partial y}(M_1N_2 - N_1M_2) - \frac{\partial}{\partial z}(P_1M_2 - M_1P_2)$

$= N_2\frac{\partial M_1}{\partial y} + M_1\frac{\partial N_2}{\partial y} - M_2\frac{\partial N_1}{\partial y} - N_1\frac{\partial M_2}{\partial y} - M_2\frac{\partial P_1}{\partial z} - P_1\frac{\partial M_2}{\partial z} + P_2\frac{\partial M_1}{\partial z} + M_1\frac{\partial P_2}{\partial z}$

$= \left(N_2\frac{\partial M_1}{\partial y} + P_2\frac{\partial M_1}{\partial z}\right) - \left(N_1\frac{\partial M_2}{\partial y} + P_1\frac{\partial M_2}{\partial z}\right) + \left(\frac{\partial N_2}{\partial y} + \frac{\partial P_2}{\partial z}\right)M_1 - \left(\frac{\partial N_1}{\partial y} + \frac{\partial P_1}{\partial z}\right)M_2$

$= \left(M_2\frac{\partial M_1}{\partial x} + N_2\frac{\partial M_1}{\partial y} + P_2\frac{\partial M_1}{\partial z}\right) - \left(M_1\frac{\partial M_2}{\partial x} + N_1\frac{\partial M_2}{\partial y} + P_1\frac{\partial M_2}{\partial z}\right) + \left(\frac{\partial M_2}{\partial x} + \frac{\partial N_2}{\partial y} + \frac{\partial P_2}{\partial z}\right)M_1$

$-\left(\frac{\partial M_1}{\partial x} + \frac{\partial N_1}{\partial y} + \frac{\partial P_1}{\partial z}\right)M_2$. Now, i-comp of $(\mathbf{F}_2 \cdot \nabla)\mathbf{F}_1 = \left(M_2\frac{\partial}{\partial x} + N_2\frac{\partial}{\partial y} + P_2\frac{\partial}{\partial z}\right)M_1$

$= \left(M_2\frac{\partial M_1}{\partial x} + N_2\frac{\partial M_1}{\partial y} + P_2\frac{\partial M_1}{\partial z}\right)$; likewise, i-comp of $(\mathbf{F}_1 \cdot \nabla)\mathbf{F}_2 = \left(M_1\frac{\partial M_2}{\partial x} + N_1\frac{\partial M_2}{\partial y} + P_1\frac{\partial M_2}{\partial z}\right)$;

i-comp of $(\nabla \cdot \mathbf{F}_2)\mathbf{F}_1 = \left(\frac{\partial M_2}{\partial x} + \frac{\partial N_2}{\partial y} + \frac{\partial P_2}{\partial z}\right)M_1$ and i-comp of $(\nabla \cdot \mathbf{F}_1)\mathbf{F}_2 = \left(\frac{\partial M_1}{\partial x} + \frac{\partial N_1}{\partial y} + \frac{\partial P_1}{\partial z}\right)M_2$.

Similar results hold for the **j** and **k** components of $\nabla \times (\mathbf{F}_1 \times \mathbf{F}_2)$. In summary, since the corresponding components are equal, we have the result

$\nabla \times (\mathbf{F}_1 \times \mathbf{F}_2) = (\mathbf{F}_2 \cdot \nabla)\mathbf{F}_1 - (\mathbf{F}_1 \cdot \nabla)\mathbf{F}_2 + (\nabla \cdot \mathbf{F}_2)\mathbf{F}_1 - (\nabla \cdot \mathbf{F}_1)\mathbf{F}_2$

(b) Here again we consider only the i-component of each expression. Thus, the i-comp of $\nabla \times (\mathbf{F}_1 \cdot \mathbf{F}_2)$

$= \frac{\partial}{\partial x}(M_1M_2 + N_1N_2 + P_1P_2) = \left(M_1\frac{\partial M_2}{\partial x} + M_2\frac{\partial M_1}{\partial x} + N_1\frac{\partial N_2}{\partial x} + N_2\frac{\partial N_1}{\partial x} + P_1\frac{\partial P_2}{\partial x} + P_2\frac{\partial P_1}{\partial x}\right)$

$=$ i-comp of $(\mathbf{F}_1 \cdot \nabla)\mathbf{F}_2 = \left(M_1\frac{\partial M_2}{\partial x} + N_1\frac{\partial M_2}{\partial y} + P_1\frac{\partial M_2}{\partial z}\right)$,

i-comp of $(\mathbf{F}_2 \cdot \nabla)\mathbf{F}_1 = \left(M_2\frac{\partial M_1}{\partial x} + N_2\frac{\partial M_1}{\partial y} + P_2\frac{\partial M_1}{\partial z}\right)$,

i-comp of $\mathbf{F}_1 \times (\nabla \times \mathbf{F}_2) = N_1\left(\frac{\partial N_2}{\partial x} - \frac{\partial M_2}{\partial y}\right) - P_1\left(\frac{\partial M_2}{\partial z} - \frac{\partial P_2}{\partial x}\right)$, and

i-comp of $\mathbf{F}_2 \times (\nabla \times \mathbf{F}_1) = N_2\left(\frac{\partial N_1}{\partial x} - \frac{\partial M_1}{\partial y}\right) - P_2\left(\frac{\partial M_1}{\partial z} - \frac{\partial P_1}{\partial x}\right)$.

Since corresponding components are equal, we see that

$\nabla \times (\mathbf{F}_1 \cdot \mathbf{F}_2) = (\mathbf{F}_1 \cdot \nabla)\mathbf{F}_2 + (\mathbf{F}_2 \cdot \nabla)\mathbf{F}_1 + \mathbf{F}_1 \times (\nabla \times \mathbf{F}_2) + \mathbf{F}_2 \times (\nabla \times \mathbf{F}_1)$, as claimed.

21. The integral's value never exceeds the surface area of S. Since $|\mathbf{F}| \le 1$, we have $|\mathbf{F} \cdot \mathbf{n}| = |\mathbf{F}||\mathbf{n}| \le (1)(1) = 1$ and

$\iiint_D \nabla \cdot \mathbf{F}\, d\sigma = \iint_S \mathbf{F} \cdot \mathbf{n}\, d\sigma$ [Divergence Theorem]

$\le \iint_S |\mathbf{F} \cdot \mathbf{n}|\, d\sigma$ [A property of integrals]

$\leq \iint_S (1)\, d\sigma \qquad [|\mathbf{F}\cdot\mathbf{n}| \leq 1]$

$= \text{Area of S.}$

22. Yes, the outward flux through the top is 5. The reason is this: Since $\nabla\cdot\mathbf{F} = \nabla\cdot(x\mathbf{i} - 2y\mathbf{j} + (z+3)\mathbf{k}$ $= 1 - 2 + 1 = 0$, the outward flux across the closed cubelike surface is 0 by the Divergence Theorem. The flux across the top is therefore the negative of the flux across the sides and base. Routine calculations show that the sum of these latter fluxes is -5. Therefore the flux across the top is 5.

23. (a) $\frac{\partial}{\partial x}(x) = 1,\ \frac{\partial}{\partial y}(y) = 1,\ \frac{\partial}{\partial z}(z) = 1 \Rightarrow \nabla\cdot\mathbf{F} = 3 \Rightarrow \text{Flux} = \iiint_D 3\, dV = 3\iiint_D dV$

$= 3(\text{Volume of the solid})$

(b) If $\mathbf{F}$ is orthogonal to $\mathbf{n}$ at every point of S, then $\mathbf{F}\cdot\mathbf{n} = 0$ everywhere $\Rightarrow \text{Flux} = \iint_S \mathbf{F}\cdot\mathbf{n}\, d\sigma = 0$.

But the flux is $3(\text{Volume of the solid}) \neq 0$, so $\mathbf{F}$ is not orthogonal to $\mathbf{n}$ at every point.

24. $\nabla\cdot\mathbf{F} = -2x - 4y - 6z + 12 \Rightarrow \text{Flux} = \int_0^a\int_0^b\int_0^1 (-2x - 4y - 6z + 12)\, dz\, dy\, dx$

$= \int_0^a\int_0^b (-2x - 4y + 9)\, dy\, dx = \int_0^a (-2xb - 2b^2 + 9b)\, dx = -a^2b - 2ab^2 + 9ab = ab(-a - 2b + 9) = f(a, b);$

$\frac{\partial f}{\partial a} = -2ab - 2b^2 + 9b$ and $\frac{\partial f}{\partial b} = -a^2 - 4ab + 9a$ so that $\frac{\partial f}{\partial a} = 0$ and $\frac{\partial f}{\partial b} = 0 \Rightarrow b(-2a - 2b + 9) = 0$ and $a(-a - 4b + 9) = 0 \Rightarrow b = 0$ or $-2a - 2b + 9 = 0$, and $a = 0$ or $-a - 4b + 9 = 0$. Now $b = 0$ or $a = 0$ $\Rightarrow \text{Flux} = 0$; $-2a - 2b + 9 = 0$ and $-a - 4b + 9 = 0 \Rightarrow 3a - 9 = 0 \Rightarrow a = 3 \Rightarrow b = \frac{3}{2}$ so that $f\left(3, \frac{3}{2}\right) = \frac{27}{2}$ is the maximum flux.

25. $\iint_S \mathbf{F}\cdot\mathbf{n}\, d\sigma = \iiint_D \nabla\cdot\mathbf{F}\, dV = \iiint_D 3\, dV \Rightarrow \frac{1}{3}\iint_S \mathbf{F}\cdot\mathbf{n}\, d\sigma = \iiint_D dV = \text{Volume of D}$

26. $\mathbf{F} = \mathbf{C} \Rightarrow \nabla\cdot\mathbf{F} = 0 \Rightarrow \text{Flux} = \iint_S \mathbf{F}\cdot\mathbf{n}\, d\sigma = \iiint_D \nabla\cdot\mathbf{F}\, dV = \iiint_D 0\, dV = 0$

27. (a) From the Divergence Theorem, $\iint_S \nabla f\cdot\mathbf{n}\, d\sigma = \iiint_D \nabla\cdot\nabla f\, dV = \iiint_D \nabla^2 f\, dV = \iiint_D 0\, dV = 0$

(b) From the Divergence Theorem, $\iint_S f\nabla f\cdot\mathbf{n}\, d\sigma = \iiint_D \nabla\cdot f\nabla f\, dV$. Now,

$$f\nabla f = \left(f\frac{\partial f}{\partial x}\right)\mathbf{i} + \left(f\frac{\partial f}{\partial y}\right)\mathbf{j} + \left(f\frac{\partial f}{\partial z}\right)\mathbf{k} \Rightarrow \nabla\cdot f\nabla f = \left[f\frac{\partial^2 f}{\partial x^2} + \left(\frac{\partial f}{\partial x}\right)^2\right] + \left[f\frac{\partial^2 f}{\partial y^2} + \left(\frac{\partial f}{\partial y}\right)^2\right] + \left[f\frac{\partial^2 f}{\partial z^2} + \left(\frac{\partial f}{\partial z}\right)^2\right]$$

$= f\nabla^2 f + |\nabla f|^2 = 0 + |\nabla f|^2$ since f is harmonic $\Rightarrow \iint_S f\nabla f \cdot \mathbf{n}\, d\sigma = \iiint_D |\nabla f|^2\, dV$, as claimed.

28. From the Divergence Theorem, $\iint_S \nabla f \cdot \mathbf{n}\, d\sigma = \iiint_D \nabla \cdot \nabla f\, dV = \iiint_D \left(\frac{\partial^2 f}{\partial x^2} + \frac{\partial^2 f}{\partial y^2} + \frac{\partial^2 f}{\partial z^2}\right) dV$. Now,

$f(x,y,z) = \ln \sqrt{x^2+y^2+z^2} = \frac{1}{2}\ln(x^2+y^2+z^2) \Rightarrow \frac{\partial f}{\partial x} = \frac{x}{x^2+y^2+z^2}, \frac{\partial f}{\partial y} = \frac{y}{x^2+y^2+z^2}, \frac{\partial f}{\partial z} = \frac{z}{x^2+y^2+z^2}$

$\Rightarrow \frac{\partial^2 f}{\partial x^2} = \frac{-x^2+y^2+z^2}{(x^2+y^2+z^2)^2}, \frac{\partial^2 f}{\partial y^2} = \frac{x^2-y^2+z^2}{(x^2+y^2+z^2)^2}, \frac{\partial^2 f}{\partial z^2} = \frac{x^2+y^2-z^2}{(x^2+y^2+z^2)^2}, \Rightarrow \frac{\partial^2 f}{\partial x^2} + \frac{\partial^2 f}{\partial y^2} + \frac{\partial^2 f}{\partial z^2}$

$= \frac{x^2+y^2+z^2}{(x^2+y^2+z^2)^2} = \frac{1}{x^2+y^2+z^2} \Rightarrow \iint_S \nabla f \cdot \mathbf{n}\, d\sigma = \iiint_D \frac{dV}{x^2+y^2+z^2} = \int_0^{\pi/2}\int_0^{\pi/2}\int_0^a \frac{\rho^2 \sin\phi}{\rho^2}\, d\rho\, d\phi\, d\theta$

$= \int_0^{\pi/2}\int_0^{\pi/2} a \sin\phi\, d\phi\, d\theta = \int_0^{\pi/2} \left[-a\cos\phi\right]_0^{\pi/2} d\theta = \int_0^{\pi/2} a\, d\theta = \frac{\pi a}{2}$

29. $\iint_S f\nabla g \cdot \mathbf{n}\, d\sigma = \iiint_D \nabla \cdot f\nabla g\, dV = \iiint_D \nabla \cdot \left(f\frac{\partial g}{\partial x}\mathbf{i} + f\frac{\partial g}{\partial y}\mathbf{j} + f\frac{\partial g}{\partial z}\mathbf{k}\right) dV$

$= \iiint_D \left(f\frac{\partial^2 g}{\partial x^2} + \frac{\partial f}{\partial x}\frac{\partial g}{\partial x} + f\frac{\partial^2 g}{\partial y^2} + \frac{\partial f}{\partial y}\frac{\partial g}{\partial y} + f\frac{\partial^2 g}{\partial z^2} + \frac{\partial f}{\partial z}\frac{\partial g}{\partial z}\right) dV$

$= \iiint_D \left[f\left(\frac{\partial^2 g}{\partial x^2} + \frac{\partial^2 g}{\partial y^2} + \frac{\partial^2 g}{\partial z^2}\right) + \left(\frac{\partial f}{\partial x}\frac{\partial g}{\partial x} + \frac{\partial f}{\partial y}\frac{\partial g}{\partial y} + \frac{\partial f}{\partial z}\frac{\partial g}{\partial z}\right)\right] dV = \iiint_D (f\nabla^2 g + \nabla f \cdot \nabla g)\, dV$

30. $\iint_S (f\nabla g - g\nabla f) \cdot \mathbf{n}\, d\sigma$

$= \iiint_D \nabla \cdot (f\nabla g - g\nabla f)\, dV$

$= \iiint_D (\nabla \cdot f\nabla g - \nabla \cdot g\nabla f)\, dV$ (Exercise 18a)

$= \iiint_D (f\nabla \cdot \nabla g + \nabla f \cdot \nabla g - g\nabla \cdot \nabla f - \nabla g \cdot \nabla f)\, dV$ (Exercise 19a)

$= \iiint_D (f\nabla^2 g - g\nabla f)\, dV$, since $\nabla f \cdot \nabla g = \nabla g \cdot \nabla f$

31. (a) The integral $\iiint_D p(t,x,y,z)\,dV$ represents the mass of the fluid at any time t. The equation says that the instantaneous rate of change of mass is flux of the fluid through the surface S enclosing the region D: the mass decreases if the flux is outward (so the fluid flows out of D), and increases if the flow is inward (interpreting **n** as the outward pointing unit normal to the surface).

(b) $\iiint_D \frac{\partial p}{\partial t}\,dV = \frac{d}{dt}\iiint_D p\,dV = -\iint_S p\mathbf{v}\cdot\mathbf{n}\,d\sigma = -\iiint_D \nabla\cdot p\mathbf{v}\,dV \Rightarrow \frac{\partial \rho}{\partial t} = -\nabla\cdot p\mathbf{v}$
$\Rightarrow \nabla\cdot p\mathbf{v} + \frac{\partial p}{\partial t} = 0$, as claimed

32. (a) ∇T points in the direction of maximum change of the temperature, so if the solid is heating up at the point the temperature is greater in a region surrounding the point $\Rightarrow \nabla T$ points away from the point $\Rightarrow -\nabla T$ points toward the point $\Rightarrow -\nabla T$ points in the direction the heat flows.

(b) Assuming the Law of Conservation of Mass (Exercise 31) with $-k\nabla T = \mathbf{v}$ and $c\rho T = p$, we have
$\frac{d}{dt}\iiint_D c\rho T\,dV = -\iint_S -k\nabla T\cdot\mathbf{n}\,d\sigma \Rightarrow$ the continuity equation, $\nabla\cdot(-k\nabla T) + \frac{\partial}{\partial t}(c\rho T) = 0$
$\Rightarrow c\rho\frac{\partial T}{\partial t} = -\nabla\cdot(-k\nabla T) = k\nabla^2 T \Rightarrow \frac{\partial T}{\partial t} = \frac{k}{c\rho}\nabla^2 T = K\nabla^2 T$, as claimed

CHAPTER 14 PRACTICE EXERCISES

1. Path 1: $\mathbf{r} = t\mathbf{i} + t\mathbf{j} + t\mathbf{k} \Rightarrow x = t,\ y = t,\ z = t,\ 0 \le t \le 1 \Rightarrow f(g(t),h(t),k(t)) = 3 - 3t^2$ and $\frac{dx}{dt} = 1,\ \frac{dy}{dt} = 1,$

$\frac{dz}{dt} = 1 \Rightarrow \sqrt{\left(\frac{dx}{dt}\right)^2 + \left(\frac{dy}{dt}\right)^2 + \left(\frac{dz}{dt}\right)^2}\,dt = \sqrt{3}\,dt \Rightarrow \int_C f(x,y,z)\,ds = \int_0^1 \sqrt{3}(3 - 3t^2)\,dt = 2\sqrt{3}$

Path 2: $\mathbf{r}_1 = t\mathbf{i} + t\mathbf{j},\ 0 \le t \le 1 \Rightarrow x = t,\ y = t,\ z = 0 \Rightarrow f(g(t),h(t),k(t)) = 2t - 3t^2 + 3$ and $\frac{dx}{dt} = 1,\ \frac{dy}{dt} = 1,$

$\frac{dz}{dt} = 0 \Rightarrow \sqrt{\left(\frac{dx}{dt}\right)^2 + \left(\frac{dy}{dt}\right)^2 + \left(\frac{dz}{dt}\right)^2}\,dt = \sqrt{2}\,dt \Rightarrow \int_{C_1} f(x,y,z)\,ds = \int_0^1 \sqrt{2}(2t - 3t^2 + 3)\,dt = 3\sqrt{2};$

$\mathbf{r}_2 = \mathbf{i} + \mathbf{j} + t\mathbf{k} \Rightarrow x = 1,\ y = 1,\ z = t \Rightarrow f(g(t),h(t),k(t)) = 2 - 2t$ and $\frac{dx}{dt} = 0,\ \frac{dy}{dt} = 0,\ \frac{dz}{dt} = 1$

$\Rightarrow \sqrt{\left(\frac{dx}{dt}\right)^2 + \left(\frac{dy}{dt}\right)^2 + \left(\frac{dz}{dt}\right)^2}\,dt = dt \Rightarrow \int_{C_2} f(x,y,z)\,ds = \int_0^1 (2 - 2t)\,dt = 1$

$\Rightarrow \int_C f(x,y,z)\,ds = \int_{C_1} f(x,y,z)\,ds + \int_{C_2} f(x,y,z) = 3\sqrt{2} + 1$

2. Path 1: $\mathbf{r}_1 = t\mathbf{i} \Rightarrow x = t,\ y = 0,\ z = 0 \Rightarrow f(g(t),h(t),k(t)) = t^2$ and $\frac{dx}{dt} = 1,\ \frac{dy}{dt} = 0,\ \frac{dz}{dt} = 0$

$$\Rightarrow \sqrt{\left(\frac{dx}{dt}\right)^2 + \left(\frac{dy}{dt}\right)^2 + \left(\frac{dz}{dt}\right)^2}\ dt = dt \Rightarrow \int_{C_1} f(x,y,z)\ ds = \int_0^1 t^2\ dt = \frac{1}{3};$$

$\mathbf{r}_2 = \mathbf{i} + t\mathbf{j} \Rightarrow x = 1,\ y = t,\ z = 0 \Rightarrow f(g(t),h(t),k(t)) = 1 + t$ and $\frac{dx}{dt} = 0,\ \frac{dy}{dt} = 1,\ \frac{dz}{dt} = 0$

$$\Rightarrow \sqrt{\left(\frac{dx}{dt}\right)^2 + \left(\frac{dy}{dt}\right)^2 + \left(\frac{dz}{dt}\right)^2}\ dt = dt \Rightarrow \int_{C_2} f(x,y,z)\ ds = \int_0^1 (1+t)\ dt = \frac{3}{2};$$

$\mathbf{r}_3 = \mathbf{i} + \mathbf{j} + t\mathbf{k} \Rightarrow x = 1,\ y = 1,\ z = t \Rightarrow f(g(t),h(t),k(t)) = 2 - t$ and $\frac{dx}{dt} = 0,\ \frac{dy}{dt} = 0,\ \frac{dz}{dt} = 1$

$$\Rightarrow \sqrt{\left(\frac{dx}{dt}\right)^2 + \left(\frac{dy}{dt}\right)^2 + \left(\frac{dz}{dt}\right)^2}\ dt = dt \Rightarrow \int_{C_3} f(x,y,z)\ ds = \int_0^1 (2-t)\ dt = \frac{3}{2}$$

$$\Rightarrow \int_{\text{Path 1}} f(x,y,z)\ ds = \int_{C_1} f(x,y,z)\ ds + \int_{C_2} f(x,y,z)\ ds + \int_{C_3} f(x,y,z)\ ds = \frac{10}{3}$$

Path 2: $\mathbf{r}_4 = t\mathbf{i} + t\mathbf{j} \Rightarrow x = t,\ y = t,\ z = 0 \Rightarrow f(g(t),h(t),k(t)) = t^2 + t$ and $\frac{dx}{dt} = 1,\ \frac{dy}{dt} = 1,\ \frac{dz}{dt} = 0$

$$\Rightarrow \sqrt{\left(\frac{dx}{dt}\right)^2 + \left(\frac{dy}{dt}\right)^2 + \left(\frac{dz}{dt}\right)^2}\ dt = \sqrt{2}\ dt \Rightarrow \int_{C_4} f(x,y,z)\ ds = \int_0^1 \sqrt{2}(t^2+t)\ dt = \frac{5}{6}\sqrt{2};$$

$$\mathbf{r}_3 = \mathbf{i} + \mathbf{j} + t\mathbf{k} \text{ (see above)} \Rightarrow \int_{C_3} f(x,y,z)\ ds = \frac{3}{2}$$

$$\Rightarrow \int_{\text{Path 2}} f(x,y,z)\ ds = \int_{C_3} f(x,y,z)\ ds + \int_{C_4} f(x,y,z)\ ds = \frac{5}{6}\sqrt{2} + \frac{3}{2} = \frac{5\sqrt{2}+9}{6}$$

Path 3: $\mathbf{r}_5 = t\mathbf{k} \Rightarrow x = 0,\ y = 0,\ z = t,\ 0 \le t \le 1 \Rightarrow f(g(t),h(t),k(t)) = -t$ and $\frac{dx}{dt} = 0,\ \frac{dy}{dt} = 0,\ \frac{dz}{dt} = 1$

$$\Rightarrow \sqrt{\left(\frac{dx}{dt}\right)^2 + \left(\frac{dy}{dt}\right)^2 + \left(\frac{dz}{dt}\right)^2}\ dt = dt \Rightarrow \int_{C_5} f(x,y,z)\ ds = \int_0^1 -t\ dt = -\frac{1}{2};$$

$\mathbf{r}_6 = t\mathbf{j} + \mathbf{k} \Rightarrow x = 0,\ y = t,\ z = 1,\ 0 \le t \le 1 \Rightarrow f(g(t),h(t),k(t)) = t - 1$ and $\frac{dx}{dt} = 0,\ \frac{dy}{dt} = 1,\ \frac{dz}{dt} = 0$

$$\Rightarrow \sqrt{\left(\frac{dx}{dt}\right)^2 + \left(\frac{dy}{dt}\right)^2 + \left(\frac{dz}{dt}\right)^2}\ dt = dt \Rightarrow \int_{C_6} f(x,y,z)\ ds = \int_0^1 (t-1)\ dt = -\frac{1}{2};$$

$\mathbf{r}_7 = t\mathbf{i} + \mathbf{j} + \mathbf{k} \Rightarrow x = t,\ y = 1,\ z = 1,\ 0 \le t \le 1 \Rightarrow f(g(t),h(t),k(t)) = t^2$ and $\frac{dx}{dt} = 1,\ \frac{dy}{dt} = 0,\ \frac{dz}{dt} = 0$

$$\Rightarrow \sqrt{\left(\frac{dx}{dt}\right)^2 + \left(\frac{dy}{dt}\right)^2 + \left(\frac{dz}{dt}\right)^2}\ dt = dt \Rightarrow \int_{C_7} f(x,y,z)\ ds = \int_0^1 t^2\ dt = \frac{1}{3}$$

$$\Rightarrow \int_{\text{Path 3}} f(x,y,z)\ ds = \int_{C_5} f(x,y,z)\ ds + \int_{C_6} f(x,y,z)\ ds + \int_{C_7} f(x,y,z)\ ds = -\frac{1}{2} - \frac{1}{2} + \frac{1}{3} = -\frac{2}{3}$$

3. $\mathbf{r} = (a \cos t)\mathbf{j} + (a \sin t)\mathbf{k} \Rightarrow x = 0,\ y = a \cos t,\ z = a \sin t \Rightarrow f(g(t), h(t), k(t)) = \sqrt{a^2 \sin^2 t} = a\,|\sin t|$ and

$\frac{dx}{dt} = 0,\ \frac{dy}{dt} = -a \sin t,\ \frac{dz}{dt} = a \cos t \Rightarrow \sqrt{\left(\frac{dx}{dt}\right)^2 + \left(\frac{dy}{dt}\right)^2 + \left(\frac{dz}{dt}\right)^2}\, dt = a\, dt$

$\Rightarrow \int_C f(x,y,z)\, ds = \int_0^{2\pi} a^2 |\sin t|\, dt = \int_0^{\pi} a^2 \sin t\, dt + \int_{\pi}^{2\pi} -a^2 \sin t\, dt = 4a^2$

4. $\mathbf{r} = (\cos t + t \sin t)\mathbf{i} + (\sin t - t \cos t)\mathbf{j} \Rightarrow x = \cos t + t \sin t,\ y = \sin t - t \cos t,\ z = 0$

$\Rightarrow f(g(t), h(t), k(t)) = \sqrt{(\cos t + t \sin t)^2 + (\sin t - t \cos t)^2} = \sqrt{1 + t^2}$ and $\frac{dx}{dt} = -\sin t + \sin t + t \cos t$

$= t \cos t,\ \frac{dy}{dt} = \cos t - \cos t + t \sin t = t \sin t,\ \frac{dz}{dt} = 0 \Rightarrow \sqrt{\left(\frac{dx}{dt}\right)^2 + \left(\frac{dy}{dt}\right)^2 + \left(\frac{dz}{dt}\right)^2}\, dt$

$= \sqrt{t^2 \cos^2 t + t^2 \sin^2 t}\, dt = |t|\, dt = t\, dt$ since $0 \le t \le \sqrt{3} \Rightarrow \int_C f(x,y,z)\, ds = \int_0^{\sqrt{3}} t\sqrt{1 + t^2}\, dt = \frac{7}{3}$

5. $\frac{\partial P}{\partial y} = -\frac{1}{2}(x + y + z)^{-3/2} = \frac{\partial N}{\partial z},\ \frac{\partial M}{\partial z} = -\frac{1}{2}(x + y + z)^{-3/2} = \frac{\partial P}{\partial x},\ \frac{\partial N}{\partial x} = -\frac{1}{2}(x + y + z)^{-3/2} = \frac{\partial M}{\partial y}$

$\Rightarrow M\, dx + N\, dy + P\, dz$ is exact; $\frac{\partial f}{\partial x} = \frac{1}{\sqrt{x + y + z}} \Rightarrow f(x,y,z) = 2\sqrt{x + y + z} + g(y,z) \Rightarrow \frac{\partial f}{\partial y} = \frac{1}{\sqrt{x + y + z}} + \frac{\partial g}{\partial y}$

$= \frac{1}{\sqrt{x + y + z}} \Rightarrow \frac{\partial g}{\partial y} = 0 \Rightarrow g(y,z) = h(z) \Rightarrow f(x,y,z) = 2\sqrt{x + y + z} + h(z) \Rightarrow \frac{\partial f}{\partial z} = \frac{1}{\sqrt{x + y + z}} + h'(z)$

$= \frac{1}{\sqrt{x + y + z}} \Rightarrow h'(x) = 0 \Rightarrow h(z) = C \Rightarrow f(x,y,z) = 2\sqrt{x + y + z} + C \Rightarrow \int_{(-1,1,1)}^{(4,-3,0)} \frac{dx + dy + dz}{\sqrt{x + y + z}}$

$= f(4,-3,0) - f(-1,1,1) = 2\sqrt{1} - 2\sqrt{1} = 0$

6. $\frac{\partial P}{\partial y} = -\frac{1}{2\sqrt{yz}} = \frac{\partial N}{\partial z},\ \frac{\partial M}{\partial z} = 0 = \frac{\partial P}{\partial x},\ \frac{\partial N}{\partial x} = 0 = \frac{\partial M}{\partial y} \Rightarrow M\, dx + N\, dy + P\, dz$ is exact; $\frac{\partial f}{\partial x} = 1 \Rightarrow f(x,y,z)$

$= x + g(y,z) \Rightarrow \frac{\partial f}{\partial y} = \frac{\partial g}{\partial y} = -\sqrt{\frac{z}{y}} \Rightarrow g(y,z) = -2\sqrt{yz} + h(z) \Rightarrow f(x,y,z) = x - 2\sqrt{yz} + h(z)$

$\Rightarrow \frac{\partial f}{\partial z} = -\sqrt{\frac{y}{z}} + h'(z) = -\sqrt{\frac{y}{z}} \Rightarrow h'(z) = 0 \Rightarrow h(z) = C \Rightarrow f(x,y,z) = x - 2\sqrt{yz} + C$

$\Rightarrow \int_{(1,1,1)}^{(10,3,3)} dx - \sqrt{\frac{z}{y}}\, dy - \sqrt{\frac{y}{z}}\, dz = f(10,3,3) - f(1,1,1) = (10 - 2\cdot 3) - (1 - 2\cdot 1) = 4 + 1 = 5$

7. $\frac{\partial P}{\partial y} = x \cos z = \frac{\partial N}{\partial z},\ \frac{\partial M}{\partial z} = y \cos z = \frac{\partial P}{\partial x},\ \frac{\partial N}{\partial x} = \sin z = \frac{\partial M}{\partial y} \Rightarrow \mathbf{F}$ is conservative $\Rightarrow \int_C \mathbf{F} \cdot d\mathbf{r} = 0$

8. $\frac{\partial P}{\partial y} = 0 = \frac{\partial N}{\partial z},\ \frac{\partial M}{\partial z} = 0 = \frac{\partial P}{\partial x},\ \frac{\partial N}{\partial x} = 3x^2 = \frac{\partial M}{\partial y} \Rightarrow \mathbf{F}$ is conservative $\Rightarrow \int_C \mathbf{F} \cdot d\mathbf{r} = 0$

9. Let $M = 8x \sin y$ and $N = -8y \cos x \Rightarrow \frac{\partial M}{\partial y} = 8x \cos y$ and $\frac{\partial N}{\partial x} = 8y \sin x \Rightarrow \int_C 8x \sin y\, dx - 8y \cos x\, dy$

$= \iint_R (8y \sin x - 8x \cos y)\, dy\, dx = \int_0^{\pi/2}\int_0^{\pi/2} (8y \sin x - 8x \cos y)\, dy\, dx = \int_0^{\pi/2} \left(\pi^2 \sin x - 8x\right) dx$

$= -\pi^2 + \pi^2 = 0$

10. Let $M = y^2$ and $N = x^2 \Rightarrow \frac{\partial M}{\partial y} = 2y$ and $\frac{\partial N}{\partial x} = 2x \Rightarrow \int_C y^2\, dx + x^2\, dy = \iint_R (2x - 2y)\, dx\, dy$

$= \int_0^{2\pi}\int_0^{2} (2r \cos \theta - 2r \sin \theta)\, r\, dr\, d\theta = \int_0^{2\pi} \frac{16}{3}(\cos \theta - \sin \theta)\, d\theta = 0$

11. Let $z = 1 - x - y \Rightarrow f_x(x,y) = -1$ and $f_y(x,y) = -1 \Rightarrow \sqrt{f_x^2 + f_y^2 + 1} = \sqrt{3} \Rightarrow$ Surface Area $= \iint_R \sqrt{3}\, dx\, dy$

$= \sqrt{3}$(Area of the circular region in the xy-plane) $= \pi\sqrt{3}$

12. $\nabla f = -3\mathbf{i} + 2y\mathbf{j} + 2z\mathbf{k}$, $\mathbf{p} = \mathbf{i} \Rightarrow |\nabla f| = \sqrt{9 + 4y^2 + 4z^2}$ and $|\nabla f \cdot \mathbf{p}| = 3$

$\Rightarrow$ Surface Area $= \iint_R \frac{\sqrt{9 + 4y^2 + 4z^2}}{3}\, dy\, dz = \int_0^{2\pi}\int_0^{\sqrt{3}} \frac{\sqrt{9 + 4r^2}}{3}\, r\, dr\, d\theta = \frac{1}{3}\int_0^{2\pi} \left(\frac{7}{4}\sqrt{21} - \frac{9}{4}\right) d\theta = \frac{\pi}{6}(7\sqrt{21} - 9)$

13. $\nabla f = 2x\mathbf{i} + 2y\mathbf{j} + 2z\mathbf{k}$, $\mathbf{p} = \mathbf{k} \Rightarrow |\nabla f| = \sqrt{4x^2 + 4y^2 + 4z^2} = 2\sqrt{x^2 + y^2 + z^2} = 2$ and $|\nabla f \cdot \mathbf{p}| = |2z| = 2z$ since

$z \ge 0 \Rightarrow$ Surface Area $= \iint_R \frac{2}{2z}\, dA = \iint_R \frac{1}{z}\, dA = \iint_R \frac{1}{\sqrt{1 - x^2 - y^2}}\, dx\, dy = \int_0^{2\pi}\int_0^{1/\sqrt{2}} \frac{1}{\sqrt{1 - r^2}}\, r\, dr\, d\theta$

$\int_0^{2\pi} \left[-\sqrt{1 - r^2}\right]_0^{1/\sqrt{2}} d\theta = \int_0^{2\pi} \left(1 - \frac{1}{\sqrt{2}}\right) d\theta = 2\pi\left(1 - \frac{1}{\sqrt{2}}\right)$

14. (a) $\nabla f = 2x\mathbf{i} + 2y\mathbf{j} + 2z\mathbf{k}$, $\mathbf{p} = \mathbf{k} \Rightarrow |\nabla f| = \sqrt{4x^2 + 4y^2 + 4z^2} = 2\sqrt{x^2 + y^2 + z^2} = 4$ and $|\nabla f \cdot \mathbf{p}| = 2z$ since

$z \ge 0 \Rightarrow$ Surface Area $= \iint_R \frac{4}{2z}\, dA = \iint_R \frac{2}{z}\, dA = 2\int_0^{\pi/2}\int_0^{2\cos\theta} \frac{2}{\sqrt{4 - r^2}}\, r\, dr\, d\theta = 4\pi - 8$

(b) $\mathbf{r} = 2 \cos \theta \Rightarrow \mathbf{dr} = -2 \sin \theta\, d\theta$; $ds^2 = r^2\, d\theta^2 + dr^2$ (Arc length in polar coordinates)

$\Rightarrow ds^2 = (2 \cos \theta)^2\, d\theta^2 + dr^2 = 4 \cos^2\theta\, d\theta^2 + 4 \sin^2\theta\, d\theta^2 = 4\, d\theta^2 \Rightarrow ds = 2\, d\theta$; the height of the cylinder is $z = \sqrt{4 - r^2} = \sqrt{4 - 4\cos^2\theta} = 2|\sin \theta| = 2 \sin \theta$ if $0 \le \theta \le \frac{\pi}{2}$

$\Rightarrow$ Surface Area $= \int_{-\pi/2}^{\pi/2} h\, ds = 2\int_0^{\pi/2} (2 \sin \theta)(2\, d\theta) = 8$

15. $f(x,y,z)=\frac{x}{a}+\frac{y}{b}+\frac{z}{c}=1 \Rightarrow \nabla f=\left(\frac{1}{a}\right)\mathbf{i}+\left(\frac{1}{b}\right)\mathbf{j}+\left(\frac{1}{c}\right)\mathbf{k} \Rightarrow |\nabla f|=\sqrt{\frac{1}{a^2}+\frac{1}{b^2}+\frac{1}{c^2}}$ and $\mathbf{p}=\mathbf{k} \Rightarrow |\nabla f\cdot\mathbf{p}|=\frac{1}{c}$

since $c>0 \Rightarrow$ Surface Area $=\iint_R \frac{\sqrt{\frac{1}{a^2}+\frac{1}{b^2}+\frac{1}{c^2}}}{\left(\frac{1}{c}\right)}\,dA = c\sqrt{\frac{1}{a^2}+\frac{1}{b^2}+\frac{1}{c^2}}\iint_R dA=\frac{1}{2}abc\sqrt{\frac{1}{a^2}+\frac{1}{b^2}+\frac{1}{c^2}}$,

since the area of the triangular region R is $\frac{1}{2}ab$

16. (a) $\nabla f=2y\mathbf{j}-\mathbf{k}$, $\mathbf{p}=\mathbf{k} \Rightarrow |\nabla f|=\sqrt{4y^2+1}$ and $|\nabla f\cdot\mathbf{p}|=1 \Rightarrow d\sigma=\sqrt{4y^2+1}\,dx\,dy$

$$\Rightarrow \iint_S g(x,y,z)\,d\sigma=\iint_R \frac{yz}{\sqrt{4y^2+1}}\sqrt{4y^2+1}\,dx\,dy=\iint_R y(y^2-1)\,dx\,dy=\int_{-1}^{1}\int_0^3 (y^3-y)\,dx\,dy$$

$$=\int_{-1}^{1} 3(y^3-y)\,dy=3\left[\frac{y^4}{4}-\frac{y^2}{2}\right]_{-1}^{1}=0$$

(b) $$\iint_S g(x,y,z)\,d\sigma=\iint_R \frac{z}{\sqrt{4y^2+1}}\sqrt{4y^2+1}\,dx\,dy=\int_{-1}^{1}\int_0^3 (y^2-1)\,dx\,dy=\int_{-1}^{1}3(y^2-1)\,dy$$

$$=3\left[\frac{y^3}{3}-y\right]_{-1}^{1}=-4$$

17. $\nabla f=2y\mathbf{j}+2z\mathbf{k}$, $\mathbf{p}=\mathbf{k} \Rightarrow |\nabla f|=\sqrt{4y^2+4z^2}=2\sqrt{y^2+z^2}=10$ and $|\nabla f\cdot\mathbf{p}|=2z$ since $z\ge 0$

$$\Rightarrow d\sigma=\frac{10}{2z}\,dx\,dy=\frac{5}{z}\,dx\,dy=\iint_S g(x,y,z)\,d\sigma=\iint_R (x^4y)(y^2+z^2)\left(\frac{5}{z}\right)dx\,dy$$

$$=\iint_R (x^4y)(25)\left(\frac{5}{\sqrt{25-y^2}}\right)dx\,dy=\int_0^4\int_0^1 \frac{125y}{\sqrt{25-y^2}}x^4\,dx\,dy=\int_0^4 \frac{25y}{\sqrt{25-y^2}}\,dy=50$$

18. Define the coordinate system so that the origin is at the center of the earth, the z-axis is the earth's axis (north is the positive z direction), and the xz-plane contains the earth's prime meridian. Let S denote the surface which is Wyoming so then S is part of the surface $z=(R^2-x^2-y^2)^{1/2}$. Let R_{xy} be the projection of S onto the xy-plane. The surface area of Wyoming is $\iint_S 1\,d\sigma=\iint_{R_{xy}}\sqrt{1+\left(\frac{\partial z}{\partial x}\right)^2+\left(\frac{\partial z}{\partial y}\right)^2}\,dA$

$$\iint_{R_{xy}}\sqrt{\frac{x^2}{R^2-x^2-y^2}+\frac{y^2}{R^2-x^2-y^2}+1}\,dA=\iint_{R_{xy}}\frac{R}{(R^2-x^2-y^2)^{1/2}}\,dA=\int_{\theta_1}^{\theta_2}\int_{R\sin 45^\circ}^{R\sin 49^\circ} R(R^2-r^2)^{-1/2}\,r\,dr\,d\theta$$

(where θ_1 and θ_2 are the radian equivalent to $104°3'$ and $111°3'$, respectively)

$$=\int_{\theta_1}^{\theta_2} -R(R^2-r^2)^{1/2}\Big|_{R\sin 45^\circ}^{R\sin 49^\circ}=\int_{\theta_1}^{\theta_2} R(R^2-R^2\sin^2 45^\circ)^{1/2}-R(R^2-R^2\sin^2 49^\circ)^{1/2}\,d\theta$$

$= (\theta_2 - \theta_1)R^2(\cos 45^\circ - \cos 49^\circ) = \frac{7\pi}{180} R^2(\cos 45^\circ - \cos 49^\circ) = \frac{7\pi}{180}(3959)^2(\cos 45^\circ - \cos 49^\circ)$
$\approx 97{,}751$ sq. mi.

19. A possible parametrization is $\mathbf{r}(\phi, \theta) = (6 \sin\phi \cos\theta)\mathbf{i} + (6 \sin\phi \sin\theta)\mathbf{j} + (6 \cos\phi)\mathbf{k}$ (spherical coordinates); now $\rho = 6$ and $z = -3 \Rightarrow -3 = 6 \cos\phi \Rightarrow \cos\phi = -\frac{1}{2} \Rightarrow \phi = \frac{2\pi}{3}$ and $z = 3\sqrt{3} \Rightarrow 3\sqrt{3} = 6 \cos\phi$ $\Rightarrow \cos\phi = \frac{\sqrt{3}}{2} \Rightarrow \phi = \frac{\pi}{6} \Rightarrow \frac{\pi}{6} \le \phi \le \frac{2\pi}{3}$; also $0 \le \theta \le 2\pi$

20. A possible parametrization is $\mathbf{r}(r, \theta) = (r \cos\theta)\mathbf{i} + (r \sin\theta)\mathbf{j} - \left(\frac{r^2}{2}\right)\mathbf{k}$ (cylindrical coordinates);

now $r = \sqrt{x^2 + y^2} \Rightarrow z = -\frac{r^2}{2}$ and $-2 \le z \le 0 \Rightarrow -2 \le -\frac{r^2}{2} \le 0 \Rightarrow 4 \ge r^2 \ge 0 \Rightarrow 0 \le r \le 2$ since $r \ge 0$; also $0 \le \theta \le 2\pi$

21. A possible parametrization is $\mathbf{r}(r, \theta) = (r \cos\theta)\mathbf{i} + (r \sin\theta)\mathbf{j} + (1 + r)\mathbf{k}$ (cylindrical coordinates);

now $r = \sqrt{x^2 + y^2} \Rightarrow z = 1 + r$ and $1 \le z \le 3 \Rightarrow 1 \le 1 + r \le 3 \Rightarrow 0 \le r \le 2$; also $0 \le \theta \le 2\pi$

22. A possible parametrization is $\mathbf{r}(x, y) = x\mathbf{i} + y\mathbf{j} + \left(3 - x - \frac{y}{2}\right)\mathbf{k}$ for $0 \le x \le 2$ and $0 \le y \le 2$

23. Let $x = u \cos v$ and $z = u \sin v$, where $u = \sqrt{x^2 + z^2}$ and v is the angle in the xz-plane with the x-axis $\Rightarrow \mathbf{r}(u, v) = (u \cos v)\mathbf{i} + 2u^2\mathbf{j} + (u \sin v)\mathbf{k}$ is a possible parametrization; $0 \le y \le 2 \Rightarrow 2u^2 \le 2 \Rightarrow u^2 \le 1$ $\Rightarrow 0 \le u \le 1$ since $u \ge 0$; also, for just the upper half of the paraboloid, $0 \le v \le \pi$

24. A possible parametrization is $\left(\sqrt{10} \sin\phi \cos\theta\right)\mathbf{i} + \left(\sqrt{10} \sin\phi \sin\theta\right)\mathbf{j} + \left(\sqrt{10} \cos\phi\right)\mathbf{k}$, $0 \le \phi \le \frac{\pi}{2}$ and $0 \le \theta \le \frac{\pi}{2}$

25. $\mathbf{r}_u = \mathbf{i} + \mathbf{j}$, $\mathbf{r}_v = \mathbf{i} - \mathbf{j} + \mathbf{k} \Rightarrow \mathbf{r}_u \times \mathbf{r}_v = \begin{vmatrix} \mathbf{i} & \mathbf{j} & \mathbf{k} \\ 1 & 1 & 0 \\ 1 & -1 & 1 \end{vmatrix} = \mathbf{i} - \mathbf{j} - 2\mathbf{k} \Rightarrow |\mathbf{r}_u \times \mathbf{r}_v| = \sqrt{6}$

$\Rightarrow$ Surface Area $= \iint_{R_{uv}} |\mathbf{r}_u \times \mathbf{r}_v| \, du \, dv = \int_0^1 \int_0^1 \sqrt{6} \, du \, dv = \sqrt{6}$

26. $\iint_S (xy - z^2) \, d\sigma = \int_0^1 \int_0^1 [(u + v)(u - v) - v^2] \sqrt{6} \, du \, dv = \sqrt{6} \int_0^1 \int_0^1 (u^2 - 2v^2) \, du \, dv$

$= \sqrt{6} \int_0^1 \left[\frac{u^3}{3} - 2uv^2\right]_0^1 dv = \sqrt{6} \int_0^1 \left(\frac{1}{3} - 2v^2\right) dv = \sqrt{6}\left[\frac{1}{3}v - \frac{2}{3}v^3\right]_0^1 = -\frac{\sqrt{6}}{3} = -\sqrt{\frac{2}{3}}$

27. $\mathbf{r}_r = (\cos\theta)\mathbf{i} + (\sin\theta)\mathbf{j}$, $\mathbf{r}_\theta = (-r\sin\theta)\mathbf{i} + (r\cos\theta)\mathbf{j} + \mathbf{k} \Rightarrow \mathbf{r}_r \times \mathbf{r}_\theta = \begin{vmatrix} \mathbf{i} & \mathbf{j} & \mathbf{k} \\ \cos\theta & \sin\theta & 0 \\ -r\sin\theta & r\cos\theta & 1 \end{vmatrix}$

$= (\sin\theta)\mathbf{i} - (\cos\theta)\mathbf{j} + r\mathbf{k} \Rightarrow |\mathbf{r}_r \times \mathbf{r}_\theta| = \sqrt{\sin^2\theta + \cos^2\theta + r^2} = \sqrt{1+r^2} \Rightarrow \text{Surface Area} = \iint_{R_{r\theta}} |\mathbf{r}_r \times \mathbf{r}_\theta|\,dr\,d\theta$

$= \int_0^{2\pi}\int_0^1 \sqrt{1+r^2}\,dr\,d\theta = \int_0^{2\pi} \left[\frac{r}{2}\sqrt{1+r^2} + \frac{1}{2}\ln\left(r+\sqrt{1+r^2}\right)\right]_0^1 d\theta = \int_0^{2\pi} \left[\frac{1}{2}\sqrt{2} + \frac{1}{2}\ln\left(1+\sqrt{2}\right)\right] d\theta$

$= \pi\left[\sqrt{2} + \ln\left(1+\sqrt{2}\right)\right]$

28. $\iint_S \sqrt{x^2+y^2+1}\,d\sigma = \int_0^{2\pi}\int_0^1 \sqrt{r^2\cos^2\theta + r^2\sin^2\theta + 1}\,\sqrt{1+r^2}\,dr\,d\theta = \int_0^{2\pi}\int_0^1 \left(1+r^2\right) dr\,d\theta$

$= \int_0^{2\pi} \left[r + \frac{r^3}{3}\right]_0^1 d\theta = \int_0^{2\pi} \frac{4}{3}\,d\theta = \frac{8}{3}\pi$

29. $\frac{\partial P}{\partial y} = 0 = \frac{\partial N}{\partial z}, \frac{\partial M}{\partial z} = 0 = \frac{\partial P}{\partial x}, \frac{\partial N}{\partial x} = 0 = \frac{\partial M}{\partial y} \Rightarrow$ Conservative

30. $\frac{\partial P}{\partial y} = \frac{-3zy}{\left(x^2+y^2+z^2\right)^{-5/2}} = \frac{\partial N}{\partial z}, \frac{\partial M}{\partial z} = \frac{-3xz}{\left(x^2+y^2+z^2\right)^{-5/2}} = \frac{\partial P}{\partial x}, \frac{\partial N}{\partial x} = \frac{-3xy}{\left(x^2+y^2+z^2\right)^{-5/2}} = \frac{\partial M}{\partial y} \Rightarrow$ Conservative

31. $\frac{\partial P}{\partial y} = 0 \neq ye^z = \frac{\partial N}{\partial z} \Rightarrow$ Not Conservative

32. $\frac{\partial P}{\partial y} = \frac{x}{(x+yz)^2} = \frac{\partial N}{\partial z}, \frac{\partial M}{\partial z} = \frac{-y}{(x+yz)^2} = \frac{\partial P}{\partial x}, \frac{\partial N}{\partial x} = \frac{-z}{(x+yz)^2} = \frac{\partial M}{\partial y} \Rightarrow$ Conservative

33. $\frac{\partial f}{\partial x} = 2 \Rightarrow f(x,y,z) = 2x + g(y,z) \Rightarrow \frac{\partial f}{\partial y} = \frac{\partial g}{\partial y} = 2y + z \Rightarrow g(y,z) = y^2 + zy + h(z)$

$\Rightarrow f(x,y,z) = 2x + y^2 + zy + h(z) \Rightarrow \frac{\partial f}{\partial z} = y + h'(z) = y + 1 \Rightarrow h'(z) = 1 \Rightarrow h(z) = z + C$

$\Rightarrow f(x,y,z) = 2x + y^2 + zy + z$

34. $\frac{\partial f}{\partial x} = z\cos xz \Rightarrow f(x,y,z) = \sin xz + g(y,z) \Rightarrow \frac{\partial f}{\partial y} = \frac{\partial g}{\partial y} = e^y \Rightarrow g(y,z) = e^y + h(z)$

$\Rightarrow f(x,y,z) = \sin xz + e^y + h(z) \Rightarrow \frac{\partial f}{\partial z} = x\cos xz + h'(z) = x\cos xz \Rightarrow h'(z) = 0 \Rightarrow h(z) = C$

$\Rightarrow f(x,y,z) = \sin xz + e^y$

35. Over Path 1: $\mathbf{r} = t\mathbf{i} + t\mathbf{j} + t\mathbf{k}$, $0 \le t \le 1 \Rightarrow x = t$, $y = t$, $z = t$ and $d\mathbf{r} = (\mathbf{i}+\mathbf{j}+\mathbf{k})\,dt \Rightarrow \mathbf{F} = 2t^2\mathbf{i} + \mathbf{j} + t^2\mathbf{k}$

$\Rightarrow \mathbf{F}\cdot d\mathbf{r} = \left(3t^2+1\right) dt \Rightarrow \text{Work} = \int_0^1 \left(3t^2+1\right) dt = 2;$

Over Path 2: $\mathbf{r}_1 = t\mathbf{i} + t\mathbf{j},\ 0 \le t \le 1 \Rightarrow x = t,\ y = t,\ z = 0$ and $d\mathbf{r}_1 = (\mathbf{i}+\mathbf{j})\,dt \Rightarrow \mathbf{F}_1 = 2t^2\mathbf{i} + \mathbf{j} + t^2\mathbf{k}$

$\Rightarrow \mathbf{F}_1 \cdot d\mathbf{r}_1 = (2t^2+1)\,dt \Rightarrow \text{Work}_1 = \int_0^1 (2t^2+1)\,dt = \frac{5}{3}$; $\mathbf{r}_2 = \mathbf{i} + \mathbf{j} + t\mathbf{k},\ 0 \le t \le 1 \Rightarrow x = 1,\ y = 1,\ z = t$ and

$d\mathbf{r}_2 = \mathbf{k}\,dt \Rightarrow \mathbf{F}_2 = 2\mathbf{i} + \mathbf{j} + \mathbf{k} \Rightarrow \mathbf{F}_2 \cdot d\mathbf{r}_2 = dt \Rightarrow \text{Work}_2 = \int_0^1 dt = 1 \Rightarrow \text{Work} = \text{Work}_1 + \text{Work}_2 = \frac{5}{3} + 1 = \frac{8}{3}$

36. Over Path 1: $\mathbf{r} = t\mathbf{i} + t\mathbf{j} + t\mathbf{k},\ 0 \le t \le 1 \Rightarrow x = t,\ y = t,\ z = t$ and $d\mathbf{r} = (\mathbf{i}+\mathbf{j}+\mathbf{k})\,dt \Rightarrow \mathbf{F} = 2t^2\mathbf{i} + t^2\mathbf{j} + \mathbf{k}$

$\Rightarrow \mathbf{F} \cdot d\mathbf{r} = (3t^2+1)\,dt \Rightarrow \text{Work} = \int_0^1 (3t^2+1)\,dt = 2$;

Over Path 2: Since f is conservative, $\oint_C \mathbf{F} \cdot d\mathbf{r} = 0$ around any simple closed curve C. Thus consider

$\int_{\text{curve}} \mathbf{F} \cdot d\mathbf{r} = \int_{C_1} \mathbf{F} \cdot d\mathbf{r} + \int_{C_2} \mathbf{F} \cdot d\mathbf{r}$, where C_1 is the path from $(0,0,0)$ to $(1,1,0)$ to $(1,1,1)$ and C_2 is the path

from $(1,1,1)$ to $(0,0,0)$. Now, from Path 1 above, $\int_{C_2} \mathbf{F} \cdot d\mathbf{r} = -2 \Rightarrow 0 = \int_{\text{curve}} \mathbf{F} \cdot d\mathbf{r} = \int_{C_1} \mathbf{F} \cdot d\mathbf{r} + (-2)$

$\Rightarrow \int_{C_1} \mathbf{F} \cdot d\mathbf{r} = 2$

37. (a) $\mathbf{r} = (e^t \cos t)\mathbf{i} + (e^t \sin t)\mathbf{j} \Rightarrow x = e^t \cos t,\ y = e^t \sin t$ from $(1,0)$ to $(e^{2\pi},0) \Rightarrow 0 \le t \le 2\pi$

$\Rightarrow \frac{d\mathbf{r}}{dt} = (e^t \cos t - e^t \sin t)\,\mathbf{i} + (e^t \sin t + e^t \cos t)\,\mathbf{j}$ and $\mathbf{F} = \frac{x\mathbf{i} + y\mathbf{j}}{(x^2+y^2)^{3/2}} = \frac{(e^t \cos t)\mathbf{i} + (e^t \sin t)\mathbf{j}}{(e^{2t}\cos^2 t + e^{2t}\sin^2 t)^{3/2}}$

$= \left(\frac{\cos t}{e^{2t}}\right)\mathbf{i} + \left(\frac{\sin t}{e^{2t}}\right)\mathbf{j} \Rightarrow \mathbf{F} \cdot \frac{d\mathbf{r}}{dt} = \left(\frac{\cos^2 t}{e^t} - \frac{\sin t \cos t}{e^t} + \frac{\sin^2 t}{e^t} + \frac{\sin t \cos t}{e^t}\right) = e^{-t}$

$\Rightarrow \text{Work} = \int_0^{2\pi} e^{-t}\,dt = 1 - e^{-2\pi}$

(b) $\mathbf{F} = \frac{x\mathbf{i} + y\mathbf{j}}{(x^2+y^2)^{3/2}} \Rightarrow \frac{\partial f}{\partial x} = \frac{x}{(x^2+y^2)^{3/2}} \Rightarrow f(x,y,z) = -(x^2+y^2)^{-1/2} + g(y,z) \Rightarrow \frac{\partial f}{\partial y} = \frac{y}{(x^2+y^2)^{3/2}} + \frac{\partial g}{\partial y}$

$= \frac{y}{(x^2+y^2)^{3/2}} \Rightarrow g(y,z) = C \Rightarrow f(x,y,z) = -(x^2+y^2)^{-1/2}$ is a potential function for $\mathbf{F} \Rightarrow \int_C \mathbf{F} \cdot d\mathbf{r}$

$= f(e^{2\pi},0) - f(1,0) = 1 - e^{-2\pi}$

38. (a) $\mathbf{F} = \nabla(x^2ze^y) \Rightarrow \mathbf{F}$ is conservative $\Rightarrow \oint_C \mathbf{F}\cdot d\mathbf{r} = 0$ for any closed path C

(b) $\int_C \mathbf{F}\cdot d\mathbf{r} = \int_{(1,0,0)}^{(1,0,2\pi)} \nabla(x^2ze^y)\cdot d\mathbf{r} = (x^2ze^y)\Big|_{(1,0,2\pi)} - (x^2ze^y)\Big|_{(1,0,0)} = 2\pi - 0 = 2\pi$

39. (a) $x^2+y^2=1 \Rightarrow \mathbf{r} = (\cos t)\mathbf{i} + (\sin t)\mathbf{j},\ 0 \le t \le \pi \Rightarrow x = \cos t$ and $y = \sin t \Rightarrow \mathbf{F} = (\cos t + \sin t)\mathbf{i} - \mathbf{j}$ and $\frac{d\mathbf{r}}{dt} = (-\sin t)\mathbf{i} + (\cos t)\mathbf{j} \Rightarrow \mathbf{F}\cdot\frac{d\mathbf{r}}{dt} = -\sin t\cos t - \sin^2 t - \cos t$

$\Rightarrow \text{Flow} = \int_0^{\pi} (-\sin t\cos t - \sin^2 t - \cos t)\,dt = \left[\frac{1}{2}\cos^2 t - \frac{t}{2} + \frac{\sin 2t}{4} - \sin t\right]_0^{\pi} = -\frac{\pi}{2}$

(b) $\mathbf{r} = -t\mathbf{i},\ -1 \le t \le 1 \Rightarrow x = -t$ and $y = 0 \Rightarrow \mathbf{F} = -t\mathbf{i} - t^2\mathbf{j}$ and $\frac{d\mathbf{r}}{dt} = -\mathbf{i} \Rightarrow \mathbf{F}\cdot\frac{d\mathbf{r}}{dt} = t \Rightarrow \text{Flow} = \int_{-1}^{1} t\,dt = 0$

(c) $\mathbf{r}_1 = (1-t)\mathbf{i} - t\mathbf{j},\ 0 \le t \le 1 \Rightarrow \mathbf{F}_1 = (1-2t)\mathbf{i} - (1-2t+2t^2)\mathbf{j}$ and $\frac{d\mathbf{r}_1}{dt} = -\mathbf{i} - \mathbf{j} \Rightarrow \mathbf{F}_1\cdot\frac{d\mathbf{r}_1}{dt} = 2t^2$

$\Rightarrow \text{Flow}_1 = \int_0^1 2t^2\,dt = \frac{2}{3};\ \mathbf{r}_2 = -t\mathbf{i} + (t-1)\mathbf{j},\ 0 \le t \le 1 \Rightarrow \mathbf{F}_2 = -\mathbf{i} - (2t^2 - 2t + 1)\mathbf{j}$ and $\frac{d\mathbf{r}_2}{dt} = -\mathbf{i} + \mathbf{j}$

$\Rightarrow \mathbf{F}_2\cdot\frac{d\mathbf{r}_2}{dt} = -2t^2 + 2t \Rightarrow \text{Flow}_2 = \int_0^1 (-2t^2 + 2t)\,dt = \frac{1}{3} \Rightarrow \text{Flow} = \text{Flow}_1 + \text{Flow}_2 = \frac{2}{3} + \frac{1}{3} = 1$

40. $\mathbf{r}_1 = (\cos t)\mathbf{i} + (\sin t)\mathbf{j} + t\mathbf{k},\ 0 \le t \le \frac{\pi}{2} \Rightarrow \mathbf{F}_1 = (2\cos t)\mathbf{i} + 2t\mathbf{j} + (2\sin t)\mathbf{k}$ and $\frac{d\mathbf{r}_1}{dt} = (-\sin t)\mathbf{i} + (\cos t)\mathbf{j} + \mathbf{k}$

$\Rightarrow \mathbf{F}_1\cdot\frac{d\mathbf{r}_1}{dt} = -2\sin t\cos t + 2t\cos t + 2\sin t \Rightarrow \text{Circ}_1 = \int_0^{\pi/2} (-2\sin t\cos t + 2t\cos t + 2\sin t)\,dt$

$= \left[\cos^2 t + 2t\sin t\right]_0^{\pi/2} = \pi - 1;\ \mathbf{r}_2 = \mathbf{j} + \left(\frac{\pi}{2}\right)(1-t)\mathbf{k},\ 0 \le t \le 1 \Rightarrow \mathbf{F}_2 = \pi(1-t)\mathbf{j} + 2\mathbf{k}$ and $\frac{d\mathbf{r}_2}{dt} = -\frac{\pi}{2}\mathbf{k}$

$\Rightarrow \mathbf{F}_2\cdot\frac{d\mathbf{r}_2}{dt} = -\pi \Rightarrow \text{Circ}_2 = \int_0^1 -\pi\,dt = -\pi;\ \mathbf{r}_3 = t\mathbf{i} + (1-t)\mathbf{j},\ 0 \le t \le 1 \Rightarrow \mathbf{F}_3 = 2t\mathbf{i} + 2(1-t)\mathbf{k}$ and

$\frac{d\mathbf{r}_3}{dt} = \mathbf{i} - \mathbf{j} \Rightarrow \mathbf{F}_3\cdot\frac{d\mathbf{r}_3}{dt} = 2t \Rightarrow \text{Circ}_3 = \int_0^1 2t\,dt = 1 \Rightarrow \text{Circ} = \text{Circ}_1 + \text{Circ}_2 + \text{Circ}_3 = 0$

41. $\nabla \times \mathbf{F} = \begin{vmatrix} \mathbf{i} & \mathbf{j} & \mathbf{k} \\ \frac{\partial}{\partial x} & \frac{\partial}{\partial y} & \frac{\partial}{\partial z} \\ y^2 & -y & 3z^2 \end{vmatrix} = -2y\mathbf{k}$; unit normal to the plane is $\mathbf{n} = \frac{2\mathbf{i} + 6\mathbf{j} - 3\mathbf{k}}{\sqrt{4+36+9}} = \frac{2}{7}\mathbf{i} + \frac{6}{7}\mathbf{j} - \frac{3}{7}\mathbf{k}$

$\Rightarrow \nabla \times \mathbf{F}\cdot\mathbf{n} = \frac{6}{7}y;\ \mathbf{p} = \mathbf{k}$ and $f(x,y,z) = 2x + 6y - 3z \Rightarrow |\nabla f\cdot\mathbf{p}| = 3 \Rightarrow d\sigma = \frac{|\nabla f|}{|\nabla f\cdot\mathbf{p}|}\,dA = \frac{7}{3}\,dA$

$$\Rightarrow \oint_C \mathbf{F}\cdot d\mathbf{r} = \iint_R \tfrac{6}{7}y\,d\sigma = \iint_R \left(\tfrac{6}{7}y\right)\left(\tfrac{7}{3}\,dA\right) = \iint_R 2y\,dA = \int_0^{2\pi}\int_0^1 2r\sin\theta\, r\,dr\,d\theta = \int_0^{2\pi} \tfrac{2}{3}\sin\theta\,d\theta = 0$$

42. $\nabla\times\mathbf{F} = \begin{vmatrix} \mathbf{i} & \mathbf{j} & \mathbf{k} \\ \frac{\partial}{\partial x} & \frac{\partial}{\partial y} & \frac{\partial}{\partial z} \\ x^2+y & x+y & 4y^2-z \end{vmatrix} = 8y\mathbf{i}$; the circle lies in the plane $f(x,y,z) = y+z = 0$ with unit normal

$$\mathbf{n} = \frac{1}{\sqrt{2}}\mathbf{j} + \frac{1}{\sqrt{2}}\mathbf{k} \Rightarrow \nabla\times\mathbf{F}\cdot\mathbf{n} = 0 \Rightarrow \oint_C \mathbf{F}\cdot d\mathbf{r} = \iint_R \nabla\times\mathbf{F}\cdot\mathbf{n}\,d\sigma = \iint_R 0\,d\sigma = 0$$

43. (a) $\mathbf{r} = \sqrt{2}t\mathbf{i} + \sqrt{2}t\mathbf{j} + (4-t^2)\mathbf{k},\ 0\le t\le 1 \Rightarrow x = \sqrt{2}t,\ y = \sqrt{2}t,\ z = 4-t^2 \Rightarrow \frac{dx}{dt} = \sqrt{2},\ \frac{dy}{dt} = \sqrt{2},\ \frac{dz}{dt} = -2t$

$$\Rightarrow \sqrt{\left(\frac{dx}{dt}\right)^2 + \left(\frac{dy}{dt}\right)^2 + \left(\frac{dz}{dt}\right)^2}\,dt = \sqrt{4+4t^2}\,dt \Rightarrow M = \int_C \delta(x,y,z)\,ds = \int_0^1 3t\sqrt{4+4t^2}\,dt = \left[\tfrac{1}{4}(4+4t)^{3/2}\right]_0^1$$

$$= 4\sqrt{2} - 2$$

(b) $M = \int_C \delta(x,y,z)\,ds = \int_0^1 \sqrt{4+4t^2}\,dt = \left[t\sqrt{1+t^2} + \ln\left(t+\sqrt{1+t^2}\right)\right]_0^1 = \sqrt{2} + \ln\left(1+\sqrt{2}\right)$

44. $\mathbf{r} = t\mathbf{i} + 2t\mathbf{j} + \frac{2}{3}t^{3/2}\mathbf{k},\ 0\le t\le 2 \Rightarrow x = t,\ y = 2t,\ z = \frac{2}{3}t^{3/2} \Rightarrow \frac{dx}{dt} = 1,\ \frac{dy}{dt} = 2,\ \frac{dz}{dt} = t^{1/2}$

$$\Rightarrow \sqrt{\left(\frac{dx}{dt}\right)^2 + \left(\frac{dy}{dt}\right)^2 + \left(\frac{dz}{dt}\right)^2}\,dt = \sqrt{t+5}\,dt \Rightarrow M = \int_C \delta(x,y,z)\,ds = \int_0^2 3\sqrt{5+t}\,\sqrt{t+5}\,dt$$

$$= \int_0^2 3(t+5)\,dt = 36;\ M_{yz} = \int_C x\delta\,ds = \int_0^2 3t(t+5)\,dt = 38;\ M_{xz} = \int_C y\delta\,ds = \int_0^2 6t(t+5)\,dt = 76;$$

$$M_{xy} = \int_C z\delta\,ds = \int_0^2 2t^{3/2}(t+5)\,dt = \frac{144}{7}\sqrt{2} \Rightarrow \bar{x} = \frac{M_{yz}}{M} = \frac{38}{36} = \frac{19}{18},\ \bar{y} = \frac{M_{xz}}{M} = \frac{76}{36} = \frac{19}{9},\ \bar{z} = \frac{M_{xy}}{M} = \frac{\left(\frac{144}{7}\sqrt{2}\right)}{36}$$

$$= \tfrac{4}{7}\sqrt{2}$$

45. $\mathbf{r} = t\mathbf{i} + \left(\frac{2\sqrt{2}}{3}t^{3/2}\right)\mathbf{j} + \left(\frac{t^2}{2}\right)\mathbf{k},\ 0\le t\le 2 \Rightarrow x = t,\ y = \frac{2\sqrt{2}}{3}t^{3/2},\ z = \frac{t^2}{2} \Rightarrow \frac{dx}{dt} = 1,\ \frac{dy}{dt} = \sqrt{2}\,t^{1/2},\ \frac{dz}{dt} = t$

$$\Rightarrow \sqrt{\left(\frac{dx}{dt}\right)^2 + \left(\frac{dy}{dt}\right)^2 + \left(\frac{dz}{dt}\right)^2}\,dt = \sqrt{1+2t+t^2}\,dt = \sqrt{(t+1)^2}\,dt = |t+1|\,dt = (t+1)\,dt \text{ on the domain given.}$$

Then $M = \int_C \delta\,ds = \int_0^2 \left(\frac{1}{t+1}\right)(t+1)\,dt = \int_0^2 dt = 2;\ M_{yz} = \int_C x\delta\,ds = \int_0^2 t\left(\frac{1}{t+1}\right)(t+1)\,dt = \int_0^2 t\,dt = 2;$

$$M_{xz} = \int_C y\delta\,ds = \int_0^2 \left(\frac{2\sqrt{2}}{3}t^{3/2}\right)\left(\frac{1}{t+1}\right)(t+1)\,dt = \int_0^2 \frac{2\sqrt{2}}{3}t^{3/2}\,dt = \frac{32}{15};\ M_{xy} = \int_C z\delta\,ds$$

$= \int_0^2 \left(\frac{t^2}{2}\right)\left(\frac{1}{t+1}\right)(t+1)\,dt = \int_0^2 \frac{t^2}{2}\,dt = \frac{4}{3} \Rightarrow \bar{x} = \frac{M_{yz}}{M} = \frac{2}{2} = 1;\ \bar{y} = \frac{M_{xz}}{M} = \frac{\left(\frac{32}{15}\right)}{2} = \frac{16}{15};\ \bar{z} = \frac{M_{xy}}{M}$

$= \frac{\left(\frac{4}{3}\right)}{2} = \frac{2}{3};\ I_x = \int_C (y^2+z^2)\delta\,ds = \int_0^2 \left(\frac{8}{9}t^3 + \frac{t^4}{4}\right)dt = \frac{232}{45};\ I_y = \int_C (x^2+z^2)\delta\,ds = \int_0^2 \left(t^2 + \frac{t^4}{4}\right)dt = \frac{64}{15};$

$I_z = \int_C (y^2+x^2)\delta\,ds = \int_0^2 \left(t^2 + \frac{8}{9}t^3\right)dt = \frac{56}{9};\ R_x = \sqrt{\frac{I_x}{M}} = \sqrt{\frac{\left(\frac{232}{45}\right)}{2}} = \sqrt{\frac{116}{45}};\ R_y = \sqrt{\frac{I_y}{M}} = \sqrt{\frac{\left(\frac{64}{15}\right)}{2}} = \sqrt{\frac{32}{15}};$

$R_z = \sqrt{\frac{I_z}{M}} = \sqrt{\frac{\left(\frac{56}{9}\right)}{2}} = \sqrt{\frac{28}{9}}$

46. $\bar{z} = 0$ because the arch is in the xy-plane, and $\bar{x} = 0$ because the mass is distributed symmetrically with respect to the y-axis; $\mathbf{r}(t) = (a \cos t)\mathbf{i} + (a \sin t)\mathbf{j},\ 0 \le t \le \pi \Rightarrow ds = \sqrt{\left(\frac{dx}{dt}\right)^2 + \left(\frac{dy}{dt}\right)^2 + \left(\frac{dz}{dt}\right)^2}\,dt$

$= \sqrt{(-a \sin t)^2 + (a \cos t)^2}\,dt = a\,dt$, since $a \ge 0$; $M = \int_C \delta\,ds = \int_C (2a - y)\,ds = \int_0^\pi (2 - a \sin t)\,a\,dt$

$= 2a\pi - 2a^2;\ M_{xz} = \int_C y\delta\,dt = \int_C y(2a-y)\,ds = \int_0^\pi (a \sin t)(2a - a \sin t)\,dt = \int_0^\pi \left(2a^2 \sin t - a^2 \sin^2 t\right)dt$

$= \left[-2a^2 \cos t - a^2\left(\frac{t}{2} - \frac{\sin 2t}{4}\right)\right]_0^\pi = 4a^2 - \frac{a^2\pi}{2} \Rightarrow \bar{y} = \frac{\left(4a^2 - \frac{a^2\pi}{2}\right)}{2a\pi - 2a^2} = \frac{8a - a\pi}{4\pi - 4a} \Rightarrow (\bar{x},\bar{y},\bar{z}) = \left(0, \frac{8a - a\pi}{4\pi - 4a}, 0\right)$

47. $\mathbf{r}(t) = (e^t \cos t)\mathbf{i} + (e^t \sin t)\mathbf{j} + e^t\mathbf{k},\ 0 \le t \le \ln 2 \Rightarrow x = e^t \cos t,\ y = e^t \sin t,\ z = e^t \Rightarrow \frac{dx}{dt} = (e^t \cos t - e^t \sin t),$

$\frac{dy}{dt} = (e^t \sin t + e^t \cos t),\ \frac{dz}{dt} = e^t \Rightarrow \sqrt{\left(\frac{dx}{dt}\right)^2 + \left(\frac{dy}{dt}\right)^2 + \left(\frac{dz}{dt}\right)^2}\,dt$

$= \sqrt{(e^t \cos t - e^t \sin t)^2 + (e^t \sin t + e^t \cos t)^2 + (e^t)^2}\,dt = \sqrt{3e^{2t}}\,dt = \sqrt{3}\,e^t\,dt;\ M = \int_C \delta\,ds = \int_0^{\ln 2} \sqrt{3}\,e^t\,dt$

$= \sqrt{3};\ M_{xy} = \int_C z\delta\,ds = \int_0^{\ln 2} \left(\sqrt{3}\,e^t\right)(e^t)\,dt = \int_0^{\ln 2} \sqrt{3}\,e^{2t}\,dt = \frac{3\sqrt{3}}{2} \Rightarrow \bar{z} = \frac{M_{xy}}{M} = \frac{\left(\frac{3\sqrt{3}}{2}\right)}{\sqrt{3}} = \frac{3}{2};$

$I_z = \int_C (x^2+y^2)\delta\,ds = \int_0^{\ln 2} \left(e^{2t} \cos^2 t + e^{2t} \sin^2 t\right)\left(\sqrt{3}\,e^t\right)dt = \int_0^{\ln 2} \sqrt{3}\,e^{3t}\,dt = \frac{7\sqrt{3}}{3} \Rightarrow R_z = \sqrt{\frac{I_z}{M}}$

$= \sqrt{\frac{7\sqrt{3}}{3\sqrt{3}}} = \sqrt{\frac{7}{3}}$

48. $\mathbf{r}(t) = (2 \sin t)\mathbf{i} + (2 \cos t)\mathbf{j} + 3t\mathbf{k}$, $0 \le t \le 2\pi \Rightarrow x = 2 \sin t$, $y = 2 \cos t$, $z = 3t \Rightarrow \frac{dx}{dt} = 2 \cos t$, $\frac{dy}{dt} = -2 \sin t$,

$\frac{dz}{dt} = 3 \Rightarrow \sqrt{\left(\frac{dx}{dt}\right)^2 + \left(\frac{dy}{dt}\right)^2 + \left(\frac{dz}{dt}\right)^2}\, dt = \sqrt{4+9}\, dt = \sqrt{13}\, dt$; $M = \int_C \delta\, ds = \int_0^{2\pi} \rho\sqrt{13}\, dt = 2\pi\rho\sqrt{13}$;

$M_{xy} = \int_C z\delta\, ds = \int_0^{2\pi} (3t)(\rho\sqrt{13})\, dt = 6\rho\pi^2\sqrt{13}$; $M_{yz} = \int_C x\delta\, ds = \int_0^{2\pi} (2 \sin t)(\rho\sqrt{13})\, dt = 0$;

$M_{xz} = \int_C y\delta\, ds = \int_0^{2\pi} (2 \cos t)(\rho\sqrt{13})\, dt = 0 \Rightarrow \bar{x} = \bar{y} = 0$ and $\bar{z} = \frac{M_{xy}}{M} = \frac{6\rho\pi^2\sqrt{13}}{2\rho\pi\sqrt{13}} = 3\pi \Rightarrow (0, 0, 3\pi)$ is the center of mass

49. Because of symmetry $\bar{x} = \bar{y} = 0$. Let $f(x,y,z) = x^2 + y^2 + z^2 = 25 \Rightarrow \nabla f = 2x\mathbf{i} + 2y\mathbf{j} + 2z\mathbf{k}$

$\Rightarrow |\nabla f| = \sqrt{4x^2 + 4y^2 + 4z^2} = 10$ and $\mathbf{p} = \mathbf{k} \Rightarrow |\nabla f \cdot \mathbf{p}| = 2z$, since $z \ge 0 \Rightarrow M = \iint_R \delta(x,y,z)\, d\sigma$

$= \iint_R z\left(\frac{10}{2z}\right) dA = \iint_R 5\, dA = 5(\text{Area of the circular region}) = 80\pi$; $M_{xy} = \iint_R z\delta\, d\sigma = \iint_R 5z\, dA$

$= \iint_R 5\sqrt{25 - x^2 - y^2}\, dx\, dy = \int_0^{2\pi}\int_0^4 \left(5\sqrt{25 - r^2}\right) r\, dr\, d\theta = \int_0^{2\pi} \frac{490}{3}\, d\theta = \frac{980}{3}\pi \Rightarrow \bar{z} = \frac{\left(\frac{980}{3}\pi\right)}{80\pi} = \frac{49}{12}$

$\Rightarrow (\bar{x}, \bar{y}, \bar{z}) = \left(0, 0, \frac{49}{12}\right)$; $I_z = \iint_R (x^2 + y^2)\delta\, d\sigma = \iint_R 5(x^2 + y^2)\, dx\, dy = \int_0^{2\pi}\int_0^4 5r^3\, dr\, d\theta = \int_0^{2\pi} 320\, d\theta = 640\pi$;

$R_z = \sqrt{\frac{I_z}{M}} = \sqrt{\frac{640\pi}{80\pi}} = 2\sqrt{2}$

50. On the face $z = 1$: $g(x,y,z) = z = 1$ and $\mathbf{p} = \mathbf{k} \Rightarrow \nabla g = \mathbf{k} \Rightarrow |\nabla g| = 1$ and $|\nabla g \cdot \mathbf{p}| = 1 \Rightarrow d\sigma = dA$

$\Rightarrow I = \iint_R (x^2 + y^2)\, dA = 2\int_0^{\pi/4}\int_0^{\sec\theta} r^3\, dr\, d\theta = \frac{2}{3}$; On the face $z = 0$: $g(x,y,z) = z = 0 \Rightarrow \nabla g = \mathbf{k}$ and $\mathbf{p} = \mathbf{k}$

$\Rightarrow |\nabla g| = 1 \Rightarrow |\nabla g \cdot \mathbf{p}| = 1 \Rightarrow d\sigma = dA \Rightarrow I = \iint_R (x^2 + y^2)\, dA = \frac{2}{3}$; On the face $y = 0$: $g(x,y,z) = y = 0$

$\Rightarrow \nabla g = \mathbf{j}$ and $\mathbf{p} = \mathbf{j} \Rightarrow |\nabla g| = 1 \Rightarrow |\nabla g \cdot \mathbf{p}| = 1 \Rightarrow d\sigma = dA \Rightarrow I = \iint_R (x^2 + 0)\, dA = \int_0^1\int_0^1 x^2\, dx\, dz = \frac{1}{3}$;

On the face $y = 1$: $g(x,y,z) = y = 1 \Rightarrow \nabla g = \mathbf{j}$ and $\mathbf{p} = \mathbf{j} \Rightarrow |\nabla g| = 1 \Rightarrow |\nabla g \cdot \mathbf{p}| = 1 \Rightarrow d\sigma = dA$

$\Rightarrow I = \iint_R (x^2 + 1^2)\, dA = \int_0^1\int_0^1 (x^2 + 1)\, dx\, dz = \frac{4}{3}$; On the face $x = 1$: $g(x,y,z) = x = 1 \Rightarrow \nabla g = \mathbf{i}$ and $\mathbf{p} = \mathbf{i}$

$\Rightarrow |\nabla g| = 1 \Rightarrow |\nabla g \cdot \mathbf{p}| = 1 \Rightarrow d\sigma = dA \Rightarrow I = \iint_R (1^2 + y^2)\, dA = \int_0^1\int_0^1 (1 + y^2)\, dy\, dz = \frac{4}{3}$; On the face $x = 0$: $g(x,y,z) = x = 0 \Rightarrow \nabla g = \mathbf{i}$ and $\mathbf{p} = \mathbf{i} \Rightarrow |\nabla g| = 1 \Rightarrow |\nabla g \cdot \mathbf{p}| = 1 \Rightarrow d\sigma = dA$

$$\Rightarrow I = \iint_R (0^2 + y^2)\, dA = \int_0^1 \int_0^1 y^2\, dy\, dz = \frac{1}{3} \Rightarrow I_z = \frac{2}{3} + \frac{2}{3} + \frac{1}{3} + \frac{4}{3} + \frac{4}{3} + \frac{1}{3} = \frac{14}{3}$$

51. $M = 2xy + x$ and $N = xy - y \Rightarrow \frac{\partial M}{\partial x} = 2y + 1,\ \frac{\partial M}{\partial y} = 2x,\ \frac{\partial N}{\partial x} = y,\ \frac{\partial N}{\partial y} = x - 1 \Rightarrow \text{Flux} = \iint_R \left(\frac{\partial M}{\partial x} + \frac{\partial N}{\partial y}\right) dx\, dy$

$$= \iint_R (2y + 1 + x - 1)\, dy\, dx = \int_0^1 \int_0^1 (2y + x)\, dy\, dx = \frac{3}{2};\ \text{Circ} = \iint_R \left(\frac{\partial N}{\partial x} - \frac{\partial M}{\partial y}\right) dx\, dy$$

$$= \iint_R (y - 2x)\, dy\, dx = \int_0^1 \int_0^1 (y - 2x)\, dy\, dx = -\frac{1}{2}$$

52. $M = y - 6x^2$ and $N = x + y^2 \Rightarrow \frac{\partial M}{\partial x} = -12x,\ \frac{\partial M}{\partial y} = 1,\ \frac{\partial N}{\partial x} = 1,\ \frac{\partial N}{\partial y} = 2y \Rightarrow \text{Flux} = \iint_R \left(\frac{\partial M}{\partial x} + \frac{\partial N}{\partial y}\right) dx\, dy$

$$= \iint_R (-12x + 2y)\, dx\, dy = \int_0^1 \int_y^1 (-12x + 2y)\, dx\, dy = \int_0^1 (4y^2 + 2y - 6)\, dy = -\frac{11}{3};$$

$$\text{Circ} = \iint_R \left(\frac{\partial N}{\partial x} - \frac{\partial M}{\partial y}\right) dx\, dy = \iint_R (1 - 1)\, dx\, dy = 0$$

53. $M = -\frac{\cos y}{x}$ and $N = \ln x \sin y \Rightarrow \frac{\partial M}{\partial y} = \frac{\sin y}{x}$ and $\frac{\partial N}{\partial x} = \frac{\sin y}{x} \Rightarrow \oint_C \ln x \sin y\, dy - \frac{\cos y}{x}\, dx$

$$= \iint_R \left(\frac{\partial N}{\partial x} - \frac{\partial M}{\partial y}\right) dx\, dy = \iint_R \left(\frac{\sin y}{x} - \frac{\sin y}{x}\right) dx\, dy = 0$$

54. (a) Let $M = x$ and $N = y \Rightarrow \frac{\partial M}{\partial x} = 1,\ \frac{\partial M}{\partial y} = 0,\ \frac{\partial N}{\partial x} = 0,\ \frac{\partial N}{\partial y} = 1 \Rightarrow \text{Flux} = \iint_R \left(\frac{\partial M}{\partial x} + \frac{\partial N}{\partial y}\right) dx\, dy$

$$= \iint_R (1 + 1)\, dx\, dy = 2 \iint_R dx\, dy = 2(\text{Area of the region})$$

(b) Let C be a closed curve to which Green's Theorem applies and let **n** be the unit normal vector to C. Let $\mathbf{F} = x\mathbf{i} + y\mathbf{j}$ and assume **F** is orthogonal to **n** at every point of C. Then the flux density of **F** at every point of C is 0 since $\mathbf{F} \cdot \mathbf{n} = 0$ at every point of $C \Rightarrow \frac{\partial M}{\partial x} + \frac{\partial N}{\partial y} = 0$ at every point of C

$\Rightarrow \text{Flux} = \iint_R \left(\frac{\partial M}{\partial x} + \frac{\partial N}{\partial y}\right) dx\, dy = \iint_R 0\, dx\, dy = 0$. But part (a) above states that the flux is 2(Area of the region) $\Rightarrow$ the area of the region would be 0 $\Rightarrow$ contradiction. Therefore, **F** cannot be orthogonal to **n** at every point of C.

55. $\frac{\partial}{\partial x}(2xy) = 2y$, $\frac{\partial}{\partial y}(2yz) = 2z$, $\frac{\partial}{\partial z}(2xz) = 2x \Rightarrow \nabla \cdot \mathbf{F} = 2y + 2z + 2x \Rightarrow \text{Flux} = \iiint_D (2x + 2y + 2z)\, dV$

$= \int_0^1 \int_0^1 \int_0^1 (2x + 2y + 2z)\, dx\, dy\, dz = \int_0^1 \int_0^1 (1 + 2y + 2z)\, dy\, dz = \int_0^1 (2 + 2z)\, dz = 3$

56. $\frac{\partial}{\partial x}(xz) = z$, $\frac{\partial}{\partial y}(yz) = z$, $\frac{\partial}{\partial z}(1) = 0 \Rightarrow \nabla \cdot \mathbf{F} = 2z \Rightarrow \text{Flux} = \iiint_D 2z\; r\, dr\, d\theta\, dz$

$= \int_0^{2\pi} \int_0^4 \int_3^{\sqrt{25-r^2}} 2z\; dz\; r\; dr\, d\theta = \int_0^{2\pi} \int_0^4 \left(16 - r^2\right) dr\, d\theta = \int_0^{2\pi} 64\; d\theta = 128\pi$

57. $\frac{\partial}{\partial x}(-2x) = -2$, $\frac{\partial}{\partial y}(-3y) = -3$, $\frac{\partial}{\partial z}(z) = 1 \Rightarrow \nabla \cdot \mathbf{F} = -4$; $x^2 + y^2 + z^2 = 2$ and $x^2 + y^2 = z \Rightarrow z = 1$

$\Rightarrow x^2 + y^2 = 1 \Rightarrow \text{Flux} = \iiint_D -4\; dV = -4 \int_0^{2\pi} \int_0^1 \int_{r^2}^{\sqrt{2-r^2}} dz\; r\; dr\, d\theta = -4 \int_0^{2\pi} \int_0^1 \left(r\sqrt{2 - r^2} - r^3\right) dr\, d\theta$

$= -4 \int_0^{2\pi} \left(-\frac{7}{12} + \frac{2}{3}\sqrt{2}\right) d\theta = \frac{2}{3}\pi\left(7 - 8\sqrt{2}\right)$

58. $\frac{\partial}{\partial x}(6x + y) = 6$, $\frac{\partial}{\partial y}(-x - z) = 0$, $\frac{\partial}{\partial z}(4yz) = 4y \Rightarrow \nabla \cdot \mathbf{F} = 6 + 4y$; $z = \sqrt{x^2 + y^2} = r$

$\Rightarrow \text{Flux} = \iiint_D (6 + 4y)\; dV = \int_0^{\pi/2} \int_0^1 \int_0^r (6 + 4r \sin\theta)\; dz\; r\; dr\, d\theta = \int_0^{\pi/2} \int_0^1 \left(6r^2 + 4r^3 \sin\theta\right) dr\, d\theta$

$= \int_0^{\pi/2} (2 + \sin\theta)\; d\theta = \pi + 1$

59. $\mathbf{F} = y\mathbf{i} + z\mathbf{j} + x\mathbf{k} \Rightarrow \nabla \cdot \mathbf{F} = 0 \Rightarrow \text{Flux} = \iint_S \mathbf{F} \cdot \mathbf{n}\; d\sigma = \iiint_D \nabla \cdot \mathbf{F}\; dV = 0$

60. $\mathbf{F} = 3xz^2\mathbf{i} + y\mathbf{j} - z^3\mathbf{k} \Rightarrow \nabla \cdot \mathbf{F} = 3z^2 + 1 - 3z^2 = 1 \Rightarrow \text{Flux} = \iint_S \mathbf{F} \cdot \mathbf{n}\; d\sigma = \iiint_D \nabla \cdot \mathbf{F}\; dV$

$= \int_0^4 \int_0^{\sqrt{16-x^2}/2} \int_0^{y/2} 1\; dz\, dy\, dx = \int_0^4 \left(\frac{16 - x^2}{16}\right) dx = \left[x - \frac{x^3}{48}\right]_0^4 = \frac{8}{3}$

61. $\mathbf{F} = xy^2\mathbf{i} + x^2y\mathbf{j} + y\mathbf{k} \Rightarrow \nabla \cdot \mathbf{F} = y^2 + x^2 + 0 \Rightarrow \text{Flux} = \iint_S \mathbf{F} \cdot \mathbf{n}\, d\sigma = \iiint_D \nabla \cdot \mathbf{F}\, dV$

$$= \iiint_D (x^2 + y^2)\, dV = \int_0^{2\pi}\int_0^1\int_{-1}^1 r^2\, dz\, r\, dr\, d\theta = \int_0^{2\pi}\int_0^1 2r^3\, dr\, d\theta = \int_0^{2\pi} \tfrac{1}{2}\, d\theta = \pi$$

62. (a) $\mathbf{F} = (3z + 1)\mathbf{k} \Rightarrow \nabla \cdot \mathbf{F} = 3 \Rightarrow$ Flux across the hemisphere $= \iint_S \mathbf{F} \cdot \mathbf{n}\, d\sigma = \iiint_D \nabla \cdot \mathbf{F}\, dV$

$$= \iiint_D 3\, dV = 3\left(\tfrac{1}{2}\right)\left(\tfrac{4}{3}\pi a^3\right) = 2\pi a^3$$

(b) $f(x, y, z) = x^2 + y^2 + z^2 - a^2 = 0 \Rightarrow \nabla f = 2x\mathbf{i} + 2y\mathbf{j} + 2z\mathbf{k} \Rightarrow |\nabla f| = \sqrt{4x^2 + 4y^2 + 4z^2} = \sqrt{4a^2} = 2a$ since $a \ge 0 \Rightarrow \mathbf{n} = \frac{2x\mathbf{i} + 2y\mathbf{j} + 2z\mathbf{k}}{2a} = \frac{x\mathbf{i} + y\mathbf{j} + z\mathbf{k}}{a} \Rightarrow \mathbf{F} \cdot \mathbf{n} = (3z + 1)\left(\frac{z}{a}\right)$; $\mathbf{p} = \mathbf{k} \Rightarrow \nabla f \cdot \mathbf{p} = \nabla f \cdot \mathbf{k} = 2z$

$\Rightarrow |\nabla f \cdot \mathbf{p}| = 2z$ since $z \ge 0 \Rightarrow d\sigma = \frac{|\nabla f|}{|\nabla f \cdot \mathbf{p}|} = \frac{2a}{2z}\, dA = \frac{a}{z}\, dA \Rightarrow \iint_S \mathbf{F} \cdot \mathbf{n}\, d\sigma = \iint_{R_{xy}} (3z + 1)\left(\frac{z}{a}\right)\left(\frac{a}{z}\right) dA$

$$= \iint_{R_{xy}} (3z + 1)\, dx\, dy = \iint_{R_{xy}} \left(3\sqrt{a^2 - x^2 - y^2} + 1\right) dx\, dy = \int_0^{2\pi}\int_0^a \left(3\sqrt{a^2 - r^2} + 1\right) r\, dr\, d\theta$$

$= \int_0^{2\pi} \left(\frac{a^2}{2} + a^3\right) d\theta = \pi a^2 + 2\pi a^3$, which is the flux across the hemisphere. Across the base we find $\mathbf{F} = [3(0) + 1]\mathbf{k} = \mathbf{k}$ since $z = 0$ in the xy-plane $\Rightarrow \mathbf{n} = -\mathbf{k}$ (outward normal) $\Rightarrow \mathbf{F} \cdot \mathbf{n} = -1 \Rightarrow$ Flux across the base $= \iint_S \mathbf{F} \cdot \mathbf{n}\, d\sigma = \iint_{R_{xy}} -1\, dx\, dy = -\pi a^2$. Therefore, the total flux across the closed surface is $(\pi a^2 + 2\pi a^3) - \pi a^2 = 2\pi a^3$.

CHAPTER 14 ADDITIONAL EXERCISES–THEORY, EXAMPLES, APPLICATIONS

1. $dx = (-2 \sin t + 2 \sin 2t)\, dt$ and $dy = (2 \cos t - 2 \cos 2t)\, dt$; Area $= \frac{1}{2}\oint_C x\, dy - y\, dx$

$$= \frac{1}{2}\int_0^{2\pi} [(2 \cos t - \cos 2t)(2 \cos t - 2 \cos 2t) - (2 \sin t - \sin 2t)(-2 \sin t + 2 \sin 2t)]\, dt$$

$$= \frac{1}{2}\int_0^{2\pi} [6 - (6 \cos t \cos 2t + 6 \sin t \sin 2t)]\, dt = \frac{1}{2}\int_0^{2\pi} (6 - 6 \cos t)\, dt = 6\pi$$

2. $dx = (-2\sin t - 2\sin 2t)\,dt$ and $dy = (2\cos t - 2\cos 2t)\,dt$; Area $= \frac{1}{2}\oint_C x\,dy - y\,dx$

$$= \frac{1}{2}\int_0^{2\pi} [(2\cos t + \cos 2t)(2\cos t - 2\cos 2t) - (2\sin t - \sin 2t)(-2\sin t - 2\sin 2t)]\,dt$$

$$= \frac{1}{2}\int_0^{2\pi} [2 - 2(\cos t\cos 2t - \sin t\sin 2t)]\,dt = \frac{1}{2}\int_0^{2\pi} (2 - 2\cos 3t)\,dt = \frac{1}{2}\left[2t - \frac{2}{3}\sin 3t\right]_0^{2\pi} = 2\pi$$

3. $dx = \cos 2t\,dt$ and $dy = \cos t\,dt$; Area $= \frac{1}{2}\oint_C x\,dy - y\,dx = \frac{1}{2}\int_0^{\pi}\left(\frac{1}{2}\sin 2t\cos t - \sin t\cos 2t\right)dt$

$$= \frac{1}{2}\int_0^{\pi}\left[\sin t\cos^2 t - (\sin t)\left(2\cos^2 t - 1\right)\right]dt = \frac{1}{2}\int_0^{\pi}\left(-\sin t\cos^2 t + \sin t\right)dt = \frac{1}{2}\left[\frac{1}{3}\cos^3 t - \cos t\right]_0^{\pi} = -\frac{1}{3} + 1 = \frac{2}{3}$$

4. $dx = (-2a\sin t - 2a\cos 2t)\,dt$ and $dy = (b\cos t)\,dt$; Area $= \frac{1}{2}\oint_C x\,dy - y\,dx$

$$= \frac{1}{2}\int_0^{2\pi}\left[\left(2ab\cos^2 t - ab\cos t\sin 2t\right) - \left(-2ab\sin^2 t - 2ab\sin t\cos 2t\right)\right]dt$$

$$= \frac{1}{2}\int_0^{2\pi}\left[-2ab\cos^2 t\sin t + 2ab(\sin t)\left(2\cos^2 t - 1\right)\right]dt = \frac{1}{2}\int_0^{2\pi}\left(2ab + 2ab\cos^2 t\sin t - 2ab\sin t\right)dt$$

$$= \frac{1}{2}\left[2abt - \frac{2}{3}ab\cos^3 t + 2ab\cos t\right]_0^{2\pi} = 2\pi ab$$

5. (a) $\mathbf{F}(x,y,z) = z\mathbf{i} + x\mathbf{j} + y\mathbf{k}$ is $\mathbf{0}$ only at the point $(0,0,0)$, and curl $\mathbf{F}(x,y,z) = \mathbf{i} + \mathbf{j} + \mathbf{k}$ is never $\mathbf{0}$.
   (b) $\mathbf{F}(x,y,z) = z\mathbf{i} + y\mathbf{k}$ is $\mathbf{0}$ only on the line $x = t$, $y = 0$, $z = 0$ and curl $\mathbf{F}(x,y,z) = \mathbf{i} + \mathbf{j}$ is never $\mathbf{0}$.
   (c) $\mathbf{F}(x,y,z) = z\mathbf{i}$ is $\mathbf{0}$ only when $z = 0$ (the xy-plane) and curl $\mathbf{F}(x,y,z) = \mathbf{j}$ is never $\mathbf{0}$.

6. $\mathbf{F} = yz^2\mathbf{i} + xz^2\mathbf{j} + 2xyz\mathbf{k}$ and $\mathbf{n} = \frac{x\mathbf{i} + y\mathbf{j} + z\mathbf{k}}{\sqrt{x^2 + y^2 + z^2}} = \frac{x\mathbf{i} + y\mathbf{j} + z\mathbf{k}}{R}$, so $\mathbf{F}$ is parallel to $\mathbf{n}$ when $yz^2 = \frac{cx}{R}$, $xz^2 = \frac{cy}{R}$,
   and $2xyz = \frac{cz}{R} \Rightarrow \frac{yz^2}{x} = \frac{xz^2}{y} = 2xy \Rightarrow y^2 = x^2 \Rightarrow y = \pm x$ and $z^2 = \pm\frac{c}{R} = 2x^2 \Rightarrow z = \pm\sqrt{2}x$. Also,
   $x^2 + y^2 + z^2 = R^2 \Rightarrow x^2 + x^2 + 2x^2 = R^2 \Rightarrow 4x^2 = R^2 \Rightarrow x = \pm\frac{R}{2}$. Thus the points are: $\left(\frac{R}{2}, \frac{R}{2}, \frac{\sqrt{2}R}{2}\right)$,
   $\left(\frac{R}{2}, \frac{R}{2}, -\frac{\sqrt{2}R}{2}\right)$, $\left(-\frac{R}{2}, -\frac{R}{2}, \frac{\sqrt{2}R}{2}\right)$, $\left(-\frac{R}{2}, -\frac{R}{2}, -\frac{\sqrt{2}R}{2}\right)$, $\left(\frac{R}{2}, -\frac{R}{2}, \frac{\sqrt{2}R}{2}\right)$, $\left(\frac{R}{2}, -\frac{R}{2}, -\frac{\sqrt{2}R}{2}\right)$,
   $\left(-\frac{R}{2}, \frac{R}{2}, \frac{\sqrt{2}R}{2}\right)$, $\left(-\frac{R}{2}, \frac{R}{2}, -\frac{\sqrt{2}R}{2}\right)$

7. Set up the coordinate system so that $(a,b,c) = (0,R,0) \Rightarrow \delta(x,y,z) = \sqrt{x^2+(y-R)^2+z^2}$ $= \sqrt{x^2+y^2+z^2-2Ry+R^2} = \sqrt{2R^2-2Ry}$; let $f(x,y,z) = x^2+y^2+z^2 = R^2$ and $\mathbf{p} = \mathbf{i}$

$\Rightarrow \nabla f = 2x\mathbf{i}+2y\mathbf{j}+2z\mathbf{k} \Rightarrow |\nabla f| = 2\sqrt{x^2+y^2+z^2} = 2R \Rightarrow d\sigma = \frac{|\nabla f|}{|\nabla f \cdot \mathbf{i}|}\,dz\,dy = \frac{2R}{2x}\,dz\,dy$

$$\Rightarrow \text{Mass} = \iint_S \delta(x,y,z)\,d\sigma = \iint_{R_{yz}} \sqrt{2R^2-2Ry}\left(\frac{R}{x}\right) dz\,dy = R\iint_{R_{yz}} \frac{\sqrt{2R^2-2Ry}}{\sqrt{R^2-y^2-z^2}}\,dz\,dy$$

$$= 4R\int_{-R}^{R}\int_0^{\sqrt{R^2-y^2}} \frac{\sqrt{2R^2-2Ry}}{\sqrt{R^2-y^2-z^2}}\,dz\,dy = \frac{16\pi R^3}{3} \quad \text{(we used a CAS integrator)}$$

8. $\mathbf{r}(r,\theta) = (r\cos\theta)\mathbf{i}+(r\sin\theta)\mathbf{j}+\theta\mathbf{k}$, $0 \le r \le 1$, $0 \le \theta \le 2\pi \Rightarrow \mathbf{r}_r \times \mathbf{r}_\theta = \begin{vmatrix} \mathbf{i} & \mathbf{j} & \mathbf{k} \\ \cos\theta & \sin\theta & 0 \\ -r\sin\theta & r\cos\theta & 1 \end{vmatrix}$

$= (\sin\theta)\mathbf{i}-(\cos\theta)\mathbf{j}+r\mathbf{k} \Rightarrow |\mathbf{r}_r \times \mathbf{r}_\theta| = \sqrt{1+r^2}$; $\delta = 2\sqrt{x^2+y^2} = 2\sqrt{r^2\cos^2\theta+r^2\sin^2\theta} = 2r$

$$\Rightarrow \text{Mass} = \iint_S \delta(x,y,z)\,d\sigma = \int_0^{2\pi}\int_0^1 2r\sqrt{1+r^2}\,dr\,d\theta = \int_0^{2\pi}\left[\tfrac{2}{3}\left(1+r^2\right)^{3/2}\right]_0^1 d\theta = \int_0^{2\pi} \tfrac{2}{3}\left(2\sqrt{2}-1\right)d\theta$$

$= \frac{4\pi}{3}\left(2\sqrt{2}-1\right)$

9. $M = x^2+4xy$ and $N = -6y \Rightarrow \frac{\partial M}{\partial x} = 2x+4y$ and $\frac{\partial N}{\partial x} = -6 \Rightarrow \text{Flux} = \int_0^b\int_0^a (2x+4y-6)\,dx\,dy$

$= \int_0^b \left(a^2+4ay-6a\right)dy = a^2b+2ab^2-6ab$. We want to minimize $f(a,b) = a^2b+2ab^2-6ab = ab(a+2b-6)$.

Thus, $f_a(a,b) = 2ab+2b^2-6b = 0$ and $f_b(a,b) = a^2+4ab-6a = 0 \Rightarrow b(2a+2b-6) = 0 \Rightarrow b = 0$ or $b = -a+3$. Now $b = 0 \Rightarrow a^2-6a = 0 \Rightarrow a = 0$ or $a = 6 \Rightarrow (0,0)$ and $(6,0)$ are critical points. On the other hand, $b = -a+3 \Rightarrow a^2+4a(-a+3)-6a = 0 \Rightarrow -3a^2+6a = 0 \Rightarrow a = 0$ or $a = 2 \Rightarrow (0,3)$ and $(2,1)$ are also critical points. The flux at $(0,0) = 0$, the flux at $(6,0) = 0$, the flux at $(0,3) = 0$ and the flux at $(2,1) = -4$. Therefore, the flux is minimized at $(2,1)$ with value -4.

10. Let the plane be given by $z = ax+by$ and let $f(x,y,z) = x^2+y^2+z^2 = 4$. Let C denote the circle of intersection of the plane with the sphere. By Stokes's Theorem, $\oint_C \mathbf{F}\cdot d\mathbf{r} = \iint_S \nabla \times \mathbf{F}\cdot\mathbf{n}\,d\sigma$, where $\mathbf{n}$ is a unit normal to the plane. Let $\mathbf{r}(x,y) = x\mathbf{i}+y\mathbf{j}+(ax+by)\mathbf{k}$ be a parametrization of the surface. Then

$\mathbf{r}_x \times \mathbf{r}_y = \begin{vmatrix} \mathbf{i} & \mathbf{j} & \mathbf{k} \\ 1 & 0 & a \\ 0 & 1 & b \end{vmatrix} = -a\mathbf{i} - b\mathbf{j} + \mathbf{k} \Rightarrow d\sigma = |\mathbf{r}_x \times \mathbf{r}_y|\,dx\,dy = \sqrt{a^2+b^2+1}\,dx\,dy$. Also,

$\nabla \times \mathbf{F} = \begin{vmatrix} \mathbf{i} & \mathbf{j} & \mathbf{k} \\ \frac{\partial}{\partial x} & \frac{\partial}{\partial y} & \frac{\partial}{\partial z} \\ z & x & y \end{vmatrix} = \mathbf{i}+\mathbf{j}+\mathbf{k}$ and $\mathbf{n} = \frac{a\mathbf{i}+b\mathbf{j}-\mathbf{k}}{\sqrt{a^2+b^2+1}} \Rightarrow \iint\limits_S \nabla \times \mathbf{F}\cdot\mathbf{n}\,d\sigma$

$= \iint\limits_{R_{xy}} \frac{a+b-1}{\sqrt{a^2+b^2+1}}\sqrt{a^2+b^2+1}\,dx\,dy = \iint\limits_{R_{xy}} (a+b-1)\,dx\,dy = (a+b-1)\iint\limits_{R_{xy}} dx\,dy$. Now

$x^2+y^2+(ax+by)^2 = 4 \Rightarrow \left(\frac{a^2+1}{4}\right)x^2 + \left(\frac{b^2+1}{4}\right)y^2 + \left(\frac{ab}{2}\right)xy = 1 \Rightarrow$ the region R_{xy} is the interior of the ellipse $Ax^2+Bxy+Cy^2 = 1$ in the xy-plane, where $A = \frac{a^2+1}{4}$, $B = \frac{ab}{2}$, and $C = \frac{b^2+1}{4}$. By Exercise 47 in Section 9.3, the area of the ellipse is $\frac{2\pi}{\sqrt{4AC-B^2}} = \frac{4\pi}{\sqrt{a^2+b^2+1}} \Rightarrow \oint\limits_C \mathbf{F}\cdot d\mathbf{r} = h(a,b) = \frac{4\pi(a+b-1)}{\sqrt{a^2+b^2+1}}$.

Thus we optimize $H(a,b) = \frac{(a+b-1)^2}{a^2+b^2+1}$: $\frac{\partial H}{\partial a} = \frac{2(a+b-1)(b^2+1+a-ab)}{a^2+b^2+1} = 0$ and

$\frac{\partial H}{\partial b} = \frac{2(a+b-1)(a^2+1+b-ab)}{a^2+b^2+1} = 0 \Rightarrow a+b-1 = 0$, or $b^2+1+a-ab = 0$ and $a^2+1+b-ab = 0$

$\Rightarrow a+b-1 = 0$, or $a^2-b^2+(b-a) = 0 \Rightarrow a+b-1 = 0$, or $(a-b)(a+b-1) = 0 \Rightarrow a+b-1 = 0$ or $a = b$.
The critical values $a+b-1 = 0$ give a saddle. If $a = b$, then $0 = b^2+1+a-ab \Rightarrow a^2+1+a-a^2 = 0$

$\Rightarrow a = -1 \Rightarrow b = -1$. Thus, the point $(a,b) = (-1,-1)$ gives a local extremum for $\oint\limits_C \mathbf{F}\cdot d\mathbf{r} \Rightarrow z = -x-y$

$\Rightarrow x+y+z = 0$ is the desired plane.

Note: Since $h(-1,-1)$ is negative, the circulation about $\mathbf{n}$ is clockwise, so $-\mathbf{n}$ is the correct pointing normal for the counterclockwise circulation. Thus $\iint\limits_S \nabla \times \mathbf{F}\cdot(-\mathbf{n})\,d\sigma$ actually gives the maximum circulation. You may wish to obtain 3D or contour plots for the surface $H(a,b)$.

11. (a) Partition the string into small pieces. Let $\Delta_i s$ be the length of the ith piece. Let (x_i, y_i) be a point in the ith piece. The work done by gravity in moving the ith piece to the x-axis is approximately $W_i = (gx_iy_i\Delta_i s)y_i$ where $x_iy_i\Delta_i s$ is approximately the mass of the ith piece. The total work done by gravity in moving the string to the x-axis is $\sum\limits_i W_i = \sum\limits_i gx_iy_i^2\Delta_i s \Rightarrow \text{Work} = \int\limits_C gxy^2\,ds$

(b) Work $= \int_C gxy^2\,ds = \int_0^{\pi/2} g(2\cos t)(4\sin^2 t)\sqrt{4\sin^2 t + 4\cos^2 t}\,dt = 16g\int_0^{\pi/2} \cos t\,\sin^2 t\,dt$

$= \left[16g\left(\frac{\sin^3 t}{3}\right)\right]_0^{\pi/2} = \frac{16}{3}g$

(c) $\overline{x} = \dfrac{\int_C x(xy)\,ds}{\int_C xy\,ds}$ and $\overline{y} = \dfrac{\int_C y(xy)\,ds}{\int_C xy\,ds}$; the mass of the string is $\int_C xy\,ds$ and the weight of the string is

$g\int_C xy\,ds$. Therefore, the work done in moving the point mass at $(\overline{x}, \overline{y})$ to the x-axis is

$W = \left(g\int_C xy\,ds\right)\overline{y} = g\int_C xy^2\,ds = \frac{16}{3}g.$

12. (a) Partition the sheet into small pieces. Let $\Delta_i\sigma$ be the length of the ith piece and select a point (x_i, y_i, z_i) in the ith piece. The mass of the ith piece is approximately $x_i y_i \Delta_i\sigma$. The work done by gravity in moving the ith piece to the xy-plane is approximately $(gx_i y_i \Delta_i\sigma)z_i = gx_i y_i z_i \Delta_i\sigma \Rightarrow$ Work $= \iint_S gxyz\,d\sigma$.

(b) $\iint_S gxyz\,d\sigma = g\iint_{R_{xy}} xy(1-x-y)\sqrt{1+(-1)^2+(-1)^2}\,dA = \sqrt{3}g\int_0^1\int_0^{1-x}(xy - x^2y - xy^2)\,dy\,dx$

$= \sqrt{3}g\int_0^1 \left[\frac{1}{2}xy^2 - \frac{1}{2}x^2y^2 - \frac{1}{3}xy^3\right]_0^{1-x} dx = \sqrt{3}g\int_0^1 \left[\frac{1}{6}x - \frac{1}{2}x^2 + \frac{1}{2}x^3 + \frac{1}{3}x^4\right] dx$

$= \sqrt{3}g\left[\frac{1}{12}x^2 - \frac{1}{6}x^3 + \frac{1}{6}x^4 + \frac{1}{15}x^5\right]_0^1 = \sqrt{3}g\left(\frac{1}{12} + \frac{1}{15}\right) = \frac{3\sqrt{3}g}{20}$

(c) The center of mass of the sheet is the point $(\overline{x}, \overline{y}, \overline{z})$ where $\overline{z} = \frac{M_{xy}}{M}$ with $M_{xy} = \iint_S xyz\,d\sigma$ and

$M = \iint_S xy\,d\sigma$. The work done by gravity in moving the point mass at $(\overline{x}, \overline{y}, \overline{z})$ to the xy-plane is

$gM\overline{z} = gM\left(\frac{M_{xy}}{M}\right) = gM_{xy} = \iint_S gxyz\,d\sigma = \frac{3\sqrt{3}g}{20}.$

13. (a) Partition the sphere $x^2 + y^2 + (z-2)^2 = 1$ into small pieces. Let $\Delta_i\sigma$ be the surface area of the ith piece and let (x_i, y_i, z_i) be a point on the ith piece. The force due to pressure on the ith piece is approximately $w(4 - z_i)\Delta_i\sigma$. The total force on S is approximately $\sum_i w(4 - z_i)\Delta_i\sigma$. This gives the actual force to be

$\iint_S w(4-z)\,d\sigma.$

(b) The upward buoyant force is a result of the **k**-component of the force on the ball due to liquid pressure. The force on the ball at (x,y,z) is $w(4-z)(-\mathbf{n}) = w(z-4)\mathbf{n}$, where $\mathbf{n}$ is the outer unit normal at (x,y,z). Hence the **k**-component of this force is $w(z-4)\mathbf{n}\cdot\mathbf{k} = w(z-4)\mathbf{k}\cdot\mathbf{n}$. The (magnitude of the) buoyant force on the ball is obtained by adding up all these **k**-components to obtain $\iint_S w(z-4)\mathbf{k}\cdot\mathbf{n}\,d\sigma$.

(c) The Divergence Theorem says $\iint_S w(z-4)\mathbf{k}\cdot\mathbf{n}\,d\sigma = \iiint_D \operatorname{div}(w(z-4)\mathbf{k})\,dV = \iiint_D w\,dV$, where D is $x^2+y^2+(z-2)^2 \le 1 \Rightarrow \iint_S w(z-4)\mathbf{k}\cdot\mathbf{n}\,d\sigma = w\iiint_D 1\,dV = \frac{4}{3}\pi w$, the weight of the fluid if it were to occupy the region D.

14. The surface S is $z = \sqrt{x^2+y^2}$ from $z=1$ to $z=2$. Partition S into small pieces and let $\Delta_i\sigma$ be the area of the ith piece. Let (x_i,y_i,z_i) be a point on the ith piece. Then the magnitude of the force on the ith piece due to liquid pressure is approximately $F_i = w(2-z_i)\Delta_i\sigma \Rightarrow$ the total force on S is approximately

$\sum_i F_i = \Sigma\, w(2-z_i)\Delta_i\sigma \Rightarrow$ the actual force is $\iint_S w(2-z)\,d\sigma = \iint_{R_{xy}} w\left(2-\sqrt{x^2+y^2}\right)\sqrt{1+\frac{x^2}{x^2+y^2}+\frac{y^2}{x^2+y^2}}\,dA$

$$= \iint_{R_{xy}} \sqrt{2}\,w\left(2-\sqrt{x^2+y^2}\right)dA = \int_0^{2\pi}\int_1^2 \sqrt{2}w(2-r)\,r\,dr\,d\theta = \int_0^{2\pi} \sqrt{2}w\left[r^2-\frac{1}{3}r^3\right]_1^2 d\theta = \int_0^{2\pi} \frac{2\sqrt{2}w}{3}\,d\theta$$

$$= \frac{4\sqrt{2}\pi w}{3}$$

15. Assume that S is a surface to which Stokes's Theorem applies. Then $\oint_C \mathbf{E}\cdot d\mathbf{r} = \iint_S (\nabla\times\mathbf{E})\cdot\mathbf{n}\,d\sigma$

$= \iint_S -\frac{\partial\mathbf{B}}{\partial t}\cdot\mathbf{n}\,d\sigma = -\frac{\partial}{\partial t}\iint_S \mathbf{B}\cdot\mathbf{n}\,d\sigma$. Thus the voltage around a loop equals the negative of the rate of change of magnetic flux through the loop.

16. According to Gauss's Law, $\iint_S \mathbf{F}\cdot\mathbf{n}\,d\sigma = 4\pi GmM$ for any surface enclosing the origin. But if $\mathbf{F} = \nabla\times\mathbf{H}$ then the integral over such a closed surface would have to be 0 by Stokes's Theorem (since **F** is conservative).

17. $\oint_C f\nabla g\cdot d\mathbf{r} = \iint_S \nabla\times(f\nabla g)\cdot\mathbf{n}\,d\sigma$ (Stokes's Theorem)

$= \iint_S (f\nabla\times\nabla g + \nabla f\times\nabla g)\cdot\mathbf{n}\,d\sigma$ (Section 14.8, Exercise 19a)

$= \iint_S [(f)(\mathbf{0}) + \nabla f \times \nabla g] \cdot \mathbf{n}\, d\sigma$ (Section 14.7, Equation 12)

$= \iint_S (\nabla f \times \nabla g) \cdot \mathbf{n}\, d\sigma$

18. $\nabla \times \mathbf{F}_1 = \nabla \times \mathbf{F}_2 \Rightarrow \nabla \times (\mathbf{F}_2 - \mathbf{F}_1) = \mathbf{0} \Rightarrow \mathbf{F}_2 - \mathbf{F}_1$ is conservative $\Rightarrow \mathbf{F}_2 - \mathbf{F}_1 = \nabla f$; also, $\nabla \cdot \mathbf{F}_1 = \nabla \cdot \mathbf{F}_2$ $\Rightarrow \nabla \cdot (\mathbf{F}_2 - \mathbf{F}_1) = 0 \Rightarrow \nabla^2 f = 0$ (so f is harmonic). Finally, on the surface S, $\nabla f \cdot \mathbf{n} = (\mathbf{F}_2 - \mathbf{F}_1) \cdot \mathbf{n}$ $= \mathbf{F}_2 \cdot \mathbf{n} - \mathbf{F}_1 \cdot \mathbf{n} = 0$. Now, $\nabla \cdot (f \nabla f) = \nabla f \cdot \nabla f + f \nabla^2 f$ so the Divergence Theorem gives $\iiint_D |\nabla f|^2\, dV + \iiint_D f \nabla^2 f\, dV = \iiint_D \nabla \cdot (f \nabla f)\, dV = \iint_S f \nabla f \cdot \mathbf{n}\, d\sigma = 0$, and since $\nabla^2 f = 0$ we have

$\iiint_D |\nabla f|^2\, dV + 0 = 0 \Rightarrow \iiint_D |\mathbf{F}_2 - \mathbf{F}_1|^2\, dV = 0 \Rightarrow \mathbf{F}_2 - \mathbf{F}_1 = \mathbf{0} \Rightarrow \mathbf{F}_2 = \mathbf{F}_1$, as claimed.

19. False; let $\mathbf{F} = y\mathbf{i} + x\mathbf{j} \neq \mathbf{0} \Rightarrow \nabla \cdot \mathbf{F} = \frac{\partial}{\partial x}(y) + \frac{\partial}{\partial y}(x) = 0$ and $\nabla \times \mathbf{F} = \begin{vmatrix} \mathbf{i} & \mathbf{j} & \mathbf{k} \\ \frac{\partial}{\partial x} & \frac{\partial}{\partial y} & \frac{\partial}{\partial z} \\ x & y & 0 \end{vmatrix} = 0\mathbf{i} + 0\mathbf{j} + 0\mathbf{k} = \mathbf{0}$

20. $|\mathbf{r}_u \times \mathbf{r}_v|^2 = |\mathbf{r}_u|^2 |\mathbf{r}_v|^2 \sin^2\theta = |\mathbf{r}_u|^2 |\mathbf{r}_v|^2 (1 - \cos^2\theta) = |\mathbf{r}_u|^2 |\mathbf{r}_v|^2 - |\mathbf{r}_u|^2 |\mathbf{r}_v|^2 \cos^2\theta = |\mathbf{r}_u|^2 |\mathbf{r}_v|^2 - (\mathbf{r}_u \cdot \mathbf{r}_v)^2$ $\Rightarrow |\mathbf{r}_u \times \mathbf{r}_v|^2 = \sqrt{EG - F^2} \Rightarrow d\sigma = |\mathbf{r}_u \times \mathbf{r}_v|\, du\, dv = \sqrt{EG - F^2}\, du\, dv$

21. $\mathbf{r} = x\mathbf{i} + y\mathbf{j} + z\mathbf{k} \Rightarrow \nabla \cdot \mathbf{r} = 1 + 1 + 1 = 3 \Rightarrow \iiint_D \nabla \cdot \mathbf{r}\, dV = 3 \iiint_D dV = 3V \Rightarrow V = \frac{1}{3} \iiint_D \nabla \cdot \mathbf{r}\, dV$ $= \frac{1}{3} \iint_S \mathbf{r} \cdot \mathbf{n}\, d\sigma$, by the Divergence Theorem

NOTES:

NOTES:

NOTES:

NOTES:

NOTES:

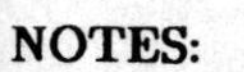
NOTES:

NOTES:

NOTES:

NOTES:

NOTES:

NOTES: